FEATURES OF THE **COMPANION WEBSITE** INCLUDE:

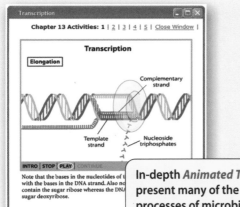

Transcription
Chapter 13 Activities: 1 | 2 | 3 | 4 | 5 | Close Window |

Transcription

Elongation

Complementary strand

Template strand

Nucleoside triphosphates

INTRO | STOP | PLAY | CONTINUE

Note that the bases in the nucleotides of the with the bases in the DNA strand. Also no contain the sugar ribose whereas the DNA sugar deoxyribose.

Done

In-depth *Animated Tutorials* present many of the dynamic processes of microbiology in an animated format, making them easier to follow and understand. All tutorials include an introduction, an animation, and a short quiz.

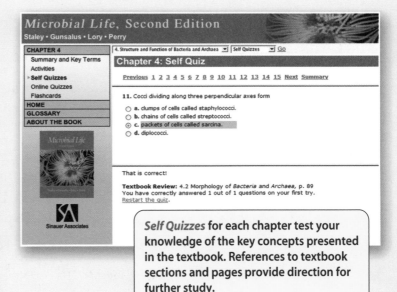

Microbial Life, Second Edition
Staley • Gunsalus • Lory • Perry

CHAPTER 4
Summary and Key Terms
Activities
▸ Self Quizzes
Online Quizzes
Flashcards
HOME
GLOSSARY
ABOUT THE BOOK

4. Structure and Function of Bacteria and Archaea ▾ | Self Quizzes ▾ | Go

Chapter 4: Self Quiz

Previous 1 2 3 4 5 6 7 8 9 10 11 12 13 14 15 Next Summary

11. Cocci dividing along three perpendicular axes form
○ **a.** clumps of cells called staphylococci.
○ **b.** chains of cells called streptococci.
◉ **c.** packets of cells called sarcina.
○ **d.** diplococci.

That is correct!

Textbook Review: 4.2 Morphology of *Bacteria* and *Archaea*, p. 89
You have correctly answered 1 out of 1 questions on your first try.
Restart the quiz.

Self Quizzes for each chapter test your knowledge of the key concepts presented in the textbook. References to textbook sections and pages provide direction for further study.

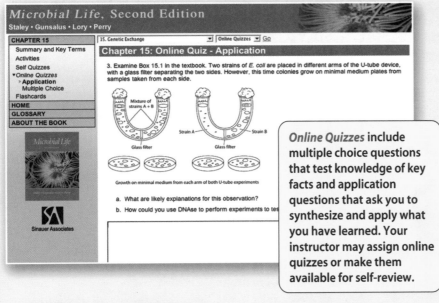

Microbial Life, Second Edition
Staley • Gunsalus • Lory • Perry

CHAPTER 15
Summary and Key Terms
Activities
Self Quizzes
▾Online Quizzes
   ▸ Application
   Multiple Choice
Flashcards
HOME
GLOSSARY
ABOUT THE BOOK

15. Genetic Exchange ▾ | Online Quizzes ▾ | Go

Chapter 15: Online Quiz - Application

3. Examine Box 15.1 in the textbook. Two strains of *E. coli* are placed in different arms of the U-tube device, with a glass filter separating the two sides. However, this time colonies grow on minimal medium plates from samples taken from each side.

Mixture of strains A + B

Strain A   Strain B

Glass filter   Glass filter

Growth on minimal medium from each arm of both U-tube experiments

a. What are likely explanations for this observation?
b. How could you use DNAse to perform experiments to tes

Online Quizzes include multiple choice questions that test knowledge of key facts and application questions that ask you to synthesize and apply what you have learned. Your instructor may assign online quizzes or make them available for self-review.

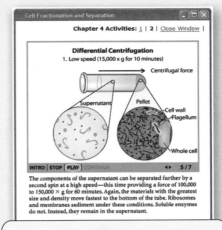

Cell Fractionation and Separation
Chapter 4 Activities: 1 | 2 | Close Window |

Differential Centrifugation
1. Low speed (15,000 × g for 10 minutes)

Centrifugal force

Supernatant   Pellet

Cell wall
Flagellum

Whole cell

INTRO | STOP | PLAY | CONTINUE   5/7

The components of the supernatant can be separated further by a second spin at a high speed—this time providing a force of 100,000 to 150,000 × g for 60 minutes. Again, the materials with the greatest size and density move fastest to the bottom of the tube. Ribosomes and membranes sediment under these conditions. Soluble enzymes do not. Instead, they remain in the supernatant.

Major Technique Tutorials provide detailed depictions of many of the tools and methods that are fundamental to the work of microbiologists.

Additional Features...

Dynamic Illustrations break complex figures down into easy-to-follow, step-by-step presentations.

Chapter Summaries provide thorough overviews of the main concepts covered in each chapter.

Flashcards and **Key Terms** lists for each chapter help you master the extensive vocabulary of microbiology.

The complete **Glossary** is included on the website, giving you quick access to definitions of terms.

COMPANION WEBSITE ACTIVITIES

 Look for this icon throughout the book. It refers to the following activities on the Companion Website, which fall into three categories: Animated Tutorials, Major Techniques, **and** Dynamic Illustrations.

Microbial Life

Microbial Life

SECOND EDITION

James T. Staley
Department of Microbiology,
University of Washington

Robert P. Gunsalus
Department of Microbiology,
University of California, Los Angeles

Stephen Lory
Department of Microbiology and Molecular Genetics,
Harvard Medical School

Jerome J. Perry
Raleigh, North Carolina

 SINAUER ASSOCIATES, INC., Publishers • *Sunderland, Massachusetts*

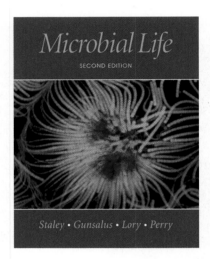

About the cover

This micrograph shows a colony of *Gloeotrichia echinulata*, a freshwater cyanobacterium in the *Rivularia* group. The colony consists of many filaments, and each filament in turn consists of many bacterial cells. Carotenoids and other red pigments give the bases of the filaments their red color; the blue and pink result from autofluorescence. ©M. I. Walker/Photo Researchers, Inc.

About the book

Editor: Andrew D. Sinauer
Project Editor: Laura Green
Market Research and Reviewing: Susan McGlew
Copy Editors: Catherine Minick and Heidi Thaens
Production Manager: Christopher Small
Book Layout and Production: Joan Gemme and Joanne Delphia
Illustrations: Elizabeth Morales
Book Design: Joan Gemme
Cover Design: Jefferson Johnson
Photo Research: David McIntyre

Address orders and editorial correspondence to:
Sinauer Associates, Inc., P.O. Box 407
23 Plumtree Road, Sunderland, MA, 01375 U.S.A
Fax: 413-549-1118
Internet: www.sinauer.com; Email: publish@sinauer.com

Library of Congress Cataloging-in-Publication Data

Microbial life / James T. Staley ... [et al.]. — 2nd ed.
    p. ; cm.
  Includes index.
  ISBN-13: 978-0-87893-685-4 (hardcover : alk. paper)
  ISBN-10: 0-87893-685-8 (hardcover : alk. paper)
  1. Microbiology. I. Staley, James T., 1938-
  [DNLM: 1. Microbiology. QW 4 M6227 2007]

QR41.2.P467 2007
579—dc22
                                    2007001581

6 5 4 3 2 1

*To Sonja, Greg, Wendy, Mark, Jack, and my colleagues
and students in science.*

J. T. STALEY

To my family, my mentors, and all my students.

R. P. GUNSALUS

*To Karlissa, for her patience, and to my scientific mentors, John Collier
and Bernard Davis.*

S. LORY

*To Elizabeth for her many contributions, including typing my
manuscript, and to Jerome, Jr., Marianne, Chuck, and Neil.*

J. J. PERRY

*The authors also wish to dedicate this book to the late Holger Jannasch
for his interest and support.*

About the Authors

JAMES T. STALEY is Professor in the Department of Microbiology at the University of Washington, Seattle. He earned a B.A. from the University of Minnesota, an M.Sc. from Ohio State University, and a Ph.D. from the University of California, Davis. Before coming to the University of Washington, he held academic appointments at Michigan State University and the University of North Carolina. Dr. Staley's major research interests are in general microbiology, with emphasis on microbial evolution, diversity, ecology, and bacterial taxonomy. He was the first Director of the cross-disciplinary NSF IGERT Astrobiology Ph.D. traineeship program at the University of Washington, which uses extreme environments on Earth as models for possible life on other planets and studies the early evolution of life on Earth. Currently he is Chair of Bergey's Manual Trust.

ROBERT P. GUNSALUS is Professor of Microbiology in the Department of Microbiology, Immunology and Molecular Genetics in the David Geffen School of Medicine at UCLA. He earned a B.S. from South Dakota State University, and M.Sc and Ph.D. degrees from the University of Illinois at Urbana–Champaign. Following postdoctoral studies at Stanford University, he has taught and has academic positions in the School of Life Sciences and the School of Medicine at UCLA. Dr. Gunsalus' major research interests are in microbial physiology and regulation with emphasis on the genetic control of electron transport pathways and anaerobic carbon flow in bacteria and archaea. He has served as a member of the editorial boards of the *Journal of Bacteriology, Archives of Microbiology,* the *Journal of Biological Chemistry,* and *Molecular Microbiology,* as well as Editor of *FEMS Microbiology Letters.*

STEPHEN LORY is Professor of Microbiology and Molecular Genetics at Harvard Medical School. He earned a B.A. in Bacteriology and a Ph.D. in Microbiology, both from the University of California, Los Angeles. Following a research fellowship in the Bacterial Physiology Unit at Harvard Medical School, he taught at the University of Washington, Seattle for several years before returning to Harvard. Dr. Lory's scientific interests are in microbial pathogenesis, and he directs research projects in microbial evolution, signal transduction, and gene regulation. The author of numerous journal articles and book chapters, Dr. Lory has served as a member of the editorial boards of the *Journal of Bacteriology, Infection and Immunity* and *Molecular Microbiology,* and as Editor of *Microbiology and Molecular Biology Reviews.*

JEROME J. PERRY, an experienced research scientist and educator, taught general microbiology to college undergraduates for more than 25 years. He received his B.S. from Pennsylvania State University and his Ph.D. from the University of Texas. Dr. Perry's major research interests included metabolism of gaseous alkanes, co-oxidation, and bioremediation. In addition to *Microbial Life,* he has written extensively on hydrocarbon metabolism in papers, chapters, and various proceedings, and co-edited *Introduction to Environmental Toxicology,* a major textbook in its field.

Contributors

 WILLIAM B. WHITMAN is Professor of Microbiology at the University of Georgia. He received his B.S. from the State University of New York at Stony Brook and his Ph.D. from The University of Texas at Austin. He has always been interested in unusual microorganisms. His current research focuses on the carbon metabolism of the methane-producing archaeon *Methanococcus* and the sulfur metabolism of the marine roseobacter *Silicibacter*. He also studies the evolution and systematics of prokaryotes from soils and seawater.

 SINA M. ADL is an Assistant Professor in the Department of Biology at Dalhousie University, Canada. He received his B.S., M.S., and Ph.D. from the University of British Columbia in Vancouver, Canada. Part of his Ph.D. was through the University of Paris VI, at the C.N.R.S. at Gif-sur-Yvette. His postdoctoral research included work at the University of Paris XI, and at the Institute of Ecology at the University of Georgia. His interest has always been in the ecology and comparative cell biology of protists. His current research is in soil ecology, with experiments in deserts, forests, pastures, and the lower-arctic habitats.

 ALASTAIR G. B. SIMPSON is an Assistant Professor in the Department of Biology at Dalhousie University, Canada, and a "Scholar" of the Canadian Institute for Advanced Research. He received his B.Sc. and Ph.D. from the University of Sydney, Australia, before completing a postdoctoral fellowship at Dalhousie University. Dr Simpson is interested in the biodiversity and evolution of protists, specializing in free-living flagellates in general, and "excavates" in particular.

 CYNTHIA L. BALDWIN is a Professor in the Department of Veterinary and Animal Sciences at the University of Massachusetts, Amherst. She has taught immunology and infectious disease there and, previously, at The Ohio State University, where she was in the Department of Microbiology for nearly 20 years. She received her Ph.D. from Cornell University in Immunology after having received her B.A. from Hartwick College. She is also a contributing author to the textbook *Infection, Resistance, and Immunity*. Her research focuses on cellular immunity to pathogens with a particular interest in gamma delta T cells and intracellular protozoa and bacteria. She has conducted research in Africa, where she did her postdoctoral studies at the International Laboratory for Research on Animal Diseases, as well as in the U.S.

 SAMUEL J. BLACK is a Professor of Immunology and Head of the Department of Veterinary and Animal Sciences at the University of Massachusetts, Amherst. He has taught immunology and infectious disease for the past 18 years, both in his current position and, previously, at The Ohio State University. Professor Black received his Ph.D. from the University of Edinburgh, UK and received postdoctoral training in Immunogenetics at the University of Cologne, Germany, and Stanford University. He then joined the faculty of the International Laboratory for Research in Animal Diseases, Nairobi, Kenya in 1979 and was Coordinator of the Immunobiology Lab there until 1989. He is a contributing author to the textbook *Infection, Resistance, and Immunity* and a co-editor of the *World Class Parasites* series. His research focuses on host and parasite interactions that affect immune responses, as well as mechanisms of pathology that cause tissue injury.

Preface

One of the most distinctive features of *Microbial Life* is its emphasis on evolution. This theme is introduced in the first chapter with this quotation from Theodosius Dobzansky: "In biology nothing makes sense except in the light of evolution." This statement is particularly relevant to microbiology, because microorganisms are inextricably tied to the origin and evolution of life and the biosphere as well as the subsequent evolution of plants and animals. Indeed, it can be said that understanding microorganisms and their evolution is key to understanding biology.

Microbiology is also of extreme practical importance. Not only are microorganisms and viruses the cause of infectious diseases of animals and plants, but they are also used in the industrial production of antibiotics, foods, and ethanol for fuel. Microbes are also responsible for food spoilage and are key players in the recycling of garbage and toxic waste. Microbiologists play an active role in the burgeoning areas of biotechnology and nanotechnology.

Microbial Life introduces students to the remarkable field of microbiology. This new edition blends fundamental concepts with the sense of excitement arising from recent discoveries regarding the vast diversity of microorganisms, pathogenic microbiology, immunology, and virology, as well as in applied and environmental microbiology.

NEW FEATURES OF THE SECOND EDITION OF MICROBIAL LIFE

Microbial Life has undergone a major overhaul in this new edition. The most significant change is the addition of a new author, Robert Gunsalus, as one of the four principal authors. Rob has taken charge of Parts II and III, which pertain to microbial physiology and metabolism, and helped integrate this material with other chapters of the book. As a result the content has undergone major revisions and includes much new, cutting-edge material.

Sina Adl and Alastair Simpson have also joined the book team as coauthors of the chapter on eukaryotic microbial life. They have provided thorough coverage of this diverse and important group of protists and fungi, which is poorly treated or even omitted entirely in some traditional textbooks.

Each chapter in this new edition commences with a clear statement of objectives, succinctly written as bullet points. The chapter then continues with an introductory overview. As a student learning aid, a brief summary referred to as "Section Highlights" follows each major section in this Second Edition of *Microbial Life*.

Another significant change is that the new edition of *Microbial Life* has increased by about 20% in content from the previous edition. This major expansion has enabled the authors to provide greater in-depth coverage of topics and to add additional information about recent and exciting developments in the field.

NEW FEATURES IN EACH PART

Part I, ***The Scope of Microbiology***, provides an introduction to microbiology with chapters on evolution, history, chemistry, and structure and function. New features include an expanded section on astrobiology, boxes to pique student interest in the history chapter, and a more comprehensive and better-illustrated treatment of the structures of bacterial and archaeal cells. The latter has been thoroughly integrated with the book's treatment of microbial physiology.

Part II, ***Microbial Physiology: Nutrition and Growth***, contains up-to-date coverage of methods for culturing anaerobic bacteria, improved coverage of bacterial growth and the effects of environmental conditions on growth and survival, and a more detailed treatment of antimicrobial agents and their mechanisms of action.

Extensive changes have been made in Part III, ***Microbial Physiology: Metabolism***. The coverage of electron transport chain components, types of respiratory pathways, and the formation and use of the proton motive force is presented in greater detail. In addition, there is expanded coverage of the electron transfer process during conversion of light energy to chemical energy, as well as more extensive treatment of the CO_2 fixation pathways. Chapter 11 contains a comprehensive treatment of chemotaxis/flagella components, associated signal reception and processing, and the control of flagellar motion. Also, the treatment of protein secretion mechanisms has been updated and expanded. This chapter also includes material on the various types, biogenesis, and roles of pili. The last chapter of the Part has an expanded treatment of aerobic biodegradative processes in microbes, and features new illustrations and boxes.

In Part IV, ***Genetics and Basic Virology***, the chapters on genetics and genetic exchange mechanisms have been fully integrated with the physiology section treated in Part III. The new edition includes an expanded discussion of the control of gene expression, including transcriptional initiation, attenuation and anti-termination, and the control of mRNA stability. The revised edition contains entirely new sections on bacterial communication (quorum sensing) and the increasingly important role of small non-coding RNAs in gene regulation. The expanded genomics chapter includes the very latest developments in genome annotation and analysis. The virology chapter has been updated with treatments of emerging viruses such as SARS.

Part V on **Microbial Evolution and Diversity** includes a revised, phylogenetic treatment of the *Bacteria* and *Archaea* that adheres to the classification used in the Second Edition of *Bergey's Manual of Systematic Bacteriology*. New findings on the ecology, evolution, and activities of the *Bacteria* and *Archaea* are explored. New material on archaeal nitrification, the *Nanoarchaea* (a new archaeal phylum), the anammox reaction, bacterial tubulins, and the physiology and ecology of the spirochetes in the termite gut are just a few examples of recent findings treated in this new edition. The topic of horizontal gene transfer and its impact on microbial speciation, phylogeny, and diversity are also discussed.

Part VI, **Microbial Ecology**, has extensively rewritten and expanded coverage of microbial biodiversity, the concept of an ecosystem, and roles of microorganisms in communities. A unique feature of this book is an entire chapter devoted to microbe–microbe, as well as microbe–eukaryote interactions that are essential to both partners. For example, a new discussion of the Hawaiian bobtail squid symbiosis illustrates the structural alterations the host makes to harbor a symbiont. In addition, newer techniques used in microbial ecology, such as stable isotope probing, expression analyses, and ecological genomics, are discussed.

Part VII, **Immunology and Medical Microbiology**, treats the mechanisms used by all animals, including humans, to ward off infectious diseases. Chapter 27 includes a thorough discussion of the barriers that prevent invasion of host cells and the immune response that destroys invasive pathogens. Diseases caused by bacteria, fungi, protists, and viruses are considered (Chapters 28, 29), with expanded coverage of emerging diseases. Extensive new material has been added on pathogenesis, immune evasion, and the mechanisms of toxin action. Epidemiology and clinical microbiology are covered in Chapter 30, including discussions of the diagnosis, treatment, and prevention of infectious disease.

Part VIII, **Applied Microbiology**, describes the application of microbes as agents to benefit humankind, and provides a major source of support for careers in microbiology. These two chapters cover industrial microbiology and biotechnology, and applied environmental microbiology. Extensive updates include a field test that illustrates how chlorinated compounds can be broken down using bioaugmentation in flow-through systems.

CONTRIBUTORS

Microbiology is a rapidly expanding field of science, too broad now to be covered by three authors. This edition introduces a fourth major contributor, Robert Gunsalus, who is an expert on microbial physiology and metabolism. In addition, we invited a few colleagues to share their expertise in writing important chapters for this book. We thank William "Barny" Whitman for the fascinating chapter on the *Archaea*, and Sina Adl and Alastair Simpson for the new and up-to-date chapter on eukaryotic microorganisms. The superb coverage of immunology and pathogenesis was written by Cynthia L. Baldwin and Samuel J. Black.

ACKNOWLEDGEMENTS

The authors could not have produced this book without the encouragement, professional support, and dedicated staff of Sinauer Associates, who have been key to making this edition a reality. We are truly grateful to Andy Sinauer for his advice and patience throughout this process. He and all of the staff members have been reliable, industrious, and extremely helpful to the authors, whose schedules were not always amenable to the publication process. We all worked closely with Laura Green, who performed superbly as Project Editor. Laura provided insight and professional skills all along, which resulted in making this new edition uniform in style and content, and highly readable. Susan McGlew provided us with exceptionally helpful reviews from professional microbiologists who are actively involved in teaching and research. Marie Scavotto led the team effort in the marketing. Finally, the resourceful David McIntyre found many fantastic photographs to help illustrate this new edition. We are also grateful to our many colleagues who took their valuable time to provide honest and helpful reviews that have greatly aided us in improving this new edition.

SPECIAL THANKS

Jim Staley expresses his thanks to Brian Oakley, a postdoc during much of the revision process, as well as to graduate students Clara Fuchsman and John Kirkpatrick. In addition, he wishes to thank all the reviewers, some of whom were anonymous, for their frank and helpful reviews of his material. Jim also thanks the Department of Microbiology and, in particular, Kari Johnstone, for help and other support for the new edition. Jim also wishes to thank Dennis Kunkel for taking some great SEMs, in particular the incredible image of *Verrucomicrobium spinosum*.

J. J. Perry expresses special thanks to Herman R. Berkhoff for his indispensable aid with the computer. Thanks also to T. J. Schneeweis and Larry F. Grand for their contributions.

JAMES T. STALEY • ROBERT P. GUNSALUS
STEPHEN LORY • JEROME J. PERRY
MARCH 2007

Reviewers

Laurie Achenbach, Southern Illinois University

Thomas Alton, Western Illinois University

O. Roger Anderson, Columbia University

Esther Angert, Cornell University

Martha Apple, Montana Tech of the University of Montana

Ann Auman, Pacific Lutheran University

Larry Baresi, California State University, Northridge

Jim Berger, University of British Columbia

Sarah Boomer, Western Oregon University

Robert A. Britton, Michigan State University

Carl E. Cerniglia, National Center for Toxicological Research

James Champoux, University of Washington

Peggy Cotter, University of California, Santa Barbara

James Courtright, Marquette University

James F. Curran, Wake Forest University

John Davis, Columbus State University

John Deighton, Rutgers Pinelands Field Station

Jeffrey DeStefano, University of Maryland

Nancy DiIulio, Case Western Reserve University

Timothy Donohue, University of Wisconsin, Madison

John Ferguson, University of Washington

Marvin S. Friedman, Hunter College

John Fuerst, University of Queensland

Daniel Gage, University of Connecticut

Douglas H. Graham, Grand Valley State University

Eileen Gregory, Rollins College

Dennis Grogan, University of Cincinnati

Brian Hedlund, University of Nevada, Las Vegas

George Hegeman, Indiana University

Michael Ibba, The Ohio State University

Thomas R. Jerrells, University of Nebraska Medical Center

Mark S. Johnson, Loma Linda University

Teresa Johnson, College of Wooster

C. Cheng Kao, Texas A&M University

Laura Katz, Smith College

Daniel Kearns, Indiana University

Dennis Kitz, Southern Illinois University, Edwardsville

Donald A. Klein, Colorado State University

Edward R. Leadbetter, University of Connecticut

Susan Leschine, University of Massachusetts

Bruce Levin, Emory University

William Lorowitz, Weber State University

Michael J. McInerney, University of Oklahoma

Bob McLean, Texas State University

Naomi Morrissette, University of California, Irvine

Scott B. Mulrooney, Michigan State University

Ivan Oresnik, University of Manitoba

Paul Orndorff, North Carolina State University

Susan Payne, Texas A&M University

Walt Ream, Oregon State University

Sabine Rech, San Jose State University

Larry Reitzer, University of Texas, Dallas

Ingrid Ruf, University of California, Irvine

Kirsty A. Salmon, University of California, Irvine

David A. Sanders, Purdue University

Wendy Schluchter, University of New Orleans

Tom Schmidt, Michigan State University

Imke Schroeder, University of California, Los Angeles

Richard Seyler, Virginia Polytechnic Institute and State University

James Shapleigh, Cornell University

Teri Shors, University of Wisconsin, Oshkosh

Patricia Siering, Humboldt State University

Kevin R. Sowers, University of Maryland Biotechnology Institute

Miriam Susskind, University of Southern California

Monica Tischler, Benedictine University

Delon E. Washo-Krupps, Arizona State University

William B. Whitman, University of Georgia

Marianne Wilkerson, Los Alamos National Laboratory

Chris Yost, University of Regina

Brief Contents

Contents

4 Structure and Function of Bacteria *and* Archaea 75

PART II *Microbial Physiology: Nutrition and Growth 123*

5 Nutrition, Cultivation, and Isolation of Microorganisms 125

PART **III** *Microbial Physiology: Metabolism 201*

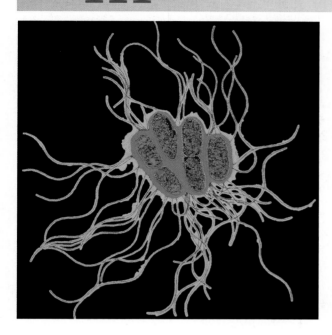

11 Assembly of Bacterial Cell Structures 287

12 Roles of Microbes in Biodegradation 317

PART IV Genetics and Basic Virology 339

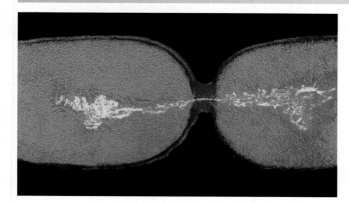

13 Basic Genetics 341

16 *Microbial Genomics* 451

PART V *Microbial Evolution and Diversity* 483

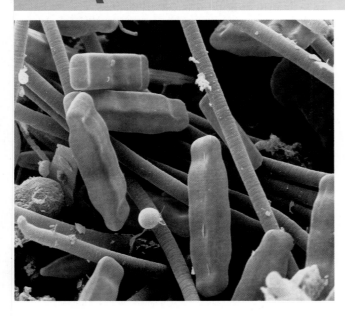

17 *Taxonomy of* Bacteria *and* Archaea 485

18 Archaea 515

19 *Nonphotosynthetic* Proteobacteria *545*

20 *Gram-Positive Bacteria:* Firmicutes *and* Actinobacteria *601*

21 *Phototrophic* Bacteria *633*

PART VI *Microbial Ecology 737*

24 *Microbial Ecology 739*

25 *Beneficial Symbiotic Associations 785*

26 *Human Host–Microbe Interaction 811*

PART VII *Immunology and Medical Microbiology 845*

27 *Immunology 847*

28 Microbial Diseases of Humans 893

PART VIII *Applied Microbiology 1007*

Boxes

Boxes throughout the text describe Milestones, Research Highlights, **or** Methods & Techniques **in the study of microbiology.**

Media and Supplements

For Students

Companion Website
(www.sinauer.com/microbial-life)

New for the Second Edition, the *Microbial Life* companion website features a wide array of study and review tools for students. The site combines all of the content from the previous edition's student CD and printed study guide into one robust online resource. Access to the site is free and requires no access code. (The online quizzes require instructor registration.)

Features of the companion website:

◆ *Chapter Summaries* for a review of major concepts

◆ *Animated Tutorials* that present key concepts using a clear, easy-to-follow narrative

◆ *Major Technique Tutorials* that describe many of the important laboratory methods used by microbiologists

◆ *Dynamic Illustrations* that present key figures in a step-by-step, animated format

◆ *Key Terms* lists for each chapter

◆ *Online Quizzing*, including two separate multiple-choice quizzes for each chapter

◆ *True/False* and *Fill-in-the-Blank Questions*

◆ *Application and Synthesis Questions* that help students synthesize the concepts and facts presented in each chapter

◆ *Flashcards* for learning terminology

◆ The complete *Glossary*

For Instructors

Instructor's Resource Library

The Second Edition Instructor's Resource Library includes a wealth of electronic resources to aid instructors in preparing courses, delivering lectures, and assessing students. Contents include:

◆ *Browser Interface*

◆ *Textbook Figures:* All of the figures, photos, and tables from the textbook in both high- and low-resolution JPEG formats.

◆ *PowerPoint Presentations:* Two presentations are provided for each chapter:
  • A *figures presentation* that includes every figure, photograph, and table from the chapter, ready to be inserted into your own presentations.
  • A new complete *lecture presentation* that includes a thorough outline of the material covered in the chapter, along with selected figures.

◆ *Clicker Questions:* For each chapter, a set of questions has been prepared specifically for use with "clicker" (classroom response) systems.

◆ *Supplemental Photos:* An additional collection of micrographs and color photos.

◆ *Animations:* All of the companion website animations and dynamic illustrations are included for use in lecture.

◆ *Instructor's Manual & Test Bank* (see details below)

◆ *Computerized Test Bank:* The entire test bank is also provided in Brownstone Diploma format, for quick exam creation (software included).

◆ *Chapter Outlines*

Instructor's Manual & Test Bank
(included in the Instructor's Resource Library)

Cheryl Keller Capone, *Pennsylvania State*, and Kirsty Salmon, *University of California, Irvine*

The *Microbial Life* Instructor's Manual provides instructors with resources for developing lectures and for assessment. Contents include the following for each chapter:

◆ *Summary* and *Outline:* A thorough summary and outline of the important concepts in the chapter.

◆ *Teaching Resources:* Useful tips on how to locate additional resources on the Web.

◆ *Test Bank:* Expanded for the Second Edition, the test bank includes multiple-choice, true/false, and matching questions, as well as the end-of-chapter review questions from the textbook, with answers.

Overhead Transparencies

This expanded set of 300 full-color transparencies includes a wide range of key figures and tables from the textbook. The figures are formatted for excellent image quality when projected.

PART

I

The Scope of Microbiology

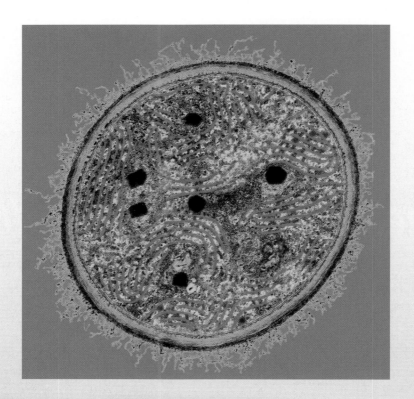

The objectives of this chapter are to:

◆ Provide evidence that microbial life was the first life on Earth and is expected to survive long after plant and animal life becomes extinct.

◆ Present the Tree of Life of all living organisms on Earth to illustrate the vast diversity of microbial life.

◆ Introduce microorganisms and differentiate them from plant and animal life.

◆ Explain how microbial life is the foundation of the biosphere from both an evolutionary as well as functional perspective.

◆ Discuss the unique and important roles that bacterial and archaeal life play in the great biogeochemical cycles of nitrogen, carbon, and sulfur.

◆ Introduce astrobiology and the importance of microbiology to its study.

1

Evolution, Microbial Life, and the Biosphere

In biology nothing makes sense, except in the light of evolution.
—*Theodosius Dobzhansky, Geneticist 1900–1975*

Our solar system originated through physical and chemical processes. After Earth formed, organisms originated and evolved. These first organisms were microorganisms, and they had a profound impact on Earth and the formation of its biosphere, the shell about Earth where life occurs. Certain bacterial groups played especially crucial roles early on in Earth's development. For example, geochemical and fossil evidence indicates that the production of oxygen in the atmosphere was due to the photosynthetic activity of cyanobacteria. The evolution of microorganisms that produced oxygen was of monumental significance because all plant and animal life that exists today requires oxygen. Thus, plants and animals could evolve only because microorganisms evolved first. In this chapter we discuss Earth's origin, the evolution of life, and the importance of microorganisms to life on Earth.

1.1 Origin of Earth and Life

The origin of Earth and the evolution of life on our planet has been a long process. The universe, which is estimated to have an age of 13 Ga (1 Ga, a giga-annum, is 10^9 years), began with a "Big Bang" that produced two principal elements, hydrogen (^1H) and helium (^4He), with smaller amounts of other light elements. Following the Big Bang, the universe expanded, as it continues to do today. At its periphery the original light elements condensed to form clouds of gases and dust. In the clouds heavier elements developed from the lighter ones.

Our solar system was formed by an accretional process in which micrometer-sized dust particles collided to form centimeter-sized bodies. These objects were located in a planar disk that orbited the sun. The accretional process continued as the dust and rock particles aggregated to form boulders and larger bodies that eventually attained the size of the planets. Therefore, ultimately by gravitational contraction, our solar system, with the Sun, Earth, and other planets, formed about 4.5 Ga ago. The final stages of accretion involved collisions between large bodies at high velocities. A major collision between early Earth and a Mars-sized object resulted in the formation of our moon and Earth.

The 600 million years following Earth's formation is called the era of "heavy bombardment" because of the high frequency of collisions between Earth and large asteroids and comets. Some of these collisions, such as the one responsible for the formation of the moon, were so violent that they heated Earth to sterilizing temperatures, that is, temperatures at which life as we know it would have perished. Even collisions with bodies only 100 km in diameter could result in sterilization within the planet to depths of several kilometers. Furthermore, the heat from these collisions would have removed volatile substances such as water.

During its first 600 million years, Earth was not a hospitable planet for life. Water was not initially available. It was brought to Earth by comets and asteroids that came from farther out in the solar system. Once water was available and the era of heavy bombardment had ended, conditions became conducive for the origin of life.

Scientists have determined the date of Earth's formation by studying slowly decaying radioactive isotopes, the decay of which occurs at a constant rate independent of temperature and pressure. Several isotopes are relied on for dating such ancient events (e.g., potassium [^{40}K], which decays to argon [^{40}Ar] with a half-life, the time required for half the radioactivity to decay, of 1.26 billion years, and uranium [^{238}U], which decays to lead [^{206}Pb] with a half-life of 4.468 billion years). Radioisotopic methods are also used for dating strata in sedimentary rocks and therefore offer a means of dating fossilized life forms in rocks.

Fossil Evidence of Microorganisms

By the nineteenth century it was known that fossils were the remains or impressions of plants and animals that had been preserved in sedimentary rocks. Accurate dating methods for the rocks that contained fossils were not yet available, so the estimated dates were only guesses. We now know via such radioisotope-based dating methods that some of the organisms that became fossilized, such as the dinosaurs, lived and became extinct millions of years ago. By examining fossils, paleontologists came to several conclusions about the evolution of life. They noted that fossils nearest the surface—that is, in the most recently deposited sedimentary rocks—were structurally more complex than those in deeper layers. Fossils found in deeper strata were increasingly simple in structure, including fossils of very simple, extinct animals such as trilobites. The gradation of complexity, from simple organisms in the most ancient rocks to more complex forms in the more recent sedimentary rocks, argued for an evolutionary process in which more complex forms of plants and animals arose from simpler organisms. Ge-

TABLE 1.1   Geological timetable on Earth

Eon/era	Period	Years Before Present (millions)	Major Events
Precambrian	Hadean	4,500	Heavy bombardment period
	Archaean	3,800	First sedimentary rocks
	Proterozoic	2,600	Appearance of O_2
Paleozoic	Cambrian	570	Animals evolve
	Ordovician	500	
	Silurian	440	Land colonization by plants, animals
	Devonian	395	Fish diversify
	Carboniferous	345	Reptiles evolve; large "fern" forests
	Permian	280	Mass extinction at end
Mesozoic	Triassic	245	Early dinosaurs; first mammals
	Jurassic	190	Plants and animals diversify
	Cretaceous	145	Mass extinction at end; 75% of species lost
Cenozoic	Tertiary	65	Plant and animal radiation
	Quaternary	1.8	Humans evolve

BOX 1.1 *Milestones*

The Discovery of Microbial Fossils

In the early twentieth century an American geologist, Charles Doolittle Walcott, was studying Precambrian sedimentary rocks in Glacier National Park in northwestern Montana. He noted that some had curious undulating wavelike structures (these are now called stromatolites) and postulated that they were fossilized forms of Precambrian reefs. Contemporary scientists doubted his theory, and it remained untested for many years.

American micropaleontologists Stanley Tyler from the University of Wisconsin and Elso Barghoorn from Harvard University were the first to test Walcott's hypothesis. They were studying stromatolites from 1 to 2 billion-year-old Precambrian Gunflint chert deposits from the Great Shield area in the Great Lakes vicinity of North America. When they examined sections of these stromatolites using the light microscope they discovered microbial fossils.

Microbiologists were incredulous when Tyler and Barghoorn first reported their observations in the 1950s and 1960s, as most microbiologists did not believe microbial fossils existed. However, Tyler and Barghoorn's clear photomicrographic evidence convinced a whole generation of skeptical microbiologists. More recently, evidence from biomarkers traces the evolution of the cyanobacteria to about 2.5 Ga BP and another report indicates that sedimentary cherts in the Pilbara Craton formation of Western Australia contain laminated structures that have been dated to 3.43 Ga BP.

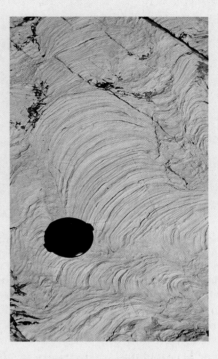

The undulating layers of this sedimentary rock of Glacier National Park are stromatolites containing fossil microorganisms. The lens cap serves as a scale marker. Courtesy of Beverly Pierson.

---

ologists and paleontologists worked hand in hand to develop time scales for sedimentary rock deposits and named the various time periods of Earth's history based upon fossil records (Table 1.1).

At the time of Charles Darwin (1809–1882) the fossil record was being carefully studied, but there was no means of assessing the ages of fossils. Today, we realize that the fossil record for plants and animals extends back to a period of 570 million years ago (mya). Rocks older than that contain no plant and animal fossils. The time before the appearance of plants and animals became known as the Precambrian era (see Table 1.1).

In the 1950s two American scientists, Stanley Tyler and Elso Barghoorn, made a startling discovery. They reported to the scientific world that they had found fossils of microorganisms in sedimentary rocks dated to the Precambrian era (Box 1.1). This important discovery provided the first convincing evidence that the earliest life forms on Earth were microorganisms.

The microbial fossils were discovered in laminated sedimentary rocks called **stromatolites** (Figure 1.1A). Many of the multilayered stromatolite structures contain calcium carbonate along with the fossils of filamentous microorganisms (Figure 1.1B). Living stromatolites still exist on Earth today. The columnar stromatolites occur in intertidal marine areas such as Shark Bay, Western Australia (Figure 1.2). These living stromatolites contain microorganisms that deposit calcium carbonate and other minerals, forming the successive layers of the stromatolite structure. Other precursors of fossil stromatolites are microbial **mat communities**, which occur extensively in intertidal marine environments throughout the world (Figure 1.3A). Photosynthetic microorganisms, including cyanobacteria and other photosynthetic bacteria, are found in distinct layers in living stromatolites (Figure 1.3B–D). The structures formed by these mat communities are flatter and broader than the columnar-shaped classical stromatolites, but they are produced in similar saline, intertidal environments by similar microorganisms. Evidently, during some major geological events, living stromatolites became fossilized and preserved in sedimentary deposits.

Fossil microorganisms have been dated at 3.5 Ga before the present (BP) and therefore are found in some of the earliest sedimentary deposits on Earth. However,

(A)

(B)

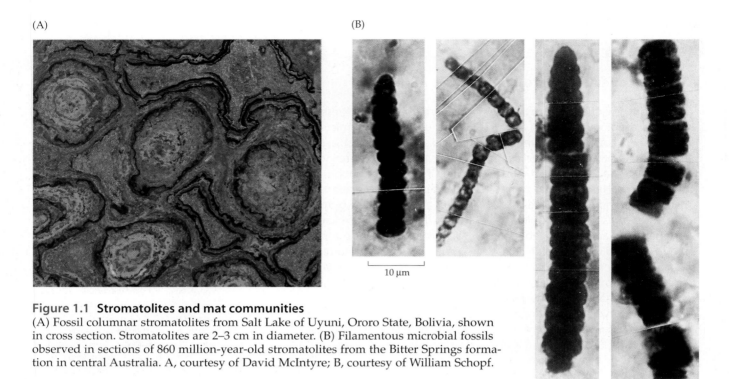

10 μm

Figure 1.1 Stromatolites and mat communities
(A) Fossil columnar stromatolites from Salt Lake of Uyuni, Ororo State, Bolivia, shown in cross section. Stromatolites are 2–3 cm in diameter. (B) Filamentous microbial fossils observed in sections of 860 million-year-old stromatolites from the Bitter Springs formation in central Australia. A, courtesy of David McIntyre; B, courtesy of William Schopf.

special conditions are necessary for the formation and preservation of fossils. Because very early sedimentary rocks were formed so long ago, finding definitive, well-preserved fossils in them is not always possible. Other evidence for early microbial life comes from studies of chemicals left by microbial activities in early sedimentary rocks. These chemicals are found in organic materials, called **kerogen**, deposited in ancient rocks.

The Isua formation in Greenland, which is more than 3.5 Ga old, is one of the oldest sedimentary deposits known. Over the long period following the deposition of organic carbon by microorganisms, the organic matter was altered considerably to form the kerogen. Geochemists who have examined the Isua kerogen note that it has a significantly higher ratio of ^{12}C to ^{13}C than does the associated inorganic carbon from the same strata. This high ratio of ^{12}C to ^{13}C in kerogen is indicative of a biological process that deposited organic material that was eventually transformed to kerogen. This dates the biological process to 3.5 Ga ago.

Figure 1.2 Living columnar stromatolites, Shark Bay, Western Australia
The largest stromatolite shown here is about 1 m in diameter. Courtesy of Beverly Pierson.

(A)

(B)

(C)

Four layers of photosynthetic organisms are visible (from top to bottom): cyanobacteria, two layers of purple sulfur bacteria (of different species), and green sulfur bacteria.

(D)

METRIC 1 2

Multiple years of bacterial buildup are visible. The green surface layer contains living cyanobacteria.

Figure 1.3 Microbial mat communities
(A) This marine intertidal community in Massachusetts, called Sippewissett Marsh, contains a microbial mat community. Some areas are sectioned off by ribbons for research purposes. (B) The mat community just beneath the surface is made visible by cutting through the upper layers of the sand using a razor blade, shown here to provide a size scale. (C) A vertical section of the mat showing the four layers of photosynthetic microorganisms. Each layer is about 1 mm thick. The Sippewissett Marsh mat forms during the summer months; winter storms disrupt it, and a mat re-forms the next summer season. Other mat communities remain stable for many years, such as this one (D) at Laguna Mormona (Laguna Figueroa), Baja California del Norte, Mexico. A–C courtesy of Beverly Pierson; D, courtesy of William Schopf.

How can the occurrence of high concentrations of ^{12}C in the kerogen be attributed to biological activity? Here is the reasoning. Some organisms, called **autotrophs**, use carbon dioxide as a carbon source for growth and from this produce organic cellular material called **biomass**. These organisms selectively use $^{12}CO_2$ in preference to its heavier, stable isotopic form, $^{13}CO_2$, which is also present in the environment. As a result, by a process called **isotopic fractionation**, the biomass becomes enriched in the lighter isotope (^{12}C) leaving behind the heavier isotope in the environment. Determination of the relative amounts of ^{12}C and ^{13}C isotopes in the kerogen and inorganic carbon deposits from a sample from the same stratum can therefore be used to determine whether biological activity is involved in geochemical processes (see Chapter 24).

Therefore, both the fossil and the geochemical evidence suggest that microorganisms originated on Earth within a billion years of its formation. In fact, during the 3 billion years between 3.5 to about 0.5 Ga ago, living mat and stromatolite communities covered vast areas of intertidal zones on the planet and were likely the dominant feature of life on Earth.

Mat communities are still common in intertidal areas, but the columnar stromatolites are much rarer. Presumably the evolution of predatory animals led to the selection of organisms that preyed on the microorganisms in stromatolite communities, and this led to the demise of these microbial communities in many areas on Earth. So, except in special environments such as Shark Bay, with its high salt concentration that is inhibitory to predators, columnar stromatolites have disappeared.

Origin of Life on Earth

Early fossils provide evidence that microbial life existed on Earth within a billion years of its formation, but we have many questions about this early period. How did life originate? What were the first forms of life? What were the conditions on Earth that permitted the origin of life? These are important and intriguing questions, but they cannot be answered by direct observation.

Nonetheless, from what we know of life and the early history of the planet, the process can be partially reconstructed. For example, we know that life cannot exist without liquid water. This means that, at the time life originated, the temperature somewhere on Earth must have been between 0°C and 100°C (at atmospheric pressure). Furthermore, we know that the atmosphere was **anoxic**, that is, without free oxygen gas (O_2). Oxygen could not have formed chemically in any great amount, and certainly it did not make up 20% of the atmosphere as it does today.

Another precondition for the origin of life is the presence of organic compounds. It is inconceivable that cells could have originated de novo in the absence of organic compounds, which are part and parcel of all living organisms and biological processes. Thus, an important question is—can organic compounds such as sugars and amino acids be produced in the absence of organisms, that is, **abiotically**? The first experiments to address this question were conducted by Stanley Miller in 1953. He constructed an apparatus for the interaction of a mixture of gases thought to be present in Earth's early atmosphere. The experimental device mimicked prebiotic conditions (Figure 1.4).

The sterile apparatus contained 500 ml of water, representing an "ocean," and an anoxic gas mixture of methane, hydrogen, and ammonia representing an "atmosphere." The water was boiled, and steam rose into the atmosphere to mix with the gases. A condenser subsequently cooled the gases to produce liquid water, that is, "rain." Miller included as a source of energy a 60,000 volt spark discharge that represented lightning in the atmosphere. The gases and water were recirculated and the anoxic process was run continuously.

In a matter of a few days of operation, Miller's apparatus yielded a dark tarry liquid. This material was analyzed

(A)

(B)

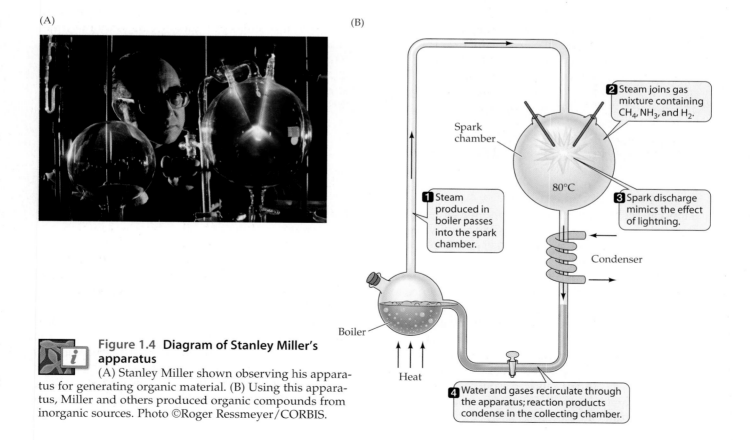

Figure 1.4 Diagram of Stanley Miller's apparatus
(A) Stanley Miller shown observing his apparatus for generating organic material. (B) Using this apparatus, Miller and others produced organic compounds from inorganic sources. Photo ©Roger Ressmeyer/CORBIS.

Spark chamber

2 Steam joins gas mixture containing CH_4, NH_3, and H_2.

1 Steam produced in boiler passes into the spark chamber.

80°C

3 Spark discharge mimics the effect of lightning.

Condenser

Boiler

Heat

4 Water and gases recirculate through the apparatus; reaction products condense in the collecting chamber.

and found to contain, in addition to tarry hydrocarbons, a variety of other organic compounds such as glycine, alanine, lactate, glycolate, acetate, and formate, as well as smaller amounts of other organic compounds. Thus, organic materials were formed under anoxic, abiotic conditions that resembled those found on the early Earth.

However, we now know that the gas mixture used by Miller does not best represent that of the early atmosphere; similar experiments have been conducted by other investigators, who used gas mixtures with compositions more closely resembling those of the atmosphere of early Earth as later understood. These are the gases, called fumarolic gases, that are released from Earth's hot mantle by volcanoes. In addition to the gases and water that Stanley Miller used, the fumarolic gases include large amounts of carbon dioxide, nitrogen, sulfur dioxide, and hydrogen sulfide. Ultraviolet light, which was intense on early Earth, has been successfully used as an alternative to Miller's spark discharge as an energy source. In addition, volcanism was more prevalent on early Earth because the nuclear reactions in its interior core produced more heat than they now do. Therefore, the heat from within Earth's crust would have influenced many of these early reactions. In all of these subsequent experiments in which conditions were anoxic, as they were on early Earth, organic compounds similar to those found by Miller were synthesized.

The overall results of these Miller-type experiments indicate that organic compounds can readily be synthesized from inorganic compounds under conditions that resemble Earth's prebiotic environment. However, we also know that organic compounds are synthesized in intergalactic space. These organic compounds, including amino acids and polycyclic aromatic hydrocarbons, would have been brought to Earth by comets and meteors. Therefore, a large variety of organic compounds would have been present on Precambrian Earth in the so-called **primordial soup**.

We now realize that it is unlikely that life originated and evolved in shallow aquatic habitats, because these habitats would have been continually susceptible to destruction during the period of heavy bombardment. Many scientists now believe that life evolved either in deep sea environments such as hydrothermal vents or in subterranean environments (see Chapters 24 and 25) because these environments were less likely to be disrupted by asteroid impacts. Furthermore, conditions are also favorable for the abiotic synthesis of organic compounds in such environments. Mineral surfaces are thought by some to have played an important role in enabling the evolution of life before cellular organisms arose.

The most difficult questions still remain unanswered. How did the first cell originate? What were its characteristics? Was the first cell a progenitor of all life?

> ### SECTION HIGHLIGHTS
> Chemical and fossil evidence allows scientists to trace the origin of life to 3.5 to 3.8 Ga ago. Although the origin of life remains a puzzle, evidence indicates that organic materials and liquid water, the essential ingredients for life, were present soon after Earth was formed 4.5 Ga ago.

1.2 Tracing Biological Evolution

How can we trace biological evolution? Two approaches have been used. The first is to look at the fossil evidence for microorganisms in sedimentary deposits. This approach, discussed earlier in the chapter, requires the examination of sedimentary rocks for evidence of fossilized microorganisms or their chemical traces, or for evidence of their geochemical activity.

The second approach is to construct an evolutionary tree based on knowledge about current living organisms. This is accomplished by analyzing the sequences of the monomers (subunits) of large molecules called macromolecules, such as deoxyribonucleic acid (DNA), the monomers of which are purines and pyrimidines, or protein, with amino acid monomers. The sequences in these macromolecules provide the necessary information to trace the evolutionary history of organisms, as discussed in greater detail in Chapter 17.

However, before further discussion of the evolution of life, we need to provide some background on the characteristics of organisms that live on Earth today. The first characteristic of all living organisms is that they are composed of one or more cells; this is the **cell theory of life**.

Cell Theory: A Definition of Life

Microorganisms can be placed into five groups on the basis of form and function: bacteria, archaea, fungi, algae, and protozoa (protists). Like plants and animals, all microorganisms consist of one or more cells. Or to put it another way, if something is living, it must be cellular. Viruses, also discussed in this book, are not cellular and therefore they are not regarded as living organisms. Nonetheless, they are important biological agents that develop only as intracellular parasites of organisms, including microorganisms.

The **cell** is the fundamental unit of living organisms and has characteristic functional and structural features. These functions include **metabolism**, the chemical reactions and physical activities by which cells obtain and transform energy and synthesize cell material for

Figure 1.5 Cell structure
This diagram shows the components of a typical cell: cytoplasm, cell membrane, and nuclear area. Some cells also have a cell wall.

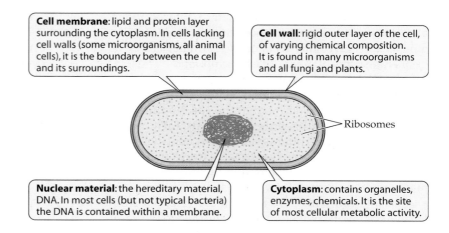

Cell membrane: lipid and protein layer surrounding the cytoplasm. In cells lacking cell walls (some microorganisms, all animal cells), it is the boundary between the cell and its surroundings.

Cell wall: rigid outer layer of the cell, of varying chemical composition. It is found in many microorganisms and all fungi and plants.

Ribosomes

Nuclear material: the hereditary material, DNA. In most cells (but not typical bacteria) the DNA is contained within a membrane.

Cytoplasm: contains organelles, enzymes, chemicals. It is the site of most cellular metabolic activity.

growth. Metabolism is accomplished by biochemical reactions catalyzed by proteins called **enzymes**. The other basic function of cells is **reproduction**, the process by which cells duplicate themselves to produce progeny.

Cells have three major groups of structural components (Figure 1.5):

- The cytoplasm, the aqueous fluid of the cell in which most of the enzymatic and metabolic activities occur. For example, **ribosomes**, small structures responsible for protein synthesis, are located in the cytoplasm.

- A central nuclear area that contains **DNA**, the hereditary material that is duplicated during reproduction.

- A cell membrane, or plasma membrane, the boundary between the cell's cytoplasm and its environment. The cell membrane consists of lipids and proteins.

Many, but not all, microorganisms also contain a fourth layer external to the cell membrane that is referred to as the **cell wall** (see Figure 1.5), a rigid structure that confers shape on the cell. All fungi have cell walls, as do most algae and bacteria. Most protozoa lack cell walls, so their bounding structure is the cell membrane. The chemical composition and structure of the cell walls of microorganisms differ from one group to another. Chapter 4 covers these structures and their functions in greater detail.

Unlike plants and animals, which are all multicellular (containing millions of cells), many microorganisms consist of a single cell and are therefore called **unicellular**. Most but not all bacteria and protozoa are unicellular. Only one group of fungi is unicellular—the yeasts. Plants and animals are macroscopic; they contain many cells organized into tissues and organs—features not found among microorganisms.

The Tree of Life

The macromolecules that have been most useful in tracing evolution are found in the ribosome. Because ribosomes are responsible for protein synthesis in all organisms, microorganisms as well as plants and animals have

TABLE 1.2	Major differentiating characteristics of the three domains of life		
	Bacteria	*Archaea*	*Eukarya*
Nuclear membrane	No	No	Yes
Plastids	No	No	Yes
Peptidoglycan cell walls	Yes[a]	No	No
Membrane lipids	Ester-linked	Ether-linked	Ester-linked
Ribosome size	70S	70S	80S

[a]Three bacterial groups, the chlamydia, planctomycetes, and mycoplasmas, lack cell wall peptidoglycan (the structure of this material is discussed in Chapter 4).

them. The ribosome is a complex structure containing **ribonucleic acid** (**RNA**) and protein (see Chapter 4). Studies of ribosomal RNA (rRNA) molecules indicate that they have changed very slowly during evolution. Because of their highly conserved nature and universal occurrence, rRNA molecules have been used in the study of the evolutionary relatedness among organisms. The 16S and 18S rRNA molecules, or the genes that encode them, are the most commonly used (where S refers to the Svedberg unit, which relates to the mass and density of a molecule).

As a consequence of these studies, three major domains of organisms are now recognized by biologists: *Bacteria*, *Archaea*, and *Eukarya* (Figure 1.6). *Bacteria* contains many of the common microorganisms encountered in typical soil and aquatic environments and includes those that are known to cause disease. *Archaea* comprises a separate group of microorganisms, some of which live in saturated salt environments or high-temperature environments. *Eukarya* contains the microbial groups fungi, algae, and protozoa, as well as plants and animals. The major differentiating characteristics among these organisms are shown in Table 1.2. It is noteworthy that the

three-domain system is the first truly scientific classification of life (see Chapter 17).

Biologists have known for a long time that the cell types of the *Eukarya* are structurally different from those of the *Bacteria* and *Archaea*; the cells of the *Eukarya* are called eukaryotic and those of the *Bacteria* and *Archaea* prokaryotic. In the next section we discuss the differences between eukaryotic and prokaryotic cell structure, or **morphology**, and compare other major features of eukaryotic and prokaryotic microorganisms.

SECTION HIGHLIGHTS

All organisms comprise cells, which carry out metabolism and undergo reproduction. All cells have DNA, RNA, protein, and lipid. Ribosomal RNA, which is found in all cells, has been used to construct the Tree of Life with three domains: *Bacteria*, *Archaea*, and *Eukarya*. All of the branches of the *Bacteria* and *Archaea* and most branches of the *Eukarya* are microbial.

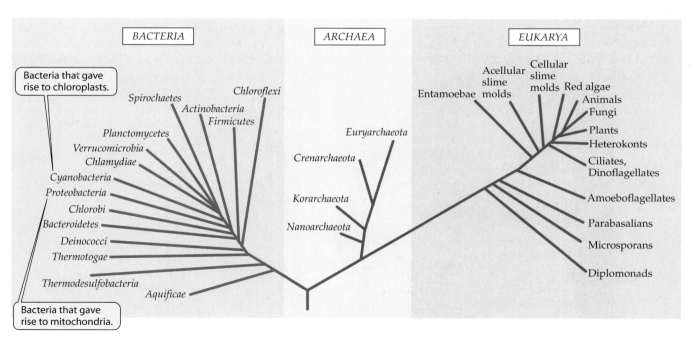

Figure 1.6 Tree of Life

This diagram shows the evolutionary tree of various groups of organisms based on 16S and 18S rRNA gene sequence analysis. The two prokaryotic domains are the *Bacteria* and *Archaea*. All eukaryotic microorganisms are placed in a separate domain, the *Eukarya*, along with the plant and animal "kingdoms." The *Eukarya* contains many "kingdoms" of microorganisms, including the fungi and various protists. The *Bacteria* and *Archaea* also contain many "kingdoms," which in this book we call phyla. The *Bacteria* alone have more than 40 phyla, many of which have never been grown and studied in the laboratory and, therefore, are not shown in this Tree of Life.

(A) Prokaryotes

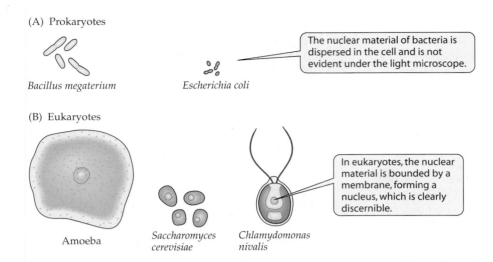

Bacillus megaterium

Escherichia coli

The nuclear material of bacteria is dispersed in the cell and is not evident under the light microscope.

(B) Eukaryotes

Amoeba

Saccharomyces cerevisiae

Chlamydomonas nivalis

In eukaryotes, the nuclear material is bounded by a membrane, forming a nucleus, which is clearly discernible.

Figure 1.7 Drawings of representative microorganisms, as they appear by light microscopy The two examples of bacteria are a large rod, *Bacillus megaterium*, and a small rod, *Escherichia coli*. The eukaryotic organisms are an amoeba (a protozoan), a yeast (*Saccharomyces cerevisiae*), and an alga (*Chlamydomonas nivalis*). Note the cup-shaped, green chloroplast and the two flagella of *C. nivalis*.

1.3 Prokaryotic versus Eukaryotic Microorganisms

When examined under the light microscope, bacteria appear different from eukaryotic microorganisms (**Figure 1.7**). Bacterial and archaeal cells are usually very small and have no apparent nucleus. In contrast, cells of algae, protozoa, fungi, plants, and animals are typically much larger and have a distinct nucleus.

These differences noted by observations with the light microscope are borne out by more detailed examination using the electron microscope. Microorganisms can be sliced into very thin sections (less than 100 nm) and examined at high magnification with the transmission electron microscope (TEM) (see Chapter 4). When viewed in this manner, the bacteria and archaea (**Figure 1.8**) exhibit striking structural differences in comparison to other microorganisms (**Figure 1.9**). Bacterial

(A)

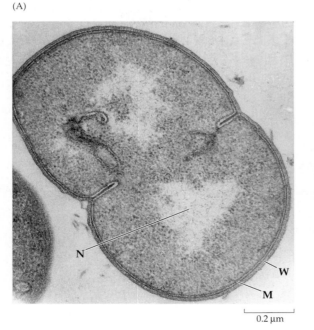

N

W

M

0.2 µm

(B)

0.1 µm

Figure 1.8 Cross section of bacterial and archaeal cells
(A) This electron micrograph of a thin section of a dividing cell of a bacterium in the genus *Sporosarcina ureae* shows the cell wall (W), cell membrane (M), and nuclear material (N), which appears as fibrous matter dispersed in the cytoplasm.

(B) A thin section of an archaeal cell of *Methanococcus voltae*. Note that the internal structure of both of these cells is similar, hence the reason for terming them prokaryotic. Photos ©T. J. Beveridge/Visuals Unlimited.

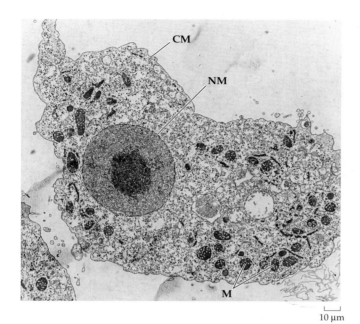

Figure 1.9 Cross section of a eukaryotic cell
A protozoan of the genus *Acanthamoeba,* showing the cell membrane (CM), nuclear membrane (NM), and mitochondria (M). Courtesy of T. Fritsche.

and archaeal cells have a much simpler structure and are referred to as **prokaryotic** (from the Greek meaning "before nucleus") organisms. In contrast, algae, fungi, and protozoa are called **eukaryotic** ("true nucleus"). Table 1.3 lists the major differences between these two basic types of cellular organization. As the terms imply, the single major difference between these two cell types is related to their nuclear material (Box 1.2). The nucleus of the cell of a eukaryotic microorganism (as well as plants and animals) is bounded by a membrane referred to as a **nuclear envelope** or **nuclear membrane** (see Figure 1.9). In prokaryotic organisms

the nuclear material, which appears as a central fibrous mass in thin sections, is not bounded by a membrane. Instead it is in direct contact with the cytoplasm (see Figure 1.8). There is one exception in the *Bacteria* in which some members of the phylum *Planctomycetes* have a membrane-bounded nucleoid.

Other differences exist between prokaryotic and eukaryotic cells, some structural, and others genetic and physiological. The nuclear material of prokaryotes typically consists of a single type of DNA molecule, called a **chromosome**. More than one copy of it may be present, depending on how fast the organism is growing.

TABLE 1.3	Major differentiating characteristics of prokaryotes and eukaryotes	
Characteristic	**Prokaryote**	**Eukaryote**
Nuclear structure and function		
Nucleus with membrane	No	Yes
Chromosomes	One (haploid)	Two or more (diploid)
Mitosis	No	Yes
Sexual reproduction	Rare; only part of genome involved	Common; all chromosomes involved
Meiosis	No	Yes
Cytoplasmic structures		
Mitochondria	No	Yes[a]
Chloroplasts	No	Yes (if photosynthetic)
Ribosomes	70S	80S[b]
Typical cell volume	$<5\ \mu m^3$	$>5\ \mu m^3$

[a]A few lack mitochondria.
[b]Some rare, primitive eukaryotic microorganisms have 70S ribosomes.

BOX 1.2 *Milestones*

Separating the Organisms of Earth into Two Categories on the Basis of Cell Structure and the Continuing Controversy of the Term "Prokaryote"

Although his views were largely ignored in the 1930s, the French biologist E. Chatton noted the differences in cellular structure between "higher" and "lower" forms of life. He coined the terms "eukaryotic" and "prokaryotic" based on his light microscopic observations of the differences between the cells of plants and animals and bacteria (see Figure 1.7). Only after the invention of the electron microscope (in the late 1930s) and the subsequent development of appropriate procedures to thin-section organisms (1950s to 1960s) did other biologists confirm the fine structural differences between these two types of cellular organization. In addition to these morphological features, a number of other differences were also discovered that permitted a clear distinction between these two types of cells. The major features that distinguish prokaryotic from eukaryotic cells were eloquently stated in an important publication by Roger Stanier and C. B. van Niel in 1962.

However, even today controversy persists about the use of the term "prokaryote" to refer to both the *Bacteria* and *Archaea*. Most microbiologists agree that the electron microscopic evidence of cell structure supports grouping these two separate domains into a single group. However, a few prominent microbiologists disdain the use of the term, as they believe that it supports the views of other biologists that the *Bacteria* and *Archaea* should be regarded as a single domain or empire of life.

Consequently, rapidly growing cells might have two or four copies of the DNA molecule, but all copies are identical. Some prokaryotic organisms also have nonchromosomal DNA called **plasmids** in their cells, discussed in greater detail in Chapter 15. Plasmids are smaller than the chromosome but often contain genes that are significant to the bacterium.

In contrast, the nuclei of eukaryotic organisms contain several to many nonidentical chromosomes, each with its own genetic material. Thus, bacterial and archaeal cells can be regarded as typically having a single chromosome and eukaryotic microorganisms as having more than one chromosome. To ensure orderly, accurate, and precise delivery of their multiple chromosomes during the process of cell division, eukaryotic organisms undergo **mitosis**. In this process, each chromosome replicates and aligns along the division axis of the cell before asexual cell division occurs (Figure 1.10). This elaborate physiological and morphological orchestration does not occur in prokaryotes.

Other Morphological Differences

Ribosomes appear as granules (approximately 5 nm in diameter) in the cytoplasm. Prokaryotic ribosomes are called 70S ribosomes. Eukaryotes, with rare exceptions in some protozoa, have slightly larger 80S ribosomes (see Chapter 4). Molecular differences in the RNA and proteins of ribosomes account for the differences in size.

One of the striking features of eukaryotes is the membrane-bounded organelles in their cytoplasm. There are several types of organelles. All are distinct compartments surrounded by one or more membranes, which, like the cell membrane, contain both protein and lipid. The most common organelle of this type, found in almost all eukaryotic cells, is the mitochondrion (Figure 1.11). The **mitochondrion** is the site of respiratory activity in eukaryotes. Mitochondria have their own internal DNA, cytoplasm, and ribosomes. One exciting fact of cell biology is that the DNA of the mitochondrion is similar to typical prokaryotic DNA, that is, it is a circular molecule that is not bounded by a nuclear membrane. Furthermore, the ribosomes of the mitochondrion are 70S, like those of prokaryotes. These features and other lines of evidence (see "Evolution of Eukaryotes" on page 21) suggest that the mitochondrion evolved from a bacterium that developed a close interdependence or **symbiotic association** with another organism more than 1 Ga ago. Indeed, based on the phylogenetic evidence from the 16S rRNA of the mitochondrion, it has descended from a member of the alpha group of the phylum *Proteobacteria*.

The chloroplast is the site of photosynthesis in eukaryotes. Like the mitochondrion, the **chloroplast** is a membrane-bounded organelle found in the cytoplasm. It also resembles the mitochondrion in that its DNA has no nuclear envelope and its ribosomes are 70S. However, unlike mitochondria, it also has internal membranes con-

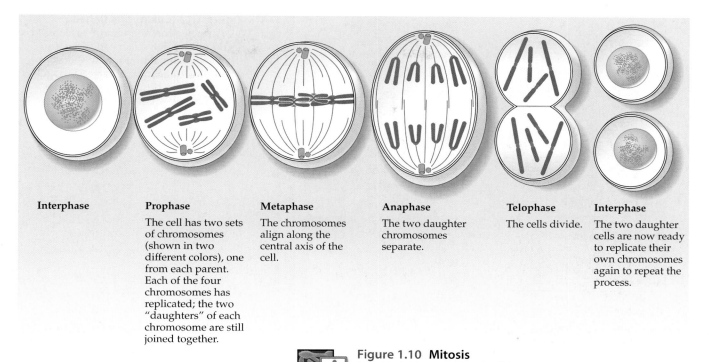

Interphase

Prophase

The cell has two sets of chromosomes (shown in two different colors), one from each parent. Each of the four chromosomes has replicated; the two "daughters" of each chromosome are still joined together.

Metaphase

The chromosomes align along the central axis of the cell.

Anaphase

The two daughter chromosomes separate.

Telophase

The cells divide.

Interphase

The two daughter cells are now ready to replicate their own chromosomes again to repeat the process.

Figure 1.10 Mitosis
Mitosis is the process of cell division for eukaryotic cells. In mitosis, a dividing eukaryotic cell duplicates its chromosomes and distributes one copy to each of the newly forming daughter cells. This particular cell has two sets of chromosomes.

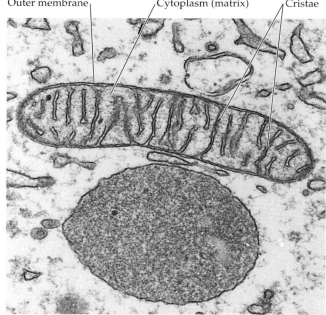

Outer membrane Cytoplasm (matrix) Cristae

Figure 1.11 Mitochondrion
Electron micrograph of a mitochondrion from a eukaryotic microorganism, showing the outer membrane, the inner membrane folded into cristae, and the enclosed cytoplasm. ©Barry F. King/Biological Photo Service.

taining the chlorophyll pigments involved in photosynthesis. Chloroplasts are found in algal and plant cells. They, too, are thought to be derived through an evolutionary process from a prokaryotic organism, in this case an organism from the photosynthetic group called the cyanobacteria (see Chapter 21).

The organelles of motility of eukaryotic cells—the flagellum and cilium—are larger and more complex than the flagellum of prokaryotes. A cross-section of the eukaryotic flagellum reveals an elaborate fibrillar system called the "9 + 2" arrangement, with fibrils called microtubules forming niñe outer doublets and an inner pair (Figure 1.12). In contrast, the prokaryotic flagellum has a single fibril when viewed in cross-section; the thread is of such a fine diameter that a single flagellum cannot be seen by light microscopy. Eukaryotic flagella and cilia, in contrast, are readily observed with the light microscope (see the alga in Figure 1.7).

Reproductive Differences

All prokaryotes reproduce by asexual cell division. Cells simply enlarge in size, replicate their DNA (i.e., produce a second identical copy of their DNA), and divide to

(A)

(B)

Inner microtubules

Outer microtubules

Plasma membrane

The arrangement of two central microtubules surrounded by nine pairs of microtubules (9 + 2) is characteristic of eukaryotic cilia and flagella.

Figure 1.12 Eukaryotic flagella
(A) This electron micrograph is a cross section through a eukaryotic flagellum. (B) A schematic representation of a eukaryotic flagellum. A bacterial flagellum has a very different structure: its single fibril is smaller than one of the microtubules shown here. Photo ©W. L. Dentler/Biological Photo Service.

form two new cells, each containing a copy of the DNA molecule (Figure 1.13). Therefore, prokaryotes have only one copy of DNA and are called **haploid**. Sexual reproduction is relatively rare in prokaryotes. Although many bacteria are able to exchange genetic material between mating types, this is not known to be a universal characteristic. As discussed in subsequent text (see Chapter

15), this rarely results in the formation of a **diploid** cell, with one copy of the DNA molecule from each of the mating cells. A diploid cell has two copies of each chromosome, that is, two copies of each DNA molecule.

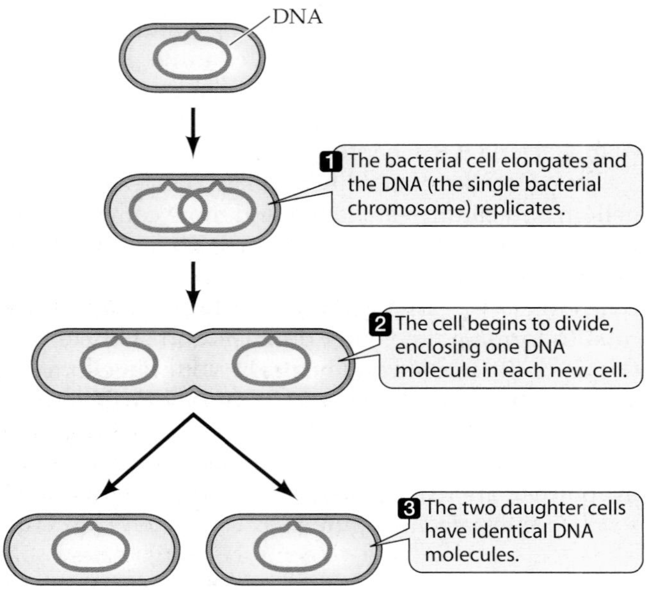

In contrast to prokaryotes, most eukaryotes exist as diploid organisms or have diploid stages in their life cycles. Thus, their cells have two sets of chromosomes: one set from the "male" and another set from the "female" mating types. For example, human body cells have 46 chromosomes. These exist as 23 paired chromosomes. Half, or one set of 23, is derived from the father and the other 23 from the mother. To generate reproductive cells, the number of chromosomes and amount of DNA are reduced by half. **Meiosis** is the process whereby, for example, the 46 human chromosomes are reduced to 23 in preparation for sexual reproduction. Meiosis results in the formation of haploid mating cells called **gametes**: the sperm and the ovum, produced by male and female mating types, respectively (Figure 1.14).

During sexual reproduction the gametes fuse during fertilization to form a diploid **zygote** (e.g., the fertilized egg). Therefore, the zygote contains a full genetic complement from each of the parental mating cells. Sexual reproduction is very common among eukaryotic organisms. Except for haploid gametes, the cells of most higher eukaryotes are diploid. Because prokaryotes contain only a single copy of each gene, genetic studies are much simplified. There are no dominant and recessive characteristics, which means that any genetic change is expressed fully and immediately in progeny cells. In contrast, mutations in eukaryotic cells may not show up in the next generation, because the diploid cells have two copies of each gene. Thus, prokaryotes are model organisms for the study of genetics.

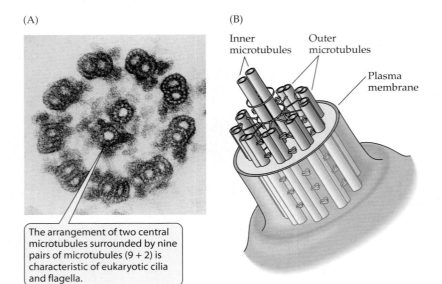

DNA

1 The bacterial cell elongates and the DNA (the single bacterial chromosome) replicates.

2 The cell begins to divide, enclosing one DNA molecule in each new cell.

3 The two daughter cells have identical DNA molecules.

Figure 1.13 Prokaryotic cell division
Although this process is analogous to mitosis in eukaryotic organisms (compare with Figure 1.10), mitosis involves complex structural features that are absent in bacteria.

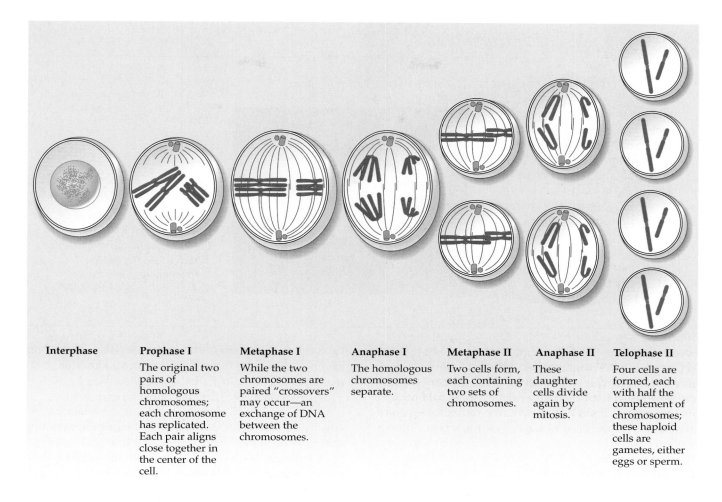

Interphase	Prophase I	Metaphase I	Anaphase I	Metaphase II	Anaphase II	Telophase II
	The original two pairs of homologous chromosomes; each chromosome has replicated. Each pair aligns close together in the center of the cell.	While the two chromosomes are paired "crossovers" may occur—an exchange of DNA between the chromosomes.	The homologous chromosomes separate.	Two cells form, each containing two sets of chromosomes.	These daughter cells divide again by mitosis.	Four cells are formed, each with half the complement of chromosomes; these haploid cells are gametes, either eggs or sperm.

Figure 1.14 Meiosis
Like mitosis, meiosis is a process that occurs only in eukaryotic organisms. Meiosis occurs in organisms that undergo sexual reproduction. A diploid cell undergoes two rounds of division to form four haploid cells, the gametes. In this example, two pairs of chromosomes are shown.

Cell Size: Volume and Surface Area

As mentioned previously, cell size is an important characteristic for an organism. Most eukaryotic organisms have larger cells than prokaryotic organisms (see Figure 1.7)—but there are some exceptions. For example, although typical bacterial cells range in diameter from 0.5 to 1.0 µm, some wider than 50 µm have been reported (Box 1.3). Likewise, some eukaryotes, such as *Ostreococcus tauri*, a small alga of the marine planktonic community, are only 1.0-µm wide but still contain mitochondria and chloroplasts. However, the cells of typical eukaryotes range in diameter from 5 to 20 µm, with most about 20 µm, although some species have larger cells. Specialized cells in multicellular organisms can be much larger. A human neuron can be as long as 1 m. Although the size ranges of eukaryotes and prokaryotes

overlap, a typical eukaryote is larger than a prokaryote by a factor of 10.

Many bacteria grow and reproduce at very rapid rates. Some can double in size or in number of cells in less than 10 minutes under optimal growth conditions. This implies that metabolic processes can be extremely rapid in these organisms. The rapid metabolic rate is due in part to the small size of bacteria. Their small size ensures that all the cytoplasm is in proximity to the surrounding environment from which bacteria derive their nutrients. The greatest distance between the cytoplasm and the growth environment is only 0.5 µm in a bacterium with a diameter of 1.0 µm, whereas it is 10.0 µm in a typical eukaryotic organism with a diameter of 20.0 µm.

Another way to consider the close spatial relationship between the cytoplasm of a cell and its environment is

BOX 1.3 *Research Highlights*

You Can't Tell a Bacterium by Its Size Alone!

Although most bacteria are very small, some are amazingly large. The largest bacterium we know of is *Thiomargarita*. Individual cells of this bacterium can be seen by the naked eye. The bacterium lives in the intertidal area off the coast of Namibia, in southwest Africa. Although it has not yet been isolated in pure culture, *Thiomargarita* is known to be a sulfur bacterium that lives by the oxidation of reduced sulfur compounds.

Three cells of *Thiomargarita*

A chain of spherical cells of *Thiomargarita* is lying next to a fruit fly, indicating their huge (for bacteria) size. Reprinted with permission from *Science*, Vol. 284, pp. 493–495 ©1999 AAAS.

6 mm

to calculate the ratio of a cell's surface area to its volume (**Figure 1.15**). Let's assume that a bacterial cell is cubical, with sides 1.0 µm in length (actually, one extreme salt-loving bacterium is a cube!); its surface area is 6.0 µm^2 and its volume is 1.0 µm^3. Therefore, the ratio of its surface area to its volume (SA/V) is 6.0. In comparison, a hypothetical eukaryotic microorganism of the same shape with 10.0 µm sides has an SA/V of 0.6. This smaller value for the eukaryote indicates that it has a tenfold greater amount of cytoplasm per unit of cell membrane surface than does the smaller prokaryote. Given that nutrients for growth must enter the cell by crossing the cell membrane, more nutrients are available per unit of cytoplasm in the

prokaryote (6.0) than in the eukaryote (0.6). The larger SA/V ratio enables faster metabolism and growth.

Microbial Nutrition

Algae and several groups of *Bacteria* are **photosynthetic**, that is, like plants, they obtain their energy from light (e.g., sunlight). Also, like plants, they use carbon dioxide as their principal source of carbon for growth. This type of nutrition, which is based entirely on inorganic compounds, is referred to as **autotrophic** (self-nourishing or self-feeding). Algae are therefore called **photoautotrophic** to indicate that they obtain their energy from sunlight and their carbon from carbon dioxide.

In contrast to algae, fungi and many prokaryotes obtain their energy directly from chemical compounds, not sunlight. This type of nutrition is referred to as **chemotrophic** (chemical feeding). Fungi require organic chemical compounds as their sources of energy and carbon. Such nutrition is termed **heterotrophic** ("other" or "different" feeding, as distinct from autotrophic) or **organotrophic** (organic compound-feeding). Therefore, all fungi are **chemoheterotrophic**—they use chemical compounds as energy sources and organic compounds as carbon sources. Only dissolved organic carbon sources can pass through the cell walls of fungal cells. Consequently, fungi are well known for their ability to use simple sugars and other dissolved substances as carbon sources. Some fungi can also degrade particulate organic materials, such as cellulose, by excreting enzymes that solubilize the organic material outside the cell; they then transport the dissolved compounds into the cell.

Like fungi, protozoa are chemoheterotrophic organisms, using organic compounds as sources of carbon and

	1.0 µm	10.0 µm
Surface area (SA)	6.0 µm^2	600 µm^2
Volume (V)	1.0 µm^3	1,000 µm^3
SA/V	6	0.6

Figure 1.15 Surface area and volume
Hypothetical cubical cells showing how the ratio of surface area to volume (SA/V) varies with cell size. The larger cell has a much smaller SA/V ratio. The text explains the implications of this difference.

energy for growth. Typical protozoa, which lack cell walls, engulf bacteria and other microorganisms in much the same way that higher animals eat food. The protozoan's source of food is particulate organic material; this type of feeding is called **phagotrophic**.

Bacteria and *Archaea* are exceedingly diverse in their nutritional capabilities. Some of the *Bacteria* are similar to algae in being **photoautotrophic**. Other *Bacteria* are **photoheterotrophic**, that is, they can obtain energy from light (sunlight) but use organic compounds as carbon sources. It is noteworthy that none of the *Archaea* are photoautotrophic. Most species of *Bacteria* and *Archaea* are **chemoheterotrophic**, deriving both energy and carbon from organic compounds. One especially interesting type of nutrition found only in some members of *Bacteria* and *Archaea* is termed **chemoautotrophic**. Chemoautotrophs, like photoautotrophs mentioned previously, obtain energy by the oxidation of inorganic compounds, such as ammonia or hydrogen sulfide, and use carbon dioxide as their principal carbon source. This nutritional category is unique to these prokaryotic microorganisms. The five principal groups of microbes and their types of nutrition are shown in Table 1.4. Microbial nutrition is discussed in more detail in Chapter 5.

With this background in microbiology, we are ready to address more specifically the early evolution of organisms.

SECTION HIGHLIGHTS

Bacteria and *Archaea* have a simple cell structure referred to as the prokaryotic cell. *Eukarya* have compartmentalized cells that contain organelles, such as mitochondria and chloroplasts, not found in prokaryotic cells. Microorganisms are remarkably diverse in terms of what they can use as carbon and energy sources for growth.

1.4 Microbial Evolution and Biogeochemical Cycles

Although we do not know which organisms were the first biological entities on Earth, various theories have been presented. Most scientists believe that the first forms of life were anaerobic (living in the absence of O_2), based on evidence that early Earth was anoxic. Many also believe that the earliest bacteria were hyperthermophilic (heat-loving), living in high-temperature environments such as hydrothermal systems or deep within Earth's crust where it is hot. Evidence supporting this hypothesis is that the earliest branches in the Tree of Life contain thermophilic *Bacteria* and *Archaea* (see Figure 1.6; see also Chapter 17). However, all theories on the origin of life and the first microorganisms are speculative and will not be addressed in great detail here.

Possible Early Metabolic Types

The Russian evolutionist A. I. Oparin argued that the initial metabolic type was likely a simple heterotrophic bacterium. He reasoned that autotrophic organisms are inherently more complex, so they would not have evolved first. As he noted, although autotrophs can live on simple nutrients, they are more complex in that they need not only metabolic pathways for the generation of energy, but also additional pathways to carry out carbon dioxide fixation, that is, the conversion of CO_2 into organic material. In contrast, simple fermentative heterotrophs require only a few enzymes for energy generation, and they could have lived on the organic compounds formed abiotically early in Earth's history.

Others have argued that early life forms might have been hydrogen bacteria, those that obtain energy from the oxidation of hydrogen gas. Both bacterial and archaeal hydrogen users are known. These organisms have simple nutritional requirements. Some grow autotrophically, generating energy from the oxidation of hydrogen gas and using carbon dioxide as a sole source of car-

---

TABLE 1.4	Principal groups of microorganisms

Microbial Group	Number of Cells per Organism	Cell Walls	Nutritional Type
Eukaryotes			
Algae	Usually one, some filamentous	Yes	Photoautotrophic
Protozoa	One	No	Chemoheterotrophic
Fungi	Filamentous, except yeasts (unicellular)	Yes	Chemoheterotrophic
Prokaryotes *Bacteria* and *Archaea*	Usually one, some multicellular	Yes[a]	Photoautotrophic, photoheterotrophic, chemoautotrophic, or chemoheterotrophic

[a]A few bacteria, namely, the mycoplasmas and thermoplasmas, lack cell walls.

bon (see Chapter 8). Furthermore, many are anaerobic and could have existed in an anoxic environment like that of early Earth.

Although photosynthetic bacteria may not have been the first organisms, it is believed that they evolved early. Several groups of the *Bacteria* carry out photosynthesis using chlorophyll-type compounds, whereas this feature is unknown among the *Archaea*. The first types of photosynthesis may have resembled the metabolisms found in photosynthetic *Proteobacteria* or *Chlorobi*, two of the major lineages of *Bacteria*. These bacteria carry out photosynthesis anaerobically using hydrogen sulfide or elemental sulfur for carbon dioxide fixation (see Chapters 9 and 21):

$$(1) \ CO_2 + H_2S \rightarrow (CH_2O)_n + S^0$$
$$(2) \ CO_2 + S^0 \rightarrow (CH_2O)_n + SO_4^{2-}$$

where $(CH_2O)_n$ represents organic material (equations are not balanced). This is an example of photoautotrophic metabolism mentioned previously. It is easy to see that in equation (1), H_2S acts as a hydrogen donor for CO_2 fixation (i.e., CO_2 reduction to nongaseous organic carbon).

The volcanism commonplace on early Earth would have been ideal for these organisms, because it provided abundant quantities of carbon dioxide and hydrogen sulfide, the essential "ingredients" for their photosynthesis. This type of photosynthesis is termed **anoxygenic photosynthesis** because it proceeds in an anoxic environment without the production of oxygen gas. Interestingly, some of the photosynthetic *Proteobacteria* can use reduced iron (Fe^{2+}) in place of hydrogen sulfide in carrying out anoxygenic photosynthesis. Like hydrogen sulfide, reduced iron would have been abundant in the early anoxic environments of Earth. These photosynthetic bacteria were likely important in producing the rock layers known as **banded iron formations** because they were first produced when conditions on Earth were still anoxic about 2.5 to 3.0 Ga ago. The bands of these formations are alternating millimeter-thick layers of quartz and iron oxides, partially oxidized forms of iron (FeO and Fe_2O_3) that can be produced by the oxidation of Fe^{2+} by the photosynthesis carried out by this group of the *Proteobacteria* (see Chapter 21). Banded iron formations are not being produced today because the concentration of oxygen in the atmosphere and in the oceans is too high. Instead, more recent iron deposits are called **red beds**, so named because they contain iron oxides known as hematites (Fe_3O_4), a more highly oxidized form of iron that gives them their red color.

Anoxic photosynthesis was likely very important early in Earth's history. Most photosynthesis that occurs today, however, generates oxygen. Cyanobacteria are the only prokaryotic organisms that can carry out this oxygenic (oxygen-producing) photosynthesis, and their first appearance was a major event in the history of life on Earth.

Cyanobacteria and the Production of Oxygen

To understand the importance of cyanobacteria in the production of oxygen, we must first review their metabolism. The metabolism of the cyanobacteria is similar to that of the anoxygenic photosynthetic bacteria (see reaction [1] above). However, there is one major difference: cyanobacteria use water in place of hydrogen sulfide as the hydrogen donor. Therefore, the overall equation for cyanobacterial photosynthesis is:

$$(3) \ CO_2 + H_2O \rightarrow (CH_2O)_n + O_2$$

This process is termed **oxygenic photosynthesis,** because oxygen is produced.

C. B. van Niel, a Dutch-born American microbiologist who studied photosynthetic bacteria, noted that the O_2 produced in reaction (3) must be derived from the water molecule rather than from the carbon dioxide. He concluded this based upon analogy to reaction (1) for anoxygenic photosynthesis in which elemental sulfur (S^0) is produced from H_2S. His hypothesis was confirmed when scientists used radiolabeled water, $H_2^{18}O$, to show that the label ended up in the oxygen produced ($^{18}O_2$), whereas the label from $C^{18}O_2$ did not. Thus, the oxygen comes from a reaction referred to as the "water-splitting" reaction. This reaction, which is the key reaction of oxygenic photosynthesis, is found in all cyanobacteria, algae, and plants.

The cyanobacteria evolved about 2.5 to 3.0 Ga ago, when Earth's atmosphere still lacked O_2. Initially, when the cyanobacteria first began producing oxygen, its concentration in the atmosphere would have remained very low. This is because oxygen is highly reactive chemically and would have combined with the large amounts of highly reduced compounds that existed on Earth at the time. These reduced compounds, such as ferrous iron and sulfides, would have reacted with the free oxygen, preventing it from accumulating rapidly in the atmosphere. For this reason the cyanobacteria may have also played a major role in the formation of the banded iron deposits because they began to produce oxygen that could have resulted in the oxidation of reduced iron and the resultant formation of some of these deposits. Gradually, the oxygen concentration in the atmosphere increased over the past 2 to 3 billion years to reach its present level of about 20% of atmospheric gases.

One of the recent exciting discoveries about cyanobacteria is that some of them can also carry out anoxygenic photosynthesis, as in reaction (1). This finding suggests that the cyanobacteria may have evolved from anoxygenic photosynthetic bacteria similar to purple or green sulfur bacteria. Indeed, evidence supporting this comes from molecular phylogenetic studies of the two photosystems of photosynthesis (see Chapter 9), one of which is thought to be derived from a member of the *Chlorobi* and the other from a member of the *Firmicutes*, a photosynthetic gram-positive group (see Chapter 21). However, some variant must have evolved that could use water in place of hydrogen sulfide as a reductant in photosynthesis and therefore could split water and carry out oxygenic photosynthesis, as in reaction (3). This important process may have evolved by natural selection when hydrogen sulfide became scarce and water for photosynthesis was abundant.

The oxygen produced by cyanobacteria would have been toxic to early life forms. Fortunately, for the reasons noted previously, free oxygen would not have been available in the atmosphere for many millions of years after the first oxygenic cyanobacterium began producing it. This lengthy period provided favorable conditions for the selection and evolution of enzymes such as peroxidases, which would protect sensitive bacteria from the oxidizing effects.

Impact of Bacteria *and* Archaea *on Biogeochemical Cycles*

Bacterial and archaeal groups carry out significant reactions, called **biogeochemical reactions**, that are crucial to the operation of Earth's biosphere. Examples of these reactions occur in the great cycles of elements such as the carbon cycle, in which autotrophic organisms fix carbon dioxide to form organic carbon that is recycled back to CO_2 by heterotrophic organisms. The ability of microorganisms to degrade all the organic cellular material of all organisms and thereby enable it to be recycled is one of C. B. Van Niel's postulates (see Chapter 12). These reactions are discussed in greater detail elsewhere in the book, so we present just a brief summary here.

Because of the early evolution of microorganisms, particularly *Bacteria* and *Archaea*, they were provided with many energy sources 3 billion years before plants and animals evolved. As a result of this long period of evolution, microorganisms have diversified into many different metabolic groups that uniquely use unusual growth substrates such as methane, ammonia, hydrogen gas, sulfur, and reduced iron. In addition, several different groups of the *Bacteria* carry out photosynthesis using light energy.

All of the biological transformations of the nitrogen cycle can be carried out by microorganisms. Likewise, all of the biological transformations of the carbon and sulfur cycles can be carried out by microorganisms. Indeed, these cycles, as we know them today, were in place about 1.5 to 2 Ga ago, at least 1 Ga before plants and animals evolved. The *Bacteria* and *Archaea* continue to carry out unique steps in these cycles—such as nitrogen fixation—that eukaryotic microorganisms, plants, and animals cannot perform. Thus, humans and other animals, as well as plants and eukaryotic microorganisms, absolutely depend on the biogeochemical services performed by the *Bacteria* and *Archaea*. More information on the biogeochemical cycles and the important roles of microorganisms in them are provided in Chapter 24.

SECTION HIGHLIGHTS

The early evolution of microorganisms means that they produced the first biosphere. Primary production, the fixation of carbon dioxide into organic material, likely began with chemosynthesis followed by photosynthesis. Because of their diverse metabolic capabilities, most of the reactions of the biogeochemical cycles of nature are primarily carried out by members of the *Bacteria* and *Archaea*.

1.5 Evolution of Eukaryotes

The origin of eukaryotic organisms is obscure at this time. However, recent evidence from Roger Buick's lab traces eukaryotes to at least 2.45 Ga ago based upon biomarkers from ancient sedimentary rocks in Australia.

The most popular hypotheses about eukaryotic evolution stem from the ideas of scientists such as Lynn Margulis and Wolfram Zillig. Zillig has proposed that eukaryotic organisms, the *Eukarya*, evolved through a fusion event between an ancestor of the *Bacteria* and an ancestor of the *Archaea* (Figure 1.16). Another scientist, Carl Woese, has proposed that an ancestral form of life, the progenote, had a permeable cell membrane (possibly protein) that allowed genetic communication with other species. As evolution proceeded, some event, perhaps physical isolation, led to the separation of two different populations—the Prebacteria and the Prearchaea. During this period of separation, the two groups of organisms developed different metabolic patterns and genetic systems. Also at this time, the organisms developed their own characteristic lipid cell membranes: ester-linked fatty acid lipids for the Prebacteria and ether-linked lipids for the Prearchaea. According to Zillig, it was at about this time that the fusion event occurred between these two cell types, giving rise to the eukaryotic cell.

Figure 1.16 Evolution of the main lines of descent
This diagram illustrates the possible origin of prokaryotic and eukaryotic organisms from a progenote. The early progenotes may have been precellular RNA forms without organic cell membranes.

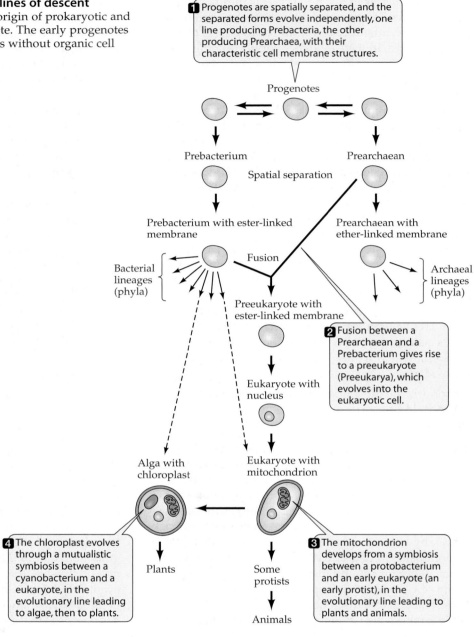

1 Progenotes are spatially separated, and the separated forms evolve independently, one line producing Prebacteria, the other producing Prearchaea, with their characteristic cell membrane structures.

Progenotes

Prebacterium   Prearchaean

Spatial separation

Prebacterium with ester-linked membrane   Prearchaean with ether-linked membrane

Bacterial lineages (phyla)   Fusion   Archaeal lineages (phyla)

Preeukaryote with ester-linked membrane

2 Fusion between a Prearchaean and a Prebacterium gives rise to a preeukaryote (Preeukarya), which evolves into the eukaryotic cell.

Eukaryote with nucleus

Alga with chloroplast   Eukaryote with mitochondrion

4 The chloroplast evolves through a mutualistic symbiosis between a cyanobacterium and a eukaryote, in the evolutionary line leading to algae, then to plants.

Plants   Some protists

3 The mitochondrion develops from a symbiosis between a protobacterium and an early eukaryote (an early protist), in the evolutionary line leading to plants and animals.

Animals

Similarities between *Eukarya* and the *Bacteria* and *Archaea* are cited as evidence supporting this hypothesis. For example, the cell membranes of *Eukarya* and *Bacteria* contain fatty acids linked to glycerol by ester linkages. In contrast, *Archaea* do not produce long-chain fatty acids and they use ether linkages in their isopranyl cell membranes (see Chapters 4 and 18). This evidence suggests that if a fusion event occurred, it occurred when the bacterial ancestor had already developed its cell membrane and this feature was thus incorporated into the *Eukarya*.

The fusion event hypothesis is also consistent with the observation that *Eukarya* genomes possess a mixture of *Archaea*-like and *Bacteria*-like genes. Some scientists have hypothesized a metabolic basis for fusion events, which may have been preceded by a close symbiotic association between a member of the *Bacteria* and *Archaea* in which the hydrogen-producing bacterium and hydrogen-utilizing archaeon produced methane.

Although much of this is hypothetical, it is interesting to note that as scientists further dissect organisms at the molecular level, they are beginning to infer likely, if not actual, evolutionary events that occurred billions of years ago.

Impact of Oxygen on the Evolution of Plants and Animals

The evolution of oxygenic photosynthetic organisms had a profound impact, not only on the chemistry of Earth, but on the evolution of animal and plant life. Virtually all plants and animals use oxygen and carry out aerobic respiration, a process that occurs in the mitochondrion. Mitochondria are not found in *Bacteria* or *Archaea*. However, the mitochondrion is bacterial, by which we mean:

- A mitochondrion has its own DNA, but its DNA is not bounded by a nuclear membrane.

- A mitochondrion has ribosomes, which are not the 80S ribosomes of eukaryotes but the smaller, 70S ribosomes of prokaryotes.

- The size and structure of a mitochondrion are reminiscent of a gram-negative bacterium without a cell wall.

- Sequence analyses of mitochondrial 16S rRNA reveal that, regardless of the eukaryotic source, mitochondria are descended from the *Proteobacteria* (see Figure 1.6).

Similarly, the chloroplast, the photosynthetic organelle of algae and higher plants, bears a striking resemblance to another prokaryotic group, the *Cyanobacteria*. Thus, the chloroplast, like the mitochondrion, is prokaryotic. It, too, has DNA but no nuclear membrane, has 70S ribosomes, and lacks a cell wall, although it is descended from one of the bacterial phyla. Sequence analysis of its 16S rRNA places the chloroplast within the cyanobacterial branch of the *Bacteria* (see Figure 1.6).

Endosymbiotic Evolution

The theory of **endosymbiotic evolution**, championed by Lynn Margulis, has been developed to explain the origin of mitochondria and chloroplasts. According to this theory, early in the evolution of eukaryotic organisms certain prokaryotic organisms (the premitochondrion and prechloroplast) developed intracellular symbioses with eukaryotic cells (see Figure 1.16). As time passed, the partners in these symbioses became more and more interdependent, until the bacterium became an organelle inside the eukaryotic cell. Strong support for endosymbiotic evolution of these organelles derives from 16S rRNA gene sequences indicating that mitochondria and chloroplasts are closely related to two prokaryotic phyla, the *Proteobacteria* and the *Cyanobacteria*.

Further support for the endosymbiotic theory is found in intracellular bacterial associations with protozoa. For example, certain protozoa harbor bacterial cells as "parasites" that provide unique features to their hosts. One type of association is the "killer paramecium." This strain of *Paramecium* contains a bacterium called a **kappa particle**. The kappa particle lives inside the protozoan and is responsible for the production of an organic compound that kills other paramecia that do not harbor these particles. The kappa particles retain their cell wall, so the symbiosis is not as highly evolved as that of the mitochondrion. Other *Paramecium*-bacterial relationships are also known (Figure 1.17). Another example of endosymbiosis occurs in certain flagellate and amoeboid protozoans that have a photosynthetic organelle called a **cyanelle**. The cyanelles appear to be cyanobacteria replete with all their bacterial features, including their cell wall.

Clearly, plants and animals, through evolution, have obtained many of their genes from microorganisms, such as those introduced with intracellular endosymbiotic organisms. In this manner, genes have been available from a vast pool of different types of organisms. The macroorganism

Lambda particles

Figure 1.17 A *Paramecium*
This protist contains numerous endosymbiotic bacteria called "lambda particles," which are analogous to the "kappa particles" described in the text. Courtesy of John Preer Jr., Louise Preer, and Artur Jurand.

can therefore be viewed, at least in part, as a chimera of different genes, some derived from microorganisms.

Origin of the Ozone Layer

Crucial to the evolution of higher life forms was the development of the ozone layer in Earth's stratosphere. Ozone is produced by a photochemical oxidation reaction of oxygen in the upper stratosphere. It is believed to have first formed as a consequence of oxygen production by early cyanobacterial photosynthesis. The ozone layer is important to life on Earth in that it strongly absorbs ultraviolet (UV) light, a very active oxidizing agent and mutagen. Some aquatic forms of life could have evolved and lived largely unaffected by UV radiation prior to the formation of an ozone layer, because water also strongly absorbs UV light. However, terrestrial plants and animals would have been unprotected and could not have evolved until the ozone shield was established, which occurred in the Silurian period (see Table 1.1).

Oxygen and the development of the ozone layer had a profound impact on the evolution of higher organisms. Especially intriguing is the diversity and complexity of multicellular eukaryotic life forms, all of which evolved during the last 600 million years, particularly in terrestrial environments. In contrast, prokaryotes, which have had about 3.5 Ga to evolve, have rather simple structures. However, the simple and largely unicellular mor-

phology of prokaryotes belies their vast genetic, metabolic, and physiological diversity, as this book so clearly demonstrates.

> **SECTION HIGHLIGHTS**
> The origin of the eukaryotic cell remains a mystery; however, most scientists agree that endosymbiotic evolution led to the acquisition of the mitochondrion and chloroplast. The evolution of oxygenic photosynthesis by cyanobacteria resulted in oxygen accumulating in the atmosphere. This paved the way for the evolution of aerobic respiration, the development of the ozone layer, and ultimately the evolution of plants and animals.

1.6 Sequence of Major Events during Biological Evolution

The timetable shown in Figure 1.18 portrays, to the best of our current knowledge, the probable major sequential events during biological evolution. The initial composition of the atmosphere was determined in large part by gaseous emissions from volcanoes. From these anaer-

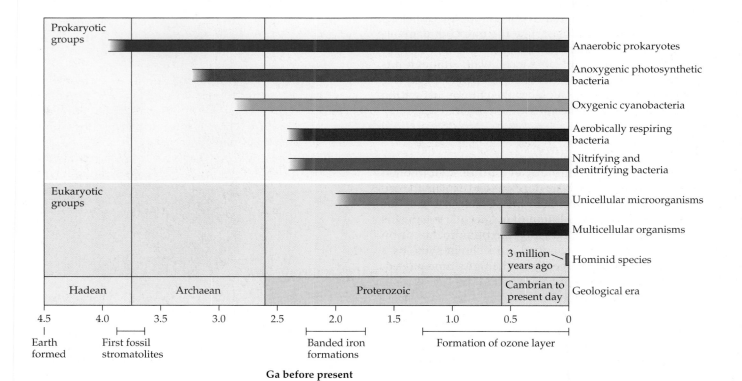

Ga before present

obic gases and water, organic compounds could have been produced on Earth and mixed with organic compounds from intergalactic space. Within a billion years of Earth's formation, the first microorganisms appeared. These were probably thermophilic, anaerobic bacteria, including heterotrophic organisms that could live by fermentative processes. Hydrogen-utilizing bacteria were also likely to have evolved early.

Anoxygenic photosynthetic bacteria likely had evolved by 3.0 to 3.5 Ga ago. These communities of organisms are thought to have colonized all aquatic habitats favorable for life. Photosynthesis had a major impact on Earth's biosphere. By producing large amounts of organic carbon, it greatly enhanced the growth of heterotrophic organisms.

The cyanobacterial branch of photosynthetic bacteria probably first appeared between 2.5 and 3.0 Ga ago. The production of oxygen would have had a major impact on many different processes in the biosphere. First, it would have caused the oxidation of reduced inorganic compounds such as iron sulfides that were present in enormous amounts early on. Indeed, it likely provided conditions that favored the evolution of new metabolic types such as aerobic, iron-oxidizing bacteria. In addition, oxygen would have been a poison to the anaerobic bacteria that had already evolved, and it would have adversely affected processes such as nitrogen fixation, which occur most efficiently under reducing conditions. Moreover, it would have led to the evolution of aerobic, respiratory bacteria that have a more efficient mode of metabolism. And, of course, it would have provided the key conditions for the evolution of plants and animals, which had obtained their mitochondria from proteobacteria and chloroplasts from the cyanobacteria.

Another consequence of oxygenic photosynthesis was the development of an ozone layer, which fostered the evolution of land plants and animals by protecting them from UV radiation. Aquatic organisms would have been less affected by UV light, because water strongly absorbs radiation of these wavelengths.

It is interesting, from the standpoint of human evolution, that hominid primates date back to only about 3 mya, and the species *Homo sapiens* to a mere 100,000 years or so ago. Humans are therefore very recent participants in the biosphere and very dependent on the other processes and organisms that occur on Earth that evolved before them.

Because humans and other animals and the plants are so recent, they evolved in a world of microorganisms, and it is interesting to note how plants and animals made use of microorganisms as they evolved. Both groups rely on microorganisms and their activities in very intimate ways. For example, the roots of plant species support associated bacteria and some have special fungi, called mycorrhizae, that the plant relies on to bring in nutrients from the soil. Likewise, almost all animals have an anoxic digestive tract in which foods are broken down by microorganisms to provide nutrients, including amino acids and vitamins, for the host animal. Without these microorganisms, animals could not survive.

> ### SECTION HIGHLIGHTS
> Much of the broad sweep of biological evolution has been dominated by microbial life, which was accompanied by the evolution of many diverse metabolisms and biogeochemical processes. Therefore, it should not seem surprising that plants and animals, which are relatively recent inhabitants of the biosphere, are completely dependent on microbial life for their survival.

1.7 Astrobiology

Astrobiology is the study of life in the universe. Of course, at this time, the only place where life is known to exist is Earth. What we have learned about the evolution of life on Earth poses interesting questions about evolution on other planetary bodies.

What type of life is most likely to be found on other potentially habitable planets? Consider the following factors:

- Earth has had microbial life for more than 3.5 Ga of its 4.5 Ga history, whereas plants and animals are less than 600 million years old.

- Microbial life can survive asteroid impacts that would kill other forms of life.

- Microbes live everywhere that organisms live on Earth including extreme environments inhospitable to plants and animals.

- Microbial life is predicted to survive on Earth several Ga after plants and animals have become extinct.

◀ **Figure 1.18 Geological and evolutionary timetable**
The timetable shows the major geological events, beginning with the formation of Earth 4.5 Ga ago, and the possible evolutionary events leading to various bacterial, archaeal, and eukaryotic groups based on microfossil, geochemical, and carbon isotopic evidence. Note that hominid primates have occupied Earth for only 3 million years or so, a minute fraction (less than one-thousandth) of the time since cellular life forms arose.

(A)

(B)

The dark red band in the ice layer indicates the presence of the microbial community, containing mainly algae and bacteria.

(C)

(D)

Bands of fungi (upper dark layer) and algae (lower greenish layer), components of a lichen, grow beneath the rock surface.

10 cm

Figure 1.19 Extreme environments
(A) Aerial view of Grand Prismatic Spring in Yellowstone National Park, containing thermophilic (heat-loving) bacteria. (B) A core taken through the sea ice of Antarctica, home to sea ice microorganisms, which are found in the dark-colored band of the sea ice microbial community. (C) On this salt farm in Thailand, sea water is evaporated and becomes more and more concentrated in salt. Only halophilic (salt-loving) microorganisms can grow in such salt-saturated water, which develops a red coloration due to the growth of halophilic microorganisms. (D) In Antarctica's cold desert area in Victoria Land, called the Dry Valleys, microbes are the only life forms. The microbes live inside the rocks, as shown by this lichen revealed in a broken sandstone rock. The dark layer near the surface contains the fungal component of the lichen whereas the green layer it is due to the growth of the algal component of the lichen. A, courtesy of Jim Peaco, NPS photo; B, courtesy of J. T. Staley; C, ©SOMBOON-UNEP/Peter Arnold, Inc.; D, courtesy of J. Robie Vestal.

Although this chapter has touched on several of these points, some explanation is needed for the last two of them. To amplify on the third point, it is noteworthy that some microorganisms grow at boiling temperatures in hot springs (Figure 1.19A) and at over 110°C in undersea volcanic hydrothermal vents. Others live in sea ice at temperatures below freezing (Figure 1.19B). Some produce sulfuric acid from sulfur compounds and live at a pH of 1.0, equivalent to 0.1 N H_2SO_4. A few microbes live in saturated salt brine solutions (Figure 1.19C), whereas others live in pristine mountain lakes as pure as distilled water. Microbes even live inside rocks in the harsh environment of Antarctica (Figure 1.19D). Thus, microorganisms can live in extreme environments not colonized by plants and animals.

The last point pertains to the hardiness of microbial life. As the Earth ages, the sun will become brighter as it emits more radiation. This will result in the Earth grad-

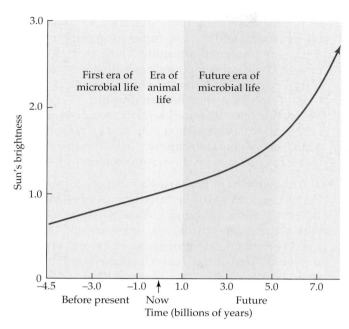

Figure 1.20 The sun's brightness and its effect on Earth's life

The sun's brightness has been and will continue to increase—first gradually, then more rapidly—as shown by the rising line. The first era of microbial life refers to the early period of their evolution before the origin of plants and animals, about 0.6 Ga BP. Plant and animal life is projected to continue for about one more Ga before becoming extinct. Of course, microorganisms are still persisting on Earth and will continue to do so for several Ga after plants and animals have disappeared.

ually warming over the next few Ga. It is estimated that the life of plants and animals can persist for another 500 million years or so. After that, conditions will be too hot and they will become extinct. However, microbial life will continue to persist for another 4 Ga (Figure 1.20).

All these factors are consistent with the hypothesis that if life is found elsewhere in the universe, it is much more likely to be microbial life rather than larger life forms such as plants and animals. This statement applies strongly to our own solar system. There is now very little doubt that "little green men" do not live on Mars or any other planets in our solar system. However, extreme environments are known to be common elsewhere in our solar system, so microorganisms are their most likely inhabitants. Indeed, Mars may still be volcanically active and appears to contain aquatic habitats beneath its surface that may be hospitable for microorganisms. Recently, European scientists detected methane on Mars, although it is not yet known whether it is of abiotic or biotic origin. Likewise, Jupiter's moon, Europa, although frozen at its surface, may teem with microbial life in its ocean and possible hydrothermal vents (Figure 1.21). Furthermore, we know that asteroid impacts on planets can disperse rocks from one planet to another. For example, several rocks from Mars have been found on Earth, having fallen as meteorites. This means that it may be possible to seed other planets within a solar system with hardy forms of life, that is, microbial life surviving within asteroid impact fragments that become meteorites.

These exciting recent developments have led the National Aeronautics and Space Administration (NASA) to

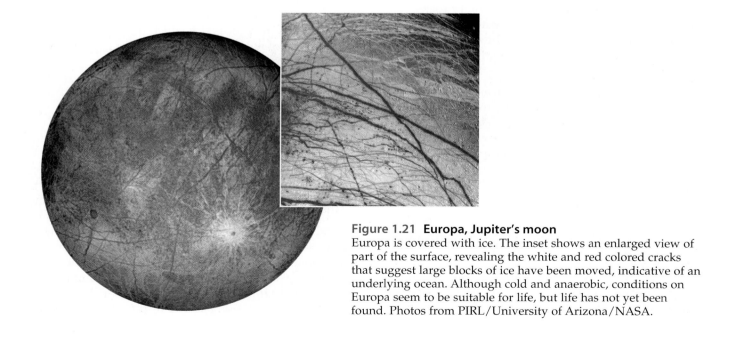

Figure 1.21 Europa, Jupiter's moon

Europa is covered with ice. The inset shows an enlarged view of part of the surface, revealing the white and red colored cracks that suggest large blocks of ice have been moved, indicative of an underlying ocean. Although cold and anaerobic, conditions on Europa seem to be suitable for life, but life has not yet been found. Photos from PIRL/University of Arizona/NASA.

launch a program on astrobiology to study the habitability of planets in the Universe as well as early life and its evolution on Earth (http://astrobiology.arc.nasa.gov). Other countries have also established astrobiology programs. Indeed, preparations are being made for expeditions to Mars and Europa for the exploration of past or present life. Pursuit of these lines of research may help humans understand some of the great questions of life, such as, where did we come from? And are we alone in the Universe?

SECTION HIGHLIGHTS

Microbiology is important to astrobiology because most regard it to be more likely to find microbial life rather than plant and animal life elsewhere in our own and other solar systems. This is because microbial life evolved prior to larger life forms on Earth, will persist after all plants and animals become extinct, and can live under the most extreme conditions for life known.

SUMMARY

- Earth and its solar system formed by physical and chemical processes about 4.5 Ga ago. Microbial fossils were first reported by micropaleontologists only recently (1950s). The oldest microbial fossils are found in **stromatolites** dated at about 3.5 Ga.

- Stromatolites are laminated sedimentary rocks, some of which were produced by microorganisms. Intertidal mat communities are the most common type of living stromatolites currently found on Earth.

- Many organic molecules can be formed abiotically from inorganic compounds in an anoxic environment with an energy source such as ultraviolet (UV) light or lightning. Some organic compounds are produced in space and brought to Earth in meteors and comets.

- Microorganisms are the smallest living things. Most can be seen only under a microscope. Microorganisms are ubiquitous in the biosphere; some grow in extreme habitats such as boiling hot springs, saturated brine solutions, or sea ice; some produce sulfuric acid.

- Microorganisms that use light as an energy source and carbon dioxide as a carbon source are said to be **photoautotrophic**. Those that use light as an energy source and organic carbon as a carbon source are **photoheterotrophic**. Those that use organic chemical compounds as both energy and carbon sources are **heterotrophic**, **chemoheterotrophic**, or **chemoorganotrophic**. Microorganisms that use inorganic compounds as energy sources and carbon dioxide as a carbon source are **chemoautotrophic**. No eukaryotic organisms are known to be chemoautotrophic.

- The two groups of prokaryotic microorganisms are *Bacteria* and *Archaea*. The three groups of eukaryotic microorganisms are fungi and the two types of protists, algae and protozoa. All organisms on Earth are classified into three domains of life and organized on a "Tree of Life," which is based on the sequence of 16S and 18S ribosomal RNAs (rRNAs). These domains are the *Bacteria*, *Archaea*, and *Eukarya* (the latter including all eukaryotic microorganisms and plants and animals).

- Viruses lack cytoplasm, and they do not produce cells. Therefore, they cannot be included in the Tree of Life.

- Prokaryotic organisms lack a nuclear membrane and do not have membrane-bounded organelles such as mitochondria and chloroplasts.

- Prokaryotes have 70S ribosomes, whereas eukaryotic organisms have larger, 80S ribosomes.

- Eukaryotic organisms have a nuclear membrane and most possess mitochondria and, in photosynthetic cells, chloroplasts. **Mitosis** is the process by which the chromosomes are duplicated and separated during asexual division of eukaryotic cells. **Meiosis** is the process by which the diploid chromosomes of eukaryotic cells are separated to form haploid gametes.

- The cyanobacteria are thought to be the first O_2-producing (oxygenic) photosynthetic organisms. The O_2 produced by cyanobacteria resulted in the formation of the protective ozone layer and led eventually to the high concentration of O_2 in the atmosphere. Through natural selection, O_2 in the atmosphere led to the evolution of **aerobic respiration**, a much more efficient type of metabolism than **fermentation** and **anaerobic respiration**.

- The theory of **endosymbiotic evolution** holds that mitochondria, chloroplasts, and possibly other organelles of higher animals and plants are derived from prokaryotic organisms that developed symbioses with higher organisms. Evidence from 16S rRNA sequence analysis of mitochondria indicates that these organelles evolved from the *Proteobacteria*. Evidence from 16S rRNA sequence analysis of chloroplasts indicates they evolved from cyanobacteria. The cyanelle is a cyanobacterial symbiont within some protozoa.

- Plants and animals could not have evolved unless microorganisms evolved first.

- Microorganisms, particularly *Bacteria* and *Archaea*, carry out unique steps in the **biogeochemical cycles** of the biosphere.

- **Astrobiology** is the study of life in the universe. Because of their early and rapid evolution and hardiness, microorganisms are much more likely to be found at other locations in the Universe than macroorganisms.

 Find more at www.sinauer.com/microbial-life

REVIEW QUESTIONS

1. What is a stromatolite and what phylum of microorganisms are necessary for their formation?

2. Compare and contrast oxygenic and anoxygenic photosynthesis.

3. What method would be used to date geological events more than one billion years ago?

4. How do you explain the existence of some eukaryotes that lack mitochondria?

5. What is the carbon source used by a photo-autotroph?

6. Why do bacteria grow in such unusual habitats?

7. Compare the nutrition of humans with that of a typical chemoheterotrophic bacterium.

8. Some bacteria can reproduce to form progeny cells in less than 10 minutes. Why is human reproduction so much slower?

9. Compare bacterial cell structure with that of a yeast.

10. Describe Earth's biosphere before and after the evolution of *Cyanobacteria*.

11. How much information can be obtained by the discovery and dating of microbial fossils? Can the type and metabolic activity of the microorganism be determined?

12. Is there life on Mars? On the moon? On Europa? Explain your answer.

SUGGESTED READING

Gest, H. 1987. *The World of Microbes.* Madison, WI: Science Tech Publishers.

Jacobson, M. C., R. J. Charlson, H. Rodhe and G. H. Orians, eds. 2000. *Earth System Science.* New York: Academic Press.

Lwoff, A. 1957. "The Concept of Virus." *Journal of General Microbiology* 17: 239–253.

Sapp, J. 2005. "The prokaryotic-eukaryotic dichotomy: meanings and mythology." *Microbiology and Molecular Biology Reviews* 69: 292-395.

Schopf, J. W. 1983. *Earth's Earliest Biosphere.* Princeton, NJ: Princeton University Press.

Staley, J. T., and A.-L. Reysenbach, eds. 2002. *Biodiversity of Microbial Life: Foundation of Earth's Biosphere.* New York: John Wiley & Sons.

Stanier, R. Y., and C. B. van Niel. 1962. "The Concept of a Bacterium." *Archiv für Mikrobiologie* 42: 17–35.

Woese, C. R., O. Kandler and M. C. Wheelis. 1990. "Towards a Natural System of Organisms: Proposal for the Domains Archaea, Bacteria, and Eucarya." *Proceedings of the National Academy of Sciences, USA* 87: 4576–4579.

The objectives of this chapter are to:

◆ Outline the history of microbiological research.

◆ Introduce some of the concepts that dominated microbiology in the past and how they shaped the progress of the science.

◆ Describe the origins of ideas and techniques that continue to influence present-day microbiology.

2

Historical Overview

*The accidents of health had more to do with the march of great
events than was ordinarily suspected.*
—H. A. L. Fisher, 1865–1940

*T*he previous chapter was devoted to a discussion of the evolution and role of microorganisms in the living world. Clearly, microbes make up a considerable part of the biosphere and are therefore studied for their own sake. However, microbiology is a multifaceted discipline that is concerned with infectious diseases; agricultural practice; sanitation; and the industrial production of food, beverages, and chemicals. Microorganisms have been and remain important model organisms for studies in nutrition, metabolism, genetics, and biochemistry. Results of studies with microorganisms were instrumental in the development of biotechnology. Microorganisms have also had a profound effect—both positive and negative—on human welfare. In this chapter we examine the contributions of pioneer scientists to our understanding of the microbe.

2.1 Effects of Infectious Diseases on Civilization

Microorganisms have markedly affected the course of human history. Disease-causing microorganisms often decided the well-being and morale of populations, the strength of armies, and the outcomes of battles. The mobilization of armies themselves, with the consequent concentration of young soldiers from across the country, created environments that were ripe for the spread of epidemics. Every soldier was both a potential carrier of infectious diseases from his home and a potential victim of disease due to lack of previous exposure. In rapidly growing urban areas, the absence of proper sanitation in crowded conditions also contributed to the spread of infectious agents. As world trade and travel increased, infectious diseases slowly but inevitably moved from region to region. They then retraced their steps back to a new generation of susceptible individuals. This section briefly presents some examples of these and other effects of microbial diseases on human populations, institutions, and wars throughout the history of civilization.

The decline of Rome under the Emperor Justinian (AD 565) was certainly hastened by epidemics of bubonic plague and smallpox. The inhabitants of Rome were decimated and demoralized by these massive epidemics and were left powerless against the barbarian hordes that destroyed the empire. Through the Middle Ages and beyond, each human generation was subject to renewed attacks by epidemics. Some of these infec-

BOX 2.1   *Milestones*

The Spread of the Black Plague in Europe

The Great Plague of the fourteenth century was also referred to as the "Black Death." The epidemic originated in China in 1331 and moved slowly across Asia, reaching the outskirts of Europe in 1347. The disease spread from caravan stop to caravan stop, moving west with the rat–flea–human community associated with the caravans. The high rate of fatalities among these hosts in thinly populated areas of Asia was probably responsible for the slow movement of the epidemic across Asia. When the infected flea–rat population reached the Mediterranean area, the disease agent (*Yersinia pestis*) spread into the black rat populations. Shipping and commerce carried the black rats and their bubonic plague–infected fleas upward across Europe. The Black Death reached Sweden by late 1350, a mere three years from when it arrived in the Mediterranean area.

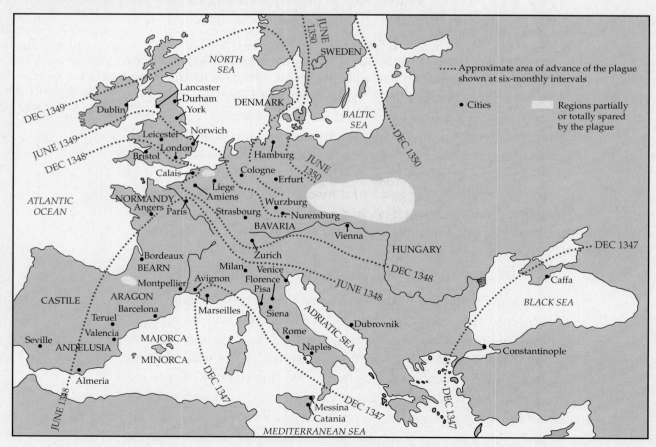

tious diseases spread across the continents, whereas others were more localized. Typhus, plague, smallpox, syphilis, and cholera were some of the diseases that caused suffering and great loss of human life.

The populations of Europe, North Africa, and the Middle East totaled about 100 million when a bubonic plague epidemic struck in AD 1346 (the Black Death). The epidemic traveled west down the "Silk Road" (the main trade route from China), bringing death to Asia; the epidemic then spread throughout Europe, resulting in the loss of 25 million lives in a few short years (Box 2.1). Recurrences of plague through the sixteenth and seventeenth centuries kept populations in check. In 1720 to 1722, one last great epidemic occurred in France that killed 60% of the population of Marseilles, 60% of Poulon, 44% of Arles, 30% of Aix, and 30% of Avignon. The most recent major plague pandemic originated in Yunnan, China, in 1892, moved across India, and arrived in Bombay in 1896. That outbreak killed an estimated 6 million people in India alone.

In wars that were fought prior to World War II, the outcome was generally decided by arms, strategy, and pestilence. Pestilence, more often than not, played the leading role, as the following examples affirm.

By the time Napoleon began his retreat from Moscow in 1812, most of his army had fallen victim to typhus, pneumonia, dysentery, and other diseases. These diseases, the cold, and deprivation all played a role in his departure. The following year (1813), the irrepressible Napoleon recruited a new army of 500,000 young soldiers. As with the previous army, their youth and the crowded, unsanitary conditions soon rendered them susceptible to infectious diseases. By the time Napoleon and his new army faced the allies at Leipzig, preliminary battles and disease had reduced his army of approximately 500,000 to about 170,000. An estimated 105,000 were casualties of earlier battles, but about 220,000 were incapacitated by illness. Therefore, microbial infections were a major factor in the ultimate defeat of Napoleon at Waterloo in 1815.

There were more than 550,000 deaths among the soldiers in the American Civil War (1861–1865), and more were victims of infectious diseases than died from battle. Of the Union Army troops lost, 93,443 were killed in action or died from wounds or wound infections, whereas 210,400 succumbed to various diseases. Infections were a major cause of death among the wounded, as field conditions were unsanitary and care of the wounded was haphazard. Of those who died of disease, records indicate that 29,336 died from typhoid fever, 15,570 from other "fevers," 44,558 from dysentery, and 26,468 from pulmonary disease (mostly tuberculosis). The remaining deaths were from other undetermined causes. Records for the army of the Confederacy are less extensive, but an estimated 90,000 were killed in action

or died from wounds. The number of deaths from disease exceeded 180,000 soldiers. It is highly probable that typhoid fever, dysentery, and pulmonary diseases were also the main causes of death among the Confederate soldiers. The approximate totals for the two armies were 183,000 deaths from combat wounds and more than 390,000 victims of infectious diseases.

SECTION HIGHLIGHTS

Diseases caused by microbes have had a tremendous impact on the history of human civilization. In many wars, the casualties resulting from disease have outnumbered those resulting from wounds received in action.

2.2 Why Study the History of a Science?

Microbiology, as with any field of endeavor, can be comprehended best if one has a reasonable understanding of the historical development of the field. The ingenious experimentation and insights that led scientists, such as Pasteur and Koch, to logical explanations for the observable manifestations of disease should be familiar to the modern student in microbiology. One cannot fully comprehend present theories and concepts without first understanding the logical steps that led to those ideas.

The remainder of this chapter presents the scientific contributions made by several intellectual giants in microbiology. This list of contributors is by no means complete, and this discussion does not do justice to the many early scientists who contributed to the foundations of microbiology. Most college libraries can provide sources for those interested in furthering their knowledge of the historical development and significance of microbiology.

2.3 Status of Microbial Science Prior to 1650

From the dawn of civilization until the middle of the nineteenth century, any success in combating disease, in fighting the microbe's destructive power or in harnessing their fermentative capabilities came about through inexact processes of trial and error. Those occurrences we now assign to the microbe were then ascribed to a supreme being, "miasmas," spontaneous generation, magic, chemical instabilities, or other interpretations limited only by the human imagination.

Human thought was influenced by the great philosophers, whose writings indicate that they were intrigued by theories supporting **spontaneous generation**, the immediate origin of living organisms from inert organic materials. It should be emphasized that spontaneous generation is defined, for our purposes in this chapter, as the creation of identifiable living creatures from the inanimate. (Note: The evolution of the earliest viable cell was discussed in Chapter 1 and is not an example of spontaneous generation as applied here.)

Aristotle (384–322 BC) and others wrote of the formation of frogs from damp earth and of mice from decaying grain. After the fall of Rome and the decline of civilization, free inquiry and acquisition of knowledge became severely limited. Through the centuries of the Dark Ages (AD 476–1000), disease epidemics and plagues were recorded, but little of scientific consequence was written. The Renaissance (1250–1550) was a period of awakening and the beginning of open inquiry into the forces that shaped human life. The first major writings on the causation of disease were by **Girolamo Fracastoro** (ca. 1478–1553), who wrote extensively on the "contagions" involved in the disease process. Based on his investigations, he wrote that disease could be spread by direct contact, handling contaminated clothing, or through the air.

It is evident from the literature of that time (prior to the Age of Enlightenment) that scholars in the sixteenth and early seventeenth centuries were seeking logical explanations for the phenomena of nature. Theory and experimentation were one thing, but acceptance of concepts contrary to the dogma of the time was quite another. It was particularly difficult to gain acceptance of biological explanations for natural phenomena, such as food decay, when many renowned scientists, particularly chemists, attributed this process to chemical instabilities and clung to theories of spontaneous generation. Typical of this group was Jan Babtista van Helmont (1580–1644), a forerunner of scientific chemistry, who published a recipe for producing mice from soiled clothing and a little wheat.

SECTION HIGHLIGHTS

Before the existence of microbes was established, many microbial phenomena, such as infectious disease and fermentation, were attributed to non-biological processes, such as spontaneous generation and chemical instability.

2.4 Microbiology from 1650 to 1850

During the seventeenth century there were individuals of a scientific bent whose contributions gave impetus to a further awakening of the human spirit. Biology benefited significantly from such developments as the microscope and the realization that living matter was composed of individual cells. This was the Age of Enlightenment, a time when people questioned traditional doctrines. Science, reason, and individualism replaced obedience to accepted dogma.

Fabrication of the original microscope is generally attributed to a Dutch spectacle maker, Zacharias Janssen, and his father, Hans, between 1590 and 1610. The first to employ a microscope extensively in the examination of biological material was the Italian scientist **Marcello Malpighi** (1628–1694), and he is considered by many to be the "father of microscopic biology." Malpighi made major contributions through his discovery of capillaries and his investigations of embryonic development. **Antony van Leeuwenhoek** (1632–1723) developed a solar microscope with high resolving power that led to the first recorded observations of bacteria in 1683. **Robert Hooke** (1635–1703) examined the structure of cork and, based on these observations, suggested that all living creatures were made up of individual cells. The following section covers Leeuwenhoek and his discovery of bacteria in some detail. Approximately 200 years passed before the nature of these organisms was generally accepted. Why 200 years? Because humans were not yet ready to disregard completely the prevailing dogma that chemical instabilities and spontaneous generation were responsible for all activities that we now ascribe to microbes.

Antony van Leeuwenhoek

Antony van Leeuwenhoek was born in Delft, Holland, in 1632 into a relatively prosperous family (Figure 2.1). At the age of 16, he was sent to Amsterdam to apprentice in a draper's shop and learn a useful trade. He was a bright lad and was appointed cashier in the business before returning to Delft to spend the remaining 70 years of his life. He did not attend a university, but was learned in mathematics, and he became a successful businessman, a surveyor, and the official wine gauger for the town of Delft. In the latter capacity, he assayed all of the wines and spirits entering the town, and it was his responsibility to calibrate the vessels in which they were transported. His lack of a university education was inconsequential because, at that time, such institutions were mostly devoted to theology, law, and philosophy. Scientific research was done by amateurs as an avocation, by the wealthy, or by individuals under the sponsorship of a patron.

Leeuwenhoek apparently had solar microscopes available and made extensive observations of bacteria prior to 1673 for, in that year, several of his studies were

Figure 2.1 Antony van Leeuwenhoek
Antony van Leeuwenhoek (1632–1723) developed a microscope and was the first to observe and accurately describe bacteria. Courtesy of National Library of Medicine.

communicated to the Royal Society in London, England, by his friend, Renier de Graaf, a noted Dutch anatomist. Fortuitously, the editor of *Philosophical Transactions* (published by the Royal Society) was Henry Oldenberg, a German-born scientist who was well-versed in several languages, including Dutch. Oldenberg firmly believed that scientific information should have no nationalistic restrictions, and he carried on an extensive correspondence with scientists throughout Europe. Leeuwenhoek was encouraged to communicate with the Royal Society. His first report, published in the *Philosophical Transactions*, dealt with molds, the mouthparts and the eye of the bee, and gross observations on the louse. Due to Oldenberg's contacts and immediate translation of Leeuwenhoek's correspondence into several languages, his work generated considerable interest in the scientific world. Leeuwenhoek continued to send letters to the Royal Society throughout his life, and in 1680, he was unanimously elected a Fellow of the Royal Society.

More than 200 of Leeuwenhoek's original letters have been preserved in the library of the Royal Society. Upon his death, Leeuwenhoek bequeathed a cabinet containing 26 microscopes to the Royal Society, and these were delivered by his daughter, Maria, with the handwritten passage, "every one of them ground by myself and mounted in silver and furthermore set in silver . . . that I extracted from the ore . . . and therewithal is writ down what object standeth before each glass." The cabinet and microscopes remained in the Royal Society collection for a century, but unfortunately have since been lost and never recovered. The microscopes were rather simple in design, and he left no description of how they were operated (Figure 2.2). The exactness of his drawings suggests that he had a keen eye, but most certainly he also had an unexplained method for using his microscope to examine bacteria. Because his microscopes could magnify by only 300 diameters (less than one-third of what modern light microscopes can do), it is probable that he achieved his detailed observations of bacteria by careful focusing of solar light.

It is evident from Leeuwenhoek's descriptions of what he called "**animalcules**" and from his size estimates by comparison with grains of sand or human red blood corpuscles (which he measured fairly accurately at 1/300

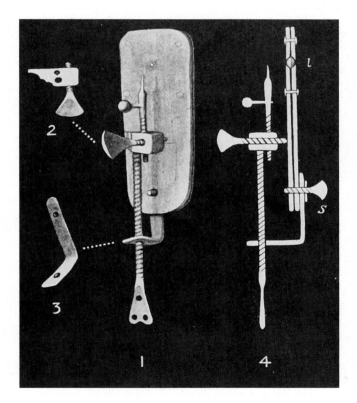

Figure 2.2 The microscope developed by Leeuwenhoek
It is very difficult to observe microbes with this crude instrument, but Leeuwenhoek held the implement to the solar light in such a way that small specimens such as bacteria could be viewed. His drawings confirm the accuracy of his observations. From *Antony van Leeuwenhoek and His "Little Animals,"* edited by Clifford Dobell, Dover Publications, 1960.

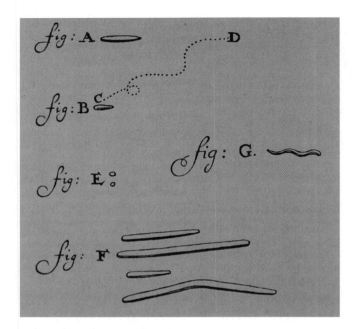

Figure 2.3 Drawings made by Antony van Leeuwenhoek
Their probable identification: (A) *Bacillus*, (B) motile *Selenomonas*, (E) *Micrococcus*, (F) *Leptothrix*, and (G) a spirochete. From *Antony van Leeuwenhoek and His "Little Animals,"* edited by Clifford Dobell, Dover Publications, 1960.

inch) that he indeed closely observed the living forms that he depicted. His writings describe some of the major bacterial forms now known—spheres, rods, and spirals—and he described motility in rod- and spiral-shaped cells (Figure 2.3). Leeuwenhoek's descriptions of bacteria were superior to those made by Louis Joblot (1645–1723) and Robert Hooke (1635–1703), who used slightly more sophisticated compound microscopes. Leeuwenhoek's extensive letters described an array of protozoa, spermatozoa, and red blood cells, and he is generally considered to be the founder of protozoology and histology.

Leeuwenhoek emphasized the abundance of protozoa, yeast, bacteria, and algae in environments as diverse as the mouth and seawater. Although Leeuwenhoek's observations were widely known, apparently no one attributed familiar processes such as fermentation and decay to these animalcules. He was firm in his belief that his animalcules arose from preexisting organisms of the same kind and did not arise by spontaneous generation.

From Leeuwenhoek to Pasteur

The studies by a number of experimentalists (now much ignored) between 1725 and 1850 laid the foundation for the great advances in bacteriology made in the latter half of the nineteenth century. These scientists did much to discredit the theories of spontaneous generation, provide a foundation for studies on the immune response, and suggest rational classification schemes for bacteria. A major factor in affirming the role of microbes in nature was disproving the generally held concept of spontaneous generation.

Francesco Redi (1626–1697), an Italian physician, published a book in 1668 attacking the doctrine of spontaneous generation. At that time, the presence of maggots in decaying meat was considered a prime example of spontaneous generation. Redi believed otherwise, and to prove this he placed meat in beaker-like containers, covered some with fine muslin, and left others exposed to invasion by blowflies. Although fly eggs were deposited on the muslin over the covered jars, maggots did not develop on the meat. Extensive growth of maggots, however, did occur in the meat that was left uncovered. When Redi placed the eggs that were deposited on the muslin onto the surface of the meat, maggots quickly appeared. Clearly maggots came from the eggs deposited by the flies. Studies of this type led to careful experimentation and clear evidence that animals and insects, discernible by the eye, could not arise spontaneously from decaying matter. The proponents of spontaneous generation turned to phenomena the causes of which were not so apparent. The inability to identify the factor responsible for fermentation and putrefaction in infusions (meat or vegetable broth) gave their theory continued life.

Lazzaro Spallanzani (1729–1799), an Italian naturalist and priest, was familiar with the studies of Redi and did a series of experiments to confirm and extend Redi's earlier work. Spallanzani hermetically (airtight) sealed meat broth in glass flasks and reported that 1 to 2 hours of heating the enclosed infusions was sufficient to render the contents incapable of supporting growth. John Needham, an English cleric and proponent of spontaneous generation, attacked these studies. Needham proposed that a "vegetative force" was responsible for spontaneous generation and that hermetically sealing and heating the flasks destroyed this vital force. Spallanzani then did a series of experiments to overcome this objection, which he considered to be conclusive and from which he affirmed that to render a broth sterile, it was necessary to seal the flask and not allow unsterile air to enter. Spallanzani also concluded that boiling for a few minutes destroyed most organisms, but that others, now known to be spore formers, withstood boiling for a half hour. Spallanzani's work was quite advanced for his time, and his deductions and conclusions were similar to those reported later by Pasteur.

During the latter part of the eighteenth century, Antoine Lavoisier and others demonstrated the indispens-

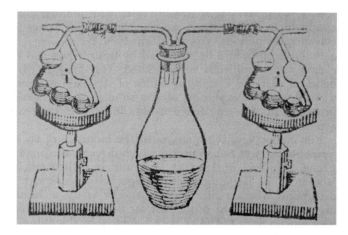

Figure 2.4 Sterilization by heat
The experiment by Franz Schulze did much to confirm that air does not contain a "vital force." From *The History of Bacteriology*, by William Bulloch, M.D., Oxford University Press, 1960.

ability of oxygen (O_2) to animals. This led to the assumption that oxygen was the mysterious element (vegetative force?) necessary for spontaneous generation, and that it was this element that had been excluded in Spallanzani's experiments.

In 1836, **Franz Schulze** (1815–1873), a German chemist, performed a crucial experiment indicating that oxygen depletion was not the sole reason for sterility in Spallanzani's flasks. Schulze took a flask that was half filled with vegetable infusion and closed it with a cork through which two bent glass tubes were fitted (**Figure 2.4**). He thoroughly boiled the infusion in a sand bath, and while steam was being emitted he attached an absorption bulb to each of the two glass tubes. One bulb contained concentrated sulfuric acid and the other a solution of potassium hydroxide (potash). Every day for several months, he drew air out through the potash bulb, and incoming air passed through the acid. The flask remained sterile, whereas a control flask without acid sterilization of incoming air had visible mold growth after a few days. When the sterile infusion was opened and exposed to the atmosphere, mold growth followed in a few days. This experiment demonstrated that infusions could be sterilized (no viable microbes) and that microbes could be introduced from the air. John Tyndall did many modifications of the Schulze experiment, and Theodor Schwann made further refinements; these finally led to refutation of the doctrine of spontaneous generation.

During this period other scientists were helping to disprove spontaneous generation by examining how microbes were involved in fermentation. The first report that gave a clear account of the yeast cell and its role in fermentation of beer and wine appeared in an 1836 report by **Charles Cagniard-Latour** (1777–1859). He asserted that yeasts were nonmotile organized globules capable of reproduction by budding and that they probably belonged to the vegetable kingdom. In 1837, he suggested that the vital activity of the yeast cell was responsible for converting a sugar solution to carbonic acid and alcohol.

Theodor Schwann (1810–1882), a German physiologist, independently discovered and described the yeast cell in 1837. Although his report also concerned the doctrine of spontaneous generation, he wrote that beer yeast consisted of granules arranged in rows, and that they resembled fungi. He believed them to be plants and observed their reproduction by budding. The relationship between the growth of yeast and the process of fermentation was clear to Schwann, and he called the organisms "zuckerpilz" (sugar fungus), from which the term *Saccharomyces* (a genus of common yeast) was derived. He also noted the indispensable requirement for nitrogenous compounds in the fermentation process.

A third independent worker who contributed to the discovery of the fermentation process was **Friedrich Kützing** (1807–1893), a German naturalist, and his major publication was also dated 1837. He described the nucleus of the cell and developed the concept that all fermentation is caused by living organisms. He was among the first to suggest that physiologically distinct organisms brought about different types of fermentation.

The role ascribed to the yeast cell in fermentation was contemptuously attacked by the chemists of that time. J. J. Berzelius, F. Wöhler, and J. von Leibig (all influential chemists) attributed a chemical character to every vital process and suggested that chemical instabilities were responsible for fermentation. Although much rhetoric ensued, the criticisms by the chemists led to more definitive experimentation and a clearer understanding of the role of yeast and fungi in many different types of fermentation.

Classification Systems

The earliest classification scheme for living organisms that included the bacteria was formulated by the Swedish botanist **Carolus Linnaeus** (1707–1778). In *Systema Naturale* (1743), he described the "animalcules" (Leeuwenhoek) as "infusoria" and placed this group, which included the bacteria, in the genus *Chaos*. A major shortcoming of the Linnaean classification scheme was the emphasis that he placed on a few known characteristics. This emphasis has permeated classification schemes until recent times; for example, we often use terms such as rod or coccus without considering the metabolic capabilities of the microbe. There were other classification schemes suggested, and the **Adansonian** classification scheme proposed by **Michael Adanson** in 1730 gave equal weight to all characteristics of a species. The Adansonian classification scheme was more amenable to modern computerized classification systems of bacteria.

The Danish naturalist **Otto F. Müller** (1730–1784) presented several works that described, arranged, and named a number of microscopic organisms. His major study, *Animalcula infusoria et Marina*, published posthumously in 1786, was 367 pages in length and described 379 species of bacteria. In his scheme, two of five genera contained bacterial forms. The genera were named *Monas* and *Vibrio*. The term *Vibrio* has been retained to the present day.

The first attempt to define bacterial forms as distinct from complex organisms was made by **Christian G. Ehrenberg** (1795–1876) in 1838. His treatise, entitled *Die Infusionsthierchen als Volkommene Organismen* (*The Infusoria as Complete Organisms*), used some of Leeuwenhoek's drawings of microbial cells. It was 547 pages in length and recognized three families that comprised forms we now recognize as bacteria. These families—*Monadina*, *Crystomonadina*, and *Vibrionia*—encompassed several genera: *Bacterium*, *Vibrio*, *Spirochaeta*, and *Spirillum*. All of these terms are still in common usage.

SECTION HIGHLIGHTS

The Enlightenment witnessed many advances in microbiology, greatly facilitated by the development of microscopes that could be used for direct observation of microbes. Over the next 200 years microbes were described and classified, and many experiments were conducted to demonstrate the involvement of microbes in decay and fermentation.

2.5 Medical Microbiology and Immunology

Several scientists merit mention for their contributions to our understanding of immunity and the cause of disease prior to 1850. In 1822, the Italian scientist **Enrico Acerbi** postulated that parasites existed that were capable of entering the body and that their multiplication caused typhus fever. This theory was advanced by the 1835 work of **Agostino Bassi** (1773–1856), who made classic observations on the diseases of silkworms.

A disease that was rampant at that time caused the death of the worms; the dead silkworms were covered with a hard, white, limy substance. At the time, the limy coat was considered to arise spontaneously from un-

known factors. However, Bassi demonstrated that aseptic transfer of subcutaneous material from sick, living worms to healthy worms resulted in the disease. He suggested that a fungus caused the disease, which was ultimately renamed *Botrytis bassiana* in his honor. In later life, although nearly blind, he developed his theory that contagion (infectious agents) in such diseases as cholera, gangrene, and plague resulted from living parasites.

During 1798, **Edward Jenner** (1749–1823) published studies on the immunization of humans against smallpox (Figure 2.5). Jenner and others observed that individuals who were routinely exposed to cows often developed pustules on their hands and arms that were similar to those caused by the dreaded smallpox. This "cowpox" was not fatal in humans, and apparently those infected with the cowpox did not contract the smallpox. Jenner inoculated an eight-year-old boy (James Phipps) with material from the infected cowpox pustules on the hands of a milkmaid. The boy became ill and had characteristic cowpox pustules at the site of inoculation but quickly recovered. A challenge dose later from an active case of smallpox did not result in illness. The boy was apparently immune to smallpox. Jenner then inoculated several healthy individuals with cowpox exudate and observed that they were made immune to smallpox. Jenner has often been criticized for the empirical nature of his experimentation, but given the context of the time, he deserves much credit for his contribution.

Figure 2.5 Effective vaccination
Edward Jenner vaccinating James Phipps, an eight-year-old boy, May 14, 1796. Pus from the hand of Sarah Nelmes, a dairy maid, was used. Print by Georges Gaston Melingue, ©Bettmann/Corbis.

2.6 Microbiology after 1850: The Beginning of Modern Microbiology

During the latter part of the nineteenth century, a number of gifted scientists working independently established the disciplines that are now encompassed by the term "microbiology." Much was accomplished during this half-century, and the foundations for immunology, medical microbiology, protozoology, systematics, fermentation, and mycology were all set in place. However, up until 1860, the doctrine of spontaneous generation of microbes remained a generally accepted concept. This was largely due to the widespread acceptance that chemists were infallible—a view they did little to discourage—and they tended to be more dogmatic than were the naturalists. They would soon meet their match in a brilliant, strong-willed Frenchman named Louis Pasteur.

A notable publication appeared in 1861 written by Louis Pasteur and entitled *Mémoire sur les corpuscules organisés qui existent dans l'atmosphère: Examen de la doctrine de générations spontanées* (*Report on the organized bodies that live in the atmosphere: Examination of the doctrine of spontaneous generation*). It marked the beginning of a new epoch in bacteriology.

Louis Pasteur (1822–1895) was a chemist, and his approach to the study of microorganisms was markedly influenced by his chemical training and analytical mind (**Figure 2.6**). Pasteur was a genius in his thinking, argument, and experimentation, but another of his unique qualities was the ability to communicate and convince scientists and laypeople alike of his views. He laid to rest forever the idea of spontaneous generation, established immunology as a science, developed the concepts of fermentation and anaerobiosis, and developed many microbiological techniques. In short, he contributed to every phase of microbiology.

The Pasteur School

Louis Pasteur entered the controversy surrounding spontaneous generation at a time when the dogma of **heterogenesis** was being used to explain the origin of

Figure 2.6 Louis Pasteur
Louis Pasteur (1822–1895), an outstanding researcher and the developer of the rabies vaccine. Pasteur is credited with disproving "spontaneous generation." From *Life of Pasteur*, by René Vallery-Radot, Doubleday Page, Garden City, NY, 1923.

living matter. This theory was avidly expounded by **Felix-Archimede Pouchet** (1800–1872), a noted French physician and naturalist and honored member of many learned societies in France. Pouchet believed that life could spring de novo from a fortuitous collection of molecules and that the "vital force" came from preexisting living matter. In contrast, Pasteur's studies on fermentation indicated that "ferments" (cause of fermentation) were actually organic living beings that reproduced and, by their vital activities, generated the observed chemical changes. In a series of brilliant experiments, Pasteur showed that "germs" present in the air were the cause of ferments and that such organisms were widely distributed in nature. Pasteur's "germs" are actually what we call bacteria and fungi today.

Starting in 1859, Pasteur dealt with the problem of microbes in air by designing an aspirator filter system to recover them. Pasteur was aware that H. G. F. Schröder and T. von Dusch had found spun cotton-wool to be an effective filter for airborne microbes. Employing this knowledge, Pasteur drew copious quantities of air through spun cotton and then dissolved the cotton in a mixture of alcohol and ether. Microscopic examination of the resulting sediment revealed a considerable number of small round or oval bodies that were indistinguishable from the "germs" previously described by others. Pasteur noted that the number of these organisms varied with the temperature, moisture, and movement of the air, and the height above the soil that the aspirator's inlet tube was placed.

We digress briefly here to present work done by others that influenced Pasteur's experimentations.

During this era, the English physicist **John Tyndall** (1820–1893), who was also opposed to the dogma of spontaneous generation, performed a series of experiments that supported the work of Pasteur. Tyndall used optics to demonstrate that microbes were present in air and that heated infusions placed in optically clear chambers remained sterile, whereas infusions placed under an ordinary atmosphere exhibited growth. A major contribution was his empirical observation that some bacteria have phases: one is a thermolabile (unstable when heated) phase during which time the bacteria are destroyed at 100°C, and the other is a thermoresistant phase that renders some microbes incredibly resistant to heat. **Ferdinand Cohn** (1828–1898) confirmed these suppositions with the demonstration that hay bacilli could form the heat-resistant bodies we now call **endospores**.

To dispel the theories of spontaneous generation, Pasteur initiated a series of experiments with long-necked flasks of various types (Figure 2.7) to improve on experiments by earlier workers such as Schwann and Spallanzani. He fashioned one flask with a horizontal neck and placed distilled water in the flask that contained 10% sugar, 0.2% to 0.7% albuminoid (soluble proteins), and the mineral matter from beer yeast. The flask was boiled for several minutes to sterilize the contents, and the neck was attached to a platinum tube that was maintained

at a red-hot temperature as the flask cooled. The air drawn into the flask was sterilized by passage through this heated tube. Pasteur noted that flasks containing various infusions treated in this manner remained clear and free of microbial growth. He also showed that swannecked flasks, which were long and bent down in a way that excluded passage of dust on cooling, but allowed a free exchange of air, also remained sterile. Microbes grew rapidly in all of the flasks if the necks were broken off or if the infusions were spilled into the necks and allowed to drain back into the flasks.

In 1864 Pasteur gave a public lecture about his experiments and demonstrated some infusions that had remained unspoiled for four years. In his address, he stated:

> And, therefore, gentlemen, I could point to that liquid and say to you, I have taken my drop of water from the immensity of creation, and I have taken it full of the elements appropriated to the development of inferior beings. And I wait, I watch, I question it, begging it to recommence for me the spectacle of the first creation. But it is dumb, dumb since these experiments were begun several years ago; it is dumb because I have kept it from the only thing that man cannot produce, from the germs that float in the air, from life, for life is a germ and a germ is life. Never will the doctrine of spontaneous generation recover from the mortal blow of this simple experiment.

(From Vallery-Radot, René. 1920. *The Life of Pasteur.* Garden City, NY: Garden City Publishing Company, Inc., p. 108–109.)

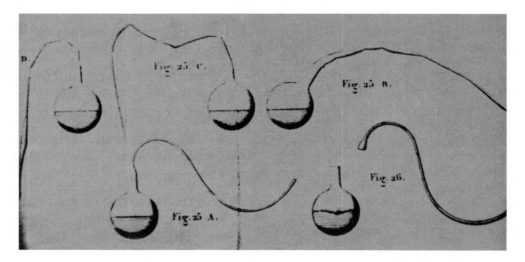

Figure 2.7 Pasteur's bent glass flasks
Pasteur's drawings of his bent neck flasks, which were employed in disproving spontaneous generation. From *The History of Bacteriology*, by William Bulloch, M.D., Oxford University Press, 1960.

Much heated controversy continued during the period from 1860 to 1880, and the heterogenesists continued to attack the experiments and writings of Pasteur, Tyndall, Lister, Cohn, and others. Despite this, the voice of the opposition was weakening. The Academy of Science in Paris and scientists everywhere were convinced of the correctness of Pasteur's experimental work.

Pasteur was aware of and continued the studies of Cagniard-Latour, Schwann, and Kützing (1837), who reported that the fermentation of sugar to an alcohol was due to the biological activities of a viable organism. Pasteur initiated studies on fermentation in about 1857 and published an extensive paper in 1861 that established a number of properties of alcohol fermentation (Box 2.2).

In 1865, Napoleon III asked Pasteur to investigate the causative agent of bad wine. Pasteur reported that one specific organism that was different from the agent of good wine caused this problem. He suggested that the juices of grapes be heated to 50 to 55°C to destroy the resident populations and that the resultant material be inoculated with a proven producer of good wine.

In addition to working with wine, Pasteur studied the fermentation of beer over several years. He felt that it was his patriotic duty to make French beer superior to the brew produced elsewhere, particularly in Germany. Whether he laid the foundation for the ultimate production of a superior brew is left to the taste of the connoisseurs. His research resulted in the publication of a book on beer fermentation in 1876.

Pasteur's studies on butyric acid fermentation (conversion of sugar to butyric acid and other products) were very successful, as in this work he confirmed that anaerobic forms of life do exist. While microscopically examining fluid from butyric acid fermentation, he noted that organisms near the edge of the cover glass placed on the slide ceased movement, whereas those in the middle swam about vigorously. He wondered whether oxygen in the air might be harmful to the organisms. To test this hypothesis, he passed a stream of oxygen through a group of cells that were actively fermenting. He found that butyric acid production was halted and the organisms died. Pasteur was the first (1863) to use the terms

---

BOX 2.2   *Milestones*

Anaerobic Growth

In 1861 Louis Pasteur published a treatise entitled *The Influence of Oxygen on the Development of Yeast and on Alcoholic fermentation*. This study was a monumental contribution to our understanding of aerobic and anaerobic modes of life. Here Pasteur discussed his observation that cells of the yeast *Mycoderma vini (cerevisiae)*, added to a sugar-containing liquid in a shallow dish and in contact with air, grew prolifically. He suggested that under these conditions the organism "practically works for itself only" and produced virtually no alcohol. The sole products were cell mass and carbonic acid (CO_2). When the yeast acted on sugar in a much deeper dish and deprived of oxygen the submerged cells "grew slowly, with difficulty and of short

duration." Copious quantities of alcohol were produced. He noted also in this writing that a source of nitrogen (NH_4 or organic) was necessary for growth to occur.

Pasteur suggested that the yeast grew anaerobically by extracting oxygen from the sugar, with the generation of small amounts of energy and production of alcohol. He was correct in assuming that the sugar provided a replacement

for molecular oxygen, but this replacement was a product of sugar fermentation and not oxygen itself.

The major contribution offered by this study was a clear understanding of the aerobic–anaerobic mode of life and Pasteur put his theory in this concise formula: "Fermentation is life without air."

Yeast cells in contact with air grow rapidly to high density and produce little alcohol.

Yeast cells deprived of air grow slowly and produce large amounts of alcohol.

"aerobe" and "anaerobe" to describe the effects of air on microorganisms.

Pasteur's studies with microbes resulted in his development of new methods to work with these organisms. Many of these methods are still in use.

Knowledge of the extraordinary sterilizing effects of superheated steam came from studies in Pasteur's laboratory. He and his coworker, Charles Chamberland, noted that when hermetically sealed flasks were placed in a bath of calcium chloride and heated above 100°C, the flask solutions were free from viable organisms. Experimental work resulted in an improvement in this process so that equally effective sterilization resulted if the flasks were plugged with wool and heated in a closed container. Heating in a closed container resulted in internal temperatures that exceeded 100°C. The modern apparatus used for this purpose is the **autoclave**, which was originally manufactured in 1884 by a Parisian engineering firm under the name Chamberland's Autoclave.

In 1865, Pasteur's chemistry professor from his earlier days at the Sorbonne (who was in later life a senator from the south of France) asked his former pupil to investigate a devastating problem in silkworms. Pasteur was reluctant to go because, as a chemist, he knew nothing of silkworms. Eventually he took his entourage to Alais, France, to initiate studies on the disease. It took Pasteur five years and generous help from his able assistant, M. Gernez, to show clearly that the agent that was destroying the extensive silkworm industry in France was a transmissible microbe, specifically, the protozoan *Nosema*. Pasteur outlined a course for eradication of the disease based on isolating healthy worms, retaining the eggs produced, and examining the progenitor for a period of time. If the parent worm remained healthy, the eggs were allowed to hatch. These robust progeny were free of the disease, and following this regimen, the French silk industry was restored to its former glory.

Most of Pasteur's remaining years were dedicated to the understanding and prevention of infectious diseases in animals and humans.

Pasteur studied the disease anthrax in farm animals and suggested that immunization to this scourge was possible. Apparently, further experimentation was unsuccessful, and Pasteur abandoned the study. He was more successful in devising methods whereby immunization could prevent cholera in chickens. Pasteur obtained the chicken cholera organism and grew it in culture. Inoculation of chickens with laboratory cultures of this organism resulted in a mild illness from which they recovered. At a later date, a virulent (infectious) strain of the cholera organism was used to inoculate these chickens, but they were completely resistant to the disease. Apparently, growth in laboratory culture yielded a bacterial strain that was less virulent but could still elicit immunity. Pasteur realized the implications and potentials of this discovery and then turned to a study that was to be the crowning achievement of his career—the use of a weakened virus to prevent hydrophobia (rabies) in humans bitten by rabid dogs.

During the late nineteenth century, rabies was a serious health problem in France, and Pasteur decided to devote his efforts to the eradication of this dreadful disease. Perhaps he was intrigued by the always-fatal consequence of a bite by a rabid dog and sensed the profound effect that a cure would have on the scientific world. Pasteur recognized that the infection settled in the brain and nervous system of animals. All previous efforts to isolate a microscopically visible agent (now known to be a virus) of this disease had been unsuccessful. He outlined an empirical method to develop a vaccine that would counteract the feared effects of rabies.

In 1885, he published a method for protecting dogs after they had been exposed to rabies. A dog so exposed was protected by inoculation with an emulsion prepared from the dried spinal cord of a rabbit that had succumbed to the disease. Because he was able to prevent symptoms of rabies in animals exposed to other rabid animals through this procedure, Pasteur was convinced that the method was also applicable to unfortunate humans who were bitten by rabid animals. An opportunity to test this belief soon presented itself.

A nine-year-old boy, Joseph Meister, was brought to Pasteur in July 1885 after he had been severely bitten by a dog that was certainly rabid. Meister was injected over several days with the emulsions prepared from animal spinal cord material. After two weeks, the boy was given an injection of virus that had maximal virulence when tested in a rabbit. The boy survived, as did thousands of others treated by the same procedure, and Pasteur received worldwide acclaim.

In 1886, a commission was appointed within the Academy of Sciences in Paris to erect a scientific institute in honor of the man who had contributed so much to world health. More than 2.5 million francs were collected from throughout the world, and the Pasteur Institute was established. This institute continues today as a major scientific research center. Although Pasteur had many noble accomplishments during his lifetime, he may have erred in some of his judgments and experimentation. However, this should not detract from his great contributions to society. The well-known microbiologist A. T. Henrici summed it up well:

> It has been hinted that Pasteur was not always willing to give credit to those who had preceded him, that his ideas and experiments were not always strictly original with him. It is one thing to discover a truth, another to get it established as an accepted fact. What-

ever criticism may be directed towards Pasteur as regards the originality of his ideas, nothing can be said to belittle his ability to put them across. His genius for quick and accurate thinking, for keen argument, and for obtaining publicity, was not less important to the development of microbiology than his ingenious experiments. He "sold" the science to the public.

(From Henrici, A. T. 1939. *The Biology of Bacteria*, 2nd ed. New York: D. C. Heath & Co., p. 10.)

The Koch School

The Koch school of thought concentrated on the isolation of pure cultures of pathogenic (disease-causing) and saprophytic (live on dead or decaying matter) bacteria. The members of Koch's laboratory were the originators of pure culture methods and made significant contributions to many specialized branches of bacteriology (see later discussions).

Robert Koch (1843–1910) was trained as a physician and became a country doctor in Wollstein, East Prussia (Figure 2.8). His wife, seeking to allay his restless curiosity, gave him a microscope as a gift, little knowing the far-reaching consequences this small instrument would have on advancing science and in alleviating human misery. With his microscope, Koch examined many specimens, including the blood of an ox that had succumbed to **anthrax**. Anthrax is an infectious disease of warm-

Figure 2.9 Causative agent of anthrax
Koch's drawings of *Bacillus anthracis* that he recognized as the causative agent of anthrax in cattle. The drawing in the lower right indicates why he described the bacterium as "sticklike." From *The History of Bacteriology*, by William Bulloch, M.D., Oxford University Press, 1960.

Figure 2.8 Robert Koch
Robert Koch (1843–1910), the great medical microbiologist; Koch confirmed the "germ theory" of disease. From *The History of Bacteriology*, by William Bulloch, M.D., Oxford University Press, 1960.

blooded animals. (See Chapter 28 for a discussion of anthrax in humans.) He noted the constant presence of stick-like bodies in diseased animal blood, which were absent in blood taken from healthy animals (Figure 2.9). He found that the disease symptoms could be transmitted by inoculating a healthy mouse with blood from an animal that had died from anthrax. His tool for inoculation was a fire-sterilized splinter.

Koch surmised that these long, cylindrical bodies might be the viable causative disease agent and proceeded to culture the organism in fluid obtained from the eye of an ox. The organism, transferred in several passages of the fluid medium, would again cause anthrax when injected into the tail of a mouse. The transfers were done to ensure that no other agent was carried along that could cause the disease. This was the first clear experimental evidence that a bacterium was an agent of disease (see Figure 2.9). From these experiments, Koch formulated his theories on the causal relationships between microbes and disease (Box 2.3).

BOX 2.3	*Research Highlights*

Koch's Postulates

Studies on anthrax disease in cattle and the isolation of the infectious organism responsible for the disease symptoms led Koch to the supposition that many of the diseases evident in humans may well result from the actions of a specific microbial species. His protocol for establishing whether there is a causal relationship between microbe and diseases followed that employed in his earlier experimentation with *Bacillus anthracis*. He proposed a logical series of experimental tests that are known as **Koch's Postulates** that could be employed in confirming a causal relationship between microbe and disease. Koch's Postulates are as follows:

1. A specific microbe must be present in all cases of the disease and not present in healthy susceptible animal species.

2. The organism should be grown in pure culture outside the host.

3. When the pure culture is inoculated into a susceptible host, symptoms must develop equivalent to those observed in the original host.

4. The microbe must be reisolated from the experimentally infected host.

Illustrated below is the procedure Koch followed in demonstrating that the stick-like bodies (*B. anthracis*) that he isolated from dead cattle were the agents of anthrax.

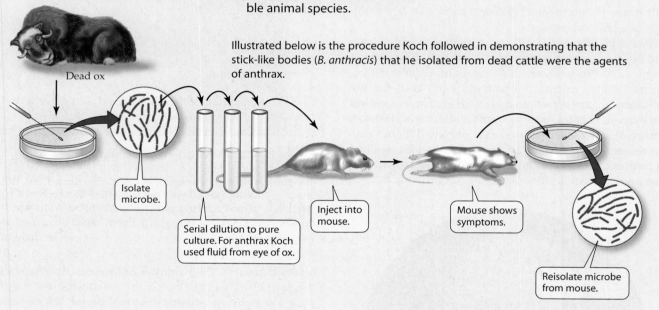

Dead ox

Isolate microbe.

Serial dilution to pure culture. For anthrax Koch used fluid from eye of ox.

Inject into mouse.

Mouse shows symptoms.

Reisolate microbe from mouse.

The magnitude of this advancement can be appreciated by considering that when Koch developed his methods in 1881, only anthrax was suspected of being caused by a microbe. During the next 20 years, Koch and his coworkers confirmed that microbes were the etiological (causative) agents of 15 significant human and animal diseases (Table 2.1).

Koch observed that the causative agent of anthrax has a life cycle involving a dormant spore, and he theorized correctly that these were the resistant bodies responsible for survival in soil. Koch (1876) wrote to Ferdinand Cohn, a noted scientist in Breslau, Germany, telling him of his investigations and was invited to Breslau to demonstrate his work. He was received with enthusiasm, and his discoveries were acknowledged with acclaim. Later he was invited to Berlin to set up a laboratory and devote his energies to studying microbes.

The origin of pure culture methods can be traced to Koch's observation that individual bacterial colonies growing on potato slices often differed in appearance (Figure 2.10). Microscopic examination of stained cells revealed that the organisms within a colony were similar, but were often unlike organisms in other colonies. Koch theorized that a colony arose from a single cell, and he developed streaking methods using a platinum loop (his invention) that enabled him to isolate organisms in pure culture. **Gelatin** was the solidifying agent for culture media during the early years of Koch's career (Box 2.4). However, gelatin had major shortcomings, and Fannie Hesse, the American-born wife of Wal-

TABLE 2.1	Diseases whose causative agents were discovered by Koch and his coworkers	
Date	**Discoverer**	**Disease**
1882	Koch	Tuberculosis
1882	Loeffler and Schutz	Glanders
1884	Koch	Asiatic cholera
1884	Loeffler	Diphtheria
1884	Gaffky	Typhoid fever
1884	Rosenbach	Staphylococcal and streptococcal infections
1885	von Bumm	Gonorrhea
1886	Fraenkel	Pneumonia
1887	Bruce	Malta fever
1887	Weichselbaum	Meningococcal infections
1889	Kitasato	Tetanus
1891	Wolff and Israel	Actinomycosis
1894	Kitasato and Yersin	Plague
1897	van Ermengen	Botulism
1898	Shiga	Acute dysentery

ter Hesse, a coworker in Koch's laboratory, suggested that **agar** could be used as a solidifying agent. At that time (1882), agar was added as a jellying agent in fruit preserves. This discovery was a considerable aid in the isolation of pathogenic microorganisms. In 1887, an assistant in Koch's laboratory, R. J. Petri, developed the

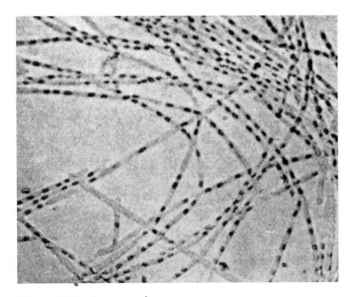

Figure 2.10 A pure culture
Koch's photomicrograph of bacteria from a single colony confirming that it was composed of one morphological type. From *The History of Bacteriology*, by William Bulloch, M.D., Oxford University Press, 1960.

dish (or plate) named in his honor that is still used in culturing bacteria. The design of the Petri dish is virtually unchanged from his original except that glass has been replaced with plastic.

Paul Ehrlich (1854–1915) was another coworker in Koch's laboratory and made far-reaching discoveries in immunology and chemotherapy. In devising staining methods for visualizing infectious microbes in tissue, he observed that bacterial cells present often absorbed selected dyes to a markedly greater extent than did surrounding tissue. Ehrlich reasoned that a toxic dye might destroy the bacterium without significant damage to the host tissue because so much more would adhere to the invading microbe. Ehrlich theorized that organic arsenicals might be synthesized that would be harmless to animals but toxic to invading parasites. **Arsenicals** are organic derivatives of arsenic, a toxic element. In his laboratory, scores of arsenicals were synthesized and tested on trypanosomal infections in mice. **Trypanosomes** are protozoan parasites that infect the blood of vertebrates. The 606th arsenical (Salvarsan) that he synthesized proved to be effective in curing these protozoan infections.

At the time, the spirochete that caused syphilis was reported to be related to the trypanosomes by its discoverer, Fritz Schaudinn. Ehrlich proceeded to test 606 as a cure for syphilis. It was remarkably successful, and thus the "magic bullet" (often mentioned in the popular press) against the dreaded disease syphilis was discovered.

Chemical antiseptics originated with **Joseph Lister** (1827–1912), an English physician, who employed car-

BOX 2.4　*Milestones*

Discovery of Agar as a Solidifying Agent

Agar has long been the universal solidifying agent used in preparing media for growing microbes, but it was not the first employed. The first solid medium was employed by Robert Koch and was an aseptically cut slice of potato. He was able to isolate colonies on this substrate and developed his theory that all organisms in a colony were of one species. Koch assumed correctly that all organisms in a discrete colony grew from a single cell. These studies were the origin of pure culture methods. The next solidifying agent employed by Koch was gelatin. Gelatin is a protein and was unsatisfactory for two principal reasons: (1) it melts at 37°C, the favored incubation temperature for most pathogens; and (2) many bacteria can digest gelatin.

In 1882 Fannie Hesse, the American-born (New Jersey) wife of Walter Hesse, suggested that a jellying agent used in making fruit preserves might be a replacement for gelatin in bacterial media. Walter Hesse (1846–1911) was an associate

Fannie Hesse, first to suggest the use of agar. Courtesy of ASM Archives, University of Maryland, Baltimore County.

in Koch's laboratory and probably discussed the problems of gelatin with his wife. Fannie Hesse's suggestion led to the adoption of agar as the choice for a solidifying agent.

The Dutch apparently brought agar from their East Indian colonies, where it was used to improve the setting quality of jam. Chemically, agar is an extract of algae that thrive in the Pacific and Indian Oceans and Japan Sea. Agar is a complex polysaccharide containing sulfated sugars. It cannot be digested by most microbial species. Laboratory grade agar is inhibitor free, and virtually all organisms grow well in its presence.

Agar melts at 100°C and remains in the liquid state down to about 45°C. The high melting point makes agar useful for growing all organisms including thermophiles at temperatures up to almost 100°C. The low solidifying temperature permits the addition of bacteria to melted agar (45°C). They can be distributed by mixing, and the isolated colonies can be observed.

Thus, an observation by Fannie Hesse on the qualities of a simple kitchen commodity became an object of worldwide utility.

---

bolic acid for antisepsis during surgery. Koch extended the work of Lister and devised a method for comparing the efficiency of chemical antiseptics. He dried cultures of bacteria, generally anthrax spores, on small pieces of silk thread, which were then immersed in the antiseptic solution. At intervals, a thread was removed from the antiseptic, washed in sterile water, and placed in growth medium to determine whether the organism remained viable. Koch found that carbolic acid was relatively weak in its disinfecting (killing) power, and of all the substances he tested, perchloride of mercury was most effective. It destroyed bacterial spores at a high dilution and in the shortest period. Mercuric compounds today remain widely used as disinfectants, although we are now concerned about the toxic effects of mercury in the environment.

Microbes as Agents of Environmental Changes

The adverse effects that microbes had on humans, food, and animals were the major considerations of the Koch and Pasteur schools. The causative agents of diseases, such as tuberculosis or rabies, caused much human suffering and challenged the scholar of scientific bent. During the latter years of the nineteenth century, knowledge of disease expanded rapidly along with a broad acceptance of the germ theory of disease. Another area that caught attention during the latter part of the nineteenth century was the potential role that microorganisms might play as a component part of all life on Earth. This leads us to the next conceptual development in microbiology—the role of microbes in nature.

Ferdinand Cohn suggested in 1872 that microbes were involved in the ultimate cycling of all living matter and that the activities of organisms in the biosphere allowed for the reutilization of cellular constituents. Our knowledge of the indispensable role that diverse microorganisms play in recycling constituents of living cells has expanded greatly since this pronouncement by Cohn. Microbial ecology, the study of the relationship between the microbe and the environment, and the scope of microbiology expanded markedly in the twentieth century.

An important technique developed by the great Dutch botanist **Martinus Beijerinck** (1851–1931) provided much of the foundation for the elucidation of the various functions of microbes in the cycles (carbon, nitrogen, and sulfur) of matter (Figure 2.11). He introduced the principles of enrichment culture, which gave clarity and rationality to microbial ecology. **Enrichment culture** is a means by which organisms that evolved to exist under any specific conditions of substrate, temperature, pH, salinity, osmolarity, and oxygen availability can be isolated. The sole limitation on isolation is the existence of an organism in the inoculum taken from a natural environment.

Figure 2.11 Martinus Beijerinck
Martinus Beijerinck (1851–1931), a major contributor to our understanding of the role of microbes in nature. From *Martinus Willem Beijerinck: His Life and His Work*, by G. van Iterson Jr., L. E. den Dooren de Jong, and A. J. Kluyver, Martinus Nijholt, The Hague, 1940.

The enrichment culture technique provided a means for the isolation of various physiological types of microorganism that exist in natural environments. An enrichment medium is prepared with a defined chemical composition and inoculated with soil or water rich in microbes; only those microbes capable of growth on that particular medium will grow. The microbes that thrive will be those best equipped by heredity to survive on that medium at the temperature, pH, and other conditions chosen. With this technique, Beijerinck and his followers readily obtained microbes with differing physiological capabilities, and they were able to assess the potential role of that microbe under natural conditions.

Beijerinck discovered free-living, nitrogen-fixing (assimilate atmospheric nitrogen) bacteria by the application of enrichment using a medium devoid of nitrogenous compounds such as ammonia or amino acids. The aerobic microbe that he obtained was given the genus name *Azotobacter* (Box 2.5). He also published extensively on the symbiotic nitrogen-fixing organisms (*Rhizobium*) that form nodules in the roots of legumes such as peanuts.

Beijerinck discovered and described many major groups of bacteria: the luminous organisms (*Photobacterium*), the sulfate reducers (*Desulfovibrio*), the methane-generating bacteria (now *Archaea*), and *Thiobacillus denitrificans*, an organism involved with denitrification (reduction of nitrates to nitrogen gas). Beijerinck did much of the early work on lactic acid bacteria and proposed the genus name *Lactobacillus*. He recognized that "soluble" living germs existed that he called "contagium vivum fluidum," which is generally accepted as the initial description of a virus (specifically, the tobacco mosaic virus). His contributions to microbiology are legion, and his perceptions of the great role of microbes established the foundation of modern approaches to microbial physiology and microbial ecology.

During the era of Koch, Pasteur, and Beijerinck, there arose another major figure in the field of general bacteriology (Figure 2.12). **Sergei Winogradsky** (1856–1953) was born in Russia in 1856, and during his long and fruitful life he witnessed the origins of the science of microbiology and survived to see the Age of Antibiotics. It is noteworthy that both Winogradsky and Beijerinck spent time in the laboratory of the great mycologist and plant pathologist **Anton de Bary** (1831–1888) in Strassburg, Germany. Winogradsky arrived in Strassburg shortly after Beijerinck left. In Strassburg, Winogradsky initiated studies on sulfur-oxidizing bacteria. He concluded that the bacterium *Beggiatoa* could utilize inorganic H_2S as a source of energy and atmospheric CO_2 for carbon in the synthesis of cellular material. He named these organisms "orgoxydants" and thus opened up the entire concept of **autotrophy**, which is the ability of bacteria to manufac-

BOX 2.5	*Research Highlights*

Atmospheric Nitrogen Fixation

Both Sergei Winogradsky and Martinus Beijerinck were aware that gaseous N_2 was released into the atmosphere in anaerobic environments. They also believed that the ongoing sustenance of life on Earth would occur only when this essential nutrient was replenished by natural means. Because microbes are so plentiful in virtually all soil and water, they proposed that they were most likely involved in this process (now called nitrogen fixation). Winogradsky assumed that N_2 fixation probably occurred via an anaerobic process,

and Beijerinck believed that it could occur aerobically. Accordingly they devised experiments that were consistent with their thinking. Both prepared enrichment cultures that were devoid of any form of fixed nitrogen and Winogradsky's were incubated anaerobically, whereas Beijerinck's were exposed to air.

Interestingly both Winogradsky and Beijerinck noted that a film developed on the surface of the liquid is the initial enrichment. Winogradsky did not transfer this but drew material from deeper in

the vessel as he sought anaerobes where as Beijerinck carried this surface material along. The organism in the film proved to be the aerobic N_2-fixing *Azotobacter* species. The anaerobe isolated by Winogradsky was identified as *Clostridium pasteurianum*. It is noteworthy that N_2 fixation is indeed an anaerobic process and strongly inhibited by molecular oxygen but the aerobic N_2 fixers have the ability to compartmentalize the process and keep O_2 away from the site of N_2 fixation.

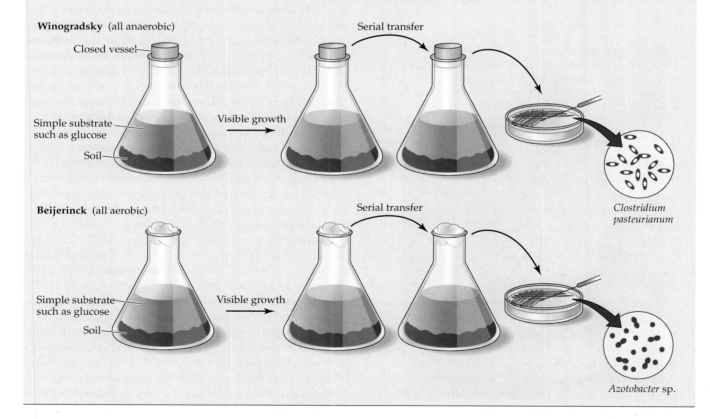

Winogradsky (all anaerobic)
Closed vessel
Simple substrate such as glucose
Soil
Serial transfer
Visible growth
Clostridium pasteurianum

Beijerinck (all aerobic)
Simple substrate such as glucose
Soil
Serial transfer
Visible growth
Azotobacter sp.

ture the componants of their cells from CO_2 and use inorganic chemicals or light as a source of energy. Prior to his study, only chlorophyll-containing plants were believed to use CO_2 as their sole carbon source.

Following the death of de Bary, Winogradsky went to Zurich, where he isolated and clarified the role of the bacteria that convert ammonia to nitrate, the autotrophic nitrifying bacteria. During this period, he also

Figure 2.12 Sergei Winogradsky
Sergei Winogradsky (1856–1953), a Russian-born microbiologist. Winogradsky was the father of autotrophy. He lived from the days of Pasteur and Koch to the modern era of microbiology. From *Sergei N. Winogradsky: His Life and Work*, by S. A. Waksman, ©1953 by the Trustees of Rutgers College. Reprinted by permission of Rutgers University Press.

showed that green and purple bacteria could oxidize hydrogen sulfide to sulfate, but he was uncertain whether this was a photosynthetic process. In addition, he isolated the nitrogen-fixing anaerobe *Clostridium pasteurianum*.

In 1891, the noted scientist and Nobel laureate **Elie Metchnikoff** (1845–1916) carried a personal letter from Pasteur to Zurich inviting Winogradsky to work at the Pasteur Institute in Paris. Metchnikoff was the discoverer of phagocytosis and cellular immunity. After much deliberation Winogradsky decided to accept a position in St. Petersburg, Russia, thus ending the first half of his illustrious scientific career. He did go to Paris in 1892 to represent Russia at the seventieth-birthday celebration for Pasteur and met Beijerinck for the first and only time. The revolution in Russia led Winogradsky to emigrate, and in 1922 he accepted an offer from the Pasteur Institute and moved to Paris. He spent the remainder of his life there studying the broad aspects of soil bacteriology. Winogradsky compiled his life's work and published it in a monumental treatise entitled *Microbi-ologie du Sol* (*Microbiology of the Soil*). He survived the deprivations of World War II and died in Paris at the age of 97.

Microbes and Plant Disease

Among the first to recognize that microbes might be directly involved in plant diseases was Anton de Bary, who in 1853 suggested that *Brandpilze* (plant rust) was caused by a parasitic fungus. He later proved experimentally that a fungus, *Phytophthora infestans*, caused late blight of the potato. This disease had caused crop failures in the 1840s in Ireland, and in 1845 M. J. Berkeley demonstrated that the blight was caused by a fungus. The potato blight continued in the 1850s and 1860s and led to a widespread famine, thousands of deaths, and the immigration of 1.5 million Irish to the United States.

Although the causal relationship between bacteria and animal disease had gained widespread acceptance by 1900, few botanists were willing to believe that bacteria could cause disease in plants. The fungi were generally acknowledged to be the agents of plant disease. J. H. Wakker, a Dutch scientist working with de Bary in 1881, isolated and identified the bacterium *Xanthomonas hyacinthi* as the pathogen causing widespread damage to hyacinth bulbs in Holland. Erwin Smith, in 1895, demonstrated that a bacterium caused a wilt in cucurbits (cucumbers and other members of the gourd family), and, in 1898, Beijerinck concluded that a virus was the cause of tobacco mosaic.

Despite these and other reports, considerable controversy remained regarding the role of bacteria in plant disease. It took two American scientists, Thomas Burrill and Erwin Smith, to convince the scientific world that bacteria did indeed cause plant disease. Burrill was the first to discover and demonstrate a bacterial disease of plants when he reported in 1877 that pear blight was caused by *Micrococcus* (*Erwinia*) *amylovorus*. Burrill investigated a number of diseases of corn, potatoes, and fruit, but he is probably best known as the first person in America to offer a laboratory course in bacteriology.

During the period from 1895 to 1900, Erwin Smith implicated bacteria in diseases of tomatoes, cabbage, and beans, and in the wilt of maize. By applying the postulates of Koch (a hero to Smith), his group at the U.S. Department of Agriculture provided definitive proof that strains of *Xanthomonas, Pseudomonas, Erwinia*, and *Corynebacterium* were the causative agents of disease in agricultural crops. Smith also demonstrated that crown gall, a plant tumor, resulted from an *Agrobacterium tumefaciens* infection. These studies established plant pathology as a field of scientific inquiry.

SECTION HIGHLIGHTS

In the mid-1800s, the work of Louis Pasteur finally laid to rest the concept of spontaneous generation. Pasteur also made many contributions to the understanding of fermentation and the prevention of infectious disease. Robert Koch established that bacteria are causative agents of infectious disease and formulated Koch's postulates, guidelines for the rigorous identification of disease-causing microbes that are used to this day. Other scientists, notably Martinus Beijerinck and Sergei Winogradsky, turned their attention to the role of microbes in environmental processes. In the late 1800s the first bacterial plant pathogens were identified.

2.7 Transition to the Modern Era

Louis Pasteur died in 1895, and Koch died in 1910. Beijerinck retired in 1921, and the first part of Winogradsky's scientific career ended at the turn of the century when he returned to Russia. Thus ended what is considered to be the "Golden Era of Microbiology," for during the 50-year period from 1870 to 1920, the discipline of microbiology was firmly established.

Albert Jan Kluyver (1888–1956) replaced Beijerinck as professor of microbiology at the Technical University at Delft in 1922. Knowledge of metabolic processes in microbes and in living cells in general was then quite limited. Most questioned whether metabolic reactions in microbes and in higher forms of life could occur in any equivalent manner. Knowledge of biochemical activities in microbes was mostly restricted to a number of unrelated transformations such as nitrification brought about by individual organisms.

During the decade following his arrival at Delft, Kluyver introduced order to chaos by presenting a simple, coordinated model for metabolic events in all living cells. His experimental approach led to our modern understanding that unity exists in biochemical reactions, a concept termed **comparative biochemistry**. This concept arose from his proposal that the basic feature of virtually all metabolic processes is a transfer of hydrogen (oxidation–reduction reactions). This transfer is universal and occurs whether the organism is aerobic or anaerobic, autotrophic or heterotrophic. He also believed that biosynthetic and biodegradative pathways in cells are highly coordinated and relatively few in number. He proposed that metabolic processes are functionally equivalent in all living cells. It is interesting to note that the general ideas behind the "Unity of Biochemistry" concept have been borne out by the recent comparative study of genomes.

Kluyver and **Cornelis B. van Niel** (1897–1985), one of his students, proposed that aerobic and anaerobic respiration could be illustrated by the following simple formula:

$$AH_2 + B \rightarrow A + BH_2$$

in which A represents a more reduced element, such as NH_4^+ or CH_4, and B represents oxygen in aerobic respiration or some less-reduced metabolic intermediate in anaerobic respiration. He was certain that oxidation–reduction reactions are basic to all life forms.

They also suggested that the general formula for all photosynthetic reactions would be as follows:

$$CO_2 + 2\,H_2A \xrightarrow{\text{light}} CH_2O\ (\text{cell material}) + H_2O + 2\,A$$

In plant photosynthesis, A would represent oxygen because the hydrogen atom for CO_2 reduction is donated by H_2O. Molecular oxygen (O_2) is released in this process. Prior to their studies, it was widely considered that light was used in photosynthesis to decompose carbonic acid and generate oxygen (O_2). For photosynthesis to occur in the anaerobic photosynthetic bacteria, A could be hydrogen, sulfide, or a reduced organic compound. As a consequence of the reduction of CO_2, an oxidized product such as sulfate would be generated.

One should not underestimate the contribution that the comparative biochemistry concept made to advances in the area of cellular metabolism. The bacteria became a major model system for study only with the general acceptance of this unity by all biologists. During the decade following World War II, a fundamental understanding of the role and structure of DNA, enzymes, metabolic regulatory mechanisms, and microbial genetics was attained. These discoveries led to a broad, integrated study of cellular growth that we now call **molecular biology**, a term coined by William Astbury in 1945.

Many characteristics of *Bacteria* and *Archaea*—rapid growth, simple nutritional requirements, growth under harsh conditions, and ease in handling—make them attractive tools for physiological studies. Because researchers generally choose to progress from simpler systems to those of greater complexity, the *Bacteria* and *Archaea* have provided an attractive model system for studying animal and plant processes. The broad array of *Bacteria* and *Archaea* species available—from those that use light as an energy source to the myxobacteria, which form fruiting bodies—makes them a system of choice for investigations on regulation, biosynthesis, gene expression, and molecular interactions. Studies of their ability to grow under extreme conditions of pH, temperature, salinity, and anaerobiosis (and combinations of these extremes) have yielded many clues to the elucidation of life strategies. Other unique characteristics that allow study at the molecular level include nitrogen fixation, antibiotic synthesis and action, parasitism (for bacteria and eucaryotes), directed motility, and development of bacteriophages (bacterial viruses).

Much might be written on the growth of microbiology as a scientific discipline over the last 50 years. The following chapters give ample evidence that much has been accomplished.

SECTION HIGHLIGHTS

During the early to mid-twentieth century the concept of comparative biochemistry—that unity exists among biochemical reactions—was developed and gained general acceptance. The realization that many cellular processes are similar in bacteria, archaea, and higher organisms led to the extensive use of microbes as model organisms.

SUMMARY

- **Epidemic diseases** played a major role in battles throughout recorded history. Napoleon was driven from Russia in 1812 by disease and deprivation and ultimately lost at Waterloo because about one-half of his army was incapacitated by illness.

- The first significant writings on contagious disease were those of **Girolamo Fracastoro** during the first half of the sixteenth century. He suggested that invisible organisms were the agents of disease.

- **Antony van Leeuwenhoek** observed and made extensive drawings of bacteria in 1683, but it was 200 years before "spontaneous generation" was disproved and the "germ theory" of disease was accepted.

- A number of scientists who preceded Louis Pasteur did much careful experimentation that disproved **spontaneous generation**. Among these were: Redi, Spallanzani, Schulze, Cagniard-Latour, Schwann, and Kützing.

- Edward Jenner developed an effective **vaccination** that protected humans against smallpox long before the germ theory of disease was established.

- When it became clear that animals could not spring from inanimate material, the proponents of spontaneous generation moved to changes caused by decay or putrefaction as proof that **chemical instabilities** were the basis for the observable changes in organic matter.

- By 1864 Pasteur had clearly proven that **spontaneous generation** did not occur, and his ideas on **fermentation** were generally accepted.

- Pasteur and his colleague, Chamberland, developed the **autoclave**.

- Pasteur was the first to observe and report that **anaerobic bacteria** existed. He also recognized that immunity to disease was possible and developed an effective rabies vaccine.

- The Academy of Sciences in Paris collected funds from around the world and established the renowned **Pasteur Institute** in honor of Louis Pasteur.

- Robert Koch was the first to clearly demonstrate that bacteria were a causative **agent of infectious disease**.

- Koch proposed a set of postulates (**Koch's postulates**) that could be employed to demonstrate that a culturable organism was the causative agent of disease symptoms.

- Koch developed **pure culture** methods. His laboratory developed the loop and the Petri dish and was the first to use agar as a solidifying agent for bacterial growth.

- Paul Ehrlich is considered the founder of **chemotherapy**, the use of chemicals to inhibit infectious organisms in vivo.

- Martinus Beijerinck was a pioneer in assessing the role of microorganisms in the **cycles of matter** (carbon, nitrogen, and sulfur) in nature. He discovered the free-living nitrogen-fixing *Azotobacter* species and the symbiotic nitrogen-fixing *Rhizobium* species.

- **Sergei Winogradsky** was the first to recognize that **autotrophic** bacteria were of widespread occurrence. He spent the last 34 years of his life at the Pasteur Institute studying the activities of bacteria in soil.

- The American scientists **Thomas Burrill** and **Erwin Smith** elucidated the role of bacteria in plant disease.

- **Albert Jan Kluyver**, a Dutch bacteriologist, proposed that basic metabolic reactions in all living cells were equivalent, a concept known as **comparative biochemistry**.

- *Bacteria* and *Archaea* have proven to be an excellent **model system** for studies in metabolism, genetics, and molecular interactions.

 Find more at www.sinauer.com/microbial-life

REVIEW QUESTIONS

1. What are some of the influences the microbe has had on human history? Discuss in terms of war, food, population control, and other factors.

2. It took 200 years, from Antony van Leeuwenhoek's observations and descriptions of bacteria (1673) until the latter part of the nineteenth century, for the role of bacteria to be elucidated. Why did it take so long?

3. Define "spontaneous generation." Why did this concept have so many proponents?

4. Spallanzi and Redi made significant contributions to disproving "spontaneous generation." What were some of their contributions, and why were their experiments not accepted as definitive proof? What did Franz Schulze add that was crucial?

5. Koch and Pasteur were interested in the effects of microbes on human health and well-being, and Winogradsky and Beijerinck in the role of microbes in nature. Give examples that support this reasoning.

6. How did Louis Pasteur contribute to conquering rabies? What experiment(s) led him to conclude that a treatment was possible?

7. Most of the basic techniques used in microbiological studies originated in Robert Koch's laboratory. What are some of them?

8. What are Koch's postulates? How did he use these in discovering the causative agents of disease? What are some of the common diseases for which Koch's laboratory discovered the causative agent?

9. Do bacteria play a role in plant disease? How did our understanding of the role of fungi and bacteria in plant disease evolve? What important historic development resulted from plant disease?

10. How did the philosophies of Albert Jan Kluyver and C. B. van Niel influence modern science?

SUGGESTED READING

Beck, R. P. 2000. *A Chronology of Microbiology in Historical Context*. Herndon, VA: ASM Press.

Brock, T. D., ed. and trans. 1999. *Milestones in Microbiology 1546–1940*. Herndon, VA: ASM Press.

Brock, T. D. 1999. *Robert Koch: A Life in Medicine and Bacteriology*. Herndon, VA: ASM Press.

Bulloch, W. 1938. *The History of Bacteriology*. London: Oxford University Press.

Dobell, C., ed. and trans. 1960. *Antony van Leeuwenhoek and His "Little Animals."* New York: Harcourt, Brace and Company.

Dubos, René. 1998. *Pasteur and Modern Science*. Herndon, VA: ASM Press.

Joklik, W. K., L. G. Ljungdahl, A. D. O'Brien, A. von Graevenitz and C. Yanofsky. 1999. *Microbiology: A Centenary Perspective*. Herndon, VA: ASM Press.

Kamp, A. F., J. W. M. La Riviere and W. Verhoeven, eds. 1959. *Albert Jan Kluyver: His Life and Work*. Amsterdam: North-Holland Publishing Co.

Vallery-Radot, R. 1920. *The Life of Pasteur* (translated by R. L. Devonshire). New York: Garden City Publishing Co.

Van Iterson, G., L. E. den Dooren de Jong and A. J. Kluyver. 1940. *Martinus Willem Beijerinck: His Life and Work*. The Hague, Netherlands: Martinus Nijhoff.

Waksman, S. A. 1953. *Sergei N. Winogradsky: His Life and Work*. New Brunswick, NJ: Rutgers University Press.

The objectives of this chapter are to:

◆ Review the structure of atoms and molecules.

◆ Review the types of chemical bonds that hold molecules together.

◆ Describe the other forces that influence intra- and inter-molecular interactions.

◆ Describe the properties of molecules particularly relevant to biological systems.

3

Fundamental Chemistry of the Cell

Life, atom of that infinite space that stretcheth,
twixt the here and there.
—Sir Richard Francis Burton, 1821–1890

*T*he basic physical principles that govern the chemical activities of all elements apply equally to both living and nonliving
systems. The living cell differs in that it has the capacity to
utilize these physical properties to generate energy, to grow, and to
reproduce. To understand viable systems, we must first have a fundamental knowledge of chemical principles and how these principles
relate to the structure and function of a cell. For example, we know
that all matter is composed of molecules that are made up of atoms.
We also know that matter has characteristics that are determined by
the way in which these atoms are joined to form molecules.

A molecule is formed when two or more atoms bond to one another. They may be like atoms and form a molecule such as O_2 or N_2,
or be unlike atoms and form a compound such as CO_2 or glucose.
Macromolecules are composed of interconnected compounds that,
working in concert, produce life. This chapter is devoted to an
overview of chemical principles underlying the formation of the molecules that collectively make up a living cell.

3.1 Atoms

The atom is the smallest unit that has all of the characteristics of an element. There are 92 different naturally occurring elements. An atom
can exist either as a single unit or in combination with other atoms.
The smallest atom is that of hydrogen, which consists of one proton
and one electron. The proton is located in the dense core of the atom
called the nucleus, and the electron orbits the nucleus at great speed.

Neutrons are present in the nuclei of all elements except hydrogen (protium lacks a neutron but deuterium—"heavy" hydrogen—has a neutron in its nucleus). A proton has a positive charge, an electron has a negative charge, and a neutron has no charge. The number of protons in the nucleus of an atom is equal to the number of electrons orbiting around it, which effectively renders an atom electrically neutral. The attraction between the positively charged nucleus and the negatively charged electrons keeps the atom intact. The nucleus contains 99.9% of the mass of one atom but occupies only one-hundred-trillionth (10^{-14}) of the volume.

The number of protons in the nucleus of any element is constant, and this is called the **atomic number**. The number of neutrons and electrons associated with an atom may vary. There is almost the same number of protons and neutrons in the nucleus, and the sum of these is the **mass number**. The average atomic mass for a naturally occurring element, expressed in atomic mass units, is called its **atomic weight**. Electrons are inconsequential to the atomic weight of an atom because an electron is only 1/1836 of the relative mass of a proton or neutron. The elements that are most abundant in living systems in nature generally contain an equal number of protons and neutrons.

Isotopes are atoms which have the same atomic number (or number of protons) but different mass numbers. Some isotopes of biological interest are listed in Table

3.1. The atomic number of an isotope does not differ from the natural element because the number of protons is invariant. The atomic weight, of course, differs because of a different number of neutrons.

Some isotopes are stable, whereas others spontaneously decay. Those that decay with the release of subatomic particles (radioactivity) are termed **radioisotopes**. Radioisotopes can occur naturally and most used in laboratory studies are produced by physical means called neutron bombardment. Thermal neutrons that are activated during nuclear fission may enter the nucleus of an atom, thereby creating an isotope.

A naturally occurring element may consist of either a single isotope, as is the case for sodium, or a mixture of two or more isotopes. For example, the most common naturally occurring isotope of oxygen is ^{16}O, but ^{17}O and ^{18}O also are found in nature. The relative abundance of these isotopes in the environment is 99.76%, 0.04%, and 0.01%, respectively. The radioisotopes of carbon (^{14}C), phosphorus (^{32}P), sulfur (^{35}S), and the radioisotope of hydrogen (^{3}H), commonly called tritium, are widely employed in research. Radiolabeled atoms can be detected in exceedingly small amounts, thus allowing scientists to follow the fate of a radioisotope in a biological system. The use of a radioisotope in studying metabolic pathways is illustrated in Box 3.1. The stable isotopes of carbon (^{13}C), nitrogen (^{15}N), and oxygen (^{18}O) are also employed in biological research, and their presence is

TABLE 3.1	Some isotopes of biological interest			
Element	**Protons**	**Neutrons**	**Isotope**	**Type**[a]
Sulfur	16	16	^{32}S	Elemental
	16	18	^{34}S	Stable isotope
	16	19	^{35}S	Radioisotope
Carbon	6	6	^{12}C	Elemental
	6	7	^{13}C	Stable isotope
	6	8	^{14}C	Radioisotope
Oxygen	8	8	^{16}O	Elemental
	8	9	^{17}O	Stable isotope
	8	10	^{18}O	Stable isotope
Phosphorus	15	16	^{31}P	Elemental
	15	17	^{32}P	Radioisotope
Nitrogen	7	7	^{14}N	Elemental
	7	8	^{15}N	Stable isotope
Hydrogen	1	0	^{1}H	Elemental
	1	2	^{3}H	Radioisotope

[a]The elemental form is most abundant in nature and has virtually the same number of protons and neutrons in the nucleus. Stable isotopes have a greater number of neutrons, but the excess neutrons do not decay. The excess neutrons in radioisotopes spontaneously decay with the release of subatomic particles.

BOX 3.1 | *Milestones*

i CO₂ Fixation and the Nuclear Age

In 1796, Jan Ingen-Housz, a Dutch scientist, wrote that plants obtain their cellular carbon by assimilating carbon dioxide from the environment. Subsequently, others discovered that selected bacteria could also obtain their cell carbon by "CO_2 fixation." Elucidating the pathway whereby CO_2 was incorporated into a cell proved to be very difficult. From a biochemical standpoint, autotrophs (CO_2 as sole carbon source) were indistinguishable from heterotrophs (organics as carbon source). Data available indicated that CO_2 entered into metabolic sequences much the same as sugars and other substrates. But how? The answer to this intriguing mystery came only when better tools were available. Ironically, the research that provided these tools occurred indirectly in the development of the atomic bomb during World War II (1941 to 1945). Research that led to atom bombs brought us into the nuclear age. The lessons learned in bomb development have been extended to provide us with nuclear energy for electrical power generation, nuclear medicine, and important research tools. In a way, the problem of CO_2 fixation ran headlong into the nuclear age.

Under proper conditions, a nuclear reactor can be manipulated to introduce extra neutrons into the nuclei of selected elements. The extra neutrons in the nucleus of an element tend to decay with the release of radioactivity (subatomic particles). This radioactivity is readily detectable and permits us to locate the site of a radioactive compound in a cell. A radioactive compound is also called a **tracer** because we can "trace" the movement of a radioactive compound in a biological system. With the knowledge that became available in 1947, it was possible to create radio-labeled carbon dioxide and follow its incorporation into an organism during photosynthesis. Radiolabeled carbon dioxide is abbreviated $^{14}CO_2$, indicating that there are eight neutrons and six protons present in the carbon atoms rather than the normal six of each. The total of 14 neutrons plus protons is indicated by the superscript 14, as this is the mass number of the isotope.

In 1949, Melvin Calvin and Andrew Benson initiated a series of experiments employing $^{14}CO_2$, and their studies led to an understanding of the biochemistry of CO_2 fixation by exposing photosynthesizing algae to $^{14}CO_2$ for varying lengths of time. They then isolated and identified the radioactive compounds present in the cells. They reasoned that the compounds that became radioactive after brief exposure were those in which the CO_2 was initially fixed. Those compounds involved in succeeding reactions would become radioactive with time. Analysis of the pattern of $^{14}CO_2$ fixation into compounds and the distribution of the radiolabel within these compounds ultimately answered the question of how a cell can fix carbon from the atmosphere. CO_2 fixation occurs in most autotrophs as follows:

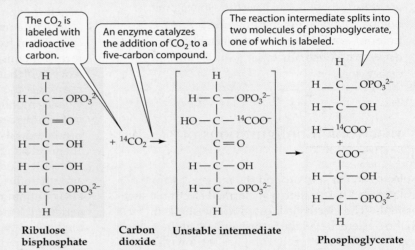

> The CO_2 is labeled with radioactive carbon.

> An enzyme catalyzes the addition of CO_2 to a five-carbon compound.

> The reaction intermediate splits into two molecules of phosphoglycerate, one of which is labeled.

Ribulose bisphosphate | Carbon dioxide | Unstable intermediate | Phosphoglycerate

Thus, a seemingly unrelated event, the development of the atomic bomb, presented the unlikely solution to a major biological question. We now know that there are other CO_2 fixation reactions in autotrophic bacteria. These reactions will be discussed in Chapter 10.

> Algae plus $^{14}CO_2$

> Light

Calvin–Benson experiment. Courtesy of the Lawrence Berkeley National Laboratory.

determined by mass spectrometry. Stable isotopes have more neutrons than the natural form of the element and therefore a larger mass.

3.2 Electronic Configurations of Molecules

The electrons that spin around the nucleus assume definite patterns. These patterns are termed electronic shells or orbitals. The electrons located in the shells farthest from the nucleus travel the fastest and are at the highest energy level. The first orbital (1s) has the lowest energy level. The other orbitals, which are farther from the nucleus, are termed the p, d, and f orbitals. Whereas the 1s orbital is spherical, the other orbitals can be either spherical or dumbbell-shaped. The electrons can be depicted as an electron cloud that surrounds the nucleus, as presented in Figure 3.1. The distribution of electrons around the nucleus may also be referred to as the electron con-

figuration. An electron will fill a lower energy orbital first, and the total number of electrons that can occupy the first orbital is two; the second shell has four orbitals (see Figure 3.1) and can contain eight electrons; the third has a maximum of 18 electrons in nine orbitals; and the fourth contains 32 electrons in 16 orbitals.

There is a tendency for atoms to achieve an electron configuration in which the outer orbit will contain the maximum allowable number of electrons. Thus, an atom may donate or gain electrons to attain this configuration. When the outer orbital is filled, a chemically stable state is attained. For example, argon, neon, and helium have filled outer shells, and these gasses are mostly inert (unreactive). Some molecules of biological interest are depicted in Figure 3.2.

The **valence** of an atom is, by definition, based on the capacity of hydrogen to donate (share) an electron to form a chemical bond. An atom may have a positive valence, which is the number of electrons it can donate in forming a chemical bond with another atom. A negative valence represents the number of electrons it may gain in forming a stable compound. Thus in the examples presented, two atoms of hydrogen (valence +1) form a chemical bond with oxygen (–2) to form water, a neutral molecule. Carbon has four electrons in its outer orbital and, therefore, can form a stable molecule by donating (+4) or attaining (–4) electrons. This can occur by interaction of 4 molecules of hydrogen (+1 each) to form methane or with 2 molecules of oxygen (–2 each) to generate carbon dioxide.

(A)   (B)

Two dumbbell-shaped orbitals, p_x and p_y, have two electrons each. The p_y orbital is perpendicular to the p_x orbital.

- ⊖ Electron
- ⊕ Proton
- ○ Neutron

The two electrons closest to the nucleus move in the spherical orbital, s.

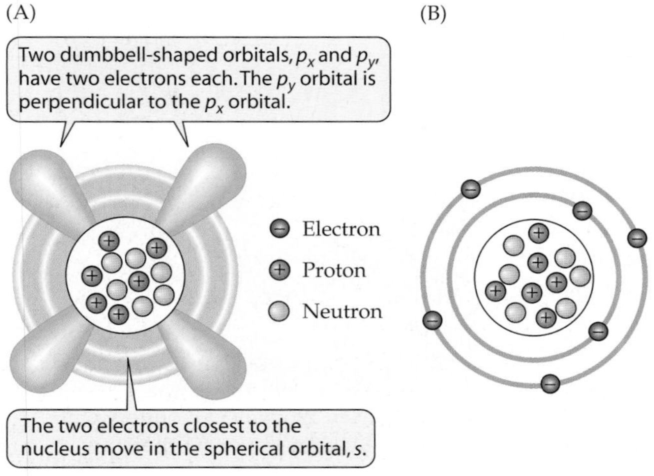

Figure 3.1 Structure of an atom
(A) Orbital model of a carbon atom, showing spherical and dumbbell-shaped orbitals. (B) The Bohr model (named for the Danish physicist Niels Bohr), a less accurate representation but commonly used for convenience. The nucleus in these representations is proportionately much larger than in an actual atom, simply to accommodate display of the protons and neutrons (this applies also to Figures 3.2, 3.3, and 3.4).

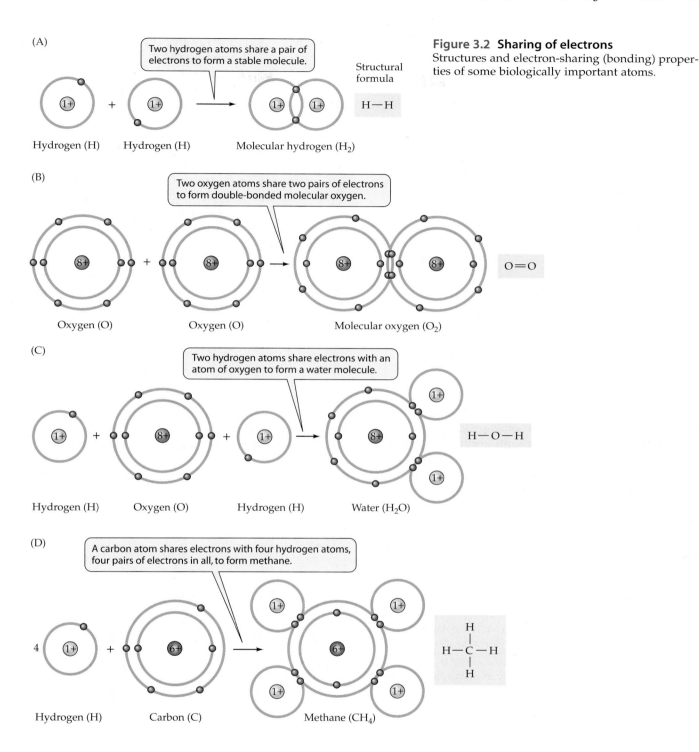

Figure 3.2 Sharing of electrons
Structures and electron-sharing (bonding) properties of some biologically important atoms.

(A)

Two hydrogen atoms share a pair of electrons to form a stable molecule.

Structural formula

H—H

Hydrogen (H)   Hydrogen (H)   Molecular hydrogen (H₂)

(B)

Two oxygen atoms share two pairs of electrons to form double-bonded molecular oxygen.

O=O

Oxygen (O)   Oxygen (O)   Molecular oxygen (O₂)

(C)

Two hydrogen atoms share electrons with an atom of oxygen to form a water molecule.

H—O—H

Hydrogen (H)   Oxygen (O)   Hydrogen (H)   Water (H₂O)

(D)

A carbon atom shares electrons with four hydrogen atoms, four pairs of electrons in all, to form methane.

H
|
H—C—H
|
H

Hydrogen (H)   Carbon (C)   Methane (CH₄)

3.3 Molecules

An atom may exist separately, in combination with dissimilar atoms, or, in some cases, combined with like atoms. For example, two atoms of hydrogen may share electrons (one from each atom) to form H_2 (see Figure 3.2A). Oxygen atoms are generally paired, and the Earth's atmosphere contains O_2 (see Figure 3.2B). However, atoms most often pair with unlike atoms, and this results in the formation of compounds. A compound that is formed by combinations of atoms other than carbon is considered an inorganic compound; those that contain carbon are generally termed organic compounds (exceptions would be compounds such as carbon monoxide or dioxide). The configurations of other simple compounds are illustrated in Figure 3.2C and D. For example, oxygen picks up one electron from each of two hydrogen atoms to fill its outer shell and form water. Methane is formed from the sharing of four electrons in the second shell with four atoms of hydrogen.

When we wish to denote the chemical composition of a molecule, we use a **chemical formula**. The chemical formulas for water, methane, and carbon dioxide are H_2O, CH_4, and CO_2, respectively.

SECTION HIGHLIGHTS

Atoms can be combined to form molecules. A molecule formed from nonidentical atoms is called a compound. Compounds containing carbon atoms are termed organic; those without carbon atoms are inorganic.

3.4 Chemical Bonds

The forces that hold molecules together are called chemical bonds. In general, chemical bonds are formed because the product(s) are in a lower energy configuration than the reactants. There are three main types of bonds that join atoms or molecules together: ionic bonds, covalent bonds, and hydrogen bonds. These bonds are formed because atoms seek a full complement of electrons in the outer shell or tend to achieve neutrality through interactions between polar (charged) molecules. There are a number of theories proposed for the variety of geometries present in molecular species. The level of complexity in the theories has grown in parallel with experimental observation. Although valence bond theory can be applied to many simple organic and inorganic compounds, there are other models, and no single theory adequately explains all observations. In fact the "ionic" and "covalent bonds," as presented here, are extreme cases and there is actually a smooth continuum of interactions that lie between them.

Ionic Bonds

Two atoms achieve a full complement of electrons in their outer orbital when one of the atoms donates an appropriate number of electrons and the other atom accepts these electrons. The atom that loses electrons becomes a **cation** (**positive ion**), and the atom that gains electrons becomes an **anion** (**negative ion**). The loss of electron(s) in the donor results in an excess of protons relative to electrons, and the gain of an electron(s) results in an excess of electrons over protons. An element with an excess number of

protons is referred to as positive (+). Some examples include Na^+, K^+, Ca^{2+}, Mg^{2+}, and Mn^{2+}. An element with an excess of electrons is negatively (−) charged; examples include Cl^-, I^-, and S^{2-}. An element with a positive charge can bond with an element that has a negative charge. The bond that joins them is considered an **ionic bond**.

The classical example of the formation of a compound by ionic bonding is the reaction between Na^+ and Cl^- to form sodium chloride—NaCl (table salt) (Figure 3.3). Sodium donates an electron and achieves a full complement of electrons (8) in the outermost shell. It is actually an ion because it carries an electrical charge due to the excess of one proton relative to electrons (Na^+). Chlorine picks up one electron, resulting in a full outer shell, but it now has one more electron than protons in the nucleus (Cl^-). The attraction between the cation Na^+ and the anion Cl^- results in the formation of a salt. Each ring surrounding the nucleus represents a shell.

Ionic bond formation has limited applicability to biological systems. However, it is important in the neutralization of DNA in a bacterial cell. DNA is acidic;

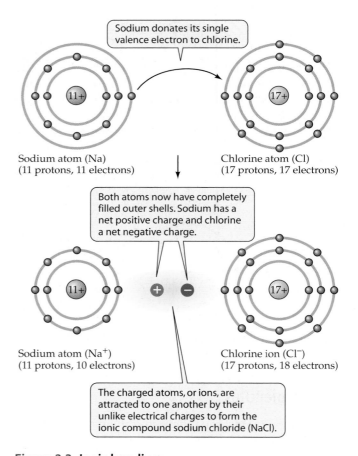

Sodium donates its single valence electron to chlorine.

Sodium atom (Na)
(11 protons, 11 electrons)

Chlorine atom (Cl)
(17 protons, 17 electrons)

Both atoms now have completely filled outer shells. Sodium has a net positive charge and chlorine a net negative charge.

Sodium atom (Na^+)
(11 protons, 10 electrons)

Chlorine ion (Cl^-)
(17 protons, 18 electrons)

The charged atoms, or ions, are attracted to one another by their unlike electrical charges to form the ionic compound sodium chloride (NaCl).

Figure 3.3 Ionic bonding
Formation of a compound by ionic bonding. Shown here is the compound sodium chloride.

TABLE 3.2	Types of covalent bonds that are an integral part of biomolecules— these are the major functional groups present in living cells

Carbon

$$-\overset{|}{\underset{|}{C}}-H \qquad -\overset{|}{\underset{|}{C}}-\overset{|}{\underset{|}{C}}- \qquad \overset{}{\underset{}{>}}C=C\overset{}{\underset{}{<}}$$

$$-\overset{|}{\underset{|}{C}}-O-\overset{|}{\underset{|}{C}}- \qquad \overset{}{\underset{}{>}}C=O \qquad -\overset{|}{\underset{|}{C}}-OH$$

Nitrogen

$$\overset{}{\underset{}{>}}N-H \qquad -\overset{}{\underset{}{>}}C-N= \qquad -\overset{|}{\underset{|}{C}}-N\overset{}{\underset{}{<}} \qquad \overset{}{\underset{}{>}}C=N-$$

Phosphorus

$$\overset{\backslash|}{\underset{/|}{P}}-O-\overset{|}{\underset{|}{P}}\overset{}{\underset{}{<}} \qquad \overset{}{\underset{}{>}}P=O$$

Sulfur

$$-S-H \qquad -\overset{|}{\underset{|}{C}}-S- \qquad \overset{\backslash|}{\underset{/|}{S}}-O-$$

$$\overset{}{\underset{}{>}}S=O \qquad -S-S-$$

therefore, divalent cations such as Mg^{2+} form ionic bonds with and neutralize the anionic phosphate PO_4^{2-} in bacterial DNA. Neutralization of DNA in eukaryotes differs in that it is neutralized through interaction with basic (positively charged) proteins termed histones.

Covalent Bonds

The **covalent bond** is more relevant than the ionic bond in biological systems because it is the major bond joining the component elements that form the molecules of the cell. Carbon-to-carbon bonds are examples of covalent bonds and are the basic backbone in cellular lipids, proteins, nucleic acids, and carbohydrates. Some of the covalent bonds that are important in cell structures are presented in Table 3.2. A covalent bond is formed when one or more pairs of electrons are shared. The simplest example of shared electrons is the hydrogen molecule where

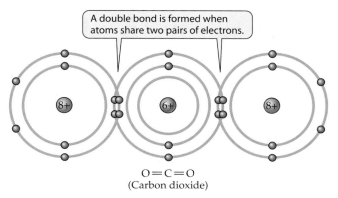

A double bond is formed when atoms share two pairs of electrons.

O=C=O
(Carbon dioxide)

Figure 3.4 Formation of double bonds
The sharing of electrons between an atom of carbon and two atoms of oxygen in carbon dioxide.

two protons share two electrons (see Figure 3.2A). These electrons orbit both nuclei. This is an example of a **single covalent bond** because one pair of electrons is shared. Other examples of the sharing of a single pair of electrons are methane and water. A **double covalent bond** is formed when two pairs of electrons are shared. An example of this is carbon dioxide (Figure 3.4). Carbon dioxide is a stable molecule because each oxygen molecule has a valence of –2 and carbon has a valence of +4. The four electrons in the outer shell of the carbon atom are shared with two electrons from each of two oxygen atoms, resulting in a full complement of electrons in the outer shell of each atom. Double bonds, the sharing of two electron pairs, occur frequently between adjacent carbons (C=C) in many of the important compounds in cells.

In some cases, a triple bond occurs because three pairs of electrons are shared. A biomolecule of this type is N≡N (N_2). *Bacteria* and some of the *Archaea* are able to cleave triple bonds, as some species utilize molecular nitrogen gas from the atmosphere as a source of organic nitrogen. This cleaving of N_2 by these organisms requires considerable expenditure of energy because the N≡N triple bond is one of the strongest of all chemical bonds (Table 3.3).

Hydrogen Bonds

The relatively large nucleus of the nitrogen or oxygen attracts the electron from hydrogen more strongly than does the small single proton in the hydrogen nucleus. In a molecule of water, the total electrons that are orbiting the two nuclei at any given moment are closer to the oxygen nucleus than to the hydrogen nucleus. This generates a partial electronegativity (δ^-) surrounding the oxygen nucleus and a positivity (δ^+) around the hydrogen nucleus, and results in an attraction between the posi-

TABLE 3.3	Bond energies between atoms of biological importance

Single Bonds	Energy consumed in breaking bond Kcal/mol
C—C	82
O—O	34
S—S	51
C—H	99
N—H	94
O—H	110
Multiple Bonds	
C=C	147
O=O	96
C=N	147
C=O	167
N≡N	226

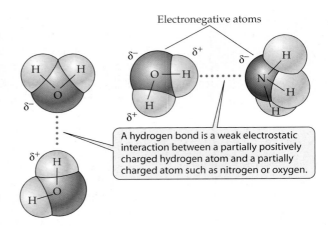

Figure 3.5 Hydrogen bond
A hydrogen bond forms between a hydrogen atom covalently bonded to a larger electronegative atom and another electronegative atom.

tive part of a water molecule and the negative part of another electronegative atom such as ammonia or another water molecule (Figure 3.5). The hydrogen of the water molecule is attracted to the nitrogen atom, and the resultant bond between the two molecules is called a **hydrogen bond**. A hydrogen bond possesses about 5% of the strength of a covalent bond.

Hydrogen bonds are of considerable importance in biological systems. The double helix in DNA and conformation of cellular proteins are dependent on hydrogen bonding. The role of hydrogen bonding in the double helix conformation of DNA is illustrated in Figure 3.6, where hydrogen bonding is involved in the planar

arrangement of base pairs. Adenine (A) and thymine (T) are always bonded via two hydrogen bonds. Guanine (G) is always paired with cytosine (C) through three hydrogen bonds. The helical core is also maintained by stacking interactions between the planes of adjacent base pairs. These stacking interactions involve dipole–dipole and van der Waals forces (discussed in subsequent text) that are equal in magnitude to the stabilizing energy of the hydrogen bonds between base pairs.

Hydrogen bonding is important in the conformation of functional proteins, where bonding can occur either between polar constituents of amino acids on separate polypeptides chains or between polar areas within a

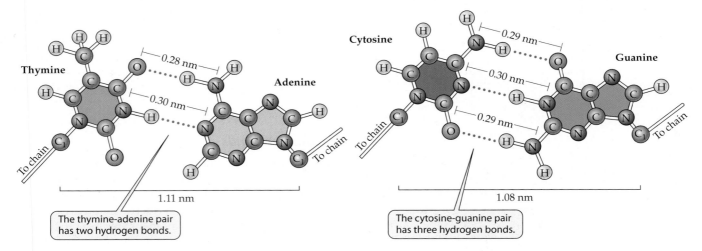

Figure 3.6 Hydrogen bonding in DNA
Hydrogen bonding between A and T and between C and G retains the planar arrangement of purines and pyrimidines in the double helix of DNA. The AT and GC pairs are equiva

lent in dimension, which is essential to the compact helical structure of DNA. Other interactions at the periphery of the helix maintain the "stacked" arrangement of base pairs one above the other.

polypeptide. Polypeptide chains with internal bonding can assume a helical shape (Figure 3.7A). In this case, the hydrogen bonds between three amino acids form a loop (helix). Two individual polypeptide chains may form sheets by bonding portions of the same polypeptide (Figure 3.7B).

SECTION HIGHLIGHTS

Molecules are held together by three types of chemical bonds: ionic bonds, covalent bonds, and hydrogen bonds. Ionic bonds are formed when electrons have been transferred from one atom to fill the outermost shell of another atom, forming a positively charged cation and a negatively charged anion. The oppositely charged atoms are attracted to one another, forming an ionic bond. A covalent bond is formed when one or more pairs of electrons are shared by two atoms, orbiting around both nuclei. A hydrogen bond is the weak electrostatic interaction that occurs between a positively charged hydrogen atom of one molecule and a negatively charged atom, such as oxygen or nitrogen, from another molecule.

3.5 van der Waals Forces

van der Waals forces are interactions that occur between adjacent nonpolar molecules and are not due to ionic, covalent, or hydrogen bonds. These interactions can be based on either an attraction or a repulsion. The attractions are due to short-lived fluctuations in the electron charge densities that surround adjacent nonbonded atoms. This fluctuation is called a dipole moment. A dipole occurs when a molecule that is electrically neutral becomes polar. This polarity results when the center of negative charge (the electrons) does not center on the site of positive charge (Figure 3.8). Consequently, there is a momentary polarity in the otherwise neutral molecule, and this can attract an opposite charge in a neighboring molecule. Some hydrophobic molecules such as hexane are liquids rather than gases because the molecules are held together by weak van der Waals forces. These van der Waals forces also play a role in enzyme substrate binding and protein interactions with nucleic acid.

Repulsion occurs when atoms that are not covalently bonded move too close together and repel one another. The electrons surrounding one molecule overlap the space occupied by the electrons of another, and the negative charges tend to push the atoms apart.

SECTION HIGHLIGHTS

Molecules can also be attracted to one another through short-lived fluctuations in their charge densities that cause them to have transient opposite charges. Also, two atoms can be repelled from one another if their electron clouds move too close together. These interactions are known as van der Waals forces.

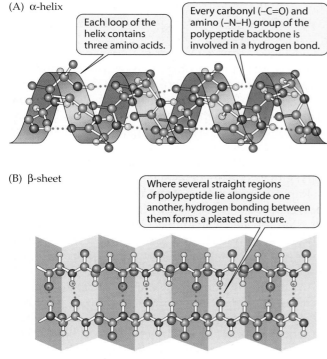

(A) α-helix

Each loop of the helix contains three amino acids.

Every carbonyl (–C=O) and amino (–N–H) group of the polypeptide backbone is involved in a hydrogen bond.

(B) β-sheet

Where several straight regions of polypeptide lie alongside one another, hydrogen bonding between them forms a pleated structure.

Figure 3.7 Hydrogen bonding in a peptide
Models of (A) α-helix and (B) β-sheet as they occur in a protein.

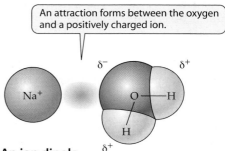

An attraction forms between the oxygen and a positively charged ion.

δ^- δ^+

Na$^+$ O — H

H

δ^+

Figure 3.8 An ion dipole
An ion dipole between a sodium ion (Na$^+$) and a polar molecule (water). As electrons orbit around the atoms of the water molecule they can "stray" toward the oxygen atom, giving it a partial negative charge. A number of water molecules orient around the ion, a process called hydration (see Figure 3.9).

3.6 Hydrophobic Forces

Hydrophobic forces are significant in biological systems and are complex. These forces essentially prevent a solute from interacting with a solvent. A hydrocarbon such as hexane is virtually insoluble in water because the water (solvent) withdraws in the area of contact with the apolar hydrophobic molecule (solute). The water molecules form a rigid hydrogen-bonded network among themselves. This network effectively restricts the orientation of water molecules at the interface. The molecules of the hydrophobic compound cannot interact with the tightly oriented water molecules. This is why oil floats on water and hydrocarbon chains in the core of cytoplasmic membranes are effective in creating a barrier that is impenetrable to most substances. Hydrophobicity is also involved in the protein–nucleic acid interactions that preserve the tertiary structure of DNA.

SECTION HIGHLIGHTS

Hydrophobic forces prevent the interaction of a non-polar solute (such as oil) with a polar solvent (such as water). Hydrophobic forces are important for protein–nucleic acid interactions and for maintaining the structure of proteins.

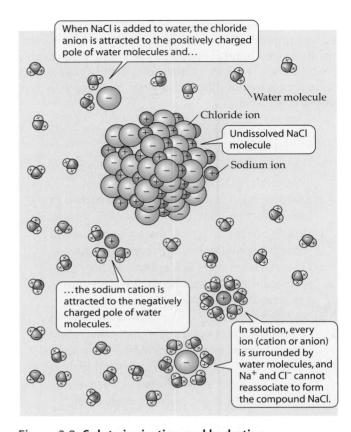

When NaCl is added to water, the chloride anion is attracted to the positively charged pole of water molecules and…

Water molecule

Chloride ion

Undissolved NaCl molecule

Sodium ion

…the sodium cation is attracted to the negatively charged pole of water molecules.

In solution, every ion (cation or anion) is surrounded by water molecules, and Na^+ and Cl^- cannot reassociate to form the compound NaCl.

Figure 3.9 Solute ionization and hydration
When an ionic compound is added to water, it dissolves as it ionizes and its component ions are hydrated.

3.7 Water

Water is the basic requirement of all living cells. Life evolved in water, and without water there is no life. If moisture is present in a microenvironment, it is probable that microorganisms (bacteria, cyanobacteria, or fungi) will be present as well. Water is the solvent in living cells, and a cell is 60% to 95% water. Even inert spores or seeds are 10% to 20% water by weight.

Water is essentially a neutral compound. The two constituent elements, oxygen and hydrogen, however, differ in electronegativity, and the distribution of charge around these two elements is asymmetrical. As a result, water actually has polarity (see Figure 3.5). This polarity allows for hydrogen bonding and an affinity of water molecules for one another. Water molecules are dynamic and are constantly involved in arrangements where hydrogen bonds are formed and broken. This and other properties of water make it an ideal solvent.

Polarity permits water molecules to surround a solute and bring it into solution. The classical example of a solvent–solute interaction is sodium chloride in water (Figure 3.9). The lone pair of electrons of the H_2O molecule may attract the positive parts of a solute and vice versa. These interactions permit the sodium and chloride ions to solubilize. Pure water has a high **dielectric constant**, which means that it has a high ability to act as an insulator of electric charges. Organic acids and amino acids are also ionizable, and this is a major factor in their water solubility. Sugars also contain many polar hydroxyl (OH) groups and are generally quite soluble in water.

Rapid temperature changes do not occur when water is exposed to heat because of the extensive hydrogen bonding between water molecules (Figure 3.10). Generally heat absorption by molecules increases their kinetic energy (both the rate of motion of molecules and their reactivity). Absorption of heat by water, however, results in the hydrogen bonds breaking apart rather than increasing rates of motion of the atoms. Much more heat is required to raise the temperature of water than would be required for a non–hydrogen-bonded liquid. This makes water an excellent insulator because it is slow to heat and slow to cool. This property is important to warm-blooded animals but of somewhat limited conse-

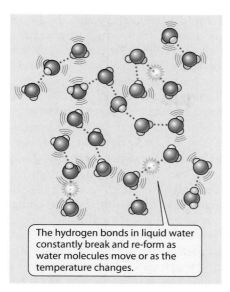

The hydrogen bonds in liquid water constantly break and re-form as water molecules move or as the temperature changes.

Figure 3.10 Hydrogen bonding of water
Hydrogen bonding between water molecules causes them to stick together. This property underlies the cohesive and adhesive properties of water, responsible for surface tension (cohesive forces) and capillary action (adhesive forces).

quence to microbes that assume the temperature of their environment.

Water can form thin layers on surfaces in part because of hydrogen bonds. This phenomenon is termed **surface tension** and is due to the attraction of water molecules for one another (see Figure 3.10). Surface tension occurs on the surfaces of bodies of water and allows insects to "walk" on water because of the adhesion between water molecules. Surface tension is also a function of the relative interaction between dissimilar liquids or a liquid–solid interface.

A thin film of water protects biological membranes. Water atoms actually bond with oppositely charged atoms at or near the surface of the membrane. As a result of this surface layer of water a membrane retains fluidity and flexibility and is slow to dry out.

Water repels hydrophobic molecules such as lipids, and, as a result, the lipids aggregate into compact molecular arrangements. This is an essential property in biological membranes where the lipid portion of the cellular membrane lines up to form a tight barrier to the flow of polar molecules in and out of the cell.

Most nutrients for microbes must be water soluble because microbes generally are unable to ingest large particles. When a microbe attacks large molecules such as proteins, lipids, or starches they are cleaved to lower-molecular-weight soluble compounds extracellularly before they can be carried into the cell.

SECTION HIGHLIGHTS
The polarity of water and its ability to form hydrogen bonds gives it many properties upon which life depends. It is an excellent solvent and insulator, forms thin films that protect membranes from drying out, and interacts with hydrophobic molecules in such way as to maintain the structure of membranes and other macromolecules.

3.8 Acids, Bases, and Buffers

The Arrhenius definition of an acid is a substance that, when dissolved in water, increases the concentration of the hydronium ion, H_3O^+ (hydrogen ions, H^+). A strong acid is one that ionizes readily in water, even in a dilute solution. Among the strong acids are hydrochloric acid, nitric acid, and sulfuric acid. Organic acids such as acetic acid or citric acid are much weaker because they ionize poorly and are only partly ionized even when in a dilute solution.

A **base** is a compound that, when ionized, releases one or more cations and negatively charged ions (OH^-). These negative ions can accept protons from solution. Thus the OH^- ions remove H^+ from solution, leading to an excess of hydroxyl groups. A solution with an excess of OH^- is basic. A strong base such as sodium hydroxide ionizes completely in dilute solution, whereas ammonium hydroxide, a weaker base, does not.

Water has a tendency to ionize—that is, to dissociate into hydrogen ions (H^+) and hydroxide ions (OH^-).

$$HOH \rightleftharpoons H^+ + OH^-$$

In pure water, the hydrogen ion concentration in solution is 0.0000001 M or 10^{-7}. The term **pH** is defined as the logarithm of the reciprocal of the hydrogen ion concentration. Pure water would have a pH of 7. When the concentration of H^+ is greater than 10^{-7}, the solution is acidic; if less than 10^{-7}, the solution is basic. The pH scale is logarithmic, so a solution of pH 6 contains a hydrogen ion concentration 10 times greater than a solution of pH 7. The pH's of some often-encountered materials are presented in Figure 3.11.

Microbial growth in laboratory media can result in the production of organic acids. These acids can lower the pH of the medium, which can result in cessation of growth or death of the cell. To prevent a significant change in pH, a **buffer** is generally added to the growth medium. A buffer can absorb or release hydrogen ions to maintain a favored pH. A buffering system that is often

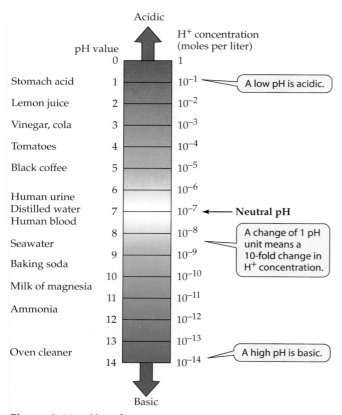

Figure 3.11 pH scale
The pH scale and the pH of some common materials.

The most abundant element in cells is hydrogen, followed by carbon, oxygen, nitrogen, phosphorus, and sulfur. Cations that are present in virtually all cells are sodium, potassium, magnesium, calcium, zinc, and iron. Some organisms require trace elements (in small amounts) such as cobalt, molybdenum, copper, vanadium, or manganese.

There are a limited number of functional groups present in microbial cells, and the major groups were presented in Table 3.2. Carbon-to-carbon bonds are the backbone of the monomers that are joined to form macromolecules. These monomers are generally joined together through nitrogen, oxygen, or phosphorus atoms (or combinations of these). Phosphorus-containing compounds are generally the medium of energy exchange and also a structural component of nucleic acids. Nitrogen is a part of many cellular components including proteins and the nucleic acid bases. Sulfur is generally a component of proteins and some of the B vitamins.

SECTION HIGHLIGHTS

Cells are composed of a limited range of elements, the most abundant of which are hydrogen, carbon, oxygen, nitrogen, phosphorus, and sulfur. These elements are combined to form a limited number of functional groups, which in turn make up the bulk of cellular components. Cells also contain various ions and trace elements.

---

employed in growth media is composed of potassium dihydrogen phosphate (KH_2PO_4) and the salt dipotassium hydrogen phosphate (K_2HPO_4). This buffering system is effective around the neutral range (pH 6 to 8) because it can accept protons if the medium becomes acidic and donate protons if the medium becomes basic.

SECTION HIGHLIGHTS

An acid is a compound that ionizes in water to produce hydrogen ions. A base releases hydroxyl ions, which can remove hydrogen ions from solution. The pH of a solution is defined as the logarithm of the reciprocal of the hydrogen ion concentration. Acidic solutions have a low pH and basic solutions have a high pH. A buffer can maintain a stable pH by absorbing or releasing hydrogen ions into solution.

3.9 Chemical Constituents of the Cell

The number of possible reactions in living cells is limited only by the restricted range of elements present in cells.

3.10 Major Monomers

This discussion is limited to the structure of typical monomeric chemicals that make up a cell. The pathways involved in the synthesis of the major monomers (building blocks) that are present in a cell will be discussed in Chapter 10. The reactions involved in assembling these building blocks into functional structures will be presented in Chapters 4 and 11.

Carbohydrates

Carbohydrates (sugars) are a primary source of energy in many microbial cells. They are also a constituent of many structural units in the microbe, including nucleic acids, cell walls, and adenosine triphosphate (ATP). The general formula for a sugar is $(CH_2O)_n$. The commonly occurring sugars contain three to seven carbons. A simple sugar—and the most abundant product of living cells—is glucose (Figure 3.12). Glucose is also called grape or blood sugar and is depicted along with fructose (fruit sugar). Other monosaccharides of biological

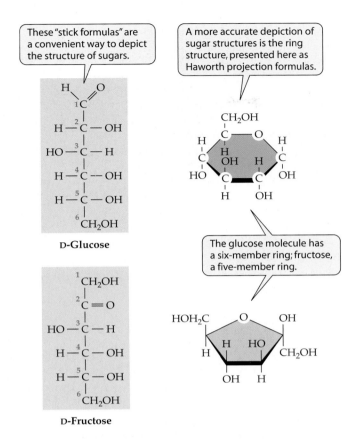

These "stick formulas" are a convenient way to depict the structure of sugars.

D-Glucose

D-Fructose

A more accurate depiction of sugar structures is the ring structure, presented here as Haworth projection formulas.

The glucose molecule has a six-member ring; fructose, a five-member ring.

Figure 3.12 Structure of glucose and fructose
Glucose and fructose are common six-carbon sugars. They have the same chemical formula ($C_6H_{12}O_6$) and thus are structural isomers.

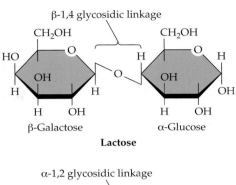

β-1,4 glycosidic linkage

β-Galactose   α-Glucose

Lactose

α-1,2 glycosidic linkage

α-Glucose   β-Fructose

Sucrose

Figure 3.14 Important disaccharides, showing typical sugar linkages
The glycosidic linkages in these molecules are common in disaccharides and polysaccharides. Specific enzymes cleave these bonds.

interest are glyceraldehyde, erythrose, ribose, and sedo-heptulose (Figure 3.13).

A common disaccharide (2-sugar molecule) is sucrose or cane sugar. It is composed of one molecule of glucose and one of fructose. Lactose (milk sugar) is also a disaccharide and is composed of glucose and galactose. The structures of these two important disaccharides are pre-sented in Figure 3.14. The disaccharide lactose has a **glycosidic** linkage between and carbons 1 and 4 of the sugars, and this configuration is called beta (β). Lactose is galactose-β-1,4-glucose. The glycosidic linkage in sucrose is between the carbons 1 and 2 as well as in the alpha (α) configuration. Sucrose is glucose-α-1,2-fructose. Glycogen, starch, and cellulose are composed of glucose units that are polymerized into large molecules called polysaccharides. Glycogen and starch are important carbon and energy reserves in bacteria, animals, and plants. Cellulose is present in fungi and plants as a structural component of the cell wall and is also present in a limited number of bacteria.

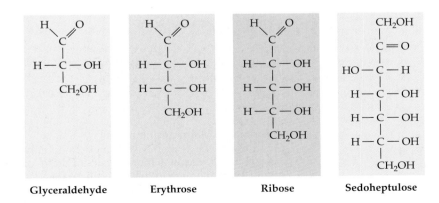

Glyceraldehyde   Erythrose   Ribose   Sedoheptulose

Figure 3.13 Important sugars
Examples of three-, four-, five-, and seven-carbon sugars important in biological systems.

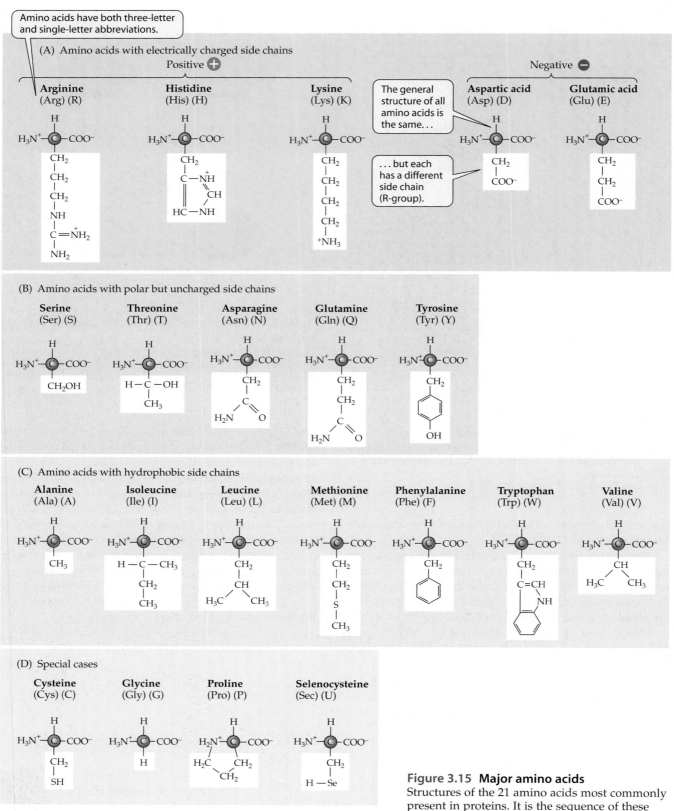

Figure 3.15 Major amino acids
Structures of the 21 amino acids most commonly present in proteins. It is the sequence of these amino acids in a protein that determines the character of that protein.

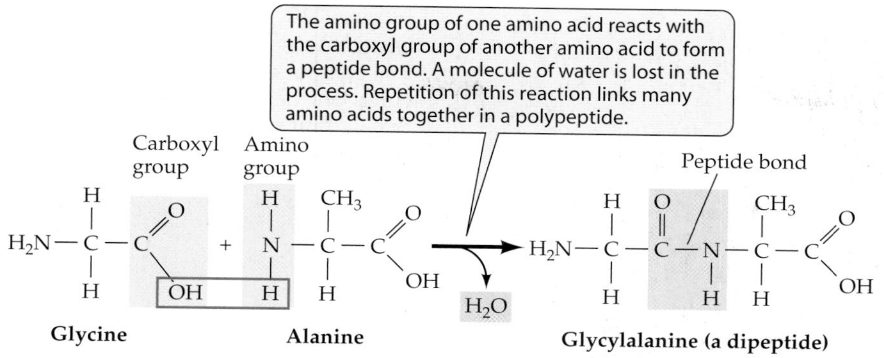

The amino group of one amino acid reacts with the carboxyl group of another amino acid to form a peptide bond. A molecule of water is lost in the process. Repetition of this reaction links many amino acids together in a polypeptide.

Figure 3.16 Formation of a peptide bond Amino acids in polypeptides are linked by peptide bonds.

Glycine   Alanine   Glycylalanine (a dipeptide)

Lipids

Lipids are composed of fatty acids covalently linked to a sugar molecule, commonly glycerol (see Figure 10.19). The fatty acids in bacteria are mostly saturated or monounsaturated, and are 16 or 18 carbons in length. A monounsaturated fatty acid has a double bond between two adjacent carbons, most often nine carbons from the carboxyl end of the long-chain fatty acid. Long-chain fatty acids that commonly occur in bacteria are depicted in Figure 10.17. Alternatively, the lipids in the *Archaea* are isoprenoid and hydroisoprenoid hydrocarbons linked to glycerol through an ether linkage (see Figure 4.47). Higher organisms (plants and animals) generally have more than one set of double bonds in their fatty acids.

Amino Acids and Proteins

Proteins are a major component of all living organisms and perform most of the work in a cell. They are involved in catabolic and anabolic reactions and fuel energy production, and they serve as structural and functional components. About 50% (dry weight) of a cell is protein, and a bacterium, such as *Escherichia coli*, can synthesize well over 1,000 different proteins. The monomeric components of a protein are amino acids. An amino acid has one or more amino groups ($-NH_3^+$) and one or more carboxylic acid groups ($-COO^-$). The general structure of an amino acid is as follows:

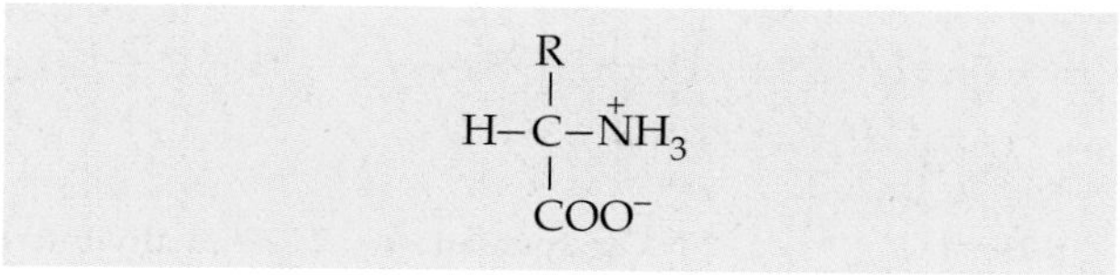

$$H-\overset{\overset{\textstyle R}{|}}{\underset{\underset{\textstyle COO^-}{|}}{C}}-\overset{+}{N}H_3$$

The R-group (side chain) differs among the major amino acids (Figure 3.15). In water, the amine gains a proton and has a positive (+) charge, whereas the carboxyl loses a proton and has a negative (−) charge.

This **zwitterionic** form of the alpha amino acids occurs at physiological pH values. Molecules that bear charged groups of opposite polarity are termed zwitterions or **dipolar ions**. At near-neutral pH both the carboxylic acid and amino groups are completely ionized—an amino acid can act as either an acid or a base. The structures of the 21 major amino acids are presented in Figure 3.15. Amino acids are joined to form an amino acid chain called a **peptide**. The bond between amino acids in a peptide chain is called a **peptide bond** (Figure 3.16).

Nucleic Acids

There are two major nucleic acid classes in all cells: **ribonucleic acid** (**RNA**) and **deoxyribonucleic acid** (**DNA**), the nucleic acid that carries the blueprint of the cell. RNA is responsible for the translation of information inherent in DNA into functional proteins. Just as amino acids are the structural units of a protein, nucleotides are the structural units of nucleic acids. There are four bases present in DNA: adenine, guanine, cytosine, and thymine. Thymine is replaced by uracil in RNA (Figure 3.17). The five-carbon sugar in DNA is deoxyribose, and in RNA it is ribose. The order of bases in DNA determines the sequence of amino acids in a protein, and, as mentioned previously, the characteristics of a protein depend on this amino acid sequence.

SECTION HIGHLIGHTS
Cells contain several types of monomers that are combined in various ways to generate cellular structures. These monomers are: carbohydrates, lipids, amino acids, and nucleic acids.

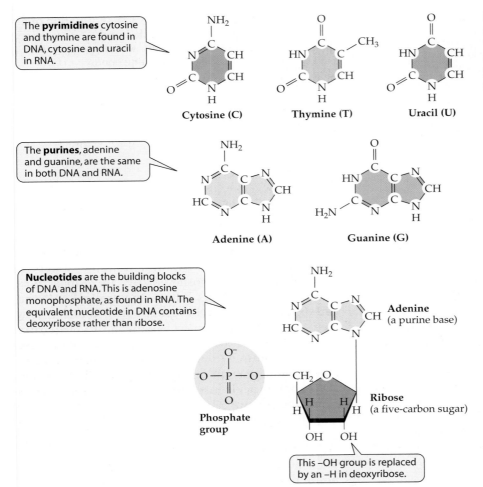

The **pyrimidines** cytosine and thymine are found in DNA, cytosine and uracil in RNA.

Cytosine (C)   Thymine (T)   Uracil (U)

The **purines**, adenine and guanine, are the same in both DNA and RNA.

Adenine (A)   Guanine (G)

Nucleotides are the building blocks of DNA and RNA. This is adenosine monophosphate, as found in RNA. The equivalent nucleotide in DNA contains deoxyribose rather than ribose.

Adenine (a purine base)

Phosphate group

Ribose (a five-carbon sugar)

This –OH group is replaced by an –H in deoxyribose.

Figure 3.17 Nucleic acid components

The components of nucleic acid molecules: purine and pyrimidine bases, a five-carbon sugar (ribose in RNA; deoxyribose in DNA), and phosphate.

3.11 Isomeric Compounds

Isomers are compounds that have identical molecular formulas but differ in the arrangement of the constituent atoms. Such isomers are termed structural isomers. Isomers composed of 6 carbons and 14 hydrogens (C_6H_{14}) are illustrated in Figure 3.18. Although these three compounds have identical molecular weights (86.18) and molecular formulas, they differ in physical properties as indicated by their boiling points. The structure would also influence biodegradability, as hexane would be the most readily degraded and 2,3-dimethylbutane the least.

Stereoisomers are isomers in which all of the bonds in the compounds are the same but the spatial arrangements of the atoms differ. Essentially, stereoisomers are mirror images of one another (Figure 3.19A,B). Some stereoisomers are also termed **enantiomers** and are mirror images of one another. In the one configuration, a stereoisomer will rotate a plane of polarized light passing through it in one direction and the mirror-image iso-

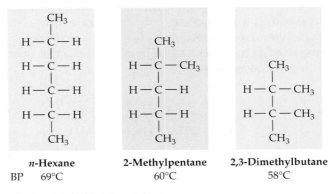

n-Hexane	2-Methylpentane	2,3-Dimethylbutane
BP 69°C	60°C	58°C

Figure 3.18 Structural isomers

Isomers of the six-carbon saturated hydrocarbon, C_6H_{14}. They have the same molecular weight (86.18) but differ in physical properties, such as boiling point (BP).

(A)

The horizontal wedges indicate bonds projecting out from the central carbon, toward the viewer…

…and the vertical wedges indicate bonds projecting behind the central carbon, away from the viewer.

$$CHO$$
$$HO - C - H$$
$$CH_2OH$$
L-Glyceraldehyde

$$CHO$$
$$H - C - OH$$
$$CH_2OH$$
D-Glyceraldehyde

$$COOH$$
$$H_3N^+ - C - H$$
$$CH_2OH$$
L-Serine

$$COOH$$
$$H - C - NH_3^+$$
$$CH_2OH$$
D-Serine

(B)

This is the carbon that is related to the central carbon of glyceraldehyde, so this is the D isomer.

$$CHO$$
$$HO - C - H$$
$$H - C - OH$$
$$HO - C - H$$
$$HO - C - H$$
$$CH_2OH$$
L-Glucose

$$CHO$$
$$H - C - OH$$
$$HO - C - H$$
$$H - C - OH$$
$$H - C - OH$$
$$CH_2OH$$
D-Glucose

Figure 3.19 Stereoisomers
Isomers, such as those of amino acids, are denoted as D or L based on their configuration relative to D-glyceraldehyde and L-glyceraldehyde (A). D-Glucose (B) is the most common form of glucose in biological systems.

mer will rotate light in the opposite direction (Box 3.2). If the polarized light passing through the compound is rotated to the right, the isomer is dextrorotatory (D), and if rotated to the left the isomer is levorotatory (L). The D,L system of nomenclature is based on the optical properties of D-glyceraldehyde and L-glyceraldehyde, respectively (see Figure 3.19A). The absolute configurations of all other carbon-based molecules are referenced to D- and L-glyceraldehyde.

Proteins are normally composed of L-amino acids. The cell wall peptidoglycan contains D-amino acids, and the D form is also present in selected antibiotics. Many isomers of sugars exist in nature and may have the same molecular and structural formula but are mirror images of one another. These sugars are stereoisomers and designated by either D or L. Sugars are generally present in nature in the D configuration (see Figure 3.19B). Enzymes that are present in bacteria (racemases) can convert the D-amino acids to the L form. Racemases are also present in bacteria that interconvert the L configuration of sugars to the D form.

BOX 3.2 *Milestones*

Louis Pasteur, the Crystallographer

Louis Pasteur began his scientific career as a chemist. He became intrigued with the observation that a chemical such as sulfur would crystallize in two distinct shapes. Earlier workers had reported that tartaric acid from grapes deflected the plane of polarized light, but sodium ammonium tartrate was optically neutral. When examining crystals of the tartrate salt, Pasteur noted that there were two types of crystals. One had "faces" that inclined to the right, and the faces of the second inclined to the left. He meticulously separated the two crystal types, dissolved each in water, and placed the respective solutions before polarized light. Pasteur found that the dissolved crystals with faces on one side rotated the polarized light in one direction, whereas the others rotated light in the opposite direction. An equal mixture of the two types was optically neutral. This was a clear proof that there can be a relationship between optical activity and molecular structure. That living things generally synthesized only one optical form was intriguing to Pasteur, and this ignited a lifelong interest in biological studies.

Courtesy of the National Library of Medicine.

SECTION HIGHLIGHTS

Compounds that have identical molecular formulas but differ in the arrangement of the chemical bonds between the constituent atoms are known as structural isomers. Stereoisomers have the same bonds as one another, but have a different spatial arrangement (they are mirror images). Cellular structures (e.g., proteins) are usually made up of one stereoisomer and not the other.

SUMMARY

- Essentially an atom is **electrically neutral** because the number of **negatively charged electrons** in orbit about the nucleus is equal to the **positively charged protons** present within the nucleus. **Neutrons** are present in all elements except hydrogen and **have no charge**.

- The **atomic number** is equal to the number of protons in the nucleus of an element. The sum of the number of neutrons and protons present is the **mass number**.

- The three major bonds that join atoms or molecules together are **ionic**, **covalent**, and **hydrogen bonds**. Carbon-to-carbon bonds are covalent bonds and vital in the formation of biological molecules.

- **Hydrogen bonds** are important in the conformation of nucleic acids and proteins.

- **Water** is essential to life, and a cell is 60% to 95% water. It is actually a polar compound and has considerable polymerization at room temperature.

- **Sugars** are essential components of living cells and are present in nucleic acids, cell walls, and ATP. They are also important sources of energy. Glucose is considered the most abundant product of living cells.

- Bacterial membranes contain **fatty acids** that are generally 16 or 18 carbons in length and linked to glycerol via an ester linkage. These fatty acids may be saturated or monounsaturated. Archaeal membrane lipids are composed of **isoprenoid** chains with an ether linkage to glycerol.

- One-half of the **dry weight** of a **cell** is **protein**, and the constituent parts of a protein are the amino acids. The bond that joins one amino acid to another is the **peptide bond**.

- The major nucleic acids in cells are **deoxyribonucleic acid** (**DNA**) and **ribonucleic acid** (**RNA**). DNA is the blueprint of the cell, and RNA is involved in **translating** the information present in DNA to functional proteins. Purines and pyrimidines are the bases present in nucleic acids.

- **Structural isomers** have equivalent molecular formulas but a different arrangement of the constituent atoms. In **stereoisomers**, the bonds between atoms are in the same place but the spatial arrangement differs. **Enantiomers** are isomers that are mirror images of one another.

 Find more at www.sinauer.com/microbial-life

REVIEW QUESTIONS

1. Draw an atom and label the parts. What part retains the structure of an atom?

2. How does atomic weight differ from atomic number?

3. Define isotope. How would you design an isotope and experimentation to determine how a microbe capable of growth on methane assimilates this hydrocarbon?

4. What structural feature is involved in the reactivity of an atom? How is this related to valence?

5. Three types of bonds occur between molecules. What are they and how do they differ? Which are most prevalent in a bacterium?

6. How are hydrogen bonds involved in the basic configuration of proteins? Nucleic acids?

7. What are van der Waals forces? Where might they play a role in biological systems?

8. Give several reasons why water is an effective solvent in biological systems.

9. How does an acid differ from a base? Why is water neutral?

10. Why are buffers often essential in bacterial culture media?

11. Name the five most abundant elements in a cell. What sorts of functions would they have in macromolecules?

12. What are the major monomers in polysaccharides, lipids, proteins, and nucleic acids?

13. How are the major monomers attached to one another?

14. Basically, how does one protein differ from another?

SUGGESTED READING

Berg, J. M., J. L. Tymoczko and L. Stryer. 2006. *Biochemistry.* 6th ed. New York: W. H. Freeman.

Brown, T. L., H. E. LeMay and B. E. Bursten. 1999. *Chemistry: The Central Science.* 8th ed. Upper Saddle River, NJ: Prentice Hall.

Cooper, G. M. and R. E. Hausman. 2006. *The Cell: A Molecular Approach.* 4th ed. Sunderland, MA: Sinauer Associates.

Nelson, D. L. and M. M. Cox. 2004. *Lehninger Principles of Biochemistry.* 4th ed. New York: W. H. Freeman.

Voet, D., J. G. Voet and C. W. Pratt. 2005. *Fundamentals of Biochemistry: Life at the Molecular Level.* 2nd ed. Hoboken, NJ: John Wiley & Sons.

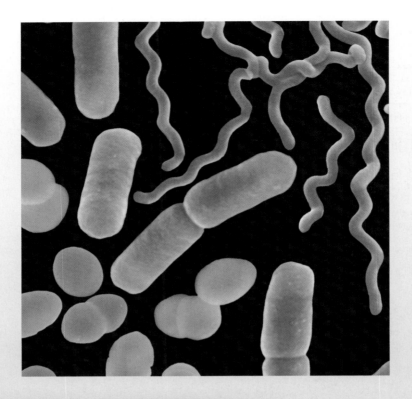

The objectives of this chapter are to:

◆ Introduce microscopy and the microscopes and methods used to examine microorganisms.

◆ Discuss the shapes and sizes of *Bacteria* and *Archaea*.

◆ Provide evidence of the structure, composition, and role of the various cell components.

4

Structure and Function of
Bacteria *and* Archaea

C'est un grand progrè, monsieur.
—Louis Pasteur, 1881*

*I*n Chapter 1 we compared and contrasted prokaryotes with eukaryotes and affirmed that the *Bacteria* and *Archaea* have simpler shapes and structures than eukaryotic organisms. Nonetheless, prokaryotes have all the necessary structural features for performing essential life functions, including metabolism, growth, and reproduction. And more complex prokaryotic organisms have additional structures and life cycles not found in the simpler species. This chapter introduces the various unicellular and multicellular shapes of prokaryotes and the types of cell division found among them. It also discusses the structure, chemical composition, and function of common morphological features found in prokaryotic organisms.

However, before discussing the properties of these microorganisms in greater detail, we need to recognize the special instruments, techniques, and procedures used to study them. Microbiology did not become a scientific discipline until appropriate instruments and techniques were developed to enable scientists to observe and study these small organisms (see Chapter 2). Of foremost importance was the development of quality microscopes. Indeed, it is not surprising that microorganisms were discovered by a lens maker, Antony van Leeuwenhoek, rather than by a biologist—an early illustration of the importance of instrumentation in microbiology. In this chapter, then, we focus first on microscopes, the instruments routinely used by microbiologists, and then discuss the structure and function of prokaryotic organisms.

---

*"That is important progress, sir." Remark made by Louis Pasteur to Robert Koch at the International Medical Congress in London, 1881, after Koch demonstrated his pure culture methods.

4.1 Microscopy

The human eye has some ability to magnify objects. Natural magnification is simply achieved by bringing the object closer to the eye. The closer the object, the larger it appears because the image on the retina is larger (Figure 4.1). The maximum enlargement depends on how close the object can be brought to the eye and still remain in focus. This **near point** is typically 250 mm for an adult human, and the distance increases as people age.

Objects smaller than about 0.1 mm (the "eye" of a needle is about 1.0 mm wide) cannot be seen distinctly because their image does not occupy a sufficiently large area on the retinal surface. Therefore, the unaided human eye cannot see small organisms such as prokaryotes, which typically have a cell diameter of only 0.001 mm (1.0 μm, or 10^{-6} m). Microscopes have been developed to aid the eye by increasing its ability to **magnify**, that is, increasing the size of a small specimen so that it can be more readily observed.

Simple Microscope

The simplest optical device that can be used to assist the eye in enlarging objects is appropriately called the **simple microscope**, which in principle is a magnifying glass. Antony van Leeuwenhoek constructed his own simple microscopes specifically to observe microorganisms (see Figure 2.2).

Magnifying glasses are not particularly powerful. They usually magnify objects about fivefold. It is a tribute to Leeuwenhoek that his simple microscopes not only could magnify 200- to 300-fold but did so with great clarity. Indeed, the images Leeuwenhoek produced starting in about 1670 (he lived from 1632 to 1723) were comparable to those obtained with light microscopes of similar magnification that are used today.

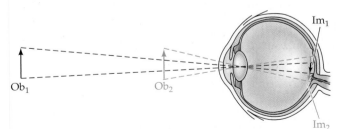

Figure 4.1 How the human eye magnifies
Diagram illustrating the magnifying capacity of the human eye. The object (Ob) is shown at two locations, Ob_1 and Ob_2. The image formed on the retina is smaller when the object is farther away from the eye (Im_1) than when it is closer (Im_2).

Compound Microscope

Another Dutchman, Zacharias Jansen (1580 to 1638), is generally given credit for the development of the compound light microscope in 1595, although the origins of this instrument are somewhat obscure. It is interesting to note that compound microscopes were available at the time Leeuwenhoek made his momentous discoveries. However, the early compound microscopes were inferior to Leeuwenhoek's simple microscope—Robert Hooke (1635–1703), the English scientist who coined the term "cell" in 1655, could not confirm Leeuwenhoek's reports of microorganisms with the early compound microscopes available to him.

The **compound light microscope** is named for the two lenses that separate the object from the eye. The objective lens is placed next to the object or specimen to be viewed; the **eyepiece** or **ocular lens** is located next to the eye. The object to be viewed is normally placed on a glass slide and illuminated with a light source.

The viewer uses the microscope to form an image by "focusing" the specimen, moving the objective lens and ocular (eyepiece) lens together relative to the specimen until the image is clear. When the specimen has been properly focused, the objective lens produces a **real image** (one that can be displayed on a screen) within the ocular diaphragm of the microscope. The viewer looking through the ocular lens sees a **virtual image**, which cannot be displayed on a screen (Figure 4.2A).

The **magnification power** of a compound light microscope is determined by multiplying the magnification of the objective lens by that of the eyepiece lens. Usually the eyepiece magnification is 10×. Most microscopes of this type have several separate objective lenses, mounted on a rotating nosepiece, that typically give magnifications of 10× (low power), 60× (high, dry power), and 100× (oil immersion). The resulting magnifications attainable from this microscope would be 100×, 600×, and 1,000×, respectively.

The typical compound microscope (Figure 4.2B) also has a third lens system. Light from a lamp or other source is focused on the specimen by a condenser lens. The **condenser lens**, which is not directly involved in image formation, is necessary to provide high-intensity light, because the brightness of the object becomes a limiting factor as the specimen is increasingly magnified. The light intensity is adjusted by opening or closing a diaphragm called the **iris diaphragm**, located in the condenser.

It might seem reasonable to assume that one could increase the magnification indefinitely using the compound light microscope simply by constructing increasingly powerful objective and eyepiece lenses. In reality, the light microscope has a useful magnification maxi-

(A)

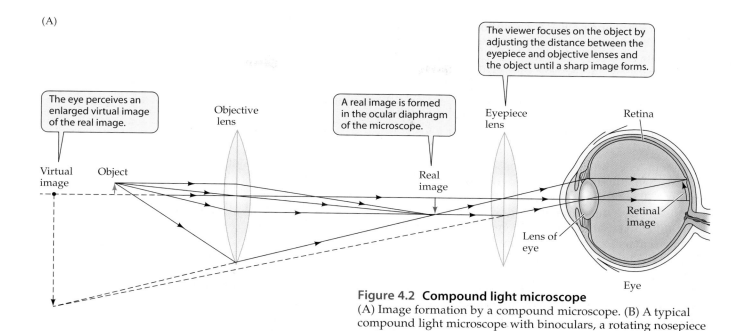

The viewer focuses on the object by adjusting the distance between the eyepiece and objective lenses and the object until a sharp image forms.

The eye perceives an enlarged virtual image of the real image.

Objective lens

A real image is formed in the ocular diaphragm of the microscope.

Eyepiece lens

Retina

Virtual image

Object

Real image

Retinal image

Lens of eye

Eye

Figure 4.2 Compound light microscope
(A) Image formation by a compound microscope. (B) A typical compound light microscope with binoculars, a rotating nosepiece with objective lenses, and a condenser lens system with a built-in light source. Special knobs are used for focusing and orienting the specimen in the *x–y* directions on the stage. Photo courtesy of Nikon Corporation.

(B)

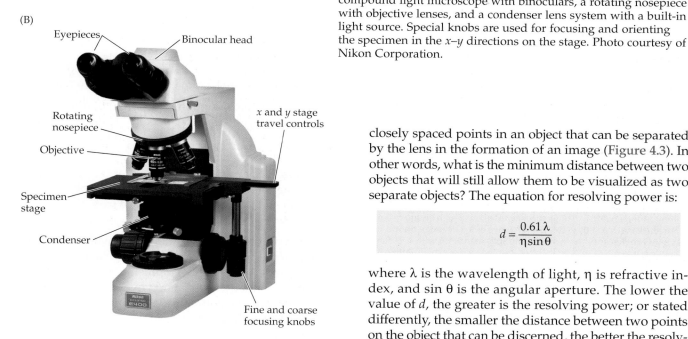

Eyepieces

Binocular head

Rotating nosepiece

x and *y* stage travel controls

Objective

Specimen stage

Condenser

Fine and coarse focusing knobs

mum of only 1,000 to 2,000×. Although magnification beyond that level is attainable, it is referred to as **empty magnification** because it does not provide any greater detail of the specimen.

IMPROVING IMAGE FORMATION IN MICROSCOPY
In magnifying, it is important that the image retains the true shape of the specimen. The ability of an optical system to produce a detailed, accurate image of the specimen or object is termed its **resolving power**. The resolving power, *d*, is defined as the distance between two closely spaced points in an object that can be separated by the lens in the formation of an image (Figure 4.3). In other words, what is the minimum distance between two objects that will still allow them to be visualized as two separate objects? The equation for resolving power is:

$$d = \frac{0.61\,\lambda}{\eta \sin \theta}$$

where λ is the wavelength of light, η is refractive index, and $\sin \theta$ is the angular aperture. The lower the value of *d*, the greater is the resolving power; or stated differently, the smaller the distance between two points on the object that can be discerned, the better the resolving power or **resolution** of the image. Therefore, to attain the best resolving power, or resolution, the following conditions are required:

• Short-wavelength light (λ is low)

• A suspending medium (for the specimen) of high refractive index (η is high)

• A high angular aperture, the angle at which light enters the objective (θ, and thus $\sin \theta$, is high)

The light microscope cannot be operated with wavelengths of less than 400 to 500 nm (violet light), as the

Figure 4.3 Resolving power or resolution
Resolution is the degree to which two separate points in an object can be distinguished. Improved lenses allow increased resolution.

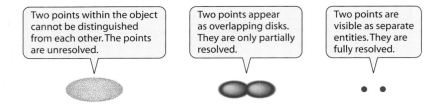

Two points within the object cannot be distinguished from each other. The points are unresolved.

Two points appear as overlapping disks. They are only partially resolved.

Two points are visible as separate entities. They are fully resolved.

eye cannot see shorter wavelengths. For maximum resolution, then, the shortest wavelength that can be used is about 500 nm (0.5 µm).

By increasing the **refractive index**, a measure of the ability of a material to bend light, more light from the specimen enters the objective lens resulting in greater resolution. The crown glass used in lenses has a high refractive index; air has a refractive index of 1.0 (essentially identical to that of a vacuum), water 1.33, and crown glass 1.5. To permit a continuous, high refractive index between the condenser and the objective, immersion oils of high refractive index are placed on the specimen and on the condenser lens system. Such **homogeneous immersion** systems allow the objective to be brought closer to the specimen, further increasing resolution (**Figure 4.4**).

The final factor affecting resolution is the aperture or opening of the **objective lens**, that is, the lens placed next to the specimen. Because the objective lens has a finite aperture, a point on the specimen is not imaged as a point on the image but as a disk (called an "Airy disk") with alternate dark and light rings. The size of this disk can be

decreased by increasing the aperture of the lens, but there is a limit to this. As a result, two closely spaced points on the object are seen as fuzzy disks on the image, and if they are too close may actually overlap and not be separate points (see Figure 4.3). The theoretical maximum value for the sine of the angular aperture is 1.0.

When these ideal values of the shortest possible wavelength (η = 0.5 µm), homogeneous oil immersion (λ = 1.5), and the angular aperture (sin θ = 1.0) are substituted into the equation for resolving power, then d = 0.61(0.5 µm)/1.5, which is about 0.2 µm. Because typical prokaryotic cells are about 1.0 µm in diameter, the compound light microscope has a satisfactory resolution for observation of these cells. It is not well suited, however, for observing internal cell structures, which are much smaller.

Early versions of the compound microscope were plagued by lens aberrations, which are of two types:

- **Chromatic aberration**, in which light of different wavelengths entering the objective from the specimen is focused at different planes in the formation of the image

- **Spherical aberration**, in which light rays from the specimen that enter the periphery of the objective lens are focused at a different place from those that enter the center of the lens

Correction of these aberrations was made possible by the use of multicomponent lens systems in the objective lens (**Figure 4.5**). These developments, and the introduction of the condenser, transpired over a period of about two centuries. The compound light microscope was finally perfected in the late nineteenth century by opticians including Ernst Abbe in Germany (1840–1905).

Because the refractive index of prokaryotic cells is similar to that of water, prokaryotes are almost invisible when viewed with an ordinary compound light microscope. Three approaches have been taken to overcome this problem:

- Staining the cells with dyes to produce higher contrast

- Using a modified compound microscope, such as the phase contrast microscope or the darkfield microscope

- Using a fluorescence microscope with fluorescent dyes

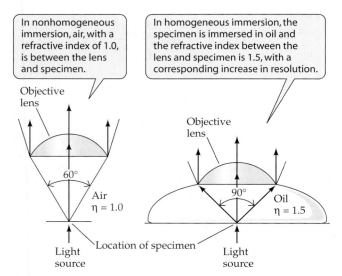

In nonhomogeneous immersion, air, with a refractive index of 1.0, is between the lens and specimen.

In homogeneous immersion, the specimen is immersed in oil and the refractive index between the lens and specimen is 1.5, with a corresponding increase in resolution.

Objective lens

60°

Air
η = 1.0

Objective lens

90°

Oil
η = 1.5

Light source

Location of specimen

Light source

Figure 4.4 Nonhomogeneous and homogeneous immersion
Immersion oil increases resolving power, not only by increasing refractive index but also by allowing the specimen to be brought closer to the lens, thus increasing the angular aperture, θ.

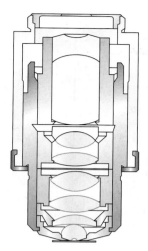

Figure 4.5 Objective lens
Cross-sectional view of a microscope objective lens showing its multicomponent lens system. Courtesy of Carl Zeiss, Inc.

DYES AND STAINS   Microorganisms to be observed in the light microscope are often stained with dyes, because dyes increase the contrast between the cells and their environment. Two types of staining procedures can be used. The most common type is **positive staining** in which the organism or a component of it takes up the dye and is stained. However, **negative staining** can also be used, although the technique is much less widely applied. Negative stains are not taken up by the organism, but stain the background.

Most dyes used for positive staining of microorganisms are aniline dyes, intensely pigmented organic salts derived from coal tar. They are called **basic dyes** if the **chromophore** (pigmented portion) of the molecule is positively charged. For example, crystal violet and methylene blue are basic dyes (Figure 4.6). Other basic dyes commonly

used to stain microorganisms are basic fuchsin, malachite green, and safranin. Under normal growth conditions, most prokaryotes have an internal pH near neutrality (pH 7.0) and a negatively charged cell surface, so basic dyes are generally the most effective staining agents. **Acid dyes** such as Congo red, eosin, and acid fuchsin have a negatively charged chromophore and are useful in staining positively charged cell components such as protein.

Simple stains of microorganisms are made by spreading a suspension of the organism on a glass slide, allowing it to dry, and then gently heating it to fix it to the slide (a **fixative**, in this case heat, allows the specimen to adhere—like frying an egg in a pan without oil). Such a preparation is called a **smear**. The stain is added, and after a brief period of exposure, the excess dye is removed by gentle rinsing. The slide is then viewed with the microscope (Figure 4.7A).

Differential stains, such as the Gram stain (Box 4.1 and Figure 4.7B), distinguish one microbial group from another, in this instance, gram-positive bacteria from gram-negative bacteria. Likewise, only acid-fast bacteria such as those in the genus *Mycobacterium* retain the color of the dye when the stained preparation is rinsed in a solution of ethanol containing hydrochloric acid at a final concentration of 3%.

Some staining procedures allow the identification of structures within the cell. Thus, specific stains exist for bacterial endospores (Figure 4.7C), flagella (Figure 4.7D), capsules, deoxyribonucleic acid (DNA), and other cell materials. Lipophilic dyes, such as Sudan black, can be used to specifically stain lipid inclusions such as poly-β-hydroxybutyric acid (PHB) (see subsequent text).

Brightfield and Darkfield Microscopes

An ordinary light microscope is called a **brightfield** microscope, because the entire field of view containing the specimen and the background is illuminated and appears bright. The brightfield microscope can be modified to produce a **darkfield microscope**, so named because the cells appear bright against a dark background (Box 4.2). One advantage of the darkfield microscope is that it can be used to view living cells. Cell suspensions are observed in **wet mounts**, prepared by placing a coverslip over the live suspension. The movement of motile cells, for example, is readily visible in these preparations.

Phase Contrast Microscope

As mentioned earlier, most microbial cells appear to be colorless, transparent objects when observed by ordinary brightfield microscopy (Figure 4.8A) analogous to looking at water in a transparent plastic bag immersed in a bathtub. A slight difference exists, however, between the refractive index of the cell (η is about 1.35) and that

$[(CH_3)_2NC_6H_4]_2C =\!\!=\!\!\overset{+}{N}(CH_3)_2Cl^-$

Crystal violet

$(CH_3)_2N$ ⟶ $\overset{N}{\underset{\underset{Cl^-}{S^+}}{}}$ ⟶ $N(CH_3)_2$

Methylene blue

Figure 4.6 Dyes
Chemical structures of crystal violet and methylene blue, two basic (positively charged) dyes, shown here as their chloride salts.

(A)

(B)

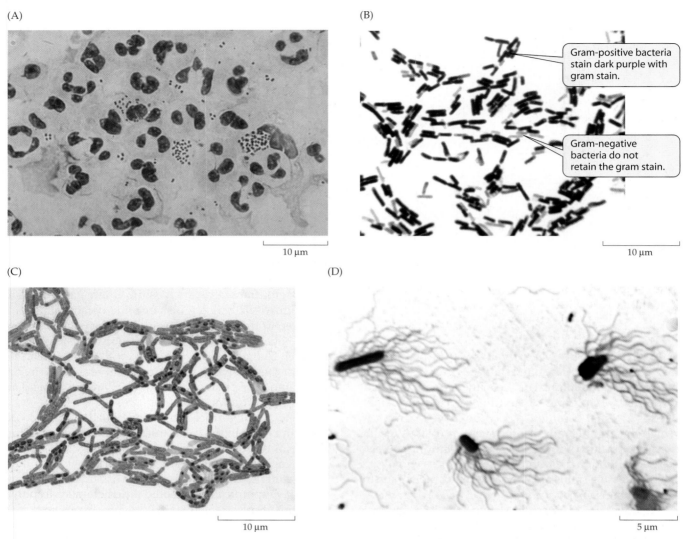

Gram-positive bacteria stain dark purple with gram stain.

Gram-negative bacteria do not retain the gram stain.

10 μm

10 μm

(C)

(D)

10 μm

5 μm

Figure 4.7 Various stains used for bacteria and archaea (A) A gram-negative, aerobic coccus, *Neisseria gonorrhoeae*, stained with methylene blue (see labels) in a sample from an infection that also contains neutrophils; (B) Gram stain for gram-negative and gram-positive bacteria; (C) *Bacillus* species stained with malachite green spore stain and safranin; and (D) *Salmonella typhosa* bacteria stained with a flagellum stain. A, ©Manfred Kage/Peter Arnold, Inc.; B, ©Dr. Jack Bostrack/ Visuals Unlimited; C, Courtesy of Larry Stauffer/CDC; D, ©Dr. John D. Cunningham/Visuals Unlimited.

(A)

(B)

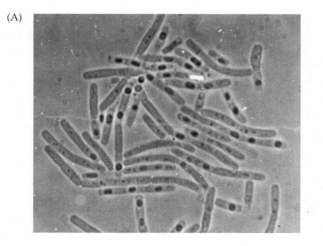

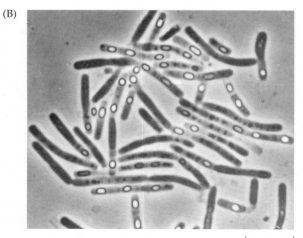

5 μm

| BOX 4.1 | *Methods & Techniques* |

The Gram Stain

The Gram stain procedure was developed unwittingly in 1888 by Christian Gram, a Danish physician studying in Berlin. He was examining lung tissues during autopsies of individuals who had died of pneumonia. He noted that *Streptococcus pneumoniae* found in the lung tissues retained the primary stain known as Bismark brown (crystal violet is now used), whereas the lung tissue did not. It was subsequently determined that certain bacteria, now termed gram-positive, like *S. pneumoniae,* retain the purple dye when stained by this method, whereas others, the gram-negative bacteria, do not. The Gram stain is called a differential stain because it distinguishes between these two groups of bacteria.

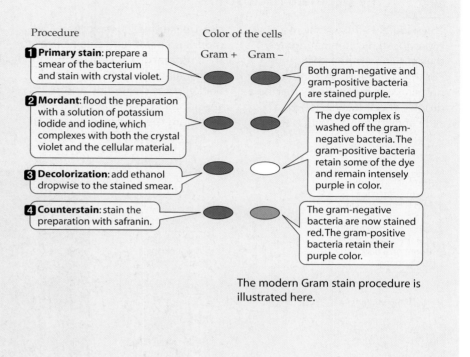

Procedure

1 **Primary stain**: prepare a smear of the bacterium and stain with crystal violet.

2 **Mordant**: flood the preparation with a solution of potassium iodide and iodine, which complexes with both the crystal violet and the cellular material.

3 **Decolorization**: add ethanol dropwise to the stained smear.

4 **Counterstain**: stain the preparation with safranin.

Color of the cells

Gram +   Gram −

Both gram-negative and gram-positive bacteria are stained purple.

The dye complex is washed off the gram-negative bacteria. The gram-positive bacteria retain some of the dye and remain intensely purple in color.

The gram-negative bacteria are now stained red. The gram-positive bacteria retain their purple color.

The modern Gram stain procedure is illustrated here.

---

of its aqueous environment (η is 1.33). The **phase contrast microscope**, or **phase microscope**, amplifies this slight difference in refractive index and converts it to a difference in contrast. The result is that the cells appear dark against a bright background (**Figure 4.8B**). The bright, highly refractive bodies within the cells are spores which have a very high refractive index relative to the cells or water. As with dark field microscopy, cells can be observed while alive, in wet mounts.

Fluorescence Microscope

Certain dyes used for staining microorganisms are called **fluorescent dyes**, because when illuminated by short-wavelength light they emit light of a longer wavelength (they fluoresce). One of the most commonly used fluorescent dyes is **acridine orange**. It specifically stains nu-

cleic acid components of cells. When stained preparations are illuminated with an ultraviolet light source, the cells fluoresce green to orange in color (**Figure 4.9**). One special advantage of fluorescence microscopy is that it

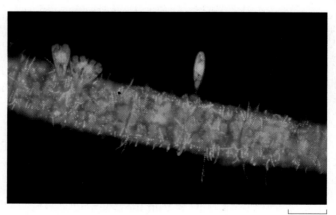

20 µm

Figure 4.9 Fluorescence microscopy
Bacteria (green) and diatoms (orange) stained with the fluorescent dye acridine orange and illuminated with ultraviolet light appear green and yellow to orange to the eye.
©Paul W. Johnson/Biological Photo Service.

◀ **Figure 4.8 Phase contrast microscopy**
Photomicrographs of an unstained bacterium, *Bacillus megaterium,* showing its appearance by (A) brightfield microscopy and (B) phase contrast microscopy. The oval bright areas in the cells are endospores, a specialized, hardy cell commonly found in *Bacillus* species. Courtesy of J. T. Staley.

BOX 4.2 *Methods & Techniques*

The Darkfield Microscope

The darkfield effect in microscopy is produced by illuminating only the cells and not the background. For this procedure, wet-mount preparations are made by placing a droplet of cell suspension on a slide and covering it with a cover-slip. Unlike smears, the cells in wet-mount preparations remain alive in their normal growth environment while being observed. The dark-field effect is created by using a special condenser that introduces light onto the specimen at such an angle that the only light entering the objective is light that is scattered into the objective lens from the cells (A). As a result, the cells appear bright against a dark background (B).

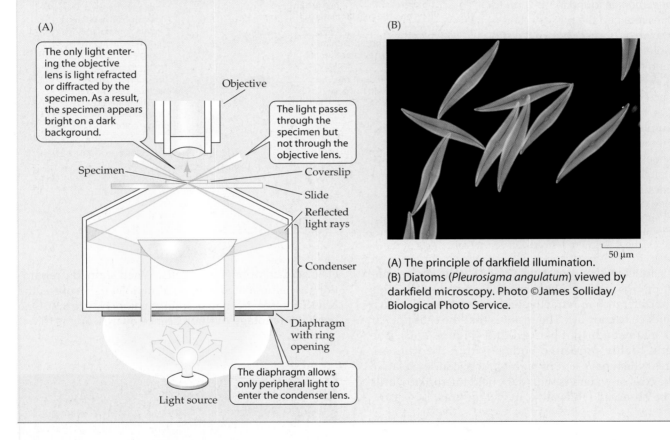

(A)

The only light enter-ing the objective lens is light refracted or diffracted by the specimen. As a result, the specimen appears bright on a dark background.

Objective

The light passes through the specimen but not through the objective lens.

Specimen

Coverslip

Slide

Reflected light rays

Condenser

Diaphragm with ring opening

The diaphragm allows only peripheral light to enter the condenser lens.

Light source

(B)

50 μm

(A) The principle of darkfield illumination.
(B) Diatoms (*Pleurosigma angulatum*) viewed by darkfield microscopy. Photo ©James Solliday/Biological Photo Service.

allows the observation of cells located on an opaque surface such as a soil particle. These would be impossible to see with an ordinary light microscope in which light must be transmitted through the specimen.

Other examples of fluorescent dyes include **DAPI**, a nucleic acid stain specific for DNA, and the lipophilic compound FM4-64 that localizes to membranes. Location of specific proteins within the cell may also be determined by attaching a fluorescent protein tag (e.g, GFP, green fluorescent protein, or RFP, red fluorescent protein). Finally, specific cell molecules can be identified using fluorescent dye–labeled antibodies (see Chapter 30).

In fluorescence microscopes, short-wavelength light is provided by either a mercury lamp (ultraviolet, UV) or a halogen lamp (near ultraviolet). The light is passed through the objective lens system onto the specimen to provide incident illumination, that is, illumination from above the specimen (not transmitted through the specimen as in an ordinary light microscope—although some fluorescence microscopes do use transmitted light). The longer-wavelength light emitted by the fluorescent dyes is visible when viewed through the ocular. Special barrier filters prevent any harmful short-wavelength UV light from reaching the eyes.

(A)

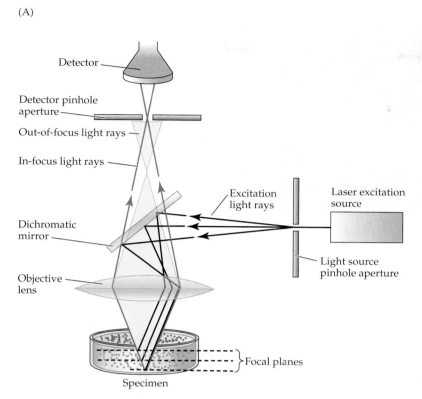

(B)

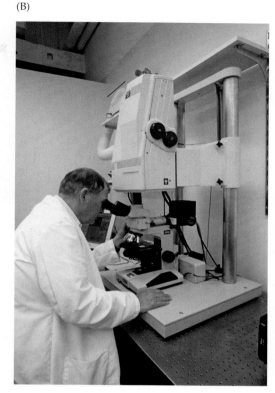

Figure 4.10 Confocal Microscope
(A) This diagram shows the light path and lenses of the confocal microscope. The laser source provides excitation light rays, which are focused at specific vertical levels of the specimen. Fluorescent light from parts of the specimen passes through the dichromatic mirror to the detector for observation. (B) A scientist using a confocal laser scanning microscope. Photo ©Inga Spence/Visuals Unlimited.

Confocal Scanning Microscope

The confocal scanning microscope is especially useful when viewing microorganisms in three-dimensional space. The specimen is illuminated with a laser beam, which is focused on one point of the specimen using an objective lens mounted between the condenser lens and the specimen (**Figure 4.10**). Mirrors are used to pass (scan) the laser beam across the specimen in the *x* and *y* directions. Only those cells that are in the plane of illumination appear on the display screen (see Figure 24.4). The objective lens used for viewing the specimen magnifies the image, which is free from diffracted light, and the image is reconstructed on a video display screen. This technique is also referred to as **epifluorescence scanning microscopy** when the light source is provided from above.

The sample to be viewed can be a pure culture of microbes (**Figure 4.11**) or a natural community containing microorganisms (**Figure 4.12**).

Transmission Electron Microscope (TEM)

The light microscope has a useful magnification of about 1,000 to 2,000×. Although greater magnification can be achieved, as noted previously, finer detail will not be visible, in part because of the properties of the illuminating source itself. It is not possible to observe objects well if they are smaller than the wavelength of the illuminating source. An analogy may help illustrate this point. Suppose you wish to make an impression of your hand and two materials are available: fine particles of wet clay (analogous to a short wavelength) or gravel (a long wavelength). You can make a much more detailed and accurate impression of your hand with the clay.

The **transmission electron microscope** can produce very short wavelengths by using a beam of electrons as the illuminating source. The wavelength is controlled by the voltage applied to an electron gun, which is the source of electrons. If the accelerating voltage is 60,000 volts, the wavelength of the electron beam is 0.005 nm,

(A)

(B)

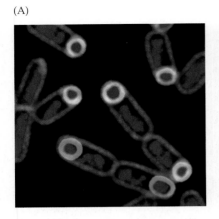

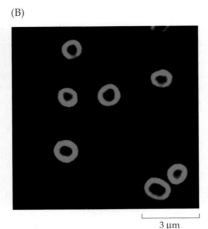

3 μm

Figure 4.11 Confocal laser microscopy
Localization of a GFP-tagged membrane protein expressed in the forespore of *Bacillus subtilis*. The endospore membrane is labeled with a membrane protein fused to GFP (green). Cellular membranes are stained with FM 4-64 (a dye that gives off red fluorescence) and the DNA is stained with DAPI (blue). (A) A medial focal plane image showing all three fluorescent signals. (B) An image showing the GFP fluorescence alone. The GFP-tagged membrane protein is randomly distributed throughout the spherical forespore membrane, and therefore appears as circles in the medial focal plane. Courtesy of Aileen Rubio and Kit Pogliano.

or 0.000005 μm. This is 100,000 times shorter than the wavelength of violet light (about 500 nm). As a result, the theoretical resolution of the electron microscope is about 2 Å (Å is an angstrom; 1 Å = 10^{-10} m), which is twice the diameter of the hydrogen atom.

In place of optical lenses, the TEM uses electromagnetic lenses to bend the electron beam for focusing. A vacuum is essential in permitting the flow of electrons through the lens system, so the entire electron microscope must have an enclosed chamber and accompanying vacuum pumps. This makes the TEM a much larger instrument than the ordinary light microscope (Figure 4.13). The real image is formed by electrons bombarding a phosphorescent screen, and the photograph, called an **electron micrograph**, is taken by a camera mounted below the screen, containing film sensitive to electron radiation.

The lenses of the TEM are equivalent to those of the compound light microscope—a condenser lens, an ob-

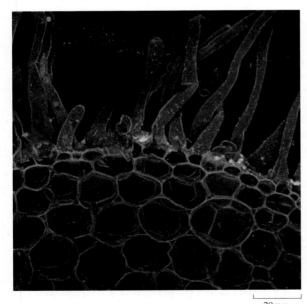

20 μm

Figure 4.12 Natural community
Cells of GFP-expressing *Burkholderia* sp. (bright yellow-green) on the surface of a *Gleditsia triacanthus* root. Courtesy of Michelle Lum and Ann Hirsch.

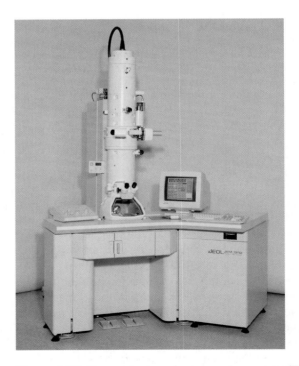

Figure 4.13 Transmission electron microscope (TEM)
The TEM is very large because the entire device is contained in a vacuum and it uses electromagnets for lenses. The illuminating source is an electron gun located at the top of the microscope. The specimen is placed between the electron gun and the phosphorescent screen that is used to view the image. Courtesy of JEOL USA, Inc.

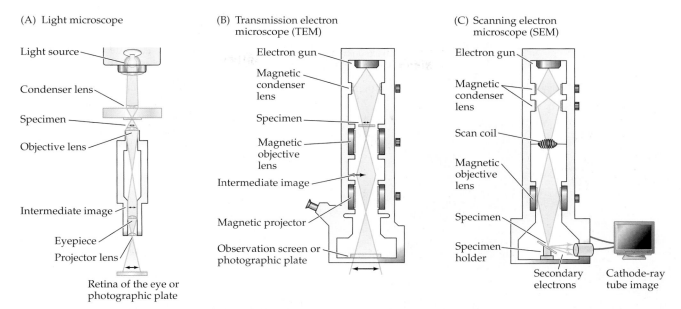

(A) Light microscope

Light source
Condenser lens
Specimen
Objective lens
Intermediate image
Eyepiece
Projector lens
Retina of the eye or photographic plate

(B) Transmission electron microscope (TEM)

Electron gun
Magnetic condenser lens
Specimen
Magnetic objective lens
Intermediate image
Magnetic projector
Observation screen or photographic plate

(C) Scanning electron microscope (SEM)

Electron gun
Magnetic condenser lens
Scan coil
Magnetic objective lens
Specimen
Specimen holder
Secondary electrons
Cathode-ray tube image

Figure 4.14 Illumination in light and electron microscopes
A comparison of the illuminating paths in the light microscope shown upside down for comparison with the electron microscopes, the transmission electron microscope and the scanning electron microscope. Note that all three have illuminating sources, condenser lenses, and objective lenses. Because a virtual image cannot be viewed, a projector lens is needed in the TEM and when cells are photographed in the light microscope. The image of the SEM specimen is viewed on a cathode ray tube.

jective lens, and a projector (ocular) lens. The fundamental differences are that electrons are the illuminating source and magnets replace optical **lenses** (Figure 4.14).

Figure 4.15 compares the appearance of bacterial cells in a light microscope and in a TEM. Note the increased detail in the electron micrograph, even though the magnification is about the same for both preparations. Because the preparation to be observed in the electron microscope is placed in a vacuum chamber, it is not possible to observe living cells with the TEM. In place of a glass slide, cell preparations are placed on a small metal (usually copper) screen (4.0 mm in diameter, about the size of this **O**) called a **grid**. A thin plastic film is placed on the grid to hold the cell preparation. In the simplest preparation, cells are placed on the plastic-coated grid and allowed to dry. The whole cell preparation is then stained with a heavy metal stain such as phosphotungstic acid or uranyl acetate, allowed to dry, and then placed in the electron microscope.

The major application of the TEM in biology is to observe internal cell structures. This requires a more elaborate procedure for preparing the specimen for observation, a procedure called **thin sectioning**. Because the specimen will ultimately be observed in a vacuum, some procedure must be used to preserve the structure of the specimen in the absence of water. This is accomplished

through a process called **dehydration and embedding,** in which the cells are taken from their aqueous environment and transferred into a plastic resin. Although more complex, this process is similar to embedding insects in plastic resins, as practiced by some hobbyists.

First, cells are harvested from their growth medium by centrifugation. They are gradually dehydrated by transferring them step by step from the aqueous medium into ethanol solutions of increasing concentration, until they are placed in 100% ethanol. The next step is to transfer the cells through an acetone–ethanol series until they are suspended in 100% acetone. Unlike ethanol and water, acetone is a plastic solvent and is miscible with plastic resins. At this point, the preparation is completely dehydrated. The next step is to embed the cells in a plastic mixture. Again, a series of transfers is made until the cells are in 100% plastic resin. Sufficient time is permitted to ensure that the resin completely displaces the acetone in the cells. The preparation is "cured" (polymerized to form a solid) by heating at 60°C in an oven. The organism is now embedded.

The embedded cells are sliced into thin sections with an **ultramicrotome**. The ultramicrotome is analogous to a meat slicer, except that it has a diamond knife and cuts extremely thin sections (about 60-nm thick). The thin sections are stained with heavy metals (e.g., lead

(A)

This bacterium, like many prokaryotes, has fimbriae, short appendages made of proteinaceous material. In this species the fimbriae are too small to be seen by light microscopy.

(B)

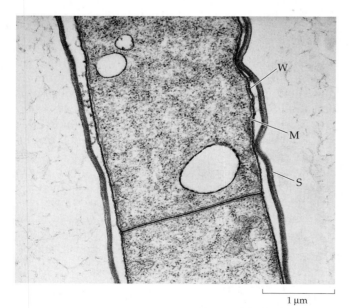

2 μm

2 μm

Figure 4.15 Resolution in light and electron microscopy
A large unidentified colonial bacterium, obtained from a lake, is viewed under (A) the electron microscope and (B) the light microscope. Fimbriae are discussed later in the chapter (see also Figure 4.69). Courtesy of J. T. Staley and Joanne Tusov.

citrate) to increase contrast. They are then placed on TEM grids and examined. A typical thin section is shown in Figure 4.16.

Scanning Electron Microscope (SEM)

As in the TEM, electrons are the illuminating source for the **scanning electron microscope** (SEM), but the electrons are not transmitted through the specimen as in transmission electron microscopy. In SEM the image is formed by incident electrons scanned across the specimen and back-scattered (reflected) from the specimen, as described for the fluorescence microscope. The reflected radiation is then observed with the microscope in the same manner that we observe objects illuminated by sunlight. Solid metal stubs are used to hold the preparations. Before the organism is viewed, it is dried by **critical point drying**, carefully controlled drying in which water is removed as vapor so that structural damage to cells is minimized. The specimen is then coated with an electron-conducting noble metal such as gold or palladium.

The SEM has a larger depth of field than the TEM so there is a greater depth through which the specimen remains in focus. Therefore, all parts of even a relatively large specimen, such as a eukaryotic cell, remain in focus when viewed by the SEM (Figure 4.17A). The result is an image that looks three-dimensional. This is the best procedure available to examine colonial forms of microorganisms, such as the fruiting structures of the myxobacteria (Figure 4.17B).

An SEM can be equipped with an x-ray analyzer, enabling the researcher to determine the elemental composition of microorganisms or their components. Elements with atomic weights greater than 20 can be assayed in a semiquantitative manner using this instrument (Figure 4.18).

1 μm

Figure 4.16 Thin section of a gram-negative bacterium
A thin section of *Thiothrix nivea* showing (from the outside) its sheath (S), cell wall (W), and cell membrane (M). Courtesy of Judith Bland and J. T. Staley.

(A)

(B)

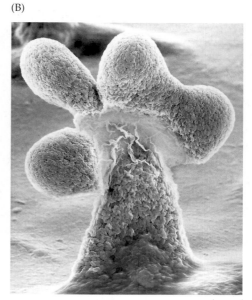

100 μm

100 μm

Figure 4.17 Scanning electron microscope (SEM)
(A) A radiolarian, a eukaryotic protist with a siliceous (silica-containing) shell. (B) The fruiting structure of the myxobacterium *Stigmatella aurantiaca*. The entire structure is about 1 mm in length. A, courtesy of Barbara Reine; B, ©K. Stephens and D. White/Biological Photo Service.

The fungal microcolony is growing on a desert rock. The x-ray data for the fungus reveal the presence of manganese, (Mn), iron (Fe), and other elements.

The x-ray analyses of four areas (1, 2, 3, and 4) of surrounding rock reveal iron and other elements but no manganese.

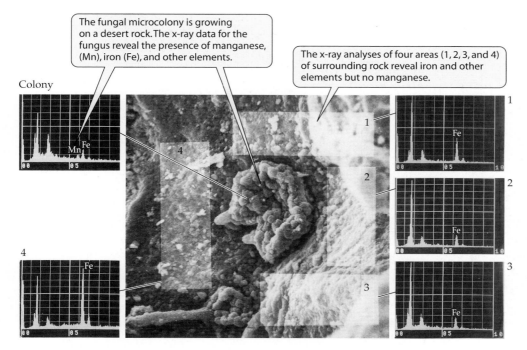

Figure 4.18 SEM-elemental analysis of manganese and iron
SEM of a fungal microcolony growing on a desert rock. The x-ray findings indicate that the colony contains manganese (Mn) as well as iron (Fe) , whereas the outlying regions (denoted by rectangular areas) contain only iron. Courtesy of F. Palmer and J. T. Staley.

Atomic Force Microscopy (AFM)

This is a scanning microscopy method developed in the mid-1980s that provides extremely high-resolution images of cells and subcellular structures. Samples can be imaged in air, in liquids, or following vacuum treatment. The image is generated by repeated scanning of the specimen surface with a cantilevered probe (Figure 4.19). Movement of the probe in the vertical dimension is monitored by a laser-coupled photo-detector, while the movement of the probe relative to the specimen in the x and y directions is controlled by a computer. The resulting recording of the probe positions in the x, y, and z directions is then processed by computer to reconstruct the three-dimensional image of the object (Figure 4.20). The theoretical resolution of the microscope approaches the atomic scale, although in practice resolution of bacterial structures in the range of 0.05 μm can be achieved. A distinct advantage of this method is that it enables scientists to directly visualize living biological materials without special processing.

The varieties of microscopes and associated procedures described previously were instrumental in the developing field of microbiology. In the next section we discuss the various morphological attributes of prokaryotes at both the organismal and the subcellular level, their chemical composition, and their functional role in the organism. Keep in mind that most microbiologists study the behavior of microorganisms in the laboratory, although their ultimate interest lies in understanding the function of a given structure in the organism's natural environment.

SECTION HIGHLIGHTS

Microscopes use lenses and light or electrons as illuminating sources to magnify microorganisms with high resolution to accurately portray cells or structures within them. Specific microscopes, stains, and fluorescent dyes are used to enhance the image of microorganisms so that they and their internal structures can be best visualized and examined. The development of microscopy has gone hand-in-glove with our understanding of microbial life.

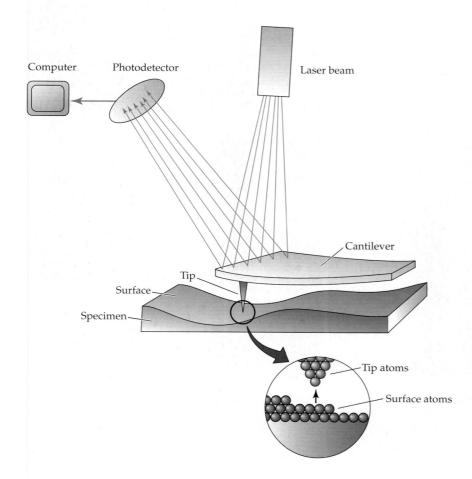

Figure 4.19 Atomic force microscope
This diagram illustrates the operation of the atomic force microscope. Tip atoms of a miniature stylus probe mounted on a cantilever are moved over the surface of a specimen. The contour (shape) of the specimen is detected by the differences in distance of a laser beam captured by a photodetector. Dimensions are recorded in the x, y, and z directions and the image reconstructed with aid of a computer.

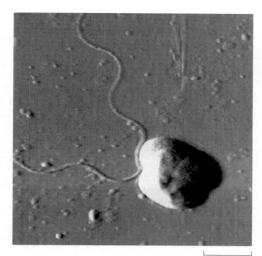

Figure 4.20 Atomic force microscopy
Nanoscale view of a bacterium, *Pseudomonas aeruginosa*, as observed by atomic force microscopy. Courtesy of J. K. Gimzewski.

1 μm

4.2 Morphology of *Bacteria* and *Archaea*

Bacteria and *Archaea* come in a variety of simple shapes, sizes, and organismal arrangements, which are components of their **morphology**. Most are single-celled but some are multicellular forms consisting of numerous cells living together. Some organisms are **pleomorphic** in that they exhibit different shapes within the culture. Compound microscopy is commonly used to determine the shape and size of microorganisms.

Unicellular Organisms

The simplest shape for a single-celled prokaryote is the sphere. Unicellular **spherical** organisms are called **cocci** (coccus, singular). Random cell division in different planes of a coccus produces a grape-like cluster of cells referred to as a staphylococcus (Figure 4.21A). Some cocci grow and divide along only one axis (Figure 4.22A; discussed in more detail in Chapter 6). If the cells remain attached after cell division, this results in the formation of chains of cells of various lengths. A diplococcus

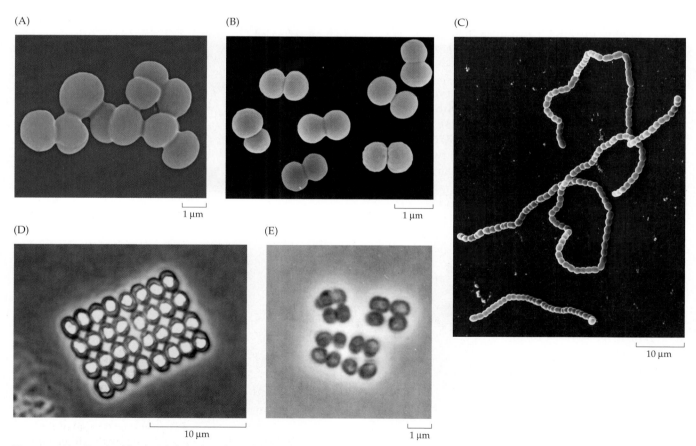

(A)

(B)

(C)

1 μm

1 μm

(D)

(E)

10 μm

10 μm

1 μm

Figure 4.21 Typical bacterial and archaeal shapes
Various formations of cocci, spherical cells, as shown by microscopy. (A) A staphylococcus, a grapelike cluster; (B) a diplococcus, *Neisseria meningitidis*; (C) streptococcus, a chain of cells; (D) sheet (internal bright areas of each cell are gas vacuoles); (E) eight-cell packet, or sarcina. A and B, ©Dennis Kunkel Microscopy, Inc.; C, ©David M. Phillips/Visuals Unlimited; D and E, courtesy of J. T. Staley and J. Dalmasso.

(A)   (B)

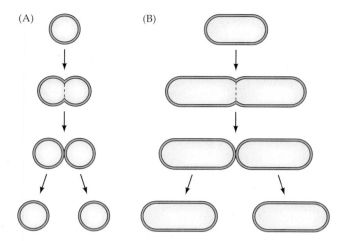

(A)

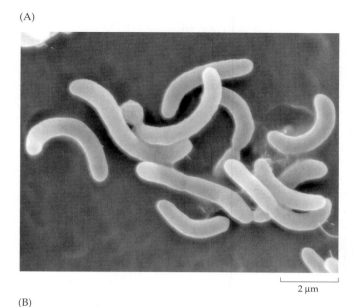

2 μm

Figure 4.22 Coccus and rod shape showing binary transverse fission
Typical binary transverse fission in (A) a coccus and (B) a rod-shaped bacterium.

(B)

is a "chain" of only two cells (Figure 4.21B); a streptococcus can contain many cells in its chain (Figure 4.21C). Some cocci divide along two perpendicular axes in a regular fashion to produce a sheet of cells (Figure 4.21D). Other cocci divide along three perpendicular axes, resulting in the formation of a packet or sarcina of cells (Figure 4.21E).

The most common shape in the prokaryotic world is not a sphere, however. It is a cylinder with blunt ends, referred to as a **rod** or **bacillus** (Figure 4.23). A cell grows and divides along one axis (Figure 4.22B; discussed in more detail in Chapter 6), to give rise to two new cells termed "**daughter cells**." If the rods remain attached to

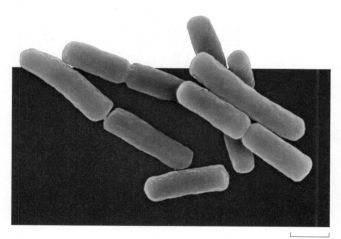

Figure 4.23 Bacilli, or rods
Rod-shaped cell of a unicellular *Bacillus anthracis*, shown by SEM. ©Dennis Kunkel Microscopy, Inc.

2 μm

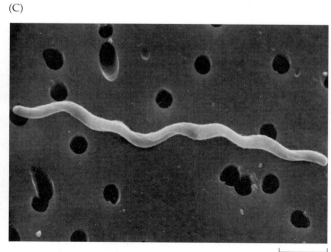

2 μm

(C)

2 μm

◄ **Figure 4.24 Curved and helical cells**
(A) Bent rod, or vibrio; (B) spirillum; (C) large spirochete, from the style of an oyster. A,B, ©Dennis Kunkel Microscopy, Inc.; C, ©Paul W. Johnson and John Sieburth/Biological Photo Service.

one another after division across the transverse (short) axis of the cell, a cell chain results, analogous to that of the cocci.

A less common shape for unicellular bacteria is a helix. A very short helix (less than one helical wavelength long) is called a bent **rod** or **vibrio** (Figure 4.24A). A longer helical cell is called a **spirillum** (Figure 4.24B) if the cell shape is rigid and unbending, or a **spirochete** (Figure 4.24C) if the organism is flexible and changes its shape during movement.

Variations of these common shapes of unicellular bacteria also exist. For example, some bacteria produce appendages that are actually extensions of the cell, called prosthecae, which give the cell a star-shaped appearance (Figure 4.25). The morphological diversity of *Bacteria* and *Archaea* is described further in Chapters 18 to 22, where individual genera are discussed.

Prokaryotes maintain their shapes during the process of **reproduction**, an asexual process whereby a single organism divides to produce two progeny cells (see Figure 4.22). For most unicellular prokaryotes there are two ways in which this may be accomplished, either by binary transverse fission or by budding (discussed in Chapter 6).

Multicellular Bacteria *and* Archaea

Numerous prokaryotic organisms exist as multicellular forms. One such group is the **actinobacteria**. These rod-shaped organisms produce long filaments containing

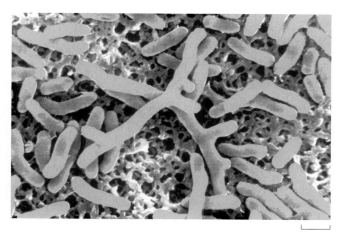

Figure 4.26 Mycelial bacterium
Streptomyces sp. illustrating the complex network of filaments with some branches, called a mycelium. ©VEM/Photo Researchers, Inc.

many cells. The filaments form branches, resulting in an extensive network comprising hundreds or thousands of cells. This network is referred to as a **mycelium** (Figure 4.26).

Another common multicellular shape is the **trichome**, which is frequently encountered in the cyanobacteria (Figure 4.27A). Although a trichome superficially resembles a chain, adjoining cells have a much closer spatial and physiological relationship than do the cells in a chain. Motility and other functions result from the concerted action of all cells of the trichome. And some cells in the trichome may have specialized functions that benefit the entire trichome. For example, the **heterocyst** (Figure 4.27B), a specialized cell produced by some filamentous cyanobacteria, is the site of nitrogen fixation (see Chapter 21).

Finally, individual cells of some bacterial species can associate with one another or with other species under certain conditions to form multicellular aggregates that perform specialized metabolic functions not associated with single cells (e.g., biofilms, developmentally programmed structures, and symbiotic relationships). These are discussed in later chapters, including Chapters 19, 24, and 25.

Size of Bacteria *and* Archaea

In addition to having characteristic cell shapes (i.e., rods, cocci, vibrios, and spirilla), the majority of bacteria have characteristic and uniform cell sizes that range from 0.2 μm to 5 μm in their dimensions. Some species such as those of *Mycoplasma* are small (ca. 0.2 μm) while others can be quite large. The largest known so far is the sulfur bacterium *Thiomargarita* (ca. 700 μm in diameter), which can be seen by the unaided eye (see Box 1.3). Another

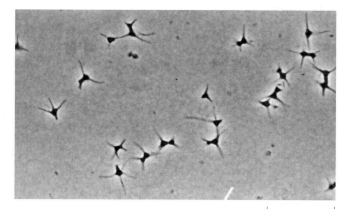

Figure 4.25 Prosthecate bacterium
A star-shaped bacterium, *Ancalomicrobium adetum*. Courtesy of J. T. Staley.

(A)

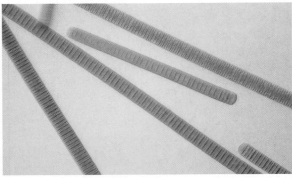

(B)

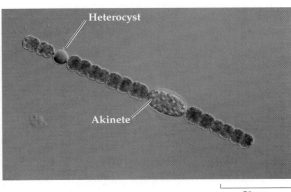

10 µm

50 µm

Figure 4.27 Filamentous bacteria
(A) *Oscillatoria*, a multicellular filamentous cyanobacterium, showing the close contact between cells in the trichome. Trichomes can be as large as 40 µm in diameter. (B) An *Anabaena* sp. filament with typical and specialized cells. (Cell diameter about 10 µm.) Heterocysts are nonpigmented cells, the site of nitrogen fixation; the akinete is a resting stage. A, ©blickwinkel/Alamy; B, ©Paul W. Johnson/Biological Photo Service.

large bacterium is *Epulopiscium*, which exceeds the size of typical protists, is found in the intestine of the surgeonfish (see Figure 20.11). Cell size and shape are predetermined by the genetic plan encoded by the organism's genome as well as by its nutritional state.

SECTION HIGHLIGHTS

Morphology refers to the shape, size, and arrangements of organisms. *Bacteria* and *Archaea* are generally very small single-celled organisms, typically about 1 µm in width, and occur as cocci, rods, vibrios, and spirilla. Some microbes produce multicellular filaments or mycelia.

4.3 Structure, Composition, and Cell Function

The remainder of this chapter is devoted to the description of various prokaryotic structures, their chemical composition, and their cellular functions. The terms "fine structure" and "ultrastructure" refer to subcellular features that are best observed using the electron microscope. Studies of the fine structure of microbial cells began in the 1950s and 1960s, when electron microscopy procedures were perfected. Scientists used a combination of procedures to "break open" or lyse cells, followed

by centrifugation to separate the various subcellular components. These components were purified and then analyzed biochemically (see Chapter 11). The electron microscope was used at various steps in the procedure to identify and assess the purity of the structures.

We begin with internal structures found in the **cytoplasm** and then consider the outer cell layers called the **cell envelope**, and, finally, certain cell materials external to the cell envelope. The discussion in this chapter is confined to structures commonly found in many *Bacteria* and *Archaea*. Thus, we do not discuss structures such as cysts, specialized developmental cell forms, and magnetite crystals, which are covered in descriptions of the particular organisms that have these structures (Chapters 18 to 22).

Cytoplasm of Bacteria and Archaea

The cell cytoplasm is defined as including all cell components bounded by the cell or cytoplasmic membrane (Figure 4.28). It is the location of the genetic material, machinery for transcription and translation, most cellular enzymes (e.g., biosynthetic, catabolic, and repair enzymes), and the pools of small molecules needed for cell metabolism. It is also the location of several types of discrete structures involved in specialized cell functions (for example, gas vesicles and storage granules).

INTRACELLULAR ENZYMES   Most biosynthetic, catabolic, and repair enzymes are located in the cytoplasm. They are responsible for synthesis of cell intermediates, proteins, DNA, ribonucleic acid (RNA), cofactors, and some macromolecular structures including ribosomes. The specific names and activities of these enzymes and enzyme complexes are described in Chapters 10, 11, and 13.

SMALL MOLECULES   The cytoplasm contains **small-molecule pools** of many types of molecules required for cell metabolism. These include amino acids needed for

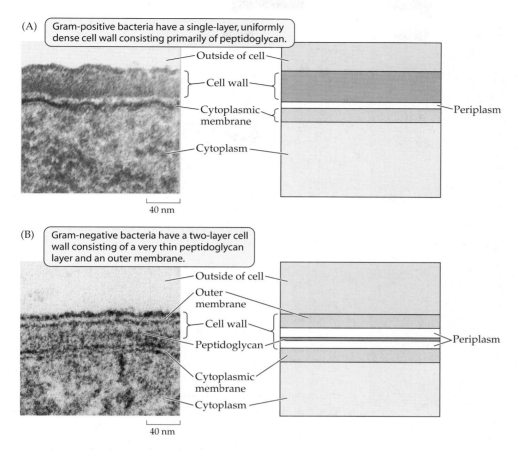

(A) Gram-positive bacteria have a single-layer, uniformly dense cell wall consisting primarily of peptidoglycan.

Outside of cell
Cell wall
Cytoplasmic membrane
Cytoplasm
Periplasm

40 nm

(B) Gram-negative bacteria have a two-layer cell wall consisting of a very thin peptidoglycan layer and an outer membrane.

Outside of cell
Outer membrane
Cell wall
Peptidoglycan
Cytoplasmic membrane
Cytoplasm
Periplasm

40 nm

Figure 4.28 Cross section of bacterial cell envelopes
Thin sections and diagrams showing the cell envelopes of (A) a gram-positive bacterium, *Bacillus megaterium*, and (B) the gram-negative bacterium *Escherichia coli*. Note the characteristic outer membrane of the gram-negative bacterium. The thin peptidoglycan layer of gram-negative bacteria is not always evident in electron micrographs. Note also the location of the periplasm in each type of organism. Photos courtesy of Peter Hirsch and Stuart Pankratz.

protein synthesis; nucleotides for DNA and RNA synthesis; precursors for lipid and carbohydrate synthesis; and many cofactors needed to produce biosynthetic, catabolic, and repair enzymes. Finally, small-molecule pools include certain ions involved in energy conservation, cellular homoeostasis, or both (e.g., H^+, K^+, Na^+, and compatible solutes).

MACROMOLECULES AND MACROMOLECULAR COMPLEXES   Foremost among the intracellular macromolecules of all cells is their DNA, the hereditary material of the cell. Other universal intracellular components of *Bacteria* and *Archaea* include proteins and RNA. In addition, ribosomes, gas vesicles, and various reserve materials are discussed individually in subsequent text.

DNA   The DNA of prokaryotes is a circular, or, more rarely, linear, double-stranded helical molecule. The two strands are held together by hydrogen bonds, the nucleotide bases of one strand forming hydrogen bonds with the bases of the opposite strand: adenine with thymine and cytosine with guanine (see Chapter 3).

The DNA appears as a fibrous material in the cytoplasm when prokaryotic cells are viewed in thin sections (**Figure 4.29**). As noted in Chapter 1, the DNA of typical prokaryotes is not surrounded by a **nuclear membrane**

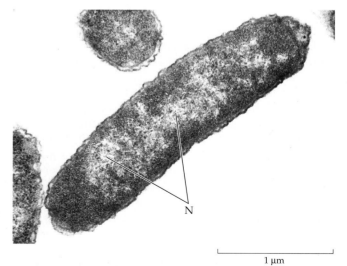

1 μm

Figure 4.29 Appearance of DNA by electron microscopy
Thin section through *Salmonella typhimurium* showing the fibrous appearance and diffuse distribution of the nuclear material (N) in a typical prokaryotic cell. Courtesy of Stuart Pankratz.

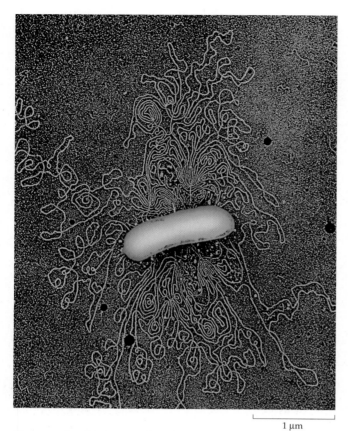

1 μm

Figure 4.30 DNA strands released from cell
Electron micrograph showing DNA strands released from a lysed bacterial cell. ©Dr. Gopal Murti/SPL/Science Source/Photo Researchers Inc.

Covalently closed circular duplex (supercoiled)

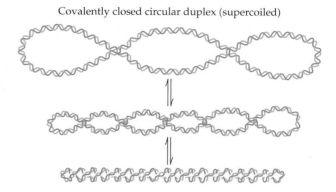

Figure 4.31 Supercoiled DNA
Increasing degrees of supercoiling of DNA produce a more tightly compacted molecule.

and thus does not appear in a confined area within the cell; rather, it appears as a somewhat diffuse, dispersed fibrous material. For this reason the region is not called a nucleus but a **nucleoid** or **nuclear area**.

If gentle conditions are used to lyse the cells (like eggs, cells can be broken carefully; the cytoplasm can be freed from the cell membrane and wall), the DNA is released and appears as a coiled structure spilled from the cell (Figure 4.30). When stretched out, the length of the DNA molecule is about 1 mm, about a 1,000 times longer than the 1 to 3 μm length of the typical prokaryotic cell! In order to package all of this material within the cell, the DNA molecule is tightly wound in **supercoils** (Figure 4.31). Special enzymes are responsible for supercoiling and for controlling the unwinding of the DNA during replication (DNA synthesis) and transcription (production of RNA from the DNA template; see Chapter 13).

The molecular weight of the chromosomal DNA molecule of prokaryotes ranges from about 10^9 to 10^{10} Da (Da is a dalton, a unit of mass approximately equal to the mass of the hydrogen atom, ^{1}H). The typical prokaryotic DNA contains about 4×10^6 base pairs (4 mega-base pairs, or 4 MB). However, some intracellular symbiotic bacteria such as *Buchnera* species have smaller genomes (0.65 MB), and some prokaryotic genomes are as large as 10 MB. These are typically smaller than the size of most eukaryotic genomes, the chromosomes of which may be ten or more times larger than prokaryotic chromosomes.

Prokaryotic cells may have one or more copies of the same chromosomal DNA molecule. For example, when the cell is growing and dividing rapidly, two, four, or more partial or complete copies may be present (see Chapter 13). *Burkholderia cenocepacia* is known to have several copies of its large genome in its cells.

In addition to the genomic DNA, the cell often contains other circular molecules of DNA called extrachromosomal elements or **plasmids**. These, too, are double-stranded DNA molecules (see Chapter 15). Plasmids do not carry genetic material that is essential to the growth of an organism, although they do contain features that may enhance the survivability of the organism in a particular environment. For example, some bacteria carry plasmids with genes that allow them to degrade antibiotics (substances produced by one organism that inhibit or kill other organisms (see Chapter 7). Other microbes possess plasmids that allow them to degrade specific carbon compounds like naphthalene—which is not useful to the bacterium unless the compound is in its immediate environment.

The primary function of the prokaryotic genome is to store its hereditary information carried in its genes. Furthermore, in some prokaryotes, under the appropriate conditions genetic material can be transferred from one organism to another. This can be accomplished by three different processes, depending on the prokaryote (see Chapter 15):

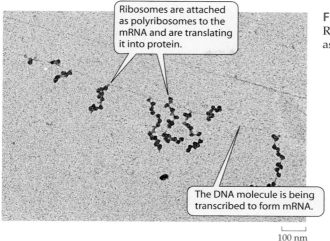

Ribosomes are attached as polyribosomes to the mRNA and are translating it into protein.

The DNA molecule is being transcribed to form mRNA.

100 nm

(A)

50 nm

Figure 4.32 Protein synthesis
Ribosomes in action in *Escherichia coli*. The DNA and RNA appear as filaments. Courtesy of Oscar L. Miller.

- **Transformation**, occurring when DNA released into the environment by lysis (cell breakage) of one organism is taken up by another organism

- **Conjugation**, in which transfer occurs during cell-to-cell contact between two closely related bacterial strains

- **Transduction**, in which prokaryotic viruses are involved in transferring DNA from one organism to another

RIBOSOMES   As mentioned in Chapter 1, ribosomes are complex macromolecular structures that carry out protein synthesis (Figure 4.32), a process referred to as translation (see Chapter 13) in which messenger RNA (mRNA) carries the message in nucleotides from the genome to the ribosome, where amino acids are linked together by peptide bonds to form protein.

At high magnification the prokaryotic ribosome can be seen to consist of two subunits, the small 30S subunit and the larger 50S subunit (Figure 4.33). Note that the Svedberg units (S), sedimentation densities, are not additive: the 30S and 50S ribosome subunits comprise a 70S ribosome. This is because the Svedberg unit is directly related not to molecular mass but to the density of particles in ultracentrifugation. (Likewise, the eukaryotic 80S ribosome consists of a small 40S subunit and a larger 60S subunit.)

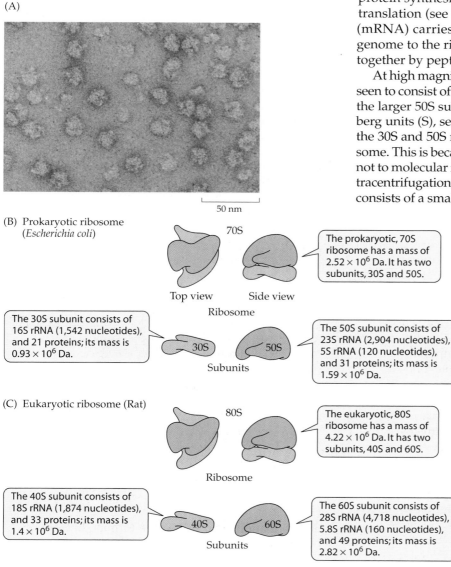

(B) Prokaryotic ribosome
(*Escherichia coli*)

70S

Top view   Side view
Ribosome

The prokaryotic, 70S ribosome has a mass of 2.52×10^6 Da. It has two subunits, 30S and 50S.

The 30S subunit consists of 16S rRNA (1,542 nucleotides), and 21 proteins; its mass is 0.93×10^6 Da.

30S   50S
Subunits

The 50S subunit consists of 23S rRNA (2,904 nucleotides), 5S rRNA (120 nucleotides), and 31 proteins; its mass is 1.59×10^6 Da.

(C) Eukaryotic ribosome (Rat)

80S

Ribosome

The eukaryotic, 80S ribosome has a mass of 4.22×10^6 Da. It has two subunits, 40S and 60S.

The 40S subunit consists of 18S rRNA (1,874 nucleotides), and 33 proteins; its mass is 1.4×10^6 Da.

40S   60S
Subunits

The 60S subunit consists of 28S rRNA (4,718 nucleotides), 5.8S rRNA (160 nucleotides), and 49 proteins; its mass is 2.82×10^6 Da.

Figure 4.33 Ribosome structure
(A) High-resolution electron micrograph of 70S ribosomes. Note the differences between the bacterial ribosome of *Escherichia coli* (B) and the eukaryotic ribosome of the rat (C). Photo courtesy of James Lake.

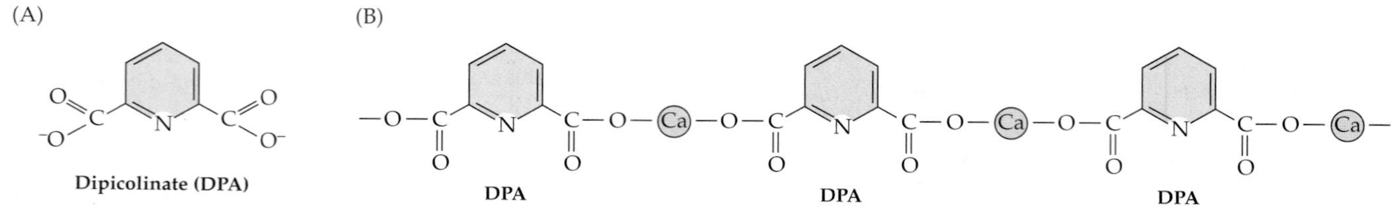

(A)

Dipicolinate (DPA)

(B)

DPA DPA DPA

Figure 4.34 Dipicolinate
(A) Chemical formula of dipicolinate and (B) proposed structure of the calcium-dipicolinate chelate.

Ribosomes consist of both protein and a type of ribonucleic acid called ribosomal RNA (rRNA). Figure 4.33 (B and C) shows the RNA and protein components of each of these subunits from the 70S and 80S ribosomes.

Even though bacterial and archaeal ribosomes have the same sedimentation coefficients, they differ somewhat in structure and composition. As a result, these organisms respond differently to the same **antibiotic** (see Chapter 7). For example, certain antibiotics, including chloramphenicol and the aminoglycosides, disrupt ribosome activity (and therefore inhibit protein synthesis) in *Bacteria*, but have no adverse effects on *Archaea*. Likewise, eukaryotic ribosomes are not sensitive to some of the antibiotics that affect bacteria.

Internal Structures Produced by Some Bacteria and Archaea

Although prokaryotes do not produce intracellular organelles, some produce internal structures that can be visualized by the light microscope. Two common types of structure are the endospore and gas vacuole.

BACTERIAL ENDOSPORE   The endospore, as the name implies, is a spore or resting stage unique to bacteria that is formed within the cell. Endospores are produced by several groups of the gram-positive bacteria, which are discussed in Chapter 20. The endospore has been the subject of many years of study, as it is an important example of cellular differentiation or morphogenesis. The dormant spore allows the bacterium to survive extended periods of desiccation and high temperature. These two conditions are often encountered in soil where these organisms are most commonly found. Not only are they more resistant to heat and desiccation than typical bacteria, but also to UV radiation and disinfection. Experiments have shown that the endospores of some species can survive storage in some environments for at least 50 years.

Endospores appear highly refractile when observed by phase contrast microscopy (see Figure 4.8). This high refractility is due to the low wa-

ter content of the spore. The endospore is rich in calcium ions and a unique compound, dipicolinic acid (Figure 4.34A). These compounds are believed to form a chelate complex because of their equal molar concentrations and chemistry (Figure 4.34B), and are thought to be responsible for the heat resistance of the endospore and other unique properties.

Endospores are very difficult to stain using ordinary staining procedures. For example, in the Gram stain, the endospore does not take up either the primary stain or the counterstain and therefore appears colorless on completion of the staining process. Endospores, however, can be stained by special staining procedures. One common procedure involves steaming the smear with the dye malachite green. After this is completed, the smear is counterstained with safranin. In this procedure the cell appears red, whereas the endospore appears green (see Figure 4.7C).

Endospore-forming bacteria have two phases of growth. **Vegetative growth**, which is the normal type of growth and reproduction, is similar to that found in other bacteria that undergo binary transverse fission. However, spore formation or **sporulation** is unique to these bacteria and is a highly regulated process (Figure 4.35). Sporulation is primarily a response to nutrient lim-

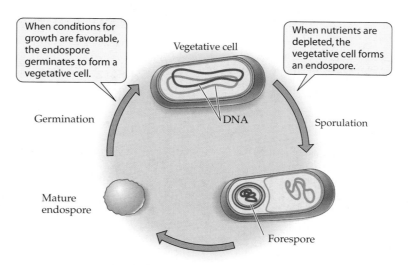

Figure 4.35 Sporulation of an endospore-forming bacterium
Although typical vegetative cells reproduce by binary transverse fission, cells that encounter nutrient deprivation may undergo sporulation as diagrammed here.

itation and usually occurs when the density of the population is high. However, the first response of the cell to exhaustion of preferred nutrients is not sporulation. The bacteria induce a range of adaptive responses such as chemotaxis, motility, and synthesis of extracellular degradative enzymes in order to procure secondary sources of nutrients. If these responses do not provide the cell with adequate nutrients for growth, the cells enter the sporulation pathway. Thus, the endospore is produced at the end of the exponential phase of growth.

A number of stages, along with specific morphological changes, are noted during sporulation (Figure 4.36).

(A)

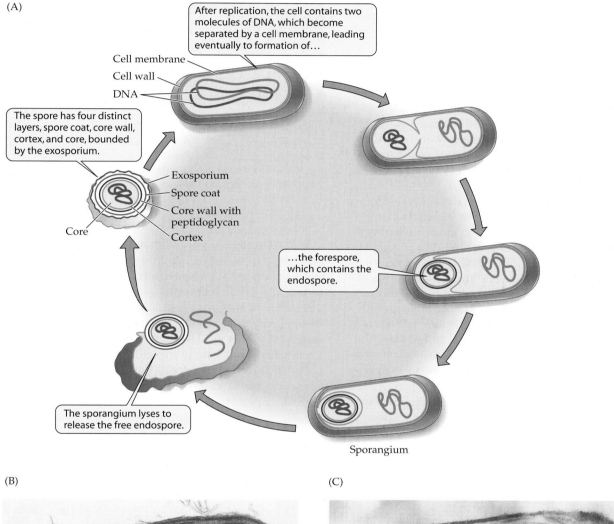

After replication, the cell contains two molecules of DNA, which become separated by a cell membrane, leading eventually to formation of…

Cell membrane
Cell wall
DNA

The spore has four distinct layers, spore coat, core wall, cortex, and core, bounded by the exosporium.

Exosporium
Spore coat
Core wall with peptidoglycan
Cortex

Core

…the forespore, which contains the endospore.

The sporangium lyses to release the free endospore.

Sporangium

(B)

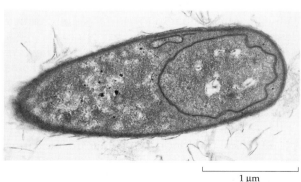

1 µm

(C)

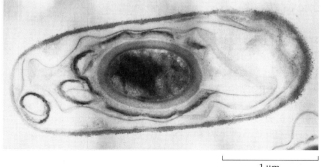

1 µm

Figure 4.36 Sporulation process
(A) Diagram of stages of sporulation as determined from electron micrographs. The thin sections through an endospore-forming bacterium show (B) the early stage of spore formation and (C) the mature spore within the sporangium. Photos courtesy of S. Pankratz.

After the period of vegetative growth ceases, two resulting molecules of DNA are present in the cell. These first coalesce and are then separated from one another by a cell membrane formed during a special differentiation process. An engulfment stage ensues in which the membrane from one incipient cell grows around the membrane of the other until it completely surrounds it. The engulfed incipient daughter cell, termed the **forespore**, eventually becomes the mature endospore. The other "cell," which contains the endospore, is called the sporangium. At this stage the process of sporulation is irreversible, regardless of the presence of nutrients.

During sporulation the spore becomes increasingly refractile and develops four distinct layers, the spore coat, the core wall with its modified peptidoglycan, the cortex, and the core wherein the calcium ions and dipicolinic acid are found along with the nuclear material. Thin sections through endospores show each of these features (see Figure 4.36B,C). A membrane called the exosporium binds these layers. Eventually, the original cell lyses to release the free endospore.

The endospore is a resting stage that can survive long periods in the environment. It remains viable, and under appropriate conditions it can undergo the process of **germination**, whereby it begins to develop into a vegetative cell again (see Figures 4.35 and 4.36A). In this process the endospore takes up water, swells, and breaks the spore coat and exosporium, releases its calcium dipicolinate, and begins multiplying as in normal vegetative growth.

Some endospores are known to be able to survive boiling temperatures for long periods. These were the heat-resistant structures that caused so many problems for the proponents and opponents of the theory of spontaneous generation discussed in Chapter 2.

The location of the endospore in the cell and its shape are important features in classification. Most species have oval-shaped endospores that are located in a central to subterminal location within the cell or sporangium. At the other extreme, some species, such as *Clostridium tetani*, produce a terminal endospore that is larger than the cell diameter, giving the sporangium a club-shaped appearance (Figure 4.37).

GAS VESICLES   Among the most unusual structures found in prokaryotes are gas vesicles, produced by some aquatic species. These special protein-shelled structures provide buoyancy to many aquatic prokaryotes, and they are not found in any other life forms. When bacteria with gas vesicles are observed in the phase contrast microscope they appear to contain bright, refractile areas, called **gas vacuoles**, with an irregular outline (Figure 4.38A). When gas vacuolate cells are viewed with the TEM, the vacuoles are found to consist of numerous subunits, called **gas vesicles** (Figure 4.38B).

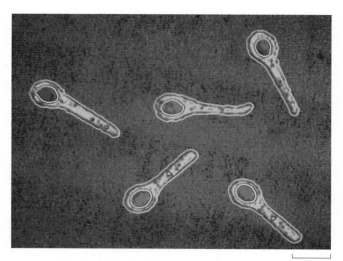

Figure 4.37 *Clostridium tetani*

2 μm

Electron micrograph (with added false color) of *Clostridium tetani*. Note that the endospores are larger than the cell, giving the cells a club-shaped appearance. ©Alfred Pasieka/Photo Researchers, Inc.

Gas vacuoles are found in widely disparate prokaryotes. They are common in photosynthetic groups such as the cyanobacteria, proteobacteria, and green sulfur bacteria (*Chlorobi*). They are also found in heterotrophic bacteria such as the genus *Ancylobacter* and in the prosthecate genera *Prosthecomicrobium* and *Ancalomicrobium*. The anaerobic gram-positive genus *Clostridium* also has gas-vacuolate strains. Finally, some *Archaea* produce gas vacuoles, including members of the genera *Methanosarcina* (methanogens) and *Halobacterium* (a salt-loving archaeal genus).

Gas vesicles have been isolated from bacteria and studied biochemically. They are obtained by gently lysing cells to release the vesicles, then separating the vesicles from cellular material by low-speed differential centrifugation. Cell material is relatively dense and is spun to the bottom of the centrifuge tube, whereas the buoyant gas vesicles float to the surface. The purified vesicles have a distinctive shape. They appear as cylindrical structures with conical end pieces (Figure 4.39). The size varies from about 30 nm in diameter in some species to about 300 nm in others. The vesicles can exceed 1,000 nm in length, depending upon the organism.

Each vesicle consists of a thin (about 2-nm thick) protein shell that surrounds a hollow space. The shell is composed of one predominant protein the repeating subunits of which have a molecular mass of about 7,500 Da. Amino acid analyses of this protein from different prokaryotes indicate that its composition is highly uniform from one organism to another. About half of the

(A)

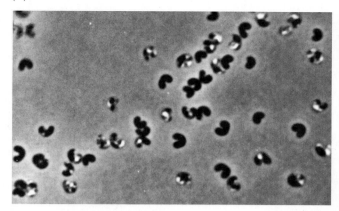

2 µm

(B)

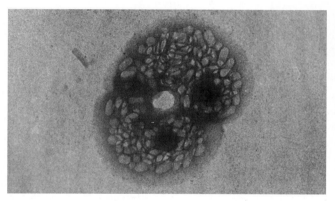

1 µm

Figure 4.38 Gas vacuoles and gas vesicles
Gas vacuoles and gas vesicles in *Ancylobacter aquaticus*, a vibrioid bacterium. (A) Phase contrast photomicrograph showing gas vacuoles and (B) electron micrograph showing the numerous transparent gas vesicles that comprise a gas vacuole. Courtesy of M. van Ert and J. T. Staley.

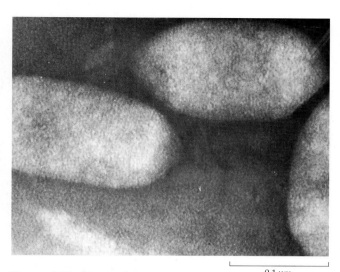

0.1 µm

Figure 4.39 Gas vesicles
Gas vesicles isolated from *Ancylobacter aquaticus*. The vesicle diameter is about 0.1 µm. Courtesy of J. T. Staley and A. E. Konopka.

protein consists of hydrophobic amino acids (such as alanine, valine, leucine, and isoleucine). It is thought that the hydrophobic amino acids are located on the inside of the shell and that their presence prevents water from entering the vesicle. Gases freely diffuse through the shell and are thus the sole constituents of the interior.

Gas vesicles do not store gases in the same way as a balloon. Gases freely diffuse through the vesicle shell, and the structure maintains its shape, not because it is inflated but because of its water impermeable, rigid protein framework. Because all gases freely diffuse in and out of the gas vesicle, the gases found in the vesicles are those present in the organism's environment.

The primary function of gas vesicles is to provide buoyancy for aquatic prokaryotes (Box 4.3). The density

of the organism is reduced when the cell contains gas vesicles, thereby permitting the organism to be buoyant in aquatic habitats. The mechanisms by which organisms regulate gas vacuole formation in nature are only poorly understood. Some cyanobacteria can descend in the environment by producing increased quantities of dense storage materials such as polysaccharides or by collapsing weaker gas vesicles. Depending upon their requirements for light, oxygen, and hydrogen sulfide, various bacterial groups are found in different strata in lakes during summer thermal stratification (see Chapter 24). The vertical position that gas vacuolate species occupy in the lake depends on their cell density, which is determined by the proportion of cell volume occupied by gas vesicles at a particular time.

INTRACELLULAR RESERVE MATERIALS   Many prokaryotes store a variety of organic and inorganic materials as nutrient reserves. Almost all these materials are stored as polymers, thereby only minimally increasing the internal osmotic pressure (see the discussion of osmotic pressure in "Function of the Cell Wall" in subsequent text).

The main organic compounds stored by bacteria are:

- Glycogen
- Starch
- Poly-β-hydroxybutyric acid
- Neutral lipids/hydrocarbons
- Cyanophycin
- Polyphosphate (volutin)
- Elemental sulfur

| BOX 4.3 | *Methods & Techniques* |

The Hammer, Cork, and Bottle Experiment

In the early 1900s, C. Klebahn conducted an important but simple experiment called the "hammer, cork, and bottle" experiment, which provided the first evidence that the gas vacuoles of cyanobacteria actually contain gas. For the experiment, cyanobacteria containing the purported gas vacuoles were taken from a bloom in a lake. Microscopic examination showed that the organisms contained bright areas indicative of gas vacuoles. Samples were placed into two bottles, one to be used as a control for the experiment.

In both bottles, the cyanobacteria initially floated to the surface. A cork was then placed in one bottle and secured in such a way that no air space was left between the cork and the water. Then, a hammer was used to strike a sharp blow on the suspension of cyanobacteria in the corked bottle. The blow briefly increased the hydrostatic pressure of the water in the experimental bottle.

The cyanobacteria in the experimental bottle soon sank to the bottom, whereas those in the control bottle remained floating on top. Furthermore, an air space formed

(A) Two bottles containing gas vacuolate cyanobacteria collected from a lake. The cell suspension with collapsed gas vesicles has a darker appearance. Gas vacuolate cells cause much greater refraction of light, so the bottle on the right appears more turbid. (B) Appearance of the bottles after a few minutes. Courtesy of A. E. Walsby.

between the water and the cork in the experimental bottle. When the cells from this bottle were subsequently examined in the microscope, no bright areas remained in the cells, showing that the cyanobacteria had lost their buoyancy and their gas vacuoles at the same time. The air space that had

collected in the experimental bottle was due to the gas released from the broken vesicles and contained by the cork.

This simple experiment showed that, indeed, gas vacuoles do contain gas and that they are essential in providing buoyancy to the cyanobacteria.

Glycogen and **starch** are common storage materials in prokaryotic organisms. They are polymers of glucose units linked together primarily by α-1,4 linkages (see Chapter 11). These storage materials cannot be seen using a light microscope, but when observed with the electron microscope they appear as either small, uniform granules, as in some cyanobacteria (see Figure 21.16), or as larger spheroidal structures, as in heterotrophic bacteria. Glycogen and starch can be degraded as energy and carbon sources. Interestingly, glycogen is also formed by animal cells, and starch is formed by eukaryotic algae and higher plants.

Unlike glycogen and starch, **poly-β-hydroxybutyric acid (PHB)** and related poly-esters appear as visible

granules in bacteria when viewed with a light microscope. In phase contrast microscopy, these lipid substances appear as bright, refractile, spherical granules (Figure 4.40). PHB granules are usually synthesized during periods of low nitrogen availability in environments that have excess utilizable organic carbon. In contrast to glycogen and starch, PHB is not found in eukaryotic organisms. PHB is a polymer of β-hydroxybutyric acid and is synthesized from acetyl-coenzyme A (see Chapter 11). The granules are enclosed within a protein membrane, which is needed for their synthesis or degradation or both. Related storage materials include longer chain fatty acid polymers (C5 to C11) called **polyhydroxyalkanoates**.

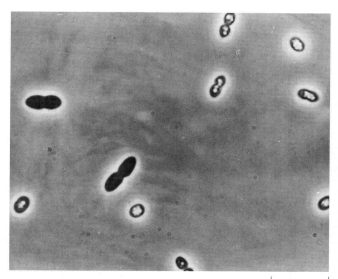

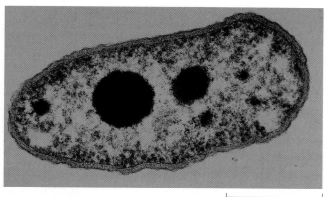

Figure 4.41 Polyphosphate granules   1 µm
Thin section of a bacterial cell (*Pseudomonas aeruginosa*) shows polyphosphate granules as dark areas. ©T. J. Beveridge/Visuals Unlimited.

Figure 4.40 PHB granules   10 µm
Poly-β-hydroxybutyrate (PHB) granules appear as bright refractile areas by phase contrast microscopy, as seen in *Azotobacter* sp. Courtesy of J. T. Staley.

NEUTRAL LIPIDS/HYDROCARBONS   Lipids or hydrocarbons like hexadec-1-ene can be accumulated by certain bacterial species (e.g., *Pseudomonas aeruginosa*, *Acinetobacter* sp.) from the cell's environment.

CYANOPHYCIN   The only nitrogen-containing polymer stored by prokaryotes is **cyanophycin**. This is a copolymer of the amino acids, aspartic acid and arginine. Cyanophycin appears as distinctive granules when viewed in thin section with an electron microscope (see Figure 21.15). As implied by the name, these granules are found only in cyanobacteria.

Some prokaryotes also store a variety of inorganic compounds, including polyphosphates and sulfur. **Volutin**, or **metachromatic granules**, consists of linear **polyphosphate** molecules, polymers of covalently linked phosphate units. Volutin appears as dark granules in phase contrast microscopy. The granules can be stained with methylene blue, which results in a red color caused by the metachromatic effect of the dye–polyphosphate complex. They appear as dense areas when viewed by the electron microscope (Figure 4.41).

Volutin is synthesized by the stepwise addition of single phosphate units from adenosine triphosphate (ATP) onto a growing chain of polyphosphate. Although its degradation has not been studied in detail, volutin may serve as both an energy source for the synthesis of ATP from ADP in some bacteria, and as a reserve material for nucleic acid and phospholipid synthesis. Volutin is apparently stored by organisms when other nutrients are limiting their growth. However, some bacteria are known to store volutin during active growth when there is no nutrient limitation, an effect called **luxurious phosphate uptake**.

Sulfur is stored as **elemental sulfur granules** by certain bacteria involved in the sulfur cycle. Sulfide-oxidizing photosynthetic proteobacteria and colorless filamentous sulfur bacteria, both of which produce sulfur by the oxidation of sulfide, store the sulfur as granules, which appear as bright, slightly yellow spherical areas in their cells (Figure 4.42). The sulfur can be further oxidized to sulfate by both groups of organisms. Thus, the granules are transitory and serve as a source of energy (when oxidized by filamentous sulfur bacteria; see Chapter 19) or a source of electrons (for carbon dioxide fixation in photosynthetic bacteria; see Chapter 21).

SECTION HIGHLIGHTS

All *Bacteria* and *Archaea* have DNA, RNA, proteins, and ribosomes in their cytoplasm where replication, transcription, and translation occur. The genomes of *Bacteria* and *Archaea* are generally small, less than 10 MB; therefore, those of many of the pathogenic and environmentally important species have been sequenced. Several internal storage polymers are produced, some of which can be used as energy sources for growth. Some organisms produce specialized structures in the cytoplasm. For example the endospore is a hardy reproductive structure produced by some gram-positive bacteria, and various aquatic bacteria produce gas vesicles as flotation devices.

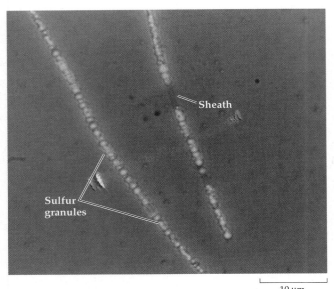

Figure 4.42 Sulfur granules
Phase contrast photomicrograph of the sulfur bacterium *Thiothrix nivea* showing many bright yellow and orange sulfur granules, and a sheath. Courtesy of Judith Bland.

4.4 The Cell Envelope

The cell envelope is defined as the cell materials that surround the cytoplasm (see Figure 4.28 and Figure 4.43). This includes the **cytoplasmic membrane** present in all

Bacteria and *Archaea*, plus other associated materials that include the **cell wall**, the **periplasmic space**, the **outer membrane**, **cell surface appendages**, and **extracellular polymeric substances** (**EPSs**). The composition of the cell envelope may differ depending on the prokaryote species. Three major categories include the gram-positive bacteria, the gram-negative bacteria, and the *Archaea*.

Cytoplasmic Membrane

The cellular membrane or **cytoplasmic membrane** serves as the primary boundary of the cell's cytoplasm. In prokaryotes it looks like two thin lines separated by a transparent area when viewed in thin section via the electron microscope (see Figures 4.28 and 4.43). At the molecular level, a model known as the **fluid mosaic** model, with a nonrigid membrane in which proteins can move, best explains the structure and composition of the cell membrane (Figure 4.44).

The cell membranes of prokaryotes are composed of **phospholipids** and **membrane proteins**. As many as seven different phospholipids and approximately 200 proteins have been identified in the typical membrane of bacteria such as *Escherichia coli*. One of the phospholipids is phosphatidyl serine (Figure 4.45). This complex lipid consists of a glycerol moiety covalently bonded to two fatty acids by an **ester linkage** and, at its third carbon, to a phosphoserine. The result is a bipolar molecule with a hydrophobic end (the hydrocarbon chains of the

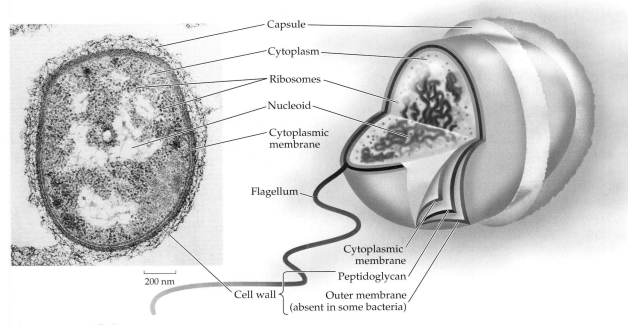

Figure 4.43 Cell diagram
A thin section of a bacterial cell is shown next to a diagram identifying the cell envelope layers. Photo ©J. J. Cardamone & B. K. Pugashetti/Biological Photo Service.

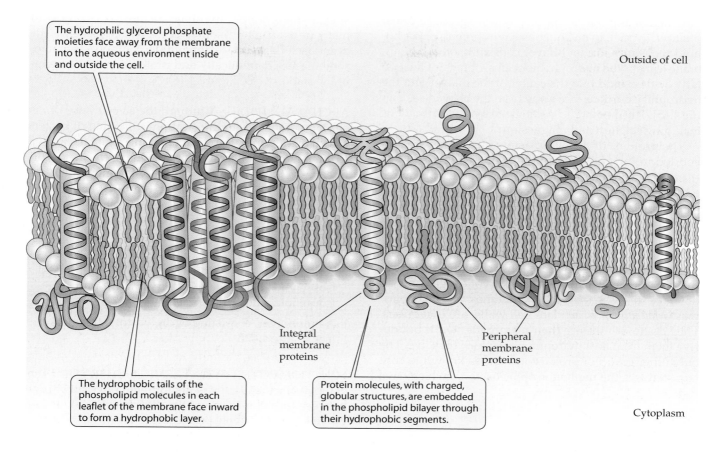

The hydrophilic glycerol phosphate moieties face away from the membrane into the aqueous environment inside and outside the cell.

Outside of cell

Integral membrane proteins

Peripheral membrane proteins

The hydrophobic tails of the phospholipid molecules in each leaflet of the membrane face inward to form a hydrophobic layer.

Protein molecules, with charged, globular structures, are embedded in the phospholipid bilayer through their hydrophobic segments.

Cytoplasm

Figure 4.44 Bacterial cell membrane structure
The fluid mosaic model of cell membrane structure. The bilayer structure consists of two phospholipid leaflets.

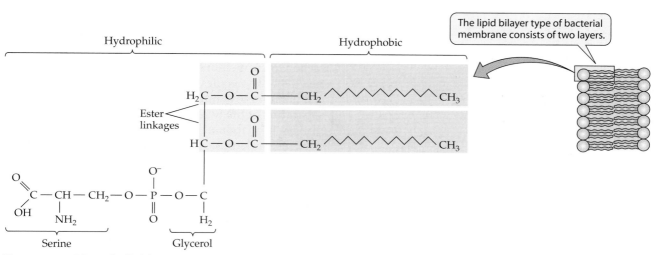

Hydrophilic

Hydrophobic

The lipid bilayer type of bacterial membrane consists of two layers.

Ester linkages

Serine

Glycerol

Figure 4.45 Phospholipid
Chemical structure of phosphatidyl serine, a typical phospholipid. The hydrocarbon side chains are fatty acids, typically containing 12 to 18, sometimes more, carbon atoms. Each is attached to glycerol by an ester linkage. Serine is attached to the phosphate moiety and forms the polar "head" group.

fatty acids) and a hydrophilic end (the glycerol phosphatidyl serine). In an aqueous environment, phospholipid molecules form a **bilayer** consisting of two membrane leaflets: the hydrophobic chains, or tails, of the two fatty acids extend into the center of the bilayer and the hydrophilic portions face away from the bilayer (see Figure 4.45). The membrane is stabilized by divalent cations including calcium and magnesium.

The proteins of the cytoplasmic membrane are embedded in the phospholipid bilayer (see Figure 4.44). Many of the proteins serve in the transport of substances into the cell. Others are involved in energy harvesting, DNA replication and cell division, and protein export to the cell envelope or beyond. Together, they may constitute up to 10% of the total cell protein.

Sterols are known to have a stabilizing effect on cell membranes because of their planar chemical structures (Figure 4.46). However, most prokaryotic organisms do not have sterols in their cell membranes. The mycoplasmas (*Mollicutes*), which lack cell walls, are exceptions. They do not synthesize their membrane sterols but derive them from the environment in which they live. The only other group of prokaryotes that is known to contain sterols is the methanotrophic bacteria, which have special membranes involved in the oxidization of methane as an energy source. Some bacteria, such as the cyanobacteria, produce **hopanoids**, compounds that resemble sterols chemically and probably provide membrane stability (see Figure 4.46).

ARCHAEAL LIPIDS   Although the appearance of archaeal membranes in thin section is like that of bacterial cell membranes, the cell membranes of archaea have a different chemical composition. One of the major differences is that the lipid moieties of the membrane are not **glycerol-linked esters** but **glycerol-linked ethers**. Furthermore, rather than fatty acids, most *Archaea* have **isoprenoid** side chains with repeating five-carbon units (Figure 4.47). Two patterns are found in archaeal cell membranes:

- A **bilayer**, with two leaflets of glycerol ether-linked isoprenoids held together by their hydrophobic isoprenoid side chains (see Figure 4.47A)

- A **monolayer**, consisting of diglycerol tetraethers (see Figure 4.47B)

FUNCTION OF THE CYTOPLASMIC MEMBRANE   The cytoplasmic membrane is the ultimate physical barrier between the cytoplasm and the external environment of the cell. It is selectively permeable to chemicals; the molecular size, charge, polarity, and chemical structure of a substance determine its permeability properties. For example, water and gases diffuse freely through cell membranes, whereas sugars and amino acids do not. These important organic compounds, which serve as substrates for growth and energy metabolism, are concentrated in the cell by special transport systems (see Chapter 5). This selective permeability of cell membranes is also important in cell energetics, as discussed in Chapter 8.

One additional function of the cytoplasmic membrane is its role in the replication of DNA, which is attached to the membrane. Following replication, the two new DNA molecules are physically separated by new cell membrane synthesis. The eventual outcome is the separation of the two newly replicated DNA molecules prior to formation of the septum and cell division.

Some prokaryotes contain membranes that extend into the cell itself. For example, phototrophic bacteria have internal membranes that are involved in photosynthesis. Some chemoautotrophs that oxidize ammonia and nitrite, and methanotrophs, which oxidize methane, possess similar intracytoplasmic membranes. These bacteria are discussed in Chapters 19 and 21.

In the remainder of the chapter we consider the layers external to the cell cytoplasm, beginning with those closest to the cytoplasm and working outward to the outer cell surface.

(A) Cholesterol

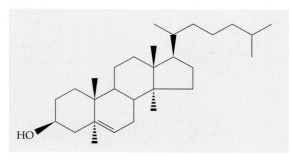

(B) A hopanoid from a cyanobacterium

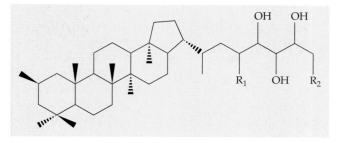

Figure 4.46 Sterols and hopanoids
(A) A sterol and (B) a hopanoid found in some bacterial cell membranes. The hopanoid shown here is from a cyanobacterium. R_1 and R_2 indicate two different hydrocarbon chains.

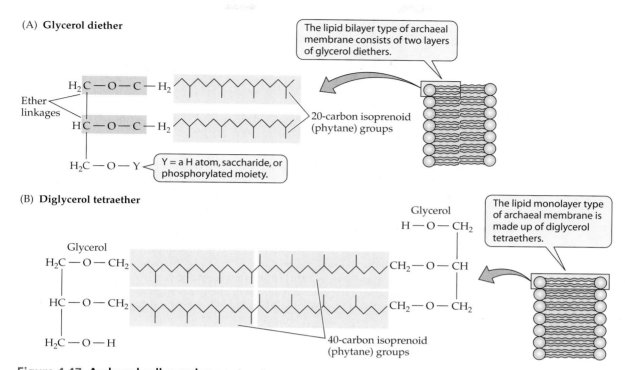

Figure 4.47 Archaeal cell membrane structure
Archaeal membranes are of two types: (A) a lipid bilayer, consisting of two leaflets of glycerol-linked isoprenoids, and (B) a lipid monolayer, consisting of a single layer of diglycerol tetraethers.

The Cell Wall

Almost all *Bacteria* and some *Archaea* have a **cell wall** composed of a polymeric mesh-like bag or sack that surrounds the cytoplasmic membrane. It is an essential component for these prokaryotic species. Unlike other components of the cell envelope, the wall material provides rigidity to prevent the cytoplasmic membrane from rupturing (i.e, to lyse or break open with loss of the cytoplasmic contents). Without this protection, many organisms could not survive in their normal habitats. The cell wall also provides the characteristic cell size and shape of the organism.

One special class of bacteria, the *Mollicutes* (commonly called the mycoplasmas) lack a cell wall entirely. As noted previously, these bacteria without cell walls survive because their cell membranes differ from those of typical bacteria. For example, their membranes contain sterols or other compounds (see preceding text) that help stabilize membrane structure, and in this respect they are similar to animal cell membranes, which also lack cell walls. These bacteria are not able to live in low-solute environments.

As mentioned in Chapter 1, prokaryotes are classified as either *Archaea* or *Bacteria* based on evolutionary stud-ies of 16S rRNA. In addition to this difference in their RNAs, the two domains also differ in the chemistry of their cell walls. In *Bacteria*, the chemical polymer of the wall is called **peptidoglycan**, or **murein** (Figure 4.48). Peptidoglycan is a complex molecule containing amino sugars and amino acids, as described in detail in subsequent text. The *Archaea* lack peptidoglycan; however, some species contain a related cell wall material called **pseudomurein**. Although functionally equivalent to peptidoglycan, this material is structurally unique (see Chapter 18).

Peptidoglycans all have the same general chemical composition and architecture, but the chemical structure varies from one bacterial species to another. Two amino sugars, glucosamine and muramic acid, are joined together by β-1,4 linkages to form a chain, or linear polymer (a glycan). The chains of amino sugars are cross-linked by peptides. Figure 4.49 shows the peptidoglycan of *Staphylococcus aureus*. Note that the amino sugars are **N-acetylglucosamine** (**NAG**) and **N-acetylmuramic acid** (**NAM**). Almost all bacteria have these *N*-substituted acetyl groups; some *Bacillus* species have unsubstituted amino groups on the sugars, and mycobacteria and some of the streptomycetes have *N*-glycolyl groups.

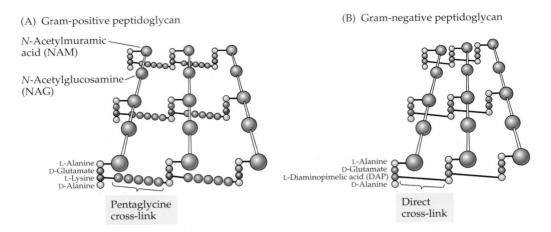

Figure 4.48 Cell walls of gram-positive and gram-negative bacteria
The diagrams show the two-dimensional network of the peptidoglycan sac surrounding (A) a gram-positive and (B) a gram-negative cell. This layer is the major structural component of bacterial cell walls. The *N*-acetylglucosamine (NAG) and *N*-acetylmuramic acid (NAM) are linked to form the amino sugar backbone (glycan). The glycan chains are held together by peptide bridges.

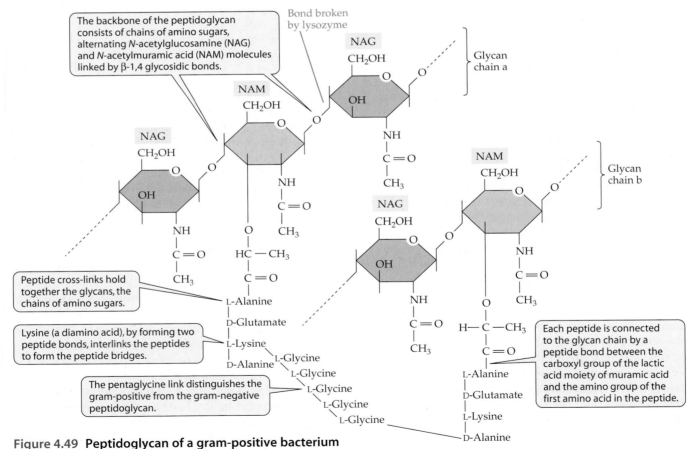

Figure 4.49 Peptidoglycan of a gram-positive bacterium
Chemical structure of the peptidoglycan layer of *Staphylococcus aureus*. Note that some amino acids in the peptide cross-links, alanine and glutamic acid, are in the D-stereoconfiguration. See also Figure 4.48.

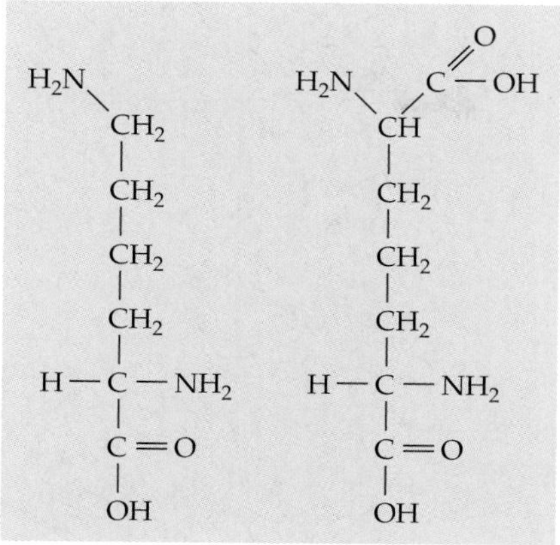

Figure 4.50 Diamino acids
Lysine and diaminopimelic acid (DAP) are diamino acids found in peptidoglycans.

The amino group of the first amino acid in the peptide cross-link is joined to the carboxyl group of the lactic acid moiety of *N*-acetylmuramic acid by a peptide bond. Some of the amino acids (e.g., alanine and glutamic acid) exist in the D stereo-configuration.

Note that one of the amino acids in the peptide—in Figure 4.49, lysine—is a **diamino acid**; that is, it contains two amino groups (Figure 4.50). It is essential that one of the amino acids in the cross-linking peptide bridge is a diamino acid, because the single amino groups of the other amino acids are tied up in peptide linkages all the way back to the muramic acid. The diamino acid can link one of its amino groups to one peptide chain and its other amino to the other chain, thus cross-linking the peptides and their amino sugar chains. The result of this cross-linking is the formation of a two-dimensional network around the cell (see Figure 4.48). It is this two-dimensional sac that provides the rigidity and strength of the cell wall.

Lysine and **diaminopimelic acid** (see Figure 4.50) are the most common diamino acids responsible for this cross-linking. Diaminopimelic acid is not found in proteins; it occurs uniquely in the peptidoglycan of virtually all gram-negative bacteria (see Figure 4.50). Considerable variation exists in the amino acids that form the cross-linking peptides of peptidoglycans. Approximately 100 types of cell wall peptidoglycan structures are known! However, certain amino acids are never found in the peptide bridge, including the sulfur-containing amino acids, aromatic amino acids, and branched-chain amino acids, as well as arginine, proline, and histidine (see Chapter 3).

GRAM-POSITIVE AND GRAM-NEGATIVE BACTERIAL CELL WALLS The domain *Bacteria* is divided into two groups based upon the cells' reaction to a staining pro-

cedure called the Gram stain (see Box 4.1). The differences between gram-positive and gram-negative bacteria relate to differences in their cell envelope structure and chemical composition. Thin sections of gram-positive bacteria reveal walls that are thick, nearly uniformly dense layers (see Figure 4.28A). In contrast, the cell walls of gram-negative bacteria are more complex, because in addition to a peptidoglycan layer they have another layer called an outer membrane (see Figure 4.28B).

The structural differences between the cell walls of gram-positive and gram-negative bacteria reflect differences in biochemical composition. When techniques were developed to permit the separation of cell walls from cytoplasmic constituents, scientists could chemically analyze the cell wall. The major constituent found in the cell wall of gram-positive bacteria was peptidoglycan. It makes up about 40% to 80% of the dry weight of the wall, depending upon the species. Peptidoglycan has a tensile strength similar to that of reinforced concrete! Gram-positive and gram-negative bacteria differ in the numbers of peptidoglycan layers: the former may contain up to 50 concentric layers, thus accounting for the much thicker wall observed by TEM relative to gram-negative cells, which may have only one or two layers of peptidoglycan.

Other constituents of gram-positive cell walls include **teichoic acids** and **teichuronic acids**. These acids are complexed with calcium in the cell wall. Teichoic acids are polyol-phosphate polymers, such as poly-glycerol phosphate and poly-ribitol phosphate (Figure 4.51). Sugars (such as glucose and galactose), amino sugars (such as glucosamine), and the amino acid D-alanine are found in some of these compounds. Teichoic acids are attached covalently to the 6-hydroxyl group of muramic acid in the peptidoglycan, and extend vertically or perpendicular to the peptidoglycan layers. The cross-links to successive layers provide additional strength and rigidity to the cell.

Teichuronic acids are polymers of two or more repeating subunits, one of which is always a uronic acid, such as glucuronic acid, or the uronic acid of an amino sugar, such as aminoglucuronic acid (Figure 4.52). The subunits are also linked covalently to peptidoglycan, but the linkage group is unknown.

Gram-negative bacteria do not contain teichoic or teichuronic acids. However, in other respects their cell wall structures are more complex than those of the gram-positive bacteria, as indicated by the presence of an outer membrane (described in subsequent text). Peptidoglycan makes up a much smaller proportion of the cell wall material and comprises only about 5% by weight of the gram-negative cell wall. Despite the reduced amount of this material, it still serves an important role as a rigid barrier outside the cell membrane, because it is covalently attached to one of the major pro-

(A)

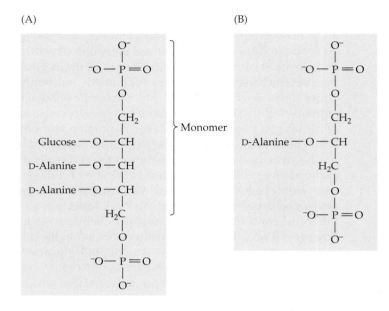

(B)

(C)

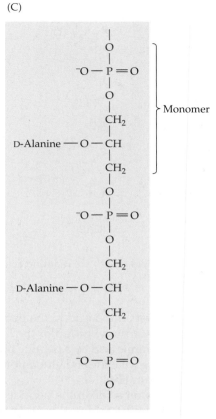

(D)

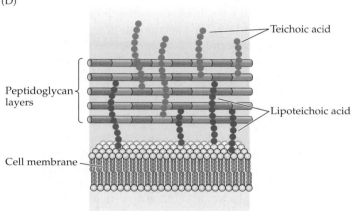

Teichoic acid

Peptidoglycan layers

Lipoteichoic acid

Cell membrane

Figure 4.51 Teichoic acids
Chemical structures of two teichoic acids found in gram-positive bacteria. Monomers of (A) the polyribitol teichoic acid of *Bacillus subtilis* and (B) the polyglycerol teichoic acid of *Lactobacillus* sp. (C) The monomers are joined by phosphate linkages. Shown here is a polyglycerol teichoic acid. (D) A diagram showing teichoic acids in relation to the cell wall and cell membrane.

teins in the outer membrane. In contrast, gram-positive bacteria that have multiple peptidoglycan layers encasing the cell cytoplasm and cytoplasmic membrane devote a considerable amount of carbon and energy to wall construction.

ARCHAEAL CELL WALLS The domain *Archaea* consists of four major phyla (see Chapter 18). These are the *Crenarchaeota* (consisting of hyperthermophiles), the *Euryarchaeota* (consisting of the methane-producing archaea and the extremely halophilic archaea), the *Nanoarchaea* (which are a group of small parasites that live on other archaea), and the *Korarchaeota* (a little understood group, since no isolates have been obtained). Only some methane-producing species have cell walls with a material similar to peptidoglycan or murein, called **pseudomurein** (Figure 4.53). The amino sugar chain of the pseudomurein contains *N*-acetylglucosamine (NAG), but **N-acetyl-talosaminouronic acid** (**TAL**) replaces the *N*-acetylmuramic acid (NAM) found in *Bacteria*. Further-

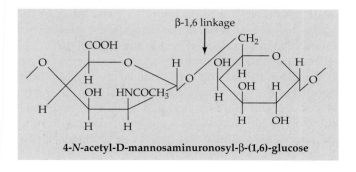

β-1,6 linkage

COOH

CH₂

OH

H
OH
HNCOCH₃

H
OH
H

H

H
OH

4-*N*-acetyl-D-mannosaminuronosyl-β-(1,6)-glucose

Figure 4.52 Teichuronic acids
A teichuronic acid found in the cell wall of *Micrococcus luteus*.

```
       |
— NAG — TAL — NAG — TAL —
        |
        Gal
        |
        Ala
        |
Glu — Lys
 |     |
Lys   Glu
 |     |
Ala   Ala
 |
Glu
 |
— TAL — NAG — TAL — NAG —
              |
```

Figure 4.53 Pseudomurein
The pseudomurein of *Methanobacterium thermautotrophicum*. The amino sugar backbone consists of alternating NAG (*N*-acetylglucosamine) and TAL (*N*-acetyltalosaminuronic acid). The amino acids are alanine (Ala), glutamate (Glu), and lysine (Lys), which is the cross-linking amino acid.

more, the two amino sugars are linked in a 1,3 rather than a 1,4 configuration, and the amino acids of the peptide cross-links are not D-stereoisomers.

Most archaeal species lack pseudomurein. The extreme halophilic and hyperthermophilic *Archaea* and some of the methanogens have glycopeptide cell walls termed the S-layer containing a variety of sugars linked covalently to protein (see Chapter 18 for more detail). The differences in cell wall composition may explain how these species are able to survive in their special environments. For example, halophilic *Archaea*, which live in high osmotic concentrations, may not need the rigid peptidoglycan cell walls typical of *Bacteria*.

FUNCTION OF THE CELL WALL   Prokaryotes generally grow in **hypotonic** aquatic environments, that is, the concentration of particles (ions and molecules) is greater inside the cell than outside the cell. The cell membrane, which is freely permeable to water, permits uptake of water by osmosis so that the concentration of particles on both sides of the cell membrane tends to equalize due to simple diffusion. If this osmotic equalization occurs, the cell swells and the membrane distends to accommodate the increase in volume. This osmotic pressure within the cell, called **turgor pressure**, is the normal consequence of living in hypotonic environments. Under normal conditions, this membrane pressure difference can be two atmospheres (28 psi). The cell wall prevents turgor pressure from stretching the membrane too far. Excessive distension of the membrane would otherwise cause it to break and result in lysis (rupture) of the cell.

Thus one function of the cell wall is to protect the cell from turgor cell lysis, or plasmoptysis. Gram-positive bacteria such as those of the genus *Bacillus* readily illustrate this function. The cell wall layer of this rod-shaped cell can be removed by treating the cell with **lysozyme**, an enzyme found in tears and other human body excretions that breaks the covalent bonds between the amino sugars (see Figure 4.49) of the cell wall peptidoglycan. If this occurs in a normal hypotonic growth environment, the cells simply lyse. However, if the environment is rendered isotonic (i.e., the osmotic pressure on both sides of the cell membrane is equalized)—for example, by suspending the cells in a 2 M sucrose solution—then the wall-less cells do not lyse. Instead, the rod-shaped cells round into spheroidal units called **protoplasts**, which lack a cell wall (Figure 4.54A versus 4.54B). A protoplast can persist in an iso-

(A)

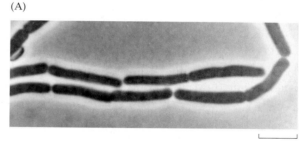

(B)
10 μm

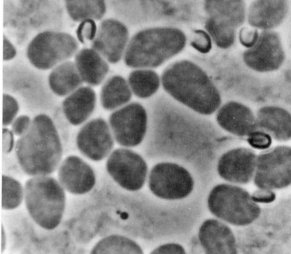

Figure 4.54 Protoplasts
10 μm
(A) Normal cells of *Bacillus megaterium*, chains of rod-shaped cells. (B) Protoplasts of the same *B. megaterium* preparation, produced by lysozyme treatment and stabilized by sucrose. Note the distension of protoplasts due to the low sucrose concentration of the surrounding medium. Courtesy of Carl Robinow.

tonic environment because the turgor pressure of the cell is balanced by the osmotic pressure of (i.e., concentration of sucrose molecules in) the medium.

Protoplasts cannot be produced from gram-negative bacteria because their cell walls cannot be completely removed in this way. Instead, removal of the peptidoglycan layer results in the formation of **spheroplasts**, which, although lacking in cell wall peptidoglycan, still possess the outer membrane.

Because peptidoglycan is a unique structure found only in members the domain *Bacteria*, it has been used as a target for specific **antibiotics** such as the penicillins, which prevent its synthesis. The result is that pathogenic bacteria can be specifically killed when these antibiotics are successfully applied therapeutically (see Chapter 7).

In summary, the cell walls of bacteria are rigid structures that not only give the cell its characteristic size and shape but also protect it from lysis. In addition, they serve a limited role in determining which molecules will pass through the cell membrane.

The Outer Membrane

Surrounding the cell wall of gram-negative cells is an **outer membrane (OM)** that appears similar to a cyto-plasmic membrane when viewed through an electron microscope (see Figure 4.28B). Lipids and proteins predominate. However, polysaccharides extend into the aqueous environment from the membrane surface. As in the cytoplasmic membrane, each of the two leaflets of the outer membrane consists of lipid molecules with their hydrophilic ends (polar head groups) facing away from the membrane and toward the aqueous environment. Likewise, the hydrophobic portions (hydrocarbon chains) face inward and hold the two leaflets together (Figure 4.55).

The outer leaflet of the membrane serves as the site of attachment for **lipid A** (see Figure 4.55). Bonded covalently to lipid A is a core polysaccharide (Figure 4.56) that has specific side-chain polysaccharides (called somatic polysaccharides, **O-polysaccharides** or **O-antigen**) that vary from one species of gram-negative bacterium to another. This lipid A–polysaccharide complex is referred to as a **lipopolysaccharide**, or simply **LPS**. The LPS that contains lipid A is called **endotoxin** because it is toxic to a variety of animals, including humans. Animals injected with endotoxin may experience fever, hemorrhage, shock, miscarriage, and a variety of other symptoms depending upon the dose and source. Many gram-negative bacteria, including *Salmonella* species

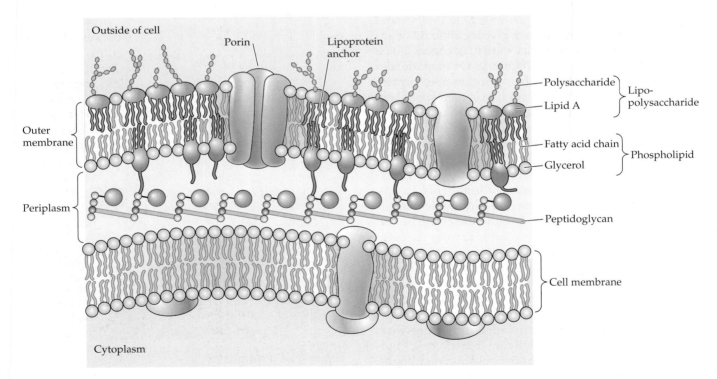

Figure 4.55 Outer membrane
Diagram of the outer membrane of a gram-negative cell. Note the location of the lipid A of LPS, the transmembrane porin proteins, and the lipoprotein anchor that attaches the outer membrane to the peptidoglycan layer. It also has a lipopolysaccharide (LPS) complex consisting of lipid A attached to the outer leaflet of the outer membrane and covalently bonded to a polysaccharide. The polysaccharide tail extends away from the cell into the external environment.

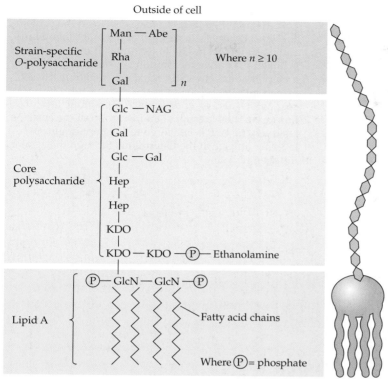

Outside of cell

Strain-specific
O-polysaccharide

Man — Abe
Rha
Gal

Where *n* ≥ 10

Core
polysaccharide

Glc — NAG
Gal
Glc — Gal
Hep
Hep
KDO
KDO — KDO —(P)— Ethanolamine

Lipid A

(P)— GlcN — GlcN —(P)
Fatty acid chains
Where (P) = phosphate

Inside outer membrane

Figure 4.56 Polysaccharide portion of LPS
Chemical composition of the polysaccharide portion of LPS in a *Salmonella* strain. The polysaccharide portion extends outside the cell, where it acts as a strain-specific antigen, and may serve to help attach the cell in its environment. Abe, abequose; Gal, galactose; Glc, glucose; GlcN, glucosamine; Hep, heptose; KDO, ketodeoxyoctonate; Man, mannose; NAG, *N*-acetylglucosamine; Rha, rhamnose.

(which cause enteric diseases, including typhoid fever) and *Yersinia pestis* (which causes bubonic plague), owe their toxic properties to this compound. Other gram-negative bacteria with LPS contain a structural version that is nontoxic because it lacks specific parts of the lipid A moiety. Note that LPS is not present in the inner leaflet of the outer membrane.

The outer membrane of gram-negative bacteria also contains many proteins, but they differ in type and amount from those in the cytoplasmic membrane. The most prevalent is a small polypeptide (~7,200 Da) called Braun's **lipoprotein** or **LPP** and it serves a structural role. It has three fatty acid groups covalently attached to one end (i.e., the N terminus) that anchors it in the inner leaflet of the outer membrane (Figure 4.57). The other end, the C terminus, is joined covalently to the cell wall via a diamino acid moiety to peptidoglycan. LPP aids in stabilizing the outer membrane by anchoring it to the rigid cell wall layer. Thus, the combined peptidoglycan-lipoprotein complex may be envisioned as a single molecule of extremely high molecular weight!

Other important outer membrane proteins serve as membrane pores called **porins** that permit the passage of small molecules through the outer membrane (see Figure 4.55). Porins are trimeric (three-subunit) structures that contain three hydrophilic channels across the hy-

drophobic lipid bilayer (Figure 4.58). Because they are water-filled, the pores allow the free diffusion of water-soluble nutrients, such as sugars and amino acids, as well as inorganic ions across the OM (for example, the *E. coli* porins called OmpC and OmpF). However, porins limit passage of molecules larger than about 600 Da. Entry of larger molecules (e.g., bile salts that function as membrane detergents) is thus prevented. Finally, some specialized porins are responsible for the entry of certain larger compounds such as vitamin B_{12}, iron chelates, disaccharides, or phosphorylated compounds across the outer membrane so they can be subsequently taken up into the cell.

Periplasm

The space between the cytoplasmic membrane and the outer membrane of gram-negative bacteria is called the **periplasmic space**. This space, although small (see Figure 4.28), is important to the physiology of the cell. The periplasm is an area of considerable enzymatic activity (e.g., phosphatases, proteases, and other hydrolases). Several steps in the synthesis and assembly of the cell wall occur in this area. Furthermore, the periplasmic **binding proteins** produced by gram-negative bacteria are located in this space and function in nutrient uptake

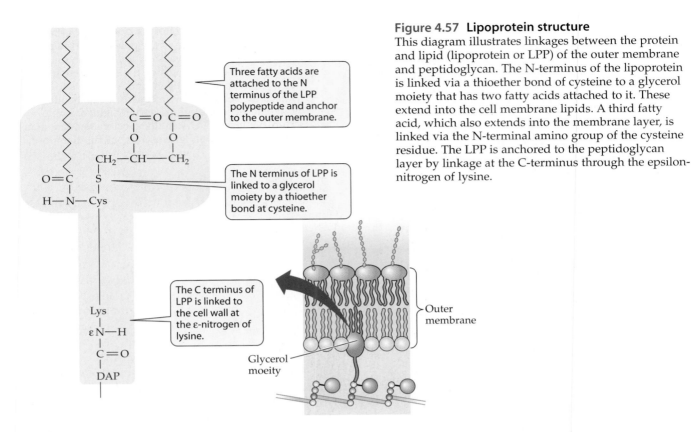

Figure 4.57 Lipoprotein structure
This diagram illustrates linkages between the protein and lipid (lipoprotein or LPP) of the outer membrane and peptidoglycan. The N-terminus of the lipoprotein is linked via a thioether bond of cysteine to a glycerol moiety that has two fatty acids attached to it. These extend into the cell membrane lipids. A third fatty acid, which also extends into the membrane layer, is linked via the N-terminal amino group of the cysteine residue. The LPP is anchored to the peptidoglycan layer by linkage at the C-terminus through the epsilon-nitrogen of lysine.

(see Chapter 5). They combine reversibly with specific molecules and deliver them to cytoplasmic membrane carrier proteins to assist in solute transport across the cytoplasmic membrane into the cytoplasm.

Some controversy exists among microbiologists as to whether gram-positive bacteria and archaea possess a periplasm. This is no doubt due to the extensive research that has focused on gram-negative bacteria, which have served as model organisms for the study of the periplasmic space. Furthermore, the periplasm of the two types of organisms undoubtedly differs, because gram-negative bacteria have the additional permeability barrier, the outer membrane. However, electron microscopy suggests the presence of a functionally equivalent space evidenced by an electron transparent region between the cytoplasmic membrane and the outer cell layer. Gram-positive bacteria and archaea also have periplasmic **binding proteins** involved in nutrient uptake. However, these proteins contain a hydrophobic amino acid helix at their N terminus that anchors them to the outer sur-

Figure 4.58 Porin
Ribbon diagrams showing the three-dimensional structure of a porin molecule, based on x-ray crystallography data. The diagram on the left is a side view and that on the right is a top view, showing the pore formed by the molecule. Data from M. S. Weiss and G. E. Schulz. 1992. "Structure of porin refined at 1.8 Å resolution." *J. Mol. Biol.* 227: 493–509.

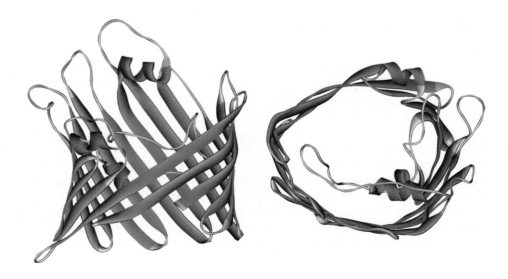

face of the cytoplasmic membrane. This presumably prevents their loss to the cell exterior.

SECTION HIGHLIGHTS

The cell membrane of *Bacteria*, which surrounds the cytoplasm, consists of a phospholipid bilayer with glycerol-linked esters and in which proteins are embedded. Archaeal membranes differ in that they have glycerol-linked ethers and isoprenoid groups. Cell membranes are selectively permeable so that they control the entry and exit of substances from the cytoplasm. The cell wall of gram-positive bacteria contains large amounts of peptidoglycan. In contrast, gram-negative bacteria have a thin layer of peptidoglycan between the cell membrane and an outer membrane, which is an outer phospholipid bilayer with porins and lipid A.

4.5 Extracellular Layers

Some prokaryotes produce discrete layers external to the cell wall or outer membrane or both. These are of several types, varying in structure and function:

- Capsules, or extracellular polymeric substances (EPSs)
- Slime layers
- Sheaths
- Protein jackets, or S-layers

If the external layer is removed from the cell, the organism does not lose its viability. Thus, these layers do not appear to be essential to the organisms that produce them, at least under some conditions of growth.

Capsules

Capsules, also called glycocalyxes are composed of **extracellular polymeric substances (EPS)** that surround the cell wall/outer membrane. They constitute a physical barrier that protects the cell from the external environment. For example, they may aid in slowing the desiccation process when the cell encounters dry conditions, or they may prevent virus attachment and infection. Capsules are also important in mediating attachment of some bacteria to solid surfaces. This is well illustrated by one of the bacteria responsible for dental cavities. In the presence of sucrose, *Streptococcus mutans* produces a polysaccharide capsule that permits the bacterium to attach to dental enamel.

As *S. mutans* grows on the surface of the tooth, acid produced by fermentation of sucrose causes etching of the tooth enamel, and a cavity may eventually develop.

In the laboratory, capsules can be removed without affecting a cell's viability, indicating they are not always essential to survival. Nevertheless, several functions have been attributed to capsules, and under certain conditions they may be crucial in determining the survival of the organism in its natural environment. For example, the capsules of some species are important virulence factors (factors that enhance the ability of the organism to cause disease). This can be illustrated by the species *Streptococcus pneumoniae*, one of the bacteria responsible for bacterial pneumonia. Unencapsulated strains of this bacterium are not pathogenic, that is, they are not capable of causing disease. The reason is that the unencapsulated cells are more easily killed by phagocytes (white blood cells) in the host organism's circulatory system. Apparently the white blood cells can ingest bacteria without capsules much more readily than those with capsules. Therefore, capsules may provide a means by which the bacteria evade the host defense system in their natural environment.

Capsules are diffuse structures that are often difficult to visualize without special techniques. This is well exemplified by the capsule of *S. pneumoniae*. Its capsule is not visible by ordinary phase contrast microscopy or by simple staining techniques because of the diffuse consistency of the polysaccharide that forms the capsule. A special technique has been developed to permit visualization of the capsule, called the **Quellung** (German for "swelling") reaction. An antiserum (see Chapter 27) is prepared from capsular material of *S. pneumoniae* and is used to "stain" *S. pneumoniae* cells, complexing with their capsules. After staining, the bacterium appears to have grown much larger or to have swollen. In reality, the antiserum has complexed with the extensive but previously transparent capsule and made it thicker and more visible (Figure 4.59).

Negative stains can also be used to stain capsules for light microscopy. These stains consist of insoluble particulate materials such as those present in India ink (nigrosin). The dark particles of the stain do not penetrate the capsule, so a large unstained halo appears around the cell (Figure 4.60).

Most capsules are composed of polysaccharides (Table 4.1). **Dextrans** (polymers of glucose) and **levans** (polymers of fructose, also known as levulose) are common **homopolymers** (polymers consisting of identical subunits) found in capsules produced by lactic acid bacteria (*Streptococcus* and *Lactobacillus* spp.). However, some bacteria produce **heteropolymers**, containing more than one type of monomeric subunit. Hyaluronic acid is an example of a heteropolymer found in capsules; it consists of *N*-acetylglucosamine and glucuronic acid. Another example is the glucose–glucuronic acid polymer, also produced by some *Streptococcus* species.

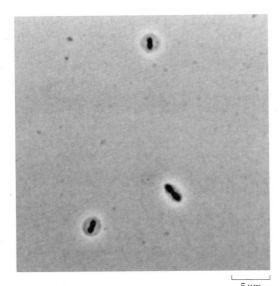

Figure 4.59 Capsule stain
Capsules of *Streptococcus pneumoniae* have been stained by the Quellung reaction. Courtesy of J. Dalmasso and J. T. Staley.

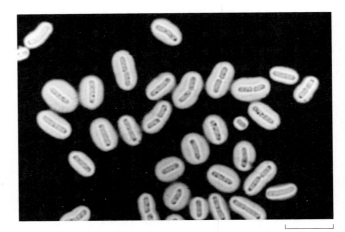

Figure 4.60 Negative stain
An India ink negative stain shows the capsule of a *Bacillus* sp. This is called a negative stain because the background around the capsule is stained, not the capsule itself. Courtesy of Carl Robinow.

Capsules can also be composed of polypeptides. For example, some members of the genus *Bacillus* have capsules made of polymers that are repeating units of D-glutamic acid (see Table 4.1). It is interesting to note that the D-stereoisomers of amino acids are found almost exclusively in the capsules and cell walls of some bacteria. Virtually all other biologically produced amino acids are of the L-stereo configuration (see Chapter 3). Thus, other than capsule and cell wall peptides, all proteins in bacteria (as well as in other organisms) contain L-amino acids. The poly-D-glutamic acid of *Bacillus* is also noteworthy because the amino acids are linked together by their γ-car-

boxyl group, not the α-carboxyl group as is characteristic of proteins. Synthesis occurs independent of ribosomes.

Slime Layers

Some prokaryotes, such as those of the *Cytophaga–Flavobacterium* group, move by a type of motility referred to as **gliding**. This type of movement is distinct from "swimming" (see Chapter 6), and requires that the organism remain in contact with a solid substrate. Gliding bacteria produce an EPS termed "slime," part of which is lost in the trail left by the organism during movement (Figure 4.61). Members of the *Cytophaga* and *Flavobacterium* are noted for their ability to degrade high-molec-

Cells gliding together   Slime trail

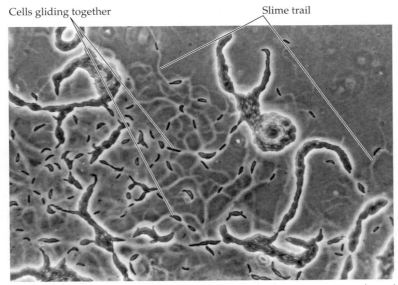

Figure 4.61 Gliding motility
The myxobacterium *Stigmatella aurantiaca* produces slime trails as the cells glide across an agar surface. Note that even the smallest units contain at least two cells gliding together. Courtesy of Hans Reichenbach.

ular-weight polysaccharides found in plant material (cellulose) and insect or arthropod shells (chitin). It is advantageous for organisms that degrade these materials for nutrition to be in physical contact with them. As the bacterium glides on the surface of the material, the cellulase and chitinase enzymes, bound to the organism's extracellular surface, are kept close to the substrate as well as to the cell. In this manner the cell can derive nutrients more effectively directly from the environment.

The chemical nature of the EPS slime material varies. In some species it is a polysaccharide. For example, *Cytophaga hutchinsonii* produces a heteropolysaccharide of

---

TABLE 4.1 Chemical composition of capsules from various bacteria

Bacterium	Capsular Material	Structure of Repeating Units
Leuconostoc mesenteroides	Dextran	α-1,6-poly-D-glucose
Streptococcus pneumoniae	Polyglucose glucuronate	Glucuronic acid / Glucose
Streptococcus spp.	Hyaluronic acid	Glucuronic acid / N-acetyl glucosamine
Bacillus anthracis	γ-poly-D-glutamic acid	γ–D-glutamic acid

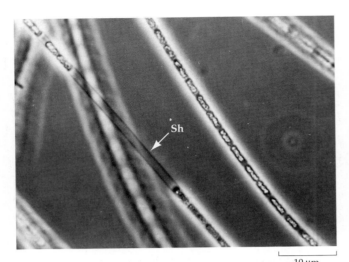

Figure 4.62 Sheathed bacteria
The filamentous gram-negative bacterium *Sphaerotilus natans* has a sheath (Sh). This species grows in flowing aquatic environments. Courtesy of J. T. Staley.

10 µm

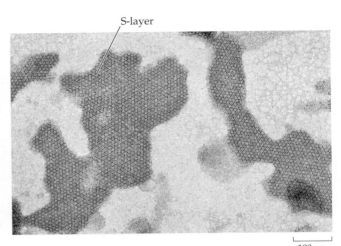

S-layer

100 nm

Figure 4.63 S-layers
Electron micrograph of negatively stained fragments of a highly textured protein jacket, or S-layer, from *Aquaspirillum serpens* strain VHA. The staining reveals the highly organized nature of the structure. Courtesy of S. F. Koval.

arabinose, glucose, mannose, xylose, and glucuronic acid. The mechanism of gliding motility is poorly understood, but recent research on *Flavobacterium johnsoniae* indicates that a special sulfur-containing lipid (sulfonolipid) found on the surface of the cell is essential for gliding in this organism. Studies with the myxobacteria indicate a role for pili in cell movement (see subsequent text).

Sheaths

Some prokaryotes, the **sheathed bacteria**, produce a much denser and highly organized external layer, or sheath. Unlike a capsule or slime layer, this structure is easily discerned with a light microscope (Figure 4.62). Sheathed bacteria form filamentous chains and grow in flowing aquatic habitats such as rivers and springs. The sheath protects the cells against disruption by turbulence in the water and ultimately against being removed from the area of the habitat that is most favorable for growth. The few chemical analyses of sheaths that have been made indicate that they are more complex than capsules, having, in addition to polysaccharides, amino sugars and amino acids.

Protein Jackets

External protein layers are produced by certain *Bacteria* and *Archaea*, such as members of the genera *Spirillum* and *Bacillus*. These **protein jackets**, also called **S-layers** are often highly textured surfaces (Figure 4.63). Their function is not well understood. The protein jackets produced in some *Archaea* make up the only layer external to the cytoplasmic membrane, and thus may function in place of a cell wall. However, *Spirillum* species and certain cyanobacteria that have these layers also have a cell wall, so the S-layers cannot serve that role in these organisms. For some prokaryotes, the S-layer serves as a site on the cell surface to which bacterial viruses adhere.

Cell Appendages

Many *Bacteria* and *Archaea* produce special appendages that extend from the surface of the cell. The two most common types, widely found among prokaryotes, are flagella and fimbriae.

FLAGELLA   Prokaryotes that move by swimming produce flagella: long, flexible appendages resembling "tails" (Figure 4.64). These proteinaceous appendages, generally about 10 to 20 nm in diameter and 5 to 20 µm in length, are found in some members of the gram-positive and gram-negative bacteria as well as in some archaea. They act like a propeller: as the cell rotates the flagella, it moves through the water. In motility, chemical energy generated from metabolism is converted into mechanical energy.

The number and location of flagella on the cell surface have been used as important features in the classification of bacteria. Some bacteria have a single polar flagellum extending from one end of the cell and thus move by what is called **monotrichous** polar flagellation. These single flagellar filaments cannot be resolved by the light microscope but are readily discerned with the electron microscope (see Figure 4.64). Other species produce polar tufts or bundles (**lophotrichous** flagellation;

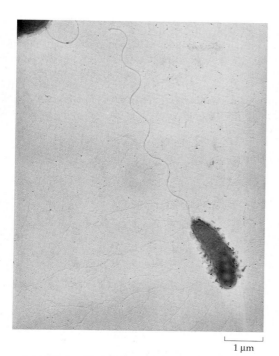

Figure 4.64 Polar flagellum (monotrichous flagellation) Electron micrograph of a whole gram-negative bacterium showing its single polar flagellum. Courtesy of J. T. Staley.

species, for example, produce polar flagella for movement in water but produce lateral flagella when grown on an agar surface.

A flagellum consists of three major parts:

- A basal structure, which anchors the flagellum into the cell wall and membrane by special disk-shaped structures called plates or rings

- A hook, a curved section connecting the filament to the cell surface

- A long filament, which extends into the surrounding environment

The **basal structure** for a typical gram-negative bacterium contains a rod structure plus L, P, S, and M rings associated with cell envelope components (Figure 4.67) that anchor it to the cell. Several of the rings function as parts of a molecular motor that spins or rotates the hook and filament assembly in a clockwise or counterclockwise direction. This action propels the cell in space in response to external signals, including chemicals, light, and magnetism. The "swimming" motion confers selective advantage to motile cells in seeking out environmental niches favorable for growth and survival.

Flagellar rotation, and hence cell movement, requires energy, which is generated by the proton motive force (pmf) (see Chapter 8). This energy is imparted to the Mot

note that "flagellation" can refer both to the arrangement of flagella and the mode of locomotion), which can be seen with light microscopy (Figure 4.65). Still other prokaryotes produce flagella from all locations on the cell surface (**peritrichous** flagellation; Figure 4.66). Motile enteric bacteria, such as *Proteus vulgaris*, exhibit this latter type of flagellation. Finally, a few bacteria are known to move by mixed flagellation. Some *Vibrio*

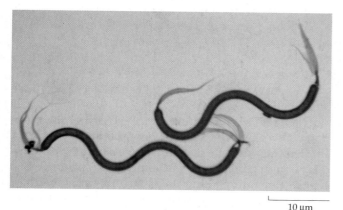

Figure 4.65 Polar flagellar tufts (lophotrichous flagellation)
Spirillum volutans bacteria with polar tufts of flagella. ©Dr. John D. Cunningham/Visuals Unlimited.

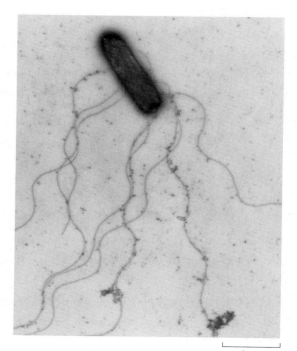

Figure 4.66 Peritrichous flagellation
Electron micrograph of a peritrichously flagellated bacterium showing many flagella about the cell surface. ©W. L. Dentler/Biological Photo Service.

(A)

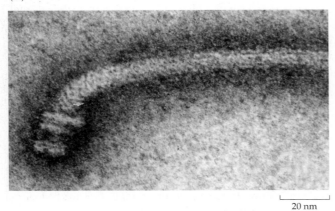

20 nm

Figure 4.67 Flagellar structure
(A) Electron micrograph showing the basal ring structure of a gram-negative flagellum that has been removed from a cell. (B) Model of a flagellum cross section showing the location of the rings and the Mot proteins in the cell envelope of a gram-negative bacterium. (C) Model of a flagellum from a gram-positive bacterium. A, ©J. Adler/Visuals Unlimited.

(B)

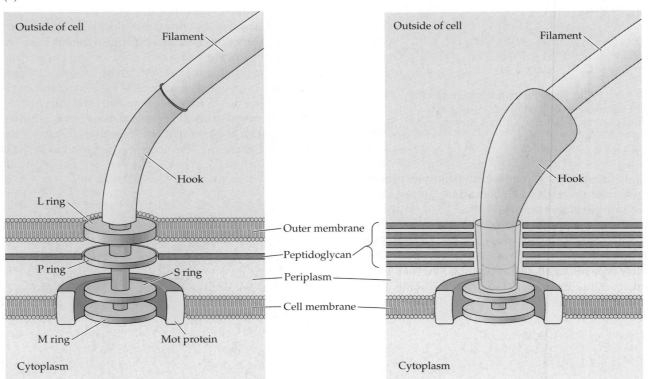

proteins, which are embedded in the cell membrane and act as a motor that drives flagellar rotation. The rate of rotation varies among species but is in the range of 200 to more than 1,000 rpm. Motility requires a major expenditure of energy, but the result is immensely useful because it allows the cell to translocate to a place where conditions for growth are more favorable.

The **hook structure** contains the FlgG protein that connects the rod of the basal structure to the flagellar filament, which is constructed of hook protein.

The **flagellar filament** is composed of a protein subunit called **flagellin**. With few exceptions, the bacteria studied thus far have flagellin with a single type of protein subunit (*Caulobacter* and some *Vibrio* spp. have two types). Repeating subunits of flagellin are joined together to form the filament, which is helically wound and has a hollow core. The filament (and thus the flagellum) is about 20 nm in diameter. The filaments are not straight. Each has a characteristic sinusoidal curvature that varies with each species.

The flagella of enteric bacteria have been studied most intensively. More than 40 genes have been identified in *Salmonella typhimurium* and *E. coli* that are involved in motility. The assembly and function of these structures is discussed in Chapter 11 as well as those genes that regulate the synthesis of the flagellum.

Tactic Responses

Bacteria and archaea use their flagella to propel themselves into environments where conditions are more favorable to the cell. Three general types of cellular response include chemotaxis, phototaxis, and magnetotaxis.

CHEMOTAXIS   Prokaryotes swim toward or away from low-molecular-weight compounds, either organic or inorganic. This phenomenon is termed chemotaxis and can be subdivided into **positive chemotaxis** and **negative chemotaxis**. Positive taxis is caused by a chemical that attracts the motile cell and the signaling molecule is termed an **attractant**. Conversely, a chemical that repels cells is termed a **repellant** and it causes cell motion away from that substance (i.e., negative taxis). For example, a variety of amino acids and sugars function as attractants to *E. coli*. Some alcohols and heavy metals function as repellants. The cell has the ability to detect numerous chemical signals in its environment; it then integrates these messages and elicits an appropriate response. Finally, the cell also has the ability to compare signal strength over time. By doing so, it can seek out higher versus lower concentrations of nutrients (attractants) or repellants (see Chapter 11). For some sugars that serve as attractant molecules, positive chemotaxis can occur at concentrations as low as 10^{-8} M.

PHOTOTAXIS   Many photosynthetic bacteria respond to light by swimming to or away from certain wavelengths of light. Like chemotaxis, cell movement is flagella driven. This light-associated phenomenon can be illustrated by projecting a spectrum of visible light onto a glass microscope slide. When a suspension of flagellated photosynthetic bacteria is placed on the illuminated slide, the organisms quickly concentrate at the wavelengths of light that are absorbed by the photosynthetic pigments of that particular species (Figure 4.68). Thus, the cells position themselves at the specific wavelengths used for photosynthesis. They also respond to light intensity and seek out conditions optimal for their growth and survival. Many phototrophs also exhibit chemotactic behavior.

MAGNETOTAXIS   Lastly, some cells respond by swimming to or away from one of the magnetic poles (N or

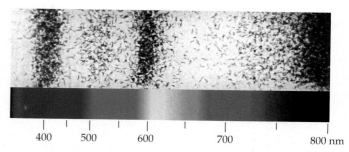

Figure 4.68 Phototaxis
A spectrum of visible light has been projected on a slide of *Thiospirillum jenense*, a purple sulfur photosynthetic bacterium. The bacteria have accumulated at those wavelengths at which their photosynthetic pigments best absorb light. Photo courtesy of Norbert Pfennig.

S), a process termed magnetotaxis. These magnetic bacteria are discussed in more detail in the microbial diversity section (see Chapters 19 to 21).

Fimbriae, Pili, and Spinae

Many prokaryotes have proteinaceous appendages—known as **fimbriae**, **pili**, or **spinae**—that are about 1 μm in length and range from about 3 to 5 nm in diameter in most bacteria (Figure 4.69) to about 100 to 200 nm in some aquatic varieties, which are sufficiently large to be seen with the light microscope. The three forms are structurally similar in that they all consist of one or a few protein subunits. The subunits assemble into a helix, leaving a pore through the center. Some forms also have a glycoprotein tip, called **adhesin**, that aids in adhesion.

The terms "fimbriae" and "pili" are often used interchangeably. Fimbriae is usually considered a more general term for the thin fibrillar appendages so commonly found on prokaryotes, whereas pili are usually considered to be a subset of fimbriae that are typical of many enteric bacteria and pathogenic species. A single bacterial cell may produce more than one type of fimbria or pilus. In fact, some bacterial viruses attach to one type of fimbria or pilus but not to the cell surface or to other types of fimbriae or pili produced by the same bacterial species.

Pili are needed for the attachment of some bacteria to other cells or surfaces. For example, one type of pilus of *E. coli*, called the fertility (F) pilus, mediates the attachment of the two types of mating cells during sexual conjugation and is therefore specifically called the "sex pilus" (see Figure 4.69B; see Chapter 15). Likewise, pili produced by *Neisseria gonorrhoeae* are responsible for the attachment of pathogenic strains to endothelial tissue of the genitourinary tract in humans. Only piliated strains cause gonorrhea.

(A)

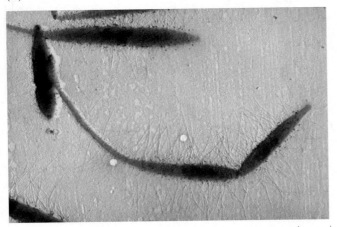

1 µm

(B)

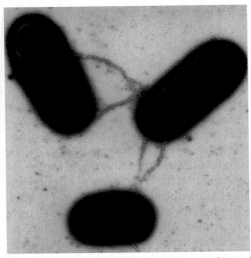

0.5 µm

(C)

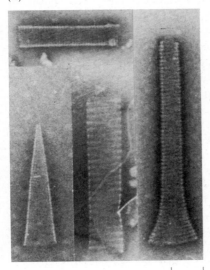

0.1 µm

Figure 4.69 Fimbriae
(A) Electron micrograph of a cell of *Prosthecobacter fusiformis* showing many fimbriae. (B) False-color transmission electron micrograph (TEM) of a male *Escherichia coli* bacterium (upper right) conjugating with two females. (C) Electron micrographs of isolated spinae. A, courtesy of J. T. Staley; B, ©Dr. L. Caro/SPL/Photo Researchers, Inc; C, courtesy of K. B. Easterbrook.

Several distinct types of pili are known. One type of pilus of *E. coli* is referred to as the type I or P pilus (Figure 4.70). This pilus contains a membrane anchor composed of several proteins that provide a platform for the protein subunits that form the rod extending from the cell surface. It may be composed of several different polypeptide types and may have a special adhesin subunit at the tip for cell attachment.

In some *Pseudomonas* species, pili are involved in the extrusion of protein from the cell. Furthermore, a type of motility termed twitching, which occurs, for instance, in the gliding bacteria referred to as myxobacteria (slime bacteria), is also caused by fimbriae. Twitching is a poorly studied process whereby cells that seem to be sessile (immobile) suddenly jerk 1 µm or so in distance.

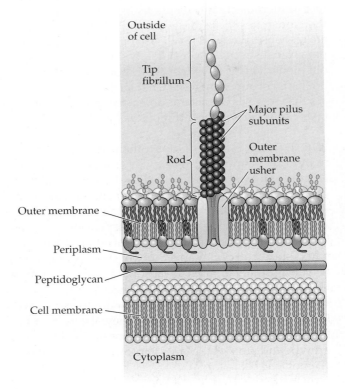

Figure 4.70 Pili
Model of a type I pilus anchored in the outer membrane of a gram-negative bacterium.

The largest appendages are spinae (see Figure 4.69C). The spinae of aquatic bacteria serve to increase the cell's surface area and, hence, their drag, enabling the organisms to remain suspended for longer periods or to be carried farther by currents. Thus, the spinae assist the organism in maintaining its position in the plankton.

SECTION HIGHLIGHTS

Many bacteria produce extracellular polymeric substances outside the cell. These can be capsules, slime layers, S-layers, or sheaths. In addition, appendages such as flagella and pili may be produced. Flagella are essential for swimming motility, whereas pili are involved in a variety of functions such as cell movement (gliding and twitching) and transport of substances to and from the cell.

SUMMARY

- A **simple microscope** is analogous to a magnifying glass with a single lens; **compound microscopes** have two lens systems, including the **eyepiece** (or ocular) lens and the **objective lens**. In addition, a **condenser lens** system is used to concentrate light on the specimen.

- The **resolution** of the light microscope is limited by the **wavelength** of light, the **refractive index** of the suspending medium, and the closeness of the objective to the specimen, which determines the **angular aperture** of the lens.

- Under most conditions of growth, bacteria have a negative surface charge and are therefore stained with **basic dyes** (positively charged). In contrast, **negative stains**, such as India ink, do not stain cell material but stain the aqueous background.

- **Phase contrast microscopy** increases the contrast of bacteria by amplifying the difference in refractive index between the cells and the aqueous environment.

- **Transmission electron microscopes** (**TEMs**) have high resolution because they use electrons as their source of illumination. **Thin sections** to be studied for intracellular details are prepared by dehydrating and embedding cells in plastic resins, which are then sliced with ultramicrotomes.

- **Scanning electron microscopes** (**SEMs**) produce images of whole bacterial cells or colonies. **Atomic force microscopes** (**AFMs**) can be used to visualize cells in their normal aqueous environments.

- Many prokaryotes are unicellular **rods, cocci, vibrios,** or **spirilla**. Some species form chains, filaments, or packets of cells. The **actinobacteria** are gram-positive bacteria that produce branching filamentous structures called **mycelia**.

- Most prokaryotes divide by **binary transverse fission**, in which a mother cell divides to produce two mirror-image daughter cells.

- The prokaryotic chromosome is a double-stranded circular molecule of DNA. Many prokaryotes harbor **plasmids**, double-stranded DNA molecules containing genes that are not essential for the livelihood of the host cell but may carry important genes for increasing its physiological capabilities.

- Some gram-positive bacteria produce **endospores**. These hardy structures can survive heat and dessication and will germinate under favorable conditions for growth. Endospores are rich in calcium and **dipicolinic acid** (**DPA**), which form a complex in the spore coat.

- **Gas vesicles** are protein-membrane structures that exclude water and therefore accumulate gases and reduce cell density, enabling cells to become buoyant in aquatic habitats.

- Organic storage compounds of prokaryotes include **glycogen**, **starch**, **poly-β-hydroxybutyrate** (**PHB**), and **cyanophycin**. Inorganic storage compounds include **volutin** (also called polyphosphate) and **elemental sulfur**.

- The cell membranes of bacteria contain proteins embedded in lipid bilayer membranes, which contain **fatty acids** linked to **glycerol** through **ester linkages**. The cell membranes of archaea contain proteins and lipids with **isoprenoid** groups that are linked to glycerol by **ether linkages**.

- The **periplasm** is an area between the cell membrane and cell wall; it is the site of a variety of enzymatic activities, including cell wall synthesis, chemoreception, and transport.

- Almost all members of the domain *Bacteria* produce **peptidoglycan** or **murein**, a unique cell wall structure that contains a backbone of *N*-acetylglucosamine and *N*-acetylmuramic acid, and peptide bridges cross-linking the amino sugar backbones. Peptidoglycan contains some unique D-amino acids, such as D-alanine and D-glutamic acid, not found in protein. Peptidoglycan contains a diamino acid, such as **diaminopimelic acid** or **lysine**, that cross-links the peptide bridge. Peptidoglycan can be hydrolyzed by **lysozyme**. Normal peptidoglycan synthesis is inhibited by penicillin and certain other antibiotics.

- **Gram-positive cell walls** may also contain **teichoic acids** and **teichuronic acids**, which are not found in gram-negative bacteria. In contrast, **gram-negative bacteria** contain, in addition to peptidoglycan, an **outer membrane** composed of lipid, protein, and polysaccharides. The polysaccharides of the outer membranes are linked to **lipid A** to form a **lipopolysaccharide** (or **LPS**), called **endotoxin**, which is toxic to vertebrates.

- **Archaeal cell walls** lack peptidoglycan; most have polysaccharide or protein cell walls.

- **Extracellular layers**, found in some prokaryotes, are **capsules**, **slime layers**, **protein jackets** or **S-layers**, and **sheaths**, depending on the species.

- **Flagella** are protein fibrils that prokaryotes use for motility.

- **Fimbriae** and **pili** are proteinaceous extracellular fibrillar appendages of some bacteria that are typically used for transport of materials to the cell, as well as attachment and in some cases motility.

- **Positive chemotaxis** is movement of a cell toward a nutrient. **Negative chemotaxis** is movement away from a toxic substance. Photosynthetic bacteria can respond to light by either positive or negative **phototaxis**.

 Find more at www.sinauer.com/microbial-life

REVIEW QUESTIONS

1. Why does the TEM have better resolution than the light microscope?

2. What types of microscopy are best suited for observing motility of microorganisms?

3. What type of microscopy would you use to view (a) ribosomes, (b) gas vesicles, (c) gas vacuoles, (d) colonies of bacteria, (e) periplasm?

4. How does fluorescence microscopy work?

5. What is the gas in a gas vesicle?

6. Which prokaryotes are not sensitive to penicillin? Why?

7. In what ways do the *Archaea* differ from the *Bacteria*?

8. Why is a diamino acid important in peptidoglycan?

9. What are the different types of extracellular structures of bacteria? What uses do they serve for the cell?

10. What are the various types of storage materials found in prokaryotes? Would you categorize them as energy sources or nutrient sources?

11. It has been hypothesized that the gas vacuole is a primitive organelle of prokaryotic motility. What evidence is consistent with this?

12. Distinguish between positive and negative chemotaxis. What is known about the molecular basis for chemotaxis?

13. Is the response of photosynthetic bacteria to light considered phototaxis? Explain your answer.

SUGGESTED READING

Beveridge, T. J. 1989. "The structure of bacteria." *In*: J. S. Poindexter and E. R. Leadbetter, eds. *Bacteria in Nature, Vol. 3.* New York: Plenum Publishing Corp.

Beveridge, T. J. and L. L. Graham. 1991. "Surface layers of bacteria." *Microbiological Reviews* 55: 684–705.

Gerhardt, P., R. G. E. Murray, W. A. Wood and N. R. Krieg. 1994. *Methods for General and Molecular Bacteriology.* 4th ed. Washington, DC: ASM Press.

Neidhardt, F. C., J. L. Ingraham and M. Schaechter. 1990. *Physiology of the Bacterial Cell: A Molecular Approach.* Sunderland, MA: Sinauer Associates, Inc.

Schatten, G. and J. B. Pawley. 1988. "Advances in Optical, Confocal, and Electron Microscopic Imaging for Biomedical Researchers." *Science* 239: 164–165.

Microbial Physiology: Nutrition and Growth

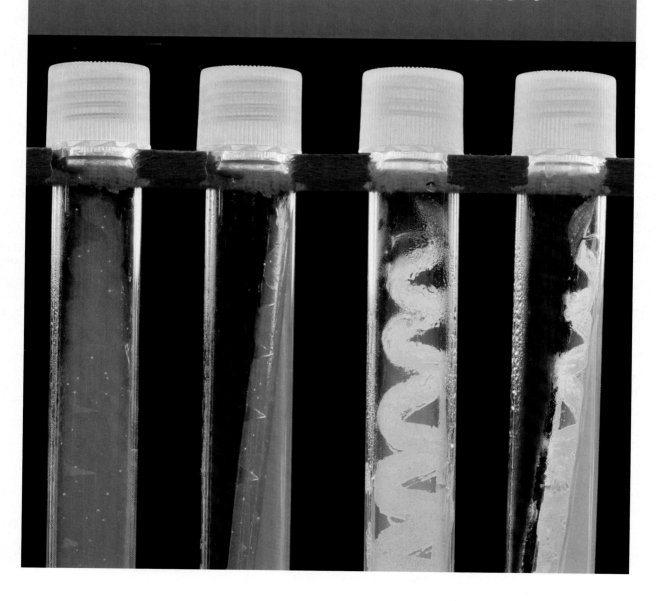

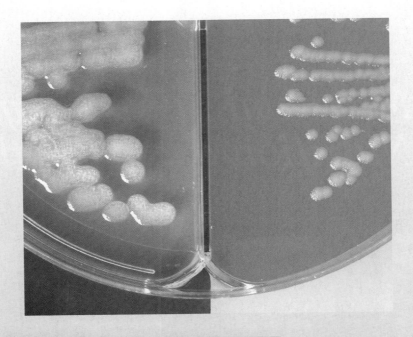

The objectives of this chapter are to:

◆ Introduce the basic nutritional needs of microbes.

◆ Discuss nutrient acquisition by *Bacteria* and *Archaea*.

◆ Outline basic cultivation methods.

◆ Describe strategies for isolation of different classes of bacteria.

5

Nutrition, Cultivation, and Isolation of Microorganisms

It appears incongruous to delineate metabolism of bacteria from that of fungi, since the two overlap in so many respects and since each group displays, more than any other group of microorganisms, a kaleidoscopic array of metabolic diversity.
—Jackson W. Foster

Green plants are photosynthetic, gaining their energy from sunlight and using CO_2 as their carbon source. In contrast, animals require preformed organic molecules for both energy and building material. These preformed molecules may be microbial, plant, or animal in origin. A plant or an animal species is generally differentiated from any other plant or animal species by observable differences in size, shape, and component structures. Mites to elephants or moss to sequoias are examples of the extremes in size and structure in the animal and plant world.

In contrast to plants and animals, prokaryotic microorganisms often do not differ markedly from one another either in size or structure. Microorganisms are generally differentiated from one another by their nutrition, which is the sum of the processes whereby the organism takes in and utilizes food for energy and building material (carbon). Among the *Bacteria* and the *Archaea*, nutritional needs vary from species to species. It is axiomatic that some microorganisms in nature can utilize, as a nutrient source, any carbon-containing constituent that is or has been a component part of living cells. Many other compounds unrelated to living cells can also be utilized as substrates (sources of food) for growth. Among such compounds that microorganisms can utilize as carbon/energy sources are carbon monoxide, methane, cyanide, and hexane. Others readily utilize aromatic compounds—such as benzene or naphthalene—that are toxic to most eukaryotes. These marked nutritional differences are referred to collectively as representing "metabolic diversity in the microbial world." This chapter examines the major nutritional differences that distinguish microorganisms, how different nutritional types can be isolated, and how they may be cultivated and preserved.

5.1 Fundamental Aspects of Diversity

Microorganisms live in virtually every environmental niche on Earth. These microbial populations can utilize the naturally occurring organic and inorganic compounds that fall into these niches as sources of carbon and/or energy. Many products of industrial chemical synthesis, such as pesticides, are also metabolized by natural microbial populations. However, it should be noted that some chemically synthesized compounds are resistant to microbial degradation, and these tend to accumulate in the environment.

Evolution has resulted in a seemingly endless array of microbial types with broad metabolic capacities. These organisms differ from one another in their nutritional capabilities and their adaptation to the physical conditions in which they live. Species of *Bacteria* and *Archaea* are adapted to growth under wide ranges of physical conditions, including pH (<1 to >11), temperature (<0°C to >100°C), salinity (freshwater to saturated saltwater), or availability of O_2 (aerobic to anaerobic). These physical conditions also constrain the nature and availability of substrates for growth of a microorganism (the forms of carbon, nitrogen, or phosphorus).

Many bacterial species utilize simple monomeric substrates such as sugars or amino acids as growth substrates, whereas others digest complex proteins and polysaccharides. Although some bacteria, such as *Pseudomonas cepacia*, can utilize any one of more than 100 different carbon-containing compounds for growth, most organisms can use only a limited number. Other microbes are restricted to growth on only one or a few chemically related substrates. For example, *Methylomonas methanooxidans* will grow only when methane or methanol is provided as growth substrate. Collectively, however, microbial populations are capable of growth on all of the unending variety of substrates encountered in nature. This broad metabolic diversity is essential in moving carbon through the biosphere and in maintaining the level of CO_2 in the atmosphere. This CO_2 is the carbon source for the photosynthetic populations that are the basis for most biological systems on this planet.

The Evolution of Metabolic Diversity

The presence of diverse nutritional types, as the quote at the beginning of the chapter affirms, is a consequence of the evolutionary role that microbes have assumed. Much of the microbial community has evolved with the ability to mineralize constituents or end products of all biota. The word *mineralize* is used in this text to emphasize that microorganisms can break down organic molecules to basic inorganic constituents (NH_4^+, CO_2, HS^-, PO_4^{2-}, etc.). Over the course of microbial evolution, mutation and selection gave rise to populations that are able to utilize every available carbon compound as a source of carbon and/or energy. Populations of microbes have also evolved that can obtain energy from light and from inorganic substrates such as hydrogen (H_2), sulfides, reduced iron, and ammonia. Other organisms evolved that can utilize gaseous carbon sources, such as CO, CO_2, methane, ethane, or propane. The ability to "fix" atmospheric nitrogen (N_2) lent a distinct advantage to selected *Bacteria* and *Archaea* occupying environments where little combined nitrogen (NH_4^+, NO_3^-, or organic nitrogen compounds) would be available. The ability to utilize a substrate more effectively than other microbes, or to use a substrate under physical conditions where others could not, gave a survival advantage. These selective pressures also gave rise to a broad array of microbial species with small but significant differences in their nutrient requirements and tolerance to physical stresses. Basically, survival advantage has been the key force during the 4 billion years of microbial evolution and a major factor in generating microbial diversity (Box 5.1).

Nutritional Types of Microorganisms

All microorganisms must have a source of carbon to provide the building blocks for synthesis of cell components, and a source of energy to drive these reactions. Based on the source of this carbon, *Bacteria* and *Archaea* can be divided into two types. When carbon is acquired exclusively from an inorganic source such as carbonate or carbon dioxide, the microbe is termed an **autotroph** (from *auto* meaning "self," and *troph* meaning "feeding"). A microbe that uses organic compounds as a source of carbon is called a **heterotroph** (from *hetero* meaning "different," and *troph* meaning "feeding").

Regarding energy acquisition, a microbe that harvests its energy from light is termed a **phototroph** (from *photo* meaning "light," and *troph* meaning "feeding"). In contrast, if either an organic or inorganic compound is used as an energy supply, the microbe is a **chemotroph** (from *chemo* meaning "chemical," and *troph* meaning "feeding"). Note that these terms clearly distinguish between the use of carbon for cell biosynthesis versus energy.

These terms thus provide a framework for separating microbes into reasonable and concise nutritional types. By combining the terms for carbon and energy supply, four nutritional groups follow:

BOX 5.1	*Milestones*

The Unique Character of *Bacteria* and *Archaea*

Bacterial and archaeal species possess a significant number of unique capabilities not found anywhere among the eukaryotes. They:

- Fix atmospheric nitrogen (N_2)
- Synthesize vitamin B_{12}
- Photosynthesize without chlorophyll

- Use H_2S, H_2, or organics as electron donors in photosynthesis
- Use inorganic energy sources (NH_4^+, H_2S, H_2, Fe^{2+}, NO_2^-) as electron donors
- Have extensive capacity for anaerobic growth
- Utilize inorganic molecules

(CO_2, NO_3^-, SO_4^{2-}, S^0, Fe^{3+}, SeO_4^{2-}, AsO_4^{3-}) as terminal electron acceptors as an alternative to O_2
- Grow at temperatures in excess of 80°C

- **Photoautotrophs** use light as their source of energy and CO_2 as their source of carbon. An obligate photoautotroph is an organism that will grow only in the presence of both light and CO_2. Obligate photoautotrophs use inorganic compounds such as H_2O, H_2, or H_2S as electron donors to reduce CO_2 to cellular carbon (CH_2O, which represents a general formula for carbohydrates).

- **Photoheterotrophs** utilize light as an energy source and use organic compounds or H_2 as an electron donor. They also assimilate organic molecules from the environment as a source of carbon. Commonly, organisms in this group require selected growth factors, such as B vitamins, and several species will grow on organic substrates if oxygen is available.

- **Chemoautotrophs** utilize reduced inorganic substrates for both the reductive assimilation of CO_2 and as a source of energy. The oxidation of H_2, NH_3, NO_2^-, H_2S, or Fe^{2+} supplies energy for these microorganisms. Aerobic chemoautotrophs utilize O_2 as a terminal electron acceptor, and some anaerobic bacteria and archaea, for example, can utilize inorganic sulfur or sulfate as a terminal electron acceptor (see Chapters 18 and 19).

- **Chemoheterotrophs** are microorganisms that assimilate organic substrates as sources of both carbon and energy. Chemoheterotrophs may use a single substrate such as glucose or succinate, for both needs, or may use one substrate for energy and another for carbon. For example, the sulfate reducers may use H_2 for energy but require an organic substrate for cellular biosynthesis. Most of the microbes that have been studied are chemoheterotrophs.

Bacteria representing each of these categories are presented in Figure 5.1. Some organisms can fit into more than one category, whereas others have exclusive lifestyles (discussed in Box 5.2). Note that this nutritional classification scheme applies equally well to all plants and animals!

Nutritional Requirements for Cell Growth

This nutritional classification of microbes based on energy and carbon supply affirms that the various bacterial and archaeal species may have significant differences in their nutritional requirements. Under laboratory conditions, microorganisms grow in an artificial environment that is rarely equivalent to that encountered in nature. However, a microorganism living in soil, water, or the intestinal tract of a warm-blooded animal survives there because its nutritional requirements are met. These requirements may be simple or they may be complex. If we wish to take a microorganism from an environmental niche and grow it in the laboratory, we must somehow mimic the conditions that permitted it to survive in the environment. These would include both physical conditions (pH, temperature, osmolarity, light) and substrate availability. A laboratory medium must provide the nutrients that are essential for growth. The following section examines in some detail how to prepare a growth medium.

Composition of Microbial Cells

Biochemical analysis of bacterial or archaeal cell mass affirms that their cellular composition is generally similar, regardless of the species or the composition of the medium on which they grew. They are composed of the same major cellular components, including the monomers that make up the major macromolecules (proteins, nucleic acids, carbohydrates, and lipids), biosynthetic intermediates, and inorganic ions. The elemental compositions of cells are likewise very similar, with little

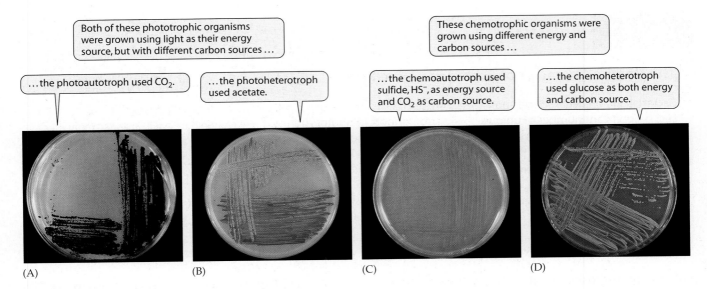

Both of these phototrophic organisms were grown using light as their energy source, but with different carbon sources …

These chemotrophic organisms were grown using different energy and carbon sources …

…the photoautotroph used CO_2.

…the photoheterotroph used acetate.

…the chemoautotroph used sulfide, HS^-, as energy source and CO_2 as carbon source.

…the chemoheterotroph used glucose as both energy and carbon source.

(A)   (B)   (C)   (D)

Figure 5.1 Nutritional types
Colonies of microorganisms representing the major nutritional groups: (A) *Anabaena* sp., a photoautotroph; (B) *Rhodomicrobium* sp., a photoheterotroph; (C) *Thiobacillus perometabolis*, a chemoautotroph; and (D) *Micrococcus luteus*, a chemoheterotroph. A, courtesy of Robert Tabita; and B–D, courtesy of Jerome Perry.

variance from the percentages of elements shown in Table 5.1. This property provides a framework to understand bacterial and archaeal nutrition.

A laboratory growth medium must provide a microorganism with all the essential elements in one form or another, and in the appropriate amounts. The macromolecules that are present in all microorganisms are composed of the first six elements listed in Table 5.1. A requirement for the abundant cations (K^+, Na^+, Mg^{2+}, and Ca^{2+}) is nearly universal, whereas the need for iron and the trace metals may vary according to species and the carbon/energy supply. These elements are discussed here as components of the growth medium.

Carbon

The most abundant element in any living cell is carbon (see Table 5.1). Carbon is the backbone of functional biological molecules. Microbial diversity reflects the ability of organisms to synthesize all of their component molecules from the carbon sources available. Marked differences exist among microbes in this capacity. For example, a photoautotrophic cyanobacterium can synthesize *de novo* all of its cellular components (protein, nucleic acids, lipids) from CO_2 if mineral salts are available (Table 5.2). Other organisms may have a very limited biosynthetic capacity and require a complex growth medium, such as that shown in Table 5.3. A growth

---

BOX 5.2   *Milestones*

Microbial Versatility: Obligate versus Facultative

A microorganism that will grow only if provided with CO_2 for carbon, and light as an energy source is called an **obligate photoautotroph**. Likewise, one microbial species that grows only with CO_2 and NH_4^+ or S^0 as energy source is an **obligate chemoautotroph**. Restriction to an obligate lifestyle can be found in the microbial world, but versatility is the more common attribute. An **obligate autotroph** can actually assimilate a very limited amount of selected organic substrates while growing with CO_2 as bulk carbon source and light (for photoautotrophs) or inorganics (for chemoautotrophs) as energy source. However, many photoautotrophs or chemoautotrophs can also grow as chemoheterotrophs. Such organisms are termed **facultative**. The term facultative is also used to describe an organism that can grow aerobically if O_2 is available or anaerobically in the absence of O_2.

TABLE 5.1	A typical analysis of the elements present in a bacterial or archaeal cell

Element	Dry Weight[a]
Carbon	50
Oxygen	20
Nitrogen	14
Hydrogen	8
Phosphorus	3
Sulfur	1
Sodium	1
Potassium	1
Calcium	0.5
Magnesium	0.5
Iron	0.2
Cu, Zn, Mn, Mo, B, Se, Cl, Ni, Cr, Co, and W	0.2
Total	100%

[a]Percentage of total dry weight (note, a microbial cell is composed of 80% to 90% water).

TABLE 5.2	A typical mineral salts medium for the isolation and growth of free-living bacteria

Constituent	Amount[a] (mg/liter H_2O)
NH_4Cl	500
$NaNO_3$	500
Na_2HPO_4	210
NaH_2PO_4	90
$MgSO_4 \cdot 7 H_2O$	200
KCl	40
$CaCl_2$	15
$FeSO_4$	1

Trace Element	(μg/liter H_2O)
$ZnSO_4 \cdot 7 H_2O$	70
H_3BO_3	10
$MnSO_4 \cdot 5 H_2O$	10
$Na_2MoO_4 \cdot 2 H_2O$	10
$NiCl_2$	10
$CoCl_2$	10
$Na_2SeO_4 \cdot 2 H_2O$	10
$CuSO_4 \cdot 5 H_2O$	5

[a]It would be necessary to add a carbon source such as glucose at 0.5 to 2.0%.

medium with all of the compounds listed in the table would be necessary for the growth of the fastidious lactic acid–producing bacterium *Streptococcus agalactiae*.

The carbon source for microbial growth can range from CO_2 or CH_4 to any naturally occurring organic compound present in the biosphere (such as carbohydrates, peptides, and organic acids). A variety of microbes also exist that can, under appropriate conditions, grow with various synthetic organic compounds as substrate. This is the basis for "bioremediation," the removal of pollutant chemicals from the environment by use of selected microbes that can utilize these pollutants as a carbon/energy source.

Oxygen

The total number of oxygen atoms present as a cellular constituent is equivalent in aerobes and anaerobes. Depending on the microbe, oxygen can be acquired from water, molecular oxygen (O_2), or organic compounds. However, molecular oxygen (O_2) is toxic to most strictly anaerobic *Bacteria* and *Archaea*, so these organisms obtain this element in a combined form from a substrate. As a rule, anaerobes utilize growth substrates that are in an oxidation–reduction state equal to or more oxidized than cellular material (CH_2O). For example, anaerobes can utilize sugars, amino acids, or CO_2 as carbon sources. Aerobic bacteria can grow on more reduced substrates (such as benzene, methane, and fatty acids) when molecular oxygen is available. These aerobic microorganisms use O_2 as a terminal electron acceptor in their electron transfer systems.

Nitrogen

Nitrogen is an integral constituent of amino acids (proteins), nucleotides (nucleic acids), and constituents of the cell wall. Most free-living microorganisms assimilate ammonia from their environment, or they assimilate nitrate and then reduce it to ammonia for subsequent incorporation into organic building blocks. One or both of these forms of nitrogen are commonly included in a growth medium.

A requirement for an organic nitrogen source—such as an amino acid—is generally confined to microorganisms that evolved in relatively nutrient-rich environments, where various amino acids, nucleic acids, and B vitamins were readily available. For example, Table 5.3 lists the nutrients in a growth medium for a microorganism that has a limited ability to synthesize nitrogen-containing intermediates. The lactic acid bacteria evolved in or on animals or plants, where organic nitrogen compounds were available.

A unique property of some bacteria and archaea is their ability to obtain cellular nitrogen by fixation (i.e., the reduction of atmospheric N_2 to ammonia). These nitrogen fixers then incorporate the fixed nitrogen, at the level of NH_4^+, into carbon compounds via the biosynthetic machinery of the cell. The fixation of N_2 is not lim-

TABLE 5.3	A defined medium for the growth of the lactic acid bacterium *Streptococcus agalactiae*

Compound	Final Concentration (μg/ml H₂0)
Carbon and Energy Source	
Glucose	10,000
Inorganic salts	
K₂HPO₄	1,000
KH₂PO₄	1,000
NaHCO₃	500
FeSO₄ • 7 H₂O	10
MnCl₂	25
NaCl	10
ZnSO₄ • 7 H₂O	10
MgSO₄ • 7 H₂O	80
Amino acids	
L-alanine	250
L-arginine	320
L-aspartic acid	500
DL-asparagine	100
L-cystine	600
L-cysteine-HCl	600
L-glutamic acid	400
L-glycine	200
L-glutamine	100
L-histidine	320
L-leucine	200
L-lysine	320
L-isoleucine	200
DL-methionine	200
L-phenylalanine	200
L-proline	200
DL-serine	400
L-tryptophan	200
L-tyrosine	200
L-valine	200
Bases	
Adenine	10
Guanine	10
Xanthine	10
Uracil	10
Vitamins	
Nicotinic acid	10
Ca pantothenate	10
Pyridoxal HCl	10
Thiamine HCl	0.6
Riboflavin	1
Biotin	0.2
Folic acid	0.02

From N. P. Willett and G. E. Morse. 1966. Long-chain fatty acid inhibition of the growth of *Streptococcus agalactiae* in a chemically defined medium. *J. Bacteriology* 91:2245.

ited to a few species, as was long believed, but occurs in an array of bacterial and archaeal types, as outlined in Table 5.4.

The growth of most heterotrophic bacteria is stimulated by adding rich nitrogenous material, such as yeast extract (the soluble portion of digested yeast cells) at 0.05%, to a basic mineral salts medium. Growth is stimulated because energy need not be expended to synthesize amino acids, purines, pyrimidines, and other nitrogen-containing compounds.

Hydrogen

Hydrogen plays a number of roles in the life of bacteria and archaea—it is a structural atom in organic molecules

TABLE 5.4	Some genera of *Bacteria* and *Archaea* that have nitrogen fixation ability

Bacteria
  Heterotrophs
    Aerobes
      Azotobacter
      Klebsiella
      Beijerinckia
    Anaerobes
      Clostridium
      Bacillus (facultative)
  Phototrophs
    Cyanobacteria
      Anabaena
      Oscillatoria
      Gloeocapsa
    Purple and green bacteria
      Chromatium
      Chlorobium
      Rhodospirillum
  Symbiotic associations
    Legumes
      Clover + *Rhizobium*
      Soybeans + *Bradyrhizobium*
      Bluebonnets + *Rhizobium*
    Nonlegumes
      Bayberry + Actinomycetes
      Alder + *Frankia*

Archaea
  Methanogens
    Anaerobes
      Methanococcus
      Methanosarcina
      Methanobacterium
      Methanothermus

and it participates in the complex process of energy generation. A source of hydrogen is essential in autotrophic microorganisms to reduce CO_2 for incorporation into cell material (CH_2O). Protons (H^+) are involved in the production of ATP via the ATP synthase system in the cytoplasmic membrane of most microorganisms (see Chapter 8). Microorganisms that respire anaerobically (CH_4 producers, denitrifiers, and sulfate reducers, among others) gain their energy by transfer of electrons from a substrate to a selected acceptor (Table 5.5). Aerobic respiration involves the transfer of electrons to O_2, a factor in establishing a proton gradient for ATP synthesis (see Chapter 8). Hydrogen is generally acquired from water or organic compounds.

Sulfur

Sulfur is a constituent of two of the amino acids that make up proteins: cysteine and methionine. It is also present in certain B vitamins (biotin and thiamine) and some other essential constituents of a cell. Sulfur is usually added to a growth medium as a sulfate salt. $MgSO_4$ added to a medium would serve as a source of both sulfur and magnesium. The capacity to reduce sulfate to the level of sulfide (HS^-), the form present in cellular constituents, is a common attribute of bacteria. If an organism cannot reduce sulfate, the addition of sulfide or the amino acid cysteine to the medium will permit growth.

Addition of yeast extract or peptone (an enzyme digest of protein) to a culture medium meets the need for reduced sulfur compounds in most microorganisms. Reduced inorganic sulfur compounds such as H_2S or pyrite (iron sulfide) can serve as the energy source for a group of bacteria called the thiobacilli. Oxidation of sulfides generates sulfate. Elemental sulfur (S^0) can serve as a terminal electron acceptor in some of the *Bacteria*, and many of the *Archaea*. Sulfate serves this purpose in sulfate-reducing *Bacteria* and *Archaea*.

Phosphorus

This element has played a major role in the evolution of life systems on Earth. Phosphorus is a constituent of high-energy compounds, the phospholipids in cell membranes, and nucleic acids. Adenosine triphosphate (ATP) is the principal medium of energy exchange in cellular metabolism and is an indispensable contributor to biosynthetic reactions involved in cell reproduction and growth. Thus, inorganic phosphates are an integral constituent of culture media. Inorganic phosphates also provide an effective buffer at near neutral pH and at concentrations that are not usually inhibitory to bacterial growth. Thus, phosphate salts are commonly added to culture media to satisfy the phosphate requirement and to provide a buffer to prevent significant changes in pH during growth.

The Abundant Cations

Several other elements present in *Bacteria* and *Archaea* are generally involved with enzymatic activity and/or in cell stability. These are mostly cations and include potassium, sodium, magnesium, and calcium (Table 5.6). They also may serve as counter-ions inside the cell to balance the negative charge of proteins, nucleic acids, and other cell molecules. The divalent cation magnesium is needed to stabilize cell surface molecules such as lipopolysaccharide (LPS) (see Chapter 4).

Iron

Iron is a constituent of electron transport chains and therefore is an absolute requirement among aerobes and most anaerobes. Iron is also present in some biosynthetic enzymes.

Trace Elements

Additional elements are present in *Bacteria* and *Archaea* and are generally involved in enzymatic reactions. These include iron, zinc, cobalt, nickel, copper, selenium,

TABLE 5.5	Examples of respiratory pathways that occur in *Bacteria* and *Archaea*
Type of Respiration	**Example**
Aerobic respiration	
Oxygen	
$O_2 \rightarrow H_2O$	*Pseudomonas fluorescens*
Anaerobic respiration	
Iron	
$Fe^{3+} \rightarrow Fe^{2+}$	*Shewanella putrefaciens*
Nitrate	
$NO_3^- \rightarrow NO_2^-, N_2O, N_2$	*Thiobacillus denitrificans*
$NO_3^- \rightarrow NO_2^-, NH_4^+$	*Escherichia coli*
Fumarate	
Fumarate $\rightarrow$ Succinate	*Proteus rettgeri*
Selenate	
Selenate $\rightarrow$ Selenite, Se^0	*Thauera selenatis*
Sulfate	
$SO_4^{2-} \rightarrow HS^-$	*Desulfovibrio desulfuricans*
Sulfur	
$S^0 \rightarrow HS^-$	*Desulfurococcus mucosus*
Carbonate	
$CO_2 \rightarrow CH_4$	*Methanosarcina barkeri*
$CO_2 \rightarrow CH_3COO^-$	*Acetobacterium woodii*

TABLE 5.6	The function of various elements in bacterial and archaeal nutrition

Element	Function
Potassium	Utilized in a number of enzymatic reactions as a cofactor and especially in protein synthesis
Sodium	Involved along with chloride in the regulation of osmotic pressure; affects activity of some enzymes; uptake of solutes in some species with Na^+-dependent transport systems
Magnesium	Integral part of chlorophyll; the cation is required in enzymatic reactions including those involved in ATP synthesis or hydrolysis; needed to stabilize DNA, membranes, and ribosomes
Calcium	Used to stabilize cell walls of some microbes; a component of endospores
Iron	Reactive center of heme-containing proteins (cytochromes, catalase, etc.); component of other proteins, including non-heme iron–sulfur proteins
Cobalt	Constituent of vitamin B_{12}; complexed to some enzymes
Copper, zinc, molybdenum, nickel, tungsten, and selenium	Essential components of certain enzymes

molybdenum, boron, and tungsten (see Table 5.6). Copper, zinc, nickel, and molybdenum and/or tungsten are required in small amounts as cofactors in certain enzymes. Selenium is required by some bacteria for incorporation into seleno-proteins or as an enzyme cofactor. Together, these nutrients are called **trace elements** due to the limited amount of each needed for cell growth.

Growth Factors

Growth factors are specific, relatively low-molecular-weight organic compounds that must be available in the growth medium of some microorganisms because they cannot synthesize them. The substances that often serve as growth factors are selected amino acids, purines, pyrimidines, and B vitamins. Microbes do not require fat-soluble vitamins, such as vitamins A and D, as they are not constituents of microorganisms.

The three types of growth factor most often required in bacterial nutrition are:

- **Vitamins**. These organic compounds serve as the prosthetic group (nonprotein catalytic part) of a number of enzymes. Small catalytic amounts of vitamins are required, as they are present in cells in low quantity (varying from a nanogram of vitamin B_{12} to 250 micrograms of nicotinic acid per gram dry weight). The vitamins most frequently required are those less susceptible to destruction by

light: thiamine, biotin, and nicotinic acid. The function of the B vitamins in nutrition is outlined in Table 5.7.

- **Amino acids**. Proteins are made from 21 amino acids (see Chapter 3), and some bacteria and archaea cannot synthesize one or more of these. They are therefore required to be present in the growth medium. For example, most strains of *Staphylococcus epidermidis*, a normal inhabitant of human skin, require six amino acids: proline, arginine, valine, tryptophan, histidine, and leucine. The lactic acid bacteria require even more amino acids, as indicated in Table 5.3. The required amount of each amino acid is proportional to the amount of that amino acid in the cellular protein. This amount can be calculated as follows: a cell is about 50% protein, and the aromatic amino acid phenylalanine (for example) constitutes about 5% of the protein on average. Therefore, to grow one gram of an organism that requires this amino acid, one should add at least 25 mg phenylalanine to the growth medium (1 gram of cells = 500 mg protein, $500 \times 0.05 = 25$ mg).

- **Purines and pyrimidines**. Requirements for the nucleotidyl bases are most often observed in lactic acid bacteria (see Table 5.3) and other fastidious organisms with many growth factor requirements. The need for added purines and pyrimidines is rare in free-living soil microbes.

TABLE 5.7	The functions of various vitamins in *Bacteria* and *Archaea*
Compound	**Function**
p-Aminobenzoic acid	Precursor of folic acid, a coenzyme involved in one-carbon unit transfer
Folic acid	Coenzyme involved in one-carbon unit transfer
Biotin	A prosthetic group for enzymes that act in carboxylation reactions
Nicotinic acid	Precursor of NAD^+ and $NADP^+$, which are coenzymes involved in hydrogen/electron transfer
Riboflavin	A component of the flavin mononucleotide (FMN) and dinucleotide (FAD) involved in hydrogen/electron transfer
Pyridoxine	A component of the coenzyme for aminotransferases and decarboxylase enzymes
Vitamin B_{12}	A coenzyme involved in molecular rearrangements
Thiamin	The prosthetic group for a number of decarboxylases, transaldolases, and transketolases
Pantothenic acid	A functional part of coenzyme A and the acyl carrier proteins
Coenzyme M	A coenzyme in methane-generating bacteria

SECTION HIGHLIGHTS

Cell growth requires a complete set of essential nutrients in sufficient amounts to support balanced cell growth. Each species of *Bacteria* and *Archaea* will exhibit a distinct nutritional profile based on its genetic makeup, where some nutrients are essential and others are nonessential.

5.2 Uptake of Nutrients into Cells

Microorganisms in nature generally live in environments where many nutrients are available at low concentrations. Effective growth can occur only if nutrients can be accumulated inside the cell at levels that exceed those present externally. Microorganisms have evolved mechanisms that permit them to concentrate nutrients such as sugars or required cations to levels inside the cell that are thousand-fold or more higher than those present in the environment.

The cytoplasmic membrane of a bacterial or archaeal cell forms a highly selective barrier between the external environment and the cytoplasm (see Chapter 4). This barrier permits or facilitates the entry of essential nutrients inward and rejects many of those that are harmful or unnecessary. The hydrophobic nature of the lipid bilayer is responsible for the high degree of selective permeability inherent in cytoplasmic membranes. Polar solutes such as amino acids or sugars cannot traverse unaided across the cytoplasmic membrane. Water can pass freely across the membrane; alcohols, fatty acids, or other fat-soluble compounds may also pass through at varying rates.

To accomplish the selective uptake of nutrients essential to growth and reproduction, a microorganism utilizes a number of distinct transport mechanisms. Among the most common transport mechanisms are **facilitated diffusion**, **active transport**, and **group translocation**. These protein-mediated processes are necessary because simple diffusion would result in an internal concentration that is no more than that present externally. Under most environmental conditions, this would not support balanced growth. The following is a discussion of the mechanisms that are important in concentration of nutrient solutes inside a bacterial cell.

Diffusion

Two types of diffusion occur across cytoplasmic membranes: **passive diffusion** and **facilitated diffusion**. The former process is protein independent, whereas the latter requires specific membrane proteins called **permeases**. Diffusion, both passive and facilitated, requires a concentration gradient, and molecules will flow across the membrane from an area of high concentration to one of low concentration until equilibrium is established. Generally, passive diffusion occurs with gasses (O_2, CO_2, H_2), water, and fat-soluble compounds. Glycerol is a compound that may enter a cell by passive diffusion, and the rate of glycerol uptake is solely dependent on the external concentration.

Facilitated diffusion is a carrier-mediated transport process using transmembrane proteins termed permeases. One end of the transport protein protrudes to the outside of the membrane, and the other end extends to the interior. Some of these permeases are highly selective in that they transport only a single type of molecule. Others recognize a class of substrates such as a group of similar amino acids or a series of related sugars.

Facilitated transport does not require energy input by the cell. Rather, it depends on a concentration gradient across the membrane to drive a conformational change in the transport protein. The solute to be transported binds to the external portion of the transport protein, then a change in protein conformation moves the solute inward, and the molecule is released to the inside of the cell (Figure 5.2). Although facilitated diffusion speeds the entry of solutes into a cell, the total internal concentration will not exceed the level in the immediate external environment. Facilitated diffusion is effective because it "speeds up" the diffusion process. The transported material is generally metabolized on entry, and this maintains a low internal concentration, thus promoting continued solute uptake.

Because facilitated transport will not function against a concentration gradient, a microbe in nature requires

an energy-driven mechanism for concentrating a nutrient that is present at a low external concentration.

Active Transport

Active transport is the direct utilization of cell energy to move a solute from the outside of a membrane where the concentration is low to establish a higher concentration on the other side. This process accumulates a solute against a concentration gradient. As with facilitated diffusion, solute-specific transmembrane proteins are involved in active transport, and the solute transported is not chemically altered as it passes into the cytoplasm. For a single solute, an organism may have alternative transport systems that may differ in the energy source required and their affinity for the solute that is transported.

The nutrients that are taken into the cell by active transport include almost all of the organic and inorganic molecules needed for cell growth: sugars, amino acids, purines, pyrimidines, organic acids, inorganic cations, vitamins, and trace metals. Some microbes may contain as many as several hundred distinct active transport systems.

The energy for active transport generally comes from ATP, or from a proton gradient established across the cytoplasmic membrane. The proton gradient can result from the metabolism of inorganic or organic substrates by the process of cellular respiration, or it can be generated from light energy by the process of photosynthesis (see Chapters 8 and 9).

Types of Active Transport

ABC TRANSPORTERS   Transport of molecules by ATP-dependent systems involves **ATP-Binding Cassette transporters** (**ABC transporters**). These systems are present in bacterial, archaeal, and eukaryotic cells. Although they may vary in composition, an ABC transporter generally has three types of proteins: a hydrophobic transmembrane protein, an ATP hydrolyzing protein, and a periplasmic solute-binding protein. The transmembrane protein forms a selective pore that spans the cytoplasmic membrane. Attached on the cytoplasmic side is a nucleotide-binding protein that, upon binding ATP, hydrolyzes it to drive solute uptake into the cell (Figure 5.3). The ATP-dependent uptake of substrate requires a solute-binding protein that is either located in the periplasmic space (gram-negative bacteria), or anchored to the outer surface of the cytoplasmic membrane by a hydrophobic membrane-spanning α-helix (gram-positive bacteria and archaea). The binding-protein selectively binds the molecule to be transported and passes it to the membrane protein for translocation across the membrane. Conformational changes in the transporter protein, promoted by energy

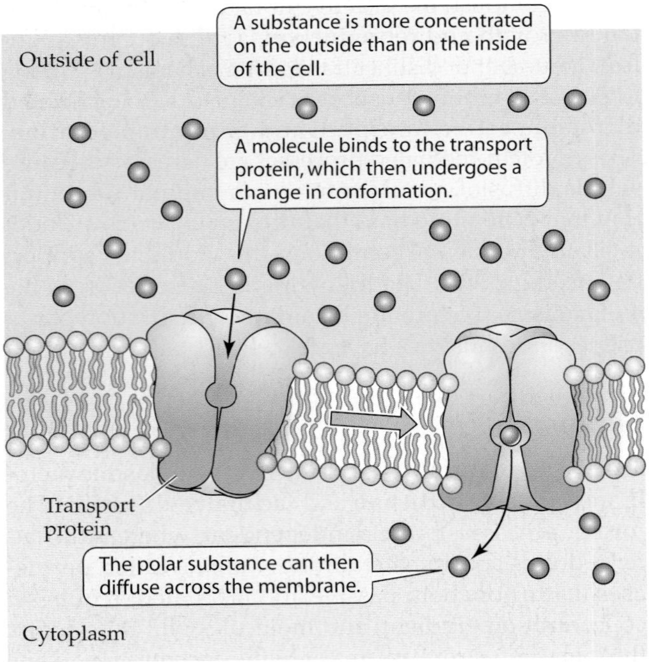

Figure 5.2 Facilitated diffusion
In facilitated diffusion, the substance passes through a transporter protein, or permease. No energy source is required. If the internal concentration exceeds the external concentration, the process can be reversed.

Labels in figure:
- Outside of cell
- A substance is more concentrated on the outside than on the inside of the cell.
- A molecule binds to the transport protein, which then undergoes a change in conformation.
- Transport protein
- The polar substance can then diffuse across the membrane.
- Cytoplasm

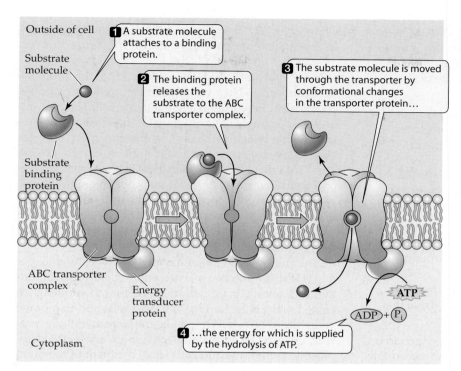

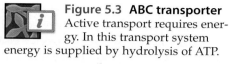

Figure 5.3 ABC transporter
Active transport requires energy. In this transport system energy is supplied by hydrolysis of ATP.

derived from ATP hydrolysis, drive the molecule into the cell against a concentration gradient. In gram-positive microorganisms, the molecules to be transported would diffuse through the cell wall to the membrane surface, where they would be available to the binding proteins. In gram-negatives, the molecules would pass through porins in the outer membrane (see Figure 4.55) and come in contact with the binding proteins located in the periplasmic space.

SECONDARY TRANSPORTERS   Bacteria also transport solutes by utilizing the energy inherent in an electrical charge separation across the membrane. These transporters, termed secondary transporters, are similar to facilitated transporters except that they use energy to drive solute movement against a concentration gradient. There are three types of these transport systems: uniporters, symporters, and antiporters (Figure 5.4). Periplasmic binding proteins are generally absent.

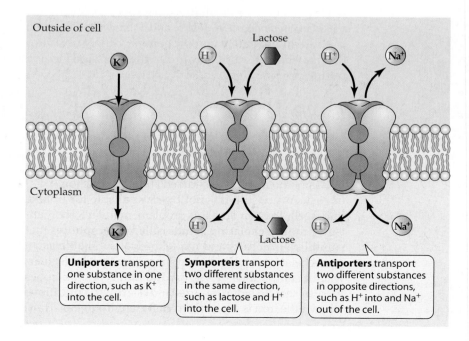

Figure 5.4 Ion-driven transport
Three examples of active transporters that use an ion gradient (protons or another ion) as the energy source to drive solute movement across the membrane: uniporter, symporter, and antiporter. The proton motive gradient is used directly or indirectly via other ions to transport substances across the membrane.

- **Uniporters** are transport proteins that take in cations such as K^+ down the electrochemical gradient established by a high positive charge on the outside and negative charge on the inside of the membrane (see Figure 5.4).

- **Symporters** carry sugars, anions, or amino acids into the cell when accompanied by a proton or a sodium ion. The binding of a proton to the transport protein, for example, drives a change in conformation that transports the solute inward. The lactose permease of *Escherichia coli* is a representative example that employs protons to drive accumulation of lactose (Figure 5.5).

- **Antiporters** use the energy of a proton gradient, for example, to move a proton inward through the transport protein and drive a cation such as Na^+ outward, thus creating a sodium concentration gradient. This sodium gradient can then drive the uptake of an amino acid or other molecule.

Most of the solute transport mechanisms depend on the generation of a proton gradient across the membrane to either directly or indirectly drive the transport processes. The charge separation, termed a **proton motive force**, is established by a proton gradient generated during photosynthetic or respiratory electron transport. The proton motive force also drives ATP synthesis (see Chapter 8); ATP hydrolysis is then used to fuel the uptake of substances through the ABC transport proteins.

GROUP TRANSLOCATION   Group translocation is a transport process in which the transported compound is chemically altered upon entry into the cell. The best understood example of group translocation is the **phosphotransferase system** (**PTS**) involved in transport of sugars such as glucose, fructose, and α-glucosides into the cell. The PTS is more complex than the ABC transport and secondary transport systems. It employs the high-energy phosphate bond energy from phosphoenolpyruvate (PEP) and the direct participation of at least four, but generally five, enzymes to transport one sugar molecule into the cell (Figure 5.6).

The first two enzymes, **Enzyme I** and a heat-stable protein (**HPr**), are soluble and present in the cytoplasm. **Enzymes IIa** and **IIb** are peripheral membrane proteins attached to the inner surface of the cytoplasmic membrane at the site of **Enzyme IIc**, an integral transmembrane protein. Enzyme IIc forms a selective pore across the cytoplasmic membrane and is the final recipient of the high-energy phosphate, which the enzyme then uses to phosphorylate the sugar and drive its entry into the cell.

The EIIa, EIIb, and EIIc enzymes are quite specific and are involved in the transport of a single sugar, such as glucose. HPr and Enzyme I are nonspecific and function in the PTS systems for other sugars. The PTS is energy conserving in that the high-energy phosphate from PEP ultimately becomes a part of the transported sugar. This process also accomplishes the first step in the breakdown of the sugar for energy generation.

The PTS is present in obligately anaerobic bacteria such as *Clostridium* and *Fusobacterium*. It is also present in many genera of facultative anaerobes including *Escherichia*, *Staphylococcus*, *Vibrio*, and *Salmonella*. However, it is rare in obligately aerobic genera. Other substrates that may be transported by group translocation include purines, pyrimidines, and fatty acids.

Iron Uptake

Iron is needed to form cytochromes and iron–sulfur proteins that function in electron transport during aerobic respiration. In aerobic habitats in nature, however, most iron is in the insoluble oxidized state as ferric ion (Fe^{3+}) or rust. As a result, it is not freely available for uptake into a cell. To overcome this problem, many microorganisms produce **chelating agents** called **siderophores** that solubilize iron salts and make the iron available as an iron–chelate complex. Siderophores are low-molecular-weight compounds of several different types. Microbes tend to secrete siderophores under growth conditions where little iron is available from the environment. Thus,

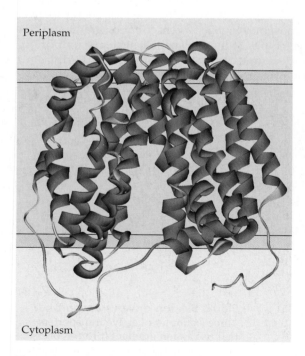

Figure 5.5   The lactose permease of *E. coli*
The three-dimensional structure of the protein is oriented in the cell membrane with the channel open on the cytoplasmic side. Coils represent transmembrane α–helices.

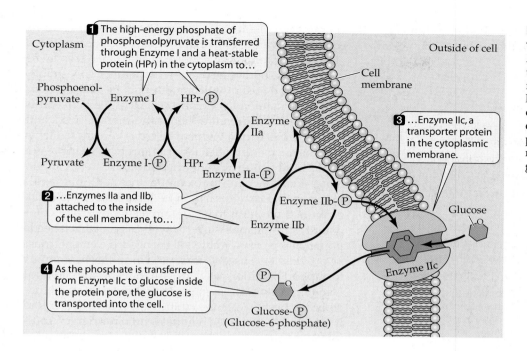

1 The high-energy phosphate of phosphoenolpyruvate is transferred through Enzyme I and a heat-stable protein (HPr) in the cytoplasm to…

Cytoplasm

Outside of cell

Cell membrane

Phosphoenol-pyruvate   Enzyme I   HPr-P

Pyruvate   Enzyme I-P HPr

Enzyme IIa

Enzyme IIa-P

3 …Enzyme IIc, a transporter protein in the cytoplasmic membrane.

2 …Enzymes IIa and IIb, attached to the inside of the cell membrane, to…

Enzyme IIb-P

Enzyme IIb

Glucose

Enzyme IIc

4 As the phosphate is transferred from Enzyme IIc to glucose inside the protein pore, the glucose is transported into the cell.

P

Glucose-P
(Glucose-6-phosphate)

Figure 5.6 The phospho-transferase system
Active transport of glucose into the cell via the phosphotrans-ferase system (PTS). HPr is a heat-stable protein; the other components of the system are enzymes. Energy is supplied by phosphoenolpyruvate (PEP), resulting in phosphorylation of glucose to glucose-6-phosphate.

even small amounts of environmental iron are made available to the cell.

A typical siderophore produced by *E. coli* is **enterobactin** (Figure 5.7), a derivative of catechol. It chelates ferric iron through the oxygen molecules on the catechol rings and thus maintains the iron in a soluble state. The siderophore–iron complex is bound by a binding protein and then transferred to a specific siderophore receptor in the cell envelope; the siderophore–iron complex is then transported across the cytoplasmic membrane by a specific

ABC transport system (Figure 5.8). The ferric iron (Fe^{3+}) is then reduced to ferrous iron (Fe^{2+}) by the enzyme **ferric reductase**, for subsequent incorporation into proteins.

Iron may be provided in a culture medium in a complex with a chelating agent such as citrate or ethylene

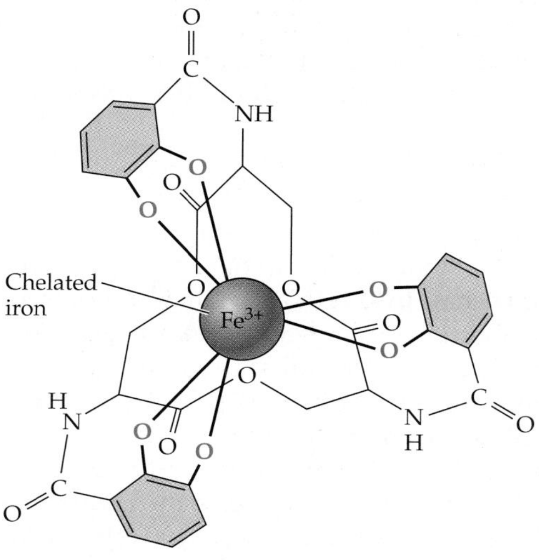

Figure 5.7 Iron transport
Enterobactin, a siderophore of *E. coli*, involved in transport of iron into the cell. This catechol derivative chelates iron through the oxygen groups on the catechol rings.

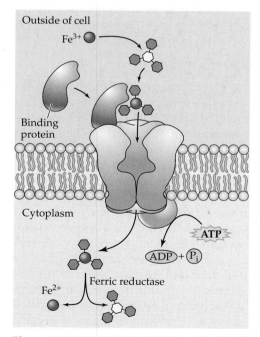

Figure 5.8 Uptake of enterobactin–iron into a cell
The iron–siderophore complex is recognized and bound by a specific binding protein, and delivered to a specific ABC transport system for uptake into the cell.

diamine tetraacetic acid (EDTA). The bound iron is released to the uptake system of the cell to supply metal for biosynthetic needs.

5.3 Cultivation of Microorganisms

Microbiologists have devised techniques that enable them to separate a single species of microorganism from a mixed environmental sample. When successfully isolated, a strain can then be cultivated in **pure culture**, also known as an **axenic culture** (*a*, without; *xenos*, strange). Some of the standard procedures for the isolation, growth, maintenance, and preservation of pure cultures are presented here.

A Culture Medium

A culture medium is composed of biologically or chemically derived materials that provide an environment conducive to growth of a particular type of microorganism. There are hundreds of described media that can be used for the growth of bacteria. A variety of ingredients—defined (the chemical composition is known) or undefined—can be used in their preparation. Components of common media may include such undefined ingredients as extracts of beef heart or brain, whole blood or serum, yeast extract (autolyzed yeast), peptone (protein digest), hydrolyzed casein (milk protein), or soil extract. A medium composed of any of these would be termed a **complex medium** because the compounds present are not defined chemically in type or amount. In addition, the various constituents may vary from batch to batch. Defined ingredients such as sugars or organic acids might also be added to a complex medium.

When a **solid medium** is desired, a gelling agent such as the complex polysaccharide **agar** is added to the liquid medium to solidify it. A typical concentration is 1.5% w/v (weight per volume; 1% w/v = 1 g per 100 ml). Agar

contains small amounts of organic matter and sulfur-containing compounds that inhibit some species. Silica gel or gelatin, or a purified form of agar (agarose), can be used for the solidifying agent when agar is undesirable.

A **defined** or **synthetic** medium can be employed for growth of many microorganisms. All of the ingredients in a defined medium are chemically characterized, and the amount of each ingredient added is known (see Table 5.2). Such a medium usually contains a buffering agent to aid in pH control. A defined medium may contain many components—for example, the medium for culture of *Streptococcus agalactiae* (see Table 5.2 and Table 5.3). A defined medium may contain a single carbon compound if the microbe is able to synthesize all that it needs from that type of molecule. All the needed elements must be provided in a useable form, and in sufficient amounts to support significant cell growth. If a specific chemical is required, it is termed an **essential nutrient**. If the microbe in question can use a nutrient, but does not require it for growth, it is called a **nonessential nutrient**. A **minimal medium** generally contains all the essential nutrients to support cell growth, plus a pH-buffering agent.

Sterilization of Media

Sterilization is the process of killing or removing *all* living things from a particular environment. Heating with live steam under pressure (in an autoclave) is commonly used for sterilization (see Chapter 7). Growth media may be autoclaved in covered or cotton-stoppered test tubes or flasks. Stainless steel or plastic caps are commonly used now, as they are less cumbersome than cotton stoppers and permit a rapid exchange of air into the culture vessel. Petri plates containing nutrient agar are prepared by first autoclaving the medium in a flask and aseptically pouring the liquified agar/nutrient into a sterile Petri plate, where the medium then gels as it cools.

Isolation in Pure Culture

In nature, most microorganisms coexist with countless other microbial species. For example, hundreds of different species live in the intestinal tracts of animals. A gram of fertile soil contains 10^7 to 10^9 bacteria and scores of species. This natural state, where many different species coexist, is referred to as a **microbial community** or a **mixed population**.

Enrichment procedures generally yield a mixture of microbial species best suited to the culture conditions. Therefore, it is necessary to apply techniques that will permit one to **isolate** individual species from this mixture. This is accomplished by use of **aseptic** isolation procedures. **Asepsis** means "in the absence of microorganisms" (its antonym is **sepsis**, which means nonsterile, or "in the presence of microorganisms"). In this context,

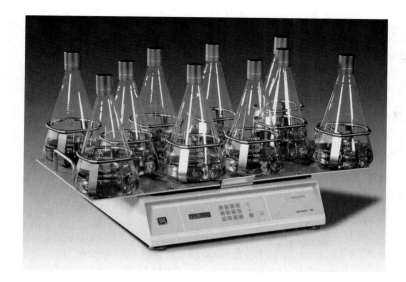

Figure 5.9 A bench-top shaker
The platform of this Certomat® RM benchtop shaker gyrates at controlled rates from 40 to 400 rpm. It can be operated at room temperature or in combination with an incubation hood or refrigerator for greater temperature control. The caps (or alternatively cotton stoppers), ensure that filtered, sterile air enters the flasks. Courtesy of Sartorius.

asepsis mandates that one must handle materials in such a way that unwanted microorganisms are not introduced during isolation. Aseptic procedures include sterilization of medium, plate streaking, and pour plating.

Incubation Conditions

Aerobic organisms require oxygen for growth and can be cultivated on solid medium in Petri plates or on agar slants (see subsequent text) by placing them in **incubators**. An incubator is an insulated chamber with a thermostatically controlled device to maintain the desired temperature. For example, common soil organisms may be incubated at 25°C or 30°C, whereas those from human or other mammalian sources may be incubated at 37°C (body temperature).

Aerobic organisms grown in broth (liquid medium) culture tend to deplete the oxygen from the medium during active respiration, leading to curtailment of growth. Thus, mechanical **shakers** are sometimes used to agitate the cultures during growth so that more oxygen is stirred in at a constant rate (Figure 5.9). Obligately anaerobic organisms, of course, cannot be grown in the presence of oxygen, and special media and media preparation procedures are needed to cultivate these microbes (see subsequent text and Chapter 6).

Streak Plate Procedure

The classic method for isolation of bacteria is the **streak plate procedure**, which was developed in Robert Koch's laboratory (see Chapter 2). In this technique, a sample of a bacterial community is picked up with a metal wire that has a circular loop at its end (Figure 5.10). The wire is first sterilized by heating it directly in a

(A)

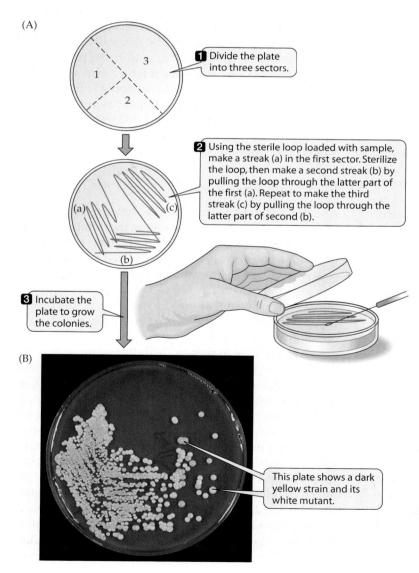

1 Divide the plate into three sectors.

2 Using the sterile loop loaded with sample, make a streak (a) in the first sector. Sterilize the loop, then make a second streak (b) by pulling the loop through the latter part of the first (a). Repeat to make the third streak (c) by pulling the loop through the latter part of second (b).

3 Incubate the plate to grow the colonies.

(B)

This plate shows a dark yellow strain and its white mutant.

Figure 5.10 Streaking procedure
(A) The proper streaking procedure for obtaining isolated colonies. (B) Results of a three-way streak of *Micrococcus luteus* strains. Courtesy of Jerome Perry.

Bunsen burner flame until it is red hot. The sterile loop is then cooled, for example, by touching it on the sterile liquid or solid medium. By touching the loop to a colony on a plate or immersing the loop in a liquid that contains microbes, such as a sample of lake water or a liquid culture, a mixture of microbes would be collected on the loop. For example, lake water typically contains about 10,000 viable bacteria per milliliter. The loop, which has an approximate volume of 0.01 ml, would thus pick up 100 or more bacterial cells. For identification, the bacteria in the droplet must be separated.

A common method of separation is by **streaking** the loop's contents on the solid surface of an **agar plate**. Streaking over the surface would be accomplished in three successive steps that separate individual cells (see Figure 5.10A).

In the first streak, a loop bearing the sample is spread back and forth in non-overlapping lines to cover about one-third of the agar surface, followed by flaming and cooling of the loop. In the second streak, the sterilized loop would be passed across the first streak to obtain some of the cells and streaked over another third of the plate. A third streak repeats the process to further ensure the separation of bacteria.

In this streaking procedure, individual microorganisms from the original sample become spaced farther and farther apart, and the numbers are reduced in successive sectors. After streaking, the plate would then be incubated at the appropriate temperature to permit growth of the organisms.

Individual microorganisms that are spaced apart by the streaking process can multiply to produce a **colony** (or **clone**) made up of cells derived from a single parent. The colony represents an **isolated strain** of the microorganism (see Figure 5.10B). To ensure purity of a desired species or colony type, however, it is necessary to repeat the streaking process one or more times on a freshly prepared plate. A pure culture is obtained when only one type of colony grows on the plate. This pure culture is referred to as a **bacterial strain**.

Spread Plate Procedures

Spread plates may also be used for isolation of bacteria (Figure 5.11). A small volume (typically about 0.1 ml) of the sample is placed on the sterile agar surface in a Petri plate. Then it is spread over the surface with a sterile bent glass rod, which resembles a "hockey stick" (see Figure 5.11A). This process distributes the cells evenly on the surface, so if they are sufficiently diluted they grow as discrete colonies (see Figure 5.11B). This procedure is less restrictive than pour plating (discussed in subsequent text) because the cells are not exposed to the elevated temperature of the molten agar.

Figure 5.11 Spread plate and pour plate procedures ▶ (A) In the spread plate method, individual colonies are obtained by spreading a liquid sample of microorganisms over the agar surface. (B) Results of a spread plate inoculated with diluted nasal mucus, showing four *Staphylococcus* species: the small gray colonies are *Staphylococcus epidermidis*; the small white colonies, *Staphylococcus capitis*; the large white colonies, *Staphylococcus hemolyticus*; and the single yellow colony, *Staphylococcus aureus*. Contrast this with results from the pour plate method (C). The colonies on, or "trapped" within, the agar can be separated after incubation. Photos courtesy of Jerome Perry.

Some organisms may grow poorly or not at all on an agar surface. There may be several reasons for this. For example, the composition or pH of the medium or the incubation conditions may not be satisfactory for their growth. Some organisms may be fragile or sensitive to drying. Some bacteria cannot grow as separate colonies, but require closely associated bacteria or other organisms to provide biochemicals that may not be provided in the medium. Organisms that require the presence of other organisms for growth are referred to as **symbiotic bacteria**. They grow very well in **consortia** (cultures containing more than one species), but not as pure cultures. For these reasons, it is estimated that 99% of bacterial species have not been isolated and grown under laboratory conditions.

Pour Plate Procedure

Another method for separating mixtures of microorganisms, and one often used in counting the number of each colony type in mixtures, is the **pour plate procedure**, in which a soil or water sample is mixed with a molten agar medium *before* it solidifies. After a thorough mixing, the molten agar (cooled to 45°C to 48°C) is poured into a Petri plate and allowed to solidify. In this procedure, the colonies are distributed uniformly throughout the solid medium (see Figure 5.11C). This procedure has one disadvantage in that the high temperature needed to keep the medium molten may actually kill some species. For example, bacteria from marine samples or temperate zone lakes rarely, if ever, encounter temperatures greater than 25°C to 30°C and may be killed by brief exposure to 45°C. Most species, however, are not harmed at this temperature. Note that cell plating and colony isolation of oxygen-sensitive anaerobic microorganisms on Petri plates may be performed inside an anaerobic glove box (Figure 5.12). The Petri plates may then be incubated within the hood or, alternatively, placed in anaerobic jars and sealed prior to removal from the box.

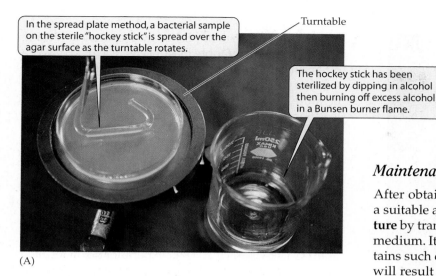

In the spread plate method, a bacterial sample on the sterile "hockey stick" is spread over the agar surface as the turntable rotates.

Turntable

The hockey stick has been sterilized by dipping in alcohol then burning off excess alcohol in a Bunsen burner flame.

(A)

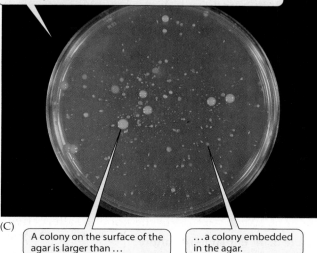

(B)

Four types of *Staphylococcus* colonies are identifiable.

This plate has been produced by the pour plate method: a culture of microorganisms in liquid was added to melted agar, and the mixture was poured onto the Petri plate, allowed to solidify, then incubated.

(C)

A colony on the surface of the agar is larger than …

…a colony embedded in the agar.

Maintenance of Stock Cultures

After obtaining a pure culture, it can be transferred to a suitable agar medium and maintained as a **stock culture** by transfer at suitable intervals to a newly prepared medium. It is of utmost importance that a researcher retains such cultures in an axenic state, as faulty research will result from **contaminated cultures** (cultures that contain a mixture of two or more organisms).

It is often convenient to store stock cultures on an agar surface in a test tube. To increase the surface area for growth in the tube, the agar medium is allowed to solidify at an angle, resulting in a **slant** (Figure 5.13). Cultures transferred to a slant would be incubated and after growth placed in a refrigerator maintained at 4°C to 5°C. At this temperature, microbial cultures are generally viable for months, but this can vary from species to species.

Longer-term maintenance is accomplished either by **lyophilization** (freeze-drying) or by cryogenic treatment. The latter method involves freezing bacterial or archaeal cells in glycerol at exceedingly low temperatures (–70°C or lower) in special freezers or in liquid

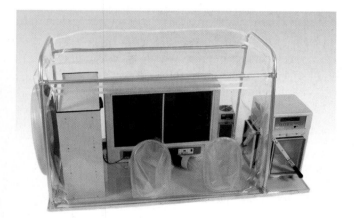

Figure 5.12 Oxygen-free glove box for manipulation of strictly anaerobic environmental samples and cell cultures
Environmental samples are introduced into the chamber through an airlock that is evacuated and flushed with oxygen-free gas. Once inside, the samples can be manipulated by access through the flexible gloves. Courtesy of Coy Labs.

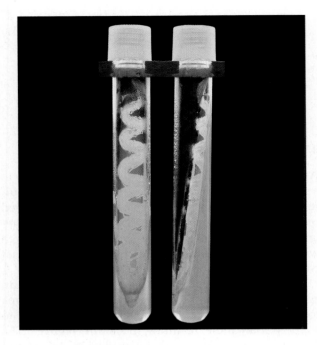

Figure 5.13 An agar slant
Two agar slants of *Kocuria rhizophila* (*Micrococcus luteus*).
Courtesy of D. McIntyre.

nitrogen. Lyophilized and frozen cultures kept in freezers retain viability for extended periods—most cultures can be stored for 10 years or longer and therefore require minimal maintenance. This approach, which keeps the cells in an inactive state, minimizes genetic drift due to the accumulation of random mutations over time—a concern after prolonged serial transfer on slants.

SECTION HIGHLIGHTS

A cell culture medium, whether it is a complex or a defined medium, must contain each essential nutrient in order to support growth of an organism. An acceptable environment with respect to pH, ions, and temperature must be provided along with a usable supply of carbon and energy. Finally, application of aseptic technique allows axenic cell culture and long-term storage.

5.4 Isolation of Selected Microbes by Enrichment Culture

The basic requirements for the growth of bacteria and archaea have been described in the previous sections. The isolation of microorganisms with selected nutritional re-

quirements can be accomplished by preparing and using growth media that supply these needs. The physical conditions for growth can be adjusted to select for microorganisms that can grow under specific conditions of temperature, pH, oxygen availability, and osmotic pressure. Isolation of specific types of microorganisms by a combination of nutrient and physical conditions is generally termed **enrichment culture**. Here we will focus on *Bacteria*; the isolation of *Archaea* will be discussed in Chapter 18.

Origins of Enrichment Culture Methodology

The first scientists to apply enrichment culture on a broad scale were Martinus Beijerinck and Sergei Winogradsky (see Chapter 2). Their experimentation resulted in the isolation of a broad array of bacterial types and led to a rational approach to microbial ecology. Beijerinck's basic method was to prepare Petri plates with a mineral salts agar medium; to this medium he added a sample of soil or water taken from diverse environments. Beijerinck then placed crystals of selected organic compounds (carbon source) on the surface of the inoculated agar. After a few days at room temperature, he noted that in virtually every case colonies developed on the agar surface where these organic compounds were added (Figure 5.14). Using this experimental approach,

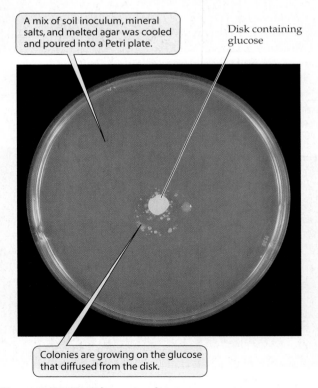

A mix of soil inoculum, mineral salts, and melted agar was cooled and poured into a Petri plate.

Disk containing glucose

Colonies are growing on the glucose that diffused from the disk.

Figure 5.14 Enrichment culture
Microorganisms grown in a culture enriched with glucose.
Courtesy of Jerome Perry.

Beijerinck observed that equivalent conditions applied to soil or water samples obtained from widely separated habitats generally yielded similar bacterial types.

Thus, Beijerinck essentially demonstrated the Darwinian concept of natural selection in a Petri plate. The organism or organisms most capable of development under the specific regimen of pH, temperature, and substrate were the ones that arose in greatest numbers from the soil or water inoculum. **Enrichment culture** is the selection of a specific organism(s) from all those in the inoculum (1 gram of fertile soil generally contains 10^7 to 10^9 microbes).

If one were to set up enrichment cultures under equivalent selected conditions by using soil from North America, eastern Asia, and northern Europe, the organisms obtained from these three soils would be physiologically similar. The characteristics of the organisms from the separate continents would probably not differ any more than two cultures obtained from a single acre of soil.

Applications of Enrichment Culture

The use of bacteria obtained by enrichment was of considerable value in studies that led to the elucidation of both biodegradative and biosynthetic pathways. **Biodegradation**, in this sense, refers to the stepwise conversion of an organic compound to $CO_2 + H_2O$. Although notable exceptions exist, biosynthetic pathways generally approximate the reversal of the biodegradative route. **Biosynthetic pathways** are the chemical reactions that transform a substrate, whether glucose or CO_2, to the various macromolecular components of a cell.

For the purposes of structural analysis and chemical identification, it is generally easier to obtain sufficient quantities of biodegradation intermediates of a molecule than it is to obtain biosynthetic intermediates. Identification of metabolic intermediates was particularly difficult when analytical methods were far less sophisticated than they are today. It should also be noted that during the period when many biosynthetic pathways were elucidated, the genetic manipulation of bacteria was also in its infancy. Genetic mutants that would accumulate intermediates were not readily available. However, the application of these tools has greatly accelerated the study of metabolism and regulation.

Often the goal in isolating microorganisms is to obtain a specific microbial type for practical purposes. For example, thousands of actinomycetes have been isolated over the last 50 or more years in the quest for those that yield useful antibacterial, antiviral, and antitumor agents. These organisms are commonly obtained from soil, compost, fresh water, and the atmosphere. A number of enrichment substrates have been devised for this purpose. One containing low levels (0.1%) of hydrolyzed

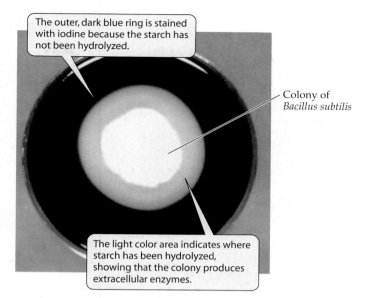

Figure 5.15 Detection of extracellular enzymes
A plate containing starch (1%) was inoculated in the center with *Bacillus subtilis*. After 3 days, an alcohol solution of iodine (which stains starch) was flooded over the surface and the excess was poured off. The technique of incorporating a polymer in the agar can also be effective in enriching cultures for microorganisms that utilize high-molecular-weight, water-insoluble compounds such as cellulose or lipids. Courtesy of Jerome Perry.

casein (animal protein), soytone (plant protein), and yeast extract (autolyzed yeast cells) has proven quite effective. The low level of sugars present in such a medium curtails overgrowth by fast-growing motile genera, such as *Bacillus* and *Pseudomonas*, and permits the slow-growing actinomycetes to form colonies.

Colonies of microbes that have the capacity for extracellular digestion of insoluble high-molecular-weight sugar polymers, lipids, and proteins can easily be selected. To do this, the polymer is incorporated into the agar medium as the carbon source. The surface of the agar is then inoculated with soil or water, and digestion of the polymer is evidenced by a clear area in the agar around the colony (Figure 5.15).

General Enrichment Methods

Both aerobic and anaerobic microorganisms can be isolated following a preliminary enrichment. Samples should be obtained from environmental niches that favor the type of organism sought—soil for aerobes and mud or sediments for anaerobes. Basic conditions for enrichment are described here.

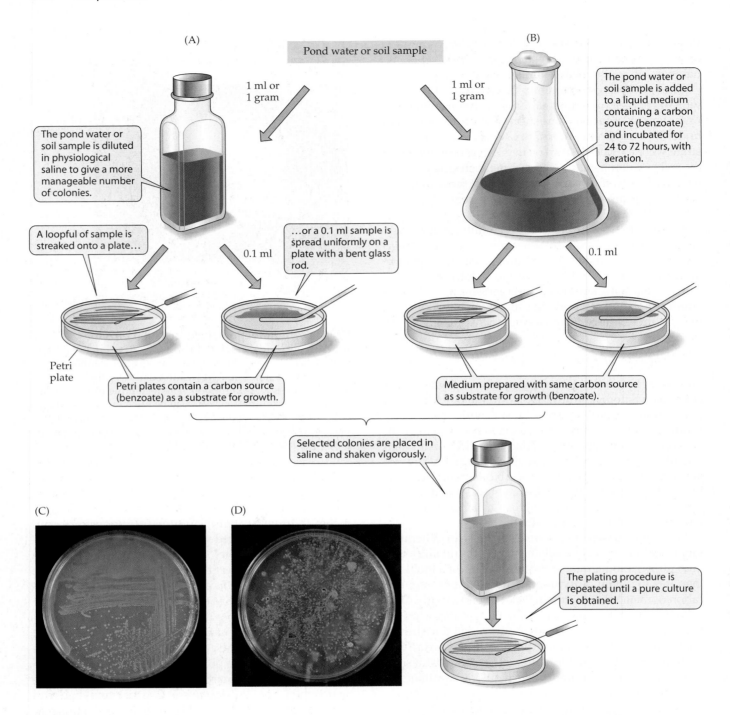

(A)

Pond water or soil sample

(B)

1 ml or 1 gram

1 ml or 1 gram

The pond water or soil sample is diluted in physiological saline to give a more manageable number of colonies.

The pond water or soil sample is added to a liquid medium containing a carbon source (benzoate) and incubated for 24 to 72 hours, with aeration.

A loopful of sample is streaked onto a plate...

...or a 0.1 ml sample is spread uniformly on a plate with a bent glass rod.

0.1 ml

0.1 ml

Petri plate

Petri plates contain a carbon source (benzoate) as a substrate for growth.

Medium prepared with same carbon source as substrate for growth (benzoate).

Selected colonies are placed in saline and shaken vigorously.

(C)

(D)

The plating procedure is repeated until a pure culture is obtained.

Aerobic Enrichment

Two general procedures that can be used to enrich for aerobic bacteria are liquid medium and solid medium (Figure 5.16). A streak from a liquid enrichment for an organism that might use a substrate such as sodium benzoate would appear as shown in Figure 5.16C. A pure culture and mixed culture can be distinguished by their colony appearance on plates (i.e., uniformly shaped and colored colonies of similar size versus diverse colony

Figure 5.16 Isolation of microorganisms with specific nutritional requirements
Isolation of a bacterium that can utilize a specific compound as carbon and energy source. (A) Enrichment on a solid medium. (B) Enrichment in a liquid medium. (C) Plate streaked from a liquid enrichment. The liquid enrichment technique tends to favor rapidly growing organisms. Note that only one type of colony has developed. (D) Spreading samples directly onto the surface of an agar plate yields a variety of microorganisms. Photos courtesy of Jerome Perry.

forms, shapes, and sizes). Direct spreading of a sample on an agar surface would result in the growth of many different colonies (see Figure 5.16D).

A shortcoming of liquid enrichment is a marked tendency for faster-growing organisms, such as pseudomonads, to be favored, and these become dominant on continued transfer to a newly prepared medium. Enrichment on a solid medium will generally yield a greater variety of organisms, provided that one has the patience to wait for slower-growing colonies to appear. Unfortunately, organisms that are present as a small percentage of the total population may be overlooked by any enrichment procedure unless the enrichment is highly specific.

Anaerobic Enrichment

Enrichment for anaerobic bacteria is technically more difficult than enrichment for aerobes. This is particularly true for strict anaerobes that are killed by exposure to even minute levels of oxygen. Samples for enrichment must be handled carefully to preclude exposure to oxygen prior to isolation procedures. Highly effective methods have been developed for handling obligate anaerobes, including the fabrication of entire rooms from which all traces of oxygen can be removed. A technique now in general use involves an oxygen-free **glove box** (see Figure 5.12) for manipulation of strictly anaerobic bacteria. Sample manipulation is performed inside the glove box, which permits one to use techniques gener-

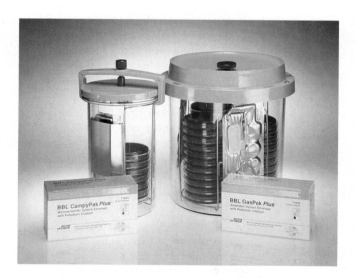

Figure 5.17 Anaerobic jars
Inoculated plates or tubes are placed in the jar along with a catalyst. After sealing the jar, the catalyst is activated to remove all atmospheric oxygen. Because transfer and manipulation of the culture occur outside the jar, this system cannot be used for strict anaerobes. Courtesy of Becton Dickinson Microbiology Systems.

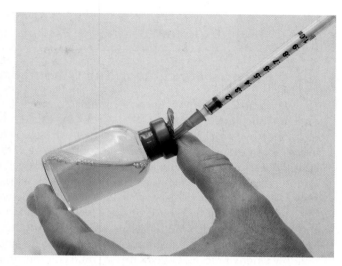

Figure 5.18 Butyl rubber–stoppered serum bottle used for culture of strict anaerobic microorganisms
The culture medium is generally de-oxygenated and then placed into the containers under a stream of oxygen-free nitrogen. An oxygen impermeable rubber stopper is then used to seal the tube or bottle via a metal crimp cap. Courtesy of T. Baskin and R. Gunsalus.

ally applicable to aerobes, for example, preparing dilutions and plating cells, or streaking cells onto Petri plates containing appropriate anaerobic media.

Microorganisms can also be grown in these specialized chambers where all traces of oxygen can be removed from the atmosphere. Alternatively, an **anaerobic jar** is a convenient apparatus for growing anaerobes in the laboratory (Figure 5.17). After tubes or plates are placed in the jar and the jar is sealed, a catalyst is activated to remove all traces of oxygen, which is then replaced with hydrogen gas. These jars are suitable for growth of air-tolerant anaerobes that can be manipulated in the atmosphere but must have an anaerobic environment for growth.

Anaerobes can be easily cultured in sealed tubes and bottles; because the isolation of many strictly anaerobic organisms is prevented by even brief exposure to air. Use of butyl rubber–stoppered serum bottles and tubes is often the method of choice for liquid enrichment and routine cell culture (Figure 5.18). The culture medium is prepared anaerobically and dispensed into tubes or bottles, which are then sealed prior to sterilization. Liquid transfers are accomplished outside a glove box by use of anaerobic gas–flushed syringes.

Many anaerobic photosynthetic bacteria, for example, may be enriched with mud from an anaerobic environment such as the shallow area of a pond. Placing a sample of this mud in a glass-stoppered bottle filled to the top with medium would ensure anaerobiosis. Oxygen diffusion from the outside would be limited, and oxygen

BOX 5.3   *Milestones*

Special Enrichment, The Winogradsky Column

The Winogradsky column is an enrichment culture technique developed by Sergei Winogradsky in the latter part of the nineteenth century (see Chapter 2). A typical Winogradsky column is a long glass tube (1½ × 24 inches), closed at one end, and with about two-thirds of the tube filled with rich mud. The column should be placed in a north window where it receives adequate but not intense light. It is a microcosm where one can follow growth and succession in an anaerobic environment.

Organic-rich mud from a shallow area of a pond is a suitable source of mud for a column. Before placing the mud in the column, a few grams of calcium carbonate and calcium sulfate can be incorporat-ed. In practice, it is preferable to mix the calcium sulfate in the mud that will occupy the lower one-fourth of the column along with some starch, cellulose powder, or shredded filter paper. These carbon sources can be varied depending on the imagination of the individual preparing the column. There should be no air pockets in the mud column, and any that form can be disrupted with a glass rod. The mud column is topped with a layer of pond water.

The appearance of the column after a few weeks would be as illustrated. Anaerobic bacteria ferment the cellulose or starch at the bottom to hydrogen, organic acids, and alcohols. These would be utilized as substrate by sulfate-reducing bacteria, and sulfate would serve as terminal electron acceptor. This would cause the formation of hydrogen sulfide, resulting in an H_2S gradient from the bottom of the column upward. An O_2 gradient would occur from the top downward. The bottom area of the column becomes black due to formation of metal sulfides. Restricting calcium sulfate to the lower area curtails an excessive production of black precipitate throughout. The green sulfur bacteria develop immediately above the dark area because of their tolerance for hydrogen sulfide. The purple sulfur bacteria are less tolerant and appear in the area above the green sulfur bacteria. The purple nonsulfur bacteria grow nearer the top in the absence of sulfide, and they may tolerate the low levels of O_2 present. Aerobic microorganisms, including cyanobacteria and algae, grow in the water layer. One would find anaerobic bacteria growing throughout and some sulfide oxidizers such as *Beggiatoa* and *Thiothrix* growing in the upper area. Samples can be removed periodically with a length of glass tubing narrowed at the end. Organisms can be isolated from these samples by aerobic or anaerobic techniques described in this chapter.

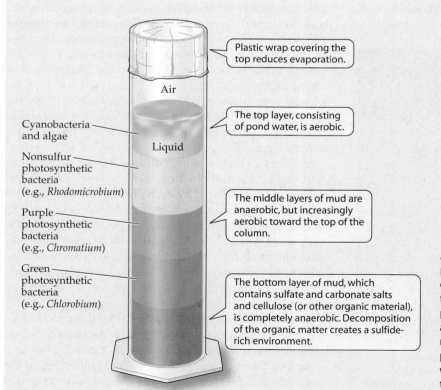

Cyanobacteria and algae

Nonsulfur photosynthetic bacteria (e.g., *Rhodomicrobium*)

Purple photosynthetic bacteria (e.g., *Chromatium*)

Green photosynthetic bacteria (e.g., *Chlorobium*)

Plastic wrap covering the top reduces evaporation.

Air

Liquid

The top layer, consisting of pond water, is aerobic.

The middle layers of mud are anaerobic, but increasingly aerobic toward the top of the column.

The bottom layer of mud, which contains sulfate and carbonate salts and cellulose (or other organic material), is completely anaerobic. Decomposition of the organic matter creates a sulfide-rich environment.

A Winogradsky column. Anaerobic decomposition of organic matter in the bottom of the cylinder creates an anaerobic sulfide-rich environment. The green and purple sulfur bacteria are present immediately above the dark sulfide area. The purple nonsulfur bacteria would be above the purple sulfur bacteria. Cyanobacteria would grow at the top in the area exposed to air.

TABLE 5.8	Enrichment conditions for chemoheterotrophic aerobic or facultative aerobic organisms		
Substrates (Carbon and Energy Source)	**Inoculum**	**Special Conditions**	**Organism Favored**
Ethanol	Soil	N_2 as nitrogen source	*Azotobacter* spp.
Uric acid	Pasteurized soil (80°C for 15 min.)		*Bacillus fastidiosus*
Glucose	Pasteurized soil		*Bacillus* spp.
Casein + thiamine	Pasteurized soil	Add urea at pH 9.0	*Bacillus pasteurii*
Glucose + yeast extract	Soil	Incubate at 60°C	*Bacillus stearothermophilus*
Nutrient broth	Pasteurized soil	10% NaCl, 0.5% $MgCl_2$	*Sporosarcina halophila*
Propane	Soil	Gas 50/50 in air	*Mycobacterium* spp.
n-Hexadecane	Soil	Incubate at 60°C	*Bacillus thermolevorans*
Filter paper	Soil		*Cytophaga* spp.
Chitin	Soil		Actinomycetes

would be removed by other organisms living in the consortium. The source of CO_2 for autotrophic growth would be bicarbonate added to the growth medium. The organism would utilize light as an energy source and glycerol as an electron donor for CO_2 assimilation. Such an enrichment could also be done by obtaining the inoculum from a Winogradsky column, as illustrated in Box 5.3.

Enrichment for Specific Metabolic Types

The foregoing discussion outlined general methods for isolating microorganisms. Proper manipulation of these conditions can lead to the isolation of an organism that will utilize a selected substrate. This was illustrated in Figure 5.16 in the isolation of a bacterium that utilized benzoic acid as substrate. More restrictive selective conditions can be established that will lead to the isolation of a microorganism of a distinct metabolic type such as chemoheterotrophs, chemoautotrophs, photoautotrophs, and photoheterotrophs. Some of these selective methods are outlined here.

Chemoheterotrophic Aerobic Bacteria

Chemoheterotrophic or facultative aerobic bacteria can be isolated from soil or water by using a variety of substrates and special conditions of atmosphere, temperature, and pH (Table 5.8). It should be emphasized that a soil or water inoculum will contain a great variety of microorganisms, and many of these can grow under primary enrichment conditions. They grow on products of, or in association with, an organism metabolizing the enrichment substrate. Obtaining a pure culture of a desired organism may require careful and continued restreaking on the appropriate medium. Such procedures are the general aspects of enrichment culture. To actually obtain specific types of microorganisms, more involved meth-

ods must be followed (see the Suggested Reading list at the end of the chapter).

Many distinctly different bacterial species can fulfill their nitrogen requirement by fixing atmospheric N_2 (see Table 5.5). For the isolation of *Azotobacter* or any nitrogen-fixing organisms, an inorganic or organic nitrogen source (such as NH_4^+ and NO_3^- or an amino acid) must be eliminated from the medium. Eliminating a fixed nitrogen source from the medium would select for microorganisms that can obtain this key element from the atmosphere. The metal molybdenum must be present in such enrichments because it is a constituent of a cofactor for the enzyme nitrogenase that is involved in nitrogen fixation.

Chemoheterotrophic Anaerobic Bacteria

Chemoheterotrophic anaerobic bacteria and archaea can be isolated from a variety of sources (Table 5.9). Many anaerobes are killed by the presence of even very low levels of oxygen, and these microorganisms require special precautions in their handling.

To isolate nitrogen-fixing anaerobes, several different substrates could be used (see Table 5.9). With starch as the enrichment substrate and pasteurized soil as inoculum, the organism most likely isolated would be *Clostridium pasteurianum*. This anaerobe forms heat-resistant endospores that readily survive the pasteurization process and germinate and grow on the starch medium.

Chemoautotrophic Bacteria and Archaea

To isolate chemoautotrophs, the carbon source is limited to carbon dioxide, but any of the following inorganics could be the source of energy: NH_4^+, NO_2^-, S^0, HS^-, Fe^{2+}, or H_2 (Table 5.10). Probably the most important factor in the isolation of the nitrifying bacteria (organisms that oxidize $NH_4^+ \rightarrow NO_2^- \rightarrow NO_3^-$) is patience. These are

TABLE 5.9	Enrichment conditions for chemoheterotrophic anaerobic *Bacteria* or *Archaea*		
Substrates (Carbon and Energy Source)	**Inoculum**	**Special Conditions**	**Type of Organism**
Sugars + yeast extract	Plant material	pH 5–6	Lactic acid bacteria
Mixed amino acids	Pasteurized soil		*Clostridium* spp.
Starch	Pasteurized soil	N_2 as nitrogen source	*Clostridium pasteurianum*
Uric acid + yeast extract	Pasteurized soil	pH 7.8	*Clostridium acidiurici*
Organic acids	Pond mud	Added SO_4^{2-}	*Desulfovibrio* spp.
Organic acids	Soil	Added NO_3^-	Denitrifying bacilli and pseudomonads
Methanol	Rumen fluid		*Methanosarcina* spp.
Lactate + yeast extract	Swiss cheese		*Propionibacterium* spp.

slow-growing organisms, and visible colonies on Petri plates can take from 1 to 4 months to appear. Enrichment and isolation of the nitrifiers is generally carried out by **serial dilution** of rich soil (for an example of serial dilution, see Figure 5.19). Growth of the ammonia (NH_4^+) oxidizers is measured by chemically analyzing for an increase in nitrite (NO_2^-) concentration in the medium. Growth of those that oxidize nitrite to nitrate (NO_3^-) is measured by adding measured amounts of nitrite and then determining the amount that disappears.

The sulfur-oxidizing chemoautotrophic bacteria play a key role in the conversion of reduced sulfur compounds (S^0, HS^-) to sulfate (SO_4^{2-}). These bacteria are present in a wide range of habitats, from pH 1 to 9, and many species are thermophilic ("heat loving").

The hydrogen-oxidizing aerobic bacteria do not form a cohesive taxonomic unit, and the ability to grow using hydrogen (H_2) as the energy source is an attribute of microorganisms in many of the bacterial phyla. Enrichment is rather straightforward because these organisms grow under an atmosphere of hydrogen, carbon dioxide, and oxygen. The hydrogen-oxidizing bacteria can be isolated by sprinkling soil on an agar surface and then incubating under the proper $CO_2/H_2/O_2$ atmosphere.

Phototrophic Bacteria

The major requirement for the isolation of phototrophic bacteria is a constant source of light at the appropriate wavelength(s). Other conditions of enrichment include an electron donor, minerals, and availability of CO_2 (Table 5.11). Provided that light is available, photosynthetic bacteria can flourish under a broad range of conditions. Cyanobacteria, for example, grow in some very harsh natural habitats: hot springs, Antarctic lakes, deserts, and areas of high salinity. They are the photosynthetic symbiont with fungi in many lichen associations and are common inhabitants of terrestrial, marine, and freshwater habitats. As the cyanobacteria produce O_2 during photosynthesis, they are all aerobic, and isolation can be accomplished in vessels exposed to air.

The green and purple bacteria do not produce O_2 during photosynthesis, and they grow under anaerobic conditions. Nonsulfur purple bacteria can be isolated from mud or water samples from ponds, ditches, and shores of eutrophic lakes (lakes rich in nutrients but low in oxygen). Nonsulfur purple bacteria utilize reduced electron donors other than sulfur compounds, such as hydrogen (H_2). The mud from the bottom of a shallow eutrophic lake may contain as many as 1 million purple nonsulfur

TABLE 5.10	Enrichment for chemoautotrophic *Bacteria* or *Archaea* using CO_2 as carbon source[a]		
Energy Source	**Inoculum**	**Special Conditions**	**Organism Obtained**
H_2	Soil or water	Aerobic	Hydrogen-utilizing bacteria
	Rumen fluid	Anaerobic	Methanogens
NH_4^+	Soil or water	Aerobic	*Nitrosomonas* spp.
NO_2^-	Soil or water	Aerobic	*Nitrobacter* spp.
$S_2O_3^{2-}$	Soil or water	Anaerobic + KNO_3	*Thiobacillus denitrificans*
Fe^{2+}	Estuarine mud	Aerobic, pH 2.5	*Thiobacillus ferrooxidans*

[a]Bicarbonate ($NaHCO_3$) may serve as source of CO_2.

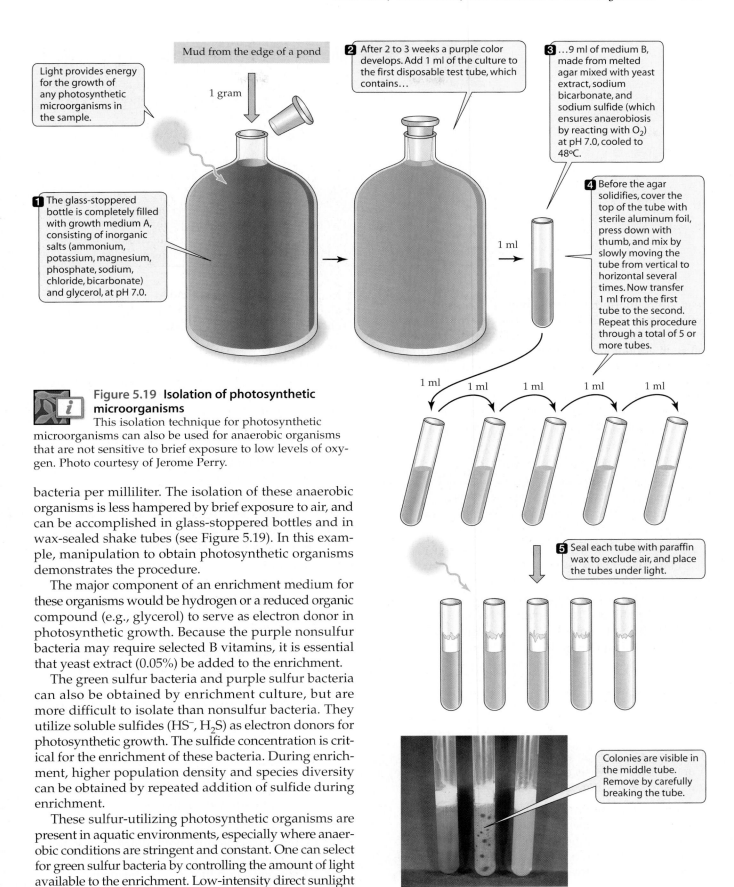

Figure 5.19 Isolation of photosynthetic microorganisms
This isolation technique for photosynthetic microorganisms can also be used for anaerobic organisms that are not sensitive to brief exposure to low levels of oxygen. Photo courtesy of Jerome Perry.

bacteria per milliliter. The isolation of these anaerobic organisms is less hampered by brief exposure to air, and can be accomplished in glass-stoppered bottles and in wax-sealed shake tubes (see Figure 5.19). In this example, manipulation to obtain photosynthetic organisms demonstrates the procedure.

The major component of an enrichment medium for these organisms would be hydrogen or a reduced organic compound (e.g., glycerol) to serve as electron donor in photosynthetic growth. Because the purple nonsulfur bacteria may require selected B vitamins, it is essential that yeast extract (0.05%) be added to the enrichment.

The green sulfur bacteria and purple sulfur bacteria can also be obtained by enrichment culture, but are more difficult to isolate than nonsulfur bacteria. They utilize soluble sulfides (HS^-, H_2S) as electron donors for photosynthetic growth. The sulfide concentration is critical for the enrichment of these bacteria. During enrichment, higher population density and species diversity can be obtained by repeated addition of sulfide during enrichment.

These sulfur-utilizing photosynthetic organisms are present in aquatic environments, especially where anaerobic conditions are stringent and constant. One can select for green sulfur bacteria by controlling the amount of light available to the enrichment. Low-intensity direct sunlight

TABLE 5.11	Enrichment for photosynthetic microorganisms that utilize light as energy source	
Special Conditions	**Source of Inoculum**	**Organism**
Aerobic		
N_2 as nitrogen source	Surface water or soil	Nitrogen-fixing cyanobacteria
NH_4^+ as nitrogen source	Surface water or soil	Cyanobacteria and eukaryotic algae
Anaerobic		
Glycerol	Mud from pond edge	Purple or green nonsulfur bacteria
H_2S (low levels)	Sulfide-rich mud	Green sulfur bacteria
H_2S (high levels)	Mud from pond edge	Purple sulfur bacteria

(e.g., from a north-facing window) or a tungsten light of low intensity (5 to 2000 lux) favors green sulfur bacteria, as these intensities are not sufficient to support growth of most other phototrophic bacteria. The purple sulfur bacteria can be enriched by higher light intensities.

SECTION HIGHLIGHTS

Different strategies can be devised for enriching and isolating different classes of microbial species based on their metabolic abilities with respect to carbon and energy acquisition. The conditions encountered at the site of the inoculum's collection, that is, the natural habitat, are considered in designing a culture medium. The successful enrichment and isolation of a particular species will also rely on proper incubation conditions. Cell culture conditions may then be optimized to study pure culture isolates (nutritional requirements, plus other environmental parameters including light, oxygen, temperature, and ionic environment).

SUMMARY

- *Bacteria* and *Archaea* generally differ from one another on the basis of the compounds they utilize as a carbon and/or energy source.

- The **substrates** utilized by *Bacteria* and *Archaea* as sources of carbon include all constituents of living cells and many synthetic chemicals. This microbial biodegradation can occur over a broad range of pH, temperature, salinity, and other physical parameters.

- Microorganisms can be divided into four nutritional groups: **photoautotrophs**, **photoheterotrophs**, **chemoautotrophs**, and **chemoheterotrophs**.

- Bacteria, regardless of species or growth substrate, are composed of the same **elements**. The monomers from which the macromolecules are synthesized are equivalent. A growth medium must, in one way or another, supply all elements that make up a microbial cell.

- Some microorganisms can synthesize all cellular components from CO_2, whereas others require a source of amino acids, vitamins, and nucleotidyl bases.

- Many bacterial and archaeal species can "**fix**" **atmospheric N_2** into cellular material. N_2 fixation is not known among the eukaryotes.

- **Phosphorus** is present in microorganisms in relatively low amounts (~3% of dry weight), but it is an essential element in nucleic acids and in energy generation.

- The **growth factors** (needed in small amounts) most commonly required by microorganisms are B vitamins, amino acids, and nucleotidyl bases.

- A **trace element** is one required in very small amounts (traces). Among these are Cu^{2+}, Fe^{2+}, Zn^{2+}, Ni^{2+}, and Co^{2+}.

- Microorganisms need specific nutrient uptake systems, as they generally live in environments where many nutrients are present at low levels. The three major systems for nutrient uptake are **diffusion** (**simple** and **facilitated**), **active transport**, and **group translocation**.

- **Facilitated diffusion** does not require energy input, and cannot concentrate nutrients internally to a greater concentration than exists outside the cell.

- **Active transport** requires energy to move a substance against a concentration gradient. The energy may come from ATP hydrolysis or an ion gradient.

- Ion gradients can drive uptake by three mechanisms—**uniport**, **symport**, and **antiport**.

- **Group translocation** results in chemical alteration of the compound transported. Sugars may be transported into the cell via this process. Group translocation is the most elaborate of the transport systems and involves a number of enzymes in uptake of a single solute.

- Iron is required by all organisms, and aerobic microorganisms may have specific chelators called **siderophores** for iron uptake.

- **Complex media** and **defined media** can be used to culture microbes. They both contain **essential nutrients** but may or may not contain **nonessential nutrients**.

- **Enrichment culture** is a method used for isolating microorganisms that grow on a specific nutrient and/or under selected physical conditions. The organism obtained from an environmental sample through this procedure will have the ability to grow under the specific enrichment conditions employed.

- Most microorganisms are maintained in collections as **pure (axenic) cultures**. All studies with pure cultures must be done under aseptic conditions.

- Pure cultures can be obtained by **streaking, pour plate**, or **spread plate** techniques.

- Microorganisms can be maintained for extended periods by **lyophilization** or by placing them at exceedingly low temperatures in freezers or in **liquid nitrogen**.

 Find more at www.sinauer.com/microbial-life

REVIEW QUESTIONS

1. What is metabolic diversity? How did it come about? How does this relate to the role of microbes in nature?

2. Microbes survive and actually thrive in virtually every environmental niche on earth. Describe how this relates to the perpetration of life on earth.

3. What are the basic requirements in the development of a growth medium? Why is each constituent or component added? How does this relate to the composition of a bacterial cell?

4. Define "growth factor." Why do some organisms require them?

5. How would enrichment culture be used to isolate an organism capable of growth with *p*-aminobenzoic acid as sole source of carbon?

6. Why are bacteria sometimes utilized in studying nutrition rather than using higher organisms such as rats or other animals? How does this relate to comparative biochemistry?

7. Where would one obtain samples to be used in isolating various photosynthetic bacteria? Would the source for cyanobacteria differ from that for green sulfur photosynthetic bacteria? Why?

8. What is a microbial community? Are microbial communities prevalent in nature?

9. What is the streak plate method, and how does one employ this to obtain a pure culture?

10. Define axenic culture, consortia, and a bacterial strain.

11. How does a pour plate differ from a streak plate? What are the advantages of each?

12. What are the two major methods for long-term storage of bacteria?

SUGGESTED READING

Atlas, R. 2004. *Handbook of Microbiological Media*. 3rd ed. Boca Raton, FL: CRC Press.

Burage, R., R. Atlas, D. Stahl and G. Geesey. 1998. *Techniques in Microbial Ecology*. New York: Oxford University Press.

Dworkin, M., S. Falkow, E. Rosenberg, K-H. Schliefer and E. Stackebrandt, eds. 2000. *The Prokaryotes: An Evolving Electronic Resource for the Microbiological Community*. 3rd ed. New York: Springer-Verlag. Release 3.7, latest update December 2001.

Gerhardt, P., ed. 2007. *Methods for General and Molecular Bacteriology*. 3rd ed. Washington, DC: ASM Press.

Gottschal, J. C., W. Harden and R. A. Prins. 1992. *Principles of Enrichment, Isolation, Cultivation, and Preservation of Bacteria*. In A. Balows, H. G. Truper, M. Dworkin, W. Harden and K. H. Schleifer, eds. *The Prokaryotes*. 2nd ed. New York: Springer-Verlag.

Neidhardt, F. C., J. L. Ingraham and M. Schaechter. 1990. *Physiology of the Bacterial Cell*. Sunderland, MA: Sinauer Associates, Inc.

Norris, J. R. and D. W. Ribbons, eds. 1969. *Methods in Microbiology*. Vol. 3B. New York: Academic Press.

Staley, J. T. and A-L. Reysenbach. 2002. *Biodiversity of Microbial Life*. New York: Wiley-Liss.

Veldkamp, H. 1970. *Enrichment Cultures of Prokaryotic Organisms*. In J. R. Norris and D. W. Ribbons, eds. *Methods in Microbiology*. Vol. 3A. New York: Academic Press.

- ◆ Introduce alternative modes of cell division.

- ◆ Discuss growth of cell populations and ways to measure growth.

- ◆ Evaluate effects of environmental conditions on cell growth and viability.

- ◆ Investigate cell growth in batch versus continuous culture.

6

Microbial Growth

*The paramount evolutionary accomplishment of bacteria as a
group is rapid, efficient cell growth in many environments.
Bacteria grow and divide as rapidly as the environment permits.*
—J. L. Ingraham, O. Maaløe, F. C. Neidhardt, Growth of the
Bacterial Cell, 1983

An overview of the nutritional requirements for the growth of representative bacterial and archaeal cultures was presented in Chapter 5. The indispensable need for carbon, oxygen, nitrogen, hydrogen, sulfur, and various cations and anions was discussed. Methods were also outlined for the isolation of specific autotrophic and heterotrophic microorganisms. When the nutrients and physical conditions required by a specific organism are met, the organism will grow (see chapter opening quote).

In microbiology, the term *growth* generally refers to an increase in the number of cells in a population, or **population growth**. An individual bacterial or archaeal cell may increase in size, and growth in this sense would be called **cell growth**. During population growth, a bacterial or archaeal cell divides to generate two progeny. Each of the progeny receives a genome that is a precise copy of the genetic information in the parent cell (see Chapter 13).

A microorganism is generally considered viable only if it is able to reproduce. In fact, the test most often used for viability is to place the organism under the growth condition for that species, incubate, and examine the culture for visible growth after a suitable time. It should be noted that growth, either as a process or a way of acquiring cells, is a major method by which microbes are studied. However, we must be aware that most of the microorganisms in nature have not yet been cultivated in the laboratory. These organisms are most assuredly viable, and our inability to grow them is due to our lack of understanding of their growth requirements.

6.1 Cell Division of *Bacteria* and *Archaea*

Prokaryotes maintain their shapes during the process of **asexual reproduction**—a process in which a single organism divides to produce two progeny. For unicellular prokaryotes there are two ways in which this may be accomplished, either by binary transverse fission or by budding.

Binary Transverse Fission

The most common type of bacterial cell division is **binary transverse fission**. In this process, the cell (which may be a coccus, rod, spirillum, or other shape) elongates as growth occurs along its longitudinal axis (Figure 6.1). When a certain length is reached, a **septum** (wall structure) is produced along the transverse axis of the cell, midway between the cell ends. When the septum has formed completely, the two resulting cells become separate entities. This process is called binary fission because two cells are produced by a division or "splitting" of one original cell. The process is described as transverse because the septum that separates the two new cells is formed along the transverse, or short, axis of the original cell.

In binary transverse fission, DNA replication precedes septum formation. The two resulting cells are mirror images of one another. Analyses of cell wall components of dividing cells indicate that the chemical constituents of the original "mother" cell wall are equally shared in the cell walls of the two "daughter" cells. The cell division (septation) process is not well understood at the molecular level. However, one key protein involved in this process is a cytoskeletal protein called FtsZ. At the onset of division, numerous FtsZ molecules assemble at the inner surface of the cytoplasmic membrane to form a ring, called the Z ring, where septation will commence (see Figure 6.1C). Localization of FtsZ to the central axis of the cell occurs by interactions with a number of cell division proteins, including MinC and MinE, that together function to identify and mark the plane of cell division. By an energy-requiring process involving GTP hydrolysis and other Fts accessory proteins (FtsA, FtsI, FtsE, and FtsK), the Z ring, along with the cell envelope to which it is attached, contracts until the membrane is sealed, giving rise to two distinct compartments. Following closure of the remaining parts of the cell envelope, the two daughter cells then may separate. The events of division are controlled in a temporal and spatial fashion whereby all molecules of the cell are doubled in number and par-

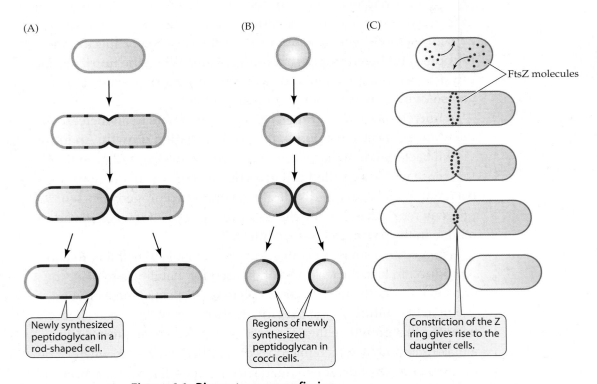

(A)   (B)   (C)

FtsZ molecules

Newly synthesized peptidoglycan in a rod-shaped cell.

Regions of newly synthesized peptidoglycan in cocci cells.

Constriction of the Z ring gives rise to the daughter cells.

Figure 6.1 Binary transverse fission
Typical binary transverse fission in (A) a rod-shaped bacterium and (B) a coccus. (C) The formation of the Z ring at the cell axis is indicated by the red FtsZ molecules.

titioned equally to the daughter cells. In multicellular prokaryotes with trichomes (see Figure 4.27), the organism divides by transverse fission and the trichome ultimately separates into two separate trichomes.

Budding

Budding, or **bud formation**, is a less common form of cell division among prokaryotic organisms. As in binary transverse fission, this is an asexual division process that results in the formation of two cells from the original cell. In the budding process, however, a small protuberance, a **bud**, is formed on the cell surface of the mother cell. The protuberance enlarges as growth proceeds and eventually becomes sufficiently large and mature to separate from the mother cell (Figure 6.2). Examples of microbes that reproduce by budding include *Nitrosomonas* sp., *Rhodobacter* sp., and *Ancalomicrobium adetum*.

Budding differs in several ways from binary transverse fission. During binary transverse fission, the symmetry of the cell with respect to the longitudinal and transverse axes is maintained throughout the entire process (see Figure 6.1). This results in the mother cell producing two daughter cells and losing its identity in the process. In the budding process, however, symmetry with respect to the transverse axis is not maintained during division (see Figure 6.2). In further contrast to binary transverse fission, most of the new cell wall components are used in the synthesis of the bud instead of being divided equally between the two progeny cells. The result is that, during budding, the mother cell produces daughter cells while retaining its identity generation after generation. Whether there is a limit to the number of buds a mother cell can produce during its existence is as yet unknown.

Polar growth, like budding, involves asymmetric synthesis and assembly of newly formed cellular materials. Therefore, these modes of cell reproduction must be unique and inherently different from the program for binary fission, where each daughter cell is composed of half old cell material and half newly synthesized material. A cell newly formed by polar growth receives a significantly larger fraction of new cell material than the mother cell (see Figure 6.2).

Fragmentation

Another type of cell division process occurs in the mycelial bacterial group, the actinobacteria or streptomycetes (see Chapter 20). These organisms have "multinucleate" filaments that lack septa between the cells; these filaments are

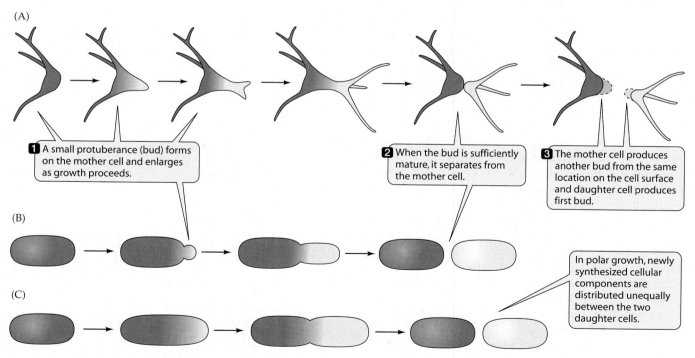

1 A small protuberance (bud) forms on the mother cell and enlarges as growth proceeds.

2 When the bud is sufficiently mature, it separates from the mother cell.

3 The mother cell produces another bud from the same location on the cell surface and daughter cell produces first bud.

In polar growth, newly synthesized cellular components are distributed unequally between the two daughter cells.

Figure 6.2 Budding and polar growth
(A) Bud formation in *Ancalomicrobium adetum*, a prosthecate bacterium. The budding process can be repeated as long as nutrients are available for growth. (B) Bud formation in *Pirella*. (C) Polar growth in *Nitrobacter* sp. Dark shading indicates pre-existing cell material, whereas light areas represent newly synthesized cell components.

referred to as **coenocytic** and are analogous to the filaments found in some fungi. Some bacteria of this type undergo a multiple fission process. Actinobacteria undergo a **fragmentation** in which the filament develops septations between the nuclear areas, resulting in the simultaneous formation of numerous unicellular rods. In an analogous fashion, some cyanobacteria produce numerous smaller daughter cells called **baeocytes** from a single large mother cell during cell division (see Chapter 21).

SECTION HIGHLIGHTS

The means of cell division varies among species of the *Bacteria* and the *Archaea*. However, the processes share many common features, and involve a series of specific temporal and spatial events. These programs coordinate the generation of progeny cells that retain the characteristic morphology of the species.

6.2 Population Growth

Population growth is the culmination of a complex series of biochemical events that are driven by light or chemical energy. This energy is utilized to synthesize or assimilate monomers, and these are in turn assembled into macromolecules. During the course of growth and division, a bacterial or archaeal cell may synthesize an estimated 1,800 different proteins, more than 400 different RNA molecules, a complete copy of the genomic DNA, new cytoplasmic membrane, and, where necessary, sufficient cell wall to surround the newly formed cell.

All the synthetic processes involved in population growth are regulated and integrated to produce duplicate cells, and this occurs in a short period called the **generation time (g)**. Generation time is defined as the time required for a population of cells to double in number.

During binary fission, a bacterial cell generally grows in size as a prelude to cellular division. Thus, this phase of the cell cycle (Box 6.1) is considered cell growth (increase in mass), whereas cellular division results in an increase in the number of cells. When a bacterium is placed in a suitable growth medium, an initial adjustment period, termed the **lag phase**, occurs, and during this time there is no increase in cell number. This phase is followed by an exponential increase in the bacterial population. This increase in cell number or cell mass per unit time occurs at a constant and reproducible rate for a given organism in a selected medium and is called the **growth rate**. Because growth of a population is **exponential** (that is, 1 cell → 2 cells → 4 → 8, etc.), it can be

depicted on a logarithmic scale as a straight line or may be depicted on an arithmetic scale as an exponential curve (Figure 6.3). The generation time for a species is the time required for the population to double during the exponential phase of growth. The generation time (also called doubling time) is usually determined under what would be considered optimal conditions for growth of that species (Box 6.2).

Typical generation times range from only 20 minutes to more than 24 hours (Table 6.1). It is apparent from these values that some organisms divide much more rapidly than others. An organism such as *Escherichia coli* that can reproduce in 20 minutes could, starting with one cell and under continuously optimal growth conditions, yield 4.7×10^{21} progeny in 24 hours (Box 6.3). In 13 hours, a microbe with a generation time of 20 minutes can produce 1×10^{12} cells, approximately the number of cells in the entire human body! A single cell of a slower-growing soil organism such as *Bacillus subtilis* (generation time 30 minutes) would generate 1×10^{12} progeny in about 20 hours. These progeny numbers would apply only if an unlimited supply of all nutrients

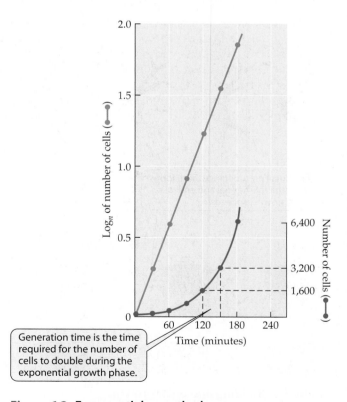

Generation time is the time required for the number of cells to double during the exponential growth phase.

Figure 6.3 Exponential growth phase
Exponential growth of a population of bacteria, plotted on logarithmic (left axis) and arithmetic (right axis) scales. The generation time for this microorganism is 30 minutes.

BOX 6.1	*Research Highlights*

The Cell Cycle in Bacteria

The cell cycle is the period in which a newly formed bacterium elongates, replicates its DNA, and divides to generate two cells. The replication of DNA and events involved in division to form two cells is under tight regulatory control.

In a newly formed cell there is a regulatory event that initiates the replication of the bacterial genome. The time required to replicate the 4.2×10^6 base pairs in the DNA of bacteria with a generation time between 20 and 60 minutes is 40 minutes. This is the C phase (see diagram). After completion of DNA replication (termination), there is a 20-minute period before division occurs. This is the D (delay) phase. During the D phase, the replicated DNA is separated into opposite ends of the elongated cell. The cytoplasmic membrane is involved in this separation. After separation of the DNA, the process of constructing cytoplasmic membrane and cell wall begins at the mid-

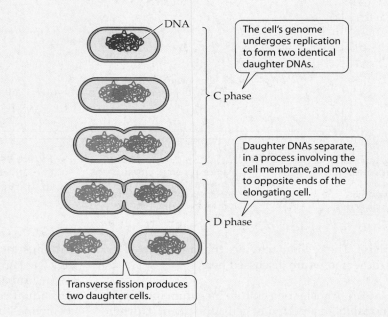

DNA

The cell's genome undergoes replication to form two identical daughter DNAs.

C phase

Daughter DNAs separate, in a process involving the cell membrane, and move to opposite ends of the elongating cell.

D phase

Transverse fission produces two daughter cells.

point of the cell. The division into two distinct daughter cells is called **transverse fission**.

You might ask how a species can have a generation time of 20 minutes when replication of DNA requires 40 minutes. The answer is that a second round of replication begins on the part of the DNA that has already replicated before the C phase of the first replication is complete. Thus, both daughter cells may receive a genome that has ongoing replication forks. This is explained further in Chapter 13.

TABLE 6.1	Approximate generation times for several organisms growing in media optimal for growth

Species	Generation Time
Escherichia coli	20 min
Bacillus subtilis	28 min
Staphylococcus aureus	30 min
Pseudomonas aeruginosa	35 min
Thermus aquaticus	50 min
Thermoproteus tenax	1 hr 40 min
Rhodobacter sphaeroides	2 hr 20 min
Sulfolobus acidocaldarius	4 hr
Thermoleophilum album	6 hr
Thermofilum pendens	10 hr
Mycobacterium tuberculosis	13 hr
Syntrophomonas wolfei	54 hr

were available and if the culture medium were free of excreted inhibitory products.

Growth Curve Characteristics

Adding a small population of bacteria (an **inoculum**) to a suitable volume of culture growth medium results in a predictable increase in cell numbers. If the number of organisms present is determined at intervals throughout the period of population growth and plotted on semilogarithmic paper (or by aid of a computer), a growth curve for that microorganism is obtained. A typical growth curve is shown in **Figure 6.4**. The numbers on the ordinate and abscissa of the graph vary quantitatively for different bacterial species, but the overall pattern does not. The growth curve reveals four distinct phases, each having a different slope. These are: (1) **lag phase**, (2) **exponential growth phase**, (3) **stationary phase**, and (4)

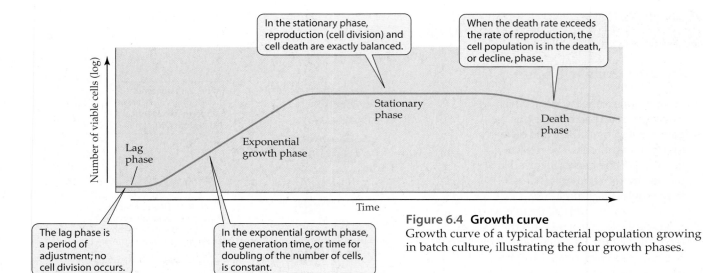

In the stationary phase, reproduction (cell division) and cell death are exactly balanced.

When the death rate exceeds the rate of reproduction, the cell population is in the death, or decline, phase.

Stationary phase

Death phase

Exponential growth phase

Lag phase

Number of viable cells (log)

Time

The lag phase is a period of adjustment; no cell division occurs.

In the exponential growth phase, the generation time, or time for doubling of the number of cells, is constant.

Figure 6.4 Growth curve
Growth curve of a typical bacterial population growing in batch culture, illustrating the four growth phases.

death phase. The cellular activities that are involved in each of these phases are discussed here.

LAG PHASE   Transfer of a culture inoculum into fresh medium results in a period of cell adjustment. During this initial lag phase, there is no increase in cell number, and in many cases there is actually a decrease. The cells, however, may grow in size, particularly if the in-

oculum is taken from a culture that is not actively growing. The lag phase is a period of unbalanced growth. During **unbalanced growth**, the various components of individual cells are synthesized to provide enzymes, coenzymes, and metabolites that are essential to begin the highly coordinated process of **balanced** growth and division. During this phase, proteins necessary for the uptake of available nutrients may also be synthesized.

BOX 6.2   *Methods & Techniques*

Calculating the Generation Time

The generation time for a microorganism that is growing under a determined set of conditions is constant. It can be calculated if three bits of information are available.

N_0 The number of bacteria present at an early stage in exponential growth

N_t The number of bacteria present after a period of exponential growth

t The time interval between N_0 and N_t

Exponential (logarithmic) growth must be understood as a

geometric progression of the number 2. Exponential growth proceeds: 2 cells $\rightarrow$ 4 $\rightarrow$ 8 $\rightarrow$ 16 $\rightarrow$ 32 ... or 2^n where n = number of generations. Exponential growth can be described mathematically by the following equation:

$$N_t = N_0 \times 2n$$

To solve for n, one would take the logarithm of the two sides:

$$\log N_t = \log N_0 + n \log 2$$

or solving for n

$$n = \frac{\log N_t - \log N_0}{\log 2}$$

Applying real numbers to the equation, one would proceed as follows:

Assume
N_0 = 6,000 cells (6×10^3)
N_t = 38,000,000 cells (3.8×10^7)

consulting a log table and filling in the equation:

$$n = 12.6 \text{ generations}$$

Assuming that the elapsed time between N_0 and N_t was 5 hours (300 minutes), the generation time would be:

$$\text{Generation time (g)} = \frac{300}{12.6} = 23.8 \text{ minutes}$$

BOX 6.3	*Research Highlights*

Growth: A Matter of Perspective

A population increase where each cell divides during a unit of time is termed *exponential growth*. If the generation time of a microorganism is 20 minutes, then each cell will reproduce every 20 minutes. This rate of population growth will continue as long as the organism remains in the exponential growth phase. During growth in a culture medium, the exponential phase of a bacterium is relatively short. It can be limited by nutrient depletion, oxygen deprivation, the accumulation of inhibitory products, or other factors.

Imagine that we devise a means for sustaining a culture of *Escherichia coli* (generation time 20 minutes) in the exponential growth phase for 24 hours. What would be the number of progeny?

A single bacterium would yield 4,722,366,478,574,681,194,496 cells.

They would weigh 4,722,366,478 grams, which is 10,401,687 pounds or 5,200 tons.

Assuming the average weight of individuals in a crowd to be 160 pounds, this biomass would equal 65,010 people.

Assuming that each cell is 2 μm long end to end, these cells would stretch 9,444,732,957,149,362 meters or 9,444,732,957,149 kilometers or 5,855,734,433,432 miles.

Placed end to end, these bacteria would circle the earth 243,988,935 times or stretch to the moon and back 12,459,009 times.

Utilizing glucose as growth substrate, these bacteria would consume about 10,000 tons in the 24-hour growth period.

Assuming the biomass to be a satisfactory food source, it would feed the entire population of North Carolina for one day.

Never underestimate the power of a microbe.

Once achieved, balanced growth is a steady-state situation, a period where every component of a cell in culture increases by the same constant factor per unit time.

The length of the lag phase can vary considerably. It can be long for an inoculum obtained from cells in late stationary phase. It can be negligible or short when the inoculum is obtained from a culture in exponential growth, or it can be quite long if the inoculum is obtained from a rich complex medium, and the microorganisms are transferred to a mineral salts/glucose medium (see Table 5.2). In the latter case, it would probably be necessary for the organism to synthesize many of the proteins that are involved in generating the various monomers that were readily available in the complex medium.

EXPONENTIAL GROWTH PHASE   During the exponential or logarithmic phase, cells are in **balanced growth**. All aspects of cell metabolism and growth are operating in unison. Each cell in the population doubles within an essentially identical unit time (generation time; see Figure 6.3). If there is one organism present at the beginning of the exponential growth phase, the total number present after a defined period of time would be 2^n, where n would be the number of generations. For example, starting with one organism with a generation of 20 minutes, after 4 hours or 12 generations there would be 2^{12}, or 4,096 progeny cells. This illustrates the explosive nature of exponential growth. The actual rate of exponential growth depends on the composition of the medium, with the most rapid growth occurring in a rich medium where microorganisms do not need to spend energy in synthesizing monomers.

Exponential growth is an important factor in the rapidity of food spoilage or the onset of an infectious disease. For example, consider a human that becomes infected with 32 disease-causing cells of an organism (pathogen) with a generation time of 30 minutes. Assume further that the infecting organisms are in the exponential growth phase. During the next 30 minutes, each bacterial cell would divide once, adding 32 more organisms for a total of 64. However, should the invading cells continue growing exponentially, after 7.5 hours there would be 1,048,576 progeny, and these would double in the next 30 minutes, yielding 2,097,152 organisms after 8 hours. This rapid growth is obviously an important consideration in the early treatment of an infectious disease. The sooner therapy is initiated, the more effective it is.

STATIONARY PHASE   Microbial populations, in general, will not maintain exponential growth indefinitely. Either the substrate for carbon and/or energy, or an essential nutrient will be exhausted. Alternatively, an inhibitory product of microbial metabolism will accumulate in the medium. Other factors may also apply, including a form of population density sensing that limits the number of cells in the population (called quorum sensing; see Chapter 13). Many species placed under conditions that are considered to be ideal will grow to a low density and then stop, whereas other species will grow to a high cell density before reproduction ceases. Any of the previously noted factors bring an end to exponential growth. The culture then enters the **stationary phase**. Cells continue to metabolize slowly during the stationary phase, although almost all biosynthesis processes have ceased. This metabolism supports cell maintenance to ensure cell viability as long as is possible. If carbon is not yet limiting, some strains synthesize storage material such as β-hydroxybutyrate or glycogen (see Chapter 4). Members of the genera *Bacillus* and *Clostridium* that form endospores do so during this period.

During stationary phase, some of the cells die and lyse when cell energy is exhausted. The materials released by lytic cells can provide nutrients for other cells, and these divide and replace the dead ones. However, most of the population in the stationary phase survives but simply does not proliferate. The individual cells in this phase differ in certain biochemical components from cells in the exponential growth phase. Generally, stationary phase cells are more resistant than exponential growth phase cells to adverse physical conditions such as increased heat, radiation, or change in pH.

The stationary phase is a period of survival and may mimic the conditions in nature when the microorganism is growing slowly or ceases to grow. There are a number of distinct genes that are expressed in microorganisms as they enter the stationary phase that are not expressed during exponential growth. Included among these are the genes for secondary metabolites such as antibiotics (see Chapter 31). Other genes expressed during the stationary phase are known as the s̲u̲r̲vival (*sur*) genes, and these are indispensable to the survival of an organism entering the stationary phase. *E. coli* mutants have been obtained that lack or have defective *sur* genes, and these mutants die rapidly as they enter the stationary phase. Microbes living in nature often face conditions that are unfavorable for growth, and it appears that evolution has given them mechanisms for protecting themselves in times of stress.

DECLINE OR DEATH PHASE   When a culture is maintained in stationary phase beyond a certain length of time (depending on the species, and on the environmental and nutritional conditions), the microorganisms die. Estimation of cell mass by turbidimetry and direct microscopic counting of cells suggests that the total number of cells remains constant. Viability counts, however, reveal otherwise. Generally, the death phase is an exponential function during which a logarithmic decrease in the number of viable cells occurs with time. The ac-

tual death rate depends on the particular organism involved and the conditions in the environment.

SECTION HIGHLIGHTS

Entry of cells into the exponential growth phase proceeds from an inactive or resting state by means of adaptive changes in cell metabolism. These events generally require RNA and protein synthesis to form the cellular machinery needed for balanced growth. When fully adapted, each cell increases in mass and initiates cell division independent of all other cells. This process continues until a nutrient becomes limiting or a waste product accumulates to an inhibitory concentration, whereupon cell growth immediately ceases. Each cell then shifts to a metabolic resting/survival mode. Death will occur unless the cell encounters new nutrients, or is displaced to another environmental niche where conditions are favorable to growth.

6.3 Ways to Measure Growth

A measure of cell number or total biomass versus time is required to accurately measure growth rate, substrate utilization, effect of inhibitors, and other parameters. In this section, we will discuss methods that can be used to determine cell mass and cell numbers. Some methods are direct and others indirect. Methods generally employed to determine mass directly are wet or dry weight of cells. Indirect measurement of cell mass can be accomplished by chemical analysis of a specific cellular component (such as nitrogen). Total number of cells can be determined by direct counting of individual cells or by a viability count. Turbidimetry is an indirect method of obtaining a relative estimate of cell mass. Each of these methods is discussed briefly.

A reasonably accurate estimation of total cell number or mass is essential in most studies involving the growth of microorganisms.

Total Weight of Cells

A liquid medium containing cultured cells can be centrifuged to concentrate the cells at the bottom of the centrifuge tube. The **wet weight** of packed centrifuged cells can be ascertained by placing them in a pan of known weight and determining the actual weight. This method is quite often employed to estimate a cell mass taken for enzyme isolation or lipid analysis.

The total **dry weight** is another reasonably accurate method whereby the cell mass is placed in a weighed pan and dried to constant weight (until no more weight is being lost). One can also filter cells onto a membrane filter with a pore size that will capture the desired organism. The filter and cells would then be dried to a constant weight. Drying is usually at 100°C to 105°C for 8 to 12 hours. Unfortunately, the drying process may render cells unsuitable for some types of biochemical analyses.

Chemical Analysis of a Cell Constituent

The total biomass can also be deduced by analytical chemical procedures designed to quantitate the amount of a cellular constituent. The constituent most widely used in laboratory research is total cell protein. It comprises about 50% of the cell dry weight, and can be easily detected by any of several analytical methods. One common method employs a dye (Brilliant Blue G) that binds specifically to proteins; the amount of color detected is proportional to the protein content of the sample. A standard curve is prepared using known amounts of protein, and the dye signal obtained from the bacterial sample is evaluated relative to the standard curve to establish protein content. This colorimetric method is rapid and relatively inexpensive to perform.

Another constituent used in laboratory research is total cell nitrogen. Because it has been established that cells are about 14% nitrogen, an analysis of the nitrogen present in a given sample will give a reasonable approximation of total biomass. In this procedure, the cells are washed free of residual medium and digested with sulfuric acid to release and convert the cellular nitrogen to ammonium, and the total NH_4^+ released is then determined by adding a reagent that combines with the ammonium to give a distinct color. The concentration can be determined colorimetrically.

Total cellular carbon can be determined by digesting a sample of cells with a strong oxidizing agent such as potassium dichromate in sulfuric acid. The amount of carbon present is proportional to the dichromate reduced. These methods rely on the assumption that the percentage of cell protein, nitrogen, or carbon does not vary throughout the growth phases.

Direct Count of Cells

The total number of microorganisms in a cellular suspension can be determined by a direct microscopic count. A sample of the suspended cells is placed in a **Petroff-Hausser counting chamber**. The Petroff-Hausser counting chamber is a thick glass slide with a grid etched onto its surface (Figure 6.5). Each square of the grid is of a known area (in square millimeters). After adding cells suspended in liquid to the chamber, it is placed under a microscope and the number of organisms in a large square (total volume of 0.02 mm^3 in this example) is determined by direct count. Several squares should be counted, and the average number present determined. This number can then be multiplied by a predetermined

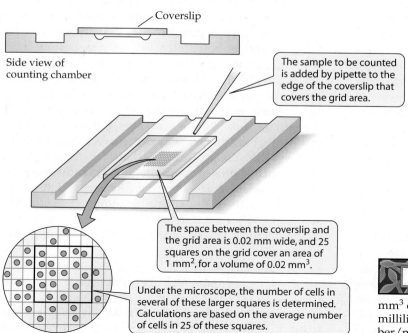

Coverslip

Side view of counting chamber

The sample to be counted is added by pipette to the edge of the coverslip that covers the grid area.

The space between the coverslip and the grid area is 0.02 mm wide, and 25 squares on the grid cover an area of 1 mm^2, for a volume of 0.02 mm^3.

Under the microscope, the number of cells in several of these larger squares is determined. Calculations are based on the average number of cells in 25 of these squares.

Figure 6.5 Cell counts using the Petroff-Hausser counter
In this example, there are 16 cells in the 0.02 mm^3 of the larger grid area, so the number of cells in a milliliter of original culture is 16 cells $\times$ 25 $\times$ 50 (number/mm^2) $\times$ 10^3 (number/mm^3) = 2 $\times$ 10^7 cells/ml.

factor based on the volume of the liquid in a large square to give the total number of cells per milliliter in the original suspension. A concern with this technique is that both viable and dead cells can have the same appearance.

A direct counting of microorganisms in suspension can also be accomplished with an electronic instrument called a **Coulter counter**. The cells to be counted are suspended in a conducting fluid (saline solution) that flows through a minute aperture in the instrument. There is an electric current across this aperture. During passage of any nonconducting particle, such as a bacterium, the electrical resistance within the aperture is increased, causing a brief spike of increased voltage across the aperture. The voltage pulse can be recorded electronically, and the number of pulses per unit volume is directly proportional to the number of particles flowing through the aperture.

Because the amplitude of the voltage pulse is proportional to the particle volume, the sizes of the cells can also be determined as they pass through the aperture. The aperture for bacteria is generally 10 to 30 µm in diameter; therefore, suspensions of cells to be counted must be diluted to minimize the probability of simultaneous passage of two or more cells, or clogging of the aperture.

Viability Count—Living Cells

A **viability count** is a determination of the number of living cells (able to reproduce) present in a suspension. A determination of the number of bacteria present in a sample can be accomplished by adaptations of the spread or pour plate methods discussed in Chapter 5. Generally, viable counts are made on samples such as soil, ground meat, and milk that contain significant numbers of bacteria and therefore must be diluted. Dilutions are tenfold or hundredfold and done in physiological saline (or seawater for marine organisms) to protect the microorganisms from the osmotic shock that might occur if diluted with tap water. To make a tenfold dilution, add 1 ml of sample to 9 ml of the diluent. If l ml of this 1/10 dilution is then added to another 9 ml of diluent, it would produce a 1/100 dilution or 1×10^{-2}. The number of dilutions necessary would be determined by the nature of the original sample.

A dilution series is illustrated in Figure 6.6. The number of colonies that are countable on a plate and would be statistically valid is 30 to 300 per standard-sized plate. Each of the dilutions should be plated in triplicate to increase statistical validity. The preferred volume of liquid that can be spread on the surface is 0.1 to 0.2 ml (more would flood the plate and result in indistinct colonies). Therefore, the dilution that would yield a countable plate should have 150 to 1,500 cells per ml. Spreading

0.2 ml of such a dilution would yield 30 to 300 colonies, a number that is readily countable. In the series illustrated in Figure 6.6, the number of cells in the original suspension would be:

$$\text{Spread 0.2 ml} \rightarrow 124 \text{ colonies} = 620 \text{ cells/ml} \times 10^5$$
$$= 6.2 \times 10^7 \text{ cells/ml}$$

Concentrating the sample, rather than diluting it, can determine the number of organisms in samples that contain fewer than 10 organisms per milliliter, such as some natural waters. Concentration can be accomplished by centrifugation of a measured volume of water. A known volume of water may also be filtered through a sterile membrane filter with a pore size that retains microorganisms. The filter may then be placed on the surface of suitable medium in an agar plate, and the number of colonies that develop can be determined after incubation (Figure 6.7).

One problem with the viable count method is in selecting a culture medium that will permit the growth of a reasonable number of the microorganisms present in the sample. An accurate estimation of the viable microorganisms in a soil sample would not be possible because of the diverse nutritional requirements of a soil population. As mentioned previously, it is clear that we are unable to grow most of the microorganisms present in soil. An alternative approach employs a selective cell dye or stain that reveals the presence of live cells. Microscopy is used to count the stained cells.

Turbidimetry: Scattered Light and Cell Density

The most convenient and rapid method for measuring cell mass, and consequently the one generally employed in the laboratory, is **turbidimetry**, which involves the use of a **spectrophotometer**. In a spectrophotometer, a beam of light is passed through a suspension of bacteria, and the light is scattered in proportion to the number of cells present. A bacterial suspension appears turbid because each cell scatters light. The more cells present, the more light is scattered, and the greater the visible turbidity. The wavelength of light passing through the suspension can be varied over a broad range, but bacterial suspensions are generally analyzed with the wavelength set between 400 and 600 nm.

Readings from the spectrophotometer are in absorbency (A), which is defined as the log of the ratio of incident light (I_o) to the light transmitted (I) through the suspension:

$$A = \log (I_O/I)$$

Because the shape and size of a bacterium affects the scattering of light, the absorbency is a relative figure. To

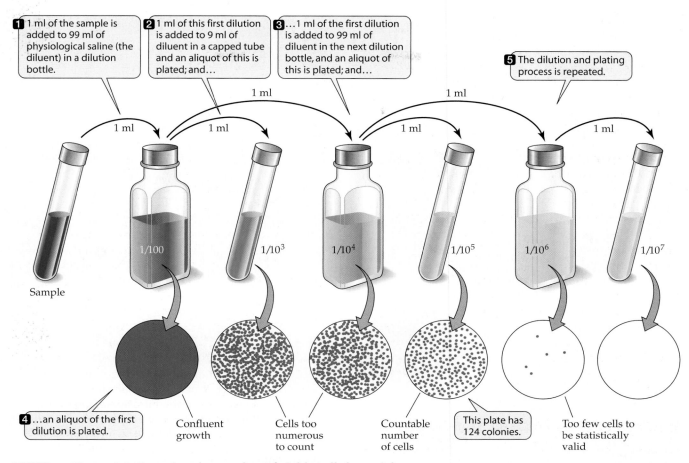

1 1 ml of the sample is added to 99 ml of physiological saline (the diluent) in a dilution bottle.

2 1 ml of this first dilution is added to 9 ml of diluent in a capped tube and an aliquot of this is plated; and…

3 …1 ml of the first dilution is added to 99 ml of diluent in the next dilution bottle, and an aliquot of this is plated; and…

5 The dilution and plating process is repeated.

1 ml   1 ml   1 ml   1 ml   1 ml

1/100 $1/10^3$ $1/10^4$ $1/10^5$ $1/10^6$ $1/10^7$

Sample

4 …an aliquot of the first dilution is plated.

Confluent growth

Cells too numerous to count

Countable number of cells

This plate has 124 colonies.

Too few cells to be statistically valid

Figure 6.6 Counting the number of viable cells by serial dilution and plate count
Aliquots of each dilution are spread on the surface of nutrient agar, as outlined in Figure 5.11. The appearance of the plates after incubation might be as depicted in bottom row.

estimate the actual cell mass for a particular species of microorganism, one must relate the absorbency to the actual cell dry weight or cell numbers that give a particular absorbency. This relationship is illustrated in Figure 6.8. Note that at high densities of cell mass, the absorbency deviates from linearity. As a consequence, one must dilute thicker suspensions to an absorbency that is in the linear range and multiply by the dilution factor to obtain a more accurate measure.

Turbidity is superior to other methods because it is quick, is readily repeatable, and does not adversely affect the cells. Turbidimetric measurements are not practical for cells that grow in clumps or for suspensions containing fewer than 10^7 cells per ml (lower limit of visible turbidity). Mycelial organisms such as actinomycetes or fungi do not form uniform suspensions. Consequently, one must resort to other methods for these organisms. Total nitrogen, total carbon, or wet/dry weight would be the methods of choice for any microbe that does not give uniform suspensions.

Synchronous Growth of Cells in Culture

The sequential events that occur during the cell cycle (see Box 6.1) are not easily measured in a single bacterial cell. Some microscopy techniques are available that are sufficiently sensitive to follow aspects of DNA replication (C phase), the delay (D phase) period before division occurs, and the macroscopic changes in morphology during cell division. However, other events at the molecular level pertaining to protein synthesis, polymerization reactions, and subsequent cell assembly reactions are not as easily measured. The sequence of events during a cell cycle can be quantitated with reasonable accuracy only when we examine a **population** in which all of the cells are at the same stage of division. This

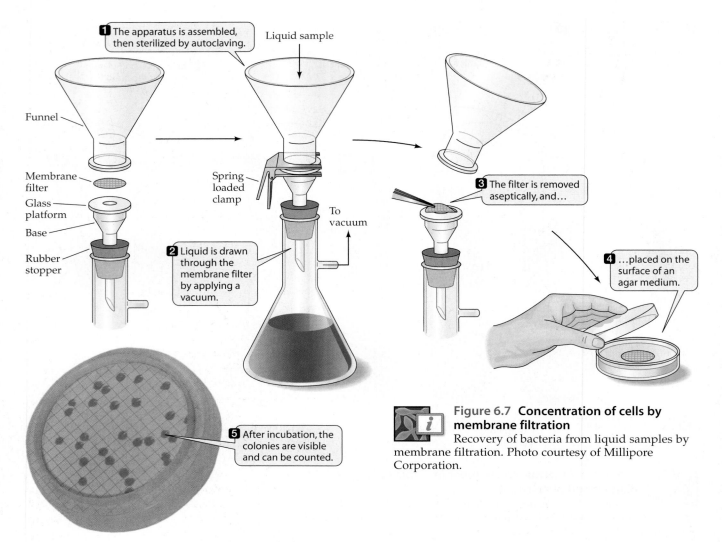

1 The apparatus is assembled, then sterilized by autoclaving.

Liquid sample

Funnel

Membrane filter

Glass platform

Base

Rubber stopper

Spring loaded clamp

To vacuum

2 Liquid is drawn through the membrane filter by applying a vacuum.

3 The filter is removed aseptically, and...

4 ...placed on the surface of an agar medium.

5 After incubation, the colonies are visible and can be counted.

Figure 6.7 Concentration of cells by membrane filtration
Recovery of bacteria from liquid samples by membrane filtration. Photo courtesy of Millipore Corporation.

means that an equal phasing of all growth processes exists in every cell in the population.

A culture of organisms where every cell is in the same stage of division is called a **synchronous culture**. Data obtained from experimentation on such a population may then be extrapolated to a single cell. A growth curve for a synchronous culture (Figure 6.9) is a periodicity in cell number versus time. This indicates that all cells in the culture divide at the same time (in this case, approximately every 60 minutes).

Bacteria do not divide simultaneously for more than two or three generations. Thus, experimentation involv-

Theoretical

Absorbency no longer proportional to cell mass

Absorbency

Cell mass (dry weight)

Figure 6.8 Relationship between light absorbency and cell mass
Measurements of absorbency in a spectrophotometer must be made using cell densities for which absorbency is proportional to cell mass (straight portion of the curve) or cell number.

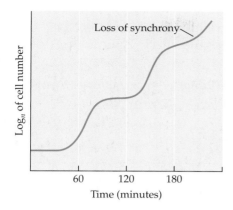

Figure 6.9 Growth curve for synchronous culture
A typical growth curve for a microorganism in synchronous culture. Cultures tend to lose synchrony after a few generations. Compare this curve with the straight line obtained in asynchronous growth (see Figure 6.3).

ing metabolic activities occurring during the cell cycle should be done in the first or second generation unless the event being studied can be examined in a single cell.

A number of techniques have been developed that permit one to establish a synchronous culture: filtration, light stimulation, altered temperature, or nutrient limitation. A synchronous culture may be obtained by selecting organisms of equivalent size from a population. Young cells are generally the best choice in synchronous culture studies, and because they tend to be the smallest, physical means of obtaining this population of small cells have been devised.

One effective method of sorting cells by size is to filter a population of microorganisms through a cellulose nitrate filter with a pore size that retains most of the cells. Many bacterial species will adhere tightly to this type of filter. The filter can be turned over and liquid passed through to dislodge all microorganisms not tightly attached to the membrane. After removing these unattached cells, a nutrient is poured through to permit the attached cells to divide. The cells released by the nutrient wash are the small progeny of the attached cells. The effluent collected over a short period will contain bacteria of equivalent size and at the same cell cycle stage.

Growth Yield (y_S and y_{ATP})

The efficiency of an organism in converting a substrate to cell mass is termed the **efficiency of growth**. This is analogous to asking how efficient a factory is in making a finished product, in this case cells, from the raw starting materials. For microbes, efficiency of growth is measured by comparing the grams biomass produced per mole substrate consumed. The ratio is defined as the **growth yield** ($y_{s \text{ or }} y_{sub}$) and can be expressed as grams cell mass obtained per mole of substrate utilized:

$$y_s = \frac{x_{max} - x_O}{\text{moles of substrate utilized}}$$

In this case the x_{max} is the cell mass obtained, and x_O is the initial weight of cell in the inoculum. The growth yield can be expressed as y_s or y_{sub} (gram dry weight of cells/mole substrate).

In organisms for which the catabolic pathway for utilization of the substrate is known, the **energy yield** (ATP formed) can also be calculated. Growth yield can then be expressed as y_{ATP}, which is the dry weight (in grams) of biomass per mole of ATP available.

Anaerobic bacteria, such as those that produce lactic acid, require a complex growth medium (see Chapter 5), and generally these anaerobic species utilize glucose solely as a source of energy according to the following reaction:

$$\text{Glucose} + 2\text{ ADP} + 2\text{ Pi} \rightarrow 2\text{ lactic acid} + 2\text{ ATP}$$

One can readily determine the mass of cells produced per mole of ATP generated in organisms such as this, where the substrate is utilized solely as energy source.

The growth yield, y_{ATP}, is generally constant among bacteria. One mole of ATP will support the growth of about 10 grams of cell mass (dry weight). Consequently, the cell yield on a substrate utilized solely as the energy source is a direct function of the number of ATP molecules produced per mole of energy source utilized (Table 6.2).

The preceding analysis indicates that despite the diversity of microorganisms, the y_{ATP} value is nearly constant, and the biosynthetic pathways that are fueled by ATP are thus similar in all species. The diversity of microbial life is therefore determined by the individual means through which microbes generate their ATP (i.e., diversity of catabolic pathways present and the type of substrates utilized).

The growth-yield in a chemoheterotrophic bacterium utilizing a substrate as both energy and carbon source is more difficult to assess in terms of ATP generated. Aerobic organisms that utilize a sugar or other substrate as both carbon and energy source will vary in their efficiency.

Experimentation indicates that the efficiency of substrate conversion to biomass by chemoheterotrophs varies from 15% to 50%. The maximum conversion of substrate to biomass for a sugar is about 50% (one-half is assimilated into cells and one-half is released as CO_2 or waste product). An organism growing on methane (CH_4) for which 1 mole of atmospheric oxygen is added

TABLE 6.2	Growth yields of anaerobic bacteria utilizing glucose as the energy source		
	Mol ATP/mol glucose	y_{max} (g of cell/mol glucose)	y_{ATP} (g of cell/mol ATP)
Lactobacillus delbrueckii[a]	2	21	10.5
Enterococcus faecalis[a]	2	20	10
Zymomonas mobilis[b]	1	9	9

[a]Homolactic fermentation, Embden–Meyerhof pathway (see Chapter 10).
[b]Alcoholic fermentation, Entner–Doudoroff pathway (see Chapter 10).

per mole of methane assimilated can have a cell yield as high as 70%.

Effect of Nutrient Concentration on Growth Rate

In the previous discussions we have considered population growth of cells cultured in a defined medium under controlled laboratory conditions. This is commonly referred to as **batch culture**. Nutrients are added to the growth medium to provide a source of building blocks (cell carbon) and energy. The amount initially present is in excess of the level necessary to attain a maximum growth rate. Following inoculation and cell adaptation during lag phase, cells rapidly establish a rate of growth that reflects their maximum capacity under the conditions encountered. Two principles are important in evaluating the potential for cell growth and the growth yield.

First, the newly acquired energy is used for two purposes: cell maintenance and cell growth. The former need is met before the second. When the source of energy available to an organism is at a very limited level, growth is not possible. Organisms require a fixed amount of energy, called **maintenance energy**, simply to stay alive and perform basic functions. Maintenance energy is used for DNA repair, accumulation of nutrients inside the cell, motility, turnover of macromolecules, maintenance of ion gradients across the cytoplasmic membrane, and other basic cellular needs. An increase in cell numbers can occur only when available energy exceeds these non–growth-related maintenance demands. This relationship is depicted in Figure 6.10. At low nutrient levels, cell yield is nil. In practice, this is not observed in batch culture due to the inability to maintain low nutrient levels. As nutrient levels are increased, cell yield becomes linear. However, for *continuous culture* conditions (described in subsequent text) this principle is easily demonstrated.

Second, a decreased growth rate at low nutrient levels results from the inability of an organism to transport substrate at a rate sufficient to maintain balanced growth. The intracellular accumulation of essential intermediates is lower than at substrate saturation, and not all of the precursors are available simultaneously to provide for a maximal rate of cell synthesis and reproduction. The nutrient concentration also affects the total number of cells produced (see Figure 6.10). As the nutrient level increases, the cell yield correspondingly increases. However, if the concentration of nutrient in a medium is well above a critical threshold, the growth yield may actually be impaired. Under these conditions the cells may shift to a less efficient mode of metabolism.

It should be emphasized that growth rates, total growth, and the concentration of substrate necessary to support growth all vary among species. Use of *continuous culture* conditions are key to understanding cell yield and growth rate due to the ease in controlling nutrient level. As mentioned previously, some organisms will not grow to high cell density, even though they are presented with a favored nutrient and what are assumed to be optimal growth conditions.

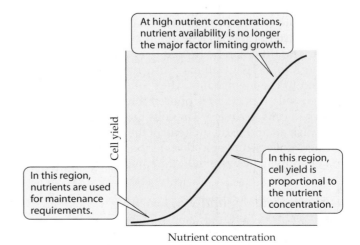

At high nutrient concentrations, nutrient availability is no longer the major factor limiting growth.

In this region, nutrients are used for maintenance requirements.

In this region, cell yield is proportional to the nutrient concentration.

Cell yield

Nutrient concentration

Figure 6.10 Relationship between nutrient concentration and total cell mass
Cell yield is directly proportional to nutrient availability only within a certain concentration range.

6.4 Growth of Cells in Batch versus Continuous Culture

In nature, batch culture conditions are encountered when cells are introduced into a new environment where the nutrient levels may be moderate or high, and relatively fixed (e.g., fresh foods, decaying plant and animal tissues, some infections). In contrast, continuous culture conditions are encountered in distinctly different environmental niches, where continual addition of nutrients and removal of waste products occur. For example, consider bacterial cells attached to a rock surface in a stream or river, bathed by the dilute "medium" flowing by. Another example is the growth of bacteria in human intestines where they are perfused by food passing by. Such habitats often have low to moderate levels of nutrients.

Batch Culture

Bacteria, yeasts, or filamentous fungi are generally cultured in the laboratory by the batch method. In **batch culture**, a nutrient liquid medium is prepared in an Erlenmeyer flask, sterilized, and inoculated. The inoculated flask is then placed on a shaker (if aerobic) at the appropriate growth temperature for a suitable growth period.

Culturing bacteria by the batch method results in a population that is in a **balanced state** of growth for only a few generations. This balanced state occurs during the early to middle exponential growth phase (see Figure 6.4). Bacteria that grow in batch culture actually go through an "aging" process as they progress through the growth cycle. This means that cells in the latter stage of exponential growth can be physiologically different from cells present during the middle exponential growth phase. Differences can be detected in enzymatic composition, presence of storage material, and membrane protein composition. Why does this happen?

Batch culture conditions approximate a true **closed system**, wherein following inoculation of the flask, nothing is added or removed. In reality, for aerobic cell growth, oxygen is added by shaking the Erlenmeyer flask. However, over the course of time as cells multiply, nutrients are consumed and end products accumulate. When any essential nutrient becomes limited, cell growth cannot be sustained and the culture enters stationary phase. Alternatively, accumulation of a toxic end product inhibits growth before nutrients are consumed. The "aging" process is a cell response to adapt to the changing conditions. For example, the cell may need to synthesize an alternative solute uptake system or switch to making a building block if it is depleted. Likewise, the cell may switch to an alternative energy pathway to avoid production of a toxic product, or it may begin make stress response proteins for protection.

Continuous Culture

Techniques have been devised that permit maintenance of a bacterial culture in the exponential growth phase for an extended period lasting weeks or months. This is accomplished by establishing a **continuous culture** in an instrument termed a *chemostat* (Figure 6.11). It represents an **open system**, since the nutrients are continually added and waste products are removed. Cells continue in balanced growth indefinitely.

For continuous culture in the laboratory, sterile medium is fed into a culture vessel in a controlled manner. The culture vessel has a mixing system to ensure that the bacterial population is uniformly exposed to substrate. An overflow line permits release of cell suspension at a rate equal to the rate of substrate addition. The **dilution rate** (*D*) in the culture vessel can be expressed as:

$$D = f/v$$

where *f* is the flow rate of the incoming medium and *v* is the volume of medium in the culture vessel.

The culture in a chemostat is controlled by the concentration of available substrate. Overall, the stability relies on this limitation of growth rate by the availability of a growth-limiting substrate such as an energy, nitrogen, or phosphate supply. As the concentration of the growth-limiting nutrient is increased, and the dilution rate remains constant, the cell concentration will increase, but growth rate will be unchanged. The growth rate increases only when the nutrients are not limiting. If the substrate level does not exceed the requirements for maintenance energy, the culture "**washes out**." When

**Figure 6.11 Continuous culture
in a chemostat**
Model of a chemostat apparatus for continuous growth of a bacterial culture.

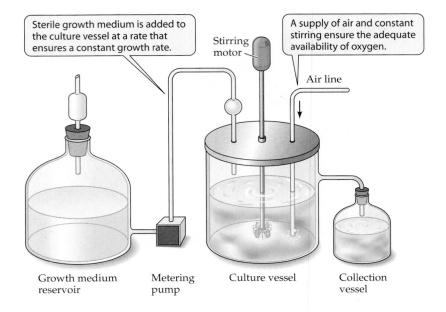

Sterile growth medium is added to the culture vessel at a rate that ensures a constant growth rate.

Stirring motor

A supply of air and constant stirring ensure the adequate availability of oxygen.

Air line

Growth medium reservoir

Metering pump

Culture vessel

Collection vessel

the total cell concentration and the nutrient supply are balanced and in equilibrium, the system is in a **steady state**. The theoretical steady-state relationship between substrate concentration and output of bacterial mass is depicted in Figure 6.12. The maximum growth yield (y_{max}) is the ratio of the cell mass to the substrate consumed and is equal to about 0.5 for virtually all bacterial species. Note that washout also occurs if the dilution rate (i.e., rate of nutrient addition) exceeds the capability of the cells to grow. *E. coli*, for example, cannot double faster than once every 20 minutes. It would washout with a dilution rate greater than 3 per hour.

A chemostat operates with a regulated flow rate such that the organism can grow at a rate equal to the dilution rate. Another instrument that functions solely by regulation of cell density is called a **turbidostat**. A turbidostat is equipped with a device that monitors the cell density in the growth chamber. If the cell density exceeds a preset level, the culture medium automatically enters the culture vessel, and the overflow is retained. If the cell

Figure 6.12 Steady-state relationship between substrate concentration and output of bacterial mass
The relationship between substrate concentration, bacterial biomass, concentration of bacteria, and generation time in a steady-state chemostat.

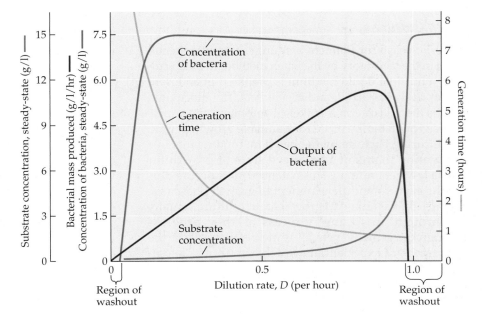

density falls below the preset level, no culture medium flows into the vessel.

6.5 Effect of Environmental Conditions on Cell Growth

A number of physical factors determine the ability of a microorganism to thrive in an environment. Among these are pH, temperature, presence of O_2, CO_2 availability, and availability of water. These also influence the growth of microbes in culture and are discussed in this section.

Acidity and Alkalinity

The relative hydrogen ion concentration in a solution is expressed as the pH. It is based on the dissociation of water, in which there would be equal levels of H^+ and OH^-. This is defined as pH 7 (the negative log of the H^+ ion concentration in water). The pH scale is from 0 (the most acidic) to 14 (the most alkaline). Because pH is a log function, a difference of one pH unit represents a tenfold difference in hydrogen ion concentration.

Among the *Bacteria* and *Archaea* are species that grow at a pH of near 1. Other species can thrive at pH 11. Generally, microorganisms grow best at pH values between these extremes. Filamentous fungi grow over a pH range of 2 to 9, with a majority favoring a pH of 4 to 6. Yeasts cover the same range, but most species require a pH above 2. Most soil bacteria grow best at a pH near neutral, or a pH of 7.

The pH range for growth of any individual species is limited. Most microorganisms can grow within a range of 2 to 3 pH units (which is actually a hundredfold to thousandfold difference in H^+ concentration) and are insensitive to small variations in pH. Growth of a microorganism in a culture medium requires that the pH be adjusted and maintained in a range near that favored by the organism.

In the course of growth, an organism utilizing a fermentable substrate such as glucose may excrete acidic intermediates (acetic, lactic, and formic acids, among others), causing a decrease in the pH of the medium. Conversely, growth on proteins or amino acids results in ammonia production, causing the culture fluid to become alkaline. To prevent significant changes in pH, a chemical **buffer** is routinely incorporated into a culture medium. Buffers retard pH changes by removing excess H^+ or by donating H^+ to balance excess acidity or alkalinity, respectively. The most commonly used buffering system for organisms that grow near a neutral pH is a mixture of KH_2PO_4 and K_2HPO_4. This also provides the organism with phosphate for the synthesis of phosphorylated cellular constituents. Carbonates may be added to control pH when excess acid production is anticipated. For example:

$$CH_3COOH + CaCO_3 \rightarrow (CH_3COO)_2Ca\ (insoluble) + CO_2$$

The ability of a typical bacterium such as the intestinal microbe *E. coli* to grow at different pH values is depicted in Figure 6.13. It is readily apparent that pH has a profound influence on whether the cell will grow, and, if so, at what growth rate. The **optimal pH** for growth is around 7, as evidenced by the growth rate peak. The speed of cell doubling is progressively slower when the pH of the medium is higher or lower than the optimum. This organism possesses the ability to grow within this limited pH range, and it cannot neither grow nor survive

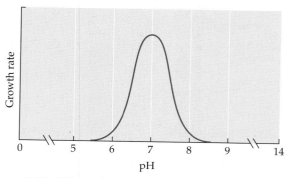

Figure 6.13 Effect of pH on cell growth
Cell growth rate is compared to the pH of the medium used for cell growth.

for long beyond these limits. However, other microbial species have evolved to colonize and thrive at this and other pH values that span the range permissible for microbial life. There is considerable diversity of life on Earth!

Microbes that grow at low pH are called **acidophiles**. The thiobacilli that are present in acid mine drainage are acidophiles and grow well at a pH of 1. *Thiobacillus thiooxidans* has a pH optimum of 2.5 and can survive at a pH of nearly 0. An interesting archaeal species is *Thermoplasma acidophilum*, which grows in coal refuse piles and can grow at 59°C and a pH of 1. This is remarkable because survivability of bacteria generally decreases at any given temperature as pH is lowered.

In contrast, **alkaliphiles** are organisms that grow optimally under very alkaline conditions. The Wadi el Natrum, a lake in Egypt, for example, has a pH ranging from 9 to 11. The water contains a rich population of halophilic (salt-loving) and phototrophic bacteria. Many of the organisms in this environment have not yet been characterized.

Presence of Oxygen (O₂)

About 20% of the earth's atmosphere is molecular oxygen (O_2). As a consequence, many environmental niches are constantly exposed to this element. Swamps and subsurface areas may have little free oxygen present and may be completely anaerobic. Virtually all eukaryotic organisms are dependent on the presence of molecular oxygen for survival. Eukaryotes evolved on Earth in an aerobic environment, and most of the animal world generates energy utilizing CH_2O (biological material) in the presence of O_2 with the concomitant production of water and CO_2. Among the eukaryotes, only a few fungi and protozoa that inhabit anaerobic environments can survive without molecular O_2.

Bacteria and *Archaea* species exhibit a mixed response to oxygen. The atmosphere of the primordial earth was completely anaerobic, and for more than 1 billion years, bacterial and archaeal populations evolved without O_2-associated respiratory mechanisms for energy generation. Thus, throughout evolution, some bacterial and archaeal populations have occupied anaerobic niches and have retained life styles that are independent of molecular oxygen. Some of these organisms are unaffected by atmospheric levels of oxygen, whereas others are killed by a brief exposure. *Bacteria* and *Archaea* can be divided into three major groupings based on their response to molecular oxygen (see Figure 6.14):

1. Aerobes: Obligate (strict) or facultative

2. Microaerophiles

3. Anaerobes: Obligate or aerotolerant

An **obligate aerobe** can grow only in the presence of molecular oxygen. It has no mechanism for energy generation other than a respiratory pathway that is geared to O_2 as a terminal electron acceptor. These microorganisms generally respond favorably to elevated oxygen availability during growth. A **facultative aerobe** follows a respiratory pathway when oxygen is available, but it can employ alternate anaerobic energy-generating systems when oxygen is not available (see Chapters 8 and 9). In facultative organisms, growth is generally better in the presence of air.

Microaerophiles are organisms that occupy niches where the atmosphere has limited levels of oxygen. Although they must have oxygen for respiration, they

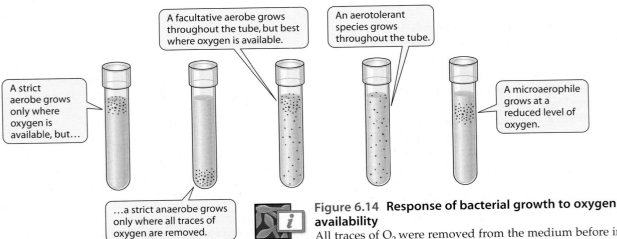

A strict aerobe grows only where oxygen is available, but...

...a strict anaerobe grows only where all traces of oxygen are removed.

A facultative aerobe grows throughout the tube, but best where oxygen is available.

An aerotolerant species grows throughout the tube.

A microaerophile grows at a reduced level of oxygen.

Figure 6.14 Response of bacterial growth to oxygen availability
All traces of O_2 were removed from the medium before inoculating a tube. A concentration gradient of available O_2—high at the top and low at the bottom—forms by diffusion of O_2 from the air space above the medium.

grow best at reduced levels that range from 2% to 10% (dry atmospheric air is 21% oxygen).

An **obligate anaerobe** cannot grow in the presence of oxygen and may be killed by minute levels. Tolerant anaerobes are oblivious to the presence of oxygen and can survive both with and without oxygen, but they do not use molecular oxygen for energy generation.

Toxicity of Oxygen

Oxygen is a highly reactive molecule, and metabolic reactions involving molecular oxygen can generate toxic byproducts. Most organisms that live in the presence of air synthesize enzymes that protect them from the toxic effects of oxygen byproducts. The major toxic products are superoxide, hydrogen peroxide, hydroxyl radical, and singlet oxygen. These products, if not removed, damage the cell by reacting with proteins, lipids, or nucleic acids. These toxic intermediates are generated as follows:

$$O_2 + e^- \rightarrow O_2^-$$ Superoxide
$$O_2^- + e^- + 2\,H^+ \rightarrow H_2O_2$$ Hydrogen peroxide
$$H_2O_2 + e^- + H^+ \rightarrow H_2O + OH\bullet$$ Hydroxyl radical
$$Chlorophyll + hv \rightarrow chlorophyll^*$$
$$Chl^* + O_2 \rightarrow {}^1O_2$$ Singlet oxygen

Superoxide ($O_2 + e^- \rightarrow O_2^-$) is a free radical that has gained an extra unpaired electron. During respiration, the reduction of oxygen to water requires the addition of four electrons. This stepwise reduction occurs by single electron transfer. An intermediate product is O_2^-, and small amounts of this toxic compound may be released during respiration. Superoxide must be removed as it is formed to protect the cell. This is accomplished by the enzyme **superoxide dismutase** ($O_2^- + O_2^- + 2\,H^+ \rightarrow H_2O_2 + O_2$).

Hydrogen peroxide ($O_2^- + e^- + 2\,H^+ \rightarrow H_2O_2$) is the product of two electron additions to form O_2^{2-}. The anion is present as H_2O_2. It is commonly formed by a two-electron reduction of oxygen. Such reactions are mediated by flavoproteins, which are normal electron carriers. Once formed, hydrogen peroxide is destroyed by the enzymes **catalase** ($H_2O_2 + H_2O_2 \rightarrow 2\,H_2O + O_2$) or **peroxidase** ($H_2O_2 + NADH \rightarrow 2\,H_2O + NAD^+$).

Hydroxyl radical ($H_2O_2 + e^- + H^+ \rightarrow H_2O + OH\bullet$) is the most reactive of the toxic products of oxygen respiration and is a very strong oxidizing agent. It is readily formed by ionizing radiation. There is no specific defense against this toxic product. However, the radical is relatively short lived and decomposes to H_2O.

Pigments such as chlorophyll are sensitive to light (photosensitizers). Exposure to light can convert these pigments to a triplet state (indicated by an asterisk, Chl*), which can result in a secondary reaction that produces **singlet state oxygen** (${}^1O_2$). Singlet state oxygen is a very powerful oxidant and thus highly lethal to cells.

Aerobic organisms contain active superoxide dismutase and either catalase or peroxidase to break down hydrogen peroxide. These enzymes are an essential defense mechanism and a constituent part of all aerobic microorganisms. Many photosynthetic organisms employ carotenoid molecules to absorb or quench the energy of singlet oxygen.

Toxicity of O₂ and Growth of Anaerobes

Growing anaerobic bacteria and archaea in culture requires specialized procedures for handling because many anaerobes are destroyed by traces of oxygen. The major obligate anaerobic bacteria include the endospore-forming clostridia (such as *Clostridium botulinum*, the causative agent of botulism), the sulfate reducers, the methanogens, and many types of microorganisms that inhabit the animal intestinal tract. Anaerobic protozoa and fungi are also present in the rumen of ruminants.

Anaerobic microorganisms can be grown in liquid media placed in glass tubes to which a chemical compound is added to remove any trace of oxygen. For example, the addition of thioglycollate is effective in removing oxygen from a culture medium. If a bacterial population is distributed in an agar culture medium prior to solidification, the growth patterns depicted in Figure 6.14 occur after suitable incubation. This affirms the various response variations microorganisms have to O_2 availability.

Strict anaerobes must be maintained in the total absence of O_2 at all times. This can be accomplished by manipulating them under a stream of nitrogen gas, an awkward procedure that is no longer in widespread use. It is much easier to carry out all manipulations of strict anaerobes in sealed chambers (see Figure 5.12). The atmosphere in these chambers can be replaced with a nonreactive gas such as nitrogen. Catalysts or O_2 scavenger chemicals are placed inside to remove all traces of O_2. Use of these anaerobic chambers permits one to perform all operations, including transfer and streaking, inside the box in the complete absence of O_2. Routine cell culture is accomplished using sealed tubes or bottles containing anaerobic medium (see Figure 5.18).

O_2 is highly toxic to anaerobes because cells often lack effective means to avoid damage. Reactive oxygen species formed from O_2 exposure may react with sensitive proteins needed for electron transfer reactions, for example, iron sulfur proteins (see Chapter 8), and are irreversibly inactivated. O_2 may also act to directly auto-oxidize certain reduced cofactors in proteins that cannot be easily recycled to their functional state.

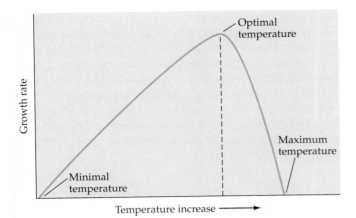

Figure 6.15 Response of bacterial growth to temperature
The cardinal temperatures for the growth of a bacterium.

Effect of Temperature

Microorganisms are of widespread occurrence in environments in which the ambient temperature can range from less than 0°C to more than 100°C. For example, saltwater beneath polar ice can reach –10°C, and volcanic areas under the sea can exceed 110°C. Microorganisms can thrive in both these environments, and those that live there have little competition from other life forms. It is clear that species diversity tends to decrease as the harshness of the environment increases.

The effect of temperature on the growth rate of a microorganism is depicted in Figure 6.15. A given species has a **minimum temperature**, a **maximum temperature**, and an **optimal temperature** for growth. These three temperatures are referred to as **cardinal** temperatures. As the temperature rises above the minimum, the metabolic reactions in the cell accelerate, and growth becomes more rapid. This increase leads to a physiological state at which all cellular reactions are operating at their optimum level. This corresponds to the optimal temperature. A little above this temperature, the cytoplasmic membrane may lose function, proteins or nucleic acids may be denatured, and the cell begins to sustain damage. The cellular reaction(s) that is least heat stable is primarily responsible for the demise of the microorganism as the temperature rises. The maximum temperature for growth represents a balance of thermal damage and the ability of the cell to repair while continuing to grow.

TEMPERATURE RANGE   Microbes are divided into four groups based on the range of temperature at which they can grow (Figure 6.16). These groups include the following: the *psychrophiles*, which grow at temperatures below 20°C; the *mesophiles*, which generally grow at temperatures between 20°C and 44°C; the moderate *thermophiles*, which grow between 45°C and 70°C; and the *hyperthermophiles*, which are microorganisms that require growth temperatures above 70°C to higher than 110°C. The optimal growth temperature for microorganisms that fall into each category are also presented. Table 6.3 presents a listing of representative organisms in each group and their ideal growth temperature ranges.

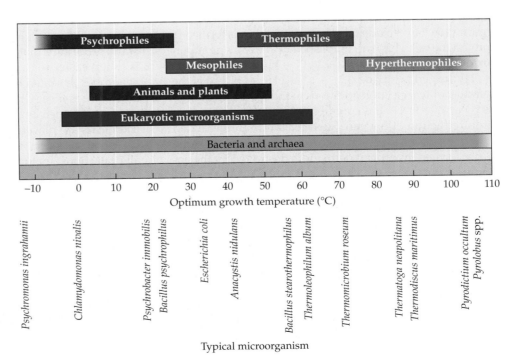

Figure 6.16 Growth temperature ranges for various life forms
The *Bacteria* and the *Archaea* are the most versatile types of organisms, with species that can grow at temperatures of below 0°C to about 115°C. The only microorganisms known to grow at temperatures greater than 92°C are *Archaea*.

TABLE 6.3	Temperature ranges for growth of selected bacteria and the archaea
Species	**Range (°C)**
Psychrophiles	
Cytophaga psychrophila	4–20
Bacillus insolitus	<0–25
Aquaspirillum psychrophilum	2–26
Mesophiles	
Escherichia coli	10–40
Lactobacillus lactis	18–42
Bacillus subtilis	22–40
Pseudomonas fluorescens	4–40
Thermophiles	
Bacillus thermoleovorans	42–75
Thermoleophilum album	45–70
Thermus aquaticus	40–79
Chloroflexus aurantiacus	45–70
Hyperthermophiles	
Hyperthermus butylicus	85–108
Methanothermus fervidus	65–97
Pyrodictium occultum	80–110
Thermococcus celer	70–95
Thermotoga maritima	75–90

PSYCHROPHILES   A psychrophile (from *psychro* meaning "cool") is defined as a microorganism that grows at or somewhat below 0°C and has an optimal growth temperature of less than 15°C. An obligate psychrophile cannot grow at a temperature above 20°C. Truly psychrophilic types are present in seawater, where the average temperature is 5°C, and these microorganisms are particularly abundant in polar ice and water at –5°C. The only requirement for survival is that liquid water be available, and this does occur in microenvironments within sea ice.

The term *psychrophile* is somewhat less precise than the other terms used to describe the temperature ranges for growth. Many microorganisms that grow optimally at 25°C to 30°C can also grow very slowly at 0°C. The ability to grow at psychrophilic temperatures is not considered of major significance in classification.

Facultative psychrophiles are sometimes present in the soil and water from temperate climates. These microorganisms grow very slowly at 0°C but grow well at 22°C. They are often present in or on food and can cause spoilage in food that is refrigerated for extended periods. Cold-tolerant fungi also cause spoilage of refrigerated foods. Food is best preserved for extended periods by drying or freezing, because these processes eliminate liquid.

MESOPHILES   The mesophiles comprise the vast group of microorganisms that are ubiquitous in soil and water and inhabit nearly all animals and plants. They generally grow between 20°C and 45°C, although some will grow at temperatures as low as 10°C. The mesophiles are mostly responsible for the cycles of matter, disease, fermentation, deterioration of material, and other activities that are important to humans. Consequently, they have been the subject of the most research. Human pathogens such as *Streptococcus pneumoniae* grow optimally at 37°C, the temperature of the human body. Soil organisms generally grow best at 26°C to 30°C.

THERMOPHILES   The thermophiles are organisms that grow optimally at temperatures greater than 45°C and below 70°C. These temperatures commonly occur in hot springs, volcanic areas, and compost heaps. Soil exposed to direct sunlight can reach temperatures that are appropriate for moderate thermophiles. As a consequence, thermophiles are present in almost every environment, including the Antarctic. Thermophilic bacteria are present in significant quantities in heated industrial water, hot water heaters, water from cooling towers at nuclear energy plants, and other artificially created thermal environments.

HYPERTHERMOPHILES   The hyperthermophiles are microorganisms that grow at temperatures higher than 70°C, and many grow at temperatures near or above the boiling point of water. The hyperthermophiles might appropriately be divided into two groups based on the cardinal growth temperature: those that grow from 70°C to 90°C in one group and those that grow from 80°C to 110°C in the other. For example, thermal vents in the ocean floor, called black smokers, may harbor a variety of chemolithotrophic bacteria and archaea (see Chapter 24, Figure 24.21).

Liquid water is a necessity for all life forms, so these microorganisms that grow at temperatures near 100°C or higher are found in habitats that are under sufficient pressure to maintain water as a liquid. These conditions occur at volcanic areas beneath the ocean surface.

Several genera of the *Archaea* have been obtained in cultures that grow at temperatures around 100°C. Not all *Archaea* are hyperthermophiles, but the microorganisms described to date that grow at temperatures in excess of 90°C are *Archaea*. Examples include *Pyrococcus furiosus*, *Archaeoglobus fulgidus*, and *Pyrodictium abyssi* (see Chapter 18).

An organism has been reported that grows under pressure at considerably higher temperatures (>200°C), and reference to this still appears in the popular press. However, there is no creditable evidence. Present evidence suggests that the maximum temperature at which molecular constituents (ATP, protein) in bacteria can function is in the range of 120°C to 130°C.

Temperature Limitations

Proteins are one of the major cellular constituents that may limit the ability of a microorganism to grow at high temperatures. Thermophiles must have heat-stable proteins, and many of the proteins in mesophiles are heat labile (unstable at high temperatures). Generally, the cytoplasmic proteins in a mesophile begin to precipitate if heated at 60°C for 8 to 10 minutes. The equivalent protein in a thermophile would withstand this temperature.

Growth temperature can also influence the fatty acid composition of the membranes in bacterial cells. This is affirmed by comparison of the cytoplasmic membrane fatty acids in a microorganism grown at a minimum temperature with those present after growth at a maximum temperature. At the minimum temperature, the relative proportion of unsaturated fatty acids in the cellular lipids is high. Growth at the maximum temperature leads to a greater relative percentage of saturated fatty acids. In some organisms the fatty acid chain length may also vary (e. g., 14–18 carbons long). The chain length and the degree of saturation of fatty acids in the cytoplasmic membrane determines the fluidity of the membrane at a given temperature. Because the proper functioning of a cytoplasmic membrane depends on fluidity, it is not surprising that membrane composition reflects the growth temperature.

Carbon Dioxide Availability

Virtually all microorganisms require carbon dioxide for growth. However, the level necessary for optimal growth varies with the microbial species. The atmospheric level of carbon dioxide (0.03% by volume) satisfies the minimal requirement for virtually all heterotrophic and autotrophic microorganisms. Optimal growth of autotrophic microorganisms, however, is attained in culture when they are provided with carbonates or with air enriched in carbon dioxide. Heterotrophs cannot utilize carbon dioxide as the sole source of carbon, but they do require it for growth. These microorganisms assimilate carbon dioxide into a number of metabolic intermediates involved in synthesis and energy generation, and complete removal often leads to cessation of growth. Some animal and human pathogens, such as *Neisseria gonorrhoeae*, the agent of the venereal disease gonorrhea, will not grow in laboratory culture unless they are incubated in the presence of an atmosphere containing 5% to 10% carbon dioxide.

Water Availability

All microorganisms—whether bacterial or archaeal—require a source of water for survival. Availability of water is a major factor in determining whether a natural environment is colonized. It also markedly influences the type of microorganisms that can inhabit an environmental niche. The term **water availability** differs from **presence of water** because even if it is present, solids or surfaces in the environment may absorb molecules of water, rendering it unavailable to support life. For example, solutes such as sugar or salts that dissolve in water have a high affinity for water. If the concentration of a solute is high enough, it can withdraw water and make it unavailable to a cell. For this reason, microbes generally do not grow in solutions such as saturated salt or undiluted honey.

Water availability is expressed as **water activity** (a_w) and is defined as the vapor pressure of air over a substance or a solution divided by the vapor pressure over pure water. If a solute such as sodium chloride absorbs water, the vapor pressure in the atmosphere above that solution becomes lower. *Bacteria* and *Archaea* strains differ in their tolerance for decreased available water (Table 6.4).

If water is placed on one side of a semipermeable membrane and concentrated saltwater is placed on the other side, the water diffuses across the membrane into the salt solution. This process is called **osmosis**. A bacterium placed in a culture medium is faced with this physical problem. If the solute concentration is higher in the medium than inside the cell, water leaves the cell, and the cytoplasmic membrane collapses inward. The organism may die of dehydration, a process called **plasmolysis**. Bacteria can prevent this from occurring by increasing the solute concentration inside the cell to one greater than in the suspending medium.

Microbes use a variety of solutes to establish this higher cytoplasmic solute concentration with normal cellular activities. Because they do not interfere with normal cellular activities, they are called **compatible**

TABLE 6.4	Tolerance of selected bacteria and archaea for decreased water activity a_w	
Type	**Organisms**	a_w
Nonhalophiles	*Aquaspirillum* and *Caulobacter* spp.	1.00
Marine forms	*Pseudomonas* and *Alteromonas* spp.	0.98
Moderate halophiles	*Vibrio* species and gram-positive cocci	0.91
Extreme halophiles	*Halobacterium* and *Halococcus* spp.	0.75

solutes. Compatible solutes may be ions pumped into the cell from the environment, or they may be organic solutes synthesized by the organism. Many organisms use potassium ions as a solute, including those as diverse as the halophilic organisms that live in highly saline environments and *E. coli*, which lives in the gastrointestinal tract of humans. The cytoplasm of some halophiles ("salt lovers") is virtually saturated with potassium ions (see Chapter 18). The gram-positive cocci synthesize the amino acid proline as solute. Many other species synthesize or accumulate glutamate and glycine–betaine. Yeasts that live in high salt or sugar concentrations synthesize polyalcohols such as sorbitol to serve as the compatible solute.

Because many microbes exist in habitats ranging from moderate to high salt, the terms **halophile** and **extreme halophile** are used to describe their growth (Figure 6.17). Marine bacteria are halophiles that thrive in seawater containing about 0.4 M NaCl. Many of these are **obligate halophiles** and cannot grow without sodium present. Extreme halophiles grow at higher salt levels that can approach saturation (4 M NaCl). Microbes found in freshwater often cannot tolerate moderate salt, and are called **nonhalophiles**. Other bacteria that can adapt well to no sodium to moderate levels are termed **halotolerant**. Again, there is considerable diversity among species in their adaptation to different niches.

Pressure

Some microbes have the ability to grow under extreme pressures found in deep ocean waters and sediments, and the intestinal tracts of deep-sea marine animals. These habitats may extend to 11,000 meters below sea level and reach 50 to 100 Mpa (one Mpa = one million Pascals, where one atmosphere ~0.1 MPa). An organism

that requires high pressure for growth is termed an **obligate barophile**. One that does not is termed a **barophile**. Although most barophilic microbes are also psychrophilic, one hyperthermophilic habitat was mentioned previously: thermal vents in the ocean floor.

SECTION HIGHLIGHTS

Environmental conditions have a profound effect on the ability of a cell to grow and survive. Because an individual species has a limited range of conditions (temperature, pH, oxygen and ion levels, and water availability) that is permissible, its natural habitat is generally restricted to niches with similar properties. These parameters are easily defined experimentally, and, once established, allow one to predict the potential for a cell to grow.

SUMMARY

- An increase in the number of cells in a culture medium is termed **population growth**. An increase in the size of a bacterial cell is **cell growth**.

- A **viable microorganism** is one that can reproduce in the laboratory. This definition is flawed, because conditions for the growth of most microorganisms present in nature have not yet been defined. Our inability to grow them results from our ignorance, not from their lack of viability.

- An increase in the number of microorganisms per unit time is the **growth rate**. The **generation time** is the time required for a population to double in number.

- The addition of a small number of microorganisms to a suitable growth medium and recording the number of microorganisms present versus time will produce a **growth curve**. The component parts of a growth curve are: **lag phase, exponential growth phase, stationary phase**, and **decline** or **death phase**.

- During **balanced growth**, every component of a cell culture increases by the same constant factor per unit time.

- During the **exponential growth phase**, the number of organisms present increases logarithmically, that is, $1 \rightarrow 2 \rightarrow 4 \rightarrow 8 \rightarrow 16 \rightarrow 32$, etc.

- The **stationary phase** is a period in which a culture does not increase in number. It is a period of survival.

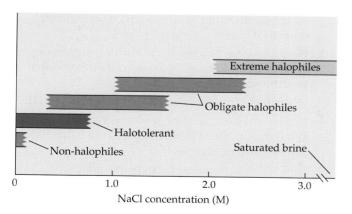

Figure 6.17 Optimal salt concentrations for different types of halotolerant and halophilic microorganisms Halotolerant and halophilic microorganisms are adapted for growth at NaCl concentrations far higher than those tolerated by non-halophiles.

- The **wet weight** of packed cells is an effective method for determining biomass when survival of the microorganism is important for subsequent analysis, such as enzyme isolation or metabolic studies. **Dry weight** of cell mass is more accurate but tends to destroy the microorganism. **Total nitrogen** and **total protein** are effective indicators of biomass, and it can be accomplished on a small amount of material.

- A **direct count** of bacterial or archaeal cells in a suspension can be performed microscopically with a **calibrated slide** or with a **Coulter counter**.

- When microorganisms grow in a clear liquid medium, they create a visible **turbidity**. This turbidity can be measured in a **spectrophotometer** and is a reliable method for determining biomass. Turbidity is evident when there are at least 10^7 microorganisms present per milliliter.

- **Maintenance energy** is that amount necessary to stay alive.

- The events in the **cell cycle** can be followed in a population with all the cells in the same physiological state. A culture in this state is called a **synchronous culture**. A number of techniques are available for establishing synchronous cultures.

- **Growth yield** is a determination of biomass produced per mole substrate utilized.

- When substrate is added below a threshold level in a chemostat, the growth rate is adversely affected and cells **washout**.

- When substrate is added above a critical dilution rate in a chemostat, the cells washout.

- A microorganism growing on a liquid nutrient in a closed system under favorable conditions for a set period is called a **batch culture**.

- A culture may be maintained in the exponential growth phase, and in balanced growth for extended periods in **continuous culture**. Continuous culture may be accomplished in a **chemostat apparatus** or a **turbidostat**.

- Selected species of the *Bacteria* and the *Archaea* can grow from a **pH** of near 1 to 11. Most soil bacteria favor a pH near neutrality (7); fungi generally grow better at an acidic pH (2 to 6).

- Virtually all eukaryotic organisms are **aerobic** and have a requirement for molecular oxygen in their energy generation. The *Bacteria* and the *Archaea* initially evolved in an **anaerobic** environment, and many species can thrive in the absence of O_2.

- Most microorganisms that live in the presence of air must have mechanisms for removal of toxic products of O_2 utilization: **superoxide**, **hydrogen peroxide**, the **hydroxyl radical**, and **singlet oxygen**.

- The most effective way to manipulate strict anaerobes in the laboratory is in a **glove box**, where all molecular O_2 can be excluded.

- The **temperature ranges** for growth of various microbial types: 0°C to 20°C, **psychrophiles**; 20°C to 44°C, **mesophiles**; 45°C to 70°C, **thermophiles**; and 70°C to higher than 110°C, **hyperthermophiles**.

- **Carbon dioxide** availability is necessary for the growth of virtually all microorganisms, and growth of autotrophs is stimulated by the presence of higher than atmospheric levels of CO_2.

- **Water activity** (a_w) or water availability is an important factor in the growth of microorganisms. Life cannot exist without available water.

 Find more at www.sinauer.com/microbial-life

REVIEW QUESTIONS

1. What is the difference between cell growth and population growth? How do these concepts fit into the cell cycle?

2. A growth curve is divided into phases designated lag, exponential growth, stationary, and decline or death. What is happening to individual cells during each of these phases? How does this relate to microorganisms in nature?

3. Total dry weight and viable count of cells are two methods employed in measuring cell mass. What are some of the pros and cons of each method? How can turbidity be employed in measuring microbial mass?

4. Why are some studies done with cells growing in synchronous culture? How would one obtain a synchronous culture?

5. Describe how a chemostat works. A turbidostat.

6. Define aerobe, anaerobe, and microaerophile. What do *strict* and *facultative* mean in reference to aerobes and anaerobes?

7. How have aerotolerant organisms adapted to tolerate the presence of toxic oxygen products?

8. What are some of the reasons that microorganisms have a fairly limited temperature range for growth? What is a cardinal temperature?

9. Water availability is an important factor in microbial growth. How can an organism grow in high concentrations of sugar or salt?

10. What metabolic/structural adaptations for extreme temperatures have psychrophiles and thermophiles made?

SUGGESTED READING

American Waterworks Association. 1998. *Standard Methods for the Examination of Water and Waste Water*. 20th ed. Washington, DC: American Waterworks Association.

Cooper, S. 1991. *Bacterial Growth and Division: Biochemistry and Regulation of Prokaryotic and Eukaryotic Division Cycles*. New York: Academic Press.

Dworkin, M., S. Falkow, E. Rosenberg, K-H. Schliefer and E. Stackebrandt, eds. 2000. *The Prokaryotes: An Evolving Electronic Resource for the Microbiological Community*. 3rd ed. New York: Springer-Verlag. Release 3.7, latest update December 2001.

Jennings, D. and S. Isaac. 2006. *Microbial Culture: Introduction to Biotechniques*. Oxford: Garland/Bios Scientific Publishers.

Gerhardt, P., ed. 1993. *Methods for General and Molecular Bacteriology*. 2nd ed. Washington, DC: ASM Press.

Neidhardt, F. C., J. L. Ingraham and M. Schaechter. 1990. *Physiology of the Bacterial Cell*. Sunderland, MA: Sinauer Associates, Inc.

Norris, J. R. and D. W. Ribbons. 1970. *Methods in Microbiology*. Vols. 2 and 3A. New York: Academic Press.

The objectives of this chapter are to:

◆ Examine the basis of microbial growth control by physical agents including temperature, pH, and ionizing radiation.

◆ Survey chemical agents used to control microbial growth.

◆ Describe biological inhibitors of cellular function that can be used as antimicrobial agents.

7

Control of Microbial Growth

In order to use chemotherapy successfully, we must search for substances which have an affinity to the cells of the parasites and a power of killing them greater than the damage such substances cause to the organism itself, so that the destruction of the parasites will be possible without seriously hurting the organism.
—Paul Ehrlich, Father of Chemotherapy, 1854–1915

The title of this chapter, "Control of Microbial Growth," is essentially a contradiction in terms. We live in a world where about one-half of the total biomass is microbial, and our bodies contain more microbial cells than human cells. In reality, we coexist with microorganisms. We attempt to keep the harmful ones at bay and foster the growth of those that are helpful. But regardless of our efforts, they are always around, and some can pose a serious threat. Their growth is never really under total "control."

During the last 130 years, humans have come to better understand the microbe, and our strategies for influencing microbial growth have thus evolved. We have developed procedures whereby, to a limited extent, microbes can be controlled. This chapter discusses many of the procedures and processes that are employed to control the microbe in areas where this is desirable. We sterilize with heat and chemicals, we filter air or liquids, and antibacterial chemicals are now available to cure disease. We have developed immunization methods and public health measures to survive in a world of the ubiquitous microbe.

In previous chapters we have discussed the composition of microbial cells, and how environmental factors may affect growth. These features provide a basis for understanding the effectiveness of antimicrobial agents in inhibiting and/or killing of cells. For example, how does temperature or other environmental factors interplay with the concentration and duration of exposure to antimicrobial agents. How do cell population size and composition affect the rate of cell death?

7.1 Historical Perspective

To counteract unwanted activities by microorganisms, humans have spent much money and effort devising preservation methods and have met with moderate success. In early times before microorganisms were known and understood, whatever anyone did was the result of an empirical observation that something worked. Human populations have historically suffered from the inability to understand or gain more than a limited amount of control over microorganisms until recent times.

Early civilizations used many methods to protect food from microbial destruction. These methods included the salting of meat, fish, and vegetables. As discussed in the previous chapter, salting removes available water and prevents bacterial growth. Certain bacterial species either alone or combined with salt proved effective in preserving vegetables, such as cabbage to produce sauerkraut. Long ago, people found that fermented milk products such as cheese, butter, and yogurt were stable for some length of time even without refrigeration. Acidification with vinegar (pickling) was also effective in preserving selected food products. In another example, American Indians dried meat, berries, and fish. In 1800, Napoleon Bonaparte offered a reward for devising new methods for food preservation in order to support his armies. By 1809 the French confectioner Nicholas Appert (1750–1841) discovered that meat and other foodstuffs could be preserved by heating in a closed container (canning); this set the stage for many advances that have radically changed how we eat and live.

The development of effective control measures for microorganisms in medicine in general parallels the evolution of microbiology as a science (see Chapter 2). Several important early workers in medicine discovered microbial control techniques that significantly decreased the number of deaths during medical procedures. It is generally acknowledged that **Ignaz Semmelweis** (1816–1865) pioneered the use of disinfectants. He insisted that all individuals working in his hospital in Vienna wash their hands in chlorinated lime before working with patients. This brought about a marked decrease in disease and death in the obstetrics ward. **Joseph Lister** (1827–1912) was able to conduct aseptic surgery by sterilizing instruments with carbolic acid (phenol) and spraying phenol around the room during surgery. As discussed in Chapter 2, Robert Koch demonstrated that mercuric chloride would kill all bacteria tested, including the highly resistant endospore.

John Tyndall (1820–1893) developed an early and effective sterilization procedure, termed "**tyndallization**" (see Chapter 2). In this procedure, a broth would be steamed for a few minutes on three or four successive occasions separated by 12- to 18-hour intervals of incubation at a favorable temperature. This would permit dormant spores to become vegetative cells and be killed by boiling.

The invention of the autoclave in Pasteur's laboratory revolutionized laboratory work, as it permitted one to sterilize media and to work under **aseptic conditions**.

SECTION HIGHLIGHTS

An understanding of the scientific basis of food preservation and human health has developed over the last 150 years. This basic and applied knowledge has radically changed human health, lifespan, and culture.

7.2 Physical Methods of Control

The control of physical parameters including temperature and radiation may be used to disrupt cells or to destroy molecules essential for survival. An assessment of these methods and their effectiveness is addressed below.

Heat Sterilization and Pasteurization

Sterilization is the destruction of all viable forms of life, including endospores. A properly packaged sterile material should be devoid of all life until opened and exposed to the outside environment. Sterilization can be attained by various procedures including fire or flame (surface sterilization), moist heat, dry heat, radiation, filtering, and toxic chemicals. Any procedure that disrupts an essential component of the cell—protein, nucleic acid, or cytoplasmic membrane—may lead to a loss of cell function. The more extensive the disruption, the greater the chance that the damaged microorganism will die. When you watch an egg fry, you are witnessing the effectiveness of heat in denaturing (breaking down) egg protein. Some proteins are more stable than others at high temperatures, but this is only a matter of time until denaturation occurs. Given sufficient heat, the structural and intrinsic function of any protein will be lost. Heat, in some form, is the agent most often employed for routine sterilization.

Studies in a microbiology laboratory are dependent on the ability to sterilize media, inoculating loops, glassware, and all material involved in culturing microorganisms. Most studies are performed with pure cultures (axenic), and this can be accomplished only when all manipulations are done under sterile conditions. Sterility is also an important requirement for surgical instruments, bandages, and in most medical procedures.

Several factors can affect any sterilization procedure. First, the presence of organic matter can protect bacterial and archaeal cells from the destructive effects of heat, chemicals, or radiation. Second, a material to be sterilized usually harbors a mixed population of microorganisms. Consequently, the regimen selected for sterilization must be designed to kill *all* of the most resistant organisms present.

The pH of the suspending medium is also a significant factor. In an acidic medium, heat is considerably more effective in killing microorganisms than in neutral medium. This has been a major factor in home canning, where acidic foods, such as fruits and tomatoes, rarely spoil. Foods that have a more neutral pH, such as corn, peas, and beans, spoil more often because on harvesting they harbor endospore-forming bacteria that may survive the canning process. On rare occasions, the anaerobic spore-forming organism *Clostridium botulinum* may survive in improperly canned foods. This organism then may germinate and grow, producing botulinum toxin, which causes botulism, a type of food poisoning (see Chapters 20 and 28). Ingestion of minute amounts of botulinum toxin can lead to death. This was a danger in the days of extensive home canning, when inefficient or improperly used pressure cookers were sometimes employed.

Heat Sterilization

Microbial death by heating occurs at an exponential rate (Figure 7.1); that is, at any given temperature a set fraction of the cells present in a heated solution will be killed per unit of time. The effect of heat on the survival of a bacterium is illustrated in Figure 7.1 and affirms that a higher temperature (95°C) kills more rapidly than a lower temperature (75°C). A survival curve can be obtained by suspending bacterial cells in water and heating to a selected temperature. At intervals, an aliquot can be removed and the number of viable cells present de-

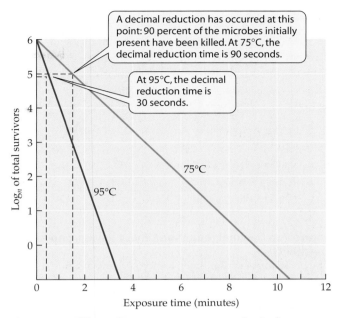

Figure 7.1 Effect of temperature on survival of bacterial cells
An exponential plot of number of surviving cells versus time exposed to heat, at two different temperatures ($D_{75} = 90$; $D_{95} = 30$), for the vegetative cells of a mesophilic bacterium.

termined by a plate count as described in Chapter 6. The length of time required to reduce a population tenfold (for example, from 10^6 to 10^5 cells) is the **decimal reduction time**, designated **D**. The temperature tested, is indicated in the *D*-value subscript, for examples D_{95} or D_{75}.

Three general techniques are available for killing microorganisms by heat: boiling, steam under pressure, and dry heat. Boiling for 15 minutes will kill virtually all vegetative bacterial and archaeal cells, viruses, and fungi, but it does not generally kill bacterial endospores. Endospores of various bacterial species have different resistances to boiling (100°C), as indicated by the D_{100} values (Table 7.1).

TABLE 7.1	Heat resistance of bacterial endospores obtained from various genera and species in the genus *Bacillus*. Decimal reduction time at 100°C
Organism	D_{100} **Value**[a] **(minutes)**
Bacillus cereus	0.8
Bacillus megaterium	2.1
Bacillus anthracis	5
Bacillus cereus var. *mycoides*	10
Bacillus licheniformis	24.1
Bacillus coagulans	270
Bacillus stearothermophilus	459

[a]Time required to reduce the population tenfold.

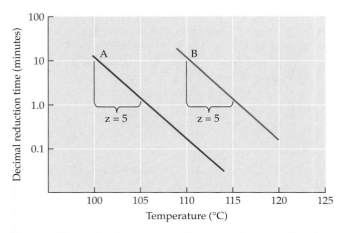

Figure 7.2 Decimal reduction time, z values, and heat sensitivity
The decimal reduction times for two organisms: one relatively heat sensitive, *B. anthracis* (A), the other more heat resistant, *C. botulinum* (B). The rate of killing for each increases at higher temperatures.

A comparison of the decimal reduction times at different temperatures can distinguish between a heat-sensitive microorganism and one that is relatively resistant (Figure 7.2). The time required to sterilize a suspension of endospores varies over a wide range, with *Bacillus anthracis* requiring less than 5 minutes. In contrast, a suspension of *C. botulinum* endospores can be heated for 5 hours or longer at 100°C, and some of the endospores will sur-

vive. If placed in an appropriate medium, the endospores can germinate and grow. It is clear that boiling water has its limitations for sterilization. When the *D* values are measured over a range of different temperatures as indicated in Figure 7.2, one may derive a **z value**, which is an empirical measure of the time and temperature relationship for cell killing of a particular microbial species. For example, the z values for both strains in Figure 7.2 indicate that a 5 degree increase in temperature achieves a tenfold increase in cell killing. This measure has practical applications in food and medicine related industries since it describes the thermal resistance properties of a particular strain under the conditions tested.

MOIST HEAT—STEAM UNDER PRESSURE   The most effective sterilization system is steam under pressure. This method is widely used in the research laboratory and throughout the food and pharmaceutical industries. The instrument available for accomplishing this is the **autoclave**. An autoclave is a closable metal vessel (Figure 7.3) that can be filled with steam at greater than atmospheric pressure.

The autoclave is designed so that the air present inside when the door is closed and sealed is driven out by steam. A valve closes automatically when the air passing outward reaches 100°C, an indication that steam now fills the entire inner space. Laboratory autoclaves operate at 15 psi (pounds per square inch) pressure, and at this pressure, the temperature rises to 121.6°C. The au-

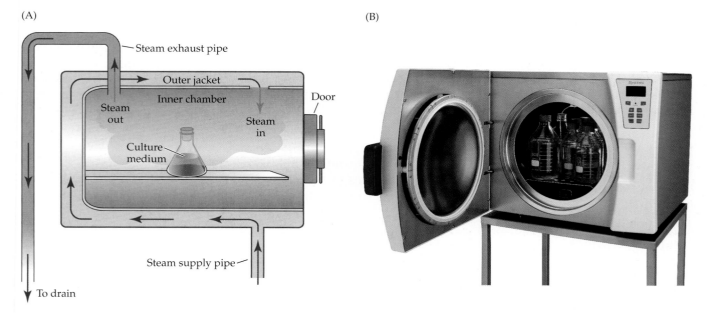

Figure 7.3 Design of an autoclave used for live steam sterilization
(A) Live steam heated to 121°C enters the inner chamber and is held at pressure for a period of time necessary to sterilize the culture medium or sample. Pressure is then slowly released to prevent the super heated liquid from boiling over. (B) An example of an autoclave. B, courtesy of Systec GmbH.

toclave must be constructed to withstand the internal pressure produced when water boils. At this temperature, sterilization of routine material can be achieved within 10 to 15 minutes.

Large batches of material must be autoclaved for a longer time-period because heat transfer to the interior can be slow. The denser the material, the slower the heat is transferred and the longer the time required for sterilization. Thus, it takes longer to sterilize a flask containing 1 liter of liquid than it does when the same flask is empty. The autoclave is effective for the sterilization of most laboratory growth media. Selected organic compounds such as vitamins and antibiotics may not be stable at autoclave temperatures due to thermal denaturation, and should therefore be sterilized by filtration or other less-destructive means, as discussed in subsequent text.

DRY HEAT   **Dry heat** is effective in sterilizing glassware, pipettes, and other heat-stable solid materials. Pipettes can be placed in metal canisters, glassware wrapped in paper, and flasks and other vessels capped with aluminum foil. This protects the objects from contamination when they are handled following sterilization. An oven that can be heated to 160°C to 180°C is used for dry heat sterilization, and the materials are retained at this temperature for 2 to 4 hours. Dry heat has been preferred for sterilizing pipettes, syringes, and other heat-stable objects, as autoclaving leaves them moist. However, most laboratory autoclaves now have a vacuum cycle that will remove moisture from such objects after sterilization.

PASTEURIZATION   **Pasteurization** is a process that employs low heat and is often applied to milk products and other heat-sensitive foods that would be destroyed if boiled or autoclaved. Pasteurization reduces the microbial population but does not sterilize the product. The method was devised by Louis Pasteur to control wine spoilage. Pasteurization was originally accomplished by heating the material at temperatures from 63°C to 66°C and holding it at this temperature for 30 minutes. This method will kill 79% to 99% of the microorganisms present in milk. A short-term "flash method" is now generally employed, whereby milk is passed through coils where it is heated quickly to a temperature of 71.6°C for 15 seconds, and then rapidly cooled to preserve food flavor and color.

Pasteurization of milk was adopted as a common practice in the dairy industry because raw milk often contained *Mycobacterium tuberculosis*, the causative agent of tuberculosis (TB), and/or *Brucella abortus*, which causes brucellosis (see Chapter 30). These organisms have now been mostly eliminated from dairy herds in the United States through mandatory testing. However, pasteurization is still employed as a precaution and to extend the shelf life of certain food products, including milk and beer. Due to the widespread consumption of dairy products, the potential spread of disease by this consumable is a constant threat. Why?

Sterilization by Radiation

Selected wavelengths of the electromagnetic spectrum (Figure 7.4) can cause the death of living organisms, thus

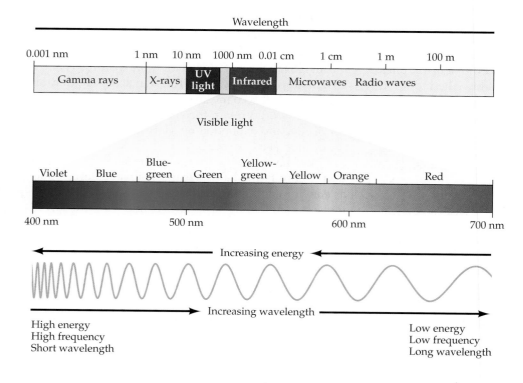

Figure 7.4 The electromagnetic spectrum
Humans can detect "visible" wavelengths of light, which range from about 400 nm (violet) to 750 nm (red). Shorter wavelengths (UV, x-rays, gamma rays) denature biological molecules.

making radiation a practical method of sterilization. The two types of radiation that are effective microbe killers are **ultraviolet (UV) light** and **ionizing radiation**. Light in the UV range can disrupt cellular DNA and some other cell constituents, whereas ionizing radiation can cause harm directly and indirectly to all constituents of a cell. Gamma (γ) rays, x-rays, alpha (α) and beta (β) particles, and neutrons are all forms of ionizing radiation.

IONIZING RADIATION   As is true for other forms of ionizing radiation, gamma rays kill cells by causing the formation of highly reactive free radicals. These free radicals, such as the hydroxyl radical (OH•), can react with and inactivate any of the various macromolecules present in a cell. Because there are multiple copies of most cellular macromolecules, damage to a limited number of the copies of a protein will not cause cell death. Generally, however, the cellular DNA will contain a single copy of a specific gene, and disruption of a gene can cause cell death. Ionizing radiation is destructive to genetic material because it causes single- and double-stranded breaks in the DNA double helix.

The sensitivity of different organisms to ionizing radiation varies considerably. Endospores of *Clostridium* or *Bacillus* species are highly resistant, as are cells of *Deinococcus radiodurans*. *D. radiodurans* is exceedingly resistant to UV or gamma radiation and also to most mutagens. Materials that have been exposed to ionizing radiation often bear this organism as the sole survivor. The ability to repair DNA damage efficiently and rapidly is a major survival factor in radiation-resistant microorganisms.

Gamma rays (see Figure 7.4) are an effective sterilizing agent because they can penetrate deeply into objects such as containers. Consequently, sterilization can be accomplished after a food or surgical product is packaged. Major sources of usable radiation for sterilization include cobalt-60 (gamma rays), x-rays, and the neutron piles that are available at nuclear plant sites. The radiation source most often used commercially is cobalt-60. Generally, items that are heat sensitive and nonfilterable are sterilized by radiation. Among the items sterilized in this way are plastics, antibiotics, sera, various medicinals, and, in some cases, food. This procedure does not generate harmful products.

ULTRAVIOLET LIGHT   Wavelengths of electromagnetic radiation in the ultraviolet range (UV) between 220 and 320 nm are particularly important biologically because cellular constituents can absorb these wavelengths. More intense, shorter wavelengths of UV (less than 220 nm) are emitted by the sun but are absorbed by the ozone layer in the Earth's upper atmosphere. This shorter wavelength UV radiation is lethal to microorganisms and harmful to many other organisms as well. These

deleterious effects are the cause of the present concern over the damage to the ozone layer over the Arctic and Antarctic areas.

Germicidal lamps and lights can be constructed to emit the short, damaging UV wavelengths. UV lamps with this potential may be placed in sterile rooms and safety cabinets to sterilize the air, but cannot be operated when people are present. These wavelengths burn skin and may damage the eyes. UV radiation does not penetrate glass, water, or nongaseous substances to a significant extent. This limits the effective use of UV radiation for sterilization of surface areas and air.

Cellular components that strongly absorb UV light are the purines and pyrimidines. Both are present in DNA and RNA and absorb UV maximally at 260 nm (see Figure 7.4). A UV lamp will produce levels of radiation that can effectively penetrate bacterial, archaeal, or other viable cells and cause alterations in cellular nucleic acids. The production of **thymine dimers** is probably the most important result of UV damage to DNA. A dimer is formed by the linkage of two adjacent thymine molecules on a single strand of DNA. Such linkages prevent the replication of DNA, and this can be lethal to a cell. Thymine-cytosine or cytosine-cytosine dimer formation occurs less frequently but also disrupts DNA replication.

Microbes have **DNA repair systems** that can excise the dimers and replace them with unaltered thymine molecules. UV light is lethal if a sufficient number of dimers are created such that all of them cannot be repaired. In some cases, the enzyme repair system will make an error, which leads to mutations or death.

Sterilization by Filtration

The ability of filters to retain bacterial and archaeal cells and allow viruses and other small particles to pass through has been recognized for more than 100 years. In 1892, Dmitri Ivanowsky, a Russian scientist, demonstrated that tobacco mosaic virus would pass through a filter that would retain the smallest bacteria. This was confirmed and expanded by Martinus Beijerinck a few years later (see Chapter 2).

Filters have proven useful for the selective sterilization of liquids and gases. Heat-sensitive materials, such as enzymes, vitamins, and sugars, can be filtered to remove contaminating bacteria. Beer and wine are now filtered in bulk to remove bacteria that would oxidize the ethanol to acetic acid and render these beverages unacceptable (see Chapter 31). Virologists and cell biologists routinely filter blood serum and other heat-sensitive culture media to prevent unwanted bacterial growth. Media for plant or animal cell cultures are generally complex, and components break down at high temperatures. They, too, are sterilized by filtration.

Basically, a filter is constructed with a pore size that permits liquids to pass, but the pores are small enough to retain *Bacteria*, *Archaea*, and eukaryotic cells. These organisms, in general, are longer than 1 μm and are greater than 0.5 μm in diameter. Because some bacterial cells are smaller than this, effective filters must be able to retain cells that are quite small. Filtration of virus particles require yet smaller pore sizes.

In practice, a filter can also be employed to separate or concentrate microbes based on their size. Bacteria and archaea can be separated from yeasts by employing a membrane with a pore diameter of approximately 3 μm. Most prokaryotic cells can pass through a filter such as this, but yeasts would be retained. Small bacteria such as *Bdellovibrio*, an organism that parasitizes other bacteria, can be concentrated by passing water or dilutions of soil through membrane filters with a pore diameter of 0.45 μm. The *Bdellovibrio* are only 0.25 μm in width and would pass through, whereas most other microorganisms would be retained. The organisms in the filtrate can then be concentrated by centrifugation.

Most early filters were composed of stacks of asbestos fibers, sintered glass (glass heated to fuse without melting), or diatomaceous earth. These materials formed circuitous passageways that retained microbial cells. Several different types of filters are used today. A **glass fiber depth filter** is a random stacking of glass fibers that forms a barrier impenetrable by bacteria and other particles (Figure 7.5A).

The filters that are now in common use are **membrane filters** made of cellulose acetate, polycarbonates, or cellulose nitrate (Figure 7.5B). These filters are quite thin (200 μm) and become relatively transparent when wet, allowing direct microscopic examination of microorganisms that are trapped on the filter. Membrane filters can be produced with pore sizes ranging from 0.1 to 0.5 μm, with 80% to 85% of the filter being open space. This construction allows a high rate of fluid passage. With this much open space available, a filter can be placed on the upper surface of an agar medium, and nutrients will diffuse to the cells through the filter by capillary action. Following incubation and growth, one can count the number of colonies that develop directly on the filter and observe their colonial morphology (see Figure 6.7).

The **isopore** membrane filter has widespread application in research and industry. It contains clearly defined cylindrical holes that pass vertically through the membrane (Figure 7.5C). These filters are manufactured by treating thin sheets of polycarbonate with nuclear radiation. The radiation results in a random distribution of minute holes in the polycarbonate that can be enlarged by chemical etching. The length of time that the membrane is etched determines the size of the pores. A major advantage of these filters is that bacteria and other objects remain on the surface of the membrane and are thus readily observable with a scanning electron microscope (see Figure 7.5C). With other types of membrane filters, the bacteria absorb to the sides of the pores

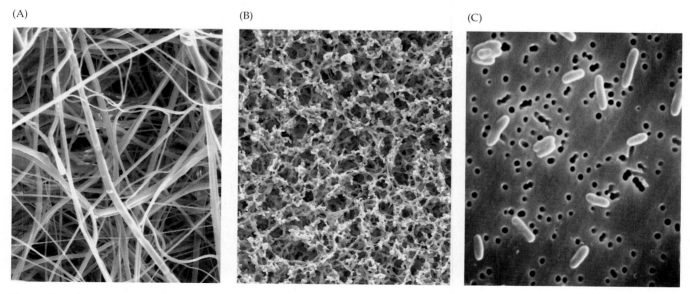

(A)   (B)   (C)

Figure 7.5 Scanning electron micrographs of filters
Scanning electron micrographs of (A) a glass fiber depth filter; (B) a membrane filter; (C) *Thermoleophilum album* trapped on an isopore filter. A,B, courtesy of Millipore Corporation; C, courtesy of Jerome Perry.

(A) (B)

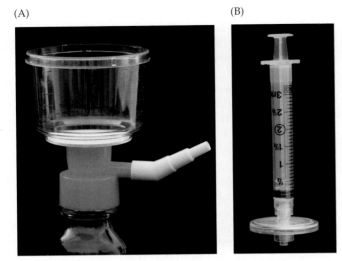

Figure 7.6 Microbiological filter unit for liquid sterilization
A liquid sample is pulled through the membrane filter and into a sterile chamber by application of a vacuum (A) or pushed through with a syringe (B). Courtesy of C. Baldwin.

and are not so readily observable. One problem with the isopore filters, however, is the limited total space on the filter that is porous. This may result in low flow rates or clogging.

Pre-sterilized disposable filter units available commercially are now used routinely in the laboratory (Figure 7.6). In industry, much larger and more sophisticated filtration systems are used for large volumes of liquid. These larger volumes are sterilized by utilizing the filters placed in large stainless-steel cartridges. As an alternative to centrifugation, particularly for large volumes, bacteria can also be harvested from culture fluids by using membrane filters.

SECTION HIGHLIGHTS

Physical agents may provide effective but non-selective means of killing microbes. The effectiveness will depend on the type of agent used, and the length of treatment. Although some methods are destructive (temperature, pH, UV light, ionizing radiation) other methods rely on physical separation of the microbe from the liquid sample (filtration). Methods have also been devised to quantitatively evaluate the action of antimicrobial agents under defined conditions.

7.3 Chemical Methods of Control

We previously mentioned that some of the nineteenth-century scientists made use of chemical disinfectants. Included were Koch (mercuric chloride), Semmelweis (chlorinated lime), and Lister (phenol). Today we have a vast number of chemical agents available for use as disinfectants or to control the growth of undesirable microorganisms. These range from toxic gases to the medically important antibiotics. Following is a discussion of the chemicals now employed to kill or control microorganisms.

Types of Antimicrobial Agents and their Properties

An array of chemical agents is now available and commonly used in the control of undesirable microorganisms. These chemicals are called **antimicrobial agents**, or antimicrobials (Table 7.2). Some of the agents are naturally occurring, some are synthetic, and others are a combination of the two.

An effective method for sterilizing surgical instruments and some heat-sensitive materials is to expose them to toxic compounds in the form of gases. Sheets and bedding in a hospital are sterilized in this manner. Toxic gases are also used to sterilize plastic Petri plates and other plastic laboratory containers. Ethylene oxide (Figure 7.7) and propylene oxide are generally the compounds of choice. Both function as alkylating agents in disrupting cellular DNA. Because these gases are exceedingly toxic to humans, their use in the sterilization process must be carefully controlled.

A much-desired property of an antimicrobial agent is **selective toxicity**. That is, the agent will eliminate a pest without doing significant harm to other species in the environment. Spraying plants with a chemical that kills invading bacteria or fungi but also harms the plant is of little value. Humans cannot be cured of disease by treating them with medicines that are overly harmful to the host (see the quote at the beginning of this chapter).

In addition, humankind has become more and more concerned with the effects of chemical agents on the total environment. Experience has taught us the harm that can be done by careless application of antimicrobial substances. For example, for years, compounds containing mercury were used in the pulping industry and other industrial processes to control microbial contamination. These mercury compounds ultimately entered streams and rivers, resulting in considerable harm to fish and other wildlife. Stringent regulations are now in place for all potentially toxic compounds that are artificially introduced into the environment. They must be tested for safety, biodegradability, carcinogenicity, or other un-

TABLE 7.2 Common antimicrobial agents and their uses

Compound	Use
Organic compounds	
Cresols	Disinfectant in laboratories
o-Phenylphenol	Disinfectant in laboratories
Phenol	Disinfectant in laboratories
Formaldehyde	Disinfect instruments
Quaternary ammonium compounds	Skin antiseptic compounds
Ethanol and isopropanol	Disinfect instruments
Ethylene oxide	Sterilize instruments
Propylene oxide	Sterilize instruments
Halogens	
Chlorine gas	Disinfect water
Iodine solution	Skin antiseptic
Hexachlorophene	Skin antiseptic
Triclocarban	Skin antiseptic
Chlorine bleach	Domestic cleaning
Heavy metals	
Mercuric chloride	Disinfectant
Silver nitrate	Eye treatment to prevent ophthalmic gonorrhea
Copper sulfate	Algicide
Sodium arsenate	fungicide
Other	
Hydrogen peroxide	Skin antiseptic
Mercurochrome	Skin antiseptic
Merthiolate	Skin antiseptic
Cetylpyridinium chloride	Mouthwash antiseptic
Ozone	Drinking water and fish tanks

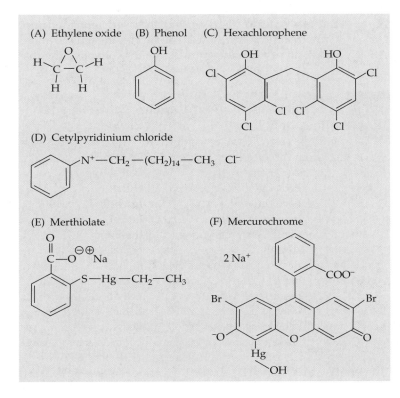

Figure 7.7 Disinfectant and antiseptic compounds
Chemical structures of (A) ethylene oxide, (B) phenol, (C) hexachlorophene, (D) cetylpyridinium chloride, (E) merthiolate, and (F) mercurochrome.

desired effects on humans. Their application is a balance between potential benefit versus potential harm.

Another example of a widely applied agent was DDT (dichlorodiphenyltrichloroethane), which was used as an insecticide. DDT was effective against insects, but it accumulated in the natural environment and was very harmful to birds and other wildlife. Accumulation resulted from the inability of microbes to mineralize DDT at rates commensurate with application rates. DDT had a half-life of about 30 years in most environments, and its use has now been curtailed in the United States. The insecticide is still used in other parts of the world.

A compound that kills a target group is identified by combining the designated organism's name with the suffix *-cide*, from the Latin *cida*, meaning "having power to kill" (Table 7.3). Other agents are available that do not kill but rather prevent or inhibit growth. These are referred to as **bacteriostatic agents** (for bacteria) or **fungistatic agents** (for fungi). **Germicide** is a general term that describes substances that kill or inhibit microbes ("germs"). Some antibacterial substances bring about **lysis**, or disintegration, of the cell and are thus termed **bacteriolytic agents**. For example, a lysozyme is a lytic enzyme that weakens the cell wall peptidoglycan backbone. When sufficient strands are broken, the cell wall no longer protects the integrity of the cytoplasmic membrane. Subsequent water uptake results in bursting of the wall-impaired membrane.

A **disinfectant** can destroy microorganisms on contact, including those that potentially cause disease. Phenol (see Figure 7.7) and mercuric chloride are disinfectants. Disinfectants are indiscriminate destroyers that, if applied to wounds, would also disrupt host tissue. An **antiseptic** is a substance—such as Merthiolate or Mercurochrome (see Figure 7.7)—that kills or prevents the growth of disease-causing pathogens. An antiseptic should not do significant harm to host tissue. As a rule, we use powerful disinfectants such as phenol on inanimate objects such as laboratory benches, and less-potent antiseptics to treat wounds.

A **chemotherapeutic** agent is a chemical compound that can be applied topically, taken orally, or injected that will inhibit or kill microbes that cause infections. The term is also applied to compounds that are effective against viruses or tumor growth. **Antibiotics** are sometimes called chemotherapeutic agents and are widely used as antimicrobials, but the term "chemotherapeutic" is generally reserved for products of chemical synthesis. The antibiotics will be discussed later.

Tests for Measuring Antimicrobial Activity

Robert Koch (1843–1910) first devised a laboratory procedure for quantifying the effectiveness of antimicrobial compounds. Koch dried endospores of *Bacillus anthracis* on threads and placed these in solutions of potential antibacterial agents. Endospores were selected because they are more resistant to harsh physical conditions than other life forms.

Koch then estimated the effectiveness of a given compound by determining how long the endospores would survive exposure to the agent. At selected time intervals, threads with attached endospores were removed from the test solution and placed in a nutrient medium. Growth indicated survival. The shorter the time required to kill all of the endospores, the more effective the dis-

TABLE 7.3	Chemical agent, their target organisms, and mode of action		
Agent	**Target Organism**	**Action**
Bactericide	Bacteria	Lethal
Bacteriolytic	Bacteria	Lethal
Bacteriostatic	Bacteria	Inhibitory
Fungicide	Fungi	Lethal
Fungistatic	Fungi	Inhibitory
Algicide	Algae	Lethal
Pesticide	Pests	Lethal
Herbicide	Plants	Lethal
Insecticide	Insects	Lethal
Germicide	"Germs"	Lethal
Disinfectant	All organisms	Lethal
Antiseptic	All organisms	Lethal to microbe but not to human
Chemotherapeutic agent	Specific type of organism	Lethal or inhibitory

infectant. Koch discovered that mercuric chloride was one of the most effective of all available compounds. Mercuric chloride has been employed as a disinfectant up to the present day. Unfortunately, the use of large quantities of this mercury compound is discouraged because of the environmental problems it causes. It is nonselective in that it irreversibly inactivates sulfhydryl-containing biological molecules.

The relative strength of an antimicrobial can be determined by serial dilution of the parent compound. The more a compound can be diluted and still inhibit the growth of a test organism, the greater its relative strength. The lowest concentration that will inhibit growth is designated as the **minimum inhibitory concentration**, or **MIC** (Figure 7.8). Many factors affect the MIC, including the test organism selected, the inoculum size, and the amount of organic material present. A strong germicide, such as ethylene oxide or formaldehyde (37% solution), will kill endospores and virtually all forms of life. Intermediate-strength antimicrobials such as phenolics can kill viruses and most vegetative cells but are less effective against endospores. Such agents are used for research and in hospital laboratories. Antimicrobials with less killing power are effective against vegetative cells (fungal and bacterial) but are not very effective against endospores, fungal spores, or some viruses. These weaker agents have less toxicity to humans and can be applied directly to the skin or can be included as a component in mouthwash. The general mode of action of various germicides is in the destruction of proteins (Table 7.4).

One standard that has been employed to compare the relative efficiency of germicides is the **phenol coefficient**. A phenol coefficient is a measure of the effectiveness of a test compound compared with the disinfecting power of phenol. For example, if a new antimicrobial at a dilution of 1:100 kills a standard population of *Salmonella typhi* that is killed by a 1:50 dilution of phenol, the phenol coefficient would be 100/50, or 2. The test antimicrobial would thus be twice as effective as phenol in destroying the test population.

However, the phenol coefficient has some limitations, as many factors can influence the effectiveness of disinfectants. Some are more affected by the physical environment including light, oxygen, organic matter, and solid substrates. Gram-positive organisms are more readily killed by some disinfectants than gram-negatives due to the properties of their cell envelopes. Members of the genus *Pseudomonas* are not only resistant to, but may also grow on, selected disinfectants. The ability of pseudomonads to thrive in the presence of antiseptics is of particular concern for burn patients.

SECTION HIGHLIGHTS

A variety of organic and inorganic chemical agents are routinely used to eliminate unwanted microbes. They do so in a relatively non-selective fashion, where the concentration of agent and the time of exposure are important variables in predicting cell death. As was described for the application of physical agents, the efficacy of a chemical agent will also depend on the types of microbes present, the population size, and the condition of application.

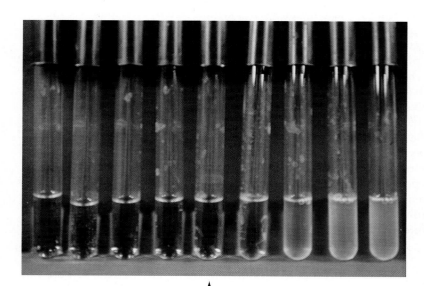

↑
MIC

Figure 7.8　Dilution series for estimating antibiotic sensitivity
A tube dilution series for estimating the sensitivity of a microorganism (in this case *E. coli*) to antibiotic (here, chloramphenicol). The minimum inhibitory concentration (MIC) is the least amount of antibiotic that completely inhibits growth. Courtesy of Jerome Perry.

TABLE 7.4	Mode of action of some antimicrobial agents

Agent	Mode of Action
Alcohols	Denature proteins and dissolve membrane lipids.
Aldehydes	Combine with proteins and denature them.
Halogens	Iodine oxidizes cellular constituents and can iodinate cell proteins.
	Chlorine and hypochlorite oxidize cell constituents.
Heavy metals	Combine with proteins generally through sulfhydryl groups.
Phenolics	Denature proteins and disrupt cell membranes.
Quaternary ammonium compounds	Disrupt cellular membranes and can denature proteins.

7.4 Biological Methods of Control

The antimicrobial agents discussed in the previous section may be used to limit the growth of undesirable organisms in the environment. For example, they may be employed as disinfectants on laboratory benches or other solid material but are not suitable for control of viral, bacterial, or fungal infections within the animal or human body. They would either be too toxic or be inactivated by the presence of organic matter.

Chemotherapeutic Agents

Prior to the development of modern chemical and biological approaches, molds (fungi) were apparently used in ancient times to treat infections (in ancient Greece, Egypt, and China). However, the underlying basis for their use was unknown. Likewise, during the American Civil War and before, soldiers' wounds that were luminous (that glowed in the dark) were an indication of healing and recovery. Only later was it known that an antibiotic-producing luminescent bacterium (*Photorhabdus luminescens*) colonized the wound. From the time of **Paul Ehrlich** (1854–1915), scientists have sought chemicals that selectively inhibit the growth of infectious bacteria without harm to the human host. Ideally, such a **chemotherapeutic agent** would be **specific** for the microbe invading the body but with little or no harmful effect on normal body function. There is a constant search for better chemotherapeutics, including antibacterials, antivirals, and antitumor agents. Following is a discussion of some of the chemotherapeutics that have been synthesized and are effective against disease-causing microorganisms.

Chemically Synthesized Chemotherapeutics

Paul Ehrlich is considered the "father" of chemotherapy (see quote at the beginning of the chapter). He originated the concept that invading pathogenic organisms might be destroyed by selective chemicals that would not seriously harm the host (see Chapter 2). Others, aware of his contributions, have sought chemical compounds that might be employed as chemotherapeutic agents.

The **sulfa drugs** are chemotherapeutic agents that were discovered by Gerhard Domagk in the 1930s. Domagk was assessing the potential use of dye substances for antibacterial activity following the approach pioneered by Paul Ehrlich. Sulfanilamide was one of the compounds tested, and Domagk found that it possessed antibacterial activity both *in vitro* (Latin—in glass [outside the body]) and *in vivo* (Latin—living [inside the body]). Sulfanilamide is a structural analog of *p*-aminobenzoic acid, a component of the vitamin folic acid (Figure 7.9). Folic acid is a coenzyme involved in nucleic acid synthesis. It is synthesized by enzymatically joining a molecule of 6-methyl pterin with *p*-aminobenzoic acid and a molecule of the amino acid glutamic acid. The presence of sulfanilamide results in the displacement of *p*-aminobenzoic acid in microorganisms that synthesize folic acid. The product thus generated cannot form a peptide bond with glutamic acid, as the inactive pteridine-sulfa conjugate predominates. Because folic acid is a mandatory catalyst in the synthesis of purines and pyrimidines, a microorganism without functional folic acid would not survive. Animals are unaffected by sulfanilamide because they obtain folic acid in their diet and do not synthesize it from the base constituents.

An understanding of the selective toxicity of sulfanilamide led to a concerted search for other chemotherapeutic agents (structural analogs of amino acids, purines, pyrimidines, and vitamins). Considerable emphasis has been placed on synthesizing analogs of the bases in DNA and RNA (adenine, thymine, guanine, uracil, and cytosine) in an effort to find chemical compounds that have antiviral or antitumor activity. Of thousands of

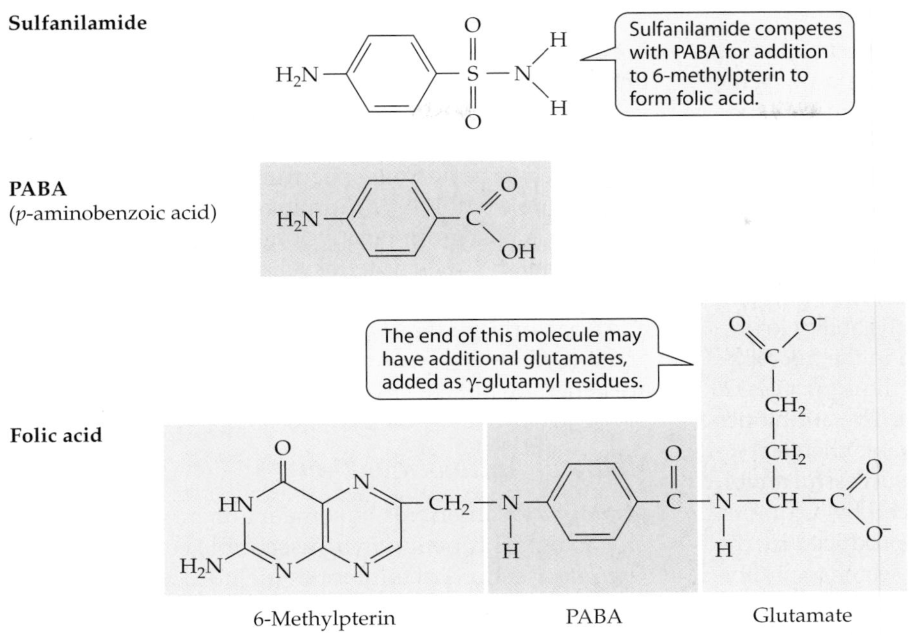

Sulfanilamide

Sulfanilamide competes with PABA for addition to 6-methylpterin to form folic acid.

PABA
(*p*-aminobenzoic acid)

The end of this molecule may have additional glutamates, added as γ-glutamyl residues.

Folic acid

6-Methylpterin   PABA   Glutamate

Figure 7.9 Sulfa drug as structural analog Sulfanilamide is a competitive inhibitor of metabolic function in microorganisms. It is an analog of, and competes with, *p*-aminobenzoic acid in the synthesis of the vitamin folic acid.

compounds that have been synthesized, a limited number have proven effective as chemotherapeutics.

Several important antiviral agents have been developed in recent years, and these include acyclovir and azidothymidine (AZT). **Acyclovir** is a structural analog of deoxyguanosine (Figure 7.10) and can be used in the treatment of herpes virus infections. When a herpes virus (a DNA virus) multiplies in a cell, the enzyme thymidine kinase is activated. This enzyme gratuitously phosphorylates acyclovir, resulting in the formation of an unnatural triphosphate, which then competes for dGTP as a substrate for the DNA polymerase. It can also act as a chain terminator of DNA synthesis. As a consequence, the assembly of DNA in viral particles is curtailed. The

nucleotide drugs valacyclovir, famciclovir, and penciclovir are effective antiviral agents for herpes simplex virus type 1 or type 2 that cause cold sores, and/or genital blisters. The mode of action is analogous to that described for acyclovir.

AZT (see Figure 7.10) is an antiviral used in the treatment of AIDS (acquired immunodeficiency syndrome). This compound inhibits the multiplication of HIV (human immunodeficiency virus), the retroviral causative agent. Unfortunately, neither the antiviral agent AZT nor acyclovir can destroy the nonreplicating intracellular virus. Both the herpes virus and HIV persist for life in selected cells within the human host from the time of infection (see Chapter 29).

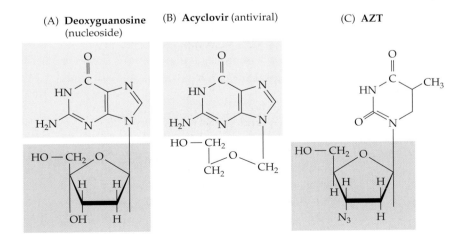

(A) **Deoxyguanosine** (nucleoside)

(B) **Acyclovir** (antiviral)

(C) **AZT**

Figure 7.10 Acyclovir and AZT as structural analogs The antiviral agent acyclovir is structurally related to the nucleoside deoxyguanosine (A). Phosphorylation of acyclovir yields an analog of deoxyguanosine triphosphate, which inhibits the viral DNA polymerase. (B) The structure of acyclovir, showing its structural similarity to deoxyguanosine. (C) The structure of AZT.

Antibiotics as Chemotherapeutics

The term "**antibiotic**," which is familiar to most of us, was coined by Selman Waksman in 1953. Waksman was the discoverer of the antibiotic **streptomycin**, which played a significant role in the control of tuberculosis. He defined an antibiotic as ". . . a chemical substance, produced by microorganisms, which has the capacity to inhibit the growth and even to destroy bacteria and other microorganisms, in dilute solutions." Antibiotics that are effective in controlling disease are produced during growth by certain bacteria, *actinomycetes*, and fungi. In fact, the ability to produce antibiotic substances is widespread in the microbial world, and the group of bacteria termed the *actinomycetes* is particularly adept at producing them. The antibiotics synthesized by *actinomycetes* have remarkable chemical diversity. However, the first commercially successful antibiotic used in chemotherapy was produced by a fungus. This antibiotic was **penicillin** and was produced by the fungus *Penicillium notatum* (Box 7.1). The search for antibiotic substances and their industrial production is presented in Chapter 31. The medical aspects of antibiotics are covered in Chapters 28 and 29. The present section provides a brief discussion of their basic **mode of action** and the manner in which infectious organisms become resistant to them.

Antibiotics are selectively toxic to certain types of living cells and less so to others. This selectivity is due to distinct differences in fundamental physiological processes in affected cells and unaffected cells. Some antibiotics are effective against a limited range of bacteria and thus are considered to be **narrow-spectrum** antibiotics. Penicillin is an example of this type of antibiotic because it is mostly effective against gram-positive organisms. A **broad-spectrum** antibiotic, such as tetracycline, inhibits both gram-positive and gram-negative bacteria. Antibiotics that inhibit bacteria generally do not inhibit eukaryotes or archaea. Conversely, antibiotics that inhibit eukaryotes are generally not effective against bacteria.

Modes of Antibiotic Action

An antibiotic interferes in some manner with a normal physiological function in a susceptible cell (Table 7.5). The major actions of antibiotics include:

- Disruption of cell wall synthesis
- Destruction of cell membranes
- Interference with some aspect of protein synthesis
- Interference with some aspect of nucleic acid synthesis

TABLE 7.5	The producing organisms and mode of action of several antibiotics

Antibiotic	Producer	Mode of Action
Cell wall synthesis inhibitor		
Penicillin	*Penicillium* sp.	Blocks transpeptidation in peptidoglycan synthesis
Ampicillin	Semisynthetic	Blocks transpeptidation in peptidoglycan synthesis
Bacitracin	*Bacillus subtilis*	Inhibits isoprenyl pyrophosphate dephosphorylation
Disruption of membrane potential		
Tyrocidine	*Bacillus brevis*	Ionophore disrupts cell membrane integrity and function
Polymyxin B	*B. polymyxa*	Disrupts membrane transport and function
Vancomycin	*Streptomyces* sp.	Potassium-specific transporter
DNA and RNA synthesis inhibitor		
Novobiocin	*S. spheroides*	Binds to a subunit of DNA gyrase and inhibits action
Metabolic inhibitor		
Trimethoprim	Synthetic	Dihydrofolate reductase inhibitor
Dapsone	Synthetic	Dihydrofolate reductase inhibitor
Sulfonamide	Synthetic	Inhibits dihydropteroate synthetase inhibitor
Protein synthesis inhibitor		
Rifampicin	*S. mediterranei*	Blocks transcribing enzyme RNA polymerase
Erythromycin	*Streptomyces erythreus*	Binds to 50S ribosomal subunit and stops peptidyltransferase
Streptomycin	*S. griseus*	Inhibits 30S ribosomes, blocks amino acid incorporation into peptides
Tetracycline	*S. aureofaciens*	Inhibits binding of aminoacyl tRNA to ribosomes
Chloramphenicol	*S. venezuelae*	Binds to ribosomes blocking peptidyltransferase
Neomycin		Binds to the 30S subunit to inhibit translocation

| BOX 7.1 | *Milestones* |

The Antibiotic Age

Fortune favors the prepared mind.

—Louis Pasteur
(Microbiologist)

The more I practice, the luckier I get.

—Lee Trevino
(Golf champion)

The first quote is a brief summation of the circumstances that led Sir Alexander Fleming (1881–1955) to the discovery of penicillin. All too often, scientific advances are attributed to chance or serendipitous good fortune, but usually they are not. More often they are the product of intuition and hard work. The discovery of penicillin resulted from Fleming's long-held and firm belief that naturally occurring substances existed that had useful antimicrobial properties.

During World War I Fleming worked in a field hospital in France. There he treated casualties of battle and experimented with better ways of treating deep wounds. The accepted treatment at that time was copious quantities of antiseptic—carbolic acid, boric acid, or peroxide. Unfortunately, in deep wounds these antiseptics were quickly neutralized by tissue, and serious infections often followed. Fleming and his colleague, Sir Almroth Wright (1861–1947; a noted developer of antityphoid vaccine), observed that wounds carefully cleansed and treated with minimal antiseptic healed faster and with fewer serious infections than wounds treated with massive amounts of antiseptic. They believed that stimulation of natural-defense mechanisms by encouraging the exudation of lymph into the wound site was particularly effective in promoting healing. They found that bathing wounds with a saline solution was effective in stimulating the exuding of lymph. The success of this experimentation convinced Alexander Fleming that naturally produced substances could be employed successfully in the treatment of infectious diseases. Fleming explained to a physician visiting his laboratory at Boulogne, "What we are looking for is some chemical substance which can be injected without danger into the blood stream for the purpose of destroying the bacilli of infection, as salvarsan destroys the spirochaetes."

Following the war, Fleming returned to his position at St. Mary's Hospital in London. As time permitted, he tested various naturally occurring substances for antibacterial activity. Among the materials he tested was nasal mucus, and he observed that addition of mucus to a suspension of *Staphylococcus aureus* led to a rapid clearing of the turbidity. He discovered that this clearing resulted from lysis of the bacterial cells. The factor involved was an enzyme termed *lysozyme* (see Chapter 4). This discovery, made in 1921, was important scientifically, but lysozyme was not useful as a chemotherapeutic agent. It did, however, offer confirmation to Fleming that his belief in natural antibacterials was warranted.

One of Fleming's research interests was *S. aureus,* and he often had cultures growing on Petri dishes lying about the lab. One day while examining some of these old cultures, he observed that one was contaminated with a mold and that bacterial colonies did not develop in the area adjacent to the mold growth. Fleming was intrigued and likened the phenomenon to the

Sir Alexander Fleming. (Photo by Sydney R. Bayne/NLM.)

action of lysozyme on *S. aureus.* He then took a picture of the plate (shown in the drawing) and wrote up the observation in his research notebook. He showed the plate to a colleague with the remark—"Take a look at that, it's interesting—the kind of thing I like; it may well turn out to be important." Important, indeed; the antibacterial substance produced by the mold (*Penicillium notatum*) was penicillin. This ushered in the "antibiotic age." Fleming's monumental discovery led eventually to the conquest of such ancient scourges as pneumonia, tuberculosis, syphilis, staphylococcal infections, typhus, and a host of other human ills.

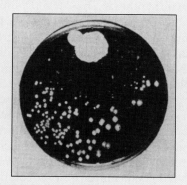

The original *Penicillium* plate. From Maurois, André. 1959. *Life of Sir Alexander Fleming,* E. P. Dutton Co.

The most useful antibiotics employed to control human infections interfere with structural or physiological reactions that are unique to the invading pathogen. This is possible because the bacteria are physiologically different from the eukaryotes. Consequently, antibacterial antibiotics can exploit these differences and destroy the invader without significant harm to the human host.

Control of fungal, protozoan, and viral invaders is much more difficult because physiological reactions in these pathogens are quite similar to that in all eukaryotes including a human host. The physiological differences among the eukaryotes are simply not so distinct that these differences can readily be exploited. Thus, effective antifungal and antiprotozoan agents are generally toxic to humans. Because viral infections are intracellular, they are also difficult to control with chemotherapeutic agents. A virus is synthesized, for the most part, by the metabolic machinery of the host, and destruction of this machinery can lead to the destruction of the host.

CELL WALL SYNTHESIS   Some antibiotics exert their effect by preventing the synthesis of normal bacterial cell walls (either gram-negative or gram-positive cells, see Chapters 4 and 11). Antibiotics in the **penicillin** and **cephalosporin** groups (Figure 7.11), both of which are produced by certain fungi, act by inhibiting the enzymes responsible for biosynthesis of peptidoglycan (i.e., the transpeptidation reaction, see Chapter 11). As a result, bacteria exposed to these antibiotics produce a weakened cell wall, which results in the eventual lysis, and destruction, of the bacterial cell. Because microorganisms generally live in osmotically unfavorable environments, one having a weakened cell wall will take up water and burst, or lyse. The *Archaea* that do not have peptidoglycan in their cell wall are generally unaffected by antibiotics such as penicillin that interfere with the synthesis of this polymer.

It is noteworthy that penicillin causes lysis and death of bacteria only under conditions that permit growth and cell wall synthesis. Thus, when antibiotics are used therapeutically, bacterial cells that are not growing and dividing can survive and may begin growing following the antibiotic treatment. These cells must be eliminated by the host's own defense activities, such as by phagocytes.

Note that the effect of **lysozyme** on the cell wall (peptidoglycan) is different from that of antibiotics previously mentioned. As an enzyme, lysozyme can continually degrade the peptidoglycan, whereas penicillin is active only on growing cells because it affects only the synthesis of peptidoglycan. Treatment of growing bacteria with penicillin in isotonic solution results in the formation of protoplasts and spheroplasts in much the same manner as lysozyme. Bacitracin, another peptidoglycan synthesis inhibitor, is a cyclic polypeptide antibiotic made by certain strains of *Bacillus subtilis*. It acts to block an early step in peptidoglycan precursor molecule production (see Chapter 11). This antibiotic is used topically on cuts and abrasions since it cannot be taken orally. Like several other cyclic antibiotics it is synthesized independent of ribosomes. Several other peptidoglycan synthesis inhibitors include phosphomycin, tunicamycin, and vancomycin. The latter is produced by *Streptomyces orientalis*, and must be administered intravenously since it cannot be absorbed orally.

Humans treated with these antibiotics are not affected (unless they have an allergy to them) because human cells do not synthesize peptidoglycan. Gram-negative bacteria tend to be less sensitive to penicillin because their outer membrane and associated lipopolysaccharide (LPS) can prevent the antibiotic from reaching the peptidoglycan layer of the cell (see Chapter 4). As one would predict, penicillin is generally much more effective against gram-positive bacteria than against gram-negative bacteria, since the former lack this permeability barrier.

For use against gram-negative bacteria, penicillin and other antibiotics can be modified to permit better solubility in the lipid outer membranes. **Ampicillin**, for example, is a penicillin with greater hydrophobicity that is used to treat infections caused by gram-negative bacteria. Penicillin is effective in the treatment of certain gram-negative infections, including syphilis, caused by *Treponema pallidum*, and gonorrhea, caused by *Neisseria gonorrhoeae*, as well as in most infections by gram-positive bacteria.

MEMBRANE DISRUPTION   **Tyrocidine, polymyxin,** and **valinomycin** are polypeptide antibiotics that disrupt cell membranes. The first two are produced by bacteria of the genus *Bacillus*. Tyrocidine is an **ionophore**, which destroys selective permeability by forming channels across the cell membrane, resulting in leakage of monovalent cations (e.g., K^+, H^+, Na^+). As a consequence, the organism cannot establish a proton motive force, and solute transport into and out of the cell is impaired. Polymyxin causes similar cytoplasmic membrane damage. Peptide antibiotics are not taken internally but are applied topically to treat skin infections. Enzymes present in the intestinal tract would digest an ingested peptide antibiotic. Valinomycin (see Figure 7.11) is produced by *Streptomyces* sp. and functions as a potassium channel that disrupts the cell's ion balance.

DNA SYNTHESIS   Certain antibiotics interfere directly with bacterial DNA replication. Among these is **novobiocin** (see Figure 7.11), which binds directly to the beta subunit of the DNA gyrase responsible for unwinding supercoiled DNA during replication. Other antibiotics, such as **rifampicin**, block the RNA polymerase involved

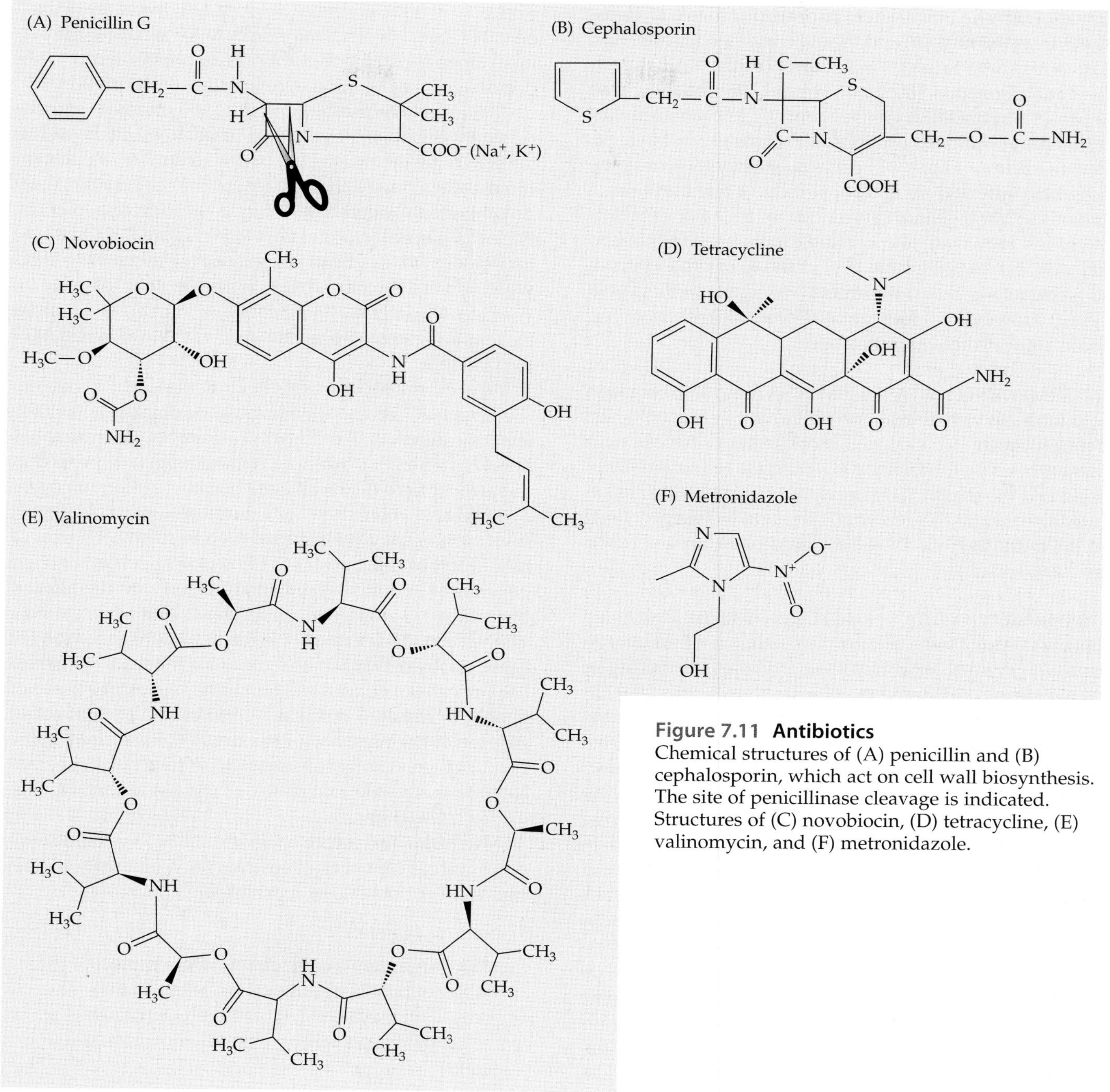

Figure 7.11 Antibiotics
Chemical structures of (A) penicillin and (B) cephalosporin, which act on cell wall biosynthesis. The site of penicillinase cleavage is indicated. Structures of (C) novobiocin, (D) tetracycline, (E) valinomycin, and (F) metronidazole.

in transcribing the information on the DNA molecule to make messenger RNA (mRNA) for protein synthesis (see Chapter 13). Most of these antibiotics are ineffective in the *Archaea*. **Metronidazole** is used to treat various anaerobic bacterial and protozoan infections. Following uptake into a susceptible strain, the nitro group is reduced by a cell protein to form an inhibitor of DNA synthesis.

PROTEIN SYNTHESIS   Many of the clinically useful antibiotics inhibit bacterial protein synthesis. Protein synthesis requires **ribosomes**, which are structures made up of subunits of unequal size. Ribosomes are involved in the synthesis of protein. Bacterial and eukaryotic ribosomes differ in relative size, protein content, and ability to bind antibiotics. On the basis of sedimentation velocity during ultracentrifugation, the smaller bacterial ribosome is designated **70S** and the larger eukaryotic ribosome size is **80S**.

Antibacterial antibiotics that inhibit protein synthesis do so by only binding to the bacterial ribosome. This occurs with the antibiotics **chloramphenicol, strepto-mycin, erythromycin,** and **tetracycline** (see Figure 7.11). The antifungal antibiotic **cycloheximide** binds to eu-karyotic ribosomes (80S) but not the 70S ribosome that is present in bacteria. Consequently, cycloheximide inhibits fungi but does not inhibit the growth of bacteria. Because humans have 80S ribosomes, they too would be adversely affected by cycloheximide. A few antibiotics, including tetracycline, can bind to both 70S and 80S ribosomes. However, tetracycline can be used in humans because it does not inhibit the synthesis of protein in eukaryotic cells at the concentrations used chemotherapeutically. However, it does impede wound healing in humans under some circumstances.

INTERMEDIARY METABOLISM   Some antibiotics interfere with enzymatic reactions involved in biosynthesis. **Trimethoprim,** for example, blocks tetrahydrofolic acid biosynthesis by inhibiting dihydrofolate reductase. **Dapsone** and the structurally unrelated **sulfonamide** antibiotics also inhibit this enzyme. Dapsone is currently used in the treatment of AIDS-associated pneumonia caused by *Pneumocystis*.

CLASSIFICATION BY STRUCTURE   **Macrolides** make up a structural class of antibiotics that contain a large lactone ring attached to a deoxy-sugar, for example, erythromycin. As noted previously these antibiotics inhibit protein synthesis by interfering with peptidyl-tRNA translocation on the ribosome. Their applications include respiratory tract infections and soft tissue infections. **Aminoglycosides** contain aminosugars as the name implies, and are often used to treat gram-negative aerobic bacterial infections. They are derived from *Streptomyces* sp. (streptomycin, kanamycin, neomycin) and either block protein synthesis by interfering with mRNA reading (misreading) or by inhibiting ribosomal peptidyl-tRNA translocation. **Quinolones** and the related fluoroquinolones (ciprofloxacin, levofloxacin, moxifloxacin) have broad-spectrum activity. Due to the presence of a nalidixic acid group in the molecule they inhibit DNA replication and transcription. Lastly, the **peptide antibiotics,** composed of various amino acids, act by disrupting the membrane ion balance needed for cell viability.

Resistance to Antibiotics

Antibiotics have been effective in controlling many of the diseases that have been a scourge to humankind. Tuberculosis, bacterial pneumonia, syphilis, and many other human infectious diseases that were once fatal only 60 years ago can now generally be treated with an-tibiotics. Antibiotics have been called "wonder drugs" because they effect a dramatic cure for what had previously been incurable. But there is a problem with "wonder drugs" that became evident after widespread use.

The extensive use of penicillin as a chemotherapeutic agent led to the evolution of pathogenic bacterial strains that were unaffected by the drug. These resistant organisms retained their potent pathogenicity but were no longer controlled by the administration of penicillin. For example, when penicillin G (see Figure 7.11) was first introduced, virtually all strains of *Staphylococcus aureus* were sensitive to the drug. Within a span of only 10 years, essentially all *Staphylococcus* infections acquired in hospitals were caused by strains that were resistant to penicillin.

Why do microorganisms become resistant to chemotherapeutics? Abundant microbial populations have existed throughout all of Earth's history because microbes are adaptable. For example, when oxygen appeared in the atmosphere or the climate became cooler, microbes evolved to survive these new environmental conditions. Environmental constraints drove natural selection of new microbial processes. Those that survived carried genes that better adapted them to serve in the altered conditions. Human control methods of all types create a challenge to bacteria that leads to **natural selection** for those organisms in the environment that can withstand the prevailing condition. Thus, the widespread use of penicillin resulted in the selection of strains that could survive in the presence of the drug. For example, some penicillin-resistant strains produce **penicillinase,** a β-lactamase enzyme that destroys the antibiotic (see Figure 7.11; Chapter 15).

Microbial resistance to an antibiotic, a chemotherapeutic agent, or other chemicals such as heavy metals can occur for several of reasons:

1. **Natural resistance**

   - The organism may lack the target molecule that the antibiotic inhibits, as occurs with mycoplasma, which lacks cell walls and is thus unaffected by penicillin (described in subsequent text).

   - The cell wall or cytoplasmic membrane of an organism may be impermeable to an antibiotic or other chemical. Some *Pseudomonas* strains are notable examples due to the structure of their outer membranes.

2. **Acquired resistance**

   - A resistant microorganism may produce a substance that inactivates the antibiotic, as occurs in strains of *S. aureus* in producing the enzyme penicillinase, which cleaves the penicillin molecule into an inactive form.

- Alterations in solute transport can occur due to reduced permeability of the cytoplasmic membrane.

- Accumulation of mutations in chromosomal DNA may result in altered cellular molecules that will not bind the antibiotic or other chemical. For example, the gene that encodes transpeptidase in staphylococci can mutate so that the enzyme does not bind penicillin.

- Antibiotic resistance may arise by the acquisition of genetic information that renders a microorganism resistant to that drug. Pieces of extrachromosomal DNA called "plasmids" can be easily transferred from cell to cell (see Chapter 15). For example, a plasmid containing a gene that encodes the enzyme penicillinase, confers on the cell a newly acquired ability to degrade penicillin. Other resistance strategies involve a gene product (enzyme) that modifies the antibiotic by covalent modification (e.g., by acetylation, methylation, or phosphorylation). By another mechanism, tetracycline resistance is conferred by the presence of a gene that encodes a membrane efflux pump that exports the drug from the cytoplasm so that it cannot effectively act at its target site inside the cell.

To maintain human health it is thus important that we address the issue of antibiotic resistance. Current research efforts focus on identifying and synthesizing new antibiotics for which no resistance mechanisms have evolved. These efforts have focused on devising new derivatives of the macrolide, aminoglycoside, and quinolone-type antibiotics, as well as identifying new classes of antibiotics that possess different modes of action.

Probiotics

The natural flora of humans, animals, and plants serve a beneficial role in the health of these organisms by preventing or reducing the ability of pathogenic bacteria to cause infection. In humans, for example, these normal inhabitants occupy environmental niches within the intestinal tract, vagina, and mouth to protect the host from colonization by newly introduced pathogens. Current research is focused on understanding the ecology of the native and foreign microbes in competition for an environmental niche. While this approach does not serve an antimicrobial role, its application avoids the selective pressures that invariably lead to antibiotic resistance due to the use of antibacterial molecules.

SECTION HIGHLIGHTS

Antibiotics are naturally occurring or man-made compounds that either inhibit or kill microbes by interfering with an essential cell function. They may be grouped by structure, mode of action, or spectrum (range) of antimicrobial activity. Use of antibiotics in the world over the last 60 years has dramatically affected human health and longevity. However, their widespread application has also resulted in resistance to specific antibiotic types. The long-term benefits of antibiotics to humans will depend on the development of new classes of antibiotics as well as their use, and controlled applications of existing antibiotics.

SUMMARY

- Human survival often depends on the ability to counteract the activities of harmful bacteria and lower eukaryotes.

- An understanding of the nature of microbial resistance has led to more direct means for controlling these organisms.

- **Sterilization** is the destruction of all viable life forms. It can be attained by fire or flame, moist heat, dry heat, radiation, filtering, or toxic chemicals.

- Microbial death by heating occurs at an **exponential rate**. The length of time required to reduce a population tenfold is termed the **decimal reduction time** (*D*).

- The most effective means for routine sterilization of media and food is steam under pressure. The common laboratory apparatus employed for this is the **autoclave**.

- Dry heat at 160°C is effective in sterilizing glassware, pipettes, and other heat-stable material.

- **Pasteurization** is a method for reducing the number of microorganisms present in milk or other products. It can destroy selected disease or spoilage-causing microorganisms.

- The wavelengths of light that can cause the death of microorganisms are **ultraviolet light** (**UV**), which disrupts RNA/DNA, and **ionizing radiation**, which can potentially harm any constituent of a cell.

- Microorganisms have DNA repair systems that can correct damage caused by radiation.

- Filters with pores small enough to retain bacteria can be employed to sterilize liquids. A number of types are available, but membrane filters are the most widely used.

- Toxic gases such as ethylene oxide and propylene oxide are used to sterilize plastic Petri plates, filters, and other disposable material.

- **Antimicrobial agents** may be products of chemical synthesis, natural products, or a combination of the two. The favored agent has selective toxicity in that it destroys the target microbe but produces limited activity toward other cells.

- A **disinfectant** is an agent such as phenol that destroys all living cells on contact. An **antiseptic** is less potent and prevents the growth of disease-causing organisms. Merthiolate that is applied to superficial wounds is an antiseptic.

- The **minimum inhibitory concentration (MIC)** is the lowest concentration of a **germicide** that will completely inhibit the growth of a test organism.

- The **phenol coefficient** is a measure of the effectiveness of a test compound as compared with that of phenol.

- The first effective chemically synthesized chemotherapeutic agent was **sulfanilamide**. An understanding of the role of this compound in blocking vitamin synthesis in bacteria encouraged the scientific world to search for other chemical agents that would be effective against viruses, protozoa, fungi, and cancer.

- By definition an **antibiotic** is a product of microorganisms. Antibiotics have proven to be effective in controlling human and animal diseases caused by bacteria. They generally interfere with metabolic activities or structures in disease-causing bacteria but do not interfere with them in a host eukaryote.

- The major **modes of action** of antibiotics include: disruption of cell wall synthesis, permeabilization of cytoplasmic membranes, and disruption of nucleic acid or protein synthesis.

- Microbial resistance mechanisms have evolved and spread among bacterial species to render them resistant to one or more commonly used antibiotics.

 Find more at www.sinauer.com/microbial-life

REVIEW QUESTIONS

1. Why are we so concerned with sterilizing things? What agent is most often used in sterilization?

2. What is the difference between sterilization and pasteurization?

3. Dry heat is employed to sterilize selected materials. Name some of the materials best sterilized by dry heat.

4. Define the terms, "*D*" and "*z*" values and explain how they may be used to characterize two distinct microbial strains.

5. What agent is most often used to sterilize air in a closed room? List what concerns there would be with other methods.

6. Give some advantages of sterilization by filtration. How is filtration applied in the laboratory? In industry?

7. How did Robert Koch determine the effectiveness of disinfectants? What compound did he find most effective with this method?

8. How would one calculate a phenol coefficient? Is this a useful number? List the pros and cons.

9. Sulfanilamide is a classical example of a competitive inhibitor. How does it function?

10. What are the basic shortcomings of antiviral agents such as AZT and acyclovir?

11. Why is it so difficult to find effective antiviral and antifungal agents when antibacterials are plentiful?

12. Discuss the mode of actions and give some examples of some common antibacterial agents.

13. Briefly describe five ways in which bacteria become resistant to drugs and give an example of each.

SUGGESTED READING

Block, S. S., ed. 1991. *Disinfection, Sterilization, and Preservation*. 4th ed. Philadelphia: Lea and Febiger.

Favero, M. S. and W. W. Bond. 1991. "Sterilization, Disinfection, and Antisepsis in the Hospital." In A. Ballows, W. J. Hausler, K. L. Herrmann, H. O. Isenberg and H. J. Shadomy, eds. *Manual of Clinical Microbiology*. 5th ed. (pp. 183–200). Washington, DC: ASM Press.

Hardman, J. G. and L. E. Limbird, eds. 1995. *The Pharmacological Basis of Therapeutics*. 9th ed. New York: Pergamon Press.

Harte, J., C. Holdren, R. Schneider and C. Shirley. 1991. *Toxics A to Z: A Guide to Everyday Pollution Hazards*. Los Angeles: University of California Press.

Levy, S. B. 1992. *The Antibiotic Paradox, How Miracle Drugs Are Destroying the Miracle*. New York: Plenum Press.

Reese, R. E., R. Betts and B. Gumustop. 2000. *Handbook of Antibiotics*. 3rd ed. Philadelphia, PA: Lippincott, Williams and Wilkins.

Walker, R. and M. Buckley. 2006. *Probiotic Microbes: The Scientific Basis*. Report from the American Academy of Microbiology. Washington DC: ASM Press.

Walsh, C. 2003. *Antibiotics: Actions, Origins, Resistance*. Washington DC: ASM Press.

Microbial Physiology: Metabolism

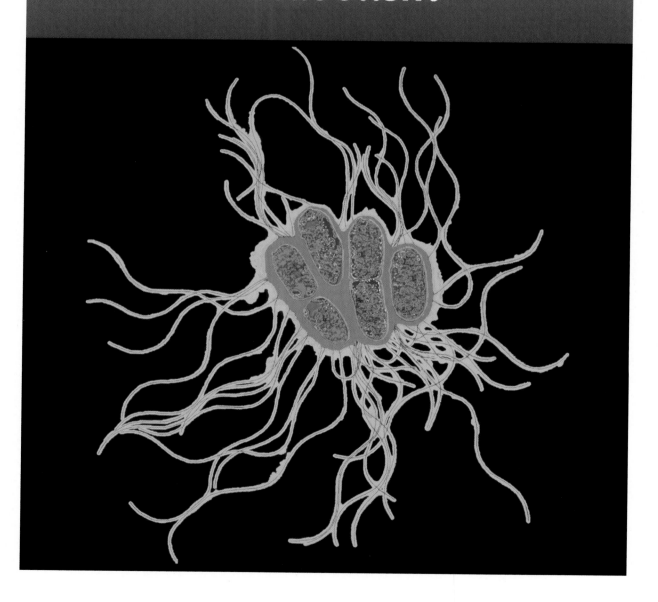

8

Cellular Energy Derived from Chemicals

Life is driven by nothing else but electrons, by the energy given off by these electrons while cascading down from the high level to which they have been boosted up by photons. An electron going around is a little current. What drives life is thus a little electric current, kept up by the sunshine. All the complexities of intermediary metabolism are but lacework around this basic fact.
—*Albert Szent-Györgyi*

This chapter introduces the principal mechanisms whereby microorganisms generate energy to fuel biological processes such as cell biosynthesis. Much of the energy for life on Earth is furnished either directly or indirectly by sunlight, and this energy is captured during photosynthesis by plants and bacteria. Among the products are ATP and potential energy–yielding compounds, such as sugars and protein, which, in turn, provide energy for heterotrophic life. Animals and many prokaryotes survive by consuming energy-yielding plant material, or by devouring/decomposing an animal that ate plants. In addition, many prokaryotes obtain energy by metabolizing inorganic compounds originating in the Earth's crust. Together there exists an amazing diversity in energy-generating strategies in microbes! In this chapter, we discuss the mechanisms whereby potential chemical energy is converted to ATP. Chapter 9 will present the processes whereby light energy is converted to ATP.

8.1 Basic Principles of Energy Generation

Among the *Bacteria* and the *Archaea*, one should be aware that the source of energy is, in many cases, different from the carbonaceous substrate used for synthesis of cell material. This is a fundamental difference between the microbial and the animal worlds. One commonality in the evolution of all living systems is that adenosine-5′-triphosphate (ATP), formed from adenosine-5′-diphosphate (ADP)

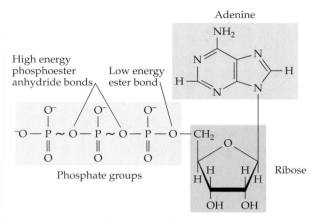

Figure 8.1 Adenosine-5′-triphosphate (ATP)
ATP is a major carrier of chemical energy in living cells. The energy-rich phosphate anhydride bonds are designated by a squiggle (~).

and inorganic phosphate (P_i), is a universal medium of biochemical energy exchange.

$$ADP + P_i \rightleftharpoons ATP$$

The energy conserved in the "energy-rich" bonds of ATP can be employed by the cell to do work. This nu-

cleotide is involved in the biosynthesis of cell constituents, transport, motility, polymerization reactions, and other activities of the cell. Two of the bonds in ATP are high-energy bonds (Figure 8.1). Hydrolysis of the high-energy phosphate bonds of ATP and certain other molecules (Box 8.1) can raise the free energy level of metabolic intermediates so that they can participate in biosynthetic reactions.

In other biosynthetic reactions, other high-energy **nucleotides** are involved. For example, uridine triphosphate (UTP) participates in the synthesis of polysaccharides, cytidine triphosphate (CTP) is involved in lipid synthesis, and guanosine triphosphate (GTP) is required for protein synthesis. These nucleotides are thermodynamically equivalent to ATP. They are formed by exchange of phosphate from ATP to G-, U-, and CDP. GTP is also a product of the tricarboxylic acid (TCA) cycle of some organisms (see Chapter 10).

The *Bacteria* and the *Archaea* employ one or more of three general mechanisms to generate ATP: respiration-linked phosphorylation, photophosphorylation, and substrate-level phosphorylation.

- **Substrate-level phosphorylation** (**SLP**) is the synthesis of ATP by direct transfer of a high-energy phosphate from a phosphorylated organic compound to convert ADP to ATP.

BOX 8.1 *Milestones*

ATP: A High Energy Molecule

The energy generated during respiration and fermentation reactions can be conserved in **ATP**, the major high-energy phosphate compound in living cells. ATP is an energy "carrier" that is formed during **exergonic reactions** (energy-yielding) and can drive **endergonic reactions** (energy-requiring). In ATP and some other phosphorylated compounds, the outer two phosphate groups are joined by an anhydride bond (see Figure 8.1). The free energy ($\Delta G_o'$) that can be released when the high-energy bonds are hydrolyzed is −31.8 kJ/mol. In contrast, the hydrolysis of the low-energy ester bond of AMP would yield only −14.2 kJ/mol. The phos-

phate bond in the glycolytic intermediate phosphoenolpyruvate (PEP) is high energy ($\Delta G_o' = -51.6$ kJ/mol), and the phosphate bond in glucose-6-phosphate is low energy ($\Delta G_o' = -13.8$ kJ/mol). The free energy available in phosphate

bonds is of value to cells only when it can be coupled to a second reaction that requires energy.

The approximate free energy values for the hydrolysis of common phosphate bonds in biological molecules are:

Hydrolysis Reaction	Approximate Free Energy
ATP → ADP + P_i	−32 kJ/mol
ADP → AMP + P_i	−32 kJ/mol
AMP → adenosine + P_i	−14 kJ/mol
PEP → pyruvate + P_i	−52 kJ/mol
1,3-Bisphosphoglycerate → 3-PG + P_i	−52 kJ/mol
Acetylphosphate → acetate + P_i	−45 kJ/mol
Butyrylphosphate → butyrate + P_i	−45 kJ/mol
Glucose-6-phosphate → glucose + P_i	−14 kJ/mol

- **Respiration-linked phosphorylation (RLP)** is the production of ATP by oxidation of a reduced organic or inorganic compound (the electron donor) coupled with the reduction of an inorganic or organic electron acceptor. These reactions are performed by many types of heterotrophic and autotrophic microorganisms and generate a membrane-associated proton motive force.

- **Photophosphorylation**, also called **photon-linked phosphorylation (PLP)**, occurs in photosynthetic microorganisms and plants that generate ATP by using light energy to create a proton motive force.

Evidence suggests that a major event in the evolution of living organisms was the advent of the ability to establish a **proton motive force (pmf)** or an **electrochemical potential** across the cytoplasmic membrane. This charge separation is a form of potential energy that may be equated to that of a car or flashlight battery that provides a current to activate a starter or produce light. The electrochemical potential established across a biological membrane can then be converted to chemical energy in the form of ATP. A general representation of a charge separation across a cytoplasmic membrane is presented in Figure 8.2. Two membrane-associated enzymes perform the oxidation of an electron donor (DH) and the reduction of an electron acceptor (A). These reactions are coupled by a transfer of electrons and the creation of a pmf. The details of these events will be discussed later.

Solar energy is a major source of biological energy on Earth, and photons from the sun can drive protons across a membrane to establish an electrochemical potential. Light energy can also extract electrons from water in plants and many bacteria. The electrons extracted from water ($2\ H_2O \rightarrow O_2 + 4\ H^+ + 4\ e^-$) may then be passed through a series of membrane-associated electron carriers to generate a pmf (see Chapter 9). Energy generated in the form of ATP during photosynthesis is utilized for the assimilation of CO_2 from the atmosphere. Products of CO_2 fixation, such as glucose, store this chemical energy, and this provides the basis for much of the heterotrophic life on Earth.

Chemotrophic respiration is a second major source of biological energy on Earth. Aerobic respiration utilizes O_2 as the terminal electron acceptor, whereas anaerobic respiration uses other compounds such as NO_3^-, SO_4^{2-},

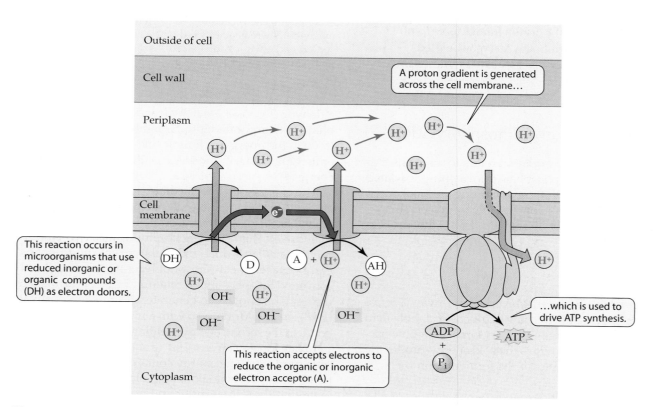

Figure 8.2 Establishing a charge separation across a cell membrane
The concentration of protons (H^+) on the outer surface and of negative ions (OH^-) inside the cell creates a charge difference across the membrane called a proton motive force (pmf). This charge difference can be a source of energy for synthesis of adenosine-5'-triphosphate (ATP) and other energy-consuming activities of the cell.

CO_2, ferric iron, or fumarate. The products generated from reduction of the electron acceptor compounds would be H_2O, N_2 or NH_4^+, H_2S, CH_4, ferrous iron (Fe^{2+}), or succinate, respectively. The electrons needed to reduce an oxidized electron acceptor are provided by an intermediate electron carrier such as NADH, or externally provided compounds such as H_2, lactate, or formate.

Both of the above photo- and chemo-energy harvesting strategies extract electrons from donor molecules and use these to establish a pmf across the cytoplasmic membrane. This electrochemical potential can be the driving force in generating ATP.

Substrate level phosphorylation occurs, as the name suggests, by an entirely different mechanism. The processes whereby ATP is generated by the respiration-linked phosphorylation reactions and substrate level phosphorylation reactions are discussed here in some detail. The photon-driven process is the subject of Chapter 9.

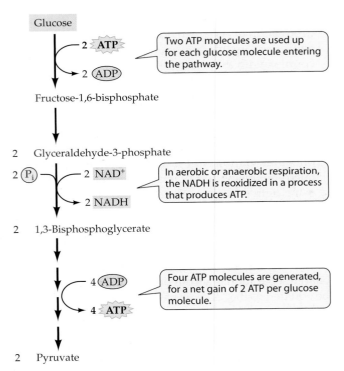

Figure 8.3 Glycolysis
The basic features of glycolysis, a pathway in which a molecule of glucose is catabolized to yield two molecules of pyruvate. There is a net yield of 2 ATP per glucose consumed and a net loss of 4 electrons. For more details of this pathway see Figure 10.5.

SECTION HIGHLIGHTS

There are three general ways organisms form ATP: chemotrophic respiration–linked phosphorylation, photophosphorylation, and substrate-level phosphorylation. The first two involve formation of a proton motive force (pmf) across the cell membrane that may be used for several cell needs including ATP synthesis.

8.2 Substrate-Level Phosphorylation

Substrate-level phosphorylation refers to reactions that involve the transfer of phosphate from a phosphorylated carbon substrate to ADP. It occurs during the breakdown of many types of sugars, for example by glycolysis, and under certain anaerobic conditions by the process termed fermentation.

Glycolysis

Also termed the Embden-Meyerhof pathway (Figure 8.3), **glycolysis** is a major route of glucose catabolism in many aerobic microorganisms. In some anaerobic bacteria, this pathway provides the sole energy-producing mechanism, for example, in the fermentation by lactic acid bacteria.

The pathway initially expends two molecules of ATP in going from glucose to fructose-1,6-bisphosphate, which is then cleaved to two molecules of glyceraldehyde-3-phosphate. In an oxidation step carried out with NAD^+ as the electron acceptor and the enzyme glycer-

aldehyde-3-phosphate dehydrogenase, inorganic phosphate is incorporated into glyceraldehyde-3-phosphate to form the energy-rich molecule 1,3-bisphosphoglycerate plus NADH. The stage is now set for harvesting energy.

The energy-rich intermediate, 1,3-bisphosphoglycerate, is converted to a molecule of 3-phosphoglycerate plus one molecule of ATP by substrate-level phosphorylation (Figure 8.4). This first energy capturing reaction of the glycolytic pathway is catalyzed by the enzyme **phosphoglycerate kinase**. The 3-phosphoglycerate is subsequently converted to 2-phosphoglycerate and then dehydrated by the enzyme **enolase** to generate phosphoenolpyruvate (PEP), an anhydride, also with a high-energy phosphate bond. In the second energy-yielding reaction, the enzyme **pyruvate kinase** converts PEP to another molecule of ATP plus pyruvate. The energy contained in the phosphate bonds of several biological molecules is shown in Box 8.1. Note that the values differ and not all phosphorylated compounds can be used to form ATP.

The late reactions in the glycolytic pathway result in the generation of four molecules of ATP from two molecules of 1,3-bisphosphoglycerate (see Figure 8.3). As

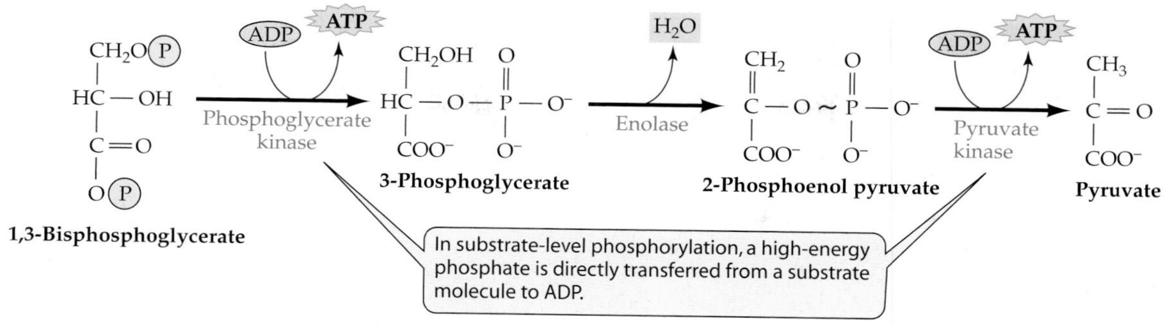

Figure 8.4 Substrate-level phosphorylation
The energy contained in one anhydride bond of 1,3-biphosphoglycerate is used to form ATP. A mutase enzyme (not shown) then forms 2-phosphoglycerate from 3-phosphoglycerate. Dehydration of 2-phosphoglycerate produces an anhydride with a high-energy bond, phosphoenolpyruvate. This phosphate can be transferred to ADP to form ATP.

two molecules of ATP are required in the early reactions, the overall pathway generates two molecules of ATP:

$$\text{Glucose} + 2\text{ ADP} + 2\text{ P}_i \rightarrow 2\text{ pyruvate} + 2\text{ ATP}$$

Pyruvate is further metabolized in all microbes, but the fate differs in the aerobic versus the anaerobic microorganisms. Aerobes respire with oxygen as an electron acceptor (discussed in subsequent text) and may completely oxidize pyruvate to carbon dioxide and water. This occurs by the TCA cycle pathway (see Chapter 10). The total energy generated by glycolytic catabolism of glucose is 2 ATPs, plus an additional 36 ATPs from aerobic respiration–linked phosphorylation reactions. This is due to the ability of the cell to perform electron transport using the 2 NADH generated during glycolysis (see Figure 8.3) plus three additional molecules of NADH or NADPH and one molecule of FADH derived from oxidation of each pyruvate to carbon dioxide via the TCA cycle (see Chapter 10). Respiratory metabolism is energetically superior to fermentation!

Without the ability to use oxygen, anaerobic microorganisms must obtain energy by substrate-level phosphorylation reactions and/or with anaerobic respiration–linked phosphorylation reactions.

Fermentation

Fermentation may be defined as the sum of the anaerobic catabolic reactions that provide for the growth of microorganisms when energy is derived solely by substrate-level phosphorylation. It occurs in the absence of O_2 or any other added electron acceptor. It involves the **incomplete oxidation** of a substrate through a series of interconversions characterized by the reduction of fermentation pathway intermediates to yield organic compounds such as alcohols and acids. This is accompanied by the regeneration of NAD$^+$ molecules needed to metabolize additional molecules of a substrate, for example, glucose. One necessity is a **fermentation balance**, meaning that the combined oxidation–reduction state of the products is equivalent to that of the substrate(s). Simply put, all electrons and protons present in the substrates are accounted for in the product(s).

The principal substrates for fermentation include carbohydrates, amino acids, purines, and pyrimidines. These are the products of anaerobic decomposition of complex plant and animal materials. Several well-studied fermentations include alcoholic fermentation by yeasts and bacteria, lactic acid fermentation by bacteria, and anaerobic dissimilation of amino acids by some species of *Clostridia*. In amino acid fermentation, one amino acid is oxidized and a second amino acid serves as electron acceptor.

Compounds produced during sugar fermentations include short-chain alcohols and organic acids, CO_2, and molecular hydrogen (H_2). Among these are homolactic, heterolactic, mixed acid, and alcoholic fermentations (Table 8.1). Some fermentations not involving sugars are listed in Table 8.2. The substrates include amino acids, purines, pyrimidines, and some organic acids, where the products are as noted previously.

There are many types of fermentations, but in all cases ATP is generated by substrate-level phosphorylation. Recall that fermentation reactions occur essentially within a **closed system** in which a substrate is oxidized and an intermediate product of the substrate is reduced.

Three examples of balanced fermentation reactions that occur under strictly anaerobic conditions are **homolactic fermentation**, **yeast alcoholic fermentation**, and the **Stickland reaction**.

HOMOLACTIC FERMENTATION BY LACTIC ACID BACTERIA One group of the lactic acid bacteria, so called by the end-product of their catabolism, use glucose to generate 2 ATP molecules solely by substrate-level phos-

TABLE 8.1	Examples of products generated during fermentation of glucose and the microorganism involved	
Fermentation Type	**Products**	**Microorganism**
Alcoholic	Ethanol + CO_2	*Zymomonas mobilis*
		Saccharomyces cerevisiae
Homolactic	Lactate	*Lactobacillus acidophilus*
Heterolactic	Lactate + ethanol + CO_2	*Lactobacillus brevis*
Butyric acid	Butyrate + 2 CO_2 + 2 H_2	*Clostridium butyricum*
Butanediol	2,3-Butanediol + ethanol + CO_2 + H_2[a]	*Enterobacter aerogenes*
Mixed acid	Ethanol + acetate + lactate + succinate + formate + CO_2 + H_2[a]	*Escherichia coli*

[a]Amounts of each product may vary.

phorylation. They are abundant in decaying carbohydrate-rich materials: for example, plants and milk products. The overall fermentation reaction is:

$$\text{Glucose} + 2 \text{ ADP} + 2 \text{ P}_i \rightarrow 2 \text{ lactate} + 2 \text{ ATP}$$

The molecules of pyruvate generated by glycolysis (see Figure 8.3) are reduced by the enzyme **lactate dehydrogenase** with NADH to form lactate that is excreted as a waste product. The final reaction, performed by lactate dehydrogenase, permits the regeneration of NAD^+.

$$2 \text{ Pyruvate} + 2 \text{ NADH} \rightarrow 2 \text{ lactate} + 2 \text{ NAD}^+$$

A key feature of fermentation involves the oxidation of an organic substrate (energy source) that serves as a donor of electrons, whereas another organic compound, often an intermediate, serves as the electron acceptor to regenerate needed cofactors (e.g., NAD^+). In this fermentation, glucose is oxidized to pyruvate, which is then reduced to lactate.

In fermentation these reactions are balanced, in that the oxidation level of the substrates and products are equivalent. There is no externally supplied electron acceptor such as O_2. The necessity for a balanced redox level in the substrate and the product limits the range of compounds that can serve as an energy source for fermentation. Most such substrates are essentially intermediate in their redox state (approximately, they have one atom of carbon to two of hydrogen or one of oxygen). They cannot be highly oxidized or highly reduced and must provide intermediates that are sufficiently oxidized so that they can serve as electron acceptors.

ALCOHOLIC FERMENTATION BY YEAST   Another common fermentation intermediate that would be an effective electron acceptor is acetaldehyde. Reduction of this compound by NADH would produce ethanol. This process occurs in yeast and in certain bacteria that thrive in fermenting fruits and grains. Glycolysis yields pyruvate (see Figure 8.3) that is in turn cleaved to acetaldehyde and carbon dioxide by the enzyme **pyruvate decarboxylase**. Acetaldehyde is then reduced by the enzyme alcohol dehydrogenase to regenerate NAD^+. Because the carbon in the organic substrate cannot be further oxidized in a fermentation, only a lim-

TABLE 8.2	Bacterial fermentations that utilize substrates other than sugars	
Type	**Overall Reaction**	**Organism**
Stickland	Glycine + alanine $\rightarrow$ acetate	*Clostridium sporogenes*
Propionic acid	Lactate $\rightarrow$ propionate + acetate	*Clostridium propionicum*
Acetylene	Acetylene + $H_2O \rightarrow$ ethanol + acetate	*Pelobacter acetylenicus*
Oxalate	Oxalate $\rightarrow$ formate + CO_2	*Oxalobacter formigenes*
Malonate	Malonate $\rightarrow$ acetate + CO_2	*Malonomonas rubra*
Homoacetic acid	4 H_2 + 2 $CO_2 \rightarrow$ acetate	*Clostridium aceticum*

ited amount of the potential energy available is released. An alcoholic fermentation by yeasts yields the following products:

$$\text{Glucose} + 2\,\text{ADP} + 2\,\text{P}_i \rightarrow 2\,\text{ethanol} + 2\,\text{CO}_2 + 2\,\text{ATP}$$

These reactions are essential for production of many foods including breads, beer, and wine (see Chapter 31). Bacterial production of ethanol proceeds by alternative fermentation pathways.

AMINO ACID FERMENTATION BY CLOSTRIDIA The strictly anaerobic clostridia (genus *Clostridium*) have

evolved an array of ATP-generating mechanisms that can employ many different substrates, including pyruvate, purines, pyrimidines, nicotinic acid, carbohydrates, and amino acids. These are generated by the decomposition of animal and plant materials.

Various species of clostridia utilize the **Stickland reaction** to anaerobically catabolize amino acids (**Figure 8.5**). In the classic Stickland reaction, one amino acid of a pair is oxidized and the other is reduced. An NAD^+-mediated oxidation generates the high-energy intermediate, acetyl-coenzyme A (acetyl-CoA) following alanine reduction and decarboxylation (**Figure 8.6**). Then a CoA–phosphate exchange reaction results in the incor-

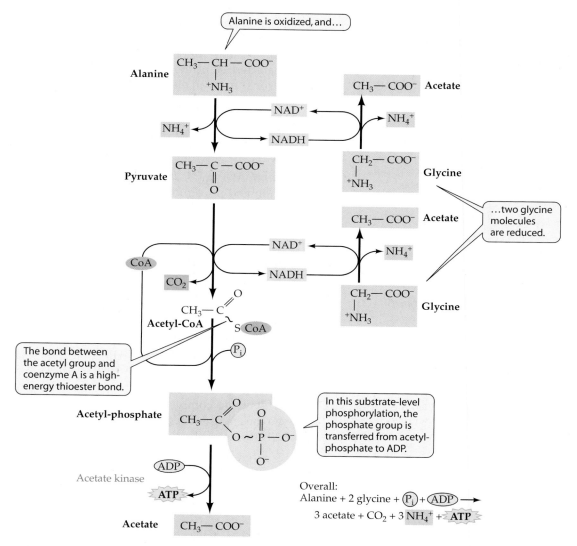

Figure 8.5 Stickland reaction
In the Stickland reaction, pairs of amino acids are used to generate ATP by substrate-level phosphorylation. This reaction occurs in anaerobic organisms such as *Clostridium sporogenes*.

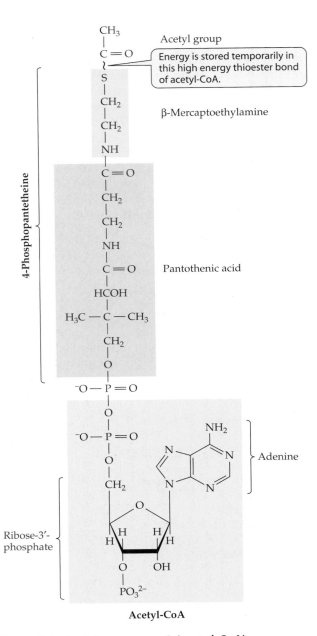

Figure 8.6 Acetyl-coenzyme A (acetyl-CoA)
The thioester bond between the β-mercaptoethylamine moiety of CoA and the acetyl groups (~) is an energy-rich bond.

poration of one P_i molecule to generate acetyl-phosphate with the concomitant release of coenzyme A. The high-energy phosphoryl group from acetyl-phosphate can then be transferred to ADP to form ATP by substrate-level phosphorylation (see Box 8.1). **Acetate kinase** is responsible for this energy-conserving step. A second amino acid—for example, glycine—is oxidized to regenerate NAD^+.

SECTION HIGHLIGHTS
Substrate-level phosphorylation reactions occur via catabolism, for example of sugars, during anaerobic conditions when no external electron acceptor is present. Different fermentation pathways are employed to degrade a variety of organic molecules, where the energy of a high-energy phospho-ester bond intermediate is captured to form ATP.

8.3 Respiration-Linked Phosphorylation

ATP is common to all forms of life and respiration-linked phosphorylation is one means to form it. The oxidation–reduction reactions involved in the synthesis of ATP are varied among bacteria. Aerobic as well as anaerobic **respiration** is characterized by the transfer of reducing equivalents (electrons) from an electron donor (DH), such as NADH, succinate, lactate, or H_2, to a terminal electron acceptor (A) (see Figure 8.2). During aerobic respiration this acceptor is O_2. This process results in a charge separation across the cytoplasmic membrane to form a proton motive force (pmf), which is then used by **ATP synthase** to generate ATP from ADP and P_i (inorganic phosphate). These respiration-linked phosphorylation processes are also termed oxidative phosphorylation.

Electron Transport Chains in Energy-Transducing Membranes

Energy-transducing membranes are composed of two distinct protein assemblies (see Figure 8.2):

1. A highly organized respiratory or *electron transport chain* that catalyzes the energy-yielding (exergonic) transfer of electrons from a primary electron donor to a terminal acceptor such as O_2. This results in formation of a pmf.

2. An ATP synthase complex that uses this pmf to catalyze the energy-requiring (endergonic) synthesis of ATP from ADP + P_i.

In microorganisms, the source of the electrons that bring about the extrusion of protons can be the oxidation of an organic or inorganic substrate. For example, NADH or H_2 may serve as an electron donor (DH):

$$NADH + H^+ \rightarrow NAD^+ + 2\,e^- + 2\,H^+$$
$$H_2 \rightarrow 2\,e^- + 2\,H^+$$

These reactions are coupled to a reduction reaction that consumes the electrons. For example, oxygen may serve as the electron acceptor (A).

$$O_2 + 4 H^+ + 4 e^- \rightarrow 2 H_2O$$

In the process, protons accumulate on the exterior of the membrane to create a positive charge (cathode), and there would be a negative charge (anode) on the interior (see Figure 8.2). The energy inherent in the pmf is captured as ATP when it passes through a specific membrane-bound enzyme and returns to the interior.

The cytoplasmic membrane of respiring and photosynthetic prokaryotic cells, the inner membrane of the mitochondrion, and the thylakoid membrane of a chloroplast are all **energy-transducing membranes**. The transfer of electrons occurs via sequential redox reactions involving highly organized **electron transport chains** that contain proteins with electron carriers such as the flavins, nonheme iron clusters, and cytochromes. They are oriented to transfer electrons in a sequential fashion from the electron donor (DH) to the terminal electron acceptor (A) (Figure 8.7). The orientations and types of proteins present in an electron transport chain may differ, and are dependent on the nature of the microorganism, the primary energy source, and growth conditions. Lipid soluble cofactors, for example ubiquinone (coenzyme Q), are freely diffusable in the membrane and function to transfer electrons and protons from one respiratory enzyme to another.

Considerable variation exists in the constituents of the electron transport chains used by various bacteria and ar-chaea, and the total ATP yield will vary according to the type of substrate and the efficiency of the system. Aerobic, facultatively anaerobic, and obligately anaerobic chemotrophs can utilize a variety of electron acceptors to derive energy from an electrochemical gradient generated by electron transport (Table 8.3). Recall that chemoorganotrophs obtain energy from organic compounds, whereas chemolithotrophs use inorganic compounds. The following section will discuss mechanisms for generating ATP via respiration-linked phosphorylation.

Oxidation–Reduction Reactions

Oxidation–reduction (or redox) reactions occur because electrons flow from a component of higher energy potential, an electron donor, to another of lower potential (an electron acceptor). Recall the analogy to a car or flashlight battery that provides a current. The donor is termed the reductant and the electron acceptor, the oxidant. In an electron transport chain, a series of intermediate oxidation–reduction reactions are coupled to join a primary electron donor and a terminal electron acceptor (see Figure 8.7, described in subsequent text).

In an individual reaction, one component in a redox reaction is oxidized and the other is reduced. The capacity to either donate or receive electrons is defined as the redox potential (or E) (Box 8.2). The redox potential of a compound is measured in volts (V) and is based on the voltage required to remove electrons from H_2 under standard

TABLE 8.3	Electron acceptors used by the *Bacteria* or the *Archaea* during aerobic and anaerobic respiration.		
Condition	**Terminal Electron Acceptor**	**Product**	**Example Microorganism**
Aerobic			
	O_2	H_2O	*Micrococcus luteus*
Anaerobic			
	SO_4^{2-}	H_2S	*Desulfovibrio desulfuricans*
	Fe^{3+}	Fe^{2+}	*Geobacter metallireducens*
	Mn^{4+}	Mn^{2+}	*Desulfuromonas acetoxidans*
	NO_3^-	NO_2^-	*Escherichia coli*
	NO_2^-	NH_4^+	*Escherichia coli*
	NO_3^-	N_2	*Thiobacillus* sp., *Paracoccus* sp.
	CO_2	CH_4	*Methanosarcina barkeri*
	CO_2	CH_3COO^-	*Clostridium aceticum*
	S^0	H_2S	*Desulfuromonas acetoxidans*
	SeO_4^{2-}	Se^0	*Thauera* sp.
	AsO_4^{3-}	AsO_3^{2-}	*Chrysiogenes arsenatis*
	Fumarate	Succinate	*Wolinella succinogenes*
	TMAO[a]	TMA	*Vibrio, Rhodospeudomonas* sp.

[a]TMAO, trimethylamine-N-oxide; TMA, trimethylamine.

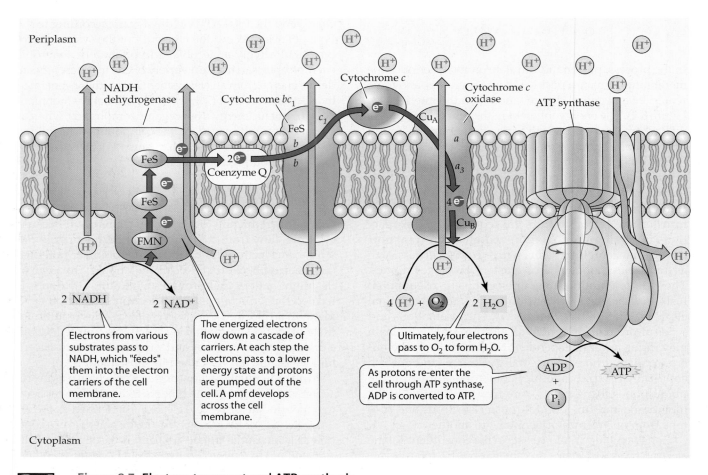

Periplasm

NADH
dehydrogenase

Cytochrome bc_1

Cytochrome *c*

Cytochrome *c*
oxidase

ATP synthase

FeS
b
b

c_1
FeS

$2e^-$
Coenzyme Q

Cu_A
a

a_3

$4e^-$
Cu_B

FeS
e^-

FeS
e^-

FMN

2 NADH → 2 NAD⁺

Electrons from various
substrates pass to
NADH, which "feeds"
them into the electron
carriers of the cell
membrane.

The energized electrons
flow down a cascade of
carriers. At each step the
electrons pass to a lower
energy state and protons
are pumped out of the
cell. A pmf develops
across the cell
membrane.

$4\ H^+ + O_2$ → $2\ H_2O$

Ultimately, four electrons
pass to O_2 to form H_2O.

As protons re-enter the
cell through ATP synthase,
ADP is converted to ATP.

ADP
+
P_i

ATP

Cytoplasm

Figure 8.7 Electron transport and ATP synthesis

The overall reaction for the electron transport chain is 2 NADH + 4 H⁺
+ O_2 → 2 NAD⁺ + 2 H_2O. Each carrier in the chain alternates between its
oxidized and reduced state. Carriers may transfer one or two electrons at a time.
Protons are consumed or moved from the cell interior to the outside of the membrane. Re-entry of protons through the ATP synthase is accompanied by ATP synthesis. The letters a, a_3, b, and c_1 denote distinct cytochrome types.

conditions with an electrode. According to this scale the
hydrogen electrode has been assigned a value of 0.0 V:

$$2\ H_2 \rightarrow 4\ H^+ + 4\ e^-$$

However, when this reaction occurs under standard
biochemical conditions, with reactants at a concentration of 1M (molar) and at a pH of 7, the redox potential
is actually –0.42 V, and thus termed E_o'. The reaction for
H_2 just presented would be termed a **half reaction** because electrons cannot exist alone in solution, and for
the preceding oxidation reaction to occur, there must be
another half reaction and a corresponding reduction. A
half reaction (an oxidation) must be coupled to another
half reaction (a reduction) such as:

$$O_2 + 4\ e^- + 4\ H^+ \rightarrow 2\ H_2O$$

Together they comprise the coupled oxidation–reduction (redox) reaction:

$$2\ H_2 + O_2 \rightarrow 2\ H_2O$$

In a biological system, the amount of free energy released in a redox reaction is the difference in redox potential between the electron donor and the electron acceptor (see Box 8.2). This energy difference is designated
by $\Delta E_o'$. The term $\Delta E_o'$ is the difference in redox potential at pH 7 (neutrality) and with 1M concentrations of
the reactants. Thus, the reduction of O_2 by H_2 is the sum
of two half reactions:

BOX 8.2 *Methods & Techniques*

Oxidation–Reduction (Redox) Reactions

Oxidation is the loss of electrons, and reduction is the gain of electrons. Because electrons cannot exist free in solution, an oxidation reaction must be coupled with a reduction reaction. Oxidation releases energy, and this can occur only if a substance is available that can be reduced. Biological reactions can involve the transfer of hydrogen atoms, which consist of an electron and a proton. When the electron is removed from NADH + H⁺, one **proton** will generally accompany each electron. For example:

$$NADH + H^+ \rightarrow NAD^+ + 2\,H^+ + 2\,e^-$$

When electrons are introduced to molecular oxygen, **protons** will generally accompany the electrons. For example:

$$O_2 + 4\,H^+ + 4\,e^- \rightarrow 2\,H_2O$$

Redox potential is a quantitative measure of the tendency for a substance to give up electrons in biological systems. It is measured in volts and generally under standard conditions at pH 7 (the pH of cytoplasm), and at 1 M concentration of the reactants (E_o'). Redox potentials of many common biological redox reactions are presented in Table 8.4. The redox potentials for the two half reactions above are:

$$NAD^+ + 2\,H^+ \rightarrow$$
$$NADH + H^+ = -0.32\text{ V}$$

$$2\,H_2O \rightarrow$$
$$4\,H^+ + O_2 = +0.82\text{ V}$$

$$\Delta E_o' = (e^-\text{ acceptor}) - (e^-\text{ donor})$$

$$\Delta E_o' = (+0.82\text{ V}) - (-0.32\text{ V})$$

$$\Delta E_o' = 1.14\text{ V}$$

The coupled oxidation–reduction-reaction is:

$$2\,NADH + 2\,H^+ + O_2 \rightarrow$$
$$2\,NAD^+ + 2\,H_2O$$

In redox reactions, electrons are transferred from one component (a donor), which is oxidized, to another (an acceptor), which is reduced. Reductions may result in the gain of one or more electrons.

The oxidized form of cytochrome *c* accepts 1 e⁻:

$$\text{cyt } c\ Fe^{+3} + e^- \rightarrow \text{cyt } c\ Fe^{+2}$$

Depending on the acceptor, it may also gain protons.

NAD⁺ accepts 2 e⁻ and gains 1 proton:

$$NAD^+ + 2\,e^- + H^+ \rightarrow NADH$$

Ubiquinone accepts 2 e⁻ and gains 2 protons:

$$Q + 2\,e^- + 2\,H^+ \rightarrow QH_2$$

When a biological oxidation is coupled to a reduction reaction, for example during aerobic respiration using NADH as the electron donor, energy is released (exergonic reaction). The release of energy for the coupled reactions is proportional to the change in mid-volt potentials. The calculation of free energy released ($\Delta G_o'$) for this coupled reaction is described in Box 8.3.

$$2\,H^+ + 2\,e^- / H_2\ (E_o' = -0.42\text{ V})$$
$$O_2 + 4\,H^+ + 4\,e^- / 2\,H_2O\ (E_o' = +0.82\text{ V})$$
Thus:
$$\Delta E_o' = E_o'(e^-\text{ acceptor}) - E_o'(e^-\text{ donor})$$
$$\Delta E_o' = (+0.82\text{ V}) - (-0.42\text{ V}) = +1.24\text{ V}$$

This change in redox potential is available to perform work (as a flashlight battery provides energy to make light). The total free energy change $\Delta G_o'$ (under standard conditions) occurring in these two half reactions can be calculated by the following equation:

$$\Delta G_o' = -nF\Delta E_o'$$

where $\Delta E_o'$ is the redox change under standard conditions, n is the number of electrons transferred, and F is the Faraday constant (96.5 kJ/mol/V). Substituting in values for each factor in the coupled reaction gives the change in free energy as:

$$\Delta G_o' = -(4)(96.5\text{ kJ/mol/V})\,[(+0.82\text{ V}) - (-0.42\text{ V})]$$
$$\Delta G_o' = -(4)(96.5\text{ kJ/mol/V})\,(1.24\text{ V})$$
$$\Delta G_o' = -478.6\text{ kJ/mol}$$

Recall that a negative $\Delta G_o'$ value represents energy release (exergonic), whereas a positive value requires energy to drive the reaction (endergonic). Note that the $\Delta G_o'$

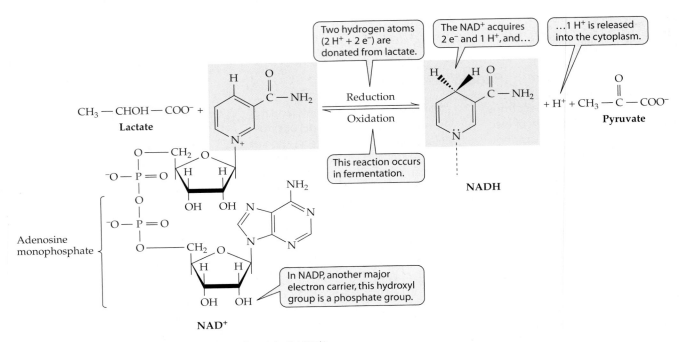

Two hydrogen atoms (2 H⁺ + 2 e⁻) are donated from lactate.

The NAD⁺ acquires 2 e⁻ and 1 H⁺, and...

...1 H⁺ is released into the cytoplasm.

Reduction

Oxidation

This reaction occurs in fermentation.

In NADP, another major electron carrier, this hydroxyl group is a phosphate group.

Adenosine monophosphate

NADH

NAD⁺

Figure 8.8 Nicotinamide adenine dinucleotide (NAD⁺)
Nicotinamide adenine dinucleotide (NAD⁺) is a major two electron carrier. When electrons are enzymatically removed from a donor (such as lactate), they pass to NAD⁺ to generate NADH.

value of –478.6 kJ/mol is based on one molecule of O_2. Per molecule of H_2 this value would be –239.3 kJ/mol.

A common redox reaction occurring in many biological systems involves enzymatic removal of an electron and accompanying proton from an organic substrate such as glyceraldehyde-3-phosphate (see Figure 8.3) and passage to an electron **carrier** such as nicotinamide adenine dinucleotide (**NAD⁺**) (Figure 8.8).

The reduction potentials for NAD⁺/NADH and NADP⁺/NADPH are equivalent at –0.32 V, and both are effective electron carriers. NAD⁺ is principally involved in energy-generating (catabolic) reactions, and NADP⁺ is involved mostly in biosynthesis (anabolism). NAD⁺ is a hydrogen atom carrier that receives one hydrogen and two electrons (H⁺ + 2 e⁻) in metabolic reactions, referred to as reduction (see Figure 8.8).

An electron carrier such as NAD⁺ is an intermediate that **carries** electrons from a **donor enzyme** to an **acceptor enzyme**. Reduction of NAD⁺ by several glycolytic intermediates would yield NADH + H⁺, which for brevity is written NADH. This reduced carrier can pass the electrons on to a series of electron carriers in a cascade of increasing E_o' (Table 8.4). In the half-reaction with a more negative E_o' the substrate tends to donate electrons and become oxidized, whereas the substrate in the half reaction with a more positive E_o' accepts electrons and becomes reduced. The energy released in a typical pair of biological half reactions in aerobic organisms is described in Box 8.3.

$$4 \text{ NADH} + 4 \text{ H} + O_2 \rightarrow 4 \text{ NAD}^+ + 2 \text{ H}_2\text{O}$$

This coupled biological reaction releases considerable energy to drive other cellular reactions (ATP formation, flagella rotation, solute uptake, and cell biosynthesis). The remaining energy is lost as heat. Other redox reactions may yield considerably less potential energy. This is described in a following section.

Components of Electron Transport Chains

The high energy electrons in NADH can be delivered to the membrane-bound electron transport chain by NADH dehydrogenase, which oxidizes NADH and reduces quinones. Reduced quinine molecules then deliver the electrons to other components of the electron transport chain (see Figure 8.7). The energy released is used to translocate protons (H⁺) to the *exterior* of the cytoplasmic membrane. This translocation creates a proton motive force (electrochemical potential), as mentioned earlier. As shown in Figure 8.7 the protons ultimately traverse back across the cytoplasmic membrane via a membrane-associated protein complex called the ATP synthase complex, and ATP is generated. ATP synthase can undergo a proton-driven conformational change that covalently bonds a molecule of ADP and inorganic phosphate (P$_i$) to form ATP. The energy-rich chemical bond between ADP and P$_i$ now bears the en-

TABLE 8.4	Half reactions, the number of electrons transferred (n), and the redox potential under standard biochemical conditions (E_o')		
Half Reaction (oxidized/reduced)		**n**	**E_o' (V)**
CO_2/formate		2	−0.43
Ferredoxin (oxidized/reduced)		2	−0.43
$2 H^+/H_2$		2	−0.42
$NAD^+ + 2 H^+/NADH + H^+$		2	−0.32
1,3-bisphosphoglycerate + 2 H^+/ glyceraldehyde-3-P + P_i		2	−0.29
S^0/HS^-		2	−0.27
$FMN + 2 H^+/FMNH_2$		2	−0.22
$FAD + 2 H^+/FADH_2$		2	−0.22
Acetaldehyde/ethanol		2	−0.20
Pyruvate/lactate		2	−0.19
Dihydroxyacetone-P/glycerol-P		2	−0.19
HSO_3^-/HS^-		6	−0.12
Menaquinone (oxidized/reduced)		2	−0.07
Glycine/acetate + NH_4^+		2	−0.01
Fumarate + 2 H^+/succinate		2	+0.03
Ubiquinone (oxidized/reduced)		2	+0.11
Cytochrome b (Fe^{3+}/Fe^{2+})		1	+0.08
Cytochrome c (Fe^{3+}/Fe^{2+})		1	+0.25
TMAO/TMA[a]		2	+0.13
DMSO/DMS[b]		2	+0.16
NO_2^-/NO		1	+0.35
$NO_3^- + 2 H^+/NO_2^- + H_2O$		2	+0.42
SeO_4^{2-}/SeO_3^{2-}		2	+0.48
Fe^{3+}/Fe^{2+}		1	+0.77
$O_2 + 4 H^+/H_2O$		4	+0.82
NO/N_2O		1	+1.17
N_2O/N_2		1	+1.35

[a]TMAO, trimethylamine oxide; TMA, trimethylamine.
[b]DMSO, dimethylsulfoxide; DMS, dimethylsulfide.

ergy available from the substrate and carried by NADH to the electron transport chain.

The enzymes and cofactors that make up the electron transport chain participate in a series of redox reactions that carry the electrons from a high-energy state to a lower-energy state, where the electrons ultimately react with the terminal acceptor. In the following discussion, we will focus on the links between electron transport and chemical structures of components of these chains in bacteria.

Note that electrons do not exist in the free state in biological systems but flow from electron donor molecules to acceptor molecules. Components of electron transport chains, such as the cytochromes, in the cytoplasmic membranes of microorganisms are efficiently organized to accept electrons and then become donors as they pass

the electrons along to an acceptor of lower redox potential (see Figure 8.7).

FLAVOPROTEINS **Flavoproteins** are a family of enzymes, each consisting of a protein (of varying molecular mass) bound to a nucleotide form of the B vitamin riboflavin. The **flavins—flavin adenine dinucleotide** (FAD) and **flavin mononucleotide** (FMN)—are two- electron (proton) carriers (Figure 8.9). Depending on the type of flavoprotein, FAD is sometimes covalently bound to an amino acid side chain, or, alternatively, is tightly bound through multiple noncovalent interactions within the protein. FMN is usually bound to the protein by its negatively charged phosphate groups. FAD and FMN have a relatively high redox potential (0.003 to 0.09 V) when bound to protein but have a redox potential of −0.22 V as free

(A)

(B)

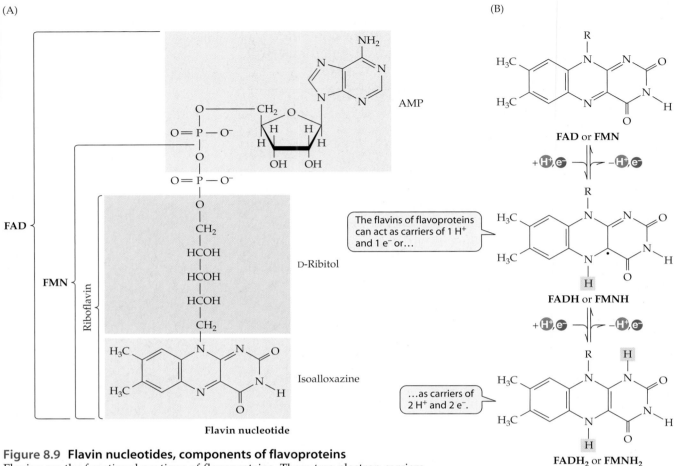

AMP

D-Ribitol

Isoalloxazine

FAD

FMN

Riboflavin

Flavin nucleotide

FAD or FMN

The flavins of flavoproteins can act as carriers of 1 H+ and 1 e− or...

FADH or FMNH

...as carriers of 2 H+ and 2 e−.

FADH₂ or FMNH₂

Figure 8.9 Flavin nucleotides, components of flavoproteins
Flavins are the functional portions of flavoproteins. These two-electron carriers are derivatives of the B vitamin riboflavin. (A) Detailed structure of the flavin nucleotides, FAD and FMN. (B) Oxidation and reduction reactions of the flavin nucleotides FAD and FMN.

coenzymes (see Table 8.4). The $NAD^+/NADH$ couple (−0.32 V) can transfer electrons to either the $FAD/FADH_2$ or $FMN/FMNH_2$ couple as electrons flow freely from a more negative carrier to a more positive one. Examples of flavoproteins include the electron-donating NADH dehydrogenase and succinate dehydrogenase enzymes, as well as the electron-accepting fumarate reductase.

IRON-SULFUR PROTEINS **Iron-sulfur proteins**, also known as nonheme iron or Fe-S proteins, contain one or more clusters of iron and sulfur atoms (Figure 8.10). These clusters are held in position by covalent linkage through the sulfur atom of the amino acid cysteine, and may contain 2, 3, or 4 atoms each of inorganic sulfur and iron. The atoms in the [3Fe-4S] type and [4Fe-4S] type form a cube, as shown, whereas the [2Fe-2S] type clusters are planar. Because the iron-sulfur clusters are redox active, they may accept one electron to become reduced, or donate one electron and become oxidized. The

redox potential for these electron carriers spans a broad range from −0.60 V to +0.35 V. A common iron-sulfur protein in biological systems is **ferredoxin**, which has a [2Fe-2S] configuration, and is present in many anaerobic microbes. Fe-S proteins can be membrane bound such as NADH dehydrogenase or succinate dehydrogenase, or soluble proteins like ferredoxins.

CYTOCHROMES **Cytochromes** are constituents of the respiratory chain in aerobic and in many anaerobic microorganisms. Cytochromes are a heterogeneous group of heme-containing proteins that are located primarily in the cytoplasmic membrane, and are characterized on the basis of their spectrophotometric properties. Those cytochromes (cyt) with similar absorption properties are designated by lowercase letters as cyt *a*, cyt *b*, cyt *c*, and so forth.

Within each of these cytochrome groups there are measurable differences in spectral and biochemical prop-

BOX 8.3 *Methods & Techniques*

Energy Efficiency in Viable Systems

The efficiency of respiration-linked phosphorylation can be calculated as follows. For the complete oxidation of 2 moles of NADH via the electron transport chain with 1 mole of oxygen (O_2) as terminal electron acceptor:

$$\Delta G_o' = -nF\Delta E_o'$$

n = number of electrons

F = Faraday constant (96.5 kJ/Mol/V)

$\Delta E_o'$ = redox potential difference between two half reactions

$\Delta E_o'$ = (electron acceptor) – (electron donor)

$NAD^+ + 2H^+$ / $NADH + H^+ = -0.32$ V

$O_2 + 4H^+$ / $2H_2O = +0.82$ V

Therefore:

$$\Delta G_o' = -4(96.5)[(0.82\text{ V}) - (-0.32\text{ V})] = -440\text{ kJ/mol}$$
(theoretical for two molecules of NADH + H$^+$

The $\Delta G_o'$ for ATP hydrolysis is:

$$\Delta G_o' = -30.5\text{ kJ/mol}$$

If 3 moles of ATP are synthesized per mole of NADH:

$$3 \times (-30.5) = -91.5\text{ kJ/mol (actual)}$$

Efficiency would be:

$$\frac{-91.5\text{ kJ/mol}}{-220\text{ kJ/mol}} = 0.42$$

or ~ 42% of the energy is conserved as ATP, with the remaining energy lost as heat to the cell surroundings.

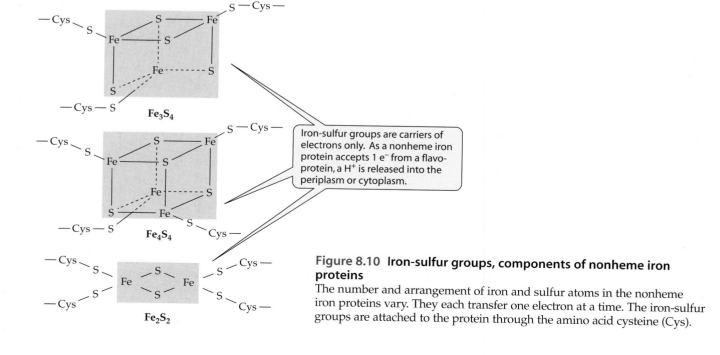

Iron-sulfur groups are carriers of electrons only. As a nonheme iron protein accepts 1 e$^-$ from a flavoprotein, a H$^+$ is released into the periplasm or cytoplasm.

Figure 8.10 Iron-sulfur groups, components of nonheme iron proteins

The number and arrangement of iron and sulfur atoms in the nonheme iron proteins vary. They each transfer one electron at a time. The iron-sulfur groups are attached to the protein through the amino acid cysteine (Cys).

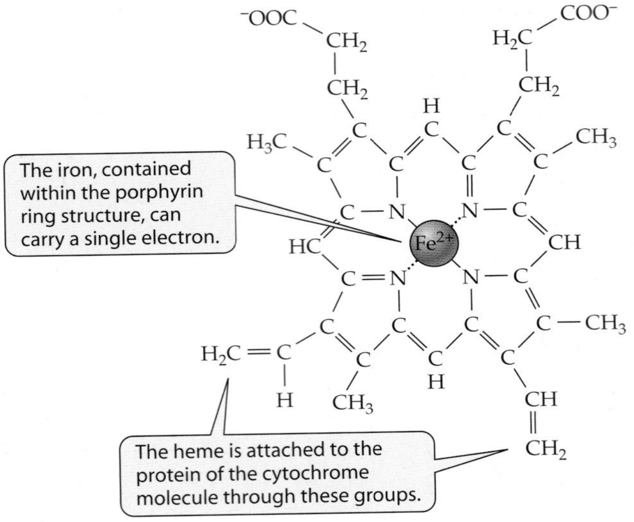

The iron, contained within the porphyrin ring structure, can carry a single electron.

The heme is attached to the protein of the cytochrome molecule through these groups.

Figure 8.11 Heme
The heme portion of a cytochrome molecule. One electron is transferred at a time.

erties. These are indicated by numerical subscripts, for example, cyt a_1, cyt a_2, and cyt a_3. In some cases, a cytochrome is identified by the wavelength of light at which it absorbs maximally (e.g., cytochrome c_{555}). The functional portion of a cytochrome molecule is an iron-porphyrin electron transport component, called **heme** (Figure 8.11). Some heme molecules are covalently attached (cyt *c*), whereas others are bound noncovalently (cyt *a*, *b*, and *d*) to the protein. The central iron atom in the porphyrin nucleus accepts a single electron on reduction.

The redox potential of the various cytochromes ranges from −100 to +400 mV, their wavelength absorption maxima from 450 to 800 nm, and their molecular masses from 12,000 to 350,000 Da. Examples of cytochromes include the membrane-bound cytochrome *c* oxidase that functions in the terminal step in oxygen reduction, and a small soluble cytochrome *c* protein that functions to transfer electrons from the bc_1 complex to the terminal cytochrome *c* oxidase (see Figure 8.7).

QUINONES   The terms **coenzyme Q**, **CoQ**, or **Q** denote a family of **quinones** that are of widespread occur-

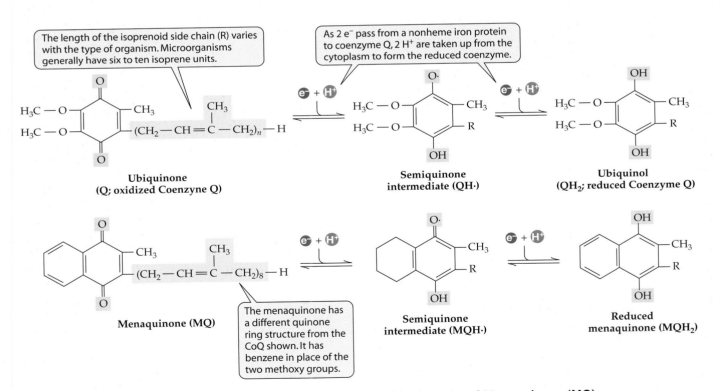

The length of the isoprenoid side chain (R) varies with the type of organism. Microorganisms generally have six to ten isoprene units.

As 2 e⁻ pass from a nonheme iron protein to coenzyme Q, 2 H⁺ are taken up from the cytoplasm to form the reduced coenzyme.

Ubiquinone
(Q; oxidized Coenzyme Q)

Semiquinone
intermediate (QH·)

Ubiquinol
(QH₂; reduced Coenzyme Q)

Menaquinone (MQ)

The menaquinone has a different quinone ring structure from the CoQ shown. It has benzene in place of the two methoxy groups.

Semiquinone
intermediate (MQH·)

Reduced
menaquinone (MQH₂)

Figure 8.12 Coenzyme Q (ubiquinone) and Menaquinone (MQ)
Coenzyme Q is a quinone (also know as ubiquinone). Its fully reduced form is ubiquinol. Menaquinone (MQ) is the form present in some microorganisms, particularly gram-positive bacteria. Both cofactors transfer two electrons and two protons.

rence in energy-transducing membranes of prokaryotes and eukaryotes. Among these are the **ubiquinones**, which are the principle quinones in membranes of eukaryotes and gram-negative bacteria (Figure 8.12). Other bacteria (generally gram-positives and many anaerobes) contain **menaquinone (MQ)**. The structures of ubiquinone and menaquinone are presented in Figure 8.12.

Quinones are small lipid-diffusible cofactors that are abundant in the cytoplasmic membrane. They mediate electron transfer between respiratory enzymes embedded in the membrane. Because quinones can also carry protons, they sometimes shuttle these across the membrane and generate a pmf. The ability of quinones to function as H^+ shuttles depends on the site of reduction and oxidation by the participating enzymes. For example, NADH dehydrogenase transfers two electrons to CoQ, which are in turn donated to the $bc1$ complex (see Figure 8.7).

SECTION HIGHLIGHTS

Respiration-linked phosphorylation reactions occur in a variety of lithotrophic and organotrophic microorganisms to form ATP from ADP and P_i. Membrane-associated electron transport chains composed of at least two oxidoreductase enzymes and diffusible cofactors act to couple the oxidation of an electron donor to an electron acceptor. The formation of a proton motive force (pmf) across the cell membrane is employed to drive ATP synthesis by the action of the ATP synthase. Considerable diversity exists among the *Bacteria* and the *Archaea* in using a large variety of electron donors and acceptor molecules to harvest useful energy.

8.4 Generation of the Proton Motive Force

Formation of the pmf occurs by one of three mechanisms:

- **Scalar consumption**—protons are consumed by the reduction reaction occurring at the inner surface of the cytoplasmic membrane. Examples include the incorporation of protons into O_2 by cytochrome *c* oxidase (see Figure 8.7).

- **Vectorial movement**—protons are pumped from the cell interior to the exterior. Examples include

the NADH dehydrogenase and cytochrome *c* oxidase (see Figure 8.7).

- **Q-loop**—protons are transferred to quinone on the cytoplasmic side of the membrane and are deposited at the periplasmic side. The cytochrome bc_1 complex, for example, performs this process (see Figure 8.7).

These processes that form a pmf require the presence of a closed membrane compartment that is impermeable to protons.

The pmf in respiring microorganisms is generated by the reactions catalyzed by the electron transport chain. We will consider variations that may occur in the types of carriers and their orientation later. In the electron transport chain shown in Figure 8.7, the oxidation of NADH, the electron donor, leads to reduction of O_2, the terminal electron acceptor. NADH donates two electrons to the quinone. This reaction is catalyzed by NADH dehydrogenase (also called complex I), which contains flavin and nine nonheme iron-sulfur redox centers that span the transmembrane protein complex (see Figures 8.9 and 8.10). This electron transfer is accompanied by movement of two protons (H^+) from the cytoplasm to Q to generate reduced cofactor, QH_2 (see Figure 8.12). Two additional protons are pumped through the NADH dehydrogenase complex and into the periplasm to form the pmf. The same process generates a proton gradient across the cytoplasmic membrane in gram-positive bacteria, although they lack a true periplasm.

The reduced quinone, freely diffusible within the membrane, transfers two electrons to the next intermediate carrier, **cytochrome bc_1 complex** (also called complex III). By a process known as the "Q-cycle," the two protons from the reduced quinone are deposited on the periplasmic side of the cytoplasmic membrane. The two electrons from QH_2, after passage through the bc_1 complex, are then transferred to a small electron carrier protein, called cytochrome *c*, located in the periplasm. In the last step of this electron transport chain, electrons are given to **cytochrome *c* oxidase** (also known as complex IV), where the reduction of O_2 results in H_2O. Like the NADH dehydrogenase complex, cytochrome *c* oxidase also pumps additional protons from the cytoplasm to the periplasm. Energy released by transfer of high energy electrons "downhill" to the terminal acceptor thus generates a positive charge (H^+) on one side of the membrane and a negative charge on the other.

The pmf is composed of chemical (pH) and electrical potential gradients (Box 8.4). The value of each can be determined experimentally, and the combined values expressed in milivolts (Δp) or as kJ/mol ($\Delta \mu H^+$). Peter Mitchell (Box 8.5) formulated the theoretical basis for these principles.

BOX 8.4	*Milestones*

Chemiosmosis and Energy Conservation

The chemiosmotic hypothesis (principle) is based on the concept of a charge separation across the bacterial cytoplasmic membrane (or membranes of mitochondria and chloroplasts in eukaryotes). The major role of the electron transport proteins in the electron transport chains is to promote the spatially oriented movement of protons across the membrane (unidirectional) whereby an electrochemical gradient is formed.

As a consequence, the pH on the outside of the cell becomes more acidic, whereas the pH in the interior of the cell becomes more alkaline. This pH difference across the cell cytoplasmic membrane is referred to as the **pH gradient** or Δ**pH**, and can be determined experimentally by use of a pH electrode.

Because protons are positively charged, a proton gradient also causes a charge difference (i.e., positive on the outside and negative on the inside of the cell). The difference in charge is referred to as the **electrical membrane potential** or $\Delta\Psi$. This can be experimentally determined by using lipophilic, positively charged, radioactive molecules that can freely diffuse into the cell. For *E. coli*, some typical values are: ΔpH = 1 pH unit, and $\Delta\psi$ = –160 mV.

The combined values of ΔpH and $\Delta\Psi$ are defined as **proton motive force** or **the chemiosmotic membrane potential** (Δ**p**). The proton motive force can be calculated by the following equation:

$$\Delta p = \Delta\Psi - Z\Delta pH$$

The term **Z** is a constant that equals 60 mV (2.3 RT/F) at 25°C and allows the conversion of ΔpH into millivolts (mV), where 1 pH unit equals 60 mV.

The term **electrochemical proton gradient**, ($\Delta\mu_H^+$), is often used instead of proton motive force. Whereas Δp is given in mV, $\Delta\mu_H^+$ is given in kJ/mol. For *E. coli*, a typical value for Δp = –240 mV, or $\Delta\mu_H^+$ = –23 kJ/mol.

$$\Delta\mu_H^+ = F\Delta p \ (F = \text{Faraday constant} = 96.5 \text{ kJ/mol/V})$$

This proton motive force drives the synthesis of ATP. The number of ATP molecules generated per transfer of two electrons from NADH through the respiratory chain is three in microbes having the cytochrome bc_1 complex and two in those lacking this component.

The coupling of the $\Delta\mu H^+$ to the production of ATP by the ATP synthase is presented in the chapter text.

Inhibitors of Respiration

Much of our understanding of the electron transport chain and respiratory ATP synthesis resulted from studies with chemical compounds that interfere with one or both of these processes. Early workers described two classes of chemicals that affect electron transport, and these were termed *uncouplers* and *inhibitors*. **Uncouplers** prevent the synthesis of ATP but do not interfere with electron transport. Among the effective uncouplers are 2,4-dinitrophenol (DNP) and carbonyl cyanide-*p*-trifluoromethoxyphenylhydrazone (FCCP). This "uncoupling" occurs because DNP and FCCP are lipophilic molecules and readily pass through a cytoplasmic membrane. They are also acidic and may bind protons on the outside of the membrane and carry them to the inside, thereby dissipating the proton gradient.

A respiratory **inhibitor** is a chemical that blocks electron transport or ATP synthesis. Inhibitors are compounds such as cyanide, carbon monoxide, or azide that can bind to the iron center of cytochromes and block electron transport. Without electron transport, there is no proton gradient formed, and therefore no ATP synthesis. The ATP synthase inhibitor DCCP (dicyclohexylcarbodiimide) binds to a membrane subunit to block proton movement.

SECTION HIGHLIGHTS

Formation of a proton motive force involves the action of electron transport chains in the cell membrane. This force, consisting of a H^+ concentration gradient, is the result of consumption of protons in the cytoplasm, or pumping of protons across the membrane, and/or the complex Q-loop mechanism. Various inhibitors can be used to define and study the nature of the proton motive force and electron transport chain components.

BOX 8.5   *Milestones*

Evolution of a Concept—Chemiosmosis

Peter Mitchell was born in Surrey, England, in 1920. He attended Cambridge University, where he completed his Ph.D. in 1951. Mitchell remained at Cambridge until 1955, when he moved to the University of Edinburgh. There he directed the chemical biology unit for the next eight years. In 1961 Peter Mitchell, with his assistant, Jennifer Moyle, published a land-mark paper that revolutionized thinking on cell energetics. This was the *chemiosmotic hypothesis,* based on his belief that enzymatic reactions in solution might differ from those that occur when enzymes are embedded in a membrane. Mitchell believed that a chemical reaction associated with a membrane could result in translocation of a chemical group. This directional (vectorial) event could cause the movement of a proton from the inner surface of the membrane to the outer surface. He theorized that this would create an electrochemical proton gradient ($\Delta\mu H^+$) across the membrane. The relatively stable $\Delta\mu H^+$ was the central energy currency in respiration. In his hypothesis, the exterior of the membrane would be positively charged, and the interior would be rendered negative. The protons would be drawn toward the interior and pass through a proton-translocating ATP synthase that would drive ATP synthesis. This hydrogen ion current would be the driving force for performance of chemical, osmotic, and, in the case of flagellar movement, mechanical work.

The scientific establishment working in the field of bioenergetics considered Mitchell's hypothesis bizarre and untenable. The central dogma of the time was that a high-energy intermediate composed of respiratory chain enzymes and an ATP synthetase must exist. Intermediates were sought; many were considered, but all failed the ulti-mate test because no such factor existed. It was not until the mid-1960s—at which time many bio-chemists had worked long and hard and failed to find a bound high-energy phosphoryl intermediate associated with respiratory phos-phorylation—that the world came around to Mitchell's ideas. There was no "coupling" factor between respiratory chain enzymes and the ATP synthase system.

In 1963 ill health led to Mitchell's retirement to a small laboratory in Cornwall. He rebuilt an eighteenth-century derelict Cornish mansion called Glynn House, and there he reestablished his research program. To the restful, calm environs of the Glynn Research Laboratory, he invited his detractors and gradually convinced all of them that the chemiosmotic hypothesis was indeed correct. In 1978 he received the Nobel Prize, and few doubted the wisdom of his theories.

Peter Mitchell, winner of the Nobel Prize for chemistry in 1978, proposed the chemiosmotic theory for genera-tion of ATP. Courtesy of *Times Union.*

Glynn House stands as an anom-aly and a viable alternative to the huge collaborative research efforts that receive funds from centralized funding sources. Is truly original research accomplished better in the hurly-burly of vast enterprises? Can original research always thrive best with centralized funding based on previous results? These are interesting questions in the age of big science, and Glynn House stands as mute testimony that big-ger is not necessarily better.

8.5   ATP Synthase

ATP synthase is present in the energy-transducing mem-branes of mitochondria, chloroplasts, and aerobic respir-ing and photosynthetic bacteria (Figure 8.13). It is also present in the *Archaea* and anaerobic bacteria that do not rely solely on substrate-level phosphorylation. The struc-ture of the complex is remarkably similar regardless of the source. Electron microscopy of energy-transducing membrane preparations reveals the characteristic mush-room-like proteins that protrude from one side of the membrane. In the *Bacteria*, the *Archaea*, and in mitochon-dria, they protrude into the internal space (cytoplasm or matrix), and in isolated thylakoid or chromatophore membranes, they project outward. Regardless of direc-

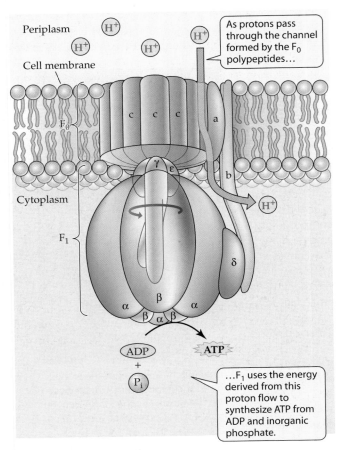

Periplasm

Cell membrane

As protons pass through the channel formed by the F_0 polypeptides…

F_0

Cytoplasm

F_1

…F_1 uses the energy derived from this proton flow to synthesize ATP from ADP and inorganic phosphate.

ADP + P_i

ATP

Figure 8.13 ATP synthase
The structure of ATP synthase, showing the F_0 and F_1 subunits and the individual polypeptide types.

tion, these projections are functionally equivalent because ATP is synthesized on the side of the membrane from which the proteins project, whereas protons enter from the opposite side.

The function of the ATP synthase is to utilize the pmf, also termed the electrochemical proton gradient ($\Delta\mu H^+$) (see Box 8.5) to maintain a mass-action ratio toward **equilibrium** and thus drive ATP synthesis. Under certain conditions the ATP synthase may function as an ATP-hydrolyzing enzyme. In fermentative bacteria, this activity utilizes ATP to maintain a $\Delta\mu H^+$ for active transport of nutrients into the cell. Many organisms, including some of the photosynthetic bacteria, use electron donors that have a redox potential more positive than that of NAD^+ or $NADP^+$ (see Chapter 9). These bacteria cannot produce NADH or NADPH directly and need a source of reduced pyridine nucleotide for synthetic reactions. If they are CO_2-fixing autotrophs, they also must obtain NADH for the reduction of 3-phosphoglycerate (they reverse glycolysis to synthesize sugars). ATP can drive a reverse $\Delta\mu H^+$-generating system that generates reduced pyridine nucleotides.

There are in reality two major forms of energy utilized by the bacterial cell; one is ATP and the other is the electrochemical potential (proton motive force). These energy forms are interconvertible by the membrane-bound F_1F_0 ATPase. The proton motive force is utilized for many purposes in the bacterial cell, including nutrient transport, membrane biogenesis, protein export, turning flagella, and maintaining a favorable intracellular pH.

The region of the energy-transducing ATP synthase that protrudes from the membrane, designated $\mathbf{F_1}$, contains the site of ATP synthesis (see Figure 8.13). The bacterial F_1 is composed of five polypeptides, and isolated F_1 can catalyze an exceedingly rapid hydrolysis of ATP: One mole of F_1 can hydrolyze 10^4 moles of ATP per minute. The reverse reaction (ADP + $P_i \rightarrow$ ATP) can only be observed in vitro when F_1 is attached to a membrane and driven by a proton motive force. The region of the ATP synthase complex, termed $\mathbf{F_0}$, is a hydrophobic protein rotor assembly that is buried in the membrane. This protein complex serves as a channel for passage of protons from the exterior of the membrane toward F_1.

The energetically favorable return of protons to the cytoplasm is coupled to ATP synthesis by the F_1 subunit, which catalyzes the synthesis of ATP from ADP and phosphate ions (P_i). Enzymatic and structural studies have established the mechanism of ATP synthase action, which involves mechanical coupling between F_0 and F_1 subunits. In reality, the flow of protons through F_0 drives the rotation of F_1, which acts as a rotary motor to drive ATP synthesis. Electrical energy is transduced into chemical energy whereby energy captured by movement of three protons across the membrane may be used to form one molecule of ATP.

SECTION HIGHLIGHTS

The membrane-associated ATP synthase is a multi-subunit protein complex composed of two domains termed F_1 and F_0. It captures energy stored in the proton motive force to drive the endergonic reaction of ATP synthesis from ADP and P_i. Mechanistic and structural studies of ATP synthase have provided considerable insight into the operation of this highly conserved energy-transducing machine.

8.6 Diversity of Bacterial Electron Transport Chains

The electron transport chains present in bacterial cytoplasmic membranes, the inner membrane of eukaryotic mitochondria, and the membranes of photosynthetic bacteria share a number of common properties. There is, however, much greater diversity in the components of the electron transport chains among the bacteria than among eukaryotes. Among the major differences between the transport chains in bacteria and eukaryotes:

- Bacteria have a greater variety of electron carriers.
- Bacteria can use different electron donors and electron acceptors.
- Many bacterial species respond to changes in growth conditions and alter their respiratory chains.

Diversity of Electron Acceptors and Electron Donors

The use of oxygen as an electron acceptor coupled to NADH oxidation was discussed previously. However, there are variety of organic and inorganic electron donors and acceptors present in nature. Bacterial electron transport chains reflect this diversity by their ability to respire one or more types of substrates. By producing distinct electron transport enzymes that use these different substrates, bacteria can extract available energy from a range compounds of markedly different redox potential (see Table 8.4).

In addition to oxygen, other inorganic acceptors include nitrate, sulfate, sulfite, elemental sulfur, selenate, arsenate, and uranium (see Table 8.3). Organic acceptors include fumarate, trimethylamine-*N*-oxide, and dimethyl sulfoxide, where the resulting products would be succinate, trimethylamine, and dimethylsulfide, respectively. Five representative respiratory enzymes (three donating types and two accepting types) produced by various bacteria to utilize different substrates are shown in Figure 8.14. Several additional examples of this diversity are briefly described below.

Oxygen Reduction (Aerobic Respiration)

The respiratory chain in *Paracoccus denitrificans* is typical of many bacteria growing aerobically (see Figure 8.7) and is remarkably similar to the respiratory chain in mitochondria. However, some bacteria lack a cytochrome

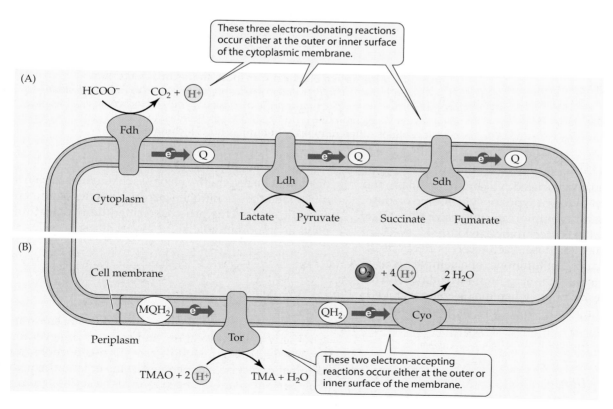

Figure 8.14 Alternative respiratory enzymes in bacterial electron transport chains
The location and orientation of several types of (A) primary dehydrogenases and (B) terminal oxidoreductases along with the substrates and products. (A) formate dehydrogenase (Fdh), lactate hydrogenase (Ldh), and succinate dehydrogenase (Sdh). (B) TMAO reductase (Tor) and cytochrome *o* oxidase (Cyo). The electron carrier cofactors are not indicated.

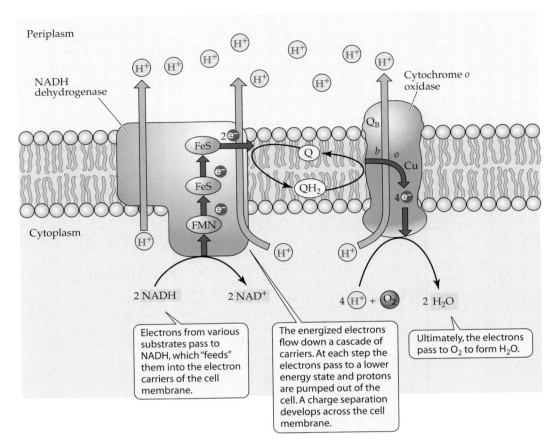

Figure 8.15 Alternative electron transport chains in a bacterium
The electron transport chain operating during aerobic growth in *Escherichia coli* by one of the two alternative cytochrome oxidase enzymes, cytochrome *o* oxidase (*b*, cyt *b*; *a*₃, cyt *a*₃; Q_B, quinone binding site). Cytochrome *d* oxidase is similarly oriented but contains different types of subunits and cytochromes.

*bc*₁ complex and an associated cytochrome *c* oxidase, and instead contain different types of cytochrome oxidase. *Escherichia coli*, for example, can synthesize two different enzymes, called cytochrome *o* oxidase (Figure 8.15) and cytochrome *d* oxidase. Each receives electrons directly from reduced ubiquinone; one complex is used when O₂ is abundant (the cytochrome *o* oxidase) and the other is synthesized only when the O₂ concentration is low. The O₂ affinity of each enzyme differs accordingly. This strategy allows the cell to adapt and thrive under changing environmental conditions.

Nitrate Reduction

When *P. denitrificans* is grown anaerobically with NO₃⁻ as the terminal electron acceptor, there is a distinctly different electron transport chain present in the cytoplasmic membrane. Cytochrome *c* oxidase (see Figure 8.7) is no longer made, and in its place the cell produces a set

of alternative respiratory enzymes that reduce nitrate to nitrite, and then to nitrogen gas (N₂) in series of four enzymatic steps. This process, termed **denitrification**, plays an important role in the global nitrogen cycle.

$$NO_3 \xrightarrow[\substack{\text{Nitrate} \\ \text{reductase} \\ (\text{Nar})}]{2e^-} NO_2 \xrightarrow[\substack{\text{Nitrite} \\ \text{reductase} \\ (\text{Nir})}]{e^-} NO \xrightarrow[\substack{\text{Nitric oxide} \\ \text{reductase} \\ (\text{Nor})}]{e^-} N_2O \xrightarrow[\substack{\text{Nitrous oxide} \\ \text{reductase} \\ (N_2\text{or})}]{e^-} N_2$$

The denitrification enzymes exhibit different substrate specificity, cellular location, and cofactor composition (represented in Figure 8.16). Each enzyme functions as a terminal reductase independent of the others, although they operate in unison. Each receives electron(s) from a primary electron donor such as NADH via one or more intermediate electron carriers. Note that the **Nar** enzyme catalyzes nitrate reduction to nitrite at the cytoplasmic surface of the cell membrane. In contrast, the **Nir** and **N₂or** (also called Nos) enzymes needed for reduction of

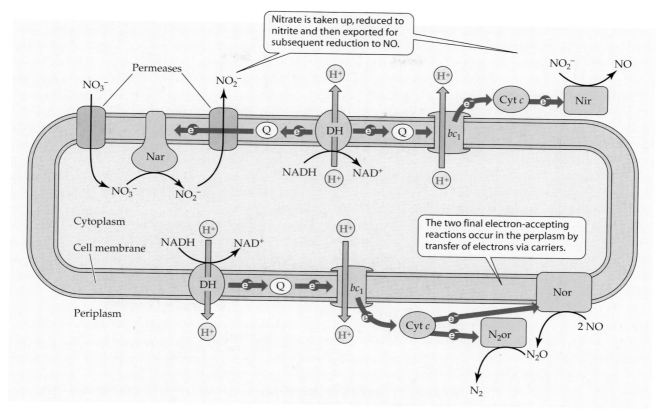

Figure 8.16 Denitrification enzymes in the *Paracoccus denitrificans* cytoplasmic membrane
The locations and reactions of the four denitrifying enzymes for reduction of nitrate to N_2 are indicated within the cell. Nar (nitrate reductase), Nir (nitrite reductase), N_2or (nitric oxide reductase), Nor (nitrous oxide reductase), bc_1 (cyt bc_1 complex), and DH (NADH dehydrogenase).

nitrite and nitrous oxide, respectively, are located in the cell periplasm and receive electrons via a soluble cytochrome *c*. The **Nor** enzyme is membrane attached and reduces nitrous oxide in the periplasm also via cytochrome *c*. Denitrification represents a complex set of electron transport reactions that work in concert to reduce a substrate in discrete steps.

Some bacteria perform nitrate reduction but produce ammonia rather than N_2 gas as the end-product. This process is called **ammonification** and involves only two enzymes (Figure 8.17), which catalyze the following reactions:

$$NO_3 \xrightarrow[\substack{\text{Nitrate} \\ \text{reductase} \\ \text{(Nar)}}]{2e^-} NO_2 \xrightarrow[\substack{\text{Nitrite} \\ \text{reductase} \\ \text{(Nrf)}}]{6e^-} NH_3$$

The **nitrate reductase** (**Nar**) is a membrane-bound, molybdenum-containing enzyme that catalyzes a 2-electron reduction of nitrate to nitrite. Like the bacterial denitrifying Nar enzyme, this nitrate reductase is localized in the cytoplasmic membrane with the active site facing the cytoplasm. However, in some ammonia-producing microorganisms an alternative type of nitrate reductase enzyme (**NapF**), is localized in the cell periplasm and receives electrons from associated Nap proteins (see Figure 8.17). The **nitrite reductase** (**Nrf**), distinct from the Nir enzyme used in denitrification, then forms ammonia from NO_2^-. The former is membrane bound, contains cytochrome *c*–type heme, and catalyzes the 6-electron reduction of nitrite to ammonia. The active site faces the periplasm.

In some bacteria the Nar enzyme may receive electrons from oxidation of formate (Figure 8.18). These coupled redox reactions form a proton motive force by a Q-loop mechanism and by scalar consumption of protons in the cytoplasm.

Note that the denitrification and ammonification pathways described previously constitute **dissimilatory reactions** (for energy-harvesting purposes) rather than **assimilatory reactions** (for acquisition of cell nitrogen). The latter process is discussed in Chapter 10.

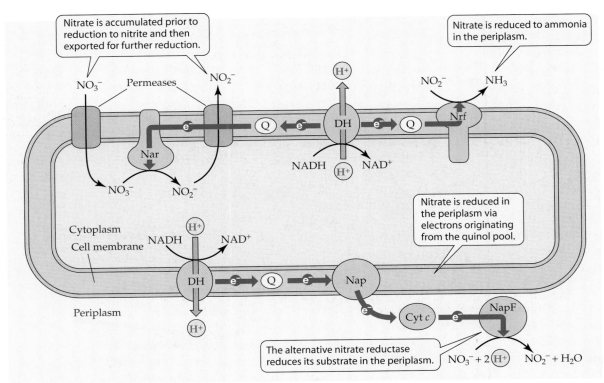

Figure 8.17 Ammonification enzymes in the *Escherichia coli* cytoplasmic membrane
The locations and reactions of the two alternative enzymes for reduction of nitrate to nitrite (Nar and Nap) and the nitrite reductase enzyme (Nrf) that forms ammonia.

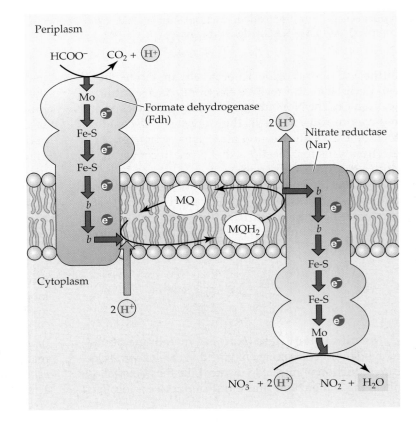

Figure 8.18 Formation of a proton motive force (pmf) by the Q-loop mechanism
The location and orientation of a primary dehydrogenase and terminal oxidoreductase that are coupled to form a pmf. Fdh, formate dehydrogenase; Nar, nitrate reductase. (Fe-S, iron sulfur centers; *b*, cyt *b*; Mo, molybdopterin cofactor.)

Alternative Electron Acceptors and Donors in Nature

Considerable microbial diversity exists with respect to the use of different electron acceptors, for example, sulfate, elemental sulfur, selenate, arsenate, and CO_2 plus O_2, NO_3^-, and NO_2^- mentioned previously (see Table 8.3). If free energy is available from oxidation–reduction of a donor and acceptor substrate pair, some organism has likely evolved to do so! These substrate pairs may have markedly different redox potentials and therefore release considerable energy in the coupled reactions (see Table 8.4). The difference may be small; for example, oxidation of reduced iron (Fe^{2+}) to (Fe^{3+}) with oxygen as an electron acceptor only provides a $\Delta E_o'$ of 0.05 V (described below).

Diverse bacteria utilize many organic or inorganic compounds as electron donors, including formate, lactate, glycerol, succinate, ethanol, H_2S, ammonia, and NO_2^-. For example, the primary dehydrogenases succinate dehydrogenase, lactate dehydrogenase, and formate dehydrogenase transfer electrons to CoQ using succinate, lactate, and formate, respectively (see Figure 8.14). These substrate-specific membrane enzymes generate the reduced products fumarate, pyruvate, and CO_2 that may be further metabolized by the organism by other reactions. Because the $\Delta E_o'$ of the above electron donors coupled to a given electron acceptor will differ (see Table 8.4), the free energy released varies accordingly.

Synthesis of many redox proteins—the primary dehydrogenase and/or the terminal oxidase—are often controlled in the bacterium depending on environmental conditions. *E. coli*, for example, can make two alternative cytochrome oxidase enzymes depending on the concentration of O_2 available to it, and two nitrate reductase enzymes and two nitrite reductase enzymes that vary depending on the nitrate availability in the environment. *E. coli* can also synthesize distinct enzymes to reduce the organic electron acceptors fumarate, trimethylamine-*N*-oxide, and dimethylsulfoxide. Synthesis of the enzymes is controlled by the cell in response to the cell growth conditions.

Iron Oxidation

The aerobic iron-oxidizing bacterium *Thiobacillus ferrooxidans* has an abbreviated electron transport chain (Figure 8.19) and is an example of a bacterium that utilizes its environment to advantage in energy generation. *T. ferrooxidans* utilizes iron as an energy source where the external pH is about 2. The pH in the cytoplasm of the

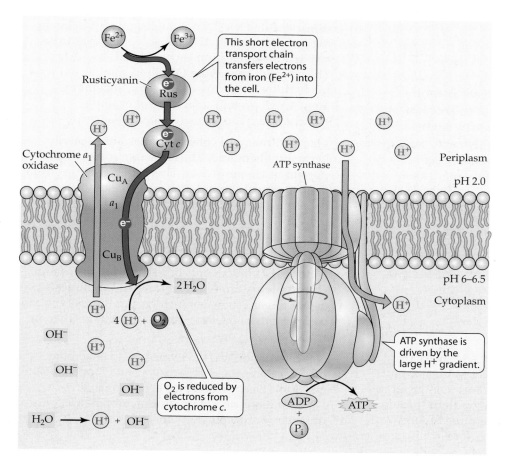

Figure 8.19 Abbreviated electron transport chain of an iron-oxidizing bacterium The electron transport chain of *Thiobacillus ferrooxidans*, which uses Fe^{2+} as an energy donor and O_2 as an electron acceptor.

microorganism growing under these conditions would be 6 to 6.5. At pH 2, iron is available as soluble Fe^{2+}, and at this acidic pH there would be an abundance of protons (H^+) in the area surrounding the outer membrane surface. At the outer surface of the cytoplasmic membrane in *T. ferrooxidans* is the copper-containing protein **rusticyanin**. Rusticyanin removes an electron from Fe^{2+}, thereby producing insoluble Fe^{3+}. These electrons flow from rusticyanin to cyt *c* and inward to cyt a_1 on the inner surface of the cytoplasmic membrane. There the electrons are transferred via a cytochrome oxidase to O_2 with the concomitant uptake of four protons from the cytoplasm to form H_2O.

Removal of protons internally results in a relatively high concentration of protons on the exterior of the cytoplasmic membrane relative to the interior (H^+ outside, OH^- inside). Protons are drawn from the acidic outer environment inward, driving the ATP synthase reaction and generating ATP. The role of the abbreviated electron transport chain in *T. ferrooxidans* is to pump electrons inward, thereby removing protons from the cytoplasm and creating a proton motive force. This is a remarkable example of the versatility in the prokaryotic world. The low $\Delta E_o'$ value indicates that little energy is available from the oxidation of one Fe^{2+} atom when compared with oxidation of one molecule of NADH. However, iron oxidation is a successful strategy for harvesting energy by some bacterial species.

SECTION HIGHLIGHTS

The *Bacteria* and the *Archaea* exhibit astonishing diversity in their ability to respire inorganic and organic molecules. They do so by synthesizing molecule-specific redox enzymes to oxidize an electron donor and to reduce an appropriate electron acceptor. The energy released by these coupled reactions is either captured as a proton motive force or lost as heat. Together, the diverse respiratory reactions serve a major role in recycling inorganic and organic molecules in nature.

SUMMARY

- A universal medium of biological energy exchange is **adenosine-5′- triphosphate (ATP)**. A major part of the ATP available to living organisms is generated by an **electrochemical proton gradient (proton motive force)** across a membrane. **Light energy** from the sun can be employed to establish this gradient in photosynthetic bacteria and plants. **Chemical energy** can be used by chemotrophic bacteria and archaea to generate this electrochemical potential.

- Products of photosynthesis and chemosynthesis are the source of food for heterotrophic life. Thus, the proton motive force provides energy for most biological systems.

- There are **three major mechanisms for ATP generation**: (1) respiration-linked phosphorylation, (2) photophosphorylation, and (3) substrate-level phosphorylation.

- **Redox** (oxidation–reduction) reactions require an **electron donor** and an **electron acceptor**. Oxidation is a loss of electrons and reduction is a gain of electrons. In coupled biological redox reactions, one molecule is oxidized and another is reduced.

- The amount of **free energy** available in a redox reaction is the difference in **redox potential (E_o')** between the donor and acceptor of electrons. The redox potential is the inherent propensity of electrons to flow from a higher energy state to a lower energy state.

- In biological systems, **electron carriers** carry electrons enzymatically removed from a substrate and pass them to an acceptor. **Nicotinamide adenine dinucleotide (NAD$^+$)** is soluble and moves freely in the cytoplasm. NAD$^+$ can accept two electrons and is a major electron carrier in cells as NADH. The cofactor, **coenzyme Q** is lipid soluble and thus freely diffusible in the cytoplasmic membrane. Some small protein cofactors, for example, **cytochrome *c***, move freely within the periplasm.

- **Electron transport chain** enzymes are generally located in membranes but in some instances are present in the periplasm. They carry electrons from the primary electron donor molecule to the terminal electron acceptor molecule and establish a proton gradient across the cytoplasmic membrane. These enzymes contain bound cofactors that aid in electron transfer.

- Among the components of respiratory chains are **flavoproteins, iron-sulfur proteins, quinones**, and **cytochromes**. There is much greater diversity in the components of bacterial respiratory chains than in those of eukaryotes.

- A **fermentation** can be defined as an anaerobic reaction that provides energy when O_2 or a suitable anaerobic electron acceptor is unavailable. Fermentations do not involve membrane electron transport or proton gradients.

- **Substrate-level phosphorylation** is the direct formation of ATP from adenosine diphosphate and an activated phosphate group on a metabolic intermediate.

- **Glycolysis** is a major route of glucose catabolism in living cells. During fermentation, glycolysis occurs in a **closed system** where the products generated have the same redox balance as the substrates. For example, in lactic acid fermentation:

$$C_6H_{12}O_6 \rightarrow 2\ C_3H_6O_3$$
$$\text{Glucose} \qquad \text{Lactic acid}$$

- **Respiration** occurs when reducing equivalents (e^-) from donors are passed to electron acceptors. During this process, electrons move from the substrate to membrane-bound electron transport chains in the membrane. Components of these chains **translocate protons** to the exterior of the membrane. Examples of the ultimate acceptors of electrons that carry the protons outward would be oxidized molecules (O_2, NO_3^-, SO_4^{2-}, or fumarate). Electron donors include H_2, formate, NADH, succinate, and lactate.

- A cytoplasmic membrane that bears membrane-bound electron carriers is called an **energy-transducing membrane**. The inner membrane of mitochondria, the thylakoid membranes in chloroplasts, and the cytoplasmic membrane of respiring and photosynthetic bacteria are energy-transducing membranes.

- According to the **chemiosmotic principle**, the proton motive force generated by electron transport is employed to drive **ATP synthase**, a transmembrane enzyme complex that synthesizes ATP from ADP and P_i.

- The energy available from an electrochemical gradient—the proton motive force—depends on the difference between the pH and electrical membrane potential on the outer surface and the inner surface of the membrane. The H^+ concentration gradient is termed **ΔpH**, whereas the charge difference across the membrane is termed **Δψ**. This proton motive force occurs because a cytoplasmic membrane is impermeable to protons, and protons can flow to the cytoplasm only through the ATP synthase system, by solute uptake systems, or by flagellar rotation.

- The ATP synthase complex can function in either direction. When protons flow inward, ATP is

formed. ATP can be **hydrolyzed** to ADP + P_i and the energy utilized to move protons outward to establish a proton gradient.

 Find more at www.sinauer.com/microbial-life

REVIEW QUESTIONS

1. What is the universal medium of energy exchange in viable cells? There are three major mechanisms for formation of this compound. How do they differ? What do they have in common?

2. Define proton motive force. What are the components that make up this energy form?

3. A redox reaction occurs with a transfer of electrons. List some components that might function in a redox reaction. Think in terms of fermentation and respiration.

4. If ferredoxin was coupled in a redox reaction with cytochrome *c*, what would be the $\Delta G_o'$? What is a half reaction? Do half reactions actually occur in biological systems?

5. Give examples of electron transport enzymes. What do they share in common? Explain how diffusible and membrane-bound electron carriers differ.

6. How do substrate-level phosphorylation and respiration-linked phosphorylation differ?

7. What is fermentation? Cite examples of fermentation reactions and how they are named.

8. What are some basic differences between fermentation and respiration? What do they have in common?

9. The electron transport chain in *Thiobacillus ferrooxidans* is unique. How does it differ functionally from that in *Escherichia coli*?

10. What is an energy-transducing membrane and why is it given this name?

11. How do electron transport chains in bacteria differ from those in eukaryotes? What do they have in common?

12. Outline the structure and operation of the ATP synthase.

13. Terminal electron acceptors in eukaryotes and bacteria differ in some respects. Explain what these differences are and what makes the prokaryotes unique.

SUGGESTED READING

Harold, F. M. 1986. *The Vital Force: A Study of Bioenergetics.* New York: W. H. Freeman and Co.

Harris, D. A. 1995. *Bioenergetics at a Glance.* Cambridge, MA: Blackwell Science.

Shively, J. M. and L. L. Barton, eds. 1991. *Variations in Autotrophic Life.* New York: Academic Press.

Nichols, D. G. and S. J. Ferguson. 2002. *Bioenergetics 3.* New York: Academic Press.

El-Mansi, E. M. T., C. F. A. Bryce, A. L. Demain and A. R. Allman. 2006. *Fermentation Microbiology and Biotechnology.* Boca Raton, FL: CRC Press.

White, D. 2006. *The Physiology and Biochemistry of Prokaryotes.* New York: Oxford University Press.

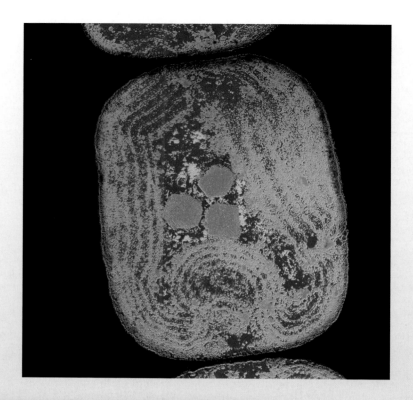

9

Cellular Energy Derived from Light

Life on planet Earth is completely dependent on the harvesting of the sun's light energy, and on the transmission and chemical storage of that energy by photosynthetic organisms and plants.
—Peter Mitchell

Photophosphorylation is the process whereby light energy is converted to chemical energy, adenosine triphosphate (ATP). This process has much in common with respiration-linked phosphorylation, which was covered in the previous chapter. The two processes, respiration-linked phosphorylation and photophosphorylation are, in a sense, mirror images of one another. In the former, the hydrogens (H^+) on a reduced substrate such as glucose or H_2 react with O_2 to release energy, generating H_2O, and CO_2. In photophosphorylation, light energy is used to extract electrons (and H^+) from H_2O (or certain other reduced substrates). The light-activated electrons are employed to generate ATP and ultimately to convert CO_2 and/or acquired organic compounds into cellular components. The reactions that harvest light energy to generate a proton motive force and ATP are termed **light-dependent reactions**; those that convert CO_2 to organic matter are called **light-independent reactions** (discussed in Chapter 10). Light is required to drive the former but not the latter. The chapter-opening quote from Peter Mitchell affirms the importance of these processes. In this chapter we will discuss the diverse processes whereby light energy is conserved in photosynthetic bacteria. After a brief look at the evolution of photosynthesis and common features among diverse photosynthetic processes, we will examine the various mechanisms utilized by diverse photosynthetic bacterial species.

9.1 Photosynthetic Organisms

Photosynthetic microorganisms arose very early in evolution and are the basis of most food chains on Earth. Whether alternative energy harvesting strategies arose earlier or later is uncertain. However, examination of microorganisms now present in the environment, based on comparative analysis of their component parts and metabolic functions, leads to firm conclusions about successful energy harvesting strategies. It is certain that photosynthetic bacteria preceded the evolution of the eukaryotes (see Chapter 1). The following discussion provides a basis to understand these light harvesting processes.

Early Evolution

The evolution of complex living organisms on planet Earth was dependent on the development of a capacity for **biosynthesis**—the capacity of early life forms to synthesize the kinds and amounts of compounds needed to sustain life and reproduce. Simple accretion of molecules present in the "primordial soup" would have yielded beings of very restricted capabilities. Only when sources of chemical energy became available in the environment could the evolution of regenerative systems occur. The singular event that ultimately allowed for a systematic and orderly evolution of viable systems was the advent of the ability to capture the inexhaustible source of earthly energy—sunlight. The coupling of this light-capturing process to the production of relatively stable, "energy-rich" compounds allowed for the formation of the biologically derived molecules that compose cells.

The nature of our planet's primitive organisms is highly speculative, and many of their characteristics remain unknown. Fermentative energy generation, akin to the substrate-level phosphorylation used by present-day *Clostridium* species (discussed in Chapter 8), may have occurred in some of the earliest organisms. However, the abiotic synthesis (not produced by living organisms) of sugar under conditions that existed in primeval times is questionable. Consequently, an early metabolism based on the fermentation of sugars is improbable.

It is possible that an early life form may have utilized anaerobic respiration in energy generation. However, the absence of abundant oxidized compounds to serve as electron acceptors would have placed limitations on the effectiveness of an anaerobically derived proton gradient for ATP generation. Evidence suggests that sulfate and nitrate can be generated abiotically, and these compounds may have served as terminal electron acceptors for some early anaerobic microorganisms.

One attractive concept is that a simple photosystem, such as that present in *Halobacterium*, might have oc-

curred early in evolution. Such photosystems have recently been found to be quite prevalent in marine environments. The generation of a proton motive force across a membrane with a simple photosensitive carotene-type compound (see later discussion) might have been used to drive ATP synthesis.

A simplistic picture of the energetics involved in light reactions can be drawn as follows:

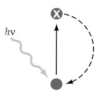

A quantum of energy from the sun, also known as a photon and designated *hv*, drives an electron in a light-absorbing compound from the ground state (●) to a higher level of electronic excitation (✕), illustrated by the vertical line. This excitation, *if untrapped,* would return to the ground state, as indicated by the dashed line. Energy for "life" would be available if excitation-capturing molecules were interposed between the two circles (solid line), thereby conserving the quantum of light energy (or **excitation**) inherent in ✕. This can be depicted as follows:

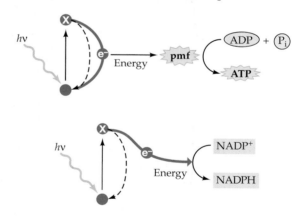

The driving force for all types of cellular processes, both autotrophic and heterotrophic, can be illustrated as components of the following scheme:

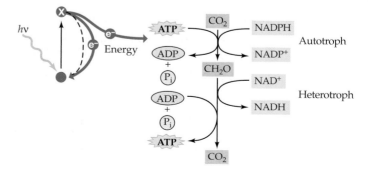

This scheme presents a simple, abbreviated illustration of the reactions that support most life forms. Photophosphorylation systematically captures light energy in a chemically stable intermediate (ATP) and generates the reductant (NADPH) to sustain biosynthetic reactions observed in photoautotrophs and photoheterotrophs (Box 9.1). The reduced products serve as source of carbon and energy for many nonphotosynthetic heterotrophs. It is important to understand the principles presented in Box 9.1 in order to comprehend this chapter.

Origin of Photosynthetic Organisms

Because *Bacteria* and *Archaea* existed for more than 3 billion years prior to the advent of eukaryotes, it seems logical for light energy–harvesting systems to have originated with these microorganisms. Three distinct photochemical energy–capturing systems have evolved in microorganisms:

- **Oxygenic** systems, which generate oxygen
- **Anoxygenic** systems, which do not produce oxygen
- **Rhodopsin** systems, which utilize carotenoid-based pigments structurally similar to the animal rhodopsin found in eye organs

The ability of bacteria known as the green and purple photosynthetic bacteria to grow anaerobically, as well as other characteristics, suggests that microorganisms of this type preceded the oxygen-generating cyanobacteria in evolution. It is not known whether the early anoxygenic photosynthetic bacterium—either a purple or green sulfur bacterium, described below—had a chlorophyll pigment similar to the chlorophyll *a* present in the oxygenic cyanobacteria. Chlorophyll *a* absorbs light of a wavelength (< 700 nm) that penetrates the deep anaerobic reaches of aquatic environments.

The occurrence of chlorophyll-based photosynthesis in bacteria that are phylogenetically dissimilar supports the concept that photosynthesis may have been an early event in evolution. The development of mechanisms for oxygenic photosynthesis was a key event because it permitted water to be used as an electron donor in addition to reduced organic and inorganic materials.

The development of this oxygen-generating system was the fundamental event in forming an O_2-containing atmosphere; it allowed eukaryotic organisms to evolve beyond the need for an aquatic habitat, thus opening up the endless possibilities of life on land (see Chapter 1). This is supported by the geological and fossil records (see Chapter 1). Based on the remarkable similarity of their photosynthetic machinery, it is clear that plant chloroplasts were derived from the oxygenic photosynthetic bacteria.

The primeval anaerobic environment bore great quantities of ferrous iron (Fe^{2+}), which served as a trap for O_2 generated by the first cyanobacterial-type organisms. Bands of ferric iron Fe_2O_3, called banded iron formations (BIFs), can be seen in geologic strata deposited during the era when oxygenic photosynthesis evolved, attesting to the probable role played by ferrous iron (Box 9.2).

Ferrous iron was gradually depleted through reactions with O_2 and, over 1.5 to 2 billion years ago, O_2 began accumulating in the atmosphere. This O_2 was subjected to the following reaction in the stratosphere: O_2 + sun (UV) → O_3. This formed an ozone (O_3) layer, which screened out ultraviolet (UV) rays below a wavelength of 290 nm—a wavelength that would be harmful to any organism constantly exposed to unfiltered sunlight. Without this ozone layer, evolution would certainly have taken a different course, because UV causes chemical changes in nucleic acids, which absorb in the 260-nm range. Without this UV trap, most living creatures would have had to remain in aquatic environments, because a thin layer of water absorbs UV.

With the ozone layer established, the lower wavelengths of UV no longer reached the planet's surface in great quantity, and life on land became feasible. The diversity of living forms that were possible under the ozone layer was virtually unlimited.

It is possible that UV-resistant microorganisms did evolve prior to the formation of the ozone layer. This resistance may have been lost after the ozone layer came into being, as it gave these microbes no selective advantage. Perhaps *Deinococcus radiodurans*, a microbe highly resistant to radiation, is a remnant from that era.

SECTION HIGHLIGHTS
Photosynthesis originated early in Earth's history and involved biological processes to capture light energy from the sun. This gave rise to the anoxygenic and oxygenic photrophs, the latter being the basis for the O_2-containing atmosphere.

9.2 The Photosynthetic Apparatus

All photosynthetic microorganisms (except the rhodopsin-based systems) have four unifying features, although they differ somewhat in the way they achieve photosynthetic growth. They all harvest light as an energy source and use this energy either to fix carbon dioxide as the source of cell carbon or to acquire organic car-

BOX 9.1   *Milestones*

The Current of Life

A singular achievement in the evolution of viable microorganisms on planet Earth occurred when chemical systems arose that could capture light energy. The primordial soup contained compounds that could absorb light energy. Absorption of a quantum of light energy by a light-sensitive molecule (M) could elevate a constituent electron to a higher energy level, as shown in the diagram (M*). The energy of excitation now inherent in M* has four possible fates:

I.   The energy might be dissipated as **heat** as the electron returns to a lower orbital.

II.   The energy of excitation might reappear as emitted light or **fluorescence.** A photon of light is emitted as the electron returns to a lower orbital.

III. The energy of excitation may be transferred by **resonance energy transfer** to a neighboring molecule, thus raising an electron in the receptor molecule to a higher energy state as the photo-excited electron in the absorbing molecule returns to the ground state.

IV.   The energy of excitation may change the reduction potential of the absorbing molecule to a level such that it can become an **electron donor.** Reacting this excited electron donor to a proper electron acceptor can lead to the transduction of light energy (photons) to chemical energy in the form of reducing power.

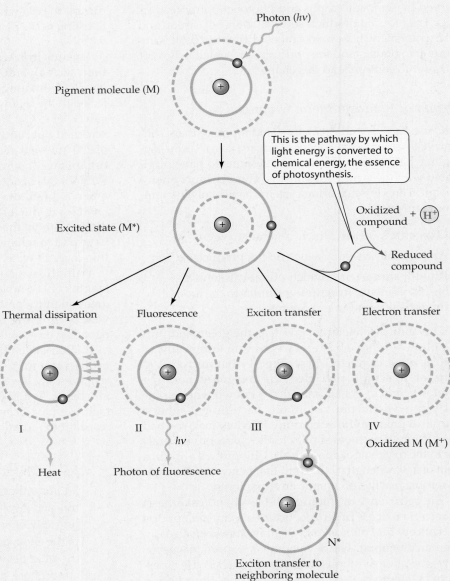

Photon ($h\nu$)

Pigment molecule (M)

This is the pathway by which light energy is converted to chemical energy, the essence of photosynthesis.

Excited state (M*)

Oxidized compound $+$ H^+

Reduced compound

Thermal dissipation   Fluorescence   Exciton transfer   Electron transfer

I   II   III   IV

$h\nu$ Oxidized M (M⁺)

Heat   Photon of fluorescence

N*

Exciton transfer to neighboring molecule

It is fate IV that permits the transduction of light energy into chemical energy, which is the essence of photosynthesis. Most life on Earth depends on this simple equation:

$$\text{Light energy} \rightarrow \text{chemical energy}$$

The purposeful use of this chemical energy for synthesis and reproduction led to the evolution of the first viable microbe.

BOX 9.2 *Research Highlights*

Banded Iron Formations

During the Archaean and Proterozoic ages (over 2 billion years ago) banded iron formations (BIFs) were laid down over extensive areas of the planet. These BIFs have been attributed to the chemical oxidation of reduced iron (Fe^{2+}) by molecular oxygen generated by the cyanobacteria. The cyanobacteria evolved with the ability to utilize water as an electron donor in photochemical energy generation. The resultant cleavage of water ($H_2O \rightarrow 2\,H^+ + \frac{1}{2}\,O_2$) was the original source of atmospheric oxygen. Recent evidence suggests that anaerobic oxidation of soluble ferrous iron could also have contributed to BIFs through the activities of the anoxygenic purple bacteria. These photosynthetic bacteria can utilize reduced iron, which is present in anoxic iron-rich sediments as an electron donor for photophosphorylation. As the soluble ferrous iron, Fe^{2+}, is oxidized

Banded Iron Formation, Dales Gorge, Wittenoom, Western Australia. The formation is 2.47 billion years old. Courtesy of Eric Cheney.

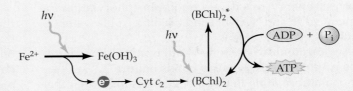

to the insoluble ferric state, $Fe(OH)_3$, a brown precipitate is formed.

This is another example of the remarkable diversity that prevails in the microbial world.

---

bon from the environment for cellular needs. These features include possession of:

1. Light-gathering systems
2. Electron transfer chains
3. An ATP-generating system
4. An electron donating system to generate NADPH

These photosynthetic components together are termed the light reactions, since they require light energy for operation. The four components are assembled into the cell membranes and allow an electrochemical gradient (proton motive force [pmf]) to be formed across it. The light energy as stored in the pmf is then used to synthesize ATP. This involves a **cyclic photosynthetic** mode of electron transfer. Finally, in a process called **noncyclic photosynthesis**, an electron donating system is used to replenish electrons at the photosynthetic reaction center that have been withdrawn to synthesize NADPH (as will be discussed later).

The Light-Gathering Systems

The primary bacterial energy-capturing pigment is **chlorophyll** (or bacteriochlorophyll) (Table 9.1); all photosynthetic bacteria also have auxiliary antenna pigments that help them to capture a broad range of light wavelengths by excitation (see Box 9.1). These include other chlorophyll types, **carotenoids**, and—in the oxygenic phototrophs—specialized light-capturing proteins called **phycocyanin** and **phycoerythrin**. Some *Archaea* and *Bacteria* contain a distinct light-gathering system consisting of a carotenoid known as "**bacterial rhodopsin**." These are all pigment molecules responsible for the absorption of light energy and the transformation of this energy of excitation into stable intermediates.

The chlorophylls are tetrapyrroles, which are quite similar to the basic structure found in the heme portion of cytochromes (see Figure 8.11). The iron present in heme is replaced by an atom of magnesium in the chlorophyll molecule. Magnesium is crucial to the capture of light because it is a "close shell" divalent cation that changes the

TABLE 9.1	Bacteriochlorophylls present in photosynthetic bacteria and primary acceptors involved in energy-conserving reactions	
Group	**Electron Donor**	**Electron Acceptor**
Purple nonsulfur bacteria	Bacteriochlorophyll *a* and *b*	Bacteriopheophytin *a*, Q_A, Q_B
Green sulfur bacteria	Bacteriochlorophyll *c*, *d*, and *e*	Bacteriopheophytin *a* and FeS-protein
Cyanobacteria (photosystem I)	Chlorophyll *a*	Chlorophyll *a* and FeS-protein
Cyanobacteria (photosystem II)	Chlorophyll *a*	Pheophytin *a*, Q_A, Q_B, and plastoquinones
Heliobacteria	Bacteriochlorophyll *g*	Bacteriochlorophyll *c*, FeS-protein

electron distribution in chlorophyll pigments and produces powerful excited states (Figure 9.1). There are a number of chlorophyll molecules, which differ slightly in their chemical structure. They are designated by the lowercase letters *a*, *b*, *c*, *d*, *e*, and *g* (see also Chapter 21).

Most of the chlorophyll molecules in a green plant or photosynthetic bacterium is involved with light gathering (Figure 9.2) and the transfer of light-generated excitation to reaction centers. The oxygenic photobacteria, like plants, possess two distinct reaction centers. Each harvests energy; one drives photophosphorylation to generate ATP and NADPH and the other splits water (2 $H_2O \rightarrow 4\ H^+ + 4\ e^- + O_2$).

The **carotenoids** are long-chain hydrocarbons with extensive conjugated double bonds (Figure 9.3; see additional information in Chapter 21). It is likely that the carotenoids evolved to provide light-harvesting capacity in the 400- to 550-nm range. Carotenoids can also intercept highly toxic singlet-state oxygen, which is gen-

erated during the photolysis of water, and convert it to a less reactive state.

Bacteriorhodopsin is a special type of carotenoid found in the halophilic (salt-loving) *Archaea* (see Chapter 18). These pigments, related to the rhodopsin proteins found in the eyes of higher organisms, function in a manner markedly different from the chlorophylls. Recent ecological experiments suggest that this mode of photosynthesis is also distributed among members of the *Bacteria* in the oceans (see Chapter 21).

The primary photochemical site is called a **photosynthetic reaction center** (**PRC**) (see Figure 9.2) and is defined as a single photochemical unit because it transfers one electron at a time. Reaction centers are protein-chlorophyll complexes within the cytoplasmic membrane of photosynthetic bacteria where quanta of light energy are converted to stable chemical energy.

The absorption of an incident photon by a molecule of chlorophyll leads to a redistribution of electrons

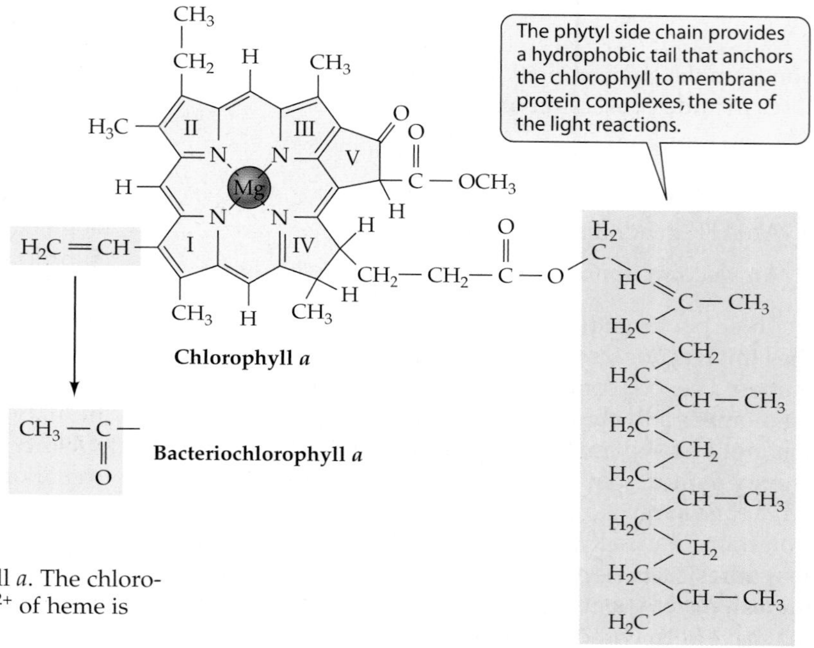

Figure 9.1 Chlorophylls
Structures of chlorophyll *a* and bacteriochlorophyll *a*. The chlorophylls are structurally related to heme, but the Fe^{2+} of heme is replaced by Mg^{2+} in the chlorophylls.

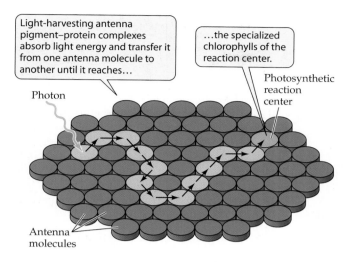

Light-harvesting antenna pigment–protein complexes absorb light energy and transfer it from one antenna molecule to another until it reaches...

...the specialized chlorophylls of the reaction center.

Photon

Photosynthetic reaction center

Antenna molecules

Figure 9.2 Photosynthetic light harvesting
Diagram of a photosynthetic unit, showing the pathway of excitation transfer from antenna molecules to the reaction center (orange).

within the basic framework of the tetrapyrrole. This state of the molecule is known as an **excited state**, and it has more energy than that of the **initial** or **ground state** (see Box 9.1). Incident radiation is absorbed only if it provides the correct energy to achieve one of the allowed activated states. Photons with intermediate energies may be absorbed by chlorophyll but do not lead to an excited state. Only when sufficient excitation energy is present, can it be transmitted to a neighboring molecule by **resonance energy transfer**, which, in turn, raises this molecule to an excited state. The process is repeated until energy is transferred to the reaction center (see Figure 9.2), where an electron may be ejected from the PRC, thus initiating the energy-harvesting process involving

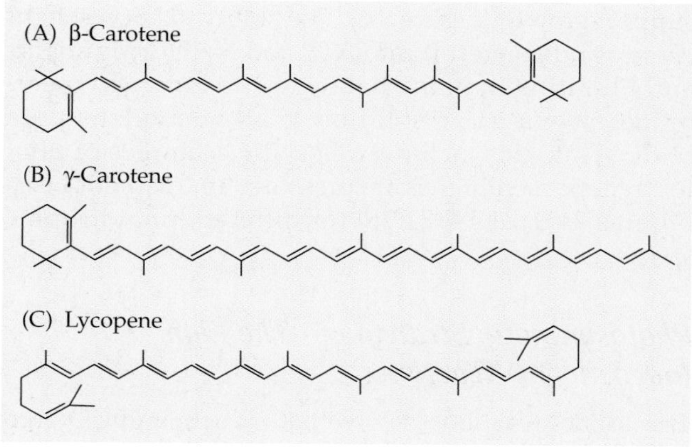

a specialized electron transport chain as will be discussed below. Note that a variety of light-harvesting pigments are used by different phototrophic species.

Bacteriochlorophyll present in the green bacteria absorbs light at a wavelength of about 700 nm. This light-harvesting capacity permits growth at lower depths in water but does not have sufficient potential as an oxidant to extract electrons from water. Thus, evolution in the anaerobic environment yielded a photosynthetic population limited to the use of reduced molecules such as hydrogen (H_2) or sulfide (H_2S) as sources of electrons. This is termed **anoxic photosynthesis**. Eventually (about 2.5 billion years ago) an organism evolved with a photosystem that had sufficiently strong oxidative power to extract electrons from H_2O, according to the following:

$$2\ H_2O \xrightarrow{\ h\nu\ } 4\ H^+ + 4\ e^- + O_2$$

This event, which distinguishes **oxygenic photosynthesis**, resulted in the formation of our oxygen atmosphere.

The Electron Transport Chains

The photosynthetic reaction center, once in the excited state, in principle could return to the ground state with the dissipation of energy. Instead, an electron is donated to a set of specialized electron acceptor molecules, which constitute the photosynthetic electron transport chain located in the cell membrane (**Figure 9.4**). The electron acceptors may include various types of bacteriopheopytin (Bph) molecules, quinones, metals, Fe-S proteins, and cytochromes similar to those found in respiratory electron transport chains (see Chapter 8). Bph is a tetrapyrrole molecule lacking a metal atom, such as the magnesium contained in the chlorophylls. Quinones include ubiquinone and menaquinone; a variety of cytochrome types may be employed, depending on the microbe. Whereas the composition of these photosynthetic systems may vary among different bacteria, they have in common the ability to generate an electrochemical gradient (proton motive force) across the photosynthetic membrane. The electron is ultimately recycled back to the PRC to restore the initial or ground state. As this cyclic process is repeated, the electrochemical membrane potential increases to "store" the captured light energy.

The ATP-Generating System

ATP is synthesized by using the electrochemical gradient (proton motive force) formed by the photosynthetic

Figure 9.3 Structures of common carotenoids in photosynthetic bacteria
(A) β-Carotene, (B) γ-carotene, and (C) lycopene.

(A) β-Carotene

(B) γ-Carotene

(C) Lycopene

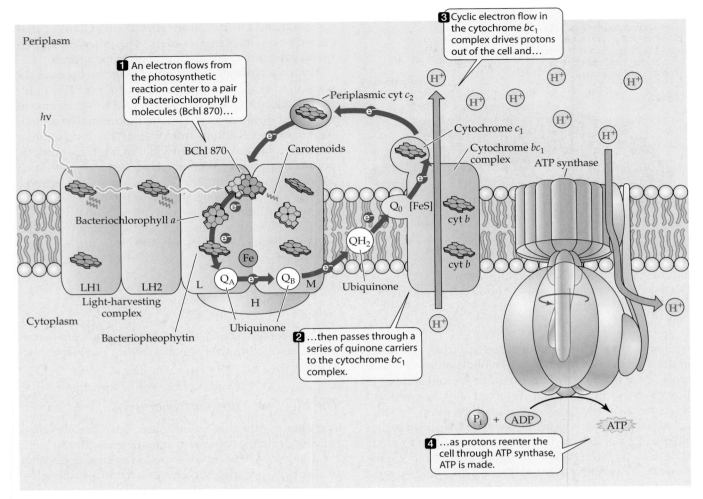

Figure 9.4 Reaction center complex of purple nonsulfur bacterium
Arrangement of electron carriers and electron flow in the reaction center of *Rhodopseudomonas sphaeroides*. The subunits (polypeptides) of the reaction center are designated light (L), medium (M), and heavy (H), based on their relative size.

electron transport chain via the membrane bound **F_1F_0 ATP synthase complex** (see Figure 9.4). Protons entering the cytoplasm drive ATP synthesis from ADP and P_i, just as ATP is formed during bacterial respiration (see Chapter 8).

The remaining and essential energy task needed for fueling cell biosynthesis is the formation of NADPH. This involves a specialized electron donating system to restore the PRC to the initial ground state (discussed below).

The photosynthetic machinery—the light-gathering, electron-transport, and ATP-generating systems—are all topologically positioned in or at the surface of the photosynthetic membranes of the cell. Different phototrophic strains possess distinctive membrane morphologies (e.g., tubes, lamellae, and vesicles) that effectively enlarge the membrane surface area to accommodate large numbers of reaction centers needed to

support growth (Figure 9.5). The ability to harvest light energy and convert it into ATP and NADPH is apparently limiting in many habitats, and enlarged membranes provide one solution to accommodate more photosynthetic machinery. Specific examples of photosynthetic membranes are presented in Chapter 21 (see Figures 21.3, 21.14, 21.16) for different phototrophic species.

Photosynthetic Strategies—The Light-Independent Reactions

The "dark" reactions involve the reduction of CO_2 into organic carbon (CO_2 fixation in plants and some photosynthetic bacteria). Related energy-consuming reactions occur in other phototrophs to incorporate organic compounds into the precursor molecules needed for the

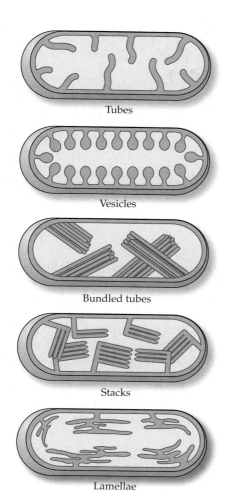

Figure 9.5 Structures of internal membranes in several types of photosynthetic bacteria
The photosynthetic membranes can take on a variety of configurations, depending on the species.

macromolecular assembly of cells. The process may be represented simply as:

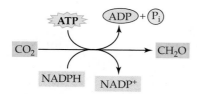

Among the photosynthetic bacteria, there is considerable diversity of biochemical pathways used to fix CO_2. These include the Calvin cycle, a major route in the oxygenic photosynthetic microbes, and two others: the reductive carboxylation pathway, also known as the reverse TCA cycle (*Chlorobium*), and the hydroxypropi-

onate pathway (*Chloroflexus*). Details of the carbon incorporation pathways are presented in Chapter 10.

The following sections focus on how the purple bacteria, green bacteria, and cyanobacteria capture energy from the sun.

SECTION HIGHLIGHTS

A variety of photosynthetic molecules contained in membrane-organized proteins capture light energy to form a proton motive force (pmf). The ATP synthase consumes this captured energy to form ATP from ADP and P_i. Distinct electron transport reactions are required for NADPH formation and the associated regeneration of the PRC ground state.

9.3 The Photosynthetic Bacterial Groups

There are five distinct groups of photosynthetic *Bacteria*; they are distinguished by their photopigments, source of electrons, cell carbon, and environmental habitat. They are commonly known as:

• The purple nonsulfur bacteria

• The purple sulfur bacteria

• The green sulfur bacteria

• The heliobacteria

• The cyanobacteria

Their general distinguishing properties are presented in Table 9.2. A major difference between these organisms is the electron donor involved in their photochemistry; for example, the purple nonsulfur bacteria generally use H_2 or reduced organics, the green and purple sulfur bacteria use reduced sulfur compounds, and the heliobacteria use reduced organics as an electron donor. The cyanobacteria use H_2O. The various photosynthetic bacteria also differ morphologically and in the composition of their electron transport chains.

Photosynthesis in Purple Nonsulfur Bacteria

The purple nonsulfur bacteria are prevalent on the surface of mud and in the water of lakes and ponds. Water that is rich in organic matter and has relatively low sulfide can have a population of these bacteria in numbers that give the water a distinct purple color. In addition to anoxygenic photosynthesis, some of these bacteria grow heterotrophically in the dark by aerobic respiration or fermentation, while most strains utilize a range of elec-

TABLE 9.2	Some general properties of the various photosynthetic bacteria				
	Purple Nonsulfur Bacteria	Purple Sulfur Bacteria	Green Sulfur Bacteria	Cyanobacteria	*Heliobacteria*
Source of reducing power (e⁻)	H_2, reduced organic	H_2S	H_2S	H_2O	Lactate, organic
Oxidized product	Oxidized organic	SO_4^{2-}	SO_4^{2-}	O_2	Oxidized organic
Source of carbon	CO_2 or organic	CO_2	CO_2	CO_2	Lactate, pyruvate
Heterotrophic growth	Common	Limited[a]	Limited[a]	Limited[a]	Required

[a]Generally limited to assimilation of low-molecular-weight organic compounds during autotrophic growth.

tron donor molecules to grow photosynthetically (see Table 9.2). They are perhaps the most metabolically versatile of all the prokaryotes.

The photosynthetic purple nonsulfur bacteria have been studied extensively because they grow readily in culture. The photosynthetic reaction centers have been isolated and crystallized from *Rhodobacter sphaeroides* and *Rhodopseudomonas viridis*. The reaction centers from *R. sphaeroides* and *R. viridis* were obtained by breaking cells by pressure treatment followed by differential centrifugation (centrifuging at speeds that separate material by size or weight) to recover the membrane components (see Box 11.1). The reaction centers were then purified by general protein separation techniques, crystallized, and studied.

These reaction center crystals yielded much data on the structural relationships between the component parts, and the mechanism whereby photosynthesis occurs (Figure 9.6A). They have also been useful as models for studies of the photosystems in higher plants.

The reaction center in the nonsulfur purple bacterium *R. spaeroides* contains:

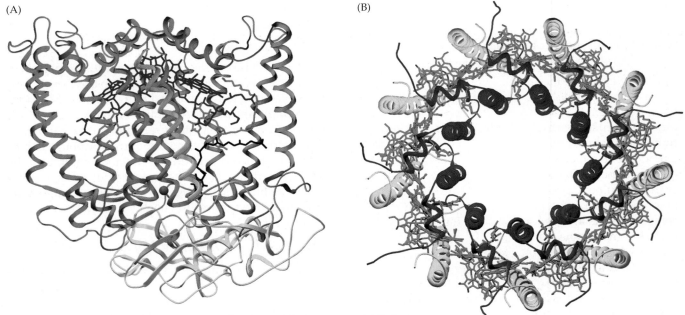

(A)   (B)

Figure 9.6 Structures of the photosynthetic reaction center and light harvesting complex LH2
(A) The reaction center structure of *Rhodobacter sphaeroides* with the L, M, and H subunits shown in different colors. (B) The LH2 light-harvesting complex of *Rhodopseudomonas acidophila*.

- Four molecules of bacteriochlorophyll *a*
- Two molecules of **bacteriopheophytin** (a molecule of bacteriochlorophyll *a* that lacks magnesium)
- A molecule of **ubiquinone** at the "Q_A site"
- A molecule of **ubiquinone** at the "Q_B site"
- A molecule of non-heme iron bound by four histidine residues
- A molecule of the carotenoid spheroidene

These components are bound to two protein subunits (L and M) of the reaction center complex that spans the membrane (see Figure 9.4). Another subunit protein (H) is attached to the L-M proteins; it also spans the membrane and faces the cytoplasm. Two of the bacteriochlorophyll molecules are paired and share the function of a photochemical electron donor (designated **Bchl 870**). One of the bacteriopheophytin molecules acts as the intermediary electron carrier between bacteriochlorophyll and ubiquinone. The roles of the second pair of bacteriochlorophyll *a* and bacteriopheophytin molecules are not known at present, although they are arranged in a nearly symmetrical fashion to the other molecules. The reaction center carotenoid is probably not involved in light gathering or excitation transport but rather in quenching singlet oxygen, a by-product that can damage the reaction center.

Adjoining the photosynthetic reaction center complex is the antenna or light-harvesting LH1 and LH2 complexes composed of 12 to 17 subunits of a single protein type (Figure 9.6B). Each subunit contains one or two molecules of bacteriochlorophyll plus several carotenoid molecules that receive light energy and relay it to the bacteriochlorophyll reaction center (BChl 870).

During **cyclic photosynthesis** in the purple nonsulfur bacteria (see Figure 9.4), the light-induced translocation of an electron from the bacteriochlorophyll BChl 870 reaction center in the LMH complex results in transfer of an electron to bacteriopheophytin. The electron is then passed to a ubiquinone molecule (Q_A), and subsequently to a second ubiquinone (Q_B). The electron(s) are then transferred to the ubiquinone pool in the membrane to reduce it to ubiquinol (QH_2). Subsequent electron flow to the **cyt bc_1 complex** results in the extrusion of protons across the cytoplasmic membrane, resulting in formation of a pmf (see Figure 9.4). Electrons are then transferred from the **cyt bc_1 complex** to a soluble periplasmic **cytochrome c_2** protein that contains a c-type heme. To complete the cycle, cytochrome c_2 then donates an electron to the oxidized bacteriochlorophyll (Bchl 870) molecule in the reaction center of the LMH complex to regenerate the initial state.

In reaction centers obtained from a strain of *R. viridis,* some of the details differ: for example, the cytochrome c_2 is attached to the reaction center LMH complex,

menaquinone replaces a molecule of ubiquinone, and bacteriochlorophyll *b* replaces bacteriochlorophyll *a*. A different molecule of carotenoid is bound to the reaction center particle. As in *R. sphaeroides*, the overall cyclic reaction system generates a proton gradient that is then used to generate ATP.

The process of **noncyclic photosynthesis** in purple nonsulfur bacteria generates the NADH (NADPH) needed for cell biosynthesis (Figure 9.7). This process proceeds differently than that seen in Figure 9.4. Photoreduction of NADP$^+$ proceeds by transfer of electrons from QH_2 by a process termed **reversed electron transfer**. A suitable organic substrate (electron donor) is oxidized at another site to provide electrons to reduce cyt c_2. Transfer of an electron from cyt c_2 then restores the reaction center to the initial state. A variety of other substrates may donate electrons, including hydrogen, ethanol, isobutanol, succinate, and malate. Recent evidence affirms that ferrous iron (Fe^{2+}) may also serve as electron donor in the purple bacteria (see Box 9.2).

During photoheterotrophic growth some of the resultant oxidized organic products (such as acetate and pyruvate) can be further metabolized, whereas others (such as isobutyric acid) are not. The overall reaction for anaerobic photoautotrophic growth of *R. sphaeroides* by the Calvin cycle can be depicted as:

Many of the purple nonsulfur bacteria are facultative chemoheterotrophs and grow aerobically on organic substrates. In this metabolic mode, aerobic respiratory chains are utilized in place of photosynthesis (see Chapter 8). If these microorganisms are placed in the dark with utilizable organic substrates, the energetics for heterotrophic aerobic growth is as outlined in Figure 9.7C. The major requirement for aerobic heterotrophic growth is the synthesis of the terminal cytochrome aa_3 oxidase complex. Both the aerobic respiratory chain and the anaerobic photosynthetic electron transport chain have some common components that function in either mode of growth. The presence of oxygen halts the synthesis of photosynthetic pigments. Thus, cultures respiring in the dark have little pigment per unit cell mass. A decrease in the oxygen supply results in the resumption of pigment synthesis as the major mechanism for gaining energy is through anaerobic photophosphorylation.

Photosynthesis in Purple Sulfur Bacteria

Purple sulfur bacteria are similar to the purple nonsulfur bacteria in some respects. However, they exhibit many distinct morphological and phylogenetic proper-

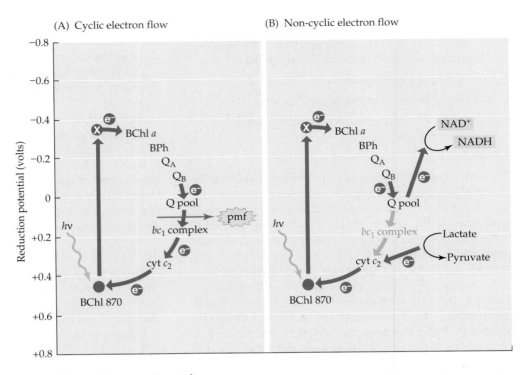

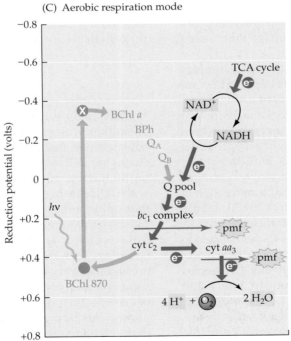

Figure 9.7 Electron flow in cyclic and non-cyclic modes of the photosynthetic cycle and in aerobic respiration in non-sulfur purple bacteria
(A) Cyclic flow of electrons in *Rhodobacter sphaeroides*. (B) Non-cyclic flow of electrons in *R. sphaeroides*. (C) Flow of electrons in aerobic respiration in *R. sphaeroides*. The pathways that are inactive are depicted in gray. BChl, bacteriochlorophyll *a*; BPh, bacteriopheophytin; Q, ubiqinone.

ties and contain a PRC distinct from that of other phototrophs. They can use hydrogen sulfide as electron donor for anoxic photosynthesis (see Table 9.2) and accumulate elemental sulfur granules within the cell. They thrive at elevated levels of H_2S that are toxic to the purple nonsulfur bacteria. Details of the properties and ecology of these microbes are presented in Chapter 21.

Photosynthesis in Green Sulfur Bacteria

Green sulfur bacteria differ significantly from purple bacteria and in some characteristics resemble the cyanobacteria. These bacteria are called green sulfurs because they appear green and generally utilize reduced sulfur compounds as an electron donor. The light reaction at the PRC (Bchl 840) in green sulfur bacteria results in the forma-

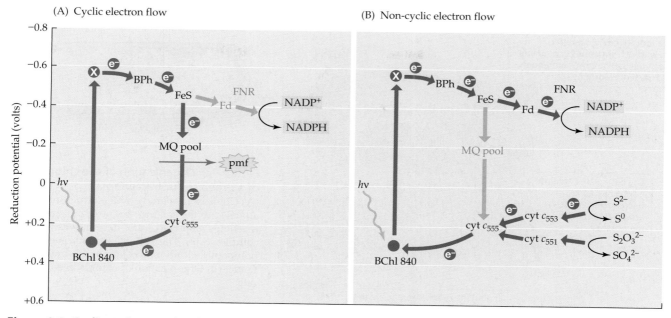

Figure 9.8 Cyclic and noncyclic electron flow in green sulfur bacteria
(A) Electron flow in the cyclic mode. (B) Electron flow in the non-cyclic mode. The P840* of these organisms has a sufficiently high reduction potential to directly reduce pyridine nucleotide. The inactive pathways are depicted in gray. BChl, bacteriochlorophyll *b*; BPh, bacteriopheophytin; FeS, iron-sulfur center; Fd, ferredoxin; FNR, ferredoxin nucleotide reductase; MQ, menaquinone.

tion of a stable reductant with a potential of about –540 mV, which is a much stronger reductant than is generated in nonsulfur purple bacteria (see Figure 9.7A). This reductant is sufficiently negative to permit a noncyclic, light-driven reduction of NADP (Figure 9.8) without the reverse electron flow essential in purple bacteria (see Figure 9.7A).

The reaction centers in green sulfur bacteria are more difficult to purify in a functional form; thus, less is known about the mechanics of photophosphorylation in these organisms. The major antenna pigments in green bacteria are two different chemical forms of bacteriochlorophyll: *c* and *a*. The reaction center itself contains bacteriochlorophyll *a* and bacteriopheophytin *a*. The ratio of antenna bacteriochlorophyll *c* to bacteriochlorophyll *a* is 1,000 to 1,500 per reaction center in green bacteria, which is unique among the photosynthetic bacteria. Plants have 200 to 400 molecules of chlorophyll per center, and other photosynthetic bacteria have 50 to 100 molecules.

The origins of green sulfur photosynthetic bacteria may offer a plausible explanation for this high level of antenna molecules that funnel energy of excitation to the reaction centers. Green sulfur bacteria are considered to have evolved in the lower depths of aquatic environments, where reduced sulfur compounds were available as electron donors. The green sulfur bacteria would

therefore have existed under swarms of cyanobacteria and purple bacteria. To absorb the weak light that filtered down through the layers of cyanobacteria and purple bacteria, the green sulfurs evolved with an extensive network of antenna pigments to capture the excitation energy of the limited light available.

The way that the light-harvesting pigments are bound to the cytoplasmic membranes in green bacteria is similar to that in which these pigments are bound to thylakoids in the cyanobacteria. The antenna pigments in green sulfur bacteria are packed into vesicles called **chlorosomes**, which lie along the inside of the cytoplasmic membrane (Figure 9.9). The chlorosomes are bound by a proteinaceous (nonlipid) membrane. Each vesicle contains up to 10,000 bacteriochlorophyll *c* molecules, and each is attached to the cellular membrane by a **subantenna** of a bacteriochlorophyll *a*. Excitation energy, generated by light and absorbed by antenna pigments, passes through this subantenna to the reaction centers.

Experimental evidence suggests that the reaction centers and probably some of the antenna bacteriochlorophyll *a* are built into a molecular structure, called a **base plate**. The protein base plate is embedded in the cytoplasmic membrane and joins the chlorosome to the cytoplasmic membrane. The high efficiency of the photosystem in green bacteria indicates that excitation energy

(A)

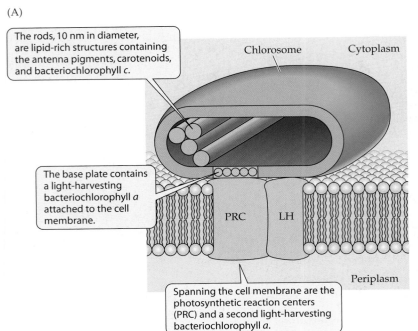

The rods, 10 nm in diameter, are lipid-rich structures containing the antenna pigments, carotenoids, and bacteriochlorophyll *c*.

The base plate contains a light-harvesting bacteriochlorophyll *a* attached to the cell membrane.

Chlorosome

Cytoplasm

PRC

LH

Periplasm

Spanning the cell membrane are the photosynthetic reaction centers (PRC) and a second light-harvesting bacteriochlorophyll *a*.

(B)

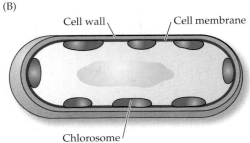

Cell wall

Cell membrane

Chlorosome

Figure 9.9 Organization of the chlorosome of a green sulfur bacterium
(A) Chlorosomes are vesicular structures attached to the cell membrane at the site of a reaction center (PRC) and light harvesting complex (LH). The vesicular membrane is a polypeptide lipid matrix, not a unit membrane. (B) Organization of chlorosomes in a cell.

captured by the antenna pigments in a saturated reaction center can readily be passed to another receptive center.

Electron transfer in green sulfur bacteria (see Figure 9.8B) begins with the primary photochemical donor (Bchl 840) passing electrons via BPh to an acceptor molecule, considered to be an Fe-S protein. The energy-enriched Fe-S protein then transfers electrons through a soluble ferredoxin (Fd) to an Fd-NADP reductase (FNR) to yield NADPH. This noncyclic pathway requires that electrons be replenished, and these are provided by reduced substrates such as sulfide, thiosulfate, or hydrogen. Electrons are transferred from substrates to bacteriochlorophyll in the reaction centers via specific types of cytochrome *c*: cytochrome c_{551} for thiosulfate and cytochrome c_{553} for sulfide. A specific reductase for each substrate acts as an intermediate carrier of electrons from the substrate to the cytochrome *c* involved.

The overall photochemistry in green sulfur bacteria is relatively straightforward and resembles photosystem I in cyanobacteria and green plants. ATP is generated through a cyclic photophosphorylation, and reduced pyridine nucleotides are produced by the noncyclic Fd-reductase FNR system. Sulfides and other reduced sulfur compounds are of widespread occurrence in the environments in which green sulfur bacteria thrive. The primary requirements for survival of an autotrophic, anaerobic photosynthetic bacterium include a ready source of reductant for CO_2 assimilation and ATP to drive synthetic reactions. These organisms have evolved with a marvelous capacity for survival in sulfurous anaerobic environments where low levels of light are available as an energy source.

Photosynthesis in Heliobacteria

The heliobacteria are distinct from other anoxygenic photosynthetic bacteria. These gram-positive microbes are of widespread occurrence in soils, particularly those that are subjected to periods of wetness and dryness. Some species form endospores, which may provide a survival advantage during dry periods. The bacteriochlorophyll in these microorganisms is bacteriochlorophyll *g*; they contain no bacteriochlorophyll *a*. Bchl *g* is unique among bacteriochlorophylls, as it contains a vinyl (—CH=CH₂) constituent on ring I, as is present in plant chl *a* (see Figure 9.1). The Bchl *g* in heliobacteria absorbs maximally at 788 nm, a wavelength not absorbed by other photosynthetic bacteria. The reaction centers of these microorganisms are embedded in the cytoplasmic membrane, and they do not form thylakoids or chlorosome-like structures. The only sources of carbon utilized by known species are simple monomers such as pyruvate, acetate, and lactate. The heliobacteria are not known to grow autotrophically (see Chapter 21).

Photosynthesis in Cyanobacteria

The cyanobacteria, formerly called the blue-green algae or blue-green bacteria, are the only photosynthetic bacteria that generate O_2 during photosynthesis (oxygenic phototrophs). The cyanobacteria are a physiologically and morphologically diverse group that appeared over 2.5 billion years ago (see Chapter 21). During evolution, they developed two independent light reactions: **photosystem I (PSI)** and **photosystem II (PSII)**; these operate in series with a redox span from H_2O/O_2 to $NADP^+/NADPH$.

The oxygen-generating photochemical system (PSII) of the cyanobacteria was instrumental in providing an O$_2$-containing atmosphere that eventually permitted the commencement of aerobic respiration, first in the *Bacteria* and *Archaea*, and then in eukaryotes.

Photosynthetic reactions in cyanobacteria are similar to and a forerunner of the reactions in green plants. However, the only relationship that the cyanobacteria have to eukaryotes is the similarity of the photosystems in these bacteria to the photosystems in the chloroplasts

of green plants. Our understanding of green plant photosynthesis has been enhanced by studies of the photoapparatus in the more readily manipulated cyanobacteria. These rapidly growing microorganisms are easily cultured under laboratory conditions and offer considerable functional diversity.

Cyanobacteria combine the salient features of photoreactions in both purple and green bacteria (see Figures 9.7, 9.8 and 9.10). The photosystem containing the reaction center P700, designated PSI, is similar to the

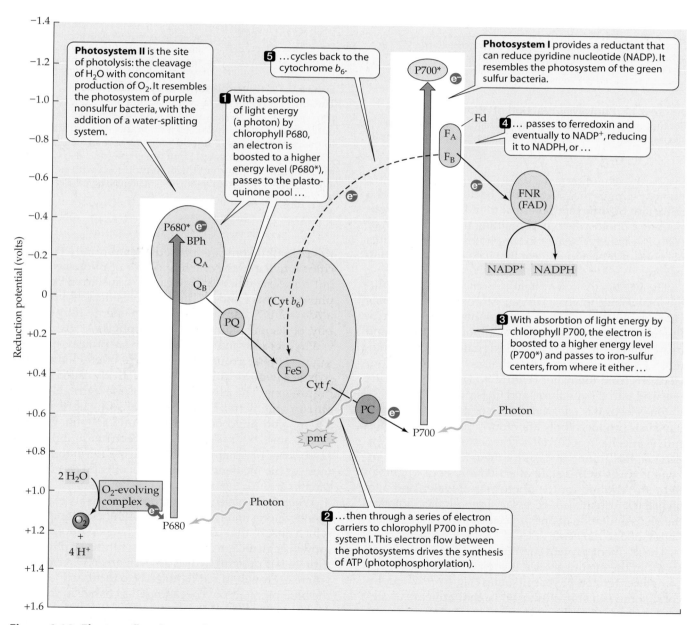

Figure 9.10 Electron flow in reaction center of a cyanobacterium

A cyanobacterium has two photosystems, designated photosystems I and II (PSI and PSII). In this diagram, electrons flow from the left (from H$_2$O) to the right. BPh, bacteriopheophytin; Q$_A$, Q$_B$, PQ, plastoquinones; FeS, iron-sulfur center; PC, plastocyanin; F$_A$, F$_B$, iron-sulfur centers; Fd, ferredoxin; FNR (FAD), the flavoprotein enzyme ferredoxin-NADP$^+$ oxidoreductase.

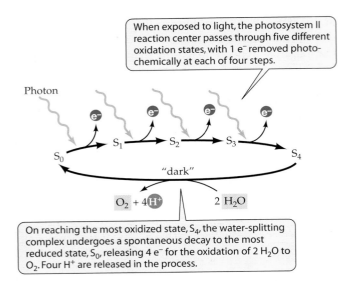

Figure 9.11 Photolysis reaction of photosystem II
Evolution of one molecule of oxygen requires the stepwise accumulation of four oxidizing equivalents in photosystem II.

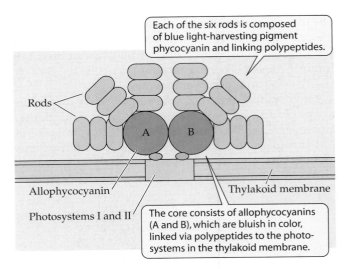

Figure 9.12 Phycobilisome of cyanobacteria
The antenna pigments of cyanobacteria are arranged in phycobilisomes. These knoblike structures project from the outer surface of the thylakoid membrane, and are attached to the photosystems (PSI, PSII). Shown here is the phycobilisome of *Synechococcus* sp. Sometimes the rods also contain the red pigment phycoerythrin. Reproduced with permission. Arthur Grossman et al. 1993. *Microbiological Reviews*, 57:725–749.

photosystem in the green sulfur bacteria. It provides a reductant of sufficient potential to reduce pyridine nucleotides. Photosystem II, containing the reaction center P680, is somewhat similar to the photosystem in the purple nonsulfur bacteria except that it has a water-splitting system.

The light-activated system responsible for the splitting of water (photolysis) to generate oxygen and donate electrons to PSII requires four quanta of light energy. Photolysis occurs when two water molecules are bound to a light-sensitive enzyme, designated the S system, which absorbs four positive charges that extract four electrons from the water (Figure 9.11). The protons are released into the medium, and O_2 is produced.

One electron is released at each of the four steps, and oxygen is produced only after four separate photon capture events have occurred on the S system. S_0 is the enzyme at the ground state; after the release of O_2 the enzyme returns to this state of activation. Four quanta of light are required for the photolysis of water, and a total of 8 to 10 quanta are necessary for both the photosynthetic dissimilation of one molecule of water and assimilation of one molecule of CO_2.

The pigment-protein complex involved in photosynthesis in the cyanobacteria is present in multi-protein complexes termed **phycobilisomes** (Figure 9.12). A phycobilisome is a subcellular body that appears as tiny knobs attached to the outer surface of photosynthetic membranes. Little if any transfer of quanta occurs between these photosynthetic units, and each is a discrete miniaturized power cell. The light energy absorbed in

the phycobilisomes is passed preferentially through chlorophyll *a* to the reaction center of photosystem II, but some excitation energy can be transferred between photosystems I and II. The system as depicted permits a "downhill" cascade of excitation energy through the phycobilisome pigments to chlorophyll *a*.

Phycocyanin and **phycoerythrin** are proteins containing prosthetic groups consisting of open-chain tetrapyrroles (Figure 9.13) and are collectively called phycobiliproteins. The phycobiliproteins are light-gathering antenna pigments. Phycocyanin absorbs light around 630 nm and phycoerythrin at around 550 nm. Chlorophylls absorb a rather narrow spectrum of light, and these accessory pigments allow the organism to capture more of the incident light that falls outside this narrow range.

When the wavelength of light available is altered, some cyanobacteria undergo an induced synthesis of pigments which absorb that particular wavelength. For example, growth in green light results in the synthesis of phycoerythrin (a red protein), which absorbs green wavelengths, whereas growth in red light leads to increased levels of the blue pigment phycocyanin. This change in the phycobilisomes is termed **chromatic adaptation** (Figure 9.14).

The primary electron acceptors in photosystem I are soluble Fe-S proteins similar in function to those in green sulfur bacteria. Electrons may flow cyclically back to

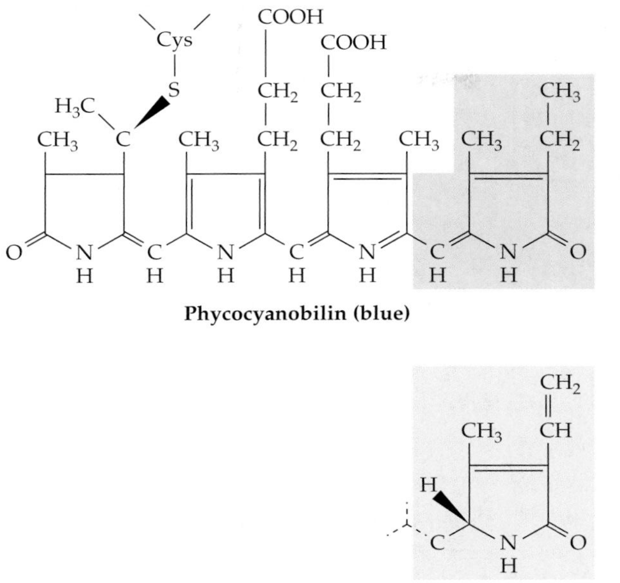

Phycocyanobilin (blue)

Phycoerythrobilin (red)

Figure 9.13 Chromophores of phycobilisomes
Structures of two of the chromophores (bilins) of the phycobilisomes of cyanobacteria. They differ only at the fourth
pyrrole ring. Each bilin is joined to protein at a cysteine
residue.

chlorophyll P700 via the cytochrome $b_6 f$ complex , analogous to the bc_1 complex, and to the copper-containing
protein **plastocyanin** (**PC**) before restoring the P700 to
its original state (see Figure 9.10). As in the purple nonsulfur bacteria (see Figure 9.4), this light-dependent
transfer of electrons results in the formation of the pmf

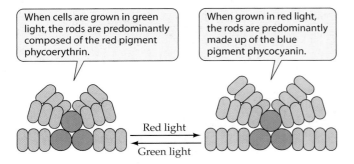

| When cells are grown in green light, the rods are predominantly composed of the red pigment phycoerythrin. | When grown in red light, the rods are predominantly made up of the blue pigment phycocyanin. |

Red light →
← Green light

Figure 9.14 Chromatic adaptation of a phycobilisome
In some bacterial species, the composition of phycobilisomes can be altered by changing the wavelength of light
provided for growth. This scheme shows chromatic adaptation in *Fremyella diplosiphon*. Reproduced with permission.
Arthur Grossman et al. 1993. *Microbiological Reviews*,
57:725–749.

and ATP synthesis (cyclic mode). Operation of PSI and
PSII in tandem is noncyclic and generates NADPH.

SECTION HIGHLIGHTS
Of the five groups of photosynthetic bacteria,
four perform anoxygenic reactions and the
fifth, the cyanobacteria, perform oxygenic reactions. Each group differs with respect to the
PRC molecule, electron transport components,
and the types of electron donors used to replenish the PRC when operating in the noncyclic mode.

9.4 Photosynthesis with Rhodopsin

The halophilic microorganisms are a remarkable group
that thrive in highly saline environments, as they require
substantial concentrations of NaCl for growth (see Chapter 18). The extreme halophiles are all in the *Archaea* domain. Of these organisms, *Halobacterium halobium* has
been studied most extensively. *H. halobium* is obligately
aerobic and operates a conventional membrane-bound
respiratory pathway for ATP generation when sufficient
O_2 is available. Because the solubility of oxygen in water
decreases as salinity increases, the availability of oxygen
beneath the surface of water with a high salt concentration is limited. When the halobacteria encounter low oxygen concentrations, they synthesize purple patches that
appear within the cytoplasmic membrane. These patches
can cover over one-half the surface of the cellular membrane and provide a light-driven proton ejection system
that is not inhibited by cyanide. Aerobic respiration in
these organisms would be inhibited by cyanide due to
its ability to bind to and inactivate cytochromes.

The purple patches are made up of flat sheets containing a crystalline array of a single protein—**bacteriorhodopsin**. This protein has a molecular mass of 26,000,
and each protein assembly is attached to a molecule of
retinal through a lysyl residue in the protein (Figure
9.15). Bacteriorhodopsin (vitamin A aldehyde) is chemically similar to the visual pigments of animals. The bacteriorhodopsin absorbs light maximally at a wavelength
of 570 nm and becomes bleached as it ejects a proton to
the exterior of the cytoplasmic membrane. The deprotonated bacteriorhodopsin then takes up a proton from the
cytoplasm of the cell (see Figure 9.15). Bacteriorhodopsin
absorbs at wavelengths at which energy is available
from sunlight, particularly when there is a water layer
over the organism. No accessory pigments are necessary
for the photochemical reaction to occur, and respiratory

(A)

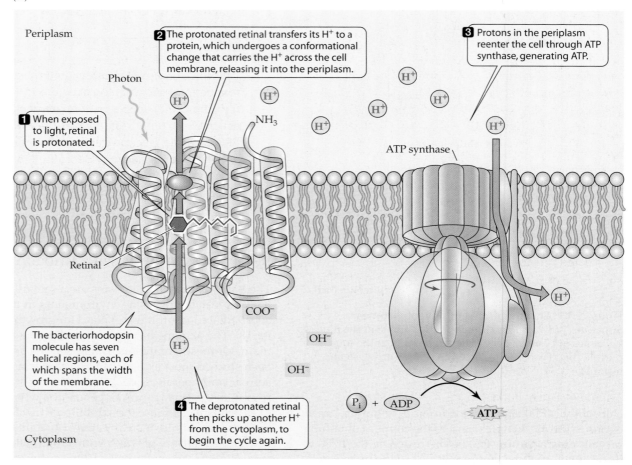

Periplasm

2 The protonated retinal transfers its H⁺ to a protein, which undergoes a conformational change that carries the H⁺ across the cell membrane, releasing it into the periplasm.

3 Protons in the periplasm reenter the cell through ATP synthase, generating ATP.

Photon

1 When exposed to light, retinal is protonated.

NH₃

ATP synthase

Retinal

The bacteriorhodopsin molecule has seven helical regions, each of which spans the width of the membrane.

COO⁻

OH⁻

OH⁻

4 The deprotonated retinal then picks up another H⁺ from the cytoplasm, to begin the cycle again.

Cytoplasm

Pᵢ + ADP

ATP

(B)

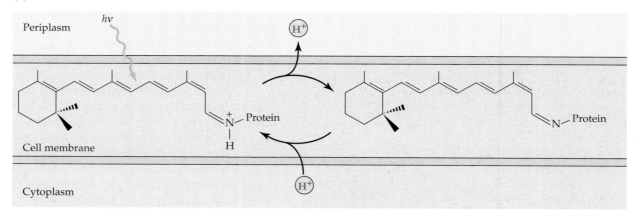

hv

Periplasm

H⁺

Cell membrane

⁺N—Protein
|
H

N—Protein

Cytoplasm

H⁺

Figure 9.15 Light-driven proton pump of halophilic archaea
(A) The light-driven proton pump in the membranes of halophilic (salt-loving) archaea consists of bacteriorhodopsin, a pigmented protein molecule, to which is attached a retinal molecule. (B) The chemical reactions of retinal underlying the pumping mechanism. No electron transport is involved in this system.

electron transport is not involved. The reaction is a simple **light-driven proton pump** that involves a *cis-trans* isomerization of the bacteriorhodopsin molecule following excitation (see Figure 9.15B). The *trans* configuration is eventually restored to the *cis-* form by a series of steps not shown.

Elucidation of this system provided considerable credibility to the chemiosmotic theory of ATP generation. The isolated purple membrane has been reconstituted in closed vesicles, which, when exposed to light, act as proton pumps. ATP synthase from beef heart has been incorporated into these vesicles, where they can perform a light-dependent synthesis of ATP (Box 9.3). It would be difficult to ascribe this phenomenon to any direct coupling reaction because no electron transfer is involved. Replacement of the halophile's ATP synthase with that from an animal confirms that no electron transfer–related phenomenon occurs in the native ATP synthase. Electron flow in respiration then serves only to transduce protons across the membrane.

The quantum requirement for the ejection of one proton by *H. halobium* is approximately 2, as compared with 0.5 for purple photosynthetic bacteria. Because light at a wavelength of 570 nm has more energy than light at 870 nm, the overall efficiency of *H. halobium* is less than 25% of the efficiency of purple bacteria.

Recent studies, employing metagenomic DNA analysis, indicate that there are bacterial species in marine habitats bearing genes that encode another form of rhodopsin. These microorganisms have not yet been grown in culture. Their rhodopsin (or rhodopsins) differs from the bacteriorhodopsin present in the *Archaea* and is designated **proteorhodopsin**. The microorganisms involved are deep-sea dwellers and absorb blue light, which penetrates to considerable depths. A number of proteorhodopsin variants appear to exist and likely rep-

---

BOX 9.3 · *Research Highlights*

Archaea, Bovines, and Chemiosmosis

In 1974, Efraim Racker, a biochemist, and Walther Stockenius, a pioneer in elucidating the role of bacteriorhodopsin in halophilic bacteria, collaborated to perform an experiment that further confirmed Mitchell's chemiosmotic hypothesis. They reconstituted vesicles from the cytoplasmic membranes of *Halobacterium halobium*, which were inverted so that the bacteriorhodopsin would pump protons inward when exposed to light. They also obtained bovine mitochondrial ATP synthase and incorporated it into the vesicle, as shown. When illuminated, the bacteriorhodopsin pumped protons into the vesicles and generated a proton gradient inside them which drove ATP synthesis by the ATP synthase. There were two distinct types of protein in the vesicles, one from a member of the *Archaea* and the other from a mitochondrion. Together they synthesized ATP, as the Mitchell hypothesis predicted. No electrons were involved in the synthesis of this ATP, thus confirming that the role of respiratory electron transport chains is the transduction of protons, not phosphorylation of ATP.

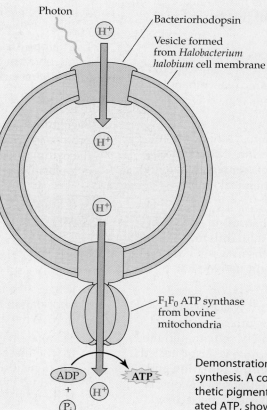

Demonstration that a proton gradient drives ATP synthesis. A combination of archaeal photosynthetic pigments with bovine ATP synthase generated ATP, showing that a proton gradient, not electron transport, drives ATP synthesis.

resent a diversity of habitats. They probably support a significant amount of biomass in these environments.

SECTION HIGHLIGHTS

Certain members of the *Archaea* and *Bacteria* perform photosynthesis without the need for electron transport chains or chlorophylls. They employ rhodopsin molecules to pump protons across the cell membrane to form an electochemical gradient, which can be used to drive ATP synthesis.

SUMMARY

- **Photophosphorylation** is the sum of the processes that generate ATP through the use of light energy.

- **All life** on the surface of the Earth **relies** either directly or indirectly **on energy from the Sun**. Photoautotrophs use photons directly, whereas heterotrophs exist by utilizing photosynthetic products.

- **Oxygenic** photosynthesis results in the release of O_2, as the source of electrons for this process is H_2O ($2 H_2O \rightarrow 4 H^+ + O_2 + 4 e^-$).

- **Anoxygenic** photosynthesis does not result in O_2 production. The source of electrons in anoxygenic photosynthesis is reduced organic (e.g., ethanol $\rightarrow$ acetate) or inorganic compounds (e.g., $S_2O_3^{2-} \rightarrow SO_4^{2-}$).

- Much of the chlorophyll in photosynthetic organisms is involved in **light gathering**. The acquired energy is transferred to **reaction centers**, where the **light energy** is converted to **chemical energy**.

- There are **several chlorophylls**—*a*, *b*, *c*, *d*, *e*, and *g*— each having a **characteristic absorption spectrum**. This permits microorganisms with differing chlorophylls to grow at various depths in water because they absorb wavelengths of light that are not absorbed by those above them.

- The **ozone layer** above the Earth was probably formed when oxygenic photosynthesis evolved. The O_2 produced reacted with UV radiation in the stratosphere to form O_3. The ozone layer protected earthly creatures from **harmful UV radiation** and was a major contributor to evolution of life on dry land. Aquatic life was not adversely affected by UV radiation because this wavelength does not penetrate into water.

- **Reaction centers** have been crystallized from purple nonsulfur photosynthetic bacteria. Much has been learned about structural arrangements in reaction centers by studying these crystals.

- Most **purple nonsulfur bacteria** grow either with light energy or as chemoheterotrophs. They use parts of the same respiratory components when grown by either mode.

- The reaction centers in **green sulfur bacteria** are complex and have considerably more **antenna pigments** than are present in the purple nonsulfur reaction centers. This permits them to gather light of much lower intensity. These antenna pigments are arranged in **chlorosomes**.

- The **photosystems** in cyanobacteria and green sulfurs can easily **reduce $NADP^+$ to NADPH**. The purple nonsulfur bacteria do not have sufficient energy to accomplish this. The latter must generate NADPH for CO_2 fixation by reversing ATP synthase to generate an electrochemical gradient across the membrane.

- The antenna pigments in cyanobacteria are arranged in projections attached to the surface of membranes termed **phycobilisomes**. The pigments in these organelles can change as the wavelength of the light available changes.

- The **halophilic *Archaea*** and some marine *Bacteria* are aerobic but can generate ATP anaerobically by utilizing **bacteriorhodopsin** and **proteorhodopsin**, respectively, in combination with retinal in a system that is similar to that found in the eyes of animals. Bacteriorhodopsin can use light energy to move protons from the cytoplasm to the outer surface of the cytoplasmic membrane. The protons reenter the cytoplasm via ATP synthase, thus producing ATP.

 Find more at www.sinauer.com/microbial-life

REVIEW QUESTIONS

1. What is the basic difference between photophosphorylation and respiration-linked phosphorylation? What are the common features?

2. There are basic differences in the electron donor for the three photosynthetic bacterial types—purple nonsulfurs, green sulfurs, and cyanobacteria. What are these differences? Are they related to habitat?

3. Which of the three bacterial types mentioned in question 2 would you consider the most ancient?

Why? Consider the atmosphere of the primordial earth and the nature of compounds available.

4. Describe the structural makeup of chlorophyll and the relationship it has to heme in blood. Why does Mg^{2+} replace iron?

5. The evolution of the cyanobacteria with oxygenic photosynthesis was of major importance in the movement of life forms to terrestrial habitats. Explain.

6 Antenna pigments are important to photosynthetic organisms. Why? Consider all types of photosynthetic bacteria.

7. Why have the purple nonsulfur photosynthetic bacteria been a model system for clarifying the events that occur in a reaction center?

8. Compare the role of electron transport in purple bacteria grown photosynthetically to that present during aerobic heterotrophic growth. How much actual difference is there?

9. How does a photosynthetic bacterium respond to increases in available light intensity? Consider chlorophyll and growth rate.

10. What is a chlorosome, and what role does it play in green sulfur bacteria? What are the major parts?

11. How have green sulfur bacteria adapted to life at the bottom of the pond?

12. List the differences in the photosynthetic apparatus among the three photosynthetic types: the cyanobacteria, the green sulfur bacteria, and the purple nonsulfur photosynthetic bacteria.

13. What is bacteriorhodopsin? How does photosynthesis with this pigment differ from that due to chlorophyll?

SUGGESTED READING

Blankenship, R. E., M. T. Madigan and C. E. Bauer, eds. 1995. *Anoxygenic Photosynthetic Bacteria*. Dordrecht, The Netherlands: Kluwer Academic Publishing.

Bryant, D. A. and N. U. Frigaard. 2006. "Prokaryotic Photosynthesis and Phototrophy Illuminated." *Trends in Microbiology* 14: 488–496.

Peschek, G. A., W. Loffelhardt and G. Schmetterer, eds. 1999. *The Phototrophic Prokaryotes*. New York: Kluwer Academic/Plenum Publishers.

Raghavendra, A. S., ed. 2000. *Photosynthesis: A Comprehensive Treatise*. Cambridge, UK: Cambridge University Press.

Staley, J. T. and A-L. Reysenbach, eds. 2002. *Biodiversity of Microbial Life; Foundation of Earth's Biosphere*. New York: Wiley-Liss.

10

Biosynthesis of Monomers

*...a superficial survey of the biochemical field is apt to fill one
with profound astonishment at the practically unlimited diversity of the chemical constituents of living organisms...We have to
accept the undeniable fact that all these substances have ultimately been derived from carbon dioxide and inorganic salts.*
—A. J. Kluyver

The previous chapter was concerned with the strategies employed by biological systems in the conservation of energy
captured from the sun. The quanta of light that reach the
Earth activate those systems that convert photons into stable, energy-
rich chemicals such as adenosine triphosphate (ATP) or reduced pyridine nucleotides (NADPH and NADH). Chemolithotrophic reactions
involved in the Earth's crust also provide an energy supply for the
formation of ATP and NADPH. These compounds then serve as key
intermediaries in cell biosynthesis.

If ATP and NADH/NADPH are considered as the exchange cur-
rencies in energy transformation, then the C—H bond might be considered a "savings bank." Generally, ATP and the electron carriers act
solely as transitory go-betweens, with the C—H and C—C bonds
present in sugars, amino acids, lipids, and other constituents of cells,
serving as a stable, life-giving energy source for cell growth. Thus,
the formation of these bonds during the photosynthetic and
chemoautotrophic processes provide stable energy units that can support the Earth's heterotrophic food chains.

The combined mechanisms whereby the major constituents of the
bacterial cell are synthesized from simple building blocks is the process
termed **biosynthesis (anabolism)**. All cellular material ultimately is
synthesized from carbon dioxide (CO_2); it is the primary substrate
supporting life and is incorporated into cells via an assortment of

mechanisms. Some of the other 1-carbon compounds utilized by bacteria include carbon monoxide (CO), cyanide (CN), formate (HCOO⁻), methanol (CH_3OH), and methane (CH_4). These substrates may be funneled through CO_2 and enter anabolic sequences through the classic Calvin cycle, or they may be assimilated through other CO_2 fixation pathways.

Certain organisms can satisfy their nitrogen requirements by the fixation of gaseous nitrogen (N_2) available from the atmosphere; others utilize ammonia (NH_3), nitrate (NO_3^-), and/or nitrogen-containing organic compounds. Sulfur needs are met by inorganic sulfate (SO_4^{2-}) or sulfide (HS⁻). Phosphate for nucleic acid synthesis and the generation of high-energy intermediates can come from inorganic phosphate sources.

This chapter considers the biosynthetic mechanisms that have evolved among the prokaryotes for the generation of the monomolecular units or **monomers** that can be polymerized to form the macromolecules of a cell. Synthesis by autotrophs is a major basis for this discussion, as these microorganisms can construct all of their cellular components from CO_2, the most rudimentary of building blocks.

10.1 Macromolecular Synthesis from Precursor Substrates

Evolution ultimately led to the selection of viable organisms that grew and reproduced by using the low-molecular-weight substrates available to them in the "primordial soup." It is possible that some of these early organisms may have utilized light as a source of energy. The use of low-molecular-weight compounds in biosynthesis permitted the survival of these organisms without relying on random abiotic (or biotic) generation of building blocks. The development of counterpart biodegradative processes—which constantly replenished the low-molecular-weight primary substrates—was essential in early evolution. Ultimately, the central substrate for biosynthesis and also the end product of biodegradative activities came to be **carbon dioxide**. Thus, CO_2 was the starting material for **anabolism** (synthesis) and the end product of **catabolism** (breakdown).

A microorganism will grow and reproduce only if it can synthesize *all essential* cellular macromolecules from substrates available to it (see Chapters 5 and 6). Autotrophic organisms that use CO_2 as their sole carbon source and N_2 as their nitrogen source represent the ultimate in biosynthetic capacity. Research has shown that the broad synthetic ability of autotrophs and other microorganisms that grow on low-molecular-weight substrates rests on the organisms' effective use of a limited number of biosynthetic pathways. In this section, anabolic sequences as they function in *Bacteria* and *Archaea* are emphasized. Eukaryotic organisms that can synthesize these cellular constituents follow identical or very similar biosynthetic routes. Indeed, studies on the metabolic pathways in model bacteria and fungi have been of fundamental importance to our understanding of the biochemistry of these systems as well as those in higher animals and plants (see Chapter 2; Box 10.1).

There are four major classes of macromolecules in a bacterial or archaeal cell: (1) proteins, (2) nucleic acids, (3) polysaccharides, and (4) lipids. These macromolecules are synthesized from a limited number of monomers. A generalized scheme of cellular synthesis is presented in Box 10.2. According to this outline, CO_2 assimilation (fixation) reactions lead into central reactions in gluconeogenesis (formation of 6-carbon sugars) in autotrophs. Other substrates lead from glycolysis and the tricarboxylic acid (TCA) cycle in heterotrophic microorganisms, as shown on the lower part of the diagram in Box 10.2. All of the major monomers—such as amino acids, purines, pyrimidines, and sugars—that are polymerized to form macromolecules are readily synthesized from intermediates derived from the various pathways present (e.g., the TCA cycle, gluconeogenesis, glycolysis, and/or related pathways).

These metabolic reactions generate the **precursor metabolites** at a continuous and appropriate level to promote orderly cellular reproduction. Twelve key precursor metabolites are involved in the biosynthesis of all building blocks, prosthetic groups, coenzymes, and the various cellular components. These key metabolites and their pathways of origin are listed in Table 10.1. Several precursors also arise directly from the CO_2 fixation pathways.

How are these key precursor metabolites generated in *Bacteria* and *Archaea*?

SECTION HIGHLIGHTS

There are four major classes of macromolecules in a bacterial or archaeal cell: (1) proteins, (2) nucleic acids, (3) polysaccharides, and (4) lipids. These macromolecules are synthesized from a limited number of monomers (e.g., amino acids, purines, pyrimidines, and sugars), which are derived from intermediates of central metabolic pathways (e.g., the TCA cycle, gluconeogenesis, and glycolysis). Most organisms, including eukaryotes, use similar pathways to synthesize the macromolecules they require if they cannot obtain them from their environment.

BOX 10.1 *Milestones*

The Microbe as a Paradigm

Back in 1930, the eminent Dutch microbiologist A. J. Kluyver gave a series of lectures on the metabolism of microorganisms at the University of London. His was a clear exposition on the status of the field at that time. These lectures were published in a classic treatise entitled *The Chemical Activities of Micro-Organisms* (1931, University of London Press, Ltd.). It was through the landmark studies of Kluyver and the Delft School of microbiologists (M. Beijerinck, C. B. van Niel, and others) that bacteria became the paradigm for experimentation in biochemistry of the cell (see Chapter 2). The following is a quote concerning colorless (nonphotosynthetic) microorganisms from one of Kluyver's elegant lectures, as it was presented to the audience in 1930:

> *...for here we find the most remarkable fact—the significance of which is often underestimated, although known since the time of Pasteur—that a single organic compound suffices to ensure a perfectly normal development of these organisms, although they are cut off from any external energy supply. Here we find the biochemical miracle in its fullest sense, for we are bound to conclude that all the widely divergent chemical constituents of the cell have been built up from the only organic food constituent, and that without any intervention of external energy sources. The chemical conversions performed by these organisms rather resemble witchcraft than chemistry!*

In 1954, Kluyver and Van Niel presented the John M. Prather Lectures at Harvard University. These lectures were published in another classic, entitled *The Microbe's Contribution to Biology* (1956, Harvard University Press, Cambridge, MA). This series of lectures confirmed that, in the years between 1931 and 1954, studies on microorganisms did indeed teach us much about the biochemistry of the cell. By this time the chemical conversions performed by microorganisms looked much less like witchcraft.

TABLE 10.1 Major precursor metabolites in biosynthetic reactions

Metabolite	Predominant Pathway of Origin[a]			
	EM	**HMS**	**ED**	**TCA**
Glucose-6-phosphate	X	X	X	
Fructose-6-phosphate	X			
Ribose-5-phosphate or ribulose-5-phosphate		X		
Erythrose-4-phosphate		X		
Glyceraldehyde-3-phosphate	X	X	X	
3-Phosphoglycerate	X			
Phosphoenolpyruvate	X			
Pyruvate	X	X	X	
Acetyl-CoA	X	X	X	
Oxaloacetate				X
α-Ketoglutarate				X
Succinyl-CoA				X

[a] EM = Embden-Meyerhof pathway; HMS = hexose monophosphate shunt; ED = Entner-Doudoroff pathway; TCA = tricarboxylic acid cycle.

BOX 10.2 *Milestones*

Economics of Growth

"Metabolic diversity" describes the seemingly endless capacity of microorganisms to utilize a vast array of naturally occurring chemicals (and products of chemical synthesis) as substrates for growth. They expeditiously convert these disparate substrates into the amino acids, purines, sugars, and fatty acids that make up the cell. How can a microorganism do this?

Evolution in microorganisms, as evident in the processes of anabolism, has proven to be exceedingly economical. Economical as defined by

Webster—"marked by careful, efficient and prudent use of resources: operating with little waste." Once Mother Nature chanced on an effective mechanism for accomplishing a task, that mechanism became de rigueur. The heart and soul of anabolism begins with CO_2 fixation and the catabolic pathways that occur in one form or another in virtually all living cells. These pathways are the main stream, and from this stream come the precursors of all the macromolecules that make up the cell. A simple depiction might be as follows:

What then of "metabolic diversity"? How does the ability to utilize so many different substrates fit into the economics of evolution? Neatly, because all of the processes of catabolism lead directly to the main stream. All substrates that can be utilized by microorganisms as a carbon source are, by a limited number of enzymatic reactions, converted to acetate, sugars, keto acids, and other intermediates that can be funneled directly into the glycolytic or TCA main stream.

Together these catabolic and anabolic reactions generate a new cell economically and expeditiously.

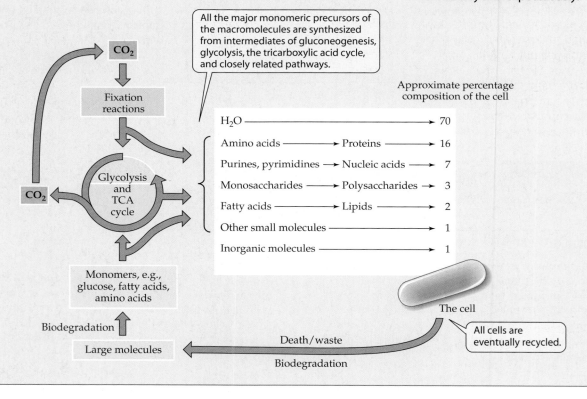

10.2 Macromolecular Synthesis from 1-Carbon Substrates

Autotrophy, by definition, involves the incorporation of inorganic carbon into biological molecules, and nature

has a variety of ways for accomplishing this essential task. Thus, microorganisms appear to have independently evolved many solutions over the past 3+ billion years (Box 10.3)! These include four pathways for CO_2 fixation:

BOX 10.3 *Milestones*

Alternative Pathways of 1-Carbon Assimilation

Elucidation of the Calvin cycle, also called the Calvin-Benson cycle, in the 1950s led to an assumption that this was the universal mechanism for autotrophic 1-carbon assimilation. These Nobel Prize–winning studies were performed using algae and plant leaves as models. However, experiments with green sulfur photosynthetic bacteria and some other autotrophs clearly indicated that some microorganisms utilize distinctly different pathways for CO_2 assimilation. When ^{14}C-radiolabeled CO_2 was placed in the atmosphere supplied for growth of green sulfur photosynthetic bacteria, the primary products were not the radiolabeled sugars associated with the Calvin cycle but rather radiolabeled fatty acids. Similar experiments with organisms that utilize methane, methanol, or other 1-carbon compounds as substrates affirmed that they too utilized alternative pathways.

Anaerobic *Bacteria* and *Archaea* that occupy niches such as swamps in which electron donors more negative than NADH (see Table 7.1) are available can fix CO_2 by reductive carboxylation reactions. Among these organisms are the acetogenic anaerobes, green sulfur photosynthetic bacteria, and methanogens (methane-producing bacteria). Despite the fact that these organisms fix CO_2 via a common pathway (reductive carboxylations), they are not phylogenetically related.

A major contribution to understanding CO_2 incorporation into heterotrophic cells was made by Harland G. Wood and C. H. Werkman in the mid-1930s. They demonstrated that the propionic acid bacteria combined one molecule of $^{13}CO_2$ with pyruvate to give oxaloacetate. Using ^{14}C-radiolabeled material, they confirmed these findings in 1941, thus overturning the dogma that only photosynthetic and adapted

Harlan Wood. Photo courtesy of the National Library of Medicine.

chemosynthetic cells were capable of fixing CO_2. They went on to demonstrate that almost all types of heterotrophic *Bacteria, Archaea,* and eukaryotes, including mammals, incorporate CO_2.

- Calvin cycle
- Reductive citric acid pathway
- Hydroxypropionate pathway
- Acetogenic or reductive acety-CoA pathway and two pathways for formate reduction:
- Glycine-serine pathway
- Ribulose monophosphate pathway

All these pathways require input of energy to drive the formation of carbon–carbon and C—H bonds. Interestingly, the pathways differ in the types and amounts of energy used. They also differ in their distribution across the bacterial and archaeal divisions.

CO_2 *Fixation: The Calvin Cycle*

The major pathway for utilization of carbon dioxide as the sole source of carbon is designated the **Calvin cycle**. This pathway is used by all plants and many types of photosynthetic bacteria, where ATP and NADPH are generated from light energy (see Chapter 9) and drive the CO_2 fixation reactions. Annually, in excess of 10^{11} tons of CO_2 are fixed, principally through the mediation of roughly 40 million tons of the enzyme ribulose-1,5-bisphosphate carboxylase. This key enzyme in the Calvin cycle provides the major mechanism for CO_2 fixation in many autotrophs and is among the most abundant proteins in the biosphere. Two key events render an organism capable of autotrophic growth via this cycle (Figure 10.1):

- A carboxylation reaction that leads to the assimilation of a molecule of CO_2

- The molecular rearrangements that regenerate ribulose-1,5-bisphosphate, which is the CO_2 acceptor molecule

A primary reaction in the Calvin cycle is the phosphorylation of ribulose-5-phosphate to produce ribulose-1,5-bisphosphate. The next enzyme involved in assimilating a molecule of CO_2 is ribulose-1,5-bisphosphate carboxylase/oxygenase (**Rubisco**) (see Figure 10.1). The Rubisco carboxylase reaction incorporates a molecule of CO_2 into ribulose-1,5-bisphosphate with the simultaneous cleavage of the unstable intermediate formed to yield two molecules of 3-phosphoglycerate. This product, 3-phospho-

glycerate, is reduced to glyceraldehyde-3-phosphate, an intermediate in gluconeogenesis (described below), and gluconeogenesis combined with the TCA cycle provides the organism with the needed precursor metabolites.

The regeneration of ribulose-5-phosphate is of utmost importance to autotrophic organisms that utilize the Calvin cycle for CO_2 fixation. The first step in the regeneration process is the conversion of 3-phosphoglycerate to glyceraldehyde-3-phosphate, which can be coupled to form hexoses for cell biosynthetic reactions. The hexoses can be rearranged to generate 4-, 5-, and 7-carbon sugars as will be described below (see Figure 10.6; Figure 12.3).

A key intermediate in regenerating ribulose-5-phosphate is the 7-carbon sugar sedoheptulose-7-phosphate, as two carbons from this sugar can be added to one mol-

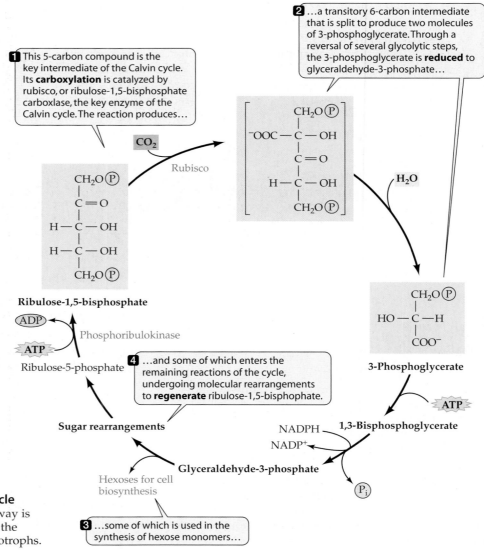

Figure 10.1 Calvin cycle
This CO_2-fixation pathway is predominant in plants, the cyanobacteria, and most other autotrophs.

ecule of glyceraldehyde-3-phosphate, resulting in the formation of two molecules of 5-carbon sugar (7 – 2 = 5; 3 + 2 = 5). The enzymes **transaldolase** and **transketolase** catalyze these interconversions of sugars (see Figure 10.6; Figure 12.3). These interconversion reactions also occur in many heterotrophic organisms in conjunction with the hexose monophosphate shunt (HMP) and supply an organism with ribulose-5-phosphate and erythrose-4-phosphate for biosynthetic purposes. The transaldolase and transketolase are also of importance in the biodegradation of various sugars (see Figure 10.6; Figure 12.3).

The Calvin cycle functions as a cycle because the products of CO_2 fixation, two molecules of 3-phosphoglycerate, can be rearranged to generate ribulose-1,5-bisphosphate. The enzyme Rubisco incorporates a CO_2 into the ribulose-1,5-bisphosphate, and thus the cycle continues.

CO_2 Fixation: The Reductive Citric Acid Pathway

Certain green sulfur photosynthetic microorganisms, the sulfate-reducing bacteria, and thermophilic hydrogen-oxidizing bacteria utilize a reductive carboxylation mechanism for CO_2 fixation known as the **reductive citric acid pathway** (Figure 10.2A). This pathway is essentially a reversal of the TCA cycle (discussed later). **Oxaloacetate** is a principal intermediate in this cycle, and one complete turn essentially results in the net production of one molecule of acetate. Several key enzymes of this cycle (ATP-dependent citrate lyase, and α-ketoglutarate synthase) differ from the oxidative cycle and are thus **diagnostic enzymes**—that is, their presence in a microbe indicates that it probably uses this cycle to fix carbon.

Acetyl-CoA can enter into the "synthetic TCA cycle" and provide the organism with the needed precursor metabolites. Decarboxylation of oxaloacetate would provide the organism with pyruvate. Gluconeogenesis then leads to formation of glucose from pyruvate (2 pyruvate → → → glucose (described later).

In certain *Chorobium* species, CO_2 fixation occurs by a slightly different route, whereby pyruvate is made by carboxylation of acetyl-CoA by the enzyme pyruvate oxidoreductase in a ferredoxin-dependent step. Oxaloacetate is then formed in a second carboxylation step (as in gluconeogenesis, described later).

CO_2 Fixation: The Hydroxypropionate Pathway

Members of genus *Chloroflexus* have a unique pathway for CO_2 fixation called the **hydroxypropionate pathway** (Figure 10.2B). Hydroxypropionyl-CoA and propionyl-CoA are key intermediates; one cycle leads to the formation of one molecule of glyoxylate via additional CoA

reactions. Glyoxylate can then enter major cycles via malate and the TCA cycle as described later.

CO_2 Fixation: The Reductive Acetyl-CoA Pathway

A number of organisms generate acetate as a major product of anaerobic respiration and do so by the **reductive acetyl-CoA pathway**, also known as the Wood–Ljungdahl pathway. These organisms are called **acetogenic bacteria**. They use H_2 as a reductant to couple two molecules of CO_2 in forming acetate. Among these organisms is *Acetogenium kivui*, a chemolithotrophic anaerobe that obtains carbon and energy according to the following equation:

$$4 H_2 + 2 CO_2 \rightarrow CH_3COO^- + H^+ + 2 H_2O$$

A general scheme for the energetics and pathway of acetate synthesis in acetogenic bacteria, as well as in some sulfate-reducing bacteria, is presented in Figure 10.3. A key enzyme complex in the acetogenic pathway is **acetyl-CoA synthase/CO dehydrogenase**. A molecule of CO_2 is reduced to carbon monoxide by this nickel-containing enzyme complex, followed by the addition of a methyl group. The methyl group is donated by the vitamin B_{12}–containing corrinoid enzyme and originates from the stepwise reduction of CO_2 in a distinct series of tetrahydrofolate–protein reactions. The methyl group then couples to a molecule of CO attached to the acetyl-CoA synthase/CO-dehydrogenase enzyme, forming the acetyl derivative. A molecule of coenzyme A is attached to the complex, and the enzyme complex catalyzes the release of acetyl-CoA. The three compounds—a molecule of coenzyme A (CoA), the methyl group, and CO—are independently bound to the enzyme. The molecule of acetate formed in this reaction series can be utilized in the synthesis of cellular components. The acetogenic bacteria are considered autotrophs because they synthesize their cellular material from CO_2 and utilize H_2 as source of energy. In methanogens the cofactors F_{420} and tetrahydromethanopterin replace ferredoxin and tetrahydrofolate, respectively.

C-1 Fixation: The Glycine-Serine and Ribulose Monophosphate Pathways

Some microorganisms can grow aerobically with the C-1 compound methane as the sole source of carbon and energy. Likewise, methanol is utilized by many types of bacteria and yeasts. The methane-utilizing microbes have the enzyme methane monooxygenase (see Chapter 19) and the methanol utilizers do not. Both methane and methanol are oxidized to the formaldehyde (HCHO)

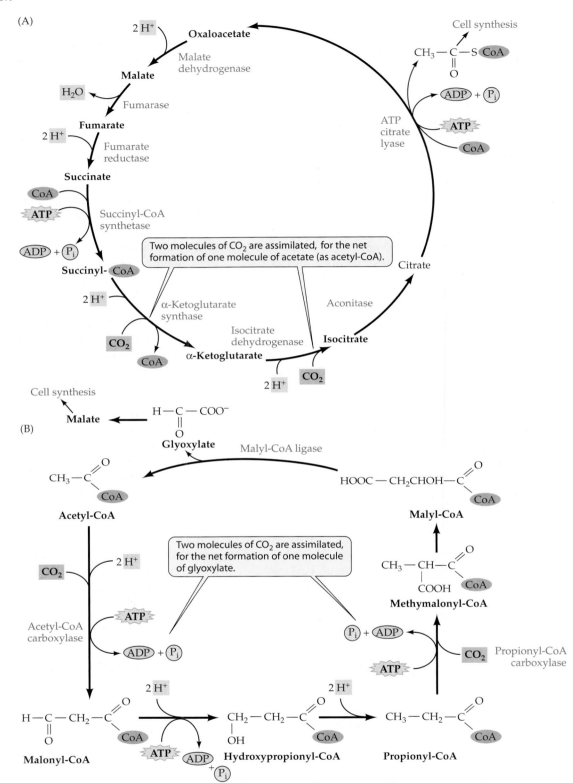

Figure 10.2 Two novel autotrophic pathways for CO$_2$ fixation

(A) The reductive citric acid pathway, which is present in green sulfur bacteria (e.g., *Chlorobium limicola*), thermophilic hydrogen-oxidizing bacteria (e.g., *Hydrogenobacter thermophilus*), and some of the sulfate-reducing bacteria (e.g., *Desulfobacter hydrogenophilus*). Note the similarity to the TCA cycle (see Figure 10.5). However, whereas the aerobic TCA cycle is a way of extracting energy from acetate by oxidizing it to CO$_2$, the reductive cycle shown here is an *anaerobic* pathway for *fixing* CO$_2$. (B) The hydroxypropionate pathway, present in *Chloroflexus*. Not all reactions are indicated.

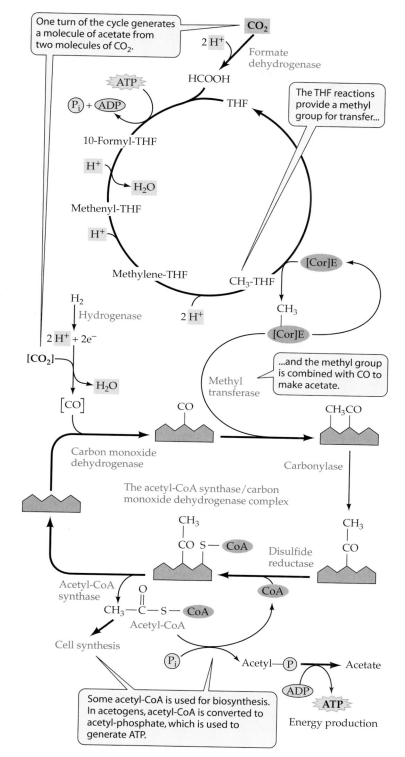

Figure 10.3 Pathway of CO$_2$ fixation in acetogenic bacteria
This pathway is present in homoacetogenic bacteria (e.g., *Clostridium thermoaceticum*), most of the sulfate-reducing bacteria (e.g., *Desulfobacterium autotrophicum*), and selected methanogenic *Archaea* (e.g., *Methanosarcina barkeri*). THF, tetrahydrofolic acid; [Cor]E, vitamin B$_{12}$ corrinoid enzyme.

level, and this molecule can be assimilated by two distinct pathways in different organisms (Figure 10.4):

1. Glycine + HCHO → serine

2. Ribulose-5-phosphate + HCHO → fructose-6-phosphate

In the **glycine-serine pathway** (see Figure 10.4A), the amino acid glycine serves as the acceptor of the 1-carbon intermediate. Energy is derived aerobically and occurs by the oxidation of CH$_4$ or CH$_3$OH to CO$_2$. In the cycle illustrated, one mole of formaldehyde via a methylene tetrahydrofolate intermediate (see Figure 10.3) and one mole of CO$_2$ would be assimilated in each turn to generate one molecule of acetyl-CoA. This product is then used to produce other intermediates for cell synthesis.

The other distinct group of methane-utilizing microorganisms assimilates the formaldehyde via a **ribulose monophosphate pathway** (see Figure 10.4B). This is somewhat similar to the ribulose-1,5-bisphosphate pathway operating in the autotrophs. Both employ a 5-carbon sugar and ultimately generate fructose-6-phosphate, which can be rearranged by the enzymes transaldolase and transketolase to regenerate ribulose-5-phosphate. Fructose-6-phosphate and modifications of the tricarboxylic acid cycle present in anaerobes would be a source of precursor metabolites for cell synthesis.

SECTION HIGHLIGHTS
Microorganisms have evolved six routes for incorporating C-1 compounds—CO$_2$ or formate—into cellular precursors needed for macromolecular synthesis. These alternative pathways share many common elements and involve a series of energy-requiring steps to form C—C and C—H bonds, the "bank" of energy storage by cells. Diagnostic enzymes present in cell types are indicative of the pathway(s) used in anabolic reactions.

10.3 Pathways that Generate Biosynthetic Precursors from Sugars

The precursor metabolites identified in Table 10.1 originate during catabolism of glucose and related sugars via four major metabolic pathways:

(A)

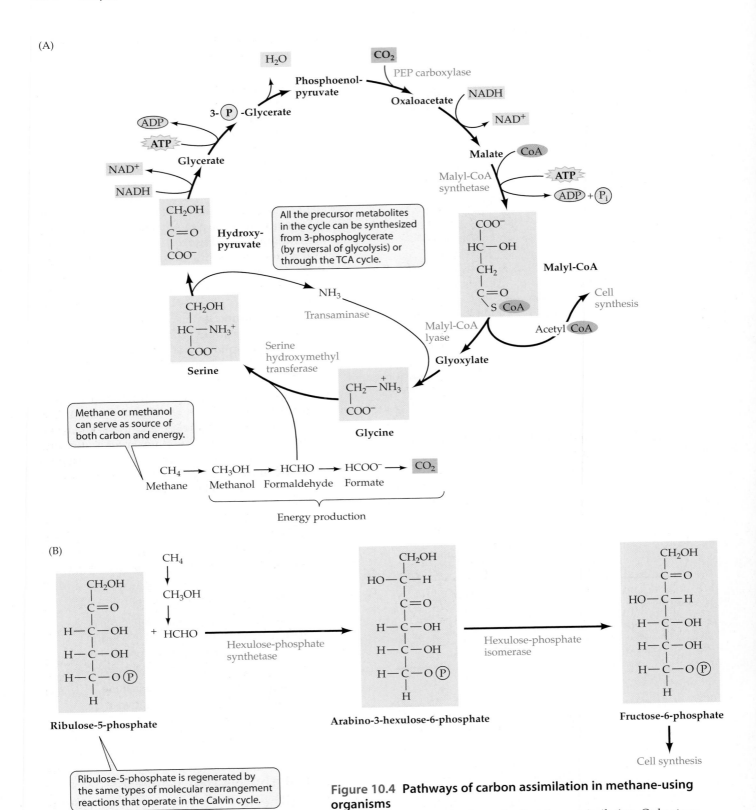

(B)

Figure 10.4 Pathways of carbon assimilation in methane-using organisms
(A) The serine-glycine pathway for 1-carbon assimilation. Only structures not previously depicted are shown (shaded), and only the enzymes specific to this pathway are named. (B) The ribulose-5-phosphate pathway for 1-carbon assimilation.

- The Embden-Meyerhof (EM) pathway, also termed glycolysis.
- The hexose monophosphate shunt (HMS).
- Some metabolites are also intermediates in the Entner-Doudoroff (ED) pathway.
- The tricarboxylic acid (TCA) cycle.

Many free-living microorganisms obtain precursor metabolites by using one or more of the above pathways or variations of them that yield equivalent products. Microorganisms that can utilize other low-molecular-weight compounds often direct the intermediates of their breakdown directly into the TCA cycle. However, considerable diversity exists with respect to what pathways are present in an organism and when they are utilized.

The **Embden-Meyerhof (EM) pathway**, also termed **glycolysis** (Figure 10.5), is the most commonly utilized pathway for glucose catabolism in microbes. It yields a net of two molecules of ATP plus two molecules of NADH per molecule of glucose utilized (see Chapter 8). Two molecules of ATP are required for the initial steps in the conversion of glucose to fructose-1,6-bisphosphate. The aldolase cleavage of the fructose 1,6-bisphosphate generates two molecules of triose (glyceraldehyde-3-phosphate). Esterification with inorganic phosphate yields two molecules of 1,3-bisphosphoglycerate. Each phosphate group (two per molecule of 1,3-bisphosphoglyceric acid) leads to the generation of a molecule of ATP by substrate-level phosphorylation, so four ATPs are produced from the two 1,2-bisphosphoglycerates. The oxidation of glyceraldehyde-3-phosphate generates NADH, which in turn is reoxidized during respiration or fermentation (see Chapter 8).

The **hexose monophosphate shunt** (HMS) and the associated **pentose phosphate** (PP) pathway reactions are important in microorganisms during growth on hexoses and pentoses (Figure 10.6). Glucose-6-phosphate is first converted to 6-phosphogluconate by a dehydrogenase. The decarboxylation of 6-phosphogluconate then provides the microorganism with the pentoses essential for the biosynthesis of nucleotides for RNA, DNA, and other cellular components. The hexose monophosphate shunt is also a ready source of NADPH, the favored nucleotide cofactor in biosynthetic reactions. Ribulose-5-phosphate is converted to ribose-5-phosphate and xylose-5-phosphate by an isomerase and epimerase, respectively. The associated oxidative pentose phosphate cycle then serves to interconvert these C5 sugars to C7, C3, C5, C4, and C6 sugars as they are needed by the cell. Note that **transaldolase** and **transketolase** play a key role in these interconversions. Generally, the HMS is a secondary pathway operative in organisms that utilize the EM or ED pathway as the major dissimilatory path-

way for glucose catabolism. Note that all reactions following ribulose-5-phosphate are energy-neutral. Pyruvate may be formed in a subsequent reaction.

The **Entner-Doudoroff (ED) pathway** yields one ATP with the formation of two molecules of pyruvate (Figure 10.7). One molecule each of NADPH and NADH is also formed per molecule of glucose. The glyceraldehyde-3-phosphate is converted to pyruvate by the reactions shown in Figure 10.5. This pathway is present in *Pseudomonas*, *Azotobacter*, and various gram-negative genera, but it does not occur to any great extent in gram-positive or anaerobic bacteria. The ED pathway is of greatest utility in microorganisms that use gluconate, mannonate, or hexuronate as substrate; some of these microorganisms use it exclusively, as they lack alternative routes. Other organisms such as *Escherichia coli* employ glycolysis for glucose metabolism; but when gluconate is provided as substrate, the key enzymes of the ED pathway are induced and utilized. Why?

Role of Pyruvate in the Synthesis of Precursor Metabolites

Pyruvate is a product of the three pathways of glucose catabolism—EM, HMS/PP, and ED (see Table 10.1). Aerobically, a microorganism can use pyruvate to generate energy and precursor metabolites. In aerobes, pyruvate dehydrogenase is the predominant enzyme used; it generates reduced purine nucleotide (NADH), acetyl-CoA and CO_2 (Table 10.2). Following the decarboxylation of pyruvate, the acetyl-CoA generated enters the tricarboxylic acid cycle by condensing with a molecule of oxaloacetate, where it is further oxidized to CO_2 and water. The TCA cycle also supplies the organism with precursor metabolites: α-ketoglutarate, oxaloacetate, and succinyl-CoA. The balance between catabolism (energy generation) versus anabolism (precursor synthesis) depends on the composition of the growth medium and other factors. Under special conditions, some aerobes, including *E. coli*, make an alternative enzyme, pyruvate oxidase, which decarboxylates pyruvate to acetate and CO_2 with the production of a molecule of reduced quinone (ubiquinol). These different end products affect the growth rate and efficiency of cells.

Anaerobically, many organisms utilize pyruvate or products of pyruvate catabolism as a terminal electron acceptor during fermentation and as a source of precursor metabolites. Some anaerobes possess the enzyme pyruvate-formate lyase, which decarboxylates pyruvate to give formate and acetyl-CoA (see Table 10.2). Formate is then excreted as a waste product or recycled as an electron donor. Other anaerobes employ pyruvate ferredoxin oxidoreductase to decarboxylate pyruvate to acetyl-CoA and reduce ferredoxin. **Ferredoxin (Fd)**, a

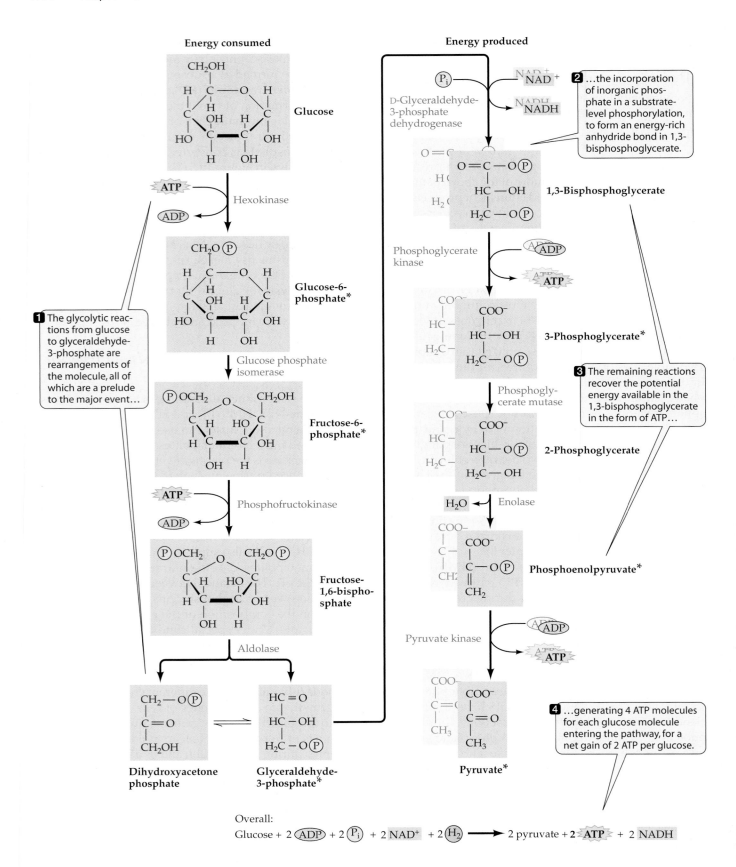

Energy consumed

Glucose

ATP

Hexokinase

ADP

1 The glycolytic reactions from glucose to glyceraldehyde-3-phosphate are rearrangements of the molecule, all of which are a prelude to the major event…

Glucose-6-phosphate*

Glucose phosphate isomerase

Fructose-6-phosphate*

ATP

Phosphofructokinase

ADP

Fructose-1,6-bisphosphate

Aldolase

Dihydroxyacetone phosphate

Glyceraldehyde-3-phosphate*

Energy produced

P_i

NAD⁺

D-Glyceraldehyde-3-phosphate dehydrogenase

NADH

2 …the incorporation of inorganic phosphate in a substrate-level phosphorylation, to form an energy-rich anhydride bond in 1,3-bisphosphoglycerate.

1,3-Bisphosphoglycerate

Phosphoglycerate kinase

ADP

ATP

3-Phosphoglycerate*

Phosphoglycerate mutase

3 The remaining reactions recover the potential energy available in the 1,3-bisphosphoglycerate in the form of ATP…

2-Phosphoglycerate

H_2O Enolase

Phosphoenolpyruvate*

Pyruvate kinase

ADP

ATP

4 …generating 4 ATP molecules for each glucose molecule entering the pathway, for a net gain of 2 ATP per glucose.

Pyruvate*

Overall:

Glucose + 2 ADP + 2 P_i + 2 NAD⁺ + 2 H_2 ⟶ 2 pyruvate + **2** ATP + 2 NADH

◀ **Figure 10.5 Embden-Meyerhof (EM) pathway, or glycolysis**
Glycolysis is the major pathway of glucose metabolism and ATP synthesis in anaerobic organisms and the first stage of glucose metabolism in aerobic organisms. This
pathway also produces a number of precursor metabolites (indicated by an asterisk*). Note that a single molecule of glucose yields two molecules each (shadow
boxes) of the energy-producing 3-carbon derivatives.

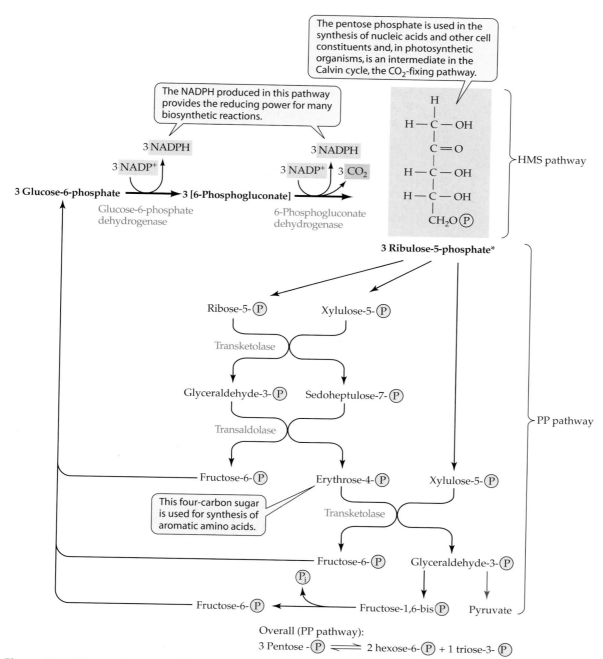

**Figure 10.6 Hexose monophosphate shunt (HMS) and the associated oxida-
tive pentose phosphate (PP) pathway**
The hexose monophosphate shunt is a major source of pentose sugars and
NADPH. These C5 units may then be interconverted to additional C7, C6, C4, and
C3 sugars for cell needs.

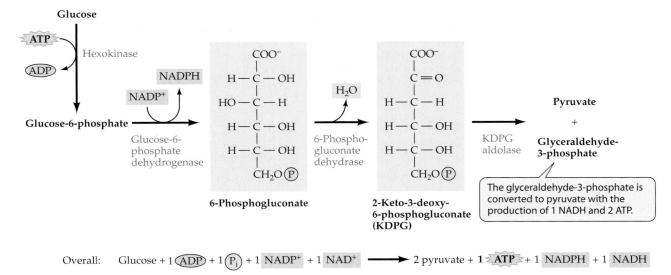

Figure 10.7 Entner-Doudoroff (ED) pathway
This pathway of glucose metabolism is common among the pseudomonads and some other gram-negative bacteria. The catabolism of one molecule of glucose to two molecules of pyruvate produces only 1 ATP (not 2 ATP as in glycolysis).

small iron-sulfur protein, then passes electrons on to a hydrogenase to make H_2 gas. Alternatively, anaerobic yeasts decarboxylate pyruvate to acetaldehyde, which is subsequently reduced to ethanol and excreted as a waste product. Although some microbes possess only one type of pyruvate-utilizing enzyme, others have several types and produce them under alternative conditions of cell growth. Why?

The Tricarboxylic Acid Cycle

The reactions involved in the tricarboxylic acid (TCA) cycle are presented in Figure 10.8. One turn of the cycle —also called the Krebs cycle or citric acid cycle—can re-

sult in the complete oxidation of one molecule of acetate, with the concomitant generation of 1 GTP (or 1 ATP), 3 NADH, 1 $FADH_2$, and 2 CO_2 molecules. Some microbes may generate a molecule of NADPH in place of NADH: *E. coli*, for example.

Aerobes can extract the maximum amount of energy from NADH oxidation by respiration and generate about 12 molecules of ATP/acetate (see Chapter 8). Anabolic reactions withdraw intermediates from the TCA cycle: these include **α-ketoglutarate** (to glutamic acid and other amino acids), **oxaloacetate** (to aspartic acid and other amino acids), and **succinyl-CoA** (to porphyrin, lysine, and methionine). These reactions deplete the 4-carbon intermediate (i.e., oxaloacetate) that combines with acetate

TABLE 10.2	Fate of pyruvate	
Condition Made	**Enzymatic Reaction**	**Example**
+O_2	Pyruvate dehydrogenase (PDH) pyruvate + CoA + NAD^+→acetyl-CoA + CO_2 + NADH	*E. coli*
+O_2	Pyruvate oxidase (POX) pyruvate + ubiquione-8→ acetate + CO_2 + ubiquionol-8	*E. coli*
–O_2	Pyruvate formate lyase (PFL) pyruvate + CoA→ acetyl-CoA + formate	*E. coli*
–O_2	Pyruvate ferredoxin oxidoreductase (PFOR) pyruvate + CoA + Fd(ox)→ acetyl-CoA + CO_2 + Fd(red)	*Clostridium* spp.
–O_2	Pyruvate decarboxylase (PDC) pyruvate + H^+→ acetaldehyde + CO_2	Brewer's yeast *Zymomonas* spp.

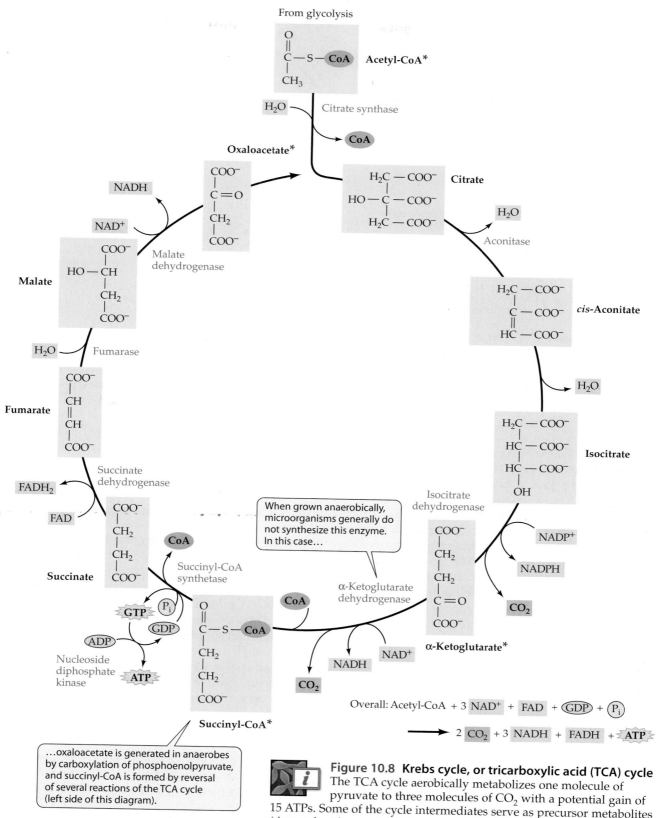

Figure 10.8 Krebs cycle, or tricarboxylic acid (TCA) cycle
The TCA cycle aerobically metabolizes one molecule of pyruvate to three molecules of CO_2 with a potential gain of 15 ATPs. Some of the cycle intermediates serve as precursor metabolites (denoted with an asterisk*).

From glycolysis

Acetyl-CoA*

H_2O — Citrate synthase

Oxaloacetate*

NADH

NAD^+

Malate dehydrogenase

Malate

H_2O — Fumarase

Fumarate

$FADH_2$

FAD

Succinate dehydrogenase

Succinate

Succinyl-CoA synthetase

GTP — P_i

GDP

ADP

ATP

Nucleoside diphosphate kinase

Succinyl-CoA*

Citrate

H_2O — Aconitase

cis-Aconitate

H_2O

Isocitrate

Isocitrate dehydrogenase

$NADP^+$

NADPH

CO_2

α-Ketoglutarate*

α-Ketoglutarate dehydrogenase

CoA

NADH — NAD^+

CO_2

When grown anaerobically, microorganisms generally do not synthesize this enzyme. In this case…

…oxaloacetate is generated in anaerobes by carboxylation of phosphoenolpyruvate, and succinyl-CoA is formed by reversal of several reactions of the TCA cycle (left side of this diagram).

Overall: Acetyl-CoA + 3 NAD^+ + FAD + GDP + P_i

⟶ 2 CO_2 + 3 NADH + $FADH$ + ATP

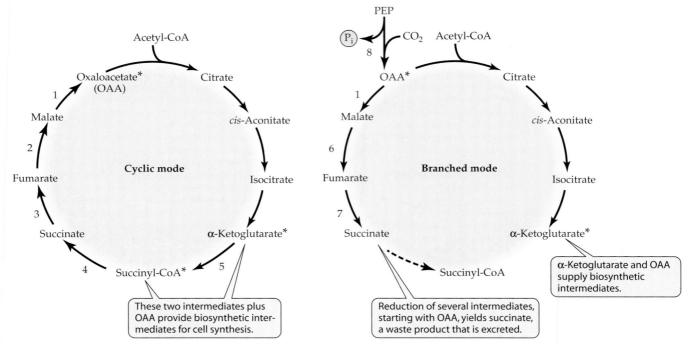

Figure 10.9 Branched TCA cycle
During anaerobic conditions in some organisms, the TCA cycle operates in a branched mode (right) to provide for the biosynthetic intermediates needed by the cell (oxaloacetate, α-ketoglutarate). The waste product, succinate is generated from malate by the malate dehydrogenase (1) operating in the reverse direction, plus a distinct fumarase enzyme (6) and fumarate reductase (7). Synthesis of three of the TCA cycle enzymes—succinate dehydrogenase (3), α-ketoglutarate dehydrogenase (4), and succinyl-CoA synthetase (5)—is suppressed. OAA is formed from PEP and CO_2 by PEP carboxylase (8).

to form citrate. For the TCA cycle to continue, the oxaloacetate must be replenished. This can be achieved in most organisms by the carboxylation of pyruvate or phosphoenolpyruvate to form a molecule of oxaloacetate (Figure 10.9). Any reaction that replenishes intermediates in a central pathway such as this is called an **anaplerotic reaction**.

In anaerobes, a modified or "branched" tricarboxylic acid cycle is employed to provide biosynthetic intermediates and as a means to dispose of reducing potential in the form of succinate (see Figure 10.9). The enzyme α-ketoglutarate dehydrogenase is not generally synthesized in anaerobes; where necessary, succinyl-CoA is formed in a pathway starting with the reduction of oxaloacetate. In a microorganism growing anaerobically, oxaloacetate can be synthesized by the carboxylation of phosphoenolpyruvate.

The Glyoxylate Shunt

The **glyoxylate shunt** is a variation of the TCA cycle; it is of considerable consequence to organisms growing on acetate, fatty acids, long-chain *n*-alkanes, or other substrates that are catabolized through 2-carbon intermediates. As anabolic reactions remove intermediates from

the TCA cycle, it is essential that organisms growing on 2-carbon substrates have a mechanism for replenishing substantial quantities of **oxaloacetate** for continued operation of the cycle. Since these microorganisms do not have a 3-carbon intermediate (such as phosphoenolpyruvate) to carboxylate to form oxaloacetate, they use the inducible glyoxylate shunt (Figure 10.10). In these shunt reactions, isocitrate is cleaved by the inducible enzyme **isocitrate lyase**, yielding succinate and glyoxylate. The glyoxylate thus formed can condense with a molecule of acetyl-CoA to form a molecule of malate via malate synthase. Thus, an organism growing on acetate would have an unlimited supply of 4-carbon intermediates to continue the TCA cycle. Precursor TCA metabolites such as α-ketoglutarate and succinyl-CoA would then be made available. Note that this process neither requires nor produces energy in the form of ATP or NADH/NADPH.

Biosynthesis of Glucose

When cells are grown in the absence of 6-carbon sugars, they must biosynthesize them for incorporation into cell components (peptidoglycan, LPS, polysaccharide capsule). Precursor molecules—for example, acetate or amino

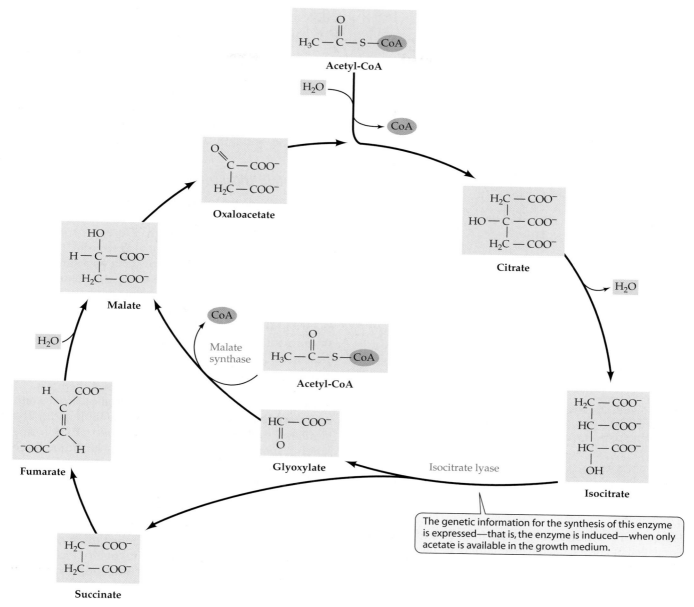

Figure 10.10 Glyoxylate cycle
The glyoxylate cycle is induced in microorganisms grown on acetate or other substrates (e.g., long-chain fatty acids) metabolized through a 2-carbon intermediate. Note the number of common intermediates in this and the TCA cycle (see Figure 10.5).

The genetic information for the synthesis of this enzyme is expressed—that is, the enzyme is induced—when only acetate is available in the growth medium.

acids—are first converted to oxaloacetate, the TCA cycle intermediate. Oxaloacetate is then decarboxylated to phosphoenolpyruvate by the enzyme PEP carboxykinase. If oxaloacetate is unavailable, pyruvate may be converted to oxaloacetate in the ATP-dependent addition of CO_2 by the enzyme pyruvate carboxylase. In some organisms pyruvate may be converted to PEP in an ATP-dependent reaction catalyzed by PEP synthase.

PEP carboxykinase, together with six of the glycolytic pathway enzymes (see Figure 10.5) running in the reverse direction, accomplishes the anabolic production of glucose-6-phosphate by the process known as **gluconeogenesis** (Figure 10.11). A distinct phosphatase enzyme removes phosphate from fructose-1,6-bisphosphate to give fructose-6-phosphate. A subsequent conversion step yields glucose-6-phosphate, a precursor molecule for the synthesis of polysaccharides, LPS, and other cell constituents. Because glucose is not directly involved in these reactions, little is actually made by the cell.

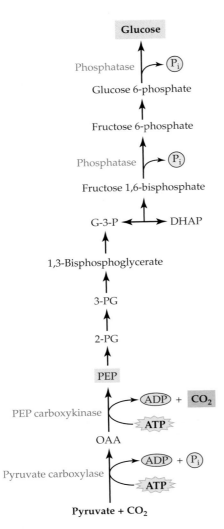

Figure 10.11 Synthesis of glucose from C3 and C4 units
Formation of glucose intermediates from PEP proceeds by
the reversal of reactions of the Embden-Meyerhof pathway,
with the exception that several phosphatase enzymes are
used to remove phosphate from fructose-1,6-bisphosphate
and glucose-6-phosphate. Thus, ATP is not generated by
the reversal of the glycolytic enzymes (see Figure 10.2).

SECTION HIGHLIGHTS
Heterotrophic microorganisms obtain the 12
precursor metabolites by one or more of the
following pathways: the Embden-Meyerhof,
Entner-Doudoroff, hexose monophosphate
shunt and associated oxidative pentose phosphate cycle, and TCA cycle. Additional pathways, including that for gluconeogenesis, permit the synthesis of biosynthetic precursors
from intermediates derived from CO_2 fixation.

10.4 Biosynthesis of Amino Acids, Purines, Pyrimidines, and Lipids

Proteins, nucleic acids, and lipids are the major constituents of a cell; their synthesis therefore requires significant production of precursors. This section briefly
addresses the origin of these precursors and outlines the
general strategy for formation of amino acid, purine,
pyrimidine, and fatty acid monomers.

Amino Acid Biosynthesis

Proteins provide the cell with structure, motility, and enzymatic capacity. One gram of dry bacterial cell mass
contains 500 to 600 mg of protein (50% to 60% of the cell
dry weight). This protein is synthesized from 21 different amino acids. An organism growing on a simple carbon substrate, whether glucose or CO_2, expends a considerable portion of its energy and biosynthetic capacity
on the synthesis of the monomeric amino acids. Much
energy is also spent organizing and assembling these
amino acids into functional proteins. The ability to synthesize many of the requisite amino acids is a trait shared
by most of the free-living *Bacteria* and *Archaea*. Obviously, the strict autotrophs can synthesize them all.

Amino acids are synthesized as "families" from a limited number of precursor metabolites (Table 10.3). The
families are the glutamate family, alanine family, serine
family, aspartate family, and aromatics. Histidine is the
sole product of its own biosynthetic pathway. The precursors include α-ketoglutarate, pyruvate, 3-phospho-
glycerate, and oxaloacetate (Figure 10.12). Selenocys-
teine, the twenty-first amino acid, is derived from serine
and is an essential constituent of certain respiratory enzymes, including formate dehydrogenase. A complete
step-by-step depiction of the biosynthetic pathways for
the amino acids from precursor metabolites may be
found in a biochemistry text.

Purine and Pyrimidine Biosynthesis

Nucleic acids are a constituent part of all living cells,
since they are the genetic determinants in all creatures
from simple, acellular viruses to complex mammals. Before the advent of life, conditions on the primeval Earth
included gaseous mixtures of hydrocarbon, hydrogen,
and ammonia as well as energy from ultraviolet radiation and lightning; a combination that can lead to the
formation of purines and pyrimidines (see Chapter 1).
Because the nucleic acid bases can be formed abiotically,
they were apparently available early in the evolution
of viable organisms. Polymerization of these bases created a structure that evolved as the key inheritable element in all life forms. Nucleic acid bases are also com-

Family	Precursor Metabolite	Amino Acid
TABLE 10.3 Precursors of the major amino acids in protein and their family designations		
Glutamate	α-Ketoglutarate	Glutamate Glutamine Proline Arginine
Alanine	Pyruvate	Alanine Valine Leucine
Serine	3-Phosphoglycerate	Serine Glycine Cysteine
Aspartate	Oxaloacetate	Aspartate Asparagine Methionine Lysine Threonine Isoleucine
Aromatics	Phosphoenolpyruvate + erythrose-4-phosphate	Phenylalanine Tyrosine Tryptophan
Histidine	5-Phosphoribosyl-1-pyrophosphate	Histidine

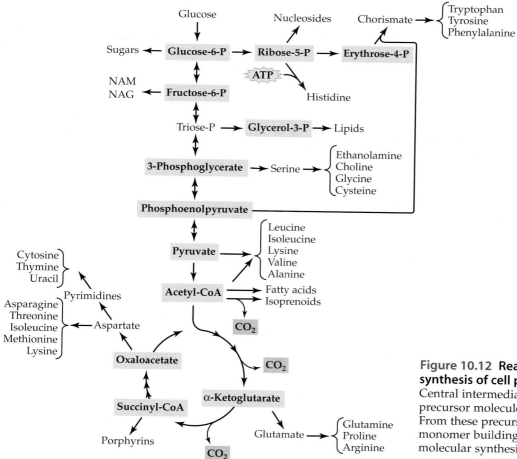

Figure 10.12 Reactions that lead to synthesis of cell precursors and monomers Central intermediates that constitute the 12 precursor molecules are indicated in bold. From these precursors, the cell forms the monomer building blocks employed in macromolecular synthesis.

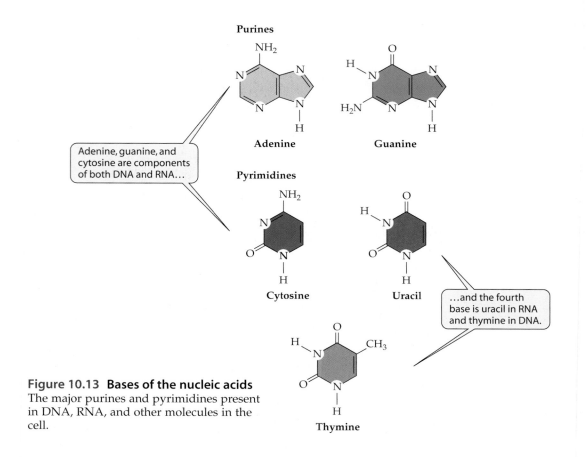

Purines

Adenine

Guanine

Adenine, guanine, and cytosine are components of both DNA and RNA...

Pyrimidines

Cytosine

Uracil

...and the fourth base is uracil in RNA and thymine in DNA.

Thymine

Figure 10.13 Bases of the nucleic acids
The major purines and pyrimidines present in DNA, RNA, and other molecules in the cell.

ponent parts of ATP, NAD$^+$, NADP$^+$, FAD$^+$, and some of the B vitamins.

The structures of the bases generally present in nucleic acids are presented in Figure 10.13. The purines adenine and guanine are constituents of both RNA and DNA. The two pyrimidines that are constituents of DNA are cytosine and thymine. Uracil replaces thymine in RNA. Recall that roughly 20% of the cell dry weight is composed of nucleic acids of one type or another. Pre-

cursors needed for purine synthesis include aspartate, glutamine, glycine, and CO_2 (see Figure 10.12). The origin of each molecule of the nine-member purine ring system is presented in Figure 10.14. Other modified bases are present in many tRNAs.

The biosynthesis of the pyrimidine molecule occurs by a more direct route than that for purines. Carbamoyl phosphate condenses with a molecule of aspartate to form the six-member ring (Figure 10.15). A detailed depiction of the biosynthetic reactions involved in the synthesis of these molecules may be found in a biochemistry textbook.

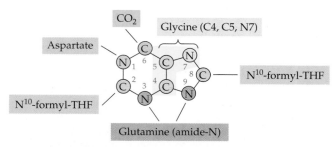

CO_2

Glycine (C4, C5, N7)

Aspartate

N^{10}-formyl-THF

N^{10}-formyl-THF

Glutamine (amide-N)

Figure 10.14 Origin of the nine atoms in the purine ring
The major contributor to the purine ring of adenine and guanine is the amino acid glycine.

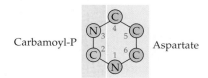

Carbamoyl-P

Aspartate

Figure 10.15 Origin of the six atoms in the pyrimidine ring
Aspartate is the main precursor of the pyrimidine ring of cytosine, thymine, and uracil.

Biosynthesis of Lipids

Phospholipids are a major constituent of the cytoplasmic membrane and outer membranes of bacterial cells and may comprise 9% or more of the dry weight (see Chapter 4). The major components of these phospholipids are the long-chain fatty acids that form the hydrophobic core of the cytoplasmic membrane (see Figure 4.45). The interior of the cytoplasmic membrane in archaeal species is made up of compounds of ether-linked long-branched hydrocarbon chains (see Figure 4.47). Both the long-chain fatty acids in bacterial cells and the branched chains of archaeal cells are synthesized by condensation of acetate via mechanisms discussed here. Precursors may include acetyl-CoA, 3-phosphoglycerate, glycerol-3-phosphate, and CO_2 (see Figure 10.12).

Fatty Acids

Bacteria generally synthesize long-chain fatty acids by the condensation of a molecule of acetyl-CoA with one of malonyl-CoA and NADPH as reductant (Figure 10.16). The malonyl-CoA is formed by carboxylation of acetyl-CoA by an ATP-driven reaction. The malonyl-CoA thus formed reacts with a protein, called the **acyl carrier protein** (**ACP**), to form the malonyl-ACP derivative. ACP is an important low-molecular-weight acidic protein of about 9,000 daltons, which is functionally equivalent to coenzyme A but activates and transfers fatty acids having chain lengths longer than those of the lower-molecular-weight groups involved with coenzyme A.

A molecule of the acetyl-ACP reacts with the enzyme β-keto-ACP synthase to form **keto-acyl-ACP synthase** (**KAS**), a derivative enzyme. The acetyl-KAS reacts with malonyl-ACP to form acetoacetyl-ACP. The decarboxylation of malonyl-ACP provides energy for the condensation of the acetyl-KAS and the methylene carbon of the malonyl-ACP molecule. Acetoacetyl is reduced stepwise to the saturated fatty acid. This fatty acid remains attached to ACP. Repetition of these reactions lengthens the chain, two carbons at a time, to the required length. Eight independent enzymatic reactions are involved in fatty acid synthesis (each involves a different protein).

The unsaturated fatty acids in bacteria generally have one double bond (monounsaturated). Those present in eukaryotic organisms commonly have more than one double bond (polyunsaturated). The cyanobacteria are the only members of the *Bacteria* that have more than a single double bond in their cellular lipids. Many bacteria can directly incorporate exogenously supplied polyunsaturated fatty acids into their cellular lipids; these apparently are not harmful. Direct incorporation bypasses the need to synthesize the molecule de novo, thus conserving energy for other purposes.

Biosynthesis of the monounsaturated fatty acids can occur by either the aerobic or the anaerobic pathway. In the anaerobic pathway, the double bond is introduced between carbons 9 and 10 during fatty acid synthesis. This occurs in strict anaerobes and may occur in various facultative aerobes, including *E. coli*, during aerobic growth. The aerobic pathway is utilized by micrococci, endospore-forming bacilli, some cyanobacteria, fungi, and animals. In the aerobic pathway, the double bond is introduced after synthesis between carbons 9 and 10 of the long-chain fatty acid.

There is considerable variety among the fatty acids present in the major lipids of bacteria (Figure 10.17). The chain length varies from C14 to C18, depending on the cell-growth temperature. They may be straight or branched, saturated or unsaturated, or they may have a cyclopropane or hydroxy constituent. Except in cyanobacteria, they rarely have more than one double bond. The hydroxy fatty acids that are constituents of the outer envelope of gram-negative bacteria are synthesized by the anaerobic pathway. This structural diversity is sometimes employed to identify certain species by their "signature" molecules through <u>f</u>atty <u>a</u>cid <u>m</u>ethyl <u>e</u>ster analysis (FAME).

Branched-Chain Lipids

The carotenoids and related long-chain branched hydrocarbons are common constituents of many bacterial and archaeal species. They are composed of isoprene subunits:

$$\left[-CH_2-\underset{\underset{CH_3}{|}}{C}=CH-CH_2-\right]_n$$

These compounds, termed isoprenoids, are synthesized by condensation of these molecules of acetyl-CoA to form mevalonate (Figure 10.18). This intermediate is then converted to isopentenyl and prenyl units, which are polymerized into the isoprenoids. The yellow and red pigments of photosynthetic and other bacteria and the phytol chain that anchors chlorophyll to the photosynthetic membrane are composed of isoprene units. Isoprenoid molecules strung together are present in *Archaea* in place of fatty acids.

The phospholipids are assembled as outlined in Figure 10.19. The precursor metabolite, glycerol 3-phosphate, provides the backbone for polar lipid synthesis; the addition of ACP-conjugated fatty acids yields phosphatidic acid. This acid is a key intermediate in the synthesis of the various membrane lipids, including phosphatidylserine, phosphatidylethanolamine, and cardiolipin. The synthesis of membrane lipids in *Archaea* is discussed in Chapter 18.

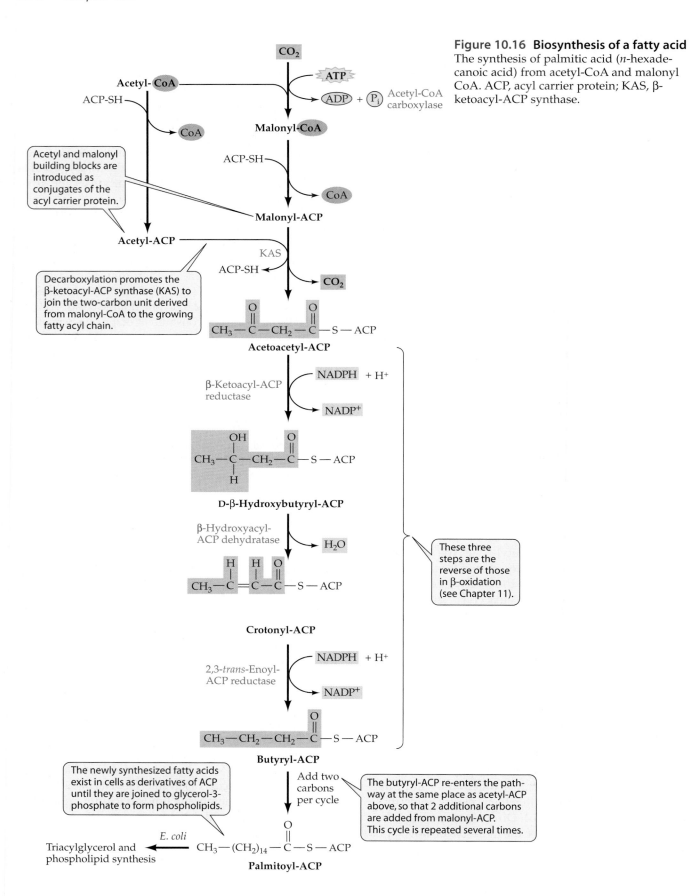

Figure 10.16 Biosynthesis of a fatty acid
The synthesis of palmitic acid (*n*-hexadecanoic acid) from acetyl-CoA and malonyl CoA. ACP, acyl carrier protein; KAS, β-ketoacyl-ACP synthase.

Acetyl and malonyl building blocks are introduced as conjugates of the acyl carrier protein.

Decarboxylation promotes the β-ketoacyl-ACP synthase (KAS) to join the two-carbon unit derived from malonyl-CoA to the growing fatty acyl chain.

These three steps are the reverse of those in β-oxidation (see Chapter 11).

The newly synthesized fatty acids exist in cells as derivatives of ACP until they are joined to glycerol-3-phosphate to form phospholipids.

The butyryl-ACP re-enters the pathway at the same place as acetyl-ACP above, so that 2 additional carbons are added from malonyl-ACP. This cycle is repeated several times.

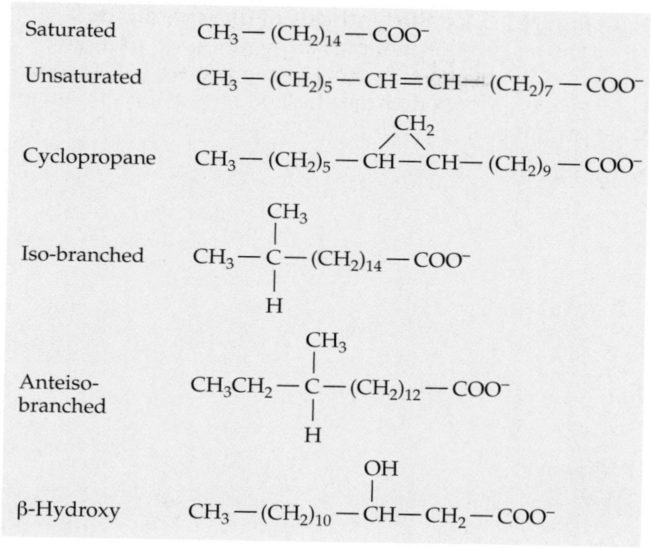

Saturated	$CH_3 - (CH_2)_{14} - COO^-$
Unsaturated	$CH_3 - (CH_2)_5 - CH = CH - (CH_2)_7 - COO^-$
Cyclopropane	$CH_3 - (CH_2)_5 - \overset{\overset{CH_2}{\diagup\diagdown}}{CH - CH} - (CH_2)_9 - COO^-$
Iso-branched	$CH_3 - \overset{\overset{CH_3}{\vert}}{\underset{\underset{H}{\vert}}{C}} - (CH_2)_{14} - COO^-$
Anteiso-branched	$CH_3CH_2 - \overset{\overset{CH_3}{\vert}}{\underset{\underset{H}{\vert}}{C}} - (CH_2)_{12} - COO^-$
β-Hydroxy	$CH_3 - (CH_2)_{10} - \overset{\overset{OH}{\vert}}{CH} - CH_2 - COO^-$

Figure 10.17 Fatty acids
Long-chain fatty acids that are commonly present in the phospholipids of bacteria.

SECTION HIGHLIGHTS
Biosynthesis of the monomer units (amino acids, purines, pyrimidines, and lipids) needed for macromolecular synthesis and cell assembly occurs from a limited number of precursor molecules. The amino acids, made in families, are derived primarily from intermediates of the glycolytic pathway and the TCA cycle. Lipids (fatty acids and branched chain lipids) are derived primarily from acetyl-CoA by a series of polymerization reactions. Lastly, the complex synthesis of pyrines and pyrimidines occurs from several amino acid precursors and carbamoyl phosphate.

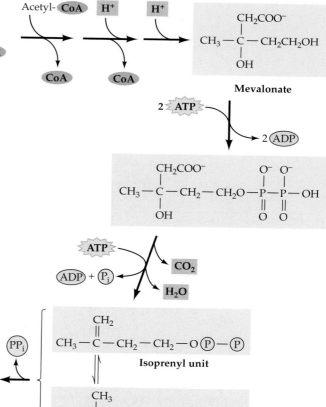

Figure 10.18 Biosynthesis of isoprenoids
The branched isoprenoids are commonly present in the membrane lipids of *Archaea* and in the carotenoids of photosynthetic and other pigmented bacteria.

The isoprenoids are branched compounds consisting of repeating isoprene units. These are formed by the condensation of acetate groups from acetyl-CoA via acetoacetyl-CoA and mevalonate.

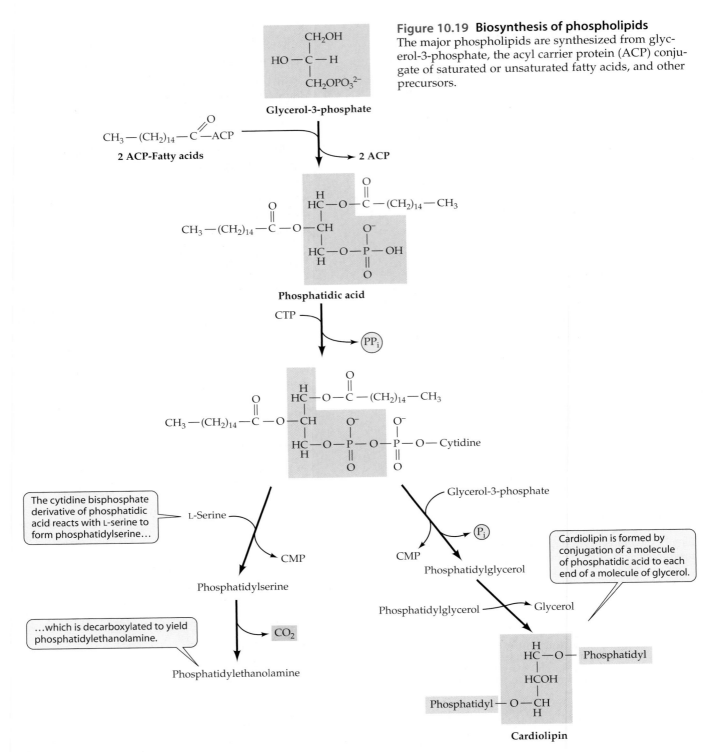

Figure 10.19 Biosynthesis of phospholipids
The major phospholipids are synthesized from glycerol-3-phosphate, the acyl carrier protein (ACP) conjugate of saturated or unsaturated fatty acids, and other precursors.

10.5 Nitrogen Assimilation

Nitrogen is a major constituent of all living cells and makes up about 14% of the dry weight of a prokaryote. The *Bacteria* and *Archaea* generally use inorganic nitrogen in the form of the ammonium ion (NH_4^+) or nitrate (NO_3^-), while a limited number of species obtain cellular nitrogen by reducing atmospheric dinitrogen (N_2). Many microbial species can also meet their nutritional requirement by uptake of one or more nitrogen-containing organic compounds from the environment (amino acids, purines, etc.). For routine bacterial growth, ammonium ion (NH_4^+) is commonly utilized (see Chapter 5), since the ability of these species to assimilate nitrate is somewhat more limited, while the ability to fix N_2 is highly restricted to certain prokaryotic species.

The general strategy for the acquisition of nitrogen by a cell (Figure 10.20) is based on minimizing the amount of energy expended to meet its biosynthetic needs. This energy cost is a combination of energy required to transport the molecule into the cell and, when necessary, to convert it to the form used for subsequent incorporation into cell material. Active transport of ammonia, nitrate, or organic nitrogen compounds typically requires one ATP or one proton per molecule of solute accumulated. Nitrogen gas is freely diffusible across the membrane and thus requires no energy input. However, nitrogen fixation is an energetically costly process relative to the reduction of nitrate to ammonia. Thus, if a cell is presented with more than one available nitrogen source, it acquires ammonia before reducing nitrate; in turn, it will fix N_2 only as a last resort. Note that some microbial species can use only one or two types of nitrogen compounds, as dictated by their genetic makeup.

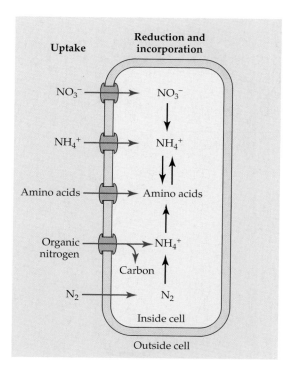

Figure 10.20 Cell acquisition of nitrogen
Cells may take up alternative nitrogen-containing compounds for cell biosynthesis. Nitrate is reduced by the assimilatory enzymes to ammonia, while N_2 gas is reduced by the nitrogenase enzyme system. Following their uptake, amino acids and other nitrogenous compounds may require deamination.

Nitrate Reduction

Two enzymatic systems are principally responsible for nitrate reduction by *Bacteria* and *Archaea*: **dissimilatory** and **assimilatory nitrate reduction** systems.

The **dissimilatory pathway** is induced in selected facultatively anaerobic bacteria during anaerobic respiratory growth in the presence of NO_3^-. Nitrate serves as the terminal electron acceptor and various gaseous intermediates are generated—including NO (nitric oxide) and N_2O (nitrous oxide)—leading to the formation of dinitrogen (N_2) in a series of membrane-associated reactions (see Chapter 8). This process is termed **denitrification** and can occur in agricultural soil that becomes waterlogged and anaerobic. Denitrification is a dissimilatory reaction, as the reduced product is not incorporated into the cell. This process provides an essential step in the Earth's nitrogen cycle (see Chapter 24).

The **assimilatory nitrate reduction pathway** is induced in those organisms that can utilize NO_3^- as the nitrogen source and then only in the absence of available NH_4^+. The pathway is composed of two soluble enzymes, an **assimilatory nitrate reductase (NAS)** and an **assimilatory nitrite reductase (NIS)**. The enzymatic reactions proceed as follows:

$$\underbrace{NO_3^- \xrightarrow{2e^-} NO_2^-}_{\text{Nitrate reductase (NAS)}}$$

$$\underbrace{NO_2^- \xrightarrow{2e^-} HNO \xrightarrow{2e^-} NH_2OH \xrightarrow{2e^-} NH_3}_{\text{Nitrite reductase (NIS)}}$$

The molybdenum cofactor-containing nitrate reductase system reduces the nitrogen in NO_3^-, with a valence of +5, to nitrite (NO_2^-, valence +3). A siroheme-containing nitrite reductase then reduces nitrite to ammonia (valence −3); no free intermediates are involved. The joint action of the two cytoplasmic assimilatory enzymes requires energy input in the form of NADPH and FADH. The end product of this reduction is then utilized in the synthesis of cellular constituents. Since considerable energy is expended in these reactions, NH_4^+ is the preferred inorganic nitrogen source.

Nitrogen Fixation

Dinitrogen (N_2) is the major constituent of the Earth's atmosphere (80%), and some bacterial and archaeal species have evolved with the capacity for **N_2 fixation** to ammonia. This N_2-fixing capacity is confined to the prokaryotic world, and these reactions ultimately provide a source of nitrogenous compounds for sustenance of all eukaryotes. Dinitrogen is a very unreactive molecule due to the triple bond N≡N; because of this unre-

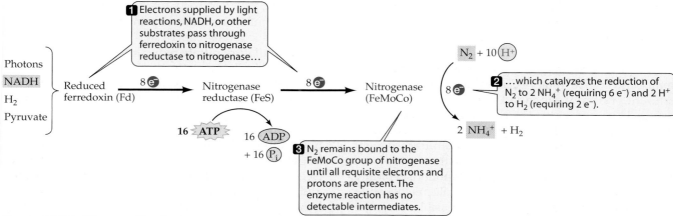

Figure 10.21 Nitrogen fixation
The primary electron donor in the nitrogenase reaction is reduced ferredoxin. When grown under iron-limited conditions, some bacteria can synthesize the flavoprotein flavodoxin, which is functionally equivalent to ferredoxin.

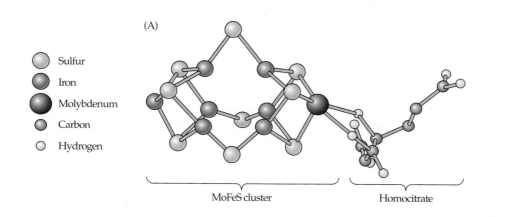

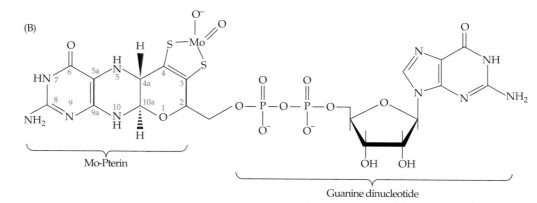

Figure 10.22 Structure of the molybdenum cofactors
(A) The nitrogenase cofactor located at the active site of the enzyme is Fe_7MoS_9 homocitrate, called FeMoCo. (B) The molybdenum containing cofactor of nitrate reductase, molybdopterin guanine dinucleotide.

active nature, considerable activation energy is required to reduce (fix) N_2 to NH_4^+.

Nitrogenase (Figure 10.21) is the enzyme complex involved in N_2 fixation (reduction). It is composed of two subunits: an iron-sulfur protein (FeS) and a molybdenum-iron-sulfur protein (FeMoCo). A total of eight electrons are sequentially transferred from a donor (for example, reduced ferredoxin) to the FeS protein and then to the FeMoCo protein, which contains a unique molybdenum-containing cofactor at the active site for N_2 reduction (Figure 10.22). Two molecules of ammonia are produced along with one molecule of hydrogen gas. The electrons required for N_2 reduction are provided by photosynthesis, respiration, or fermentation. A considerable expenditure of ATP (16 per mole N_2 fixed) is also necessary to drive electron transfer for N_2 reduction. Nitrogenase is inactivated by molecular oxygen (O_2) due to the lability (instability) of the Fe-Mo cofactor. Aerobic microbes have evolved cellular strategies to deal with this problem.

Nitrogen fixation occurs in a range of anaerobic, aerobic, and photosynthetic prokaryotes. All N_2 fixers that grow in the presence of air have a mechanism for keeping O_2 away from their nitrogenase system. For example, *Azotobacter vinelandii*, an obligate aerobic soil microbe, is somehow able to maintain low oxygen levels at the site of N_2 fixation. The O_2-producing cyanobacteria that fix N_2 develop a specialized anaerobic cell type (called a **heterocyst**) that becomes the site of nitrogen fixation within their multicellular chains (see Chapter 21). Many facultative anaerobes—such as *Bacillus polymyxa* and *Klebsiella oxytoca*—fix N_2 only when growing anaerobically. Strict anaerobes that fix N_2 do not routinely face this challenge.

In addition to the molybdenum-containing nitrogenase, two related nitrogenase enzymes exist in some bacterial and archaeal species. They contain either a vanadium or an iron atom in place of the molybdenum, but are otherwise similar in enzyme operation. The vanadium nitrogenase is synthesized when molybdenum is limiting in the cell environment. Likewise, the iron nitrogenase is made only when both molybdenum and vanadium are limiting: the regulatory basis is not understood. The ability to fix N_2 defines life in many environmental niches. A variety of interesting microbe–plant interactions have evolved that allow symbiotic nitrogen fixation (see Chapter 25).

Fixed nitrogen can often be limiting for the growth of many types of prokaryotes and plants. In the early twentieth century, Fritz Haber devised a chemical method for the synthesis of ammonia from atmospheric N_2. Named after him, this revolutionary process—the Haber-process—has resulted in increased food production through the application of man-made fertilizers worldwide. He was awarded a Nobel Prize in 1918 for this contribution.

Utilization of Nitrogenous Compounds

Many microbes accumulate amino acids and/or a variety of other nitrogenous organic molecules from the environment (see Chapter 4). These amino acids can be directly incorporated into cell material or converted to ammonia (see Figure 10.20). Other nitrogenous compounds are disassembled to ammonia and the carbon backbone, whereupon the ammonia is then incorporated into cell material. The remaining carbon is either utilized or discarded, depending on the capability of the organism.

Incorporation of Ammonia

Bacteria generally link NH_4^+ to organic intermediates via three major reactions. The enzymes mediating these reactions are glutamate dehydrogenase, glutamine synthetase, and alanine dehyrogenase.

The enzyme **glutamate dehydrogenase (GDH)** catalyzes the reductive amination of α-ketoglutarate to form glutamate (Figure 10.23). The reductant for this reaction is either NADH or NADPH. **Glutamine synthetase (GS)**

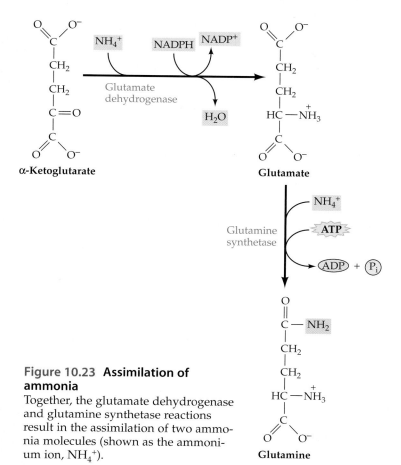

Figure 10.23 Assimilation of ammonia
Together, the glutamate dehydrogenase and glutamine synthetase reactions result in the assimilation of two ammonia molecules (shown as the ammonium ion, NH_4^+).

then catalyzes an ATP-dependent amidation of the γ-carboxyl of glutamate to form glutamine (see Figure 10.23). Glutamine is a major amine donor for biosynthesis of other amino acids, purines, and pyrimidines. Combined, the GDH and GS enzymes are responsible for most of the ammonium assimilated into prokaryotic cell material.

When a microorganism is grown in an environment with an ample supply of NH_4^+, the sequence of reactions is that shown in Figure 10.23. *Escherichia coli* grown on a mineral salts medium with ammonia as the sole nitrogen source, for example, derives about 75% of its nitrogen from glutamate and 25% from glutamine.

However, when the availability of NH_4^+ is limited, the glutamate dehydrogenase (GDH) reaction is not effective and glutamine synthetase (GS) becomes the major NH_4^+ assimilation reaction. This occurs because glutamine synthetase has a much higher affinity for ammonia. Because the glutamine synthetase reaction reduces the level of glutamate, there is a need for an alternate mechanism for glutamate production. The enzyme **glutamate:oxoglutarate aminotranferase (GOGAT)** carries out the reductive amination of α-ketoglutarate with the amide-N of glutamine as N-donor according to the following reaction:

$$NADPH + α\text{–ketoglutarate} + glutamine \xrightarrow{\text{GOGAT}}$$
$$2\,glutamate + NADP^+$$

Two molecules of glutamate are generated—one from amination of α-ketoglutarate and the other from the deamidation of glutamine. Together the two enzymes GS and GOGAT constitute a significant pathway for the assimilation of NH_4^+ when nitrogen is limiting:

$$2\,NH_4 + 2\,ATP + 2\,glutamate \xrightarrow{\text{GS}}$$
$$2\,glutamine + 2\,P_i$$

$$NADPH + α\text{–ketoglutarate} + glutamine \xrightarrow{\text{GOGAT}}$$
$$2\,glutamate + 2\,NADP^+$$

Sum:

$$2\,NH_4^+ + α\text{-ketoglutarate} + NADPH + 2\,ATP \rightarrow$$
$$glutamine + NADP^+ + 2\,P_i$$

The amino group can be transferred from glutamate to other keto acids via enzymes termed **aminotransferases** (Figure 10.24).

Distinct from the above processes, **alanine dehydrogenase** is used—notably during certain rhizobia–legume symbioses—to incorporate ammonia by the following reaction:

$$Pyruvate + NH_4^+ + NADPH + H^+ \xrightarrow{\text{Alanine dehydrogenase}}$$
$$alanine + NADP^+$$

Alanine is then secreted from the bacterium and supplied to the plant as a source of fixed nitrogen. The distribution of the ammonia incorporation reactions in other organisms is not well known.

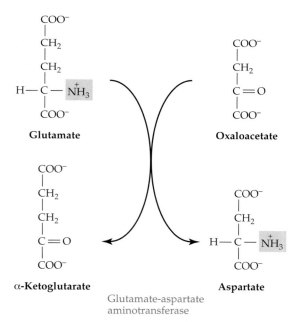

Figure 10.24 Transamination reaction
The glutamate-dependent transamination of OAA (an α-keto acid) to give aspartate is a fundamental reaction of amino acid synthesis.

> **SECTION HIGHLIGHTS**
> Microorganisms accumulate nitrogen (N) for cell biosynthetic needs from one or more environmental sources. Besides the nitrogen-containing organic molecules derived from living cells, other molecules include inorganic oxides of nitrogen (nitrate or nitrite), ammonia, and nitrogen gas (N_2). The ability of an organism to use a specific type of N will depend on its genetic potential. When alternative nitrogen sources may be used, the cell will, as a rule, minimize the energy expended and incorporate the least costly one into cell precursor molecules.

10.6 Sulfur Assimilation

Sulfur is an indispensable component of amino acids (cysteine, methionine), water-soluble vitamins (biotin, thiamine), and other key cellular constituents, including coenzyme A. Many species of *Bacteria*, *Archaea*, and fungi can use sulfate as their source of sulfur, but it must first be reduced to the sulfide HS$^-$ level for incorporation into the constituents of cells. A few organisms, such as the methanogens, lack the capacity for sulfate reduction and require H_2S as a sole source of sulfur. The key reaction in sulfur assimilation is as follows:

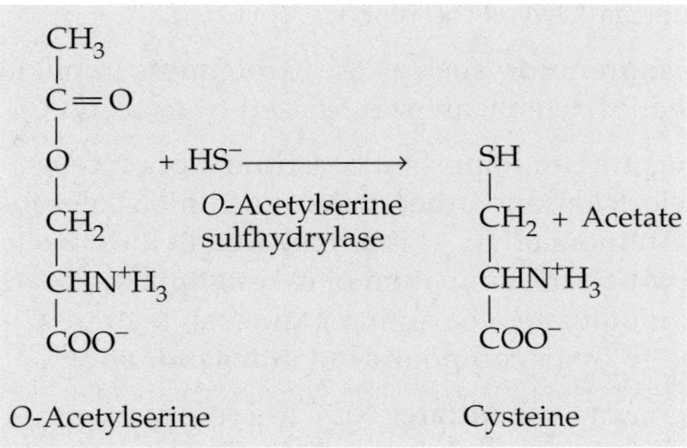

The assimilatory reduction of sulfate (SO_4^{2-}) to sulfide occurs in a series of reactions involving ATP, NADPH, and the sulfur-containing protein thioredoxin. In the initial reaction (Figure 10.25), sulfate reacts with ATP to form adenosine-5'-phosphosulfate (APS) and inorganic pyrophosphate. Hydrolysis of PP_i drives this otherwise endergonic reaction. APS reacts with another molecule of ATP to yield 3'-phosphoadenosine-5'-phosphosulfate (PAPS). PAPS is reduced by thioredoxin, which releases a molecule of sulfite (SO_3^{2-}), which is then reduced by NADPH to the sulfide (HS$^-$) level by the assimilatory sulfite reductase. Sulfide is then incorporated into biomolecules through *O*-acetylserine.

There are limited amounts of sulfate in some soils, and some microorganisms in the environment can assimilate other sulfur compounds, such as sulfur esters (R-CH$_2$OSO$_3^-$), sulfonates (R-CH$_2$SO$_3^-$), sulfur-containing amino acids, or other carbon–sulfur compounds. The role of sulfur-containing compounds in the sulfur cycle is discussed in Chapter 24.

SECTION HIGHLIGHTS

Assimilation of sulfur into precursor molecules within the cell occurs at the level of sulfide. However, sulfate and other sulfur-containing compounds may be utilized by many microbial species. As described for the assimilation of nitrogen, cells also minimize energy expenditure when alternative sources of sulfur are present.

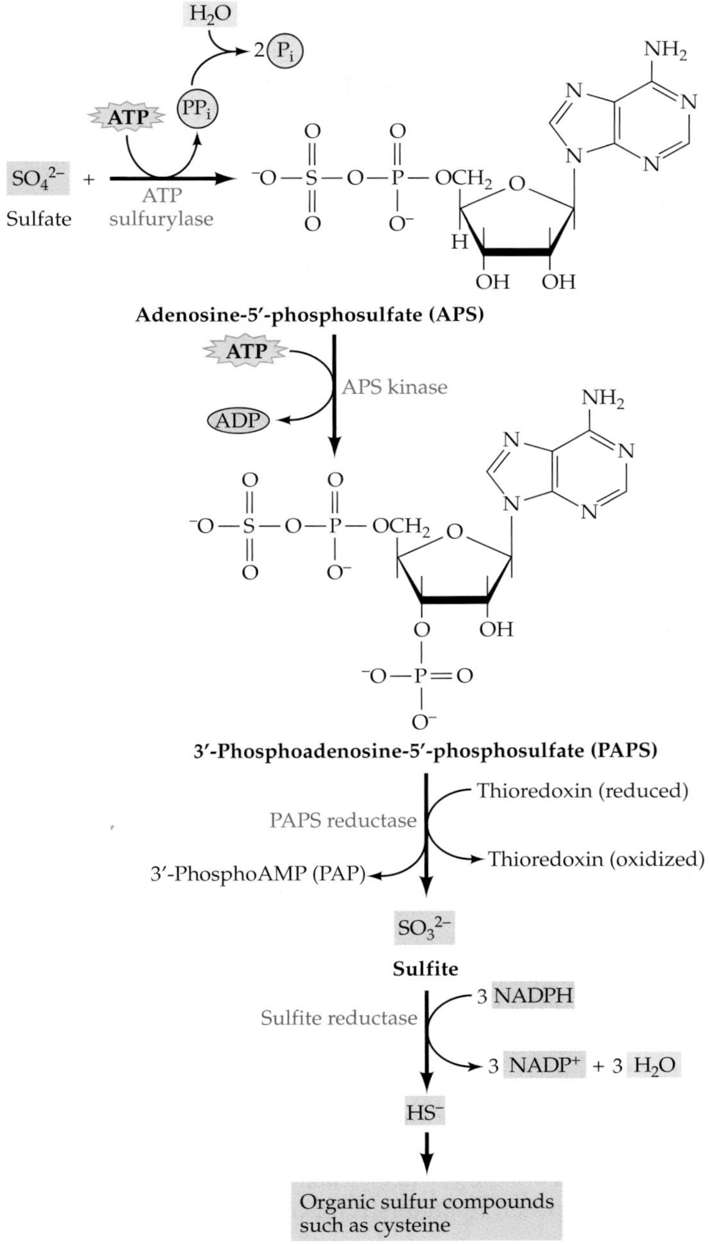

Figure 10.25 Sulfur assimilation
Sulfate is assimilated through the production of sulfide (HS$^-$), which is then used in the synthesis of organic sulfur-containing compounds via *O*-acetylserine.

SUMMARY

- The **autotrophs** have the most thorough biosynthetic capacities, as they are able to synthesize all constituents of a living cell from CO_2 and inorganic salts. Many can also obtain cellular nitrogen from atmospheric N_2.

- The **macromolecules** that make up a cell are of four types: proteins, nucleic acids, polysaccharides, and lipids.

- Twelve monomers, termed **precursor metabolites**, are involved in the biosynthesis of all the constituent parts of a cell's macromolecules. These metabolites are mostly products of **glycolysis**, **gluconeogenesis**, and the **tricarboxylic acid cycle**.

- Glucose is catabolized in bacteria by any of three pathways—**glycolysis** (the **Embden-Meyerhof** pathway), the **Entner-Doudoroff** pathway, **hexose monophosphate shunt**, or pentose phosphate pathway.

- The **tricarboxylic acid (TCA) cycle** may be used by **aerobic** organisms to completely oxidize one molecule of acetyl-CoA to two molecules of carbon dioxide. These organisms also utilize the TCA cycle to provide **precursor metabolites**. **Anaerobically**, the cycle is employed to generate **precursor metabolites** and the reduced compound, succinate.

- An **anaplerotic reaction** is one that permits metabolic sequences such as the TCA cycle to continue by replacing intermediates removed for biosynthetic purposes.

- **Carbon dioxide** is generally assimilated into autotrophs via the **Calvin cycle**. In this cycle, a molecule of CO_2 is added to a 5-carbon sugar (ribulose-1,5-bisphosphate), resulting in the formation of two molecules of 3-phosphoglyceric acid, which is also an intermediate in glycolysis.

- The green sulfur photosynthetic bacteria assimilate CO_2 via a mechanism that is essentially a reversal of the tricarboxylic acid cycle. This is termed a **reductive citric acid cycle** for CO_2 fixation.

- **Acetogenic bacteria** can combine two molecules of CO_2 using H_2 as the energy supply to form **acetic acid** by the reductive acetyl-CoA pathway. This is another mechanism for autotrophic CO_2 assimilation.

- **Amino acids** are synthesized as "**families**" from precursor metabolites. An example of a family of amino acids would be those synthesized from 3-phosphoglycerate: serine, glycine, and cysteine.

- The nucleic acid bases are synthesized de novo from simple substrates. The **pyrimidines** are synthesized from aspartate and carbamoyl phosphate, whereas the **purines** are synthesized from CO_2, glycine, two formyl groups, an amine nitrogen from aspartic acid, and two amide nitrogens from glutamine.

- **Fatty acids** in bacteria are synthesized from **acetyl-CoA** and **malonyl-CoA**. During the joining of acetyl- CoA and malonyl-CoA, a decarboxylation occurs that drives the reaction forward.

- The **isoprenoids**, such as the **carotenoids** in photosynthetic bacteria, are synthesized from acetyl-CoA.

- **Nitrogen** is an important constituent of proteins, nucleic acids, and other major macromolecules in a cell. Ammonia (NH_4^+) is assimilated in most bacteria through the amination of α-ketoglutarate to form glutamate. The amino group can be transferred to other compounds by **transamination**.

- Bacteria can use **nitrate** (NO_3^-) as source of nitrogen and must reduce it to ammonia. The **assimilatory enzymes** involved are the assimilatory **nitrate** and **nitrite reductase enzymes**.

- **Nitrogen fixation** (use of dinitrogen as a nitrogen source) is confined to the bacterial and archaeal world. An array of free-living, commensal, and photosynthetic bacteria can supply their need for nitrogen by assimilating it from the air.

- The key reaction in the incorporation of **sulfur** into bacteria and fungi is the formation of **cysteine** from the serine derivative *O*-acetylserine.

 Find more at www.sinauer.com/microbial-life

REVIEW QUESTIONS

1. In discussing macromolecule synthesis, CO_2 was selected as the starting substrate. Why?

2. What are the four major macromolecular types that are present in a bacterial cell? What are the monomers that make up these macromolecules? Draw typical monomers that are assembled to form these macromolecules and show how are they linked.

3. A precursor metabolite is a molecule that is directly involved in biosynthesis. Why are these metabolites called precursors? What does the existence of these monomers tell us about biological systems?

4. How does precursor synthesis relate to metabolic diversity?

5. Review the three major pathways for glucose catabolism in microorganisms. What are some of the significant differences between them? What role might each play?

6. What is an anaplerotic reaction? What is the function of these reactions? How is CO_2 involved?

7. What are the key enzymes in the Calvin cycle?

8. How different enzymatically is an organism utilizing CO_2 as substrate from one growing on glucose? How does the initial product of CO_2 fixation fit into the concept of the precursor metabolite?

9. The green sulfur bacteria are probably ancient organisms. How do they fix CO_2, and in what way is this related to aerobic metabolism?

10. Ammonia is "funneled" into the metabolic machinery through two intermediates. Draw these reactions.

11. Sulfate is "funneled" into the metabolic machinery through several intermediates. Describe this process.

12. What is a family of amino acids? How many families are there?

13. What low-molecular-weight compounds are utilized in the synthesis of a purine? A pyrimidine?

14. What is a fatty acid? List different types. Outline the mechanism of synthesis of phospholipids and isoprenoids.

SUGGESTED READING

Cooper, G. M. and R. E. Hausman. 2007. *The Cell: A Molecular Approach.* 4th ed. Washington, DC and Sunderland, MA: ASM Press and Sinauer Associates, Inc.

Garrett, R. H. and C. M. Grisham. 2006. Biochemistry. 3rd ed. Belmont, CA: Brooks/Cole.

Lawlor, D. W. 1993. *Photosynthesis: Molecular, Physiological, and Environmental Processes.* 2nd ed. Essex, England: Longman Scientific and Technical.

Lengeler, J. W., G. Drews, and H. G. Schlegel eds. 1999. *Biology of the Prokaryotes.* Malden, MA: Blackwell Science.

Neidhardt, F. C., R. Curtiss III, J. L. Ingraham et al. 1996. *Escherichia coli and Salmonella: Cellular and Molecular Biology.* Washington, DC: ASM Press.

Neidhardt, F. C., J. L. Ingraham, and M. Schaechter. 1990. *Physiology of the Bacterial Cell: A Molecular Approach.* Sunderland, MA: Sinauer Associates, Inc.

Schlegel, H. G., and B. Bowien, eds. 1989. *Autotrophic Bacteria.* Madison, WI: Science Tech Publishers.

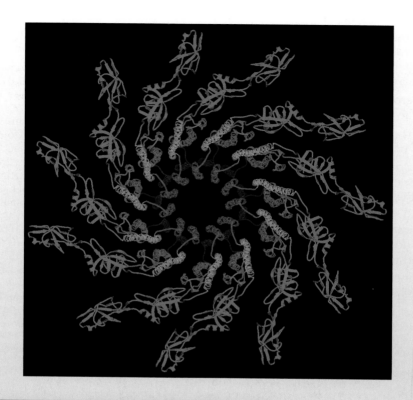

The objectives of this chapter are to:

◆ Describe the steps of membrane synthesis.

◆ Outline the process of peptidoglycan synthesis.

◆ Introduce the concepts of protein structure and function.

◆ Highlight important aspects of protein synthesis and export.

◆ Describe the structure and function of cell appendages.

11

Assembly of Bacterial Cell Structures

...Topologically, all the layers of the envelope are closed surfaces and must be physically continuous for cell integrity and viability to be maintained. ...All the constituents of the envelope must grow coordinately and with special regard to their location... The overall process is similar in gram-positive and gram-negative bacteria but differs in detail...
—Neidhardt, Ingraham and Schaechter, Physiology of the Bacterial Cell: A Molecular Approach, *1990*

The generation of **precursor metabolites** and the reactions that convert these to the monomeric building blocks of a cell were discussed in Chapter 10. The precursor metabolites originate either from CO_2 fixation, gluconeogenesis, the tricarboxylic acid cycle, glycolysis, or allied pathways. Through a limited number of well-integrated reactions, the major monomers—including the amino acids, nucleotides, sugars, and fatty acids—are synthesized from these precursors. Rapid and orderly growth depends on the **polymerization** or assembly of monomeric building blocks to form macromolecules. Among the essential polymerization reactions are the formation of proteins from amino acids, polysaccharides from sugars, and nucleic acids from nucleotides. Phospholipids are derived from fatty acids. Once formed, macromolecules (DNA, RNA, proteins, and phospholipids) are assembled to generate a cell (**Figure 11.1**). Note that proteins and RNA make up the major part of a living cell. The actual number of molecules of each macromolecular component present in an *Escherichia coli* cell is listed in Table 11.1.

Replication of DNA, RNA synthesis, and the role of the nucleic acids in protein synthesis will be discussed in Chapter 13. The following is a brief discussion of the synthesis and assembly of cell membranes and protein structures and the assembly and export of constituents to the cell envelope (peptidoglycan layers), outer membrane (in gram-negative bacteria), plus cell appendages.

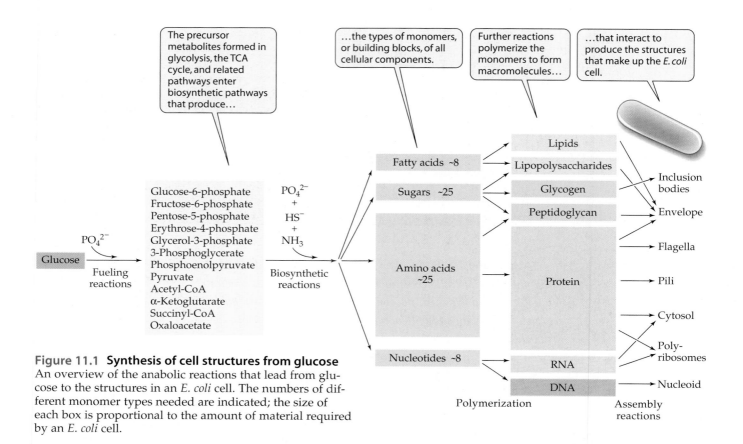

Figure 11.1 Synthesis of cell structures from glucose
An overview of the anabolic reactions that lead from glucose to the structures in an *E. coli* cell. The numbers of different monomer types needed are indicated; the size of each box is proportional to the amount of material required by an *E. coli* cell.

TABLE 11.1	The overall macromolecular composition of an *Escherichia coli* cell		
Molecule	**Percentage of Total Dry Weight of Cell**	**Number of Molecules per Cell**	**Types of Molecules Possible**
Protein	55.0	2,360,000	~4200
RNA	20.5		
23S rRNA		18,700	1
16S rRNA		18,700	1
5S rRNA		18,700	1
transfer RNA		205,000	~60
messenger RNA		Variable	~1,380
DNA	3.1	~2.1	1
Lipid[a]	9.1	22,000,000	4
Lipopolysaccharide	3.4	1,200,000	1
Peptidoglycan	2.5	1	1
Glycogen	2.5	4,360	1
Soluble organic pool[b]	2.9	Large	~850
Inorganic pool[c]	1.0	Large	~20

[a]The phospholipids are of four general classes, which may exist in a variety of types based on fatty acid chain.
[b]Metabolites, vitamins, and precursors.
[c]Anions, cations.
Adapted from Neidhardt, Ingraham, and Schaechter, 1990.

11.1 Membrane Synthesis

One structure that is present in all *Bacteria* and *Archaea* is the cytoplasmic membrane, also called the cell membrane. The basic structure of a cytoplasmic membrane was described in Chapter 4 (see Figures 4.44 and 4.45); it serves as the primary boundary of the cell's cytoplasm. The bacterial cytoplasmic membrane is similar in its basic structure to other membranes—for example, those surrounding eukaryotic cells, nuclei, mitochondria, and other organelles. The membranes of *Archaea* are quite different, as will be discussed in Chapter 18.

A cytoplasmic membrane in a growing cell must assimilate new membrane components (phospholipids and proteins) and integrate these into the membrane, which is increasing in surface area. During integration of the newly synthesized phospholipids, the permeability and barrier functions of the membrane must be maintained.

Synthesis of Lipids

Membrane lipids compose about 10% of the cell dry weight (see Table 11.1) and thus represent a considerable expenditure of cell energy. Enzymes for lipid synthesis are generally located in the cytoplasmic membrane with the exception of those needed for precursor biosynthesis (glycerol-3-phosphate) and fatty acid synthesis; these are discussed in Chapter 10. Since lipid types differ among bacteria with respect to fatty acid chain length, saturation, and polar head group, the reader is directed to advanced textbooks and scientific literature for detailed information. As a general plan, long-chain fatty acids (C14–18) are assembled in the cytoplasm from acetyl-CoA precursor molecules via small acyl-carrier protein (ACP) intermediates. The degree of fatty acid saturation may vary depending on the cell growth condition; the fatty acids are then esterified with glycerol at the inner surface of the cytoplasmic membrane. A polar head group is then added to join two fatty acids, yielding the mature phospholipid (see Figure 4.45). *E. coli*, for example contains three types of phospholipids (phosphatidylglycerol, phosphatidylethanolamine, and cardiolipin (diphosphatidylglycerol), where the polar head groups are glycerol, ethanolamine, and phosphatidylglycerol, respectively. Other microbes, for example *Bacillus subtilis*, may contain different head groups: glucose and glucose—O—glucose in addition to ethanolamine and glycerol.

The newly synthesized phospholipids are incorporated into the inner leaflet of the cytoplasmic membrane bilayer, and in a relatively short time appear in the outer leaflet. This movement from the inner to the outer membrane leaflet occurs at a more rapid rate in actual cell membranes than it does in model bilayers, suggesting that rotation from the inner to the outer leaflet in a viable cell may be controlled by proteins in the membrane. This process may require ATP, but that is uncertain at the present time.

Modification of bacterial membranes can occur independently of cell growth and division. A constant turnover of membrane proteins or phospholipids (or both) occurs as organisms adjust to changing environmental conditions. Cells increase the proportion of membrane unsaturated fatty acids as a response to a decrease in growth temperature. This occurs because a functional membrane must be fluid and because the double bonds in unsaturated fatty acids are more fluid at lower temperatures. A cell may also vary the fatty acid chain length. In *E. coli*, C18 fatty acids are more abundant at high growth temperatures, while C16 predominates at lower temperatures. Newly formed solute transport proteins or respiratory enzymes may also be inserted into the cytoplasmic membrane as a microorganism encounters a different substrate or other changes in growth conditions.

SECTION HIGHLIGHTS

Assembly of phospholipids occurs at the surface of the cytoplasmic membrane and requires the carrier molecule ACP. The lipid and protein composition of a cell may be adjusted independently of cell growth in order to adapt to temperature and nutritional changes.

11.2 Peptidoglycan Synthesis

Peptidoglycan (murein) is the main structural component of the bacterial cell wall. It is the source of strength and provides the characteristic cell size and shape. The composition, structure, and function of peptidoglycan in both gram-positive and gram-negative bacteria were discussed in Chapter 4. The position beyond the cytoplasmic membrane (Figure 11.2) presents interesting questions regarding its synthesis and assembly.

The assembly of peptidoglycan precursor units occurs in the cytoplasm and the cytoplasmic membrane (Figure 11.3). At the inner surface of the cell membrane, both *N*-acetylglucosamine (NAG) and *N*-acetylmuramic acid (NAM) are synthesized and then coupled to bactoprenol (undecaprenol phosphate, Udc; Figure 11.4). **Bactoprenol** is a long-chain hydrocarbon that can enter the hydrophobic core of the cytoplasmic membrane and facilitate the movement of attached hydrophobic molecules into or through the lipid bilayer.

In an initial membrane-associated step in peptidoglycan synthesis, bactoprenol displaces the uridine triphos-

(A) Gram-positive cell envelope

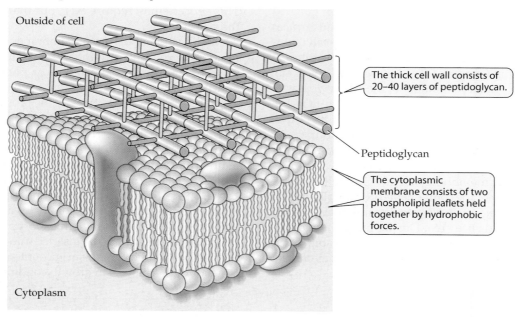

Outside of cell

The thick cell wall consists of
20–40 layers of peptidoglycan.

Peptidoglycan

The cytoplasmic
membrane consists of two
phospholipid leaflets held
together by hydrophobic
forces.

Cytoplasm

(B) Gram-negative cell envelope

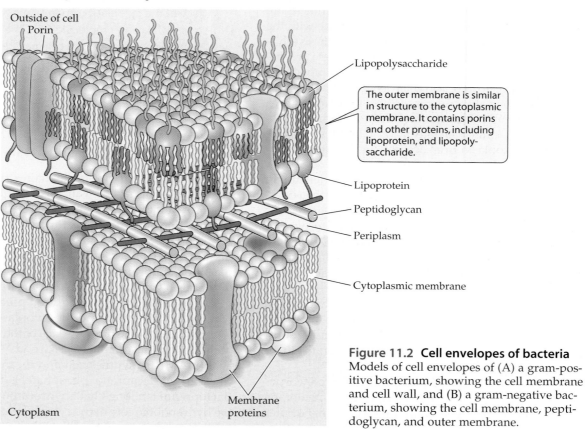

Outside of cell
Porin

Lipopolysaccharide

The outer membrane is similar
in structure to the cytoplasmic
membrane. It contains porins
and other proteins, including
lipoprotein, and lipopolysaccharide.

Lipoprotein

Peptidoglycan

Periplasm

Cytoplasmic membrane

Membrane
proteins

Cytoplasm

Figure 11.2 Cell envelopes of bacteria
Models of cell envelopes of (A) a gram-positive bacterium, showing the cell membrane
and cell wall, and (B) a gram-negative bacterium, showing the cell membrane, peptidoglycan, and outer membrane.

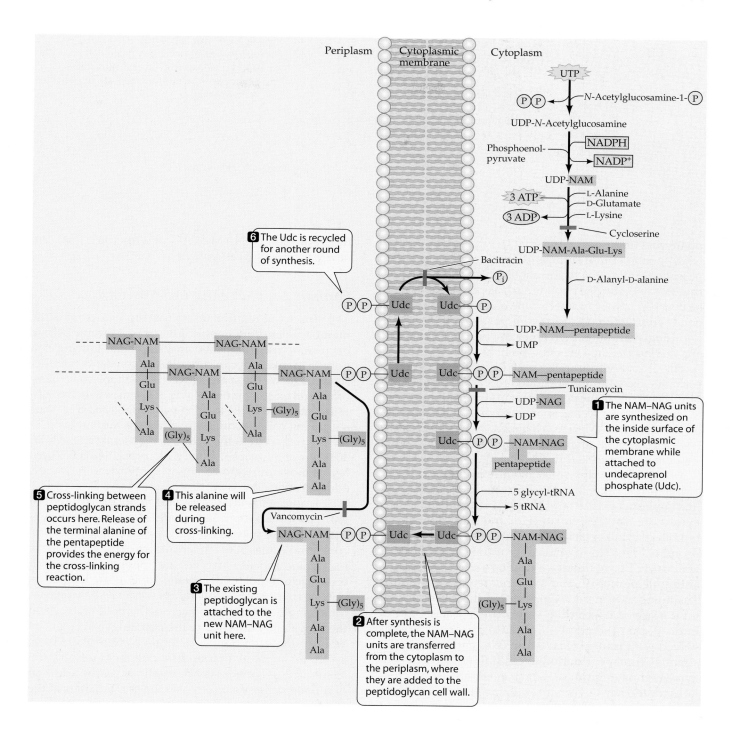

Figure 11.3 Synthesis of peptidoglycan
Steps in the synthesis of peptidoglycan. Gram-positive peptidoglycan is depicted here; gram-negative peptidoglycan has DAP in place of lysine and pentaglycine (see Figure 4.48) and DAP is directly cross-linked to the D-alanine of the adjacent strand. UTP, uridine triphosphate; UDP, uridine diphosphate; UMP, uridine monophosphate; NAG, N-acetylglucosamine; NAM, N-acetylmuramic acid; Udc, undecaprenol phosphate. Steps blocked by antibiotics are indicated.

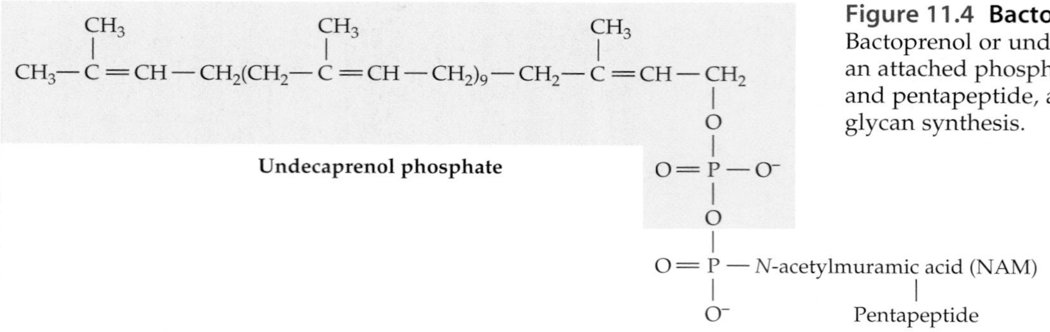

Figure 11.4 Bactoprenol
Bactoprenol or undecaprenol phosphate with an attached phospho-*N*-acetylmuramic acid and pentapeptide, an intermediate in peptidoglycan synthesis.

phate precursor (see Figure 11.3), which renders the NAM-pentapeptide sufficiently hydrophobic to allow passage through the hydrophobic membrane. During this passage, a bridge peptide such as pentaglycine may be attached to the terminal amine of lysine via a peptide bond. This would occur in gram-positive bacteria. The peptidoglycan units then enter the periplasmic space and are inserted at a growing point in the cell wall.

The *N*-acetylmuramic acid (NAM)–*N*-acetyglu-cosamine (NAG) units are added to the existing NAM–NAG backbone of the peptidoglycan. The peptide bridge, in this case pentaglycine, is joined to an alanine molecule on an adjacent NAM–NAG chain. The NAM–NAG backbone and the peptide cross-linking are responsible for the strength of the peptidoglycan cell wall.

A battery of periplasmic enzymes is involved in the covalent reactions that result in extension and cross-linking between the peptidoglycan strands. The enzymes are also responsible for the septation of the murein cell wall, which occurs during cell division. The proteins involved in extension of the peptidoglycan wall during growth have a unique ability to bind the antibiotic **penicillin** and some related antibacterials. The number of these "penicillin-binding" proteins on the surface of a bacterium varies with species. Studies indicate that the penicillin-binding proteins are involved with **transglycosylation** (elongation of glycan strands), **transpeptidation** (cross-linking), and the enzyme **carboxypeptidase**, which cuts preexisting cross links for the new glycan insertion needed for cell elongation and division (see Figure 6.1). Binding of penicillin to these proteins inhibits murein biosynthesis and can destroy the integrity of the cell.

A variety of other inhibitors also interfere with peptidoglycan synthesis and assembly. **Bacitracin** inhibits the removal of phosphate from undecaprenol-pyrophosphate, thus preventing recycling of undecaprenol-phosphate. **Tunicamycin** blocks addition of NAM-pentapeptide to undecaprenol-phosphate, whereas **cycloserine** inhibits addition of D-alanine to the elongating peptide chain on UDP-NAM. Finally, **vancomycin** inhibits the

incorporation of NAM–NAG-pentapeptide into the newly forming glycan polymer.

Forming Gram-Positive Cell Envelopes

The structural relationship between the cytoplasmic membrane and the peptidoglycan in a gram-positive bacterium was described in Chapter 4. During cell growth, peptidoglycan subunits are constantly synthesized and enter the space between the cytoplasmic membrane and existing murein layer at points where the subunits are linked to the existing peptidoglycan layer. Thus a growing cell is continually adding a murein layer in the area adjacent to the cytoplasmic membrane. About 40 layers of murein surround a gram-positive cell. These layers move outward as newly synthesized murein is added to the inner layer.

Teichoic acids are the second major constituent of gram-positive bacterial envelopes (see Chapter 4). Synthesis of these polymers occurs by assembly from precursor molecules at the inner surface of the cytoplasmic membrane and also involves the lipophilic carrier molecule **bactoprenol** (undecaprenol). A built-up long-chain polymer of ribitol phosphate, as in *Staphylococcus aureus*, for example, is transferred across the lipid bilayer and then covalently attached to the peptidogycan layers by a phosphodiester bond. Bactoprenol-phosphate is recycled for subsequent rounds of polymer assembly. Teichoic acid polymers may be modified by the addition of monosaccharide or oligosaccharide moieties at the –OH group of ribitol by specific glycosyltransferases. There is considerable diversity among the gram-positive bacteria with respect to the type of polymer (see Chapter 4) and side-chain modification. Details of teichoic acid synthesis as well as production of lipoteichoic acids are provided in advanced textbooks.

Forming Gram-Negative Cell Envelopes

The cell wall structures of gram-negative bacteria are considerably more complex than those of gram-positives (see

Figure 11.2). The gram-negative bacteria have a complex outer membrane that surrounds the cell envelope. Growth and expansion of this envelope is consequently more elaborate. It is apparent that fatty acids, sugars, phospholipids, and proteins must move from the cytoplasm to the periplasm during wall expansion or during environmental changes in gram-negative bacteria.

Expansion has been viewed as occurring in minute openings in the cell wall, and these have been termed **zones of adhesion** because the inner and outer membranes actually make direct contact. The junctions occur in the peptidoglycan layer, and all components for the construction of the outer cell envelope pass through these gaps. However, more recent studies implicate specific protein secretion pathways for cell assembly, very likely not involving these junctions.

The components of the lipopolysaccharide (see Chapter 4) present in the outer membrane are synthesized on the inner surface of the cytoplasmic membrane and are carried outward with the assistance of **bactoprenol**. Assembly then occurs on the outer surface of the cell by interactions between the envelope components. The phospholipid components, including lipid A of the outer membrane, are likewise synthesized and translocated to form the outer membrane layer.

Methods to characterize the various components that make up the cell envelope are described in Box 11.1. They are routinely applied in the molecular study of model organisms and can be adapted to study of newly isolated strains from the environment.

SECTION HIGHLIGHTS

Synthesis of peptidoglycan, teichoic acids, and LPS initiates in the cyptoplasmic membrane and involves the carrier molecule, molecule bactoprenol. Translocation and insertion of these cell precursor molecules generates the mature cell envelope characteristic of a gram positive or gram negative bacterium.

11.3 Protein Assembly, Structure, and Function

The term "protein" broadly defines molecules composed of one or more polypeptide chains. A polypeptide is a polymer that generally exceeds several dozen amino acids in length. Some proteins are present in the cytoplasm; others are associated with the cytoplasmic membrane, outer envelope, or periplasm of a bacterium (Table 11.2).

TABLE 11.2	Location of major proteins associated with a gram-negative bacterium
Location	**Proteins**
Cytoplasm	Enzymes involved in catabolism of soluble substrates
	Enzymes involved in anabolism
	Enzymes involved in DNA replication
	Ribosomes and enzymes for protein synthesis
	Proteins involved in gene regulation
Cytoplasmic membrane	Components of electron-transport chain
	Proton translocating ATP synthase
	Solute uptake proteins
	Lipid biosynthesis enzymes
	Cell wall and outer-envelope biosynthetic enzymes
	Secretory machinery
	Chemotactic receptors
Periplasm	Binding proteins involved in transport and chemotaxis
	Enzymes for peptidoglycan assembly
	Certain secretory proteins
Outer envelope	Porin proteins
	Receptors for bacteriophage, etc.
External surface	Fimbriae/pili
	Flagella
Extracellular	Hydrolases
	Proteases
	Lipases
	Nucleases
	Protein toxins
	Capsule assembly proteins

For example, proteins in the cytoplasm are involved with catabolic functions and with the synthesis of DNA, RNA, and cellular components. The electron transport systems in microorganisms capable of respiration are located in the cytoplasmic membranes, as are the proteins of the ATP synthase system. The outer envelopes of gram-negative bacteria contain the protein porins involved in passage of molecules into the periplasmic space, where binding proteins can retain them. Transport of these molecules to the cytoplasm occurs through the action of proteins in the cytoplasmic membrane. Finally, microorganisms that can digest polysaccharides, fats, or other large molecules secrete enzymes (proteins) to the cell exterior that can hydrolyze large molecules to low-molecular-weight compounds. These compounds can then be transported into the cell.

BOX 11.1   *Methods & Techniques*

Cell Fractionation/Separation and Biochemical Analyses of Cell Structures

In order to determine the composition of various components of the cell, scientists separate these components from the rest of the cell, purify them, and analyze them biochemically. The initial step is to break open the cells. Either chemical or physical procedures can be used to break open small, prokaryotic cells. For example, chemical procedures include lysis of the cells by enzymes or detergents. Physical

methods include ultrasound (called **sonication** by biologists), in which high-frequency sound waves vibrate cells until they break. A sonicator probe is inserted into a cell suspension for this purpose, as shown in the illustration. Alternatively, cells can be broken by passing thick suspensions of cells through a small orifice ("French pressure" cell) at very high pressure.

Once the cells have been broken, the various structural fractions are separated, usually by centrifugation. Two types of centrifugation can be used. In **differential** or **velocity centrifugation**, fractions are separated by the length of time they are centrifuged at different gravitational forces. Denser structures—such as unbroken cells or bacterial endospores, cell membranes, or cell walls—sediment at

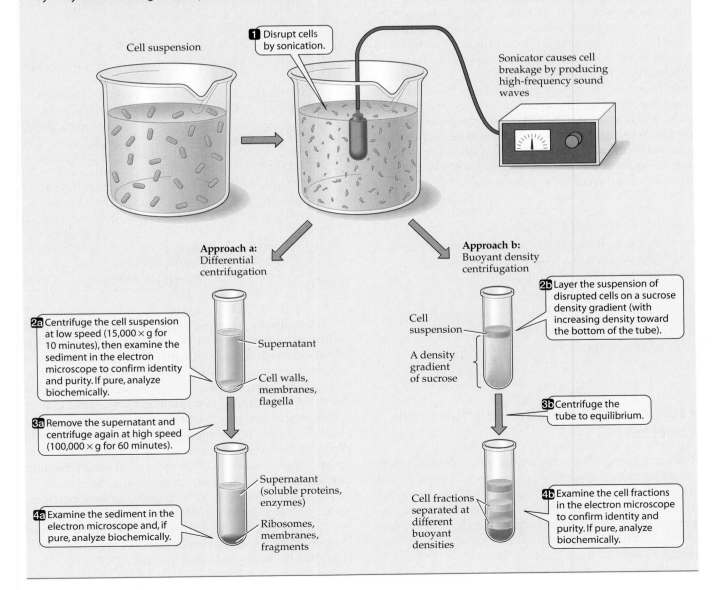

BOX 11.1 *Continued*

low speeds (15,000 × g for 10 minutes). The supernatant is removed and centrifuged at higher speed to spin out less dense structures. For example, ribosomes sediment only after centrifugation at higher speeds (100,000 × g for 60 minutes). The remaining material which does not sediment in the centrifuge tube contains soluble constituents such as cytoplasmic enzymes.

Alternatively, **buoyant density** or **density gradient** centrifugation can be used to separate the various cell fractions. In this procedure, a density gradient is set up in the centrifuge using different concentrations of a solute, such as sucrose. The sample is layered on the surface and centrifuged at moderate speed until the cellular fractions equilibrate with the layer in the gradient that has the same buoyant density. They can then be removed with a pipette and studied as purified fractions. For example, the cytoplasmic membranes of *E. coli* are less dense than the outer membrane because of differences in the types of molecules present.

When the cell fraction of interest to the microbiologist is separated and purified by the procedures outlined above, it can be analyzed chemically. The electron microscope may be used to check the identity and purity of the material at each step in the process. Proteins may be separated on polyacrylamide gels (native or denaturing) to characterize those present. Biochemical assays can be performed using marker enzymes as an indication of fraction purity versus contamination by other cell fractions.

Protein Structure

Amino acids are joined together by **peptide bond** formation on ribosomes through a series of peptidyl transfer reactions from amino acid-tRNA molecules. When a series of amino acids are joined together, the result is a polypeptide chain referred to as the **primary structure** (Figure 11.5). The nature and character of a protein is determined to a considerable extent by the total number and sequence of amino acids in this chain. Functional proteins are generally composed of polypeptides that are folded or coiled into a three-dimensional structure; this is the **conformation** that a functional polypeptide ultimately assumes.

The folding pattern is determined largely by the amino acid sequence, and the formation of a functional protein can potentially occur by unassisted self-assembly. However, there is evidence suggesting that random unassisted assembly can result in a structure that is nonfunctional (Box 11.2). Consequently there are preexisting proteins in the growing cell, termed **chaperones**, which act to prevent incorrect molecular interactions that would lead to nonfunctional secondary or tertiary structures. These chaperones assist in forming but are not a part of the final functional protein. Generally the folding of peptides to a functional secondary structure is energetically favorable; that is, it occurs without energy input.

Structural Arrangements of Proteins

The types of bonding that occur between internal areas (domains) of a polypeptide, giving the protein a secondary and tertiary structure, are outlined in Figure 11.6. The variability in the side chains of the 21 major amino acids (see Figure 3.15), the total number incorporated, and their sequence in a polypeptide are all factors that contribute to the biochemical characteristics of the protein. Amino acids with nonpolar hydrocarbon side chains (valine, leucine, phenylalanine, and tryptophan) are hydrophobic (do not interact with water). These side chains point outward on the external surface of the proteins embedded in hydrophobic membrane lipids. In water-soluble proteins present in cytoplasm, these hydrophobic groups extend inward. Correspondingly, the side chains of hydrophilic amino acids (aspartic acid, glutamic acid, lysine, arginine, and histidine) extend outward to the aqueous environment.

Typically, the twisting and coiling of polypeptide chains results in the formation of a **secondary structure** that is either helical in form, called an **α-helix**, or a flat arrangement, designated a **β-sheet**. The α-helix is a linear polypeptide that is wound like a spiral staircase (Figure 11.7). It is held together by hydrogen bonding between the amine hydrogen of one amino acid and an oxygen from another amino acid. The hydrogen bonding occurs as the polypeptide is formed and leads to a stable helical structure. Glutamic acid, methionine, and alanine are the strong formers of the α-helix.

β-sheets (see Figure 11.7) are formed when two or more extended polypeptide chains come together side by side so that regular hydrogen bonding can occur between the amide (NH) and the carbonyl (C═O) of an adjacent-chain peptide backbone. Addition of more polypeptides results in a multistranded structure. The

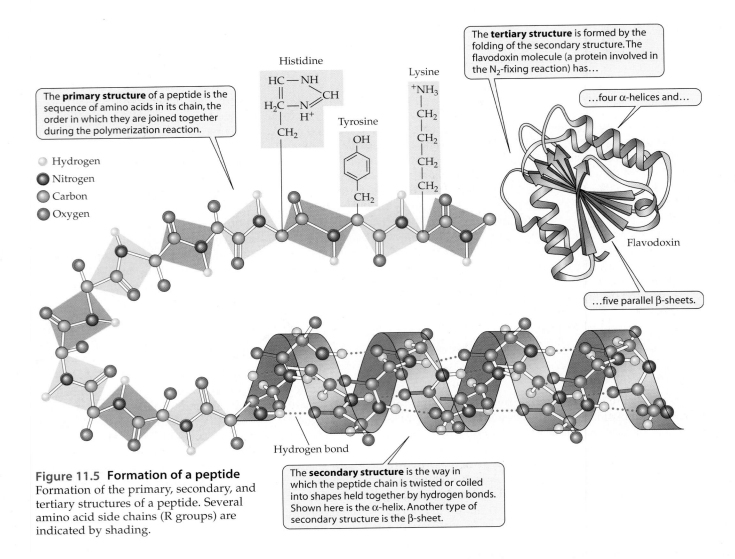

The **primary structure** of a peptide is the sequence of amino acids in its chain, the order in which they are joined together during the polymerization reaction.

Histidine

Tyrosine

Lysine

The **tertiary structure** is formed by the folding of the secondary structure. The flavodoxin molecule (a protein involved in the N₂-fixing reaction) has…

…four α-helices and…

…five parallel β-sheets.

Flavodoxin

○ Hydrogen
● Nitrogen
○ Carbon
○ Oxygen

Hydrogen bond

The **secondary structure** is the way in which the peptide chain is twisted or coiled into shapes held together by hydrogen bonds. Shown here is the α-helix. Another type of secondary structure is the β-sheet.

Figure 11.5 Formation of a peptide
Formation of the primary, secondary, and tertiary structures of a peptide. Several amino acid side chains (R groups) are indicated by shading.

amino acids valine, tyrosine, and isoleucine are common β-sheet formers.

In many proteins, the polypeptide chain is bent at specific sites and folded back and forth, resulting in the tertiary structure (see Figure 11.5). Although the α-helices and β-pleated sheets contribute to the tertiary structure, only parts of the macromolecule usually have these secondary structures; large regions consist of structures unique to a particular protein. The protein shown is flavodoxin, an electron transport protein in sulfate reducing bacteria. The folding of a polypeptide chain to form a discrete compact protein molecule requires bends in the chain of amino acids that reverse the direction;

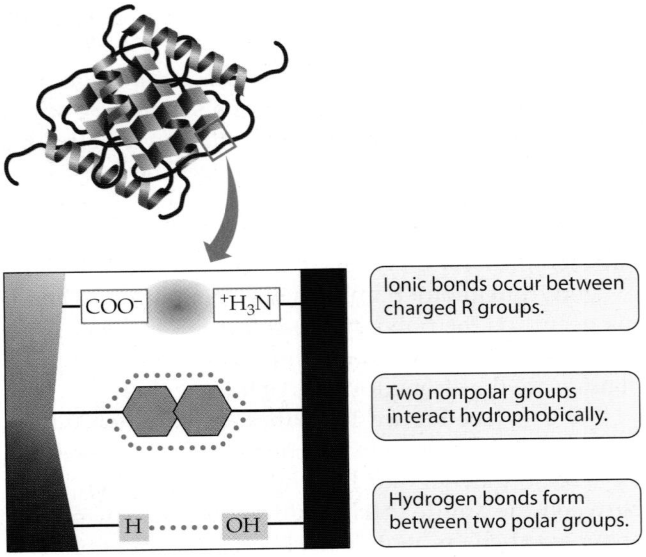

Ionic bonds occur between charged R groups.

Two nonpolar groups interact hydrophobically.

Hydrogen bonds form between two polar groups.

Figure 11.6 Types of bonds between areas of a peptide chain
Several types of bonds and interactions form between the domains of a polypeptide chain. These determine the shape of the protein.

(A) α-helix

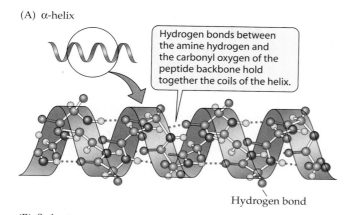

Hydrogen bonds between the amine hydrogen and the carbonyl oxygen of the peptide backbone hold together the coils of the helix.

Hydrogen bond

(B) β-sheet

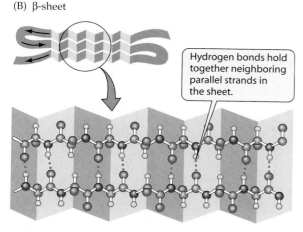

Hydrogen bonds hold together neighboring parallel strands in the sheet.

Figure 11.7 Secondary structures
Hydrogen bonds hold a polypeptide chain in an α-helix or β-sheet configuration. These are important determinants of secondary structure.

these are called **β-bends**. Glycine and proline are frequently present in β-bends. A β-bend is a tight loop that results when a carbonyl group of one amino acid forms a hydrogen bond with the NH_2 group of another amino acid three positions down the polypeptide chain. This results in the polypeptide folding back on itself.

Many proteins are composed of more than one polypeptide chain. These subunits are adjoined by non-covalent bonding (hydrogen bonding, hydrophobic interaction, or ionic bonding). The joining of these subunits forms what is termed a **quaternary structure** (Figure 11.8). The individual polypeptide subunits that make up a quaternary structure can be either the same or a polypeptide of different composition or size. A classic example of a quaternary structure is the oxygen-carrying protein hemoglobin. This protein is composed of four polypeptides—of two different kinds—and is termed an $\alpha_2\beta_2$ tetramer. The individual subunits in a quaternary structure are proteins which themselves have typical secondary and tertiary structures.

α Subunits

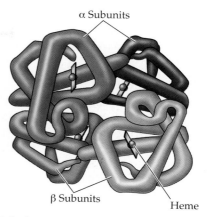

β Subunits   Heme

Figure 11.8 Quaternary structure
The quaternary structure of hemoglobin is composed of two α– and two β–polypeptide chains or subunits. Note the compact symmetry of the structure.

The **relative mass (M_r)** of a protein is the total molecular mass of the assembled subunits. Some representative proteins, their relative masses, and the number of subunits that form each protein are listed in Table 11.3.

SECTION HIGHLIGHTS

The characteristic properties of a protein are dictated in large part by its primary amino acid sequence. Folding of the polypeptide chain into the native or mature state may proceed unaided, or may require protein chaperones to aid in forming the correct secondary, tertiary, and quaternary structures.

TABLE 11.3	Relative mass and number of subunits in selected proteins[a]	
Protein	**Relative Mass (M_r)**	**Subunits**
Ferredoxin	6,500	1
Ribonuclease	13,700	1
Lysozyme	14,388	1
Luciferase	80,000	2
Hexokinase	100,000	2
Lactate dehydrogenase	223,000	4
Urease	483,000	5
NADH dehydrogenase	550,000	13

[a]The average molecular mass of an amino acid residue in a protein is 110 Da, or daltons (average molecular weight less a molecule of H_2O), and an "average" protein molecule has about 330 amino acid residues, or M_r 36,000. A dalton is defined as one-twelfth of the mass of a carbon atom.

BOX 11.2   *Milestones*

Protein Conformation

Classic experiments in the early 1960s by Christian B. Anfinsen led to the hypothesis that the sequence and nature of the amino acids in a polypeptide chain were the major contributing factors to the folding and final conformation of a protein. This was based on experimentation with bovine pancreatic ribonuclease, a small protein made up of 124 amino acids. The secondary structure of this protein is maintained by four disulfide bridges. Denaturation can be accomplished by cleaving these covalent disulfide linkages, which reduces the —S—S— linkage to —SH HS— as follows:

agents and exposure to air results in a random reassembly and an inactive enzyme. However, exposure of the denatured protein to trace levels of β-mercaptoethanol promotes the proper rearrangement of the —S—S— cross links and results in a reformation of the active enzyme. This and other in vitro experiments suggest that proper folding might be an inherent property of the primary structure of polypeptides. Thus folding to secondary and tertiary structures would be promoted by the proper spacing of amino acids whose side chains could interact with counterpart amino acids

trated in Figure 11.6. This sort of in vitro study apparently does not completely reflect the conditions in the cytoplasm of a cell. The levels of polypeptide required for test tube refolding are much higher than would occur in vivo, and the physiochemical conditions of the in vitro experiments do not exist in the living cell. In reality, studies in vitro that mimic in vivo physiological conditions result in misfolding or aggregation of the polypeptide. This misfolding is uncommon in vivo except with mutant proteins or protein folding at elevated temperatures.

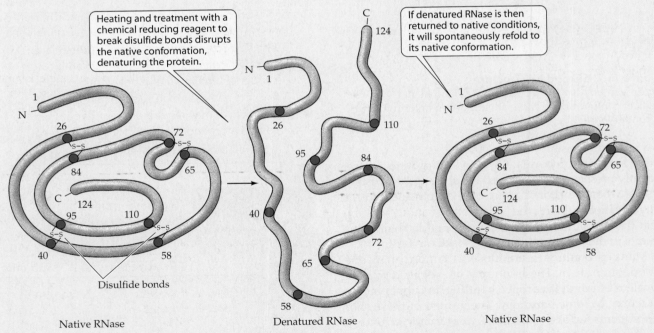

Heating and treatment with a chemical reducing reagent to break disulfide bonds disrupts the native conformation, denaturing the protein.

If denatured RNase is then returned to native conditions, it will spontaneously refold to its native conformation.

Disulfide bonds

Native RNase

Denatured RNase

Native RNase

The secondary structure, and thus the activity, of bovine pancreatic ribonuclease is maintained by disulfide bonds between cysteine residues.

Disruption of the sulfide alters the conformation of the protein and leads to loss of enzymatic activity. Removal of the denaturing

elsewhere in the polypeptide chain. The bonding producing the functional structures would be formed by the side chain interactions illus-

Studies in recent years affirm that there are two classes of proteins associated with the proper folding of polypeptides in a cell.

BOX 11.2	*Continued*

One class consists of the conventional enzymes that catalyze specific isomerization reactions that result in proper polypeptide conformations for proper folding.

The second class of proteins comprises the chaperones that stabilize unfolded or partially folded structures and preclude the formation of inappropriate intra- or interchain interactions. A chaperone may also interact with individual proteins to promote protein-protein arrangements that yield quaternary functional structures. Among the roles that protein chaperones play are the following:

- They prevent folding of secretory proteins before translocation.
- They are involved with the assembly of bacterial viruses.
- They promote the assembly of pili and of the carboxysome in photosynthetic organisms.
- They prevent protein denaturation during environmental stress, such as elevated temperature.

The secondary structure, and thus the activity, of bovine pancreatic ribonuclease is maintained by disulfide bonds between cysteine residues.

11.4 Export of Proteins

Approximately 20% of the polypeptides synthesized by bacteria are inserted into the cytoplasmic membrane or translocated across the membrane and into the cell envelope or beyond. In *E. coli*, this number is estimated to be about 800 distinct proteins. In most cases, the secretion or targeting of a protein to its extracytoplasmic destination requires the activity of one of several complex secretion machineries. Secreted proteins move into or through the cytoplasmic membrane via two distinct pathways, referred to as the Sec and TAT pathways. Each of these pathways will be discussed separately. Once transferred to the external side of the cytoplasmic membrane, the proteins can be released to the cell exterior, as is the case for secreted proteins of gram-positive bacteria, or into the periplasm, in the case of gram-negative bacteria. The presence of a second membrane (the outer membrane) in gram-negative bacteria means that additional machinery is required to move proteins that have been released into the periplasm to the outer membrane or exterior of the cell. Alternatively, several proteins from gram-negative bacteria can be secreted by a one-step process, in which proteins synthesized in the cytoplasm are translocated to the exterior by machineries (types I to V, see subsequent text) that span the inner membrane, periplasm, and outer membrane.

Signal Sequences

The distinguishing characteristic of a subclass of secretory proteins is the presence of a **signal sequence** (sometimes called a **signal peptide**) of approximately 17 to 22 amino acids at the N-terminus of the pre-secretory protein. The signal peptide functions as a recognition signal for the Sec or Tat machinery, directing the secretory proteins into or across the cytoplasmic membrane. Signal peptides are usually cleaved off during or following the membrane translocation process. There are two distinct types of signal sequence found on pre-secretory proteins, depending on whether they are recognized by the Sec or the TAT pathway. A somewhat different signal sequence is found on pre-lipoproteins, a class of outer membrane proteins that are modified by covalent lipid attachment during export.

Each signal sequence consists of three distinct regions: the n, h, and c regions. The leading or N-terminal region of the signal sequence is termed "n" and is polar, with a net positive charge (Figure 11.9). The middle region, termed "h," has at least 10 hydrophobic amino acids and is inserted into the membrane. The "c" region is also hydrophobic and is recognized by one of two peptidases, called **signal peptidases**, which cleave the signal sequences from the secretory proteins. Signal peptidase I (SPase I) is responsible for cleaving signal sequences from the majority of secreted preproteins utilizing the Sec and TAT pathways, whereas signal peptidase II (SPase II) removes signal sequences from precursors of lipoproteins. A significant fraction of bacterial membrane proteins are not translocated, but instead are integrated into the cytoplasmic membrane via the Sec pathway. These integral membrane proteins contain a sequence at their amino terminus that resembles a signal peptide and is not cleaved off. Once in the membrane these sequences have a second function in anchoring the proteins in the lipid bilayer and thus preventing further translocation.

The Sec-Dependent Protein Secretion Pathway

Secretion of most Sec-dependent pathway proteins in bacteria is mediated by membrane-associated **translo-**

(A) Sec signal peptide

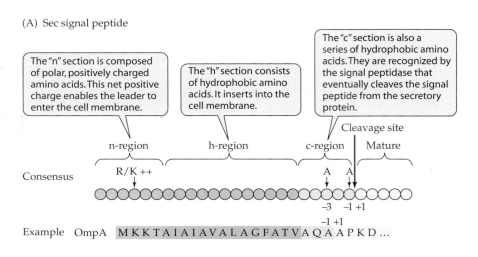

The "n" section is composed of polar, positively charged amino acids. This net positive charge enables the leader to enter the cell membrane.

The "h" section consists of hydrophobic amino acids. It inserts into the cell membrane.

The "c" section is also a series of hydrophobic amino acids. They are recognized by the signal peptidase that eventually cleaves the signal peptide from the secretory protein.

Cleavage site

n-region   h-region   c-region   Mature

Consensus   R/K ++   A   A

−3   −1 +1

Example   OmpA   M K K T A I A I A V A L A G F A T V A Q A A P K D ...

−1 +1

(B) TAT signal peptide

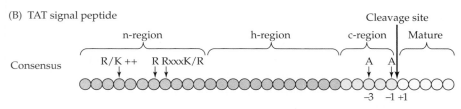

Cleavage site

n-region   h-region   c-region   Mature

Consensus   R/K ++   R RxxxK/R   A   A

−3   −1 +1

Example   DmsA   M K T K I P D A V L A A E V S R R G L V K T T A I G G L A M A S S A L T L P F S R I A H A V D S A ...

*   −1 +1

(C) Lipoprotein signal peptide

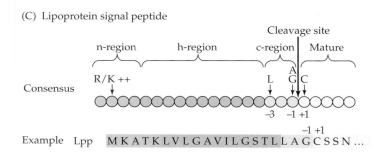

Cleavage site

n-region   h-region   c-region   Mature

Consensus   R/K ++   L   A
G   C

−3   −1 +1

Example   Lpp   M K A T K L V L G A V I L G S T L L A G C S S N ...

−1 +1

Figure 11.9 The tripartite structure of signal peptides
Generic structures of signal peptides showing the n, h, and c regions. The sites cleaved by signal peptidases are between −1 and +1. The amino acid at position +1 is the first amino acid of the mature protein. One example is provided from *E. coli* for each class of secreted proteins utilizing the Sec (A), TAT (B), and lipoprotein (C) export pathways. Note that the lengths of the n, h, and c regions are variable, so the specific examples don't necessarily match the consensus as depicted. OmpA, the outer membrane protein A; DmsA, the A subunit of dimethyl sulfoxide reductase; and Lpp, the Braun lipoprotein. For the TAT-secreted DmsA, the twin arginines are underlined and the asterisk points to the basic amino acid in the c region that is responsible for preventing the protein precursor from entering the Sec pathway.

case complexes. In *E. coli,* this complex is composed of at least three proteins (SecY, SecE, and SecG) that form a hydrophobic channel spanning the cytoplasmic membrane (**Figure 11.10A**). The complex also contains a bound protein, SecA, which provides energy for the translocation by hydrolyzing ATP.

Depending on the amount of delay following the synthesis of the secreted protein, two general routes for protein translocation via the Sec-type translocase complex are followed: **cotranslational** and **posttranslational**. Insertion of integral membrane proteins occurs almost exclusively cotranslationally.

During **cotranslational protein translocation**, the signal sequence (or the N-terminal anchor sequence) emerging from the ribosome attaches to a cytoplasmic chaperone called **signal recognition protein**, or **SRP** (see Figure

11.10A). The mRNA–ribosome–nascent polypeptide–SRP complex then docks to its receptor, FtsY, and perhaps to another, as yet unidentified membrane-bound receptor. These events facilitate the transfer of the pre-protein to the translocase complex. As protein synthesis proceeds on the ribosome, the pre-protein is exported though the translocase complex in an ATP-dependent process assisted by the SecA protein. Many of the newly synthesized proteins remain in the membrane and are not further processed. However, several types of membrane proteins with cleavable signal sequences are routed

Figure 11.10 Secretion of proteins across the cytoplasmic membrane
(A) Post- and cotranslational secretion by the Sec pathway. (B) The TAT translocation pathway.

(A) Sec secretion pathway

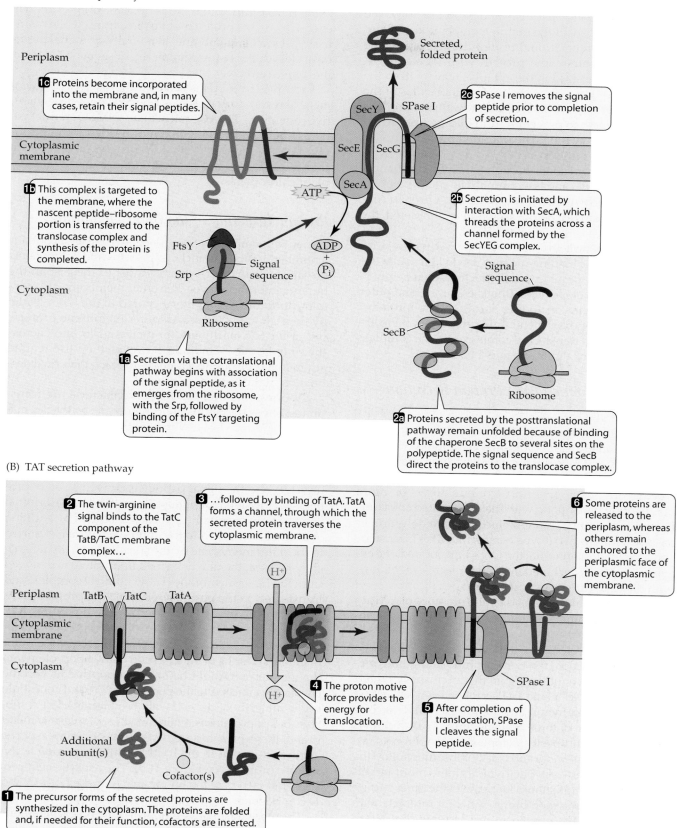

1c Proteins become incorporated into the membrane and, in many cases, retain their signal peptides.

Periplasm

Secreted, folded protein

2c SPase I removes the signal peptide prior to completion of secretion.

SPase I

SecY

Cytoplasmic membrane

SecE   SecG

SecA

ATP

1b This complex is targeted to the membrane, where the nascent peptide–ribosome portion is transferred to the translocase complex and synthesis of the protein is completed.

ADP + P$_i$

2b Secretion is initiated by interaction with SecA, which threads the proteins across a channel formed by the SecYEG complex.

FtsY

Srp

Signal sequence

Signal sequence

Cytoplasm

Ribosome

SecB

Ribosome

1a Secretion via the cotranslational pathway begins with association of the signal peptide, as it emerges from the ribosome, with the Srp, followed by binding of the FtsY targeting protein.

2a Proteins secreted by the posttranslational pathway remain unfolded because of binding of the chaperone SecB to several sites on the polypeptide. The signal sequence and SecB direct the proteins to the translocase complex.

(B) TAT secretion pathway

2 The twin-arginine signal binds to the TatC component of the TatB/TatC membrane complex…

3 …followed by binding of TatA. TatA forms a channel, through which the secreted protein traverses the cytoplasmic membrane.

6 Some proteins are released to the periplasm, whereas others remain anchored to the periplasmic face of the cytoplasmic membrane.

H$^+$

Periplasm   TatB   TatC   TatA

Cytoplasmic membrane

Cytoplasm

H$^+$

SPase I

4 The proton motive force provides the energy for translocation.

5 After completion of translocation, SPase I cleaves the signal peptide.

Additional subunit(s)

Cofactor(s)

1 The precursor forms of the secreted proteins are synthesized in the cytoplasm. The proteins are folded and, if needed for their function, cofactors are inserted.

through the Srp/FtsY pathway and SPaseI removes the signal peptide to yield the mature polypeptide.

During **posttranslational protein translocation**, the secreted pre-protein (the protein with its signal sequence) is synthesized in the cytoplasm away from the SecYEG translocase complex. Several molecules of the soluble chaperone protein SecB bind to different portions of the pre-protein to prevent it from folding prior to export (see Figure 11.10A). The unfolded protein binds to the translocase complex, initially by an interaction with SecA. It is then threaded through a membrane channel formed by SecYEG and is exported. The pre-protein export is driven by ATP hydrolysis via SecA protein conformational changes. Before secretion is completed, the signal sequence is removed by SPaseI.

Examples of proteins secreted by the Sec pathway include periplasmic binding proteins needed for solute uptake, periplasm-located proteins such as alkaline phosphatase, cell envelope biosynthetic enzymes, and outer membrane proteins, including porins and lipoproteins. Given the demand for membrane processing, there are an estimated 500 translocase complexes in the cytoplasmic membrane of a gram-negative bacterium.

The Twin-Arginine (TAT) Protein Secretion Pathway

TAT secretion systems export fully assembled protein complexes across the cytoplasmic membrane. Such systems are present in many but not all bacteria and mainly export enzymes involved in cell respiration. These secreted complexes generally contain three or more polypeptide types along with their associated cofactors. For example, the *E. coli* enzyme, formate dehydrogenase-N, contains two *b*-type hemes, four 4Fe-4S clusters, and a molybdopterin cofactor. Once translocated, the enzyme complex is anchored in the cytoplasmic membrane with the catalytic site exposed to the periplasm.

TAT systems are so named by the presence of a cleavable "twin arginine"–containing motif, RRxxxK/R, located at the N-terminus of one of the secreted polypeptides (see Figure 11.9). Like the signal sequence involved in Sec-type secretion, the twin arginine–containing motif provides a recognition signal that exclusively targets the protein complex to the TAT secretion complex located in the cytoplasmic membrane (Figure 11.10B). This complex is composed of three proteins, TatA, TatB, and TatC. Prior to secretion, specific cytoplasmic chaperone(s) assist in enzyme assembly and cofactor insertion following polypeptide synthesis. The twin-arginine motif targets the assembled multisubunit enzyme to the inner surface of the cytoplasmic membrane, where it interacts with TatC and TatB. This pre-secretory enzyme–TatC–TatB complex then combines with TatA. Following this interaction, a channel opens in TatA, through which the en-

zyme is transported into the periplasm. Translocation is energized by the proton motive force (pmf) while maintaining a sealed membrane. In a final step, signal peptidase cleaves the twin-arginine signal sequence to yield the mature enzyme.

Examples of proteins secreted by the TAT pathway include respiratory enzymes (formate dehydrogenase, hydrogenase, TMAO reductase, DMSO reductase [DmsABC]) and certain enzymes needed for peptidoglycan assembly. Related TAT secretion pathways are found in *Archaea* and in the thylakoid membranes of plant chloroplasts.

Protein Targeting to the Outer Membrane

The outer membrane of gram-negative bacteria contains a number of proteins, including porins and components of various transport systems; most of these span the outer membrane bilayer. The outer membrane also contains lipoproteins that are anchored in the inner leaflet by their attached lipids. All outer membrane proteins utilize the Sec machinery for their translocation across the cytoplasmic membrane; however, they also require additional transport machineries to reach their final destination in the outer membrane.

Lipoproteins in gram-negative bacteria are transported across the cytoplasmic membrane by the Sec machinery (Figure 11.11). When the pre-protein is translocated to the periplasmic face of the inner membrane, two fatty acids are attached via glycerol to the cysteine thiol group located at the signal peptide cleavage site (see Figure 11.9), and the signal peptide is cleaved by a lipoprotein-specific signal peptidase (SpaseII). The N-terminal amino group of cysteine is further modified by addition of another fatty acid and the lipoprotein is then exported to the outer membrane by the so-called LolABCDE machinery. First, the newly formed lipoprotein is removed from the inner membrane by the LolBCD complex, and transferred to the periplasmic shuttling protein LolA. This is an energy-dependent process and requires ATP hydrolysis by LolD. Interaction of the LolA–lipoprotein complex with LolB at the periplasmic face of the outer membrane results in the insertion of the lipoprotein into the membrane. About half of the lipoproteins become covalently attached to the peptidoglycan, and contribute to the integrity of the cell wall (see Figure 4.55). A subclass of lipoproteins is retained in the cytoplasmic membrane; these proteins are distinguished by an aspartic acid residue next to the lipid-modified cysteine in the mature protein (see Figure 11.9). The aspartic acid prevents interaction of the lipoprotein with the LolB, C, and D components of the transport machinery.

The mechanism of assembly of other outer membrane proteins, many of which function as multimeric channels, is not completely understood. The newly synthe-

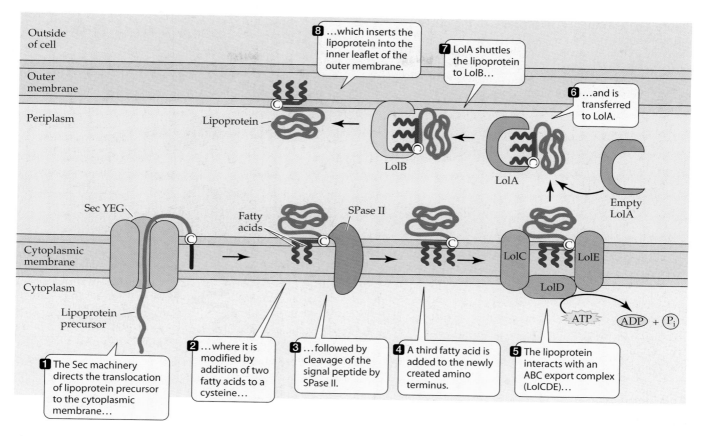

Figure 11.11 Transport of lipoproteins to the outer membrane
Transport of lipoproteins to the outer membrane involves the Lol machinery and special modifications of the amino terminus.

sized pre-proteins are first translocated through the Sec machinery, followed by removal of their signal peptide by SPase I (**Figure 11.12**). It is likely that two chaperones, Skp and SurA, are responsible for preventing aggregation of these proteins and shuttling them to the outer membrane insertion complex, which consists of YaeT (sometimes called Omp85), YfgL, YfiO, and NlpB proteins. Here the proteins assemble into functional integral outer membrane proteins.

Extracellular Secretion by Gram-Negative Bacteria

A number of bacterial proteins are not retained in any of the cellular compartments but are instead secreted from the cell. For gram-positive bacteria, which have a single cytoplasmic membrane, secretion of a protein across this membrane is sufficient for its release into the surrounding medium, unless the protein is modified such that it becomes attached to cell wall components. In contrast, secretion from gram-negative bacteria requires transfer across two membranes. Various gram-negative bacteria

have evolved specialized machineries of varying complexity to accomplish extracellular secretion. Several different secretion routes can coexist in a single bacterial cell and each protein must be targeted to the correct one. For proteins to be recognized by their cognate secretion machinery they possess a signature (an "address")—typically a short sequence or a structural motif that is part of the polypeptide chain.

Here we review the major extracellular secretion pathways of gram-negative bacteria. They have been numbered type I-V, based on the order of their discovery.

TYPE I SECRETION PATHWAY   This secretion pathway utilizes the so called **ABC machinery**—named after its ABC (for $\underline{A}$TP-$\underline{b}$inding $\underline{c}$assette) components (**Figure 11.13**). The type I machinery consists of a complex of three different proteins, each located in a different compartment of the cell. The ABC component is located within the cytoplasmic membrane, and is linked to an outer membrane protein (OMP) through a periplasmic bridge—the so-called membrane fusion protein (MFP). The type I secretion complex assembles only after initial

Figure 11.12 **Transport of proteins to the outer membrane**

Outer membrane proteins enter the periplasm via SecYEG. Chaperones (SurA, Skp) transfer the newly synthesized protein to an outer membrane protein complex, which inserts the protein into the outer membrane.

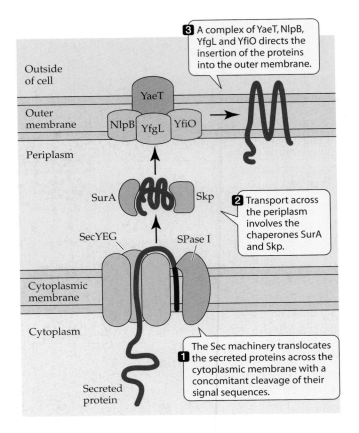

3 A complex of YaeT, NlpB, YfgL and YfiO directs the insertion of the proteins into the outer membrane.

2 Transport across the periplasm involves the chaperones SurA and Skp.

1 The Sec machinery translocates the secreted proteins across the cytoplasmic membrane with a concomitant cleavage of their signal sequences.

Figure 11.13 **The type I, IV, and V secretion pathways**

Three of the five pathways used by gram-negative bacteria to secrete proteins to the outside of the cell.

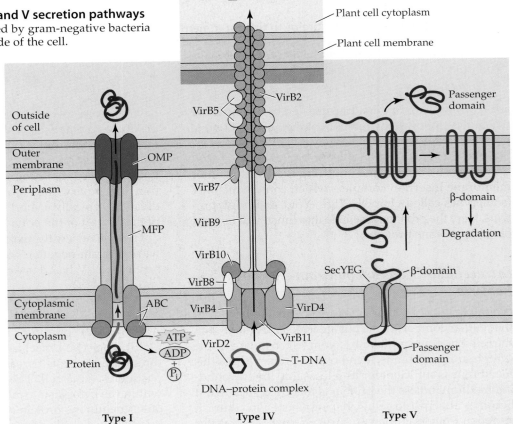

interaction of the secreted protein with the ABC component, which then triggers the assembly of the rest of the machinery. An active type I secretion apparatus consists of a dimer of the ABC component plus two trimers: one of Mfp and one of Omp. The energy for protein export is provided by ATP hydrolysis, which is carried out by the ABC component.

The proteins secreted by this pathway can be as large as 1,000 amino acids. Because they do not utilize either the Sec or TAT pathway, they lack a cleavable N-terminal signal peptide. The signal for routing these proteins through the type I secretion pathway is usually located at their C-terminal end and is not cleaved during secretion. Many of the secreted proteins have distinctive glycine-rich repeats—G-G-x-G-X-D-x-x-x (where x is any amino acid)—that specifically bind calcium. Given the large size of the secreted proteins that utilize this pathway, they are transported in an unfolded state and then are folded immediately following their exit. It is likely that calcium is required for their correct refolding.

Among the many secreted proteins that utilize the type I pathway are the hemolytic protein of *E. coli*, the secreted proteases of *Pseudomonas aeruginosa* and *Erwinia chrysanthemi*, the adenylate cyclase toxin of *Bordetella pertussis*, and the iron-sequestering protein of *Serratia marcescens*.

TYPE II SECRETION PATHWAY   Secretion via the **type II secretion pathway** is a two-step process. Proteins are initially synthesized with an N-terminal signal peptide, which targets them for secretion via the Sec pathway and results in a transient periplasmic localization (**Figure 11.14**). In the periplasm, these proteins fold into their final conformation and enter the secretion pathway that will transport them across the outer membrane. The machinery of type II secretion is referred to as the **general secretory apparatus** (GSP). The precise mechanism of protein transport by this apparatus is not well understood. One of the proteins of the GSP machinery, GspE, has a nucleotide-binding domain and probably provides energy for secretion by hydrolysis of a nucleotide, pos-

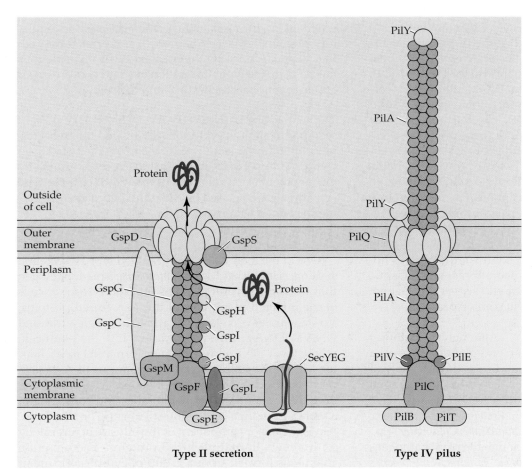

Figure 11.14 The type II secretion pathway and the type IV pilus
A comparison of the structure and components of the type II secretion apparatus and the type IV pilus.

sibly ATP. GspE is found in the cytoplasmic membrane, where it associates with two other membrane proteins, namely GspL and GspF. A filamentous structure—consisting largely of GspG plus a few molecules of GspJ, GspI, and GspH—spans both the inner and outer membranes and connects the cytoplasmic membrane components with a ring-like structure of assembled GspD monomers in the outer membrane. This filament is referred to as pseudopilin, because it resembles the type IV class of pili (see subsequent text).

The secreted proteins fold into their final conformation in the periplasm, and in some cases, such as cholera toxin, they assemble into their final multimeric complexes prior to release from the cell. Because the type II secretion machinery must differentiate between proteins that are destined to remain in the periplasm and those that are to be secreted, the latter proteins must contain specific extracellular secretion signals. However, the precise sequence or structural motifs that define these signals have not yet been identified.

The type II secretion pathway is one of the major mechanisms utilized by bacterial pathogens to secrete protein toxins, which can affect organs that are located far from the site of colonization (see Chapter 26).

One class of appendages, the type IV pili on the surface of certain gram-negative bacteria, resembles the assembled type II secretion apparatus (see Figure 11.14). The type IV pili are used primarily as adhesins by certain pathogenic bacteria to attach themselves to solid surfaces, including tissues of infected humans (see Chapter 26). The main feature of these pili is the presence of a long filament—a multisubunit structure—anchored in the cell envelope. Many of the components of type IV pili are similar in amino acid sequence to those of the type II secretion apparatus, and therefore they probably function in an analogous manner (see Figure 11.14). The type IV pili do not secrete proteins but instead direct the export and polymerization of the pilus subunit (the PilA protein). In contrast, the protein homolog to PilA in the type II secretion system, GspG, forms the pseudopilin structure and serves as a channel for the secretion of a number of soluble proteins.

TYPE III SECRETION PATHWAY   Although this pathway is common among mammalian and plant pathogens, a closely related mechanism that directs the assembly of bacterial flagella has a much wider distribution among gram-negative bacteria. Type III secretion systems have been identified in *Yersinia* spp., *Pseudomonas aeruginosa*, *Shigella flexneri*, *Salmonella typhimurium*, *Chlamydia* spp., and certain pathogenic *E. coli*, and in the plant pathogens or symbionts *Pseudomonas syringae*, *Erwinia*, *Xanthomonas* spp., and rhizobia.

The unique feature of the type III pathway for host/pathogen interactions is the ability of the secretion machinery to target the secreted proteins directly into eukaryotic cells. Typically, bacteria adhere to the target cells, bringing their surfaces into close proximity. The external tip of the secretion apparatus is inserted into the host plasma membrane, allowing the secreted protein to be delivered directly into the eukaryotic cell without ever being released into the surrounding medium. The proteins are "injected" into the target cell cytoplasm, with the bacterium being the syringe and the secretion apparatus the needle. The complete type III secretion machinery, sometimes referred to as an "injectosome," consists of 20 components that span the cytoplasmic membrane, the periplasm, and the outer membrane. The secretion signals that target proteins to the type III secretion pathway are approximately 20 amino acids long and are located at the N-termini of the secreted proteins. They consist of an equal mix of hydrophobic and charged amino acids and include several serines. The signals are not cleaved during secretion.

Several components of the type III secretion apparatus are similar to components of a bacterial flagellum, and the flagellum plays a limited role in secretion. The flagellum monomers (FliC) that are added to the tip of the filament during flagellum assembly and the regulatory protein FlgM (see Chapter 13) are secreted through the channel formed by the basal body. By an analogous mechanism, the type III secretion machinery of a *Yersinia* species exports its toxins to a targeted cell.

TYPE IV SECRETION PATHWAY   The type IV protein secretion system is evolutionarily related to the bacterial conjugation machinery. It resembles the type III secretion pathway in its ability to transfer proteins directly into another bacterium or into a eukaryotic cell. The type IV pathway is also used by a handful of proteins, including the *Bordetella pertussis* toxin, which are secreted into the medium instead of being transferred into the cytoplasm of another cell.

One of the most extensively studied type IV secretion systems is in *Agrobacterium tumefaciens*, a plant pathogen that transfers several proteins into the cytosol of cells of the infected plant. One of the transferred proteins, VirD2, carries a covalently attached single-stranded DNA (T-DNA) from *A. tumefaciens* into the plant cell, where the T-DNA induces tumor formation (see Chapter 16). The assembled type IV apparatus of *A. tumefaciens* (see Figure 11.13) is a multisubunit structure spanning the inner and outer membranes and consisting of a secretion channel and a pilus filament. The VirB2 protein is the major component of the pilus structure, along with a few alternating VirB5 proteins. The secretion channel is formed from a group of cytoplasmic membrane proteins, VirB4, VirB6, VirB8, and VirB11, and two additional proteins, VirB7 and VirB9, which form an anchor for the pilus in the outer membrane and connect it to the cytoplasmic components of the apparatus. VirB4 and VirB11 are inner membrane proteins that are ATPases and provide the energy for the transfer process.

In addition to *A. tumefaciens* and *B. pertussis*, a number of intracellular parasites (bacteria that grow inside infected host cells), including *Legionella*, *Brucella*, and *Bartonella* species, utilize the type IV secretion pathway to deliver proteins into the cytoplasm of their host cell. These proteins facilitate bacterial survival and growth within the host cell.

TYPE V SECRETION PATHWAY   This is the simplest of the gram-negative extracellular secretion pathways and does not use any energy for protein transport. The type V secretion pathway is sometimes referred to as **autotransporter** secretion. As can be inferred from this name, the proteins promote their own transfer across the outer membrane. Proteins secreted by this mechanism are relatively large, and share a similar overall domain organization. They are synthesized with an N-terminal signal peptide followed by the **passenger domain**, which specifies the protein's extracellular function. At their carboxy-termini, these proteins contain another domain, the **β-domain**, which forms a pore in the outer membrane. The secretion of these proteins is initiated by the Sec pathway, which cleaves the signal peptide and releases the protein into the periplasm. During the next stage of export, the β-domain inserts into the outer membrane and forms a barrel-like structure, through which the passenger domain is then threaded. Some proteins secreted by this pathway remain associated with the outer surface of the cells, whereas others are released following cleavage of the bond between the passenger and β-domains.

Several important secreted proteins of pathogenic gram-negative organisms utilize the type IV secretion pathway. These include the antibody-specific protease of *N. gonnorhoeae*, the protease of *H. influenzae*, and the vacuolating toxin produced by *H. pylori*. A number of surface adhesins—proteins that mediate binding of bacteria to tissues, such as the protractin of *B. pertussis* and the BabA adhesin of *H. pylori*—are also autotransporters.

Protein Secretion in Lower Eukaryotes

In fungi, the production and secretion of enzymes and other proteins is fundamentally different from that in bacteria. Fungi, of course, are eukaryotes, and fungal proteins are synthesized at the rough endoplasmic reticulum; they are then transported in vesicles and enter the Golgi apparatus. Vesicles then bud from the Golgi and are transported to the inner surface of the cell membrane. At the site of secretion, the membrane of the protein-containing vesicle fuses with the cytoplasmic membrane and the protein is released to the outside. Interestingly, SRP-type proteins are employed, as are TAT-like systems, in certain eukaryotic processes. The origin of these processes is arguably prokaryotic.

SECTION HIGHLIGHTS

The synthesis and folding of cytoplasmic proteins generally occurs spontaneously. However, the synthesis of proteins destined to the cell envelope often requires chaperone proteins. Movement of these proteins into or across the cytoplasmic membrane generally requires the assistance of specialized protein secretion machinery, usually the Sec or TAT apparatus. Proteins destined for the outer membrane or beyond require additional specialized machinery; these are termed type I to V secretion pathways. Regardless of the mechanism used, protein secretion involves the recognition of specific targeting signals on the secreted protein and energy in the form of ATP or ion gradients to drive the movement.

11.5 Structure and Function of Cell Appendages

Flagella and pili are composed of multiple protein subunits assembled into macromolecular structures outside the cytoplasmic membrane. Their synthesis, assembly, and function are described below.

Bacterial Flagella

Flagella are specialized cellular appendages that allow directed cell movement under appropriate environmental conditions. They are multiprotein assemblies that span the cell envelope and extend several cell lengths beyond the cell surface. Some types of microbes possess a single flagellum located at one pole of the cell (*Pseudomonas*, for instance), while others, *E. coli*, for example, possess multiple flagella distributed uniformly about the cell (see Chapter 4). Flagellar movement is controlled by the cell in response to environmental stimuli that allow the cell to seek out nutrients and to avoid toxic materials that create unfavorable conditions for survival.

STRUCTURE   A flagellum consists of three major parts:

- A long filament, which extends into the surrounding environment

- A hook, a curved section connecting the filament to the cell surface

Figure 11.15 Flagellar structure
(A) Model of a flagellar cross section showing the location of the rings and the Mot proteins in the cell envelope of a gram-negative bacterium. (B) Electron micrograph showing the basal ring structure of a gram-negative flagellum. B, courtesy of David DeRosier, Brandeis University.

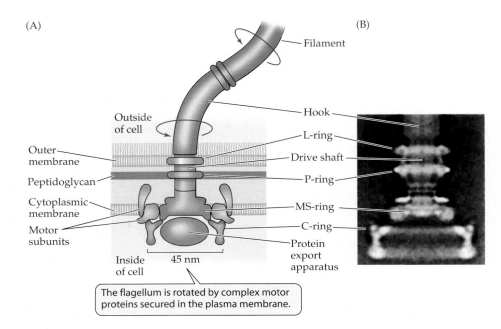

(A)

(B)

Filament

Outside of cell

Hook

L-ring

Drive shaft

P-ring

MS-ring

C-ring

Protein export apparatus

Outer membrane

Peptidoglycan

Cytoplasmic membrane

Motor subunits

Inside of cell

45 nm

The flagellum is rotated by complex motor proteins secured in the plasma membrane.

• A basal structure, which anchors the flagellum and hook into the cell wall and membrane by special disk-shaped structures called plates or rings

The basal structure for a typical gram-negative bacterium contains a rod structure plus L, P, MS, and C rings that associate with cell envelope components (**Figure 11.15**) to anchor it to the cell. Several of the rings (MS and C) function as parts of a molecular motor that, along with the motor shaft protein (FlgG), spins or rotates the hook and filament assembly in a clockwise or counterclockwise direction. This action propels the cell in three-dimensional space in response to external signals including chemicals, light, and magnetism. The motion, sometimes called "swimming," confers selective advantage to motile cells in seeking out environmental niches favorable for growth/survival.

The **hook** is composed of a single protein type (FlgE) that transmits the rotary movement of the motor shaft to the filament (see Figure 11.15). Several additional proteins are involved in joining it to the filament.

The **flagellar filament** is composed of a protein (FliC) called flagellin. With exceptions, bacteria generally have a single type of flagellin subunit (*Caulobacter* and some *Vibrio* spp. have two subunit types). Repeating subunits of flagellin are joined together to form the filament, which is helically wound and has a hollow core that is capped by FliD (**Figure 11.16**). The filament (and thus the flagellum) is about 20 nm in diameter. The filaments are not straight, and each has a characteristic sinusoidal curvature that varies with each species.

MOVEMENT   Flagellar rotation, and hence cell movement, requires energy; it is fueled by the proton motive force (pmf; see Chapter 8). Some bacterial species use protons, while others, especially marine microorganisms, use sodium ions to drive rotation. This energy is imparted to the **Mot proteins (MotA and MotB)**, which are embedded in the cell membrane and act as a motor that drives flagellar rotation. The rate of rotation varies among species but is in the range of 200 to more than 1000 rpm. Motility requires a major expenditure of energy, but the result is immensely useful because it allows the cell to "swim" to a place where conditions for growth are more favorable.

The flagella of enteric bacteria have been studied intensively. More than 50 genes involved in motility have been identified in *Salmonella typhimurium* and *E. coli*. The assembly and function of these structures include the

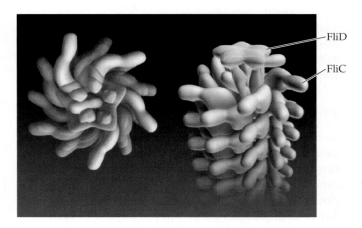

FliD

FliC

Figure 11.16 Flagellar filament and flagellin subunits
Top (left) and side (right) views of the flagellar filament with flagellin subunits and cap structure. Courtesy of Keiichi Namba.

genes responsible for producing the flagellin subunit, hook, basal body structure with associated motor proteins, and additional genes that regulate flagellar biogenesis. Another set of genes encode proteins making up the sensory system that detects environmental signals and, in response, directs the flagellar motor to change the direction of rotation. The functions of these structures in signal gathering and processing are described below.

ASSEMBLY   The synthesis and assembly of flagella is controlled temporally (time of event) and spatially (location in the cell); these processes are initiated by the induction of gene expression in response to environmental conditions. The 50 or so flagellar (*fla*) genes needed for this to occur are divided into three groups called the "early," "middle," and "late" genes based on the timing of their expression. The early genes encode key regulatory proteins which, upon induction, direct transcription of the middle genes, followed by the late genes. Control of gene expression allows for the synthesis of certain proteins before others, so that the assembly of polypeptides into a mature flagellum occurs in the proper sequence. The middle genes encode parts of the basal body structure and hook. The late genes encode additional proteins needed for motor assembly, filament assembly, and chemosensory proteins that control flagellar movement. Several additional *fla* genes encode chaperone proteins needed to assist the structural proteins to their destinations.

Assembly of a new flagellum occurs in several highly regulated stages. The process begins with synthesis and assembly of the MS-ring proteins at the cytoplasmic membrane surface (see Figure 11.15). This is followed by the ordered synthesis and insertion of other proteins composing the basal body, including the motor shaft proteins; the C-, P-, and L-ring proteins; and the hook. Products of the late genes are then synthesized and assembled. These include the addition of motor stator proteins (MotA and B) and motor switch proteins (FliM, N), and the buildup of the filament (FliC). A specialized transport apparatus composed of at least eight proteins directs protein secretion and assembly of the mature structure; this occurs by **type III secretion**. To reach their final destination at the growing tip of the filament, FliC subunits are transferred through the hollow core of the basal body shaft, hook, and growing filament, where they self-assemble into the helically wound structure. By a mechanism that is not well understood, filament growth, once it extends to several cell lengths, is ultimately terminated by the action of another filament protein (FliD), which caps the tip (see Figure 11.16).

TACTIC RESPONSES: CHEMOTAXIS, AEROTAXIS, AND PHOTOTAXIS   Heterotrophic bacteria use their flagella

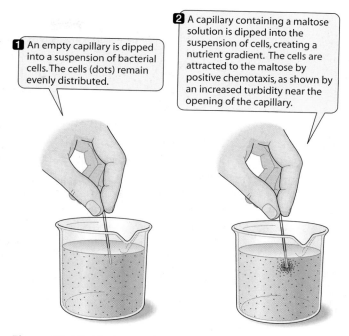

1 An empty capillary is dipped into a suspension of bacterial cells. The cells (dots) remain evenly distributed.

2 A capillary containing a maltose solution is dipped into the suspension of cells, creating a nutrient gradient. The cells are attracted to the maltose by positive chemotaxis, as shown by an increased turbidity near the opening of the capillary.

Figure 11.17 Chemotactic capillary assay
An illustration of the capillary chemotactic effect.

to propel themselves into areas where organic nutrients are concentrated. This phenomenon, termed **chemotaxis**, can be illustrated with a capillary technique. If a glass capillary containing an organic energy source such as maltose is placed in a suspension of heterotrophic bacteria, the bacteria will accumulate at the tip of the capillary from which the nutrient is diffusing (Figure 11.17).

Chemotaxis has been thoroughly studied in enteric bacteria such as *E. coli*. These peritrichously flagellated bacteria orient their flagella in a polar bundle during motility. The bundle is rotated counterclockwise during forward movement, called running or smooth swimming. Running is disrupted when the flagella reverse to the clockwise direction: the cell stops and then somersaults—a movement referred to as tumbling. After a moment, the flagella reverse again to rotate in a counterclockwise direction, and the organism begins another run for an extended period. However, the direction of cell movement is completely random. When the organism is in the gradient of increasing nutrient, fewer tumbling events occur, allowing the organism to spend more time in the favorable area (Figure 11.18). This phenomenon, in which an organism effectively becomes directed toward a utilizable nutrient—an attractant—is **positive chemotaxis**. **Negative chemotaxis**—movement away from toxic materials, called repellants—is also commonly exhibited by bacteria.

Many utilizable organic nutrients serve as **attractant molecules**, causing positive chemotaxis. In *E. coli*, these would be various sugars and amino acids (Figure 11.19).

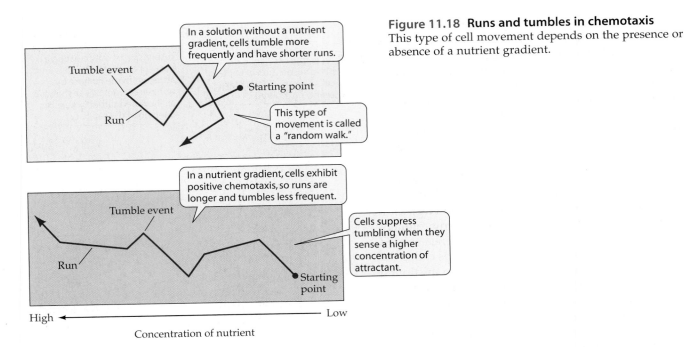

Figure 11.18 Runs and tumbles in chemotaxis
This type of cell movement depends on the presence or absence of a nutrient gradient.

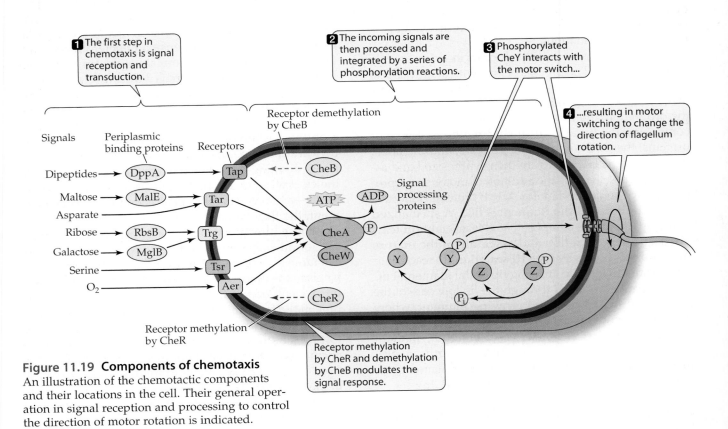

Figure 11.19 Components of chemotaxis
An illustration of the chemotactic components and their locations in the cell. Their general operation in signal reception and processing to control the direction of motor rotation is indicated.

The molecules are bound to specific receptor proteins located either in the periplasmic space or the cell membrane. For the amino acids serine and aspartate, binding occurs at the specific membrane-bound chemoreceptors called Tsr and Tar, respectively. Other attractants are bound to specific binding proteins, located in the periplasmic space, that interact at the chemoreceptor (for example the maltose-MalE complex binds to Tar). *E. coli* possesses five different chemoreceptors; each exhibits specificity toward a subset of the attractants and repellants. In the presence of their signal molecules, each receptor transfers a message across the membrane to the CheA-CheW proteins, which, in turn, process and integrate all the newly received information from outside the cell (i.e., the periplasm). Based on the information accumulated, the flagella are then instructed to either rotate counterclockwise or clockwise (motor switching), thus directing cell movement toward attractants and away from repellants. For *E. coli*, **repellant molecules** include ethanol, phenol, and certain heavy metals. The chemical signaling events that lead to directed cell movement are described in Box 11.3. Note that each chemoreceptor molecule exists in the cell as a dimer complexed with two molecules each of CheA and CheW. Depending on growth conditions, there may be over 10,000 receptor complexes per cell!

It is noteworthy that in chemotaxis, both spatial gradients and temporal changes in the chemical environment have an effect on cell movement. Adaptation of the organism to its environment also occurs in chemotaxis. In the prolonged presence of the attractant substrate, the rate of tumbling is less than it would be in the absence of the attractant (see Box 11.3). This may be a mechanism whereby the organism can conserve energy, when it is not needed, that would be needlessly spent on changing direction.

The simple behavior exhibited by bacteria during motility is of great interest to biologists, because an understanding of this behavior at the molecular level may lead to a fuller understanding of more complex behavioral responses of eukaryotic organisms, such as the sense of smell. Bacteria like *E. coli* and *S. typhimurium* continue to provide excellent models for unraveling the highly sophisticated mechanisms needed for protein secretion, assembly, and cell behavior.

Flagella are also involved in the phenomena of aerotaxis and phototaxis. The former process involves a special receptor protein called Aer (see Figure 11.19), which detects oxygen and transmits a signal through the chemotactic machinery to control flagellar motor rotation. Depending on the bacterial species, it may seek oxygen at low, intermediate, or high levels or avoid it completely. Phototaxis occurs in many types of photosynthetic bacteria. This light-associated phenomenon was described in Chapter 4 (see Figure 4.68); it allows a cell to position itself at the specific wavelengths used for photosynthesis.

Bacterial Pili

Pili are hair-like appendages on many gram-negative bacteria and are involved in attachment, motility, and genetic exchange (see Chapter 4). Structures with similar appearance can also be found as components of a number of extracellular secretion machineries (see Figure 11.14). The various classes of pili have somewhat different structures, roles, and routes of assembly. Many of the functions of pili involve interactions with solid surfaces in a cell's environment, including other cells. These include conjugation pili involved in DNA transfer between cells (see Chapter 15), in cell adhesion, or in cell motility, where the pilin subunits form a hair-like structure that pulls the bacteria along a solid surface or towards other bacteria. Regardless of type, pilus synthesis is an energy consuming process and involves chaperone proteins to prevent misfolding of subunits prior to translocation and assembly. Several examples follow.

One type of pilus involved in bacterial adhesion—termed the **type I pilus**—was described in Chapter 4 (see Figure 4.71). *E. coli* strains that infect the human urinary track use pili to attach to the surface of eukaryotic cells in order to colonize this niche (see Chapter 26). These pili are assembled as shown in Figure 11.20. Although the P pili and type I pili (as well as related pili on *E. coli* and other *Enterobacteriaceae*) are similar in their structure and assembly mechanisms, the primary determinants of their specificity for different human tissues are the unique proteins located on the tip of the filament (PapG in Figure 11.20), which recognize specific receptors on the surface of the mammalian cells (see Chapter 26).

Certain types of pili are involved in cell movement by a mechanism termed **gliding**. This process is distinct from chemotaxis because the cell employs a retractable pilus to propel itself across a solid surface (Figure 11.21). *Myxobacter xanthus*, for example, assembles multiple **type IV pili** at the poles of the cell. The extended pili at one pole attach to the solid surface in front of the cell via an adhesion protein at the tip. Retraction of the pili then pulls the cell forward. This retraction occurs by depolymerization and removal of the pilin subunits that make up each pilus filament. Pilus extension sets the stage for another cycle of cell movement. The cell can reverse direction using the pili at the opposite pole, although the control of these gliding events is not understood. Type IV pili are related to the type II secretion apparatus (see Figure 11.14) and are used by many bacterial pathogens to attach to mammalian cells during infection.

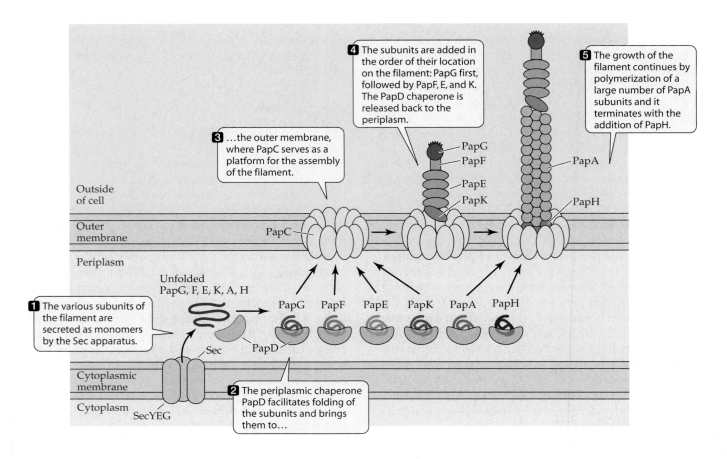

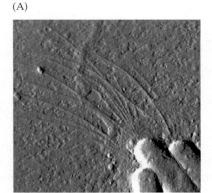

(A)

Figure 11.21 *Myxococcus xanthus* **type IV retraction pilus involved in cell motility.**
(A) Atomic force micrograph of *Myxococcus xanthus*. Note the polar location of retraction pili needed for gliding. (B) Diagram showing the cycle of extension and retraction of pili during gliding. A, courtesy of J. K. Gimzewski.

(B)

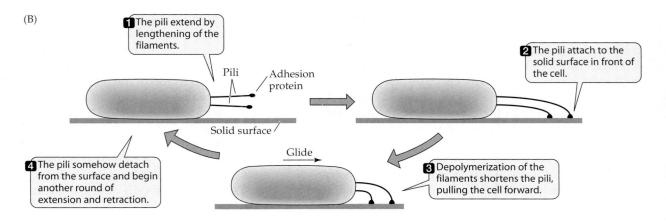

◀ **Figure 11.20 Assembly of the P pilus**
The P pilus is assembled from subunits secreted into the periplasm. PapC, the outer membrane component of the pilus, is assembled first and it serves as a platform for polymerization of the remaining subunits that make up the filament. Components that are at the tip of the filament are added first, followed by the components that are closest to the cell surface. A periplasmic chaperone, PapD, plays an important role in the assembly process; it prevents misfolding and premature aggregation of the major (PapA) and minor (PapG, F, K, and H) subunits of the filament.

BOX 11.3 | *Milestones*

Chemotaxis Signaling

How does the chemotaxis signaling process operate at the molecular level? If the concentration of an attractant molecule is *lower* than a prior cell signal determination, the chemoreceptors are unable to efficiently interact with the CheA–CheW proteins. This leads to CheA autophosphorylation, where ATP is the phosphate donor. CheA-phosphate then donates this phosphate to the CheY messenger protein to give the activated form, CheY-phosphate. Since CheY-phosphate (and CheY) is a soluble cytoplasmic protein, it is free to diffuse throughout the cell. If CheY-phosphate levels are sufficiently high, it binds to the CheY-phosphate receptor protein complex (FliM–FliN) located at the base of the flagellar motor (see Figure 11.15). This causes the motor to reverse direction to clockwise rotation (see the figure), thus inducing a cell tumble event. However, the flagella quickly reverse direction again (counterclockwise) and initiate a period of smooth swimming (see the figure). Since the cell is continually sampling its environment, the length of the smooth run is dependent on signal intensity. Short runs report decreasing levels of attractant, whereas longer swimming runs report increasing attractant level(s).

Let us examine the chemosensory process when the attractant concentration is *higher* than a prior cell

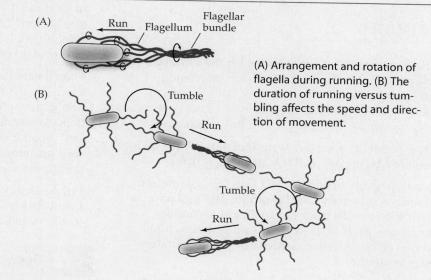

(A) Arrangement and rotation of flagella during running. (B) The duration of running versus tumbling affects the speed and direction of movement.

signal determination. The chemoreceptors report the current signal level to CheA–CheW as before. However, the elevated attractant signal suppresses CheA autophosphorylation, and this in turn results in a low CheY-phosphate level that is insufficient to trigger motor reversal. This permits a longer period of smooth swimming (i.e., suppresses cell tumbling). For some sugars serving as attractant molecules, positive chemotaxis can occur at concentrations as low as 10^{-8} M. Recall that a reverse signaling relationship applies for repellants.

An additional level of chemotaxis control involves chemoreceptor methylation and demethylation. This highly complex process allows for the modulation of signal transfer from the Tap, Tar, Trg, and Tsr chemoreceptors to the CheA–CheW

processing proteins and results in the ability of the cell to momentarily remember attractant and repellant conditions. This chemical memory device allows temporal (time) and spatial (space) comparisons. Methylation is accomplished by a methyltransferase (CheR), an enzyme that uses *S*-adenosylmethionine as a substrate. The process of chemoreceptor binding and subsequent methylation of the **methyl-accepting chemoreceptor** (termed **methyl-accepting chemotaxis protein**, or **MCP**) results in an attenuated signal. Demethylation is accomplished by a methylesterase (CheB), an enzyme that relieves the attenuated signaling by the chemoreceptor, thus increasing signal response. Aer signaling is independent of CheB and CheR.

SECTION HIGHLIGHTS

Formation of cell appendages requires specialized protein secretion and assembly machinery. Flagella and many pili are generated using a type III secretion process. The functioning of these assemblies differs considerably, and range from taxis to cell attachment, DNA transfer, and toxin delivery.

SUMMARY

- The **monomers** that are essential for the synthesis of cell matter are **polymerized** to form macromolecules, and these are the functional components of the cell. Major macromolecules in a cell are proteins, phospholipids, DNA, RNA, and polysaccharides.

- Over one-half of the dry weight of a bacterial cell is **protein**. About 20% is **RNA** and 10% is **lipid**.

- A **polypeptide** may become a functional **protein** by assuming a **tertiary configuration**. The tertiary configuration results from the joining of **secondary structures** such as the α-**helix** and β-**sheets**. Joining to two or more tertiary structures yields a **quaternary structure**.

- Cytoplasmic membranes are formed from bilayers of phospholipid.

- About 20% of the polypeptides synthesized by bacteria are **integrated** into the cytoplasmic membrane or **translocated** across the membrane.

- Most proteins are translocated across the membrane as they are **translated** on a ribosome. This process is termed **cotranslational translocation**.

- A **chaperone** is a protein that assists in the folding of a polypeptide chain to assume the proper functional configuration.

- The **Sec-dependent secretion pathway** proteins identify a specific protein export signal, termed a **signal sequence**, located at the N-terminus of a secreted protein. Several alternative pathways exist that allow membrane targeting.

- The **TAT-dependent protein secretion pathway** targets fully folded and assembled protein complexes to the cytoplasmic membrane. This secretion system recognizes a distinct targeting signal located on one of the secreted polypeptides.

- The structural component of a bacterial cell wall is **peptidoglycan**. It is composed of sugar amines that are complexed with peptides. Synthesis of peptidoglycan units occurs in the cytoplasm and within the cytoplasmic membrane. Final assembly occurs outside the cytoplasmic membrane.

- Five different extracellular secretion pathways function in bacteria to deliver proteins to their exterior.

- **Gram-positive** microorganisms have a thick peptidoglycan layer whereas a **gram-negative** bacterium has a thinner peptidoglycan layer. The gram-negative bacteria have an elaborate cell envelope outside the peptidoglycan layer that requires specialized protein export machines for its construction.

- **Flagella** are highly specialized appendages that confer on the cell the ability "swim." Their formation requires the synthesis of numerous proteins that are assembled in a temporal and spatial fashion. When instructed by the associated **chemosensory proteins**, the cell exhibits the ability to seek attractants and avoid repellants.

- **Pili** perform varied roles and, as a result, contain different structural proteins and are assembled by different means, aided by distinct chaperone proteins.

 Find more at www.sinauer.com/microbial-life

REVIEW QUESTIONS

1. Proteins make up over half of the dry weight of a cell. What are the diverse roles that proteins play and where are the various types located?

2. Define primary, secondary, tertiary, and quaternary protein structural information. Draw a peptide bond of a peptide composed of several amino acids.

3. What are the two major secondary structures that a protein may assume? How do they differ, and what is our shorthand for depicting them? What is the major bonding that holds polypeptides in these configurations?

4. What are some structures (see question 3) that are involved in the functional configuration of proteins?

5. Draw a phospholipid. Which is the hydrophobic and hydrophilic end of the molecule? What happens when a cell is shifted to grow at an acceptably higher temperature?

6. How can a peptide pass through the hydrophobic cytoplasmic membrane? What is the difference between cotranslational insertion versus post-translational insertion? What role might chaperones play in secretory protein translocation?

7. What types of proteins are exported by the TAT secretory pathway and why?

8. Where are the major components of a gram-negative outer envelope assembled? Discuss the processes involved.

9. Outline several distinct routes for export of proteins to the cytoplasmic membrane, and to the outer membrane.

10. How do pili types differ in structure and function?

SUGGESTED READING

Berg, H. C. 2003. "The rotary motor of bacterial flagella." *Annu Rev Biochem* 72:19–54.

Garrett, R. H. and C. M. Grisham. 1999. *Biochemistry*. 2nd ed. Fort Worth, TX: Saunders College/Harcourt Brace Publishing.

Gerhardt, P. R., G. E. Murray, W. A. Wood and N. R. Krieg. 2007. *Methods for General and Molecular Bacteriology*. 3rd ed. Washington, DC: ASM Press.

Neidhardt, F. C., J. L. Ingraham and M. Schaechter. 1990. *Physiology of the Bacterial Cell: A Molecular Approach*. Sunderland, MA: Sinauer Associates Inc.

Neidhardt, F. C., R. Curtiss III, J. L. Ingraham, C. C. Lin, K. B. Low, B. Magasanik, W. S. Reznikoff, M. Riley, M. Schaechter and H. E. Umbarger. 1996. Escherichia coli *and* Salmonella: *Cellular and Molecular Biology*. Washington, DC: ASM Press.

Purves, W. K., D. Sadava, G. H. Orians and H. C. Heller. 2007. *Life: The Science of Biology*. 8th ed. Sunderland, MA: Sinauer Associates Inc.

Saier, M. H. 2006. "Protein secretion systems in gram-negative bacteria." *Microbe* 1:414–419.

The objectives of this chapter are to:

◆ Present an overview of the carbon cycle.

◆ Discuss how major biopolymers from living cells are degraded and recycled.

◆ Illustrate how some polymers are useful for humanity.

◆ Provide information on the degradation of some compounds that are produced chemically, which are harmful to the environment.

12

Roles of Microbes in Biodegradation

All these tidal gatherings, growth and decay, shining and darkening, are forever renewed; and the whole cycle impenitently revolves; and all the past is future.
—Robinson Jeffers

The previous chapters in Part 3 dealt with the major processes microorganisms use to generate energy and the biosynthetic reactions involved in the construction of a living cell. It is axiomatic that survival, growth, and perpetuation of species depend on the continued generation of progeny. A source of building material and the assembly of this material into viable cells is an indispensable component of survival and growth. But building materials are limited. How long can nature generate new life without renewing the ultimate source of building blocks? Not long!

The ultimate source of building blocks for new cell synthesis is through the recycling of dead and dying organisms. Such turnover replenishes the ultimate source of structural material—carbon—by generating CO_2. **Microbial biodegradation** is nature's way of recycling the organic and inorganic components of the biosphere. A major step in the evolution of complex biological systems was the appearance, more than 2.5 billion years ago, of viable organisms that could assimilate simple compounds as substrates during oxygen-generating (oxygenic) photosynthesis. These microorganisms were the cyanobacteria, a group that evolved with a synthetic system that utilized CO_2 as a carbon source and a photosystem that yielded molecular oxygen:

$$(1)\ CO_2 + H_2O \rightarrow CH_2O\ (\text{cellular mass}) + O_2$$

Proliferation of the cyanobacteria, with the inexhaustible supply of solar energy, resulted in an atmosphere containing O_2, an oxidant suitable for the later evolution of aerobic heterotrophs. The broad diversity of microbes that evolved as a consequence of O_2 production was instrumental in the evolution of the multicelled eukaryotes.

Life on Earth can be sustained only if the cellular material produced either directly or indirectly by photosynthetic reactions is ultimately recycled back to CO_2, which constitutes only about 0.03% of the gases in the atmosphere. The success in meeting this requirement is apparent from the general balance and stability that is evident in nature. If it were not balanced by respiration, the massive biological productivity resulting from photosynthesis would markedly lower the CO_2 levels in the atmosphere within about 30 years. Conversely, the burning of geological carbon reserves (coal, oil, and natural gas) has the opposite effect of increasing atmospheric CO_2 levels.

Various microorganisms (*Archaea*, *Bacteria*, and some eukaryotes), through their metabolic activities, account for greater than 85% of the biologically derived CO_2 released into the atmosphere. Without this constant renewal of atmospheric CO_2 plants would cease photosynthesis, the pH of the ocean would increase, and much of life on Earth would be altered. Animals and plants would ultimately be doomed.

In a schematic depiction of the biosphere (Figure 12.1), photosynthetic reactions are those appearing at or above the surface of the soil, and biodegradative reactions are those that occur at or below the surface of the soil. In aquatic environments photosynthesis occurs mainly at or near the surface and to depths where light is available (about 100 meters). Biodegradative activities in oceans predominate beneath this zone. In soils and sediments, the biodegradative processes may extend deep into the subsurface (to more than 3,000 meters). Microbial activities are also responsible for the cycling of organic nitrogen and sulfur to their inorganic forms. The release of phosphate and other minerals present in living cells is also a result of microbial mineralization (breakdown of an organic compound to its component parts CO_2, NH_4^+, SO_4^{2-}).

The prominent role of microorganisms in perpetual recycling of the elements of life can be summed up in two statements, termed **Van Niel's postulates:**

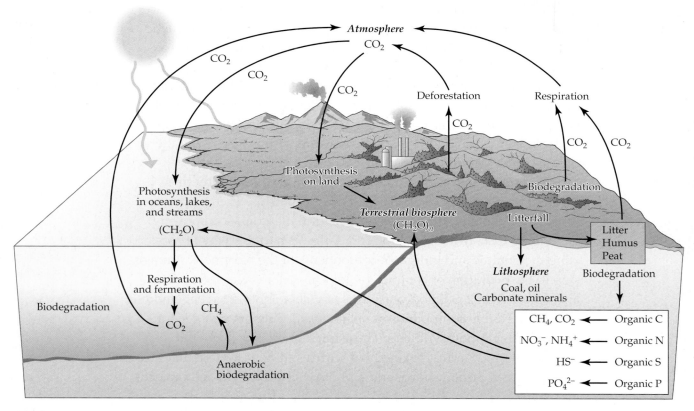

Figure 12.1 Carbon cycle
Photosynthesis uses light energy and fixes carbon dioxide into organic material. The organic material is an energy and carbon source for other microorganisms and animals. Organisms recycle carbon dioxide by respiration and decomposition of organic materials. (See text for details.)

- A microorganism is present in the biosphere that can utilize every constituent part or product of the living cell as a source of carbon and/or energy.

- Microbes with this capacity are present in every environmental habitat on Earth.

These postulates are based on an extensive body of knowledge about the metabolism of microorganisms. The postulates are a clear statement of the role that microbes play in the balance that exists between biosynthesis and biodegradation. It is also evident that as eukaryotes evolved with complex molecular components, such as sterols or cerebrosides, a microbial population also evolved in due course that mineralized these unique molecules.

This chapter discusses some of the mechanisms involved in the biodegradation of biological material, selected xenobiotics (products of chemical synthesis), and other compounds. Chapter 32 provides additional information on the role of microorganisms in biodegradation. The examples presented in this chapter are but a few representatives of the vast array of biodegradative pathways that occur in nature. Following an overview of the carbon cycle, we will examine the catabolism of individual carbonaceous compounds including polysaccharides, proteins, nucleic acids, fatty acids, and aromatic and aliphatic hydrocarbons.

12.1 Overview of the Carbon Cycle

The carbon cycle is described in greater detail in Chapter 24. Here we provide a brief overview. Most organic material synthesized on Earth comes from carbon dioxide fixation during photosynthesis. Some is also provided by bacterial and archaeal chemosynthesis, in which autotrophs derive their energy from reduced inorganic materials such as hydrogen gas. In photosynthesis, the sun's light is harvested by photosynthetic organisms including anoxygenic photosynthetic bacteria, cyanobacteria, algae, and plants. These photosynthetic organisms convert the light energy into chemical bond energy in the form of organic materials produced by carbon dioxide fixation. Part of this organic material is excreted into the environment, but the majority of it, along with inorganic nitrogen, phosphate, sulfate, calcium, and trace elements required by all organisms, is used to produce cells for growth of the photosynthetic species.

This organic material is recycled back into inorganic carbon as illustrated by the carbon cycle (see Figure 12.1), and is accomplished by microbial and animal consumers, which feed on the photosynthetic microorganisms and plants. The organic material is converted back into carbon

dioxide (i.e., recycled) through two major processes: (a) **respiration**, which is the reversal of the carbon dioxide fixation process shown in equation (1) above and is carried out by all organisms, and (b) **anaerobic decomposition**, which is the process in which microorganisms degrade the organic material of dead organisms in the absence of O_2.

The process of photosynthesis in which carbon dioxide is fixed and the processes of respiration and decomposition, along with forest and grass fires that produce carbon dioxide, are balanced so that the normal concentration of carbon dioxide on Earth is kept steady at about 0.03%. Balancing these processes is a major challenge in closed artificial biosystems such as that in Biosphere 2 (Box 12.1).

The remainder of this chapter focuses primarily on the degradation of organic materials by aerobic microorganisms, which play a major role in the carbon cycle. The primary pathways involved in the synthesis of cellular constituents from precursor metabolites are presented in Chapter 10. Recall that the precursor metabolites utilized in biosynthesis are derived from glycolysis, the tricarboxylic acid (TCA) cycle, the hexose monophosphate shunt, and related pathways. The precursor metabolites are converted to building blocks (amino acids, nucleotides, etc.), and these are then polymerized to form macromolecules. These macromolecules are then assembled to generate a cell. All of these processes constitute **anabolism**.

Archaea, *Bacteria*, and eukaryotes (especially fungi) can reverse anabolism, and the resulting catabolic reactions can break down macromolecules (catabolic reactions) and release monomers that are in turn utilized to generate adenosine triphosphate (ATP) and ion gradients. In many instances, such biodegradation reverses the pathways involved in synthesis. However, there are notable exceptions, and an example of this is fatty acid synthesis (see Figure 10.16) and biodegradation (see Section 12.4) that utilize somewhat different pathways and enzymes. Regardless of the route taken, macromolecules are disassembled to produce monomers similar to those from which the macromolecules were assembled. Thus, the reversal of anabolic processes or the use of additional specialized catabolic reactions can transform the constituents of the dead cell or the discarded products of living cells into low-molecular-weight compounds that can be completely mineralized (Figure 12.2).

The total mineralization of sugars and other metabolites under aerobic conditions is depicted in Figure 12.3. Note that each carbon in a sugar, from a three-carbon sugar to one with seven carbons, can become a molecule of CO_2. Acetate, pyruvate, or glyceraldehyde-3-phosphate can enter into the TCA cycle and can be catabolized completely to CO_2 (see Figure 10.8). The microorganisms that are involved in these catabolic reactions

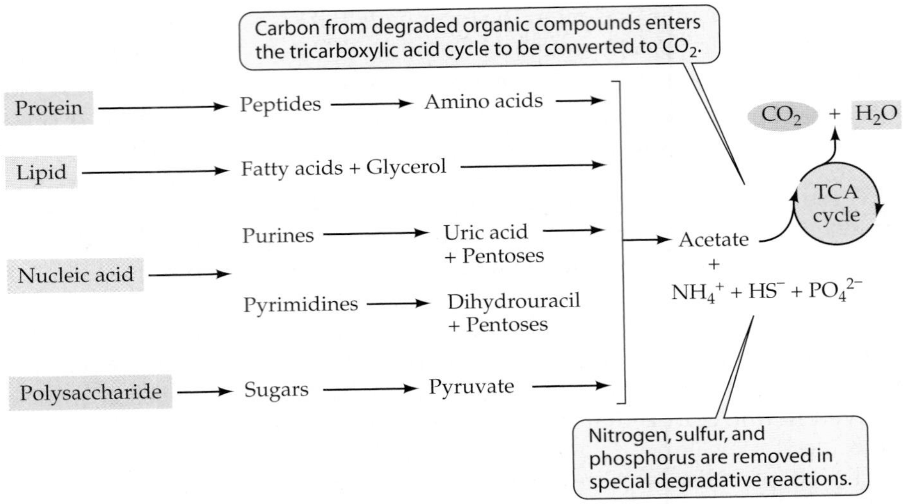

Figure 12.2 Fate of major biomolecules
Cell components and waste products are ultimately hydrolyzed by microorganisms to their constituent parts. Most aerobic biodegradative pathways eventually lead to the tricarboxylic acid cycle or related cycles.

BOX 12.1 *Research Highlights*

Biosphere 2

In the 1990s a grand experiment began in a special "greenhouse-like" structure called "Biosphere 2" in the Sonoran Desert near Tuscon, Arizona. The facility was designed to serve as a miniature biosphere in which people could live with plants and microorganisms. As a contained structure, the aim was to replicate the processes that occur in nature including the carbon cycle.

The experiment lasted about 2 years. The major problem encountered was directly related to the carbon cycle. Biosphere 2 had many plants to carry out photosynthesis to provide food for humans. However, there was insufficient oxygen for the decomposition and mineralization of waste organic material that was generated. As a result, the oxygen concentration became too low for human habitation. This experiment demonstrated that the homeostatic mechanisms we rely on in our biosphere are finely attuned to the levels that are needed to sustain life and may be difficult to replicate elsewhere.

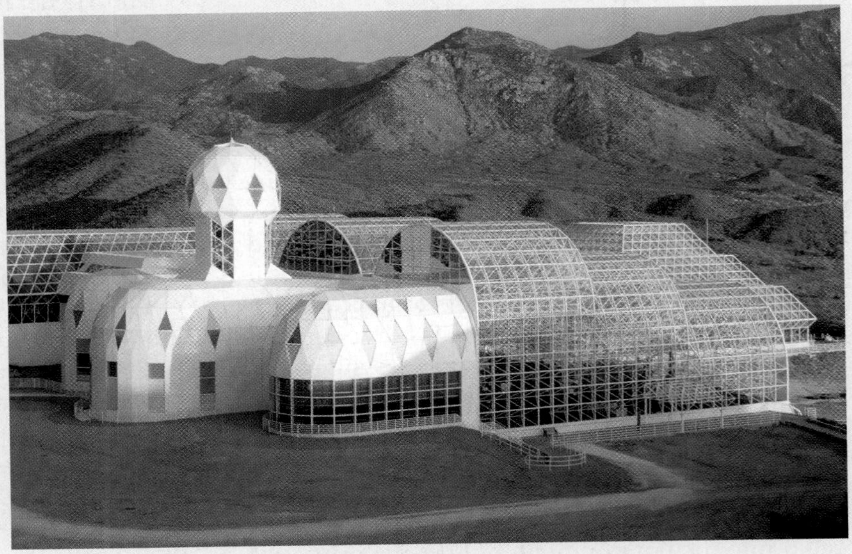

Biosphere 2 is located near Tuscon, Arizona. Note its resemblance to a large greenhouse structure. ©Joe Sohm/Jupiter Images.

Biosphere 2 is currently operated by Columbia University, which uses it to study plant growth under carefully controlled conditions.

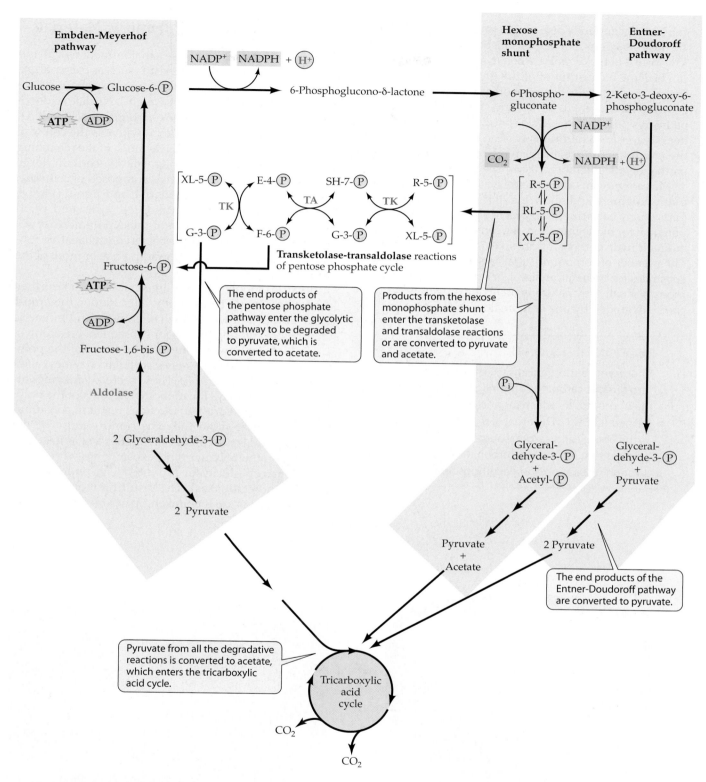

Figure 12.3 Catabolism of sugars
Transketolases (TKs) and transaldolases (TAs) are key enzymes in transforming sugars with three, four, five, or six carbons into intermediates that are further catabolized and enter the tricarboxylic acid cycle. Note that ATP and NADPH/NADH, both of which are used in biosynthesis, are generated during these catabolic reactions. In the pentose phosphate cycle, the sugars are: R, ribose; RL, ribulose; XL, xylulose; E, erythrose; SH, sedoheptulose; G, glyceraldehyde. Details of the other pathways are given in Figures 10.5 through 10.8.

gain energy and may also obtain precursor metabolites via these reactions. For aerobic microbes, approximately 35% to 45% of the organic carbon in a utilizable substrate can become a constituent of the growing bacterial, archaeal, or fungal cell. For anaerobic microbes, this value is lower (about 5% to 10%), since the biodegradative pathways provide less usable energy. Mineralization under oxygen-free conditions is often complex, requiring the combined efforts of fermenting, syntrophic, and methanogenic microorganisms (see Chapters 8 and 18).

Microorganisms can mineralize such compounds as plant polymers and alkaloids that are not constituent parts of any bacterium. They have also evolved with the potential for biodegradation of a considerable number of aromatic hydrocarbons, alkanes, ketones, and xenobiotics. The number of compounds that can potentially serve as substrates for bacterial, archaeal, or fungal growth is enormous. Virtually any naturally occurring organic molecule can be degraded by one or more microbial types.

SECTION HIGHLIGHTS

Photosynthetic organisms fix carbon dioxide into organic carbon using sunlight. This is the major process by which organic materials are formed on Earth. The organic material serves as the energy and carbon source for other organisms that recycle the carbon dioxide through respiration and decomposition.

12.2 Catabolism of Polymeric Materials

Polymeric materials comprise a major fraction of all living cells. For unicellular organisms such as bacterial, archaeal, and fungal cells, proteins comprise about 50% to 60% of the total cell mass (dry weight). Nucleic acids (ribonucleic acid [RNA] and deoxyribonucleic acid [DNA]) account for another 20% to 25%, whereas the cell wall components and lipids comprise much of the remaining polymeric materials. Woody plants contain large amounts of polymeric materials composed of cellulose, hemicellulose (including xylans), and lignin (Figure 12.4), which provide the rigid cell walls and plant shapes. Breakdown of each of these polymeric types presents special tasks for the bacterial, archaeal, or fungal cells that specialize in degrading one or more of the polymers.

Because biopolymers are large in molecular size, they cannot be directly taken-up by cells. Rather, they must first be broken down to small oligomeric and momeric units before entering the cell for subsequent breakdown. The initial step involves hydrolytic attack of the polymer by specialized types of extracellular enzymes called **exoenzymes** (Figure 12.5). These secreted enzymes attack at bonds joining the monometric units of the polymer, where the hydrolytic reactions result in consumption of a water molecule. The bond energy of the polymer bond is not conserved, but it is lost as heat. Depending on the polymer type, exoenzymes with different polymer specificity are produced (e.g., proteases, nucleases, and cellulases). The resulting monomer and small oligomers are then taken up across the cell enve-

(A)

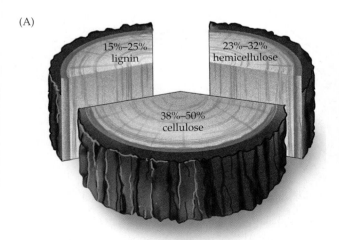

(B)

Figure 12.4 Composition of woody plants
(A) The three major constituents of wood and their proportions are indicated. (B) Micrograph of cellulose fibers. B, ©Biophoto Associates/Photo Researchers, Inc.

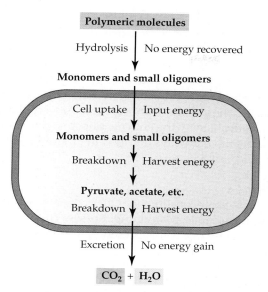

Figure 12.5 Schematic of polymer degradation by microbes.
The initial degradation of polymers occurs outside of the cell. Cells must use energy to import the resulting monomers and oligomers, but harvest more energy than expended from the further breakdown of these molecules inside the cell. The products of degradation are excreted into the environment.

lope via active transport systems (see Chapter 5). This process requires energy input in the form of ATP or ion gradients. Once inside the cell, the molecules are then degraded with the harvesting of energy (see Figure 12.5).

Polysaccharides

Simple as well as complex polysaccharide polymers are abundant in nature and are a ready food source for microorganisms. Polysaccharides are long chains of monosaccharides (sugars) connected by glycosidic linkages. Starch, glycogen, and cellulose are high-molecular-weight polysaccharides (in excess of 1,000,000 MW) composed of glucose, but they differ in the linkage between sugar units. Starch and glycogen are not chemically stable and are readily degraded by enzymes. They are therefore a good storage material. Cellulose is much more stable and is used by plants as a structural material. Xylans, too, are structural components of plants and are considered heterogeneous polymers because they contain small amounts of hexose and are quite complex. Chitin is a tough polymer of *N*-acetylglucosamine with hydrogen bonding between the long chains.

The following section presents pathways involved in the microbial catabolism of selected polysaccharide substrates.

Starch

Starch is a major storage product in plants. There are two distinct polymers present in starch granules—amylose and amylopectin. Amylose consists of unbranched D-glucose chains in an α-1,4-glucosidic linkage with 200 to 500 sugar units per chain (**Figure 12.6**). The other polymer, amylopectin, is composed of the same glucosidic linkages but with branching in the 1,6-position every 25 to 30 sugar residues. Starch is generally hy-

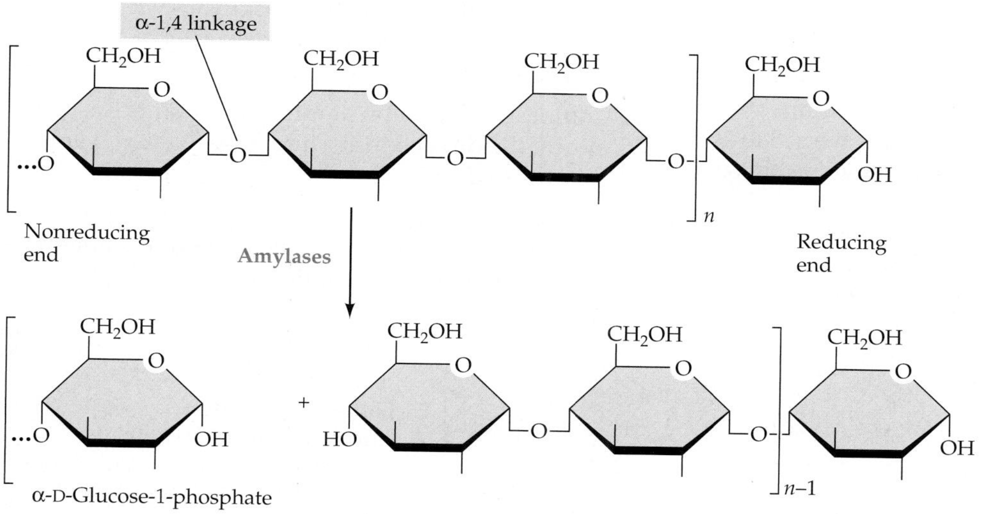

Figure 12.6 Starch
Starch consists of α-1,4-linked glucose units. Glycogen has a similar structure. The major difference between starch and glycogen is in the number of branches, which are formed through 1,6 linkages. The α-1,4 linkages result in a bending between glucose units and a helical structure to the starch molecule. Amylases cleave the linkage between the glucose units and release α-D-glucose and maltose units.

drolyzed by the enzymes, α-amylase or β-amylase. α-Amylase hydrolyzes the α-1,4 linkages to a mixture of glucose and maltose. Maltose is a disaccharide made up of two molecules of glucose in an α-1,4 linkage. β-Amylase hydrolyzes the α linkages of amylose at alternating positions to yield mostly maltose.

These two enzymes, α- and β-amylase, also cleave amylopectin to a mixture of glucose, maltose, and a branched core. A different enzyme, amylo-1,6-glucosidase, cleaves the 1,6 linkage responsible for branching. This enzyme, in combination with α-amylase or β-amylase, can lead to the complete degradation of starch to maltose or glucose. The enzyme maltase is present in organisms that degrade starch and can hydrolyze one molecule of maltose to two molecules of glucose.

Amylases, for example, are present in many *Bacillus* species, including the obligate thermophiles *Bacillus acidocaldarius* and *Bacillus stearothermophilus*. The glucoamylases are produced by *Aspergillus niger*, *Rhizopus niveus*, and other fungi. Anaerobic thermophiles produce amylases that are stable at a range of temperatures and are of potential value in the production of commercial ethanol from starch. Starch digestion in the rumen of cattle and other ruminants is essential for the efficient utilization of plant material and grain. Rumen bacteria, such as *Streptococcus bovis* and *Bacteroides amylophilus*, produce α-amylase. *Entodinum caudatum*, a rumen ciliate, can hydrolyze starch to maltose and glucose. Maltose and maltose dextrans generally enter microbial cells by ABC (<u>A</u>TP-<u>b</u>inding-<u>c</u>assette)-type uptake systems, whereas glucose is acquired by PTS (<u>p</u>hospho<u>t</u>ransferase <u>s</u>ystem) uptake machinery (see Chapter 5).

Glycogen

Glycogen is a carbon and energy storage polysaccharide present in animals and some bacteria. It is also called *animal starch* and is more highly branched than plant starch. Branching in glycogen occurs once every 8 to 10 glucose residues, whereas branching in amylopectin occurs once every 25 to 30 residues.

Glycogen is hydrolyzed by a glycogen phosphorylase that removes one glucose and simultaneously incorporates an inorganic phosphate to yield a molecule of glucose-1-phosphate (see Figure 12.6). The enzyme glycogen phosphorylase has been isolated from the yeast *Saccharomyces cerevisiae* and the cellular slime mold *Dictyostelium discoideum*. Because enteric bacteria, spore formers, and cyanobacteria store glycogen, it is apparent that these organisms also have the glycogen phosphorylase. The glucose-1-phosphate is converted to glucose-6-phosphate by the enzyme phosphoglucomutase and enters the pathways outlined in Figure 12.3.

Chitin

Chitin is a major constituent of the cell walls of selected plants and animals. Many fungi possess chitin as a cell wall component, and it is present in the shells of crustaceans (lobsters, shrimp, and crabs). Chitin also forms the tough, outer integument of grasshoppers, beetles, cockroaches, and other insects. It is a chemically stable polymer of *N*-acetylglucosamine (NAG) units linked by β-1,4-glycosidic bonds. Its stability is derived from hydrogen bonding between adjacent chains through the *N*-acetyl moiety (Figure 12.7). The widespread occurrence of chitin in soil and aquatic environments has led to the evolution of a wide variety of microorganisms that can decompose this substance. The soil actinobacteria (filamentous bacteria, see Chapter 20) are major utilizers of chitin. The addition of chitin as a substrate in a soil enrichment medium will consistently result in the isolation of diverse actinobacteria. In aquatic habitats other bacteria are involved in chitin degradation. For example, the genus *Vibrio*, with its species, *Vibrio cholerae* is a well-known pathogen that resides on the exoskeleton of cladocerans. Many species in this and related marine bacteria are capable of chitin degradation, which occurs in the water column as well as in marine sediments that receive the crustaceans when they die as well as when they molt their exoskeletons.

Chitin is cleaved by two enzymes:

1. Chitinase splits chitin at numerous points to mostly the disaccharide chitobiose ($NAG_{n=2}$) and chitotriose ($NAG_{n=3}$).

2. Chitobiase hydrolyzes chitobiose and chitriose to monomers that are readily biodegraded inside the cell following removal of the *N*-acetyl side chain.

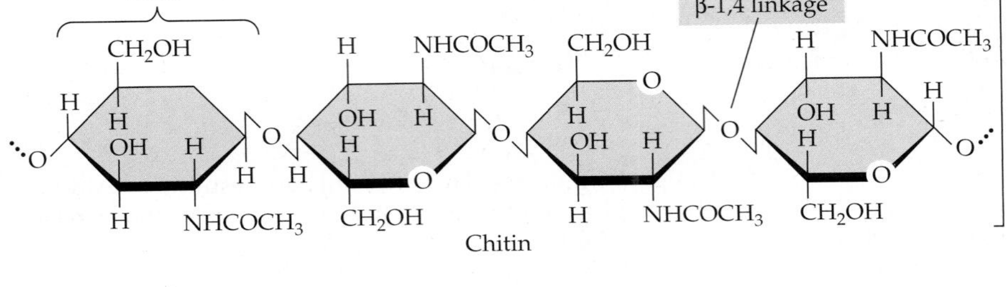

Figure 12.7 Chitin
Chitin consists of *N*-acetylglucosamine (NAG) units. As in cellulose, the sugar units are joined by β-1,4 linkages. Chitin occurs as long ribbons in the cell walls of fungi and the exoskeletons of crustaceans, insects, and spiders.

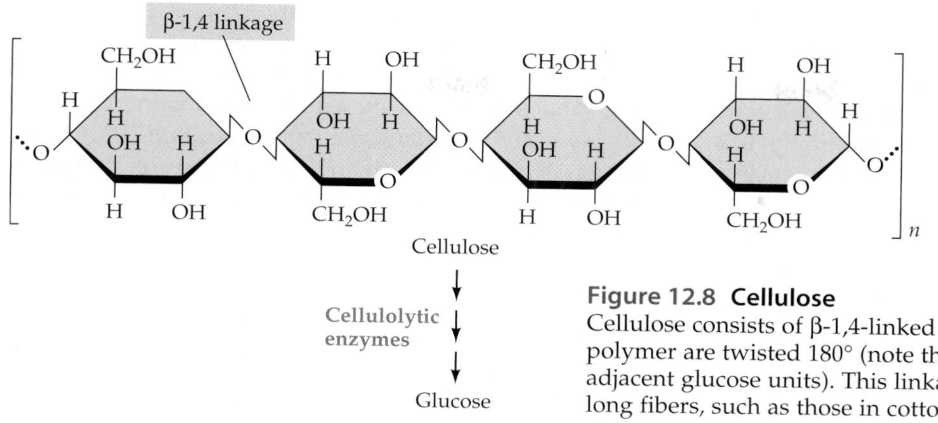

β-1,4 linkage

Cellulose
↓
Cellulolytic enzymes
↓
Glucose

Figure 12.8 Cellulose
Cellulose consists of β-1,4-linked glucose units. The linkages in the polymer are twisted 180° (note the positions of the —CH₂OH groups in adjacent glucose units). This linkage permits the polymer to grow into long fibers, such as those in cotton.

Cellulose

Cellulose is considered to be the most abundant product of biosynthesis present on Earth. It is a major constituent of plants and is also synthesized by a limited number of bacterial species (*Acetobacter xylinum* and *Sarcina ventriculi*). Fungi and bacteria are the major decomposers of cellulose in nature. Fungi function mostly in acidic soils and in woody tissue, where the cellulose present is protected from enzymatic attack by **lignin**, whose composition and properties will be discussed later. Both wood-rotting and soil fungi are instrumental in the depolymerization and mineralization of cellulose.

The aerobic bacteria that decompose cellulose include myxobacteria, cytophaga, sporocytophaga, and some other species. Characteristics of these organisms are discussed in Chapters 20 and 22. Cellulose is fermented anaerobically in soil by clostridia and in the rumen of cattle (see Chapter 25) and other ruminants by an array of anaerobic bacterial species.

The cellulose polymer is composed of glucose molecules that are joined by a β-1,4 linkage. An unbranched cellulose fibril can be composed of up to 14,000 glucose molecules (**Figure 12.8**). Decomposition of cellulose occurs by the concerted action of several enzymes:

- Endo-β-1,4-glucanase hydrolyzes bonds in the interior of chains, releasing long chain fragments.

- Exo-β-1,4-glucanase removes two glucose units (called *cellobiose*) from the long fragments.

- α,β-Glucosidases hydrolyze cellobiose to glucose.

Some cellulose degrading microbes attack the high-molecular-weight polymer in plant cell walls using a specialized extracellular enzyme complex called a **cellulosome** (**Figure 12.9**). This multiprotein assembly is composed of a scaffold protein complex called scaffoldin, which is attached to the outer surface of the bacterial cell by an anchoring protein. Attached to the extended scaffold proteins are various cellulose-attacking glycosidase

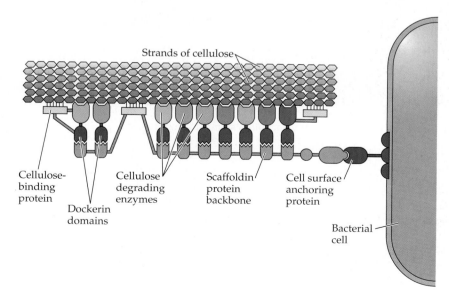

Strands of cellulose

Cellulose-binding protein

Dockerin domains

Cellulose degrading enzymes

Scaffoldin protein backbone

Cell surface anchoring protein

Bacterial cell

Figure 12.9 The cellulosome
The cellulosome–scaffold complex is attached to the bacterial cell surface by an anchoring protein. Cellulolytic enzymes of differing enzyme specificity are displayed on the scaffoldin base by dockerin domains. Cellulose-binding proteins link the large multiprotein assembly to the cellulose fibrils.

enzymes (endoglucanases and cellobiohydrolases) plus additional specialized proteins with cellulose-binding domains that adhere to the cellulose fiber. This multi-enzyme cellulose-degrading complex provides increased efficiency in plant cell wall polymer degradation, while improving uptake of monomers and small oligomers into the cell due to their release close to the cell.

Endonucleases responsible for degradation of cellulose produce cellobiose and glucose. These products repress the synthesis of the endonucleases, thereby slowing the process. Industries that have an interest in the large-scale conversion of cellulose to glucose must overcome this repression to increase their efficiency. Production of glucose from the abundant supplies of cellulose available is of considerable interest, as the glucose released can be fermented to ethanol (Box 12.2) and other industrial chemicals.

Hemicellulose, Xylans, and Pectins

Xylans, which are a major component of hemicellulose, are heterogeneous polysaccharides composed mainly of D-xylose, a pentose sugar (Figure 12.10). Hemicellulose is second in abundance to cellulose among the sugar-based polymers that occur in nature. These random amorphous polymers are a major constituent of mature woody tissue of all land plants (see Figure 12.4). In contrast to cellulose, which confers cell rigidity due to its crystalline structure, xylans have little tensile strength and are more easily hydrolyzed.

Xylans are degraded by bacteria and fungi with two key enzymes:

- β-1,4-Xylanases hydrolyze the internal bonds that link the xylose molecules into polymers, yielding lower molecular weight xylo-oligosaccharides. An oligosaccharide contains a small number of monosaccharide units.

- β-Xylosidases release xylosyl residues through an endwise attack on the xylo-oligosaccharides.

When free xylose is released, it is then taken up into the cell and phosphorylated to xylose-5-phosphate before it enters the pentose cycle (see Figure 12.3).

BOX 12.2 *Research Highlights*

Alternate Fuels to Reduce the World's Reliance on Fossil Oil

Because oil is a fossil resource, it will eventually be depleted. Brazil, which lacks oil resources, recently began using yeast to convert sucrose from the country's large sugar cane crop into ethanol. In 2006 the Brazilian government announced that the country was no longer dependent on the importation of expensive oil for use as fuel for automobiles.

Other countries, such as the United States, use starch to produce ethanol as a supplement to gasoline. However, starch from corn is an expensive source of carbon for ethanol. For this reason, some are looking at plant cellulose and hemicellulose as an abundant, inexpensive, and recyclable source of carbon to produce ethanol. The primary barrier for the production of ethanol from cellulose lies in identifying appropriate processes to inexpensively hasten its depolymerization to sugars.

An ethanol plant next to a cornfield. ©Jim Parkin/istockphoto.com.

Of course, all carbon sources of fuel, from coal to organic materials, result in the production of carbon dioxide that is released into the atmosphere. The increased CO_2 in the atmosphere due to human activities is directly related to global warming. For this reason, other fuels, such as hydrogen gas, which produce only water when they are combusted, can also be obtained by microbial fermentation and are becoming the focus of future fuels.

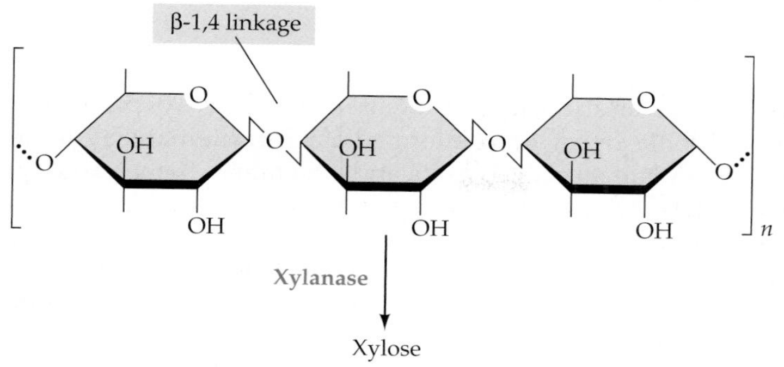

β-1,4 linkage

Figure 12.10 Xylan
Xylan is a polymer of the pentose D-xylan. The xylan chain consists of β-1,4-linked xylose units.

Xylanase → Xylose

Pectin is a highly branched heteropolysaccharide component of plant cell walls that forms a matrix with hemicellulose. It is also highly abundant in fruits and contains a variety of C-6 and C-5 sugars including rhamnose, arabinose, and galacturonic acid. Pectin is broken down by a class of hydrolase enzymes called pectinases. The release of bond energy to yield mono- and disaccharides is lost as heat. Molecules enter the cell by active transport mechanisms, and the C-6 and C-5 sugars enter one or more of the sugar catabolic pathways (see Figure 12.3).

Catabolism of Proteins

Proteins are a major constituent of all cell types and constantly enter into the environment, either from dead cells or as waste materials. Microorganisms (*Bacteria*, *Archaea*, and fungi) that biodegrade proteins are widespread in nature. Microorganisms initiate digestion of these large molecules by exporting protein-degrading enzymes to the cell surface and beyond; these enzymes hydrolyze the protein to smaller molecules (Figure 12.11A). These small molecules are then taken up into the cell by molecule-specific active transport systems where they are either utilized for energy production, or serve as substrates for cell biosynthesis.

Extracellular **proteolytic enzymes** (**proteases**) hydrolyze proteins to polypeptides, and these in turn are cleaved by endopeptidases to oligopeptides. Endopeptidases (also called proteinases) hydrolyze peptide bonds within polypeptide chains. The endopeptidases are specific and hydrolyze the peptide bond only if specific amino acids are present at the cleavage site. For example, the endopeptidase **thermolysin** from *Bacillus thermoproteolyticus* cleaves peptide linkages where leucine, phenylalanine, tryptophan, or tyrosine is one of the amino acids present. As for other types of exoenzymes, the energy released upon hydrolysis of the peptide bond is lost as heat.

An **exopeptidase** is an enzyme that hydrolyzes amino acids from the end of a peptide chain (Figure 12.11B). A **carboxypeptidase** attacks the C-terminal end of the chain, and an **aminopeptidase** removes an amino acid from the N-terminal end of a peptide chain. The enzymes involved in proteolysis act in concert and with or without high substrate specificity.

The amino acids released by the hydrolysis of a protein are transported into the bacterial or fungal cells by specific active transport systems (see Chapter 5). The oligopeptides may also be transported into cells where they are hydrolyzed by other peptidases. The free amino acids can be incorporated into the proteins of the cell or deaminated and catabolized to provide carbon and/or energy for growth. For example, the amino group of glycine and alanine is removed to yield acetate or pyruvate plus ammonia. Excess ammonia is released into the surrounding medium.

Catabolism of Nucleic Acids

Nucleic acids (RNA and DNA) comprise a significant fraction of most types of living cells, and are released upon cell death and lysis. These macromolecules are readily degraded by a variety of extracellular enzymes

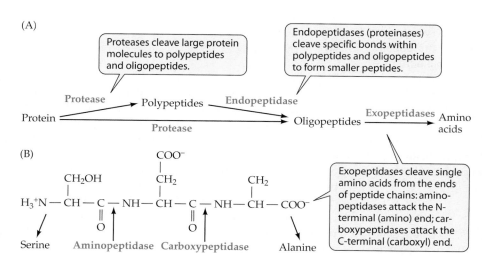

(A)

Proteases cleave large protein molecules to polypeptides and oligopeptides.

Endopeptidases (proteinases) cleave specific bonds within polypeptides and oligopeptides to form smaller peptides.

Protein — Protease → Polypeptides — Endopeptidase → Oligopeptides — Exopeptidases → Amino acids
Protease

(B)

Exopeptidases cleave single amino acids from the ends of peptide chains: aminopeptidases attack the N-terminal (amino) end; carboxypeptidases attack the C-terminal (carboxyl) end.

Serine — Aminopeptidase — Carboxypeptidase — Alanine

Figure 12.11 Protein biodegradation
Proteases, endopeptidases, and exopeptidases act in concert to break down proteins to amino acids. (A) Overview of the degradation process. (B) Detailed view of the action of exopeptidases.

called **nucleases**. By analogy to the endopeptidases that act to hydrolyze proteins, RNA- and DNA-dependent endonucleases attack the respective macromolecules to generate oligonucleotides, which in turn are further degraded to short oligonucleotides and mononucleotides by the action of endo- and exonucleases. The nucleotides are then further cleaved to the constituent bases (purines and pyrimidines) plus the ribose and deoxyribose sugars. Following cell entry by specific active transporters, the sugar moieties are rearranged to xylose and enter the pentose cycle (see Figure 12.3). The purines (adenine and guanine) and pyrimidines (thymine, cytosine, and uracil) are also recycled or degraded for energy production.

> **SECTION HIGHLIGHTS**
> Polymeric organic compounds are first broken down into lower molecular weight subunits by enzymes that are active at the outer surface of the microorganism. The subunits are taken into the cell and metabolized as an energy source. Some polymers can only be broken down by specialized microorganisms.

12.3 Catabolism of Low-Molecular-Weight Compounds

Many types of low–molecular-weight compounds in plants, animals, and microbes are released upon cell death, or are produced as waste products of metabolism. In general, living cells contain only 5% of their carbon in the form of small organic molecules. Many of these molecules, when available, are readily taken up from the environment as a source of carbon and/or energy. Several examples follow.

Uric Acid

Birds, scaly reptiles, and carnivorous animals secrete uric acid rather than urea as a product of purine catabolism. Selected soil microorganisms catabolize uric acid as a source of carbon and energy (Figure 12.12). One of these organisms is the unique endospore-forming *Bacillus fastidiosus,* which cannot grow with any substrate except uric acid and its degradation products, allantoic acid and allantoin. Glyoxylate, CO_2, and urea are the products of uric acid degradation. Urea is degraded to ammonia and CO_2 by urease, an enzyme present in many species of bacteria.

Organic Acids

Many mono-, di-, and tri-carboxylic acids, such as acetate, citrate, malate, succinate, α-ketoglutarate, oxaloacetate,

and tartaric acid are present in many types of cells as anabolic or catabolic intermediates, or, alternatively, accumulate in mature fruits and vegetables. Some carboxylates are also excreted as waste products (e.g., lactate and acetate). These molecules are readily taken up by many types of bacteria, archaea, and fungi as a source of carbon and or energy. They may enter the TCA cycle (see Figure 10.8) either directly or indirectly via accessory reactions, and provide for ATP generation via cell respiration.

> **SECTION HIGHLIGHTS**
> A variety of small molecules, such as uric acid and organic acids, are products of polymers. Uric acid is degraded by a specialized group of bacteria, whereas organic acids are readily degraded by many microorganisms.

12.4 Catabolism of Lipids

An indispensable structure in all cells, prokaryotic and eukaryotic, is the cytoplasmic membrane. The membrane is composed mainly of polar phospholipids (see Figure 4.45), with the exception of the ether-linked lipids in the *Archaea* (see Figure 4.47). The internal organelles of eukaryotes (mitochondria, nucleus, among others) are also enclosed in phospholipid membranes. The major components of the phospholipids are the long-chain fatty acids (or branched-chain alcohols in the *Archaea*) that are attached to a glycerol moiety. Long-chain n-alkanes ($>C_{12}$) are also present in most living cells. The role of these hydrocarbons is not known.

For convenience in this section, we will first discuss the disassembly of the phospholipids to their constituent parts, followed by the stepwise breakdown of the fatty acids and phospholipid moieties. The attack on n-alkanes is discussed in a following section.

Phospholipids

Phospholipids account for approximately 10% of the dry weight of a typical prokaryotic or eukaryotic cell. They are constantly introduced into the environment by the decay of dead cells. Bacteria and fungi are commonly present in the environment that can disassemble phospholipids and utilize each of the constituent parts. The biodegradation occurs by the disassembly of the phospholipid to yield two long-chain fatty acid molecules, one glycerol molecule, and one phosphoserine molecule (Figure 12.13). This is accomplished by the concerted action of specific **phospholipases** that hydrolyze the ester linkages. The fatty acids are then catabolized by β-oxidation (Figure 12.14),

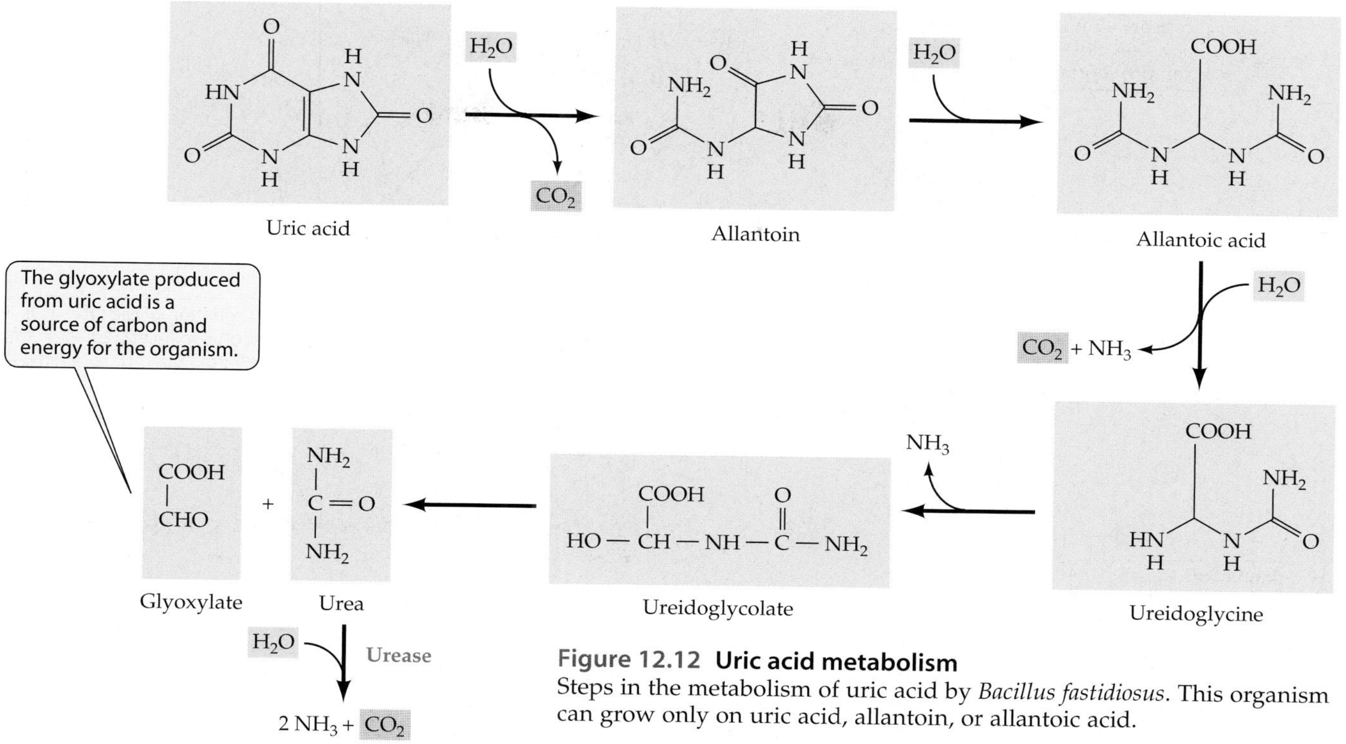

Figure 12.12 Uric acid metabolism
Steps in the metabolism of uric acid by *Bacillus fastidiosus*. This organism can grow only on uric acid, allantoin, or allantoic acid.

whereas the glycerol and phosphoserine units are recycled or degraded via the TCA cycle (see Figure 12.2).

β-Oxidation of Fatty Acids

The long-chain fatty acids commonly present in phospholipids, such as hexadecanoic (C_{16}) or octadecanoic acid (C_{18}), are catabolized by the **β-oxidation** pathway (see Figure 12.14). The initial step involves the ATP dependent incorporation of CoASH to form a fatty acyl-CoA intermediate by the enzyme acetyl CoA synthase. The fatty acyl-CoA molecule is then disassembled in a series of reactions termed *β-oxidation* because the β-carbon is oxidized to a keto group ($C=O$) prior to the cleavage of an acetyl-CoA unit from the chain. Each cycle generates 1 NADH and 1 FADH, which provide reductant for electron transport–derived ATP production. Depending on their origin, the fatty acid side-chains may range from 12 to 18 or more carbons in length.

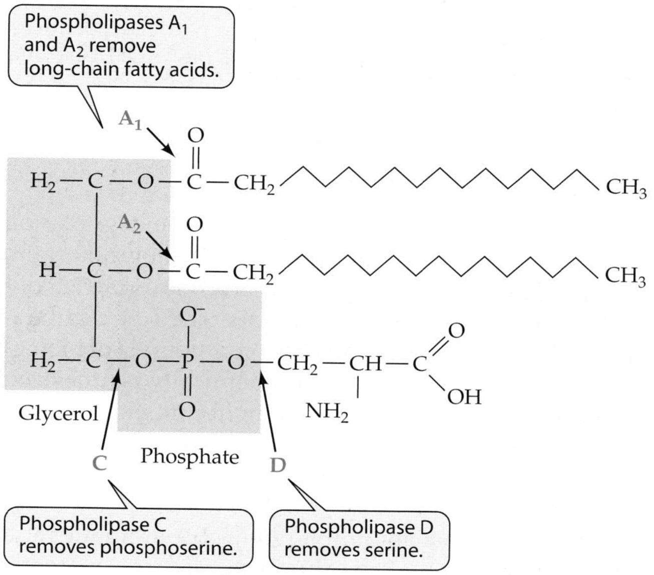

SECTION HIGHLIGHTS
Membrane phospholipids and their subunit fatty acids are degraded by phospholipases and β-oxidation, respectively.

Figure 12.13 Phospholipase specificity
Phospholipases are the hydrolytic enzymes that cleave polar phospholipids. The various bonds are cleaved by specific enzymes. Shown here are the effects of these enzymes on phosphatidylglycerol.

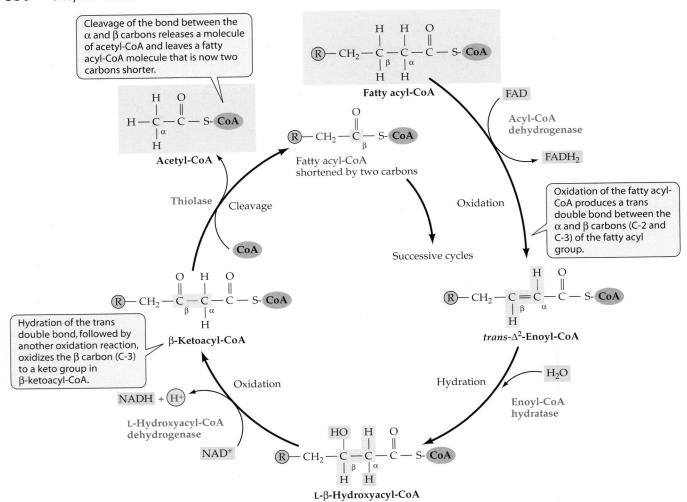

Figure 12.14 β-Oxidation of a fatty acyl-CoA
Each cycle of the β-oxidation pathway produces
one molecule of FADH₂, one of NADH, and one

of acetyl-CoA. The fatty acid is shortened by two carbons.
Delta (Δ) indicates a double bond, and the superscript (2)
indicates its position (between C-2 and C-3).

12.5 Catabolism of Aromatic and Aliphatic Compounds

In keeping with van Niel's postulates, the degradation
of relatively refractory compounds of biological origin,
such as lignin and aromatic hydrocarbons, is carried out
by microorganisms. This section discusses the degradative pathways for these and other compounds of biological origin, as well as for some chemically synthesized
compounds that may enter the environment.

Lignin

Lignin is the most abundant renewable aromatic molecule on Earth. Its natural abundance may be exceeded
only by cellulose. Biodegradation of lignin, therefore, is
an essential part of the carbon cycle. Lignin surrounds
and protects the xylans and cellulose that are components of wood from destruction by microorganisms. Be-

tween 15% and 25% of wood and plant vascular tissue
is composed of lignin (see Figure 12.4). All higher
plants, including ferns, contain lignin, but lower plants,
such as mosses and liverworts, do not. Lignin is relatively indestructible, and it is probably not metabolized
by any single microbial species as a sole source of carbon or energy.

The resistance of lignin to microbial destruction is
founded in the reactions whereby it is synthesized.
Lignin is synthesized from three precursor alcohols by
free-radical copolymerization (Figure 12.15A). The complex large polymer of lignin (Figure 12.15B) would present a microbial species with an insurmountable task in
any attempt to break it down. The molecule is a heterogeneous polymer assembled in a haphazard way that
would be attacked only by enzymes with very high levels of nonspecificity. Such enzyme specificity has apparently not evolved in nature.

(A)

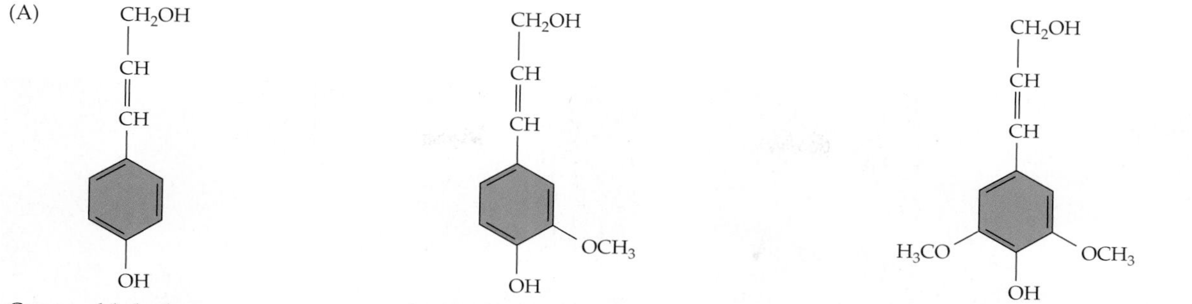

Coumaryl (*p*-hydroxy cinnamyl) alcohol

Coniferyl alcohol (4-hydroxy-3-methoxy cinnamyl) alcohol

Sinapyl alcohol (3,5-dimethoxy cinnamyl) alcohol

(B)

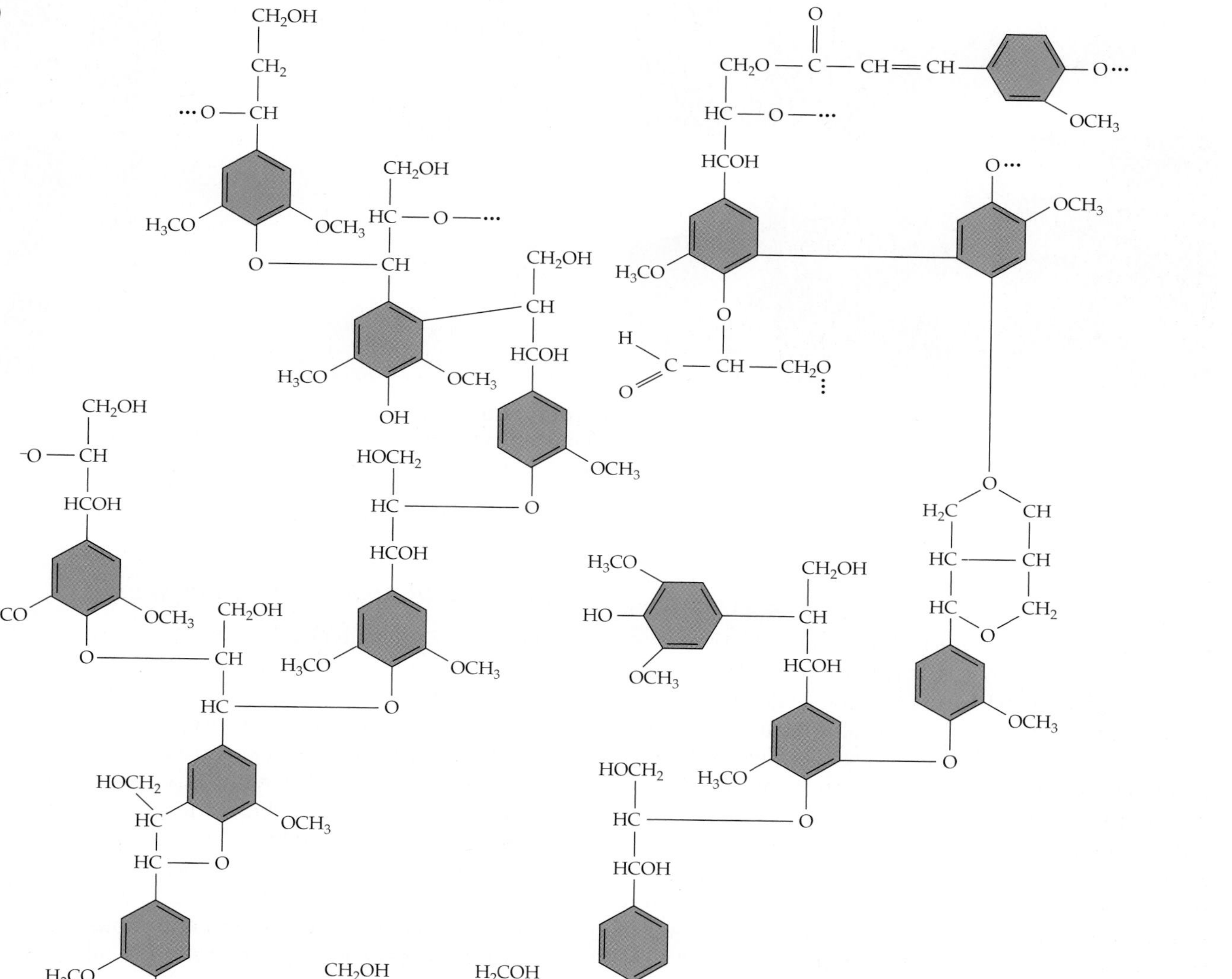

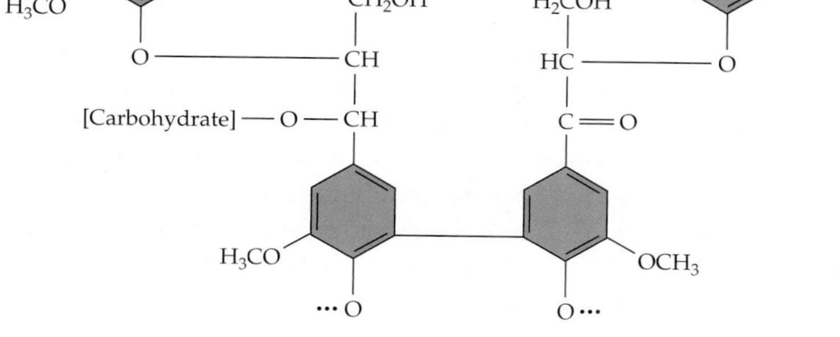

[Carbohydrate]

Figure 12.15 Lignin
The structure of lignin. (A) The alcohols that are polymerized to form lignin. (B) A generalized structure of lignin, formed by a random polymerization of the alcohols shown in (A). The constituents of lignin are oxidized to various phenolic compounds by the concerted action of peroxides and the enzyme ligninase. Many microorganisms can use these phenolics as substrates.

(A)

(B)

Figure 12.16 White-rot fungus
(A) Fruiting bodies of a white-rot fungus (*Armillaria* sp.) that co-metabolizes the lignin and cellulose components deposited in the plant cell wall. (B) A close-up view of white-rot fungus on white-rotted wood. The white cellulose strands remain after breakdown of the lignin and hemicellulose. Courtesy of D. McIntyre.

Lignin is biodegraded in nature by white-rot basidiomycetes (fungi) (Figure 12.16) via a mechanism befitting its synthesis, that is, a random oxidative attack. Lignin is not known to serve as growth substrate for this or any other fungus, so an alternate carbon or energy source, termed a cosubstrate, must be available. Xylans, cellulose, and carbohydrates can serve as this cosubstrate.

Experiments with a white-rot fungus, *Phanerochaete chrysosporium*, have affirmed that lignin is degraded by enzymatic "combustion." During growth on a cosubstrate, *P. chrysosporium* synthesizes enzymes that require H_2O_2 to function and catalyze the oxidative cleavage of both β-O-4 ether bonds and carbon-to-carbon bonds. The hydrogen peroxide is supplied by peroxidases in the fungus. Lignin is not degraded in the absence of O_2. The breakdown of lignin is initiated by the action of single electrons that produce aromatic cationic radicals in the lignin structure. These, then, function in nonenzymatic radical-promoted reactions that disrupt the lignin molecule. This process leads to a nonspecific oxidation of lignin to random, low-molecular-weight products that are then available for mineralization by fungi or bacteria.

Alkanes and Aliphatic Hydrocarbons

Alkanes are saturated hydrocarbons of 12 to 20 carbons that are a minor constituent of most living cells, and *n*-alkanes of greater carbon chain length are present in many plant tissues as **oils** and **waxes**. Crude oil contains **paraffins** ranging from gaseous methane (CH_4) to solid paraffins of considerable length ($>C_{20}$). Thus, significant amounts of these hydrocarbons are released into the environment on a regular basis by natural seepage and by spillage from commercial hydrocarbon extraction. In addition, major oil tanker spills, such as the *Exxon Valdez* spill in Alaska, have resulted in enormous damage to marine environments, especially to birds, other wildlife, and commercial fisheries (Figure 12.17).

Microorganisms that can utilize alkanes and paraffins as growth substrate are abundant in soil and water. The initial site of enzymatic attack on the *n*-alkane series from *n*-butane (C_4) to eicosane (C_{20}) is at a terminal methyl group that ultimately yields the homologous carboxylic acid. There is substantial evidence that *n*-alkanes more than 30 carbons long may be cleaved at sites in the interior of the chain to yield more manageable, smaller mol-

Figure 12.17 Illustration of effects of an oil spill
Workers clean up a beach contaminated by an oil spill. ©Vanessa Vick/Photo Researchers, Inc.

ecules. Many bacteria, including pseudomonads, mycobacteria, actinomycetes, and corynebacteria, can utilize *n*-alkanes as a sole substrate for carbon and energy. Yeasts and filamentous fungi are also capable of growth on these compounds.

A general scheme for the degradation of a molecule of *n*-alkane to the homologous alcohol and fatty acid is presented in Figure 12.18. The initial step requires the action of a special class of enzymes called **oxygenases** that are of two types: monooxygenases and dioxygenases. In the example illustrated, an alkane monooxyge-

nase oxidizes the *n*-alkane to the corresponding alcohol to prepare the molecule for subsequent enzymatic attack. Energy input is required in the form of NADH, whereby one atom of oxygen is incorporated into the alkane and the other oxygen atom is reduced to water. Subsequent steps of reduction by alcohol dehydrogenase and aldehyde dehydrogenase generate NADH and the carboxylic acid. The hydrolysis of one molecule of ATP drives the covalent addition of CoASH to form the fatty acyl-CoA derivative. Finally, the fatty acyl-CoA molecule is catabolized by β-oxidation (see Figure 12.14).

This series of reactions is termed "β-oxidation" because the β-carbon is oxidized to a keto group (C=O) prior to the cleavage of an acetyl-CoA from the chain.

Oxygenase enzymes (monooxygenases and dioxygenases) serve critical roles in oxidation of many types of molecules in nature. In contrast to a monooxygenase that inserts one atom of oxygen into the substrate, a dioxygenase inserts both atoms of molecular oxygen into its substrate (Figure 12.19). There is considerable diversity in enzymes and substrate specificity, which accounts for the vast range of molecules that are attacked and degraded by microbes in nature. A brief description of the

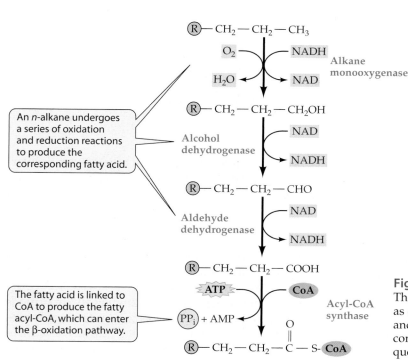

An *n*-alkane undergoes a series of oxidation and reduction reactions to produce the corresponding fatty acid.

The fatty acid is linked to CoA to produce the fatty acyl-CoA, which can enter the β-oxidation pathway.

Figure 12.18 *n*-Alkane oxidation
The stepwise oxidation of a long-chain *n*-alkane such as octane (C_8) to the homologous alcohol, aldehyde, and fatty acid. The fatty acid is then converted to the corresponding fatty acyl-CoA derivative for subsequent metabolism.

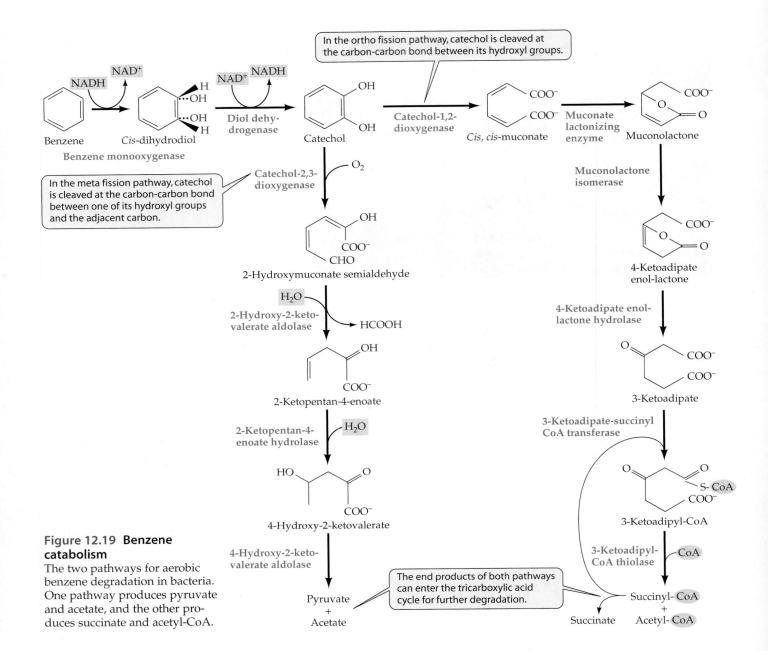

Figure 12.19 Benzene catabolism
The two pathways for aerobic benzene degradation in bacteria. One pathway produces pyruvate and acetate, and the other produces succinate and acetyl-CoA.

role of oxygenase enzymes in the degradation of aromatic compounds follows.

Aromatic Hydrocarbons

Aromatic and polyaromatic hydrocarbons constitute a large fraction of oils and tars present in the Earth. Their extraction provides a rich source of starting material for commercial use. For example, phenol or pyridine are industrial chemicals that occur both as products of, or intermediates in, chemical syntheses. Polyaromatic hydrocarbons are also formed during combustion of fossil fuels and are natural constituents of unaltered fossil fu-

els. Benzene, toluene, xylenes, and other low-molecular-weight aromatics serve as commercial solvents and are constituents of paint. Various other aromatic compounds are intermediates in the synthesis of dyes, medicines, plastics, and other chemicals that have become a part of modern life. In addition, aromatic compounds are a source of considerable environmental pollution. Aromatic compounds such as phenylalanine and tyrosine are constituents of living cells.

The aerobic biodegradation of aromatic hydrocarbons has been studied extensively. Benzene is one that has received much study, and research has affirmed that there are two major pathways for the biodegradation of this compound (see Figure 12.19). The pathway followed depends on the bacterial strain involved. Note that dioxygenase and monooxygenase enzymes are used to drive the endothermic (energy requiring) reactions that allow subsequent molecule rearrangements and eventual mineralization to CO_2. Both benzene pathways result in products (pyruvate, succinate, and acetate) that are further metabolized via the TCA cycle. Energy harvesting reactions involving dehydrogenase enzymes provide NADH for subsequent cell respiration.

Xenobiotic Compounds

Xenobiotic compounds are chemically synthesized molecules that do not otherwise exist in nature. Among these are derivatives of aromatic hydrocarbons, particularly chlorinated compounds that are widely used as pesticides, insulators, and preservatives. However, the accumulation of these chloroaromatics has become a serious environmental problem when they are either toxic and/or not recycled. For example, the cumulative annual worldwide production of **polychlorinated biphenyls (PCBs)** (Figure 12.20) is presently estimated to exceed 750,000 tons. Ultimately, a considerable portion of this amount ends up in the environment. There are 210 possible PCBs, all having the same biphenyl carbon skeleton and differing only in the number and position of chlorine atoms. PCBs are used in capacitors, transformers, plasticizers, printing ink, paint, and a host of other materials and products. The use of PCBs has now been curtailed in the

United States. **Pentachlorophenol** (see Figure 12.20A) is another broadly applied compound that has become an environmental hazard. It is used in termite control, as a preharvest defoliant, and in wood preservation.

The biodegradation of chlorinated hydrocarbons occurs via dehalogenation reactions that yield compounds that are biodegradable (Box 12.3). The removal of chlorines from an organic compound may be a slow process, and cosubstrates often must be available to provide a source of energy and carbon for the dehalogenating microorganism. A pathway for mineralization of trichloroethylene is outlined in Figure 12.20B.

SECTION HIGHLIGHTS

Lignin is perhaps the most refractory organic compound in nature. Specialized white-rot fungi are able to degrade lignin by producing hydrogen peroxide, which causes the cleavage of the bonds of the complex structure. Aliphatic hydrocarbons are broken down by monooxygenases, whereas dioxygenases are required to break the ring structures of aromatic hydrocarbon compounds. Many xenobiotic compounds have halogen atoms and are persistent in the environment. However, some bacteria can break down these compounds by dehalogenation.

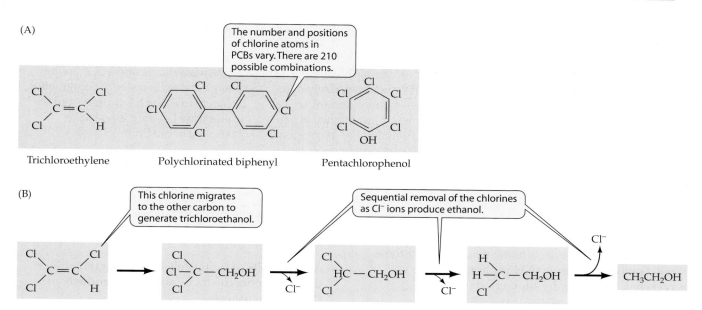

Figure 12.20 Chlorinated hydrocarbons
Typical chlorinated hydrocarbons and their bacterial degradation. (A) Some significant environmental pollutants. Trichloroethylene (TCE) is a degreasing agent; polychlorinat-ed biphenyls (PCBs) have been employed as insulators; and pentachlorophenol (PCP) is a wood preservative. (B) The biodegradation of trichloroethylene by *Mycobacterium vaccae*, following growth of this organism on propane.

BOX 12.3 *Milestones*

Recalcitrant Molecules

Recalcitrant—obstinately defiant of authority or restraint; difficult to handle or operate; not responsive to treatment.

Webster's New Collegiate Dictionary

Molecular recalcitrance measures the relative resistance a molecule has to microbial attack. Many recalcitrant compounds are halogenated aromatic or aliphatic compounds such as DDT and trichloroethylene (TCE). These compounds are of considerable concern because they are harmful environmental pollutants. The persistence of many halogenated chemicals and their toxicity renders them hazardous to health and to the environment. Chlorinated compounds are widely applied as pesticides, solvents, and hydraulic fluids, and thus eventually enter the environment. For example, pentachlorophenol (PCP) (see Figure 12.15) is used as a wood preservative (telephone poles, fence posts) and is an effective long-term protector against rot.

Halogenated compounds have been considered impervious to microbial assault because they are the products of chemical synthesis, and microorganisms may not have evolved with enzyme systems capa-

Lindane	Mirex®	Dalapon®	DBCP
180 days	10+ years	30 days	180 days

The half-life of some compounds that have been used as pesticides. DBCP is dibromochloropropane.

ble of their mineralization. One may ask whether the inability of microbes to dehalogenate is due to the presence of halogenated compounds in nature solely as products of chemical synthesis. Clearly this is not the case. There are more than 700 naturally occurring halogenated compounds that have been identified thus far and at least seven distinct enzyme systems that can cleave the carbon–halogen bond.

Recalcitrance is not solely related to the presence of a —Cl, —Br, —F, or —I atom. The number, position, and electronegativity of the halogen–carbon(s) bonds are of fundamental importance in determining whether the dehalogenating enzymes can function. It is

apparent that some halogenated compounds are not recalcitrant, whereas others are impervious to microbial attack. The structures and half-life of some pesticides are presented in the figure above.

The half-life in soil is for temperate regions such as the United States and is somewhat shorter in tropical regions. Lindane and Mirex® are insecticides, Dalapon® is a herbicide, and DBCP is a nematocide. Dalapon does not accumulate in nature regardless of usage, whereas lindane and DBCP gradually build up in soil with continued use. Mirex, which has been used to eradicate fire ants, is no longer applied in the United States.

SUMMARY

- Sunlight supports biosynthetic reactions, and these reactions can continue only if photosynthate is **recycled** constantly by **biodegradation**.

- Without replenishment, the level of CO_2 in the atmosphere would support the present rate of photosynthesis on Earth for only **30 years**.

- **Microbial decomposition** is the source of 85% of the CO_2 in the atmosphere.

- Some **biodegradative pathways** are a **reverse** of **synthetic** reactions, whereas others follow distinctly **different** pathways.

- **All products** of living cells are biodegraded by some microorganism, **as are many products of chemical synthesis**.

- **Cellulose** is the most abundant product of **biosynthesis** present on earth. It is biodegraded by **both bacteria** and **fungi**.

- Starch is a major storage product in plants. Enzymes termed **amylases** can biodegrade starch to glucose.

- **Chitin**, the **tough integument** of insects and crustaceans, is mineralized by the **actinomycetes** in soils and *Vibrio sp.* in marine environments.

- The **alkanes** are of widespread occurrence in nature and are oxidized by a **monooxygenase** that ultimately produces the homologous carboxylic acid.

- **Phospholipids** are biodegraded by the **sequential removal of the fatty acids** and polar group from the glycerol backbone by a series of phospholipases of differing specificity.

- Initial attack on large molecules such as proteins occurs by the action of **extracellular** enzymes termed **proteases**. These enzymes release lower-molecular-weight polypeptides that are cleaved to **oligopeptides**. There are specific and nonspecific enzymes (**exopeptidases**) that hydrolyze amino acids from the oligopeptides.

- **Uric acid**, a product of purine catabolism, is utilized by *Bacillus fastidiosus* as sole source of carbon and energy.

- **Lignin** is an aromatic complex that makes up 20% to 30% of woody vascular tissue. It is **resistant** to **enzymatic attack** and is depolymerized by the peroxides generated by wood-rotting fungi. Both bacteria and fungi utilize the **products** of this **oxidative attack** as a carbon source.

- **Aromatic hydrocarbons** occur widely as **environmental pollutants**. Many are degraded by bacteria and fungi.

- **Xenobiotic** compounds such as the **chlorinated** hydrocarbons are generally resistant to enzymatic attack. Removal of the **chlorine atoms** through incidental attack while metabolizing **cosubstrates** is a key element in ridding the environment of these compounds.

 Find more at www.sinauer.com/microbial-life

REVIEW QUESTIONS

1. Why is recycling of biological materials important in the perpetuation of life on Earth? What geochemical event was of significant importance in the evolution of heterotrophic microorganisms and multicelled eukaryotes?

2. What percentage of atmospheric CO_2 comes from microbial degradation? How long would the present supply last without replenishment? What would be the consequence of lowered levels of CO_2 in the atmosphere?

3. What are Van Niel's postulates? How do they relate to the role of microbes on Earth?

4. Biodegradation is sometimes a reversal of biosynthesis. In some cases, biodegradation follows a pathway different from that involved in biosynthesis. Explain and consider examples of both.

5. How would a microbe attack a large molecule like cellulose? Like protein? What are the similarities and differences between cellulose and protein degradation?

6. Why are amylases important to the well-being of some animals? What types of organisms produce these enzymes?

7. Where does commonality occur in the biodegradation of long-chain *n*-alkanes and long-chain fatty acids?

8. What are the unique features of lignin biosynthesis? Biodegradation?

9. How can xenobiotic compounds be degraded by organisms that have not previously been exposed to them?

SUGGESTED READING

Atlas, R. M. and J. C. Philp, eds. 2005. *Bioremediation: Applied Microbial Solutions for Real-World Environmental Cleanup.* Washington, DC: ASM Press.

Alexander, M. 1999. *Biodegradation and Bioremediation.* 2nd ed. New York: Academic Press.

King, R. B., J. K. Sheldon and G. M. Long. 1998. *Practical Environmental Bioremediation: The Field Guide.* Boca Raton, FL: CRC Press.

Kirk, T. K. and R. L. Farrell. 1987. "Enzymatic 'Combustion': The Microbial Degradation of Lignin." *Annual Review of Microbiology* 41: 465–505.

Young, L. and C. Cerniglia. 1995. *Microbial Transformation and Degradation of Toxic Organic Chemicals.* New York: John Wiley & Sons.

Wackett, L. P. and C. D. Hershberger. 2001. *Biocatalysis and Biodegradation: Microbial Transformation of Organic Compounds.* Washington, DC: ASM Press.

Zehnder, A. J. B., ed. 1988. *Biology of Anaerobic Bacteria.* New York: John Wiley & Sons.

PART

IV

Genetics and Basic Virology

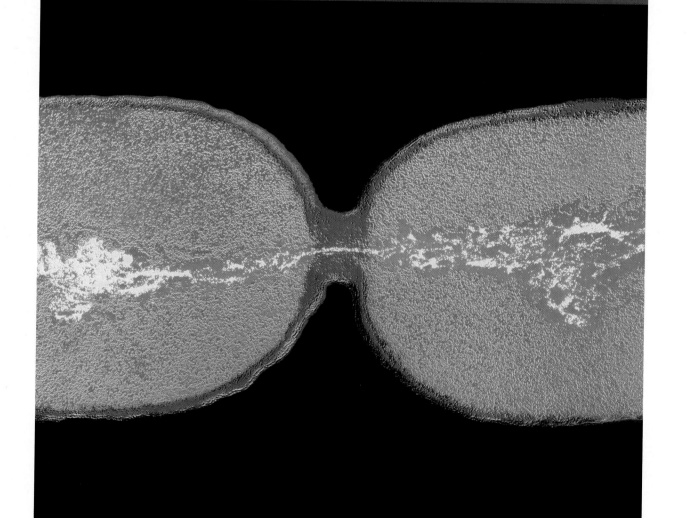

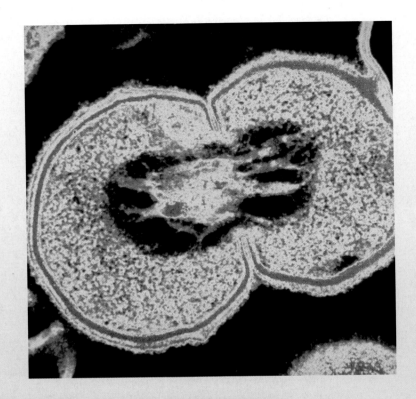

The objectives of this chapter are to:

◆ Discuss the importance of DNA structure in storing the hereditary information of microorganisms.

◆ Explain the details of DNA replication and repair.

◆ Present the process of transmission of genetic information from genes to proteins through messenger RNAs.

◆ Discuss the various systems used by bacteria to control gene expression using transcriptional and post-transcriptional regulatory mechanisms.

◆ Discuss the types and consequences of mutations.

13

Basic Genetics

A cell is DNA's way of making more DNA.
—Author Unknown

All the members of a bacterial or archaeal species are remarkably similar in their structure and ability to survive in a given environment. The thousands of biochemical reactions carried out within the cell—synthesizing cellular constituents, utilizing available nutrients, and responding to environmental stimuli—are nearly identical in all members of a species living under equivalent conditions. How is it that this species identity is maintained as each cell moves through a predetermined life cycle and divides, producing daughter cells that are identical to the parent? During replication, a copy of genetic instructions written in chemical code is transferred to each daughter cell. It is this genetic code—unique to each bacterial species—that safeguards the identity of the species from generation to generation.

This chapter defines the chemical nature of the genetic code, delineates the basic principles of informational transfer during cell division, and describes the cellular machinery responsible for translating the "words" of the code into the "actions" of the cell, such as enzymes and structures. We also examine how the coded instructions carried by the genes can be altered and investigate the consequences of such alterations on traits that bacteria possess and pass on to their progeny or, in some cases, to their siblings.

13.1 The Nature of the Hereditary Information

The hereditary information of all living cells is stored in ordered segments of deoxyribonucleic acid (DNA) called **genes**; genes are found strung together in large molecules called chromosomes. The discovery of DNA as the hereditary material and determination of its structure were two seminal discoveries that provided the basis for understanding the transmission of genetic information in molecular detail and explained the nature of similarities within species and differences among individuals.

The complete set of an organism's genes is called its **genome**. Whereas *Eukarya* carry their genes in multiple, linear chromosomes, the genome of most bacteria consists of a single circular chromosome, with a few bacterial species possessing linear chromosomes or combination of both. Some bacteria contain extrachromosomal molecules of DNA that carry information often necessary for survival in special environments. These molecules, called plasmids, are chemically identical to chromosomal DNA. The smallest genomes are found in viruses; viral genomes can be circular DNA (double- or single-stranded), linear DNA, or double- or single-stranded ribonucleic acid (RNA).

The flow of information within the cell may be viewed as ideas (DNA), transcribed to words (RNA), and translated to actions (proteins). Maintenance of the store of DNA guarantees that the process can be repeated. These

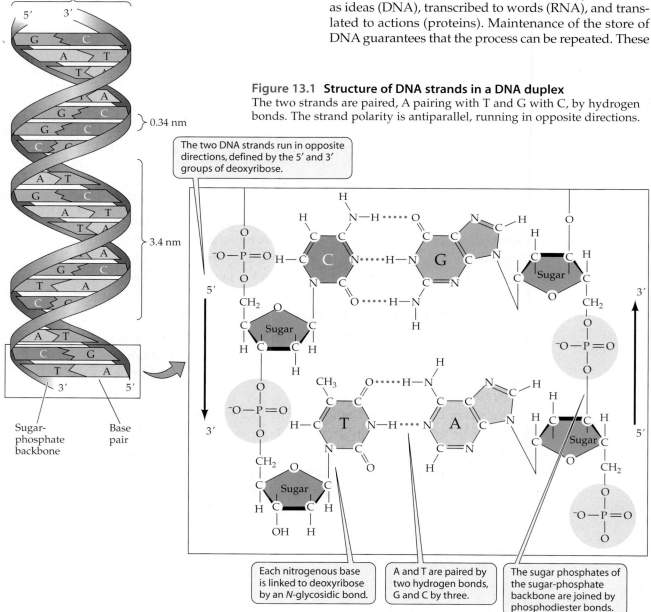

Figure 13.1 Structure of DNA strands in a DNA duplex
The two strands are paired, A pairing with T and G with C, by hydrogen bonds. The strand polarity is antiparallel, running in opposite directions.

The two DNA strands run in opposite directions, defined by the 5' and 3' groups of deoxyribose.

Each nitrogenous base is linked to deoxyribose by an *N*-glycosidic bond.

A and T are paired by two hydrogen bonds, G and C by three.

The sugar phosphates of the sugar-phosphate backbone are joined by phosphodiester bonds.

2 nm

0.34 nm

3.4 nm

Sugar-phosphate backbone

Base pair

processes require energy and are therefore regulated at many levels. The remainder of this chapter reviews the mechanics that make this regulated flow possible.

Structure of DNA

Of all the reactions in a cell, the consistently accurate copying of genetic information can be considered one of the most important. Species identity is written in DNA, and reliable replication is the guardian of this identity. The key to this accuracy is found in base-pairing of the double-stranded structure of DNA, the high fidelity of the machinery responsible for DNA replication, and the proofreading mechanism that corrects even the rarest of the mistakes.

Chemically, DNA is a relatively simple molecule: deoxyribonucleotides linked by phosphodiester bonds. All of these deoxyribonucleotides consist of the same sugar phosphate, **2-deoxyribose**, but they differ with respect to their nitrogenous bases, as shown in Figure 13.1. **Purines** (**adenine** and **guanine**) and **pyrimidines** (**cytosine** and **thymine**) are attached to the sugar by specific *N*-glycosidic linkages. The long DNA chain is the result of links made by a phosphodiester bond between the hydroxyl group attached to the 3'– and the phosphate group on the 5'–carbon atoms of each sugar. It is the order of nucleotides in a gene that carries its unique meaning, just as the order of the letters in a word conveys its meaning. Again, the accurate copying of this sequence during bacterial division allows a bacterium to transmit its specific characteristics from the parent cell to the daughter cells.

Chromosomal and plasmid DNA are always present in a **double-stranded form**, with hydrogen bonds holding the two strands together. **Base-pairing** takes place between adenine and thymine and between guanine and cytosine, with a purine bonding to a pyrimidine in each case (see Figure 13.1). The number of hydrogen bonds between the bases is not identical: two hydrogens participate in a bond between adenine and thymine, whereas guanine and cytosine form hydrogen bonds with three participating hydrogens. Nevertheless, this bonding maintains a distance of just under 0.3 nm between the bases of each strand, giving the DNA molecule a uniform width.

As a result of base pairing, the sequence in each strand is **complementary**; that is, a specific sequence in one strand determines the sequence in the other. Moreover, because the two polynucleotide chains run in opposite directions with respect to their 5'-to-3' links, they are said to be **antiparallel**. This is depicted in Figure 13.1, where the phosphodiester linkages in one strand run in the 5'-to-3' direction, and the complementary strand runs 3' to 5'.

The linear arrangement of nucleotide bases, together with strict base-pairing between specific purines and pyrimidines, puts additional constraints on the DNA molecule. Hydrophobic interactions between adjacent bases result in stacking, forcing the phosphate backbone into a helical configuration. The predominant form of double-stranded DNA is a **right-handed helix**, turning in a clockwise direction, with a regular periodicity of 10.6 base pairs (bp) every 3.4 nm.

An "average" bacterial cell contains about 5×10^6 bp of the four nucleotides, giving a total (unwound) bacterial chromosome length of approximately 1 mm. This single huge molecule fits into a 1- to 2-μm-long bacterial cell, a feat comparable to a 10-m string folding into a 1-cm box. DNA is packaged into the cell, taking on several additional higher-order structural features, most important of which is supercoiling. Like any circular object, if a circular DNA molecule is twisted, it will cross itself. When a right-handed, clockwise helix is twisted in a left-handed direction, it becomes underwound. DNA twisted in the same right-handed direction as the helix becomes overwound. The torsional stress is relieved by **supercoiling**, in which the double-stranded DNA twists onto itself. The condensed, supercoiled form of DNA brings together the negatively charged phosphate backbones, which are naturally repulsing. Neutralizing these charges is accomplished by a coating on the DNA that contains small, basic proteins and a variety of cationic molecules. These factors allow packaging of the genomic DNA into a compact space without disrupting its ability to provide genetic information to the cell.

SECTION HIGHLIGHTS

All living cells carry hereditary information in deoxyribonucleic acid (DNA), which is organized into genes. In bacteria, genes are usually found in a single circular chromosome. DNA is composed of four deoxyribonucleotides: deoxyadenosine monophosphate (dAMP), deoxythymidine monophosphate (dTMP), deoxyguanosine monophosphate (dGMP), and deoxycytidine monophosphate (dCMP). The deoxyribonucleotides are linked together via phosphodiester linkages between 5' phosphoric acid and 3' hydroxyls on neighboring deoxyribose sugars. Bacterial chromosomal DNA is **double-stranded**; opposite, **antiparallel** strands are held together by hydrogen bonds formed between adenine and thymine and between guanine and cytosine bases of the deoxyribonucleotides.

13.2 DNA Replication

Bacterial division is an asexual process in which a single cell divides into more or less equal parts. This mode of reproduction is called binary fission and results in the formation of two daughter cells that are indistinguishable from the single parent cell. Constituents within the cell cytoplasm and cell envelope are divided approximately equally, and, if necessary, minor differences in cytoplasmic content, such as enzymes and cofactors, can be made up by synthesis in the daughter cell immediately following cell division. The exact partition of the genetic material, however, is absolutely necessary. Even the smallest error in segregation of the genome can result in daughter cells permanently losing all ability to further replicate, which in biological terms is equivalent to death.

The replication of the genome results in two virtually identical double-helical molecules. One strand in each molecule is the parent DNA, whereas the other is a complementary, newly synthesized strand. The strand separation of a parent molecule, its use as a template, and its incorporation into the daughter genome is termed **semiconservative replication** and is shown in Figure 13.2.

Overview of DNA Replication

DNA replication in bacteria is rapid and continues throughout the cycle of cell division, in contrast to replication of DNA by eukaryotic cells, which takes place during one distinct stage of the cell cycle. The replication of DNA within the circular bacterial chromosome begins at a single specific site on the bacterial chromosome, termed *oriC* for the ***ori**gin of replication*. The DNA is unwound, the double-stranded DNA is separated into two single-stranded templates, and synthesis proceeds from this site in opposing directions. This bidirectional replication mechanism requires the presence of two sites of synthesis of the new DNA. Each of these two sites, termed **replication forks** (Figure 13.3), comprises a complex of proteins that incorporate complementary deoxynucleotides, accurately copying the template of the parent cell. When the replication forks meet at approximately the opposite side of the circular DNA duplex, replication terminates and each strand has been copied into a complete double-stranded helix, which can then be partitioned into daughter cells.

The replication of this enormous molecule is very rapid; replication forks proceed at a rate of 1,000 bp per second, resulting in repli-

cation of the total 1-mm-long chromosome in about 40 minutes. This, however, does not explain how certain rapidly growing bacteria can divide in 20 minutes. To take advantage of optimal growing conditions, DNA replication is reinitiated before one round of replication is completed. The genomes partitioned into the new daughter cells will already be partially replicated, enabling them to divide again more quickly. A concomitant effect is that in these rapidly dividing cells, genes near *oriC* will be present in more than one copy.

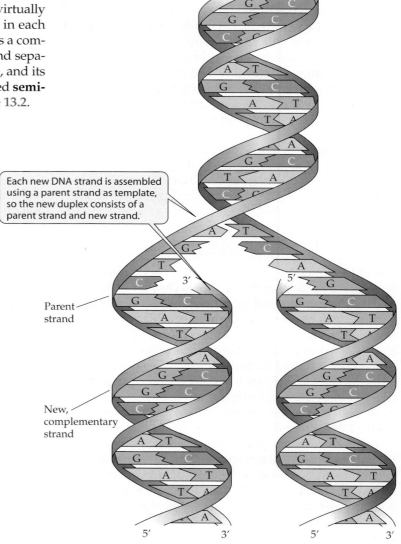

Each new DNA strand is assembled using a parent strand as template, so the new duplex consists of a parent strand and new strand.

Parent strand

New, complementary strand

Figure 13.2 Semiconservative DNA replication
Each new DNA molecule consists of a parent strand and a new, complementary daughter strand.

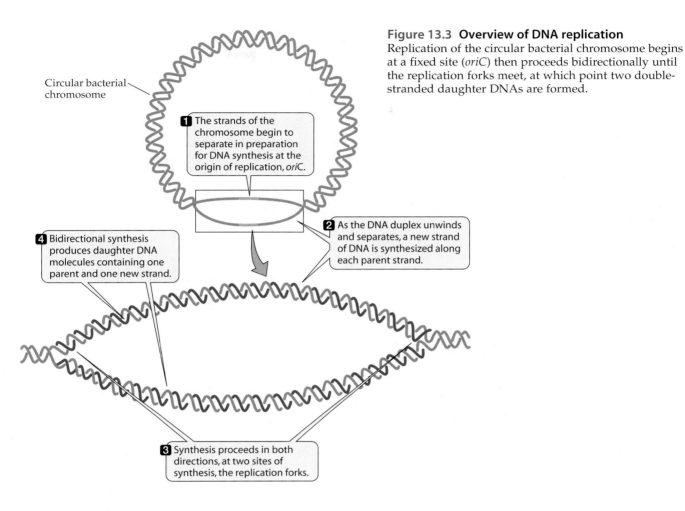

Figure 13.3 Overview of DNA replication
Replication of the circular bacterial chromosome begins at a fixed site (*oriC*) then proceeds bidirectionally until the replication forks meet, at which point two double-stranded daughter DNAs are formed.

Circular bacterial chromosome

1 The strands of the chromosome begin to separate in preparation for DNA synthesis at the origin of replication, *oriC*.

2 As the DNA duplex unwinds and separates, a new strand of DNA is synthesized along each parent strand.

4 Bidirectional synthesis produces daughter DNA molecules containing one parent and one new strand.

3 Synthesis proceeds in both directions, at two sites of synthesis, the replication forks.

Details of the Replication Cycle

Let's now examine in greater detail the events that establish the replication fork and subsequent replication. The bacterial replication apparatus is quite complex. It has been dissected with the aid of mutations in genes that specify its components and with meticulous biochemical reconstruction of the replicative events using purified enzymes. The group of proteins involved in DNA replication is shown in Table 13.1.

Replication of the chromosome is initiated at a site called *oriC*, a 245-bp sequence that is highly conserved in gram-negative bacteria (Figure 13.4). There are four 9-bp repeats within *oriC* and another 3 repeats of a 13-base AT-rich sequence. The replication process is initiated by the binding of up to 40 molecules of the **DnaA** protein to the 9-bp repeats, causing the separation of the duplex DNA to single strands at the adjacent AT-rich 13-mers. A protein complex, designated **DnaB** hexamer,

TABLE 13.1	Proteins required for replication of DNA in *Escherichia coli*
Protein	**Function**
DNA gyrase and helicase	Unwinding of DNA
SSB	Single-stranded DNA binding
DnaA	Initiation factor
DnaB	DNA unwinding
HU	Histone-like (DNA binding)
PriA	Primosome assembly, $3' \rightarrow 5'$ helicase
PriB	Primosome assembly
PriC	DNA unwinding, $5' \rightarrow 3'$ helicase
DnaC	DnaB chaperone
DnaT	Assists DnaC in the delivery of DnaB
DnaG	Primase, synthesis of RNA primers
DNA Pol III	Strand elongation during DNA synthesis
DNA Pol I	Excises RNA primers, fills in with DNA
DNA ligase	Covalently links Okazaki fragments
Tus	Termination of replication

Figure 13.4 Initiation of replication
Details of the events occurring during initiation of *oriC* replication in *E. coli*.

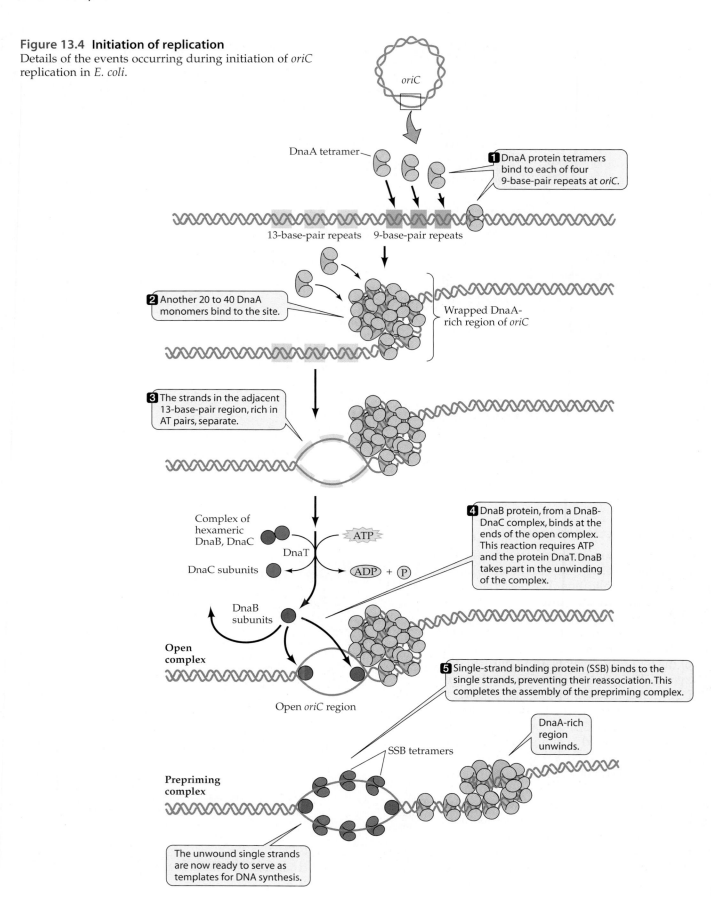

DnaA tetramer

1 DnaA protein tetramers bind to each of four 9-base-pair repeats at *oriC*.

13-base-pair repeats   9-base-pair repeats

2 Another 20 to 40 DnaA monomers bind to the site.

Wrapped DnaA-rich region of *oriC*

3 The strands in the adjacent 13-base-pair region, rich in AT pairs, separate.

Complex of hexameric DnaB, DnaC

ATP

DnaT

DnaC subunits

ADP + P

4 DnaB protein, from a DnaB-DnaC complex, binds at the ends of the open complex. This reaction requires ATP and the protein DnaT. DnaB takes part in the unwinding of the complex.

DnaB subunits

Open complex

5 Single-strand binding protein (SSB) binds to the single strands, preventing their reassociation. This completes the assembly of the prepriming complex.

Open *oriC* region

DnaA-rich region unwinds.

SSB tetramers

Prepriming complex

The unwound single strands are now ready to serve as templates for DNA synthesis.

binds to the two replication forks with the assistance of a protein designated **DnaC**. The addition of the DnaB complex, which forms a ring around the DNA, completes the assembly of the prepriming complex. DnaB has helicase activity and, assisted by **DNA gyrase**, promotes unwinding of the DNA duplex. The **single-strand binding pro-**

tein (**SSB**) tetramers attach to the single strands as they arise. Unwinding exposes the base sequences in the single strands to serve as templates in replication.

Further additions to the pre-priming complex generate a replisome assembly that can replicate the genome directionally (Figure 13.5). The key addition is the **primase**

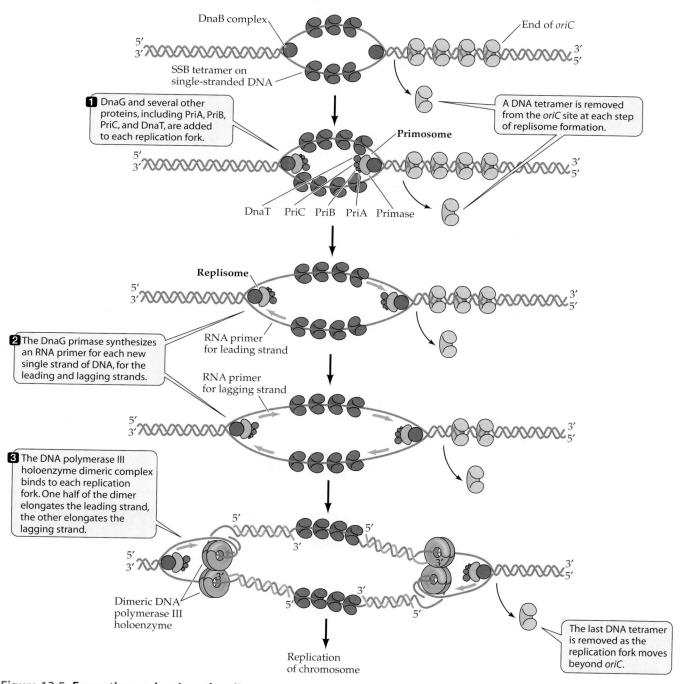

Figure 13.5 Formation and action of replisomes
A replisome assembly forms at each fork in the prepriming complex. Steps 1 through 3 occur at each replication fork. When the forks meet at the opposite side of the chromosome, replication is complete.

(**DnaG**), which synthesizes the RNA primer necessary for DNA synthesis and elongation of the DNA strand.

DNA polymerase is the enzyme that polymerizes deoxynucleoside 5′-triphosphate into DNA. Bacteria possess three distinct DNA polymerases, named **DNA Pol I, II,** and **III**. The enzyme termed **DNA Pol III** carries out the elongation stage of DNA synthesis as the replication fork moves forward (Figure 13.6). DNA Pol I, in concert with DNA ligase, plays a crucial role in filling in gaps and removing RNA primers, annealing short gaps, and in DNA repair. The role of DNA Pol II in replication of the chromosome is not known, and mutations in the gene encoding this enzyme have no effect on bacterial cell division.

DNA Pol III can add only nucleotides to a preexisting 3′-hydroxyl; consequently the template must be "primed" with a nucleotide from which the polymerase can start. For this purpose, a short **RNA primer**, complementary to the exposed template, is synthesized by **DnaG**, an RNA polymerase called **primase**. This short RNA segment provides the necessary 3′-hydroxyl end to which single nucleotides are added. The polymerization reaction takes place in the 5′-to-3′ direction; that is, the incoming 5′-nucleoside triphosphate is linked by its α-phosphate to the 3′-hydroxyl group of the ribose on the primer, with the release of pyrophosphate The accuracy of replication is dictated by the same hydrogen-bonding rules that hold together the nucleoside triphosphates of the original duplex: a guanine (G) is paired with a cytosine (C), and a thymine (T) with an adenine (A).

The fact that DNA Pol III can only polymerize in the 5′-to-3′ direction presents a special problem regarding one of the two strands of DNA. The strand depicted above the replication fork in Figure 13.5 contains constantly available primer, and synthesis is continuous in the 5′-to-3′ direction. This so-called **leading strand** is synthesized as a single molecule complementary to the template. The same 5′-to-3′ direction of synthesis on the opposite DNA template strand, called the **lagging strand**, however, requires discontinuous synthesis involving several steps. First, RNA primase synthesizes a short, 10-nt RNA primer about 1,000 to 2,000 nucleotides upstream (that is, in the 5′ direction) from *oriC*. This process is analogous, although not identical, to the initiation of DNA synthesis at *oriC*. The 3′-hydroxyl group then serves as a primer for polymerization of DNA from the primer back toward the origin of replication. The DNA segments polymerized from such an RNA primer are called **Okazaki fragments**, named for the researcher who first characterized them. In leapfrog fashion, the primase continues laying primers on the lagging strand upstream of the last Okazaki fragment as DNA synthesis proceeds. The resulting complementary strand on the lagging strand is discontinuous, consisting of a series of RNA primers linked to Okazaki fragments. The primers are removed by DNA Pol I via its 5′-to-3′ exonuclease activity (discussed later in the section on DNA proofreading), and the gaps formerly occupied by RNA are filled in by polymerization of deoxyribonucleoside triphos-

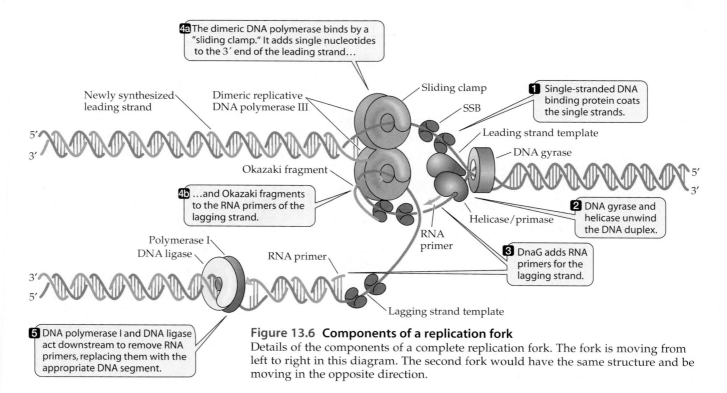

Figure 13.6 Components of a replication fork
Details of the components of a complete replication fork. The fork is moving from left to right in this diagram. The second fork would have the same structure and be moving in the opposite direction.

phates into DNA. The final stage of synthesis is sealing of the nicks between fragments by another enzyme, DNA ligase.

The replication of the circular bacterial chromosome terminates when two replication forks meet. Termination of replication is facilitated by the protein **Tus**, which binds to the *ter* sites. These sites are located on the circular bacterial chromosome opposite to the origin of replication *oriC*. The *Escherichia coli* chromosome contains 10 tandem *ter* sites, each capable of binding a single molecule of Tus. The *ter*-Tus complex coordinates termination of replication by blocking the progress of the replication fork from one direction, until the fork from other direction arrives; the fusion of the forks leads to formation of two double-stranded DNA molecules. Being helical in nature, however, the DNA undergoes rotation to accommodate the moving replication fork. This leads to excessive supercoiling and tension on the DNA molecule. As mentioned previously, partial relief of this tight winding generated during synthesis, as well as the eventual untangling of the daughter strands, is directed by **DNA topoisomerases**. Two types of this enzyme have been isolated from bacteria. **DNA topoisomerase I** is an enzyme that creates a nick on one of the DNA strands, allowing free rotation around the phosphodiester bond opposite the nick, thereby relieving the tension. This break persists only transiently and is rapidly closed. **DNA topoisomerase II** (also called **gyrase**) causes double-stranded breaks, allowing two helices that cross each other to pass through the break on one of the strands before the break is resealed. The reaction catalyzed by DNA topoisomerase II is especially important in freeing the concatamer (two interlocked circles) on completion of replication, which allows free separation of each chromosome into each daughter cell during cell division.

Proofreading During DNA Replication

The specificity of base pairing (AT, GC) is the major mechanism by which reliable copying of the parent strand is guaranteed. Errors during replication can occur when an incorrect, mispaired base is added to the growing chain, creating a transient mismatch. When these strands separate and are partitioned into daughter cells, the chromosome of the daughter receiving the mismatched base will be different from the parent's. This change could be deleterious if it is in an important gene. The process of DNA replication is, however, remarkably efficient and has a very low error rate. It is estimated that in *E. coli*, the frequency of replication error is less than 10^{-9} per replicated base pair—that is, less than one error per billion base pairs. This low frequency of mismatched bases is due to the proofreading and repair activities of DNA polymerases. One system for corrections is the **mismatch repair system** present in *E. coli* and other bacteria, which monitors newly replicated DNA for mispaired bases, removes the mismatched base, and replaces it with the correct one. Repair is by a DNA polymerase–mediated local replication. To replace a mismatched base, the enzyme must identify which base in the pair should be replaced. A mismatch repair mechanism that can identify and remove an incorrect base is the methyl-directed mismatch repair system consisting of the **MutS**, **MutH**, and **MutL** proteins. The activity of this repair system is shown in Figure 13.7. The parental strand in a duplex is methylated, whereas the newly formed strand is not, because methylation of bases occurs after replication. One of the components of this repair system, the MutS protein, binds to a mismatched base pair, whereas MutH continues along the DNA duplex until it encounters a methylated base. The third protein, MutL, binds simultaneously to both MutS and MutH. This bridging results in "nicking" of the unmethylated, newly synthesized strand by MutH. The portion of the DNA strand between MutS and MutH is removed, including the wrong base that caused the mismatch. The final stage of repair process is the resynthesis of the repaired strand.

All three bacterial polymerases are capable of both 5′-to-3′ polymerizing and 3′-to-5′ exonucleolytic activity. The immediate repair of mismatched bases formed during replication is carried out by the 3′-to-5′ exonuclease activity of DNA Pol III, the major polymerizing enzyme in the replication fork. Filling in small gaps and repairing damaged DNA is accomplished by continuous degradation and synthesis in the 5′-to-3′ direction, carried out by DNA Pol I. Additional enzymes are involved in the repair of DNA damaged by physical or chemical agents; these types of repair are discussed later in this chapter.

Alternative DNA Replication Mechanisms

Many bacteria carry accessory genetic elements, called plasmids (see Chapter 15). Because plasmids exist free of the chromosome, they replicate independently and often utilize specialized replicative mechanisms. Two common types of mechanisms are utilized by many plasmids and by certain bacterial viruses.

In general, plasmid replication is analogous to the replication of circular chromosomes. The replication is initiated at a specific origin (*oriV*), and at these single-stranded regions replication begins at annealed RNA primers. In certain instances, the replication is bi-directional. Two replication forks move around the entire plasmid until they meet at a site that is approximately opposite the origin, creating two daughter molecules that are separated by the action of topoisomerases. Some

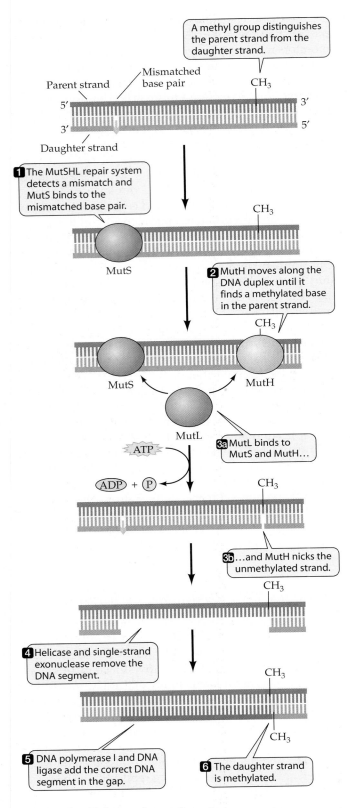

A methyl group distinguishes the parent strand from the daughter strand.

Parent strand
Mismatched base pair
CH_3
5′ 3′
3′ 5′
Daughter strand

1 The MutSHL repair system detects a mismatch and MutS binds to the mismatched base pair.

CH_3

MutS

2 MutH moves along the DNA duplex until it finds a methylated base in the parent strand.

CH_3

MutS MutH

MutL

ATP

ADP + P

3a MutL binds to MutS and MutH...

CH_3

3b ...and MutH nicks the unmethylated strand.

CH_3

4 Helicase and single-strand exonuclease remove the DNA segment.

CH_3

5 DNA polymerase I and DNA ligase add the correct DNA segment in the gap.

6 The daughter strand is methylated.

CH_3

Figure 13.7 Mismatch repair system
Errors encountered during DNA synthesis are repaired by the methyl-directed mismatch repair system, MutSHL.

plasmids replicate by utilizing a single replication fork, and replication terminates when the replication fork returns to the origin. The term **theta replication** refers to the intermediate structures (resembling the Greek letter θ) formed during replication.

Plasmids and certain bacterial viruses with single-stranded genomes can also replicate by a so-called **rolling-circle mechanism**, shown schematically in Figure 13.8. These plasmids contain two sites for replication. One of these, *dso* (double-stranded origin), is used during synthesis of a complementary strand of the "parental" plasmid, whereas the *sso* (for single-stranded origin), is used as the origin for the replication of the single-stranded molecule, generated after one of the strands has been copied. The replication commences after the initiator dimeric **Rep** protein binds to the plasmid double-stranded origin *dso*, resulting in the formation of a hairpin that is immediately nicked by the bound Rep protein. One of the subunits of the Rep protein becomes covalently attached to the 5′-phosphate at the nick site. A replication machinery then assembles at this site, consisting of DNA helicase, the single-stranded DNA-binding protein, and DNA Pol III. Replication commences when one of the un-nicked templates of the duplex is copied and then begins to utilize the free 3′-hydroxyl end as a primer, with a concomitant displacement of the strand that is not being copied.

Replication terminates when the replication fork returns to the *dso*. The replication apparatus then proceeds about 10 nucleotides beyond this site, and, following several nicking and sealing reactions, the displaced strand is released as a single-stranded circular molecule. Then one of the Rep proteins nicks once more at the original *dso* site, releasing the enzyme complex with a short bound oligonucleotide. The gap is filled in and closed to regenerate the original double-stranded molecule. At the same time, the single-stranded DNA is converted to a double-stranded form after bacterial RNA polymerase forms a short RNA primer at the *sso* site. DNA Pol I then extends this primer, followed by DNA Pol III–directed replication, removal of the RNA primer, and sealing of the newly synthesized strand, forming a second double-stranded DNA molecule.

Certain bacterial viruses, such as lambda, use a variation of the rolling circle mechanism to generate long **concatemers** (long tandem repeats; see Figure 14.10) of their genomes. In this case, the displaced single strand is not closed into a circular molecule when the first round of replication is complete. Instead the circular genome continues replicating, generating concatemers of the displaced strand. During this process, RNA primers initiate the synthesis of the complementary strand, resulting in a long double-stranded molecule consisting of multiple copies of the genome. This linear

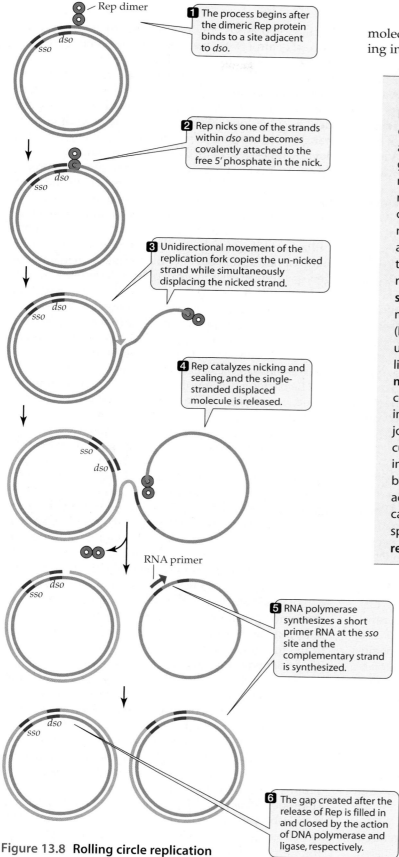

1 The process begins after the dimeric Rep protein binds to a site adjacent to *dso*.

2 Rep nicks one of the strands within *dso* and becomes covalently attached to the free 5' phosphate in the nick.

3 Unidirectional movement of the replication fork copies the un-nicked strand while simultaneously displacing the nicked strand.

4 Rep catalyzes nicking and sealing, and the single-stranded displaced molecule is released.

RNA primer

5 RNA polymerase synthesizes a short primer RNA at the *sso* site and the complementary strand is synthesized.

6 The gap created after the release of Rep is filled in and closed by the action of DNA polymerase and ligase, respectively.

Figure 13.8 Rolling circle replication

molecule is then cut into genome-size units for packaging into bacteriophage capsids.

SECTION HIGHLIGHTS

DNA replicates by a **semiconservative** mode of replication, where the strands are separated and each strand is copied, with the daughter genome receiving one parental strand and a newly synthesized complementary copy. DNA replication begins at a single, relatively well-conserved site called the **origin** (*ori*). Two replication forks move away from the origin, as nucleotides are incorporated with the assistance of a number of proteins. Within each replication fork, one strand (the **leading strand**) is synthesized as a single contiguous molecule, whereas the complementary strand (**lagging strand**) is synthesized in a discontinuous fashion, primed by short RNA segments linked to DNA fragments called **Okazaki fragments**. The newly synthesized DNA strand is completed by removal of RNA primers and filling in the gaps with DNA, followed by the joining of the fragments into a single molecule. The low frequency of mistakes made during DNA replication is due to the efficient—but not perfect—proofreading and repair activities associated with enzymes that replicate DNA. Certain plasmids replicate using a specialized mechanism called **rolling-circle replication**.

13.3 Transcription and its Regulation

There are two occasions when the cell copies its genes. In the first, replication, the entire genome is copied from DNA to DNA for the purpose of continuing the species. In the second, **transcription**, genes are copied from DNA to RNA for the purpose of expressing those genes, that is, to make proteins and RNA needed for sustaining the life of the individual cell.

The genetic information encoded in the DNA almost always specifies proteins. The intermediary in the transfer of genetic information from DNA to proteins is a molecule of **messenger ribonucleic acid (mRNA)**. Because the process of synthesizing RNA from DNA is called transcription, molecules of RNA are sometimes referred to as **transcripts**. Separate classes of RNA, not translated into protein, make up the protein synthesis machinery; these are ribosomal and transfer RNA (rRNA and tRNA) and several small noncoding RNAs with various functions. All types of RNA are synthesized by an identical enzymatic reaction. One significant difference between the types of transcripts is that whereas rRNA and tRNA are produced from larger RNA precursor molecules by enzymatic processing, bacterial mRNA is not processed after transcription. (This difference is not found in eukaryotes, where all types of RNAs are extensively processed, and the final form of the mRNA is the result of cutting and splicing from a large precursor molecule.)

Bacterial mRNA is rather labile, the average half-life being only 1.5 to 2 minutes, unlike eukaryotic mRNA, which can persist and be translated for hours. Short-lived mRNA enables bacteria not only to respond rapidly to environmental changes but also to conserve energy. Within minutes of a new signal from the environment, bacteria can stop putting their resources into transcription of genes that are no longer needed. Nucleotides from the degraded RNA become available for new transcripts. More important, the extremely energy-costly process of protein synthesis is not wasted on unnecessary proteins.

Steps in RNA Synthesis

The process of RNA synthesis is a chemical polymerization, which in principle resembles the copying of one strand of DNA during replication: **initiation**, **elongation**, and **termination**. Transcription, however, is site-specific and involves copying only a short segment of the DNA (genes or clusters of linked genes), and the product is built of ribonucleotides rather than deoxyribonucleotides. The steps in initiation of transcription are shown in Figure 13.9.

The enzyme **RNA polymerase** carries out the synthesis of RNA. Using a DNA template, it recognizes and binds to a specific sequence, which indicates the beginning of a gene. This site is termed the **promoter**; the transcription initiates 10 to 15 bp downstream from the middle of the promoter sequence. The site of initiation is usually T or C, hence all mRNAs start with an A or G. Because only one strand (called the template strand) of the double-stranded helix is copied, it is necessary for the RNA polymerase to displace the complementary copy as it moves along incorporating free nucleoside triphosphates into a polymerized RNA strand. However, the DNA duplex is immediately re-formed following the passage of the transcriptional apparatus.

One method used most frequently to control gene expression is to regulate how often transcription is initiated.

Regulation of RNA polymerase binding allows bacteria to exert control over an entire cluster of genes that specify related functions, utilizing a common regulatory mechanism. A promoter serves as a start for transcription of a single gene, or that of an operon resulting in **polycistronic mRNA**. Because bacteria do not have a nucleus as eukaryotic cells do, the nascent transcript can be translated into proteins from its free 5′ end while it is yet being synthesized. In further contrast to eukaryotic transcription, bacteria frequently synthesize polycistronic mRNA, that is, mRNA that specifies more than one gene. Bacterial genes are frequently arranged in **operons**, sets of adjacent genes that are transcribed and translated together and are functionally related. Genes in an operon utilize a single promoter, preceding the first gene, and transcription terminates after the last gene. Therefore, the genes within an operon are expressed together, and control of their expression usually involves the regulatory sites near the single promoter (see following text). Operons generally code for enzymes that function together in a common biosynthetic or catabolic pathway. This enables the bacterial cell to coordinately regulate expression of related genes, increasing its efficiency.

The bacterial RNA polymerase is a multisubunit enzyme, consisting of five subunits: two **alpha** (α), one **beta** (β), one **beta′** (β′), and one **sigma** (σ). The four subunits, except σ, assemble into a multimeric structure (called the **core enzyme**), which is joined by the σ subunit to form the complete RNA polymerase (the so-called **holoenzyme**), capable of binding to the promoter region. It is the σ subunit within the holoenzyme that makes contacts with specific bases within the promoter—the site from which the transcription is initiated. The subsequent steps in the initiation of transcription are melting of the double-stranded DNA (i.e., separating the two strands) and polymerizing the first few nucleotides, after which the σ factor is not needed. At the end of this stage of initiation of transcription, referred to as promoter clearance, the σ subunit dissociates from the holoenzyme, leaving the core enzyme (2α, β, and β′ subunits) moving along the DNA strand and polymerizing nucleotides into RNA.

Signals built into the DNA at the end of a gene allow the RNA polymerase to terminate transcription. One type of termination sequence is identified as an inverted repeat followed by a stretch of T's. An inverted repeat (sometimes referred to as a palindrome) is a sequence of nucleotides that is the reversed complement of an adjacent sequence. As an RNA copy is made of this section, same-strand pairing of the inverted repeats forms a stem-loop structure (by pairing of the inverted repeats transcribed into RNA), causing the RNA polymerase to be released from the DNA template, thereby terminating transcription.

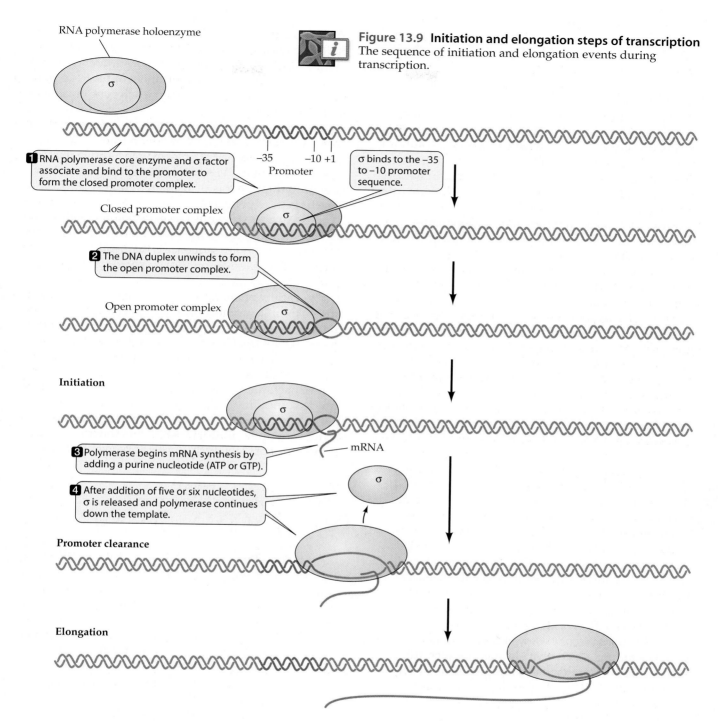

RNA polymerase holoenzyme

σ

Figure 13.9 Initiation and elongation steps of transcription The sequence of initiation and elongation events during transcription.

1 RNA polymerase core enzyme and σ factor associate and bind to the promoter to form the closed promoter complex.

−35   −10 +1
Promoter

σ binds to the −35 to −10 promoter sequence.

Closed promoter complex

σ

2 The DNA duplex unwinds to form the open promoter complex.

Open promoter complex

σ

Initiation

σ

3 Polymerase begins mRNA synthesis by adding a purine nucleotide (ATP or GTP).

— mRNA

4 After addition of five or six nucleotides, σ is released and polymerase continues down the template.

σ

Promoter clearance

Elongation

Another mechanism of termination of transcription involves both a signal on the template that causes RNA polymerase to pause and a specific termination protein called rho (ρ). Rho factor is an ATP-dependent helicase that promotes unwinding of RNA–DNA hybrid duplexes. Rho factor binds to C-rich regions in the RNA transcript and advances in the 5′-to-3′ direction until it reaches the RNA polymerase. At the GC-rich termination site where the RNA polymerase stalls, the ρ factor unwinds the transcript from the template, and the mRNA is released.

The difference between the rho-dependent and rho-independent termination mechanisms is the type of sequence information encoded at the termination site (i.e., whether it leads to pausing of the RNA polymerase, or whether it leads to an RNA stem-loop). Following release of the RNA polymerase, the core enzyme associates with a σ subunit present in the cytoplasm, thereby forming a holoenzyme capable of recognizing another promoter and initiating transcription of another gene.

Promoters and Sigma Factors

The key step in gene expression is the selective transcription of segments of the genome, accomplished by recognition of specific promoter sequences by the σ **subunit** of the RNA polymerase holoenzyme. This interaction determines which gene is to be expressed. In bacteria there exist several types of σ subunits that can form complexes with the RNA polymerase core enzyme. Each species of σ subunit in different RNA polymerase complexes recognizes different promoter sequences, guiding the polymerase to specific genes.

σ^{70} (named for its molecular weight of about 70,000 daltons in *E. coli*) is sometimes referred to as σ^A, and it recognizes the majority of genes encoding essential functions in the bacterial cell. The promoter sequences of these genes are relatively conserved, although they are not identical. These promoters often have two regions of particular similarity: a consensus sequence TATAAT approximately 10 bp upstream from the site of initiation of transcription and a sequence TTGACA, 35 bp upstream, termed the –35 region. (The first base pair copied into RNA is referred to as +1; hence, the TATAAT sequence is at position –10, TTGACA is –35.) The TATAAT is termed a Pribnow box and is named for David Pribnow, the first to recognize the importance of this sequence in transcription. These sequences were deduced by examination of a large number of different bacterial promoters. They represent what is called a consensus promoter for recognition by σ^{70}.

Bacteria contain **alternative σ factors**, which also form complexes with the core RNA polymerase and lead to transcription of genes that contain their cognate promoter sequences. Some of these are summarized in Figure 13.10. The genes requiring alternative σ factors for transcription are involved in cellular responses to a variety of specialized conditions, including survival at elevated temperature, and synthesis of proteins for nitrogen metabolism, motility, and sporulation. The alternative σ factors are a minority and do not effectively compete with σ^{70} under normal conditions.

One of the important features of all promoters, including those that are recognized by the alternative σ factors, is that they contain specific sequences relative to +1. Therefore, promoters give RNA polymerase a direction for transcription; once RNA polymerase binds to a promoter, it will move unidirectionally, as determined by the relative position of the bases in the promoter (e.g., in the case of σ^{70} by the –10 and –35 sequences). Occasionally, two genes

are transcribed from promoters that overlap, and the RNA is copied from opposite strands, but the position of the –10 and –35 sequence (or similar sequences for alternative σ factors) ensure movement of the RNA polymerase in the correct direction for each gene on a given strand.

The term "promoter strength" is used to describe the efficiency of transcriptional initiation from a given promoter. High frequency of holoenzyme binding to a promoter results in high levels of RNA synthesis, and consequently high levels of protein synthesized from a specific gene. A perfect match to the consensus sequence at both –10 and –35 would be expected to result in the highest level of gene expression. However, agreement of a specific promoter sequence is frequently not the sole determinant of the amount of a specific gene product made at any one time.

The variety of alterative σ factors and their recognition of specific promoter sequences act as a type of regulatory mechanism for gene expression that is part of the transcriptional machinery. Conditions that control the synthesis of these σ factors simultaneously control the transcription from their target promoters. Transcription in bacteria can be also regulated by proteins that interfere with the association of σ factors with the core RNA polymerase. These are the so-called **anti-sigma factors** and they control (antagonize) the transcription of groups of promoters that utilize alternative σ factors.

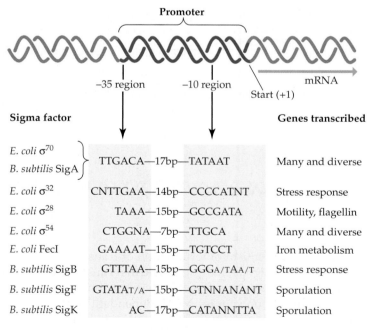

Figure 13.10 Alternative sigma (σ) factors of *Escherichia coli* and *Bacillus subtilis* and their cognate promoter sequences Small letters and slashes indicate where alternative bases are possible.

Anti-sigma factors can function during rapid responses and adaptation to environmental stress, the control of synthesis of polysaccharides, or the expression of virulence genes by pathogens. A number of alternative σ factors have been shown to play an important role in regulating the expression of genes that need to be temporarily or spatially organized, such as those encoding determinants of spore development in certain gram-positive organisms, or the assembly of complex cellular machineries.

One of the best known mechanisms for transcriptional regulation by an anti-sigma factor occurs during the coordinated assembly of the flagella (see Chapter 11). In motile bacteria, the formation and function of the flagellum requires more than 50 gene products. These genes are expressed in an ordered fashion, depending on the need of a particular protein while the flagellum is being assembled. The genes encoding the components of the flagellar motor, consisting of the basal body and hook, are expressed first, followed by the genes for the major (flagellin) and minor subunits of the filament, followed by the genes encoding various chemotactic functions. Any significant accumulation of flagellins prior to their incorporation into the flagellum would result in their premature self-assembly into filaments within the cytoplasm or the periplasm and would likely affect bacterial viability. Therefore, the flagellin gene cannot be expressed before the platform containing the hook-basal body is completed. The coupling of timing of transcription to assembly is regulated by the interaction between a specific σ factor (σ^{28}) controlling the expression of late genes such as those for flagellin subunits, and the anti-sigma factor FlgM (Figure 13.11). Prior to the completion of the basal body and hook structure, the intracellular level of FlgM is high and this protein reversibly binds σ^{28}, sequestering it from the RNA polymerase core enzyme. The transcription of genes for flagellin and other late-acting functions is effectively blocked. Once the basal body and hook are assembled in the cell envelope, FlgM is secreted through the channel formed by this partially assembled organelle, resulting in the release of σ^{28}, which can now associate with RNA polymerase and direct it toward the transcription of the flagellin gene. Flagellin is transported though the channel formed by basal body–hook and the assembly of the organelle is completed.

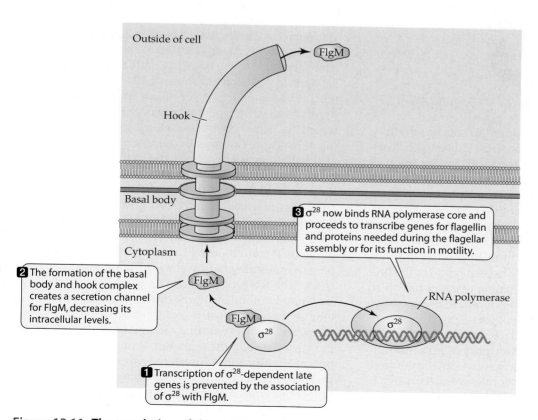

Figure 13.11 The regulation of the activity of the anti-sigma factor FlgM during the assembly of the bacterial flagellum

13.4 Regulation of Gene Expression

Bacteria have the capacity to adapt to specific environmental conditions, and they do so by altering the levels of mRNA available for translation. From the point of view of energetics, this is most efficiently accomplished by controlling the initiation of transcription to prevent the unnecessary use of nucleotide triphosphates and consequent energy-costly synthesis of protein. Although there are many genes expressed constitutively (i.e., without any control of the levels of mRNA other than the affinity of the RNA polymerase holoenzyme for the promoter), there are also a large number of genes for which the ability of RNA polymerase to initiate transcription is controlled by other proteins that respond to signals from the environment of the bacterial cell.

Control at the level of transcriptional initiation, responding to physical or chemical cues from the environment, can be divided into two classes, according to whether a particular gene is repressed by regulatory factors that interfere with its transcription. Environmental signals can be required for these regulatory factors to function. However, different signals from the environment can also reverse repression, allowing a previously repressed gene to be expressed. In contrast, positive regulation centers on the activity of regulatory proteins that are required to facilitate transcription in response to environmental signals. In certain instances, a gene or an operon can be regulated negatively under one set of conditions and positively under another.

Negative Regulation

There are several modes employed to decrease the capacity of RNA polymerase to transcribe a specific gene or operon. The most common mechanism involves a regulatory protein called a **repressor**. Repressors usually bind to specific sites, termed **operators**, near or within the promoter region. The operator sequence is usually located between the promoter and +1 position on the DNA, and frequently overlaps the −10 region of the promoters. The binding of the repressor, therefore, prevents transcription by physically blocking the polymerase from either binding its operator or, if bound, from moving forward.

One of the best-studied examples of negatively regulated genes is the lactose utilization operon (*lac*) that controls catabolism of lactose by *E. coli*. The regulatory mechanism is shown in Figure 13.12. The genes of the *lac* operon are expressed only when bacteria are growing in the presence of lactose as the sole carbon source. The operon includes three genes: *lacZ* (encoding the enzyme β-galactosidase), *lacY* (lactose permease), and *lacA* (transacetylase). The negative regulatory aspect of the *lac* operon comes from the ability of a repressor protein, called the lactose repressor (product of *lacI*), to tightly bind to the operator site near the promoter from which the lactose operon is transcribed, preventing binding of RNA polymerase. In the absence of lactose, none of the three genes necessary for lactose catabolism are transcribed. When *E. coli* finds itself in a medium where lactose is the only carbon and energy source, small amounts of lactose enter the cell, and it is isomerized to the actual inducer allolactose and binds to the repressor protein. The interaction of allolactose with the repressor prevents it from binding to the operator, and those repressor molecules that are already bound dissociate from DNA. The RNA polymerase now can bind to the promoter and initiate the transcription of the *lac* operon. As long as there is lactose present in the bacterial environment (and a corresponding amount of allolactose inside the bacterial cell), the repressor is unable to bind to the operator and cannot interfere with transcription. It is precisely these conditions that necessitate the production of enzymes that are involved in lactose utilization. It is of interest to note that other synthetic molecules that resemble lactose, but are not metabolized by *E. coli*, can cause dissociation of the repressor from the operator.

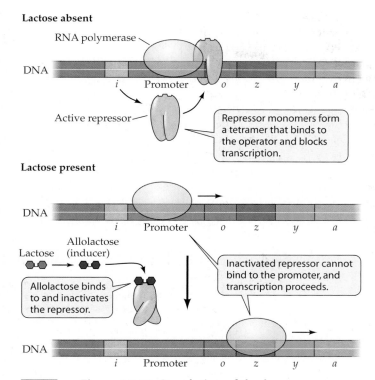

Lactose absent

RNA polymerase

DNA

i   Promoter   *o*   *z*   *y*   *a*

Active repressor

Repressor monomers form a tetramer that binds to the operator and blocks transcription.

Lactose present

DNA

i   Promoter   *o*   *z*   *y*   *a*

Lactose   Allolactose (inducer)

Allolactose binds to and inactivates the repressor.

Inactivated repressor cannot bind to the promoter, and transcription proceeds.

DNA

i   Promoter   *o*   *z*   *y*   *a*

Figure 13.12 Regulation of the *lac* operon
Regulation of gene expression in the *lac* operon. In the absence of lactose, repressor blocks the operator; when lactose is the sole substrate, its isomer (allolactose) inactivates the repressor and the genes are transcribed.

These small molecules and similar signaling molecules that interfere with the activities of repressor proteins are called **inducers**. The common mechanism of all inducers is that they bind to the repressor, causing it to lose its ability to interact with the operator sequences.

Although all negatively regulated systems rely on the ability of repressor proteins to block transcriptional initiation, the precise interaction with small signaling molecules varies. The inducer (allolactose or its structural analogs) causes dissociation of the repressor from the operator. Similarly, a regulatory protein (AraC), which controls genes of arabinose metabolism, acts as a true repressor, binding to a specific operator sequence and preventing the initiation of transcription in the absence of arabinose. In the presence of the inducer arabinose, the AraC–arabinose complex dissociates from the repressor, analogous to the behavior of the lactose repressor–allolactose complex. The situation with the AraC protein is a little more complex, because the AraC–arabinose complex also participates in a positive regulatory mechanism, facilitating transcription for arabinose uptake and utilization.

A slightly different mechanism of negative regulation involves interaction of a signaling molecule with a free, unbound repressor protein, in which this complex interacts with the operators. One of the most extensively studied systems of this type is the regulatory mechanism involved in transcription of the tryptophan biosynthetic genes (Figure 13.13). The inactive "empty" repressor (called the aporepressor), which is not complexed with tryptophan, is unable to bind to the operator of the *trp* operon, thus tryptophan biosynthetic genes are transcribed and bacteria synthesize their own tryptophan. In the presence of tryptophan in the growth medium, there is no need to synthesize the enzymes of the tryptophan biosynthetic pathway because external tryptophan is taken up and utilized; therefore, the *trp* operon is not transcribed. The shutoff of the *trp* operon is due to formation of an active repressor following its binding of tryptophan. In this system, the signal molecule is called a **co-repressor**, as it actively participates with the repressor in the negative regulation of transcriptional initiation.

Positive Regulation

Positive regulation involves direct or indirect action of signaling molecules with **activator** proteins, thereby stimulating transcription. Although the mechanism of negative repression is obvious (i.e., interference with binding of RNA polymerase), the molecular basis of activation is less clear. The activator proteins interact with sequences usually upstream of the promoter, and these regulatory sites can be several hundred base pairs from the affected gene. The interaction of the activator protein with its cognate regulatory sequences results in a direct contact between the regulatory protein and RNA polymerase bound to the promoter. If the binding site for the regulatory protein is far from the promoter, this requires extensive looping of the DNA between the two sequences to allow for the regulatory protein to contact RNA polymerase. Some activator proteins can recognize sequences that are found near or within promoters. In case of the promoters of positively regulated genes, which are recognized by the σ^{70} RNA polymerase, they have –10 regions that are well conserved with other promoters, but there is a considerable variation in the sequence at their –35 regions. The most plausible explanation for the stimulatory effect of the proteins involved in the positive regulation of gene expression is that they stabilize the RNA polymerase complex with the promoter sequence. Alternatively, they may help RNA polymerase in melting of the DNA duplex into single-stranded regions during the critical incorporation of the first few nucleotides at the initiation of transcription.

One of the well studied examples of positively regulated genes is the maltose **regulon**. The term "regulon" denotes a set of operons under the control of a common regulatory element. In the case of the maltose reg-

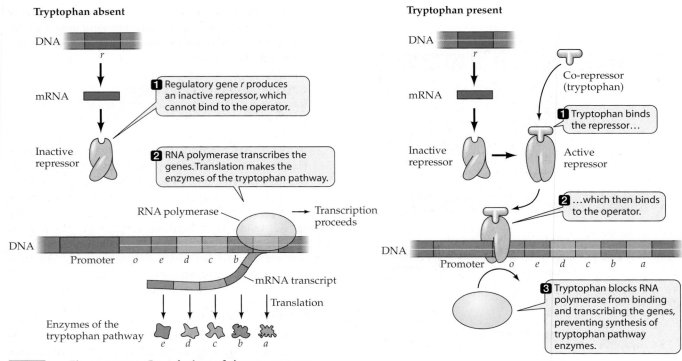

Figure 13.13 Regulation of the *trp* operon
In the absence of tryptophan in the growth medium, the genes are read constitutively and the enzymes required for formation of tryptophan are synthesized. When tryptophan is present in the medium, the promoter is blocked and transcription halts.

ulon, the positive regulatory protein MalT stimulates transcription of at least four different operons involved in uptake and metabolism of maltose or related sugars (**Figure 13.14**). The *malT* gene product is initially inactive (MalT$_i$), but becomes active (i.e., capable of binding to the DNA recognition sequences) after complexing with maltose and this form is referred to as MalT$_a$. The

promoters of the maltose regulon lack typical −35 sequences but contain the binding site for MalT$_a$. The sequence of this so-called MalT box within the promoter regions of MalT$_a$-regulated genes or operons is GGA(T/G)GA. Binding of MalT$_a$ to the MalT box facilities the RNA polymerase's σ factor association with the −10 region.

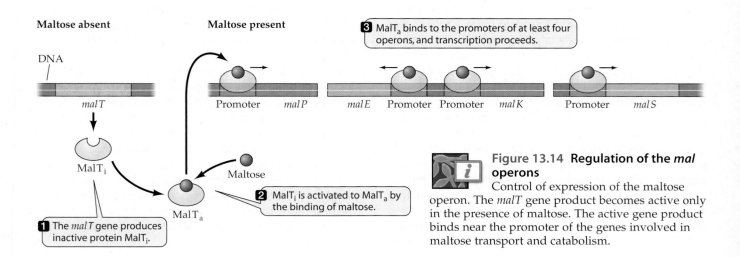

Figure 13.14 Regulation of the *mal* operons
Control of expression of the maltose operon. The *malT* gene product becomes active only in the presence of maltose. The active gene product binds near the promoter of the genes involved in maltose transport and catabolism.

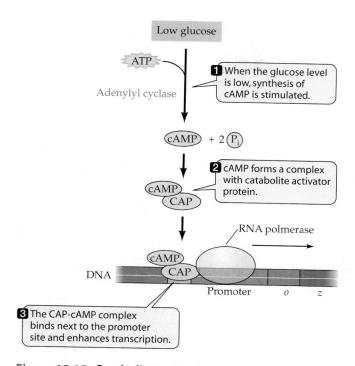

Figure 13.15 Catabolite activation
Catabolite activation occurs when products of catabolism regulate the rate at which genes are transcribed, as in the *lac* operon. Low glucose levels enhance transcription of the *lac* genes.

Several negatively regulated genes are positively controlled as well, whereas some genes are regulated only by activator proteins. In *E. coli* the *lac* operon, for example, is regulated by the levels of cyclic AMP (cAMP) in the bacterial cell, through the activity of the catabolite activator protein (CAP, also called CRP, for cAMP receptor protein), as shown in Figure 13.15. Catabolites are complex carbohydrates (including maltose or lactose) that require energy-consuming breakdown into monomers prior to their utilization in metabolism, and are considered a relatively poor source of energy when compared to monosaccharide such as glucose. cAMP is an adenosine monophosphate molecule with an internal phosphodiester bond between the 5' and 3' carbon atoms of adenosine. It is synthesized from ATP by an enzyme adenylate cycles, whose activity, and, therefore the intracellular level of cAMP, is increased when capabilities are the sole energy source, whereas it is repressed in the presence of high-energy sources such as glucose. In the context of transcriptional regulation of the *lac* operon, cAMP is a co-inducer, because it binds directly to the regulatory protein. In the presence of elevated levels of cAMP (which takes place when preferred carbohydrate sources are depleted—another condition that allows *E. coli* to take full advantage of lactose metabolism), the cAMP-CAP complex binds to a site adjacent to the promoter, and if the operator site is unoccupied by the repressor, the transcription of the *lac* operon is enhanced severalfold above that which would result in the presence of the inducer alone.

Regulation by Two-Component Signal Transduction Pathways

There are several cases of positive control of transcription, where the environmental signal is sensed by activator proteins through a covalent modification, which converts it into a form capable of binding to regulatory sequences and helps RNA polymerase in initiating transcription. The conversion of an inactive regulatory protein (response regulator) into its active form involves also a second protein, a sensor; therefore, these systems have been referred to as **two-component signal transduction systems**. The signal leading to transcriptional regulation starts with autophosphorylation of the **sensor**, a protein kinase. The signal-induced conformational change in the sensor activates its latent self-modifying enzymatic activity (hence the term **autophosphorylation**), using ATP as a donor of the phosphate moiety. Transfer of the phosphate follows this to the **response regulator**. The phosphorylated regulatory protein then binds to a specific regulatory sequence, which is located on the 5' (upstream) of the gene and facilitates transcription of a promoter from an adjacent site. The phosphorylation of the response regulator is reversible, and specific enzymes (phosphatases) can remove the phosphate from the regulatory proteins. In certain instances, the sensor kinase has the ability to carry out the removal of the phosphate, thus performing dual and opposite reactions, depending on the availability of the environmental signal.

The mechanism of sensing high osmolarity by the membrane-embedded protein EnvZ (sensor) and the transcriptional activator OmpR (response regulator) of *E. coli* is outlined in Figure 13.16. Similar phosphorylation mechanisms have also been shown to function in controlling selective expression of genes that respond to nitrogen limitation in enteric bacteria, nitrogen fixation in rhizobia, signaling for transcription of the apparatus necessary for DNA transfer from *Agrobacterium* to plants, and biosynthesis of pili and flagella in an assortment of bacteria. Two-component systems are one of the more common sensory mechanisms used by a variety of bacterial pathogens to respond to conditions within infected hosts, and they activate signal transduction pathways leading to expression of virulence factors. Although most two-component regulators function in transcriptional activation, in some regulated systems binding of phosphorylated response regulators leads to interference with RNA polymerase binding and a negative regulatory effect.

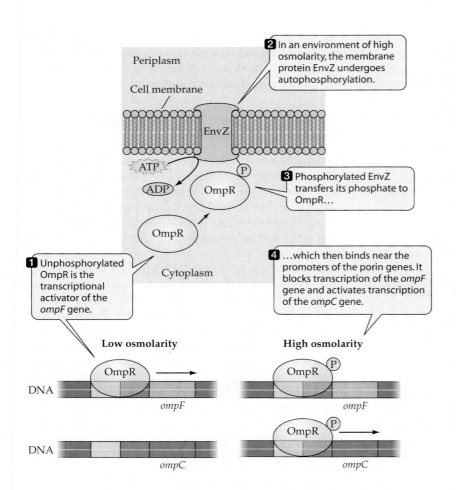

Figure 13.16 Osmoregulation of gene expression
Osmoregulation of porin gene expression in *E. coli* is controlled by a two-component regulatory system consisting of EnvZ, which senses external osmolarity, and OmpR, a transcriptional activator. *E. coli* can express two porins, OmpF and OmpC, which differentially regulate flux of solutes from the environment due to their different pore sizes. In environments of low osmolarity, OmpF is the major porin in the outer membrane, but OmpC becomes the predominant porin at high osmolarity. By altering the ratios of OmpC and OmpF, cells can adjust to environmental changes in various solutes. At low osmolarity, OmpR is mostly unphosphorylated, and it binds efficiently to regulatory sites adjacent to the promoter of the *ompF* gene, resulting in its expression. At high osmolarity, the autokinase activity of EnvZ sensor is stimulated and this leads to transfer of phosphate to the response regulator OmpR. The phosphorylated form of OmpR now has an increased affinity for the *ompC* regulatory region, stimulating expression of high levels of OmpC, while simultaneously repressing the *ompF* gene.

SECTION HIGHLIGHTS

When the product of a gene is not needed, bacteria utilize **negative regulation** to prevent the gene's transcription. This often involves binding of a **repressor** protein to an **operator** site located adjacent to a promoter sequence. Tight binding of the repressor to a sequence that often overlaps the protein interferes with σ-factor–directed binding of the RNA polymerase, and thus prevents transcription of the negatively regulated gene. Alternatively, protein factors (**activators**) stimulate transcription by binding to DNA sites upstream of promoters and facilitate the rate-limiting incorporation of the first bases into RNA. A variety of bacterial species control gene expression through the use of a two-component signal transduction pathway, consisting of a sensor protein kinase and its substrate, a response regulator whose phosphorylation state controls its activity as a transcription factor.

13.5 Protein Synthesis

Depending on the genome size of a particular bacterium, between 1,000 and 7,000 different polypeptides (the biochemical term for proteins) can be found in a cell; most of these are the cellular structural components or enzymes. Chemically, proteins are relatively simple molecules, consisting of linearly linked amino acids. The order of the assembled amino acids—the total number and nature—distinguishes one protein from another. There are 20 different amino acids that can be used as building blocks. Amino acids are similar to one another, each consisting of a central carbon (called the α-carbon) with an amino group, a carboxyl group, a hydrogen, and a side group. Of the 20 amino acids, 19 differ only in the structure of the side groups. The twentieth amino acid, proline, differs from the others by having its side group linked to the α-carbon, resulting in a ring structure. The structures of the 20 amino acids present in proteins were presented in Figure 3.15.

Each amino acid is linked to its adjacent amino acid by a peptide bond (see Figure 3.16), formed when the primary amino group of one amino acid condenses with the carboxyl group of the adjacent one. Short chains of bonded amino acids are therefore called peptides. Because of the directional bond, one end of the peptide is the amino terminus and the other is the carboxyl terminus. The sequence of linked amino acids makes up the primary structure of a polypeptide. Proteins, however, are rarely found in these extended configurations, but rather assume additional conformations. The secondary structure of a polypeptide refers to the shape of the backbone of each α-carbon and the positioning of the side groups (i.e., whether the backbone has a spiral shape with the side chains pointing outward, or there are regions that are relatively straight with intermittent turns). Furthermore, the polypeptide chain can fold, and this tertiary conformation can bring together amino acids that are distant from one another in the linear polypeptide. Moreover, independent polypeptide chains can aggregate to bring together amino acids that are present on separate polypeptide chains (see Figures 11.6, 11.7, and 11.8).

Different polypeptides consist of differing amounts of the 20 amino acids. The differences among proteins are due to the characteristics (polar, nonpolar, or electrically charged) and order of the amino acids in their primary structure. The order of the specific amino acids along the polypeptide chain is determined by the specific order of the nucleotides in the gene coding that polypeptide. The genetic information is first transcribed from DNA into linear RNA called messenger RNA (mRNA), as it serves as the messenger of genetic information between the genes and the protein synthesizing apparatus. Distinct triplet nucleotide sequences in mRNA, called codons, encode specific amino acids. The order of the codons, from a fixed point, determines the order of the amino acids in the polypeptide chain. Hence, the process of protein synthesis serves two main functions simultaneously: (1) it chemically catalyzes the linking of amino acids into a protein, and (2) it decodes the triplet codons on mRNA and converts this information to the primary structure of a protein. The process of protein synthesis is therefore called **translation**. One of the major breakthroughs in biological science took place in the early 1960s, when the genetic code was determined and the specific assignment of triplet nucleotides to each amino acid was made. There are 61 codons specifying all 20 amino acids that are components of proteins. Codon usage is relatively redundant. For example, any one of six different triplets (UUA, UUG, CUU, CUC, CUA, or CUG) in mRNA can specify leucine in the polypeptide chain. However, two amino acids have only one codon each: methionine (AUG), and tryptophan (UGG) (Table 13.2).

In addition to codons specifying amino acids, several signaling codons determine the beginning and the end

| TABLE 13.2 | The genetic code: the codons specify the amino acids incorporated into protein |||||

First Position (5′ end)	Second Position				Third Position (3′ end)
	U	C	A	G	
U	UUU Phe UUC Phe UUA Leu UUG Leu	UCU Ser UCC Ser UCA Ser UCG Ser	UAU Tyr UAC Tyr UAA Stop UAG Stop	UGU Cys UGC Cys UGA Stop UGG Trp	U C A G
C	CUU Leu CUC Leu CUA Leu CUG Leu	CCU Pro CCC Pro CCA Pro CCG Pro	CAU His CAC His CAA Gln CAG Gln	CGU Arg CGC Arg CGA Arg CGG Arg	U C A G
A	AUU Ile AUC Ile AUA Ile AUG Met[a]	ACU Thr ACC Thr ACA Thr ACG Thr	AAU Asn AAC Asn AAA Lys AAG Lys	AGU Ser AGC Ser AGA Arg AGG Arg	U C A G
G	GUU Val GUC Val GUA Val GUG Val	GCU Ala GCC Ala GCA Ala GCG Ala	GAU Asp GAC Asp GAA Glu GAG Glu	GGU Gly GGC Gly GGA Gly GGG Gly	U C A G

[a]AUG signals translation initiation as well as coding for Met residues.

of polypeptide chains. **Initiation codons** signal the start (the amino terminus) of the polypeptide chain by coding for a methionine (AUG) or, rarely, valine (GUG). The end of the carboxyl terminus of the polypeptide is determined by termination codons, also called **stop codons**. Stop codons (UAA, UAG, or UGA) do not specify any amino acids; one or several of these on the mRNA will stop the ribosomes from continuing. The sequence of codons between the initiation and termination signals is the **reading frame**. Given that a codon is a triplet, any one nucleotide sequence of mRNA contains three possible reading frames, depending on which of three positions is used as the beginning point, as illustrated by the following. The one translated is determined by the location of the initiation codon upstream of this sequence:

mRNA Reading frame #1	...GUC Ala	UUU Phe	AUA Ile	ACA Thr	CAC His	CCU Pro	
mRNA Reading frame #2	...G	UCU Ser	UUA Leu	UAA Stop	CAC	ACC	CU
mRNA Reading frame #3	...GU	CUU Leu	UAU Tyr	AAC Asn	ACA Thr	CCC Pro	U

If the sequence of a functional protein is Ala-Phe-Ile-Thr-His-Pro, the initiation codon must be located upstream, in multiples of three and in reading frame #1. If the initiation codon were in frame with either the second or third reading frames, proteins with totally different primary sequences would result. Note that reading frame #2 contains a stop codon, which would result in a short polypeptide chain having Leu as its carboxyl terminus.

Components of the Protein Synthesis Machinery

Protein synthesis takes place on complex particles called **ribosomes**. Ribosomes, together with a number of RNA molecules and polypeptides, recognize the coded information on mRNA and translate it into a polypeptide. Ribosomes in bacterial cells are relatively large, multisubunit complexes of RNA and proteins. In its functional stage, an individual ribosome consists of two subunits, one large and one small. As mentioned in Chapter 4, the physical properties of ribosomes were originally studied by their behavior during high-speed centrifugation. Hence, the size of a ribosome and its subunits are in Svedberg units (S), the units of velocity of travel in a centrifugal field. The complete ribosome has a size of 70S, whereas the small subunit is 30S and the large one is 50S.

A comparison of the size and the RNA–protein composition of individual subunits of the *E. coli* ribosome is presented in Table 13.3. Operons encoding ribosomal RNA genes are generally present in more than one copy in the bacterial chromosome. This redundancy of genes is rare in prokaryotes and enables bacteria to respond to sudden changes of growth conditions by synthesizing a large quantity of the translational machinery. Ribosomes are also targets of a number of antibiotics used in the treatment of infectious diseases (Box 13.1).

The most critical components of the translational apparatus are small RNAs that carry specific amino acids and are responsible for the conversion of genetic information into protein sequences. These **tRNAs** are the "trans-

TABLE 13.3	The structural components of *E. coli* ribosomes		
	Ribosome	**Small Subunit**	**Large Subunit**
Sedimentation coefficient	70S	30S	50S
Mass (kD)	2520	930	1590
Major RNAs		16S = 1542 nt	23S = 2904 nt
Minor RNAs			5S = 120 nt
RNA mass (kD)	1664	560	1104
RNA proportion	66%	60%	70%
Protein number		21 polypeptides	31 polypeptides
Protein mass (kD)	857	370	487
Protein proportion	34%	40%	30%

From: Garrett and Grisham, *Biochemistry*, 1995. Figure 32.1, page 1041.

BOX 13.1 *Milestones*

Antibiotics as Inhibitors of Bacterial Protein Synthesis

The ability to synthesize proteins is essential for bacterial survival and growth in all environments. Interference with this process leads to bacterial death. A number of antibiotics, useful in antibacterial therapy for infectious diseases, are directed against ribosomes and result in the inability to grow or to synthesize proteins. Because bacterial ribosomes differ substantially from eukaryotic ribosomes, these antibiotics show significant specificity and can thus be used against infectious diseases, selectively inhibiting prokaryotic pathogens without causing any damage to protein synthesis in the human cells. A number of the commonly used antibiotics and their mechanisms of action are listed:

Antibiotic	Mode of Action
Streptomycin	Binds to 30S subunit; causes misreading and irreversible inhibition of protein synthesis
Other aminoglycosides (e.g., gentamicin)	Similar to streptomycin, although have different targets on 30S and 50S subunits
Tetracyclines	Reversibly bind to 30S subunit, inhibit binding of aminoacyl-tRNA
Chloramphenicol	Reversibly binds to 50S subunit; inhibits peptidyl transferase and peptide-bond formation
Erythromycin and lincomycin	Bind to 50S subunit; inhibit peptidyl transferase and translocation reactions

lators" of the translation process. The tRNA molecule is about 70 nt long and folds into the configuration shown in Figure 13.17. Four extensive regions of complementarity are formed, and three open loops are present. One of these is the anticodon loop, containing a region precisely complementary to the codon sequence on mRNA. For example, the tRNA that is involved in adding tryptophan to the polypeptide chain contains an anticodon sequence ACC, which is complementary to the codon UGG in the mRNA. There are more than 20 but fewer tRNAs than there are possible codons, because incorrect base pairing in the third position of the codon can still result in incorporation of the correct amino acid into the polypeptide. This is referred to as third-base wobble. The tRNA genes, as with genes for rRNA, are redundant, and tRNA is often cut out from a larger precursor RNA molecules.

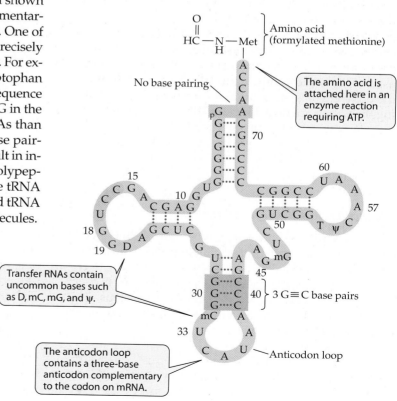

Figure 13.17 The structure of *E. coli* *N*-formyl-methionyl-tRNA^fMet
The base areas that differ from those in non-initiator tRNAs are highlighted. This is the initiator of peptide synthesis. Several unusual bases can be also found in tRNAs; tRNA^fMet contains D, mC, mG, and ψ, which are dihydrouridine, methyl cytidine, methyl guanosine, and ribofuranosyl uracil, respectively.

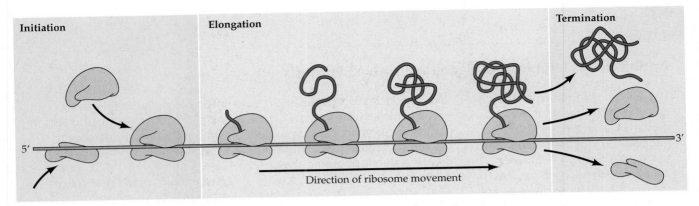

Figure 13.18 Overview of the three stages of protein synthesis

The attachment of the amino acid to tRNA is an enzymatic step involving recognition of a tRNA molecule by the enzyme **aminoacyl-tRNA synthetase** and attachment of the amino acid (aa) to the end of the tRNA:

$$aa + tRNA + ATP \rightarrow tRNAaa + ADP$$

A tRNA that is attached to its specific amino acid is said to be charged. Each tRNA is specific for a codon, and it is indicated by a superscript. For example, the tRNA for the tryptophan codon is referred to as tRNA^{Trp}. Once it is charged with the correct amino acid (tryptophan in this example) by the cognate amino acyl-tRNA synthase it now becomes Trp-tRNA^{Trp}.

The enzyme aminoacyl-tRNA synthetase has a specific recognition for both the tRNA and the amino acid corresponding to the codon. For example, the amino acid asparagine is added only to the tRNA containing the sequence UUA or UUG in the anticodon loop. Errors (placing the wrong amino acid onto a tRNA) are exceedingly rare. They would result in an incorrect translation of the genetic code, thereby synthesizing a polypeptide with different amino acids than those specified by the genetic information.

Mechanism of Protein Synthesis

The process of assembling a polypeptide chain is an orderly reaction consisting of the stepwise transfer of amino acids directed by an mRNA sequence. The overall scheme of protein synthesis is outlined in Figure 13.18. The process can be divided into three steps: **initiation**, **elongation**, and **termination**. Each of these steps is discussed here separately.

INITIATION   The initiation of protein synthesis involves the interaction of ribosomes with the site on mRNA that includes the codon for the first amino acid

(the amino terminus) of the polypeptide chain. The precise placement of the N-terminal amino acid is important because the reading frame of the remaining polypeptide is determined from this point.

A series of soluble factors and the free 50S and 30S subunits are joined with mRNA and the aminoacyl-tRNA to form what is called the initiation complex. Initiation of protein synthesis involves binding the small ribosomal subunit to mRNA with the aid of several polypeptides called **initiation factors** (**IFs**). There are three IFs in the bacterial cytoplasm: IF1, IF2, and IF3. IF3 is required to stabilize the interaction of the 30S subunit with mRNA binding. IF1 is also part of the 30S–mRNA complex and assists IF3 although its precise role in the initiation of protein synthesis is not known. IF2 functions during binding of the initiator tRNA and guanine triphosphate GTP to the ribosome.

The association of the 30S subunit with a precise site on mRNA that encodes the first amino acid of the polypeptide is accomplished by pairing of a specific sequence on the mRNA to a complementary sequence at the 3' end of the 16S ribosomal RNA. This sequence on mRNA is the so-called **ribosome-binding site** (**RBS**) or the **Shine-Dalgarno sequence**, named after its discoverers. It is approximately 5 to 8 nucleotides (nt) from the initiation codon. As shown in Figure 13.19, the sequence at the 3' end of the *E. coli* 16S rRNA (3'-AUUCCUC-CACUAG....5') is complementary to a portion of each Shine-Dalgarno sequence, base-pairing over a sequence ranging from 3 to 8 nt. In general, longer complementary sequence results in higher affinity for mRNA of the 30S ribosome and more efficient initiation of translation. The interaction of the small subunit with a sequence near the start of translation allows the ribosomes to find the beginning of the protein coding sequence anywhere on the transcript, including polycistronic mRNAs. This positioning of the nearest start codon also determines the reading frame of the polypeptide.

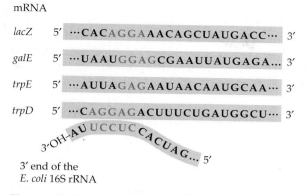

mRNA

lacZ 5′ ···CACAGGAAACAGCUAUGACC··· 3′

galE 5′ ···UAAUGGAGCGAAUUAUGAGA... 3′

trpE 5′ ···AUUAGAGAAUAACAAUGCAA··· 3′

trpD 5′ ···CAGGAGACUUUCUGAUGGCU··· 3′

3′OH-AUUCCUCCACUAG··· 5′

3′ end of the
E. coli 16S rRNA

Figure 13.19 Location of Shine-Dalgarno sequences (ribosome binding sites) near the start codons on selected mRNAs
The base-pairing of the *trpD* Shine-Dalgarno sequence of the *trpD* mRNA with the commentary sequence at the 3′ end of the 16S rRNA is also shown.

Following binding of the 30S ribosomal subunit to mRNA, the first tRNA, carrying methionine, binds to this complex. In *E. coli*, the first amino acid is almost always **N-formylated methionine (fMet)**, which is generated by first linking methionine to a specialized tRNA by aminoacyl-tRNA synthetase, and subsequently attaching a formyl group to the free amino group. The interaction of fMet-tRNA^fMet with ribosome involves interaction with IF2 and formation of a hydrogen-paired region between the anticodon sequence UAC and the complementary codon AUG. Occasionally initiation involves a mismatch, where the anticodon UAC pairs with GUG codon on mRNA. At the same time, a molecule of GTP binds to a specific site on the 30S subunit. The final stage in the assembly of the initiation complex involves association of the 50S subunit, hydrolysis of GTP to GDP, and release of all of the initiation factors. The ribosome is now ready to move along the mRNA and progressively synthesize a polypeptide chain.

ELONGATION The detailed steps for this phase of protein synthesis are shown in Figure 13.20. Following completion of the assembly of the initiation complex, the ribosome contains two sites that can bind aminoacyl-tRNA. The peptidyl site (or P site) contains the fMet-tRNA, whereas the adjacent acceptor site (or A site) is unoccupied, although it does contain the exposed triplet codon of the second amino acid

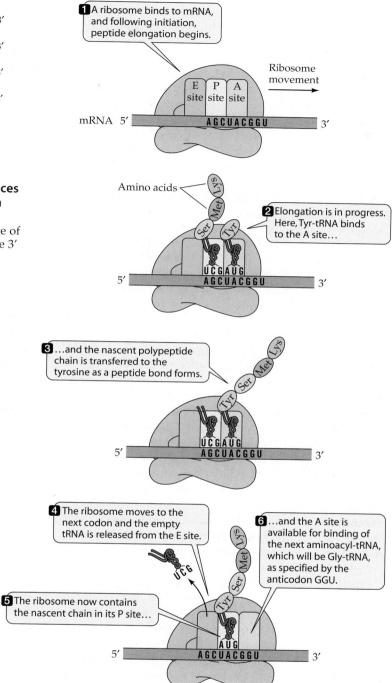

1 A ribosome binds to mRNA, and following initiation, peptide elongation begins.

Ribosome movement →

E site P site A site

mRNA 5′ AGCUACGGU 3′

Amino acids

2 Elongation is in progress. Here, Tyr-tRNA binds to the A site…

5′ UCGAUG / AGCUACGGU 3′

3 …and the nascent polypeptide chain is transferred to the tyrosine as a peptide bond forms.

5′ UCGAUG / AGCUACGGU 3′

4 The ribosome moves to the next codon and the empty tRNA is released from the E site.

6 …and the A site is available for binding of the next aminoacyl-tRNA, which will be Gly-tRNA, as specified by the anticodon GGU.

5 The ribosome now contains the nascent chain in its P site…

5′ AUG / AGCUACGGU 3′

Figure 13.20 Protein synthesis
The basic steps in protein synthesis. A ribosome has three sites that interact with tRNA during protein synthesis: the acceptor (A) site, the peptidyl (P) site, and the exit (E) site. During the continuous synthesis of a polypeptide, the P site is occupied by the tRNA with its nascent polypeptide chain, whereas the A site contains the next charged aminoacyl-tRNA. After addition of this amino acid, the ribosome moves along one codon and the empty tRNA is released from the E site.

in the reading frame. The formation of the first peptide bond between fMet and the second amino acid of the polypeptide chain takes place following the interaction of the aminoacyl-tRNA charged with the second amino acid and the A site. This is directed by the correct hydrogen-bonded, double-stranded region formed between the codon and anticodon sequence of tRNA. The binding of aminoacyl-tRNA to the A site is a stepwise process, involving several elongation factors (EFs) and GTP hydrolysis. The aminoacyl-tRNA first binds to a complex of **elongation factor Tu** (**EF-Tu**) and GTP. The EF-Tu places the aminoacyl-tRNA into the A site, with the concomitant hydrolysis of GTP to GDP. The EF-Tu–GDP complex is released. The regeneration of fresh EF-Tu–GTP complex is necessary for the next step of the elongation process; another **elongation factor, EF-Ts**, releases GDP from EF-Tu and reloads it with fresh GTP.

The formation of the actual peptide bond between the two amino acids within the ribosomes is accomplished by the peptidyl transferase activity of the ribosome. This reaction is also unusual because it is catalyzed by the RNA component of the ribosome, unlike most biochemical reactions, which are catalyzed by proteins. The bond formed between the amino acids results in transfer of fMet to the second aminoacyl-tRNA occupying the A site. The final step in elongation is called translocation, in which the ribosome moves three nucleotides and moves the free tRNA to a region of the ribosome called the exit (E) site, from which it is immediately discharged. The translocation reaction is catalyzed by another **elongation factor (EF-G)** and requires hydrolysis of GTP. The translocation leaves the A site open for the next incoming aminoacyl-tRNA, specified by the third codon of the mRNA. The subsequent elongation steps proceed in a manner entirely identical to the reaction between fMet and the second amino acid. In this process, all ribosomal elongation factors are reused, and the elongation of the polypeptide requires energy in the form of GTP, one to charge EF-Tu and one to fuel the EF-G–mediated translocation. Together with the requirement of ATP to charge a tRNA, the elongation stage of polypeptide chain formation is an energy-demanding process.

TERMINATION The termination of the polypeptide chain is directed by the location of one of the termination codons immediately following the codon for the carboxyl-terminal amino acid. The termination codons do not code for any amino acid, so when they occupy the A site no tRNA has the corresponding anticodon in its loop structure. Termination codons are recognized by release factors (RFs) that bind to the A site. The termination codons UAA and UAG are recognized by RF1, whereas RF2 recognizes UAA or UGA. The third release factor, RF3, complexed with GTP, then binds to the ribosome and catalyzes the cleavage of the peptide chain from the last tRNA. The hydrolysis of the GTP in RF3 ends the protein synthesis cycle and provides the energy for the dissociation of the ribosomes and the release of the polypeptide and the RFs. The polypeptide is now free to fold into its native tertiary structure.

When the Ribosomal Machinery Stalls...

One of the key features of successful translation is the uninterrupted movement of the ribosomes along mRNA, from initiation to termination codons, allowing for their recycling. What happens when ribosomes stall? This is apparently not an infrequent occurrence and occurs most often when mRNA, already engaged in translation, is damaged, losing a 3' portion that contains termination signals following the ribonuclease cleavage. Occasionally, ribosomes can read through the termination codon, particularly if cells have a small amount of tRNA suppressors (see subsequent text), or when the RNA forms a secondary structure that is stable enough to block the progress of a ribosome. These events can lead to ribosome depletion and the synthesis of a short defective protein. Bacteria utilize a simple mechanism that releases stalled ribosomes and tags incomplete proteins for degradation, utilizing a hybrid molecule consisting of a tRNA portion and a short reading frame, specifying a 10–amino acid polypeptide. This RNA molecule is referred to as **tmRNA**, (**t**rans-**m**essenger RNA), and has also been called **SsrA** and 10sRNA.

When tmRNAs are folded, the 5' and 3' ends form a structure resembling tRNA^{Ala} (Figure 13.21). The remainder of the molecule forms a series of hydrogen-bonded complementary strands of various lengths. The middle part, tmRNA, contains a single-stranded segment, encoding a reading frame for a peptide, but lacks an initiation codon, although it contains an in-frame termination codon.

The rescue mechanism of stalled ribosomes is depicted in Figure 13.22. Alanyl-tRNA synthetase charges the tmRNA with alanine after which EF-Tu directs it to the unoccupied acceptor (A)-site. At this point another protein, called SmpB, associates with the tmRNA. The function of SmpB is to facilitate the stabilization of the tmRNA within the ribosome, in the absence of a cognate anticodon sequence. Because of a standard transpeptidation reaction, the alanine is then transferred from tmRNA to the nascent polypeptide chain, followed by the translocation of the ribosome releasing the damaged mRNA and replacing it with the portion of tmRNA that encodes the 10–amino acid peptide. This mRNA portion of tmRNA is then used as a template for translation, adding the 10–amino acids, (the so-called "tag"), to the carboxy terminus of the originally translated protein.

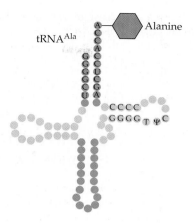

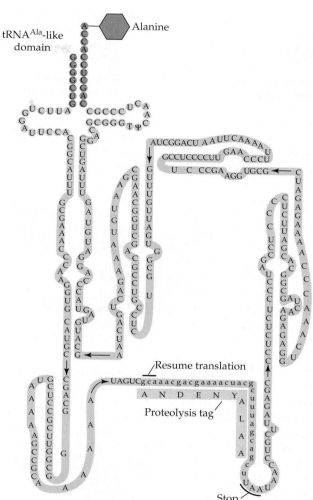

Termination codons and release factors cause the ribosome to dissociate and release the tagged protein. The added sequence serves as a recognition signal for proteases that degrade the truncated protein. This simple and efficient quality control mechanism is important during times when cells are subject to stress, and ribosomal depletion can influence viability and damaged proteins can exhibit undesirable activities or be toxic.

Figure 13.21 Structure of tmRNA
Comparison of folded *E. coli* tRNA^Ala and tmRNA.

SECTION HIGHLIGHTS

The information in mRNA is converted into proteins by a process of translation, which involves ribosome-assisted polymerization of 20 amino acids into a sequence specified by the triplet code of mRNA, copied from its complementary DNA strand. The specific order of such codons is called the reading frame. After amino acids are covalently attached to tRNA, they are incorporated into polypeptides. Each tRNA recognizes a sequence of three bases on mRNA (called codons) via a complementary sequence. The ribosomes recognize and bind to a specific site at the beginning of the mRNA. This so-called ribosome-binding site is 8 to 10 nt from the codon for the first amino acid of the polypeptide, which is always methionine. Ribosomes terminate translation at specific codons (termination, or stop codons), for which there are no corresponding tRNAs. Stalled ribosomes due to mRNA breaks are released though the activities of a specialized small RNA (tmRNA).

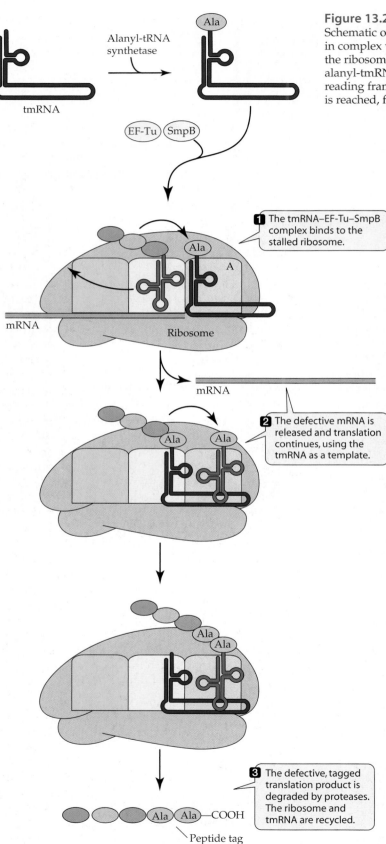

Alanyl-tRNA synthetase

tmRNA

Ala

EF-Tu SmpB

1 The tmRNA–EF-Tu–SmpB complex binds to the stalled ribosome.

Ala

A

mRNA

Ribosome

mRNA

2 The defective mRNA is released and translation continues, using the tmRNA as a template.

Ala Ala

Ala
Ala

3 The defective, tagged translation product is degraded by proteases. The ribosome and tmRNA are recycled.

Ala Ala —COOH

Peptide tag

Figure 13.22 Rescue of stalled ribosomes
Schematic of tmRNA action during ribosome rescue. The tmRNA, in complex with Ef-Tu and SmpB, binds to the empty A site on the ribosome. The nascent polypeptide chain is transferred to the alanyl-tmRNA. The defective mRNA is replaced by the open reading frame of tmRNA, which is translated until a stop codon is reached, followed by the release of the tagged protein.

13.6 Regulation by RNA Secondary Structure

A number of additional mechanisms allow microorganisms to fine-tune gene expression based on the need of the cells by controlling the ability of the core RNA polymerases to continue the elongation phase of transcription. These regulatory checkpoints also influence the translation or degradation of mRNA. Here we review several examples of regulation of gene expression that exploit the coupling of transcription and translation.

Once RNA polymerase has cleared the promoter, transcription can be further regulated by secondary RNA structures that are formed in the 5′ leader region of mRNA as it emerges from the RNA polymerase. A type of regulatory mechanism, referred to as **attenuation**, is defined as a process whereby RNA polymerase temporarily stops ("pauses") and in response to a specific signal terminates the transcription of the downstream genes. A mechanistically closely related regulatory process, often used interchangeably with attenuation is called **antitermination**. In this latter process the action of regulatory signals allow RNA polymerase to continue transcription by overcoming possible terminator structures.

The attenuation and termination events respond to RNA **hairpin structures** (double-stranded structures formed by the complementary pairing of inverted repeats) within the newly synthesized nascent transcript. The hairpin loop, followed by a string of U's that causes the termination of transcription, is called the transcription **terminator**, and it effectively halts the progress of RNA polymerase. The terminator structure can be eliminated when an alternative secondary hairpin structure is formed (the **antiterminator**), which sequesters part of the terminator sequence. Figure 13.23 illustrates the role of the alternating terminator and antiterminator structures on the ability of RNA polymerase to proceed into structural genes and complete transcription. The signals for attenuation or antitermi-

Attenuation

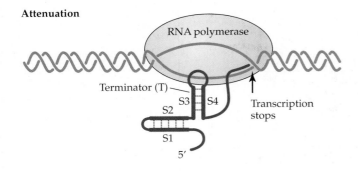

Antitermination

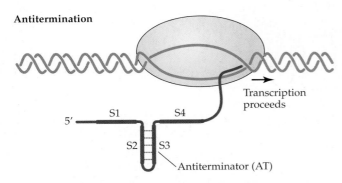

Figure 13.23 General model for attenuation or antitermination of transcription
The process is directed by a series of four RNA segments located at the 5' end of the transcript. As these segments emerge from RNA polymerase, they can form secondary double-stranded, hydrogen-bonded hairpin structures because their sequences are inverted repeats of each other. In case of attenuation, the first inverted repeat (consisting of segments S1 and S2) forms a base-paired duplex, which allows the next pairing of inverted repeats (S2 and S3) to form another hairpin structure, called the terminator (T). The top portion of the hairpin interacts with RNA polymerase, and blocks further transcription. Antitermination occurs when conditions that favor formation of a different hairpin structure, called the antiterminator (AT) which, lacks the sequences necessary to interact with RNA polymerase; its function is to block the formation of the terminator by sequestering the sequence of the inverted repeats into the S2–S3 hairpin. The formation of terminator or antiterminator RNA structure is controlled by a variety of cellular conditions, and several examples are provided in the text.

nation are defined as any conditions that affect the formation or disruption of secondary structures within the nascent mRNA chain, capable of influencing the progress of RNA polymerase. For enzymes involved in amino acid biosynthesis, the regulatory signals that control formation of terminator or antiterminator mRNA structures are charged tRNA molecules. Here we review several mechanisms that allow cells to assess the availability of amino acids, as reflected by the ratio of uncharged versus charged tRNAs, and modulate the transcription of a particular gene or an operon.

In *E. coli* and other gram-negative bacteria, the transcription of tryptophan biosynthetic operon (the *trp* operon) is under the positive control of the TrpR regulator, which is inactive in the presence of tryptophan. However, the transcription is not entirely shut off in the presence of excess tryptophan, and it does not proceed to maximum rate under the conditions of tryptophan limitation. A fine-tuning of the regulatory effect is accomplished by an attenuation mechanism (shown schematically in Figure 13.24).

In *E. coli*, when RNA polymerase clears the *trp* operon promoter, but before it reaches *trpE* (the first genes of the tryptophan biosynthetic cluster), it has to traverse though and enter the 140-bp region between the promoter of the *trp* operon and *trpE*. The corresponding 141-bp leader transcript has the potential to form three different overlapping hairpins—referred to as pause, antiterminator, and terminator structures. After transcribing the region containing the sequence specifying the first hairpin (resulting from the pairing between S1 and S2 complementary segments), RNA polymerase pauses, presumably as a result of an interaction with this mRNA structure. At the same time, ribosomes initiate the translation of a short, 14–amino acid peptide, encoded in the sequence that includes the S1 portion of the S1:S2 hairpin. Therefore, ribosomes that make sufficient advance into the coding sequence of this peptide will disrupt the complementarity of the S1:S2 hairpin and simultaneously relieve the transcriptional block that leads to RNA polymerase pausing. Transcription (and translation of the newly synthesized mRNA) resumes. The continuation of transcription at this point is determined by a sensing mechanism that "measures" the levels of available tryptophan. If the level of tryptophan is low, then the fraction of tRNA^{Trp} charged with this amino acid is low. The 14–amino acid nascent peptide contains two adjacent tryptophans near the end of this peptide. When Trp-tRNA^{Trp} is low, the ribosomes stall at the codons for this tryptophan pair. As a consequence of this interaction, a new hairpin structure, the so called antiterminator can form. The antiterminators consist of the S2 segment and the adjacent complementary S3 sequence. Under these conditions, the RNA polymerase is free to complete the transcription of the genes of the tryptophan operon, providing the bacteria with the necessary enzymes for tryptophan biosynthesis.

When cellular levels of tryptophan increase, this condition results in the corresponding rise in the level of charged tRNA^{Trp}. The ribosomes have no difficulty completing the synthesis of the 14 amino acid peptide, and thus disrupt the S2:S3 structure. The S3 part of the hairpin is now free to form a secondary structure with its adjacent complementary sequence. This new hairpin now acts as a transcriptional terminator interacting with RNA polymerase and preventing its entry into the coding region for the enzymes of the *trp* biosynthetic operon.

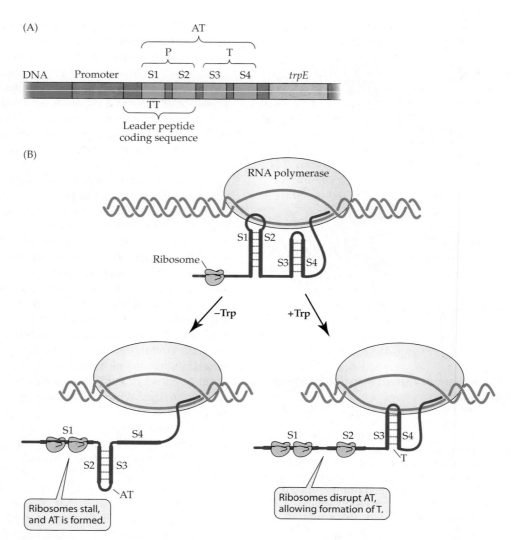

Figure 13.24 Regulation by transcriptional attenuation of the *trp* operon of *E. coli*
(A) Organization of the various regulatory segments in the 5′ region of the *trp* operon. (B) Coupling transcription and translation to form antiterminator (AT) or terminator (T) structures based on the availability of tryptophan. When RNA polymerase encounters the pause site (consisting of segments S1 and S2) the ribosomes begin the simultaneous synthesis of the short peptide and release the RNA polymerase from pausing by disrupting the S1:S2 structure. Low intracellular tryptophan concentration causes ribosomes to stall at the first UGG codon of tryptophan mRNA. This allows for the formation of an antiterminator structure (consisting of S2 and S3) and consequent continued transcription of the *trp* operon by RNA polymerase. When tryptophan is abundant, ribosomes complete the synthesis of the peptide that allows the ribosomes to occupy a portion of the S3 segment, disrupting the antiterminator structure. Thus segment S3 can pair with S4, forming a terminator structure, leading to a halt in the progress of RNA polymerase.

The attenuation functions as an additional block preventing the synthesis of unnecessary enzymes (or assuring maximal synthesis of enzymes when necessary). The combined effect of repression by low-level transcription by RNA polymerase in the absence of TrpR is several hundred, which is almost in order of magnitude more effective than relying entirely on regulatory signals provided at the level of transcriptional initiation through the activity of the TrpR.

Regulatory Mechanisms Controlling Transcription and Translation

Attenuation or antitermination relies on the formation of adjacent hairpin structures from direct repeats on mRNAs to control the progress of RNA polymerase. In many instances, the terminator structure includes a segment of mRNA that contains the RBS and the initiation codon. The consequences of formation of terminator and antiterminator loops allows cells to simultaneously regulate not only transcription but also translation of mRNAs, by making the sites involved in initiation of translation unavailable for ribosomes. Translational regulatory mechanisms allow cells to assimilate additional signals informing it about its need for a particular protein.

Expression of genes specifying amino acid biosynthetic enzymes or tRNA synthetases in gram-positive bacteria are controlled at the translational level through a mechanism that detects the ratio of charged to uncharged tRNA for a particular amino acid. Because the extent of charged tRNAs is determined by the abundance of the particular amino acid, the proportion of charged tRNAs is an accurate indication whether the cell should or should not commit toward additional synthesis of this amino acid. This termination mechanism depicted in Figure 13.25, is used as an example of the control of biosynthetic genes for a number of amino acids, including glycine. The abundance of glycine in the cell results in a decreased need for additional synthesis. This is reflected by the elevated levels of charged tRNA^{Gly}. The tRNA^{Gly} is unable to exert its regulatory effect on the terminator hairpin loop structure formed in the newly transcribed 5'-leader sequence of the *glyQS* genes, causing RNA polymerase pausing. However, when the intracellular levels of glycine drop, with concomitant decrease in the frac-

tion of tRNA charged with glycine, the terminator structure is disrupted by the formation of the alternative antiterminator hairpin structure, utilizing some of the segments of mRNA that were previously involved in formation of the terminator. The antiterminator hairpin loops are formed with the aid of a stabilizing effect of hydrogen bonds formed between the tRNA and the 5' mRNA. The specificity of the mechanism that controls the expression of downstream loci is in correct pairing of the anticodon of the tRNA with a single codon, designated the "specifier sequence," in the first stem-loop structure leader sequence. For the example in Figure 13.24, the anticodon sequence 5'-GCC-3' of tRNA^{Gly} interacts by hydrogen-bond base-pairing with the specifier sequence 3'-CGG-5' which is a part of the *glyQglyS* transcript. For another amino acid, such as threonine, the sequence 5'-GGU-3' is the anticodon sequence for tRNA^{Thr} and would pair with 3'-CCA-5', found in the leader region of the *thrS* (threonyl-tRNA synthetase) transcript.

The transition between antiterminator and terminator loops found within the 5' untranslated regions of certain mRNAs can be mediated directly by small-molecule effectors, typically metabolites. In each case, the binding of the metabolites rearranges the stem-loop structure within the leader region, resulting in the formation of a termi-

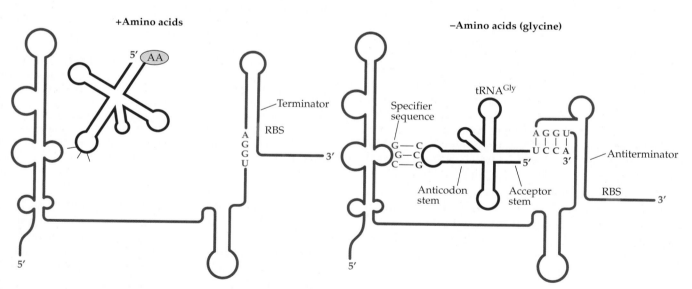

Figure 13.25 Schematic model for formation of terminator/antiterminator structures directed by the availability of a particular amino acid
When intracellular levels of the amino acid are high, a majority of their cognate tRNAs are charged. This form of tRNA is unable to hydrogen-bond with a complementary sequence within the antiterminator structure. Instead a terminator loop is formed, preventing RNA polymerase from further movement along the DNA template, and, simultaneously, blocking access of ribosomes to the ribosome-binding site (RBS)

sequestered within the terminator structure. The low level of an amino acid is reflected in a high proportion of uncharged tRNAs. This form of tRNA stabilizes an antiterminator structure by forming two short hydrogen-bonded, double-stranded segments, one with a complementary sequence in the specifier sequence and another with a sequence within the free portion of the antiterminator. This interaction frees RNA polymerase and the RBS. The specific interactions between tRNA^{Gly} and the segment at the 5' portion of the *glyQglyS* mRNA is also shown.

nator that blocks further elongation of the transcript by RNA polymerase. The mechanism for controlling elevated levels of metabolites is called "**riboswitching**," and it is the most direct way that a nascent transcript can sense the need for the early termination of transcription. Although the molecular details of the various structures formed between the small molecule and the 5′ untranslated leader sequence is not known, it is very likely that the metabolites stabilize a "competing" structure (anti-antiterminator) that now allows the formation of the terminator duplex. Because these regions also contain the ribosome-binding site, riboswitches may play an equally important role in translational regulation. For example, in gram-positive bacteria the thiamine (vitamin B_1) biosynthetic genes are regulated by transcriptional attenuation utilizing the riboswitch mechanism, where the activated form of thiamine (thiamine pyrophosphate) binds to the leader RNA and stabilizes the terminator structure (Figure 13.26). In many gram-negative organisms, thiamine pyrophosphate regulates gene expression at the level of translation, where it promotes formation of secondary structures that sequester the RBSs and interfere with translational initiation.

Control of Gene Expression by Regulatory Noncoding RNAs

Bacteria transcribe certain RNAs that do not code for proteins but have specific biological functions on their own. Most abundant of this class of RNAs are the components of the translational machinery, namely the rRNAs and tRNAs. In addition, bacteria produce another class of relatively short RNAs (50 nt to 400 nt) that are not transcribed, and the primary role of which is regulating translation and degradation or both of specific mRNA. They will be referred to as **ncRNAs** (<u>n</u>on<u>c</u>oding RNA) but they are sometimes called <u>s</u>mall <u>r</u>egulatory RNAs (srRNAs). The majority of these ncRNAs can either stimulate or repress translation of their target mRNAs with a concomitant effect on transcript stability. As described previously for control of translation by tRNAs or riboswitches, any block in translation can have a consequence of affecting transcript stability, facilitating ribonuclease-mediated degradation of unprotected mRNA. Certain ncRNAs can have an opposite affect on translation, relieving a translational block imposed on an mRNA by a regulatory protein or the secondary structure of the transcript, allowing ribosomes an unimpaired access to initiate protein synthesis and concomitantly, stabilize the mRNA.

The two well-characterized major translational regulatory mechanisms that utilize ncRNAs are illustrated in Figure 13.27. In one case, the ncRNA is complementary to the 5′ end of the mRNA and forms a double-stranded structure that includes the Shine-Dalgarno sequence (the RBS) and often the start codon (AUG). This

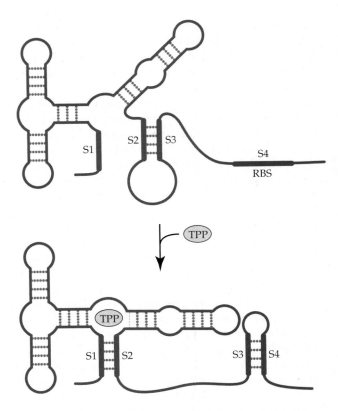

Figure 13.26 Attenuation control of the *B. subtilis thi* operon by a riboswitch mechanism
Thiamine pyrophosphate (TPP) is present at elevated intracellular levels when the cells grow in thiamine-replete media and stabilizes several secondary structures in the 5′ leader region of the transcript. The most significant of these in terms of regulation is the formation of an antiterminator structure (consisting of mRNA segments S1 and S2) stabilized by TTP, that not only prevents the formation of an antiterminator structure with S2 and S3 segments, but allows for creation of a transcriptional terminator, a hairpin consisting of the S3 and S4 segments. The terminator simultaneously sequesters the RBSs and prevents translation of the *thi* genes.

interaction effectively blocks initiation of translation by occluding the ribosomes. In the absence of translating ribosome, the mRNA is degraded. The interaction of the ncRNA with its complementary RNA is facilitated and stabilized by a protein factor called **Hfq**. Hfq plays a particularly important role when the ncRNA and its cognate target mRNA sequence show only partial complementarity. In *E. coli*, where this system was originally discovered, the Hfq-ncRNA regulatory mechanism controls translation of a number of genes, including those encoding iron-binding proteins, porins, and the cell-division proteins. However, analogous mechanisms involving ncRNAs and stabilizing proteins appear to be widely distributed among various bacterial species.

The ncRNAs can also stimulate translation. Certain mRNAs are transcribed with relatively long untranslated sequences at their 5′ end, which are also known as the 5′ untranslated regions (5′ UTR). Certain sequences within the 5′ UTRs can form secondary structures by base-pairing, which sequester the RBS and prevents translational

(A) Regulation mediated by Hfq and ncRNA

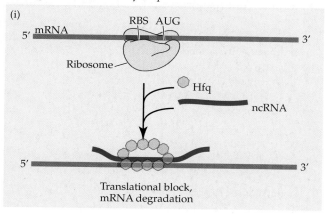

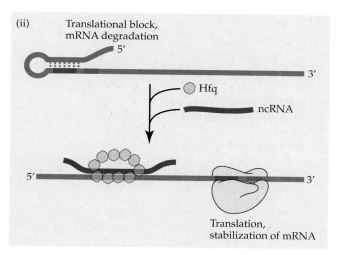

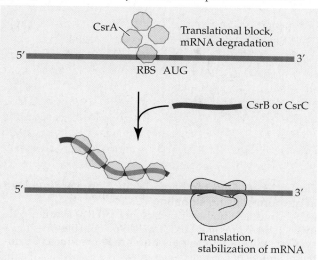

(B) Regulation mediated by translational repressors and ncRNA

initiation. In such cases, the translational block can be relieved by disruption of this intrinsic inhibitory structure by a complementary ncRNA and with the cooperation of the Hfq protein. One of the best-studied examples of translation regulation by this mechanism involves the activities of two related ncRNAs, DsrA and RprA, which regulate the synthesis of the σ factor RpoS from its mRNA. The 5′ UTR of the *rpoS* mRNA forms a hairpin structure occluding the RBS, thereby preventing translational initiation. Pairing of the ncRNAs, DsrA, or RprA, each complementary to a slightly different segment of the *rpoS* 5′ UTR, blocks the intrinsic inhibitory pairing with the *rpoS* ribosome-binding region of the mRNA, thereby allowing unimpaired translation of this transcript when cells enter stationary phase of growth.

Certain bacterial ncRNAs are capable of binding to translational regulatory proteins and reverse their inhibitory activity. In most instances, a regulatory protein (a translational repressor) binds to the 5′ UTR, including the RBS, and interferes with translational initiation and accelerates mRNA degradation. The translational block can be reversed by the activities of specific ncRNA, the sequence of which resembles that of the translational repressor-recognition site. For example, in *E. coli*, translation of several mRNAs encoding the glycogen metabolism is controlled by a repressor protein CsrA (or carbon storage regulator). CsrA repressor protein binds to a sequence in the 5′ end of the transcripts that overlaps the RBS, which interferes with the mRNA translation and degradation. Two small ncRNAs of similar sequences (CsrB and CsrC) relieve this translation block by binding to as many as 18 CsrA molecules per regulatory RNA. The CsrB and CsrC ncRNAs contain a total of 18 repeats

Figure 13.27 Regulation of translation and mRNA stability by noncoding RNAs (ncRNAs) through complementary base-pairing
(A) Activity of ncRNAs and the Hfq protein (i) in interfering with translation (top) or (ii) in facilitating translation (bottom). In each case, the ncRNA recognizes a sequence in the 5′ end of mRNA and, together with Hfq multimers, forms a stable base-paired structure. In cases in which translation of mRNAs is repressed by the ncRNAs, the RNA-Hfq complex blocks access of translational machinery to the RBS and the initiation codon (AUG). In contrast, translation of certain mRNAs is naturally blocked by formation of inhibitory secondary structures by their 5′ untranslated regions (5′ UTR), sequestering RBS and occluding the translational machinery. The class of ncNRAs that functions as translational activators does so with Hfq-assisted base-pairing with portions of this inhibitory sequence, freeing the RBS for ribosomes to initiate translation. (B) Binding of ncRNAs to translational repressors. A regulatory protein (CsrA) inhibits translation by binding to the RBS of mRNA. Regulatory ncRNAs (CsrB, CsrC, or RsmY, RsmZ) sequester the translational repressor into a multimeric complex, freeing the RBS for the ribosomes.

of sequences similar to that of RBS and additional short sequences surrounding it in certain mRNAs, which explains the ability of these two ncRNAs to sequester CsrA away from its mRNA target. When CsrB or CsrC levels are high, there is not enough free CsrA to bind to the mRNA targets, allowing unhindered initiation of translation. An analogous system regulates translation of a number of genes in various *Pseudomonas* species, where the action of a translational repressor RsmA is antagonized by two ncRNAs, RsmZ and RsmY. Although CsrA has been found to be primarily a repressor of translation, at least one mRNA encoding flagellar regulators has been shown to be positively regulated by CsrA.

> ### SECTION HIGHLIGHTS
> Various regulatory mechanisms control progress of RNA polymerase by attenuation and antitermination mechanisms, some of which simultaneously regulate translation. Small RNAs that do not code for proteins (ncRNAs) regulate gene expression by interfering or facilitating translation and mRNA degradation.

13.7 Quorum Sensing

Bacteria are unicellular organisms, and they usually assimilate chemical or physical stimuli from their environment as individual cells. However, it has become increasingly apparent that in certain habitats microorganisms live in single- or multiple-species communities, and, consequently, they have the ability to communicate with each other. A number of bacterial species utilize chemical signals, allowing them to monitor the population density of other closely related bacteria that occupy the same niche. In each cell, the signals allow bacteria to respond in a coordinated fashion with their close relatives and neighbors by expressing genes that encode determinants for adaptation that would effectively benefit the entire community. The mechanism allowing bacteria to determine whether they are present in sufficient or critical numbers (a quorum) is called **quorum sensing**. A variety of chemical signals are produced and secreted by bacteria that inform them about each other's presence. These molecules affect equally the producer and recipient cells, and are sometimes referred to as **autoinducers**, (also referred to as **quoromones**). The function of autoinducers is to modulate the activities of regulatory proteins. The major group of signaling molecules involved in quorum sensing by gram-negative bacteria are related to **acyl-homoserine lactones** (**AHLs**), whereas most gram-positive organisms communicate using **peptides** that perform similar cell-density signaling functions.

Quorum-sensing systems are particularly common among bacteria that form symbiotic or parasitic relationships with their host. Quorum sensing was first described in the luminescent marine bacterium *Vibrio fischeri*. These organisms produce light only when they reach relatively high densities in the light organs of certain luminescent fish or squid (see Box 25.1). Lumines-

(A) Acyl-homoserine lactones (AHL)

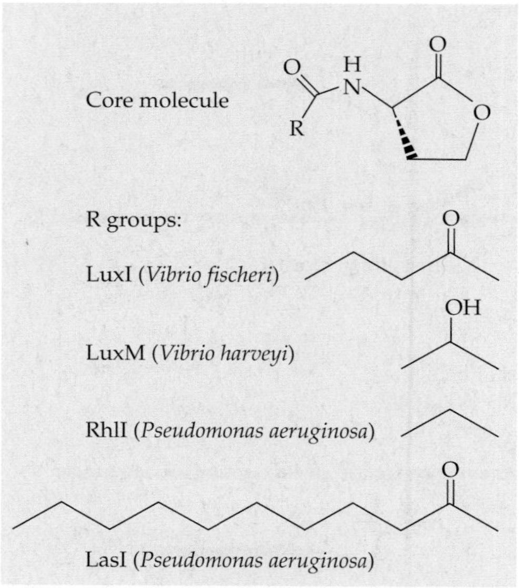

(B) *Staphylococcus aureus* oligopeptide autoinducers

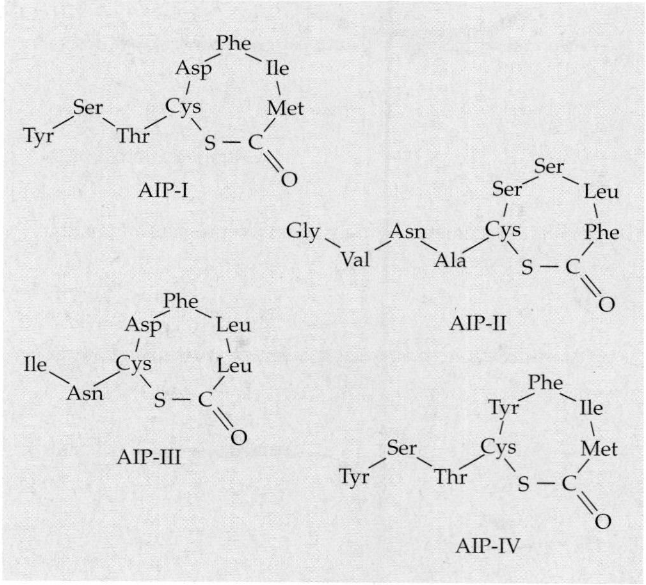

Figure 13.28 Selected bacterial autoinducers involved in quorum-sensing signaling
(A) Various acyl-homoserine lactones (AHLs) synthesized by bacteria expressing LuxI/LuxR-type quorum-sensing systems. (B) Autoinducer peptides (AIPs) produced by different strains of *Staphylococcus aureus*.

cence is the result of density-dependent expression of *lux* genes, which encode enzymes responsible for production of compounds used for light production. The *V. fischeri* AHL autoinducer (specifically, 3-oxo-C6-homoserine lactone) is synthesized by an enzyme, LuxI, in a two-step reaction involving formation of the homoser-ine lactone ring from S-adenosylmethionine (SAM) and the attachment of a fatty acyl moiety from an acyl carrier. A number of related AHLs are shown in Figure 13.28A. Figure 13.29A shows schematically the quorum-sensing regulatory circuit using AHLs. As the cell density increases, so too does the production of AHL and

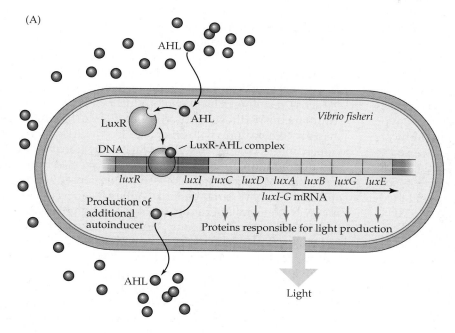

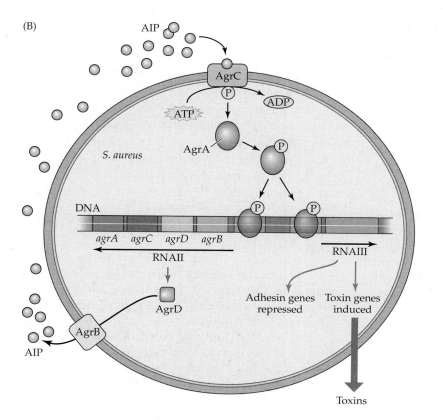

Figure 13.29 Quorum-sensing systems utilizing acyl-homoserine lactones (AHLs) or autoinducer peptides (AIPs) (A) Density-dependent light production by *Vibrio fischeri* colonizing the light organs of fish or squid. An increase in cell number results in accumulation of AHLs, namely the 3-oxo-C6-homoserine lactone, which enters cells and interacts with the transcriptional regulator LuxR. The LuxR-AHL complex activates transcription of the *luxI-G* operon, resulting in production of proteins responsible for light production. Because the *luxI-G* polycistronic mRNA also encodes the LuxI protein, the level of this enzyme also rises, resulting in the production of additional autoinducer, thereby amplifying the group response of all *V. fischeri* in the light organ, and leading to the production of sufficient quantities of light for the host. (B) Regulation of virulence factor expression by a peptide-based quorum-sensing system in *S. aureus*. Following initial colonization of the host, bacteria increase in density, resulting in accumulation of the autoinducer peptide (AIP). The membrane protein AgrB is responsible for cleaving AgrD into the AIP, and then modifying and secreting it into the media. The extracellular AIP activates AgrC, the sensory component of a two-component response-regulatory system, by binding to its external domain. Activated AgrC phosphorylates AgrA, which binds to two promoters and directs the transcription of RNAII and RNAIII. RNAII is the mRNA for the components of the Agr system leading to the synthesis and assimilation of more AIP signal. RNAIII itself encodes a secreted hemolysin. RNAIII controls the expression of virulence genes that encode attachment factors (so-called adhesins) and secreted toxins. The density-dependent quorum sensing promotes dissemination of *S. aureus* by blocking any further synthesis of attachment factors and stimulating production of secreted toxins that are used by the invading bacteria to spread from the site of infection into deeper tissues and blood.

the presence of extracellular AHL. The coordinated response in the producer cells as well as their neighbors is triggered when their density (and consequently, the AHL levels) reaches a critical concentration (a quorum) and the autoinducer enters cells to activate transcription by acting as a co-activator for the transcription factor LuxR. The AHL-LuxR complex binds to regulatory sequences that control transcription of the *lux* genes involved in bioluminescence. The response is amplified by production of additional LuxI (and consequently more AHL), since it is also encoded within the *luxI-G* operon, as well as more LuxR, the genes of which is also autoregulated.

Different bacteria can produce AHLs that vary only in the length of their acyl-chains. Certain bacteria such as *Pseudomonas aeruginosa* can produce more than one type of AHL (see Figure 13.28A). Each of these autoinducers carries out a specific response, even when exposed to similar AHLs. The specificity determined by the LuxR-type regulators, each of which can only recognize its cognate AHL. Therefore, in *P. aeruginosa* two different LuxI-type enzymes (LasI and RhlI) synthesize the two AHLs and each binds its own LuxR-type receptor. This allows the response to be species-specific and the "crosstalk" between different autoinducers, produced by a mixed community of bacteria, is kept to a minimum. However, several interspecies signaling systems have been recently described, wherein coordinate expression of genes of a mixed community is controlled by unique types of diffusible signaling molecules that vary in their structure. Moreover, as novel quorum-sensing systems are discovered, it is becoming clear that other chemical moieties in addition to AHLs can serve as autoinducers, provided that a cell has the machinery for their synthesis and specific receptor proteins to assimilate the signal and direct an appropriate transcriptional response.

A somewhat different mechanism coordinates the group response in several gram-positive bacteria. The pathogen *Staphylococcus aureus* uses quorum sensing to differentially express genes required for infection at different cell densities. When bacteria colonize the infection site and are present in small numbers, they preferentially express a number of surface proteins necessary for attachment to tissues (so-called adhesins). However, as their number increases, production of various secreted toxins is induced and these factors facilitate dissemination of the bacteria into various organs and blood. The change in the expression of these virulence factors is cell-density dependent and is regulated by a quorum-sensing system called Agr acting through an **autoinducer peptide (AIP)**. Structures of AIPs made by different strains of *S. aureus* are shown in Figure 13.28B. The regulatory circuitry activated by an increase in cell density is shown in Figure 13.29B. The AIP is made as a large protein precursor, AgrD, which is processed into a short

peptide by the AgrB protein. AgrB is also responsible for further modification of the AIP (formation of a cyclic thiolactone bond between an internal cysteine and the carboxyl terminus) and its subsequent secretion from the cell. Extracellular AIP, when present at sufficiently high levels, binds to an external site on the membrane protein AgrC, which is a sensor protein (a kinase of a two-component response-regulatory system). AIP binding stimulates AgrC to autophosphorylate, followed by transfer of the phosphate onto its cognate response regulator, AgrA. Phosphorylated AgrR activates transcription from two adjacent promoters. One of these promoters controls the synthesis of additional *agr* genes (*agrB*, *agrD*, *agrC*, and *agrA*), resulting in increased production of AIP. AgrA simultaneously directs transcription of RNAIII, which shares several features with other ncRNAs discussed previously in this chapter. RNAIII regulates the translation and degradation, or both, of mRNAs for several transcriptional activators and repressors. The consequence of RNAIII activity is repression of adhesin-gene expression with concomitant induction of expression of genes that encode secreted toxins. Unlike other ncRNAs, RNAIII itself encodes a secreted protein, a toxin with hemolytic activity. Analogous to various AHLs of gram-negative quorum-sensing systems discussed previously, various *S. aureus* strains synthesize slightly different AIPs (see Figure 13.28B), which are recognized by cognate domains of AgrC proteins, thereby allowing utilization of the AIP only by the same strains of *S. aureus* that were responsible for its production.

> ### SECTION HIGHLIGHTS
> Quorum sensing is a mechanism utilized by bacteria to communicate with each other and coordinate gene expression as a group. In general, gram-negative bacteria produce quorum-sensing metabolites, the most common of which are acyl-homoserine lactones. Gram-positive bacteria communicate via short signaling peptides.

13.8 Mutations

A heritable change in the DNA sequence is an event called a mutation. One consequence of having a haploid chromosome is that a change in a specific gene is immediately expressed by the daughter cells because the bacterial genome lacks the duplicated copy of each gene that, in diploid organisms, often masks the effect of a mutation in one of the chromosomes. This section discusses the types of mutations that occur in bacteria, the predicted consequences of mutations, and the effect of environmental conditions on the frequency of mutations in bacteria.

Fluctuation Test

When a small amount of streptomycin (an antibiotic made by *Streptomyces*) is added to a culture of bacteria, almost all bacteria are killed within a few minutes. Similarly, bacteria can be killed by exposure to bacterial viruses (also called bacteriophages) that adsorb to their cell walls, inject their genomes, and, following a cycle of replication, release new viruses, with the concomitant death of the host bacterium. A small fraction, however—fewer than one in a million—can survive the exposure to potential killing agents. When these bacteria are propagated, they and their progeny are completely resistant to killing by the same agents that devastated the members of the previous, sensitive population. The new, resistant population of bacteria owes its survival to mutations: in this case changes in a gene that encodes a ribosomal protein such that the ribosomes no longer bind streptomycin. This allows the streptomycin-resistant bacteria to synthesize protein even in the presence of streptomycin. Similarly, resistant bacteria that are isolated from a population of cells killed by bacteriophages have altered genes

that encode new cell-wall components that do not allow bacterial viruses to attach to or enter the cell.

Because these mutant bacteria appear to arise only following exposure to the appropriate agent, it had been debated for many years whether mutations are adaptive changes, wherein a specific agent would induce the required alteration in the genetic material. An alternative hypothesis, supported by the Darwinian theory of evolution, proposed that changes occur spontaneously and continuously in a culture, and it is the selective process (such as exposure of bacteria to streptomycin and bacteriophages used as examples) that allows us merely to identify such bacteria that carry any one specific mutation.

The controversy was resolved in the early 1940s when M. Delbruck and S. Luria devised the fluctuation test, which demonstrated that mutations are spontaneous and independent of a selective environment. The original test was based on the ability of *E. coli* to mutate into a form that is resistant to the killing action of bacteriophage (phage) T1. When exposed to T1, a population of 10^8 bacteria gives rise to about 10 to 50 mutants that are T1-resistant; that is, they can grow in the presence of this phage.

The fluctuation test is outlined in Figure 13.30. A culture of *E. coli* propagated in a flask was divided into two

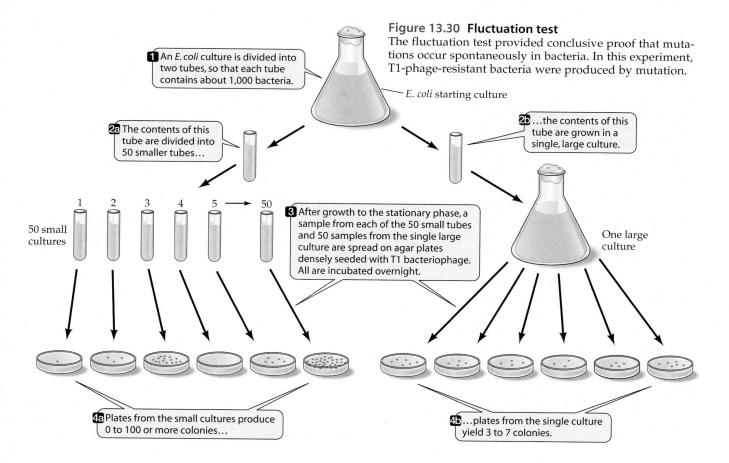

Figure 13.30 Fluctuation test
The fluctuation test provided conclusive proof that mutations occur spontaneously in bacteria. In this experiment, T1-phage-resistant bacteria were produced by mutation.

1 An *E. coli* culture is divided into two tubes, so that each tube contains about 1,000 bacteria.

E. coli starting culture

2a The contents of this tube are divided into 50 smaller tubes...

2b ...the contents of this tube are grown in a single, large culture.

50 small cultures

1 2 3 4 5 → 50

3 After growth to the stationary phase, a sample from each of the 50 small tubes and 50 samples from the single large culture are spread on agar plates densely seeded with T1 bacteriophage. All are incubated overnight.

One large culture

4a Plates from the small cultures produce 0 to 100 or more colonies...

4b ...plates from the single culture yield 3 to 7 colonies.

tubes, each containing about 10^3 bacteria. The contents of one of the tubes were further subdivided into 50 smaller tubes. All of the cultures were grown to stationary phase. From the single large culture, as well as from the 50 individual tubes, small aliquots were spread over an agar surface densely seeded with T1 bacteriophage, which will kill most susceptible bacteria, leaving only those that have become resistant to T1 by virtue of a mutation. These resistant bacterial mutants can give rise to colonies even in the presence of large numbers of bacteriophage. After overnight incubation, the number of phage-resistant colonies was determined. The culture that was incubated in a single tube yielded about three to seven resistant colonies. The range of phage-resistant colonies obtained from the individually grown tubes was from none to 100 or more.

What was the reason for this variation in the number of resistant colonies? If the mutations arise as a result of contact with the phage on the plate, aliquots from both large and small cultures should give rise to the

ing in duplication or elimination of the looped-out region in the newly synthesized strand. For any form of mutation to take effect, the change also has to escape the repair activity of DNA Pol, namely the ability to back up in the 3′-to-5′ direction, remove the incorrect base, and resynthesize the correct strand. Because base-pairing errors, loop-outs, and failure of the repair system are all very rare, mutations resulting from errors of this type arise infrequently.

Consequence of Mutations

Changes in the DNA sequence can have a number of effects in the bacterial cells that inherit the mutant version of a specific gene. As an example, examine the effect of substitutions as well as deletions or insertions at the highlighted codon on the following polypeptide coding sequence (in this example, DNA strand 1 is the template used by RNA polymerase to generate the mRNA for this protein):

DNA strand 1	TAC	TTT	GCT	AAT	ACG	CTC	TAG	TTG	ACC	ATT
DNA strand 2	ATG	AAA	CGA	TTA	TGC	GAG	ATC	AAC	TGG	TAA
mRNA		AUG	AAA	CGA	UUA	UGC	GAG	AUC	AAC	UGG	UAA
Protein		Met	Lys	Arg	Leu	Cys	Glu	Ile	Asn	Trp	Stop

same number of resistant colonies. If phage-resistant mutants are arising at all times, the individual cultures should have a great variation in number, depending on the time the mutation takes place during growth. That is, if mutation takes place immediately after subdivision, as many as 100 resistant bacteria will be identified in that tube following spreading on a bacteriophage-containing agar plate. If mutation takes place long after subdivision or not at all, small numbers of phage-resistant bacteria will be identified. Clearly, results support the second hypothesis because of the great fluctuation in the number of resistant bacteria that arose from the small cultures.

Types of Mutation

A mutation involving a single base-pair substitution, in which a purine (G or A) is replaced by another purine, or a pyrimidine (T or C) is replaced by another pyrimidine, is called a **transition**. A **transversion** is a type of mutation wherein a purine is replaced by a pyrimidine or a pyrimidine is replaced by a purine (see Figure 10.13). These **substitutions** become part of a newly synthesized chain as a consequence of incorrect base-pairing during the polymerization of the chain; for example, a G pairs with a T instead of a C. Deletions and insertions arise when the template or the newly synthesized strand loops out and the DNA Pol makes a slip, result-

Now, consider the following consequence of several changes in the DNA encoding this hypothetical nine amino acid peptide, all involving substitutions that change the mRNA from specifying the codon UUA, which codes for the amino acid, leucine.

1. A transversion of the second AT base pair for GC results in the codon UGA in the mRNA. This is a termination codon, resulting in synthesis of a shortened three amino acid peptide.

2. A transition of the TA base pair for a GC pair changes the codon on mRNA into UUG. This is a so-called silent mutation and has no effect on the polypeptide because UUG, like UUA, encodes leucine.

3. Another transversion of the first AT to TA alters the codon into one specifying isoleucine (AUA). The peptide now has a leucine-for-isoleucine substitution. The outcome of this mutation is unknown and depends on the role this particular leucine played in the function of the peptide.

In addition to substitutions, **insertions** or **deletions** can take place. For example, insertion of one or two base pairs results in the alteration of the order of the triplet codons relative to the initiation codon. A new reading frame will originate from the site of insertion, encoding a completely different polypeptide. These so-called

frameshift mutations often have the same result as a chain termination mutation at the same site because the new polypeptide, however long it is, usually does not specify a functional protein. Frameshift mutations terminate when ribosomes encounter a termination codon in the new reading frame. Deletion of one or two nucleotides also results in frameshift mutations.

A frameshift mutation due to insertion of a T paired with A would look like this:

DNA strand 1	TAC TTT GCT TAA TAC GCT CTA GTT GAC CAT T
DNA strand 2	ATG AAA CGA ATT ATG CGA GAT CAA CTG GTA A
mRNA	AUG AAA CGA AUU AUG CGA GAU CAA CUG GUA A
Protein	Met Lys Arg Ile Met Arg Asp Gln Leu Val

Note that as a result of the insertion, the new reading frame continues beyond the stop codon of the wild-type (that is, original) mRNA until a new termination codon is reached. The polypeptide sequence after the site of mutation is completely different from that of the original peptide. Insertions or deletions of three nucleotides (or multiples of three nucleotides) yield true insertions or deletions of amino acids, without altering the reading frame around the site of the mutation; however, these could still have a profound impact on the function of the protein.

Changes in a specific base sequence can have variable effects on the polypeptide that is encoded by a specific gene. Alteration of the length of the protein by a mutation to a termination codon can abolish the function of the protein. Mutations that result in substitution of one amino acid for another may have only minor effects, especially if the amino acids are similar in size, such as the example in which leucine is substituted by isoleucine. However, if the substitution is in an important part of the molecule, it can also alter the functional area of the polypeptide. Mutations that have a more dramatic affect on a protein's function or stability are those that result in substitution of amino acids having opposite charges or very different side chains. Occasionally the change will cause altered physical properties in the polypeptide. Certain substitutions render enzymes temperature sensitive, such that they can function only at lower temperatures. This is presumably due to incorrect folding of the polypeptide chain, making it more susceptible to thermal denaturation. The effect of the mutation would therefore be noticeable only under nonpermissive conditions—that is, when the protein is denatured. Under permissive conditions, however, less-than-perfect folding may still result in a protein with only slightly altered activity.

Restoration by a counteractive mutation is called **suppression**. For mutations that result in alteration of cel-

lular metabolism, such that the cell does not function optimally, there is sufficient selective pressure for secondary compensatory mutations that repair the original defect. Suppression of the mutation can occur by several mechanisms, the simplest of which is exact reversal of the mutation, that is, a mutation that restores the original codon, or, when possible, creates a new codon for the same amino acid. However, other mechanisms of suppression involve genetic changes at adjacent or distal sites on the DNA. Suppression of a base-substitution mutation can be accomplished by a mutation in a different amino acid, which would nullify the disruptive effect of the first mutation in the polypeptide chain. Frameshift mutations can sometimes be suppressed by a genetic change that mirrors the original mutation. If the mutation occurs by an insertion or deletion of a base, it can be suppressed by a reciprocal deletion or insertion, respectively. This second mutation will restore the original reading frame from the site of the second mutation. The resultant protein would have an amino acid sequence that is identical to the original wild-type sequence, with the exception of the stretch of amino acids between the two mutations. The suppressor mutation would conceivably restore the protein's function, provided it occurred in the same or adjacent codon relative to the original mutation.

A specialized category of suppression deals with nonsense mutations (Figure 13.31). Nonsense mutations occur when a base substitution changes a codon for an amino acid into one of the chain-termination codons. Nonsense suppressors are mutations in genes encoding the tRNA anticodon loop corresponding to an mRNA stop codon, such that the tRNA adds an amino acid to the growing peptide rather than terminating it. Figure 13.31 presents an example of a chain-termination mutation suppressed by a mutant suppressor tRNA. Because cells still need the correct tRNA genes for normal protein synthesis, only those tRNAs that are present in more than one copy per cell can mutate to nonsense codon suppressors without killing the cell. Moreover, suppressor tRNA can interfere with normal termination of protein synthesis by a single termination codon, resulting in added carboxyl-terminal tails extending into the adjacent gene up to the next chain termination codon. This happens relatively infrequently, because most coding sequences of polypeptides terminate with more than one in-frame termination codon.

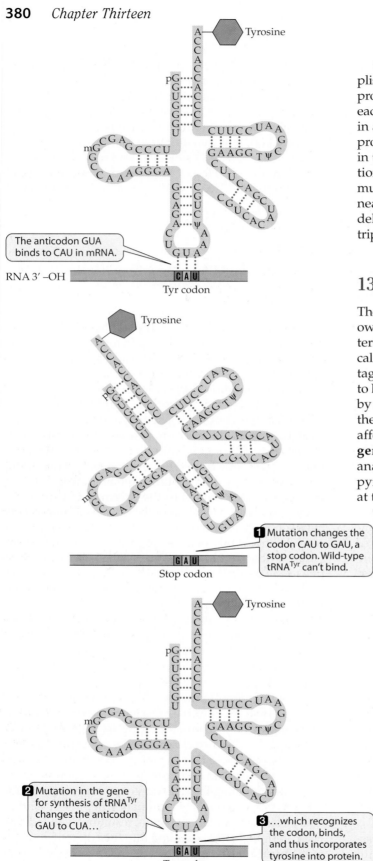

The anticodon GUA binds to CAU in mRNA.

RNA 3'—OH

Tyr codon

1 Mutation changes the codon CAU to GAU, a stop codon. Wild-type tRNA^Tyr can't bind.

Stop codon

2 Mutation in the gene for synthesis of tRNA^Tyr changes the anticodon GAU to CUA...

3 ...which recognizes the codon, binds, and thus incorporates tyrosine into protein.

Tyr codon

Figure 13.31 Suppressor mutation

Often, two proteins associating with each other accomplish a cell function, and these complexes are formed by protein–protein interactions involving distinct regions of each polypeptide called interactive domains. A mutation in a gene encoding the interactive domain of one such protein can be suppressed by a compensatory mutation in the second protein's interactive domain. Some mutations are more difficult to suppress, such as frameshift mutations, in which the reading frame has to be corrected near the site of the insertion or deletion by insertion or deletion of one or two base pairs to restore the original triplet reading frame.

13.9 Mutagens and DNA Repair

The rate of spontaneous change in DNA is relatively low, owing to the effective repair mechanisms present in bacterial cells. Exposing cells to several physical or chemical agents can increase the rate of mutation. Physical mutagenesis almost exclusively involves radiation damage to DNA. A variety of chemicals can also cause mutations by interacting with DNA directly or by interference with the DNA replication and repair processes. Compounds affecting mutational rates are termed **chemical mutagens** and fall into three broad categories: structural analogs of nucleotides, chemicals that alter purine or pyrimidine bases, and agents that intercalate the DNA at the replication fork (Box 13.2).

Mutagenesis with Base Analogs

5-Bromouracil (BU) is an example of a base analog because of its similarity to thymine. During replication, it can substitute for thymine, resulting in a BU:A pair. (The colon indicates a hydrogen bond.) However, thymine can occasionally pair with guanine in a relatively unstable transient bond. When BU substitutes for thymine, the BU:G bond is relatively stable, and during subsequent replication the stabilized G pairs with C, resulting in a permanent transition of an AT pair to GC (Figure 13.32). Treatment with 5-bromouracil can also result in the opposite transition, in which an existing GC pair is replaced by an AT pair in daughter cells.

Mutagenesis with Agents That Modify DNA

Agents that modify DNA include a large number of compounds with nucleotide bases that readily incorporate into DNA. Most of these cause transition mutations.

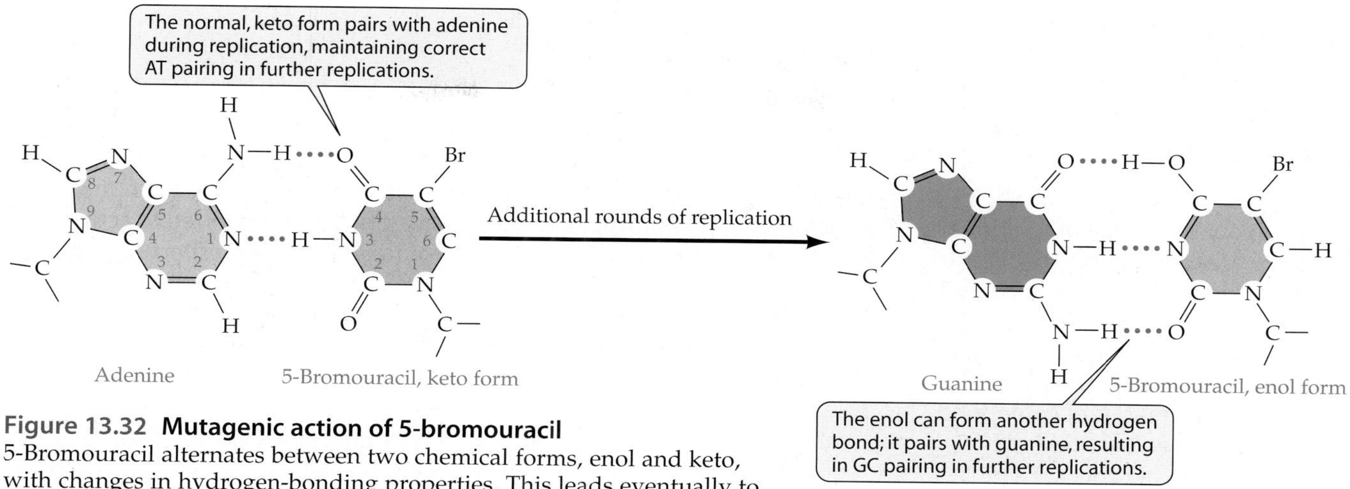

Figure 13.32 Mutagenic action of 5-bromouracil
5-Bromouracil alternates between two chemical forms, enol and keto, with changes in hydrogen-bonding properties. This leads eventually to an AT pair being replaced by a GC pair.

1. *Nitrous acid* deaminates cytosine and adenine, converting them to uracil and hypoxanthine, respectively. During the subsequent rounds of replication, uracil pairs with adenine, whereas hypoxanthine pairs with cytosine. The net effect of the reaction of nitrous acid with DNA is the transition of AT to GC and GC to AT.

2. *Hydroxylamine* and *nitrous acid* specifically modify cytosine so it can pair with adenine, resulting in the transition of a GC pair to an AT pair.

3. *Ethylmethane sulfonate* is an alkylating agent that modifies the oxygen at position 6 of guanine by adding an alkyl group (a methyl group), converting it to O^6-methylguanine. This modified base can now weakly pair with thymine during subsequent replication. This modification leads to a GC-to-AT transition in one of the daughter cells. This mispairing can also lead to an AT-to-GC transition.

4. *Nitrosoguanidine* also alkylates guanines, giving a GC-to-AT transition. This often leads to small deletions.

Mutagens That Intercalate with DNA

Several complex organic molecules—such as acridine orange, proflavin, and ethidium bromide—intercalate with DNA bases and result in distortion of DNA. Distortion leads to DNA Pol slippage, resulting in small insertions or deletions. Frameshift mutations are the most common types of lesions in DNA that occur as a result of interaction with intercalating compounds. The structures of all of these compounds are different, yet they all have conjugated rings of a size similar to a base pair in a DNA duplex, and they most likely insert between hydrogen-bonded adjacent bases in the duplex, resulting in the bending of DNA.

Radiation Damage

Ultraviolet (UV) radiation is capable of causing DNA damage by covalently cross-linking adjacent pyrimidines. The most common form of cross-link is between two adjacent thymines, forming a thymine dimer by linking carbons at the 5 and 6 positions of the stacked pyrimidine rings. However, T-C and C-C dimers and 6,4 photoproducts (bonds formed between the carbons at positions 6 and 4 of adjacent pyrimidines) are also often seen following UV treatment of DNA. Errors in repair may cause deletions, but pyrimidine dimers can also lead to transition and transversion of bases during subsequent replication. Ionizing radiation causes hydrolysis of phosphodiester linkages forming the phosphate backbone of DNA; subsequent errors in repair can lead to deletions and insertions.

DNA Repair

Bacteria possess a number of DNA repair systems that can correct damaged DNA. The mutations described in the previous section therefore become part of the genetic repertoire of the bacterium only if they are not repaired. Enzymes present in many bacterial and archaeal species accomplish direct reversal of DNA damage. Methylated guanines, such as those caused by exposure to ethylmethane sulfonate or other alkylating mutagens, are repaired by methylguanine transferase, an enzyme that recognizes the modified guanine in the DNA duplex and transfers the methyl group onto the enzyme itself, thus restoring guanine. A process called **photoreactivation**

BOX 13.2 *Methods & Techniques*

Ames Test

It is well known that many chemicals in our environment, including foods, are carcinogens; that is, they can induce cancers in animals and humans. Testing these chemicals is very complex, requiring the exposure of experimental animals to high concentrations of compounds over prolonged periods and noting development of cancer. A simplified and sensitive test has been developed to examine carcinogens, based on the ability of most of these compounds to induce mutations. The mutagenic ability of a chemical is tested as outlined in the accompanying figure. The test compound is applied to a strain of *Salmonella* that has a single mutation in an enzyme of histidine biosynthesis, designated His⁻ (unable to grow on media without histidine). If the compound is mutagenic, it can increase the frequency of mutations in the *Salmonella* genome and, at a certain frequency, will result in back-mutations in the original gene responsible for the histidine requirement. Those bacteria can then synthesize their own histidine and grow on media that is not supplemented by this nutrient. In the Ames test, a pure culture of His⁻ *Salmonella* is mixed with the suspected carcinogen and plated on minimal media that lacks histidine. As a control, the same culture is plated without the carcinogen. After the plates are incubated, the number of colonies is counted. A significant increase in the number of bacteria on the test sample, resulting from an increase in back-mutations from His⁻ to His⁺, compared to the control, indicates mutagenic activity of the compound. This test can be adapted for those compounds that are not mutagenic by themselves, but are metabolically converted to active mutagens after ingestion. Usually, the compound is mixed with an extract from rat liver, to allow conversion to the mutagenic form.

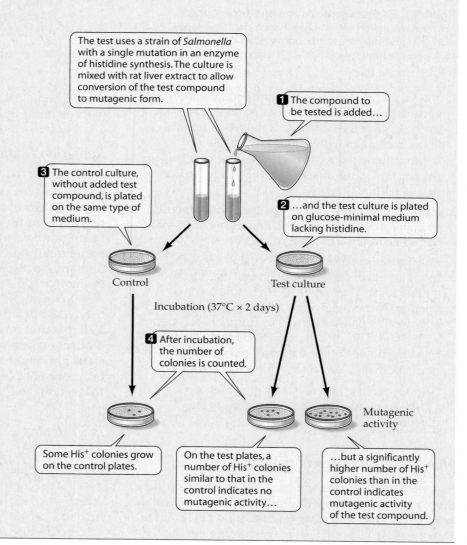

---

repairs adjacent pyrimidines that become cross-linked as a result of the mutagenic action of UV light. The resulting lesion is recognized by the enzyme photolyase, which, on absorption of visible light, breaks the bond between the dimerized pyrimidines, thereby restoring their original structure.

In addition to direct reversal of mutations by enzymes, a common repair mechanism involves excision repair, in which the damaged DNA strand is enzymatically removed and the gap correctly resynthesized using the complementary strand as a template. For example, a mutation created following deamination of cytosine to uracil can be repaired by the enzyme DNA glycosylase, which removes uracil from the sugar–phosphate backbone. This region lacks bases and is called the apyrimidinic (AP) site. This defective lesion is removed by the combined action of two enzymes: AP endonuclease and deoxyribophosphodiesterase, which leave behind a one base gap. DNA Pol then repairs the nick by incorporating the correct base (dC) and DNA ligase seals the nick.

Bulky distortions in DNA resulting from gross mispairing or UV light–induced pyrimidine dimers can be repaired by an **excision mechanism** (Figure 13.33) that relies on the ability of a multisubunit endonuclease (in *E. coli*, encoded by genes *uvrA*, *uvrB*, and *uvrC*) to hydrolyze the phosphate backbone at two places flanking the damaged site. The action of DNA helicase results in the release of a 12-base oligonucleotide containing the damaged bases; the resulting gap is filled in by DNA Pol I, and the backbone is closed by the action of DNA ligase.

Another mechanism by which thymine dimers can be removed is called **recombination repair**. When DNA Pol III reaches a thymine dimer, it skips it and continues synthesis at the other side of the dimer, leaving a gap in the newly synthesized duplex. The opposite strand is replicated normally. The gap that is created around the damaged DNA is filled by excision of the corresponding fragment from the second strand and insertion into the gap by a mechanism similar to recombination between two related DNA molecules, as described in Chapter 15. This repair, just like recombination between DNA segments of similar sequence, is also dependent on the RecA protein, as well as a multisubunit nuclease (exonuclease V), which contains products of genes *recB*, *recC*, and *recD*. The newly created gap in the parental strand is then filled in by repair synthesis, because it involves copying an undamaged strand.

The ability of bacteria to sense the presence of damaged DNA involves synthesis of a number of specialized proteins, in addition to those that function in normal repair. In *E. coli*, a complex regulatory system is activated, which directs synthesis of many of these proteins. This regulatory network, called the **SOS regulon**, is negatively controlled by a repressor protein, LexA, which recognizes operator sequences for a number of genes for repair enzymes, including *umuC*, *uvrA*, *uvrB*, and *uvrC*, as well as the *recA* gene. All of these genes express proteins

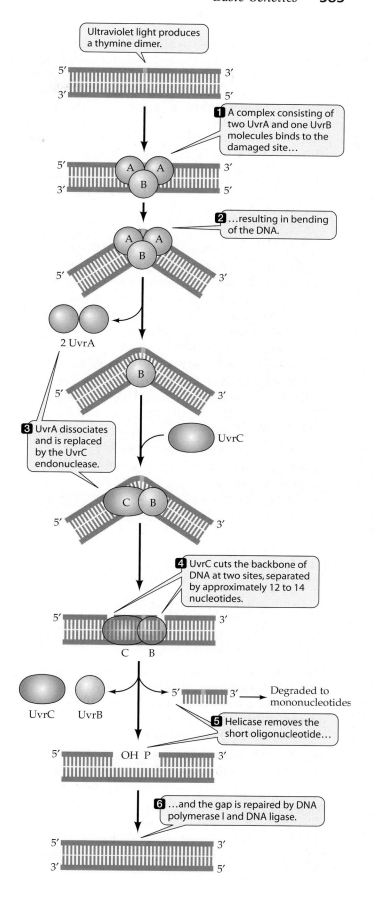

Figure 13.33 Excision repair
Thymine dimers, created by exposure of DNA to UV light, are repaired by the UvrABC system.

that function at one stage or another in the previously mentioned repair systems. It is the sensing of damaged DNA by the RecA protein, resulting in cleavage of the LexA repressor, which initiates the various DNA repair mechanisms by relieving the repression of the repair enzymes' genes. The SOS response is activated when the replication fork stalls and RecA protein binds to exposed ssDNA. The RecA–DNA complex binds LexA, and this binding induces a conformational change that causes LexA to cleave itself.

SECTION HIGHLIGHTS

Changes in the base-pair composition of the DNA are called mutations. Depending on the specific change in the coding sequence of a gene, a mutation can have a different consequence on the activity of the protein. Certain chain-termination mutations can be suppressed by mutated tRNAs that recognize the termination codon on the mRNA and facilitate insertion of an amino acid into the polypeptide during translation. Various chemicals can affect the frequency of mutations in a specific gene, either by altering the DNA or by interfering with DNA repair during replication; these are called mutagens. There are several mechanisms through which cells repair DNA and eliminate the deleterious effects of mutations.

SUMMARY

- All living cells carry hereditary information in deoxyribonucleic acid (DNA), which is organized into **genes**. In bacteria, genes are usually found in a single circular **chromosome**.

- DNA is composed of four deoxyribonucleotides: deoxyadenosine monophosphate (**dAMP**), deoxythymidine monophosphate (**dTMP**), deoxyguanosine monophosphate (**dGMP**), and deoxycytidine monophosphate (**dCMP**).

- The deoxyribonucleotides are linked together via **phosphodiester linkages** between phosphoric acid and hydroxyls on neighboring deoxyribose sugars.

- Bacterial chromosomal DNA is **double stranded**; opposite, **antiparallel** strands are held together by hydrogen bonds formed between adenine and thymine and between guanine and cytosine bases of the deoxyribonucleotides.

- DNA replicates by a **semiconservative** mode of replication, wherein the strands are separated and each strand is copied, with the daughter genome receiving one parental strand and a newly synthesized complementary copy.

- DNA replication begins at a single, relatively well-conserved site called the origin (*ori*). Two **replication forks** move away from the origin as nucleotides are incorporated with the assistance of a number of proteins.

- Within each replication fork, one strand (the **leading strand**) is synthesized as a single contiguous molecule, whereas the complementary strand (**lagging strand**) is synthesized in a discontinuous fashion, involving short RNA segments that serve as primers for DNA segments (**Okazaki fragments**). The newly synthesized DNA strand is completed by removal of RNA primers and filling in the gaps with DNA, followed by the joining of the fragment into a single molecule.

- The low frequency of mistakes made during DNA replication is due to the **proofreading** and **repair** activities associated with enzymes that replicate DNA.

- **RNA** is a copy of one of the DNA strands. It consists of polymerized ribonucleotides with the same base composition as DNA, except that thymine is replaced by uracil. The complementarity of the RNA copy is ensured by base-pairing of the nitrogenous bases by hydrogen bonding during synthesis of the RNA strand.

- RNA that specifies proteins is called messenger RNA (**mRNA**). If a single RNA can specify several proteins, it is called a polycistronic mRNA.

- Other forms of RNA specify components of ribosomes, the ribosomal RNA (**rRNA**), or the adapter molecules for protein synthesis, called transfer RNAs (**tRNA**).

- The **promoter**—the beginning of the gene, where transcription initiates—is recognized by the σ subunit of RNA polymerase. After binding, σ dissociates from the RNA polymerase complex, leaving a form called the **core enzyme**, which continues elongating the RNA strand. RNA polymerase recognizes the end of a gene after formation of a stem-loop structure or by termination assisted by the so-called **rho** protein.

- When the product of a gene is not needed, bacteria utilize **negative regulation** to prevent transcription of the gene. This often involves binding of a **repressor** protein to an **operator** site located adjacent to a promoter sequence. Tight binding of the repressor to a sequence that often overlaps the promoter interferes with σ-factor–directed binding of the RNA polymerase, and thus prevents transcription of the negatively regulated gene.

- Alternatively, protein factors (**activators**) stimulate transcription by binding to DNA sites upstream of promoters and facilitating the rate-limiting incorporation of the first bases into RNA.

- The information in mRNA is converted into proteins by a **translation**, which involves ribosome-assisted polymerization of amino acids into a sequence specified by the triplet code of the mRNA, copied from its complementary DNA strand. The specific order of such codons is called the **reading frame**.

- Amino acids are incorporated into polypeptides after they are covalently attached to tRNA. Each tRNA recognizes a sequence of three bases on mRNA (called codons) via a complementary sequence.

- The ribosomes recognize and bind to a specific site at the beginning of the mRNA. This so-called **ribosome binding site** is adjacent to the codon for the first amino acid of the polypeptide, which is always methionine. Ribosomes terminate translation at specific codons (**termination**, or **stop codons**), for which there are no corresponding tRNAs.

- Various regulatory mechanisms control progress of RNA polymerase by **attenuation** and **antitermination mechanisms**, some of which simultaneously regulate translation.

- Changes in the base-pair composition of the DNA are called **mutations**.

- Various chemicals can affect the frequency of mutations in a specific gene, either by altering the DNA or by interfering with DNA repair during replication. These are called **mutagens**.

- There are several mechanisms through which cells repair DNA and eliminate the deleterious effects of smutations.

 Find more at www.sinauer.com/microbial-life

REVIEW QUESTIONS

1. Contrast the basic structural differences between deoxyribonucleotides and ribonucleotides.

2. Draw the structure of deoxycytidine-deoxyguanosine (dC-dG) and deoxyadenosine-deoxythymidine (dA-dT) base pairs as they appear in the antiparallel form in double-stranded DNA. Indicate the location of the hydrogen bonds holding the strands together.

3. Distinguish between DNA synthesis on a leading and on a lagging strand, addressing the different mechanisms of initiation and elongation of the DNA strand.

4. What is the mechanism for relieving the tension in DNA that is caused by supercoiling during replication?

5. What would be the consequences for *E. coli* in its ability to synthesize the enzyme β-galactosidase of the following mutations in the regulatory components of the lactose utilization operon? Explain you answer by considering bacteria that are present in medium with or without lactose.

   (a) A mutation in the operator sequence of the *lac* operon, such that it cannot bind the repressor protein.

   (b) A mutation in the DNA sequence of the *lacI* gene, such that the mutant repressor cannot bind an inducer (allolactose or its structural analog).

   (c) A two–base-pair mutation in the *lacI* gene, changing, in mRNA, the AUG-initiating codon to UAG.

   (d) A double mutant containing both types of mutations described in (b) and (c).

   (e) Would a mutation in the gene coding for cyclic AMP-binding protein (CAP), resulting in the inability of bacteria to synthesize this protein, offset the effect of the mutation described in (b), such that the bacteria synthesize wild-type levels of β-galactosidase?

6. Consider a polycistronic operon consisting of genes *X*, *Y*, and *Z*. The promoter located upstream of gene *X* undergoes a mutation, such that three of the bases in the –10 region are changed. How would this affect the synthesis of proteins encoded by genes *X*, *Y*, and *Z*? What would be the effect of a three–base-pair mutation in the ribosome-binding site of gene *X* on synthesis of proteins *X*, *Y*, and *Z*?

7. When bacteria are exposed to antibiotics, they are killed, except for a small population of mutants that are resistant, due to their ability to undergo mutations in the target of the antibiotic. Devise a set of experiments (analogous to the fluctuation test described in the chapter) that demonstrate that mutations arise spontaneously in the bacterial population and are not induced by exposure of the bacterial culture to antibiotics.

8. A bacterial culture was exposed to a mutagen, and some mutants were isolated that were unable to synthesize an essential biosynthetic enzyme. When these mutants were grown in pure culture and reexposed to the mutagen, a small fraction of bacteria regained their ability to synthesize the biosynthetic enzyme. Describe several possible mechanisms for this phenomenon, explaining how the second exposure to a mutagen "corrected" the original mutation.

9. What are the mechanisms of removal of cross-linked thymine dimers in DNA?

10. What is the mechanism of induction of DNA repair processes, when the DNA replication fork stalls at the site of DNA damage, exposing single-stranded regions?

11. Find the most likely small regulatory RNA from the selection (a–c) that would inhibit the translation of the following mRNA (only the 5' portion is shown).

    mRNA: 5'-UUGAGCUUUCCUG<u>GAAGA</u>ACAAA<u>AUG</u>UCAAUAUUAGCCA...3'

    (a) 5'-GACAUUUUGUUCUUCCAGGAAGCCGG-3'
    (b) 5'-GCAAUUUAGUUGGUUUGGAGAAUUAG-3'
    (c) 5'-GGGAAUUUAGGCCUUCCAGGAAGGAU-3'

12. The partial DNA sequence (below) from the *E. coli* genome encodes a gene for a short polypeptide. The transcriptional start site is highlighted.

    (a) Identify the most likely promoter for this gene. Is it recognized by the major RNA polymerase?

    (b) Write out the sequence of the mRNA transcribed from this promoter.

    (c) Translate the sequence into the polypeptide according to the genetic code.

    (d) Identify the position in the DNA sequence where, by a single base-pair substitution or a frameshift mutation, the result would be synthesis of a peptide that would have the identical sequence but half the size of the wild-type peptide.

    (e) Identify the position in the nucleotide sequence where a mutation involving a single base-pair change (insertion, deletion, or substitution) would result in synthesis of a peptide identical to the wild type throughout the sequence, but having four additional arginine residues at its carboxyl terminus.

SUGGESTED READING

Maloy, S. R., J. E. Cronan and D. Freifelder. 1994. *Microbial Genetics.* 2nd ed. Boston: Jones and Bartlett.

Primrose, S. B. and R. Twyman. 2002. *Principles of Genome Analysis and Genomics.* 3rd ed. Cambridge, MA: Blackwell Science.

Baunberg, S. 1999. *Prokaryotic Gene Expression.* Oxford: Oxford University Press.

Higgins, N. P. 2004. *The Bacterial Chromosome.* Washington, DC: American Society for Microbiology.

Snyder, L. and W. Champness. 2003. *Molecular Genetics of Bacteria.* 2nd ed. Washington, DC: ASM Press.

Gottesman, S. 2004. "The small RNA regulators of *Escherichia coli*: roles and mechanisms." *Annual Review of Microbiology* 58: 303–328.

Waters, C. M. and B. L. Bassler. 2005. "Quorum sensing: cell-to-cell communication in bacteria." *Annual Review of Cell and Developmental Biology* 21: 319–346.

+1

GGCTTGACACCCGGCTAGCGTAGTTGTATAATGGTCAGGCTTTGAGGAGGTCTAGGCATGAAAGCTCAACGTAAGGGGGATCCGGAGTAGGAGACGGAGGTAAG
CCGAACTGTGGGCCGATCGCATCAACATATTACCAGTCCGAAACTCCTCCAGATCCGTACTTTCGACTTGCATTCCCCCTAGGCCTCATCCTCTGCCTCCATTC

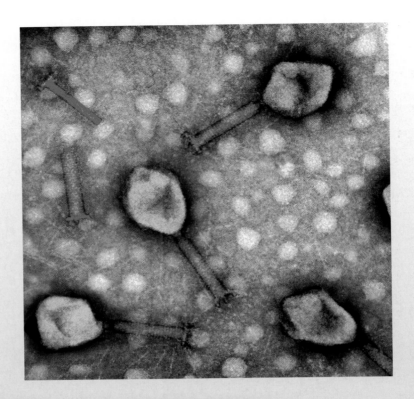

The objectives of this chapter are to:

◆ Examine current theories about the evolution of viruses.

◆ Introduce the genome structures, morphologies, and reproductive phases generally found among viruses.

◆ Describe some of the laboratory methods used to cultivate, quantitate, and purify viruses.

◆ Explain how viral genomes are replicated.

◆ Survey the wide variety of viruses found in nature and the strategies they use to infect bacteria, animals, plants, fungi, and algae.

◆ Illustrate how these relatively simple infectious agents exploit the cellular functions of their hosts to survive and reproduce.

14

Viruses

*It is, indeed, an RNA world, not a DNA world. In present-day
life, DNA acts entirely by way of RNA, whereas RNA can act on
its own. Viruses are a case in point.*
—*Christian de Duve,* Blueprint for a Cell: The Nature and
Origin of Life

S elf-perpetuation is the key to survival of a species, and as discussed previously, the blueprint for "self" is present in nucleic
acids (genome). Therefore, a bit of nucleic acid that can generate a like nucleic acid may be considered to have some properties of
life. Thus we define a **virus**: a bit of nucleic acid—RNA or DNA—
surrounded by a protective coat; and this small bit of nucleic acid can
regenerate by commandeering the metabolic machinery of an appropriate prokaryotic or eukaryotic host. It is certain, however, that
viruses are neither alive nor have the characteristics associated with a
viable cell.

Viral nucleic acid can gain entry into a suitable host, where it mobi-
lizes the essential cellular synthetic capacities and compels the host to
generate viral particles. Viral mRNA (message) can be synthesized by
polymerases available in the host or furnished by the virus. However,
viruses do not contain translational enzymes, and they depend on the
translational machinery of a host to synthesize viral proteins.

An understanding of how viruses function begins with an
overview of the basic structure of the virus (Figure 14.1). At the core
of the virus is the viral genome, which can be composed of either
RNA or DNA. Unlike the cells of animal, plant, or bacteria, a virus
rarely contains both. The nucleic acid is bounded by a protective protein coat termed a **capsid**. The capsid and genome together form a
nucleocapsid. In some viruses, a membrane envelope surrounds the

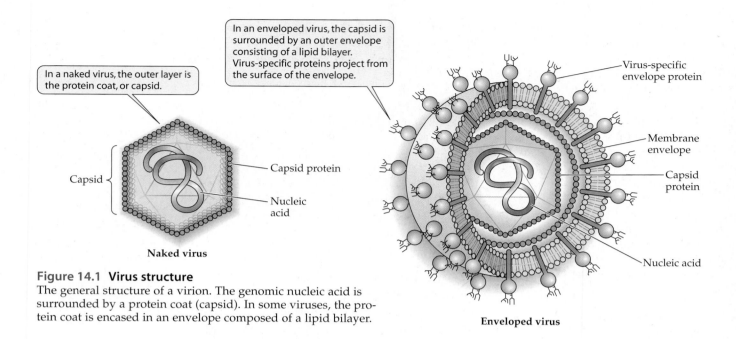

Figure 14.1 Virus structure
The general structure of a virion. The genomic nucleic acid is surrounded by a protein coat (capsid). In some viruses, the protein coat is encased in an envelope composed of a lipid bilayer.

nucleocapsid, and these viruses are considered to be **enveloped viruses**. For these viruses, host cellular membrane is the source of the viral envelope and surrounds the nucleocapsid as it exits from an infected cell. Virus-specific proteins that are present in the envelope are embedded in the cellular membrane prior to virus envelopment.

All viruses exist in two states: extracellular and intracellular. In the **extracellular state**, the **virion** or **virus** particle is inert. The virion is the form in which a virus passes from one host cell to another host. In the **intracellular state**, the viral nucleic acids induce the host cell to synthesize the virus. Such viral reproduction is called infection. A viral infection can be productive, which leads to viral progeny, or nonproductive, whereby a virus may remain dormant in the cell, causing no apparent manifestations of infection. With some animal viral infections, multiplication is always accompanied by disease symptoms, as would occur with a case of measles. There are other infections where the inapparent infection rate may be 100 or more cases of viral multiplication for each clinical case. This chapter is concerned with the mechanisms whereby prokaryotic and eukaryotic viruses are propagated. The first half of the chapter presents a general overview of viruses: their size, their structure, how they are cultivated and quantitated, and how they reproduce. The second half considers selected viruses in some detail.

14.1 Origin of Viruses

The origin of viruses is an intriguing question and has been the subject of considerable speculation in the sci-

entific community. There is general agreement that viruses (bacteriophage) probably infected bacteria during those millennia when *Archaea* and *Bacteria* were the sole life forms on Earth. As eukaryotes evolved, they too became subject to viral attack. There are three major theories that have been advanced for the origin of viruses:

1. Viruses originated in the primordial soup and co-evolved with the more complex life forms.

2. They evolved from free-living organisms that invaded other life forms and gradually lost functions by a process called "retrograde evolution."

3. Viruses are "escaped" pieces of nucleic acid no longer under the control of a cell—also termed the escaped gene theory.

The great variety of viruses present in the living world affirms that they originated independently many, many times during evolution. Viruses also originated from other viral types through mutation. A brief discussion of theories of viral evolution follows.

Coevolution

We will consider first the theory of coevolution. The earliest self-replicating genetic system was probably composed of RNA. RNA can promote RNA polymerization, although this would be a slow process and proteins present in the primordial soup may have become promoters of RNA replication. The DNA template is much more effective as the genetic material and originated early in evolution. RNA then became the messenger between the DNA template and protein synthesis. Therefore, the genetic code came into being and permitted orderly replication.

Gradually the early replicative forms gained in complexity and became encased in a lipid sac, the purpose of which was to separate its metabolic machinery from the environment. This may have been the ancestor of the progenote and later the *Archaea* and *Bacteria*. Another replicative form may have retained simplicity and may have been composed mainly of self-replicating nucleic acid surrounded by a protein coat. This entity may have been the forerunner of the virus and evolved to depend for duplication on its ability to invade and take over the genetic machinery of a host. Thus, there was coevolution of the host and virus. This hypothesis has few supporters, but it does offer an explanation for the evolution of viruses with genomic RNA.

Retrograde Evolution

The theory that viruses originated as free-living or parasitic microorganisms is based on the concept that a microorganism became a predator and gradually lost unused genetic information. Genes for biosynthesis of intermediates supplied by the host could be lost by mutation without harm to the invader. Eventually this prokaryote may have evolved to nothing more than a group of genes, now termed a virus. Prior to the advent of the eukaryote, bacterial and archaeal life forms were mostly free-living, but predators similar to *Bdellovibrio* may also have existed. An intracellular parasitic microorganism could have become more and more dependent on the host cell and would only need to retain the ability to replicate nucleic acids and a mechanism for traveling from cell to cell.

An equivalent occurrence in eukaryotic cells might account for the evolution of plant and animal viruses. An intracellular parasitic bacterial type such as the *Chlamydiae* would be an example of a bacterium that potentially could regress to the viral state. The *Chlamydiae* are bacteria that have no cell wall and cannot live outside of living cells.

A bacterial origin would be more likely for complex viruses such as the poxviruses, but the genetic information in these viruses differs so markedly from that in prokaryotes that the retrograde evolution theory has little support. The absence of any life form between intracellular bacterial pathogens and viruses is also cited as a concern, and it is difficult to explain how RNA viruses came into existence through loss of genetic information.

"Escaped" Gene Theory

This is the most plausible theory and has considerable support. It proposes that pieces of host-cell RNA or

DNA "escaped" and gained independence from cellular control. Living organisms make duplicates of their genetic information by initiating replication at a specific site called the origin of replication (*ori*). The replication cycle ends when a full complement of the genome is synthesized. If initiation of replication begins elsewhere on the genome, this duplication would occur independently of host control. An entity that could recognize nucleotide sequences at sites other than the origin and carried the proper polymerase could have the capacity to produce RNA or DNA without interference from normal control mechanisms.

The origin of viruses may have been with episomes (plasmids) or transposons, DNA molecules that can be integrated into or excised from various sites in the host chromosome. Plasmids can also move from cell to cell carrying information such as fertility or antibiotic resistance. Transposons are bits of DNA present in both prokaryotic and eukaryotic cells that can move from one site in a chromosome to another site, carrying genetic information.

There is actually one type of transposon known that directs the assembly of an RNA copy of its own DNA. This transposon can use this messenger RNA to synthesize DNA via reverse transcriptase. This copy may then be inserted into the chromosome. The DNA of the transposon carries a gene for synthesis of the reverse transcriptase, and transposon elements with such properties have some similarity to retroviruses. Analysis of nucleotide sequences in viruses indicates that they are much the same as specific sequences in the host cell. Evidence accumulated to date strongly supports the proposal that viruses originated from "escaped" host nucleic acid.

SECTION HIGHLIGHTS

Three major theories have been put forward to explain the origin of viruses: coevolution, retrograde evolution, and the escaped gene theory. The coevolution theory proposes that viruses arose at the same time as the earliest ancestors of *Archaea* and *Bacteria* and coevolved with more complex life forms. According to the retrograde evolution theory, viruses originated from predatory prokaryotes that gradually lost functions that were provided by the host. The escaped gene theory, which proposes that viruses arose from pieces of host cell RNA or DNA that escaped cellular control, is currently the most widely accepted explanation.

(A) Icosahedron   (B)   (C)

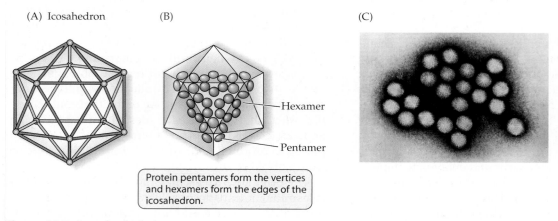

Hexamer

Pentamer

Protein pentamers form the vertices
and hexamers form the edges of the
icosahedron.

Figure 14.2 Icosahedral virus
Graphic representation of (A) an icosahedron and (B) the arrangement of proteins
in an icosahedral virus. (C) Electron micrograph of icosahedral viral particles. C,
courtesy of the Centers for Disease Control and Prevention.

14.2 Size and Structure of Viruses

Electron microscopy has affirmed that viruses vary widely
in size and structure. A very small virus, the poliovirus,
is about 20 nm (a nanometer is 1/1,000 of a micrometer)
in diameter. This virus actually approaches the size of a ribosome. The larger viruses, such as the poxviruses, are 400
nm long and 200 nm in width, and are large enough to
be visible by phase contrast microscopy. The genomes of
RNA or DNA viruses can be single- or double-stranded.
The smallest RNA virus has a single-stranded genome
with about 3,600 nucleotides. The size of the genome for
DNA viruses varies over a broad range from about 5,000
to 200,000 bases or base pairs (bp). Simian virus 40, a small
virus, has a genome makeup of 5,224 bp, and vaccinia, a
large virus, has 190,000 bp. A bacterial genome would
range from one million to nine million bp, depending on
species. The smallest bacterial genomes are those of intracellular parasites and are in the 600,000-bp range.

Electron microscopy, x-ray diffraction, biochemical
analysis, and immunological techniques have provided
much information on the structure and composition of
viruses. Viruses are of several morphologic types, which
are described below.

ICOSAHEDRAL VIRUSES   The capsids in these viruses
are constructed from protein subunits termed **protomers**. In an icosahedral capsid, these protomers are
arranged in capsomers composed of pentamers (five
sides) and hexamers (six protein subunits). The bonding
between capsomers is generally noncovalent and
weaker than the noncovalent bonding within a cap-
somer. The hexamers form the edges and faces of the
icosahedron and pentamers are at the vertices. The 12
vertices in an icosahedron are shown in Figure 14.2A, a

structure that has 20 equilateral triangles as faces. The
capsid of a small virus such as poliovirus is made up of
12 pentamers, and the capsid of the larger adenovirus is
composed of 252 capsomers (12 pentamers and 240
hexamers). Generally the capsids follow the laws of crystallography. The pentamers and hexamers may be composed of one type of protein subunit or the proteins in
the pentamers may differ from those in the hexamers
(Figure 14.2B). The procapsid is assembled and then

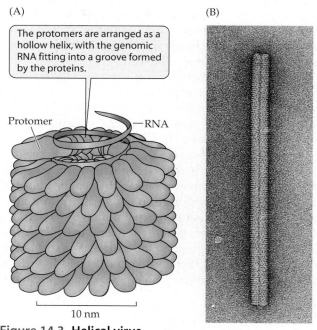

(A)

The protomers are arranged as a
hollow helix, with the genomic
RNA fitting into a groove formed
by the proteins.

Protomer   RNA

10 nm

(B)

Figure 14.3 Helical virus
(A) Graphic representation and (B) photomicrograph of
tobacco mosaic virus. Photo ©R. C. Williams/Biological
Photo Service.

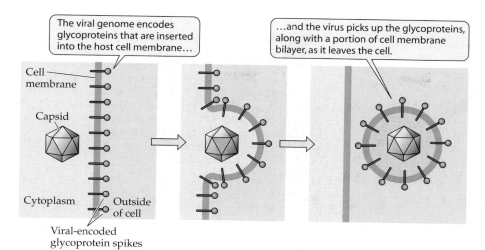

The viral genome encodes glycoproteins that are inserted into the host cell membrane…

…and the virus picks up the glycoproteins, along with a portion of cell membrane bilayer, as it leaves the cell.

Cell membrane

Capsid

Cytoplasm

Outside of cell

Viral-encoded glycoprotein spikes

Figure 14.4   Enveloped virus
An enveloped virus acquires its lipid membrane coat as it exits through the cell membrane or nuclear membrane of the host cell.

filled with nucleic acid. In some eukaryotic viruses the nucleic acid may be incorporated during assembly. Viruses have specific mechanisms to ensure that an entire genome is incorporated.

HELICAL VIRUSES   These are long, thin, hollow cylinders composed of distinct protomers. The helical tobacco mosaic virus (TMV) is depicted in Figure 14.3. The protein subunits are arranged in a helical spiral with the genome fitted into a groove on the inner portion of the protein. The nucleic acid is enclosed in the capsid as the protomers are assembled. The helical viruses are narrow (15 to 20 nm) and may be quite long (300 to 400 nm). These viruses may be rigid or flexible, depending on the nature of the constituent protomers. The helix of TMV has one type of self-assembling protomer composed of 158 amino acids. There are 2,130 copies of this one protein subunit in the capsid of a mature virus particle.

ENVELOPED VIRUSES   These viruses have an outer, membranous structure that surrounds the nucleocapsid (see Figure 14.1). The membranous envelope consists of a lipid bilayer that has specific glycoproteins (spikes) embedded in the lipid. The glycoprotein spikes and other surface proteins are incorporated into the host cell membrane prior to exit of the virus and are picked up by the virus as it leaves the infected cell by budding through the membrane (Figure 14.4). The glycoproteins are encoded by the virus genome and are not normal constituents of the host cell membrane. These glycoproteins are generally responsible for infection, as they attach to specific sites on host cells.

COMPLEX VIRUSES   These have capsid symmetry, but often the virus is assembled from parts synthesized separately (e.g., head, tail). The *Escherichia coli* virus T4 (Figure 14.5) is a complex virus. The virus is composed of a

(A)

(B)

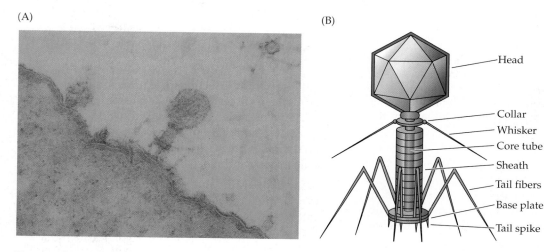

Head

Collar
Whisker
Core tube
Sheath
Tail fibers
Base plate
Tail spike

Figure 14.5 Complex virus
(A) T4 virus (phage) attacking an *Escherichia coli* cell. (B) Graphic representation of T4. Photo ©J. Broek/SPL/Science Source/Photo Researchers, Inc.

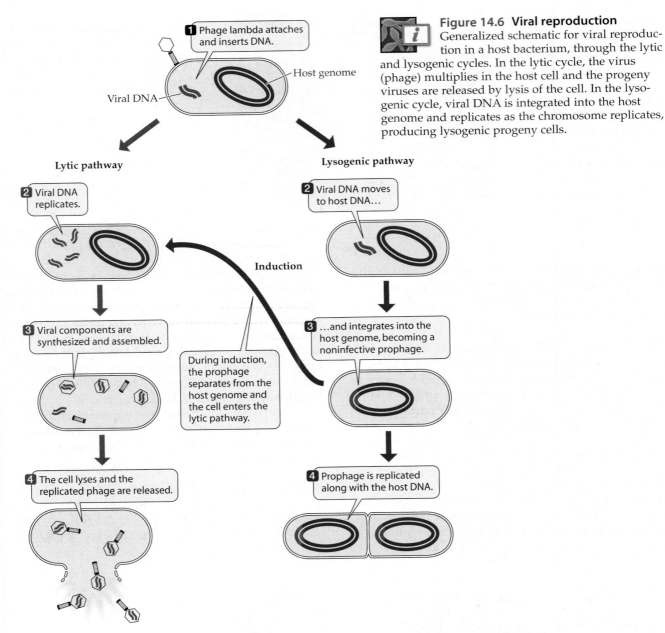

Figure 14.6 Viral reproduction
Generalized schematic for viral reproduction in a host bacterium, through the lytic and lysogenic cycles. In the lytic cycle, the virus (phage) multiplies in the host cell and the progeny viruses are released by lysis of the cell. In the lysogenic cycle, viral DNA is integrated into the host genome and replicates as the chromosome replicates, producing lysogenic progeny cells.

polyhedron head, a tail, whiskers, tail fibers, and other parts. This virus is discussed in some detail later in the chapter.

Viral Propagation

There are four rather distinct phases in viral reproduction: (1) attachment, (2) penetration, (3) gene expression and genome replication, and (4) assembly. These are as follows:

1. **Attachment or adsorption**. There are specific sites (virus receptors) on the surface of a cell where the virion attaches.

2. **Penetration or injection**. The nucleic acid of the virion gains access to the cytoplasm of the host. Bacterial and plant viruses initiating an infection must gain entry through the cell wall, whereas animal viruses enter through the cell membrane. This process influences virus structure.

3. **Gene expression and genome replication**. Once inside, the viral genome is expressed and duplicated. This leads to the formation of gene products that control the replication of viral nucleic acid and capsid synthesis.

4. **Assembly and release**. The viral nucleic acid is packaged into the capsid and the mature virus is

released. A virus needs a host for replication, since ATP, ribosomes, and other building blocks are supplied by the host cell. Enzymes involved in viral replication are often also supplied by the host with or without modification. Generally, virally encoded lytic enzymes cause host-cell dissolution.

These phases are illustrated in (Figure 14.6) for a bacteriophage. Entry of the viral genome generally results in an immediate cessation of many normal host-cell functions and the replication of viral nucleic acid. Later in the infectious process, capsid and other proteins are synthesized that become part of the mature viral particle.

SECTION HIGHLIGHTS

Viruses range from 20 nm to 400 nm in their longest dimension. Viral genomes, the largest of which contain about 200,000 bases or base pairs, are much smaller than the typical bacterial genome. Generally, viruses may be icosahedral, helical, enveloped, or complex in morphology. Viral reproduction has four phases: attachment, penetration, gene expression, genome replication, and assembly.

14.3 Cultivation of Viruses

Viruses can be grown only within living cells. This is because a virus has no biosynthetic capacity of its own, but can manipulate a cell genetically to synthesize viral particles. In early experimentation, viruses were cultivated by transfer from plant to plant, animal to animal, or bacterial culture to bacterial culture. The symptoms or manifestations of disease that resulted from inoculation were the principle means of identification. Animal viruses such as smallpox, polio, and mumps were routinely propagated in embryonated eggs. A fertile egg was incubated for 6 to 8 days at a temperature suitable for hatching and then inoculated with a viral suspension. Egg inoculation was often suitable for diagnostics, as the lesion produced in the embryo was characteristic of a given virus. More exacting cell-culturing techniques have been developed for use in research and diagnostic laboratories today. Some viruses are still grown in chicken eggs for selected purposes such as antigen production.

Viruses that infect animal cells are now generally grown in monolayers of cells obtained from animal tissue. Plant cells can be cultivated as clusters of cells called calli (callus, singular). Cells removed aseptically from a plant or animal can be placed in a sterile vessel made of glass or plastic. Generally the cells used in viral propagation are from established cell lines. The animal cells may attach to the solid surface, and when fed properly they can form a monolayer. This monolayer is called a **cell culture**; likewise, tissue can be maintained in culture and is termed **tissue culture**. A virus inoculum can be spread over a cell monolayer or introduced into tissue culture, and if the cells in the culture are susceptible, the virus will replicate. After viral attachment to a monolayer, a layer of agar can be spread over the surface to limit the spread of virions to cells that are adjacent to the infected cell. In this way, areas of cell destruction can be observed.

Bacterial viruses can be cultivated by inoculating a suitable susceptible bacterial culture with a virus. Liquid cultures or cultures growing in an agar medium are suitable for this purpose.

Plant viruses can be cultivated in masses of plant tissue or in cultures of separated cells in suspension, in protoplast cultures, or in intact plants. The leaf may be abraded with a mixture of virus and an abrasive material, and localized necrotic lesions can be observed. Viral suspensions may be injected into a plant with a syringe, thus emulating transfer by insects in nature.

SECTION HIGHLIGHTS

Viruses must be propagated inside living cells. Bacterial viruses are generally cultivated on bacterial cultures in liquid media or on agar plates. Animal viruses are propagated in fertilized chicken eggs or in monolayers of tissue culture cells. Plant viruses may be grown in various types of plant tissue culture or in intact plants.

14.4 Quantitation of Viruses

To understand the nature of viruses, their reproduction, and their properties, it is essential that methods be available to quantify the virus particles present in a suspension. This can be accomplished by directly counting the virions or quantifying viruses as infectious units. A direct count can be accomplished with an electron microscope (EM). Dilutions of virions are generally stained with heavy metals and mixed with a predetermined concentration of small latex beads, and the mixture examined by EM. The ratio of beads to viruses in a field would give a reasonable estimate of viral particles present.

Many viruses agglutinate red blood cells (RBCs), a phenomenon termed **hemagglutination**. This has been adapted as an effective method for quantitating selected viruses such as the influenza virus. Serially diluted suspensions containing virus are added to a standardized number of RBCs. The hemagglutination titer is the highest dilution of virus that causes visible aggregates.

Viruses are most often quantified as **virus infectious units**. A virus infectious unit is the smallest unit that causes a detectable effect on a specific host. The number of units present in a suspension can be determined by a plaque assay or by the direct effect on a host animal or plant.

Plaque assays are an effective means of measuring the number of infectious particles in a viral suspension and can be routinely applied to bacterial viruses. This can be accomplished by pouring a sterile agar medium that can support growth of the host into a Petri dish. Dilutions of the virus suspension are added to a suspension of host bacteria in molten agar and, after mixing, the viral–bacterial suspension poured over the agar surface. A dilution of virus with significantly fewer virions than there are bacteria present will result in clear visible areas in the soft agar layer that lack intact bacteria, and these are termed **plaques** (Figure 14.7). The plaque results from the destruction of a bacterium followed by destruction of those adjacent to it, and each plaque is assumed to have resulted from an initial infection with one virion. Counting the number of plaques on a plate and adjusting for the dilution factor permits one to estimate the number of **plaque forming units** (**PFUs**) in the original suspension.

The plaque assay may also be adapted to animal viruses. However, some animal viruses cause lethal cytopathic effects that do not result in discernible plaques. For these viruses, degenerative changes in host cells must be quantitated by microscopic examination.

The inoculation of a suitable host with serially diluted virus and then determination of the highest dilution that causes symptoms of disease or death can be used to quantitate animal viruses that cannot be quantitated by plaque assays.

Leaf assays are useful in estimating the number of plant viruses present in a suspension. A dilution of virus particles would be combined with an abrasive and rubbed into a suitable host leaf. The number of necrotic lesions evident after incubation would be an estimate of the number of viruses present in the original suspension. Abrasion is necessary because plant viruses generally cannot penetrate intact cells.

SECTION HIGHLIGHTS

Viruses can be counted directly under an electron microscope. For viruses that cause red blood cells to aggregate, a hemagglutination assay can be used for quantitation. However, viruses are most commonly quantified as virus infectious units—measured by plaque assays (for bacterial viruses and some animal viruses), by leaf assays (for plant viruses), or by inoculating suitable hosts with serial dilutions of the virus (for animal viruses that can't be measured with a plaque assay).

(A)

Uninfected bacteria form a lawn on the plate.

Plaques are cleared spots where infected bacteria have lysed.

(B)

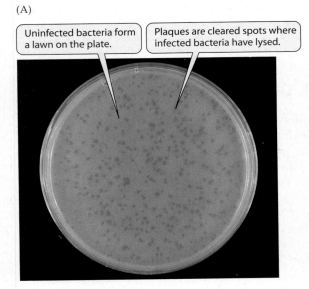
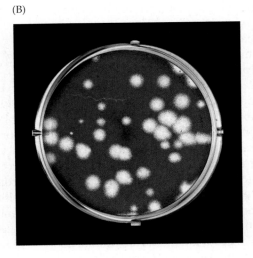

Figure 14.7 Viral plaques
(A) Plaques formed in a lawn of *Escherichia coli* by phage T4. (B) Plaque assay of coronaviruses inoculated onto a human cell line. A, ©E. Chan/Visuals Unlimited; B, courtesy of Sean Whelan.

14.5 Purification of Viral Particles

Virus particles must be separated from host cells to determine their chemical nature, for preparation of vaccines, and for other purposes. The particular procedure adopted depends on the nature of the virus and the host cell. No single technique is effective for all viral particles, but there are purification methods that are useful, including differential centrifugation, precipitation, denaturation, and enzymatic digestion.

Differential centrifugation. The infected host cells are lysed, followed by differential centrifugation. The large host-cell parts form a pellet at a relatively low speed, and the supernatant contains the virus. The virus is then pelleted with use of high-speed centrifugation.

Precipitation. Proteins from cell debris and viruses precipitate at different salt concentrations. Salts compete with protein for available water, and specific proteins tend to associate at a particular salt concentration. Each protein has a specific salting-out point.

Denaturation. Some viruses are tolerant of organic solvents such as butanol, which denatures host-cell debris. The virus enters the aqueous phase, the denatured cell debris is present at the interface, and lipids enter the solvent phase.

Enzymatic digestion. Selected proteases and nucleases digest cell components, and a specific virus may be resistant to these enzymes. The cell components are solubilized and the virus obtained by differential centrifugation.

SECTION HIGHLIGHTS

The particular procedure used to purify a virus depends on the nature of the virus and its host, but may include differential centrifugation, precipitation, denaturation, or enzymatic digestion steps.

14.6 Viral Genomes and Reproduction

The genomes of all cells are composed of double-stranded DNA. A virus genome differs in that it can be either DNA or RNA. Viruses have not been characterized that contain both RNA and DNA, although certain DNA viruses contain RNA primers. The genomes of viruses that infect plants, animals, or bacteria can be composed of a single strand (ss) or double strand (ds) of RNA or DNA. An exception would be the hepadnaviruses, which are partially double-stranded and partially single-stranded DNA. By convention, the negative (–) strand of dsDNA is the strand transcribed into mRNA. Consequently, the mRNA translated into protein is considered plus (+) and is a complementary copy of one strand of DNA. However, the ssDNA geminiviruses that infect plants make a complementary DNA strand and transcribe mRNA from both strands. An RNA viral genome may be a single strand of RNA that is mRNA (plus strand) or an RNA (minus strand) that must be transcribed to produce mRNA. Some RNA viruses have an RNA genome that is part (+) and part (–) strand. There are no ssRNA (minus strand) viruses yet characterized that infect bacteria.

The pathways followed in production of mRNA from the genomic nucleic acids of various viruses are presented in Table 14.1. Capsid size limits the amount of nucleic acid a virus may contain. Thus, certain viruses have mechanisms for producing mRNAs that lead to genetic economy not known to occur in eukaryotic and prokaryotic life forms. For example, $\phi \chi 174$, an ssDNA bacteriophage, contains enough DNA to code for four to five proteins (1,795 base triplets; 400 amino acids/protein). The virus uses overlapping reading frames, different initiation sites, and frameshifts within a reading frame to produce 11 proteins. Parvoviruses, small ssDNA animal viruses, geminiviruses, ssDNA plant viruses, and Leviviridae, or (+) ssRNA bacteriophages, also employ overlapping reading frames.

In most cases a (+) ssRNA viral genome would actually be mRNA. However, the (+) ssRNA genome of retroviruses, such as the virus that causes acquired immunodeficiency syndrome (AIDS), is reverse-transcribed to make dsDNA that is integrated into the host genome. From there it can direct the transcription to viral mRNA and the original viral genome.

Some viruses produce early and late mRNA, and some of the early proteins resulting from translation of the early mRNA may direct the host cell to cease normal function. Other early mRNAs direct the synthesis of viral-specific enzymes that result in the replication of the

TABLE 14.1	Pathways to mRNA for viruses with various types of genomes
Genome type	**Pathway for mRNA production**[a]
dsDNA	→ mRNA
ssDNA	→ dsDNA → mRNA
(+) ssRNA	acts as mRNA
(+) ssRNA (retroviruses)	→ (–) ssRNA → dsDNA → mRNA
(–) ssRNA[b]	→ mRNA
ds RNA	→ mRNA

[a]Arrows indicate transcription, reverse transcription, and DNA synthesis steps.
[b]There are plant and animal viruses of this type, but bacterial (–) ssRNA viruses have not been characterized. See text for definitions of (+) and (–) strands.

viral genome. The late proteins synthesized from late mRNA are involved with the synthesis of the capsid and other proteins needed for the assembly of the mature virus. A bacterial virus may direct the production of lytic enzymes that lyse the host cell and free the mature viral particles.

DNA viruses can transcribe and replicate their genomes using host cell enzymes without modification, although some code for their own enzymes or modify host enzymes or both. Replication of viral DNA parallels host DNA replication by requiring a small RNA primer to provide the 3′–OH. The ssDNA of parvoviruses self-primes by using palindromes to create a hairpin at the 3′ terminus (Figure 14.8A). Adenovirus replication occurs without the involvement of the lagging strand as illustrated in (Figure 14.8B). Poxviruses also self-prime by nicking one of their dsDNA covalently closed ends to give a free 3′–OH. Eukaryotic and prokaryotic cells do not contain the RNA-dependent RNA polymerase required to transcribe and replicate the genomes of RNA viruses. The (+) strand RNA viruses code for an RNA polymerase, which is made early in the infection process, whereas an RNA polymerase is present in the virion of the (–) strand RNA viruses and dsRNA viruses and is used to make mRNAs. Hepadnaviruses make genomic DNA from an RNA template using a reverse transcriptase (RNA-dependent DNA polymerase), whereas retroviruses employ a reverse transcriptase to replicate their RNA genome using a DNA intermediate.

SECTION HIGHLIGHTS

Viral genomes are structurally diverse; they may be made of DNA or RNA, and be single-stranded or double-stranded. Some viral genomes even have a combination of these features. Because viral genomes are generally small, viruses use a variety of mechanisms (overlapping reading frames, variable initiation sites, and frameshifts) to maximize the protein coding capacity of their genetic material. Some viruses produce early and late mRNA; the former encodes proteins necessary for replication of the viral genome and proteins that may interrupt the host cell's normal functions, and the latter encodes proteins required for the assembly and release of progeny viruses. Viruses may use host cell enzymes, with or without modification, or their own enzymes for genome replication.

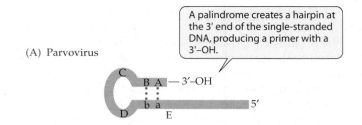

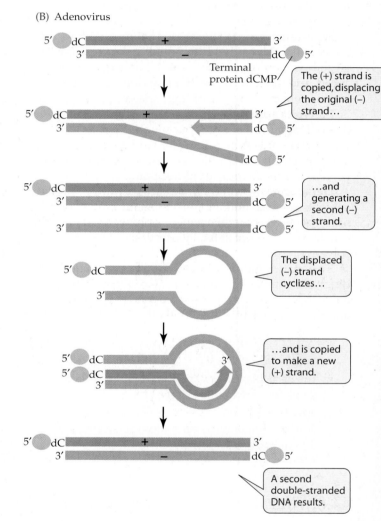

Figure 14.8 Priming of viral replication
Mechanisms used by single-stranded DNA viruses to prime replication of their genomes. (A) Parvovirus; (B) Adenovirus.

14.7 Bacterial Viruses

The first scientist to observe and report the manifestation of a viral attack on bacteria was Frederick W. Twort of England in 1915. Twort noted that an agent that caused dissolution of bacterial colonies could be transferred to intact colonies and that these also were lysed. Twort also realized that the agent was active at high dilution, was heat labile, and passed through a bacterial filter. He suggested that it might be a virus. In 1917, Felix D'Herelle of the Pasteur Institute in Paris observed the lytic phenomenon and called the agent a bacteriophage, which means "bacteria-eating." Today, the name *bacteriophage*, or simply phage, can be used interchangeably with bacterial virus, as virions that attack bacteria have essentially the same attributes of animal or plant viruses—they simply have a different host specificity.

Bacterial viruses have proved to be excellent models for the study of viral infection and propagation. Some of the viral types and their size, structure, host, and nucleic acid components are presented in Table 14.2. The diversity of viral types is too great to allow a discussion of more than a limited number.

Bacterial viruses have been widely employed for the study of host–parasite interactions in viral pathogenesis. The viruses that attack enteric bacteria, particularly *Escherichia coli*, have been studied most. There is a broad array of viral types that attack *E. coli*, and *E. coli* is useful for study, as we know much about the genetics of the organism. The value of bacteria as a model is apparent when you consider that *E. coli* has a generation time of about 20 minutes, which means that considerable cell mass can be obtained overnight. A virus that attacks *E. coli* can produce mature progeny in approximately 22 minutes. Therefore, one can follow the entire sequence of molecular events—from attachment of virus to the bacterium to lysis and release of virus—over a very short period.

There are viruses that attack virtually all microorganisms, including thermophiles and cyanobacteria. The number of viral types is broad, and the mechanism for viral reproduction is varied.

Taxonomy

Bacterial viruses are generally identified by a Greek symbol or a combination of strain number and symbol. Because viruses apparently are polyphyletic (arose multiple times), a grouping based on phylogeny would not be possible. There are more than 2,000 different bacterial viruses described in the literature. Unfortunately, the guidelines for describing a new virus were not established until recent years, and most of those described are

---

TABLE 14.2 The families of some major bacterial viruses

Genome Type/Virus	Size (nm)	Example	Host	
dsDNA, linear				
Siphoviridae	Head 90 Tail 200 × 15	λ	*E. coli*	
Myoviridae	Head 80 × 110 Tail 110 × 25	T4	*E. coli*	
dsDNA, linear				
Corticoviridae	60	PM2	*Pseudomonas*	
Plasmaviridae	50 to 120	MV-L2	*Mycoplasma*	
ssDNA, circular				
Microviridae	30	φΧ174	*E. coli*	
Inoviridae	9 × 890	M13	*E. coli*	
dsRNA, linear				
Cystoviridae	85	φ6	*Pseudomonas phaseolicola*	
ssRNA, linear				
Leviviridae	30	MS2	*E. coli*	

inadequately characterized. Relatedness among viruses is determined by host range, virion morphology, type of nucleic acid present in the genome, and other factors. Immunological identification of viral components is now employed and is of value when characterizing a specific viral group and relating it to others. A classification scheme for bacterial viruses is presented in Table 14.2.

Lysogenic and Temperate Viruses

There are two potential outcomes of a viral attack on a susceptible bacterium. The virus may attach to gain entry, reproduce, and destroy the host. A virus can also attach, gain entry, and establish a mutually beneficial interaction with a bacterium. This interaction allows both viral propagation and survival of the host in an association termed **lysogeny**. Those viruses that can establish this relationship are called **temperate** viruses. The nucleic acid of a temperate virus enters the host, and may become integrated into the host genome. The progeny of a lysogenized cell generally are lysogenic (see Figure 14.6). In nature, many bacteria may be lysogenic for one or more viruses. The lysogenic state can cause an alteration in the host outer envelope that can modify attachment sites for other bacterial viruses. Invariably there are specific sites on a bacterium where the virus attaches, and infection cannot occur unless there is attachment. Lysogeny, then, may be beneficial in that it can protect the host bacterium from attack by more virulent viruses that are related to the temperate one. Lysis of all hosts could lead to extinction of a virus, whereas a lysogenized bacterium could retain the virus indefinitely.

In a lysogenic culture, 1/1,000 or less of the infected cells become lytic and produce virions. Proteins that repress the expression of genes for a lytic infection are produced by the integrated prophage (provirus). It is these repressor proteins that maintain the lysogenic state. Under certain conditions the lysogenic state is lost, the prophage DNA is activated, and lysis occurs. This is called **induction of the lytic state**. Induction occurs at a low spontaneous rate but is accelerated in the presence of mutagens such as ultraviolet (UV) light, x-rays, and nitrogen mustards. Much research has been devoted to the lysogenic state because it has implications for gene regulatory mechanisms in all organisms and for dormancy in human viruses. The integration of viral nucleic acid into the host genome is also of considerable interest in diseases such as AIDS and cancer.

Lysogenic Conversion

As mentioned previously, lysogeny can alter surface components of the host bacterium and render it immune to attack by other bacteriophages. Lysogenic cells can also undergo **lysogenic conversion**, thereby gaining other properties that nonlysogenized counterparts do not have. These conversions result from expression of genes on the phage chromosome that are not repressed during lysogeny and will continue to be expressed as long as the bacterium retains the prophage. Lysogenic conversion changes the phenotype of a cell and can lead to an increase in pathogenicity in some disease-causing bacteria.

Lysogenic strains of *Corynebacterium diphtheriae* produce a potent toxin that is the cause of diphtheria (see Chapter 28). Scarlet fever results from a toxin released by lysogenized *Streptococcus pyogenes*, and the botulinum toxin is a product of prophage-bearing *Clostridium botulinum*. The epsilon phage of *Salmonella* sp. alters surface lipopolysaccharides by altering the activity of enzymes involved in their synthesis. This changes the antigenic character of the surface and susceptibility of the pathogen to specific antibodies. Elimination of the prophage from the aforementioned bacteria removes the virus-induced pathogenicity.

Lambda (λ) Phage

Lambda (λ) has long been a model system for understanding lysogeny and gene regulation mechanisms in its host, *E. coli*. In the lysogenic state, λ DNA is integrated into the host genome at a specific site (*attB*) located between the *gal* operon (for catabolism of galactose) and the *bio* operon (for biotin biosynthesis). The integration of λ phage is effected by the protein integrase, a gene product that is encoded by the *int* gene in the viral genome. The genome of λ has a region of DNA (*attP*) that is homologous to the *attB* site of the host genome. Immediately after λ infects the host, its linear dsDNA is circularized. The circularized DNA integrates by a single reciprocal crossover between *attP* and *attB*, leading to the lysogenic state.

Whether lysis or lysogeny occurs with λ is determined by a number of host and phage factors, but ultimately depends on the concentration of two viral encoded proteins: Cro and cI (the λ repressor). The *cro* and *cI* genes are adjacent in the λ genome and are transcribed in opposing directions (Figure 14.9). The adjacent promoters are designated P_{RM} ("repressor maintenance") for *cI* transcription and P_R ("rightward") for *cro* transcription. cI and Cro proteins bind to the same transcriptional operators (O_{R1}, O_{R2}, and O_{R3}) adjacent to P_{RM} and P_R, although with different affinities. When cI binds to O_{R1} and O_{R2}, it represses rightward transcription from P_R, but activates transcription from P_{RM}. Therefore, more cI repressor is made, and the lysogenic state is maintained. The cI repressor prevents expression of virtually all phage genes, including those transcribed from P_L (the

Figure 14.9 Lambda phage lysogeny
The genome of λ phage, with details of the critical area that determines whether the phage enters a lytic cycle or remains in the lysogenic state. The determining factor is the relative amounts of the two repressor proteins, cI and Cro.

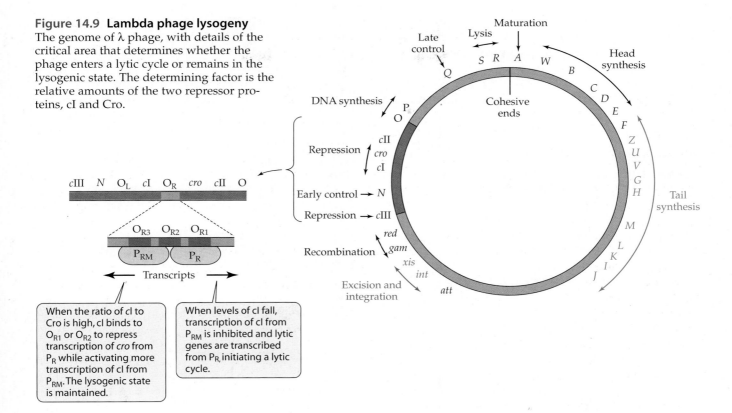

When the ratio of *cl* to Cro is high, cl binds to O_{R1} or O_{R2} to repress transcription of *cro* from P_R while activating more transcription of *cl* from P_{RM}. The lysogenic state is maintained.

When levels of cl fall, transcription of cl from P_{RM} is inhibited and lytic genes are transcribed from P_R, initiating a lytic cycle.

leftward promoter). Therefore, the cI protein is both a repressor and an activator of transcription. The level of the cII protein, an activator of λ *int* gene transcription, is also important for promoting lysogeny. The cII protein is unstable and affected by the physiological state of the bacterium. Induction of the lytic cycle depends on the relative amounts of the two repressor proteins cI and Cro. Although these repressors bind to the same operators (O_{R1}, O_{R2}, and O_{R3}) that regulate P_R and P_{RM}, they bind them in a different order. The cI repressor has the highest affinity for O_{R1} and lowest for O_{R3}, whereas Cro has the highest affinity for O_{R3}. A high cI-to-Cro ratio will inhibit P_R and maintain lysogeny as described. If cI level is low after infection or the level drops in a lysogen (by UV-induced cleavage, for example), Cro binds first to O_{R3}, repressor synthesis from P_{RM} is inhibited, and the lytic genes are transcribed from P_R and P_L. Expression of lytic genes in a lysogen results in excision of the prophage from the bacterial genome *att* site as a covalently closed circle.

The λ DNA replication cycle is presented in **Figure 14.10.** Early after infection the linear genome is circularized using the cohesive ends. The circularized genome becomes supercoiled, and early replication via "theta structures" yields additional copies of the circularized λ phage genome. Later, rolling circle replication leads to synthesis of linear **concatemers** (a chain composed of a number of repeated units of duplex genomic DNA). The concatemers are cleaved to form the individual λ phage genomes that are packaged in a separately synthesized capsid. Mature viruses then escape by lysing the host cell.

Mutagens and other agents that disrupt bacterial DNA and affect the integrity of the cI repressor in a lysogen are ultimately responsible for inducing the lytic cycle and virus release. Lysis can therefore be considered a survival mechanism for the phage.

We will now examine in some detail two other bacterial viruses that have been studied extensively and that serve as examples of alternative mechanisms by which bacterial viruses propagate: M13 and T4.

M13

The filamentous phage designated M13 has an unusual property in that it can invade a cell (*E. coli*), reproduce, and exit without serious harm to the host. The M13 virion is a thin filament about 895 nm in length and 9 nm in diameter (**Figure 14.11**). The inner diameter of the protein capsid is 2.5 nm. This encloses a single-stranded circle of DNA that extends the entire length of the filament. Stretched out, the viral genome would be nearly 2 µm in length. The capsid coat consists of about 2,700 individual protein subunits that overlap one another. At one end of the virion are four pilot proteins that guide

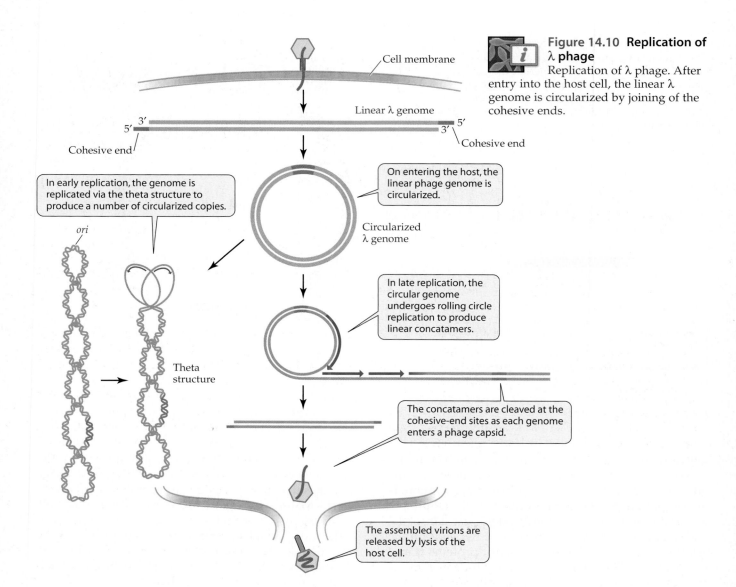

Cell membrane

Linear λ genome

3′
5′
3′
5′

Cohesive end

Cohesive end

Figure 14.10 Replication of λ phage
Replication of λ phage. After entry into the host cell, the linear λ genome is circularized by joining of the cohesive ends.

On entering the host, the linear phage genome is circularized.

In early replication, the genome is replicated via the theta structure to produce a number of circularized copies.

Circularized λ genome

ori

Theta structure

In late replication, the circular genome undergoes rolling circle replication to produce linear concatamers.

The concatamers are cleaved at the cohesive-end sites as each genome enters a phage capsid.

The assembled virions are released by lysis of the host cell.

the virus into and out of the host. In one generation, *E. coli* infected with M13 can produce 1,000 intact progeny virions. A considerable mass of virions can readily be collected (100 mg/L growth culture) free of the host-cell debris that would be associated with a lytic infection.

GENOME   The M13 genome is composed of 6,407 nucleotides, and there are ten defined genes (see Figure 14.11). The genome has an intergenic space (about 8% of the genome) that is the origin of DNA replication. Single-stranded foreign DNA can be inserted into the intergenic space of the M13 genome to generate DNA fragments for sequencing and other studies.

COAT   The M13 coat protein is composed of subunits that contain 50 amino acid residues. The capsid protein from gene VIII accumulates on the plasma membrane of the infected cell and is assembled on the genome as the

DNA exits from the cell. The binding proteins that protect the M13 genome are removed by an endopeptidase as the pilot protein guides the genome to the cell membrane. These binding proteins are then recycled by the virus.

PENETRATION AND INFECTION   The DNA of M13 is not injected into the host cell, as occurs with most bacterial viruses. The virion attaches at or near the F pilus of *E. coli* (organisms lacking an F pilus are not susceptible to attack). Two theories have been advanced for viral penetration: (1) the virion follows a groove in the F pilus to a receptor site on the cell surface where it enters or (2) the virion stimulates a progressive depolymerization of the pilus that retracts the tip and brings the M13 to the surface.

Entry of the intact virion occurs, and decapsidation occurs during entry into the cell. Replication of the (+) single strand of viral DNA occurs simultaneously with coat

(A)

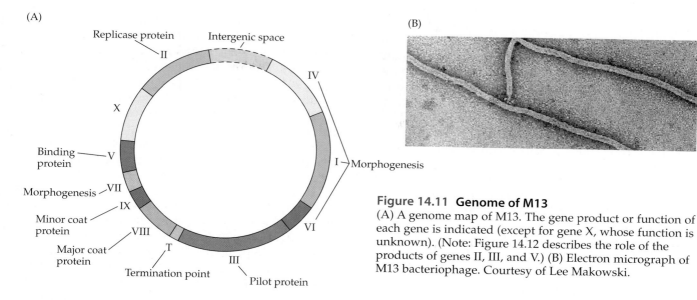

Figure 14.11 Genome of M13
(A) A genome map of M13. The gene product or function of each gene is indicated (except for gene X, whose function is unknown). (Note: Figure 14.12 describes the role of the products of genes II, III, and V.) (B) Electron micrograph of M13 bacteriophage. Courtesy of Lee Makowski.

removal. Most of the capsid protein is conserved and deposited on the plasma membrane and reappears as capsid on progeny DNA along with newly synthesized capsids. The gene III pilot protein present at the lead end of the virus remains bound to the genome, as it is initially replicated to the duplex form. The enzymes involved in initial replication are provided by the host and are normally associated with synthesis of extrachromosomal el-

ements in the bacterium. M13 does not take over the genetic apparatus of the host. The growth rate of *E. coli* producing M13 virions is reduced by about one-third.

REPLICATION   The overall events involved in the replication of M13 viral DNA are presented in Figure 14.12. The process can be divided into three stages. In the first stage (see Step 2 in Figure 14.12), the viral (+) ssDNA is

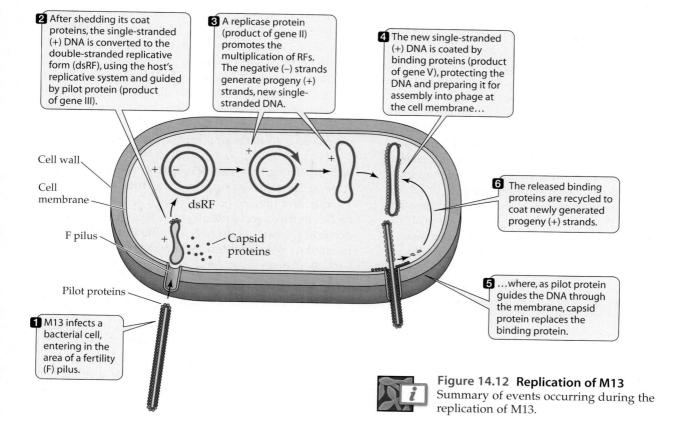

2 After shedding its coat proteins, the single-stranded (+) DNA is converted to the double-stranded replicative form (dsRF), using the host's replicative system and guided by pilot protein (product of gene III).

3 A replicase protein (product of gene II) promotes the multiplication of RFs. The negative (−) strands generate progeny (+) strands, new single-stranded DNA.

4 The new single-stranded (+) DNA is coated by binding proteins (product of gene V), protecting the DNA and preparing it for assembly into phage at the cell membrane…

6 The released binding proteins are recycled to coat newly generated progeny (+) strands.

5 …where, as pilot protein guides the DNA through the membrane, capsid protein replaces the binding protein.

1 M13 infects a bacterial cell, entering in the area of a fertility (F) pilus.

Figure 14.12 Replication of M13
Summary of events occurring during the replication of M13.

converted to the double-stranded replicative form (RF). There are no viral-coded proteins synthesized for this step. The phage relies on the host replicative system, and the gene III–encoded pilot protein that entered with the DNA apparently is involved in directing this synthesis. Remember that in M13 infections the host synthetic machinery is not disrupted and the bacterial genome remains intact and functional.

In the second stage, a virus-encoded gene (II) is expressed and leads to synthesis of a replicase protein that promotes the multiplication of RFs. The negative strands of these RFs generate progeny (+) strands and serve as a template for continued production of viral genomes (see Step 3 in Figure 14.12).

In the third stage, the synthesized viral (+) ssDNA is coated by binding proteins (see Step 4 in Figure 14.12). This envelops the DNA in a form for assembly at the membrane. The function of the gene V binding protein is to facilitate the processes that generate (+) ssDNA. Binding also prevents the (+) strands from being employed as templates for synthesis of duplexes and protects against nucleases. As the pilot protein guides the DNA through the cytoplasmic membrane, the capsid protein replaces the binding protein. The mature virus is extruded through the multilayered gram-negative bacterial cell wall without disrupting the host cell.

T4

The *E. coli* phage T4 is a complex phage with several distinct functional parts including a head, tail, whiskers, and tail fibers (see Figure 14.5). T4, along with T2 and T6, is one in a series of bacterial viruses called the T-even phages, which share about 85% DNA homology. The most studied has been T4 because it is large, complex, and has many capabilities, including the ability to propagate in a short period. As discussed previously, the host for T4, *E. coli*, has a doubling time of about 20 minutes under nutrient-rich conditions. The genome of *E. coli* has about 4,300 genes and a function has been assigned to about 2,700 of these. During normal growth, *E. coli* can generate a duplicate genome in about 40 minutes. The organism can divide in 20 minutes because multiple rounds of DNA replication are ongoing in a rapidly growing cell. However, when *E. coli* becomes infected

with T4, the virus assumes complete control of the cell's reproductive machinery. Within about 22 minutes after infection, the *E. coli* host lyses and releases over 100 viral particles. These viral particles each contain a genome bearing 200 genes. Essentially, a normally dividing *E. coli* produces over 4,000 genes in 40 minutes, whereas the machinery of an infected bacterium can generate 60,000 genes in an equivalent length of time. The ability to take over a cell and accomplish this feat in so short a time makes the T4 virus a prime object for study. The events that occur during a lytic infection of *E. coli* follow.

GENOME   The T4 genome is among the largest known in viruses. It encompasses about 200 genes and is synthesized as a concatemer, by rolling-circle replication. As the genome is packaged in the head of the virion, somewhat more than one genome (105%) enters. This ensures that the same gene will be present at or near the end of each genome (Figure 14.13), a phenomenon called **terminal redundancy**. The total complement of genes in each viral head is equivalent, but the genome may start and end with a different gene. Therefore, each viral particle has a complete linear genome plus two sets of at least one gene at the ends.

The genome of T4 is linear as packed in the phage head and enters a bacterial host as a linear duplex. All markers or genes in the genome are linked. Given that the gene at the end of each genome may differ, no single gene would consistently appear at the end of the linear chain. The genes as mapped would, therefore, be linked on both sides with other known genes. Each gene would be in the middle of a linear genome with the same frequency. This phenomenon is called **circular permutation**.

The expression of T4 genes is tightly regulated. The organization of the genome is such that expressing the genes in a defined order would follow. Early genes are transcribed in a counterclockwise direction, and late genes tend to be read in the opposite or clockwise direction.

ADSORPTION AND PENETRATION   There are distinct sites on the outer envelope of *E. coli* that act as receptors for the fiber tips of the T4 virion. The viral particle attaches at these recognition sites. This attachment may occur at a fusion point between the inner cytoplasmic membrane and the outer cell envelope. After attachment

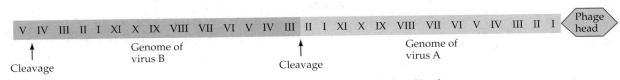

Figure 14.13 Terminal redundancy
A concatemer of T4, as it might be cleaved by endonuclease. Virus A would receive duplicates of genes I and II.

of the base plate, the sheath reorganizes, and the inner core of the tail penetrates through the cell envelope and cytoplasmic membrane. The sheath contains 24 protein rings, but during penetration the proteins slide downward to form 12 rings. The tail sheath contains ATP, and this contraction has been compared with muscle activity. The core is unplugged, and the DNA genome is injected into the host.

EFFECT ON HOST   Within one minute after injection of viral DNA, the synthesis of host-specific macromolecules ceases. This stoppage occurs because the bacterial RNA polymerase has a strong affinity for the promoters of T4 DNA, so T4 DNA transcription gets higher priority. Transcription of selected phage genes is initiated within 2 minutes after infection (Figure 14.14). This early mRNA is translated to form early proteins before viral-specific DNA is synthesized. These early proteins and enzymes take over various processes in the host cell and effect changes in the host genome. The synthesis of viral nucleic acid is initiated by these modifications. The host genome is broken up, and fragments of the DNA move to the cytoplasmic membrane. Degradation of the host chromosome to produce nucleotides that can be as-

sembled to form viral DNA occurs within 5 minutes after infection. In less than 10 minutes, the viral attack turns the host cell into a virus-synthesizing machine.

REPLICATION   To induce the host to immediately express the T4 genome, the virus must take control of transcription and translation. Shortly after infection, proteins are synthesized, employing viral genome information that modifies the specificity of host RNA polymerase. Many of the enzymes actually involved in replication of viral DNA and encoded by the virus genome have catalytic functions that are equivalent to the enzymes synthesized by the host. The virus stimulates synthesis of enzymes leading to a more rapid synthesis of T4-specific DNA. The replication rate for phage DNA is ten times the rate of host DNA synthesis in an uninfected, normally dividing bacterium. Phage-induced enzymes are active in promoting this increase. There are at least 20 viral-encoded proteins synthesized immediately after infection, and some of these effect the synthesis of a new tRNA. This tRNA reads the T4 mRNA more efficiently and rapidly than occurs in a normal growing cell. The phage also induces the synthesis of about 30 replication proteins. These proteins

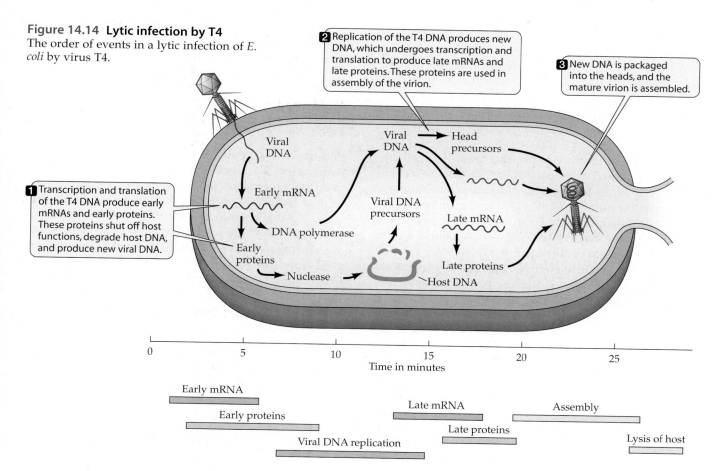

Figure 14.14 Lytic infection by T4
The order of events in a lytic infection of *E. coli* by virus T4.

2 Replication of the T4 DNA produces new DNA, which undergoes transcription and translation to produce late mRNAs and late proteins. These proteins are used in assembly of the virion.

3 New DNA is packaged into the heads, and the mature virion is assembled.

1 Transcription and translation of the T4 DNA produce early mRNAs and early proteins. These proteins shut off host functions, degrade host DNA, and produce new viral DNA.

Viral DNA

Viral DNA → Head precursors

Early mRNA

Viral DNA precursors

DNA polymerase

Late mRNA

Early proteins

Nuclease

Host DNA

Late proteins

Time in minutes
0   5   10   15   20   25

Early mRNA

Early proteins

Late mRNA

Assembly

Late proteins

Viral DNA replication

Lysis of host

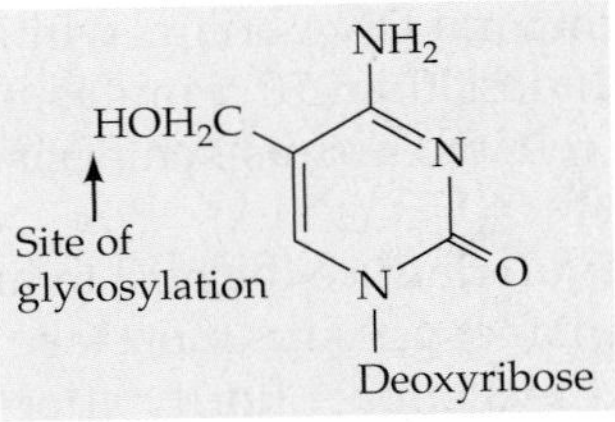

Figure 14.15 5-Hydroxymethylcytosine
The unique pyrimidine present in the DNA of T4, 5-hydroxymethylcytosine. This base is often glycosylated (has a glucose attached) at the hydroxyl, which renders it resistant to host restriction enzymes. The structure of cytosine is shown in Figure 3.17.

are involved in nucleotide biosynthesis, DNA synthesis, and modifications of DNA.

The pyrimidine cytosine is the nucleic acid base generally paired with guanine in cellular DNA. In T4 DNA, the cytosine is replaced by the unique base 5-hydroxymethylcytosine (Figure 14.15). Endonucleases encoded by the viral genome hydrolyze the host DNA. To prevent the inclusion of unaltered deoxycytidine triphosphate (dCTP) into viral DNA as dCTP is formed, it is converted to dCMP. The dCMP reacts with hydroxymethylate to form 5-hydroxymethyl dCMP. The 5-hydroxymethyl dCMP is then phosphorylated to form 5-hydroxymethyl dCTP, which is incorporated into the viral genome. A glucosylation reaction adds a glucose molecule to the hydroxy group after the dCTP derivative is incorporated into viral DNA, thereby protecting the DNA of the virion from attack by host restriction enzymes in subsequent attacks on *E. coli*.

ASSEMBLY   After the polyhedron head of T4 is completed, it is packed with viral DNA. A few low-molecular-weight core proteins and phage-induced enzymes are also incorporated into the T4 head. Magnesium and polyamines are also included to neutralize the genome DNA. It is remarkable that the 0.085 μm-wide capsid of T4 can incorporate a DNA genome that is 500 μm in length.

Translation of the late mRNA results in the synthesis of tail fibers and whiskers. The base plate is assembled as a unit and contains the T4-encoded lysozyme that facilitates subsequent entry events. The core or tube through which the DNA is injected is assembled on this base plate. The core is capped at the distal end and encircled by the sheath proteins. The head and tail are then joined together. Whiskers are then put in place, and these probably direct the placement of the long tail fibers on the base plate.

RELEASE   Lysis of *E. coli* and release of 100 to 300 viral particles occurs about 25 minutes after infection. The T4 genes direct the synthesis of a protein that disrupts the cytoplasmic membrane. A phage-encoded lysozyme

moves through the disrupted membrane to breach the peptidoglycan cell wall, and the cell disintegrates, releasing viral particles.

SECTION HIGHLIGHTS
Bacterial viruses, especially those that attack *E. coli*, are useful models for studying viral biology. A bacterial virus typically enters its host, replicates, and lyses the host cell to release its progeny. However, temperate viruses may also enter into a mutually beneficial relationship, termed lysogeny, with the host bacterium. Lysogenic bacterial cells may express characteristics not found in uninfected cells, such as increased resistance to other viruses or increased pathogenicity (lysogenic conversion). Lambda (λ) phage is a well-studied example of a temperate phage. M13 phage is a non-temperate *E. coli* phage that has the unusual ability to enter the host, replicate, and release progeny without killing the host cell. T4, another *E. coli* phage, is remarkable for the speed with which it takes over the host cell; in less than ten minutes after viral entry host cell functions are completely turned over to the synthesis of new viruses.

14.8 Viral Restriction and Modification

Bacteria have several mechanisms to protect against viral invaders. The outer surface may be modified to prevent attachment or a bacterium may produce enzymes, termed restriction enzymes, that cleave the DNA of an invader at one or more sites. Restriction enzymes are specific and attack only selected nucleotide sequences that are generally 4 to 6 bp in length. Restriction enzymes are used extensively in recombinant DNA technology (see Chapter 16). The host is protected against its own restriction enzymes by chemical alteration of its DNA at sites where the restriction enzyme might function. An example of chemical modification is methylation of bases at potentially susceptible sequences.

Viruses can counteract restriction enzymes by adding methyl or sugar groups (glycosylating) to the nucleic acid bases at susceptible sites. The T-even phages (T2, T4, and T6) that infect *E. coli* can glycosylate selected bases in DNA, rendering the genome resistant to restriction enzymes. The λ phages methylate adenine and cytosine bases in their DNA. Methylation and glycosylation occur after replication of the DNA.

14.9 Eukaryotic Viruses

Prokaryotic and eukaryotic cells differ significantly in structure. The strategy of viral attack on these two distinct cell types reflects these differences. The nuclear material, ribosomes, and other cell machinery of a prokaryote are encased in the cytoplasmic membrane. There are no internal membrane-bound structures or separate functional parts in these organisms. The eukaryotic cell is more complex, with discrete internal organelles including the nucleus, mitochondrion, and other structures that are membrane-bound.

When a virus attacks a susceptible bacterial host, the genome of the virus must traverse the cell wall and cytoplasmic membrane. Once the viral genetic information enters the cytoplasm, it has immediate access to the transcription–translation machinery and energy-producing systems of the cell all in one compartment. In infecting a eukaryotic cell, the virus must pass through the cytoplasmic membrane, and in most DNA viruses and some RNA viruses, the viral genome must also gain access to the membrane-bound nucleus.

The attachment protein present on the surface of a virus binds to a specific receptor site on the surface of the host cell. This determines which host the virus can attack, and, in multicellular hosts, the specific cells that will be infected. The cells in an animal are highly differentiated, and the surface components of these cells differ. Potential receptors on the surface of the various organs in an animal differ, and viruses that could attach to one cell type would not necessarily affect another. This is considered tissue specificity.

Classification, Size, and Structure

Some major animal and plant viruses are illustrated in Tables 14.3 and 14.4. These viruses, as with bacterial viral particles, are classified by the nature of the nucleic acids in the genomes, although the presence or absence of an envelope is also considered. The animal viruses range from the tiny parvoviruses (20 nm) to very large poxviruses (250×350 nm). Some of the plant viruses are very long, narrow filaments. For example, the potato γ virus is 10 nm across, but 700 nm in length.

Viruses can differ in their effects on the host cell. An **acute infection** has a rapid onset and short duration but results in the destruction of the host cell. *Picornaviruses* such as the polio and common cold viruses cause acute infections. In a **persistent infection** the virus replicates actively, but there may be few symptoms, and the host cell retains viability and viral production continues for an extended period. The mature virus can leave the host cell by budding without disrupting the integrity of the cell. Measles and hepatitis B are examples of persistent or chronic infections. A **latent infection** is one in which the virus is generally not actively replicating and remains dormant within the host. During latency, there are no symptoms and antibodies are not produced against the virus. When herpesvirus (Herpes simplex type I), the cause of cold sores, goes dormant, this is an example of a latent viral infection. Most plant viruses cause persistent infections, and a few are involved in latent infections.

Some viral infections can cause **transformation** of the host cell. Transforming viruses can change a normal cell into a cancer cell with fewer growth factor requirements than for the normal cell. As a consequence, the transformed cell reproduces more rapidly, resulting in a mass of cells called a tumor. Some tumors are self-limiting, do not spread, and are called **benign**. Others are **malignant** tumors, and these spread and grow in other tissue and organs, causing dysfunction or death of that tissue. Malignant tumors initiated by viruses result from genetic changes, in either expression or structure of the genes, that lead to a loss of growth regulation. Abnormal masses of cells result, and these cancers or neoplasms can grow unchecked. Phosphorylation, transcription, and enzymes involved in DNA or RNA synthesis are among the functions altered by tumor viruses, and these same functions can also be altered by other physical factors such as mutations. Chemicals, diet, and environmental factors can cause genetic alterations that ultimately lead to cancer.

TABLE 14.3	Some of the major animal virus genera

Genome Type/Virus	Size (nm)	Example of Disease	
dsDNA, linear			
Poxvirus	250×350	Smallpox,	
Herpesvirus	$180 - 120$	Genital herpes, chicken pox	
Adenovirus	75	Common cold, conjunctivitis	
ssDNA, linear			
Parvoviruses	20	Erythema infectiosum	
dsDNA, circular			
Papovavirus	50	Warts	
Baculovirus	40×400	Polyhedrosis (lepidopteran insects)	
ssRNA, (+) strand			
Picornavirus	27	Polio, common cold	
Togavirus	50	Rubella	
Flavivirus	40×50	Yellow fever	
Retrovirus	80	AIDS	
Coronavirus	25	SARS	
ssRNA, (–) strand			
Orthomyxovirus	110	Influenza	
Paramyxovirus	200	Measles, mumps	
Rhabdovirus	70×170	Rabies	
Bunyavirus	90	Encephalitis	
Filovirus	50×1000	Hemorrhagic fever (Ebola)	
dsRNA			
Reovirus	65	Diarrhea	

TABLE 14.4	Some of the major plant virus genera		
Genome Type/Virus	**Size (nm)**	**Disease Symptom/Host**	
dsDNA, linear			
Caulimovirus (pararetrovirus)	50	Cauliflower mosaic	
ssDNA, circular			
Geminivirus (paired genome segments)	15×30	Maize streak	
ssRNA, (+) strand, linear			
Tobamovirus	15×300	Tobacco mosaic	
Comovirus (2 genome segments)	30	Cowpea mosaic	
Carlavirus	15×165	Carnation latent	
Cucumovirus (3 genome segments)	30	Cucumber mosaic	
Potyvirus	10×70	Potato	
dsRNA, linear			
Phytoreovirus (mutiple genome segments)	80	Wound tumor	
ssRNA, (–) strand			
Tospovirus	90	Tomato spotted wilt	
Rhabdovirus	70×170	Lettuce necrotic yellows	

14.10 Animal Viruses

As with bacterial viruses, the genomes of the viruses that infect animals are composed of either DNA or RNA (see Table 14.3). The genomes of DNA viruses are mostly double-stranded, but a few, such as the parvoviruses, are single-stranded. The genome of RNA viruses is generally single-stranded RNA (reoviruses and birnaviruses are double-stranded). The genome may be surrounded by a nucleocapsid, or an envelope of varying complexity may in turn surround a nucleocapsid. Attachment to the host cell may involve the nucleocapsid, the envelope, or the spikes that extend out from the surface of the virus. The unenveloped viruses enter a host cell by mechanisms that are not clearly understood. For example, when the poliovirus attaches to a susceptible cell, the protein capsid loses structural integrity, and the RNA protein complex is translocated into the cytoplasm. It is not clear whether the capsid enters the cytoplasm in all cases with naked capsid viruses. The attachment and penetration of the genome of enveloped viruses into the host can occur by fusion of the viral envelope with the cell membrane or by endocytosis (Figure 14.16). A limited number of representative viruses will be discussed in some detail.

DNA Viruses

Most DNA viruses replicate in the nucleus of the host cell. Poxviruses are an exception, as they replicate in the cytoplasm. Herpesviruses are dsDNA viruses that replicate in the nucleus of the host cell and are the causative agent of several human ailments, including fever blisters, shingles, genital herpes, chicken pox, and mononucleosis (see Chapter 29). The Herpes simplex virus is presented in some detail as a typical DNA virus.

Herpesviruses

The herpesvirus particle is complex, with an icosahedral capsid enclosed by an envelope bearing elaborate glycoprotein spikes that protrude outward (Figure 14.17). The herpesvirus nucleocapsid is about 100 by 200 nm in the greatest dimension. Between the envelope

Figure 14.16 Viral penetration of an animal cell

Viruses attach to and enter a host cell by (A) membrane fusion between the host cell membrane and the viral envelope or (B) endocytosis, with invagination of the host cell membrane.

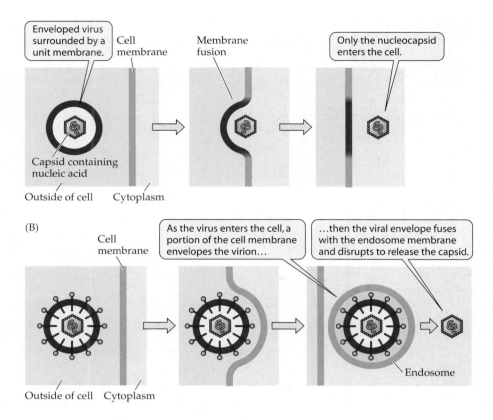

and nucleocapsid is an electron-dense amorphous material. The icosahedral nucleocapsid is composed of 162 capsomers made up of several distinctly different proteins. The genome is a dsDNA that carries genetic information for the synthesis of more than 100 different polypeptides.

The glycoprotein spikes that protrude outward from the envelope are the means of attachment to specific host-cell receptors. After fusion to the host cytoplasmic membrane, the nucleocapsid is extruded into the cytoplasm. The viral particle is uncoated, and the viral dsDNA enters the host-cell nucleus. Initially the viral genome directs the synthesis of mRNA that is translated into regulatory proteins, followed by mRNA that generates the proteins involved in DNA replication. Later mRNAs are transcribed from the viral genome and code for viral structural proteins and envelope proteins that are an integral part of the mature viral particle.

The herpesvirus DNA is synthesized in the host-cell nucleus as long concatemers. These long concatemers are processed to viral genome length as they are incorporated into the nucleocapsid. The viral nucleocapsid capsomers are synthesized in the cytoplasm, and the nucleocapsid is assembled in the host nucleus. The amorphous fibrous coat that surrounds the nucleocapsid and the envelope are acquired as the virus buds through the host nuclear membrane into the cytoplasm. The virus particle is translocated to the endoplasmic reticulum of the host cell and exits

the cell from this site. In many cases the accumulation of viruses or expression of viral proteins may cause disruption of the cytoplasmic membrane.

Some of the herpesviruses, such as Herpes simplex I, can be latent for extended periods and become active when the host is under stress (e.g., fever, sunburn). The virus remains latent in neurons of sensory ganglia and is not integrated into the host genome.

Poxviruses

The poxviruses are complex viruses that have some features of a free-living cell. One disease caused by this group, smallpox, was important historically (see Chapter 2), but it has been eradicated by worldwide vaccination.

The poxviruses are large, and one species, the vaccinia virus, is a box-like structure ($400 \times 240 \times 200$ nm) with an outer coat composed of protein filaments in an array that resembles a membrane (see Figure 29.4). The vaccinia virus is taken up by host cells via a process that resembles phagocytosis. The plasma membrane of the host cell actually extends around the virus, resulting in viral entry. The core of vaccinia virus contains the DNA genome and several viral-encoded proteins. The poxvirus replicates in the cytoplasm. As the virus does not penetrate the nucleus of the host, there are no proteins generated on infection that are transcribed from host DNA. Host genes are not expressed after infection, indicating that a molecular mes-

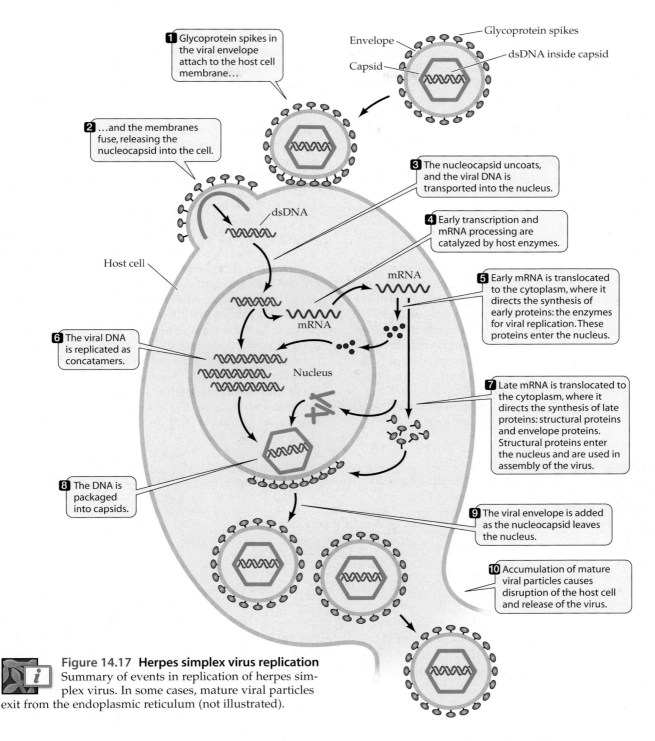

Figure 14.17 Herpes simplex virus replication Summary of events in replication of herpes simplex virus. In some cases, mature viral particles exit from the endoplasmic reticulum (not illustrated).

sage does enter the host nucleus to turn off the host genetic machinery. All of the proteins expressed postinfection are encoded in the vaccinia genome. Enzymes and DNA polymerase are also specified by the viral genome. All of the enzymes necessary to transcribe early viral DNA are brought in by the virus upon infection. The mRNA transcripts of viral origin are capped and polyadenylated even before the viral coat is completely removed. The synthesis of poxvirus DNA in the cytoplasm is unusual, as

such extranuclear DNA synthesis generally occurs only in specialized organelles such as mitochondria.

After a viral particle enters the host cytoplasm, an inclusion body is formed. If several viruses infect the host cell, there will be an inclusion body formed around each virus particle. It is in these inclusion bodies that transcription, replication, and encapsidation of mature viral progeny occur. Disintegration of the host cell leads to release of virus particles.

RNA Viruses

The genome of most RNA viruses is a single strand of RNA (ssRNA), and this strand may be the equivalent of mRNA ([+] strand) or the ssRNA may be complementary to mRNA ([–] strand) and in this case it must be copied to generate mRNA (Table 14.5). Major positive-strand RNA viruses that infect humans are polio and rhinoviruses that cause the common cold. Negative-strand viruses are the causative agents of rabies, influenza, and other human ailments. Two major diseases caused by RNA viruses— polio and HIV—will be discussed.

Poliovirus

The poliovirus has a genome composed of a single-strand of RNA. Polio was once a dreaded paralytic disease, but widespread use of a vaccine has led to its virtual elimination in the United States. The poliovirus multiplies in the intestinal tract of human beings, and infection is generally asymptomatic or causes only mild symptoms. The paralytic form, poliomyelitis, results from passage of the virus to motor neurons in the spinal cord. Destruction of these nerve cells leads to paralysis. How a limited number of infections progress from the mild intestinal tract infection to a paralytic form is not known.

The poliovirus attaches to specific receptors on cells of the host intestine and the RNA enters host cells as described in the preceding text. Following entry of the virus, all host-cell protein synthesis ceases. The viral genome is mRNA ([+] strand) and about 7,500 bases in length. The 5′ end of the RNA codes for capsid proteins, and replication proteins are coded at the 3′ end. At the 5′ terminus of the linear viral genome is a small protein attached covalently, and at the 3′ end is a polyadenylate chain. The small 5′ protein (22 amino acids) is involved in the priming of RNA synthesis during genome replication. The (+) strand is copied by an RNA-dependent RNA polymerase. The (–) strand generated then serves as a template for the production of multiple copies of the (+) strand by the same RNA polymerase. The viral mRNA is translated into a single, large protein molecule. This polyprotein contains all of the proteins involved in poliovirus synthesis. After formation, the large protein is cleaved into about 20 individual proteins. Some of these proteins are structural; others are the RNA polymerase involved with the synthesis of (–) strand templates and a protease that cleaves the polyprotein. Replication of viral RNA occurs a short time after infection by generating (+) strands from the (–) strand templates. An infected cell can contain hundreds of (–) strands that can generate up to a million viral (+) strands. The individual proteins

TABLE 14.5 Basic mechanisms used by animal RNA viruses for production of viral proteins and genomes

Genome Type	Examples of Diseases	Mechanism of Production
(+) ssRNA	Polio, common cold	Protein: —Processing→ mRNA —→ Protein Genome: —RNA replicase→ (±) dsRNA —→ (+) ssRNA
(+) ssRNA (retroviruses)	AIDS, Rous sarcoma	Protein: —Reverse transcriptase→ (–) ssDNA —Reverse transcriptase→ (±) dsDNA →mRNA→ Protein Genome: —Reverse transcriptase→ (–) ssDNA —Reverse transcriptase→ (±) dsDNA —→ (+) ssRNA
(–) ssRNA	Influenza, encephalitis, mumps	Protein: —RNA-dependent polymerase→ mRNA —→ Protein Genome: —RNA replicase→ (±) dsRNA[a]
(±) dsRNA	Encephalitis, diarrhea	Protein: —Transcriptase→ mRNA —→ Protein Genome: —RNA replicase→ (±) dsRNA

[a]The (+) strand is mRNA, the (–) strand is the genome.

cleaved from the polyprotein direct the synthesis of structural capsid proteins and these are assembled to form the mature viral particle.

Retroviruses

The retroviruses can cause important diseases in animals, and Rous sarcoma, a form of chicken cancer, was the first virally induced cancer described. The avian leukemia virus and murine leukemia virus are also retroviruses. In humans only a rare T-cell lymphoma is caused by a retrovirus. HIV is another retrovirus and is the causative agent of AIDS. HIV will be discussed in this section as a representative of these unique viruses.

The retroviruses are defined by their ability to reverse-transcribe a single-stranded RNA genome to form dsDNA. The (+) ssRNA genome is first copied into ssDNA, and this ssDNA is replicated to form dsDNA. The viral dsDNA is an intermediate that is integrated into the host genome and resides there as a **provirus**. The dsDNA can be transcribed to generate mRNA and new genome RNAs for continued virus production.

Retroviruses are enveloped viruses with a diploid single-stranded RNA genome. They are considered diploid because they carry two copies of the single-stranded RNA genome (Figure 14.18). The inner core contains the ssRNA genomes and a number of enzymes that are responsible for the early replication events. Among the enzymes present are integrase, protease, and a protein that has both reverse-transcriptase and ribonuclease H activities. This inner core is surrounded by a protein capsid. There is a protein matrix present on the inner surface of the lipid membrane, and this stabilizes the viral particle. The envelope is a lipid bilayer with spikes composed of glyco-

proteins encoded by the virus, plus numerous cellular proteins that protrude from the surface of the virion.

Infection occurs after direct exposure to the retrovirus HIV-1 through body fluids—blood, breast milk, semen, or vaginal fluids. The virus eventually encounters a cell with a high-affinity receptor for the virus surface glycoprotein. There is a highly specific binding between the glycoprotein protruding from the exterior of the virus and a surface molecule on these selected host cells. The host cell receptor is a glycoprotein present in considerable quantity on the outer surface of helper T lymphocytes, and this glycoprotein is designated CD4. The glycoprotein is also present but in smaller amounts on the surface of monocytes and macrophages. Any cell bearing the specific glycoprotein is termed CD4$^+$. Additional co-receptors, including CCR5 and CXCR4, plus minor co-receptors are also required for entry into a host cell. If the required receptors are present on a cell, HIV-1 attaches and enters the cell by fusion with the host-cell cytoplasmic membrane (Figure 14.19).

After entry into a host cell, the viral RNA is reverse-transcribed to DNA by the action of a specific RNA-dependent DNA polymerase (reverse transcriptase). The enzymes responsible are contained in the virus particle. The order of the genetic information in the HIV-1 genome is outlined in Table 14.6. Synthesis of DNA occurs in the cytoplasm, usually within the first 6 hours of infection. After reverse transcription of the ssRNA viral genome to cDNA, the RNA is destroyed by the viral ribonuclease H. The viral reverse transcriptase that reverse-transcribes RNA into dsDNA is inaccurate, and there is no mechanism for the correction of errors. Therefore, variants occur continuously, and progeny have a DNA template that varies in base sequence from that of the parent virus. This

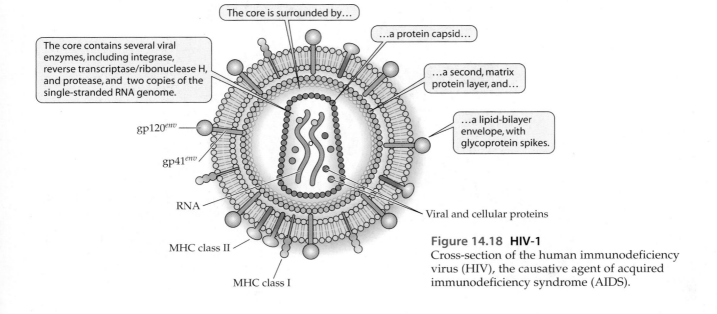

The core contains several viral enzymes, including integrase, reverse transcriptase/ribonuclease H, and protease, and two copies of the single-stranded RNA genome.

The core is surrounded by…

…a protein capsid…

…a second, matrix protein layer, and…

…a lipid-bilayer envelope, with glycoprotein spikes.

gp120^{env}

gp41^{env}

RNA

Viral and cellular proteins

MHC class II

MHC class I

Figure 14.18 HIV-1
Cross-section of the human immunodeficiency virus (HIV), the causative agent of acquired immunodeficiency syndrome (AIDS).

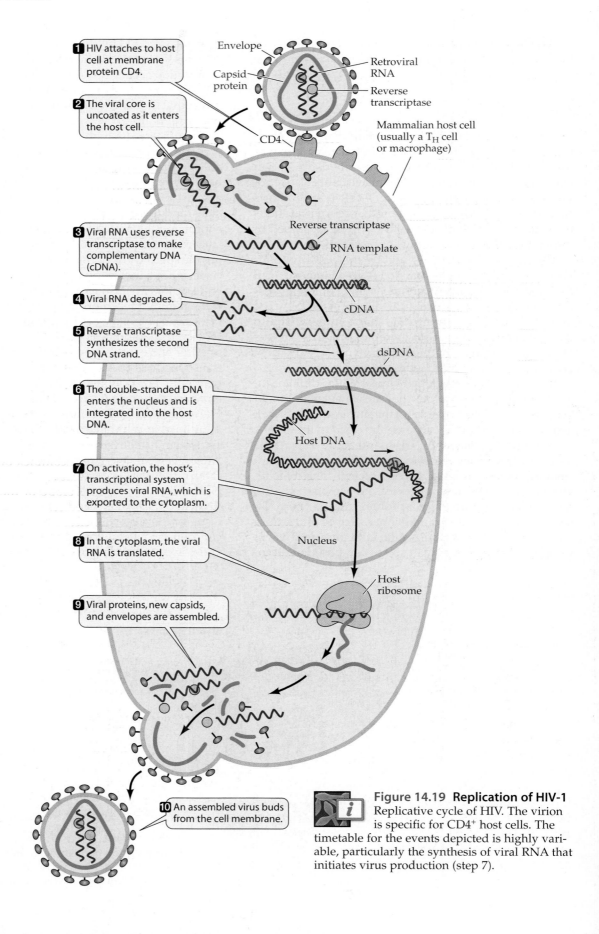

1 HIV attaches to host cell at membrane protein CD4.

2 The viral core is uncoated as it enters the host cell.

3 Viral RNA uses reverse transcriptase to make complementary DNA (cDNA).

4 Viral RNA degrades.

5 Reverse transcriptase synthesizes the second DNA strand.

6 The double-stranded DNA enters the nucleus and is integrated into the host DNA.

7 On activation, the host's transcriptional system produces viral RNA, which is exported to the cytoplasm.

8 In the cytoplasm, the viral RNA is translated.

9 Viral proteins, new capsids, and envelopes are assembled.

10 An assembled virus buds from the cell membrane.

Envelope
Capsid protein
CD4
Retroviral RNA
Reverse transcriptase
Mammalian host cell (usually a T_H cell or macrophage)
Reverse transcriptase
RNA template
cDNA
dsDNA
Host DNA
Nucleus
Host ribosome

Figure 14.19 Replication of HIV-1 Replicative cycle of HIV. The virion is specific for CD4+ host cells. The timetable for the events depicted is highly variable, particularly the synthesis of viral RNA that initiates virus production (step 7).

TABLE 14.6	HIV-1 genes and their functions

Gene Name	Function
gag	Provides structural elements of the virus: p24, the viral capsid, p6 and p7, the nucleocapsid, and p17, the matrix.
pol	Codes for enzymes: Reverse transcriptase, which transcribes the viral RNA to dsDNA Integrase, which integrates the dsDNA into host genome Protease, which processes proteins produced by *gag* and *pol* into functional units (protease inhibitors block this step)
env	Codes for the precursor glycoprotein gp160. Host enzymes cleave this precursor into two viral envelope proteins: gp120, which binds to CD4 receptor, and gp41, which assists in fusion with host cell.
tat	Compensates for defect in HIV RNA and allows transcription to occur at a rapid rate. Tat is released by infected cells and is absorbed by cells not infected with HIV, causing cell death.
rev	Provides for regulation of replication.
vif	Essential for viral replication.
vpr	Required for virus replication in nondividing cells. Induces cell cycle arrest in proliferating cells, causing immune dysfunction.
nef	Expressed early in viral infection, causing T cell activation and establishment of persistent state of infection.
vpu	Involved in viral budding and release from host cell.

variation generates progeny that may be unaffected by antibodies and the cellular arm of the immune system that would otherwise act against the progenitor virus. This would confer a survival advantage.

Although the genome of HIV-1 is RNA, the ultimate genetic information that generates viral progeny is dsDNA. This dsDNA is integrated into the host genome by the HIV integrase, where it remains as a provirus. The viral integrase serves three major functions: (1) it trims the ends of the viral dsDNA, (2) it cleaves host DNA, and (3) the integrase covalently links the termini of the trimmed DNA to host DNA. The inserted viral DNA (provirus) has all of the characteristics of a cellular gene. A treatment for HIV-1 infection that might excise the provirus is improbable.

The synthesis of the viral genome begins with an RNA transcript formed from the viral DNA template present in the host genome. Transcription rates are regulated in part by host proteins that are involved with the transcription of other host genes. Consequently, production of viral RNA transcripts can result from normal activities that occur in the host. An inactive T cell that contains the viral genome would not produce viral RNA; however, macrophages that are not dividing but contain the provirus do transcribe viral RNA and produce HIV-1.

The viral mRNA transcribed from the provirus DNA is translated to produce the capsid proteins and enzymes needed for viral assembly, and, in a spliced form, the RNA also encodes for the envelope proteins and other auxiliary proteins. The capsid proteins are synthesized as a polyprotein. The protein moves to the inner surface of the host cytoplasmic membrane, and viral RNA binds to the capsid protein precursor. The precursor then surrounds the replicative enzymes and viral RNA, and the spherical particle buds through the host cell membrane. The glycoprotein surface complex present on HIV-1 is formed in the endoplasmic reticulum of the host and is added to the outer envelope of the virus as it buds through the host cell surface.

It is not completely clear how the HIV-1 virus curtails the immune response or causes the destruction of T cells. The CD4$^+$ T cells are an essential part of the immunological response. Despite this, an HIV-1 infection will ultimately result in loss of virtually all functional T cells in the immune system. Destruction of T cells may occur by apoptosis (programmed cell death) initiated by cross-linking of molecules by gp120 (envelope protein) of the HIV virus and binding of antigen to the T-cell receptor. It is probable that a number of factors combined are involved in the destruction of T cells by infection with HIV-1, and no single factor is responsible for AIDS pathogenesis.

SECTION HIGHLIGHTS

Animal viruses may have RNA or DNA genomes and may be enveloped or unenveloped. Examples of DNA viruses are the herpesviruses and poxviruses. RNA viruses include poliovirus and retroviruses (e.g., Rous sarcoma virus, murine leukemia virus, and human immunodeficiency virus [HIV]).

14.11   Insect Viruses

The baculoviruses are one of the most studied of the insect viruses, as they are virulent and kill susceptible insect populations. Eastern tent caterpillars and the tussock moth that defoliates Douglas fir trees are highly susceptible to baculoviruses. The East African armyworm that devastates extensive grassland areas, forest sawfly, orchard codling moth, and cotton bollworm are all potentially controllable by baculovirus. Baculoviruses sprayed on crops may not kill infected insects immediately, thus allowing them to mate and spread the infection to the general population. Killing can be accelerated by introducing genes for insect-specific toxins into the virus. In theory, baculoviruses can be distributed among humans safely, as the virus has a narrow host range and is confined to insects.

Baculoviruses are excellent vectors for carrying foreign genes into insect larval cells for the creation of insect-manufactured products. Insects are an effective system for gene cloning through introduction of a gene into cells cultured from adult insects or into their larvae. These insect systems have many advantages over the use of engineered mammalian cells. Genes from plants, animals, and humans expressed in insect cells generate an amount of product far in excess of that obtained from a mammalian cell. The insect system also does posttranslational processing, as in animal cells, to yield functional proteins. The insect systems can be employed in the production of pharmaceutics, pesticides, and various proteins.

> ### SECTION HIGHLIGHTS
> Some insect viruses, for example baculoviruses, are potentially useful as insecticides, because they do not infect other animals. Baculovirus-infected insects are also important for producing recombinant eukaryotic proteins for pharmaceutical and agricultural applications.

14.12   Plant Viruses

Although the first virus described was the tobacco mosaic virus (TMV), the plant viruses have received less attention than bacterial or animal viruses. Plant viruses may well receive increased attention in the future as the world food supply dwindles and populations grow, because viruses adversely affect many crops important in human nutrition. Plant viruses have proved to be rather difficult to study, but in recent years methods have been developed for growing many of them in plant cell culture.

The plant virus structures are essentially equivalent to the bacterial and animal viruses. Some are long, flexible helices that are very narrow, but of a length approaching that of a bacterium such as *E. coli*. The clostervirus, called beet yellow, is 1.3 μm in length but exceedingly narrow (<10 nm) and cannot be visualized through a light microscope. The capsid of plant viruses is generally composed of a single protein.

The genome of most plant viruses is composed of either single- or double-stranded RNA. There are exceptions, however, and the caulmovirus genome is composed of dsDNA and that in geminivirus is ssDNA. Caulmovirus, the causative agent of cauliflower mosaic, is a pararetrovirus, as it generates mRNA that is reverse-transcribed to dsDNA by a reverse transcriptase. The geminivirus has a single (+) strand of DNA and employs host enzymes to form a dsDNA. The (–) strand produces multiple (+) strands via the rolling-circle method. These (+) strands are encapsulated to form a mature virion.

A difficulty that a plant virus must overcome in nature is in passing through the plant cell surface barriers. Plant viruses are disseminated by wind or vectors such as insects or nematodes. The insects that transmit viruses include aphids, leafhoppers, whiteflies, and mealy bugs. Viruses infect the mouthparts of these insects during feeding and are transferred to uninfected hosts during normal feeding as the insects penetrate plant cells to withdraw sap. Some plant viruses are stored in the foregut of aphids and inoculated into the plant by regurgitation during feeding. Plant viruses also enter through breaks or abrasions on plant surfaces. Seeds or pollen transmits some plant viruses. Budding or grafting to nursery stock is also a concern, as parent stock must be virus-free to ensure propagation of healthy progeny.

A select number of plant viruses propagate in both the plant and in the insects that transmit them, and in these cases the plant and insect are both host to the pathogen. The wound tumor viruses multiply in the tissue of leafhoppers before they move to the salivary glands. The virus is then transmitted by the bite of the insect. The RNA viruses that can reproduce in both plant and insect vectors produce viral genomes via a virus-carried RNA-dependent replicase. Tospoviruses (tomato spotted wilt) and rhabdoviruses (lettuce necrotic yellow) are two other examples of viruses that infect the insects that transmit them.

Virus diseases of plants are classified according to apparent manifestations of disease symptoms. These manifestations include the following:

- Mosaic diseases cause mottling of leaves, yellow spots, blotches, and necrotic lesions. These symptoms sometimes are observed on flowers or fruit. The mosaic viruses produce variegations in leaves or flowers of ornamental plants. The variegated petals observed in tulip flowers are a result of viral infection.

- A number of viruses cause leaves to curl and/or turn yellow. Dwarfing of leaves or excessive branching can result from viral infections.

- Wound tumors can occur on roots or stems.

- Wilt diseases can result in the complete wilting and death of a plant.

SECTION HIGHLIGHTS

Most plant viruses have single- or double-stranded RNA genomes, but a few have genomes composed of DNA. To infect a plant, viruses have to breach physical barriers at the plant surface, either through breaks or abrasions in the cell wall or via insect vectors that penetrate the plant during feeding. Viruses may also be spread through pollen and seeds, or by grafting. Viral diseases of plants are generally classified according to the symptoms produced.

14.13 Fungal and Algal Viruses

Plant, animal, bacterial, and archaeal viruses are transmitted horizontally as they multiply in one cell, exit, and enter another susceptible cell. Some animal viruses such as herpesvirus or HIV may be transferred to adjacent cells by cell fusion. Fungal viruses are unique in that they are spread by cell-to-cell fusion, and extracellular virions have not been observed. Latent viral infections are apparently common in fungi, and the majority of *Saccharomyces cerevisiae* strains carry dsRNA viruses. Cytoplasmic mixing during cell fusion transfers these viruses.

Killer strains occur among various genera of yeasts including *Saccharomyces*, *Hansenula*, and *Kluyveromyces*. These killer strains secrete a protein toxin that is encoded in specific regions of the genome of a latent dsRNA virus. Resistance to the toxin is also encoded in the viral genome, thereby rendering the host immune to the toxin. The killer toxin binds to specific glucans in the cell wall of nonimmune yeasts and causes cell death by creating cation-permeable pores in the cytoplasmic membrane. The killer trait (virus) can be introduced into yeast strains employed in the fermentation industry. The presence of the killer trait in a yeast that is immune

to the toxin does not affect the fermentative ability of the strain.

Filamentous fungi, such as *Penicillium chrysogenum* are also susceptible to infection by extracellular viruses. An ssRNA virus (Barnavirus) can infect the cultivated mushroom *Agaricus bisporus* and is a concern in commercial production.

Algal viruses such as *Phycodnavirus* (dsDNA) are ubiquitous in fresh water. These viruses are host-specific and attach to cell walls of susceptible unicellular eukaryotic algae. The virus causes dissolution of the cell wall at the site of attachment, and viral DNA enters the algal cell. Following virus multiplication, the viral particles are released by cell lysis.

It is apparent from this brief discussion that viruses have evolved that can infect all types of living cells. There are limited chemotherapeutic agents available for treatment of viral infections and resistance induced by vaccination is the major control measure now available (see Chapter 29).

SECTION HIGHLIGHTS

A unique feature of fungal viruses is that they are spread by cell-to-cell fusion, with no observed extracellular virions. Among the yeasts, latent viral infections can produce killer strains that secrete virus-encoded protein toxins that kill noninfected cells. Such strains are sometimes used in the fermentation industry. There are also viruses that have evolved to infect filamentous fungi and algae.

14.14 Viroids

Viroids are the smallest nucleic acid–containing infectious agents known. They were first described by T. O. Diener, a plant physiologist, in 1971. Viroids have been identified only in plants and are composed solely of RNA. The RNA is made up of 270 to 380 nucleotides (Figure 14.20). This amount of RNA would be about one-tenth the genetic material present in the smallest of the viruses that have been characterized. This number of nucleotides could theoretically produce a 100-amino-acid protein if every base were used in a single reading frame, or two proteins if a double (frameshift) reading frame were used. However, much evidence suggests that viroids do not produce any proteins. The viroid is a circular single strand of RNA collapsed into a rod by intrastrand base pairing. Viroids are not surrounded by a protein capsid. Viroids appear in the nucleus of the plant

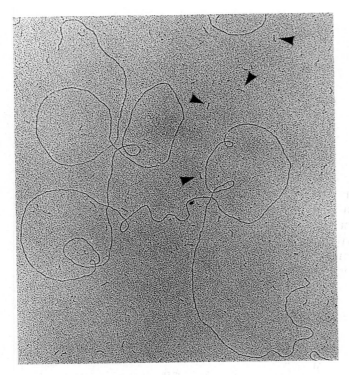

Figure 14.20 A viroid
Comparison of PSTD, a low-molecular-weight RNA viroid (arrowheads), with the DNA of T7, an *E. coli* phage. Courtesy of T. O. Diener, U. S. Department of Agriculture.

cells and apparently do not function as mRNA. A viroid is synthesized by host RNA polymerases by employing the viroid particle as a template. The polymerase of the host cell recognizes the RNA as it would a piece of DNA.

The viroid most studied is called the potato spindle-tuber disease agent (PSTD). It is 359 nucleotides in length and may function by interfering with host gene regulation. Comparison of nucleotide sequences in viroids suggests that they are similar to the RNAs of introns. They may have evolved from introns and may function by interfering with the normal splicing of introns in cells. A number of strains of PSTD have been isolated, and some cause mild symptoms in plants, whereas others cause lethal infections. The difference in virulence is due to alterations in the sequence of the nucleotides.

Viroids are the causative agent of diseases in agricultural crops, including cucumbers, potatoes, and citrus fruit, and these diseases cause losses in the millions of dollars each year. The spread of viroids by grafting in fruit trees and through the use of modern harvesting machinery may be involved in viroid transmission. Although viroids have not been identified in animal infections, hepatitis D exhibits many of the properties of a viroid. A viroid is not apparent in infected tissue without employing special techniques that involve nucleotide sequencing. Application of these techniques may lead to the identification of viroid-caused infections in humans.

SECTION HIGHLIGHTS

Viroids are composed solely of a single RNA molecule and cause many diseases in plants. The nucleotide sequences of viroids suggest that they may have evolved from introns and may cause disease by interfering with normal mRNA splicing within the host cell. No animal viroids have been found, but hepatitis D shows similarities to viroids.

14.15 Prions

Prion is an acronym for proteinaceous infectious particles. Prions have no nucleic acid, they are heat- and UV-resistant, and in the active form are somewhat resistant to proteases. Interestingly, the prion protein does not elicit an immune response during typical infections. This has hampered efforts to develop vaccines.

Prions are known to cause a number of animal diseases including: mad cow disease, scrapie in sheep, wasting disease in deer and elk, and kuru and Creutzfeldt-Jacob disease in humans (Box 14.1). These diseases are collectively termed transmissible spongiform encephalopathies (TSEs).

Evidence currently available indicates that prions cause disease by modifying a glycoprotein normally present in healthy animals. Most of this glycoprotein (PrPc) is associated with neurons and closely resembles the disease-causing prion protein (PrPsc). When PrPsc infects a susceptible animal and encounters the PrPc associated with neurons, the infectious agent apparently alters the conformation of PrPc. The altered PrPc then is self-propagating, forms aggregates, and disrupts the normal function of the neurons. Brain and nerve tissue are destroyed and death is the ultimate result.

SECTION HIGHLIGHTS

Prions are proteinaceous infectious particles that contain no nucleic acid and do not elicit an immune response. They cause a variety of neurological diseases of animals, which are collectively known as transmissible spongiform encephalopathies.

BOX 14.1 *Milestones*

Other Infectious Particles

D. Carleton Gajdusek, of the National Institutes of Health, received the Nobel Prize in 1976 for studies on the human disease kuru. Kuru was then attributed to a "slow virus" and occurred mostly in New Guinea. The disease was prevalent among women and small children. Gajdusek went to New Guinea and followed the course of kuru in the population. It was apparent that the disease was transmitted to women, as it was their task to prepare bodies of the dead for disposal. In the ritual of preparation, the women rubbed their bodies with raw brain tissue of the deceased. Gajdusek attributed the disease to protein particles that were introduced through breaks in the skin and traveled to the brain. Children were infected because they accompanied the women during the ritual. Kuru symptoms occur 1 to 15 years after infection and appear as a severe headache, loss of coordination, and a tendency to giggle at inappropriate times. There is a loss of the ability to walk and swallow, and death occurs within a year of the appearance of symptoms.

Creutzfeldt-Jakob disease (CJD) is heritable in humans and appar-ently caused by a heritable genetic defect. The prion involved in CJD also is now known to be transmitted by corneal transplantation and through the injection of human growth hormone into children who fail to grow normally. The cornea and growth hormone had been obtained from human cadavers. In recent years human growth hormone has been obtained from genetically engineered bacteria (see Chapter 31).

During the mid-1990s it became clear that a disease resembling Creutzfeldt-Jacob disease was occurring in the British Isles. The disease was ultimately traced to cattle infected with Bovine Spongiform Encephalopathy (BSE) also called "mad cow disease." Studies on the source of infection indicated that beef products were responsible. At the time, cattle were fed supplements that contained rendered cattle and sheep protein. Diseased animals, unfit for human consumption, may well have been the source of these supplements. Use of such products has been banned. Export and import of beef products were adversely affected by this discovery. Banning

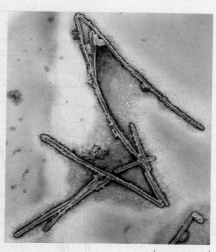

0.24 µm

Colored transmission electron micrograph of fibrils associated with scrapie, a transmissible spongiform encephalopathy which affects sheep and goats. Scrapie-associated fibrils are believed to be aggregations of a misfolded form of a protein normally present in the brain. ©Vla/Photo Researchers, Inc.

of animal products in supplements fed to cattle led to a marked decline in cases. Cattle, apparently infected before the ban, are occasionally encountered.

SUMMARY

- Viruses vary in **size** from those that are close to the size of a ribosome (polio) to viruses that approach free-living bacteria (poxviruses). Smallpox is a large virus that is visible in phase-contrast microscopy.

- A viral genome is surrounded by protein **capsid** that protects the virus when outside a host cell. Many viruses also have a membranous **envelope** that surrounds the capsid, with specific glycopro-teins (spikes) embedded in this envelope. These spikes are the site of attachment to host cells.

- Viruses grow **only within living cells**, including the cells of plants, animals, *Bacteria,* and *Archaea.* Plant and animal viruses can be grown in cell or tissue culture.

- **Quantitation** of bacteriophage can be accomplished by placing dilutions of a viral suspension on a "lawn" of the susceptible host growing on an agar surface. The phage will lyse organisms in a confined area, forming clear areas called plaques. Plant or animal viruses can be quantitated in cell culture or by treating the host with dilutions of virus and determining the highest dilution that produces manifestation of disease.

- Viruses are generally **purified** by differential centrifugation, enzyme treatment, and other procedures commonly applied to protein purification.

- The viral genome is composed of either **DNA** or **RNA**. This nucleic acid may be a single strand (ss) or a double strand (ds). A viral genome may be mRNA or transcribed to produce an early message for the proteins that initiate the infectious process.

- Some animal **tumor viruses** and the HIV-1 virus that causes **AIDS** have ssRNA genomes. There are also DNA tumor viruses. The ssRNA retroviruses are transcribed in reverse to produce dsDNA. This dsDNA is integrated into the genome of the host.

- **All types of bacteria** are subject to viral attack. There are more than 2,000 described bacterial viruses. A bacteriophage infection can result in lysis of the host, or in some the genome can be **integrated** into the host genome. The viral genome would be replicated along with host DNA and passed along to progeny without harm to the host. This is called **lysogeny**.

- A very elaborate bacterial virus is **T4**, which infects *Escherichia coli*. This virus can enter a host cell and produce up to 300 viral progeny in about 25 minutes. *E. coli* can be infected by a unique virus, **M13**, that can enter the cell, reproduce, and exit without lysis of the host.

- Viruses can cause a **lytic** infection that destroys the host cell, a **persistent** infection where the host remains viable and sheds virus for an extended time, or a **latent** infection where the virus exists in a **dormant** state.

- Viruses that infect eukaryotes transfer genetic information into the cytoplasm, and, for most DNA viruses, viral nucleic acid transcription occurs in the **nucleus**. RNA viruses and some DNA viruses (e.g., poxviruses) are transcribed in the cytoplasm. Some eukaryotic virus particles contain viral-encoded proteins that are involved in the infectious process. Enveloped animal viruses enter a cell by **fusing** to the cytoplasmic membrane or by **endocytosis**.

- The **retrovirus** that is the causative agent of AIDS attaches to cells (macrophages, T lymphocytes, and monocytes) that have a specific glycoprotein on their surface. These cells are called **CD4+** cells. The **T cells** are ultimately destroyed by the AIDS virus, resulting in a loss of immune function in the infected human.

- Herpesvirus is a **DNA virus** that can remain dormant in a host and cause an active infection when the host is under stress.

- Poliomyelitis is an infection caused by a **ssRNA virus**. The genome is actually mRNA that is copied to form an RNA template that is involved in production of viral genomes.

- Viruses enter plants through abrasions or via insects that suck sap from plants. Some plant viruses can infect the insects that transmit them as well as plant hosts. Plants resistant to viruses can be obtained through selective plant breeding.

 Find more at www.sinauer.com/microbial-life

REVIEW QUESTIONS

1. What are the extremes in the sizes of viruses and how does this compare with sizes of bacteria?

2. What are the general morphologies of viruses? Can one predict the morphology of a virus based on the host it attacks—bacteria, animals, or plants?

3. Discuss cell culture and how it might be employed to propagate viruses.

4. Why are bacterial viruses employed as a model system for the study of viral replication?

5. What are the four phases involved in viral replication?

6. How can a potential host cell protect itself from viral invasion? How can the virus overcome these barriers?

7. From an evolutionary standpoint, why would the lysogenic state be favored? Why is lambda (λ) of interest?

8. Outline the steps involved in reproduction of M13. What is the unique attribute of M13? Why is M13 an attractive virus for recombinant DNA technologies?

9. Outline the steps involved in reproduction of the virus T4.

10. Compare the mechanism whereby an animal virus and a bacterial virus attach to and enter a host cell. What difference in structure of the two cell types is important in these processes?

11. Define *lytic*, *persistent*, *latent*, and *transforming* as these terms apply to animal viruses.

12. What is a retrovirus and why have these viruses been given this name?

13. Outline the life cycle of a virus such as HIV-1. How does it attack, become integrated, and cause disease?

14. How do plant viruses spread? What difficulties do they encounter and how are these overcome?

SUGGESTED READING

Calendar, R. L. 2005. *The Bacteriophages*. 2nd ed. New York: Oxford University Press.

Collier, L. and J. Oxford. 2006. *Human Virology*. New York: Oxford University Press.

Dimmock, N. J., A. Easton and K. Leppard. 2006. *Introduction to Modern Virology*. 6th ed. Malden, MA: Blackwell Publishing, Inc.

Flint, S. J., L. W. Enquist, R. M. Krug, V. R. Racaniello and A. M. Skalka. 2003. *Principles of Virology: Molecular Biology, Pathogenesis and Control*. 2nd ed. Washington, DC: ASM Press.

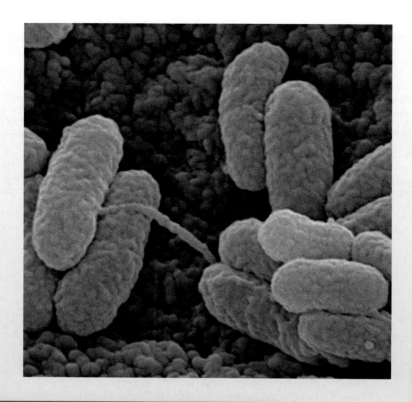

The objectives of this chapter are to:

◆ Explain the properties of accessory genetic elements such as plasmids and bacteriophages.

◆ Discuss the various mechanisms of gene transfer between bacteria.

◆ Discuss the various recombination mechanisms.

◆ Discuss the key features of mobile genetic elements such as insertion sequences and transposons.

15

Genetic Exchange

If it is ultimately proved . . . that the transforming activity . . . is actually an inherent property of the nucleic acid, one must still account on a chemical basis for the biological specificity of its action...although the constituent units and general pattern of the nucleic acid molecule have been defined, there is as yet relatively little known of the possible effect that subtle differences in molecular configuration may exert on the biological specificity of these substances.
—O. T. Avery, C. M. McLeod, and M. McCarty
(1944, original paper on DNA as the transforming principle,
Journal of Experimental Medicine *79: 137–158)*

*T*racing the effects of a mutation introduced into the genes of higher animals and plants would take a long time, as the reproductive rates of these organisms are measured in years and their numbers are comparatively sparse. Countering this inability to make quick changes in their genetic makeup, animals and plants have large and complex genomes that enable them to display an array of behaviors for coping with changes, opportunities, and threats in their environment.

By contrast, the effects of a new mutation in a bacterium can be seen in a matter of hours in the millions of cells that arise from the original. If the environment changes beyond the ability of bacteria to respond by expression of genes already in their genomes, new traits may arise that meet the needs for adaptation to the new environment through preferential growth of mutant variants present in a relatively homogeneous population.

Random mutations in a bacterial population occur at a low but constant measurable rate. Some mutations can be deleterious and lead to the loss of a functional gene, leading ultimately to the death of the cell. Even so, mutations are one of the main mechanisms by which bacteria acquire new traits and adapt to life in changing environmental niches. However, the evolution of bacteria is greatly accelerated by their ability to acquire and express entire genes that they obtain from other bacteria. This chapter discusses the mechanisms of this genetic exchange. New traits that bacteria gain by genetic exchange allow them to adapt to new environments much more rapidly than would occur by mutation alone. This adaptability is important in circumstances where newly acquired genetic information permits bacteria to survive in what previously was a potentially lethal environment. Further, the new traits may augment the metabolic versatility of the bacterium, allowing survival in an environment in which the cell previously lacked the ability to utilize available nutrients.

15.1 Plasmids and Bacteriophage

Bacterial plasmids and bacteriophage (bacterial viruses) are small genetic elements that are usually not part of the chromosome. They replicate in the bacterial cytoplasm and hence utilize the metabolic machinery and replicative apparatus of the host bacterial cell. Genes found on these extrachromosomal elements can specify functions that may influence the life of the bacterial host. In general, however, the genes encoded by these elements are not essential; that is, they are useful only when the bacterium finds itself in a specialized environment.

Plasmids

Plasmids present in bacteria are generally circular, double-stranded DNA molecules. A single bacterial cell can be completely devoid of plasmids, or it can carry many different plasmids. Plasmids are variable in size, ranging from several hundred to many thousands of base pairs. In fact, some large plasmids that are present in certain species of the genus *Rhizobium* can be considered to be mini-chromosomes encoding hundreds of genes. Although many plasmids exist, as in a single copy within a bacterial cell, some plasmids are present in several to as many as 20 or 30 copies. Many but not all plasmids can be transferred to other bacteria by a process called **conjugation** (see discussion in subsequent text); therefore, bacteria that are plasmid free can acquire plasmid-encoded genes. Usually, only a closely related member of the genus can receive plasmids, but some plasmids, called **promiscuous**, can be transferred to bacteria that are unrelated.

Plasmids Specifying Resistance to Antibiotics

Bacterial plasmids are classified on the basis of the information that is encoded in the genes of these plasmids. One very important group of plasmids is the **R factors**

(or **R plasmids**) that encode antibiotic-resistance determinants. A bacterium carrying one of these plasmids can be resistant to a specific antibiotic, whereas the member of the same species lacking the plasmid can be readily killed by the antibiotic. A number of different mechanisms exist for plasmid-encoded antibiotic resistance, the most common of which is synthesis of enzymes that destroy the antibiotic. Another example is plasmid-encoded membrane proteins that mediate active expulsion (efflux) of antibiotics from the cell. Occasionally, a plasmid can encode a product that modifies the target of the antibiotic, rendering it resistant to the antibiotic. Table 15.1 summarizes some common mechanisms of plasmid-encoded resistance.

The prevalence of R factors among pathogenic (disease-causing) bacteria is of concern to physicians who are treating infectious diseases, because it can limit the range of antibiotics that can be administered to treat a particular infection. The presence of a plasmid encoding β-lactamase in *Neisseria gonorrhoeae* requires the use of costly and less-effective antibiotics for the treatment of gonorrhea, even though previously susceptible strains had been killed by administration of a single large dose of penicillin to the infected patient. Many R plasmids can move from one bacterium to another by conjugation, thereby allowing a pathogen that is normally susceptible to therapy with an antibiotic to become resistant. A plasmid can acquire additional resistance determinants from other R factors; therefore, plasmids carrying multiple antibiotic resistance determinants are common. Figure 15.1A is a schematic representation of RK2, a plasmid that encodes resistance to tetracycline, kanamycin, and ampicillin. This plasmid is also transferable to other bacterial hosts, including most of the gram-negative bacteria. This promiscuous behavior of a plasmid implies that the plasmid or the bacterial host carrying the plasmid has the capability of transferring it to a recipient, and once in the new host, the plasmid can stably replicate.

Virulence Plasmids

Plasmids of pathogenic bacteria may encode genes that are required for virulence. Plasmids conferring pathogenic abilities on *Escherichia coli* have been well characterized. Certain strains of *E. coli* responsible for diarrheal diseases in humans and animals carry large plasmids containing genes for two types of **toxins**. The same plasmid can also carry genes for **adhesins**, surface molecules that are required for mucosal colonization. A diagram of plasmid pCG86 that encodes toxin production as well as resistance determinants to three different antimicrobial agents is presented in Figure 15.1B. Plasmids of *E. coli* that are important in virulence can encode both hemolytic factors (that is, proteins that destroy membranes of a variety of cells, including red blood cells) as well

TABLE 15.1	Common mechanisms of plasmid-encoded antibiotic resistance
Antibiotic	**Mechanism of Resistance**
β-Lactams	Synthesis of β-lactamases, enzymes that hydrolytically destroy the antibiotic.
Chloramphenicol	Synthesis of an enzyme that acylates chloramphenicol, rendering it inactive.
Aminoglycosides	Synthesis of one of several enzymes that inactivate the antibiotic by acetylation, phosphorylation, or adenylation.
Tetracycline	Synthesis of a membrane protein capable of pumping the antibiotic from the cell before it can act on the ribosomes.
Erythromycin	Synthesis of an enzyme that methylates bacterial 23S ribosomal RNA; methylated ribosomes cannot bind the antibiotic.
Trimethoprim	Synthesis of a mutant, trimethoprim-insensitive form of dihydrofolate reductase.

(A)

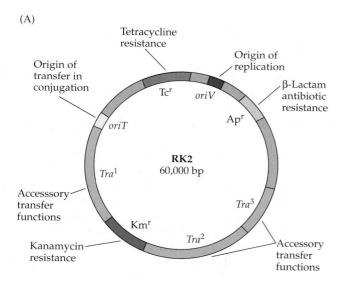

(B)

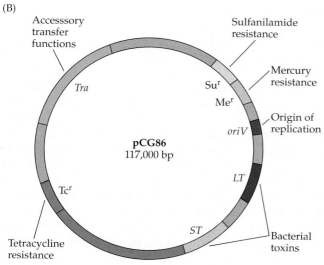

as **siderophores** (polypeptides responsible for efficient iron uptake, a metal often limiting to growth in infected tissues). Pathogenic gram-positive bacteria also carry plasmids that encode toxins; the insecticidal toxins of *Bacillus thuringiensis* are plasmid encoded, as are *Clostridium* neurotoxins. The major virulence determinants of *Bacillus anthracis* are encoded on two large plasmids: pXO1 (182,000 bp) carries the genes for the production of the multi-component anthrax toxins, whereas pXO2 (96,000 bp) contains the genes for the synthesis of a poly-D-glutamic acid anti-phagocytic capsule.

Acquisition of plasmids that carry virulence genes is considered to represent a major step in the evolution of pathogenic species from nonpathogenic, or relatively less virulent, members of closely related bacteria. *Yersinia pestis*, the causative agent of plague, is one of the most virulent human pathogens, and as few as ten microorganisms can kill an infected human. *Y. pestis* is a close relative of *Yersinia pseudotuberculosis*, a pathogen responsible for minor enteric infections. All *Yersinia* species are pathogens and cause infections of varied severity; they carry a common plasmid that encodes several secreted toxins and a dedicated secretion system. *Y. pestis* is believed to have evolved from *Y. pseudotuberculosis* approximately 10,000 years ago by acquiring two plasmids from an unknown source. One of these plasmids encodes a toxin and a capsule; both of these virulence determinants are important during infection of mammals (rodents and humans) and for dissemination via the insect (flea) vector. The other plasmid specifies proteins that promote dissemination of the bacteria from the site of infection. Yet it was the acquisition of the additional plasmids, and mutations in a handful of chromosomal genes that allowed *Y. pestis* to acquire new properties, including the ability to effectively transmit using fleas as vectors and increase its virulence in humans.

Infection of certain plants with the pathogen *Agrobacterium tumefaciens*, which carries specialized tumor-inducing (Ti) plasmids, results

Figure 15.1 Plasmids of pathogenic bacteria
Genome maps of some plasmids found in pathogenic bacteria. (A) A widely distributed R plasmid, RK2. (B) Plasmid pCG86 of pathogenic *E. coli*. The gene product or function of the genes is indicated. β-Lactam antibiotics include ampicillin.

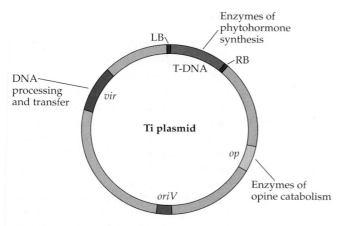

Figure 15.2 Ti plasmid
Genome map of the Ti plasmid of *Agrobacterium tumefaciens*, showing the gene product or function of the genes. The *vir* region encodes the machinery responsible for the processing of T-DNA and its transfer from the bacterium into the plant nucleus. The T-DNA encodes the phytohormone-producing enzymes. Also shown is *oriV*, the *Agrobacterium* origin of replication, and LB and RB, the two 25-bp repeats at the borders of the T-DNA.

in formation of crown gall tumors. The plant tumor results from the expression in the plant of genes encoded on the bacterial plasmid, a portion of which is transferred into the nucleus of the host where it becomes incorporated into the plant genome. A schematic representation of a Ti plasmid is presented in Figure 15.2. Two regions of this plasmid are important for virulence: the T-DNA (so named because it carries the genes required for tumor-induction in the host plant) and the *vir* genes (for *virulence*) that mediate the various stages of transfer of T-DNA to the plant by a process analogous to transfer of DNA between bacteria by the process of conjugation (see subsequent text). The T-DNA is about 30 kilobases (kb) in length and is bordered on each side by 25-bp direct repeats (LB, RB). The T-DNA is excised from the plasmid by the action of the *vir* genes and is transferred into the host genome. T-DNA genes encode enzymes responsible for the synthesis of opines, some of which are analogs of the amino acid arginine, which can be used by *Agrobacterium* as a source of energy and nitrogen. In the plant cell, T-DNA genes also direct synthesis of plant hormones (phytohormones), resulting in uncontrolled growth of the plant tissue and formation of a tumor. The *vir* genes are not transferred into the plant but nevertheless are responsible for transfer of T-DNA. They include, in addition to genes encoding the DNA transfer apparatus, regulatory genes that respond to signals from the plant, as well as genes involved in processing T-DNA prior to its transfer to the plant cell.

Another group of plasmids can direct synthesis of proteins that are toxic to other bacteria. These bactericidal proteins, called **bacteriocins**, are encoded by specialized plasmids termed **bacteriocinogenic plasmids**. Commonly, the names of the plasmid and its toxic protein are derived from the genus or species name of the host bacterium. For example, *E. coli* can carry **colicinogenic plasmids**, which encode a variety of **colicins**. Colicins kill other bacteria through a variety of mechanisms, ranging from direct damage to the cell membrane to the destruction of ribosomes. Bacteria carrying bacteriocinogenic plasmids protect themselves from the lethal bacteriocins by expressing **immunity determinants**, which are coded on the same bacteriocinogenic plasmids. The bacteriocin-immunity system thus allows bacteria to survive while killing other bacteria that may compete with them in an environment of limited nutrients.

Plasmids Encoding Genes for Specialized Metabolism

Certain metabolic functions can also be encoded by plasmids, including the biodegradation of complex organic molecules by some species in the genus *Pseudomonas*. Large plasmids that encode catabolic pathways for aliphatic or aromatic compounds, such as toluene, naphthalene, chlorobenzoic acid, octanes, and decanes have been described in recent years. The organization of genes on one such degradative plasmid is presented in Figure 15.3 together with the enzymes of naphthalene and salicylate oxidation. This plasmid codes for the enzymes that catabolize naphthalene, leading to pyruvate and acetyl aldehyde. Such degradative functions encoded on plasmids allow the bacterium to utilize unusual organic compounds as their sole carbon and energy source. Because many environmental pollutants are complex organic molecules, bacteria that carry degradative plasmids are prime candidates for use in bioremediation (see Chapter 31).

Plasmid Replication, Maintenance, and Partitioning During Cell Division

Plasmids can replicate by one of three mechanisms. Starting from the origin of plasmid replication (*oriV*), the two replicative forks can move bidirectionally, until they meet at a site opposite to the origin, analogous to the replication of circular chromosomes. The other two mechanisms, described in Chapter 13, are represented by the unidirectional movement of the fork around the plasmid, terminating at the origin, and the rolling circle mechanism. These latter two replication mechanisms are unique to bacteria and the genomes of certain bacterial viruses. In addition to the host replicative appara-

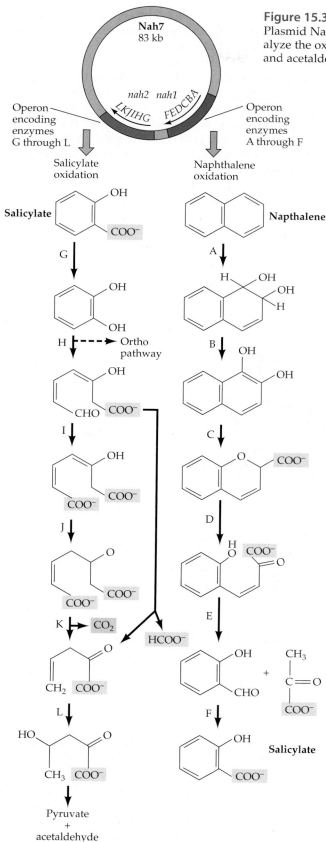

Figure 15.3 Plasmid-encoded degradation of aromatic compounds
Plasmid Nah7, harbored by *Pseudomonas putida*, encodes enzymes that catalyze the oxidation of naphthalene to salicylate, and salicylate to pyruvate and acetaldehyde.

tus, most plasmids also require expression of plasmid-encoded genes. These genes code for various enzymes that assist in DNA replication, such as unwinding and separating strands and ensuring correct partition of replicated plasmids into daughter cells.

Although a bacterial cell can carry several plasmids, not all plasmids can coexist in the same cell. Plasmids that use identical or closely related replication mechanisms are mutually incompatible in the same cell and, therefore, belong to the same **incompatibility group**. If a cell receives two plasmids of the same incompatibility group by any one of the genetic exchange mechanisms, these two plasmids will segregate during cell division such that each daughter cell will have only one of the plasmids. If there are multiple copies of each incompatible plasmid, it may take several cycles of cell division before the plasmids are completely segregated—that is, before each daughter cell contains only one type of plasmid from the incompatibility group. This property allows yet another method of classifying plasmids—on the basis of plasmid compatibility. The mechanism of incompatibility is not well understood, but for a number of plasmids it appears to be directed by a control system that coordinates plasmid replication with cell division.

Occasionally, plasmids can be lost by a bacterium, and this may occur naturally during cell division when both copies of a replicated plasmid end up in one daughter cell and the second daughter cell does not receive a plasmid. The daughter cell without a plasmid is said to be **cured**. Environmental stresses, such as exposure to extreme temperatures, nutritional limitation, or treatment with chemicals that interact with DNA, greatly stimulate curing. If the bacterium is in an environment where it depends on the expression of a plasmid-encoded gene, such as in the presence of antibiotics, the cured daughter cell will not survive.

However, loss of plasmids during cell division by imprecise distribution following their replication is usually a rare event. The accuracy of this process is assured by the various partitioning systems that ensure placement of at least one copy of the replicated plasmid into daughter cells prior to formation of the division septum. These partitioning systems have been studied most extensively in low-copy number plasmids; there is evidence that even those plasmids that exist in tens of copies per cell do not portion their plasmids randomly, but their localization into daughter cells is directed by plasmid-encoded determinants and host factors.

Although there are subtle differences between the various plasmid partitioning systems, their components (the Par proteins) are highly conserved among a wide range of plasmids. The components of the partitioning complex are encoded by a pair of linked *par* genes and an adjacent DNA site. The most extensively studied portioning system is that of plasmid R1 and the model for partitioning of this plasmid is shown in Figure 15.4. One of the Par proteins, ParM, is an ATPase related to the actin family of eukaryotic proteins, capable of forming polymerized helical filaments. It functions in conjunction with ParR, a DNA-binding protein that binds specifically the adjacent *parC* sequence. Typically, between 6 and 8 ParR molecules bind to an equal number of repeated sequences within *parC*. Following completion of plasmid replica-

tion, the plasmids are joined by interactions between some of the dimeric ParR molecules and *parC* sequences located in different plasmids, in what is called the partition complex. This complex triggers polymerization of ParM, and it binds one end of the ParM-containing filament. ParM polymerization continues at the site of the ParR–*parC* complex, the so called "head" of the filament, and entry of a ParM monomer requires hydrolysis of ATP within the ParM molecule bound to ParR–*parC*. Polymerization of the filaments leads to the separation of the plasmids to the opposite poles of the cell. At the distal "tail" end of the filament, ATP is also hydrolyzed within the ParM protein; however, since no new ParM monomers are added, the filament progressively dissociates. Competition for the *parC*-like sites between Par proteins from

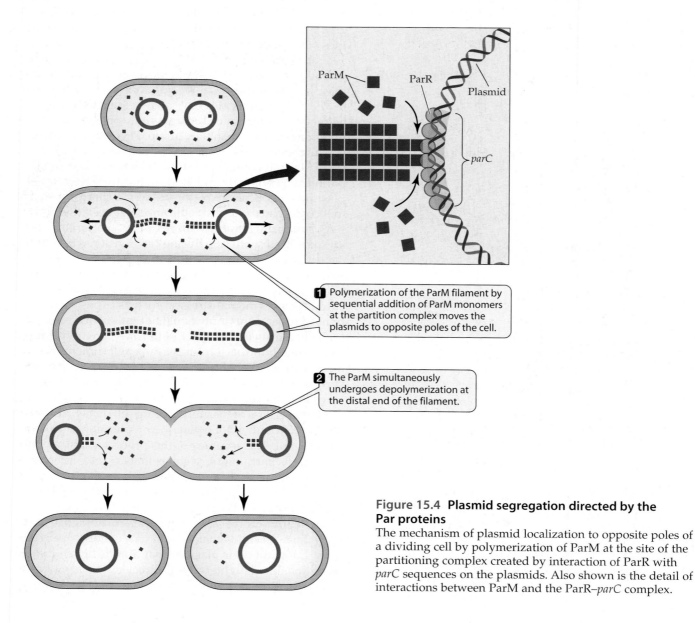

1 Polymerization of the ParM filament by sequential addition of ParM monomers at the partition complex moves the plasmids to opposite poles of the cell.

2 The ParM simultaneously undergoes depolymerization at the distal end of the filament.

Figure 15.4 Plasmid segregation directed by the Par proteins
The mechanism of plasmid localization to opposite poles of a dividing cell by polymerization of ParM at the site of the partitioning complex created by interaction of ParR with *parC* sequences on the plasmids. Also shown is the detail of interactions between ParM and the ParR–*parC* complex.

related plasmids is believed to be one of the bases of plasmid incompatibility.

Bacteriophages

Viruses are obligate parasites that can propagate only inside host cells. Virtually every type of living cell can be infected by one or several kinds of viruses, and bacteria are no exception. Although viruses can self-replicate, they do so only when inside a host cell, parasitizing the host cell's metabolic machinery. Outside a host, viruses are inert molecules of nucleic acids surrounded by protein or lipid envelopes. For these reasons, viruses are not considered living organisms, although they are generally discussed in the context of their host. Bacterial viruses utilize the metabolic and biosynthetic machinery of their host, having replication and gene regulation patterns similar to those of *Bacteria* or *Archaea*, whereas viruses that infect eukaryotic cells follow the patterns of gene expression that are seen in eukaryotic cells. The replication of viral nucleic acid and assembly of a virus is considered in Chapter 14.

SECTION HIGHLIGHTS

Although most of the essential genes are found on the bacterial chromosome, additional genes for a variety of ancillary cellular functions are encoded by extrachromosomal genetic elements called plasmids. R factors (R plasmids) carry antibiotic resistance genes, and virulence plasmids carry genes that encode virulence determinants, such as adhesins and toxins. The plant pathogen *Agrobacterium tumefaciens* causes tumors in plants by transferring a portion of its plasmid (called tumor-inducing, or Ti plasmid) to the plant, where the Ti-encoded genes induce synthesis of plant hormones. Plasmids replicate independently from the chromosome, and their correct portioning into daughter cells is directed by the action of two proteins (ParM and ParR) acting on a specific site on the plasmid called *parC*.

15.2 Recombination

The replication of genetic material is controlled by a set of enzymes that, by minimizing changes in the specific DNA bases and the order of blocks of genes in the chromosome, ensures the accurate transfer of genetic information to daughter cells. Errors or mutations can have potentially lethal consequences, yet they are also responsible for adaptive evolution to novel functions. A gene can acquire new properties by mutation and selection. An altered gene can arise by duplication of a related gene and extensive mutation in one of the two genes. Another very important mechanism in bacterial evolution is the acquisition of novel genetic information from another bacterium. Rearrangement of blocks of DNA within the bacterial genome and integration of newly acquired genetic information following genetic exchange is a complex process. These processes involve specialized enzymes as well as components of the DNA replication and repair machinery. Orderly rearrangement of DNA, called genetic recombination, is a process that occurs in both prokaryotic and eukaryotic cells. Recombination involves cutting and ligating DNA molecules, replacing one block of DNA with another. Recombination can take place between any two separate DNA molecules, such as plasmids, plasmids and chromosomes, and genomes of bacterial viruses. In addition, two segments of DNA within the same molecule can recombine.

Genetic recombination can be divided into two classes: general and site-specific. The former process refers to recombination between any two DNA molecules, provided they share relatively long homologous (that is, matching or very similar) base sequences at the site of recombination. The latter process involves recombination between highly specific and relatively short sequences. The two classes of recombination also differ in their utilization of specific enzymes that catalyze cutting and ligation of the DNA strands undergoing recombination.

General (Homologous) Recombination

General (homologous) recombination is a mechanism of genetic rearrangement involving DNA sequences that share significant sequence similarity. The key protein that controls general recombination is encoded by the *recA* gene. This type of recombination is also called **crossover recombination**, because the event appears to be a simple one-step breakage of two double-stranded DNA molecules and ligation across each strand as illustrated in Figure 15.5. At the molecular level, this process is much more complex. It can be divided into two stages: formation of a hybrid structure between two DNA molecules followed by their resolution into free DNA molecules. These steps are illustrated in Figure 15.6.

The recombination process between two double-stranded DNA molecules requires a nick, or a single-stranded gap, in one of the two DNA duplexes. The nick can be formed by physical damage or by an enzyme complex consisting of RecB, RecC, and RecD, which un-

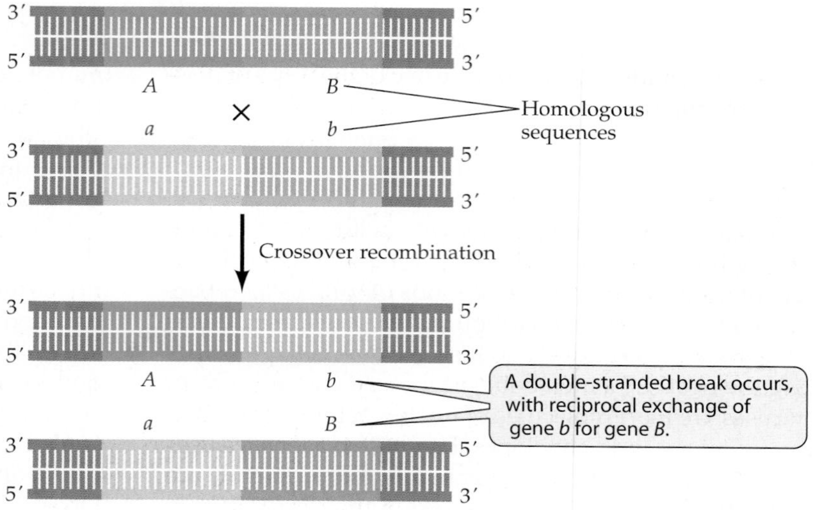
3′
5′
5′
3′
A
B
×
a
b
Homologous sequences
3′
5′
5′
3′
Crossover recombination
3′
5′
5′
3′
A
b
a
B
3′
5′
5′
3′
A double-stranded break occurs, with reciprocal exchange of gene *b* for gene *B*.

Figure 15.5 Homologous recombination Exchange of two DNA segments by homologous recombination.

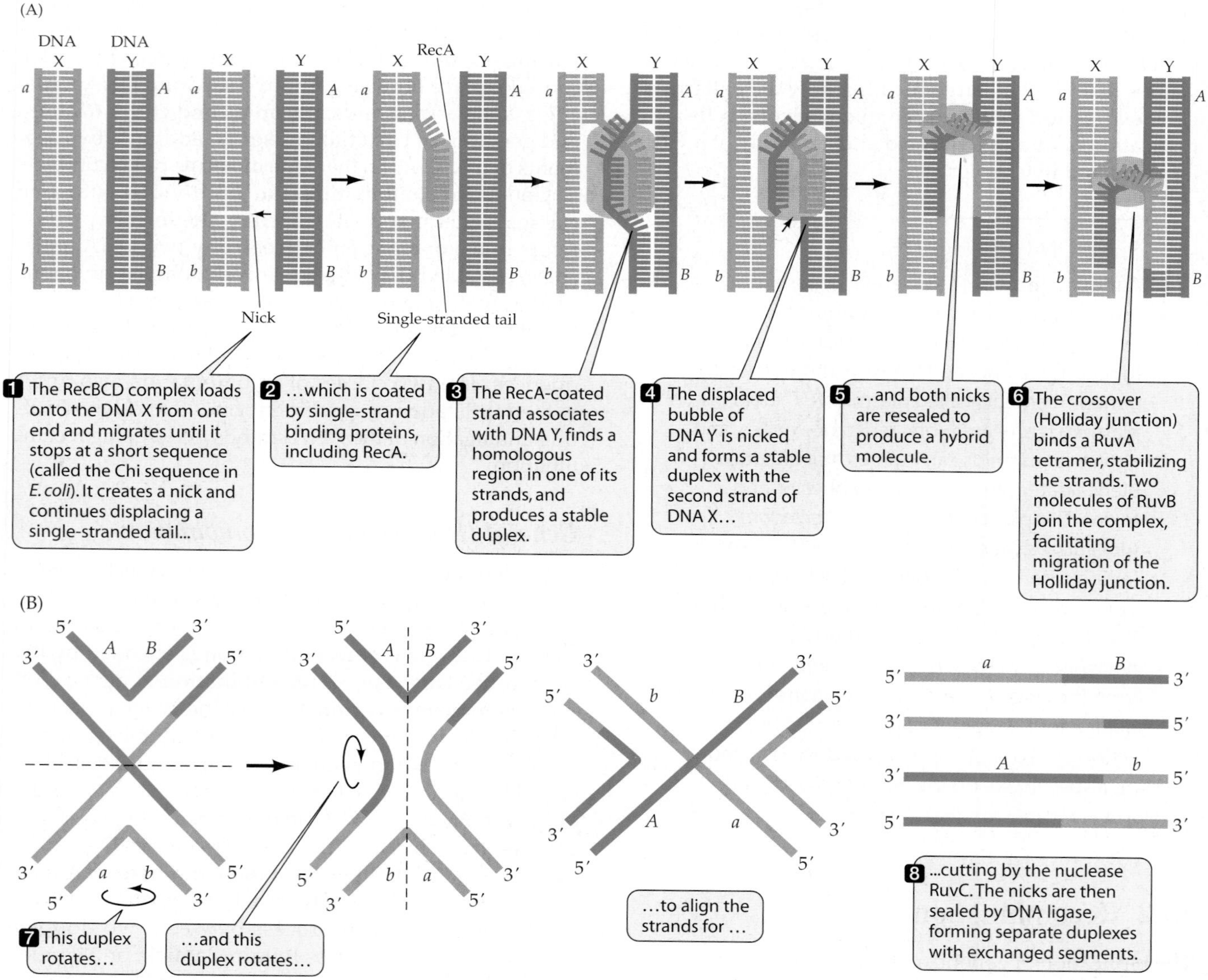
(A)
DNA X
DNA Y
a
A
b
B
Nick
RecA
Single-stranded tail
X
Y
1 The RecBCD complex loads onto the DNA X from one end and migrates until it stops at a short sequence (called the Chi sequence in *E. coli*). It creates a nick and continues displacing a single-stranded tail...
2 …which is coated by single-strand binding proteins, including RecA.
3 The RecA-coated strand associates with DNA Y, finds a homologous region in one of its strands, and produces a stable duplex.
4 The displaced bubble of DNA Y is nicked and forms a stable duplex with the second strand of DNA X…
5 …and both nicks are resealed to produce a hybrid molecule.
6 The crossover (Holliday junction) binds a RuvA tetramer, stabilizing the strands. Two molecules of RuvB join the complex, facilitating migration of the Holliday junction.
(B)
5′
3′
A
B
a
b
7 This duplex rotates…
…and this duplex rotates…
…to align the strands for …
8 ...cutting by the nuclease RuvC. The nicks are then sealed by DNA ligase, forming separate duplexes with exchanged segments.

Figure 15.6 Details of homologous recombination Molecular details of homologous recombination between two double-stranded DNA molecules, designated here as X and Y. (A) Formation of the hybrid molecule and (B) its resolution.

winds DNA and cuts one strand at a specific short sequence. In *E. coli*, it is the sequence GCTGGTCG, which is called the Chi site. A single-stranded tail is formed, which becomes coated by single-stranded DNA-binding proteins, including RecA. The bound form of RecA is now capable of aggregating with double-stranded DNA, searching for homology and invading the duplex of the second DNA molecule, thereby allowing formation of a stable duplex from the two homologous strands. The displaced "bubble" of the invaded strand is then nicked and forms a stable double-stranded structure with the other strand, resulting in a transient hybrid

molecule. The nicks in both of the strands are resealed. The final stage in recombination is **isomerization**, which leads to alignment of the two remaining outside strands across each other. Isomerization involves simultaneous rotation of the two strands, followed by cutting and ligation of the crossed strands. This process effectively exchanges a segment of two DNA strands.

Now consider some of the DNA rearrangements that can result from homologous recombination. When two plasmids, each containing similar DNA sequences, recombine, the result is a single, large, composite plasmid (Figure 15.7A). A similar process involving chromoso-

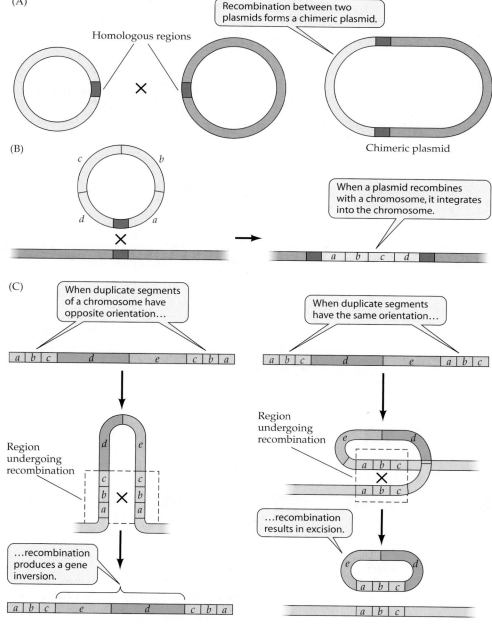

Figure 15.7 Types of homologous recombination in bacteria
Homologous recombination between (A) two plasmids, (B) a plasmid and a chromosome, and (C) two segments of a chromosome containing duplicated regions *a, b, c.*

mal sequences with sequences on a plasmid leads to incorporation of the plasmid into the bacterial genome (Figure 15.7B). The outcome of homologous recombination within a contiguous sequence (chromosome or plasmid) depends on the relative orientation of the homologous sequences (Figure 15.7C). If such sequences are in an opposite orientation, the reciprocal exchange results in inversion of the intervening region. If the sequences are in the same orientation, the result of general recombination is excision of a circular DNA consisting of the intervening region and potential loss of its genetic information during subsequent cell division.

Site-Specific Recombination

Site-specific recombination between two DNA segments requires relatively short regions of homology between the base pairs involved in the exchange of DNA strands.

The process is called site-specific because it takes place between specific sites on the DNA, as defined by their nucleotide sequence. Moreover, site-specific recombination requires a specialized set of enzymes that recognizes these sites and mediates the recombination process. This form of recombination mediates a number of important cellular processes, including integration of viral genomes into the bacterial chromosome, insertion of specific genes by a transposition mechanism (see following), and expression of genes by inversion of promoter or regulatory sequences.

The mechanism of inversion of DNA resulting in expression, or lack of expression, of specific genes has been extensively studied in a number of microorganisms. One of the best-understood examples is flagellar **phase variation** in the flagellin subunit proteins by *Salmonella typhimurium* (Figure 15.8). This bacterium alternates between expression of either one of the two possible

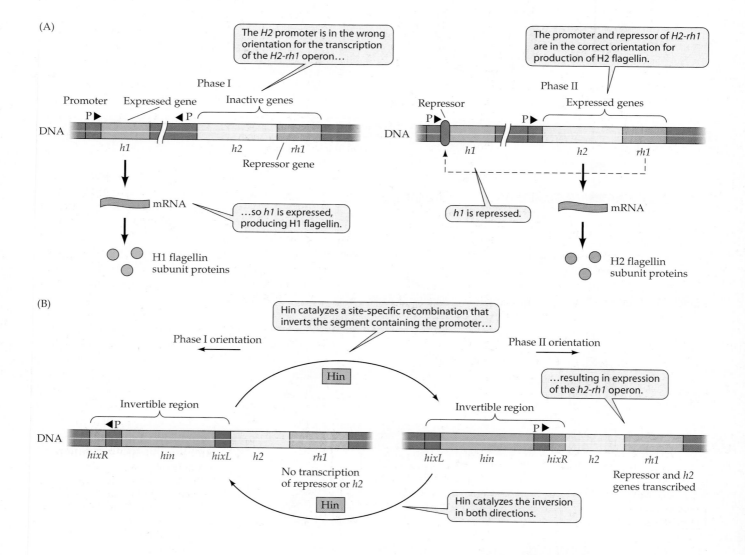

immunologically distinct flagellin proteins (H1 and H2) encoded by two different genes, thus avoiding an immune response to the flagellin expressed by the infecting organism. This cleverly designed molecular circuit not only ensures that one of the flagellin genes is expressed, but also prevents simultaneous expression of the other gene. The expression of the gene for H2, and repression of the gene specifying H1, is accomplished because expression of H2 is linked to the expression of another protein, a repressor, which prevents expression of H1. The repressor gene *rh1* (repressor of *h1*) is a second gene of the two-gene operon, transcribed from a single promoter preceding the gene for H2. The expression of the *h2–rh1* operon is controlled by an invertible 995-bp DNA segment, flanked by short, 26-bp inverted repeat sequences (*hixL* and *hixR*); the end nearest to the operon contains its promoter. At a regular frequency, the 995-bp segment is inverted, resulting in the promoter being oriented away from the genes for H2 and the repressor. The consequence of this rearrangement is concomitant loss of expression of *h2* and lack of synthesis of the repressor. In the absence of the repressor protein, the gene for H1 is now expressed, and these new subunits are assembled into flagella. The inversion process can be reversed (going from *h1* expression to *h2*) by the reversal of the same rearrangement, allowing for synthesis of H2 as well as the repressor of *h1*.

The molecular details of this type of site-specific rearrangement have been worked out (see Figure 15.8). The key mediator of this process is Hin, a recombinase enzyme the gene of which (*hin*) is encoded within the 995-bp invertible segment. Hin mediates this type of site-specific recombination by binding to the ends of the invertible segment, within *hixR* and *hixL*, where with the aid of several ancillary proteins, the recombination complex, together with bound DNA, assembles. Within this protein complex, the DNA is cleaved and the ends that form the opposite ends of the segments are ligated together. The consequence of inversion of the segment is a reversal of the sequence between the *hixR* and *hixL* sequences, thereby moving the promoter away from its site adjacent to the *h2* gene. Mechanistically similar reactions are responsible for expression of genes that encode receptors for certain bacterial viruses, and those specifying pili in a number of bacterial species.

◀ **Figure 15.8 Flagellar phase variation**
Phase variation during expression of *Salmonella typhimurium* flagellar proteins. (A) Overall scheme for expression of H1 and H2 flagellins and control of the *h1* gene by the repressor. (B) Details of the inversion process. Hin mediates inversion in both directions, so cells alternate between H1 and H2 flagellin production.

SECTION HIGHLIGHTS

Homologous recombination involves breakage and joining of two DNA strands along sequences of identity, or a high degree of similarity. Homologous recombination utilizes a DNA site that possesses a nick, generated by the **RecB**, **RecC**, and **RecD** nuclease complex, at a sequence called the **chi site**. Another key component, required for homologous recombination, is the **RecA** protein, which binds to single-stranded DNA and then identifies homologous regions in the opposite strand. Another complex of proteins, **RuvA** and **RuvB**, stabilizes a critical intermediate called the Holliday structure. These proteins also promote migration of Holliday junctions and alignment of the DNA strands for cutting by the **RuvC** protein, separating the two strands. Recombination is site-specific because it takes place between two specific DNA sequences and requires enzymes that promote recombination between these particular sequences.

15.3 Genetic Exchange among Bacteria

Bacterial recombination via a general or site-specific mechanism is most easily manifested following acquisition of foreign DNA by a bacterium. This is a normal process in higher eukaryotes in which sexual reproduction is initiated by fertilization of an egg, which brings together genetic material from two different sources. Bacteria also have the capacity to acquire DNA from other bacteria, but it is not tied to cell division. Genetic exchange in bacteria is an important vehicle for evolution and rapid adaptation to a new environment.

Bacteria can acquire novel genetic information, that is, new segments of DNA, by three different mechanisms: **conjugation**, which is a plasmid-mediated transfer of DNA from one bacterium to another; **transformation**, in which a bacterial cell takes up free DNA from its environment; and **transduction**, in which a bacteriophage carries bacterial DNA from one cell to another. The ability to detect microorganisms that have received new genetic information by any one of these methods depends on expression of the gene in the new host. Often the genes become part of the bacterial genetic repertoire, and the trait thus gained is inherited by daughter cells after cell division.

BOX 15.1   *Milestones*

Sex Among Bacteria

The discovery of transformation by Avery and his colleagues, as a means of bacterial acquisition of genetic information, raised the possibility that bacteria can acquire traits by more conventional (i.e., sexual) means. In 1946, Tatum and Lederberg demonstrated the exchange of genetic information among bacteria. This simple experiment, outlined in part A of the figure, demonstrated transfer of genes between two strains of *E. coli,* each carrying two mutations in biosynthetic genes. One of the strains required both biotin and methionine for growth on minimal medium, whereas the second was a threonine- and proline-requiring double mutant of *E. coli.* When a culture of these two mutant strains was mixed and plated onto minimal medium, a few bacteria grew that did not require any of the supplements that characterized the mutations of the parental strains. When a comparable number of each of the *E. coli* mutants was plated onto minimal medium, no wild-type bacteria were ever recovered. This was the first demonstration that recombina-

Demonstration of (A) recombination among bacteria and (B) the need for intimate contact for genetic recombination to occur.

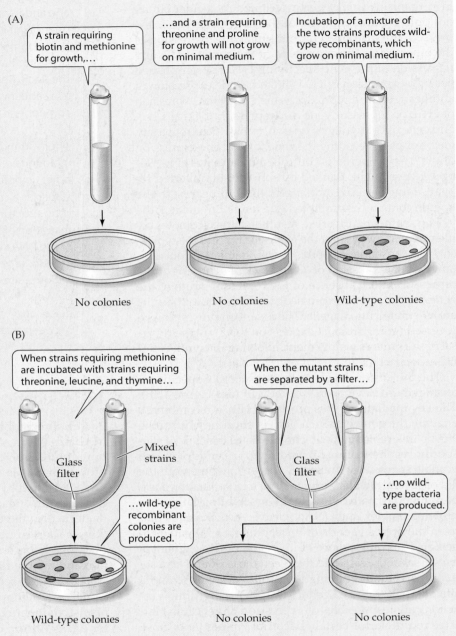

(A)

A strain requiring biotin and methionine for growth,…

…and a strain requiring threonine and proline for growth will not grow on minimal medium.

Incubation of a mixture of the two strains produces wild-type recombinants, which grow on minimal medium.

No colonies   No colonies   Wild-type colonies

(B)

When strains requiring methionine are incubated with strains requiring threonine, leucine, and thymine…

When the mutant strains are separated by a filter…

Glass filter   Mixed strains

Glass filter

…wild-type recombinant colonies are produced.

…no wild-type bacteria are produced.

Wild-type colonies   No colonies   No colonies

Conjugation

Conjugation is a mechanism of DNA exchange mediated by plasmids or genetic elements referred to as conjugative transposons. Some but not all plasmids are **transmissible**, which means they can promote their own transfer as well as mediate the transfer of other plasmids and even portions of chromosomal DNA. Bacteria that contain transmissible plasmids are called **donor** cells;

those cells that receive plasmids are the **recipients**. Bacterial conjugation is superficially analogous to sexual transfer of genetic information in higher eukaryotes from male to female members of the species; the donor bacteria are often called male cells, whereas recipients are called females (Box 15.1).

Plasmids—such as the F plasmid of *E. coli* or the large R plasmids of a variety of other bacteria—are circular

BOX 15.1   *Continued*

tion can occur between two bacteria, even in the absence of apparent lysis and release of DNA, because DNase present during the coincubation of the two bacterial strains failed to block recombination.

The additional evidence that the observed recombination requires cell-to-cell contact was provided by Davis in 1950. His experiment, shown in part B of the figure, utilized a U-shaped tube that was separated by a fritted glass filter, the pores of which were large enough

to allow passage of liquids but not of bacteria. One side of the tube was inoculated with an *E. coli* strain that required threonine, leucine, and thymine, whereas the other side contained a methionine-requiring mutant of *E. coli*. In a separate experiment, the two bacterial mutants were mixed in one of the arms of the tube. The medium was flushed back and forth and incubated for several hours. Bacteria were recovered from the tube and plated onto minimal media.

The tube that contained a mixture of bacteria yielded several wild-type recombinants, but the tube in which bacteria were separated failed to give rise to any recombinants capable of growing on media lacking the appropriate nutritional supplements. Hence, genetic exchange by intimate contact was unambiguously demonstrated, which was subsequently shown to be mediated by plasmids mobilizing DNA from donor (male) to recipient (female) cells.

---

molecules of DNA. These extrachromosomal genetic elements encode a number of genes responsible for their own replication and maintenance within the bacterial cell. A significant number of cellular genes are dedicated to the promotion of DNA transfer via conjugation. These transfer functions include several genes that encode the pilus components as well as proteins involved in the regulation of expression and biogenesis of the pilus. The pilus is an organelle formed on the surface of a plasmid-bearing donor cell that recognizes the recipient and makes the initial contact. Each given plasmid typically expresses a distinct pilus type. Another key component of the transfer process is the actual conjugative pore—through which DNA is transferred—formed at the junctions of the two bacterial cell envelopes. For many conjugative plasmids, the conjugative pore is a structure resembling, and very likely functioning, as a specialized protein secretion apparatus (the so called type IV secretion system, see Chapter 11).

The process of conjugation can be divided into several steps, which are outlined in Figure 15.9. Conjugation is initiated by recognition of the recipient cell by the pilus of the donor, resulting in pilus retraction, which brings the two bacterial cells into a tight contact, resulting in formation of a stable mating pair and the establishment of a conjugative pore. DNA transfer begins with the interaction between the origin of transfer (*oriT*), and several proteins, most prominent of which is **relaxase**. This DNA-protein complex is called the relaxosome. One of the functions of relaxase is to nick the DNA within *oriT*, attach covalently to the free 5'-phosphate of the nicked strand, and guide the single-stranded DNA molecule to the **coupling protein** (the DNA pump) at

the cytoplasmic membrane of the donor's side of the conjugative pore. The DNA–relaxase complex is then transferred by the coupling protein to the pore, possibly on the recipient's side, with concomitant replacement of the transferred strand in the donor. This is highly similar to rolling circle replication (see Chapter 13 for details), with relaxase performing many of the functions of the Rep protein. Given the similarity of the conjugative pore to the type IV secretion apparatus, it is conceivable that the transfer machinery functions exactly as a protein-secretion apparatus that transports relaxase (and the attached DNA) from the donor cell into the recipient. Within the recipient cell, relaxase permits formation of a single-stranded circular molecule with a release of the primase. During the transfer of DNA, some plasmids also transfer a specific primase, which is responsible for the synthesis of RNA primers for initiation of synthesis of the complimentary strand of the transfer DNA molecule.

The conjugative transfer described for F plasmids or certain R plasmids requires that the plasmid contain all of the information for its own transfer. These plasmids are called **self-transmissible** because they can enter a recipient using entirely their own transfer machinery. The transfer of a self-transmissible plasmid into a recipient converts that recipient to a donor because by inheriting a self-transmissible plasmid, it acquires the ability to conjugate with other recipients. Not all plasmids are self-transmissible. Some plasmids lack a number of genes that encode the transfer functions, rendering them incapable of transfer to another cell. These plasmids, however, need not be restricted to a permanent residence in a bacterial cell. Because a bacterium can contain more

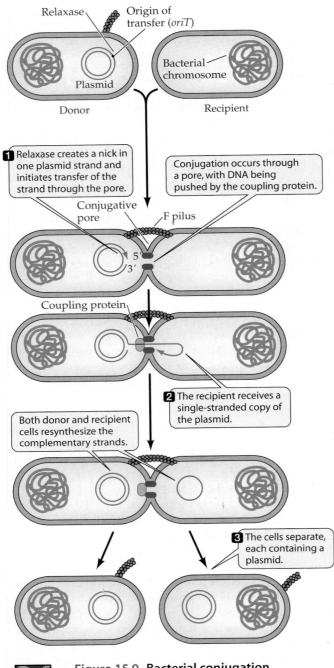

Figure 15.9 Bacterial conjugation
Transfer of F plasmid from donor to recipient cells by conjugation. Once transfer is complete, both cells have an intact copy of F plasmid and can act as donors.

1 Relaxase creates a nick in one plasmid strand and initiates transfer of the strand through the pore.

Conjugation occurs through a pore, with DNA being pushed by the coupling protein.

2 The recipient receives a single-stranded copy of the plasmid.

Both donor and recipient cells resynthesize the complementary strands.

3 The cells separate, each containing a plasmid.

than one type of plasmid, a non–self-transmissible plasmid may be able to move to a recipient cell using conjugal functions provided by other plasmids in the donor cell. For example, a plasmid lacking a certain transfer function, such as pili, can still transfer to a recipient cell,

provided the donor cell has another plasmid that encodes pili. This "helper" plasmid thus is capable of mobilizing another plasmid for transfer. Note that, unlike the case of self-transmissible plasmids, the recipient receiving such an incomplete plasmid cannot be a donor unless it already contains a helper plasmid.

Plasmids lacking *oriT* are not transmissible, even with helper plasmids. However, these plasmids can encode transfer functions and can act as helper plasmids for other transfer-deficient plasmids. One way a plasmid lacking *oriT* or any additional transfer functions can transfer from one cell to another is by "hitchhiking" a ride on a transmissible element. This can be accomplished by having some homologous sequence on both plasmids such that a homologous recombination event fuses the two plasmids into a **chimeric** plasmid (that is, from two sources). The transfer of the chimeric plasmid is originated from the *oriT* of the transmissible plasmid component.

Plasmid-Mediated Exchange of Chromosomal Genes

Transfer of genes among bacteria, mediated by self-transmissible plasmids, is not restricted to genes that are located on plasmids themselves. A number of conjugative plasmids have the ability to transfer chromosomal genes as well. This property was first discovered in the *E. coli* F plasmid. However, it is now apparent that similar mechanisms of chromosomal transfer exist among several R plasmids. The ability to mediate the transfer of chromosomal genes is referred to as the **chromosome mobilizing ability**.

The F plasmid can mediate transfer of chromosomal genes by two related mechanisms, both of which require formation of **high-frequency recombinant cells (Hfr)** in which the F plasmid has integrated, by recombination, into the chromosome (Figure 15.10). Integration is not random; a few preferred sites exist in the *E. coli* chromosome because recombination requires that the plasmid and the chromosomal site share the same DNA sequence. The most common integration site in *E. coli* is a sequence referred to as insertion sequence IS2 (see discussion of insertion sequences later in this chapter) and a copy of IS2 is found also on the F plasmid. Recombination between the two identical IS2 sequences leads to integration of F into the chromosome, where it is flanked by two IS2 sequences. The integration of the F plasmid has no effect on the cell per se because it replicates along with the chromosome. The integration is not absolutely stable; in a population of cells, an equilibrium exists between bacteria with integrated F plasmid (Hfr) and cells where the F plasmid is not part of the bacterial chromosome.

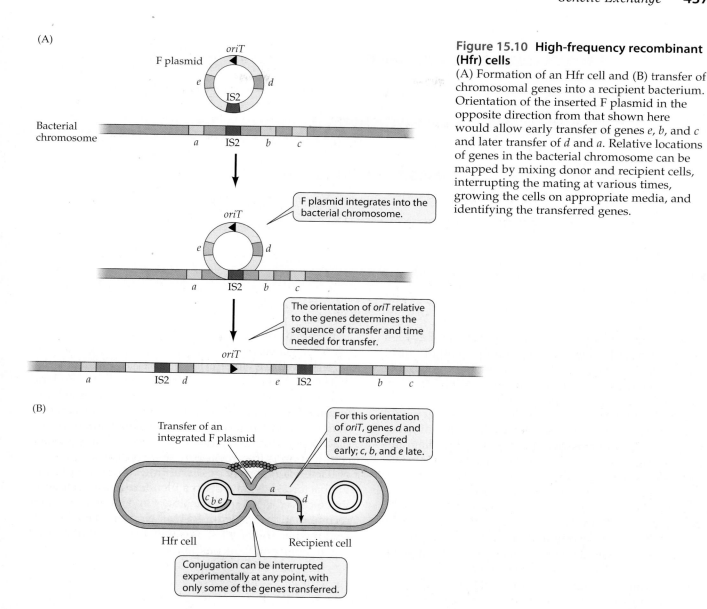

Figure 15.10 High-frequency recombinant (Hfr) cells
(A) Formation of an Hfr cell and (B) transfer of chromosomal genes into a recipient bacterium. Orientation of the inserted F plasmid in the opposite direction from that shown here would allow early transfer of genes *e*, *b*, and *c* and later transfer of *d* and *a*. Relative locations of genes in the bacterial chromosome can be mapped by mixing donor and recipient cells, interrupting the mating at various times, growing the cells on appropriate media, and identifying the transferred genes.

Once the F plasmid has integrated, the cells can initiate a transfer process virtually identical to the one described for transfer of the F plasmid itself (see Figure 15.10B). Following contact of an Hfr cell with a recipient, a single-stranded nick is created in the integrated F plasmid, and transfer begins by a process analogous to the transfer of the free, nonintegrated F plasmid, in the 5'-to-3' direction. The process continues as long as the cells remain in contact. The entire chromosome of a donor cell can be transferred into a recipient, resulting in a diploid cell containing a reconstituted Hfr chromosome. However, the association of donor and recipient is usually not very strong, and the amount of transferred donor chromosomal DNA is directly proportional to the duration of contact between the bacterial mating pair.

The recipient cell usually receives only a small portion of the chromosome of a donor, containing only a fraction of the integrated F plasmid. Therefore, the recipient cell is unable to donate DNA to another recipient. In complete matings in which the entire chromosome is transferred, however, the recipients are Hfr; that is, they are capable of further chromosomal transfer.

Once a DNA segment is taken up by a cell and the complementary strand is synthesized, the genes transferred by this process can undergo general (homologous) recombination with the chromosome of the recipient, not unlike the fragments of DNA that a cell receives by transformation. Because homologous recombination requires significant regions of DNA sequence similarity or identity between the acquired DNA and the site of chromo-

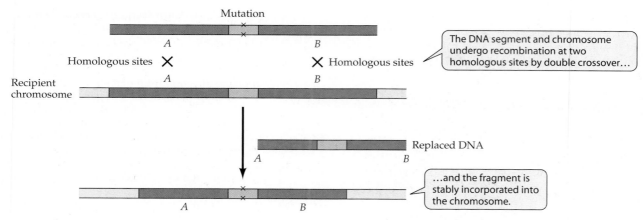

Figure 15.11   Double recombination
Introduction of a mutation into a bacterial chromosome from a piece of DNA acquired by transformation.

somal integration, this process favors exchange between similar or identical genes and could potentially limit the acquisition of new traits to changes in the genes that are small enough to not interfere with recombination. Moreover, because incorporation of linear DNA into a bacterial chromosome by a single-crossover event would disrupt the chromosome, stable incorporation of genes requires a **double-crossover** event. As shown diagrammatically in Figure 15.11, such double recombination effectively replaces the DNA sequence from the bacterial chromosome with that of the transforming DNA.

For double-crossover, the requirement for a high degree of sequence similarity or identity is limited to the sites of the actual recombination event; the sequences flanked by these homologous sequences can be quite different. Entire genes can be introduced or eliminated by transforming DNA, provided the DNA participating in the crossovers flanks the DNA containing the mutation. As illustrated in Figure 15.12, a newly acquired gene or even several linked genes can be inserted into the chromosome by recombination, given that sufficient homology exists on either side of the acquired gene to allow for efficient general recombination. Analogous events can lead to loss of the gene. The double crossover, when applied to the introduction of new or modified genetic material, is sometimes referred to as **gene replacement**. The replaced DNA fragment is generally degraded by bacterial nucleases in the cytoplasm.

Bacterial conjugation via Hfr is a useful method for mapping genes in a bacterial cell. A series of Hfr strains is generated, carrying the integrated plasmid at any one of its preferred or secondary chromosomal sites, and the distance of a gene from a particular integration site is determined. For each site of integration of F, one can mix donor and recipient cells and allow conjugation to take place in a controlled manner. At various time intervals, conjugation is interrupted by shearing the suspension of bacteria, generally by disrupting contact by placing the bacterial suspension in a blender. Those recipients that received DNA from the donor and have undergone recombination are identified by their growth on specialized media. Therefore, if a recipient is a mutant requiring an amino acid, recombination of the defective gene

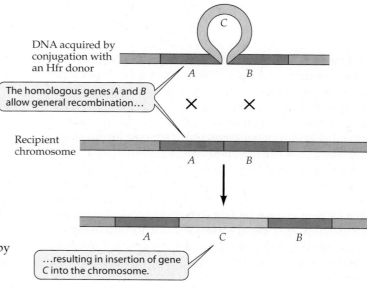

Figure 15.12   Gene replacement
Introduction of a new gene into a bacterial chromosome by transformation. General recombination occurs between homologous sequences that flank the gene.

with DNA from a wild-type donor cell allows identification of such genes. The farther a gene is located from the site of integration of F, and hence, from the origin of transfer, the lower the frequency of transfer of that particular gene to a recipient.

F plasmids or conjugative R plasmids can mediate transfer of chromosomal genes by yet another mechanism. As alluded to earlier, integration of the F plasmid into the bacterial chromosome is not permanent, and excision does take place occasionally. When the F plasmid excises from the chromosome by reversal of the reciprocal recombination, the site of integration is usually perfectly restored, yielding a complete F plasmid and uninterrupted chromosomal DNA. However, occasionally imperfect excision leads to removal of a small portion of the chromosome, such that genes adjacent to the site of recombination are now part of the excised plasmid. These plasmids are called **F prime** (**F′**), or if the excision involved an R plasmid, they are called **R′**. Transfer of DNA from an F′-donor (or an R′-donor) to a recipient proceeds by a mechanism identical to the transfer of the F plasmid (Figure 15.13). In the case of plasmids that contain most of the F plasmid's genome intact, the recipient becomes a donor. Because the recipient can contain the same genes as are on the larger F plasmid transferred during mating with F′, the recipient may become a partial diploid (or merodiploid); that is, it contains two copies of a gene in a single cell. Moreover, regions of extensive homology exist between the chromosome and the plasmid, allowing general recombination to take place along these sequences. Insertion of an F-like plasmid into the bacterial chromosome is one way new Hfr strains can be generated, thereby allowing chromosomal gene transfer from a specific location.

Transformation

The process of acquisition of heritable traits by exposure to a solution of DNA was a true landmark in the history of genetics, as it was the first demonstration that genes are located on DNA. In 1944, O. T. Avery, C. M. MacLeod, and M. McCarty reported that avirulent (non–disease-producing) *Streptococcus pneumoniae*, lacking a protective capsule, regained the ability to synthesize capsules and become virulent when exposed to crude DNA from virulent strains (Figure 15.14). They further showed that the ability of the DNA preparation to transform an avirulent *S. pneumoniae* into a virulent form was lost when the DNA preparation was exposed to DNAses (enzymes that destroy DNA), but not when exposed to proteinases or RNases (enzymes that destroy proteins or RNA). This proved that the transforming genetic information was carried in cellular DNA, not in proteins or RNA.

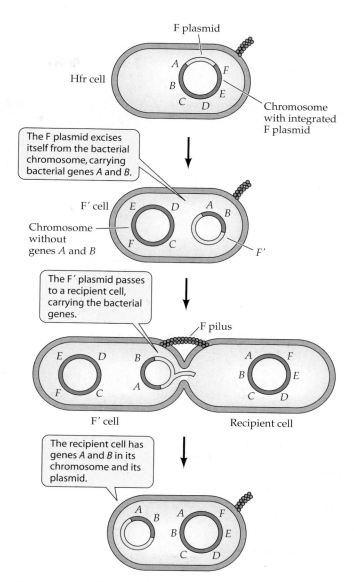

Figure 15.13 F′ cells
Formation of an F′ cell from an Hfr cell, and transfer of a bacterial chromosome segment to a recipient cell.

A significant fraction of bacteria have the ability to take up DNA from the environment, provided that they are **competent**. **Transformation competence** is a state of a bacterial cell during which the usually rigid cell wall can transport a relatively large DNA macromolecule. This is a highly unusual process for bacteria, for they normally lack the ability to transport large macromolecules across the rigid cell wall and through the cytoplasmic membrane. Competence in some bacteria, such as *Haemophilus*, *Streptococcus*, or *Neisseria*, is expressed during a certain stage of cell division and most likely represents a stage when the reforming cell wall can allow passage of DNA. These bacterial genera possess **natural compe-**

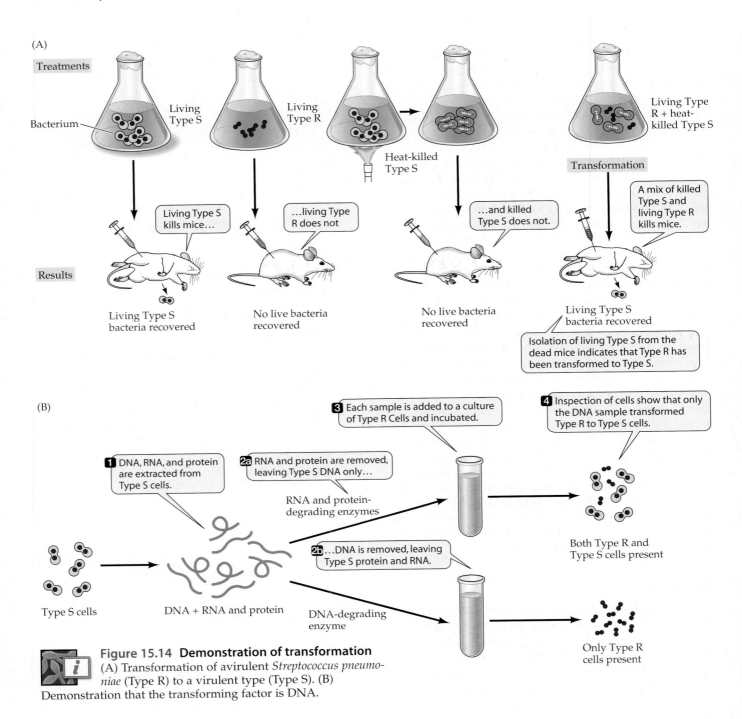

Figure 15.14 Demonstration of transformation
(A) Transformation of avirulent *Streptococcus pneumoniae* (Type R) to a virulent type (Type S). (B) Demonstration that the transforming factor is DNA.

tence because their cells do not require any special treatment to increase their ability to take up DNA. Not all bacteria, however, are naturally competent to take up DNA, but treatment of such cells with calcium or rubidium chloride alters their envelope, and they become competent. This form of competence, termed **artificial competence**, is one of the more important techniques of introducing recombinant DNA molecules into bacteria.

Genetic transformation of naturally competent cells is a multistep process. Exogenous double-stranded DNA binds to specific protein receptors that are expressed on the surface of the recipient cells. Efficient uptake of DNA by a number of bacteria occurs only if the transforming DNA contains short, species-specific DNA sequences distributed in many copies in the genome. The requirement for this sequence during transformation ensures that only closely related species are able to exchange DNA using the transformation mechanism. Once bound to the surface of a gram-positive organism, the DNA is fragmented by nucleases and transported into the cytoplasm using a multicomponent transport machinery. In contrast, gram-negative bacteria first import the fragmented DNA into their periplasm. The transport across the cytoplasmic membrane into the cell is accompanied by degradation of one of the strands, and, consequently, the transforming DNA enters as a single-stranded molecule. Once inside the cell the DNA it is immediately bound by many copies of the single-stranded (ss) DNA-binding protein and RecA, thereby initiating the homologous recombination process.

When cells are chemically permeabilized to become artificially competent, they are capable of taking up double-stranded (ds) DNA. If the transformed DNA is a plasmid capable of replicating within that particular species of bacterium, the cell becomes plasmid-bearing and plasmid DNA replicates as an extrachromosomal element. Genes encoded on the plasmid are expressed, and synthesis of plasmid-encoded proteins allows the bacterial host to take advantage of some additional traits that its siblings without plasmids lack. For example, if the bacterium took up a plasmid encoding a gene that specified aminoglycoside resistance, the transformant would then be resistant to aminoglycosides, a trait certainly useful if the bacterium is in an environment containing this class of antibiotic.

It is not clear to what extent transformation contributes to natural genetic exchange, although recent evidence suggests that for some bacteria that are naturally competent, individual genes can be acquired even from genomes of different species. However, transformation requires the presence of free DNA in the bacterial environment, and there it can be readily attacked by nucleases. The availability of DNA in the environment of a bacterium would be limited. However, in certain environments where bacteria live in large numbers and at high density, some bacteria die and lyse, and it is not unreasonable to expect that some DNA escapes digestion by extracellular nucleases. As is the case with all mechanisms of DNA exchange, if the acquired trait gives the recipient bacterium significant survival advantage, the progeny of the rare transformant will outnumber the bacteria lacking such a trait within several generations, even if the DNA acquisition mechanism is inefficient.

Transduction

The process of transferring genes between bacteria as part of bacteriophage particles is called **transduction**. Two fundamentally distinct transductional mechanisms operate among bacteriophage. One system, called **general transduction**, results in transfer of all bacterial genes at low but identical frequencies. During **specialized transduction**, certain genes are transferred at high frequencies, whereas others are transferred at low rates or not at all.

Generalized Transduction

Generalized transduction is likely carried out by most bacteriophage, but certain ones are more efficient than others. This process is outlined in **Figure 15.15**. Following infection of the cell, these bacteriophage undergo a complete replication cycle. During the infection process, the host chromosome is disrupted and used as a source of nucleotide building blocks for the phage genome.

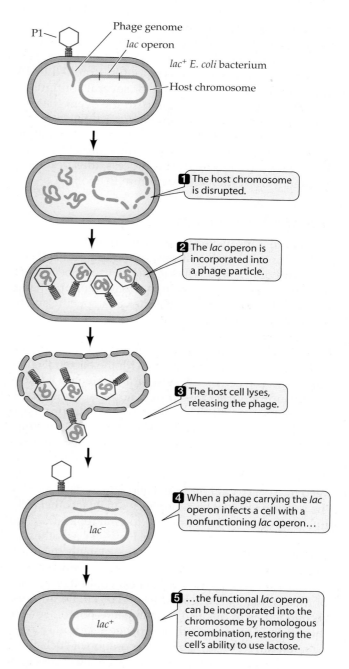

1 The host chromosome is disrupted.

2 The *lac* operon is incorporated into a phage particle.

3 The host cell lyses, releasing the phage.

4 When a phage carrying the *lac* operon infects a cell with a nonfunctioning *lac* operon…

5 …the functional *lac* operon can be incorporated into the chromosome by homologous recombination, restoring the cell's ability to use lactose.

Figure 15.15 Generalized transduction
Generalized transduction carried out by bacteriophage P1.

Occasionally, fairly large host DNA fragments remain in the cytoplasm at the time of viral DNA–particle packaging. Infrequent packaging errors result in bacterial DNA being packaged into the capsid instead of the phage genome. Upon completion of the infection cycle, the host cell lyses, thereby releasing the progeny virus. Although most capsids contain viral DNA and are fully capable of binding to new host cells and thus repeating the infectious cycle, the rare capsid containing the host DNA will also adsorb to another host and inject its DNA into the bacterium. This interaction of a cell with a viral particle does not result in the death of the host, because the introduced DNA does not specify viral particles. Rather, the cell now contains a double-stranded DNA fragment that may contain one or several contiguous genes from the donor's genome. In the event that the transduced DNA shares homology with the chromosome of the recipient, double reciprocal recombination can result in incorporation of the transduced DNA into the bacterial chromosome. This process, thus, results in transfer of DNA from one bacterium to another.

Specialized Transduction

Specialized transduction requires incorporation of the viral DNA into the bacterial chromosome, and it is therefore carried out only by lysogenic viruses. Specialized transduction occurs following formation of prophage, whereby the viral DNA integrates at a specific site in the host genome by recombination between a site on the phage DNA (*attP*) and the corresponding site on the bacterial chromosome (*attB*). Although the prophage replicates for many generations with the bacterial chromosome, it can occasionally excise and initiate a lytic cycle, including viral DNA replication, packaging, and lysing of the host. Occasionally excision of the prophage is imprecise, resulting in removal of some DNA on either the left, right, or both sides of the site of integration, as shown schematically in Figure 15.16. This excised DNA now is part of the viral genome; it replicates and is packaged into every viral particle. Following lysis of the cell, the released virulent bacteriophage can infect other cells. Infection of a new host can lead to viral replication and continuation of the virulent cycle. However, as is the case with the original infecting phage, occasionally the transducing virus can enter into a lysogenic relationship with its new host. The transduced genes would then be part of the genome of the second bacterial cell. Because the second recipient can also contain genes homologous to those carried by the transducing phage, reciprocal recombination can result in stable incorporation of such genes into the bacterial chromosome, without the necessity of establishing a bacterial lysogen.

Specialized transduction therefore results in phage-mediated transfer of genes that are near the attachment site of the lysogenic prophage. Although this process is limited to such genes, it is extremely efficient because once the error in excision has occurred, such genes are part of the bacteriophage genome and are efficiently transferred.

SECTION HIGHLIGHTS

There are three types of genetic exchange mechanisms among bacteria: (1) direct transfer of DNA between donor and recipient bacteria, called conjugation; (2) uptake of free DNA from the media, called transformation; and (3) bacteriophage-mediated transfer of DNA, called transduction. DNA exchange by conjugation requires that the donor bacterium carry a transmissible plasmid, which encodes the entire transfer machinery, including conjugal (sex) pili. Plasmid transfer by conjugation is replicative; that is, a copy of the plasmid is transferred from the donor to a recipient without the loss of the plasmid in the donor bacterium. High-frequency recombinant (Hfr) *E. coli*, which carry a chromosomally integrated form of the F plasmid, can transfer chromosomal sequences starting from the *oriT* of the integrated F plasmid. Upon entry into the recipient *E. coli*, the partial genome is incorporated into the chromosome by homologous recombination. Occasionally, a copy of the entire genome is transferred into a recipient, in which case the recipient can become a donor. Uptake of DNA by transformation requires that the recipient bacterium be in a physiological state called natural competence. In some bacteria, artificial competence can be induced by chemical treatment. Specialized transduction results from imprecise excision of a lysogenic bacteriophage; chromosomal DNA from the region flanking the attachment site is excised along with the phage genome and packaged into bacteriophage capsids. Generalized transduction involves random packaging of chromosomal DNA fragments, instead of bacteriophage genomes, into bacteriophage capsids during the final stages of phage maturation.

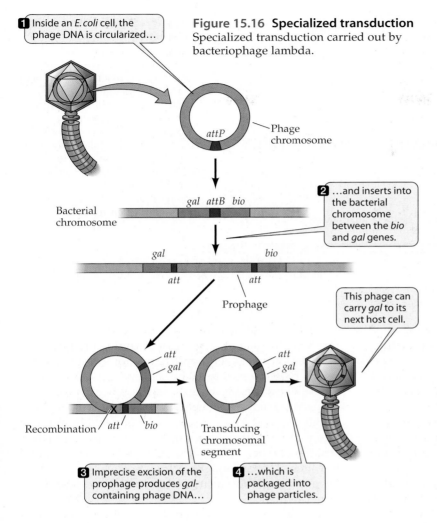

1 Inside an *E. coli* cell, the phage DNA is circularized...

attP

Phage chromosome

Bacterial chromosome

gal attB bio

2 ...and inserts into the bacterial chromosome between the *bio* and *gal* genes.

gal   *bio*

att   *att*

Prophage

This phage can carry *gal* to its next host cell.

att
gal

att
gal

Recombination   *att*   *bio*

X

Transducing chromosomal segment

3 Imprecise excision of the prophage produces *gal*-containing phage DNA...

4 ...which is packaged into phage particles.

Figure 15.16 Specialized transduction Specialized transduction carried out by bacteriophage lambda.

15.4 Mobile Genetic Elements and Transposons

This chapter has considered genetic exchange in terms of blocks of DNA that move from cell to cell by being integrated into defined, self-replicating units, such as viruses or plasmids. Occasionally, a gene can move from one place on the chromosome to another in that cell, or a gene can move between a chromosomal site and a plasmid site. This process is called transposition. Movement from one location to another by the transpositional mechanism requires recombination to take place between the transposing DNA and the site where it integrates, by distinct mechanisms that are similar to site-specific recombination. This occurs because the DNA exchange leading to integration of a transposon does not require homologous sequences as required in general recombination. The site-specific recombination utilizes DNA sequences that define the site of transposition, and

specialized enzymatic machinery that catalyzes the transpositional event.

The simplest form of such a mobile gene is an **insertion sequence**, abbreviated IS. A number of different insertion sequences have been characterized in bacteria, ranging in size from 700 to 5,000 bp. They encode no functions other than those involved in their movement from one DNA site to another. All IS elements have several common features. They contain short (16 to 41 bp) inverted repeat sequences at their ends. Each IS also specifies an enzyme, called **transposase**—encoded by a single gene, or, occasionally, two genes—that specifically mediates the site-specific recombination event during transposition. A schematic diagram of a well-characterized insertion sequence, IS1, is shown in Figure 15.17A.

Transposition of an IS element into a new site can create insertion mutations by physically disrupting a sequence that encodes a polypeptide. The presence of this large block of DNA in an **operon** (a set of genes transcribed from a single promoter) can cause **polar mutations**; that is, insertion in the first gene of an operon can block expression of genes downstream from the site of insertion. This is due to the inability of RNA polymerase to traverse through an IS element, presumably because of the presence of transcriptional termination signals within the element.

Transposons are more complex forms of mobile genetic elements and are characterized by the presence of genes beyond those needed for transposition. The most common genes found within transposons are those specifying antibiotic-resistance determinants. However, other genes, such as those encoding toxins, have also been identified on these transposable elements.

Classification of Transposons

The simplest type of transposon is a **composite transposon**, which is a mobile genetic element in which two IS elements surround a gene. Transposition then involves translocation of a copy of the entire element (two ISs and the internal gene) into a new site, using the transposase gene of one of the IS elements. Composite transposons can serve as a source of IS sequence transposition, because as long as the IS element contains a functional transposase it can transpose independently from the transposon. An example of such a transposon is shown

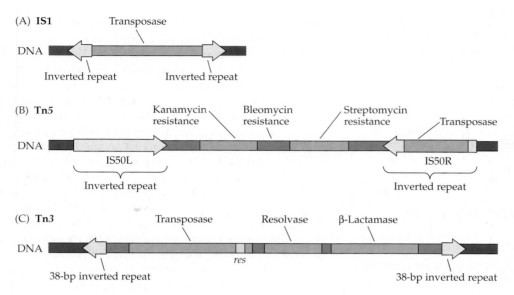

Figure 15.17 Transposable elements
Structures of some bacterial insertion sequences and transposons. (A) Insertion sequence IS1. (B) A composite transposon contains antibiotic genes flanked by two insertion sequences as direct or inverted repeats. Shown here is the Tn5 transposon, with inverted repeats. (C) The Tn3 transposon. It is not flanked by insertion sequences, but by short, 38-bp inverted repeats.

in Figure 15.17B. As can be seen in the figure, some composite transposons, such as Tn5, contain multiple drug-resistance genes within the IS boundaries. Composite transposons represent a simple way for any bacterial gene to become mobile. Two identical IS sequences can surround a bacterial gene; under some circumstances that gene becomes a transposon. Every active transposon requires expression of a functional transposase gene, which is generally found within at least one of the insertion sequences.

A distinct class of transposons comprises those modeled after transposon Tn3, shown in Figure 15.17C. This family of transposons does not contain repeated IS elements at its ends. However, the transposon is a single DNA segment that contains small (38-bp) inverted repeats at its ends. Within the element, there are genes for antibiotic-resistance determinants (β-lactamase in Tn3), as well as two genes encoding proteins involved in transposition: transposase and resolvase. Resolvase is an enzyme that catalyzes recombination between two integrated transposons, an essential step in transposition by the replicative mechanism, using the *res* sites as sites of recombination (see subsequent text).

Mechanism of Transposition

The most distinguishing feature of insertion sequences or transposons is their mobility, the ability to move (transpose, hop, or jump) from a **donor** site (bacteriophage, plasmid, or chromosome) into **target** sites, which can be another region of the bacterial chromosome, or a site on another plasmid or a bacteriophage. The target sequence can vary, depending on the transposon, and can include variables in addition to a specific sequence.

Some transposons require specific sites that occur only rarely within a chromosome or plasmid, and target sequences vary between 2 and 12 bp. For example, transposon Tn10 inserts into a 9-bp sequence NGCTNAGCN, where N can be any base, whereas other transposons or insertion sequences such as IS1 prefer AT-rich regions. Because not all AT-rich regions are targeted by IS1, it is likely that additional parameters such as local DNA structure (supercoiling) also play a role in determining the sites of insertion. Other transposons, such as Tn5, have low target specificity and can insert into almost any target sequence.

Transposons utilize two distinct recombinational mechanisms for transposition into a new site, which are differentiated by the ability of the transposon donor to be duplicated in the process, called **replicative transposition**. Alternatively the donor transposon can be lost or eliminated, and this mechanism is called **nonreplicative transposition**, sometimes referred to as cut-and-paste transposition. The basic steps that accompany replicative and nonreplicative mechanisms of transposition are shown in Figures 15.18 and 15.19.

Integrons

Conjugative plasmids, transducing bacteriophage, and transposons provide the means for disseminating genes—including a number of determinants of antibiotic resistance—between bacteria. A distinct class of genetic elements, called **integrons**, is often found on plasmids and within or near transposons or insertion sequences. Integrons consist of an **integrase** gene and an adjacent **recombination site** (*attI*). An integron is capable of capturing **antibiotic-resistance gene cassettes**

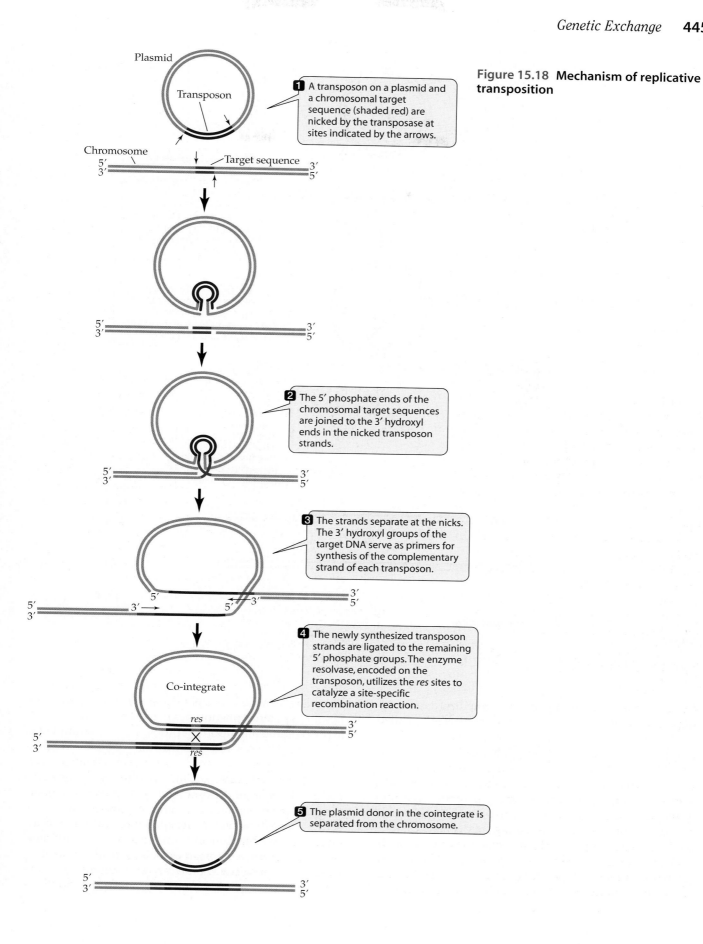

Figure 15.18 Mechanism of replicative transposition

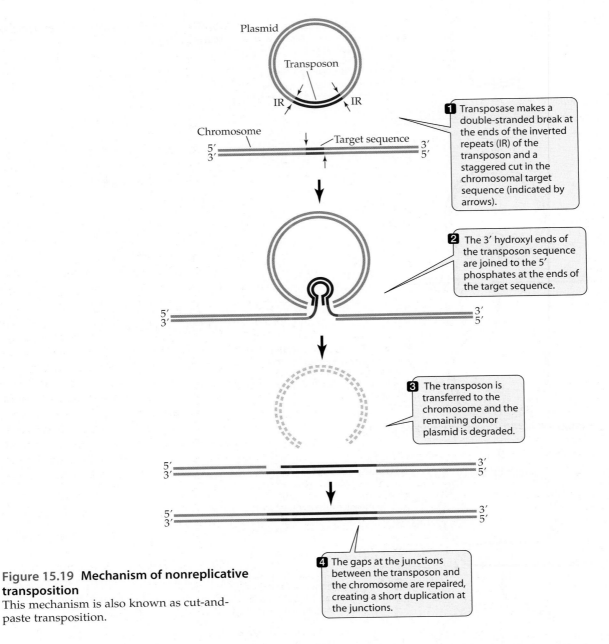

Plasmid

Transposon

IR   IR

Chromosome

5′
3′   Target sequence   3′
5′

1 Transposase makes a double-stranded break at the ends of the inverted repeats (IR) of the transposon and a staggered cut in the chromosomal target sequence (indicated by arrows).

5′
3′   3′
5′

2 The 3′ hydroxyl ends of the transposon sequence are joined to the 5′ phosphates at the ends of the target sequence.

3 The transposon is transferred to the chromosome and the remaining donor plasmid is degraded.

5′
3′   3′
5′

5′
3′   3′
5′

4 The gaps at the junctions between the transposon and the chromosome are repaired, creating a short duplication at the junctions.

Figure 15.19 Mechanism of nonreplicative transposition
This mechanism is also known as cut-and-paste transposition.

from circular structures by integrase-mediated recombination between the *attI* site and the *attC* site (also called the 59-bp element) located next to the gene within the cassette (Figure 15.20). Although the gene is within the circular cassette, it lacks a promoter. However, the gene can be expressed following integration into the *attI* site; it is now transcribed from a strong promoter reading outward (P_{out}) located at the 3′ end of the integrase gene on the opposite strand. The capture of the gene cassette by a resident integron can be repeated many times, with each new cassette inserting into the *attI* site. In addition, the proximity of the P_{out} ensures high-level expression of the newly captured gene. The buildup of different an-

tibiotic-resistance genes is responsible for multiple-antibiotic-resistance phenotypes of certain pathogens, rapidly becoming an important problem in the treatment of infectious diseases. Integrase can also cause the excision of the antibiotic-resistance gene cassette that can now move to a new location. If the cassette integrates into a mobile element (transposon, plasmid, or bacteriophage) the genes can move into another organism; these cassettes are responsible for the spread of antibiotic resistance to previously sensitive organisms. Although most of the integrons have been associated with acquisition of antibiotic-resistance gene cassettes, certain genomes contain integrons with gene cassettes for metabolic or

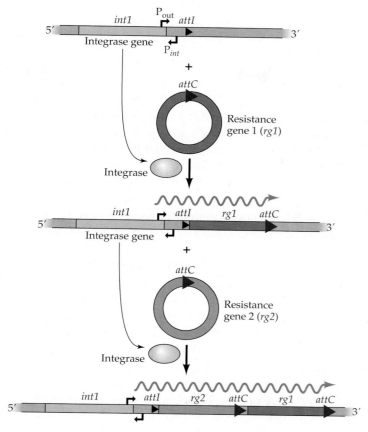

Figure 15.20 Capture of antibiotic-resistance cassettes by integrons

A resident integron contains an integrase gene and an adjacent *attI* site. While the integrase is transcribed from its own promoter (P_{int}), a second promoter (P_{out}) is active only after capture of gene cassettes, in this case, antibiotic-resistance genes (*rg*). A gene cassette is a circular DNA molecule, containing a resistance gene and the *attC* sequence, and it is inserted into the *attI* site by integrase-mediated site-specific recombination. Sequential acquisition of an *n* number of antibiotic-resistance genes (*rg1–n*), all transcribed from the (P_{out}) promoter, results in a strain carrying multiple-resistance genes.

virulence functions. These integrons often accumulate a large number of cassettes (more than 100 in some cases), and they are referred to as **super-integrons**.

Conjugative Transposons

One class of conjugative elements encompasses several features of transposons and plasmids. These are the so-called **conjugative transposons**, initially described in certain gram-positive bacteria; however, they appear to be rather widely distributed among bacterial species. Although conjugative transposons reside in the chromosome, they can excise and assume covalently closed circular forms that can transfer to recipient cells by the

process of conjugation. The distinction between conjugative transposons and plasmids is that the former does not have the ability to replicate autonomously; conjugative transposons are only copied during transfer from the donor cell to the recipient. Once in the recipient, they immediately integrate into the chromosome.

Conjugative transposons contain many genes functionally related to those found on conjugative plasmids and primarily specify transfer determinants. Although they lack the ability to replicate, they form a transient circular intermediate, they are nicked at *oriT*, and they are transferred as single-stranded molecules into the recipient, where they re-form a double-stranded covalently closed circle prior to integration. These transposons encode all of the enzymes necessary for DNA processing, including excision, transfer, and integration, as well as the determinants for the formation of a conjugative pore. However, there are several distinctions between conjugative transposons and classical transposons. The movement of a conjugative transposon from the donor site is accomplished by precise or near-precise excision from the chromosome. Similarly, the integration occurs by a site-specific recombinational insertion of the circular element into specific locations in the chromosome. Moreover, they do not utilize either the replicative or nonreplicative integration mechanisms. The majority of conjugative transposons studied to date carry antibiotic-resistance genes; however, genes encoding metabolic enzymes were identified on similar elements in *Lactobacillus* and certain *Pseudomonas* species.

SUMMARY

- Although most of the essential genes are found on the bacterial chromosome, additional genes for a variety of cellular functions are encoded by extra-chromosomal genetic elements called plasmids. A class of plasmids carrying antibiotic-resistance genes is called **R factors** or **R plasmids**. A special class of plasmids, called **virulence plasmids**, carries genes that encode essential virulence determinants, such as adhesins and toxins.

- A plant pathogen, *Agrobacterium tumefaciens*, causes tumors in plants by transferring a portion of its plasmid (called tumor-inducing, or **Ti plasmid**) to the plant, where the Ti-encoded genes induce synthesis of plant hormones.

- **Homologous recombination** involves breakage and joining of two DNA strands along identical or highly similar sequences. Homologous recombination utilizes a DNA site that possesses a nick, generated by the **RecB**, **RecC**, and **RecD** nuclease complex, at a sequence called the **Chi site**. Another key component required for homologous recombination is the **RecA** protein, which binds to single-stranded DNA and then identifies homologous regions in the opposite strand.

- There are three types of genetic exchange mechanisms among bacteria: (1) uptake of free DNA from the media, called **transformation**; (2) direct transfer of DNA from donor to recipient bacteria, called **conjugation**; and (3) bacteriophage-mediated transfer of DNA, called **transduction**.

- Uptake of DNA by transformation requires that the recipient bacterium be in a physiological state called **natural competence**.

- In some bacteria, chemical treatment can induce **artificial competence**, such that they can take up DNA fragments or plasmids.

- DNA exchange by conjugation requires that the donor bacterium carry a **transmissible plasmid**, which encodes the complete transfer machinery, including **conjugal (sex) pili**. Plasmid transfer by conjugation is **replicative**—that is, a copy of the plasmid is transferred from the donor to a recipient without loss of the plasmid from the donor bacterium.

- *E. coli* carrying a chromosomally integrated form of a well-studied F plasmid, called **high-frequency recombinant (Hfr)**, can transfer chromosomal sequences adjacent to this integrated form, starting from the *oriT* of the F plasmid. Upon entry into the recipient *E. coli*, the partial donor genome is incorporated into the recipient's chromosome. Occasionally, a copy of the entire genome (including a complete F plasmid) can be transferred into a recipient, in which case the recipient can become a donor.

- **Specialized transduction** is the consequence of imprecise excision of a lysogenic bacteriophage from the chromosome, such that portions of the host chromosomal DNA from the region flanking the attachment site is packaged into bacteriophage capsids.

- **Generalized transduction** involves random packaging of chromosomal DNA fragments—instead of bacteriophage genomes—into bacteriophage capsids during the final stages of phage maturation.

- Movement of genes between various locations on the chromosome or plasmids is referred to as **transposition**. Insertion of a transposon into a gene results in a mutation in the gene due to the disruption of its coding sequence.

- Simple mobile elements are **insertion sequences**, which contain short **inverted repeats** at their ends and encode one enzyme, **transposase**. Some genes can transpose when they are flanked by two insertion sequences. These mobile genetic elements are called **composite transposons**. Certain transposons, such as Tn3, lack flanking insertion sequences. However, they still contain inverted repeats at their ends and, within the transposon, encode two essential enzymes, **transposase** and **resolvase**. Depending on the type of transposon, the site of insertion—the **target site**—varies from a defined short sequence to any random DNA sequence.

 Find more at www.sinauer.com/microbial-life

REVIEW QUESTIONS

1. Does a bacterium bearing a self-transmissible plasmid require a functional DNA synthesizing machinery for it to transfer the plasmid into a recipient by conjugation? Does a recipient need functional DNA synthesis for it to receive the plasmid from a donor?

2. How does the fate of a chromosomal DNA fragment differ from that of a plasmid when it is introduced into bacteria by transformation?

3. When bacteria contain more than one compatible plasmid, some plasmids, which are not self-transmissible, can transfer to recipients. Describe the

mechanism that makes that possible. If the same plasmid lacks an origin of transfer, can it ever transfer to a recipient by conjugation?

4. When a chromosomal gene is encoding an antibiotic-resistance determinant and is flanked by two identical DNA sequences, homologous recombination can lead to the loss of the resistance gene such that some bacteria become susceptible to the antibiotic. Explain.

5. Which method of DNA transfer between bacteria would not take place if the donor and recipient were separated by a filter with a pore size of 0.45 μm? Which method of transfer would be blocked by the presence of high concentrations of DNase (enzymes capable of degrading DNA)?

6. A transposon insertion into the *lacZ* gene of the lactose utilization operon results in inability of the bacterium to utilize lactose as a sole carbon and energy source. When this bacterium acquires an F′ *lac* (an F plasmid with the intact *lacZ* gene), it is still unable to utilize lactose. Provide a reasonable explanation for this phenomenon.

7. Chromosome mobilization by the integrated F plasmid (Hfr) usually does not lead to the recipient being capable of acquiring donor functions. In contrast, F′-mediated transfer usually converts the recipient to a donor. Explain this difference in transfer mechanisms. What condition of DNA transfer results in the appearance of rare recipients that have acquired donor capabilities after receiving DNA by the Hfr-mediated transfer?

8. How can the Hfr strain be used to map the relative order of genes?

9. A recipient *Bacillus subtilis*, with mutations in the genes encoding enzymes of histidine and arginine biosynthesis, is transformed with DNA that came from *B. subtilis* carrying a mutation in a ribosomal protein, rendering it resistant to streptomycin. The transformants were plated onto streptomycin, and subsequently replica plated onto minimal media containing either arginine or histidine. Growth (or lack of growth) on these supplemented minimal plates revealed that among the original streptomycin-resistant transformants, 12% did not require histidine for growth and 6% did not require arginine for growth, whereas the remaining 82% was streptomycin-resistant and retained the original requirement for both histidine and arginine. No streptomycin-resistant isolates that lost both of the arginine and histidine requirements were obtained. What is the location of the arginine and histidine

biosynthetic genes, relative to the streptomycin-resistance-determining ribosomal protein gene?

10. Compare and contrast specialized and generalized transduction.

11. A plasmid carrying a kanamycin-resistance gene and containing a composite transposon (two insertion sequences flanking a tetracycline-resistance gene) was introduced into a recipient by conjugation. When the bacteria were passaged for many generations in an antibiotic-free medium, a fraction of these bacteria did not contain any plasmid. When these bacteria were individually analyzed for antibiotic resistance, it was found that some were resistant to tetracycline because they contained a chromosomal copy of the transposon. However, some were resistant to kanamycin, and the kanamycin-resistance gene was inserted into the chromosome. No isolates that contained both of the resistance genes were recovered. Explain the events in the recipient bacterial cell following introduction of the original plasmid.

12. A wild-type strain of *E. coli* is transformed by DNA from *E. coli* that has a Tn5 transposon in a gene that encodes an enzyme of the tryptophan biosynthetic pathway. After plating transformants on rich medium containing kanamycin (the resistance gene carried on Tn5), the colonies are individually transferred onto minimal medium. Of 100 colonies tested, 18 required tryptophan for growth, 70 did not require any supplemental nutrients, and 12 could not grow on minimal media, even when it was supplemented with tryptophan. Provide an explanation for the three types of transformants. If the recipient were a *recA* mutant of *E. coli*, would you expect a different distribution of the three classes of transformants based on their nutritional requirements?

SUGGESTED READING

Burrus, V. and M. K. Waldor. 2004. "Shaping Bacterial Genomes with Integrative and Conjugative Elements." *Research in Microbiology* 55: 376-386.

Craig, N. L., R. Cragie, M. Gellert and A. M. Lambowitz, eds. 2002. *Mobile DNA II*. Washington, DC: ASM Press.

Ebersbach, G. and K. Gerdes. 2005. "Plasmid Segregation Mechanisms." *Annual Review of Genetics* 39: 453-479.

Funnell, B. E. and G. J. Phillips, eds. 2004. *Plasmid Biology*. Washington, DC: ASM Press.

Mazel, D. 2006. "Integrons: Agents of Bacterial Evolution." *Nature Reviews in Microbiology* 4: 608-620.

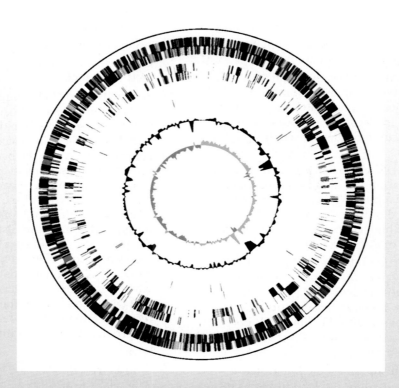

The objectives of this chapter are to:

◆ Introduce the basic tools, technologies, and applications of gene cloning.

◆ Describe the methods of genome sequencing and use of genome sequence information in microbiological research.

◆ Explain the principles of horizontal gene transfer.

◆ Describe the use of microarrays in studies of gene expression and gene content.

16

Microbial Genomics

If we should succeed in helping ourselves through applied
genetics before vengefully or accidentally exterminating
ourselves, then there will have to be a new definition of
evolution, one that recognizes a process no longer directed
by blind selection but by choice.
—M. A. Edey and D. C. Johanson, Blueprints, 1989

*P*revious chapters in this book have dealt with the mechanisms for gene transfer and recombination that allow bacteria to acquire genes from other microorganisms and express traits encoded by these genes. Our growing understanding of DNA replication, genetic variation, and transfer of genetic information among microorganisms, as well as of the molecular mechanisms that control bacterial gene expression, has led to the development of a new industry: biotechnology. Biotechnological applications, based on laboratory manipulation of genetic information, have had a major impact on all aspects of human health, ranging from diagnosis of disease to the production of useful therapeutics and vaccines. Furthermore, the advanced technologies resulting in rapid determination of sequences of entire bacterial genomes have enormously enhanced our understanding of the potential of these organisms to carry out a multitude of known as well as unexpected biological processes.

In this chapter, we discuss the rationale, methodology, and outcome of deliberate laboratory manipulations of genes, a process called **genetic engineering**. The aim of genetic engineering, generally, is to isolate genes from eukaryotic, bacterial, or archaeal genomes and to introduce these genes into a different species, where they will be propagated in a stable manner, thereby resulting in a **transgenic** organism.

Moreover, the identification of the entire genetic repertoire of a bacterium or higher eukaryote makes it possible to manipulate these organisms on a genome-wide basis and, not surprisingly, provides unparalleled opportunities to engineer entirely new species with potentially beneficial, as well as harmful, effects on humankind.

16.1 Basis of Genetic Engineering

Early work by bacterial geneticists with transducing phages and conjugative plasmids (see Chapter 15) that carry fragments of the bacterial chromosome suggested a possible means of isolating and manipulating individual genes. This work has promoted the study of genetic organization and expression of genes and has been used successfully to produce large amounts of specific proteins for research and for use in agriculture, industry, and medicine.

The isolation of genes by classical bacterial genetic techniques such as F′ plasmids or transducing phages has been feasible for decades. However, it took the discovery of **restriction endonucleases** to launch the field of genetic engineering. By employing these specialized enzymes from bacteria, researchers gained a powerful tool for manipulating DNA containing information. Simultaneously, increased knowledge of the replication requirements for plasmids and phages provided vehicles for isolating identical copies of individual genes and propagating them in pure form—that is, **cloning** them. Finally, development of readily applied methods for determining the precise nucleotide sequence of DNA allowed studies of genetic organization, genetic structure, and the function of gene products. Recombinant DNA technology has affected virtually every aspect of biomedical science, ranging from the production of human hormones in bacteria to the study of evolution. This chapter reviews some principles of genetic engineering and summarizes some of its more common applications.

Restriction and Modification

One of the most effective barriers to free genetic exchange of DNA between unrelated bacteria is the presence of **restriction–modification** systems. These are enzymes that bacteria employ in nature to maintain the integrity of their genomes and to degrade DNA that they may accidentally acquire from other bacteria by any of the genetic exchange mechanisms discussed in Chapter 15. **Restriction enzymes** are also useful in protecting bacteria from DNA viruses by degrading the viral genetic material after it enters the bacterial cytoplasm. Restriction enzymes (or **restriction endonucleases**) introduce double-stranded breaks in foreign DNA at specific base sequences. However, it is likely that the chromosomes of bacteria producing these enzymes also contain such sequences. So the

bacteria prevent destruction of self-DNA by using **modification enzymes** to alter (modify) the recognition sites of restriction enzymes in their own genomes, such that they are no longer degraded by these enzymes. Each bacterial species has characteristic restriction–modification enzymes that recognize unique sequences. For example, certain strains of *Escherichia coli* produce an enzyme, *Eco*RI, which cleaves DNA wherever the sequence GAATTC occurs. The genome of *E. coli* itself has such sequences, and these are modified by methylation of the second adenine of this sequence on both strands, rendering the sequence resistant to *Eco*RI cleavage.

The best example of the advantage conferred by a restriction–modification system is observed in infections by bacterial viruses. The efficiency of infection of a particular host by bacterial viruses is inverse to the extent of damage to the viral DNA by the host's restriction endonucleases, following entry of the viral genome into the bacterial cell. An efficient restriction system can reduce the successful infections to an exceedingly low level, such that most of the bacteria survive the infecting virus, and only 1 in 1,000 bacteria infected will ever yield viral progeny. However, when these rare infective virions that infect that 1 in 1,000 are isolated and used to infect the same bacterial strain again, the viral infection will be very successful, with hundreds of virions obtained from each host. The reason for the high level of infectivity of the second generation virions is that they survived in the first host's cytoplasm long enough to have their DNA modified by the host's modification enzymes. Thus, during the subsequent infection, this modified DNA is not recognized as foreign by the host's restriction enzymes.

This same restriction barrier is effective against the uptake of foreign DNA during transformation or conjugation (see Chapter 15). Therefore, the fate of a plasmid or a DNA fragment depends in part on whether it has been protected by modification and whether it can be cleaved by the host's particular restriction enzymes. Overall, restriction–modification systems, along with other barriers to DNA exchange, favor recombination of DNA within identical or closely related species.

Analysis of DNA Treated with Restriction Endonucleases

In practice, restriction enzymes are powerful tools for genetic engineering because they permit hydrolysis of a DNA molecule into defined fragments containing individual genes or clusters of genes. Thousands of restriction enzymes with unique recognition sequences have been identified, and many have been purified for this purpose, providing an array of enzymes to fragment DNA to a desired size. Representative examples of restriction enzymes are shown in Table 16.1. A restriction enzyme binds to specific sequences and hydrolyzes the phosphodiester bond in each DNA strand. These sequences are usually symmetrical and range between 4 and 16 bp in length. A large fraction of restriction enzymes cleave within their recognition sequences, which are continuous, but the recognition sequence also can be discontinuous (i.e., the two halves are separated by a short stretch of base pairs that do not affect the enzyme's specificity). Some enzymes cleave at a specific distance outside of their recognition sequence. If the cut is staggered, the cleaved DNA molecule has either a protruding 3' or 5' single-stranded sequence. However, some enzymes create "flush" (blunt) ends by hydrolyzing the phosphodiester bonds that are opposite to each other. The types of ends created by three commonly-used restriction enzymes *Eco*RI, *Pst*I, and *Sma*I are shown in Figure 16.1.

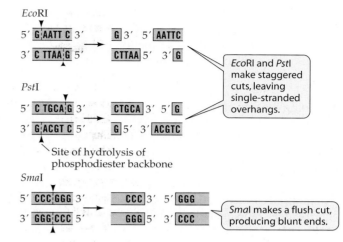

Figure 16.1 Restriction enzymes
Cleavage of DNA strands by restriction enzymes. *Eco*R1 creates a 5' overhang; *Pst*I creates a 3' overhang; *Sma*I produces blunt ends.

TABLE 16.1	Common types and sources of restriction endonucleases (enzymes) and their DNA recognition sequences

Microorganism	Endonuclease	Target Sequences, Showing Axis of Symmetry (|) and Cleavage Sites (▼)
Recognize and cleave within continuous sequence		
Bacillus amyloliquefaciens H	*Bam*HI	5' G G A|T C C 3' 3' C G T|A G G 5'
Streptomyces albus	*Sal*I	5' G T C|G A C 3' 3' C A G|C T G 5
Xanthomonas badrii	*Xba*I	5' T C T|A G A 3' 3' A G A|T C T 5'
Thermus aquaticus	*Taq*I	5' T C|G A 3' 3' A G|C T 5'
Haemophilus aegyptius	*Hae*III	5' G G|C C 3' 3' C C|G G 5'
Recognize and cleave discontinuous sequences		
Pseudomonas fluorescens F	*Pfl*FI	5' G A C N N N G T C 3' 3' C T G N N N C A G 5'
Bacillus globigii	*Bgl*I	5' G C C N N N N N G G C 3' 3' C G G N N N N N C C G 5'
Cleave outside of their recognition sequences[a]		
Pseudomonas lemoingei	*Ple*I	5' G A G T C N N N N N N 3' 3' C T C A G N N N N N N 5'
Saccharopolyspora species	*Sap*I	5' G C T C T T C N N N N N 3' 3' C G A G A A G N N N N N 5'

[a]Recognition sequences are underlined.

A simple method of DNA analysis is **agarose gel electrophoresis**, a procedure that separates DNA fragments by differential migration, as a function of size and conformation, in an electrical field applied across a hydrated porous matrix of agarose (a highly purified polysaccharide from seaweed). Among the critical parameters determining the rate of migration are the size and the topological form of the DNA. Agarose gel electrophoresis can thus separate plasmids based on size, and the plasmid content of independent isolates can be compared to see whether they contain identically sized plasmids. The extent of supercoiling of circular DNA molecules such as a plasmids, can influence the shape and migration in agarose gels. However, the mobility of linear DNA molecules, generated by treatment with restriction enzymes, is primarily determined by the length of each fragment. As illustrated in Figure 16.2A (left panel), different clinical isolates of the same organism can contain unique plasmids, and their respective electrophoretic mobilities can establish the size relationship among them.

The technique of comparing plasmids based on migration in an electrical field is rather unreliable, mainly because two plasmids can have similar sizes and yet be totally unrelated in sequence. Treatment of plasmids with restriction enzymes and subsequent separation by agarose gel electrophoresis allows a finer matching of sequences. If two plasmids are identical, the location of the restriction-site recognition sequences is the same. Therefore, following treatment with a particular restriction enzyme, an identical or similar pattern of fragments is obtained. Examples of such analyses are shown in Figure 16.2A (middle panel), which illustrates a case in which some of the fragments in strains A, B, C, and D are the same size and strains A and B may be identical.

To determine the relationship among any set of sequences, a method of DNA hybridization was developed by Edward Southern in 1975. This technique, called **Southern blotting** (Figure 16.2B), relies on the formation of stable duplexes (hybrids) between electrophoretically separated restriction fragments and radioactively labeled, denatured DNA probes. The restriction fragments are denatured into single-stranded DNA by treating the gel with an alkali. The gel is then overlaid with a nitrocellulose sheet. Capillary action (blotting) moves the DNA onto the nitrocellulose sheet, where it binds firmly. Specially treated nylon membranes may be used instead of nitrocellulose. In this way a permanent copy of the gel is obtained, with the fragments immobilized in the exact position where they migrated on the gel. This blot can then be soaked in a solution of a denatured radiolabeled probe. If the probe is one of the plasmids (for example, plasmid A), it will anneal to any complementary sequences immobilized on the solid support of the nitrocellulose. By use of this method, similar sequences can

be radioactively "tagged" and visualized following exposure of the blot to photographic film. Results presented in Figure 16.2A (right panel) indicate that plasmid A is probably identical to plasmid B, but only partially similar to plasmid C. Plasmid D is completely different from plasmid A, despite having an identical size and sharing some common-sized restriction fragments. Extra sequences found in plasmid C are probably due to recombination or transposition from other plasmids or chromosomal sites having sequences similar to plasmid A.

This powerful technique can be applied to the analysis of sequences found on specific fragments of bacterial or even eukaryotic chromosomes. DNA is extracted from cells and treated with restriction enzymes. As with plasmid DNA, fragments are separated according to size by using agarose gel electrophoresis. Unlike fractionation of plasmids, which yields only a handful of fragments, chromosomal DNA contains hundreds or thousands of recognition sites. Following gel electrophoresis, many bands appear, often preventing clear differentiation of one band from closely migrating bands of equivalent size. Nevertheless, following Southern blotting, a radiolabeled **probe** —a single-stranded piece of DNA with a sequence complimentary to a sequence present in the chromosomal DNA sample—can clearly identify related sequences in one or more of the restriction fragments, as shown in Figure 16.3. This method can detect proviruses in bacterial genomes (proviruses are viruses that have integrated their DNA into the host genome), as well as specific genes that encode determinants of antibiotic resistance or of bacterial pathogenicity, such as those that encode toxins.

Probes may be obtained from a variety of sources. Usually they are cloned pieces of DNA from different species that share some sequence similarity. Alternatively, a probe may be a short piece of chemically synthesized DNA or the DNA product from the polymerase chain reaction (PCR) (Box 16.1). The synthetic oligonucleotide probe can be made on the basis of a DNA sequence deduced from a known amino acid sequence of a protein. For example, an enzyme, X, has been purified and has an amino terminal sequence of Met-Trp-Asp-Trp. Based on the codons for these four amino acids (see Table 13.2), oligonucleotides ATGTGGGATTGG and ATGTGGGACTGG can be synthesized chemically and can be used as probes in a Southern blot hybridization to bind to fragments containing the complementary sequence. (Note that codons GAT and GAC can both encode aspartic acid, so two probes must be made.) Labeled with radioactive phosphorus or with fluorescent dyes, these probes hybridize with their complementary sequence on one of the DNA strands within the gene that encodes enzyme X. This probe can also be used directly to identify recombinant plasmids containing the cloned gene for enzyme X.

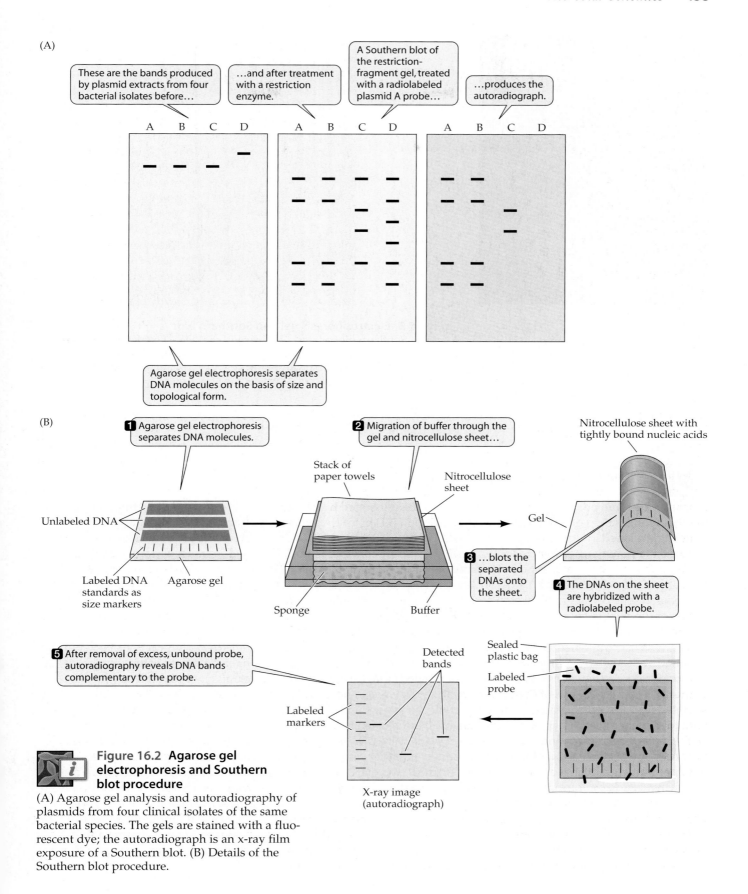

Figure 16.2 Agarose gel electrophoresis and Southern blot procedure
(A) Agarose gel analysis and autoradiography of plasmids from four clinical isolates of the same bacterial species. The gels are stained with a fluorescent dye; the autoradiograph is an x-ray film exposure of a Southern blot. (B) Details of the Southern blot procedure.

(A)

The photograph includes a ruler, so the size of the sample bands can be determined.

DNA fragments of known size (markers) are loaded on both sides of the gel.

(B)

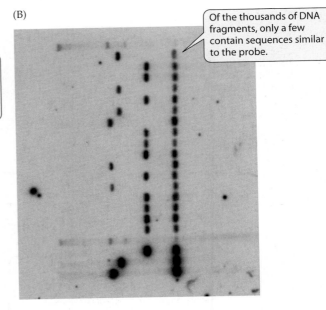

Of the thousands of DNA fragments, only a few contain sequences similar to the probe.

Figure 16.3 Electrophoresis gel and Southern blot
Results of Southern blot analysis of chromosomal DNA restriction fragments from different clinical isolates of the same bacterial species. (A) Agarose gel after electrophoresis, stained with fluorescent dye; (B) autoradiograph of the Southern blot. Courtesy of Steve Lory.

BOX 16.1   *Methods & Techniques*

The Polymerase Chain Reaction

One of the more powerful techniques for isolating genes that has been developed in recent years is called the **polymerase chain reaction** (PCR). PCR generates copies of a target DNA sequence by a repeated series of steps: denaturation of the double-stranded target DNA is followed by primer-directed synthesis of complementary strands, and then by additional rounds of denaturation and synthesis. Because synthesis of DNA strands by DNA polymerase requires primers (see Chapter 13), this technique also depends on knowledge of at least part of the target sequence, such that the polymerization can be initiated from short oligonucleotides that hybridize to the ends of the target. The PCR usually starts from a double-stranded DNA (called the **template**). However, the first DNA mol-

ecule can be generated by reverse transcription (RT) of mRNA to give a copy of complementary DNA (cDNA), which is converted into a double-stranded template. This modification of PCR is called RT-PCR. As illustrated in the figure, PCR is divided into cycles, each cycle consisting of three steps. Raising the temperature of the DNA solution denatures the double-stranded DNA. Next, the temperature is lowered, and complementary oligonucleotides (the primers) are allowed to anneal to the DNA. Finally, DNA polymerase is added, together with dATP, dCTP, dTTP, and dGTP, and the complementary strand is synthesized. The denaturation–primer annealing–complementary strand synthesis is repeated 30 to 40 times. Because each newly synthesized complementary strand can

serve as a template for the next round of PCR, the amount of DNA synthesis increases exponentially, and from a few starting DNA molecules, 20 cycles can generate more than a million copies. The PCR-generated DNA can be used for many techniques, among them DNA sequencing and preparation of DNA probes for Southern hybridizations, or they can be cloned into plasmid or viral vectors. It is noteworthy that the repeated denaturation exposes the DNA polymerase to temperatures that usually inactivate most enzymes. However, enzymes utilized in PCR are isolated from thermophilic bacteria (such as *Thermus aquaticus*), which usually replicate at high temperatures, and they possess thermostable DNA polymerases.

BOX 16.1 *Continued*

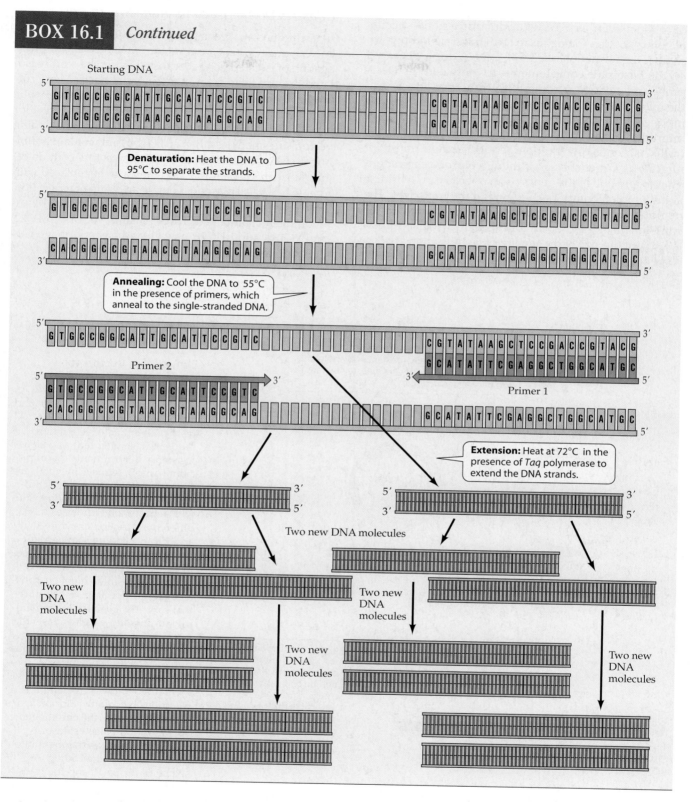

A technique called **northern blotting** is closely related to Southern blotting and is used to identify the presence of specific RNAs expressed by a cell. Following specialized nucleic acid isolation procedures that separate RNA from DNA, the total RNA is size-fractionated by agarose gel electrophoresis. It is then transferred onto a nylon sheet by blotting and is probed using labeled single-stranded DNA probe consisting of a gene or a portion of a gene, such as a PCR product or a gene-specific oligonucleotide. Analogous to the Southern blotting procedure, the formation of an RNA–DNA duplex is based on the base-pair complementarity between the DNA probe and a corresponding RNA transcript attached to the nylon sheet at a location determined by its size, giving the appearance of a labeled band at the position of migration of the RNA. Polycistronic messenger RNAs (mRNAs) containing multiple genes in a single transcript migrate as larger species, although a probe corresponding to any one of the genes in an operon can hybridize to its complementary portion within the transcript. The migration of RNA, as determined by the location of the band to which the probe hybridized, provides information about the size of the transcript, and its intensity corresponds to the abundance of the RNA in the cell.

Construction of Recombinant Plasmids

Extrachromosomal genetic elements, such as plasmids and bacteriophages, have become the logical choice for propagation of isolated genes because they can be readily isolated in pure form from their bacterial hosts. Many different plasmids and bacteriophages are used in genetic engineering as vectors, that is, vehicles of transmission for cloning isolated genes. Most of the useful plasmid cloning vectors are small (less than 10,000 bp) and contain antibiotic-resistance genes. Bacteriophage vectors contain regions that are dispensable, and these can be replaced with foreign DNA. All plasmids contain an origin of replication, which is a piece of DNA that allows the plasmid to be autonomously maintained. Three plasmid vectors, pBR322, pACYC184, and pUC18, as well as a bacteriophage vector, lambda, are shown schematically in Figure 16.4.

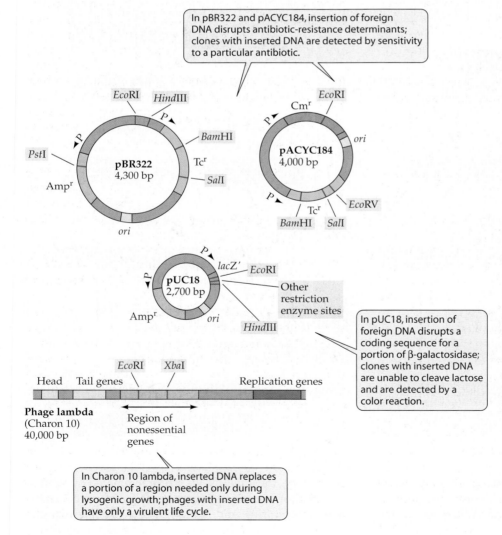

In pBR322 and pACYC184, insertion of foreign DNA disrupts antibiotic-resistance determinants; clones with inserted DNA are detected by sensitivity to a particular antibiotic.

In pUC18, insertion of foreign DNA disrupts a coding sequence for a portion of β-galactosidase; clones with inserted DNA are unable to cleave lactose and are detected by a color reaction.

In Charon 10 lambda, inserted DNA replaces a portion of a region needed only during lysogenic growth; phages with inserted DNA have only a virulent life cycle.

Figure 16.4 Vectors for construction of recombinant molecules
Plasmids pBR322, pACYC184, and pUC18 contain several unique restriction enzyme recognition sequences (shaded red) as well as resistance determinants for ampicillin (Amp^r), tetracycline (Tc^r), and chloramphenicol (Cm^r). Plasmid pUC18 has a multiple cloning site with recognition sequences for 13 different restriction enzymes that are within a coding sequence for a portion (LacZ′) of β-galactosidase. In pUC18, to screen for inserted DNA, the plasmids must be transformed into special *E. coli* strains that carry the gene for the remaining portion of β-galactosidase; the two portions of the protein form the active enzyme. Artificial formation of an active enzyme from two fragments (called alpha complementation) allows screening for insertions by colony color on agar containing X-gal, the chromogenic indicator of β-galactosidase. Insertion of a DNA fragment into pUC18 disrupts the coding sequence for LacZ, and the transformants are unable to form an active enzyme. Bacterial colonies with intact plasmids have a blue appearance; colonies with plasmids that carry inserts are white.

Construction of a recombinant plasmid begins by treating the appropriate vector with a restriction enzyme with a known recognition sequence and treating the DNA to be cloned (the donor DNA) with the same enzyme. The two populations of DNA molecules are mixed, and recombinants are generated by the addition of DNA ligases, which covalently ligate the foreign DNA into the plasmid vector. DNA ligase also recloses the vector. The mixture of ligated DNA is introduced into bacteria (usually *E. coli*) by transformation of artificially competent cells. The bacteria are plated onto selective media containing an antibiotic to which the plasmid encodes a resistance determinant. Only the cells that have taken up a plasmid expressing the antibiotic-resistance–specifying gene will grow on such a medium. Because, in general, the number of competent cells exceeds the number of reassembled plasmid molecules, each colony that arises from plating of transformants is expected to be a clone of a single-cell transformant that took up a single plasmid with a DNA insert. Individual colonies can be isolated and plasmids can be extracted. After treatment with a restriction enzyme, these can be analyzed by agarose gel electrophoresis. Cloning of fragments from a large plasmid into vector pBR322 is outlined in Figure 16.5.

The mixture of bacteria or phage containing the various inserts is called a **library** of DNA. A simple library, such as that outlined in Figure 16.5, is represented by four different clones plus one vector. Thus, the hundreds of possible transformants contain five populations of plasmids: four with inserts and one containing only the vector without an insert DNA. The identification of a specific clone, therefore, requires screening of at least four randomly selected clones by agarose gel electrophoresis of the plasmid digests. A more complex library, prepared by restriction enzyme treatment of bacterial or eukaryotic chromosomal DNA ligated with a vector, contains thousands of bacteria, each carrying a recombinant plasmid with a single fragment of the chromosome. This is called a **genomic library** and is a collection of plasmids carrying all of the chromosomal DNA as fragments in recombinant plasmids. The task of isolating genes, then, is reduced to identification of a desired clone from a library of DNA fragments in a vector.

Identification of Recombinant Clones

In construction of recombinant plasmids, one often deals with the technically more difficult problem of identifying a specific clone from the collection (library) of recombinant plasmids. A number of different techniques are available to investigators that allow them to identify a bacterium with a specific recombinant plasmid.

EXPRESSION OF A GENE IN A HETEROLOGOUS HOST A library of DNA fragments carried by thousands of bacteria can easily be screened for a desired cloned gene, provided that gene endows the host with a unique characteristic that distinguishes its particular host from others carrying different fragments of the donor chromosome. For example, to identify a clone encoding a hemolysin, a protein toxin made by *Pseudomonas aeruginosa*, a collection of *E. coli* carrying a library of *P. aeruginosa* DNA is plated on agar-based medium containing sheep blood. The clone containing the hemolysin gene can be identified by a zone of clearing around the colony, as shown in Figure 16.6. Similarly, hydrolysis of starch or casein can be used as a test for expression of amylases or proteases, respectively. These assays rely on expression of the gene in the cloning host, and this is not always guaranteed, especially if the gene in its natural host is regulated. These regulatory signals may be absent in the host carrying the recombinant plasmid.

High-level expression of a cloned gene product can be ensured by using specialized cloning vectors containing regulatory sequences adjacent to the cloning sites. These sequences are usually promoters of highly expressed genes, and the vectors containing these promoters are called **expression vectors**. Modified strong promoters from the lactose operon, bacteriophage lambda, or T7, have been incorporated into some vectors (Figure 16.7). Expression vectors allow efficient transcription, provided the cloned gene is in the same orientation as the promoter.

SCREENING WITH DNA OR ANTIBODY PROBES One of the simplest and most direct methods for the identification of a clone that carries a recombinant plasmid with a specific DNA sequence is to use DNA hybridization techniques. This approach does not depend on expression of the cloned gene. The method for identifying such a clone is yet another application of a localized DNA hybridization technique, **colony hybridization**, which is analogous to Southern blotting (Figure 16.8). The oligonucleotide used in the previous example to identify enzyme X can be used to screen a gene library of DNA fragments. Bacteria growing as colonies on a plate are allowed to stick to a nylon or nitrocellulose support, which is placed on the agar surface. Once the support is removed, these bacteria, which are bound to the support as an imprint of the original plate (hence called the **replica**), are lysed, and the DNA is denatured. The DNA becomes immobilized, and the radioactive probe hybridizes to the site on the support that corresponds to the colony carrying a plasmid with the gene coding for the enzyme.

Another technique based on colony hybridization uses an antibody as a probe. This necessitates the use

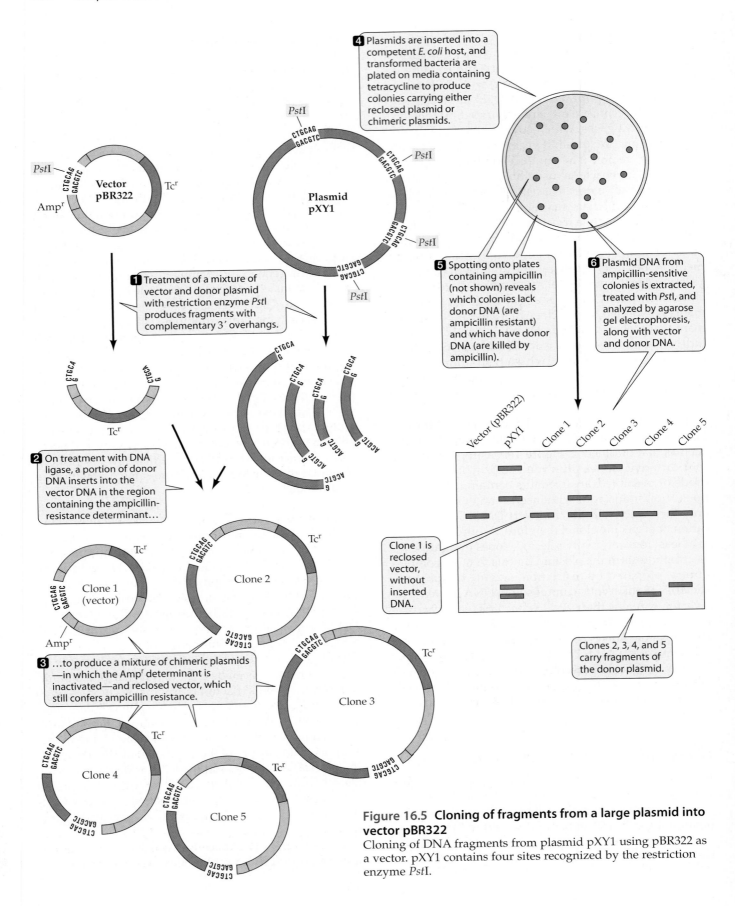

Figure 16.5 Cloning of fragments from a large plasmid into vector pBR322
Cloning of DNA fragments from plasmid pXY1 using pBR322 as a vector. pXY1 contains four sites recognized by the restriction enzyme *Pst*I.

Figure 16.6 Identification of hemolysin-containing clones from a library of chromosomal DNA
Hemolysin is an enzyme that breaks down red blood cells. DNA from a hemolytic microorganism, *Pseudomonas aeruginosa*, was treated with restriction enzyme *Bam*HI, and a library of this DNA was prepared in pBR322. The ligation mixture was plated onto blood agar plates to detect clones containing the hemolysin gene. Courtesy of Steve Lory.

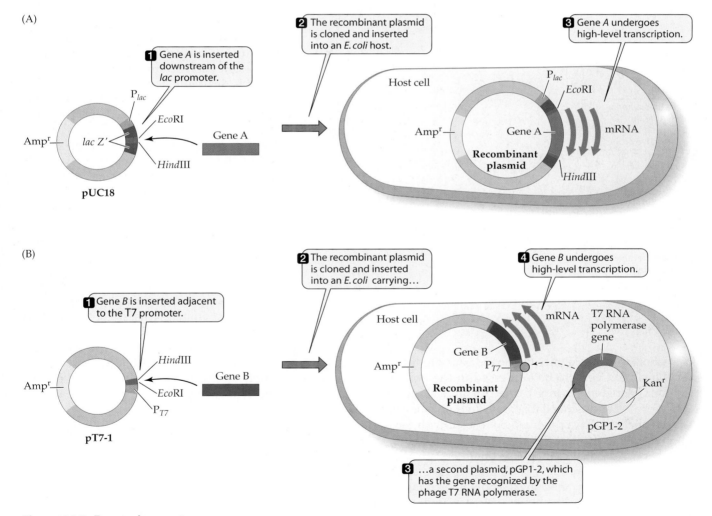

Figure 16.7 Expression vectors
(A) Plasmid pUC18 contains the promoter for the *lac* operon of *E. coli* (P_{lac}) adjacent to a cluster of restriction sites, which can be used to insert foreign genes. Transcription from P_{lac} results in high-level expression of the cloned gene. (B) An expression system based on components of a bacteriophage. The cloning vector pT7-1 contains a promoter from bacteriophage T7 (P_{T7}), which is recognized by a specialized RNA polymerase from the same phage.

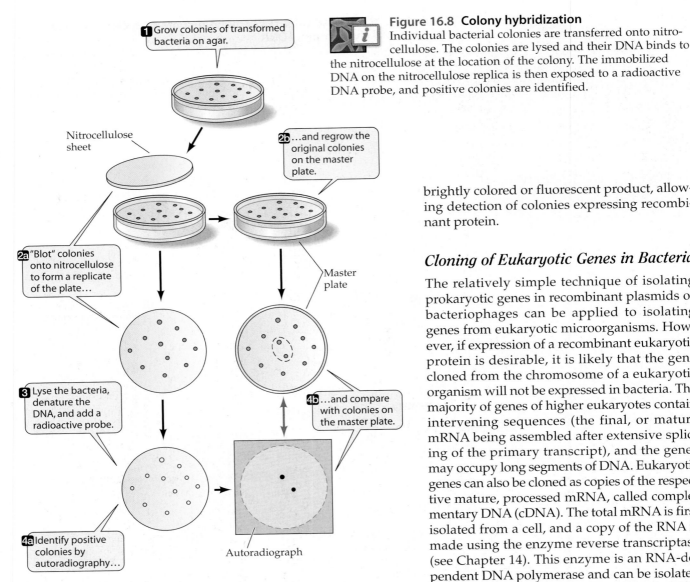

1 Grow colonies of transformed bacteria on agar.

Nitrocellulose sheet

2b ...and regrow the original colonies on the master plate.

2a "Blot" colonies onto nitrocellulose to form a replicate of the plate...

Master plate

3 Lyse the bacteria, denature the DNA, and add a radioactive probe.

4b ...and compare with colonies on the master plate.

4a Identify positive colonies by autoradiography...

Autoradiograph

Figure 16.8 Colony hybridization
Individual bacterial colonies are transferred onto nitrocellulose. The colonies are lysed and their DNA binds to the nitrocellulose at the location of the colony. The immobilized DNA on the nitrocellulose replica is then exposed to a radioactive DNA probe, and positive colonies are identified.

brightly colored or fluorescent product, allowing detection of colonies expressing recombinant protein.

Cloning of Eukaryotic Genes in Bacteria

The relatively simple technique of isolating prokaryotic genes in recombinant plasmids or bacteriophages can be applied to isolating genes from eukaryotic microorganisms. However, if expression of a recombinant eukaryotic protein is desirable, it is likely that the gene cloned from the chromosome of a eukaryotic organism will not be expressed in bacteria. The majority of genes of higher eukaryotes contain intervening sequences (the final, or mature mRNA being assembled after extensive splicing of the primary transcript), and the genes may occupy long segments of DNA. Eukaryotic genes can also be cloned as copies of the respective mature, processed mRNA, called complementary DNA (cDNA). The total mRNA is first isolated from a cell, and a copy of the RNA is made using the enzyme reverse transcriptase (see Chapter 14). This enzyme is an RNA-dependent DNA polymerase and can be isolated from selected animal viruses. The RNA–DNA hybrid is denatured, the RNA component is removed by hydrolysis, and a complementary copy of the DNA is synthesized with DNA polymerase. The final double-stranded product can be cloned into phage and plasmid vectors.

Practical Applications of Recombinant DNA Technology

Recombinant DNA technology has revolutionized biological and medical research by providing scientists with the ability to obtain large quantities of products that were previously available from their natural sources only in minute quantities. For example, rare enzymes from both bacterial and eukaryotic sources can now be obtained in gram quantities by cloning the appropriate

of antibodies that recognize the cloned gene product and for the cloned gene to be expressed in the host bacterium. Colonies of bacteria carried on nitrocellulose filters, prepared as described for colony hybridization with a radiolabeled DNA probe, can be reacted with the antibody probe. To detect the binding of the antibody to proteins within the lysed colony, the filters are soaked in a solution of the primary antibody that has been "labeled" by chemically attaching a highly fluorescing compound. Alternatively, the bound primary antibody can be detected using a secondary antibody—also called anti-antibody—that recognizes the protein-specific primary antibody. The secondary antibody is covalently attached to an enzyme that converts a substrate to a

genes in expression vectors and isolating the recombinant products from bacterial extracts. Bacteria, as the source of recombinant products, offer an additional economic advantage because they can be propagated on relatively inexpensive media and grow rapidly.

Bacterial, and, more recently, yeast-based, expression systems have been extensively employed in industrial-scale production of various hormones, such as insulin and human growth hormone, as well as in vaccines against hepatitis B and herpes (see Chapter 31). An added benefit of recombinant products, recently realized, is that they are often safer than their natural counterparts, being free of potentially allergenic protein contaminants. Finally, many microorganisms, particularly the members of the genus *Streptomyces*, produce metabolites and antibiotics that are useful in therapy for infectious diseases, and some have been shown to possess anti-cancer activities. Through genetic engineering of the genes encoding the components of the biosynthetic pathways, and by combining enzymes or catalytic domains of enzymes from different species of microorganisms, recombinant organisms producing various useful metabolites can be constructed. Often these organisms produce novel drugs that can be obtained by relatively inexpensive fermentation processes. Recently, crop plants, many of which can be transformed (genetically engineered) by the process outlined in Box 16.2, have been explored as a possible system for synthesizing pharmaceutical products.

Currently research is devoted to development of genetically engineered vaccines for a variety of infectious diseases. This includes introducing specific mutations into the proteins that make up vaccine components, such that they retain their immunogenicity but do not have harmful effects on the vaccinated patients. This approach has been used to engineer **toxoids**, detoxified bacterial protein toxins. Genetically attenuated bacterial and viral hosts are also being developed to provide a safe and efficient way to deliver such vaccines.

An integral part of the identification and study of genes is the knowledge of the precise nucleotide sequence that allows deduction of the amino acid sequence of the protein product. Moreover, sequence information has been useful in studies of molecular evolution, by comparing sequences between common genes and thus establishing relationships among genes and organisms (Box 16.3; see Chapter 17). The ability to detect the presence of certain genes in bacteria has been exploited in diagnostic microbiology as well. Agarose gel electrophoresis of plasmid DNA, coupled with Southern blot analysis, often allows tracking of bacteria responsible for dissemination of antibiotic-resistance genes. This type of plasmid fingerprinting allows detection of specific size categories of plasmids, as well as specific DNA sequences that reside on such plasmids as a result of transposition from the chromosome or another plasmid. Another molecular method, PCR, is becoming a widely accepted tool for identification of pathogens in diagnostic laboratories. Moreover, PCR detection of pathogens based on "signature" DNA sequences can facilitate their identification in tissue samples, eliminating the need for time-consuming laboratory culture.

SECTION HIGHLIGHTS

Classical bacterial genetics was the foundation for recombinant DNA research, which is based on the ability of scientists to **isolate individual genes** from more complex genomes. Restriction endonucleases, made by bacteria, introduce double-stranded breaks at specific DNA sequences. These are used to digest DNA into pieces that contain parts of genes or, if fragments are sufficiently large, intact genes. Restriction endonucleases, together with their cognate **modification enzymes**, serve as barriers to unrestricted DNA exchange among different bacterial species. Bacteria carrying a specific gene—a **clone**—are identified by colony hybridization using a single-stranded complementary DNA probe, or by immunological or activity screens to detect expression of the gene's protein product. Specialized cloning vectors, called **expression vectors**, include specific promoter sequences to maximize expression of the cloned genes. Eukaryotic genes encoding proteins are often cloned from mRNA sequences. First, a copy of mRNA, called **cDNA**, is made by the enzyme **reverse transcriptase**, and the second DNA strand, complementary to cDNA, is made by DNA polymerase. This double-stranded form is then inserted into the appropriate plasmid or bacteriophage vectors. Any sequence from a genome, or from a mixture of various DNA species, can be isolated by repeated denaturation–complementary strand synthesis–denaturation cycles with sequence-specific primers—a technique called the **polymerase chain reaction** (**PCR**). Specific DNA sequences within a DNA fragment fractionated by agarose gel electrophoresis can be identified by membrane **hybridization** using a known single-stranded DNA probe with a complementary sequence. This procedure is known as **Southern blotting**.

BOX 16.2 *Research Highlights*

Genetic Engineering of Plants Using *Agrobacterium tumefaciens* and Ti Plasmids

Crown gall disease in plants, described in Chapter 15, is caused by integration into a plant's chromosome of genes from the bacterium *Agrobacterium tumefaciens*. Recall that the genes that are transferred are on the bacterium's Ti plasmid, within the T-DNA region, which is flanked by short repeats, called the left and right borders (LB and RB, respectively). *Agrobacterium*'s *vir* genes encode proteins that promote the transfer of T-DNA from *Agrobacterium* into the plant and across the nuclear membrane, where it integrates into the chromosome. Only the T-DNA region (as defined by LB and RB) is integrated.

Genetic engineers have exploited this natural system by replacing the tumor-causing genes between LB and RB with genes that may produce new, desirable traits in the plant. In fact, the plasmids carrying the foreign genes between LB and RB need not be part of the Ti-plasmid. The *Agrobacterium* strains used for genetic engineering of plants carry two plasmids. One is a "disarmed" Ti-plasmid that has been modified by removing the genes involved in tumor formation but carries functional *vir* genes. The strains also carry a second plasmid, the so-called binary vector, where the foreign gene intended for transfer into the plant nucleus can be inserted between the LB and RB. Following infection of cultured plant cells, and under the direction of the *vir* gene products expressed from the disarmed Ti-plasmid, the region flanked by LB and RB is

excised from the binary vector and is transferred into the plant, where it integrates into the plant chromosome. The steps involved are outlined in the accompanying figure.

When the plant cells carrying T-DNA divide, each daughter cell receives a copy of the cloned DNA. This can be verified by including a drug-resistance marker gene among the cloned genes. The cells are hormonally induced to produce roots and shoots, and grow into plants in which every cell contains a copy of the cloned DNA. Such plants will now express the new trait introduced by the T-DNA–mediated gene transfer. *Arabidopsis thaliana*, a plant used extensively for research, can be genetically engineered by dipping its flowers in a suspension of *Agrobacterium*. Some of the resulting seeds produce plants carrying the T-DNA. The foreign genes used for genetically engineering plants include those that improve growth characteristics of the plant; speed up ripening of fruit;

render the plant resistant to insects, viruses, or disease-causing bacteria; or increase tolerance to herbicides or to bacterial toxins used for insect control.

1 For modification of plants, a strain of *Agrobacterium* carries a disarmed Ti-plasmid and a second plasmid, a binary vector carrying a foreign gene inserted between the LB and RB repeats. This region also contains a drug-resistance gene.

Foreign gene · Drug-resistance gene

LB · RB

vir

Disarmed Ti-plasmid

Agrobacterium

Plant cell

2 The *vir* genes of the Ti-plasmid promote transfer of the region between the LB and RB into the plant nucleus…

3 …where it inserts into the chromosome.

4 Transformed cultured plant cells are propagated on medium containing antibiotics.

5 Plants are regenerated. The mature plant exhibits new traits provided by the foreign gene.

BOX 16.3 *Methods & Techniques*

Determination of a Specific DNA Sequence of a Gene

There are several methods available for the determination of a specific nucleotide sequence of a gene, primarily by sequencing of DNA fragments cloned in various cloning vectors. The technique most commonly employed for sequencing large fragments of DNA is based on a method devised by Frederick Sanger and colleagues in 1977. This method relies on synthesis of complementary DNA in a test tube that contains modified nucleoside triphosphates, called 2′,3′-dideoxynucleoside triphosphates (ddNTPs), along with a slight excess of natural deoxynucleoside triphos-

phates (dNTPs). These dideoxynucleotides can be incorporated into the growing chain; however, they lack the necessary 3′ hydroxyl group to form a phosphodiester bond with the next dNTP. The consequence of incorporation of ddNTP bases is termination of DNA synthesis at a precise place. DNA sequencing by this method is carried out in four reactions, each containing a mixture with a single-stranded template, containing the DNA sequence of interest, and a short radiolabeled oligonucleotide primer, which anneals to its complementary sequence. The location

of the primer determines the starting point of the sequence analysis. The four reactions are started by adding DNA polymerase I, all four dNTPs in excess, and a small amount of one ddNTP to each reaction. The chain elongation reaction proceeds using dNTPs, but incorporation of a ddNTP terminates synthesis. For example, the reaction using ddATP will terminate at positions on the template containing a thymine (illustrated in the accompanying figure, A). The other three ddNTPs will terminate reactions corresponding to the positions of their complementary bases on the

Continued on next page

(A)

DNA sequence determination using the dideoxynucleotide (ddNTP) chain termination method. Shown here are two techniques for analyzing the products of reactions using individual ddNTPs. (A) Analysis of radioactive fragments by acrylamide gel electrophoresis. Part (B) is on next page.

BOX 16.3 *Continued*

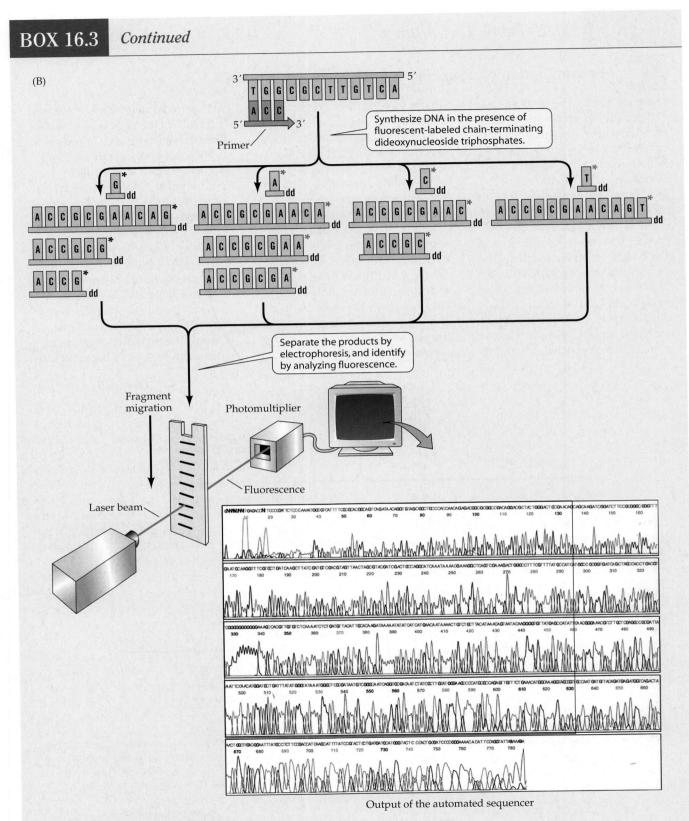

Output of the automated sequencer

Sequence analysis by incorporation of fluorescent ddNTPs and automated sequencing.

| BOX 16.3 | *Continued* |

their complementary bases on the DNA. These reactions give radioactively labeled products of different length, which are size-fractionated by electrophoresis on acrylamide gels. The relative size of the fragments and the source of the ddNTP will allow determination of a corresponding sequence in the 5′ direction from the primer. This method allows sequence analysis of a region of DNA up to 1,000 bp from the site of the primer hybridization. Once this sequence is known, a new primer can be chemically synthesized, radioactively labeled, and annealed to the same DNA template or a new template carrying an additional portion of the gene; this will serve as a starting point for the next round of sequencing.

An innovation in DNA sequencing technology was made possible through the use of dideoxynucleotides that are labeled with fluorescent dyes (see accompanying figure, B). Each of the four dyes emits fluorescence at a different wavelength, and the bands migrating in the gel can be detected by a laser scanner. The four reactions are combined and analyzed in a single lane of a gel. This process can be automated, and a sequencing machine can analyze up to a hundred sequencing reactions at a time.

16.2 Microbial Genomics

One of the youngest disciplines of microbiology, microbial genomics, is based on increasing the efficiency of the simple method of DNA sequencing, which can then be applied to entire genomes of cells. The era of microbial genomics refers to the start of a flurry of sequencing activities and completion of genomic sequences of over one hundred different bacterial species. The contribution of genomic sequencing to the understanding of cellular processes will undoubtedly be experienced for a long time. One of the most exciting results of the genomic sequencing projects is that a significant fraction (in some bacteria, nearly one-half) of all proteins deduced from the nucleotide sequence of corresponding genes cannot be assigned a biological function. This remarkable finding, in the face of the impressive body of knowledge in prokaryotic and eukaryotic genetics, biochemistry, and physiology, indicates that many biological processes remain to be discovered.

Rapid automation of DNA sequencing technology enabled the sequencing of large segments of DNA in relatively short periods. The most important accomplishment in this regard took place in 1995, when the entire genome of a living organism was sequenced. Although the genome of *Haemophilus influenzae* is relatively small (only 1,830 kb), it represented one of the major scientific accomplishments in biology. The strategy used to achieve this, called **whole-genome shotgun sequencing** (Figure 16.9), became a model for all sequencing projects involving bacteria, and it has also been used in various parts of the project to sequence the entire human genome. This approach takes advantage of the high efficiency of automated DNA sequencing machines and computer-assisted tools, which are used to assemble the final sequence of the genome.

The process of generating the sequence of an entire bacterial genome begins with the isolation of chromosomal DNA from a culture of the organism and preparation of a DNA fragment library in a cloning vector. This is analogous to the procedure described earlier in this chapter (illustrated in Figure 16.5) for random cloning of DNA fragments into plasmids. Once a large collection of plasmids carrying chromosomal DNA inserts is generated, individual plasmids are isolated, and the ends of each insert are sequenced using dideoxy-termination methods. Approximately 500 to 800 bp of DNA sequence are routinely obtained from each end. These sequences are then entered into a computer, which identifies overlaps of these partial sequences and aligns these overlaps. This stage is called sequence assembly. Once a sufficient number of short sequences has been generated (up to 60,000 sequences of 500 to 800 bp each for a genome of 4,000,000 bp) and assembled, the genome sequence is now represented as a series of long DNA segments, called contigs (short for contiguous), which may be separated by regions where insufficient sequence information was generated. These assembly gaps often represent "toxic" sequences that are not clonable and that are therefore absent from the original insert library. Alternatively, these gaps may represent repetitive sequences in the genome, which cannot be readily assembled. The gap closure and final stage of assembly of the genome by contig linking involves PCR amplification across each gap, followed by sequencing the entire PCR product without cloning it into a vector.

The final stage of a sequencing project consists of the identification of sequences that code for proteins or ribosomal or transfer RNA (tRNA). This process, called **annotation**, is carried out with the extensive use of computer-assisted tools. From the linear nucleotide sequence, the coding sequence for each gene is determined. Because

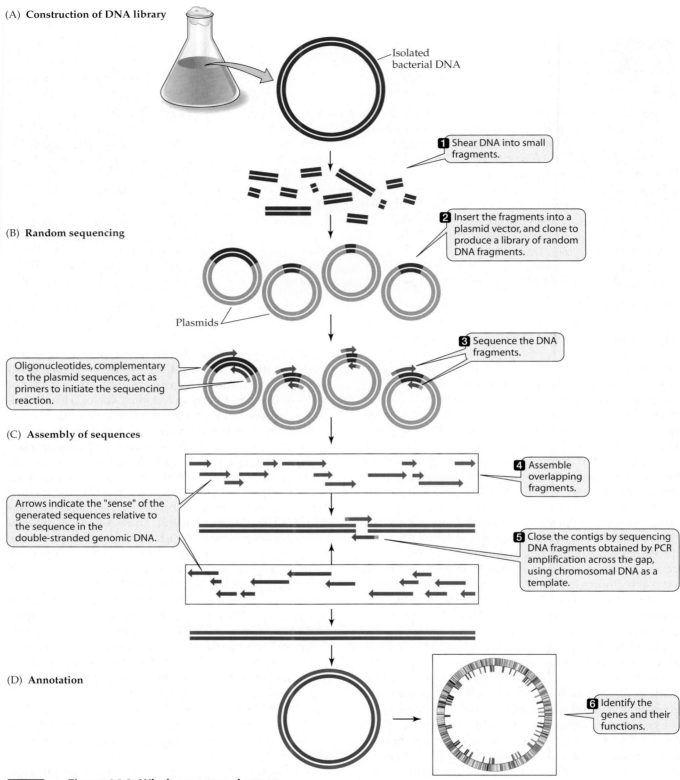

(A) Construction of DNA library

Isolated bacterial DNA

1 Shear DNA into small fragments.

(B) Random sequencing

2 Insert the fragments into a plasmid vector, and clone to produce a library of random DNA fragments.

Plasmids

3 Sequence the DNA fragments.

Oligonucleotides, complementary to the plasmid sequences, act as primers to initiate the sequencing reaction.

(C) Assembly of sequences

4 Assemble overlapping fragments.

Arrows indicate the "sense" of the generated sequences relative to the sequence in the double-stranded genomic DNA.

5 Close the contigs by sequencing DNA fragments obtained by PCR amplification across the gap, using chromosomal DNA as a template.

(D) Annotation

6 Identify the genes and their functions.

Figure 16.9 Whole-genome shotgun sequencing
Determination of a microbial genome sequence. (A) Construction of a random library of DNA fragments in a cloning vector. (B) Random sequencing of clones. Short sequences are obtained from each end of the cloned DNA, and thousands of clones are sequenced (see Box 16.3 for a detailed discussion of DNA sequencing methods).

(C) Assembly of the final sequence of the bacterial genome. The thousands of short sequences are assembled into long segments called contigs, which are cloned. The assembled bacterial genome is usually a continuous circle. (D) Completion of the sequencing project. Annotation involves the identification of genes and, where possible, designation of a biological function for the gene products.

there are three possible reading frames for a protein within each DNA strand (see Chapter 13), several criteria are used to assign the correct sequence for a gene. These include the recognition of initiation and termination codons, the presence of a ribosome-binding site (Shine-Dalgarno sequence) preceding the initiation codons, and the codon usage pattern known to be employed by the protein synthesis machinery of the organism. Genes, as characterized by a contiguous coding sequence between initiation and termination codons, are also called **open reading frames (ORFs)**. Once these are identified, predictions of function for a product of the gene can be made. The process of annotation is perhaps the most important part of a genome-sequencing project, because it generates information about the possible function of all recognizable genes encoded in the genome, and it defines the genetic repertoire of the organism (Box 16.4).

Most bacteria have a single circular chromosome, but notable exceptions can be found, such as the two chromosomes for *Vibrio cholerae* and a circular and linear chromosome in *Agrobacterium tumefaciences*. As shown in **Figure 16.10**, the DNA sequence is densely packed with genes, and only a small percentage of the DNA consists of intergenic regions, which are often binding sites for RNA polymerase and transcriptional regulators.

The annotation of the genome serves as a blueprint of the hundreds of interactive processes in the living cell. Even cursory examination of the various genes that can be tentatively assigned functions can provide some interesting insights into the lifestyle of a particular bacterium. For example, the fraction of genes dedicated to transcriptional regulation (Table 16.2) varies greatly among the different bacteria. In *P. aeruginosa*, a free-living prototroph capable of colonizing a wide range of environmental niches, 8.4% of its genome encodes putative transcriptional regulators. This may reflect the range of environmental signals that this microorganism is capable of responding to, and other prototrophic bacteria similarly dedicate a high percentage of their genomes to transcriptional regulation. At the other extreme is *Helicobacter pylori*, a human pathogen that thrives in a very specific environment (the human stomach) with a small range of environmental variables, and this is reflected in a very limited regulatory circuitry. Only 1.1% of the *H. pylori* genome encodes regulatory proteins.

One of the striking findings is the large number of genes that encode proteins involved in unknown cellular processes (Table 16.3). These proteins of unknown function can be further subdivided into two groups. One

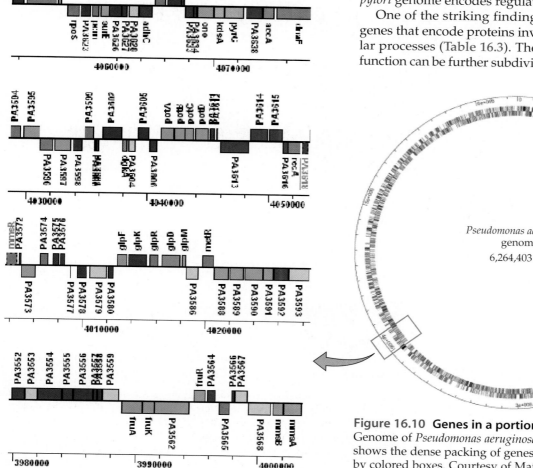

Figure 16.10 Genes in a portion of bacterial genome Genome of *Pseudomonas aeruginosa*. The enlarged area shows the dense packing of genes, which are represented by colored boxes. Courtesy of Mathew Wolfgang.

BOX 16.4 | *Methods & Techniques*

Alignment of Protein Sequences Deduced from Genome Sequencing

A key feature of annotation is computer-assisted homology assignment, based on sequence similarity with a known gene in a different organism. This example shows the amino acid sequence alignment of several proteins—related to the *murB* gene product of *E. coli*—that were predicted from the genomic sequences of *P. aeruginosa, V. cholerae, H. influenzae, M. tuberculosis,* and *H. pylori.* Boxed amino acids represent those that are identical or similar at each position of the different MurB proteins. All of these proteins have been assigned a function: UDP-*N*-acetylglucosamine enolpyruvyl reductase, which catalyzes a reaction in peptidoglycan synthesis. Although this enzymatic activity has been characterized only in *E. coli*, it is very likely that all of these proteins catalyze the same biosynthetic reaction in all bacteria that contain peptidoglycan.

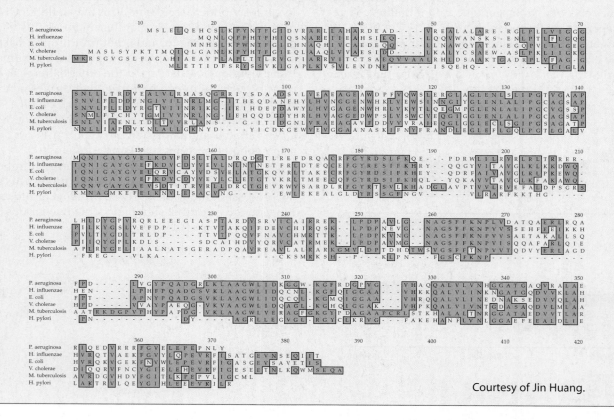

Courtesy of Jin Huang.

TABLE 16.2	Comparison of regulatory genes in selected bacterial genomes		
Microorganism	**# Genes in the Genome**	**# Regulatory Proteins**	**% of Total**
Pseudomonas aeruginosa	5570	468	8.4
Escherichia coli	4289	250	5.8
Bacillus subtilis	4100	217	5.3
Mycobacterium tuberculosis	3918	117	3.0
Helicobacter pylori	1566	18	1.1

category includes proteins of unknown function that have similar proteins present in more than one bacterial genome. The other group contains proteins of unknown function that are unique to a particular bacterium. Given that only a small fraction of bacterial genomes have been sequenced, the complexity of life represented by bacteria is truly amazing. A survey of 24 genomes, encompassing 51,627 genes, shows that 21,248 of them encode products of unknown function, and 11,083 of these products are unique.

Functional assignments can also provide a bird's-eye view of metabolic activities in the cell. Such reconstructions of the potential cellular processes are based on predicted functions deduced from the annotated genome sequence. They can provide comprehensive insights into the ability of a particular bacterium to function in specific environments. A transport and metabolic reconstruction, such as one shown for *Neisseria meningitidis* in Figure 16.11, can suggest various pathways of nutrient uptake, energy generation, biosynthesis, and survival in the environment of the infected host.

Horizontal Gene Transfer

The availability of complete sequences of genomes from taxonomically related as well as unrelated bacteria, including sequences of multiple strains of the same species, provided many new insights into prokaryotic evolution. The most striking result that arose from the analysis of gene content in sequenced genomes was the recognition of the plasticity of bacterial genomes, whereby new genetic traits are often not acquired by vertical (i.e., from "mother cell to daughter cells") transmission. Transmission of genetic material from one organism to another which is not its offspring is referred to as **horizontal (lateral) gene transfer**. The amount of horizontally acquired genes varies for different species. For example, the percentage of foreign genes in various *E. coli* genomes has been estimated to be as high as 20%, whereas many bacterial strains, particularly those with small genomes, show no evidence of horizontal gene transfer. A major factor in the rapid evolution and adaptation of bacteria to new environments is the acquisition of novel traits by horizontal gene transfer in the form

TABLE 16.3	Distribution of genes of unknown function among selected bacterial genomes					
Organism	**Genome Size (Million bp)**	**Number of ORFs**	**Genes of Unknown Function (% of total)**		**Genes Unique to the Organism (% of total)**	
A. aeolicus VF5	1.50	1,749	663	(44%)	407	(27%)
A. fulgidus	2.18	2,437	1,315	(54%)	641	(26%)
B. subtilis	4.20	4,779	1,722	(42%)	1,053	(26%)
B. burgdorferi	1.44	1,738	1,132	(65%)	682	(39%)
Chlamydia pneumoniae AR39	1.23	1,134	543	(48%)	262	(23%)
Chlamydia trachomatis MoPn	1.07	936	353	(38%)	77	(8%)
C. trachomatis serovar D	1.04	928	290	(32%)	255	(29%)
Deinococcus radiodurans	3.28	3,187	1,715	(54%)	1,001	(31%)
E. coli K-12-MG1655	4.60	5,295	1,632	(38%)	1,114	(26%)
H. influenzae	1.83	1,738	595	(35%)	237	(14%)
H. pylori 26695	1.66	1,589	744	(45%)	539	(33%)
Methanobacterium thermotautotrophicum	1.75	2,008	1,010	(54%)	496	(27%)
Methanococcus jannaschii	1.66	1,783	1,076	(62%)	525	(30%)
M. tuberculosis CSU#93	4.41	4,275	1,521	(39%)	606	(15%)
M. genitalium	0.58	483	173	(37%)	7	(2%)
M. pneumoniae	0.81	680	248	(37%)	67	(10%)
N. meningitidis MC58	2.24	2,155	856	(40%)	517	(24%)
Pyrococcus horikoshii OT3	1.74	1,994	589	(42%)	453	(22%)
Rickettsia prowazekii	1.11	878	311	(37%)	209	(25%)
Synechocystis sp.	3.57	4,003	2,384	(75%)	1,426	(45%)
T. maritima MSB8	1.86	1,879	863	(46%)	373	(26%)
T. pallidum	1.14	1,039	461	(44%)	280	(27%)
Vibrio cholerae	4.03	3,890	1,806	(46%)	934	(24%)
Total	**48.93**	**50,577**	**22,358**	**(43%)**	**12,161**	**(23%)**

From Fraser, et al. 2000. *Nature* 406: 800.

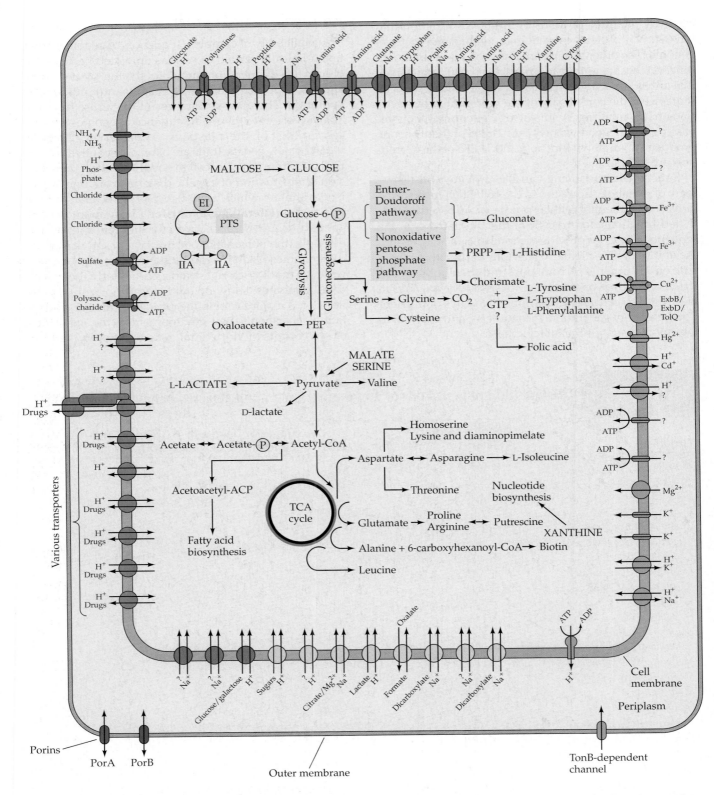

Figure 16.11 Cellular functions based on an annotated genome

Reconstruction of transport and metabolism of *Neisseria meningitidis*, based on the annotated genome. The reconstruction shows the potential pathways for the generation of energy and metabolism of organic compounds. Question marks indicate that a transporter's substrates are unknown, and functional assignment is based on its overall similarity to other transporters. After Nelson, K. T., I. T. Paulson and C. M. Fraser. 2001. *ASM News* 67: 310–317.

of complete genes encoding fully functioning products. A large number of genes can be acquired in a single genetic exchange event. These genes can have an immediate impact in the recipient. In contrast, the evolution of genes through mutations is slow and may take many generations before a new phenotype emerges.

For genes to be transmitted by horizontal gene transfer, several genetic and molecular mechanisms must be functional before the genetic information of one organism becomes part of the permanent genetic repertoire of the recipient. First, there has to be a gene transfer mechanism operating in the context of the donor, the recipient, and the particular environment where these two cell types encounter each other. Any one of the three mechanisms of genetic exchange (transformation, conjugation, and transduction) discussed in Chapter 15, could serve as vehicles for moving DNA into a recipient cell. This acquired DNA must have a specific means of replication, which is typically gained by incorporation into the chromosome. This could occur by homologous recombination or by site-specific recombination events. Finally, the genes have to be expressed under conditions where this particular physiological trait is useful.

Not only individual genes or portions of genes can be transferred by horizontal gene transfer, but also large segments, ranging from ten to hundreds of kilobase pairs, can be acquired by bacteria through horizontal gene transfer. These large blocks of foreign DNA, incorporated in bacterial genomes, are called **genomic islands**. The islands often specify functions that are associated with a specialized lifestyle, including virulence, plant–bacterial symbiosis, unusual metabolic capabilities, and antibiotic resistance. For example, genomic islands encoding virulence determinants are called pathogenicity islands. They have been described in a wide variety of bacterial pathogens, and in many instances have been implicated as genetic determinants responsible for conferring a non-pathogenic species with virulence traits.

Because genomic islands acquired by an organism did not co-evolve with the rest of the genome, they can be recognized in sequenced bacterial genomes by several signature features. A typical genomic island is depicted in Figure 16.12. The shared features of these islands include atypical nucleotide composition (% G + C), a unique pattern of usage of the genetic code that is different from the genes comprising the conserved portions of the chromosome, and, often, a similar location in the host chromosome. If the mechanism of transmission of a particular genomic island involves specialized transducing bacteriophages or conjugative plasmids, the islands will also include genes specifying determinants of intercellular movement and stable integration into the recipient's genome. Genomic islands are often adjacent to tRNA genes, which serve in the recipient cells as targets for the integration of the newly acquired DNA. The islands are flanked by specific sequences (direct repeats) that are created by recombination between identical sites on the chromosome and the foreign DNA, which has circularized following its entry into the cytoplasm of the bacterial host. The mechanism of recombination that causes the integration of genomic islands into the chromosome resembles the generation of prophage by the lambda bacteriophage or an Hfr-like structure by the F plasmid. This process is discussed in Chapter 15. Moreover, the genomic islands frequently contain transposons and insertion sequences. This suggests that these mobile genetic elements are capable of "hitchhiking" and exploiting horizontal gene transfer to gain access to new bacterial hosts.

Metagenomics

Molecular tools have been also exploited in microbial taxonomy, where sequences of portions of 16S rRNA genes are used to identify individual bacteria and establish evolutionary relationships among different species

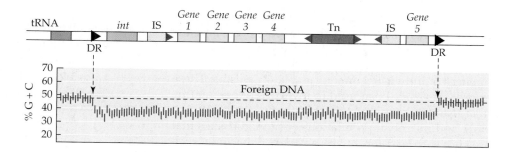

Figure 16.12 A schematic representation of a genomic island
The islands are typically adjacent to tRNA genes and are flanked by direct repeats (DRs). The analysis of the nucleotide composition (% G + C) usually reveals a significant deviation from the rest of the chromosome. A number of genes are carried on the islands and they occasionally acquire transposons (Tn) and insertion sequences (IS). The integrase gene (*int*) encodes the enzyme mediating insertion of the island into the chromosome.

(see Chapter 17). This approach also can be applied toward the analysis of bacterial diversity in the environment. Moreover, the 16S rRNA gene sequence information that identifies individual bacteria can be obtained without the need to culture the microorganisms. Overcoming the technical difficulties associated with the analysis of DNA isolated from microbial communities in environmental samples ("environmental DNA") was greatly facilitated by the application of PCR, using primers corresponding to conserved sequences flanking species-specific segments of ribosomal RNA genes. The PCR-generated fragments are cloned into suitable vectors and are subjected to DNA sequence analysis. This work led to the remarkable conclusion that more than 99% of prokaryotes in the environment are previously unrecognized species and the majority of them are not culturable in the laboratory.

The field of metagenomics takes the study of the environmental genetic repertoire even further. Metagenomics is defined as the genetic analysis of a total environmental sample, including nonculturable organisms. Following extraction of total DNA from an environmental sample, it is directly amplified by PCR, or it is fragmented by restriction endonuclease treatment. Libraries of these fragments are generated and propagated in surrogate bacterial hosts such as *E. coli*. Sequencing of these plasmid inserts, or the occasional successful expression of the cloned genes, can be used to identify proteins encoded in the genomes of the organisms present in the original sample. This work led to the discovery of novel enzymes and entire metabolic pathways that were previously unrecognized because the bacteria carrying the corresponding genes could not be cultured in the laboratory. The use of environmental DNA has generated much interest as a potential means for obtaining useful novel metabolites, such as antibiotics, that may be produced by nonculturable organisms. Simple cloning of DNA fragments that are large enough to carry the genetic information for a particular biosynthetic pathway may be sufficient to gain access to these important natural products.

Libraries of recombinant plasmids carrying fragments of environmental DNA can also be sequenced, and, using overlapping sequences, they can be assembled into contigs and, occasionally, even into entire genomic sequences of individual nonculturable organisms. These approaches have been used to investigate total marine microbial communities, inhabitants of various polluted sites, and microbial populations of human tissues that are known to be colonized by a variety of bacteria, such as the oral cavity and the intestinal tract. In practice, the successful assembly of contigs into complete genomes is difficult and depends on the limited diversity of the organisms that were the source of the original DNA sample. The most promising application of metagenomic approaches to human health focuses on determination of

causative agents of certain diseases where the symptoms suggest microbial infections, yet no obvious candidate pathogens have been identified. It is conceivable that bacteria that escaped detection—because of lack of information about media and conditions that are necessary for their successful propagation in the laboratory—may be responsible for these diseases.

SECTION HIGHLIGHTS

The sequence of a microbial genome is determined by a process called whole-genome shotgun sequencing. The first step of this process is to construct a library of DNA fragments in cloning vectors. Thousands of randomly selected clones are sequenced, each providing up to 1,000 bp of sequence from a portion of the genome. The short sequences are assembled into longer segments, called contigs, by aligning the short overlapping sequences. Gaps between the contigs are closed by specialized methods, such as PCR amplification across the gap, followed by sequencing of the PCR products. The final stage of genome sequencing is **annotation** of the genome. First, the positions of genes are localized within the sequenced DNA by identifying open reading frames, which define the most likely coding sequences of proteins. The function of an individual protein, the amino acid sequence of which is derived from the DNA sequence, can sometimes be deduced from its similarity (homology) to proteins with a well-characterized biological function in other organisms. A large number of genes in the genome encode proteins that have no known function, and a substantial fraction of these are unique to each organism. Comparison of genomes of related organisms allows the identification of genes or blocks of genes that were acquired by horizontal gene transfer. Large blocks of horizontally acquired genes are called genomic islands.

16.3 Functional Genomics

The major challenge faced by the research community, in light of the availability of complete genome sequences of a variety of organisms, is to understand the biological function of each gene product, specifically within the context of the parallel activities of the thousands of other components that make up a living cell. This postgenomic phase of research is called functional genomics. This field is driven by technological developments in many different fields, including classical genetics, physics, engineering, molecular biology, and biochemistry. Most notable is the rise of computational biology as a central component of modern biomedical research. Computational tools played an important role in genome-sequencing efforts and will continue to be the dominant technology supporting all disciplines of postgenomic research.

DNA Microarrays and Global Transcriptional Analysis

Of all the postgenomic technologies developed during the last 10 years, the most significant impact has been provided by the use of DNA arrays (also called microarrays or DNA chips) to measure changes in mRNA levels in cells, which are the result of changes in the rates of transcription and mRNA degradation. For any biological process, it is important to understand the flow of genetic information encoded in genes, and microarray technology provides a truly comprehensive view of gene expression in the entire cell by simultaneously measuring the levels of mRNA transcribed from every gene in the bacterial genome.

Most applications of microarrays can be divided into three stages: preparation of microarrays, hybridization of nucleic acid probes (derived from RNA or DNA of cells), and computer-assisted data analysis. First, a complete collection of PCR-amplified genes, or oligonucleotides consisting of segments unique to each gene, are generated. The guide for this process is the annotated genome, which is the source of sequence information for every gene. Usually, specific primers are designed for each gene, and the corresponding sequences are amplified and placed into multiwell dishes. If oligonucleotides are used, they are prepared by chemical synthesis. The number of PCR products or oligonucleotides that are needed for a comprehensive, all-inclusive array depends on the size of the genome. For example, a complete gene array for *H. pylori* would require advance preparation of 1,566 unique clones, PCR products, or oligonucleotides, whereas for *P. aeruginosa* this number would be considerably higher (5,570).

The spotting or "arraying" (sometimes referred to as printing) is carried out by robotic devices adapted from tools developed for industrial production of computer chips. These robots have movable arms that deposit very small (1 to 2 nl) aliquots at high density on a solid support, in a process called array printing. The arrays are often printed onto coated microscope slides. Once the droplets of DNA solution are spotted; the slides are exposed to ultraviolet light to covalently link the DNA to the slide. Each PCR product (representing a gene) is therefore immobilized at a specific location on the array. The schematic diagram of a robotic arrayer is shown in Figure 16.13, and the overall procedure for microarray analysis is outlined in Figure 16.14.

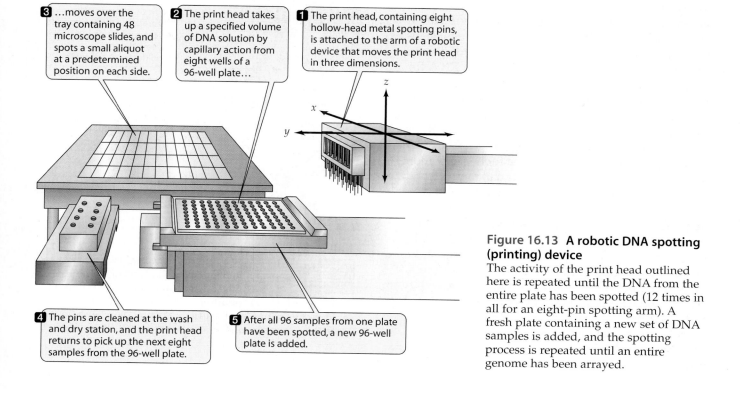

3 ...moves over the tray containing 48 microscope slides, and spots a small aliquot at a predetermined position on each side.

2 The print head takes up a specified volume of DNA solution by capillary action from eight wells of a 96-well plate...

1 The print head, containing eight hollow-head metal spotting pins, is attached to the arm of a robotic device that moves the print head in three dimensions.

4 The pins are cleaned at the wash and dry station, and the print head returns to pick up the next eight samples from the 96-well plate.

5 After all 96 samples from one plate have been spotted, a new 96-well plate is added.

Figure 16.13 A robotic DNA spotting (printing) device
The activity of the print head outlined here is repeated until the DNA from the entire plate has been spotted (12 times in all for an eight-pin spotting arm). A fresh plate containing a new set of DNA samples is added, and the spotting process is repeated until an entire genome has been arrayed.

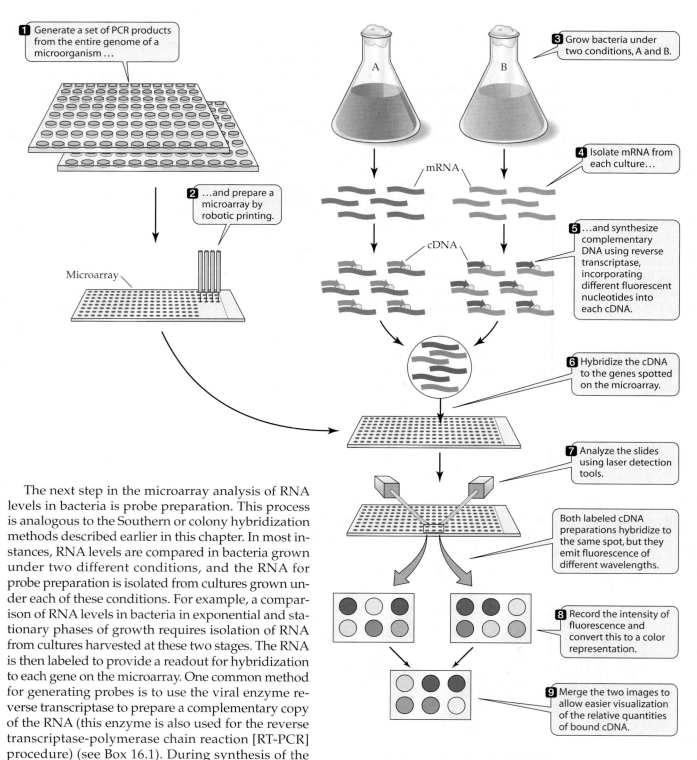

1 Generate a set of PCR products from the entire genome of a microorganism …

2 …and prepare a microarray by robotic printing.

Microarray

3 Grow bacteria under two conditions, A and B.

mRNA

4 Isolate mRNA from each culture…

cDNA

5 …and synthesize complementary DNA using reverse transcriptase, incorporating different fluorescent nucleotides into each cDNA.

6 Hybridize the cDNA to the genes spotted on the microarray.

7 Analyze the slides using laser detection tools.

Both labeled cDNA preparations hybridize to the same spot, but they emit fluorescence of different wavelengths.

8 Record the intensity of fluorescence and convert this to a color representation.

9 Merge the two images to allow easier visualization of the relative quantities of bound cDNA.

The next step in the microarray analysis of RNA levels in bacteria is probe preparation. This process is analogous to the Southern or colony hybridization methods described earlier in this chapter. In most instances, RNA levels are compared in bacteria grown under two different conditions, and the RNA for probe preparation is isolated from cultures grown under each of these conditions. For example, a comparison of RNA levels in bacteria in exponential and stationary phases of growth requires isolation of RNA from cultures harvested at these two stages. The RNA is then labeled to provide a readout for hybridization to each gene on the microarray. One common method for generating probes is to use the viral enzyme reverse transcriptase to prepare a complementary copy of the RNA (this enzyme is also used for the reverse transcriptase-polymerase chain reaction [RT-PCR] procedure) (see Box 16.1). During synthesis of the DNA strand complementary to mRNA, fluorescently labeled nucleotides are incorporated into the cDNA strand. These labeled nucleotides can be synthesized with slightly different chemical tags that fluoresce at different wavelengths, so that cDNA prepared

Figure 16.14 Microarray analysis
Steps in the microarray analysis of all genes expressed in a bacterium grown under two different experimental conditions, designated A and B. These could be variations in physical environment or nutritional content, or comparisons of a mutant to a wild-type cell.

from RNA from different organisms can be readily differentiated. For example, when mRNA in bacteria at two different growth stages is compared, probes synthesized from RNA isolated from cells at the exponential phase can be labeled with modified deoxynucleotides that emit at a wavelength that is detectable in the "red channel" of the fluorescence detector. Similarly, the probe from cells at the stationary phase can be labeled with deoxynucleotides that emit at a wavelength detectable in the "green channel."

The final stage of the microarray analysis is to determine the proportion of each hybridized probe. This stage is carried out by mixing the two probes and hybridizing them to the PCR products printed on the slide. Each probe will hybridize to the region of the array that contains the complementary DNA sequence (i.e., the gene that encodes the original transcript). Because the DNA on the microarray is in excess, the two fluorescent probes do not compete with each other, and the amount of bound probe is proportional to the amount of template mRNA in the original culture.

The images generated from the laser scanner can now be analyzed. As shown in Figure 16.15, the intensity of

fluorescence emission from each spot on the array reflects the relative ratio and quantity of transcripts for the gene represented by that particular spot. In the example shown, gene A is highly expressed during the exponential phase of growth, whereas gene B is predominantly transcribed when the bacteria are in the stationary phase, assuming that their mRNAs are degraded at approximately the same rate. Gene C is not growth-phase regulated, and its mRNA accumulates equally at all phases of growth. Other genes show differential levels and ratios of expression. Because microarrays allow comprehensive transcriptional analysis, quantification of the relative mRNA abundance in the cell can be carried out on a genome-wide basis. The total body of all RNA transcripts in a cell is called the **transcriptome**. Unlike a genome, which is a fixed entity consisting of all genes, the transcriptome reflects in part the environmental conditions of the bacterium, and many different RNA profiles can be generated from a single genome. Moreover, perhaps only a fraction of all genes is expressed at any one time, reflecting the basic metabolic activities of the organism in that particular environment. Figure 16.16 shows the *P. aeruginosa* transcriptome in media with

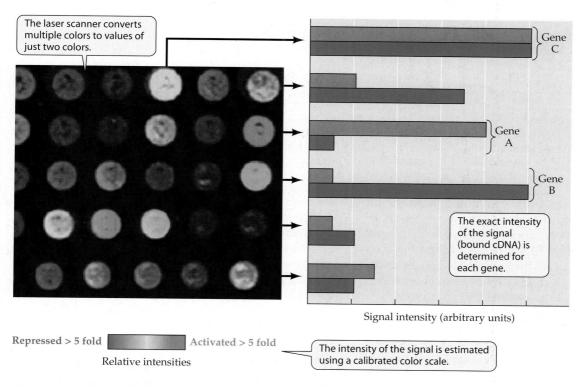

Figure 16.15 Analysis of data from a microarray experiment
The merged color representation (see Figure 16.14) reveals the relative quantities of bound cDNA. On the bar graph, the red bars represent the transcript level during exponential growth and the green bars the levels during stationary phase.

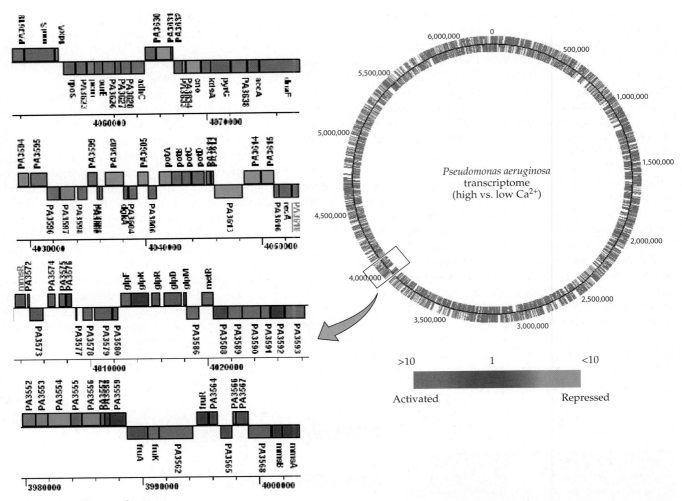

Figure 16.16 Transcriptome
The transcriptome of *Pseudomonas aeruginosa* grown in media of varying calcium concentration. The enlarged section shows the relative expression of each gene within a 7-kb segment of the chromosome. Courtesy of Mathew Wolfgang.

varying amounts of calcium. The highlighted region shows the coordinated expression of genes under different calcium-limiting conditions, as they are displayed on the genetic map (see Figure 16.10). This provides an overview of changes in the genome-wide levels of mRNA, and it serves as one starting point for dissection of the complex regulatory circuits that control the bacterial response to environmental changes.

The use of robotic devices to spot DNA solutions onto slides is only one of several methods currently available for producing DNA microarrays. An example of an alternative method is direct synthesis of short oligonucleotides at predetermined locations on solid supports. There are also at least a dozen different methods for preparing probes, involving covalent modification of RNA, direct labeling of RNA, or other ways of preparing cDNA probes. New versions of fluorescent nucleotides are being developed that allow simultaneous detection of more than two conditions in a single experiment. Yet another exciting application, derived from DNA microarray technology, is the development of protein arrays, where all the individual proteins expressed by a microorganism are deposited on a support. Protein arrays have a large number of applications, ranging from detection of antibodies against pathogens, to studies of interactions between the proteins encoded in the genome.

DNA Microarrays as Tools of Comparative Genomics

DNA microarrays are also useful in large-scale comparisons of genomes of related bacteria. Hybridization (or lack of hybridization) of probes derived from the genomic DNA of a test strain is compared to the reference

Figure 16.17 Comparison of genomes of different isolates of *Helicobacter pylori*
DNA was extracted from seven different isolates (indicated at bottom), and hybridized to a *H. pylori* microarray. A strong hybridization signal corresponding to a specific gene sequence was scored as present (blue) and lack of hybridization as absent (yellow). The genes are organized in a linear format (1–1,156) based on their position on the chromosome of the reference strain. Also shown are the locations of a common pathogenicity island (cagPAI), and two regions (PZ1 and PZ2) that in certain strains contain DNA segments acquired by horizontal gene transfer. Adapted from Joyce, E. A., et al. 2002. *Nature Reviews Genetics* 3: 462–473.

strain's genome, and the presence or absence of genes in the genome of the test strain is determined. An example of one such analysis, comparing different *H. pylori* strains is shown in Figure 16.17. This application of microarrays to the analysis of genome contents cannot identify genes that are present in the genomes of the test strains, but absent from the reference strain's genome; these genes would not be represented on the microarray. The availability of genome sequences of different strains of the same genus makes it possible to construct more comprehensive arrays that include genes that are present in more than one strain.

Surveys of strains by this method can provide useful information about the evolution of the genome of the particular species. The overall hybridization pattern is a "signature" of the organism's genome, and it can be used to establish the closeness of the relationship between any two strains. This is useful in epidemiological studies where the sources of the pathogenic organisms and route of their dissemination are being traced. Specific segments of DNA that are acquired by horizontal gene transfer can be readily identified as blocks of DNA, found in some but not all members of the species. Surveys of genome content of a significant number of different taxonomically related strains can be used to identify the so-called core genome, the shared repertoire of genes that is found in all members of the particular species. The core genome defines the set of genes that encode products essential for survival in environments most commonly occupied by these organisms. With the development of more advanced array manufacturing technologies, several sequence variants of individual genes that are subject to mutations can be placed on the same microarray. These tools are then applicable toward finer probing of sequence polymorphisms in individual genes, often at a single nucleotide level. For example, specific patterns of mutations in genes that encode proteins targeted by the immune system or by antibiotics—changes which might lead to resistance—could be monitored using microarrays containing different variants of the same protein-coding sequences.

One of the fastest-growing disciplines of functional genomics is **proteomics**, the study of the structure and function of proteins, including their relative abundance under certain conditions of growth. Comparisons of mRNA levels revealed by DNA microarrays provide a good approximation of the response of a cell to environmental stimuli. However, it is the proteins that carry out the vast majority of cellular processes, and protein levels do not necessarily correlate with transcript levels. Some of the post-transcriptional regulatory mechanisms of bacteria are discussed in Chapter 13. In addition, some proteins are modified after their synthesis and before they become fully active; these changes are impossible to detect by gene expression studies alone. Therefore, other methods are used to determine what forms of a protein are present and active within a cell.

A major tool for analyzing protein changes during environmental or genetic perturbation is **two-dimensional gel electrophoresis**. Cells grown under different conditions are lysed and their proteins are fractionated in one dimension by protein charge, followed by a size-based separation in the second dimension, giving a two-dimensional pattern of protein spots. Because the amino acid composition of each protein is unique with regard to one or both of these parameters (charge and size), an extremely high resolution can be obtained. Comparison of the sizes of equivalent protein spots in two gels—each containing an extract from cells propagated under one of two different conditions—can be used to quantify the relative abundance of proteins. If the identity of a particular protein is not known, the protein present in the corresponding spot in the gel can be excised and its identity determined by a technique called **mass spectrometry**. This analytical technique is used to identify proteins based on determination of their amino acid sequence, following fragmentation of the proteins in the mass spectrometer. The instrument is able to determine the sequence of each fragment based on their precise mass. These short sequences are matched to a databases of all open reading frames of the organism (deduced from the genome sequence), resulting in identification of the protein present in the spot from the two-dimensional gel. Mass spectrometry can also detect post-translational modifications of proteins, such as phosphorylation or glycosylation. Currently used mass spectrometry instruments allow the analysis of complex protein mixtures by including a fractionation step prior to protein fragmentation, bypassing the need for two-dimensional gel electrophoresis. It is certain that the development of more sophisticated mass spectrometers and corresponding computational tools to analyze the massive amount of data will make it possible to rapidly determine the entire complement of proteins in a cell.

SECTION HIGHLIGHTS

Functional genomics is the study of the biological functions of genes and their products, using technology enabled by the availability of whole genome sequences. For example, **DNA microarrays** can be used to determine the levels of mRNA for each gene in the genome under a given set of conditions. Microarrays contain immobilized DNA fragments corresponding to a complete or partial sequence of each gene in the genome. They are usually prepared using robots that apply aliquots of DNA to a solid support at high density. For example, the DNA may be spotted onto microscope slides that have been chemically treated to allow cross-linking and immobilization of the DNA. An entire genome of a microorganism can be represented on a single slide. Microarrays are analyzed by hybridization of labeled RNA or fluorescently labeled cDNA. A matrix describing the abundance of RNA for each gene in a genome under a specified set of conditions is called a **transcriptome**. Microarrays can also be used to compare genomes of related bacteria or to study genetic variation within a species, an area of research called **comparative genomics**. **Proteomics** is the study of the amount and activity of the proteins present in a cell under a given set of conditions.

SUMMARY

- Classical bacterial genetics was the foundation for recombinant DNA research, which is based on the ability of scientists to **isolate individual genes** from more complex genomes.

- Bacteria make **restriction endonucleases**, which introduce double-stranded breaks at specific DNA sequences. These are purified and used to fragment DNA into pieces that contain intact genes or parts of genes. Restriction endonucleases, together with their cognate **modification enzymes**, serve as barriers to unrestricted DNA exchange among different bacterial species.

- DNA fragments generated by treatment with restriction endonucleases are usually analyzed by

agarose gel electrophoresis, which separates DNA fragments according to their size.

- Specific DNA sequences found within a DNA fragment fractionated by agarose gel electrophoresis can be identified by a method of in situ **hybridization** of a single-stranded DNA probe with its complementary sequence in the gel. This procedure is known as **Southern blotting**.

- Construction of recombinants involves insertion of a DNA fragment into a **vector**, which can be either a **plasmid** or **virus**, followed by introduction of the vector into the bacterial **host** by transformation. If the vector is a bacteriophage (usually bacteriophage lambda), the recombinant DNA is packaged into the bacteriophage head and is introduced into the bacteria by infection.

- Plasmid vectors contain sites where a DNA fragment may be inserted following treatment of the insert DNA and the vector with the same restriction endonuclease, which generates compatible ends. The insert DNA is joined with the plasmid, to give a chimeric circular molecule, by the action of **DNA ligase**.

- Recombinant plasmids can be propagated in pure culture as **clones**. They are identified by colony hybridization or by expression of protein product using either immunological or activity screens.

- Specialized cloning vectors, called **expression vectors**, include specific sequences to maximize expression of the cloned genes.

- Cloned genes can be altered in the laboratory, and introduced into bacterial genomes via either plasmids or bacteriophage vectors. Even plants can be genetically engineered, using *Agrobacterium*-mediated transfer of recombinant DNA cloned within its **Ti plasmid**.

- Eukaryotic genes encoding proteins are often cloned from mRNA sequences. First, a copy of mRNA, called **cDNA**, is made by the enzyme **reverse transcriptase**. Then, a second DNA strand, complementary to the cDNA, is made by DNA polymerase. This double-stranded form is then inserted into the appropriate plasmid or bacteriophage vectors.

- Rare sequences from complex genomes, or from a mixture of different RNA species, can be isolated by repeated cycles of denaturation–complementary strand synthesis–denaturation, a process called the **polymerase chain reaction (PCR)**.

- **Functional genomics** is a term applied to the study of biological processes in the entire cell, using information generated from genome sequencing.

- The determination of the sequence of a microbial genome involves a process called **whole-genome shotgun sequencing**.

- There are several distinct stages in the whole-genome sequencing process. The first step is the construction a library of fragments of DNA in cloning vectors. Thousands of randomly selected clones are sequenced, each providing a short 500 to 800 bp sequence from a portion of the genome. The short sequences are assembled into longer segments, called **contigs**, by aligning the short overlapping sequences. The genome sequence still contains **gaps** between the contigs, which are closed by specialized methods, such as PCR amplification across the gap and sequencing the PCR products.

- The final stage of genome sequencing is **annotation** of the genome. First the location of genes within the sequenced DNA is determined. This involves the identification of open reading frames, which define the most likely coding sequence of proteins.

- The function of an individual protein, the sequence of which was deduced from the DNA sequence, can be sometimes determined by the extent of similarity (homology) to proteins in other organisms that have a well-characterized biological function.

- A large number of genes in the genome encode proteins that have no known function, and a substantial fraction of these are unique to each organism.

- **DNA microarrays** are the tools to determine the levels of mRNA for each gene in the genome and can also be used to compare bacterial genomes or genes. Microarrays are immobilized DNA fragments corresponding to a complete or partial sequence of each gene in the genome. They are usually prepared using robots that apply aliquots of DNA to solid supports at a very high density. DNA is spotted onto chemically treated microscope slides that allow cross-linking and immobilization of the DNA. An entire genome of a microorganism can be represented on a single slide.

- Microarrays are probed with labeled RNA or a fluorescently labeled cDNA, and the amount of probe bound to each spot of DNA is then quantitated.

- A set of data describing the abundance of RNA for each gene in a genome under a specified set of conditions is called a **transcriptome**.

 Find more at www.sinauer.com/microbial-life

REVIEW QUESTIONS

1. What are the most common reasons for failure to obtain a complete genome from the assemblies of thousands of short sequences in whole-genome shotgun sequencing?

2. Why do genomes of bacteria vary so much in size?

3. An annotation of a genome of a microorganism revealed the presence of 4,000 open reading frames. Does this imply that there are 4,000 different proteins present in the cell at any one time?

4. Describe the process of a genome-based microarray preparation.

5. Which genes bind the most to a probe in a genome microarray of a bacterium?

6. What is the relationship of a genome to a transcriptome?

7. Why don't bacteria that produce restriction enzymes destroy their own DNA?

8. Below is a portion of the *E. coli* chromosome containing the *lacZ* gene (coding for β-galactosidase).

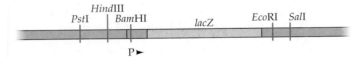

(a) Design a strategy to clone the shortest fragment that still carries the full coding sequence for the enzyme, into plasmid pBR322.

(b) Describe several methods to identify the recombinant *E. coli* carrying the cloned *lacZ* gene.

(c) How would your cloning strategy change if you wanted to ensure that the *lacZ* gene is expressed in a host strain other than *E. coli*?

9. Some *E. coli* strains, used as cloning hosts, carry a mutation in their *recA* gene. Why is it sometimes important to use such a mutant strain when cloning certain genes?

10. A solution of an oligonucleotide of the sequence 5'-CGGCCCGGGATCCGAATTCCTAGGCA-3' is heated and mixed with the oligonucleotide 5'-TGGGATGGAATTCGGGATCCCGCGCCG-3', and then allowed to cool.

(a) Identify all the sites recognized by restriction enzymes (use Table 16.1).

(b) Would a solution of the first oligonucleotide alone be recognized and cut by any restriction enzyme?

11. A patient was admitted to a hospital with a urinary tract infection, and antibiotic-sensitive *Klebsiella pneumoniae* were recovered. During hospitalization the patient became nonresponsive to antibiotic therapy, and the new isolates of *K. pneumoniae* were resistant to ampicillin and kanamycin. Stool samples of all of the patients on the same floor revealed that several of them carried *E. coli* that was also resistant to ampicillin and kanamycin. You are in charge of determining if the *K. pneumoniae* somehow acquired an R plasmid from *E. coli*. What molecular tools would you use to test the plasmid-transfer hypothesis?

12. You wish to isolate a gene from a chromosome, which is flanked by recognition sites for either *Bgl*II or *Sma*I restriction endonucleases. Is it possible to use either vector pBR322 or pACYC184, despite the fact that neither of these plasmids has a recognition sequence for *Bgl*II or *Sma*I?

13. A procedure analogous to Southern blotting is called northern blotting, where total RNA is extracted from bacteria and size-fractionated on agarose gels. Specific mRNA sequences are identified by hybridization with radiolabeled probes, the sequences of which can be deduced from protein sequence. If you wish to analyze for the presence of mRNA of gene X (encoding a protein with the N-terminal sequence Met-Trp-Asp-Trp), could you use, in the northern blot, the same oligonucleotide probes

5'-ATGTGGGATTGG-3'
5'-ATGTGGGACTGG-3'

used to detect gene X in Southern blot?

14. Describe the differences between colony hybridization aimed at detecting specific DNA sequences and the related immunological techniques for the identification of clones producing a specific protein.

15. Explain the difference in the initial steps of the polymerase chain reaction (PCR) when starting from mRNA, rather than DNA.

SUGGESTED READING

Blalock, E. M, ed. 2003. *A Beginner's Guide to Microarrays.* Norwell, MA: Kluwer Academic Publishers.

Fraser, C., T. Read and K. Nelson. 2004. *Microbial Genomes.* Totowa, NJ: Humana Press Inc.

Watson, J., J. Witkowski, M. Richard and A. Caudy. 2006. *Recombinant DNA: Genes and Genomics: A Short Course.* Washington, DC: ASM Press.

Microbial Evolution
and Diversity

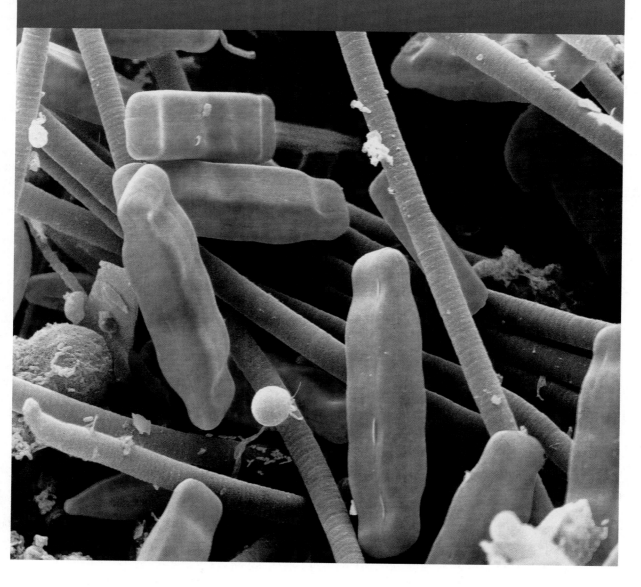

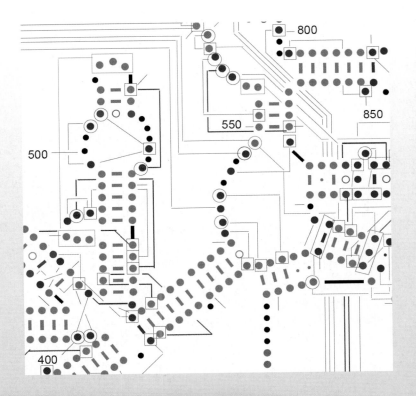

The objectives of this chapter are to:

◆ Provide information on how bacteria are named and what is meant by a validly named species.

◆ Discuss the classification of *Bacteria* and *Archaea* and the recent move toward an evolutionarily based, phylogenetic classification.

◆ Describe the ways in which the *Bacteria* and *Archaea* are identified in the laboratory.

17

Taxonomy of Bacteria *and* Archaea

It's just astounding to see how constant, how conserved, certain sequence motifs—proteins, genes—have been over enormous expanses of time. You can see sequence patterns that have persisted probably for over three billion years. That's far longer than mountain ranges last, than continents retain their shape.
—Carl Woese, 1997 (in Perry and Staley, Microbiology)

This part of the book discusses the variety of microorganisms that exist on Earth and what is known about their characteristics and evolution. Most of the material pertains to the *Bacteria* and *Archaea* because there is a special chapter dedicated to eukaryotic microorganisms. Therefore, this first chapter discusses how the *Bacteria* and *Archaea* are named and classified and is followed by several chapters (Chapters 18–22) that discuss the properties and diversity of the *Bacteria* and *Archaea*.

When scientists encounter a large number of related items—such as the chemical elements, plants, or animals—they characterize, name, and organize them into groups. Thousands of species of plants, animals, and bacteria have been named, and many more will be named in the future as more are discovered. Not even the most brilliant biologist knows all of the species. Organizing the species into groups of similar types aids the scientist not only in remembering them but also in comparing them to their closest relatives, some of which the scientist would know very well. In addition, biologists are interested in evolution, because this is the process through which organisms became diverse. Unraveling the route of evolution leads to an understanding of how one species is related to another. As discussed in subsequent text of this chapter, evolutionary relationships are assessed by **molecular phylogeny**, the analysis of gene and protein sequences to determine the relatedness among organisms. To date, approximately 5,000 bacterial and archaeal species have been named and, based on their characteristics, placed within the existing framework of other known species. The branch of bacteriology that is responsible for characterizing and naming organisms and organizing them into groups is called **taxonomy** or **systematics**.

Taxonomy can be separated into three major areas of activity. One is **nomenclature**, which is the naming of bacteria. The second is **classification**, which entails the ordering of bacteria into groups based on common properties. In **identification**, the third area, an unknown bacterium, for example, from a clinical or soil sample, is characterized to determine its species. This chapter covers all three of these areas.

17.1 Nomenclature

Bacteriologists throughout the world have agreed on a set of rules for naming *Bacteria* and *Archaea*. These rules, called the "International Code for the Nomenclature of Bacteria" (1992), state what a scientist must do to describe a new species or other taxon (taxa, pl.), which is a unit of classification, such as a species, genus, or family. Each bacterium is placed in a genus and given a species name in the same manner as are plants and animals. For example, humans are *Homo sapiens* (genus name first, followed by species), and a common intestinal bacterium is named *Escherichia coli*. This binomial system of names follows that proposed for plants and animals by the Swedish taxonomist Carl von Linné (Linnaeus; 1707–1778).

According to the rules of bacterial nomenclature, the root for the name of a species or other taxon can be derived from any language, but it must be given a Latin ending so that the genus and species names agree in gender. For example, consider the species name *Staphylococcus aureus*. The first letter in the genus name is capitalized, the species name is lowercase, and they are both italicized to indicate that they are Latinized. When writing species names in longhand, as for a laboratory notebook, they should be underlined to denote that they are italicized. The genus name *Staphylococcus* is derived from the Greek *Staphyl* from *staphyle*, which means a "bunch of grapes," and *coccus,* from the Greek, meaning "a berry." The *o* ("oh") between the two words is a joining vowel used to connect two Greek words together. The figurative meaning of the genus name is "a cluster of cocci," which describes the overall morphology of members of the genus. The species name *aureus* is from the Latin and means "golden," the pigmentation of members of this species. The *-us* ending of the genus and species names is the Latin masculine ending for a noun (*Staphylococcus* in this case) and its adjective (*aureus*). Successively higher taxonomic categories are family, order, class, phylum, and domain (Table 17.1)

TABLE 17.1	Hierarchical classification of the bacterium *Spirochaeta plicatilis*

Taxon	Name
Domain	*Bacteria*
Phylum	*Spirochaetes* (vernacular name: spirochetes)
Class	*Spirochaetes*
Order	*Spirochaetales*
Family	*Spirochaetaceae*
Genus	*Spirochaeta*
Species	*plicatilis*

The *International Journal of Systematic and Evolutionary Microbiology* (IJSEM) is a journal devoted to the taxonomy of bacteria that is published by the Society for General Microbiology. IJSEM publishes papers that describe and name new bacterial taxa and contains an updated listing of all new bacteria whose names have been validly published. Thus, although bacterial species may be described in other scientific journals, they are not considered validly published until they have been included on a validation list in IJSEM.

IJSEM also provides a forum to debate specific controversies in nomenclature by allowing a scientist to challenge the current nomenclature of an organism or group of organisms. Such challenges, if accepted by peer review, are then published as a question in the IJSEM. The question is then evaluated by the Judicial Commission of the International Union of Microbiological Societies, which subsequently publishes a ruling in the journal. One typical example of a problem considered by the Judicial Commission was the question about *Yersinia pestis*, the causative agent of bubonic plague. Scientific evidence indicates that *Y. pestis* is really just a subspecies of *Yersinia pseudotuberculosis*, a species name that has precedence over *Y. pestis* because of its earlier publication. Because of the potential confusion and possible public health issues that could arise by renaming *Y. pestis*, *Y. pseudotuberculosis* subspecies *pestis*, the Judicial Commission ruled against renaming the bacterium despite its scientific justification.

SECTION HIGHLIGHTS

Nomenclature is concerned with naming organisms. For *Bacteria* and *Archaea*, specific rules must be followed in order to name and describe new species. Organisms that are placed on the approved or validated lists are officially recognized species.

17.2 Classification

Classification is that part of taxonomy concerned with the grouping of bacteria into taxa based on common characteristics. The earliest classifications did not consider microorganisms. There were two kingdoms of life: Plants and Animals. In 1868, Ernst Haeckel, a German scientist, proposed a third kingdom specifically for microorganisms. Approximately a century later, in 1969, Robert Whittaker proposed a five-kingdom system of classification. His classification included Plants (*Plantae*), Animals (*Animalia*), *Fungi*, *Protista*, and *Monera*. In this system, the eukaryotic microorganisms were placed in the Protista kingdom and the fungi had their own special kingdom. The bacteria and archaea were placed, as prokaryotes, in the kingdom Monera. Organisms were separated from one another on the basis of nutrition and cell structure. Therefore, plants are photosynthetic eukaryotes, fungi are heterotrophs that use dissolved nutrients, and animals are heterotrophs that ingest their food. The five-kingdom classification remained popular until recently. Then, in 1990, Carl Woese and colleagues proposed an entirely new classification, the Tree of Life (see Chapter 1). Unlike all previous classifications this new classification uses a molecular phylogenic approach. The Tree of Life is based on the sequence analysis of a common macromolecule that all organisms share, the RNA in the small subunit of the ribosome (see Chapter 4). This RNA was used to separate all organisms on Earth into three different domains, the *Bacteria*, the *Archaea,* and the *Eukarya*. Viruses, which are not organisms because they are not cellular (see Chapter 1), cannot be classified by this system.

Classification systems can be either artificial or natural. **Artificial systems** of classification are based on *expressed characteristics* of the organisms, or the **phenotype** of the organism. In contrast, **natural** or **phylogenetic systems** are based on the purported *evolution* of the organism. Until recently, all bacterial classifications were artificial because there was no meaningful basis for determining their evolution. In contrast, plants and animals have a fairly extensive fossil record on which to base an evolutionary classification system. Although fossils of microorganisms do exist (see Chapter 1), the simple structures of microorganisms do not permit their identification into a taxonomic group by morphological criteria. Furthermore, the stable morphology of and ontology of plants and animals have been useful in developing natural systems of classification. This type of information is rarely available for microorganisms.

When considering bacterial classification, it is important to keep in mind that bacteria have been evolving on Earth for the past 3.5 to 4 billion years. Therefore, it should not seem surprising that two separate domains of prokaryotic organisms exist—the *Bacteria* and the *Archaea*—versus only one domain for eukaryotes, *Eukarya*. In addition, the *Eukarya* may have evolved more recently as the result of symbiotic events between different early prokaryotic forms of life (see Chapter 1). Because of the long period of evolution of bacteria and archaea, the various groups within these domains exhibit considerable diversity, particularly metabolic and physiological. In contrast, the metabolic diversity of the *Eukarya* is limited, especially with respect to energy generation. The vast diversity of metabolic types of prokaryotes is discussed more fully in Chapter 5 and in Chapters 18–22. This chapter first discusses the traditional system of classification and then covers what is being done to make it phylogenetic.

Artificial versus Phylogenetic Classifications

Conventional artificial taxonomy uses phenotypic tests to determine differences between strains and species. These tests are typically weighted so that characteristics that are considered to be more important are given higher priority. For example, in traditional taxonomy, the Gram stain has been given more weight in determining the classification of an organism than whether the organism uses glucose as a carbon source. Therefore, all gram-positive strains would be ascribed to one family or genus, and, within that group, certain species or strains would use glucose and others would not.

Most bacteriologists favor a phylogenetic system for the classification of bacteria, and with the advent of molecular phylogeny, this hope is now being realized. The current accepted treatise that contains a complete listing of prokaryotic species and their classification is *Bergey's Manual of Systematic Bacteriology* (2001–2008), published by Springer, and its more condensed edition, *Bergey's Manual of Determinative Bacteriology* (1994). In addition to containing a complete classification of bacteria and archaea, the more comprehensive version of *Bergey's Manual* contains a description of all known validly described bacterial species. Therefore, it is the "encyclopedia" of the bacteria and archaea that is widely used by bacteriologists (Box 17.1). *Bergey's Manual of Systematic Bacteriology* is now in its second edition. The first edition was based on an artificial classification because too little phylogenetic information was available. However, the second edition is based on the Tree of Life that uses a phylogenetic framework as discussed in subsequent text of this chapter.

Phenotypic Properties and Artificial Classifications

Phenotypic properties are those that are expressed by an organism, and they have always played a major role in

Bergey's Manual Trust

David Bergey was a professor of bacteriology at the University of Pennsylvania in the early 1900s. As a taxonomist he was a member of a committee of the Society of American Bacteriologists (SAB—now called the American Society for Microbiology), which was interested in formulating a classification of the bacteria that could be used for identification of species. In 1923, he and four others published the first edition of *Bergey's Manual of Determinative Bacteriology*. This was followed by new editions every few years. Royalties collected by the publication activities of the committee were held in SAB. When David Bergey and his co-editor, Robert Breed, requested money from the account to be used for preparation of the fifth edition, the leadership in SAB refused. After a long and bitter fight, the SAB relented and turned the total proceeds over to Bergey, who promptly put the money into a nonprofit trust with a board of trustees to oversee the publication of manuals on bacterial systematics. The Trust, now named in his honor as Bergey's Manual Trust, is responsible for the publication of *Bergey's Manual of Determinative Bacteriology*, which is now in its ninth edition, as well as other taxonomic books such as *Bergey's Manual of Systematic Bacteriology*. The Trust is headquartered at University of Georgia and has a nine-member international board of trustees, as well as associate members from many countries.

David Bergey. Courtesy of the National Library of Medicine.

microbial taxonomy. Indeed, early classifications had to be based entirely on phenotypic properties because they were the only properties that could be studied.

In classifications, it is important to select phenotypic characteristics that allow one to group organisms together and others that enable one to distinguish organisms from one another. Furthermore, it is important that the identifying characteristics selected be easy to determine. Two examples of simple phenotypic characteristics that have been widely used in artificial classification schemes are the Gram stain and cell shape. It turns out that each of these has some utility in classifications. For example, the Gram stain tells something about the nature of the cell wall (see Chapter 4). Furthermore, the Gram stain happens to be important phylogenetically because two of the phylogenetic groups of *Bacteria* are gram-positive (i.e., *Firmicutes* and *Actinobacteria*) and the other 20 or more are gram-negative. However, there are some drawbacks to phenotypic properties. For example, it is noteworthy that some species of gram-positive bacteria stain as gram-negative bacteria. Likewise, the mycoplasmas, which stain as gram-negative organisms, have been found to be members of the *Firmicutes* through 16S rRNA analyses (see Chapter 20). The rea-son the mycoplasmas stain as gram-negative is that they lack cell walls altogether. Therefore, they have apparently evolved from a group of gram-positive bacteria that lost their peptidoglycan wall during evolution.

In contrast to the gram-positive bacteria, gram-negative bacteria and archaea fall into many different phylogenetic groups, including peptidoglycan-containing types and non–peptidoglycan-containing types that are bacteria as well as archaea. Therefore, gram-negative organisms are very diverse phylogenetically.

At one time, some bacteriologists proposed that the simplest, and purportedly the most stable, cell shape—the sphere—must have been the shape of the earliest bacteria. They then developed an evolutionary scheme based on this theory, in which all of the coccus-shaped bacteria were included in the same phylogenetic group. The validity of this classification has not been borne out by molecular phylogeny. For example, there are both gram-negative as well as gram-positive cocci. Some cocci are photosynthetic *Proteobacteria*, others are nonphotosynthetic, some are highly resistant to ultraviolet (UV) light (*Deinococcus*), some are *Archaea*, and others are *Bacteria*.

Nonetheless, cell shape is important for some groups. For example, the spirochetes phylum contains all of the helically shaped bacteria with periplasmic flagella (see Table 17.1 and Chapter 22). Because of their morphology, these bacteria were correctly classified with one another in the order *Spirochaetales* and now in the phylum *Spirochaetes* long before phylogenetic data confirmed the grouping. Apart from this group, overall cell shape has little meaning at higher taxonomic levels. Nonetheless, it can still be significant at the species, genus, and even family levels.

Other phenotypic properties have also proved useful in both artificial and phylogenetic classifications. For example, because of their unique ability to produce methane gas, the methanogenic microorganisms have always been classified together in artificial classifications. Likewise, from a phylogenetic standpoint, all the methanogenic organisms are members of the *Euryarchaeota* of the *Archaea*.

Of course, phenotypic properties have a special significance not found in gene sequence analyses in that these features provide information about what the organism is capable of doing. One cannot directly conclude from a sequence that an organism is or is not a methanogen, for example, unless that particular feature has been tested for and determined. Thus, phenotypic tests provide valuable information about the capabilities of the organism that may help explain its role in the environment in which it lives.

Numerical Taxonomy

When a large number of similar bacteria are being compared, computers are very useful in the analysis of the data. This aspect of taxonomy, which has been used in artificial classifications, is referred to as **numerical taxonomy**. Numerical taxonomy is most useful at the species and strain level, where phylogenetic relatedness has already been established by ribosomal RNA (rRNA) sequencing and DNA–DNA reassociation.

In numerical taxonomy *all characteristics are given equal weight*. Therefore, metabolism of a particular carbon source is considered to be as important as the Gram stain or the presence of a flagellum. In characterizing strains in this manner, a large number of characteristics are determined, and the similarity between strains is then compared by a similarity coefficient. Each strain is compared with every other strain. The **similarity coefficient**, S_{AB}, between two strains, A and B, is defined as follows:

$$S_{AB} = \frac{a}{a+b+c}$$

where a represents the number of properties shared in common by strains A and B; b represents the number of properties positive for A and negative for B; and c represents the number of properties positive for B and negative for A.

Characteristics for which both strains A and B are negative are considered irrelevant because there would be many such features that would have no bearing on their similarity. For example, endospore formation is an uncommon characteristic for bacteria. It is not significant to incorporate this characteristic when comparing two species within a genus that do not produce endospores. It would, however, be of value in comparing endospore-forming organisms to closely related organisms. It should be noted that the similarity coefficient can be used to relate not only phenotypic features of one organism to another, but also to relate the sequence similarity of macromolecules of different organisms.

In numerical taxonomy, it is best to have as many tests of phenotypic characters as possible. Typically at least 50 *independent* characters are used, and many strains are usually compared simultaneously. Ideally, each characteristic should represent a single and separate gene. The same gene should not be assessed more than once, and, therefore, overlapping phenotypic tests must be avoided. Typically, S_{AB} values are greater than or equal to 70% within a species and greater than 50% within a genus. An example of a numerical analysis is shown in Box 17.2.

In artificial classifications, bacteria are grouped into a hierarchy based on phenotypic properties. For example, the chemoautotrophic bacteria that obtain energy from the oxidation of inorganic nitrogen compounds, such as ammonia and nitrite, would be classified in the group of nitrifying bacteria. This group would be regarded as an order. One subgroup, termed the family, would contain ammonia oxidizers and another would contain nitrite oxidizers. Within each of these groups, features such as cell shape would be further used to define differences among the various genera and species. However, this artificial classification does not take into account the evolutionary relatedness among the members of the nitrifying bacteria.

Phylogenetic Classification and Molecular Phylogeny

An important article published by Zuckerkandl and Pauling in 1962 suggested that the evolution of organisms might be recorded in the sequences of their macromolecules. Subsequent research in the late 1960s and 1970s has supported this concept, resulting in a revolution in bacterial taxonomy. In particular, molecules such as rRNA and some proteins have changed at a very slow rate during evolution; therefore, their sequences provide important clues to the relatedness among the various bacterial taxa and their relatedness to plants and animals

BOX 17.2   *Methods & Techniques*

Determination of Similarity Coefficients

In this example, eight strains, A through H, are compared to one another by ten phenotypic tests. The results are shown in the first table:

Table 1 Results of phenotypic tests for eight strains, A–H

Tests	Strains Tested							
	A	**B**	**C**	**D**	**E**	**F**	**G**	**H**
1	+	–	+	+	+	+	–	+
2	–	+	+	–	+	–	+	+
3	+	+	–	+	–	+	+	+
4	+	–	–	–	+	+	–	+
5	+	–	–	+	+	+	–	–
6	+	+	–	+	–	–	+	–
7	–	+	+	+	+	–	–	+
8	+	–	–	+	–	+	+	–
9	+	–	+	+	+	+	–	+
10	–	+	+	–	+	–	+	+

Similarity coefficients are then determined by comparing the results of the tests for each of the strains against one another, using the formula given in the text. The results are shown in Table 2.

Table 2 Similarity coefficients (×100 to give percent similarity) for the eight strains

Strains	A	B	C	D	E	F	G	H
A	100							
B	20	100						
C	20	43	100					
D	75	33	33	100				
E	40	38	71	40	100			
F	86	10	22	63	44	100		
G	33	67	25	33	20	22	100	
H	40	50	71	40	75	44	33	100

The information from this matrix is then used to group the strains into similar types, as shown in the following matrix:

Table 3 Similarity matrix of grouped strains

Strain	A	F	D	E	H	C	G	B
A	100							
F	86	100						
D	75	63	100					
E	40	40	40	100				
H	40	44	40	75	100			
C	20	22	33	71	71	100		
G	33	22	33	20	33	25	100	
B	20	10	33	38	50	43	67	100

According to these tests, strains A, F, and D are very similar to one another and probably compose a single species. Likewise, E, C, and H are very similar and appear to be a separate species. Strains B and G may also be a different species, although more tests should probably be performed to substantiate this. As mentioned earlier in the chapter, phenotypic data such as this can provide an indication of relatedness at the species level, but if new species are being described, DNA/DNA reassociation tests should be performed.

as well. The result is that a major breakthrough has occurred in the classification of prokaryotic organisms, which has rapidly become phylogenetic through the study of molecular phylogeny as discussed in subsequent text. The fact that the two domains *Bacteria* and *Archaea* exist was not at all appreciated until molecular phylogenetic studies were performed. Bacteriologists are now trying to determine when the split occurred between the *Bacteria, Archaea,* and *Eukarya* and what the nature was of their last common ancestor.

By the 1970s, data had accumulated indicating that a true phylogenetic classification of bacteria was possible. What made it possible was, first of all, acceptance of the evidence that some of the macromolecules of organisms were highly conserved, that is, changed very slowly during evolution, and that their sequences held the key to unlocking the relatedness of bacteria to one another and to plants and animals. Second, sequencing techniques were developed and improved so that it became easy to conduct sequence analyses of rRNA and other macromolecules.

In this section, emphasis will be given to rRNA sequencing, in particular the RNA of the small subunit of the ribosome (see Chapter 4), 16S rRNA (or 18S rRNA of *Eukarya*), as it is the most common conserved molecule used to study the phylogeny of microorganisms (Figure 17.1). In actual practice, the 16S rRNA gene, or 16S rDNA, is sequenced because the polymerase chain reaction (PCR) procedures are simple, and both strands of the DNA can be used to confirm the actual sequence.

Several reasons justify the choice of rRNA as an evolutionary marker. First, the ribosome is found in *all* cellular organisms (see Chapter 4) and therefore allows one to compare all organisms. Second, the function of the ribosome as *the* structure responsible for protein synthesis holds true for all classes of life. Therefore, it is possible to compare the phylogeny of all organisms by analysis of a single structure with an important cellular function. The third advantage of using the ribosome in phylogeny is that ribosomes are highly conserved and therefore have changed very slowly over many millions of years. This is true because it is a very complex structure that carries out a specific function—protein synthesis (see Chapter 11). Keep in mind that the 16S rRNA is interacting in a three-dimensional structure with protein and other rRNA molecules as well messenger RNA (mRNA). Therefore, a high rate of evolutionary change in ribosome structure has been selected against during evolution. Mutant organisms with dysfunctional ribosomes would be unable to compete with existing types and have therefore not survived. In this manner, evolution has selected against *major* changes in the ribosome. Nonetheless, incremental modifications have occurred over the billions of years of biological evolution and these differences are used to construct evolutionary trees.

Ribosomal RNAs are not the only macromolecules that have been considered in determining relatedness at higher taxonomic levels. Proteins such as cytochrome *c* and ribulose bisphosphate carboxylase are just two examples of other molecules that have been used. However, not all organisms synthesize these macromolecules. Furthermore, not all cytochrome *c*–like molecules found in different organisms have the same physiological function. Because these other molecules are not universally distributed among all organisms, they cannot be used to compare distantly related taxa. However, the sequences of highly conserved proteins, such as cytochromes and nitrogenase, are very useful in constructing phylogenies of their origin and evolution within disparate microbial groups. It should be noted that because proteins are difficult and expensive to sequence, their sequences are normally deduced from their gene sequences.

Within the biological world, ribosomes share many similarities, indicating the conservative nature of the structure. Prokaryotic ribosomes contain three types of RNA: 5S, 16S, and 23S. Both 5S and 16S rRNA have been used to determine relatedness among organisms. Because the 16S molecule is larger (with about 1,500 bases), it contains more information (see Figure 17.1) than the smaller 5S molecule with only about 120 bases (Figure 17.2). Less work has been done on the 23S molecule because it is longer (about 3,000 nucleotides) and therefore not as easy to study. Therefore, scientists interested in the classification and evolution of bacteria have concentrated on 16S rRNA. The method of evaluation that provides the most information is sequence determination, especially for the complementary DNA of 16S rRNA, that is, the double-stranded 16S rRNA gene or 16S rDNA. It has been found that some regions of these molecules are more highly conserved than others. The more highly conserved regions permit the comparison of distantly related organisms (Figure 17.3), and the more variable domains are used for comparing more closely related organisms. Figure 17.3 shows the "stem-loop" secondary structure of a model 16S rRNA that allows for comparison among all organisms. The stem and loop portions, designated by the blue color, contain the most highly variable regions. For example, the blue area designated with a red star has been expanded for three species—*E. coli* (a bacterium), *Methanococcus vannielii* (an archaean), and *Saccharomyces cerevisiae* (a eukaryote)—to illustrate the variation among these three species. The regions that are unique to a given taxon are termed signature sequences. They can be used for the design of specific hybridization probes for identification (see later section on Identification).

It appears that an analysis of 16S rDNA sequences provides important information on the evolution of prokaryotes. However, before concluding that the se-

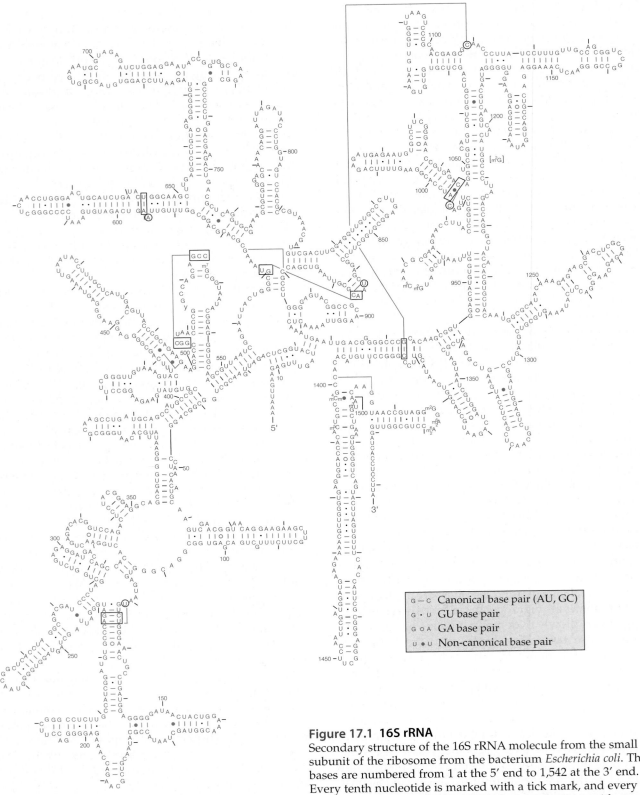

Figure 17.1 16S rRNA

Secondary structure of the 16S rRNA molecule from the small subunit of the ribosome from the bacterium *Escherichia coli*. The bases are numbered from 1 at the 5′ end to 1,542 at the 3′ end. Every tenth nucleotide is marked with a tick mark, and every fiftieth nucleotide is numbered. Tertiary interactions with strong comparative data are connected by solid lines. From the Comparative RNA Web Site, www.rna.icmb.utexas.edu (courtesy of Robin Gutell).

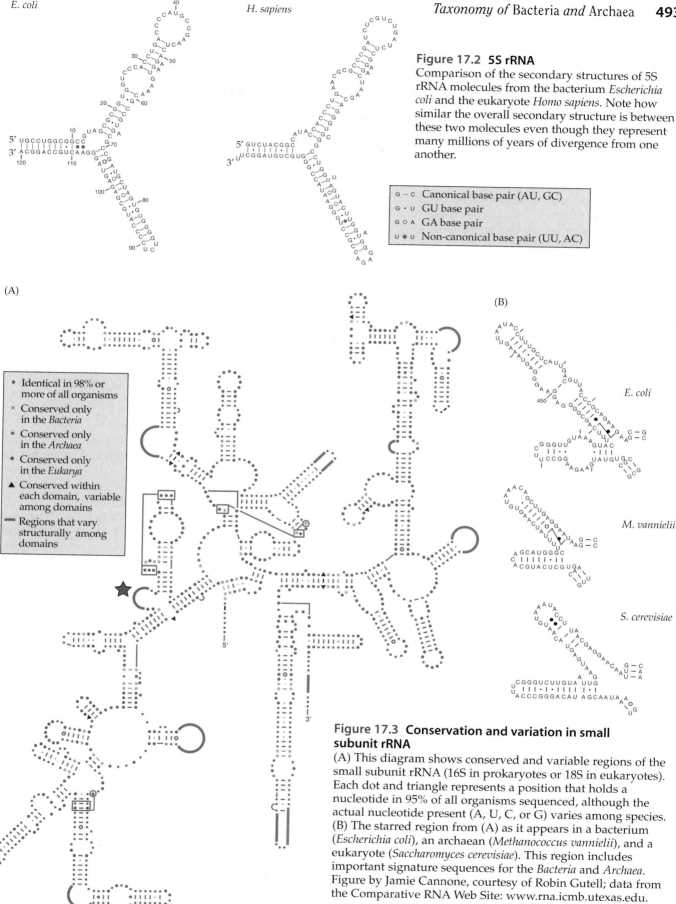

E. coli

H. sapiens

Figure 17.2 5S rRNA
Comparison of the secondary structures of 5S rRNA molecules from the bacterium *Escherichia coli* and the eukaryote *Homo sapiens*. Note how similar the overall secondary structure is between these two molecules even though they represent many millions of years of divergence from one another.

G — C	Canonical base pair (AU, GC)
G • U	GU base pair
G ○ A	GA base pair
U • U	Non-canonical base pair (UU, AC)

(A)

- Identical in 98% or more of all organisms
- Conserved only in the *Bacteria*
- Conserved only in the *Archaea*
- Conserved only in the *Eukarya*
- ▲ Conserved within each domain, variable among domains
- Regions that vary structurally among domains

(B)

E. coli

M. vannielii

S. cerevisiae

Figure 17.3 Conservation and variation in small subunit rRNA
(A) This diagram shows conserved and variable regions of the small subunit rRNA (16S in prokaryotes or 18S in eukaryotes). Each dot and triangle represents a position that holds a nucleotide in 95% of all organisms sequenced, although the actual nucleotide present (A, U, C, or G) varies among species. (B) The starred region from (A) as it appears in a bacterium (*Escherichia coli*), an archaean (*Methanococcus vannielii*), and a eukaryote (*Saccharomyces cerevisiae*). This region includes important signature sequences for the *Bacteria* and *Archaea*. Figure by Jamie Cannone, courtesy of Robin Gutell; data from the Comparative RNA Web Site: www.rna.icmb.utexas.edu.

quence of bases in rDNA accurately reflects the phylogeny of organisms, it is important to find totally separate and independent evolutionary markers to confirm the classification. Some work has been performed with sequencing of ATP synthases, elongation factors, RNA polymerases, and other conserved macromolecules. The outcome of this research, which represents one of the most exciting areas of biology, is leading to the development of a complete phylogenetic classification of the *Bacteria*, *Archaea*, and other microorganisms. The 16S rDNA–based phylogeny will face its most stringent test as additional genomes are sequenced and compared.

Before we discuss the use of 16S rDNA sequences in arriving at the current phylogeny of *Bacteria* and *Archaea*, it is first necessary to consider phylogenetic trees.

PHYLOGENETIC TREES   Like a family tree, a phylogenetic tree contains the tree of descendants of a biological family or group. However, whereas the family tree traces the genealogy of a family of humans, phylogenetic trees trace the lineage of a variety of different species. Thus, **phylogenetic trees** reflect the purported evolutionary relationships among a group of species, usually through the use of some molecular attribute they possess, such as the sequence of their rRNA. In this particular section we will discuss molecular phylogeny based on a comparison of the 16S rRNA sequences of organisms.

Phylogenetic trees have two features—**branches** and **nodes** (Figure 17.4). Each node represents an individual species. External nodes (usually drawn to the extreme right of the tree) represent living species, and internal

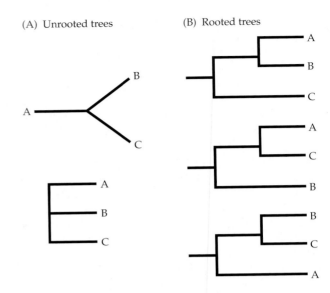

(A) Unrooted trees   **(B)** Rooted trees

Figure 17.5 Unrooted and rooted trees
Representations of the possible relatedness between three species: A, B, and C. (A) A single unrooted tree (shown in both formats; see Figure 17.4). (B) Three possible rooted trees (in one format).

nodes represent ancestors. A branch is a length that represents the distance between or degree of separation of the species (nodes).

Trees may be rooted or unrooted. Figure 17.5A shows unrooted trees containing three different species: A, B, and C. **Unrooted trees** typically compare one feature of a group of related organisms, such as the sequence of their 16S rRNA gene. There is only one shape to this particular tree with three species.

In contrast, **rooted trees** provide more information. Typically, rooted trees use the same gene from a distantly related organism for the root. This allows one to compare the relatedness of the more closely related species—A, B and C—to one another. A rooted tree containing three species has three possible shapes (Figure 17.5B). In the examples shown, A and B are more closely related to one another in the uppermost example, whereas A and C and B and C, respectively, are more closely related in the two lower examples.

Alternatively, an additional gene can be used to compare the species. When using this approach, a different gene such as the sequence of a macromolecule that underwent a gene duplication event prior to when the taxa that are being studied, diverged from one another. As an example, a comparison of the elongation factors Tu and G (see Chapter 13) was used to root the 16S rRNA Tree of Life (see Figure 1.6).

For bacterial phylogeny, trees are constructed from information based on the sequence of subunits in macro-

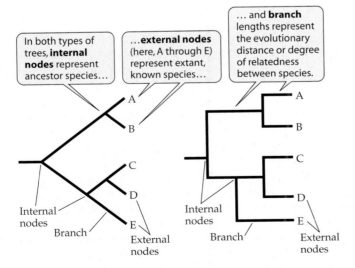

In both types of trees, **internal nodes** represent ancestor species…

…**external nodes** (here, A through E) represent extant, known species…

… and **branch** lengths represent the evolutionary distance or degree of relatedness between species.

Internal nodes   Branch   External nodes

Internal nodes   Branch   External nodes

Figure 17.4 Phylogenetic trees
Two different formats of phylogenetic trees used to show relatedness among genes or species.

molecules. As mentioned previously, rRNA, in particular the small subunit rRNA molecule, 16S rRNA, has been selected as the molecule of choice because of its conserved nature and length.

SEQUENCING 16S rDNA   If one isolates a new bacterium and wishes to determine its phylogenetic position among the known bacteria, it is necessary to determine the sequence of its 16S rRNA gene. This can be accomplished in a number of ways. One of the most common ways is to use polymerase chain reaction (PCR) to amplify the 16S rDNA from genomic DNA from the bacterium (see Chapter 16). The amplified 16S rDNA may be sequenced directly or ligated into a cloning vector and cloned into *E. coli*. This latter step allows a large quantity of 16S rDNA to be produced through growth prior to sequence analysis. The entire 16S rDNA can be sequenced using a standard set of oligonucleotide primers and standard sequencing techniques.

ALIGNMENT WITH KNOWN SEQUENCES   The next step is to incorporate the determined linear DNA sequence into an alignment with the sequences of other known organisms. Two internet databases are of special importance in this regard. One is called the Ribosome Database Project, located at Michigan State University. It contains the 16S rRNA sequences for those bacteria that have been sequenced. Sequences from this database can be retrieved electronically via the Internet (http://rdp.cme.msu.edu/). Another frequently used database for sequence analysis is the National Center for Biotechnology Information (NCBI) (www.ncbi.nlm.nih.gov/Genbank), which contains a comprehensive listing for 16S rRNA genes as well as other genes.

By using a series of computer programs and careful manual examination, one can align and compare the sequence with those retrieved from the database.

PHYLOGENETIC ANALYSIS   Having the sequence is only half of the story. It is next necessary to compare the sequence of the unknown bacterium to that of other bacteria from the database. The determination of the evolutionary relatedness among organisms can be accomplished by one of a number of phylogenetic methods. There are several types of analysis that can be used. **Distance matrix** methods are one type of approach. In distance matrix methods, the evolutionary distances, based on the number of nucleic acid or amino acid monomers that differ in a sequence, are determined among the strains being compared. A second approach is to use **maximum parsimony** methods. In maximum parsimony, the goal is to find the simplest or most parsimonious phylogenetic tree that could explain the relatedness between different sequences. In both approaches,

the sequence of nucleic acid subunits among different strains of bacteria or other genes is compared.

Figure 17.6 illustrates the use of two different methods to create a tree from an aligned hypothetical sequence region of rRNA from four different strains (shown in Figure 17.6A). The first approach is a distance matrix method called the "unweighted pair group method with arithmetic mean," or UPGMA (see Figure 17.6B). This is one of the simplest analytical methods that can be used. Figure 17.6C shows an analysis of the same sequences by maximum parsimony.

In the UPGMA method, a distance matrix is set up to compare the differences in sequence or "distance," d, between each of the four strains. There are four nucleotide differences between the sequence of organism a and the sequence of organism b; therefore, d_{ab} is determined to be 4. Likewise, d_{ac} is 5, d_{bc} is 5, and so forth. In this instance, the shortest distance, 2, is between strains c and d. From these two strains, which show the closest relationship to one another, a simple tree is constructed that shows c and d connected by a node that is half the distance between the two, that is, 1 unit. This is expressed in the actual tree as a horizontal branch length of one unit from each of the organisms to a common ancestral node (see Figure 17.6B).

The next step is to construct another matrix in which c and d are considered as a single composite unit (cd) and compared with a and b. From this matrix, the two most similar organisms are a and b, and the length of this branch is calculated as the distance, $d_{ab}/2$, which is equal to 2 units. From this an intermediate second tree is formed. Finally, a is different from (cd) by $(d_{ac} + d_{ad})/2$, or $(5 + 6)/2 = 5.5$. Likewise, b is different from (cd) by $(d_{bc} + d_{bd})/2 = 4.5$. To determine the connection between the composite branch cd and ab, the distance is calculated as the average of $d_{(cd)(ab)}$, or $5.5 + 4.5/2$, which is 5.0. This value is then divided by 2 to give the average distance between the two composites, cd and ab. Using this reasoning, a final tree (see Figure 17.6B) is produced showing the relationship among the four different strains.

The UPGMA method is the simplest of the distance methods used. More sophisticated distance methods include transformed distance and neighbor-joining methods, which will not be discussed here.

As mentioned earlier, in maximum parsimony the goal is to identify the simplest tree that could explain the difference between two different sequences or species. This approach has its philosophical basis in **Occam's razor**, commonly used in the sciences, which states that *the likely solution to a problem is the simplest one.* In this case, to explain the evolutionary difference between two species, one looks at the tree that has the fewest changes (mutational events) that could explain their differences. This is accomplished with a computer that, in theory, con-

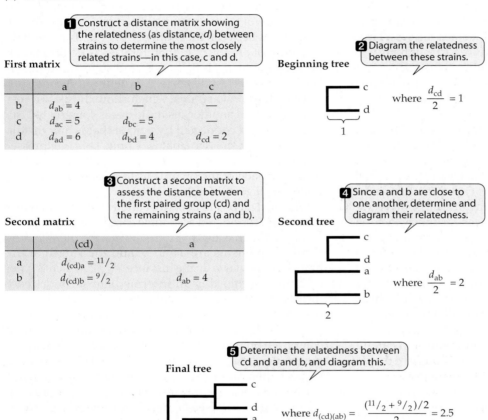

(A) 16S rRNA sequences

This table shows the sequence of a nine-base region of the 16S rRNAs of the four strains.

Organism (strain)	Site number								
	1	2	3	4	5	6	7	8	9
a	G	C	G	G	A	C	A	A	A
b	G	A	C	G	C	C	A	A	G
c	G	A	A	A	U	C	U	A	A
d	G	A	A	A	G	C	U	A	G

Figure 17.6 Phylogenetic analysis
Phylogenetic analysis of four different strains—a, b, c, and d—using a hypothetical region of their 16S rRNA that contains nine bases (A). (B) The UPGMA method of determining a phylogenetic tree. (C) The maximum parsimony method (see text for details).

(B) UPGMA method

1 Construct a distance matrix showing the relatedness (as distance, d) between strains to determine the most closely related strains—in this case, c and d.

First matrix

	a	b	c
b	$d_{ab} = 4$	—	—
c	$d_{ac} = 5$	$d_{bc} = 5$	—
d	$d_{ad} = 6$	$d_{bd} = 4$	$d_{cd} = 2$

2 Diagram the relatedness between these strains.

Beginning tree

where $\dfrac{d_{cd}}{2} = 1$

3 Construct a second matrix to assess the distance between the first paired group (cd) and the remaining strains (a and b).

Second matrix

	(cd)	a
a	$d_{(cd)a} = {}^{11}/_2$	—
b	$d_{(cd)b} = {}^{9}/_2$	$d_{ab} = 4$

4 Since a and b are close to one another, determine and diagram their relatedness.

Second tree

where $\dfrac{d_{ab}}{2} = 2$

5 Determine the relatedness between cd and a and b, and diagram this.

Final tree

where $d_{(cd)(ab)} = \dfrac{({}^{11}/_2 + {}^{9}/_2)/2}{2} = 2.5$

siders all possible trees and then identifies the simplest one (the one with the fewest assumed mutational events). As for all phylogenetic analyses, the alignment of the sequences must be accurate.

In maximum parsimony it is important to recognize sites in the sequences that are useful for a comparison between organisms. These are termed **informative sites**. These sites are then used to determine the most parsimonious tree. For example, Site 1 in the example given (see Figure 17.6C) is not informative because all the bases are identical. Site 2 is not informative either, because three of the strains have A and one has C, suggesting that a single mutational event has occurred. Site 3 is not informative, because Trees 1 and 2, which have two changes, are equally parsimonious, and Tree 3 differs from Tree 2 only in the inferred ancestor. Site 5 is not informative because all trees constructed from the information at this site differ from one another by three mutations. In contrast, Site 4 is informative, because one tree (Tree 1) is more parsimonious than the other two. Sites 6 and 8 are

(C) Maximum parsimony method

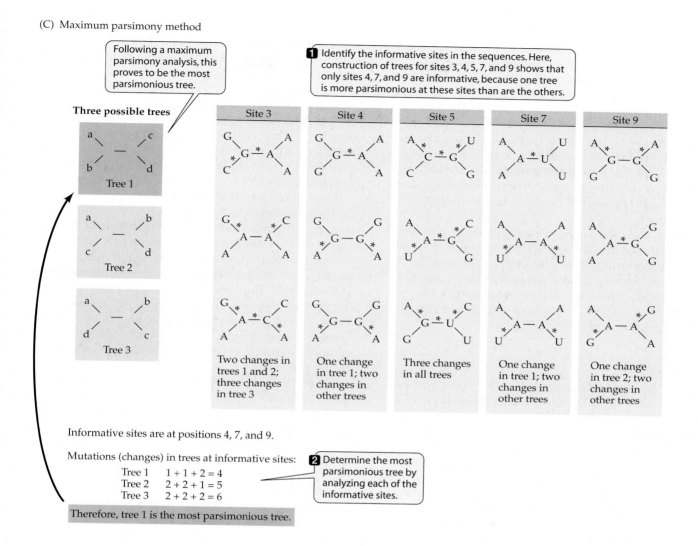

Following a maximum parsimony analysis, this proves to be the most parsimonious tree.

1 Identify the informative sites in the sequences. Here, construction of trees for sites 3, 4, 5, 7, and 9 shows that only sites 4, 7, and 9 are informative, because one tree is more parsimonious at these sites than are the others.

Three possible trees

Informative sites are at positions 4, 7, and 9.

Mutations (changes) in trees at informative sites:

Tree 1	1 + 1 + 2 = 4
Tree 2	2 + 2 + 1 = 5
Tree 3	2 + 2 + 2 = 6

2 Determine the most parsimonious tree by analyzing each of the informative sites.

Therefore, tree 1 is the most parsimonious tree.

not informative because all bases are identical, but both Sites 7 and 9 are informative. Site 7 favors Tree 1, whereas Site 9 favors Tree 2. Thus, for this set of data, Tree 1 is favored two of three times, Tree 2 is favored one of three times, and Tree 3 is not favored at any time. Adding the changes at those three sites gives the following data: Tree 1 is the most parsimonious because a total of only four changes (1 + 1 + 2 = 4) would explain its phylogeny, whereas in Tree 2, five changes are required (2 + 2 + 1 = 5), and in Tree 3 six changes are required (2 + 2 + 2 = 6).

Note that the two trees that were constructed from the distance matrix and maximum parsimony methods are identical in shape. Both trees indicate that two of the strains, a and b, are more closely related to one another than they are to c and d. Likewise, c and d are more closely related to one another than they are to a and b. As you can imagine, some rather sophisticated computer programs have been developed to handle the immense amount of information inherent in longer sequences such

as 16S rDNA, which contains about 1,500 bp. Moreover, when such large sets of data are being analyzed, it is often not possible to determine that a proposed tree is, in fact, the true tree. Indeed, *all* trees should be considered as hypotheses until additional information has been analyzed. For example, the inclusion of a sequence from a newly discovered, closely related organism may alter the shape of a tree when it is included in the analysis.

To help support the reliability of a given tree, other techniques are used. For example, in "bootstrap" analyses, random portions of the sequence are selected by the computer, and the trees formed from them are compared with the proposed tree to assess the statistical significance of the proposed tree. In this statistical approach, some 100 or 1,000 different bootstrap comparisons might be made and provided as evidence that the proposed tree is indeed the most parsimonious one. Bootstrap analyses are applicable for all phylogenetic treatment procedures.

As mentioned previously, other analytical methods can be used to analyze sequence information and construct phylogenetic trees. A common one used by microbiologists is the **maximum likelihood** method, which involves selecting trees that have the greatest likelihood of accounting for the observed data. This is accomplished by assigning a probability to the mutation of any one base to any other base at each possible sequence position. From this, all possible topological trees are constructed. By integrating the probabilities for each mutation over each tree, a degree of improbability for a tree is assessed. The least improbable tree is chosen as the "true" tree.

A final note on studying microbial phylogeny: It should be emphasized that many other genes besides 16S rDNA can be used for phylogenetic analysis. An appropriate gene must have the degree of conservation necessary for the analysis desired and must have a **homolog** in the other organisms of interest. A homologous gene is a gene that shares a common ancestry.

Speciation

The process by which organisms evolve is termed **speciation**. As with plants and animals, bacteria evolve in habitats. However, unlike plants and animals, bacteria can evolve very quickly because of their rapid growth rates, high population sizes, and haploid genomes that allow for the rapid expression of favorable mutations through natural selection. Thus, a lineage of bacteria is determined in large part through **vertical inheritance**, the process by which the parental genotype is transferred to the progeny cells following DNA replication and asexual reproduction.

However, bacteria can also acquire genetic material from other different organisms through their various genetic exchange mechanisms: conjugation, transformation, or transduction (see Chapter 15). This phenomenon is referred to as **horizontal** (or **lateral**) **gene transfer** (HGT) to distinguish it from vertical inheritance. As a result, prokaryotic organisms may undergo dramatic changes in their population structure in a relatively short period. For example, we know that multiple-drug resistance can be rapidly acquired by a bacterial species that is sensitive to antibiotics if it is exposed to antibiotics in the presence of other antibiotic-resistant bacteria.

Consider the situation of a bacterium that is a member of the normal microbiota of the intestinal tract that is exposed to an antibiotic to which it is sensitive. The bacterium may either perish or—if a gene is available in the environment that confers resistance and the bacterium has the capability—acquire the resistance gene through an HGT process and survive. This example of a strong selective pressure likely explains how it is pos-

sible for sensitive bacteria to quickly become resistant to an antibiotic. This scenario applies equally to other environmental pressures that confront bacteria, such as exposure to potentially toxic hydrocarbons that are used as an energy source by other species in the environment.

Many of the known examples of rapid genetic change occur through the acquisition of plasmids from related organisms. Thus, in the preceding examples, some plasmids are known that carry multiple antibiotic resistant genes and others are known that carry hydrocarbon-degrading genes.

In addition, we do know that genes can be acquired from distantly related organisms. For example, it has been recently reported that some members of the *Proteobacteria* and other phyla of the *Bacteria* have been found to contain a gene responsible for bacteriorhodopsin synthesis, which was only known previously from members of the *Archaea*. Thus, this example appears to represent the transfer of genetic material across two different domains as well as phyla within the *Bacteria*. Another example is the bacterium, *Agrobacterium tumefaciens*, which naturally transfers genetic material to plants (Chapters 16 and 19).

From an analysis of genomes that have been sequenced, genes that have been derived from other organisms have been identified in some prokaryotic genomes (Figure 17.7). The transfers between phyla appear to be relatively rare events. In addition, it should be noted that each species has hundreds of core genes that can be used to compare its relatedness to other organisms. If HGT occurred too extensively, it could confuse phylogenetic classifications based on vertical inheritance to such an extent that it would render them utterly useless. Fortunately, this does not appear to pose a major problem for 16S rRNA gene trees.

SECTION HIGHLIGHTS

Both phenotypic and genotypic properties have been used to describe microorganisms, and both are important in describing and naming new prokaryotic species. Artificial taxonomy entails the use of phenotypic tests, whereas a taxonomy based on evolutionary processes relies on molecular phylogeny. Molecular phylogeny uses 16S rRNA gene sequence and protein sequence analyses, which have become very important in the classification of *Bacteria* and *Archaea*. Several different methods, such as distance, parsimony, and maximum likelihood methods, are used to construct phylogenetic trees.

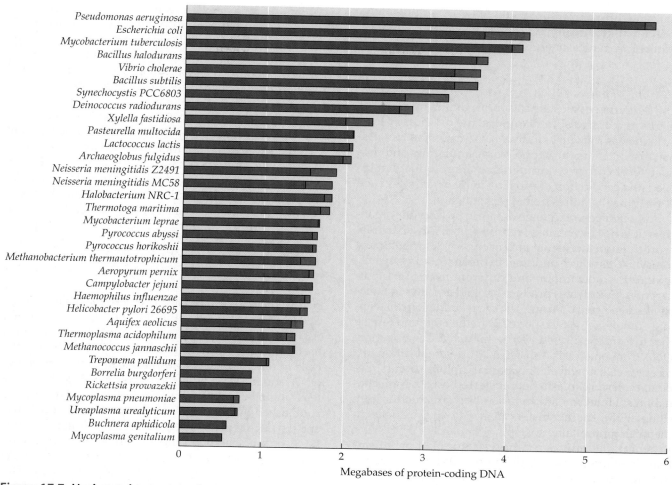

Figure 17.7 Horizontal gene transfer
Analyses of sequenced bacterial genomes indicate that a significant proportion of
their genes can be traced to other phylogenetic groups, indicating the importance of
horizontal gene transfer (HGT) in bacterial speciation. This diagram shows the proportion of DNA that was acquired by HGT (in red) in some microbial genomes.
Courtesy of Jeffrey Lawrence.

17.3 Taxonomic Units

The basic taxonomic unit is the species, although as mentioned earlier, some species have subspecies categories
as well. The categories above the species are (sequentially) genus, family, order, class, phylum, and domain
(see Table 17.1). It should be noted that uncertainties exist in bacteriology about the meaning of the higher taxonomic categories such as kingdom because oftentimes
the phylogenetic markers that have been used (primarily 16S rDNA sequences) cannot definitively resolve the
earliest branching points in the Tree of Life. Thus, although we know that each of these major branches is
equivalent to the plant and animal "kingdoms," how the
microbial "kingdoms" or phyla are related to one another is only poorly understood.

Each colony or culture of an organism represents an
individual **strain** or **clone** in which all of the cells are descended from one single organism. In a somewhat different sense of meaning, a strain can also refer to a mutant of a species that has changed characteristics (for
example, lacks a particular gene). The strain, however,
is not considered a formal taxonomic unit, and Latin
names are therefore not ascribed to strains; they have
only informal designations, such as *E. coli* strain K12.
There can also be **varieties** within species that exhibit
differences. These are called **biovars** (i.e., biological
varieties). For example, a serological variety such as *E.
coli* O157:H7 is a pathogenic **serovar** that causes hemolytic uremic syndrome and can be lethal to children
who become infected by eating contaminated food. Likewise, pathogenic varieties are termed **pathovars**, ecolog-

ical types **ecovars**, and so forth. Now let us look at the individual taxa beginning with the species to see what features are typical at each taxonomic level.

The Species

The definition of a bacterial species differs from that of plants and animals. In mammals, the classical species is defined as a group of individuals (males and females) that exhibit evident morphological similarities and produce fertile progeny through sexual reproduction. Indeed, the production of progeny in many animals such as mammals requires sexual reproduction.

Although gene exchange occurs in prokaryotic organisms, it is not essential for reproduction. Most bacterial reproduction is asexual and occurs by simple binary transverse fission or budding. In prokaryotic organisms, sexuality is uncommon and different from that of eukaryotes. Eukaryotes produce haploid gametes in meiosis (see Chapter 1). During sexual reproduction, the haploid gametes (egg and sperm) from the male and female fuse to form the diploid zygote. In bacterial conjugation, DNA from one cell is transferred during replication to a receptor cell; however, only partial diploidy occurs. Genetic material can also be transferred by other mechanisms such as transformation and transduction (see Chapter 15). These transfers are not always restricted to members of the same species.

A bacterial species comprises a group of organisms that share many phenotypic properties and a common evolutionary history and are therefore much more closely related to one another than to other species. This definition, which is very subjective, has been interpreted differently by bacteriologists in describing species. For example, at one extreme some taxonomists are called **lumpers** because they group (or "lump") fairly diverse organisms into a single species or genus. An example of a lumper is F. Drouet, who has proposed reducing the number of cyanobacteria from 2,000 species to only 62! At the opposite pole are **splitters**. These are taxonomists who consider even the slightest differences sufficient for a new species. For example, many years ago it was proposed that the genus *Salmonella* be "split" into hundreds of different species, a separate species for each of the hundreds of different serotypes (or serovars) that are recognized based on specific cell-surface antigens of their lipopolysaccharides and flagella. However, the views of lumpers and splitters illustrated here are considered to be extreme and are not accepted by the majority of microbiologists.

In fact, more recently, a less arbitrary, quantitative basis has been proposed to define a bacterial species. Agreement was reached by a group of prominent bacterial taxonomists to define a bacterial species based on

genomic similarity between strains. Accordingly, a species is defined as follows: two strains of the same species must have a similar mole percent guanine plus cytosine content (mol % G + C) and must exhibit 70% or greater DNA–DNA reassociation. The procedures used to determine these features are described here.

MOLE PERCENT GUANINE PLUS CYTOSINE (MOL % G + C)   The **mol % G + C** refers to the proportion of guanine and cytosine to total bases (guanine, cytosine, adenine, and thymine) in the DNA. Recall that because G and C are paired in the double-stranded DNA molecule by hydrogen bonds, as are A and T, they occur in equal concentrations. The formula is given as:

$$\text{mol} \% \, G + C = \frac{\text{moles}\,(G + C)}{\text{moles}\,(G + C + A + T)} \times 100$$

Several methods can be used to determine the mol % G + C, sometimes also called the "**GC ratio**" of a bacterium. All of them require that the DNA be first isolated from a bacterium and purified. Thus, it is necessary to lyse the cells to release the cytoplasmic constituents including DNA, and the DNA must then be purified to remove proteins and other cellular material. Cell lysis is typically accomplished by treatment with lysozyme and detergents, and the DNA is precipitated with ethanol. When the DNA has been sufficiently purified, it can be analyzed chemically to determine the content of each of the bases. Several different procedures can be used to determine the GC ratio of the purified DNA. We will describe two of them here.

A common procedure to determine GC ratios is by **thermal denaturation**. The principle behind this method is that the hydrogen bonds between the double strands can be broken by heating dissolved DNA. As the hydrogen bonds are broken and the two strands separate, the absorbance of the DNA increases. This procedure, called "melting the DNA," is conducted with a spectrophotometer set at 260 nm, a wavelength at which DNA absorbs strongly. The hydrogen bonding of the GC base pair is stronger than the AT pair in the double-stranded DNA molecule. Therefore, a higher temperature is required to melt DNA that has a high content of GC pairs, that is, a high GC ratio. Figure 17.8 shows a graph of the melting of a double-stranded DNA molecule. This process is accomplished by gradually increasing the temperature of a solution of the DNA in an appropriate buffer (ionic strength is important). As the temperature is increased, the melting process begins and continues until the double-stranded DNA molecule is completely converted to the single-stranded form. The absorbance increases during this melting process. The midpoint temperature (T_m) is directly related to the GC ratio of the DNA. Thus, the

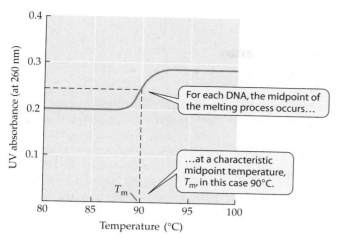

Figure 17.8 DNA melting curve
Melting curve for a double-stranded molecule of DNA. As the temperature is increased during the experiment, the double-stranded DNA is converted to the single-stranded form and the UV absorbance of the solution increases. The midpoint temperature, T_m, can be calculated from the curve. This process is reversible if the temperature of the solution is slowly decreased to allow the single strands to reanneal. The T_m of this species, *Escherichia coli*, can be used to determine its mol % G + C content (see Figure 17.9).

GC ratio can be read from a chart showing the relationship between T_m and GC content (Figure 17.9). Once the DNA has been melted, it will reanneal if the temperature is slowly lowered. Thus, the process shown in Figure 17.8 is reversible. However, if the solution is cooled rapidly, hybrid formation does not recur, and the molecules of DNA are left in the single-stranded state.

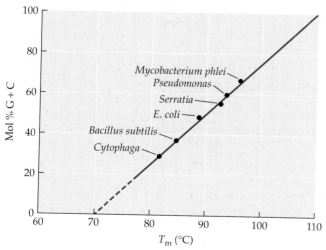

Figure 17.9 T_m and DNA base composition
Graph showing the direct relationship between mol % G + C and midpoint temperature (T_m) of purified DNA in thermal denaturation experiments.

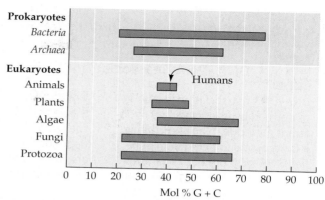

Figure 17.10 DNA base composition range
Range of mol % G + C content among various groups of organisms. Note the broad range of GC ratios for bacteria, archaea, and the lower eukaryotes in comparison to plants and animals.

Another method is to use the "readout" from a genome sequence, which contains all of the genetic information of a species.

Figure 17.10 shows the range of GC ratios in various groups of organisms. On this basis alone, one can see that bacteria, which have GC ratios ranging from approximately 20 to greater than 70, are truly a very diverse group. In contrast, higher organisms such as animals have a very restricted GC ratio range.

The GC ratio provides only the relative amount of guanine and cytosine compared to total bases in the DNA of an organism and says nothing about the inherent characteristics of the organisms or what genes are present. Indeed, *two very different organisms can have similar or even identical GC ratios*. For example, the DNA of *Streptococcus pneumoniae* and humans have the same mol % G + C content.

DNA–DNA REASSOCIATION OR HYBRIDIZATION Although the determination of GC ratio is useful in bacterial taxonomy, it does not tell us anything about the linear arrangement of the bases in the DNA. It is the arrangement of the DNA subunits that codes for specific genes and proteins and therefore determines the features of an organism. DNA–DNA reassociation or hybridization is one method used to compare the linear order of bases in two different organisms (Box 17.3). It is important to recognize that the = 70% level of reassociation used for the species definition does not indicate that the two DNAs are 70% homologous or identical.

In DNA–DNA reassociation, the actual order of bases in the DNA is not determined, but rather the extent of reannealing between the DNAs of two different strains is assessed. Ideally one would like to know the actual

BOX 17.3   *Methods & Techniques*

DNA–DNA Reassociation

DNA–DNA reassociation can be performed by using a variety of different methods. In all approaches, it is necessary to begin with purified DNA from the two organisms that are being compared. The DNA is first cut into smaller segments (i.e., sheared) and then denatured by melting. DNA from the two different strains are mixed and allowed to cool together to allow reannealing to occur. This reannealing will occur both between DNA strands of the same species and between strands of the comparison species. The degree of reannealing depends on how similar the DNAs are to one another. If two strains are very similar, their DNAs will reanneal to a high degree. In contrast, if two strains are very different, then the extent of reannealing will be much less. One way to perform DNA–DNA reassociation is to radiolabel the DNA by growing the bacterium with tritiated thymidine or ^{14}C-labeled thymidine (other DNA bases or ^{32}P labeling can be used as well). If the bacterium takes up this labeled substrate and incorporates it into DNA, then the DNA becomes labeled. Alternatively, the DNA can be purified from the bacterium and labeled enzymatically in the laboratory.

After the DNA has been labeled and purified, it is sheared to an appropriate length by sonication. It is then ready for the hybridization

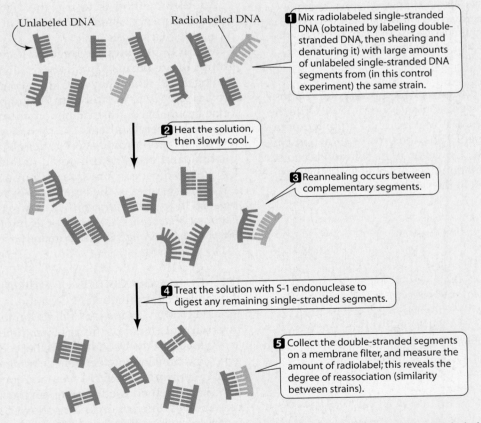

DNA–DNA reassociation. In this example, which is a control experiment (the radiolabeled sample is reannealed with unlabeled DNA from the same strain), the degree of reassociation is highest and treated as 100%. If a different strain is reannealed with the radiolabeled DNA, it will show a lower degree of reannealing (compared with the 100% attributed to the control), indicative of the similarity between the two strains being tested. Strains with reannealing values of 70% or greater are considered to be the same species.

BOX 17.3   *Continued*

experiments. First, single-stranded DNA is prepared. This is accomplished by heating the isolated DNA molecules to render them single-stranded and then cooling them rapidly to prevent reannealing.

First, let's look at the control assay for the DNA reassociation experiment. In this case, a small amount of sheared radiolabeled DNA is rendered single-stranded. This is then mixed with a much larger amount of unlabeled DNA obtained from the same bacterial strain. These are heated together and cooled slowly to allow the two single-stranded groups to reanneal to form hybrid double strands. Because the amount of labeled DNA relative to the unlabeled DNA is small, there is a very low probability that it will reanneal with other labeled strands. Most of the reassociations will occur between unlabeled strands, and most of the remainder will be between the labeled and unlabeled strands. The

single-stranded fragments that did not reanneal are removed by enzyme digestion using S-1 endonuclease, which specifically degrades only single-stranded DNA, and the double-stranded fragments are collected on a membrane filter or in a column. The amount of radioactivity remaining on the filter or on the column, after washing to remove low-molecular-weight material, represents the amount of hybrid formation between the labeled and unlabeled DNA for this identical strain. This is the *control reaction,* and the amount of radiolabel (the extent of hybridization) is considered to be 100%.

To determine the extent of reassociation between the strain described and an unknown strain, similar experiments need to be performed. In this instance, unlabeled single-stranded DNA from the unknown strain is prepared and mixed with the known strain for which we have labeled the DNA. As

indicated previously, those strains that show 70% or greater reassociation or hybrid formation with the labeled strain (determined by the amount of hybrid DNA that is radiolabeled compared with the same strain control of 100% shown in the figure) are considered to be the same species. Anything less is considered a different species.

The temperature and salt concentration at which the DNA–DNA reannealing occurs will influence the degree of reassociation between single strands. Scientists conducting DNA–DNA reassociation experiments typically use a reannealing temperature that is 25°C lower than the average midpoint temperature (T_m) of the DNAs being compared.

This temperature is sufficiently high that only those sequences that are most complementary will reanneal. Therefore, this is considered a **stringent** condition for reannealing.

sequence of genes of a species. Indeed, sequencing entire bacterial genomes has become quite common (see Chapter 16). It is worthwhile noting that one could consider that the actual DNA base sequence of a strain is the ultimate definition of a strain—analogous to the chemical formula for a compound. However, it is important to recognize that, unlike chemical compounds, bacterial strains and species are not static; they continue to evolve.

Interestingly, the bacterial species definition appears to be much broader than that used for animals and plants when one considers DNA–DNA reassociation and other molecular criteria. For example, the bacterial species *E. coli* can be compared to its host mammalian species using a variety of molecular features, including the range in GC ratio, 16S rDNA sequence (versus 18S rDNA sequence), and DNA–DNA reassociation (Table 17.2). Therefore, although there is essentially no variation in the range in GC ratio for the human species, it is about 4 mol % within the *E. coli* species. Indeed, there is less variation in GC ratio in the *Primate* order than

there is in the species *E. coli.* Likewise, the 16S rDNA sequences of *E. coli* possess more than 15 substitutions, whereas the difference between 18S rDNA of the mouse (order *Rodentia*) and the human is less than 16. Finally, DNA–DNA reassociation data, which are used to define the bacterial species at 70% or greater, indicate that humans are much more highly similar to one another in comparison with *E. coli.* Therefore, it is evident that the typical bacterial species is equivalent to a genus or family of mammals based on molecular divergence, indicating that a bacterial species is defined much differently from its eukaryotic counterparts. This difference is further evidenced by the biological species definition for animals, which requires that within a species, mating between sexes produces fertile progeny.

The Genus

All species belong to a genus, the next higher taxonomic unit. When DNA–DNA reassociation is performed

TABLE 17.2	Comparison of *E. coli* and its host species[a]			
		Property		
Comparison	Mole % G + C	16S or 18S rRNA Substitutions	DNA/DNA Reassociation	
Among *E. coli*	48–52	>15 bases	>70%	
Among *Homo sapiens*	42	—	—	
Among all primates	42	—	—	
Between *H. sapiens* and mouse	—	16 bases	—	
Between *H. sapiens* and chimpanzee	—	—	98.6%	
Between *H. sapiens* and lemurs	—	—	>70%	

[a]Adapted from J. T. Staley, *ASM News*, 1999.

within a genus, some species of the genus may show little or no significant reassociation with other species. This does not indicate that they are unrelated to one another, only that this technique is too specific to identify outlying members of the same genus. Therefore, DNA–DNA hybridization has limited utility for determining whether a species is a member of a known bacterial genus.

The definition of the genus is based on one or more prominent phenotypic characteristics that permit it to be distinguished from its closest relatives. Oftentimes some striking physiological or morphological feature is present that permits the genus to be differentiated from closely related taxa. For example, the genus *Nitrosomonas* is a group of rod-shaped bacteria that grow as chemoautotrophs, gaining energy from the oxidation of ammonia. Other ammonia oxidizers with coccus-shaped and helical cells are placed in other genera. Of course, all strains of each of these genera need to be more closely related to one another phylogenetically than to strains of other genera in a phylogenetic classification. Ideally then, the genus makes up a **monophyletic** lineage (i.e., one in which all are members of the same phylogenetic cluster or clade).

Higher Taxa

Odd bedfellows are sometimes found in phylogenetic trees; therefore, photosynthetic and nonphotosynthetic members of some closely related groups have been reported. Of course, loss of a key gene or two may result in converting a formerly photosynthetic organism to one that is not photosynthetic. Consequently, although the plant and animal kingdoms are differentiated on the basis of whether or not they are photosynthetic, both features have been reported in two closely related bacterial genera. However, because phenotype is so important at the genus level, important features such as photosyn-

thesis are sufficient to proclaim a separate genus, even though two groups are otherwise very closely related.

As mentioned previously, the taxonomy of prokaryotes is undergoing major changes. Although according to classical taxonomy each genus belongs to a family of similar genera, relatedness at the familial and higher level is often uncertain for bacteria. Bacteriologists have been reluctant to ascribe organisms to formal Latinized families and orders. However, as more becomes known about bacterial phylogeny, it is increasingly apparent that higher taxonomic levels do have meaning and can be distinguished from one another by comparing the sequences of certain macromolecules, as is reflected in the new edition of *Bergey's Manual of Systematic Bacteriology*.

SECTION HIGHLIGHTS

Bacteria and *Archaea* are classified in a hierarchical structure from the domain level to the phylum, class, order, family, genus, and finally species. Species are described based on both phenotypic and genotypic properties including GC ratio, 16S rRNA sequence, and DNA–DNA hybridization.

17.4 Major Groups of *Archaea* and *Bacteria*

Bacteriologists have begun to construct classifications using phylogenetic information from rRNA analyses. As mentioned in Chapter 1, some prokaryotes are very different from others, a revelation that came through an analysis of rRNA (Box 17.4). Ribosomal RNA data allow the division of all organisms on Earth, prokaryotic and eukaryotic, into three domains: *Bacteria, Archaea,* and *Eukarya* (see Figure 1.6). These domains can also be dis-

BOX 17.4   *Research Highlights*

The Discovery of *Archaea*

Clearly, one of the most exciting developments in bacterial classification of the twentieth century was the discovery that there are two major groups called Domains, named the *Bacteria* and the *Archaea*. The appreciation of the difference between the *Bacteria* and the *Archaea* was the culmination of years of research by microbiologists throughout the world. However, the final piece of evidence that convinced microbiologists of this dichotomy was the discovery that these organisms had very different 16S rRNAs. This research was performed in Carl Woese's laboratory at the University of Illinois. At that time, rRNA sequencing was not done routinely in laboratories. Instead, 16S rRNA was purified and digested by ribonucleases that cut between specific nucleotide pairs. The oligonucleotide fragments produced were subjected to two-dimensional electrophoresis. The pattern of spots on a two-dimensional chromatogram represented the various rRNA oligonucleotides that were typical of each species. Studies from Woese's laboratory demonstrated that the patterns were very different for *Bacteria* and *Archaea*. Indeed, their studies of 18S rRNA from eukaryotic organisms indicated that the *Archaea* are as different from *Bacteria* as they are from eukaryotes. The seminal findings of this work have forever changed the way microbiologists view taxonomy and phylogeny.

Carl Woese. Courtesy of Jason Lindsey.

---

tinguished from one another by phenotypic testing. For example, consider the cell envelope composition of the organisms. Peptidoglycan is found only in *Bacteria*, although two groups—the mycoplasmas and the *Planctomycetales*—lack it. Furthermore, the lipids of the *Bacteria* and *Eukarya* are ester-linked, whereas they are ether-linked in the *Archaea* (see Chapters 4 and 18).

At this time, 28 different phylogenetic groups, referred to here as **phyla**, are known. Included are 24 phyla of *Bacteria* and four phyla of *Archaea*. Each of these phyla has specific signature sequences in their ribosomes that are distinctive to them. The prokaryotic phyla are listed here along with a brief description of their major features. They are treated in more detail in subsequent chapters (Chapters 18 through 22), and the groups in each chapter are indicated below. It is noteworthy that many new phyla of the *Bacteria*, in particular, have been discovered in natural environments using clone library approaches, but have not yet been isolated in pure culture (Box 17.5). Thus, at least ten to 15 additional major prokaryotic groups are very poorly understood.

The second edition of *Bergey's Manual of Systematic Bacteriology* has been largely followed in organizing the bacterial groups treated in this book. The *Archaea* are described in Chapter 18 and the *Bacteria* in Chapters 19–22 (Table 17.3).

Domain: Archaea

The *Archaea* are divided into the following four phylogenetic groups or phyla.

CRENARCHAEOTA   This phylum contains the most thermophilic organisms known. Some of these organisms grow at temperatures higher than the boiling point of water. Most rely on sulfur metabolism either as an energy source or as an electron sink. For example, some oxidize reduced sulfur compounds aerobically to produce sulfuric acid. Others reduce elemental sulfur and use it as an electron acceptor to form hydrogen sulfide. Some are iron and manganese reducers. Not all organisms are thermophilic. They are also significant in deep-sea environments as well as in polar seas.

EURYARCHEOTA   The methanogens (methane producers) are noted for their ability to produce methane gas

BOX 17.5 Research Highlights

Novel Phyla Discovered by Molecular Analyses of Natural Habitats

One of the major advances in exploring the diversity of microorganisms in natural environments has been the application of molecular approaches developed in Norman Pace's laboratory. In the most recent variation of this approach, DNA is extracted from the environment of interest. Then PCR is used to amplify genes of interest from the DNA. For phylogenetic (i.e., diversity) information, 16S rDNA primers for the *Bacteria* or *Archaea* or universal primers that amplify both groups are commonly used. The segments retrieved, typically about 500 bp, can then be sequenced to identify the phylogenetic groups. Although PCR approaches are not quantitative, the various 16S rDNA types retrieved from a natural sample can provide important information about the diversity of prokaryotes that occur in that environment.

Using these approaches, it has recently been determined that more than 50 major phyla of *Bacteria* exist, yet isolates have been obtained of only 24.

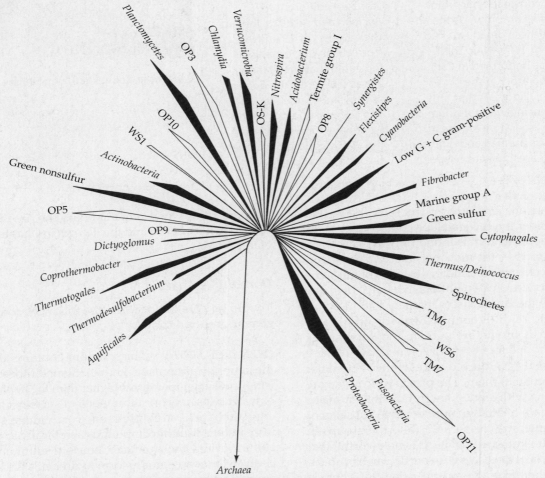

A phylogenetic tree of 16S rDNA sequences of *Bacteria,* based on pure cultures and clonal libraries from natural samples. Note the existence of many phyla (shown in outline rather than as solid black lines) that have not yet been cultivated. Courtesy of Phil Hugenholz and ASM Publications (Hugenholz, P., B. M. Goebel and N. R. Pace. 1998. *J. Bacteriol.* 180: 4765–4774).

| TABLE 17.3 | Overall organization for treatment of *Bacteria* and *Archaea* |

Bacterial Group	This Book	*Bergey's Manual of Systematic Bacteriology*[a]
Archaea	Chapter 18	Volume 1
Proteobacteria	Chapter 19	Volume 2
Gram-positive bacteria	Chapter 20	Volumes 3 and 5
Phototrophic bacteria	Chapter 21	Volumes 1, 2, and 3
Other bacterial phyla	Chapter 22	Volume 4

[a]For more complete treatment of organization of taxa see *Bergey's Manual of Systematic Bacteriology* (2nd ed.).

from simple carbon sources. Some use carbon dioxide and hydrogen gas, whereas others use methanol or acetic acid. These archaea are anaerobes, some of which grow at the lowest oxidation–reduction potentials of all prokaryotes. Some of these archaea fix carbon dioxide, but they use neither the Calvin cycle nor the reductive tricarboxylic acid (TCA) cycle. Some of the *Euryarcheota* are hyperthermophilic. Extreme halophiles make up another phenotypic subgroup. These extremely halophilic archaea grow only in saturated salt-brine solutions. They lyse when placed in distilled water.

It should be noted that there is an overlap between the extreme halophiles and methanogens. Thus, some species of methanogens grow in high-salt environments.

NANOARCHEOTA   This recently discovered group of the *Archaea* comprises obligate parasites of other members of the *Archaea*. They are among the smallest of organisms, hence their name.

KORARCHEOTA   These archaeal microorganisms have been found in hot springs, but no strains have yet been isolated in pure culture, so little is known about their phenotypic properties.

Domain: Bacteria

The *Bacteria* are divided into a number of phyla, which are described in the following text.

PROTEOBACTERIA   The *Proteobacteria* comprise a very large and diverse group of organisms. All four of the major bacterial nutritional types are represented within this group. Some of these organisms are photosynthetic (treated in Chapter 21), whereas some are heterotrophic and others are chemolithotrophic. The chemolithotrophic bacteria include the nitrifiers, the thiobacilli, the filamentous sulfur oxidizers (*Beggiatoa* and related genera), and many species that grow as hydrogen autotrophs. Carbon dioxide fixation, when

present, is via the Calvin cycle in all members of this phylum.

This phylogenetic group contains many of the well-known gram-negative heterotrophic bacteria such as *Pseudomonas*, the enteric bacteria including *E. coli, Vibrio* and luminescent bacteria, and the more morphologically unusual bacteria such as the prosthecate bacteria. In addition, many symbiotic genera such as *Agrobacterium, Rickettsia*, and *Rhizobium* are members of this group.

The gram-negative bacterial sulfate reducers such as *Desulfovibrio* are also found in this group. Also included in this phylum are the exotic myxobacteria that form fruiting structures as well as unicellular, nongliding forms such as the bacterial predator *Bdellovibrio*.

Finally, the mitochondria present in almost all eukaryotes evolved from this group of bacteria.

FIRMICUTES   The bacteria in this group are all gram-positive, although the *Mycoplasma* group lacks a cell wall altogether and therefore stains as gram-negative. All other members of the group contain large amounts of peptidoglycan in their cell wall structure.

The *Firmicutes* are unicellular organisms that have a low mol % G + C content. Most are cocci or rods, and some produce endospores. *Bacillus* are aerobic or facultative spore formers, whereas *Clostridium* species are anaerobic fermenters. Some are sulfate reducers. One group, the heliobacteria, are photosynthetic, and members produce a unique form of bacteriochlorophyll, Bchl *g*.

ACTINOBACTERIA   These gram-positive bacteria range in shape from unicellular organisms to branching, filamentous, mycelial organisms. Most are common soil organisms, some of which produce specialized dissemination stages called conidiospores, which enable them to survive during dry periods.

CHLOROFLEXI   This group contains the genus *Chloroflexus*, a green gliding bacterium that is metabolically versatile. Members can grow as heterotrophs or photosynthetically. Carbon dioxide is not fixed by either the Calvin cycle or the reductive TCA cycle but by a special pathway known only for this group of bacteria.

CHLOROBI The green sulfur bacteria are anoxygenic photosynthetic bacteria. Some are unicellular forms, and others produce networks of cells. None are motile by flagella or gliding motility. Some have gas vacuoles. They use the reductive TCA cycle rather than the Calvin cycle to fix carbon dioxide.

CYANOBACTERIA The cyanobacteria are the only bacteria that carry out oxygenic photosynthesis. This is a diverse group of bacteria ranging from unicellular to multicellular filamentous and colonial types. Some grow in association with higher plants and animals. All cyanobacteria use the Calvin cycle for carbon dioxide fixation. The chloroplast found in all eukaryotic photosynthetic organisms evolved from this group of bacteria.

AQUIFICAE These bacteria are hydrogen autotrophs. This phylogenetic group contains the most thermophilic member of the *Bacteria* known and makes up one of the deepest branches of the *Bacteria*.

THERMOTOGAE This is a fermentative genus that contains some of the most thermophilic members of the *Bacteria* known. The term "toga" refers to the outer extracellular material that surrounds the cells. They grow at temperatures from 55°C to 90°C. Their cell lipids are unusual.

THERMOMICROBIA The genus *Thermomicrobium* contains small, rod-shaped thermophiles that grow as heterotrophs in hot springs with an optimal temperature for growth of 70°C to 75°C. The cell wall contains very low amounts of diaminopimelic acid.

THERMODESULFOBACTERIA This is a group of thermophilic sulfur-reducing bacteria.

DEINOCOCCUS–THERMUS This is a very small group of organisms currently represented by very few genera. The genus *Deinococcus* contains gram-positive bacteria. However, they differ from other gram-positive bacteria in showing strong resistance to gamma radiation and UV light. *Thermus* contains thermophilic, rod-shaped bacteria. Ornithine is the diamino acid in the cell walls of both *Thermus* and *Deinococcus*.

BACTEROIDETES This is a diverse group containing heterotrophic aerobes and anaerobes. Some are gliding heterotrophic bacteria that have a low DNA base composition (about 30 to 40 mol % G + C).

PLANCTOMYCETES The *Planctomycetes* group of *Bacteria* are budding, unicellular, or filamentous bacteria. These bacteria lack peptidoglycan.

CHLAMYDIAE The *Chlamydiae* are a group of obligately intracellular parasites and pathogens whose closest relatives are the *Planctomycetes*. They also lack peptidoglycan.

VERRUCOMICROBIA These bacteria are unusual in that some members have bacterial tubulin genes. Very few representatives of this phylum have been isolated in pure culture, although they comprise up to 3% of the microbiota from soils.

LENTISPHAERA This is a newly discovered phylum. Some isolates are marine, others are intestinal symbionts of mammals.

SPIROCHAETES The spirochetes are morphologically distinct from other bacteria. Their flexible cells are helical. All are motile due to a special flagellum-like structure, the axial filament, not found in other bacteria.

FIBROBACTERES These are anaerobic bacteria, some of which live in the gastrointestinal tracts of animals. Some are cellulose degraders.

ACIDOBACTERIA These bacteria are commonly found in soils and sediments, but few strains have been cultivated. In addition to the aerobic genus *Acidobacterium*, this phylum contains homoacetogenic bacteria, *Holophaga*, and iron-reducing bacteria in the genus *Geothrix*.

FUSOBACTERIA These obligately anaerobic bacteria are commonly found in the oral cavities and intestinal tracts of animals.

DICTYOGLOMI Species of this group are thermophilic, obligately anaerobic fermentative bacteria.

SECTION HIGHLIGHTS
Currently four phyla have been described in the domain *Archaea* and 24 have been described in the domain *Bacteria*.

17.5 Identification

The final area of taxonomy is identification. Bacteriologists are often confronted with determining to which species a newly isolated organism belongs. Clinical microbiologists need to know whether a specific pathogenic bacterium is present so they can properly diagnose a disease. Food microbiologists need to determine whether *Salmonella*, *Listeria*, or other potentially pathogenic bacteria are present in foods. Dairy microbiologists need to keep their important lactic acid bacteria in culture to produce a uniform variety of cheese. Brewers and wine makers need to keep their cultures pure to inoculate the proper strains for quality control of their fermentations. Analysts at water treatment plants need to make sure that their chlorination treatment is effective in killing coliform bacteria in the treated water and distribution systems. Microbial ecologists need to identify bacteria that are responsible for important processes such as pesticide breakdown and nitrogen fixation.

The process of identification first assumes that the bacterium of interest is one that has already been described and named. This is the usual case for most clinical specimens or specimens from known fermentations. However, microbial ecologists often find that the organism they are interested in is new. It is estimated that less than 1% of prokaryotic species have been isolated, studied in the laboratory, and named. Therefore, it is not always possible to identify a bacterium that has been isolated from an environmental sample.

Phenotypic Tests

Phenotypic tests based on readily determined characteristics are often used to identify a species. Most are simple to perform and inexpensive. Furthermore, the amount of time and equipment required for conducting genotypic tests such as DNA–DNA reassociation preclude the use of these tests in routine diagnosis. By performing a battery of some 10 to 20 simple phenotypic tests, it is often possible to determine the genus, and perhaps even the species, of a clinically important bacterium, although some taxa are much more difficult to identify than others.

Traditional methods for identification require growing the organism in question in pure culture and performing a number of phenotypic tests. For example, if one wishes to identify a rod-shaped bacterium, the first test would be a Gram stain. If the organism is determined to be a gram-negative rod, the next questions to ask include the following: Is it motile? Is it an obligate aerobe? Is it fermentative? Does it have catalase? Can it grow using acetate as a sole carbon source? The answers to these questions will direct the investigator toward the next step in identification. Conversely, if the organism is a gram-positive rod, then a completely different set of tests would need to be performed such as a test for endospore formation.

These tests take time to perform, and the appropriate tests for one genus of bacteria differ from that of oth-

ers. Therefore, the results of one set of tests will determine which tests will need to be performed for further clarification of the taxon. Sometimes several weeks might be required to conduct all the tests needed to identify a strain. Rapid tests are extremely helpful, especially in the medical, food, and water-testing areas. Fortunately, standardized, routine tests can be performed for most clinically important bacteria that allow for their rapid identification. Several companies have now produced commercial kits that are helpful in assisting in the identification of unknowns (see Figure 30.4).

An increasingly popular approach to identification of bacterial unknowns involves characterization of their fatty acids. The fatty acids are found in membrane lipids and are readily extracted from the cells and analyzed. Different species of *Bacteria* produce different types and ratios of fatty acids. The *Archaea*, of course, do not produce fatty acids, so this procedure is not of value for their identification. However, they do produce characteristic lipids that are useful taxonomically, especially for the halobacteria.

The fatty acid analysis procedure involves hydrolyzing a small quantity of cell material (about 40 mg is all that is needed) and saponifying it in sodium hydroxide. This is acidified with hydrochloric acid in methanol so that the fatty acids can be methylated to form methyl esters. The fatty acid methylated esters (FAME) are then extracted with an organic solvent and injected into a gas chromatograph. The resulting chromatogram (Figure 17.11) can be used to identify the fatty acids that are indicative of a species. Commercial firms have developed databases of fatty acid profiles that can be used for the identification of species. The advantage of this procedure is that many samples can be analyzed quickly and without great effort. However, all organisms must be grown under controlled conditions of temperature and length of incubation and on the same medium.

Nucleic Acid Probes and Fluorescent Antisera

One exciting current area of research and commercial application involves the development of DNA or RNA "probes" that are specific for the signature sequences of rRNA or some other appropriate gene such as an enzyme that is characteristic of a species of interest. The probes are labeled in some manner so that the hybridization can be visualized. This is accomplished either by making the probes radioactive or by tagging them to a fluorescent dye or an enzyme that gives a colorimetric reaction.

By the proper selection or design of probes, it is possible to identify an organism to a domain, genus, or species by demonstrating specific hybridization to the probe (see Chapter 25).

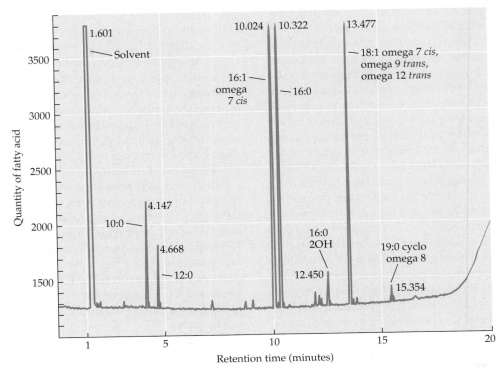

Figure 17.11 Fatty acid analysis
Fatty acid methyl ester (FAME) chromatogram of an unknown species, showing chromatographic column retention times and peak heights. Note: 10:0, 12:0, 16:0, and 19:0 indicate saturated fatty acids with 10, 12, 16, and 19 carbons; 16:1 and 18:1, monounsaturated 16-carbon and 18-carbon fatty acids; omega number, the position of the double bond relative to the omega end—that is, the hydrocarbon end (not the carboxyl end)—of the fatty acid chain; *cis* and *trans*, the configuration of the double bond. For example, omega 7 *cis* indicates a *cis* double bond between the seventh and eighth carbons from the omega end of the fatty acid. Also, 2OH indicates a hydroxyl group at the second carbon from the omega end; cyclo omega 8, a cyclo-carbon at the eighth position from the omega end. The 18:1 omega 7 *cis*, omega 9 *trans*, and omega 12 *trans* peak results from either one fatty acid or a mixture of fatty acids with double bonds at the three positions indicated (the chromatographic column does not separate these three fatty acids). Courtesy of MIDI (Microbial Identification, Incorporated, Delaware).

Potential applications for probe technology are considerable. A number of commercial firms are already marketing probes to identify pathogenic bacteria from clinical samples. A goal of this technology is to enable the rapid identification of organisms directly from clinical or environmental samples without actually growing them in culture first. Fluorescent antiserum tests are also useful. For example, *Legionella* spp., the causative agents of Legionnaires' disease, are very difficult to cultivate. However, good fluorescent antisera are available for the identification of these species directly from clinical samples or even from environmental samples (Figure 17.12).

Another approach in the identification of species is through the use of multiple locus sequence typing (MLST). In this approach, several core genes (typically seven or eight) of an organism are sequenced and used to compare the organism to known species. In the clinical setting, if the organism has been isolated from a patient, its MLST "type" can be compared to that of a large database to determine whether it belongs to a known pathogenic species. The MLST approach is rapidly gaining acceptance as a means of identification of pathogenic strains and species.

Culture Collections

Unlike plants and animals, many bacteria and archaea can be easily grown in pure culture and preserved by **freeze-drying (lyophilization)**, handled in small test tubes and vials, and readily sent anywhere in the world. As lyophils, many of these organisms remain viable for 10 to 20 years or more and can be revived and studied by anyone anywhere. Cultures can also be frozen at −80°C in vials containing a suspension medium amended with 15% glycerol. These remain viable for many years. Thus,

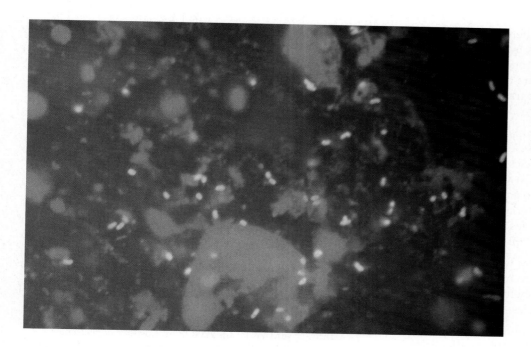

Figure17.12 *Legionella* **fluorescence**
Legionella species are difficult to grow, but can be identified to species by fluorescence microscopy using tagged and specifically labeled antisera. In this image the bacteria are yellow-green. ©Michael Abbey/Visuals Unlimited.

unlike plants, for which an herbarium is used to preserve the specimens collected of an original species, bacteria are preserved as **type cultures** that are clones of the original viable **type strain** of a species. *The type strain of a species is the one on which the species definition has been based.* Through **culture collections** type strains are made available to professional microbiologists throughout the world. Many countries maintain national collections of microorganisms, such as the American Type Culture Collection (ATCC) in the United States (www.atcc.org). If a microbiologist from India wishes to determine if he or she has a new species, the original type strain can be obtained from a culture collection and used to conduct DNA–DNA reassociation assays and other tests to compare it with his or her isolates.

Because of the importance of biological materials to science and industry, culture collections have become biological resource centers. Thus, strains that have been patented are also deposited in culture collections so that they are accessible. Clones of genomes that have been sequenced are also deposited as are viruses and other biological materials.

SECTION HIGHLIGHTS

A number of tests are used to identify bacteria that have been isolated from clinical specimens and natural sources. Some tests are phenotypic, whereas others use molecular sequence information.

SUMMARY

- Bacterial taxonomy or systematics consists of three areas: nomenclature, classification, and identification. **Nomenclature** is the naming of an organism. An International Code for the Nomenclature of Bacteria has been published containing the rules for naming *Bacteria* and *Archaea*.

- **Classification** is the organization of *Bacteria* and *Archaea* into groups of similar species. *Bacteria* and *Archaea* are classified in increasing hierarchical rank from species, genus, family, order, class, phylum, and domain. **Artificial classifications** are not based on the evolution of organisms but on expressed features or the phenotype of an organism that includes properties such as cell shape and nutritional patterns. **Phylogenetic classifications** are based on the evolution of a group of organisms. Bacterial phylogeny is now based on the sequence information from the highly conserved macromolecule, 16S rRNA, as well as the sequences of other genes and proteins.

- **Phylogenetic trees**, with branches and nodes, can be constructed based on the sequence of macromolecules such as rRNA. The length of the branch represents the inferred difference (number of changes) between organisms. **External nodes** represent extant species, whereas **internal nodes** represent ancestor species. **Rooted** trees are based on a comparison of related species and an out-group.

- **Speciation** is the process whereby organisms evolve. Typically genetic material is transferred from the parental bacterium to the progeny through **vertical inheritance**. **Horizontal gene transfer** (HGT) occurs in which genetic material is transferred among bacteria that may or may not be closely related.

- A **bacterial species** is a group of similar strains that show at least 70% DNA–DNA hybridization. Organisms of the same species will have similar if not identical DNA mol % G + C content. Organisms that have the same GC ratio are not necessarily similar. Based on molecular criteria, such as DNA–DNA reassociation, bacterial species are much more broadly defined than are plant and animal species.

- **Identification** is the process whereby unknown cultures can be compared to existing species to determine if they are sufficiently similar to be members of the same species.

- **Type strains** of all species must be deposited in at least two different types of **culture collections**, repositories where strains are preserved by lyophilization and deep freezing. The culture collections provide cultures to microbiologists worldwide so they can compare unidentified strains to the official type strains.

 Find more at www.sinauer.com/microbial-life

REVIEW QUESTIONS

1. Is it important to name and classify bacteria?

2. What procedure(s) are necessary to identify a bacterial isolate as a species?

3. Differentiate between an artificial and a phylogenetic classification.

4. In what ways does the classification of *Bacteria* differ from that of eukaryotic organisms?

5. How do the *Archaea* differ from *Bacteria*? From eukaryotes?

6. How is DNA melted and reannealed, and why is this useful in bacterial taxonomy?

7. How would you go about identifying a bacterium that you isolated from a soil habitat?

8. Why is morphology of little use in bacterial classification? Is it of any use?

9. What is *weighting* and should phenotypic features be weighted in a bacterial classification scheme?

10. Distinguish between lumpers and splitters.

11. If you were working in a clinical laboratory, outline the types of procedures you would use to identify isolates. Why do you recommend using the procedures you suggest?

12. Why is rRNA of use in bacterial classification?

13. Compare the information obtained from determining the DNA base composition (GC ratio) with that obtained by DNA reassociation experiments.

SUGGESTED READING

Boone, D., R. Castenholz and G. Garrity, eds. 2001. *Bergey's Manual of Systematic Bacteriology.* 2nd ed., Vol. 1. New York: Springer-Verlag.

Brenner, D. J., N. R. Krieg, J. T. Staley and G. Garrity, eds. 2005. *Bergey's Manual of Systematic Bacteriology.* 2nd ed., Vol. 2. New York: Springer-Verlag.

Gerhardt, P., ed. 1993. *Methods for General and Molecular Microbiology.* Washington, DC: ASM Press.

Graur, D. and W. H. Li. 2000. *Fundamentals of Molecular Evolution.* 2nd ed. Sunderland, MA: Sinauer Associates, Inc.

Hall, B. G. 2004. *Phylogenetic Trees Made Easy.* 2nd ed. Sunderland, MA: Sinauer Associates, Inc.

COMPUTER INTERNET RESOURCES

American Type Culture Collection: http://www.atcc.org/. This site has a listing of all the bacterial strains deposited in the American Type Culture Collection, as well as growth media and conditions.

Bergey's Manual Trust. Headquarters at the University of Georgia. This website has information on the current classification of *Bacteria* and *Archaea*: http://www.bergeys.org/.

National Center for Biotechnology Information (NCBI): http://www.ncbi.nlm.nih.gov/. This center contains information that allows for comparison of genes from different organisms through BLAST (Basic Local Alignment Search Tool), Genbank, which contains a huge collection of gene sequences that can be used for comparative analyses and a section on taxonomy.

Ribosome Database Project (RDP), Michigan State University: http://rdp.cme.msu.edu/. This site has information on the 16S rRNA sequences of more than 250,000 bacterial and archaeal sequences. The database allows one to conduct phylogenetic analyses of unknown strains whose 16S rRNA sequence has been determined and to compare the sequence with those already reported.

Comparative RNA Web Site: www.rna.icmb.utexas.edu. A remarkable collection of RNA sequence information presented with secondary structure models, conservation diagrams, and more. Published as Cannone, J. J., et al. 2002. "The Comparative RNA Web Site: An online database of comparative sequence and structure information for ribosomal, intron, and other RNAs." *BioMed Central Bioinformatics* 3: 2.

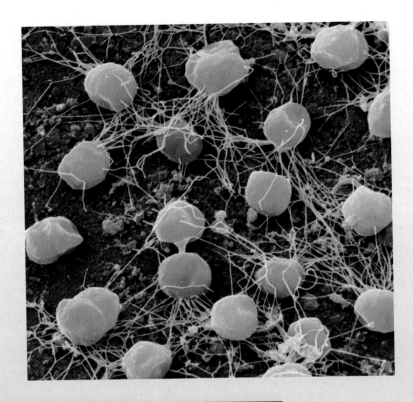

The objectives of this chapter are to:

◆ Describe the evidence for the *Archaea*, a group of prokaryotes more closely related to the eukaryotes than the *Bacteria*.

◆ Elaborate on the importance of *Archaea* to our understanding of fundamental questions concerning the early evolution of life and biodiversity.

◆ Demonstrate the importance of the *Archaea* to our ordinary lives, including global warming and industrial enzymes.

◆ Describe the major phenotypic groups of the *Archaea*, including their physiology, ecology, and systematics.

18

Archaea

Hence without parents, by spontaneous birth,
Rise the first specks of animated earth . . .
Organic life beneath the shoreless waves
Was born and nursed in ocean's pearly caves;
First, forms minute, unseen by spheric glass,
Move in the mud or pierce the watery mass;
These, as successive generations bloom,
New powers acquire, and large limbs assume;
Whence countless groups of vegetation spring,
And breathing realms of fin and feet and wing.
—*Erasmus Darwin, from* The Temple of Nature, *1802*

When Carl Woese and his collaborators at the University of Illinois began to study the phylogenetic relationships among prokaryotes during the mid-1970s, they discovered that one group was very different from all other prokaryotes. According to sequence analyses, the 16S rRNAs of these prokaryotes were no more closely related to the bacterial 16S rRNAs than to the 18S rRNAs of eukaryotes. They called these different prokaryotes archaebacteria because they seemed to resemble the primitive organisms believed to exist in Archean times 3 to 4 billion years ago. This discovery piqued the interest of microbiologists around the world. After much debate, microbiologists now generally agree that these organisms are very different from *Bacteria*. In recognition of this fact, they are currently classified as **Archaea**, one of the three primary domains of organisms.

18.1 Why Study *Archaea*?

Although the study of unusual organisms is inherently interesting and needs no special justification, the study of *Archaea* has the potential to provide special insights into four areas of basic knowledge. First, many of the currently cultivated *Archaea* are also extremophiles, and they flourish under conditions that are lethal to most *Bacteria* and

Eukarya. In fact, the most extreme thermophiles are *Archaea*. As such they set one of the boundaries of the biosphere, especially in the deep earth where temperature increases with depth. Thus, the potential of microorganisms to transform the deep subsurface depends upon the upper temperature limit for life and the physiology of these extreme organisms. Studies of the extremely thermophilic and extremely halophilic *Archaea* also contribute to our understanding of growth mechanisms in these harsh environments. A question of special interest is how the macromolecules of extremophiles maintain their structure and activity under conditions that denature the macromolecules of most organisms. This question is also of practical importance for the design of commercially valuable thermostable enzymes. The optimal growth temperature of many extremely thermophilic *Archaea* is also above melting temperature of their DNA. How these organisms maintain a functional chromosome is not fully understood.

Second, *Archaea* offer a valuable model system because some of their features closely resemble those of the eukaryotes. Their simpler cell structure and smaller size may prove to be a valuable research tool in understanding the function of the eukaryotic-type of RNA polymerase, DNA replication, and other processes common to both domains.

Third, as one of the most ancient lineages of living organisms, the *Archaea* represent one of the extremes of evolutionary diversity. Most of our knowledge of prokaryotes comes from studies of a relatively small number of model organisms, especially proteobacteria such as *Escherichia coli*. Comparison with the *Archaea* has illustrated the enormous diversity of prokaryotes, even in mechanisms once thought to be universal. For instance, when the first genomic sequence of an archaeon, *Methanocaldococcus jannaschii*, was completed, only 16 of the 20 aminoacyl-tRNA synthetases believed to be essential for all living organisms were detected. Subsequent studies showed that many *Archaea* possess alternative synthetases and that these synthetases were common in the *Bacteria* as well. Previous studies limited to the proteobacteria and a few eukaryotes had underestimated the extent of aminoacyl-tRNA synthetase diversity. Similarly, this first archaeal genomic sequence led to the discovery of a large number of genes that had never been encountered before. Although some of these genes have since been found widely distributed in living organisms, many have so far been found only in *Archaea*. These uniquely archaeal genes may encode enzymes and other proteins involved in processes unique to the *Archaea*, such as methanogenesis. The discovery of novel genes associated with the unique aspects of the archaeal lifestyle is a further goal of archaeal research.

Fourth, the *Archaea* offer valuable insights into the nature of the earliest organisms. Because most biochemical features of microorganisms are absent in the fossil record,

these properties must be deduced from modern organisms. By studying modern organisms, we also know that form changes rapidly throughout evolution but that basic biochemical processes are inherited by even very remote ancestors. Because the *Archaea* are an extreme in prokaryotic evolution, they offer the opportunity to obtain unique insights into the nature of the earliest prokaryotes. As extremes, they delineate one of the boundaries within which all modern organisms must lie.

Models of Early Life

The importance of the discovery of the *Archaea* to the study of early life may be illustrated by the following example. First, consider that all modern organisms evolved from two lines of descent, a proposition which we know to be false (Figure 18.1, model A). Properties that are the same in both lines of descent, such as the genetic code, were probably inherited from the common ancestor; however, for different properties, such as the struc-

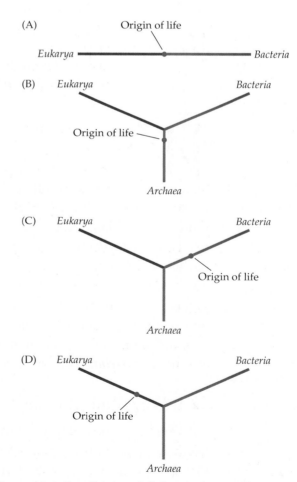

Figure 18.1 Possible models for ancient evolution
Different possible positions for the origin of life are indicated.

ture of the DNA-dependent RNA polymerase, very little may be concluded. For instance, did the complex RNA polymerase of eukaryotes evolve by the addition of complexity to the simpler bacterial RNA polymerase, or did the bacterial RNA polymerase evolve by subtraction from, and a more efficient construction of, the complex eukaryotic enzyme? Either scenario is plausible, and this problem cannot be solved without additional evidence.

Now consider that all modern organisms evolved from three lines of descent that arose very early in life's history, a proposition which is probably true. The location of the origin of life is not known in this scenario, but it is probably close to the point where all three lines intersect and on one of the three branches (that is not at the center). This scenario may be represented by the three models B, C, and D (see Figure 18.1). Only one of these can be correct, but the correct model is not known. Again, properties that are common to all three lines of descent, like the genetic code, were probably inherited from the common ancestor. But for properties that are different, it is now possible to draw firmer conclusions. For example, both the *Archaea* and the *Eukarya* possess the complex RNA polymerase, and only the *Bacteria* possess the simple enzyme. Did the common ancestor possess the simple or complex enzyme? If either model B or D is correct, then it is very likely that the common ancestor contained the complex RNA polymerase because it is unlikely that this level of complexity evolved more than once. If model C is correct, then it is likely that the ancestor of the *Eukarya* and the *Archaea* contained a complex enzyme, but the properties of the common ancestor with the *Bacteria* are still not determined. Thus, even without detailed knowledge of the early events, the presence of three lineages introduces new constraints on our models of early life. This example also illustrates the importance of the location of the origin of life on our interpretation of the properties of modern organisms. Although current evidence seems to suggest that the origin of life was on the bacterial lineage (model C), this evidence is by no means conclusive.

Abundance of Archaea

Although their unique evolution was not recognized until the 1970s, many of the organisms now known to comprise the *Archaea* were first described much earlier. Although not named as such until the 1950s, the first isolates of the modern genus of the extremely halophilic archaea, *Halobacterium*, were described as early as 1919. A number of genera of the methanogens were described in 1936. Thus, many of these organisms have long been familiar to microbiologists. However, it was not until the application of molecular methods to microbial ecology in the 1990s that it was possible to fully appreciate the abundance of these organisms. Prior to that time, surveys based upon

cultured organisms seemed to indicate that the *Archaea* were found in only a few specialized environments. Therefore, it was surprising when surveys of natural populations that did not rely on culture techniques discovered large populations of *Archaea* in the oceans, soil, and subsurface. For instance, mesophilic *Archaea* represent about 1% of the prokaryotic cells in soil and 10% to 30% of the prokaryotic cells in seawater. Thus, it is clear that *Archaea* are a significant fraction of the prokaryotes on earth.

Diversity of Archaea

In addition to being very different from other organisms, the *Archaea* are very different from each other. Some of this diversity is illustrated by the large number of the morphological types common in *Archaea* (Figure 18.2) and the large differences in metabolism (see subsequent text). In fact, the phyla or deepest phylogenetic groups within the *Archaea* are at least as different from each other as the phyla within the *Bacteria*, even though many fewer archaeal phyla have been identified. A likely explanation for the large diversity is that the *Archaea* are an ancient line of descent. Thus, there has been adequate time for their ancestors to specialize and form novel lineages.

> ### SECTION HIGHLIGHTS
> Distantly related to *Bacteria*, the *Archaea* are abundant prokaryotes that offer unique insights into the nature of the earliest organisms and the diversity of life.

18.2 Biochemical Differences and Similarities to Other Organisms

The *Archaea* are composed of three major phenotypic groups. These include the methane-producing archaea, the extreme halophiles, and the extreme thermophiles. In addition, a representative of the uncultured crenarchaeotes in temperate environments has recently been shown to nitrify, and this physiological type may also prove to be a major phenotypic group (see subsequent text). Although the physiology of these archaea is very different, some features are common to most if not all archaea (Table 18.1). The cell membranes are composed of isoprenoid-based glycerol lipids. Murein is absent in their cell walls, and it is usually replaced by a protein envelope. The enzyme DNA-dependent RNA polymerase, which copies DNA to form cellular RNA, is different from the bacterial type but very similar to the eukaryotic RNA polymerase II. The proteins in DNA replication are more similar to the eukaryotic than bac-

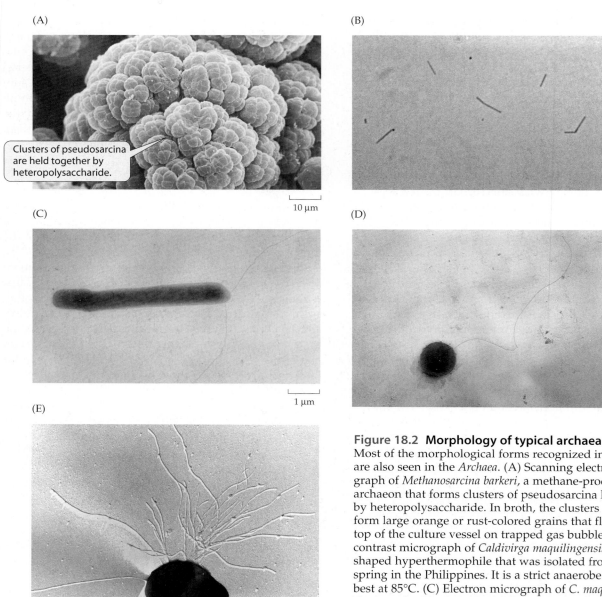

(A)

Clusters of pseudosarcina are held together by heteropolysaccharide.

10 μm

(B)

10 μm

(C)

1 μm

(D)

1 μm

(E)

1.5 μm

Figure 18.2 Morphology of typical archaea
Most of the morphological forms recognized in the *Bacteria* are also seen in the *Archaea*. (A) Scanning electron micrograph of *Methanosarcina barkeri*, a methane-producing archaeon that forms clusters of pseudosarcina held together by heteropolysaccharide. In broth, the clusters aggregate to form large orange or rust-colored grains that float to the top of the culture vessel on trapped gas bubbles. (B) Phase contrast micrograph of *Caldivirga maquilingensis*, a rod-shaped hyperthermophile that was isolated from a hot spring in the Philippines. It is a strict anaerobe that grows best at 85°C. (C) Electron micrograph of *C. maquilingensis* shadowed with platinum/palladium; pili are attached to one end. (D) Electron micrograph of a novel extremely thermophilic archaeon belonging to the family *Thermoproteaceae* that was isolated from a hot spring in Japan. (E) Platinum-shadowed electron micrograph of *Pyrococcus furiosus*. This heterotrophic coccus has a temperature optimum near 100°C and possesses a flagellar bundle. The cell diameter is about 1.5 μm. A, ©Ralph Robinson/Visuals Unlimited; B–D, courtesy of Dr. T. Itoh, Japanese Culture Collection; E, courtesy of Prof. Dr. K. O. Stetter and Dr. R. Rachel, University of Regensburg, Germany.

terial homologs. Lastly, *Archaea* are insensitive to many common antibiotics that are potent inhibitors of *Bacteria* or eukaryotes.

Archaeal Membranes

Like the membrane lipids of the *Bacteria*, the archaeal lipids are composed of hydrophobic side chains linked to glycerol. However, many of the details are different (Figure 18.3). Instead of fatty acids joined by an ester linkage of glycerol, archaeal lipids are composed of iso-

prenoid side-chains joined by an ether linkage to glycerol. Isoprenoids are branched-chain alkyl polymers based upon a five-carbon unit synthesized from mevalonate. In bacteria and eukaryotes, they are found in the side-chains of quinones and chlorophyll, intermediates in the biosynthesis of sterols like cholesterol, and in rubber. Isoprenoids

TABLE 18.1   Comparison of *Archaea* to *Bacteria* and *Eukarya*

	Archaea	Bacteria	Eukarya
Typical organisms	Methane-producing archaea, haloarchaea, extreme thermophiles	Enteric bacteria, cyanobacteria, *Bacillus*, etc.	Fungi, plants, animals, algae, protozoa
Size	1–4 µm	1–4 µm	>5 µm
Physiological features	Aerobic and anaerobic; mesophiles, extremely thermophilic and halophilic	Aerobic and anaerobic, chlorophyll-based photosynthesis	Largely aerobic, chlorophyll-based photosynthesis
Genetic material	Small circular chromosome, plasmids and viruses, genome associated with histones	Small circular chromosome, plasmids and viruses, no histones	Complex nucleus with more than one large linear chromosome, viruses, genome associated with histones
Differentiation	Frequently unicellular, cellular differentiation infrequent	Frequently unicellular, cellular differentiation infrequent	Unicellular and multicellular, cellular differentiation common
Cell wall	Protein, glycoprotein, pseudomurein, wall-less	Murein and lipopolysaccharide (LPS), protein and wall-less forms rare	Great variety, peptidoglycan absent
Cytoplasmic membrane	Glycerol ethers of isoprenoids, site of energy biosynthesis	Glycerol esters of fatty acids, site of energy biosynthesis	Glycerol esters of fatty acids, sterols common
Intracytoplasmic membranes	Generally absent	Generally absent, when present they frequently contain large amounts of protein	Common in organelles like mitochondria and chloroplasts, nucleus, Golgi apparatus, endoplasmic reticulum and vacuoles, site of energy biosynthesis
Protein synthesis	70S ribosome, insensitive to chloramphenicol and cycloheximide, diphthamide present in elongation factor	70S ribosome sensitive to chloramphenicol, insensitive to cycloheximide, diphthamide absent in elongation factor	80S and 70S (organelle) ribosomes, insensitive to chloramphenicol (80S), sensitive to cycloheximide (80S), diphthamide present in elongation factor
Locomotion	Simple flagella	Simple flagella, gliding, twitching	Complex flagella, cilia, legs, fins, wings
RNA polymerase	Complex	Simple	Complex

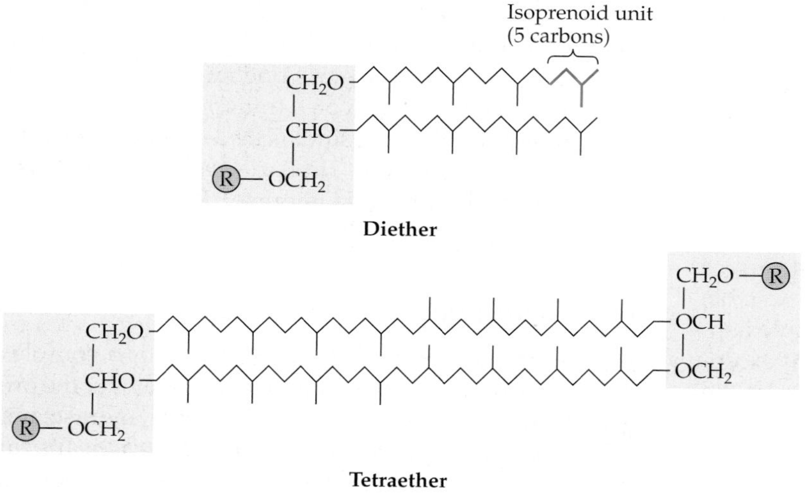

Figure 18.3 Archaeal glycerol lipids Typical archaeal glycerol lipids contain isoprenoid side chains linked to glycerol by an ether bond. A single C-5 isoprenoid unit is indicated in the diether. The tetraether is believed to span the membrane. One glycerol moiety is on the cytoplasmic side of the membrane and the second is on the external face.

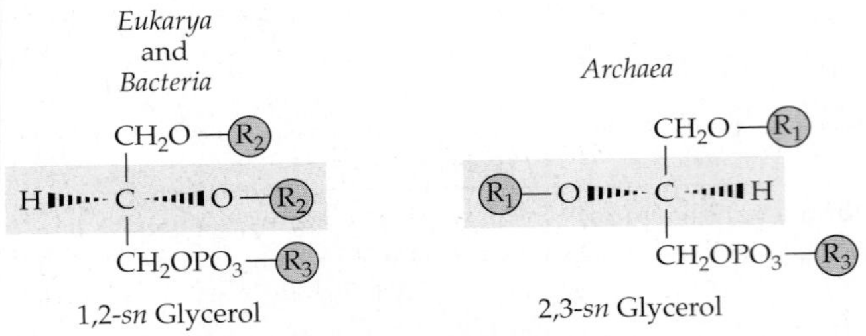

Figure 18.4 Stereochemistry of the glycerol lipids in the *Bacteria* (1,2-*sn* glycerol) and the *Archaea* (2,3-*sn* glycerol)
R_1 = ether-linked isoprenoid side chains, R_2 = ester-linked fatty acids, and R_3 = hydrophilic amino acids or sugars.

are never major components of glycerol lipids. Although branched-chain fatty acids are sometimes found in bacterial lipids, they are not synthesized from mevalonate. Instead, they are synthesized from the branched-chain amino acids or volatile fatty acids. In *Archaea*, the ether linkage of the isoprenoid side-chains to glycerol is also distinctive. The substituted glycerol contains one asymmetrical carbon. Therefore, it has two stereoisomers (Figure 18.4). In the *Archaea*, only the stereoisomer called 2,3-*sn* glycerol is found. The glycerol lipids of the *Bacteria* and the *Eukarya* contain only the other stereoisomer of glycerol, 1,2-*sn* glycerol. Although a few bacteria contain ether-linked fatty acids, they have the typical bacterial stereochemistry. This difference in stereochemistry is evidence for a different mechanism for biosynthesis of the linkages between glycerol and the side chains in *Bacteria* and *Archaea*.

Archaeal Cell Walls

The cell walls of the *Archaea* are different from the gram-negative and gram-positive cell wall types common in the *Bacteria*. Cell walls composed of murein are never found in the *Archaea*. Instead, the cell walls are usually composed of S-layers: protein subunits arranged in a regular array on the cell surface. Frequently, polysaccharides are also found associated with the envelope, although in many cases their structures are not known. The S-layers from many *Archaea* are very sensitive to detergents like SDS (sodium dodecyl sulfate), which solubilizes proteins. The cells of these archaea also lyse rapidly in solutions containing low concentrations of detergents. Although S-layers are common in *Bacteria*, they are usually only one component of a complex envelope and found outside the murein layer. Some exceptions to this generality are the budding bacteria of the genera *Planctomyces* and *Pasteuria* and the thermophilic anaerobe *Thermomicrobium roseum*. In these bacteria, the S-layer is the major cell wall component.

The cell envelopes of some *Archaea* are also similar to the cell envelopes of *Bacteria*. *Thermoplasma*, *Ferroplasma*, and *Picrophilus* are archaea that lack cell walls as do the bacterial mycoplasmas. Some methanogens, like *Methanobacterium*, contain a cell wall polymer called pseudomurein, which is strikingly similar to murein, the peptidoglycan in *Bacteria*. Pseudomurein is composed of polysaccharides crosslinked by amino acids much like murein (Figure 18.5). However, the polysaccharide is composed of *N*-acetylglucosamine (or *N*-acetylgalactosamine) and *N*-acetyltalosaminuronic acid. Muramic acid, the common saccharide component of murein, is never found. Moreover, D-amino acids, which are common in bacterial peptidoglycan, are not found. In other respects, many of the chemical and physical properties of pseudomurein and murein are similar. Both are resistant to proteases, or enzymes which hydrolyze peptide bonds. Both provide the cell with a rigid sacculus. In addition, many of the archaea that contain pseudomurein also stain gram-positive. The archaeal and bacterial peptidoglycans appear to be an example of convergent evolution, where two prokaryotic lineages have independently developed similar solutions to the problem of how to make an enzyme-resistant sacculus.

Archaeal RNA Polymerase

The enzyme DNA-dependent RNA polymerase synthesizes messenger RNA (mRNA), which is complementary to the DNA sequence of genes. Because this function is essential to all living cells, this enzyme is believed to be very ancient. The bacterial enzyme is relatively simple in structure, and it consists of a core enzyme with three subunits, β, β', and α. Binding of the core enzyme to DNA also requires an additional subunit, called the sigma factor (σ). In contrast, the eukaryotic polymerase is much more complex and contains 9 to 12 subunits. The archaeal enzyme more closely resembles the eukaryotic polymerase than the bacterial enzyme. It has a complex subunit structure, and the amino acid sequence of some of the subunits closely resembles that of the eukaryotic enzyme. As one might expect, in the *Archaea* as well as eukaryotes transcription is initiated at promoters containing similar DNA sequences called TATA-boxes and require similar accessory proteins in addition to the RNA polymerase. The promoters of *Bacteria* are very different.

DNA Replication

In all living organisms, DNA replication is a complex process requiring binding of initiation proteins at the origin prior to the recruitment of DNA polymerase and other proteins required for elongation. *Archaea* appear to follow the same general pathway found in the *Eukarya*

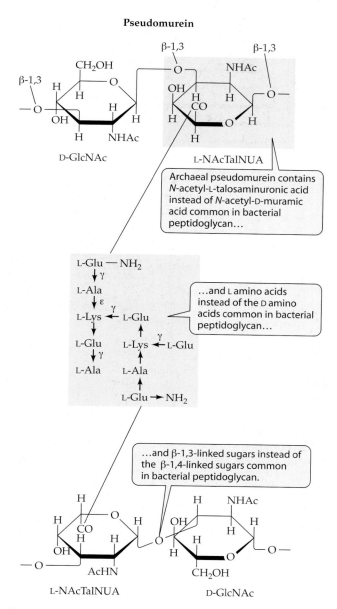

Figure 18.5 Pseudomurein of *Archaea*
Comparison of the structure of the peptidoglycan (murein) in *Bacteria* to the structure of pseudomurein in the archaeon *Methanobacterium*. Major differences include the substitution of N-acetyl-L-talosaminuronic acid (L-NAc TalNUA) for N-acetyl-D-muramic acid (D-MurNAc) and L-amino acids for D-amino acids. Note also the differences in the bonds in the polysaccharide chain. N-acetyl-D-glucosamine (D-GlcNAc) is present in both polymers. DA is a diamino acid like diaminopimelic acid, lysine, or ornithine. Modified from O. Kandler and H. Konig. 1985. "Cell envelopes of *Archaea*." In C. R. Woese and R. S. Wolfe, ed. *The Bacteria*, Vol. 8. pp. 413–457.

and *Bacteria*. However, many of the proteins involved are clearly homologous to the eukaryotic and not the bacterial proteins. In addition, many *Euryarchaeota* also possess histones, which compact and fold DNA in the eukaryotic nucleosome. The *Crenarchaeota* have unrelated DNA-compacting proteins.

Antibiotic Sensitivity of Archaea

Because the biochemistry of the *Archaea* is very different from that of other organisms (see Table 18.1), it might be expected that the sensitivity to antibiotics would also be different. This is true. For instance, because the *Archaea* do not contain murein, they are not sensitive to most antibiotics that inhibit bacterial cell wall synthesis. Thus, most of the *Archaea* are not affected by very high concentrations of penicillin, cycloserine, vancomycin, and cephalosporin, all of which are inhibitors of murein synthesis. Because of this selectivity, these antibiotics are also useful additions to enrichment cultures for *Archaea*, where they prevent the growth of bacterial competitors. Likewise, the DNA-dependent RNA polymerase from *Archaea* is not inhibited by rifampicin, which inhibits the bacterial enzyme at low concentrations. Archaeal protein synthesis is also not affected by the common antibiotics chloramphenicol, cycloheximide, and streptomycin, although neomycin is inhibitory at high concentrations. Tetracycline is also a very poor inhibitor even though it inhibits protein synthesis in both *Bacteria* and *Eukarya*. These results suggest that the structure of the archaeal ribosome is very different from that of the bacterial and eukaryotic ribosomes.

Similarities to Other Organisms

Equally interesting are the similarities between the *Archaea* and other organisms. For instance, even though the difference is more pronounced between the archaeal and bacterial 16S rRNAs than between any two bacterial 16S RNAs, archaeal and bacterial 16S rRNAs are still identical at 60% of their positions. Likewise, the initiator tRNAs in both *Archaea* and *Eukarya* add methionine, rather than formyl-methionine as in *Bacteria*, as the first amino acid in protein biosynthesis. Translation in both *Archaea* and *Eukarya* is also sensitive to diphtheria toxin, whereas bacterial translation is not. Moreover, the genetic code is essentially the same in all organisms, many of the major anabolic pathways are the same, and many of the same coenzymes and vitamins are found in all organisms.

In conclusion, the differences found between the *Archaea* and members of the other domains are a matter of degree, and a major challenge facing modern microbiologists is to understand the significance and evolution of this diversity. An illustration of the types of questions this understanding will enable us to address is the significance of "horizontal" evolution, or gene transfer between distantly related organisms. In eukaryotes, the mitochondrion and chloroplast are two well-documented cases of horizontal evolution. Did similar events occur in prokaryotic evolution? Currently, the evidence is mixed. Certain properties like methanogenesis and chlorophyll-based photosynthesis are found solely in the archaeal and the bacterial domains, respectively. For these properties, horizontal evolution probably did not occur between domains. However, with the availability of genomic sequences in prokaryotes, many examples of horizontal gene transfer of small numbers of genes have been observed between the *Archaea* and the *Bacteria* as well as within the *Bacteria*. At least one example of a potentially massive horizontal gene transfer, as massive as may have occurred in the formation of the mitochondrion or chloroplast, has been observed. In this case, the transfer seemed to be from a hyperthermophilic archaeon related to *Pyrococcus* to the ancestor of the hyperthermophilic bacterium *Thermotoga*.

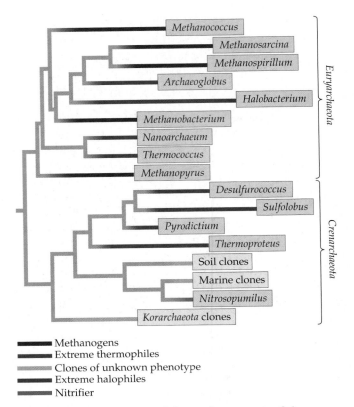

Methanogens
Extreme thermophiles
Clones of unknown phenotype
Extreme halophiles
Nitrifier

Figure 18.6 Phylogeny of the major groups of the *Archaea* based upon the sequences of 16S rRNA The major groups of the methanogens are related. The extreme halophile *Halobacterium* and *Archaeglobus* are also related to the methanogens in the orders *Methanosarcinales* and *Methanomicrobiales*. In contrast, the extreme thermophiles are found in both of the two major lines of descent in the *Archaea*. There is also evidence for many groups that have never been cultivated, including marine and soil clones and a possible phylum *Korarchaeota*. Because they have never been cultivated, their phenotypes are unknown.

SECTION HIGHLIGHTS

Archaea possess features that are uniquely archaeal, such as the composition of their cellular lipids and pseudomurein, as well as features distinctly eukaryotic, such as their RNA polymerase and DNA replication system, or distinctly bacterial, such as S-layers in their cell walls. However, these differences are frequently a matter of degree, and many features, such as the genetic code and many major metabolic pathways, are shared with all living organisms.

18.3 Major Groups of *Archaea*

The *Archaea* are divided into three phyla on the basis of their rRNA structures (Figure 18.6). One phylum, the *Korarchaeota*, has no members that are currently in pure culture. Because these organisms are abundant in hot springs, they are believed to be extreme thermophiles. In the remaining two phyla, only some of the phylogenetic groups identified by rRNA sequences have been cultured. The phylum *Euryarchaeota* contains all the methanoarchaea. The extreme halophiles and sulfate-reducing archaea are closely related to each other and to one branch of the methanogens within this group. Some of the extreme thermophiles also appear as deep branches. Thus, both the methanogens and the extreme halophiles are related phylogenetically as well as phenotypically. In contrast, the extreme thermophiles are not a closely knit phylogenetic group. In addition to the extremely thermophilic *Euryarchaeota*, the cultivated members of the phylum *Crenarchaeota* include mostly extreme thermophiles.

Other lineages of special interest include a large group of *Crenarchaeota* that have been detected in marine and soil environments but not cultured. Because of their habitats, these are expected to be mesophiles. Recently, a representative of these organisms, *Nitrosopumilus*, was cultivated and shown to be a nitrifier. However, it is still not known if this property is widespread among the remaining members of this group.

Methane-Producing Archaea: The Methanoarchaea

These archaea are truly cosmopolitan. By this we mean that these strict anaerobes are found in most anaerobic environments on earth, including water-logged soils, rice paddies, lake sediments, marshes, marine sediments, and the gastrointestinal tracts of animals. They are also found over the full temperature range of life, from psychrophilic to mesophilic to hyperthermophilic. Usually they are found in association with anaerobic bacteria and eukaryotes, and they participate in anaerobic food chains that degrade complex organic polymers to CH_4 and CO_2.

Methanogenesis is a significant component of the carbon cycle on earth. About half of the methane produced by these archaea is oxidized before it reaches the atmosphere, and the total biological production on earth is estimated to be about 8×10^{14} grams of methane C per year. Given that microbially produced methane also results in the production of an equal amount of CO_2, the *process* of methanogenesis results in the mineralization of about 1.6×10^{15} grams of the 100×10^{15} grams of C fixed every year during photosynthesis, or about 1.6% of the net primary productivity. Large amounts of methane produced by the methanoarchaea are released into the earth's atmosphere each year. Because methane is a greenhouse gas, it contributes to the warming of the planet. The present atmospheric concentration of methane is about 1.7 ppm. However, this value is more than twice the concentration found in air frozen in ice cores before the industrial revolution. Currently, the concentration of atmospheric methane is increasing at a rate of about 1% per year. Because methane is a much stronger greenhouse gas than CO_2, it is making a significant contribution to increases in global warming even though its concentration is low.

A summary of the properties of the major genera of methanoarchaea is given in Table 18.2. All methane-producing archaea have the ability to obtain their energy for growth from the process of methane biosynthesis. So far, no methanogens have been identified that can grow using other energy sources. Thus, these archaea are obligate methane producers. In prokaryotic nomenclature, the names of genera of methanoarchaea contain the prefix "methan-." This prefix distinguishes them from an unrelated group of *Bacteria*, the methylotrophic bacteria, which consume methane. The names of genera of methylotrophic bacteria contain the prefix "methyl-."

The substrates for methane synthesis are limited to a few types of compounds (Table 18.3). The first type is used by the methanogens that reduce CO_2 to methane. Because most of these organisms use H_2 as an electron donor, they are called **hydrogenotrophic** or "hydrogen-gas eating." Another common electron donor for this reduction is formate. In addition, some hydrogenotrophic methanogens can use alcohols like 2-propanol, 2-butanol, 2-pentanol, cyclo-pentanol, and ethanol as electron donors. Most of the alcohols are oxidized to the ketone level. For instance, in the oxidation of 2-propanol, acetone is the product. Because eight electrons are required to reduce CO_2 to methane, four molecules of H_2, formate, or 2-propanol are consumed. An exception is ethanol,

which undergoes a four electron oxidation to acetate. The limited ability to utilize even these alcohols illustrates the narrow substrate range of these organisms. Even though formate is a reduced C-1 compound, it is oxidized to CO_2 before CO_2 is reduced to methane. The second type of substrate for methanogenesis includes C-1 compounds containing a methyl carbon bonded to O, N, or S. Because they "eat methyl-groups," these organism are referred to as **methylotrophic**. Compounds of this type include methanol, monomethylamine, dimethylamine, trimethylamine, methanethiol, and dimethylsulfide. The methyl group is reduced to methane. The electrons for this reduction are usually obtained from the oxidation of an additional methyl group to CO_2. Because six electrons can be obtained from this oxidation and only two are required to reduce a methyl group to methane, the stoichiometry of this reaction is three molecules of methane are formed for every molecule of CO_2 formed. This process is termed a **disproportionation** because some of the substrate is oxidized and some is reduced. An alternative methylotrophic metabolism is found in a few organisms, such as *Methanosphaera stadtmaniae*, which can only use H_2 to reduce methanol to methane. Apparently, this archaeon lacks the ability to oxidize methanol. The third type of substrate is acetate. In this reaction, the methyl (C-2) carbon of acetate is reduced to methane using electrons obtained from the oxidation of the carboxyl (C-1) carbon of acetate. Organisms that grow by this metabolism are called **aceticlastic** or "acetate breaking" because they split acetate into methane and CO_2. Interestingly, the ethanol-utilizing methanoarchaea are not aceticlastic and cannot use the acetate they form as a substrate for methanogenesis.

Biochemistry of Methanogenesis

Elucidation of the pathway of methane synthesis has been a major challenge to microbial physiologists. Despite simple substrates, the complexity of the pathway is comparable to bacterial photosynthesis. In addition, many of the enzymes of methanogenesis are extremely sensitive to oxygen, and their study requires specialized equipment and techniques. Even after extensive research, many features of archaeal methanogenesis are still not well understood.

The reduction of CO_2 to methane requires five novel coenzymes or vitamins (Figure 18.7). Elucidation of the structure and function of these coenzymes was achieved primarily in the laboratories of R. S. Wolfe at the University of Illinois at Urbana, G. D. Vogels at the University of Nijmegen, and R. K. Thauer at the University of Marburg. The structures of these coenzymes represent a fascinating blend of novel and familiar themes in biochemistry. Methanofuran is an unusual molecule containing an aminomethylfuran moiety. During methanogenesis,

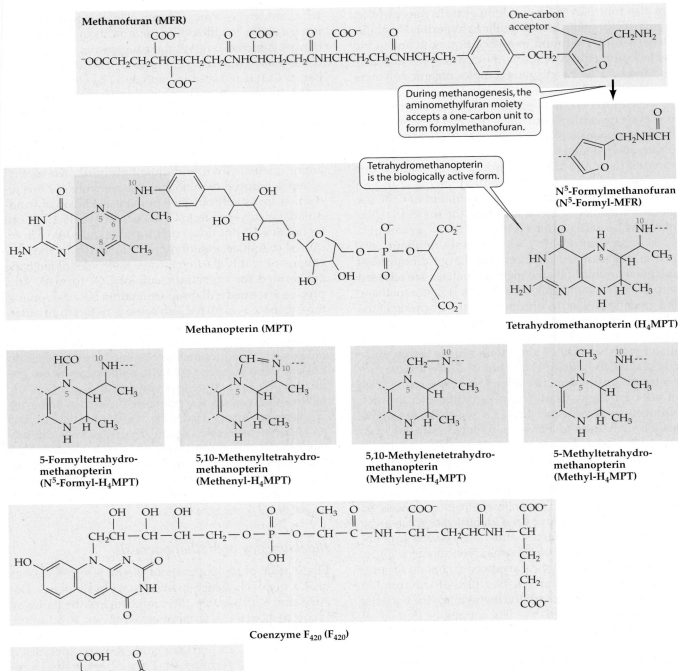

Methanofuran (MFR)

One-carbon acceptor

During methanogenesis, the aminomethylfuran moiety accepts a one-carbon unit to form formylmethanofuran.

N^5-Formylmethanofuran (N^5-Formyl-MFR)

Tetrahydromethanopterin is the biologically active form.

Methanopterin (MPT)

Tetrahydromethanopterin (H$_4$MPT)

5-Formyltetrahydro-methanopterin (N^5-Formyl-H$_4$MPT)

5,10-Methenyltetrahydro-methanopterin (Methenyl-H$_4$MPT)

5,10-Methylenetetrahydro-methanopterin (Methylene-H$_4$MPT)

5-Methyltetrahydro-methanopterin (Methyl-H$_4$MPT)

Coenzyme F$_{420}$ (F$_{420}$)

Coenzyme F$_{430}$ (F$_{430}$)

Coenzyme M (HS-CoM)

Methyl-coenzyme M (methyl-CoM)

7-Mercaptoheptanoylthreonine phosphate (HS-HTP)

◀ **Figure 18.7 Coenzymes of methanogenesis**
Although discovered in the methanoarchaea and closely related *Archaea*, many of these coenzymes have also been found in some of the *Bacteria*.

TABLE 18.2	Summary of properties of the methane-producing *Archaea*[a]			
	Morphology	**Major Energy Substrates**[b]	**Temperature Optimum (°C)**	**Cell Wall**[c]
Order *Methanobacteriales*				
Family *Methanobacteriaceae*				
Genus *Methanobacterium*	Rod	H_2, (formate)	37–45	Pseudomurein
Methanobrevibacter	Short rod	H_2, formate	37–40	Pseudomurein
Methanosphaera	Coccus	H_2 + methanol	37	Pseudomurein
Methanothermobacter	Rod	H_2, (formate)	55–65	Pseudomurein
Family *Methanothermaceae*				
Genus *Methanothermus*	Rod	H_2	80–88	Pseudomurein + protein
Order *Methanococcales*				
Family *Methanococcaceae*				
Genus *Methanococcus*	Coccus	H_2, formate	35–40	Protein
Methanothermococcus	Coccus	H_2, formate	60–65	Protein
Family *Methanocaldococcaceae*				
Genus *Methanocaldococcus*	Coccus	H_2	80–85	Protein
Methanotorris	Coccus	H_2	88	Protein
Order *Methanomicrobiales*				
Family *Methanomicrobiaceae*				
Genus *Methanomicrobium*	Rod	H_2, formate	40	Protein
Methanoculleus	Irregular coccus	H_2, formate	20–55	Glycoprotein
Methanofollis	Irregular coccus	H_2, formate	37–40	Glycoprotein
Methanogenium	Irregular coccus	H_2, formate	15–57	Protein
Methanolacinia	Rod	H_2	40	Glycoprotein
Methanoplanus	Plate or disc	H_2, formate	32–40	Glycoprotein
Family *Methanospirillaceae*				
Methanospirillum	Spirillum	H_2, formate	30–37	Protein + sheath
Family *Methanocorpusculaceae*				
Genus *Methanocorpusculum*	Small coccus	H_2, formate	30–40	Glycoprotein
Methanocalculus[d]	Irregular coccus	H_2, formate	30–40	ND
Order *Methanosarcinales*				
Family *Methanosarcinaceae*				
Genus *Methanosarcina*	Coccus, packets	(H_2), MeNH$_2$, Ac	35–60	Protein + HPS
Methanococcoides	Coccus	MeNH$_2$	23–35	Protein
Methanohalobium	Flat polygons	MeNH$_2$	40–55	ND
Methanohalophilus	Irregular coccus	MeNH$_2$	35–40	Protein
Methanolobus	Irregular coccus	MeNH$_2$	37	Glycoprotein
Methanomethylovorans	Coccus, clusters	MeNH$_2$	20–50	ND
Methanimicrococcus	Irregular coccus	H_2 + MeNH$_2$	39	ND
Methanosalsum	Irregular coccus	MeNH$_2$	35–45	ND
Family *Methanosaetaceae*				
Genus *Methanosaeta* (*Methanothrix*)	Rod	Ac	35–60	Protein + sheath
Order *Methanopyroles*				
Family *Methanopyraceae*				
Genus *Methanopyrus*	Rod	H_2	98	Pseudomurein

[a]All of the methanoarchaea are members of the phylum *Euryarchaeota*.
[b]Major energy substrates for methane synthesis. MeNH$_2$ is methylamines, Ac is acetate. Parentheses means utilized by some but not all species or strains.
[c]Cell wall components include HPS for heteropolysaccharide. ND is not determined.
[d]Placement in higher taxon is tentative.

TABLE 18.3	Free energies for typical methanogenic reactions	
Reaction		$\Delta G^{\circ\prime}$ (kJ/mol of CH_4)
Type 1: CO_2-reducing		
$CO_2 + 4\,H_2 \rightarrow CH_4 + 2\,H_2O$		−130
$4\,HCOOH \rightarrow CH_4 + 3\,CO_2 + 2\,H_2O$		−120
$CO_2 + 4\,(\text{isopropanol}) \rightarrow CH_4 + 4\,(\text{acetone}) + 2\,H_2O$		−37
Type 2: methyl-reducing		
$CH_3OH + H_2 \rightarrow CH_4 + H_2O$		−113
$4\,CH_3OH \rightarrow 3\,CH_4 + CO_2 + 2\,H_2O$		−103
$4\,CH_3NH_3Cl + 2\,H_2O \rightarrow 3\,CH_4 + CO_2 + 4\,NH_4Cl$		−74
$2\,(CH_3)_2S + 2\,H_2O \rightarrow 3\,CH_4 + CO_2 + 2\,H_2S$		−49
Type 3: aceticlastic		
$CH_3COOH \rightarrow CH_4 + CO_2$		−33

the aminomethylfuran is a C-1 acceptor, forming formyl-methanofuran. The use of a furan moiety as a reactive center is unique. In contrast, methanopterin is very similar to folate in structure and function. In both molecules, the reactive center is the tetrahydro form and composed of a pterin and *p*-aminobenzoic acid (PABA). However, the side-chains are strikingly different. Interestingly, unusual pterins are common in other archaea as well, and methanopterin has also been found in the methylotrophic bacteria. Coenzyme F_{430} is also familiar. Named because it absorbs light at 430 nm, it is a tetrapyrrole with Ni at the active center. Although the ring structure is more reduced than other common tetrapyrroles containing Fe (heme), Mg (chlorophyll), and Co (cobamide), the biosynthesis and function of coenzyme F_{430} is probably similar. Coenzyme M and 7-mercaptoheptanoylthreonine phosphate (or HS-HTP) are two thiol-containing coenzymes. Although thiols are common functional groups in other compounds in the cell, such as coenzyme A and cysteine, the coenzymes from methanogens have some very unusual features. HS-HTP contains a threonyl phosphate residue, which has only been previously found in phosphorylated proteins. A phosphorylated amino acid in a coenzyme is extremely unusual. Likewise, coenzyme M is a sulfonate, whereas most highly polar coenzymes are phosphate derivatives.

The requirement for a large number of unusual coenzymes in methanogenesis has profound implications. The ability to utilize methanogenesis as an energy source requires a large amount of genetic information for coenzyme biosynthesis in addition to the enzymes for the pathway of methanogenesis. For this reason, methanogenesis probably evolved only once and all methanogens are related.

Methanogenesis from CO_2 Reduction Occurs Stepwise

CO_2 is bound to carriers and reduced successively through the formyl, methylene, and methyl oxidation–reduction levels to methane (Figure 18.8). Initially, CO_2 binds methanofuran (MFR) and is reduced to the formyl level. The formyl group is then transferred to tetrahydromethanopterin (H_4MPT), which is the C-1 carrier for the next two reductions. Thus, after dehydration of formyl-H_4MPT to methenyl-H_4MPT, the C-1 moiety is reduced to methylene-H_4MPT and then to methyl-H_4MPT. The methyl group is then transferred to coenzyme M. The last reaction forms methane from the reduction of methyl coenzyme M (methyl-CoM). The methylreductase, the enzyme which catalyzes this reaction, requires two additional coenzymes, coenzyme F_{430} and HS-HTP. HS-HTP reduces methyl-CoM to form methane plus the heterodisulfide of coenzyme M and HS-HTP (see Figure 18.7). This mixed disulfide must then be reduced to regenerate HS-HTP.

The electrons for the reduction of CO_2 to methane are obtained from H_2, formate, or alcohols. For some of these reductions, the electron carriers are not known with certainty. However, for at least the reduction of methenyl-H_4MPT to methylene-H_4MPT and then methyl-H_4MPT, the electron carrier coenzyme F_{420} is utilized. Coenzyme F_{420} is named because of its absorption maximum at 420 nm. Although it was discovered in methanogens, it is also present in other organisms. In *Streptomyces*, it functions in antibiotic synthesis. In many bacteria and eukaryotes, it is also essential for photoreactivation of ultraviolet (UV)-damaged DNA. In methanogens, coenzyme F_{420} is present in very high levels.

(A)

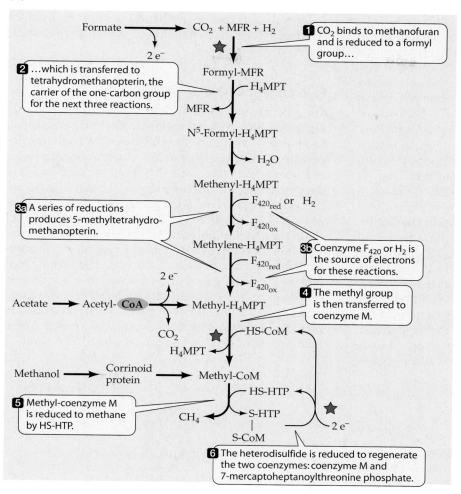

(B)

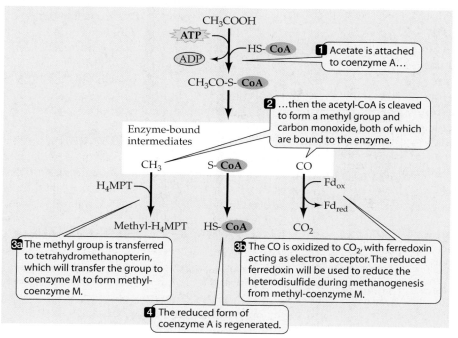

Figure 18.8 Pathways of methanogenesis
(A) Methyl coenzyme M (methyl-CoM) is a central intermediate in methanogenesis from CO_2, acetate, and methanol or methylamines. For some of the reactions, either F_{420} or H_2 is the electron donor. In cases designated by "2 e⁻," the immediate electron donor is not known. Reactions marked with stars indicate potential points of coupling with the proton motive force. MFR = methanofuran, H_4MPT = tetrahydromethanopterin, HS-HTP = 7-mercaptoheptanoylthreonine phosphate. (B) The aceticlastic reaction for biosynthesis of methyl-H_4MPT from acetate. Fd_{ox} and Fd_{red} are the oxidized and reduced forms of ferredoxin, respectively.

Methanogenesis from Other Substrates

Methyl coenzyme M is also a central intermediate in the pathway of methane synthesis from methyl compounds and acetate. For methyl compounds, the C-1 group is first transferred to a cobamide-containing protein and then to coenzyme M. The methyl group is then either reduced to methane or oxidized by the reverse of the pathway of CO_2 reduction. Acetate must first be activated with ATP to form acetyl-CoA (Figure 18.8B). Acetyl-CoA is then cleaved to form an enzyme-bound methyl group, an enzyme-bound carbon monoxide (CO), and HS-CoA. The methyl group is transferred to coenzyme M via tetrahydromethanopterin. The enzyme-bound CO is oxidized to CO_2, and the electron carrier ferredoxin is reduced. The ATP used for acetyl-CoA biosynthesis is recovered at this step. The ferredoxin is then used to reduce methyl-CoM to methane. The enzyme complex, which oxidizes acetyl-CoA, is also called CO dehydrogenase because it oxidizes CO to CO_2. A similar enzyme is found in the homoacetogenic clostridia and many anaerobic autotrophs, including the hydrogenotrophic methanogens. In the autotrophs, the enzyme works to make acetyl-CoA, which is the first step in CO_2 fixation, and all the organic molecules in the cell are derived from it.

Bioenergetics of Methanogenesis

Methanogenesis is a type of anaerobic respiration and serves as the primary means of energy generation in these archaea. Three coupling sites to the proton motive force have been identified: the reduction of CO_2 to formylmethanofuran, the methyl transfer from methyl-H_4MPT to HS-CoM, and the reduction of the heterodisulfide to HS-HPT and HS-CoM (see Figure 18.8). The first of these reactions is endergonic in the direction of methane synthesis, and the proton motive force is required to drive the formation of formylmethanofuran. At the remaining sites, the reactions are exergonic, and either a sodium or proton motive force

is generated. These ion gradients may then be used for ATP biosynthesis, CO_2 reduction, or other energy requiring reactions in the cell.

Ecology of Methanogenesis

The methanoarchaea flourish in anaerobic environments where sulfate, oxidized metals, and nitrate are absent. In these environments, the substrates for methanogenesis are readily available as the fermentation products of bacteria and eukaryotes. The methanogens catalyze the terminal step in the anaerobic food chain where complex polymers are converted to methane and CO_2. These principles are illustrated in a typical methanogenic food chain for a thermophilic bioreactor as shown in Figure 18.9. Polymers are first degraded by specialized microorganisms, like the cellulolytic bacteria. The major products are simple sugars such as glucose, the disaccharide cellobiose, lactate, volatile fatty acids (VFAs) such as acetate, and alcohols such as ethanol. These products are further metabolized by the intermetabolic group. These microorganisms convert simple sugars to VFAs and al-

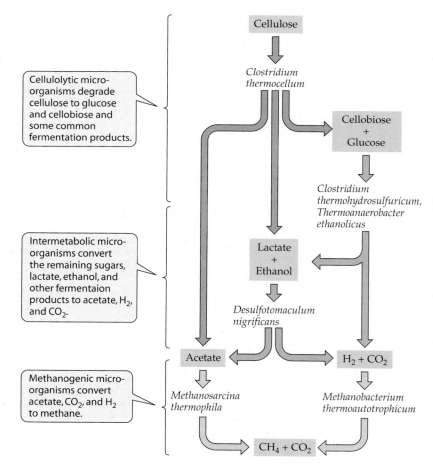

Figure 18.9 Typical anaerobic food chain found in a thermophilic bioreactor
Cellulose is converted to methane and CO_2 by the action of the cellulolytic, intermetabolic, and methanogenic groups of microorganisms. In this fermentation, about 95% of the combustion energy of the cellulose is released as methane. Modified from J. Wiegel and L. G. Ljungdahl. 1986. *CRC Critical Reviews in Biotechnology* 3: 39–108.

cohols as well as the VFAs and alcohols to acetate and H_2, which are major substrates for the methanoarchaea.

Interspecies Hydrogen Transfer

Molecular hydrogen (H_2) is a key intermediate in this process. For many of the fermentations catalyzed by the intermetabolic group, it is one of the major products. When it accumulates, it inhibits the further fermentation. For instance, under standard conditions when the H_2 partial pressure is 1 atmosphere, the fermentations of VFAs and alcohols to acetate and H_2 are thermodynamically unfavorable (Figure 18.10). Therefore, the microorganisms that catalyze these reactions cannot grow. The VFAs then accumulate to concentrations that are toxic to the cellulolytic and intermetabolic groups of microorganisms, and the fermentation of cellulose ceases. However, if the methanoarchaea are present, H_2 is rapidly metabolized, and its partial pressure is maintained below 10^{-3} to 10^{-4} atmospheres.

Under these conditions, the fermentations of VFAs and alcohols are thermodynamically favorable (see Figure 18.10). Because these compounds are rapidly metabolized, their concentrations are maintained below the toxic levels. This interaction between the H_2-producing intermetabolic organisms and the H_2-consuming methanogens is an example of interspecies hydrogen transfer. In many anaerobic environments, it is a key regulatory mechanism. Because the H_2-consuming methanogens play a critical role, they are said to "pull" the fermentation of complex organic polymers to methane and CO_2.

Interspecies hydrogen transfer is not limited to methanogenic food chains. In environments rich in sulfate or nitrate, anaerobic bacteria oxidize H_2, VFAs, and alcohols. When sulfate is present, this activity is catalyzed by the sulfate-reducing bacteria. Because the oxidation of H_2 with sulfate as the electron acceptor is thermodynamically more favorable than when CO_2 is the electron acceptor (as in methanogenesis), the sul-

(A) 1 atmosphere of H_2

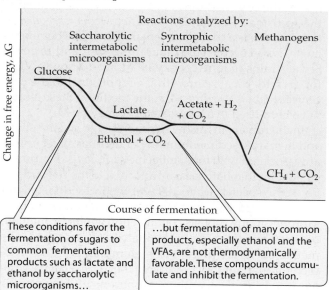

(B) 10^{-5} atmospheres of H_2

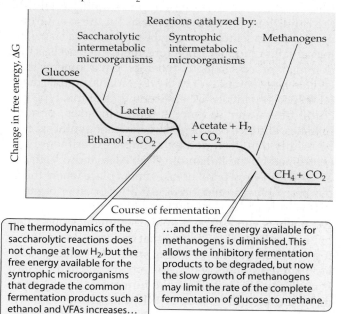

Figure 18.10 Free energy changes during fermentation of glucose to methane
Changes in free energy (ΔG) during a typical fermentation of glucose to methane under standard conditions (1 atmosphere of H_2) and low partial pressures of H_2 (10^{-5} atmospheres of H_2). The reactions catalyzed by each of the three groups of microorganisms required for the fermentation are the saccharolytic intermetabolic group, the syntrophic intermetabolic group, and the methanogen group. (A) Under high partial pressures of H_2, the thermodynamics of the fermentation of sugars to lactate and ethanol is very favorable. The microorganisms that can catalyze these reactions grow and these

products are formed. However, the thermodynamics of the fermentations of lactate and ethanol are unfavorable or only slightly favorable. Therefore, the bacteria that catalyze these reactions cannot grow, these compounds accumulate, and the fermentation ceases. (B) In contrast, under low concentrations of H_2, the fermentations of lactate and ethanol are favorable, and the intermetabolic group of organisms grows well. Therefore, lactate and ethanol are rapidly converted to acetate, H_2, and CO_2. Notice that the free energy available for the growth of the methanogens is now diminished, and growth of the methanogens may limit the rate of the complete fermentation of glucose to methane.

fate-reducing bacteria outcompete the methanogens for H_2. For similar reasons, the sulfate-reducing bacteria also outcompete the methanogens for other important substrates such as acetate and formate. Therefore, methanogenesis is greatly limited in marine sediments that are rich in sulfate. The oxidation of H_2 with nitrate, Fe^{2+}, and Mn^{2+} as electron acceptors is also thermodynamically more favorable than methanogenesis. The denitrifying and iron-and magnesium-reducing bacteria also outcompete the methanogens when these electron acceptors are present. In conclusion, with few exceptions, methanogenesis only dominates in habitats where CO_2 is the only abundant electron acceptor for anaerobic respiration.

Methanogenic Habitats

Just as aerobic microorganisms rapidly deplete the O_2 in environments rich in organic matter to establish anaerobic conditions, sulfate-reducing bacteria, iron- and magnesium-reducing bacteria, and denitrifying bacteria frequently consume all the sulfate, Fe^{3+}, Mn^{4+} and nitrate in anaerobic environments and rapidly establish the conditions for methanogenesis. In these environments, CO_2 is seldom limiting because it is also a major fermentation product. Therefore, methanogenesis is the dominant process in many anaerobic environments that contain large amounts of easily degradable organic matter. Especially important environments of this type include freshwater sediments found in lakes, ponds, marshes, and rice paddies. Here methane synthesis can be readily demonstrated by a simple experiment first performed by the Italian physicist Alessandro Volta in 1776. In a shallow pond or lake, find a place where there has been an accumulation of leaf litter or other organic debris. When this material is stirred with a long stick, gas bubbles will escape from the sediment. Collect the bubbles in an inverted funnel that has been closed at the top with a short piece of tubing and a pinch clamp. Collect about four liters of gas. To demonstrate that the gas is methane, ignite it. First, the funnel is partially submerged to pressurize the gas. When the clamp is slowly opened, a match is touched to the escaping gas, which will burn with a blue flame. Care must be taken to keep your hair and eyebrows clear of the opening and to avoid be burned.

Methanogenesis also occurs in other habitats. Methane is formed by the methanoarchaea in the anaerobic microflora of the large bowel. About one-third of healthy adult humans excrete methane gas. Some methane is also absorbed in the blood and excreted from the lungs. The most numerous methanogen in humans is *Methanobrevibacter smithii*, which is a gram-positive coccobacillus that utilizes H_2 or formate to reduce CO_2 to methane. In people who excrete methane, it is found in numbers of 10^7 to 10^{10} cells per gram dry weight of feces, or between 0.001 and 12% of the total number of viable anaerobic microorganisms. Why the numbers of *M. smithii* fluctuates so greatly in apparently healthy individuals remains a mystery. *Methanosphaera stadtmaniae* is also present in the human large bowel, but in much lower numbers. This interesting methanogen will grow only by the reduction of methanol with H_2. It cannot reduce CO_2 to methane or utilize acetate and methylamine. Like *M. smithii*, it is also gram-positive and contains pseudomurien in its cell walls.

The rumen is another major habitat for methanogens, and approximately 10% to 20% of the total methane emitted to the earth's atmosphere originates in the rumen of cows, sheep, and other mammals. In the rumen, complex polymers from grass and other forages are degraded to volatile fatty acids, H_2, and CO_2, by the cellulolytic and intermetabolic groups of bacteria, fungi, and protists. The VFAs are absorbed by the animal and are a major energy source. Therefore, little methane is produced from acetate. The H_2 is used to reduce CO_2 to methane, which is emitted. Methanogenesis represents a significant energy loss to the cow, and up to 10% of the caloric content of the feed may be lost as methane. *Methanobrevibacter ruminantium* is a common methanogen in the bovine rumen. It is a coccobacillus that utilizes H_2 and formate. Some strains require coenzyme M in addition to VFAs for growth. Because growth is proportional to the concentration of coenzyme M in the medium, these strains have been used in a bioassay for coenzyme M.

In the rumen, in freshwater sapropel or swamp slime, and in marine sediments, many of the anaerobic protists are associated with methanogenic symbionts. For rumen ciliates, methanoarchaea are attached to the cell surface. For some species, up to 100% of the individual ciliates are associated with methanogens. In aquatic sludge or sapropel, many protists contain endosymbiotic methanogens. These include both ciliates and amoebae. The methanogens are present in high numbers and distributed throughout the cytoplasm. The marine ciliate *Metopus contortus* also contains numerous endosymbiotic methanogens of the species *Methanoplanus endosymbiosus*. This disc-shaped methanogen has a protein cell wall and lyses rapidly in 0.001% sodium dodecyl sulfate. It is located in parallel rows on the cytoplasmic side of the pellicle and on the surface of the nuclear membrane, and it is present in very high numbers, greater than 10^{10} methanogens per milliliter of cytoplasm. *M. endosymbiosus* is probably associated with the hydrogenosomes, which are microbodies that convert pyruvate to acetate, H_2, and CO_2. The methanogen is then the electron sink for H_2. Conceivably, the ciliate may use this mechanism to divert reducing equivalents away from the sulfate-reducing bacteria and limit the production of H_2S, which is very toxic.

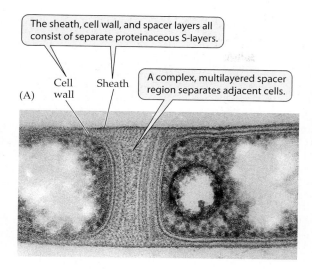

The sheath, cell wall, and spacer layers all consist of separate proteinaceous S-layers.

(A) Cell wall | Sheath | A complex, multilayered spacer region separates adjacent cells.

(B)

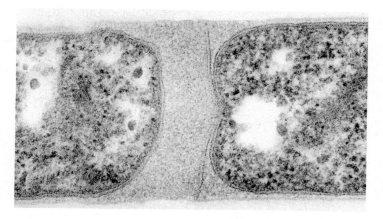

Figure 18.11 Sheaths of methanoarchaea
Some methanoarchaea have complex cell walls. (A) *Methanospirillum hungatei* is common in bioreactors and fresh water sediments. The rod-shaped cells are enclosed in a sheath and often form long filaments. Between each cell is a complex, multilayered spacer region seen in the thin section. The sheath, cell wall, and spacer layers all consist of separate proteinaceous layers. (B) Sheaths are also observed on *Methanosaeta concilii*, an aceticlastic methanogen. In thin section, the spacer region appears to be less complex. A, ©T. J. Beveridge/Visuals Unlimited; B, courtesy of T. J. Beveridge, University of Guelph, Canada.

habitats. A hydrogenotrophic methanogen, it has a distinctive spirillum morphology. It forms long filaments, where the individual cells are encased in a sheath (Figure 18.11). The cells are separated by a complex spacer region composed of protein layers. *Methanosaeta concilii* is another sheathed archaeon. This obligately aceticlastic methanogen has a very high affinity for acetate, and it is common in bioreactors where the acetate concentrations are low. Because *M. concilii* has a generation time of one day, it is frequently observed only after the bioreactor has been running for an extended period.

Anaerobic Methane Oxidation

In marine sediments near methane seeps or other sources of methane, methane is oxidized to CO_2 under anaerobic conditions. Although aerobic methane-oxidizing bacteria are well known, the nature of organisms capable of anaerobic methane oxidation is just becoming understood. Pure cultures have yet to be isolated; however, culture-independent methods suggest that anaerobic methane oxidation is performed by a consortium of archaea related to the methylotrophic methanogens and sulfate-reducing bacteria. Microbial mats that contain numerous archaea also contain high levels of a novel form of the nickel tetrapyrrole coenzyme F_{430}. This circumstantial evidence suggests that the pathway of anaerobic methane oxidation may resemble the reverse of the pathway of methane synthesis.

Extreme Halophiles

The extremely halophilic archaea or haloarchaea are all closely related, and they comprise a single family of 14 genera (Table 18.4). They are all halophilic, and most grow only at concentrations of NaCl greater than 1.8 M. Most species will also grow at concentrations greater than 4 M, which is close to the saturation point for NaCl in water. In addition, a number of genera are alkalinophilic and only grow at basic pH. Some genera of the *Methanosarcinales* are also extremely halophilic, such as *Methanohalophilus* and *Methanohalobium*. In addition, these methanogens and the haloarchaea possess similarities in lipids, the presence of

Although methanogens are most frequently found at the top of anaerobic food chains associated with the intermetabolic microorganisms, in some ecosystems they are the primary consumers of geochemically produced H_2 and CO_2. The submarine hydrothermal vents found on the ocean floor expel large volumes of very hot water containing H_2 and H_2S. As the water cools from several hundred degrees Celsius to the temperature of the ocean, zones suitable for the growth of thermophilic methanogens are established. *Methanocaldococcus jannaschii* is found in such an environment. It has an optimum temperature of 85°C and a protein cell wall. Methane produced by this bacterium escapes into the surrounding water, where it is utilized by symbiotic methylotrophic bacteria living in the gills of mussels. In this ecosystem, the methanogen is the primary producer at the bottom of the food chain.

Methanoarchaea are common in bioreactors that convert organic wastes to methane gas, which can then be used a fuel. *Methanospirillum hungatei* is a common methanoarchaea found in this and other freshwater

TABLE 18.4   Summary of properties of extremely halophilic *Archaea*[a]

		Morphology	Substrates[b]	pH Optimum	Optimum NaCl (M)
Family *Halobacteriaceae*					
Genus	*Halobacterium*	Long rod	Amino acids	5–8	3.5–4.5
	Halalkalicoccus	Coccus	CH_2O	9.5–10	3.4
	Haloarcula	Short pleomorphic rods	Amino acids, CH_2O	7.0–7.5	2.5–3.0
	Halobaculum	Rod	Organic	6–7	1.0–2.5
	Halobiforma	Pleomorphic	Amino acids, CH_2O	7.5–8.9	2.6–3.4
	Halococcus	Coccus	Amino acids, CH_2O	6.8–9.5	3.5–4.
	Haloferax	Pleomorphic	Amino acids, CH_2O	7	2.5
	Halogeometricum	Pleomorphic	Amino acids, CH_2O	6–8	3.5–4
	Halomicrobium	Short rod	CH_2O	6.2–8	3–3.5
	Halorhabdus	Pleomorphic	CH_2O	6.7–7.1	4.6
	Halorubrum	Rod	Amino acids, CH_2O	ND[c]	1.7–4.5
	Halosimplex	Rod	CH_2O	7–8	4.3
	Haloterrigena	Rod or oval	CH_2O	ND	2.5–4.3
	Halovivax	Pleomorphic	CH_2O	7–7.5	3.4
	Natrialba	Rod	Organic, CH_2O	9.5	3.5
	Natrinema	Rod	Amino acids, CH_2O	7.2–7.8	3.4–4.3
	Natronobacterium	Rod	CH_2O, organic acids	9.5–10.0	3.0
	Natronococcus	Coccus	CH_2O	9.5–10.0	2.5–3.0
	Natronolimnibius	Rod, pleomorphic	CH_2O	9–9.5	2.6–4.3
	Natronomonas	Short rod	Amino acids	8.5–9.5	3.5
	Natronorubrum	Pleomorphic	Organic	9.0–9.5	3.4–3.8

[a]All of the haloarchaea are members of the phylum *Euryarchaeota* and the order *Halobacteriales*.
[b]CH_2O is carbohydrate. "Organic" includes complex C sources, the exact nature of the components used are unknown. Oxygen is the electron acceptor for all the extremely halophilic archaea. Some species may also grow anaerobically by fermentation or nitrate reduction or denitrification.
[c]ND, not determined.

gas vesicles, and phylogenies of the 16S rRNA and other genes not shared with other archaea. This evidence suggests that the aerobic haloarchaea are specifically related to this group of methanoarchaea despite their other great differences in phenotype.

In addition to the haloarchaea, some bacteria, algae, and fungi can also grow in very high concentrations of NaCl. Examples of extremely halophilic bacteria include the purple photosynthetic bacterium *Ectothiorhodospira* and the actinomycete *Actinopolyspora*. In addition, *Halomonas* is a halotolerant bacterium that grows optimally in concentrations near 1 M NaCl, but it also grows slowly at 4 M NaCl. Although these bacteria share with the haloarchaea the ability to grow in high concentrations of salt, they are unrelated and their biochemistry and physiology are quite different.

The haloarchaea are obligate or facultative aerobes. Most utilize amino acids, carbohydrates, or organic acids as their principal energy sources. *Halobacterium* oxidizes glucose by a modification of the Entner-Doudoroff pathway (Figure 18.12). In this modification, 2-keto-3-deoxygluconate-6-phosphate is the first phosphorylated intermediate, and glucose-6-phosphate is not found. A further modification in which none of the intermediates are phosphorylated is also found in the extremely thermophilic archaea (see Figure 18.12). Carbohydrates are further oxidized by pyruvate oxidoreductase and the complete tricarboxylic acid (TCA) cycle. NADH generated by the TCA cycle is oxidized by an electron transport chain very similar to that found in the bacteria. In the absence of oxygen, *Haloferax denitrificans* is also capable of growth by the anaerobic respiration of nitrate. *Halobacterium salinarium* can ferment arginine in the absence of oxygen or nitrate.

Bacteriorhodopsin and Archaeal Photophosphorylation

Many haloarchaea are also capable of an interesting type of photophosphorylation. Photophosphorylation is that portion of photosynthesis involved in converting light energy to the proton motive force and hence chemical energy, typically ATP biosynthesis. It does not include the CO_2 reduction steps associated with photosynthesis. Under low partial pressures of oxygen, haloarchaea insert large amounts of a protein called bacteriorhodopsin into their cytoplasmic membrane. This protein forms patches that are called purple mem-

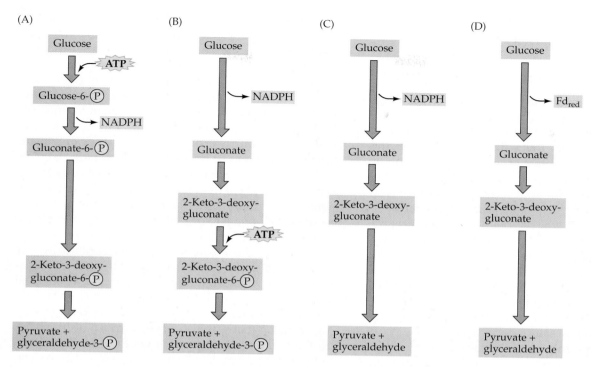

Figure 18.12 Modifications of the Entner-Doudoroff pathway in the *Archaea*
(A) Pathway common in *Pseudomonas* and other bacteria. (B) Pathway found in *Halobacterium* and *Clostridium aceticum*. (C) Pathway found in the extremely thermophilic archaea *Sulfolobus* and *Thermoplasma*. (D) Pyroglycolytic pathway found in the hyperthermophilic archaeon *Pyrococcus*. Fd is ferredoxin.

branes. The color is due to the presence of the pigment retinal, which is also the common visual pigment in eyes. The retinal is covalently bonded to a lysyl residue of bacteriorhodopsin via a Schiff base (see Figure 9.15). In the dark, the retinal is in the all-*trans* configuration and the Schiff base is protonated. Upon the absorption of light, the retinal isomerizes to the 13-*cis* configuration, and the Schiff base is deprotonated. A proton is also expelled outside the cell. The Schiff base then returns to the protonated form by removing a proton from the cytoplasm, and the retinal reverts to the all-*trans* isomer. In this fashion, bacteriorhodopsin functions as a light-driven proton pump. The proton motive force it generates is then used for ATP synthesis and transport. The haloarchaea have at least two additional retinal-based photosystems. One is a chloride pump called halorhodopsin. The second is important in phototaxis and is called "slow rhodopsin."

Although long believed to be unique to the *Archaea*, bacteriorhodopsin has also been found in marine *Bacteria*. First, a gene homologous to the archaeal bacteriorhodopsin gene was found in DNA cloned from an uncultured γ-proteobacterium, SAR86. This gene was expressed in *E. coli*, and the recombinant protein formed a functional proton pump. Although currently uncul-

tured, proteobacteria containing this bacteriorhodopsin gene are now known to be widely distributed in the ocean.

Retinal is also the visual pigment in the eye, and the function of the animal rhodopsin is different from that of the bacteriorhodopsin. In rhodopsin, the retinal is not covalently bonded to the protein opsin. In the dark, retinal is in the 11-*cis* configuration. Exposure to light causes an isomerization to the all-*trans* configuration. Proton movement does not occur. Instead, the change in retinal causes a conformational change in the protein, which leads to the activation of a second messenger in the visual response. Because the mechanisms of the two retinal-based photosystems are fundamentally different, they may be an example of convergent evolution.

Photophosphorylation in the haloarchaea is also very different from that in the chlorophyll-based system found in the *Bacteria* and chloroplasts. The electron transport chain does not participate in retinal-based photophosphorylation, and it is not possible to directly reduce $NADP^+$ from electron donors that have unfavorable redox potentials, like water or elemental sulfur. Moreover, extensive light-trapping pigments like the light-harvesting chlorophylls, which are elaborated by photosynthetic bacteria in dim light, have not been described in the haloarchaea. Therefore, the retinal-based system is more suitable for growth at high light intensities.

Adaptations to High Concentrations of Salt

The haloarchaea are adapted specifically for growth in high concentrations of salt. The water activity (a_w) of the cytoplasm of all microorganisms is equal to or less than the a_w of their medium. To the first approximation, this means that the concentration of solutes in the cytoplasm equals or exceeds the concentration outside the cell. For halophilic microbes, the concentration of intracellular solutes must be very high. In the haloarchaea, the intracellular solute is KCl, which is often found at concentrations greater than 5 M. High concentrations of salts also tend to weaken the ionic bonds necessary to maintain the conformation of proteins and the activity of enzymes. To maintain their structure, the proteins from the haloarchaea have an increased number of acidic and hydrophobic residues, and their enzymes frequently require high concentrations of salts for activity. In the absence of salts, the acidic residues on the surface of their proteins are no longer shielded by cations, and many of their enzymes denature. Below NaCl concentrations of 1 M, the protein cell walls of many haloarchaea lose their structural integrity for the same reasons, and the cells lyse.

The haloarchaea are common in salt lakes and salterns, or ponds used to prepare solar salt by the evaporation of sea water. Frequently, they are abundant and impart a red color to the brine. The red color is due to the presence of C_{50} carotenoids called bacterioruberins in the cell envelope. Although nonpathogenic, the haloarchaea can spoil fish and animal hides treated with solar salt.

In salt lakes, the total concentration of salts is generally between 300 and 400 grams per liter, but the specific ions present may vary greatly. For instance, the Dead Sea in Palestine contains abundant chloride, magnesium, and sodium ions. In the Great Salt Lake in Utah, chloride, sodium and sulfate ions predominate. The pH of these lakes are near neutrality. In contrast, the Wadi Natrun in Egypt contains high concentrations of bicarbonate and carbonate ions, which maintain the pH near 11. The alkalinophilic haloarchaea, *Natronobacterium* and *Natronococcus*, have been found in these lakes. In the salt lakes, the haloarchaea mineralize organic carbon to CO_2. The organic carbon is produced by other halophilic microorganisms including the green algae *Dunaliella*, cyanobacteria, and *Ectothiorhodospira*.

Extreme Thermophiles

This name refers to a disparate group of the *Archaea* that are not particularly related, either phylogenetically or physiologically. Moreover, the term extreme thermophile

is somewhat of a misnomer. Although most species are extremely thermophilic, some are only moderate thermophiles. Moreover, some methanogens are extremely thermophilic but are not included in this group.

The name is justified because extreme thermophily is almost exclusively a property of *Archaea*. Extremely thermophilic organisms were first discovered independently by J. A. Brierley and T. D. Brock in the late 1960s. Although very controversial at the time, these organisms are now known to be common in thermophilic habitats worldwide. Although many more species have been described for *Bacteria* than for *Archaea*, only two bacterial genera, *Thermotoga* and *Aquifex*, have optimal growth temperatures equal to or greater than 80°C. In contrast, many *Archaea* have optimal temperatures greater than 80°C, and the most extreme hyperthermophiles are *Archaea*. *Pyrolobus fumarii* is a facultatively aerobic chemolithoautotroph isolated from a deep sea hydrothermal vent. It has a temperature optimum of 106°C and a temperature maximum of 113°C. Strain 121 is a Fe(III)-reducing archaeon that was also isolated from a deep sea hydrothermal vent. It has a maximum growth temperature of 121°C. Organisms like *Pyrolobus* and strain 121 that can grow at 90°C or above are also called **hyperthermophiles**. Interestingly, extremely thermophilic archaea become abundant just above the temperature where thermophilic bacteria become rare (Figure 18.13). The cultured archaea with optimal temperatures below 80°C are either methanogenic, extremely halophilic, nitrifying, or both moderately thermophilic and acidophilic. Therefore,

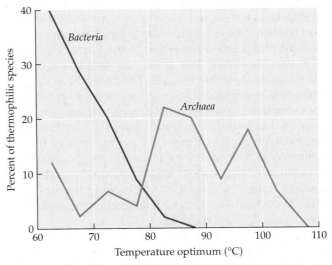

Figure 18.13 Optimum temperature for growth of thermophilic *Archaea* and *Bacteria*
The percentage of 45 thermophilic species from each domain is given. Modified from T. D. Brock. 1985. *Science* 230: 132–138.

they occupy niches in which bacteria are absent or rare. A possible explanation for this distribution is that bacteria exclude the archaea from temperate environments and that archaea are only abundant either in habitats where bacteria grow poorly such as at high temperatures or high salt or if they utilize a means of producing energy unavailable to the bacteria such as methanogenesis. Further studies into the nature of the abundant uncultured temperate crenarchaeotes is a test of this hypothesis. If bacteria systematically outcompete archaea in common temperate environments, the physiologies of the temperate *Crenarchaeota* should be highly specialized, such as the methanogens (see subsequent text).

The extremely thermophilic archaea may be further subdivided into the obligate and facultative aerobes and the obligate anaerobes (Table 18.5). The aerobes are usually acidophilic, and have pH optima close to 2.0. The obligate anaerobes are mostly neutrophilic, and have pH optima greater than 5.0. Despite the extreme conditions in which these organisms are usually found, the adaptations to their environment are of the same complexity as found in mesophiles. For instance, the hyperthermophile *Pyrodictium abyssi* forms tiny white balls in broth at 97°C composed of a complex three-dimensional network of tubules (Figure 18.14). The physiological rationale for this unique structure is not known. Because the cells obviously dedicate a major proportion of their resources to its construction, we must assume that this structure is important to their survival in the environment.

It is important to note that the extremely thermophilic archaea are a recent discovery. More than 90% of the species currently known were isolated after 1980, largely due to the pioneering work of Karl Stetter at the University of Regensburg and Wilfram Zillig at the Max-Planck-Institüt in Martinsried and their colleagues in Germany. Because of their recent discovery, many aspects of their biology, physiology, and biochemistry are not yet known. However, even at this early stage in their study, the extremely thermophilic archaea have challenged many of the basic tenets of modern biology.

Habitats and Physiology

The extremely thermophilic archaea grow above the temperature limit of photosynthesis. Therefore, they are generally restricted to environments where geothermal energy is available. These areas include hot springs, solfatara fields, geothermally heated marine sediments, and submarine hydrothermal vents. Here, H_2, H_2S and S^0 are abundant energy sources. Many of these archaea are chemolithotrophic and are capable of some form of sulfur metabolism. *Acidianus infernus* is a particularly interesting example. It is an obligate autotroph and uses CO_2 as its sole carbon source. During aerobic growth, S^0

is oxidized to H_2SO_4 (Figure 18.15). Acid production maintains the pH at very low levels, and this organism is also an acidophile. Under anaerobic conditions, H_2 is oxidized and S^0 is now reduced to H_2S. Therefore, S^0 functions as either the electron donor in the presence of O_2 or the electron acceptor in the presence of H_2. Other extremely thermophilic archaea are capable of either one of these two modes of sulfur metabolism. The aerobe *Sulfolobus* oxidizes S^0 and organic compounds. It is also an acidophile. The anaerobes in the order *Thermoproteales* oxidize H_2, sugars, or amino acids using S^0 as an electron acceptor. In addition, growth by fermentation or other types of anaerobic respiration may also be possible.

Thermoplasma is an unusual archaeon that was first isolated from burning coal refuse piles. It has since been found in hot springs as well, and two species have been identified, *Thermoplasma acidophilus* and *Thermoplasma volcanium*. *Thermoplasma* species are facultative aerobes and grow on the peptide components of yeast extract; these microorganisms have not been grown in defined medium. During aerobic growth, *Thermoplasma* species utilize an electron transport chain consisting of cytochrome *b* and a quinone. Nevertheless, its bioenergetics are poorly understood, and ATP is probably synthesized by substrate level phosphorylation. *Thermoplasma* lacks a cell wall, yet it is osmotically stable and contains glycoprotein and an unusual lipopolysaccharide in its cell membrane. Interestingly, *Thermoplasma* also contain a basic DNA-binding protein that is functionally very similar to the histones found in eukaryotes. However, a comparison of the amino acid sequence reveals that it is more closely related to the bacterial DNA-binding proteins than to histones.

Archaeoglobus is another exceptional extreme thermophile and the only archaeon capable of sulfate reduction. Phylogenetically, it is a close relative of the methanogens and contains tetrahydromethanopterin, methanofuran, and coenzyme F_{420}. However, coenzyme F_{430} and coenzyme M are absent. During growth on lactate, *Archaeoglobus* utilizes an aceticlastic system similar to that found in the aceticlastic methanogens to oxidize acetyl-CoA (Figure 18.16). Methyltetrahydromethanopterin is then oxidized by the reverse of the pathway of CO_2 reduction.

Nanoarchaeum is an obligate symbiont that was discovered in cultures of another extremely thermophilic archaeon *Ignicoccus*. Never grown in pure culture, it appears as buds on the cells of *Ignicoccus* (Figure 18.17). In stationary phase cultures, cells of *Nanoarchaeum* detach and can be isolated free in the culture medium. Remarkably, optimal growth of this coculture occurs near 90°C, illustrating that life at high temperatures may be as complex and varied as that found at mesophilic temperatures. Although little is known about the physiology of this un-

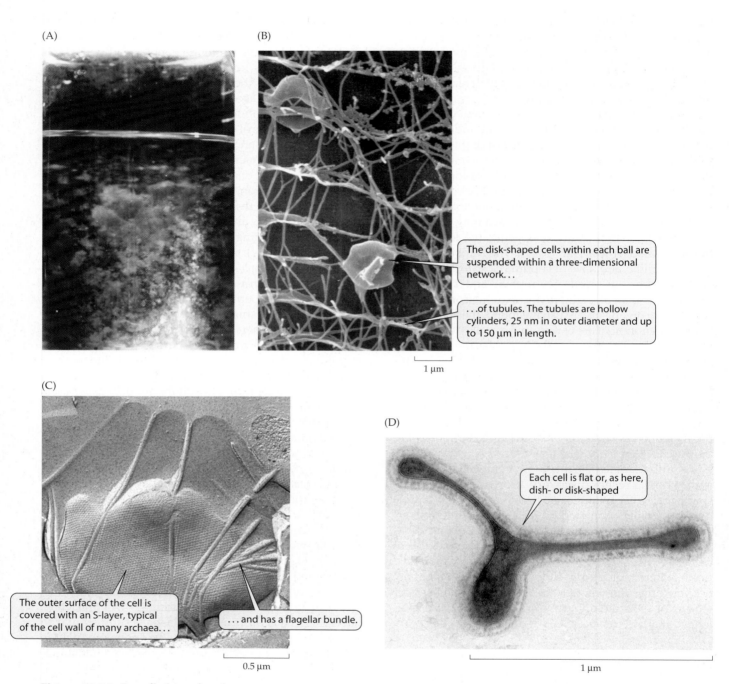

(A)

(B)

The disk-shaped cells within each ball are suspended within a three-dimensional network. . .

. . .of tubules. The tubules are hollow cylinders, 25 nm in outer diameter and up to 150 μm in length.

1 μm

(C)

(D)

Each cell is flat or, as here, dish- or disk-shaped

The outer surface of the cell is covered with an S-layer, typical of the cell wall of many archaea. . .

. . . and has a flagellar bundle.

0.5 μm

1 μm

Figure 18.14 *Pyrodictium abyssi*
Ultrastructure of the hyperthermophilic crenarchaeote *Pyrodictium abyssi*. This strictly anaerobic heterotroph has a temperature optimum for growth of 97°C and was isolated from a submarine "black smoker" off Guaymas, Mexico. (A) In broth culture, these cells form tiny white balls, about 1 mm in diameter. (B) Upon higher magnification by scanning electron microscopy, these balls are composed of a three-dimensional network of cells and tubules. The tubules appear to be hollow cylinders, with an outer diameter of 25 nm and a length up to 150 μm. (C) The outer surface of the cells are covered with an S-layer, typical of the cell wall of many archaea, as well as a flagellar bundle (freeze etch electron microscopy). (D) An ultrathin section of the cells showing that they are flat and dish or disk-shaped. Courtesy of Prof. Dr. K. O. Stetter and Dr. R. Rachel, University of Regensburg, Germany.

Figure 18.15   Sulfur metabolism of the facultative aerobe *Acidianus infernus*

During aerobic growth, elemental sulfur (S_8) is oxidized to sulfuric acid. During anaerobic growth, elemental sulfur is reduced to hydrogen sulfide. The aerobic metabolism of *Acidianus* is typical of the obligately aerobic extreme thermophiles. The anaerobic metabolism is typical of the obligately anaerobic extreme thermophiles.

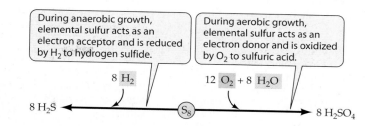

During anaerobic growth, elemental sulfur acts as an electron acceptor and is reduced by H_2 to hydrogen sulfide.

During aerobic growth, elemental sulfur acts as an electron donor and is oxidized by O_2 to sulfuric acid.

$$8\ H_2S \longleftarrow \quad 8\ H_2 \quad \boxed{S_8} \quad 12\ O_2 + 8\ H_2O \quad \longrightarrow 8\ H_2SO_4$$

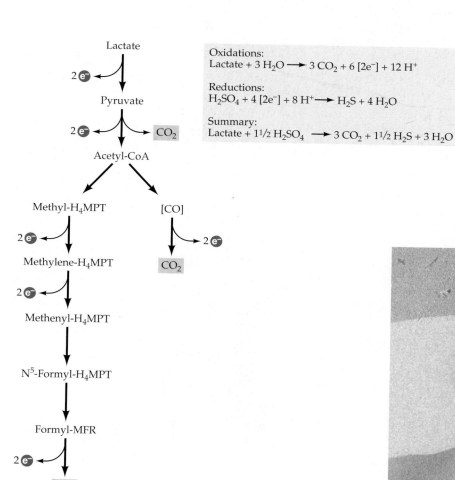

Oxidations:
$$\text{Lactate} + 3\ H_2O \longrightarrow 3\ CO_2 + 6\ [2e^-] + 12\ H^+$$

Reductions:
$$H_2SO_4 + 4\ [2e^-] + 8\ H^+ \longrightarrow H_2S + 4\ H_2O$$

Summary:
$$\text{Lactate} + 1^1/_2\ H_2SO_4 \longrightarrow 3\ CO_2 + 1^1/_2\ H_2S + 3\ H_2O$$

Figure 18.16   Lactate oxidation by the sulfate-reducing archaeon *Archaeoglobus*

Oxidation of lactate by *Archaeoglobus* proceeds through pyruvate to acetyl-CoA. Using an enzyme similar to that found in the aceticlastic methanogens, acetyl-CoA is oxidized to CO_2 and methyltetrahydromethanopterin (methyl-H_4MPT; see Figures 18.7 and 18.8). Methyl-H_4MPT is then oxidized by the same pathway used by the methylotrophic methanogens (see Figure 18.8). MFR = methanofuran, H_4MPT = tetrahydromethanopterin.

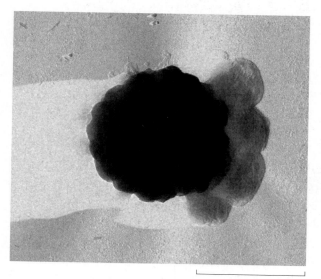

1 µm

Figure 18.17   The extremely thermophilic *Ignicoccus* infected with the symbiont *Nanoarchaeum equitans*

The large cell is *Ignicoccus*. Four smaller cells of *Nanoarchaeum equitans* are attached. In this electron micrograph, cells were fixed by glutaraldehyde and then shadowed with platinum-carbon. Courtesy of Prof. Dr. K. O. Stetter and Dr. R. Rachel, University of Regensburg, Germany.

TABLE 18.5 Summary of properties of extremely thermophilic *Archaea*

	Morphology	Electron Donor[a]	Electron Acceptor	Temperature Optimum (°C)	pH Optimum
Phylum *Crenarchaeota*					
Order *Sulfolobales*					
Family *Sulfolobaceae*					
Genus *Sulfolobus*	Irregular coccus	Organic, S^0, H_2S, Fe^{2+}, H_2	O_2	65–85	2.0–4.5
Acidianus	Irregular coccus	H_2, S^0, organic	S^0, O_2	70–90	1.5–2.5
Metallosphaera	Coccus	S^0, FeS_2, organic	O_2	75	1.0–4.5
Stygiolobus	Irregular coccus	H_2	S^0	80	1.0–5.5
Sulfurisphaera	Irregular coccus	Peptone	S^0	84	2.0
Sulfurococcus	Coccus	S^0, FeS_2, organic	O_2	60–75	2.0–2.6
Order *Thermoproteales*					
Family *Thermoprotaceae*					
Genus *Thermoproteus*	Thin rod	Organic, H_2	S^0	85–88	5.0–6.8
Caldivirga	Thin rod	Organic	S^0	85	3.7–4.2
Pyrobaculum	Thin rod	H_2	S^0	100	6.0
Thermocladium	Thin rod	Organic, H_2	S^0	75	4.2
Vulcanisaeta	Thin rod	Organic	S^0, $S_2O_3^{2-}$	85–90	7.0–4.5
Family *Thermofilaceae*					
Thermofilum	Thin rod	Organic	S^0	85–90	5.0–6.0
Order *Desulfurococcales*					
Family *Desulfurococcaceae*					
Genus *Desulfurococcus*	Coccus	Organic	S^0	85–92	6.0
Aeropyrum	Coccus	Organic	S^0	90–95	7
Ignicoccus	Coccus	H_2	S^0	90	6
Staphylothermus	Coccus	Organic	S^0	92	ND[b]
Stetteria	Disk	Organic	S^0	95	6
Sulfophobococcus	Coccus	Organic		87	7.5
Thermodiscus	Disk	Organic	S^0	90	5.5
Thermosphaera	Disk, aggregates	Organic		85	6.5–7.2
Family *Pyrodictiaceae*					
Genus *Pyrodictium*	Disk	H_2, organic	S^0	97–105	5.5
Hyperthermus	Irregular coccus	Peptides, H_2	S^0	95–106	7.0
Pyrolobus	Irregular coccus	H_2	NO_3^-, $S_2O_3^{2-}$, O_2	106	5.5
Family unassigned					
Genus *Acidilobus*	Coccus	Organic	S^0	85	3.8
Caldisphaera	Coccus	Organic	S^0	70–78	3.5–4.5
Ignisphaera	Coccus	Organic	None	92–95	6.4
Phylum *Euryarchaeota*					
Order *Archaeoglobales*					
Family *Archaeoglobaceae*					
Genus *Archaeoglobus*	Coccus	H_2, organic	SO_4^{2-}, $S_2O_3^{2-}$	80–85	6.0–7.0
Ferroglobus	Irregular coccus	Fe^{2+}, H_2S, H_2	NO_3^-, $S_2O_3^{2-}$	85	7.0
Order *Thermoplasmatales*					
Family *Thermoplasmataceae*					
Genus *Thermoplasma*	Pleomorphic	Organic	O_2, S^0	55–59	1.0–2.0
Family *Picrophilaceae*					
Genus *Picrophilus*	Irregular coccus	Organic	O_2	60	0.7
Family *Ferroplasmaceae*					
Genus *Ferroplasma*	Irregular coccus	Fe^{2+}, Fe_2S, Mn^{2+}	O_2	35	1.7
Order *Thermococcales*					
Family *Thermococcaceae*					
Genus *Thermococcus*	Coccus	Organic	S^0	75–88	6.0–8.0
Palaeococcus[c]	Coccus	Organic	S^0	83	6.0
Pyrococcus	Coccus	Organic		100	7.0
Phylum *Nanoarchaeota*					
Order and family unassigned					
Genus *Nanoarchaeum*	Coccus	Obligate symbiont	ND	90	6

[a]Organic substrates include amino acids, sugars or yeast extract.

[b]ND, not determined.

[c]Placement in this family is tentative.

usual symbiont, the genome of *Nanoarchaeum* is among the smallest known for a prokaryote. It lacks most of the genes for amino acid, nucleoside, coenzyme, and lipid biosynthesis. It also lacks genes for the major catabolic pathways including glycolysis and the TCA cycle. Thus, it must be nearly completely dependent upon the host *Ignicoccus* for nutrients. *Ignicoccus* can be grown by itself, and it is not clear whether or not *Nanoarchaeum* is detrimental, since the coculture grows as well as the monoculture.

Based upon its unusual physiology and rRNA genes, *Nanoarchaeum* is classified within a novel phylum (see Table 18.5). However, its origins may be more recent than this suggests. Most of its protein encoding genes are closely related to those of another hyperthermophile *Thermococcus*. This euryarchaeote is an anaerobic heterotroph that respires elemental S^0. Thus, *Nanoarchaeum* may have evolved from an ancestor of *Thermococcus*, losing the ability to grow independently as it became an obligate symbiont. Its unusual rRNA and some other genes may have then evolved as adaptations to its very unusual life style.

Viruses

As is true for the *Bacteria* and *Eukarya*, viruses specific to *Archaea* are common. The viruses of *Sulfolobus* have been studied extensively, and four morphological types are recognized (Figure 18.18). These viruses are ubiquitous in acidic hot springs, environments where *Sulfolobus* is common. So far all the *Sulfolobus* viruses found are temperate, that is, they replicate and are shed from the host without killing it. This may be an adaptation to the extreme habitat of high temperature and low pH. Under these conditions, the viruses are quickly inactivated.

Perspectives and Applications

The extremely thermophilic archaea are especially interesting from at least four very different points of view. First, these microorganisms grow at the upper temperature limit for life. Although the upper limit is not known with certainty, its precise value is of importance because it sets one boundary on the biosphere. Therefore, earth scientists studying geothermally heated aquifers, hot springs, deep vents, and similar geological formations need to know this boundary to know where biological transformations can be significant. In addition, in the search for life on other planets, the upper temperature limit determines where life as we know it can exist. Currently, the upper temperature limit for life appears to be about 121°C, which is the temperature maximum for growth of strain 121. By studying the growth conditions and physiology of these organisms as well as isolating new species from geothermal environments, it may be possible to increase the known temperature limit.

(A)

(B)

(C)   (D)

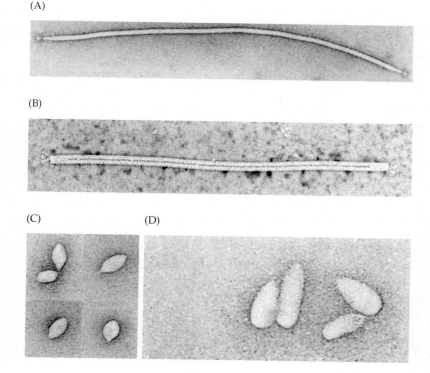

Figure 18.18 Viruses of the extremely thermophilic acidophile *Sulfolobus*
(A) Electron micrograph of Lipothrixvirus SIFV, a flexible virion with a linear DNA genome. (B) Rudivirus SIRV2 forms a stiff virion also with a linear DNA genome encased by plugs at the ends. (C) Fusellovirus SSV1, a spindle-shaped virion with a circular DNA genome. (D) Guttavirus SNDV, a droplet-shaped virion with a dense beard of long thin filaments. From D. Prangishvili, K. Stedman and W. Zillig. 2001. *Trends in Microbiology* 9: 39–43.

Thermoadaptation

A second area of interest is how these archaea cope with high temperatures. Most proteins and nucleic acids from mesophilic organisms denature at temperatures well below 100°C. At temperatures greater than 150°C, the macromolecules that form most of the cell rapidly hydrolyze. Although cells are known to use a variety of mechanisms for coping and even thriving at high temperatures, there does not seem to be any single adaptation that is universal among all (hyper)thermophiles.

One important mechanism of thermoadaptation is the biosynthesis of specialized proteins and enzymes that are both thermostable and maximally active at high temperatures. For example, many enzymes from *Pyrococcus* are most active at temperatures greater than 95°C and stable for several days at those temperatures. These enzymes are virtually inactive at room temperature. A variety of mechanisms contribute to the biochemical basis for this thermoadaptation. These include increases in the number of ionic and hydrophobic bonds that determine the tertiary structure, decreases in the interactions of the amino acids with water, and changes in the amino acid content.

However, not all biomolecules from extreme thermophiles are stable at high temperatures. Double-stranded DNA and the secondary structure of ribosomal RNA denature well below the temperature optimal for many hyperthermophiles. In some archaea, special DNA-binding proteins enhance the thermostability of doubled-stranded DNA in vitro. Similarly, the extremely thermophilic archaea possess an enzyme reverse gyrase that introduces positive supercoils into DNA. This enzyme is unique to thermophiles and may play a role in thermoadaptation. Hyperthermophiles also tend to possess very efficient DNA repair systems to compensate for the increased rates of DNA decomposition at high temperature. Ribosomal RNAs from thermophiles possess higher GC and lower U content than those found in mesophiles, have extensive modifications of their nucleotides, and have increased affinity for ribosomal proteins. These factors may all contribute to their thermostability. Lastly, many hyperthermophiles have high concentrations of intracellular salts to further increase the melting temperatures of nucleic acids.

Some small molecules are also thermolabile. NADH and NADPH are very unstable at temperatures above 90°C. During growth on sugars, *Pyrococcus* utilizes the pyroglycolytic pathway, where glucose is oxidized by a ferredoxin-dependent glucose oxidoreductase instead of the conventional NAD(P)$^+$-dependent glucose dehydrogenase. *Pyrococcus* also contains a ferredoxin-dependent aldehyde oxidoreductase instead of the conventional glyceraldehyde dehydrogenase. This ferredoxin-dependent pathway is less temperature sensitive than the NAD(P)$^+$-dependent pathway. A similar rationale may explain why other extreme thermophiles replace some ATP-dependent reactions with ADP-reactions. ADP is more thermostable than ATP. By minimizing the role of ATP, cells could be come less sensitive to high temperatures.

At high temperatures, the lipid bilayer of the cell becomes more flexible and permeable to protons. Hyperthermophilic archaea compensate by increasing the rigidity of the membranes. Two mechanisms are common. Lipid tetraethers are biosynthesized that span the membrane (see Figure 18.3). These tetraethers are formed by a tail-to-tail condensation of diethers to form a monolayer. Further increases in rigidity are achieved by internal cyclization of the isoprenoid side chains to form cyclopentane rings. Increases in proton permeability of the membrane with temperature affect the ability of cells to generate a proton motive force, which is necessary for coupling ATP synthesis with respiration, transport of solutes, and a variety of other bioenergetic processes. Hyperthermophiles may compensate by coupling these processes to Na$^+$ ions, which are less permeable.

Some extreme thermophiles also contain high concentrations of unusual intracellular anions, which may contribute to thermostability. Two compounds, called "thermoprotectants," which may have this role are di-inositol-1,1'-phosphate from *Pyrococcus* and cyclic 2,3-diphosphoglycerate (or cDPG) from some methanogens. However, thermoprotectants have not been identified in all extreme thermophiles, so this mechanism may not be general.

Industrial Applications

Because of their high temperature optimum and thermostability, enzymes from extreme thermophiles may have important commercial applications. Activity at high temperatures is important because most industrial processes operate between 50°C and 100°C. Enzymes with a temperature optimum in this range are cheaper to use because less protein is required for the same amount of activity. Thermostability is important because enzymes are expensive, and thermostable enzymes last longer. Types of enzymes being characterized from extreme thermophiles include esterases and lipases (which are widely used in organic syntheses), proteases and peptidases, saccharolytic enzymes, and alcohol dehydrogenases.

Thermostability is also correlated with greater stability at mesophilic temperatures and resistance to denaturing chemicals. One of the major uses of moderately thermophilic enzymes is alkaline proteases in laundry detergent. These enzymes have high activity at 50°C, the temperature of wash water. They are very stable at room temperature, which extends their shelf-life in the supermarket. They are also resistant to the other components of laundry detergent that can denature enzymes, like surfactants and metal chelators. Although proteases from extreme thermophiles have not yet been used in laundry detergents, many have been tested in this capacity.

Another example of extremely thermophilic enzymes currently used industrially is the DNA polymerase from *Pyrococcus* and *Thermococcus*. The biotechnology industry uses these enzymes for DNA sequencing and the polymerase chain reaction. In addition to their high activity and thermostability, these enzymes are especially attractive because they have a very low error rate. For instance, the popular enzyme Pfu was obtained from *Pyrococcus furiosus*. Because of its proofreading activity, its error rate is much lower than that of the common Taq enzyme, which was discovered in the thermophilic bacterium *Thermus aquaticus*.

Early Life

The extreme thermophiles represent very deep branches of the *Archaea*. Similarly, the extreme thermophiles, *Thermotoga* and *Aquifex*, represent the deepest branches of the *Bacteria*. Based in part upon these observations, it has been proposed that the earliest organisms were extreme thermophiles. If true, extant extreme thermophiles may have retained some ancient characteristics that were lost in the evolution of mesophiles. Thus, the unusual enzymes and pathways common in extreme thermophiles may be remnants of the earliest organisms. These microorganisms have been described as "living fossils" for this reason. Even if this hypothesis is not true, it is a refreshing perspective on the origin of life.

Uncultured and Nitrifying Archaea

Molecular methods have been used to clone 16S rRNA genes directly from the environment. These studies have discovered an enormous diversity not represented among the cultured organisms. In hot springs, one new phylum, the *Korarchaeota* is abundantly represented in rRNA clones (see Figure 18.6). Similarly, 16S rRNA genes with very low similarity to cultured archaea have been detected in samples from the deep ocean.

Of particular interest is a large group of uncultivated crenarchaeotes found in the ocean and soil (see Figure 18.6). In the oceans, these organisms are widespread, representing 1% to 2% of the total number of prokaryotes in the surface layer and up to 30% in the deeper water. Phylogenetically similar organisms are also present in many soils, where they typically represent 1% of the total number of prokaryotes. Because they are found in temperate environments, it is unlikely that they are extreme thermophiles like the cultured crenarchaeotes.

Although little is known about the physiology and growth properties of the temperate crenarchaeotes, the recent isolation of one representative of this group is providing new insights. Although *Nitrosopumilus* was isolated from a marine aquarium, the sequence of its 16S rRNA gene is very similar to many of the clones found in sea water. This organism is a nitrifier, and it grows by oxidizing ammonia using O_2 as an electron acceptor (Table 18.6). Like many of the nitrifying bacteria, it is an obligate autotroph and obtains all its carbon from CO_2 fixation. It is closely related to *Cenarchaeum symbiosum*, which was one of the first temperate crenarchaeotes characterized. Never grown in pure culture, this organism is a symbiont of the marine sponge *Axinella mexicana*, where it may represent up to 60% of the microorganisms present in the body cavity. It remains to be discovered if this and other uncultivated organisms in this group are also nitrifiers.

SECTION HIGHLIGHTS

Three major phenotypic groups, the methanogens, halophiles, and extreme thermophiles, represent most of the cultivated *Archaea*. Each of these groups is uniquely adapted and highly specialized, and they all play important roles in the environment. In addition, many phylogenetic groups of the *Archaea* have never been cultivated and their physiological properties are unknown.

TABLE 18.6	Summary of properties of the nitrifying *Crenarchaeota*			
	Morphology	Electron Donor	Electron Acceptor	Autotroph

Order   *Nitrosopumilales*
  Family   *Nitrosopumilaceae*
    Genus   *Nitrosopumilus*[a]
	Thin rod	NH_3	O_2	yes

[a]Taxa are currently tentative.

SUMMARY

- The *Archaea* are a diverse group of prokaryotes more closely related to the *Eukarya* than to the familiar *Bacteria*.

- Unusual biochemical features of the *Archaea* are the presence of isoprenoid-based glycerol lipids, protein cell walls, a eukaryotic type of RNA polymerase and DNA replication system, and insensitivity to many common antibiotics.

- The major physiological groups of the cultivated archaea are the methanogens, the extreme halophiles, and the extreme thermophiles.

- The *Archaea* are widely distributed in aerobic and anaerobic, mesophilic and thermophilic, and freshwater and extremely halophilic environments.

- Methanogenesis is a form of anaerobic respiration and requires a complex biochemical pathway that contains unusual coenzymes: methanofuran, methanopterin, coenzyme M, coenzyme F_{420}, coenzyme F_{430}, and 7-mercaptoheptanoylthreonine phosphate.

- Methanogens catalyze the last step in the anaerobic food chain in which organic matter is degraded to CH_4 and CO_2. This food chain is the basis of many symbiotic associations of methanogens with eukaryotes and bacteria.

- The extreme halophiles among the *Archaea* contain bacteriorhodopsin, a light-driven proton pump that contains retinal. This type of photophosphorylation differs fundamentally from chlorophyll-based photophosphorylation, common in plants and bacteria, because it is not coupled to an electron transport chain.

- The enzymes from the extreme halophiles are unusually salt tolerant and many also require salt for activity and stability.

- Organisms with growth optimum temperatures greater than 80°C are called extreme thermophiles. Organisms that grow at temperatures greater than 90°C are called hyperthermophiles. Most extreme and hyperthermophiles are archaea. These organisms set the upper temperature limit for life.

- The extreme thermophiles are common in geothermal habitats where H_2 and reduced sulfur compounds are abundant.

- The extreme thermophiles are a rich source of thermostable enzymes, which may have valuable applications in biotechnology.

 Find more at www.sinauer.com/microbial-life

REVIEW QUESTIONS

1. What was the basis for establishing the *Archaea* as a domain? On what is the three-domain concept of life based?

2. Give the major reasons for studying *Archaea*. What can be learned from studying the *Archaea* that cannot be learned from studying the *Bacteria*?

3. What are some distinguishing features of *Archaea*?

4. Where would one find the methanogens in nature?

5. In what ways might one consider the methane-producing archaea specialists?

6. Methanogens possess some unusual coenzymes: To what common bacterial coenzymes are the methanogen coenzymes related and what are their functions? Give examples of coenzymes that are unique to methanogens.

7. How does the presence of sulfate or nitrate affect methanogenesis? For respiration, why is CO_2 considered the electron acceptor of last resort?

8. Are the haloarchaea autotrophs or heterotrophs? Give reasons for your answer.

9. What is bacteriorhodopsin? How does photosynthesis with this compound differ from chlorophyll-based photosynthesis?

10. Define an extreme thermophile. What habitats do they occupy?

11. What can we learn from the hyperthermophiles? How might they contribute to industry?

12. Many of the *Archaea* have adapted to growth under extreme conditions of high salt or high temperature. What adaptations are necessary for growth under these conditions?

SUGGESTED READING

Balows, A., H. G. Trüper, M. Dworkin, W. Harder and K. H. Schleifer. 1992. *The Prokaryotes*. 2nd ed. New York: Springer-Verlag.

Dworkin, M., ed. 2001. *The Prokaryotes: An Evolving Electronic Resource for the Microbiological Community*. New York: Springer-Verlag.

Ferry, J. G., ed. 1993. *Methanogenesis: Ecology, Physiology, Biochemistry and Genetics*. New York: Chapman & Hall.

Huber, H., M. J. Hohn, R. Rachel, T. Fuchs, V. C. Wimmer and K. O. Stetter. 2002. "A new phylum of *Archaea* represented by a nanosized hyperthermophilic symbiont." *Nature* 417: 63–67.

Könneke, M., A. E. Bernhard, J. R. de la Torre, C. B. Walker, J. B. Waterbury and D. A. Stahl. 2005. "Isolation of an autotrophic ammonia-oxidizing marine archaeon." *Nature* 437: 543–546.

Woese, C. R. and R. S. Wolfe. 1985. *Archaebacteria*. New York: Academic Press, Inc.

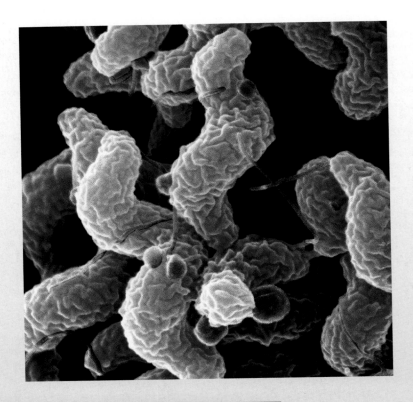

The objectives of this chapter are to:

◆ Introduce the five different classes of *Proteobacteria*.

◆ Describe important genera and species of each of the classes and orders.

◆ Discuss the diverse array of heterotrophic species that range from plant and animal pathogens to those that play important roles in biogeochemical cycles.

19

Nonphotosynthetic Proteobacteria

*The extraordinary diversity . . . among the bacteria is shown by
a simple enumeration of the properties which they can possess.
They can be unicellular, multicellular, or coenocytic; permanently immotile, or motile by any one of three distinct mechanisms;
able to reproduce by binary fission, by budding, or by the formation of special reproductive cells, such as the conidia of actinomycetes; photosynthetic or nonphotosynthetic.*
—R. Y. Stanier, *in* The Bacteria, *1964*

The ***Proteobacteria***, also called the "purple bacteria" because
of its photosynthetic species, makes up the largest and most
heterogeneous phylogenetic groups of the ***Bacteria***. There are
more than 450 genera in this phylum of gram-negative bacteria. This
phylum was appropriately named after the Greek god Proteus, who
could change his shape. Encompassed within this diverse phylum are
photosynthetic, chemoautotrophic, and heterotrophic bacteria, indicative of a complex, extensive evolutionary history. Most species
are unicellular rods, vibrios, and cocci that divide by binary transverse fission. However, some have life cycles with two or more types
of cells that divide by budding, and one group, the myxobacteria,
have a complex developmental cycle culminating in the formation of
an elaborate fruiting structure. Many are flagellated, but others move
by gliding motility or gas vacuoles. Some produce stalks. Some are
obligate aerobes, whereas others are facultative or obligate anaerobes.
Some are well known pathogens including *Vibrio cholerae*, *Brucella
abortus*, and *Salmonella typhi*.

The current evolutionary chart, based on 16S rRNA analyses, indicates that there are five major subgroups of the phylum *Proteobacteria*,
designated as the *Alphaproteobacteria*, *Betaproteobacteria*, *Gammaproteobacteria*, *Deltaproteobacteria*, and *Epsilonproteobacteria* classes (Table
19.1). Many species of common, well-known heterotrophic bacteria
are included in these groups. Please note that all of the photosynthetic

| TABLE 19.1 | Classes, Orders and Characteristics of the Phylum *Proteobacteria* |

Class	Representative Orders	Characteristics of Selected Members of the Order
Alphaproteobacteria		
	Rhizobiales	Plant symbionts, pathogens; prosthecate, and nitrogen-fixing bacteria
	Rhodospirillales	Diverse array of phototrophic, acetic acid–producing, magnetotactic, and nitrogen-fixing bacteria
	Rickettsiales	Pathogenic and symbiotic species
	Rhodobacteriales	Some phototrophs; some heterotrophic prosthecate bacteria
	Sphingomonadales	Common heterotrophs
	Caulobacterales	Prosthecate bacteria
Betaproteobacteria		
	Burkholderiales	Sheathed and other heterotrophic bacteria
	Hydrogenophilales	Unicellular hydrogen- and sulfur-oxidizing bacteria
	Methylophilales	Methane and methanol oxidizers
	Neisseriales	Commensal and pathogenic bacteria
	Nitrosomonadales	Chemolithotrophic and heterotrophic bacteria
	Rhodocyclales	Phototrophic and propionic acid–producing species
Gammaproteobacteria		
	Chromatiales	Phototrophic and nitrifying bacteria
	Xanthomonadales	Plant symbionts and pathogens
	Thiotrichales	Sulfur-oxidizing filamentous species
	Legionellales	Symbiotic and pathogenic species
	Methylococcales	Methane- and methanol-oxidizing bacteria
	Oceanospirales	Marine genera
	Pseudomonadales	Common heterotrophic bacteria
	Alteromonadales	Diverse marine genera
	Vibrionales	Important animal symbionts and pathogens
	Aeromonadales	Anaerobic succinic acid producers
	Enterobacteriales	Large, diverse and important order containing many well-known genera
	Pasteurellales	Animal pathogens
Deltaproteobacteria		
	Desulfovibrionales	Important sulfate-reducing bacteria
	Desulfomonadales	Sulfur and iron bacteria
	Bdellovibrionales	Bacterial parasites of other bacteria
	Myxococcales	Bacteria that form fruiting structures
Epsilonproteobacteria		
	Campylobacterales	Animal symbionts and pathogens

members of the *Proteobacteria* will be treated in Chapter 21. In this chapter, the major representatives of the heterotrophic and chemolithotrophic *Proteobacteria* are discussed.

19.1 Class *Alphaproteobacteria*

This class contains a diverse array of taxa including phototrophs, nitrogen fixers, methane oxidizers, acetic acid bacteria, and prosthecate bacteria (Table 19.2). The phylogeny is shown in Figure 19.1. Important genera are discussed in this section. For a complete listing of the genera, please see the Appendix.

Order Rhizobiales

This order contains many nitrogen-fixing species, important plant and animal symbionts and pathogens, many prosthecate bacteria, and some of the methane-

TABLE 19.2	Characteristics of Genera of the Class *Alphaproteobacteria*[a]	
Orders	**Representative Genera**	**Characteristics**
Rhizobiales	*Beijerinckia*	Free-living nitrogen fixer
	Derxia	Free-living nitrogen fixer
	Rhizobium	Nitrogen-fixing symbiont
	Agrobacterium	Plant pathogens
	Bartonella	Animal pathogens
	Brucella	Animal pathogen
	Methylocystis	Methane oxidizers
	Methylosinus	Methanol oxidizers
	Ancylobacter	Ring-forming shape
	Bradyrhizobium	Nitrogen-fixing symbiont
	Nitrobacter	Nitrifiers
	Hyphomicrobium	Prosthecae, buds
	Ancalomicrobium	Prosthecae, buds
	Prosthecomicrobium	Prosthecae, buds
	Pedomicrobium	Prosthecae, buds
Rhodospirillales	*Rhodospirillum*	Photoheterotroph
	Azospirillum	Nitrogen-fixing grain symbiont
	Azomonas	Free-living soil nitrogen fixer
	Magnetospirillum	Magnetotaxis
	Acetobacter	Vinegar producer
	Gluconobacter	Acetic acid producer
	Stella	Prosthecae
Rickettsiales	*Rickettsia*	Animal pathogen
	Wolbachia	Insect parasite
Rhodobacteriales	*Rhodobacter*	Photoheterotroph
	Hyphomonas	Prosthecae, buds
Sphingomonadales	*Sphingomonas*	Important heterotroph
	Zymomonas	Ethanol producer
Caulobacterales	*Caulobacter*	Prosthecate bacteria
	Asticcacaulis	Prosthecate bacteria

[a] A complete listing of genera is provided in the Appendix.

oxidizing bacteria. We will begin with the nitrogen-fixing bacteria.

NITROGEN-FIXING *ALPHAPROTEOBACTERIA* The ability to utilize atmospheric N_2 as a sole source of cellular nitrogen is widespread among prokaryotic genera. Interestingly, the genes for this process are scattered among phylogenetic groups like many other genes (Box 19.1). For example, virtually all anoxygenic and some oxygenic phototrophic bacteria are able to fix nitrogen. Some of the *Archaea* are also nitrogen fixers. Some but not all members of the *Alpha-*, *Beta-* and *Gammaproteobacteria* are nitrogen fixers. How can the seemingly helter-skelter dis-

tribution of this activity be explained? It appears that it may be a combination of vertical inheritance and loss of characteristics as well as horizontal gene transfer (see Box 19.1). Because many fixed forms of nitrogen (e.g., nitrate and ammonium) become limiting to growth in natural environments, such as the ocean, soils, and freshwater habitats, those organisms that have the capacity to carry out this energetically expensive process have an advantage (see Chapter 25).

FREE-LIVING NITROGEN-FIXING BACTERIA Several otherwise unrelated aerobic gram-negative nitrogen-fixing bacteria are characterized in Table 19.3. Although

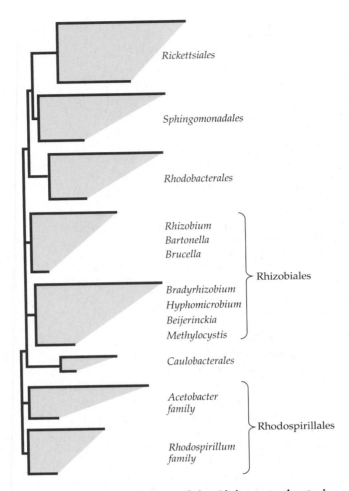

Figure 19.1 16S rRNA Tree of the *Alphaproteobacteria*
This tree shows the various orders and genera within the *Alphaproteobacteria*. The upper and lower boundaries of the four-sided blocks show the divergence within each of the designated phylogenetic orders with some genera noted. Adapted by permission from *Bergey's Manual of Systematic Bacteriology*.

these organisms are genetically distinct, they share a number of characteristics and are thus considered here as a group. All are free-living bacteria found in soil and waters. Most strains are motile. These aerobic organisms have an oxygen-sensitive, nitrogen-fixing nitrogenase system (see Chapter 10). Under favorable conditions, they fix about 10 mg of nitrogen per gram of carbohydrate consumed. Because nitrogenase is sensitive to oxygen, it must be protected from excessive O_2 in order for the fixation process to be induced and function. All of the organisms can utilize ammonia as their nitrogen source, and nitrogenase is induced as a response to the depletion of ammonia and other forms of "fixed" nitrogen.

Beijerinckia is a genus of nitrogen fixers found in acidic soils, particularly of tropical regions. *Beijerinckia* species are straight or curved rods with rounded ends. The cells are often misshapen with polar lipid bodies surrounded by a membrane. The lipid bodies are composed of poly-β-hydroxybutyric acid. Nitrogen-fixing colonies produce copious quantities of a tenacious elastic slime. Fixation is enhanced by reduced O_2 tension. Cysts containing one cell may occur in *Beijerinckia*; however, during growth on a nitrogen-free medium, several individual cells will be enclosed in a common capsule. *Beijerinckia* species can use glucose, fructose, or sucrose as the sole source of carbon for growth.

Derxia species are also important nitrogen fixers. They can grow on sugars, alcohols, and organic acids and are differentiated from *Beijerinckia* by their ability to grow as hydrogen-utilizing chemoautotrophs. *Xanthobacter* species can also grow as facultative hydrogen-utilizing chemoautotrophs. They use a variety of carbon sources for growth including butanol, propanol, and tricarboxylic acid (TCA) cycle intermediates.

TABLE 19.3	Free-living nitrogen-fixing bacteria				
Class	**Genus**	**Form Cysts or Resting Bodies**	**Autotrophic Growth**	**Mol % G + C**	**Habitat**
Alphaproteobacteria					
	Beijerinckia	+	−	55–61	Tropical soil; acid pH
	Azospirillum	+	±	69–71	Free-living in soil or with roots
	Derxia	−	+	69–73	Tropical soil
	Xanthobacter	−	+	65–70	Wet soil and water
Gammaproteobacteria					
	Azotobacter	+	−	63–67	Rich soil, high phosphate, slight acid to alkaline pH
	Azomonas	−	−	52–59	Soil and water

BOX 19.1 *Research Highlights*

"Odd Bedfellows" of Bacterial Phylogeny

The Phylum *Proteobacteria* provides many of the most vivid examples of what is best interpreted as past occurrences of **horizontal gene transfer** (HGT) (see Chapters 16 and 17). For example, members of the nitrifying, nitrogen-fixing, methane-oxidizing, and prosthecate bacteria are found scattered about among phylogenetically disparate genera and classes, yet they contain similar morphological features and enzymatic pathways. Furthermore, the sequences of some of their genes display strong homologies, indicating they have likely shared a common ancestry. So, why is the distribution of these genes so spotty and seemingly helter-skelter? Part of the explanation may be HGT and another part could be vertical inheritance accompanied by loss of the genes in some taxa through deletion events.

To illustrate examples of this within the phylum consider the nitrifying bacteria. Ammonia oxidizing nitrifying genera of the phylum *Proteobacteria* are found in both the *Betaproteobacteria* (*Nitrosomonas* and *Nitrosospira*) as well as in the *Gammaproteobacteria* (*Nitrosococcus*). Furthermore, *Nitrosococcus* is found in a family, *Chromatiaceae* that contains more than 20 genera, none of the others of which are nitrifiers. Likewise, one of the nitrite oxidizing genera, *Nitrobacter*, is a member of the

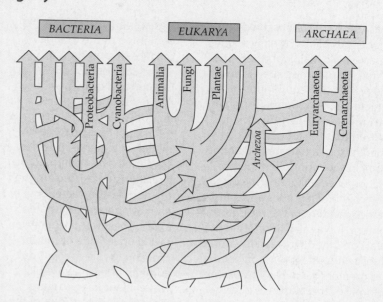

Figure shows a 16S rRNA "tree of life" that illustrates how phylogenetic groups can be difficult to resolve from one another if horizontal gene transfer is rampant. The result is not a tree but a web. Adapted from W. Ford Doolittle. 1999. "Phylogenetic Classification and the Universal Tree." *Science* 284: 2124–2128.

Betaproteobacteria, whereas another, *Nitrococcus*, is a member of the *Gammaproteobacteria*. One could argue that these similar metabolisms evolved independently in these two different groups. But, a general rule that applies to most distinct and complex enzymatic systems in which some enzymes are highly conserved is that it is less likely that a separate evolution has occurred than a sharing of the genes through HGT. This view has been supported by genomic analyses for most complex processes, such as nitrogen fixation. For this reason, some scientists have argued that while it is possible to construct a valid phylogeny of highly conserved genes, it is not possible to construct a valid phylogeny of the organisms. However, most scientists believe that there are a sufficient number of core genes within a taxon and that it is possible to use these for constructing the phylogeny of the organisms. Because scientists now have full genomes to study, this dilemma is likely to be resolved in the foreseeable future.

SYMBIOTIC NITROGEN-FIXING BACTERIA Rhizobial species may be free living or they may develop a symbiotic nitrogen-fixing association with leguminous plants (Box 19.2). Legumes are plants that bear seeds in pods and include soybeans, clover, alfalfa, peas, vetch, lupines, beans, and peanuts. The bacteria penetrate the roots or, in some cases, the stems of the plant and form a nodule. The bacteria reside in this nodule and, under appropriate conditions, fix atmospheric nitrogen, thereby providing a source of nitrogen for plants growing in soil deficient in

BOX 19.2 *Milestones*

History of Symbiotic Nitrogen Fixation

Root nodules were observed on leguminous plants by early scientists. Marcello Malpighi (1628–1694), an Italian anatomist, drew elaborate pictures of leguminous plants depicting root nodules. During the early nineteenth century, scientists realized that the amount of combined nitrogen (nitrate and ammonium) in soils controlled the production of cereal grains. In contrast, the growth of legumes appeared to be independent of the nitrogen content. It was also noted that the total combined nitrogen measured in soils after the growth of legumes was greater than what was present before. For this reason, it became common practice to grow legumes in nitrogen-poor soil and plow the resultant crop back into the soil. The soil was then fit for growth of cereal grains. This alternation of legumes and grains was the beginning of the routine practice of crop rotation.

In 1888, Martinus Beijerinck isolated and cultivated bacteria from root nodules. He demonstrated that these bacteria could cause nodule formation in legumes grown in sterile soil. This confirmed that a relationship existed among the nodule, the bacterium, and nitrogen replenishment in soil. The potential utility in taking advantage of this for growing food crops without addition of nitrogen fertilizer was evident in agriculture. Since that time, studies of the symbiosis between legumes and rhizobia has become an important and exciting research area.

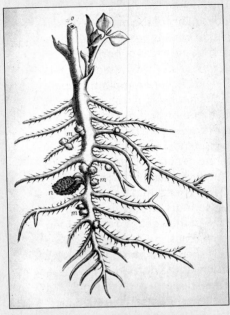

Nodules on the root of *Vicia fabia*, from a woodcut by Marcello Malpighi.

this nutrient. The symbiotic association of these bacteria and legumes is discussed more fully in Chapter 25.

Some of the characteristics of rhizobia are listed in Table 19.4. The genera *Rhizobium, Sinorhizobium,* and *Mesorhizobium* are generally involved in nodule formation in plants in the temperate zone. They are faster growing (with a generation time of 4 hours) and produce colonies 2 to 4 mm in diameter within 3 to 5 days on a yeast-mannitol-mineral salts agar medium. *Bradyrhizo-*

bium grows slowly with a generation time of 8 hours. *Bradyrhizobium* is especially effective in nodulation of tropical leguminous plants, although it can nodulate some temperate zone plants. *Bradyrhizobium* produces colonies on the yeast-mannitol medium that do not exceed 1 mm in diameter in 5 to 7 days.

Importantly, a plant legume species can only be successfully nodulated by a particular rhizobial species or strain. The leguminous plants infected by a particular strain are considered to be a cross-inoculation group (Table 19.5). The three major described species of rhizobia are *Rhizobium leguminosarum, Sinorhizobium meliloti,* and *Mesorhizobium loti.* The three major strains of *R. leguminosarum* are *trifolii, phaseoli,* and *viceae.*

AGROBACTERIUM: PLANT PATHOGENS   Members of the genus *Agrobacterium* naturally induce tumors in dicotyledonous (flowering plants in which the seed contains two halves, e.g., peas), but not monocotyledonous (flowering plants such as grains in which the seed is a single unit) plants. The disease is commonly called **crown gall**, as it frequently occurs at the soil-stem interface, the crown of the plant (Figure 19.2). Other varieties

TABLE 19.4	Characteristics of free-living rhizobial species

Gram-negative rods (0.5–0.9 ×1.2–3.0 μm)
Aerobic
Motile
Produce copious amounts of extracellular slime
Mol % G + C = 59–64
Metabolize glucose via Entner-Duodoroff pathway
Fix N_2 under microaerophilic conditions
Some species can grow as H_2 chemoautotrophs

Figure 19.2 Crown galls
Photograph of crown galls on a plant. This plant was artificially wounded by cutting with a knife, then inoculated with *Agrobacterium tumefaciens*. Courtesy of E. W. Nester.

TABLE 19.5	Cross-inoculation groups nodulated by rhizobial species[a]	
Bacterial Species	**Plant Genus**	**Plant Common Name**
Rhizobium leguminosarum		
	Pisum	Field pea
	Lathyrus	Pea
	Vicia	Vetch
	Lens	Lentil
	Phaseolus	Bean
	Trifolium	Clover
Sinorhizobium meliloti		
	Melilotus	Sweet clover
	Medicago	Alfalfa
Mesorhizobium loti		
	Lotus	Trefoil
	Lupinus	Lupines
	Mimosa	Mimosa
Bradyrhizobium japonicum		
	Glycine	Soybean
	Arachis	Peanut

[a]Species listed normally cause root nodules on some, but not necessarily all, genera of legumes listed.

of the disease are hairy root and cane gall. The major species involved in gall formation is *Agrobacterium tumefaciens*, whereas *Agrobacterium rhizogenes* causes hairy root disease.

The genus *Agrobacterium* is closely related to the rhizobia; indeed, the primary differences are due to the plasmids that are found in *Agrobacterium* spp. *Agrobacterium* are gram-negative rods (0.6–1.0 μm × 1.5–3 μm) that are motile by peritrichous flagellation. They are aerobes with a mol % G + C of 57 to 63. Agrobacteria utilize a variety of carbohydrates, organic acids, and amino acids as carbon sources. Some species use ammonia or nitrate as nitrogen sources, whereas others require amino acids and other growth factors. *Agrobacterium* species are common in soil and are often abundant in the rhizosphere, or root zone, of plants. In the past, *A. tumefaciens* received considerable attention as a model of tumor induction with possible implications for tumor formation in humans. Now it is of commercial importance as a mechanism for genetic engineering of plants.

A. tumefaciens generally attacks plants at a wound site at the root stem interface. It is believed that the bacterium attaches at the wound site by fimbriae that ad-

here to the plant cells. The entire bacterium does not enter the plant. Instead, all oncogenic (tumor-producing) strains have a large conjugative plasmid, designated Ti for tumor inducing, that is released into the site of infection. This plasmid carries several important genes:

1. Virulence genes that induce tumor formation

2. Genes for substances that regulate the production of plant growth hormones, auxins, and cytokinins

3. Genes that direct the plant to synthesize **opines** (Figure 19.3), special amino acids that serve as specific substrates for *Agrobacterium* species in their soil environment

4. Genes for enzymes that degrade opines

Only a specific part of the Ti plasmid enters the plant cell. This portion, designated T-DNA (where T stands for transforming) is a series of genes residing in the Ti plasmid between two 23-bp direct repeat sequences. The T-DNA carries the information listed previously, is inserted into the plant chromosome at various sites, and is maintained in the plant cell nucleus. Once a plant cell is transformed, its progeny continue to exhibit the tumor-producing characteristic in the absence of bacteria. There is considerable evidence that T-DNA gene expression is controlled by plant regulatory elements; that is, the plant actually assists in the disease process.

Octopine

$$NH_2 - \overset{\overset{\displaystyle NH}{\|}}{C} - NH - (CH_2)_3 - \underset{\underset{\displaystyle CH_3 - CH - COO^-}{\overset{\displaystyle |}{NH}}}{CH} - COO^-$$

Nopaline

$$NH_2 - \overset{\overset{\displaystyle NH}{\|}}{C} - NH - (CH_2)_3 - \underset{\underset{\displaystyle {}^-OOC - (CH_2)_2CH - COO^-}{\overset{\displaystyle |}{NH}}}{CH} - COO^-$$

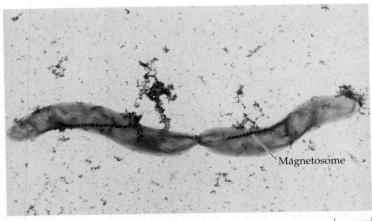

Magnetosome

0.5 μm

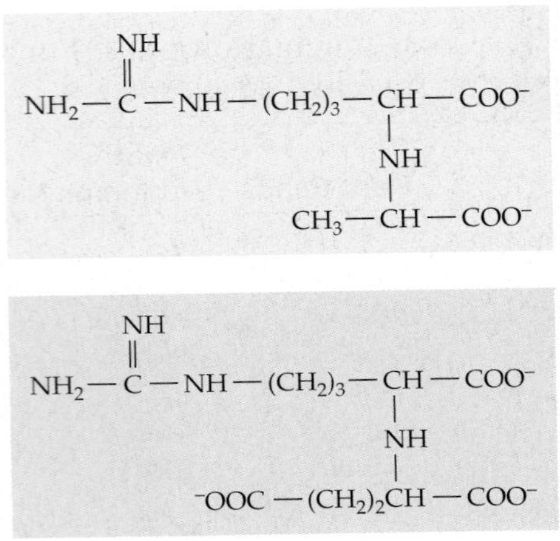

Figure 19.3 Opines
Chemical structures of two unusual amino acids, octopine and nopaline, opines produced by plants infected with *Agrobacterium tumefaciens*. The genes for producing these opines are transferred from the Ti plasmid to the plant genome, the plant excretes the opines, and they are degraded by the same *Agrobacterium* strain, living in the soil. This suggests one way in which *Agrobacterium* benefits from the crown gall association.

Figure 19.4 *Aquaspirillum magnetotacticum*
Aquaspirillum magnetotacticum cell, showing its linear magnetosome and magnetite particles. Courtesy of M. Sarikaya.

The Ti plasmid is an excellent vector for introducing genetic information into a plant. For example, genes can be introduced into tobacco and other plants by genetic engineering (see Box 16.2). Some traits that are beneficial to establish in crop plants include herbicide resistance, resistance to plant pathogens, resistance to adverse conditions (drought, heat, and salinity), and improved nutritional features. Although dicotyledonous plants are normally infected by *Agrobacterium*, researchers have introduced the Ti plasmid into monocotyledonous plants such as corn, wheat, rice, and other cereal grains by artificial means. As a result, these plants, which are staple grains for humans, are also being successfully engineered genetically.

SPIRILLA AND *MAGNETOSPIRILLUM*   The spirillum is one of the more common bacterial cell shapes. Like the rod and coccus, different phylogenetic groups contain spirilla. Many of the heterotrophic spirilla are members of the *Proteobacteria*, but spirilla are also found in other phyla, too, such as the *Cyanobacteria* (genus *Spirulina*) and even in the *Archaea* (methanogenic genus *Methanospirillum*).

Previously in this chapter we discussed *Azospirillum*, a nitrogen-fixing bacterium associated with monocotyledonous plants. A phototrophic genus in this class, *Rhodospirillum* will be discussed in Chapter 21. Another spirillum genus of the *Alphaproteobacteria* is the unusual magnetite-containing *Magnetospirillum*.

Magnetospirillum magnetotacticum is one of the most fascinating bacteria known. Each cell contains small crystals of **magnetite** (Fe_3O_4) that are aligned along the long axis of the cell (Figure 19.4). The cells actually orient themselves and move in a magnetic gradient. Therefore, if a magnet is placed against a test tube containing the cells, they will move toward and accumulate where the magnet is located. The magnetite crystals are located inside a membrane called the magnetosome. It is unclear whether this membrane is in direct contact with the cell membrane.

These bacteria, which are microaerophilic, live in the sediment of freshwater and marine habitats poised at a location where oxygen is available, albeit in low concentration. The magnetic forces of the Earth provide a direction for the cell to move down and toward the sediments where the concentration of oxygen is low and iron is readily available to the cell. This type of movement is termed **magnetotaxis** (see Chapter 4). The magnets allow the cells to orient themselves toward either the North or South Pole. It should be noted that the magnetic force of Earth is not sufficient to "pull" the organism, but only to orient the cell. The flagella provide the force that permits cell movement.

Materials engineers and scientists are very interested in knowing how bacteria can make such small and perfect magnets. For this reason, understanding the molecular biology of this process is an exciting area of research.

It is interesting to note that many other species of bacteria produce magnetite, but very few others have been isolated in pure culture. Some of them, such as *Magnetospirillum*, are members of the *Proteobacteria*, but recent studies indicate that some magnetotactic bacteria such as *Magnetobacterium bavaricum* are found in the phylum *Nitrospira* (see Chapter 22).

Ancylobacter is unusual in that it forms ring shapes (see Figure 4.38). Most strains are gas-vacuolate. It is quite versatile physiologically. In addition to growing as an ordinary heterotroph using sugars and amino acids as carbon sources, it is able to degrade methanol and also grows as a hydrogen autotroph. This is a common genus found in freshwater environments.

Brucella abortus is an important animal pathogen that causes abortion in cattle. Humans can be a secondary host causing a disease referred to as **brucellosis**, undulant fever, or Bang's disease after Frederick Bang.

Order Rhodospirillales

This order contains phototrophs, vinegar bacteria, and some nitrogen fixing and magnetotactic bacteria. We will begin with the vinegar bacteria.

VINEGAR BACTERIA: *ACETOBACTER* AND *GLUCONOBACTER*   Vinegar has been in use as a preservative and flavoring agent throughout human history. As early as 6000 B.C., the Babylonians depicted methods for converting ethanol from beer to vinegar (acetic acid). Two genera of bacteria, *Acetobacter* and *Gluconobacter*, are the major microbes responsible for vinegar production (Table 19.6). Both of these genera are natural inhabitants of flowers, fruit, sake, grape wine, brewers yeast, honey, and cider. *Gluconobacter* is found on the beech wood shavings of vinegar generators (see Chapter 31). *Gluconobacter* and *Acetobacter* can be differentiated from other gram-negative genera by their ability to oxidize ethanol to acetic acid at low pH (4.0 to 5.0). Both genera also cause pink disease of pineapple and the rotting of apples and pears. They are pests in the brewing and wine industry because they convert ethanol to acetic acid. Besides acetification, they cause ropiness, turbidity, and off flavors in beverage products, including soft drinks.

Gluconobacter cells are also ellipsoidal to rod shaped and often form enlarged, irregular forms. Some strains are motile by polar flagella. Extensive analysis of about 100 strains indicates there is only one species in this genus, designated *Gluconobacter oxydans*. *Gluconobacter* is a strict aerobe, but it lacks a complete TCA cycle and is therefore called an **underoxidizer**. As a result, it produces acetic acid as a terminal product. It is **ketogenic**, meaning that it oxidizes glucose to 1-ketogluconate, and some strains produce 5-ketogluconic acid. *G. oxydans* also can produce intermediates such as dihydroxyacetone from glycerol.

Acetobacter cells are also ellipsoidal to rod-shaped gram-negative organisms, but they are motile by peritrichous flagellation. These are obligately aerobic bacteria that prefer an acid pH for growth. *Acetobacter* is called an **overoxidizer** because it can oxidize acetic acid all the way to carbon dioxide via its complete TCA cycle. *Acetobacter* is also ketogenic. It converts sorbitol to 2-keto sorbitol and glucose to 2-keto- and 5-ketogluconic acids. For this reason, this genus is used commercially for the production of ketonic acids such as ascorbic acid (vitamin C).

Some strains of *Acetobacter* form extracellular cellulose microfibrils that surround the dividing cell mass. Therefore, the cells become embedded in a large mass of cellulose microfibrils. The cellulose is formed internally and secreted through pores in the lipopolysaccharide outer envelope. In static culture, cellulose-synthesizing cells are favored, and in shake culture, cellulose-free mutants predominate. Cellulose production is rare among bacteria, and its purpose is unknown. It has been postulated that cellulose production aids in maintaining the organisms on the surface of liquids so they have oxygen available for growth. However, many aerobic organisms survive in equivalent environments without such an elaborate mechanism for flotation.

AZOSPIRILLUM: A NITROGEN-FIXING GENUS   This is a unique genus in that it is a nitrogen fixer that grows in association with the roots of grains such as corn. *Azospirillum* species are plump curved rods that grow either as free-living bacteria in soils or may be associated with the roots of monocotyledonous plants including cereal grains, grasses, and tuberous plants. They do not produce root nodules. *Azospirillum* species form distinct capsules around the cell that provide some resistance to desiccation. Sugars and organic acids such as malate are used as substrates; some strains are also hydrogen chemoautotrophs. Nitrogen is fixed only under microaerophilic conditions.

TABLE 19.6	Differences between *Acetobacter* and *Gluconobacter*	
Characteristic	***Acetobacter***	***Gluconobacter***
Cellulose production	+ or −	−
Complete tricarboxylic acid cycle	+	−
Flagellation	Peritrichous	Polar
Mol % G + C	56–60	56–64

Order Rickettsiales

Rickettsia species are unable to grow outside a living host cell. Instead, they live as intracellular parasites of animals and are therefore called **obligate intracellular parasites**. *Rickettsia* species are parasitic in humans and other vertebrates. Members of this genus are the causative agents of typhus (*Rickettsia prowazeki*), Rocky Mountain spotted fever (*Rickettsia rickettsii*), and scrub typhus (*Rickettsia tsutsugamushi*). *Rickettsia* species are often associated with arthropods, usually intracellularly. Arthropods such as ticks, lice, and fleas are vectors that introduce the *Rickettsia* species into humans and other mammals through a bite (the medical aspects are considered in more detail in Chapter 28).

Rickettsia species are short rods (0.3–0.5 × 0.8–2.0 µm) that have not yet been cultivated from host tissues (Figure 19.5). They multiply readily in embryonated eggs or metazoan cell culture, provided that the host cells are viable. They have a generation time in vivo of 8 to 9 hours. Their cell walls are typical of gram-negative bacteria, with muramic acid and diaminopimelic acid in their peptidoglycan.

The rickettsias derive energy by oxidation of glutamic acid via the TCA cycle. The generation of NADH is coupled to an electron transport system for adenosine triphosphate (ATP) synthesis. They require a source of adenosine monophosphate (AMP) and other nucleoside monophosphates. *Rickettsia* species have limited synthetic capacity but can synthesize low-molecular-weight proteins and lipids. Monomers such as amino acids must be provided by the host cell.

R. prowazeki multiplies to a high density in the cytoplasm of a host, causing disruption of host cells. The released bacteria move on and infect other cells. *R. rickettsii* produces small numbers of cells in the cytoplasm of infected cells that escape to extracellular spaces. These then go on to infect other host cells.

Order Caulobacterales

The namesake of this order is the genus *Caulobacter*, which are prosthecate bacteria. Prosthecate bacteria are unicellular bacteria that have appendages extending from their cells that give them remarkable and distinctive cell shapes (Figure 19.6A). The appendages are

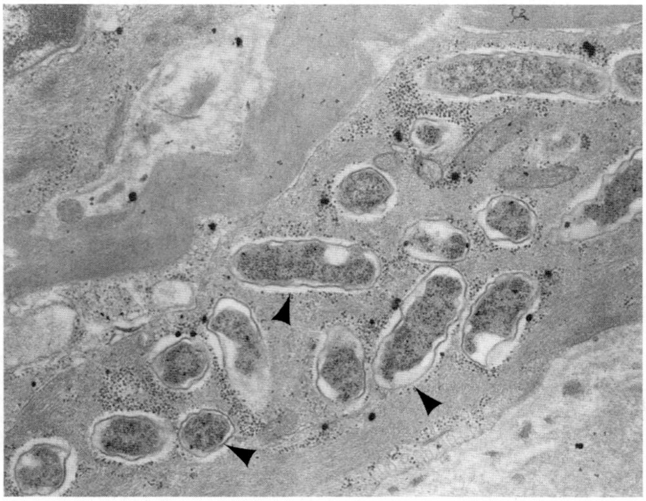

(A)

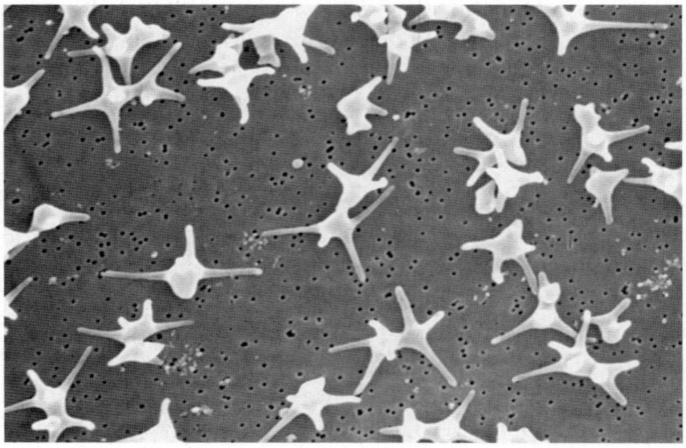

5 µm

(B)

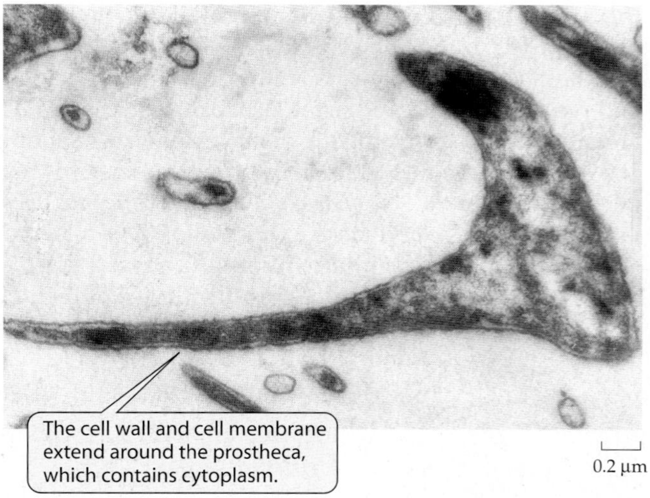

The cell wall and cell membrane extend around the prostheca, which contains cytoplasm.

0.2 µm

Figure 19.5 *Rickettsia*
Electron micrograph showing *Rickettsia* cells (arrowheads) infecting animal cells. Photo ©Science VU/Visuals Unlimited.

1 µm

Figure 19.6 *Ancalomicrobium*
(A) Scanning electron micrograph of *Ancalomicrobium adetum*. Each cell has several long prosthecae extending from it. (B) Thin section through *A. adetum* cell and prostheca. Courtesy of J. T. Staley.

TABLE 19.7	Groups of heterotrophic prosthecate bacteria		
Group	**Budding**	**Prosthecae (No. per Cell; Location)**	**Reproductive Role of Prosthecae**
Caulobacters	–	One (rarely two); polar	–
Hyphomicrobia	+	One to several; polar[a]	+
Polyprosthecate bacteria	+	Several to many	–

[a]The genus *Pedomicrobium* can produce prosthecae at other locations on the cell surface.

called **prosthecae**, which, by definition, are cellular appendages, or extensions of the cell that contains cytoplasm (Figure 19.6B). There are two important consequences of having these appendages. First, they increase the surface area of the cell, resulting in a higher surface area to volume ratio, thereby allowing the cells increased access to the nutrients in the environment. Because these organisms live in aquatic environments that have low nutrient concentrations, the prosthecae provide greater exposure to the nutrients they require for growth. Studies have shown that these organisms have very high affinity nutrient uptake systems, which enables them to take up nutrients at very low concentrations. They are therefore examples of **oligotrophic** bacteria. Second, the increased surface area is also important for these planktonic organisms because it provides greater friction, or drag, and therefore slows their settling out from the plankton.

There are three major groups of prosthecate *Proteobacteria*: the caulobacters, which are in the order *Caulobacterales*, the hyphomicrobia and the polyprosthecate bacteria both of which are in the order *Hyphomicrobiales* (Table 19.7). Almost all heterotrophic prosthecate bacteria are members of the *Alphaproteobacteria*. Other heterotrophic prosthecate bacteria are found in the phylum *Verrucomicrobia* (see Chapter 22). Next, each of the prosthecate groups is considered individually.

CAULOBACTERS   There are several genera of caulobacters (Table 19.8). The best-studied genus is *Caulobacter* with a single polar prostheca, which in this genus is commonly called a "stalk" (Figure 19.7). *Caulobacter* species, and certain other prosthecate and or budding bacteria, have a distinctive life cycle (Figure 19.8). The life cycle of *Caulobacter* species consists of two stages (see Figure 19.8A): (1) a motile stage and (2) a sessile, or prosthecate, stage. In the motile stage, the cell is called a **swarmer cell**, which has a single polar flagellum and, at the same site, bears a **holdfast**, a special adhesive organelle that mediates attachment. The holdfast allows the cell to attach to detritus or other particulate material in the environment. Before the swarmer cell divides, it loses motility and develops a polar prostheca at the same site as the flagellum. Normally by this time in its development, the cell would be attached to a detritus particle or a dead algal cell in the environment by virtue of its sticky holdfast. The *Caulobacter* cell then elongates and produces a daughter swarmer cell at the nonprosthecate pole. Next, the daughter cell becomes motile and separates from the mother cell to become free in the planktonic environment and ready to repeat the cycle. The mother cell continues to produce daughter cells from the same position on its cell surface as long as conditions are favorable for its growth. A life cycle such as this, with two separate morphological forms (a flagelled nonprosthecate cell and a nonflagellated prosthecate cell), is termed a **dimorphic life cycle**. Because of their life cycles and their simple prokaryotic nature, *Caulobacter* species have become model organisms in the study of **cellular differentiation** or **morphogenesis** (Box 19.3).

In pure cultures, the holdfasts of *Caulobacter* enable several cells to attach together to form a rosette structure (see Figure 19.7C), something that is not found in the natural environment.

TABLE 19.8	Genera of caulobacters		
Order	**Genus**	**Prostheca Position**	**Habitat**
Caulobacterales			
	Caulobacter	Polar	Freshwater
	Asticcacaulis	Subpolar	Freshwater
Rhodobacterales			
	Maricaulis	Polar	Marine

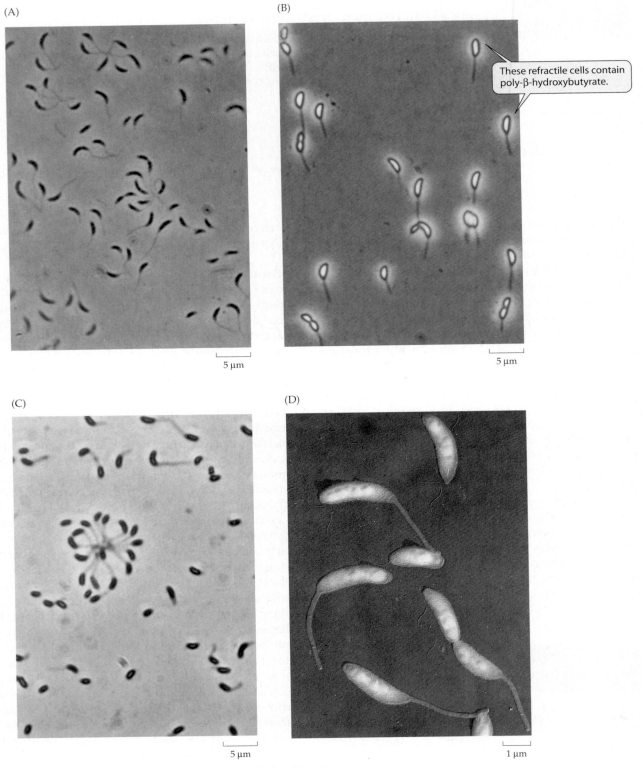

(A)

(B)

These refractile cells contain poly-β-hydroxybutyrate.

5 μm

5 μm

(C)

(D)

5 μm

1 μm

Figure 19.7 *Caulobacter*
(A–C) Phase contrast photomicrographs of *Caulobacter crescentus*, showing the polar prostheca. (D) Electron micrograph of a shadowed preparation of *C. crescentus*. Note single crossbands in each of the prosthecae. Courtesy of J. Poindexter.

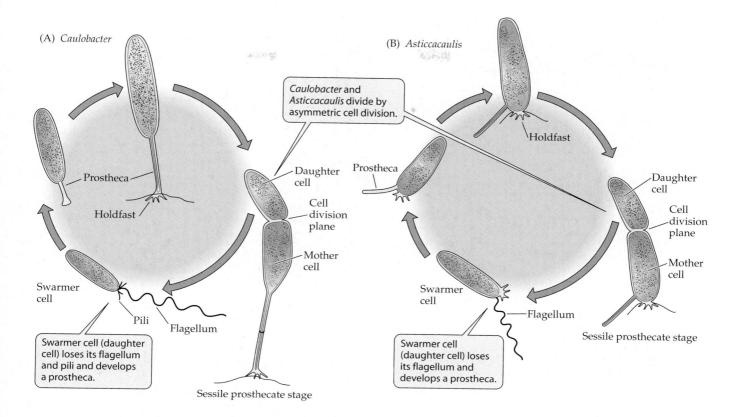

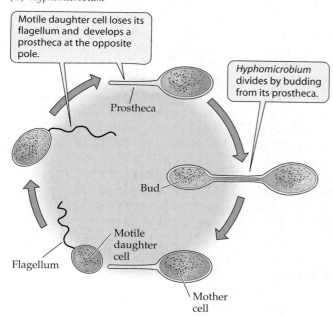

Figure 19.8 Life cycles of prosthecate bacteria
Diagrams of life cycles of (A) *Caulobacter*, (B) *Asticcacaulis*, and (C) *Hyphomicrobium*. Note that all produce a flagellated swarmer cell.

These organisms also produce structures called **crossbands** in their prosthecae (see Figure 19.7D and **Figure 19.9**). Studies have shown that these are produced at the time of cell division. Therefore, it is possible to determine the "age" of a mother cell by counting the number of crossbands (hence, the number of times it has divided), much like counting tree rings.

These bacteria are metabolically similar to *Pseudomonas* species. They utilize a wide variety of soluble organic carbon sources, including sugars, amino acids, and organic acids, which are found in low concentrations in aquatic habitats (see Chapter 24). They use the Entner-Duodoroff pathway and TCA cycle and are active in aerobic respiration. They occur in both freshwater and marine habitats as well as in soils.

Interestingly, there is a morphologically homologous marine genus—named *Maricaulis*—that is in the order *Rhodobacterales* of the *Alphaproteobacteria* (see Table 19.1).

The genus *Asticcacaulis* is less frequently encountered in most habitats. This genus is very similar to *Caulobacter* in all respects, except that its flagellum and prostheca are in a subpolar position on the cell surface and one species produces two lateral prosthecae (see Figure 19.8B). However, the holdfast remains in a polar position. Thus, rosettes formed by pure cultures hold the cells together so that the prosthecae extend away from the center.

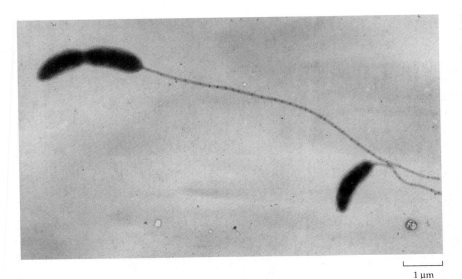

Figure 19.9 Crossbands in prosthecae
Electron micrograph of a *Caulobacter* cell, showing 31 crossbands in its stalk, indicating that it has undergone 31 cell divisions. Courtesy of J. T. Staley and T. Jordan.

1 µm

BOX 19.3   *Research Highlights*

Caulobacter crescentus, a Model Prokaryote for Studying Cell Morphogenesis or Differentiation

Caulobacter crescentus is ideally suited for studying morphogenesis because it undergoes a transformation from a flagellated swarmer cell to an immotile, prosthecate cell during its life cycle (see figure). In the cell cycle, the cell must be programmed to "switch on" and "switch off" various biosynthetic processes at specific times. Furthermore, the cell has the capability of localizing these events at one pole of the cell or the other.

Let us consider initially the flagellated swarmer cell. This stage of the life cycle not only has a polar

flagellum and is active chemotactically, but the flagellated pole of the cell also contains the holdfast, a specific pilus structure, and DNA phage receptor sites. Thus, there is strong polar orientation of functions. Then, at some time during the differentiation process, the swarmer cell loses its flagellum and pili and begins to develop a prostheca (called a "stalk") and a holdfast at the same pole of the cell.

Only after the stalk forms does the cell switch on DNA replication. The stalked cell elongates, DNA replication occurs, and the polar

differentiation process whereby the new swarmer cell is formed occurs.

Not surprisingly, the length of time it takes for a swarmer cell to undergo cell division is longer (typically an additional 30 minutes) than it takes a cell with a stalk to divide (90 minutes). This is due to the length of time required for the swarmer cell differentiation process. The molecular biological events controlling the life cycle of *C. crescentus* are being analyzed genetically and biochemically by several research groups.

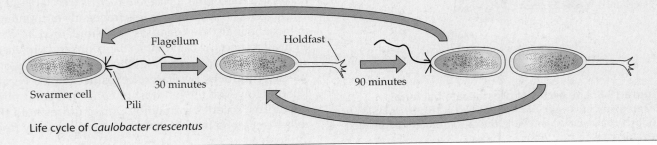

Life cycle of *Caulobacter crescentus*

Caulobacter is widespread in aquatic habitats. Studies indicate that they are found in lakes of all trophic states from oligotrophic to eutrophic. They comprise a significant proportion of the total heterotrophs in oligotrophic and mesotrophic habitats, making up as much as 50% of the viable heterotrophic count during their periods of greatest abundance (see Chapter 24). Although the natural function of the *Caulobacter* prostheca is not yet well understood, it is known that it lengthens during conditions of phosphate limitation. Because phosphate is a common limiting nutrient in aquatic habitats, especially during summer algal blooms, this appendage may enhance phosphate uptake and improve their ability to survive when this nutrient becomes limiting during the summer periods.

Order Hyphomicrobiales

HYPHOMICROBIA   Like the caulobacters, hyphomicrobia make up a group of several genera with similar morphological traits (Table 19.9). They all have polar prosthecae and divide by forming a bud at the tip of it (Figure 19.10A,B). Buds are usually motile by a single flagellum; some species produce a holdfast for attachment. In this group, the holdfast is located on the surface of the cell, not at the tip of the prostheca.

These organisms also have a dimorphic life cycle consisting of a nonprosthecate motile daughter cell and a prosthecate mother cell (see Figure 19.8C). The motile daughter cell must first form a prostheca. It then divides by a budding process in which a bud develops from the tip of the prostheca. The daughter cell enlarges, becomes motile, separates from the mother cell, and repeats the cycle. Meanwhile, the appendaged mother cell develops

a new bud, either at the same site as the earlier bud or on a newly formed polar appendage. Apparently, if conditions are favorable for growth, an unlimited number of buds can be produced from the mother cell, in the same manner as caulobacters.

Hyphomicrobium species are primarily methanol utilizers, although they can also use some other one-carbon compounds such as methylamine as their carbon source. They use the serine-glycine pathway for methanol degradation (see Chapter 8). Some carry out denitrification using nitrate as an electron acceptor in anaerobic respiration and oxidizing methanol as the energy source.

Hyphomonas and *Hirschia*, which are members of the order *Alphaproteobacteria* termed the *Rhodobacterales*, look identical to *Hyphomicrobium*, but they do not use methanol as a carbon source. Instead, they use amino acids and organic acids as carbon sources. Amino acids are required. The life cycles of these bacteria are similar to that of *Hyphomicrobium*, despite their metabolic differences.

Pedomicrobium species are different from the other hyphomicrobia. They are soil and aquatic bacteria that grow on acetate and pyruvate as well as more complex organic compounds such as fulvic acid. They are best known for their ability to oxidize reduced iron and manganese compounds and form deposits of the oxides on the cell surface. Their life cycle is similar to that of the other hyphomicrobia. However, they commonly produce more than one prostheca per cell, and some of these are in nonpolar positions (Figure 19.11).

The hyphomicrobia are widespread in soils and aquatic environments. *Hyphomicrobium* are commonly found in freshwater and soil habitats. Most *Hyphomonas* and *Hirschia* species have been isolated from marine habitats. *Pedomicrobium* was first reported from soils, but they

(A)

(B)

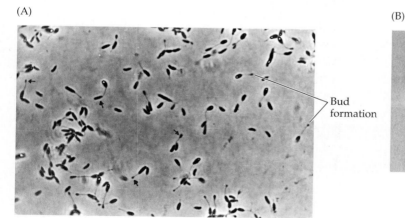

Bud formation

5 μm

1 μm

Figure 19.10 *Hyphomicrobium*
(A) Phase photomicrograph of cells of *Hyphomicrobium zavarzinii*, showing bud formation at the tips of the prosthecae. The bright area in some cells is poly-β-hydroxybutyrate. (B) Electron micrograph of a budding cell of *Hyphomicrobium facilis*. Courtesy of P. Hirsch.

			Carbon	Special Features
Order	Genus	Mol % G + C	Source	and Habitat
Hyphomicrobiales				
	Hyphomicrobium	59–65	Methanol	Polar prosthecae; soil, freshwater
	Pedomicrobium	62–67	Acetate, pyruvate	Lateral and polar prosthecae; soil, lakes
Rhodobacterales				
	Hyphomonas	57–62	Amino acids	Polar prosthecae; marine
	Hirschia	45–47	Amino acids, organic acids	Polar prosthecae; marine

TABLE 19.9 Genera of hyphomicrobia

have also been obtained from freshwater lakes and water pipelines where they form a coating of manganic oxides if the water contains reduced manganese ions (Mn^{2+}).

POLYPROSTHECATE BACTERIA Polyprosthecate bacteria are a diverse group of bacteria represented by several genera (Table 19.10). Unlike other prosthecate bacteria, they have several prosthecae per cell, which gives their cells starlike shapes. In fact, the genus *Stella* produces six prosthecae all in one plane, so they resemble a perfect six-pointed star (Figure 19.12). This genus is a member of the *Alphaproteobacteria*.

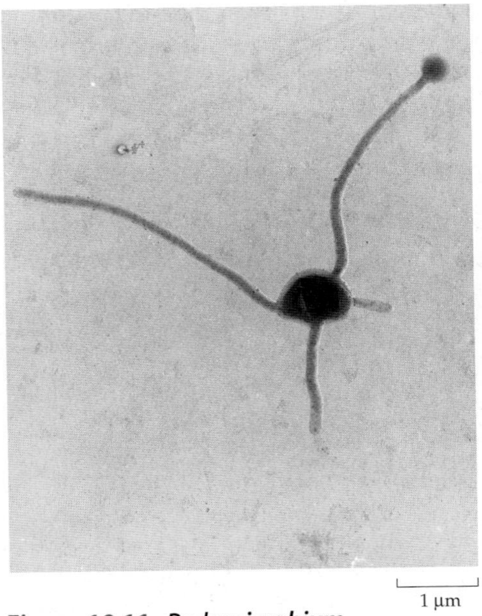

Figure 19.11 *Pedomicrobium*
Electron micrograph of a *Pedomicrobium* cell, showing a bud at the tip of one prostheca. Courtesy of R. Gebers.

1 μm

The genus *Prosthecomicrobium* is the most common member of this group. They are found in freshwater, marine, and soil environments. Some species are motile by a single subpolar flagellum, whereas others are immotile or produce gas vacuoles (Figure 19.13).

Unlike all other heterotrophic prosthecate bacteria, the genus *Ancalomicrobium* (see Figure 19.6) is a fermentative facultative aerobe. Indeed, it is a mixed-acid fermenter whose products are identical to that of *Escherichia coli* (discussed later in this chapter). However, unlike *E. coli*, *Ancalomicrobium* produces prosthecae and gas vacuoles and is well adapted to live and compete successfully in aquatic habitats where nutrients are in much lower concentrations than in the intestinal tract of animals. *Ancalomicrobium adetum* uses a variety of sugars and other organic carbon sources for growth and requires one or more B vitamins such as thiamine, biotin, riboflavin, or B_{12}.

Like the caulobacters, these bacteria live in aquatic habitats and soils, particularly those that are oligotrophic to mesotrophic. Their appendages enhance the uptake of nutrients in habitats where nutrients occur in low concentrations. Each genus and species specializes in the uptake of a group of low-molecular-weight dissolved

TABLE 19.10 Genera of polyprosthecate bacteria

Genus	Mol % G + C	Special Features
Prosthecomicrobium	64–70	>10 Short prosthecae around cell; aerobic
Ancalomicrobium	70–71	<10 Long prosthecae; facultative aerobe; fermentative
Stella	69–74	Star-shaped; prosthecae in one plane

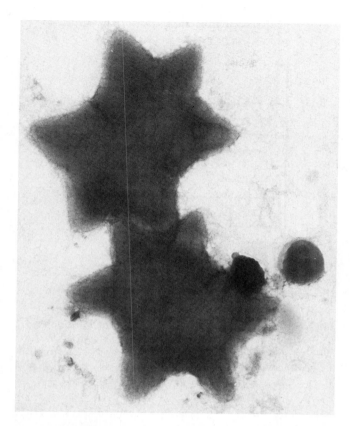

Figure 19.12 *Stella*
Two cells of *Stella humosa*, a six-pointed, star-shaped budding bacterium. Cells are about 1 μm in width. Courtesy of J. T. Staley.

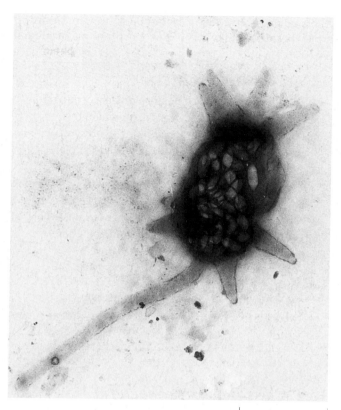

Figure 19.13 *Prosthecomicrobium*   1 μm
Electron micrograph of a gas vacuolate cell of *Prosthecomicrobium pneumaticum*. Some individual gas vesicles can be seen. Courtesy of J. T. Staley.

organic nutrients, such as sugars and organic acids, and is therefore responsible for keeping these nutrients in low concentrations in oligotrophic environments. In this manner, they maintain a competitive edge against eutrophic bacteria that require organic nutrients in much higher concentrations and are therefore not competitive in oligotrophic environments.

Conversely, the low maximum growth rates of the prosthecate bacteria (generation times are longer than 2 hours in laboratory culture) mean that they cannot outgrow fast-growing eutrophic bacteria such as the enteric bacteria in more enriched habitats, such as the intestinal tract, which has much higher concentrations of organic compounds.

Order Sphingomonadales

This order contains 12 genera of heterotrophic bacteria that are widespread in the environment.

SPHINGOMOMONAS   This common genus is typically found in soil and aquatic environments. This genus of aerobes contains strains that are known for their ability to de-grade a variety of toxic aromatic compounds such as dibenzo-*p*-dioxins and dibenzofurans. This versatile genus also contains strains that are opportunistic pathogens that can cause meningitis, peritonitis, and septicemia.

ZYMOMONAS   *Zymomonas* is a gram-negative non-motile rod that is facultatively anaerobic and oxidase negative. The mol % G + C is 47 to 50. Strains are occasionally microaerophilic, and some are strictly anaerobic. *Zymomonas* is often a spoiler of beer and cider in which it produces a heavy turbidity and unpleasant odor due to formation of acetaldehyde and sulfide. The organism ferments glucose anaerobically via the Entner-Duodoroff pathway (see Chapter 8):

$$(1) \text{ Glucose} \rightarrow \text{ethanol} + \text{lactate} + CO_2$$

Small amounts of acetaldehyde and acetyl methyl carbinol are also produced during this fermentation. The palm wines of the Far East and Africa use *Zymomonas mobilis* as the fermenting agent. The organism is also involved in transforming the sugary sap of various agaves to pulque, a fermented beverage produced in Mexico.

Zymomonas is found on honeybees and in ripening honey because it tolerates high concentrations of sugars.

19.2 Class *Betaproteobacteria*

Like the *Alphaproteobacteria* the *Betaproteobacteria* have considerable diversity including phototrophs, methane-oxidizers, nitrifiers, and pathogens (Figure 19.14; Table 19.11). Representative members of several of the orders are discussed in this section.

Order Neisseriales

The *Neisseria* group comprises gram-negative, aerobic bacteria of varying morphology (Table 19.12). Most are nonmotile. *Neisseria* is the major genus of the group. The best-known species of *Neisseria* is *Neisseria gonorrhoeae*, the causative agent of the common venereal disease, **gonorrhea** (see Chapter 28). Another pathogenic member of this genus is *Neisseria meningitidis*, a causative agent of **cerebrospinal meningitis**. Other species are typically found in the nasopharynx and respiratory passages of warm-blooded animals. These rarely cause disease and are considered part of the normal microflora. *Neisseria* is best cultivated on chocolate-blood agar at 37°C in a 3% to 10% atmosphere of CO_2.

Simonsiella is a unique filamentous, gliding bacterium. It resembles a watchband (Figure 19.15). One side of the

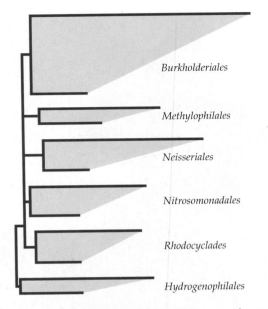

Figure 19.14   16S rRNA tree of the *Betaproteobacteria* This phylogenetic grouping contains the *Betaproteobacteria* with several orders noted. Adapted by permission from *Bergey's Manual of Systematic Bacteriology*.

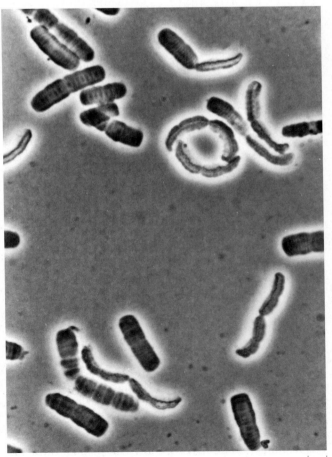

Figure 19.15   *Simonsiella* Phase contrast photomicrograph of *Simonsiella* cells. Note that the filaments appear as watchbands. The flat, ventral side is the surface used for gliding. The dorsal side is rounded and shows the cell segments. Courtesy of J. T. Staley.

5 µm

TABLE 19.11	Characteristics of genera of the class *Betaproteobacteria*[a]	
Orders	**Representative Genera**	**Characteristics**
Burkholderiales		
	Burkholderia	Common, versatile heterotroph
	Alcaligenes	Heterotroph
	Bordetella	Whooping cough
	Comamonas	Small rod
	Leptothrix	Sheathed iron oxidizers
	Sphaerotilus	Sheathed
Hydrogenophilales		
	Hydrogenophilus	Hydrogen oxidizers
	Thiobacillus	Unicellular sulfur oxidizer
Methylophilales		
	Methylophilus	Methane and methanol oxidizers
	Methylobacillus	Methane and methanol oxidizers
	Methylovorus	Methane and methanol oxidizers
Neisseriales		
	Neisseria	Commensal and pathogenic bacteria
	Kingella	Oral commensal
	Simonsiella	Oral commensal
Nitrosomonadales		
	Nitrosomonas	Chemolithotrophic nitrifiers
	Nitrosospira	Chemolithotrophic nitrifiers
	Spirillum	Heterotrophic spirilla
Rhodocyclales		
	Rhodocyclus	Phototroph
	Zoogloea	Floc former

[a]A complete listing of genera is provided in the Appendix.

filament, the ventral side, which is in contact with the surface on which it glides, is flattened. The other, or dorsal, side of the filament is rounded.

This genus has a very distinctive habitat. The organism is found in the oral cavity of mammals including humans, dogs, cats, and sheep, each animal species harboring its own distinctive species. These organisms are aerobic and respire using a variety of sugars as substrates for growth. They are not harmful to humans or other animals but are instead part of the normal commensal microbiota.

Kingella, another genus of this group, is a common organism in the mucous membrane of humans as part of the normal microbiota.

TABLE 19.12	*Neisseria* group of gram-negative bacteria			
Genus	**Cell Shape (Size)**	**Mol % G + C**	**Oxidase Reaction**	**Habitat**
Neisseria	Cocci (0.6–1.0 μm)	47–54	+	Mucous membranes of mammals
Simonsiella	Filamentous (2–8 × 10–50 μm)	41–55	+	Mucous membranes of mammals
Kingella	Rods (1 × 2–3 μm)	47–55	+	Mucous membranes of humans

(A)

(B)

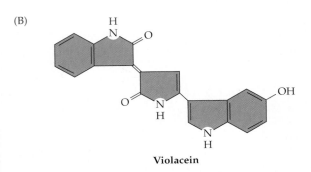

Violacein

Figure 19.16 *Chromobacterium violaceum*
(A) Colony of *Chromobacterium violaceum* showing purple pigmentation. (B) Chemical structure of violacein, the purple pigment of *Chromobacterium violaceum*. A, ©Dr. Elmer Koneman/Visuals Unlimited.

Chromobacterium violaceum is one of the most striking of all bacteria because it produces deep-violet colonies on a solid medium (Figure 19.16A). The pigment, called **violacein** (Figure 19.16B), is produced in significant amounts during growth on tryptophan, and it bears a structural relationship to this amino acid. Members of the genus are motile by polar flagella and are rod shaped. *C. violaceum* occurs primarily in soil and water and is especially common in tropical soils.

Order Burkholderiales

This order contains a large variety of heterotrophic bacteria that are found in freshwater, soil, and marine environments.

BURKHOLDERIA AND *THERMOTHRIX* The namesake genus of this order, *Burkholderia* is remarkable for its metabolic capabilities. In particular, it can utilize more

than 100 different carbon compounds for its growth. To illustrate this physiological diversity, some members are plant pathogens, others are implicated in cystic fibrosis, and some species are capable of nitrogen fixation. The genus is very important in the bioremediation of toxic organic compounds. Several genomes of this genus have been subjected to genome analysis.

Thermothrix is a sulfur-oxidizing genus. It is a polarly flagellated rod that grows at temperatures from 63°C to 86°C. In the low oxygen conditions of the hot springs where it lives as a facultative chemoautotroph, *Thermothrix* produces filaments.

SHEATHED BACTERIA: *SPHAEROTILUS* AND *LEPTO-THRIX* Some bacteria produce **sheaths**, which are distinctive layers formed external to the cell wall (Figure 19.17). The three major genera in this group are *Sphaerotilus*, *Leptothrix*, and *Crenothrix*. They are differentiated from one another on the basis of their deposition of fer-

(A)

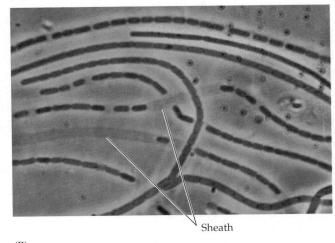

Sheath

(B)

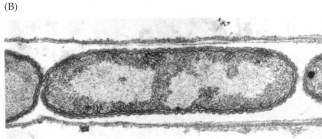

Figure 19.17 *Sphaerotilus*
(A) Micrograph of *Sphaerotilus* filaments, showing the cells in chains within a sheath. The sheath appears as a transparent tube about 1.5 μm in diameter. (B) TEM of *Sphaerotilus natans* cells in sheath. A, ©Science VU/Visuals Unlimited; B, ©Judith F. M. Hoeniger/Biological Photo Service.

TABLE 19.13	Differences among *Sphaerotilus*, *Leptothrix*, and *Crenothrix*		
Genus	**Mol % G + C**	**Flagellation**	**Distinguishing Features**
Sphaerotilus	70	Polar tuft	May deposit iron oxide
Leptothrix	69–71	Single polar	Deposits iron oxide; oxidizes Mn^{2+}
Crenothrix	Unknown	Unknown	Tapering sheath

ric hydroxide and manganese oxide as well as other features (Table 19.13).

Sphaerotilus and *Leptothrix* are large rod-shaped bacteria ($1-2 \times \sim 10$ µm) that are motile by polar flagella. They live in flowing aquatic habitats, where they attach to inanimate objects such as rocks and sticks. They have a distinctive life cycle (Figure 19.18). Single motile cells have a holdfast structure that allows them to attach to inanimate substrata such as sand grains. When they are attached, they become sessile. These cells then reproduce by binary transverse fission to form a chain of cells enclosed in the sheath. As the chain grows and elon-

gates, motile cells are produced and released from the unattached end. These motile swarmer cells repeat the life cycle and provide the organism with a means of dispersal to other habitats.

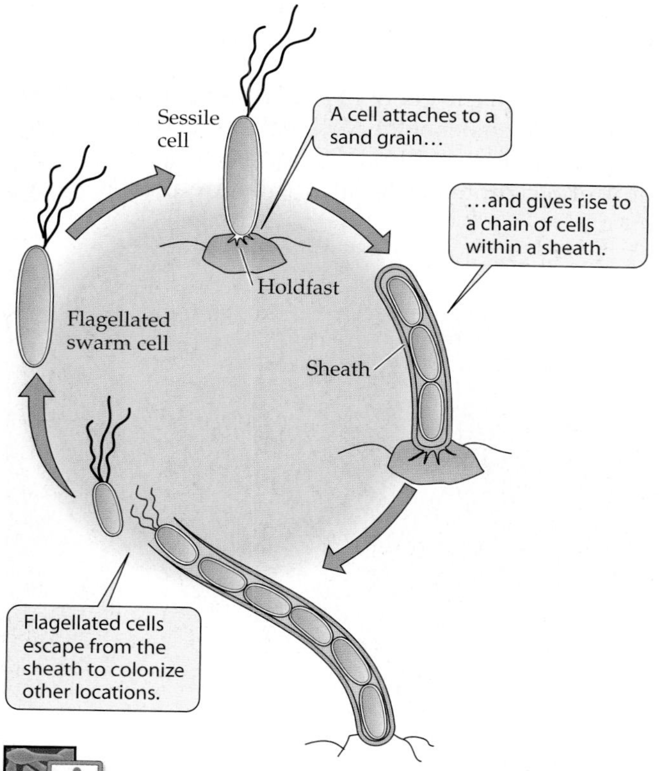

These bacteria are aerobic heterotrophs that use a variety of organic substrates for growth, including sugars, alcohols, and organic acids. All species known require vitamin B_{12} for growth. They are common in flowing aquatic habitats where they utilize the nutrients that pass by them in the water. They are particularly common in the Pacific Northwest, where they grow downstream of pulp mills and sewage treatment facilities. They can be a nuisance because when they grow profusely in enriched habitats, clumps can break off and clog fishermen's nets and water treatment inlets. Their growth has been controlled considerably since pulp mills have been required by the Environmental Protection Agency (EPA) to treat pulping-plant effluent prior to discharge into receiving streams.

The sheaths of *Sphaerotilus natans* have been analyzed chemically. They have a complex composition similar to that of the outer membrane of gram-negative bacteria consisting of protein, carbohydrate, and lipid.

Leptothrix species appear dark brown or black due to the deposition of iron and manganese oxides in their sheaths (Figure 19.19). These organisms are commonly found in iron springs. Samples from such springs are filled with brown, encrusted sheaths that are devoid of cells. Conditions away from the mouth of the spring are not conducive to the growth of these bacteria, and the cells leave the sheaths or are lysed. However, samples taken close to the mouth of the spring where conditions are more reduced contain cells within the sheath.

Some controversy exists over the ability of these bacteria to obtain energy from the oxidation of reduced iron and manganese ions. As yet there has been no conclusive demonstration of chemolithotrophy using these inorganic substrates. The pH of the environment in which these organisms grow is near neutral, and virtually all iron would be already oxidized to the ferric state. How-

Figure 19.18 Life cycle of *Sphaerotilus natans*
Diagram of the life cycle of *Sphaerotilus natans*.

Figure 19.19 *Leptothrix*
Phase contrast photomicrograph of a filament of a *Leptothrix* sp. from a natural sample. The FeO(OH) and MnO$_2$ accumulated in its sheath have imparted a brownish color. The sheath is about 1.5 µm in diameter. Courtesy of J. T. Staley.

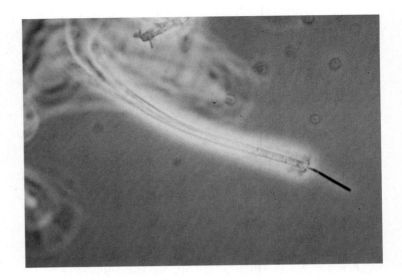

ever, manganese is available in part in the Mn^{2+} state and could be used as a source of energy, as suggested early on by Winogradsky. However, although enzymes have been found that enable the oxidation of Mn^{2+} to the Mn^{4+} state in *Leptothrix*, as yet there has been no demonstration of ATP production in this process.

Crenothrix is a genus that has been known for more than a century, but has not yet been successfully grown in pure culture. Although it resembles the other sheathed bacteria, it has a tapered sheath resembling a cornucopia (Figure 19.20) and produces small cells from the larger unattached end of the sheath.

BORDETELLA   *Bordetella* is a genus that harbors the pathogen *Bordetella pertussis,* the causative agent of

whooping cough (see Chapter 28). This disease has been largely eradicated, but because many children have not received vaccines to prevent it, there have been recent outbreaks.

Order Nitrosomonadales

This order contains many nitrifying genera as well as the classical genera *Spirillum* and the iron-oxidizing genus *Gallionella*.

SPIRILLUM   The genera *Spirillum* and *Aquaspirillum*, which is a member of the *Neisseriales* order of the *Betaproteobacteria*, are spirillum-shaped bacteria. These heterotrophic proteobacterial spirilla are primarily mi-

(A)

(B)

(C)

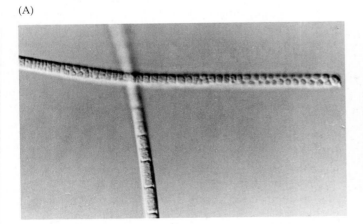

Figure 19.20 *Crenothrix*
Photomicrographs of *Crenothrix polyspora* from a freshwater source. Note the slightly tapered sheaths with the smallest cells near the tips in (A) and (B). Cells are being released from the sheath in (C). These are Nomarski interference photomicrographs. Courtesy of P. Hirsch.

croaerophilic and grow by the oxidation of organic acids. All are motile with either a single polar flagellum at each pole or with polar flagellar tufts (lophotrichous flagellation). The genus *Spirillum* contains the largest species, *Spirillum volutans*, which move by a flagellar tuft at each pole, and cells that are healthy are in constant motion. These bacteria live in natural aquatic habitats and are particularly successful at growing when nutrients are in low concentration.

NITRIFYING CHEMOLITHOTROPHS   The **chemolithotrophic bacteria** obtain energy from the oxidation of reduced inorganic compounds (Box 19.4). Several different groups exist and are differentiated from one another based on the inorganic compounds they oxidize as energy sources (Table 19.14). **Nitrifying bacteria** oxidize reduced nitrogen compounds, **sulfur oxidizers** oxidize reduced sulfur compounds, **iron bacteria** oxidize reduced iron, and **hydrogen bacteria** oxidize hydrogen gas.

By and large, these chemolithotrophic activities are uniquely associated with the *Bacteria* and the *Archaea*. Only rarely have eukaryotic organisms been found that can carry out any of these processes, and when this association has been found, as for example, nitrification by

TABLE 19.14	Groups of chemolithotrophic *Proteobacteria*	
Group	**Energy Source(s)**	**Product(s)**
Nitrifiers	NH_4^+ or NO_2^-	NO_2^- or NO_3^-
Sulfur oxidizers	H_2S, S^0, $S_2O_3^{2-}$	SO_4^{2-}
Iron oxidizers	Fe^{2+}	Fe^{3+}
Hydrogen bacteria	H_2	H_2O
Methane oxidizers[a]	CH_4	$CO_2 + H_2O$

[a]Methane is the simplest organic compound; however, the metabolism of the methane oxidizers is very similar to that of the nitrifiers, therefore they are considered here.

some fungi, their activity is not considered to be significant in the environment. In contrast, these processes carried out by bacteria are well known and considered to be extremely important in the biogeochemical cycles of nature (see Chapter 24). Each of the chemolithotrophic groups is discussed individually in greater detail. The chemolithotrophic *Proteobacteria* carry out carbon dioxide fixation using the Calvin cycle (see Chapter 10).

BOX 19.4   *Milestones*

Winogradsky and the Concept of Chemolithotrophy

Sergei Winogradsky, a scientist who was born and raised in Czarist Russia, became one of the most remarkable figures in microbiology (see Chapter 2). His most important discoveries relate to the phototrophic bacteria and the chemolithotrophic bacteria. For example, he described by direct microscopic observation several genera of the purple sulfur bacteria. During the last century, many microbiologists have worked on these bacteria, and his classification of them still holds largely unchanged today.

But we are most indebted to Sergei Winogradsky for his novel contributions to our understanding of bacterial chemolithotrophy. These studies began while he was a student in Switzerland. It was when he was in the Alps that Sergei

Winogradsky first observed *Beggiatoa* in sulfur springs. Based on his careful observations of their growth and sulfur chemistry, he concluded that these filamentous bacteria must obtain energy by oxidizing the sulfide emanating from the spring. He concluded this because he noted that the cells contained the more oxidized form of sulfur, elemental sulfur, in granules located inside their cells (see Figure 19.38). Therefore, he proposed that these bacteria were able to capture some of the chemical energy from the oxidation of sulfide to sulfur and to use this for their metabolic processes. Thus, the novel concept of chemolithotrophy was born.

Sergei Winogradsky was also the first to grow nitrifying bacteria successfully. He cultivated them on a

silica gel medium because they do not form colonies on agar media. Most isolates of nitrifiers are now obtained by dilution of enrichment samples to extinction using liquid inorganic media. To accomplish this, active enrichments containing the bacteria are diluted serially with many replicates. By chance, some of the highest dilution tubes with growth will receive only one cell of a nitrifier of interest, and from this a pure culture can be obtained. In this tedious manner representatives of several new genera have been isolated in pure culture. Many of these strains were first isolated in pure culture using this technique in the late Stanley Watson's laboratory at Wood's Hole Oceanographic Institute. Most pure cultures must be maintained in liquid media to keep them viable.

Ribulose 1,5-bisphosphate carboxylase (Rubisco) is the key enzyme in this pathway.

As their name implies, the nitrifying bacteria use reduced inorganic nitrogen compounds, ammonia and nitrite, as their sources of energy for growth. It should be noted that nitrifying members of the *Crenarchaeota* group of the *Archaea* have recently been discovered and are treated in Chapter 18. In the nitrifying *Proteobacteria*, ammonia is oxidized to nitrite by one group of nitrifiers, called the **ammonia oxidizers**:

$$(2)\ 2\,NH_3 + 3\tfrac{1}{2}\,O_2 \rightarrow 2\,NO_2^- + 3\,H_2O$$

Nitrite is oxidized to nitrate by another group of *Proteobacteria* called the **nitrite oxidizers**:

$$(3)\ NO_2^- + \tfrac{1}{2}\,O_2 \rightarrow NO_3^-$$

No single nitrifying bacterium is able to oxidize ammonia all the way to nitrate, although pure cultures of some fungi have been reported to carry out this process. These fungi, however, are not considered important in the environment. Normally the two groups of nitrifying bacteria grow in close association in the environment to carry out the two-step sequential oxidation. The two processes appear to be tightly coupled so that nitrite, which can be toxic to plants and animals, does not accumulate in high concentration.

Nitrifying bacteria are found in all soil and aquatic habitats and are especially common in alkaline or neutral pH environments where ammonia rather than ammonium ion is abundant, consistent with ammonia being the form of nitrogen used as the energy source. Also, more carbonate is available for CO_2 fixation in moderately alkaline environments.

Nitrifiers are superb examples of chemolithotrophs. They grow in pure culture on completely inorganic media. Furthermore, they are primary producers (i.e., chemoautotrophs) that can fix carbon dioxide into organic material. However, most species are also able to use organic carbon sources such as acetate for growth and thus are facultative chemolithotrophs. They grow poorly in culture even under the best of conditions, having generation times of 24 hours or more. Rarely do they grow to sufficient cell densities to produce turbidity in media, even though their activities in the medium can be readily demonstrated. The reason for their poor growth in pure culture is not well understood; however, it may be due in part to the toxicity of nitrite. We first discuss the ammonia oxidizers, which are sometimes also called the nitrosofying bacteria.

AMMONIA OXIDIZERS The oxidation of ammonia to nitrous acid involves two major steps with hydroxyl amine, NH_2OH, as an intermediate (see also Chapter 8):

$$(4)\ NH_3 + O_2 + XH_2 \xrightarrow{AMO} NH_2OH + H_2O + X$$

$$(5)\ NH_2OH \xrightarrow{HAO} HNO + H_2O \xrightarrow{HAO} HNO_2 + 2\,H^+ + 2\,e^-$$

where AMO = ammonia monooxygenase
HAO = hydroxylamine oxidoreductase
X = a hydrogen carrier such as NADH
$\Delta G^{o\prime}$ (reactions 4 and 5 combined) = -275 kJ/mol

Several enzymes are involved in the oxidation pathway to form nitrous acid. The first enzyme in this pathway is the key enzyme, ammonia monooxygenase (AMO), which carries out the oxidation of ammonia to hydroxyl amine. Ammonia monooxygenase shows rather broad specificity. Thus, it can also oxidize methane, making these bacteria resemble methane-oxidizing bacteria, which have a methane monooxygenase of similar broad specificity (see Methanotrophic Bacteria section that follows). A variety of inhibitors affect the ammonia monooxygenase, including acetylene, which is used as a blocking inhibitor in ecological studies, as well as 2-chloro-6-trichloromethylpyridine (nitrapyrin).

Energy is generated in the second reaction in which hydroxylamine is oxidized to HNO and subsequently to nitrous acid. This reaction is catalyzed by the enzyme hydroxylamine oxidoreductase (see reaction 5). HNO spontaneously degrades to produce N_2O, nitrous oxide (laughing gas). Electrons from the oxidations are passed through an extensive membrane-bound system of cytochromes (these give the cultures a red appearance). The free energy of this reaction is sufficiently high to generate ATP. Acid produced by oxidation of the inorganic nitrogen compounds results in the formation of protons that are passed through the membrane, and ATP is formed by a cell membrane ATPase. Reducing power generated during the oxidation process is used for the carbon dioxide fixation reactions. Most ammonia oxidizers are members of the *Betaproteobacteria*, but some are *Gammaproteobacteria*.

NITRITE OXIDIZERS Nitrite oxidation is a one-step oxidation process carried out by the enzyme nitrite oxidoreductase. Water is the actual electron donor for the oxidation process:

$$(6)\ NO_2^- + H_2O \xrightarrow{Nor} NO_3^- + 2\,e^- + 2\,H^+$$

where Nor = nitrite oxidoreductase
ΔG^o = -76 kJ/mole

In reaction 6, the protons are produced in the periplasmic space, thereby generating a proton motive force across the cell membrane. The electrons from reaction 6 are passed onto oxygen through the cytochrome system,

TABLE 19.15	Important genera of nitrifying bacteria		
Group/Class or Phylum	**Genus**	**Mol % G+C**	**Special Characteristics**
Ammonia oxidizers			
Betaproteobacteria			
	Nitrosomonas	45–54	Common rod
	Nitrosospira	53–55	Tightly coiled helix
	Nitrosolobus	53–56	Multilobed coccus
Gammaproteobacteria			
	Nitrosococcus	48–51	Coccus
Nitrite oxidizers			
Alphaproteobacteria			
	Nitrobacter	60–62	Common rod, divides by budding
Betaproteobacteria			
	Nitrospina	58	Thin rod
Deltaproteobacteria			
	Nitrococcus	51	Marine genus
Phylum *Nitrospira*[a]			
	Nitrospira	50	Separate phylum: helix, common marine genus

[a]See Chapter 22.

resulting in the formation of water in the cytoplasm (see Chapter 8). Under standard conditions, this reaction provides less energy per mole of substrate oxidized than ammonia oxidation. However, as with ammonia oxidizers, there is sufficient energy available in this process to generate ATP, and this is accomplished by chemiosmotic means. However, unlike the ammonia oxidizers, the nitrite oxidizers must produce NADPH by reverse electron transport (see Chapter 9).

Table 19.15 lists some of the genera of ammonia and nitrite-oxidizing bacteria, and representatives of selected genera are illustrated in Figure 19.21. Most nitrifying bacteria have a complex internal membrane system that contains cytochromes (see Figure 19.21B,D,F, and H). These intracytoplasmic membranes are thought to contain the cytochrome systems that act as the site of ammonia or nitrite oxidation as well as the site of generation of NADH. *Nitrococcus* is a member of the *Gammaproteobacteria*, whereas all others, with the exception of *Nitrospira*, which is in a separate phylum, are members of the *Betaproteobacteria*.

ECOLOGICAL IMPORTANCE OF NITRIFIERS Although nitrifiers occur in relatively low concentrations in habitats, they are very active metabolically. Nitrite does not accumulate in most environments because of the tight coupling between ammonia oxidation and nitrite oxidation. Therefore, nitrite, which is toxic (and mutagenic) to plants and animals, typically occurs in very low concentrations in environments because of its rapid oxidization to nitrate.

The activities of nitrifying bacteria are especially important in soil environments because of the role they play in nitrogen cycling (see Chapter 24). Ammonium nitrogen is a preferred source of nitrogen as a crop fertilizer. The principal reason for this is that ammonium is more readily retained in soils because it is positively charged. Conversely, nitrate is highly soluble and readily leached from soils. Moreover, nitrate formed by nitrifiers can be converted to nitrogen gas by denitrifiers, and therefore lost from soils. For these reasons, the inhibitor nitrapyrin (see preceding text) is produced commercially as a "fertilizer" to inhibit nitrification and thereby prevent nitrogen loss from soils.

It is also noteworthy that N_2O, an intermediate formed in nitrification as well as denitrification, is important environmentally in that it can react photochemically with ozone in the upper atmosphere (see Chapter 24) and destroy it.

Sewage contains high concentrations of ammonia due to the decomposition of proteins and amino acids (see Chapter 32). As a result, ammonia is discharged in large quantities in sewage outfalls. Like organic compounds, ammonia has a high biochemical oxygen demand (BOD, see Chapter 32) because the process of nitrification requires oxygen. Therefore, even treated sewage, which has low organic carbon concentrations, can cause low oxygen concentrations that reduce the quality of receiv-

(A)

(B)

(C)

(D)

Nitrosolobus is unusual in having compartmentalized cells.

(E)

Nitrosospira is so tightly coiled that it appears as rods.

(F)

(G)

(H)

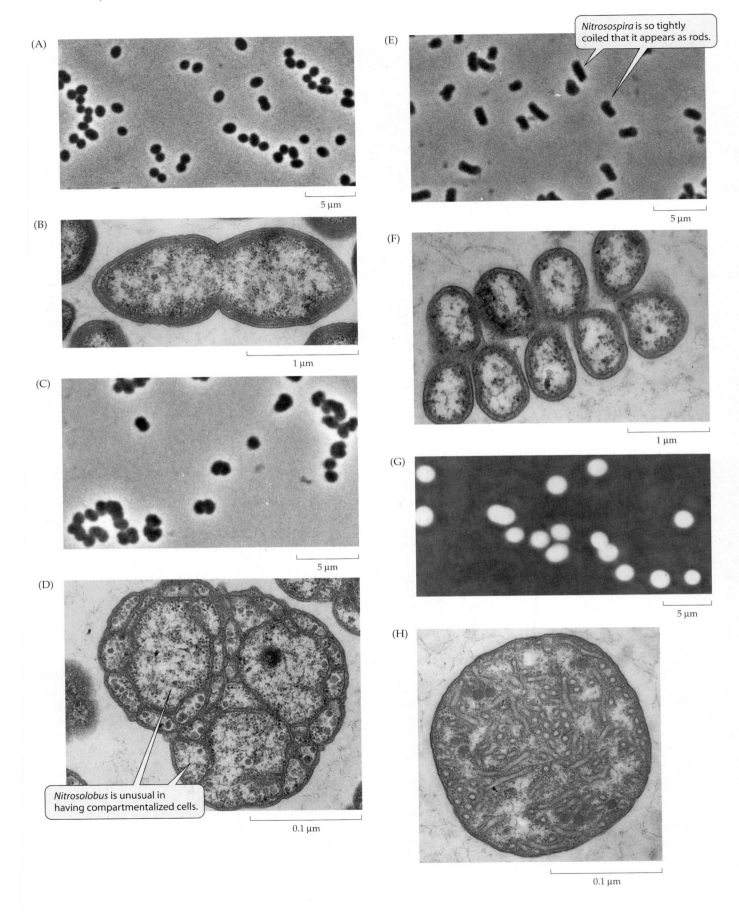

◀ **Figure 19.21 Nitrifying bacteria**
Phase contrast photomicrographs and electron micrographs of thin sections of nitrifying bacteria. (A,B) *Nitrosomonas europaea*; (C,D) *Nitrosolobus multiformis*; (E,F) *Nitrosospira briensis*; and (G,H) *Nitrococcus mobilis*. Courtesy of S. Watson, J. Waterbury, and C. Remsen.

ing waters because of nitrification. For this reason, modern wastewater treatment systems are designed to encourage the growth of nitrifying bacteria *within* the treatment plant so that downstream waters will be unaffected by ammonia (see Chapter 32). Otherwise, receiving waters with little organic material could become anaerobic by the activity of these bacteria.

GALLIONELLA: AN IRON-OXIDIZING CHEMOAUTO-TROPH   Some bacteria are able to obtain energy by the oxidation of ferrous iron. Because iron oxidation at typical physiological pHs, that is, near neutrality (pH 4 to 8), occurs spontaneously in the presence of oxygen, the bacteria that carry out this process either live in environments with very low pH or very low oxygen concentrations where ferrous ion remains reduced. Two different iron-oxidizers are discussed in more detail.

Iron oxidation by *Thiobacillus ferrooxidans* occurs at low pH. These bacteria grow in association with acidophilic sulfur oxidizers and are capable of acidophilic sulfur oxidation themselves. In this environment, ferrous ion is abundant, and these bacteria have been shown to be able to use it as an energy source (see Figure 8.19).

In contrast, *Gallionella ferruginea* grows at neutral pH in an environment that is very low in oxygen. This bacterium lives in iron springs, where it appears as masses of twisted filaments that are reddish brown from the oxidized iron, FeO(OH). These are the stalks of the bacterium (Figure 19.22A). The cell appears as a small vibrio that produces the inorganic stalk of ferric hydroxide from the concave side of the cell (Figure 19.22B). As the organism grows, the stalk elongates and twists to form a helix. When the cells divide, the stalk bifurcates. Meanwhile, the cells remain small and almost indiscernible but continue to produce more stalk material as they grow. This iron-oxidizing bacterium is better known for its stalk than for itself, because when one observes it in the microscope, one primarily sees the massive amounts of iron oxide-encrusted stalks.

G. ferruginea is a chemoautotrophic iron oxidizer. Although it has been difficult to obtain *G. ferruginea* in pure

(A)

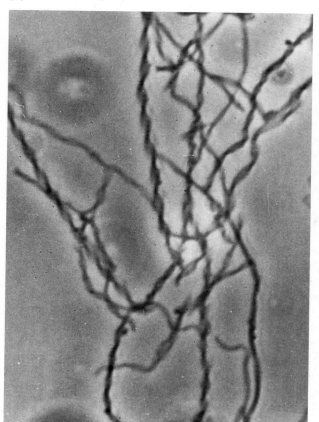

(B)

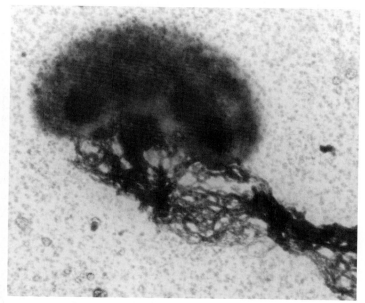

Figure 19.22 *Gallionella*
(A) Twisted stalks of *Gallionella ferruginea* as they appear when obtained from natural spring sources. (B) Electron micrograph of *G. ferruginea*, showing the vibrioid appearance of the cell and the stalk arising from the concave side of the cell. Courtesy of H. Hanert.

culture, some strains have been isolated. Results of physiological studies indicate that this bacterium is a true chemoautotroph because it has the enzymes of the Calvin cycle.

Order Hydrogenophilales

This order contains some of the hydrogen gas–oxidizing bacteria as well as unicellular sulfur-oxidizing bacteria.

HYDROGEN BACTERIA   Hydrogen gas (H_2) is an excellent energy source for bacterial growth. Among the *Bacteria,* several types of hydrogen utilizers are found. One group is the mesophilic, facultative hydrogen chemolithotrophs that can also use organic carbon sources for growth. Another group, the *Aquificales,* contains thermophilic bacteria. A third group also is involved in the oxidation of carbon monoxide. Still another bacterial group is anaerobic bacteria, called acetogenic bacteria because they produce acetic acid from carbon dioxide. Each of these four groups is discussed individually in this chapter.

The ability to use hydrogen gas as an energy source is widespread among the *Bacteria* and the *Archaea.* At one time all hydrogen-oxidizing *Bacteria* were placed in the genus "*Hydrogenomonas.*" However, many of these bacteria were found to be closely related to other heterotrophic bacteria from a variety of different genera. Furthermore, with the exception of the thermophilic genera *Hydrogenobacter* and *Calderobacterium* in the phylum *Aquificae,* all aerobic hydrogen bacteria are facultative hydrogen bacteria that can use organic compounds as energy sources as well as hydrogen gas. Therefore, most, such as the genus *Hydrogenophilus,* a member of the order *Hydrogenophilales,* are facultative chemolithotrophs or mixotrophs, not obligate chemolithotrophs. Another example is *Ralstonia eutropha,* which is one of the most thoroughly studied hydrogen bacteria and is a member of the Order *Burkholderiales* of the *Betaproteobacteria.* Typical hydrogen oxidizers obtain energy from the oxidation of hydrogen gas using a membrane-bound hydrogenase:

$$(7)\ H_2 \rightarrow 2\,H^+ + 2\,e^-$$

The electrons generated are passed through an electron transport chain, and ATP is generated by proton pumping and membrane-bound ATPases. Only a few genes are needed for the hydrogen oxidation process. In some species, these have been found on plasmids. This suggests that the ability to generate energy from hydrogen can be genetically transferred from one species to another. We call this ability lateral gene transfer. However, in addition to having a hydrogenase and an electron transport system for energy generation, these bacteria also need to

have the enzymes for carbon dioxide fixation in order to grow autotrophically. All of the hydrogen-oxidizing *Bacteria* studied so far that can grow on carbon dioxide use the Calvin cycle for carbon dioxide fixation.

SULFUR AND IRON OXIDIZERS   Two different groups of lithotrophic *Proteobacteria* obtain energy by the oxidation of reduced inorganic sulfur compounds, the "unicellular sulfur oxidizers" and the "filamentous sulfur oxidizers." In addition, one proteobacterium that is a sulfur oxidizer, *Thiobacillus ferrooxidans,* can also carry out the oxidation of pyrite, FeS_2, with the formation of ferric oxide and sulfate and obtain energy both from iron oxidation as well as the oxidation of sulfur. Thus, it is also considered in this section. Other sulfur oxidizers, such as the family *Sulfolobaceae,* are *Archaea* (see Chapter 18).

Some bacterial sulfur oxidizers use sulfide as an energy source. The sulfide can be supplied as hydrogen sulfide, or as a metal sulfide such as iron or copper sulfide. Some use elemental sulfur, S^0, whereas others use thiosulfate, $S_2O_3^{2-}$. Some can use all three forms of sulfur. The final product formed in all cases is sulfuric acid. The overall reactions for the complete oxidation of these various sulfur forms are shown following:

$$(8)\ S^{2-} + 4\,O_2 \rightarrow 2\,SO_4^{2-}$$

$$(9)\ 2\,S^0 + 3\,O_2 + 2\,H_2O \rightarrow 2\,H_2SO_4$$

$$(10)\ S_2O_3^{2-} + 2\,O_2 + H_2O \rightarrow 2\,SO_4^{2-} + 2\,H^+$$

The production of sulfuric acid (as sulfate or sulfuric acid) leads to the lowering of pH during the growth of these bacteria. Indeed, the acidophilic members of the sulfur oxidizers, as exemplified by *Thiobacillus thiooxidans,* can produce enough acid to lower the pH to 1.0, equivalent to 0.1 N H_2SO_4 and still remain viable. Clearly, these are examples of ultimate proton pumpers! These oxidations yield energy for the cells as shown in Figure 19.23.

UNICELLULAR SULFUR OXIDIZING PROTEOBACTERIA   *Thiobacillus* is a genus of polarly flagellated rods that is widespread in soil and aquatic habitats. *Thiobacillus* is especially diverse metabolically (Table 19.16). At one extreme are obligate chemolithotrophs, such as *T. thiooxidans,* which do not use organic compounds at all. These obligate chemolithotrophs are also called chemolithoautotrophs because they use (1) chemicals as energy source (hence, *chemo*-), (2) inorganic sources of electrons (that is, **litho**-, meaning rock), and (3) carbon dioxide as a sole carbon source (hence, **auto**trophic). Many autotrophic species produce carboxysomes in their cells, which con-

(A)

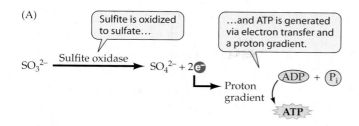

Sulfite is oxidized to sulfate…

…and ATP is generated via electron transfer and a proton gradient.

$$SO_3^{2-} \xrightarrow{\text{Sulfite oxidase}} SO_4^{2-} + 2e^-$$

(B)

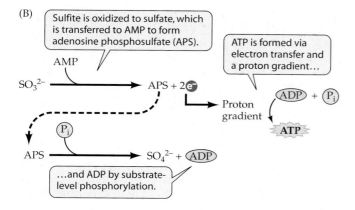

Sulfite is oxidized to sulfate, which is transferred to AMP to form adenosine phosphosulfate (APS).

ATP is formed via electron transfer and a proton gradient…

…and ADP by substrate-level phosphorylation.

Adenosine-5′-phosphosulfate (APS)

Figure 19.23 Oxidation of sulfur compounds by *Thiobacillus*
Pathways for oxidation of reduced sulfur compounds and generation of energy by *Thiobacillus* species. Sulfide, elemental sulfur, and thiosulfate are oxidized by the enzyme sulfide reductase to form sulfite. The sulfite is metabolized by one of two different pathways, (A) or (B). (A) shows the principal pathway.

tain high concentrations of Rubisco (Figure 19.24). *Thiobacillus novellus* is an intermediate group of so-called facultative chemolithotrophs—or mixotrophs—that fix carbon dioxide but can also assimilate organic carbon sources for anabolic purposes. *Thiobacillus denitrificans* can utilize nitrate as an electron acceptor.

Thermothrix, which is another sulfur-oxidizing genus of the *Betaproteobacteria*, is found in hot springs where it grows as a filamentous organism.

Thiomicrospira is an important sulfur oxidizer in the marine environment and is a member of the *Gammaproteobacteria*.

Finally, at the other extreme are some ordinary heterotrophs including some *Pseudomonas* species that oxidize reduced sulfur compounds but do not obtain energy from this process and cannot utilize carbon dioxide as a carbon source.

Order Rhodocyclales

The genus Zoogloea forms masses of cells, called **zoogloea**—for which the genus was named—enclosed in a polysac-

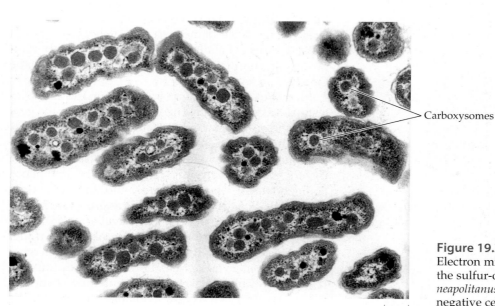

Carboxysomes

1 μm

Figure 19.24 *Thiobacillus*
Electron micrograph of rod-shaped cells of the sulfur-oxidizing bacterium *Thiobacillus neapolitanus*, showing the typical gram-negative cell structure. Note the carboxysomes in the cells. Courtesy of J. Shively.

TABLE 19.16	Characteristics of unicellular sulfur-oxidizing bacteria			
Class/Genus/Species	Nutrition Group	pH Optimum	Sulfur Source	Mol % G + C
Betaproteobacteria				
Thiobacillus thiooxidans	Chemolithoautotroph	2–4	$S^0, S_2O_3^{2-}$	51–53
T. ferrooxidans	Chemolithoautotroph	2–4	S^0, S^{2-} (incl. metal sulfides)	53–65
T. thioparus	Chemolithoautotroph	6–8	$S^0, S^{2-}, S_2O_3^{2-}$	63–66
T. acidophilus	Mixotroph	2–4	$S^0, S_2O_3^{2-}$	61–64
T. novellus	Mixotroph	6–8	$S_2O_3^{2-}$	67–68
T. denitrificans	Chemolithoautotroph	6–8	$S^{2-}, S^0, S_2O_3^{2-}$	63–68
Thermothrix thiopara	Mixotroph	6–8	$S^{2-}, S_2O_3^{2-}$	40
Gammaproteobacteria				
Thiomicrospira pelophila	Chemolithoautotroph	6.5–7.5	$S_2^-, S^0, S_2O_3^{2-}$	44

charide gelatinous matrix (Figure 19.25). *Zoogloea* occur in organically polluted freshwater and in wastewaters such as activated sludge in sewage treatment plants. In addition *Zoogloea ramigera* carries out denitrification.

SECTION HIGHLIGHTS

The *Betaproteobacteria* contains some pathogenic bacteria such as *Neisseria gonorrhoeae*, the causative agent of gonorrhoea, and *B. pertussis*, which causes whooping cough. In addition, many of the chemolithotrophic nitrifying bacteria are found in this class as well as others that obtain energy from the oxidation of hydrogen gas and reduced iron and sulfur compounds.

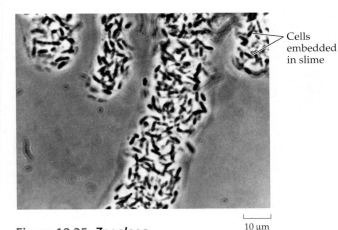

Figure 19.25 *Zoogloea*
Zoogloeal cell mass. Each finger-like projection contains numerous cells embedded in an extracellular slime matrix, which appears transparent. Courtesy of R. Unz.

Cells embedded in slime

10 μm

19.3 Class *Gammaproteobacteria*

The class *Gammaproteobacteria* is the largest class in the phylum *Proteobacteria*. It contains more than 150 genera, some of which are listed in Table 19.17. Included in the class are important pathogens, important environmental genera, phototrophs, and chemoautotrophs. As shown in Figure 19.26 the phylogeny shows that the *Betaproteobacteria* are actually embedded within this class, another illustration that the phylogeny of the *Proteobacteria* is complex and still poorly understood (see Box 19.1). Selected orders and their important species are discussed in this section.

Order Enterobacteriales

Many of the important members of the *Proteobacteria* are fermentative. None of these are obligately anaerobic, but instead grow as facultative aerobes. Several important groups are known, all of which are members of the *Gammaproteobacteria*.

Many but not all bacteria in this group live in the intestinal tracts of animals, hence accounting for the general name of "enteric bacteria" for these organisms. The enteric bacteria are small, nonsporeforming rods, typically about 0.5 μm in width and from 1 to 5 μm in length. In general, they have simple nutritional requirements. Some, however, require vitamins and/or amino acids for growth. Most of the bacteria in this group are motile, and, if so, they have peritrichous flagella.

Enteric bacteria are facultative aerobes that, under anaerobic conditions, ferment glucose and certain other sugars to form an array of end products. They are catalase positive and oxidase negative, and most reduce nitrate to nitrite when oxygen availability is limited. The DNA from the various species exhibits considerable homology, based on hybridization tests (Figure 19.27).

TABLE 19.17	Characteristics of genera of the class *Gammaproteobacteria*[a]	

Orders	Representative Genera	Characteristics
Chromatiales	*Chromatium*	Phototroph
	Lamprocystis	Phototroph
	Thiocapsa	Phototroph
	Thiocystis	Phototroph
	Thiodictyon	Phototroph
	Thiopedia	Phototroph
	Thiospirillum	Phototroph
	Ectothiorhodospira	Phototroph
	Nitrosococcus	Nitrifier
Xanthomonadales	*Xanthomonas*	Plant pathogen
	Xylella	Plant pathogen
Thiotrichales	*Thiothrix*	Filamentous sulfur oxidizer
	Beggiatoa	Filamentous sulfur oxidizer
	Thioploca	Filamentous sulfur oxidizer
	Leucothrix	Filamentous sulfur oxidizer
	Francisella	Animal pathogen
Legionellales	*Legionella*	Animal pathogen
	Coxiella	Q fever
Methylococcales	*Methylococcus*	Methylotroph
	Methylobacter	Methylotroph
	Methylomicrobium	Methylotroph
	Methylomonas	Methylotroph
Oceanospirillales	*Oceanospirillum*	Marine heterotroph
	Halomonas	Halophile
Pseudomonadales	*Pseudomonas*	Common heterotroph
	Azomonas	Nitrogen fixer
	Azobacter	Nitrogen fixer
	Moraxella	Heterotroph
	Acinetobacter	Heterotroph
Alteromonadales	*Alteromonas*	Marine heterotroph
	Marinobacter	Marine heterotroph
	Shewanella	Marine heterotroph
Vibrionales	*Vibrio*	Animal pathogens
	Photobacterium	Luminescent bacteria
Aeromonadales	*Aeromonas*	Common heterotroph
	Ruminobacter	Rumen anaerobe
	Succinomonas	Succinic acid producers
Enterobacteriales	*Enterobacter*	Common in soil
	Escherichia	Colon enteric
	Buchnera	Aphid symbiont
	Klebsiella	Pathogen
	Erwinia	Plant pathogen
	Proteus	Common heterotroph
	Salmonella	Animal pathogen
	Serratia	Enteric bacterium
	Yersinia	Bubonic plague
Pasteurellales	*Pasteurella*	Animal pathogen
	Haemophilus	Human diseases

[a]A complete listing of genera is provided in the Appendix.

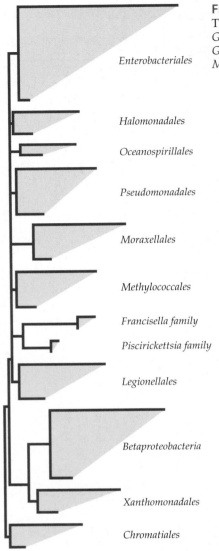

Figure 19.26 16S rRNA tree of the *Gammaproteobacteria*
This tree shows the various orders and selected genera of the *Gammaproteobacteria*. Note that the *Betaproteobacteria* are a subgroup of the *Gammaproteobacteria* by this analysis. Adapted by permission from *Bergey's Manual of Systematic Bacteriology*.

Enterobacteriales

Halomonadales

Oceanospirillales

Pseudomonadales

Moraxellales

Methylococcales

Francisella family

Piscirickettsia family

Legionellales

Betaproteobacteria

Xanthomonadales

Chromatiales

These organisms are metabolically and genetically similar but demonstrate considerable diversity in ecology and pathogenic potential for humans and other vertebrate animals (cold and warm blooded), insects, and plants. The intimate association of many enteric bacteria with higher eukaryotes suggests a long evolutionary relationship. Major genera in the enteric group and their habitats are shown in Table 19.18.

The enteric bacteria have had considerable influence on human history (see Chapter 2). Their medical importance is evident, as members of this group are the causative agents of diseases such as plague (*Yersinia pestis*), typhoid fever (*Salmonella typhi*), and bacillary dysentery (*Shigella dysenteriae*). Those that are not pathogenic in healthy individuals can cause disease under appropriate conditions and are therefore called opportunistic or secondary pathogens. Because they are so firmly associated with humans, a huge number of enterobacteria have been isolated and characterized. For example, 1,464 different serogroups have been defined as serovar subspecies in the genus *Salmonella*. The considerable medical and epidemiological importance of the enterics is discussed in Chapters 28 and 30.

The enteric bacteria are separated into two physiological groups, based on the products they generate during glucose fermentation. The **mixed-acid fermenters** (Figure 19.28) produce significant amounts of organic acids. In contrast, the **2,3-butanediol fermenters** (Figure 19.29) produce mostly neutral compounds. *E. coli* is

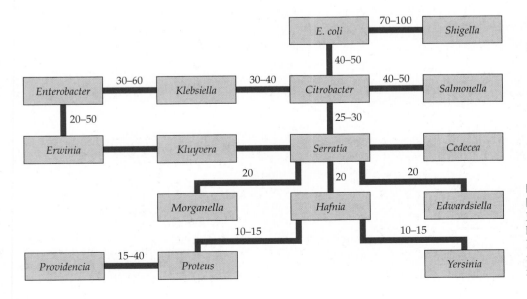

Figure 19.27 Enteric bacterial genera
Relatedness among the enteric bacteria. Numbers indicate DNA–DNA reassociation values, a measure of the relatedness among the various genera.

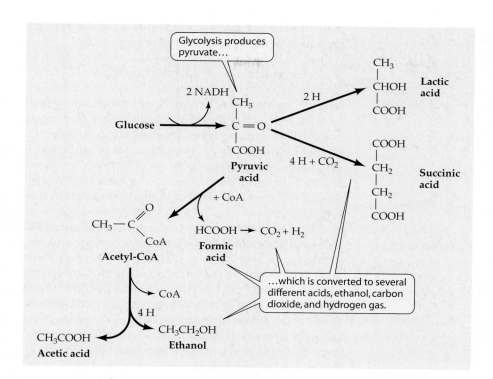

Figure 19.28 Mixed-acid fermentation
Mixed-acid fermentation of glucose by *Escherichia coli*. Pyruvate produced by glycolysis (the Embden-Meyerhof pathway) is catabolized to a mixture of products, high in acids. Acids are shown in their nonionized form.

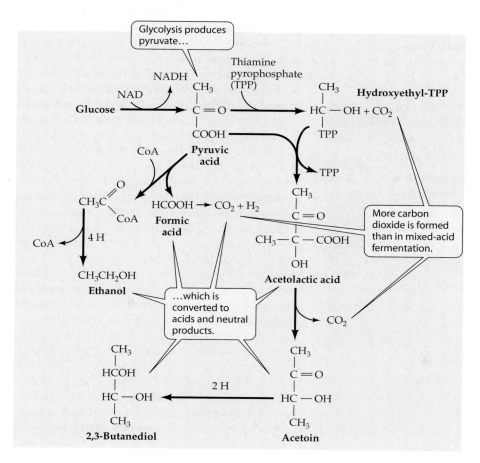

Figure 19.29 2,3-Butanediol fermentation
In the butanediol variation of mixed-acid fermentation, more neutral products are formed, as well as a higher ratio of CO_2 to H_2.

TABLE 19.18 Normal habitat of the enteric bacteria

Genus	Habitat
Escherichia	Normal microbiota of intestinal tract of warm-blooded animals
Salmonella, Shigella, and *Providencia*	Intestinal pathogens of humans and other primates; also associated with turtles
Hafnia	Feces of humans, other animals, and birds
Edwardsiella	Cold-blooded animals; may be pathogenic for eels and catfish
Proteus	Intestinal tracts of humans and other animals; also soil and polluted water
Morganella	Mammal and reptile feces
Yersinia	Humans, rats, and other animals
Klebsiella, Citrobacter, Enterobacter, and *Serratia*	Human intestine; also soil and water
Erwinia	Plants as pathogenic, saprophytic, or epiphytic microflora

a mixed-acid fermenter. It metabolizes glucose via the Embden-Meyerhof pathway to pyruvate. The reduced NADH generated by glyceraldehyde-3-phosphate dehydrogenase is reoxidized with the formation of three major acids: acetic, lactic, and succinic acids. Formic acid is catabolized by the enzyme formic hydrogen lyase to CO_2 and H_2 in a ratio of one to one (Table 19.19). Fermentation of 100 moles of glucose by *E. coli* leads to the formation of about 128 moles of organic acids.

Butanediol fermentation results in the formation of neutral products and considerable quantities of CO_2 (see Table 19.3). *Enterobacter aerogenes* is a typical butanediol-fermenting enteric bacterium. Formic acid is cleaved to form $H_2 + CO_2$, but the significantly higher proportion of CO_2/H_2 occurs because other reactions produce CO_2 but not hydrogen. Only about 20 moles of acidic product are generated from 100 moles of glucose. A distinctive and therefore diagnostically important intermediate called acetoin is formed in this fermentation (see Figure 19.29).

The precise identification of enterics is of considerable importance in public health microbiology and epidemiology. Consequently, a variety of diagnostic tests have been devised to identify genera in this group (Tables 19.20 and 19.21). Identification of a strain from a clinical specimen entails a large variety of diagnostic tests, which are set up systematically as illustrated by a dichotomous tree (Figure 19.30). Typically, the clinical microbiologist performs these tests routinely, along with whole cell fatty acid methyl ester analysis (see Chapter 17), using prepackaged automated or semiautomated formats that are available from commercial manufacturing firms. The results of these tests can be subjected to computer analysis to find the best-fit species identification for the unknown strain.

E. coli, the best-known species of the enteric bacteria and arguably the most widely known of all the bacteria, has been more thoroughly studied than any other living organism. This common bacterium is a normal inhabitant of the intestinal tract of humans and most other warm-blooded animals. As a facultative aerobe of the intestinal tract, it is well equipped to survive anaerobically and consume oxygen that enters its habitat. This activity is essential for maintaining anaerobic conditions in the large intestine. Although it is an important species in the human intestinal tract, it is not the dominant one.

Because *E. coli* is found in the intestinal tract of warm-blooded animals, it is an important test organism for fecal contamination of food and drinking water. Because *E. coli* does not grow or survive for long periods in food, water, or soil, its presence in these environments is indicative of contamination by fairly recent fecal material and, therefore, should not be consumed. Thus, *E. coli* is termed an **indicator bacterium**, whose presence in the environment indicates fecal contamination from warm-blooded animals (see Chapter 32).

Shigella dysenteriae is the causative agent of **bacillary dysentery**, a severe type of gastroenteritis (see Chapter 28). The species of this genus are

TABLE 19.19 Relative amounts of product from a mixed-acid and a neutral (butanediol) fermentation

	Moles of Product per 100 Moles of Glucose Fermented	
	Mixed-Acid Fermentation (*Escherichia coli*)	Neutral Fermentation (*Enterobacter aerogenes*)
Acetic acid	36	0.5
2,3-Butanediol	0	66
Ethanol	50	70
Lactic acid	79	3
Succinic acid	11	0
Formic acid	2.5	17
H_2	75	35
CO_2	88	172
Total moles of acid	128.5	20.5
Ratio CO_2/H_2 produced	1.2	4.9

TABLE 19.20	Types of tests used to differentiate enteric bacteria from one another
Test	**Description**
Urease	Streak the organism on urea agar base. Urea agar contains high levels (2.0%) of urea and phenol red indicator. Organism producing urease liberates ammonia that turns indicator red near streak.
Indole	Grow organism in a peptone medium with high tryptophan content. Test for indole production after growth.
Motility	Tubes contain tryptose in a soft agar medium (0.5% agar). Inoculate by stabbing down the center of the agar. Diffuse growth out from the inoculum indicates motility.
Methyl red	A buffered glucose-peptone medium containing methyl red indicator. Turns red if large amount of acid is produced, as in mixed-acid fermentation.
Acetoin (Voges-Proskauer)	A liquid medium containing glucose is inoculated and, after growth, assayed for acetoin.
Citrate	Organism is tested for its ability to grow with citrate as sole carbon source.
H_2S production	Inoculate agar medium containing ferrous sulfate by stabbing. Blackening of the agar indicates sulfide production.
Phenylalanine	After growth on a medium containing phenylalanine deaminase (0.1%), assay for production of phenylpyruvic acid by adding ferric chloride. A green color indicates deaminase activity.
Ornithine decarboxylase	Measured manometrically by following CO_2 release.
Gas from glucose	Gas production from glucose or other sugars is measured by placing a Durham tube in a broth containing sugar. A Durham tube is a small test tube that is placed inverted in the broth before autoclaving. Autoclaving causes the tube to be filled with broth. Gas production is indicated by a gas bubble within the Durham tube.
β-Galactosidase	Grow organism on a lactose medium. Add a disc containing O-nitrophenyl-β-D-galactoside to the surface. After 15 to 20 minutes the yellow O-nitrophenol is released if the enzyme is present.

so closely related to *E. coli*, as determined by DNA–DNA reassociation, that they could be considered the same species. However, for historical reasons and because of their distinctive pathogenesis, a separate genus has been maintained.

Salmonella typhi causes **typhoid fever** and gastroenteritis. As mentioned earlier, there are hundreds of serovars in this genus. The antigens for the serovars are determined from three types of surface polymers, including the outer cell membrane lipopolysaccharides ("O" antigens), the fla-

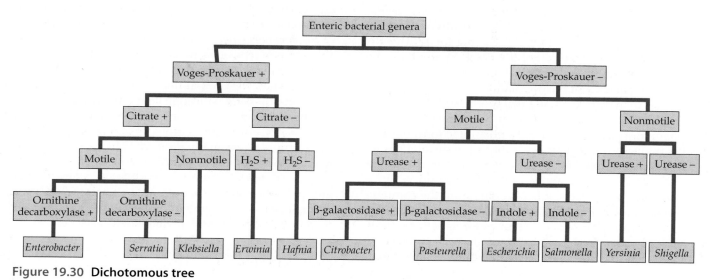

Figure 19.30 Dichotomous tree
This dichotomous tree illustrates how phenotypic tests can be used to separate and identify various genera of enteric bacteria. (Please see text for details.)

TABLE 19.21	Characteristics used to distinguish the common genera of enteric bacteria

Genera	Methyl Red	Voges-Proskauer	Citrate Utilization	Urease	Indole Production	Motility	H₂S Production	Gas from Glucose	β-Galactosidase	Ornithine Decarboxylase	Phenylalanine Deaminase	Mol % G + C
Mixed-Acid Fermenters												
Eschericia	+^{a}	–	–	–	+	+	–	+	+	+	–	48–52
Salmonella	+	–	V	–	–	+	–	+	V	+	–	50–53
Shigella	+	–	–	–	V	–	–	–	V	V	–	49–53
Edwardsiella	+	–	–	–	+	+	–	+	–	+	–	53–59
Citrobacter	+	–	+	+	V	+	–	+	+	+	–	50–52
Proteus	+	V	V	+	V	+	V	+	–	V	+	38–40
Morganella	+	–	–	+	+	+	–	+	–	+	+	50
Providencia	+	–	+	V	+	+	–	+	–	–	+	39–42
Yersinia	+	–	–	+	–	–	–	–	+	V	–	46–50
2,3-Butanediol Producers												
Klebsiella	+	+	+	V	V	–	–	+	+	V	–	56–58
Enterobacter	V	+	+	–	–	+	–	+	+	+	–	52–60
Serratia	V	+	+	–	–	+	–	V	+	–	+	52–60
Erwinia	+	+	–	–	–	+	+	–	V	–	–	50–58
Hafnia	V	V	–	–	–	+	–	+	+	+	–	48–49

^{a}Symbols: + = positive for most strains; – = negative for most strains; V = variable within genus.

gella ("H" antigens), and the outer layer polysaccharides ("Vi" antigens). These serovars are helpful in tracing the source of organisms during an outbreak of typhoid fever.

Another important pathogenic genus is *Yersinia*, which contains the dreaded agent of **bubonic plague**, *Yersinia pestis* (see Chapter 28). During the fourteenth century, in a scourge called the Black Death, bubonic plague spread throughout Europe resulting in the death of more than 25% of the population, a proportion much greater than caused by any war (see Chapter 2).

Klebsiella pneumoniae can cause a type of bacterial pneumonia as well as urinary tract infections, but it is not normally pathogenic. As a soil and freshwater group, this genus is noted for its lack of motility, its formation of a capsule, and the ability of many strains to fix nitrogen. In other respects, it is very similar to the common soil bacterium, *Enterobacter aerogenes*.

The genus *Proteus* is well known for its active **urease**, which breaks down urea to ammonia and carbon dioxide. It is also noted for its active motility, called swarming, which occurs when cells align themselves in the long axis and rapidly spread over the surface of an agar plate.

Serratia marcescens is a distinctive enteric bacterium because of its red pigment. The pigment of this bacterium is a bright red, pyrrole-based compound called **prodigiosin**, which is produced at temperatures less than 37°C. Prodigiosin is a tripyrrole (**Figure 19.31**) akin to the tetrapyrroles of chlorophyll, cytochrome, and heme. The function of prodigiosin in *S. marcescens* is still unknown.

Plesiomonas shigelloides causes a disease similar to shigellosis, a type of bacterial dysentery. Although it has been isolated from fish and land mammals, it is not a constituent of the normal flora of humans. The disease is apparently spread through contaminated water.

One of the most recent isolates of the enteric group is *Xenorhabdus*, which grows in association with nematodes. Some strains can actually produce light, a phenomenon referred to as **bioluminescence**. This fascinating capability is more widely found in the *Vibrio-Photobacterium* group discussed next. Bioluminescent bacteria are usually associated with symbiotic associations, and they are therefore discussed in more detail in Chapter 25.

Order Vibrionales

This order contains important marine bacteria, some of which are bioluminescent. In addition, several species are pathogens of humans.

(A)

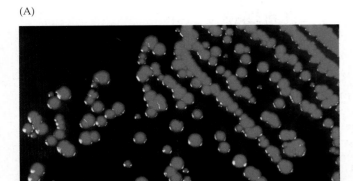

(B)

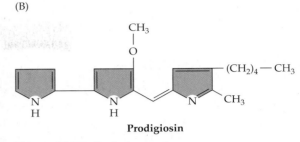

Prodigiosin

Figure 19.31 Red pigment
(A) A culture of *Serratia marcescens* showing red pigmentation. (B) Chemical structure of the tripyrrole prodigiosin, the bright red pigment of *Serratia marcescens*. A, ©Christine Case/Visuals Unlimited.

VIBRIO AND *PHOTOBACTERIUM*   *Vibrio* and *Photobacterium* species are gram-negative facultatively anaerobic straight or curved rods. All are motile via polar flagella. All are capable of respiration or fermentation. None can use nitrate as the terminal electron acceptor in respiration. Most are oxidase positive and utilize glucose as the sole source of carbon and energy. These three genera are primarily aquatic organisms; those that live in seawater require 2% to 3% NaCl or a seawater-based medium for optimal growth. Some species are pathogenic to humans and other vertebrates. Some are pathogenic to fish, frogs, and certain invertebrates.

Certain species of *Photobacterium* and *Vibrio* are bioluminescent and emit a blue-green light. The bacteria live as symbionts in the animal light organs, special areas on the animals where these bacteria are maintained in high concentrations (see Chapter 25). The mechanism of bioluminescence has been extensively studied in *Vibrio harveyi* and *Vibrio fischeri*.

The polar flagellum of *Vibrio* species is enclosed in a sheath. The sheath is continuous with the outer envelope of this gram-negative bacterium. Some species also produce lateral flagella when they are grown on solid media. These species lack a sheath and their lateral flagella have a shorter wavelength than the polar flagellum. Such flagellation, which is unique to this group, is termed **mixed flagellation**. Apparently, synthesis of the lateral flagella occurs when the microenvironment is particularly viscous, such as on an agar surface. In this manner, the additional flagella aid in the cell's movement.

Vibrio species live primarily in marine and estuarine environments (Box 19.5). Many live in the intestines or on the outer surfaces of marine animals. Almost all grow best with added sodium ions in the medium and therefore are typically grown in media with marine salts. Most species grow on a variety of organic compounds (sugars, amino acids, and organic acids) and do not require growth factors. Extracellular hydrolases are produced by many

species and include amylase, lipase, chitinase, and alginase. Also, some are agar digesters, which produce depressions or cavities on the surface of agar media where their colonies grow. One species, *Vibrio natriegens*, has the shortest known doubling time, 6 minutes, of any organism.

Vibrio cholerae is the most thoroughly studied species in the group because it is the causative agent of **cholera** (see Chapter 28). Cholera is an epidemic human disease that occurs most commonly in densely populated coastal areas, such as Bangladesh during periods of flooding. It is transmitted by fecal contamination of water and food. Effective drinking water treatment has eliminated the disease from most developed countries.

Photobacterium species, like *Vibrio*, are widespread in marine habitats and are also found associated with fish. The luminescent species, *Photobacterium phosphorum*, can be isolated by incubating marine fish, squid, or octopus partially submerged in seawater at 10°C to 15°C. After 15 to 20 hours, luminescent areas develop, and material from them can be streaked on plates for isolation of strains. See the symbiosis chapter for more information on their bioluminescence (see Chapter 25).

PASTEURELLA AND *HAEMOPHILUS*   *Pasteurella* and *Hemophilus* are parasites of invertebrates and are often pathogens for both mammals and birds. All are nonmotile rods that are oxidase positive. They are generally fastidious microbes requiring organic nitrogen sources, B vitamins, amino acids, and hematin for growth. Characteristics of this group are shown in Table 19.22.

Pasteurella species are the causative agents of diseases in cattle and of "fowl cholera" in poultry. They are parasites of the mucous membranes of the respiratory tracts of mammals and birds. Some species are found in the digestive tracts of animals, but they rarely cause human disease.

Haemophilus species are obligate parasites of the mucous membranes of humans and other animals. Blood or blood derivatives are required for growth. *Haemophilus*

BOX 19.5   *Research Highlights*

Are Marine Bacteria Unique?

The oceans began forming very early in the evolution of Earth. Their saltiness increased as land was eroded by rainfall and the rivers carried the dissolved salts into the sea. The total concentration of salts now equals a salinity of about 3.5% or, as oceanographers state, 35‰ (parts per thousand). Thus, marine bacteria must be able to grow in a saline environment with a lower water activity than their freshwater relatives.

Some microbiologists have defined marine bacteria as those that grow optimally when the salinity of the medium, or growth environment, is at about 3.5%. To test this, special media are prepared that are amended with the appropriate concentrations of sea salts—this is referred to as artificial seawater, or sterile seawater itself may be used.

In addition to the preference to grow best at 3.5% salinity, some microbiologists have proposed that true marine bacteria must also have an absolute requirement for sodium ions. This is tested by preparing a medium in which the sodium ions are replaced entirely by potassium ions. Using this definition, if the organism cannot grow under these conditions, it is not regarded as a marine organism.

Recently it has been discovered that strains of *Caulobacter* isolated from seawater environments differ phylogenetically from those isolated from freshwater, giving rise to a separate marine genus, *Maricaulis*. This is consistent with the view that, millions of years ago, these freshwater and marine groups diverged to form separate genera within the *Alphaproteobacteria*.

---

influenzae was originally isolated from patients with viral influenza and was thought to be the causative agent of the disease, hence the name. The organism is present in the nasopharynx of healthy individuals and is probably a secondary invader in infections of the respiratory tract. This is the first organism whose entire genome was sequenced in 1995.

Order Pseudomonadales

Many of the *Proteobacteria* are obligately aerobic, nonfermentative rods and cocci that obtain their energy through aerobic respiration. They use a large variety of organic compounds as energy sources, depending on the genus. For example, some species of *Pseudomonas* can use more than 100 different organic monomers, including amino acids, sugars, and organic acids. Others in this group are nitrogen fixers. Some members of this group are widespread in aquatic and soil environments, whereas others are obligate parasites of humans and other animals.

The "pseudomonads" are *Gammaproteobacteria* that utilize oxygen as a terminal electron acceptor. Some species also utilize nitrate as an alternate electron acceptor in anaerobic respiration, and therefore can grow anaerobically. The pseudomonads, however, do not ferment, which is a hallmark of the enteric and *Vibrio* groups. The pseudomonads are gram-negative rods (generally 0.5×1–4 µm), and virtually all species are motile by one or more polar flagella (Figure 19.32). The pseudomonads are almost entirely oxidase and catalase

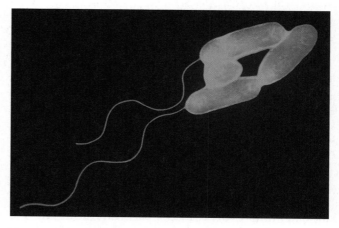

Figure 19.32 *Pseudomonas*
Electron micrograph (with false color) of typical *Pseudomonas* cells with single polar flagella. Photo ©Kwangshin Kim/Photo Researchers, Inc.

TABLE 19.22	Characteristics of the *Pasteurella* group	
Genus	**Morphology**	**Mol % G + C**
Pasteurella	Ovoid or rod (0.3–1.0×1.0–2.0 µm)	40–45
Haemophilus	Coccobacillus (0.3–0.5×0.5–3.0 µm)	37–44

TABLE 19.23	Differentiation of the pseudomonads from the vibrios and enteric bacteria		
Characteristic	Pseudo-monads	Vibrios	Enteric Bacteria
Flagella	Polar	Polar or mixed	Peritrichous
Oxidase[a]	+[b]	+	–
Glucose fermentation	–	+	+

[a]The oxidase test is a measure of the enzyme, cytochrome oxidase. It is assayed by picking up a colony of the test organism with a platinum loop and placing a smear directly on filter paper containing two or three drops of a 1% solution of tetramethyl-*p*-phenylenediamine dihydrochloride. If the smear turns violet, the organism is considered oxidase positive.

[b]Some *Xanthomonas* strains are oxidase negative.

positive. **Oxidase-positive** organisms produce large amounts of cytochrome oxidase. The pseudomonads can be differentiated from the *Vibrio* and enteric groups by a few simple diagnostic tests (Table 19.23).

The genus *Pseudomonas* is the most thoroughly studied and characterized of the group. It is of widespread occurrence in nature. Some species reside in soil, in water, and on surfaces such as human skin. Species are defined on the basis of physiological characteristics. Some species produce water-soluble fluorescent pigments. These are yellow-green pigments called **pyocyanin** and **pyoverdin** that diffuse into the medium and fluoresce under ultraviolet light. The pyocyanins are blue-colored phenazines (Figure 19.33). The chemical structure of the pyoverdins is not completely known because they are unstable. However, they are believed to be **siderophores**, which are iron-binding compounds that bring ferric ion from the external environment into the cytoplasm (see Chapter 5).

Members of the genus *Pseudomonas* like the betaproteobacterial genus *Burkholderia* exhibit remarkable nutritional versatility. Most can grow on 50 or more different substrates, and some can use more than 100 different organic compounds. Because of this metabolic versatility, *Pseudomonas* species are very important in the degradation of organic compounds in soil and aquatic environments. Among the substrates utilized are sugars, fatty acids, di- and tricarboxylic acids, alcohols, aliphatic hydrocarbons, aromatic hydrocarbons, amino acids, various amines, and other naturally occurring compounds as well as many manufactured chemicals such as chlorinated hydrocarbons that are toxic to animals and plants. Glucose is catabolized by many of the pseudomonads via the Entner-Duodoroff pathway (see Chapter 8). *Pseudomonas* species also have a limited capacity to hydrolyze polymeric compounds.

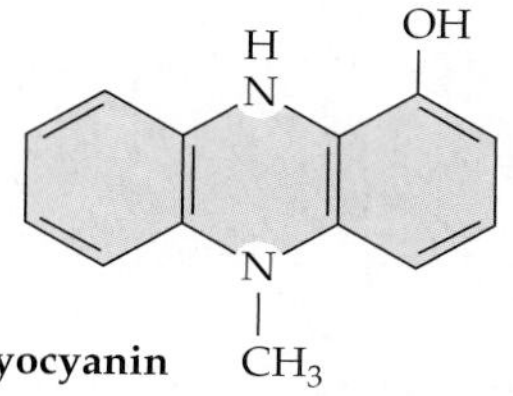

Pseudomonas species are used extensively as biochemical tools in the elucidation of catabolic pathways. They have been particularly useful in ascertaining pathways in the catabolism of aromatic compounds (see Figure 12.19). The genetics of many fluorescent strains have been clarified. Both conjugational and transduction systems have been established. The genetic information for synthesis of the enzymes involved in catabolism of many uncommon organics occurs on transferable plasmids. Among these are genes for catabolism of salicylate, camphor, octane, and naphthalene. It should be noted that the genus *Pseudomonas* has been modified greatly since its taxonomy was evaluated by rRNA analysis. The original genus has now been split into several additional genera, including *Comamonas* (an alpha subdivision genus that does not produce fluorescent pigments); *Deleya* and *Halomonas*, which are marine; *Stenotrophomonas* and *Hydrogenophaga* (H_2-oxidizing facultative chemolithotrophs); and *Burkholderia*.

Pseudomonas aeruginosa is an opportunistic human pathogen. For example, it causes serious skin infections of burn victims and grows in the lungs of patients with cystic fibrosis. However, it can also cause urinary tract and lung infections in healthy individuals. Normally, it is found in soils and is an important denitrifying genus. Certain other *Pseudomonas* species are animal pathogens such as *Pseudomonas mallei*, which causes glanders in horses.

A number of species of *Pseudomonas*, including *Pseudomonas syringae*, are plant pathogens, which can cause diseases such as brown spot and halo blight on bean plants. *P. syringae* owes its invasiveness to its ability to form ice crystals, which damage the plant tissue (Box 19.6).

Moraxella is found in the eyes and upper respiratory tracts of humans and other warm-blooded animals. Some species cause conjunctivitis (eye inflammation) in humans and bovines. *Moraxella catarrhalis* is a normal inhabitant of the nasal cavity of humans, but it is an opportunistic pathogen associated with disease in unhealthy patients.

Most strains of *Acinetobacter* have simple nutritional requirements, and as the genus name implies, are im-

Figure 19.33 Fluorescent pigment
Chemical structure of pyocyanin, the blue phenazine pigment of *Pseudomonas aeruginosa*.

BOX 19.6   *Research Highlights*

Ice Nucleation

Certain gram-negative plant pathogenic bacteria, including some species of *Pseudomonas*, *Xanthomonas*, and *Erwinia,* cause ice nucleation. These bacteria normally colonize plant leaves in the spring and summer. In the fall, when the temperature begins to dip toward freezing, they cause frost damage by forming ice crystals at temperatures somewhat higher than normal freezing temperatures. The ice crystals damage the plant leaf cells, which then exude nutrients that these bacteria utilize. Ice nucleation is initiated by a specific protein located in the outer cell membrane of these bacteria. This protein has a distinctive repeating amino acid region, which is the site of ice nucleation.

Pseudomonas syringae, one of the ice-nucleating species, is also used to prevent ice nucleation damage to crops. To accomplish this, a mutant strain that has a deletion of the ice nucleation gene, an "ice-minus" strain, is sprayed on crop plants early in the spring. The concept is to enable the mutant bacterium to colonize the leaves before a wild-type, "ice-plus" strain reaches the leaves. Then, later on when a pathogenic, "ice-plus" strain arrives at the plant leaf, it will be prevented from colonizing because the ice-minus mutant is already there. This ecological concept is termed "pre-empting the niche."

The ice-nucleating bacterium, *Pseudomonas syringae*, is used commercially in snowmaking. The strain is added to water in snow-making machines used in ski areas to raise the temperature at which the water freezes to make snow. In this manner, less energy is required to cool the water to sufficiently low temperatures.

Ice nucleation. A snow-making machine in operation. ©Robert Cocquyt/istock-photo.com.

motile. *Acinetobacter* comprise a group of nutritionally diverse organisms that can utilize a wide array of substrates, including selected hydrocarbons. They cannot use hexoses as the sole source of carbon and energy, although they can oxidize aldose sugars to the corresponding sugar acid: for example, glucose + O_2 → gluconic acid. *Acinetobacter calcoaceticus* is a prominent member of this genus and is readily isolated from soil and water.

Azotobacter is a genus of free-living nitrogen-fixing bacteria that grow as bulbous rods (Figure 19.34). The organisms produce a capsular material that may aid in nitrogen fixation by keeping the oxygen concentration low in the vicinity of the organism. Cells may form resting cells called cysts that are resistant to dessication.

Order Xanthomonadales

Xanthomonas is a genus of yellow-pigmented plant pathogens. The yellow pigment is a brominated aryl polyene of unknown function (Figure 19.35) called **xanthomonadin** that is located in the cell membrane. Dis-

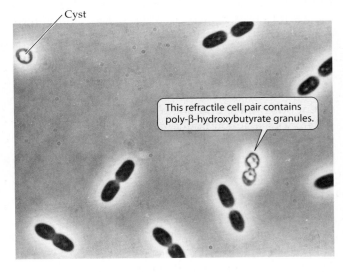

Cyst

This refractile cell pair contains poly-β-hydroxybutyrate granules.

Figure 19.34 *Azotobacter*
Cells of an *Azotobacter* sp. Courtesy of J. T. Staley.

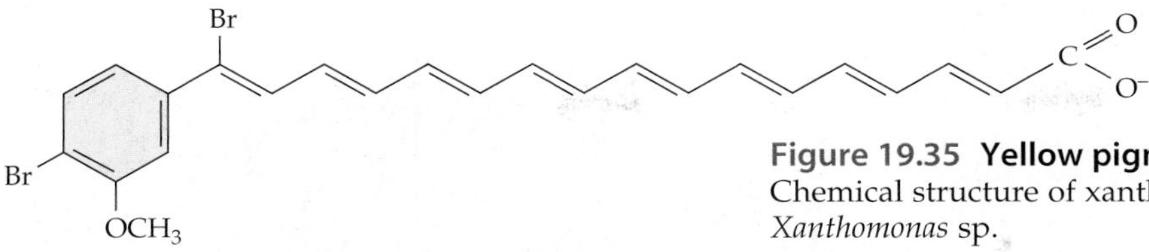

Figure 19.35 Yellow pigment
Chemical structure of xanthomonadin, the brominated pigment from *Xanthomonas* sp.

Xanthomonadin

ease symptoms caused by *Xanthomonas* vary with plant species and the strain involved. Stem wilt, leaf necrosis, and other manifestations of disease are observed. These are ascribed to toxins, enzymes, ice nucleation activity, and other metabolic products that accumulate during the growth of the bacterium in its host tissue.

Order Alteromonadales

The genus *Shewanella* is remarkable in the variety of alternative electron acceptors it can use in respiration. As a facultative aerobe, it is capable of using oxygen as its terminal electron acceptor in aerobic respiration. However, it can also use a variety of other electron acceptors in anaerobic respiration including oxidized iron compounds, manganese, nitrate, nitrite and various sulfur compounds. Therefore, like *Pseudomonas* and *Burkholderia*, it has metabolic diversity, but its diversity is especially directed at the large variety of electron acceptors it is capable of utilizing. The genus is widespread in fresh waters, estuarine environments, and in the marine environment as well.

Order Legionellales

This order contains two important genera of human pathogens, *Legionella* and *Coxiella*.

LEGIONELLA   DNA–DNA hybridization and 16S rRNA sequencing confirm that *Legionella* species and related organisms form a distinct group among the aerobic, gram-negative rods. *Legionella pneumophila*, which is the causative agent of Legionnaires' disease, is the type species (see Chapter 28). All species in the genus have been implicated in human respiratory disease. Morphologically, they are relatively small bacteria (0.5 × 2 μm). All strains are motile by polar or lateral flagella. Members of this genus have complex nutritional requirements, and L-cysteine and iron salts must be provided for growth. All are strict aerobes that utilize amino acids as their source of carbon and energy. They cannot oxidize or ferment carbohydrates.

Strains of *Legionella* and associated genera are commonly found in ponds, lakes, and wet soil. They also thrive in the warm water associated with evaporative cooling towers. It is probable that aerosols from this source caused an outbreak of the severe respiratory infection among the attendees at an American Legion Convention in Philadelphia in 1976. Several fatalities resulted from this outbreak. Difficulties in culturing the organism and assessing its nature pose a problem in prevention and treatment of the disease. It is not transferred from one human to another by contact. Sensitive immunofluorescence techniques indicate that *Legionella* strains are of widespread occurrence in the environment. Thus, they must either be of low pathogenicity or transferred to humans only rarely.

COXIELLA   *Coxiella burnetii* is the causative agent of the human disease Q fever (see Chapter 28). This organism also lives in close associations with arthropods as well as other vertebrate hosts.

Order Aeromonadales

Aeromonas species live both in fresh and marine waters as well as in sewage. *Aeromonas salmonicida* is a strict parasite that causes severe diseases of salmon and trout and can also infect humans. It lives in the blood and kidneys of fish.

Order Oceanospirales

The *Gammaproteobacteria* also contains spirilla such as the genus *Oceanospirillum*, which is widespread in the oceans. This is a member of a family of largely marine bacteria many species of which are spiral in shape. This order also contains the photosynthetic genus, *Thiospirillum*, which is discussed in Chapter 21.

Order Methylococcales: *Methylotrophic Bacteria*

Methane and methanol are considered by some to be two of the simplest organic compounds. However, it could be argued that true organic compounds have carbon–carbon bonds. Of course, microorganisms are not mindful of human definitions. The methylotrophic bacteria treat methane and methanol from a purely physical and biochemical standpoint—as substrates that can provide energy for growth. **Methanotrophic bacteria** use methane as a carbon source and oxidize it to carbon dioxide. These bacteria can also use methanol and a variety of other one-carbon and a few two-carbon compounds as carbon sources. Another separate group of bacteria use methanol, but not methane gas, as a carbon source for growth. Therefore, these are not methanotrophic bacteria, but are instead called **methylotrophic bacteria**. It should be noted that methanotrophic bacteria are also methylotrophic because they can also use methanol and other one-carbon compounds for growth. For this reason, this section is entitled the methylotrophic bacteria. In addition to these bacteria that use methanol, some of the *Archaea*, in particular the methanogens in the genus *Methanosarcina*, can use methanol as a carbon and energy source for growth.

METHANOTROPHS   There are two groups of methane-oxidizing bacteria, both of which have internal membrane systems similar to those of the nitrifying bacteria. The **type I methanotrophs** have an internal membrane system that lies perpendicular to the long axis of the cell (Figure 19.36A,B), whereas the **type II methanotrophs** have membranes that lie parallel to the cell membrane (Figure 19.37). The genera that have been described are listed in Table 19.24. The type I methanotrophs are members of the *Gammaproteobacteria*, whereas the type II methanotrophs belong to the *Alphaproteobacteria*. They are treated in this section on the *Gammaproteobacteria*.

Some of the methanotrophs produce a resting stage, called cysts or exospores. The cysts are analogous to those produced by *Azotobacter* species in that they have complex cell wall layers and may store poly-β-hydroxybutyrate. They are formed during conditions of nutrient depletion and can germinate during periods of nutrient sufficiency.

ENERGY GENERATION FROM METHANE   The methanotrophs are all gram-negative obligate aerobes that derive energy from the oxidation of methane to carbon dioxide. The first reaction is unique to these bacteria and involves

(A)

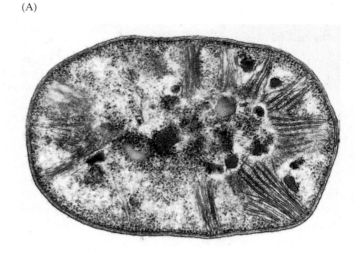

0.2 μm

(B)

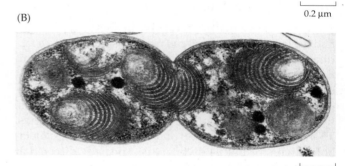

0.3 μm

Figure 19.36 Type I methanotrophs
Thin sections through type I methanotrophs, (A) *Methylomonas agile* and (B) *Methylomonas methanica*, showing the typical transverse membrane systems. Courtesy of C. Murrell.

TABLE 19.24	Genera of methanotrophic bacteria		
Group/Class	**Genus**	**Mol % G + C**	**Special Features**
Type I methanotrophs			
(*Gammaproteobacteria*)	*Methylobacter*	50–54	Rod, polar flagellum, cyst
	Methylococcus	62–64	Coccus, nonmotile, cyst
	Methylomonas	50–54	Rod, polar flagellum, cyst
Type II methanotrophs			
(*Alphaproteobacteria*)	*Methylocystis*	62-63	Rod or vibrio, nonmotile
	Methylosinus	62–63	Curved rod, polar flagellum, exospore

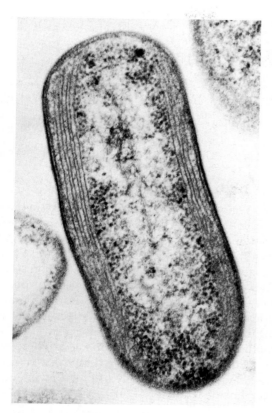

Figure 19.37 Type II methanotroph
Thin section of the type II methanotroph *Methylocystis parvus*, showing the membranes lying along the periphery of the cell. Courtesy of C. Murrell.

an enzyme called methane monooxygenase (MMO), which converts methane to methanol as shown following:

$$(11) \quad CH_4 + O_2 + XH_2 \rightarrow CH_3OH + H_2O + X$$

where X = hydrogen carrier (cytochrome carrying electrons).

The MMO is a complex copper-containing enzyme that can be either membrane-bound or free in the cytoplasm. The second reaction involves a methanol dehydrogenase that converts the methanol to formaldehyde as follows:

$$(12) \quad CH_3OH \rightarrow HCHO + 2\,e^- + 2\,H^+$$

The final reactions involved in carbon dissimilation for energy generation include either an oxidation via formate to carbon dioxide (see Chapter 10), or a more complex route to carbon dioxide called the ribulose monophosphate (RMP) pathway (see Chapter 10). Many type I methanotrophs use the RMP pathway for energy generation, but some of these can also use the serine-glycine pathway. Some of the species using the serine-glycine pathway have a distinctive quinone, pyrrolo-quinoline quinone (PQQ), as a hydrogen acceptor.

CARBON ASSIMILATION BY METHANOTROPHIC BACTERIA   To assimilate carbon, the type I methanotrophs use the RMP pathway (see Figure 10.4B), whereas the type II methanotrophs use the serine-glycine pathway (see Figure 10.4A). In both pathways, formaldehyde serves as the substrate for organic carbon synthesis. Thus, they are not autotrophs in that they cannot use carbon dioxide as their sole source of carbon.

ECOLOGICAL AND ENVIRONMENTAL IMPORTANCE OF METHANOTROPHIC BACTERIA   The methanotrophic bacteria are very important environmentally. They are responsible for the oxidation of methane produced by methanogenesis (see Chapter 18) and natural gas seeps in the biosphere. Methane is a greenhouse gas that has been increasing in the atmosphere at a rate of about 1% per year during the last decade. If it were not for the activity of these bacteria, the gas would increase at much more rapid rates, with deleterious environmental effects. These bacteria reside in soil and aquatic habitats where oxygen is available as well as methane. Therefore, they live at the interface between the reduced zones of the environment in which methane is produced and oxygenated zones exposed to air. Although methane gas escapes in part from shallow ponds and marshes, in deeper water bodies it is largely oxidized by the activity of methanotrophs before it reaches the surface of the water column. Methane also reaches the atmosphere through the activities of termites as well as ruminant animals that harbor methanogenic bacteria.

The MMO of methanotrophs can also remove halogen ions from toxic halogenated compounds. This process is called dehalogenation or dechlorination if the compound is chlorinated. Brominated compounds can be similarly dehalogenated. The enzyme is most effective at dehalogenating low-molecular-weight, one- and two-carbon, aliphatic compounds such as chloroform and trichloroethene. The methanotrophs do not derive energy from these dechlorination reactions. This is an example of co-metabolism (see Chapter 24) in that an energy source, in this instance methane, must be provided in order for these organisms to carry out the dechlorination process. They cannot obtain energy from the dehalogenation reactions. These bacteria are important agents in dehalogenation of wastes containing low-molecular-weight halogenated compounds and are therefore of considerable importance in bioremediation (see Chapter 32).

The methanotrophic bacteria share similarities with ammonia-oxidizing bacteria. Members of both groups have been shown to be able to carry out the oxidation of each others' substrates, ammonia, and methane, using their characteristic monooxygenases. In addition, both are members of the *Proteobacteria*, and therefore are phylogenetically related to one another. However, neither group obtains energy from the oxidation of the other's substrate and therefore is not considered to be ecologically important in the other's niche.

Other Methylotrophs

As mentioned heretofore, the term **methylotroph** refers to a microorganism that can utilize one-carbon compounds such as methane, methanol, or methylamine as sole carbon and energy sources for growth. This is a general term that encompasses methanotrophs discussed previously as well as bacteria such as *Hyphomicrobium* species that can grow on methanol as sole carbon source. A variety of heterotrophic genera are known to be methanol or methylamine users. All of these bacteria that have been studied use the serine pathway for carbon assimilation and grow by aerobic or nitrate respiration. It should be emphasized that none of them are known to use methane gas, and none of them produce the enzyme methane monooxygenase. In addition, all of them are facultative methylotrophs that can use other organic compounds for growth including sugars, amino acids, and organic acids, depending on the taxon.

One interesting newly discovered habitat for methylotrophic bacteria is on the surface of leaves. Methanol is derived from the degradation of methoxylated compounds produced by the plant. Strains especially adapted to growth on the plant leaf surface use the volatile methanol that is released. These strains are all pink in pigmentation.

Order Thiotrichales

This order is best known for its filamentous sulfur-oxidizing bacteria. However, in addition, it contains an important marine unicellular sulfur oxidizing genus, *Microspira*, the pathogen *Francisella* and a remarkable aromatic hydrocarbon degrading marine genus, *Cycloclasticus*.

FILAMENTOUS SULFUR OXIDIZERS   The filamentous sulfur oxidizers are found at the interface of sulfide-containing muds and the aerobic environment. For example, they are common on the surface of marine and freshwater sediments and in sulfur springs. These types of habitats, which have an active sulfur cycle, are called **sulfureta**. There are several important genera as presented in Table 19.25.

The genus *Beggiatoa* is the best-known member of the filamentous sulfur bacteria (Figure 19.38). Studies of this organism by Sergei Winogradksy led to his proposal of

TABLE 19.25	Filamentous sulfur-oxidizing bacteria	
Genus	**Mol % G + C**	**Distinctive features**
Beggiatoa	37–51	Long gliding filaments with sulfur granules
Thiothrix	52	Rosette-forming; gliding gonidia
Thioploca	Unknown	Gliding filaments in sheath

the concept of chemolithotrophy (see Box 19.4). Although Winogradsky did not study *Beggiatoa* in pure culture, several strains of *Beggiatoa* have now been isolated. All isolated strains produce sulfur granules and are known to oxidize sulfides, but some are facultative chemoautotrophs that use acetate as a carbon source.

The classical metabolic type of *Beggiatoa*, as proposed by Winogradsky, is a strict chemolithotroph that uses sulfide as an energy source and oxidizes it completely to sulfuric acid. Furthermore, it also grows chemoautotrophically using carbon dioxide as a sole source of carbon. Strains of this type have been isolated from the marine environment and they use the Calvin cycle for carbon dioxide fixation.

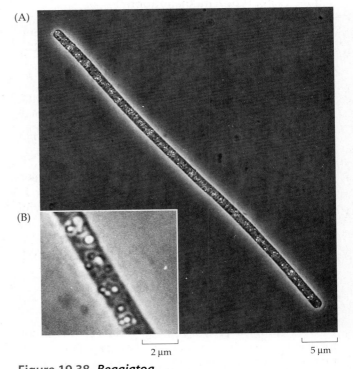

(A)

(B)

2 μm   5 μm

Figure 19.38 *Beggiatoa*
(A) Phase contrast photomicrograph of the filamentous gliding sulfur-oxidizing bacterium *Beggiatoa*. (B) Inset shows enlarged section of *Beggiatoa* with bright intracellular sulfur granules. Courtesy of J. T. Staley.

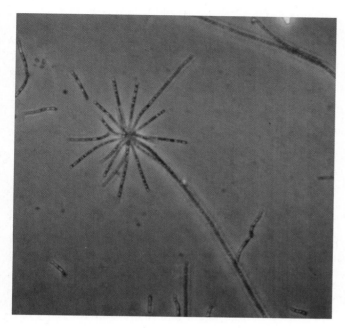

Figure 19.39 ***Thiothrix***

A rosette of *Thiothrix* from a sulfur spring. Note bright internal sulfur granules. Courtesy of J. Bland and J. T. Staley.

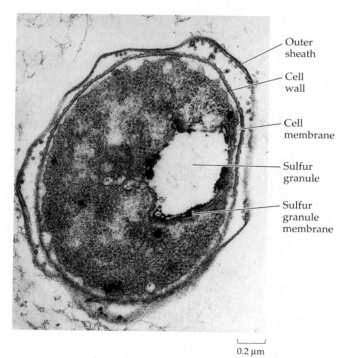

Outer sheath

Cell wall

Cell membrane

Sulfur granule

Sulfur granule membrane

0.2 μm

Figure 19.40 ***Thiothrix* fine structure**

Thin section through a *Thiothrix* cell, showing the sheath, septum, and typical gram-negative structure. Note that the sulfur granules are bound by a membrane. Courtesy of J. Bland and J. T. Staley.

However, heterotrophic types are currently known and have been studied in pure culture. The heterotrophic strains are common in freshwater habitats and sulfur springs. This type oxidizes sulfide only to sulfur, not sulfate. They use acetate as a carbon source and obtain energy by its oxidation. These strains may also be able to fix carbon dioxide by the Calvin cycle, because low levels of (Rubisco), a key enzyme in this pathway, have been detected.

Thiothrix species are found not only in sulfur springs, but also in sewage treatment facilities as bacterium that causes bulking (see Chapter 32). The filaments contain sulfur granules and closely resemble *Beggiatoa* sp.; however, they produce holdfasts, form rosettes (Figure 19.39), and divide by generating gliding gonidia in a manner analogous to the genus *Leucothrix*, a genus of *Proteobacteria* that attach to seaweeds in intertidal marine areas. They also produce a thin sheath (Figure 19.40). Little is yet known about the physiology and metabolism of this group, although some strains have been cultivated in pure culture.

Another important filamentous sulfur bacterium is the genus *Thioploca*. It is found in lake sediments and in the sediments of certain marine intertidal zones, such as in the Gulf of Mexico and off the coast of Chile, where they serve as an important food source for marine animals. Pure cultures have not yet been obtained of these bacteria. They resemble *Beggiatoa* sp. in that they produce filaments with sulfur granules; however, in addition, they produce a sheath in which one or more filaments reside (Figure 19.41A,B). One of the unique attributes of this genus is that the cells appear "empty" when they have been examined by thin section in the electron microscope (Figure 19.42). Recent studies of these bacteria indicate that these vacuoles are used to store nitrate. These unusual bacteria have a unique capability. In the natural sediment habitat, the sheathed filaments are oriented vertically so that the trichomes can glide up and down through the aerobic–anoxic interface. They are able to migrate from anoxic zones in the sediments into aerobic zones, where they concentrate the nitrate from the environment. They can then glide down into the sediments, which are rich in sulfide, and carry out denitrification, thereby obtaining energy for growth. Thus, they have evolved to take advantage of both the oxidized and reduced zones in sediment habitats.

Thiomargarita, which has the largest cells of any known bacterium, also carries out this process (Figure 19.43). In addition, some strains of *Beggiatoa* may also be denitrifiers and store nitrate.

OTHER MEMBERS OF *THIOTRICALES* *Thiomicrospira* is a unicellular group of sulfur oxidizers that is common in the marine environment.

(A)

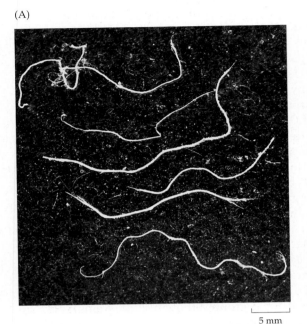

5 mm

(B)

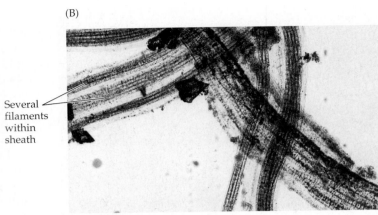

Several filaments within sheath

0.2 mm

Figure 19.41 *Thioploca*

(A) Sheathed filaments of *Thioploca* washed from marine intertidal sediments. These filaments are visible to the unaided eye. (B) Ensheathed filaments of *Thioploca* as they appear by bright-field light microscopy. Note that each sheath contains several filaments. Courtesy of S. Maier.

Cycloclasticus is also a marine genus that is found universally in marine environments. This genus of small rod-shaped bacteria is surprisingly active in the degradation of polycyclic aromatic hydrocarbon (PAH) compounds such as naphthalene and pyrene, that occur in crude oil and creosote. This genus degrades many of the toxic compounds aerobically. It carries two separate genes for naphthalene dioxygenase, which is the key enzyme that causes a break in the aromatic ring.

Another species in this order is *Francisella tularensis*, which is the causative agent of tularemia (see Chapter 28).

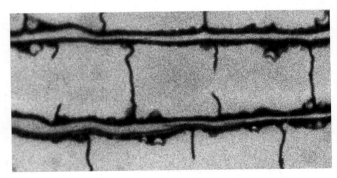

Figure 19.42 *Thioploca* **fine structure**

Thin section through *Thioploca* filaments, showing their empty appearance. Apparently the cytoplasm is pushed against the cell membrane by internal vacuoles. Courtesy of S. Maier.

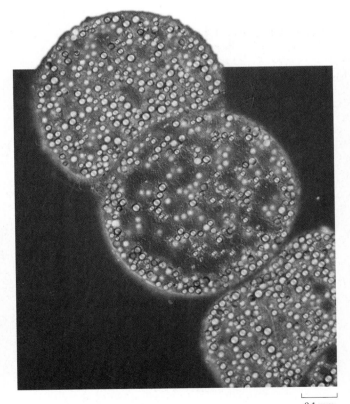

0.1 mm

Figure 19.43 *Thiomargarita*

This is the largest bacterium known. Bright yellow granules are sulfur. Photo by Ferran Garcia Pichel, from the cover of *Science*, Vol. 284, No. 5413. ©1999 AAAS.

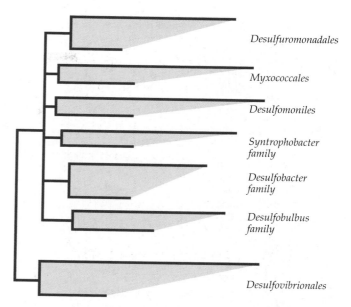

Figure 19.44 Phylogenetic tree of *Deltaproteobacteria* This 16S rRNA tree shows various orders and genera of the *Deltaproteobacteria*. Adapted by permission from *Bergey's Manual of Systematic Bacteriology*.

19.4   Class *Deltaproteobacteria*

This class contains some of the most unusual members of the *Proteobacteria* (Figure 19.44; Table 19.26). Herein resides a group of bacterial predators, the remarkable fruiting myxobacteria and a major group of sulfate reducers.

Order Bdellovibrionales

The most important genus in this order is *Bdellovibrio*, an unusual predator of other bacteria.

BDELLOVIBRIO: THE BACTERIAL PREDATOR   The genus *Bdellovibrio* contains unique gram-negative bacteria that actually parasitize other gram-negative bacteria.

They are small vibrios, only about 0.25 μm in diameter. The *Bdellovibrio* cell has a polar flagellum that rapidly propels the organism through the environment until it collides with an appropriate host cell. It hits the host cell with such force that the host is moved several micrometers. The *Bdellovibrio* cell then attaches to the gram-negative host at its nonflagellated pole and rotates rapidly at the site of attachment (Figure 19.45). After a few min-

TABLE 19.26	Characteristics of genera of the *Deltaproteobacteria* and *Epsilonproteobacteria* classes[a]	
Class Order	**Representative Genera**	**Characteristics**
Deltaproteobacteria *Desulfovibrionales*		
	Desulfovibrio	Sulfate reducers
Desulfomonadales		
	Desulfuromonas	Sulfate reducers
	Geobacter	Iron respiration
Bdellovibrionales		
	Bdellovibrio	Parasites of other bacteria
Myxococcales		
	Myxococcus	Form fruiting structures
	Chondromyces	Form fruiting structures
Epsilonproteobacteria *Campylobacterales*		
	Campylobacter	Animal pathogen
	Helicobacter	Stomach symbiont and pathogen

[a]A complete listing of genera is provided in the Appendix.

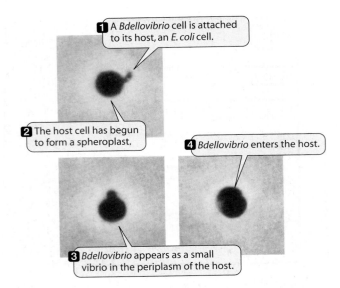

Figure 19.45 Stages of *Bdellovibrio* life cycle
Stages in the life cycle of *Bdellovibrio bacteriovorus*, shown by light microscopy. Courtesy of R. J. Seidler.

utes during which it produces enzymes that are responsible for the breakdown of the cell wall, it enters the cell envelope of the host and moves into the periplasmic space. This process of penetration requires the production of enzymes such as proteases, lipases, and a type of lysozyme, all of which enable the *Bdellovibrio* cell to penetrate the outer lipopolysaccharide and peptidoglycan cell wall layers so that it can enter the cell. This entire process takes several minutes. The *Bdellovibrio* cell does not break the cell membrane and lyse the cell but instead

resides within the periplasmic space during its parasitism. The host cell becomes rounded during this process, due to the destruction of its cell wall, and is called a **bdelloplast**. Once the *Bdellovibrio* cell enters the periplasm, it loses its flagellum and becomes immotile.

While in the periplasm, the *Bdellovibrio* cell produces enzymes that degrade the host cell macromolecules including its DNA, RNA, and protein. The degradation products are then used by *Bdellovibrio* in making its own protein, DNA, and RNA. The *Bdellovibrio* cell enlarges as growth proceeds to form a long, helical cell during the next 4 hours or so (Figure 19.46). When mature, this cell divides by multiple fission to form up to six motile daughter cells in a typical unicellular gram-negative bacterium, such as *E. coli* or *Pseudomonas* species. *Bdellovibrio* progeny then swim away from the dead host cell and are free to repeat the life cycle.

One of the remarkable features of this organism is that it actually produces plaques such as phage plaques on plates of host bacteria. It was this feature that enabled the German microbiologist Heinz Stolp to discover the bacterium when he was isolating phage from soil and sewage samples. The plaques form after most phage plaques have already developed, and they keep on enlarging, making them unusual and suspicious. When wet mount preparations from such plaques were examined in the light mi-

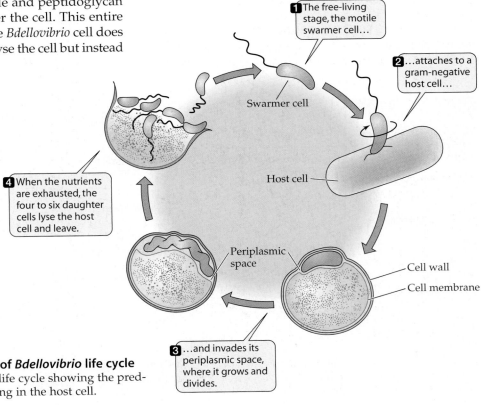

Figure 19.46 Diagram of *Bdellovibrio* life cycle
Bdellovibrio bacteriovorus life cycle showing the predator invading and growing in the host cell.

croscope, the small bacterium was observed quickly darting about among the cells of the host species.

Bdellovibrio cells are obligately aerobic with a TCA cycle. They do not use sugars but rather degrade proteins and use the amino acids as their organic carbon source. Studies of the physiology and metabolism of this bacterium have been conducted with some mutant strains, the so-called host-independent strains that can be grown in the absence of host cells.

Bdellovibrio species are widespread in aquatic environments as well as in soils and sewage. Their major role appears to be analogous to that of virulent bacteriophages in controlling the numbers of gram-negative bacteria. However, they have a broader host range than do typical phages and carry out a much more active type of parasitism. It is curious that they have not yet been found to be chemotactic, an ability that would seem to be of great significance in locating prey in the dilute environments in which they reside.

Order Myxobacteriales: *The Fruiting, Gliding Bacteria*

The **myxobacteria** (slime bacteria) are named for the production of extracellular polysaccharides and other material they excrete as they glide on surfaces. Myxobacteria are gram-negative, rod-shaped bacteria (Figure 19.47) that have complex life cycles. The cells live together in close association as they grow. Most species require complex nutrients such as amino acids and protein for growth and are therefore commonly found on animal dung in soil environments. They produce extracellular hydrolytic enzymes to degrade protein, nucleic acids, and lipids. Using these enzymes, they are able to lyse

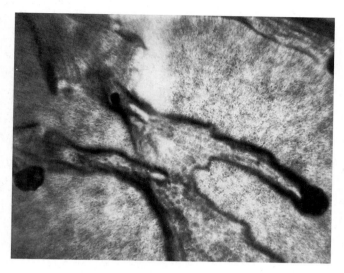

Figure 19.48 Swarming of myxobacteria
Swarm of myxobacteria moving across an agar surface. Each finger-like projection contains hundreds of cells. Courtesy of H. McCurdy.

and degrade bacteria and other microorganisms as their principal nutrient source. These bacteriolytic species do not use sugars as carbon sources for growth. Although most species have an active TCA cycle, they do not have a complete Embden-Meyerhof pathway, which explains their inability to use sugars.

One genus of myxobacteria, *Polyangium*, lives on tree bark and degrades cellulose as its organic energy and carbon source and otherwise uses inorganic nutrients for growth. Therefore, it does not require the complex organic nitrogen sources needed by others. Typical myxobacteria, including *Polyangium*, are aerobic respiratory bacteria, but some are anaerobic.

The most remarkable aspect of the developmental process of these bacteria is their ability to aggregate together to form a swarm (Figure 19.48) and then, as nutrients or water become depleted, to develop a complex **fruiting body** (Figure 19.49). The fruiting body is a dormant stage in which reproductive structures called **microspores** or **microcysts** (Figure 19.50) are produced. Each cell of the swarm is converted into a microcyst in the fruiting structure. The microcysts are resistant to desiccation, thereby enabling the organism to survive until conditions are again favorable for growth. The microcysts can be dispersed much like the spores of fungi that form analogous fruiting structures for reproduction and dispersal. In some species, the microcysts are formed in larger structures called **sporangia**. These sporangia may be formed within the cellular mass or borne on stalks. Stalked varieties such as *Chondromyces crocatus* have an elaborate morphology (see Figure 19.49). Table

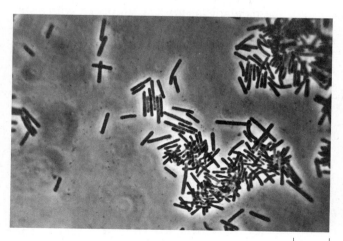

Figure 19.47 Myxobacterial cells   5 µm
Typical rod-shaped vegetative cells of the myxobacterium *Chondromyces crocatus*. Courtesy of H. McCurdy.

(A)

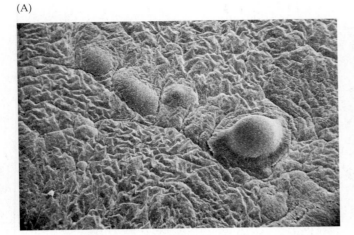

(B)

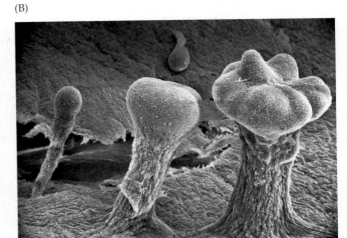

(C)

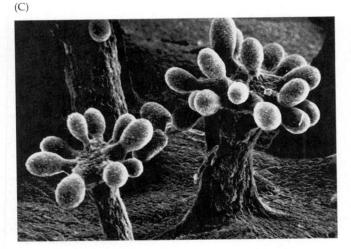

Figure 19.49 Fruiting body formation in a myxobacterium species
Scanning electron micrographs showing development of a fruiting structure in *Chondromyces crocatus*. (A) Cells have aggregated and the structure has begun to rise. (B) Rising fruiting structures at various stages of differentiation. (C) The mature fruiting structure is about 1 mm in height. Each "fruit" is a sporangium filled with microcysts. Courtesy of Patricia Grilione and Jack Pangborn.

19.27 lists the important genera of these bacteria. Because they are simple, one-celled organisms with complex life cycles, these bacteria have become model organisms for the study of developmental processes. The species that has been most thoroughly studied in the laboratory is *Myxococcus xanthus*.

Gliding motility in these bacteria is poorly understood. As with other gliding bacteria, these organisms must be in contact with a surface to be able to move. They produce extracellular slime as they glide, leaving a trail of slime as well as an indentation or actual groove, called a slime track or trail, if they are on agar. Once a track has been formed by one cell, others follow it in preference to making their own. The cells maintain a "herd" or "pack" instinct while gliding. Only rarely do they venture away from the pack except at the advancing edge of growth, and even then it is only for a brief moment. During this period of growth, myxobacteria are actively involved in the degradation of particulate organic material that includes other microor-

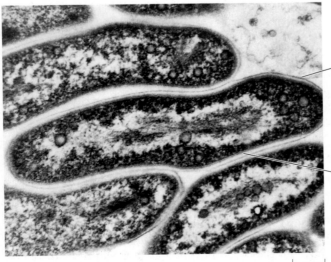

Spore coat

Cell membrane

0.5 μm

Figure 19.50 Microcysts of myxobacteria
Thin section through microcysts of *Cystobacter fuscus*. Courtesy of H. McCurdy.

Genus	Mol % G + C	Distinguishing Features
Cells tapered		
Myxococcus	68–71	Spherical microcysts; no sporangium
Archangium	67–69	Rod-shaped microcysts; no sporangium
Cystobacter	68–69	Rod-shaped microcysts in nonstalked sporangia
Mellitangium	Unknown	Rod-shaped microcysts in sporangium on single stalk
Stigmatella	68–69	Rod-shaped microcysts in sporangium borne singly or in clusters on stalk that may be branched
Cells have blunt, rounded ends, microcysts resemble vegetative cells		
Polyangium	69	Rod-shaped cells; form nonstalked sporangium; may degrade cellulose
Nannocystis	70–72	Coccoidal cells; form nonstalked sporangium
Chondromyces	69–70	Rod-shaped cells; sporangia formed singly or in clusters on stalks

TABLE 19.27 Important genera of myxobacteria

ganisms. They elaborate several lytic enzymes as they move, and if a host bacterium is approached by the pack, its cells are lysed even before physical contact is made. Therefore, although they resemble *Bdellovibrio* sp. in that they parasitize other bacteria, they do not actually enter individual host cells. Furthermore, they have a broader host range and can parasitize gram-positive as well as gram-negative bacteria.

Typically, the myxobacteria cells come together to aggregate at the onset of drying or nutrient depletion, although in some instances they merely enter a stationary growth phase like other bacteria. Aggregation is thought to occur in response to the "A signal," the accumulation of a specific group of amino acids in the environment. This results in drawing of the cells together. They pile on top of one another to form a mound, which eventually becomes the fruiting structure. The fruiting structure containing the microcysts develops into a distinctive shape, characteristic for each species. These fruiting structures, as well as the vegetative cell masses, are pigmented in vivid yellow, orange, or red colors, depending on the species. A-signal mutants cannot aggregate or differentiate into microcysts.

In the simplest of the fruiting structures, such as the one formed by *M. xanthus*, the cells aggregate together to form a small, raised mound about 1 mm in diameter and height. Each of the cells in the fruiting structure is converted into a microcyst. The microcyst then, is a modified cell that shows greater resistance to drying and also ultraviolet exposure than the vegetative cell. Although they also exhibit somewhat greater resistance to heating, they are not similar morphologically or physiologically to the more inert, heat-resistant endospores of the gram-positive bacteria.

The myxobacteria with the most elaborate fruiting structures form their microcysts in a special structure called a sporangium. Sporangia may contain hundreds of microcysts. In some species, the sporangium lies on the same surface as the vegetative growth. However, in the more exotic forms, the cells aggregate, climb on top of one another, and produce a stalk. The stalk consists of hardened, extracellular slime. The sporangia then develop on top of the stalk (see Figure 19.49). Clearly, these are among the most complex prokaryotic organisms.

Most myxobacteria are bacteriolytic and require complex organic materials for growth. They obtain these nutrients by killing and lysing other bacteria. To perform this activity, the myxobacteria produce lysozyme and some of their own antibiotics, most of which have not yet been characterized. Once they lyse the host bacterial cells, they degrade the cells proteins, lipids, and nucleic acids. Most myxobacteria have been isolated from animal dung, which contains other bacteria in high concentrations. However, *Polyangium* are cellulolytic and grow by degrading wood cellulose. Thus, they are found on logs and bark in forested areas.

Orders Desulfovibrionales and Desulfobacteriales: *Sulfate- and Sulfur-Reducing Bacteria*

The sulfate- and sulfur-reducing bacteria are a physiologically and ecologically related group of bacteria. They are strict anaerobes that generate energy by anaerobic respiration of a variety of organic compounds. Some can use hydrogen gas as an energy source. Elemental sul-

TABLE 19.28	Representative genera of sulfate- and sulfur-reducing bacteria[a]		
Class/Genus	**Cell Shape**	**Carbon Source**	**Electron Acceptor**
Desulfovibrio	Vibrio	Lactate, ethanol	SO_4^{2-}; SO_3^{2-}; $S_2O_3^{2-}$
Desulfococcus	Coccus	Lactate, ethanol	SO_4^{2-}; SO_3^{2-}; $S_2O_3^{2-}$
Desulfosarcina	Ovoid; aggregates	Lactate, benzoate	SO_4^{2-}; SO_3^{2-}; $S_2O_3^{2-}$
Desulfobacter	Rod; vibrio	Acetate	SO_4^{2-}; SO_3^{2-}; $S_2O_3^{2-}$
Desulfobulbus	Oval	Lactate, acetate	SO_4^{2-}
Desulfomonile	Rod	Pyruvate, benzoate	SO_4^{2-}; SO_3^{2-}; $S_2O_3^{2-}$
Desulfomicrobium	Rod	Lactate	SO_4^{2-}; SO_3^{2-}; $S_2O_3^{2-}$
Desulfobacterium	Rod; vibrio	Lactate; acetate, longer-chain fatty acids	SO_4^{2-}; SO_3^{2-}; $S_2O_3^{2-}$
Desulfurella	Rod	Acetate	S^0
Desulfuromonas	Oval, curved rod	Acetate, propionate	S^0

[a]Some gram-positive bacteria in the genus *Desulfotomaculum* and some thermophilic *Bacteria* and *Archaea* are also dissimilatory sulfur- and sulfate-reducing organisms (see Chapters 18, 20, and 22).

fur (S^0) or oxidized sulfur compounds such as sulfite and sulfate act as terminal electron acceptors that are converted to H_2S in the process of **dissimilatory sulfur reduction**. Some of the common genera of the sulfate and sulfur reducers and their characteristics are listed in Table 19.28. It should be noted that many ordinary heterotrophic bacteria, such as some enteric bacteria and pseudomonads, can reduce sulfate to sulfide; however, these organisms cannot use sulfate as the sole terminal electron acceptor in anaerobic respiration, the hallmark of the dissimilatory sulfate reducers.

Not all members of this group are closely related to one another phylogenetically. Most bacterial sulfate reducers, however, such as the common genus *Desulfovibrio*, are members of the delta subphylum of the *Proteobacteria*. Although *Desulfonema* differs from the others in that it is filamentous, moves by gliding, and stains as a gram-positive bacterium, phylogenetic analysis indicates it is also a member of the delta *Proteobacteria*. Its closest relative is *Desulfosarcina variabilis*, a packet-forming (sarcina) coccus.

In contrast, *Desulfotomaculum* is not a proteobacterium. It is a genus of endospore-forming bacteria with a broad range of DNA base composition (37–50 mol % G + C). 16S rRNA sequence analysis confirms that this endospore-forming genus is closely related to *Clostridium* and other gram-positive bacteria. Therefore, it is treated in Chapter 20. Likewise, some *Archaea* (see Chapter 18) and other phyla of the *Bacteria* have members that also carry out the dissimilatory reduction of sulfur compounds (see Chapter 22).

Sulfur- and sulfate-reducers are abundant in anaerobic soils as well as freshwater, estuarine, and marine sediments. They are especially common in estuarine and marine sediments, which are in constant contact with seawater that is rich in sulfate. Aquatic and terrestrial habitats that become anaerobic through decomposition of organic matter are rich in sulfate and sulfur reducers. Sulfate-reducing bacteria also live in the intestinal tracts of some animals.

The organic energy sources used by sulfur- and sulfate-reducing bacteria (such as acetate, lactate, ethanol, pyruvate, and butyrate) are end products of fermentation produced by other bacteria. As noted in Table 19.28, many sulfur reducers cannot catabolize acetate because they have an incomplete TCA cycle. Those with a complete TCA cycle oxidize acetate completely to CO_2. In addition to the respiratory generation of ATP with sulfate as the terminal electron acceptor, some genera carry out fermentations. Thus, *Desulfococcus*, *Desulfosarcina*, and *Desulfobulbus* ferment pyruvate or lactate to propionate and acetate. This provides an electron flow that can be utilized in ATP generation.

SECTION HIGHLIGHTS

The *Deltaproteobacteria* contain the unusual gliding and fruiting myxobacteria, which are able to assemble into cell aggregates and differentiate into a remarkable fruiting structure. In addition, this class contains a bacterium, *Bdellovibrio bacteriovorus*, a predator of gram-negative bacteria. Another important group in this class is the sulfate-reducing bacteria, which can use sulfate as an electron acceptor in anaerobic respiration.

19.5 Class *Epsilonproteobacteria*

Campylobacter and *Helicobacter* are curved microaerophilic bacteria that can cause disease in humans and other animals. Both are members of the epsilon subphylum of the *Proteobacteria* (see Table 19.26). *Campylobacter fetus* causes enteritis in humans.

Helicobacter pylori is the causative agent of peptic ulcer disease, which may progress to stomach cancer. *H. pylori* is especially interesting in its global distribution. Each subgroup of humans harbors its own distinctive strains. Thus, strains of *H. pylori* from Asian, African, and European populations have diverged from one another as the human species dispersed from Africa in the last 40,000 to 50,000 years. Evidence for this comes from specific genetic markers in the *H. pylori* strains.

SUMMARY

- The **Proteobacteria** is the most diverse phylum of the *Bacteria* and contains heterotrophic, chemoautotrophic, and photosynthetic bacteria as well as important symbionts and pathogens. The *Proteobacteria* consists of five classes, the *Alpha-, Beta-, Gamma-, Delta-,* and *Epsilonproteobacteria*.

- The class *Alphaproteobacteria* contains several important orders including the *Rhizobiales* with its **symbiotic nitrogen-fixing genera *Rhizobium*, *Sinorhizobium*, *Mesorhizobium*,** and ***Bradyrhizobium***. These genera are associated with legumes, where they form root nodules. Rhizobial cells multiply in the root nodule, and some become specialized cells, called **bacteroids**, which produce nitrogenase for nitrogen fixation. Nitrogen fixation in the legume is a truly symbiotic association—for example, **leghemoglobin**, an oxygen-scavenging compound is formed in part by the plant (globin portion) and in part by the bacterium (heme portion).

- *Agrobacterium tumefaciens* causes **crown gall** disease in dicotyledonous plants. *A. tumefaciens* carries a plasmid, called the **Ti (tumor-inducing) plasmid**, the DNA of which is incorporated into the plant host DNA upon infection—normally this plasmid contains genes that cause gall formation and plant disease. The Ti plasmid of *A. tumefaciens* carries genes for the production of unusual amino acids, called **opines**, which the plant then synthesizes. The opines synthesized by infected plants can be used by the same strain of *A. tumefaciens* that has caused the plant infection. The Ti plasmid can be **genetically engineered** to carry specific desirable genes into the plant.

- The *Alphaproteobacteria* also contains *Acetobacter* and *Gluconobacter*, which are aerobes that produce acetic acid and are therefore called the **vinegar bacteria**. *Zymomonas* ferments sugars to ethanol, lactic acid, and carbon dioxide. *Magnetospirillum magnetotacticum* is a **magnetotactic** bacterium that has magnetite crystals that allow it to orient in a magnetic field.

- **Prosthecae** are cellular appendages produced by some aquatic bacteria. Three major groups of prosthecate bacteria in the *Alphaproteobacteria* are: the caulobacters, the hyphomicrobia, and the polyprosthecate bacteria. The caulobacters, including the genus *Caulobacter*, live in freshwater and marine habitats. The hyphomicrobia produce **buds** from their appendages and include the genera *Hyphomicrobium*, a group of denitrifying bacteria that use methanol as a carbon source; *Hyphomonas*, which use amino acids and organic acids as carbon sources; and *Pedmicrobium*, which are common in soils and freshwater habitats and are involved in the deposition of iron and manganese oxides. The polyprosthecate bacteria include *Prosthecomicrobium* and *Stella*, which are aerobic water and soil bacteria, and the genus *Ancalomicrobium*, which lives in aquatic habitats and ferments sugars by a mixed-acid fermentation identical to that of *E. coli*.

- Two important pathogenic genera in the *Alphaproteobacteria* are *Rickettsia* (cause of Rocky Mountain Spotted fever) and *Brucella*.

- The *Betaproteobacteria* contain some pathogenic species. *Neisseria* is a genus of gram-negative cocci that cause **gonorrhea** (*N. gonorrhoeae*) and **bacterial meningitis** (*N. meningitidis*).

- **Sheathed bacteria**, including *Sphaerotilus* and *Leptothrix*, live in flowing freshwater habitats—some deposit iron oxides and manganese oxides in their sheaths.

- Several groups of chemolithotrophic bacteria exist, including the **nitrifiers**, the **sulfur oxidizers**, the **iron oxidizers**, and the **hydrogen bacteria**. The nitrifying bacteria comprise two separate groups, the **ammonia oxidizers** and the **nitrite oxidizers**, most genera of which are found in the *Alpha-, Beta-* and *Gammaproteobacteria*.

- The sulfur-oxidizing bacteria also consist of two unicellular genera: *Thiobacillus* and *Thiomicrospira*, a marine group. Many sulfur-oxidizing bacteria grow as **facultative chemolithotrophs** or **mixotrophs** by requiring or using various organic carbon sources. **Hydrogen bacteria** in the *Alphaproteobacteria* also grow as facultative chemolithotrophs.

- The *Gammaproteobacteria* contain several important orders such as the *Enterobacteriales*. The order includes *E. coli, Enterobacter, Erwinia, Salmonella,* and *Shigella*; all are facultative aerobes that ferment sugars and belong to the gamma subgroup. Some enteric bacteria, such as *E. coli*, have a **mixed-acid fermentation** and produce acetic, succinic, and lactic acids as well as hydrogen and carbon dioxide. Some enteric bacteria, such as *Enterobacter aerogenes*, have a neutral or **butanediol fermentation** in which 2,3-butanediol, ethanol, carbon dioxide, and hydrogen are produced. Some enteric bacteria cause human disease, for example, **typhoid fever** (*Salmonella typhi*), **plague** (*Yersinia pestis*), and **bacterial dysentery** (*Shigella dysenteriae*).

- *Vibrio* species are typically marine or estuarine members of the *Gammaproteobacteria*—some are pathogenic to fish and humans. *Vibrio cholerae* is the causative agent of **human cholera**. The *Vibrio* species are facultative aerobic bacteria that ferment sugars. Some *Vibrio* species and *Photobacterium* species produce light by **bioluminescence**, using bacterial luciferase—many luminescent bacteria live as symbionts in the light organs of fish.

- The *Pseudomonadales*, another order of the *Gammaproteobacteria*, includes the genera that are obligately aerobic, oxidase-positive, gram-negative rods that are motile by polar flagella. Most species of *Pseudomonas* are not pathogenic; however, some are opportunistic pathogens of humans (*P. aeruginosa*) or plant pathogens (*P. syringae*). *Pseudomonas* species are versatile in their degradative abilities— many can use more than 100 different sugars, organic acids, amino acids, and other monomers as sole carbon sources for growth. Some *Pseudomonas* species can degrade toxic compounds such as toluene and naphthalene, and the genes for this degradative ability are borne on plasmids. *Shewanella* species are known for their ability to use several different electron acceptors in aerobic and anaerobic respiration. Some *Pseudomonas* species and *Xanthomonas* species are plant pathogens that cause ice nucleation, which damages plant cells.

- The **filamentous sulfur-oxidizing** genera, such as *Beggiatoa, Thiothrix,* and *Thioploca* are members of the *Gammaproteobacteria*.

- *Azotobacter* is a genus of free-living soil bacteria capable of **nitrogen fixation**. *Legionella pneumophila* is the pathogen that causes **Legionnaires' disease**.

- **Methylotrophic bacteria** use one-carbon compounds such as methanol or methane as a sole carbon and energy source for growth. **Methanotrophic bacteria** can use methane as a sole source of carbon; they comprise two groups, the type I and type II methanotrophs, which have internal membranes where the enzyme methane monooxygenase resides. Type I methanotrophs use the ribulose monophosphate pathway for carbon assimilation. Type II methanotrophs use the serine pathway for carbon assimilation. Methane-oxidizing bacteria are all aerobic and oxidize methane produced by methanogens in natural environments.

- The *Deltaproteobacteria* contain many sulfate-reducing bacteria, which use sulfate in anaerobic respiration.

- *Bdellovibrio* is a bacterial genus whose species are predators of other gram-negative bacteria.

- The **myxobacteria** are gliding bacteria that have a complex life cycle. Cells move in "packs" when nutrients are plentiful. When nutrients are depleted, they aggregate and undergo a complex fruiting process in which the cells are converted to a more resistant stage called a **microcyst**, which are formed in a **fruiting body**. Microcysts can germinate when conditions for growth again become favorable.

- The *Epsilonproteobacteria* have two important pathogenic genera: *Helicobacter* and *Campylobacter*.

 Find more at www.sinauer.com/microbial-life

REVIEW QUESTIONS

1. Briefly describe the morphology and life cycle of *Caulobacter*.

2. Some bacteria move by gliding, some have flagella, and others are nonmotile. What are the advantages or disadvantages of each type of existence?

3. Give the major characteristics of the nitrifying bacteria and discuss their ecological significance.

4. What is the role(s) of the prostheca? Why do some bacteria have more than one?

5. When you enter a winery you smell vinegar. Why?

6. What are the possible desirable features that could be engineered into plants by use of the *Agrobacterium* Ti plasmid? Could undesirable features be engineered, too?

7. What advantages does the sheath provide for an organism?

8. Compare the motility of spirochetes to that of flagellated and gliding bacteria.

9. Why is it that only prokaryotic organisms can carry out nitrogen fixation?

10. List several positive and negative impacts that sulfur-oxidizing bacteria have on the environment and human activities.

11. Why are the pseudomonads such important bacteria?

12. How do methanotrophic and methylotrophic bacteria live?

13. How do hydrogen bacteria obtain their energy?

14. How are the large vacuoles in *Thioploca* used in their environment?

15. Characterize the genus *Bdellovibrio* and outline its life cycle.

16. Briefly describe the myxobacterial life cycle. What are fruiting bodies and microcysts?

SUGGESTED READING

Balows, A., H. G. Trüper, M. Dworkin, W. Harder and K. H. Schleifer, eds. 1992. *The Prokaryotes*. 2nd ed. Berlin: Springer-Verlag.

Holt, J. G., editor-in-chief. 1994. *Bergey's Manual of Determinative Bacteriology*. 9th ed. Baltimore, MD: Williams and Wilkins.

Krieg, N., D. Brenner, J. T. Staley and G. Garrity, eds. 2005. *Bergey's Manual of Systematic Bacteriology*. 2nd ed. Vol II. New York: Springer.

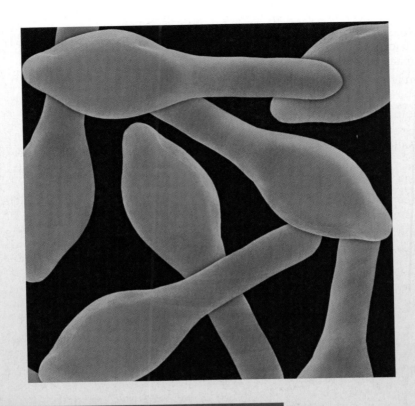

The objectives of this chapter are to:

◆ Introduce the two phyla of gram-positive bacteria.

◆ Discuss the distinctive fermentations and other metabolic pathways of gram-positive bacteria.

◆ Present examples to illustrate the diversity of gram-positive bacteria.

◆ Describe the medical, commercial, agricultural, and ecological significance of the gram-positive bacteria.

20
Gram-Positive Bacteria: Firmicutes *and* Actinobacteria

The most important discoveries of the laws, methods and progress of Nature have nearly always sprung from the examination of the smallest objects which she contains.
—J. B. Lamarck, Philosophie Zoologique, 1809

The common gram-positive *Bacteria* comprise a spectrum of morphological types ranging from unicellular organisms to branching, filamentous, multicellular organisms. Most are heterotrophs that grow as saprophytes living off nonliving organic materials. Gram-positive bacteria are particularly abundant in soil and sediment environments. A number of them are important plant and animal pathogens. This phylogenetic group contains some of the simplest heterotrophic organisms from a metabolic standpoint, such as the lactic acid bacteria. A few gram-positive bacteria are hydrogen autotrophs that make acetic acid from carbon dioxide and H_2. One phototrophic group, the heliobacteria, has also been described. Gram-positive bacteria contain significant amounts of peptidoglycan in their cell walls (see Chapter 4). Their cell walls, however, lack lipopolysaccharide, a characteristic component of the cell walls of gram-negative bacteria. They also have a simpler cell wall architecture. However, one subgroup, the *Mollicutes* or **mycoplasmas**, lack a cell wall altogether. Nonetheless, they are members of the gram-positive bacteria based on 16S rDNA sequence analyses.

The gram-positive *Bacteria* are classified into two separate phylogenetic groups, the *Firmicutes* and the *Actinobacteria*, based on 16S rDNA sequences. The phylum *Firmicutes* contains unicellular gram-positive *Bacteria* that have a low mol % G + C content (less than 50%). In contrast, the phylum *Actinobacteria* contains many filamentous, multicellular species, the DNA of which has a high mol % G + C (greater than 50%).

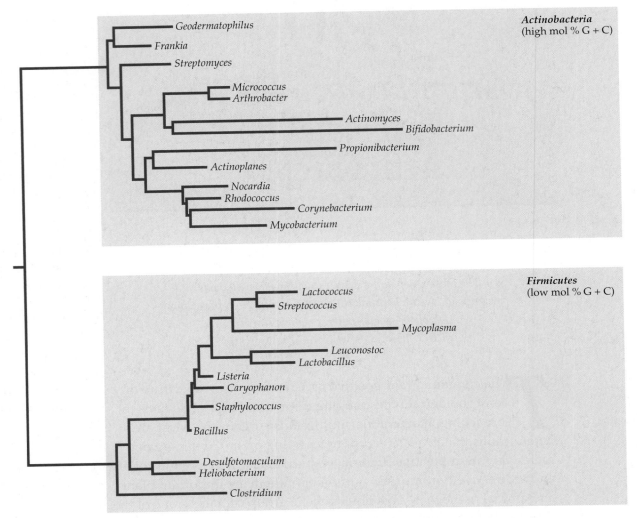

Figure 20.1 Phylogeny of gram-positive bacteria
Diagram showing the two phyla of gram-positive bacteria based on 16S rRNA analyses. Note that the unicellular, low mol % G + C lactic acid bacteria and endospore-forming bacteria are found in the *Firmicutes* branch, whereas the mycelial, high mol % G + C *Actinobacteria* form a separate branch. Adapted from Ribosomal Database Project.

The phylogenetic relatedness among the gram-positive bacteria is shown in Figure 20.1 and to the other bacterial phyla in Figure 1.6. Note that the low mol % G + C group branches off separately from the high mol % G + C group. Each phylum is discussed individually, beginning with the *Firmicutes*.

20.1 Phylum: *Firmicutes*

The *Firmicutes* phylum is very diverse. As shown in Appendix I, there are three major classes within this phylum—*Bacilli*, *Clostridia*, and *Mollicutes*—which include 10 orders and 34 families. The phylum contains many important pathogenic bacteria of diseases with household names, such as tetanus and anthrax. In addition, members of this phylum are widely distributed in the environment where they carry out important processes such as nitrogen fixation, sulfate reduction, degradation of cellulose, and acetogenesis. Some members are even phototrophic. We begin by considering the class *Bacilli*.

Class Bacilli

This class contains two major orders, the *Lactobacillales* and the *Bacillales*. As the name *Lactobacillales* implies, this is an important group of genera that produce lactic acid as an end product of their metabolism. They are discussed first.

Domain *Bacteria*

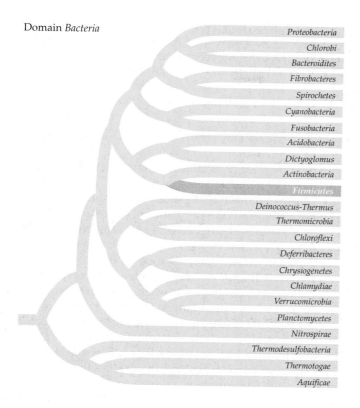

Proteobacteria
Chlorobi
Bacteroidites
Fibrobacteres
Spirochetes
Cyanobacteria
Fusobacteria
Acidobacteria
Dictyoglomus
Actinobacteria
Firmicutes
Deinococcus-Thermus
Thermomicrobia
Chloroflexi
Deferribacteres
Chrysiogenetes
Chlamydiae
Verrucomicrobia
Planctomycetes
Nitrospirae
Thermodesulfobacteria
Thermotogae
Aquificae

Order Lactobacillales*: Lactic Acid Bacteria*

As their name implies, the **lactic acid bacteria** are noted primarily for their ability to ferment sugars to produce lactic acid. This group contains several key genera—*Streptococcus, Lactobacillus,* and *Enterococcus*—the activities of which in human health and the environment are familiar to all of us. Most of us have had sore throats, called "strep" throat, caused by *Streptococcus* species. The genus *Streptococcus* contains several important pathogens, such as *Streptococcus pyogenes,* which causes scarlet fever and rheumatic fever.

The lactic acid bacteria are also significant in food and dairy microbiology as well as in agriculture. The fermentation of vegetable materials is used in the manufacture of pickles and sauerkraut. Milk fermentation by lactic acid bacteria leads to the manufacture of commercial products such as yogurt, buttermilk, and cheeses. Some of the lactic acid species are known to cause tooth decay.

All lactic acid bacteria are **aerotolerant anaerobes**; that is, they grow in the presence of oxygen but do not use it in respiration (see Chapter 6). Instead, they produce energy by fermentation of sugars (see Chapter 8). In addition, most of these bacteria have complex nutritional requirements. Many require amino acids and vitamins, and some even need purines and pyrimidines to grow. Not surprisingly, they are found in organic-rich environments such as decaying plant and animal materials. Table 20.1 lists the major genera of lactic acid bacteria as well as some of their common features.

PHYSIOLOGY AND METABOLISM   Although all lactic acid bacteria carry out lactic acid fermentation, they can be separated into two different groups based on the type of fermentation process. The **homofermentative** bacteria carry out a simple fermentation in which lactic acid is the sole product from sugar fermentations:

$$\text{Glucose} \rightarrow 2 \text{ lactic acid}$$

The Embden-Meyerhof or glycolytic pathway is used in this process (see Chapter 8). The pyruvic acid formed is reduced by the enzyme lactate dehydrogenase to produce the characteristic end product, lactic acid (Figure 20.2).

In contrast, the **heterofermentative** lactic acid bacteria produce ethanol and carbon dioxide as well as lactic acid:

$$\text{Glucose} \rightarrow \text{lactic acid} + \text{ethanol} + CO_2$$

The heterofermentative pathway (Figure 20.3) lacks the key enzyme aldolase that is present in the glycolytic pathway of the homofermentative bacteria. Aldolase cleaves fructose-1,6-bisphosphate to form the two phosphorylated trioses that ultimately lead to the production of two ATPs. Therefore, although homofermenters obtain 2 moles of ATP for each mole of glucose fermented,

TABLE 20.1	Principal genera of lactic acid bacteria		
Genus	**Morphology**	**Fermentation Type**	**Lactic Acid Form**
Streptococcus	Cocci in chains	Homofermentative	L-
Leuconostoc	Cocci in chains	Heterofermentative	D-
Pediococcus	Cocci in tetrads	Homofermentative	DL-
Lactobacillus	Rods	Homo- or heterofermentative	Varies with species

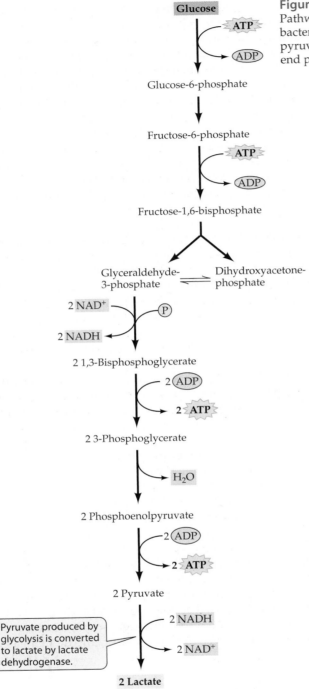

Figure 20.2 Homofermentation pathway
Pathway for dissimilation of glucose by homofermentative lactic acid bacteria. These bacteria use the glycolytic pathway for formation of pyruvate and lactate dehydrogenase to produce lactate as the final end product.

(b) acetyl-phosphate, which is reduced to acetaldehyde, and then to ethanol.

The only energy available to lactic acid bacteria is through ATP generated by substrate-level phosphorylation. The metabolism of these bacteria is among the simplest of the various energy-yielding processes found in bacteria or other organisms. The resulting energy yield per mole of glucose (Y_{ATP}) is very low compared to the yield of many other bacteria, especially those that can respire (see Chapter 6). The homofermenters that produce 2 mole of ATP per mole of glucose are twice as efficient as the heterofermenters, and they capture about 27% of the total energy available in this process (Table 20.2). Although very little energy is harvested by these bacteria by such fermentations, they have been remarkably successful in establishing themselves in important niches in the environment, as discussed later.

The optical form of lactic acid produced by lactic acid bacteria varies among the different genera. Some produce one stereoisomer, the D-form, which rotates light toward the right (*dextro* rotary), whereas others produce the L-form (left or *levo* rotary). The reason for these differences lies in the stereospecificities of the lactate dehydrogenase enzyme itself. Some species produce enzymes that make both D- and L-forms, resulting in a racemic mixture of the two. This example illustrates the narrow specificity of some enzymes in comparison to typical chemical reactions.

Lactic acid bacteria can grow in the presence of oxygen but are unable to use it metabolically. They lack cytochrome enzymes (which have iron-containing heme groups that they cannot synthesize) and an electron transport system with which to generate ATP by electron transport phosphorylation. Lactic acid bacteria do have flavoproteins, however, and when exposed to oxygen they produce hydrogen peroxide that can be toxic to the cells.

It is interesting to note that most lactic acid bacteria are **catalase-negative**; that is, they are unable to make the enzyme catalase, an iron-containing heme-protein (like cytochromes) that degrades peroxides in the following manner:

$$H_2O_2 \rightarrow 2\,H_2O + O_2$$

A few lactic acid bacteria make a special manganese-containing catalase and are therefore phenotypically cata-

the heterofermenters obtain only 1 mole, as can be seen by examining their pathway. This pathway involves an initial oxidation of glucose to 6-phosphogluconic acid, which is decarboxylated to form CO_2 and ribulose-5-phosphate. The pentose formed is converted to (a) the three-carbon intermediate, glyceraldehyde-3-phosphate, which gives rise to one molecule of ATP by substrate-level phosphorylation in the formation of lactic acid, and

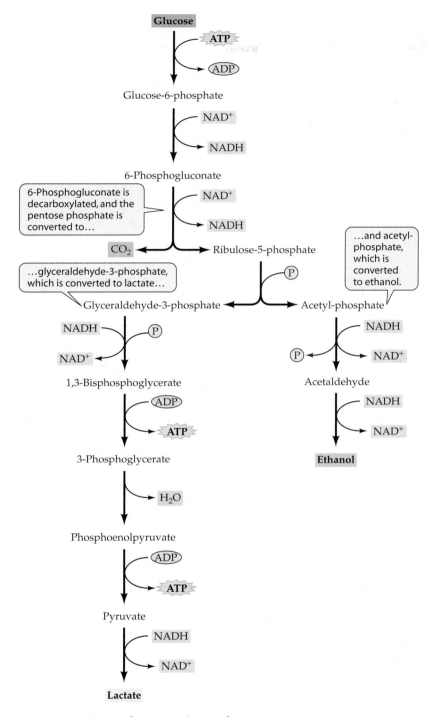

Figure 20.3 Heterofermentation pathway
Pathway for dissimilation of glucose by heterofermentative lactic acid bacteria. These bacteria lack aldolase, a key glycolytic enzyme, and therefore do not use the Embden-Meyerhof fermentation pathway.

Reactive oxygen compounds, such as hydrogen peroxide, can lead to damage of almost all cell components, including nucleic acids, membrane lipids, and proteins. The toxic effect on the cell is referred to as oxidative stress. Enzymes such as peroxidases degrade the toxic oxygen forms, thereby providing a protective effect.

Another toxic form of oxygen is the superoxide anion, O_2^-, which is formed in various physical and biochemical ways by single electron reductions of O_2. The lactic acid bacteria, as well as most aerobic and facultative aerobic bacteria, protect themselves from this strong oxidizing agent by producing the enzyme **superoxide dismutase**, which decomposes superoxide as follows:

$$2\,O_2^- + 2\,H^+ \rightarrow O_2 + H_2O_2$$

Note that hydrogen peroxide is formed by the enzyme. The concerted action of the superoxide-dismutase and either catalases or peroxidases results in the destruction of these toxic oxidizing agents. In contrast to the lactic acid bacteria, many obligate anaerobes do not produce these enzymes; therefore, these anaerobes can be exposed to lethal oxidative stress if they encounter aerobic conditions.

NUTRITION AND ECOLOGY   The lactic acid bacteria are regarded as **fastidious** organisms because almost all species have complex nutritional requirements. With the exception of *Streptococcus bovis* from the intestine of cattle, virtually all require preformed amino acids and vitamins, and some even require purines and pyrimidines. Thus, although these bacteria do not live as intracellular pathogens of other organisms, they live in habitats with other microorganisms, animal or plant tissues, or decaying organic substances that serve as sources of these materials. Because they require these compounds presynthesized, these bacteria conserve energy that would otherwise be required for their synthesis.

Despite their limited metabolic capabilities and their complex nutritional requirements, the lactic acid bacteria have survived well in selected environments due to their specialization in sugar fermentation. The principal habitats of groups of the lactic acid bacteria are shown in Table 20.3.

lase-positive. Those species lacking catalase have other enzymes, **peroxidases**, which degrade hydrogen peroxide through organic compound–mediated reductions.

TABLE 20.2	Comparison of the efficiency of the homofermentative versus the heterofermentative lactic acid bacteria[a]

Energy Sources and Efficiency	Energy Available (kjoules)
Glucose (1 mol)	2,870
Pyruvate (1,326 kjoules/mol × 2 mols)	− 2,652
Theoretical energy available in oxidation of glucose to pyruvate	218
Energy captured by ATP synthesis (29 kjoules/mol ATP):	
Homofermenters (2 ATP/mol glucose)	58
Heterofermenters (1 ATP/mol glucose)	29
Efficiency of process	
Homofermenters (58/218 ×100%)	27%
Heterofermenters (29/218 ×100%)	13%

[a]Because pyruvate is more oxidized than glucose, the two moles of pyruvate formed from glucose contain 218 kjoules less energy than the initial mole of glucose. Two molecules of ATP are synthesized in the homofermentative process, and each of these is equivalent to 29 kjoules of energy per mole of glucose. Thus, 58 kjoules of energy have been captured in the formation of ATP by homofermenters, representing an efficiency of 27%. In contrast, heterofermentative lactic acid bacteria are only half as efficient because they produce only one molecule of ATP per molecule of glucose fermented. Also note that some 92% of the total energy available in the glucose molecule is still available in the two molecules of pyruvic acid formed (2652/2870 ×100 = 92%), indicating that much of the original energy still remains in pyruvate.

One of the major habitats of lactic acid bacteria is decomposing plant materials. These bacteria ferment the hexoses and pentoses that are the predominant decomposition products. The lactic acid formed prevents the growth of many other organisms, especially if the pH is lowered to 5.0 or so. For example, in pickle fermentations, the cucumbers are placed in a vat with a small amount of salt and seasonings. They are covered with water and incubated at or about room temperature. Initially, aerobic and facultative aerobes grow and through aerobic respiration remove the free O_2 that is dissolved in the water in the vat. In a short time, the vat becomes anaerobic, and the plant sugars begin fermenting. The initial fermentation is accomplished by the genus *Streptococcus* because it grows at higher pH values than does the genus *Lactobacillus*. The pH is lowered by the former group until it reaches about 5.5 to 6.0, at which point *Lactobacillus* sp. begin to grow. These organisms ferment the remaining sugars and lower the pH even further, to about 4.5. As long as conditions remain anaerobic, the

TABLE 20.3	Habitats of lactic acid bacteria	
Habitat	**Predominant Groups**	**Activity or Product**
Decomposing plant material	*Streptococcus* spp. and *Lactobacillus plantarum*	Pickles, kim chee, silage, and sauerkraut
Dairy	*Streptococcus lactis*, *Lactobacillus casei*, *L. acidophilus*, *L. delbrueckii*, *Leuconostoc mesenteroides*, and *L. lactis*	Cheeses, yogurt, etc.
GI tract of animals (oral)	*Streptococcus salivarius*, *S. mutans*, and *Lactobacillus salivarius*	Normal microbiota, dental caries
GI tract of animals (intestinal)	*Enterococcus faecalis*	Intestine; some urinary tract pathogens
Mammal vagina	*Streptococcus* spp. and *Lactobacillus* spp.	Normal microbiota

plant materials are preserved very well due to the low pH of the lactic acid. Therefore, the pickles keep for many weeks or for many months if refrigerated or canned.

The same general protocol is followed for the manufacture of sauerkraut and kim chee, the Korean fermented bok choy product. The characteristic flavor of sourdough bread also results from the fermentation of lactic acid bacteria. Likewise, in agriculture, silage is made by placing hay of the right moisture content in silos or, more recently, in large plastic bags to exclude air and to allow the fermentation of plant materials. The resulting silage can be used during the winter months when fresh hay is not available.

Lactic acid bacteria are a critical component of the dairy industry as well. These bacteria are the natural souring agents of milk. They gain access to the milk through plant materials such as the feed or, in some cases, from the cattle themselves. If milk is permitted to sour naturally, these bacteria become established, with first the *Streptococcus* group, followed by the *Lactobacillus* group, as discussed in pickle fermentation. They ferment the milk sugar lactose and form lactic acid in the process. When the pH drops sufficiently, the principal milk protein, casein, is precipitated. Thus, the **curd**, or the precipitated casein, is separated from the **whey**, the remaining fluid. It is this same process that is used for the commercial manufacture of cheeses. Special strains called **starter cultures** of lactic acid bacteria are used by dairies for the manufacture of cottage cheese, yogurt, acidophilus milk, butter, buttermilk, and various cheeses. The distinctive flavor and aroma of butter is due to diacetyl formed spontaneously from acetoin produced by some lactic acid bacteria in butter production.

Another habitat of the lactic acid bacteria is the gastrointestinal tract of animals. These organisms grow in the oral cavity, where they ferment sugars such as sucrose with the formation of lactic acid. Species of both *Streptococcus* and *Lactobacillus* occur in the mouth, where the lactic acid they produce can cause decay of tooth enamel, resulting in cavities. These bacteria also live in the small intestine of humans and other animals.

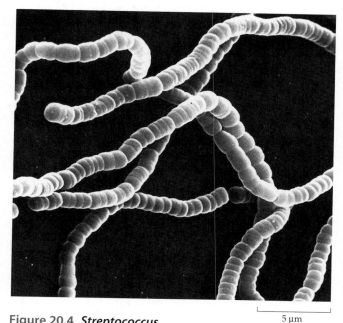

Figure 20.4 *Streptococcus*
Typical appearance of a *Streptococcus* sp. in chains of cocci, as seen by scanning electron microscopy. Photo ©David M. Phillips/Visuals Unlimited.

A final habitat of significance is the vagina of female mammals. Both *Streptococcus* and *Lactobacillus* ferment glucose from the glycogen normally secreted by the vagina of adult females. The resulting lactic acid serves as a barrier to vaginal infection by other bacteria.

Some of the more common genera of lactic acid bacteria are discussed individually next.

STREPTOCOCCUS This genus of cocci (Figure 20.4) contains several important species (Table 20.4). They all grow poorly on lab media and form only small, pinpoint colonies during growth. *Streptococcus lactis* is a common dairy organism found in milk and other dairy products. It does not cause disease, but is of industrial significance. Note that it is incapable of causing hemolysis of red blood cells (see Table 20.4). This is because it lacks the ability to produce **hemolysins**, special enzymes that are

		Hemolysis of Red Blood Cells	
Species	**Group Name**	**Alpha- (greening)**	**Beta- (clearing)**
S. lactis	Dairy group	–	–
S. salivarius	Viridans group	+	–
S. pyogenes	Pyogenic	–	+
S. pneumoniae	Pneumococcus	+	–

TABLE 20.4 Differential characteristics of important species of the genus *Streptococcus*

responsible for this activity. Hemolysis is regarded as a virulence factor, a characteristic that enhances the pathogenicity of species.

Streptococcus salivarius is one of many bacteria that grow in the upper part of the gastrointestinal tract, the oral cavity of humans. This organism lives off the sugars supplied in the diet or produced from starches by amylases secreted by the salivary glands. A close relative of this species, *Streptococcus mutans*, produces a capsule in the presence of disaccharides that enables it to attach to enamel and cause tooth decay through the production of lactic acid. Oral streptococci hydrolyze table sugar—the disaccharide sucrose (see Figure 3.14)—to the monomers fructose and glucose. Fructose is fermented to produce lactic acid, and the glucose is used to produce a polymer called a **glucan** (alpha-1,6 and alpha-1,3-linked glucose), which serves as an adhesive that allows the streptococci to attach to teeth.

Sucrose → glucan + fructose

Other oral species that produce dental caries are *Streptococcus gordonii*, *Streptococcus oralis*, and *Streptococcus mitis*.

Members of this genus, the so-called **viridans group** streptococci (from Greek meaning green), cause a partial clearing or greening effect on blood agar plates (Figure 20.5A). This phenomenon is referred to as **alpha-hemolysis**, but in fact red blood cells are not actually lysed in this process. In alpha-hemolysis hemoglobin is converted to methemoglobin in vivo—this form of hemoglobin has a reduced capacity to carry oxygen and loses its red color. Most of the members of the viridans group streptococci are not pathogenic, but are implicated in tooth decay.

The **pyogenic** (pus-forming) group of *Streptococcus* characteristically produces hemolysins that completely lyse red blood cells, resulting in a clearing on a blood agar plate prepared with mammalian blood. This type of hemolysis is termed **beta-hemolysis** (Figure 20.5B). The members of the pyogenes group of streptococci are serious human pathogens (see Chapter 28). In addition to hemolysins, they produce other virulence factors that are important for pathogenicity such as fibrinolysin, an enzyme that dissolves blood clots; an erythrogenic toxin, which causes the characteristic rash of scarlet fever; and leucocidin, which destroys white blood cells. Furthermore, they produce the enzyme hyaluronidase, which attacks hyaluronic acid in connective tissue. These organisms are ideally adapted for growth in warm-blooded animals because their growth temperature range is between 10°C and 45°C. *S. pyogenes* is the causative agent of several diseases including scarlet fever, puerperal fever of postdelivery mothers, and endocarditis (an inflammation of heart tissue).

S. pneumoniae, one of the bacterial species that causes pneumonia, comprises its own group and is commonly referred to as the "**pneumococcus**." Unlike the other members of the genus, this species does not typically form long chains of cells under most culture conditions, but rather grows as pairs of cocci, so it is also sometimes

(A)

(B)

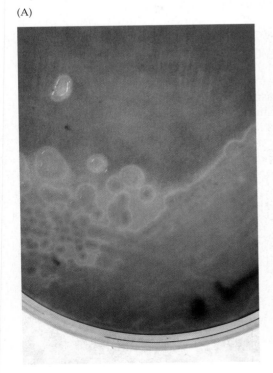

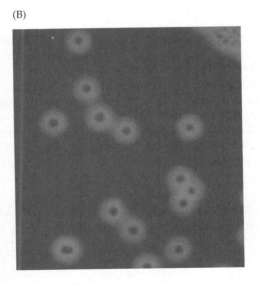

Figure 20.5 Blood cell hemolysis by viridans group
(A) Alpha-hemolysis, or "greening" hemolysis, due to the growth of a member of the viridans group of *Streptococcus* on blood agar plates. (B) Beta-hemolysis by *Streptococcus pyogenes* colonies results in the lysis and clearing of the red blood cells on blood agar plates. A, ©L. M. Pope and D. R. Grote/Biological Photo Service; B, ©G. W. Willis/Visuals Unlimited.

called the "**diplococcus**." Virulent members of this species have a capsule that is important for their pathogenicity (see Figure 4.59). If the capsule is removed, the cells are readily phagocytized. Only strains with capsules can cause pneumonia. Encapsulated strains form colonies that are described as "smooth" in texture and are distinguished from unencapsulated strains that form "rough" colonies. As discussed in Chapter 15, these bacteria served as important experimental organisms in Avery and MacCleod's demonstration that genetic material can be passed from one organism to another, even if the donor organism is dead. This was the first demonstration of the phenomenon of genetic transformation.

Streptococcus is also known for the devastating tissue damage caused by some strains that infect humans. This rare "flesh-eating" disease has a rapid onset and, if not treated quickly, may result in loss of limb or life.

LACTOBACILLUS   This genus of rod-shaped lactic acid bacteria is also widespread in the same habitats in which the genus *Streptococcus* resides, including plant materials and the oral and genital tracts of humans. However, unlike *Streptococcus*, the genus *Lactobacillus* does not harbor any pathogenic strains.

These bacteria can grow at lower pH values than the streptococci. As mentioned previously, they do not grow during the initial stages of fermentations, but only when the pH is lowered to 5.5 to 6.0. In turn, they continue to produce lactic acid and lower the pH of the environment to values below 5.0. This effectively eliminates the genus *Streptococcus*. This is a bacterial example of an ecological succession.

ENTEROCOCCUS   The genus *Enterococcus* is known to inhabit the lower part of the human digestive tract. *Enterococcus faecalis* is a normal and therefore nonpathogenic inhabitant of the human intestine—this species is particularly abundant in the large intestine. This environment is anaerobic and contains ample nutrients, making it an ideal habitat for growth. Some *E. faecalis* strains cause infections of the urinary tract.

OTHER IMPORTANT GENERA   Two other important genera of *Lactobacillales* are *Leuconostoc* and *Pediococcus*. *Leuconostoc* spp. are common in dairy products as well as in the oral cavity. They are especially well known as producers of **dextrans** (alpha-1,6-glucans), which they form as capsular materials in the fermentation of sucrose, a disaccharide consisting of glucose and fructose. *Leuconostoc mesenteroides* produces an extracellular enzyme called dextran sucrase that converts the glucose of the sucrose molecule to dextran and releases the fructose into the environment.

Sucrose → dextran + fructose

Although dextran is a polymer of glucose subunits linked in the alpha-1,6 position, the glucose subunits are added to an initial primer molecule of sucrose. Thus, each dextran molecule has one subunit of fructose as its terminal sugar. Dextran is used commercially as a blood plasma extender.

Pediococcus species also carry out important lactic acid fermentations. They can be troublesome in the brewing industry, as the name *Pediococcus damnosus* implies, in that they can interfere with the normal yeast alcoholic fermentation by producing undesirable products of lactic acid fermentation.

Order Bacillales

The second order in the class *Bacilli* is *Bacillales*, which includes several endospore-forming genera.

ENDOSPORE-FORMING BACTERIA   The bacterial endospore is a very distinctive structure produced by relatively few genera (Table 20.5). The two principal genera are the aerobic to facultatively aerobic genus *Bacillus* (see

TABLE 20.5	Important genera of endospore-forming bacteria		
Genus	Oxygen Requirement	Mol % G + C	Nutrition/ Physiology
Bacillus	Aerobe or facultative aerobe	25–69	Aerobic respiration or fermentation
Clostridium	Obligate anaerobe	22–55	Fermentation
Desulfotomaculum[a]	Obligate anaerobe	37–50	Anaerobic respiration (sulfate reduction)
Thermoactinomyces	Obligate aerobe	52–55	Aerobic thermophile; compost
Sporosarcina	Obligate aerobe	40–42	Urea degrader

[a]See gram-negative dissimilatory sulfate reducers in Chapter 19.

Figure 4.8) and the obligately anaerobic genus *Clostridium*, treated subsequently in the class *Clostridia*. Four additional genera of ecological importance include *Desulfotomaculum*, a group of anaerobic sulfate-reducing bacteria discussed in Chapter 19; *Sporosarcina*, a soil coccus; and *Thermoactinomyces*, a thermophile. In addition, some species of *Sarcina*, for example *Sarcina ventriculi* from the stomach, are coccus-shaped endospore formers.

With the exception of *Sporosarcina* and *Sarcina*, which are cocci, all endospore formers are rod-shaped bacteria that have the typical gram-positive cell wall ultrastructure and stain gram-positive to variable. Many species are motile with peritrichous flagella.

HABITATS OF ENDOSPORE-FORMING BACTERIA   The endospore-forming bacteria are found primarily in soils, aquatic sediments, and muds. Like other gram-positive bacteria, they are only occasionally reported in planktonic habitats such as oligotrophic lakes and the ocean. The endospore is an ideal survival device for soil bacteria. Organisms that can convert to resting stages during dry periods in soil environments can survive periods of prolonged desiccation. These bacteria are also found on weathered rock surfaces in deserts, where hot, dry periods are common.

Endospore-forming bacteria can be selectively isolated from habitats by first pasteurizing the soil or mud sample at 80°C for 10 minutes. Although this kills normal soil bacteria, the heat-resistant endospores of most of these bacteria will not be affected. In fact, some endospore formers do not germinate unless they are first heat shocked, that is, first exposed to high temperature.

BACILLUS   This genus contains aerobes to facultative anaerobes. Unlike the strictly anaerobic genus *Clostridium*, *Bacillus* species produce catalase, which may explain the difference in the survival of these two genera in the presence of oxygen. The various species of this genus are classified into groups based on their cell morphology, using in particular the shape of the endospore and its location. Table 20.6 provides a grouping of some of the major species based on cell morphology.

Bacillus subtilis is a small, obligately aerobic spore former that is a common soil inhabitant. It is noted for its ability to degrade plant polysaccharides and pectin, and some strains are even known to produce a rot in potato tubers. It can cause a condition known as "ropy bread" caused by polymer production during growth after baking contaminated bread.

Bacillus licheniformis resembles *B. subtilis*, but it is a facultative anaerobe that ferments sugars. It carries out a characteristic fermentation:

$$3 \text{ Glucose} \rightarrow 2 \text{ glycerol} + 2 \text{ 2,3-butanediol} + 4 \text{ CO}_2$$

Most strains of *B. licheniformis* carry out denitrification, which is a relatively unusual attribute within this genus.

Bacillus megaterium (from the Greek *megaterium* meaning "big beast"), as the species name implies, is very large, about 2.5 μm in width and up to 10 μm in length. Because it is large, bacteriologists use it as a teaching organism in introductory laboratories such as for Gram staining, and in research laboratories interested in morphology and ultrastructure. Like *B. subtilis*, this species is an obligate aerobe with very simple nutritional re-

TABLE 20.6	Some major species of *Bacillus* and their properties	
Group	**Major Species**	**Distinctive Properties**
I. Oval spores		
A. Sporangium not swollen	*B. subtilis*	Common soil form; aerobic gramicidin producer
	B. licheniformis	Denitrifier, fermentative; bacitracin producer
	B. megaterium	Large rod
	B. cereus	Common soil form, fermentative
	B. mycoides	Like *B. cereus* but distinctive hair-like colonies
	B. anthracis	Like *B. cereus*, but causes anthrax
	B. thuringiensis	Insect pathogen
B. Sporangium swollen	*B. stearothermophilus*	Thermophile
	B. circulans	Colony rotates on agar media
	B. polymyxa	Nitrogen fixer, polymyxin producer
	B. popilliae	Insect pathogen
II. Spherical spores, nonfermentative		
	B. pasteurii	Degrades urea, grows at high pH

quirements (ammonium or nitrate can serve as the sole nitrogen source).

Bacillus cereus and *Bacillus mycoides* are also quite large (about twice the size of *Escherichia coli*) and are among the most common soil bacteria that grow on decaying plant materials. They grow anaerobically by sugar fermentation and require amino acids for growth. *B. mycoides* is commonly dispersed in air and therefore often contaminates plates in microbiology laboratories.

Closely related to *B. cereus* are two important pathogens. *Bacillus anthracis* causes anthrax in animals as well as humans. Unlike *B. cereus,* it is immotile. The source of human infection is largely through domestic farm animals, although many other animals, including horses, mink, dogs, and even birds, fish, and reptiles can acquire the disease. In humans, the disease occurs most commonly in sheep and cattle tanners and meat workers. The organism normally invades through open cuts in the skin but can also infect through the lungs (see Chapter 28). The organism produces a gummy D-glutamic acid polypeptide capsule, an important virulence factor during invasion. The disease symptoms are produced by a protein exotoxin formed during growth in the host tissues. The toxin acts by causing physiological shock and ultimately kidney failure. The toxin is heat labile. The spores survive well in the environment, so animals that have died of the disease present a problem for disposal; they are usually buried. *B. anthracis* is also one of the bacteria favored as an agent for biological warfare (Box 20.1).

The best-known insect pathogen in this genus is *Bacillus thuringiensis.* This bacterium produces a large amount of a crystalline protein during the sporulation process. The protein crystal can be seen both with the light microscope and the electron microscope (Figure 20.6). It is formed in the sporangium along with the spore and is referred to as a parasporal body. This protein crystal is toxic to insects. Vegetative cells of *B. thuringiensis* and its spores are found on plant leaves and are ingested by insects that feed on the leaves. The most common hosts are the larval stages of moths and butterflies (class *Lepidoptera*). These insects have an alkaline gut that dissolves the toxin, a neurotoxin that causes paralysis and death of the caterpillar. Presumably, the bacterium benefits from its host by growing on the carcass (Box 20.2).

Bacillus stearothermophilus is a thermophile with a temperature growth range between 45°C and 65°C. This particular thermophile is found in hot springs, deserts, and temperate zone soils.

Bacillus polymyxa is a facultative anaerobe that, like *B. licheniformis,* ferments sugars with the production of 2,3-butanediol, ethanol, acetic acid, and CO_2. It is also distinctive in being one of the few members of the genus to carry out nitrogen fixation, which it does when growing anaerobically.

Figure 20.6 *Bacillus thuringiensis* **crystal**
Thin-section electron micrograph of the protein crystal of *Bacillus thuringiensis.* Courtesy of S. Pankratz.

Bacillus circulans is similar to *B. polymyxa* but is peculiar in its formation of motile colonies. The colonies of this organism actually rotate and can move across a Petri dish due to the concerted action of the flagella.

Another member of the genus that is particularly noteworthy is *Bacillus pasteurii.* This bacterium is especially suited for growth on urea. For example, it does not grow on ordinary nutrient broth unless it is supplemented with urea. As the urea is degraded, ammonia is produced and the pH rises:

$$CO(NH_2)_2 + H_2O \rightarrow CO_2 + 2 NH_3$$

The organism can tolerate pH values between 8.0 and 9.5, much higher than tolerated by typical bacteria. Because of its ureolytic ability, *B. pasteurii* is commonly found in urinals and can be easily isolated from these sources or from soils by the addition of urea to nutrient broth. Although other bacteria can degrade urea, they cannot do so at the high pH values that this bacterium can. Thus, other bacteria are eliminated from environments where this is a major activity because the resulting high pH is so selective. Some *Bacillus* species, such as *Bacillus alcalophilus,* grow at even higher pH values. Such microorganisms that grow at high pH are termed

BOX 20.1	*Milestones*

Biological (or Germ) Warfare and Bioterrorism

Some pathogenic microorganisms have been studied as agents of germ warfare by many technologically advanced countries. *Bacillus anthracis* has been studied extensively because of the hardiness of the endospore, the ability to infect populations by aerial dispersal, and the deadliness of pulmonary anthrax.

At the international Biological Weapons Convention in 1972 many nations, including the United States and the Soviet Union, signed an agreement to ban research on the development and use of biological weapons. However, some countries continue to study biological warfare even today.

In 1979 a tragic incident occurred in Sverdlovsk, a city of 1.2 million inhabitants located 1,400 kilometers east of Moscow in the former Soviet Union. This city was the site of a biological weapons research facility. Apparently, due to an accidental release from the military research plant, 77 individuals were infected with anthrax, and 66 subsequently died. This epidemic occurred during the cold war between the Soviet Union and the United States after they were signatories to the 1972 Convention. Official information released by the Soviet Union regarding the outbreak stated that individuals developed gastrointestinal, not pulmonary, anthrax after eating contaminated meat.

Following the collapse of the Soviet Union, the real cause of the epidemic was determined in a study led by a U.S. scientist, Matthew Meselson (*Science* 266: 1202–1208, 1994). This group of scientists discovered that, in fact, most individuals died of pulmonary anthrax, indicating that the outbreak was due to airborne dispersal and inhalation of spores, not from ingestion of meat as officially claimed. This man-made epidemic provides some insight into the deadliness of *B. anthracis* as a biological warfare agent.

During 2001, an incident occurred in the United States in which letters that had been heavily inoculated with anthrax spores were mailed through the U.S. Postal Service. The deadly letters were mailed to U.S. senators, congressional representatives, a private company, and the U.S. Department of Justice. Several postal workers and citizens died from pulmonary anthrax, even though they had not even opened the envelopes, indicating that the spores passed through the paper, contaminated other mail, and became aerosolized. These were some of the individuals who contracted pulmonary anthrax.

Others contracted the less-deadly cutaneous form in which they developed skin lesions on their hands and arms. Once the disease is recognized, it can be readily treated with antibiotics such as penicillins. However, those who contracted the pulmonary form died unless their illness was detected quickly. This incident indicates how vulnerable modern societies are to malevolent individuals and terrorist organizations that use weapons of mass destruction with the intent of killing innocent people.

In the past, the United States has also studied the offensive and defensive aspects of biological weapons. In the 1960s, airplanes released an aerosol of *Serratia marcescens* over the San Francisco metropolitan area with the intent of monitoring the effectiveness of its aerial dispersal, an important aspect of bacterial dissemination. Because of its distinctive red pigment, prodigiosin (see Chapter 19), it is easy to determine the concentration of this species when air samples are plated on agar media. At that time *S. marcescens* was not regarded as a pathogen. However, several individuals who were being treated in hospitals in the Bay area acquired previously unknown *S. marcescens* lung infections due to the aerial spraying, which resulted in one death.

alkaliphiles. They are typically found in alkaline soils such as those found in deserts including the Sahara Desert in Egypt. Some alkaliphiles have been reported to grow at pH 10 and above.

SPOROSARCINA AND *SARCINA*   These are two other genera that are able to produce endospores. Some of them are coccus shaped. They divide to form groups of cells in tetrads and packets of eight. They are obligately aerobic and motile. The two species of *Sporosarcina* are *S. ureae*, from garden soils, and *S. halophila*, from salt marshes. Media supplemented with 3% urea select against most other bacteria due to the resulting high pH produced by the growth of this species. *Sarcina ventriculi* is found in the mammalian gastrointestinal tract.

THERMOACTINOMYCES   As the name implies, *Thermoactinomyces* is a genus of thermophilic bacteria. Like *Bacillus*, the members of this genus produce endospores and are aerobic. The organism is a septated, filamen-

BOX 20.2   *Research Highlights*

Biological Pesticides

Bacillus thuringiensis toxin, often referred to simply as **Bt**, is currently being used as a biological insecticide. Commercial preparations are available that are used to eradicate tussock moths and other troublesome insects. The advantage of using biological agents against such pests is that other animals and plants are not affected because of the high specificity of its action. Moreover, because the toxin is of biological origin, it is readily degraded in the environment. In contrast, DDT, which was used for control of these same pests, had devastating effects on other members of the food chain. Furthermore, DDT undergoes biomagnification (see Chapter 24) and persists in the environment for a long time. It was for this reason that DDT was banned in the United States and many other countries.

Some strains of *B. thuringiensis* as well as other species of *Bacillus, B. popillae* and *B. lentimorbis,* produce toxins that are effective against other insects. The commercial development of the mosquito toxin promises to be useful in the control of malaria.

In a quite different approach aimed at controlling plant insect infestation, these toxin genes are now being incorporated into plants by genetic engineering using *Agrobacterium tumefaciens.* Thus, when insects eat plant leaves, they will ingest the toxin and be killed by the genetically engineered bioinsecticide.

tous bacterium. Many endospores are formed within the same filament. The organism has been found in decaying compost piles where temperatures can become so hot from decomposition that they begin burning spontaneously.

LISTERIA   This is a food-borne pathogen that causes **listeriosis**, a type of gastrointestinal disease. Although it is usually not fatal, it can cause fatalities in infants and debilitated individuals. *Listeria* is a gram-positive rod that is catalase positive and oxidase negative. It ferments sugars to form L-lactic acid. However, unlike *Lactobacillus*, it is motile (peritrichous flagella), produces cytochromes, and grows aerobically by respiration. In addition, it grows at low temperatures and therefore multiplies in the refrigerator. For this reason it has been a problem with refrigerated foodstuffs such as cheese and fish.

STAPHYLOCOCCUS   Like the lactic acid bacteria, members of the genus *Staphylococcus* grow anaerobically and produce lactic acid from sugar fermentation. These bacteria are also gram-positive cocci, but they occur in grapelike clusters of cells (Figure 20.7) rather than chains. The genus differs from the lactic acid bacteria in nutrition and physiology. For example, staphylococci do not have the complex nutritional requirements typical of lactic acid bacteria. Furthermore, they produce heme pigments and, as facultative aerobes, carry out aerobic respiration with an electron transport system containing cytochromes.

Staphylococci are associated with the skin and mucosal membranes of animals. Indeed, all humans carry *Staphylococcus epidermidis*, a nonpigmented coccus, as a normal inhabitant on the surface of their skin. *Staphylococcus aureus* is a human pathogen capable of causing a variety of health problems. Unlike *S. epidermidis*, *S. aureus* is pigmented gold. *S. aureus* also produces several virulence factors that enhance its ability to cause disease. One distinctive factor, called **coagulase**, is an enzyme that causes fibrin to clot. Another factor, leukocidin, attacks leukocytes. Several other factors can be produced by pathogenic strains, including β-hemolysin, lipase, fibrinolysin, hyaluronidase, deoxyribonuclease, and ribonuclease.

S. aureus is a normal inhabitant of the nasopharyngeal tissues of humans. Some humans harbor a specific strain for many years and are therefore called **carriers**. Infants come into contact with the organism during their first

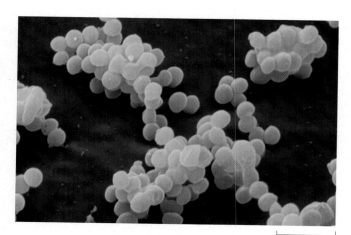

Figure 20.7 *Staphylococcus*
Typical grapelike clusters of *Staphylococcus aureus*. Photo ©Oliver Meckes/Photo Researchers, Inc.

5 μm

week of life. The strain they receive may come from the mother or other close relatives or hospital personnel. Normally these strains do not result in any infection. However, if the infant is unhealthy, it is susceptible to the colonizing strain, and serious skin infections can result. Treatment of such infections can be difficult because many *S. aureus* strains are antibiotic resistant.

Some *S. aureus* strains can also cause a common type of food poisoning. Such strains produce an **enterotoxin**, a type of **exotoxin** (exotoxins are to be distinguished from the endotoxin produced by gram-negative bacteria, which is their lipopolysaccharide—see Chapter 4). Foods that have been prepared and not refrigerated for extended periods allow for the growth of the toxin producer. Problem foods include cooked meats, potato salads, and cream desserts such as eclairs. During summer months, when the weather is warm and families have picnics, conditions are ideal for potential problems. Leftover foods that are not refrigerated properly and are served on subsequent days are especially hazardous. The symptoms of *Staphylococcus* food poisoning are acute and memorable. The toxin, which is produced while the bacterium grows in the food, takes effect within a few hours of ingestion and causes headache and dizziness, fever, diarrhea, and nausea that persists for 24 hours or so. During this period, the afflicted individual feels extremely uncomfortable and is totally incapacitated. Fortunately, the poisoning is self-limiting, and individuals recover fully after 24 to 48 hours.

Toxic shock syndrome (TSS) is a serious malady caused by *S. aureus*. Women who use high-absorbency tampons for prolonged periods during menstruation are susceptible to TSS. This is because *S. aureus* grows and produces toxins in the vagina under these conditions. If unchecked, TSS can be fatal.

Considerable concern has developed about the antibiotic resistance of some strains of *S. aureus*. The major antibiotic used to treat it is methicillin; however, methicillin-resistant strains, referred to as MRSA, have arisen and pose a serious health issue.

Class Clostridia

The clostridia comprise a group of anaerobic gram-positive bacterial genera. These bacteria carry out various types of fermentations, some of which allow them to break down polymers (e.g., cellulose) to monomers such as sugars, purines, and amino acids. The common endospore-forming genus *Clostridium* is discussed first.

CLOSTRIDIUM This genus of anaerobic spore formers is even more diverse than *Bacillus* and contains some ecologically important species as well as human pathogens. *Clostridium* spp. are grouped according to their fermentative abilities into several major subgroups (Table 20.7).

These bacteria are obligate anaerobes that lack cytochrome and an electron transport system. Therefore, they rely solely on the formation of ATP by substrate-level phosphorylation during fermentation of various carbon sources. Although these organisms are obligately anaerobic, they are not as difficult to grow as methanogenic bacteria (see Chapter 18). Unlike methanogens, most can be grown easily in ordinary anaerobic jars in which oxygen has been removed by reaction with H_2. Most species can be grown in liquid media in tubes that have been supplemented with sodium thioglycollate. Therefore, they can be maintained in tubes on the lab bench without additional precautions.

The genus is metabolically diverse. Two species, *Clostridium cellobioparum* and *Clostridium thermocellum*, carry out an anaerobic fermentation of cellulose to form cellodextrins, which are ultimately fermented to produce

TABLE 20.7	Major groups and representatives of *Clostridium*		
Group or Substrate	**Fermentation Type**	**Representative Species**	**Distinctive Products or Features**
Cellulolytic	Acetate, lactate ethanol, H_2, CO_2	C. cellobioparum C. thermocellum	Rumen organism Soil and sewage
Saccharolytic	Butyrate (or butanol-acetone); proteolytic	C. butyricum C. pasteurianum C. acetobutylicum C. perfringens	 Nitrogen fixer Wound infections (gas gangrene)
Amino acid pairs	Stickland reaction	C. sporogenes C. tetani C. botulinum	 Causes tetanus Causes botulism
Purines	Uric acid and other purines fermented to acetic acid, NH_3, CO_2	C. acidurici C. fastidiosus	
H_2 and CO_2	Acetogen	C. aceticum	Acetic acid formed

acetic and lactic acids, ethanol, H_2, and CO_2 as the major end products. *C. cellobioparum* is found in the rumen of cattle and sheep and enables these higher animals to utilize cellulose in their diet (see Chapter 25). Interestingly, this organism is inhibited when too much H_2 accumulates as a result of the fermentation. These bacteria live in close association with other ruminant microbes, the methanogenic bacteria, which remove the hydrogen by forming methane gas. The removal of the hydrogen gas lowers its concentration to a sufficiently low level that

the formation of more hydrogen gas is favored. This interaction involving two different groups is referred to as an **interspecies hydrogen transfer** and is an example of synergism (see Chapter 25). *Clostridium thermocellum* is common in decaying soils containing cellulose.

A major group of the clostridia ferments sugars and occasionally starch and pectin to form butyric acid, acetic acid, CO_2, and H_2 as the principal end products. The pathway for this fermentation is shown in Figure 20.8. In this **butyric acid fermentation**, glucose is fermented

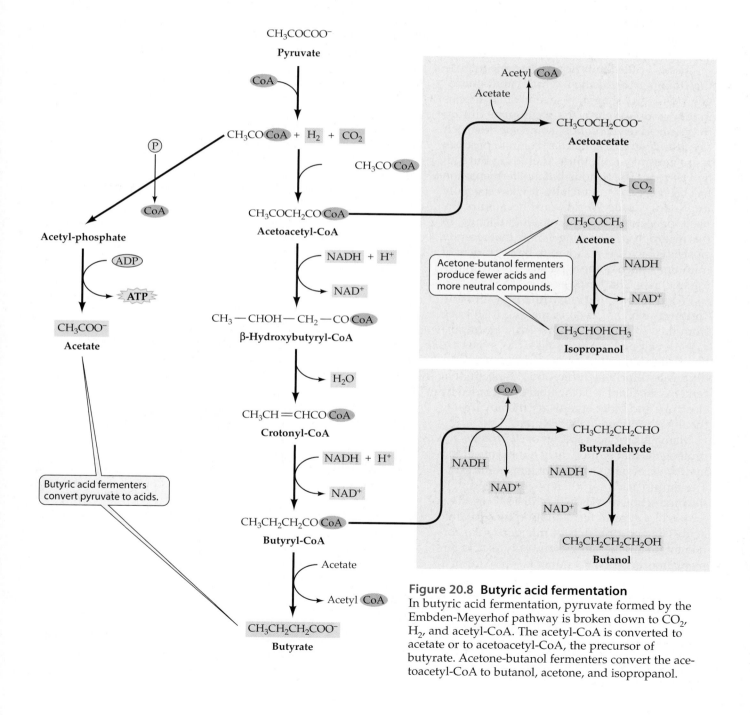

Figure 20.8 Butyric acid fermentation
In butyric acid fermentation, pyruvate formed by the Embden-Meyerhof pathway is broken down to CO_2, H_2, and acetyl-CoA. The acetyl-CoA is converted to acetate or to acetoacetyl-CoA, the precursor of butyrate. Acetone-butanol fermenters convert the acetoacetyl-CoA to butanol, acetone, and isopropanol.

to pyruvic acid via the Embden-Meyerhof pathway. The pyruvate is split into carbon dioxide and hydrogen gas in the formation of acetyl-coenzyme A (CoA). Some of the acetyl-CoA is used for ATP generation in the formation of acetic acid via acetyl-phosphate. Also, the acetyl-CoA can be condensed with another molecule of acetyl-CoA to form acetoacetyl-CoA, which is the precursor of butyric acid.

Some of the butyric acid bacteria produce fewer acids and more neutral products in prolonged fermentations. These are the so-called acetone-butanol fermenters. Butanol is formed from butyryl-CoA via butyrylaldehyde (see Figure 20.8). Acetone and isopropanol are formed from acetoacetyl-CoA by decarboxylation and subsequent reduction, respectively.

These so-called **butyric acid bacteria** include *Clostridium pasteurianum,* which can fix nitrogen, a property shared by several other species in this group. As the acids accumulate during butyric acid fermentation, some species, such as *Clostridium acetobutylicum,* begin to produce more neutral compounds, including butanol and acetone. The acetone-butanol fermentation has been used commercially. It was especially important during World War I because of the need for acetone in munitions manufacture. In this regard, it is interesting to note that recently clostridia have been shown to be degraders of munitions, such as the explosive TNT (trinitrotoluene) (see Chapter 32).

Pectin-fermenting clostridia play a role in "retting," which is used in making Irish linen from flax. In this process, which was empirically developed centuries ago, flax stems are bundled together and immersed in water. Conditions become anaerobic, and the pectin that cements the plant cells together is degraded by clostridia and other organisms, thereby freeing the fibers for linen manufacture.

Many butyric acid species of *Clostridium* are proteolytic; that is, they carry out the anaerobic hydrolysis of proteins, resulting in amino acids. The amino acids can then be fermented with the production of ATP. **Figure 20.9** shows an example of a typical fermentation for glutamic acid. The end products of this particular fermentation are the same as the butyric acid fermentation, except that ammonia is also formed.

Some of the proteolytic clostridia carry out unique fermentation reactions called **Stickland reactions,** in which two amino acids are catabolized. For example, L-alanine can be oxidized and L-glycine reduced by some species, leading to the formation of the final products: acetic acid, CO_2, and NH_3, along with ATP (see Figure 8.5).

Some disease-producing amino acid fermenters include *Clostridium tetani,* the causative agent of tetanus, and *C. botulinum,* which is responsible for botulism (see Chapter 28). *C. tetani* can grow in wounds exposed to soil, which is the normal environment of the organism. The bacterium does not need to grow very much in the tissue because the toxin it produces is extremely potent. The disease is best handled by preventive medicine: children are given tetanus vaccine to build up a natural an-

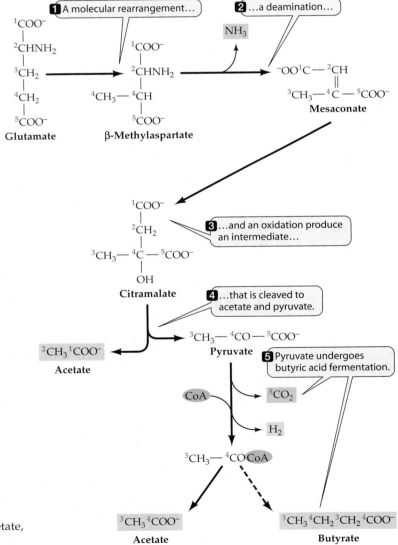

Figure 20.9 Glutamic acid fermentation
Many clostridia ferment glutamate to ammonia, acetate, and butyrate.

tiserum immunity to the organism's toxin before any exposure to the bacterium through wound infections. However, if the skin of an immunized person is cut or punctured and exposed to dirt or soil, it is recommended that the individual receives a tetanus booster shot and perhaps receive antitoxin as well.

Clostridium botulinum is another common soil bacterium and is also a pathogen. It causes a fatal food poisoning called **botulism**. Botulinum toxin is a protein exotoxin that is excreted from the bacterium as it grows. It affects the normal release of acetylcholine from motor nerve junctions, thereby causing paralysis (see Chapter 28). It is one of the most potent toxins known (Box 20.3). Despite its toxicity, the toxin (commonly referred to as "Botox") is actually used therapeutically. Its primary application is for treatment of dystonia, a condition in which muscles move involuntarily, resulting in tremors. Upper body muscles such as those in the neck or throat are frequently affected. Botox is injected into the affected muscles, thereby blocking the aberrant nerve pulses that cause the muscles to contract. Botox injections are also widely used in cosmetic applications as a means to control wrinkles (see Box 20.3).

Clostridium perfringens is another important pathogen and food-poisoning organism. Like *S. aureus,* this species produces an enterotoxin. The *C. perfringens* enterotoxin is formed during sporulation and can be seen as a protein crystal in the sporangium (Figure 20.10). This bacterium is widely distributed in soils and commonly contaminates foods. In addition, it is particularly trouble-some to foot soldiers during times of war. Soldiers' wounds that are exposed to soil can become infected by the bacterium. It grows profusely in the tissues and produces gas that clogs blood vessels and can lead to compromised blood circulation and eventually "gas gangrene," which may necessitate amputation.

BOX 20.3 *Milestones*

Food (or a face) to Die For!

The toxin produced by *Clostridium botulinum* is among the most deadly toxins known to humans. Most incidents of botulism occur from eating improperly sterilized home-canned foods, such as vegetables that are contaminated with soil containing the bacterium. If canning does not kill the bacteria, they can grow in the canned food and release the toxin into the contents of the can. As the toxin is readily denatured by boiling, it is primarily a problem in foods such as canned peas that are often used cold in salads without prior cooking.

When botulism is detected in commercially canned food, it can have a disastrous financial impact on manufacturers. A few years ago, a handful of individuals from around the United States were diagnosed with botulism. By examining the food consumption of infected individuals, epidemiologists quickly identified the source of the problem

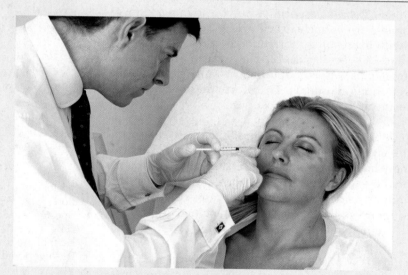

A patient being treated with Botox. Credit to come, to come, to come, to come. ©Lauren Shear/Photo Researchers, Inc.

as canned salmon. Ultimately, the cans were traced to a defective canning machine in one plant in Alaska. When the news broke, sales of canned salmon plummeted, and the canned salmon market took years to recover, despite assurances by canners that the problem had been corrected.

The toxin produced by *C. botulinum*, familiar as Botox, has become popular as a cosmetic treatment to eliminate wrinkles. Tiny amounts of the toxin are injected into facial muscles, paralyzing them and relaxing the overlying skin.

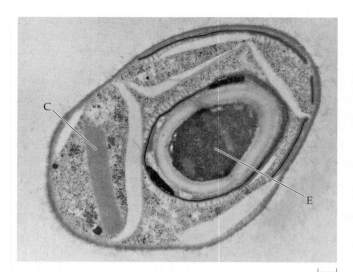

Figure 20.10 *Clostridium perfringens* **toxin**
Thin section showing the endospore (E) and enterotoxin
crystal (C) of *Clostridium perfringens*. Courtesy of Judith
Bland.

Other *Clostridium* species are involved in the fermentation of purines such as uric acid (*Clostridium fastidiosus*) with the resulting formation of acetic acid, ammonia, and CO_2 (see Figure 12.12). Uric acid is excreted as
a nitrogenous waste product of birds, analogous to urine
excretion by mammals. Therefore, these bacteria are
widespread in soils and rookeries.

EPULOPISCIUM   One of the most remarkable members
of the clostridia is the genus *Epulopiscium*. This extremely large bacterium (Figure 20.11) lives in the intestinal tract of surgeonfish, a tropical marine fish. Although it has not been isolated in pure culture, it has
been identified as a member of the clostridia by 16S
rRNA gene sequences.

DESULFOTOMACULUM   Although this is also a genus
of anaerobic spore formers, it differs from the genus
Clostridium in two important ways. First, it carries out
an anaerobic respiration involving sulfate and is therefore a sulfate-reducing bacterium (see Chapter 19 for
more information on sulfate reducers). Second, it contains some cytochrome enzymes (cytochrome *b* but not
c_3), and therefore it has a limited electron transport system that is needed as a means of passing the electrons
generated from the anaerobic oxidation of organic compounds onto sulfate, resulting in the formation of hydrogen sulfide. However, ATP is not generated by chemiosmotic processes, but only by substrate-level phosphorylation. Species using lactic or pyruvic acid as carbon sources oxidize these incompletely to produce acetic

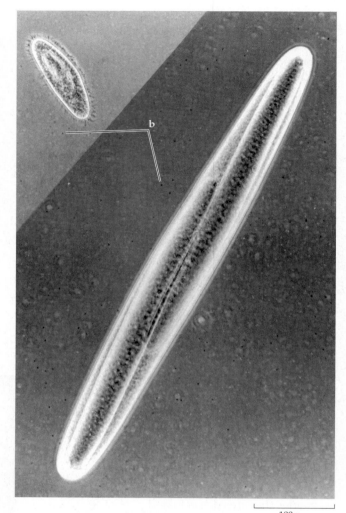

Figure 20.11 *Epulopiscium*
Compare this large bacterium with the protist, a ciliated
protozoan, shown in the inset taken at the same magnification. *Epulopiscium*, like *Thiomargarita* (Box 1.3) is the exception to the rule that bacteria are smaller than eukaryotic
cells. The small black specks, marked "b," are typical bacteria that are dwarfed by these larger organisms. Courtesy of
E. Angert and N. Pace.

acid and carbon dioxide. Species using acetic acid oxidize it completely to carbon dioxide.

Desulfotomaculum is found in soils, geothermal regions, sediments, and anaerobic muds. It is also known
to produce off-flavors in canned foods, referred to as
"sulfur stinker." One species, *Desulfotomaculum nigrifi-
cans,* is thermophilic and grows at temperatures from
45°C to 70°C, whereas the others are mesophilic.

Clostridium aceticum is a representative of a group of
gram-positive bacteria that can grow as chemolithotrophs. This bacterium is a hydrogen autotroph that
gains its energy by oxidation of hydrogen. This is an ex-

TABLE 20.8	Acetogenic gram-positive bacteria
Taxon	**Special Features**
Clostridium aceticum	Endospore-former
Acetobacterium spp.	Non–spore-forming rods; mesophilic; also ferments sugars to produce acetate
Acetogenium kivui[a]	Non–spore-forming rod; thermophilic (opt. temp. 66°C); also ferments sugars to form acetate

[a]Stain as gram-negative, but have a gram-positive cell wall type.

ample of an acetogenic bacterium in which carbon dioxide is reduced to form acetic acid in the following overall reaction:

$$2\ CO_2 + 4\ H_2 \rightarrow CH_3COOH + 2\ H_2O$$

A number of non–spore-forming gram-positive bacteria are also acetogenic. These include the genera *Acetobacterium* and *Acetogenium* (Table 20.8). All of these bacteria fix carbon dioxide by the acetyl-CoA pathway (see Chapter 10), a much different mechanism of carbon dioxide fixation than that of the Calvin cycle. These bacteria can also grow as ordinary heterotrophs by utilization of sugars in fermentations.

HELIOBACTERIA   This group is one of the most recently discovered bacterial groups. These bacteria are the only phototrophic gram-positive bacteria known. They are photoheterotrophs that require organic compounds as carbon sources and use light for energy generation. They have a unique type of bacteriochlorophyll called bacteriochlorophyll *g* (see Chapter 21). Although they stain as gram-negative bacteria, an analysis of their 16S rRNA places them with the gram-positive bacteria. Furthermore, some members of the group produce heat-resistant bacterial endospores similar to those of *Clostridium* sp.

Two common genera in this group are *Heliobacterium* and *Heliobacillus*. *Heliobacterium* sp. are gliding bacteria, whereas members of *Heliobacillus* are motile by use of peritrichous flagella. The heliobacteria grow as obligate anaerobes and carry out anoxygenic photosynthesis using organic compounds such as pyruvate as carbon sources.

The finding of phototrophic bacteria in the gram-positive phylum is consistent with photosynthesis being a primitive characteristic that is shared by many of the other phyla of *Bacteria*. This group is discussed further, along with the other photosynthetic bacteria, in Chapter 21.

Class Mollicutes: *Cell Wall–Less Bacteria*

The *Mollicutes* are a major taxonomic group of cell wall–less bacteria that stain as gram-negative. However, studies of their 16S rRNA have clarified their relatedness to other bacteria and indicate that they are most closely allied with gram-positive bacteria in the genus *Clostridium*.

Because they completely lack a cell wall, the organisms have unusual shapes (Figure 20.12). They are also among the smallest of the bacteria and the simplest in structure. Some have a diameter of about 0.25 μm, which approaches the theoretical minimum for the size of an organism, in other words, a structure large enough to contain the DNA, ribosomes, and necessary enzymes with which to perform the functions of life. Their genomes are also small, about 0.75 to 1.0 MB, which is comparable to the size of the obligately parasitic *Chlamydia* sp. (see Chapter 28).

Their lack of a cell wall makes the *Mollicutes* particularly fragile osmotically. Many species must be grown on complex media, and some require sterols to stabilize their plasma membranes (Table 20.9). Because they cannot synthesize sterols, they must obtain them from the medium in the form of cholesterol, often added as serum. It should be noted, however, that some species do not require sterols.

MYCOPLASMA   One important genus of the *Mollicutes* is *Mycoplasma*. Colonies of *Mycoplasma* species appear as "fried eggs" with dense, granular centers and more transparent outer regions. Some members of this genus

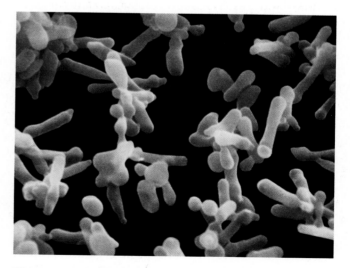

Figure 20.12 *Mycoplasma*
Electron micrograph of cells of *Mycoplasma gallisepticum*, a member of the *Mollicutes*. Note how the cells vary in shape. Photo ©Michael Gabridge/Visuals Unlimited.

Genus	Mol % G + C	Sterol Requirement	Habitat	Distinctive Properties
Mycoplasma	23–40	+	Animals	
Ureaplasma	27–30	+	Animals	Degrade urea
Acholeplasma	27–36	–	Animals	
Spiroplasma	25–31	+	Plants; insects	Helical shape
Anaeroplasma	29–34	+	Cattle rumen	Obligate anaerobes

TABLE 20.9   Important genera of *Mollicutes*

obtain energy by fermentation of sugars through the Embden-Meyerhof pathway to form lactic, pyruvic, and acetic acids. Others degrade arginine as an energy source or derive energy from acetyl-CoA by phosphoacetyl transferase and acetate kinase.

Many species of this genus are parasitic or pathogenic to animals. They are found particularly associated with mucoid epithelial tissues. *Mycoplasma pneumoniae* causes a type of bacterial pneumonia. Unlike pneumococcal pneumonia, however, it is not possible to use penicillin as an antibiotic in its treatment because *M. pneumoniae* lacks peptidoglycan. However, mycoplasmas are inhibited by tetracycline and chloramphenicol.

OTHER IMPORTANT GENERA   *Ureaplasma* spp. obtain energy through the degradation of urea to ammonium ions and the subsequent conversion of ammonium to ammonia plus protons, from which they produce ATP chemiosmotically. These are parasites in animal genitourinary tracts and respiratory tracts.

Acholeplasma spp. do not require sterols for growth. They produce **lipoglycan**, which may help stabilize their cell membranes. Unlike the lipopolysaccharide of gram-negative bacteria, lipoglycan is not linked to a lipid A backbone. They resemble *Mycoplasma* species in that they obtain energy from sugar fermentation. None are known to be pathogens, but they are common parasites of animals.

Spiroplasma spp. have a characteristic helical shape (Figure 20.13). They obtain energy by sugar fermentation. They are found primarily as commensals of plants and insects. Insects feeding on infected plants can obtain the bacterium from plant phloem and carry it to another plant.

Anaeroplasma are strict anaerobic organisms found in the rumen of cattle and sheep. They ferment starch and other carbohydrates to produce acetic, formic, lactic, and propionic acids, as well as ethanol and carbon dioxide. They lyse other bacteria as well, but are not known to be pathogenic to their hosts.

SECTION HIGHLIGHTS

The phylum *Firmicutes* contains three classes the *Bacilli*, *Clostridia*, and *Mollicutes*. The *Bacilli* includes the lactic acid bacteria and the aerobic or facultatively aerobic endospore-forming bacteria. It also includes several important pathogenic genera such as *Streptococcus*, *Staphylococcus*, *Listeria*, and *Bacillus*. The *Clostridia* are anaerobic endospore-forming bacteria, some of which are pathogens that cause tetanus, botulism, or gas gangrene. Other members of this class are sulfate reducers, acetogens, phototrophs, and important fermenters of cellulose and other substances. Bacteria in the class *Mollicutes* lack cell walls. Many are pathogens of animals and plants.

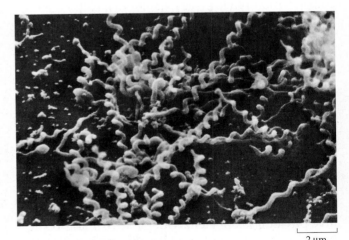

2 μm

Figure 20.13   *Spiroplasma*
Cells of the cell wall–less genus *Spiroplasma* have a characteristic helical shape. Photo ©David M. Phillips/Visuals Unlimited.

20.2 Phylum: *Actinobacteria*

The "true" *Actinobacteria* (also referred to as the "actinomycetes") are mycelial organisms; that is, the cells produce branches as well as filaments. Therefore, they have a mycelial growth habit resembling that of the fungi. Interestingly, although they are gram positive, the *Actinobacteria* are high mol % G + C bacteria that are phylogenetically separate from the *Firmicutes* in the Tree of Life (see Chapter 1). This is a large and diverse group of bacteria noted for their degradation of plant materials and production of antibiotics, which contains only a few animal pathogens. The first organisms studied in this group were species of the genus *Actinomyces*, after which the group is named.

Some *Actinobacteria* produce only a substrate mycelium in which the growth is either within or on the surface of the agar or other growth medium. The substrate mycelium is **coenocytic**, lacking cell septa. Some genera produce an aerial mycelium as well as a substrate mycelium. The aerial mycelium can form special reproductive spores called conidia or conidiospores. Conidia are more resistant than the mycelium to ultraviolet light and can survive well under dry conditions. They are disseminated by the wind as a means of dispersing the organism from soil environments. However, unlike endospores, they are not particularly resistant to high temperatures. Almost all of the actinomycetes are nonmotile; however, some types produce flagellated spores that permit dispersal in aquatic habitats.

Unicellular Actinobacteria

Although many members of the *Actinobacteria* produce multicellular filaments or mycelia, some, such as *Micrococcus*, are unicellular. We begin by discussing *Micrococ-*

Domain *Bacteria*

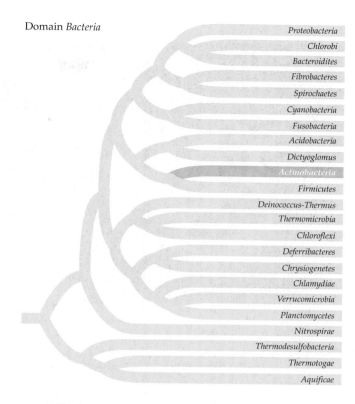

cus, Bifidobacterium, Arthrobacter, and *Geodermatophilus,* which are common soil genera, and *Frankia,* which is a plant symbiont (Table 20.10).

MICROCOCCUS   Although *Staphylococcus* and *Micrococcus* resemble each another superficially in both being nonmotile, gram-positive cocci, they are considerably different. The genus *Micrococcus* is a group of obligately aerobic organisms with a high mol % G + C (66–72%) and is phylogenetically classified with the *Actinobacteria*. In

TABLE 20.10	Important genera of unicellular, filamentous, and coryneform bacteria			
Genus	**Shape**	**Mol % G + C**	**O₂ Requirement**	**Habitat**
Micrococcus	Cocci in clusters	66–72	Obligate aerobe	Soil
Bifidobacterium	Rods	55–67	Anaerobe	Animal intestine
Corynebacterium	Irregular rods, V-shapes	51–60	Facultative aerobe	Soil; animal pathogen
Propionibacterium	Club-shaped rods	57–67	Facultative aerobe	Animal intestine, cheese
Mycobacterium	Rods, some branching	62–70	Aerobe	Animal pathogen
Nocardia	Slight to extensive branching	64–69	Aerobe	Soil
Rhodococcus	Coccus to filamentous mycelium	60–69	Aerobe	Soil
Renibacterium	Rod	53	Aerobe	Fish pathogen, salmonids
Geodermatophilis	Filamentous	73-75	Aerobe	Soil and rocks
Frankia	Filamentous	66–71	Aerobe	Alder N₂-fixing symbiont

(A)

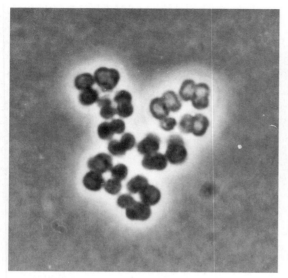

(B)

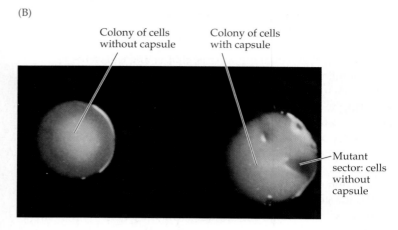

Colony of cells without capsule

Colony of cells with capsule

Mutant sector: cells without capsule

Figure 20.14 *Micrococcus*
(A) *Micrococcus luteus* as it appears in the phase microscope. Note how the cells occur in packets. (B) Colonies of *Micrococcus roseus*, without and with capsules. Note the larger, gummy appearance of the colony of encapsulated cells. A, courtesy of J. T. Staley; B, courtesy of Wesley Kloos.

contrast, the genus *Staphylococcus* is a group of facultative anaerobes with a low DNA base ratio (30 to 35 mol % G + C), which are members of the class *Bacilli*.

The genus *Micrococcus* is a common soil organism that is dispersed in air. Some species also reside on the skin of humans and other mammals. Most strains produce carotenoid pigments that give their colonies a yellow to red pigmentation. The pigments may protect these organisms from ultraviolet light during their dispersal in air by absorbing lethal ultraviolet radiation. Some of them produce packets of cells (Figure 20.14A). Some *Micrococcus* spp. produce capsules that give colonies a characteristic gummy appearance (Figure 20.14B).

RENIBACTERIUM This is the genus of a slow-growing, rod-shaped bacterium noted as the causative agent of bacterial kidney disease (BKD) in salmonid fish (see Table 20.10). It has been cultivated in pure culture only recently. It is an obligate aerobe that uses sugars as carbon sources for growth and requires cysteine.

BIFIDOBACTERIUM *Bifidobacterium* is a genus of anaerobic irregular rod-shaped bacteria that ferment sugars to acetic and lactic acids (Figure 20.15). These nonmotile bacteria are found primarily in the intestinal tracts of animals. One species, *B. bifidus*, is commonly found in the intestines of humans that are breast fed and is therefore a pioneer colonizer of the human intestinal tract. It is particularly well adapted to growing on human breast milk, which contains an amino sugar disaccharide not found in cow's milk. This species is unusual in that it requires amino sugars for growth.

ARTHROBACTER The species of this genus are commonly found as inhabitants of soil. Sergei Winogradsky, the microbiologist who discovered biological nitrification (see Chapter 2), was the first to note that small coccoid cells were abundant in soils. Actually, the coccoid cells are one of two cell types exhibited by this genus. When cells are actively growing, they grow as irregular rods (Figure 20.16A). However, as the cells enter stationary phase, they become shorter and rounded in appearance (Figure 20.16B). In fact, some studies have shown that the length of the rod is directly related to the growth rate of the bacterium. If they are growing slowly, they are shorter than if they are growing more rapidly. These bacteria are gram-positive, but some may stain as gram-negative cells. Some strains produce motile cells.

Arthrobacter spp. are obligate aerobes that use a variety of sugars as carbon sources. Many grow on a simple medium with ammonium as a nitrogen source, although some strains require biotin or other vitamins.

GEODERMATOPHILUS This unusual genus, with the aptly named species, *Geodermatophilus obscurus*, is a common inhabitant of soils, particularly desert soils and rocks. It has been found on Mt. Everest as well as in Antarctica. Like *Micrococcus*, it is an obligate aerobe that utilizes a variety of sugars as carbon sources for growth. Colonies produce a greenish black pigment, probably a melanin. The organism grows as masses of cocci to form a gummy colony that rises above the agar surface. Motile cells, which are termed the "R-form," are rod-shaped, whereas nonmotile cells are termed the "C-form" and are coccoid-shaped.

Figure 20.15 *Bifidobacterium* **fermentation**
The bifidobacteria carry out an unusual lactic acid fermentation, as shown here.

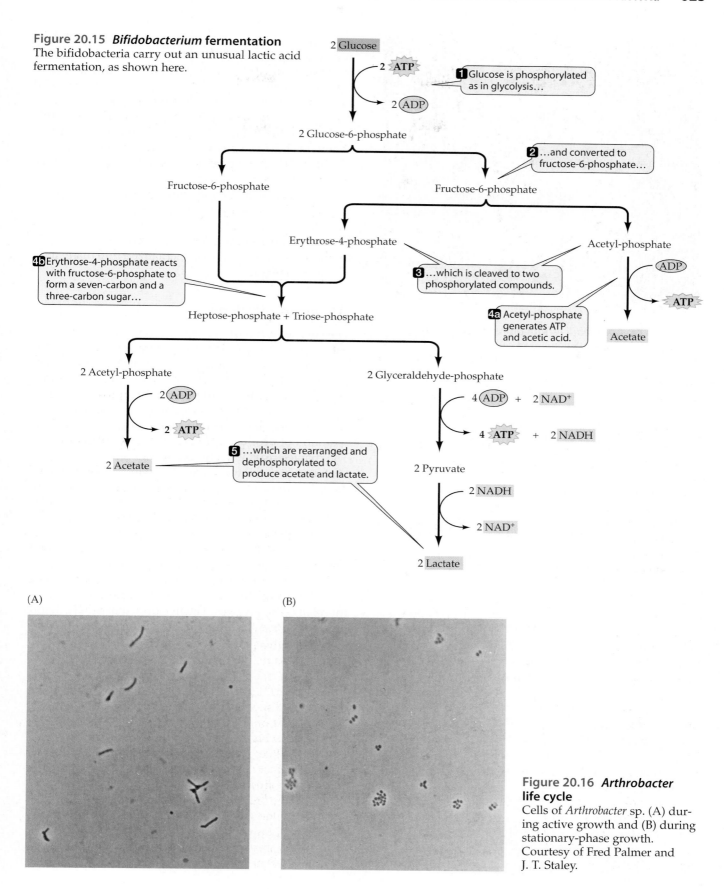

2 Glucose

2 **ATP**

1 Glucose is phosphorylated as in glycolysis…

2 (ADP)

2 Glucose-6-phosphate

2 …and converted to fructose-6-phosphate…

Fructose-6-phosphate

Fructose-6-phosphate

Erythrose-4-phosphate

Acetyl-phosphate

4b Erythrose-4-phosphate reacts with fructose-6-phosphate to form a seven-carbon and a three-carbon sugar…

3 …which is cleaved to two phosphorylated compounds.

(ADP)

ATP

4a Acetyl-phosphate generates ATP and acetic acid.

Acetate

Heptose-phosphate + Triose-phosphate

2 Acetyl-phosphate

2 (ADP)

2 **ATP**

2 Glyceraldehyde-phosphate

4 (ADP) + 2 NAD^+

4 **ATP** + 2 NADH

5 …which are rearranged and dephosphorylated to produce acetate and lactate.

2 Acetate

2 Pyruvate

2 NADH

2 NAD^+

2 Lactate

(A)

(B)

Figure 20.16 *Arthrobacter* **life cycle**
Cells of *Arthrobacter* sp. (A) during active growth and (B) during stationary-phase growth. Courtesy of Fred Palmer and J. T. Staley.

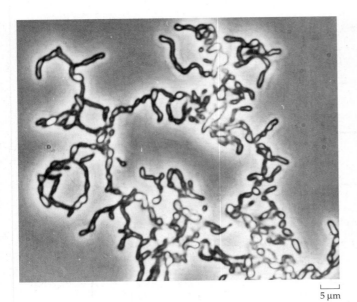

5 μm

Figure 20.17 *Corynebacterium diphtheriae*
Typical orthogonal arrangement of cells of the coryneform bacterium *Corynebacterium diphtheriae*. Courtesy of J. T. Staley.

FRANKIA This genus is an important plant symbiont. It produces root nodules analogous to those formed by *Rhizobium*; however, it associates with a variety of non-leguminous plants such as *Alnus* (alder), *Ceanothus* (wild lilac), and *Casuarina* (Australian pine or she-wood). Like *Rhizobium*, *Frankia* carries out nitrogen fixation while growing in the plant as a symbiont, thereby benefiting the plant. Like the nitrogen-fixing cyanobacteria (see Chapter 21), *Frankia* produces specialized cells in which the nitrogen fixation occurs.

Frankia is difficult to cultivate away from its host plant. The organism is microaerophilic and produces an aerial as well as substrate mycelium. The aerial mycelium develops a sac or **sporangium** that is referred to as multi-locular because it is compartmentalized into many individual spores (see Figure 25.8).

Coryneform Bacteria

The coryneform bacteria are a group of organisms, primarily soil forms, which show characteristic branching cells with the formation of Y, V, or orthogonal shapes (Figure 20.17). The Y shapes are due to rudimentary branch formation, a common characteristic of the filamentous gram-positive bacteria, the actinobacteria (except for some of the cyanobacteria, gram-negative bacteria do not produce true branches in their filaments). The characteristic V shapes occur after cell division. Cells have an outer cell wall layer not shared by other bacteria. When the cells have completed division, turgor pressure is exerted on the space between the dividing cells, and they separate incompletely. This is called a postfission snapping movement. After cell division, the two cells do not separate completely from one another and thereby produce a V-shaped configuration.

The coryneform bacteria appear to be intermediate morphological forms between unicellular and true mycelial actinomycetes. The five principal genera in this group are *Propionibacterium*, *Corynebacterium*, *Mycobacterium*, *Nocardia*, and *Rhodococcus* (see Table 20.10).

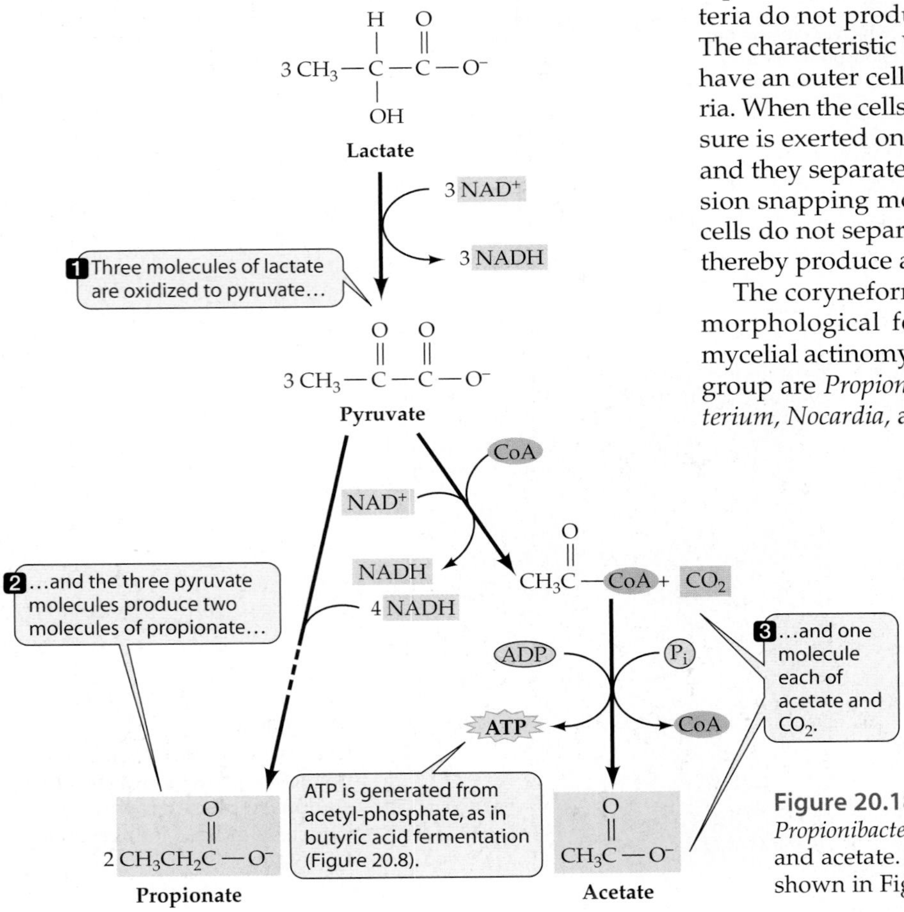

$$3 CH_3 - \underset{\underset{OH}{|}}{\overset{\overset{H}{|}}{C}} - \overset{\overset{O}{||}}{C} - O^-$$

Lactate

3 NAD$^+$

3 NADH

1 Three molecules of lactate are oxidized to pyruvate...

$$3 CH_3 - \overset{\overset{O}{||}}{C} - \overset{\overset{O}{||}}{C} - O^-$$

Pyruvate

CoA

NAD$^+$

NADH

4 NADH

$$CH_3\overset{\overset{O}{||}}{C} - CoA + CO_2$$

ADP

P$_i$

ATP

CoA

2 ...and the three pyruvate molecules produce two molecules of propionate...

3 ...and one molecule each of acetate and CO$_2$.

ATP is generated from acetyl-phosphate, as in butyric acid fermentation (Figure 20.8).

$$2 CH_3CH_2C - O^-$$

Propionate

$$CH_3\overset{\overset{O}{||}}{C} - O^-$$

Acetate

Figure 20.18 **Propionic acid fermentation**
Propionibacterium spp. ferment lactate to propionate and acetate. The route from pyruvate to propionate is shown in Figure 20.19.

PROPIONIBACTERIUM   This genus is so named because its members produce propionic acid as a principal product of their fermentation. They are gram-positive aerotolerant fermentative bacteria that are found in two different habitats. One habitat occupied by the classic type of *Propionibacterium* is the intestinal tract of animals. An allied habitat is cheese. Indeed, the characteristics of Swiss (Ementhaler) cheese are due to the growth of propionibacters. These organisms take the lactic acid formed by the lactic acid bacteria in cheese fermentation and further metabolize it to propionic and acetic acids as well as CO_2 (Figure 20.18). These reactions are more complex, comprising a series of steps beginning with the oxidation of lactic acid to pyruvate and then transformation of the pyruvate (Figure 20.19). The concomitant gas production results in the characteristic holes, called "eyes," of Swiss cheese (if gas is not produced in adequate amounts, holes are not formed, and the cheese is said to be "blind"). Although cheese was the initial source of these bacteria, their true natural habitat has been traced to the rumen of

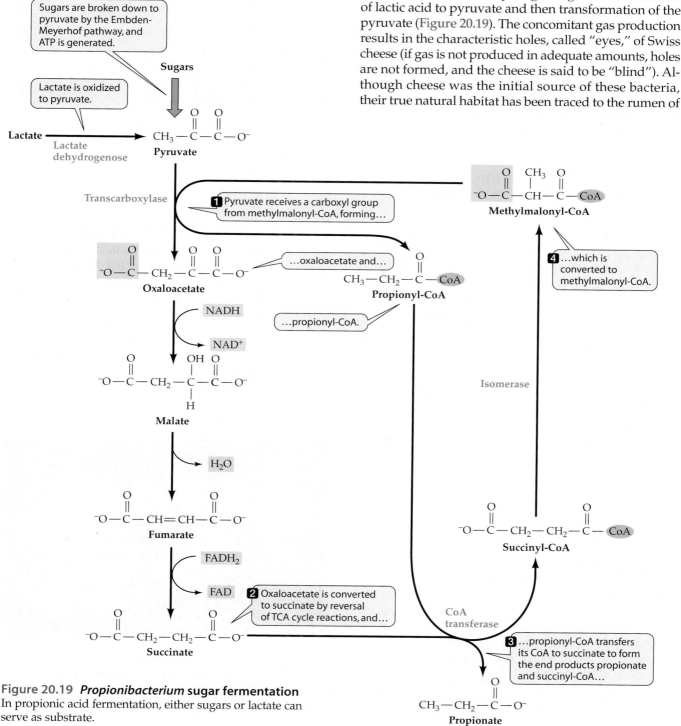

Figure 20.19 *Propionibacterium* **sugar fermentation**
In propionic acid fermentation, either sugars or lactate can serve as substrate.

cattle and other animals. Rennin, an enzyme used for the curdling of milk in the production of cheeses, is obtained from the stomachs of calves. Therefore, rennin contaminated with propionibacteria was the likely original source of these organisms in the cheese-making industry. Sugars are also fermented by propionibacteria. They use the Embden-Meyerhof pathway to produce pyruvate, which is then fermented as shown in Figure 20.19.

The other major habitat of propionibacteria is the skin of mammals. One species, *Propionibacterium acnes*, is found on the skin of all humans. It grows in the sebaceous gland (not sweat glands), where it produces propionic acid in abundance. It ferments the lactic acid produced by *Staphylococcus epidermidis* to form propionic and acetic acid, as previously discussed. Humans fall into two groups based upon the numbers of *P. acnes* they harbor on their skin. Some individuals have more than 1 million per cm^2 of skin surface, whereas others have fewer than 10,000 per cm^2. Because propionic and acetic acids are volatile fatty acids with distinctive smells, they give animals, including humans, a characteristic natural scent.

CORYNEBACTERIUM   This genus is a group of common aerobic soil organisms. The first isolate of the genus was *Corynebacterium diphtheriae*, which is a normal inhabitant of the oral cavity of animals, where it lives as a parasite. Pathogenic strains of this species carry a piece of DNA that they have received from a phage by lysogenic conversion (see Chapter 14). This genetic material is responsible for production of the protein diphtheria toxin, an exotoxin, which is the principal virulence factor for diphtheria (see Chapter 28). The disease is rare now, having been controlled largely through vaccination with the DPT (diphtheria, pertussis, tetanus) vaccine.

Soil "diphtheroids," which appear as club-shaped rods (Figure 20.20), are similar nutritionally and metabolically to the animal parasites. All are immotile and most are facultative aerobes that produce propionic acid during fermentation of sugars. Some species cause diseases of plants.

MYCOBACTERIUM   The mycobacteria are aerobic, nonmotile rod-shaped bacteria that may show true branching and typically bundle together to form cord-like groups (Figure 20.21). They are rods or curved rods that occur naturally as saprophytes in soils. Unlike other bacteria, they produce a distinctive group of waxy substances called **mycolic acids** (Figure 20.22) that are linked covalently to the peptidoglycan. In addition to glucosamine and muramic acid, they have a polymer of arabinose and galactose (called an arabinogalactan) bound to the peptidoglycan. The mycolic acids make the organisms difficult to stain using ordinary simple stain-

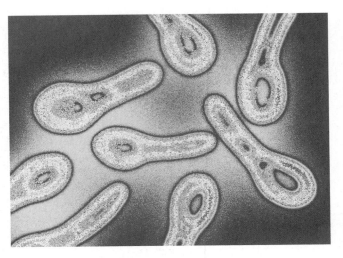

Figure 20.20 *Corynebacterium diphtheriae*
False-colored TEM showing club-shaped diphtheroids of *Corynebacterium diphtheriae*. Photo ©Alfred Pasieka/SPL/ Photo Researchers, Inc.

ing procedures; nonetheless, they are regarded as being gram-positive. These same mycolic acids make these bacteria **acid-alcohol fast**, which refers to the ability of these organisms, once stained with a solution of basic fuchsin in phenol, to withstand decolorization with acidified ethanol during a staining procedure called the Ziehl-Neelson stain. The basic fuchsin binds strongly to the mycolic acids. This acid-alcohol–fast property is not found in any other bacterial group, making it an excellent differential property for distinguishing these from all other bacteria.

Mycobacteria grow on simple inorganic media with ammonium as the sole nitrogen source and glycerol or

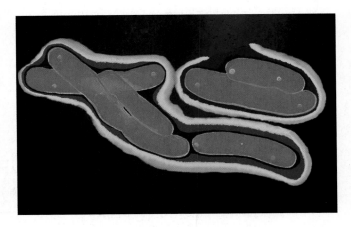

Figure 20.21 *Mycobacterium tuberculosis*
Cells of *Mycobacterium tuberculosis*, showing its tendency to aggregate into cord-like structures. Photo ©Dr. Linda Stannard/SPL/Photo Researchers, Inc.

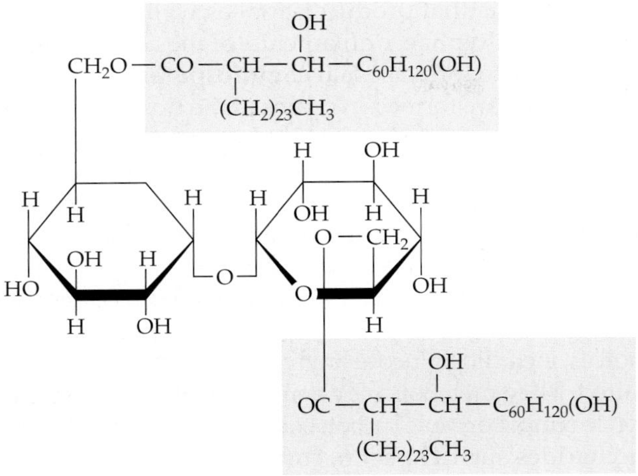

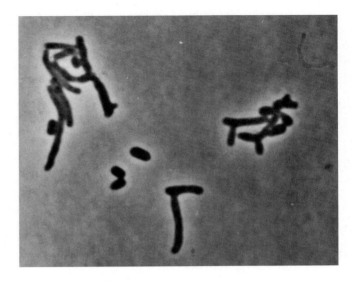

Figure 20.22 Mycolic acids
General formula for mycolic acids.

Figure 20.23 *Nocardia* sp.
Phase contrast photomicrograph of a *Nocardia* sp., showing branching. Courtesy of J. T. Staley.

acetate as a carbon source. The incorporation of lipids into growth media tends to enhance growth of some species. Their slow growth is in part due to the hydrophobicity of the organisms, which prevents rapid uptake of dissolved nutrients.

Two species are important pathogens of humans: *Mycobacterium tuberculosis* is the causative agent of tuberculosis, and *Mycobacterium leprae* causes leprosy. See Chapter 28 for more detail on these diseases.

NOCARDIA   The nocardia are separated into two groups. One group is acid-fast and therefore similar to the mycobacteria, whereas the other group is non–acid-fast. These nonmotile, rod-shaped to filamentous organisms show somewhat greater branching than the mycobacteria (Figure 20.23). Some strains even show extensive mycelial development and may even produce an aerial mycelium. One feature characteristic of this group is that the mycelium, if formed, tends to fragment into small units during the stationary growth phase.

Nocardia are common soil bacteria. They are obligately aerobic heterotrophs the metabolism of which has not

been well studied. As a group and individually, they utilize a wide variety of sugars and organic acids as carbon sources for growth.

RHODOCOCCUS   The genus *Rhodococcus* is widely distributed in soils and aquatic environments. It is particularly well known for its ability to degrade hydrocarbon compounds. It produces capsular materials that are excellent dispersants and emulsifiers of hydrocarbon compounds. The cells grow at the interface between oil and water and are particularly adept at degrading aliphatic hydrocarbons. Some species are pathogenic to animals.

Highly Branched and Mycelial Actinobacteria

The more highly branched *Actinobacteria* are common soil organisms that are prokaryotic counterparts of the fungi (Table 20.11). These organisms are not acid-fast. Some of them produce an aerial mycelium that has a fruiting structure that bears spores called conidiospores.

TABLE 20.11	Important genera of branching and mycelial *Actinobacteria*			
Genus	**Shape**	**Mol % G + C**	**Habitat**	**Special Features**
Actinomyces	Rods; some branching	58–63	Oral cavity; some pathogens	Only facultative aerobes in group
Micromonospora	Substrate and aerial mycelium	71–73	Soil	Single conidium
Streptomyces	Substrate and aerial mycelium	69–78	Soil	Many conidia
Actinoplanes	Substrate and aerial mycelium	72–73	Soil	Motile sporangiospores

Figure 20.24   *Streptomyces*
The aerial mycelium of a *Streptomyces* sp., showing conidiospores. Photo ©Frederick P. Mertz/Visuals Unlimited.

These are asexual spores, sometimes referred to as exospores to differentiate them from endospores.

ACTINOMYCES   This genus, for which the entire group was named, is atypical of other mycelial members, primarily because it is a group of anaerobic or facultative anaerobic bacteria. These organisms ferment sugars, such as glucose, to produce formic, acetic, lactic, and succinic acids and do not carry out aerobic respiration. Organic nitrogen compounds are required for growth, and supplemental carbon dioxide greatly enhances it. Though these organisms are mycelial, they do not produce an aerial mycelium. They grow in the oral cavities of animals and can cause serious infections such as lumpy jaw, as caused by *Actinomyces bovis*.

STREPTOMYCES   The foremost genus of the mycelial actinobacteria is *Streptomyces*. This is a diverse group of soil bacteria that produce aerial as well as substrate mycelia. The **hyphae**, or filaments, of the aerial mycelia differentiate to form asexual **conidiospores**. The spores of this genus are formed in chains at the tips of aerial hyphae (Figure 20.24). Species differences are based partially on the morphology of the spores. For example, some have a warty or spiny appearance, whereas others are smooth. Physiological differences, especially antibiotic production, are also important.

Members of this genus grow on simple inorganic media supplemented with a variety of organic carbon sources, including glucose or glycerol; vitamins are not required. They metabolize by aerobic respiration. In addition to simple organic carbon sources, some can use polysaccharides such as pectin, chitin, and, even, latex. If a culture is started with conidiospores, they first germinate to produce the vegetative or substrate mycelium. After this, the aerial mycelium is formed. If the medium has a sufficiently high carbon-to-nitrogen ratio, the aerial mycelium differentiates to produce conidiospores when the nutrients have been depleted. The conidia are actually formed within the outer wall of the conidiophore; however, they do not at all resemble endospores. The conidia become pigmented blue, gray, green, red, violet, or yellow, but the color can be influenced by medium composition. The substrate mycelium can also be pigmented.

Streptomyces occur in viable concentrations of 10^6 to 10^7 per gram in soil environments. They are responsible for imparting the characteristic "earthy" odor to soil. This is due to the **geosmins**, a group of volatile organic compounds that they produce during growth. Geosmins cause odors and flavors in drinking water supplies as well. However, cyanobacteria also produce geosmins and are more likely to be responsible for this problem in water supplies.

This group is noted for the commercially important antibiotics they produce including streptomycin, chloramphenicol, and tetracycline (Table 20.12). Approximately half of the commercially produced antibiotics are

TABLE 20.12	Important antibiotics produced by *Streptomyces* species		
Antibiotic Group	**Species**	**Common Name**	**Effective Against**
Chloramphenicol	*S. venezuelae*	Chloramphenicol	Broad spectrum
Tetracycline	*S. aureofaciens*	Tetracycline	Broad spectrum
Chlortetracycline	*S. aureofaciens*	Chlortetracycline	Broad spectrum
Polyenes	*S. noursei*	Nystatin	Fungi
Macrolides	*S. erythreus*	Erythromcin	Most gram-positives
	S. lincolnensis	Clindamycin	Obligate anaerobes
Aminoglycosides	*S. griseus*	Streptomycin	Most gram-negative
	S. fradiae	Neomycin	Broad spectrum

derived from this genus. See Chapters 7 and 31 for more detail. In addition, they are sources of anticancer drugs used in chemotherapy (see Table 31.9).

Although this is an extremely important genus of soil microorganisms, little is known of their ecological roles. Intriguingly, it is unclear whether *Streptomyces* species produce the antibiotics while growing in their natural habitat. Certainly, this would appear to favor their activities because they could use the antibiotics to inhibit competitors.

The proliferation of great numbers of species in this genus (more than 500 species have been proposed in the past) largely reflects the need for pharmaceutical firms who study these bacteria to propose a new species when they wish to patent and manufacture a new antibiotic.

Other genera of mycelial actinomycetes are differentiated from this genus, primarily by their cell wall composition and the morphology of their conidiospore-bearing structure.

ACTINOPLANES This is a genus of actinomycetes that produces a flagellated spore. Like *Streptomyces*, this organism has both a substrate and an aerial mycelium. However, the spores are produced within a sac or sporangium, a stage that can survive desiccation. When conditions in the environment are moist and favorable for germination and growth, the sporangium ruptures and releases the motile spores.

SECTION HIGHLIGHTS

The *Actinobacteria*, also called the actinomycetes, range from unicellular cocci and rods to mycelial organisms, which are coenocytic. Some genera have pathogenic species, such as *Mycobacterium*, *Corynebacterium*, and *Actinomyces*, but most are common soil organisms. *Rhodococcus* is an important degrader of hydrocarbons. Some of the mycelial species produce antibiotics including streptomycin, chloramphenicol, and tetracycline.

SUMMARY

- **Lactic acid bacteria** ferment sugars to produce lactic acid either by the **homofermentative** or **heterofermentative** pathway. All energy generated by the lactic acid bacteria is via substrate-level phosphorylation. More energy is available to the homofermentative (2 ATP/mol glucose) compared to the heterofermentative (1 ATP/mol glucose) lactic acid bacteria.

- Lactic acid bacteria are **aerotolerant anaerobes** that are indifferent to the presence of oxygen in their growth environment. Most lactic acid bacteria are nutritionally **fastidious** in that they require complex nutrients for growth, including vitamins, amino acids, and purines or pyrimidines.

- Several groups of lactic acid bacteria exist, including those associated with plants, dairy products, and in animals in the mouth, intestines, and vagina. In natural fermentations of plant materials and dairy products, the fermentation undergoes a succession in which the *Streptococcus* group predominates first and lowers the pH to 5.5 or so, then the *Lactobacillus* group succeeds and lowers the pH even further, to about 4.5—the low pH acts as a preservative for the foodstuff.

- Some *Streptococcus* spp. cause dental caries, others cause **scarlet fever**, **puerperal fever**, **pneumonia**, **tissue necrosis**, and **endocarditis**. *Streptococcus* spp. may cause **alpha-** or **beta-hemolysis** of red blood cells.

- *Bacillus* spp. are aerobic to facultatively aerobic endospore-forming rods commonly found in soils. *Clostridium* spp. are obligately anaerobic spore-forming rods. **Endospores** are specially modified cells that are heat-resistant resting stages of various genera including *Bacillus, Clostridium, Sporosarcina, Desulfotomaculum,* and *Thermoactinomyces.*

- Some *Bacillus* spp., such as *Bacillus thuringiensis*, are insect pathogens that produce a characteristic protein crystal that is toxic to moths. *Bacillus anthracis* is the causative agent of anthrax in cattle and humans.

- *Staphylococcus epidermidis* is part of the normal microbiota of human skin—all humans are colonized by this bacterial species. *Staphylococcus aureus* is a normal resident of the nasopharynx of humans, but it can also be a human pathogen—it is noted for causing skin infections, food poisoning, and toxic shock syndrome.

- *Clostridium* spp. ferment cellulose, sugars, amino acids, or purines. In the **Stickland fermentation**, two amino acids are catabolized: one amino acid is fermented and the other is oxidized. One species, *Clostridium aceticum,* is an acetogen that obtains energy from oxidation of H_2 and fixes CO_2 via the acetyl-coenzyme A pathway. *Clostridium botulinum* is the causative agent of botulism; *Clostridium tetani* is the causative agent of tetanus.

- *Heliobacterium* and *Heliobacillus* are photoheterotrophic bacteria that use organic compounds such as pyruvate as carbon sources for growth and derive energy from sunlight.

- The **mycoplasmas**, or *Mollicutes*, lack a cell wall, but are close relatives of *Clostridium*—many species, such as *Mycoplasma pneumoniae*—are pathogenic to humans.

- *Micrococcus* spp. are obligate aerobes that are common soil inhabitants and have a higher mol % G + C content.

- *Propionibacterium* is noted for its production of **propionic acid** from lactic acid; it is responsible for Swiss cheese production. *Bifidobacterium bifidus*, which ferments sugars to acetic and lactic acids, is part of the normal intestinal microbiota of breast-fed infants.

- *Corynebacterium* is the formally named genus of the **coryneform bacteria** that are common soil organisms. *Corynebacterium diphtheriae* lives as a parasite of the oral cavities of humans—strains that cause diphtheria have received a piece of DNA from a phage that encodes for the diphtheria protein exotoxin.

- The **actinobacteria** are a group of branching filamentous organisms, some of which produce both a **substrate** and an **aerial mycelium**. *Mycobacterium* spp. are acid-fast and produce **mycolic acids**; two species, *Mycobacterium tuberculosis* and *Mycobacterium leprae* (leprosy), are well-known pathogens.

- *Streptomyces* is a common soil actinomycete noted for its production of **antibiotics** such as tetracyclines, chloramphenicol, streptomycin, and erythromycin. *Streptomyces* impart an "earthy" smell to soil due to the production of geosmins.

 Find more at www.sinauer.com/microbial-life

REVIEW QUESTIONS

1. What evidence indicates that the Gram stain has phylogenetic significance? Name groups that are exceptions.

2. The primitive Earth was anaerobic and contained anaerobic bacteria. The modern world is aerobic (with anaerobic sediments), but it has highly evolved exotic aerobic organisms, such as humans. Do you believe there is a parallel to this in the evolution of gram-positive bacteria?

3. What bacterial groups do we ingest in large numbers with certain foods?

4. What is the significance of the bacterial endospore?

5. In what ways are gram-positive bacteria known to be commercially important?

6. List several genera of coccus-shaped gram-positive bacteria. How do they differ from one another?

7. What are the photosynthetic members of the gram-positive bacteria?

8. Compare the genus *Bacillus* to the genus *Clostridium*.

9. Identify pathogenic species of the gram-positive bacteria. What is the basis for their pathogenesis?

10. Identify agriculturally useful gram-positive bacteria and discuss their value.

11. What are some examples of fermentations carried out by gram-positive bacteria? Are they commercially important?

12. Why are some bacteria acid-fast?

SUGGESTED READING

Balows, A., H. G. Trüper, M. Dworkin, W. Harder and K. H. Schleifer, eds. 1992. *The Prokaryotes*. 2nd ed. Berlin: Springer-Verlag.

Bergey's Manual of Systematic Bacteriology. 2nd ed., Vol. III. 2007. New York: Springer-Verlag.

Holt, J. G., editor-in-chief. 1993. *Bergey's Manual of Determinative Bacteriology*. Baltimore, MD: Williams and Wilkins.

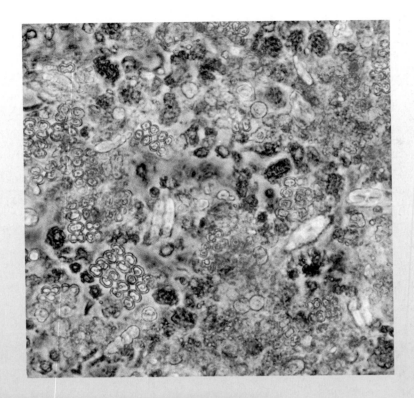

The objectives of this chapter are to:

◆ Introduce the five major groups of photosynthetic *Bacteria*.

◆ Distinguish between anoxygenic and oxygenic photosynthesis.

◆ Describe the various carbon dioxide fixation pathways and energy generation systems (photosystems) of photosynthetic bacteria.

◆ Discuss the major carbon dioxide fixation pathways of the photosynthetic bacteria.

◆ Provide information on the importance of bacterial photosynthesis to primary production on Earth.

21

Phototrophic Bacteria

The evolution of photosynthesis, particularly oxygenic photosynthesis, by ancestral Bacteria was perhaps the most significant metabolic event in Earth's history.
—*Anonymous*

The evolution of photosynthesis heralded an important event on Earth. Prior to that, organic materials were much less abundant. In addition, the energy generated by photosynthesis set the stage for the evolution of metabolisms that resulted in the production of scarce oxidants such as sulfate, oxygen, and nitrate, which are important now in sulfate reduction, aerobic respiration, and denitrification, respectively.

This chapter discusses *Bacteria* that obtain energy from light (**phototrophic metabolism**). The phyla we discuss include the *Proteobacteria, Chlorobi, Chloroflexi, Firmicutes,* and *Cyanobacteria*. Some of these bacteria are **photoautotrophic**; that is, they are able to use carbon dioxide as their sole source of carbon for growth and to carry out photosynthesis. However, some of the phototrophic bacteria require organic compounds and are therefore termed **photoheterotrophic** (Table 21.1). Some members of the *Bacteria* and *Archaea* produce **bacteriorhodopsin**, which is used in the generation of ATP from sunlight, but they do not produce the chlorophyll pigments that are the hallmark of all true photosynthetic organisms.

All photosynthetic organisms, from phototrophic bacteria to higher plants, obtain energy from light. The light source is typically sunlight, but for some phototrophs may also include invisible infrared radiation from geothermal sources. The reactions used to capture the energy from sunlight and transform it into chemical energy are referred to as the **light reactions** of photosynthesis (see Chapter 9). The energy

TABLE 21.1	Phyla and properties of photosynthetic bacteria	
Phylum	**Carbon Source**	**Carbon Metabolism**
Proteobacteria		
Purple sulfur	CO_2 and organics	Calvin cycle
Purple nonsulfur	CO_2 and organics	Photoheterotrophic
Chlorobi: Green sulfur	CO_2 and organics	Reductive TCA cycle
Chloroflexi: Green filamentous[a]	CO_2 or organic	3-Hydroxypropionate pathway
Cyanobacteria	CO_2	Calvin cycle
Firmicutes: Heliobacteria	Organics	Photoheterotrophic

[a]The *Chloroflexi* are very versatile metabolically. See text for details.

generated from photosynthesis, as well as the reducing power produced as NADPH or NADH, is required for the carbon dioxide fixation reactions. Because the actual carbon dioxide fixation reactions do not directly involve sunlight, they are referred to as the **dark reactions**. Before a discussion of the various photosynthetic bacteria, the light and dark reactions that were treated in Chapter 9 and 10 are reviewed briefly.

21.1 Light and Dark Reactions

The light reactions initially involve absorption of radiant energy by pigments in the cells. The pigments fall into three groups. **Reaction center** chlorophyll pigments, located in the photosynthetic membranes, are chlorophyll molecules that play a direct role in generating ATP by photophosphorylation. The type of reaction center chlorophyll varies from one group of organisms to another (Table 21.2). Second, **antenna** or **light-harvesting** chlorophyll molecules are associated with the reaction center and are involved in harvesting light radiation for the reaction center chlorophyll. Finally, photosynthetic microorganisms also have **accessory** pigments, such as carotenoids, that collect light energy and transfer it to the chlorophyll molecules. The nature of these accessory pigments varies from one bacterial group to another (see Table 21.2).

ATP is synthesized by a proton gradient established across the photosynthetic membranes and associated with the reaction center where the electron transfer reactions occur (see Chapter 9). Thus, like typical heterotrophic bacteria, photosynthetic bacteria generate ATP by use of proton pumps and their associated ATP synthases. The unique feature of photosynthesis is the generation of electron gradients using pigments that harvest sunlight.

A common biochemical pathway used by phototrophic bacteria for carbon dioxide fixation is the Calvin cycle (see Chapter 10). Not all photosynthetic bacteria use this pathway. The green sulfur and green filamentous bacteria use different pathways for carbon dioxide fixation (see the following discussion).

SECTION HIGHLIGHTS

The unique feature of photosynthesis is the conversion of light energy to ATP by photophosphorylation. This process involves pigments for harvesting photons and a reaction center where ATP is formed.

TABLE 21.2	Characteristics of photosynthetic bacteria	
Group	**Reaction Center Chlorophyll**	**Antenna (Accessory Pigments)**
Proteobacteria	Bchl *a*	Bchl *b* (carotenoids)
Chlorobi	Bchl *a*	Bchl *c*; Bchl *d*; Bchl *e* (carotenoids)
Chloroflexi	Bchl *a*	Bchl *c* (carotenoids)
Cyanobacteria	Chl *a*	Phycobilins or Chl *b* (in prochlorophytes)
Heliobacteria	Bchl *g*	Carotenoids

21.2 Phototrophic Phyla

First of all, it is noteworthy that several phyla of the *Bacteria* are either exclusively phototrophic or contain some phototrophic members (see Table 21.1). In contrast, none of the *Archaea* is truly photosynthetic. Although it is true that some halophilic members of the *Archaea* produce bacteriorhodopsin and can generate ATP through a light-mediated reaction, the metabolism of these organisms is different from that of the phototrophic bacteria, which have complex light-harvesting centers involving bacteriochlorophyll and accessory pigments such as carotenoids and phycobiliproteins.

Domain *Bacteria*

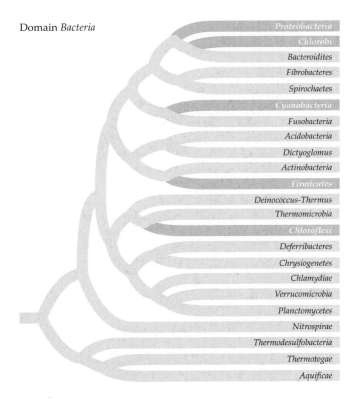

The photosynthetic bacteria that can use light energy for growth have a distinct metabolic advantage over those that rely on organic or inorganic chemical energy sources, because of the much greater amount of energy available for their metabolism.

The phototrophic bacteria fall into five major phylogenetic groups: (1) photosynthetic *Proteobacteria* (purple sulfur bacteria and purple nonsulfur bacteria), (2) *Chlorobi* (i.e., green sulfur bacteria), (3) *Chloroflexi* (green filamentous bacteria), (4) photosynthetic *Firmicutes* (heliobacteria), and (5) *Cyanobacteria*. These groups are differentiated from one another on the basis of phylogeny, carbon dioxide fixation pathways, and pigments (see Tables 21.1 and 21.2).

Photosynthetic bacteria produce a variety of chlorophyll pigments (see Table 21.2; Figure 21.1). Bacteriochlorophyll *a* (Bchl *a*) is a common reaction center chlorophyll used by both the purple and green bacteria. In contrast, Bchl *b* is produced only by purple bacteria, whereas Bchl *c*, Bchl *d*, and Bchl *e* are produced by various members of the *Chlorobi*, and Bchl *g* is produced by the heliobacteria. Unlike other prokaryotes, the *Cyanobacteria* produce chlorophyll *a* as their reaction center chlorophyll. It is interesting to note that the bacteriochlorophylls absorb very long wavelengths of light, some of which are in the infrared region (wavelength >800 nm) not detected by human eyesight (see Figure 21.1). In contrast, chlorophyll *a* and *b* absorb much shorter wavelengths of light in the visible range. Because each photosynthetic group has its own characteristic absorption spectrum for light, these groups can coexist in the same habitat without directly competing for available light.

It is noteworthy that some members of the *Bacteria* are able to produce rhodopsins and use them to generate ATP without carrying out photosynthesis (Box 21.1). Although bacteriorhodopsins were first reported in the halophilic *Archaea*, rhodopsins appear to be very common in various bacterial groups especially in marine planktonic habitats.

Each of the photosynthetic bacterial groups is discussed in subsequent text of this chapter in greater detail.

> ### SECTION HIGHLIGHTS
> Phototrophic bacteria, which obtain energy from sunlight, occur in five different bacterial phyla, each of which produces its own distinctive photosynthetic pigments. The carbon dioxide fixation pathway varies from one phylogenetic group to another.

21.3 *Proteobacteria*

The photosynthetic *Proteobacteria* are commonly called the purple bacteria because of their reddish or purplish coloration. The photosynthetic *Proteobacteria* comprise two separate subgroups, the **purple sulfur bacteria** and the **purple nonsulfur bacteria**. Like all *Proteobacteria* that fix carbon dioxide, both of these subgroups use the Calvin cycle, although they utilize different compounds as reducing power for the dark reactions. The purple sulfur bacteria use reduced sulfur compounds such as sulfide or elemental sulfur as their source of electrons. In contrast, the purple nonsulfur bacteria typically use or-

(A)

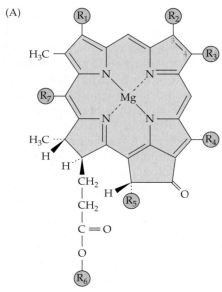

Figure 21.1 Bacterial chlorophylls
(A) General structure of the chlorophyll molecule. (B) Structures of the R groups and the absorption maximum for each chlorophyll type. P, phytyl ester ($C_{20}H_{39}$—); G, geranylgeraniol ester ($C_{10}H_{17}$—); F, farnesyl ester ($C_{15}H_{25}$—). Where two or more structures are shown for an R group, the native chlorophyll contains a mixture. Adapted from A. Gloe, N. Pfennig, H. Brockmann and W. Trowitsch. 1975. *Archives of Microbiology* 102:103-109; and A. Gloe and N. Risch. 1978. *Archives of Microbiology* 118:153-156.

(B)

Chlorophyll	R1	R2	R3	R4	R5	R6	R7	Absorption maximum of cells (nm)
Bacterio-chlorophyll a	$-\overset{\overset{\displaystyle O}{\|\|}}{C}-CH_3$	$-CH_3$	$-CH_2-CH_3$	$-CH_3$	$-\overset{\overset{\displaystyle O}{\|\|}}{C}-O-CH_3$	P/G	$-H$	805 830–890
Bacterio-chlorophyll b	$-\overset{\overset{\displaystyle O}{\|\|}}{C}-CH_3$	$-CH_3$	$=\underset{\underset{\displaystyle H}{\|}}{C}-CH_3$	$-CH_3$	$-\overset{\overset{\displaystyle O}{\|\|}}{C}-O-CH_3$	P	$-H$	835–850 1020–1040
Bacterio-chlorophyll c	$-\overset{\overset{\displaystyle H}{\|}}{\underset{\underset{\displaystyle OH}{\|}}{C}}-CH_3$	$-CH_3$	$-C_2H_5$ $-C_3H_7$ $-C_4H_9$	$-C_2H_5$ $-CH_3$	$-H$	F	$-CH_3$	745–755
Bacterio-chlorophyll d	$-\overset{\overset{\displaystyle H}{\|}}{\underset{\underset{\displaystyle OH}{\|}}{C}}-CH_3$	$-CH_3$	$-C_2H_5$ $-C_3H_7$ $-C_4H_9$	$-C_2H_5$ $-CH_3$	$-H$	F	$-H$	705–740
Bacterio-chlorophyll e	$-\overset{\overset{\displaystyle H}{\|}}{\underset{\underset{\displaystyle OH}{\|}}{C}}-CH_3$	$-\overset{\overset{\displaystyle O}{\|\|}}{C}-H$	$-C_2H_5$ $-C_3H_7$ $-C_4H_9$	$-C_2H_5$	$-H$	F	$-CH_3$	719–726
Bacterio-chlorophyll g	$-\overset{\overset{\displaystyle H}{\|}}{C}=CH_2$	$-CH_3$	$-C_2H_5$	$-CH_3$	$-\overset{\overset{\displaystyle O}{\|\|}}{C}-O-CH_3$	F	$-H$	670, 788
Chlorophyll a	$-\overset{\overset{\displaystyle H}{\|}}{C}=CH_2$	$-CH_3$	$-C_2H_5$	$-CH_3$	$-\overset{\overset{\displaystyle O}{\|\|}}{C}-O-CH_3$	P	$-H$	440, 680
Chlorophyll b	$-\overset{\overset{\displaystyle H}{\|}}{C}=CH_2$	$-\overset{\overset{\displaystyle O}{\|\|}}{C}-H$	$-C_2H_5$	$-CH_3$	$-\overset{\overset{\displaystyle O}{\|\|}}{C}-O-CH_3$	P	$-H$	645

BOX 21.1   *Research Highlights*

Aerobic Bacteriochlorophyll *a*– and Bacteriorhodopsin-Producing *Bacteria*

Photosynthetic *Proteobacteria* and *Chlorobi* produce bacteriochlorophyll *a* only under anaerobic conditions. In contrast, several genera of aquatic bacteria have been discovered recently that produce Bchl *a* only under *aerobic* conditions. Many of these bacteria can make ATP by photophosphorylation, but all so far studied require organic compounds as carbon sources. Some of these bacteria have recently been isolated from marine habitats, including the genera *Erythrobacter* and *Roseobacter*, whereas others such as *Porphyrobacter* are from freshwater habitats. All are members of the *Alphaproteobacteria*.

Similarly, bacteriorhodopsin-producing members of the *Bacteria* were first discovered in the *Proteobacteria* and were referred to as proteorhodopsin-producing bacteria. They were discovered in Ed DeLong's laboratory during an environmental genomics study in which a proteobacterial 16S rDNA sequence was found that was in the same segment along with the rhodopsin gene. This is an example of one of the remarkable discoveries made by the use of genomic approaches in microbial ecology.

Now, bacterial rhodopsins have been found in several other bacterial phyla. Furthermore, recent studies show that these organisms are very abundant in the aerobic water column of the oceans. For example, studies of the Mediterranean Sea indicate that 13% of the planktonic bacteria produce bacterial rhodopsin. The common occurrence and widespread distribution of these bacteria indicate that they comprise an important part of the normal microbiota of aquatic habitats. Like Bchl *a*–producing bacteria, these organisms use these pigments as a means of generating ATP to supplement the ATP produced by their normal heterotrophic metabolism.

ganic compounds, especially organic acids and alcohols, as their source of reducing power. Photosynthesis in purple sulfur bacteria is represented by the following two reactions (reactions are not balanced):

$$(1) \quad H_2S + CO_2 \rightarrow (CH_2O)_n + S^0$$

$$(2) \quad S^0 + CO_2 + H_2O \rightarrow (CH_2O)_n + H_2SO_4$$

where $(CH_2O)_n$ represents organic carbohydrate carbon.

Photosynthesis in purple nonsulfur bacteria is represented by:

$$(3) \quad H_2A + CO_2 \rightarrow (CH_2O)_n + A$$

where H_2A = an organic compound such as acetic acid.

As can be seen by the preceding three reactions for carbon dioxide fixation, oxygen is *not* formed during photosynthesis by these bacteria. This type of photosynthesis is therefore referred to as **anoxygenic photosynthesis**. As discussed in Chapter 9, the electron flow in anoxygenic photosynthesis is noncyclic, involving only one photosystem, photosystem I, with Bchl *a*. In addition, a characteristic of this type of photosynthesis is that it occurs at a relatively high redox potential, approximately –0.15 volts. Thus, NADPH cannot be formed directly from the electron flow; instead energy must be expended by reverse

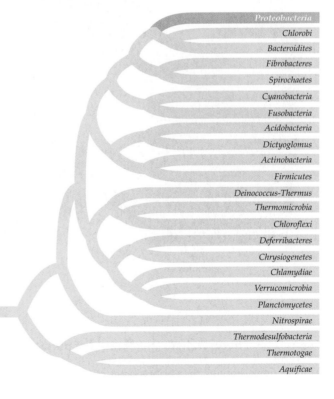

Proteobacteria
Chlorobi
Bacteroidites
Fibrobacteres
Spirochaetes
Cyanobacteria
Fusobacteria
Acidobacteria
Dictyoglomus
Actinobacteria
Firmicutes
Deinococcus-Thermus
Thermomicrobia
Chloroflexi
Deferribacteres
Chrysiogenetes
Chlamydiae
Verrucomicrobia
Planctomycetes
Nitrospirae
Thermodesulfobacteria
Thermotogae
Aquificae

electron flow to obtain the NADPH needed for CO_2 reduction (see Chapter 9 for more detail).

The two proteobacterial photosynthetic groups are discussed individually.

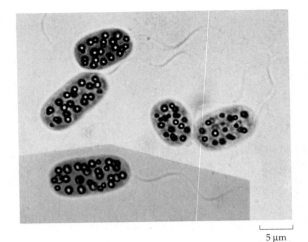

5 μm

(A)

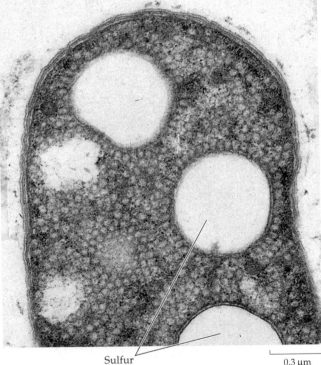

Sulfur
granules

0.3 μm

Figure 21.2 Purple sulfur bacterium
Phase contrast photomicrograph of the purple sulfur bacterium *Chromatium okenii*. Note the polar flagellum (which is actually a tuft) and the internal sulfur granules. Courtesy of N. Pfennig.

Purple Sulfur Bacteria

The purple sulfur bacteria are large unicellular bacteria (Figure 21.2) that may reach more than 6.0 μm in diameter. Many are motile with polar flagella. Some have gas vacuoles, and all can utilize hydrogen sulfide in photosynthesis and form elemental sulfur granules inside their cells (Figure 21.3).

The photosynthetic *Proteobacteria* have extensive intracellular membrane systems that contain the photopig-

(B)

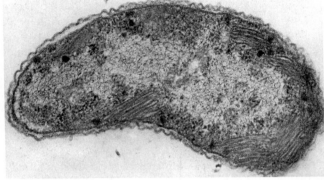

0.25 μm

(C)

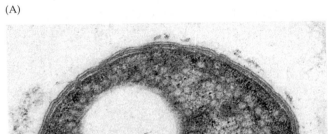

Figure 21.3 Photosynthetic membranes of purple bacteria
Thin sections showing the various types of intracellular photosynthetic membranes of purple bacteria. (A) Vesicular membranes in *Chromatium vinosum*; (B) lamellar membranes in *Rhodospirillum molischianum*; (C) tubular membranes in *Thiocapsa pfennigii*. A and C, courtesy of S. Watson, J. Waterbury, and C. Remsen; B, courtesy of G. Drews.

0.25 μm

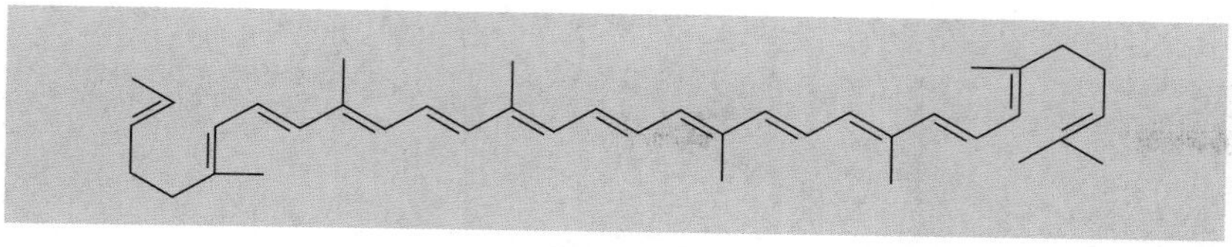

Red-brown

Lycopene

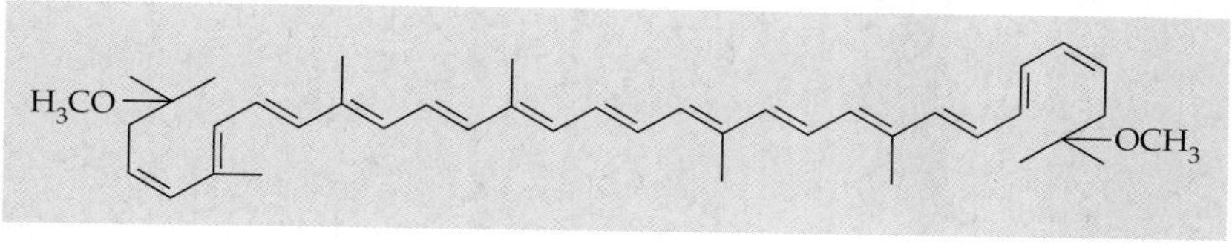

Pink-purple

Spirilloxanthin

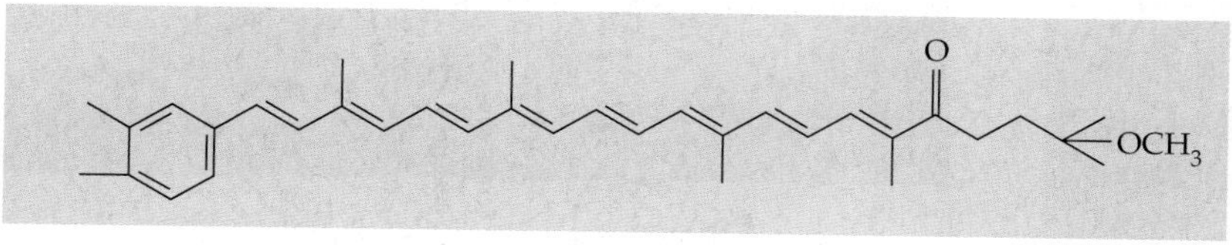

Red-purple

Okenone

Figure 21.4 Carotenoids of purple bacteria Structures of common carotenoid pigments of purple bacteria and the colors they impart to cells.

ments and photoreaction centers. These membranes are extensions of the cytoplasmic membrane and may be vesicular, lamellar, or even tubular (see Figure 21.3). The photosynthetic pigments are all bacteriochlorophylls, with Bchl *a* found in all but one species, *Thiocapsa pfennigii*, which contains Bchl *b*. They also contain carotenoid pigments (Figure 21.4). These carotenoid pigments actually mask the green to blue bacterial chlorophyll pigments, thereby giving the organisms their characteristic red, orange, and purple colors (Figure 21.5). A variety of carotenoid pigments may occur, depending on the species. The pigmentation of an organism is related to the amount of the various carotenoid pigments produced. For example, lycopene, which is brown, is the biochemical precursor of spirilloxanthin, which is purple. The amounts of these two pigments and their intermediates and, hence, the color of an organism, will vary depending on the environment and growth state of the organism.

Table 21.3 lists important genera and their characteristics. *Chromatium* is one of the most common genera. This is a genus of large, polarly flagellated rods (see Figure 21.2). The genus *Thiospirillum*, as the name implies, contains helical bacteria with polar flagellar tufts (see Figure 21.5A). *Thiodictyon* is a genus of net-forming rods (see Figure 21.5B), and *Thiopedia* species form flat plates of coccoid cells with gas vacuoles (see Figure 21.5C). All purple sulfur bacteria are members of the gamma group of the *Proteobacteria*.

Under anaerobic conditions in light, purple sulfur bacteria grow as photolithoautotrophs, using sulfide or ele-

TABLE 21.3 Characteristics of genera of purple sulfur bacteria

Genus	Flagella	Gas Vacuoles	Mol % G + C	Morphology
Chromatium	+	–	48–70	Rod
Thiocystis	+	–	61–68	Coccus
Thiospirillum	+	–	45–46	Spirillum
Thiocapsa	–	–	63–70	Coccus
Lamprobacter	+	+	64	Ovoid cells
Lamprocystis	+	+	64	Coccus
Thiodictyon	–	+	65–67	Rod; forms network of cells
Amoebobacter	–	+	64–66	Coccus; single or clusters
Thiopedia	–	+	62–64	Cocci in one plane

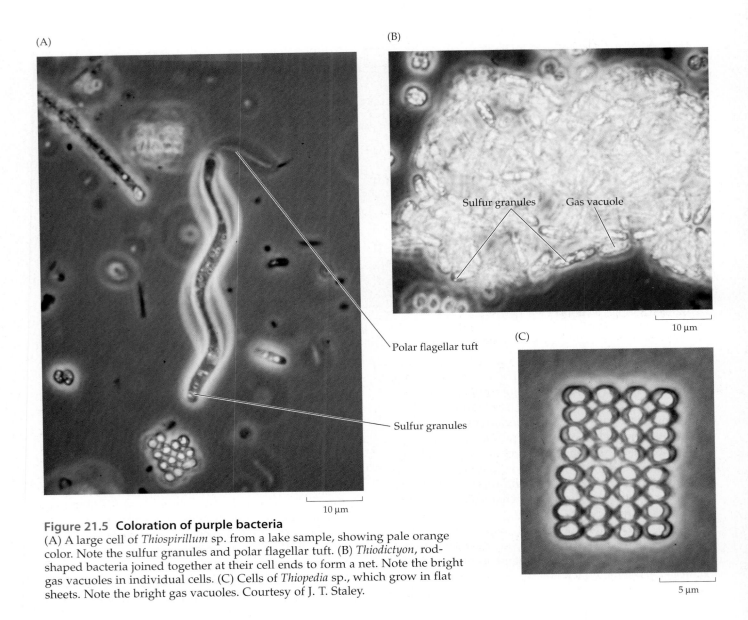

(A)

(B)

Sulfur granules   Gas vacuole

10 μm

(C)

Polar flagellar tuft

Sulfur granules

10 μm

5 μm

Figure 21.5 Coloration of purple bacteria
(A) A large cell of *Thiospirillum* sp. from a lake sample, showing pale orange color. Note the sulfur granules and polar flagellar tuft. (B) *Thiodictyon*, rod-shaped bacteria joined together at their cell ends to form a net. Note the bright gas vacuoles in individual cells. (C) Cells of *Thiopedia* sp., which grow in flat sheets. Note the bright gas vacuoles. Courtesy of J. T. Staley.

mental sulfur as an electron donor and carbon dioxide as carbon source. They are best known for this physiological activity. In addition, many species can also use organic compounds, such as acetate or pyruvate, as a reductant as shown in Reaction 3, which is typical of the purple non-sulfur bacteria. When organic carbon sources are used, these bacteria still require sulfide, however, because they cannot carry out assimilatory sulfate reduction.

Many species can also use hydrogen gas as a reducing agent for photosynthesis under anaerobic conditions. Virtually all species are nitrogen fixers. Although most are obligate anaerobes, some can grow under microaerobic to aerobic conditions in the dark. When doing this, they obtain energy from the oxidation of inorganic compounds

such as hydrogen gas or sulfide, or even organic carbon sources, in the same manner as some chemolithotrophs and heterotrophs, respectively.

The purple sulfur bacteria are most commonly found in anaerobic environments where sulfide is abundant and light is available (see Chapter 24). Such conditions occur in the anaerobic hypolimnion of many eutrophic lakes, at the surface of marine and freshwater muds, or in sulfur springs (Figure 21.6). A characteristic vertical layering of phototrophic bacteria occurs in such environments. The *Cyanobacteria* are found nearest the surface, the purple sulfur bacteria are beneath them, and the green sulfur bacteria occur at the lowest depth. This pattern of vertical stratification is identical to that found

(A)

(B)

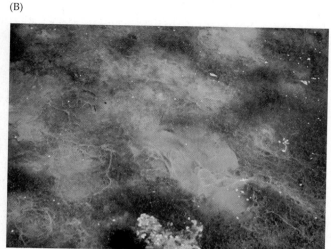

Figure 21.6 Sulfur springs
(A) Yankee Springs, a sulfur spring in central Michigan. The darker green growth
on the surface in some areas is due to cyanobacteria. (B) Close-up of the "veils"
in Yankee Springs, containing several different species of purple sulfur bacteria.
Courtesy of Peter Hirsch.

in microbial mat communities as discussed earlier in
Chapter 1.

The genus *Ectothiorhodospira* is closely related to the purple sulfur bacteria. It resembles the purple sulfur bacteria
in its metabolism, pigmentation, and internal membrane
systems. Like the purple sulfur bacteria, *Ectothiorhodospira*
sp. are members of the gamma *Proteobacteria*. Their cells
are small, motile, and vibrioid to rod-shaped. Depending
on the species, they live in marine habitats, alkaline soda
lakes, or hypersaline lakes. Some species are gas vacuolate. Unlike the purple sulfur bacteria, however, the sulfur
granules they produce are deposited *outside* of the cell
(hence the name *ectothio-*, which means *extracellular sulfur*).

Purple Nonsulfur Bacteria

The purple nonsulfur bacteria share characteristics of
both the heterotrophic proteobacteria and the purple sulfur photosynthetic bacteria. Like many bacteria, they can
grow aerobically (in the absence of light) as heterotrophs
using certain organic substrates. However, if they are
grown anaerobically in the light, they carry out photosynthesis, much like the purple sulfur bacteria. They differ from purple sulfur bacteria in that organic carbon
sources, such as organic acids (for example, acetate,
pyruvate, or lactate) or ethanol, rather than sulfide, are
preferred as electron donors for carbon dioxide fixation.
Some can also grow as photolithoautotrophs, using hydrogen gas or sulfide as reducing agents in the same
manner as the purple sulfur bacteria. However, they
do not tolerate high concentrations of sulfide and for this

reason are called purple nonsulfur bacteria. When sulfide is available in low concentrations, those that use it
as a reductant for photosynthesis do not form sulfur
granules inside the cells. Elemental sulfur is either
formed outside of the cell, or the sulfide is oxidized completely to sulfate.

The cells of most purple nonsulfur bacteria are
smaller than those of the purple sulfur bacteria. Representatives of several common genera, *Rhodobacter, Rhodopseudomonas,* and *Rhodospirillum,* are shown in Figure
21.7. The cells of the nonsulfur bacteria also differ from
the purple sulfur bacteria in that individual cells do not
appear pigmented when observed with the light microscope. This is because the smaller cells are not thick
enough to contain sufficient pigments to reveal their true
color.

Like the purple sulfur bacteria, the purple nonsulfur
bacteria produce intracytoplasmic photosynthetic membranes when growing photosynthetically in the light under anaerobic conditions. These membranes are extensions of the cell membrane and may be either vesicles or
lamellae. They produce more of these membranes and
photosynthetic pigments under lower light intensity as
a means of compensating for the decrease in available
light. These organisms also contain the same carotenoid
pigments as produced by the purple sulfur bacteria, with
characteristic types for each species. Many of the species
in this group are motile by flagella. There are eight major genera (Table 21.4).

Most of the nonsulfur purple bacteria are members
of the alpha purple group of *Proteobacteria* and are there-

(A)

(B)

(C)

(D)

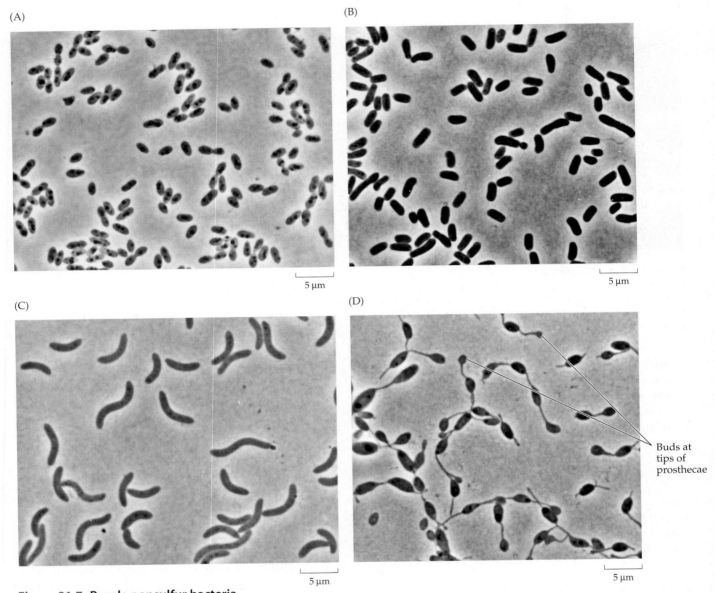

Buds at
tips of
prosthecae

Figure 21.7 Purple nonsulfur bacteria
Phase contrast photomicrographs of four representative purple nonsulfur bacteria. (A) *Rhodobacter spheroides*, (B) *Rhodopseudomonas acidophila*, (C) *Rhodospirillum rubrum*, and (D) *Rhodomicrobium vannielii*. Courtesy of N. Pfennig.

fore phylogenetically related to the prosthecate, budding bacteria. Indeed, *Rhodomicrobium* sp. (see Figure 21.7D) and some *Rhodopseudomonas* sp. produce prosthecae and divide by budding, confirming their close phylogenetic relatedness to the heterotrophic prosthecate *Proteobacteria*. Three genera, *Rhodocyclus*, *Rhodovivax*, and *Rhodoferax*, are members of the *Betaproteobacteria*. Unlike the purple sulfur bacteria, none are members of the *Gammaproteobacteria*.

When discussing the purple bacteria, it is important to note that several bacteriochlorophyll *a*–producing

genera that superficially resemble purple nonsulfur bacteria have been discovered recently. However, unlike the purple nonsulfur bacteria, these bacteria produce Bchl *a* while growing under *aerobic* conditions. The role of these bacteria in nature is not yet understood (see Box 21.1).

Another group of the *Proteobacteria* contain rhodopsin pigments that are related to the rhodopsin found in obligate halophilic members of the *Archaea* such as the genus *Halobacterium*. These, too, have been found to be abundant in marine environments based upon genetic evidence from DNA extracted from the water column

TABLE 21.4	Genera of purple nonsulfur bacteria	
Genus	**Mol % G + C**	**Cell Morphology and Features**
Alphaproteobacteria		
Rhodospirillum	60–66	Helical cells
Rhodobacter	64–73	Ovoid to rod-shaped; neutrophilic
Rhodopila	66	Ovoid to rod-shaped; acidophilic
Rhodopseudomonas	61–72	Ovoid to rod-shaped; budding
Rhodomicrobium	61–64	Prosthecate, budding
Betaproteobacteria		
Rhodocyclus	64–73	Curved rod to ring-shaped
Rhodoferax	59–61	Curved rod
Rhodovivax	70–72	Curved rod

and analyzed by community genomics and polymerase chain reaction (PCR). Like *Halobacterium* sp. and the bacteriochlorophyll *a*–producing bacteria, these bacteria are able to obtain ATP by photophosphorylation using their rhodopsin pigments (see Box 21.1).

SECTION HIGHLIGHTS

The photosynthetic *Proteobacteria*, called purple bacteria because of their pigments, carry out anoxygenic photosynthesis. They are separated into two groups: the photoautotrophic group, which are also termed the purple sulfur bacteria, and the photoheterotrophic group, called the purple nonsulfur bacteria. They are typically found in aquatic habitats and sediments and fix carbon dioxide by use of the Calvin cycle.

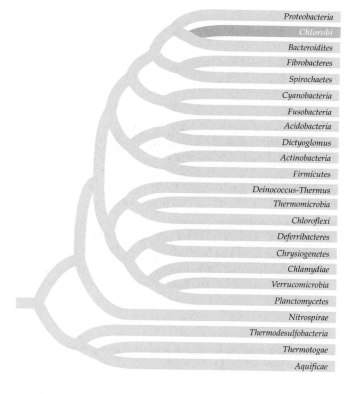

21.4 *Chlorobi*

The *Chlorobi*, or green sulfur bacteria, and the *Chloroflexi* (green filamentous bacteria) resemble one another in that they both have the **chlorosome** as their intracellular membrane system for photosynthesis. The chlorosome is a cigar-shaped structure situated next to the cell membrane (**Figure 21.8**). It is here and in the cytoplasmic membrane that the photopigments and photoreaction center of these bacteria are located. These two groups of bacteria are considered individually because they differ in their phylogeny, are found in different habitats, and have their own carbon dioxide fixation reactions.

The *Chlorobi*, or green sulfur bacteria, comprise a separate group of photosynthetic bacteria with an independent phylogeny. They are so named because most species are green due to the presence of characteristic bacteriochlorophylls (see Figure 21.1), predominantly Bchl *c*, Bchl *d*, or Bchl *e*. Others appear brown in color, principally due to carotenoid pigments (**Figure 21.9**). In addition, all strains contain small amounts of Bchl *a* as their reaction center chlorophyll. They are termed "sulfur" bacteria because they carry out anoxygenic photosynthesis using H_2S as the reductant for carbon dioxide fixation in the same manner as do the purple sulfur bacteria. However, they grow at a reduced redox potential and, because of their electron transport chains, are

Figure 21.8 *Chlorobium* sp., a green sulfur bacterium

Thin section through a green sulfur bacterium, showing the oval chlorosomes just inside the cell membrane. ©T. J. Beveridge/Biological Photo Service.

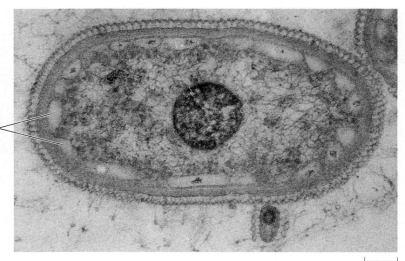

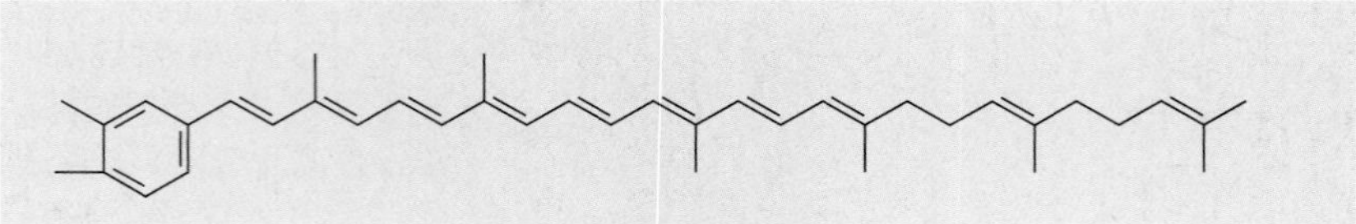

Chlorosomes

0.15 µm

Green

Chlorobactene

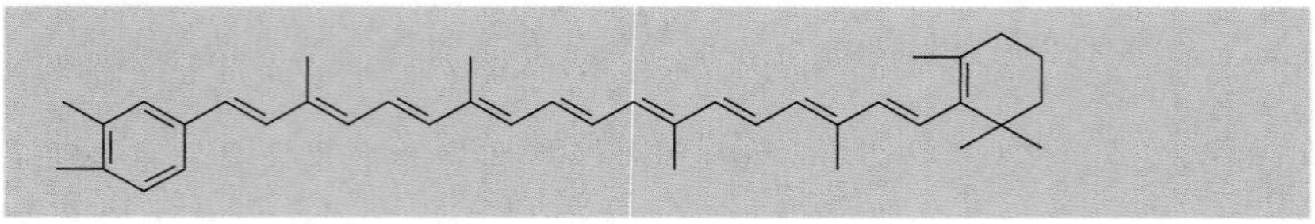

Brown

β-Isorenieratene

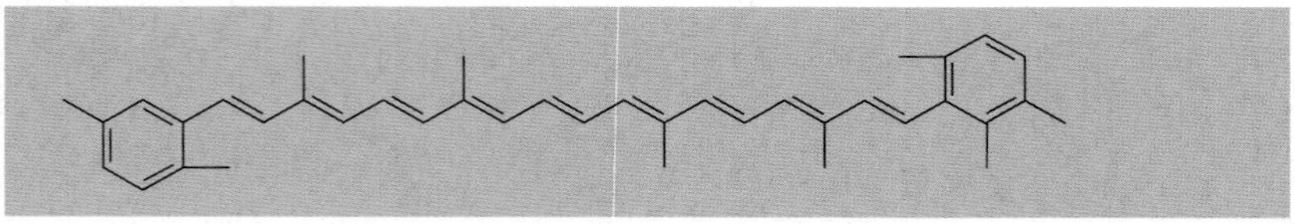

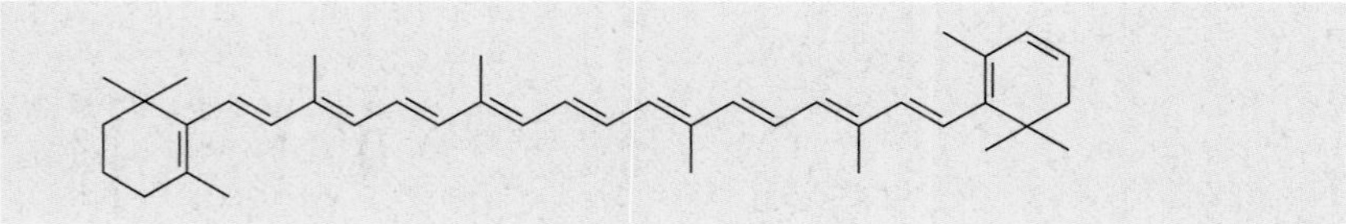

Brown

Isorenieratene

Orange

β-Carotene

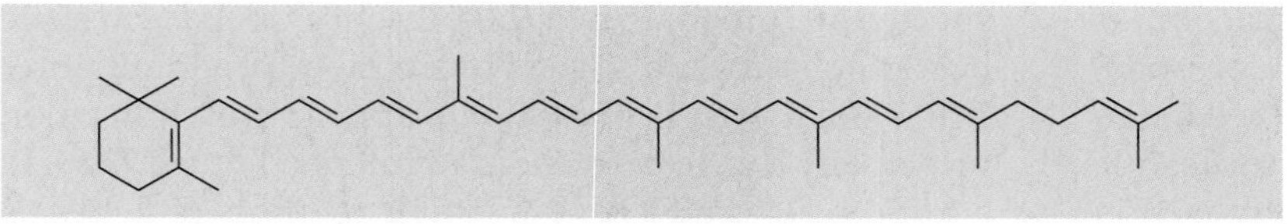

Green

γ-Carotene

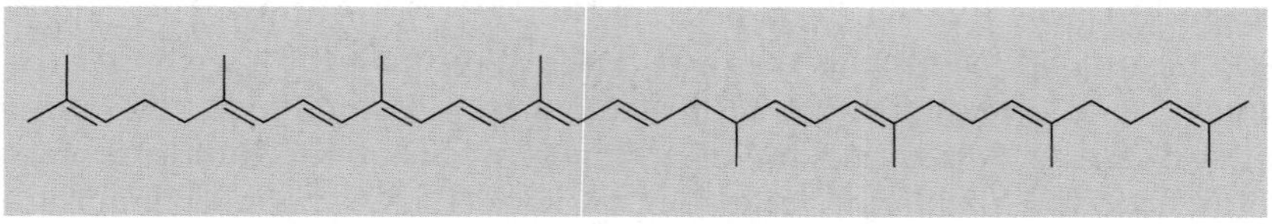

Red-brown

Neurosporene

Figure 21.9 Carotenoids of green sulfur bacteria

Structures of carotenoid pigments of green bacteria and heliobacteria and the colors they impart to cells.

able to reduce NADPH directly without resorting to reverse electron transport, thereby conserving energy.

The *Chlorobi* are small organisms, much smaller than the purple sulfur *Proteobacteria*. Like typical bacteria, their cell diameter is about 0.5 to 1.0 μm. None are motile by flagella, although many have gas vacuoles and use them to regulate their vertical position in stratified habitats (see Chapters 4 and 24). One genus, *Chloroherpeton*, moves by gliding motility.

Like the purple sulfur bacteria, the green sulfur bacteria use sulfide as an electron donor in photosynthesis and form elemental sulfur as an intermediate in this process. However, the sulfur granules are deposited *outside* the cells in the same manner as in *Ectothiorhodospira* (Figure 21.10). The *Chlorobi* can continue to oxidize the sulfur granules and ultimately produce sulfate. Therefore, the overall photosynthetic reactions are identical to that of the purple sulfur bacteria shown previously (Reactions 1 and 2). However, these bacteria do not use the Calvin cycle for carbon dioxide fixation. Instead, they use the reductive tricarboxylic acid (TCA) cycle (see Figure 10.2A). Also, unlike the purple bacteria, the green bacteria are strict anaerobes. Although most use carbon dioxide as their sole carbon source, some can also use simple organic acids in the presence of sulfide and carbon dioxide. Again, as with the purple sulfur bacteria, these bacteria must have sulfide when using organic compounds because they cannot carry out assimilatory sulfate reduction.

Several genera of *Chlorobi* are known, each with its own characteristic morphology (Table 21.5). *Chlorobium* is a common genus of unicellular rods, the species of which are found in many anaerobic, sulfide-rich habitats

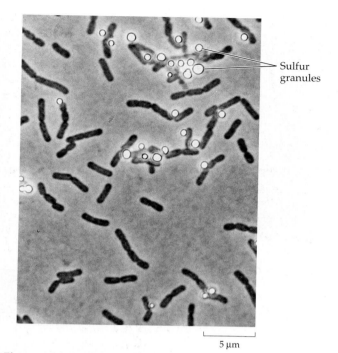

Sulfur granules

5 μm

Figure 21.10 *Chlorobium*
Phase contrast micrograph of a pure culture of *Chlorobium limicola*, showing external deposition of sulfur granules, which appear as bright spheres. Courtesy of N. Pfennig.

(see Figure 21.10). *Pelodictyon* is also very common. It is a genus of rod-shaped, gas-vacuolate bacteria, the cells of which divide to produce a net-like structure similar to that of the purple sulfur genus *Thiodictyon*. Some species appear green in color (Figure 21.11A), whereas others appear brown (Figure 21.11B), due to differences in pig-

	Mol %	Gas	
Group/Genus	**G + C**	**Vacuoles**	**Morphology**
Chlorobi (Green Sulfur Genera)			
Chlorobium	49–58	–	Single cells, rods
Prosthecochloris	50–56	–	Prosthecate; marine or saline habitats
Pelodictyon	48–58	+	Rods in netlike clusters
Ancalochloris	Unknown	+	Prosthecate; freshwater
Chloroherpeton	45–48	+	Long rods; gliding
Consortium	Unknown	+/–	Microcolonial aggregates of two separate species
Chloroflexi (Green Filamentous Genera)			
Chloroflexus	55	–	Gliding; moderate thermophile
Roseiflexus	62	–	Gliding; moderate thermophile
Chloronema	Unknown	+	Gliding; mesophile

TABLE 21.5 Important genera of green bacteria

(A)

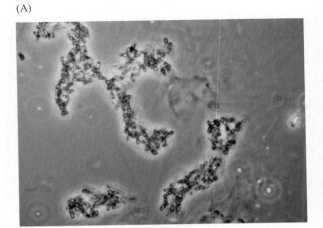

(B)

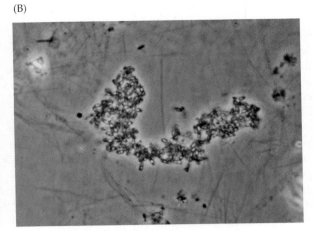

(C)

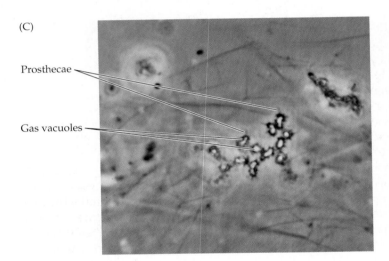

Prosthecae

Gas vacuoles

Figure 21.11 *Pelodictyon*
(A) Cells of *Pelodictyon clathratiforme*, showing net formation and small individual cells with gas vacuoles. Note the green color. (B) *Pelodictyon phaeum*, showing the brownish color of cells in the net. (C) *Ancalochloris*, a net-forming prosthecate bacterium found in lakes. Note the prosthecae and gas vacuoles. As yet, no *Ancalochloris* sp. have been isolated in pure culture. Courtesy of J. T. Staley.

ment composition. *Ancalochloris* species, which have not yet been isolated in pure culture, resemble *Pelodictyon* in that they form net-like microcolonies, but they also produce prosthecate cells (Figure 21.11C). These bacteria are found in the plankton of the anoxic hypolimnion of lakes, in freshwater and marine muds, and in marine or saline anoxic habitats. *Prosthecochloris*, as the name implies, is prosthecate and forms nets and is therefore similar to *Ancalochloris*, but it is found in saline habitats and does not produce gas vacuoles. The genus *Chloroherpeton* contains rod-shaped gliding bacteria with gas vacuoles and is found in marine habitats (Figure 21.12).

The green sulfur bacteria grow at lower redox potentials than the purple sulfur bacteria and therefore can be found in very low light conditions beneath all the other photosynthetic microorganisms in thermally stratified lakes and mat communities.

The consortium species, such as "*Chlorochromatium aggregatum*," is one of the most unusual forms of bacterial life. This consortium is found in

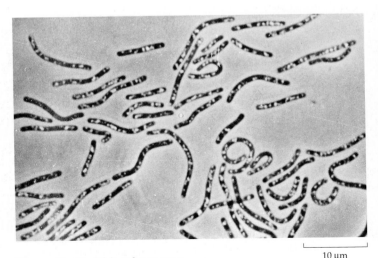

10 µm

Figure 21.12 *Chloroherpeton*
Phase contrast photomicrograph of *Chloroherpeton* cells. Note the gas vacuoles. Courtesy of J. Gibson and J. Waterbury.

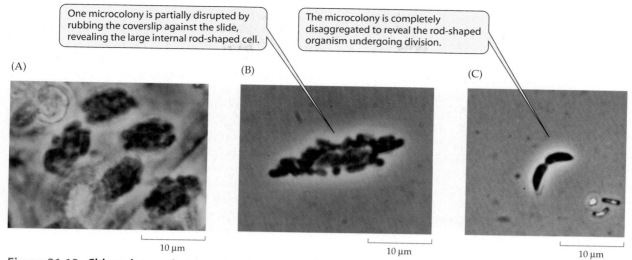

One microcolony is partially disrupted by rubbing the coverslip against the slide, revealing the large internal rod-shaped cell.

The microcolony is completely disaggregated to reveal the rod-shaped organism undergoing division.

(A) (B) (C)

10 µm   10 µm   10 µm

Figure 21.13 *Chlorochromatium aggregatum*
(A) Phase contrast micrograph of several microcolonies of the consortium species *"Chlorochromatium aggregatum"* in a lake sample. (B) One microcolony has been partially disrupted by rubbing the coverslip against the slide. (C) After further treatment, the internal rod-shaped organism is revealed. Note the flagellar tuft. Courtesy of J. T. Staley.

the same anaerobic sulfide-rich habitats as the green sulfur bacteria. It is unique among the phototrophic bacteria in that it consists of a symbiotic association between two separate organisms. For this reason, they cannot be officially named as a species, but instead quotation marks are used to designate this association between these two different species. They appear as multicellular aggregates in the anoxic, sulfide-rich lower depths of lakes (Figure 21.13A). At the center of the aggregate, or microcolony, is a large, nonpigmented motile rod, usually seen in the process of division, which propels the consortium (Figure 21.13B,C). Attached to the large rod and forming a shell around it are numerous rod-shaped green sulfur bacteria. Members of the consortium have not been studied in pure culture, so we do not have a clear understanding of the interactions between the two species. However, it appears that the large rod is an anaerobic heterotroph that obtains organic nutrients excreted from the green sulfur bacteria, and the green sulfur bacteria gain motility from the large rod.

SECTION HIGHLIGHTS

The *Chlorobi*, or green sulfur bacteria, are anoxygenic photosynthetic bacteria. Like the purple sulfur bacteria, they reside in sulfide-rich, anoxic waters and sediments that receive light. They fix carbon dioxide by the reductive TCA cycle.

21.5 *Chloroflexi*

The *Chloroflexi* resemble the green sulfur bacteria only in that they contain chlorosomes and are phototrophic. They resemble the green sulfur genus *Chloroherpeton* most closely in that they move by gliding motility. However, they are filamentous and multicellular organisms that evolved as a separate phylogenetic group. Three genera are currently known including *Chloroflexus*, *Heliothrix*, and *Oscillochloris*. The latter two genera have not been isolated in pure culture. In addition, this phylum contains a genus of heterotrophic gliding bacteria, *Herpetosiphon*. Interestingly, this phylum is one of the more commonly encountered phyla found in environmental clone libraries, indicating that it contains some of the most challenging bacteria to be grown in pure culture.

The most thoroughly studied genus is *Chloroflexus*. This long, thin (0.5 to 1.0 µm diameter) filamentous gliding bacterium (Figure 21.14) was first isolated from hot springs where it forms distinct layers in mats; however, some strains have also been found in mesophilic habitats. The temperature range for growth of the thermophilic strains is 45°C to 70°C.

The nutrition of *Chloroflexus aurantiacus* is complex. It grows best as a photoheterotroph, anaerobically in the light. It utilizes a variety of organic carbon sources including sugars, amino acids, and organic acids during photoheterotrophic growth. *C. aurantiacus* can also grow as a chemoheterotroph aerobically in the dark. Finally, it grows as a photoautotroph anaerobically in the

Proteobacteria
Chlorobi
Bacteroidites
Fibrobacteres
Spirochaetes
Cyanobacteria
Fusobacteria
Acidobacteria
Dictyoglomus
Actinobacteria
Firmicutes
Deinococcus-Thermus
Thermomicrobia
Chloroflexi
Deferribacteres
Chrysiogenetes
Chlamydiae
Verrucomicrobia
Planctomycetes
Nitrospirae
Thermodesulfobacteria
Thermotogae
Aquificae

for carbon dioxide fixation (see Figure 10.2B). This pathway is also used by some of the chemoautotrophic *Crenarchaeota*.

Like *Chloroflexus*, *Roseiflexus* is a filamentous photosynthetic genus that is found in alkaline hot springs and grows best at 50°C. It uses Bchl *a* as its primary photosynthetic pigment, but is colored orange due to the large amount of carotenoid pigments produced. *Roseiflexus* grows both as a photoheterotroph and as a heterotroph. *Chloronema* species have not been isolated in pure culture. This is a filamentous, gas-vacuolate genus with chlorosomes and a sheath, and is found in sulfide-rich lake habitats.

Herpetosiphon is a little-studied genus of filamentous gliding heterotrophic bacteria that produce a sheath and is mentioned here only because it has been shown to be closely related to *Cloroflexus* by 16S rRNA sequence analysis. Thus, as in many bacterial phylogenetic groups, both phototrophic and heterotrophic members occur. *Herpetosiphon* sp. are aerobic heterotrophs that live on a variety of organic carbon sources. Some species degrade cellulose and chitin. Both marine and freshwater species occur.

light, using sulfide or hydrogen gas as an electron donor.

It is interesting to note that *C. aurantiacus* uses neither the Calvin cycle nor the reductive TCA cycle for carbon dioxide fixation. The best evidence available suggests that instead it uses the 3-hydroxypropionate pathway

SECTION HIGHLIGHTS

Many of the filamentous green photosynthetic members of the phylum *Chloroflexi* live in hot springs, some at temperatures as high as 70°C. They are photoheterotrophic and fix carbon dioxide by the 3-hydroxypropionate pathway.

(A)

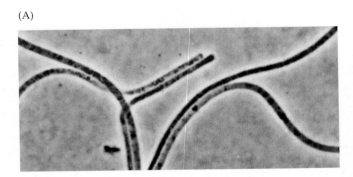

(B)   Chlorosomes

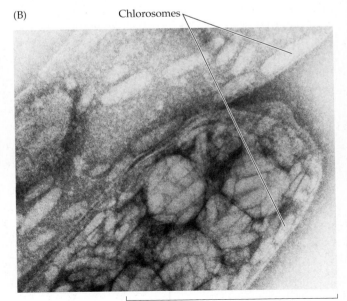

1 μm

Figure 21.14 *Chloroflexus*
Filaments of *Chloroflexus aurantiacus* as seen by (A) light microscopy and (B) electron microscopy. Note the chlorosomes in the electron micrograph. A, courtesy of R. Castenholz; B, courtesy of M. Broch-Due.

21.6 *Firmicutes*—The Heliobacteria

The heliobacteria are the most recently discovered photosynthetic bacteria. They are unique because they comprise the first known group of gram-positive phototrophic prokaryotes and have therefore been discussed in Chapter 20. The genera *Heliobacterium* and *Heliobacillus* appear to be closely related to *Clostridium,* and some strains even produce true endospores that are rich in calcium and dipicolinic acid. They are all photoheterotrophic, requiring an organic carbon source for growth, and in this respect resemble the photoheterotrophic purple nonsulfur bacteria. Acceptable carbon sources include organic acids such as acetate or pyruvate.

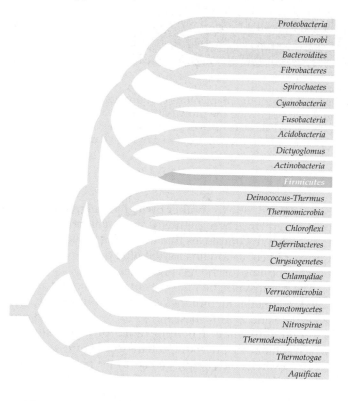

They are also unique in that their bacteriochlorophyll, Bchl *g,* is not found in other organisms and most closely resembles chlorophyll *a* (see Figure 21.1B). Photosynthesis in the heliobacteria is reminiscent of the *Chlorobi* in that the primary reduced electron acceptor is sufficiently reduced (−0.5 V) in its reduced form that NAD$^+$ can be reduced directly without the use of reverse electron flow, as required by the photosynthetic proteobacteria and chloroflexi. Cells appear red-brown due to their characteristic carotenoid pigment, neurosporene (see Figure 21.9). Some strains are capable of nitrogen fixation. The heliobacteria are found in alkaline soils and soda lakes. In addition, some live in rice paddies, where their nitrogen fixation activities are likely important.

SECTION HIGHLIGHTS

The heliobacteria is the only known group of gram-positive photosynthetic bacteria. All members known are photoheterotrophic. Some produce endospores like their close relatives in the genus *Clostridium*.

21.7 *Cyanobacteria*

This group of photosynthetic bacteria differs from all others in that they carry out **oxygenic photosynthesis** in which oxygen is evolved as a waste product:

$$(4) \quad CO_2 + H_2O \rightarrow (CH_2O)_n + O_2$$

Oxygen is derived from water by "water splitting," a process characteristic of all oxygenic photosynthetic organisms including algae and plants. All oxygenic photosynthetic organisms have photosystem II and chlorophyll *a* (see Chapter 9). It is in photosystem II that the water-splitting reaction occurs, resulting in oxygen formation. Recent evidence suggests that the *Cyanobacteria* evolved from two anoxygenic photosynthetic groups: the *Chlorobi* and the *Proteobacteria* (see Chapter 9). Thus, their photosystems I and II are thought to have descended from the photosystems of each of these two groups, respectively.

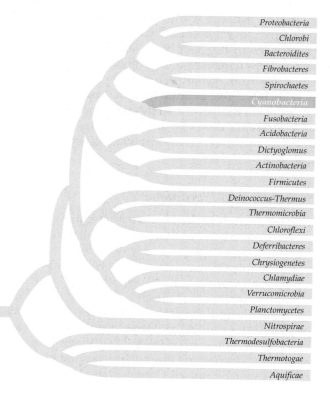

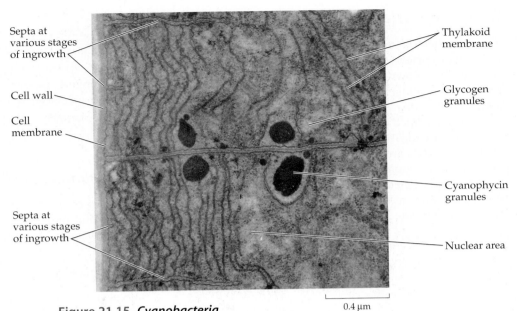

Septa at various stages of ingrowth

Cell wall

Cell membrane

Septa at various stages of ingrowth

Thylakoid membrane

Glycogen granules

Cyanophycin granules

Nuclear area

0.4 µm

Figure 21.15 *Cyanobacteria*
Thin section through adjoining cells of the filamentous cyanobacterium *Symploca muscorum*, an oscillatorian species, showing the prokaryotic nature of the cell (note the nuclear area) and the characteristic structures of cyanobacteria. The thylakoid lamellar membranes extend throughout the cell. Courtesy of H. S. Pankratz and C. C. Bowen.

The *Cyanobacteria* were traditionally classified with the algae as the blue-green "algae" because they have chlorophyll *a* and carry out oxygenic photosynthesis. However, because it was discovered by electron microscopy that their cell structure is truly prokaryotic (Figure 21.15), they have been claimed by bacteriologists and are now named by the bacteriological code and classified accordingly. Furthermore, phylogenetic analyses of their 16S rRNA indicate that they are a separate line of descent of the *Bacteria*. At this time, there is a dual system of classification of *Cyanobacteria*, because they are also named and classified by botanists using the botanical code.

Structure and Physiology

The fine structure of *Cyanobacteria* is typical of other gram-negative bacteria in that they have a multilayered cell wall containing peptidoglycan and an outer membrane. However, not only do they have these layers, but their cell walls usually contain other layers as well (Figure 21.16).

In addition to chlorophyll *a* and cytochromes, the *Cyanobacteria* have characteristic, unique pigments called **phycobilins**, including **allophycocyanin**, **phycocyanin**, and **phycoerythrin** (see Figure 9.13). The phycocyanins give the *Cyanobacteria* their characteristic blue-green color. However, not all *Cyanobacteria* are blue-green.

Some are red, due to another phycobilin pigment called phycoerythrin. These pigments are covalently bonded to proteins in complexes that are called **phycobiliproteins**—phycocyanobilins for the blue pigments or phycoerythrobilins for the red pigments.

The photosynthetic pigments of *Cyanobacteria* are located in lamellar membranes called **thylakoids** (see Figures 21.15 and 21.16). These intracellular membranes are studded with small structures called **phycobilisomes**, which contain the phycobiliproteins (see Figures 9.12 and 9.14). The thylakoid complex serves as the photosynthetic reaction center for photosynthesis in *Cyanobacteria*. Thus, both the physical process of light-harvesting as well as the chemical process of electron transfer occurs here.

The *Cyanobacteria* fix carbon dioxide via the Calvin cycle. One of the key enzymes in this cycle is ribulose-bisphosphate carboxylase (Rubisco). Some *Cyanobacteria* store this enzyme in structures in the cell in **carboxysomes**, a feature shared by some of the thio-bacilli discussed in Chapter 19. Most *Cyanobacteria* do not require vitamins and do not utilize organic compounds and are therefore excellent examples of photolithoautotrophs. Some species, however, can photoassimilate simple organic compounds such as acetate. They do not use these as energy sources, but simply as carbon sources for growth, thereby relying on photo-phosphorylation as their sole means of formation of ATP. The organic mate-

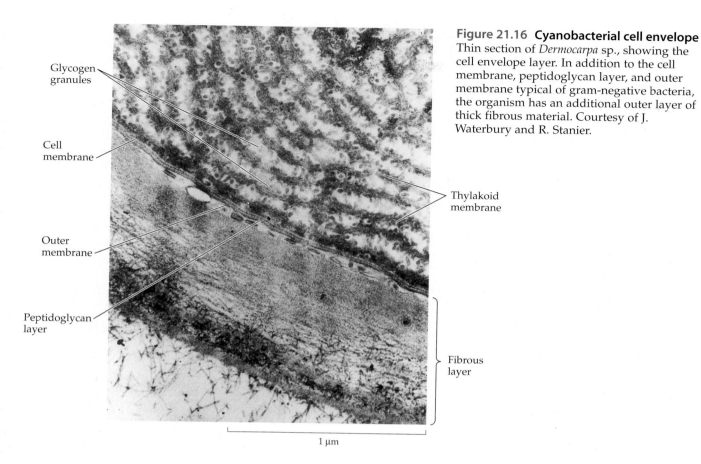

Glycogen granules

Cell membrane

Outer membrane

Peptidoglycan layer

Thylakoid membrane

Fibrous layer

1 μm

Figure 21.16 Cyanobacterial cell envelope
Thin section of *Dermocarpa* sp., showing the cell envelope layer. In addition to the cell membrane, peptidoglycan layer, and outer membrane typical of gram-negative bacteria, the organism has an additional outer layer of thick fibrous material. Courtesy of J. Waterbury and R. Stanier.

rial produced by photosynthesis during daylight is respired in part during the night.

One of the recent exciting discoveries regarding *Cyanobacteria* is that some of them can perform anoxygenic photosynthesis as well as oxygenic photosynthesis. When they carry out anoxygenic photosynthesis, sulfide, but not sulfur, serves as the electron donor. Sulfur granules are, in fact, formed outside the cells. This discovery has major implications for evolution, suggesting that the *Cyanobacteria* may have evolved early on from anoxygenic photosynthetic bacteria, which is consistent with their purported descent from green sulfur bacteria.

One of the unique features of *Cyanobacteria* is that they can store a compound called **cyanophycin** (see Figure 21.15). This is unique in that it is a polymer of aspartic acid, with each residue of aspartic acid containing a side-group of arginine:

$$
\begin{array}{cccc}
- \text{asp} & \text{asp} & \text{asp} & \text{asp} - \\
| & | & | & | \\
\text{arg} & \text{arg} & \text{arg} & \text{arg}
\end{array}
$$

Cyanophycin is unique, as it is the only nitrogen-containing storage granule of prokaryotic organisms. Because many environments contain low concentrations of nitro-

gen, it is often a limiting nutrient; therefore, cyanophycin is a useful storage product. In addition, cyanophycin can be used as an energy source during breakdown. Arginine dihydrolase leads to the synthesis of ATP:

$$
\text{Arginine} + \text{ADP} + \text{P}_i + \text{H}_2\text{O} \rightarrow
$$
$$
\text{ornithine} + \text{ATP} + \text{NH}_3 + \text{CO}_2
$$

Ecological and Environmental Significance of Cyanobacteria

The *Cyanobacteria* are important ecologically. Perhaps their major significance is that they are important contributors to carbon dioxide fixation and, hence, primary production in aquatic environments. Whereas the anoxygenic purple and green photosynthetic bacteria contribute at best about 10% of the annual primary production of the habitats in which they occur, the *Cyanobacteria* are responsible for more than half. Moreover, the *Cyanobacteria* are common in all aquatic habitats, whereas the anoxygenic photosynthetic bacteria are restricted to special anaerobic habitats that are rich in sulfide. Therefore, in the oceans where algae and *Cyanobacteria* are the dominant primary producers, up to almost one-half of the primary production is due to *Cyanobacteria*. Similarly high rates of primary production are attributable to these organisms in freshwater habitats (Box 21.2).

Cyanobacteria are also important members of contemporary mat communities. They form the uppermost layer in mats where they carry out oxygenic photosynthesis. The mat communities in which they dominate are common in thermal habitats, polar lakes, and hypersaline environments throughout Earth. Some species also grow well on the surface of soil.

BOX 21.2 · *Research Highlights*

Cellular Absorption Spectra of Photosynthetic Bacterial Groups

As mentioned in the text, the various photosynthetic bacterial groups occupy the same aquatic habitats as mat communities in springs and intertidal zones or the water columns of lakes, ponds, and marine ecosystems. The pattern in the vertical distribution of these photosynthetic groups is always the same: the *Cyanobacteria* reside at the surface, nearest the oxygen, the purple bacteria lie beneath them, and the green bacteria occupy the deepest layer, farthest from the oxygen and closest to the sulfide. This same pattern is found in Winogradsky columns (see Chapter 5). Each of these groups has its own characteristic bacteriochlorophyll or chlorophyll pigments and accessory pigments, such as the phycobiliproteins and carotenoids, and each of these pigments has its own characteristic absorption spectrum (see figure). Each group absorbs light of a different wavelength; therefore, each is able to obtain some of the light radiating from the sun because the overlying groups filter out only some of the wavelengths. It is also interesting to note that the ratio of light-harvesting pigments in the reaction centers is about 30:1 for purple bacteria, but 1,000:1 for green bacteria, because they lie beneath the purple bacteria where the light intensity is lower.

It should be noted that the photosynthetic bacteria that reside below the *Cyanobacteria* in these vertically stratified communities

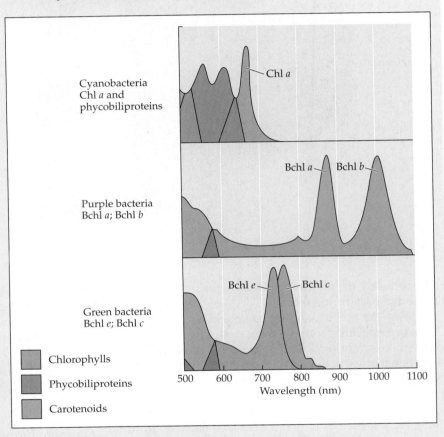

Absorption spectra of bacteriochlorophyll and chlorophyll from various photosynthetic bacterial groups shown in the vertical stratification in which they occur in natural communities. The cyanobacteria are shown at the surface and the purple bacteria underneath; the green bacteria occupy the lowest layer.

receive less light for photosynthesis. Furthermore, in planktonic communities, the little light that is available for the purple and green sulfur bacteria that reside in the deepest zones of the water column is energetically less favorable. This is because the longer wavelength light that these photosynthetic groups require is absorbed more strongly by the overlying water than is the shorter wavelength light used by the *Cyanobacteria*. Nonetheless, these less efficient photosynthetic bacteria have survived for hundreds of millions of years.

Because *Cyanobacteria* require sunlight and oxygen for their metabolism, they grow at the surface of the aquatic habitats and mat communities where they reside. Therefore, they are exposed to especially high solar radiation. To protect their cells from mutations due to UV radiation, many species produce pigments that absorb UV light. One such pigment is called **scytonemin** (Box 21.3).

Many *Cyanobacteria* are nitrogen fixers. Because nitrogen is a common limiting nutrient in many marine and some freshwater environments, their contribution to the nitrogen budget can be extremely important.

It should be noted, however, that some *Cyanobacteria* carry out undesirable activities in the environment. For example, certain gas-vacuolate species are responsible for **nuisance blooms** in freshwater habitats (Figure 21.17). Nuisance blooms occur in the midsummer or fall. These gas-vacuolate *Cyanobacteria* float to the surface of the lake and are carried onshore by the wind. They accumulate and subsequently decay onshore, where

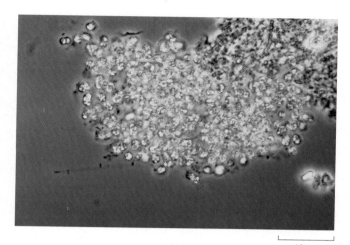

10 μm

Figure 21.17 Gas-vacuolate *Cyanobacteria*
Phase contrast photomicrograph of *Microcystis aeruginosa*. Bright areas are the gas vacuoles. Courtesy of R. W. Castenholz.

BOX 21.3 *Research Highlights*

Scytonemin: An Ancient Prokaryotic Sunscreen Compound

All organisms are susceptible to mutation caused by UV light. Humans can develop skin cancer if exposed to excess solar radiation. Animals produce melanin pigments to absorb UV radiation, and many plants produce flavonoid compounds. However, to further retard incoming radiation for humans, a variety of chemicals are commercially produced that are used as effective sunscreens to absorb damaging UV radiation. For example, *p*-aminobenzoic acid (PABA), a precursor in the synthesis of the vitamin folic acid, is commonly used in sunscreen and sunblocking lotions to absorb UV radiation. The reason for selecting these compounds is that PABA and other aromatic compounds are excellent absorbers of UV light.

Many *Cyanobacteria* produce their own remarkable sunscreen compound called **scytonemin**. Although this substance was named about 150 years ago, only recently has its chemical formula

The chemical structure of scytonemin.

Scytonemin

been determined by the *Cyanobacteria* expert Richard Castenholz and his colleagues at the University of Oregon.

Considering its complex aromatic structure, it is not surprising that this compound is an excellent UV blocking agent. Scytonemin is especially effective at absorbing UV-A (325 to 425 nm) but is also effective at absorbing UV-B (280 to 325 nm) and UV-C (maximum absorption at 250 nm).

Scytonemin is deposited in the extracellular sheath of the *Cyanobacteria* that produce it. There it can intercept the UV light

before it reaches the cytoplasm and DNA of the cell. It has been estimated that at the normal concentrations of scytonemin found in a trichome, 85% to 90% of the incident UV will be absorbed by this biological sunscreen before it enters the cell.

Because sheathed *Cyanobacteria* are found in early fossil stromatolites, it is likely that scytonemin evolved early on through these *Cyanobacteria*. The UV radiation would have been much more intense earlier in Earth's history than it is now.

they cause odors. Under normal conditions, the gas vacuoles allow the cells to float at or near the surface of lakes and ponds, where they grow at a location where photosynthesis is most active.

Some species, such as *Microcystis aeruginosa*, may produce toxic compounds that can actually kill animals, including cattle and dogs that eat them. Humans who ingest contaminated animals in the food chain have also died.

Cyanobacteria can cause odor and off-flavor tastes in drinking waters as well. This is due to the production of geosmin compounds, which are similar chemically to the geosmins produced by some *Actinobacteria*. This problem normally occurs in the late summer or fall when these *Cyanobacteria* are common in lakes and reservoirs that may be used for the water supply in municipalities.

Major Taxa of Cyanobacteria

Several orders of *Cyanobacteria* have been named and classified on the basis of their morphological characteristics (Table 21.6). They range in morphology from unicellular rods and cocci to filamentous types, some of which show branching division. Many species have not yet been isolated in pure culture. This is due to the difficulty that microbiologists have in cultivating them. Part of the problem is that many *Cyanobacteria* produce an external capsule or sheath to which other bacteria attach, making isolation especially difficult. Consequently, the cultivation of these bacteria in pure culture is a major area of research for bacteriologists. A brief description of each order and representative groups follows.

Chroococcales are common *Cyanobacteria* that are widely distributed in fresh and marine waters. Table 21.7 provides a listing of the important genera of the *Chroococcales*, some of which are discussed next.

Synechococcus species (Figure 21.18), which are unicellular coccoid to rod-shaped *Cyanobacteria*, are among the most important photosynthetic organisms in marine habitats. It has been estimated that they account for approximately 25% of the primary production that occurs

TABLE 21.7	Important genera of *Chroococcales*	
Genus	**Mol % G + C**	**Characteristics**
Chaemosiphon	46–47	Divide by budding
Synechococcus	47–56	Rods of freshwater, marine, and hot spring origin
Gloeothece	40–43	Cocci in common sheath; nitrogen fixer
Microcystis	45	Cocci; gas vacuolate; may cause toxic water blooms

in typical marine habitats (see Chapter 24). *Synechococcus* sp. are also found in hot springs and are among the most thermophilic phototrophs known, capable of growth in temperatures up to about 73°C.

The genus *Gloeothece* is remarkable in being able to fix atmospheric nitrogen in the presence of oxygen. Perhaps the capsular material of this organism (Figure 21.19) is somehow involved in keeping the oxygen concentration low within the cells. However, it is known also that some unicellular *Cyanobacteria* carry out nitrogen fixation during the dark period when oxygen is not formed by photosynthesis, and this may account for the ability of this genus to carry out this process.

Another notable genus in this order is *Microcystis*. Members of this genus are gas vacuolate, and this is one of the principal genera responsible for nuisance water blooms (see Figure 21.17). Furthermore, as mentioned previously, these organisms may produce toxins that can be deadly to animals that ingest them. *Chaemosiphon*, another genus in this order, is one of the few unicellular *Cyanobacteria* that divides by budding.

TABLE 21.6	Cyanobacterial orders	
Order	**Distinctive Features**	
Chroococcales	Unicellular or multicellular, nonfilamentous	
Pleurocapsales	Baeocytes formed	
Oscillatoriales	Straight filaments without specialized cells	
Nostocales	Straight filaments with heterocysts	
Stigonematales	Branching filaments	

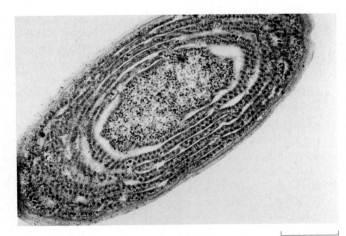

0.5 μm

Figure 21.18 *Synechococcus*
Electron micrograph of a thin section of *Synechococcus*, a common marine cyanobacterium. ©Science VU/Dr. Elizabeth Gentt/Visuals Unlimited.

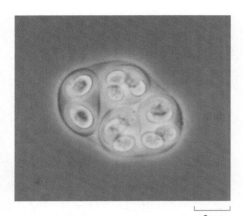

Figure 21.19 *Gloeothece*
Phase contrast micrograph of *Gloeothece* sp., showing the cells within a sheath. Courtesy of J. T. Staley.

Pleurocapsales have an unusual morphology. They grow as unicellular organisms that enlarge in size and undergo internal cell divisions to form **baeocytes**, small spherical reproductive cells that are released when the division cycle is complete. The baeocytes then develop into larger cells and repeat the division cycle. Some members of this order form multicellular filaments, but baeocyte formation also occurs in these organisms.

Oscillatoriales are common filamentous gliding bacteria. Because the cells in the filaments are in very close contact to one another and act in a concerted fashion to effect motility, they are regarded as being truly multicellular. The name for this gliding multicellular filament, which may contain 50 to 100 or more cells, is the **trichome** (Figure 21.20A,B,C). The trichome may divide to form groupings of 5 to 15 cells called **hormogonia**,

Figure 21.20 *Oscillatoriales*
(A) *Oscillatoria formosa*, a typical oscillatorian species, showing its filamentous growth habit. Each multicellular filament is a trichome. (B) Gas-vacuolate oscillatorian from a lake sample. (C) *Lyngbya* sp., showing its dark sheath. (D) *Spirulina*, a helical, filamentous cyanobacterium. A–C, courtesy of J. T. Staley; D, ©M. Abbey/Visuals Unlimited.

TABLE 21.8	Common genera of *Oscillatoriales*	
Genus	**Mol % G + C**	**Description**
Oscillatoria	40–50	Straight trichomes; some gas vacuolate; freshwater or hot springs
Trichodesmium	Unknown	Resemble *Oscillatoria*, but marine origin; fix nitrogen
Lyngbya	42–49	Sheathed; resembles *Oscillatoria*
Spirulina	54	Helical trichomes

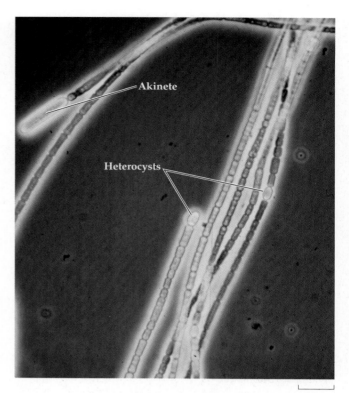

Figure 21.21 ***Anabaena***
Anabaena sp., showing typical filaments with heterocysts and akinetes. Courtesy of J. T. Staley.

which are short, gliding filaments used for dispersal and reproduction. These are commonly formed during asexual reproduction in all types of trichome-producing *Cyanobacteria*. A list of the common genera of this order is given in Table 21.8.

Oscillatoria sp. (see Figure 21.20A and B) are common in freshwater habitats and hot springs and can be found growing on the surface of soils. They vary in pigmentation from light green to almost black. Some are even red due to phycoerythrin pigments. Lake forms may be gas vacuolate (see Figure 21.20B). Their marine counterpart is the genus *Trichodesmium*. These bacteria are gas vacuolate and grow in clusters of filaments. *Trichodesmium* sp. are able to carry out nitrogen fixation even though they do not have specialized cells, that is, heterocysts (see *Nostocales*). *Lyngbya* sp. are similar to *Oscillatoria*; however, in addition, they also have a prominent sheath (see Figure 21.20C).

Spirulina species are unusual in that they have a helical trichome (Figure 21.20D). They are used as a food source by Africans in Chad, and more recently *Spirulina* has been sold as a dietary supplement in health food stores.

Nostocales is an order that represents an important group of nitrogen-fixing filamentous bacteria (Table 21.9). *Anabaena* species (Figure 21.21) are widely distributed in freshwater and saline habitats. They, like other members of the order *Nostocales,* have specialized cells called **heterocysts** that are the sites of nitrogen fixation (see Figure 21.21). Heterocysts are formed in the filaments from ordinary cells. Their location within the filament is distinctive for the species. Some are located at the end of the filament, whereas others are located within it. When ammonia or other forms of fixed nitrogen are depleted from the environment, heterocyst formation is induced. During the process in which a cell is

TABLE 21.9	Important genera of *Nostocales*	
Genus	**Mol % G + C**	**Distinctive Features**
Anabaena	35–47	Untapered trichomes, most gliding; many gas vacuolate
Aphanizomenon	Unknown	Like *Anabaena*; however, trichomes aggregate to form colonies; gas vacuolate
Nostoc	39–45	Like *Anabaena*, but sheathed
Rivularia	Unknown	Tapering filaments with polar heterocysts
Gloeotrichia	Unknown	Tapering filaments with polar heterocysts in microcolonies

converted to a heterocyst, the synthesis of phycobilins, the antennae pigments for photosynthesis, is stopped. Therefore, the heterocyst is no longer able to photosynthesize, so the production of oxygen ceases. Because oxygen is inhibitory to nitrogen fixation, this enables the heterocyst to carry out fixation when nitrogenase is synthesized. A considerable amount is known about heterocyst differentiation and the process of nitrogen fixation in *Anabaena* sp. (Box 21.4).

Some species produce **akinetes**, which are special cyst-like resting cells. These are usually somewhat larger than the normal vegetative cells in filaments (see Figure 21.21). They are typically formed after the heterocysts have been produced. Generally, they do not have gas vesicles, even though normal vegetative cells may have them. They can survive better under conditions of starvation and desiccation than the vegetative cells.

One species of *Anabaena*, *Anabaena azollae*, grows in a symbiotic association with the small water fern *Azolla*. The bacterium is harbored in a special sac of the fern (see Chapter 25). The water fern obtains fixed nitrogen from *A. azollae*, and, in return, the bacterium has a protected place where it can grow. Water ferns are widely distributed in aquatic habitats such as rice paddies, and this association is important in ensuring the fertility of such habitats.

Aphanizomenon flos-aquae as well as *Anabaena flos-aquae* are common gas-vacuolate organisms that cause water blooms. *A. flos-aquae* occur in characteristic colonies or "rafts" of trichomes that glide back and forth against one another.

Nostoc species are also common in aquatic habitats. They closely resemble *Anabaena* sp.; however, their filaments are enclosed within a sheath. Indeed, the sheathed structures can be as large as golf balls, or even baseballs, in habitats where they occur (Figure 21.22). Like *Anabaena* sp., *Nostoc* sp. have heterocysts and therefore carry out nitrogen fixation.

The *Rivularia* group differs from the other heterocystous filamentous genera in that they have tapering filaments. However, as shown by the genus *Gloeotrichia* (Figure 21.23), they also have heterocysts, which are located at the base of the filament and are responsible for nitrogen fixation.

Stigonematales is an order of *Cyanobacteria* that contains branching clusters of filamentous organisms. They have the most complex morphology of any of the *Cyanobacteria*. Some produce heterocysts as well as hormogonia and akinetes.

The Prochlorophytes: Chlorophyll a– and Chlorophyll b–containing Cyanobacteria

Prochloron, Prochlorococcus, and *Prochlorothrix* are *Cyanobacteria* that are referred to as prochloro-

Figure 21.22 *Nostoc*
A large colony of *Nostoc pruniforme* from Mare's Egg Spring in Klamath County, Oregon. Note the new, smaller colonies forming on its sides. Courtesy of W. K. Dodds and R. W. Castenholz.

phytes. They differ from typical *Cyanobacteria* in that they possess chlorophyll *b* as well as chlorophyll *a*, and they also lack phycobilins. At one time they were postulated as a possible evolutionary intermediate between *Cyanobacteria* and some of the algae. However, 16S rRNA phylogenetic analyses indicate they are members of the *Cyanobacteria*. The genus *Prochloron*

Figure 21.23 *Gloeotrichia*
Fluorescence micrograph of colonies of the filamentous cyanobacterium *Gloeotrichia echinulata.* ©M. I. Walker/Photo Researchers, Inc.

BOX 21.4 *Research Highlights*

The Cyanobacterial Heterocyst

The heterocyst is a specialized cell produced by some filamentous *Cyanobacteria*. The heterocyst is formed from a typical vegetative cell through a developmental process that is induced when the concentration of fixed nitrogen forms such as ammonia and nitrate are depleted from the environment. Once the heterocyst has formed, it is unable to undergo cell division and therefore resembles a resting cell such as a cyst. A differentiation process occurs whereby the heterocyst develops a complex coat of three predominant layers—an outer fibrous layer, a homogeneous layer, and an inner laminated layer—which are thought to restrict the passage of oxygen and other gases, including nitrogen, into the heterocyst. Nitrogen, which is needed by nitrogenase, enters the heterocysts through the ends of the heterocysts, called the microplasmodesmata.

The preheterocyst cell loses its ability to produce phycobilins, the antennae pigments for photosynthesis. This shuts off photosystem II, and, therefore, the production of oxygen as well as fixation of carbon dioxide cease. Photosystem I, however, remains intact, allowing the heterocyst to produce ATP by photophosphorylation. This ATP is needed by nitrogenase for nitrogen fixation. The reducing power needed for nitrogen fixation is produced through metabolism of carbohydrates, which are produced by adjacent photosynthesizing cells.

The decrease in oxygen concentration coupled with the decrease in fixed nitrogen available to the organism is a signal for the induction of nitrogenase and the process of nitrogen fixation. The ammonia formed

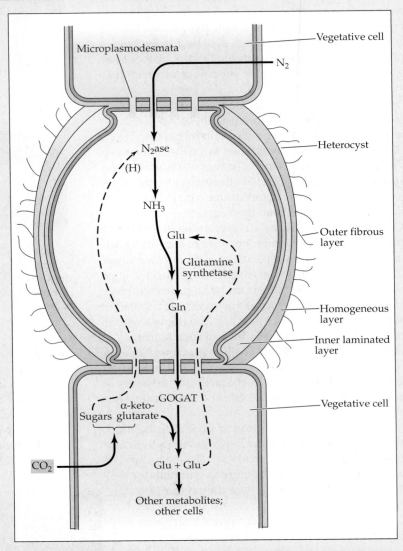

Diagram of the heterocyst showing its structure and the manner in which fixed nitrogen is formed and transferred from the heterocyst.

by nitrogenase (N_2ase) is carried to adjoining cells through the formation of glutamine as follows: The heterocyst has a high level of the enzyme glutamine synthetase, which carries out the following reaction:

$$\text{Glutamate} + NH_3 \rightarrow \text{glutamine}$$

The glutamine (Gln) formed from the heterocyst is converted in the adjoining vegetative cells to glutamate (Glu) by an enzyme called glutamine oxoglutarate amino transferase (GOGAT), which couples glutamine to α-ketoglutarate to form two molecules of glutamate. Part of the glutamate is recycled to the heterocyst (see figure), and part is transferred to adjacent cells as a form of utilizable fixed nitrogen. Thus, organic, fixed nitrogen is produced that can be used by the vegetative cells in anabolic processes.

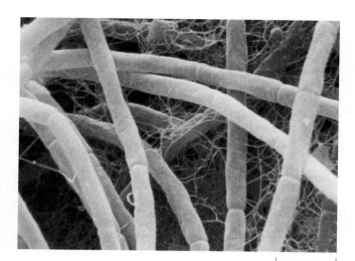

Figure 21.24 *Prochlorothrix*

3 μm

Scanning electron micrograph of *Prochlorothrix*, a marine prochlorophyte. Courtesy of W. Krumbein.

grows in symbiotic association with marine animals called didemnids and has not yet been cultivated in pure culture. *Prochlorothrix* (Figure 21.24) is a free-living marine bacterium.

Prochlorococcus is a genus of marine *Cyanobacteria*. They are small cocci, about 1 μm in diameter, that are common in the marine water column of temperate and tropical zones. Interestingly, they produce an altered form of chlorophyll *a* referred to as divinyl chlorophyll *a*. Like other *Cyanobacteria*, prochlorophytes, in particular *Prochlorococcus* sp., are significant contributors to the process of carbon dioxide fixation in the open oceans, where it is estimated that they fix about 25% of the carbon in marine habitats. Two different genetic types of *Prochlorococcus* are found in the marine water column. One type

lives closer to the surface and the other at greater depth. Apparently there was an evolutionary divergence, an example of an adaptive radiation that occurred to give rise to the two separate physiological types (Box 21.5).

The Chloroplast

Studies of 16S rRNA isolated from chloroplasts from plants and from algae indicate that they have a common origin. Furthermore, their ancestors all come from the *Cyanobacteria* branch of the *Bacteria,* and they would be placed with this group of the prokaryotes if they were free-living microorganisms. These results strongly support the endosymbiotic theory of evolution and argue that the progenitor of the chloroplast was a cyanobacterium that evolved into a close association with eukaryotic cells early in evolution (see Chapters 1 and 17).

SECTION HIGHLIGHTS

The *Cyanobacteria* are among the most important photosynthetic organisms on Earth. They are the only bacterial group that carries out oxygenic photosynthesis to produce oxygen as an end product. This diverse group of bacteria is widely distributed in aquatic habitats and soils. They produce blooms in freshwater habitats, and their marine species are responsible for half of the oceans' primary production. An ancestral cyanobacterium evolved to become the chloroplast, the photosynthetic organelle of plants. Carbon dioxide is fixed using the Calvin cycle.

BOX 21.5 *Research Highlights*

High and Low Light *Prochlorococcus*

Two major lineages of *Prochlorococcus* are found in marine water columns. The high-light strains reside near the surface of the water column, whereas the low-light strains live beneath them. Based upon 16S rDNA trees, both are descended from a common low-light ancestor. The high-light–adapted strains are markedly different from the low-light strains.

High-light strains have a smaller sized genome of only 1.65 Mb whereas the low-light strains have a genome size of 2.41 Mb. However, most surprisingly, they differ in mol % G + C content with the high-light strains at 30.8 versus 50.7 for the low light strains. This marked difference in genome size and DNA base composition is totally unexpected for members of the

same genus. The lower base composition may be explained by the loss of a gene, *Muty*, from the high-light strain. *Muty* is responsible for removing incorrectly paired adenosine from DNA. Thus, over many millennia this single genetic change may have led to the remarkable divergence in DNA base composition.

SUMMARY

- There are several major phylogenetic phyla that contain photosynthetic *Bacteria*, including the photosynthetic *Proteobacteria*, the *Chlorobi*, the *Chloroflexi*, the photosynthetic *Firmicutes* (heliobacteria), and the *Cyanobacteria*.

- The **phototrophic *Proteobacteria***, which comprise two groups, the **purple sulfur** bacteria and the **purple nonsulfur** bacteria, carry out **anoxygenic photosynthesis** in which oxygen is not produced, and either reduced sulfur compounds or organic compounds are used, respectively, as electron donors for carbon dioxide fixation, which occurs anaerobically. The purple bacteria, like other *Proteobacteria*, fix carbon dioxide by the Calvin cycle. The coloration of the purple bacteria is determined by **carotenoid pigments** that mask the **bacteriochlorophyll** *a* in the reaction center where **photophosphorylation** occurs.

- The *Chlorobi* and the *Chloroflexi* have a special structure called the **chlorosome**, which contains the photosynthetic pigments and reaction center. In addition to Bchl *a*, they produce other green or brown bacteriochlorophylls, Bchl *c*, Bchl *d*, and Bchl *e*. The *Chlorobi* use the **reductive TCA cycle** for carbon dioxide fixation. The **consortium species** consist of two separate organisms, a green sulfur bacterium and a heterotroph, living in close symbiotic association. *Chloroflexus aurantiaca*, a green filamentous member of the *Chloroflexi*, has its own pathway for carbon dioxide fixation, termed the **3-hydroxypropionate pathway**.

- The **heliobacteria** are members of the *Firmicutes* and are photoheterotrophic. They are the only known gram-positive phototrophs. Some form endospores.

- The oxygenic photosynthetic bacteria are members of the *Cyanobacteria*. *Cyanobacteria* have photosystem I and photosystem II and carry out the water-splitting reaction to form oxygen from water. The accessory photopigments of the *Cyanobacteria* are phycobilins, **phycocyanin**, and **phycoerythrin**. These are located in phycobilisomes, which are structures on the photosynthetic lamellae termed **thylakoids**. Some *Cyanobacteria* store an amino acid polymer called **cyanophycin**, which is composed of repeating units of aspartic acid and arginine. The **prochlorophytes** are *Cyanobacteria* that lack phycobilin pigments and contain chlorophyll *b*.

- The multicellular structure typical of many filamentous *Cyanobacteria* is termed a **trichome**. Most *Cyanobacteria* that are nitrogen-fixing carry out the process in a specialized modified vegetative cell called the **heterocyst**.

 Find more at www.sinauer.com/microbial-life

REVIEW QUESTIONS

1. Distinguish among the terms chemoautotroph, chemolithoautotroph, photoautotroph, photolithoautotroph, photolithoheterotroph, and mixotroph.

2. List the various carbon dioxide fixation pathways used by photosynthetic *Bacteria*. Which organisms use which pathways?

3. What distinguishes anoxygenic from oxygenic photosynthetic bacteria? What groups fall into each category?

4. What is the evidence that prochlorophytes are or are not intermediates in the evolution of algae?

5. Compare the metabolisms of *Escherichia coli*, *Methylococcus* sp., *Beggiatoa* sp., and *Chromatium okenii*. How do you explain that they are all members of the *Gamma proteobacteria*?

6. Compare the morphological and metabolic diversity of the various cyanobacterial orders, including the prochlorophytes.

7. Why are the *Cyanobacteria* so important in the evolution of life and on the biosphere of Earth?

8. Why do two different genetic variants of prochlorophytes exist in the marine water column? How did they arise?

SUGGESTED READING

Boone, D. R., R. W. Castenholz and G. M. Garrity, eds. 2001. *Bergey's Manual of Systematic Bacteriology*, 2nd ed., Vol. 1. New York: Springer.

Brenner, D. J., N. R. Krieg, J. T. Staley and G. M. Garrity, eds. 2005. *Bergey's Manual of Systematic Bacteriology*. 2nd ed., Vol II. Baltimore: Lippincott, Williams & Wilkins.

Dworkin, M., S. Falkow, K. Schleifer, E. Rosenberg, K-H. Schleifer and E. Stackebrandt. 2006. *The Prokaryotes*. 3rd ed. Heidelberg, Germany: Springer.

Staley, J. T. and A-L. Reysenbach. *Biodiversity of Microbial Life: Foundation of Earth's Biosphere*. 2002. New York: John Wiley & Sons.

22

Exploring Bacterial Diversity: Other Phyla, Other Microbial Worlds

There are wide areas of the bacteriological landscape in which we have so far detected only some of the highest peaks, while the rest of the beautiful mountain range is still hidden in the clouds and the morning fogs of ignorance. The gold is still lying on the ground, but we have to bend down to grasp it.
—*Preface*, The Prokaryotes

The domain *Bacteria* contains many other phylogenetic groups than the six discussed in Chapters 19 through 21. It has been estimated that at least 50 phyla of *Bacteria* exist altogether. As surprising as it may seem, bacteriologists do not even have representatives of some of these in pure culture. They are known to exist only from analyses of the diversity of 16S rRNA gene sequences of the DNA extracted from natural communities (see Box 17.5). This chapter contains information about the remaining 17 phyla of *Bacteria* that have representatives that have been isolated in pure cultures. Phyla for which pure cultures are not yet available are not described, because virtually nothing is known about their phenotypic attributes. Thus, much about the vast diversity of the bacterial world still remains largely unknown (Box 22.1).

For some phyla in which representatives have been isolated, the few known strains represent only a small fraction of the diversity of the group. In fact, several phyla are represented by only a single species because they are so poorly understood. For those phyla, only brief thumbnail sketches of their descriptions are provided.

The chapter begins by discussing the various phyla that contain thermophilic members of the *Bacteria*. These phyla are thought to represent some of the deepest branches in the domain *Bacteria* on the 16S rRNA tree of life.

BOX 22.1	*Research Highlights*

Billions of Years of Evolution and Only 5,000 Species?

Although bacteria have lived on Earth for more than 3.5 billion years, fewer than 5,000 species have been described and named. In contrast, about 750,000 species of insects have been named, and they have lived on Earth less than one billion years. How can this be? There are a number of reasons. First of all, as discussed in Chapter 17, bacterial species are defined very differently from plants and animals. But there is also another major reason. For the most part, bacteria must be isolated in pure culture and studied in the laboratory before they can be officially named and described, and it is not always simple or inexpensive to do this. For example, consider the *Plancto-mycetes* phylum of *Bacteria*.

The *Planctomycetes* are among the most unusual members of the *Bacteria*. Although they were first observed in aquatic habitats in the early 1900s, none were reported in pure culture until 1971. Even now, some of the most morphologically striking species of this group have not been cultivated in pure culture, although they grow in ordinary habitats we have all visited. One example of this is *Planctomyces guttaeformis*, commonly found in freshwater lakes (see figure). This organism can be readily identified by its morphology using the light microscope. Why then, hasn't it

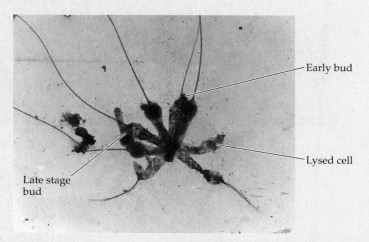

Transmission electron micrograph of *P. guttaeformis* from a lake in North Carolina. Note the long spikes, unique to this uncultivated species, extending from the pole of each ovoid cell in the rosette. This rosette contains some cells that are lysed. Courtesy of J. T. Staley.

been grown in the laboratory? We simply do not know how to grow it, although microbiologists have tried.

One of the reasons that this organism is so difficult to grow is that we do not know its requirements for growth, and there is no obvious way to determine these requirements. One approach is to simply attempt to grow it under a variety of laboratory conditions using various carbon and nitrogen sources. This approach is not particularly scientific and has not yet proved successful.

Strange as it may seem, the inability to cultivate bacteria is commonplace in the microbial realm. Thus, there are many bacteria in natural water and soil environments that have never been grown in pure culture. As discussed in Chapter 24, it is estimated that less than 1% of all bacteria have been isolated and studied in pure culture in the laboratory. A similar situation exists for eukaryotic microorganisms. The good news is that microbiology will continue to be an exciting field of discovery for many years to come for microbiologists who have the imagination and patience to investigate these organisms.

22.1 Phyla Containing Thermophilic Bacteria

Table 22.1 contains a listing of some of the genera of bacterial thermophiles that have been isolated from hot springs and other geothermal environments. All are gram-negative rods of various sizes with optimal growth temperatures in the range of 40° to 100°C. Most would be considered moderate thermophiles, in contrast to some of the hyperthermophilic *Archaea* that grow optimally at temperatures greater than 75°C (see Chapter 6). However, some, such as *Aquifex pyrophilus*, are regarded as hyperthermophiles because they are capable of growth at temperatures in excess of 80°C.

Some of the genera are aerobes, and others are anaerobes. It is interesting to consider that some of these

TABLE 22.1	Characteristics of bacterial phyla containing thermophiles			
Phylum/Genus	**O_2 Requirement**	**Temperature Range (°C)**	**Mol % G + C**	**Habitat (Energy Source)**
Aquificae				
Aquifex	Aerobe	60–95	40	Hot springs (H_2 autotroph)
Hydrogenobacter	Aerobe	60–80	38–44	Hot springs
Desulfurobacterium	Anaerobe	40–75	35	Hydrothermal vents
Thermodesulfobacteria				
Thermodesulfobacterium	Anaerobe	60–70	31–38	Hot springs (H_2; organic acids)
Thermotogae				
Thermotoga	Anaerobe	55–90	46–51	Hydrothermal vents (sugars)
Fervidobacterium	Anaerobe	40–80	33–40	Hot springs (sugars)
Thermosipho	Anaerobe	35–77	60–63	Hydrothermal vents (yeast extract; peptone) (H_2 autotroph)
Nitrospirae				
Thermodesulfovibrio	Anaerobe	40–70	30–38	Hot springs (H_2 + organic acids)
Deinococcus-Thermus				
Thermus	Aerobe	40–85	60–67	Hot spring (glucose; acetate)
Thermomicrobia				
Thermomicrobium	Aerobe	45–80	64	Hot springs (tryptone; yeast extract)
Dictyoglomus				
Dictyoglomus	Anaerobe	50–80	29	Hot springs (carbohydrates)

phyla—*Thermodesulfobacteria*, *Thermotogae*, and *Aquificae*—appear to represent rather deep branches in the bacterial lineage and are not closely related to any other known mesophilic bacterial groups.

Phylum Aquificae

The *Aquificae* contains the most thermophilic species of the *Bacteria* known. The maximum growth temperatures of some species exceed 95°C, and therefore they can be regarded as hyperthermophiles. All cultured strains that do not grow on organic compounds are obligate hydrogen autotrophs.

Aquifex is the most thoroughly studied genus. The 1.55 MB genome of *A. aeolicus* has been sequenced. These gram-negative bacteria are true hyperthermophiles that grow at an optimum temperature of 85°C. These bacteria grow as chemolithotrophs that typically grow as aerobes or microaerophiles. In addition, they can grow as denitrifiers, whereby they use nitrate as an electron acceptor and produce nitrite and N_2 gas. They fix carbon dioxide through the reductive citric acid cycle. In addition to using H_2 as an energy source, they can also use thiosulfate and sulfur, which they oxidize to sulfuric acid. The genus *Hydrogenobacter* is similar metabolically in that it also uses the reductive tricarboxylic acid (TCA) cycle.

Desulfurobacterium grows chemolithotrophically by oxidizing hydrogen gas as an energy source and reducing thiosulfate, S^0, or sulfite to H_2S. This is a gram-negative, obligate anaerobe that was isolated from deep-sea hydrothermal vents (**Figure 22.1**).

Phylum Thermodesulfobacteria

Thermodesulfobacterium is the sole genus in this phylum. These gram-negative rod-shaped organisms are heterotrophic sulfate-reducing bacteria that use lactate and pyruvate as energy sources and thiosulfate or sulfate as electron acceptors. H_2S is formed by their sulfate-reducing metabolism. The organic acids are incompletely oxidized to acetic acid and carbon dioxide. These bacteria live in hot springs and hot subterranean oil reservoirs. The upper temperature range for growth of members of this group is from 60°C to 80°C.

Phylum Thermotogae

The genera *Thermotoga* and *Thermosipho* are closely related anaerobic organisms isolated from submarine thermal environments. In many phylogenetic trees these gram-negative bacteria represent one of the deepest branches of any known lineage of the *Bacteria*. The cells

Figure 22.1 Deep sea hydrothermal vent
Hydrothermal vents are the source of many thermophilic bacteria, including *Desulfurobacterium*. Courtesy of OAR/National Undersea Research Program/NOAA.

of both genera are surrounded by a proteinaceous sheath-like structure, a so-called **toga** that balloons over the ends of the rods (Figure 22.2). A number of rods in a chain may be surrounded by one sheath. The function of the sheath is unknown. The optimum temperature for this genus is 66°C to 80°C.

Thermotoga species are chemoheterotrophs that ferment sugars such as glucose to lactate, acetate, CO_2, and H_2. A genome sequence of *Thermatoga maritima* indicates that it is 1.86 MB and has a mol % G + C of 46. *Thermosipho* grows as an anaerobic chemoheterotroph on rich media such as yeast extract and requires the amino acid cysteine. Some strains of *Thermotoga* are motile by monotrichous flagellation.

Phylum Nitrospirae

This phylum contains a variety of bacteria, most of which are mesophilic. However, in keeping with our

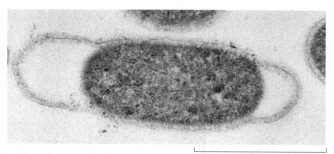

Figure 22.2 Thermotoga
The "toga" of *Thermotoga* is shown as the outer faint layer extending beyond the cells shown. Courtesy of K. O. Stetter and R. Rachel.

theme of thermophiles, we begin by discussing the single known genus in the phylum that is a thermophile, *Thermodesulfovibrio*.

THERMODESULFOVIBRIO   *Thermodesulfovibrio* strains have been isolated from thermal hot springs in Yellowstone National Park and Iceland. As their name implies, these are anaerobic chemoheterotrophic sulfate-reducing bacteria that use organic carbon sources as energy sources and reduce sulfate, thiosulfate, and sulfite to H_2S. Lactate and pyruvate are used as energy sources. Their optimal growth temperature is 65°C.

NITROSPIRA   The phylum *Nitrospirae* is the only phylum apart from the *Proteobacteria* and some members of the archaeal *Crenarchaeota* in which nitrification has been reported. The namesake genus for this phylum is *Nitrospira*. These are spiral-shaped chemoautotrophic nitrifying bacteria that oxidize nitrite to nitrate as an energy source. They are obligate aerobes that use oxygen as their electron acceptor. *Nitrospira* strains have been isolated from freshwater and marine sources as well as from soil and activated sludge.

MAGNETOBACTERIUM   This is a genus of magnetotactic bacteria that has not yet been isolated in pure culture (Figure 22.3). However, the 16S rRNA gene sequence has been determined, and fluorescent probes have been used to identify the organism in the environment. This magnetic bacterium is a large rod that contains oval-shaped magnetite granules. It has been found at the oxic–anoxic transition zone in freshwater lake sediments.

Phylum Deferribacteres

This phylum contains heterotrophic bacteria that respire anaerobically using oxidized inorganic electron acceptors such as Fe^{3+}, Mn^{4+}, Co^{3+}, S^0, or nitrate.

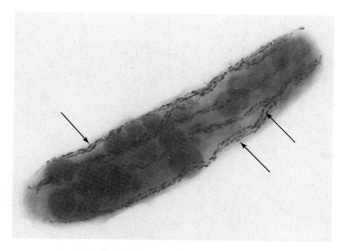

Figure 22.3 Magnetic bacterium
Electron micrograph of *Magnetobacterium bavaricum*, showing its intracellular oval-shaped magnetite particles. Arrows indicate magnetite particles. Courtesy of S. Spring.

DEFERRIBACTER AND *GEOVIBRIO*   The genus *Deferribacter* contains anaerobic, rod-shaped bacteria that oxidize organic acids and transfer the electrons to Fe^{3+}, Mn^{4+}, or nitrate. Complex organic compounds can also be used as substrates, none of which are fermented. This moderately thermophilic genus grows over a temperature range of 50°C to 65°C. The original strain was isolated from a thermal petroleum reservoir in the North Sea.

Geovibrio is a genus of anaerobic vibrioid bacteria, whose metabolism is similar to that of *Deferribacter*. These organisms are mesophiles that oxidize acetate using Fe^{3+}, S^0, or Co^{3+} as electron acceptors. Fe^{3+} can be used as an electron acceptor for oxidation of other energy sources, including H_2, proline, and a variety of organic acids and other amino acids. Strains have been isolated from soils.

Phylum Thermomicrobia

Thermomicrobium is the sole known genus. *Thermomicrobium roseum* is an aerobic, pink-pigmented, nonmotile species originally isolated from Toadstool Spring in Yellowstone National Park. The temperature at this site was 74°C. *Thermomicrobium* cells are short gram-negative pleomorphic rods and dumbbell shapes. They grow best chemoheterotrophically on a complex medium containing low concentrations of nutrients.

Phylum Dictyoglomi

This phylum is named for the single thermophilic genus *Dictyoglomus*. These anaerobic, elongated, rod-shaped

bacteria grow in alkaline hot springs. The gram-negative cells are about 0.5 μm in diameter and 5 to 20 μm in length. They form bundles containing several cells in a distinctive spherical structure. The temperature range of growth is from 50°C to 80°C. These are chemoheterotrophic fermentative bacteria that use a variety of sugars as energy sources for growth.

Phylum Deinococcus–Thermus

This phylum contains two major groups of organisms that differ from one another phenotypically. We begin by discussing the radiation-resistant group.

DEINOCOCCUS AND *DEINOBACTER*   Members of the gram-positive genus *Deinococcus* look superficially like *Micrococcus* species in their morphology, in that they stain as gram-positive, nonmotile cocci that form tetrads (Figure 22.4A). However, they differ from the micrococci in their complex cell wall structure (Figure 22.4B), which is more reminiscent of gram-negative bacteria, and their high resistance to ultraviolet and gamma radiation. These bacteria can survive exposure to blasts of gamma radiation thousands of times greater than the level that would kill a human being. The genus also differs because the diamino acid of its peptidoglycan is L-ornithine, and it lacks phosphatidylglycerol in its cell membrane. *Deinococcus radiodurans* (from Latin meaning "strange berry that withstands radiation") was first isolated from meat that had been sterilized by gamma irradiation.

The radiation resistance of this organism is in part attributable to its ability to rapidly repair DNA that has been damaged, not only by gamma but by ultraviolet radiation as well. Recent evidence indicates there is a two-step repair mechanism that allows irradiated or desiccated *D. radiodurans* to completely reassemble its radiation- or dessication-shattered chromosome from hundreds of short DNA fragments in just a few hours. The genome structure suggests that it may contribute to the radiation resistance, because it has two chromosomes, as well as one megaplasmid and a smaller plasmid that encode the genes for DNA repair. Because of their radiation resistance, these bacteria are being considered as candidate organisms for the treatment of radioactive wastes (see Chapter 32). This is one example of how different and novel are some of the bacteria from the poorly studied phyla described in this chapter.

The genus is also noted for its pink to red pigments, which are carotenoids. The membrane carotenoids absorb light and thereby aid the cells in their ability to survive exposure to ultraviolet radiation. The optimal temperature for growth is between 25°C and 35°C. These bacteria have been isolated from soil in temperate zones

(A)

(B)

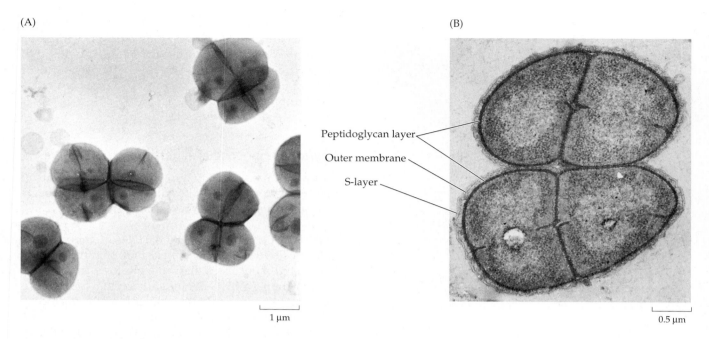

Peptidoglycan layer

Outer membrane

S-layer

1 µm

0.5 µm

Figure 22.4 *Deinococcus radiodurans*
(A) Electron micrograph of actively growing cells, showing their tendency to grow as four-celled packets, or tetrads. Note that the septa are forming as "curtains." (B) Section showing the multilaminated cell wall structure, consisting of an outer single S-layer, outer membrane, and thick peptidoglycan layer. Courtesy of R. G. E. Murray.

and rocks from Antarctica where they may be exposed to higher than normal doses of ultraviolet and gamma radiation. Another genus of the deinococci, *Deinobacter grandis*, contains rod-shaped cells (Figure 22.5).

THERMUS   Although they are phylogenetically close relatives, the genus *Thermus* has phenotypic properties

Outer membrane   Peptidoglycan layer

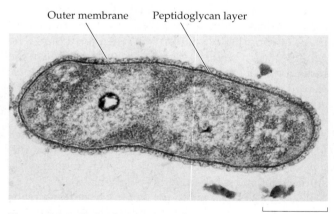

Figure 22.5 *Deinobacter*   0.5 µm
Thin section through *Deinobacter grandis*, a rod-shaped member of the deinococci group. The outer membrane appears as a looped layer surrounding the darker peptidoglycan layer. Courtesy of R. G. E. Murray.

that are different from those of *Deinococcus*. *Thermus* contains aerobic nonmotile bacteria that form filaments from 5 to up to 200 µm in length. Unlike *Deinococcus*, it stains as a gram-negative bacterium, but like *Deinococcus*, it has ornithine as the diamino acid in its cell wall peptidoglycan. Colonies are often pigmented yellow, orange, or red due to the production of carotenoids. *Thermus* strains have been isolated from thermal environments throughout the world. Some have postulated that *Deinococcus* and *Thermus* evolved from a common ancestor to occupy very different niches, high radiation or thermophily, respectively.

Thermus species are heterotrophs that use a variety of carbohydrates, amino acids, carboxylic acids and peptides for growth. Some species grow anaerobically with nitrate or nitrite as electron acceptors but none are fermentative.

One species, *Thermus aquaticus*, is noted in molecular biology for its heat-stable DNA polymerase, the Taq (for *T. aquaticus*) polymerase, which is used in the polymerase chain reaction (PCR) technology. The enzyme withstands the multiple heating and cooling cycles required in PCR. The genome is 1.9 MB and exists in the cell along with a 232 KB megaplasmid. Genes for the Embden-Meyerhof pathway and the TCA cycle, as well as those for gluconeogenesis have been found. The optimal growth temperature of *T. aquaticus* ranges from 70°C to 75°C. It is an obligate aerobe that grows as a heterotroph with ammonium salts as the nitrogen source and sugars as the carbon source.

T. aquaticus is one of the most common thermophilic bacteria known. It grows in most neutral-pH hot springs, but it is also found in ordinary home water heaters. The usual way to isolate it is to inoculate a broth medium from hot tap water and incubate at 70°C.

SECTION HIGHLIGHTS

The thermophilic members of the *Bacteria* are thought to be among the most ancient bacterial lineages. Many of these species are chemoheterotrophic; however, some are chemoautotrophs, including nitrifiers and hydrogen bacteria. The phylum also contains gamma radiation–resistant bacteria and anaerobes that can use sulfate, oxidized metals, and nitrate as electron acceptors in anaerobic respiration.

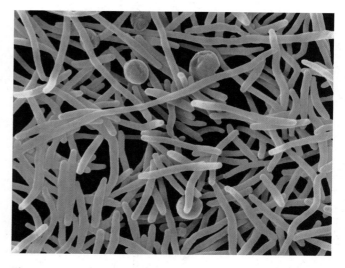

Figure 22.6 *Cytophaga* group
False-colored scanning electron micrograph of *Capnocytophaga ochracea* cells, showing morphology typical of the *Cytophaga* group. Photo ©Dennis Kunkel Microscopy, Inc.

22.2 Phyla that May Contain Thermophiles

It is noteworthy that although we discussed those bacterial phyla that contain thermophilic genera that have been isolated in pure culture, it is known from DNA extracted from thermal environments that some of the remaining phyla discussed next appear to contain thermophilic members. However, apart from *Isosphaera*, which is a moderately thermophilic genus of the *Planctomyetes*, none of these have been grown in culture, so virtually nothing is known about them at this time.

Phylum Bacteroidetes

This diverse collection of bacteria contains heterotrophic, nonphotosynthetic organisms. Some are aerobic or fac-

ultative aerobic and move by gliding motility, as represented by the *Cytophaga–Flavobacterium* group, whereas others are anaerobic nonmotile fermenters represented by the *Bacteroides* group. These bacteria and their relatives are discussed here individually.

CYTOPHAGA–FLAVOBACTERIUM GROUP   These are rod-shaped to filamentous bacteria (Figure 22.6), some of which are more than 100 μm in length, depending on the genus (Table 22.2). They move by **gliding motility** or are immotile. Unlike the fruiting myxobacteria, which are also gliding heterotrophs, these bacteria have very low mol % G + C (30–48%). Most are obligately aerobic, but some are fermentative. Because of their gliding motility, their colonies are thin and spreading, almost always with a yellow to orange (or rarely red) pigmentation. The pigments are cell-bound carotenoids and flexirubins (Figure 22.7), some of which are unusual biologically because they are chlorinated. One diagnos-

TABLE 22.2	Principal genera of the *Cytophaga–Flavobacterium* group		
Genus	**Mol % G + C**	**Cell Length (μm)**	**Special Features**
Cytophaga	30–45	1.5–15	Some degrade cellulose
Capnocytophaga	33–41	2.5–6	Oral cavity
Flavobacterium	30–45	1–6	Some degrade agar or chitin
Flexibacter	37–47	15–50	Some degrade chitin or starch; freshwater
Microscilla	37–44	10– >100	Like *Flexibacter*, but marine, and do not degrade chitin
Saprospira	30–48	1.5–5.5	Helical filament; saprophyte

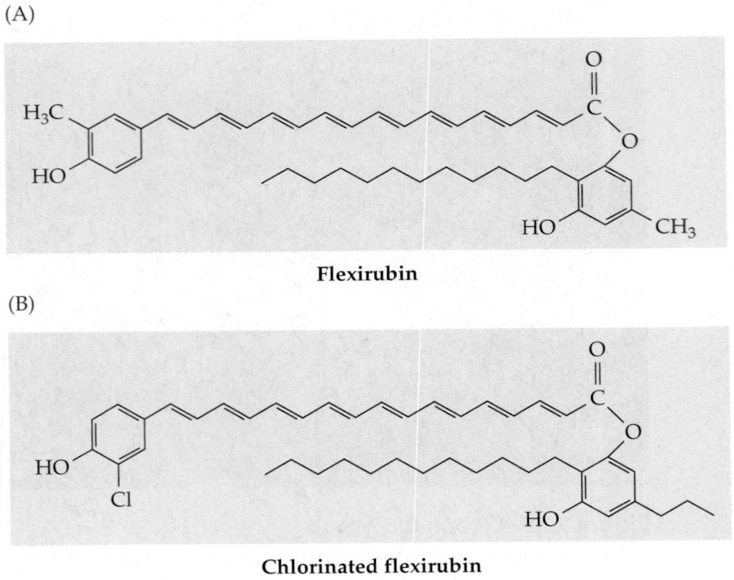

(A)

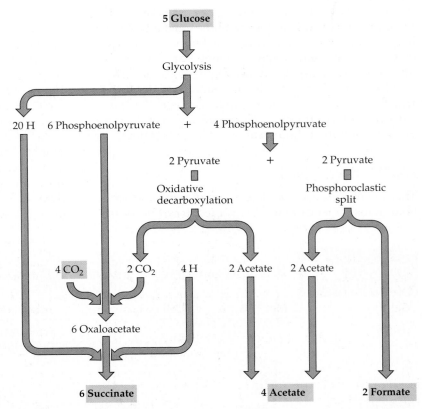

Flexirubin

(B)

Chlorinated flexirubin

Figure 22.7 Flexirubin pigments
(A) The parent compound, flexirubin, from *Flexibacter filiformis*; (B) this chlorinated flexirubin compound is typical of those produced by *Flavobacterium* species.

tic feature of flexirubin is the "flexirubin reaction" in which the color of the colony changes from yellow to purple or red-brown when it is flooded with alkali (20% KOH).

These bacteria are aerobic respirers or facultative anaerobic fermenters. Therefore, they are thought to have an Embden-Meyerhof glycolytic pathway and a TCA cycle. Electron transport systems contain menaquinones as the respiratory quinone.

A typical fermentation of glucose by *Flavobacterium succinicans* produces succinate, acetate, and formate (Figure 22.8). *Flavobacterium succinicans* requires carbon dioxide for anaerobic growth, and it is thought that they produce succinate through condensation of phosphoenolpyruvate with carbon dioxide, followed by a reduction to succinate.

The genus *Capnocytophaga* is also a fermentative facultative aerobe. *Capnocytophaga* species are

Figure 22.8 Fermentation pathway
Proposed pathway for glucose fermentation by *Flavobacterium succinicans*.

Net reaction: 5 glucose + 4 CO_2 ⟶ 6 succinate + 4 acetate + 2 formate

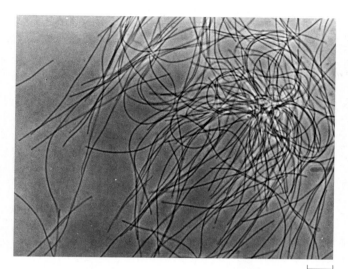

Figure 22.9 *Flexibacter*
Flexibacter elegans grows as long, sinuous, gliding filaments.
Courtesy of H. Reichenbach.

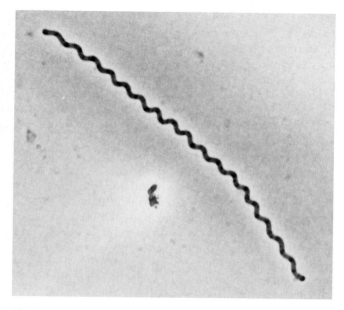

0.5 μm

Figure 22.10 *Saprospira*
Filament of the gliding helical bacterium *Saprospira*.
Courtesy of J. T. Staley.

found in the oral cavities of humans and other animals. Organisms from this genus require carbon dioxide on primary isolation and initial cultivation. Acetate and succinate are the major acid byproducts of the fermentation of carbohydrates.

Several genera in this group, including *Cytophaga*, *Flavobacterium*, *Flexibacter* (Figure 22.9), and *Microscilla*, are best known for their ability to degrade macromolecules. Indeed, Sergei Winogradsky described the first member of *Cytophaga* as a cellulose-degrading organism similar to *Cytophaga hutchinsonii*. Cellulose is degraded while these bacteria grow and glide in direct contact with the cellulose fibers. These organisms produce a cell membrane–bound endoglucanase as well as soluble periplasmic exoglucanases that are responsible for cellulose degradation. *Sporocytophaga*, which is an active cellulolytic bacterium, is differentiated from *Cytophaga* because it produces a microcyst structure analogous to that produced by some of the myxobacteria.

Other species from marine, freshwater, and soil habitats degrade agar or chitin. In addition, some species degrade starch and others degrade proteins such as gelatin, casein, and keratin from hair or feathers. Furthermore, many produce extracellular RNases and DNases.

Saprospira cells grow in long, multicelled, helical filaments (Figure 22.10). The filaments can be up to 500 μm in length and are composed of individual cells that are generally 0.8 to 1 μm in diameter and 1.5 to 5 μm in length. Most species produce pink, yellow, orange, or red carotenoid pigments. Filaments of some strains form coils, and others are uncoiled. Although the uncoiled filaments are nonmotile, the coiled strains glide via a

screw-like motion along their long axis. *Saprospira* species are all strictly aerobic and are common inhabitants of sand, mud, and decaying organic materials that collect along seacoasts in temperate and warm climates around the world. Most strains are marine, but some live in freshwater.

The type species is *Saprospira grandis*, a marine organism with complex nutritional requirements. It grows in culture on peptones and amino acid mixtures. Unknown growth factors in yeast extract are also required. Glucose and other sugars stimulate growth of some strains. The generation time of *S. grandis* is 2 to 2.8 hours at 30°C. The mol % G + C is 30 to 37. The presence of these bacteria in decaying organic matter and as inhabitants of oxic areas in sewage treatment plants indicates that they may play a significant role in the degradation of organic material in various environments.

BACTEROIDES GROUP   Members of the genus *Bacteroides* are all obligately anaerobic gram-negative heterotrophs. *Bacteroides* are rod shaped, often with terminal or central swelling of the cells. *Bacteroides* are the predominant microbe in the lower digestive tracts of mammals.

The *Bacteroides* group inhabits areas of the gastrointestinal tract, where limited digestion occurs. They are predominant in the human cecum and colon. These areas are essentially fermentation chambers where food that is either not digested or is indigestible by the host is fermented by *Bacteroides*.

The *Bacteroides* group consists of oxygen-sensitive bacteria, but they can withstand exposure to air for 6 to 8 hours, particularly if blood and hemin are available. Many species grow best under increased CO_2 tension. *Bacteroides fragilis* is a species found in the human alimentary tract. It ferments glucose to fumarate, malate, and lactate. The cells are 0.8 to 1.3 μm in diameter and 1.6 to 8 μm in length and have rounded ends, occur singly or paired, and are usually encapsulated. Growth of *B. fragilis* is enhanced by 20% bile. Some species of *Bacteroides*, such as *B. fragilis*, chemically transform various bile acids. *B. fragilis* has a generation time of 8 hours without added hemin and 1 hour in a glucose-rich medium with added hemin.

As much as 30% of the fecal mass in humans consists of bacteria, and there are 10^{10} *Bacteroides* species per gram versus only 10^6 *Escherichia coli* per gram. *Bacteroides* may cause disease in humans, generally in immune-compromised individuals or those with other predisposing factors that make them susceptible to infection. Lowered oxygen levels in tissue caused by trauma, vascular constriction, necrosis, or concomitant infection by other bacteria all increase the risk of infection from these bacteria. Because *Bacteroides* are the most numerous bacterial genus in the human colon, they are found in significant numbers in fecal material and domestic sewage. They have been considered as potential markers of water quality to trace animal fecal material.

Phylum Fibrobacteres

The genus *Fibrobacter* is the only genus so far described in this entire phylum. *Fibrobacter* species are anaerobic gram-negative, rod-shaped bacteria. *Fibrobacter succinogenes*, which live in the rumen of cattle, ferment sugars to produce volatile fatty acids, including acetic and propionic acids, which are used as an energy source by the animal (see Chapter 25). The volatile fatty acids are absorbed from the lining of the rumen and are used as a source of carbon and energy for the ruminant. In addition, the protein-rich microorganisms subsequently pass into the small intestine, where they are degraded as an organic, nitrogenous food source for the animal.

Phylum Spirochaetes

The spirochetes are noted for their distinct helical shape and unique mode of motility. Electron microscopy has revealed that the spirochetes have an unusual motility due to their **periplasmic flagella**. These flagella make up a structure called the **axial filament**, which extends around the helical body of these gram-negative bacteria within the outer membrane or outer envelope (Figure 22.11). The axial filament is anchored in the cytoplasmic membrane in the same manner as are typical bacterial flagella with the characteristic flagellar plate. However,

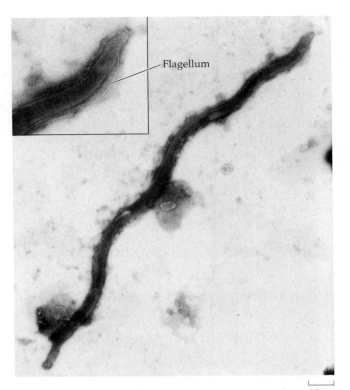

Flagellum

0.5 μm

Figure 22.11 Spirochete morphology
Electron micrograph of a spirochete from the hindgut of a termite, showing its characteristic helical morphology. Note the periplasmic flagella at each pole (see inset). Courtesy of D. A. Odelson and J. A. Breznak.

the hooks and filaments of the flagella remain in the periplasmic space (Figure 22.12) and for this reason they are also called **endoflagella** because they are located inside the cell membrane. The axial filament is formed by the overlapping and apposition of two sets of flagella (called fibrils), which are attached at each pole of the cell. The number of periplasmic flagella varies from 2 to more than 100 per cell, depending on the genus and species. Because of their helical shape and axial filament, the spirochetes are especially well adapted to movement through viscous liquids. Flagella movement is coordinated at both poles. By their concerted action, the periplasmic flagella produce rotational movements that propel the organism forward in a corkscrew fashion, resulting in ordinary translocation. These are analogous to runs in typical bacteria undergoing positive chemotaxis. Unusual flexing movements of the cell are analogous to tumbles and are caused by uncoordinated activity of the two separate polar flagellar tufts, thereby interrupting translocation. They can also creep and crawl on solid media due to their unusual flexing movements, which may also distort the cell shape. Many species are so slender that, despite their great length, the cells can-

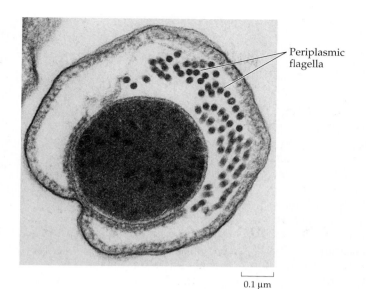

Periplasmic
flagella

0.1 μm

Figure 22.12 Spirochete in cross-section
Electron micrograph of thin section through a termite
spirochete. Note the location of the numerous periplasmic
flagella of the axial filament. Courtesy of J. A. Breznak and
H. S. Pankratz.

not be seen when observed unstained with the light microscope.

The cytoplasm and cell membrane of spirochetes are analogous to those of other gram-negative bacteria. However, apart from the genus *Leptospira*, their peptidoglycan contains ornithine—not diaminopimelic acid—as the diamino acid. The flexible outer membrane of spirochetes is referred to as an outer sheath or outer envelope. Its composition varies from genus to genus, but all contain protein, lipid, and polysaccharide. The outer sheath is indispensable—its removal results in cell death.

Most spirochetes are heterotrophic, although some of the termite spirochetes are exceptions because they carry out acetogenesis (Box 22.2). Spirochetes may be obligate anaerobes, facultative aerobes, microaerophiles, or aerobes. They divide by binary transverse fission and can utilize carbohydrates, amino acids, long-chain alcohols,

or fatty acids as carbon and energy sources. Many spirochetes placed in unfavorable conditions form spherical bodies, which appear as bulbous, swollen areas, usually at the tips of the cells. These represent cells undergoing early stages of lysis.

The spirochetes range from free-living bacteria in soil and aquatic habitats to obligate parasites and pathogens of eukaryotes. Some grow in symbiotic associations with eukaryotes, including the termite symbionts (see Box 22.2). There are several genera of *Spirochaetes* (Table 22.3). The five most important genera are discussed individually.

SPIROCHAETA   *Spirochaeta* is a genus of anaerobes and facultative anaerobes that live in freshwater and marine environments. As anaerobes these organisms thrive in muds, ponds, and marshes. They are not pathogenic. They utilize carbohydrates as the source of carbon and energy, and their fermentation products are ethanol, acetate, CO_2, and H_2. The species *Spirochaeta plicatilis* is found in H_2S-containing habitats. This species has not been obtained in pure culture, although Ehrenberg first described it more than 150 years ago. Many other species in this genus have been isolated. During cell division, a new set of flagella appears at the middle of a cell, a septum is laid down, and the two cells separate. Some strains are multicellular and grow to 250 μm or longer.

This free-living genus is the best understood of all the spirochetes. The most thorough studies on the metabolism, physiology, and motility have been conducted with members of this genus.

CRISTISPIRA   This genus was assigned the name *Cristispira* because they produce a bundle of 100 or more periplasmic flagella. When these flagella are intertwined on the protoplasmic cylinder, they distend the outer sheath, forming a ridge or crest. *Crista* is the Latin word for *crest*. The genus is widely distributed in both marine and freshwater molluscs (clams, oysters, and mussels). They inhabit the crystalline style that is a part of the digestive tracts of these invertebrates. They can also be found in other organs in molluscs. The *Cristispira* are not pathogenic and occur

TABLE 22.3	Important genera of *Spirochaetes*			
Genus	**Cell Length (μm)**	**Mol % G + C**	**Disease**	**Habitat**
Spirochaeta	5–250	51–65	None	Sediments, mud, ponds, marshes
Cristispira	30–180	Unknown	None	Crystalline style and digestive tracts of molluscs
Treponema	5–20	25–53	Syphilis; yaws	Mouths, intestinal tracts, genital areas of animals
Borrelia	3–20	Unknown	Relapsing fever; Lyme disease	Mammals and arthropods
Leptospira	1–2	35–53	Leptospirosis	Free-living, warm-blooded animals

BOX 22.2   *Research Highlights*

Spirochetes in Termites

Microscopic examination of the microbiota of the hindgut of termites and wood-eating cockroaches reveals that spirochetes are part of the normal flora. Indeed, about half of the bacteria from the termite hindgut are spirochetes. When they were first reported in 1877 by J. Leidy, they were incorrectly regarded as being spirilla. However, their distinct morphology as revealed by electron microscopy indicates they are true spirochetes (see Figure 22.11). One termite species, *Pterotermes occidentis*, carries at least 15 types of spirochetes, based on size of the cell body, wavelength and amplitude of primary coils, and number of periplasmic flagella. The spirochetes vary in size from 0.2 μm by 3 μm to as large as 1μm by 100 μm. The number of periplasmic flagella ranges from a few to more than 100.

The metabolism that occurs in the hindgut of lower termites is similar to that of the rumen. Cellulose is a primary foodstuff that the termite itself cannot digest. Therefore, it relies on its hindgut microbiota to carry out this and other important metabolic processes that ultimately provide the energy and organic compounds required by the termite. Cellulose is ingested by the termites and, as the first step in its metabolism in the hindgut, it is broken down to sugars by cellulolytic protozoa.

The spirochetes carry out several important roles in the termite. Although most of the termite spirochetes are free-living in the gut fluids, one group of the spirochetes is associated with some of the resident protozoa as **ectosymbionts**. These spirochetes are attached to the protozoa and by their movement actually propel the protists around the gut.

Until recently none of the termite spirochetes had been grown in culture. However, pure cultures that have been isolated in John Breznak's lab are known to carry out novel processes not previously reported in the *Spirochaetes* phylum. One group of the spirochetes is acetogenic. They compete with methanogens for hydrogen gas and carbon dioxide produced in the sugar fermentations. Interestingly, the spirochetes are oriented closer than the methanogens to the fermenting bacteria that produce these gases, thereby enabling them to utilize most of these two substrates. The methanogens live near the periphery of the hindgut where the concentration of hydrogen and carbon dioxide is low.

Another group of the spirochetes in termites carries out nitrogen fixation, an important means of obtaining nitrogen for production of amino acids and other organic nitrogen sources for the bacteria and its host. Nitrogen fixation is especially important for animals that rely on plant foodstuffs such as wood, which has a high ratio of organic carbon to nitrogen.

(A)

(B)   Spirochetes

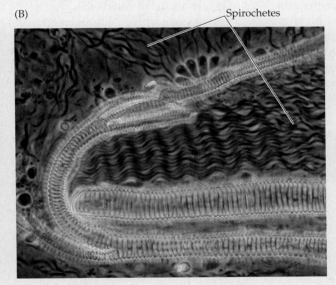

(A) A termite. (B) Spirochetes from a termite gut. A, courtesy of Scott Bauer/USDA; B, ©L. Amaral-Zettler, L. Olendzenski and D. J. Patterson/micro*scope.

in healthy univalve and bivalve molluscs. None, however, has ever been isolated in pure culture.

TREPONEMA Members of the genus *Treponema* live in the mouths, digestive tracts, and genital areas of humans and other animals. One of the ancient scourges of humans is due to a member of this genus, *Treponema pallidum*, the causative agent of the sexually transmitted disease syphilis. *T. pallidum*, subspecies *petunae*, causes yaws, a contagious disease in tropical countries (see Chapter 28). The treponemes are strictly anaerobic or microaerophilic. The human pathogens are probably microaerophiles, but have not been grown on artificial media. The treponemes that have been cultured are anaerobic and ferment carbohydrates or amino acids. Some require long-chain fatty acids for growth. Others require the short-chain volatile fatty acids present in rumen fluid.

BORRELIA Some species of *Borrelia* have been grown in culture. All are microaerophilic with complex nutritional requirements. Two noted pathogenic types occur in the genus. *Borrelia burgdorferi* is the causative agent of Lyme disease. This disease is transmitted to humans by the black-legged (deer) tick *Ixodes scapularis*. Other species are tick-borne pathogens of relapsing fever. Both diseases are discussed more fully in Chapter 28. Genomes of these organisms are not circular but linear, which is an unusual feature for bacteria.

LEPTOSPIRA *Leptospira* is the simplest genus morphologically of the spirochetes. Species in this genus have only two periplasmic flagella in their axial filament. The cells are tightly coiled, and the fibrils of the axial filament rarely overlap. Frequently, the cells end with a hook at both ends. Most species are free-living organisms that occur in soil, freshwater, and marine environments. Some are parasites of humans and other animals, whereas others are pathogenic. *Leptospira* species are obligate aerobes with a DNA base composition of 35 to 41 mol % G + C. The major cell wall diamino acid is diaminopimelic acid.

The source of carbon and energy for members of this genus is long-chain fatty acids (15 carbons or more) or long-chain fatty alcohols. They cannot synthesize fatty acids and directly incorporate substrates into cellular lipids. Among the species in the genus are two important species: the nonpathogen *Leptospira biflexa*, and the pathogen *Leptospira interrogens*. The latter is the causative agent of leptospirosis, a disease known worldwide. The natural hosts for leptospirosis are rodents and other mammals (see Chapter 28) including rats and domestic animals such as cattle. For this reason, leptospirosis poses an occupational hazard for sewer workers, miners, fish farmers, sugarcane cutters, rice-field workers, and dairy farmers.

Phylum Planctomycetes

The *Planctomycetes* comprise a phylum of unusual bacteria whose species lack most or all of the genes for producing peptidoglycan. They are also remarkable for their mode of cell division—budding—which may be due to their lack of FtsZ, the cell division protein that is produced by typical members of the *Bacteria* as well as the methanogens and halophilic members of the *Euryarchaeota*.

The *Planctomycetes* are also known for an unusual structural feature called the **crateriform structures**, which appear as distinctive pits on the cell surface (Figure 22.13). In addition, some species have unique noncellular appendages called **stalks** (see Figure 22.13 and Figure 22.14). These stalks are not prosthecate and thus differ from those of *Caulobacter*, a member of the *Alphaproteobacteria* (see Chapter 19).

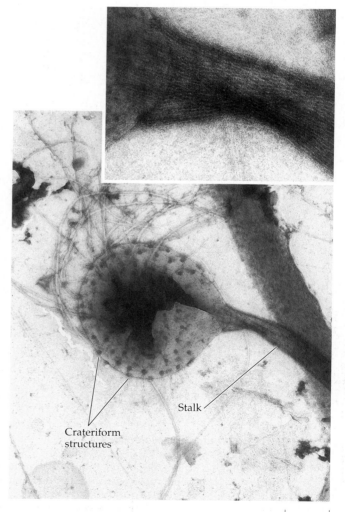

Figure 22.13 *Planctomyces bekefii*
0.5 μm
The "pits," or crateriform structures, on the surface of *Planctomyces bekefii* are unique to these bacteria. Note also the fibrillar nature of the stalk, shown best in the inset. Courtesy of J. Fuerst and J. T. Staley.

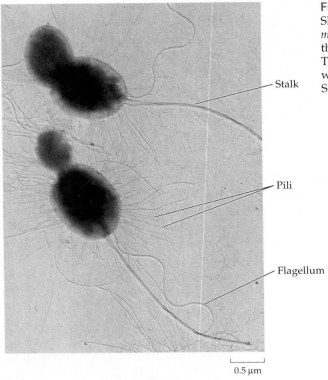

Stalk

Pili

Flagellum

0.5 µm

Figure 22.14 *Planctomyces maris*
Shadowed transmission electron micrograph of *Planctomyces maris*. This is a marine budding species. Note that in addition to the stalks, some cells have flagella and all cells have many pili. The stalk consists of several fibrils bundled together in a fascicle with a holdfast at its distal tip. Courtesy of J. Bauld and J. T. Staley.

Most striking of all, the *Planctomycetes* contain intracellular membrane-bound compartments of functional importance, not found in other members of the *Bacteria* or *Archaea*. For example, the nuclear material of *Gemmata obscuriglobus* is located within a nuclear body compartment that bears a strong resemblance to the nuclear compartment of eukaryotic organisms, since it is surrounded by an envelope comprising two membranes (Figure 22.15). Likewise, the unique anammox (anaerobic ammonia oxidation) reaction (discussed later on in this section) that is carried out by some members of this group occurs within an intracellular compartment called the **anammoxosome** (Figure 22.16), a region bounded by a single membrane.

Some of these intracellular compartments contain ribosomes (e.g., the **riboplasm**), whereas others lack ribosomes (e.g., the **paryphoplasm**, which lies at the periphery of the cytoplasm). Therefore, the structural organization is more

(A)

(B)

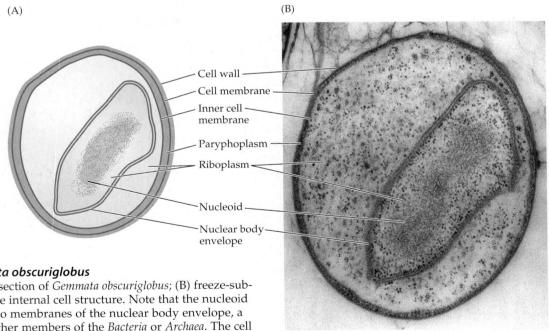

Cell wall
Cell membrane
Inner cell membrane
Paryphoplasm
Riboplasm
Nucleoid
Nuclear body envelope

Figure 22.15 *Gemmata obscuriglobus*
(A) Diagram of a cross section of *Gemmata obscuriglobus*; (B) freeze-substituted cell showing the internal cell structure. Note that the nucleoid is located within the two membranes of the nuclear body envelope, a feature not known in other members of the *Bacteria* or *Archaea*. The cell wall is proteinaceous and lies outside of the cell membrane. Inside the cell membrane lies the paryphoplasm, which is bounded by another membrane referred to as the inner cell membrane. Ribosome-like particles lined up on the inner nuclear envelope membrane are indicated by arrowheads. Photo courtesy of John Fuerst.

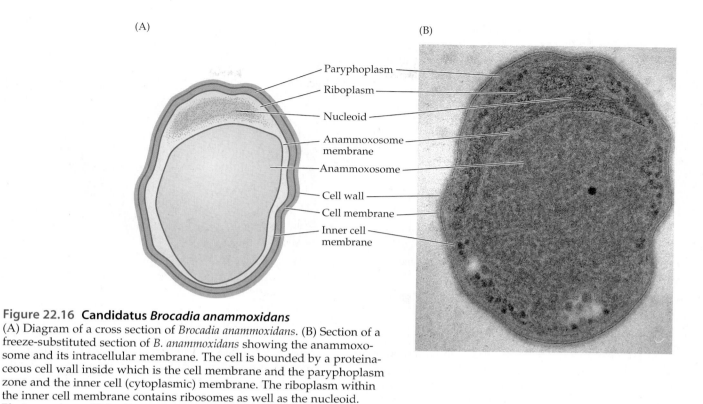

(A) (B)

Figure 22.16 Candidatus *Brocadia anammoxidans*
(A) Diagram of a cross section of *Brocadia anammoxidans*. (B) Section of a freeze-substituted section of *B. anammoxidans* showing the anammoxosome and its intracellular membrane. The cell is bounded by a proteinaceous cell wall inside which is the cell membrane and the paryphoplasm zone and the inner cell (cytoplasmic) membrane. The riboplasm within the inner cell membrane contains ribosomes as well as the nucleoid. Photo courtesy of John Fuerst.

complex than that found in typical members of the *Bacteria* discussed in Chapter 4.

Most of the described planctomycetes are heterotrophic, and many grow by either fermentation or respiration of sugars as carbon sources. However, the anammox bacteria can live autotrophically using ammonia as an energy source and nitrite as an electron acceptor, generating N_2 (dinitrogen) as a result (see following section). The planctomycetes are a diverse group that is very large and poorly studied. Several genera have been described and are listed in Table 22.4. Recent information on the group indicates they are one of the dom-

inant bacterial phyla in typical soils and in one of the most extreme soil habitats, the Atacama Desert in Chile, which is the driest desert on Earth.

PIRELLULA, GEMMATA, AND *PLANCTOMYCES* *Pirellula, Gemmata,* and *Planctomyces* are genera that have life cycles of alternating motile and sessile cells (Figure 22.17). The sessile cells have a holdfast by which they can attach, much like *Caulobacter* species, to other organisms or detritus in the habitat. They are commonly found on sheathed organisms in aquatic habitats (see Figure 22.17). In all three genera, the sessile mother cells divide by bud formation at the unattached pole to produce a motile daughter cell with subpolar flagella. In *Pirellula,* at least, it has been found that the motile daughter cell subsequently attaches to particulate matter and becomes immotile. It can then produce a bud and repeat the cycle. The mother cell remains sessile and periodically forms new buds from the budding (or reproductive) pole of the cell as growth proceeds.

Planctomyces species have a stalk (see Figures 22.13, 22.14, and 22.17), which consists of a bundle of fibrils that resemble fimbriae, except that they have a noticeable holdfast at their tip. The holdfast allows them to attach to particulate material. Typically, unicellular species produce a stalk after the cell attaches to particulate ma-

TABLE 22.4		Genera of *Planctomycetes*
Genus	**Mol % G + C**	**Morphology**
Pirellula	54–57	Lack stalks; most have polar flagellum
Gemmata	64	Lacks stalks; polar bundle
Planctomyces	50–58	Stalks; polar flagellum
Isosphaera	62	Lack stalks; filamentous gliding bacterium

Figure 22.17 Life cycle of *Planctomyces*
Diagram of life cycle of a *Planctomyces* sp.

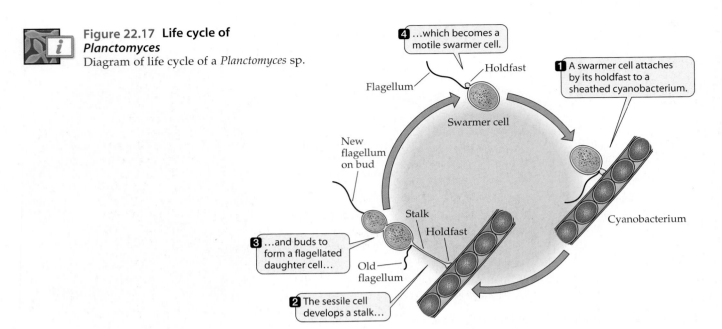

4 ...which becomes a motile swarmer cell.

Holdfast

Flagellum

1 A swarmer cell attaches by its holdfast to a sheathed cyanobacterium.

Swarmer cell

New flagellum on bud

Stalk

Holdfast

Cyanobacterium

3 ...and buds to form a flagellated daughter cell...

Old flagellum

2 The sessile cell develops a stalk...

terial in the environment. In contrast, the microcolonial species aggregate to form rosettes in their freshwater environment (**Figure 22.18**). The rosettes are strikingly complex and beautiful, but none of these rosette formers has yet been isolated in pure culture. Some of the rosette formers, such as *Planctomyces bekefii*, deposit iron and manganese oxides in their stalks.

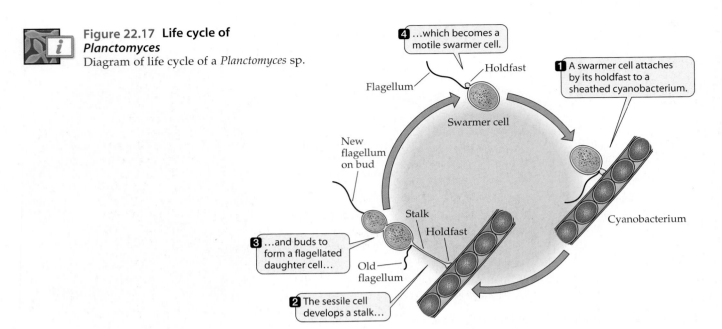

— Holdfast

— Stalk

Figure 22.18 *Planctomyces bekefii* colony
Electron micrograph of *Planctomyces bekefii*, showing the rosette structure as it appears when collected from lake waters. Cells appear as balls with stalks extending toward the center of rosette. Courtesy of J. T. Staley.

Some genomes of the *Planctomycetes* have been sequenced. *Pirellula marina* has a large genome: 7.35 Mb. The genome sequence of an enriched consortium containing the anammox bacterium *Kuenenia stuttgartensis*, discussed next, has also been reported.

"*CANDIDATUS ANAMMOX*" GENERA: *KUENENIA* AND *SCALINDUA*   These genera are considered "Candidatus genera" because none of them have been isolated in pure culture. However, sufficient information is known about them that they can be quite well described (see Chapter 17). In fact, a genome sequence of one of them, *Kuenenia stuttgartensis*, a freshwater organism, has been published. Marine candidatus genera, such as *Scalindua* are also known. Recently, this group of the planctomycetes has been discovered to carry out a unique biochemical reaction, called the **anammox reaction**, in which dinitrogen is produced from ammonia and nitrite:

$$NH_4^+ + NO_2^- \rightarrow N_2 + 2\,H_2O$$

In this reaction, one of the nitrogen atoms of the N_2 they produce is derived from the energy source, ammonia, whereas the other is from the electron acceptor, nitrite. None of the anammox species have been isolated in pure culture, but they can be grown as a consortium with other species in enrichment cultures in which they comprise about 95% of the biomass. These highly enriched cultures are used to study physiology and biochemical reactions. The anammox reaction occurs within an intracellular compartment called the **anammoxosome** using a unique biochemical pathway (**Figure 22.19**). Note that one of the intermediates in the process is hy-

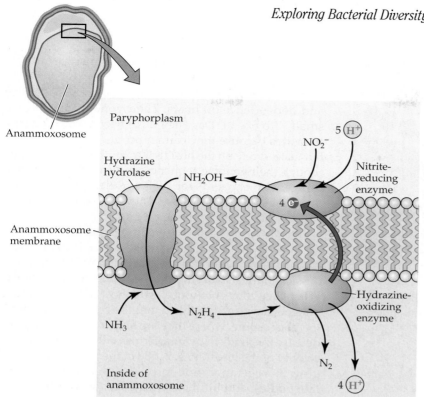

Anammoxosome

Paryphorplasm

Hydrazine hydrolase

Anammoxosome membrane

NO_2^-

$5\,H^+$

NH_2OH

Nitrite-reducing enzyme

$4\,e^-$

NH_3

N_2H_4

Hydrazine-oxidizing enzyme

N_2

Inside of anammoxosome

$4\,H^+$

Figure 22.19 Metabolism in anammoxosome
Diagram shows a postulated pathway of the anammox reaction thought to occur in the anammoxosome membrane. Nitrite is reduced to hydroxylamine, which in the presence of ammonia and the enzyme hydrazine hydrolase forms hydrazine, N_2H_4. This is then oxidized to produce N_2 gas and protons that generate a proton gradient used to produce ATP through membrane-embedded ATP synthase.

drazine, a chemical that is so reactive it is used as rocket fuel. A special type of membrane lipid termed a ladderane lipid (Figure 22.20) is found in the anammoxosome membrane. These lipids, which are very tightly packed together, are thought to contain the volatile intermediates of the reaction. Interestingly, these ladderane lipids contain not only the typical ester linkages found in other bacterial lipids, but some ether linkages as well. Apparently they can have either or both ether- and ester-linkages in the same ladderane lipid moiety. This is the only report to date of a bacterium that contains ether-linked lipids, although they are found exclusively in all members of the *Archaea*.

It is also noteworthy that the planctomycetes that carry out the anammox reaction are autotrophic. A recent genome sequence of a consortium enriched for one of these species indicates that it uses the acetyl-coenzyme A pathway for carbon dioxide fixation.

The anammox reaction has changed our conception of the nitrogen cycle. In this process ammonia is oxidized as an energy source and nitrite is reduced. The importance of this reaction is still not well understood. Some oceanographers have hypothesized that the reaction is favored over denitrification in environments that have low concentrations of organic material. Research is underway to test this hypothesis. Some studies indicate

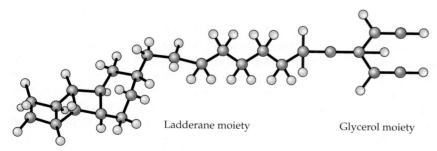

Ladderane moiety

Glycerol moiety

Figure 22.20 Ladderane lipids
The ladderane lipids of the anammoxosome membrane are unique. The example shown is derived from a family of different structures. Note that the hydrocarbon chains are attached to a glycerol moiety on the right by ether linkages previously reported only in members of the *Archaea*. These unique lipid structures pack tightly together and are thought to retain the volatile intermediates of the anammox reaction. Carbon atoms are gray, hydrogen atoms white, and oxygen atoms red.

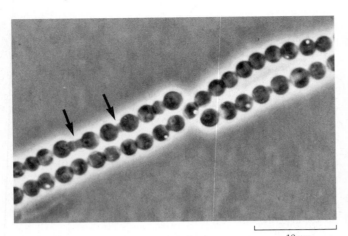

Figure 22.21 *Isosphaera pallida*

Isosphaera pallida is a filamentous planctomycete found in hot springs. Note the buds (arrows) and the gas vacuoles (bright areas) in the cells. Courtesy of Steve Giovannoni.

10 µm

that massive N_2 losses from anoxic marine waters may be effected via the anammox bacteria.

ISOSPHAERA *Isosphaera pallida* is quite different from the other genera of this order. It occurs in alkaline hot springs, where it grows as a filamentous gliding bacterium (Figure 22.21). It grows at temperatures between 35°C and 55°C. The organism is also gas vacuolate.

Phylum Chlamydiae

The genus *Chlamydia* is the best known genus in this phylum. *Chlamydia* are small gram-negative rods (0.2 µm in diameter and 0.4 µm in length) that are obligate intracellular parasites of animals, including humans. Like the planctomyces group, they lack peptidoglycan in their cell walls. Nonetheless, for some unknown reason, they are susceptible to penicillin and other peptidoglycan-acting antibiotics, and their genomes contain the genes for the enzymes necessary for peptidoglycan synthesis. Perhaps peptidoglycan serves some purpose in these bacteria other than conferring cell wall strength. Analyses of 16S rRNA gene sequences confirm that the *Verrucomicrobia* and *Planctomycetes* phyla are their closest relatives. These three phyla are considered to comprise a superphylum.

Chlamydia trachomatis causes a severe conjunctivitis called trachoma, an eye disease of humans and other animals, and also causes one of the most common sexually transmitted diseases of humans, non-gonococcal urethritis (NGU) as well as lymphogranuloma venereum (see Chapter 28).

Chlamydophila psittaci causes diseases of parrots and other birds as well as psittacosis in humans who become exposed to avian fecal material. *Chlamydophila pneumoniae* has not only been associated with pneumonia, but

has been implicated in cardiovascular disease (atherosclerosis) as well.

The chlamydias closely resemble the rickettsias in size and dependence on hosts. Their genome size is also small, 4 to 6×10^8 Da. They have been termed **energy parasites** because they cannot produce their own ATP but instead rely on the host tissues for this function, even though their genome sequences indicate that they have the genetic capacity for ATP synthesis with proton pumping and ATP synthase. They are cultivated on the yolk sac of eggs or in tissue culture.

Chlamydia species have a complex life cycle (Chapter 28). They are transmitted to the host epithelial tissues (in the eye or genitourinary tract) at a nonmultiplying dessication-resistant stage termed an **elementary body**. When they come into contact with host cells, they induce phagocytosis and are taken into the cell. They reside in the phagosome, where they are able to counter the host defense mechanism by interfering with the normal lytic activity of the lysosomes. While in the phagosome, they grow as vegetative, non-infectious cells, called **reticulate bodies**, and ultimately produce the infectious elementary body cysts that are released into the environment when the host cells disintegrate.

Phylum Verrucomicrobia

Members of the *Verrucomicrobia* are estimated to comprise between 1.2% and 10.9% of the total bacteria in soil and represent about 5% of all 16S rRNA genes from environmental surveys. The phylum is one of the most widely distributed, having been reported in soil and aquatic habitats and from the termite hindgut. However, only about a dozen species have been isolated in pure culture. Unfortunately, until strains are obtained from all six of the major subgroups of the phylum, it will not be possible to understand the complete phenotypic and genetic diversity of this group, because pure cultures are necessary for physiological studies.

Although the phylum *Verrucomicrobia* is closely related to the *Planctomycetes* and the *Chlamydiae*, they differ from both in that they produce cell wall peptidoglycan.

Prosthecobacter and *Verrucomicrobium* are two genera of heterotrophic, aquatic bacteria. Species of both of these genera have prosthecae. *Prosthecobacter* (Figure 22.22) attaches by a holdfast at the tip of its prostheca to algae and sheathed bacteria. Its life cycle resembles that of *Caulobacter* sp. (see Chapter 19) except it has no motile stage. In contrast, *Verrucomicrobium spinosum* produces several prosthecae that extend from its cell surface (Figure 22.23) and in this respect, like *Caulobacter*, resembles another genus of the *Alphaproteobacteria*, *Prosthecomicrobium* (see Chapter 19).

Recent studies have indicated that *Prosthecobacter* species contain two genes for tubulin synthesis. This is

Figure 22.22 *Prosthecobacter*
Electron micrograph of *Prosthecobacter fusiformis*, showing several cells in the process of division. Courtesy of J. T. Staley.

remarkable because, until now, no prokaryotic organisms had been reported to contain tubulin genes. Alpha- and beta-tubulin gene homologs have been found in all four of the *Prosthecobacter* species (Figure 22.24). All eukaryotes have tubulin genes coding for the subunits

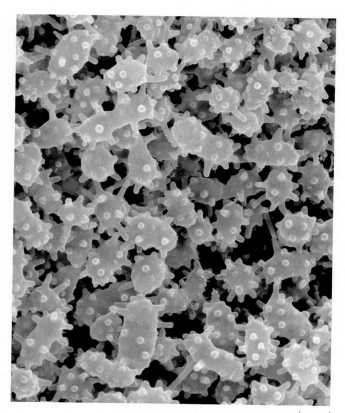

1 µm

Figure 22.23 *Verrucomicrobium spinosum*
Note the numerous prosthecae that extend from the cell of *Verrucomicrobium spinosum*. In this respect it resembles *Prosthecomicrobium* sp. ©Dennis Kunkel Microscopy, Inc.

of microtubules that are responsible for movement of cell organelles, chromosomes, and flagella. Tubulin genes are found in all eukaryotic organisms. Although evidence has been reported that they are being expressed in *Prosthecobacter* sp., their role in these bacteria remains unclear. These bacterial tubulins, termed BtubA and BtubB for bacterial tubulin A and B, are homologs for alpha- and beta-tubulin, respectively. The tubulin genes have been cloned into *E. coli*, and the expressed proteins form protofilaments analogous to microtubules in the presence of the GTP (guanosine triphosphate), an energy source for microtubule synthesis.

The origin of these tubulin genes is uncertain. They may have been derived from a eukaryotic organism by horizontal gene transfer. However, if this occurred, it happened tens to thousands of millions of years ago, since they are not closely related to the genes of any eukaryotic group.

Another interesting member of this phylum may also make microtubules. This is a bacterium that is a symbiont of a marine ciliate protist. The bacterium, referred to as the epixenosome ("surface stranger"), grows attached to the outer layer of the protist (Figure 22.25A). When the protist is threatened by another protist, the bacterium produces a proteinaceous "harpoon" that is shot from its cells outward toward the invader (Figure 22.25B and Figure 22.26). This remarkable activity appears to be novel in the microbial world. As yet, tubulin genes for this bacterial symbiont have not been identified. However, reactivity to anti–eukaryote tubulin antibody has been reported for the microtubule-like tubules of the verrucomicrobial epixenosomes.

Other important genera in this group include the anaerobic genus *Opitutus*, which has been cultivated from rice paddies, and *Chthonobacter*, which is a genus of aerobic soil organisms.

Phylum Chrysiogenetes

This phylum contains only a single genus and species at this time, *Chrysiogenes arsenatis*, an anaerobic, motile bent rod that grows at mesophilic temperatures. Acetate is used as an energy source and is oxidized to carbon dioxide using arsenate as the electron acceptor. Arsenate is reduced to arsenite in the process. This bacterium was isolated from arsenic-contaminated muds of a wastewater purification system associated with a commercial gold-mining facility in Australia.

Phylum Acidobacteria

The *Acidobacteria* are gram-negative bacteria commonly found in 16S rRNA gene libraries from DNA extracted from natural samples. The original culture was obtained from an acidic mineral soil. However, clonal libraries

Figure 22.24 Tubulin tree
Phylogenetic tree showing relatedness between α-
and β-tubulins of representative eukaryotes and
the BtubB and BtubA of *Prosthecobacter*. Courtesy of
C. Jenkins and J. T. Staley.

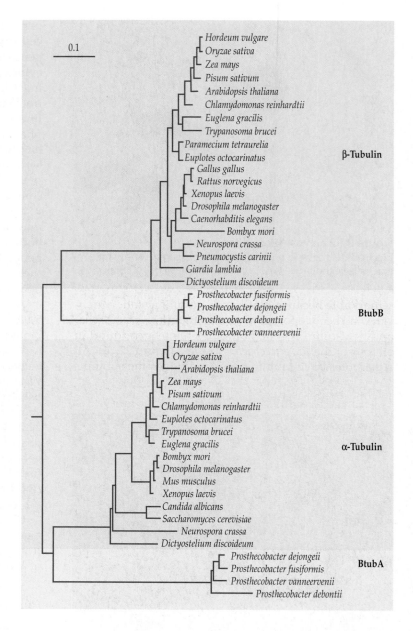

indicate that these bacteria are found in many other soil habitats including forests, deserts, grasslands, and agricultural soils. Yet, to date, only one species, *Acidobacterium capsulatus*, has been obtained in pure culture. This species of rod-shaped bacteria is orange pigmented. Cells contain menaquinones as their sole membrane quinones.

Phylum Fusobacteria

The genus *Fusobacterium* contains unicellular anaerobic organisms that are rod-shaped and sometimes spindle-shaped cells. They are nonmotile and fermentative. The major fermentation end product is butyrate along with

acetate, lactate, and smaller amounts of propionate, succinate, and formate. These bacteria live in the gastrointestinal and genital tracts of animals. *Fusobacterium nucleatum* is an oral commensal bacterium. It lives in consortia with other bacteria, such as *Porphyromonas gingivalis*, which plays a role in periodontal disease. Its 2.17-Mb genome has been sequenced. Interestingly, it can synthesize only three amino acids.

This phylum contains several other genera that are fermentative, such as *Propionigenium* and *Leptotrichia*, which produce propionic acid and lactic acid, respectively, as major fermentation end products. These bacteria have been found in the oral cavities of animals.

(A)

Cells of *X. brevicoli*

Figure 22.25 Verrucomicrobial protist symbiont
(A) Electron micrograph showing cells of the epixenosome member of the *Verrucomicrobia* attached to the pole of its host protist, *Euplotidium arenarium*. (B) Epixenosome remnant and the harpoon-like structure that has been extruded. Courtesy of G. Petroni and G. Rosati.

(B)

Epixenosome
remnant

Harpoon

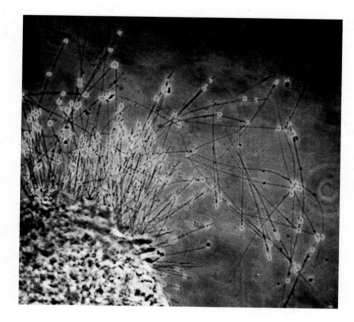

Figure 22.26 Extruded "harpoons"
Electron micrograph of host protist, showing extruded "harpoons" from epixenosome symbionts. DNA from the bacterial cell is found at the apical tip of the harpoon. Courtesy of G. Petroni and G. Rosati.

SECTION HIGHLIGHTS

Several unusual phyla are described in this section. Two of them, the *Spirochaetes* and *Chlamydiae*, contain important animal pathogens. Some contain atypical bacteria that do not produce peptidoglycan or do not contain FtsZ. *Prosthecobactor* species contain homologs for alpha- and beta-tubulin, previously reported only in eukaryotes. Many species are abundant and widely distributed in soil and aquatic habitats.

SUMMARY

- The deepest phylogenetic branches of the *Bacteria* contain thermophiles, including *Thermodesulfobacterium*, *Thermotoga*, and *Aquifex*—some of which grow at temperatures as high as 95°C. Cells of the *Thermotoga* group (*Thermotoga* and *Thermosipho*) produce a characteristic external sheath called a **toga**.

- Some species of the genera *Aquifex, Thermodesulfobacterium*, and *Hydrogenobacter* grow as **aerobic, obligate hydrogen autotrophs** that fix carbon dioxide by the reductive carboxylic acid cycle.

- The *Bacteroidetes* phylum contains aerobic gliding bacteria and anaerobic fermenters, respectively. *Cytophaga* and *Flavobacterium* are aerobic or facultatively aerobic—most species are noted for **degradation** of **polymeric substances**. They are frequently pigmented yellow to orange, and many contain, in addition to carotenoids, special pigments called flexirubins—some of which may be chlorinated. *Bacteroides* species are common **fermentative bacteria** that are found in the digestive tracts of animals, where they occur in very high concentrations.

- The *Spirochaetes* comprise their own phylogenetic group of **helical bacteria** with unique **periplasmic flagella**, forming a structure called an **axial filament**, that imparts a characteristic motility that is especially well adapted to highly viscous environments. Several spirochetes are **pathogenic**, including *Treponema pallidum* (**syphilis**), *Borrelia burgdorferi* (**Lyme disease**), and *Leptospira* (**leptospirosis**).

- The *Planctomycetales* comprise a separate phylogenetic group containing **budding bacteria** that are widespread in **aquatic habitats** and soils. *Planctomyces* species have a multifibrillar stalk with a holdfast at its tip. *Isosphaera* is a gliding planctomycete found in alkaline hot springs at 35°C to 55°C. *Chlamydiae*, which are the closest relatives of the *Planctomycetes*, are obligate **intracellular parasites**. The *Verrucomicrobia* are unique among prokaryotes in that at least some species carry homologs for **alpha-** and **beta-tubulin genes**.

- Many additional phyla of the *Bacteria* are known from clonal libraries based on DNA extracted from natural samples. Some of these have cultured representatives, but much more remains to be learned about the extent and true diversity of the *Bacteria*.

 Find more at www.sinauer.com/microbial-life

REVIEW QUESTIONS

1. Why do microbiologists know so little about some of the major phyla of living organisms?

2. Do you believe that thermophilic *Bacteria* that branch deeply in the Tree of Life are among the earliest phyla of *Bacteria* to evolve? What reasons can you give? What methods are used to analyze such depth of branching?

3. The phylum *Verrucomicrobia* contains some prosthecate bacteria. What other phyla contain them and how would you explain the evolution of this feature?

4. Why is it advantageous for bacteria that degrade particulate organic materials, like cellulose in plant tissues and chitin in the exoskeleton of insects and other animals, to have gliding motility?

5. Bioprospecting refers to the search for novel biological materials that might have commercial significance, such as antibiotics from *Actinobacteria*. What example(s) can you cite from the phyla in this chapter?

6. Explain why the discovery of tubulin genes in *Bacteria* might be important in understanding the evolution of eukaryotic organisms.

7. Compare the motility of spirochetes to that of flagellated and gliding bacteria.

8. How do you explain the occurrence of nitrifying bacteria in the *Nitrospirae* and in the *Proteobacteria*?

9. The antibiotic ampicillin is commonly added to media to selectively isolate *Planctomyetes* strains from enrichment cultures. Why is this so effective?

SUGGESTED READING

Balows, A., H. G. Trüper, M. Dworkin, W. Harder and K. H. Schleifer eds. 1992. *The Prokaryotes*. 2nd ed. Berlin: Springer-Verlag.

Boone, D. R., R. W. Castenholz and G. M. Garrity, eds. 2001. *Bergey's Manual of Systematic Bacteriology*. 2nd ed., Vol. I. New York: Springer-Verlag.

Krieg, N. R., J. T. Staley, B. Hedlund, B. Paster, N. Ward and W. Whitman, eds. In press. *Bergey's Manual of Systematic Bacteriology*. 2nd ed., Vol. IV. New York: Springer-Verlag.

Holt, J. G., editor-in-chief. 1994. *Bergey's Manual of Determinative Bacteriology*, 9th ed. Baltimore, MD: Williams and Wilkins.

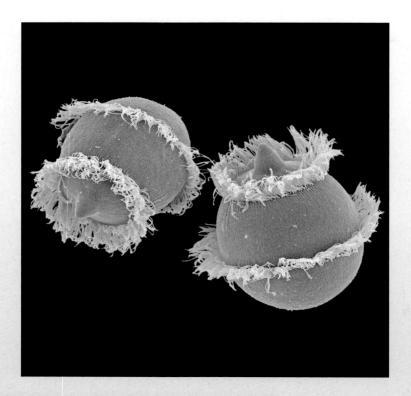

The objectives of this chapter are to:

◆ To describe the roles and diversity of organelles found in protists.

◆ To introduce the Opisthokonta as including protists that are at the base of the fungi and animals, and to introduce the diversity of fungi.

◆ To introduce the Excavata as an example of a diverse group with species that are heterotrophic or phototrophic, flagellated or amoeboid or both, parasitic or symbiotic or free-living.

◆ To introduce the Rhizaria as a diverse group that have few common morphological features besides fine pseudopodia.

◆ To introduce the Archaeplastida as grouping the red algae and the green algae, including the ancestors of higher plants.

◆ To introduce Chromalveolata, whose origins are believed to be from endosymbiosis between a protist-feeding cell and a red algal photosynthetic ancestor. This large and diverse group includes some of the algae of greatest ecological importance as well as major parasitic lineages.

◆ To introduce the Amoebozoa as a group that contains amoeboid species, most of which have broad pseudopodia.

23

Eukaryotic Microorganisms

And in the summer, when I feel disposed to look at all manner of little animals, I just take the water that has been standing a few days in the leaden gutter up on my roof, or the water out of stagnant shallow ditches: and in this I discover marvellous creatures.
—From a letter of Anthony van Leeuwenhoek to Constatijn Huygens, Dec. 26, 1678

Of the three domains of life, we have so far covered the *Bacteria* and the *Archaea*, which are prokaryotes. The third domain consists of the eukaryotes, which are distinguished from the prokaryotes by several morphological and biochemical features. This chapter describes the microbial eukaryotes, generally referred to as protists.

The main distinctions between prokaryotes and eukaryotes are described in Chapter 1. They are summarized briefly here. The words "prokaryote" and "eukaryote" refer to the absence of a nucleus in the prokaryotes and the presence of a true nucleus in the eukaryotes. Some of the other distinguishing morphological features of eukaryotes are the endomembrane network in the cytoplasm; the cytoskeleton, which provides structural support and locomotion; and the mitochondria. Many protists have plastids for photosynthesis, albeit different types of plastids are found in different protists. Other organelles—such as flagella, extrusomes, hydrogenosomes, mineral skeletons, and external scales—are common. Besides the morphological features, eukaryotes have biochemical differences from prokaryotes. These include differences in the structure of chromosomes and ribosomes, RNA transcription, DNA repair, glycoprotein synthesis, the lipid composition of membranes, and the types of polysaccharides synthesized.

Figure 23.1 Eukaryotic cell
Diagram showing a generalized eukaryotic cell to illustrate various organelles.

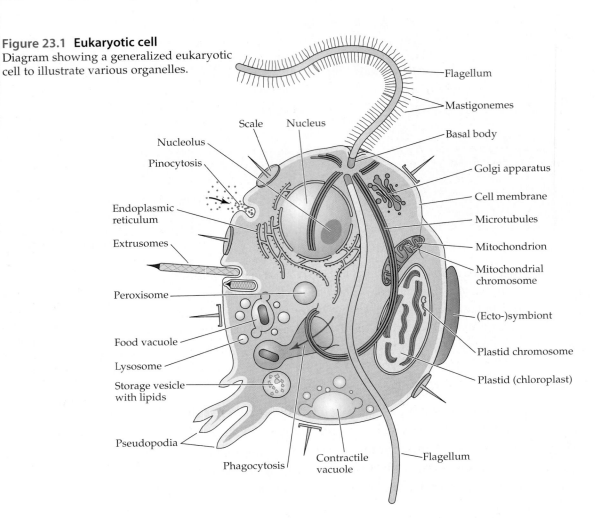

23.1 Eukaryotic Organelles

The diversity of organelles found in protists provides useful morphological markers to identify different groups (**Figure 23.1**). There are variations in the ultrastructure of nuclei, mitochondria, flagella, plastids, cytoskeleton components, and other organelles such as extrusomes and mineral elements. However, on their own, morphological characters are insufficient to elucidate the evolutionary relatedness among groups. Comparisons of gene sequences, especially of rRNA genes, have been very useful in explaining phylogenetic relationships among protists (see Chapter 17).

The Cell Membrane and Cortex

The cell membrane of eukaryotes is similar to that of prokaryotes in that it consists of a lipid bilayer and proteins that provide a semipermeable barrier to the environment, as described by the fluid mosaic model (see Chapter 4). There are several important differences, however. Eukaryote cell membranes contain **sterols** among the phospholipids, which increase the rigidity of the membrane. Typical sterols include cholesterol, found in animals; stigmasterol, found in plants; ergosterol, found in fungi; and tetrahymenol, found in ciliates. Phospholipids are diverse but include phosphatidylserine, phosphatidylcholine, and sphingomyelin. The exact lipid composition of the cell membrane is adjusted with temperature fluctuations, and it is different for different groups of eukaryotes. For example, with increasing temperature, some protists incorporate more saturated lipids (lacking double bonds) in their membranes. These lipids pack closer together than those with double bonds and thus increase the stability of the membrane against thermal disruption by the increased heat.

The cortex consists of the cell membrane with additional components connected to it on the cytoplasmic side and the outside. Those on the outside may include one or more protective elements. These may be thick or strong, such as mineral or organic scales, a mineral or organic vase-shaped enclosure called a test or lorica, and cell walls made of polysaccharides such as cellulose or chitin. Alternatively, the outside may be coated with mucus or glycoproteins. Mucus consists of glycoproteins and polysaccharides secreted by the cell. Amoeboid cells need flexibility to change shape and tend to have a covering consisting only of glycoproteins. The inside of the cell membrane is usually covered with filaments that link membrane proteins to the cytoskeleton. In some cells, such as those of the Ciliophora, the cytoskeleton beneath the cell membrane can be extensive. In other cells, such as the Euglenida, proteinaceous strips support the membrane. A variety of combinations are observed in the cortical areas of different protists, as we shall see later.

The Endomembrane Network

The endomembrane network is visualized in transmission electron micrographs and consists of the membranes and vesicles in the cytoplasm (**Figure 23.2**). The membranes consist of a single lipid bilayer that can bud into spherical vesicles. The endomembrane network is a dynamic network without fixed form. It is responsible for coordinating the formation of vesicles in the cytoplasm and secretion out of the cell, the synthesis of glycoproteins and lipids, and the compartmentalization of different cellular processes; it also forms the nuclear envelope. For example, the endomembrane network separates genetic transcription and nucleic acid–related processes in the nucleus from protein synthesis, organelle synthesis, and the secretion related processes in the cytoplasm. The endomembrane network provides a variety of different membrane-bound compartments where these various processes occur. Eukaryotic cells tend to be much larger than prokaryotic cells. The separation of different cellular processes into specialized compartments in eukaryotic cells increases their efficiency as larger cells (see Figure 23.2).

SECRETION   The endoplasmic reticulum (ER) and Golgi apparatus are part of the protein synthesis and secretory network. These membranous organelles are flattened sac-like structures enclosing an internal space called a lumen or cisterna. Ribosomes are the site of protein synthesis, and some are found lining part of the cytosolic surface of the endoplasmic reticulum. The ER associated with ribosomes is called the rough ER. The

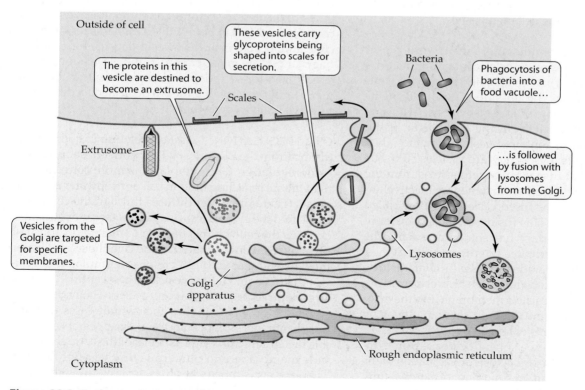

Figure 23.2 Endomembranes and secretion
The rough endoplasmic reticulum and Golgi membranes synthesize glycoproteins and enzymes, which are packaged into vesicles destined for different membranes and compartments.

proteins being synthesized on the ribosomes are passed into the lumen of the ER. Inside the ER lumen, the proteins are modified by the addition of sugar groups, which help fold proteins into their functional shape. The proteins are processed further in the internal spaces of the Golgi network, where additional sugar moieties are added to the proteins, yielding mature glycoproteins. Other processes in the Golgi include the assembly of different ER-modified proteins into extrusomes, scales, or small secretory vesicles.

Secretory vesicles from the Golgi are destined to fuse with the cell membrane in order to release the contents to the outside. The contents of the vesicles can be digestive enzymes serving to dissolve substrates outside of the cell, as in saprotrophs. In some species, the vesicles may contain mucus released to the cortex. In other species, the vesicles may contain cell wall material, such as the siliceous vesicles of Bacillariophyta. In yet other species, surface scales for the cortex are synthesized and shaped in the Golgi before being secreted to the surface.

Not all the ER is associated with ribosomes. Some of the ER is free of ribosomes and called smooth ER, the site of lipid synthesis. Newly synthesized lipids are released as numerous tiny membrane vesicles (50 to 500 nm) into the cytoplasm. In Fungi, for example, the growth and extension of cells requires fusion of these vesicles with the existing cell membrane.

ENDOCYTOSIS   The endomembrane network is also responsible for processing vesicles containing material from outside the cell; it can invaginate into a vesicle containing solution from the outside. These cell membrane–derived vesicles are processed by fusing with specific Golgi-derived vesicles in the cytoplasm. If the Golgi-derived vesicles contain digestive enzymes, they are called **lysosomes**. Lysosomes are vesicles 50 to 500 nm in diameter that fuse with target vesicles from the cell membrane; their digestive enzymes are inactive until this fusion occurs. The fused vesicles are now called a secondary lysosome. This fusion initiates transfer of protons into the secondary lysosome, increasing the acidity. The acid pH activates the digestive enzymes, which solubilize and degrade the vesicle's contents. The soluble molecules produced by digestive enzymes are transferred across the vesicle's membrane into the cytoplasm for cell metabolism. This is the main pathway by which dissolved nutrients enter cells.

There are two main types of endocytosis: **pinocytosis** and **phagocytosis**. In pinocytosis, invagination of the cell membrane results in a small amount of external fluid being internalized in 50- to 500-nm vesicles. In many cells, this is a continuous process that helps bring in dissolved nutrients as well as recycling some of the cell membrane. In addition, a different kind of pinocytosis

is found in many protists. It is called receptor-mediated endocytosis and occurs specifically at coated pits, which are membrane pockets coated with various receptors on the external side and clathrin proteins on the cytoplasmic side. The receptors bind scarce molecules such as hormones and signal molecules, essential nutrients, or minerals such as iron. The invaginated vesicles are enriched in the molecules bound to the receptors. The clathrin coat on the cytoplasmic side is required for invagination and vesicle formation and is released once the vesicle has been formed.

In phagocytosis, cell membrane invagination engulfs larger particles such as organic material, bacteria, or protists. These larger membrane-bound structures are called food vacuoles or **phagosomes**. Lysosomes fuse with phagosomes to digest their contents. Vacuoles containing undigested food material from different lysosomes fuse into a **residual body**. Protists without a cell wall are able to excrete the contents of the residual body from the cell. This occurs by fusion with the cell membrane, releasing the material outside. In some ciliates, this excretion occurs at a specific location called the **cytoproct**.

SITE OF SECRETION AND ENDOCYTOSIS   The cell cortex can be a barrier to secretion or endocytosis, especially if a cell wall is present. Many cells have specialized regions dedicated to secretion or endocytosis where the cortex is modified to permit this. Often this region is discrete and located near the base of flagella; for example, flagellar pits are invaginations of the plasma membrane specialized for secretion or endocytosis; they occur in many protist groups. If there is a dedicated feeding structure, it is called a **cytostome**.

OSMOREGULATION   A set of membranes and vesicles distinct from the ER and Golgi network is responsible for osmoregulation. The cytoplasm is more concentrated than the surrounding fluid in soil or freshwater environments. Therefore water diffuses through the cell membrane into the cytoplasm. This dilutes the cytoplasm and expands the cell volume. In order to maintain a constant physiological environment in the cell, the excess water must be eliminated. It is therefore accumulated in vesicles that fuse into larger vesicles and eventually form a large **contractile vacuole** (Figure 23.3). The water in the vacuole is eliminated when the vacuole fuses with the cell membrane. Its appearance and disappearance are visible by light microscopy. Dispersal of the contractile vacuole membrane into numerous tiny vesicles follows. Then the cycle repeats itself as vesicles fill with excess water and fuse again. In a small number of species, the collecting vesicles do not fully disperse at each cycle.

In marine environments the solute concentration is higher outside of the cell; therefore, water will tend to

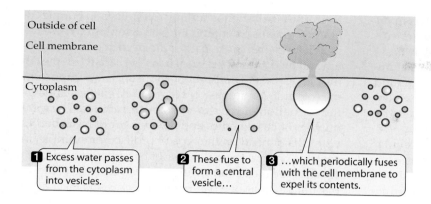

Figure 23.3 Contractile vacuoles
Contractile vacuoles expel excess water from the cell. There is a regular cycle of filling and emptying that varies with osmotic potential.

Outside of cell

Cell membrane

Cytoplasm

1 Excess water passes from the cytoplasm into vesicles.

2 These fuse to form a central vesicle…

3 …which periodically fuses with the cell membrane to expel its contents.

diffuse out of the cytoplasm. In this situation, cells increase the cytoplasm's osmotic potential by accumulating certain ions or producing soluble organic molecules, called **osmotic balancers**, that do not interfere with metabolism. Another situation where osmotic balancers are needed is in drying soil—as water evaporates or in preparation for freezing. Osmotic balancers include the carbohydrate trehalose and amino acids or derivatives such as proline and glycine-betaine.

The Cytoskeleton

The cytoskeleton is perhaps the component of protists most responsible for their morphological and functional diversity. It is visualized in micrographs obtained with the transmission electron microscope as a variety of arrays of microtubules, fibrous filaments, actin microfilaments, and fine filaments that support the cell membrane, crisscross the cytoplasm, anchor the cilium (also known as the eukaryotic flagellum), and support cytoplasmic extensions. The cytoskeleton is responsible for the locomotion of eukaryotic cells, which move by swimming with flagella (or cilia), gliding on surfaces with flagella, and amoeboid crawling or wriggle motions. The cytoskeleton is also responsible for many of the cell projections used for feeding and food capture. The cytoskeleton is also a key component of eukaryotic cell division. The microtubules form a mitotic spindle that separates the chromosomes into sister nuclei, and the cytoskeleton is involved in cell fission.

MICROTUBULES   These are observed in transmission electron micrographs as hollow cylinders 25 nm in diameter. The wall of the cylinder is composed of two types of globular proteins called α- and β-tubulin. The α- and β-tubulin monomers are not stable in isolation but exist as α-β heterodimers. Microtubules—which are assembled from α-β heterodimer subunits to form a long cylinder—are found singly, in loose bundles, or in tight organized patterns. The microtubules play a role in (1)

supporting the cell membrane and cell shape, (2) directional transport of vesicles and organelles in the cytoplasm, (3) supporting extensions of the cell, (4) cell locomotion, and (5) separating chromosomes at cell division.

Microtubules under the cell membrane, in the cortex, are responsible for supporting cell shape and providing additional rigidity to the cell (Figure 23.4). They often extend from the basal bodies at the base of flagella and spiral around the cell along the cortex. Some may extend from the flagella toward the nucleus or into the cytoplasm in fixed patterns of bundles of microtubules, such as the axostyle of Parabasalia. These microtubules with a supporting or structural role are stable and are part of the permanent cytoskeleton. In contrast, the microtubules involved in vesicle traffic and the transport of organelles are part of the dynamic cytoskeleton. They are partly responsible for the streaming of organelles in the cytoplasm, visible by light microscopy. These elements form and disassemble as required to direct traffic in one direction or another. Transport along microtubules requires motor proteins called dynein and ki-

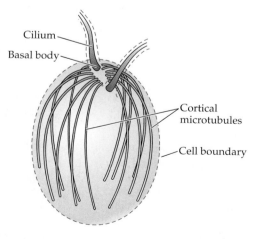

Figure 23.4 Microtubules under the cell membrane
In many protists, cortical microtubules run from the basal bodies along the cell membrane, providing support.

Cilium

Basal body

Cortical microtubules

Cell boundary

nesin. Kinesin has the ability to attach to proteins in organelle or vesicle membranes and moves along microtubules, carrying the vesicle or organelle with it. Kinesin is also involved in the movement of microtubules at mitosis. Dyneins attach to vesicle membrane proteins and are important in vesicle traffic. Dyneins are also responsible for the undulating beat of eukaryotic flagella by causing microtubules to slide against each other.

Organized patterns of microtubules occur in cellular extensions such as the cilium, the haptonema of Haptophyta, or the axopodia of Radiolaria. The microtubules in these extensions provide structural support or retraction by microtubule disassembly, enable vesicle transport into and out of the extension, and facilitate bending by using motor proteins.

CILIA OR EUKARYOTIC FLAGELLA   Many protists use flexible extensions called cilia (sing., cilium) for locomotion; these are also referred to as eukaryotic flagella. The two terms are equivalent and may be used interchangeably. The particular fixed arrangement of microtubules in the cilium is called the **axoneme** (Figure 23.5), and consists of nine pairs of microtubule doublets arranged in a circle around one pair in the middle. This is called the 9 + 2 pattern. The base of each cilium has a different structure called a kinetosome or **basal body**, which is similar in structure to the centrioles. It consists of nine triplets of microtubules arranged in a circle, called the 9 + 0 pattern. Basal bodies initiate the assembly of cilia and of the stable microtubular arrays. The transition zone between the basal body and the axoneme is called a basal plate. Basal bodies also serve as anchors for other elements of the cytoskeleton that hold the cilium in place. Details of the arrangement of these cytoskeletal elements about the basal body are useful characteristics in identifying protists.

Cilia are enclosed by the cell membrane, supported by the 9 + 2 axoneme. The basal body–associated cytoskeleton consists of various short microtubules and bundles of fibrous proteins extending into the cytoplasm. These cytoskeletal elements are part of the permanent cytoskeleton and help to hold the basal body in place when the cilium is beating. The basal body itself is an important microtubule organizing center (MTOC) which initiates the formation of the axoneme microtubules. The total number of basal bodies and the number of basal bodies with cilia depend on the type of protist. For example, in chytrids (Fungi) there is a pair of basal bodies, only one of which bears a cilium; but in *Chlamydomonas* (Chloroplastida) there is one pair of basal bodies and each has a cilium. The bending of the cilium is caused by dynein motors moving along adjacent microtubules in the axoneme, resulting in a sliding motion of the microtubules. The axoneme bends because the sliding microtubules are not free but held in place by the basal body at the basal plate. Repeated cycles of bending and relaxing cause the cilium to beat. The process is ATP- and calcium-dependent and can generate about 50 beats per second. The frequency and direction of the beat can be modulated by the basal body and axonemal physiology.

In some groups, cilia have one of a number of accessory appendages. For example, many Stramenopiles have hair-like structures attached to the ciliary membrane. These hairs are called **mastigonemes**. Different types of mastigonemes are encountered in other protists, some with branching fibrous proteins or with several parts. In some groups, the cilium is also supported by an internal rod of proteins, called the **paraxonemal rod**, adjacent to the axoneme. In some protists the cell membrane is connected to the ciliary membrane along part of its length, resulting in a thin connection; this is called an **undulating membrane**. Some protists have a layer of fine scales covering the outside of the cilium.

Numerous thin fibrous proteins link different elements of the cytoskeleton. Called **fine filaments**, they are found in the cortex, under the cell membrane, and

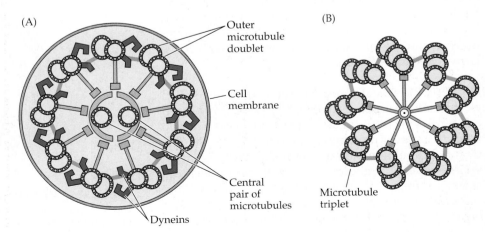

Figure 23.5 Cilium or "eukaryotic flagellum" and basal body
(A) Diagram of a cross section of a cilium showing the characteristic 9 + 2 arrangement of the axoneme. (B) Diagram of a cross section through a basal body, showing the characteristic 9 + 0 arrangement of microtubule triplets, as also found in centrioles.

(A)

Outer microtubule doublet

Cell membrane

Central pair of microtubules

Dyneins

(B)

Microtubule triplet

in complex cytoskeletal structures of cells. Some contribute to cell contractions. Others regulate the assembly and disassembly of complex cytoskeletal structures or of the stable microtubular network. There are many other fibrous cytoskeletal elements in protists that can be visualized by electron microscopy. Most have not been studied but are found in the basal body–associated cytoskeleton and in other complex structures of the stable cytoskeleton. **Microfilaments** are 4- to 7-nm filaments consisting of actin assembled as two parallel filaments wound together in a helix. They often assemble into bundles and form contractile filaments associated with other elements of the cytoskeleton. They are particularly important in the amoeboid locomotion of some protists. They are also responsible for cytoplasmic streaming in slime molds and have been implicated in the formation of food vacuoles in some protists.

MICROTUBULE ORGANIZING CENTERS   The microtubule organizing centers (MTOCs) are responsible for initiating the polymerization of microtubules. Two main types of MTOCs are found in protists. Their common feature is probably the presence of γ-tubulin, required for initiating microtubular assembly. The first kind are basal bodies and centrioles. Except in species without cilia, basal bodies are the most common MTOC in protists. Centrioles consist of two orthogonal cylinders of microtubules, each with the 9+0 pattern otherwise characteristic of basal bodies. The other kind of MTOC is seen as an amorphous region near the nucleus, as in many Radiolaria and Fungi, where it forms a spindle plaque at mitosis. It can be located on electron micrographs by tracing the origin of microtubules.

AMOEBOID LOCOMOTION   Many protists have a flexible cortex that permits changes in shape. A temporary deformation or extension of the cytoplasm is called a pseudopodium and it is used both in feeding and locomotion. Amoeboid locomotion has arisen independently many times throughout protist lineages. Thus, details of force generation for the formation of pseudopodia vary. In some protists, the tip of the extending cytoplasm is clear of organelles and called hyaline. Extension of the cytoplasm involves the contraction of a network of actin microfilaments and myosin below the cell membrane. The network contracts, generating a force that causes the cytoplasm in the center of the cell to flow forward, somewhat like squeezing toothpaste from a tube. Our understanding of amoeboid locomotion is incomplete and largely based on observations in the Mycetozoa slime molds *Physarum* and *Dictyostelium*, and the Amoebozoa *Amoeba proteus*.

Pseudopodia can be used to engulf large prey, such as another protist, by phagocytosis. Extending pseudopo-

dia wrap around a prey, which is then ingested into a food vacuole. Most phagocytosis, however, involves smaller membrane invaginations that occur along most amoeboid surfaces or at the cytostome of nonamoeboid protists. Phagocytosis involves contraction of a network of actin microfilaments and myosin to invaginate the food vacuole. The microfilaments attach to the cell membrane in the cortex.

FILOPODIA AND AXOPODIA   Another type of pseudopodial extension is called a filopodium—a long, fine extension of the cytoplasm less than 1 μm in diameter that is supported by actin microfilaments. Filopodia are common in the Rhizaria, especially those that live in soils and sediments. They are used to extend into fine pores and crevices to find bacteria for phagocytosis. Similarly, axopodia are used by Radiolaria and certain other amoebas to capture prey by phagocytosis (Figure 23.6). Axopodia are supported by microtubules, which are

(A)

(B)

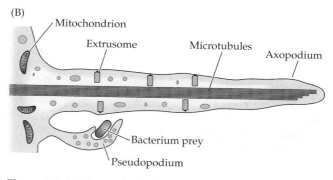

Figure 23.6 Microtubules in axopodia
(A) Some amoeboid organisms use axopodia to capture prey. Organisms that bump into the axopodia stick to them and are ingested by phagocytosis. (B) Diagram showing the bundle of microtubules supporting an axopodium. A, ©J. Solliday/Biological Photo Service.

arranged in different patterns according to taxonomic group.

The Nucleus

This is the defining organelle of eukaryotes and is present in all species. It consists of a **nuclear envelope** surrounding eukaryotic chromosomes (see Figure 23.1). The nuclear envelope appears as two layers of membrane in electron micrographs. The nuclear envelope is derived from the endomembrane network and is connected to the ER. The outer layer may be associated with ribosomes, while the inner layer is lined with fibrous cytoskeletal proteins that are related to the lamin and vimentin intermediate filaments found in animals. The space between both membranes is called the perinuclear space. The two membrane layers are continuous with each other, so they are really one membrane 'folded over' on itself. Nuclear pores traverse both layers, forming tunnels or pores that connect the cytoplasm with the nuclear matrix. The pores consist of a complex of proteins that regulate traffic into and out of the nucleus.

Inside the nuclear envelope, the chromosomes are kept isolated from the rest of the cytoplasm. This compartmentalizes metabolic processes and protein synthesis in the cytoplasm from gene transcription and DNA replication processes in the nucleus. The chromosomes are structurally different from prokaryotic chromosomes. Eukaryotic chromosomes are linear, with **telomeres** at each end and a **centromere**. Telomeres consist of short repeated sequences of nucleotides. The exact repeat pattern varies with different protists. The 3' end of the chromosome has an extended single strand of telomere that folds back on itself. This is called a hairpin loop and caps the chromosome, protecting the end from enzymatic degradation or ligation. The enzyme responsible for this is telomerase, which was first discovered in the ciliates *Tetrahymena* and *Oxytricha*. It requires an RNA component and works much like reverse transcriptases. The centromere consists of short repeated DNA sequences that bind a complex of proteins. The assembled protein complex is called a **kinetochore**. At cell division, the kinetochore becomes the site of microtubular attachment on the chromosome. Eukaryotic chromosomes are not supercoiled DNA, as in prokaryotes. Instead, the DNA is tightly packaged and wound around specialized proteins called **histones**.

One region of the nucleus is called the **nucleolus** and is usually visible by light microscopy. It is the site of transcription of DNA coding for ribosomes (rDNA) to ribosomal RNA (rRNA). Chromosomal regions with ribosomal genes aggregate at the nucleolus. The initial assembly of rRNA with ribosomal proteins also occurs in the nucleolus. The eukaryotic ribosome consists of a 40S small subunit and a 60S large subunit, which are assembled separately and exported through the nuclear pores to the cytoplasm. Final assembly of a ribosome from the two subunits occurs in the cytoplasm.

MITOSIS AND CELL DIVISION   Eukaryotic cell division is more complicated than binary transverse fission in prokaryotes (see Chapters 4 and 6). The two main events, nuclear division and cell division, are coordinated. Nuclear division in eukaryotes is called **mitosis** and involves the replication of chromosomes, attachment of chromosomes on a mitotic spindle, separation of the chromosomes by the mitotic spindle, and nuclear division. Replication of chromosomes involves DNA replication followed by condensation of the chromosomes. Chromosome condensation involves a tighter packing of the DNA on the histones. When DNA synthesis along each chromosome is complete, there is one pair of each chromosome side by side (see Chapter 13). At the kinetochore, the chromosomes attach to the **mitotic spindle**, which consists of microtubules extending from one end of the nucleus to the kinetochore of chromosomes. Chromosomes are separated as microtubules slide past each other and are disassembled at the cytoplasmic end. Finally, separated chromosomes at opposite ends of the cell must be repackaged inside a complete nuclear envelope. Many variations of mitosis and nuclear structure are encountered in protists (Figure 23.7). For example, in dinoflagellates, euglenids, many hypermastigotes, and ciliate micronuclei, chromosomes remain condensed all the time. The nuclear envelope in some protists remains intact (closed mitosis), with the mitotic spindle forming inside the nucleus (intranuclear spindle) or outside the nucleus (extranuclear spindle). In some protists, the nuclear envelope remains mostly intact, but the extranuclear spindle penetrates the nuclear envelope through openings. This, called semi-open mitosis, is encountered in many Chloroplastida, some brown algae, and chytrids among other protists. In other protists—for example, in the Amoebozoa, Cryptophyta, and Chrysophyceae—the nuclear envelope disintegrates and nuclear division occurs in the cytoplasm (open mitosis). Last, there is variation in the formation of the mitotic spindle. In some cases it forms from opposite ends of the nucleus and the microtubules extend through the nuclear region. In other cases, the spindle halves form at the same place and separate by moving apart and around the nuclear envelope, as in many Apicomplexa, Foraminifera, and Radiolaria.

Cell division involves the separation of replicated organelles and other cell structures. For example, the basal bodies must be duplicated along with the associated cytoskeleton. The endomembrane network is disrupted at division and must rearrange itself in two new cells. Cells

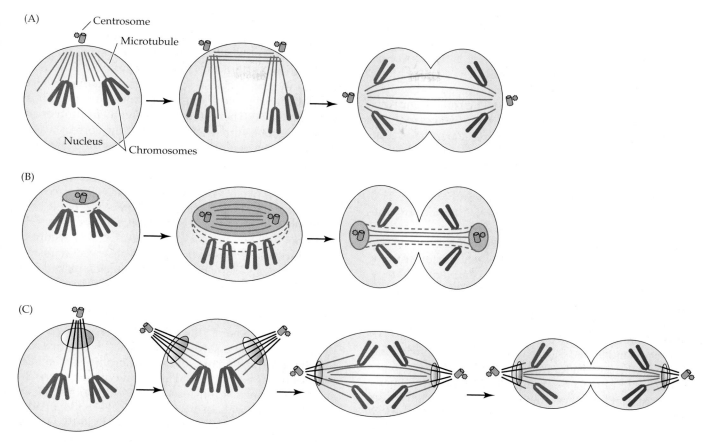

Figure 23.7 Variations of mitosis
Many types of mitosis occur in protists, such as (A) closed intranuclear mitosis,
(B) closed extranuclear mitosis, and (C) semi-open mitosis.

with a cytostome must partially disassemble the structure in order to form two new ones, one for each cell. Many vacuoles and vesicles are separated and moved to opposite poles in the new cells. The process of separation of two new cells from the parent cell is called cytokinesis. The whole process of cellular and nuclear division is called the cell cycle, which is coordinated by a set of specialized regulatory enzymes. They are responsible for the maintenance of cell size and the timing of cell division.

The Mitochondrion

Most eukaryotes have a membrane-bound organelle called a mitochondrion, which is responsible for the production of ATP in aerobic species. Small cells may have only one mitochondrion, whereas larger cells may have hundreds distributed throughout the cytoplasm. The mitochondrion consists of an outer membrane and an inner membrane which folds inward into the mitochondrial matrix. The folds of the inner membrane are called

cristae; there are different types based on their appearance in electron micrographs. The three main types are tubular, flat, and discoid. It is a morphological feature that has been useful in elucidating the relationships among protists. There is a prokaryotic chromosome in the matrix which codes for 3 to 67 protein genes, depending on the species. The presence of this prokaryote DNA in mitochondria has supported the theory that mitochondria are derived from a bacterium that was symbiotic in the cells of early eukaryotes. Other features of the mitochondria resemble bacteria—for example, 70S ribosomes in the matrix and proteins related to bacterial homologues. Genetic sequence analysis has shown that mitochondria are related to the alpha proteobacteria. The origin of the mitochondrion is from a single symbiosis event of an ancestral cell with an alpha proteobacterium. The bacterial chromosome in the mitochondrion is a highly reduced genome, with many of the ancestral genes transferred to the host nucleus or lost. A typical alpha-proteobacterial genome codes for more than a thousand genes, while all mitochondria encode less than a hundred. The bacterial symbiont contributed the tricarboxylic acid cycle (TCA cycle; see Chapter 10) to the host metabolism. The TCA cycle is the source of most of the reduced molecules that drive ATP synthesis in eukaryotes; the mitochondrial cristae and matrix remain the site of these reactions. The mitochondrion uses pyruvate and acetyl Co-A derived from glycolysis in the cytoplasm as substrates for the TCA reactions. The mitochondrial membranes are energy-transducing membranes (see Chapter 8). They hold the enzymes for the electron transport chain that leads to ATP synthesis, using oxygen as the terminal electron acceptor. The chemistry of ATP generation is explained by the chemiosmotic theory (see Chapter 8).

Many anaerobic species and some that tolerate anaerobic environments have lost the mitochondrion entirely or possess modified mitochondria. For example, many intestinal parasites and symbionts no longer have mitochondria because the anaerobic habitat does not provide sufficient oxygen for the electron transport chain reactions. In anaerobic chytrids, Parabasalia such as *Trichomonas*, and anaerobic ciliates, the mitochondrion has been modified into a **hydrogenosome**. Hydrogenosomes usually do not have a chromosome and have lost many of the typical mitochondrial reactions. Instead, they are involved in the anaerobic production of about one ATP molecule per molecule of pyruvate metabolized. Others—such as the intestinal parasite *Giardia*, a diplomonad, and the amoebozoan *Entamoeba*— have a degenerate mitochondrion called a **mitosome**, which has lost the chromosome and does not have the ability to produce ATP, but retains some metabolic functions for the host.

Plastids

Another membrane-bound organelle is found in photosynthetic eukaryotes and is called a plastid. The main functional plastid is the chloroplast, which contains chlorophylls and is responsible for photosynthetic reactions (see Chapter 9). Specialized forms of plastids sometimes lack the usual chlorophyll pigments, and are used for starch storage, oil storage, or phototaxis. In some species, the plastids are also used to compartmentalize some metabolic reactions away from the cytoplasm. These include nucleotide synthesis, amino acid synthesis, and fatty acid synthesis. In protists without plastids, these reactions occur in the cytoplasm. The plastid consists of an outer membrane and an inner membrane surrounding an internal cytoplasm and a prokaryotic chromosome. Photosynthetic plastids also contain membranes called thylakoids, which hold the light-harvesting photosynthetic pigments. The morphology of thylakoids varies among groups. The pyrenoid is the site of carbohydrate (e.g., starch) synthesis in the plastid. As

for the mitochondrion, plastids also have prokaryotic ribosomes, certain features of prokaryotic metabolism, and RNA translation reactions. Genetic analysis of the plastid chromosome indicates that the chloroplast is derived from an ancient symbiosis with a cyanobacterium, which contributed oxygenic photosynthetic reactions to the host (see Chapter 21). Therefore all photosynthetic protists are obligate aerobes that require light of the correct wavelengths to carry out photosynthesis. Over time, many of the plastid genes were transferred to the chromosomes in the nucleus. The glaucophyta, a group of Archaeplastida, retain a plastid that most resembles the ancestral cyanobacterial symbiont; it even retains some of the bacterial cell wall. The initial acquisition of a photosynthetic symbiont is called *primary endosymbiosis*. This occurred in the ancestral Archaeplastida and is observed in the red and green algae. Many protists have lost their plastid and are no longer photosynthetic. This is called secondary loss of the plastid. In these organisms, some of the prokaryotic genes that were transferred from the plastid to the nucleus remain. Other protists have acquired their plastids not from a primary symbiosis event; instead, they have acquired a photosynthetic protist—an archaeplastidan—as a symbiont. This is called a *secondary endosymbiosis*, since the enclosed photosynthetic organism is itself a protist with an internal plastid. Secondary endosymbioses result in plastids with more than two (three or four) bounding membranes. When four membranes are retained, the outer two are the descendants of the host food vacuole membrane and the symbiont cell membrane. Secondary endosymbiosis has happened several times—including in Chromalveolata, Euglenida, and Rhizaria (chlorarachniophytes).

Other Organelles

A variety of other organelles and structures occur in many but not all protists (see Figure 23.1). Some of the more important ones are described here. **Extrusomes** are synthesized in the Golgi and attach beneath the cell membrane. They are large exocytotic vesicles destined to discharge their content outside of the cell. Their role is defensive to ward off a predator or offensive to capture prey. Several types of extrusomes are encountered. Some contain a protein ribbon or fibrous proteins that discharge by extending into a long spear-like structure (trichocysts). Others (mucocysts) contain mucopolysaccharides that are released as a sticky mucus, while still others contain toxic substances (toxicysts). **Peroxisomes** are small vesicles that contain oxidative enzymes. They are important sites of oxygen utilization and have been implicated in fatty acid degradation, photorespiration in photosynthetic species, and detoxification of certain organic molecules.

Mineral skeletons and scales are formed in many protists. The mineral skeleton of Radiolaria, for example, is formed inside the ER membranes. It is not known how the ER forms the diversity of shapes observed in radiolarian mineral skeletons. In other protists the Golgi provides the chemical environment for the deposition of dissolved minerals, often inside a protein matrix also formed inside the Golgi. Similarly, the external **mineral scales** and **organic scales** of protists are formed inside the Golgi and secreted by exocytosis to the cortex outside of the cell. The glass case (frustule) of diatoms is also constructed within vesicles derived from the endomembrane system. This provides a controlled environment for the deposition of silica to form elaborately sculpted structures.

Terrestrial species as well as many freshwater and marine species are able to form **cysts**, or dormant cells that are resistant to adverse conditions. Cysts are initiated when food resources run out or environmental conditions are unfavorable to growth and survival. The process involves the synthesis of a cell wall, usually made of protein or polysaccharide. In many terrestrial species, the cyst is also dehydrated and is much more resistant to harsh conditions. These cysts require specialized proteins to protect the chromosomes, some RNA, and proteins within the dehydrated cytoplasm until growth can resume. The process of cyst formation is elaborate and called **encystment**. When cysts are stimulated to become active again, the process of reactivating the cytoplasm and reforming the active cell shape is called **excystment**. Many parasitic species form resting cysts that can be dispersed or ingested by hosts.

Diversity of Feeding Mechanisms

Protists obtain nutrients in diverse ways, often by more than one mechanism. Probably all protists are capable of **osmotrophy**, which is the uptake of dissolved organic and inorganic nutrients, including minerals, through the cell membrane. Osmotrophy occurs by diffusion of solutes through the membrane, transport through membrane proteins in the membrane, and pinocytosis. **Saprotrophy** is a special case of osmotrophy that includes both the secretion of digestive enzymes to dissolve complex substrates as well as the uptake of the dissolved nutrients. Common examples of saprotrophs are in the Fungi and the Myxogastria. Ingestion of larger particles by phagocytosis is usually divided into separate categories, depending on what is ingested. Thus, those that ingest bacteria are **bacterivores** and those that ingest protists are **cytotrophs**. Both bacterivores and cytotrophs are consumers of other organisms. The term "predation" is used when a cell ingests one prey cell at a time, while "grazing" is used when many cells are ingested and packaged within a single food vacuole. For example, many ciliates ingest hundreds of bacteria into a single food vacuole through a cytostome and are typical grazers. Many species with plastids complement photosynthesis by osmotrophy or phagocytosis. Some that feed both by phagotrophy and photosynthesis are common in the marine plankton and are called *mixotrophic*.

Sex

Sex has been described in many protists; however, many others are not known to reproduce sexually. Sexual reproduction requires several key steps. First, cells of the same species must be able to express different **mating types** or sexes. This is achieved by expressing surface cortical proteins that are different from, but complementary to, those of other mating types. Cells of complementary mating types attach in pairs by forming protein-protein connections. This step is called cell-cell recognition by mating type. Second, complementary fused cells must be able to transfer and merge nuclei. This step requires opening the cell connection between the mating pair to allow passage of a nucleus or the fusion of both cells into a common cytoplasm. This is called **nuclear fusion** and requires cytoskeletal elements to move the nuclei together. The result is a nucleus that has one chromosome set from each member of the mating pair. It is therefore called a **diploid nucleus** and the cell a **zygote**. The whole process of cell-cell recognition, cell fusion, and nuclear fusion to form the zygote is called sex, or conjugation. The diploid nucleus may divide by mitosis or "asexually." This form of cell division leads to a clonal population. The diploid nucleus may divide by **meiosis** to yield four nuclei, each with one set of chromosomes, called **haploid nuclei**. Meiosis involves the replication of the two sets of chromosomes followed by a first nuclear division to separate the chromosomes into two sets. This is immediately followed by a second separation of the chromosomes without DNA replication yielding four haploid nuclei, each with one chromosome set. Often the haploid cells will feed, grow, and then divide asexually by mitosis. These haploid cells can be stimulated by environmental factors to become sexually reactive, and the cycle begins again. Haploid cells that become sexually reactive are called gametes. The fusion of gametes leading to nuclear fusion and a diploid cell is called sex. In some organisms, sexually reactive cells secrete molecules called **pheromones** that initiate sexual reactivity in other cells of the same species. It is a form of cell-cell communication. Organisms that are not able to undergo sex are called asexual. They remain haploid or diploid and divide only by mitosis. It is unclear whether sex evolved once and was lost in many protists or whether it has evolved several times independently in different protists.

> **SECTION HIGHLIGHTS**
> The endomembrane network, the cytoskeleton, and the nucleus provide eukaryotic cells with organelles and a variety of cellular functions that distinguish them from prokaryotes. Eukaryotic cells may also have mitochondria or plastids, which are of endosymbiotic origin.

23.2 Diversity of Protists

Eukaryotic microbes are often divided into a few simple types based on their nutritional capabilities. Photosynthetic forms are often referred to as algae, while most nonphotosynthetic forms are commonly called protozoa. Traditionally, algae have been considered to be primitive forms of plants and protozoa as primitive forms of animals. It is now clear that these distinctions are artificial—they do not reflect actual evolutionary relationships. For example, photosynthesis has been acquired by several different eukaryotic groups by separate events of secondary endosymbiosis. Therefore many "algae" are much more closely related to particular "protozoan" groups than they are to land plants. The terms "algae" and "protozoa" remain useful informally in some contexts, but they no longer have taxonomic significance. Until recently it was popular to refer to all eukaryotes that were not animals, plants, or fungi as "Kingdom Protista" or "protists." The term "protist," too, is excluded from modern classification schemes because it is not monophyletic (for example, some protists are more closely related to animals than they are to any other protists). Other useful but artificial names for types of microbial eukaryotes include "flagellates" (cells that move by using flagella) and "amoebas" (nonflagellated cells that produce pseudopodia); however, many flagellates can be amoeboid and produce pseudopodia for feeding.

In this section, we will explore the diversity of microbial eukaryotes by dividing them into major groups based on evolutionary relatedness. As with prokaryotes, the advent of gene sequencing technologies has had a tremendous impact on our understanding of the diversity and evolution of eukaryotes. Several important groupings have been identified by analysis of ribosomal RNA genes or of multiple protein-coding genes. In the case of eukaryotes, many groups are identified because they share distinctive cellular features that are too small to be resolved by the light microscope and were first rec-

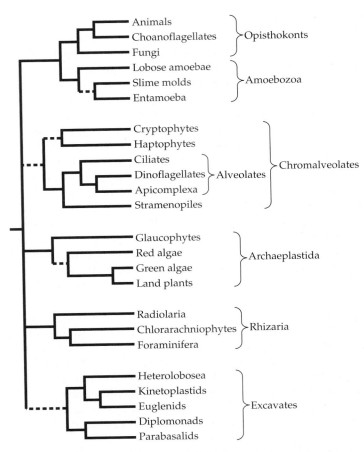

Figure 23.8 Phylogenetic relationships among eukaryote groups
Protist phylogenies are reconstructed based on cell ultrastructure, biochemistry, life cycles, and statistical analyses of DNA and protein sequence variations. Broken lines indicate uncertain branch positions.

ognized by transmission election microscopy. Our picture of the evolution of eukaryotic microbes has changed rapidly in recent years and is still not complete. The system of "supergroups" and subgroups presented here (Figure 23.8) may require further revision in the future. Also, this survey will cover only the more important (or better-studied) groups of microbial eukaryotes and is not exhaustive.

23.3 Opisthokonta

Opisthokonts includes both animals and fungi and several protist groups. The best known of the protists are the choanoflagellates, which are discussed further below. Opisthokonts also include Mesomycetozoa, which are a diverse collection of parasites, mostly of aquatic animals, and nucleariids, an obscure group of filopodia-producing amoebas that turn out to be the closest relatives of fungi.

Opisthokonts share a particular arrangement of the flagellar apparatus; each cell has two basal bodies but only a single flagellum, which emerges from the posterior end of the swimming cell (rather than from the anterior end or laterally, as in most other eukaryote groups). This organization can be seen, for example, in animal sperm. Most opisthokonts also have flattened mitochondrial cristae.

Choanoflagellata

Choanoflagellates are small unicellular or colonial flagellates (Figure 23.9). The single flagellum is surrounded by a collar of thin microvilli supported by actin microfilaments. During feeding, the beating of the flagellum produces a feeding current from behind the cell, and prokaryotes are trapped against the collar as the water currents stream along the cell. The prokaryotes are then phagocytosed by small pseudopodia that extend up the microvilli. Many choanoflagellates attach to surfaces by a short stalk; however, some marine choanoflagellates—Acanthoecidae—are free-floating and produce elaborate openwork loricas made of silica that can be ten times larger than the cell itself. The loricas may function to increase drag, such that the cell can produce a feeding current while remaining almost stationary. Acanthoecid choanoflagellates are among the most important bacterivores in the ocean.

Over a hundred years ago it was noted that choanoflagellates closely resemble the feeding cells (choanocytes) of sponges—the deepest-diverging animals. Choanocytes employ a food-capturing collar very similar to that of

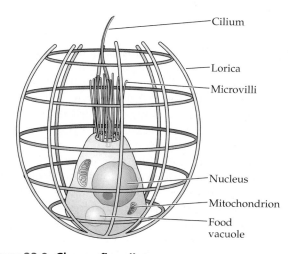

Figure 23.9 Choanoflagellates
A choanoflagellate inside a silica lorica. The microvilli are involved in accumulating bacteria for phagocytosis.

choanoflagellates. Molecular phylogenies have since confirmed that choanoflagellates are very closely related to animals.

Fungi

Fungi occur primarily in terrestrial environments, where they are important in the decomposition of organic matter. Some also occur in aquatic habitats, where they also contribute to decomposition. Most fungi are saprotrophs that feed exclusively by osmotrophy. Some fungi are important parasites of plants and animals. Many crop pathologies are caused by fungi known as powdery mildews, rusts, and smuts. Some are symbiotic with photosynthetic organisms and are called lichens. Some form a symbiotic association with plant roots, called **mycorrhizae**, and contribute to plant nutrition and survival.

The fungal cell consists of an outer cell wall made mostly of a polysaccharide polymer called chitin. Chitin also occurs in the exoskeleton of other groups of opisthokonts, the invertebrate animals. The chitin polymer is cross-linked by hydrogen bonds into a protective matrix that forms a barrier about 0.2 μm thick. The cell wall is composed of 80% to 90% carbohydrates. However, there are numerous important proteins that must be exposed to the environment. These include recognition proteins for mating types as well as structural proteins and enzymes for substrate digestion. In saprotrophic species, secretory vesicles containing digestive enzymes fuse with the cell membrane to release enzymes. Enzymes up to 20 kDa may pass through the cell wall into the substrate. Common fungal exoenzymes are diverse and include proteases, amylases, cellulase (cellobiohydrolase), xylanases, pectin-degrading enzymes (lyase, esterase, pectate lyase, polygalacturonase), ligninases (lignin peroxidases and Mn-peroxidases) and other secreted enzymes. The cell membrane must include substrate receptors and carrier proteins to take up the digested substrates. The role of the cell wall in supporting the cell is important during periods of water stress, as in plant cells, but it also serves as a defensive barrier to predation. The main reserve materials are glycogen and lipids. Mitochondria have flat cristae, and peroxisomes are present in the Ascomycetes and Basidiomycetes. Mitosis is closed, with a persistent nucleolus that divides. Chromosomes do not align along the spindle in a typical metaphase, as there are often too many chromosomes. The endomembrane network is prominent in saprotrophic species, especially where secretory enzymes are synthesized and accumulated. Some vacuoles also participate in osmoregulation and storage of amino acids, ions, or other soluble nutrients.

Three forms of growth are recognized in Fungi. One form is called **yeast growth** and consists of cell division and separation of cells into independent or loosely attached cells. A second form of growth occurs in Chytridiomycetes, where one cell extends cytoplasmic branches into the substrate and forms a mononuclear **thallus**. The third mode of growth is called **hyphal growth**, which, from successive mitoses, produces a long extending filament with nuclei at regular intervals. In these filaments, growth is by tip elongation. The repeated branching of growing hyphae forms a three-dimensional mass called the **mycelium**. In many groups, hyphae of the same species, strain, or individual are able to fuse. This is called **anastomosis** and requires a growing tip to fuse with a compatible hyphal filament. Some species of Fungi may grow as hyphal filaments under some conditions and as yeasts under different conditions. The hyphal cytoplasm is often separated by a cross wall called a **septum**. The septa separate the cytoplasm along the hyphae into cellular compartments with one or more nuclei. Pores in the septum permit the translocation of cytoplasm to other parts of the mycelium but keep nuclei separated. The role of septa is crucial in containing damage after a hypha is broken or invaded by a predator or parasite. Septal pores become blocked by cell wall deposition or a protein plug, and a new branch initiates a new growing tip. In parasitic and symbiotic fungi, the end of a growing tip that contacts a host cell may expand and enlarge. This is called a **haustoria** and is used to make good contact with the host and to penetrate into the host cell.

Species identification in Fungi involves description of the morphology of spores and reproductive structures. The morphology of asexual hyphae, especially if conidia are absent, does not provide sufficient characters for identification. Therefore many isolates cannot be assigned to a taxonomic group without sequencing of DNA regions for molecular phylogenic analysis. When reproductive or dispersal structures are known, identification may be possible; otherwise one is forced to rely on analysis of DNA sequences.

Chytridiomycetes

Chytrids are motile cells with one posterior cilium and no mastigonemes. Cells can be partially amoeboid on a surface or thin water film. Chytrids are found from polar regions to tropical climates in soil and freshwater habitats, including bogs. They have been most studied in aquatic habitats. Most soil and freshwater species are saprotrophs on chitin, keratin, pollen, cellulose, and substrates that are difficult to digest. It is believed that they facilitate the entry of other organisms into the decomposing litter. Many soil species are predatory on microinvertebrates (nematodes, rotifers, tardigrades, or their eggs), larger invertebrates, active protists, fungal spores, and other chytrids. Many species also grow on microinvertebrates and aerial parts of plants. Some species are parasitic on a variety of animals, fungi, and plants, including commercially important crops and domestic animals.

The cell wall of chytrids consists of chitin and β-glucan polymers, as in other Fungi. The mitochondria have flattened cristae, and the storage products are glycogen and lipid vesicles. Several organelles are particular to this group. There is a nuclear cap of dense ribosomes with endoplasmic reticulum closely associated. There is a lipid globule complex, called a microbody, at the cell posterior, which may be involved in lipid metabolism. Adjacent to this microbody, the **rumposome** consists of flattened vesicles, which have been implicated in sensory orientation and taxis in certain species. The rumposome often has connections to the cell membrane and microtubules associated with the basal bodies. It stores Ca^{2+} ions, which may mediate the direction of taxis. Sex probably occurs in most species and involves male and female gametes produced in separate antheridia and oogonia on terminal hyphae (Figure 23.10). Motile gametes are released; their conjugation involves pheromones.

Chytrids find their prey or adequate substrate by chemotaxis. Upon making contact, the motile cell (zoospore) encysts and loses the cilium, then grows cytoplasmic extensions into the substrate or prey. The cytoplasmic extensions, together called a thallus (pl., thalli), can be extensively branched. The growing thallus is not the result of cell divisions, does not have a cell wall as fungal hyphae do, and is capable of phagocytosis, at least in some species. Extensive secretion of extracellular digestive enzymes from the thallus dissolves nutrients for osmotrophy. When the food resources are exhausted, dispersal spores (or resistant cysts) are formed by repeated mitotic divisions. Release of the spores occurs at an opening away from the substrate and thallus. The cysts of chytrids are very resistant to long periods of desiccation, chemical attack, and anaerobiosis. For this reason, parasitic species can reappear after decades from infected agricultural soil.

The Chytridiales and Monoblepharidales contain mostly aquatic genera, whereas the Spizellomycetales and Blastocladiales are primarily soil species. For example, *Chytriomyces* (Chytridiales) can be isolated in soil and freshwater samples, where they digest chitin. The aquatic Monopblepharidales are found in decomposing insect parts, plant debris, and woody debris. The Spizellomycetales are particularly adapted to soil. The motile stage lacks the ribosome cluster, the rumposome is absent, and microtubules from the basal bodies are associated with the nucleus. The Blastocladiales (such as *Allomyces*) occur actively in fresh water and moist soils. The metabolism of many saprotrophic Blastocladiales

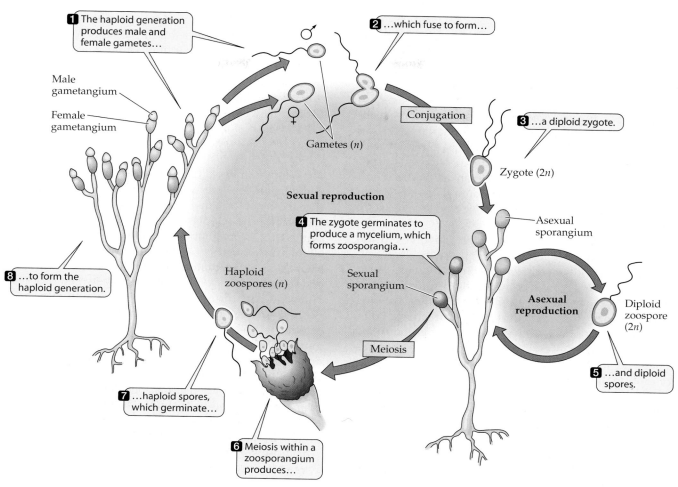

1 The haploid generation produces male and female gametes…

2 …which fuse to form…

3 …a diploid zygote.

Male gametangium

Female gametangium

Conjugation

Gametes (*n*)

Zygote (2*n*)

Asexual sporangium

Sexual reproduction

4 The zygote germinates to produce a mycelium, which forms zoosporangia…

8 …to form the haploid generation.

Haploid zoospores (*n*)

Sexual sporangium

Asexual reproduction

Diploid zoospore (2*n*)

Meiosis

5 …and diploid spores.

7 …haploid spores, which germinate…

6 Meiosis within a zoosporangium produces…

Figure 23.10 Life cycle of an aquatic chytrid
Allomyces has haploid and diploid growth forms.

species is fermentative and carboxyphilic, producing lactic acid as an end product. Characteristically, in motile cells the organelles aggregate in one part of the cytoplasm. Germination from the spores is bipolar, with each emerging hypha repeatedly branching into finer hyphae to produce the mycelial thallus. Some chytrids are predatory on invertebrates. For example, *Sorochytrium milnesiophthora* is saprotrophic on dead rotifers, nematodes and tardigrades, but it is also predatory and capable of penetrating the cuticle of living individuals. The prey eventually dies and is digested; thereafter, the release of motile zoospores propagates the chrytrid to other prey or substrate in the microhabitat. The Enteromycetes are endosymbiotic chytrids in the intestinal tracts of ruminant animals. They are anaerobic, with a broad spectrum of enzymes with which to digest the walls of plant cells. The chytrid thallus grows into the chewed particles ingested by the animal.

Zygomycota

The Zygomycetes are filamentous species that lack any form of cilium, including centrioles. Hyphal filaments contain mitochondria, peroxisomes, and many vesicles. The filaments are coenocytic (without septa) or have septa at irregular intervals. Haploid nuclei occur along the length of hyphae. Septate species have a plugged pore in the septum called a lenticular cavity, as in the Kickxellales. Many Zygomycetes are saprotrophs in soil, where they grow on decaying substrates rich in amino acids, peptides, and sugars. The better-known species belong to the Mucorales, which are ubiquitous in soils and on herbivore dung, moldy bread, decaying fruit, and decomposing mushrooms. The mucors, sometime called the sugar fungi, are early colonizers of soluble substrates such as sugars and amino acids. These species tend to lack digestive enzymes for substrates such as cellulose and chitin, which are more difficult to break down. Spores are stimulated to germinate by moisture, adequate solution, and abiotic conditions. The spore swells with water and numerous hyphae may emerge over several hours. Hyphae of mucors grow rapidly into the substrate, ahead of other fungi. Under anaerobic conditions, some species grow as yeasts, but others remain hyphal while still others are strict aerobes. In anaerobic conditions, sugars are metabolized to ethanol. Dispersal occurs from aerial or vertical hyphae that differentiate at the ends into about 10^5 sporangiospores, each with several nuclei. The wall of sporangiospores consists of sporopollenin, an oxidized and polymerized form of β-carotene. Some species also produce chlamydospores, which are thick-walled, resistant vegetative cysts inside terminal hyphae.

Sexual conjugation in Zygomycetes involves fusion of hyphae from two mating types (Figure 23.11). Terminal aerial hyphae specialized for conjugation are called zygophores. Conjugation between complementary mating types occurs when zygophores fuse, after growing toward each other by autotropism or chemotaxis. Growth of zygophores toward each other is controlled by chemical substances. Zygophore growth comes to a halt as they bend toward each other to fuse. The main zygophore hormone is trisporic acid, which occurs in several varieties. The molecule is derived from the breakdown of β-carotene. The production of trisporic acid requires the metabolic cooperation of both mating types close together because the intermediate products must diffuse from one mycelium to the other. As the zygophores touch, a thick wall forms that separates the several nuclei at the tip from the rest of the mycelium. Fusion of the complementary zygophores into a common cytoplasm forms a multinucleate zygospore with nuclei from both mycelia. The zygospore wall contains sporopollenin and a black pigment called melanin. In some cases, the zygospore wall is elaborate and ornate. Germination of the zygospores requires suitable environmental conditions and includes meiosis prior to growth of new hyphae.

The Dimargaritales are parasites of other fungi, especially mucors, and cannot be cultured without the host. The Zoopagales are encountered in soils, where many are predatory or parasitic on amoebas, ciliates, nematodes, and other small invertebrates. The hyphae are syncytial with septa at intervals. The Entomophthorales are insect parasites that usually enter the host through the cuticle. Spores attached to the cuticle germinate into the insect, where the hyphae proliferate on host tissues. Eventually, as the insect dies, hyphae emerge from its tissues, releasing propulsive conidia. These species do not produce sporangiospores for dispersal but release spores by violently expulsing conidiospores. Some species are sexual and form structures similar to the zygospore of mucors. The other orders consist of species

that are symbiotic with, or parasites of, mandibulate arthropods, either in the gut or outside on the cuticle. The Harpellales consist of filamentous species that are symbiotic with freshwater arthropods. The Endogonales are endomycorrhizal with many plants, forming arbuscular mycorrhizae. They are not known to grow without the plant symbiont.

Glomeromycota

The Glomeromycota form close symbiotic associations with plant roots. Only a small number of terrestrial plant species do not form mycorrhizal associations with Glomeromycota. The cytoplasm of hyphae contains mitochondria, peroxisomes, and simple Golgi cisternae. The Glomeromycota do not have centrioles or cilium; their hyphae are not septate; and they do not produce conidia or aerial spores. Hyphae extend between root cells and send haustoria into root cells, where they branch profusely. It is this branched morphology in the plant cell that is referred to as the arbuscule. Species are obligate symbionts that depend on their plant host for sugars and amino acids. However, this statement has not been completely verified physiologically for all species. Most hyphae are intercellular between root cells of the primary cortex and epithelium. The hyphae do not penetrate the endodermis, vascular tissues, or aerial plant organs. Some hyphae extend into the soil from the root. Anastomosis of hyphae in the root and soil occurs. Species are probably asexual and produce large dispersal spores less than 800 µm in diameter called chlamydospores. The Glomales form arbuscules and vesicles, whereas the Gigasporineae form only arbuscules. Vesicles and spores both store lipids. It is uncertain how much specificity there is between the fungal and plant species because an association is possible between a single fungus and many plant species capable of forming arbuscular mycorrhizae.

Ascomycota

Ascomycetes are by far the largest group of fungi, representing a very diverse group with many varieties of forms. They include species that form inconspicuous mycelia, some that form large mushrooms, the lichens, many yeasts, and some important parasites. The defining characteristic is the production of a terminal hypha, where meiosis leads to the production of eight spores inside the parent hypha. The terminal hypha where meiosis occurs is called an **ascus** (asci), and the haploid spores are called **ascospores**. In yeasts, the single cell becomes the ascus. Development of the ascospores requires numerous steps (Figure 23.12). In most species, the ascospores are forcefully ejected from the

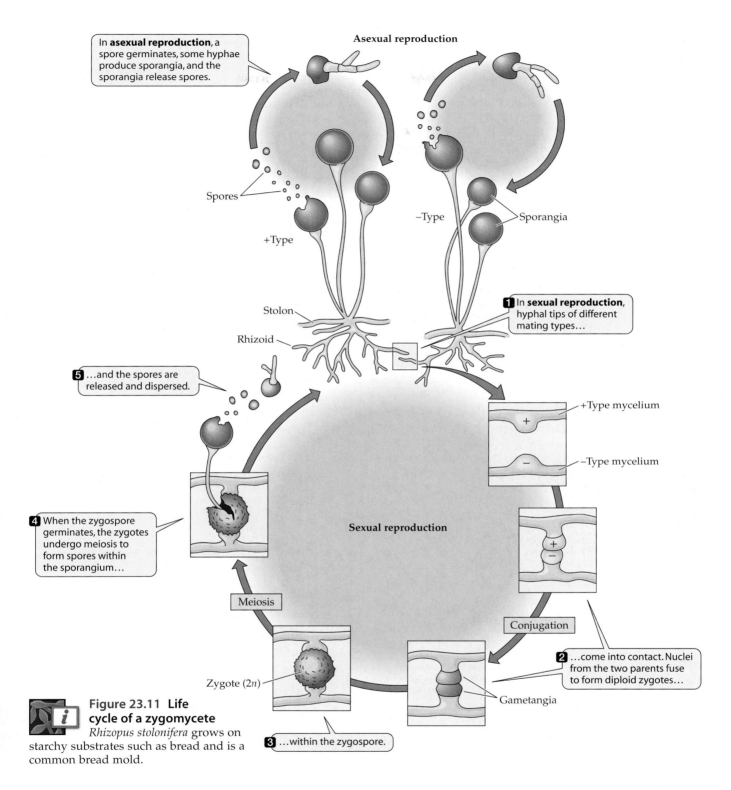

Asexual reproduction

In **asexual reproduction**, a spore germinates, some hyphae produce sporangia, and the sporangia release spores.

Spores

+Type

−Type

Sporangia

Stolon

Rhizoid

1 In **sexual reproduction**, hyphal tips of different mating types…

5 …and the spores are released and dispersed.

4 When the zygospore germinates, the zygotes undergo meiosis to form spores within the sporangium…

Meiosis

Sexual reproduction

+Type mycelium

−Type mycelium

Conjugation

2 …come into contact. Nuclei from the two parents fuse to form diploid zygotes…

Gametangia

Zygote (2*n*)

3 …within the zygospore.

Figure 23.11 Life cycle of a zygomycete
Rhizopus stolonifera grows on starchy substrates such as bread and is a common bread mold.

mature ascus. The mechanism of propulsion as well as the shape and color of the spores and of the ascus are important characters in identification. Spores germinate into branched mycelia with septa at regular intervals. The septa have simple pores that allow passage of cy-

toplasm and organelles such as mitochondria. Nuclei do not usually pass through septa with cytoplasmic streaming but can do so with sufficient force. Septa separate one or more nuclei along hyphae. Anastomosis of hyphae occurs within a single mycelium as well as among

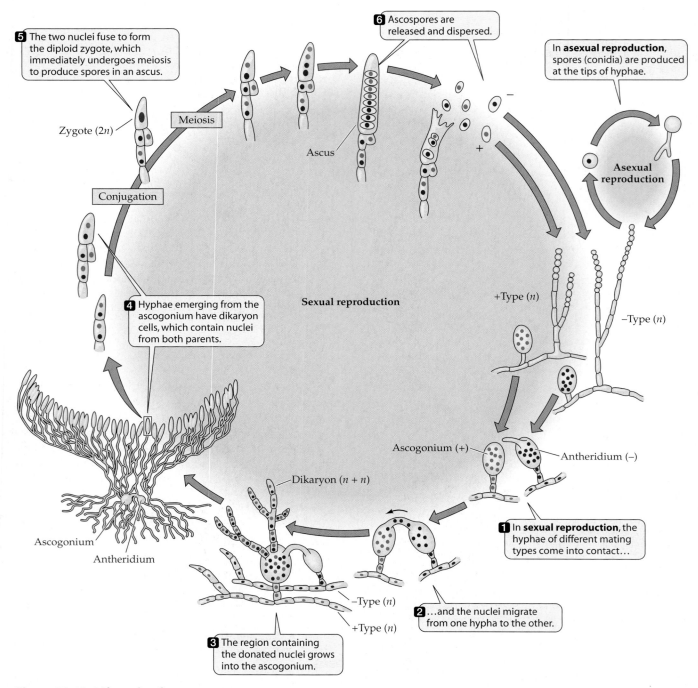

Figure 23.12 Life cycle of an ascomycete
Production of ascospores inside the ascus is characteristic of ascomycetes. Both monokaryotic and dikaryotic haploid mycelia occur prior to conjugation.

neighboring hyphae of compatible mycelia. Vegetative propagation occurs through **conidia**. Terminal hyphae of monokaryotic haploid mycelia form loosely attached cells that can easily be dispersed by wind or water. Morphology of the conidia is useful in species identification. Some species form more than one type of conidia. These asexual vegetative growth forms are referred to as **anamorphs**. Sexual mycelia also occur; they are called **teleomorphs**. As species were traditionally identified by the anamorph or teleomorph morphology, anamorphs and teleomorphs were given different names. However, as we work more with these species, anamorphs and teleomorphs are being recognized as different life cycle stages of the same species. Sexual reproduction in ascomycetes requires the conjugation of two complementary mycelia through **antheridia** (sing. antheridium) and **ascogonia** (sing. ascogonium). Each antheridium and ascogonium consists of a terminal hypha with an enlarged bulb at the apex where several haploid nuclei accumulate by mitosis. The antheridium extends a projection toward an ascogonium and all the nuclei are transferred to the ascogonium. From the ascogonium grow several branched dikaryotic septate filaments. The dikaryotic hyphae grow together into a mass of filaments called the **ascocarp** or ascomata. Terminal hyphae of the ascocarp that become the site of conjugation and ascus formation are called the **crozier**. The terminal end folds back on itself and further growth stops. Fusion of both nuclei in the apical cell forms a zygote that immediately proceeds with meiosis, forming four haploid nuclei. Each undergoes one mitosis to form eight nuclei which separate into distinct ascospores inside the ascus. Fusion of the two penultimate monokaryotic cells restores the dikaryotic state. The ascocarp is called an **apothecium** if it is disk-shaped, a **perithecium** if it is flask-shaped, or **cleistothecium** if it is spherical. In cleistothecia, the asci discharge inside the sphere and ascospores are released through an ascocarp opening. Several groups of fungi are described below. The ascomycetes also include lichens, described in Chapter 25.

PEZIZOMYCETES   Species in this group are saprotrophs that occur in forest soil organic horizons, woody debris, and animal dung. In some species, ascocarps are phototropic, bending toward light prior to spore discharge. They include the morels and the truffles, which are sought after culinary mushrooms. Morels have a stalked ascocarp that rises above the soil. Truffles, such as the highly prized *Tuber melanosporum*, form underground ascocarps, and the mycelia form mycorrhizae with beech and oak trees. This group also includes cup fungi, with their brightly colored apothecia.

SORDARIOMYCETES   These species produce large numbers of wind-dispersed ascospores that are good colonizers of organic matter such as animal dung. *Neurospora* is common on burnt wood in the tropics. *Podospora* can be found on herbivore cellulosic dung and *Sordaria* on animal dung. They include species that are cellulolytic, such as *Chaetomium* in plant debris. Some species are saprotrophic on decaying fungi and lichens. A few can be parasitic on plants. The ascomata occur in a perithecium, which is phototropic and orients toward light prior to spore discharge. Ascospore discharge is usually regulated by drying.

One species, *Neuropora crassa*, has been used extensively in genetic and biochemical studies. It was particularly useful in the early days of genetic mutant analysis and remains a useful model organism. Its ascospores can survive many years because they have thick, pigmented walls that provide resistance to ultraviolet light, heat, and drying. Germination requires heat treatment to 60°C to break dormancy. Spores grow rapidly into a branched mycelium by repeated mitotic divisions with septa between nuclei. Aerial hyphae above the substratum can produce many branched conidia with a pinkish carotenoid pigment. Two types of conidia can form. One forms larger, loosely attached cells along a branched filament, called macroconidia. Under different conditions, much smaller cells form in grape-like clusters, called microconidia. Microconidia have much less nutrients stored within them and do not survive or germinate as readily as macroconidia. Anastomosis with compatible mycelia results in a dikaryotic heterokaryon. For sexual reproduction, hyphae from the haploid monokaryotic cell form ascogonia. There are two mating types, called "A" and "a". Compatible hyphae, one from each mating type, are required for conjugation. Macroconidia or microconidia must be air-blown onto ascogonia of complementary mating type to form the dikaryotic cell. The two nuclei of complementary mating type fuse and proceed with meiosis and one mitotic division to form eight haploid ascospores. The ascospores in the long and narrow ascus remain in order, allowing the genetic analysis of mutants, because the origin of a mutation can be traced to meiosis I or II or the subsequent mitosis, and each ascospore can be separated and cultured. Further mitotic divisions in each ascospore produce a multinucleate spore.

EUROTYOMYCETES   Several genera in this group are well known and economically important. For example, there are at least 250 species of *Penicillium*, which are used to obtain antibiotics such as penicillin. These molecules are secondary metabolites that are toxic to certain organisms. For example, penicillin is effective against many bacteria. *Aspergillus flavus* secretes aflatoxins that are toxic to humans. The toxin causes liver breakdown in humans and is lethal. These secondary metabolites are generally called mycotoxins and include many produced by poisonous mushrooms in the Basidiomycetes. Other species of *Aspergillus* are used industrially to produce fermented goods. For example, *A. oryzae* and *A. soyae* are used to ferment soybeans into soy sauce. *A. nidulans* was used extensively in research and contributed a great deal to our understanding of eukaryotic genetics.

Most species in this group produce cleistothecia, although the ascocarp is entirely absent in some species. In general, the anamorph stage with conidia is more common and the teleomorph is rarely seen. Both *Aspergillus* and *Penicillium* represent the anamorph stage. Their teleomorph in both cases is known as a species within the genus *Eurotium*. In *A. herbarium*, the anamorph is observed at 15 to 25°C on media. If the sugar concentration is increased and the temperature increased to 25 to 30°C, then the teleomorph can be observed. Very small cleistothecia form; these are 150 to 200 µm in diameter and hold a number of fine asci, each with eight ascospores. In the environment, ascospores germinate in moist, decaying plant matter. However, many are able to germinate in dry or osmotically unfavorable conditions, as on the surface of fruit jams.

LEOTIOMYCETES   This group of ascomycetes includes saprophytic mushrooms called cup fungi, earth tongues, and the parasitic Erysiphales. The Erysiphales are parasites of flowering plants, fruit trees, and cereal crops. They can impart significant commercial damage to agricultural and horticultural operations. The disease is called powdery mildew, and different species affect different plants. The fungus grows on plant surfaces as a branched mycelium with septa separating uninucleate cells. The mycelium spreads on the surface only, but haustoria extend into the plant's epidermal cells. Close contact between the plant's cell membrane and the thin fungal cell wall allows passage of nutrients to the parasite. On the surface, the mycelium produces conidia that disperse to infect new surfaces. The conidia are unlike most fungal spores because they germinate without water under very dry conditions. This is an adaptation for growing on exposed dry plant surfaces. The epidemic infection continues through the summer growing season. In late summer, plant senescence begins and fewer nutrients are available to the parasite. Cleistothecia form after conjugation of compatible mycelia. Each ascocarp is only 100 µm in diameter and holds an ascus with eight ascospores. The morphology of the ascocarp and ascus is useful in species identification. The ascospores remain dormant throughout the winter. In spring, uptake of water swells the ascus and the ascospores, which break through the ascocarp and the ascus. The ascospores are ejected into the wind and begin a new infection cycle.

LABOULBENIOMYCETES   This group consists of about 1,500 species that grow on insects as ectoparasites. Some continue to grow on the dead insect and contribute to its decomposition. Filaments grow as a cellular thallus into the host but do not appear to cause harm. The parasite is inconspicuous and hard to find on the insect. The ascospores that are produced are two celled, one being modified for attachment to a new host.

SACCHAROMYCETES   These are osmotrophic species, such as *Saccharomyces*, found on surfaces of fruit and leaves, which proliferate in sugary solutions. In anaerobic conditions, they produce ethanol as an end product of metabolism. The metabolic pathway is called fermentation. Species form a small, simple mycelium or grow only as yeasts. Wild yeasts were domesticated about a thousand years ago for the production of fermented beverages. Beer is produced from the fermentation of barley malt, which contains maltose. Wine is made by fermenting fruit juices, usually grapes. Mead is made by fermenting honey. Many strains of *S. cerevisiae* have been developed over the past thousand years that impart characteristic flavors to beer or that tolerate different amounts of ethanol or degrees of temperature during fermentation. Another species, *S. sake*, is economically important for the production of sake from rice wine. *S. cerevisiae* is also used in cell biology as a model organism. It is an important species in understanding the regulation of cell-cycle in eukaryotes. *S. cerevisiae* divides by budding a small cell from the parent cell after mitosis. After fusion of complementary mating types, meiosis proceeds to form four haploid nuclei in the parent cell. Each nucleus becomes a new haploid cell inside the parent cell. This represents the ascus, with four ascospores. Another important species is *Candida albicans*, which causes opportunistic infections in humans. Cells normally grow as yeasts on mucosal membranes. Under certain conditions, the filamentous stage is initiated and the mycelium causes an infection into tissues. Affected areas include the skin, oral cavity, digestive tract, and vagina.

TAPHRINOMYCOTINA   This group also contains small cells that grow as yeasts, but some can be filamentous. The well-studied genus *Schizosaccharomyces* consists of small rectangular cells 4 to 8 µm in diameter that elongate at one end and divide by mitosis. Cells grow on dissolved nutrients by osmotrophy and are found in decomposing organic matter in soil and composts. Conjugation of two cells of complementary mating types forms a dikaryotic cell, which undergoes nuclear fusion and meiosis. The four haploid nuclei separate as four ascospores, which is equivalent to the ascus. *S. pombe* was used extensively in cell biology to understand cell-cycle progression in eukaryotes. Another genus with similar characteristics is *Taphrina*, which forms a simple dikaryotic mycelium that infects plant tissues, forming galls or lesions. A third important genus in this group is an important extracellular parasite of mammalian lungs that causes pneumonia. There are many cryptic species that are host-specific, but they were all described as *Pneumocystis carinii*. Its entire life cycle occurs inside the host lung. In humans, *Pneumocystis* has become the predominant infection causing death in HIV patients.

Basidiomycota

Basidiomycetes are terrestrial fungi involved in the recycling of organic matter. There are about 22,000 known species, which include edible mushrooms. Most are filamentous species that grow in soil on organic substrates. They release digestive enzymes in the substrate and absorb the dissolved nutrients by osmotrophy. Cell walls consist of chitin and xylose polymers. The mycelium forms from branched hyphae with septa separating nuclei. Septa have a central pore surrounded by endoplasmic reticulum. The structure is called a dolipore septum and prevents the passage of organelles and nuclei but permits cytoplasmic flow between cells. Growth of a mycelium begins with germination of a spore that has one haploid nucleus (Figure 23.13). The hyphae are about

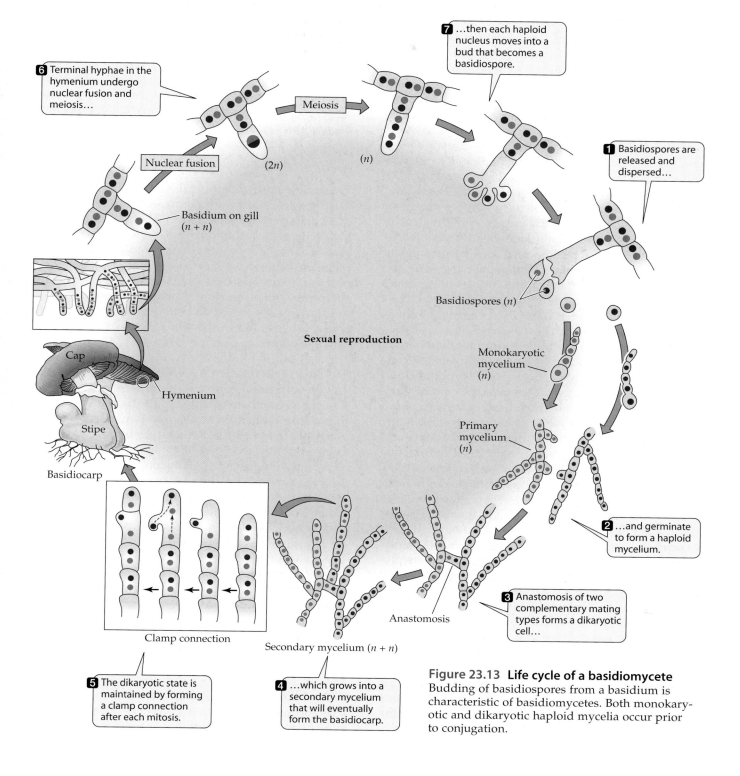

6 Terminal hyphae in the hymenium undergo nuclear fusion and meiosis…

7 …then each haploid nucleus moves into a bud that becomes a basidiospore.

Meiosis

Nuclear fusion

(2*n*)

(*n*)

1 Basidiospores are released and dispersed…

Basidium on gill (*n* + *n*)

Basidiospores (*n*)

Monokaryotic mycelium (*n*)

Sexual reproduction

Primary mycelium (*n*)

Cap

Hymenium

Stipe

2 …and germinate to form a haploid mycelium.

Basidiocarp

3 Anastomosis of two complementary mating types forms a dikaryotic cell…

Anastomosis

Clamp connection

Secondary mycelium (*n* + *n*)

5 The dikaryotic state is maintained by forming a clamp connection after each mitosis.

4 …which grows into a secondary mycelium that will eventually form the basidiocarp.

Figure 23.13 Life cycle of a basidiomycete Budding of basidiospores from a basidium is characteristic of basidiomycetes. Both monokaryotic and dikaryotic haploid mycelia occur prior to conjugation.

4 μm in diameter and grow on suitable substrates. The branching hyphae grow as a monokaryotic mycelium, with one nucleus per cell. There are often several mating types in each species. Hyphae of mycelia of complementary mating types merge. The process is called anastomosis, and the fused cells become dikaryotic, possessing one haploid nucleus from each mycelium. The dikaryotic cells continue growth as a dikaryotic mycelium. The mechanism for maintaining two distinct nuclei in the same cell is particular to Basidiomycetes. The process requires a clamp connection, which is a small backward outgrowth. One nucleus migrates into this space and both nuclei begin mitosis. New septa form to separate the dikaryote pair at the tip from the other two nuclei. This results in two cells with one nucleus. Fusion of the monokaryotes at the clamp recovers the dikaryote as one nucleus migrates into the cell. The dikaryotic mycelium differs from the monokaryotic type by having faster growth, larger diameter hyphae (about 7 μm), and branches forming at a more acute angle. Sexual reproduction is often seasonal and governed by changes in temperature, moisture, or substrate chemistry. Reproductive structures, called basidiocarps, are often prominent and aerial. The basidiocarp's hyphae are nonfeeding and structural. A button of mycelium in the soil begins to extend upward, first toward light by positive phototropism, then by negative gravitropism. The mycelium extends using water and reaches full length in a few hours. The button of mycelium initially looks like a miniature version of the basidiocarp. It contains glycogen reserves, which are used during the extension to produce chitin and new cytoplasm. The fully extended basidiocarp consists of a rigid stipe that supports a cap containing reproductive hyphae called hymenia (sing. hymenium). Many terminal hyphae of the hymenium are reproductive and called basidia (sing. basidium). In the basidium, the two nuclei fuse to form a diploid nucleus, which immediately undergoes meiosis. The resulting four haploid nuclei migrate to the tip of the basidium, where they bud off as four haploid basidiospores. The basidiospores are ejected 0.1 to 1 mm into the air for dispersal. After a few days of producing spores, the basidiocarp undergoes autolysis and the cell walls disintegrate. During the spore production period, as many as 100,000 spores can be released every minute. Each basidiospore that reaches adequate substrate can germinate and grow into a new homokaryotic mycelium.

Additional asexual dispersal structures occur. Those that occur in the homokaryon form conidia and conidiospores at terminal hyphae. In both the dikaryote and monokaryote state, hyphae can fragment at septa into independent yeast-like cells. These are called arthroconidia, and the fragmentation is usually caused by poor nutrient resources.

Typical Basidiomycetes include those commonly known as the agarics, such as *Agaricus*, the common cultivated mushroom (and related fungi such as *Boletus, Coprinus, Laccaria*), bird's nest fungi, bracket fungi, coral fungi, jelly fungi, puffballs and stink horns. Some groups include predatory species, such as the oyster mushroom (*Pleurotus ostreatus*), which capture small invertebrates and protists. Many Boletales (*Rhizopogon, Suillus*) and Agaricales (*Cortinarius, Russula*) are mycorrhizal symbionts, and many other Basidiomycetes decompose wood. In general, Basidiomycetes (and to a large extent Ascomycetes) constitute the bulk of the hyphal biomass in soil horizons and, when they are present and active, are responsible for most of the primary decomposition of organic matter.

Urediniomycetes

These fungi are commonly called rusts. There are about 7,000 species, which infect ferns, conifers, monocotyledons, and other flowering plants. They are economically important because they can devastate cereal crops across broad regions in one growing season. The disease will recur in successive years. Many rusts have two host plants that are very different. This way, if one plant is absent, the other functions as a refuge. Species that require two hosts are called heteroecious. Here we will consider wheat black-stem rust disease caused by *Puccinia graminis* as an example (Figure 23.14).

FORMING THE DIKARYOTIC INFECTIVE SPORES   The growth cycle begins with a wind-dispersed basidiospore landing on a barberry plant leaf. The basidiospore has a thin cell wall of chitin and xylose polymers. It germinates into the leaf by secreting digestive enzymes and extending branched intercellular hyphae. This results in a small mycelium a few millimeters across. The hyphae have regular septa isolating one haploid nucleus per cell. The septa have pores with a plug that regulates passage of material between cells. The mycelium obtains its nutrients by extending haustoria into plant cells. The mycelium also stimulates plant cells to divide locally. This causes a small hump of cells at the site of infection. The mycelium soon forms a pycnium on the upper surface of the leaf, which contains terminal hyphae that form uninucleate spores called pycniospores. These spores are dispersed by insects only. They are covered in a sticky, sugary solution from the pycnium that attaches to passing insects. The pycniospores are for conjugation, not infection. There are two types of morphologically identical mycelia representing two mating types. Each pycnium is of one or the other mating type. Pycniospores that reach a pycnium of opposite mating type fuse with the hyphae. This results in a dikaryotic

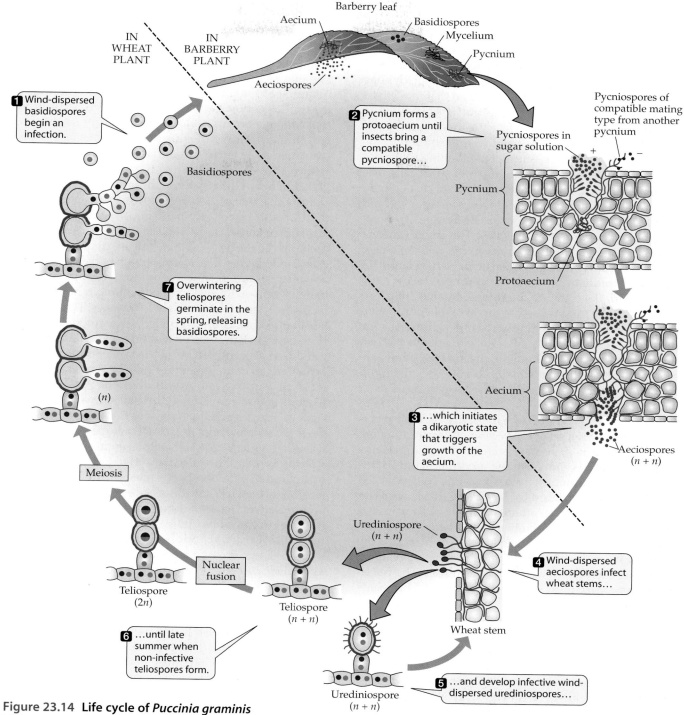

Figure 23.14 Life cycle of *Puccinia graminis*
Many rusts are heteroecious, requiring two host to complete their life cycle.

mycelium through successive mitotic divisions. The dikaryotic state initiates growth of mycelium to the lower leaf surface. The mycelium emerges through the lower surface as a spore-forming structure called an ae-cium. The aecium's terminal hyphae form dikaryotic ae-

ciospores that are propelled into the air. They are wind dispersed and infective.

THE INFECTION CYCLE The aeciospores can infect a variety of grasses including wheat, depending on the strain. The spores have a thick yellow-orange wall and contain a lipid droplet as reserve. Unlike the basidiospores, the germinating aeciospores enter the leaf through the stomata openings. The dikaryotic hyphae branch into a small mycelium similar to that in the barberry and also feed by haustoria. Eventually the mycelium breaks through the leaf as a uredinium, which contains erect terminal hyphae that are exposed to the wind outside of the leaf tissue. Repeated mitotic divisions in these hyphae form a large number of small cells. The apex of each terminal hypha buds off as a rust-colored dikaryotic urediniospore. These spores also have a lipid reserve material and a thick wall. The lipid reserves enable the cell to survive as it is blown around over long distances for some time. Each spore can reinfect a new individual plant or a new location on the same plant. This causes an epidemic, and the infection produces many rust-colored plants, hence the common name "rust." At this stage of the infection, plants are weakened by the infection and become susceptible to other diseases. One further side effect is water loss, since the plants have so many little punctures. Therefore the infected plants are more susceptible to wilting in dry conditions. Uredinium formation continues throughout the growing season.

PREPARING FOR WINTER   Gradually, as the plant is weakened and the growing season ends, uredinia are is no longer formed; instead, each new mycelium forms a telium, which is similar to the uredinium, but its terminal hyphae form only two dikaryotic cells. The cell pair is enveloped inside a thick cell wall that remains attached to the plant even after it dies. The cell pair, called a teliospore, must go through a winter before it can germinate. Germination is initiated in the spring on the decaying leaf tissues. It begins with a short hyphal filament extending out of each cell with both nuclei. The nuclei fuse and proceed immediately with meiosis. This yields four haploid nuclei in each of the two germinating hyphae. Each nucleus buds off as a basidiospore, bearing one mating type, to infect a barberry leaf.

There are a variety of noteworthy variations. Some species have only one plant host and are called monoecious, such as the blackberry rust. Other species are missing some of the life-cycle stages, such as the hollyhock rust, which lacks teliospores and basidiospores. In some rusts, such as the thistle rust, instead of forming small mycelia, the hyphae spread to other parts of the plant also. Many affect trees; for example, *Gymnosporangium juniperi virginianae* infects apples and red cedar.

Ustilaginomycetes

This is a group of plant parasites often called smuts. The name refers to infected parts of the plant, which often have a dry or dusty soot-black color caused by spore production. The hyphae grow inside host tissues as intercellular filamentous endophytes. The dispersal spores emerge on stems, leaves, and reproductive organs. Sometimes they appear as galls (plant tissue enlarged because of fungal mycelial growth inside) or tumors in the roots. Economically serious infections occur in the reproductive organs of flowering plants or seeds of cereal crops. Not all species are obligate parasites because many can be cultured through their life cycle on media without the plant. Cells do not have characteristic organelles, but xylose is absent from the cell wall.

In maize (corn) *Ustilago maydis* forms galls where there would be corn kernels (Figure 23.15). These infected corn kernels are a Mexican delicacy called huitlacoche. The galls break open to release dikaryotic spores called teliospores. The spores are dispersed with wind and are resistant to winter conditions. In spring, germination begins with fusion of both haploid nuclei into a diploid nucleus, which undergoes meiosis. The four meiotic haploid nuclei undergo successive mitotic divisions. Each new nucleus buds off as a separate cell. These cells grow as yeasts and can be cultured on simple media on agar. Cells of complementary mating types can fuse to form a dikaryotic cell. At this point the yeast cell is infectious and will no longer grow on simple media. Parasitic species require plant tissues to continue growing. On appropriate plants the dikaryote will penetrate host tissues and grow as a filamentous endophyte. At intervals the hyphae emerge to the surface, exposing dikaryotic conidia. Conidiospores released to the wind can infect new hosts. When the seed begins to form, hyphae accumulate in the seed, forming a gall. These terminal hyphae produce teliospores for dispersal. The mating type in *U. maydis* is controlled by two genetic loci called *a* and *b*. There are two alleles at the *a* locus, *a1* and *a2*. Each allele codes for a pheromone and a receptor for the complementary pheromone. Mating occurs between two haploid yeasts of complementary mating type, *a1* with *a2*. Thus nearby cells are attracted to the pheromones of the other, for which they have a receptor. They reach each other by extending toward each other. There are about 30 *b* alleles that regulate pathogenicity and sexual reproduction. Dikaryotic yeast cells must have two different *b* alleles to be infectious.

In the more typical situation there is only one mating locus, as in *U. violacea*, which affects carnations. In this example, the hyphae form teliospores in the antheridia. The teliospores are released in the wind instead of pollen. In another example, *U. avenae* infects oat grass. The teliospores form in the inflorescence where the cereal grain would be, just before the uninfected inflorescence emerges. Wind dispersal of teliospores causes infections in the healthy plants, as the teliospores stick to their stigmas and ovaries. The teliospores germinate into yeasts, which conjugate and infect the new seeds as they

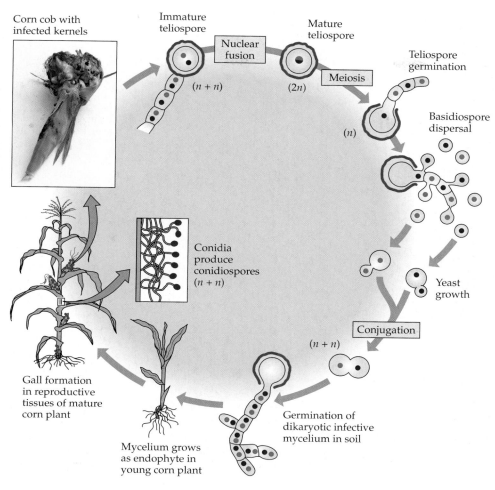

Corn cob with infected kernels

Immature teliospore

Nuclear fusion

(*n* + *n*)

Mature teliospore

Meiosis

(2*n*)

Teliospore germination

(*n*)

Basidiospore dispersal

Yeast growth

Conjugation

(*n* + *n*)

Germination of dikaryotic infective mycelium in soil

Mycelium grows as endophyte in young corn plant

Gall formation in reproductive tissues of mature corn plant

Conidia produce conidiospores (*n* + *n*)

Figure 23.15 **Life cycle of *Ustilago maydis*** Smuts are economically important pathogens that infect plant reproductive structures. Photo courtesy of J. Gemme.

form. The fungus remains in the seed through the winter for protection. In the spring, as the seed germinates, the hyphae grow as an endophyte. When the oat produces inflorescence, terminal hyphae form massive numbers of teliospores in what would be seed tissue.

Microsporidia

All Microsporidia are intracellular parasites found in invertebrates and vertebrate animals as well as in some Alveolata. Infection occurs by ingestion of contaminated food containing Microsporidia spores. The spores are very small, ranging from 1 to 20 µm in diameter, but typically about 5 µm. There is an inner spore wall of chitin called the endospore wall and an outer layer of protein called the exospore wall. At the posterior of the spore cytoplasm, there is a vacuole. There is no cilium or centriole at any stage, and mitochondria are extremely reduced and probably play no role in energy metabolism. The ribosomes are also secondarily reduced into more compact structures. The genome is very small and contains mostly gene coding sequences. The dominant features of the spore cytoplasm are one pair of nuclei (one nu-

cleus in some genera), and the extrusion organelle for penetrating host cells (**Figure 23.16**). The extrusion organelle occupies most of the volume of the spore and consists of three parts. The polar tube is a protein filament coiled around the nuclei, extending straight toward the spore's anterior. The tip of the filament has a cap, called the anchoring disk, for attachment to the host cell. Several layers of membranes surround the polar tube. The polar cap membranes cover the polar filament below the anchoring disk. The membranes below the polar cap are called the polaroplast. All these membranes are believed to be the invaginated and folded extension of the cell membrane. When a spore is ingested by a host, the alkaline pH of the intestines and substances in the mucosa stimulate the infection. Rapid inflow of water causes swelling of the spore and mechanically forces the polar tube into the host cell. The polar tube can penetrate through cell walls and cysts. The spore cytoplasm is ejected into the host cytoplasm by a poorly understood mechanism. The whole infection process is completed in 5 to 30 seconds. The parasite cell is often free in the host cytoplasm, but in some genera it is surrounded by an additional membrane provided by the parasite called the

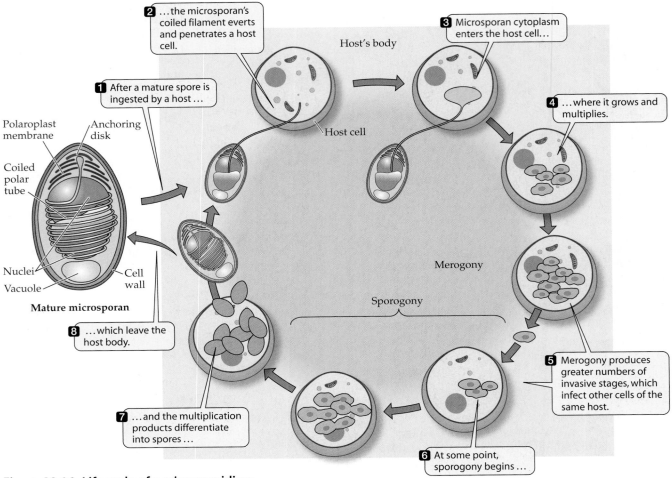

Figure 23.16 Life cycle of a microsporidium
Infection cycle of the intracellular parasite in a eukaryote host cell.

sporophorous membrane. In the host, the parasite derives all its energy requirements and nutrients from the host. Since the parasite has no functional mitochondria or respiratory organelles, it is completely dependent on the host for growth. As the parasite grows, it divides repeatedly, forming new infective spores that can spread to neighboring cells and the circulatory system. Mitosis involves closed mitosis with an amorphous centrosome outside the nuclear envelope. For some species, meiosis has been observed in the host, but it is unclear how it is regulated and how frequent it is. The shape of the parasite in the host and the mechanisms of reproduction vary with species. The current taxonomy and diagnosis is based on morphological details of the extrusion apparatus, the arrangement of the nuclei in the spore and host, and the shape of the spore and growing form in the host. Some have a simple extrusion apparatus, as in *Metchnikovella,* which occur in the gregarines (Apicomplexa) that parasitize the gut of annelid worms. Most mi-

crosporidia have a complex extrusion apparatus, as in *Nosema, Glugea,* and *Encephalitozoon*. More recently, several genera have been found to cause opportunistic infections in immunosuppressed individuals.

SECTION HIGHLIGHTS

Opisthokonts include Animalia, Fungi, and several protist groups, with the ecologically important bacterivorous choanoflagellates being closely related to animals. Chytrids are flagellated and retain fungal ancestral characters. Despite the scarcity of morphological characters in filamentous fungi, they have diversified greatly into saprophytes, symbionts, lichens, and parasites of plants and animals. Many fungi have complex sexual life cycles.

23.4 Excavata

Most members of Excavata (or "excavates") are heterotrophic unicellular flagellates, although one group (Heterolobosea) are mostly amoebas, while another—euglenids—contains many photosynthetic forms. While many excavates are free-living organisms, others are parasites that cause important human diseases. Most free-living species feed by phagocytosis of prokaryotes, which are swept by the beating flagella into a distinctive "feeding groove" that makes up much of one side of the cell. It often appears as if this groove had been dug out of the side of the cell—thus the name "excavates" for whole group.

Diplomonadida

Diplomonads are a remarkable group of eukaryotes. Each diplomonad cell is symmetrical—each half includes a nucleus associated with four flagella. Thus there are two nuclei and two flagellar apparatuses per cell (Figure 23.17). This doubled appearance gives the group its name (*diplo* = double). It is thought that diplomonads evolved from an ancestor with one nucleus and flagellar apparatus that underwent mitosis and replication of the flagellar apparatus but then failed to carry out actual cell division (cytokinesis).

Diplomonads are one of several groups of eukaryotic microbes that live in oxygen-poor environments and lack normal mitochondria with oxidative phosphorylation machinery. They instead have highly reduced organelles with no remaining role in energy production. Although there are free-living diplomonads, most species are parasites or commensals of vertebrate animal hosts.

One diplomonad, *Giardia intestinalis* (also known as *Giardia lamblia*), is a widespread parasite of humans. The infection is acquired by ingesting cysts that activate in the intestinal tract, releasing cells that divide to produce trophozoites (i.e., normal, active cells). These attach to the wall of the small intestine by means of a large adhesive disk. Trophozoites feed by osmotrophy rather than by phagocytosis. They divide by mitotic cell division and sometimes transform into new infective cysts that are shed in the feces. The disease caused by *Giardia* is called giardiasis or "beaver fever"; the symptoms include diarrhea, flatulence, cramps, and weight loss. In the developed world, giardiasis is often associated with drinking from wilderness streams, which contain cysts shed by animals that also serve as hosts for the parasite. In much of the developing world, *Giardia* is routinely passed from human to human by contaminated drinking water, and there are hundreds of millions of infections each year.

Parabasala

Almost all parabasalids are commensals or parasites of animals. Like diplomonads, parabasalids lack normal mitochondria. Instead they have organelles called hydrogenosomes that house an anaerobic energy-generation pathway that liberates hydrogen gas as a waste product. The simplest parabasalids are small or medium-sized cells with four to six flagella (Figure 23.18). One fla-

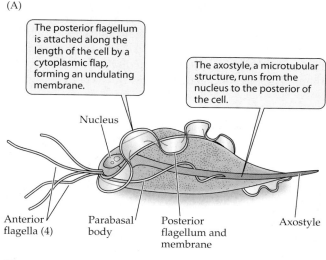

(A)

The posterior flagellum is attached along the length of the cell by a cytoplasmic flap, forming an undulating membrane.

The axostyle, a microtubular structure, runs from the nucleus to the posterior of the cell.

Nucleus

Anterior flagella (4)   Parabasal body   Posterior flagellum and membrane   Axostyle

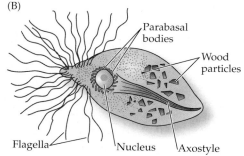

(B)

Parabasal bodies

Wood particles

Flagella   Nucleus   Axostyle

Figure 23.18 Parabasalids
(A) A simple trichomonad parabasalid with five flagella. (B) A hypermastigote with many flagella found in termite guts, where they help digest wood particles.

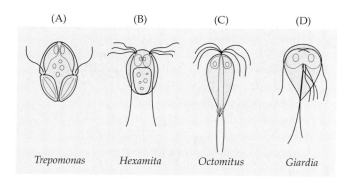

(A)	(B)	(C)	(D)
Trepomonas	*Hexamita*	*Octomitus*	*Giardia*

Figure 23.17 Diversity of diplomonads
(A) *Trepomonas*. (B) *Hexamita*. (C) *Octomitus*. (D) *Giardia*.

gellum is usually attached to the main cell body, forming an undulating membrane. The cell is supported by an axostyle—a hollow rod of microtubules that extends down the center of the cell. Many of these small parabasalids are parasitic; trichomoniasis, the most common sexually transmitted disease (STD) of humans, is caused by a small parabasalid.

Some parabasalids are huge cells with hundreds of flagella. These species live in the hindguts of certain termites and wood-eating cockroaches, along with a complex prokaryotic biota. Many phagocytose wood particles, which are partly digested within the parabasalids. This is important because the termite does not produce its own cellulases (enzymes that break down cellulose). It, in effect, feeds off the waste products of wood digestion by microbes.

Heterolobosea

Most Heterolobosea have two distinct active forms—an amoeba, which is usually the feeding stage, and a flagellate, which is used in dispersal to new habitats (Figure 23.19). Heterolobosean amoebas produce broad pseudopodia, which are used for movement and to phagocytose bacteria. Many Heterolobosea can also form distinctive cysts.

Heterolobose amoebas are found in diverse habitats but are especially common in soil. Most species are free-living, but *Naegleria fowleri* is a facultative parasite of humans. The infection is usually contracted in very warm lakes with disturbed sediment, hot springs, or warm, dirty swimming pools. After entering through the nasal passages, the amoebas migrate into the brain, causing a fatal condition called primary amoebic meningoencephalitis (PAM). Fortunately PAM is extremely rare—only a few hundred cases have ever been confirmed worldwide.

Euglenozoa

This diverse group of excavates includes two major subgroups—euglenids (many of which are photosynthetic) and kinetoplastids (a "protozoan" group including many important parasites). Most euglenozoan cells are supported by parallel microtubules lying beneath the cell membrane. In many species these microtubules can slide relative to each other, allowing the cell to change shape markedly. The flagella (there are usually two) insert into a pocket in the cell surface and have distinctive paraxonemal rods. The feeding apparatus is a tubular structure supported by microtubules. Heterotrophic species typically feed by actively moving through the environment and capturing individual prey particles.

EUGLENIDA Euglenids have a distinctive euglenid pellicle—a series of abutting proteinaceous strips that lie immediately underneath the cell membrane and, in turn, are supported by microtubules. The pellicular strips may be fused to each other, making the cell rigid,

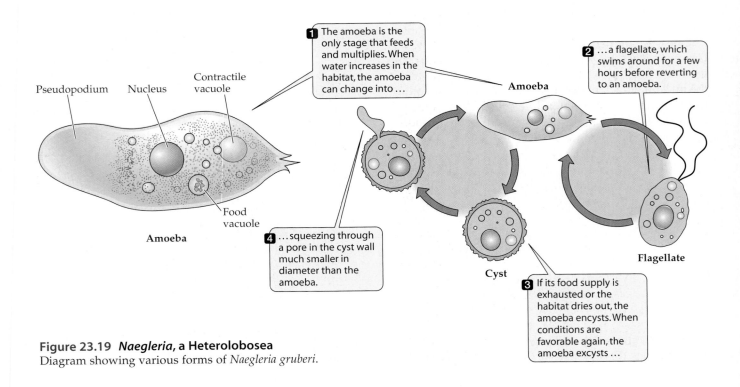

Figure 23.19 *Naegleria,* **a Heterolobosea**
Diagram showing various forms of *Naegleria gruberi.*

(A)

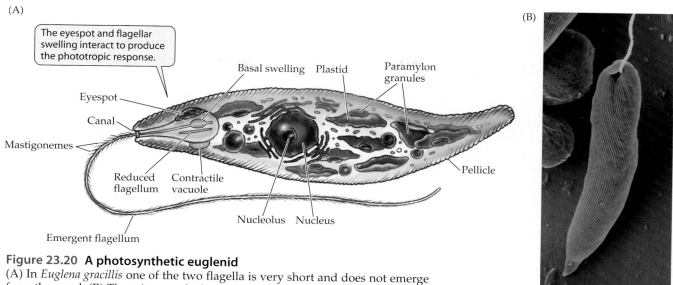

The eyespot and flagellar swelling interact to produce the phototropic response.

Eyespot

Canal

Mastigonemes

Reduced flagellum

Contractile vacuole

Emergent flagellum

Basal swelling

Plastid

Paramylon granules

Pellicle

Nucleolus   Nucleus

(B)

Figure 23.20 A photosynthetic euglenid
(A) In *Euglena gracillis* one of the two flagella is very short and does not emerge from the canal. (B) The micrograph shows the outline of pellicle strips which support the cell membrane. B, ©Biophoto Associates/Photo Researchers, Inc.

or the strips may slide relative to each other, allowing the cell to actively alter its shape. Some euglenids are large cells—more the 100 μm long.

The ancestral euglenids were phagotrophs. Some living phagotrophic euglenids with sliding pellicular strips are spectacular predators that engulf other eukaryotic cells close to their own size. The most conspicuous euglenids, however, are the photosynthetic forms (Figure 23.20), which have a plastid that is surrounded by three membranes and contains chlorophylls *a* and *b*. This plastid is the product of a secondary endosymbiosis between a phagotrophic euglenid host and a photosynthetic eukaryote symbiont, which was some kind of "green alga" (see section 23.1 above). Most photosynthetic euglenids swim using a single functioning flagellum, which is associated with a pigmented eyespot involved in phototaxis (migration towards or away from light). The photosynthetic euglenid *Euglena* has been used widely as a laboratory model organism and as a paradoxical "plant-like animal" or "animal-like plant" in teaching, although it is, of course, neither!

KINETOPLASTEA   Kinetoplastids derive their name from their kinetoplast—a huge mass of DNA within the mitochondrion. The kinetoplast is one of the strangest genomes known, mitochondrial or otherwise (Box 23.1). Many kinetoplastids are small (usually less than 10 μm), free-living cells with two flagella, called bodonids (Figure 23.21). They feed mostly on prokaryotes and are ubiquitous and abundant in many benthic systems. Several groups of kinetoplastids have evolved into parasites.

The most important of these is the trypanosomatids. In contrast to other kinetoplastids, trypanosomatids have only one flagellum, which is usually attached to the cell surface to form an undulating membrane (Figure 23.22).

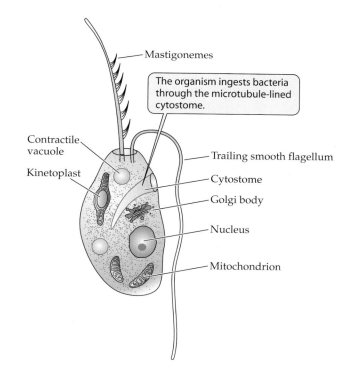

Mastigonemes

The organism ingests bacteria through the microtubule-lined cytostome.

Contractile vacuole

Kinetoplast

Trailing smooth flagellum

Cytostome

Golgi body

Nucleus

Mitochondrion

Figure 23.21 A free-living kinetosplastid
Bodonids are common small bacterivores found in soil and aquatic habitats.

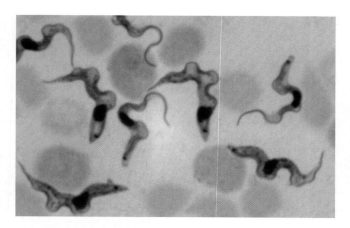

Figure 23.22 A parasitic kinetoplastid
Many trypanosomatids are blood parasites, such as these examples of *Trypanosoma lewisi,* a pathogen of rats. ©Ed Reschke/Peter Arnold, Inc.

Some trypanosomatids infect only one host species, while others alternate between two different hosts—generally a biting insect and a vertebrate. The parasite inhabits the intestinal tract or salivary glands of the insect and is transmitted to the vertebrate by biting, either due to injection of saliva containing the parasite or contamination of the wound by the insect's feces. Three serious diseases of humans are caused by trypanosomatid parasites—sleeping sickness, Chagas' disease, and the various forms of leishmaniasis (Table 23.1).

Sleeping sickness or "African trypanosomiasis" is caused by *Trypanosoma brucei.* The parasites are transmitted by biting tsetse flies. The infection begins in the bloodstream and typically causes relatively mild illness. However, the parasite eventually crosses the blood-brain barrier and causes neurological problems, including disruption of the sleep cycle. Many drugs cannot cross the blood-brain barrier, so treatment has been difficult. This stage of the disease is fatal without treatment, and the main therapy used itself carries a significant risk of death. Entire regions within central Africa have been rendered uninhabitable in some periods due to a high risk of infection.

Trypanosoma brucei is an extracellular parasite, so it is particularly exposed to the human immune system. The parasite has a surface coat of glycoproteins embedded in the cell membrane. There are numerous different

BOX 23.1 *Research Highlights*

The Kinetoplast DNA

The mitochondrial DNA of kinetoplastids forms a large mass that is visible in stained samples under the light microscope. Early scientists were not clear what this mass of DNA was and where it was located. Transmission electron microscopy resolved the puzzle by showing that it was located inside the mitochondrion. Further molecular analysis revealed this kinetoplast DNA to be unusual in two ways. First, the mitochondrial genome is structurally different because it is not a simple circular chromosome, in contrast to the mitochondrial genomes of most other eukaryotes. The kinetoplast DNA instead consists of two kinds of DNA molecules—"maxicircles" and "minicircles." The minicircles can be concatenated together like intercrossing loops. Second, many RNA molecules transcribed from the gene coding regions must be "edited" before they can be used.

Maxicircles can be compared to the mitochondrial genomes of other eukaryotes; they encode several mitochondrial genes. There are about 25 to 50 maxicircle chromosomes of 20 to 38 kb each. However, the original RNA transcripts of some of their protein-coding genes are incomplete. The genes on the maxicircles often lack conventional translation initiation codons and contain frameshifts. These mRNAs must be "edited" enzymatically after transcription by the insertion (and sometimes deletion) of uracils (U) before they will encode a protein correctly. The minicircles are small (usually about 1 kilobase) and do not encode protein-coding genes. Instead, they encode tiny RNAs that "guide" the editing of maxicircle transcripts, instructing the editing machinery of the mitochondrion exactly where to insert and delete uracils from the maxicircle transcripts. Therefore different minicircles code for different guide-RNAs, each of which hybridizes specifically to a particular maxicircle RNA region for editing. Most of the mass of the kinetoplast is composed of a huge number (5,000 to 20,000) of minicircles.

The large amount of RNA editing seen in kinetoplastids is unique, but similar processes have been described in other eukaryotes too. For example, forms of RNA editing are known to occur in the mitochondria of the amoebozoan slime mold *Physarum,* in plant mitochondria, some viruses, and in the nuclei of mammals.

TABLE 23.1	Representative Excavata	
Major Group	**Organism**	**Distinctive Properties**
Diplomonadida	*Giardia intestinalis*	Causes human giardiasis
	Spironucleus spp.	Causes disease in fish
Parabasala	*Tritrichomonas foetus*	Causes spontaneous abortion in cattle
	Trichomonas vaginalis	Causes trichomoniasis
Heterolobosea	*Naegleria fowleri*	Causes primary amoebic meningoencephalitis (PAM)
Euglenida	*Euglena*	Photosynthetic
Kinetoplastea	*Trypanosoma brucei*	Causes sleeping sickness (African trypanosomiasis)
	Trypanosoma cruzi	Causes Chagas' disease (American trypanosomiasis)
	Leishmania spp.	Cause leishmaniasis (kala-azar, etc.)

surface glycoprotein genes in the genome but only one is expressed at a time. The expressed surface glycoprotein gene is switched periodically; thus the parasite is constantly changing the surface it shows to the host immune system. This keeps the parasite one step ahead of adaptive immune responses. Furthermore, frequent recombination within the genome results in the continual creation of novel surface glycoprotein genes in the population.

SECTION HIGHLIGHTS

Most excavates are heterotrophic flagellates, and many are parasites. Diplomonads and parabasalids live in low-oxygen environments and have highly modified mitochondrial organelles. Trypanosomatid parasites cause several major human diseases, including insect-transmitted sleeping sickness. They are surprisingly closely related to euglenids, which include many photosynthetic species.

23.5. Rhizaria

Rhizaria is a grouping that was identified relatively recently by molecular sequence analyses. Most Rhizaria are free-living heterotrophic protists, two major groups of which are highlighted below (Foraminifera and Radiolaria). Rhizaria also includes several parasitic lineages and one small group of algae, called "chlorarachniophytes," which have green plastids that were acquired through secondary endosymbiosis (as in euglenids—see above).

There is no one distinctive morphological feature that unites Rhizaria (Table 23.2), although many produce fine pseudopodia (e.g., filopodia or axopodia) used to capture food and in some cases to move over surfaces. Many Rhizaria are amoeboflagellates with filopodia (e.g., cercomonads) or are amoebas (e.g., filose testate amoebas). Both types are extremely abundant in benthic and soil systems and are of considerable ecological importance.

Foraminifera

Almost all foraminifers inhabit marine environments and have a mineralized external shell, or "test." About 4,000

TABLE 23.2	Representative Rhizaria	
Major Group	**Organism**	**Distinctive Properties**
Foraminifera	Globeriginids	Planktonic marine amoebas
	Soritids	Example of large benthic marine amoebas
Radiolaria	Acantharea	Planktonic amoebas with strontium sulfate skeleton
	Polycistinea	Planktonic amoebas with silica skeleton
Cercomonads	*Cercomonas*	Abundant soil flagellates
Filose testate amoebas	*Euglypha*	Abundant soil amoebas
Chlorarachniophytes	*Chlorarachnion*	Photosynthetic amoebas
Haplosporidia	*Haplosporidium nelsonae*	Oyster parasite (MSX disease)
Plasmodiophorids	*Plasmodiophora brassicae*	Causes "club root" in cabbages

(A)

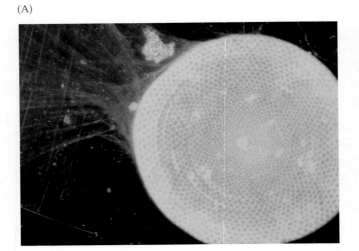

(B)

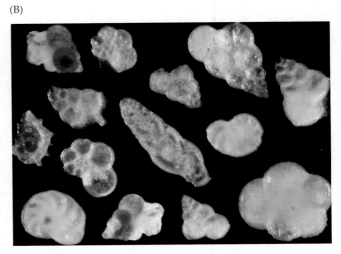

Figure 23.23 Foraminifera
(A) Micrograph of a living foraminifer showing the fine pseudopodia extending from the test apertures and forming a cytoplasmic network outside of the test. (B) Micrograph of various foraminifers showing test diversity. A, ©Biophoto Associates/Photo Researchers, Inc.; B, courtesy of D. McIntyre.

living species have been described, but their fossil record goes back more than 500 million years and includes some 40,000 species. Most foraminifers are relatively large cells—many species have tests that are several millimeters or even centimeters across (Figure 23.23).

Foraminifers have fine pseudopodia supported by microtubules; they can branch and rejoin to form a complex network. In addition to capturing food and allowing movement along surfaces, this network acts as a transport system, with organelles and vesicles shuttling rapidly in both directions by using intracellular transport mechanisms (i.e., dyneins and kinesins acting against the internal microtubules).

The tests of foraminifers are very diverse. In some species the test is composed primarily of mineral particles gleaned from the environment. In most, however, the test is "calcareous"—composed of calcium carbonate. Calcareous tests can be elaborately ornamented and pierced by numerous pores. In most species, individuals grow by adding larger and larger chambers to the test. The cell inhabits the larger chambers; however, some cytoplasm spreads over the surface of the test, so that the pseudopodial network can extend from all around the cell. The life cycles of foraminifers typically involve long periods of growth followed by a large burst of reproduction by multiple fission; it can take months or years to complete. In some, there is an alternation of haploid and diploid generations similar to that seen in some macroalgae (e.g., *Porphyra*, see section 23.6). The mature haploid and diploid individuals can look so different that they may be mistaken for different species.

Most foraminifers are benthic, exploring large areas of sediment with their pseudopodial network to find food. They live in habitats ranging from tropical reefs to the Antarctic and the deep sea and can be extremely abundant, forming a significant percentage of the total biomass in the sediment. One group of foraminifers, globerigerinids, are planktonic cells that float in the water column and trap prey which collides with their pseudopodia. Typical prey includes other large unicellular eukaryotes or even small animals. Many planktonic and shallow-water benthic species maintain symbiotic eukaryotic algae within their cytoplasm and are therefore mixotrophs.

Foraminifers play a major role in the global carbon cycle. The production of calcium carbonate by foraminifers can reach several kilograms per square meter per year, and entire sections of the ocean floor are covered by foraminiferan tests that have sunk from the water column. Some major limestone formations are essentially giant deposits of foraminiferan tests. (The Egyptian pyramids are mostly made from foraminiferan shells!) Being so abundant, fossil foraminiferan tests are a mainstay of stratigraphy (dating sedimentary rocks by their fossils) and are crucial indicators of sediment types and ages in fossil fuel exploration.

Radiolaria

Radiolaria are planktonic marine amoebas with radiating axopodia. Most have an internal mineral skeleton that takes many beautiful shapes and often has radiating spines (Figure 23.24). The mineral skeletons are composed

Figure 23.24 Radiolaria
A variety of radiolarian mineral skeletons observed with a scanning electron microscope and colored with image analysis software. ©Dennis Kunkel Microscopy, Inc.

either of strontium sulfate ($SrSO_4$) or of silica (partly hydroxylated SiO_2). The radiolarian cell itself is organized into two distinct zones—a "central capsule" and a peripheral "ectoplasm." The central capsule houses most of the major cell organelles, including nuclei and mitochondria. The ectoplasm is very diffuse, with most of its volume taken up by large vacuoles that likely control buoyancy. The ectoplasm is also the site of prey digestion. Radiolaria are ecologically similar to planktonic foraminifers. They capture other microbial organisms that encounter their axopodia. Many also host algal symbionts.

SECTION HIGHLIGHTS

Most Rhizaria produce fine pseudopodia (e.g., filopodia or axopodia). Most are free-living amoeboid and/or flagellated cells but a few are parasitic. Foraminifers are shelled amoebas that are extremely abundant and active in and on marine sediments and have a long fossil record. Radiolaria and globerigerinid foraminifers are important microbial consumers in the marine plankton.

23.6 Archaeplastida

Archaeplastida includes three distinct groups of photosynthetic eukaryotes—Chloroplastida, Rhodophyta, and Glaucophyta (Table 23.3). Chloroplastida includes the familiar land plants (Embryophyta) plus a wide diversity of macroscopic and microscopic organisms informally called "green algae." Rhodophyta, or red algae, are mostly marine macroalgae. Glaucophytes are freshwater algae of evolutionary interest; uniquely, their plastids have retained a bacterial-type peptidoglycan layer. The feature uniting Archaeplastida is the presence of plastids of primary endosymbiotic origin (Figure 23.25). Most evidence indicates that a common ancestor of all Archaeplastida engulfed a cyanobacterium, which was converted into a plastid with two bounding membranes.

Rhodophyta

Most Rhodophyta (red algae) are composed of many cells. There are 6,000 or more species, most of which are marine and attach to surfaces in shallow water or the intertidal zone. Nonetheless, some red algae live under extremely low light conditions at considerable depth (250 m), while the unicellular red alga *Cyanidium* is among the most extremophilic of eukaryotes, living in acidic hot springs at 57°C and pH 2.

The plastids of rhodophytes have retained a strong resemblance to cyanobacteria. First, the major chlorophyll present is chlorophyll *a*, whereas other algae usually also have chlorophylls *b* or *c*. Second, the thylakoid membranes have phycobilisomes attached along their surfaces. These phycobilisomes generally contain large amounts of phycoerythrins, imparting the red color typical of rhodophyte plastids.

(A) Rhodophyta

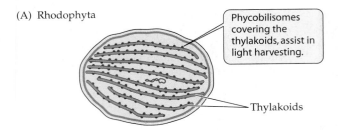

Phycobilisomes covering the thylakoids, assist in light harvesting.

Thylakoids

(B) Chloroplastida

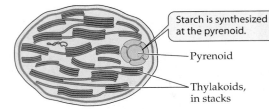

Starch is synthesized at the pyrenoid.

Pyrenoid

Thylakoids, in stacks

(C) Stramenopiles

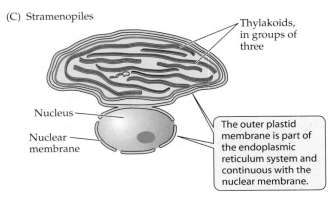

Thylakoids, in groups of three

Nucleus

Nuclear membrane

The outer plastid membrane is part of the endoplasmic reticulum system and continuous with the nuclear membrane.

Figure 23.25 Photosynthetic plastids
Diagram of plastids found in (A) red algae, (B) Chloroplastida, and (C) stramenopiles.

The cell biology of red algae is striking in other respects. They are one of the few major groups of eukaryotes to completely lack flagella and basal bodies (and centrioles). Red algae have cell walls/extracellular matrices that contain large amounts of complex, modified polysaccharides such as sulfated polygalactans (see below). In many red algae, cell division is followed by the formation of a proteinaceous pit plug that forms a physical connection between the cell membranes of the two daughter cells. Many red algae are essentially composed of filaments of cells connected end to end by pit plugs.

Most red algae are sexual. Some, including the edible nori seaweed *Porphyra*, have a two-phase life cycle alternating between haploid (gametophyte) and diploid (sporophyte) generations (**Figure 23.26**). In *Porphyra*, the gametophyte and sporophyte are completely different in appearance. Most red algae, however, have a more complex triphasic life cycle, in which there are actually two

consecutive diploid generations between each haploid generation. In a typical triphasic life cycle, the female gamete remains attached to the parental plant after it is fertilized. The zygote develops into a small diploid adult called a carposporophyte, which also remains attached. The carposporophyte then produces unicellular diploid spores, which disperse and develop into a second diploid form called the tetrasporophyte. It is the tetrasporophyte that finally undergoes meiosis to produce haploid spores, which develop into new gametophytes and complete the life cycle.

Red algae are an important part of the marine benthic algal biota, and some can reach over a meter in length (though not the astounding sizes achieved by some brown algae—see section 23.7). Several species are harvested as sources of agars and carrageenans, which are derived from the modified polysaccharides of the extracellular matrix. These are used as stiffeners in foodstuffs, paints, and other products. Agar, of course, is also essential to microbiology, being used to prepare solid media. The sheet-like red alga *Porphyra* is farmed on a large scale in much of East Asia. It is minced and dried in sheets of nori—the wraps for sushi rolls. This industry is worth well over $1 billion per year worldwide. Some red algae—corallines—deposit large amounts of calcium car-

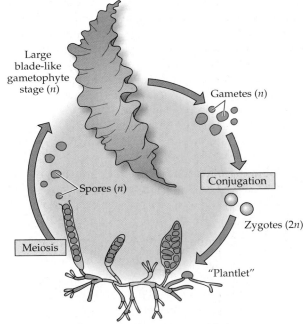

Large blade-like gametophyte stage (*n*)

Gametes (*n*)

Spores (*n*)

Conjugation

Zygotes (2*n*)

Meiosis

"Plantlet"

Sporophyte or "Conchocelis" stage (2*n*)

Figure 23.26 Life cycle of a red alga
The haploid form in *Porphyra* has a large leaf-like shape; the diploid stage is an inconspicuous branched filamentous form that grows on mussel shells. Each form was initially described as a different genus; the diploid form was once called *Conchocelis*.

TABLE 23.3	Representative Archaeplastida	
Major Group	**Organism**	**Distinctive Properties**
Glaucophyta	*Cyanophora*	Unicellular, plastid has peptidoglycan layer
Rhodophyta	*Cyanidium caldarum*	Unicellular, thermoacidophilic
	Porphyra	Cultivated for production of nori
	Gelidium	Harvested for agar (one of many species)
	Corallines	Deposit calcium carbonate
Chloroplastida	*Chlamydomonas reinhardtii*	Unicellular flagellate, laboratory research organism
	Dunaliella salina	Halophilic unicellular flagellate
	Scenedesmus	Small immobile colonies
	Volvox globator	Large motile colonies
	Ulva	Macroscopic seaweeds
	Codium	Siphonous seaweeds
	Charales (stoneworts)	Freshwater tufted algae, related to land plants

bonate in their extracellular matrix. Some form slow-growing stony crusts on existing rocks, and this deposition of calcium carbonate is important in the maintenance of coral reef formations.

Chloroplastida

Chloroplastidans (green algae) range from some of the smallest unicellular eukaryotes known (e.g., the marine flagellate *Micromonas*, which is less than 2 μm long) to complex multicellular macroalgae (e.g., Charales). There are several thousand algal species. The 200,000-plus species of land plants are descended from within the green algae and are therefore also members of the Chloroplastida.

The plastids of Chloroplastida contain chlorophylls *a* and *b* and lack phycobilisomes. The comparative dearth of accessory pigment in most species allows the green pigmentation of chlorophylls *a* and *b* to be observed; thus the plastids tend to be grass-green. Uniquely among algae, the thylakoid membranes themselves are frequently organized into thick stacks called grana. Unusually, the main energy store (starch) is manufactured and accumulated inside the plastid rather than in the cytoplasm.

Chloroplastidan cells are usually supported by an external cell wall or by numerous scales composed largely of complex carbohydrates, such as cellulose. Scales are produced within the endomembrane system and exported to the cell surface. By contrast, the cellulose fibers that form much of the cell wall are synthesized at the cell surface by membrane-bound protein complexes.

Many groups have flagella at some point in the life cycle. The flagella are usually organized into pairs that beat with an oar-like beat pattern. There is usually an eyespot located within the plastid; it interacts with the flagella to permit phototaxis.

Unicellular chloroplastids are common in marine and freshwater systems and even on land (for example, many of the phycobionts—photosynthetic partners—in lichen symbioses are unicellular chloroplastids). The unicellular chloroplastid *Chlamydomonas reinhardtii* is used as a model organism, particularly for studies of the composition, biogenesis, and function of eukaryotic flagella. *Chlamydomonas reinhardtii* is normally haploid and reproduces asexually, but cultures of opposite mating types can be induced to pair up at their anterior ends and then fuse to form a diploid zygote (Figure 23.27). This zygote then encysts and undergoes meiosis to produce new haploid cells.

Several groups of Chloroplastida are colonial. Some produce small colonies with a defined number of cells (often 4, 8, 16, or 32). Others species form filaments that are many cells long. A series of chloroplastidans form motile colonies of flagellated cells. The most complex form is *Volvox*, whose colonies contain hundreds or thousands of *Chlamydomonas*-like cells embedded into the surface of a hollow ball of mucus. Asexual reproduction in *Volvox* involves the growth of spherical daughter colonies inside the parental sphere, which eventually ruptures to release the daughter colonies.

A few groups of Chloroplastida have independently evolved into large plant-like macroalgae with complex morphologies. The ulvophyceans are the major group of green seaweeds. *Ulva*, or sea lettuce, forms large sheets two cells thick and is a common intertidal alga. Many ulvophyceans are siphonous, meaning that they are composed of one or a few extremely elongated, branching multinucleate cells yet are surprisingly plant-like in appearance.

Figure 23.27 Life cycle of *Chlamydomonas*
In *Chlamydomonas*, only the zygote spore is diploid.

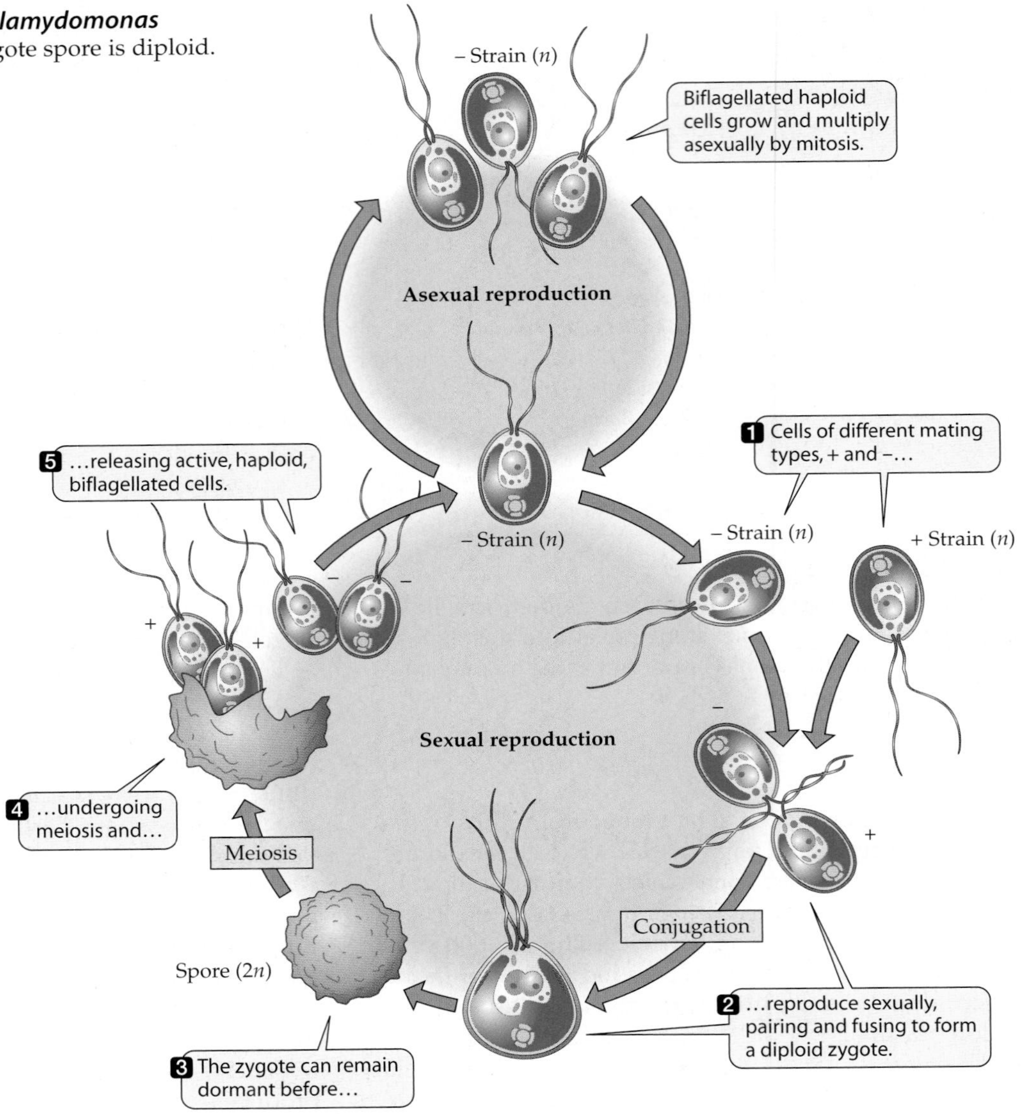

A second multicellular group, the Charales, is made up predominantly of freshwater algae with a central erect shoot supporting tufts of feathery branches that emerge from regularly spaced nodes. They often accumulate calcium carbonate in their cell walls, earning the colloquial name "stoneworts." Charales and land plants share several significant features not found in other major groups of Chloroplastida. They have large immobile female gametes (ova) as well as channels called **plasmodesmata**, which allow communication between the cytoplasm of adjacent cells; in addition, they form a structure called the **phragmoplast** between the dividing cells during cytokinesis. These features and evolutionary trees estimated from molecular sequences confirm that Charales are the closest living relatives of land plants and they closely resemble the aquatic ancestor from which land plants descended over 450 million years ago.

SECTION HIGHLIGHTS

Almost all archaeplastidans are photosynthetic species that are unicellular, filamentous, colonial, or develop as large algae. They stem from a common endosymbiotic event that gave rise to their primary plastids. Rhodophytes are important marine macroalgae that are exploited commercially in several ways. Within Chloroplastida, the Charales are the closest relatives of land plants.

TABLE 23.4	Representative Chromalveolata		
Major Group	**Subgroup**	**Organism**	**Distinctive Properities**
Cryptophyta		*Cryptomonas*	Plastid retains a nucleomorph
Haptophyta	Coccolithophorids	*Emiliania huxleyi*	Produces calcium carbonate scales (coccoliths); forms huge blooms
Stramenopiles	Chrysophyceae	*Ochromonas*	Freshwater mixotroph
	Diatoms	*Thalassiosira*	Planktonic marine centric diatom
		Chaetoceros	Spine-forming, colonial centric diatom
		Pseudonitzshia	Toxic pennate diatom (unusually, planktonic)
	Phaeophyceae	*Laminaria*	Intertidal seaweed, moderate-sized
		Macrocystis	Giant kelp, branched organism up to 50 m long
		Fucus	Common intertidal seaweed
		Ascophyllum	Common intertidal seaweed
	Oomycetes	*Phytophthora*	Filamentous plant pathogens
Alveolates	Dinoflagellates	*Ceratium*	Armored, with elongated spines
		Alexandrium tamarense	Toxic alga, causes paralytic shellfish poisoning (PSP)
		Symbiodinium	Dominant symbiotic alga of corals
		Noctiluca	Large (1-mm) planktonic predator
	Apicomplexa	*Plasmodium*	Causes malaria
		Cryptosporidium	Causes cryptosporidiosis diarrhea
		Toxoplasma	Causes toxoplasmosis
	Ciliophora	*Paramecium*	Large model ciliate
		Tetrahymena	Small model ciliate
		Tintinnids	Important planktonic spirotrich ciliates; Loricate

23.7 Chromalveolates

The Chromalveolates include four major groups, each of which includes protists with plastids that contain chlorophyll *c* in addition to chlorophyll *a*. These are (1) Cryptophyta, (2) Haptophyta, (3) stramenopiles, and (4) Alveolata, although many members of the latter two groups lack plastids. These plastids originated through secondary endosymbiosis involving red algae (in contrast to photosynthetic euglenids and chlorarachniophytes, whose secondary plastids are of green algal origin).

There are competing ideas about the origins of chromalveolates. According to one hypothesis, all chromalveolates stem from a single event of secondary endosymbiosis and are all closely related. A second hypothesis holds that the plastids were acquired in several separate endosymbiotic events and that chromalveolates do not represent one monophyletic group. While awaiting a final resolution of this controversy, it is still convenient to introduce these four groups together (Table 23.4).

In secondary endosymbiosis, the outer membrane surrounding the plastid was originally the membrane of a phagosome and is, in essence, still part of the endomembrane system of the host cell (see Figure 23.25). In three chromalveolate groups—cryptophytes, haptophytes, and stramenopiles—there are four plastid membranes, the outermost of which has attached ribosomes and closely resembles endoplasmic reticulum. It is sometimes called periplastidic endoplasmic reticulum (PER). The PER is often continuous with the nuclear envelope. The plastids of photosynthetic alveolates (dinoflagellates) are bounded by only three membranes, the outermost of which is not PER-like. Chromalveolates typically have thylakoids arranged in stacks of three (although cryptophytes have stacks of two).

In addition to their chlorophylls, chromalveolate plastids have a spectrum of accessory pigments, many of which are carotenoids. Example pigments include β-carotene, fucoxanthins, and, in the case of dinoflagellates, peridinin. In concert, these pigments typically impart a yellow-brown color to the plastids.

Cryptophyta

Cryptophytes (or cryptomonads) have two flagella that are associated with a depression in the cell often called the gullet (Figure 23.28). Most cryptophytes are photo-

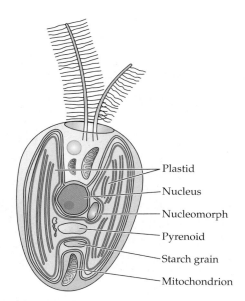

Figure 23.28 Cryptophytes
A photosynthetic cryptophyte (= cryptomonad). Some are mixotrophic or lack plastids.

synthetic and their plastids show some strange ancestral features. Cryptophytes are the only chromalveolate group that has retained red algal/cyanobacterial phycobilin pigments; as a consequence, cryptophyte plastids are often red or olive green rather than yellow-brown. Curiously, the phycobilins are not organized into phycobilisomes but instead are located within the lumen (interior space) of the thylakoids. More remarkably, the original red algal nucleus is still present in reduced form as a "nucleomorph." It sits within the plastid between the inner pair and outer pair of plastid membranes. This remnant nucleus retains a few hundred genes on three tiny linear chromosomes. A few of these genes are essential for the functioning of the plastid. Interestingly, another unrelated algal group—the chlorarachniophytes—also retain a nucleomorph (see section 23.5, above).

Haptophyta

Haptophytes are small algae, most of which are marine species that have two flagella during some or all of their life cycle. Haptophytes usually also possess a slender microtubule-supported filament called the haptonema, which can be ten times the length of the cell. In some species, one of its functions is to collect prokaryotes and small eukaryotes during feeding.

Haptophyte cells are often covered in organic scales that are synthesized in the Golgi and secreted to the cell surface. In many species the scales are massively reinforced with calcium carbonate. These mineralized scales are called coccoliths, and the haptophytes that produce them are known as coccolithophorids. Coccoliths are typically large (half as wide as the cell) and have elaborate sculpting.

The importance of coccolithophorids can be illustrated by considering just one species. *Emiliania huxleyi* exists either as a swimming haploid flagellate with carbohydrate scales or as a floating diploid cell with coccoliths. Found around the world, *E. huxleyi* can form huge blooms with up to 10^5 cells per milliliter over thousands of square kilometers of ocean. The coccoliths are reflective, and thus the blooms are easily imaged from space. When these blooms collapse, most of the coccoliths sink out of the upper ocean water and into the deep ocean. Coccolith sinking is perhaps the single most important route by which carbon is transported to the deep ocean; it is therefore tremendously important in the global carbon cycle. Many of the earth's chalk and limestone deposits were originally accumulations of coccoliths on the ocean floor.

Stramenopiles

Stramenopiles are a large and extremely diverse group that includes algae, protozoan forms, and even some superficially fungus-like organisms. Most stramenopiles have one or two flagella at some stage in their life cycle. One flagellum is directed anteriorly and has two rows of hollow, stiff mastigonemes. These mastigonemes reverse the effect of the flagellar beat, so that the flagellum seems to pull rather than push the cell when a normal flagellar beat is employed. These "thrust-reversing" mastigonemes are unique to stramenopiles and have been useful in revealing the evolutionary affinities of organisms with very different life histories.

Photosynthetic stramenopiles range from small unicellular organisms through to the largest macroalgae. Small cells generally have one plastid that is connected to the nucleus, while large cells have many nonconnected plastids. In many flagellated cells, the plastid contains a highly pigmented eyespot, which is closely associated with the nonhairy (smooth) flagellum and is involved in phototaxis. The typical storage polysaccharide is chrysolaminarin, which is synthesized and stored in the cytoplasm. Phycologists (scientists who study algae) currently divide photosynthetic stramenopiles into about a dozen distinct groups. We will consider only a couple of important groups here.

CHRYSOPHYCEAE   Most chrysophyceans are unicellular or colonial organisms with two flagella (Figure 23.29). Some chrysophyceans are covered in delicate organic or silica scales, while others secrete an organic lorica. Although most photosynthetic stramenopiles are in-

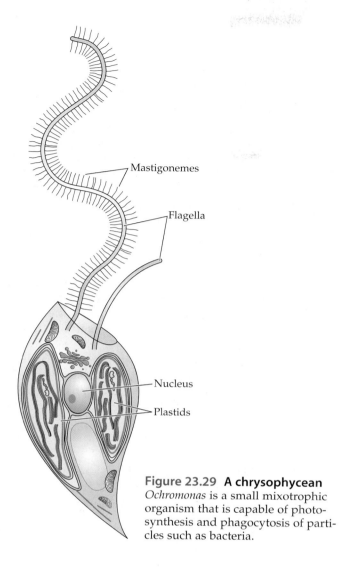

Figure 23.29 A chrysophycean
Ochromonas is a small mixotrophic organism that is capable of photosynthesis and phagocytosis of particles such as bacteria.

Figure 23.30 Diatoms
Diatoms are important in marine and freshwater food webs, where they can be identified by the shape of their glass cell walls. ©R. Brons/Biological Photo Service.

capable of phagotrophy, many chrysophyceae are mixotrophs. Typically, the hair-bearing flagellum creates a feeding current, and small prey (mostly prokaryotes) are captured and phagocytosed using a hoop-shaped feeding apparatus. Some species have reduced plastids and require phagotrophy to survive. Chrysophyceans are common in freshwater habitats, especially in oligotrophic (low nutrient) waters, perhaps because their phagotrophic abilities give them an advantage over other algae in acquiring nitrogen and phosphorous.

BACILLARIOPHYTA (DIATOMS) Diatoms are found in most environments with light and water and dominate the "spring bloom" of microphytoplankton (algal cells larger than 20 μm in diameter) in the ocean. Diatoms are unicellular or colonial phototrophs without flagella (Figure 23.30). The cells are enclosed by a characteristic frustule, in effect a "box with a lid" made of silica. The bottom portion of the box is called the hypotheca, while the lid is the epitheca. Both include a large valve (which forms the top

of the lid or bottom of the box) and several "girdle bands" (forming the side walls). The valves are often elaborately sculpted and pierced by small pores, which allow communication between the cell and the environment.

The frustule profoundly affects the life cycle of diatoms. During normal mitotic cell division, one half of the frustule—either the epitheca or the hypotheca—is inherited by each daughter cell. However, both parental thecae become epithecae in the daughters, and a new hypotheca is produced for each daughter. Since the hypotheca fits inside the epitheca, the average size of cells in a population falls slowly over time. Once the cells reach a critical size (typically 30% the maximum cell size), they can become sexual and undergo meiosis to form haploid gametes. The gametes fuse and form a large diploid cyst-like cell called an auxospore. Eventually a new diploid cell with a full-size frustule develops within the auxospore; it "resets" the original size for the next population.

There are two basic types of diatoms, centric and pennate. **Centric diatoms** generally have valves with radial symmetry and a rounded profile. In some species the valves bear elongated spines. Most centric diatoms are planktonic and move only by regulating their buoyancy. Many connect valve-to-valve to form colonies. In the sexual phase, centric diatoms produce either one or two large, immobile female gametes or numerous flagellated male gametes. **Pennate diatoms** are mostly benthic and generally have valves with bilateral symmetry and a boat-like profile. In most species one or both valves have a longitudinal slit-like structure called the raphe, through which the cell secretes mucus; it allows the cell to adhere to surfaces and, usually, to glide forward and backward. The gametes of diatoms are equal in size and lack flagella.

PHAEOPHYCEAE   The best-known group of photosynthetic stramenopiles are the Phaeophyceae or brown algae. Most browns are marine algae, ranging from microscopic, one-cell-wide filaments through to huge multicellular organisms (Figure 23.31). The largest brown algae rival land plants in size and in their degree of differentiation into distinct tissues. The cell walls of phaeophyceans are primarily composed of compounds called alginates. As in land plants and Charaleans, adjacent cells communicate via narrow plasmodesmata.

Most Phaeophyceae show a classic alternation between haploid (gametophyte) and diploid (sporophyte) generations. The generations are often highly dissimilar—in kelps the sporophyte is huge while the gametophyte is microscopic. The unicellular spores produced by sporophytes are usually flagellated, as are one or both sexes of gametes.

The kelps (Laminariales) are the largest and most complex algae on earth. Each organism is composed of a stem-like stipe that is attached to rocks or the sea floor by a holdfast and is surmounted by leaf-like blades, which are the primary photosynthetic structures. The internal tissues of the blades and stipe include a network of large colorless cells called trumpet hyphae or sieve elements that are specialized for transporting solutes such as sugars. Species that live in deeper water often have gas-filled floats called pneumatocysts or bladders at the base of each blade. These help the blades to remain near the surface, where light levels are highest. The giant kelp *Macrocystis* can be 50 m tall and has a branched stipe that bears numerous blades. *Macrocystis* and other large kelp form extensive forests in coastal waters in the Pacific and the Southern Hemisphere. *Macrocystis* is harvested as a source of alginates, which are used as thickeners and stiffeners in many commercial products. Another group of complex brown algae, the "wracks" (Fucules, e.g., *Fucus*), are the dominant intertidal seaweeds in much of the Northern Hemisphere.

Nonphotosynthetic stramenopile groups are very diverse. Some, such as bicosoecids, are mostly small, free-living flagellates that capture bacteria. Opalinids, by con-

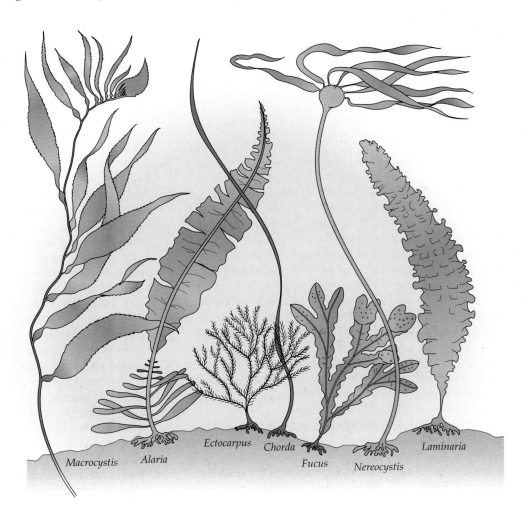

Figure 23.31 Phaeophyta
Drawings representing a diversity of large brown algae. Not drawn to scale.

Macrocystis   *Alaria*   *Ectocarpus*   *Chorda*   *Fucus*   *Nereocystis*   *Laminaria*

trast, are large multiflagellated cells that feed by pinocytosis and live in the hindguts of certain poikilothermic vertebrates, especially frogs. Finally there are several groups of stramenopiles that are fungus-like saprotrophs or pathogens. We will discuss one such group—the oomycetes.

PERONOSPOROMYCETES (OOMYCETES)   Oomycetes are important because many are pathogens that have devastating effects on crops. For example, *Phytophthora infestans* causes the late blight disease of potatoes that led to the great famine in Ireland in the 1840s. Other parasitic oomycetes attack grapevines, legumes, various trees, and even fish.

Oomycetes are saprotrophic and grow by forming a network of branching filaments. The cell wall is predominantly composed of cellulose and cellulose-like polysaccharides. Nutrients are acquired by the uptake of small organic molecules by osmotrophy. As saprotrophs, oomycetes mobilize these soluble molecules by secreting enzymes that break down complex organic compounds. The filaments are mostly coenocytic—that is, not separated into distinct cells by cross walls. Free-living oomycetes are important in freshwater systems, often fulfilling a decomposer role similar to that of fungi on land.

Oomycetes are predominantly diploid. In the sexual phase, differentiated male filaments (antheridia) and female filaments (oogonia) both undergo meiotic divisions to produce haploid gametes, and the male gamete penetrates the oogonium to fuse with a female gamete. The resulting diploid zygote (oospore) develops into a large sporangium, which produces many diploid spores, which are often flagellated. Oomycetes also reproduce asexually by producing diploid spores.

Alveolata

Alveolata includes three very different types of eukaryotes: (1) dinoflagellates, an important group of heterotrophic and photosynthetic protists in marine and freshwater environments; (2) Apicomplexa, arguably the most successful microbial eukaryotic parasites on earth; and (3) ciliates, one of the protist groups of greatest ecological significance. Their close evolutionary affinities were demonstrated by molecular studies, but alveolates also share one key cellular feature: in all three groups, there is a layer of flattened membrane sacs, called alveoli, which lie immediately beneath the cell membrane. The alveoli are supported, in turn, by cytoskeletal elements and provide structural strength to the cell.

DINOFLAGELLATA (DINOFLAGELLATES)   Dinoflagellates are the only group within Alveolata with their own

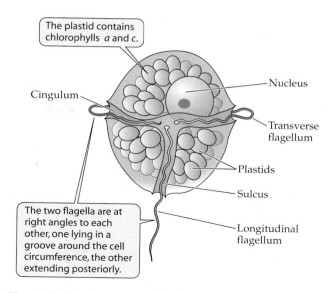

The plastid contains chlorophylls *a* and *c*.

Cingulum

Nucleus

Transverse flagellum

Plastids

Sulcus

The two flagella are at right angles to each other, one lying in a groove around the cell circumference, the other extending posteriorly.

Longitudinal flagellum

Figure 23.32 A common dinoflagellate
Species of *Gyrodinium* are common in marine coastal regions.

photosynthetic plastids. There are about 3,000 living species, ranging in size from a few micrometers up to a millimeter or more. Most are free-swimming cells with two flagella (Figure 23.32). In general, one flagellum is directed posteriorly within a channel called the sulcus. The second flagellum circles the cell in a groove called the cingulum. Both flagella usually have distinctive paraxonemal rods and/or mastigonemes. Many dinoflagellates are "armored," meaning that each alveolus contains a thecal plate of cellulose. These abut against one another, giving the cell a rigid structure. In some species, the thecal plates form elaborate spines and wings. Many dinoflagellates can transform into cysts that persist in the sediment for long periods. The nucleus of most dinoflagellates is unique. Unlike other eukaryotes, the DNA is not packaged around histones, but appears nearly naked within the chromosomes, which are condensed throughout the cell cycle.

About half of all dinoflagellate species are photosynthetic, while nonphotosynthetic dinoflagellates are generally phagotrophs or parasites. Many photosynthetic dinoflagellates are actually mixotrophs, which will also phagocytose particulate food. Phagotrophic dinoflagellates generally prey on other protists. Many use an extending, tubular peduncle to pierce the membrane of another eukaryotic cell and then suck up the cell contents. The huge (greater than 1 mm) heterotrophic dinoflagellate *Noctiluca* uses a flexible feeding tentacle to capture prey, including small planktonic invertebrates.

Dinoflagellates are of major importance in the environment. As algae, they are second only to diatoms in importance in the marine microphytoplankton. As protozoa, they are significant consumers of other protists. Virtually all reef-building corals harbor symbiotic photosynthetic dinoflagellates of the genus *Symbiodinium.* These symbionts are essential for the long-term survival of the corals and hence of coral reefs. Dinoflagellates are also the main agents of toxic algal blooms, an increasing environmental and health problem.

APICOMPLEXA Apicomplexa are a diverse group of animal parasites, most of which are intracellular and have complex sexual life cycles. There are several apicomplexans that directly affect humans. *Plasmodium* causes malaria, one of the three worst infectious diseases in the world (along with HIV/AIDS and tuberculosis). Cryptosporidiosis, toxoplasmosis, and several important livestock diseases are also caused by apicomplexan parasites.

In introducing the cell organization of Apicomplexa, we will consider the main invasive stages: sporozoites and merozoites (see below). These cells lack flagella but may move using gliding motility (Figure 23.33). Cells feed by osmotrophy and by pinocytosis at a structure called a **micropore**, a coated pit that passes through a hole in the alveolar system. The cell is dominated by a group of organelles called the **apical complex**, involved in host cell invasion. The apical complex has four main components: (1) one or two polar rings, (2) a tubulin-based truncated cone-like structure called the conoid (absent in some groups, e.g., *Plasmodium*), (3) many small secretory vesicles called micronemes, and (4) a few large secretory vesicles called rhoptries. During invasion, the apical complex of the sporozoite or merozoite is brought into contact with the host cell. Microneme proteins are involved in attachment to the host cell and in pushing the parasite into the host cell. The apicomplexan cell induces invagination of the host cell membrane, forming a parasitophorous vacuole around the parasite. Lipids and proteins discharged from the rhoptries modify the contents of the vacuole and the vacuolar membrane to suit the parasite. Remarkably, many apicomplexan parasites have a small nonphotosynthetic plastid (Box 23.2).

Plasmodium, the malarial parasite, is one of many apicomplexans that are blood-borne parasites of terrestrial vertebrates and are transmitted by biting insects. In the case of human malaria, the insects are *Anopheles* mosquitoes. Four different species of *Plasmodium* infect humans, causing somewhat different disease scenarios; most deaths are caused by *Plasmodium falciparum*. Like that of most apicomplexan parasites, the life cycle of *Plasmodium* is sexual and involves three distinct events of reproduction, each potentially involving production of a large number of progeny by multiple fission (Figure 23.34). These events are (1) merogony (or schizogony), which produces merozoites; (2) gamogony, which produces male

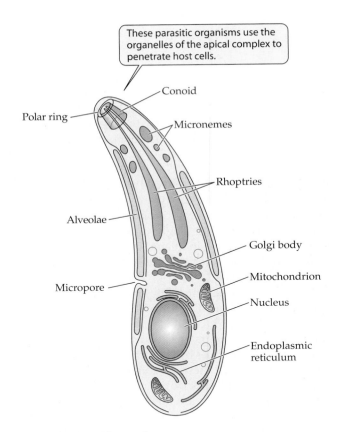

These parasitic organisms use the organelles of the apical complex to penetrate host cells.

Conoid
Polar ring
Micronemes
Rhoptries
Alveolae
Golgi body
Mitochondrion
Micropore
Nucleus
Endoplasmic reticulum

Figure 23.33 Apicomplexa
Diagram of a sporozoite. The main components of the apical complex are shown (polar ring, conoid, micronemes, and rhoptries). Micronemes and rhoptries are different types of secretory vesicles whose contents are involved in interaction with the host cell during invasion.

gametes; and (3) sporogony, which is meiotic and produces sporozoites. The stage transmitted by the mosquito is the sporozoite, a slender, scythe-shaped cell. Sporozoites enter the bloodstream and invade liver cells. There they undergo a round of merogony to produce somewhat rounded merozoites. Merozoites released from liver cells then infect red blood cells, and there undergo merogony again. Merozoites released from bursting red blood cells invade other red blood cells and this cycle then repeats itself. Some invasions result in the production of gametocytes. When an infected human is bitten by another mosquito, gametocytes are taken up into the insect gut, and develop into female gametes, or undergo gamogony and produce male gametes. Gametes there fuse, forming a zygotic cell that is the only diploid stage in the life cycle. These invade intestinal cells, encyst, and undergo sporogony to produce haploid sporozoites, which migrate to the salivary glands, ready to infect another human.

In contrast to *Plasmodium*, *Toxoplasma* and *Cyptosporidium* infections are acquired by ingesting cysts containing sporozoites. The infections are spread by fecal contamination (from cats rather than humans in the case of *Toxoplasma*).

BOX 23.2   *Research Highlights*

Can Herbicides Cure Malaria?

Some apicomplexan parasites of direct importance to humans include *Plasmodium, Cryptosporidium,* and *Toxoplasma. Plasmodium* in particular causes malaria in animals, and four species (*P. vivax, P. ovale, P. malariae,* and *P. falciparum*) affect humans. Over 2 million people die each year from malaria, and between 300 and 500 million people are affected by this parasite.

Although members of Apicomplexa are not photosynthetic, studies over the last 15 years

have shown than many species actually have a small degenerate plastid. This organelle is surrounded by four membranes and is the product of secondary endosymbiosis, most likely the same symbiosis that gave rise to the dinoflagellate plastid. The apicomplexan plastid is involved in several biosynthesis pathways, such as the production of fatty acids for the cell. Some of these essential pathways use enzymes that are found only in other plastids and prokaryotes. It is therefore possible that drugs or

compounds that inhibit these plastid enzymes specifically would harm the parasite without affecting the animal host cells (since animal cells definitely do not have plastids!) Far from being merely a cellular biological peculiarity, the apicomplexan plastid is now considered an important potential target for novel drugs to combat a major human disease, malaria. It is hypothesized that herbicides designed to interfere with plant metabolism could be used against apicomplexan parasites.

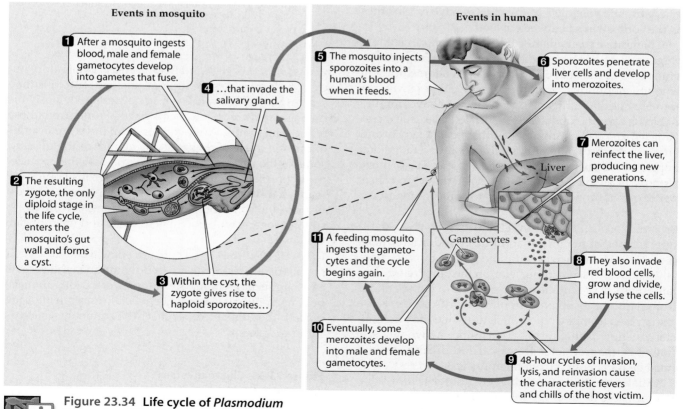

Figure 23.34 Life cycle of *Plasmodium*
Life cycle of *Plasmodium vivax*, one of the species that cause malaria in humans. The vector is an *Anopheles* mosquito. Merozoites have the same structure and function as sporozoites, but they are produced in the human host instead of in the mosquito.

CILIOPHORA   Ciliates are important consumers of other protists and prokaryotes in almost all habitats. There are two conspicuous defining features of ciliates (Figure 23.35). First is the presence of numerous cilia that move in coordination and with a distinctive ciliary beat. The second feature is nuclear dualism—the presence of two distinct types of nuclei, the macronucleus and the micronucleus, within each cell. Ciliates vary in size from 10 μm to several millimeters and have a complex and relatively robust cortex that imparts a well-defined shape (although most ciliates are also flexible or contractile). In many ciliates, the cilia are distributed all over the cell in ordered longitudinal rows called **kineties**. Cytoskeletal elements connect the basal bodies of the cilia within a kinety and in adjacent kineties. Thus there is a cytoskeletal meshwork around most of the cell, adding further strength. The continuity of the cortical cytoskeleton is usually interrupted in two places. Most ciliates have a distinct cytostome where prey is collected and packaged into food vacuoles. The cytostome is usually part of a larger oral apparatus that includes differentiated ciliary structures (see below). In some species, there is a cytoproct, which is a permanent site for exocytosis of undigested wastes. In addition, freshwater ciliates have distinct pores for the expulsion of contractile vacuoles. Many ciliates have extrusomes in the cortex that sit between alveoli.

Ciliates use their cilia for locomotion and in feeding (Figure 23.36). At any one time, there are cilia at different stages in the beat cycle across the cell, resulting in smooth continuous swimming. Some ciliates swim extremely rapidly (mm sec^{-1}). Many ciliates have bundles or blocks of numerous closely associated cilia, called cirri (sing. cirrus) and membranelles, that beat in synchrony. Organized arrays of membranelles are found in the oral apparatus, where they generate feeding currents to move food particles toward the cell and act as a filter to collect particles and funnel them to the cytostome for ingestion.

SEX AND NUCLEI IN CILIATES   Ciliates have two different types of nuclei—macronuclei and micronuclei (Box 23.3). Micronuclei are diploid but are transcriptionally inactive. The macronucleus, on the other hand, contains many copies of the active genome of the cell and is transcriptionally active—in other words, mRNAs are transcribed from the macronucleus, while the micronuclei function as the dormant germline. As we will see, however, macronuclei are discarded during sex and must be regenerated from micronuclei.

Ciliates normally reproduce by mitotic cell division but occasionally undergo sexual conjugation, when two cells of compatible mating types will partially fuse. They are then called conjugants. Their macronuclei degrade while their micronuclei undergo meiosis to form haploid nuclei. One haploid nucleus from each conjugant mi-

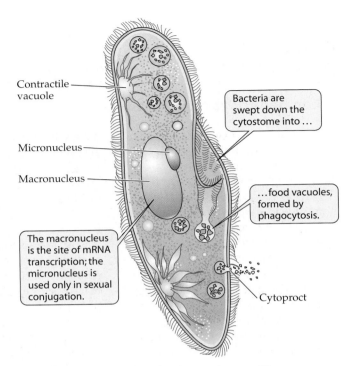

Figure 23.35 Diagram of *Paramecium*, a ciliate
Paramecium has a cytostome for collecting bacteria, defensive extrusomes, and large contractile vacuoles. Undigested food remaining in vacuoles is expelled through a cytoproct.

Contractile vacuole

Micronucleus

Macronucleus

Bacteria are swept down the cytostome into …

…food vacuoles, formed by phagocytosis.

The macronucleus is the site of mRNA transcription; the micronucleus is used only in sexual conjugation.

Cytoproct

grates to the other cell and fuses with a nonmigratory nucleus to form a zygotic diploid nucleus. The cells then separate and are now called exconjugants. Obviously the exconjugants now have genotypes that differ from those of the conjugants. After conjugation, there are a series of mitotic nuclear divisions in each exconjugant, and some of the micronuclei develop into new macronuclei by a complex process in which numerous copies are made of only part of the micronuclear genome. Cell divisions (and some more nuclear divisions) eventually result in progeny with both macronuclei and micronuclei.

CILIATE DIVERSITY   There are some 8,000 described species of ciliates. Some are free-swimming while others attach to surfaces. Some are parasites of various animals, but anaerobic ciliates are a beneficial component of the biota in the rumen (forestomach) of ruminants, such as cattle and sheep.

Figure 23.36 Ciliate diversity   ▶
Ciliates have very diverse morphologies. Here are a few examples: (A) *Tetrahymena*. (B) *Euplotes*, a hypotrich; its oral cilia are arranged in a similar way to those in *Tetrahymena* but cover a greater area. (C) *Didinium* feeding on a *Paramecium*. (D) *Tracheloraphis*, a karyorelictid that lives in the sand of seashores.

BOX 23.3	*Research Highlights*

The Strange Genomes of Ciliates

Ciliates have drawn the attention of biologists many times over the past 150 years. They were used in early studies of meiosis and, at the end of the nineteenth century, to discover the role of sex in rejuvenating the genetic material. They were used by Tracy Sonneborn to discover mating types in eukaryotes and became important study organisms in the early days of genetics and cell biology. Ciliates have also been used to describe epigenetic inheritance—that is, nongenetic heritable traits—as in the maintenance of cell shape. With the advent of molecular biology, ciliates became known for being different from other standard eukaryotic model organisms.

Ciliates have two types of nuclei, the micronucleus and the macronucleus. Initially both nuclei are the result of a mitotic division. Then one becomes inactive as the micronucleus. Its DNA becomes condensed around small modified histone proteins. The other nucleus develops into the macronucleus through a series of transformations

to the DNA. In some ciliates, only the ribosomal DNA is amplified multiple times in the macronucleus. In many ciliates, the micronuclear genome is replicated many times in the macronucleus, but many noncoding DNA regions are eliminated at the same time. The hypotrich ciliates, such as *Oxytricha*, serve as an example. During macronucleus development, the chromosomes are replicated many times in a way similar to that of the *Drosophila* polytene chromosomes. Then noncoding DNA regions between genes are eliminated. The introns are also removed, and other short noncoding regions (called internal eliminated sequences) are removed. The result is that each chromosome is trimmed to hold only coding and regulatory sequences. Furthermore, the original chromosomes become highly fragmented, such that each protein coding region is found on a single small DNA molecule, each with a short telomere. The resulting macronucleus of hypotrichs, then, holds only short DNA molecules,

with the noncoding DNA eliminated, and each molecule is amplified about 10,000 times. This allows the macronucleus to produce a large amount of RNA from the DNA template.

More oddly, some of the hypotrich genes are scrambled. That is, the introns and the exons are not in the right order or in the right direction in the micronuclear genome. During the generation of a new macronucleus, a complex DNA unscrambling mechanism is required to splice together the exons in the correct order and the correct orientation.

Last, some ciliates have evolved nonstandard genetic codes. That is, the standard nucleotides coding for amino acids and stop codons in eukaryotes and prokaryotes are modified. For example, with few exceptions, the nucleotides TAG, TAA, and TGA are recognized as stop codons in eukaryotes and prokaryotes. Instead, in many ciliates, the TAG and TAA codons code for the amino acid glutamine.

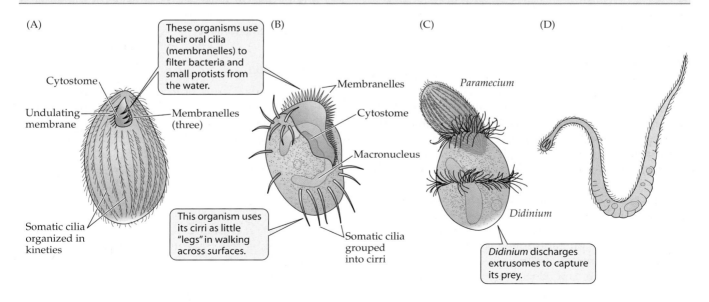

(A)

Cytostome

Undulating membrane

Membranelles (three)

These organisms use their oral cilia (membranelles) to filter bacteria and small protists from the water.

Somatic cilia organized in kineties

(B)

Membranelles

Cytostome

Macronucleus

This organism uses its cirri as little "legs" in walking across surfaces.

Somatic cilia grouped into cirri

(C)

Paramecium

Didinium

Didinium discharges extrusomes to capture its prey.

(D)

Paramecium and *Tetrahymena* are examples of ciliates with cilia over the entire cell (see Figure 23.36A). Species of both *Paramecium* and *Tetrahymena* can be grown readily in the laboratory and are important model organisms for general cell biology and molecular biology. They are especially important in studies of flagella/cilia and the cytoskeleton. *Tetrahymena* was also employed in the discovery of ribozymes (catalytic RNA molecules).

The most important group of ciliates in nature is probably the spirotrichs. In contrast to *Tetrahymena* and *Paramecium*, many spirotrichs have cilia and cirri arranged in particular positions on the cell surface. Certain spirotrichs are the dominant ciliates in the ocean plankton. They typically feed on small eukaryotes, although many also have algal symbionts or temporarily retain functional plastids from ingested algae, a phenomenon known as kleptoplasty. Some other spirotrichs associate with surfaces and use their cirri like legs (see Figure 23.36B).

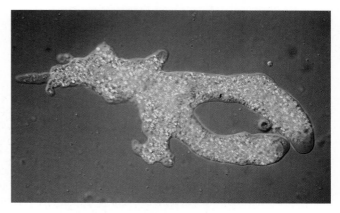

Figure 23.37 Lobose amoeba
Many amoebas move by using their pseudopodia; phagocytosis occurs by invagination of the cell membrane around prey items. ©M. Abbey/Visuals Unlimited.

SECTION HIGHLIGHTS

The origin of Chromalveolata is hypothesized to have been a secondary endosymbiosis event involving a heterotrophic eukaryote host and a red alga. Many groups are ecologically important algae, especially haptophytes (including coccolithophorids), diatoms, and dinoflagellates. Some groups lack plastids and are phagotrophic, saprotrophic, or parasitic. Apicomplexan parasites are among the most important parasites on earth. Ciliates are extremely important microbial predators.

23.8 Amoebozoa

Amoebozoa includes many of the eukaryotes that produce broad ("lobose") pseudopodia. These are different from the thinner and often microtubule-supported pseudopodia typical of Rhizaria, for example. Few amoebozoans have flagella at any point in their life cycle. Most use pseudopodia both for motility and phagocytosis. Amoeboid cells take characteristic shapes depending on the form of the pseudopodia. Most large amoebozoans can form several lobose pseudopodia simultaneously. In others, the whole cell body is effectively a single large pseudopodium. Some can produce numerous finer subpseudopodia emanating from the pseudopodia. There is a bulb-shaped uroid at the posterior end of many lobose amoebas which is probably involved in moving the contractile network material. There are three basic types of organisms within Amoebozoa: (1) lobose amoebas, (2)

Eumycetozoa or true slime molds, and (3) pelobionts and entamoebas (Table 23.5).

Lobose Amoebas

Lobose amoebas range from less than 5 μm up to more than 1 mm in length (Figure 23.37). They never have flagella. They are abundant both in nutrient-rich sediments and in soil, where their numbers can reach 10^6 cells g^{-1}. They are also common in marine environments on suspended particles and at the air-water interface. One subgroup, the "lobose testate amoebas," produces an organic test that sometimes incorporates mineral objects from the environment. The test confers desiccation resistance, and testate amoebas are common in environments in which water is limited or ephemeral, especially soils. Some lobose amoebas are parasites of animals, but those causing disease in humans generally represent opportunistic infections.

Eumycetozoa

Eumycetozoa are amoeboid organisms that produce stalked structures called fruiting bodies, which release spores. The fruiting body is an adaptation for long-range dispersal in the terrestrial environment, and most eumycetozoa inhabit forest soil, bark, rotting timber, or animal dung. There are two very different kinds of eumycetozoans—acellular slime molds (Myxogastria) and cellular slime molds (Dictyostelia).

MYXOGASTRIA (ACELLULAR SLIME MOLDS)   Acellular slime molds are saprotrophic species that feed on decaying organic matter, especially woody debris. They have a complex sexual life cycle (Figure 23.38), begin-

TABLE 23.5	Representative Amoebozoa	
Major Group	**Organism**	**Distinctive Properties**
Lobose amoebas	*Amoeba*	Large lobose amoeba
	Arcella	Testate amoeba
	Acanthamoeba	Common soil amoebas, occasionally infect the cornea
Mycetozoa	*Physarum*	Acellular slime mold with conspicuous yellow plasmodium
	Dictyostelium	Cellular slime mold with migratory pseudoplasmodium
Pelobionts and entamoebas	*Entamoeba histolytica*	Causes amoebic dysentery and liver abscesses

ning life as small haploid amoebas or flagellates that feed and undergo cell division. Eventually two cells will fuse to form a diploid cell, which grows without cell division to form a large multinucleate plasmodium. When mature, the plasmodium produces numerous fruiting bodies. Meiotic division takes place within the fruiting bodies, producing numerous haploid spores. After dispersal (by wind, rain, or passing animals), each spore releases a new amoeba or flagellate. The name "acellular slime mold" refers to the stalk of the fruiting body, which is secreted by the plasmodium and not constructed from cells.

DICTYOSTELIA (CELLULAR SLIME MOLDS) In cellular slime molds there is an amoeba stage ("myxamoebas") but no plasmodium stage or flagellates (Figure 23.39). The formation of fruiting bodies in cellular slime molds is very different from that in acellular slime molds—it does not involve cell division, nor is it part of the sexual cycle. Instead, the myxamoebas feed and divide until the local food supply is exhausted, whereupon starving individuals releases a chemical attractant—for example, pulses of cyclic AMP. This induces other nearby myxamoebas to move toward the starving cells and also begin to release the attractant, amplifying the

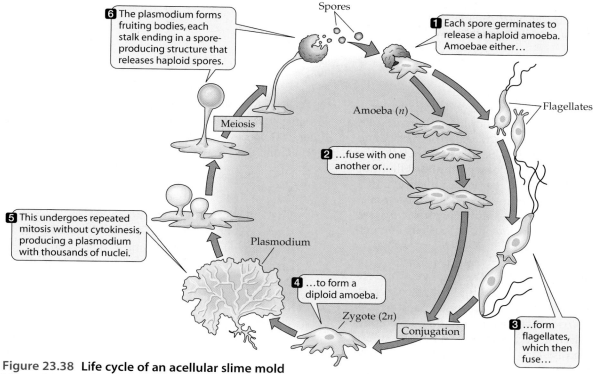

6 The plasmodium forms fruiting bodies, each stalk ending in a spore-producing structure that releases haploid spores.

Spores

1 Each spore germinates to release a haploid amoeba. Amoebae either...

Meiosis

Flagellates

Amoeba (*n*)

2 ...fuse with one another or...

5 This undergoes repeated mitosis without cytokinesis, producing a plasmodium with thousands of nuclei.

Plasmodium

4 ...to form a diploid amoeba.

Zygote (2*n*)

Conjugation

3 ...form flagellates, which then fuse...

Figure 23.38 Life cycle of an acellular slime mold
Life cycle stages in *Physarum*. After conjugation, the diploid zygote undergoes mitotic divisions without cell division, forming a large multinucleate plasmodium.

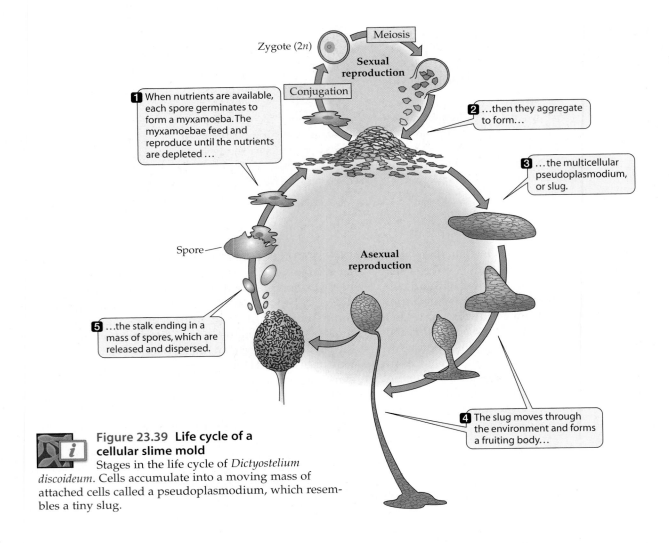

Zygote (2*n*)

Meiosis

Sexual reproduction

Conjugation

1 When nutrients are available, each spore germinates to form a myxamoeba. The myxamoebae feed and reproduce until the nutrients are depleted …

2 …then they aggregate to form…

3 …the multicellular pseudoplasmodium, or slug.

Spore

Asexual reproduction

5 …the stalk ending in a mass of spores, which are released and dispersed.

4 The slug moves through the environment and forms a fruiting body…

Figure 23.39 Life cycle of a cellular slime mold
Stages in the life cycle of *Dictyostelium discoideum*. Cells accumulate into a moving mass of attached cells called a pseudoplasmodium, which resembles a tiny slug.

signal. Eventually thousands of amoebas aggregate to form a mass of cells called a pseudoplasmodium or grex. The pseudoplasmodium can migrate along a surface and then forms into a fruiting body. Some (about 30%) of the cells are sacrificed to form the baseplate and stalk (they expand and harden, dying in the process). The expanding stalk lifts up the rest of the cells to form the sorus of the fruiting body, where they simply encyst to form the spores. The name "cellular slime mold" refers to the cellular nature of the stalk.

PELOBIONTS AND ENTAMOEBAS   Pelobionts and entamoebas live in environments with little free oxygen. Their mitochondrial organelles are small, lack cristae, and lack normal oxidative phosphorylation. Pelobionts are usually free-living, and most have a flagellum. Entamoebas, by contrast, lack flagella and are mostly commensals or parasites of animals. *Entamoeba histolytica*

causes serious diarrheal disease and also occasionally invades organs other than the intestine, in particular the liver. This species is responsible for an estimated 500 million infections and 100,000 deaths per year.

SECTION HIGHLIGHTS

The Amoebozoa consist of groups that are amoeboid, but some have flagellated stages. Most are free-living and are most common in sediments and soils, but a small number are pathogenic. There are two types of slime molds that produce fruiting bodies as a dispersal strategy, but their life cycles differ dramatically.

SUMMARY

- Eukaryotes have organelles, including mitochondria and plastids, which distinguish them from prokaryotes.

- The endomembrane network provides compartments for protein and lipid synthesis. It is responsible for the secretory pathway and for processing endocytosis vesicles and vacuoles; it also contributes to osmoregulation.

- The cytoskeleton consists of specialized proteins that provide shape and structure to the cell. It is responsible for amoeboid locomotion, supporting cellular extensions and cilia (also called eukaryotic flagella).

- The nucleus provides a compartment for gene regulation and RNA transcription by segregating the chromosomes from the cytoplasm. There are a variety of ways to separate replicated chromosomes and divide nuclei by mitosis.

- The mitochondria and plastids are organelles derived from ancient endosymbioses with bacteria. They each retain prokaryotic features and part of the bacterial chromosome. In some lineages, the mitochondria or the plastids have been lost or modified.

- Sex is known to occur in many protists but not all. It requires meiosis, which yields haploid cells that can conjugate. It involves mating type recognition as well as cell fusion and nuclear fusion, which yields a diploid nucleus.

- Opisthokonts include animals, choanoflagellates, and fungi. Two obscure groups—Mesomycetozoa and nucleariids—are amoeboid or parasitic species near the base of the animal and fungal lineages.

- The chytrids, which retain a flagellated stage, represent an ancestral state of fungi. Other fungi form mostly filamentous hyphae or yeasts and have lost their flagellated or amoeboid stages. The fungi have diversified into many lineages that function as free-living saprotrophs, symbionts, or parasites. The main form of dispersal is by spore formation.

- Rusts and smuts are fungal plant parasites with complex life cycles. Some damage commercially important plant species.

- Microsporidia are highly derived fungi, and are intracellular parasites of animals. Infection of host cells is through the complex extrusion organelle.

- Excavata includes diplomonads, parabasalids, Heterolobosea, and Euglenozoa. The Euglenozoa consist of euglenids and kinetoplastids. The euglenids include species that are photosynthetic and others that are not. The kinetoplastids are heterotrophic and contain species that are dangerous parasites of animals including humans.

- The Rhizaria consist of many flagellated, amoeboid, and amoebo-flagellated species. Most produce fine pseudopodia, for example, filopodia or axopodia.

- Foraminifera and Radiolaria, two subgroups of Rhizaria, are marine planktonic or benthic organisms. Foraminifers form tests with many small pores while radiolarians produce elaborate mineral skeletons. Many have photosynthetic symbionts.

- Archaeplastida are photosynthetic organisms derived from an endosymbiosis with a cyanobacterium. The red algae often have a complex triphasic life cycle, and some provide products of commercial value. The Chloroplastida include the green algae and have diverse morphologies; one group within Chloroplastida, the Charales, is closely related to land plants.

- Chromalveolates are believed to be derived from one or more endosymbiosis events in which a phagotrophic eukaryote ingested a photosynthetic red algal eukaryote. Many groups have secondarily lost the plastid. Chromalveolates include cryptophytes, haptophytes, stramenopiles and alveolates. The cryptophytes retain a nucleomorph—a remnant of the nucleus of the red algal endosymbiont that gave rise to the plastid.

- Many haptophytes are covered by large calcium carbonate scales. They are important in the cycling of carbon in the ocean.

- Stramenopiles are very diverse; they include photosynthetic and nonphotosynthetic groups. Important photosynthetic groups include the diatoms, which are important primary producers in the oceans, and the phaeophyceans (brown algae).

- Nonphotosynthetic stramenopiles include many heterotrophic flagellates, such as the Bicosoecida, and plant parasites, such as the Peronosporomycetes (oomycetes). A parasitic oomycete was responsible for the great Irish famine.

- The Alveolata consist of dinoflagellates, apicomplexa, and ciliates. A common feature is the presence of alveoli.

- Dinoflagellates consist of photosynthetic and heterotrophic species with diverse feeding mechanisms.

- Apicomplexa are a very successful group of animal parasites. They are intracellular parasites that often have complex life cycles. Malaria is caused by an apicomplexan.

- Ciliates have two types of nuclei as well as coordinated cilia that beat in synchrony.

- The Amoebozoa consist of free-living amoebas with lobose pseudopodia as well the eumycetozoan slime molds and parasitic species such as the entamoebas. The Eumycetozoa (slime molds) are terrestrial amoeboid organisms that produce fruiting bodies for dispersal. They may have flagellated stages in their life cycles.

 Find more at www.sinauer.com/microbial-life

REVIEW QUESTIONS

1. What are the functions of the endomembrane network?

2. What is secondary endosymbiosis? Which groups of photosynthetic eukaryotes originated by secondary endosymbiosis?

3. Name some fungi used to obtain commercial products. Which groups contain edible fungi?

4. Which groups of fungi contain plant parasites?

5. What are the main differences between the life cycles of the rusts and the smuts?

6. Name some of the organisms that are intracellular parasites of animals. Name some that are extracellular parasites of animals.

7. Which groups consist of both photosynthetic species and non-photosynthetic species?

8. What characteristics of Cryptophyta make members of this group important in understanding secondary endosymbiosis?

9. What are the roles of filopodia and axopodia in the Rhizaria?

10. What are some of the differences between the plastids of Rhodophyta, Chloroplastida, and chromalveolates?

11. How do cellular and acellular slime molds differ when forming fruiting bodies?

SUGGESTED READING

Alexopoulos, C. J., C. W. Mims and M. Blackwell, 1996. *Introduction to Mycology.* 4th ed. New York: John Wiley & Sons.

Carlile, M. J., S. C. Watkinson and G. W. Gooday. 2001. *The Fungi.* 2nd ed. New York: Academic Press.

Fuller, M. S. and A. Jaworski. 1987. *Zoosporic Fungi in Teaching and Research.* Athens, GA: Southeastern Publishing Corporation.

Graham, L. E. and L. W. Wilcox. 1999. *Algae.* Upper Saddle River, NJ: Prentice Hall.

Hausmann, K., N. Hülsmann and R. Radek. 2003. *Protistology.* 3rd ed. Berlin: E. Scweizerbart'sche Verlagsbuchhandlung.

Lee, J. L., G. F. Leedale and P. Bradbury. 2000. *The Illustrated Guide to the Protozoa.* 2nd ed. 2 vols. Lawrence, KS: Society of Protozoologists.

Microbial Ecology

The objectives of this chapter are to:

◆ Introduce fundamental concepts of microbial ecology.

◆ Present autecology and the techniques for the counting, identification, and assessment of activity of individual species in the environment.

◆ Discuss community ecology and the methods for analysis of the distribution and activities of different transformations.

◆ Present the ecological aspects of various soil, aquatic, and extreme habitats in which microbial life occurs.

◆ Discuss the great biogeochemical cycles of nature and the role of microbial groups in them.

◆ Introduce concepts in microbial biodiversity and biogeography and methods to assess them.

24

Microbial Ecology

The ultimate aim of ecology is to understand the relationships
of all organisms to their environment.
—R. E. Hungate, in The Bacteria, 1962

*E*cology is derived from the Greek word *oikos*, which means "house," that is, the place where beings live. **Microbial ecology** is the study of the interactions between microorganisms and the abiotic and biotic components of the environment in which they live. Whereas the microbial physiologist or taxonomist is interested in the *capabilities* of the organism, the microbial ecologist is interested in the *actual activities* and *livelihood* of the organism in its natural habitat.

Microbial ecology is a young field of investigation, so there are tremendous gaps in our knowledge of the diversity of microorganisms, their distribution in natural environments, and their activities in natural communities. In fact, many important questions about microbial activities remain unanswered or have not yet been addressed due to, at least in part, the difficulties of studying such small organisms. Birds and mammals can be readily observed in their natural environment and studied as they go about their daily activities. In contrast, most microorganisms are too small to be seen without the use of a microscope, and because they are so simple structurally, most cannot be identified solely by microscopic observation.

Why is the study of microbial ecology so important? Microbial activities in ecosystems are important because:

- Microorganisms live in *all* ecosystems.

- Microorganisms carry out unique and important activities not performed by other organisms.

- Microorganisms formed the first ecosystems and biosphere, and their activities are essential for all life on Earth.

24.1 The Concept of an Ecosystem

An **ecosystem** is a subunit of nature in which organisms live and interact with one another and the nonliving parts of the environment, resulting in an exchange of materials. Ecosystems are geographic areas that contain two types of organisms: autotrophs, which fix carbon dioxide to produce organic materials, and heterotrophs, which consume the organic materials produced by the autotrophs.

The classical ecosystem concept includes the following four components:

- **Abiotic environment**, which provides the chemical substrates and physical conditions required by the organisms

- **Primary producers**, which produce the cellular and other organic materials used by the consumers

- **Consumers**, which ingest the primary producers as their energy foodstuff

- **Decomposers**, which degrade the organic cellular and noncellular materials

A lake is an example of an ecosystem because it contains both biotic and the associated abiotic components of an environment. Algae and cyanobacteria serve as the primary producers in lakes and other aquatic ecosystems. A **food chain** involves the interaction between the different trophic (nutrient) levels in which the organic carbonaceous material of the primary producer is consumed and degraded, as illustrated by a marine food chain (Figure 24.1A). The primary producers occupy the first trophic level. They receive their energy from sunlight and convert carbon dioxide into organic cell material. They are therefore the base of the food chain, and all other members of the food chain rely upon them for their organic carbon source from which they obtain their energy and nutrients for growth. The primary producers are ingested by **primary consumers**, which are protists such as amoebae, flagellate and ciliate protozoa, and invertebrate animals such as copepods. The protists and small invertebrates are in turn consumed by **secondary consumers** including small fish. At the **tertiary consumer** level are larger fish and birds. Finally at the **quaternary level** are mammals.

The food chain can also be regarded as a food pyramid in which the amount of organismal material produced at each trophic level is considered (Figure 24.1B). Typically about 10% of the organic material from a lower trophic level is converted to organisms at each successively higher level in the food chain. This reduction in cellular material at each successive level is related to the efficiency of conversion of the organic material.

Figure 24.1 Diagram of a marine food chain
(A) The organisms at the base of the food chain are the primary producers, which are algae and cyanobacteria. The consumers are protists and animals that occupy successively higher levels in the chain that rely on the ingestion of organisms at the trophic level beneath them. *Bacteria, Archaea* and fungi serve as decomposers that help recycle the organic materials. All organisms produce carbon dioxide through respiration and this is recycled back to the primary producers. (B) A food pyramid showing that the amount of organismal material decreases about ten-fold at each successively higher trophic level.

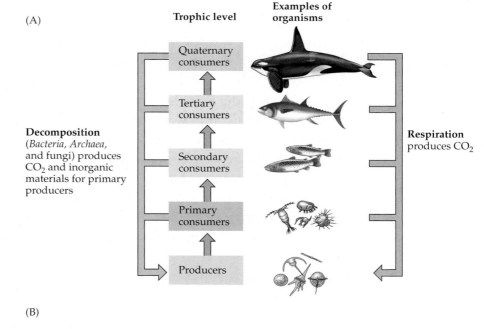

(A)

Decomposition (*Bacteria, Archaea,* and fungi) produces CO_2 and inorganic materials for primary producers

Trophic level

Examples of organisms

Quaternary consumers

Tertiary consumers

Secondary consumers

Primary consumers

Producers

Respiration produces CO_2

(B)

Dry weight (g/m^2)

	Trophic level
1	Tertiary consumers
10	Secondary consumers
65	Primary consumers
800	Producers

The "decomposers" are the heterotrophic microorganisms, including bacteria, archaea, and fungi, whose concerted action results in an almost complete recycling of the organic material from dying or dead organisms and dissolved organic carbon back into inorganic materials.

The food chain is also involved in the **recycling** of carbon. The organic material is utilized as a carbon and energy source by each of the consumer trophic levels, which oxidize it through aerobic respiration. This results in the production of carbon dioxide, which is thereby recycled back into the environment and made available again to the primary producers. In addition, the food chain also illustrates that, *whereas nutrients are recycled, energy is not*—it is transferred unidirectionally. The sun's energy that is captured by photosynthetic organisms is converted into the organic carbon of the primary producers, which in turn serves as an energy source for consumers and decomposers.

In reality, food chains are an oversimplification of the actual carbon transformations, because some secondary consumers also serve as primary consumers, and others serve as tertiary consumers and so forth. A **food web** more accurately portrays the carbon cycle.

This classical view of an ecosystem focuses on carbon cycling, which is most relevant to plants and animals. However, microorganisms are immensely important, not only in the carbon cycle, but also in other cycles in which they are responsible for carrying out many fundamental biologically mediated geochemical transformations that are essential to all forms of life in the biosphere. Furthermore, many of the transformations of the elemental cycles, termed **biogeochemical processes**, are uniquely carried out by members of the *Bacteria* and the *Archaea*. Most of these reactions began billions of years ago, long before plants and animals evolved. Therefore, plants and animals are dependent on the transformations that microorganisms perform that are fundamental to the operation of the biosphere.

Microorganisms also live in their own ecosystems, termed **microbial ecosystems**. These exist in areas such as thermal, hypersaline, acidic, alkaline, and anoxic environments that are inhospitable for macroorganisms. These microbial ecosystems may rely on chemoautotrophs rather than photoautotrophs for their primary production. Furthermore, microorganisms live in the intestinal tracts in microbial ecosystems, such as cattle rumen (see Chapter 25), which contain hundreds of microbial species, and they form close associations with the plant rhizosphere and are therefore found wherever animals and plants live. For these reasons, the **microbial biosphere** is considerably more extensive than the biosphere of macroorganisms.

In addition, all organisms, including microorganisms have viruses. These make up the **virosphere**, which is as extensive as the biosphere of microorganisms. Indeed, it is estimated that in marine environments there are tenfold more viruses than there are *Bacteria* and *Archaea*. Viruses play important roles in limiting the abundance of all organisms in ecosystems.

The chemical transformations that are carried out by microorganisms are often unapparent. For example, the process of nitrification and the bacteria that carry it out are imperceptible to us, so it is natural that most people are simply unaware that this important process even occurs. These and many other microbial processes that occur in nature cannot be readily differentiated from chemical processes, certainly not without careful scientific investigation.

Although it is true that the study of the ecology of macroorganisms and microorganisms shares many common principles and ecosystems, microbial ecology stands apart in several ways. For example, as a group, microorganisms are more widely distributed on Earth than plants and animals. Because of the small sizes of microorganisms, their rapid growth rates, their large population sizes, the many unique geochemical transformations they perform, their ability to use a large variety of redox reactions for generating energy, and the advantages of studying them in pure culture, microbial ecology is a special area of investigation. Due to their small sizes, they create their own microscale environment as discussed next.

SECTION HIGHLIGHTS

Ecosystems are subunits of the biosphere in which organisms interact with one another and the abiotic environment, which provides them with the physical conditions and chemical substrates required by organisms to metabolize and grow. Algae and cyanobacteria serve as primary producers in ecosystems along with plants. Some members of the *Bacteria*, *Archaea*, and fungi are decomposers, whereas others are chemoautotrophic or anoxygenic photoautotrophic primary producers, and still others carry out special transformations in the biosphere such as methanogenesis.

24.2 The Microenvironment

The typical bacterial cell is about one-millionth the length of a human being. Because microorganisms are so small, the natural environment, or **microenvironment**, in which they live and perform their activities, is correspondingly small. More than other organisms, microorganisms live in intimate contact with the chemicals of their environment with which they directly interact. They have uptake systems that allow them to rapidly utilize small and large molecules that cannot by utilized by plants and animals. For example, bacteria and archaea are the principal utilizers of organic molecules in the environment because these microorganisms are widely dispersed in the environment where these chemicals exist and they have the enzymes necessary for their uptake and metabolism. Furthermore, collectively they can utilize many molecules such as nitrogen gas that cannot be used by plants and animals.

For these reasons microorganisms carry out important biochemical reactions that influence the physical and chemical conditions of the microenvironment. As a result, the concentration of substrates and products is different in the vicinity of the cells as compared to the bulk environment that is measured with ordinary electrodes and by chemical analyses. Therefore, in order to study what is happening in the microenvironment, it is necessary to amplify these activities with appropriate instruments, a challenge for scientists.

The small size of the microenvironment requires that special consideration be given to designing instruments to measure conditions in the vicinity of the organisms. **Microelectrodes** for the measurement of oxygen and pH are examples of such instrumentation. These fine probes come equipped with tips as small as about 5 to 10 μm in diameter. With these instruments, microbial ecologists have demonstrated that the oxygen concentration and pH near colonized areas can be quite different within a span of a few micrometers (Box 24.1).

Microbial ecologists need to be constantly aware that the growth and activities of microorganisms occur in the environment at a microscale. Nonetheless, the concerted actions of enormous numbers of microorganisms are responsible for producing metabolic products and chem-

ical gradients that are apparent at macroscopic scales in soils, aquatic environments, and the atmosphere.

Three basic issues that microbial ecologists address are the identification, abundance, and location of microorganisms in the natural environment of interest. We begin by discussing **autecology**, the study of the individual microorganism in the microenvironment as well as population studies of the species in the macroenvironment.

SECTION HIGHLIGHTS

Most microorganisms live in correspondingly small habitats, or microenvironments, which they transform due to their metabolic activities. Some of the activities of the microenvironment can be detected using microelectrodes.

24.3 Autecology: Ecology of the Species

Autecologists focus their attention on a single species and its life and activities in its habitat. Typical issues the autecologist considers concern the abundance, distribution, and activities of the species of interest. In order to address these issues, several properties relating to the autecology of a species are of utmost importance:

- **Identification** and **enumeration** of the species in the environment
- Sampling of the environment
- Determination of the **habitat** (the *location* or "address") of the species
- Assessment of the **ecological role(s)** for which the species is responsible

---

BOX 24.1   *Research Highlights*

Thinking Small: A Microbial World within a Marine Snow Particle

Marine snow is found in the water column of marine habitats. It is called marine snow because as these particles descend in the water column, they are reminiscent of falling snowflakes to scuba divers. The marine snow particles range from 1 to about 10 mm in diameter and may contain photosynthetic algae and cyanobacteria as well as heterotrophic bacteria and protozoa.

Microprobes, with tips that are only 5 to 10 μm in diameter, were used by A. L. Alleridge and Y. Cohen to measure the pH and oxygen concentrations at the surface and at various points in the interior of marine snow particles. Micrometer gradients of oxygen and pH were found from the surface to the interior of the particles. For example, the oxygen concentration in the interior of the particle can be half that of the bulk water in which the marine snow particle is found. The concentration of oxygen is lower inside the particle because of aerobic respiration that depletes the oxygen more quickly than it can be replenished by diffusion.

Identification of Species in the Environment

Of course, if the species of interest is morphologically distinct, it can be recognized and identified by direct microscopic observation in the environment. However, as mentioned previously, it is often not possible to identify a typical unicellular bacterium at the level of species, genus, or perhaps even kingdom by observing it in the microscope. This is because many bacteria and archaea have very simple shapes. For example, members of the genus *Nitrosomonas* appear very similar to *Escherichia coli*, although the former is a genus of chemolithotrophic ammonia-oxidizing bacteria, whereas the latter is an ordinary heterotroph. Two common approaches for identification are:

- Fluorescent antibody approach
- Fluorescent in situ hybridization (FISH)

FLUORESCENT ANTIBODY APPROACH   The problem of in situ identification of microorganisms can be solved by the use of **fluorescent antibodies** that are used to microscopically identify a species. A fluorescent dye, such as fluorescein, is covalently tagged to a whole-cell antiserum or to a monoclonal antiserum prepared from the organism of interest. An environmental sample that contains the organism can be stained with the fluorescent antiserum and examined microscopically using an ultraviolet (UV) or halogen light source. Cells of the organism of interest will fluoresce because they have been labeled by the specific antiserum (Figure 24.2). In fact, this is a preferred procedure for detection and identification of *Legionella* species in natural samples, because these bacteria are so difficult to grow.

FLUORESCENT IN SITU HYBRIDIZATION (FISH)   Nucleic acid probes (either labeled RNA or DNA) are most commonly used for the identification of microorganisms from an environment. In one application of this procedure, an oligonucleotide specific for a species or other taxonomic group is tagged with a fluorescent dye. Then the hybridization reaction is prepared directly on a glass microscope slide with organisms from a natural sample and the fluorescent oligonucleotide. The preparation is then observed using a fluorescent microscope (Figure 24.3). The FISH label is usually targeted against ribosomal RNA (rRNA) because there are more ribosomes than DNA molecule sites, so the ribosome-labeled cell will fluoresce more brightly. Fluorescent rRNA probes require that the organism of interest have a sufficient number of ribosomes to serve as targets for the cell. The FISH approach has been greatly aided by the introduction of stronger fluorescing dyes.

Enumeration of Microorganisms in the Environment

In this section, we will concentrate on counting or enumerating bacteria, in particular heterotrophic bacteria, in natural habitats. When enumerating bacteria in natural environments, it is important to recognize that not all small organisms in the habitat are heterotrophic bac-

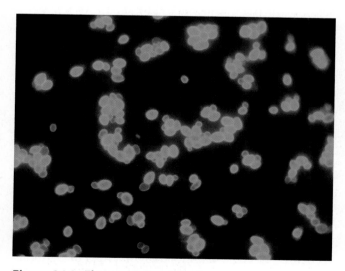

Figure 24.2 Fluorescent antibody
Oval budding yeast cells of *Candida albicans* stained with a specific fluorescent antibody. Courtesy of Maxine Jalbert and Dr. Leo Kaufman/CDC.

Figure 24.3 FISH (fluorescent in situ hybridization) labeling
Fluorescence in situ hybridization (FISH) image of a biofilm found in the Richmond Mine at Iron Mountain in northern California. Three FISH probes were used: one targeted the bacteria (EUBmix; green), another the archaea (ARC915; Cy5 Blue), and a third the bacterial genus *Leptospirillum* (LF; Cy3 Red). The extensive overlap of red and green (resulting in yellow) indicates that *Leptospirillum* cells predominate in the biofilm. Courtesy of Gene Tyson, MIT.

teria. Depending on the environment, large numbers of cyanobacteria, small algae, or autotrophic bacteria may also be found. These other microorganisms can usually be distinguished from heterotrophic bacteria.

The fundamental problem of counting individual bacteria in natural environments is challenging for two primary reasons. First, because bacteria are so small, they must be visualized with a microscope. Second, even if we can see and count them, determining whether they are dead or alive is not a simple matter. Also, as we discussed previously, it is usually impossible to distinguish one species from another. Initially, we will address how bacteria are counted and then consider how it is possible to determine whether or not they are alive or metabolically active.

TOTAL MICROSCOPIC COUNT   One of the best ways to determine the total number of bacteria in a sample is to count them microscopically by **fluorescent dye staining** such as acridine orange or DAPI (4',6-diamidino-2-phenylindole). These dyes bind to the RNA and DNA of a cell. In this staining procedure, the sample containing the bacteria is first fixed immediately after collection with formaldehyde or glutaraldehyde to preserve the organisms. They are then stained with acridine orange or DAPI. Cells that have taken up the dye appear green, blue, or orange when observed by fluorescence microscopy. This procedure is one of the best known for accurate counting of bacterial cells (Figure 24.4). Furthermore, because this is an epifluorescence procedure that uses incident light (see Chapter 4), it is possible to observe cells attached to soil particles and other opaque materials. It should be noted that it is not possible to distinguish living from dead cells

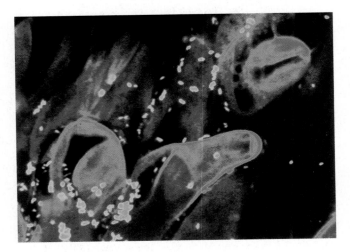

Figure 24.4 Fluorescing cells
Confocal laser photomicrograph (false color) of acridine orange–stained bacteria growing in the rhizosphere of a legume. Courtesy of Frank Dazzo.

through this technique—*this provides a total count of both living and dead cells.*

Cyanobacteria that might be present in the sample can be distinguished from other bacteria because the cyanobacteria have chlorophyll *a*, which produces a natural red fluorescence without acridine orange staining. Thus, control samples are prepared to determine whether chlorophyll *a*–containing cyanobacteria and algae are present in a sample. It is not possible to distinguish lithotrophic bacteria from heterotrophic bacteria, although in many environments chemolithotrophs would be expected to occur in low numbers relative to the heterotrophs. Likewise, archaea are enumerated by total microscopic counting methods, but are not distinguishable from bacteria by this procedure. Therefore, other techniques, such as FISH are needed in order to distinguish archaea from bacteria. This is very important to consider, especially in environments where archaea are numerous.

FLOW CYTOMETRY   An alternative to microscopic counting is the use of **cell sorters** or **flow cytometers**, instruments developed for separation of blood cells. Flow cytometers are now also used to separate microbial cells based on size. Furthermore, with special detectors it is possible to separate fluorescent particles (such as acridine orange–stained or FISH-labeled cells) from other particles and examine that group specifically. Therefore, it is possible to obtain a large number of cells of an organism that has not yet been isolated in pure culture so that it can be studied in greater detail.

VIABLE COUNTS   If one wishes to know how many bacteria are alive in a sample, it is necessary to use a viable counting procedure in which the bacteria are actually grown. The traditional method used for this is the spread plating procedure, in which a medium prepared to grow the heterotrophic bacteria is first poured into the Petri dish (see Chapter 5), and a dilution of bacteria is spread over the surface of the plate with a sterile glass rod. This procedure is superior to pour plating because the high temperature required to keep the agar molten while mixing with the bacteria kills many psychrophilic and mesophilic bacteria.

EXTINCTION DILUTION   An alternative to plate counts is the use of liquid media. For example, samples can be quantitatively diluted and inoculated into a liquid growth medium for viable enumeration such as the most probable number (MPN) procedure (see Chapter 32). This procedure has been found to typically allow higher recoveries of viable bacteria than plating. Furthermore, a selective medium can be used that will permit the growth of a group of organisms of interest, for example, those that can degrade a special carbon source such as a hydrocarbon compound.

MICROENCAPSULATION A recent novel approach to growing microorganisms from natural samples is microencapsulation. In this procedure, cells from an environment are purified and encapsulated in agarose capsules about 50 to 80 μm in diameter. The microcapsules are then placed in a column through which nutrients can be passed to allow for growth. Individual microcapsules, some of which contain only a single species, are then placed in microwell plates with medium to permit further growth. This procedure can be automated for high throughput and shows great promise for growing organisms that have been difficult to cultivate.

The Bacterial Enumeration Anomaly

Major discrepancies are found between the total microscopic counts and viable counts of bacteria from many habitats. For example, in **oligotrophic** (low concentrations of nutrients) or **mesotrophic** (moderate concentrations of nutrients) aquatic environments, less than 1% of the total acridine orange–staining bacteria can be grown on the best of media. In contrast, in **eutrophic** environments (rich in nutrients) such as wastewaters, the recovery of bacteria can approach 100% of the total count (Table 24.1). The inability to recover viable bacteria from oligotrophic and mesotrophic environments is called the **bacterial enumeration anomaly**, or the "Great Plate Count" anomaly. There are three possible explanations for this result. First, there is no such thing as a "universal medium" that will permit the growth of all bacteria, so not all viable bacteria are capable of growth on the medium that is used. Second, many bacteria may not grow well enough on artificial media using different incubation conditions to produce visible colonies on plates or turbidity in liquid media. And finally some, perhaps most, of these bacteria are dead. To address the enumeration anomaly issue, three alternative microscopic procedures have been developed to identify metabolically active ("living") bacteria from natural samples:

- Microautoradiography
- INT-reduction technique
- Nalidixic acid growth technique

Each of these procedures is described briefly next.

MICROAUTORADIOGRAPHY Microautoradiography entails the use of a radiolabeled substrate such as tritiated acetate or tritiated thymidine. If a microorganism can use acetate or thymidine as a substrate, it will incorporate the radiolabeled material into its cells as it would a normal nonradioactive substrate. When an organism takes up the radiolabeled substrate, the cells become radioactive and can then be identified by autoradiography (Figures 24.5 and 24.6). In using this approach the pro-

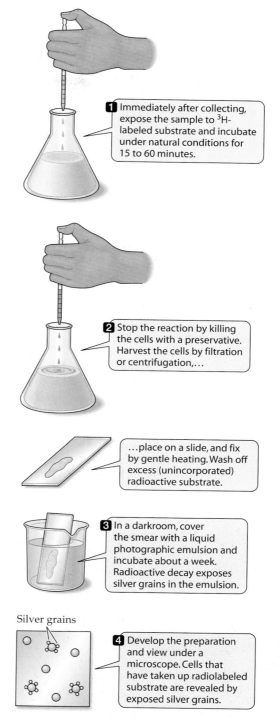

1 Immediately after collecting, expose the sample to ^{3}H-labeled substrate and incubate under natural conditions for 15 to 60 minutes.

2 Stop the reaction by killing the cells with a preservative. Harvest the cells by filtration or centrifugation,…

…place on a slide, and fix by gentle heating. Wash off excess (unincorporated) radioactive substrate.

3 In a darkroom, cover the smear with a liquid photographic emulsion and incubate about a week. Radioactive decay exposes silver grains in the emulsion.

Silver grains

4 Develop the preparation and view under a microscope. Cells that have taken up radiolabeled substrate are revealed by exposed silver grains.

Figure 24.5 Microautoradiography
Steps in microautoradiography. A radiolabeled substrate (or mixture of substrates) is added to a sample, and the sample is incubated for 15 to 60 minutes under the same conditions as in the natural habitat. The reaction is stopped, the sample fixed, and an autoradiogram prepared (see Figure 24.6) to determine what proportion of the organisms incorporated radiolabeled substrate, that is, what proportion were metabolically active. Radioactive cells are counted, and the percentage of active cells is determined by dividing the number of radioactive cells by the total cell count (determined by acridine orange counting).

TABLE 24.1	Bacterial enumeration anomaly illustrated for a mesotrophic lake, Lake Washington, compared with a eutrophic pulp mill aeration lagoon		
	Bacterial Cells/ml		
Trophic Status	**Total Count**	**Viable Count**	**% Recovery**[a]
Mesotrophic			
Lake Washington	3.0×10^6	2.0×10^3	0.067
Eutrophic			
Pulp mill oxidation lagoon	2.1×10^7	3.1×10^7	~100[b]

[a]This refers to the ability to cultivate the bacteria on a plating medium; it is the ratio of the viable count divided by the total microscopic count.
[b]In this particular instance, the viable count was actually somewhat higher than the total count, due to minor uncertainties in measurement. Therefore, the recovery is considered to be 100%.

portion of metabolically active organisms (that is, those that incorporate the label into their cells) to total cells can be readily determined.

INT REDUCTION   In the INT reduction procedure, a tetrazolium dye (INT) is used as the substrate. It is commonly metabolized by aerobic respiring bacteria if they are alive and metabolically active. They use it as an electron acceptor and reduce it to form a precipitate called formazan, which appears as a black deposit in the cell. When combined with acridine orange total counting, the organisms that are metabolically active can be distinguished from those that are not.

NALIDIXIC ACID CELL ENLARGEMENT   In the **nalidixic acid cell enlargement procedure**, a small amount of yeast extract is added to a sample along with nalidixic acid, which inhibits DNA replication. Metabolically active cells that are sensitive to this antibiotic cannot divide, but their cells continue to enlarge in size as they grow. Therefore, the proportion of cells from a sample that become very large compared to those that have not enlarged provides a measure of the percentage of metabolically active cells.

All of these procedures have their own pitfalls and drawbacks; however, they all provide similar results when used to assess viability in typical mesotrophic and

(A)

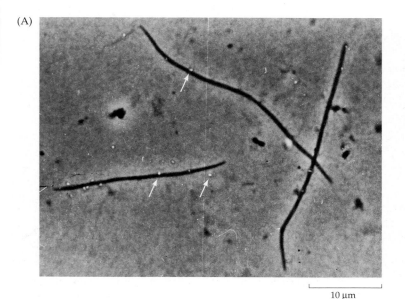

10 µm

(B)

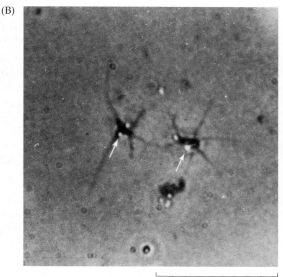

10 µm

Figure 24.6 Autoradiogram
Photomicrographs showing (A) a filamentous bacterium and (B) *Ancalomicrobium adetum* from a pulp-mill treatment lagoon. The exposed silver grains (arrows) in the emulsion overlying the cells reveal which cells are metabolically active. The labeled substrate was tritiated acetate. Courtesy of P. Stanley and J. T. Staley.

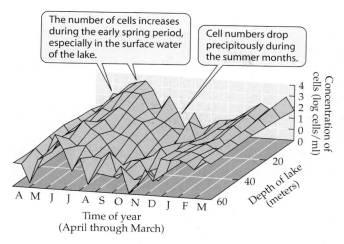

The number of cells increases during the early spring period, especially in the surface water of the lake.

Cell numbers drop precipitously during the summer months.

Figure 24.7 Seasonal distribution study
The seasonal and spatial distribution of *Caulobacter* sp. in Lake Washington. Courtesy of J. T. Staley, A. E. Konopka, and J. Dalmasso.

oligotrophic environments. Unlike viable plating in which less than 1% of the cells can be recovered, about half of the cells from most oligotrophic and mesotrophic habitats are metabolically active. Thus, either these organisms are different physiologically from known bacteria and therefore do not grow on typical plating media, or the cells are too debilitated to actually multiply on plating medium, or a combination of both. Therefore, *most of the bacteria found in environments, including oligotrophic habitats, are alive and many of these are metabolically active.*

Although there are drawbacks to using plating and other typical viable counting procedures to determine total viable counts of heterotrophs, viable counting procedures work well if one is enumerating a specific group

of organisms that are known to grow under certain laboratory conditions. For example, as mentioned previously, it is possible to determine the seasonal distribution of the genus *Caulobacter* in lakes using viable counting procedures (Figure 24.7).

Sampling Environments and Assessing the Habitat of a Species

By combining a sampling program of an environment with a means of identification of a microorganism, it is possible to conduct distributional studies of species to assist in determining its natural habitat. The first concern facing the microbial ecologist is, "How can I obtain samples from the environment?"

SAMPLING THE ENVIRONMENT   Sampling environments is not a trivial problem. Imagine the difference between sampling for the bacteria from the intestine of an animal compared to sampling for marine planktonic bacteria from the Sargasso Sea. Completely different approaches are used to obtain samples of microorganisms from each habitat. For intestinal samples, the microbiologist has some difficulty in obtaining samples—stool samples could be used for lower intestinal bacteria, but more elaborate techniques would be required to sample the small intestine. Furthermore, the microbiologist needs to maintain anaerobic, aseptic conditions during the collection process. The marine microbiologist needs to find an appropriate location to obtain samples and uses entirely different collection gear (Figure 24.8). This text cannot describe all of the techniques used for sampling for microorganisms; however, some general guidelines for sampling are provided in Table 24.2.

TABLE 24.2	Guidelines for microbial sample collection and processing[a]
Purpose of Sampling	**Preferred Handling Technique**
Total count (direct microscopic count)	Fix sample immediately after collection with formaldehyde or glutaraldehyde. Store at refrigerator temperatures in the dark. Aseptic collection is preferred, but is not always essential as long as vessel is clean and microbial counts are high.
Viable count	Collect aseptically. Enumerate as soon as possible after collection. Chill sample and keep under conditions of the habitat until processed. Incubate at or near the in situ conditions of the environment (i.e., pH, temperature, redox potential, etc.).
Activity measurements	Perform assay as soon as possible. Perform in situ if at all possible. May have to use a container for certain assays (e.g., radioisotopes), but these should be incubated under conditions of the environment, if not right in the environment.
Chemical analyses	Collection container should be chemically inert. Analyses should be performed soon after collection, especially for reactive compounds such as hydrogen sulfide, dissolved oxygen, etc. Special chemical fixation techniques are used for more reactive chemicals.

[a]As soon as a sample of microorganisms is taken from its natural habitat, conditions will change. Part of the change will be caused by the microbes themselves as they metabolize. However, physical and chemical changes also occur. It is important to minimize changes, as they can affect the results of a study. The actual sampling procedure will vary from one habitat to another.

(A)

(B)

Figure 24.8 Sampling gear
Collection gear and instruments used by marine microbiologists. (A) Plankton net for sampling larger marine organisms. (B) A sampling rosette for sampling marine microbes. A, courtesy of NOAA Ocean Explorer; B, courtesy of the U.S. Environmental Protection Agency.

In addition, it is desirable to measure the location and certain physical and chemical properties of the habitat at the time of sample collection. The location is determined from global positioning satellite (GPS) measurements. Also, in the intestinal and marine habitats described previously, it is important to measure the temperature at the time of collection as well as the pH. In addition, in the marine habitat it is important to record the time of sample collection and the hydrostatic pressure and depth. These parameters as well as others provide valuable information that is useful in interpreting what is happening in the environment.

ASSESSING THE DISTRIBUTION OF A BACTERIAL SPECIES   Autecological studies may be designed to assess seasonal changes in the distribution of a microbial species as well as its habitat. For example, nitrifying bacteria have been enumerated in marine environments and in soils by use of fluorescent antibody techniques. This procedure is species-specific, at least for this group, so it provides a means of assessing the temporal and spatial distributions of particular nitrifying species.

Morphologically distinctive microorganisms such as many of the *Cyanobacteria* and anoxygenic photosynthetic bacteria can be enumerated microscopically. Thus, photosynthetic bacteria can be counted in lakes by identification using the light microscope (Figure 24.9). As with all microorganisms, the distribution of these groups

is confined to certain strata in the lakes in which they reside. The lower depth of each photosynthetic bacterial group is determined by its need for sulfide obtained from the sediments of the lake, whereas its upper depth is determined by its need for light obtained from the surface of the lake.

Viable counting procedures can also be used. For example, as mentioned previously, the seasonal and spatial distribution of *Caulobacter* sp. has been determined in freshwater lakes (see Figure 24.7). Note that this genus is most abundant in the surface waters of the lake, and, furthermore, that it is most numerous in the spring when diatom blooms occur in the lake. These results suggest that the caulobacters may derive organic nutrients from the diatoms.

IDENTIFYING THE HABITAT OF A SPECIES   Merely finding an organism in a sample does not mean that it lives in the environment from which the sample was taken. It could be a transient or allochthonous organism that is being passively dispersed in the environment. The ideal means of determining whether an organism is indigenous (or autochthonous) to the habitat is to assess its in situ growth.

One method that can be used to assess growth is to visualize organisms growing in a habitat microscopically. Some scientists have placed glass slides in a habitat and allowed microorganisms to attach and grow on

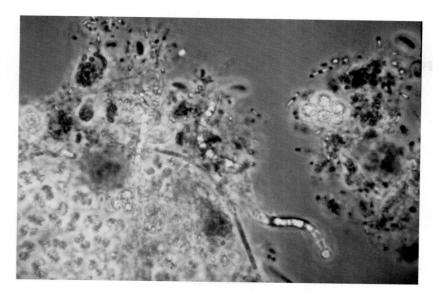

Figure 24.9 Anoxic hypolimnion sample
Sample from the anoxic hypolimnion of a lake, showing a variety of photosynthetic bacteria identifiable by their characteristic morphology. Bright areas in cells are gas vacuoles. Courtesy of J. T. Staley.

the slide surfaces (Figure 24.10). In the example shown, an alga can be observed growing on a glass slide in a pond. Note that the initial organism shown in Figure 24.10A contains two cells that were photographed in the evening. By the next morning, these two cells had divided to produce a total of four cells (see Figure 24.10B). During the next day these four cells dramatically increased in size (see Figure 24.10C) and again divided during the next night to produce eight cells (see Figure 24.10D). From these results it can be concluded that this alga is indigenous to this habitat.

An alternative method that can be used to verify whether an organism is indigenous to a habitat is to demonstrate that it is metabolically active in the habitat. Techniques already mentioned such as microautoradiography can be used for this. Although these approaches are not as definitive as growth studies, they provide evidence to support their being indigenous.

Of course, even indigenous organisms may be dormant or otherwise inactive during certain periods of the year. For example bacteria that are closely associated with algae that produce spring or summer blooms in lakes may be largely inactive during the remainder of the year, as shown by the seasonal distribution of *Caulobacter* sp. in Lake Washington (see Figure 24.7).

Evaluation of the Role of a Species

To assess the role(s) of a species in the environment, microbiologists may follow the "ecological" Koch's postulates approach. The approach can be summarized by paraphrasing the postulates used for studies of pathogenesis; however, in this instance the microbiologist is not attempting to verify that the bacterium causes a disease but that it is responsible for a particular process or chemical transformation. These modified Koch's postulates for microbial ecology are as follows:

1. The species believed responsible for a transformation must be found in environments in which the process is occurring.

2. The species must be isolated from the environment where the process occurs.

3. The species must carry out the process in the laboratory in pure culture.

4. The species that carries out the process in the laboratory must be shown to carry out the process in the environment in which the process occurs.

Of course, unlike infections that are typically mediated by a single, specific pathogen or parasite, ecological processes in the environment are frequently mediated by groups of organisms growing in concert with one another. Such mixed groups containing two or more species are referred to as a **consortium**. An example of a consortium-mediated environmental process is the process of formation of stromatolites that are produced by a variety of organisms (see Chapter 1). This structure could not be formed in the laboratory using a single pure culture of an alga or cyanobacterium. However, it is formed in nature by the concerted action of several different species. Another example is the anammox (anaerobic ammonia oxidation) process. A pure culture of an anammox organism has not yet been obtained so it appears another organism must be playing an essential role in the anammox process. Therefore, it may not always be possible to use Koch's postulates with pure cultures to show that a specific microorganism causes a specific process. How-

(A)

(B)

(C)

(D)

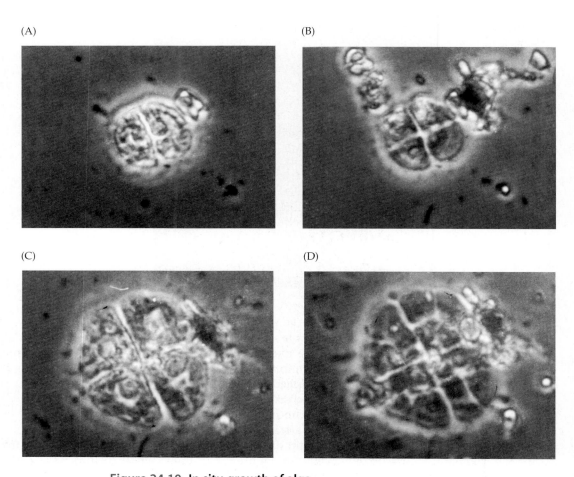

Figure 24.10 In situ growth of alga
This alga has attached to and grown on a glass slide submerged in a pond. The growth of the alga was followed by use of a microscope immersed near the slide in the pond. Note that the initial colony consists of two small cells (A) in a photo taken at 7:38 p.m. The next morning (B) at 8:27 a.m. the two cells had divided to form four. By 7:05 p.m. that evening (C) the cells had grown substantially but not divided. However, the next morning at 8:33 a.m. each of the four cells had undergone division (D). Courtesy of J. T. Staley.

ever, what is true for the species would also be true for the consortium. So, the word *consortium* can substituted for *species* in the environmental Koch's postulates. A microbial ecologist studying such a consortium should attempt to identify the members of the consortium and, ideally, also determine what each member of the consortium contributes to the overall activity.

Generally speaking, testing the first three postulates is straightforward: An organism suspected of causing a transformation can be found in various environments in which the process occurs and can be isolated in pure culture. Once in the laboratory, it can be shown to be able to carry out the process. The final step is much more challenging, however, because it is incumbent on the microbiologist to demonstrate that the species or consor-

tium is carrying out this process in the environment. One approach in assessing this last postulate is to combine microautoradiography with an identification procedure such as the fluorescent antibody procedure. For example, if one wonders whether a specific heterotrophic bacterium is using acetate in the environment, then the species can be identified by fluorescent antibody, and the uptake of radiolabeled acetate associated with the chemical transformation is noted by the exposed silver grains on the same autoradiogram.

Microautoradiography can also be used with morphologically identifiable bacteria, such as *Ancalomicrobium adetum* (see Figure 24.6B). Indeed, quantitative autoradiography can be achieved in some cases as with *A. adetum*. This species was found to comprise a minor

BOX 24.2 *Methods and Techniques*

Stable Isotope Probing

This recent procedure takes advantage of stable isotopes, especially carbon (^{13}C) and nitrogen (^{15}N). One application is to provide these as labeled substrates, for example, $^{13}CO_2$ to a community that contains primary producers. The actively growing primary producers will incorporate some of the label into their cell material including their DNA. After incubation of the community in the environment for a relatively short period of time the microorganisms can be harvested and their DNA extracted. The purified DNA is then placed in an ultracentrifuge and the heavy DNA is separated from the lighter DNA. This heavy DNA band is then analyzed to determine which organisms are the predominant primary producers. This is accomplished by using PCR to amplify the 16S rDNA sequences for analysis of the types of organisms. One caution is that if the incubation proceeds for too long, the ^{13}C-label will be transferred to heterotrophic organisms and thus provide misleading results.

component of the acetate-utilizing community in this study, in comparison with a filamentous organism (see Figure 24.6A—data not shown). This does not mean that *A. adetum* is insignificant in this habitat. It is more likely that it has a different role. For example, from pure culture studies in the laboratory, it is known that this organism preferentially uses sugars as carbon sources. Sugars are also abundant in pulp mill effluents and are likely being utilized as the primary carbon sources by this species.

Molecular approaches are also possible. For example, it may be possible to detect the expression of a particular gene that codes for a specific enzyme of the organism of interest using reverse transcription PCR (RT-PCR) (see Chapter 16). Because messenger RNA (mRNA) remains stable for a relatively short period of time, the detection of mRNA is indicative of an active process that is occurring at the time of sampling of the environment. Likewise, new procedures such as stable isotope probing (SIP) have been applied to assess active organisms in a community (Box 24.2).

Finally, it is very important that the organism that has been shown to be responsible for a process be kept in pure culture and deposited in a culture collection. If it comprises a new species, then it should also be described and named.

SECTION HIGHLIGHTS

Autecological studies are focused on the study of the individual species in its natural habitat. Special techniques are available for enumerating, identifying, and assessing the in situ activities of bacterial species.

24.4 Microbial Communities

The community ecologist is not concerned with the activities or distribution of a single species but rather is interested in the processes of a group of species or all the species of the habitat. Thus, the community ecologist might be interested in all the sulfur-oxidizing species (i.e., the guild of sulfur oxidizers) in the lake and the rates at which this process is occurring, rather than the activity of a single species of *Thiobacillus*.

Some communities consist primarily or exclusively of microorganisms. Microbial mats and biofilms are examples of microbial communities.

Microbial Mat Communities and Biofilms

Chapter 1 describes microbial mats as the dominant manifestation of life on Earth from about 3.5 Ga (giga-annum, or 10^9 years) ago until the rise of land plants about 0.5 Ga ago. Today's microbial mats are found in hot spring environments and intertidal zones of the marine environment (see Chapter 1).

Microbial mats are communities that consist of a vertical stratification of species in a layer that is typically about 1 to 4 cm in depth. Mats in hot spring environments have been studied thoroughly. At higher temperatures near the orifice of hot springs, the primary producers are nonphotosynthetic archaea such as *Pyrococcus,* or bacteria such as *Aquifex.* These organisms obtain their energy from reduced sulfur compounds or H_2, not light. Downstream, at temperatures of about 70°C, cyanobacteria become the major primary producers, using light as their energy source. Eukaryotic consumers of the microbial mats cannot grow at the highest temperatures but begin to colonize the mats when the temperatures decrease. Microbial decomposition occurs through heterotrophic bacteria and the sulfur- and sulfate-reducing bacteria (some of which are also autotrophic).

Not only is there a vertical layering of the species in hot spring communities, but also as the water flows away from the thermal source, the temperature drops. This gradient allows for the development of organisms with lower and lower temperature optima. Furthermore, day and night changes often occur in the community structure of mats. In hot springs with sulfide, *Cyanobacteria* dominate the surface layers of the mat during daylight hours. But at night, gliding sulfur oxidizers such as *Beggiatoa* migrate to the surface of the mat. *Beggiatoa* species do not require light but gain greater access to oxygen by moving to the surface of the mat where they may oxidize their stored sulfur granules for energy.

A typical mat community in a hot spring may contain a guild of several different species of primary producers. For example, in Yellowstone National Park hot springs, several different cyanobacterial strains or species are known to occur in the community (Figure 24.11). Each type has its own optimal temperature, but several types may overlap at a particular location.

Biofilms are similar to mats in that they consist of a buildup of microorganisms on a surface but are not as thick as mats. Often, like microbial mats, they exist in flowing environments like on rocks in a stream. In such environments, the biofilm may not be visible to the eye, but the rock surface is slippery, indicating its presence. Another example of a biofilm is the layer of bacteria that develops on our teeth. Heterotrophic bacteria, such as *Streptococcus mutans,* grow on the teeth and obtain their organic nutrients from the food we eat. We attempt to remove this particular biofilm by brushing our teeth; however, it builds up again after every meal.

Unlike microbial mats, biofilms may be caused by a single species of bacteria. For example, human pathogenic bacteria may form biofilms on heart valves or stents in the circulatory system. Biofilms do not necessarily have all the biotic components of a typical ecosystem. Thus, the biofilms on teeth or in a water pipeline are communities of bacteria, but they do not have primary producers that are found in major ecosystems such as ponds and lakes nor do they have macrobial consumers.

Biofilms are very important economically. Biofilms that develop on the hulls of ships not only lead to corrosion but also increase the drag of the ship as it moves through the water. Likewise, microorganisms living in pipelines carrying water, natural gas, or oil can produce biofilms that impede the delivery of the fluid, corrode

Octopus Spring

Roseiflexus

Synechococcus

Figure 24.11 Hot springs in Yellowstone National Park
This hot spring has a mat dominated by the cyanobacterium *Synecococcus* and *Roseiflexus*, a member of the *Chloroflexi* (see Chapter 21). The section (top inset) taken through the mat shows the vertical stratification of microorganisms in the mat community. The filamentous *Roseiflexus* (bottom inset) produces a reddish carotenoid pigment. Courtesy of David Ward.

the pipes, and ultimately destroy the pipe itself. For example, sulfide produced in sewage pipes can be oxidized to sulfuric acid by sulfur-oxidizing bacteria. The acid can actually destroy metal and ceramic sewer lines, necessitating their costly replacement.

Biomass and Biomarkers

How do microbial ecologists study microbial communities and assess their activities? We begin by discussing biomass and biomarkers.

BIOMASS   In the same sense that autecologists are interested in knowing the numbers or quantities of a particular species in the habitat, the community ecologist is interested in knowing the quantity of all microbial cytoplasm. This is the **biomass**, or concentration of living microbial cell material in the environment. A variety of techniques have been used for the measurement of biomass, all of which have limitations. The ATP biomass procedure is briefly discussed herein. ATP is found in the cytoplasm of all living organisms but not in dead organisms. The approach used in this procedure is to concentrate organisms from the environment and extract the ATP, the amount of which is directly related to the amount of living cytoplasm. Microorganisms are separated from other organisms in a natural sample by screening or filtration. The ATP is extracted from the harvested cells on the filter, and the ATP content is determined with an ATP photometer. This assay can provide important information to the microbial ecologist, as illustrated in Box 24.3.

BIOMARKERS   Biomarkers are used to assess the presence and concentrations of microbial groups in natural communities. A **biomarker** is a chemical substance that is uniquely produced by a group of organisms. Ideally, the biomarker selected should be simple to analyze, otherwise one might as well use other approaches such as the fluorescent antiserum procedure to identify specific groups.

A popular group of biomarkers currently used for bacteria is phospholipids. Different groups of bacteria have different membrane phospholipids and therefore can be differentiated from one another in natural communities by assessing the type of phospholipid present.

It should be noted that although biomarkers have also been used as biomass indicators for certain groups, biomarkers may survive in the environment even though the bacterium that produced them is no longer alive. Therefore, caution should be exercised in assessing the significance of finding a biomarker from a natural sample.

Biomarkers have useful applications. For example, some *Bacteroides* species that grow in the human intestine produce coprostanol, a product of cholesterol degradation. Because of its association with the human intestinal tract, coprostanol has been used as an indicator of human sewage contamination in natural environments. Therefore, if a water sample contains this biomarker, it is suggestive evidence for contamination of water supply by human fecal material. However, a recent study has shown that this particular biomarker assay has some limitations. Coprostanol was found in marine sediments near Antarctica, far from human habitation and influence. Subsequent research determined that the source was from certain marine mammals including whales and some seals! So, coprostanol is not as much a marker for human feces as it is for the bacteria that live in the intestinal tracts of mammals that transform cholesterol to coprostanol.

Is It a Microbial or Chemical Process?

When examining a transformation in nature, the first question the microbiologist asks is, "Is this transformation a chemical or biological process?" It is not always clear whether a particular activity is performed by bacteria and other microorganisms or by a chemical mechanism. Consider for example, the weathering of rocks, buildings, and monuments. It is well known that a variety of microorganisms grow on the surfaces of rocks and can be involved in weathering of the surfaces. They do this by producing organic acids and chelating agents that gradually dissolve away minerals. However, the chemical effects of acid rain and the physical effects of freezing and thawing are examples of nonbiological processes that cause weathering. For this reason, it is not always easy to determine if a specific rock is being eroded by microorganisms or by physical and chemical processes. Furthermore, both chemical and biological processes often work in concert with one another in effecting a transformation. Because these processes are so slow, they may be difficult to study in the field.

An early example in which a process was identified as a biological process is nitrification, which was studied by Schloesing and Müntz (see Chapter 19). They added soil to a cylinder with a drain and trickled sewage effluent containing ammonia through the column. In a few days, the effluent from the column contained nitrate, not ammonia. This indicated to them that the process of nitrification was occurring in the soil in the column. To determine whether this process was a biological process, they added boiling water to the column and noted that the process abruptly ceased. They reasoned that this is likely a biological process because, if it were chemical, the increased heat would speed up the process, not stop it. They then treated another nitrifying column with chloroform, which is toxic to most organisms. Again the process stopped. Because chloroform would not be

BOX 24.3 | *Research Highlights*

Effect of Mt. St. Helens' Eruption on Lakes in the Blast Zone

On May 18, 1980, Mt. St. Helens, a volcano in Washington State, erupted. The cataclysmic event devastated the forest in the blast zone area north of the mountain (Figures A and B). All of the trees and other vegetation in the blast zone were killed due to the heat and force of the blast, the ensuing mud slides, and the fall of ash and other pyroclastic materials.

Ecologists set out to determine the effect of the eruption on lakes in the blast zone on the north side of the volcano. Although it was difficult to obtain samples immediately following the eruption, after a short time helicopters were flown in, and small volume samples were collected that could be used for analysis of total and viable counts of bacteria, and analysis of chlorophyll *a* and ATP. Fortunately, some samples had been obtained from Spirit Lake prior to the eruption so that the chlorophyll *a* and ATP levels could be compared with those collected after the eruption (Table 1). Curiously those results indicat-

ed that, following the eruption, there was a decrease in chlorophyll *a*, but an increase in ATP. This was surprising because generally the most significant biomass in a lake is due to algae, and, yet, in this instance, the concentration of chlorophyll *a* actually decreased. Thus, the increase in ATP could not be explained by an increase in phytoplankton growth.

The explanation for the increase in biomass, as indicated by the increase in ATP concentration, was that there was tremendous growth of bacteria. This result was first indicated because the initial plates used for viable counting of the bacteria showed that there were much higher concentrations of bacteria in the blast zone lakes north of the volcano than were found in control lakes located south of the volcano (Merrill Lake and Blue Lake) that were not affected by the blast (Figure C). In fact, there were so many bacteria that higher dilutions were needed to accurately count them. The con-

centrations of viable heterotrophic bacteria in blast zone lakes north of the volcano (for example, Coldwater Lake), where the effects of the eruption were most severe, increased dramatically to more than 10^6 per ml in the first summer, and this pattern continued into the next year after the eruption (Table 2). In contrast, the control lakes south of the mountain (Merrill Lake) were unaffected by the eruption, with a concentration of about 10^3 viable cells per ml (compare these numbers with Lake Washington and the pulp mill oxidation lagoon in Table 24.1). It was also interesting to note that not only did the viable counts of bacteria increase in the blast zone lakes, but they also approached the total microscopic cell count (acridine orange counts), thereby indicating that the plating procedures for viable counts was able to recover most of the bacteria in the blast zone lakes. As mentioned in the text, the phenomenon of high recoveries of bacteria (>1% of total microscopic count) is found in lakes that are eutrophic (high in nutrient concentration) but not oligotrophic (low in nutrients). This suggested that the lakes in the Mt. St. Helens blast zone, which had previously been oligotrophic, became eutrophic due to the effects of the eruption. Indeed, the lakes were converted from oligotrophic mountain lakes to eutrophic lakes following the eruption (Table 2).

So, why were there such high concentrations of bacteria in the lakes? The dramatic increase in the concentration of heterotrophic bacteria in the blast zone lakes

Table 1 Characteristics of Spirit Lake prior to and following May 18, 1980, volcanic eruption

Feature	April 4 1980	July 15 1980	Enrichment
Chlorophyll *a* (μg/l)	2.5	0.3	—
ATP (μg/l)	0.25	4.3	17-fold
DOC (μg/l)	0.83	39.9	48-fold
Dissolved oxygen (mg/l)	Unknown	2.35[a]	—
Temperature	15°C	22°–24°C[b]	—

[a]Although oxygen was not measured on April 4, it would have been expected to be saturated at 4°C and therefore greater than 12 mg/l, close to the value normally expected on July 15.

[b]A temperature of 35°C was actually measured in Spirit Lake on May 19, the day after the eruption. On July 15 the temperature was still much higher than it would have been normally at this time of year, due to hot water entering the lake from the volcano. Data from R. C. Wissmar, et al. 1982. *Science* 216:178–181.

BOX 24.3 *Continued*

Table 2 Bacterial concentrations (per ml) and recovery from a blast zone lake compared to a control lake following Mt. St. Helens' eruption[a]

Date (mo/day) 1980–1981	Control Lake: Merrill			Blast Zone Lake: Coldwater		
	Viable	Total	% Recovery	Viable	Total	% Recovery
6/30/80	1.0×10^3	5.4×10^5	0.19	$>10^4$	7.4×10^6	—
9/11/80	2.0×10^2	1.4×10^6	0.01	2.5×10^6	6.4×10^6	39.1
4/30/81	ND[b]	—	—	2.2×10^6	2.7×10^6	81.5
6/29/81	ND	—	—	1.8×10^6	2.5×10^6	72.0

[a]Viable counts were on CPS (caseinate, peptone, starch) plates, and total counts were determined by acridine orange epifluorescence counting.
[b]Not determined.
Data from J. T. Staley, et. al. 1982. *Appl. Environ. Microbiol.* 43:664–670.

occurred because of the tremendous increase in nonliving organic matter (dead vegetation and animals) that became available when the plants and animals were killed following the eruption. The organic matter leached from the watersheds into the rivers and lakes where the bacteria thrived and made the lakes naturally eutrophic.

(A)

(B)

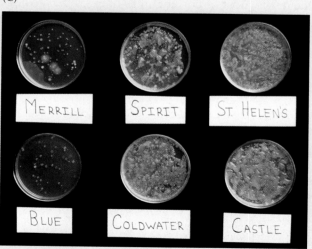

(C)

Spirit Lake, north of Mount St. Helens (A) before and (B) after eruption of the volcano. Note the dramatic impact of the eruption on the biological communities. (C) All plates contain 0.1-ml samples of water taken from lakes near Mount St. Helens shortly after the eruption. A, ©Pat and Tom Leeson/Photo Researchers, Inc.; B,C, courtesy of J. T. Staley.

Compared with the control lakes...

...all lakes in the blast zone have an overgrowth of bacteria.

expected to inhibit most chemical processes, they concluded that the result also supported the hypothesis that this is a biologically mediated process. Subsequently, nitrifying bacteria have been isolated from such columns, indicating that biological agents are indeed responsible for the transformation.

As discussed previously, Koch's postulates have been adapted for use in microbial ecology to verify that a microorganism is responsible for the process. It is increasingly found that microorganisms are responsible for many reactions that were previously thought to occur only by chemical means.

Stable isotopes can also provide important information about whether a process is biological because of isotope discrimination. For example, consider the sulfur cycle. There are nine isotopes of sulfur including four that are stable (nonradioactive): the normal isotope ^{32}S, and several others including ^{33}S, ^{34}S, and ^{36}S; the natural abundance of each in Earth's crust is, respectively, 95.0%, 0.76%, 4.22%, and 0.014%. Biological processes exhibit **isotope** discrimination, that is, organisms preferentially use lighter isotopes. For example, sulfate-reducing bacteria preferentially reduce $^{32}SO_4^{2-}$ rather than the heavier isotopic form of the ion, $^{34}SO_4^{2-}$. As a result, they produce a sulfide that is lighter (contains more ^{32}S than ^{34}S) than the substrate. From this information, it has been concluded that elemental S (S^0) deposits on Earth that have a lighter sulfur isotope content than that of Earth's crust are derived from biological sources. Likewise, the sulfur cycle itself has been dated to at least 3.2 Ga BP (before the present) on the basis of observing lighter than expected sulfur compounds in sedimentary deposits.

Rate Measurements

One of the most important areas of microbial ecology is measuring the rates of microbial processes. Knowing this allows scientists to assess the significance of the contribution of the microorganisms to a process. The following section describes some of the various techniques for rate measurements and their applications.

DIRECT CHEMICAL ANALYSES   In some cases, rate measurements can be determined by simply analyzing for the changing concentrations of chemicals (substrates and products). An example of this is the measurement of primary productivity by measuring the oxygen evolved during photosynthesis. As discussed in Chapter 1, the overall reaction for primary production is given as:

$$(1)\ 6\ CO_2 + 6\ H_2O \rightarrow (C_6H_{12}O_6)_n + 6\ O_2$$

where $(C_6H_{12}O_6)_n$ = organic carbon formed.

Reaction 1 indicates that the amount of carbon dioxide fixed (that is, the primary production) is directly re-

lated to the oxygen evolved on a mole-to-mole basis. Therefore, the rate of primary production can be determined simply by measuring the amount of oxygen produced for a 24-hour period during photosynthesis. In aquatic habitats, this is accomplished in glass-stoppered bottles containing water collected from a lake or marine habitat (Box 24.4).

USE OF CHEMICAL ISOTOPES   Chemical isotopes can also be used for rate measurements. The major advantage of using them is the extreme sensitivity that they provide for rate analysis. Two types of isotopes are available, depending on the substrate. As mentioned previously, stable isotopes are not radioactive, but their atomic weight differs from that of the common form of the element. **Radioisotopes** exhibit radioactive decay. These isotopes occur normally in nature, but they are found in reduced amounts compared to the normal or common isotope. Both stable and radioactive isotopes are used to determine the rates of ecological transformations. Whether a stable or radioactive isotope is used is determined by several factors, such as the equipment available for analysis and the half-life (the length of time required for half of the radioactivity to be lost due to decay) of the radioisotope available. For example, for carbon isotope analysis, most scientists determining rates of processes use carbon-14 (^{14}C) because it can be simply quantified using a scintillation counter, and it has a long half-life (more than 5,000 years). Conversely, ^{13}C, a stable isotope, requires that a mass spectrometer be available for analysis: this is much more expensive than a scintillation counter and the analysis is much more labor intensive. However, with nitrogen, the choice is simple: ^{15}N is used even though a mass spectrometer is required, because the half-life of ^{13}N is only a few minutes, making it impracticable for virtually all ecological experiments.

The most common way to measure **primary productivity** is by the use of radioisotopic procedures, namely using ^{14}C-labeled sodium bicarbonate. The utility of this can be seen by referring to Reaction 1 for primary production. So, rather than measuring oxygen production, which is an indirect measure of primary production (see Box 24.4), one measures carbon dioxide fixation into biomass using ^{14}C-labeled bicarbonate (Box 24.5). As mentioned earlier, this radioisotopic method is much more sensitive than ordinary chemical methods and is therefore the preferred technique.

The measurement of **heterotrophic activities** is more complicated than that of primary production. This is because in primary production, a single substrate, carbon dioxide, is being metabolized by primary producers, whereas for heterotrophic activities, there are hundreds of different organic substances that serve as substrates in the environment. Furthermore, some of these are soluble—termed dissolved organic carbon (DOC)—and

BOX 24.4 *Methods & Techniques*

Chemical Measurement of Primary Production

The chemical measurement of primary production in an aquatic habitat is performed using freshly collected water from the depth(s) of interest. This is used to fill three bottles that are incubated at the depth (Figure A). One bottle is called the "dark bottle" because it is made opaque to prevent light from entering. Although photosynthesis cannot occur in this bottle, respiration, which is the reverse of the photosynthesis reaction, does occur. The second bottle, called the "fixed control," is fixed with formalin at zero time so that no biological activity can occur in it. The third bottle, called the "light bottle," or experimental bottle, is allowed to incubate and photosynthesize as it would in nature except that the process is now occurring in an enclosed container. These bottles are incubated usually by submerging at the depth of collection, preferably for an entire 24-hour period. The reaction is then stopped by addition of formalin to the experimental bottle and the dark bottle.

Primary production is determined from measurements of the concentration of oxygen in the different bottles (Figure B). At zero time the oxygen concentration would be expected to be the same in all bottles. After a 24-hour incubation period, the oxygen concentration would be expected to be highest in the experimental bottle in which photosynthesis occurs. The difference in oxygen concentration in this bottle during the time of incubation is a measure of the net primary production that occurred in the volume of water tested and is calculated as follows:

$$\text{Net primary production/day} = [O_2]_{L2} - [O_2]_{L1}$$

where $[O_2]_{L1}$ is the moles of oxygen per liter at zero time in the light bottle and $[O_2]_{L2}$ is the moles of oxygen per liter after the incubation period in the light bottle.

The gross primary production is calculated as the net primary production plus that amount of oxygen that was used in respiration (Figure B). This latter value can be calculated as the amount of oxygen used during incubation in the dark bottle where only respiration could occur. So, gross photosynthesis is:

$$\text{Gross primary production/day} = \text{net primary production} - [O_2]_{D2}$$

The oxygen concentration in the fixed control should not have changed during the incubation. If it did, this would indicate either a leak in the bottle or some other problem with the assay that would indicate to the researchers that the experiment was flawed.

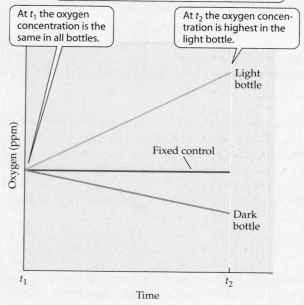

(A) Water from a given depth is divided into three samples in glass-stoppered bottles…

…one sample, untreated, in a clear glass bottle…

…one sample in an opaque bottle (which halts all photosynthetic primary production)…

…one sample fixed with preservative (which halts all activity).

Light bottle   Dark bottle   Fixed control

(B) Oxygen is measured at the start (t_1), the samples are incubated for 24 hours at the depth under study, preservative is added to the light and dark bottles, then oxygen is measured again (t_2).

At t_1 the oxygen concentration is the same in all bottles.

At t_2 the oxygen concentration is highest in the light bottle.

Light bottle

Fixed control

Dark bottle

Oxygen (ppm)

t_1 t_2

Time

Measurement of primary production in aquatic habitats by measuring oxygen production. (A) Experimental treatments of water samples from the depth at which photosynthesis is to be measured. (B) Graph showing the change in concentration of oxygen.

BOX 24.5	*Methods & Techniques*

Radioisotopic Method of Measurement of Primary Production: An Example of the Tracer Technique

The experiment is set up in the same manner as the chemical analysis experiment for primary production—that is, with both light and dark bottles as well as a fixed control (see Box 24.4, Figure A). To initiate the experiment, a small amount (trace) of ^{14}C-labeled bicarbonate is added to each bottle. The amount of bicarbonate added is small relative to the total amount already in the water, which is measured chemically. Therefore, the concentration of the substrate, both unlabeled and labeled, in the sample is virtually unaffected by the addition.

This is called the tracer approach, because the pool size or natural substrate concentration is not affected by the amount of label added because so little is added relative to the total amount of natural substrate. By following what happens to the label, the biological transformation can be assessed.

In this procedure, the amount of carbon dioxide fixed is determined by the amount of label taken into the cells during photosynthesis. Thus, on completion of the incubation, the algae are harvested from the experimental and control bottles by filtration, and the amount of label incorporated is determined by scintillation counting of the filters. The net photosynthesis is the difference between the amount of label taken up in the experimental bottle and the amount taken up in the dark bottle. The fixed control is used to assess whether chemical effects such as adsorption might have caused artifacts in the experiment.

The photosynthetic rate (PR) is calculated as:

$$\text{Photosynthetic rate (mg c/l/h)} =$$
$$\frac{\dfrac{cpm_L - cpm_D}{CPM_{added}} \times}{\dfrac{\text{Total } ^{12}C\,(mg/l) \times 1.06}{\text{Incubation time (hours)}}}$$

where cpm_L and cpm_D equal the counts per minute found on filters in the light bottle and dark bottle, respectively, and cpm_{added} is the amount of label added to each bottle in the form of ^{14}C-bicarbonate. Total ^{12}C is the amount of carbon in the bicarbonate pool in the habitat; 1.06 is the isotope discrimination factor.

A major assumption of these tracer techniques is that the labeled substrate acts just as the normal substrate would for carbon dioxide fixation. In the case of carbon dioxide, the ^{14}C label is not only radioactive, but it is heavier than the naturally occurring ^{12}C. Therefore, the factor 1.06 is multiplied by the entire photosynthetic rate to correct for isotope discrimination (6% in this case) and to determine the actual rate. As discussed previously, isotope discrimination occurs with both stable as well as radioactive isotopes, whose masses are different from the normal isotope.

The measurement of primary production using the $^{14}CO_2$ tracer procedure uses the same type of bottles as for the oxygen procedure. However, prior to incubation a trace amount of radiolabeled $NaH^{14}CO_3$ is added to initiate the reaction. For the analysis of primary production, the amount of label taken into the cells as fixed organic matter is measured using a scintillation counter. Therefore, after incubation, the cells must be concentrated on a filter before placing them in the scintillation vial. Primary production is determined as the amount of carbon dioxide fixed by the algae in the experimental bottle compared with that fixed in the dark bottle.

---

others are insoluble—termed particulate organic carbon (POC). Nonetheless, radioisotopes are available for many DOC compounds, such as sugars, amino acids, and organic acids, and either tritium (^3H)-labeled or ^{14}C-labeled forms can be used.

The general reaction that occurs aerobically in respiration is:

$$(2)\ (C_6H_{12}O_6)_n + 6\,O_2 \rightarrow 6\,CO_2 + 6\,H_2O$$

where $(C_6H_{12}O_6)_n$ represents organic carbon.

Thus, overall, this is the reverse reaction of primary production (Reaction 1). One advantage of using ^{14}C-labeled organic compounds is that it is possible to collect the carbon dioxide that is respired and to quantify it. This is not possible if the tritiated form of the organic substrate is used. Furthermore, the radioactive water formed with tritium is difficult to separate from the soluble organic substrate and products produced in the reaction. Therefore, ^{14}C-labeled substrates are most commonly used, except for microautoradiography.

Another major advantage of using isotopes is that it is possible to determine rates in systems in which a **steady state** or **approximate steady state** occurs. This is a condition in which the concentrations of substrates and products remain constant from one time to the next. This commonly occurs for extended periods in nature because the rate of formation of a product is often very similar to the rate at which the product is removed or metabolized. Therefore, it is meaningless to measure the chemical concentration of the substrate and product at time zero and some time later, because they will be approximately the same, and if they are somewhat different, it is due to the combined activity of formation and utilization. Nonetheless, utilization is occurring in the natural habitat. By adding a labeled substrate to the natural substrate pool as a tracer (see Box 24.4), it is possible to then identify (label) that part

of the substrate pool and follow it in the formation of the product.

This tracer approach is most commonly used in measuring, not only the rate of primary production but also the rate of the heterotrophic activity or metabolism of dissolved organic compounds in environments. This is usually accomplished with short-term incubations using sealed flasks (Box 24.6).

In the 1960s, Richard Wright and John Hobbie developed a procedure for the measurement of the uptake of dissolved organic compounds that is now named for them. They discovered that when ^{14}C-labeled glucose

---

BOX 24.6 *Methods & Techniques*

Radioisotopic Procedure for Assessing Heterotrophic Rates of Metabolism in Aquatic Habitats

Initially, the water sample is collected from the environment and distributed into flasks. Both experimental and fixed control flasks are set up, and the reaction is initiated by adding the radiolabeled substrate. If the actual concentration of the substrate in the natural habitat can be measured, then the reaction can be quantified by labeling the pool with a tracer label of that organic compound.

By knowing the concentrations of the substrate, (C_o), the amount that is radiolabeled at zero time (cpm_s), and the amount of label in the product (or that which is taken up by the metabolizing bacteria), cpm_p, it is possible to calculate the total amount of substrate that has been incorporated into the product, X, during a certain period of time by a simple proportion:

$$C_o/cpm_s = X/cpm_p$$

where C_o is the concentration of the substrate; cpm_s is the counts per minute of substrate; cpm_p is the concentration of the product; and X is the amount of substrate taken up by metabolizing cells.

In these experiments, some of the substrate will be converted into

Apparatus for assessing rate of use of dissolved organic carbon by microorganisms in aquatic habitats using the tracer label approach.

carbon dioxide, which is collected in a separate vial in the experimental vessel (above). Most of the remaining label in the product will have been incorporated into cells. This is determined by harvesting cell material on a membrane filter and counting its radioactivity along with that of the mineralized carbon dioxide using a scintillation counter. It is the sum of the ^{14}C-CO_2 and ^{14}C-cellular material that is determined to provide the value for cpm_p.

In addition to the aspect of the decay of radioisotopes, there is the question of specific activity of the

label. This refers to how "hot" the substrate is. Because dissolved organic compounds occur at extremely low concentrations in natural environments, if we are to use the tracer approach, it is important to add small amounts of the radiolabeled substrate to the pool. This requires that the radiolabeled substrate received from the manufacturer contain as many labeled molecules as possible. Otherwise, the natural substrate will be diluted too much to provide a sufficiently accurate and sensitive assay.

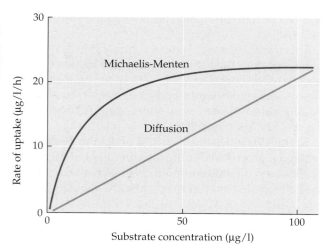

Figure 24.12 Diffusion versus Michaelis-Menten kinetics
This diagram illustrates the rate of uptake of a substrate at various substrate concentrations by diffusion in comparison with Michaelis-Menten kinetics. Note that when Michaelis-Menten uptake occurs, the high affinity enzyme system rapidly takes in the substrate until the enzyme system is saturated with substrate. In contrast, the diffusion system is much less effective at taking in the substrate at low concentrations.

was added at varying concentrations to cell suspensions of an alga and a bacterium, the results of uptake were different. The alga took up the glucose by diffusion, and, therefore, as the amount of glucose increased, there was a proportionally greater increase in glucose incorporated in the alga suspension (Figure 24.12). In contrast, the bacterium took up the glucose by a high affinity process that appeared to follow Michaelis–Menten kinetics, which characterize the rate of metabolism of a substrate in a typical enzymatic reaction. The rate of enzymatic reactions is very high at low substrate concentrations and then levels off as the enzyme becomes saturated at higher concentrations of substrate. Therefore, bacteria have high affinity uptake systems that are much more effective at the uptake of glucose and other dissolved organic compounds than algae in natural habitats where the DOC concentrations are typically very low. The **Wright–Hobbie kinetic technique** entails the addition of labeled substrates at several concentrations, some of which exceed the natural concentration of the substrate in the environment. These experiments need to be short-term in nature, ideally with incubations less than one hour, to prevent an unnatural adapted response from the community. The term **heterotrophic potential** refers to the maximum rate of uptake that occurs when the substrate is added at high levels. This represents the "potential rate" that that particular community could metabolize a substrate when the substrate is available in

high concentrations. The heterotrophic potential can be measured even though the natural substrate concentration is unknown. This is an advantage because it is often difficult to measure the in situ concentration of individual organic compounds in habitats because they are in such low concentration. Indeed, the low concentration of DOC in many natural environments is maintained by their efficient removal from the substrate pool by heterotrophic bacteria and archaea. The Wright–Hobbie technique is excellent for most freshwater and some marine habitats. However, in the mid-oceanic water column it is often difficult to detect any response of the natural community to the added substrate, indicating that the concentration of DOC is extremely low.

Radioisotopic methods can also be complemented with autoradiography as discussed previously in this chapter. This permits the scientist to associate the uptake of a specific compound with a specific microorganism. Although the organism might not be identifiable morphologically, it can be rendered identifiable by labeling it with a fluorescent antiserum or by use of the FISH approach and then examining it by both fluorescence and by light microscopy.

ALTERNATIVE SUBSTRATES   For some activities, it is possible to use an alternative substrate in place of the natural substrate for determining the reaction rate. Perhaps the best example of this is the use of acetylene in place of nitrogen for the measurement of nitrogen fixation. The overall reaction for nitrogen fixation involves three reductive steps in which N_2 ($N \equiv N$) is reduced first to $HN = NH$, then to $H_2N — NH_2$, and then ultimately to $2 NH_3$. Acetylene, which is also a triple-bonded compound, occupies the active site of nitrogenase and is reduced to ethylene gas as follows:

$$(3)\ HC \equiv CH \rightarrow H_2C = CH_2$$

Remarkably, acetylene is actually preferred to N_2 by nitrogenase, so the reaction can be carried out even in the presence of the natural substrate, N_2. The main advantage of measuring nitrogen fixation this way is that both acetylene and ethylene can be easily quantified using a gas chromatograph, whereas the only stable nitrogen isotope that is useable is the stable $^{15}N_2$ isotope and its analysis is tedious and requires a mass spectrometer.

A disadvantage of the acetylene reduction assay is that the actual rate of the reaction is influenced by the substrate provided. Furthermore, acetylene is reduced to ethylene (which escapes as a gas) by only one reductive step rather than the three required for complete nitrogen fixation. For these reasons, it is not possible to directly determine in situ rates with these procedures. Nonetheless, they provide excellent relative measures of nitrogen fixation in systems where nitrogen fixation is known to occur. Furthermore, assays can be standardized using $^{15}N_2$.

INHIBITORS   Inhibitors are sometimes used to measure the rates of reactions. If the inhibitor specifically stops the process of interest and has no other effects, then the rate of the process can be determined by the rate at which the substrate accumulates. For example, molybdate ion is a known inhibitor of sulfate reduction. Therefore, it can be incorporated into assays to inhibit sulfate reducers. One example where this has been used is in the determination of how effectively sulfate reducers compete with methanogens for hydrogen in sediment communities. The sulfate reducers are inhibited with molybdate to assess for methanogen uptake of hydrogen in the environment, and the methanogens can be inhibited by the use of chloroform to assess the uptake of hydrogen by sulfate reducers. However, the use of such inhibitors may cause other effects on the community that may adversely affect the results, so they should be used with caution.

SECTION HIGHLIGHTS

The community ecologist is interested in studying the natural community with its biotic and abiotic components. Assessment of activities such as primary production and heterotrophic activities can be undertaken by the use of specialized procedures. Microbial mats and biofilms are examples of natural microbial communities.

24.5 Major Environments

As previously mentioned, bacteria grow in some of the most unusual and extreme environments on Earth. In this section, the more common natural habitats are considered. These include freshwater, marine, and terrestrial environments.

Freshwater Habitats

Aquatic habitats comprise some of the most important habitats for microorganisms. Even waters of the most pristine lake or in the open ocean contain hundreds of thousands or millions of bacteria per milliliter. Bacteria can be easily concentrated on filters of samples from aquatic habitats and therefore counted without interference from debris, as is found in the soil environment. Freshwater habitats include lakes, ponds, rivers, and

streams. Most of these inland waters have very low concentrations of salt.

LAKES   Many freshwater lakes of sufficient depth that are located in temperate zones become thermally stratified during the summer months. In such lakes, the surface water becomes warmer than the underlying water. This occurs because the surface is increasingly heated by the Sun's radiation during the spring and summer months. Summer winds are not strong enough to mix the lighter warm surface layers with the more dense cold water that lies beneath. Therefore, the warm surface water layer becomes separated from the colder bottom water, resulting in **thermal stratification** of the lake. In contrast to deep temperate zone lakes, shallow lakes and ponds do not stratify because the winds keep them intermittently mixed.

In lakes where thermal stratification occurs, the surface layer of the lake is called the **epilimnion** and the lower body of water is called the **hypolimnion** (Figure 24.13). The zone of transition in temperature between

(A)

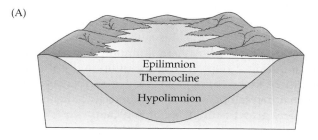

(B)

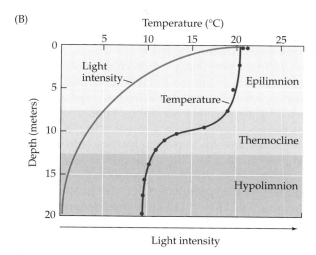

Figure 24.13 Thermal stratification
(A) Cross-section through a thermally stratified lake, showing the epilimnion, thermocline, and hypolimnion. (B) A limnological graph is a plot of the depth of a lake versus the concentration or intensity of a parameter—in this case, temperature and light intensity in a typical thermally stratified lake during the summer months.

the surface and the lower depths is called the **metal-imnion**, or **thermocline**.

The period of thermal stratification persists throughout the summer season and into early fall. However, as the air temperature cools in the fall, the surface water temperature decreases, and the water becomes denser. At some point, the phenomenon of **fall turnover** occurs in which the density of the surface water and the underlying water are almost identical. Winds can then readily mix the lake thoroughly from top to bottom. This period of mixing continues throughout the wintertime, unless the temperatures become so cold that the lake actually freezes over. If freezing occurs, then the spring warming must first melt the ice and warm the melted surface water to 4°C (water is most dense at 4°C) before it will mix with the underlying water. In either case, at least one period of mixing occurs during the winter months in typical temperate zone lakes.

These physical features of lakes make them ideally suited for spring and summer blooms of algae and the resultant growth of bacteria. In the early spring months, the nutrients in the lake water are high in concentration because they have been brought up to the surface from the lower depths and sediments during the mixing period. At this time, the concentration of dissolved nutrients such as phosphate, ammonium, and nitrate is constant at all depths. Then, as the daylight period lengthens in the spring, it provides both increased light and warmer temperatures, conditions that are favorable for algal growth and other biological processes. Thus, spring and summer blooms of algae occur. At this same time, the increased warmth of the surface waters results in a decrease in surface water density and combined with wind, it enables the separation of the epilimnion from the hypolimnion.

As discussed previously, a food chain exists in aquatic environments (see Figure 24.1). As decomposers, heterotrophic bacteria degrade the excess organic material excreted by algae as well as the remains of dead algae and other organisms that inhabit the lake. These bacteria are called decomposers because they convert the organic material back into inorganic material so that it can be recycled. If the complete degradation of an organic substance occurs to produce entirely inorganic end products including CO_2 and H_2O, the process is called **mineralization**. Most heterotrophic bacteria that are indigenous to planktonic aquatic habitats are referred to as **oligotrophs**, because they can grow using very low concentrations of organic nutrients. They are to be distinguished from **eutrophs**, which grow in organic-rich environments such as the intestinal tracts of animals.

Oligotrophs exhibit low growth rates and have high affinity systems in their cell membranes that allow for the uptake of carbon sources that occur in low concentrations as found in planktonic habitats. By their activi-

ties, the concentrations of soluble organic substrates are kept at very low levels (microgram concentrations) in mesotrophic and oligotrophic environments.

In contrast, eutrophic bacteria have rapid growth rates when they are provided with high concentrations of nutrients. In lake water, nutrients are available only in low concentrations, so the eutrophic bacteria are not competitive with oligotrophic species from these natural communities. However, the intestinal tracts of all of the consumer animals in aquatic food chains contain eutrophic heterotrophs, which play a major role in the decomposition of foodstuffs eaten by the consumers. These foods are converted into amino acids, vitamins, and organic acids that are absorbed in the intestinal tracts of the animals and used as their principal energy sources and sources of organic building materials. It is important to note that almost as much of the food consumed by the animal goes to produce microorganisms in the intestine (which are ultimately defecated) as goes to the animal's nutrition and energetics. Therefore, symbiotic microorganisms of the animal intestine play a major, often overlooked, role in all food chains and food webs.

The **microbial loop** model best explains what occurs in aquatic habitats (Figure 24.14) to account for the mineralization of organic material. In addition to the bacteria, protozoa and small invertebrate animals are involved in the degradation of organic materials. The heterotrophic bacteria utilize dissolved organic material excreted by algae and *Cyanobacteria*. Because they have high affinity uptake systems for organic compounds, bacteria grow principally on dissolved organic compounds (although through their extracellular cellulases and chitinases they can also degrade particulate organic materials). The bacteria are then ingested by small protozoa, which also ingest small algae. The bacteria are commonly attached to organic **detritus** particles (nonliving organic particulate material), which they are degrading (Table 24.3). Unlike bacteria, protozoa and other invertebrates have mouths and therefore ingest the smaller bacterial cells and small detritus particles through **phagotrophic** nutrition. These protozoa, which are microbial **grazers** (consumers that ingest bacteria, algae, and bacterially coated particles), are in turn eaten by larger protozoa and invertebrate animals. Ultimately, marine animals including fish and mammals at higher levels in the food chain ingest the invertebrates to complete the microbial loop.

Although microbial and macrobial grazers do not degrade all the particulate material during its passage through their digestive tract, their fecal pellets can be reingested by other grazers after defecation, so the process is repeated until the matter is ultimately completely degraded. Therefore, the combined efforts of bacteria and the grazing animals result in the ultimate decomposition of organic materials in aquatic ecosystems.

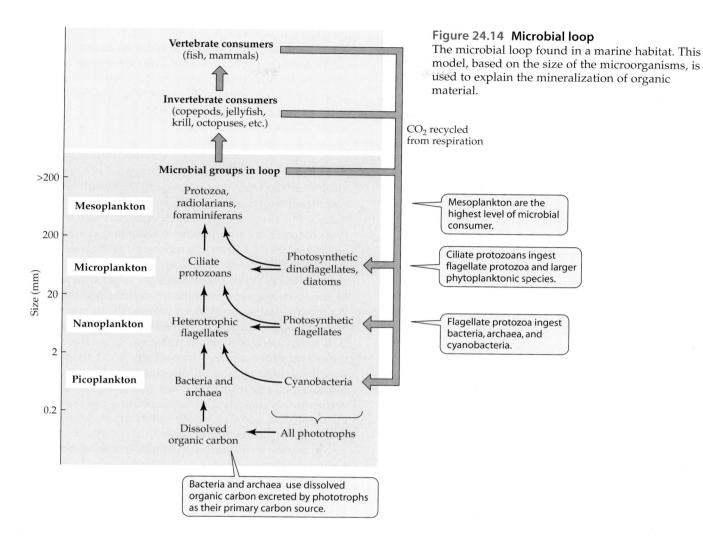

Figure 24.14 Microbial loop
The microbial loop found in a marine habitat. This model, based on the size of the microorganisms, is used to explain the mineralization of organic material.

TABLE 24.3	Composition of a natural water sample
Constituent	**Characteristics**
I. Dissolved material	Obtained by either centrifugation or filtration through a 0.2-μm pore size filter to remove particulates
A. Organic components	
1. DOC	Dissolved organic carbon
2. DON	Dissolved organic nitrogen
B. Inorganic materials	
II. Particulate materials	Obtained as centrifuged material or collected on 0.2-μm pore size filter.
A. Living material	Living cellular material is called biomass
B. Nonliving materials	
1. Organic	Nonliving organic material is called detritus
2. Inorganic	

Of course, small amounts of particulate organic materials are not completely degraded and deposited in the sediments.

Eutrophic lakes that are sufficiently enriched with nutrients may develop an anaerobic hypolimnion in the summer months during the period of thermal stratification. This occurs because the hypolimnion is physically separated from the surface layers of the lake during stratification, so the amount of oxygen available in the hypolimnion is limited to what was initially there at the time of thermal stratification and the small amount that diffuses to it from the epilimnion. Anaerobic conditions develop in the hypolimnion in the following manner. The high concentration of organic nutrients in the eutrophic lake will be degraded in the hypolimnion initially by aerobes because aerobic respiration is such an energetically advantageous process. This results in oxygen depletion, and the hypolimnion becomes anaerobic.

Anaerobic conditions in the hypolimnion permit the growth of many other bacterial groups during the summer months. For example, if light penetrates to the hypolimnion and sufficient hydrogen sulfide is produced in the sediments, then photosynthetic bacteria can grow and produce hypolimnetic blooms. Common photosynthetic bacteria that may produce anaerobic blooms in this habitat include the purple sulfur bacteria and the green sulfur bacteria (see Figure 24.8). Other anaerobic processes such as fermentation, denitrification, sulfate reduction, and methanogenesis occur at successively deeper depths in the hypolimnion or in the sediments of the lake.

Gas-vacuolate organisms are especially common in thermally stratified lakes (see Figure 24.8). These organisms readily stratify in the lake at a light intensity or nutrient concentration level that is satisfactory for them. Thus, the gas-vacuolate aerobic photosynthetic cyanobacteria are found at the surface of the lake where high light intensities occur, whereas the gas-vacuolate purple and green sulfur bacteria are found at greater depths in the lake, where there is both sulfide and at least some light.

RIVERS AND STREAMS   Flowing water habitats are quite different from lakes. These habitats are shallower and usually remain aerobic except in nonflowing backwaters and in their sediments. Microorganisms that inhabit these environments, such as *Sphaerotilus* sp., attach to rocks and sediments so that they can take advantage of the nutrients that flow by them. Streams in natural areas receive large amounts of leaf fall and other dead organic material in the autumn and winter and are therefore important habitats for leaf and litter decomposition.

Flowing habitats receive runoff from agricultural and urban areas and are greatly influenced by the nature of the material they receive. In urban areas, rivers are frequently used for sewage disposal and may receive varying loads of organic material and bacteria, depending on the degree of sewage treatment. Industrial effluents are also extremely important in larger cities and commercial waterways. In agricultural areas, excess fertilizers and pesticides reach the rivers through runoff and may pose serious pollution problems. These inputs provide nutrients for bacterial growth that may lead to anoxic conditions (see Chapter 32).

Marine Environments

Marine environments are extremely important because about two-thirds of the Earth's surface is covered by oceans. The predominant primary producers in the marine environment are planktonic algae and cyanobacteria (called phytoplankton), which serve as the base of the food chain (see Figure 24.1). The consumers in the ocean that are highest in the food chain are larger fish, sharks, toothed whales, and some other mammals. It is noteworthy that the krill-eating baleen whales, the largest mammals, occupy a rather low position in the food chain.

Because oceans cover so much of Earth, the role of marine microorganisms in primary production and other biogeochemical activities greatly surpasses that of microorganisms in inland lakes and rivers. Biological oceanographers study the biogeochemical processes that are occurring, some of which have important implications for the global climate.

Measuring primary productivity in the ocean is no small feat. Because of its vast size and seasonal and temporal variability, a few measurements here and there do not adequately assess activities. More recently, satellite-mapping procedures have been used to assess global productivity in the oceans (Figure 24.15). These procedures rely on analyzing photos that show chlorophylls in the phytoplankton. Although the quantitative information of these photos is limited due to the inability to observe phytoplankton beneath the surface, they provide valuable information on the extent and intensity of blooms over large scales heretofore impossible to study. It is important to note that the cyanobacteria in the oceans account for more than half of all the marine primary productivity, as noted in Chapter 21.

Other marine microbial activities are also important. For example, it has recently been discovered that dimethyl sulfide (DMS) is produced as a product of metabolism by some phytoplankton. DMS is volatile and enters the atmosphere, where it reacts photochemically with oxygen to form dimethyl sulfoxide (DMSO). This in turn serves as a raindrop nucleator, which results in cloud formation.

Cloud formation affects the intensity of solar radiation on an area in the ocean, thereby allowing for a decrease in primary production and a decrease in DMS production. Thus, there is an internal feedback mechanism regulating primary production and temperature as well (Figure 24.16).

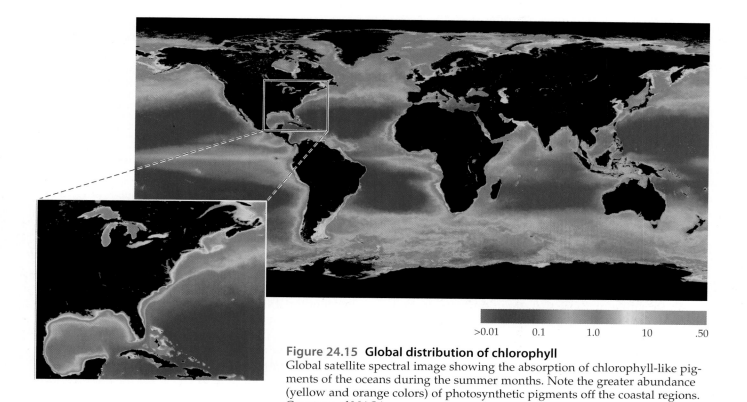

Figure 24.15 Global distribution of chlorophyll
Global satellite spectral image showing the absorption of chlorophyll-like pigments of the oceans during the summer months. Note the greater abundance (yellow and orange colors) of photosynthetic pigments off the coastal regions. Courtesy of NASA.

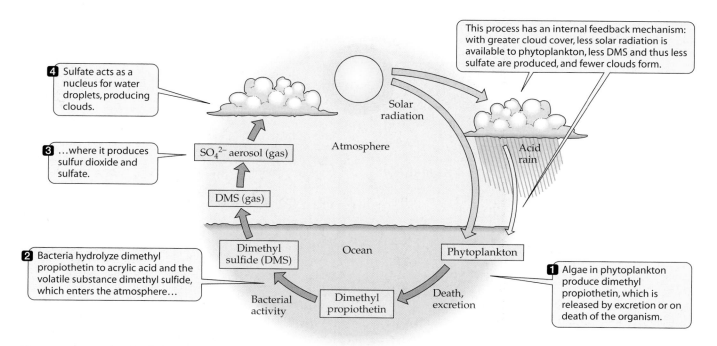

4 Sulfate acts as a nucleus for water droplets, producing clouds.

This process has an internal feedback mechanism: with greater cloud cover, less solar radiation is available to phytoplankton, less DMS and thus less sulfate are produced, and fewer clouds form.

3 ...where it produces sulfur dioxide and sulfate.

SO_4^{2-} aerosol (gas)

DMS (gas)

Solar radiation

Atmosphere

Acid rain

2 Bacteria hydrolyze dimethyl propiothetin to acrylic acid and the volatile substance dimethyl sulfide, which enters the atmosphere...

Dimethyl sulfide (DMS)

Ocean

Phytoplankton

Bacterial activity

Dimethyl propiothetin

Death, excretion

1 Algae in phytoplankton produce dimethyl propiothetin, which is released by excretion or on death of the organism.

Figure 24.16 Microbial cloud formation
Proposed pathway for formation of clouds in the marine environment. It is interesting to note that the rainfall is acidic because of the sulfate, so this is an example of a process that causes a natural type of acid rain.

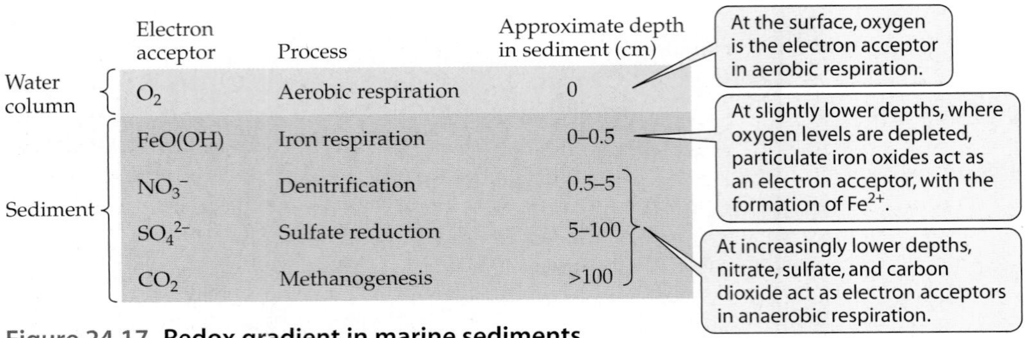

Figure 24.17 Redox gradient in marine sediments
Several types of electron acceptors are used in the biodegradation of organic substances in marine sediments, varying with depth. The actual depths at which these electron acceptors function will differ from one location in the sediment to another, depending on the amount of organic material, temperature, and other factors.

The anaerobic processes that occur in marine sediments are similar to those that occur in freshwater sediments. From the sediment surface going downward, the following are sequentially encountered: aerobic respiration, iron oxide respiration, denitrification, anammox, sulfate reduction, and finally methanogenesis (Figure 24.17). However, there are some major differences. For example, in marine systems, there is more sulfate than in freshwater habitats, so sulfate reduction is a more important process in this habitat.

Typical marine bacteria are different from freshwater bacteria. The higher salt concentration (about 3.5%) makes marine bacteria moderate halophiles (see Box 19.5). Thus, one of the characteristics of a typical marine bacterium is that it will have an optimum salinity of about 3.5%, whereas a freshwater bacterium exhibits a lower optimal salt concentration. Also, most marine bacteria have an absolute requirement for Na^+, a feature not found in typical freshwater or terrestrial bacteria.

Viruses are also very abundant in the oceans. It is estimated that there are about ten times as many viruses as *Bacteria* and *Archaea* in marine environments. They, along with grazers (e.g., protozoa and copepods), play an important role in limiting the cell densities of microorganisms.

Furthermore, great depth at some locations produces very high hydrostatic pressures. Only **barophilic** or **piezophilic** bacteria are known to grow at such depths. These organisms have been isolated from deep trench sites (Table 24.4).

TABLE 24.4	Extreme conditions for microbial growth[a]		
Environmental Parameter		**Microbial Group**	**Habitat Source**
Temperature			
	High: 116°C	Hyperthermophilic *Archaea*	Hot springs, marine hydrothermal vents
	Low: −12°C	Psychrophilic *Bacteria*	Polar sea ice
pH			
	High: >12	Alkaliphilic *Bacteria, Archaea*	Desert soils, soda lakes
	Low: 0–1	Acidophilic *Bacteria, Archaea*	Acid mine drainages, sulfur springs
Saturated salts		Halophilic *Archaea*	Salt lakes, brines
High hydrostatic pressure		Barophilic or piezophilic *Bacteria*	Deep sea
High radioactivity		Radioduric *Bacteria*, e.g., *Deinococcus radiodurans*	Radioactive sites, soils
Low relative humidity (Low water activity)		Xerophilic fungi	Deserts, saps, brines

[a]Some environments have more than one condition that is considered extreme (e.g., acidic hot springs, which contain organisms such as *Sulfolobus*).

Terrestrial Environments

Terrestrial environments are extremely important for human civilization. Modern agriculture relies on good soils for plant and animal productivity. Soils contain a mineral component, which is derived from rock weathering, as well as an organic component, which is derived from organisms that previously resided in the habitat.

SOILS The soil environment is very complex. It is made up of inorganic minerals, the weathered remains of rocks, and organic material, called **humus**. Humus comprises the partial decomposition products of plant organic material and as such contains many of the less-degradable or **refractory** organic plant substances such as lignin and humic acids. Soils vary greatly in composition and fertility that varies with climate, geology, and vegetation. Depending on the composition of the soil, the climate, and the types of vegetation, soils are classified into various groups. Soil texture, one important feature of classification, is determined by the relative proportion of clay, silt, and sand particles of the soil (Figure 24.18).

The soil profile is another important feature used in classification of soils (Figure 24.19). The profile of a soil is determined by making a vertical cross-section or cut into the soil. Each layer, called a **horizon**, has a characteristic composition and color due to the activities that occur there. For example, the A horizon at the surface of the soil is dark in color due to the large amounts of organic matter (humus) that accumulate there. Below this is the E (for eluviation or "wash out") horizon, which is lighter in color and contains resistant minerals such as quartz. The deeper B horizon is a zone of accumulation (illuviation or "wash into"), which contains materials such as Fe, Al, and silicate minerals. Each soil type has its own characteristic soil profile.

Unlike aquatic environments, water is not always available in soils. Nonetheless, water is absolutely required for the growth of all organisms. Thus, microorganisms that occur in soils grow intermittently, that is, only when there is sufficient soil moisture available for them. Microorganisms have made special adaptations to withstand these intermittent periods when water is unavailable. Some produce cysts (*Azotobacter* and myxobacteria); endospores (*Clostridium* and *Bacillus*); or conidiospores (the actinobacteria), specialized cells that survive periods of desiccation. Then, when moisture is available through rain or dew, these cells germinate to produce metabolically active vegetative cells.

In soils, the predominant primary producers are plants, not algae, although algae and cyanobacteria may grow on the surface of soils. Plants synthesize complex organic constituents such as cellulose, hemicellulose, and lignin. These types of organic material are insoluble in water and generally more refractory to decomposi-

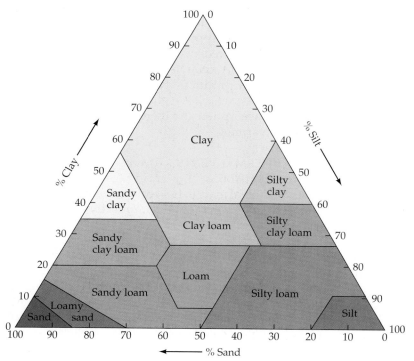

Figure 24.18 Soil textural triangle
Diagram illustrating how soils are named and described based on the proportion of sand, silt, and clay particles.

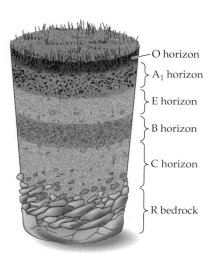

Figure 24.19 Soil profile
Each layer in this soil profile that is distinguishable by chemical composition, color, and microbial activities is referred to as a horizon.

tion than algal polysaccharides and starch. Nonetheless, heterotrophic bacteria and fungi are known to degrade these substances. The rate and extent of degradation of these substances varies. For example, peat bogs are low pH environments that have complex organic constituents such as humic acids and lignin that degrade more slowly than they accumulate. Therefore, over many centuries the refractory organic substances may be converted to coal or petroleum deposits under the appropriate geological conditions. However, almost all of the organic material that is synthesized by plants is degraded biologically and goes back into the soil and air and thereby is made available for recycling.

The "health" of soils, called **soil tilth**, is extremely important for agriculture. Microorganisms play major roles in soil nutrient cycles, making nutrients available on the one hand and removing nutrients on the other hand, depending on the process. For example, nutrients are made available through the recycling of organic compounds as previously discussed. Likewise, important transformations in the nitrogen cycle occur in soils (see discussion later in the section on Nitrogen Cycle). Among the beneficial nitrogen transformations are nitrogen fixation, both symbiotic and asymbiotic, as well as ammonification, the removal of ammonia from organic sources. Less obviously beneficial is the process of nitrification that converts ammonia (normally a good source of nitrogen for plants) to nitrite and nitrate, which is an energetically less-favorable form of nitrogen for plants. The process of denitrification is actually harmful to soil fertility because it results in a conversion of nitrate nitrogen back into atmospheric nitrogen gas, in effect reversing the nitrogen fixation process.

Extreme Environments

As mentioned previously, some habitats have such extremes of temperature, salinity, low water activity, and pH that only microorganisms can grow in them (see Table 24.4). Examples of some extreme environments are discussed herein, and some are alluded to in Chapters 18 to 21, where bacterial diversity is treated, as well as in Chapter 25, which considers symbiotic associations. Acid mine environments are discussed in Chapter 32. Keep in mind that environments may have more than one parameter that makes them extreme. For example, some hot springs are also acidic. Nonetheless, certain bacteria such as *Sulfolobus* can grow in them.

SALT LAKES   Some lakes, such as the Great Salt Lake and the Dead Sea, are very salty because they do not have outlets to rivers or the sea. The saltiness occurs because the water evaporates from the lake, leaving behind the salt, which becomes more and more concentrated

over time. After hundreds or thousands of years, these lakes can become saltier than the sea. Table salt is obtained from salt lakes or from marine habitats, using special evaporation ponds (Figure 1.19C). The high salinity in such habitats favors the growth of halophilic bacteria, including *Halobacterium* sp., and algae such as *Dunaliella*.

SEA ICE   One special microbial environment found in oceanic polar regions is the sea ice microbial community (SIMCO). This habitat develops during the early spring when the sunlight reaches the polar region. The annual sea ice, which can be more than 2 meters deep, is colonized by algae (mostly diatoms), protozoa, krill, and bacteria (Figure 24.20). Therefore, there are primary producers, consumers, and decomposers all together. It is interesting to note that the krill graze on the SIMCO, which typically lies at the interface between the ice and the underlying seawater. Also, some of the bacteria have been found to produce rhodopsins and therefore may be able to obtain part of their energy from sunlight. Others produce gas vacuoles that may permit them to rise up from the water column when the SIMCO develops in the early spring.

Typically the SIMCO grows in the lower 10 to 20 cm of the ice, just above the seawater (see Figure 1.19B). The SIMCO organisms grow at temperatures below the freezing point of seawater ($-1.8°C$). They do this by growing in the brine pockets between ice crystals. One sea ice bacterium, *Psychromonas ingrahamii* has been isolated that grows at $-12°C$.

The sea ice community is important on Earth because it is so extensive. Approximately 10% of the surface of the oceans is covered by ice on an annual basis. Furthermore, the SIMCO accounts for approximately one-third of the primary productivity of polar marine environments. Also, because it is highly reflective it serves to reduce solar absorption and thereby moderate the effects of global warming.

HYDROTHERMAL VENTS   Two major types of thermal environments occur: hot springs and hydrothermal vents. Microbial mats in hot springs were considered as an example of a microbial community earlier in this chapter.

Hydrothermal vents are among the most fascinating natural wonders of the world. They are a result of plate tectonic movements. Very briefly, the Earth's crust consists of large plates that are floating on the underlying hot mantle. The plate movements occur both in the Atlantic and Pacific Oceans. As the plates separate from one another, hot fluids from the mantle rise to produce the vents.

The most dramatic examples of the hydrothermal vents are the black smokers that rise from the ocean floor, spewing reduced gases and iron and manganese

(A)

(B)

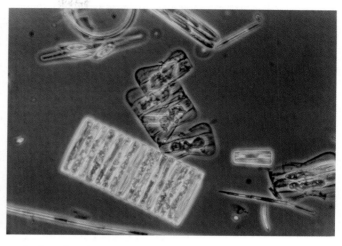

Figure 24.20 Sea ice profile
(A) Diagram showing the sea ice overlying the water column at a North Pole location. The ice is approximately 2 meters thick, and the microbial community is shown as a brown layer at the interface between the ice and the water column. A core has been cut from the sea ice (see also Figure 1.19B). (B) Diatoms predominate as primary producers in the sea-ice community. A, courtesy of John Gosink; B, courtesy of J. T. Staley.

sulfides, which impart the black appearance of a smoking chimney to the structures (Figure 24.21). These black fluids contain reduced inorganic nutrients that serve as substrates for the growth of chemolithotrophic bacteria and archaea, which serve as the primary producers of this community.

What is especially fascinating about the hydrothermal vents is the unique biota that develops in association with them. Among the most fantastic are large tube worms that extend to greater than 2 meters in height. In addition, clams and shrimp as well as other animals live in the vicinity of the vents.

The vents occur in many deep-sea environments that are 2,000 meters below the surface. Thus, they are too deep to receive sunlight. In these environments, photosynthesis cannot occur, and the amount of organic carbon raining on the ocean floor from surface photosynthesis is insignificant. Therefore, the biota in these vents is entirely dependent on the bacterial primary producers that obtain energy for their growth from the oxidation of the reduced chemicals such as hydrogen, hydrogen sulfide, iron, and manganese that are emitted from the vents. The result is that the animals have developed unique and special symbiotic associations with the *Bacteria* and the *Archaea*. Chapter 25 discusses some of these fascinating symbioses.

Both hydrothermal vents and subsurface environments are strong candidates for the origin of life on Earth. Darwin's "warm little pond" model for evolution seems highly unlikely, considering that the period when life evolved on Earth closely followed the period of heavy bombardment by asteroids and comets (see Chapter 1). Although many of these impacts could have sterilized the surface of our planet, the hydrothermal vents and subsurface environments were much better protected from their destructive effects.

SUBTERRANEAN AND SUB-SEAFLOOR ENVIRONMENTS Two of the major environments on Earth that are very poorly understood are the subterranean environment and the deep marine sediments. These habitats are aphotic and extend beneath the surface to more than a kilometer in depth. It has been proposed that, like the hydrothermal vents, the primary production that occurs in these environments is derived from geochemical energy sources such as sulfide and hydrogen gas generated from thermal sources beneath the surface. These energy sources are used by sulfate reducers and methanogens, respectively, which fix carbon dioxide to form a cellular biomass that serves as an energy source for fermenting microorganisms. Because these environments are so difficult to study, little is known about the primary production activities that occur therein. Nonetheless, the vast extent of these habitats suggests that they could be major and likely ancient environments dominated by primary producing bacteria and archaea. Recent deep sub–sea-floor drilling has been conducted at Pacific and Atlantic sites to study the numbers, types, and activities of *Bacteria* and *Archaea* that live in the sub–sea-floor, where it has been estimated that about one-half of Earth's microbial biomass resides.

ROCKS Although rocks are not normally thought of as an environment for living organisms, they are, in fact,

(A)

Figure 24.21 Hydrothermal vents
(A) A "black smoker." (B) Diagram showing the flow of seawater from the seafloor into the sediments, where it is heated and chemicals are reduced to produce fluid containing sulfide, Fe^{2+}, and Mn^{2+}. Iron sulfides give the vent emissions their black color. These reduced substances serve as energy sources for chemosynthesis. A, ©D. Foster, WHO/Visuals Unlimited.

(B)

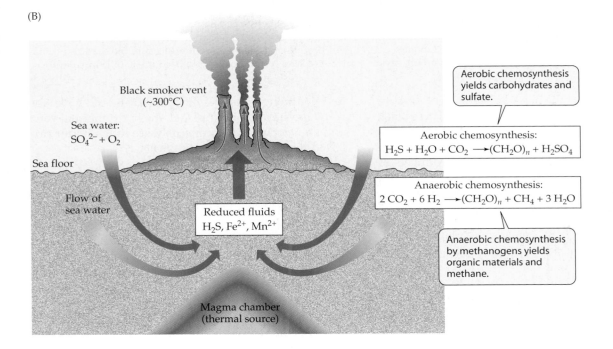

Black smoker vent
(~300°C)

Sea water:
$SO_4^{2-} + O_2$

Sea floor

Flow of
sea water

Reduced fluids
H_2S, Fe^{2+}, Mn^{2+}

Magma chamber
(thermal source)

Aerobic chemosynthesis yields carbohydrates and sulfate.

Aerobic chemosynthesis:
$H_2S + H_2O + CO_2 \longrightarrow (CH_2O)_n + H_2SO_4$

Anaerobic chemosynthesis:
$2\,CO_2 + 6\,H_2 \longrightarrow (CH_2O)_n + CH_4 + 3\,H_2O$

Anaerobic chemosynthesis by methanogens yields organic materials and methane.

common microbial habitats. Most desert rocks harbor microorganisms, with the exception of some in the driest desert on Earth, the Atacama in Peru and Chile. The growth of microorganisms on rocks in conjunction with physical and chemical processes (such as freezing of water, wind action, and acid rainfall) causes rock weathering. As mentioned previously, these weathering processes result in soil formation.

Algae, cyanobacteria, and lichens are common primary producers on rocks (Figure 24.22A). They may grow on the surface of rocks (**epilithic**) (Figure 24.22B), in cracks (**chasmolithic**), or even on the under surface (**hypolithic**) if the rocks are translucent. Lichens are particularly well suited for growth on rocks because the fungal component

that dominates in biomass can withstand severe desiccation and provide inorganic nutrients for the algal component. In turn, the algae produce organic nutrients for the growth of the fungus symbiont (see Chapter 25). Some fungi can also grow as microcolonies on rock surfaces in desert and arid areas in environments too severe for the growth of lichens (see Figure 4.18). Apparently they obtain their nutrients from windblown dust.

One of the most remarkable findings has been that some lichens and algae can actually grow inside rocks (see Figure 1.19D). Such growth is called endolithic. Endolithic microorganisms that grow in rocks in Antarctica are among the most extreme forms of terrestrial life on Earth (Box 24.7).

BOX 24.7   *Research Highlights*

Endolithic Microorganisms of Antarctica

The Victoria Land dry valleys near the U.S. Antarctic base at McMurdo Sound are one of the most extreme environments on Earth. The two principal features that limit biological growth and activity are the low temperature and low water activity (humidity). As a result, "soils" taken from these environments appear sterile, because they have such extremely low counts of bacteria. For this reason, the dry valleys have been compared to the surface of Mars.

However, some microorganisms do live in the terrestrial environ-ments of Antarctica. They reside inside the rocks in the Dry Valleys. The rocks are porous sandstones that obtain some moisture from snowfall. During brief periods of midsummer, conditions allow for melting. During these periods the endolithic lichens obtain water and are able to photosynthesize and grow. For the remainder of the year, they exist in a natural freeze-dried or lyophilized state. Carbon dating by E. I. Friedmann, the microbial ecologist who discov-ered this remarkable community of microbial life, indicates that these lichens are thousands of years old.

(A)

(A) View of the Victoria Land dry valleys. Rocks in fore-ground with a reddish color contain endolithic lichens. (B) Rocks being studied by microbial ecologists from the laboratory of the late J. Robie Vestal. (C) Broken sandstone rock showing the endolithic lichen commu-nity, which appears as a dark band beneath the surface. See also Figure 1.19D for a close-up of the algal and fungal layers of this community. Courtesy of J. R. Vestal.

(B)

(C)

(A)

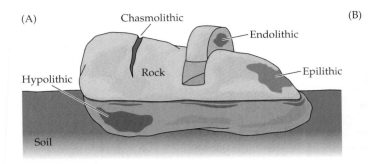

Figure 24.22 Microbial growth on rocks
(A) Diagram showing microbial colonization of rocks. Epilithic organisms grow on the surface, chasmolithics in cracks, endolithics inside, and hypolithics on the buried side of the rock. (B) Epilithic and chasmolithic lichens colonizing rocks in Massachusetts. B, courtesy of D. McIntyre.

(B)

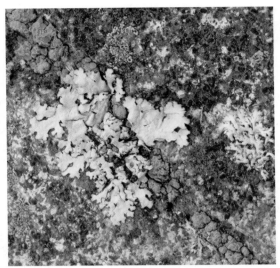

SECTION HIGHLIGHTS
Lakes, rivers, and marine habitats are major aquatic environments that have food chains with microorganisms at the levels of primary production (algae, anoxygenic photosynthetic bacteria, and cyanobacteria), consumers (protists), as well as decomposers and mineralizers. Soils, particularly agricultural soils, are important examples of terrestrial environments. Extreme environments include those with high or low temperature and pH, high salt, and aridity. Microorganisms live at the most extreme conditions in which life occurs.

24.6 Biogeochemical Cycles

Because microorganisms, particularly bacteria, evolved billions of years before higher organisms, they were largely responsible for the origin and evolution of the biogeochemical cycles and continue to play major roles in these transformations. Of course, the cycles have continued to change since the evolution of higher organisms and are also influenced by human industrial and military activities.

Atmospheric Effects of Nutrient Cycles

From a global perspective, the effects of the nutrient cycles are most keenly felt in the atmosphere. This is because all animals are dependent on oxygen that is provided by primary producers. Likewise, respiration ex- pels carbon dioxide, an essential nutrient for plants and other primary producers. Animals and plants, which require air, are profoundly affected by noxious airborne pollutants. Furthermore, the chemical composition of the atmosphere affects global temperatures and incident UV radiation. Two major current concerns are the buildup of greenhouse gases that cause global warming and the release of anthropogenic chemicals such as the chlorofluorocarbons, which destroy the ozone layer that protects all living organisms from UV radiation.

Microorganisms are involved in many of the transformations that affect the atmosphere. For example, nitrous oxide, N_2O, an intermediate product formed in two different bacterial processes, nitrification and denitrification, reacts photochemically with ozone, O_3, in the atmosphere, thereby depleting it (see the section, Nitrogen Cycle).

Both methane and carbon dioxide, other important constituents of the atmosphere, are greenhouse gases produced in part through microbial activities. Methane is produced largely through the activities of methanogenic bacteria, although it is also released from natural gas seeps and during oil drilling operations. Its concentration has been increasing in the atmosphere at a rate of about 1% per year for the last 15 to 20 years. The biological sources of methane include ruminant animals, termites, and bogs and swampy areas, all of which contain methanogenic bacteria. In the northern hemisphere an important nonbiological source comes from oil drilling operations, which release lighter hydrocarbons such as methane into the atmosphere. Methane from biological sources also exists in frozen methane hydrates or clathrates in cold, deep-sea environments. Global warming may result in an increase in the release of this methane into the atmosphere.

Carbon Cycle

Microorganisms play major roles in the carbon cycle (Figure 24.23); carbon is an element that is a key constituent of living organisms. This example shows there are four separate **reservoirs** for carbon: the ocean, the atmosphere, the lithosphere, and the terrestrial biosphere. The amount of carbon in each reservoir, termed the *mass*, is given in PgC (petagrams of carbon) (where $1\ Pg = 10^{15}\ g$). Thus, the atmosphere of Earth contains 725 Pg carbon dioxide, 3 Pg methane, and 0.2 Pg carbon monoxide. Arrows in figure designate the **flux**, or rate of movement, of carbon from one reservoir to another in terms of PgC per year. Therefore, in the ocean, a balance occurs between the flux of carbon to and from the atmosphere (80 PgC/year). The mass of the marine biota in the surface water of the ocean (3 PgC) is responsible for the processes of carbon **assimilation** (largely due to carbon dioxide fixation by primary producers), **respiration**, and **decomposition**.

Note that the fluxes to and from a box are balanced. In this example, 50 PgC per year are assimilated by the marine biota, 45 PgC per year are respired, and the final 5 PgC per year are particulate materials (largely nonliving phytoplankton and other organisms) that are lost by sedimentation. The deeper water contains large amounts of inorganic carbon in the form of carbonate, bicarbonate, and carbonic acid. It also contains substantial amounts of DOC and POC. Only a small portion of the carbon, 0.2 Pg/year, reaches the sediments.

This model also accounts for inputs of carbon to the atmosphere through volcanic eruptions (methane, carbon dioxide, and carbon monoxide) as well as fossil fuel burning and deforestation, all of which occur in the terrestrial compartment. Although this model is not very detailed, it provides important information about the major reservoirs of carbon and the most important fluxes that occur between the reservoirs. More and more complex models are being generated as more detailed experimental data become available.

Because this text is primarily interested in the microbial aspects of biogeochemical cycles, particular emphasis is given to primary production, decomposition, and methane formation and oxidation.

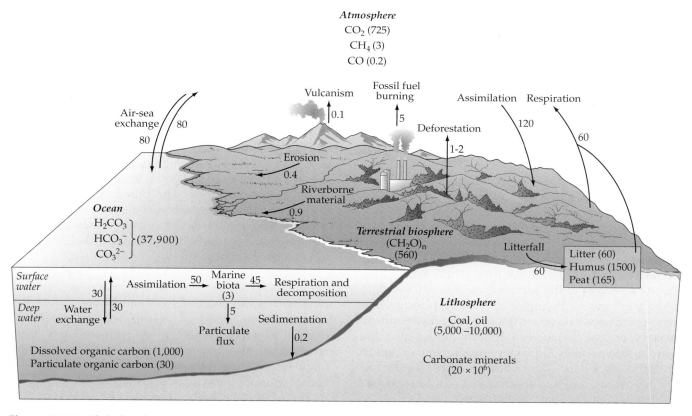

Figure 24.23 Global carbon cycle
Cycling of carbon through the atmosphere, biosphere, oceans, and lithosphere. Reservoir amounts are in pentagrams of carbon (PgC) and fluxes (shown by arrows) are in PgC per year. Freshwater environments are included in the terrestrial compartment. Adapted from K. Holmén, *Global Biogeochemical Cycles*, Academic Press.

TABLE 24.5	Primary production (carbon dioxide fixation) among the *Bacteria* and *Archaea*[a]	
I. Photosynthetic Primary Producers[b]		
A. *Bacteria*	Cyanobacteria	
	Proteobacteria	
	Chlorobi	
	Chloroflexi	
B. *Archaea*	None	
II. Chemosynthetic Primary Producers		
A. *Bacteria*	Nitrifiers	
	Sulfur oxidizers	
	Iron oxidizers (*Thiobacillus ferrooxidans*)	
	Aquificales (hydrogen oxidizers)	
	Acetogens	
B. *Archaea*	Methanogens	
	Sulfur oxidizers	

[a]Not all members of each group are primary producers.
[b]Some photosynthetic groups use light for photophosphorylation but do not fix carbon dioxide; these include the heliobacteria, the purple nonsulfur bacteria, and the halobacteria.

PRIMARY PRODUCTION   Primary producers fix carbon dioxide and convert it to organic material (Table 24.5). One group of primary producers comprises the photosynthetic organisms that derive their energy directly from sunlight. Another group is the chemolithotrophic bacteria that require chemical sources of energy (reduced inorganic compounds such as hydrogen sulfide, ammonia, or hydrogen). These energy sources are provided through geochemical activities or other biological processes, some of which may depend on sunlight. For example, chemolithotrophic sulfur oxidizers require reduced forms of sulfur derived either from geochemical sources or from reduced sulfur sources provided by sulfate reducers, which obtain their energy from organic materials ultimately produced by photosynthesis.

The principal terrestrial primary producers are plants. In freshwater and marine habitats, which account for about two-thirds of the surface area of Earth, the algal and cyanobacterial phytoplankton are the principal primary producers. In contrast, anoxygenic photosynthetic prokaryotes and chemolithotrophic bacteria play much more restricted roles in primary production, except in specialized habitats like meromictic lakes (see the following discussion) and thermal habitats such as hydrothermal marine vents, respectively.

DECOMPOSITION OF ORGANIC MATERIAL   Organic material derived from primary producers resides in living organisms and the nonliving organic material derived from them. Plants and animals and most microorganisms carry out **respiration** to form carbon dioxide and water. For animals and aerobic, heterotrophic microorganisms, respiration is the mechanism by which they obtain ATP for their metabolism.

Bacteria and fungi are the ultimate recyclers of nonliving organic material. They live as saprophytes on organic material from dead plants and animals as well as other microorganisms. They are aided in this process by higher animals that ingest particulate organic materials (herbivores and carnivores) that contain bacteria. The process is chemically analogous to respiration, but involves the degradation of nonliving organic material to obtain energy for growth. This process is called organic decomposition or degradation. If the organic compound is degraded completely to inorganic products such as carbon dioxide, ammonia, and water, the process is called mineralization.

Bacteria and fungi are especially well suited for the degradation of polymeric organic compounds that are refractory to degradation by higher organisms. Cellulose, chitin, and lignin are examples of this. Cellulose, derived from plants, and chitin, derived largely from crustaceans, insects, and some fungi, are degraded by many bacteria and fungi that produce extracellular enzymes to attack these particulate substrates.

The white rot fungi are the major known degraders of lignin. A wider variety of microorganisms can degrade soluble organic compounds including organic acids, amino acids, and sugars.

Almost all organic material produced on Earth is degraded by microbial activities. However, small amounts accumulate in sediments, as shown in Figure 24.23. Over time, these accumulations have resulted in the formation of coal and petroleum deposits.

METHANOGENESIS AND METHANE OXIDATION

Major anaerobic processes of the carbon cycle result in the fermentation of organic compounds to organic acids and gases such as hydrogen and carbon dioxide. Additional degradation by methanogens results in the formation of methane gas from highly reduced sediments and the digestive tracts of ruminant animals and termites. Although many methanogens use carbon dioxide and hydrogen gas as substrates for methane formation, some use other fermentation products, including methanol and acetic acid to produce methane (see Chapter 18). Methane-oxidizing bacteria and certain yeasts degrade methane in the biosphere, but some escapes to the atmosphere, where it becomes a greenhouse gas.

Nitrogen Cycle

All living organisms require nitrogen because it is an essential element in protein and nucleic acids. Animals require organic nitrogen sources that they obtain through digestion of plant or animal tissues. Plants use inorganic nitrogen sources such as ammonia or nitrate. Most bacteria can use ammonia or nitrate as nitrogen for growth, but some, such as the lactic acid bacteria, may require one or more of the amino acids in their diet.

Microorganisms play several important and some unique roles in the nitrogen cycle (**Figure 24.24**). A discussion of the major processes follows.

TABLE 24.6	Representative prokaryotic groups containing species that carry out nitrogen fixation[a]
Type	**Example**
I. *Bacteria*	
A. Nonsymbiotic	
1. Photosynthetic	*Proteobacteria*
	Chlorobi
	Cyanobacteria
2. Heterotrophic	*Azotobacter*
	Clostridium
	Some spirochetes
B. Bacterial Symbionts	*Rhizobium* (legumes)
	Frankia (alder trees)
	Cyanobacteria (lichens; *Azolla*)
II. *Archaea*	
	Methanogens

[a]This list is illustrative and does not contain all taxa that are capable of nitrogen fixation; not all members of each group are nitrogen fixers.

NITROGEN FIXATION Only prokaryotic organisms carry out nitrogen fixation (see Chapter 10). Some members of the *Bacteria* and the *Archaea* produce the enzyme nitrogenase, which is responsible for this process. Nitrogenase is found in both photosynthetic prokaryotes and heterotrophic prokaryotes (Table 24.6). Because nitrogen

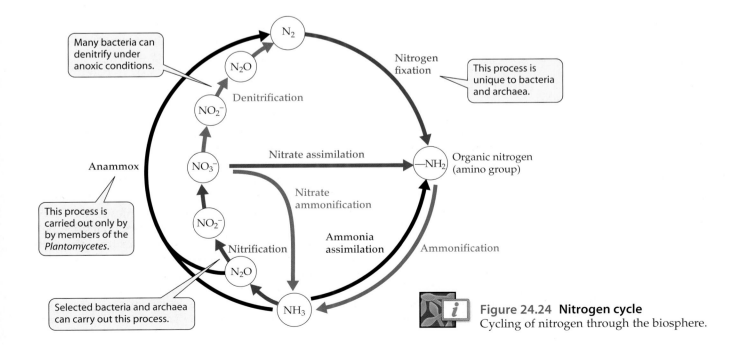

Figure 24.24 Nitrogen cycle
Cycling of nitrogen through the biosphere.

is often a limiting nutrient in terrestrial and aquatic habitats, this process is important because it recycles nitrogen back into the biosphere.

Nitrogen is fixed into organic nitrogen (as the amino group in the amino acid glutamine). This can be transformed into other amino acids using various enzymes available in the organism that fix the nitrogen or make it available to the organism's symbiotic partner (if it is a symbiotic fixer). In this manner, N_2, a rather inert gas, is made available to organisms directly from the atmosphere.

AMMONIFICATION   Organisms release nitrogen from their cells or tissues in the form of ammonia, a common decomposition product. This process, called **ammonification**, is hastened by the activity of some bacteria that have deaminases that remove amino groups from organic nitrogenous compounds to form ammonia.

NITRIFICATION   Ammonia produced by ammonification can be used directly by many plants as a source of nitrogen for synthesis of amino acids and other nitrogen-containing organic compounds.

However, ammonia can be oxidized by a special group of the *Bacteria* called nitrifiers (see Chapter 19) and a more recently discovered novel group of the *Archaea* (see Chapter 18). Nitrification is a two-step process in which ammonia is first oxidized to nitrite, and the nitrite is subsequently oxidized to nitrate (see Chapter 19). Like nitrogen fixation, this process is uniquely associated with the *Bacteria* and the *Archaea*. During this process, some nitrous oxide (laughing gas), N_2O, is produced. This is an important gas because it reacts photochemically with ozone according to the following reactions:

$$(4)\ N_2O + hv \rightarrow N_2 + O$$

$$(5)\ N_2O + O \rightarrow 2\ NO$$

$$(6)\ NO + O_3 \rightarrow NO_2 + O_2$$

where hv represents a photon.

The result of these reactions is that the ozone, O_3, in the protective ozone layer becomes depleted.

Nitrate is much more readily leached from soils than is ammonia. If excessive amounts of nitrate are leached from soils, it can accumulate in runoff water and in wells. When the concentrations become high enough, the water becomes unfit as a drinking source for humans. This happens because the nitrite formed in the intestinal tract by nitrate-reducing bacteria can have an extremely adverse effect by interacting with hemoglobin in the bloodstream to produce methemoglobin. This causes **methemoglobinemia** and, if it is not properly diagnosed, may result in the death of infants because of its effect on respiration. The color of infants becomes blue (hence the vernacular name for the ailment, "blue babies"). This rare malady occurs primarily in agricultural areas that receive excess nitrogen fertilizer.

DENITRIFICATION   A number of bacteria carry out nitrate respiration. In this process, which occurs preferentially in an anaerobic environment, nitrate is ultimately converted into nitrogen gas. This process is called denitrification. It occurs predominantly in waterlogged areas that have become anaerobic. From an agricultural perspective, this is an undesirable process because it results in a loss of fixed nitrogen back into the atmosphere. The intermediates of this process are similar to those of nitrification, including the formation of N_2O. Some of the bacteria that carry out this process include various *Pseudomonas* species, *Thiobacillus denitrificans*, and *Paracoccus denitrificans*.

ANAMMOX   Recently, the anammox (**an**aerobic **ammo**nia **ox**idation) process has been discovered, which is carried out by members of the *Planctomycetes* phylum. In this overall process, ammonia is oxidized as the energy source and nitrite is the electron acceptor for the oxidation. The end product is N_2 gas with one atom of nitrogen from ammonia and the other from nitrite. The responsible organism has not yet been isolated in pure culture; however a genome sequence of one of these organisms is now available.

The anammox organisms have also been shown to use nitrate and reduce it to ammonia. Therefore, they appear to compete with denitrifiers, at least in the marine environments where this process has been discovered.

NITRATE AMMONIFICATION   In some anaerobic environments, such as cattle rumen, nitrate is not converted to nitrogen gas by denitrification, but is reduced to ammonia by the resident bacteria. This process is called nitrate ammonification.

Sulfur Cycle

The **sulfur cycle** is also an important biogeochemical cycle. Sulfur, like carbon and nitrogen, is needed by living organisms, where it is a constituent of protein. However, it is not required in such large amounts, as carbon and nitrogen are, and has not been reported as a limiting nutrient in environments. Higher plants usually obtain sulfur in the form of sulfate, whereas animals obtain it through amino acids (cysteine, cystine, and methionine) in their diet, either from plant protein or from the protein of other animals or microorganisms (as in the ruminant animals). In contrast, microorganisms use sulfur compounds in other ways. For example, some use sulfur-containing substances as energy sources, others use

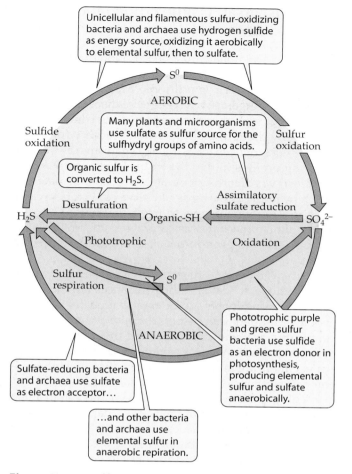

Figure 24.25 Sulfur cycle
Cycling of sulfur through the biosphere. S^0 indicates elemental sulfur.

it as electron acceptors in anaerobic respiration, and some use it as hydrogen donors for photosynthesis. Because of these diverse activities, the sulfur cycle is dominated by microbial processes and is one of the most fascinating of the elemental cycles (Figure 24.25). The oxidation state of inorganic sulfur ranges in its cycling from –2 in sulfide to +6 in sulfate.

SULFUR OXIDATION   Reduced inorganic forms of sulfur, including not only sulfide and elemental sulfur, but thiosulfate and other ions as well, can be oxidized by several different groups of organisms, as has been discussed previously (Table 24.7). This process is called **sulfur oxidation**. Photosynthetic bacteria use hydrogen sulfide produced by sulfate reducers in anaerobic environments as electron acceptors for the ultimate reduction of carbon dioxide for the synthesis of organic materials. Included in this group of organisms are the purple sulfur and green sulfur bacteria. These organisms oxidize the sulfide to elemental sulfur and ultimately to sulfate (see Chapter 21).

In addition, certain nonphotosynthetic bacteria and archaea also oxidize reduced sulfur compounds. Some of these are chemosynthetic and therefore use the reduced sulfur compounds as energy sources and inorganic carbon as a carbon source for growth. Others are heterotrophic. For the most part, these are obligately aerobic organisms, although a few can use nitrate as an electron acceptor in nitrate respiration. Both filamentous and unicellular bacteria are involved in this process, as well as some acidophilic, thermophilic archaea.

SULFUR REDUCTION   The groups of sulfur-reducing bacteria are less well studied. The best-studied group is the sulfate-reducing bacteria, which obtain carbon

TABLE 24.7	Bacterial groups responsible for the oxidation of reduced sulfur compounds	
Microbe type	**Oxidative Activity**
Photosynthetic *Bacteria*	
Purple Sulfur	$H_2S \rightarrow S^0 \rightarrow SO_4^{2-}$
Green Sulfur	$H_2S \rightarrow S^0 \rightarrow SO_4^{2-}$
Some *Cyanobacteria*	$H_2S \rightarrow S^0$
Chemosynthetic *Bacteria*	
Filamentous Sulfur Oxidizers (e.g., *Beggiatoa*)	$H_2S \rightarrow S^0 \rightarrow SO_4^{2-}$
Unicellular Sulfur Oxidizers (e.g., *Thiobacillus, Microspira*)	$H_2S \rightarrow S^0 \rightarrow SO_4^{2-}$
Heterotrophic *Bacteria*	
Filamentous Sulfur Oxidizers (e.g., *Beggiatoa*)	$H_2S \rightarrow S^0 \rightarrow SO_4^{2-}$
Unicellular Sulfur Oxidizers (e.g., some *Pseudomonas* spp.)	$H_2S \rightarrow S^0 \rightarrow SO_4^{2-}$
Archaea	
Acidianus, Sulfolobus	$H_2S \rightarrow S^0 \rightarrow SO_4^{2-}$

from organic compounds and use sulfate as an electron acceptor in sulfate respiration. Some of these also use hydrogen gas as an energy source and grow autotrophically by carbon dioxide fixation. This process of sulfate reduction is referred to as **dissimilatory sulfate reduction** to distinguish it from **assimilatory sulfate reduction**, the process by which plants, algae, and many aerobic bacteria obtain sulfur for synthesis of amino acids. The dissimilatory sulfate reduction process (see Chapter 19) requires large amounts of sulfate and occurs in anaerobic muds. It is the sulfate-reducing bacteria that produce the sulfide smell typical of black muds from lake and marine sediments.

A large number of thermophilic archaea such as *Pyrodictium* sp. respire with sulfur (see Chapter 18). Elemental sulfur is used by most of them as an electron acceptor, although a few use sulfate and thiosulfate.

Other Cycles

Microorganisms play important roles in many other elemental cycles. For example, all of the metallic elements have their own cycles, and bacteria are involved in all of them. More common microbial transformations occur with iron, manganese, calcium, mercury, zinc, cobalt, and the other metals. Ionized forms of the heavy metals precipitate protein and other macromolecules, explaining their toxicity. Many bacteria carry plasmids that have enzymes that carry out the oxidation and reduction or other transformations of these heavy metal ions, thereby detoxifying them. The metal cycles do not have an atmospheric stage, so they are not nearly as mobile as elements such as nitrogen and sulfur.

Iron and manganese oxides can serve as electron acceptors in anaerobic systems. These elements are accordingly reduced to their Fe^{2+} and Mn^{2+} states, which are soluble and therefore more mobile than their oxidized forms. These transformations are especially important in sediments. Because iron is needed by almost all organisms, its availability is restricted due to its insolubility under most aerobic conditions. Thus, bacteria have developed special iron transport compounds termed *siderophores* to bring Fe^{3+} into the cells (see Chapter 5).

SECTION HIGHLIGHTS

Microorganisms play dominant and sometimes unique roles in the biogeochemical cycles of carbon, nitrogen, and sulfur. Novel microbial roles include nitrogen fixation, anammox, denitrification, nitrification, sulfate reduction, sulfur oxidation, and various transformations of metals.

24.7 Dispersal, Colonization, and Succession

In order to disseminate from one location to another, microorganisms need to be able to survive during transit under nongrowing conditions. Many bacteria and higher microorganisms have special dispersal and survival stages such as endospores, cysts, or conidiospores that allow them to maintain their viability for many days, weeks, or even years. Other organisms can survive for long periods in the dirt on birds' claws, in the digestive tracts of birds, or on plant and animal materials eaten by birds. These birds can serve as vectors to carry a variety of microorganisms from one continent to another. Similar mechanisms occur with other animals, although their migration patterns are more limited in scale.

Perhaps the microorganisms that are best known for their abilities to traverse long distances are the fungi. Their airborne spores can be carried globally under favorable conditions for dispersal. Some of these microorganisms cause plant diseases, and their long-distance dissemination is therefore a matter of concern to agriculture. Others are important allergens that affect the pulmonary systems of sensitive individuals.

A different illustration of dispersal and colonization is that of the ruminant animals (see Chapter 25). Ruminant animals maintain a dense community of anaerobic bacteria, including methanogens that are very sensitive to oxygen, in their forestomach. However, newborn calves have a sterile intestinal tract. After they are weaned and begin eating plant materials, they develop the typical anaerobic microbiota of adult cattle. Because of the sensitivity of this community to oxygen, long-distance dissemination of the organisms is highly unlikely. The anaerobic microbiota is most likely transmitted from the mother through her saliva, which contains small numbers of these bacteria that are brought to the mouth through regurgitation and cud chewing.

Of course, newly exposed habitats, such as a new volcanic landmass, will not be colonized immediately. The process of colonization takes time. Furthermore, the colonizing or **pioneer** species may not necessarily be one of the species that survives in the new habitat. Colonization leads to the phenomenon of **succession**, in which the species composition of the habitat changes over time. An example of a microbial succession is one that occurs in the colonization of the intestinal tracts of newborn mice (Figure 24.26). In this example, the initial colonizing organisms include aerobic *Flavobacterium* species, and aerotolerant, anaerobic *Lactobacillus* and enterococcal species. Over a few days' time, the succession proceeds, and colonization by the obligately anaerobic genus *Bacteroides* occurs. Eventually, the *Flavobacterium*,

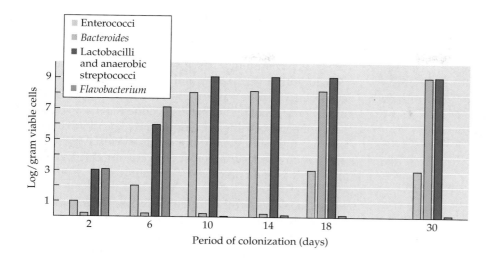

Figure 24.26 Succession
Colonization of the intestinal tract of a newborn mouse by various bacterial species. Not all early colonizers are successful. As the intestinal tract becomes anaerobic, obligate aerobes, such as *Flavobacterium* sp., can no longer survive. Adapted from R. W. Schaedler, R. Dubois and R. Costello. 1965. *The Journal of Experimental Medicine* 122: 59–66.

one of the pioneer species and an obligate aerobe, disappears. Thus, changes have occurred in the intestinal tract that are brought about in part by the colonizing organisms including the *Flavobacterium*. These changes have resulted in creating a habitat with new features (among them, anoxia) that no longer allow the *Flavobacterium* to survive. In contrast the others, which are facultative anaerobes, remain, although the concentrations of all are affected by the changed conditions they helped bring about. Eventually a **climax state** is attained in which the types and concentrations of microorganisms attain an equilibrium in the intestine. This equilibrium can be disrupted by intestinal infections or antibiotic therapy.

SECTION HIGHLIGHTS

Because of their small sizes and hardiness, microorganisms are globally dispersed through air and water currents and the use of animal vectors. If conditions are favorable for growth, microorganisms can colonize new or previously colonized habitats. Successions in microbial communities occur in natural habitats as the conditions of habitat change or are changed by the activities of the organisms.

24.8 Biodiversity of Microbial Life

One of the most remarkable features of microbial life is its vast diversity. This is illustrated nicely by the Tree of Life, which shows that two of the three Domains of life, the *Bacteria* and the *Archaea*, are entirely microbial; furthermore, apart from the plants and animals in the Domain *Eukarya*, the other main branches (i.e., Phyla or Kingdoms) are microorganisms as well. Therefore, almost all of the major divisions (Kingdoms and Phyla) of living organisms are microbial. Furthermore, to complement the biosphere of life, is the virosphere. Because all organisms have viruses, the virosphere is similarly diverse, and extends to all areas of the biosphere of organisms. A similar argument could be made about prions, although their extent and diversity are still poorly understood.

Considering that most of the life on Earth is microbial, it is ironic that most of the species that have been named are animals. For example, the insects comprise more than 750,000 known species in comparison to about 6,000 for the *Bacteria* and the *Archaea* combined.

The paucity of named microbial species is due to several reasons. First, it is not easy to distinguish one microorganism from another. Members of the *Bacteria* and the *Archaea* must be isolated in pure culture and characterized by various phenotypic and genotypic features before they can be named (see Chapter 17). This involves many hours of work and is expensive. In contrast, animals are primarily named on the basis of their morphological features, which can be seen without the use of high-powered microscopes.

Another reason is based on the Great Plate Count Anomaly. Most bacteria found in environments have not yet been grown in the laboratory. It is estimated that about 10,000 species of the *Bacteria* and the *Archaea* can be found in a gram of soil. Only a small fraction of these and those from many other environments have been grown in pure culture.

A third reason that so few species of microorganisms have been named is due to the species definition (see

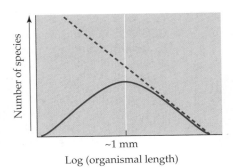

Figure 24.27 Species diversity based on the size of organisms
Most of the known species on the planet are insects whose measurement is greater than 1 mm. The solid line on this graph shows that as the size, that is, the length, of organisms decreases, the number of actual named biological species increases until it reaches its greatest height at about 1 mm before decreasing. However, some have argued that the relationship between the numbers of species may actually continue to increase to much smaller sizes as shown by the dashed line. If this is true, then many millions of unnamed species of microorganisms remain to be discovered.

Chapter 17). The *Bacteria*, the *Archaea*, and microbial eukaryotes are defined much more broadly than animal species.

How many species of the *Bacteria* and the *Archaea* exist? No one knows for sure; however, some theoretical approaches have been used to make estimates. Thus, if one looks at the relationship between organism size and the number of animal species, the numbers of species increase dramatically as the size of the organisms decreases. This is obvious, as we know that there are about 750,000 species of insects but only 5,000 species of mammals, a group of large animals. If this trend continues into the microscopic world, it would appear that the numbers of microbial species could be in the millions (Figure 24.27).

Assessing Microbial Diversity

The biodiversity of microbial life can be measured in some environments. Here we consider some of the theoretical and practical aspects involved.

THEORETICAL CONSIDERATIONS   There are two components to diversity measurements. The first is the **variety component**, which is simply the total number of species in a particular environment. The second component is that of **evenness**. These two components are best illustrated by the following example.

Consider two different communities in which there are 100 individuals and 10 species altogether. In Community A there are 91 individuals of species number one and one species each of the nine other species. In Community B there are 10 individuals of each of the 10 species. Because each community has 10 different species, one could conclude that they are equally diverse. However, this considers only the **variety** component of diversity. The other component of diversity is **evenness**. If the evenness component is considered, Community B is more diverse because there is a greater opportunity for the individuals of each species to interact with one another in Community B as compared with Community A.

SHANNON–WEAVER DIVERSITY INDEX   Some indices of microbial diversity such as the Shannon–Weaver diversity index take both the variety and evenness components into consideration. The Shannon–Weaver formula is given below:

$$H' = -3.3 \sum_{i=1}^{s} n_i / N \times \log_{10} n_i / N$$

Where H' is the species diversity; s is the total number of species; N is the total number of individuals; n_i is the number of individuals of the ith species.

With use of this formula, the diversity of Community A is 0.72, whereas that of Community B is 3.3, thereby indicating that Community B is more diverse.

PRAGMATIC CONSIDERATIONS   Microbial diversity can be assessed in a number of ways. In some instances it is possible to use plate counts to estimate diversity if all of the bacteria and archaea in an environment can be cultivated. As mentioned previously, complete recoveries of bacteria have been obtained in some eutrophic habitats such as wastewater and pulp mill effluent treatment lagoons (see Table 24.2) or in the sea-ice microbial community. In such habitats one can use colony morphology or whole cell shape as determined by electron microscopy as markers to identify "species," and thereby understand diversity. Because pure cultures can be obtained of all organisms from such environments, the actual species can be grown and identified by classical microbiological procedures.

However, as discussed previously in this chapter, in most environments the recovery of bacteria and archaea is much poorer. For such environments, molecular approaches have been used to assess the diversity of types. The most commonly used procedure to assess the molecular diversity of bacteria and archaea from these environments is 16S rRNA gene analyses. This popular approach entails the extraction of DNA directly from the environment of interest (see Box 17.5). With use of this approach many novel phyla have been discovered. In addition, many novel bacterial genera and other groups have been discovered in natural environments. However, these surveys also have their drawbacks. First, not all organisms are equally amplified and second, the 16S rRNA gene does not permit one to identify organisms at the level of the species.

Metagenomic or ecogenomic approaches are now being used to assess microbial biodiversity as well as genetic diversity. In this approach the microbial DNA is extracted from the community of an environment of interest and subjected to sequencing and analysis. Although this approach is expensive, it offers a new approach to assessing the genetic, deduced metabolic and organismal diversity in habitats (Box 24.8).

SECTION HIGHLIGHTS

Microbial biodiversity is vast and challenging to study because so little is known about the species. However, techniques are now available to study the diversity using molecular approaches that range from the construction of DNA libraries for species genes, such as the 16S rRNA gene, to metagenomic analyses.

24.9 Biogeography of Bacteria

L. M. G. Baas-Becking, a Dutch microbiologist, stated that "Everything is everywhere, the environment selects." By this he meant, for example, that if a particular soil type in Europe is very similar to a soil type in North America with similar climatic conditions, the same species of bacteria, such as *Bacillus licheniformis*, is expected to live in both places. Thus, this hypothesis argues for most bacteria having a cosmopolitan, or worldwide, distribution. The basis for this hypothesis is that microorganisms are readily dispersed throughout Earth—they are carried by the wind across the ocean, as well as by animals, especially birds and humans, that travel long distances. When they find a favorable habitat, they will colonize it and grow. Therefore, over relatively short periods of time a bacterium from Europe could be brought to North America or vice versa. The rapid exchange of bacteria with other places on Earth, coupled with the ability of the best-fit strains to be selected and survive, is consistent with the cosmopolitan hypothesis.

Although this hypothesis is widely accepted by microbiologists, it has not been thoroughly tested. Recently, however, microbiologists are using molecular techniques to analyze samples from various habitats such as soil and sea ice to determine the validity of the hypothesis. Some recent data taken from diverse geographic regions suggest that perhaps some bacteria are endemic or indigenous to specific locales. However, it is difficult to preclude the possibility that the same organism might be found in much lower concentrations in another habitat. The answer to this question is important because it will help provide information on the total number of bacterial species on Earth. If there are no endemic bacteria, then fewer bacterial species would be predicted.

BOX 24.8   *Research Highlights*

The Metagenomics Approach to Microbial Ecology

Recently, metagenomic or ecogenomic approaches have been developed to assess the genetic, biochemical, and organismal diversity of samples from natural habitats. In this approach, microbial DNA from the habitat of interest is extracted, cloned into an appropriate vector, and then subjected to sequencing by genomic approaches. Jo Handlesman's laboratory at the University of Wisconsin has pioneered this approach for soil habitats, which have enormous microbial diversity.

Ecogenomic approaches are providing a novel perspective of biodiversity and a wealth of information about the genetic diversity of soil and marine communities. Genomic studies of the microorganisms associated with humans, termed the human microbiome, are also just beginning. Among the foci of the human microbiome are the intestinal and oral microbiota.

A recent study from Craig Venter's laboratory focused on the Sargasso Sea. 1.2 million novel genes were identified among the 1,800 genomic species that were detected based upon 16S rRNA genes. A total of 782 new bacterial rhodopsin varieties were also discovered.

SECTION HIGHLIGHTS

Biogeography is one of the most challenging areas of current study in microbial ecology. However, it is well known that microbial species can live only where there are favorable chemical conditions for their growth; therefore, not all bacteria are everywhere, they are only where they can grow. What is not so apparent is whether speciation has given rise to different organisms at different locations in which the habitats are very similar.

SUMMARY

- **Microbial ecology** is the study of the interactions between microorganisms and the **abiotic** and **biotic** components of their environments. Microbial ecologists seek to understand the distribution and activities of microorganisms in their habitats, both natural and artificial. **Ecosystems** are physical areas on Earth, such as lakes, where organisms live. The microbial biosphere extends to areas on Earth not occupied by other organisms.

- **Total counts** of bacteria, including dead and living organisms, are usually conducted by using fluorescent dyes and a fluorescence microscope. Viable counts are made using growth media. The **bacterial enumeration anomaly** refers to the fact that the recovery of culturable bacteria from typical mesotrophic and oligotrophic aquatic and soil habitats is typically less than 1% of the total count.

- **Autecology** is the study of an individual species in its environment. Species can be identified in natural samples by use of fluorescent antibody techniques or by nucleic acid probes. By using these procedures, it is possible to determine the spatial and temporal distribution of a species in its environment. The activity of an individual microorganism can be assessed in the environment by microautoradiography.

- **Biomarkers** are substances that are uniquely associated with a species or other group of organisms and can therefore be used for identification.

- The **biomass** of microorganisms is the sum total of all living microbial protoplasm in a habitat at a given time. ATP measurements are often used to assess biomass in microbial communities.

- The rates of microbial processes can be measured by direct chemical analyses or by use of radioiso-

topes. Radioisotopic measurements of rates typically use the **tracer approach** in which the substrate **pool** is radioactively labeled by low concentrations of the substrate. The amount of label that is taken up by the community during a time-period is determined to provide an assessment of the rate of activity.

- Microorganisms live in all habitats known including freshwaters, marine waters, soils, and rocks. These may be aerobic or anaerobic, acidic or alkaline, or cold or hot. Microbial life is found in some of the most extreme environments known in terms of temperature, pH, and salt concentration.

- Microorganisms, particularly the *Bacteria* and the *Archaea*, play critical and unique roles in **biogeochemical cycles**. For example, they are solely responsible for such transformations as nitrogen **fixation**, **denitrification**, and **nitrification**. Bacteria and archaea are also critical to sulfur cycling, where they are responsible for sulfate and **sulfur reduction** as well as **aerobic** and **anaerobic sulfide oxidation**. In the global carbon cycle, microorganisms are very important in decomposition of organic matter, **methanogenesis**, and **methane oxidation**. However, some microorganisms, such as the algae and cyanobacteria, are primary producers and others, primarily the protozoa, are consumers.

- The **biodiversity of microorganisms** is vast and best illustrated by their unparalleled genetic and metabolic diversity. Because of the difficulty in growing all the bacteria and archaea from most habitats, molecular community approaches, such as 16S rRNA gene sequencing is used to estimate the diversity for many environments. **Metagenomic** approaches are now being used to examine the genetic and organismal diversity of some communities.

- Microorganisms are readily dispersed from one habitat on Earth to another. Therefore, it has been postulated that all species are **cosmopolitan** in their **biogeography**. However, recent studies indicate that **endemic species** may exist.

 Find more at www.sinauer.com/microbial-life

REVIEW QUESTIONS

1. What is meant by biomass and biomarkers, and how are these terms used in microbial ecology?

2. How would you go about determining the total and viable numbers of nitrifying bacteria in a lake? Do you believe probes would be of any use?

3. Discuss how the tracer approach is used to measure rates of microbial processes in nature.

4. Why does the hypolimnion of a eutrophic lake often become anaerobic during the summer months? Does this affect the temperature of the hypolimnion?

5. If you were counting only the total and viable heterotrophic bacteria in a mesotrophic lake and saw that the viable numbers increased 100-fold in a matter of 2 weeks, what would you believe was the most likely cause?

6. Define succession and provide an example. Some successions are seasonal and are repeated each year. Why?

7. In what ways are microorganisms important to the atmosphere? To the greenhouse gases? To the ozone layer?

8. How can microorganisms live on rocks?

9. Compare the sulfur cycle and the nitrogen cycle. Which do you believe is more important to the biosphere? Why?

10. What is the acetylene reduction technique and what is its use?

SUGGESTED READING

Hurst, C. J., G. R. Knudsen, M. J. McInerney, L. D. Stetzenbach and R. L. Crawford, eds. 2001. *Manual of Environmental Microbiology.* Washington, DC: ASM Press.

Jacobson, M. C., R. J. Charlson, H. Rodhe and G. H. Orians, eds. 2000. *Earth System Science.* New York: Academic Press.

Leadbetter, J. R., ed. 2005. *Environmental Microbiology. Methods in Enzymology* Vol. 397. New York: Academic Press.

Lynch, J. M. and J. E. Hobbie, eds. 1988. *Microorganisms in Action: Concepts and Applications in Microbial Ecology.* Oxford: Blackwell Scientific Publications.

Staley, J. T. and A.–L. Reysenback, eds. 2002. *Biodiversity of Microbial Life: Foundation of Earth's Biosphere.* New York: John Wiley & Sons.

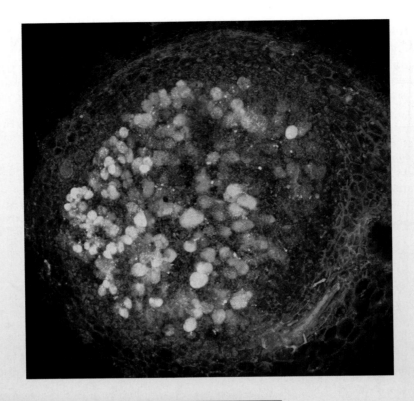

◆ Describe the types of benefits that organisms can derive from symbiotic associations with microbes.

◆ Introduce the ways in which symbiotic associations are maintained from generation to generation.

◆ Define the basic terms used to describe and classify the wide variety of symbiotic associations found in nature.

◆ Illustrate the recurring themes of symbiotic associations with specific examples of microbe–microbe, microbe–plant, and microbe–animal symbioses.

25

Beneficial Symbiotic Associations

*In symbiosis two dissimilar organisms become intimately associated.
This type of vital relationship was apparently first observed by Reinke
in 1872, who named the phenomenon consortism. In 1879, de Bary
observed and recorded the same phenomenon, and introduced the term
symbiosis, which has since been adopted into common usage.*
—I. E. Wallin, Symbionticism and the Origin of the Species, 1927

The term **symbiosis** refers to a more or less intimate association that occurs between two dissimilar organisms that live together. These interactions are outlined in Table 25.1 and range from those that are beneficial to both species (mutualism) to those associations that are harmful to both partners (competition). Except for commensalism, which is discussed in Chapter 26, this chapter considers all the symbiotic associations involving microorganisms in which the interaction is beneficial to at least one of the participants. Parasitism, antagonism (amensalism), and competition are not discussed, and harmful interactions among humans, microbial species, and viruses are discussed in Chapters 26, 28, and 29.

Keep in mind that prokaryotic life forms have lived on Earth for more than 3.0 Ga. This length of time is more than sufficient for the development of myriad, complex interspecies interactions or symbioses among prokaryotes. In addition, eukaryotes evolved in a prokaryotic world. As they evolved, eukaryotes also developed symbioses with prokaryotes. Some types, such as the endosymbiotic evolution of the mitochondrion, are so intimate that we no longer think of the prokaryote as a separate organism. Of course, most symbioses have not become that close or have not had such a long period in which to evolve.

In this chapter we cannot present all the types of important symbioses that occur between microorganisms and other organisms. However, we provide examples to illustrate the types and ranges of

TABLE 25.1	Types of symbiotic associations	
Interaction	**Effect on Interacting Species**[a]	
	Species A	Species B
Mutualism	+	+
Syntrophy	+	+
Commensalism	+	0
Parasitism	+	−
Antagonism (amensalism)	−	0
Competition	−	−

[a]Symbols: + organism gains; 0 not affected; − organism harmed.

interactions that occur and how the two different organisms interact with one another (Table 25.2). Many of these symbiotic interactions are of critical importance to life on Earth. For example, the microbiota in the digestive system of humans and other animals are **obligatory** to the animal's survival in nature.

25.1 Functions of Symbiotic Relationships

Every symbiotic association is in itself rather unique, and the benefits or harm that may accrue to each of the partners are likely to be as varied as the number of such relationships. Symbionts often provide the partner with increased access to some physical or chemical element that a free-living organism might not normally gain from its environment. To ascertain the roles of the members

in a symbiotic relationship, it is often helpful to separate the partners and determine the needs of each. This is impractical in many cases because some symbionts do not survive in the absence of a partner. Organisms in these associations are termed obligate symbionts. Even those that can survive in pure culture when separated may not perform as they would in their association.

Four functions are generally attributed to beneficial symbiotic associations. They are:

- Protection
- Access to new habitats
- Recognition aids
- Nutrition

Protection

Microbes in nature are often subjected to adverse physical conditions. Varying availability of water, pH, and temperature extremes are among the physical conditions that microbes encounter. An endosymbiont that resides within its partner is generally protected from adverse environmental conditions. The partner can prevent desiccation and variations in osmotic pressure, predation, and, in a warm-blooded animal, extremes of temperature. These symbionts in turn might protect the partner from invasion by pathogens (see Chapter 26).

Access to New Habitats

Photosynthetic organisms need access to light but cannot grow on rocks that are exposed to high light intensity, primarily because of the periodic lack of water and

TABLE 25.2	Examples of beneficial symbiotic relationships involving microorganisms
	Symbiotic Partners
Microbe–Microbe	
Bacteria–Bacteria	*"Chlorochromatium aggregatum"*
Archaea–Protist	Methanogen–Protist
Lichens	*Cyanobacteria* or algae–Fungus
Plant–Microbe	Symbiotic Partners
Rhizosphere	*Bacteria*–Plant
Mycorrhizae	Fungus–Plant
Actinorrhizae	Actinomycetes–Plant
Symbiotic N-fixers	*Rhizobium*–Leguminous plant, *Cyanobacteria–Azolla* (water fern)
Animal–Microbe	
Marine invertebrates	Sulfur-oxidizing bacterium–Tube worm
Microbe–Insects	*Bacteria, Archaea,* and protists–Termites
Microbe–Birds	*Bacteria* and *Archaea*–Leaf eaters
Microbes–Ruminants	Prokaryotes and protists–Cattle

unavailability of nutrients. However, lichens, which grow on rocks, are a symbiotic association that consists of a photosynthetic microorganism, either a cyanobacterium or eukaryotic alga, and a fungus. The fungal partner adheres to a rock and provides water and nutrients to its photosynthetic partner. The photosynthetic partner is thereby enabled to photosynthesize and provide organic nutrients for the fungal partner.

Recognition Aids

Many marine invertebrates and fish have bioluminescent bacteria either on their surface or as endosymbionts in special organs. These bioluminescent bacteria emit light that is involved in the host organism's schooling, mating, attraction of prey, evasion of predators, or massing in moon light.

Nutrition

The involvement of one symbiont or partner in providing nutrients for the other partner is common in favorable symbiotic associations. For example, the symbiotic nitrogen-fixing bacteria that infect alder, legumes, and other plants provide the eukaryote with fixed nitrogen in exchange for the plants' photosynthesis-derived nutrients and a place to reproduce.

As will be evident from the examples presented in this chapter, a considerable amount of overlap exists among these functions in many associations.

> **SECTION HIGHLIGHTS**
>
> Symbiotic associations can confer a wide range of advantages on one or both partners. The benefits of symbiosis fall into four functional categories: protection, access to new habitats, recognition aids, and nutrition.

25.2 Establishment of Symbioses

The evolution of symbiotic relationships occurred through a progressively greater interdependence of two different species. As the interdependence became more pronounced, it was essential that mechanisms evolved to ensure the continuity of the symbiosis from generation to generation. For example, in the lichen, special reproductive units containing both algae and fungal cells are dispersed through the air to allow for the colonization of new habitats. In some cases one of the partners transmits the other symbiont directly to its progeny, a common occurrence in insect–microbe symbioses. Other associations require that each generation re-establish the interaction within the environment.

Direct transmission occurs in many endosymbiotic associations. For example, the algal symbionts that are associated with protists would have no invasive ability and would have difficulty in gaining access to the protist partner. To ensure that each of the protist progeny receives its algal symbiont, a cell controls division of the symbiont. When the symbiont is present in a daughter cell, it may then proceed to division. This ensures perpetuation of a favorable symbiotic interaction.

Sexually reproducing animals, generally insects, can transmit symbionts by infection of the egg cytoplasm. Some insects carry microbial symbionts in specialized cells called **bacteriocytes**. Symbionts released from bacteriocytes move to the reproductive tract of the insect and by various mechanisms are transferred to progeny.

Adult mammals have a distinct microbiota in their digestive tract that is essential to their well-being. At birth, the intestinal tract of the neonate is sterile, and an infant gains a normal population of microorganisms via their diet and by association with adults.

The remainder of this chapter is devoted to a discussion of several symbiotic associations that serve as examples of these interactions.

> **SECTION HIGHLIGHTS**
>
> As symbiotic relationships evolved, so did mechanisms to ensure continuity of each symbiosis from one generation to the next. Some symbioses are maintained by direct transmission from parent to progeny, whereas others must be established between new partners in each generation.

25.3 Types of Symbioses

Symbiotic interactions generally benefit at least one of the partners, and the terms *mutualism* and *symbiosis* are sometimes considered to be synonymous. This is not actually the case, as the interaction of species can be of several types, including mutualism, in which both partners benefit from the association, or commensalism, in which one partner benefits from the association and the other partner is neither harmed nor benefited. Parasitism is an association in which one partner benefits and the other is harmed; antagonism, where one partner is harmed and the other unaffected, and competition, which generally harms both partners, will not be discussed here. Usually, when one of the organisms is much larger than the other, the smaller organism in a symbiotic association is termed the symbiont and the larger one the host.

An **ectosymbiont** lives outside but in proximity to the host, whereas an **endosymbiont** exists within the cells of the host. A cooperative interaction in which two or more organisms combine to synthesize a required growth factor or to catabolize a substrate, which neither could utilize independently, or to exchange nutrients is termed **syntrophy**.

We will begin by discussing microbe–microbe symbioses, then consider microbe–plant, and finally microbe–animal symbioses.

SECTION HIGHLIGHTS

Symbiotic associations can be classified according to their effect—beneficial, harmful, or neutral—on the partners involved. In symbioses that involve one partner that is much larger than the other, the former is called the host and the latter the symbiont. Symbionts may either live outside (ectosymbiont) or inside (endosymbiont) host cells.

25.4 Microbe–Microbe Symbioses

"Chlorochromatium aggregatum"

A remarkable association occurs between two prokaryotic partners, one of which is photosynthetic and the other of which is nonphotosynthetic. This mutualistic symbiosis is so intimate that it has been considered as one species and termed *"Chlorochromatium aggregatum."* The name is derived from the "aggregates" that are commonly present in anaerobic aquatic environments (see Figure 21.13). The aggregate or consortium consists of a central bacterium that is thought to be a sulfate reducer, which produces hydrogen sulfide. The sulfate reducer is surrounded by cells of an anoxygenic green sulfur photosynthetic bacterium that utilize sulfide as an electron donor in photosynthesis. The green bacterium has recently been identified as *Chlorobium chlorochromatii*. The photosynthetic partner may provide simple substrates such as photosynthesis-derived acetate for the growth of the sulfate reducer. Like other green sulfur bacteria, the photosynthetic partner is immotile. However, its symbiotic partner is motile and has been shown to be chemotactic to sulfate. Interestingly, the consortium also exhibits tactic responses to light, suggesting that signal transduction occurs between the two partners. *C. aggregatum* is just one example of several similar related consortial species that have been described.

Bacteria–Archaea

The term *syntrophy* usually refers to a relationship where two or more species living together can utilize a substrate that neither can utilize alone. An example of this would be *Syntrophus aciditrophicus*, an anaerobic bacterium that grows in pure culture solely on crotonate. *Methanospirillum hungatei* uses hydrogen-formate for methanogenesis when it grows in pure culture. When the two species grow in a syntrophic relationship, the combined culture can utilize benzoate, butyrate, hexanoate, or heptanoate as substrates. Therefore, the coculture is able to utilize an array of substrates unavailable in axenic culture to either species alone. This interaction fits the classical definition of syntrophy.

Archaea–*Protist*

Sediments at the bottom of stagnant waters are rich in organic matter. These sediments are anaerobic but support the life of various ciliates, amoebae, and flagellates. These heterotrophic protists generate energy through oxidation of organic compounds and must have an electron sink to rid themselves of protons (H^+) generated during respiration. They accomplish this through the endosymbiotic methanogens that are spread throughout their cytoplasm. One amoeba can play host to more than 10,000 methanogens. Electron micrographs of protozoa from the surface of sediments indicate that many species that harbor methanogens lack mitochondria. An amoeba such as *Pelomyxa* excretes measurable quantities of CH_4. Much of the methane generated in the upper layer of anaerobic sediments originates from the methanogens that inhabit protozoa, and not from free-living methanogens.

Lichens

Lichens comprise a classical case of mutualistic symbiosis. Lichens are commonly observed as encrustations on rocks, tree bark, and the soil surface. A lichen is composed of a heterotrophic fungus and a photosynthetic cyanobacterium or a eukaryotic alga. Typical lichens are depicted in Figure 25.1. The fungus–cyanobacterium or the fungus–alga associations are so close that they are considered a unitary vegetative body and have been classified taxonomically as a distinct biological entity. The morphology and metabolic relationship of any particular lichen is so constant and reproducible that it can be assigned to both a genus and a species. More than 20,000 species of lichen have been described.

The fungus in most lichens is an ascomycete. Basidiomycetes are found in some lichens from tropical regions. One fungal species may associate with several different algae. Each resultant lichen is considered a separate species, differing in morphology and metabolic

(A)

(B)

Figure 25.1 Lichen
Lichens (cyanobacterial–fungal symbiotic associations)
vary in shape, color, and appearance. (A) Crustose lichen,
commonly growing on rocks and tree trunks. (B) Foliose
(leaf-like; bottom) and fructicose (top) lichens. A, ©G.
Meszaros/Visuals Unlimited; B, courtesy of D. McIntyre.

such as phosphate for the association. In addition, it synthesizes sugar alcohols, such as mannitol or sorbitol,
which, as compatible solutes, absorb moisture from the
atmosphere. This water serves not only to prevent plasmolysis of cells, but also as the reductant for photosynthetic CO_2 assimilation. Many lichens grow at low temperatures in high altitudes or in polar environments. The
so-called reindeer moss in the tundra of the arctic regions is actually a lichen. Lichens grow very slowly. For
example, some of those from arctic tundra have been car-
bon-dated at more than 1,000 years of age.

Lichens can be used as indicators of air pollution, as
they are highly sensitive to sulfur dioxide, ozone, and
toxic metals. Apparently lichens absorb air pollutants,
and the toxic components are harmful to the photosynthetic partner. Cities with air pollution problems are devoid of lichens.

The color of lichens can be gray, black, blue, yellow,
green, or various shades of red or orange. The pigments
have long been extracted and employed as textile dyes.
The dyes that give the distinct color of Harris tweed, for
example, have traditionally come from lichens. Harris
tweed is woven on the Isle of Harris off the west coast
of Scotland. Other compounds produced by lichens are
litmus, the acid-base indicator employed in chemistry,
and some of the essential oils used in perfumes.

> **SECTION HIGHLIGHTS**
> Microbe–microbe symbioses allow the part-
> ners to carry out metabolism and exploit re-
> sources that would be unavailable to them as
> separate individuals. Syntrophy, for example,
> refers to relationships where two or more
> species living together can use a substrate
> they would not be able to use on their own.
> Microbe–microbe symbioses may consist of
> microbes living as consortia (e.g., "*Chlorochro-
> matium aggregatum*"), or of one microbe liv-
> ing inside the cells of another (e.g., *Pelomyxa-
> methanogen* symbioses and lichens).

interrelationship. Some of the fungal and phototrophic
partners have been separated and grown in axenic culture. The phototrophic partner is generally one that occurs as a free-living organism in nature, and these grow
readily when separated. Those fungal components that
can be grown in axenic culture grow poorly and generally require complex carbohydrates. Recombining the
separated partners is very difficult.

In a lichen association, the fungus produces a mycelial
structure that adheres to hard surfaces and allows the
lichen to inhabit rocks and tree bark. The photosynthetic
partner supplies the fungus with organic nutrients (photosynthate). The fungus scavenges inorganic nutrients

25.5 Microbe–Plant Symbioses

Various symbiotic relationships occur between plants
and microbes. Symbionts may be present on a leaf surface (the phyllosphere) or in the soil surrounding the
roots (rhizosphere). Microbes in such a relationship are
considered ectosymbionts. Fungi may grow on the surface of or invade plant roots and form a relationship
known as mycorrhizae. Single-cell bacteria and filamen-

tous actinomycetes may also invade roots and form colonies with the formation of distinct nitrogen-fixing nodules. Microbes involved in these more intimate interactions are endosymbionts.

Rhizosphere

The **rhizosphere** is the thin layer of soil remaining on plant roots after taking the plant from its environment and shaking it. The microbial population in the rhizosphere is generally one to two orders of magnitude higher than in surrounding, root-free soil, and the number in the rhizosphere frequently reaches 10^9 per gram of soil. The root system of plants can be quite extensive in area. For example, a typical cereal-grain root system can be 160 to 225 meters in length with an average diameter of 0.1 mm. The total root surface area would be 62.8 cm^2, providing ample opportunity for microorganism–root interaction.

The rhizosphere has a higher proportion of gram-negative, nonsporulating, rod-shaped bacteria than would be present in adjacent soil. Gram-positive organisms are scarce in the rhizosphere. Ammonifying and denitrifying bacteria are present, and many of these require growth factors, such as B vitamins and amino acids, that can be supplied by root exudates.

Ectorhizosphere organisms are those that grow in the soil immediately surrounding the root, whereas endorhizosphere organisms penetrate the root itself where they feed directly on the plant. The ectorhizosphere dwellers consume root exudates, lysates, and sloughed root cells. The total carbon released through the root system (including CO$_2$) can be as much as 40% of the total plant photosynthate. Root exudates include sugars, amino acids, vitamins, alkaloids, and phosphatides. The release of root exudates is in many cases stimulated by the presence of select microorganisms.

Here are some of the direct plant–microbe interactions that have been described:

- The organic fraction of wheat and barley exudate stimulates the growth of *Azotobacter chroococcum*. This bacterium is a nitrogen fixer that supplies nitrogen compounds for use by the plant. Many other species of the nitrogen-fixing genus *Azotobacter* are associated with root surfaces in the rhizosphere.

- Anaerobic clostridia occupy the *rhizosphere* of submerged seawater plants *Zostera* (eelgrass) and *Thalasia* (turtle grass). The root exudates released by these plants supply the clostridia with a carbon and energy source, and the bacteria fix nitrogen for the benefit of the plant and the bacterium.

- *Azospirillum* species inhabit the cortical layer of tropical grasses and feed on plant-generated carbon compounds. The bacteria in turn fix abundant quantities of nitrogen for the growth of both.

- *Desulfovibrio* are abundant around the roots of rice, cattails, and other swamp-dwelling plants. The sulfate-reducing members of the genus *Desulfovibrio* generate H$_2$S in quantities that would harm developing plants. The H$_2$S-oxidizing bacterium, *Beggiatoa*, lives in the rhizosphere of these swamp-dwelling plants. Oxidation of sulfide by *Beggiatoa* under limited oxygen conditions reduces the sulfide concentration but results in the production of potentially toxic levels of H$_2$O$_2$. *Beggiatoa* does not produce catalase, the enzyme that decomposes H$_2$O$_2$. A catalase-like activity from the root tips breaks down the peroxide produced by the bacterium to water and oxygen (Figure 25.2). Thus, *Beggiatoa* removes sulfide that could limit plant growth, and the plant provides catalase to prevent peroxide autointoxication of the bacterium.

- *Pseudomonas fluorescens* living in the rhizosphere of some plant species produces antifungal agents, and these ward off potential pathogens.

Mycorrhizae

Mycorrhizal plant root interactions are common in nature. The mycorrhizal relationship is beneficial to the plant in several ways:

- Provides longevity to feeder roots

- Improves nutrient absorption

- Enhances selective absorption of ions

- In some cases, increases resistance to plant pathogens

Plants growing in wet environments have mycorrhizal symbionts that increase availability of nutrients such as phosphorus. In arid environments, the fungi aid in water uptake, and, as a result, mycorrhizal plants often thrive in poor soil where nonmycorrhizal plants cannot. Most of the fungi involved in ectomycorrhizal associations are not obligate symbionts, and some of those that form endomycorrizal associations can grow independently. An estimated 80% of vascular plants have some type of mycorrhizal involvement, and more than 5,000 species of fungi form mycorrhizal associations.

The dramatic effect of mycorrhizae on the growth of a citrus seedling is illustrated in Figure 25.3. The fungi obtain nutrients from the plant and, in turn, provide nutrients (especially phosphorus) and water for plant growth. In many cases, mycorrhizal fungi can be grown

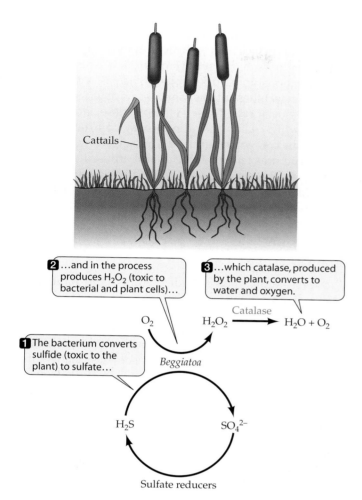

Figure 25.2 Plant–bacterium symbiosis in rhizosphere
Growth of cattails (*Typha latifolia*) in an anaerobic swamp environment is sustained by a symbiotic association with the sulfur-oxidizing bacterium *Beggiatoa*. In the low-oxygen environment, the bacterium produces toxic H_2O_2 but lacks the enzyme (catalase) to remove it; the enzyme is supplied by the plant. Sulfide that would be toxic to the plant is converted to sulfate by the bacterium.

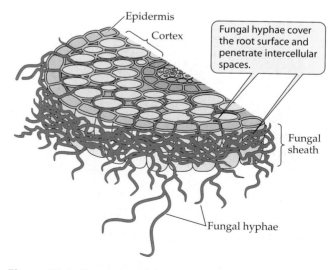

Figure 25.3 Effect of mycorrhizal organisms on growth of citrus seedlings
Citrus seedlings grown with and without inoculation with mycorrhizal organisms. Note the significant difference in root and leaf structures. Courtesy of L. F. Grand.

in culture and mixed with rooting soil resulting in better growth of the plant.

Two major types of mycorrhizal associations occur:

- **Ectomycorrhizae**, where fungi grow externally
- **Endomycorrhizae**, in which fungi grow inside root tissue

Ectomycorrhizae

These fungi grow as an external sheath approximately 40-μm thick around the root tip, as illustrated in Figure 25.4, and are present mainly on roots of forest trees, such

Figure 25.4 Ectomycorrhizae
Cross section of a root with ectomycorrhizae. The hyphae penetrate between, but not into, root cells.

as conifers and oaks. The fungi penetrate intercellular spaces of the epidermis and cortical regions but not into root cells. The relative growth of loblolly pine seedlings uninoculated and inoculated with the fungus *Pisolithus tinctorius* is illustrated in Figure 25.5. Note the superior and more uniform growth in the plants inoculated with the fungus (Figure 25.6). A marked contrast in the development of the uninoculated (see Figure 25.6A) and inoculated (see Figure 25.6B) roots is apparent. A close-up (see Figure 25.6C and D) confirms the benefit of the fungus in root development.

(A)

(B)

Figure 25.5 Effect of mycorrhizal fungus on growth of loblolly pine seedlings
Growth of loblolly pine seedlings (A) without and (B) with the mycorrhizal organism *Pisolithus tinctorius*. Courtesy of L. F. Grand.

Endomycorrhizae

In the endomycorrhizal association, there is significant invasion of cortical cells with some of the fungal hyphae extending outside the root (Figure 25.7). This fungal–plant relationship occurs in plants such as orchids and azaleas. In fact these plants cannot survive without these fungal associations. These symbiotic associations also occur in other plants, including wheat, corn, beans, and pasture and rangeland grasses. The fungi form intracellular structures termed vesicles and arbuscules, as depicted, and the association is designated a vesicular–arbuscular mycorrhiza. The fungi involved in these relationships are difficult to grow or may not grow separated from the plant.

Frankia–*Plant*

Pioneer plants, such as alder (*Alnus*) and bayberry (*Myrica*), grow in moist, nitrogen-poor environments. These woody plants and shrubs are commonly seen in bogs, dredge spoil, and abandoned open pit mines. The ability to establish a symbiotic relationship with nitrogen-fixing bacteria is, in part, the reason these plants survive in such inhospitable environments. The nitrogen fixers in these plants are filamentous actinomycetes in the genus *Frankia*. The symbiotic relationship between plant and *Frankia* is functionally equivalent to that present in the legume–*Rhizobium* symbiosis (see later discussion). The striking difference is that *Rhizobium* sp. will interact with only one family, the legumes. *Frankia* can colonize the roots of plants that are phylogenetically distinct, and symbiotic relationships are known to occur between *Frankia* and 7 different plant orders, 8 families, and 14 genera. More relationships will undoubtedly be discovered.

The endosymbiotic relationship formed between plant and *Frankia* apparently occurs through root hairs. *Frankia* detects a host and prepares for infection by responding to defense molecules secreted by the plant. Development of nodules leads to densely packed, coral-like branching roots that cease growth (Figure 25.8A), and the nodules can be quite large (Figure 25.8B). The microorganism grows slowly in axenic culture, and intact cells can fix nitrogen in atmospheric levels of oxygen. Cell extracts from *Frankia*, however, are sensitive to oxygen. The development of nitrogenase activity coincides with a differentiation of terminal areas of the filaments as depicted in Figure 25.8C.

Rhizobium–*Legume*

Two major bacterial genera that establish a nitrogen-fixing endosymbiotic association with leguminous plants are *Rhizobium* and *Bradyrhizobium*. The general proper-

(A)

(B)

(C)

(D)

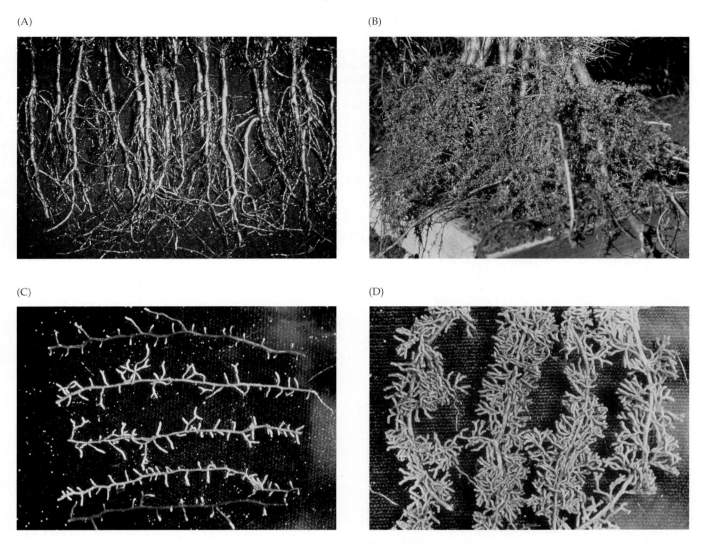

Figure 25.6 Effect of mycorrhizal fungus on development of loblolly pine roots
Development of loblolly pine roots (A) without and (B) with *Pisolithus tinctorius*.
Close-up photographs show details of the root structure of seedlings grown (C) without and (D) with the mycorrhizal fungus. Courtesy of L. F. Grand.

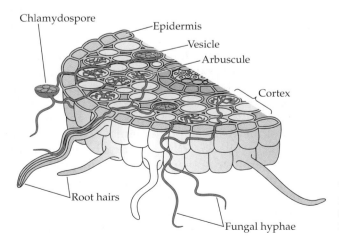

Figure 25.7 Endomycorrhizae
Cross section of root with endomycorrhizae. Intracellular growth of the fungal hyphae is evident, along with "tree-like" structures (arbuscules) and "vesicle-like" structures (vesicles) inside root cells.

(A)

Figure 25.8 Actinomycete–plant symbiosis
Growth of the nitrogen-fixing actinomycete *Frankia* on roots of alder
(*Alnus*), a pioneering plant. (A) Large nodules on alder root caused by the
symbiotic actinomycete. (B) Large nodule detached from an alder root. (C)
Electron micrograph of thick-walled terminal bodies of *Frankia*. These bodies are involved in nitrogen fixation. A, courtesy of D. McIntyre; B, courtesy of J. M. Ligon; C, ©R. H. Berg/Visuals Unlimited.

(B)

(C)

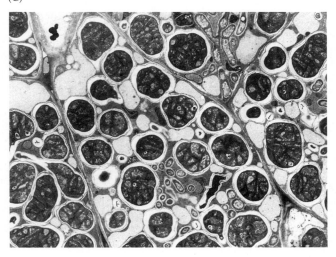

ties of these organisms were presented in Chapter 19.
Nitrogen fixation by these microbes occurs within
nodules that develop on the roots of leguminous
plants. The development of the symbiotic association
in legumes has been studied extensively, and the
events leading to successful nodulation are now well
understood. Both *Rhizobium* and *Bradyrhizobium* are
free-living microbes that can move to, and specifically
attach to, the root hairs of a specific legume. The rhizobia are very selective in the plant species they infect,
and the narrow range of plant species a given bacterial strain will infect is called a cross-inoculation group.

The establishment of the *Rhizobium*–plant symbiotic association requires a complex series of steps. The
bacterium and the plant set up a "cross talk" using
chemical signaling molecules. First, the root releases
flavonoids. These not only attract the *Rhizobium* to the
vicinity of the root, but they also induce bacterial *Nod*
genes, which encode Nod (nodulation) factors. These
factors, secreted by the *Rhizobium*, stimulate cell division in the root cortex. This leads to the formation of
a primary meristem (actively dividing cells).

The attachment of the bacterium to the root hairs
results from a specific adhesion protein, rhicadhesin,
found on the cell surface of *Bradyrhizobium* and *Rhizobium* species. Rhicadhesin is a calcium-binding protein and may function by binding calcium complexes
on the root-hair surface. Lectins were once considered
responsible for attachment but it is now known that
they are less important than rhicadhesin.

Following binding to the root hair, the root curls
back in response to substances secreted, and the bacterium enters via an invagination process (Figure
25.9). The bacterium forms an **infection thread** as it
grows and moves down the root hair. The root cells
surrounding the infection thread become infected as
growth of the bacterium continues. The bacterium divides rapidly, resulting in the formation of a nodule.
Most of the bacterial population is transformed to
branched, club-shaped, or spherical **bacteroids**. The
peribacteroid membrane surrounds the bacteroids.
The irregularly shaped bacteroids have a larger volume than their free-living counterparts do and are incapable of cell division. A few dormant, normal rod-
shaped bacteria live in nodules. These are the bacteria

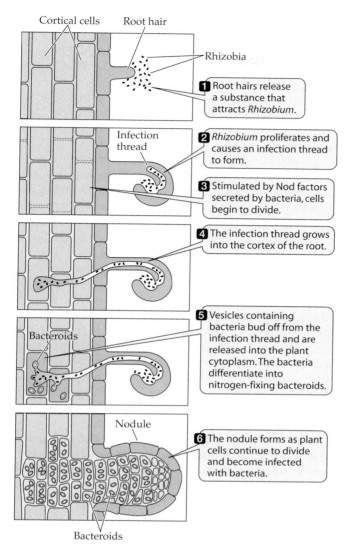

Figure 25.9 captions within diagram:

Cortical cells   Root hair

Rhizobia

1 Root hairs release a substance that attracts *Rhizobium*.

Infection thread

2 *Rhizobium* proliferates and causes an infection thread to form.

3 Stimulated by Nod factors secreted by bacteria, cells begin to divide.

4 The infection thread grows into the cortex of the root.

Bacteroids

5 Vesicles containing bacteria bud off from the infection thread and are released into the plant cytoplasm. The bacteria differentiate into nitrogen-fixing bacteroids.

Nodule

6 The nodule forms as plant cells continue to divide and become infected with bacteria.

Bacteroids

Figure 25.9 Development of root nodules in *Rhizobium*–plant symbiosis
(A) Diagram showing attachment and invasion in development of a root nodule in a leguminous plant. *Rhizobium* attaches to the root hair of a susceptible plant, enters, and forms an infection thread as it moves into the root cells. Factors contributed by both the bacterium and the root result in nodule formation.

(A)

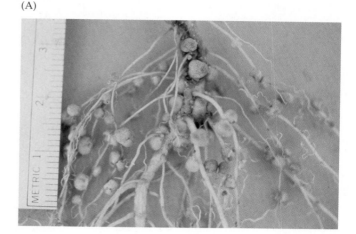

(B)

Figure 25.10 Root-nodulating bacteria in peanut plants
(A) Effective nodulation of a peanut plant (*Arachis hypogeal*) by *Bradyrhizobium* sp. (B) Peanut plants grown with and without (center two rows) inoculation by root-nodulating rhizobia strains. Courtesy of T. J. Schneeweis.

that survive to reproduce in the soil when the plant dies. They can infect legume roots in the vicinity or the roots of other plants at a later date.

A number of biochemical events occur during the infection process. More than 20 polypeptides are synthesized either by the bacterium, the plant, or both, and these substances play a role in the development of the symbiotic association. The nitrogenase and associated genes are entirely contributed by the bacterium. Under tightly controlled microaerophilic conditions bacteroids isolated from nodules can continue to fix nitrogen.

The *nod* genes are highly conserved among *Rhizobium* species. These genes are on the large **Sym** plasmids. The Sym plasmids bear the **specificity genes** that are responsible for the restricted host range of a *Rhizobium* strain. Host specificity can be transferred from one member of a cross-inoculation group to another by the Sym plasmid.

The consequence of a well-established association in the root of a peanut plant is illustrated in Figure 25.10, where nodulation of the roots is quite extensive (see Figure 25.10A). Field-grown plants that have been inoculated have superior pigmentation, total growth, and development, as depicted in Figure 25.10B.

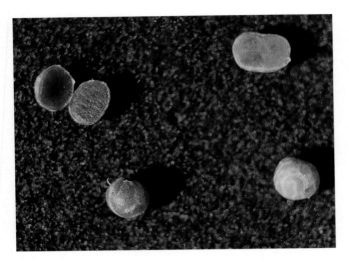

Figure 25.11 Nitrogen-fixing nodules
Effective nitrogen-fixing nodules (left) have a reddish coloration resulting from the presence of the oxygen-carrying protein leghemoglobin. Ineffective nodules (right) are unpigmented. Courtesy of T. J. Schneeweis.

Leghemoglobin surrounds the bacteroids in the nodules. The synthesis of leghemoglobin is induced by genetic information from the plant and bacterium, as neither is able to synthesize it alone. Leghemoglobin is a red-pigmented iron-containing heme protein, which is similar to animal hemoglobin. Leghemoglobin binds oxygen in the nodule and maintains an oxygen tension sufficient to allow the aerobic bacteroids to respire and generate ATP, but restricts the oxygen available to a level that does not inactivate the oxygen-sensitive nitrogenase

system. The ratio of oxygen bound to leghemoglobin to free oxygen in the nodule is in the order of 10,000:1. Nodules that are effective in nitrogen fixation are red (containing leghemoglobin) in the interior of the nodule (Figure 25.11).

The bacteroids (Figure 25.12) receive photosynthate from the plant in the form of tricarboxylic acid cycle intermediates, mainly malate, succinate, and fumarate, which they oxidize to generate ATP.

Nitrogen fixation is a stepwise addition of three pairs of hydrogen atoms to nitrogen gas. This process requires considerable energy as 15 to 20 moles of ATP are expended per mole of N_2 fixed. The product of fixation is ammonia, which is converted into the organic form, glutamine, primarily by the plant. Glutamine is a major donor of amines in amino acid synthesis.

Azorhizobium–*Legume*

Rhizobia form stem nodules on leguminous plants that grow in tropical areas. These nodules are functionally equivalent to root nodules. The nodules are generally formed on the submerged portion of stems or near the water surface. The characteristics of the rhizobia that form stem nodules are different from those that form root nodules and are placed in a separate genus, *Azorhizobium*.

Sesbania rostrata, a legume that grows in Senegal and Madagascar, forms stem nodules in association with *Azorhizobium caulinodans*. The establishment of this nitrogen-fixing system occurs through steps similar to those involved in root nodulation. The stem-nodulating rhizobia apparently produce bacteriochlorophyll *a* and may

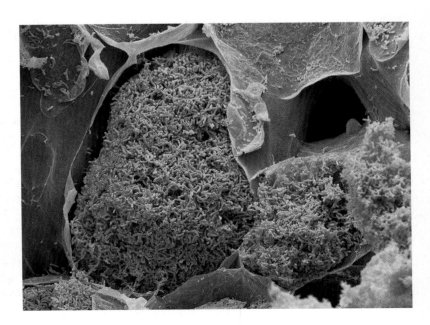

Figure 25.12 Bacteroids
Scanning electron micrograph of bacteroids inside a nodule. The cells lack a cell wall and are no longer able to reproduce. Photo ©A. Syred/SPL/Photo Researchers, Inc.

carry out anoxygenic photosynthesis. Because N_2 fixation requires considerable energy, photophosphorylation would be an effective driving force for these reactions.

Cyanobacteria–*Plant*

Azolla is an aquatic fern that grows on the surface of still waters in temperate and tropical regions and also commonly grows on the surface of rice paddies after the harvesting of the rice crop. An ancient practice in southeast Asia was to maintain the water level after harvesting to allow the fern to grow; tilling the decaying plants into the soil ensured a nitrogen supply for the subsequent growing season. The success of this practice resulted from the growth of nitrogen-fixing cyanobacteria within the tissue of the fern leaves (Figure 25.13A). The

cyanobacterium involved is *Anabaena azollae,* and during symbiotic growth, 15% to 20% of the cells in a trichome are converted to heterocysts (Figure 25.13B), specialized anaerobic nitrogen-fixing cells.

The leaf surface of *Azolla* contains cavities filled with mucilage. During development, these are open to the outside, allowing free-living cyanobacterial cells that grow in the water to attach. The cavities then close as the *Azolla* develops, thereby entrapping filaments of *Anabaena* within the leaf structure.

Bacteria–*Plant Leaf*

Bacterial populations associated with leaf surfaces (phyllosphere) often reach 100 million per gram of fresh leaf material. Under the humid conditions of the tropics, the phyllosphere is constantly moist, and this moisture provides a habitat for a diverse community of bacteria. The species present on leaves are significantly different from those that inhabit the soil in areas adjacent to the plant. Nutrients for the phyllosphere community come from the organic acids, sugars, and methoxyl compounds that leach from the leaf. Some of the organisms present on these leaves are nitrogen fixers of the following genera: *Klebsiella, Beijerinckia,* and *Azotobacter.* There is evidence that these nitrogen fixers provide the host plant with nitrogenous compounds, but the interactions involved are not yet clear. In addition, methanol-oxidizing bacteria are abundant in these environments, but any role they may play in growth of the plant has not been defined.

(A)

(B)

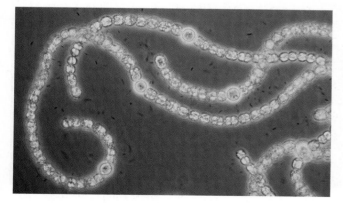

Figure 25.13 Cyanobacterium–plant symbiosis
(A) Fronds of the aquatic fern *Azolla*, which can harbor nitrogen-fixing cyanobacteria. (B) Trichomes of *Anabaena azollae*, showing the enlarged round heterocysts in which nitrogen fixation occurs. A, courtesy of D. McIntyre; B, ©J. R. Waaland/Biological Photo Service.

SECTION HIGHLIGHTS

Plants participate in many symbiotic associations with microbes. The rhizosphere (the soil in close contact with plant roots) is an environment highly enriched with microbes that engage in more or less intimate metabolic interactions with the plant. Most plant species also form mycorrhizal associations with fungi, which help the plants obtain nutrients or water from their environment. Other microbe–plant symbioses provide plants with fixed (reduced) nitrogen, which is often a limiting nutrient for plant growth. Nitrogen-fixing symbioses typically involve complex signaling events and extensive metabolic and morphological changes in both partners. Plant leaves also sustain microbial communities, but it is still unclear whether these associations benefit the plant.

25.6 Microbe–Animal Symbioses

A better understanding of the microbiota of healthy animals has led to the realization that this resident microbial community plays more than a passive role. It is clear that the microbes living in the intestinal tracts of animals are indispensable to the health of the host. Some of the beneficial roles that microbial symbionts play in animal health have already been defined and will be discussed in this section.

Symbiotic Invertebrates

In 1977, during oceanographic studies on plate tectonics and related volcanic activities on the seafloor, dense and thriving populations of novel marine invertebrates were discovered at depths from 1,000 to 3,700 meters with biomass orders of magnitudes higher than could possibly be supported by a photosynthetic food supply. How was this possible? These complex animal communities were found near "hydrothermal vents," which emit hot water containing highly reduced molecules. Some of its major chemical constituents, H_2S, H_2, and CH_4, are known to serve as reductants for chemosynthetic growth of microorganisms. A variety of feeding mechanisms among the various planktonic or sessile animals make use of this unusual microbial base of the food chain in these ecosystems. However, the largest portion of the biomass appears to be produced by symbioses between chemolithoautotrophic bacteria and certain marine invertebrates.

Among the symbiotic invertebrates from deep-sea hydrothermal vent ecosystems in the Pacific Ocean are the vestimentiferan tube worms (*Riftia pachyptila*), (Figure 25.14A), mussels (*Bathymodiolus thermophilus*), (Figure 25.14B), and the "giant" white clams (*Calyptogena magnifica*), specimens of which are up to 32 cm in length.

The vestimentiferan tube worms reach lengths of more than 2 meters and can occur at densities that appear to exceed the highest concentration of biomass per area known anywhere else in the biosphere. They may have the highest known rates of growth of any animal. Detailed anatomical studies revealed the absence of a mouth, stomach, and intestinal tract, in short, the absence of any ingestive and digestive system. Instead, the animals contain a **trophosome** (Figure 25.15) in their body cavity, which accounts for about 50% of the weight of the animal. The tissue in the trophosome consists of large coccoid prokaryotic cells interspersed with the animal's blood vessels. There are about 4 billion bacterial cells per gram of tissue. The bacteria have not yet been characterized but resemble a large, marine sulfur-oxidizing bacterium.

A "chemoautotrophic potential" in these worms is evident from the presence of enzymes catalyzing the syn-

(A)

(B)

Figure 25.14 Symbiotic worms and mussels
(A) A population of vestimentiferan tube worms (*Riftia pachyptila*), mid-Ocean ridge, Pacific Ocean. (B) Mussels (*Bathymodiolus thermophilus*) found at hydrothermal vent sites of the Mariana Arc, Pacific Ocean. A, courtesy of Cindy Lee Van Dover; B, courtesy of NOAA.

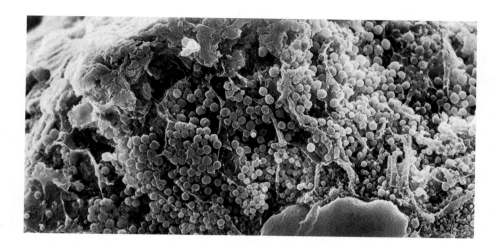

Figure 25.15 Internal tissue of tube worms
Trophosome tissue of *Riftia pachyptila* consisting of coccoid prokaryotic cells interspersed among blood vessels of the animal tissue. Courtesy of H. W. Jannasch.

thesis of ATP via sulfur oxidation (rhodanese, adenosine 5' phosphosulfate reductase [APSR], and ATP sulfurylase) as well as the Calvin cycle enzymes ribulose bisphosphate carboxylase (Rubisco) and ribulose 5' phosphate kinase (APK). ADP sulfurylase (ADPS) and phosphoenolpyruvate carboxylase (PEPC) were also found. None of these enzymes were detected within the worm tissue proper. The necessary simultaneous transport of oxygen and hydrogen sulfide from the retractable plume of gill tissue to the trophosome is carried out by an extracellular hemoglobin of the annelid-type blood system (Figure 25.16).

Electron microscopy and enzymatic studies of the bivalves have shown that the gill tissues contain prokaryotic endosymbionts that are involved in the production of ATP through the oxidation of reduced inorganic sulfur compounds (hydrogen sulfide and thiosulfate). Furthermore, they can reduce CO_2 to organic carbon. Specifically indicative of microbial metabolism (Figure 25.17) are the enzymes Rubisco and APSR. From the analyses of 5S and 16S rRNA nucleotide sequences, it is apparent that the prokaryotic symbionts of all chemolithoautotrophic marine invertebrates so far investigated belong to different, albeit closely related, groups.

Therefore, the bivalves have chemosynthetic symbioses as well, although it should be noted that the mussels are also capable of filter feeding in environments that contain high concentrations of bacteria.

The discovery of chemosynthetic symbioses from the hydrothermal environments has led to the search for symbioses among invertebrates from other marine environments. Indeed, the phenomenon is more widespread than previously imagined. Thus, symbiotic methanotrophy has been observed in mussels similar to *Bathymodiolus* from anoxic, shallow hydrocarbon seepages in the Gulf of Mexico, as well as from so-called cold seeps at the bottom of the west Florida continental slope at a depth of 3,200 meters.

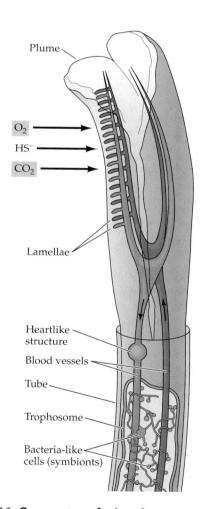

Plume

O_2

HS^-

CO_2

Lamellae

Heartlike structure

Blood vessels

Tube

Trophosome

Bacteria-like cells (symbionts)

Figure 25.16 Oxygen transfer in tube worms
Scheme of the dissolved oxygen transport system in *Riftia pachyptila* from the gill "plume" to the symbiotic trophosome tissue.

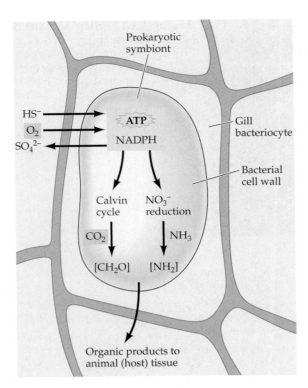

Figure 25.17 Symbiont metabolism in bivalve gills
Schematic of metabolic processes in prokaryotic symbionts within gill "bacteriocytes" of vent bivalves; the presence of enzymes catalyzing ATP production, the Calvin cycle, and nitrate reduction have been demonstrated. Adapted from Felbeck et al. 1983. *The Mollusca,* Chapter 2. Academic Press.

Chemosynthetic symbioses based on chemoautotrophic microorganisms occur in environments that are dependent on an energy source other than sunlight. A number of these environments have been discovered in the deep sea. Light does not penetrate to deeper than 100 meters into the ocean, so most of the ocean is permanently dark. We know that more than 75% of our planet's biosphere lies below 1,000 meters of seawater. At this depth, it is obvious that no photosynthesis can occur. The temperature ranges from 2°C to 4°C. The photosynthetic organic material produced by phytoplankton in the oceans' surface waters is largely recycled in the upper 300 meter layer, and only a fraction, about 5%, reaches deeper waters in the form of sinking particulate matter. About 1% is estimated to reach the deep-sea floor at 3,000 to 4,000 meters. This limited food supply controls the scarce animal populations and their life strategies in this desert-like deep-sea environment.

The most surprising characteristic of the chemolithotrophic-symbiotic sustenance of the copious deep-sea communities is its efficiency. Considering the point source of the geothermally provided energy for chemoautotrophic bacterial growth in the normal food chain, filter feeding on bacterial cells from the quickly dispersing vent plumes appears highly wasteful. This problem was overcome by transferring the chemosynthetic production of organic carbon to a site within the animal where the electron donor, as well as the acceptor, is made available with the aid of the respiratory system. This symbiotic association combined the metabolic versatility of the prokaryotes with the genetic and differentiative capabilities of the eukaryotes and takes advantage of a most direct and efficient transfer of the chemosynthate to the animal tissue. These transfer processes and the biochemical interactions between the microbial and animal metabolisms are presently the focus of research in this area.

Symbioses are also found in tectonic areas in the Atlantic Ocean. Although the large tube worms do not occur, smaller tube worms have been found at cold seeps. In addition, shrimp occur in these environments. The shrimp live in association with epsilon-proteobacteria, which have not been reported in such abundance in other environments. Clearly, much exciting research awaits microbiologists who study the microbial communities of tectonic areas in the oceans.

Termites–Microbes

The termite is an interesting social insect that relies on intestinal symbionts for its survival. The wood that termites eat cannot serve as a food for the insect unless it is predigested by its symbiotic partners. The enzymes essential for the degradation of the cellulosic compounds are supplied by microorganisms. In some cases, the cellulases are produced by microbes living in the termite gut; in other cases, the termite acquires digestive enzymes from the externally grown fungi that they eat.

The "higher" termites are those that obtain digestive enzymes by ingesting fungi that are cultivated in the termite nest. The fungi are members of the genus *Termitomyces,* a basidiomycete. Termites moving from one habitat to another carry fungal spores to the new site, where they establish a fungal garden anew. Growth of the fungi on the cellulosic material they are fed by termites ensures that cellulases will be present on the fungal mycelium. These fungi can digest lignin, which usually protects cellulose from microbial attack. How this occurs is not now known. The termites that cultivate and eat the fungi do not have internal cellulase-producing protists or bacteria.

The gut of "lower" termites has a population of protozoa that digest cellulose and release monomers and various nutrients for the termite. The gut protozoa are anaerobic and ferment cellulose to acetate, carbon dioxide, and hydrogen. The acetate is absorbed through the hindgut of the termite and is then metabolized aerobically to provide the termite with ATP and cell carbon. The termites may also harbor *Enterobacter agglomerans*, a nitrogen-fixing bacterium. Recently, nitrogen-fixing spirochetes have been reported in some termites, the first reported example of nitrogen fixation in this group of *Bacteria*. Thus, the microbial symbionts provide the termite with utilizable nitrogen as well as carbon sources.

The hindgut of lower termites also has significant populations of hydrogen-utilizing methanogens and acetogens. Recent studies indicate that the methanogens and acetogens compete for the hydrogen and carbon dioxide. The acetogens are located closer to the hydrogen-producing bacteria in the hindgut. The methanogens are farther away, near the outside wall of the gut where the concentration of hydrogen is lower. The acetate that the acetogens produce provides more than one-third of the energy requirements of the termite. The methane generated in termites is released into the environment.

Aphid–Microbe

A fascinating mutualistic symbiosis occurs between the gamma-proteobacterium *Buchnera* and the aphids that harbor them. These bacteria are endosymbionts that reside in a specialized cell, termed a bacteriocyte, inside the aphid. The bacterium provides essential amino acids (tryptophan, leucine, etc.) for the aphid, and the host aphid meets all of the nutritional needs of *Buchnera*. To enhance the production of amino acids, it is estimated

that each *Buchnera* cell produces 50 to 200 copies of its chromosome. Neither host nor symbiont can survive without the other.

What is most remarkable about the *Buchnera*–aphid association is the length of time that it has persisted. The aphid lineage is about 150 to 250 million years old. Fossil evidence has been used to date when the different host species diverged from one another. By comparing the 16S rDNA sequence of the *Buchnera* with its host 18S rDNA sequences, phylogenetic trees have been constructed. These reveal (Figure 25.18) that the *Buchnera* species have evolved in concert with their aphid host species. This is seen by the mirror image of the two phylogenetic trees. This phenomenon indicates that as the aphid species diverged from one another, so did the *Buchnera* species, a process referred to as **coevolution**.

The long-term symbiotic association has led to a *Buchnera* chromosome now composed of 650 kb, whereas the ancestral chromosome was more than 6 times larger. Apparently most of the genes involved with survival of the ancestral *Buchnera* strain have been lost over the millennia because they are no longer needed. Furthermore, isolation of the *Buchnera* in the host organ has prevented any acquisition of genetic information from other bacterial species through horizontal gene transfer.

Microbe–Mealybug

Mealybugs are insects that feed on plant sap. Although plant sap is available in abundance, it does not provide these bugs with all of their required nutrients. These essential nutrients are supplied by bacteria that are natural endosymbionts of the mealybugs and other sap-sucking insects. This close association has evolved over

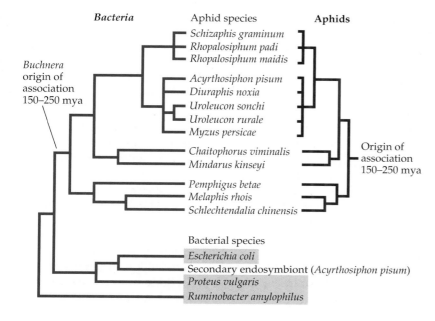

Figure 25.18 Phylogenetic trees of *Buchnera*–aphid associations
Congruence of the evolutionary relationships of *Buchnera* and its aphid hosts. The endosymbiont tree is based on 16S rRNA; the aphid tree is based on 18S rRNA and the fossil record (mya, million years ago). Shading indicates free-living bacteria. Courtesy of Paul Baumann.

millions of years, and elaborate mechanisms are available to ensure that, in each generation, the endosymbionts are passed on to insect progeny. At an early stage of the mealybug embryo development, specialized cells are produced that re-engulf symbionts that are released by maternal cells.

Recent studies on the citrus mealybug (*Planococcus citri*) affirm that this insect has two distinctly different endosymbionts: one a gamma proteobacterium and the other a beta proteobacterium. These endosymbionts are packaged in "symbiotic spheres." The symbiotic spheres are composed of insect cells that form an oval organ that takes up about one-half of the insect's abdomen. Within the symbiotic sphere of the citrus mealybug are the beta proteobacteria and *within* the beta proteobacteria are the alpha proteobacteria. There are 5 to 8 beta proteobacteria per specialized insect cell, and each of these beta proteobacterial cells contains 5 or more alpha proteobacterial cells. Analysis of RNA gene sequences indicates that the beta-types are more ancient and probably the original symbionts, whereas the alpha-type are of more recent origin. The beta-type endosymbionts apparently contribute to the nutritional needs of the mealybug, and the alpha-types serve some needs of the beta-types. Thus, we have within the mealybug a symbiont within a symbiont, a fascinating interrelationship.

Wolbachia–*Animals*

Wolbachia, a bacterium discovered in 1924, might be the most common infectious bacterium on Earth. For more than 100 million years, this alpha proteobacterium has cospeciated with invertebrate hosts, including, among others, fruit flies, shrimp, spiders, and parasitic worms. The organism is not known to infect vertebrates and has not yet been cultivated. *Wolbachia* resides in the ovaries or testes of insect hosts and manipulates the reproductive capacity of the host for its evolutionary benefit. When present in host females, the infection is passed on to succeeding generations. In the male, as sperm develops, the microbes are generally squeezed out and are not always transferred to progeny. As a consequence, the female line is favored by the symbiont. For example, if an uninfected female mates with an infected male, some or all fertilized eggs die, but a female carrying *Wolbachia* can mate with an infected or uninfected male and produce fertile eggs. As a result, infected females outcompete those that are parasite-free, and *Wolbachia* carriers increase in the overall population. There is evidence that when present in males, the bacterium produces a toxin that inactivates sperm, whereas *Wolbachia* strains present in females produce an antidote that restores sperm to full viability. Although *Wolbachia* controls the evolution of insects for survival of the bacterial lineage, it does not cause apparent curtailment of survival in most insect species.

In some cases, *Wolbachia* is essential for survival of the host. River blindness, a disease caused by a nematode, bears a *Wolbachia* symbiont, and antibiotic treatment that kills the symbiont results in death of the nematode.

Bioluminescent Bacteria–*Fish*

The capacity for light emission (bioluminescence) is widespread in the biological world. Bioluminescence has been observed in bacteria, fishes, insects, mollusks, and annelids. In some instances, the animal itself produces light, as in fireflies. However, in other animals, the light is produced by symbiotic luminescent bacteria living in or on the animal. Luminescence is employed by the animal host as a recognition device and may be involved in mating and schooling, and used for prey attraction or in predator avoidance.

A number of bacterial species in the genus *Photobacterium* emit light, as do some members of the genus *Vibrio*. Both genera are abundant in marine environments and are often associated with indigenous fish. The microbes emit light only when grown to a dense suspension. The enzyme luciferase, a flavin (FMN), O_2, and a long-chain aldehyde (RCHO) are necessary for luminescence to occur. Luciferase is induced in *Vibrio fischeri* when a metabolic product, the autoinducer (N-α-keto-caproyl homoserine lactone), accumulates to a critical level in the growth medium. This is referred to as "quorum sensing," as induction occurs only when a dense bacterial population has accumulated that produces a sufficiently high concentration of the autoinducer (see Chapter 13). This critical level is attainable in a rich growth medium or in association with a host, but does not occur among free-living organisms in the marine environment. The reaction is:

$$FMNH_2 + O_2 + RCHO \xrightarrow{\text{luciferase}}$$
$$FMN + RCOOH + H_2O + Light$$

Some marine fish have luminous glands that are an integral part of the animal. These glands are actually pouches that contain cultures of bioluminescent bacteria. The Atlantic flashlight fish has special pouches under the eyes where *Photobacterium leiognathi* grows as an endosymbiont (Figure 25.19). The fish can control the emission of light by drawing a fold of dark tissue over the gland. Some fish have luminescent bacteria in open glands, where they are nourished directly by the fish.

The squid (*Doryteuthis kensaki*) has luminous glands containing *P. leiognathi* embedded in the ink sac, which are partly enclosed in tissue.

(A)

(B)

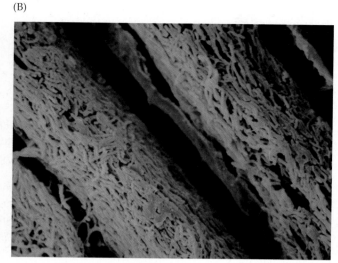

Figure 25.19 Bioluminescent bacteria in fish
Bioluminescence in marine fish. (A) The Atlantic flashlight fish *Kryptophanaron alfredi*, with a luminous organ under the eye. (B) Scanning electron micrograph of *Photobacterium leiognathi* inside the luminous organ. A, ©F. R. McConnaughey/ Photo Researchers, Inc.; B, ©J. G. Morin/Visuals Unlimited.

In all of these animals, it is believed that the bacterial symbionts are a result of selective infection, as in squid, and not through egg passage or parental transfer. The symbiotic association between the squid (*Euprymna scolopes*) and the luminous bacterium (*V. fischeri*) has been studied extensively. The formation of this remarkable association is presented in Box 25.1.

Bacteria–*Bird*

Two fascinating cases of beneficial symbiotic associations occur in specialized species of birds. One of these bird species is the beeswax-eating honey guide and the other is the leaf-eating hoatzin. The honey guides (genus *Indicator*; Figure 25.20) are small, sparrow-sized birds that live in India and on the African continent. Honey guides seek out cracks and crevices in trees and cliffs where swarms of honeybees have established residence. The birds attract the attention of sweet-toothed honey badgers (or humans), who are aware that the presence of these birds affirms that honeycombs can be found nearby, and the honey guide instinctively knows that these animals will rip open the bees' habitation to obtain honey. The bird can then feed on the exposed honeycomb that remains.

The primary source of food for the honey guide is beeswax. The digestive tract of the bird does not produce esterases and other enzymes necessary for the utilization of the wax esters. However, the intestinal tract

Figure 25.20 Symbiosis in honey guides
A black-throated honey guide, perched on a honeycomb. These birds (*Indicator* sp.) can use beeswax as a food source because they are host to bacteria that break down wax to yield long-chain fatty acids, and a yeast that cleaves the long-chain to short-chain fatty acids that the bird can metabolize. Photo ©N. Dennis/Photo Researchers, Inc.

BOX 25.1 *Research Highlights*

Squid–bacterium

Bioluminescence transmitted from the internal light organ of the Hawaiian bobtail squid, *Euprymno scolopes*, protects the animal from predators during its nocturnal foraging. The light emitted by the light organ matches that shining down from the moon and stars, thereby camouflaging the squid. The source of this luminescence is *Vibrio fischeri*, a symbiotic bacterium that dwells within the squid. This symbiosis has been studied extensively, because both partners in the squid–bacterium association can be cultured independently, thereby enabling researchers to follow development of the relationship in the laboratory.

Each day at dawn a mature squid expels 90% to 95% of its light organ symbionts into the environment. During the day, the remaining symbionts grow, and the light organ is repopulated. When the squid emerges from the sand for nocturnal feeding, it once again has a full complement of luminous bacteria.

Newly hatched squid obtain their endosymbionts from the environment, rather than from their parents. A female squid produces hundreds of eggs that remain in the vicinity of the parent colony. After embryogenesis, the eggs hatch synchronously at dusk. Thus, the newly hatched squid enter an environment enriched in the *V. fischeri* cells that were expelled by nearby mature squid. Even so, the number of *V. fischeri* cells is relatively low, and it is therefore essential that the squid have some mechanism for selectively gathering *V. fischeri* from the environment.

During embryogenesis, the squid develops two areas of ciliated epithelia on the lateroventral surfaces of the nascent light organ. Each of these consists of two protruding appendages, with the tips forming a ring and the base surrounded by three pores. Bacteria enter through these pores and travel down ciliated ducts to a crypt space inside the light organ. At first, the pores are nonselective; they allow any bacteria or particles less

than 2 μm in diameter to enter the crypts. However, all this material is subsequently eliminated by a poorly understood process. Following this removal, the host squid starts making mucus around the pores. This mucus is shed in response to peptidoglycans in the surrounding seawater. Nearby *V. fischeri* form aggregates on this mucus, although other gram-negative bacteria may also attach. The latter are outcompeted by *V. fischeri*, which becomes the dominant microbe present. After aggregating over a period of 2 to 5 hours, the *V. fischeri* cells migrate through the pores, move down the ciliated ducts, and colonize the crypt space.

After colonization, the epithelial cells involved in gathering *V. fischeri* degenerate, the ducts through which the bacterium entered constrict, and mucus shedding ceases. Cells surrounding the crypt space enlarge and microvilli in the light organ proliferate. Interestingly, the bacterial lipopolysaccharide and peptido-

---

of the honey guide has two inhabitants that can digest beeswax: *Micrococcus cerolyticus* (cero-wax lyticus-splitting) and a fatty-acid-cleaving yeast, *Candida albicans*. The micrococcus requires a growth factor that is available in the digestive tract of the honey guide. Thus, we have a remarkable symbiotic association between a bird species and two specific microorganisms. The microbes have a home, and the bird has a source of food.

The hoatzin (*Opisthocomus hoazin*) is a leaf-eating (folivorous) bird for which leaves are the major source of food. The hoatzin is a large bird, weighing about 750 grams, whose habitat ranges from the Guianas to Brazil (Figure 25.21). It is found mostly in riverine swamps, forests, and oxbow lakes.

Cellulosic components of leaves are the major food source for hoatzins. They have, therefore, evolved with

a rumen-like crop. The crop and esophagus are the major digestive organs in this species, and the pH and physical environment in these organs support the growth of bacteria at concentrations equivalent to those in the ruminant (see following section). Volatile fatty acids, including acetic, propionic, butyric, and isobutyric acids, are produced by the bacteria in the crop and esophagus. The crop and esophagus make up about 75% of the total digestive system. The volatile fatty acids are absorbed by the small intestine.

The hoatzin is by far the smallest warm-blooded animal that has a foregut that functions as a fermentation vat (rumen). All other animals that do this are the mammals, as discussed in subsequent text. There are some significant anatomical and behavioral modifications in the hoatzin to make room for the relatively large crop.

BOX 25.1   *Continued*

(A)

(B)

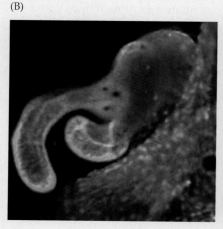

Aposymbiotic

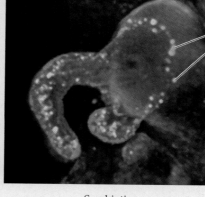

Symbiotic

Condensed
chromatin

glycan that control these developments are not specific to *V. fischeri*, but are effective only if present within the crypt space. Through these processes, *V. fischeri* becomes the sole occupant of the light organ and remains with the host throughout its lifetime.

(A) The Hawaiian bobtail squid, *Euprymno scolopes*. (B) Confocal images of juvenile light organs, showing degeneration of the epithelial cells after colonization by *V. fischeri*. The light organs have been stained with acridine orange, which fluoresces bright yellow-green when bound to DNA. The image on the left shows a light organ exposed to all other bacteria in the environment except *V. fischeri* (aposymbiotic). The diffuse fluorescence of the epithelial cells indicates they are alive. The image on the right shows a light organ after exposure to *V. fischeri* for 12 hours (symbiotic). The fluorescence in the epithelial cells is punctate, indicating chromatin condensation and cell death. Courtesy of Margaret McFall-Ngai.

Figure 25.21 Symbiosis in hoatzins
The foliage-eating hoatzin (*Opisthocomus hoazin*). The hoatzin is the smallest animal—and the only bird species—with a rumen-like digestive system. Photo ©F. Gohier/Photo Researchers, Inc.

They have a reduced sternum area for flight-muscle attachment, and the bird is a poor flyer. The young take more than 60 days before they are able to fly, and survival mandates that the juveniles have some means for protection against predation. Hoatzins have wing claws that are used in rapid crawling movements to escape from predators. They scramble to and dive into water when threatened.

Ruminant Symbioses

Ruminants are herbivorous mammals that have a complex digestive system mainly composed of a large specialized fermentation chamber called the **rumen**. Among the ruminants are domestic animals such as cattle and sheep, as well as giraffes, buffalo, and elk.

Ruminants feed on grasses and other plant materials composed mainly of insoluble polysaccharides including cellulose, hemicellulose, and pectin. These animals lack the enzymes essential for the digestion of such refractory food sources, but their rumen harbors a complex community of anaerobic bacteria, archaea, and protozoa that convert the polysaccharides to useable food for the animal and the microbe.

When a ruminant feeds on grass or other plant material, the food enters the mouth, where it is mixed thoroughly with saliva rich in bicarbonate. The material then passes down the esophagus to the rumen (Figure 25.22). A rumen in a mature cow will hold 100 liters of material and has a microbial population approaching 10^{12} per milliliter. The temperature is 30°C, at a pH of 6.5, both highly favorable for microbial fermentation. Because the rumen is anoxic, the microbial population is composed of anaerobes. The ingested food remains in the rumen for 10 to 12 hours, where the peristaltic action rotates the material, breaking up the cellulosic mass to a fine suspension. Cellulolytic bacteria and protozoa attach to the pulpy mass and free up glucose and cellobiose, which are fermented to volatile fatty acids, mainly acetic, propionic, and butyric acids (Figure 25.23). These fatty acids pass through the epithelium of the rumen wall and are oxidized by the ruminant as energy source. The low redox potential (–0.30 mV) in the rumen permits the growth of methanogenic archaea that produce methane gas from the H_2 and CO_2 generated during the sugar fermentation, thereby serving as a proton sink. A mature cow belches copious quantities of methane daily.

Worldwide the ruminants generate 80 to 100×12^{12} grams of methane each year, whereas wetlands produce about two times this amount. This far exceeds the abiogenic production of methane which is about 20% of the total.

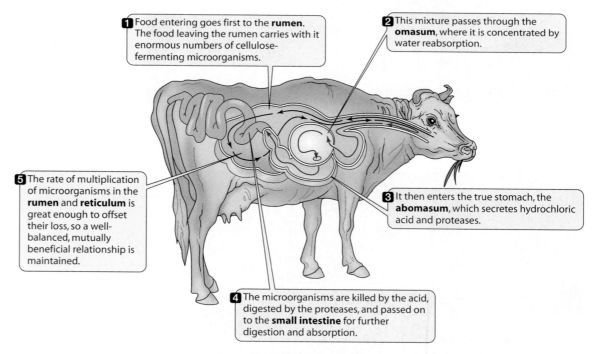

1 Food entering goes first to the **rumen**. The food leaving the rumen carries with it enormous numbers of cellulose-fermenting microorganisms.

2 This mixture passes through the **omasum**, where it is concentrated by water reabsorption.

5 The rate of multiplication of microorganisms in the **rumen** and **reticulum** is great enough to offset their loss, so a well-balanced, mutually beneficial relationship is maintained.

3 It then enters the true stomach, the **abomasum**, which secretes hydrochloric acid and proteases.

4 The microorganisms are killed by the acid, digested by the proteases, and passed on to the **small intestine** for further digestion and absorption.

Figure 25.22 Ruminant digestive system
Diagram of the rumen and gastrointestinal system of a ruminant. Food enters the rumen and is partially digested and fermented. It moves to the reticulum; cuds are formed, regurgitated, and move back through the digestive system via the omasum.

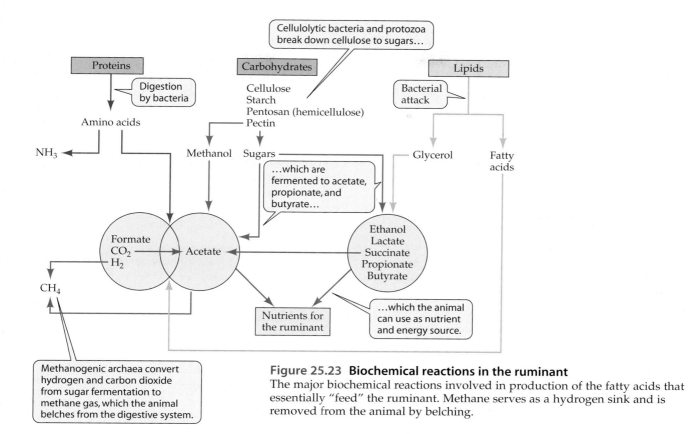

Figure 25.23 Biochemical reactions in the ruminant
The major biochemical reactions involved in production of the fatty acids that essentially "feed" the ruminant. Methane serves as a hydrogen sink and is removed from the animal by belching.

The pulpy mass then moves gradually from the rumen to the reticulum, where small clumps called cuds are formed. These cuds move up the esophagus to the mouth where they are rechewed to smaller particles and mixed with saliva. This mixture, which is liquid, is swallowed and moves to the omasum and thence to the abomasum, a more acidic area where digestive enzymes much like those of other mammals digest the material. The microbe-containing mass is digested to release amino acids, vitamins, and other nutrients. Absorption of these nutrients occurs as the material moves through the small intestine.

From a microbial ecology perspective, the rumen fermentation has been a classic source of information on microbial interactions. For example, the first methanogenic microorganisms that were cultivated were obtained from the rumen. Moreover, it was the study of ruminant metabolism that led to an understanding of the tight coupling in metabolism that can occur between different species. The concept of "interspecies hydrogen transfer," in which one species (a fermentative bacterium) produces hydrogen gas and another species (a methanogen) uses it quickly for energy with the formation of methane gas, originated from a study of rumen metabolism. The combined activity of these two species maintains an extremely low concentration of hydrogen gas. If hydrogen were to accumulate, it would inhibit the activity of the fermenter, and the fermentation could not proceed. This same tight coupling of interspecies hydrogen transfer also occurs in many anaerobic environments where methanogens proliferate.

SECTION HIGHLIGHTS

Chemolithoautotrophic symbioses between marine invertebrates and bacteria support rich communities near hydrothermal vents in the deep ocean, where darkness precludes photosynthesis. Many herbivores rely on symbioses with microbes to digest cellulose (termites, the hoatzin, ruminants) or beeswax (honey guides) in their diet. Other animals, such as aphids and mealybugs, obtain essential nutrients from microbial symbionts. Luminescent bacteria are used by their hosts for mating, schooling, prey attraction, and predator avoidance.

SUMMARY

- Eukaryotes evolved in a prokaryotic world. Alliances between the eukaryotes and prokaryotes are of common occurrence. Two or more **dissimilar** organisms may form an alliance that is termed a **symbiosis**.

- Symbiotic associations can **benefit** the **symbionts**, among the advantages: **protection**, **access to new environments** in the biosphere, **recognition aids**, and **nutrition**. Microbe–microbe, microbe–plant, and microbe–animal symbioses are common in nature.

- Symbiotic relationships are often beneficial to at least one partner. **Mutualism** is an interrelationship where both partners benefit. Nutritional cooperation, in which the concerted activity of two species results in the metabolism of a substrate, is termed **syntrophy**, a form of **mutualism**.

- **Commensalism** is an association in which one partner benefits and the other is neither harmed nor benefited.

- **Heterotrophic protists** survive in anaerobic sediments because they have methane-producing endosymbionts that serve as an electron sink. Many of these protozoa lack mitochondria, and their archaea serve the general function of this organelle.

- A **lichen** is a symbiotic association between a fungus and a cyanobacterium or eukaryotic alga.

- Fungi develop in or on roots, and these are termed **mycorrhizal relationships**. Bacteria may invade roots or stems of selected plants and form **nodules** in which nitrogen fixation occurs.

- Bacterial associations with plants that result in the formation of nitrogen-fixing nodules generally occur with **legumes** (peas, clover, etc.). The microorganisms involved are members of the genus *Rhizobium* or *Bradyrhizobium*. Formation of nitrogen-fixing nodules on roots of alder, a woody plant, is a result of invasion by *Frankia* sp., a filamentous actinomycete.

- Invertebrate animals survive at depths greater than 1,000 meters in tectonically active areas of the ocean because they have developed beneficial symbiotic associations with **chemosynthetic** prokaryotes that live within their tissues. Among those animals are tube worms, mussels, shrimp, and clams.

- **Termites** depend on bacterial and protozoan populations in their gut to **digest** the **wood** they consume.

- **Bioluminescence** is common in the biological world. Many **animals** in **marine environments** that emit light do so because they are hosts to **luminescent bacteria**.

- Symbiotic relationships in the **avian** world permit birds to utilize **beeswax** or **foliage** as foodstuff. The **honey guide** is a beeswax eater and is host to bacteria that cleave the wax to utilizable fatty acids. The **hoatzin** can survive on foliage because it has a **rumen**, much like that of mammals (cows and deer).

- **Herbivorous** animals survive on vegetation because they have a bacterial population in their digestive tract that converts **plant material** to utilizable foodstuff. This occurs in a large internal **fermentation vat** termed the **rumen**.

 Find more at www.sinauer.com/microbial-life

REVIEW QUESTIONS

1. What is symbiosis? What are the major functions of symbiotic associations? How does mutualism or commensalism differ from parasitism? Could many of these associations originate through parasitism?

2. How does a symbiotic association become established?

3. Define: ectosymbiont, endosymbiont, syntrophy, mutualism, and consortium. What are some examples of each?

4. Microbiologically, what is a rhizosphere and how does this environment differ from surrounding soil? What are benefits that the plant and microorganism receive in symbiotic associations of this type?

5. How do ectomycorrhizae and endomycorrhizae differ? Think of some of the ways that these relationships benefit a plant. What type of plant is involved in each of these?

6. What is *Frankia*? Where is it found? What does it do?

7. Outline the steps involved in the establishment of root nodules on a leguminous plant. What is the function of leghemoglobin?

8. How does a deep-sea chemosynthetic ecosystem differ from the ecosystems on the surface of the earth? Which animal systems on the earth's surface resemble those in the deep sea?

9. Outline the "feeding" of a tube worm. What are the unique features of a tube worm?

10. What is the chemical reaction involved in bacterial luminescence? Why have animals established relationships with luminescent bacteria?

11. Microbiologically, how do "higher" and "lower" termites differ?

12. What symbiotic associations occur in birds? Discuss the role of the bacterial symbionts in nutrition of the animal.

13. Outline the functions of microbes in a ruminant. What are the benefits to the animal? To the bacterium?

14. What function is served by methanogens present in the protozoa that reside in anaerobic environments? Is this similar to ruminants? How?

SUGGESTED READING

Margulis, L. and D. Sagan. 1986. *Microcosmos: Four Billion Years of Evolution from Our Microbial Ancestors.* New York: Summit Books.

Margulis, L. and R. Fester, eds. 1991. *Symbiosis as a Source of Evolutionary Innovation: Specialization & Morphogenesis.* Cambridge, MA: MIT Press.

Paracer, S. and V. Ahmadjian. 2000. *Symbiosis.* 2nd ed. New York: Oxford University Press.

Smith, S. E. and D. J. Read, eds. 1997. *Mycorrhizal Symbiosis.* 2nd ed. San Diego, CA: Academic Press.

Staley, J. T. and A-L. Reysenbach. 2002. *Biodiversity of Microbial Life.* New York: Wiley-Liss.

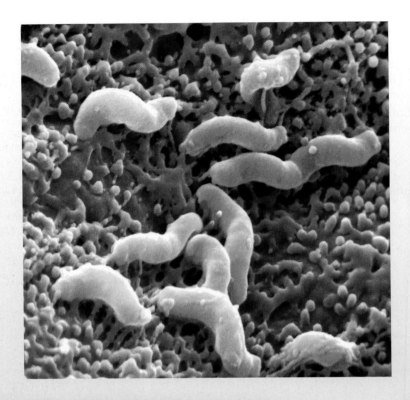

The objectives of this chapter are to:

◆ Show how humans have an ongoing relationship with a group of microbes that generally do not cause infections but actually protect them from infection.

◆ Introduce the barrier defense, the first layer of host protection against infections.

◆ Describe how pathogens (disease-causing microbes and parasites) use adherence and invasion strategies to overcome the barrier defense.

◆ Describe how the host fights back with phagocytic cells that engulf (eat) the invaders and usually kill them.

◆ Survey the mechanisms used by pathogens to cause disease, including toxins, induction of inflammation and host cell death, and avoidance of immune responses.

◆ Describe how some pathogens have evolved clever ways to invade and replicate inside host cells.

◆ Illustrate how some viruses and intracellular bacteria and protozoa cause chronic, sometimes lifelong, infections of their host.

26

Human Host–Microbe Interaction

Healing is a matter of time,
but it is sometimes also a matter of opportunity.
—*Hippocrates, 460–377 BC*

Microbes (bacteria, fungi, and viruses) and parasites (protozoa and parasitic worms) that coexist with human hosts can be divided into three groups on the basis of their relationship with the host. That is, whether they:

- Coexist with the host in a non-disease-inducing manner (termed normal microbiota)
- Cause chronic infection of the host
- Cause acute infection of the host

The **normal microbiota** is a distinct microbial community to which every healthy human plays host. The organisms making up this group reside on skin or mucosal surfaces throughout the body. The normal microbiota includes billions of bacteria along with a few fungi. The alliance between the human **host** and the normal microbiota is generally **mutualistic**, meaning that both benefit. However, some members of the normal microbiota are **commensals**, meaning that one partner benefits and the other is unaffected (see Table 25.1). For example, most of the bacteria that live in our mouths benefit by growing on the food we eat and, while they may not be disease agents, they also are not known to provide any particular benefit to us. The normal microbiota includes **opportunist pathogens**. These are normally harmless organisms that can cause infection when the host immune system is compromised or the natural barrier defenses (discussed below) are breached. Humans also have a **transient** microbiota, composed of microbes that do not establish long-term residency.

It is largely composed of bacteria, although some protozoa and fungi are included.

The second group comprises disease-causing agents, including bacteria, viruses, protozoa, parasitic worms and fungi that cause long-term infection. Such an infection is referred to as a **chronic infection**. Generally these organisms affect only a few individuals within the population at any particular time. They usually do not kill their human host immediately and sometimes never do so, although they cause intermittent episodes of disease. Maintaining the second type of relationship depends upon a balance between the infective agent and the host. That is, the ability of the infectious agent to cause injury to the host (also known as disease or **pathology**) must be matched by the ability of the host to control the infection and level of injury. (The host immune system, which conveys resistance to infection, is discussed in detail in Chapter 27.) **Virulence** is the term used to describe the capacity of an infecting agent to cause disease (Box 26.1). Virulence varies among species of organisms and viruses. Even within a species, some strains can be more virulent than others and some can be **avirulent** (unable to cause disease). Virulent strains or species produce unique molecules or **virulence factors**, some of which are not produced by the nonpathogenic organisms. This second group of disease-causing agents includes organisms that healthy humans occasionally encounter in daily life but which often fail to colonize the host. Thus, the relationship is transient, causing no apparent harm. This outcome depends upon the host's ability to resist the infection.

The third group is composed of pathogens that frequently cause **acute infections**. In that case, the relationship between the pathogen and host can be quite short, since either the host dies from the infection or the pathogen is eliminated by the host's immune response. This group includes bacteria, viruses, and protozoa. They can be highly virulent organisms and kill the host quickly. However, if the host has the time to produce a protective immune response before death, it can eliminate the organism. Although vaccines are not discussed in this chapter, it is important to note that their role is to prime the host immune system so that the immune response to a particular infectious disease agent can be activated rapidly. This may prevent serious disease or even death. Additionally, drug therapy—most commonly with antibiotics—can be used to eliminate pathogens.

Finally, some microorganisms can be categorized as **accidental pathogens**. It is only because of their presence in many parts of the environment that these microorganisms occasionally enter a human host and cause disease. This category includes members of the normal microbiota that can be opportunistic pathogens (Box 26.2). Also included are organisms that have no partic-

---

BOX 26.1 | *Milestones*

Evolution of Virulence

English rabbits were introduced into Australia in 1859 by British sportsmen interested in game hunting. There were no carnivorous animals on the Australian continent that were natural enemies of the rabbit, so the population increased at an incredible rate. The agricultural interests in Australia became alarmed at the loss of grazing lands to the rapidly growing rabbit community, and they pressured the government to find a solution.

In 1950, a myxoma virus that was prevalent in South American rabbits was introduced into the Australian rabbit population. It had low virulence in the South American rabbit population but a much higher virulence in the European rabbit strains. The virus spread rapidly in Australia, killing more than 99% of the rabbit population within a few years. Death rates were highest in the summer, as mosquito vectors spread the virus. The scientific community in Australia watched this daring experiment in pest control closely.

Two interesting phenomena became evident over time—one in the rabbit population and the other in the virus itself. Controlled laboratory studies confirmed that over a 6-year span, the virus had become less virulent than the original strain. Also the rabbits became more resistant to the virus and less susceptible to the disease it caused. Apparently, survivors of the initial epidemic were selected in some way for resistance to infection. The rabbit population in Australia now stays at about 20% of the previous high number. Experimental evidence indicated that the increased resistance in rabbits was not related to an improved immunological defense system but rather to a physiologically altered rabbit. This study indicates that both the susceptible host and the virulent virus evolved in a way that allowed both to survive. It is axiomatic that any pathogenic agent that kills all of its hosts will quickly become extinct. Evolution favors survival of both the pathogen and the host.

BOX 26.2 *Research Highlights*

Skin Staphylococci

Staphylococci are among the major bacterial colonizers of human skin. Of the bacteria present on the body surface and the nostrils, over 50% are typically staphylococci. *Staphylococcus aureus* was the first member of this genus to be isolated; it was described in 1884 by F. J. Rosenbach, who isolated the organism from an infected wound. The second species described was *S. epidermidis,* which was isolated by C. E. A. Winslow in 1908. Both *S. aureus* and *S. epidermidis* are gram-positive cocci, and are facultative anaerobes. They are generally differentiated from one another by the coagulase test. Most *S. aureus* strains produce coagulase (a blood-clotting enzyme), whereas strains of *S. epidermidis* do not. The common habitat for staphylococci is humans and other animals; their occurrence in nature is generally attributed to human or animal sources. *S. aureus* is a serious opportunistic pathogen in humans and the causative agent of boils, impetigo, pneumonia, osteomyelitis, endocarditis, menin-gitis, toxic shock syndrome, and several other infections. *S. epidermidis* strains are a clinical problem in patients with prosthetic heart valves or other implanted devices. Antibiotic-resistant strains of the staphylococci pose a persistent and acute dilemma, particularly in nosocomial (hospital-acquired) infections. Despite the early recognition that staphylococci are a common organism on the skin of humans, there had been relatively little research on the characterization and infectivity of these human parasites. This changed markedly in the 1970s.

In the early 1970s, Wesley Kloos of the United States, in collaboration with Karl-Heinz Schleifer of Germany, initiated a systematic study on the staphylococci. Over the ensuing years, they and others determined that there were at least 12 species of the genus *Staphylococcus* inhabiting the human skin. *Staphylococcus* is the dominant organism on skin, with members of the genus *Micrococcus* also being present. At least two or three species of micrococci live on most humans, although some individuals may harbor as many as seven. Coryneforms, *Acinetobacter* spp., enterobacteria, *Bacillus* spp., and *Streptomyces* spp. may also be present in significant numbers.

These studies affirm that several diverse staphylococci are common colonizers of the human skin. Other species are adapted to life on other animals: *S. intermedius,* on the domestic dog and other carnivores; *S. hyicus,* on the pig and other ungulates; *S. caseolyticus,* on the cow, hoofed mammals, and the whale; *S. felis,* on the cat; and *S. caprae,* on humans and goats. There are now 32 named species of *Staphylococcus* specifically adapted to life on animal hosts. Despite the potential pathogenicity of many of these parasitic organisms, the host and parasite have evolved with a reasonable tolerance for one another. Unfortunately, this stand-off is broken at times, and staphylococci can become a serious and deadly enemy.

---

ular relationship with the host, but, if introduced into the host, cause disease. An example is the bacterium that causes tetanus.

26.1 Normal Human Microbiota

The human fetus is virtually free of microorganisms, but during birth and immediately thereafter, the newborn is exposed to and colonized by microbes. These colonizers are picked up from the environment, including other humans. The microbes gradually establish an "ongoing" residence in exposed areas—the skin, oral cavity, respiratory tract, gastrointestinal tract, and genitourinary tract. These areas of the human body provide a favorable environment for the microorganisms that make up the normal microbiota. A favorable environment includes physical characteristics such as pH, moisture, osmotic pressure, temperature, and nutrient availability. The physical conditions and available nutrients tend to be relatively constant at a given site, but vary among sites. At permissive sites, colonization by microbes can be quite rapid; for example, the intestinal tract is colonized within 24 hours after birth. Ultimately, an estimated 10^{13} (10 trillion) bacteria live in or on an adult human, along with a few fungi.

The variability of conditions among body sites results in the formation of a distinct microbiota in each area of the body. For example, although a variety of different microorganisms can survive on human skin, internal mucous membranes have a much larger and diverse population. The distribution of microbial groups that reside in humans is presented in Figure 26.1. The genera shown make up

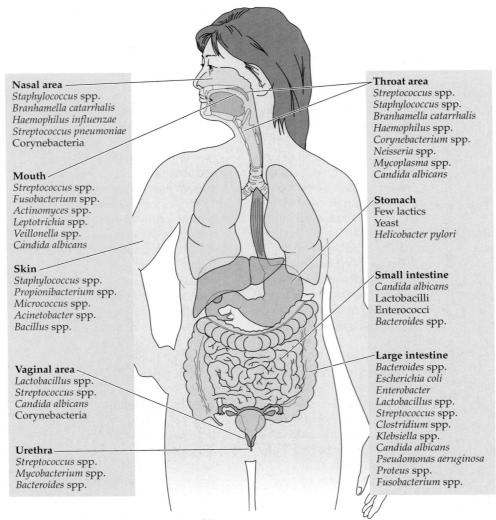

Nasal area
Staphylococcus spp.
Branhamella catarrhalis
Haemophilus influenzae
Streptococcus pneumoniae
Corynebacteria

Mouth
Streptococcus spp.
Fusobacterium spp.
Actinomyces spp.
Leptotrichia spp.
Veillonella spp.
Candida albicans

Skin
Staphylococcus spp.
Propionibacterium spp.
Micrococcus spp.
Acinetobacter spp.
Bacillus spp.

Vaginal area
Lactobacillus spp.
Streptococcus spp.
Candida albicans
Corynebacteria

Urethra
Streptococcus spp.
Mycobacterium spp.
Bacteroides spp.

Throat area
Streptococcus spp.
Staphylococcus spp.
Branhamella catarrhalis
Haemophilus spp.
Corynebacterium spp.
Neisseria spp.
Mycoplasma spp.
Candida albicans

Stomach
Few lactics
Yeast
Helicobacter pylori

Small intestine
Candida albicans
Lactobacilli
Enterococci
Bacteroides spp.

Large intestine
Bacteroides spp.
Escherichia coli
Enterobacter
Lactobacillus spp.
Streptococcus spp.
Clostridium spp.
Klebsiella spp.
Candida albicans
Pseudomonas aeruginosa
Proteus spp.
Fusobacterium spp.

Figure 26.1 Normal human microbiota
Microorganisms identified as "normal microbiota" on the skin and mucous membranes of the human body. (The abbreviation "spp." means species, plural.)

only a part of the overall picture. For example, well over 350 named species have been identified in the large intestine, and countless others have not yet been characterized. These organisms are not **pathogenic** (disease-causing) when they are on the skin and mucous membranes of the body (all of which are actually "outside" the body, if you visualize the body as a tube, with the gastrointestinal tract from the mouth to the anus being the inside surface of the tube). Nevertheless, it is important to note that when any of these organisms of the normal microbiota are found in organs, internal tissues, or blood, this generally indicates an ongoing infectious disease (see Box 26.2).

Human health depends on the establishment and maintenance of an indigenous microbiota. This is known because prolonged antibiotic therapy that alters or suppresses the normal microbiota can lead to serious health problems. For example, oral antibiotics alter the normal microbiota in the gastrointestinal tract and on the mucous membranes of the oral cavity and upper respiratory system. When this happens, yeasts or other fungal infections can occur in the oral cavity and/or lungs. Despite the many examples of the adverse effects resulting from the destruction of the normal microbiota, confirming that the indigenous microbiota of humans and animals in general are a primary defense against invasion by harmful pathogens, antibiotic supplementation of the diet is used to increase yield in some livestock systems. Low-dose dietary antibiotics have been used to promote growth in the swine industry, and their effectiveness is documented. The reasons for their effectiveness include protection of animal feed from microbial breakdown (and thus increased availability to the host) and decreased loads of microbial toxins. Despite its apparent benefits in cases such as this, chronic use of low-dose antibiotics is considered of questionable value, since it could increase the incidence of antibiotic resistance in bacterial strains. This practice is now prohibited in some countries.

Skin

Members of the normal microbiota reside on human skin, particularly in moist areas such as the armpits, scalp, and feet. Two key nutrient sources for the microbes associated with the moist areas of the human skin are the sebaceous and sweat glands (Figure 26.2). The sebaceous glands are associated with hair follicles. They secrete lipid materials (sebum) that prevent overwetting, overdrying, and abrupt changes in temperature of the epidermis layer. The relatively high concentration of sebum, lipids, free fatty acids, wax alcohols, glycerol, and hydrocarbons produced by the sebaceous glands provides nutrients for a microbiota consisting of about 10^6 bacteria per square centimeter. The sweat glands include the eccrine glands, which generate perspiration, and the apocrine glands, which generate perspiration and nutrients. The sweat glands also secrete urea, amino acids, lactic acid, and salts, which provide sustenance. The moist areas of the skin have a constant population of mostly grampositive bacteria, including species of *Staphylococcus*, *Micrococcus*, *Corynebacterium,* and *Propionibacterium* (see Box 26.2). The apocrine glands secrete lactic acid, which lowers the local pH to 3 to 5, an acidity which discour-

ages excessive microbial colonization. However, much of the skin, such as the palms of the hands, is dry because it has few or no secretory glands; in these areas the microbial densities are much lower, ranging from 10^2 to 10^4 per square centimeter.

The bacterial community of the skin changes at puberty. The concentration of *Propionibacterium acnes* increases and certain species of the staphylococci appear, including *Staphylococcus capitis* and *Staphylococcus auricularis*. Diet has little measurable influence on the skin microbiota, but antibiotic therapy can alter the skin populations and is used to treat acne.

Respiratory Tract

The respiratory tract consists of the mouth, nasopharynx, throat, trachea, bronchia, and lungs. Some of these moist locales, including the mouth, nasopharynx and throat, are potential sites for microbial colonization. For example, several hundred different species of bacteria are present in the oral cavity. In contrast, the **lower respiratory tract**, which includes the trachea, bronchi, and lungs, is typically devoid of bacteria as a result of barrier defenses, discussed in the next section.

The **mouth** is a favorable environment for bacteria, since it is relatively nutrient-rich and the temperature and pH are stable. About 10^8 bacteria live in each milliliter of saliva. Many of these microorganisms originate from colonies that adhere to the surfaces of the tongue, gums, and teeth. For example, *Streptococcus salivarius*

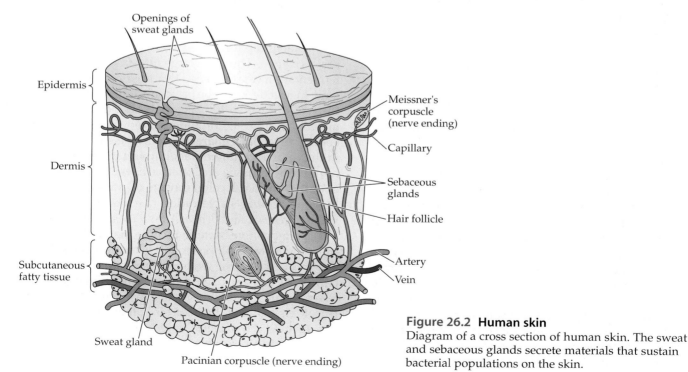

Figure 26.2 Human skin
Diagram of a cross section of human skin. The sweat and sebaceous glands secrete materials that sustain bacterial populations on the skin.

proliferates by adhering to the surface of the tongue. The surface of the **teeth** (the enamel) is made of a calcium phosphate crystalline material that readily becomes colonized by bacteria. In the oral cavity of the newborn, before the teeth develop, lactobacilli and streptococci predominate. As teeth develop, oral microorganisms such as *Streptococcus mutans* invade and attach to the enamel by producing glycans (polysaccharides) that bind the bacteria together on the tooth surface. *Streptococcus sanguis* and anaerobic *Actinomyces* use sucrose to produce capsules used for adherence to teeth. These aggregates of bacteria and organic matter are known as **plaques**.

Some species of the normal microbiota in the mouth cause disease in some hosts. The microbes in plaque ferment sugars to produce lactic acid and other organic acids that etch teeth, causing dental caries (cavities). *Actinobacillus actinomycetemcomitans*, *Porphyromonas gingivalis*, *Tannerella forsythia*, and *Treponema denticola* cause periodontal disease. Recently, high concentrations of these species in the oral cavity has been found to correlate with thickening of the walls of arteries by deposition of fats and bacteria, suggesting a possible association between peridontal disease and "hardening of the arteries." Oral bacteria are also linked to other pathologic processes. For example, the presence in maternal saliva of *Actinomyces naeslundii* Genospecies2 is associated with low birth weight and premature delivery.

The **nasopharynx** and **nose** are constantly inoculated with microorganisms as one breathes. The **nostrils** have a microbiota similar to that of facial skin, with *Staphylococcus* species and other gram-positive bacteria predominating. The nasopharynx (above the soft palate) is colonized by strains of *Streptococcus pneumoniae*, *Neisseria meningitidis*, *Branhamella*, and *Haemophilus*, although they are not necessarily disease-causing in this environment. The **oropharynx** (between the soft palate and larynx) has a microbial population that is similar to that of the nasopharynx, but which also includes some micrococci and anaerobes of the genera *Prevotella* and *Porphyromonas*.

Gastrointestinal Tract

The newborn infant has a sterile gut that quickly becomes populated by bacteria. If the infant is breast-fed, the bacteria are mostly of the genus *Bifidobacterium*, obtained from the skin of the mother, because human milk contains an amino disaccharide that is utilized by this bacterium. Bottle- and breast-fed infants differ in the species of *Lactobacillus* that dominate in their colons. This difference can be recognized by a difference in the stool color and odor. However, as a child moves to a solid diet, the microbial population of the gut gradually changes to resemble that of the adults with whom the child lives most closely and from whom the child obtains his or her normal microbiota.

In adults, a large number of resident microbes inhabit the gastrointestinal tract, including numerous species of bacteria and eukaryotic microbes. A genomic analysis of the human large intestine (colon) microbiota has shown that it contains at least 100-fold more genes than the human genome. Technological advances have opened the way to characterizing organisms in the gut microbiota that cannot be cultured. That is, profiles of these microbes have been developed by analyzing sequences of conserved and variable regions of small ribosomal RNA, expressed genes, and genomic DNA; combined with more conventional techniques such as gross morphology.

The species, number, and concentration of resident organisms varies along the length the gastrointestinal tract. DNA analyses indicate the presence of at least 128 different microbial species in the stomach, although it is not certain how many of these are actual residents and how many are transitory. In the **small intestine**, the number of bacteria present increases in concert with the digestion of food as the food moves downward through the small intestine (the upper part is the **duodenum**, the middle part the **jejunum**, and the lower part the **ileum**). The duodenum has a limited bacterial population (fewer than 10^3 per milliliter) owing to stomach acids, bile secreted by the gallbladder, and pancreatic secretions. The microbiota consists mainly of gram-positive cocci and bacilli. The jejunum's population consists of enterococci, lactobacilli, and corynebacteria, along with the yeast *Candida albicans*. The ileum's microbiota resembles that of the **large intestine** (also called the **colon**), with large numbers of *Bacteroides* species and a limited number of facultative anaerobes such as *Escherichia coli*.

The facultative aerobes in the colon function to remove oxygen. Thus, the large intestine is essentially a fermentation vat populated by masses of anaerobic bacteria. Ingested food is the basic substrate for these organisms. About 10^{10} to 10^{11} bacterial cells per gram of fecal mass live in the colon, and more than 350 different species have been characterized. The strict anaerobes in the colon outnumber facultative aerobes by several hundred-fold (none are strict aerobes). The microbiota of the healthy adult colon consists of gram-negative bacteria, including bacteroides (*Bacteroides fragilis*, *B. melaninogenicus*, and *B. oralis*) and *Fusobacterium* species. The gram-positive bacteria include members of the following genera: *Bifidobacterium*, *Eubacterium*, *Lactobacillus*, and *Clostridium*. We often hear about *E. coli* in feces, since it can be a pathogen, but, surprisingly, less than 1% of the microbiota consists of *E. coli* and species of *Proteus*, *Klebsiella*, and *Enterobacter*. A few harmless protozoa such as *Trichomonas hominis* may also live in the colon. An adult excretes 3×10^{13} bacteria per day, and 25% to 35% of fecal matter is bacterial mass.

The resident microorganisms of the gastrointestinal tract contribute to human health in many ways, including the ability to obtain nutrients. The resident microbiota digests much of what we eat, breaking complex polymers down to amino acids, simple sugars, purines, and pyrimidines, which are then absorbed by the gut epithelia. Studies in mice have shown that gut bacteria have a profound effect on utilization of food. For example, the inclusion of *Methanobrevibacter smithii* in mixes of bacteria inoculated into the intestine of mice elevated their ability to extract calories from their food. *M. smithii* is a member of the *Archaea* and grows on the products of other bacteria, including methane. Recently, related studies have shown that the relative abundance of *Bacteroidetes* versus *Firmicutes* also differs between obese and lean individuals and may contribute to weight gain (Box 26.3).

The gut bacteria also have a profound effect on the development of the gut-associated lymphoid system (see Chapter 27) and other aspects of gut physiology. Mice born and subsequently maintained in sterile ("germ-free") environments, and which therefore lack gut bacteria, have decreased gut motility and peristalsis, as well as lower mucosal and muscle cell turnover and vascularity; the associated lymphoid tissue is only rudimentary. Conversely, inappropriate activation of the gut-associated lymphoid system by the gut microbiota may contribute to inflammatory bowel disease, which results from an overactive host immune system in the gut. Inflammatory bowel disease does not arise in animals kept under germ-free conditions and, in humans, the lesions of inflammatory bowel disease are most severe in areas with the highest bacterial numbers. Some scientists argue that the improvements in public health during the last century, together with use of drugs to suppress infestations of parasitic worms in livestock and companion animals, underlies the increased incidence of inflammatory bowel disease. It has been suggested that parasitic worms provoke anti-inflammatory responses, balancing the inflammation-inducing signals provided by other microbes. With regard to this, it should be noted that bacteria also vary with respect to their inflammation-inducing qualities, with some far exceeding others. For example, lactobacilli and bifidobacteria have little or no inflammatory activity; for this reason they are being evaluated for their capacity to act as **probiotics** (that is, their ability to confer a health effect on the host when consumed in adequate quantity). This approach and the study of **prebiotics** (agents that selectively expand non-inflammatory gut bacteria already present) may contribute to the therapeutic management of inflammatory gut and bowel disorders.

It is important to note that tolerance of a given microbial species varies among individuals. For example, about half of the human population harbors *Helicobacter pylori*, which causes gastric or duodenal ulcers and gastric cancer in some. *H. pylori* is transmitted among people mouth-to-mouth and via fecal contamination of food or water. It anchors itself to the stomach wall by binding to a specific carbohydrate (sugar) attached to proteins or lipids, known as sialyl-di-Lewis x (sLex). Although about half of the human population carries *H. pylori*, only a few people develop peptic ulcers, duodenal ulcers, or gastric cancer as a result, for reasons that are unknown.

BOX 26.3 *Research Highlights*

Can Your Gut Microbes Make You Fat?

It was recently shown by Jeffrey Gordon and colleagues at Washington University School of Medicine that obese mice and obese humans have a high proportion of *Firmicutes* bacteria in the gut, while lean humans have more *Bacteroidetes* (R. E. Ley, et al. 2006. *Nature* 444: 1022–1023). When germ-free mice were inoculated with microbiota of normal human guts, they had an increase in body fat without a change in food consumption. This indicates that the bacteria allow the mice to extract more energy from their food. Even more interesting was the observation that gut bacteria from genetically obese mice (*ob/ob* mice), which had a higher proportion of *Firmicutes*, were more efficient than normal mice at releasing calories from food. This means that the mutant mice deposited more fat than wild-type mice when on the same diet.

The same observation was then extended to humans. Moreover, it was found that the bacterial communities changed when a person lost weight. Those individuals who lost 6% of their body weight on a restricted-fat diet or 2% of their weight on a restricted-carbohydrate diet had a decrease in the *Firmicutes* in their gut microflora. It is possible that this finding may provide another approach to the treatment of obesity.

Genitourinary Tract

Bacteria are present in the lower part of the urethra in both males and females, whereas the upper part of the urethra, near the bladder, is sterile. The kidneys and urinary bladder of the healthy adult are also essentially free of microorganisms. In contrast, the adult female genital tract has a complex normal microbiota which changes during the menstrual cycle. The adult vagina is colonized by acid-tolerant lactobacilli that convert the glycogen produced by vaginal epithelia to lactic acid. This maintains the pH at 4.4 to 4.6. Microbes that can tolerate this pH—the lactic acid bacteria, enterobacteria, coryneforms, the yeast *Candida albicans*, and various other anaerobic bacteria—are commonly present. Glycogen production, and consequently the pH of the vagina, changes with aging; thus the microbiota of the vagina also changes.

SECTION HIGHLIGHTS

The human body has many surfaces that are colonized by bacteria and a few fungi. Together, these organisms are termed the normal or resident microbiota. These organisms generally do not cause disease and often play beneficial roles for the host. The types of organisms in the microbiota differ among the various systems in the body; for example, that on the skin differs from that in the gastrointestinal tract.

26.2 Host Barrier Defense

Infectious agents are first encountered at body surfaces. Here, a combination of mechanisms categorized as physical, mechanical, chemical, or microbial barriers generally destroy or repel disease-causing (**pathogenic**) bacteria, viruses, fungi, protozoa, and parasites (Figure 26.3). Together these mechanisms make up the first layer of defense against infection and establish local environments suitable for survival of only the most adapted microorganisms.

Physical Barriers

The skin and the mucous membranes lining the mouth, nasal passages, upper respiratory tract, gastrointestinal tract, and vagina are obvious physical barriers to infectious diseases. These physical barriers protect us from the invasive microorganisms we encounter in our daily lives, since—under normal conditions—they are quite impenetrable to microbes. Other aspects of the physi-

cal barrier include the saliva and mucus produced on several of these surfaces, including those lining the respiratory and the gastrointestinal tracts. The viscous nature of mucus makes it difficult for microbes to attach to host cells lying beneath it. In addition, mucus itself traps bacteria very effectively.

Because pathogenic microorganisms are generally quite tissue-specific and attach and colonize particular cell types, they are readily removed from nontarget areas when the physical barriers are intact. For example, by washing your hands, you remove *Salmonella* and other intestinal pathogens. *Shigella* or *Vibrio cholerae* are also readily washed away, as they have no mechanism for attaching to skin. However, when physical barriers are broken—such as through a cut or burn—the risk of infection increases significantly. For example, infection by *Pseudomonas aeruginosa*, a common organism carried on airborne dust, is a constant threat to burn patients.

Mechanical Defenses

Mechanical mechanisms of defense can be energetic or violent—such as coughing, sneezing, vomiting, and diarrhea—or may not even be noticeable. That is, most dust and other inhaled particles settle out on mucous membranes of the upper respiratory tract before reaching the lower regions. Turbulence directs them onto the tract's mucus-covered walls, where they are trapped. Turbulence is created by the shape of the turbinate bones, the trachea, and the bronchi. The cilia lining the tract continually wave upward, pushing the trapped particles or material outward. This effect is referred to as the **mucociliary escalator**. The beating cilia push the mucus layer upward, where the material enters the throat and is subconsciously swallowed. Constant saliva production and swallowing also clears microbes that have entered by the mouth or nasal passages and sends them to the more hostile environment of the stomach via peristalsis in the esophagus. Peristalsis in the intestinal tract continues the journey for the ill-fated microbes. Both the intestines and lungs are also constantly washed by fluids that move over their surfaces, thus removing foreign matter that is not tightly attached. Blinking and tears are mechanical mechanisms that wash and clear the eyes. Similarly, microbes that enter the body through the urethra can be removed by the cleansing action of urination.

Microbial Barriers

The microbial barrier is composed of the normal microbiota described in the previous section. For example, the skin and mucosal membranes are populated by bacteria that are well adapted to survive there. These resident bacteria are unlikely to cause disease when confined to these primary niches. They also serve an

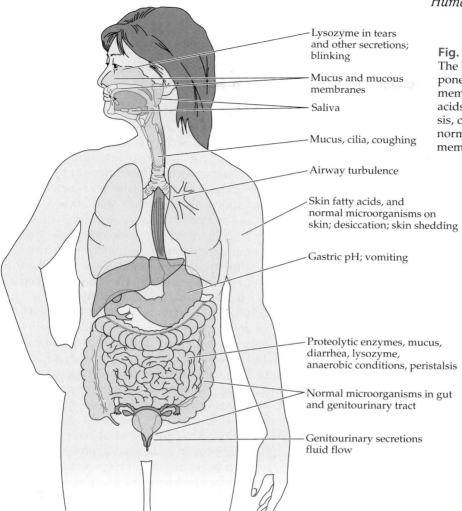

Lysozyme in tears
and other secretions;
blinking

Mucus and mucous
membranes

Saliva

Mucus, cilia, coughing

Airway turbulence

Skin fatty acids, and
normal microorganisms on
skin; desiccation; skin shedding

Gastric pH; vomiting

Proteolytic enzymes, mucus,
diarrhea, lysozyme,
anaerobic conditions, peristalsis

Normal microorganisms in gut
and genitourinary tract

Genitourinary secretions
fluid flow

Fig. 26.3 Barrier defenses
The barrier immune system has four components: physical (e.g., skin and mucous membranes), chemical (e.g., low pH, fatty acids, lysozyme), mechanical (e.g., peristalsis, coughing, blinking), and microbial (e.g., normal microbiota of skin and mucous membranes).

important function, since they prevent colonization by other organisms that are potentially pathogenic. Resident microbiota of the digestive tract compete with pathogens for nutrients, occupy epithelial binding sites, produce antimicrobial factors, and maintain the mucosal immune response in a state of "controlled physiological inflammation." The importance of the microbiota in the gut is illustrated by the overgrowth of pathogens when normal microorganisms are either eliminated or altered by aggressive antibiotic treatment. On some surfaces, the resident microbiota also decreases the pH. For example, catabolism of lipids in sebum by the resident propionic acid bacteria in the skin results in the release of fatty acids such as oleic acid. The decrease in pH can be inhibitory to organisms that are not members of the resident microbiota, and thus it is part of the chemical barrier defense system.

Occasionally, as mentioned above, some members of the normal microbiota may gain access to the bloodstream, internal tissues, and organs through a scratch, an abrasion, or minor trauma such as those that occur in brushing one's teeth or chewing food. From tissues, microorganisms can enter the bloodstream. Generally, these organisms are of little consequence, as they are removed quickly by specialized cells called **phagocytes** that engulf foreign material, including infectious disease agents; phagocytes are discussed later in this chapter. However, in some cases, disease does occur and the organism is then an **opportunistic pathogen**. For example, when the intestine is punctured, it allows the resident *E. coli* from the colon to enter the peritoneal cavity, causing peritonitis, which may be fatal.

Chemical Defenses

The body maintains an array of chemical defenses against invasion by microorganisms. Low pH is a familiar chemical barrier that precludes growth of organisms other than the resident lactic acid bacteria or other acid-tolerant species:

- Low pH in the stomach is due to stomach acid.

- The intestine is also kept acidic and anaerobic by the resident microbiota.

- The pH of urine is low because of its physiological composition (uric acid).

- Sebaceous glands in skin release oils and waxes that become organic acids, which together lower the pH of the skin.

- Catabolism of the lipids in sebum by the resident propionic acid bacteria results in the release of fatty acids such as oleic acid, which can be inhibitory to other species.

- Lactic acid–producing bacteria are found on the skin and in the vagina, where they keep the pH low, at around 5.0.

Antibacterial enzymes are also associated with barriers. **Lysozyme** is found in tears and saliva and is synthesized by the gastric mucosa. Its role is to break down the cell envelope of bacteria by disrupting the peptidoglycan layer. Lysozyme is also present in blood, sweat, nasal secretions, and other body fluids. It is a particularly effective lytic agent for gram-positive bacteria (see Figure 4.49). Gram-negative bacteria are more resistant to lysozyme because their peptidoglycan layer is located beneath the outer cell membrane. **Lactoperoxidase**, another enzyme present in saliva and milk, catalyzes a reaction between chloride ions and peroxide (H_2O_2), resulting in the generation of toxic singlet oxygen, which can kill bacteria.

In addition, a basic polypeptide, termed α-lysin, is released from the blood platelets that aggregate in cuts and abrasions. α-lysin is effective for killing gram-positive bacteria and thus serves to defend breaches in the physical barrier of the host.

Mammalian hosts also have molecules that bind and sequester iron. Many microbes must obtain iron from the host in order to colonize. Thus lactoferrin and transferrin, two iron-binding molecules found in host milk and tissues, limit the amount of iron available to microbes. The microbe must, therefore, make its own iron-binding molecules to wrest the iron from the host molecules (discussed in subsequent text).

SECTION HIGHLIGHTS

The first level of host resistance to protect against disease-causing organisms is called the barrier defense system. It has four categories: the physical barrier (for example, skin), the chemical barrier (for example, stomach acid), the microbial barrier (the normal microbiota that prevents colonization by pathogenic species), and the mechanical defense (for example, peristalsis and coughing).

26.3 How Infectious Agents Overcome Barrier Defense: Adherence and Penetration by Pathogens

Hosts are exposed to infectious disease agents in the air, food, and water and, thus, despite the barrier defense system outlined above, these organisms do occasionally invade and cause disease. Routes of infection can include breaches in the skin caused by lacerations or by penetration by ectoparasites. Some ectoparasites—including biting flies, ticks, lice, and leeches—are disease vectors. Other microbes and parasites can enter even when there is no breach in the barrier by accessing the body through the natural openings serving the respiratory system (nose, mouth), the digestive system (mouth, anus), or the urogenital system (vagina, urethra); they may also gain access through the eyes and ears. They have the capacity to attach specifically to certain membranes or epithelia.

To cause an infection, the pathogen must not only **adhere** but **colonize** (replicate). To colonize, pathogenic organisms need soluble nutrients such as sugars, amino acids, fatty acids, and/or some minerals. These are available within tissues or host cells. In the case of viral infections, the virus must also gain access to the host cell machinery needed for replication of the viral genome (RNA or DNA) and to produce viral structural molecules, including proteins and glycoproteins. Similarly, several protozoan parasites and bacteria can replicate only when they are inside host cells. Thus, pathogens must reach the appropriate tissue site or intracellular location before colonization (replication) can take place.

Disease-causing organisms can be categorized as **obligate**, **accidental**, or **opportunistic** pathogens. **Obligate pathogens** include bacteria that do not live outside a host (such as *Neisseria gonorrhoeae* and *Streptococcus pyogenes*) and protozoa that do not replicate outside a host or its insect vector (such as *Trypanosoma cruzi* and *Plasmodium*). The survival of this type of pathogen depends on its ability to move from one host to another. An **accidental pathogen** is an organism such as *Clostridium tetani* that is ubiquitous in nature and causes disease only under unusual circumstances. If this microorganism accidentally enters the body through a deep wound, it causes tetanus, which can be fatal. In contrast to an obligate pathogen, the ability of an accidental pathogen to cause disease does not play a significant role in its survival, since it survives outside the host in the environment. An **opportunistic pathogen** is an organism that does not affect a healthy host but can cause disease in an unhealthy individual or one whose immune system is not operating optimally. For example, patients suffering from acquired immunodeficiency syndrome (AIDS) are susceptible to tuberculosis caused by *Mycobacterium avium* and to many other infections that rarely occur in immune-competent individuals, even though they are regularly exposed to those organisms. Opportunistic pathogens also include some normal microbiota, such as *Staphylococcus* and *Candida* on the skin or *E. coli* in the gut. They may cause infection if the host tissue on which they reside is injured or the host's immune system is deficient.

Vector-Borne Infections

Some infectious microbes are spread by insect and arthropod vectors when they bite the host or defecate on the host's skin (see Chapter 30). Examples include bacteria, viruses, and protozoa. The protozoan that causes malaria (*Plasmodium*) is introduced in saliva from an infected mosquito when it takes a blood meal. (Only females take blood meals, because the feeding apparatus of males lacks the specialized microdissection structures needed to penetrate the human skin.) West Nile virus is a serious human pathogen that causes neurological damage and is also spread by mosquitoes. Protozoan parasites transmitted in the saliva of biting flies include the causative agent of leishmaniasis (*Leishmania*), which is transmitted by sand flies, and of African sleeping sickness (African trypanosomes), which is transmitted by tsetse flies. Ticks spread the rickettsial bacterium that causes Rocky Mountain spotted fever and also the bacterium that causes Lyme disease (*Borrelia burgdorferi*). (The saliva of mosquitoes, biting flies, and ticks contains anticoagulants that prevent blood clotting and consequently preserve the insect's proboscis and gut function.) Finally, the protozoan that causes Chagas' disease, *Trypanosoma cruzi* (South American trypanosome), is spread via the feces of kissing bugs deposited near the host's mouth and enters the host through scratches in the skin or when the host rubs the feces into the mucosa of the eye.

Some vectors of pathogens establish long-term associations with people under certain circumstances. For example, body lice, *Pedunculus humanus corporis*, colonize people living in damp, dirty places without access to bathing facilities. The lice transmit several pathogens including *Bartonella quinatana*, which is the causative agent of trench fever, a disease that affected over 1 million soldiers in World War I and that is reemerging among homeless people in urban areas of developed countries. Body lice also transmit *Rickettsia prowazekii*, which causes typhus, and *Yersinia pestis*, which cause plague.

Colonization by Bacteria

Most acute bacterial infections originate on mucosal surfaces of the respiratory, gastrointestinal, or genitourinary tracts, which have portals into the body. The bacteria generally depend on various **adherence factors** that permit them to attach to selected tissues and thus to colonize these

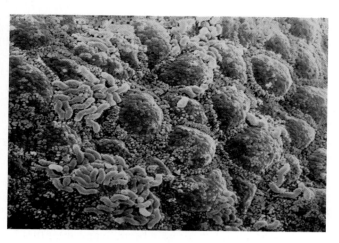

Figure 26.4 Adherence of bacteria
Specific adherence of microorganisms to an epithelium, in this case in the small intestine. Glycoproteins on the bacterial cell surface interact with specific sites on epithelial cells to cause the adherence of bacteria as a result of the receptor-ligand interaction. Pink indicates the rod-shaped bacteria, blue indicates microvilli, and brown indicates the M cells. Photo ©P. Motta and F. Carpino/SPL/Photo Researchers, Inc.

tissues or surfaces (**Figure 26.4**). These adherence factors are quite specific and do not interact with epithelial cells indiscriminately (some of the adherence factors are listed in **Table 26.1**). For example, those that attach to throat mucosal cells probably would not attach to intestinal epithelia and vice versa. Bacterial adherence factors include **fimbriae** and other surface molecules. Because of the tissue or cell specificity, different types of fimbriae occur on different bacterial strains and vary according to the type of tissue to which that bacterium attaches. The fimbriae of *Neisseria gonorrhoeae* attach to specific gangliosides on the surface of cheek mucosa in the mouth or epithelial cells in the genitourinary tract. Human pathogenic *E. coli* strains that cause urinary tract infections have P (pyelonephritic)-type fimbriae, whereas S-type fimbriae occur on *E. coli* strains that infect the central nervous system—causing, for example, meningitis in the newborn. Genetic alteration resulting in the inability to synthesize the specific fimbriae leads to a loss of virulence in the bacteria. The specificity of fimbriae for host molecules also explains why bacteria that infect one animal species generally do not infect other host species. In both cases, the bacteria cannot adhere and are swept away by the action of cilia or flushing. In most cases, if the pathogen cannot adhere to host cells, it cannot infect and colonize the host.

Some pathogens colonize or replicate on mucosal surfaces without penetrating the tissues. Bacterial diseases such as whooping cough (*Bordetella pertussis*), diphtheria (*Corynebacterium diphtheriae*), and cholera (*Vibrio cholerae*) are examples of diseases caused by pathogens that colo-

TABLE 26.1	Adherence factors involved in attachment of bacteria to host cells
Adherence Factor	**Example**
Fimbriae (adhesion proteins)	*Proteus mirabilis*—urinary tract infections
	Neisseria gonorrhoeae—attach to urinary epithelia
	Salmonella—attach to intestinal epithelia
	Streptococcus pyogenes—M protein attaches to epithelia of the pharyngeal soft palate
Capsule (glycocalyx)	*Streptococcus mutans*—dextrans attach to teeth
	Streptococcus salivarius and *S. sanguis*—attach to tongue epithelia
Teichoic acids; lipoteichoic acids	*Staphylococcus aureus*—attach to nasal epithelia

nize the surface. It has recently become apparent that in some cases the bacteria actually provide the host cell with a suitable molecule, to which the bacteria attach in order to colonize the surface. Enterohaemorrhagic *E. coli*, which causes diarrhea in children, is able to do this. The bacterium "injects" a protein called the **translocated intimin receptor (TIR)** into the host cell membrane. The TIR receptor is a second *E. coli* molecule (**intimin**) that stays associated with the bacterium. When the bacterium injects TIR into the host cell, it causes the host cell to form a pedestal-like structure to which the bacterium adheres by its intimin (**Figure 26.5**). As a result, the bacteria are not washed away by the mechanical flushing of the gut. Since the bacteria are able to receive nutrients from the gut, they are able to replicate on the intestinal epithelium.

Other pathogens need to invade tissues to replicate. To do this, disruption of host cells and tissues is needed. This can occur as a result of the activity of enzymes produced by invading pathogenic bacteria (Table 26.2). Collagenase, elastase, hyaluronidase, and lecithinase disrupt connective tissues (the intercellular cement) and host plasma membranes. *Staphylococcus aureus* and *Streptococcus pyogenes* can also synthesize streptokinase, which dissolves blood clots, permitting the pathogen to move farther into tissue after it has adhered to a place where a breach of the skin or mucosa accompanied by bleeding has occurred. Once in the tissues, some microbes are carried by the lymphatic system or bloodstream to other areas of the body. *Neisseria meningitidis*, the causative agent of a type of meningitis, is one such

(A)

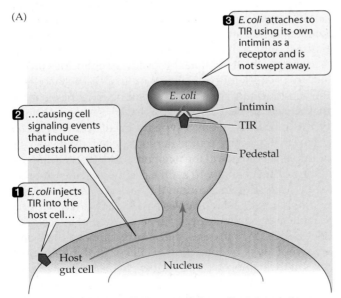

3 *E. coli* attaches to TIR using its own intimin as a receptor and is not swept away.

2 ...causing cell signaling events that induce pedestal formation.

1 *E. coli* injects TIR into the host cell...

(B)

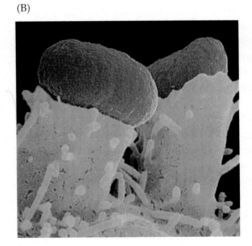

Figure 26.5 Pedestal formation for adherence of enterohaemorrhagic *E. coli*
E. coli injects a molecule known as translocated intimin receptor (TIR) into the host gut cell, which causes changes in that cell so that a pedestal is formed. *E. coli* then binds to the pedestal via the bacterial cell surface molecule intimin—a "receptor" for TIR—allowing the bacterium to adhere to the gut wall. (A) Diagram of the interaction; (B) view of the interaction by scanning electron microscope (SEM; false-colored). B, courtesy of B. Brett Finlay.

TABLE 26.2	Examples of enzymes produced by pathogenic bacteria that promote colonization and/or invasion of the host	
Enzyme	**Organism**	**Function**
Collagenase	*Clostridium*	Breaks down collagen in connective tissue
Coagulase	*Staphylococcus aureus*	Clot formation around point of entry protects from host defenses
Elastase	*Pseudomonas aeruginosa*	Disrupts membranes
Hyaluronidase	*Streptococcus, Staphylococcus, Clostridium*	Hydrolyzes hyaluronic acid—intercellular matrix
Lecithinase	*Clostridium*	Disrupts phosphatidylcholine in host cell membranes
Streptokinase	*Streptococcus*	Digests fibrin (blood) clots so bacteria can invade
Staphylokinase	*Staphylococcus*	Digests fibrin (blood) clots so bacteria can invade

bacterium. A localized infection caused by another bacterial species disrupts epithelial cells of the throat, which permits *Neisseria meningitidis* to enter and be transported to the meninges (the membranes surrounding the central nervous system), where it adheres and reproduces (colonizes), resulting in meningitis.

Colonization by Viruses

Viruses gain access to the host through virtually all surfaces of the barrier defense except the skin. They enter through the skin only when they are introduced by an insect vector that penetrates the skin (e.g., West Nile virus) or when they gain access through a cut or abrasion. In contrast, many viruses enter the respiratory system; this entry is particularly efficient when the virus is contained within an aerosol—for example, when the viral particles are expelled as part of a cough or sneeze from an infected person (see Figure 29.8). Such viruses include orthomyxoviruses and coronaviruses. Viruses also enter through the gastrointestinal tract in food or from the ingestion of fecal material. In order to infect the tissues of the gastrointestinal tract, these viruses must be able to resist the acid of the stomach. Such viruses include the rotaviruses and enteroviruses, which have acid-resistant **capsids**. The capsid is the structure of the virus particle that surrounds its genetic material (either RNA or DNA, depending upon the virus; see Figure 14.1). Other viruses such as the human immunodeficiency virus (HIV) and human papilloma virus enter through the genital tract by sexual transmission. Some viruses (such as those that cause warts) stay at the site where the initial infection occurred and replicate there. Others are carried deeper into the body by the bloodstream or lymphatic system and, in the case of the rabies virus, by the nervous system.

Regardless of the tissue through which the virus enters and where it is then carried, viral colonization requires attachment to host cells through specific receptors. The expression of host cell receptors can differ among cell types and tissues, which accounts for some of the tissue-specificity of viruses. Examples of host cell molecules used as receptors include neuraminic acid on the epithelial cells to which the virus causing the common cold attaches. The virus that causes polio specifically attaches to neuronal cells, while the HIV virus, the cause of AIDS, has surface structures that bind to the CD4 protein on T lymphocytes. Although a particular type of host molecule that serves as a virus receptor may be common to all mammals, slight variations in the molecule's composition among species can account for susceptibility to viral colonization (infection) in one species of mammal and resistance to infection in another. This is similar to the variation in host cell receptors for bacterial fimbriae discussed above. In contrast, other viruses that infect humans (for example, the rabies virus) can also infect several other mammalian species.

Viral entry into the host cell requires enzymatic reactions, with the enzymes provided either by the virus or in some cases by the host cell (Figure 26.6). Infection of host cells occurs either by endocytosis, as in the case of the influenza virus, or by fusion of the virus to the host cell's membrane, as occurs in the case of HIV. Phagocytosis of the influenza virus by the host cell is promoted by binding of host cell membrane proteins to viral molecules called hemagglutinin and neuraminidase. As a result of this binding, the host cell membrane gradually wraps itself around the viral particle (zippering) until it is inside the host cell yet still surrounded by the host cell's now inside-out membrane (see Figure 14.4). This new vesicle in the host cell is called a phagosome. Once inside the host cell and within the phagosome, the viral membrane fuses with the host cell's membrane and is thus introduced into the cytoplasm of the host cell. In the other process of entry, direct fusion of HIV with the host cell membrane is expedited by binding events that result in reorganization of the membrane lipids in the attachment zone. Once a virus is inside the host cell and uncoated, its genetic material (viral DNA or RNA) can be accessed and replicated for incorporation into the daughter viral particles. It is

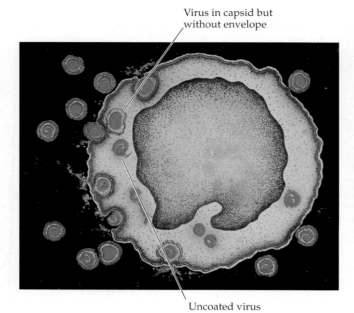

Figure 26.6 shows viral particles labeled "Virus in capsid but without envelope" and "Uncoated virus"

Figure 26.6 Adherence of virus to host cell
This transmission electron micrograph shows viral particles attaching to, entering, and uncoating within the host cell. ©Chris Bjornberg/Photo Researchers, Inc.

also used as the blueprint for producing the viral components, such as the hemagglutinin and neuraminidase in the case of the influenza virus, as well as other molecules needed to form a capsid surrounding the virus's genetic material.

Colonization by Fungi

Fungi are generally associated with the external surfaces of human hosts, although *Candida* is found in the mouth as well as the throat and vagina. In babies, infection in the throat with an overgrowth of *Candida albicans* is referred to as thrush, while women may suffer from vaginal infections. Men too may suffer from infections in the urethra. Some immunodeficient persons are infected throughout life with *Candida* on the skin, resulting in raised nodules that may cover large areas of the body, including the face. Occasionally *Candida* spp. and other fungi found in soil (e.g., *Histoplasma capsulatum* and *Cryptococcus neoformans*) cause internal infections in humans, some of which can be fatal, but generally only in people with deficient immune systems.

Very few of the adhesion molecules have been identified for fungal pathogens because of the difficulty of performing the genetic validation tests in these organisms (deleting all genes involved in one function). However, recent work on *Candida* spp. indicates there are at least four adhesion molecules that are glycosylphosphatidylinositol (GPI)-anchored into the fungal cell wall. *Candida*

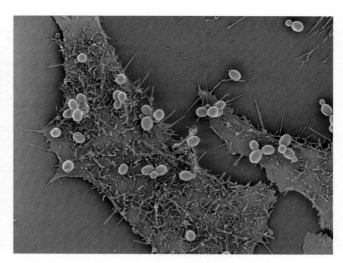

Figure 26.7 Adherence of fungi to host cells
A false-colored scanning electron micrograph of *Candida* adhering to human epithelial cells. The fungi are green. Courtesy of B. Cormack and M. Delannoy, Johns Hopkins University.

glabrata is also a human pathogen, responsible for about 15% of *Candida* infections. Using this species it was shown that the adhesion molecules are coded for by the *EPA* genes. These cell wall fungal proteins bind to various host epithelial cell sugars (Figure 26.7). Notably, *Candida* is also proficient at sticking to artificial materials used as medical devices, including urinary catheters; people can acquire severe infections by this route. The *EPA* gene products are also involved in the binding to plastic. Moreover, urine may promote adhesion (Box 26.4).

Like bacteria, fungi also produce enzymes that contribute to host tissue invasion. These enzymes function by digesting tissue and destroying host cell membranes. While they are perhaps less well characterized than those of bacteria, one group of enzymes that has been characterized comprises the secreted aspartic proteinases of *Candida albicans*. Proteinases are a type of enzyme that digests proteins.

Colonization by Protozoa

Many protozoa form **cysts**, which are primarily a means of transfer from one host to the next, often in water or food. The cysts enable the organisms to survive harsh environmental conditions and thus to infect a new host and multiply. For example, *Giardia lamblia* and *Entamoeba histolytica* cysts infect many people worldwide, including in North America, when they are acquired from contaminated water. These protozoa are especially a problem among children or individuals in communities with poor sanitation, since they can enter the water supply through feces. (Water filtration and other water treatments reduce

BOX 26.4   *Research Highlights*

Does Urine Promote Urinary Tract Infections by Yeast?

Normally we think of the barrier defense system as just that, a group of mechanisms to defend against infection. In the urinary tract the components expected to repel infection include the epithelial cells that line the tract (physical defense), the flow of urine (mechanical defense), and the composition of the urine, including its acidity (chemical defense). Urinary tract infection (UTI) can be caused by the fungus (or yeast) *Candida glabrata*, which hangs on to the epithelial cells in the urinary tract. There it replicates or colonizes the host without further invading the host tissues. Because the yeast attach to the host tissues, the flow of urine does not wash them away. Surprisingly, recent studies have shown that urine can actually promote the infection by causing the yeast to turn on genes that code for adherence molecules.

Using mice, a research team led by Dr. Brendan Cormack at Johns Hopkins Medical Center was able to show that when the yeast was in the bladder it began to colonize and spread the infection to the kidney (R. Domergue, et al., 2005, *Science* 308: 866–870). In contrast, when it was in the blood the yeast did not replicate and colonize. This finding led the researchers to hypothesize that some of the yeast's adherence genes were activated in the urinary tract but not the blood.

To test this hypothesis, the researchers used yeast genetically engineered to have the gene for green fluorescent protein (GFP) under the control of the promoter for the adherence gene known as *EPA6*. These yeast cells would fluoresce whenever the adherence gene was turned on. The researchers also made synthetic urine so that they could selectively add or remove components. They found that it was the low level of the vitamin niacin in urine that stimulated the yeast to turn on the *EPA6* gene. As a result, the yeast could produce more adherence molecules and adhere to human epithelial cells more tightly (see Figure 26.7). Because of this, the yeast is able to colonize tissues in contact with urine, as well as medical urinary catheters in hospital patients. This research illustrates the adaptation of a pathogen for colonizing a specific host tissue and may provide clues on how to reduce UTIs. This is important because such infections can be serious in humans.

the chance of infection.) These organisms cause a gastroenteritis characterized by diarrhea, among other symptoms. **Excystment** (the emergence of active cells) is initiated by exposure of cysts to stomach acids. *Giardia* then attaches to the upper small intestine by a sucker (Figure 26.8). If the parasite detaches, it is exposed to unfavorable conditions, including the presence of bile salts and alkaline pH in the lower small intestine. This leads to their encystment and mechanical expulsion from the body. *Entamoeba histolytica* colonizes the intestinal wall, using physical and lytic digestion of host cells. It starts with a small lesion where the shedding of intestinal cells occurs, forming a small cavity. As this abscess enlarges, it can make a channel into the intestinal mucosa. If it erodes the musosa enough, it can gain access to and be carried by the lymph system to the liver and sometimes the lung, where it continues to colonize, forming additional abscesses.

Cryptosporidium parvum infects a number of different mammals, including humans and livestock, by its eggs (known as **oocysts**). In humans, *C. parvum* cause diarrhea, which may last for months in people with deficient immune systems. The infection is controlled within a few weeks in healthy individuals. This organ-

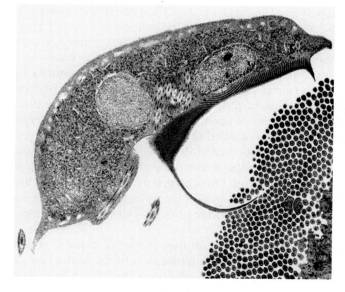

Figure 26.8   Adherence of protozoa to host gut
Giardia (in blue) adhering to gut mucosa microvilli (in green). This TEM shows a cross-section of the sucking disc (the "hooks") made of short cilia (dark blue lines within the disk). ©SPL/Photo Researchers, Inc.

ism may be a contaminant of water close to agricultural facilities as a result of fecal runoff into the water supply. The oocyst is encapsulated in a novel protein sheath, making a cyst with an impervious outer layer that is highly resistant to harsh environmental conditions. Excystment may be triggered by a variety of factors including low pH, exposure to bile salts, a raised level of carbon dioxide, and elevated temperatures. Once released from the cyst, the protozoan attaches to the microvilli of epithelial cells in the gut and becomes enveloped by host cells, forming the parasitophorous vacuole; this is composed of both host and parasite membranes and the parasite begins to replicate in this membrane "pouch."

Colonization by Parasitic Worms

Although parasites known commonly as parasitic worms are not microbes, their infective stages may be microbial. Moreover, they established long-term relationships with humans. Most parasitic **cestodes** (tapeworms) that infect humans have an intermediate host that is ingested by humans. That is, they are contained within another substance, such as meat. When this is digested, it frees the parasite to initiate infection. For example, the infective stage of *Diphylidium caninum* is called a **cysticercoid**, which lives within the dog flea; it is digested in the stomach and small intestine of humans and dogs. The freed infective cysticercoid attaches to the wall of the small intestine via its **hook-studded scolex**, where it matures into an adult within 3 weeks. The pig tapeworm, *Taenia solium*, has a juvenile form (**cysticerci**), which is passed to humans in undercooked pork. Digestion of the meat results in freeing of the cysticerci, which then can attach to the wall of the intestine by hooks and suckers (**Figure 26.9**). If *T. solium* grows to maturity in the intestine, it produces eggs. When humans ingest the eggs of *T. solium*, they cause more severe disease than that acquired from cysticerci in uncooked meat. The eggs contain larvae (called onchospheres) that are released upon interaction with bile salts. These bind to the gut wall and penetrate through the muscle with the help of parasite enzymes. From here, they can be taken into the bloodstream and, once in the tissues, develop into cysticerci, which may lodge in various sites including in the brain.

All parasitic **trematodes** of the subclass Digenea (flukes) use snails as intermediate hosts. In the case of *Schistosoma mansoni*, which causes schistosomiasis, eggs released in the feces of infected people contain larvae. These hatch from the eggs in freshwater and swim about seeking a snail host, into which they penetrate. The snail host, *Biomphalaria glabrata*, is prevalent in South and Central America, while other permissive *Bio-*

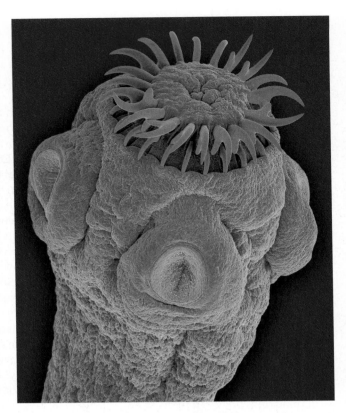

Figure 26.9 Adherence apparatus of parasitic worm
Taenia scolex showing hooks and sucker in false colors. The hooks are in bright red and are used to attach to the intestinal wall. ©Dennis Kunkel Microscopy, Inc.

mphalaria species are present in Africa. Following maturation in the snail, these organisms are infective for humans. They penetrate the skin of their human hosts through hair follicles. Thus, this infectious cycle requires the deposition of feces from infected humans near their water supply, the presence of the snail hosts, and the use of the water supply by humans. The adult stage is found in the blood supply of the liver, and eggs are found in the liver and lungs.

Parasitic **nematodes** (roundworms and whipworms) have a variety of infection strategies. *Necator americanus* was a prevalent infectious agent in rural areas of North America in earlier times, when children and adults frequently did not wear shoes. *N. americanus* enters through unbroken skin, typically through a blood capillary supplying a hair shaft. They are borne by the blood to the heart and then the lungs, where they exit the circulation into the alveolar space (spaces in lung tissue between the alveoli). The larvae are carried up the mucociliary ladder of the respiratory tract to reach the pharynx and are swallowed. Once in the small intestine, the larvae attach to epithelia and develop into adults.

26.4 Inflammation and Phagocytosis: The Second Level of Host Defense

When host cells are killed, their DNA is exposed. This can occur when a pathogen is invading and colonizing the host. DNA is a principal inducer of inflammation. **Inflammation** is a chain reaction of events whose major symptoms include redness, pain, swelling, and increased localized temperature. White blood cells (**leukocytes**), known functionally as phagocytes, respond to inflammation by moving from the blood to the site of injury to play a protective role. However, inflammation can cause tissue and organ damage if prolonged.

The two main types of phagocytes in the blood are known as **neutrophils** and **monocytes**. Phagocytes engulf and internalize foreign invaders. The local increase in temperature due to the inflammatory response can accelerate the rate of phagocytosis. Within the host cell, the invader becomes contained in a **phagosome**, which is es-

sentially an inside-out membrane-bound compartment (Figure 26.10). The phagosome fuses with other membrane vesicles—the **lysosomes**—to generate a **phagolysosome**. Fusion with the lysosome results in the introduction of many lysosomal enzymes that break down proteins, lipids, and carbohydrates. The pH in the phagolysosome is also very acidic. Antimicrobial molecules are introduced into or generated in the phagosome and phagolysosome. These include small peptides called **defensins** as well as **reactive oxygen intermediates** or **radicals** such as hypochloride (which is bleach), hydrogen peroxide, and hydroxyl radicals. As a result, most phagocytosed organisms or cells are killed rapidly.

The first phagocytes to enter tissue by crawling through the blood vessel wall are neutrophils (see Figure 27.4). The neutrophils are generally short-lived cells (half-life of 6 to 8 hours). If an infection is ongoing, there is an increased rate of release of neutrophils from bone marrow, a response termed **leukocytosis**. Thus, a relatively high number of neutrophils in the blood may indicate an active infection. Within 48 hours, neutrophils begin to be replaced by monocytes at the site of infection and tissue injury. These are long-lived cells that will remain until the infection and inflammation are resolved.

If inflammation is initiated because of an injury and bleeding occurs, a fibrinous blood clot forms. This can prevent further invasion of the host. (Some bacteria have enzymes that dissolve the clot [see Table 26.2]; such enzymes are virulence factors.) Eventually the wound will be repaired by a proliferation of epithelial and endothelial cells; in the meantime, however, phagocytic cells are of considerable importance in eliminating the invading microbes.

In addition to the role played by phagocytes recruited to the site of inflammation from the blood, **fixed tissue macrophages** are also phagocytes that defend against invaders. These arise from monocytes that have previously passed through blood vessels into tissues, where they mature and become resident in the tissues. As a result, they have limited mobility but can be very long-

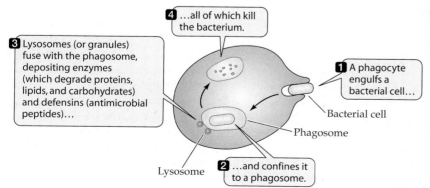

3 Lysosomes (or granules) fuse with the phagosome, depositing enzymes (which degrade proteins, lipids, and carbohydrates) and defensins (antimicrobial peptides)...

4 ...all of which kill the bacterium.

1 A phagocyte engulfs a bacterial cell...

Bacterial cell

Phagosome

Lysosome

2 ...and confines it to a phagosome.

Figure 26.10 Phagocytosis: Events occurring in phagocytosis of microbes
Neutrophils and macrophages are phagocytic cells that engulf microorganisms and confine them to a membrane-bound intracellular compartment called a phagosome. The antimicrobial reactive oxygen intermediates produced in the phagosome include hydrogen peroxide, hypohalides, and hydroxyl radicals. The lysosome fuses with the phagosome, delivering digestive enzymes. This usually results in killing of the phagocytosed organisms, unless they are specifically adapted to live inside the phagocyte.

lived at these sites. These macrophages occur in tissues that comprise the epithelial surfaces exposed to the external environment and also those lining vessels carrying blood or lymph. Tissues rich in fixed macrophages include liver, spleen, lungs, and lymph nodes. These fixed tissue macrophages provide continual surveillance of tissues for pathogenic invaders. Macrophages and monocytes collectively are sometimes referred to as the **mononuclear phagocyte system** or, previously, the **reticuloendothelial system**.

Avoiding Destruction by Phagocytes

Some bacteria have mechanisms enabling them to avoid control by host phagocytes. They include mechanisms to avoid phagocytosis altogether, while others involve killing the phagocyte. For example, capsular material surrounding pneumococci and meningococci protects these organisms from phagocytosis. A pneumococcus without a capsule is not very effective in colonization, as it will be readily captured and destroyed by phagocytes. Examples of antiphagocytic factors are listed in Table 26.3. **Leukocidins**—produced by pneumococci, streptococci, and staphylococci, and some gram-negative bacteria such as *Pasteurella*—attach to the membrane of leukocytes and promote disruption of the lysosomes. This releases hydrolytic enzymes from the lysosomes, which destroy the phagocytic cell. **Hemolysins** act by destroying the plasma membrane of host cells, including monocytes and neutrophils. They do this by forming small pores in the host cells, which cause cell leakage and uptake of water and, ultimately, cell lysis.

Some pathogens have mechanisms to survive within the phagocyte without killing the cell. In the case of the fungus *Candida*, it can switch from its usual yeast-like form to a hyphal form to avoid being killed within the phagolysosomes of neutrophils. Many organisms not only survive but replicate inside macrophages and monocytes. They use a variety of mechanisms to do so and are discussed below under "Life in Host Cells." They may rely partially upon pathogen-derived enzymes that neutralize the harsh environment within the phagocyte. These are virulence factors such as catalase, which breaks down the antimicrobial molecule hydrogen peroxide into water and oxygen. Pathogens can also neutralize the pH in the phagosome, escape into the host cell cytoplasm, or prevent fusion of the phagosome and lysosome.

SECTION HIGHLIGHTS

Once the barrier defense has been breached by an infectious disease agent, the next layer of host defense is the innate immune response, including inflammation. Inflammation is a complex response to damaged host cells and tissues. As part of inflammation, phagocytic cells are brought to the site of infection. These are cells that engulf (eat) and usually kill microbial invaders.

26.5 Virulence and Pathology

It is convenient to distinguish pathology that results from processes initiated by the pathogen from those resulting from the host response to the infection. The products of the infectious organism that contribute to pathology are referred to as **virulence factors**. These include:

- Molecules that allow a pathogen to colonize, including adhesion molecules and enzymes needed to penetrate deeper into host tissues or cells (as discussed above)
- Molecules that allow the organism to obtain nutrients efficiently
- Toxins
- Mechanisms to avoid host immune defense, thus prolonging survival of the pathogen

TABLE 26.3	Antiphagocytic factors produced by bacteria and their modes of action
Factor	**Action**
Leukocidins	Specific lytic agent for leukocytes, including phagocytes
Hemolysins	Form pores in host cells, including macrophages; streptolysin O affects sterols in membranes; streptolysin S is a phospholipase
Capsules (glycocalyx)	Long polymers of carbohydrate; physically prevent engulfment
Fimbriae	(1) Bind to surface components of phagocytes, prevent close contact, and may inhibit phagocytosis
	(2) Phase variation—a change in expression of fimbriae

TABLE 26.4	Mechanisms of virulence and pathogenicity	
Mechanism	**Origin**	**How Pathology is Caused**
Adherence to host tissues	Pathogen	Does not specifically cause pathology but is needed for the organism to colonize the host
Invasion of host tissues	Pathogen	Enzymes that kill host cells and disrupt tissues
Nutrition	Pathogen	Obtained by: • killing host cells to gain nutrients • iron-binding molecules (siderophores)
Toxins	Pathogen	Killing or disabling host cells, tissues, and organs
Immune evasion	Pathogen virulence factors to avoid host response	Pathogen survives by: • avoiding phagocytosis • killing host phagocytes • surviving in the phagosome or escaping from the phagosome into the host cell cytoplasm • changing antigens to evade antibody and T-lymphocyte responses • becoming non-replicative (latent)
Immunopathology	Host immune system	Immune response to pathogen and its products (e.g., toxins) results in tissue damage, organ failure, and in some cases host death: • endotoxin shock • toxic shock syndrome • nephritis due to complexes of pathogens and antibodies • granuloma formation • inflammation resulting in pneumonia or other organ damage

Collectively, these virulence factors result in host cell death or inhibition of host cell or organ functions. For example, bacterial infections cause pathology as a result of tissue invasion and toxins; viral infections use the host cells to propagate themselves; parasites disrupt tissue and organ architecture and function. The second aspect of pathology originates with the host. That is, the host's immune responses can be a major factor in injury to the host. A summary of the virulence mechanisms and causes of pathology covered in various parts of this chapter is provided in Table 26.4.

Bacteria

Bacterial virulence factors related to acquiring nutrition include the bacterial molecules that bind iron. Iron is essential for the synthesis of cytochromes, the electron carriers involved in the generation of energy. Vertebrates retain iron in their systems in a soluble state by high-affinity binding of reduced iron to glycoproteins. For example, iron is bound by lactoferrin in milk, tears, saliva, mucus, and intestinal fluids, and by transferrin in plasma. The level of free iron in body fluids is less than 10^{-8} molar. Invading bacteria and other pathogens, all of which require iron to survive and replicate, must acquire

it from the lactoferrin or transferrin of the host in order to colonize effectively. For this purpose, many pathogens synthesize low-molecular-weight siderophores (see Figures 5.7 and 5.8), which have a much higher affinity for iron than the host lactoferrin or transferrin. These siderophores are soluble molecules that bind receptors on the pathogen's cytoplasmic membrane and transfer the iron pulled away from the host molecules to the bacteria. The role of iron in bacterial colonization can be confirmed by feeding soluble iron to animals infected with selected bacteria and observing a marked increase in the severity of the disease.

Most virulent bacteria are highly invasive, causing inflammation and host cell death even in the absence of toxins. For example, *Streptococcus pneumoniae* can spread from the upper respiratory tract to the lower tract, killing cells in the process, even though it does not produce a specific toxin. Before the advent of antibiotics, this organism was a leading cause of death because it clogged the lungs of victims.

In contrast, some organisms, such as *Clostridium tetani* and *Clostridium botulinum,* have virtually no invasive ability but produce highly potent exotoxins. That is, while *C. tetani* is an opportunistic or accidental pathogen, its introduction into a puncture wound can

TABLE 26.5	Characteristics of exotoxins and endotoxins

Exotoxins	Endotoxins
Heat-labile at 60°C to 80°C	Heat-stable
Immunogenic	Immunogenic
Cause no fever	Cause fever
Can be lethal at low concentrations	Toxic at high doses; can be lethal
Different genera produce different toxins	Similar structure for all gram-negative bacteria
Secreted by live bacterium or released by lysis	Active when attached to bacterial outer membrane
Inactivated by chemicals that affect proteins	Not generally harmed by chemicals that affect proteins

be fatal. Ingestion of botulinum toxin, even in small amounts, can also be fatal. Moreover, all gram-negative bacteria have endotoxin in their cell walls (Table 26.5). Toxins are virulence factors, since they cause pathology, and are discussed here individually.

BACTERIAL ENDOTOXINS   Lipopolysaccharide (LPS) is another name for endotoxin and is a component of the outer envelope of most gram-negative bacteria. It is composed of lipid A, core polysaccharides, and O-polysaccharide side chains (see Figure 4.55). The polysaccharide component has little apparent toxicity but can also play a role in virulence, since when the O- side chains are eliminated by genetic mutation, the disease-causing ability of an otherwise virulent pathogen such as *Brucella abortus* is very much reduced. Lipid A exhibits virtually all of the toxic properties of the intact lipopolysaccharide complex. However, endotoxins can vary considerably in their toxic effects. Those from *Enterobacteriaceae* (*E. coli* and *Salmonella*) are most potent, while those from *Brucella* are among the least potent.

Although endotoxin generally remains associated with the bacterial organism, it can be released when the organism is lysed or, in some instances, during cell division. Release of endotoxin is not necessary, because it is on the outer surface of the bacterium and stimulates host cells through host-cell receptors; thus its toxic effects are exerted even when bound to the bacterium—hence the prefix "endo." (In contrast, exotoxins exert their effects away from the organism that makes them.)

Endotoxins damage the endothelial lining of blood vessels, thus activating them, and also bind to mononuclear phagocytes. These events lead to the release of the soluble immune mediators called **cytokines** (interleukin-1 [IL-1] and tumor necrosis factor-α [TNF-α]) as well as Hageman factor. IL-1 and TNF-α are associated with events in the immune response and are discussed in Chapter 27. The Hageman factor is a blood-clotting factor (factor XII) that initiates a clotting cascade through a series of reactions "triggered" by a single event. The clotting cascade leads to the development of blood clots.

In the absence of adequate control mechanisms, thrombosis (blockage of blood vessels) can occur, leading to widespread coagulation of blood in the vascular system. The clotting removes platelets from the blood system at rates that exceed their replacement and may cause hemorrhaging elsewhere in the body, which can lead to the failure of essential organs. High levels of endotoxin can bring about uncontrolled clotting/fibrinolytic cascades leading to circulatory collapse and potentially death. The Hageman factor also causes activation of the kallikrein system, promoting **vasodilation** (widening of the blood vessels) and drainage of blood from vessels, thus decreasing blood pressure.

Because endotoxins cause an increase in human body temperature, a condition referred to as **fever**, we call them **pyrogens** or fever inducers. It is actually the cytokines released by phagocytes and other leukocytes that cause the fever response. Because many pathogens grow optimally at 37°C, an increase of 2°C to 3°C could curtail microbial growth. "Chills" that occur with a fever are a response by the body to increase muscular activity and drive the temperature higher. There is also some evidence that an increase in temperature may decrease the amount of iron available in the system, thereby interfering with microbial growth. However, a temperature much above 41°C (105.8°F) can lead to systemic damage and be fatal to the host.

Since endotoxins are rather stable to heat and can be potent toxins, it is important to be able to detect them in our food, water, and medical solutions. An exceedingly sensitive test for trace amounts of endotoxin is the *Limulus* amoebocyte lysate assay. This test employs amoebocytes of the horseshoe crab, *Limulus polyphemus*. An endotoxin, even in trace amounts, reacts with the clot protein from these circulating amoebocytes and causes clotting. Spectrophotometric measurement of this clotting is an effective, sensitive, and very specific assay for endotoxins.

BACTERIAL EXOTOXINS   An exotoxin (Table 26.6) is a toxic protein released into the surrounding medium by a microorganism that is growing or dying. The exotoxin

TABLE 26.6	Some pathogen toxins		
Toxin	**Producing organism**	**Disease**	**Actions or Effects**
Bacterial exotoxins			
Cytotoxin	*Corynebacterium diphtheriae*	Diphtheria	Inhibits protein synthesis; affects heart, nerve tissue, liver
Botulinum toxin	*Clostridium botulinum*	Botulism	Neurotoxin; flaccid paralysis
Perfringolysin O toxin	*Clostridium perfringens*	Gas gangrene	Hemolysin, collagenase, and phospholipase activities; causes stomach pain and diarrhea
Erythrogenic toxin	*Streptococcus pyogenes*	Scarlet fever	Capillary destruction
Pyrogenic toxin TSST-1	*Staphylococcus aureus*	Toxic shock syndrome	Fever, shock
Exfoliative toxin	*Staphylococcus aureus*	Scalded skin	Massive skin peeling
Exotoxin A	*Pseudomonas aeruginosa*	Burn infections	Inhibits protein synthesis
Pertussis toxin	*Bordetella pertussis*	Whooping cough	Stimulates adenyl cyclase
Anthrax toxin	*Bacillus anthracis*	Anthrax	Pustules; blood poisoning
Enterotoxin	*Escherichia coli*	Diarrhea	Water and electrolyte loss
Enterotoxin	*Vibrio cholerae*	Cholera	Water and electrolyte loss
Enterotoxin	*Staphylococcus aureus*	"Staph" food poisoning	Diarrhea, nausea
Enterotoxin	*Clostridium perfringens*	Food poisoning	Permeability of intestinal epithelia
Neurotoxin	*Clostridium tetani*	Tetanus	Rigid paralysis
Bacterial endotoxin			
Lipopolysaccharide (LPS)	Gram-negative bacteria	Endotoxin shock (endotoxemia)	Intravascular coagulation of blood; fever
Viral			
gp41	HIV	AIDS	Destabilizes membranes
PB1-C2	Influenza virus	Flu	Host cell death
Fungal			
Gliotoxin	*Candida*	Thrush	Kills macrophages and T lymphocytes
Protozoan			
Variable surface glycoprotein	*Trypanosoma brucei*	Sleeping sickness	Weight loss (severe wasting)
Saponin-like molecule	*Entamoeba histolytica*	Dysentery	Cell death (apoptosis)

produced by a species is generally unique, differing in both structure and function from other exotoxins. If the bacterium releases an exotoxin within the human body, it may travel from the site of infection to other areas of the body. It is apparent from the list in Table 26.6 that the effects of exotoxins vary widely, from diarrheal diseases to paralysis. The effects of exotoxins on tissues can be classified as follows:

- Enterotoxins—cause dysentery and other intestinal distress (an example is cholera toxin)

- Neurotoxins—affect nerve impulse transmission (examples are tetanus and botulinum toxins)

- Cytotoxins—destroy cells by inhibiting the synthesis of proteins and also disrupt or disorder membranes, causing harm to the heart, liver, and other organs (an example is the cytotoxin causing diphtheria)

- Pyrogenic toxins—stimulate the release of cytokines, leading to fever and shock (an example is the exotoxin known as toxic shock syndrome toxin type 1 [TSST-1] produced by *S. aureus*)

The mode of action of specific exotoxins is discussed in Chapter 28.

Some exotoxins, such as the enterotoxin of *E. coli* and the staphylococcal enterotoxins, are heat-stable. Thus the activity of heat-stable enterotoxins in food is not destroyed by cooking, although the organisms that produce them are likely to be killed. Despite their heat-stability, exotoxins can be inactivated by other methods such as chemical fixation with formalin. Toxins are referred to as **toxoids** following inactivation and can be used as vaccines to immunize infants (and adults) against the diseases caused by exotoxin-producing microorganisms. Among the diseases that can be prevented by vaccina-

TABLE 26.7	Bacterial virulence factors generally encoded in plasmids	
Organism	**Factor**	**Disease**
Escherichia coli	Enterotoxin	Diarrhea
Clostridium tetani	Neurotoxin	Tetanus
Staphylococcus aureus	Coagulase enterotoxin	Diarrhea; food poisoning
Streptococcus mutans	Dextransucrase	Tooth decay
Agrobacterium tumefaciens	Tumor	Crown gall
Corynebacterium diphtheriae	Diphtheria toxin	Diphtheria

tion with toxoids are diphtheria, tetanus, and whooping cough. This is because exotoxins are effective in eliciting host adaptive immune responses (see Chapter 27) in the form of soluble molecules known as antibodies.

PLASMIDS   Some bacterial virulence factors, including toxins, are encoded in the plasmids of pathogens (Table 26.7). This permits a rapid passage of genetic information from a limited number of virulent pathogenic microorganisms to a population with lower virulence (see Chapter 15). The transfer of plasmids enables the organisms that acquire them to generate toxins and additional enzymes involved in tissue invasion and avoidance of host defenses. An infection caused by a population that had a limited number of members carrying particular plasmid-encoded virulence factors can rapidly become a more virulent population composed largely of members carrying the plasmid. Another example of transfer of **plasmid-encoded factors** is acquisition of antibiotic resistance.

Viruses

The virulence of a virus depends upon its ability to obtain the raw materials (nutrients) and machinery needed to replicate itself. Therefore it must be able to enter a host's cell. Virulence also depends upon the virus's ability to prevent host cell replication of its own DNA and production of host gene transcripts (mRNA) and, thus, host cell proteins. This causes normal host cell function to lapse into a nonproductive state. As a consequence, the host's tissue function is disrupted, which in some cases leads to organ failure. Further host cell damage can result when the assembled viral particles exit the host cell. They do this either through lysis or by a process known as **budding**. In budding viruses, the assembled virus particle buds through the host cell membrane without lysing the cells (see Chapter 14); in lytic viruses, the host cell is obviously destroyed to release the assembled viral particles, thus adding to the pathology and increasing inflammation.

Viruses can also produce molecules referred to as toxins (see Table 26.6). This is a relatively new observation, but such viral toxins have been shown to be products of both HIV and influenza virus. In the case of HIV, the toxin is a viral glycoprotein named gp41, and is a membrane-destabilizing molecule. For the influenza virus, the toxin is a protein named PB1-F2, believed to destabilize mitochondrial membranes. Such destabilization can lead to leakage of the cell and its organelles and therefore to programmed cell death (known as **apoptosis**). Such molecules may aid the survival and replication of the virus by destroying host cells—such as phagocytes and lymphocytes—involved in mediating protective immune responses.

Another mechanism of inducing pathology, which is unique to viruses among the infectious disease agents, is their ability in some cases to integrate the viral genetic material into the host cell genome. When this happens, some viral genes can cause the host cell to become transformed into a continually replicating or cancerous cell. This is a rare occurrence in humans, although some members of the papilloma virus complex cause cervical cancer. In 2006, a new vaccine was released for use in humans; it has been shown to be very effective at preventing cervical cancer as well as genital warts caused by strains of the papilloma virus.

Fungi

Like other pathogens, fungi must obtain nutrients from the host. The fungus *Candida albicans* secretes the hydrolytic enzyme group known as aspartic proteinases (SAPs). While these are used for tissue invasion, as discussed previously, another role is to digest proteins to provide nutrition. Some fungi also produce siderophores, as do bacteria, to obtain iron (see Chapter 5). Because these fungal products contribute to survival and colonization, they are considered virulence factors at least in the context of some species. For example, SAPs are virulence factors for *C. albicans* but not for *C. parapsilosis*, which is not a human pathogen. Other nonpathogenic species of *Candida* have lower proteinase (protein-digesting) activities.

Fungi also produce toxins. One that is produced by several species of human pathogenic fungi, including *C. albicans*, is gliotoxin. It has a role in immune evasion, since it kills macrophages as well as T lymphocytes, cells important in host immunity. Also, *C. albicans* produces a mannoprotein-α-glucan complex, which is secreted. Its biological effects include a shock-like syndrome that can be lethal within 15 minutes after injection, similar to the effects of bacterial endotoxin.

Parasites

Parasites gain nutrition in some cases by living inside host cells (*Toxoplasma gondii* and *Trypanosoma cruzi*), living free in the blood (*Trypanosoma brucei*), or in the lumen of the intestine (parasitic worms) or the intestinal wall (*Entomeoba histolytica*), whereas others live deeper in tissues. In the case of the parasitic worms and *E. histolytica*, they cause tissue damage at the site of colonization in the intestine, but pathology is also caused by migrating tissue-dwelling worms, which leave a trail of damage and inflammation. *Wucheria bancroftii* and *Brugia malay,* which are thread-like filarial nematodes, can also affect lymph circulation. These organisms inhabit lymphatic vessels, and chronic infections can lead to blockage of lymph ducts, resulting in lymphedema (swelling of the lymphatic system) and, in extreme cases, elephantiasis. Other filarial nematodes live in the skin, subcutaneous tissues, and ocular tissue, forming tubercles and erupting from these at intervals. Protozoan parasites that live inside host cells cause unique pathologies. *Trypanosoma cruzi,* the causative agent of Chagas' disease, which is endemic in much of South America, lives free in the host cell cytoplasm, where it replicates until a critical number of parasites is reached. At that time, the parasites differentiate to infective forms and lyse the host cell, so that they are free to invade additional cells and continue to propagate themselves, but only after they have migrated into deeper tissues and/or circulated in the blood, eventually infecting heart muscle. In contrast, *Plasmodium falciparum,* the causative agent of cerebral malaria, infects red blood cells. The infected red blood cells develop adhesive knobs that cause the cells to aggregate, clogging the capillaries of the brain. As a result, blood flow is occluded and death can ensue.

Pathology due to infections by protozoa can also be the result of parasite-derived toxigenic-like properties. The protozoan pathogens that cause sleeping sickness, *Trypanosoma brucei rhodesiense* and *T. b. gambiense,* have coats composed of a glycoprotein that completely sheathes the parasites; it is known as the variable surface glycoprotein (VSG), and is held on the surface of the parasite by a glycolipid structure called the glycosylphosphatidylinositol (GPI) anchor. However, it can be cleaved from the surface by a parasite enzyme that cuts this anchor. Released VSG retains the glycosyl group of the GPI anchor and has endotoxin-like properties in that it interacts with phagocytes and causes release of TNF-α. This ultimately causes the host to undergo weight loss or wasting. *E. histolytica* transfers a galactose-specific lectin to the enterocytes in the gut wall, followed by a parasite molecule that is a polypeptide of 77 amino acids. The latter component is similar to eukaryotic saponins and results in cell death.

Avoiding Adaptive Immunity by Antigenic Variation

Some infectious agents that cause disease periodically vary their structural component against which the host mounts a protective immune response. By doing so, they avoid the host immune response or "evade" it—a process known as **immune evasion**—and thus sustain an infection. The ability of a pathogen to do this is a virulence mechanism and is a characteristic of a wide variety of intracellular and extracellular pathogens, including viruses, bacteria, and protozoa. This type of immune evasion is known as **antigenic variation**, and results in evasion of the immune response mediated by antibodies and T lymphocytes. Another form of immune evasion to avoid phagocyte killing was described in the previous section.

There are numerous examples of antigenic variation of bacterial surface structures used for adherence. Host antibodies are often directed at fimbriae of invasive pathogens. (We previously discussed the role of fimbriae in adherence to host mucosa.) Those pathogens that display **antigenic variation** have more than one gene governing production of fimbriae, and each such gene codes for fimbriae of different amino acid sequences. Recombination between these coding genes can change part or all of a gene directing the synthesis of fimbriae. As a result, fimbriae are generated that are not affected by antibodies directed toward the original fimbriae. A virulent bacterium with altered fimbriae would therefore be able to avoid host defenses at least temporarily, until the host mounted a new immune response against the "altered fimbriae." Similarly, *Neisseria gonorrhoeae,* the causative agent of gonorrhea, a leading sexually transmitted disease, is a master of antigenic variation. It evades immune elimination through antigenic variation of its surface antigens, especially its pili, which have a repertoire of up to a million variants. Variation is generated at the DNA level and reflects recombination between pili genes as well as phase shifts brought about by adding or cleaving repeated sequences within pili genes. *Borrelia recurrentis,* the causative agent of relapsing fever, also has multiple genes for cell surface proteins. One mechanism that antibodies use to protect the host is by promoting uptake by phagocytes (known as **opsonization**; see Chapter 27); but this is avoided in the varying *Borrelia* organisms. Thus, in this case, antigenic variation enables *B. recurrentis* to avoid phagocytosis; thus it can survive in infected individuals for long periods. This results in periodic relapses of disease.

Related to bacterial antigenic variation is another mechanism known as **phase variation** (see Figure 15.8). For example, some bacteria have chromosomal genes for fimbriae synthesis whose expression is repeatedly switched on and off. This variable expression is important for pathogen survival if the fimbriae are a prime target of the immune response.

Immune evasion is also an important component of viral infections. Influenza virus uses two important molecules, hemagglutinin and neuraminidase, to attach to and enter host cells; these are referred to in shorthand fashion as "H" and "N," respectively. Strains of the influenza virus are described by the form of their H and N molecules. For example, the avian influenza virus that is currently spreading worldwide is referred to as "H1/N5." These H and N molecules are remarkable because of their ability to continually mutate or change, making the influenza virus seem continually "new" to antibodies and T lymphocytes (see Chapters 27 and 29). Because of this, the immune system is unable to block the binding and entry of the "new" influenza virus strain. In contrast to influenza, HIV antigenic variation results from mutations in the HIV genome during replication. Enzymes involved in replicating the genetic materials of HIV lack proofreading ability. This includes the reverse transcriptase of HIV and host RNA polymerase II. As a result, the variability among HIV strains is so large that each viral sample taken from a patient is given a unique isolate name, as well as being tagged with the group and subtype it resembles.

Variations of the surface antigens of protozoan parasites are exemplified by *Trypanosoma brucei,* the causative agent of human sleeping sickness and *Plasmodium falciparum,* a protozoan that causes malaria. *T. brucei* variation has been extensively studied and is one of the best-known examples of pathogen antigenic variation. The tsetse-transmitted protozoan lives in the host's blood, lymph, and tissue fluids. Each parasite is covered in a sheath composed of about 10 million copies of a single glycoprotein, called the variable surface glycoprotein (VSG). Each of the VSG molecules is embedded in the outer layer of the parasite's plasma membrane, with the packing so dense that other membrane molecules are

protected from binding by antibodies. Trypanosomes contain 1,000 different VSG genes and additional variations of the VSV type is created by VSG gene mutations and recombinations. With a few special exceptions, each trypanosome expresses only a single VSG on its surface at any given time, which is encoded by a single VSG gene. However the switching occurs just at the point when the host is beginning to make a protective antibody response to the currently expressed VSG type. For *Plamodium,* antigenic variation generates new variants within an infected individual by switching among a diverse repertoire of *var* genes, which encode a membrane protein on infected red blood cells. As a result of genetic recombination, *P. falciparum* also undergoes antigenic diversification during its sexual cycle in infected mosquitoes. Consequently it has been very difficult to produce protective vaccines against either of these important pathogens of humans in Africa, despite a large and sustained international effort.

Immunopathology

Although it is necessary to understand the immune system (covered in Chapter 27) to appreciate how an immune response to a pathogen or its products can actually contribute to the disease, it is worth noting here that this does occur. This is known as **immunopathology**; in some cases it can result in death of the host, despite the fact that we generally think of immune responses protecting the host from infection. A few examples are presented here.

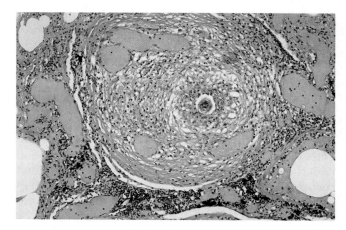

Figure 26.11 Granuloma formation
Macrophages and T lymphocytes surround eggs of *Schistosoma mansoni* to wall them off into a granuloma. This protects the surrounding host tissue. ©Dr. Frederick Skvara/Visuals Unlimited.

GRANULOMAS   *Schistosoma mansoni* eggs are released by adult parasitic worms that live in the blood system. The eggs accumulate in the liver and lungs of the host. There they provoke a response by macrophages and T lymphocytes. The responding cells surround the eggs and seal them off from the host, much like the formation of a pearl around a grain of sand. This is known as a **granuloma** (Figure 26.11). It is important because it prevents the spread of toxins released from the egg. However, it also has deleterious effects on the host, leading to fibrosis in the liver and an increase in blood pressure.

Infection of humans with *Mycobacterium tuberculosis,* the acid-fast bacterium that causes tuberculosis, results in granuloma formation in the lungs. The process is the same as for the eggs of *Schistosoma* and results in loss of function in that part of the lung. To diagnose tuberculosis, a chest x-ray is often used, since the granulomas are apparent. In some cases, the lung tissue containing the granulomas is removed in order to lessen the pathology associated with the infection.

TOXINS   The host response to bacterial endotoxin (lipopolysaccharide) is another example of immunopathology. This can be a lethal response. Although the endotoxin is a pyrogen, increasing the body temperature, it does this through its interaction with phagocytes. Phagocytes release soluble molecules known as cytokines, including TNF-α and IL-1. These are **endogenous pyrogens** that travel to the hypothalamus and interfere with the thermostat controlling body temperature. As a result of the cytokines and host inflammation, the endotoxin shock response described above ensues.

Similarly, the *Staphylococcus* exotoxin TSST-1 has its effect by stimulating host T lymphocytes. It acts as a **superantigen**, meaning that it stimulates large numbers of T lymphocytes to secrete cytokines by binding directly to their specific receptors for antigen. The effect can also be severe and even fatal. This is covered in more detail in Chapter 27.

IMMUNE COMPLEXES   Another example of immunopathology results from infections by viruses that do not become dormant or latent and cause chronic infections. As a result, the host continues to produce an immune response to the virus. In this example, antibodies are part of the immune response. These soluble molecules combine with the virus but are unsuccessful in clearing the infection. Thus, large complexes of the host antibodies and virus particles are produced. When the kidneys try to clear this "complex" from the blood, they become clogged. As a result, kidney function is compromised. Immune responses characterized by antibodies to hepatitis C virus cause this type of pathology, even though the virus itself is in the liver.

> **SECTION HIGHLIGHTS**
>
> The mere presence of a microorganism or parasite is rarely the cause of disease (pathology). Rather, multiple aspects of the infection contribute to the pathology. *Virulence* refers to the relative ability to cause pathology. Virulence factors include not only pathogenic products that facilitate colonization of the host but also toxins produced by the pathogen and the pathogen's ability to avoid the host immune response. The host immune response to the pathogen can also contribute to pathology.

26.6 Life in Host Cells

Some pathogens have been selected through evolution to live within host cells. Some survive and replicate in macrophages, whereas others do so in nonprofessional phagocytes and still others do so in both. Pathogens use a variety of mechanisms to enter and survive in these intracellular homes, as is described here. Some cannot replicate anywhere but inside a host cell and are called **obligate intracellular pathogens**, because they have an "obligation" to be inside (Table 26.8). The rest are "**facultative**" intracellular pathogens. Whether the infection is acute or chronic depends upon the organism. Residing in host cells helps avoid host immune responses and elimination, as does becoming dormant (nonreplicative) within the host cell. This is particularly relevant to viruses, but also some bacteria.

How Viruses Live in Host Cells

As stated several times in this chapter, viruses can replicate only inside host cells. Therefore, by definition, all are obligate intracellular pathogens. Since this is a large group which infects a diverse set of cells, only general principles are described in this chapter. Viruses are more fully discussed in Chapter 14. Viruses bind to and enter host cells by phagocytosis or injection, as described above. Once they enter host cells and shed their coats (the capsids), their genetic material—which can be either single- or double-stranded RNA or DNA, depending upon the virus—is available for duplication and as a template to produce the viral proteins and glycoproteins. In some cases the viral genetic material can integrate into the host's DNA, resulting in either transformation of the host cell into a cancerous cell or in a silent infection. To generate new viral particles, the genetic material is enclosed in a capsid, which either buds through

TABLE 26.8	Obligate intracellular pathogens	
Type of Pathogen	**Organism's Name**	**Disease**
Bacteria	*Chlamydia*	Infertility, trachoma, community pneumonia
	Mycobacterium leprae	Leprosy
	Rickettsia rickettsii	Rocky Mountain spotted fever
Protozoa	*Toxoplasma gondii*	Toxoplasmosis
	Trypanosoma cruzi	Chagas' disease
	Leishmania donovani	Visceral leishmaniasis
	Leishmania tropica	Cutaneous leishmaniasis
	Plasmodium spp.	Malaria
Viruses	All	Varied

the host cell membrane or is released when the host cell is lysed. Inside cells, viruses can avoid detection by the host immune system's antibodies and also its T lymphocytes by decreasing the display of viral molecules on the surface of infected cells.

How Bacteria and Protozoa Live in Phagocytes

Despite the hostile conditions within phagocytes, there are pathogens that survive and replicate inside macrophages. Because macrophages are long-lived cells, they can play a significant role in maintaining chronic infections by becoming a "home" for pathogenic organisms. In contrast, neutrophils are short-lived; thus, even if a microbe can resist being killed, the neutrophil cannot provide a long-term habitat. Pathogens that live in host cells are known collectively as **intracellular pathogens** or **intracellular microbes**.

Microbes use several mechanisms to survive in macrophages (Figure 26.12):

- Prevent fusion of the lysosme with the phagosome
- Survive in the phagolysosome
- Escape from the phagosome into the cytoplasm of the host cell

There are bacteria, protozoa, and fungi that take advantage of these possibilities, which are summarized in Table 26.9.

Brucella abortus and *B. suis,* the causative agents of undulant fever in humans, prevent fusion of the lysosome with the phagosome following phagocytosis. They survive in the human body for extended periods—often lifelong. In contrast, the vaccine strains of *B. abortus* are less virulent; in keeping with this, they are much less able to prevent phagolysosomal fusion. The ability of *Brucella* to prevent fusion of the phagosome and lysosome is linked to the type IV secretion system of the bacterium. Other bacteria that grow in macrophages in phagosomes include *Legionella pneumophila* (which causes Legionnaire's disease) and *Mycobacterium tuberculosis* (which causes tuberculosis). Protozoa, such as *Toxoplasma gondii* (which causes toxoplasmosis), can also use this mechanism.

Other protozoa, fungi, and bacteria survive in the fused phagolysosome—a situation presenting even harsher conditions. They include *Salmonella typhi,* which causes typhoid fever, and *Leishmania* spp., which cause leishmaniasis. Once the organisms are in the intracellular compartments, considerable "remodeling" may take place to make the conditions conducive to replication. This may include raising the pH to be more neutral.

A diverse group of pathogens escape from the host cell vacuole into the cytoplasm. The rickettsias, which cause diseases such as Rocky Mountain spotted fever and typhus; the bacteria *Listeria monocytogenes* and *Shigella,* which produce food-borne infections, and the protozoan *Trypanosoma cruzi* (which causes South American trypanosomiasis) escape from the phagosome and grow in the cytoplasm (see Table 26.9). All of these organisms use pore-forming molecules to escape. Although the pore-forming enzymes from the various organisms have similar functions, the enzymes are not related among the species. One example is *L. monocytogenes'* pore-forming molecule—a hemolysin called listeriolysin-O. Thus these enzymes are also virulence factors that promote invasion; if they are absent from the organism, it fails to parasitize the host cell. After escaping into the cytoplasm of the host cell, *Listeria, Shigella,* and *Rickettsia* polymerize actin so that they acquire "tails," which propel them through the cytoplasmic membrane of the cell they are in and into the neighboring cell (Figure 26.13). They do not disrupt either cell in this process but use enzymes to escape from the host membranes in the newly infected cell, continuing to replicate there, free in the cytoplasm. Vaccinia virus also uses this mechanism to propel itself, although not in macrophages.

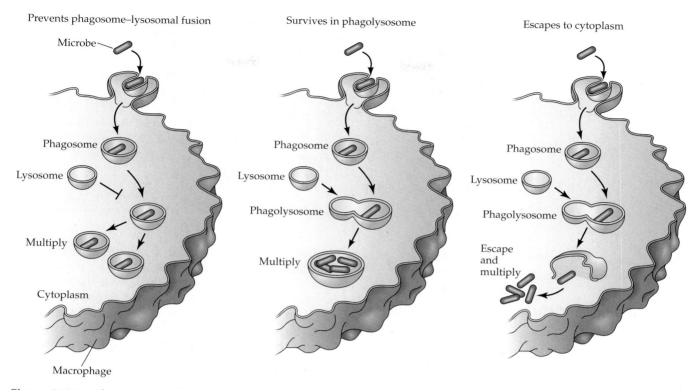

Figure 26.12 Life in host cells
Some bacteria, protozoa, and fungi survive and replicate inside host cells, using different mechanisms to achieve this, as shown here. See Table 26.9 for specific pathogens that use these mechanisms.

How Bacteria and Protozoa Live in Nonphagocytic Host Cells

Some intracellular bacteria and protozoa have the ability to invade cells that are not professional phagocytes. Most can also survive in macrophages; some of them are listed in Table 26.10. As you can see from the table, they parasitize a number of different types of host cells. Inside host cells, they may live in phagosome-like vacuoles; but these intracellular compartments are outside of the normal endosomal trafficking routes. This means that the vacuoles do not fuse with lysosomes, making the intracellular compartment more hospitable. The organisms that live in this type of compartment include

(A)

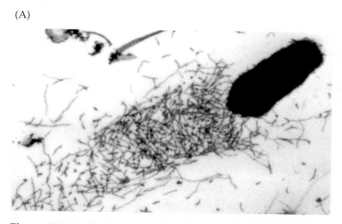

(B)

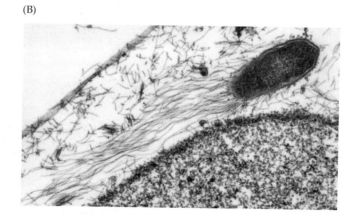

Figure 26.13 Actin propulsion of bacteria inside host cells
Once bacteria are established in the cytoplasm of the host cell, they form actin tails. These tails propel the bacteria through the membrane of the host cell and the neighboring cell, allowing them to begin to establish residency.
(A) *Listeria*, (B) *Rickettsia*. From Gouin, E., et al. 1999. *Journal of Cell Science 112*: 1697–1708. Courtesy of Pascale Cossart.

TABLE 26.9	Survival mechanisms of various pathogens in mononuclear phagocytes		
Pathogen Group	**Prevent Phagolysosomal Fusion**	**Survive in Phagolysosome**	**Escape to Cytoplasm**
Bacteria	*Brucella abortus* *Legionella pneumophila* *Mycobacterium tuberculosis*	*Salmonella typhi* *Coxiella burnetii*	*Listeria monocytogenes* *Shigella* *Rickettsia rickettsii*
Protozoa	*Cryptosporidium* *Toxoplasma gondii*	*Leishmania* spp.	*Trypanosoma cruzi*
Fungi		*Histoplasma capsulatum*	

the bacteria *Chlamydia* and *Legionella* and the protozoa *Toxoplasma*, *Plasmodium*, and *Cryptosporidium*.

Entry of Non-Viral Pathogens into Host Cells

Pathogens enter macrophages and nonprofessional phagocytes by one of three mechanisms, which are shown in Figure 26.14. The most common is zippering phagocytosis, which requires the expenditure of energy by the host cell but not by the pathogen and is used by bacteria, fungi, and protozoa. An example is *Listeria monocytogenes*, an enteroinvasive bacterium; it produces a protein called internalin, which binds E-cadherin on the host cell and is then "zippered" into the host cell.

The second method is referred to as **induced phagocytosis** but also has several alternative names (see Figure 26.14A). Examples are found for several bacteria, including *Shigella* and *Salmonella*. The pathogen actually uses its type III secretion system to "inject" the host cell,

resulting in ruffling of the host cell membrane around the bacteria (see Figure 26.14B,C). These pathogens are able to survive and replicate once inside the host cell.

Induced phagocytosis is also used by bacteria that are not intracellular microbes but that use this as a mechanism to pass from the gut lumen into the underlying submucosa by way of **M cells** (Figure 26.15), thus breaking barrier immunity. An example is the enteropathogenic bacterium *Yersinia*, one species of which is the causative agent of plague. *Yersinia* does this by producing a surface protein called invasin, which binds to molecules on the host cells; the latter are part of the α-integrin family of adhesion molecules. The invasion protein not only mediates adhesion of the bacterium to the host cell but also promotes internalization. This has been shown by coating inert beads with the invasion protein. They are then taken up by epithelial cells, which are normally not phagocytic. Thus, in the case of the M-cell invasion, it is only a transient intracellular life to access greener pastures beyond,

TABLE 26.10	Organisms living in nonprofessional phagocytes	
Organism	**Host Cell**	**Also in Macrophages**
Bacteria		
Shigella flexneri	Enterocytes	+
Brucella abortus	Trophoblasts	+
Chlamydia	Epithelial, endothelial, fibroblasts	+
Bartonella quintana	Endothelial	−
Francisella tularensis	Hepatocytes	+
Listeria monocytogenes	Hepatocytes	+
Salmonella	Epithelial	+
Protozoa		
Plasmodium spp.	Hepatocytes and RBCs	−
Trypanosoma cruzi	Heart muscle	+
Toxoplasma gondii	Broad range	+

(A)

Zippering phagocytosis	Induced phagocytosis	Active invasion

Pathogen

Host cell

- Most common entry mechanism
- Tight phagosome formed by zippering around organism
- *Leishmania* spp.
- *Mycobacterium* spp.
- *Histoplasma capsulatum*

- Also called "facilitated endocytosis," "macropinocytosis," and "trigger mechanism"
- Ruffling of host membrane so pathogen appears to SPLASH onto cell.
- Signal transduction in host cell resulting from pathogen
- *Shigella*
- *Salmonella*

- Pathogen "sinks" into host cell
- Used for entry into nonprofessional phagocytes
- No actin polymerization or host cell contraction
- No host cell signaling involved
- *Toxoplasma gondii*
- *Plasmodium* spp.

Figure 26.14 Mechanisms of entry into host cells
(A) Diagram of the three mechanisms used to enter cells. Cells colored red expend energy during the process; cells colored blue do not. (B) For induced phagocytosis, the bacterium attaches to its type III secretion system to the host cell, causing it to put up ruffles in an unguided manner. (C) These ruffles engulf the bacteria on the cell's surface, as is shown in this scanning electron micrograph. C, from P. Cossart and P. J. Sansonetti. 2004. *Science* 304: 242–248. Courtesy of Philippe Sansonetti.

(B)

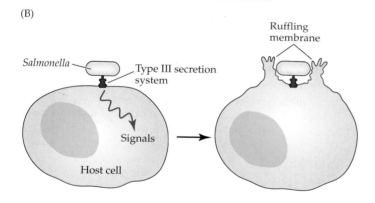

(C)

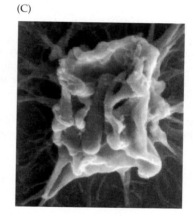

so to speak. This provides another example of how microbes overcome barrier defense, in this case of the intestine. *Shigella* and *Salmonella* also use induced phagocytosis to pass through M cells, but in their cases they can also use it to enter gut enterocytes or macrophages, where they survive and replicate intracellularly.

The third method of entering host cells is by **active invasion**. This is used by protozoan parasites and requires energy to be expended only by the pathogen. The pathogen appears to "sink" into the host cell without perturbing the host cell's actin cytoskeleton or inducing signals in the host cell.

Avoiding Adaptive Immune Responses inside Cells

When they are living in macrophages or other host cells, pathogens are protected from antibodies, which are prod-

ucts of the highest level of host defense called **adaptive immunity**. As indicated above, some intracellular microbes—such as *Listeria*, *Shigella*, and vaccinia virus particles—can transit from cell to cell without being exposed to the extracellular environment. This makes them even more efficient at avoiding antibodies. However, the infected cells may still be detected by T lymphocytes, and in some cases the infected host cells are killed by these T lymphocytes, as discussed in Chapter 27. T cells can also make the intracellular habitat less desirable by producing soluble molecules called **cytokines**, which interact with mononuclear phagocytes. This causes the phagocyte to become more effective at preventing or reducing the intracellular survival and replication of the pathogen; such a macrophage is referred to as an **activated** macrophage. This type of immune response by the T lymphocytes and mononuclear phagocytes is called **cellular immunity**.

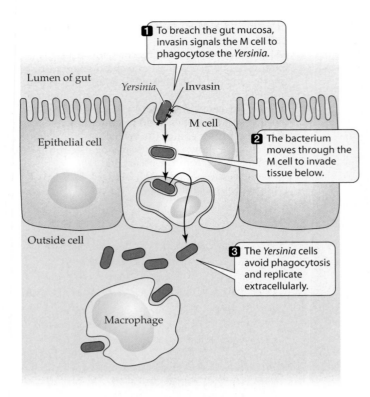

Figure 26.15   Passage through M cells in the gut
Nonintracellular bacteria pass through M cells lining the gut to reach the submucosa. They do not colonize the M cells but do replicate extracellularly after passing through. The example shown is for *Yersinia*, which also has the ability to avoid phagocytosis. It uses a bacterial molecule called "invasin" to adhere to the M cell and cause phagocytosis.

1 To breach the gut mucosa, invasin signals the M cell to phagocytose the *Yersinia*.

2 The bacterium moves through the M cell to invade tissue below.

3 The *Yersinia* cells avoid phagocytosis and replicate extracellularly.

Lumen of gut

Yersinia   Invasin

M cell

Epithelial cell

Outside cell

Macrophage

Latent Infections of Host Cells

A number of viruses that give rise to persistent infections in the host undergo mutations to generate nonreplicative forms, which have the capacity to interfere with the replication of wild-type viruses. These **defective interfering** particles have been demonstrated in influenza infections as well as in infections caused by Sendai virus, measles virus, and rubella virus. The defective interfering particles help to maintain a low level of virus in cells jointly infected with wild-type virus.

Another example is herpes viruses. These give rise to persistent infections, although at times no infectious virus particle is generated because the viral replicative cycle is completely arrested. This is called **latency** and is poorly understood. Latency can be short- or long-lived; the conditions that induce and reverse it are under investigation but are likely to include suppression of the host immune system. In the case of herpes simplex virus 1, which causes cold sores, latency develops in infected neurons and sensory ganglia. When viral transcription is reactivated, the virus spreads to the lip

epithelium, where it continues to replicate, causing a "cold sore"; the virus is then shed from these cells as infective viral particles. Similarly the herpes virus that causes chicken pox in young people can become latent and then recur much later in life in the same individual, resulting in a disease called shingles.

There is also evidence that chronic infections caused by bacteria have periods of dormancy during which they do not cause clinical symptoms. Examples include tuberculosis and brucellosis, both of which are caused by intracellular pathogens. It is presumed that these may be replicating very slowly or not at all inside host cells, sometimes within granulomas. If the host becomes immunosuppressed, including just due to aging, clinical episodes again become apparent.

SECTION HIGHLIGHTS

Although all viruses replicate in host cells, some bacteria, fungi, and protozoa also do so. Some survive and replicate in macrophages, while others do so in nonprofessional phagocytes. Microorganisms use a variety of mechanisms to enter and survive in these intracellular homes. Some become dormant (nonreplicative) in host cells. Since living inside host cells provides a mechanism to avoid host immune responses, some of these pathogens cause chronic infections.

SUMMARY

- The **normal** or **resident microbiota** is a distinct community of bacteria and a few fungi to which every human plays host.

- Most members of the normal microbiota have an alliance with the human host that is either **mutualistic** or **commensalistic**, meaning the organisms contained within the microbiota do not harm the host.

- Other organisms—including bacteria, viruses, fungi, protozoa, and parasitic worms—can cause an **infection** of the human resulting in **pathology** (**disease**). An infectious agent that causes measurable injury to the host is a **pathogen**. An infection occurs when the pathogen colonizes (grows or replicates) on or in host tissues or organs.

- **Virulence** measures the ability of a pathogen to inflict damage. Pathogens often express or use molecules and mechanisms that are not associated with nonpathogenic strains or species. These are called **virulence factors**.

- An infection can be **acute**, meaning either that it is cleared or the host dies in a limited period of time; an infection can also be **chronic**, meaning that it lasts a long time, sometimes throughout the life of the host.

- **Pathogens** can be divided into three general categories: (1) those that live only in a host and infect it, or **obligate** pathogens; (2) those that can cause infection if given access by trauma, or **accidental** pathogens; and (3) those that infect debilitated individuals, or **opportunistic** pathogens.

- The **barrier defense** is the first layer of host protection against infection. It includes physical, mechanical, microbial, and chemical components.

- The first components of barriers defense encountered are the **skin** and **mucous membranes**, which cover the surfaces of the body and are exposed to the outside environment. This is the **physical barrier defense**.

- Coughing, sneezing, and peristalsis are examples of **mechanical barrier defense**.

- **Acidity** in the stomach and vagina are examples of **chemical barrier defense**.

- Colonization of host skin and mucous membranes by nonpathogens can **preempt** potential invaders; this is the **microbial barrier defense**.

- The **skin** has physical, chemical, and microbial barriers. The skin cannot be penetrated by pathogens except for a few parasitic worms. The skin has a resident bacterial population comprising mainly **gram-positive bacteria**. The concentration on drier areas is 10^2 to 10^4 per square centimeter and about 10^6 per square centimeter in moist areas. Natural skin secretions include amino acids, fatty acids, urea, lactic acids, lipids, and mineral salts, providing nutrients for the normal microbiota and some level of chemical defense to prevent overcolonization.

- **Saliva** contains about 10^8 bacteria per milliliter; these are the normal or resident microbiota. The bacteria **adhere** to the teeth, tongue, and other surfaces and are not dislodged by swallowing. The **nasopharynx** has adequate moisture and nutrients for the growth of bacteria. The **lower respiratory tract** is mostly devoid of bacteria.

- Dust and other particles that enter this region are moved upward to the throat by action of **cilia** and swallowed. This effect is the **mucociliary escalator**, which is part of the mechanical defense system.

- Organisms in the **stomach** are usually limited to some acid-tolerant lactobacilli and yeasts. There are fewer than **10 microbes** per milliliter of stomach fluid. Colonization of the stomach by *H. pylori* can cause peptic ulcers in some individuals.

- The **small intestine** has an increasing number of bacteria as it proceeds downward. The **large intestine** is a fermentation vat, with 10^{10} to 10^{11} bacterial cells per gram of mass and high **species diversity**.

- The upper **genitourinary** tract of a healthy human is relatively free of bacteria, but the lower part of the urethra may harbor numerous species of bacteria and even fungi.

- To colonize a host, **adherence** of the pathogen to a specific tissue is needed. Pathogens have specialized molecules on their surfaces for this purpose. Bacteria have **pili** and **fimbriae**, while some protozoa can use **suckers**, parasitic worms can use **hooks**, and viruses use specialized molecule such as **hemagglutinin** in their capsids.

- **Pathology** is sometimes caused by **toxic substances** produced by the pathogen. In some cases, the toxin alone is enough to cause the disease even without the presence of the pathogen.

- The **toxins** produced by bacteria include **endotoxins**, which are part of the bacterial cell, and **exotoxins**, which are released into the surrounding medium and can work at a distance from the bacterial colonization site. Viruses also produce molecules referred to as toxins, which cause host cell death, while some protozoa and fungi have molecules with activities similar to those of bacterial endotoxins.

- Pathology is also a result of substantial **cell death** in the host (caused by most pathogens) and/or obstruction of organs or tissue functions (for example, by parasitic worms).

- Some pathology results from the host immune responses to the infectious agent.

- Virulence factors include mechanisms and molecules involved in gaining nutrients and iron.

- The **inflammatory response** is an important nonspecific defense against entry of pathogens. This response brings **phagocytes** and other protective cells to the site of injury. Exposure of DNA from dead cells initiates the response.

- **Phagocytes** are host defense cells that **engulf and destroy invaders**. The invader is brought into the phagocyte in a membrane-bound compartment called a **phagosome**. Phagocytic cells have **lysosomes**, compartments containing an array of

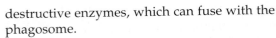

destructive enzymes, which can fuse with the phagosome.

- **Neutrophils** and **macrophages** are important phagocytic cells. Macrophages can circulate in the blood as **monocytes** or be permanently associated with host tissues.

- **Intracellular pathogens** include bacteria, protozoa, and fungi that survive and replicate inside host cells. All **viruses** are intracellular pathogens, since they need the host cell machinery to replicate themselves. Some bacterial, protozoan, and fungal pathogens can survive and replicate in **macrophages**, while others do so in **nonprofessional phagocytes**. They can protect the pathogen from the immune system.

- Intracellular pathogens enter host cells by one of three mechanisms: **zippering** phagocytosis, the most common; **induced phagocytosis**; and **invasion**. In induced phagocytosis, the pathogen forces the host cell to undergo phagocytosis by injecting pathogen-derived molecules into the host cell membrane. In the invasion method, the host cell is neutral and the pathogen forces its way in.

- Some intracellular pathogens can replicate only in host cells and are called **obligate intracellular pathogens**.

- Some intracellular pathogens move from cell to cell by making **actin tails**, which propel them through the neighboring cell's membrane.

- Some pathogens cause chronic infections or reinfections of the host by undergoing **antigenic variation**, which means that the host's adaptive immune components no longer recognize the pathogen.

 Find more at www.sinauer.com/microbial-life

REVIEW QUESTIONS

1. What is virulence?
2. Define: host, commensalism, infection, and normal microbiota.
3. What are some factors on human skin that promote the growth of bacteria? Some that inhibit or limit bacterial colonization?
4. Why is brushing one's teeth important in preventing dental caries? How do microorganisms contribute to caries?
5. How do the bacterial populations of the stomach and large intestine differ qualitatively and quantitatively? What role do the facultative aerobes play in the human gastrointestinal tract?
6. Why do germ-free mice have poorly developed gut-associated lymphoid tissue?
7. What are probiotics and prebiotics, and how might they assist in the maintenance of health?
8. What is the source of the acidic pH in the genitourinary tract of females? How does the microbiota in this area prevent invasion by pathogens?
9. Is inflammation a positive or a negative in defending against infection? Give reasons for your answer.
10. What are phagocytes? How do neutrophils differ from macrophages? What does a high neutrophil count in human blood signify?
11. How does an obligate pathogen differ from an accidental pathogen? What is an opportunist? Cite examples.
12. Adherence is important in infection. Why? Cite diseases where adherence is not essential.
13. What are some of the virulence factors, and how do they function in promoting the survival of pathogens? What are some antiphagocytic factors?
14. How do endotoxins and exotoxins differ chemically as well as in toxicity, heat stability, and origin?
15. How do endotoxins and exotoxins differ in their mode of action? What are the reasons for this? How would you determine whether an intravenous fluid might contain endotoxin?
16. What are the three ways by which microbes can survive and replicate inside host cells?
17. What are the three ways by which microbes can enter host cells?
18. How do some microbes use actin to avoid the host immune system?
19. Why does antigenic variation make it more difficult for the host to control an infection?
20. Do bacteria, viruses, and parasitic worms all adhere to host tissues by the same mechanism? If not, discuss differences.

SUGGESTED READING

Cossart, P., P. Boquet, S. Normark and R. R. Appuoli. 2000. *Cellular Microbiology*. Washington, DC: ASM Press.

Cox, F. E. G., D. Wakelin, S. H. Gillespie and D. D. Despommier, eds. 2005. *Topley & Wilson's Microbiology and Microbial Infections: Parasitology*. 10th ed. London: Hodder Arnold.

DeFranco, A. L., R. M. Locksley and M. Robertson. 2007. *Immunity: The Immune Response in Infectious and Inflammatory Disease*. Sunderland, MA: Sinauer Associates, Inc.

Galan, J. and P. Cossart, eds. 2005. *Current Opinion in Microbiology*, Vol. 8, Issue 1, *Host–Microbe Interactions: Bacteria*. Oxford, UK: Elsevier.

Groisman, E. A. 2000. *Principles of Bacterial Pathogenesis*. San Diego, CA: Academic Press.

Kreier, J. P. 2002. *Infection, Resistance and Immunity*. 2nd ed. New York: Taylor & Francis.

Minion, F. C. and M. J. Wannemuehler. 2000. *Virulence Mechanisms of Bacterial Pathogens*. 3rd ed. Washington, DC: ASM Press.

Ryan, K. J. and C. G. Ray. 2003. *Sherris Medical Microbiology: An Introduction to Infectious Diseases*. 4th ed. New York: McGraw-Hill.

Salyers, A. A. and D. D. Whitt. 2002. *Bacterial Pathogenesis*. 2nd ed. Washington, DC: ASM Press.

Immunology and Medical Microbiology

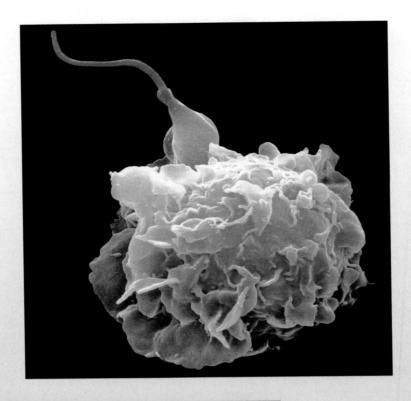

The objectives of this chapter are to:

◆ Present a comprehensive overview of the physiological processes that protect mammals from pathogens that have entered the body.

◆ Describe the cells and molecules that mediate immune responses, their origin from hematopoietic stem cells, and their distribution in the body.

◆ Explain how inflammation and innate immunity limit the establishment of infections.

◆ Explain how a class of white blood cells, called lymphocytes, discriminates among invaders using highly specific recognition structures (receptors), and how specific clones of lymphocytes are expanded to mount adaptive immune responses.

◆ Explain how adaptive immune responses generate immunological memory and can be pre-programmed by vaccination.

◆ Describe additional processes, called bridging immune responses, that can be rapidly deployed like innate immune responses and have some of the functional features of adaptive immune responses, but in most cases lack immunological memory.

27

Immunology

The immune system is complex; if it was simple it would be practically useless."
—Anonymous professor

The living body of mammals is warm, moist, and full of nutrients. Because of this, cells and tissues of mammals provide a habitat for many creatures including viruses, bacteria, fungi, protozoa, and parasitic worms. Consequently, the survival of mammals depends on the successful defense of their bodies against invasion and colonization. Healthy mammals resist invasion using an array of defense mechanisms that are encompassed within the discipline of immunology. The essential nature of immunity is exclusion of disease-causing agents (pathogens) from host tissues by barrier defenses (discussed in Chapter 26), or the control and elimination of any organisms that breach barrier immunity and enter the body. The latter processes are discussed in this chapter.

The immune defenses comprise:

- *Barrier Immunity, which prevents infectious agents from getting into the body.* This aspect of immune defense is mediated by the skin and mucosa, including their biota and secretions, and is outlined in Chapter 26.

- *Innate immunity, which kills or limits the replication of microorganisms that have breached barrier immunity.* The innate immune response is mediated by a variety of cells and molecules that are present in an inactive state in the blood and tissues of healthy individuals and that become activated in the presence of disease causing organisms (**pathogens**) or at sites of tissue injury. The innate defenses can be rapidly deployed and are relatively nonspecific, hence their activation can result in damage to tissues in which the responses are induced in addition to damage to **pathogens**. The cells that mediate the innate defenses include **polymorphonuclear leukocytes**, **monocytes**, and **macrophages** (see Figure 27.1), all of which are called **phagocytes** because they can bind, take up, and degrade particulate material, including pathogens. The molecules that mediate innate immunity include a family of enzymes, called **complement factors**, which are present in the **blood plasma**, that is, the fluid compartment of blood. In the presence of pathogens, complement factors can assemble into a complex, called the **membrane attack complex** or **MAC**. MAC can make holes in the surface of pathogens, thereby killing them. During initial steps in the MAC pathway, fragments of complement are generated that stick to pathogens, tagging them for uptake by phagocytic cells. Other complement fragments cause the directional migration of phagocytes toward the site where they are being generated. This latter process is called **chemotaxis**, and molecules that induce this response are called **chemotactic factors**.

- *Adaptive immunity, which is highly efficient at killing invaders and neutralizing their products by pathogen-specific mechanisms.* Adaptive immunity is mediated by white blood cells called lymphocytes. These cells belong to two major classes: B cells and T cells. B cells are specialized to secrete antibodies, which are soluble proteins that can neutralize toxins, block surface structures on pathogens that are required for successful invasion and colonization, and interact with complement factors of the innate immune system to more effectively kill invaders, as well as tag them for phagocytosis and destruction. T cells are specialized for cell-contact–dependent reactions including killing of target cells and the short-range delivery of cytokines. Cytokines control the functions of other cells. Certain cytokines increase the pathogen-killing power of phagocytes, others increase the efficiency of B-cell responses, and yet others increase the proliferation and functional specialization of T-cell responses. Adaptive immune responses are initiated by the interaction of B cells and T cells with **antigens**, the generic name of substances that stimulate an immune response. Antigens include proteins, carbohydrates, and lipids. The antigenic makeup of microorganisms is complex and distinct for each organism, hence, the diversity of antigens present among pathogens is enormous. Lymphocytes interact with antigens via antigen-specific receptors, which are clonally expressed. Because each lymphocyte has a particular antigen-specific receptor capable of binding only its complementary anti-

gen, a legion of distinct antigen-specific lymphocytes is required to recognize the variety of pathogens an individual might encounter. During the development of an adaptive immune response, lymphocytes with a receptor that binds an antigen expressed by the pathogen are induced to replicate, eventually reaching a level where they can contribute to killing the pathogen. The development of a critical mass of lymphocytes that express a particular antigen-specific receptor takes time, and during this period pathogens are constrained by the innate immune response. Adaptive immunity is often more efficient on second exposure to the same pathogen, a state described as immunological memory. This is exploited by vaccines.

- *Bridging immunity, which comprises a group of rapidly deployed immune responses that kill or limit the replication of microorganisms by processes that are somewhat more discriminating than those of the innate immune system but less discriminating than those operative in adaptive immunity.* The bridging immune responses have some characteristics of innate and others of adaptive responses. Bridging immune responses are mediated by particular types of lymphocytes and lymphocyte-like cells, including γδ T cells, natural killer cells, marginal zone B cells, B-1 B cells, and their products. Cytokines produced by cells of the bridging immune system can regulate (enhance, suppress, and direct) innate and adaptive immune responses.

Each of these aspects of the immune system has an important role to play in host defense. A person or animal that lacks any one of the defense processes will likely be more susceptible to infection. This is illustrated by: (1) the susceptibility of burn patients to infection as a result of a breach of barrier immunity, (2) development of recurrent infections in people or animals with a defect in the ability of cells of the innate immune system to adhere to blood vessels and enter sites of inflammation and infection, and (3) the inability of people and animals with genetic defects in lymphocyte development to survive for prolonged periods outside of sterile environments.

27.1 Cells and General Organization of the Immune System

The cells of the innate, bridging, and adaptive defense systems derive from **hematopoietic stem cells (HSCs)** that reside in the bone marrow. These stem cells have the capacity for long-term self renewal and the ability to repopulate the body with all hematopoietic lineages (Figure 27.1): red blood cells, platelets, and white blood cells.

Figure 27.1 Origin of blood cells
Blood cells arise from stem cells, and go through successive stages of differentiation before reaching their mature state.

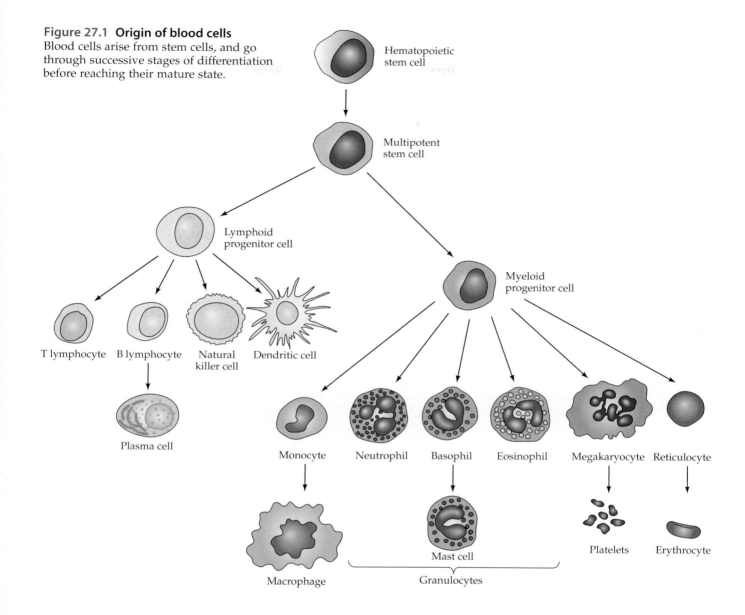

White blood cells include myeloid cells such as granulocytes and monocytes; lymphoid cells of both the T and B lineages; and lymphocyte-like cells such as natural killer cells. The paradox presented by long-term self renewal of HSCs and development of differentiated progeny cells is not satisfactorily resolved. One possibility is that it results from an asymmetric HSC division, in which one daughter cell retains full HSC potential, while the other gives rise to progenitor cells that support development of the different lineages of white blood cells. Another possible explanation is that HSCs undergo symmetric self-renewing divisions until their support niche is filled, after which divisions outside of the niche yield stem cells with lesser capability to sustain self renewal and greater tendency to undergo a differentiation division. Characteri-

zation of HSC niches and the signals that regulate HSC division and differentiation are an active area of research.

The downstream progeny of HSCs are not uniformly distributed. Platelets, red blood cells, monocytes, neutrophils, and other granulocytes are found primarily in the blood and in blood vessels of the spleen and liver. Neutrophils and monocytes that circulate in the blood are rapidly recruited to sites of inflammation, as discussed in subsequent text, where they participate in innate immune responses. T lymphocytes, also called T cells, develop in the thymus from a bone marrow–derived stem cell, whereas B lymphocytes, also called B cells, develop in the bone marrow in both mice and humans. As discussed previously, B cells and T cells are equipped with clonally restricted receptors that enable

Figure 27.2 Distribution of lymphoid organs and organs with phagocytic activity
Antigen circulating in the blood may be trapped by phagocytic cells in the liver, lungs, and spleen. Antigen deposited in the skin may be carried by lymph to draining lymph nodes. Antigens in the gut may be trapped in the gut-associated lymphoid tissues, including the Peyer's patches.

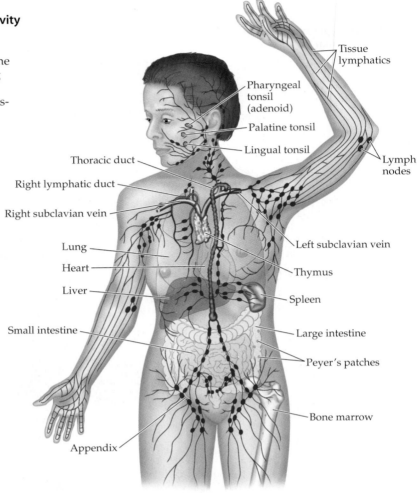

these cells to be stimulated by components of specific pathogens. The development of clonally restricted receptors and regulation of responses by B and T cells is discussed in the following text.

Mature B cells and T cells are present in large numbers in anatomically discrete sites called lymphoid organs (Figure 27.2). These fall into two categories: the **primary lymphoid organs**, namely the thymus and bone marrow, where, respectively, T cells and B cells develop; and the **secondary lymphoid organs**, namely the spleen and lymph nodes, where the lymphocytes express their functions. The secondary lymphoid organs play a major role in defense against disease. In the case of the spleen, pathogens or pieces of these are carried into the organ in the blood **plasma**, that is, the fluid component of blood, as well as in and on cells within the blood. Once in the spleen, these materials stimulate adaptive immune responses. In the case of lymph nodes, the stimulatory materials are transported into a local node via **lymph**, a fluid that is similar to the blood. The lymph

is formed from fluid that leaks from the capillaries into surrounding tissues (Figure 27.3) and is particularly copious at sites of inflammation, where it will also contain materials released from inflammatory cells and damaged pathogens as discussed in subsequent text.

Blind-ended lymphatic capillaries originate in the tissues and have flap cells that open when local fluid pressure builds up, admitting fluid which is now known as lymph as well as antigens and cells into the vessels. Once the pressure has been relieved, the flap cells close. The lymph is pumped through the lymphatic vessels toward **lymph nodes** by the contractile action of muscles surrounding the vessels. Lymph follows a unidirectional circuit that is specified by one-way flow valves (see Figure 27.3). Lymph nodes are designed like filters and are specialized to remove foreign bodies from lymph and to support the development of immune responses against this material. Lymph nodes are found clustered at the base of the skull, where they drain the tissues of the head; in the throat, where they drain the nose, mouth,

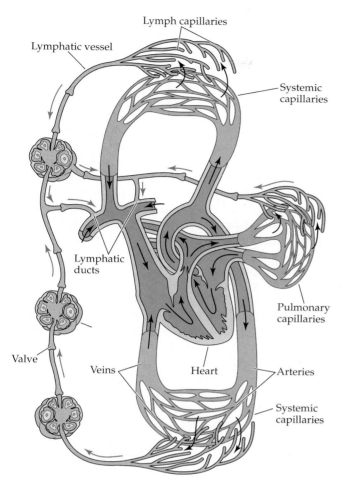

Figure 27.3 Human cardiovascular and lymphatic systems
Fluid that leaks from the blood system into tissues collects in lymphatic vessels and is transported to draining lymph nodes. Lymph drains back into blood at two ducts.

and throat; in the armpit, where they drain the lymphatic vessels of the arm; on the flank, where they drain the body wall; and in the groin and behind the knees, where they drain the legs. Lymph eventually flows back into the blood via two ducts: the right lymphatic duct and the thoracic duct. The right lymphatic duct drains lymph from the right side of the head, neck, thorax, and right upper extremity into the right subclavian vein. Lymph from the rest of the body flows into the thoracic duct that empties into the left subclavian vein. Lymph flow does not occur when one is inactive, which accounts for the swelling (**edema**) that occurs around the feet and ankles when one is immobile for a long period, such as on an airplane or in a hospital bed.

Lymphocytes are brought into the spleen in the blood. Lymphocytes are also carried to the lymph nodes in

blood and can enter lymph nodes by crossing the endothelium of blood vessels called **high endothelial venules**. Lymphocytes also circulate in lymph. The trapping of pathogens and their fragments in the spleen and lymph nodes, together with the movement of lymphocytes into and out of these organs increases the likelihood that a lymphocyte will encounter its target antigen, that is, the antigen that forms a stable interaction with its antigen-specific receptor causing the development of an adaptive immune response. Indeed cell migration through lymphoid organs is required for induction of immune responses.

SECTION HIGHLIGHTS

Many types of functionally distinct white blood cells make up the immune system. These cells can be distinguished one from the other on the basis of size, shape, nuclear structure, molecules they express on their surface, and their immune function. The many different white blood cells all derive from a common precursor cell called the hematopoietic stem cell or HSC. Progeny of the HSCs become committed to different white blood cell lineages and their more differentiated progeny give rise to, and circulate among, the organs that make up the immune system.

27.2 Inflammation and Innate Immunity

Any process that breaches barrier immunity and causes tissue damage (a burn, trauma, surgery, an insect bite, local infection, or the passage of a parasitic worm through tissues) evokes inflammation. Inflammation is a general term that describes the biochemical and cellular responses that result from tissue damage. The inflammatory response functions to:

- Seal lacerations and plug broken blood vessels

- Remove damaged cells from sites of tissue injury and repair damaged tissue

- Restrain the population growth rate of pathogens that may have invaded or have been introduced at the site of tissue damage by killing them or creating conditions that limit their replication

These diverse functions are mediated by a combination of molecules that are present in normal plasma, and to a lesser extent in tissue fluids, and by white blood cells (leukocytes). The white blood cells circulate in the blood and move to the site of inflammation, or reside in the tissues and are activated by inflammatory events. Two important leukocytes involved in the inflammatory response are polymorphonuclear leukocytes, sometimes called **PMNLs** or **neutrophils**, and monocytes (also called **macrophages** when in tissues). These cells are part of the innate immune system. Both cells are specialized to engulf cells and other particles, such as bacteria and viral particles, and subsequently destroy them. The processes involved in phagocytosis and destruction of phagocytosed material are discussed later in the text.

Because the molecules and cells that participate in the inflammatory and innate immune responses are typically present in sufficient concentration or number (albeit in an inactive state) in the blood and tissues, the inflammatory response can be rapidly deployed through a series of interdependent steps following tissue injury.

Early Events in Inflammation

Inflammation of the skin causes redness, heat, swelling, and pain (Figure 27.4), and is detected within minutes after injury of that tissue. Redness and heat result from increased local blood flow caused by an increase in the diameter of blood vessels (**vasodilation**). Swelling results from local edema brought on by the leakage of plasma from the bloodstream into tissues at the site of injury and subsequently by the infiltration of cells into this region. Pain results from tissue destruction by enzymes released by infiltrating phagocytes and from a reduced threshold of activation of nerve endings, which is a consequence of local production of a fatty acid derivative called **prostaglandin E$_2$**. Prostaglandin E$_2$ is made by damaged cells and by infiltrating phagocytes. Pain focuses the attention of the afflicted individual on the site of inflammation. Wounded animals lick and cleanse damaged areas. Humans use additional strategies to cleanse wounds and expedite healing. Pain, therefore, plays a significant role in the control of infection.

Vasodilation results from relaxation of smooth muscle of the blood vessel walls. This is brought on by the interaction of the muscle cells with a number of molecules produced at the site of inflammation. The most important molecules involved in this process are bradykinin, fibrinopeptides, complement-factor C5a, and histamine. Generation of these **vasoactive** agents is initiated by the shearing of cells from their extracellular matrix at the site of tissue damage and the concomitant rupture of local capillary beds, which results in the influx of plasma and cells. Bradykinin is generated by the kinin system, an enzyme cascade that begins with a plasma clotting factor called Hageman factor. Hageman factor is activated upon contact with collagen exposed

at sites of tissue damage. In its active form it is called Factor XIIa. Activation of the blood clotting cascade by Factor XIIa results in the generation of thrombin, which converts plasma fibrinogen to fibrin. This creates a mesh that promotes clot formation and seals breaches in the barrier, allowing repair to begin. The clot is dissolved by plasmin, which is a potent enzyme generated by the fibrinolytic system. This generates fibrinopeptides.

Mast cells, another type of leukocyte in the innate immune system, also play an important role in inflammation. Plasmin acts on a blood protein called complement factor 5 (C5), breaking it to yield complement factors C5a and C5b. C5a is one of a group of complement fragments called **anaphylatoxins**, which are small molecules that promote inflammation. (Other anaphylatoxins are complement fragments C3a and C4a.) The C5a fragment binds to receptors on mast cells. The mast cells then release histamine, which they store in granules. The released histamine causes local vasodilation (see Figure 27.4B). Mast cells also have a receptor for a type of antibody called immunoglobulin E (IgE) (discussed on pp. 874, 878, 887). The mast cells become coated with IgE, which when cross-linked by a specific antigen causes release of histamine. This is one of the ways that the adaptive immune response (which includes production of antibodies) is tied into the innate immune response.

Platelets play a central role in repair of damaged tissues. These small particles circulate in the blood. They are saucer-shaped disks, about one-fifteenth the size of a red blood cell, generated by fragmentation of a cell called a megakaryocyte (see Figure 27.1). Platelets are riddled with channels and filled with granules. They adhere to thrombin, which forms at the site of damaged blood vessels, and spread to form a plug that rapidly seals torn blood vessels. When activated, they contract vigorously and release these granules, which contain many biologically active molecules, notably:

- Platelet activating factor, which recruits more platelets

- Fibroblast growth factor, which facilitates repair of damaged extracellular components of tissues by inducing fibroblasts to replicate and to produce collagen

- Platelet-derived growth factor, which helps recruit neutrophils and macrophages

- Serotonin, which can cause vasodilation or vasoconstriction, depending on the receptors expressed on the target blood vessels, and also activates neutrophils, macrophages, and fibroblasts

- Decay accelerating factor (DAF) and factor H, both of which limit activation of complement (a major source of pro-inflammatory agents)

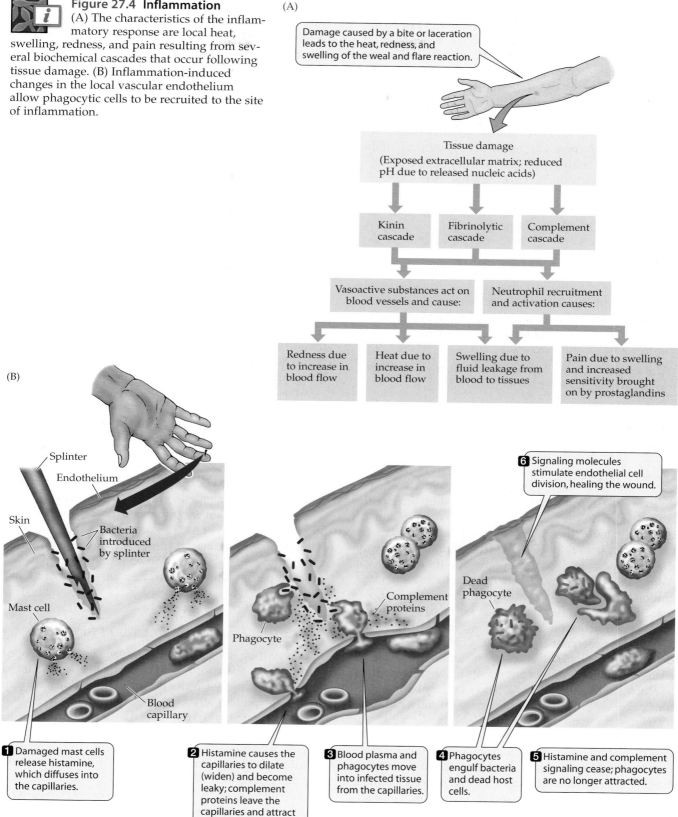

Figure 27.4 Inflammation (A) The characteristics of the inflammatory response are local heat, swelling, redness, and pain resulting from several biochemical cascades that occur following tissue damage. (B) Inflammation-induced changes in the local vascular endothelium allow phagocytic cells to be recruited to the site of inflammation.

(A)

Damage caused by a bite or laceration leads to the heat, redness, and swelling of the weal and flare reaction.

Tissue damage

(Exposed extracellular matrix; reduced pH due to released nucleic acids)

Kinin cascade

Fibrinolytic cascade

Complement cascade

Vasoactive substances act on blood vessels and cause:

Neutrophil recruitment and activation causes:

Redness due to increase in blood flow

Heat due to increase in blood flow

Swelling due to fluid leakage from blood to tissues

Pain due to swelling and increased sensitivity brought on by prostaglandins

(B)

Splinter

Endothelium

Skin

Bacteria introduced by splinter

Mast cell

Blood capillary

Phagocyte

Complement proteins

6 Signaling molecules stimulate endothelial cell division, healing the wound.

Dead phagocyte

1 Damaged mast cells release histamine, which diffuses into the capillaries.

2 Histamine causes the capillaries to dilate (widen) and become leaky; complement proteins leave the capillaries and attract phagocytes.

3 Blood plasma and phagocytes move into infected tissue from the capillaries.

4 Phagocytes engulf bacteria and dead host cells.

5 Histamine and complement signaling cease; phagocytes are no longer attracted.

In addition to causing vasodilation, bradykinin, fibrinolytic peptides, C5a, and histamine react with receptors on endothelial cells that line blood vessels, causing these cells to swell and contract, thereby allowing plasma fluids (including large molecules) to leak into surrounding tissues and to pool there, causing edema. So long as collagen is exposed at the site of plasma leakage, vasodilation, edema, clot formation, and tissue repair will proceed. To arrest inflammation, damaged tissues must be repaired. This requires both removal of the dead cells and their replacement. Phagocytes remove damaged tissue in addition to infectious agents. Recruitment and activation of the phagocytic cells for both purposes involves their interaction with the complement fragments C3a and C5a, discussed next.

Activating Complement

The key player in the inflammatory process is the **alternative pathway** of complement activation. The complement system consists of some 20 interacting proteins that are present in the blood plasma. Once activated, the complement system can:

- Kill foreign cells by rupturing their surface lipid bilayer, also called the plasma membrane
- Stimulate phagocytosis by tagging cells for recognition and uptake by phagocytic cells
- Cause inflammation that isolates infectious agents and repairs tissue damage
- Attract phagocytes

The complement proteins are activated by three pathways (Figure 27.5): the classical pathway and mannan-binding protein–dependent pathway, which are discussed later in the chapter (p. 876) and the alternative pathway discussed here. Each of these pathways involves several complement components, many of which are shared among the different pathways. In all pathways, complement components that are normally present in blood acquire new functions as a result of enzymatic processing. As each complement component is cleaved, fragments with distinct biologic functions are generated. Cleavage of each complement component results in the generation of a soluble fragment, labeled "**a**" and a surface bound fragment labeled "**b**." Table 27.1 provides a partial list of complement fragments and their main biological activities.

The alternative pathway of complement activation occurs during inflammation and infection without involvement of the adaptive immune system. Complement factor 3 (C3) plays a central role in this process. C3 is a protein that is produced by macrophages and is constitutively present in plasma. Plasma C3 is subject to slow breakdown, yielding the fragments called C3a and C3b. This goes on whether or not there is inflammation, and the concentrations of C3a and C3b generated are low and cause no adverse physiological response. However, the ongoing supply of C3b allows for the massive amplification of C3b and C3a, as well as other biologically active complement fragments, in the presence of many pathogens. C3b that is generated in plasma has two fates. It can be broken down or it can be converted to an enzyme complex, called the **alternative pathway C3 convertase**, which converts C3 to C3a and C3b in the presence of pathogens that facilitate this reaction (see Figure 27.5). These include bacteria, yeasts, some protozoa, and parasitic worms. Generation of the alternative pathway C3 convertase occurs at sites on pathogens that favor the binding of factor B to C3b over processes that degrade C3b. The binding of factor B to C3b creates the alternative C3 convertase, which then cleaves C3 into C3a and C3b, thereby increasing availability of C3b and, so long as more pathogens are available, promoting exponential cleavage of C3. Co-selection of hosts and pathogens through evolution has equipped pathogens with such sites and masked them on host cells. In addition, host cells are endowed with a molecule that both inhibits formation of, and promotes disassembly of, the alternative pathway C3 convertase. Selection through evolution has also equipped some pathogens with this capability.

Although some pathogens can inactivate complement, many do not and instead promote formation of the alternative pathway C3 convertase. The pathogens, therefore, increase the generation of products of the alternative pathway of complement activation, namely, C3b, C3a, C5a, and MAC, a donut-shaped complex that inserts into lipid membranes (Figure 27.6) and is capable of lysing cells as well as enveloped viral particles (see Chapter 14). The soluble complement fragments that are formed during the generation of MAC, namely C3a and C5a, have several pro-inflammatory properties. For example, by binding to receptors on endothelial cells, C5a increases vascular permeability and leakage of plasma proteins into the inflamed tissue as discussed in the preceding text.

Recruiting Phagocytes to Sites of Infection

When pathogens activate the alternative pathway of complement and continue to replicate unchecked at the site of tissue injury, products of the alternative pathway increase exponentially. However, in most instances that does not occur. One of the ways that activation of the alternative pathway creates conditions that limit the capacity of pathogens to replicate is through the recruitment of phagocytic cells that engulf, kill, and degrade the pathogens (see Table 27.1).

Recruitment of the phagocytes involves two distinct processes. First, phagocytes circulating in the blood adhere to the walls of inflamed blood vessels as a result of binding to a group of molecules that are expressed on

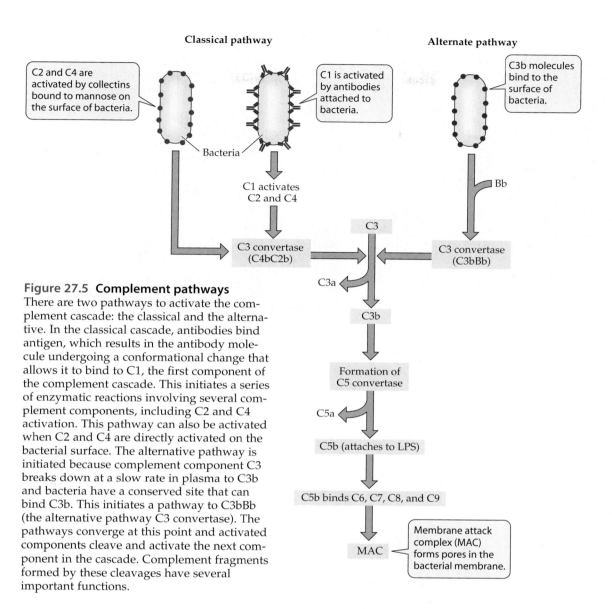

Figure 27.5 Complement pathways
There are two pathways to activate the complement cascade: the classical and the alternative. In the classical cascade, antibodies bind antigen, which results in the antibody molecule undergoing a conformational change that allows it to bind to C1, the first component of the complement cascade. This initiates a series of enzymatic reactions involving several complement components, including C2 and C4 activation. This pathway can also be activated when C2 and C4 are directly activated on the bacterial surface. The alternative pathway is initiated because complement component C3 breaks down at a slow rate in plasma to C3b and bacteria have a conserved site that can bind C3b. This initiates a pathway to C3bBb (the alternative pathway C3 convertase). The pathways converge at this point and activated components cleave and activate the next component in the cascade. Complement fragments formed by these cleavages have several important functions.

TABLE 27.1	Complement fragments and their biological activities

		Activity				
Fragment	Vasodilation	Recruitment: Neutrophils, Monocytes	Activation: Neutrophils, Monocytes	Opsonization	B Cell Activation	Membrane Attack Complex
C3a	+	+	+	–	–	–
C4a	–	+	+	–	–	–
C5a	+	+	+	–	–	–
C3b	–	–	–	+	–	–
iC3b	–	–	–	+	–	–
C3dg	–	–	–	–	+	–
C3d	–	–	–	–	+	–
C5b678(9)$_n$	–	–	–	–	–	+

(A)

(B)

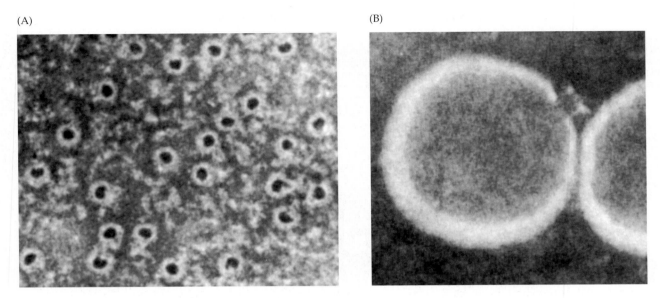

Figure 27.6 **Membrane attack complex (MAC)**

(A) Electron micrograph showing membrane attack complexes. Activation of complement leads to the assembly of a donut-shaped structure $[C5b,6,7,8(9)_n]$ that inserts into and causes holes in lipid membranes. This kills the target cell. (B) A side view of a membrane attack complex. Courtesy of Sucharit Bhakdi and Jorgen Tranum-Jensen.

endothelial cells within minutes after exposure to C5a and histamine. Second, the phagocytes migrate from the bloodstream to the damaged tissue because they are attracted to chemotactic molecules generated at the inflammatory site (Figure 27.7).

The neutrophils are normally swept along in the center of the stream of blood flowing through blood vessels but **marginate** (move to the outside of the stream) in dilated blood vessels, where blood flow becomes sluggish because of the increased diameter of the af-

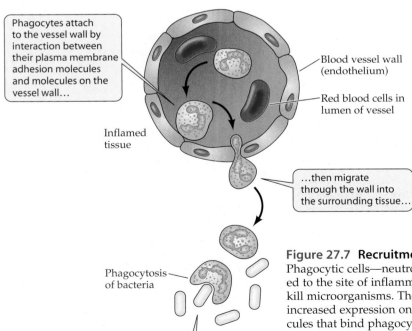

Phagocytes attach to the vessel wall by interaction between their plasma membrane adhesion molecules and molecules on the vessel wall...

Blood vessel wall (endothelium)

Red blood cells in lumen of vessel

Inflamed tissue

...then migrate through the wall into the surrounding tissue...

Phagocytosis of bacteria

...where they encounter and engulf microorganisms.

Figure 27.7 **Recruitment of phagocytes from the blood**

Phagocytic cells—neutrophils and macrophages—are attracted to the site of inflammation, where they may encounter and kill microorganisms. The inflammatory response results in an increased expression on local blood vessels of adhesion molecules that bind phagocytes, allowing these cells to pass into the surrounding tissue in a process known as diapedesis. At the site of the inflammatory response, the phagocytes are likely to encounter microorganisms.

fected vessel. Membrane proteins called **selectins**, which are expressed on endothelial cells at the site of inflammation, bind specific carbohydrate groups on blood neutrophils. At first the PMNLs do not make a firm attachment to the endothelial cells and are said to "roll" along the vessel membrane. As the phagocytes roll slowly along the region of the affected vessel wall, they are exposed to platelet activating factor, which causes them to express molecules called **integrins**. Expression of integrins on neutrophils causes these cells to bind tightly to the endothelium and cease rolling. Neutrophils that bind in this way have receptors for C3a and C5a and other chemotactic molecules, notably a cytokine called interleukin 8 (IL-8), which are produced at sites of inflammation. Binding of these chemotactic agents to their receptors on the phagocytes induces changes in the cells that allow them to traverse the blood vessel wall. This process is called **diapedesis** and is diagrammed in Figure 27.7. Once the phagocytes have squeezed between endothelial cells, they release protein-digesting enzymes that break down the extracellular matrix and escape into the tissue fluids. Neutrophils have a life span of only about 24 hours and their death at sites of inflammation gives rise to much of the composition of **pus**. Neutrophils that die in the circulation are removed by phagocytic cells in the liver without yielding pus.

Monocytes are recruited to the inflammatory site about 24 hours later than neutrophils. Their attachment to inflamed blood vessels is by molecules different from those used by neutrophils. Once they migrate into the inflamed tissue, monocytes are referred to as macrophages. Activated macrophages secrete cytokines that increase inflammation and thus are referred to as pro-inflammatory cytokines. Cytokines bind to specific receptors on target cells and elicit specific responses. **Interferon** alpha (**IFN-α**) and beta (**IFN-β**), which are two of the cytokines produced by macrophages, have the ability to inhibit replication of some viruses. When IFN-α and IFN-β bind to their receptors on the virus-infected cell, a series of events are initiated that ultimately result in activation of a ribonuclease that degrades viral RNA. Macrophages can live for years in the tissues.

Neutrophils and macrophages engulf and destroy pathogens, but they can also release microbicidal compounds. As neutrophils and macrophages migrate toward areas of increasingly high concentration of soluble complement fragments (C3a, C4a, and C5a) and other activating compounds, the binding of these materials enhances phagocytic and microbicidal activity. At high concentrations of the extracellular activators, phagocytes release microbicidal and tissue-destructive materials into the extracellular fluids. Although this process limits pathogen replication, it does so at the expense of collateral damage to host tissue.

Molecular Tags Facilitate Phagocytosis

Complement factor C3b sticks to surfaces on which it is generated, typically pathogens, tagging these cells and other molecules for **phagocytosis** by neutrophils and macrophages both of which express a receptor, **complement receptor 1 (CR1)**, for this fragment. C3b can be broken down into a number of other fragments. The first product of C3b cleavage is iC3b, which remains attached covalently to the surface on which it forms. The iC3b binds to **complement receptor 3 (CR3)** on neutrophils and macrophages. Engagement of CR1 or CR3 is one of the signals needed for the phagocytic process to proceed. Molecules that tag cells and other molecules for phagocytosis are called **opsonins**. Other opsonins are C-reactive protein and mannan-binding protein, produced by liver cells during acute inflammation.

Phagocytosis and Killing of Infectious Agents

Bacteria, protozoa, viruses, and yeasts that are engulfed by PMNLs, macrophages, and monocytes end up in vesicles known as **phagosomes** (Figure 27.8). Following phagosome formation, several events occur that create a microbicidal environment. The first of these is an increase in metabolic activity of the phagocyte, called the **respiratory burst**, which occurs as stored glycogen is metabolized, yielding glucose. Glucose is processed anaerobically yielding NADPH, which is in turn oxidized by NADPH oxidase. NADPH oxidase is assembled on the phagocyte plasma membrane from inactive sub-units as a result of signaling events that occur during phagocytosis and is included in the phagosome. Electrons captured from NADPH by the oxidase are donated to oxygen yielding superoxide anion, which spontaneously converts to hydrogen peroxide, a process that is accelerated by superoxide dismutase. If iron or copper is present in the phagosome, they will mediate the subsequent generation of hydroxyl radicals from the superoxide anion and hydrogen peroxide. These reactive oxygen intermediates are antimicrobial, particularly the hydroxyl radicals, and may damage or kill microorganisms.

Following its formation, the phagosome's pH is slowly decreased from neutral to an acidic pH of 3 to 4. This low pH is detrimental to the survival of many microorganisms. Preformed bags of enzymes—known as **lysosomes** and granules—that are present in the cytoplasm of the phagocytes, also fuse with the phagosome, depositing enzymes that degrade lipids, proteins, and carbohydrates, as well as lysozyme, which damages the cell wall of bacteria, as discussed in the section on barrier defense (see Chapter 26). In some phagocytes, particularly in neutrophils and newly recruited macrophages, the lysosomes contain the enzyme myeloperoxidase, which catalyzes the generation of hypohalides such as hypochloride (commonly known as chlorine

(A)

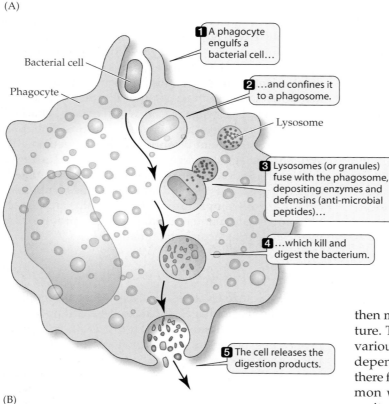

1 A phagocyte engulfs a bacterial cell...

Bacterial cell

Phagocyte

2 ...and confines it to a phagosome.

Lysosome

3 Lysosomes (or granules) fuse with the phagosome, depositing enzymes and defensins (anti-microbial peptides)...

4 ...which kill and digest the bacterium.

5 The cell releases the digestion products.

(B)

Figure 27.8 Phagocytosis
Neutrophils and macrophages are phagocytic cells that engulf microorganisms and confine them to an intracellular compartment known as a phagosome. The antimicrobial reactive oxygen intermediates produced in the phagosome include hydrogen peroxide, hypochloride, hydroxyl radicals, and nitric oxide. The process is diagrammed in (A) and shown by a scanning electron micrograph in (B), which depicts a macrophage (red) phagocytosing *Mycobacterium tuberculosis* (yellow). B, ©S. H. E. Kaufman and J. R. Golecki/SPL/Photo Researchers, Inc.

then migrate by diapedesis into tissues where they mature. These resident macrophages of tissues, called by various names such as Kupffer cell and microglial cell, depending upon the tissue (Table 27.2), may persist there for many years. They have many functions in common with the monocytes/macrophages that are recruited from blood as a result of inflammation, except that they are permanently available in organs and tissues, usually at positions of entry point for pathogens.

Macrophage Pattern Receptors and Nonphagocytic Defenses

Macrophages have molecules on their cell membranes that interact with conserved molecules on microbes. This interaction sends an alarm to the macrophage and activates it to increase transcription of genes that encode proinflammatory molecules—that is, that enhance inflammation as described in the preceding section. The receptors

bleach, which is an effective disinfectant) from hydrogen peroxide. Other components of lysosomes and granules of neutrophils include defensins and lytic peptides, which circularize and insert themselves into bacterial cell walls, resulting in bacterial killing.

Tissue Macrophages Originate from Monocytes

Monocytes may undergo an additional round of replication in the blood once they have left the bone marrow where they arise from stem cells (see Figure 27.1). They

TABLE 27.2	Macrophage names

Name of Macrophage	Tissue or Organ Where Found
Monocyte	Blood
Kupffer cell	Liver
Microglial cell	Brain
Splenic macrophage	Spleen
Alveolar macrophage	Lung
Peritoneal macrophage	Peritoneum
Bone marrow macrophage	Bone marrow

TABLE 27.3	Pattern receptors

Receptor Designation	Target Molecules
CD14	LPS and yeast
Toll 2	Lipoproteins, lipoteichoic acid of gram-positive bacteria, muramyl dipeptide, bacterial heat-shock proteins
Toll 4	LPS, bacterial heat-shock proteins
Toll 9	Bacterial CpG DNA

on the macrophages that react with conserved molecules on microbe surfaces are referred to as **molecular pattern receptors**. These include a receptor that is known as the lipopolysaccharide (LPS) receptor, which binds LPS alone but with higher affinity when associated with serum LPS-binding protein. Macrophages also express several additional pattern receptors known as **Toll-like receptors (TLRs)** because of homology with fruitfly TOLL proteins, which are involved in defense of the fly against fungal infection. Bacterial molecules that interact with pattern receptors on macrophages include LPS, lipoteichoic acid, muramyl dipeptide (all of which are part of the bacterial cell envelope), flagellin, bacterial heat shock proteins, and the dinucleotide CpG motif present on DNA released by dying bacteria. The interactions of ligands and pattern receptors on macrophages are summarized in Table 27.3.

Macrophages can be classified into several subpopulations based on their function. Two broad categories are the M1 cells, or classically activated macrophages, and the M2 cells, or alternatively activated macrophages. In the case of mice, macrophages of the M1 type respond to LPS and INF-γ by producing nitric oxide (NO), a diffusible antimicrobial agent. Whether human macrophages also produce microcidal NO is controversial; however, it appears to be likely that they do because these cells express the gene for inducible NO synthase. M2 macrophages produce arginase, which catabolizes arginine, the substrate from which NO is generated, and also produce a cytokine called transforming growth factor beta (TGF-β), which inhibits NO production. Macrophages can be driven to become M1 or M2 cells by different stimuli. The stimulatory materials, macrophage receptors, and signal pathways are active areas of research at present and beyond the scope of this chapter.

Surviving Phagocytosis—Intracellular Microbes

Some protozoa, bacteria, and fungi can survive inside monocytes or macrophages following phagocytosis. They are referred to as **intracellular organisms** because of this ability. The survival mechanisms of intracellular organisms include expression of genes that allow resistance to the low pH, the oxidative conditions, and the proteolytic environment of the phagolysosome. An example of this is protozoans of the genus *Leishmania,* which have a surface lipophosphoglycan that mediates many of these survival mechanisms. They even use C3b and iC3b to gain entry into the macrophage through phagocytosis. Other pathogens remodel the phagosome to suit their needs, prevent phagosome and lysosome fusion, or escape from the phagosome into the cytoplasm of the host cell. Some of these pathogenic survival strategies are listed in Table 27.4. To combat these microbes, the im-

TABLE 27.4	Mechanisms of survival in macrophages

Type of Organism	Name of Organisms	Disease Caused	Mechanism by Which Pathogen Survives Intracellularly
Bacteria	*Shigella* sp.	Food poisoning	Escapes into cytoplasm
	Mycobacterium tuberculosis	Tuberculosis	Remodels phagosome with proton pump to neutralize the pH
	Brucella sp.	Brucellosis (Bang's disease, undulant fever)	Survives in phagolysosome by activation of specific acid-response genes
	Listeria monocytogenes	Listeriosis	Escapes to cytoplasm
	Legionella pneumophila	Legionnaire's disease	Transits to ribosome-studded autophagosomes
Protozoa	*Leishmania major*	Leishmaniasis	Survives in phagolysosome
	Leishmania donovani	Visceral leishmaniasis	Survives in phagolysosome
	Toxoplasma gondii	Toxoplasmosis	Prevents phagolysosomal fusion
	Trypanosoma cruzi	Chagas disease or South American trypanosomiasis	Escapes into cytoplasm
Fungi	*Candida albicans*	Candidiasis	Survives in phagosome

mune system must rely on additional cells that produce a cytokine called IFN-γ, which acts on macrophages to elevate their killing mechanisms. IFN-γ is secreted by many T-cell subpopulations of the adaptive and bridging immune systems. Although IFN-γ is by far the most powerful macrophage-activating cytokine, other cytokines also activate macrophages. These include tumor necrosis factor alpha (TNF-α), which is produced by the macrophages themselves as a result of signaling through TLRs, after they contact microbes, and a related molecule lymphotoxin (TNF-β) that is produced by subpopulations of lymphocytes.

Macrophages activated by INF-γ and TNF-α have increased antimicrobial properties as a result of increased generation of reactive oxygen intermediates and production of antimicrobial NO. Nitric oxide is particularly effective against protozoan parasites and diffuses from the macrophage, as well as acting inside it. However, NO is very short lived in blood, where it rapidly reacts with hemoglobin. In addition, activated macrophages decrease the availability of intracellular tryptophan by increasing the enzyme indoleamine 2,3-dioxygenase. Tryptophan is required by many pathogens. This mechanism is particularly effective against *Chlamydia* within macrophages. *Chlamydia* is a bacterium that causes infertility in women as well as blindness. Activated macrophages also have decreased expression of transferrin receptors on the cell surface. Transferrin is an iron-transporting plasma protein. The reduced expression of transferrin receptors on macrophages results in decreased intracellular iron. Iron is needed for bacterial growth. As indicated previously, iron is also needed for generation of antimicrobial hydroxyl radicals, so a balance needs to be struck regarding iron levels to prevent growth of intracellular organisms.

Wasting, Fever, and the Acute Phase Response

An intense inflammatory response has systemic effects, all of which act to control infectious agents, but which may also have an adverse effect on the host. Many of these effects are caused by cytokines. The cytokines released by activated macrophages include IL-1, IL-6, and tumor necrosis factor alpha (TNF-α). TNF-α is responsible for the dramatic loss in body fat that occurs during chronic infections. It binds to receptors on fat cells, preventing production of an enzyme, lipoprotein lipase, that is required for the restoration of fat reserves. In the absence of lipoprotein lipase, the fat cells lose their lipid reserves. This form of wasting is called **cachexia**.

Interleukin-1, IL-6, and TNF-α induce liver cells to mount an **acute phase response**, which results in the production of C-reactive protein. C-reactive protein binds to phosphorylcholine on bacteria and promotes their phagocytosis. The acute phase response also results in the secretion of haptoglobin into plasma. Haptoglobin is a hemoglobin-binding protein that limits loss of iron from the body in the event of **hemolysis** (lysis of red blood cells) and also limits the availability of iron for invading microorganisms. The acute phase response also results in production of mannan-binding protein, which attaches to mannose residues on the surface of many bacteria, thereby facilitating their phagocytosis. Like C3b, iC3b, C-reactive protein, and IgG, mannan-binding protein is an opsonin.

Mannan-binding protein adheres to mannose and *N*-acetylglucosamine on microorganisms. It does not react with normal host cells because on these cells mannose and *N*-acetylglucosamine are covered by other carbohydrates. Mannan-binding protein that has bound to microorganisms acquires the capacity to bind and activate mannan-binding protein–associated serine proteases and activates the complement cascade in a manner similar to the classical pathway of complement activation (see Figure 27.5). Anaphylatoxins and MAC are generated, and target cells are coated with C3b and iC3b, which facilitates their phagocytosis. Therefore, generation of mannan-binding protein during the acute phase response amplifies complement activation and the clearance from the blood of pathogens with exposed target sugars. Mannan-binding protein is implicated in the efficient killing of several microorganisms, including *Neisseria meningitidis*. Deficiencies in mannan-binding protein are associated with chronic diarrhea in children and may enhance susceptibility to protozoal infections in patients with acquired immunodeficiency syndrome.

Interleukin 1, TNF-α, and prostaglandin E$_2$ react with receptors expressed on cells in the brain's thermoregulatory center, resulting in vasoconstriction and shivering, which conserve and generate heat, leading to an elevation in body temperature (i.e., fever). The fever response is detrimental to growth of pathogens, as it creates conditions that are not optimal for function of enzymes that regulate bacterial cell division. The goal here is to kill the pathogen before killing the host, although this is not always achieved. The possibility that fever enhances immune responsiveness thus facilitating pathogen elimination is still under investigation.

Inflammatory responses that result from trauma are self-limiting. In the event that a pathogen is present at the site of tissue injury, the inflammatory response will typically continue to amplify until a specific adaptive immune response develops that clears the pathogen, allowing conditions that provoke the inflammatory response to subside. When the specific immune response that develops is sufficient to limit pathogen growth but not to cause elimination, chronic inflammation can ensue. This can also occur when a pathogen induces a local self-sustaining autoimmune reaction and when inflammation is induced by a substance that is not biodegradable (e.g., asbestos fibers). When infections are not controlled, inflammatory responses continue to amplify until, racked with fever and wasted by cachexia, the afflicted individual dies.

SECTION HIGHLIGHTS

The warm, moist, nutrient-rich tissues of mammals provide potentially excellent growth environments for pathogens. Selection through evolution has equipped mammals with rapidly deployable defenses that limit pathogen survival and growth in tissues. These innate immune responses are mediated by pre-formed cells and molecules, which switch from an inactive to an active state in response to local tissue damage and to stimuli provided by pathogens. Innate responses include inflammation leading to the puncturing of pathogens by enzymes called complement factors, their uptake and degradation by phagocytic cells including polymorphonuclear leukocytes (PMNLs) and macrophages, and eventually wound healing. Pathogens have also been selected through evolution to both regulate and evade these host-protective innate immune responses.

27.3 Adaptive Immunity

Lymphocytes have receptors on their plasma membranes that allow them to interact with specific portions (**epitopes**) of molecules referred to as antigens. These antigens are typically foreign (non-self) molecules of infectious disease agents. There are millions of different T-cell and B-cell receptors expressed, which give the specificity to the response even though all of the antigen-binding receptors on a single lymphocyte have the same specificity. The genetic basis of receptor diversity is touched on only briefly in this text. Antigens recognized by the antigen-specific receptors of B cells (**B-cell receptor**) are present on the intact pathogens or their products. T-cell antigen–specific receptors recognize peptides (a small piece of a protein) derived from pathogen molecules and presented on the surface of host cells that have taken up pathogens or their products, for example, macrophages and dendritic cells that have phagocytosed the pathogen, or host cells that are infected by pathogens. Dendritic

cells are phagocytes that are specialized for presenting antigen to and stimulating lymphocytes to mount an adaptive immune response. The host cell molecules with which the antigenic peptides are displayed to the T cells are known as **major histocompatibility complex** (MHC) molecules and this process of display is known as **antigen presentation**.

The second hallmark of the adaptive immune response is memory, which enables more rapid and heightened responses of antigen-specific lymphocytes when they encounter a particular antigen the second time. Previously, memory was thought to result mainly from the replication of lymphocytes that had reacted with a particular antigenic epitope, consequently resulting in an expanded population of antigen-specific cells. Indeed there is an increase in the number of antigen-specific cells during an infection that can react with particular antigens. However, most of these cells subsequently die. Mammals would be unable to accommodate the large number of lymphocytes that would be acquired after each antigen encounter if this did not occur. Some of the antigen-stimulated lymphocytes do not die and have a less stringent requirement for activation when they encounter those antigens the next time, meaning they respond more rapidly. These are referred to as memory cells. Immunological memory is the principle behind vaccination.

To initiate an adaptive immune response, antigens from infectious agents must come in contact with lymphocytes that bear receptors for that antigen. The following sections discuss the nature of the antigen-specific receptors on T cells and B cells, how the cells develop, how they are activated, and the types of responses that they mount.

T-Cell Development

T cells and B cells develop from lymphoid progenitor cells (Figure 27.9), which derive from hematopoietic stem cells as discussed in the introduction to this chapter. T cells develop in the thymus, and their development is associated with the assembly of an antigen-specific receptor and the selection of cells that have the antigen-specific receptor of the appropriate specificity. The antigen-specific receptors of T cells are designed to recognize fragment of proteins, called peptides, which are expressed on the surface of cells. A group of host molecules has evolved to transport peptide fragments of pathogens onto the cell surface. These are called major histocompatibility complex (MHC) molecules. Antigen-specific receptors on T cells actually bind to the peptides and some amino acid residues of MHC close to where the peptide is located. There are two classes of MHC molecules: MHC class I and MHC class II. These different

Figure 27.9 Development of cells of the adaptive immune system
Cells of the adaptive immune system are known as lymphocytes; these include B cells, CD4 T cells, and CD8 T cells. They arise from a common precursor in the bone marrow and differentiate into major subpopulations in primary lymphoid organs: T cells differentiate in the thymus; and B cells differentiate in the bursa in birds and in the bone marrow, spleen, or gut-associated lymphoid tissue in mammals. Lymphocytes mature into effector or memory cells once they have encountered antigen.

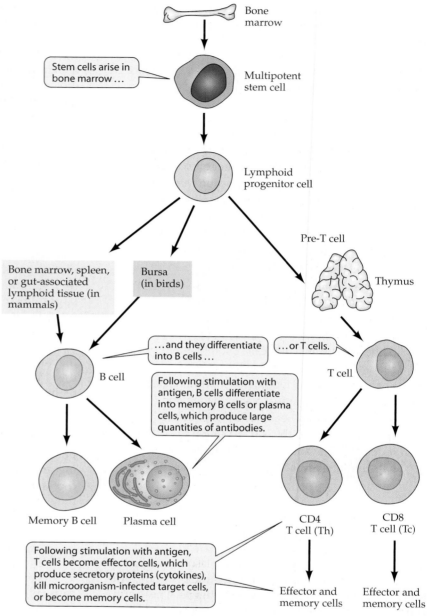

types of MHC molecule direct responses of two functionally different T-cell subpopulations called **CD4 T cells** and **CD8 T cells**.

The MHC molecules do not discriminate between peptides of pathogen origin and of host origin and consequently can display a comprehensive sampling of these on the cell surface. This is important for educating the developing immune system to selectively respond to largely non–self-antigens, that is, to antigens from pathogens and not to antigens from normal self-molecules.

Immature T cells that enter the thymus after leaving the bone marrow express an antigen-specific **T-cell receptor**, and both the CD4 and the CD8 proteins on their surface and thus are referred to as double-positive T cells. These cells do not have **CD3**, which refers to a set of plasma membrane glycoproteins associated with the T-cell receptor in mature T cells. The CD3 molecules are responsible for signal transduction after the T-cell receptor binds antigen (Figure 27.10). T cells mature in the cortex of the thymus and then transit to the medulla. By the time they enter the medulla, they are single positive, expressing either CD4 or CD8 molecules, but not both. They also express CD3 and are ready to encounter antigen. However, T cells undergo programmed cell death

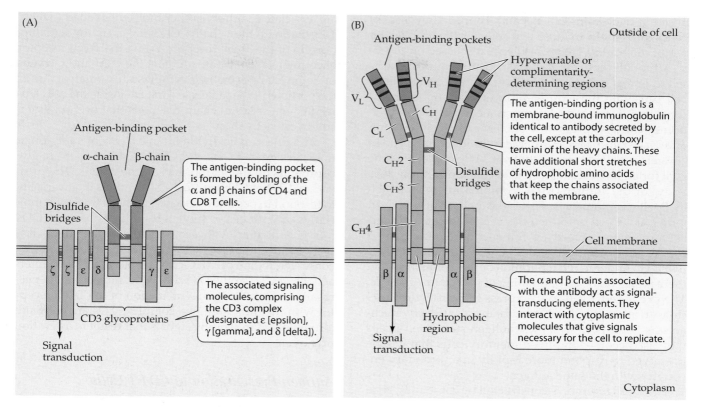

Figure 27.10 Structure of lymphocyte receptors for antigen
(A) The antigen-binding portion of the antigen-specific T-cell receptor is composed of an α- and a β-chain for both CD4 and CD8 T cells. The antigen-binding portion is associated in the cell membrane with a complex of molecules called CD3 and ζ-chains that signal the cell when the receptor has engaged antigen. (B) The B cell reacts with antigen through its plasma membrane–bound immunoglobulin and then signals the cell that it has found the appropriate antigen through associated molecules called α- and β-chains.

in the medulla of the thymus unless spared from this fate by a threshold interaction with self MHC molecules. Only those that react strongly enough receive the appropriate signal to survive; that is, they undergo **positive selection**. Subsequently, if their T-cell receptor reacts too strongly with MHC molecules that are presenting self-peptides, they are depleted or negatively selected to avoid immune responses to self-antigens or autoimmunity. This would be very detrimental to the host.

CD4 T cells react with the combination of peptide and MHC class II molecules, whereas CD8 T cells react with the combination of peptide and MHC class I molecules. The interaction of these cells with peptide MHC complexes is directed both by the antigen-specific receptors and by the accessory molecules CD4 and CD8. The T-cell receptors on CD4 T cells bind peptide and amino acid residues of the peptide-binding groove of the MHC class II molecule, whereas CD4 itself binds a conserved component of the MHC class II molecule. Antigen-spe-

cific receptors on CD8 T cells bind peptide and amino acid residues of the peptide-binding groove of the MHC molecule class I, whereas CD8 itself binds a conserved component of the MHC class I molecule. CD4 T cells produce cytokines that assist or help both B cells and CD8 T cells to proliferate and mature into fully functional effector cells, and hence CD4 T cells are referred to as helper T cells, or Th. CD8 T cells bind to and kill other host cells that express the appropriate MHC and antigenic peptide. The CD8 cells are therefore referred to as cytotoxic T cells, or Tc.

Receptors of αβ T Cells

The T cells discussed in this section on adaptive immune responses have a T-cell receptor composed of two chains called α and β. Therefore these T cells are known as **αβ T cells**. The other major subpopulation of T cells is known as **γδ** T cells because their T-cell receptor is composed of two chains called γ and δ encoded by a set of genes different from those that code for the αβ T-cell receptor molecules. The γδ T cells are discussed on p. 883.

Although there are millions of different possible T-cell receptors that an αβ T cell can express, and millions of T cells with those different receptors, an individual T cell and its progeny, known as a **clonotypic population** or **clone**, express only one type of T-cell receptor. The T-cell receptor can interact with only one type of antigen presented on one type of MHC molecule. The goal for an individual T cell is to constantly survey the body for the antigen/MHC combination with which its T-cell receptor reacts. This is the method by which an individual T cell is notified that it is needed to counteract an infection. Once a T cell encounters the correct combination of antigen and MHC, it is stimulated or activated to undergo cell replication or proliferation so that there is an expanded population of that clone of T cells. The expanded T cells will express their **effector functions** to attempt to limit infection by a variety of mechanisms. Some portion of these antigen-stimulated T cells will also become long-lived **memory cells** that will be preserved to respond more rapidly in a secondary encounter with the same antigen.

The T-cell receptor is constructed by combining gene segments selected from a large array of possibilities, very similar to the way in which the diversity of B-cell antigen-specific receptors is created. For example, the β chain is a protein coded for by recombination of a variable gene segment (called a *V* gene), a diversity segment (called a *D* gene), and a joining gene segment (called a *J* gene) with a constant region segment (called a *C-β* gene). The α chain is a protein coded for by recombination of a *V* gene, a *D* gene, and a *J* gene with an α constant region segment (*C-α* gene). The *V*, *D*, and *J* genes that code for the α chain are different from those that code for the β chain. The DNA coding for *V*, *D*, *J*, or *C* genes that were not selected is looped out and permanently removed so that the chosen gene segments can be fused together, or **recombined**, to make a single gene. The *V*, *D*, and *J* gene segments give rise to the portion of the protein that is the antigen-binding site (see Figure 27.10A), which interacts with the antigen and MHC molecules. The many millions of different T-cell receptors arise because of the high number of gene segments from which to choose. It is this diversity of T-cell receptors that allows the adaptive immune system to be so specific for a particular antigen and therefore the adaptive immune system to be so carefully regulated.

The second important component of the T-cell receptor complex is the set of associated glycoproteins called CD3 (see Figure 27.10A). The CD3 molecules are identical on all T cells and among all members of the same species. The CD3 molecules are named γ, δ, and ε and

are associated with two other molecules called ζ (zeta). Together with the ζ chains, CD3 makes up the signaling portion of the T-cell receptor. When the T-cell receptor encounters appropriate antigen, the CD3 and ζ chains undergo changes on their **cytoplasmic domains**, that is, the portions of the molecules that are inside the cell. This results in a series of chain reactions that eventually result in cell activation.

αβ T cells can be divided into two major subpopulations, designated CD4 and CD8. CD4 and CD8 are glycoproteins on the membranes of the T cells. The presence of CD4 or CD8 on T cells determines their capacity to respond to peptide associated with MHC class II molecules or with class I molecules. This is because CD4 molecules interact with a conserved region of MHC class II molecules, whereas CD8 molecules interact with a conserved region of MHC class I molecules. These interactions of CD4 and CD8 with MHC molecules are necessary for stabilizing the interaction of the T-cell receptor with the MHC–antigen complex. CD4 and CD8 are also involved in sending signals that result in activation of the cell.

Antigen Presentation to CD4 T Cells

Presentation of antigen with MHC class II molecules occurs by a mechanism known as the **exogenous pathway** (Figure 27.11). This is because the presented antigen may be synthesized outside the presenting cells (i.e., may be produced exogenously) and taken into the presenting cell by phagocytosis. Materials taken up in this manner include killed bacteria and viruses, fragments of these, or molecules derived from them. Once in the phagolysosome, the phagocytosed material is degraded into fragments. Some vesicles containing these fragments fuse with other vesicles in the macrophage that contain membrane-bound MHC class II molecules. Peptide fragments associate with the antigen-binding groove of the class II molecules and are transported to the cell surface where they are displayed in association with MHC class II to CD4 T cells.

Only select populations of host cells have MHC class II molecules and thus are equipped to present antigen to CD4 T cells. This is in contrast to the MHC class I molecules, which are on all nucleated host cells. The cells with MHC class II molecules include macrophages, B cells, and dendritic cells. However, because B cells are not phagocytes as are macrophages and dendritic cells, the antigens they present must be soluble antigens that have been endocytosed.

The stimulation of the T cell that occurs as a result of interaction of the T-cell receptor with the antigen–MHC complex is sometimes referred to as signal 1. For a CD4 T cell to become activated to proliferate (clonally ex-

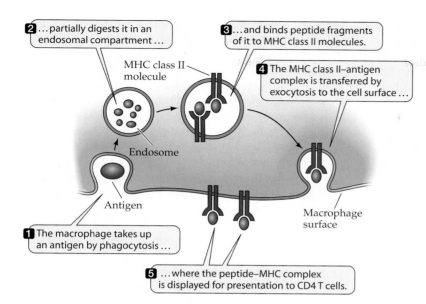

2 ...partially digests it in an endosomal compartment ...

3 ...and binds peptide fragments of it to MHC class II molecules.

MHC class II molecule

4 The MHC class II–antigen complex is transferred by exocytosis to the cell surface ...

Endosome

Antigen

1 The macrophage takes up an antigen by phagocytosis ...

Macrophage surface

5 ...where the peptide–MHC complex is displayed for presentation to CD4 T cells.

Figure 27.11 Presentation of antigen with MHC class II molecules
Following internalization of a foreign antigen (for example, a bacterium or virus) by the macrophage, the microbe is killed and its proteins degraded into small antigen peptides. These antigen peptides associate in specialized intracellular compartments with MHC class II molecules and the complex is transported and displayed on the surface of the antigen-presenting cell. For purposes of illustration, the MHC class II molecules have been greatly enlarged relative to the rest of the cell.

pand) and express its effector functions, it needs to receive a second signal that is referred to as **co-stimulation**. This second signal is received by the T cells through a co-stimulator receptor called **CD28**. T cells that have not previously encountered antigen may undergo programmed cell death as a result of signal 1 alone.

CD4 T-Cell Effector Molecules: Cytokines

When a CD4 T cell encounters the antigen–MHC complex that interacts with its T-cell receptor, it is known as a T-helper zero (**Th0**) cell and may produce a variety of cytokines. Th0 cells subsequently mature into one of two functional subsets called **type 1** (**Th1**) or **type 2** (**Th2**), distinguished from one and another by the types of cytokines they produce (Figure 27.12). This depends not on their T-cell receptor but on the cytokines they receive when encountering antigen. For example, IL-12 produced by dendritic cells and macrophages that have been stimulated through TLRs, biases T cells toward a Th1 response. In contrast, IL-4, perhaps produced by mast cells during an inflammatory response and by neighboring Th2 cells, biases T cells toward a Th2 response. It is also worthy of note that macrophages and dendritic cells are functionally heterogeneous with respect to their capacity to elicit Th1 and Th2 responses.

The main Th1 cytokines produced are IFN-γ and IL-2. The general function of Th1 cytokines is to promote **cell-mediated immunity** (also called cellular immunity). IFN-γ activates macrophages for increased control of intracellular microbes and inhibits viral replication by stimulating production of a cellular endoribonuclease that cleaves viral messenger RNA (mRNA). As discussed on pp. 857–859, activated macrophages have de-

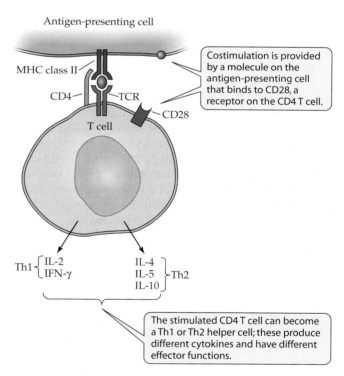

Antigen-presenting cell

MHC class II

CD4

TCR

CD28

T cell

Costimulation is provided by a molecule on the antigen-presenting cell that binds to CD28, a receptor on the CD4 T cell.

Th1 { IL-2, IFN-γ }

Th2 { IL-4, IL-5, IL-10 }

The stimulated CD4 T cell can become a Th1 or Th2 helper cell; these produce different cytokines and have different effector functions.

Figure 27.12 Interaction of CD4 T cell with antigen-presenting cell
The T-cell receptor (TCR) of CD4 T cells recognizes pieces of antigen presented in conjunction with MHC class II molecules on the surface of antigen-presenting cells. Thus, the antigen-binding pocket interacts not only with the foreign antigen fragment (or protein peptide) but with the MHC molecule as well. In addition, antigen-presenting cells such as macrophages have molecules on their surface, called co-stimulator molecules, which react with the CD28 receptor on the T cells, thereby contributing to their activation. The CD4 molecule interacts with MHC class II molecules.

creased iron availability and increased production of reactive oxygen intermediates and nitric oxide, which are toxic to intracellular microbes. IL-2 produced by Th1 CD4 T cells helps CD8 T cells to replicate. Subsequently, CD8 T cells become **cytotoxic**. This means they kill or lyse host cells that are infected with viruses or with intracellular pathogens. Thus, Th1 cells are particularly effective against infectious agents that live within host cells. However, IFN-γ also promotes B cells to develop into cells that produce a particular IgG type of antibody. IFN-γ produced by Th1 cells also influences development of other T cells into Th1 cells.

Th1 cells mediate two other responses: delayed type hypersensitivity and granuloma formation. These involve recruitment of macrophages to sites of infection. Dermal delayed type hypersensitivity is a response to antigens that are presented in the skin. It takes approximately 48 hours to develop and involves the recruitment of CD4 T cells and macrophages to the site. This response is the basis of a skin test that is used to determine whether a person or animal is sensitized to a particular antigen, such as the skin test for tuberculosis. The Th1 cell response that occurs in the skin is similar to those that occur in other tissues and organs. A granuloma is similar to a delayed type hypersensitivity response, but it is a long-term response that occurs when the influx of macrophages is unable to clear the antigenic stimulus. The macrophages adhere to one another, fusing to form multinucleate giant cells. Examples of this are the walling off of *Mycobacterium* in the lungs, the reason chest x-rays are used for individuals who potentially have tuberculosis, the walling off of eggs of the parasitic worm *Schistosomula mansoni* in the liver, resulting in enlarged livers, and the localized retention of tattoo pigments.

The main cytokines produced by Th2 cells are IL-4 and IL-10. These Th2 cytokines, often in association with other cytokines, for example, IL-5 and IL-6, promote **humoral immunity**. Promotion of humoral immunity by Th2 cytokines means they promote B-cell growth and differentiation into cells producing antibodies of the IgA, IgE, and IgG isotypes as discussed in subsequent text. These classes of antibodies are particularly effective against pathogens that are present in the blood plasma and tissue fluids and are not hidden inside host cells. In addition to expediting the development of B-cell responses, the Th2 cytokines activate eosinophils and other cells that are involved in immunity to parasitic worms (discussed on p. 877), and influence the development of other T cells into Th2 cells (Figure 27.13).

Antigen Presentation to CD8 T Cells

The association of antigen with MHC class I molecules occurs within the endoplasmic reticulum of cells and is referred to as the **endogenous pathway of antigen presentation**. This is because the presented antigen is synthesized within the cells (produced endogenously). The association of antigens of infectious agents with MHC class I molecules first requires that these molecules be free in the cell cytoplasm. For example, viruses produce viral proteins only inside host cells because they do not contain the needed cellular machinery to do this independently. Thus, to make new viral particles, a virus must attach to a host cell, enter, and uncoat to expose its genetic code (DNA or RNA) to serve as a template for producing new viral proteins (see Chapter 14). Similarly, intracellular bacteria and protozoa that replicate either in a phagosome, phagolysosomes, or within the cytoplasm of host cells can have their protein peptides presented with MHC class I molecules only if they are first transported from the vesicle into the cell cytoplasm. In addition to those intracellular microbes mentioned previously in this chapter that live within macrophages (see Table 27.3), protozoa and bacteria may also live within host cells that are not professional phagocytes. Examples include the protozoan *Plasmodium falciparum* that initially lives in hepatocytes (liver cells). How the proteins from the infectious agents living within intracellular endosomal compartments gain access to the cytoplasm is not yet entirely clear. There may be specific transport mechanisms to remove pathogen components from the compartment. It is also likely that some intracellular organisms die within the compartments, and it may be proteins from these that gain access to the cytoplasm.

Once the proteins are in the cytoplasm, they are degraded by an enzyme complex, called the **proteosome**, into peptides. Resulting peptides are moved by **chaperonins** to the endoplasmic reticulum and moved across the membrane of the endoplasmic reticulum (ER) into its cisterna by the Transporter associated with Antigen Processing (TAP transporter). Within the ER, the antigen peptides come in contact with and bind in the antigen-binding groove of MHC class I molecules, as do peptides derived from self-molecules. Together, the MHC class I molecules and the antigen or self-peptides are transported to the surface of the cell, where they are displayed to CD8 T cells (Figure 27.14). Because all nucleated cells of the host express MHC class I molecules, the antigen-specific CD8 T cells can receive signal 1 for activation from any of them.

Some viruses have developed methods to foil the immune system by decreasing the level of MHC class I molecules expressed on the cell surface or causing retrograde transport of viral peptides back into the cytoplasm of the infected cell. In this case, the immune CD8 T cells do not recognize the virally infected cells. However, the natural killer (NK) cells in the bridging immune system are alerted as a result of decreased MHC class I expression. NK cells are discussed on pp. 880, 881.

Like CD4 T cells, CD8 T cells express CD28. However unlike CD4 T cells, CD8 T cells can respond to antigen

(A)

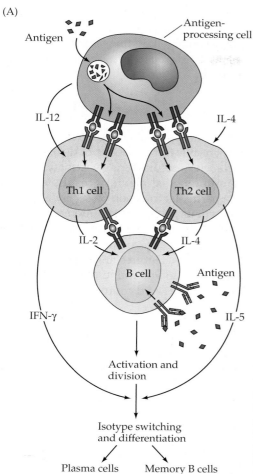

(B)

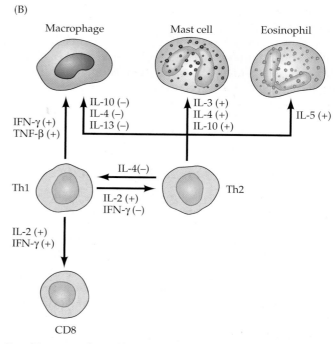

Figure 27.13 Cytokines produced by CD4 type 1 and type 2 T cells
(A) Following activation, CD4 T cells mature into type 1 or type 2 subpopulations, distinguished by the types of cytokines they produce. This depends not on their T-cell receptor (TCR) but on the cytokines they receive, such as IL-12 and IL-4, when encountering antigen. CD4 T cells that produce type 1 cytokines are known as Th1 cells; those that produce type 2 cytokines, as Th2 cells. Both type 1 and type 2 cytokines act on B cells. Type 2 cytokines such as IL-4 and IL-5 promote B-cell differentiation and growth and the generation of specific antibody isotypes. The type 1 cytokine IFN-γ promotes generation of specific IgG subclasses not promoted by type 2 cytokines. (B) Cytokines produced by CD4 T cells also act on macrophages to activate (+) or deactivate (–) them for antimicrobial activity and presentation of antigen. The cytokines IL-4 and IFN-γ affect development of other T cells into type 1 or type 2 cells. IL-2 helps other T cells proliferate. Finally, the type 2 cytokines activate mast cells and eosinophils.

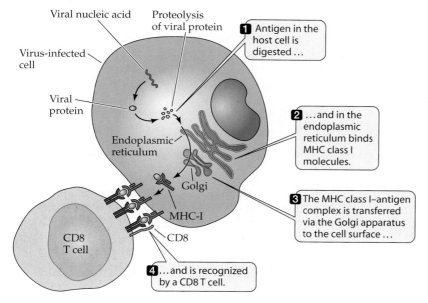

Figure 27.14 Antigen presentation with MHC class I molecules
Antigens produced in a host cell (whether as a result of a viral infection or of parasitism by a bacterium or protozoan that survives within phagocytes) are degraded into small antigen peptides that associate in the endoplasmic reticulum with MHC class I molecules. Together, the MHC class I molecules and antigen peptides are displayed on the cell surface in such a way that they can be recognized by the T-cell receptor and CD8 of CD8 T cells.

in the absence of co-stimulation. When stimulation through TCR and CD8 is weak, co-stimulation through CD28 can enhance the CD8 T-cell response. It is logical that CD8 T cells may not require co-stimulation, because CD8 T cells need to be able to respond to antigen on any infected cell of the host, and not all cells express co-stimulatory ligands for CD28. CD8 T cells also require IL-2 from CD4 Th cells, although occasionally a CD8 T-cell clone has been reported to produce its own IL-2.

CD8 T-Cell Effector Function: Cytotoxicity and Cytokines

The main effector functions of CD8 T cells are cytotoxicity and production of IFN-γ (Table 27.5). Because of the presentation of endogenously produced peptides with MHC class I molecules, CD8 T cells are particularly important in immunity to viral infections. The mechanism by which CD8 T cells kill target cells (i.e., those expressing the appropriate antigen peptide with MHC class I molecules) is diagrammed in Figure 27.15. At the apposition point of the cell membranes, the CD8 T cells insert perforin molecules into the target cell, thereby forming pores. The perforin is stored in granules within the CD8 T cells. Granzyme is then deposited into the target cells through the pores, which results in the target cell undergoing programmed cell death (**apoptosis**). An additional mechanism not unique to CD8 T cells is the production of Fas ligand that interacts with its receptor called Fas or CD95 on the target cell (see Figure 27.15), causing the target cells to undergo apoptosis.

Memory Responses by T Cells

The basic definition of immunological memory is a long-lived response that results from infection or deliberate exposure to antigen such as occurs during vaccination. Most importantly, immunological memory results in more rapid control of a particular infection on secondary challenge with the infectious agent. The requirements for activation of memory cells are less than those for activating naïve cells, that is, less antigen is required and no co-stimulation is necessary. These features, in conjunction with the clonal expansion in the first round of response to an antigen, result in the more rapid control of infection.

B Cells: Basic Structure of an Immunoglobulin

Antibodies, which are produced by B cells, contribute to adaptive immunity. Antibodies comprise a group of molecules called soluble immunoglobulins. Prior to discussing B-cell development, antibody production, and

---

TABLE 27.5 Summary of αβ T-cell effector functions for control of infectious disease

T-Cell Cytokine or Action	T Cell Type Involved	Effector Function	Infectious Agents Targeted
IFN-γ	Type 1 CD4 T cells mainly but some by type 1 CD8 T cells	Activates macrophages to kill or control replication of intracellular protozoa, bacteria, and fungi	Intracellular bacteria Intracellular protozoa Intracellular fungi
		Inhibits virus replication in host cells	All viruses after they have entered host cells
		Promotes class switching of B cells to specific IgG subclasses	Extracellular infectious agents in internal tissues and organs
TNF-α	Type 1 CD4 T cells	Assists in macrophage activation	All intracellular infectious agents (see above in this column)
IL-2	Type 1 CD4 T cells	Assists in T cell replication	
IL-4, IL-5, IL-6	Type 2 CD4 T cells	Promotes class switching of B cells to IgA, IgE, and other IgG subclasses	Immunity at body surfaces by IgA against entry of viruses and bacteria; engagement of mast and eosinophils by IgE to kill worms; IgG subclasses to promote immunity to all extracellular viral particles, bacteria, and protozoa by complement activation, agglutination and opsonization
Cytotoxicity	CD8 T cells	Kills infected cells	Virus-infected cells; cells infected with intracellular bacteria, protozoa, and fungi

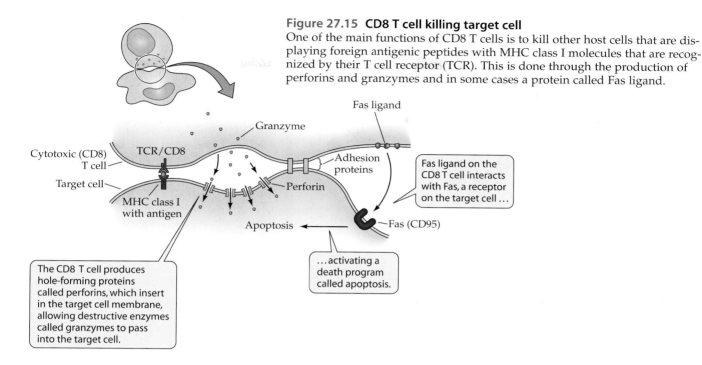

Figure 27.15 CD8 T cell killing target cell
One of the main functions of CD8 T cells is to kill other host cells that are displaying foreign antigenic peptides with MHC class I molecules that are recognized by their T cell receptor (TCR). This is done through the production of perforins and granzymes and in some cases a protein called Fas ligand.

function, it is useful to know the basic structure of a soluble immunoglobulin and the names used to describe its parts. All soluble immunoglobulins irrespective of class have the same subunit structure (Figure 27.16A). Each immunoglobulin subunit is composed of two identical polypeptide chains, called heavy or H-chains, and two identical smaller polypeptide chains, called light or L-chains. The heavy chains are covalently attached to each other by one or more disulfide bridges. Each light chain is also covalently bonded to a heavy chain by a disulfide bridge. The basic immunoglobulin structural unit is the same in most mammals with the exception of camelids, which in addition to standard immunoglobulins also have some unique immunoglobulins that are composed of dimers of novel H-chains only. There are several classes of soluble immunoglobulins, which differ in size and biological function. Immunoglobulins of the IgG and IgE classes have a single subunit, whereas soluble immunoglobulin of the IgM class is composed of five subunits. The IgM subunits are attached to each other by disulfide bonds, a process initiated by a small peptide called the J-chain. IgA is composed of two subunits held together by a J-chain (Figure 27.16B).

The variable domains of a heavy and a light chain fold to create a small pocket, called the antigen-binding pocket, capable of accommodating a structure that is equivalent in size to a small peptide of 5 to 8 amino acid residues (Figure 27.17). The antigen-binding pocket is located in the fraction-antibody, or **Fab region**, of the antibody molecule. The amino acid residues of the anti-gen-binding pocket that make contact with the antigenic epitope lie within hypervariable regions of the variable domain of H- and L-chains. There are three such hypervariable regions, also called complementarity-determining regions (see Figure 27.17). These are present in the variable region of both the heavy and light chains, in each case representing about 15% of the variable regions. The remaining stretches of amino acids in the variable region are called the framework regions. The wide range of specificities of antibody molecules reflects variation in the amino acid sequences and lengths of the six hypervariable regions.

Immunoglobulins can also be synthesized in forms that remain associated with the B-cell plasma membrane (see Figure 27.10). Cell membrane immunoglobulin has the basic immunoglobulin subunit structure irrespective of the class. Thus B-cell membrane associated IgM, IgG, IgE, and IgA in all cases comprise two heavy chains and two light chains. B-cell membrane immunoglobulin differs from secretory immunoglobulin that is made by the same B cell by having an additional short stretch of hydrophobic amino acids at the carboxy terminus of each heavy chain. The heavy chains of secretory and B-cell surface immunoglobulin are transcribed from the same gene, and the resultant RNA is differentially processed to generate two distinct mRNAs: one that includes sequence coding for the *trans*-membrane domain of the B-cell surface immunoglobulin, and the other that lacks this and codes for secretory immunoglobulin. The enzyme that removes coding sequence for the heavy chain

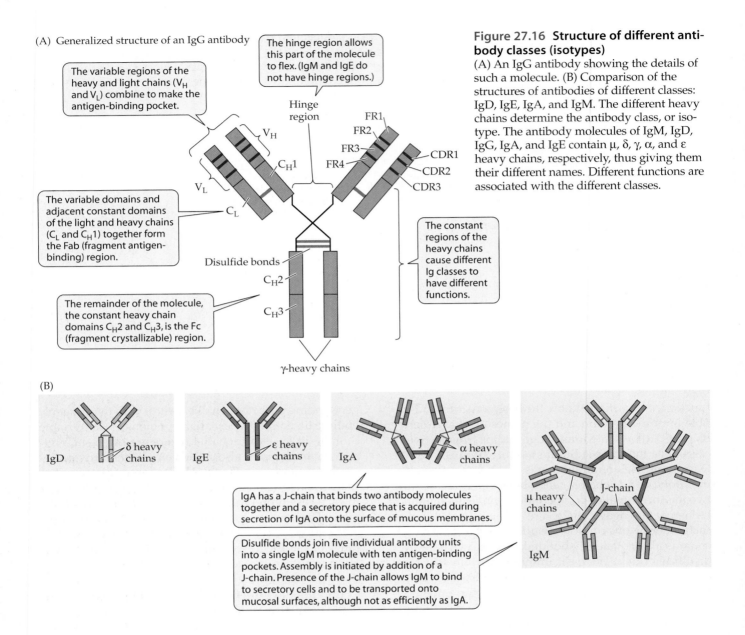

(A) Generalized structure of an IgG antibody

The variable regions of the heavy and light chains (V_H and V_L) combine to make the antigen-binding pocket.

The hinge region allows this part of the molecule to flex. (IgM and IgE do not have hinge regions.)

The variable domains and adjacent constant domains of the light and heavy chains (C_L and C_H1) together form the Fab (fragment antigen-binding) region.

The remainder of the molecule, the constant heavy chain domains C_H2 and C_H3, is the Fc (fragment crystallizable) region.

Hinge region

V_H

C_H1

V_L

C_L

Disulfide bonds

C_H2

C_H3

γ-heavy chains

FR1
FR2
FR3
FR4
CDR1
CDR2
CDR3

The constant regions of the heavy chains cause different Ig classes to have different functions.

Figure 27.16 Structure of different antibody classes (isotypes)
(A) An IgG antibody showing the details of such a molecule. (B) Comparison of the structures of antibodies of different classes: IgD, IgE, IgA, and IgM. The different heavy chains determine the antibody class, or isotype. The antibody molecules of IgM, IgD, IgG, IgA, and IgE contain μ, δ, γ, α, and ε heavy chains, respectively, thus giving them their different names. Different functions are associated with the different classes.

(B)

IgD — δ heavy chains

IgE — ε heavy chains

IgA — α heavy chains, J

IgA has a J-chain that binds two antibody molecules together and a secretory piece that is acquired during secretion of IgA onto the surface of mucous membranes.

Disulfide bonds join five individual antibody units into a single IgM molecule with ten antigen-binding pockets. Assembly is initiated by addition of a J-chain. Presence of the J-chain allows IgM to bind to secretory cells and to be transported onto mucosal surfaces, although not as efficiently as IgA.

μ heavy chains, J-chain

IgM

trans-membrane domain is expressed differentially at different stages of B-cell development, being absent from resting B cells and present in plasma cells, which are the end stage of B-cell differentiation and specialized to secrete large amounts of soluble immunoglobulin.

B-Cell Development

B cells first arise from hemopoietic stem cells in the yolk sac and liver of the embryo. After birth, they arise from hemopoietic stem cells that are mainly present in the bone marrow (see Figure 27.9). Development of B cells is associated with development of their antigen-specific receptors, which is cell-membrane associated immunoglobulin. Each mature B cell has on its surface between 100,000 and 150,000 such receptors, and for any given B cell all 100,000 or so receptors are the same and therefore have identical antigen-binding sites. The capability of the B-cell system to respond to the huge number of antigens that may be encountered on pathogens results from the presence of several million different antigen-binding pockets on several million different clonal populations of B cells. Development of this legion of antigen-specific B cells results from events that take place during early B-cell development, as well as somatic mutation events that take place in the antigen binding molecules of specific B cells during immune responses to their target antigen.

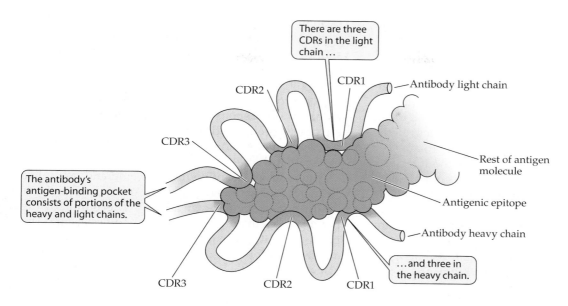

Figure 27.17 Interaction between antigen and antibody
Interaction of the antigen-binding pocket of an antibody molecule with epitopes of an antigen. The portions of the antibody molecule that interact with the antigen epitopes are hypervariable regions called complementarity-determining regions (CDRs).

There are two families of immunoglobulin light chain genes called the kappa (κ) and lambda (λ) families. Each of the families is on a different chromosome. The light chain genes are also on chromosomes different from that of the heavy chain gene. It is most useful to think of all of the germ line immunoglobulin genes as gene segments and to think of B-cell development as a process through which individual gene segments are moved together in such a way as to generate genes encoding a complete κ or λ chain gene and a complete heavy chain gene. This movement of immunoglobulin gene segments together to form mature heavy and light chain genes is referred to as immunoglobulin gene rearrangement and the mechanism of rearrangement is similar to that described previously for the generation of the T-cell antigen receptor.

Immunoglobulin gene segment rearrangements are a property of B cells only, and do not occur in other cells of the body. Rearrangement of immunoglobulin gene segments occurs independently in all precursor B cells (pre–B cells). The first immunoglobulin gene to be assembled during B-cell development is the heavy chain gene. This is made by the moving together of a heavy chain variable gene segment (V_H), a diversity segment (D_H), and a joining segment (J_H) so that they can be transcribed as a single messenger RNA, together with a gene segment encoding a heavy chain constant region, specifically that for μ. There are several heavy-chain constant-region gene segments that are adjacent to each other in

all mammal species with the segment encoding the C_Hμ being the closest to the J_H gene segment. The number of V_H, D_H, and J_H segments varies among species.

Expression of a μ heavy chain by a developing B cell starts rearrangements that lead to assembly of a light chain gene from variable light (V_L) and joining light (J_L) segments. This is first attempted with segments encoding a κ light chain, and successful rearrangement (meaning it gives rise to a gene that can successfully code for a protein) stops further light-chain gene-segment rearrangements. If rearrangement is unsuccessful for the κ gene (on either of the parental chromosomes), rearrangement continues for the λ gene. If L-chain gene rearrangement is unsuccessful for both κ and λ light-chain genes, the developing B cell dies. Successful rearrangement of a μ H-chain and an L-chain gene allows assembly of an IgM molecule and heralds the appearance of the immature B cell. This process of immunoglobulin gene assembly is common to humans and mice, irreversible in any B cell, and generates millions of different B-cell antigen-binding receptors. Other mammals, for example, cattle, also generate complete immunoglobulin genes from gene segments, but only a few complete genes are assembled. The B cells that arise subsequently diversify this basic immunoglobulin gene by a process called gene conversion, which occurs in the Peyer's patches of the gut most likely under the influence of the gut microbiota.

Some B-cell antigen receptors that are generated are capable of binding self-antigens with high efficiency. Fortunately, immature B cells are negatively regulated by antigen. When surface immunoglobulin is cross-linked by binding antigen at this stage of B-cell development, the cell dies or undergoes a cycle of receptor editing in which further rearrangements occur in L-chain genes that can alter the antigen-binding specificity of the expressed receptor. In humans and mice approximately 90% of B cells generated each day die without leaving the bone marrow. This selection process is modified by B-cell survival factors, the discussion of which is outside the scope of this chapter.

The joint processes of negative selection and receptor editing eliminate a significant portion of the self-reactive B cells that are generated. However, many self-antigens are present at only very low concentration, are absent from the bone marrow, or are developmentally regulated; therefore, not all self-reactive B cells are eliminated. Any immature B cell that is not eliminated matures to a resting or virgin B cell, which can be stimulated to proliferate by cross-linking of surface immunoglobulin as discussed on p. 873. If the mature B cell does not encounter antigen it eventually dies. How self-reactive B cells that escape elimination during their early development and encounter antigen when mature are regulated is still under investigation.

B Cells and Antibodies

Antibodies are water-soluble copies of the surface immunoglobulin expressed by the B cell that makes them. Antibodies are secreted in large quantities by plasma cells. **Plasma cells** are the end-differentiative stage of antigen-activated B cells. Differentiation of resting B cells to plasma cells is stimulated by antigen. B-cell ac-

tivation by antigen is discussed in subsequent text. Molecules present in the body of an organism are called **self-antigens** and typically stimulate no or only low-titer and harmless antibody responses in the organism. When self-antigens do elicit a high-titer antibody response, or harmful immune response, this is called an **autoimmune response**. All non-self antigenic materials are called **foreign antigens**. Some material is not capable of stimulating an immune response, and is referred to as **nonantigenic**.

The purpose of antibodies is to bind and neutralize (render harmless) antigens. Binding occurs in an antigen-binding pocket present near the N terminus of the polypeptides that make up an antibody molecule. The part of an antigen that binds in the pocket is called an antigenic **epitope** (Figure 27.18). Neutralization may be achieved by binding the antigen, as is the case with toxins; however, it is generally facilitated by additional properties of antibodies that are latent until antibody has bound to antigen. These are discussed later in this chapter. There are several classes and subclasses of antibodies. Not all antibody classes and subclasses are present in all species of mammals, although the main classes are represented (i.e., IgM, IgD, IgG, IgA, and IgE). The different immunoglobulin classes have different molecular weights and different biological properties dictated by how the non-antigen binding parts of the antibody molecule (**Fc region**) interact with cells and molecules of the immune system. The genetic processes that lead to generation of the different immunoglobulin classes and their biological functions are discussed in preceding text.

When size, shape, charge, and phobicity permit, an antigenic epitope that enters an immunoglobulin antigen-binding pocket can form a noncovalent association with amino acid residues that line the pocket, in which case the antigen and antibody now form a complex (see Figure

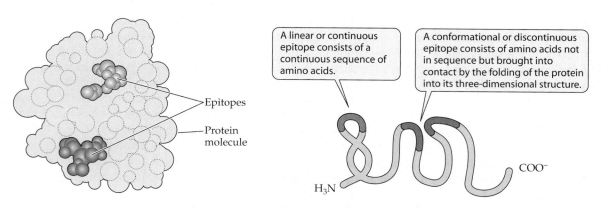

Figure 27.18 Antigenic determinants on a protein molecule
Specific portions of antigens, called the antigenic determinants or epitopes, are exposed on the intact antigen molecule and recognized directly by antibodies or B-cell receptors in their native form.

27.17). This complex is held together by four forces, each weak, but in combination adequate to sustain a stable association that can be broken only by proteolytic enzymes or chemical conditions that break the noncovalent associations. The forces are (1) hydrogen bonds in which a hydrogen atom is shared between two electronegative atoms; (2) ionic bonds between oppositely charged residues; (3) hydrophobic interactions in which the absence of polar groups prevents interaction with water, which consequently forces hydrophobic groups together; and (4) van der Waals interactions, which form between the outer clouds of two or more atoms.

An antigenic epitope could be a linear stretch of amino acid residues on a large polypeptide (called a linear or continuous epitope), a group of non-adjacent amino acids brought together by folding of a protein (called a discontinuous or conformational epitope) (see Figure 27.18), or it might be a carbohydrate, a stretch of nucleic acid, a lipid, a glycolipid, a glycoprotein, or even a chemical group called a hapten. A hapten does not induce an immune response on its own, but can do so when bound to a larger molecule.

B-Cell Activation and Immunoglobulin Heavy Chain Switching

Antigens that activate B cells fall into two categories: those that require the involvement of helper T cells, called **T cell–dependent (TD) antigens** and those that do not or require only minimal T-cell help, called **T cell–independent (TI) antigens**. The TI antigens fall into two classes, called **TI-1** and **TI-2** antigens. Activation of B cells by TI-1 antigens occurs in the complete absence of T cells. Activation of B cells by TI-2 antigens has a minimal requirement for T-cell help.

TI-1 antigens, such as bacterial lipopolysaccharide, can induce B-cell proliferation without binding to cell-surface immunoglobulin. They do so by reacting with members of a family of receptors, called **Toll-like receptors** or **TLRs** that have been selected through evolution to recognize common molecular patterns on pathogens. A high concentration of LPS activates many clones of B cells, including cells that do not bind the TI-1 antigen by their surface immunoglobulin. TI-1 antigens are therefore **polyclonal B-cell activators**. At lower concentrations, LPS induces highly specific B-cell responses. This results from the antigen being enriched on specific B-cell surfaces, consequent to being bound by LPS-specific immunoglobulin on those cells. Stimulation of B cells by TI-1 antigens typically induces only IgM production and can result in highly elevated levels of serum IgM. TI-1 antigens do not elicit immunological memory.

TI-2 antigens are highly repetitive molecules, such as polymeric proteins, bacterial cell wall material with re-

peating polysaccharide units, and bacterial flagellin. They activate B cells by extensive cross-linking of the surface immunoglobulin. Development of B-cell responses to TI-2 antigens requires low concentrations of cytokines supplied by T cells and does not result in polyclonal B-cell activation at any concentration. TI-2 antigens mainly induce IgM antibodies, but a low amount of IgG may also be made. The TI-2 antigens do not elicit immunological memory.

TD antigens are typically soluble proteins with few repeating antigenic epitopes. In the case of TD antigens, B-cell antigen-specific receptor cross-linking on its own is inadequate to stimulate cell division. Additional signals from T cells are required to elicit this response and interaction of the B cells and T cells is facilitated by presentation of antigenic peptides on B-cell MHC class II molecules, as discussed on pp. 864, 865. A co-receptor complex on B cells modifies their activation. This co-receptor complex includes complement receptor 2 (CR2, which binds C3dg and C3d as discussed in the section on Inflammation). Binding of C3d or C3dg to this receptor is required for activation of T cell–dependent B cells.

Antigen that binds to the B-cell receptor is endocytosed by the B cell. The antigen is processed, and peptides are presented on the B-cell surface in association with MHC class II antigens, leading to interactions with CD4 helper T cells, the T cell receptor of which is specific for the antigen. Activation of the B cell leads to their elevated expression of MHC class II antigens and the expression of the T-cell co-stimulator molecule called B7. Recognition of the peptide-MHC class II complex by CD4 T cells together with co-stimulation through B7 causes the T cells to express CD40 ligand. This binds CD40 on B cells, providing the second signal required for their activation. At the same time, the associated T cells release IL-2 and IL-4, which support progression of the B cells to the DNA synthesis stage of their cell division cycle and eventually to differentiate to memory cells and to high-rate antibody-secreting cells called plasma cells (see Figure 27.13).

Thus, stimulation with TD antigens in conjunction with signals provided by CD4 T cells causes B cells to replicate and give rise to plasma cells that secrete antibodies. The antibodies have the specificity of the B-cell receptor because they result from transcription of the same immunoglobulin heavy chain and immunoglobulin light chain genes. Whether the immunoglobulin heavy chain is made as a secreted (antibody) versus a surface form (B-cell receptor) is determined by differential processing of the same primary mRNA transcript. In the case of secreted immunoglobulin, two exons of the gene segment encoding the C terminal of the heavy chain are spliced from the mRNA.

Antigen binding and stimulation of B cells with T-cell cytokines can result in a process called **Ig class switch-**

TABLE 27.6	Cytokine-induced Ig class switching
Cytokine	**Ig Class Switch**
IFN-γ	IgG2a or IgG3
TGF-β	IgA or IgG2b
IL-4	IgE or IgG1
IL-2, IL-4, IL-5	IgM

ing, in which the V_H-D_H-J_H gene segment assembly, which encodes the variable region of the immunoglobulin heavy chain, is spliced onto a heavy-chain constant-region gene segment lying downstream of the immunoglobulin heavy-chain δ gene segment. In this way, the same variable region gene sequence can be joined to either an immunoglobulin heavy chain γ gene segment to generate an IgG heavy chain, or to an immunoglobulin heavy chain α gene segment to generate an IgA heavy, or to an ε segment to generate an IgE heavy chain. These rearrangements occurring in progeny of an activated mature B cell result in the generation of daughter cells producing IgM, or IgG, IgA, or IgE with the same antigen-binding specificity.

Immunoglobulin heavy-chain class switching is not reversible. It involves excision of DNA lying between flanking sequences of DNA, called switch regions. Switch regions are located upstream of each immunoglobulin heavy-chain constant-region gene segment. It is hypothesized that various cytokines make these switch regions accessible to enzymes that loop out and excise intervening sequences between the chosen switch and variable regions. This may account for the ability of different cytokines to bias B cells to make antibodies of different immunoglobulin classes (Table 27.6). The different immunoglobulin heavy-chain constant-region gene sequences imbue antibodies with different secondary functions. The functions of different antibodies are discussed on p. 878.

B-Cell Maturation, Memory, and Secondary Immune Responses

T cell–dependent antigens on infectious agents, toxins, and venoms stimulate B-cell responses that expedite clearance of these agents from the body and prepare the body for the rapid elimination of the same agents should they be reintroduced. The B-cell responses have three components:

• The development of a primary immune response that involves production of antibodies by short-lived plasma cells

• Immunoglobulin class switching, somatic mutation of the variable region of the expressed immunoglobulin genes, and selection of cells with an improved capacity to bind the specific antigen (a process called affinity maturation), and the generation of memory B cells, for example, B cells that rapidly give rise to antibody-secreting cells on re-exposure to antigen, and the development of long-lived antibody-secreting plasma cells

The development of these responses occurs in the spleen or in lymph nodes to which antigen that is coated with the complement components C3b and iC3b (and therefore C3dg and C3d) is drained. Purified proteins typically do not provoke inflammation and also do not invoke primary immune responses unless administered with material that elicits the inflammatory response. Primary antibody responses take several days to arise and result in an increase in serum IgM that is specific for the stimulatory antigens (Figure 27.19). IgM has a half-life of about 7 days in plasma and often disappears fairly rapidly from the blood, indicating that it is produced for a short time only. Infectious agents that take some time to be eliminated from the body may continue to stimulate the production of IgM, a process that requires continuous recruitment of resting B cells into T cell–independent (TI) responses, or expression of new antigens as a result of antigenic variation, or both.

Within the secondary lymphoid organs (i.e., the spleen, lymph nodes, and mucosa-associated lymphoid tissue), lymphocytes segregate into T-cell and B-cell areas, the latter in ovoid groups called **follicles**, which are regions where B cells develop into memory cells. Follicles also contain **follicular dendritic cells**, which are phagocytes that are specialized for antigen capture and retention and that form the scaffolding of an antigen-induced body called a **germinal center**. In addition, follicles contain a type of macrophage, called a tingible body macrophage, which endocytoses apoptotic cells that arise in germinal centers. Follicles also contain a few CD4 T cells that, together with TD antigens, are required for germinal center formation.

Follicles are divided into two types, called primary and secondary follicles. **Primary follicles** mainly contain small resting B cells and a loose network of follicular dendritic cells and are antigen independent, whereas **secondary follicles** contain an inner region of activated B cells associated with a denser network of follicular dendritic cells, called a germinal center, and are formed in response to antigenic stimulation. T cell–independent antigens do not induce germinal center formation.

When C3b/iC3b-coated antigen is transported into a secondary lymphoid organ, a portion encounters dendritic cells, B cells, and T cells in the interfollicular areas,

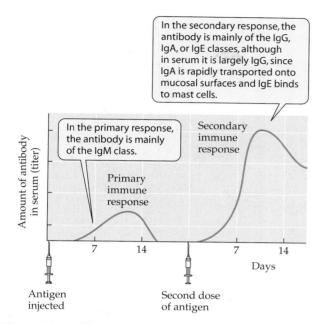

In the secondary response, the antibody is mainly of the IgG, IgA, or IgE classes, although in serum it is largely IgG, since IgA is rapidly transported onto mucosal surfaces and IgE binds to mast cells.

In the primary response, the antibody is mainly of the IgM class.

Figure 27.19 Time-course of antibody response
A hallmark of the adaptive immune system is memory: On second exposure to an antigen, the immune response develops more rapidly and reaches a higher magnitude than at the first exposure. This results from clonal expansion during primary exposure to the antigen and less stringent requirements for memory lymphocytes to respond; as a consequence, more antibody is made. Antibody circulates in the fluid component of blood and is usually measured in serum obtained by collecting the cell-free components from clotted blood. The amount of antibody present for a specific antigen is measured and is typically expressed as the titer, the reciprocal of the greatest dilution of serum that gives a measurable reaction with the antigen.

causing T-cell and B-cell activation, antibody formation, and immunoglobulin heavy chain class switching, as discussed earlier. A portion of the complement-tagged antigen also encounters follicular dendritic cells, is endocytosed, and is subsequently presented on the surface of these cells, as are antibody–antigen complexes. Activated B cells migrate from the interfollicular areas into follicles, in regions called germinal centers, where they are driven through several divisions and at the same time subjected to modifications in their immunoglobulin variable region gene segments that result in modification of the antigen-binding sites of expressed antigen-binding receptors.

This process of somatic mutation has two possible consequences with respect to antigen binding. It can be improved or diminished. This is critical to the fate of the germinal center B cells, because continued cross-linking of their antigen binding receptors is required for the expression of anti-apoptosis molecules by the B cells. In the ensuing competition among antibodies and anti-

gen binding receptors on germinal center B cells, only those B cells whose receptors bind to available antigen survive. This process is called **affinity maturation** and results in B cells whose antigen-specific receptors form very strong interactions, called high affinity interactions, with the selecting antigen. B cells that do not bind antigen after somatic mutation die, are consumed by tingible body macrophages, and their components are broken down and recycled.

B-cell activation and selection in secondary follicles also results in the development of long-lived plasma cells producing IgG, with high affinity for the selecting antigen. Many of the long-lived plasma cells relocate to the bone marrow, which provides survival niches. These cells continue to produce antibody for periods in excess of a year. The conditions that ensure survival of the long-lived plasma cells in the bone marrow and the kinetics of niche occupation and turnover are still under study. This research may result in the development of vaccines that give rise to long-term antibody responses.

In addition, B-cell activation and selection in secondary follicles gives rise to memory B cells, which are long lived and responsible for secondary immune responses that arise on re-exposure to the priming antigen (see Figure 27.19). Secondary immune responses develop more rapidly, are of higher magnitude and affinity compared to primary responses, and involve B cells that have undergone switching of the immunoglobulin heavy chain class. Immunological memory can last for many years. The longevity of the memory cell response reflects the life span of individual memory B cells and T cells, the retention of pockets of stimulatory antigen and interactions of memory cells with antibodies and T cells whose receptors react specifically with the antigen-binding portion of receptors of the memory cells (called anti-idiotypic antibodies and lymphocytes).

Orderly development of primary antibody responses, immunoglobulin heavy chain class switching, germinal center formation, B-cell affinity maturation, and memory cell formation are features of response to dead antigens and some pathogens that are rapidly cleared from the body. These processes can be disrupted, or not occur at all, during acute and chronic infections. For example, in mice that are infected with African trypanosomes, the spleen becomes massively enlarged, the distinctions between areas containing red blood cells (called red pulp) and leukocytes (called white pulp) blur, follicles do not develop, there is massive activation of splenic phagocytes, plasma cells are spread throughout the organ, memory development is poor, and necrotic areas develop. Furthermore, with time, the capacity of the spleen and lymph nodes to develop immune responses is diminished or lost as a result of immunosuppressive products released by activated phagocytes. These gross manifestations of immunopathology are common in hosts infected with disease causing organisms and are the subject of ongoing investigations aimed at elucidating mechanisms of resistance and susceptibility to disease.

Functions of Antibody Classes: IgG, IgA, IgM, and IgE

IgG is primarily viewed as an opsonin. IgG that has bound antigen is in turn bound by IgG-Fc-receptors on macrophages, and the complex is phagocytosed/endocytosed and degraded; the site on IgG that engages the macrophage receptor is exposed only after the IgG has bound antigen. Sites at the C-terminal domains of the IgG molecule also facilitate its transport across the placenta in some species to help protect the fetus and across the gut of the neonate to help protect the newborn before full development of its immune system. IgG is one of the most abundant serum proteins. There are about 80 mg of protein per milliliter of serum in an adult cow, and about 10% of this protein is IgG. There is about 10 times less serum IgM than IgG and little or no free IgE (most IgE is bound to mast cells). Serum IgA is usually present at an even lower concentration than serum IgM because of its efficient transport out of the plasma and onto mucosal surfaces (Figure 27.20).

IgM is renowned for its capacity to activate complement. IgM that has bound antigen activates complement by the classical pathway (see Figure 27.5), leading to the generation of pro-inflammatory complement fragments (C3a, C4a, C5a), fragments that promote phagocytosis of materials to which they are attached (C3b, iC3b), fragments that are required for activation of T cell–dependent B cells (C3d, C3dg), and assembly of the membrane attack complex. The site on IgM that activates complement is exposed only after IgM has bound antigen. IgM is much more efficient at initiating the classical pathway of complement activation than IgG, whereas IgA and IgE do not activate complement at all. Activation of complement by antibody requires the binding of a plasma complement component called C1q, a large protein with six globular heads, two of which must be bound for activation of C1. Each globular head group of C1q engages a site that is exposed on the Fc region of IgM or IgG when it is in complex with antigen (Figure 27.21). Because IgM is a pentamer, it potentially has five C1q binding sites, whereas IgG, being a single subunit, has only one. In order for IgG to elicit complement activation, two IgG molecules are required to attach to antigen within 30 to 40 nanometers of each other to provide two attachment sites for C1q. Once two sites of C1q are engaged, C1r and C1s are recruited into the complex, C1 is activated, and the pathway proceeds, as shown in Figures 27.21 and 27.5. Thus, a single antigen-associated IgM can activate the classical pathway of complement activation, whereas

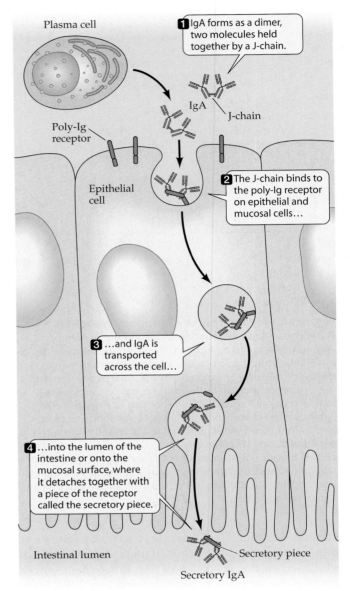

Figure 27.20 Secretion of IgA into the intestinal lumen
Plasma cells that produce IgA are enriched in the lymphoid tissues associated with the digestive system. A specialized transport system allows the IgA to get into the lumen of the intestine where it can act against infectious agents that are swallowed and survive the barrier defense systems.

IgG molecules must be bound in proximity to one another to achieve this. Although IgM is not generally considered to be an opsonin, IgM that is bound to antigen is in turn bound by IgM-Fc receptors that are expressed on most B cells and some T cells. The immunological significance of this is still under study.

IgA is specialized for transport across mucosa, where its presence on the luminal surface helps prevent pathogens with target antigenic epitopes from crossing the mucosa as well as to bind and inactivate toxins.

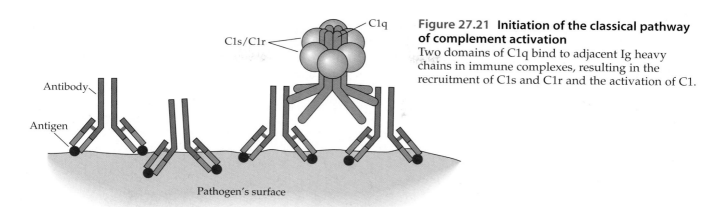

Figure 27.21 Initiation of the classical pathway of complement activation Two domains of C1q bind to adjacent Ig heavy chains in immune complexes, resulting in the recruitment of C1s and C1r and the activation of C1.

Transport of IgA across the mucosal epithelial cells involves binding to a receptor on these cells called the poly-Ig receptor. This receptor binds to the J-chain of dimeric IgA. It can also bind to the J-chain of IgM but less efficiently than to that of IgA, even though these have the same J-chain. IgA is transported through the mucosal epithelial cell in association with the poly-Ig receptor, the piece of the receptor that binds J-chain is enzymatically cleaved in the cell. The dimeric IgA is released on the luminal side of the cell with its associated section of the poly-Ig receptor, now called the secretory piece. IgA is commonly detected in tears, saliva, feces, and the fluids bathing the urogenital tract.

IgE binds to IgE-Fc-receptors on mast cells, and when the mast cell–bound IgE is engaged by antigen, a signal is propagated that causes mast cell degranulation; histamine released in this process is responsible for many of the symptoms of hay fever, which is an allergic response to pollen. IgE also binds directly to antigen, resulting in changes in its Fc region that causes it to bind to Fc ε receptors on eosinophils. Cross-linking of these receptors causes eosinophils to degranulate, releasing materials that are toxic for parasitic organisms (Figure 27.22). This is an important component of defense against parasitic worms.

Figure 27.23 presents a summary of the roles played by antibodies of different classes in host defense. To get a pure population of antibodies of one particular isotype reactive with a single antigen epitope, hybridomas are made that secrete monoclonal antibodies (Box 27.1).

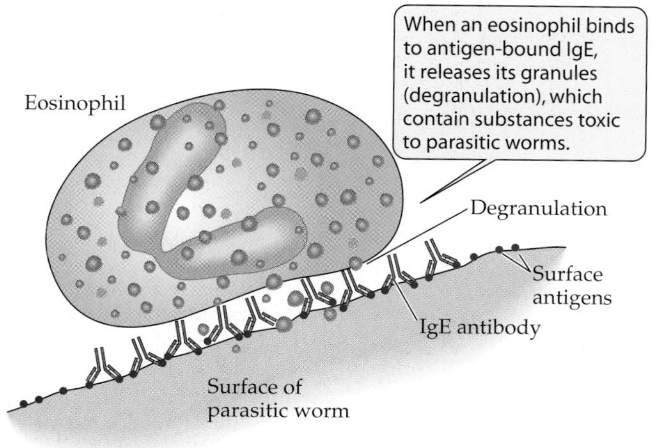

When an eosinophil binds to antigen-bound IgE, it releases its granules (degranulation), which contain substances toxic to parasitic worms.

Figure 27.22 Interaction of an eosinophil with a parasitic worm An eosinophil interacts with a parasitic worm through antibody bridging. Eosinophil granules contain several substances that are toxic to parasitic worms.

SECTION HIGHLIGHTS

Inflammation and the phagocytic cells of the innate immune system slow pathogen population growth rates but typically do not eliminate the pathogens. For this to occur, highly specific adaptive immune responses are required. The leukocytes involved in the adaptive immune system are subpopulations of lymphocytes known as B cells and T cells. B cells that respond to pathogens make and secrete water-soluble proteins called antibodies. Antibodies can neutralize toxins, coat pathogens so that they are phagocytosed, or activate complement causing the membrane attack complex to form on the surface of pathogens. T cells are specialized to kill cells that are infected by pathogens, to enhance the capacity of macrophages to kill pathogens, and to enhance responses by B cells. The hallmarks of responses by lymphocytes are specificity and immunological memory.

Antibody interactions	Effect	Types of infectious agent targeted	Examples of infectious pathogen component or structure targeted	Isotype of Ab involved	Where antibody is found
Bacterium / Host cell	Prevents attachment to host cells for colonization of the body	Bacteria	Fimbriae, pili, flagella, LPS, outer membrane proteins of bacteria	IgA	Mucous membranes in upper respiratory tract, intestinal tract, vagina
Virus / Host cell	Prevents attachment and entry into host cells for infection	Viruses	Influenza virus hemagglutinin for attachment and neuraminidase for entry into host cell	IgA IgG	Depends on the target cells for virus infection: IgA for cells associated closely with mucous membranes and IgG for internal tissues and organs
	Agglutination	Bacteria, viruses, fungi, protozoa	Any external structures with antigenic epitopes	IgM IgG	Body tissue
Toxin / Host cell	Neutralizes toxins	Produced by bacteria	Secreted toxins, such as cholera toxin, tetanus toxin, diphtheria toxin	IgA IgG	Gut lumen and tissues
Complement component	Kills by activating complement onto microbial surface	Bacteria, viruses, protozoa	Any external surface	IgM IgG	Internal tissues and organs
Phagocyte	Opsonization for phagocytosis	Bacteria, viruses, protozoa	Any external surface	IgG	Anywhere phagocytic cells are found
See Figure 27.22	Tag for recognition by eosinophils	Helminths	Tegument	IgE	Intestinal tract
See Figure 27.24	Tag for antibody-dependent cellular cytoxicity	Cells infected with enveloped virus	Influenza virus	IgG	Where it comes in contact with NK cells in internal tissues and organs
See Figure 27.5 and Figure 27.6	Kills by activating complement on surface of host cells infected with enveloped virus	Cells infected with enveloped virus	Influenza virus	IgG IgM	Internal tissues and organs

◀ **Figure 27.23 Antimicrobial activities mediated by antibodies**
This diagram summarizes the various activities of antibodies that help to prevent infection.

BOX 27.1 *Research Highlights*

ⓘ Monoclonal Antibody Production

Antigens typically have several different epitopes, each of which stimulates B cells with a distinct B-cell receptor. The resulting immune sera recognize many different epitopes, even when a single purified polypeptide is used as antigen. These polyvalent sera are not suitable to be used for diagnostic reagents because they often show cross-reactions among different antigens and are difficult to reproduce exactly. Monoclonal antibody (mAb) technology has solved this problem. mAb technology has made it possible to immortalize a clone of antibody-secreting B cells and thus to generate an unlimited amount of its secreted antibody. Because all of the progeny cells make the same antibody, there is only one type of antigen-binding site represented. Thus, the immunoglobulin product of a clone of B cells is a monoclonal antibody.

mAb technology is simple. Antigen-stimulated B cells are fused to a tumor cell, and the hybrid has the transformed phenotype of the tumor but still makes and secretes the immunoglobulin specified by the stimulated B cell. A B-cell tumor called a myeloma is used as the fusion partner, and the most commonly used myeloma has been modified so that its own

Continued on next page

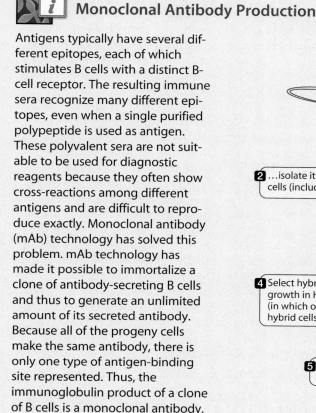

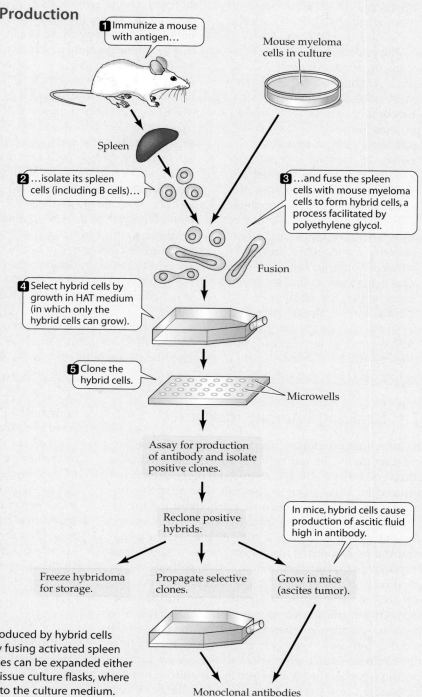

1 Immunize a mouse with antigen...

Mouse myeloma cells in culture

Spleen

2 ...isolate its spleen cells (including B cells)...

3 ...and fuse the spleen cells with mouse myeloma cells to form hybrid cells, a process facilitated by polyethylene glycol.

Fusion

4 Select hybrid cells by growth in HAT medium (in which only the hybrid cells can grow).

5 Clone the hybrid cells.

Microwells

Assay for production of antibody and isolate positive clones.

Reclone positive hybrids.

In mice, hybrid cells cause production of ascitic fluid high in antibody.

Freeze hybridoma for storage.

Propagate selective clones.

Grow in mice (ascites tumor).

Monoclonal antibodies

Monoclonal antibodies are produced by hybrid cells (thus the name *hybridoma*) by fusing activated spleen cells with myeloma cells. Clones can be expanded either in mice, as shown here, or in tissue culture flasks, where the antibodies are secreted into the culture medium.

BOX 27.1   *Continued*

Ig genes are no longer functional. Köhler and Milstein developed the method to make and select hybridomas in 1975. It is based on the way that cells obtain purines for synthesis of DNA and RNA.

The genetic material of cells is made up of purines and pyrimidines. Each time a cell replicates, it must make or obtain these essential components. Cells can make purines by the de novo synthetic pathway or obtain preformed purines from medium or extracellular fluids by the salvage pathway. The de novo synthetic pathway is sensitive to the drug aminopterin, a folic acid antagonist that blocks de novo purine synthesis. Tumor cells that are incubated in vitro with aminopterin continue to replicate in medium containing purines because they can obtain the purines by the salvage pathway.

In the salvage pathway, preformed purines are imported into the cell and remodeled.

Hypoxanthine guanine phosphoribosyl transferase (HGPRT) is a key enzyme in the purine salvage pathway. This enzyme adds a phosphoribosyl group to the purine base, thereby converting it to a purine nucleotide, which is the form of purine present in DNA. Köhler and Milstein developed a myeloma cell line that lacked HGPRT by culturing mutagenized cells with a toxic purine analog called 8-azaguanine. Cells with an active purine salvage pathway incorporated 8-azaguanine and died. 8-Azaguanine resistant cells lack a purine salvage pathway and rely solely on their de novo purine synthetic pathway for purines required for DNA and RNA synthesis. 8-Azaguanine–resistant cells are sensitive to aminopterin, which blocks de novo purine synthesis.

Aminopterin-sensitive myeloma cells were fused to antigen-stimulated B cells to generate hybrids.

Fusion was encouraged using polyethylene glycol (PEG), which causes the membranes of closely juxtaposed cells to meld into a single lipid bilayer, incorporating the content of both cells including their nuclei. Fused cells randomly assort chromosomes at the next metaphase. The resulting progeny cells that arise at cell division contain a mix of chromosomes. In a medium containing aminopterin and hypoxanthine, only those cells that have obtained an HGPRT gene from the B cell, the transforming gene from the myeloma, and a full complement of relevant housekeeping genes multiply, and all other cells die. The growing hybridomas are cloned by placing single cells in a liquid medium or by growing single cells as colonies on a medium that contains gelling agents, and antibodies produced by the hybridoma cells are screened for antigen-binding activity.

27.4 Bridging Immune Responses

The inflammatory/innate response is rapid because participating molecules and cells are already present in significant quantity in the blood, and cell replication is not required for this response to occur. The adaptive immune response is slow because participating antigen-specific lymphocytes are initially few in number and have to replicate to achieve a critical number. The bridging immune response falls between these and involves lymphocytes that can respond rapidly to pathogens. The functions of cells that make up the bridging immune system are the same as those of cells in the antigen-specific adaptive immune response (Table 27.7). However, there are significant differences in how the cells of the bridging immune system are triggered to display their functional activities, even when they have an antigen-specific receptor that is structurally very similar to that of classical T and B cells. These include a class of B cells that arises early in ontogeny and responds to TI-2 antigens called B-1 B cells, a B-cell population called marginal zone B cells that is restricted to the spleen and specialized for rapid responses to bloodborne pathogens, as well as subpopulations of T cells called γδ T cells, and CD1-restricted T cells. In most instances, these cells do not seem to display memory responses, constitute only a low percentage of the total leukocyte population of the blood, and have limited diversity of their antigen receptors. This limited diversity may allow a significant response by an otherwise small population of cells and in this way precludes the need for development into memory cells.

Primitive T Cells: NK Cells

Natural killer (NK) cells are in the lymphoid lineage and are generally considered to be primitive T cells because they have a progenitor cell in common with other T cells. However, NK cells do not mature in the thymus, unlike T cells that are involved in adaptive immune re-

TABLE 27.7	Cells of the bridging immune system		
Type of Cells	**Functions**	**Types of Antigen Interaction**	**Receptors Involved in Activation**
NK cells	IFN-γ production; cytotoxicity; ADCC	Microbial antigens via antibody; altered self-cells	NCR NKG2A NKG2D KIR Fc
γδ T cells	IFN-γ production; cytotoxicity	Soluble antigens; nonproteinaceous self-antigens	γδ T-cell receptor NKG2D
NKT cells	IL-4 production IFN-γ production	Lipid-containing bacterial proteins	Invariant αβ T-cell receptor
CD5 B cells (B-1 B cells)	Make IgM antibodies	Common carbohydrates of bacteria, self-antigens	B-cell receptor

sponses, and NK cells do not express an antigen-specific T-cell receptor. Nevertheless they share two functions in common with T cells of the adaptive immune system, that is, production of the cytokine IFN-γ and cytotoxicity. NK cells are important during the first few days of an infection before the adaptive immune response has developed. For example, NK-cell production of IFN-γ is important in protection against the intracellular bacteria *Listeria monocytogenes*. In the absence of NK cells, some infections have been shown to be lethal in mouse models. NK cells are also important in controlling virus infections by IFN-γ secretion.

NK-cell cytotoxicity (Figure 27.24) is mediated by a mechanism similar, if not identical, to that used by CD8 T cells. When activated, the NK cells have granules in their cytoplasm that contain perforin. These are visible under the light microscopy, and the NK cells are there-fore known as large granular lymphocytes. NK cells kill target cells that do not express normal levels of MHC class I molecules. They have receptors for class I MHC molecules that, when engaged, signal the NK cells not to kill the target cell. The reader may recall that one way viruses try to thwart the adaptive immune system is by decreasing host cell MHC class I expression, thereby de-creasing the presentation of antigen on class I. However this backfires to an extent, since reduced levels of class I MHC on these cells makes them vulnerable to killing by NK cells. In this way the NK cells act as a backup for the adaptive immune system.

NK cells also have other receptors. Fcγ receptors al-low them to bind to IgG and antigen complexes. These receptors can link NK cells to target cells bearing anti-gen with IgG bound, allowing the NK cell to kill the tar-get cell. This process is known as **antibody-dependent**

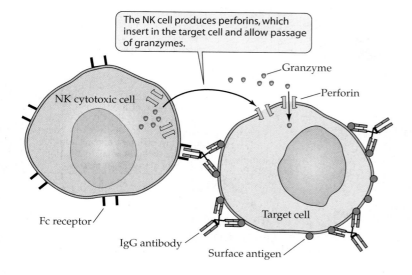

Figure 27.24 Antibody-dependent cellular cytotoxicity (ADCC) by natural killer (NK) cells Cytotoxic NK cells have Fc receptors for IgG anti-bodies, which allow the cells to bind antibody-coat-ed cells on which the antibody Fc piece is accessible, and kill them. This is known as antibody-dependent cellular cytotoxicity (ADCC). An example is a host cell infected with an enveloped virus; the enveloped virus inserts viral proteins into the host cell mem-brane before budding out of the host cell, and these viral proteins bind specific antibodies.

cellular cytotoxicity (**ADCC**) (see Figure 27.24) and might be effective against host cells infected with enveloped viruses that display viral proteins in the host cell membrane. NK cells also have a receptor called NKG2A, for MHC-class I-like molecules that are expressed by mycobacterial-infected cells. When this receptor is stimulated, the NK cells display cytotoxic activity. A third type of receptor, called the NK cell receptors, or NCR, also provides activation signals. NCR are only on NK cells and mediate cytotoxicity against Epstein-Barr and other virus-infected cells and against tumor cells.

Lymphoid Cells with Antigen-Specific Receptors: B-1 and Marginal Zone B Cells

B cells can be divided into subpopulations based on the immunoglobulin classes expressed on their surface and on whether they have an additional differentiation antigen called CD5. Mature B cells that have IgM and CD5 on their surface are called B-1 B cells (Figure 27.25) and are richly present in the peritoneal cavity. The antigen-binding receptors of B-1 B cells, and the antibodies they produce, are skewed toward reactivity with common pathogen-associated carbohydrate antigens and toward weak (and harmless) reactivity with self-antigens. B-1 B cells produce **natural antibodies**, which are antibodies of the IgM class that are sustained in plasma without obvious antigenic stimulation. B-1 B cells spontaneously produce these antibodies when placed in tissue culture or moved to an immunoglobulin-free host animal. The B-1 B cells most likely help protect against pathogens that leak into the peritoneal cavity and blood system as a result of damage to the intestinal tract. Natural antibodies produced by B-1 B cells are reactive with bacterial LPS, the archetype TI-2 antigen discussed ear-

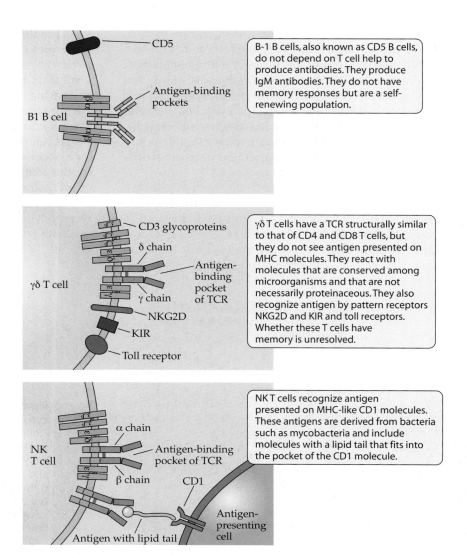

Figure 27.25 Lymphoid cells of the bridging immune system
Bridging lymphoid cells are stimulated via antigen-specific receptors.

lier, and consequently limit nonspecific B-cell activation and endotoxin shock induced by LPS. They also contribute to resistance to influenza infection and resistance to infection with *Streptococcus pneumoniae*. In addition, there is good evidence that a significant portion of IgA-secreting plasma cells in the lamina propria of the gut derive from B-1 B cells, suggesting a role in maintaining gut microbiota homeostasis and possibly controlling enteric infections.

Mature B cells that lack CD5 and have surface IgD as well as IgM fall into two categories: the marginal zone B cells and the follicular or B-2 B cells. The marginal zone B cells are restricted to the spleen and are specialized to mount rapid T cell–independent antibody responses against bloodborne pathogens, although they can also mount T cell–dependent responses. Marginal zone B cells express a MHC-like molecule called CD1, which presents glycolipid antigens to a group of T cells called CD1-restricted T cells, which are discussed in subsequent text. The marginal zone B cell and CD1-restricted T-cell interaction may be of importance in immunity to bloodborne protozoa including the causative agents of malaria, African sleeping sickness, babesiosis, and Chagas disease (*Trypanosoma cruzi*), as well as immunity to opportunistic infections.

B-2 B cells, also called follicular B cells, are responsible for adaptive immune responses against T cell–dependent antigens and were discussed previously.

Lymphoid Cells with Antigen-Specific Receptors: γδ T Cells

The antigen-binding receptors on γδ T cells (see Figure 27.25) are similar to those on αβ T cells, and the generation of diversity occurs by the same mechanism as used by αβ T cells (i.e., rearrangement of specific gene segments). However, the genes that code for the γδ T-cell receptor are largely distinct from those coding for the αβ T-cell receptor, and, overall, the T-cell receptor diversity of γδ T cells is limited. Within a tissue or organ there may be only a few γδ T-cell receptor gene combinations expressed. The CD3 signaling complex of molecules associated with the T-cell receptor is identical for the αβ and γδ T cells. However, γδ T cells do not mature in the thymus in the same manner as αβ T cells because they do not undergo positive and negative selection, and the γδ T cells in the gut do not go to the thymus at all.

Although very few molecules have been defined that stimulate γδ T cells through the T-cell receptor, it is clear that they differ significantly from those that stimulate αβ T cells. The molecules that stimulate γδ T cells do not require presentation on MHC molecules, nor are they necessarily derived from proteins or even foreign. For example, they include nonproteinaceous phospholi-

gands, which are produced by species of *Mycobacterium* and the malaria-causing parasite *Plasmodium falciparum* and alkyl amines from bacteria. A number of self-molecules that stimulate γδ T cells have been identified, including MHC class I–like molecules called MICA and MICB and heat-shock proteins that cross-react with mycobacterial heat-shock proteins.

The limited T-cell receptor diversity means that within an organ, relatively large numbers of γδ T cells have identical T-cell receptors. The only time that the frequency of αβ T cells with identical T-cell receptors is as high is at the height of an infection after they have undergone tremendous clonal expansion. This peculiarity may enable γδ T cells to have an immediate effective response to conserved molecular components of microbial pathogens, abnormal host cells such as tumor cells, and even normal self-components under special circumstances, such as during an inflammatory response, without needing to expand to establish sufficient numbers. In support of this, an influx or increase in the number of γδ T cells has been reported for a large number of infectious diseases, including those mediated by viruses, bacteria, and protozoa.

γδ T cells are most often reported to produce IFN-γ, and thus it has been suggested that they may generally cause a type 1 cytokine response. This is consistent with their role in limiting viral infections and playing a crucial role in bacterial infections, particularly the development of granulomas in mice with tuberculosis. However, it has also been shown that they may make the type 2 cytokine IL-4 in animals with a parasitic worm infection. γδ T cells also have been shown to have cytolytic activity against foreign cells, although cytotoxicity is not necessarily a common feature of all γδ T cells. Also, because of their potential ability to produce type 1 and type 2 cytokines and to respond immediately, it has been suggested that they may direct development of adaptive immune responses by αβ T cells and B cells by virtue of the cytokines they secrete.

Based on current evidence, γδ T cells may respond as part of the adaptive immune response framework as well as to respond similarly to innate/nonadaptive immune responses, bridging the two systems. The latter is supported by the fact that they share elements with NK cells, clearly cells of the nonadaptive system. That is, they express the receptor molecule NKG2D, which stimulates cytotoxic responses by γδ T cells. A role in adaptive immune responses for γδ T cells is supported by studies showing that they may have a memory response to antigens of the bacteria *Leptospira*, *Mycobacteria*, and *Listeria*. In *Listeria* infections, γδ T cells play a perceptible role in resistance to secondary challenge, although it is less than that of αβ T cells and only marginally greater than in primary infections.

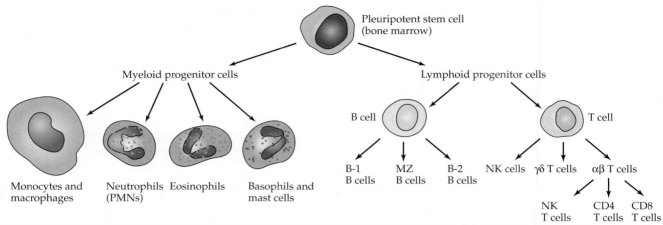

Cytokine	Produced by (cell type)	Cytokine function	Acts on (cell type)
IL-1	Macrophages, monocytes	Costimulatory molecule; fever inducer	CD4 T cells
IL-2	Th1 CD4 T cells	T cell growth factor	CD4, CD8 and γδ T cells; NK cells
IL-3	Th1 CD4 T cells	Progenitor cell growth factor	Myeloid progenitor cells
IL-4	Th2 CD4 T cells, NKT cells, dendritic cells, macrophages	B cell growth factor; drives T cells toward Type 2 and induces proliferation	B cells; CD4 T cells
IL-5	Th2 CD4 T cells	Promotes growth and differentiation	B cells; eosinophils
IL-6	Th2 CD4 T cells, macrophages	Promotes terminal differentiation into plasma cells	B cells
IL-8	Activated macrophages	Recruits phagocytes to site of inflammation	Macrophages and neutrophils
IL-10	Th2 CD4 T cells, B cells	Decreases antigen presentation; decreases macrophage activation; decreases IFN-γ production by T cells	Macrophages
IL-12	Macrophages and dendritic cells	Induces IFN-γ production	T cells, NK cells
IFN-γ	Th1 T cells, γδ T cells, NK cells	Activates macrophages; inhibits virus replication; directs T cells toward Type 1	Macrophages, T cells
IFN-α and β	Macrophages, fibroblasts, virus-infected cells	Inhibits virus replication; activates NK cells for cytolytic activity	Virus-infected cells; NK cells
TNF-α	Th1 T cells, macrophages	Activates macrophages; fever inducer; cachexia (physical wasting)	Macrophages
TGF-β	Macrophages, lymphocytes, mast cells, platelets	Antibody class switching	B cells

Figure 27.26 Relationship of all cell types involved in immunity and the cytokines they produce
All the immune system cells discussed in this chapter arise from a common leukocyte precursor in the bone marrow and differentiate into two main lineages: myeloid and lymphoid. They interact through a variety of soluble molecules called cytokines that act on other cells of the immune system.

Lymphoid Cells with Antigen-Specific Receptors: CD1-Reactive T Cells

CD1 is a nonclassical MHC molecule that does not vary among individuals of a species and that presents certain microbe antigens to specific populations of T cells. In humans there is a population of αβ T cells that recognize CD1-presented antigens, appear to have memory responses, and produce IFN-γ and kill mycobacterial-infected cells. They respond to lipid-based antigens presented in the deep groove of the CD1 molecule, including those from mycobacteria, and thus may play a role in protection against the bacteria. The lipid-containing antigens can be obtained from various endosomal compartments in macrophages.

A second type of MHC CD1–reactive T cell, called NK T cells, expresses the NK cell marker NK1.1. These cells have only one type of T-cell receptor gene expressed. They respond quickly to infection, and make either IFN-γ or IL-4 in response to glycolipids presented on CD1 molecules. The NK T cells have been shown to be important in preventing growth of the malaria-causing protozoan *Plasmodium* in liver cells and participate in granuloma formation to *M. tuberculosis*.

The relationship of the cells of the innate, adaptive, and bridging immune systems and their derivation from stem cells as well as the cytokines produced by these cells are summarized in Figure 27.26.

SECTION HIGHLIGHTS

Several important defense processes cannot be easily categorized as innate or adaptive because they are mediated by lymphocytes or lymphocyte-like cells that can be rapidly deployed but lack immunological memory.

27.5 Vaccines

In addition to invoking an adaptive immune response, antigenic stimulation of B cells and αβ T cells induces the development of progeny cells that are capable of mounting a "memory (or recall) response" when they encounter the priming antigen again. Memory lymphocytes retain the same antigen-binding specificity as their parent cells but their response is more rapid and of greater magnitude, accounting for the benefit of vaccination in preparing for defense against known pathogens. Because B cell and T cell antigen-specific receptors are clonally distributed, priming of the immune system against one antigen does not diminish its capability to mount a response against other different antigens. The vaccinologist focuses on identifying relevant antigens to be included in vaccines as well as on developing meth-

ods of delivery that induce appropriate responses without undesired side effects.

Passive and Active Immunity

Immune protection may be induced by immunization (called active immunity) or passively conferred on an individual by transfer of antibodies (called passive immunity) (Figure 27.27). Passive immunity occurs naturally in humans where antibodies may cross the placenta. Antibodies are also delivered to newborns in the colostrum of mother's milk. Before birth, the antibodies become concentrated in the mother's breast ready to impart an instant immunity to the newborn, which will require considerable time to build up active adaptive immune responses to infectious disease agents.

Passive immunity is also sometimes given to people and animals in the form of γ globulin, which contains antibodies. In more refined procedures, antibodies produced against a particular infectious agent in one individual are purified and given to another individual. This is used in situations where the infection is acute and may kill the individual before a protective immune response can develop, for example, to treat exposure to rabies and tetanus. In some cases, it has been used as a temporary prophylactic, for example, against hepatitis. The disadvantage of passive immunity is that it is short lived because IgG antibodies have a half-life in the blood of less than 3 weeks.

Active immunity is induced as a result of survival of a natural infection or through vaccination. Vaccines may be live organisms that have been attenuated so that they do not cause full-blown disease, dead organisms, or portions of organisms that contain the antigenic determinants needed to induce a protective immune response (see Figure 27.27). The use of vaccines to control infectious diseases such as smallpox, rabies, tetanus, anthrax, cholera, and diphtheria has been one of the great success stories of modern medicine. In the United States, the administration of effective vaccines has reduced the number of reported cases of diphtheria, measles, mumps, pertussis, poliomyelitis, rubella, and tetanus by at least 97% relative to prevaccine levels of disease. No other form of disease control has had such an effect on the reduction of mortality. When vaccinated correctly, the immunity induced is long lasting and may be extended by booster vaccines given at intervals, such as for the tetanus vaccine.

The advantage of living vaccines is that the organisms that make up the vaccine will increase in number following immunization, thereby increasing and prolonging the exposure of the host to antigen and minimizing the need for multiple doses or boosters. In addition, if engagement of a CD8 T-cell response is necessary for protective immunity, for example, to a virus or intracellular bacteria or protozoa, a living vaccine is needed. How-

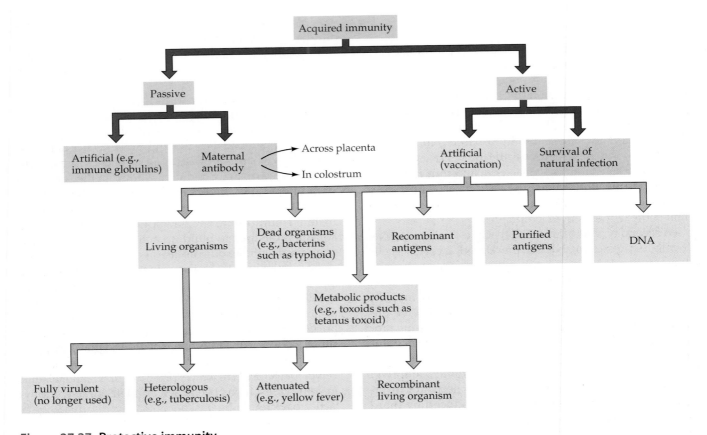

Figure 27.27 Protective immunity
Immunity can result from transfer of antibodies from an immune individual to a nonimmunized individual, such as occurs when maternal antibody passes to the fetus across the placenta or to the newborn in colostrum. This is referred to as passive immunity. Active immunity results from survival of a naturally acquired infection or as a result of vaccination with appropriate antigens.

ever, if living vaccines are to be used, the organism must be an **avirulent** strain that will induce cross-protective immunity to the **virulent** (or disease-causing) strain. This can be achieved by passaging the organism in unusual culture conditions. For example, the intracellular bacteria *Mycobacterium bovis* was attenuated by growing it for 13 years on a bile-saturated medium and is now used in many countries as a vaccine against tuberculosis, caused by the heterologous bacteria *M. tuberculosis*. The living poliovirus vaccine is obtained by growing the virus in monkey cells, whereas the rubella virus vaccine is grown in duck-embryo cells.

A more modern way of attenuating organisms for vaccine development is to disrupt genes responsible for virulence, while retaining those that encode host-protective antigens. Although these types of vaccines are theoretically possible, they are less easy to achieve than was predicted at the time of conception. For example, delet-

ing genes to attenuate *Brucella* sp. has revealed that the loss of a single gene product can usually be compensated for and thus the virulence not reduced significantly. Alternatively, depletion of some genes may result in such severe attenuation that the organism does not survive long enough in the host to induce a protective immune response. Another option is to transfer genes that code for host-protective antigens of a disease-causing infectious agent into an unrelated non–disease-causing organism. The recipient organism will then express the gene from the virulent organism and thereby produce the antigens of the pathogen needed to stimulate an immune response. This is known as a recombinant vaccine.

The use of dead organisms (for example, intact killed bacteria, viruses, protozoa) or portions of those infectious agents has an advantage over use of living organisms as vaccines. That is, the chance for causing disease is minimal. For example, the vaccine against tetanus consists only of the toxin of tetanus that has been inactivated (referred to as a toxoid). Antibodies to the toxin are sufficient to protect the recipient of the vaccine from disease. Specific antigens to be used as a vaccine can either be purified from virulent organisms using biochemical techniques or made artificially as recombinant proteins. Recombinant proteins are proteins made from genes of one organism in another harmless organism. The disadvantage of killed vaccines is that they often require more doses or boosters to achieve sufficient antigen load, and they do not usually engage responses by CD8 T cells. However, if the protective immune response is one mediated by antibodies, these vaccines may be appropriate.

One of the latest innovations in vaccinology is the **DNA vaccine**. Here DNA from an infectious agent is injected into muscle with a needle or stuck onto a gold particle and shot into a tissue by a gene gun. Either way, the DNA enters a host cell, allowing the encoded proteins to be made. The vaccinated host then makes an immune response to these foreign molecules, even though the host's own cells are making the antigens. The usefulness of this type of vaccine is that only DNA that is needed to code for antigens needs to be given. Moreover, because those antigens are produced within the host's cells, the antigen can be presented by the endogenous pathway to CD8 T cells, even though the host never received the living infectious agent. In addition, DNA is less affected by temperature and by long-term storage than proteins and its use as material for vaccination may remove a need for a cold-chain (i.e., refrigeration units) for transport and storage. This has obvious advantages for vaccination to be used in developing countries.

SECTION HIGHLIGHTS

Immunological memory is a characteristic of the adaptive immune response by both $\alpha\beta$ T cells and B-2 B cells, and is the keystone of vaccination against infectious diseases. Components of pathogens that evoke protective immune responses by $\alpha\beta$ T cells and B-2 B cells can be exploited to artificially prepare mammals to defend themselves against these pathogens.

27.6 Dysfunctions of the Immune System

An overzealous immune response can actually result in death of an individual. This occurs with allergies, as well as in some responses to bacterial infections. The latter re-

sult in debilitating conditions called toxic shock syndrome and endotoxemia. The immune system may also respond to self-molecules, recognizing these as appropriate antigens. This is known as autoimmunity and can result in a chronic debilitating state or even death. Current ideas about autoimmune responses suggest that they may actually be initiated as a response to antigens of infectious agents and that subsequently these same lymphocytes cross-react with self- or autologous molecules. The flip side of the coin is an inability to mount a sufficient immune response to control infections. Some individuals or animals are born with a defect in the immune system, known as an **immunodeficiency**, which interrupts one layer of the immune system and leaves the individual vulnerable to infection. In other cases, immunodeficiencies may be transient, resulting from poor health, poor nutrition, or stress. Infections themselves can also result in immunodeficiencies, such as human immunodeficiency virus (HIV), which causes acquired immunodeficiency disease syndrome (AIDS) primarily by infecting CD4 T cells. Because of this immunodeficiency, AIDS patients often die as a result of an uncontrolled secondary infection caused by other infectious agents.

These aberrations of an effective immune response are briefly discussed.

Allergy

Immune responses that are against antigens that do not pose a threat to the host but are persistent and annoying are often referred to as allergies. Allergies are actually immune responses that serve little useful purpose for the host, at least in the context of modern hygiene-conscious societies. Allergic responses fall within two categories of immune responses known as **hypersensitivities**: type I and type IV. Type I hypersensitivity is known as **immediate hypersensitivity**, because it takes minutes to occur, whereas type IV hypersensitivity is known as **delayed type hypersensitivity**, because this type of response takes 2 days to become apparent after contact with the allergy-causing agent.

Type I hypersensitivity responses are mediated by the IgE class of antibodies and mast cells. The reader will recall that in preceding sections, the IgE–mast cell collaborative response was discussed with respect to vasodilation, and the development of inflammation and edema. The mechanism is the same for an allergic response that falls under the type I hypersensitivity category. However, in this case, the antigen that cross-links the IgE molecules on armed mast cells is pollen, mold, insect or snake venom, food, antibiotics, or insect saliva. The mast cells degranulate, and the symptoms are those common among hay fever sufferers, that is, itchy eyes, running nose, and congestion, which are annoying but rarely life threatening. Treatment usually involves taking antihistamines to counteract histamine released by mast cells.

The response may progress to **anaphylactic shock** if enough mast cells are armed with the appropriate IgE and sufficient quantities of antigen are introduced to cause a generalized degranulation of mast cells throughout the body. Anaphylactic shock consists of itching, redness, hives, swelling of the throat, pulmonary edema, and heart failure. The hives represent areas of local mast cell degranulation, whereas the swelling and edema are the result of mast cell release of histamine that causes vascular permeability. The heart failure results from the difficulty presented to the heart of pumping blood through lungs filled with fluid. Death due to full-blown anaphylactic shock can occur in less than one half hour.

People prone to these severe allergies may carry adrenaline (norepinephrine) syringes to counteract the anaphylactic shock and seek a treatment of desensitization. The principle behind desensitization is that injection of small, incrementally increasing doses of the allergen (i.e., the molecule to which the person is allergic) intradermally will stimulate production of IgG antibodies that will mop up the antigen when it is introduced, thereby preventing it from cross-linking IgE on mast cells. The norepinephrine works by stimulating the β-receptor on mast cells that prevents their degranulation.

Allergies associated with type IV hypersensitivities are referred to as **allergic contact dermatitis** and are commonly associated with responses to poison ivy, nickel, formaldehyde, and neomycin. In this type of allergy the molecule that induces the response does so by complexing onto the surface of dendritic cells in the skin. CD4 T cells respond to this complex by producing cytokines that recruit macrophages. This can result in symptoms such as extreme itchiness and is treated with steroids that inhibit T-cell responses. The type IV hypersensitivity reaction is used as a diagnostic test for tuberculosis. Injection of tuberculin (killed preparation of mycobacterial antigens) into the skin results in development of a small white bump at the site if the individual has T cells that have been sensitized to these antigens.

Toxic Shock Syndrome and Endotoxemia

Some microbial proteins are unique in that they can stimulate as much as 20% of the total T-cell population. These proteins are called **superantigens** and result in a life-threatening condition called toxic shock syndrome. Normally, a limited number of clonal populations of T cells respond to any foreign antigen, which results in a controlled and coordinated immune response (Figure 27.28). Generally the proportion of T cells that would respond to an individual antigen would be much less than

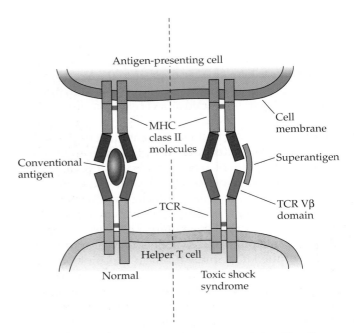

Figure 27.28 Bacterial superantigens trigger T-cell responses
Bacterial superantigens directly link the TCR with MHC class II molecules. They bind to a region of the TCR Vβ domain outside the antigen-binding groove and tie this to a region of MHC class II molecules that lies outside the antigen-binding groove.

0.01% of those in the body. Superantigens stimulate T cells by binding to those with a particular type of β-chain in their T-cell receptor and linking it to MHC class II molecules on other cells by binding outside of the antigen-binding grooves of both the T-cell receptor and the MHC molecule (see Figure 27.28). Because of the large number of T cells activated, a powerful immune response is elicited that results in overproduction of cytokines such as IFN-γ by the T cells and thus overproduction of IL-1 and TNF-α by activated macrophages. This results in toxic shock syndrome due to fever, widespread blood clotting, and shock. Superantigens made by streptococci and staphylococci bacteria include the toxic shock syndrome enterotoxins from *Staphylococcus aureus* and *Streptococcus pyogenes*.

A second type of life-threatening immune response to bacteria is known as endotoxemia, a condition where endotoxin is found in the blood. Endotoxin, also called LPS, is a major component of the outer cell envelope of gram-negative bacteria. Bacterial endotoxin stimulates macrophages through pattern receptors (see pp. 858, 859), and overproduction of macrophage cytokines including IL-1 and TNF-α results in toxic shock. The cytokines cause fever, a drop in blood pressure, and widespread blood clotting, often resulting in death. Bacteria with endotoxin molecules that mediate this violent response include *Escherichia coli*, *Klebsiella pneumoniae*, *Pseudomonas aeruginosa*, and *Neisseria meningitidis*.

Autoimmunity

Three processes contribute to the presence of lymphocytes that react with our own antigens (self-reactive lymphocytes) in the peripheral circulation, which is the cause of **autoimmunity**. These processes are (1) failure of clonal deletion of thymocytes and pre–B cells to eliminate all self-reactive cells, because not all self-antigens are accessible in the thymus or in the circulation; (2) positive selection of weakly self-reactive B-1 B cells; and (3) development of B cells with a self-reactive antigen-specific receptor as a result of somatic mutation in the germinal center and positive selection by endogenous or acquired antigens present on follicular dendritic cells. Although all of us have self-reactive lymphocytes; only 5% to 7% of us will develop debilitating autoimmune disease. The rest of us are spared by good fortune (i.e., we do not encounter infections or other environmental conditions that stimulate these self-reactive lymphocytes) and are adequately protected by regulatory responses that limit activation of self-reactive lymphocytes. The study of these regulatory responses is an exciting area of immunology.

Although some self-reactive peripheral CD4 T cells encounter their target antigens in the absence of co-stimulator activation and enter an inactive state, other self-reactive lymphocytes encounter their target antigen under activating conditions. These are usually associated with infections and inflammation. Infectious agents with molecules that are cross-reactive with self, called **molecular mimicry**, induce co-stimulator activity on antigen-presenting cells and can stimulate autoimmune responses. For example, the immunodominant antigen epitope of *Borrelia burgdorferi* outer surface protein A (amino acids 165–173) is both homologous to a sequence in human lymphocyte function associated antigen-1 and predicted to bind in the groove of the human MHC class II allele HLA-DRB1*0401 and related alleles. Thus, in people with these MHC alleles, the response of CD4 lymphocytes to the bacterial antigenic epitope may result in an autoimmune condition that can be sustained in the absence of the provoking pathogen as a result of autostimulation of memory cells. The high frequency of treatment-resistant Lyme arthritis in people with the HLA-DRB1*0401 allele is taken as evidence of this condition.

Many other instances of molecular mimicry have been noted, including between poliovirus VP2 and the receptor for acetylcholine, between papilloma virus E2 and insulin receptor, between HIV-1 envelope proteins and HLA-DR4 and DR2, variable regions of T-cell receptors, Fas protein and IgG, and between an antigen of *On-chocerca volvulus* and a 44-kDa protein of ocular tissue. Immune responses against the self-like epitopes on pathogens may contribute to autoimmune pathology. In this regard, cross-reacting B-cell epitopes may account for infection-associated pathology; for example, antibodies against *Mycobacterium leprae* react with human skin.

Autoantibodies that arise in infections may not result from cross-reacting pathogen and self-antigens. Many different autoantibodies are detected in people infected with African and South American trypanosomes and arise as a result of nonspecific B-cell activation due to the infection. Other situations where pathogens elicit autoimmune responses by processes other than molecular mimicry include T-cell responses to superantigens. These can lead to the activation of T cells with self-reactive T-cell receptors (for example, staphylococcal enterotoxin A superantigen stimulates T cells that cause thyroiditis), and human endogenous retroviral superantigen is induced by IFN-α, an inflammatory cytokine and activates T cells that cause Type 1 diabetes. Infection can also break self-tolerance by a process called epitope spreading. Here, an ongoing cytolytic response results in release of self-antigens that stimulate an autoimmune response. This is implicated in a model of multiple sclerosis in mice, brought on by Theiler's murine encephalomyelitis virus.

Whether or not a pathogen induces an autoimmune response may reflect host regulatory processes. For example, the development of autoimmunity may require strong Th1 T-cell help and does not occur if a strong Th2 response is stimulated, or the response may be suppressed by natural regulatory T cells either through antigen binding receptor (idiotypic) recognition pathways, or through natural CD4$^+$ CD25$^+$ regulatory T cells.

Immunodeficiencies

Although there are many types of inherited immunodeficiencies, an example of one that affects the innate immune system and one that affects the adaptive immune response is described here. **Leukocyte adherence deficiency (LAD)** is a defect of the innate immune system that occurs in humans and animal species, including cattle and dogs. The defect is in expression of the molecule required for neutrophils to adhere to the wall of blood vessels in inflamed tissue. As a result, neutrophils cannot be recruited from the blood to the site of infection. Animals and people with such a deficiency have recurrent bacterial infections because the ability to control them at the level of the innate immune response is severely impaired.

Severe combined immunodeficiency (SCID) affects the adaptive immune response and occurs in humans and animal species such as horses. Neither the T cells nor the B cells function, and therefore there is no adaptive immune response. Infants with SCID generally have recurrent infections during the first weeks of life, such as oral candidiasis, and pneumonias caused by organisms that are not normally significant pathogens, such as *Pneumocystis carinii*, and diarrhea. Without isolation from infectious microbes or treatment, these children will not survive. The cause of the immunodeficiency in humans is traced to a deficiency in an enzyme. Lack of the enzyme eventually results in the death of T cells and thus loss of any cell-mediated functions by these cells as well as the provision of help to B cells. As a consequence, there is a defect in antibody production as well.

SECTION HIGHLIGHTS

There are several situations in which immune responses cause more harm to the host than benefit. Sometimes the immune system responds overzealously to antigens that are not actually a threat to the health of the individual, resulting in allergy or autoimmunity. In other cases an individual may be unable to mount fully effective immune responses against infectious agents either because of an inherited or acquired immunodeficiency.

SUMMARY

- The primary role of the **immune system** is to combat invasion by infectious agents by excluding disease-causing agents (pathogens) from host tissues or controlling and eliminating them.

- Several layers of sequential, overlapping defense mechanisms accomplish this. **Barrier defense** or immunity excludes infectious agents by physical, chemical, mechanical, and microbial mechanisms (discussed in Chapter 26).

- **Inflammation** and **innate immunity** limit the replication of breakthrough microorganisms by relatively nonspecific responses.

- Adaptive **immunity** eliminates invaders by highly pathogen-specific mechanisms that are more efficient on second exposure to the same pathogen and hence is considered to have memory.

- Each of the overlapping systems has an important role to play in host defense because in the absence of any one of them, the person or animal will be susceptible to infection.

- The molecules and cells that mediate inflammation and innate immunity are constitutively present in a sufficient quantity or number to provide significant defense upon activation, and can be mobilized immediately.

- Monocytes (or macrophages) and neutrophils are **phagocytic cells** that are called into areas of inflammation. They phagocytose and kill infectious agents present.

- Phagocytes use a variety of mechanisms to kill infectious agents. These include lowering the pH in the phagosome, generating toxic oxygen radicals and nitric oxide, and exposing the organisms to lysosomal enzymes and microbicidal peptides called **defensins**.

- The adaptive immune system mediates its protective responses through soluble molecules called **antibodies** produced by B cells and **cytokines** produced by T cells. T cells are able to kill microbe-infected host cells by a process called **cytotoxicity**.

- B cells and T cells respond to foreign molecules called **antigens** that may be parts of infectious agents.

- The adaptive immune system has **specificity** and **memory**. It is a carefully regulated response to infection that takes time to develop and is more effective upon the second encounter with the same antigen.

- **Cell-mediated immunity** refers to activation of macrophages by **IFN-γ** for more efficient killing of phagocytosed microbes and generation of cytotoxic **CD8 T cells**. This type of immune response is directed by type 1 or **Th1 CD4 T cells**.

- Cytotoxic CD8 T cells kill cells infected with viruses and also cells infected with **intracellular microbial pathogens**.

- **Humoral immunity** refers to antibody-mediated control of pathogens, especially antibodies of the IgA, IgE, and several IgG subclasses. This is directed by type 2 or **Th2 CD4 T cells**.

- Extracellular bacteria and protozoa are controlled by antibody-mediated humoral immunity. These include blocking entry into the host, **opsonization** for phagocytosis, **complement**-mediated killing, and neutralization of their toxins.

- **Antibodies** are also important in the control of viral infections by blocking adhesion to and entry into host cells and mediating **antibody-dependent cellular cytotoxicity** against cells infected with enveloped viruses.

- Antibodies are important for control of parasitic worms through IgE-mediated mechanisms that result in degranulation of **mast cells** and **eosinophils**.

- Other cells in the immune system are part of the **bridging immune system** and include NK cells, γδ T cells, B-1 B cells, and NK T cells. These cells share function with the T and B cells involved in adaptive immunity.

- **Allergy**, **toxic shock syndrome**, **endotoxemia**, and **autoimmunity** are dysfunctions of the immune system, since these types of responses are detrimental to the host and provide no protective value.

- **Vaccination** is a way to exploit the properties of the adaptive immune system, since it has immunological memory.

 Find more at www.sinauer.com/microbial-life

REVIEW QUESTIONS

1. How does breach of the barrier defense and exposure of collagen initiate an inflammatory response?

2. How is it possible that individuals with an inherited deficiency in C1q synthesis are still able to generate the opsonins C3b and iC3b, and the anaphylatoxins C3a, C4a, and C5a during inflammation?

3. What are the mechanisms by which phagocytic cells control infectious disease?

4. How does IFN-γ contribute to host control of bacteria, protozoa, and viruses?

5. What are the hallmarks of the adaptive immune response?

6. What are the endogenous and exogenous pathways of antigen presentation; what roles do class I and class II MHC antigens play in this?

7. Th1 and Th2 CD4 T cells make distinct contributions to the control of pathogens. What are they?

8. How is diversity of B-cell receptors and antibodies generated?

9. Why are live vaccines against intracellular pathogens particularly able to engage responses by CD8 T cells?

10. What are the distinct functions of antibodies of the IgM, IgG, IgA, and IgE subclasses?

SUGGESTED READING

DeFranco, A. L., R. M. Locksley and M. Robertson. 2007. *Immunity: The Immune Response in Infectious and Inflammatory Disease.* Sunderland, MA: Sinauer Associates, Inc.

Goldsby, R. A., T. J. Kindt and B. A. Osborne. 2006. *Kuby Immunology.* 6th ed. New York: W. H. Freeman and Co.

Mims, C. A., A. Nash and J. Stephen. 2000. *The Pathogenesis of Infectious Disease.* 5th ed. New York: Taylor and Francis.

Paul, W. E. 2003. *Fundamental Immunology.* 5th ed. Philadelphia: Lippincott Williams & Wilkins.

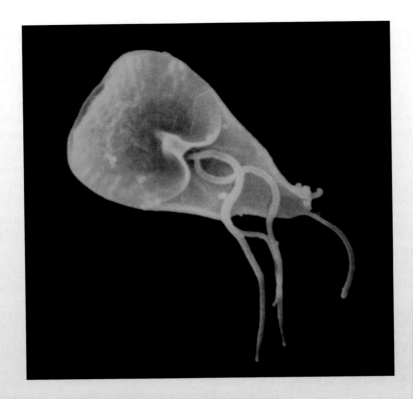

28

Microbial Diseases of Humans

Throughout nature infection without disease is the rule rather than the exception.
—*René Dubos, in* Man Adapting

The symptoms and ravages of infectious diseases have been prominently recorded in the annals of human history. Biblical and ancient writings describe the crippling and suffering caused by various scourges, and the words "plague" and "disease" have 17 citations in the concordance of the King James Bible. Ancient drawings show humans scarred by smallpox or crippled by polio. The epidemic of syphilis that spread through Europe in the latter part of the fifteenth century killed thousands. The devastations of "Black Death," the plagues that periodically ravaged Europe, are well recorded (see Chapter 2). Typhus was often the final arbiter in war; the disease foiled Napoleon's offensive at Moscow and killed 3 million Russian soldiers during World War I.

Our understanding of the relationship between a pathogenic microbe and disease symptoms has increased markedly since the pioneering studies of Robert Koch (see Chapter 2). It was Koch who clearly defined the causal relation between infectious disease and a specific bacterium, and he and his coworkers suggested definitive procedures for confirming this relationship. The minor limitation imposed by Koch's methodology (a microbe must be culturable outside the animal) has now been largely overcome by improved identification techniques.

This chapter discusses some of the significant human bacterial infectious diseases and their manifestations. A limited number of fungal and protozoan diseases are also described. The presentation is by route of entry or site of infection such as respiratory, gastrointestinal tract, and so on. Presentation in this order is both convenient and appropriate because most infectious microbes are selective in the portal of entry. Most of the tissues that could serve as sites of colonization or systemic invasion are protected by various innate host defense mechanisms. Route of entry often represents a specialized niche for individual pathogens that prefer it and is often the primary determinant of the outcome of the infection. However, some bacteria such as *Staphylococcus aureus*, *Streptococcus pyogenes*, *Escherichia coli*, and *Pseudomonas aeruginosa* can cause disease at various sites in or on the human body.

28.1 Skin Infections

The intact skin forms an effective physical barrier against pathogenic organisms, unless it is disrupted, such as by injury or penetration by catheters or surgical incisions. The skin barrier can also be breached by animal or insect bites. Infections of the skin are often considered together with infections of underlying soft tissue. These infections can be mild and self-limiting; however, in certain instances the causative bacteria can spread into deeper tissues and cause more serious, life-threatening infections. Skin and external infections that are often encountered by humans include impetigo, wound infections, acne, ringworm, and athlete's foot (Table 28.1).

Impetigo is a skin infection that occurs frequently in child-care centers and schools where youngsters are exposed to one another under confined conditions. It is caused by *Streptococcus pyogenes* or *S. aureus* and is transmitted by scratching or other interpersonal contacts. It generally occurs around the nose and spreads about the face. Streptococcal impetigo is characterized by the appearance on the skin of pus-containing vesicles that are eventually covered by a reddish crust (Figure 28.1). Impetigo caused by *S. aureus* is characterized by water-filled blisters that of-

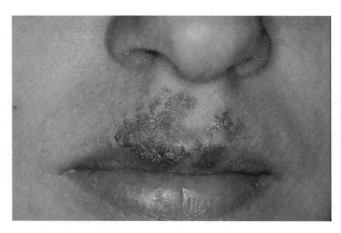

Figure 28.1 Impetigo
Impetigo on the face of a child. Note the inflammatory response (redness and swelling) around the lesions. Impetigo is caused by *Staphylococcus aureus, Streptococcus pyogenes*, or a combination of these microorganisms. Photo ©Biophoto Associates/Photo Researchers, Inc.

ten rupture. The infection caused by *S. pyogenes* can be more serious, as it spreads rapidly and can cause scarring. Impetigo should be treated without delay and can be controlled by topical application of antibiotic ointments.

Another common skin infection is acne. **Acne** occurs frequently during adolescence when the endocrine system (hormone-secreting glands) is most active. It can last for years and is usually resolved by the time an affected individual reaches adulthood, but it can result in permanent scaring. Hormonal activity stimulates sebaceous glands in the skin to overproduce sebum. The sebum can accumulate within the gland and become infected with a normal flora anaerobic organism, *Propionibacterium acnes*. This organism proliferates and produces factors that result in an inflammatory response manifested by swelling and reddening in the area. Mild acne can be treated topically by application of erythromycin, clin-

TABLE 28.1	Major diseases of the skin[a]	
Disease	**Organism**	**Symptoms**
Impetigo	*Staphylococcus aureus* *Streptococcus pyogenes*	Skin lesions
Wound infections (burns)	*Pseudomonas aeruginosa*	Bacteremia, shock, hypotension
Acne	*Propionibacterium acnes*	Redness and swelling of sebaceous glands
Ringworm (tinea corporis)	*Trichophyton mentagrophytes*	Fungal invasion of skin; ring of inflammation
Scalp ringworm	*Trichophyton tonsurans*	Hair follicle infection and loss of hair
Athlete's foot (tinea pedis)	*Trichophyton rubrum* *Trichophyton mentagrophytes* *Epidermophyton floccosum*	Peeling and cracking of skin between toes
Groin ringworm	*Epidermophyton floccosum*	Fungal infection of the groin

[a]The list is not exhaustive and other diseases occur. References to some of these are presented at the end of the chapter.

damycin, metronidazole, or benzoyl peroxide. Systemic antibiotic therapy for severe or chronic acne includes tetracycline, erythromycin, and doxycycline.

Particularly susceptible to serious skin infections are patients with extensive burn wounds. Thermal injury results not only in a breach of the normal skin barrier, but the extensive tissue damage at the burn site provides various exudates that serve as a source of nutrients for the infecting organisms. Open wounds, such as extensive burns, can be infected with bacteria found in normal flora as well as those present in the environment. Not surprisingly, *S. aureus* is a common colonizer of burn sites, as are *Candida* and *Aspergillus* species. Another microorganism that is a concern is *Pseudomonas aeruginosa*. This gram-negative bacterium is commonly present in air and in water and is an opportunistic pathogen in burns. *P. aeruginosa* can produce toxins and proteases that contribute to pathogenicity. It is naturally resistant to many antibiotics and is difficult to control. A common complication of burn infections is further dissemination including invasion of the bloodstream (**bacteremia**). Bacteremia caused by *P. aeruginosa* is associated with a very high mortality.

S. aureus is probably the most versatile pathogen that afflicts human beings (Table 28.2). Infections caused by *S. aureus* are difficult to prevent, as virulent strains of the organism are carried in the nasopharynx by up to 50% of the human population. These pathogens also normally inhabit the skin, intestine, and vagina. The organism is readily spread from asymptomatic carriers by touching, sneezing, coughing, or passage on inanimate materials. Overall, there are at least 14 species of *Staphylococcus* that colonize humans, with coagulase-positive *S. aureus* being most often associated with disease. *Staphylococcus epidermidis* and *Staphylococcus saprophyticus* can be infectious if resistance of the host is compromised by immunosuppressants.

S. epidermidis is the most prevalent of the staphylococci that infect in the vicinity of devices such as prosthetic joints, artificial heart valves, and catheters. The organism produces a polysaccharide capsule that adheres tightly to artificial materials. *S. saprophyticus* can cause cystitis, a urinary tract infection that occurs primarily in women. Cystitis is an inflammation of the urinary bladder resulting from microorganisms traveling from the urethral opening. It occurs less frequently in males, as their urethra is longer.

Infections caused by *S. aureus* are listed in Table 28.2. Generally, these infections result from a breakdown in the natural defense system. Serious staphylococcal infections occur most often in individuals with a decreased ability of their phagocytic cells to destroy the microbes. In addition to the above-mentioned surgical and burn patients, neonates with underdeveloped immune systems, individuals on immunosuppressants, or those lacking a normal immune response are all susceptible to infections caused by *S. aureus*. Scratches and blocked pores can lead to pimples, impetigo, and various superficial or deep-seated skin infections. Deep *S. aureus* infections, such as carbuncles, can lead to infections of the lymph nodes.

S. aureus is a well-equipped pathogen with an armament of virulence factors. Most strains have a cell wall component, termed **protein A**, which binds to immunoglobulins and interferes with the normal phagocytic response. Infected lymph nodes and localized internal infections can result in rapid multiplication and dissemination of the organism into the bloodstream. The organism produces a number of toxins and enzymes that aid in the invasive process (see Chapter 26). One enzyme, **coagulase**, is particularly effective at inducing the host to produce a fibrin clot that surrounds the bacterial cell, thereby preventing the normal host defenses from phagocytizing the invader. Secreted staphylococcal proteases play a role in harvesting of nutrients and degrade proteins of various tissues in the process, whereas the various nonspecific membrane-damaging toxins target membranes of epithelial and phagocytic cells. *S. aureus* also produces **leukocidin**, an antiphagocytic factor that destroys leukocytes. Production of leukocidin and coagulase permits the organism to overcome host defenses.

Viral diseases can also predispose an individual to invasion by staphylococci. Influenza can be followed by a staphylococcal pneumonia that may be fatal. It is probable that many deaths ascribed to viral influenza can be ascribed to staphylococci. Other viral infections of the

TABLE 28.2	Some infections caused by *Staphylococcus aureus*
Disease	**Site**
Pimples	Anywhere on the body
Impetigo	Generally on face
Boils	Anywhere
Carbuncles	Anywhere
Lymph nodes	Near site of infection
Septicemia	Blood
Osteomyelitis	Bones
Endocarditis	Heart
Meningitis	Brain
Enteritis (food poisoning)	Gastrointestinal tract
Nephritis	Kidneys
Toxic shock syndrome	Hypotension
Staphylococcal scalded skin syndrome	Skin peeling
Wound infections	Skin
Inner ear infections	Ear
Respiratory infections	Larynx, bronchi, lungs

respiratory system can cause tissue trauma that creates an opening for invasion by *S. aureus*, resulting in meningitis or laryngitis (see later discussion). Staphylococcal infections can be treated with cephalosporin, penicillin, cloxacillin, and other antibiotics.

Toxic shock syndrome occurs in surgical patients of both sexes where there is an instance of internal colonization by *S. aureus*. In women, toxic shock syndrome also results from the overcolonization of the vagina by *S. aureus*. This disease, first described in 1978, is a syndrome that occurs most frequently in menstruating women who use tampons. The disease symptoms are a sudden fever, diarrhea, vomiting, red skin rash, and low blood pressure. Toxic shock is caused by certain strains of *S. aureus* that carry a pathogenicity island, disseminated between various *S. aureus* by a transducing bacteriophage (see Chapter 14), which encodes a pyrogenic toxin called the toxic shock syndrome toxin (TSST). This toxin, and the enterotoxins produced by *S. aureus* during infection of the gastrointestinal tract (see subsequent text), is a **superantigen** causing uncontrolled release of cytokines by immune cells (see Chapter 27). The drop in blood pressure can lead to an irreversible state of shock and death. The death rate in confirmed cases is between 5% and 12%. Newer superabsorbent brands of tampons that bind magnesium and a lowered level of magnesium in the environment apparently stimulate the *S. aureus* strains to produce TSST. Women can reduce the risk of

contracting the disease by not using tampons, by not using the superabsorbent tampons, or by not using them continuously during the menstrual period. Other forms of toxic shock syndrome are also related to vaginal colonization, and are the consequence of antibiotic use, use of contraceptive devices, or childbirth. Because *S. aureus* often produces β-lactamase, chemical variants of β-lactam antibiotics that are not destroyed by this enzyme, such as oxacillin or nafcillin, are used for treatment of toxic shock syndrome.

Fungal diseases of the skin are a widespread occurrence in humans. Fungi that invade the keratinized part of skin are called dermatophytes, and the disease they cause is called **dermatomycosis** (or dermatophytosis), sometimes referred to as tinea or ringworm (Figure 28.2). These fungal infections are classified according to location on the body. Infections of the scalp (tinea capitis) and body (tinea corporis) are common skin infections of young children, whereas infections of groin (jock itch) and feet (athlete's foot) are common among young adults.

The causative agents of these diseases are presented in Table 28.1. The infections are caused mostly by fungi in three different genera—*Epidermophyton*, *Microsporum*, and *Trichophyton*. The sources of infections are other humans, animals, or the natural environment (soil and water). The fungi are transmitted through direct contact with infected lesions; animals that are carriers; or contaminated objects such as clothing, towels, or shoes that

(A)

(B)

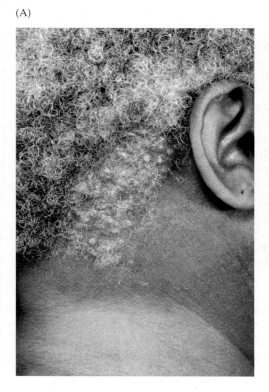

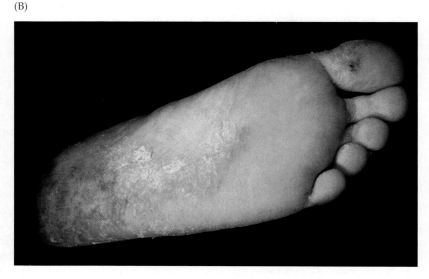

Figure 28.2 Fungal infection of the skin
(A) Ringworm on the head (tinea capitis) is caused by *Microsporum audouinii*. (B) Ringworm on the foot (athlete's foot; tinea pedis) is caused by *Trichophyton rubrum*, *Trichophyton mentagrophytes*, or *Epidermophyton floccosum*. A, ©Medical-on-line/Alamy; B, courtesy of the Centers for Disease Control and Prevention.

encourage fungus growth. Some of the fungi involved in these conditions primarily infect animals, but they may also be transmitted from animals to humans. In general, dermatomycoses are relatively mild and often self-limiting. However, they can lead to additional complications, mainly superinfection by bacteria or viruses. Some deep infections by dermatophytes can be life-threatening, particularly in immunocompromised individuals, and involve draining lymph nodes, brain, and liver. Because the infections remain localized to the surface of first layer of the skin, they are usually treated with ointments or creams containing drugs, usually belonging to the azole or allylamine/benzylamine families of compounds: terbinafine, naftifine, and butenafine or miconazole, clotrimazole, and ketoconazole, respectively. Because of poor accessibility of the infection site, dermatophytic infections of scalp or nails are treated with oral antifungal drugs.

SECTION HIGHLIGHTS

Infections of skin and the underlying soft tissues occur after disruption of the skin barrier by injury or by medical or surgical procedures. Impetigo is a form of skin infection, characterized by the appearance of pus-filled or water-filled blisters, is caused by *S. pyogenes* or *S. aureus*, respectively. The staphylococcal impetigo can lead to scarring, even after antibiotic treatment. *P. acnes* is responsible for infections of sebaceous glands and hair follicles resulting in a disease called acne. Skin infections caused by *S. aureus*, *P. aeruginosa*, or *Candida* and *Aspergillus* species that sometimes follow serious burns can be very serious and these organisms are often resistant to a number of antibiotics. *S. aureus* produces a number of virulence factors, including an immunoglobulin-binding surface protein A, and secreted enzymes such as coagulase and leukocidin, which promote dissemination and avoidance of host defense mechanisms. The major symptoms of staphylococcal toxic shock syndrome are elicited by a pyrogenic toxin called the toxic shock syndrome toxin (TSST). This toxin is a superantigen causing uncontrolled release of cytokines by immune cells. Pathogenic fungi cause dermatomycoses; these conditions are usually self limiting but can lead to secondary infections by more virulent bacterial pathogens.

28.2 Respiratory Infections

Next to skin, the organ that is most extensively exposed to the environment is the respiratory tract. Its function is to continuously provide oxygen to the organism, and, therefore, breathing ensures uninterrupted exposure of this organ to microorganisms present in the air, either suspended in droplets or found on particles such as dust or constituents of smoke. Moreover, these microorganisms are readily communicated from human to human via air droplets created by sneezing, coughing, or face-to-face talking, and by kissing or other forms of human contact. Sneezing and coughing are significant mechanisms for expelling infectious droplets into the air. A sneeze droplet can move at 100 meters per second for short distances, and a single sneeze can expel 10,000 to 100,000 or more bacteria into the air. Although microorganisms do not survive well in outdoor air, the atmosphere is constantly inoculated by organisms from soil, plants, and animals, but mostly by other humans. Indoor air has higher numbers of microbes, and the residents occupying the space generally disseminate these. If large numbers of humans are crowded into a room, microbes common in the respiratory tract will populate the air. The gram-positive organisms such as those of the genera *Micrococcus, Staphylococcus,* and *Streptococcus* can survive in air, as they are quite resistant to drying and relatively resistant to ultraviolet light. Infectious diseases spread predominantly via the respiratory tract, and the organisms that cause these diseases are summarized in Table 28.3.

The respiratory tract can be divided into upper and lower sections and the infections are categorized accordingly. Viruses cause the majority of upper respiratory tract infections, whereas infections of the lower respiratory tract have both bacterial and viral origin. The agents of bacterial respiratory tract infections and their primary sites of colonization are shown in Figure 28.3.

The anatomy of the respiratory tract contributes significantly toward the defense of this organ against pathogens. Hair that lines the nose serves as a filter for particles as small as 10 μm. The nasal cavity is lined with ciliated cells that are covered by a thin layer of hydrated mucus, a complex protein/polysaccharide matrix. Particles or bacteria are readily trapped in mucus and are expelled by the action of the cilia. The lower respiratory tract is also lined with ciliated cells and mucus. The ciliary action of these cells in the upward direction, sometimes referred to as the ciliary elevator, moves the trapped bacteria toward the throat, where it can be swallowed. Mucus-producing cells found in glands readily replace mucus removed as part of the host defense response. Moreover, the liquid that bathes the large portion of the repository tract is rich in antimicrobial factors

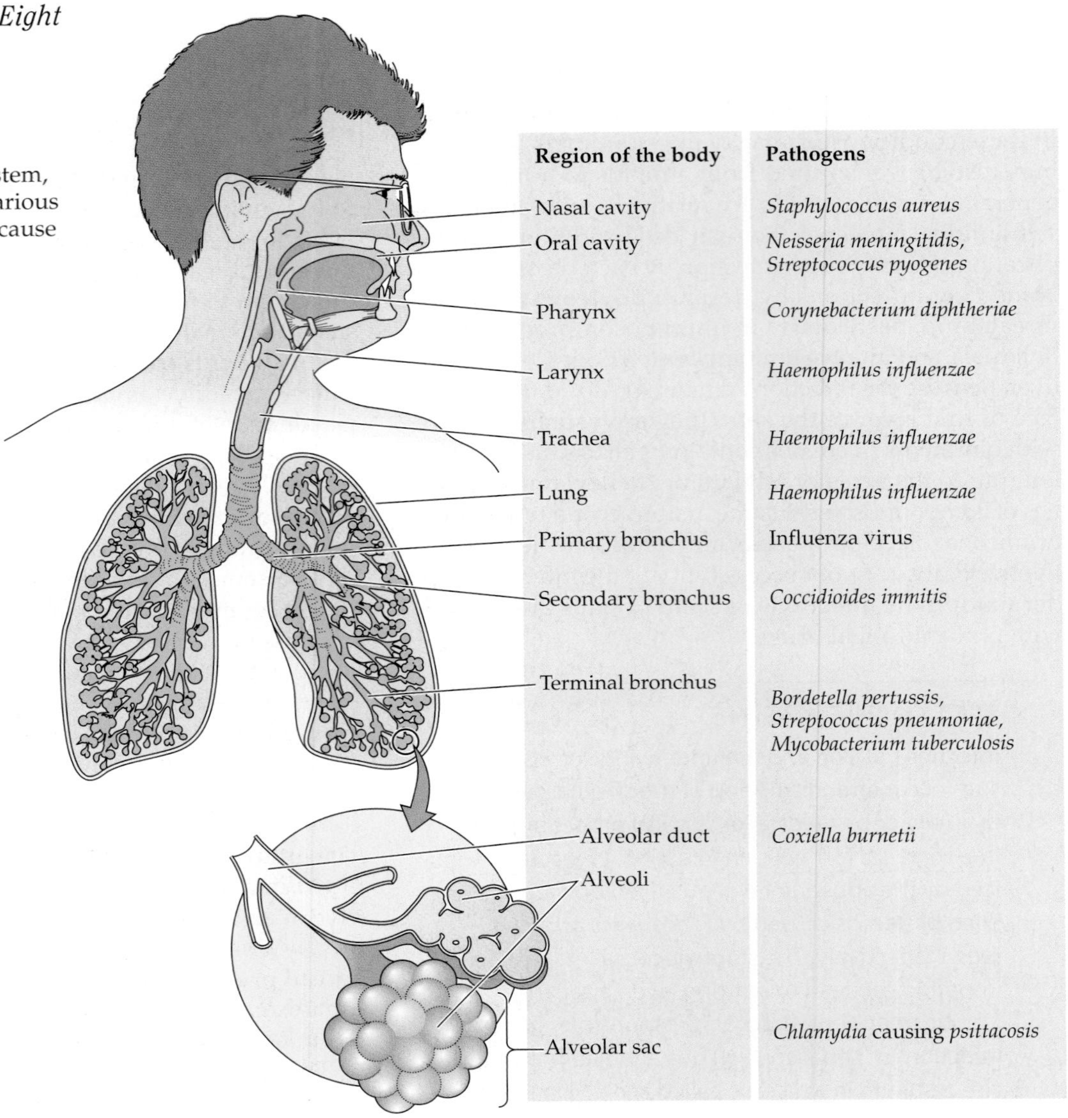

Figure 28.3 Respiratory infections
The human respiratory system, showing the sites where various pathogens may occur and cause infections.

Region of the body	Pathogens
Nasal cavity	*Staphylococcus aureus*
Oral cavity	*Neisseria meningitidis, Streptococcus pyogenes*
Pharynx	*Corynebacterium diphtheriae*
Larynx	*Haemophilus influenzae*
Trachea	*Haemophilus influenzae*
Lung	*Haemophilus influenzae*
Primary bronchus	Influenza virus
Secondary bronchus	*Coccidioides immitis*
Terminal bronchus	*Bordetella pertussis, Streptococcus pneumoniae, Mycobacterium tuberculosis*
Alveolar duct	*Coxiella burnetii*
Alveoli	
Alveolar sac	*Chlamydia* causing *psittacosis*

including **lysozyme** capable of digesting the bacterial peptidoglycan and small toxic peptides called **defensins**. Other factors also interfere with the survival of inhaled bacteria. For example, respiratory secretions contain large amounts of an iron-sequestering protein lactoferrin and limit access of iron to bacteria, severely impeding their ability to proliferate.

A significant area of the upper respiratory tract including the nasal and oral cavities and pharynx are occupied by bacterial normal flora that also participate in the defense of the respiratory tract against invading pathogens. They colonize preferred niches and interfere with colonization by pathogens. These bacteria are typically not pathogenic; however, their displacement into the lower portion of the respiratory tract can initiate a pathological process and respiratory disease. However, these bacteria often represent less virulent strains of the pathogenic species. The nose is predominantly occupied by *S. aureus* and *S. epidermidis*, and, occasionally,

Streptococcus pneumoniae, Neisseria meningitidis, and *Haemophilus influenzae* can be found in the nasal cavity and in the nasopharynx region. The oral cavity and the oropharynx area contain a large variety of resident bacteria. These include—in addition to *S. aureus* and *S. epidermis* and certain streptococci (also called viridans) that produce a specific form of a membrane-damaging protein, α-hemolysin—*Streptococcus salivarius S. mitis, S. mutans, S. milleri.* The ability of these microorganisms to exclude pathogens is based on competition for colonizable surfaces and for nutrients, and, in some instances, production of factors lethal to other bacteria such as bacteriocins.

Immunization has reduced the scourge of many respiratory diseases, particularly diphtheria and whooping cough. Antibiotics are effective against causative agents of most of the others, including tuberculosis, but strains resistant to antibiotics are a concern. A discussion of some of these respiratory infections follows.

TABLE 28.3	Diseases transmitted predominately via the respiratory tract and the organisms involved
Disease	**Organism**
Diphtheria	*Corynebacterium diphtheriae*
Whooping cough	*Bordetella pertussis*
Pneumonia	*Streptococcus pneumoniae*
	Staphylococcus aureus
	Legionella pneumophila
	Mycoplasma pneumoniae
	Klebsiella pneumoniae
	Pneumocystis carinii
Streptoccocal infections	*Streptococcus pyogenes*
Meningitidis	*Haemophilus influenzae*
	Neisseria meningitidis
	Streptococcus pneumoniae
Tuberculosis	*Mycobacterium tuberculosis*
Psittacosis	*Chlamydia psittaci*
Leprosy	*Mycobacterium leprae*
Coccidioidomycosis	*Coccidioides immitis*
Aspergillosis	*Aspergillus fumigatus*
	Aspergillus flavus

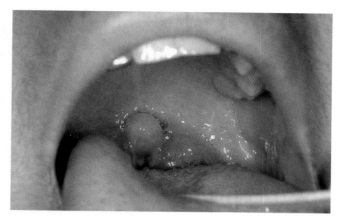

Figure 28.4 Diphtheria
Diphtheria results from colonization of the oropharynx by *Corynebacterium diphtheriae*. In an active case, as here, a pseudomembrane develops in the throat. ©Medical-on-line/Alamy.

Diphtheria

Diphtheria is caused by *Corynebacterium diphtheriae*, a gram-positive, nonmotile pleomorphic bacillus that may appear to be club shaped (see Figure 20.20). They appear as black colonies on media containing tellurite, which is used in clinical laboratories for diagnosis of diphtheria.

Diphtheria was once a leading cause of death in children in the United States. In the early part of the twentieth century, there were on the average 150,000 cases per year with up to 15,000 deaths. Effective immunization has reduced the number of cases worldwide. Currently, the incidence of diphtheria in the United States is fewer than five per year. Diphtheria is still a concern in economically deprived urban areas of the world.

C. diphtheriae enters the body through the respiratory route and adheres to tissue in the throat. There the bacteria multiply and produce a potent toxin causing localized tissue damage. It is possible to become infected through contact, for example, by wearing the clothing of an infected person. Bacterial mass, mixed with dead cells and leukocytes, appears as a dull gray **pseudomembrane** on the throat mucosa (Figure 28.4) that aids in the early diagnosis of diphtheria, differentiating it from other respiratory infections. The pseudomembrane appears first on the tonsils but often extends over the tracheal opening, blocking the passage of air into the lungs, resulting in suffocation and death. Other symptoms of diphtheria are caused by dispersal of the toxin into sites far from the site of bacterial colonization, leading to lesions in the kidney and heart.

The diphtheria toxin has been studied extensively, and its mode of action is well understood. It was the first extracellular toxin (an **exotoxin**) described. Two French scientists, Pierre Roux and Alexandre Yersin, discovered that culture filtrates of the bacterium *C. diphtheriae* were lethal to test animals, affirming that lethality in diphtheria infections was largely due to the toxic material released by the bacterium and less so by the direct tissue damage caused by these organisms growing in the throat.

The toxin is produced only by those strains of *C. diphtheriae* that are lysogenized by β-bacteriophage, and the *tox* gene, encoding the diphtheria toxin is part of the bacteriophage genome. The process of lysogenization of *C. diphtheriae* by this bacteriophage is similar to that described for bacteriophage lambda in Chapter 14. Toxin is produced at low iron concentrations, since the transcription of the phage-encoded *tox* gene is under negative control (see Chapter 13 for the discussion of negative regulation of gene expression) of the chromosomally encoded diphtheria toxin repressor (DtxR). For its binding to the operator site of the *tox* gene, DtxR requires iron as a corepressor. Therefore, it becomes inactive when iron is limiting, allowing unimpeded expression of the *tox* gene.

C. diphtheriae secretes diphtheria toxin into circulation in the form of a larger polypeptide that is cleaved into two fragments (A and B), held together by a disulfide bond (Figure 28.5). The cleavage can take place either in the circulation or after toxin binding to the receptor by host cell proteases. The toxin is taken up by **receptor-mediated endocytosis**, a process where invagination of the cytoplasmic membrane creates a so-called endosomal

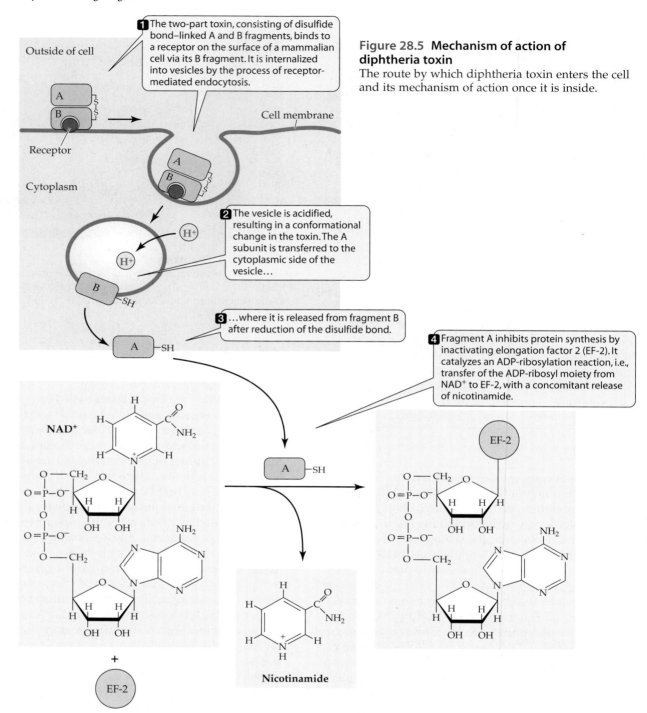

Figure 28.5 Mechanism of action of diphtheria toxin
The route by which diphtheria toxin enters the cell and its mechanism of action once it is inside.

1 The two-part toxin, consisting of disulfide bond–linked A and B fragments, binds to a receptor on the surface of a mammalian cell via its B fragment. It is internalized into vesicles by the process of receptor-mediated endocytosis.

Outside of cell

Cell membrane

Receptor

Cytoplasm

2 The vesicle is acidified, resulting in a conformational change in the toxin. The A subunit is transferred to the cytoplasmic side of the vesicle…

3 …where it is released from fragment B after reduction of the disulfide bond.

4 Fragment A inhibits protein synthesis by inactivating elongation factor 2 (EF-2). It catalyzes an ADP-ribosylation reaction, i.e., transfer of the ADP-ribosyl moiety from NAD^+ to EF-2, with a concomitant release of nicotinamide.

NAD^+

EF-2

Nicotinamide

vesicle, thus internalizing proteins that are bound to the cell surface. Acidification of the vesicle causes the toxin to unfold, with the B fragment creating a pore in the membrane facilitating the release of the A fragment into the cytoplasm. The diphtheria toxin A fragment is an enzyme, capable of catalyzing the transfer of the ADP-ribose moiety from nicotinamide adenine dinucleotide (NAD^+) onto a modified histidine on elongation factor-2 (EF-2), an essential component of the mammalian pro-

tein synthesis machinery. The ADP-ribosylated EF-2 becomes inactivated, resulting in a complete inhibition of protein synthesis and subsequent cell death. The A fragment is a highly efficient enzyme; once in the cytoplasm, one molecule can inactivate all cellular EF-2 in minutes and a single molecule of diphtheria toxin will kill a single host cell it enters. It is interesting to note that *P. aeruginosa*, a common organism in water and soil, secretes a functionally similar toxin.

Administration of antiserum (produced in horses by vaccination with diphtheria toxin) has been used since the late nineteenth century and has proven to be effective in controlling the symptoms of the disease. Nevertheless complete eradication of the bacteria is also necessary. Antibiotics, such as penicillin and erythromycin are effective against *C. diphtheriae*, but the antitoxin should also be given, as the antibiotic does not prevent the action of the toxin once it is produced.

Whooping Cough

Prior to the introduction of an effective vaccine, whooping cough (pertussis) afflicted more than 95% of children in the United States and resulted in about 4,000 deaths per year. The etiologic agent is *Bordetella pertussis*, a fragile gram-negative aerobic coccobacillus that is nonmotile and often encapsulated. This microorganism is very fragile and has a number of specific nutritional requirements, and, consequently, it survives very poorly outside the human body. Vaccination, begun in the 1940s, decreased the rate of infection from 150 reported cases per 100,000 to fewer than 1 per 100,000. However, the rate of infection has increased significantly in recent years. The disease can be deadly, particularly in infants, with 20 to 30 deaths nationwide each year. The resurgence of whooping cough is as much the consequence of the imperfect vaccine (85% protection) as well as lapses in vaccination of children because of safety concerns by parents. In addition, the vaccine probably protects for only about 10 years, leaving most adolescents and adults unprotected. Worldwide, whooping cough infects 40 to 50 million individuals each year and is responsible for nearly 300,000 deaths, mostly among young children. The highest incidence of pertussis is in countries that do not carry out active vaccination campaigns.

Whooping cough is highly contagious in young children and is communicated by respiratory discharge from infected individuals. The incubation period is 7 to 14 days. The organism binds to ciliated epithelial cells of the bronchi and trachea. The adherence factor that binds to cilia is called **filamentous hemagglutinin** because the purified factor will agglutinate erythrocytes. Other adhesive factors, such as **pili** and a surface protein called **pertactin** also mediate binding of the bacteria to ciliated as well as nonciliated cells in the upper respiratory tract. *B. pertussis* does not invade deeper tissue, and the symptoms of whooping cough are largely due to production of several secreted toxins by this organism. Three toxins, the **pertussis toxin**, the **adenylate cyclase toxin**, and the **tracheal cytotoxins** have been shown to contribute to the virulence of *B. pertussis* during infection.

Pertussis toxin and a number of other exotoxins produced by a variety of bacterial pathogens, superficially resemble diphtheria toxin in the functional organization of its domains, where one portion of the toxin is devoted to receptor binding and the other to intercellular toxic activity, analogs to the functions of fragments B and A of diphtheria toxin, respectively. Unlike diphtheria toxin pertussis toxin is a hexameric complex, assembled from coordinately expressed subunits. The receptor-binding B moiety is a complex of five proteins: one each of the S2, S3, and S5 subunits and two S4 subunits. The S1 subunit is the intracellular activity-associated A moiety of the pertussis toxin. After binding of the toxin to the receptors on target cells, the entire toxin is taken up by endocytosis and is shuttled along a poorly characterized pathway through various cellular compartments until it reaches the endoplasmic reticulum, where the S1 subunit dissociates and enters the cytoplasm, as depicted in Figure 28.6A. Pertussis toxin interferes with the function of the trimeric protein G_i, a regulator of adenylate cyclase, the enzyme responsible for the synthesis of an important intracellular signaling molecule cyclic AMP (cAMP). The intracellular activity of the S1 subunit of pertussis toxin is shown in Figure 28.6B. The S1 subunit catalyzes an ADP-ribosylation reaction, which is similar to that carried out by that diphtheria toxin A fragment, using NAD^+ as the ADP-ribose donor and a cellular protein as the acceptor. The target of S1 is the α subunit of G_i ($G_{i\alpha}$) an inhibitor of cellular adenylate cyclase. The consequence of ADP-ribosylation of the $G_{i\alpha}$ is a complete loss of the ability of G_i to regulate adenylate cyclase, leading to an increase in cytoplasmic levels of cAMP. Unnaturally high levels of cAMP perturb a number of signaling pathways regulated by this nucleotide, leading to increased insulin production, hypersensitivity to histamine, and hypertension. cAMP also regulates a number of immune functions, including phagocytosis and the oxidative burst, and these are also impaired in cells exposed to pertussis toxin.

The adenylate cyclase toxin (CyaA) of *B. pertussis* also raises intracellular levels of cAMP. The primary targets of CyaA in the host are the various cells of the immune system, mobilized as the consequence of *B. pertussis* infection. CyaA is a large, 1,700-amino acid bipartite protein, with distinct functions associated with each portion of the molecule. The adenylate cyclase catalytic domain is located within the amino terminal quarter of the protein, whereas the rest of the molecule has hemolytic activity because of its ability to lyse a variety of cells, including red blood cells. Therefore, it also fits into a general model of toxin, consisting of distinct A (activity) and B (receptor binding) moieties. The hemolytic domain (B moiety) accounts for binding of the toxin to target cells and for translocation of the adenylate cyclase domain (A moiety) into the cell cytosol. However, the hemolytic domain appears to have additional cytotoxic functions associated with its ability to create pores in mammalian cell membranes. The toxin is not taken up

(A)

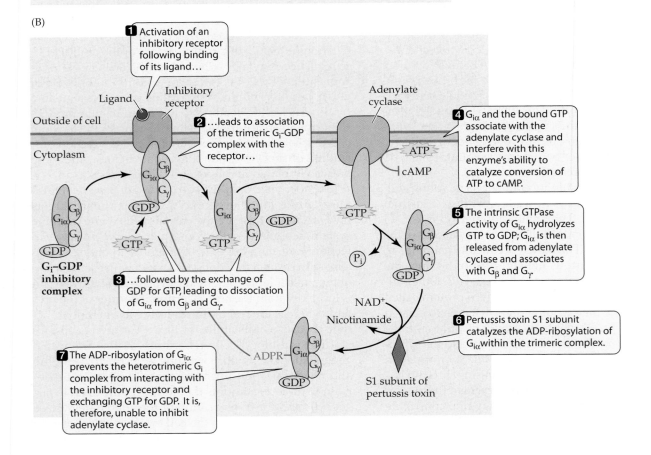

Figure 28.6 Transport and activities of pertussis toxin and cholera toxin
(A) Both toxins bind to distinct receptors on the surfaces of mammalian cells. They are then taken up into endocytic vesicles, and shuttled to the endoplasmic reticulum by a process called retrograde transport. From the Golgi compartment their respective enzymatically active subunits, S1 subunit of pertussis toxin and the A1 subunit of cholera toxin, are released into the cytosol, where they modify different regulatory subunits of adenylate cyclase. (B) Modulation of cAMP levels by pertussis toxin. Although depicted as cytoplasmic proteins, the subunits of the G proteins are anchored to the membrane and their activities (binding to receptors or to the adenylate cyclase) take place within the plane of the membrane. (C) Modulation of cAMP levels by cholera toxin.

(B)

(C)

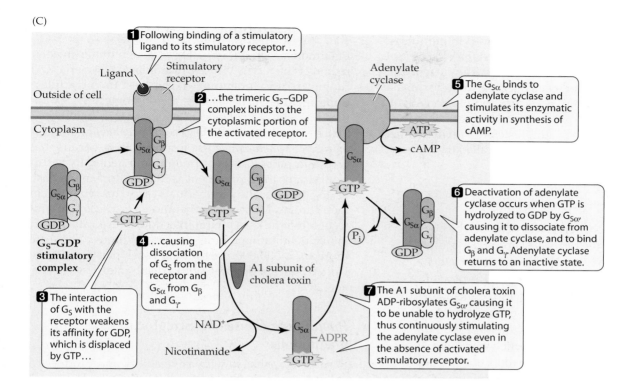

by endocytosis but it acts directly from the cell surface, and the translocated adenylate cyclase domain remains attached to the hemolytic domain. It is first activated by a cellular protein calmodulin and then proceeds to covert ATP into cAMP at a very high rate. It is the sharp rise in cAMP, together with the membrane disruptive activity of the hemolytic moiety, that kills mammalian cells exposed to CyA. Because these activities are most dramatically manifested in killing of macrophages, CyA plays an important role in the ability of *B. pertussis* to overcome host defense mobilized to the site of infection.

Tracheal cytotoxin (TCT) is a fragment of the *B. pertussis* peptidoglycan (see Chapter 4 for the discussion of peptidoglycan structure) consisting of *N*-acetyl glucosamine and *N*-acetyl muramic acid with its four nascent amino acids (glutamic acid, diaminopimelic acid, and two alanines). It is released from the bacteria during recycling of peptidoglycan that occurs during peptidoglycan remodeling during cell division. What fraction of peptidoglycan in *B. pertussis* and other bacteria is turned over and the precise steps in this process is not well understood. The major pathological effect of TCT is the destruction of ciliated respiratory epithelial cells during *B. pertussis* infection. TCT is involved in other activities, particularly when *B. pertussis* lipopolysaccharide is released simultaneously from the site of infection. Together, these two bioactive molecules cause additional epithelial cell damage accompanied by accumulation of

mucus, bacteria, and host-cell debris that are responsible for the manifestation of the whooping cough. The precise molecular mechanisms that mediate these broad effects of TCT are unknown.

Infection with *B. pertussis* begins with a catarrhal stage, which causes the inflammation of the mucous membrane, sneezing, and relatively mild coughing. After one to two weeks, the paroxysmal stage is reached and is marked by violent coughing. These violent coughs are the body's effort to rid the lungs of accumulated debris. The violent cough is followed by a "whoop," the sound of incoming air; hence, the name. Coughing can be so violent that the victim suffers **cyanosis** (turns blue from O_2 insufficiency), vomiting, and convulsions. The catarrhal and coughing stage can last six weeks with the convalescent period lasting even longer.

Whooping cough is diagnosed by plating throat swabs on Bordet-Gengou agar, where cultures have a typical identifiable appearance. However, B. *pertussis* grows very slowly on this media and 5 to 7 days are needed for the development of mature colonies. Therefore, alternative diagnostic procedures have been developed. Smears of throat swabs stained directly with a fluorescent antibody are useful for initial diagnosis. The disease can be treated with erythromycin and other antibiotics.

Vaccination begins at two months of age, since the disease can occur in infants and is associated with a high mortality in infants less than one year old. Until 1991, a whole killed-cell vaccine was employed in immunization against pertussis. In rare instances, the vaccine itself caused severe systemic toxic reactions. The current acellular vaccine consists of inactivated pertussis toxin, filamentous hemagglutinin, pertactin, and pili and should be free of side effects; toxic reactions are exceedingly rare. Pertussis vaccine is often administered with inactivated diphtheria and tetanus toxins and is called the DPT vaccine.

SECTION HIGHLIGHTS

The respiratory tract is well defended by physical barriers, the ciliary elevator, and mucosal secretions that contain antimicrobial factors such as lysozyme and defensins. The growth of organisms in the lungs is also restricted because of limited availability of iron, which is effectively sequestered by lactoferrin. Diphtheria is a respiratory tract infection with more systemic effects, due to the reduction of a protein exotoxin by the infesting bacterium *C. diphtheriae*. The diphtheria toxin's active subunit is an enzyme that transfers the ADP-ribose moiety from NAD onto a component of the eukaryotic protein synthesis machinery (elongation factor 2, or EF-2). This ADP-ribosylation reaction inhibits proteins synthesis and results in cell death. Diphtheria toxin is a prototype of an A-B toxin, where the A moiety (fragment or subunit) is responsible for the toxin's intracellular activity, whereas the B moiety recognizes receptors on target cells and facilitates the translocation of A moiety into the cytoplasm. *B. pertussis*, the causative agent of whooping cough produces two secreted toxins that perturb cells by increasing the levels of the mammalian signaling molecule cAMP by utilizing two different mechanisms. The S1 subunit of pertussis toxin (the "A" moiety) is targeted into the cell cytoplasm by the complex of one each of subunits S2, S2, and S5, and two S4 subunits (the "B" moiety). The S1 subunit catalyzes the ADP-ribosylation of Gla a component of an inhibitor of mammalian adenylate cyclase, causing a rise in cAMP. *B. pertussis* also produces an adenylate cyclase toxin that enters the cell and converts ATP to cAMP.

28.3 Pneumonia

Pneumonia is an inflammatory reaction in the alveolar region of the lungs. Viruses, mycoplasma, bacteria, and fungi can cause pneumonia. The most frequent infectious agent is *S. pneumoniae*, the etiologic agent of approximately 70% of pneumonia cases in the United States. Formerly called *Diplococcus pneumoniae* or *Pneumococci*, *S. pneumoniae* is a gram-positive coccus that grows in pairs. Cases of pneumonia generally result from strains that are present as normal flora in the host; up to 50% of healthy individuals carry this organism in their throats. Most often, pneumonia is a secondary infection that follows a viral infection, a respiratory tract injury, debilitating injuries, or chronic illness. There are more than 300,000 pneumonia cases per year in the United States, associated with a death rate between 10% and 20%.

Pneumonia Caused by Streptococcus pneumoniae

S. pneumoniae has low invasive ability and produces a cytolytic toxin that binds to cell membranes. The actual function of this toxin in pathogenesis is not clearly understood, and the capsule is considered to be more important in the development of the disease. The capsule protects the organism from phagocytosis. The loss of the capsule renders *S. pneumoniae* avirulent and phagocytes readily clear them. The bacterium grows rapidly in alveolar spaces, and the alveoli become filled with blood, bacteria, and phagocytic cells. Acute lung inflammation is a consequence of this fluid buildup.

Symptoms of pneumonia are a sudden onset of chills, labored breathing, and pleural pain. If untreated, the bacterium may escape from the lungs and cause a **bacteremia** (presence of bacteria in blood) and can be carried to the middle ear and sinuses where acute infections can occur. The organism can travel to the heart and cause **endocarditis** (inflammation of the heart valves) or **pericarditis** (inflammation of the pericardium, the membrane surrounding the heart), both of which can have long-lasting effects on health. Diagnosis of pneumonia is by chest radiograph and identification of the bacterium in the laboratory. The laboratory diagnosis includes microscopic examination of gram-stained expectorated sputum smears, where in addition to gram-positive cocci present in pairs, an increased number of leukocytes can be readily observed. This is then followed up by culturing and positive diagnosis of *S. pneumoniae* by a variety of tests, including detection of bacterial antigens using specific immunological reagents. Penicillin has been effective against *S. pneumoniae*, but resistant strains have evolved, and these cases are treated with other antibiotics such as erythromycin. A vaccine composed of purified *S. pneumoniae* polysaccharide from more than 20 pneumonococcal types is available and is recommended for adults older than age 50. A pneumococcal vaccine, consisting of a cocktail of capsular polysaccharides from the seven serotypes of *S. pneumoniae* most commonly isolated from young children, coupled to a nontoxic form of diphtheria toxin, was approved for children in the year 2000.

Pneumonia Caused by Staphylococcus aureus

S. aureus pneumonia often occurs as a secondary infection following an initial infection with the influenza virus (**Figure 28.7**). *S. aureus* accounted for a significant fraction of pneumonia acquired in the community, or in a hospital setting. Emphysema, the accumulation of pus and fluid in the lining of the lung cavity, is a more serious complication of pneumonia. Therapy for *S. aureus* pneumonia has become more difficult since the emergence of resistance to the previously effective antibiotic methicillin. Vancomycin and synthetic oxazolidine derivates have shown efficacy in therapy for staphylococcal pneumonia. A severe form of pneumonia caused by *S. aureus* is associated with the production of a specific toxin by this organism called Panton-Valentine leukocidin (PVL), a protein toxin capable of forming pores in mammalian cell membranes and inducing cell death by apoptosis.

Legionellosis

Another pneumonia that is of relatively recent description is **legionellosis (Legionnaires' disease)**, so named

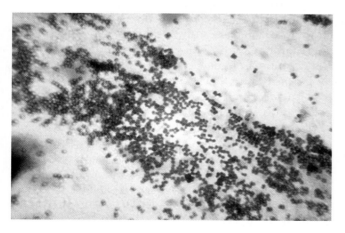

Figure 28.7 Staphylococcal pneumonia
Staphylococcus aureus in a sputum sample from a case of staphylococcal pneumonia. Courtesy of Dr. Thomas F. Sellers/Emory University/CDC.

because fatal cases of the disease occurred in attendees at a 1976 American Legion Convention in Philadelphia afflicting more than 200 individuals and causing 34 deaths. At that time the disease could not be assigned to any known organism. Since then, *Legionella pneumophila* has been identified as the causative organism.

Isolation and characterization have confirmed that *L. pneumophila* is unrelated to any of the other organisms that cause respiratory disease. It is a gram-negative pleomorphic rod that stains poorly and does not grow on media employed in diagnosing pneumonia caused by staphylococci and streptococci. Hence, earlier isolation protocols did not recover *L. pneumophila,* and the disease may have erroneously been attributed to a virus.

Legionellosis is spread by inhaling mist from air-conditioning cooling towers or being in contact with water from areas where epidemics have occurred. There is no evidence that the organism is spread by human contact. It can be readily isolated from a range of natural aquatic environments, including those that are not associated with human habitation. Despite the fact that the organism has complex nutritional requirements, including a need for high iron availability, it apparently survives in cooling-water towers, on vegetable misters, and on showerheads. *L. pneumophila* can infect and replicate in amoebae such as *Hartmannella vermiformis,* and this may be a source of virulent organisms in nature.

The organism produces and secretes a number of proteins that may be associated with virulence including several phospholipases, a ribonuclease, acid phosphatase, lipase, and a protease, but the exact role of these invasive agents is unknown. The life cycle of *L. pneumophila* in the infected macrophages (and presumably, also in the amoeba) requires the expression of factors that prevent macrophages from killing the internalized bacteria. Normal phagocytosis by macrophages involves the attachment of the bacteria to the cell surface, their internalization into the cytoplasmic within vesicles called phagosomes, followed by fusion of these vesicles with lysosomes, another type of specialized vesicle, containing microbicidal and degradative enzymes that kill and break down the internalized microorganisms (see Chapter 27). However *L. pneumophila* survives phagocytosis by the macrophages because from the phagosomal vesicle it expresses a number of protein factors that are secreted into the macrophage cytoplasm by the **type IV secretion system** (see Chapter 11). These proteins modulate several cellular functions; most importantly, they prevent fusion of lysosomes with the bacteria-containing phagosomal vesicles. *L. pneumophila* can therefore replicate within these vesicles and once released it can re-infect other cells, including macrophages.

Legionellosis occurs most often in elderly or debilitated individuals including cigarette smokers and those

(A)

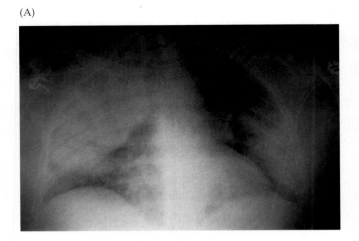

(B)

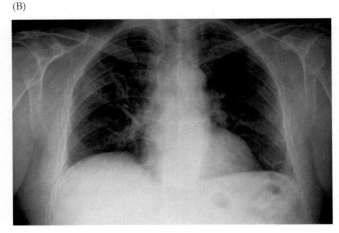

Figure 28.8 Legionnaires' disease
(A) A chest x-ray of a patient with legionellosis pneumonia. (B) An x-ray of the same patient after ten days of treatment for the disease. A,B © Hubert Raguet/Photo Researchers, Inc.

undergoing immunosuppressive therapy. Symptoms include fever, headache, chest pain, muscle aches, and bronchopneumonia (Figure 28.8). *L. pneumophila* has also been implicated in skin abscesses and inflammation of internal organs such as the heart and kidneys. In the laboratory the organism grows slowly and it may take up to a week to detect colonies even on nutritionally moist favorable media. Diagnosis therefore requires the use of fluorescent antibody or is determined by the presence of antibody in the serum of the host. Treatment is with erythromycin or rifampin.

Another disease caused by strains of *L. pneumophila* is Pontiac fever. The first incidence of this illness occurred in the county health department in Pontiac, Michigan. The symptoms of Pontiac fever resemble an allergic reaction rather than the pneumonia associated with legionellosis. Abrupt-onset fever, headache, dizziness, and muscle pain are the major symptoms. These symptoms generally resolve in 2 to 5 days and no deaths have been reported. There is a concern that in debilitated individuals the infection may progress to pneumonia. Identification of the infectious agent and treatment would be the same as with legionellosis.

Other Types of Pneumonia

Mycoplasma pneumoniae is the etiological agent for a disease called primary atypical pneumonia. These organisms are short rods (150–300 nm) and often display an irregular structure. This is very likely due to the absence of peptidoglycan, which makes them unique among the prokaryotes. To maintain rigidity, a cell envelope con-

sisting of a triple-layered membrane surrounds them. Unlike other bacteria, the *Mycoplasma* membranes contain sterols acquired from the breakdown of host cells in infected tissues or provided in the laboratory media.

The infection spreads from person to person among schoolchildren, men and women in the military service, and others living in close quarters. The symptoms include chills, fever, and a general malaise, but these symptoms are often mild and go virtually unnoticed. The mild nature of the disease and its insidious (slow) onset is the reason for calling it atypical. Diagnosis is by chest radiograph. *M. pneumoniae* can be isolated from the sputum of patients and identified by staining with fluorescent antibodies. However, the organism grows slowly, and isolation is rarely of clinical significance. Treatment is with tetracycline, erythromycin, and rifampin.

Pneumonia can be also caused by *Klebsiella pneumoniae*, a bacterium that is present as part of the respiratory tract microbiota in about 5% of the population. It is a gram-negative nonmotile rod. The organism can cause severe pneumonia that may result in chronic ulcerative lesions in the lungs. Less than 5% of pneumonia cases can be ascribed to *K. pneumoniae*, although in recent years, together with *E. coli* and *P. aeruginosa*, it has become an important agent of hospital-acquired respiratory infections. The infection can be treated with gentamicin.

Chlamydia pneumoniae is the etiological agent of a generally mild type of pneumonia. The illness is spread by contact with respiratory secretions of individuals infected. More than one-half of adults tested have antibody to this organism; however, in most individuals a subclinical infection is apparently responsible for the immune response. The symptoms of chlamydial pneumonia are a mild fever, cough, sore throat, bronchitis, and sinusitis. The symptoms disappear after a few days without treatment but erythromycin or tetracycline can be administered to prevent further complications.

Electron microscopic examination of lesions involved with coronary artery disease indicates that *Chlamydia*-like microorganisms may be present. Further testing suggests that genes and antigens of *C. pneumoniae* are present in coronary lesions and in lesions elsewhere in the vascular system. Although this type of association of a pathogen with a disease may be suggestive, definitive proof connecting *C. pneumoniae* with heart disease has not been obtained to date.

There is growing evidence that a number of mycobacterial species commonly present in soil may be agents of pulmonary infections in humans. These organisms, including *Mycobacterium avium*, *Mycobacterium intracellulare*, and *Mycobacterium* sp. are genetically related and termed the *M. avium* complex or MAC. The MAC microbes are also known to infect insects, birds, and other animals. They gain entry to the pulmonary area of humans via respiratory droplets from infected individuals and through inhalation of dust particles.

The symptoms of the disease caused by MAC in humans are similar to the symptoms of tuberculosis. The infections are most prevalent in debilitated individuals, particularly those with a deficient immune system. Members of MAC are among the major opportunistic pathogens in patients with acquired immunodeficiency syndrome (AIDS). They bring about disseminated infection in these patients, particularly when resistance levels drop in the latter stages of the disease. Symptoms include fever, weight loss, diarrhea, and night sweats. Treatment is with a combination of antibacterial antibiotics such as azithromycin, rifampin, and ciprofloxacin.

A fungus that was previously considered a protozoan causes another form of pneumonia. This organism, *Pneumocystis carinii*, inhabits many animal species, including humans. Serological examination suggests that all children have been infected with this organism. The organism remains dormant unless the host becomes immunocompromised, at which point it can cause a fatal pneumonia. In recent years, *P. carinii* has been identified frequently in cases of fatal pneumonia in immunocompromised individuals. *P. carinii* occurs in about 50% of AIDS patients and results in a fatal infection in about one-half of those infected. It is diagnosed by fluorescent antibody staining. Sulfamethoxazole and trimethoprim are given as an oral prophylactic, although novel, more effective formulations, allowing aerosol delivery, are under development.

SECTION HIGHLIGHTS

Pneumonia is often caused by normal residents of the throat or nasal cavity. The most significant causative agent of pneumonia is encapsulated *S. pneumoniae*. *S. aureus* can also cause pneumonia, usually following infection with the influenza virus. *L. pneumophila* cause pneumonia in debilitated and immunocompromised individuals. During infection, the bacteria is taken up by macrophages, and they not only survive phagocytosis but replicate within these cells. Pneumonia caused by *K. pneumoniae*, *E. coli*, *P. aeruginosa*, *Pneumocystis carinii*, and *M. avium* complex is frequently associated with a breach in host defense mechanisms including immune deficiency caused by human immunodeficiency virus (HIV) infection.

28.4 Streptococcal Infections

In addition to *S. pneumoniae*, other members of the genus *Streptococcus* can cause infections in humans. *S. pyogenes* (also called group A streptococcus, after an earlier serological classification scheme) and *Streptococcus agalactiae* (group B streptococcus) are important human pathogens, although they can be found in the normal flora of the nasal cavity or in the intestinal tract. Streptococci are gram-positive cocci that divide in one plane, and the individual cocci tend to remain together and form long chains. The organism is the causative agent of skin infections (impetigo) as described previously.

The streptococci are **aerotolerant fermenters**, so called because most grow in the presence of air but obtain energy by a lactic acid fermentation of sugar. The two most important pathogenic species of *Streptococcus* (*S. pyogenes* and *S. agalactiae*) secrete an active hemolysin that can lyse erythrocytes. The streaking of some strains of streptococci on blood-agar plates results in a greenish discoloration around colonies, indicating that there has been an incomplete destruction of red blood cells (α-hemolysis)—these strains are termed **α-hemolytic**. The **β-hemolytic** streptococci, which include all major human pathogenic members of this genus, affect a complete lysis of red blood cells with a clear zone around the colony.

Infections Caused by Streptococcus pyogenes

The major infections caused by *S. pyogenes* are **pharyngitis**, **scarlet fever**, and **necrotizing fasciitis**. However, infections caused by *S. pyogenes* can be sometimes followed by even more serious complications, namely **rheumatic fever** and **glomerulonephritis**.

S. pyogenes produces a number of virulence-promoting factors that play a role in the ability of this organism to cause disease in humans. Prominent among them are factors located on the cell surface, which play a role in colonization and evasion of host cell defense mechanisms. These include **M protein**, a **capsular polysaccharide**, and **lipoteichoic acid**. Although lipoteichoic acid (a modified form of teichoic acid described in Chapter 4) protrudes from the surface and spans the entire peptidoglycan layer, it is anchored to the cell via a covalent linkage to two fatty acids, embedded in the cytoplasmic membrane. The M protein is a long fibrous structure whose N terminus projects from the surface, covalently but very tightly with the peptidoglycan through its hydrophobic C-terminal anchor sequence. Both M protein and lipoteichoic acid mediate attachment of the bacteria to various tissue surfaces. In addition, M protein protects bacteria from phagocytosis by preventing functional binding of complement fragments to the bacterial surface. Therefore, antibodies to M protein bind to *S. pyogenes*, promote phagocytosis of the microorganisms and therefore, they could provide protection during subsequent infections. However, M protein shows extensive sequence variability in its exposed amino terminus, and more than 200 types of M proteins have been identified among natural isolates, making re-infection with the strain expressing the same type of M protein unlikely.

Many strains of *S. pyogenes* are surrounded by a capsule composed of hyaluronic acid and alternating residues of *N*-acetylglucosamine and glucuronic acid. Because hyaluronic acid is a normal host cell constituent, this gives the bacterium protection from phagocytosis and is non-immunogenic as these components are considered "self" antigens. The ability of bacterial pathogens to avoid the host immune defenses by producing components similar to those found in human tissues is referred to as **molecular mimicry** of host epitopes. The hemolytic activity is the result of the action of two proteins: **streptolysin O**, an oxygen-labile toxin, and **streptolysin S**, an oxygen-stable toxin. Both of these toxins damage the membranes of polymorphonuclear leukocytes (PMNLs) and platelets. Other toxins are produced by strains that cause specific disease and will be discussed later.

S. pyogenes is the causative agent of several familiar acute infections, including **streptococcal pharyngitis**, scarlet fever, rheumatic fever, and glomerulonephritis. Streptococcal pharyngitis is the most common infection caused by *S. pyogenes*, an acute infection with reddened tonsils, a swollen pharynx, and the production of purulent exudate. Often the local lymph nodes become enlarged and there is a high fever. The infection is commonly known as "**strep throat**" and may spread to the middle ear, and an acute infection there can cause a loss

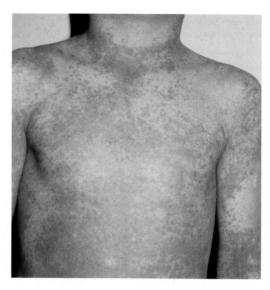

Figure 28.9 Scarlet fever
The rash associated with scarlet fever. The toxin produced by *Streptococcus pyogenes* causes a hypersensitivity reaction that results in the pink-red rash. Photo ©Biophoto Associates/Photo Researchers, Inc.

of hearing. On rare occasions, the infection breaks through the throat epithelia and is carried by the bloodstream to the meninges, causing a form of meningitis. Epidemics of strep throat are the result of personal contact with infected persons or carriers. Control with antibiotics, primarily penicillins, is essential, as streptococcal sore throat can be followed by scarlet or rheumatic fever and glomerulonephritis.

Scarlet fever is caused by strains of *S. pyogenes* lysogenized by a specific bacteriophage. The phage carries the genetic information for the production of one of the forms of the **pyrogenic** (fever-inducing) **exotoxin** or **erythrogenic toxin**. This toxin is a **superantigen** and acts by a mechanism similar to that in staphylococcal TSST, described earlier in this chapter and in Chapter 27. The erythrogenic toxin induces the skin rash that is typical of scarlet fever (Figure 28.9). Strains expressing the streptococcal pyrogenic toxins, like TSST-producing strains of *S. aureus*, can cause toxic shock syndrome. Scarlet fever can be controlled with antibiotics. It is very important to halt the infection at an early stage for, if untreated, the infection can lead to rheumatic fever.

Rheumatic fever can follow streptococcal sore throat caused by certain strains of *S. pyogenes*. This disease occurs in 3% to 5% of pharyngitis patients that fail to receive proper treatment. Rheumatic fever can result in permanent damage to the heart. The exact mechanism by which *S. pyogenes* induces rheumatic fever is unknown. The most likely theory is based on the concept of **autoimmunity**. This theory is supported by observations that several constituents of *S. pyogenes* share antigenic similarity with human tissues, including heart and neurons of the basal ganglia of the brain. Taken together, these immunological cross-reactions could theoretically account for the tissue damage seen in rheumatic fever, particularly involving heart tissue.

Glomerulonephritis is an inflammation of glomeruli in the kidney. It is often the consequence of infection by an *S. pyogenes* expressing a limited number of M-protein types. It may be an autoimmune response, as the streptococcal toxin and kidney cells share a common antigen. Antibodies to streptococcal M protein, fragments of streptococcal membrane, and a pyrogenic exotoxin were all shown to cross-react with kidney tissue. The incidence of this disease is low, and most patients undergo a slow healing of the damaged glomeruli. Others (less than 10%) may have chronic kidney disease.

Necrotizing fasciitis can be cause by several bacterial pathogens; however, the most common causative agent is *S. pyogenes*. Because of its symptoms, massive destruction of subcutaneous tissues, the infecting organisms are referred to as "flesh-eating bacteria." This disease is also likely the sequela of pharyngitis; however, direct infection of traumatized skin tissue can also lead to necrotizing fasciitis. The onset of disease is very rapid and is associated with a high death rate. The dramatic symptoms of necrotizing fasciitis are very likely the consequence of production of one of the pyrogenic exotoxin superantigens by *S. pyogenes,* resulting in hyperstimulation of macrophages that are then responsible for the massive tissue damage. Successful therapy depends on early diagnosis and rapid response, including surgical removal of dead tissue and administration of one or a combination of broad spectrum antimicrobials such as penicillin, clindamycin, gentamicin, vancomycin, or metronidazole.

Streptococcus agalactiae

The major infections caused by *S. agalactiae* (group B streptococcus) are puerperal infections (bacterial infection following childbirth), neonatal septicemia, and meningitis. These organisms can also cause skin infections and endocarditis.

S. agalactiae is a frequent colonizer of the vagina and intestinal tract. It expresses a number of adhesive factors (including lipoteichoic acid and a fibrinogen-binding protein) that allow the bacteria to successfully compete with other colonizers of these tissues. However, a key virulence attribute of this organism is the ability to breach the epithelial barriers and evade host defenses. The invasive process involves uptake of *S. agalactiae* by epithelial cells. The bacteria survive within these cells but do not replicate. In some cases, they cross the epithelial barrier without disruption of the invaded cells; however, it is equally likely that invaded cells die and release the organisms into the amniotic fluid or, during sep-

ticemia, into blood. The polysaccharide capsule produced by *S. agalactiae* appears to facilitate the invasion process. Another important virulence factor of this organism is the secreted enzyme hyaluronate lyase. Hyaluronic acid is a major component of the connective tissue and is found in high concentrations in placenta and amniotic fluid; its destruction permits this organism to spread into deeper tissues.

The surface polysaccharide capsule also plays a prominent role in immune evasion. Virtually all *S. agalactiae* associated with human disease are encapsulated. The capsule contains terminal sialic acid that limits detection by the host immune system. Because many components of the human tissues are also sialylated, the sialic acid–containing capsule shields the organism from recognition. The ability of these bacteria to surround themselves with a polymeric structure that resembles material produced by the host represents another instance of molecular mimicry, analogous to the function of the hyaluronic acid capsule of *S. pyogenes* discussed earlier. This structure also interferes directly with opsonic phagocytosis by blocking deposition of components of complement on the cell surface (see Chapter 27). The β-hemolysin of this organism, another toxin with the ability to form pores in mammalian membranes, is responsible for killing of a substantial fraction of phagocytic cells encountered by this organism during infection.

The genital and gastrointestinal flora of adults often includes *S. agalactiae* but these sites remain free of disease symptoms. However, during pregnancy, colonized women are at an increased risk of premature delivery and transmission of the organism to infants during their passage though the birth canal. Complications during labor and subsequent infections by *S. agalactiae* can affect both mother and the infant, often within days of delivery. In mothers this organism cause infections of placental membranes and that can lead to their rupture. It can also cause bloodstream infections and, less frequently, meningitis, both in mothers and infants. Detection of *S. agalactiae* in pregnant women, and administration of antibiotics (usually penicillin) during labor has proved to be effective in reducing infections by *S. agalactiae* during delivery. Nevertheless, incidence of serious infections remains at 0.5 per 1,000 live births, with 5% fatality. Several vaccines are currently under development.

Meningitis

Meningitis is an inflammation of the meninges, which are the three membranous layers that envelop the brain and spinal cord. When meningitis is suspected, the cerebrospinal fluid is tapped and examined for the presence of microorganisms.

A number of organisms involved in respiratory infections may be responsible for meningitis. These include

H. influenzae, which is responsible for nearly half of the cases of meningitis in the United States. The other organisms are *N. meningitidis, S. pyogenes, S. pneumoniae,* and *S. agalactiae.*

H. influenzae meningitis is an exceedingly dangerous disease, and survivors can have neurological disorders. Ampicillin has been an effective therapy, but ampicillin-resistant strains of *H. influenzae* are increasing in number, requiring the use of chloramphenicol as an alternative therapy. Vaccination for *H. influenzae* is now available. Meningitis caused by *S. pneumoniae* can also be treated with chloramphenicol.

Meningitis symptoms are sudden fever, stiffness in the neck, headaches, and often delirium, followed by convulsions and coma. *N. meningitidis* is called the meningococcus and is a gram-negative diplococcus.

N. meningitidis often causes a mild nasopharyngeal infection. In a limited number of cases, however, the organism enters the bloodstream and causes a systemic infection. This can be a rapidly fatal stage of the disease, as the infectious agent may cross the blood–brain barrier and infect the meninges. The disease is more prevalent in younger individuals. Treatment of meningococcal infections is with penicillin. Erythromycin and chloramphenicol are also effective, and treatment must be initiated early, as the disease can be fatal otherwise. Meningitis is such a severe, fatal disease that anyone in contact with a victim is generally treated with antibiotics as a precautionary measure.

SECTION HIGHLIGHTS

M protein and a lipoteichoic acid coat the surface of *S. pyogenes* and play a role in colonization and evasion of host cell defense mechanisms. Many strains of *S. pyogenes* are surrounded by a hyaluronic acid capsule. This substance is also found in human connective tissue and therefore the bacterial product is nonimmunogenic. This is an example of a pathogenic strategy referred to as molecular mimicry. Rheumatic fever, an autoimmune reaction directed at cardiac tissue and glomerulonephritis an inflammation of glomeruli in the kidney are two serious consequences that follow a fraction of cases of streptococcal pharyngitis. The major infections caused by *S. agalactiae* (group B streptococcus) are those that follow childbirth and can affect the mother and the child. The capsule of *S. agalactiae* also contains sialic acid, a sugar found frequently in human polysaccharides and on glycoproteins; therefore, it shields the organism from recognition by the host.

28.5 Tuberculosis

The etiologic agent of **tuberculosis** (TB) is *Mycobacterium tuberculosis*, commonly referred to as the tubercle bacillus. It was isolated in 1882 by Robert Koch, who confirmed the relationship between the organism and the disease. Historically and currently, tuberculosis is a devastating disease that continues as a leading cause of human suffering and death. Worldwide, there are 10 million active new cases of tuberculosis each year and 1.7 million deaths. This death rate is higher than that for any infectious disease except HIV. It is estimated that one-third of the world's population has either an active or latent form of tuberculosis. Despite available therapy in the second half of the twentieth century, it is estimated that tuberculosis has claimed the lives of approximately 100 million people over the last 100 years. In the United States alone, there are more than 10,000 new or reactivated cases per year. Prior to the availability of antibiotics effective against *M. tuberculosis*, victims of the disease were practically bedridden, as rest was the only known treatment and patients were isolated from the general population.

M. tuberculosis is a thin rod, sometimes bent and club shaped. The organism is acid-fast, a staining property where cells stained with hot carbolfuchsin are not decolorized by acid alcohol, and this can be used as a diagnostic test. *M. tuberculosis* is aerobic with relatively simple nutritional requirements, although it grows very slowly on most laboratory media and it takes two weeks for a colony to become visible. The organism has a high lipid content (greater than 40% cell dry-weight), and the lipid contains unique fatty acids called mycolic acids (Figure 28.10). Mycolic acids vary somewhat in size and in amount of hydroxylation and are frequently linked to sugars to form glycolipids called mycosides. A mycoside associated with certain mycobacterial strains is depicted in Figure 28.11. Here two molecules of mycolic acid are attached to the disaccharide trehalose forming a molecule called **cord factor**. Only virulent mycobacteria express cord factor. They grow in parallel and small colonies have the appearance of twisted cords. Cord factor appears to be an important virulence factor of *M. tuberculosis*. When purified cord factor is injected into mice, they undergo dramatic weight loss. This lipid is therefore considered to be responsible for the severe weight loss (cachexia) observed in tuberculosis patients.

Tuberculosis is transmitted from person to person by aerosol, via small droplets introduced into the air by infected individuals. The greater the proximity to a source, the higher is the risk of infection. A single active case can infect many bystanders, and crowded conditions favor epidemics. Inhalation of droplets containing the tubercle bacillus can lead to propagation within the lungs. Ninety

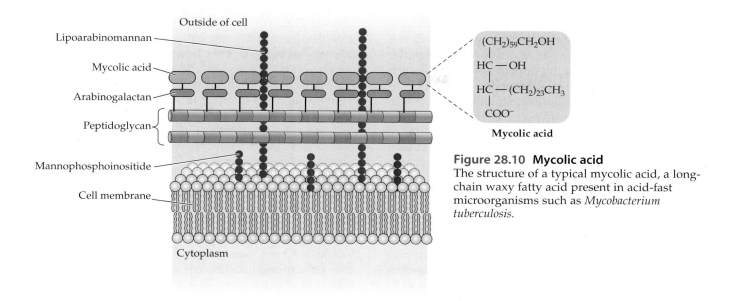

Outside of cell

Lipoarabinomannan

Mycolic acid

Arabinogalactan

Peptidoglycan

Mannophosphoinositide

Cell membrane

Cytoplasm

$(CH_2)_{59}CH_2OH$

$HC-OH$

$HC-(CH_2)_{23}CH_3$

COO^-

Mycolic acid

Figure 28.10 Mycolic acid
The structure of a typical mycolic acid, a long-chain waxy fatty acid present in acid-fast microorganisms such as *Mycobacterium tuberculosis.*

percent of infected people do not progress to clinical disease, but the bacillus remains latent. Should the patient become immunocompromised, for example, due to AIDS, the clinical disease will appear.

Growth of *M. tuberculosis* in the lungs results in inflammation and the development of lesions. The bacteria are ingested by macrophages but are not destroyed by these phagocytic cells, and instead they replicate and destroy the macrophages. Two surface lipids of *M. tuberculosis*, the cord factor and a lipopolysaccharide called lipoarabinomannan (LAM), prevent phagosome–lyso-

some fusion, thus protecting the bacteria from killing and contributing to their survival within macrophages. Infected macrophages are the vehicles for dissemination of the bacteria to lymph, blood, and to other organs and tissues. As the bacteria propagate in lesions, they produce a mass consisting of the waxy bacterium, lymphocytes, and macrophages. The lesion becomes enmeshed in fibrous connective tissue forming a nodule called a **tubercle**. A tubercle contains viable bacteria that can be dormant indefinitely. During dormancy, the infected individual is asymptomatic, and this would be considered a primary infection. The dormant primary infection may be activated by malnutrition, stress, hormonal imbalance, or decrease in immune function. Symptoms of the disease—fatigue, weight loss, and fever—are only obvious after extensive lesions are formed.

The formation of tubercles is mostly a delayed-type hypersensitivity reaction (see Chapter 27). Consequently, individuals in the primary stages of the disease are hypersensitive to protein fractions from *M. tuberculosis*. The **tuberculin test** given to children and to others during physical examinations is based on this hypersensitivity. A small amount of protein from *M. tuberculosis* is placed intradermally (beneath the skin) and observed for redness and swelling. Reddening after a few days (tuberculin positive) indicates that the individual has had an inapparent infection.

Diagnosis of active tuberculosis is by the presence of *M. tuberculosis* in sputum. Presence of lung lesions results in the expectoration of bloody sputum laden with *M. tuberculosis*. X-ray examination of the chest is employed to reveal lung damage and tubercles. There are ten different antibiotics available for the treatment of tu-

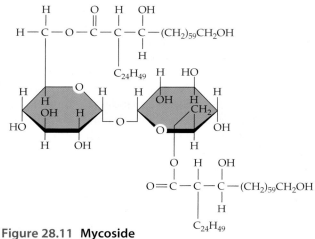

Figure 28.11 Mycoside
Structure of a cord factor, a mycoside present in virulent strains of *Mycobacterium tuberculosis*. Colonies of bacteria that synthesize this mycoside appear as pieces of cord sitting side by side.

berculosis including rifampin, isoniazid (INH), pyrazinamide (PZA), ethambutol (EMB), and streptomycin. The most common treatment regimen uses a combination of isoniazid and rifampin; however, the precise drug combination depends on the susceptibility of the infecting organisms to the drugs. Treatment must last at least 6 months to ensure that all intracellular and intranodular organisms are destroyed because the bacteria grow extremely slowly. Unfortunately, these drugs all have nasty side effects, and patients are reluctant to continue to take them while they are feeling well. This is not only bad for the patient but has serious public health consequences. The solution is DOT (directly observed therapy), when a public-health nurse observes the patient take their medication three times a week.

In recent years, strains of *M. tuberculosis* have evolved that are resistant to antibiotics commonly employed to treat tuberculosis. These resistant strains are prevalent among patients with AIDS and will ultimately spread to the general population. Poverty and inadequate efforts at eradication worldwide also contribute to the potential pool of resistant strains.

Psittacosis (Ornithosis)

Psittacosis is an avian disease that afflicts all species of birds and can infect humans. The causative agent is *Chlamydia psittaci*, an obligate intracellular prokaryotic parasite related to gram-negative bacteria. Individuals who, in certain seasons, extensively handle birds (poultry farmers, employees of poultry processing plants, and pet store employees) are at a higher risk of infection than the general population. The disease is generally latent and asymptomatic in birds. Crowding and unsanitary conditions, such as those encountered in transporting birds, can lead to active infections. The organism is present in virtually every organ of an infected bird, and considerable numbers are excreted in the feces. The consequent inhalation of dried feces by bird handlers can cause human infections. The inhaled organism travels to the liver, spleen, and lungs. Lungs become inflamed, hemorrhagic, and with symptoms of pneumonia. The pneumonia can be fatal.

A case of fatal meningitis can also occur in a limited number of psittacosis patients. The disease can be diagnosed by isolation of *C. psittaci* from blood or sputum and serologically. Treatment with tetracycline is effective, although a long regimen (up to three weeks) is required for full eradication of this organism.

28.6 Leprosy

Leprosy, also known as Hansen's disease, is a dreaded disease that afflicts about 14 million people worldwide,

mostly in tropical countries. About 100 new cases of leprosy are diagnosed in the United States each year. The organism responsible for leprosy is *Mycobacterium leprae*, a rod-shaped, acid-fast organism. *M. leprae* can be obtained from lesions but has not been grown on artificial media. The organism will grow in nude mice that have an impaired immune function. The armadillo is a natural host, where a low body temperature permits a systemic multiorgan infection. Death of the animal occurs within 2 years. Attempts to infect human volunteers have not been successful.

The epidemiology of leprosy is somewhat of a mystery. Most humans apparently are not susceptible to infection, although it afflicts twice as many men as women. Children are more susceptible than adults. Transfer by human contact is possible, because individuals with the more serious lepromatous form of leprosy have more than 10^8 cells of the infectious agent in their nasal discharge. This may be the mechanism whereby the organism is transmitted to susceptible individuals, but epidemics do not occur. There are two major forms of leprosy in humans: tuberculoid leprosy and lepromatous leprosy.

Tuberculoid leprosy is often a mild self-limiting disease. It is nonprogressive and occurs as a delayed hypersensitivity reaction to proteins in *M. leprae* or as a normal cellular immune response. Damaged nerves and loss of sensation occur in areas where lesions exist. Rarely are intact organisms isolated from material taken from these lesions. Spontaneous recovery from tuberculoid leprosy frequently occurs.

Lepromatous leprosy, on the other hand, is a progressive, nasty disease that ultimately leads to death. *M. leprae* is found in virtually every organ of the body. Skin lesions are common, often pigmented, and organisms can be obtained for examination by skin scraping. Skin nodules infected with *M. leprae* appear on the surface of the body. Nerves are damaged, and mucosal lesions are abundant. Lesions in the mucous membranes of the nose lead to the destruction of cartilage and to nasal deformities. The eyes also can be infected, and this can lead to blindness. Loss of fingers or toes is the consequence of destruction of cartilage in the appendages (Figure 28.12).

Diagnosis of leprosy in the early stages is difficult. The occurrence of numbness and the isolation of the acid-fast bacterium are the major diagnostic criteria. It has been suggested that a material (lepromin) purified from infected armadillo tissue might be employed as a diagnostic tool. This would be analogous to the use of the tuberculin test for tuberculosis in that infected individuals would be expected to be hypersensitive to lepromin. In practice it has not been very effective.

Limited information is available regarding the pathogenic mechanisms used by *M. leprae* to cause the ex-

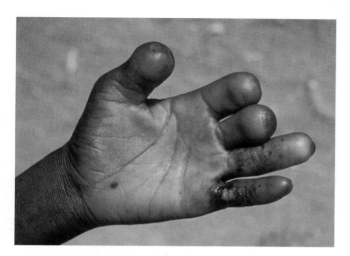

Figure 28.12 Leprosy
Hand of an individual with leprosy. The infection progresses from a loss of sensation in the skin to deformity and lesions and can result in the loss of fingers and toes. ©Phototake, Inc./Alamy.

tensive tissue damage and death seen in certain forms of leprosy. *M. leprae*, like *M. tuberculosis*, can survive in macrophages, and is not effectively eliminated by innate host defenses. However, the nerve damage and the subsequent deformities are the result of the ability of this organism to invade Schwann cells, which are components of the peripheral nervous system. These cells contain a fatty substance called myelin, and they wrap around a nerve fiber to form a protective sheath. The affinity of *M. leprae* for Schwann cells is based on the presence of a specific protein (laminin-2) on the surface of these cells, which serves as a receptor for a histone-like protein, LBP21, produced by *M. leprae* and displayed on its surface. Following this initial interaction with the Schwann cells, the bacteria are taken up by and proliferate within these cells. Eventually, Schwann cells are destroyed by the combined action of the bacteria themselves and attacks by T cells that recognize mycobacterial antigens processed and presented by the invaded cells. Affected nerve cells lose their protective myelin sheath, resulting in a severe impairment of their function.

Treatment of leprosy is also difficult. The incubation period for tuberculoid leprosy is 3 to 6 years and for lepromatous leprosy 3 to 10 years. Without a reliable diagnostic tool, treatment is typically delayed until the full onset of visible symptoms. The antibiotic Dapsone (4,4'-sulfonylbisbenzamine) alone or in combination with rifampin has been proven effective. In order to prevent relapse, therapeutic regimens for leprosy last for months or even for several years. A variant of dapsone, diacetyl dapsone, is less toxic and appears to be more effective. Vaccines that might be employed in tropical areas where leprosy is an ongoing problem are under development.

SECTION HIGHLIGHTS

The agent of tuberculosis, *Mycobacterium tuberculosis*, grows within macrophages and uses them to disseminate into a variety of tissues. The infectious agent propagates in lesions that become surrounded by fibrous connective tissue, developing into a nodule called a tubercle. In this nodule, the bacteria are dormant, but they can reactivate many years later. Psittacosis is a disease of birds and can be transmitted to infect humans. The causative agent, *Chlamydia psittaci*, is an obligate intracellular bacterium. It can infect a number of organs; however, the most serious—and occasionally fatal—infection is pneumonia. Destruction of the nerves during leprosy follows invasion of Schwann cells that surround neurons by *Mycobacterium leprae*. The damage is very likely caused by an immune reaction to the infected Schwann cells.

28.7 Anthrax

Bacillus anthracis is a saprophytic, aerobic, spore-forming gram-positive soil organism. *B. anthracis* is essentially an accidental pathogen that can cause a generally limited cutaneous infection, or inhalation of spores can cause an infection that is nearly 100% fatal. The ready dispersal of *B. anthracis* spores and their lethality when inhaled has led to the potential use of this organism as a weapon in bioterrorism (see Box 20.1). Many strains have a capsule composed of poly-D-glutamic acid that renders them resistant to phagocytosis. *B. anthracis* played a significant role in defining the microbe–disease relationship, especially in the formation of Koch's postulates (see Chapter 2).

Grazing animals are constantly exposed to spores of *B. anthracis* and may become infected, but the number of infections is unknown. Spores may be present on animal hides, and individuals working with these or otherwise exposed can develop skin infections. Historically the symptoms have been described as "wool sorters disease."

Most of the symptoms of anthrax are due to the production of lethal toxin and edema toxin, a pair of toxins by *B. anthracis* that are encoded on the plasmid pOX1. These toxins are also organized into two functionally distinct A (activity) and B (receptor binding) moieties, as described earlier for diphtheria and pertussis toxins. The steps during intoxication of cells by the anthrax lethal and edema toxins are shown in Figure 28.13. Each of the anthrax toxins is assembled by a specific combination of two of the three nontoxic proteins secreted by the bacterium. One of these proteins is called the protective antigen (PA) and it is the B moiety of the toxin PA that binds to cell surface receptors and transports the other two subunits, the edema factor (EF) and lethal factor (LF), to the cytosol. LF and EF are the "A" moieties of these multi-subunit toxins. Following the secretion of the subunits by *B. anthracis*, the monomeric PA binds to the receptors on the targeted cell. A cellular protease cleaves off a small fragment of PA, while the larger fragment remains bound to its receptor. The processed PA oligomerizes into a ring-shaped heptamer that binds three molecules of LF or EF. The resulting complex is taken up by endocytosis into a vesicle. The lumen of the vesicle is acidified and the A moieties (LF or EF) are translocated across the membrane and released into the cytosol. PA is a protease that cleaves several related signal transduction proteins (MAP kinase kinases), leading to death of the host cell by an unknown mechanism. EF is an adenylate cyclase and it interrupts signaling within the cell by producing unnaturally high levels of cAMP, similar to the action of the diptheria and pertussis toxins. EF is also responsible for the edema seen in the disease. Both enzymes are believed to facilitate bacterial survival during infection by interfering with the cells of the host innate immune system.

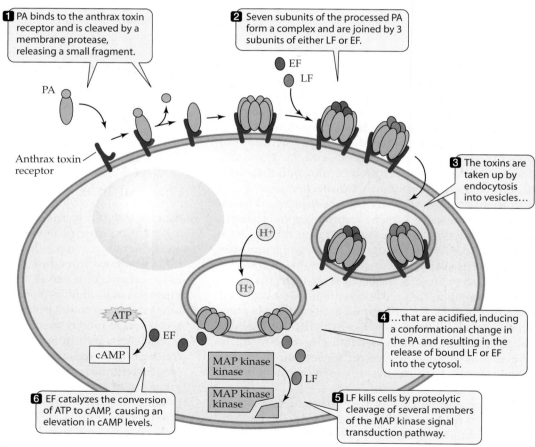

1 PA binds to the anthrax toxin receptor and is cleaved by a membrane protease, releasing a small fragment.

2 Seven subunits of the processed PA form a complex and are joined by 3 subunits of either LF or EF.

EF
LF

PA

Anthrax toxin receptor

3 The toxins are taken up by endocytosis into vesicles…

H+

H+

ATP

EF

cAMP

MAP kinase kinase

MAP kinase kinase

LF

4 …that are acidified, inducing a conformational change in the PA and resulting in the release of bound LF or EF into the cytosol.

6 EF catalyzes the conversion of ATP to cAMP, causing an elevation in cAMP levels.

5 LF kills cells by proteolytic cleavage of several members of the MAP kinase signal transduction pathway.

Figure 28.13 Steps in the activities of the lethal and edema toxins of anthrax
The three toxin subunits PA, LF, and EF are secreted separately by *B. anthracis* and are assembled into two toxins on the cell membrane before they are taken up into the cell and exert their activities in the cytosol.

B. anthracis infections can take any of three forms:

- Cutaneous anthrax occurs when *B. anthracis* on animal hides, fur, or other sites gains entry through superficial wounds. Usually this results in a localized tissue necrosis and is not lethal. It can be a serious problem if the organism passes into the bloodstream. The necrotic area in cases of cutaneous anthrax is termed an eschar.

- Pulmonary anthrax results from the inhalation of *B. anthracis* spores. Germination and growth of the organism causes severe pulmonary hemorrhaging. Unless treated promptly this infection is nearly 100% fatal. The symptoms initially resemble cases of flu and progress to difficulty in breathing and death from systemic shock.

- Gastrointestinal anthrax is caused by ingestion of spores but is rare in the United States.

An anthrax vaccine has been developed for individuals at risk of exposure to bioweapons/bioterrorism. The current vaccine is a cell-free preparation of *B. anthracis*, but more specific protein subunit vaccines are under development. A prophylactic treatment for individuals exposed to the organism is administration of antibiotics—either penicillin or ciprofloxacin. Because many bioterrorism strains are penicillin resistant, ciprofloxacin would be the antibiotic of choice in such cases. Diagnosis is by microscopic examination and culturing of clinical specimens.

Gas Gangrene

Several members of the genus *Clostridium* can cause a necrotizing (death to local tissue) infection of muscle, termed gas gangrene. *Clostridium perfringens* is most frequently recovered from infected tissue; however *Clostridium novyi, Clostridium histolyticum,* and *Clostridium septicum* can also be the causative agents of gas gangrene. These organisms are strictly anaerobic, gram positive endospore formers. They occur in most soils, beds of lakes, and are inhabitants of the human intestinal tract. These clostridia readily utilize an array of sugars and amino acids and generate gases during growth.

Entry of the organisms, usually as spores, is by deep wounds or often by gunshot wounds. Because they are killed by oxygen, the spores germinate only after penetrating into deep tissue. Depending on the strain, they produce a combination of up to 15 toxins, most of which are degradative enzymes. The **α toxin** (a phospholipase C) and another membrane-damaging toxin, perfingolysin O, are believed to play a highly significant role in tissue necrosis. Other secreted enzymes, such as collagenase, proteases, and hyaluronidase, and a number of necrotizing toxins with poorly defined biochemical

activities have also been described, and these very likely contribute to tissue disruption leading to necrosis.

Treatment of gas gangrene is by surgical removal of dead tissue and administration of antibiotics such as penicillin. When resistant organisms are encountered, or in case of allergy to penicillin, chloramphenicol or metronidazole are also effective. Introduction of increased oxygen levels may curtail growth of these anaerobes. Amputations associated with those wounded in warfare are common.

SECTION HIGHLIGHTS

Bacillus anthracis is responsible for a serious, often fatal disease (anthrax). Following inhalation or ingestion of spores, the germinating cells produce two multi-subunit toxins: the lethal toxin and the edema toxin. Several members of the genus *Clostridium* cause deeps tissue infections leading to necrosis. The most important agent of gas gangrene is *C. perfringens.* The tissue damage is due to the combined action of several secreted toxins produced by the pathogenic clostridia.

28.8 Fungal Diseases

There are a number of human respiratory diseases caused by fungi. Fungal spores are inhaled in normal respiration, and in some cases these fungi can propagate in the lungs. Among these diseases, we will consider histoplasmosis, coccidioidomycosis, and aspergillosis.

Histoplasmosis

Histoplasma capsulatum is the etiologic agent of histoplasmosis, the most common fungal disease in humans. An estimated 40 million Americans have been infected with histoplasmosis, and 250,000 new cases occur each year. *H. capsulatum* is widely distributed in soil and is generally associated with birds and bats. Abundant numbers of the bacteria are present in chicken coop litter, bat caves, and bird roosts. The disease is endemic in the Ohio and Mississippi River valleys. *H. capsulatum* is dimorphic, producing hyphae in soil and budding yeasts in the disease state. Infection occurs by inhalation of the reproductive spores (microconidia) that are present on hyphal filaments. In most cases, the normal host immune response is sufficient to rid the body of the infectious organism. A small number of infected individuals may develop a systemic histoplasmosis of variable severity. The symptoms vary from none to coughing, fever, and pain in joints. Lesions may appear in the lungs and become calcified. Generally the disease is resolved by host defenses but may become chronic, with bouts of lung infection occurring periodically. Diagnosis is by examination of sputum or by a test similar to the tuberculin test described earlier but employing a fungal protein. The disease can be treated with amphotericin B or ketoconazole, but relapses may occur.

Coccidioidomycosis

The etiologic agent of coccidioidomycosis is *Coccidioides immitis*, a filamentous fungus present in arid soils. Most cases in the United States occur in the Southwest. Infection of the lungs occurs by inhalation of the reproductive arthrospores. More than half of primary infections are asymptomatic, and most of the remainder produce a mild-to-severe pulmonary disease. Chronic or acute disseminated infections occur in less than 0.5% of those infected. Symptoms of the disease are a mild cough, fever, chest pain, and headache. Chronic cases occur if lung tissue develops localized cavities filled with the spherules (cylindrical bodies) of *C. immitis*. Amphotericin B is the treatment of choice for coccidioidomycosis.

Aspergillosis

Aspergillosis occurs in individuals frequently exposed to conidia (spores) of filamentous fungi in the genus *Aspergillus*, with *Aspergillus fumigatus* the most common etiological agent of the disease. *Aspergillus* species are commonly found in the environment, and disease occurs more often in immunocompromised individuals including patients with HIV/AIDS. Conidia inhaled into the lungs cause a delayed-type hypersensitivity reaction that appears as a typical asthma attack. Once conidia enter the lungs they germinate and form colonies. These colonies may remain small and confined or can spread to other organs, resulting in a serious disease called invasive aspergillosis. In patients with severely limited immune function, the mycelia can completely fill the lungs. In addition to respiratory infections, *Aspergillus* can also infect cardiac valves, particularly if they are damaged or patients have prosthetic valves. Treatment of aspergillosis is difficult. Amphotericin B is the antibiotic of choice; however, this agent shows poor efficacy in severely immunocompromised patients.

Candidiasis

Candida albicans is a part of the normal microbiota of the gastrointestinal tract, mouth, and vaginal area. The organism does not usually cause disease because the bacterial populations in these areas suppress the growth of *C. albicans*. In the absence of the normal microbiota, the fungus can proliferate and cause disease involving the skin or mucous membranes. This organism has become an important nosocomial (hospital-acquired) pathogen.

Oral candidiasis (thrush) occurs in neonates born to mothers with *Candida* infections of the vagina. In cases of candidal vaginitis, the infection can be transmitted to the male during sexual intercourse. The resultant infection of the penis is termed balanitis. Thrush also occurs in immunocompromised individuals. Systemic candidiasis can occur in patients with AIDS. Individuals whose hands are immersed in water for long periods, such as cannery workers, often develop candidiasis of the hands or around the nails. Candidiasis is best treated by relieving the underlying conditions that result in infection. But several effective drugs are now available for treatment including amphotericin B, clotrimazole, and fluconazole.

Vaginal infection (**vaginitis**) caused by *C. albicans* (commonly referred to as "yeast infection") occur commonly among women with diabetes, during pregnancy, following antibiotic treatment, or without any obvious predisposing conditions. The sharp increase in the use of anti-bacterial drugs (antibiotics) is most likely responsible of the increase seen in the incidence of candidal vaginitis, presumably by displacement of normal vaginal flora, allowing for the outgrowth of *Candida*. Although not always reported, the number of case in the United States is very likely about several million per year. Yeast infections can be treated topically with creams that include butoconazole, clotrimazole, miconazole, and terconazole, or orally, with fluconazole.

Cryptococcosis

Cryptococcosis is a yeast infection caused by *Cryptococcus neoformans*, a common soil inhabitant. Most human infections result from inhalation of the organism on dried pigeon feces. *C. neoformans* does not infect pigeons but grows prolifically on the droppings. The primary infection is in the lungs, and most cases are asymptomatic or undiagnosed, and cure is spontaneous in healthy individuals. However, cryptococcosis is a serious problem in patients with AIDS, where it causes infections in the brain and meninges. Cryptococci are present in the spinal fluid of 10% to 15% of these patients. In addition, cryptococcosis is fatal in immunocompromised individuals if untreated. Cryptococcosis can be diagnosed with latex beads coated with rabbit antibodies to the polysaccharide capsule of the yeast. Presence of *C. neoformans* is indicated by aggregation of the beads. Amphotericin B and 5-fluorocytosine are the drugs of choice for treatment.

SECTION HIGHLIGHTS

Inhalation of spores (conidia) of pathogenic *Aspergillus* species leads to superficial asthma-like reactions; however, it can develop into a more serious systemic disease called invasive aspergillosis. *Candida albicans* is part of normal human flora, but under certain conditions, this organism can be invasive, particularly where anatomical barriers have been disrupted or in immunocompromised patients. Oral candidiasis and vaginitis are the two most common infections caused by this fungal pathogen. Pigeons are the vehicle for spreading *Cryptococcus neoformans*. Although largely asymptomatic in healthy individual, infections of immunocompromised individuals, including those with AIDS, are more serious, and can be fatal.

28.9 Bacterial Diseases of the Gastrointestinal Tract

Diseases of the gastrointestinal tract occur frequently and range from a mild upset stomach to fatal cases of botulism. Proper sanitation, clean water, refrigeration, and public health measures have been effective in lowering the incidence of these diseases in developed nations. Because many gastrointestinal diseases are spread by fecal contamination of food and water, they are a serious concern in developing countries where sewage treatment is often inadequate. Typhoid fever and cholera are endemic in many parts of Southeast Asia and South America due to a lack of proper sewage treatment.

The intestinal tract diseases are of two major types:

- Food poisoning results from ingestion of food on which bacteria have grown. In such cases, bacteria release toxins that cause the disease symptoms. These toxins are then ingested along with the food.
- Food-borne infections occur if one ingests food or drinks liquids that are contaminated with disease-causing bacteria. The bacteria propagate in the intestinal tract at which point a toxin could be released by the bacteria causing disease. In some cases, the infection occurs directly from hand to mouth after touching or handling fecal-contaminated objects.

Food poisoning symptoms generally occur within a few hours after ingestion of contaminated food (Table 28.4). Duration of the poisoning caused by species of *Staphylococcus* or *Bacillus*, or by *C. perfringens* is relatively short, and recovery is generally rapid. The exception, of course, is botulism, which, although rare, can be a fatal disease. The incubation period for food-borne infections is generally longer than that for food-poisoning cases, and the duration of symptoms can also be considerably longer. The major route of transmission for food-borne infections is directly or indirectly related to fecal contamination. Most of the organisms listed in Table 28.5 are inhabitants of the intestinal tract of various animals and do not survive in nature for any extended period. Consequently, they survive by passing from one animal to another through fecal contamination of food or water.

In this section we consider some of the food- and water-borne infections and the causal agents of food poisoning.

Ulcers

Peptic ulcers occur in the stomach and duodenum of humans. These ulcers result from erosion of the highly alkaline mucus that forms a protective layer over the gastric epithelia. This mucus protects the epithelial cells from the

TABLE 28.4	Gastrointestinal diseases caused by food poisoning bacteria[a]	
Organism	**Time to Onset of Symptoms in Hours**	**Symptoms**
Staphylococcus aureus	1–6	Nausea, vomiting, and diarrhea
Clostridium botulinum	18–36	Dizziness, double vision, swallowing, and breathing problems
Bacillus cereus	1–6 10–12	Vomiting, abdominal pain, diarrhea, and nausea

[a]These organisms grow and produce toxins in the food prior to ingestion, except *B. cereus* may also grow for a short time and produce enterotoxin in vivo.

| | TABLE 28.5 | Food-borne and water-borne infections of the gastrointestinal (GI) tract |

Organism[a]	Infection	Source of Infection
Salmonella typhimurium	Salmonellosis	Poultry, eggs, animals
Salmonella typhi	Typhoid fever	Water, food
Vibrio cholerae	Cholera	Water, food
Shigella dysenteriae	Shigellosis	Water, food
Campylobacter jejuni	Campylobacteriosis	Poultry, shellfish
Escherichia coli	Travelers' diarrhea	Uncooked vegetables, salads, water
Yersinia enterocolitica	Enterocolitis	Milk
Listeria monocytogenes	Listeriosis	Milk, cheese
Vibrio parahaemolyticus	Hypersecretion	Shellfish
Clostridium perfringens	Hypersecretion	Meats, gravy, stews
Clostridium difficile	Pseudomembranous	Natural inhabitant of GI tract, colitis

[a]These organisms survive in the gastrointestinal tract and produce toxins.

acid and pepsin (proteolytic enzyme) secretions that regularly enter the stomach to aid in digestion of food. Ulcers have been ascribed to several factors including genetic predisposition, excess secretion of acids and pepsin, and diet. It has long been the prevailing dogma that acidity and other factors rendered the stomach an inhospitable environment and that it was free of bacterial colonization. However, in 1977, a microaerophilic bacterium, later named *Helicobacter pylori*, was discovered that could colonize the human stomach. Studies conducted since 1985, namely that aggressive antibiotic therapy is effective in treating peptic ulcers, have confirmed that this bacterium is a causative agent of peptic ulcers and may cause gastric cancer. However, it appears that gastric cancer is caused by *H. pylori* strains different from those that are responsible for peptic ulcer disease. The bacterial factors that contribute to such different outcomes of stomach colonization by *H. pylori* have not yet been identified.

Infections caused by *H. pylori* occur throughout the world and are more prevalent in developing countries than developed countries. The organism synthesizes adhesins that recognize sialic acid–containing glycolipids and glycoproteins and mediates the attachment of the organisms to stomach epithelial cells. *H. pylori* also produces a urease that degrades urea, thereby generating ammonia that may protect against acidity. Many strains produce a cytotoxin that promotes infection. The organism can be eliminated from infected patients by combinations of tetracycline, metronidazole, and bismuth.

Food Poisoning

Food poisoning describes gastrointestinal disturbances that are the consequence of consuming organisms or toxins in contaminated food. In general, thorough cooking of food kills contaminating pathogens and inactivates most toxins; however, some forms of toxins are heat resistant and some organisms responsible for food poisoning form heat-resistant endospores. Improper refrigeration of food contributes significantly to outbreaks of food poisoning. Although a number of pathogens can infect humans via the oral route, here we will discuss agents that gain access to the intestinal tract primarily through the consumption of contaminated food.

STAPHYLOCOCCUS AUREUS *S. aureus* is the agent most frequently encountered in cases of **food poisoning**. It is present in the nasopharynx of 10% to 50% of adult populations. The incidence is much higher in the nasal passages of children. It is inevitable that through negligence, unsanitary practice, or accident some *S. aureus* cells will get into food during preparation. Two conditions must be met, however, before food poisoning can occur:

- The contaminated food must be suitable for growth and toxin production by the bacterium.

- The food must stand at a temperature that will permit growth of the bacterium.

For *S. aureus*, this would be room temperature or warmer. *S. aureus* effectively produces **enterotoxin** during growth on such foods as ham or chicken salad, cream-filled pastry, custards, salad dressing, and mayonnaise. To prevent enterotoxin production, food should be refrigerated before preparation, kept cold during preparation, and refrigerated immediately afterward. Any foods that contain creams or mayonnaise and are exposed to room temperature for any length of time should be discarded.

Staphylococcal food poisoning generally occurs within six hours after ingestion of the toxin-containing food. The toxin is heat and acid stable. Severe nausea, vomiting, and diarrhea are the symptoms, and they rarely last longer than 24 hours. However, food poisoning can be a serious problem for individuals weakened by other health problems. Treatment of severe cases is by administration of intravenous fluids to prevent dehydration. Antibiotic therapy is of no value because the causative agent is the toxin.

Staphylococcal enterotoxins are carried on mobile genetic elements (plasmids and bacteriophages). They are related to the family of so-called pyrogenic toxins, with superantigen activities that include the staphylococcal and streptococcal toxic shock syndrome, as discussed previously in this chapter. The staphylococcal enterotoxins consist of distinct domains, one associated with the superantigen activity and the other responsible for the symptoms of food poisoning because of its emeticity (the ability to induce vomiting) in animals. It is conceivable that in order to produce the symptoms of staphylococcal food poisoning the two activities work synergistically; the domain associated with emeticity promotes translocation of the superantigen domain from the intestinal lumen into the bloodstream, thereby enhancing its activity as a superantigen.

BACILLUS CEREUS   *B. cereus* is an endospore-forming, gram-positive bacillary organism that is a significant source of food poisoning. It is commonly present in soil and can be introduced into food on dust particles. Food poisoning by this organism occurs with various foods but particularly with rice. Rice is often cooked gently and left at room temperature. The heating induces endospores to germinate, and while sitting at room temperature the organisms grow extensively and produce enterotoxin. Ingestion of large numbers of bacteria may also result in growth of the *B. cereus* in the gut, where it produces enterotoxin. When this occurs, it might also be considered a food-borne infection. Creamy foods and meat have also been implicated in *B. cereus* food poisoning. Diagnosis is according to symptoms and examination of suspected food. A count of 10^5 *B. cereus* cells per gram of food is a strong indicator that this organism is involved. *B. cereus* produces two different enterotoxins:

- Vomiting occurs in 1 to 6 hours after eating contaminated food and is caused by a cholera-like enterotoxin that stimulates adenylate cyclase.

- Abdominal pain and profuse diarrhea 4 to 16 hours after ingestion of contaminated food is indicative of yet another enterotoxin, but the action of the toxin is not clear.

CLOSTRIDIUM BOTULINUM   *C. botulinum* is the etiological agent of the dreaded disease **botulism**. It is a gram-positive, spore-forming bacterium. The organism and its spores are widely distributed in soil, on lake bottoms, and in decaying vegetation. The endospores can contaminate vegetables, meat, and fish. Historically, botulism occurred following the ingestion of home-canned, low-acid vegetables such as peas, beans, corn, mushrooms, and meats. The endospores of *C. botulinum* are heat resistant and survive unless the processing is complete (120°C for 15 minutes). Different strains of *C. botulinum* produce several types of toxins, but the symptoms of botulism are associated with the so-called **neurotoxin**. The botulinum neurotoxin has a high affinity for receptors at neurons and is among the most potent toxin known. Its activity is almost identical to that of the neurotoxin of *Clostridium tetani,* although the physiological effects are opposite (see subsequent text). *C. botulinum* neurotoxin binds to the nerve endings at their junctions with muscles, and blocks the release by the nerve of the neurotransmitter acetylcholine, thereby preventing the muscles from contracting. The neurotoxin is synthesized as a larger precursor and is cleaved into two chains H (heavy) and L (light) held together by a disulfide bond. Similar to the other toxins with A-B moieties, the H chain is responsible for binding to neurons and translocation, whereas the L chain is a proteolytic enzyme. Inside the neuron, the A chain cleaves proteins that are required for docking of acetylcholine-containing vesicles to the surface. The release of acetylcholine from the nerve cell is blocked; resulting in failure to stimulate muscle contraction and leads to flaccid paralysis. The toxin is heat labile and is destroyed by boiling of foods for 10 minutes.

Purified botulinum neurotoxin is used clinically to treat crossed eyes, facial spasms, and other neurological disorders characterized by abnormal muscle contractions, such as tremors. A formulation of botulinum toxin (Botox) is also used to smooth frown lines (wrinkles), although the effect of the toxin is relatively short lasting (see Box 20.3).

CLOSTRIDIUM TETANI   *C. tetani* is present in the soil and in the feces of many animals. It has no invasive ability and enters tissue through punctures or other deep wounds.

The tetanus toxin is structurally related to the botulinum neurotoxin. It is composed of two disulfide-linked H and L chains. Tetanus toxin is also internalized at neuromuscular junctions, but it is transported to the spinal cord, where it is translocated into inhibitory interneurons. The proteolytic activity of the L chain blocks the release of neurotransmitter-containing vesicles. As a result, opposing muscles that permit movement by alternating contraction/relaxation are in a constant state of contraction. This results in a painful spastic paralysis. The spasms can involve muscles of the jaws, and, consequently, the disease has been termed *lockjaw*. The toxin can also cause spasms in the respiratory muscles that can result in suffocation.

SECTION HIGHLIGHTS

The causative agent of peptic ulcers, *H. pylori*, survives the low pH of the stomach by producing an enzyme urease, which converts urea to ammonia. Symptoms of staphylococcal food poisoning are the consequence of actions of enterotoxins that are also superantigens. Ingestion of food contaminated by *C. botulinum* leads to serious, occasionally fatal consequences, because of the activities of a potent neurotoxin produced by this organism. The clostridial neurotoxin binds to the nerve endings at their junctions with muscles, and blocks the release of vesicles containing the neurotransmitter acetylcholine, thereby preventing the muscles from contracting. A neurotoxin produced by *C. tetani* also interferes with the release of vesicles from neurons, those that contain inhibitory transmitters responsible for muscle relaxation.

28.10 Food- and Water-Borne Infections

There are a number of bacterial species causing disease that are transmitted by contaminated food or water. Among these diseases are salmonellosis, cholera, shigellosis, travelers' diarrhea, and various forms of enterocolitis. The agents of these diseases, and the disease symptoms, diagnosis, and possible treatment are discussed.

Salmonellosis

Salmonellosis is an infection fostered by microorganisms in the genus *Salmonella.* There are more than 2,000 serological types within this genus. These serological types differ by one or more cellular (O) or flagellar (H) antigens. The species most commonly associated with the human infection is *Salmonella typhimurium.*

Various *Salmonella* species can be present in the intestinal tracts of animals and birds. As a result, poultry, eggs, and beef can be contaminated with the organism. People who handle these foods, as well as asymptomatic carriers, are a source of food contamination. Given the widespread distribution of salmonellae, all foods, particularly hamburger and poultry, should be treated as contaminated. Hands contaminated by handling an infected animal such as a dog or cat can be a source of disease. Water polluted with human or animal waste is another source of the disease. Some years ago small pet turtles were available in variety stores and pet shops. These turtles were *Salmonella* carriers and were responsible for an estimated 300,000 cases of human salmonellosis per year. Import and sale of these animals is now banned.

There are several million cases of salmonellosis in the United States each year. Most cases of salmonellosis follow ingestion of uncooked egg products such as custards, meringues, and eggnog. Undercooked meat, particularly chicken, is also a source of infection.

Enterocolitis is the most often observed manifestation of salmonellosis. The symptoms occur 10 to 30 hours after infection and include abdominal pain, nausea, vomiting, and diarrhea. The symptoms usually last from 2 to 5 days but may last longer. The organism multiplies in the intestine and produces cytotoxins that destroy epithelial cells in the intestinal tract, and with the aid of several proteins secreted by the type III secretion system, the organism invades cells. Feces from a patient with enterocolitis can contain one billion salmonellae per gram. Diagnosis is by isolation and identification of the infectious agent. Salmonellosis can induce a loss of fluid that may be a serious complication in children, the elderly, and debilitated individuals. Treatment is by replacement of lost electrolytes through intravenous fluids.

Salmonella typhi is the etiological agent of **typhoid fever**. There are about 500 cases of typhoid fever in the United States each year and an estimated 2,000 carriers. Typhoid fever epidemics have occurred throughout human history. The sole reservoir for *S. typhi* infections is humans; therefore, carriers and improperly treated sewage are the sources of infection. *S. typhi* is passed from feces to hands to food, and public health measures have been established to prevent carriers from being food handlers. Typhoid infections are caused by ingesting contaminated food, by putting contaminated objects in one's mouth, or by drinking contaminated water. After infection, the onset of symptoms occurs after a 10- to 14-day incubation period, and the disease is marked by headaches, abdominal pain, a high temperature, and rose-colored spots on the abdomen. *S. typhi* cells can penetrate the small intestine and spread to the lymphoid system. During the active stages of the disease, *S. typhi* is present in the blood and is disseminated to the spleen, liver, bone marrow, and gallbladder. Victims shed the bacterium for 3 months or more, whereas carriers often have an infected gallbladder that constantly sheds *S. typhi* into the bile duct and on into the intestines.

Diagnosis is by isolation of the organism from blood, urine, or feces and through serological identification. Patients are treated with chloramphenicol, which must be continued for at least 2 weeks to clear the organism from the gallbladder.

Cholera

The etiological agent for **cholera** is *Vibrio cholerae*, a gram-negative, curved rod with a single polar flagellum. The organism adheres to epithelia in the small and large intestine and secretes **cholera toxin**.

Cholera is a fearsome disease spread by fecal–oral transmission through water and food. It is inevitable that cholera will occur in areas that have breakdowns in sewage disposal systems or where human waste pollutes water supplies. Cholera is endemic in India, Pakistan, Bangladesh, and other Asian countries where sewage disposal is inadequate. Serious outbreaks of cholera have occurred in Peru and Brazil in recent years, killing more than 10,000 people.

The incubation time from ingestion to disease symptoms is only hours to 3 or more days and is dose dependent. One must ingest 108 to 109 organisms in water to acquire an infection because the stomach acidity can wipe out smaller doses. Ingestion of fewer organisms with contaminated food can result in the disease, because food tends to protect *V. cholerae* from stomach acidity.

Cholera has one major symptom: an acute diarrhea with excretion of 8 to 15 liters of liquid per day. This liquid contains *V. cholerae*, epithelial cells, and mucus. If untreated, 60% of the patients succumb to dehydration. Treatment by intravenous electrolyte and liquid replacement lowers the death rate to less than 1%.

Cholera is diagnosed symptomatically and by the isolation of the organism from feces. Serology is employed to determine the strain of organism involved. Treatment with tetracycline or other antibiotics may shorten the course of infection. Cholera can be prevented by implementing adequate sewage treatment procedures and by purifying drinking water.

The cholera toxin (and its closely related *E. coli* heat-labile enterotoxin) is responsible for the major symptoms of cholera—influx of water into the intestinal lumen and subsequent diarrhea. The structure of cholera toxin fits into the model of A-B toxins. The toxin is composed of the disulfide bond–linked A1-A2 subunit, generated from a larger precursor by a single proteolytic cut. The A subunits are joined noncovalently to a ring-like structure consisting of five B subunits. The pathway of internalization of cholera toxin is shown in Figure 28.6A and its intracellular action in Figure 28.6C. The B subunit's function is to bind the toxin to a specific receptor, consisting of ganglioside glycolipid (GM1) on the epithelia of the ileum and large intestine. The toxin is transported to the endoplasmic reticulum, where the A1 subunit is released into the cytosol. Like pertussis toxin, it induces a rise in cAMP levels through ADP-ribosylation of one of the subunits of the adenylate cyclase regulatory complex. Although the pertussis toxin modifies the inhibitory subunit ($G_{i\alpha}$) preventing inhibition of adenylate cyclase activity in response to binding of inhibitory ligands to their receptors, the target of cholera toxin is $G_{s\alpha}$, the protein responsible for stimulation of adenylate cyclase following the engagement of stimulatory ligands by their cognate receptors.

The increased levels of cAMP stimulate electrolyte transporters in intestinal cells, resulting in the release of chloride ion and bicarbonate from mucosal cells that line the intestine. These inorganic ions accumulate in the lumen (inner space) of the intestine, and the normal passage of sodium ion into the mucosal cells is blocked. This creates an osmotic imbalance such that water passes through the intestinal epithelia into the lumen (Figure 28.14). This influx of water to the lumen causes the severe diarrhea associated with the disease.

Dysentery

Another organism responsible for intestinal disease is *Shigella dysenteriae*, a gram-negative facultative anaerobe. It is the causative agent of **shigellosis**, a form of dysentery. Humans are the only known hosts for *S. dysenteriae*, and the infectious organism is transmitted via the fecal–oral route. Food and water are the major routes of transmission, whereas direct feces-to-mouth transmission is probably responsible for the prevalence of this disease in preschool children. It is highly infectious and 200 to 400 organisms can cause a fatal infection in humans. Most cases of the disease are found among children in day-care centers.

S. dysenteriae remains localized in intestinal epithelial cells and effects a loss of fluid and ulceration of the colon wall. The symptoms include abdominal cramps, nausea, vomiting, fever, and bloody and mucus-containing diarrhea.

The tissue damage is largely due to the activity of a potent toxin called **shiga toxin**, secreted by the microorganisms. Shiga toxin resembles cholera toxin not only in its structure but also in its intracellular routing to the cytosol. Shiga toxin consists of a B pentamer and the associated A moiety, cleaved into A1 and A2 subunits. The B subunit binds to specific glycolipid (globotriaosylceramide, Gb3) on the host cell, and follows the same pathway as cholera and pertussis toxins, from endosomes through the endoplasmic reticulum until its release into the cytosol. Within the cell, the A1 subunit inhibits protein synthesis following an *N*-glycosidic cleavage, which releases a specific adenine base from the sugar-phosphate backbone of 28S ribosomal RNA. The shiga toxin kills absorptive epithelial cells, and diarrhea results from this interference with liquid absorption. *Shigella* can also invade the colonic epithelium, where it induces an intense inflammatory response. Invasion of colonic cells

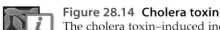

Figure 28.14 Cholera toxin
The cholera toxin–induced increase in cAMP in intestinal cells leads to a massive increase in fluid secretion into the intestinal lumen.

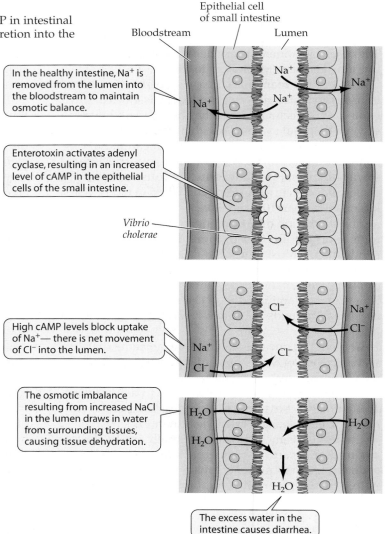

Epithelial cell of small intestine

Bloodstream   Lumen

In the healthy intestine, Na⁺ is removed from the lumen into the bloodstream to maintain osmotic balance.

Enterotoxin activates adenyl cyclase, resulting in an increased level of cAMP in the epithelial cells of the small intestine.

Vibrio cholerae

High cAMP levels block uptake of Cl⁻— there is net movement of Cl⁻ into the lumen.

The osmotic imbalance resulting from increased NaCl in the lumen draws in water from surrounding tissues, causing tissue dehydration.

The excess water in the intestine causes diarrhea.

requires secretion of various proteins through the type III secretion system (see Chapter 11). These proteins are primarily involved in modulation of various host cell functions to dampen the innate defense mechanisms.

Isolation and identification of *S. dysenteriae* in feces is one means of diagnosis, but a direct swab of lesions in the colon and subsequent identification of the organism serologically is better. Treatment is by intravenous fluid replacement and with the antibacterials sulfamethoxazole and trimethoprim.

Campylobacteriosis

The etiological agent for a major diarrheal disease termed **campylobacteriosis** is *Campylobacter jejuni*. *C. jejuni* is a microaerophilic, gram-negative rod with a single polar flagellum. *C. jejuni* is an inhabitant of the intestinal tracts of wild and domestic animals. The organism can cause sterility and abortion in cattle and sheep. It is transmitted to humans via the fecal–oral route in contaminated food and water, and underprocessed chicken and raw milk are common sources of campylobacteriosis. Children can get the infection from domestic dogs. The incubation period after ingestion (fewer than ten organisms can cause infection) is 2 to 4 days. Symptoms include a high fever, nausea, abdominal cramps, and watery, bloody feces. The exact factor responsible for these symptoms has not been identified, although they suggest activities of toxins. So far, only one toxin, known as **cytolethal distending toxin** has been described. It kills cells by blocking DNA replication and induces an inflammatory response. Diagnosis is by isolation of the organism from feces and by serological identification. The disease can be treated with fluoroquinolone.

Escherichia coli

A normal inhabitant of the human digestive tract is *E. coli*. There are, however highly virulent strains, and depending on their specific repertoire of virulence factors, each can cause distinct infections. Enterotoxigenic strains of *E. coli* cause diarrheal disease, particularly evident in individuals traveling between countries—hence the

name "travelers' diarrhea" (among others). These strains colonize the intestinal tract and they rarely invade tissues. They attach to the intestinal mucosa by **pili** or fimbriae and produce **enterotoxins**. One such toxin is the **labile toxin**, or LT, because of is sensitivity to heating. LT strongly resembles cholera toxin A1 in its structure and mechanism of action in activating adenylate cyclase. Another toxin produced by enterotoxigenic *E. coli* is **ST** (for **heat stable toxin**). This toxin bids to and stimulates guanylate cyclases that span mammalian cell membranes. Increase in cGMP levels also stimulates secretion of electrolytes into the intestinal lumen.

Another group of strains is referred to as **enteropathogenic *E. coli***. When ingested, these strains produce diarrhea by destruction of intestinal villi and induced inflammation. They also invade cells with the aid of proteins secreted using the type III secretion system. **Enterohemorrhagic** *E. coli* are similar to enteropathogenic *E. coli*; however, they possess an additional virulence factor in form of a toxin encoded on a lysogenic bacteriophage. This toxin is referred to as **shiga-like toxin**, because it resembles shiga toxin in its subunit composition and acts by an identical mechanism of action, inhibiting host cell protein synthesis. A particular enterohemorrhagic *E. coli*, serotype O157:H7, was responsible for an outbreak of a food-borne *E. coli* infection in 1993 that killed four children in the Seattle area. *E. coli* O157:H7 can be transmitted in meat products, particularly ground beef. Several outbreaks of infections with *E. coli* O157:H7 were attributed to consumption of raw vegetables. Presumably, the source of the bacteria was cow manure from an infected animal that had been used as a fertilizer. Most *E. coli* infections can be treated with trimethoprim-sulfamethoxazole or ciprofloxacin.

Enterocolitis

The etiologic agent for **enterocolitis** in humans is *Yersinia enterocolitica*. The organism is present in the intestinal tracts of cats, dogs, rodents, and domestic farm animals. It can also be isolated from lake and well water. As with many of the previously mentioned enteric diseases, enterocolitis is a fecal–oral transmitted infection. *Y. enterocolitica* adheres to and invades intestinal epithelial cells. The symptoms include fever, diarrhea, and abdominal pain. A form of arthritis can occur several days after onset of acute enteritis. The drug of choice is trimethoprim-sulfamethoxazole.

Listeriosis

An infection called **listeriosis** is caused by *Listeria monocytogenes*, a gram-positive bacillus. This bacterium can also cause disseminated infections frequently involving fetuses, newborns, and immunocompromised individuals. *L. monocytogenes* is mainly an animal pathogen and is transmitted to humans through dairy products. Milk that has been improperly pasteurized and the cheese manufactured from such milk are the major causes of infection.

Following ingestion, *L. monocytogenes* invades and grows within cells. The bacterium avoids immune recognition because of its ability to spread without leaving the infected cell, moving via membrane protrusions between two adjacent cells. Most of the virulence factors of this organism are related to its ability to invade and survive within cells and move from cell to cell. Listeriolysin O (LLO) is a secreted pore-forming protein essential for the escape of *L. monocytogenes* from the vacuole following initial internalization. To facilitate the movement of the organism within the host cell cytoplasm and

to propel it into another cell, *Listeria* express a protein (ActA) on the surface at one of their poles and this protein serves as a nucleation site for mammalian actin. Actin polymerization moves the bacterium in the cytoplasm and also pushes it against the cell membrane, creating protrusions that become inserted into the adjacent cell. Here, another *L. monocytogenes* enzyme, phospholipase C, hydrolyzes phospholipids in the host membrane, releasing the bacterium into the cytoplasm of the newly invaded cell.

L. monocytogenes may infect pregnant women, and the organism may pass through the placenta to the fetus. In the newborn, the disease is generally pneumonia at birth or shortly thereafter. About 2 to 4 weeks after birth, the listeriosis infection appears as meningitis. This disease also occurs in immunocompromised persons (such as AIDS patients). Listeriosis can be treated with ampicillin and gentamicin.

Clostridium perfringens

C. perfringens is a major agent of food poisoning in the United States. The disease is most often caused by ingestion of meat products, particularly rewarmed meat that has been cooked in bulk at fast-food restaurants. Quite high numbers ($>10^8$) are required for infection, as low numbers of *C. perfringens* are a normal inhabitant of the human intestinal tract. Strains of *C. perfringens* produce a number of toxins that contribute to intestinal damage. Diarrhea is caused by the action of a specific enterotoxin, which impairs absorption of water by cells lining the small intestine, leading to an increase in stool mass. Symptoms occur in 8 to 16 hours and include diarrhea, cramps, nausea, and vomiting. The symptoms last for about 24 hours.

Clostridium difficile

An anaerobic microorganism that forms endospores, *Clostridium difficile* is present in the intestinal tracts of some adults: the number of organisms is quite low, and *C. difficile* causes no apparent infection. If antibiotics such as clindamycin or ampicillin are administered to an individual in an amount sufficient to destroy the normal intestinal flora, *C. difficile* can proliferate to significant numbers. Overgrowth by the organism can cause a severe necrotizing (death of tissue) process. The infection is termed **pseudomembranous colitis**.

Vibrio parahaemolyticus

A marine bacterium, *V. parahaemolyticus*, grows only in the presence of moderate salt concentrations. It is the etiologic agent of a food poisoning that follows ingestion of uncooked seafood including raw oysters. The microorganism produces a membrane-damaging toxin, called thermostable direct hemolysin, which may be responsible for rapid onset of diarrhea.

SECTION HIGHLIGHTS

Ingestion of certain serotypes of *Salmonella* can lead to enterocolitis, accompanied by nausea, vomiting, and profuse diarrhea. The disease is usually self-limiting within a few days. Typhoid fever is an invasive disease, caused by *Salmonella typhi* that penetrates the small intestine and disseminates into various organs, where it causes damage by local proliferation. *V. cholerae*, enterotoxigenic *E. coli*, and *C. perfringens* infect the intestinal tract and utilize different mechanisms of colonization; however, the diarrheal symptoms are due to the production of toxin. Cholera is a diarrheal disease that follows ingestion of *V. cholerae*. While growing in the intestinal tract these organisms produce a toxin. The active subunit of the toxin acts within the intestinal epithelial cells, where it causes a rise in cAMP levels by ADP-ribosylation of a regulatory subunit of adenylate cyclase. Enterotoxigenic *E. coli* produce a so-called heat-labile toxin that is highly similar in its subunit composition to cholera toxin and has the identical intracytoplasmic mechanism of action. The enteric disease shigellosis is caused by *S. dysenteriae*. It produces a toxin that cleaves an adenine from the eukaryotic ribosomal RNA. The organism can also invade colonic epithelial cells and induce intestinal inflammation. *L. monocytogenes* has evolved sophisticated virulence mechanisms to invade cells, replicate within them, and move into neighboring cells via membrane protrusions.

28.11 Urinary Tract Infections

Infections of the urinary tract are second only to respiratory infections in number of cases annually. Urinary tract infections are by far the most prevalent nosocomial (hospital-acquired) infection, causing about 40% of all nosocomial infections. These infections can occur in the kidneys (pyelonephritis), in the bladder (cystitis), or as an inflammation of the urethra (urethritis). Infection of the kidneys generally will lead to infection of the entire urinary tract, but more often infections spread by ascending from the urethra. Proper treatment of urethritis with antibiotics (amoxicillin or trimethoprim) will prevent the spread of the infectious agent to the bladder and kidneys. Because the urethra in females is much shorter than that in males, the incidence of urinary tract infections in females is much higher.

The organism most often associated with urinary tract infections is *E. coli*, the causative agent in 80% of the cases. Other organisms involved are *Proteus mirabilis, P. aeruginosa,* and other facultative aerobic gram-negative rods and cocci. Usually a single strain is involved in an infection, and diagnosis is made by determining the number of organisms present in urine.

The bladder is generally sterile but bacteria colonize epithelial cells that line the lower end of the urethra. Urine commonly has 1,000 to 10,000 bacteria organisms per milliliter. Urine is a good growth medium, and accurate counts require prompt processing of the urine specimens. The presence of more than 10,000 organisms per milliliter implies an infection. However, in some cases of kidney infection the number of organisms present in urine is relatively low. Diagnosis depends on the presence of irritation and pain, especially during urination. Catheterization of hospital patients is a leading cause of infections by opportunistic pathogens generally present in the urethra or on the skin.

28.12 Sexually Transmitted Diseases

Sexually transmitted diseases (STDs) are a growing health problem throughout the world. These diseases affect an alarming portion of the young adult population and may be transmitted to the newborn by transfer during fetal development or during birth. The etiological agents for most STDs are bacteria and viruses. We will discuss the major bacterial diseases: gonorrhea, syphilis, and chlamydial infections. All three bacterial STDs are eminently curable and have been for the last 40 years. Genital herpes and HIV are the most prevalent viral infections (see Chapter 29). STDs have become as much a social problem as a medical concern. The greatest incidence of STDs is in young people ages 15 to 35, the age group that is least likely to seek treatment. Increased sexual activity and multiple partners make control of the diseases particularly difficult.

Gonorrhea

The etiological agent for gonorrhea is *Neisseria gonorrhoeae,* (sometimes called a gonococcus), a gram-nega-

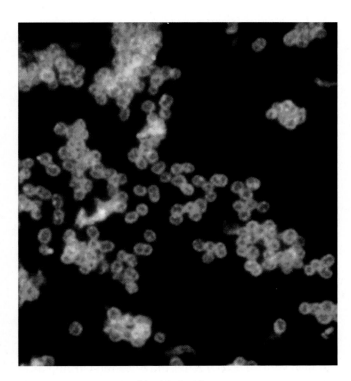

Figure 28.15 Gonorrheal infection
A positive fluorescent antibody test for *Neisseria gonorrhoeae*. Courtesy of the Centers for Disease Control and Prevention.

tive, generally diplococcal bacterium (Figure 28.15). Drying, sunlight, and UV light readily kill the organism. *N. gonorrhoeae* is transmitted in virtually every case by direct contact. Sexual activity between infected and uninfected individuals leads to dissemination of the pathogen. The pili on pathogenic strains of *N. gonorrhoeae* attach to microvilli of mucosal cells of the genitourinary tract. The bacterium is then phagocytosed by cells on the mucosal surface and transported to intercellular spaces and submucosal tissue. An outer membrane porin facilitates the uptake of *N. gonorrhoeae* by cells utilizing as yet an uncharacterized mechanism. Gonococcal porins also protect the bacteria by binding and inactivating components of the complement, thereby effectively neutralizing this important host defense mechanism.

　N. gonorrhoeae can infect the eyes of newborns during birth. Untreated, this may lead to serious eye infections and is responsible for blindness in much of the developing world. Tetracycline, erythromycin, or silver nitrate is applied to the eyes of neonates at birth as a preventive measure in the United States. This has essentially eliminated gonorrhea-related eye infections and subsequent blindness.

　The incidence of gonorrhea in the United States is alarming, with perhaps as many as one million new cases occurring annually. This figure would indicate that 1 of every 250 Americans is infected each year, and most are in the 15 to 35 age-group. Considering that individuals in this group represent about one-third of the population, the incidence would be about 1 in 80, a public health concern. The probability of infection per contact is about 25% for males and 85% for females. The incidence of gonorrhea increased over fivefold in the period following the widespread use of oral contraceptives. A higher level of sexual activity and decreased use of condoms in the period from 1975 to 1978 is considered responsible for this marked increase. Fear of AIDS has now led to an increased awareness of the dangers of "unsafe" sex and has caused some lowering of the incidence of gonorrhea over the last several years.

　Infection with *N. gonorrhoeae* does not confer immunity, and many of the cases observed are re-infections. This is due to the large repertoires of sequence (and therefore antigenic) variants of certain highly antigenic surface proteins, such as pili or porins expressed by different strains. For example, the genome of *N. gonorrhoeae* carries 20 copies of the type IV pilin gene; however, only one gene is intact and is expressed. The remaining copies of the pilin gene are partial (silent) sequences lacking 5′ segments that contain the transcriptional and translational start sites. Recombination between the sequences of expressed and silent genes results in variation in the amino acid sequence of the pilin protein expressed by each gonococcus. Theoretically, gonococci have the potential to produce over a million antigenically distinct pilin proteins, utilizing not only the silent gene sequences encoded in the genome of an individual strain, but genes from other gonococcal strains, acquired by natural transformation of DNA. It is not surprising that during re-infection, novel sequence variants are selected from the population of the infecting gonococci and these are not recognized by antibodies generated during previous infection by a bacterium expressing an antigenically different pilin variant.

　The high incidence of gonorrhea is primarily due to the absence of distinct symptoms in most females. These females are unaware of the infection, and the symptoms of mild vaginitis and minor discharge are attributed to other causes. These chronic carriers serve as a significant reservoir of infection. The gonococcus can infect the urethra, vagina, cervix, and fallopian tubes of females before it is detected and cause pelvic inflammatory disease leading to sterility.

　Infected males are rarely asymptomatic. Two to eight days after exposure they experience a painful urethritis and discharge of pus that contains numerous leukocytes. These leukocytes bear intracellular gonococci, and a microscopic examination of stained exudate is an effective diagnostic tool.

A major problem in treating gonorrhea is the increasing incidence of antibiotic resistance. Penicillin has been the treatment of choice, but β-lactamase (penicillinase)–producing resistant organisms have appeared. Tetracycline and erythromycin have been employed as alternatives. Recently, penicillin-resistant strains have appeared that have a chromosomal-mediated resistance to penicillin that also confers some resistance to tetracycline and erythromycin. Ceftriaxone is the drug of choice in this case. However, unless public health measures or individual responsibility lowers the incidence of this disease, it is likely that antibiotic-resistant strains will become more of a threat in the future.

Syphilis

Treponema pallidum is the causative agent of the historically significant disease, syphilis, which reached epidemic proportions in Europe in the sixteenth century as it spread rapidly and caused much suffering. Apparently *T. pallidum* was considerably more contagious in earlier times than it is today. About one in ten people who have a onetime exposure to syphilis actually will contract the disease. There are 25,000 to 35,000 new cases diagnosed each year in the United States. The lower incidence as compared with gonorrhea is attributed to the absence of asymptomatic female carriers. Although *T. pallidum* can be cultured in animals, under natural conditions it is exclusively a human pathogen.

Syphilis is transmitted by sexual contact. The organism enters through mucous membranes or through minor breaks in the skin during sex. About one case in ten occurs in the oral region. Pregnant women can pass the spirochete in utero to the fetus, a terrible disease that affects 300 children per year.

There are three recognizable stages in a syphilitic infection: primary, secondary, and tertiary.

PRIMARY STAGE   Within 10 days to 3 weeks after exposure, a small, painless reddened ulcer occurs at the site of infection. This ulcer is called a **chancre** and may appear on the penis or in the vaginal or oral regions (Figure 28.16A). The chancre contains spirochetes, and at this stage the infection is readily transmissible through intimate contact. In about one-third of the cases, the symptoms disappear and the disease progresses no further. In others, the spirochete enters the bloodstream and is distributed throughout the body. The primary stage lasts from 2 to 10 weeks. The disease then enters the secondary stage.

SECONDARY STAGE   After the chancre(s) heal(s), skin lesions occur about the body including the soles of the feet and palms of the hand (Figure 28.16B). They also appear on mucous membranes. The patient is generally feverish. Individuals in this stage of the infection are all serologically positive and can transmit the disease. After 4 to 8 weeks, the disease enters a latent period and

(A)

(B)

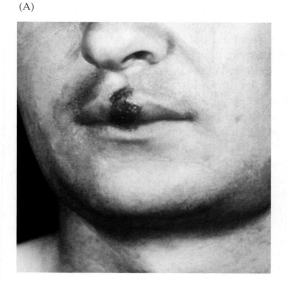
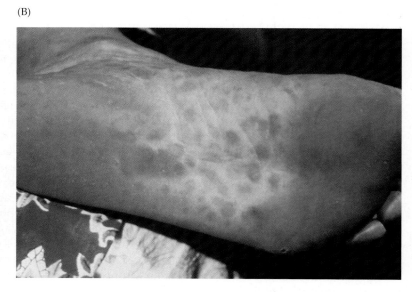

Figure 28.16 Syphilis infection
(A) A primary lesion from a syphilis infection. These lesions are known as chancres and occur at the site of infection. (B) Secondary lesions can occur on the soles of the feet and elsewhere on the body. Courtesy of Susan Lindsley/Centers for Disease Control and Prevention.

is not normally infectious except maternally to the fetus. At this stage, about one-fourth of the cases are essentially cured, one-fourth remain latent with relapses to the secondary stage, and about one-half of these enter the tertiary stage.

TERTIARY STAGE   Untreated tertiary syphilis is a degenerative disease with lesions occurring on skin, in bones, and in nervous tissue. Lesions frequently occur in the central nervous system, leading to mental retardation, blindness, and insanity. Paralysis often occurs, and those afflicted with tertiary syphilis walk with a stumbling gait. Infection of vascular tissue, and the subsequent inflammatory response, injures and weakens the walls of the blood vessels. There are actually few spirochetes in the lesions associated with tertiary syphilis, and most of the damage is attributed to hypersensitivity to the organism and its products.

Diagnosis can be made when chancrous lesions are examined by darkfield microscopy of the exudate where the spirochetes can be observed. Serological tests are widely employed for diagnosis and depend on the presence of antitreponemal antibodies to indicate the presence of *T. pallidum*. Penicillin is an effective treatment and is particularly effective in primary stages of the disease. More drugs and prolonged treatment are necessary in the later stages. No vaccine is available at this time. There is no cure for symptoms of tertiary syphilis.

Chlamydiae

Chlamydia trachomatis is an obligate intracellular parasite that causes several different infections in humans. The organisms are coccoid or short rods; they are gram negative but take up stains very poorly. They have an exceedingly small genome, reflecting their limited synthetic ability, and, so far, it has not been possible to grow any *Chlamydia* species outside of infected cells or animals. The incidence of these sexually transmitted infections greatly exceeds that of gonorrhea, and there are an estimated 3 to 10 million new cases in the United States each year.

The life cycle of *C. trachomatis*, depicted in Figure 28.17, consists of two stages: an extracellular, infectious particle, and an intracellular, replicating bacterium, referred to as elementary body and reticulate body, respectively. The elementary body is small, 0.3 μm in diameter, and has many properties of an endospore; it is metabolically inactive but extreme hardy, capable of surviving for prolonged periods outside of the host. When an elementary body encounters a host cell, it is taken up into an endosomal vesicle, where it differentiates into the vegetative form, the reticulate body, and it begins to divide by binary fission with a generation time of 2 to 3

hours. At the same time the host cell's metabolic activities decline. Because *Chlamydia* lack many metabolic pathways, they essentially feed on many constituents of the host cell. Replication continues until the large vesicle is filled with several hundred reticulate bodies, at which point they begin to differentiate into elementary bodies. Within 48 hours from the initial infection, the elementary bodies are released as the result of host cell lysis and the *Chlamydia* can start the infection process in other cells.

C. trachomatis is also the etiological agent of **trachoma**, an eye infection that occurs worldwide. More than 10 million cases of blindness, principally in developing countries, are attributed to this disease. The organism is passed to the newborn during birth and causes conjunctivitis (eye infection) and respiratory distress. Between 8% and 12% of pregnant women in the United States have chlamydial infections of the cervix, and one-half of babies born to these infected individuals have conjunctivitis. Application of tetracycline to the eyes prevents development of the disease. This organism causes much of the pneumonia in infants.

Chlamydial nongonococcal urethritis (NGU) is probably the most often acquired sexually transmitted disease. In males, *C. trachomatis* causes urethritis, and in females it causes urethritis, cervicitis, and pelvic inflammatory disease. An estimated 50,000 women become sterile each year from NGU. Inapparent infections are common in sexually active adults. Diagnosis is difficult, and until recently it was diagnosed by excluding other diseases. Now the infection is identified by staining material, taken by swab, with fluorescent antibody. Treatment with tetracycline and erythromycin is effective. Penicillin is ineffective, as *C. trachomatis* lacks a peptidoglycan cell wall. Because a significant number of patients with gonorrhea are also infected with chlamydia, it is essential that both diseases be treated.

Another strain of *C. trachomatis* is responsible for a disease in males called **lymphogranuloma venereum**. This is a sexually transmitted infection that causes swelling of lymph nodes about the groin. It is diagnosed by the presence of swollen lymph nodes and treated with tetracycline or sulfonamides.

Chancroid

Chancroid is a sexually transmitted disease that occurs commonly in the tropics and is of increasing incidence in the United States. The etiological agent for chancroid is a gram-negative rod-shaped organism termed *Haemophilus ducreyi*. This pathogen is transmitted through breaks in the epithelium in the genital area. After an incubation period of 4 to 7 days, swelling and white cell infiltration occur in the area where the bac-

(A)

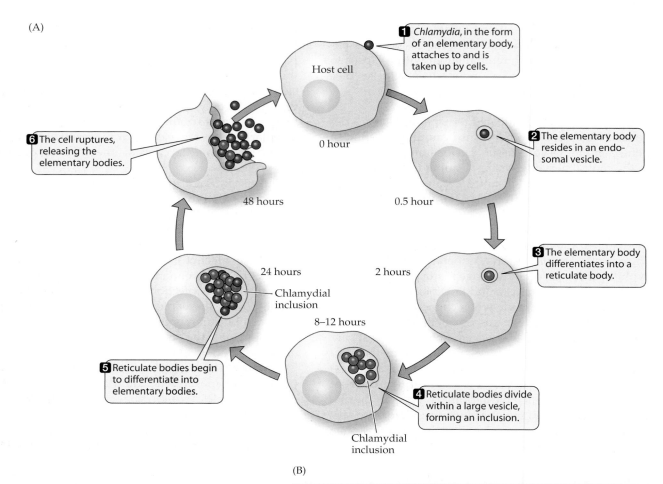

(B)

Figure 28.17 *Chlamydia* replication
(A) Life cycle of *Chlamydia*. (B) Electron micrograph showing a late stage *Chlamydia trachomatis* inclusion, with reticulate and elementary bodies. B, ©D. M. Phillips/Visuals Unlimited.

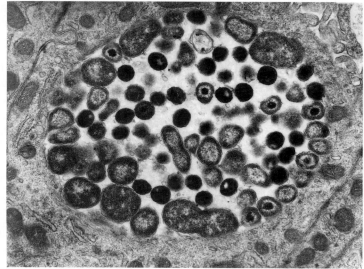

terium has gained entry. *H. ducreyi* produces a cell-destroying toxin resulting in a pustule. These pustules appear on the penis in males and the outer vaginal area in females. The pustule ruptures forming a painful ulcer.

Chancroid may be involved in transmission of HIV, as the virus can gain entry through the ulcerous lesion during intercourse. Erythromycin or azithromycin is effective in treating the disease.

serve as reservoirs. This occurs in cases where the infectious bacterium is passed transovarially (by infected eggs) from parent to progeny.

The bacteria involved in vector-borne diseases are **obligate parasites** that cannot reproduce or survive in nature outside a living host. Those that cause a fatal infection in the reservoir animal must be carried by the vector to another living host. The etiological agents for many of these vector-borne diseases are in the order *Rickettsiales,* and in the genera *Rickettsia, Bartonella,* and *Coxiella.* Fleas, lice, or ticks transmit these organisms. The reservoir is generally a mammal.

Tick-borne diseases can be prevented by proper sanitation and by ridding the human environment of rats and other vermin that carry fleas and ticks. For diseases such as Lyme diseases, ehrlichiosis, and Rocky Mountain spotted fever, one should dress properly when entering outdoor areas potentially infected with ticks. These are summertime diseases of high incidence when ticks are feeding.

28.13 Vector-Borne Diseases

Vector-borne diseases are transmitted from one infected animal to another by means of an intermediary host (Table 28.6). This intermediary host is termed a vector and is most often a biting insect (arthropod or louse) that transmits the pathogen while feeding on an uninfected host. The animal that is a constant source of the infectious agent is called a reservoir. Most reservoirs are mammals that may or may not be adversely affected by the infectious organism. Some tick and mite vectors also

Plague

The etiological agent of plague is *Yersinia pestis,* a gram-negative nonmotile coccobacillus (**Figure 28.18**). Historically, the plague has killed more humans than any other bacterial disease. In the sixth century millions died from the disease, and in the fourteenth century a plague epidemic killed more than one-fourth of the people in Europe (**Box 28.1**).

Y. pestis normally afflicts rodents and is endemic in the southeastern United States in prairie dogs, ground

TABLE 28.6	Vector-borne bacterial infections of humans			
Disease	**Etiological Agent**	**Vector**	**Reservoir**	**Transmission**
Rocky Mountain spotted fever	*Rickettsia rickettsii*	Wood tick (western U.S.) Dog tick (eastern U.S.)	Mammals	Tick to human
Q fever	*Coxiella burnetii*	Animal tick	Ticks, cattle	Tick to human, carcass to human
Tularemia	*Francisella tularensis*	Wood tick	Ticks, rabbits	Tick to human, animal; other mammals to human
Relapsing fever	*Borrelia recurrentis*	Body louse	Humans	Human to louse to human
Lyme disease	*Borrelia burgdorferi*	Deer tick	Deer, mice, voles	Flea to human
Typhus (epidemic)	*Rickettsia prowazekii*	Human louse	Humans, flying squirrels	Louse to human
Typhus (endemic)	*Rickettsia typhi*	Rat flea	Rats, ground squirrels	Flea to human
Plague	*Yersinia pestis*	Rat flea	Rats	Flea to human, human to human
Scrub typhus	*Rickettsia tsutsugamushi*	Mite	Mites	Mite to human
Trench fever	*Bartonella quintana*	Body louse	Lice, humans	Louse to human

(A)

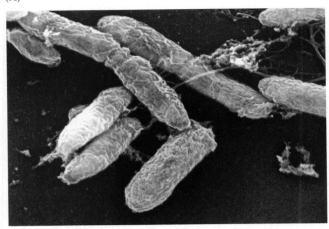

(B)

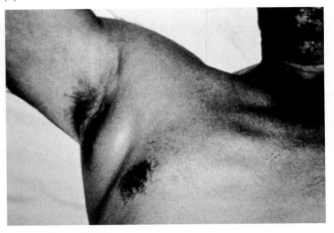

Figure 28.18 Plague
Manifestations of plague infection in humans. (A) A scanning electron micrograph of *Yersinia pestis*. (B) A bubo (swelling) in the armpit area. (C) Gangrene in hand tissue. A, ©Dr. Gary Gaugler/Photo Researchers, Inc.; B, courtesy of the Centers for Disease Control and Prevention; C, ©Biophoto Associates/Photo Researchers, Inc.

(C)

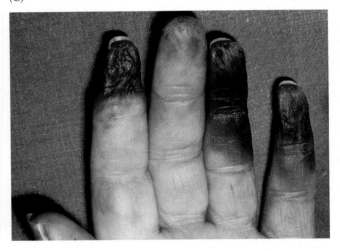

squirrels, wood rats, chipmunks, and mice. These animals are relatively resistant to plague, and the infectious agent resides in them during interepidemic periods. If domestic rats interact with these reservoir animals, the domestic rat becomes infected (Figure 28.19). Fleas infesting these infected animals can cause an epidemic of plague in the domestic rat populations. Plague is a fatal disease in domestic rats. When the rats die, the fleas travel to other rats. If rats are not available, the fleas move to and feed on humans, thereby transmitting the disease. Killing rats after the local population of fleas is infected can actually drive the fleas to humans and increase the incidence of plague.

The blood of a domestic rat infected with plague contains a large number of the *Y. pestis* organisms. The flea ingests this blood, and the bacterial coagulase induces a clot in the proventriculus of the flea. This prevents food from passing on to the stomach of the flea. The flea be-

comes hungry and feeds ravenously but regurgitates due to the gut blockage. The regurgitated blood contains large numbers of infectious bacteria that enter the bite site, thus infecting the host.

The plague bacillus moves from the entrance site to regional lymph nodes. These lymph nodes become enlarged and tender due to growth of the bacterium. The enlarged node is called a **bubo** and is the origin of the name **bubonic plague** (see Figure 28.18B). The organism can be engulfed by PMNLs and destroyed, whereas those phagocytosed by macrophages are not. Within the macrophage, *Y. pestis* is maintained at a favorable temperature (37°C) and in a low Ca^{2+} environment. Low Ca^{2+} level promotes the expression of virulence factors. The bacilli thrive in macrophages and travel to regional lymph nodes. From lymph nodes they are carried via the bloodstream to the spleen, liver, and other organs. They cause subcutaneous (below the skin) hemorrhages that appear as dark areas on the body surface. These darkened areas and the high level of mortality are the reason for calling plague "Black Death." Movement of the bacilli to the lungs results in **pneumonic plague**, a disease that is transmitted via the respiratory route. It causes a serious pneumonia associated with extensive necrosis of the lung tissue. Untreated, pneumonic plague has 100% fatality rate, and bubonic plague has a fatality rate of 75%.

The plague bacillus produces a number of toxins, and among these is a protein/lipoprotein complex that pre-

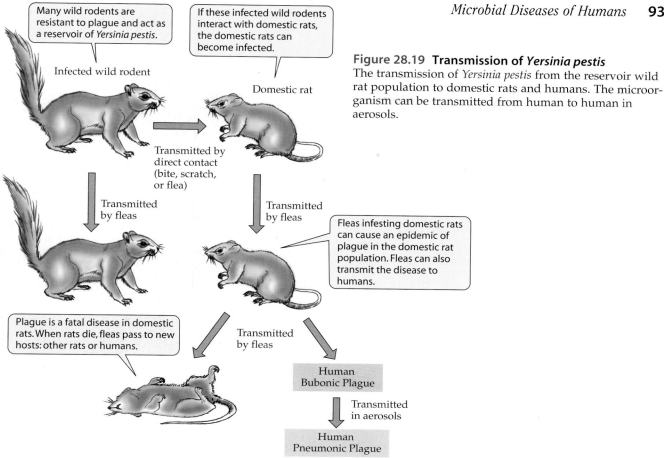

Many wild rodents are resistant to plague and act as a reservoir of *Yersinia pestis*.

If these infected wild rodents interact with domestic rats, the domestic rats can become infected.

Infected wild rodent

Domestic rat

Transmitted by direct contact (bite, scratch, or flea)

Transmitted by fleas

Transmitted by fleas

Fleas infesting domestic rats can cause an epidemic of plague in the domestic rat population. Fleas can also transmit the disease to humans.

Plague is a fatal disease in domestic rats. When rats die, fleas pass to new hosts: other rats or humans.

Transmitted by fleas

Human Bubonic Plague

Transmitted in aerosols

Human Pneumonic Plague

Figure 28.19 Transmission of *Yersinia pestis*
The transmission of *Yersinia pestis* from the reservoir wild rat population to domestic rats and humans. The microorganism can be transmitted from human to human in aerosols.

BOX 28.1 *Milestones*

A Sad Game

There is a children's game where the youngsters stand in a circle holding hands and sing:

> *Ring-a-ring o'rosie*
> *A pocketful of posie*
> *A-tishoo! A-tishoo!*
> *We all fall down.*

When the last line is recited the children drop to the ground. There are a number of versions of this poem, which may have originated in the days of the Great Plague of London (1665). Thus, this pleasant little diversion may well have had a macabre origin, as evidenced by the following:

"Ring-o-rosie"—referred to the rosy rash associated with the symptoms of plague.

"Posies of herbs"—were carried during epidemics to ward off the plague.

"A-tishoo"—referred to sneezing, which was considered a final fatal symptom.

"All fall down"—meant that everyone dropped over dead.

Illustration from *Mother Goose*, by Kate Greenaway (1881). The Granger Collection, New York.

Ring-a-ring-a-roses,
A pocket full of posies;
Hush! hush! hush! hush!
We're all tumbled down.

vents phagocytosis. There is an exotoxin termed **murine toxin** produced by virulent strains of *Y. pestis* that blocks cellular respiration in laboratory mice and probably has a similar activity in human plague. Another important surface-localized virulence factor of *Y. pestis* is protein Pla, a plasminogen-specific protease that plays an important role early during infection, releasing bacteria from the entrapment of fibrin clots.

Plague is diagnosed by symptoms and by isolation of the organism from aspirates of the buboes. Sputum can be the source of isolates in pneumonic plague, or the organism from either source can be identified serologically. Treatment with streptomycin, chloramphenicol, or tetracycline is effective if given early in the infectious process.

Lyme Disease

Lyme disease results from infection by the spirochete *Borrelia burgdorferi*. The tick-borne disease was first recognized in individuals in Lyme, Connecticut, hence its name. The disease first appeared in 1975 as cases of arthritis, and in 1982 the etiologic agent was identified. Lyme disease is the most rapidly spreading of the tick-borne diseases, with nearly 20,000 cases reported each year. Although Lyme disease is primarily diagnosed in the Northeastern and Middle Atlantic states, it has now spread to virtually every state in the United States.

The major reservoirs for *B. burgdorferi* are the deer and white-footed mouse. The tick involved is mainly the black-legged (deer) tick, *Ixodes scapularis* (Figure 28.20), although other tick species may also be infected with the spirochete. The deer tick feeds on birds and other animals, rendering them potential sources of Lyme disease. The deer tick can be a reservoir as well as the vector. A higher percentage of deer ticks are infested with *B. burgdorferi* than is the case for other ticks carrying tick-borne diseases.

The deer tick is quite small and not readily seen on human skin. The first manifestation of infection is a skin lesion at the site of the insect bite, which spreads to form a large rash area (Figure 28.21). At this stage of the disease it is readily treatable with tetracycline. Disease symptoms include headaches, chills, backache, and general malaise. The ability of *B. burgdorferi* to survive in blood is due to their resistance to killing by complement. The organisms adsorb from serum a natural inhibitor of complement preventing complement activation on their surface. Untreated Lyme disease can progress to a chronic stage, and the symptoms at this stage are arthritis and numbness in the limbs, believed to be the consequences of an immune response to the infection. The organism may attack the central nervous system, affecting vision and causing paralysis. The best preventive measure is to avoid tick habitats, and early removal of ticks is effective.

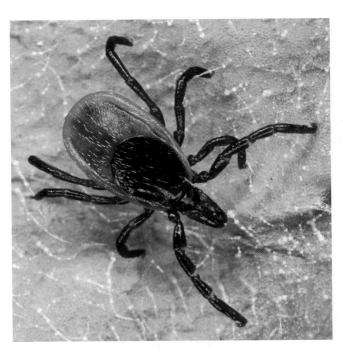

Figure 28.20 Vector of Lyme disease
The black-legged (or deer) tick (*Ixodes scapularis*) is the major vector of Lyme disease in the United States. Shown here is a female (usually about 3 mm long). Courtesy of Scott Bauer/USDA.

Rocky Mountain Spotted Fever

The etiological agent for Rocky Mountain spotted fever (RMSF) is *Rickettsia rickettsii*, a disease first recognized in Bitterroot Valley, Montana. The organism is named for the microbiologist H. T. Ricketts, who died of typhus while investigating that disease, which is also caused by a rickettsia. Rickettsia are gram-negative, highly pleiomorphic bacteria that can present as cocci, thread-like or rod-shaped. As obligate intracellular parasites, the *Rickettsia* must invade eukaryotic host cells (typically endothelial cells) for replication within their cytoplasm. Because of this, *Rickettsia* cannot be cultured on artificial media and are grown either in tissue or embryonic culture, typically in chicken embryos.

RMSF is caused by the bite of an infected tick and is acquired in humans from feces deposited by the tick in and about the bite site. In the western United States the rickettsia is carried by the wood tick (*Dermacentor andersoni*) and in the east by the dog tick (*Dermacentor variabilis*). *R. rickettsii* is passed from parent to tick progeny by transovarial passage. *D. andersoni* is a slow-feeding, hard-shelled tick, quite distinct from the fast-feeding ticks that are the vectors of Lyme disease. Consequently, removal of the RMSF tick up to 3 hours after it initiates feeding can prevent infection. There are about 1,000 cases of RMSF in the United States each year, with more

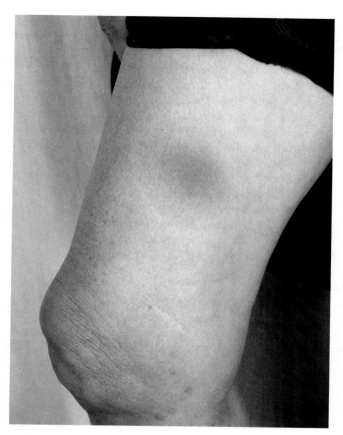

Figure 28.21 Lyme disease
The characteristic rash of Lyme disease at the site of the tick bite. The rash area increases over several days, and effective treatment should start at this stage. ©Larry Mulvehill/ Photo Researchers, Inc.

cases in North Carolina and adjacent states than in the Rockies. Dogs are readily infected with *R. rickettsii*, and organisms are present in the blood for 2 weeks after infection. A dog may serve as a reservoir.

The symptoms of RMSF include a fever, severe headache, and a rash that appears first on the palms and soles of the feet. The organism proliferates in the endothelial linings of small vessels and capillaries. The resultant hemorrhage and necrosis are responsible for the observed rash. Gastrointestinal problems can also occur. RMSF is generally diagnosed via symptoms and by establishing a history of tick bite(s). Tetracycline is an effective treatment, provided it is administered early in the infection.

Typhus

There are two types of the disease typhus, and they are caused by different rickettsial species that share common antigens. The two types are epidemic and endemic typhus. The etiological agent for epidemic typhus is *Rickettsia prowazekii*, and that for endemic typhus is *Rickettsia typhi*.

Epidemic typhus occurs in areas of substandard living conditions and poor sanitation. The major reservoir for *R. prowazekii* is the human, and the vector is the louse *Pediculus humanus corporis*. Flying squirrels can be infested with lice carrying the rickettsia. The flying squirrel can be a reservoir of *R. prowazekii*, leading to an occasional transmission to humans. There is no transovarial transfer of *R. prowazekii* in the louse, and the infected insect will die within several weeks.

All *Rickettsia*, including *R. prowazekii* are obligate intracellular bacteria that grow within the cytoplasm of their host cells. The organisms are internalized following attachment to host cell receptors, and are initially restricted to phagocytic vacuoles. They express an enzyme (phospholipase A2) that causes disruption of the host cell vacuole, and the subsequent replication of the bacteria takes place in the cytoplasm. Infection of the endothelial cell is believed to have the most significant consequences for the disease. Deposition of fibrin and platelets near infected sites causes occlusion of blood vessels and can impair heart function. Disruption of the endothelial barrier leads to increased vascular permeability, characterized by leakage of fluid from the bloodstream and its accumulation (edema) in the surrounding tissues, including the lungs. The reduced blood volume causes poor perfusion of organs such as the kidneys, which can lead to renal failure.

Infection of humans occurs while the louse is feeding, as the infectious organisms are present in feces or regurgitated blood of the insect. Symptoms include a high fever (104°F) with headaches, chills, and delirium. During the fifth or sixth day of illness, a rash develops. Recovery or death occurs 2 to 3 weeks after the onset of symptoms.

Typhus is diagnosed during epidemics by symptoms and, in sporadic cases, by the presence of a circulating antibody. The Weil-Felix test can be employed in diagnosis. This test is based on the fact that strains of the genus *Proteus* share common antigens with *R. prowazekii*. If antibody against *R. prowazekii* is present in serum, it will agglutinate *Proteus* cells. Tetracycline and chloramphenicol are effective in treating typhus. Vaccines are also available for use in areas where typhus is widespread.

Endemic typhus, also called **murine typhus**, occurs sporadically in areas throughout the world, including the Gulf Coast states in the United States, but, rarely in epidemics. Immunity to epidemic typhus (*R. prowazekii*) will confer immunity to endemic typhus (*R. typhi*). The primary reservoir for *R. typhi* is the rat or ground squirrel. Neither the rat nor the rat louse is adversely affected by the infection. The disease symptoms, diagnosis, and treatment are the same as for epidemic typhus.

Scrub Typhus

Scrub typhus is also a rickettsial disease, and the etiological agent is *Rickettsia tsutsugamushi*. This disease was of some consequence during World War II and the Vietnam War. The reservoir is the rodent and the vector is a mite. The mite transmits the infectious agent from rodent to rodent. It is transferred transovarially, and the mites lay eggs in soil that hatch into larvae (chiggers) that feed on animals, including humans. The disease symptoms are similar to those encountered with typhus. A primary lesion, called an eschar, occurs at the location of the bite. Diagnosis is either by the Weil-Felix reactions, which test the ability of sera from infected patients to agglutinate *Proteus vulgaris* cells, or by detection of the infected organism's DNA by use of polymerase chain reaction (PCR). Tetracycline and chloramphenicol are effective, but treatment should be prolonged because relapses do occur. A successful vaccine has not been developed against scrub typhus.

Q Fever

Q fever received the name Q (Query) because the etiological agent for the disease was long unknown. We now attribute the disease to *Coxiella burnetii*, a rickettsial organism remarkable for its ability to survive outside a host. Arthropods are both a reservoir and vector of Q fever and transmit it from animal to animal. The organism forms an endospore-like body that permits survival on animal carcasses and hides, and in feces. It is also unusually heat resistant. Outbreaks in humans occur among animal handlers and through consumption of unpasteurized milk. Q fever causes an influenza-like illness with prolonged fever, headaches, and chills. Pneumonia and chest pains are also symptoms. The disease is insidious in that an endocarditis (inflammation of the heart) can occur months or years after infection. Immunological tests are employed in diagnosing Q fever, and treatment is with tetracycline.

Ehrlichiosis

Ehrlichiosis is an emerging rickettsial disease and making inroads in the United States. The first case observed in the United States occurred in 1986. The causative agents are *Ehrlichia chaffeensis* and *Ehrlichia equi* and are transmitted to humans by ticks. The reservoir may be rodents, rabbits, or deer. These illnesses are relatively rare in the United States, and infections are often not properly diagnosed as the symptoms are quite nonspe-

cific. The disease may be subclinical and is rarely fatal, although long-term untreated cases may result in respiratory and renal problems.

E. chaffeensis infection results in a fever, and the early symptoms resemble influenza with malaise and headache. The organism invades monocytes and has been described as human monocyte ehrlichiosis (HME). *E. equi* infection results in a rapid onset of fever, chills, and muscle pain. Granulocytes are affected and the disease is termed human granulocytic ehrlichiosis (HGE).

There are more than 600 cases of ehrlichiosis per year and more than half are HGE. Diagnosis of the infections is based on fluorescence antibody assay or detection of *Ehrlichia* DNA in patient serum by PCR. Treatment is with doxycycline or other antibiotics.

Relapsing Fever

Humans are the reservoir for *Borrelia recurrentis*, the causative agent of relapsing fever. *B. recurrentis* is a gram-negative spirochete that forms coarse irregular coils that are 0.5 μm wide and more than 20 μm long. The disease is transferred from human to human by the body louse. A bite will transfer the spirochete into the bloodstream, which carries it to the spleen, liver, kidneys, and gastrointestinal tract. Multiple lesions occur in these organs. After a fever and 4 to 5 days of illness, the disease symptoms disappear. In most cases, a relapse (hence, relapsing fever) occurs after about 10 days, with a return of the manifestations of disease. These bouts of wellness and relapse occur three to ten times before full recovery occurs. The relapse is attributed to the development of strains with different surface proteins on the outer membrane that are unaffected by the original immune response. A period of time is necessary before the host can generate antibodies to combat each new strain.

The disease can be diagnosed by clinical symptoms and by the presence of *B. recurrentis* spirochetes in the blood. Dark-field microscopy is effective in visualizing spirochetes. The treatment of choice is penicillin.

Tularemia

The etiological agent for tularemia is *Francisella tularensis*, an extremely pleomorphic, small, gram-negative bacterium. The wood tick is both a reservoir and vector of *F. tularensis* in the Northwest and Midwest and is transmitted in the Southwest by a deer fly. It occurs worldwide mostly from handling of infected animals, particularly rabbits.

F. tularensis is an exceedingly virulent organism, but the reason for this is unclear. The organism grows intracellularly in PMNLs. There are three major manifestations of tularemia:

- **Ulceroglandular**—occurs from the bite of an infected tick or from contact with an infected animal through a scratch, and a lesion occurs at the point of entry.
- **Pneumonic**—generally spread by the respiratory system during skinning or handling of infected animals.
- **Typhoidal**—occurs after ingestion of contaminated food or water.

Tularemia can be diagnosed by microscopy of smears obtained from lesions and visualized with fluorescent antibodies; treatment is with tetracycline.

Trench Fever

Trench fever is a disease that apparently occurs mostly in unsanitary conditions associated with war. Trench fever first occurred in 1915 (World War I), and more than 1 million military personnel contracted the disease. The etiological agent is *Bartonella quintana*, a fastidious, gram-negative bacillus. The primary reservoir is the human, and body lice are the vectors. Disease symptoms include chills, fever, headache, and muscle pain. These symptoms last for 5 days; hence, the species name *quintana*. Patients are incapacitated for up to 2 months. Relapses often occur, and antibiotics such as chloramphenicol must be given for an extended period to eliminate the disease.

Cat Scratch Disease

There are more than 24,000 cases of cat scratch disease in the United States each year. The bite or a scratch, generally by a kitten, results in cat scratch fever and occurs most often in young children. Papules (red swellings) are evident 7 to 12 days after injury, and lymph nodes in the vicinity become tender. As the disease progresses, the lymph nodes swell and become pus filled. Enlarged nodes may require surgical drainage, but the disease symptoms are generally mild and localized. Recovery requires a few weeks. The etiological agent is *Bartonella henselae*, a small, curved gram-negative rod that can be isolated on blood agar under a CO_2 atmosphere. There may be other causes of this human disease, but no other agent has been isolated.

Polymicrobial Diseases

There is a growing awareness that many disease states previously attributed to a single microbial species actually occur through the concerted action of two or more individual species or strains. These diseases have been termed polymicrobial, dual, mixed, secondary, or synergistic infections. The organisms involved may have different kingdoms, genera, or different species within a genus. They may be different strains or substrains within a species.

Polymicrobial diseases are now subdivided into polyviral, polybacterial, combined viral and bacterial, polymycotic, and parasitic diseases. The ability of several types to invade simultaneously is often abetted by microbe-induced immunosuppression.

Lyme disease is a prime example of a polymicrobial disease. The ixodid tick that carries *B. burgdorferi* to human hosts has been described as a sewer of infectious diseases. This tick may simultaneously carry the causative agents of ehrlichiosis, tularemia, and cat scratch fever. Consequently polymicrobial infections are relatively common in cases of Lyme disease. Diagnosis is therefore complicated, and treatment of these mixed populations of different antimicrobial susceptibility is difficult.

SECTION HIGHLIGHTS

A number of pathogens are transmitted to humans through intermediary hosts called vectors. Vectors can be other mammals or insects. Bacteria that infect humans through a vector are usually obligate parasites, and they cannot survive outside a living host. *Y. pestis* is transmitted from infected rodents to humans by a flea bite. In bubonic plague, bacteria replicate in a lymph node that swells into a bubo. Pneumonic plague occurs in a fraction of patients with bubonic plague, where bacteria invade the bloodstream and reach the respiratory tract. It can be also acquired by person-to-person transmission via aerosol, and, if untreated, the mortality associated with this form of plague is nearly 100%. Lyme disease, an infection with a spirochete *B. burgdorferi*, can manifest itself as a mild, treatable infection, or as a chronic illness, including such symptoms as arthritis and numbness in the limbs. Other vector-borne diseases include typhus, scrub typhus, Q fever, ehrlichiosis, relapsing fever, tularemia, and cat scratch disease. Vectors may carry more than one type of infectious agent and cause polymicrobial infections.

28.14 Protists as Agents of Disease

Protozoa are the causative agents of many diseases worldwide. Tropical areas near the equator are particularly subject to protozoan diseases, and malaria is considered to be the world's leading health problem. Malaria infects 150 million humans each year and is responsible for more than 2 million deaths. Although malaria has been essentially eradicated in the United States, conditions exist for the reintroduction of the disease. Two other protozoan diseases, amebiasis and giardiasis, are a concern in the United States, and these and malaria will be discussed briefly. Another pathogenic protozoan, *Cryptosporidium*, recently caused a health crisis in Milwaukee, Wisconsin (Box 28.2).

Amebiasis

Entamoeba histolytica is the etiologic agent for amebiasis, also known as amebic dysentery. Worldwide there are an estimated 400 million active cases that cause more than 100 thousand deaths each year. There are 4,000 to 5,000 new cases in the United States annually. The disease is most prevalent in warmer climates with inadequate water treatment and sanitation.

The infection occurs when the mature cysts of *E. histolytica* are ingested. The cyst moves through the stomach, and the cyst coat is removed in the small intestine yielding the amoeboid form. The amoeboid cell, called a **trophozoite**, moves to the large intestine and penetrates the intestinal mucosa. In acute amoebiasis, lesions are formed in the mucosa, and these lead to severe ulceration. The trophozoites can multiply and invade other intestinal mucosa while feeding on erythrocytes and bacteria. The lesions become infected with bacteria present in the intestine. *E. histolytica* may penetrate the large intestine and move to and produce lesions in the liver, lungs, or brain.

The trophozoites may establish a mild or asymptomatic form of the disease called **chronic amebiasis**. In-

---

BOX 28.2 *Research Highlights*

The Water We Drink—Safe or Not?

A severe outbreak of a diarrheal disease occurred in Milwaukee, Wisconsin, in April of 1993. This outbreak affected one-fourth of the population of Milwaukee and served as a warning to all Americans that:

1. Our water supplies are under great stress.

2. Disease is lurking and can create havoc with the least breakdown in our defenses.

3. We must not take our water supplies for granted.

The pathogen involved in Milwaukee was a protozoan of the genus *Cryptosporidium*. *Cryptosporidium* is a common protozoan in mammals and birds that should be filtered from drinking water during routine water treatment. Ingestion of *Cryptosporidium* or cysts (cryptosporidia) in drinking water gave 370,000 citizens of Milwaukee acute diarrhea, nausea,

and stomach cramps. Undoubtedly there were fatalities from dehydration and other complications, but the exact number of these will never be known.

Why did this happen? The epidemic occurred in early spring, and runoff from the spring thaw may have overstressed the water treatment plant. Dairy farms abound around Milwaukee, and cattle harbor the protozoan. Water from feedlots may have carried the parasites into catchments that supply the city with potable water.

In routine treatment of water, coagulants are added, prior to sand filtration, which remove particles from the water. Alum (aluminum potassium sulfate dodecahydrate) was the coagulant employed in Milwaukee until the fall of 1992, when it was replaced with polyaluminum hydroxychloride. This replacement was due to an EPA concern that alum can leach lead from water pipes, and Milwaukee

water supplies contained unacceptable levels of lead.

During March of 1993, employees at one of the water treatment facilities noted that the treated water coming through the plant was unusually turbid. The spring thaw runoff water was evidently not clarified by the replacement coagulant employed to settle particulates. Chlorination does not kill the cryptosporidia cysts of those of other pathogenic protozoa. The water treatment plant in question was closed down, and the city returned to using alum as coagulant.

A report on drinking water quality was released in April of 1993 by the U.S. General Accounting Office. The report concluded that water quality assurance programs across the United States are "flawed and underfunded." The conclusion was that 198,000 water supplies are inadequately monitored. This is a problem we cannot ignore.

dividuals with this form of the disease become carriers and constantly shed cysts into the environment. They are a concern because they serve as a constant reservoir for infective cysts. Amebiasis is a health problem among homosexual males.

The symptoms of amebiasis are severe dysentery or colitis. The diarrhea is accompanied by blood and mucus shed from the extensive intestinal lesions. Amebiasis can be diagnosed by the presence of trophozoites in the feces filled with ingested red blood cells. Treatment is different for carriers than for those with active infections. Iodoquinol is the drug of choice for asymptomatic carriers, and Aralen phosphate is given in acute amebiasis. The best preventive measure is quality water, although chlorination of water does not destroy the cysts of *E. histolytica*.

Giardiasis

Giardia lamblia (also known as *G. intestinalis*) is the agent of an intestinal disease called giardiasis. *G. lamblia* is an ancient eukaryote that lacks a mitochondrion, and its rRNA has many bacteria-like features. Many domains in *G. lamblia* rRNA resemble that in the *Archaea*. Giardiasis is the most prevalent water-borne diarrheal disease in the United States. It occurs most often in young children, and there are about 30,000 cases per year. Giardiasis can be endemic in child day-care centers.

About 7% of the population in the United States are healthy carriers of *G. lamblia* and shed the cysts in their feces. Transmission of the cysts to humans generally occurs through ingestion of contaminated food or water. Outbreaks often occur in campers or backpackers in wilderness areas when they drink seemingly clear, fresh water that has been contaminated by domestic or wild animals, including beavers, rodents, and deer.

Acute cases of giardiasis are marked by severe diarrhea, cramps, flatulence, and dehydration. The asymptomatic carriers may have intermittent mild cases of diarrhea. Ingestion of the cysts, the infectious form of the parasite, is followed by uncoating and release of trophozoites. The trophozoites inhabit the lower small intestine where they attach to the mucosal wall through small disks and apparently feed on mucosal material. Here trophozoites undergo asexual reproduction. Some of the trophozoites are carried into the large intestine, where they develop into thick-walled cysts, followed by their excretion. Cysts are hardy and they remain viable in water for many months. Diagnosis of giardiasis is by examining for cysts in mucosal material shed in feces. The mucous sheath in the intestinal tract is shed normally, and the trophozoites or cysts would be present in this material. A commercial enzyme-linked immunosorbent assay (ELISA) (see Chapter 30) is available for de-

tection of the *G. lamblia* antigens in fecal material. Treatment is with quinacrine hydrochloride (Atabrine) or metronidazole. As with *E. histolytica*, the cysts of *G. lamblia* are resistant to chlorination. Campers should boil water to prevent infection. The disease can be prevented in populations at large by proper water treatment.

Malaria

The etiologic agents of malaria are several species of protozoa in the genus *Plasmodium*. Nearly all adults in the middle regions of Africa and in India have been infected. Malaria sickens up to half a billion people a year and kills more than a million. Many of these deaths occur in young children. Although the disease has essentially been eradicated in the United States since 1945, it has made a comeback through infected immigrants. There are about 1,000 cases per year recorded in the United States. Of the species that infect humans, *Plasmodium vivax* is found in areas with temperate and subtropical climates, whereas *Plasmodium falciparum* is common in the tropics.

Malaria results from the bite of a female mosquito of the genus *Anopheles*. The female injects saliva containing anticoagulants while sucking blood, and this saliva also transmits the malarial parasite into the human bloodstream. Male mosquitoes do not suck blood and do not carry the protozoan.

There are two phases in the life cycle of the protozoan *Plasmodium*: asexual phase in human cells and a sexual stage in the mosquito. Ingestion of male and female *Plasmodium* **gametocytes** from a human blood meal by the mosquito is followed by their reproduction and formation of thousands of **sporozoites** in the insect's salivary glands. When *Plasmodium* sporozoites are injected into the bloodstream by the feeding mosquito, they pass to the liver, where they undergo an asexual fission; in about 7 days there is the release of **merozoites**. These merozoites invade red blood cells where they reproduce. The red blood cells rupture and release more merozoites that infect other red blood cells in several cycles. Rupture of the red cells at each cycle triggers the symptoms of malaria. Some parasites do not cause red cell rupture, they develop into gametocytes and these can be taken into the stomach of a female mosquito while feeding on an infected human.

Within the mosquito, the microgametocytes form macrogametocytes and microgametocytes that fuse to form a diploid zygote. Sexual reproduction in the midgut of the mosquito leads to the production of sporozoites that travel to the salivary glands of the mosquito. The sporozoites and anticoagulants are then injected into another victim during feeding.

The symptoms of malaria—chills followed by fever and sweating—are caused by the presence of merozoites

TABLE 28.7	Zoonoses that are transmitted to humans			
Disease	**Organism**	**Transmitted by**	**Host**	**Symptoms**
Anthrax	*Bacillus anthracis*	Endospore inhalation or animal contact	Cattle, goat, sheep	Cutaneous lesions
Bovine tuberculosis	*Mycobacterium bovis*	Milk	Cattle	Lesions in bone marrow of hip, knee
Brucellosis	*Brucella abortus*	Milk, skin/eye contact or bite	Cattle	Fever, headache
	B. melitensis		Goat	
Leptospirosis	*Leptospira interrogans*	Urine, contaminated water, skin contact	Rodent, dogs, cattle	Meningitis, headaches, fever
Listeriosis	*Listeria monocytogenes*	Raw milk	Cattle	Meningitis
Yaws	*Treponema pallidum* subsp. *Pertenue*	Bite	Fly	Skin ulcers, lesions

and debris from the synchronized rupture of blood cells. A victim may have several attacks, and these are followed by remission that can last for days to months. Relapses occur as dormant parasites become activated and emerge from the liver. Malaria cannot be spread by human-to-human contact, as it requires the mosquito vector for transfer, a condition somewhat unique in infectious diseases.

Malaria is diagnosed by observing microscopically the parasites in red cells. Serological tests are also available. Treatment is with quinine derivatives. Malaria can be treated with artemisinin, a drug developed in China that is extracted from the wormwood plant. The favored treatment is a combination of artemisinin and quinine derivatives. Chloroquine is administered to potential victims as a prophylactic treatment. The only preventive measure is to destroy the breeding grounds for *Anopheles* mosquitoes. DDT was employed extensively for this purpose during the period from 1946 to the 1960s, but the use of this pesticide has been curtailed in the United States.

28.15 Zoonoses

Zoonoses are infections that normally occur in animals but can infect humans through contact with infected animals, animal products, or bites. A number of vector-borne diseases have already been discussed. There are a number of others, and some of these are presented in Table 28.7. Many of these diseases occur rarely in the United States, but outbreaks of diseases such as listeriosis do occur when milk is improperly handled. Brucellosis has been a concern in the past, but widespread testing of cattle has eliminated this disease from dairy herds in the United States. Anthrax is mainly a concern at the present time as a germ warfare agent (see Box 20.1).

Giardiasis is a common water-borne diarrheal disease. *G. lamblia* is found as cysts in the environment. Following the ingestion of the cyst form of the parasite, they become trophozoites, and feed and reproduce before reverting to cysts and returning to the environment.

The agents of malaria are various species of protozoa belonging to the genus *Plasmodium*. They replicate by a sexual cycle in the gut of the mosquito vector and by asexual replication in the infected red blood cells of the host.

SUMMARY

- Humans are subject to an array of **skin infections**, of both bacterial and fungal origin. Common cutaneous infections caused by bacteria are staphylococcal infections and by fungi, athlete's foot, or ringworm.

- **Respiratory infections** are difficult to control because they are generally transmitted in minute air droplets.

- **Diphtheria** was once a leading cause of death in children but has been virtually eliminated in the United States by a massive immunization program. It continues to be a concern in economically depressed areas of the world. Whooping cough is another disease that has been controlled in the United States by vaccination. Worldwide it is responsible for 500,000 deaths annually in young children.

- **Pneumonia** is an inflammation of the lungs that results from infection by viruses, mycoplasma, bacteria, or fungi. The leading etiologic agent *S. pneumoniae* is often a secondary infection that follows a viral infection. *S. pyogenes* is a potent pathogen and can infect the meninges or cause scarlet or rheumatic fever.

- **Meningitis** is an inflammation of the meninges that can be caused by *S. pyogenes, H. influenzae,* or *N. meningitidis.* Meningitis is a severe, often fatal, disease.

- **Tuberculosis** is an infectious disease, generally occurring in the lungs. It can be cured with antibacterial agents, but new antibiotic resistant strains are developing at an alarming rate, especially among patients with AIDS. An estimated 10 million new cases of tuberculosis occur worldwide each year.

- **Leprosy** is an ancient disease that has not yet been eliminated, but treatment is possible.

- **Fungal diseases** are a problem in part because there are few compounds that will cure infections, and those available are toxic. Histoplasmosis, coccidiomycosis, and aspergillosis are lung diseases caused by fungi.

- **Gastrointestinal tract diseases** are of two major types—food poisoning from ingestion of food on which toxin-producing microorganisms have grown and food infections from drinking liquids or ingesting food contaminated with disease-causing bacteria.

- Recent evidence indicates that **peptic ulcers** in humans are caused by growth of a bacterium, *H. pylori,* in the stomach.

- **Botulism** is caused by ingesting food on which *C. botulinum* has grown. This microorganism produces one of the most potent toxins known.

- A common food infection is **salmonellosis**, caused by various strains of the genus *Salmonella. S. typhi* is the etiological agent of typhoid, a serious human infection. Cholera is a disease caused by drinking water or eating food that has been contaminated with fecal waste.

- The incidence of **urinary tract infections** is second only to respiratory infections in the United States in total cases. These are of frequent occurrence in hospital patients.

- **Sexually transmitted diseases** (STDs) are a major health concern. These diseases are preventable/curable and their continuing incidence is mainly a societal problem.

- **Gonorrhea** is an STD that occurs most often in the 15 to 35 age-group and is of considerable concern because the organism responsible for the disease is gaining resistance to the antibiotics that are employed in treatment.

- **Syphilis** is an STD that occurs less frequently than gonorrhea but is more of a menace because of the severity of the disease.

- **Vector-borne diseases** are transmitted from one infected animal to another via an intermediate host. Most of these diseases are caused by obligate parasites such as the rickettsia. Lyme disease and Rocky Mountain spotted fever are vector-borne diseases that are rapidly increasing in incidence. Among the vector-borne diseases are typhus, Q fever, relapsing fever, tularemia, and trench fever.

- **Plague** has historically been a major killer of humans. It is carried to humans by rat fleas.

- The protozoan disease **malaria** is considered to be the leading health problem in the world. The disease infects 150 million humans each year and is responsible for 2 million deaths. Malaria is passed to humans by the bite of the *Anopheles* mosquito.

- **Amebiasis** and **giardiasis** are protozoan diseases of the intestinal tract. Inadequate treatment of water is responsible for the transmission of these diseases.

- **Zoonoses** are infections that are passed to humans through contact with infected animals.

 Find more at www.sinauer.com/microbial-life

REVIEW QUESTIONS

1. *Staphylococcus* is a versatile pathogen. What are some of the infections caused by this organism? How can they be controlled?

2. Why are respiratory diseases so difficult to control? What influence did this have on the development of vaccines to prevent these diseases? Consider some of the organisms that attack the different areas of the respiratory system.

3. What is pneumonia? Which organisms cause this disease?

4. *Streptococcus pyogenes* is the etiological agent for a number of infections. Cite some of these and the symptoms. What factors give this organism invasive ability?

5. What is meningitis? It is caused by several organisms; which is the most dangerous?

6. What is meant by an A-B toxin, using the anthrax toxins as an example?

7. How does tuberculosis affect an individual? How is it diagnosed? Why is it making a comeback?

8. What are the symptoms of leprosy? Why has it been difficult to study and treat?

9. Fungi are the causative agents of respiratory diseases in humans. Cite examples.

10. What factors are important in lowering the incidence of gastrointestinal tract infections?

11. Compare the intracellular action of pertussis and cholera toxins.

12. How does "food poisoning" differ from "food infection"? Give examples. How can either of these be prevented?

13. What are the major sexually transmitted diseases (STDs) that are caused by bacteria? Are they curable? Are they more a societal problem or a medical problem—give reasons for your answer. What is a major health concern with gonorrhea?

14. Describe the life cycle of *Chlamydia*.

15. Do STDs affect the fetus or neonate? If so, how?

16. Define vector and reservoir. Cite examples. Can an organism be both?

17. What is the disease cycle in plague? How can it be prevented? Is plague a concern in the United States?

18. Where did Lyme disease get its name? Why is this disease a danger?

19. What are the symptoms of Rocky Mountain spotted fever? If one is bitten by a dog tick and later has such symptoms; what course should be followed?

20. The remark has been made that General Typhus defeated General Napoleon. Why?

21. How does one get tularemia? How might it be prevented?

22. Protozoan pathogens are a problem in many areas of the world. Where is amebiasis of greatest concern? Giardiasis?

23. Malaria is one of the world's greatest health problems. Why? How can it be prevented?

SUGGESTED READING

Brogden, K. A., J. A. Roth, T. B. Stanton, C. A. Bolin, F. C. Minion and M. J. Wannemuehler. 2000. *Virulence Mechanisms of Bacterial Pathogens.* Washington, DC: ASM Press.

Cossart, P., P. Boquet, S. Normark and R. Rappuoli. 2005. *Cellular Microbiology.* 2nd ed. Washington, DC: ASM Press.

Fischetti, V. A., R. P. Novik, J. J. Feretti, D. A. Portnoy and J. I. Rood. 1999. *Gram-positive Pathogens.* Washington, DC: ASM Press.

Miller, V. L. (ed.). 1994. *Molecular Genetics of Bacterial Pathogenesis.* Washington, DC: ASM Press.

Roth, J. A., C. A. Bolin, K. A. Brogden, F. C. Minion and M. J. Wannemuehler. 1995. *Virulence Mechanisms of Bacterial Pathogens.* Washington, DC: ASM Press.

Salyers, A. A. and D. D. Whitt. 2002. *Bacterial Pathogenesis: A Molecular Approach.* 2nd ed. Washington, DC: ASM Press.

Sheld, W. M., W. C. Craig and J. M. Hughes. 2001. *Emerging Infections 5.* Washington, DC: ASM Press.

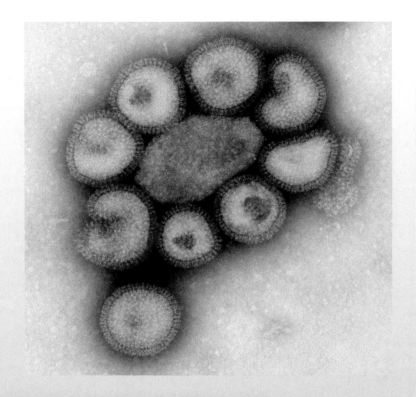

- Survey the major viral diseases that afflict humans.

- Describe the principal modes of viral transmission.

- Describe the symptoms of viral diseases and methods for treatment.

- Discuss the strategies used to prevent or control disease transmission.

29

Viral Diseases of Humans

. . . and hell itself breathes out contagions to this world.
—*Shakespeare,* Hamlet

A virus (or virion) is a bit of DNA or RNA that can enter and take over the normal functions of a specific host cell. The virus can selfishly supersede the host-cell genomic DNA, forcing the synthetic machinery of the host to function solely in the synthesis of viral particles. The viral genetic material is considered to have originated from ancestors of the organism that they attack (see Chapter 14) and therefore is attuned to the synthetic and genetic machinery of that host. Viral assault on a specific host cell generally leads to its disintegration. However, there are temperate viruses that do establish a latent or dormant infection in a eukaryotic cell (see Chapter 14). Active proliferation of viruses, the infectious cycle, may result in cell destruction. This is the mechanism that has evolved to disseminate viruses, and even latent viruses must periodically have an infectious cycle for survival.

To date there are no chemotherapeutic agents that can cure most of the viral infections that afflict humans and the favored preventive measure is vaccination. The nature of viral replication is such that administering an agent to rid the body of a viral infection, as we do routinely for bacterial infections, is not feasible. The physiological explanation for this limitation rests on the marked disparity between the components of a bacterium and a eukaryotic cell in contrast to the similarities between a virus and its host. Bacteria generally have peptidoglycan in their cell wall, and an antibiotic (such as penicillin or vancomycin) can interfere with cell wall synthesis without harm to the eukaryotic host, which lacks peptidoglycan. In addition, differences between the protein-synthesizing machinery in prokaryotes and eukaryotes can be exploited, and inhibition of protein synthesis in a bacterial invader with antibiotics (such as tetracycline or streptomycin) will not so adversely affect the host. Thus, antibiotics are effective in destroying select bacterial invaders and "curing" disease.

Viruses, on the other hand, present a different scenario—their reproductive machinery is similar to that of the cells they infect. A virion can attach to, invade, and commandeer the synthetic functions of the host, and exploitable differences between virus and host are not available. A chemotherapeutic agent that would interfere with virus replication would likely have a comparable adverse effect on equivalent reactions in a healthy cell. Experience has affirmed that this is essentially the case. There are a limited number of compounds that are administered for specific viral infections: adamantine or Tamiflu® for influenza A, acyclovir for herpesvirus infections, and combined therapy to treat AIDS; but only adamantine has a curative effect.

This chapter is devoted to a discussion of major viral infections in humans. The transmission, entry, and symptoms of viral infections are discussed along with mechanisms for prevention and/or control of the viral pathogen. For convenience, these infections are discussed according to portal of entry.

29.1 Viral Infections Transmitted via the Respiratory System

Respiratory diseases are readily transmitted from person to person via airborne droplets and are therefore dif-ficult to control. Normal, everyday human contacts can lead to a broad distribution of infectious particles. Highly contagious diseases, such as influenza, that are spread via the respiratory route can be transmitted rapidly and can reach localized epidemic proportions among human populations that have not previously been exposed.

Because many of the infections transmitted via the respiratory system are so contagious, much emphasis has been placed on developing immunization programs for them. Immunization is favored because quarantine or isolation of the ill has not proved effective. In some cases (smallpox), immunization has been totally effective, and in other cases (HIV), it has so far been a failure. In the case of influenza, vaccination has been somewhat successful by using a combined vaccine made up of antigens from previous flu outbreaks. Consequently, the development of immunity to several strains that have passed through the population previously confers limited immunity to a strain that might be introduced at a later date.

Respiratory viral infections affect 85% to 90% of the population of the United States each year. The two major types of these infections are the flu (caused by influenza virus) and the common cold (caused by rhinoviruses). All other infectious diseases *combined* affect fewer than 20% of the population. However, there was a marked surge in infection from herpes simplex virus (HSV) in the 1970s and in human immunodeficiency virus (HIV) in the early 1980s as a result of increased unprotected sexual activity. This section presents some of the major viral infections that are spread via the respiratory route: chicken pox, rubella, measles, smallpox, mumps, influenza, and common colds (Table 29.1).

TABLE 29.1	Droplet or airborne viral diseases of humans		
Infectious Disease	**Viral Agent**	**Control by Vaccination**	**Type**[a]
Chicken pox	Herpesvirus	+	E dsDNA
Rubella	Togavirus	+	E ssRNA
Measles	Paramyxovirus	+	E ssRNA
Smallpox	Vacciniavirus	Eradicated	E ssRNA
Mumps	Paramyxovirus	+	E ssRNA
Influenza	Orthomyxovirus	±	E ssRNA
Common colds[b]	Rhinovirus	–	N ssRNA
	Coronavirus	–	E ssRNA
	Adenovirus	–	N dsDNA

[a]E, enveloped; N, naked capsid.
[b]Also spread by contact and fomites.

Chicken Pox

Chicken pox is caused by the varicella-zoster virus (VZV), a member of the **herpesvirus** group, and is a highly contagious infection that occurs mainly in children. Prior to the development of an effective vaccine, there were more than 4 million cases of chicken pox in the United States annually. Eruptions on the skin appear after an incubation period of about 10 to 20 days in the mucosa of the upper respiratory tract. During an infection the virus is carried to the bloodstream and lymphatics, where replication continues. Chicken pox is characterized by a fever and the small reddish vesicles that erupt on the skin (Figure 29.1A). The rash may occur over much of the body but is more severe on the trunk and scalp than on the extremities. The contagious period begins a few days before the appearance of lesions and subsides a few days after the fever ends. The vesicles are painful and itchy and occur over 2 to 4 days as the virus replicates. Scratching by young victims of the disease can cause bacterial infection and minor scarring. The virus can move to the lower respiratory system and gastrointestinal tract with more severe consequences. Damage to lungs and to the blood vessels in essential organs can be fatal.

A primary infection confers lifetime immunity to chicken pox but can lead to the presence of a dormant virus in the nuclei of sensory nerve roots. These dormant viruses can be activated by physiological or mental stress or by immunosuppressants, and this activation leads to an active infection commonly called **shingles** (Figure 29.1B). When activated, the virus moves to sensory nerve axons, where it replicates and damages sensory nerves. Skin eruptions occur in the area of nerve damage and are painful and itchy. Generally, these lesions appear about the trunk, hence the name *zoster* (Greek for *girdle*), and occur more frequently in adults older than 60 years of age.

A vaccine has been employed in Japan for many years, and in 1995 a vaccine was approved for distribution in the United States. The chicken pox vaccine is a live attenuated varicella virus and is administered to children on the same schedule as the MMR (measles, mumps, and rubella) series. Adults who have not been exposed to chicken pox (or have not been vaccinated) should be vaccinated, as the disease can be severe in adults and lead to complications. A vaccine for shingles prevention has recently been approved for individuals over 60 years of age.

Rubella

The infectious agent for **rubella (German measles)** is a **togavirus** disseminated by respiratory secretions from infected individuals. It is moderately contagious, and infection results in a rash of red spots (Figure 29.2) and a mild fever. The illness is of short duration, and complete recovery occurs within 3 to 4 days.

The rubella virus infects the upper respiratory tract and spreads to local lymph nodes. Replicated viruses move

(A)

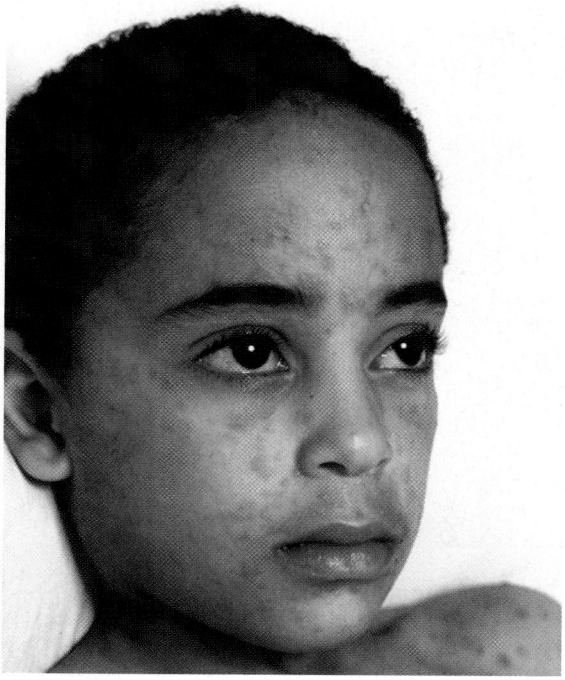

(B)

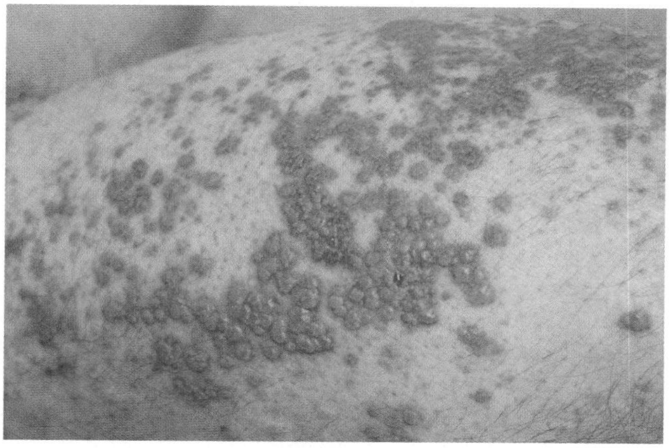

Figure 29.1 Herpesvirus infection
Symptoms of herpesvirus infection. Vesicular lesions associated with (A) chicken pox (varicella) and (B) shingles. Shingles lesions generally occur in sensory nerves around the trunk of adult humans. A, photo ©SPL/Photo Researchers, Inc.; B, ©SIU/Visuals Unlimited.

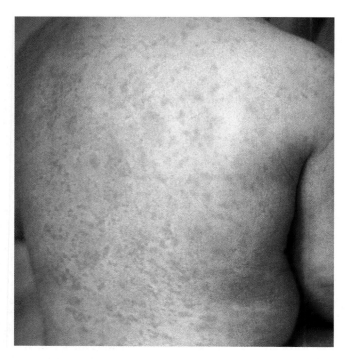

Figure 29.2 Symptoms of rubella
Rash of red spots caused by rubella (German measles). The spots are not raised. Courtesy of the Centers for Disease Control and Prevention (CDC).

throughout the body via the bloodstream. The incubation period varies from 2 to 3 weeks, but the virus is present in the upper respiratory tract for more than a week before the onset of the rash. Transmittable virus is present in the respiratory tract for a period of at least 7 days after the disappearance of the rash. This is another case where an infected individual can transmit a disease prior to the onset of symptoms and after symptoms disappear.

Rubella can be tragic for pregnant females who become infected with the virus. The virus can cross the placenta and infect the fetus during the first few months of pregnancy. This fetal infection may result in congenital defects in the newborn. These defects include hearing loss, microcephaly, cerebral palsy, heart defects, and other abnormalities. Fetal death or premature birth may also occur. A neonate that has been exposed to rubella in utero can shed (transmit) the virus for 1 to 2 years after birth. An estimated 20% of infected babies do not survive the first year.

A live attenuated virus vaccine is administered in the MMR series to preschool children in the United States. It is about 95% effective, and widespread use may develop herd immunity (see Chapter 30) that will ultimately eliminate the disease. This would substantially reduce the possibility of rubella infections in at-risk pregnant women. There were 152 cases of rubella in the United States in the year 2000 and 7 cases of congenital

rubella syndrome. Health officials are concerned that outbreaks will occur among immigrants who were not vaccinated as infants.

Measles

Measles, also known as "common measles" or **rubeola**, is a highly contagious viral disease caused by a **paramyxovirus**. Measles is an acute systemic infection marked by nasal discharge, cough, delirium, eye pain, and high fever. A rash (Figure 29.3) can appear over the entire body, and lesions in the oral cavity are common. The measles virus enters through the respiratory tract or conjunctiva (eye). The incubation period is about 11 days after exposure, and the rash appears 3 days after the onset of fever, cough, and other symptoms. These symptoms persist for 3 to 5 days and are followed by the rash, which spreads from the head downward over a subsequent 3- to 4-day period.

Measles is a severe disease and is frequently followed by secondary infections. Pneumonia, inner ear infections, and, infrequently, encephalitis may occur. This encephalitis has nearly a 25% mortality rate and can cause neurological damage in survivors.

Today there are few cases of common measles in the United States due to a comprehensive vaccination program (MMR) that was initiated in the mid-1960s. The vaccine employed is an attenuated live virus and induces immunity in more than 95% of recipients. Infants from 12 to 18 months old are vaccinated, a mandatory

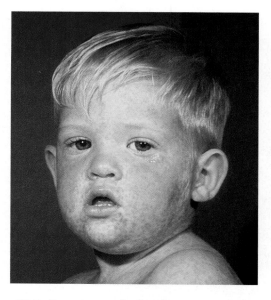

Figure 29.3 Symptoms of rubeola
Characteristic rash associated with rubeola (measles) in a young child. The rash begins on the face and moves downward to the trunk. Courtesy of the Centers for Disease Control and Prevention (CDC).

practice in most states. There have been some outbreaks in recent years among poor inner-city children, immigrants, and illegal aliens who have not been immunized. Each year there are 30 to 40 million cases of measles worldwide, causing the death of more than 700,000 people. Occasional outbreaks occur on college campuses and are a concern because adults generally have a more severe infection than that observed in children.

There has been a decline in childhood vaccination in many countries due to reports that the vaccine contains thimerosal, a mercury-based preservative. There is no thimerosal in MMR vaccine, as the vaccine is composed of weakened live virus and the preservative would destroy the virus and inactivate the vaccine. The rumor of an autism link persists, however, leading to a decline in vaccination and thus an increased incidence of measles in other countries.

Smallpox

Smallpox, a deadly disease, has caused untold misery throughout recorded human history (Box 29.1). The smallpox virus (variola) entered a potential victim through the respiratory tract with an incubation period of 12 to 16 days. Regional lymph nodes became infected. The virus then entered the bloodstream (viremia) and was disseminated throughout the body. The first symptoms were fever, chills, headache, and exhaustion. The virus continued multiplying in mucous membranes, in internal organs, and on the skin. Lesions occurred over the entire body and were filled with fluid. The lesions often erupted and became infected by bacteria. These lesions were a considerable concern given the unsanitary conditions that existed in much of the world.

There were two strains of the smallpox virus: (1) variola major, which produced a severe infection, and (2) variola minor, which caused a mild infection. The death rate from variola major infections approached 50%; whereas, variola minor had less than a 1% mortality rate. Infection with variola minor resulted in a lifelong immunity to both strains. In fact, during the years when smallpox was rampant, individuals exposed themselves to patients infected with the mild form in an effort to gain immunity to the more deadly strain.

The World Health Organization (WHO) initiated a worldwide program for smallpox immunization in 1966. The effort was funded by the United States, Russia, and other industrialized nations. All available humans were inoculated. Because humans are the sole host for the virus, the virus was eradicated, and the program was a great success. The last known naturally occurring case of smallpox occurred in 1977 in Somalia (Figure 29.4A; the virus is depicted in Figure 29.4B); the individual received treatment and survived. The disease was declared eradicated in 1979. There are concerns that terrorists might obtain access to remaining stocks of smallpox and spread the disease to susceptible populations (Box 29.2).

(A)

(B)

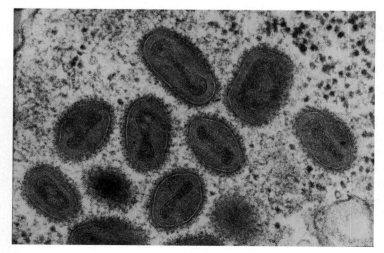

Figure 29.4 Human smallpox
(A) The last known victim of naturally acquired smallpox was Ali Maow Maalin, in Somalia in October 1977. The pustules are characteristic of smallpox. Maalin received treatment and survived. (B) Electron micrograph of the smallpox virus. A, courtesy of WHO/Centers for Disease Control and Prevention; B, ©H. Gelderblom/Visuals Unlimited.

BOX 29.1 | *Milestones*

Smallpox Epidemics in the New World

Infectious diseases have played a major role in shaping the course of human evolution. Viral and bacterial infections have destroyed armies and devastated entire civilian populations. In reality, prior to large-scale vaccination and the use of antibiotics, infectious diseases played a major role in the control of human populations, and many of these deadly diseases were caused by viruses. One such virus was **smallpox,** a constant threat to Old World inhabitants. Smallpox became a threat to New World inhabitants when explorers, fortune seekers, and immigrants brought the disease to the Americas. Because native populations had no previous exposure to smallpox, they had no resistance to the infection. As a consequence, it was exceedingly deadly to the indigenous peoples of the Americas.

The effect of smallpox epidemics on the Native American populations is well documented by historical records. Two examples are presented to illustrate this point. One was the purposeful employment of the smallpox virus against Native Americans and is a cruel story of "germ warfare" to destroy a perceived enemy. Another is the effect of an accidental exposure to the Aztecs and their subsequent inability to defend themselves against the conquistadors.

During pre-Revolutionary days, Sir Jeffery Amherst was the commander-in-chief of the British forces in North America. At the time, General Amherst was preoccupied with containing a coalition of Native American tribes that had been harassing the western frontiers of Pennsylvania, Maryland, and Virginia. This coalition was under the leadership of Chief Pontiac of the Ottawa tribe. The natives, under Chief Pontiac, had captured several forts along the western frontier in defiance of the colonists who wanted to move westward. To counter this feat, Sir Jeffery Amherst proposed the use of "germ warfare." In a 1763 letter to Colonel Henry Bouquet, commander of the western forces, he offered the following advice:

Mumps

Mumps was a common disease in children prior to the development of an effective vaccine. Mumps is caused by a **paramyxovirus** and is spread via the respiratory route. Disease symptoms occur 18 to 20 days after exposure. The mumps virus multiplies in the upper respiratory tract and progresses to local lymph nodes. The virus moves from the local lymph nodes to the bloodstream, and the resultant viremia disseminates the infection throughout the body. The obvious symptom of the disease is inflammation and swelling of the parotid glands on one or both sides of the neck (Figure 29.5). The

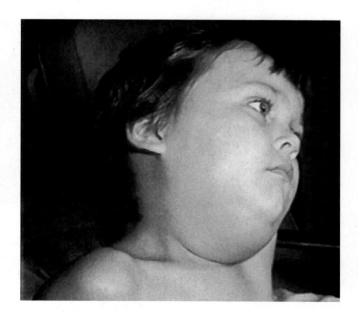

Figure 29.5 Mumps
Enlargement of the parotid gland, typically present in an individual with mumps. Courtesy of Barbara Rice/NIP/Centers for Disease Control and Prevention (CDC).

BOX 29.1   *Continued*

...could it not be contrived to send the smallpox among those disaffected tribes of Indians? We must on this occasion use every stratagem in our power to reduce them.

During July of that year, Colonel Bouquet replied to Amherst,

I will try to inoculate the _____ with some blankets that may fall in their hands, and take care not to get the disease myself.

The results of their diabolical plan are unknown, but it is a disgraceful episode in our historical dealings with Native Americans.

On another occasion, a Spanish conquistador inadvertently brought smallpox to Mexico. In June of 1520, Hernando Cortes (1485–1547) and his army marched into the Tenochtitlan, which later became Mexico City. The Aztec emperor was Montezuma, an ineffective and weak leader. Cortes arrested Montezuma and demanded that he deliver treasures to the Spaniards to ensure their safety. Montezuma subsequently collected the treasures to satisfy Cortes's demands. Due to intrigues and treachery by Cortes's men, the Aztecs rallied around Montezuma's brother, Cuitlahuac, to drive out the Spaniards. Cortes attempted to gather his loot and escape the island with his men.

As they were attempting their escape, the Aztecs attacked and blocked the Spaniards' passage to the mainland. Many Aztecs and Spaniards were killed. Unbeknownst to the Aztecs, one Spaniard killed in battle was infected by the smallpox virus. The Aztecs looted the dead, including the victim with smallpox. Within 2 weeks, smallpox infected the Aztecs and killed more than one-fourth of the native population. Because Aztecs had never been exposed to smallpox prior to arrival of the Spaniards, they had no resistance to the virus. Many in the Aztec army succumbed, including the leader, Cuitlahuac.

A few months later, Cortes returned and easily defeated the weak and demoralized Aztec army, looting the fortress city of Tenochtitlan.

Drawings by Elizabeth Perry

---

parotids are salivary glands located below and in front of the ear. Enlargement of these glands can lead to difficulty in swallowing. Infection can develop in the meninges, pancreas, ovaries, testes, and heart. Infection of the pancreas may lead to juvenile diabetes. In males who have reached puberty the viral infection can become localized in the testes, a painful condition that in some cases results in sterility. Encephalitis can also occur as an aftermath of mumps infections, but it is quite rare.

There is only one antigenic type of the mumps virus. Once the virus is contracted, immunity is apparently life long. A live, attenuated virus vaccine was developed in 1967 that has proved to be at least 95% effective. This vaccine is part of the MMR series and has reduced the number of cases in the United States to fewer than 500 per year. A mumps vaccination prior to entering school is mandatory in most states. There have been mumps outbreaks in states that do not have a mandatory vaccination program.

Influenza

Viral influenza is caused by an **orthomyxovirus**. It is an enveloped virus with an RNA genome surrounded by a matrix protein, a lipid bilayer, and glycoprotein (Figure 29.6). Influenza epidemics have occurred throughout recorded history, and **pandemics** (worldwide epidemics) have occurred with some frequency. The 1918–1919 **pandemic**, which was caused by a particularly severe and deadly strain of the influenza virus, killed at least 20 million people. The disease was termed the Spanish flu and was the H1N1 strain of influenza A. In the 1918 pandemic, the United States lost more than a half million military personnel; mostly these were otherwise healthy young men. The 1957 pandemic, which was the most recent, is referred to as the Asian flu. This pandemic was caused by the H2N2 strain of influenza A, a recombinant virus of considerable virulence. The viral strain originated in central China in February 1957 and moved to

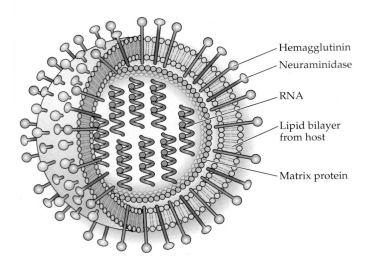

Hemagglutinin

Neuraminidase

RNA

Lipid bilayer from host

Matrix protein

Figure 29.6 Influenza virus
Diagram of the influenza virus, depicting the location of the hemagglutinin and neuraminidase involved in invasion.

Hong Kong in April. Air and naval traffic carried the infectious agent worldwide, as depicted in Figure 29.7. During 2 weeks in October of that year, a peak of 22 million cases was reported. There was also a Hong Kong flu scare in the late 1990s that moved from avians to humans. This was termed the "bird flu" and this outbreak led to the slaughter of thousands of poultry to prevent spread of the disease. This virus was the H5N1 strain of influenza A and resulted from an antigenic shift. Domestic chickens, ducks, and wild birds are potential reservoirs for this virus. There is evidence that the virus has

---

BOX 29.2 *Milestones*

History to Preserve or to Destroy?

In 1798 Edward Jenner, an English physician, developed an effective vaccination procedure to protect against smallpox (see Chapter 2). Despite this discovery, smallpox continued to kill humans throughout the nineteenth and well into the twentieth century. A worldwide immunization program initiated in 1966 led to the eradication of the disease by 1977. There has not been a documented case of smallpox in the last 24 years.

Smallpox virus survives at two known sites in the world: the Centers for Disease Control and Prevention in Atlanta, Georgia, and the Institute for Viral Preparations in Moscow, Russia. There are three major strains of the smallpox virus, and the Russian scientists were to establish a genetic map of two strains and the United States the other. All strains were to be destroyed when this task was accomplished.

Scientists are divided on the issue of destroying all of the viral strains now maintained. Some favor destruction; others favor preservation. Those in favor of preservation point out that genetic maps have not revealed some of the information that only the intact virus can reveal, and so questions remain. Among these questions— How did smallpox kill people? Can the virus yield clues about other diseases? Are there similar diseases that could potentially emerge? Has the virus been fully exploited?

The CDC maintains about 400 vials of smallpox virus and the Russians about 200 vials. These vials are maintained in liquid nitrogen and guarded closely. The fear of biological warfare has led to smallpox vaccination of the Russian, Canadian, Israeli, and U.S. armies. The rest of the world is essentially unprotected.

Should all remaining cultures of the virus be destroyed? This would achieve the goal of complete eradication of the dreaded disease. Should the cultures be maintained for scientific reasons? Once destroyed, the virus would not be available to the scientific world or to potential terrorists. The WHO in Geneva has recommended that the smallpox stocks should be eradicated, but this has not yet occurred.

During the Cold War period, from the early 1950s to the late 1970s, there was an active development of biological warfare weapons in Russia. One suspected germ warfare agent was anthrax (see Chapter 20), and there is speculation that smallpox might have also been developed as a military weapon. Since September 11, 2001, and the anthrax scare in the United States, there is understandable fear that the smallpox virus may have fallen into the hands of rogue nations. Release of this virus would be catastrophic, as vaccination against this dreaded disease ended about 30 years ago, and the level of protection available even to those vaccinated prior to the cutoff date is an unknown. Certainly there is an array of viral and bacterial agents that an amoral despot might loose onto the world. This is a reality in the twenty-first century.

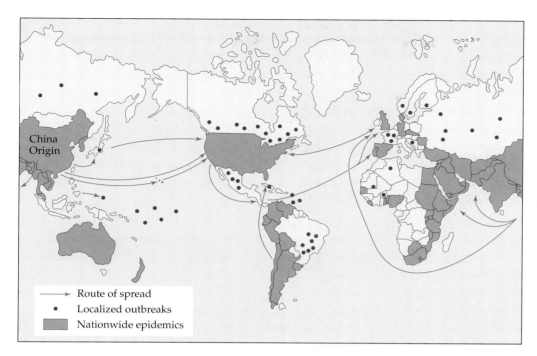

Figure 29.7 Influenza pandemic
Route followed by the influenza pandemic that originated in the interior of China in 1957 and spread throughout the world.

moved to pigs, where any reassortment of genes that occurred would potentially lead to a lethal strain.

Once the influenza virus enters human populations, it does not affect any other host. It is mainly transmitted through the air on droplets expelled by sneezing or coughing (Figure 29.8). The flu symptoms (chills, fever, headache, malaise, and general muscular aches) appear after a 1- to 3-day incubation period and last for 3 to 7 days.

Figure 29.8 A sneeze
High-speed photograph of a sneeze. ©M. R. Grapes/Michaud/Photo Researchers, Inc.

There are two major proteins on the surface of the influenza virus and these are neuraminidase and hemagglutinin. Once inhaled, the hemagglutinin attaches to sialic acid, a component of mucous membranes in the upper respiratory tract. Penetration occurs by endocytosis. The virus moves from the upper to the lower respiratory tract, destroying epithelial cells.

The destruction of epithelial cells in the respiratory system can lead to secondary infections such as bacterial pneumonia caused by *Staphylococcus aureus*, *Streptococcus pneumoniae*, or *Haemophilus influenzae*. These infections are a significant problem in infants, elderly, and otherwise debilitated individuals. Secondary infections are the leading cause of death during influenza epidemics, and prior to availability of antibiotics these deaths generally occurred from pneumonia.

The genome of the influenza virus is segmented into eight distinct fragments. When more than one virus infects a cell, there can be a reassortment of genes, which yields progeny with recombinant genomes. These progeny may then synthesize an altered surface glycoprotein that is unaffected by any antibodies that are present from a previous bout with the influenza virus. This change in the antigenic character of the viral particle is called **antigenic shift** and occurred in the virus that caused Hong Kong flu. Antigenic drift occurs when the structures of neuraminidase and hemagglutinin proteins (the N and H in the strain designations) are altered by mutation of the genes that encode them. This would change the surface properties of the virus such that antibodies to previous structures would be ineffective.

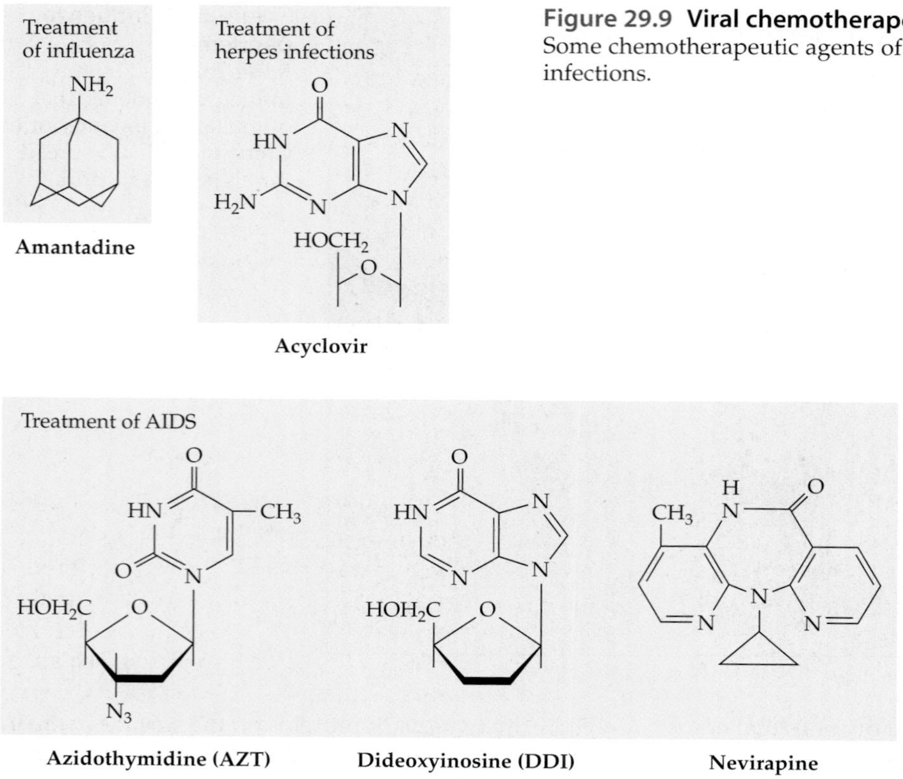

Treatment of influenza

NH$_2$

Amantadine

Treatment of herpes infections

Acyclovir

Figure 29.9 Viral chemotherapeutics
Some chemotherapeutic agents of value in preventing or treating viral infections.

Treatment of AIDS

Azidothymidine (AZT)   **Dideoxyinosine (DDI)**   **Nevirapine**

Vaccination against influenza is difficult because the antigenic shift or drift yields novel strains. Polyvalent vaccines are generally administered that are composed of a number of viral strains obtained from previous epidemics. Presently, worldwide influenza outbreaks are monitored so that the virus involved can be isolated and vaccines to new strains prepared and made available. It is the goal of these efforts to forestall a pandemic. The drug amantadine (Figure 29.9) is effective in treating influenza A if administered after exposure or early in the infection. Amantadine, given before symptoms appear, can reduce the incidence of infections by 50% to 75%. The drug blocks penetration by an uncoating of influenza virus particles. It is generally given to high-risk individuals (elderly, immunosuppressed).

Reye's syndrome is an acute encephalitis that can occur in children following some viral infections. Reye's syndrome is marked by brain swelling that results in injury to neuronal mitochondria. Liver damage can also occur. Chicken pox and influenza infections treated with salicylates, such as aspirin, increase the incidence of this syndrome. For this reason, children experiencing flu-like symptoms should not be given aspirin or related medicinals. Reye's syndrome has about a 40% fatality rate, and mental deficiency can occur in survivors.

Guillain-Barré syndrome is a delayed reaction to influenza infection that generally occurs within 8 weeks. Guillain-Barré can be a reaction to the virus or to a vac-

cine. This disease damages the cells that myelinate peripheral nerves. This demyelination results in limb weakness and sensory loss. Recovery is complete in virtually every case.

Cold Viruses

The common cold affects 40% to 45% of Americans each year and with varying severity. It is estimated that Americans miss more than 200 million work or school days annually due to symptoms of the common cold. Colds are caused by several different viruses. **Rhinoviruses** are involved in about 75% of all cases and are the most common etiological agent (Figure 29.10). Colds caused by the rhinoviruses generally occur in spring and fall.

Another group of viruses associated with the common cold are the **coronaviruses** (Figure 29.11), so named because the envelope projections resemble a solar corona. These viruses generally are involved with colds that occur in midwinter and constitute about 15% of all cases. About 10% of all colds are caused by adenoviruses and various other viruses. More cases of common cold are usually observed in winter, as people tend to congregate indoors.

The cold viruses may be transmitted in airborne droplets or by direct contact with infected individuals. Airborne transmission in volunteers has been difficult to demonstrate. The cold viruses can survive on inanimate objects for hours, and fomites (objects that harbor

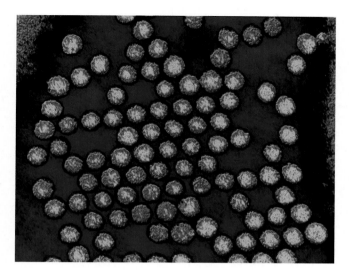

Figure 29.10 Rhinovirus, a cold virus
Electron micrograph of a rhinovirus, an etiologic agent of
the common cold. ©O. Meckes/c.o.s./Gelderblom/
Photo Researchers, Inc.

pathogens) are considered to be involved in transmission. Touching inanimate objects such as doorknobs that are contaminated with viral particles through handling by an infected person is considered a major route of transmission.

Cold symptoms include inflammation of the mucous membranes, nasal stuffiness, "runny" nose, sneezing, and a scratchy throat. The symptoms last for about 1

week, although cough-symptoms can last up to 2 weeks. Clinical symptoms such as these develop in only about one-half of infected individuals, and those without apparent symptoms are a constant source of infection to the general population. Treatment includes aspirin, antihistamines, nasal sprays, and plenty of fluids.

There are more than 110 known serotypes of the rhinovirus, effective antibodies are not generally produced, and any immunity developed is short lived. There is little hope for a successful immunization program for the common cold, since there are too many distinctly different viruses (rhinovirus, coronavirus, adenovirus, and others) that cause colds and too many serotypes.

Severe Acute Respiratory Syndrome (SARS)

SARS is a widely employed acronym for **Severe Acute Respiratory Syndrome**, a deadly viral disease that was first observed in Southeast Asia. The virus apparently originated in animal hosts that inhabited the Guangdong province in China. A strain of the virus evolved in animal species that crossed over to humans, and a surge in infections occurred in late 2002 and early 2003. During the initial stages of infection transmission rates are low between humans, but a week to 2 weeks after onset an active viremia is evident and transmission rates surge. During this stage, health care workers in contact with patients are at a considerable risk of contracting the disease. The disease is difficult to diagnose in the early stages and may progress to the actively communicable stage before safety measures are employed.

A potential source of the human strain is the Civet cat (Figure 29.12), as it has been established that these animals harbor the SARS virus. The Chinese government

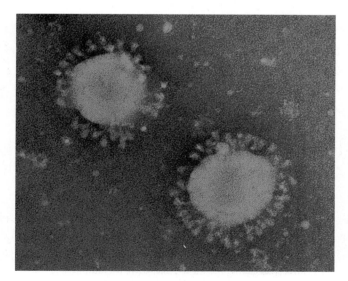

Figure 29.11 Coronavirus, a cold virus
Electron micrograph of a coronavirus, another causative
agent of human colds. Note the projections from the viral
envelope that give the virus its name. ©R. Dourmashkin/
SPL/Photo Researchers, Inc.

Figure 29.12 A Civet cat
The Civet cat, an animal eaten as a delicacy in China, is
suspected to be a reservoir for the SARS virus. ©Pete
Oxford/Nature Picture Library.

has initiated a campaign to destroy these animals, particularly in the Guangdong area. This has been difficult as the Civet cat is a delicacy and there has been considerable opposition to their destruction.

The etiological agent for SARS is a coronavirus, a single-stranded RNA virus (see Figure 29.11), which is 125 nm in diameter. The genome is composed of about 30,000 nucleotides and was sequenced within a few months after isolation of the virus. The SARS genome has little homology with other single-stranded RNA viruses, and analysis of several genes in the genome, including those encoding the replicase, surface spikes, matrix, and nucleocapsid proteins, support a mammalian origin. It is unlike any of the known coronaviruses.

The major symptom of SARS is a respiratory inflammation that often progresses to pneumonia. The mortality rate is about 10% and increases to 50% in elderly patients. There is no effective treatment. Isolation of patients to control human-to-human passage is essential, but the high level of virulence in the virus is a problem for those attending the infected. Surgical masks and general covering of the nose and mouth for those in contact with SARS patients is a must. A WHO physician and expert in communicable diseases, Carlo Urbani, who recognized the disease as a new one while working in Vietnam, became infected and succumbed to SARS.

SARS is prime example of the effects of rapid global travel on the spread of emerging infectious diseases. Controlling a disease that moves person to person respiratorily is virtually impossible. SARS is particularly a problem because diagnosis is practical only after antibodies have been generated, and this precludes quarantine before potential dissemination.

Respiratory Syncytial Virus

Respiratory Syncytial Virus (RSV) is the causative agent of a potentially dangerous infection in children under 2 years of age. It is particularly risky for premature babies, or infants with weak lungs or an impaired immune system. Contracting RSV can result in a lower respiratory tract infection, and the number of deaths that occur from the disease is significant. It is spread by hand contact and respiratory secretions. Dissemination of RSV is generally seasonal, occurring mostly from November to March. The incidence of the disease in the United States is on the increase.

RSV is a negative single-stranded RNA virus and a member of the *Paramyxoviridae* family (Figure 29.13). The virus has two major structural glycoproteins termed G and F. The G glycoprotein binds the virus to the host cell and F (fusion) brings about the attachment of the viral particles to the plasma membrane and entry of the virus into the host cell. The fusion of F to the plasma membrane

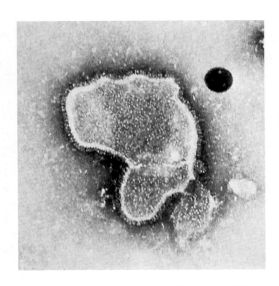

Figure 29.13 Respiratory Syncytial Virus (RSV)
RSV is a paramyxovirus that is the cause of dangerous infections in young children. Courtesy of the Centers for Disease Control and Prevention (CDC).

of a cell effects a further merging with the plasma membranes of adjacent cells. This results in a syncytium—a mass of fused cells, hence the name of the virus. An inflammatory process follows, resulting in the accumulation of fluid in the alveolar spaces. The major symptoms of an RSV infection are fever, nasal congestion, cough, and bronchitis. This may progress to an active case of viral pneumonia. Diagnosis is by clinical symptoms, and a rapid test applied to nasal swabs. Infected babies should be isolated from those that are well. Inhaled ribavirin (Virazole) can relieve symptoms.

SECTION HIGHLIGHTS

Respiratory diseases are readily transmitted from person to person via airborne droplets and are therefore difficult to control. Because these diseases tend to be very contagious, immunization programs have generally been the most effective approach to preventing infection. Immunization has been successful against some viral diseases, such as smallpox, rubella, and mumps, but less so against other diseases, such as influenza and the common cold, which are caused by viruses that frequently shift their antigenic properties or raise only a weak and short-lived immunity. In addition to their primary symptoms, many respiratory viruses can lead to life-threatening complications.

TABLE 29.2	Viral diseases with humans as reservoir and spread directly or indirectly through human contact
Disease	**Mechanism of Transmission**
Human T cell leukemia, serum hepatitis, AIDS	Contaminated blood, exchange of needles between IV drug abusers, sexual contact
Genital herpes, venereal warts	Sexual contact
Infectious mononucleosis, Herpes simplex	Mouth-to-mouth contact
Infectious hepatitis A, gastroenteritis	Fecal-to-oral route, raw shellfish
Polio	Water, food, swimming pools
Warts	Direct contact

29.2 Viral Pathogens with Human Reservoirs

There are a number of highly contagious difficult-to-control viral diseases that are transmitted directly from human to human (Table 29.2). The human serves as reservoir, and the viruses are transmitted in various ways, including through contaminated blood, by sexual contact, or intrauterine transmission to a fetus. Some of these viruses are also transmitted through kissing, touching, or via the fecal-oral route through contaminated water. Ingestion of raw or poorly cooked shellfish that has been exposed to raw sewage is a danger. Swimming in septic water can also promote the transmission of selected viruses. Common warts are caused by a virus that is passed from human to human by direct contact, not by touching frogs.

This section presents information on major viral diseases that are transmitted via human contact. These diseases are common in the United States, and these and others are prevalent in other areas of the world as well.

Acute T-Cell Lymphocytic Leukemia

There is a growing awareness that viruses can transform mammalian cells, and this can lead to sarcoma, leukemia, lymphoma, or other cancers (Table 29.3). One virus that can transform cells is the **human T-cell lymphotropic virus** (**HTLV**), a retrovirus (see Chapter 14). There are two types of HTLV: HTLV-1 and HTLV-2 (Figure 29.14).

These HTL viruses are transmitted through transfusion with contaminated blood, between intravenous (IV) drug users who exchange needles, or via sexual contact. Breast-feeding may also transmit the oncogene-bearing viruses. The HTLV-1 infects the CD4$^+$ lymphocytes (T cells) and is not cytolytic.

The leukemia virus (HTLV-1) causes cancer after long latent periods by promoting uncontrolled growth of infected cells in two ways. The virus may (1) integrate into the host genome near gene sequences that are involved in growth control and activate their expression or (2) ac-

TABLE 29.3	Some viruses known to cause tumors in humans
Virus	**Cancer**
Hepatitis B virus	Liver cancer
Human T cell lymphotrophic virus	T cell leukemia
Epstein-Barr virus	Burkitt's lymphoma Nasopharyngeal carcinoma
Papilloma virus	Cervical cancer

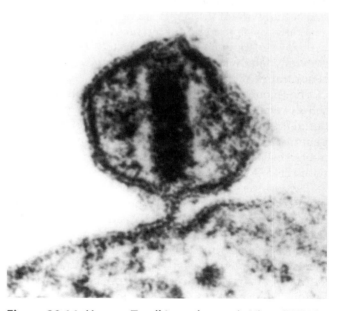

Figure 29.14 Human T-cell Lymphotropic Virus (HTLV)
HTLV transforms mammalian cells, leading to various forms of cancer. Courtesy of the National Cancer Institute.

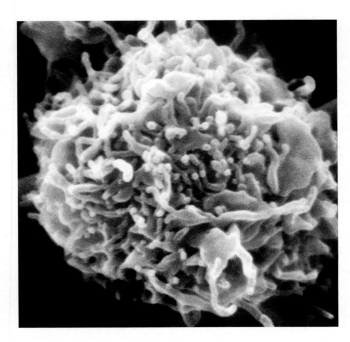

Figure 29.15 Hairy-cell leukemia
Affected lymphocytes become covered with hair-like projections of the membrane. ©Aaron Polliack/Photo Researchers, Inc.

tivate promoters that stimulate overgrowth of cells. Whichever the cause, uncontrolled cell growth can transform the lymphocytes, inducing the proliferation of a clone of aberrant T cells. This clone of HTLV growth-stimulated lymphocytes reproduces with the accumulation of the aberrant (leukemia) cells. HTLV-1 infection is generally asymptomatic but can progress to acute T-cell leukemia in about 5% of those infected. The disease is generally fatal within a year after diagnosis, and there is no treatment.

The second of these, HTLV-2, is responsible for hairy-cell leukemia. The name is derived from the membrane-derived projections that give leukocytes a "hairy" appearance (Figure 29.15). The development of this malignancy follows a pathway similar to that occurring with HTLV-1 disease.

The viruses known to cause cancer in humans may have exceedingly long latency between infection and development of the disease. This latency can be 30 years or more. Because the viruses will not transform cells in vitro, experimental investigation is difficult. The presence of the virus can be detected by the identification of virus-specific antibodies in the patient's serum. There is no cure, and prevention is by preventing transmission of the viral particle. At the present time, human blood supplies are being examined for both HTLV and HIV.

Acquired Immune Deficiency Syndrome (AIDS)

Acquired immune deficiency syndrome is caused by an RNA retrovirus, the **human immunodeficiency virus** (HIV). A brief discussion of HIV and the clinical effects of the AIDS infection follows. For a discussion of the effect of HIV on the immune system, see Chapter 27; for a discussion on transmission and development of the virus, see Chapter 14.

The HIV virus is transmitted when the body fluids (blood, saliva, semen, or vaginal secretions) of an infected individual come into contact with an open blood vessel (e.g., an open cut or a needle puncture) in an uninfected person. This passage of contaminated body fluids to the bloodstream of an uninfected individual inevitably leads to infection.

In the United States, the populations most at risk of acquiring HIV are homosexual and bisexual males, IV drug users, and heterosexuals who have sexual intercourse with IV drug users, prostitutes, or bisexual males. Transmission between heterosexuals who have multiple partners is a growing concern. HIV can also be transmitted by blood transfusions or from infected mothers to the newborn, either directly from the mother's blood to that of the fetus or via mother's milk to the infant.

More than one million individuals in the United States are HIV positive. In other countries, AIDS is not limited to discrete groups but affects men, women, and children throughout the population. Worldwide, more than 22 million people have died of AIDS since 1981, and 40 million people now alive are infected with HIV. It is estimated that each day the virus spreads to another 15,000 individuals in the world.

There is a stage in the infection, often spanning many years, during which antibodies to HIV are present in blood, but symptoms of AIDS are minimal and generally go unnoticed. The antibodies do not eliminate the virus because the viral genetic information integrates into the genome of $CD4^+$ T lymphocytes in an infected individual. During this period, diagnosis is accomplished by screening a blood sample for the presence of antibodies against HIV, and the viral infection at this point is known as ARC (AIDS-related complex). Symptoms of ARC include fever, malaise, headache, and weight loss. The symptoms generally occur within weeks of infection and last 1 to 3 weeks. Untreated, the ARC form of infection develops into full-blown AIDS within a 10-year period. It is rare for an HIV-positive individual to survive for more than 10 years without chemotherapy.

Within 5 to 6 months of an active HIV infection, the adaptive immune response in an individual is shut down. This leaves an individual vulnerable to other infectious agents (Figure 29.16; Table 29.4). The presence of increasing numbers of virus leads to the release of ad-

(A)

(B)

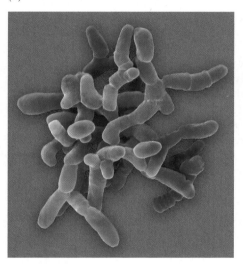

(C)

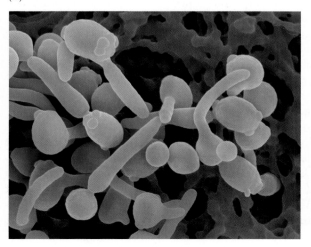

TABLE 29.4	Major opportunistic pathogens associated with active cases of AIDS
Organism	**Symptoms/Affected Organs**
Eukaryotic	
Protozoan	
Cryptosporidium	Intestinal infections
Toxoplasma gondii	Brain, central nervous system
Fungi	
Pneumocystis carinii	Pulmonary pneumocystosis
Cryptococcus neoformans	Meningitis, liver damage
Candida albicans	Oral thrush, systemic infections
Histoplasma capsulatum	Systemic infections
Coccidioides immitis	Pneumonia
Helminths	
Strongyloides stercoralis	Enteric infections
Viral	
Herpes	Ulcerative lesions
Cytomegalovirus	Pneumonia, blurred vision
Prokaryotic	
Mycobacterium tuberculosis	Tuberculosis
Mycobacterium sp.	Intestine, bone marrow
Salmonella sp.	Septicemia
Streptococcus pneumoniae	Pneumonia

ditional gp120, a protein that is a normal component of the viral envelope (see Figure 14.18). The gp120 molecules in the bloodstream are carried to the brain where they can attach to neurons, causing dementia. They may also damage other organs including the heart, kidney, and liver.

An active case of AIDS can be diagnosed by identification of an array of opportunistic secondary infections, including *Pneumocystis carinii* pneumonia. *Candida albicans* is present in the mouth and alimentary tract of healthy individuals. In the immune-compromised AIDS patient this organism can cause severe mouth infections. Another yeast infection, cryptococcosis (caused by *Cryptococcus neoformans*) can cause a damaging infection in meninges and brain. Toxoplasmosis (*Toxoplasma gondii*) and cryptosporidiosis (*Cryptosporidium*) are protozoan diseases that afflict AIDS patients. The former causes brain damage and the latter a severe form of long-last-

Figure 29.16 Causative agents of diseases in AIDS patients
Microorganisms causing the diseases associated with a fully developed case of AIDS. (A) *Toxoplasma gondii*, affecting the brain and central nervous system. (B) *Mycobacterium*, cause of intestinal and lung infections. (C) *Candida albicans*, cause of mouth and systemic infections. Photos ©Dennis Kunkel Microscopy, Inc.

ing diarrhea. Most of these secondary infections are caused by eukaryotic microbes and are difficult to bring under control. *Toxoplasma* is a protozoan; *Cryptococcus*, *Candida*, and *Pneumocystis* are fungi.

Viral infections are also a concern, including the cytomegalovirus that can cause retinal damage and blindness. The herpes simplex virus can cause chronic ulcers or bronchitis. Other viral infections can be carried to the brain tissue by macrophages, causing overproduction of nerve cells that surround the neurons and resulting in a loss of mobility and brain function.

Another symptom of AIDS is Kaposi's sarcoma, a form of cancer that causes purplish blotches on the skin of the legs and feet. In immunocompetent individuals, this cancer is not aggressive and is rarely fatal. In AIDS patients the cancer can spread to the lungs, lymph nodes, and brain. Other malignancies common in AIDS are Burkitt's lymphoma, primary brain lymphoma, and immunoblastic lymphoma. Non-Hodgkin lymphoma and other malignant lesions that are normally removed by immune host defenses can occur in the HIV-infected.

At the present time, treatment of AIDS is limited to efforts to delay symptoms; there is no cure for the disease once it is contracted. The drugs now available are of three types:

- Nucleoside analogs that interfere with the transcription of the single-stranded RNA; genetic information that otherwise leads to viral double-stranded DNA

- Reverse-transcriptase inhibitors that are not nucleoside analogs

- Protease inhibitors that prevent processing of viral polypeptides

The first **nucleoside analog** that was an effective inhibitor of transcription was azidothymidine or AZT (see Figure 29.9). Dideoxyinosine (see Figure 29.9) has also been employed but is less effective. AZT resembles thymidine and once incorporated into viral DNA by reverse transcriptase the analog prevents attachment of the next base. Inability to attach a base to the nucleotide chain results in **chain termination**. Unfortunately, AZT becomes ineffective rapidly in a patient because the high mutation rate results in resistance to the drug. A limited number of mutations are sufficient to render HIV resistant to many of the nucleoside analogs. Other nucleoside analogs that have been employed are tenofovir and lamivudine.

Reverse-transcriptase inhibitors are drugs designed to alter the conformation of the catalytic site on reverse transcriptase. Nevirapine (see Figure 29.9) is a widely administered reverse-transcriptase inhibitor. Even fewer mutations in the viral genome are necessary to overcome the nonnucleoside reverse-transcriptase inhibitors. Nevirapine is given to pregnant HIV-positive women during labor, since it reduces the transfer of HIV to the neonate in about 50% of the cases.

The **protease inhibitors** are computer-designed peptide analogs that bind to the active site on the protease in HIV that processes the viral polypeptides. This prevents virus maturation. Among the inhibitors are saquinavir and indinavir. HIV is able to generate mutants that retain virulence and are unaffected by the protease inhibitors.

A new class of drugs that has considerable promise is the **integrase inhibitors**. These drugs inhibit HIV replication by blocking the ability of the virus to patch its DNA onto that of a host cell. Combination of integrase inhibitors with tenofovir and lamivudine has been effective in both new patients and those receiving drug therapy for an extended period.

The standard treatment for AIDS is the administration of at least three drugs at the same time. This is termed HAART (highly active antiretroviral therapy). Under such a multidrug protocol, the virus would need to develop several types of resistance simultaneously. This is unlikely even in a virus as mutable as HIV. HAART usually includes a protease inhibitor combined with nucleoside analogs and/or a reverse-transcriptase inhibitor. The patient is monitored periodically for viral load by the sensitive reverse transcriptase–polymerase chain reaction (RT-PCR) assay. Changes in the viral load indicate that the combination of drugs used in treating the patient should be adjusted.

There is a concerted effort underway to prevent HIV infections by immunizing healthy individuals. In addition, chemotherapeutics are being developed that would combat HIV and prevent it from progressing to full-blown AIDS.

Indeed, parts of the HIV retrovirus appear to be vulnerable, including surface glycoproteins (gp120) that could be incorporated into a vaccine. Antibodies formed against these envelope proteins could prevent the CD4$^+$ glycoprotein interaction and prevent infection. Unfortunately, the gene encoding gp120 is very mutable, so an antibody against one form of the glycoprotein may not recognize another.

Vaccines that induce antibody against other envelope proteins have been more promising. The genes for the envelope proteins are being incorporated into harmless viruses, where they are expressed and serve as a vehicle for the HIV envelope protein antigen. Several of these antigen-producing systems have been developed and are in clinical trial.

The detection of anti-HIV antibodies in blood is an effective diagnostic procedure. It can be applied to blood from individuals or for blood donated by volunteers.

The immunological procedure that has proved effective for large-scale screening of blood or in diagnosis of potential victims is the ELISA (enzyme-linked immunosorbent assay) (see Chapter 30). However, a positive test in any individual would be confirmed by a highly specific immunoblot procedure.

Development of rapid diagnostic tests to screen individuals for HIV infection is underway. Generally these tests are rapid and simple to apply but are neither as sensitive nor as accurate as tests that are more time-consuming.

Cold Sores (Fever Blisters)

Herpes simplex virus type 1 (HSV-1) is the causative agent of **cold sores** or **fever blisters**. This virus is a member of the **herpesvirus**, a group that includes common animal pathogens. The herpes viruses are noted for their ability to cause latent infections that may be activated days or years after the initial infection. Zoster, the etiological agent of shingles, is one such virus and was discussed previously (see Chicken Pox). Herpes simplex is a double-stranded DNA virus with a genome that is surrounded by an icosahedral nucleocapsid. The capsid is enclosed in a membrane envelope with glycoproteins projecting from this envelope.

Active HSV-1 infection results in blisters around the mouth, lips, and face (**Figure 29.17**). The blister results from host- and virus-mediated tissue destruction. The lesions heal in 1 to 3 weeks without treatment, although the virus is normally not eliminated from the host. It is estimated that up to 90% of adults in the United States possess antibodies against HSV-1.

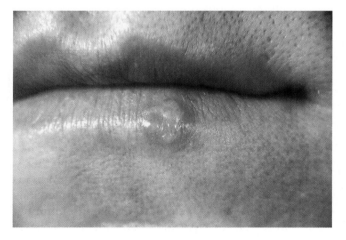

Figure 29.17 Cold sores
Cold sores, also known as fever blisters, are caused by Herpes simplex type 1. Courtesy of Dr. Herrmann/Centers for Disease Control and Prevention.

The incubation period is 3 to 5 days. After a primary infection the virus moves to trigeminal nerve ganglia and remains in the dormant state. An active viral infection can occur sporadically throughout the life of an infected individual. Clinical symptoms are triggered by stresses such as excessive sunlight, emotional upset, fever, trauma, immune suppression, or hormonal changes. When the latent virus is activated, it travels down peripheral nerves to the lips, mouth, or facial epidermis, resulting in a recurrence of the fever blister(s).

Acyclovir (see Figure 29.9), a guanine analog, is effective in treating herpes infections. When acyclovir is phosphorylated in cells, it resembles deoxyguanine triphosphate and inhibits the viral DNA polymerase. Acyclovir would be selectively activated by the thymidine kinase in an infected cell, as the virus encodes this enzyme. This would not occur in a normal uninfected cell. Dormant viruses would not be affected by acyclovir, and consequently it is not a cure.

Genital Herpes

Genital herpes is generally caused by herpes simplex virus type 2 (HSV-2), with about 10% of clinical cases the result of HSV-1 infection. Genital herpes is associated with the anogenital region and is transmitted by direct sexual contact. A primary infection occurs after an incubation period of 1 week, but many primary infections are asymptomatic. The virus causes painful blisters on the penis of a male. In females, the lesions appear on the cervix, vulva, and vagina. The blisters are a result of cell lysis and a local inflammatory response. They contain fluid and infectious viral particles. During this stage, the virus is readily transmitted to unprotected sex partners. The blisters resulting from a primary infection heal spontaneously in 2 to 4 weeks. Recurrent genital herpes infections are generally shorter and less severe than the initial episode. These lesions that occur after primary infection heal in 7 to 8 days. As with HSV-1, the HSV-2 virus travels to nerve cells and remains dormant. Genital herpes can reoccur every few weeks, a few times per year, or not at all. An active case can be activated by a fever. An estimated 50% of those infected with HSV-2 never have symptoms and are unaware of a primary infection. These individuals may shed the virus and can be significant transmitters of the disease.

The virus can be transmitted to an infant during birth, and more than 2,000 babies are born in the United States annually with HSV-2. The disease in the neonate ranges from a latent, inapparent infection to brain damage and death. To avoid infecting the newborn, cesarean birth is advised if there are virus-infected lesions in the genital area. Unfortunately, the virus can be present in genital secretions without overt symptoms, thereby exposing a vaginally delivered infant.

There is no known cure for herpes virus infections. Oral and topical administration of acyclovir can lessen the symptoms of the disease and can prevent recurrences of HSV if taken prophylactically. Genital herpes has become a significant sexually transmitted disease, with one in five teens infected. According to American Social Health statistics, 50 million Americans have genital herpes, and there are about 1 million newly diagnosed cases each year.

Infectious Mononucleosis

Infectious mononucleosis is caused by a human herpes virus called **Epstein–Barr virus (EBV)**. The virus is named for its codiscoverers and was originally isolated when T. Epstein and Y. Barr were seeking the etiologic agent for a malignancy of lymph nodes called Burkitt's lymphoma.

The EBV can be present in the mouth and throat area and is commonly called "kissing disease" because it occurs most frequently among those of college age. The symptoms include enlarged lymph nodes and spleen, sore throat, nausea, and general malaise. These symptoms may be accompanied by a mild fever. Fatigue is a common complaint of mononucleosis patients. The symptoms last for 1 to 6 weeks and rest is the only treatment.

Once contracted, EBV enters lymphatics where the virus multiplies and infects **B lymphocytes (B cells)**. The B cells proliferate rapidly and are altered in appearance. These atypical lymphocytes are the earliest indication of a mononucleosis infection. The presence of antibody against EBV is another, more reliable, indicator of the disease.

Burkitt's lymphoma is an EBV-associated malignancy that affects children in Uganda and other central African countries. EBV virus has also been implicated in a nasopharyngeal carcinoma that occurs more frequently among the Chinese than people in the Western Hemisphere. EBV is associated with Burkitt's lymphoma but is not the causative agent. Geographic areas where the malignancy occurs have a high incidence of malaria, a disease known to suppress the immune system. Suppression of the immune system might permit EBV to proliferate in B cells. Individuals that have AIDS, and consequently are immunosuppressed, are often victims of EBV infection.

Poliomyelitis

Poliomyelitis has virtually been eliminated in the United States by an intensive vaccination program. Efforts are being made by WHO to eradicate the disease worldwide, but it continues to be a concern in developing countries.

Poliomyelitis, also known as **polio** or **infantile paralysis**, is caused by a picornavirus that may survive in food or water for lengthy periods. The virus is transmitted by ingestion of contaminated food or water. Public swimming pool water was considered a source of polio infection when the disease was of widespread occurrence in the United States (prior to vaccine development, polio affected 15 of every 1,000 in the United States annually) (Box 29.3).

The incubation period after exposure is generally 7 to 14 days. The poliovirus can multiply in the oropharynx and may be transmitted through respiratory droplets during the early stages of infection. Multiplication occurs in tonsils and intestinal mucosa, and the virus moves from the intestinal tract to lymph nodes. It then invades through the bloodstream and is disseminated throughout the body.

There are three forms of clinical illness that can result from invasion by the poliovirus: abortive infection, aseptic infection, or paralytic polio. The most common form is an **abortive infection**. This infection is asymptomatic or causes some fever, sore throat, nausea, and vomiting. Recovery is rapid, and the symptoms are rarely attributed to the poliovirus. **Aseptic** infection results in similar symptoms, except that some of the virus enters the central nervous system, resulting in a stiff neck and back. In these infections, full recovery occurs in about one week. **Paralytic polio** is marked by destruction of motor neurons in the anterior horn of the spinal cord. Because these cells transmit impulses to the motor fibers of the peripheral nerves, their destruction results in paralysis. If the infection involves neurons in the medulla, bulbar poliomyelitis or loss of respiratory function results. Destruction of other neurons can result in loss of motor function and muscle paralysis.

Some unfortunate polio victims lost the ability to breathe and were forced to spend their lives in an iron lung (Figure 29.18). Many Americans are familiar with pictures of Franklin D. Roosevelt, the 32nd President of the United States, confined to a wheel chair as he contracted polio as an adult.

The Salk vaccine for polio was introduced in 1954 and was an inactivated virus administered by injection. Albert Sabin developed an attenuated viral vaccine in 1962 that is given orally. The oral vaccine is more effective than the inactivated virus because it induces an infectious cycle that establishes antibody protection in the intestinal mucosa. As a consequence, vaccinated individuals apparently do not serve as a reservoir of infectious particles. The Salk vaccine is administered to immune-compromised persons to avoid opportunistic infections.

It is now feared that the poliovirus might become established among immigrants and poor children in the inner city because many of these youngsters are not properly immunized. Because we depend on herd im-

BOX 29.3 | *Milestones*

Conquering Polio

During the first half of the twentieth century, paralytic polio was among the most dreaded diseases in the industrialized world. It struck randomly in human populations, and the epidemiology was poorly understood. We now know that improved sanitation conditions in the more developed countries were actually a significant factor in the spread of paralytic polio. The virus normally survived in the intestinal tracts of infants and conferred immunity. Exposure at a very early age was the rule in populations where sanitation was less favorable. If a child was not exposed at an early age but randomly contracted the disease at an older age, paralysis was more likely. Consequently, the chance of contracting paralytic polio increased in populations with an improved standard of living. During the period when polio was rampant, the March of Dimes was organized. This was an organization that collected dimes to support research on this feared disease.

The development of viral vaccines was stymied by the inability to obtain quantities of viral particles. Until the middle of the twentieth century, there was no available technique for propagating virus outside a living host. It was then (1946) that John Enders, Thomas Weller, and Frederich Robbins began a collaboration at the Infectious Disease Laboratory at Boston Children's Hospital. Together they developed techniques for propagating poliomyelitis virus in tissue culture. This was the essential step in the development of the polio vaccine and a major breakthrough in the study of viruses in the laboratory. These three scientists worked together from 1946 to 1952, and the tissue culture methods they developed could be used for selection of strains with altered pathogenicity—a prerequisite for developing effective vaccines for any viral disease. Polio, measles, and mumps vaccines resulted from the techniques developed in John Enders's laboratory.

Dr. John Enders. Courtesy of the National Library of Medicine.

Enders, Weller, and Robbins received the 1954 Nobel Prize for Medicine for their contributions. Although the polio vaccine has been named for the developers, Salk and Sabin, the Nobel went to the Enders group because they made the conceptual contribution that led to a successful vaccine.

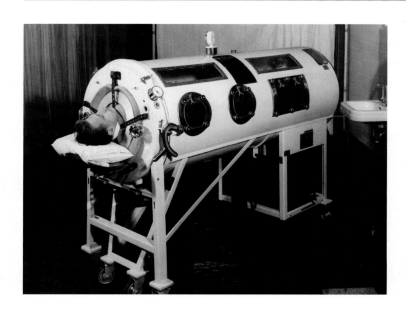

Figure 29.18 An iron lung
Iron lungs were used to maintain breathing of paralyzed polio victims in the 1940s and 1950s.
©SuperStock/Jupiter Images.

munity in most immunization programs, it is important that all individuals be immunized. No vaccine is effective in 100% of the individuals inoculated, and the goal is to have 90% to 95% of the potential victims immune and thus remove reservoirs. Without sufficient reservoirs, the disease will disappear. In the absence of this herd immunity, a new outbreak of polio could occur and have decidedly disastrous effects on those that did not gain immunity by vaccination.

Hepatitis

The word **hepatitis** means inflammation of the liver, and five distinct viruses have been implicated in hepatitis infections. It is probable that other viral agents of hepatitis are yet to be discovered (Table 29.5). These diseases are generally transmitted by exchange of blood or body fluids or via the oral–fecal route.

Hepatitis A (infectious hepatitis) infections are acquired by ingesting raw oysters, clams, or mussels harvested from fecally contaminated water or handled by an infected carrier. Person-to-person transmission can occur in daycare centers and other institutions in which sanitary conditions are difficult to maintain. The infectious hepatitis virus is exceedingly stable and can tolerate heating at 56°C for 30 minutes. It is also unaffected by many disinfectants. An estimated 40% of all acute cases of hepatitis are caused by hepatitis A.

Hepatitis A has an incubation period of 15 to 50 days. The virus multiplies in the intestinal epithelia and moves via the bloodstream to the liver. It generally causes a mild, self-limiting infection. In severe infections, nausea, vomiting, and fever are the symptoms. Hepatitis A causes a degenerative inflammation of the liver that can lead to liver enlargement and possible blockage of biliary excretions. This blockage leads to jaundice (release of bile pigments into the bloodstream giving a yellowish color to the skin). Damage to the liver is immune-mediated in that antibodies to the virus attack certain hepatic cells.

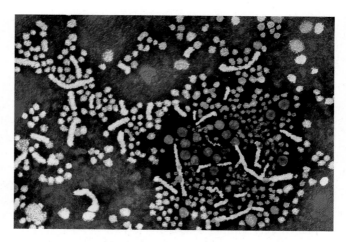

Figure 29.19 Hepatitis B virus particles
TEM of hepatitis B virus, showing tubular forms, Dane particles, and spherical particles. ©CDC/Peter Arnold, Inc.

Recovery takes about 12 weeks, and permanent liver damage is possible. Infections in children are often asymptomatic. Hepatitis A virus appears in the feces of infected individuals 10 days before onset of symptoms and for several weeks after recovery. An active case results in immunity, and about 40% of the people in the United States have antibodies to infectious hepatitis. Because symptoms can be more severe in adults than in children, a prophylactic treatment (immunoglobulin) is available to individuals exposed to the virus. A vaccine is now available for at-risk individuals.

Hepatitis B virus (HBV) (serum hepatitis) is a more serious threat to human populations than is hepatitis A (Figure 29.19). An estimated 300 million individuals worldwide are carriers of the virus. Of these, 40% will probably succumb to liver disease. There are about 300,000 humans in the United States infected each year with HBV, with about 5,000 deaths annually.

Individuals infected with hepatitis B have three different antigenic particles present in their serum: Dane

TABLE 29.5	Characteristics of hepatitis viruses				
Hepatitis Virus	**Common Name**	**Virus Type**	**Envelope**	**Transmission**	**Incubation Period (Days)**
Hepatitis A	Infectious	Picornavirus (RNA)	No	Fecal–oral	15–50
Hepatitis B	Serum	Hepadnavirus (DNA)	Yes	Blood, sexual	45–160
Hepatitis C	Post-transfusion non-A, non-B	Flavivirus (RNA)	Yes	Blood, sexual	14–180
Hepatitis D	Delta	Undefined (RNA)	Yes	Parenteral, sexual	15–64
Hepatitis E	Enteric non-A, non-B	Calicivirus (RNA)	No	Fecal–oral	15–50

particles, spherical particles, and tubular particles. The Dane particle is the infective form of the virus. The two incomplete particles (spherical and tubular) have the hepatitis B surface antigen as part of the particle and are present in greater quantity than is the infective Dane particle. The presence of these incomplete antigenic particles in serum is an effective indicator of an HBV infection. Their presence is also important in screening for hepatitis B in blood that might be used in transfusions.

The hepatitis B virus is transmitted by needle sharing, acupuncture, tattooing, and potentially by body or ear piercing. The virus can also be transferred during blood transfusions and in semen and is present in saliva and sweat. The virus traverses the placenta, and infants born to infected mothers generally become carriers.

The incubation period for HBV virus varies from 45 to 160 days. The first symptoms include fever, anorexia, and general malaise. This is followed by nausea, vomiting, abdominal pain, and chills. Hepatitis B viral infections can cause extensive liver degeneration resulting in jaundice and release of liver enzymes into the bloodstream. HBV is also associated with primary hepatocellular carcinoma, and this disease accounts for many of the deaths that occur worldwide.

There is no cure for HBV infections, but vaccines for prevention of the disease have been developed. These vaccines are recommended for administration to infants at birth with follow-up doses at 2, 4, and 6 months of age. Hospital workers and others who work in high-risk situations might also benefit from vaccination. Passive immunity with HBV immunoglobulin can be of value in individuals exposed to HBV, but it must be administered within a week of exposure.

The **hepatitis C** virus was formerly called non-A, non-B because all known cases not caused by hepatitis A or B were believed to result from infection by this virus. Hepatitis C virus is a concern in blood transfusion because 5% to 10% of transfusion recipients are victims of this form of hepatitis. Intravenous drug users who share needles also have a high incidence of hepatitis C infections. Generally the epidemiology of hepatitis C is similar to that of hepatitis B virus, and the symptoms are equivalent to those of hepatitis A or B but generally milder. Chronic hepatitis C can lead to cirrhosis of the liver. Prevention is by careful screening of blood used in transfusions.

The **hepatitis D** virus is a defective satellite virus that was formerly known as the **delta agent**. Hepatitis D virus infections occur only in individuals infected with the hepatitis B virus. Hepatitis D virus can be cotransmitted with HBV or acquired after HBV infection. HBV must be actively replicating for the hepatitis D virus to generate complete virions. A coinfection with hepatitis D and HBV leads to more severe liver damage and a higher mortality rate than occurs with an HBV infection alone. The coinfection occurs mostly in high-risk individuals, such as intravenous drug abusers. The vaccine for HBV can protect against this disease.

The **hepatitis E** virus is a problem in developing countries, and the symptoms and course of infection are similar to that for hepatitis A virus. However, the mortality rate for hepatitis E is about 10 times higher than that for hepatitis A viral infections. The mortality rate in pregnant women infected with hepatitis E virus is about 20%, and the reason for this high rate of mortality is unclear.

Warts

Papilloma viruses cause a variety of skin warts and epithelial lesions in humans. Most common warts are benign and disappear or are readily removed by minor surgery or a chemical agent. There are 100 different strains of human papilloma virus (HPV), and 15 of these cause genital warts. Genital warts can cause cervical carcinoma, and two strains (strains 16 and 18) account for 95% of these cancers. Cervical cancer kills about 4,000 women in the United States each year. Transmission of genital warts among human populations is a growing concern. It is estimated that between 10 and 30 million individuals in the United States may be infected with genital papilloma virus or carry the virus in a latent state. The number of new cases each year is estimated to be in excess of 6 million with a high prevalence among teenagers. By the age of 50 about 80% of all women have been infected. A PAP smear is used to detect this virus in sexually active women as it can diagnose early signs of cervical cancer. A vaccine to elicit immunity to HPV has been approved, and it has been suggested that the vaccine be administered to all females prior to puberty.

The human papilloma viruses are nonenveloped, double-stranded DNA viruses. The nucleocapsid is composed of protein. Warts are transmitted by direct contact and occur mainly on skin and mucous membranes. They are not highly contagious. Virtually everyone has had or seen a wart. They are spread by scratching and most often occur in children or young adults. The projections from the skin are a result of skin cell proliferation and/or a longer lifespan for infected skin cells, which then accumulate. Common warts can be treated by applying dry ice or liquid nitrogen to the wart or destroying it with a laser beam. Application of podophyllin over an extended time is effective in removing most warts. Podophyllin is a resin extracted from the dried roots of the mayapple.

Viral Gastroenteritis

Several virus types are the causative agents of **viral enteritis**, an inflammation of the stomach or intestines. **Rotaviruses**, **caliciviruses**, and **astroviruses** are among the viruses involved and are transmitted via the fecal–oral route. These diarrheal diseases are the leading cause of childhood deaths in developing countries. Rotaviruses are the cause of more than 4 million cases of infectious diarrhea in the United States annually, with several hundred deaths. The Norwalk virus has caused outbreaks of gastroenteritis, especially on cruise ships.

SECTION HIGHLIGHTS

A number of highly contagious viruses are transmitted directly from human to human, through, for example, contaminated blood, sexual contact, kissing, touching, or ingestion of contaminated water or food. Among these diseases are acute T-cell lymphocytic leukemia, AIDS, cold sores, mononucleosis, polio, hepatitis, warts, and gastroenteritis. Such infections cannot be cured, but in some cases the disease symptoms can be lessened by drugs that target viral-specific processes or enzymes within infected cells. There are also effective vaccines for some of these diseases (e.g., polio and hepatitis A and B).

29.3 Viral Diseases with Nonhuman Reservoirs

There are a considerable number of viruses that infect humans that reside primarily in nonhuman reservoirs. Most are vector-borne and transmitted by the bite of an arthropod, such as a tick or mosquito. The arthropod carrier that bites an infected primary reservoir animal and then bites a human host transfers the infectious virus in the process. Some viruses that reside in nonhuman reservoirs are contracted directly through animal bites, dust, or skin abrasions. A list of natural hosts and vectors is presented in Table 29.6. It is evident from the number and diversity of primary hosts involved that prevention of these diseases through the control of the reservoir would be virtually impossible. Reservoirs such as squirrels, rodents, and birds cannot and should not be exterminated. Controlling the infectious agents in a wild animal population is highly unlikely. Ridding the environment of mosquitoes is a worthy goal but not attainable. The practical solution to this problem would be vaccination of domesticated animals and/or human hosts. Unfortunately, effective vaccines are not available for many of these infectious diseases.

This section discusses some of the many viral infections that are transmitted to humans from nonhuman reservoirs.

Colorado Tick Fever

Colorado tick fever is one of the more common tick-borne diseases in the United States. The disease is caused by a reovirus (orbivirus), and the vector is *Dermacentor andersoni*. Colorado tick fever affects several hundred people each year, mostly in the western and northwestern United States and in western Canada. The number of cases is considered much higher, but most go unreported.

Symptoms include chills, headache, muscle aches, and fever. The symptoms appear after a 3- to 6-day incubation period and last for 7 to 10 days. The virus infects red blood cells and may persist for 20 weeks after symptoms subside. Fatalities are rare but can occur in young children. The virus apparently traverses the placenta in pregnant females and can cause abnormalities in or death of a fetus. A viral vaccine has been developed but is impractical for general use.

TABLE 29.6	Viral Infections with nonhuman primary hosts and transmitted to humans by vectors		
Disease	**Viral Agent**	**Natural Host(s)**	**Vector**
Colorado tick fever	Arbovirus	Squirrels, chipmunks	Tick (*Dermacentor andersoni*)
Encephalitis			
California	Arbovirus	Rats, squirrels, horses, deer	Mosquitoes (*Aedes* sp.)
St. Louis	Arbovirus	Birds, cattle, horses	Mosquitoes (*Culex* sp.)
Eastern equine	Arbovirus	Birds, fowl	Mosquitoes (*Aedes* sp.)
Venezuelan	Arbovirus	Rodents	Mosquitoes (*Aedes* sp., *Culex* sp.)
Western equine	Arbovirus	Birds, snakes	Mosquitoes (*Culex* sp.)
West Nile	Arbovirus	Birds	Mosquitoes (*Culex* sp.)
Lymphocytic choriomeningitis	Arbovirus	Mice, rats, dogs	Dust, food
Rabies	Rhabdovirus	Skunks, raccoons, bats	Domestic dogs and cats
Yellow fever	Togavirus	Monkeys, lemurs	Mosquitoes (*Aedes* sp.)

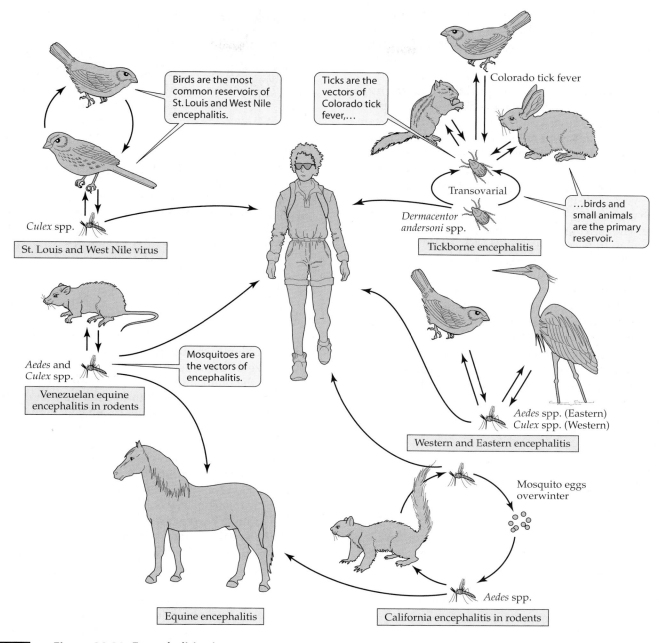

Figure 29.20 Encephalitic viruses
Reservoirs and vectors of viruses that are the etiologic agents of human encephalitis infections. The vectors are *Aedes* or *Culex* mosquitoes and the tick *Dermacentor andersoni*.

Small animals, including rodents, are the primary reservoir for Colorado tick fever (Figure 29.20). The virus is passed transovarially, and the tick can be both reservoir and vector. It is a reservoir because it passes the virus to other ticks, and it is a vector when it passes the virus to humans. The disease occurs most often among campers in tick-infested areas. Prevention is by avoiding areas where ticks might be present and by using tick repellants.

Encephalitis

The term **encephalitis** describes several diseases that are characterized by inflammation of the brain and can lead to neurological damage or fatalities. The major etiological agents of encephalitis in the United States are **arboviruses** that are transmitted by arthropod vectors. Transmission of the virus to humans is illustrated in Figure 29.20.

St. Louis and West Nile encephalitis are caused by flaviviruses, and epidemics of these diseases occur sporadically in the southern and southeastern United States. Birds are the most common reservoirs. Western, Eastern, and Venezuelan equine encephalitis are caused by an **alphavirus** and are transmitted to humans and horses by mosquitoes, although the horse is not an important reservoir.

Symptoms of infection occur 3 to 7 days after exposure. The first bout of infection results in a systemic infection with fever, chills, headache, and flu-like symptoms. This is followed by a second phase where the virus moves to the central nervous system and multiplies in the brain. Eastern equine encephalitis can be a severe infection in humans and has a high mortality rate (greater than 50%). Neurological disorders are a potential consequence for survivors of the disease.

The elimination of breeding sites for vectors might be an effective control measure for these diseases. Vaccines are available for individuals working with the Eastern and Western equine encephalitis virus. A live vaccine against the Venezuelan equine virus is available for use in domestic animals.

Rabies

The causative agent of **rabies** is a bullet-shaped virus of the **rhabdovirus** group (see Table 14.3). The virus is maintained in nature in carnivorous animals. Skunks and raccoons are the principal reservoir, with significant occurrences in bat and fox populations (Table 29.7). Feral dogs and cats or wild animals carry the disease to domestic animals. Humans are generally exposed through bites or via contact with diseased vectors. The rabies virus is present in the saliva of rabid animals and consequently aerosols in bat caves, where rabid bats may nest, are a potential source of human infection.

When transmitted in aerosols, the virus infects the epithelial tissue that lines the upper respiratory system. Generally, rabies is passed among animal reservoirs through a bite or exposure to virus-laden saliva. Although a young skunk infected with rabies will eventually succumb to the disease, it may survive to reproductive age. The progeny then become infected and pass the disease along to their offspring. Foxes do not survive long after a rabies infection, and though not an important reservoir, they tend to be aggressive and will attack humans.

Rabies is transmitted to humans through aerosols or entry of virus-laden saliva into an open wound, and this

TABLE 29.7	Major sources of the rabies virus
Animal Group	**Percent Carriers**
Wild Animals	
Skunk	43.0%
Raccoon	27.7%
Bat	13.3%
Fox	2.5%
Others	1.7%
Domestic Animals	
Dog	3.6%
Cat	3.5%
Cattle	3.7%
Others	1.0%

can occur by a bite or during handling of an infected animal. The virus multiplies in tissue at the site of inoculation and can remain localized for days or months. The length of time required for the virus to move from the site of infection to the brain depends on several factors: (1) the size of the inoculum, (2) the proximity of the wound to the brain, and (3) the host's age and immune status. The rabies virus travels from the wound site to the peripheral nerve system, farther to dorsal ganglia, and on through the spinal cord to the brain. Infection occurs in the spinal cord, brainstem, and cerebellum. The virus can move from the brain to the eye, salivary glands, and other organs. Viral replication in the brain results in the presence of cytoplasmic inclusions in the affected neurons. These inclusions are called **Negri bodies** (Figure 29.21) and have been a target of a principal diagnostic test for the disease in animals. However, diagnosis is now based on a fluorescent antibody test on nerve tissue.

Symptoms of rabies do not appear until the virus reaches the brain. The initial symptoms include fever, malaise, headache, gastrointestinal upset, and anorexia. After 2 to 10 days, neurological symptoms associated with rabies appear. **Hydrophobia** occurs in 25% to 50% of patients and is characterized by jerky contractions of the diaphragm when the victim attempts to swallow water. Generalized seizures and hallucinations follow, ending in coma and death.

Humans bitten or otherwise exposed to **potential rabies** carriers are given postexposure prophylactic treatment. These treatments are initiated unless it can be confirmed that the suspected animal involved **does not** have a rabies infection. The victim is immunized with a vaccine plus a dose of equine antirabies serum or human rabies immunoglobulin. This passive immunization provides protective antibodies until antibodies are produced in response to the vaccine.

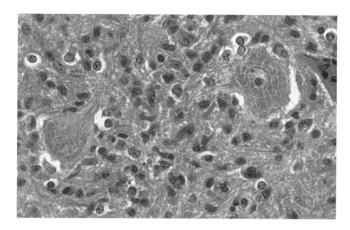

Figure 29.21 Negri bodies
Rabies victims develop cytoplasmic inclusion bodies in certain nerve cells in their brains. Courtesy of the Centers for Disease Control and Prevention.

The control of human rabies depends on the effective elimination of the disease in the wild animal populations that serve as reservoirs. Domestic dogs and cats are routinely vaccinated against the disease. Vaccination of wild animals by capture/release is out of the question. Administration of an oral vaccine by placing it on bait is one possible mechanism for widespread inoculation of wild animal populations.

Yellow Fever

Yellow fever, caused by an **arbovirus**, was the first human infection shown to be caused by a virus (1901). It is also the first viral infection demonstrated to be vector-borne. Yellow fever is transmitted by a mosquito, *Aedes aegypti*. The last epidemic in the United States occurred in 1878 and killed more than 13,000 people. There has not been a reported case in the United States in the last 70 years, but the potential for reintroduction is real because the vector is present in the southern United States. The reservoir for the disease in Central and South America is the monkey. There are 30,000 to 40,000 deaths worldwide each year from this disease.

The yellow fever virus enters the human body through a mosquito bite and is carried by the lymph to local lymph nodes. The virus multiplies in lymph nodes and the spleen and results in a systemic infection. Damage to liver, kidneys, and heart and hemorrhaging of bloody vessels follows the systemic infection. The name *yellow fever* originated because the liver damage results in jaundice. Fever, chills, headache, and nausea are also symptoms of the disease.

Where yellow fever is endemic, it is spread by distinct epidemiological patterns that are either urban or sylvatic. The urban cycle involves human-to-human transmission, with the mosquito serving as vector. In the sylvatic cycle, mosquitoes transmit the virus between monkeys and occasionally to humans.

Elimination of breeding grounds for the vector is an important and effective control measure. A vaccine is available that, when administered intradermally, leads to lifetime immunity.

Bird Flu

An emerging threat to humans worldwide is a newly evolved strain of the influenza virus. It arose when different strains of the flu virus infected, incubated in, and reassorted genetic information in domestic poultry and wild fowl. The bird flu strain, designated **H5N1**, is genetically distinct from the Hong Kong flu of 1997. The proximity of humans to poultry in Southeast Asia may have been a contributing factor in the spread of this influenza strain to humans. Although it appears, at this time, that the virus is not spread from human to human, there is no assurance that such strains will not evolve. A further concern is the transmission of the virus by migrant bird populations. The flyways for bird populations overlap over the globe and a potential bird flu pandemic is a frightening prospect.

There have been confirmed cases of the disease in Asia and the Middle East with a 50% mortality rate. All cases, to date, appear to have been transmitted through contact with fowl. Huge numbers of domestic fowl have been destroyed to prevent spread of the virus.

Ebola and Marburg

Outbreaks of **hemorrhagic fevers** have occurred in recent years in Central and West Africa. Two viruses that are known to be causative agents of these diseases are termed **Ebola** and **Marburg** (Figure 29.22). Both are single-stranded RNA, enveloped viruses (*Filoviridae*). The viruses are passed among wild vertebrates, including monkeys, and these are the apparent reservoir. Arthropods are believed to be the vector that carries the viruses animal to animal. Transmission mechanisms for these viruses are uncertain at the present time.

There was an outbreak of hemorrhagic fever among scientists in Germany in the late 1960s, with several deaths. The scientists were experimenting on monkeys imported from Uganda that were infected with the Marburg virus. The Marburg virus was transmitted to the victims via saliva or mucus.

In 1976, there was an outbreak of Ebola in Zaire and Sudan that killed more than 500 people, including Belgian physicians and nurses that were treating patients. Another Ebola outbreak occurred in Uganda in the year 2000 and

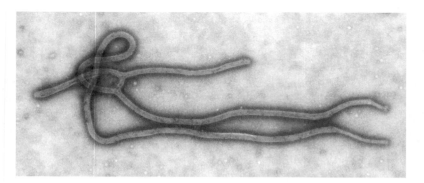

Figure 29.22 Ebola virus
This member of the *Filoviridae* has a unique structure in that it is composed of very long slender filaments. Courtesy of Cynthia Goldsmith/Centers for Disease Control and Prevention.

caused the death of 224, including health workers. Additional outbreaks occurred in 2002 in Gabon, in 2003 in the Republic of Congo, and in 2004 in south Sudan.

The incubation period for Ebola is 4 to 10 days before flu-like symptoms occur, followed by vomiting, diarrhea, and severe internal bleeding. Ebola is a highly virulent disease causing death in 50% to 90% of all clinical cases. The disease kills victims faster than it can spread and tends to "burn out" before it reaches epidemic proportions.

Because Ebola is so readily transmitted by bodily fluids, such as mucus, saliva, and blood, it is a deadly threat. For this reason, the WHO monitors occurrence of the disease and makes every effort to isolate and treat victims.

West Nile

West Nile fever is caused by the **West Nile virus (WNV)**, a mosquito borne disease that occurs widely in Africa, East Asia, and Middle East. It was first diagnosed in New York in 1999 and in a few years had spread across the United States (Figure 29.23). It now is of concern in virtually all states, with the highest incidence in the Midwest. The number of cases is increasing annually. The WNV is transmitted to humans, mammals, and an array of bird species by several species of mosquitoes. Horses are particularly susceptible to the disease and have a mortality rate approaching 50%.

The WNV causes an active infection in birds, and those that survive have acquired an active immunity to West Nile fever. Mosquitoes attacking birds that have an active infection (viremia) pick up the WNV and transmit it to susceptible birds, humans, and other animals. There are more than 5,000 diagnosed cases of West Nile fever annually but only about 20% of infected individuals show clinical symptoms of the disease.

Symptoms of West Nile fever include an increase in temperature, headache, nausea, and enlarged lymph nodes. The incubation period for the virus is 5 to 14 days and noticeable symptoms last for 3 to 6 days. The death rate for diagnosed cases is 2% to 3%. A low percentage

of infected individuals develop a form of meningitis or encephalitis, and adults older than 50 seem to be more susceptible to those clinical symptoms. The fever can be confirmed by ELISA for antibodies active against WNV. There is no effective specific treatment for West Nile fever.

Emerging Viruses

HIV, the etiological agent of AIDS, was first described in 1981. Over the last 20 years, this newly emerged virus has had a devastating effect on humans worldwide. There are other viral infections that have emerged in the last 20 years, and these may also be a serious threat to human populations. A hantavirus outbreak in the Southwest in the 1990s caused more than 30 fatalities, and this disease has a 70% to 75% death rate among victims. The virus is apparently disseminated in dust that contains rodent urine. Whether this virus may gain other mechanisms of transmission is unknown. Other viral diseases that have emerged include mosquito-borne infections such as **dengue fever** and **chikungunya**. How much of a threat these viruses will be to humans is difficult to assess (see Chapter 14).

SECTION HIGHLIGHTS

Many viruses that cause disease in humans reside primarily in nonhuman reservoirs. These viruses can be transmitted from the reservoir animal indirectly through bites from intermediate vectors, such as ticks or mosquitoes, or, directly, through animal bites, skin abrasions, or inhalation of contaminated dust. The large number and variety of primary hosts makes it impractical and undesirable to eliminate the nonhuman reservoirs of these diseases. Control measures include reducing vector populations, using insect repellants, and vaccination of humans or the animal reservoirs.

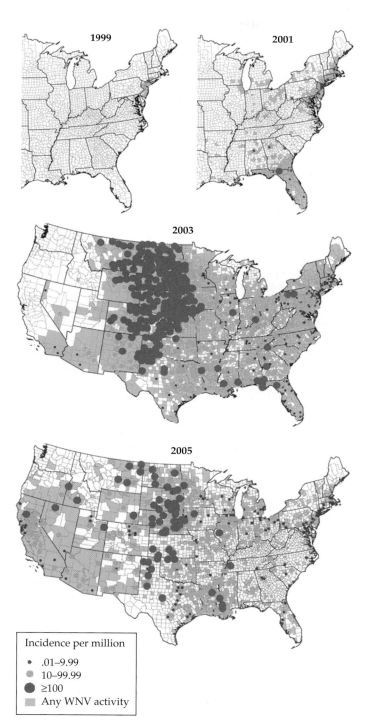

1999   2001

2003

2005

Incidence per million
- .01–9.99
- 10–99.99
- ≥100
- Any WNV activity

Figure 29.23 West Nile virus in the United States
From the first cases diagnosed in New York in 1999, WNV has now spread throughout the United States. The maps show the incidence of WNV human disease by county in 1999, 2001, 2003, and 2005. Maps from the Centers for Disease Control and Prevention.

SUMMARY

- **Viruses** can infect bacteria, plants, and animals. Infections caused by viruses are more difficult to **control** than infections caused by bacteria because the metabolic machinery of the virus is essentially equivalent to that of healthy cells.

- Infections spread in **airborne droplets** such as colds and influenza are difficult to control because the **daily** life of humans involves person-to-person contact.

- **Chicken pox** is a highly contagious disease that generally affects children. It can be a serious infection, and a vaccine is available to prevent the disease. **Shingles** is the result of a dormant viral infection and can occur in later life.

- **Rubella (German measles)** is a viral disease that was particularly dangerous if infections occurred in **pregnant** women, as the virus causes considerable harm to the fetus. Rubella can be controlled by **vaccination** of children. Another measles, **rubeola**, was a concern because it caused permanent neurological damage to some children. This disease can also be prevented by vaccinating infants.

- A deadly disease that has been **eliminated** throughout the world is **smallpox**.

- **Mumps** was a common childhood disease that caused **inflammation** of the **parotid glands** and spread to other areas of the body with unfortunate consequences. Vaccinating young children before they enter school can control this disease.

- **Influenza** can occur in **worldwide epidemics** (pandemics). It is a respiratory disease that can lead to pneumonia. Vaccines lower the risk and/or severity of the disease.

- **Common colds** affect 40% to 45% of Americans each year and result in an estimated **200 million lost school/workdays annually**. They can be neither prevented nor cured.

- Some viruses can **transform** normal human cells to cancerous cells. **Human T cell leukemia** virus is one such and is transmitted human to human in contaminated blood, exchange of needles by IV drug users, or sexual contact.

- **Acquired immune deficiency syndrome** (AIDS) has become a leading killer of young adults. Transmission of HIV, the causative agent of AIDS, can occur through transfusions with contaminated blood but most often is transmitted by **sexual contact**. There is **no cure** or prevention except "**safe sex**."

- **Herpes simplex virus type 1** is the etiological agent of **cold sores** that appear on lips and mouth. Another type of herpes (type 2) causes **genital herpes**, a growing problem in the United States. **Acyclovir** is a chemotherapeutic agent that can relieve symptoms of herpes infections, but it is not a cure.

- **Infectious mononucleosis** is a disease caused by herpesvirus (EBV) that occurs often in college age individuals. It causes fatigue, and rest is the only cure.

- Another viral disease that can be prevented by vaccination is **poliomyelitis**. It has not been eliminated because there has not been worldwide use of the vaccine. **Polio** can cause **paralysis** by destruction of motor neurons.

- There are a number of **encephalitic** diseases that are characterized by **inflammation** of the **brain**. They are transmitted to humans by vectors that include **arthropods** and **mosquitoes**. Elimination of vector breeding sites is the only preventive measure now available.

- **Rabies** is generally spread by the bite of a domestic animal and wild animals such as **skunks** and **raccoons** that serve as **reservoir**. Feral dogs and cats can carry the virus from the reservoir to domestic animals. Treatment is with antiserum and vaccination of exposed individuals.

- **Yellow fever** is caused by an arbovirus carried by a **mosquito** (*Aedes aegypti*). The disease causes severe liver damage but has been eliminated in the United States.

- Dissemination of emerging infections, such as **SARS**, by global travel is a threat to humans everywhere.

 Find more at www.sinauer.com/microbial-life

REVIEW QUESTIONS

1. Why are viral infections so difficult to control? How might this relate to the origin of virus? In considering this, would one assume that fungal or protozoan infections might also be a problem?

2. Why were the viral diseases transmitted via the respiratory route of such interest to the vaccine producers?

3. What are the symptoms of "shingles," and how does one get this disease?

4. Cite examples of viral diseases that can be transmitted by individuals that show no obvious symptoms.

5. Where have measles outbreaks occurred in recent years?

6. How was smallpox eradicated? Why was this possible from an economic standpoint?

7. What is "herd" immunity? How does this relate to mandatory vaccination? What are some dangers when there are not adequate immunization programs for preschool children?

8. Why is the influenza virus such a problem? Periodically there are epidemics and pandemics of this disease. Why?

9. One might say that the common cold will always be with us. Why is this true?

10. Why would an individual, once infected with HIV, have it for life despite chemotherapy? Elimination of a retrovirus from an infected individual is not feasible. Why?

11. AIDS patients have a major problem with infections by viruses and eukaryotic opportunists. What are some of these, and why is an AIDS victim so vulnerable to them?

12. Herpesviruses cause latent infections. Discuss this process.

13. Genital herpes is a disease that is increasing in frequency. What are the dangers?

14. The Epstein-Barr virus is of some concern in the United States. Why? Why is it more of a problem in other countries?

15. Polio may be eliminated from the world, as was smallpox. What are some problems in achieving this goal?

16. What are some of the diseases caused by hepatitis viruses? How can they be controlled?

17. Can warts cause cancer?

18. What are some of the major viral diseases in humans that have nonhuman reservoirs? How can these diseases be controlled?

19. What animals serve as major reservoirs for rabies?

SUGGESTED READING

Cann, A. J. 2005. *Principles of Molecular Virology.* 4th ed. San Diego: Academic Press.

Dalgleish, A. G. and R. A. Weiss. 1999. *HIV and the New Viruses.* 2nd ed. San Diego: Academic Press.

Granoff, A. and R. G. Webster. 1999. *Encyclopedia of Virology.* 2nd ed. San Diego: Academic Press.

Specter, S., R. L. Hodinka and S. A. Young. 2000. *Clinical Virology Manual.* 3rd ed. Washington, DC: ASM Press.

Strauss, J. and E. G. Strauss. 2002. *Viruses and Human Disease.* San Diego: Academic Press.

White, D. and F. Fenner. 1994. *Medical Virology.* 4th ed. San Diego: Academic Press.

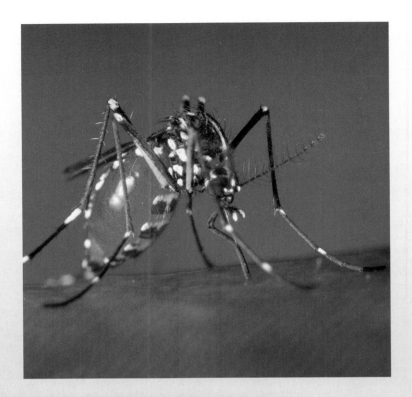

The objectives of this chapter are to:

◆ Introduce the basic terms and concepts of epidemiology.

◆ Describe the various sources and modes of infection.

◆ Outline the various types of public health measures used to control infectious disease.

◆ Highlight the differences in the public health challenges faced by developing versus developed nations.

◆ Explain the importance of accurate diagnosis to the control and treatment of infectious disease.

◆ Describe the methods used by clinical microbiologists to identify pathogens.

30

Epidemiology and Clinical Microbiology

Alas! regardless of their doom
The little victims play;
No sense have they of ills to come
Nor care beyond today.
—*Thomas Gray, 1716–1771*

The health and well-being of human populations relies, in large measure, on the control of communicable infectious diseases. Human history is a record of devastation by infectious diseases, and up to the early years of the twentieth century such diseases were the major cause of death worldwide (Box 30.1). Infectious diseases continue to take a toll. In the developing nations of Asia, Africa, and South America, they account for nearly 50% of all deaths. In the developed countries, including the United States, Canada, and Japan, and in Western Europe, less than 8% of all deaths are directly from infectious disease. The decreased death rate in the developed nations was brought about by controlling both individual infections and communitywide factors that would otherwise contribute to the spread of disease. Vaccination and potable water are primary examples of effective control.

Worldwide, very few diseases have been truly defeated, but one major affliction that caused much human misery has been eradicated, and that is smallpox. The periodic spread of infectious diseases such as influenza to virtually every nation is a reminder that communicable diseases are a concern to human populations and health professionals worldwide. The current catastrophic spread of the human immunodeficiency virus (HIV) confirms that uncontrolled highly communicable diseases can still leave their mark on societies. Emerging infectious diseases such as avian derived influenza and re-emerging infections (antibiotic resistant staphylococcus infections

Milestones

The Major Causes of Death in the United States in the Twentieth Century

In 1900 there were 318 deaths per 100,000 population caused by various infectious diseases and 215 due to heart disease and cancer. Public health measures instituted in the first half of the twentieth century lowered the death rate from these infectious diseases dramatically, and the death rate is now even lower. Public health measures, antibiotics, and immunization have combined to protect people in the United States from the scourge of infectious diseases. Heart disease and cancer continue to be leading causes of death.

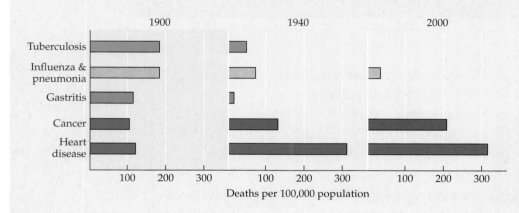

Leading causes of death in the United States in 1900; in 1940, before antibiotics were widely used; and in 2000, by which time antibiotics were widely available.

and tuberculosis, TB) are a concern. The rapid transfer of infectious agents from the unvaccinated and those afflicted with subclinical infectious diseases worldwide can disseminate diseases among continents within hours. The effect of societal changes and global warming on disease transmission is mostly unexplored.

Two major aspects of infectious disease control are (1) having an accurate determination of the etiological agent involved in infections and (2) the incidence of this infectious disease in the population. This is where epidemiology and the clinical laboratory come together. Epidemiology is the science concerned with the distribution and prevalence of communicable diseases in populations. The modern clinical laboratory has diagnostic capabilities that can determine precisely the strain of an agent involved in a disease outbreak and accomplish this in a short period. This aids the epidemiologist in pinpointing the origin and course of an epidemic and in taking measures that might curtail the spread of the infection.

30.1 Epidemiology

Epidemiologists work worldwide to monitor infectious diseases to institute measures for their control. The role of epidemiologists in pinpointing the outbreak of infectious disease is illustrated by the discovery of AIDS. In June of 1981, the Centers for Disease Control and Prevention (CDC) in Atlanta compiled data indicating that there were an unusually large number of deaths among men in the Los Angeles area. These deaths were caused by opportunistic pathogens that usually infected people who were immunodeficient, and the deaths involved homosexual males. This discovery was announced a full two years before the retrovirus responsible for the disease was isolated and identified. More important, the relation of the disease to the exchange of bodily fluids was almost fully defined by this time. This is a prime example of the effectiveness of epidemiologists in determining the frequency of a disease even before its etiological agent is identified. Since the identification of HIV, the virus has spread rapidly and is now a concern worldwide.

Following is a discussion of some of the terminology that is commonly used in epidemiology, including types of epidemics, incidences, prevalence, mortality, and morbidity.

Terminology

The word **epidemic** appears ominous, as it suggests that a disease may be spreading that will affect a massive number of people. In reality, it means that a disease is occurring in a population at a higher than normal frequency (Box 30.2). A significant increase in the frequency

The Great Plague in London

The Great Plague in London (1665) was a reasonably well-documented epidemic. The population of London at the time of the Great Plague was estimated to be approximately 400,000 inhabitants. The number that succumbed in the plague epidemic was somewhere between 25% and 40% of the population.

However, the number of deaths reported is a conservative figure, since many went unrecorded. The numbers for this drawing are mostly from Burial Registers of Churches. Many, including Quakers, were buried in their gardens, and massive numbers of others were buried without record.

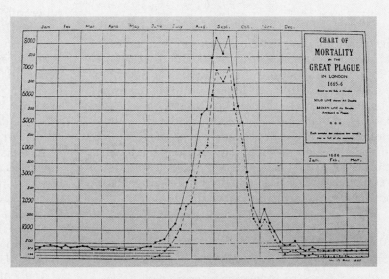

The dramatic increase in death rate in London during the plague epidemic of 1665–1666. From Bell, *The Great Plague in London* (1665), p. 238.

---

of a disease over continents or throughout the world is termed a **pandemic**. Major pandemics caused by the influenza virus occurred in 1918 and 1957, and HIV has clearly reached pandemic proportions.

Epidemics fall into three categories: sporadic, seasonal, and contact. A **sporadic** disease affects a certain percentage of the population throughout the year. **Endemic** diseases occur sporadically. The incidence may increase slightly during the year, but overall the number of cases is relatively constant. A **seasonal** disease, such as Lyme disease or Rocky Mountain spotted fever, is transmitted only in those months when the arthropod vectors are active. A **contact** disease, such as the common cold, is spread via contact with an infected individual.

The **incidence** of an infectious disease is calculated by the following formula:

$$\text{Incidence} = \frac{\text{Number of newly reported cases per given unit of time}}{100{,}000 \text{ people at risk}}$$

The **prevalence** of an infectious disease is the percentage of a total population that has the disease at a given time. For example, a seasonal disease is more prevalent during a limited number of months of the year. The prevalence of influenza among college students in the winter can reach 30%, whereas the prevalence in September is less than 1%. The term **outbreak** is also used and indicates that there is a relatively large number of cases in a limited area at a certain time.

Mortality and Morbidity

There are infectious diseases that cause a limited number of fatalities and others that are fatal to a considerable percentage of those infected. Botulism is generally fatal, but few succumb to staphylococcal food poisoning. The coming of the antibiotic age resulted in a marked change in the number of deaths from diseases such as pneumonia and TB. The **mortality rate** is the number of deaths from an infectious disease relative to the total number that contract it:

$$\text{Mortality rate} = \frac{\text{Deaths due to a disease}}{\text{Total infected}}$$

The mortality rate generally varies with the availability of health care. Cholera has a low mortality rate when adequate medical care is available but increases significantly if intravenous fluids and other health care measures are unavailable.

Morbidity is the incidence of infectious disease in a population:

$$\text{Morbidity} = \frac{\text{New cases in a selected period}}{\text{Total population}}$$

The morbidity rate is a definitive figure and serves as an accurate measure of the general health of a population. In developed nations, morbidity is considerably higher than the fatality rate, although both are relatively low for most diseases. The low fatality rate reflects the availability of better health care, antibiotics, and other chemotherapeutics in developed versus developing countries, although in developing nations the lack of health care leads to incurred fatality rates from otherwise controllable diseases. Fatality figures, in general, are not an accurate measure of health.

> **SECTION HIGHLIGHTS**
>
> The distribution and prevalence of communicable diseases are monitored by epidemiologists. An epidemic—a higher than normal frequency of a disease in a population—can be sporadic, seasonal, or contact. Incidence, prevalence, mortality rate, and morbidity rate are specific quantitative measures used by epidemiologists to describe the impact of diseases on populations.

30.2 Carriers and Reservoirs

A significant problem in controlling communicable infectious diseases is the presence of carriers in the population. Carriers are individuals with asymptomatic or subclinical infections that are not sufficiently serious to require treatment or curtail a person's activities. Often carriers with subclinical infections continue normal activities while "toughing it out" with their disease symptoms, and they tend to deny to themselves and others that they are ill with a transmissible disease. Unfortunately, these diseased people can expose everyone they encounter to the infectious microorganism they bear. Carriers can also be those convalescing from infectious diseases who return to work or school while harboring and shedding large numbers of infectious organisms. This is particularly a concern in enteric-type diseases, where convalescent patients continue to shed infectious organisms for a considerable period after symptoms disappear. In the case of typhoid, some patients (2% to 5%) become chronic carriers. A classical case of a chronic

typhoid carrier was Mary Mallon, known as Typhoid Mary (Box 30.3), who was well enough to continue working after bouts with typhoid fever.

There are readily used diagnostic tests that can be employed to identify carriers of selected diseases. The tuberculin skin test is commonly used to determine whether humans have been exposed to the etiological agent of tuberculosis. A protein derived from *Mycobacterium tuberculosis* is injected intracutaneously, and a reddening in the area within 48 hours suggests exposure to or contraction of the organism. Food handlers are often tested for the presence of *Salmonella typhi* in fecal specimens, to determine whether they might be chronic typhoid carriers.

A **reservoir** is the animate or inanimate site where an infectious disease is maintained in nature between outbreaks. If a disease agent survives only in living hosts, it must have a means of escaping from one host and of traveling and gaining entry into another host. Humans would be the reservoir for diseases transmitted via fecal contamination with food or drinking water as the intermediate (Box 30.4).

There are a number of diseases termed *zoonoses* that cause either human or animal infections. The animal is the reservoir and occasionally transmits the disease to humans; a number of these diseases are discussed in Chapter 28.

> **SECTION HIGHLIGHTS**
>
> A carrier is an individual with an asymptomatic or subclinical infection that spreads the disease to others in a population. A reservoir harbors an infectious disease agent between outbreaks, and may be human, animal, or inanimate.

30.3 Modes of Transmission of Infectious Diseases

Survival of infectious diseases depends on their transmission from host to host. In most cases, disease-causing organisms do not propagate outside a host or a reservoir. As discussed previously, many infectious diseases are endemic in the general population. There are two general types of localized epidemics (Figure 30.1). A common source epidemic (disease transmitted from a single source) occurs if a considerable number of humans eat from a large batch of contaminated food. An epidemic propagated by person-to-person transmission occurs, for example, when susceptible children are brought together during the first days of school. Major bacterial diseases that occur in humans are pre-

BOX 30.3 *Milestones*

Typhoid Mary

Mary Mallon, known as Typhoid Mary, has attained the dubious honor of having much of her life story discussed in virtually every microbiology text. Hers is a classical case of the chronic carrier of infectious disease wreaking havoc wherever she traveled. Tracking her down as the source of a typhoid outbreak tested the epidemiological investigatory abilities of the Health Department in New York City in the early 1900s. Yet, the Health Department's handling of the case parallels issues we face today in the AIDS dilemma, where the rights of individuals are pitted against the perceived welfare of the populace.

Mary Mallon was an Irish immigrant who had a serious case of typhoid fever in 1901 that resulted in a gallbladder permanently infected with *Salmonella typhi*. The infection generated large numbers of the typhoid bacillus that entered her gastrointestinal tract and were shed in her feces. Unfortunately, Ms. Mallon was a cook and housekeeper in New York City and worked in several homes. As she moved from

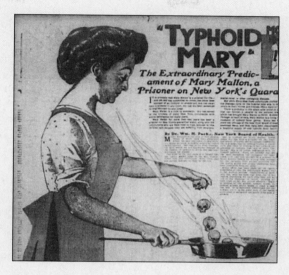

Mary Mallon, as depicted in the *New York American* June 20, 1909.

position to position, she left behind 28 cases of typhoid fever.

The New York Health Department, headed by Dr. George Soper, tracked down Mary Mallon and had her arrested. The authorities offered to remove her gallbladder to effect a cure. She refused and was released after 3 years of imprisonment on a pledge to never cook or handle food and to report periodically to the Health Department. Mary Mallon immediately disappeared, changed her name, and became a cook in hotels, hospitals, and sanitaria. She

apparently recognized that the disease was caused by her presence, as she quit her job when an outbreak occurred and moved on in order to evade authorities.

After 5 years, Mary Mallon was intercepted during a typhoid outbreak at a hospital. She spent her remaining 23 years on North Brother Island in New York City's East River, where she died in 1938. Typhoid Mary was the source of an estimated 200 cases of typhoid fever, much suffering, and several deaths.

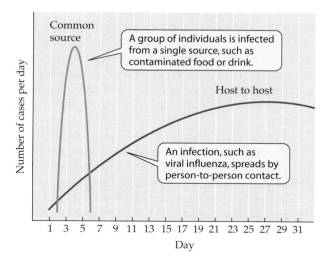

Figure 30.1 Types of epidemics
The different time-courses of common-source and host-to-host epidemics.

BOX 30.4 *Milestones*

John Snow and Cholera

John Snow (1813–1858), a British physician, was the first to recognize that humans were a reservoir for cholera. He realized that the feces of cholera patients were highly infectious and surmised correctly that human waste in drinking water could transmit the disease. Snow followed the incidence of cholera from 1853 to 1855 in a wide area of London. At that time, there were two major water systems supplying water to homes in that area: the Southwark & Vauxhall Company and the Lambeth Company. John Snow followed the cholera epidemic in homes of equivalent living standards but supplied by one or the other of these companies. It was obvious that inhabitants of houses supplied by Southwark & Vauxhall had a markedly higher incidence of cholera than those supplied by the Lambeth Company.

In the first 7 weeks of the epidemic, 315 people per 10,000 had died in houses whose water was supplied by the Southwark &

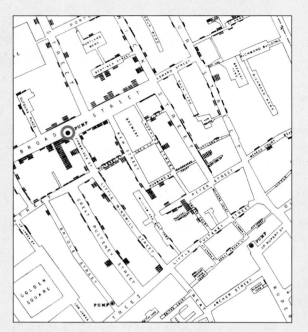

A contemporary map showing the incidence of cholera in the vicinity of the Broad Street pump (red circle) among a population that drew water from the Thames in an area contaminated by human waste. Each black bar represents one death.

Vauxhall Company compared to 37 per 10,000 people of those whose water was supplied by the Lambeth Company. Snow looked at the water supply for the two companies and found that the Lambeth Company obtained its water from the Thames above the city, where-

as the Southwark & Vauxhall Company drew water from the Thames in an area where untreated sewage from London entered the river. He concluded that the source of the cholera epidemic was the sewage-contaminated water.

---

sented in Table 30.1, along with potential measures for their control. Many of these diseases are discussed in Chapter 28.

Diseases Spread by Human Contact and Airborne Transmission

Human contact diseases and respiratory diseases are passed directly by human interaction. Transfer may occur by touching, kissing, coughing, sneezing, or, in some cases, on dust particles. Streptococcal and staphylococcal skin infections, such as impetigo, are often transferred among children by scratching. An unimpeded sneeze expels infectious droplets that are about 10 μm in diameter and contain one to several infectious agents (see Figure 29.8). The number of droplets per sneeze can be in the thousands. A hearty sneeze can travel at 100

m/sec and contain up to 100,000 bacteria or virus particles. Tuberculosis (TB) is a disease that is transmitted in respiratory droplets and is a constant threat.

There are 10,000 to 15,000 new cases of TB each year in the United States, and the mortality rate is about 5%. Clinical tuberculosis is generally a slowly progressing, chronic disease, and infected individuals can be asymptomatic for months or even years. These individuals are carriers and may transmit the disease to those they encounter.

Someone is newly infected with TB every second, and the largest number of these cases occurs in Southeast Asia. Overall about one-third of the world's population is infected with TB and approximately 2 million die every year. This percentage is greater than that for death from any infectious disease except HIV. The rapid growth of TB cases in sub-Saharan Africa has been augmented by

| TABLE 30.1 | Major bacterial diseases of humans, sources of infection, and potential control |

Disease	Primary Reservoir	Potential Means for Control
Human Contact and Respiratorily Contracted		
Streptococcal infections	Humans	Antibiotics; vaccine for pneumonia
Staphylococcal infections	Humans	Antibiotics; antiseptics
Meningitis	Humans	Specific antibiotics
Tuberculosis	Humans	Test and treat infected persons
Whooping cough	Humans	Vaccinate infants
Diphtheria	Humans	Vaccinate infants
Leprosy	Humans	Obtain proper treatment; vaccinate in endemic areas
Pneumonic plague	Humans	Eliminate rats and fleas
Water-, Food-, and Soil-borne		
Cholera	Humans	Treat sewage and water; observe proper sanitation
Typhoid fever	Humans	Pasteurize milk; proper treatment of sewage; inspect food handlers
Shigellosis (dysentery)	Humans	Observe proper sanitation
Salmonellosis	Beef, poultry	Cook meat and eggs properly
Campylobacter	Animals, poultry	Pasteurize milk; thorough cooking of food and water
Tetanus	Soil	Vaccinate
Brucellosis	Cattle	Immunize cattle and pasteurize milk
Botulism	Soil	Properly can and cook food
Staph food poisoning	Humans	Refrigerate food
Legionnaires' disease	Aquatic environments	Clean misting equipment or do not use
Pseudomonas infections	Dust	Clean air in burn wards
Sexually Transmitted		
Gonorrhea	Humans	Eliminate carriers; practice safe sex
Syphilis	Humans	Eliminate carriers; practice safe sex
Chlamydia	Humans	Eliminate carriers; practice safe sex
Louse-borne, Human to Human		
Trench fever	Humans	Proper sanitation; control lice
Relapsing fever	Humans	Control ticks and lice
Typhus (epidemic)	Humans	Proper sanitation; vaccinate
Vector-borne		
Rocky Mountain spotted fever	Mammals, birds	Wear protective clothing and examine body for ticks
Tularemia	Rodents, rabbits	Observe proper care when cleaning wild rabbits
Lyme disease	Deer	Wear protective clothing
Bubonic plague	Rats	Control rats; proper sanitation
Typhus (endemic)	Rodents	Control rats; vaccinate
Scrub typhus	Mites	Control mites
Animal Contact		
Leptospira	Vertebrates	Control rodents; vaccinate domestic animals
Anthrax	Soil	Sterilize wool, hair, other animal products
Psittacosis	Birds	Control bird imports
Q fever	Cattle	Vaccinate animal handlers

the AIDS epidemic. Weakening of the immune system by HIV promotes the ravages of TB, leading to death. In contrast to TB, leprosy is a poorly transmitted contact disease, and individuals that are immune-compromised are predisposed to contract it. There are 200 to 300 new cases annually in the United States. Vaccination for leprosy is not available, and treatment of infected individuals is the only known method for control. Vaccination is an effective control for some of the respiratorily transmitted bacterial diseases, such as whooping cough, diphtheria, and viral diseases such as mumps, chicken pox, rubeola, and measles (see Chapter 29).

Water-, Food-, and Soil-Borne Infections

Many water- and food-borne diseases are caused by pathogens that survive solely in the gastrointestinal tract. Humans are a natural host for many of these infectious microbes, including those that cause cholera, typhoid fever, and shigellosis. Because these microorganisms can survive for a limited time in raw sewage, they can be transmitted through contaminated water supplies. Proper sewage disposal and effective water treatment are important in control of gastrointestinal diseases. Cases of cholera and typhoid fever are rare in the United States. Shigellosis, a form of dysentery, is quite common and can be transmitted from infant to infant in childcare centers. Outbreaks also occur in elementary schools. Proper sanitation is the best mechanism for control of shigellosis. Pasteurization of milk, proper cooking practices, and immunization of cattle are important in preventing the transmission of food-borne diseases. As a result of such measures, brucellosis in cattle and bovine tuberculosis are of reduced concern in developed nations.

Legionnaires' disease is associated with mists created by cooling towers for air-conditioning systems and vegetable misters in grocery stores and in various aquatic environments. The control of the bacterium involved, *Legionella pneumophila,* an amoeba parasite, is difficult, and the disease occurs sporadically across the United States.

Anthrax has been an important soil-borne disease because the causative organism (*Bacillus anthracis*) can form endospores that survive in soil. Cattle killed by anthrax have been a source of the endospores. Better health practices in the animal industry have curtailed the spread of this disease. Probably the greatest threat from anthrax is its use by a terrorist as a germ warfare agent.

Sexually Transmitted Diseases

The sexually transmitted bacterial diseases are a significant societal problem that can be addressed only when

Figure 30.2 Worldwide distribution of HIV infection ▶
The estimated distribution of adults and children infected with HIV in 2004. Data from Joint United Nations Programme on HIV/AIDS.

people assume responsibility for their behavior (see Table 30.1). All can be treated and cured if caught early, yet millions of new cases of gonorrhea and chlamydial infection occur each year. An increasing incidence of antibiotic-resistant strains, particularly with gonococcus, poses a substantial threat to public health. These diseases can be controlled only by public education, public responsibility, and a greater awareness of the problem. The use of condoms and other safe sex practices are the only control now available. Immunization to prevent most sexually transmitted diseases including HIV is not on the horizon. The U.S. Food and Drug Administration (FDA) recently approved a vaccine that is effective in preventing human papilloma virus (HPV) infections. HPV is the causative agent of cervical cancer in human females.

Three sexually transmitted viral infections are epidemic in the United States. These are genital herpes (herpes simplex virus type 2, HSV-2), genital warts, and HIV. There are an estimated 30 million individuals in the United States infected with HSV-2, and about 1,000,000 new cases are diagnosed each year. Herpes in the neonate is one of the most common life-threatening infections, as there are about 2,000 babies born each year afflicted with this disease. The incidence of genital warts is increasing at an alarming rate and is now one of the more prevalent sexually transmitted diseases, particularly in promiscuous young adults.

HIV has reached pandemic proportions with more than 1,000,000 HIV-positive individuals in North America and an estimated 40 million infected worldwide (Figure 30.2). Since the epidemic began about 20 years ago, 57 million people have been infected with HIV, and 21 million have died because of these infections. HIV infection in sub-Saharan Africa is of great concern, and the United Nations has instituted programs for controlling the disease and treatment of victims. South and Southeast Asia are also experiencing a troubling increase in the incidence of HIV.

The difficulties encountered in controlling the spread of infectious diseases in developing countries are amply illustrated in Figure 30.3. The lack of therapeutic agents to treat individuals with active cases of AIDS results in those individuals being a reservoir for other contagious diseases. As mentioned previously, TB is inadequately treated in these locales, leading to the infection of increasing numbers. Consequently, there is an unending cycle of misery in vast areas of the world.

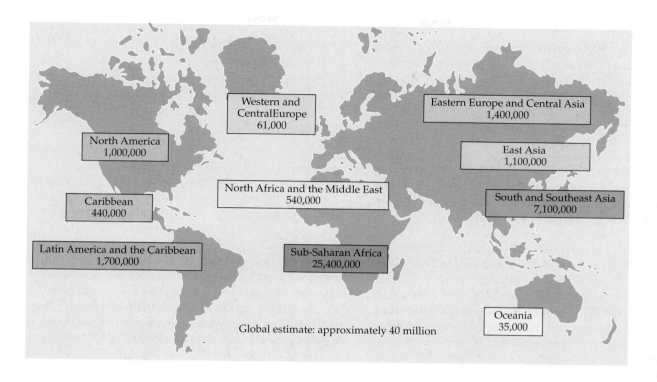

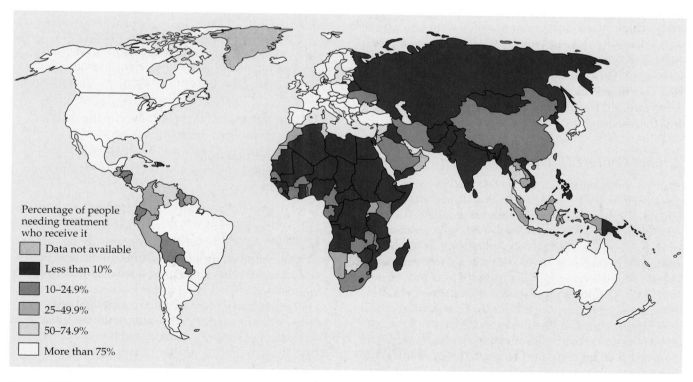

Figure 30.3 Areas of the world with unmet AIDS treatment needs
AIDS treatments are much less available in Africa and parts of Asia than they are in North America and Western Europe. Data are from June, 2005. After World Health Organization (WHO) report.

Vector-Borne Infections

A vector is an agent (insect or animal) that carries a pathogen from one host to another. Two vector-borne infections of concern in the United States are Lyme disease and Rocky Mountain spotted fever. The number of cases of both are increasing in some areas of the United States, and the sole control measure is avoidance of the ticks that serve as vectors. Protective clothing and insect repellent are essential when entering areas potentially occupied by ticks. One should check one's body for ticks after leaving such areas. Rat control has been effective in the control of vector-borne diseases such as **plague** and **endemic typhus**. **Tularemia** cannot be controlled in wild animal populations. The only way humans can avoid infection by this organism is to not dress freshly killed wild rabbits or, if doing so, to wear rubber gloves and proceed with care.

Mosquito-borne infections that are of concern in the United States are the various forms of **viral encephalitis** (see Chapter 29). **Yellow fever** and **malaria** are mosquito-borne diseases and a concern in various parts of the world. Malaria affects about 500 million people worldwide and kills someone on the average of every 10 to 15 seconds. West Nile Fever is now the greatest mosquito borne concern in the United States.

Louse-borne diseases are those that are passed among human populations by lice bites and are best controlled by proper sanitation. **Epidemic typhus** and **trench fever** are louse-borne diseases and are generally associated with substandard living conditions. Sanitation and elimination of lice can effectively control the spread of these diseases. **Relapsing fever** is a louse or tick-borne disease that occurs mainly among campers who spend time in areas infested with rodents. Campsites in locales where this disease occurs should be selected with care.

Animal Contact Diseases

Animal contact diseases (zoonoses), such as brucellosis, are of decreasing incidence in the United States. The spread of **brucellosis** and **bovine tuberculosis** has been controlled by vaccination of cattle and/or pasteurization of milk. Rabies is a problem in wild animals but can be controlled in domestic animals by vaccination. **Anthrax**, a disease found among workers exposed to wool and goat hair, is difficult to control because the causative agent, *B. anthracis*, is an endospore former that can survive in soil for decades. Sterilization of wool and goat hair is the only known control measure. Eliminating rats and vaccinating domestic animals can control **leptospirosis**. Animal handlers can and should be vaccinated against Q fever. **Psittacosis** is an ever-present danger in domestic birds and is a potential hazard for workers in poultry slaughterhouses. Pigeons are also a reservoir for the disease.

> ### SECTION HIGHLIGHTS
> Transmission refers to the means by which new individuals become infected with pathogens. Infections may be transmitted by direct contact with an infected individual, by exposure to contaminated air, water, food, or soil, by disease-carrying vectors, by sexual contact, and by contact with infected animals.

30.4 Nosocomial (Hospital-Acquired) Infections

To effectively invade a host, the pathogen must counteract the host's defenses. These normal defenses (see Chapter 26) are generally quite effective in combating invading organisms. Unfortunately, hospitals bring together patients whose normal defenses are compromised by one circumstance or another, and infections by opportunistic or accidental invaders can occur. Infections acquired in a hospital are called nosocomial infections.

Nosocomial infections occur for several reasons:

1. Illness or chemotherapeutics may compromise the immune system.

2. Abrasions or openings in epithelial barriers caused by surgery, catheters, syringes, respirators, or instruments used by physicians to examine the inner body offer the opportunistic pathogen a site to establish infection.

3. Exposed tissues of burn or wound patients offer access to airborne microorganisms.

4. Cross infection from patient to patient in crowded wards as well as from hospital workers or physicians is a concern.

5. The overuse of antibiotics in hospitals has resulted in an environment where drug-resistant microorganisms are ubiquitous.

6. The hospital environment selects for pathogens, as few hospitals have isolation wards; a virulent organism can find a reservoir in patients. This pathogen can be transmitted to other patients. Stepwise transfer of pathogens generally leads to increasing levels of virulence.

7. Hospital pathogens may bear plasmids that carry information for multidrug resistance. As a result, drug resistance can spread rapidly among a population of pathogenic microorganisms.

Major Nosocomial Infections

Each year about 2 million patients become infected during hospitalization, and the resultant fatalities number over 100,000 patients. Bacteria that are part of the normal human flora frequently cause the nosocomial infections. In a healthy nonhospitalized individual these microorganisms would not be invasive. The most prevalent organism in nosocomial infections is *Escherichia coli*, a normal inhabitant of human intestines.

Infections of the urinary tract are the nosocomial infection most often encountered in the hospital environment (Table 30.2). Of the total urinary tract infections, one-third is caused by *E. coli* and another third by *Pseudomonas aeruginosa, Enterococcus faecalis,* or *Streptococcus epidermis*. Most of the remainder of infections are caused by other gram-negative bacteria. Infection is generally a consequence of urinary catheterization of immobilized patients. Respiratory infections that appear as a form of pneumonia are often encountered and may be caused by *P. aeruginosa, Staphylococcus aureus,* or *Klebsiella* sp. These nosocomial infections, which can cause death, result from respiratory devices and the inability of the patient to clear the lungs.

S. aureus and pyogenic streptococci are a major concern in surgical patients. An estimated 7% to 12% of all surgical patients have postoperative infections. When the gastrointestinal or genitourinary tracts are involved in the surgery, the number of postoperative infections is two to three times higher.

TABLE 30.2	The relative frequency of nosocomial infections by body site
Site of Infection	**Percentage of Total Infections**
Urinary tract	40%–42%
Respiratory	16%–18%
Surgical wound	17%–20%
Bacteremias	6%–7%
Skin infections	6%–7%
Other	12%

> **SECTION HIGHLIGHTS**
>
> Hospital patients are particularly vulnerable to infections because their medical conditions and the treatments they receive weaken the body's usual defenses. The problem is compounded by the concentration of many sick people in a single environment, and the prevalence of antibiotic-resistant and highly virulent pathogens in hospitals.

30.5 Public Health Measures

The general health of people in the United States has improved dramatically since the time of our Founding Fathers. Life expectancy for a baby born today is about double that in the late eighteenth century. Nutrition has improved markedly, with fresh fruits and vegetables available year around. Housing has also improved, and working conditions are generally less stressful. Potable water is available to virtually all people in the United States, and we have effective treatment systems for human waste. We have eliminated the threat of many deadly diseases such as yellow fever, and antibiotics have alleviated dreaded diseases such as TB. A combination of these factors has tended to make our lives longer and healthier. How has this happened? By diversified efforts termed public health, which refers to the overall health of populations and the efforts of local, state, and federal public health officials to maintain reasonable health standards (Box 30.5).

Each state in the United States has a publicly funded organization generally known as the state health department. The name may vary, but the responsibilities are much the same, and a major function of this organization is to monitor the incidence of infectious diseases in the population. In many states, the health department provides a diagnostic laboratory for infectious organisms unidentifiable by local clinical laboratories. State agencies follow the incidence of disease in the state and communicate this information to the national agency, the Centers for Disease Control and Prevention (CDC) in Atlanta, Georgia. The CDC is a subunit of the U.S. Public Health Service (USPHS) within the Department of Health and Human Services.

The epidemiology unit of each state health department requires that licensed physicians report cases of selected communicable diseases that occur in patients under their care. Listed in Table 30.3 are the communicable diseases reported to the North Carolina Division of Epidemiology. Other states have similar requirements. Many of these diseases are then reported to CDC. In addition, restaurants and other food or drink establishments are required to report outbreaks of food-borne disease in employees and customers to the local health department. Many diseases, such as cholera, plague, and hepatitis, must be reported within 24 hours so that immediate control measures can be implemented.

The effectiveness of our public health institutions is very evident. Potential epidemics are tracked, and, in

BOX 30.5 | *Milestones*

Public Health Measures and Human Well-Being

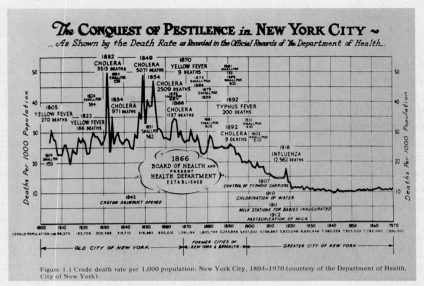

Figure 1.1 Crude death rate per 1,000 population: New York City, 1804–1970 (courtesy of the Department of Health, City of New York)

The effectiveness of public health measures on the curtailment of infectious diseases in New York City is apparent from this chart. Cholera, a disease transmitted by fecal contamination, was a considerable cause of death prior to the establishment of the Board of Health and Health Department in 1866. Smallpox, yellow fever, and typhus were also significant health problems. Chlorination of water was initiated in 1910 and pasteurization of milk in 1912, and these measures resulted in better control of infectious diseases. Note the number of deaths from the influenza pandemic in 1918.

The effect of public health measures on the incidence of disease. Courtesy of the Department of Health, New York City.

TABLE 30.3 | Communicable diseases reported to the epidemiology division of a state health department (North Carolina) and to the Centers for Disease Control and Prevention

AIDS	Granuloma inguinale	Q fever
Anthrax	*Haemophilus influenza* (invasive)	Rabies, human
Botulism	Hemolytic-uremic syndrome	Rocky Mountain spotted fever
Brucellosis	Hepatitis A	Rubella
Campylobacter infection	Hepatitis B	Rubella congenital syndrome
Chancroid	Hepatitis B carriage	Salmonellosis
Chlamydial infection	Hepatitis C	Shigellosis
Cholera	HIV infection	Streptococcal infection, group A
Cryptosporidiosis	Legionellosis	Syphilis
Cyclosporiasis	Leptospirosis	Tetanus
Diphtheria	Lyme disease	Toxic shock syndrome
Dengue	Lymphogranuloma venereum	Toxoplasmosis
E. coli O157:H7 infection	Malaria	Trichinosis
Ehrlichiosis	Measles (rubeola)	Tuberculosis
Encephalitis	Meningococcal disease	Tularemia
Enterococci (vancomycin resistant)	Meningitis, pneumococcal	Typhoid
Food-borne diseases	Mumps	Typhoid, carrier
Clostridium perfringens	Nongonococcal urethritis	Typhus, epidemic
Staphylococcal	Plague	Vibrio infection
Bacillus cereus	Polio, paralytic	Whooping cough
Gonorrhea	Psittacosis	Yellow fever

Source: Information provided by the North Carolina Health Services.

many cases, spread of disease is curtailed. An example of this was the potential for measles (rubeola) outbreaks on college campuses during recent years. The vaccine employed during the infancy of most college students at that time did not confer a long-lasting immunity. There was the risk of an epidemic in this population, and after a few cases were documented it was made mandatory that all students lacking proof of the proper immunization report for vaccination. This prevented a potential harmful outbreak, as rubeola can be a serious infection in young adults.

There is an active program designed to minimize the potential deadly consequences of the next flu epidemic by stockpiling antivirals, primarily Tamiflu®, for prophylaxis and/or treatment of those afflicted. The role of public health measures in reducing morbidity in the United States over the last century is well documented, but the bulk of money spent in the United States for health care goes to medical intervention rather than prevention.

> ### SECTION HIGHLIGHTS
>
> In the United States, state and federal agencies have been established to improve and maintain public health. Their activities include monitoring the incidence of infectious disease, assisting with the diagnosis of diseases, and developing strategies to control outbreaks.

30.6 Control Measures for Communicable Diseases

As we examine diseases and their reservoirs (see Table 30.1), it is evident that some diseases are more readily controlled than others. Diseases that can be prevented by immunization during infancy are controllable, but only when adequate vaccination programs are followed. The same applies to most water-borne infectious diseases, for which adequate water treatment can virtually eliminate the disease. Diseases transmitted in milk can be controlled by maintaining healthy dairy herds and the pasteurization of milk before consumption or prior to conversion to cheese or other milk products.

Many of the respiratory infections, particularly those passed orally, are difficult to control. A viral influenza outbreak may spread unchecked around the world. Cold viruses pass rapidly through human populations. Humans are gregarious, and infectious droplets resulting from talking, coughing, or a partly stifled sneeze are hazardous to everyone coming in contact with the infected individual. The influenza virus itself can be exceedingly virulent and constantly changes such that immunity to past epidemics does not ensure protection against the

next. There is no immunization or treatment for the common cold.

In the following section we will discuss some potential procedures for the control of infectious diseases. Difficulties that might be encountered are also considered.

Reservoir Control

Control of the reservoirs that sustain infectious microorganisms is effective in some cases but not in others. Domestic cattle have been a reservoir for human infections, but this source has been substantially eliminated by rigorous control measures. *Brucella abortus* causes spontaneous abortion in cattle, and vaccination prevents the spread of the disease among animals and to humans. Cattle are also immunized for bovine tuberculosis to prevent transmission of this infectious microorganism to humans through milk. There are strict rules on the extermination of all cattle that may become infected with lumpy jaw, a disease caused by *Actinomyces bovis*. The elimination of diseased sources is a practical mechanism for preventing the spread of an infectious disease. It is, however, essential that **all infected sources** in the potential reservoir be eliminated, otherwise the disease can erupt and travel rapidly through a nonimmune, and therefore susceptible, population.

The reservoir for a number of significant diseases is rodents, so domestic rats must be destroyed or driven from populated areas to control diseases spread by them. The rat is one wild animal that has evolved with a unique ability to live among humans, and the destructive habits, appetite for grain, and disease-causing potential of the rat are all a threat to human welfare. During 1993, a viral infection caused a number of deaths in the southwestern United States that were traced to a hantavirus that was spread by inhalation of aerosols of mouse urine and feces. The living conditions of the victims were a primary contributor to the infection.

Quarantine

Animals are frequently quarantined before they are transported from region to region or country to country. Great Britain has eliminated rabies from the isles and has imposed strict regulations on the import of animals to ensure that the disease will not return. The United States enforces a quarantine on cattle and other animals before import to ensure that they are free of infectious diseases.

Transfer of cattle from country to country has become a major concern because of the dangers of bovine spongiform encephalopathy ("mad cow disease") to human health. Sale of meat internationally has also been curtailed as the prion involved in the disease may follow this route. An effort is underway in the United States to put tracking devices in all livestock in order to identify the origin of animals that might carry this disease.

Quarantine was once commonly practiced in the United States at the local level. If a family had a member with scarlet fever or certain other transmissible diseases, a notice would be posted outside the door of their home. The notice warned visitors that an infected individual resided there, and they should not visit lest they carry the disease elsewhere. In cases where the patient was a youngster, all siblings were barred from attending school.

Prior to the availability of antibiotics, patients with TB were confined to a state-controlled sanitarium. Hawaii and other tropical areas had leprosy colonies where individuals infected with the disease were confined. Quarantine of humans in the United States, except in isolation wards of hospitals, has been discontinued. However, by international agreement, some diseases are quarantinable. These diseases include cholera, plague, yellow fever, typhoid, and relapsing fever. Those infected with one of these diseases may be barred from moving from country to country.

Food and Water Measures

Food inspection is a common practice, and among the foodstuffs subject to government inspection are meats coming from abattoirs and meat-processing plants. Milk pasteurization follows mandated procedures and has curtailed the spread of brucellosis, bovine tuberculosis, typhoid, listeriosis, and other diseases. Breakdowns in the pasteurization process have led to common source epidemics. In recent years, there were two listeriosis (*Listeria monocytogenes*) epidemics: one in Massachusetts (1983) resulting from ingestion of insufficiently pasteurized milk and the other in California (1985) from ingestion of Mexican-style cheese, which was apparently made from contaminated milk. There were 104 deaths in these two episodes. There was an outbreak in 2002 that caused 8 deaths. Cases such as this are testimony to the need for strict quality control in production and handling of the food we consume. Foods sold to the general public rightfully must meet reasonable public health standards for quality and sanitation. Despite these precautions, we continue to have periodic incidents that cause considerable human misery.

Water treatment and purification (in the developed nations) have virtually eliminated the threat of cholera, typhoid, and other waterborne diseases. Sewage treatment plants in our municipalities have been effective in processing domestic sewage and in preventing the spread of diseases from this source. A major improvement has been in separating storm runoff that enters street sewers from the domestic sewage systems. In years past, raw domestic sewage and storm water bypassed the sewage disposal plant during heavy storms. This resulted in significant local pollution from raw sewage.

Human and Animal Vaccination

Vaccination has been successfully employed in controlling a number of feared diseases. There are mandatory laws among the various states that require proof of vaccination before a child can enter school. The basic immunization program is initiated in an infant at about 2 months of age, and children in the United States are immunized against 11 different diseases (Table 30.4). These immunizations continue for the first 6 years of life.

There is growing concern that immunization programs are not begun at a sufficiently early age, particularly among the economically disadvantaged. The concern is that this delay might establish a significant pool of children up to 5 years of age who are at risk of contracting and disseminating communicable diseases. Extensive immunization of children has been successful to date in virtually eliminating many childhood diseases, such as smallpox, diphtheria, pertussis, and poliomyelitis, and this may be giving the adult population a false sense of security. Some adults are inadequately immunized and potentially can contract childhood diseases. Outbreaks in uninoculated preschoolers can potentially cause limited outbreaks among adults.

In recent years, there has been an increased incidence of autism among young children. Autism is a mental disorder that originates in infancy and is characterized by an inability to develop socially, language dysfunction, and other problems. The basis for the condition is not

TABLE 30.4	Recommended immunization schedule for infants and young children in the United States
Age	**Vaccine**
Birth	Hepatitis B
2 months	Diphtheria; pertussis; tetanus (DPT)
	Haemophilus influenzae type b (Hib)
	Poliomyelitis (OPV)
4 months	DPT; OPV; Hib
	Hepatitis B
6 months	Hepatitis B
	DPT; OPV; Hib
12–15 months	DPT; Hib; chicken pox, measles, mumps, rubella (MMR)
4–6 years	OPV; DTP; MMR

known. An English physician has made the claim that autism is due to the immunization programs established for infants. He maintained that the measles, mumps, and rubella (MMR) series overburdens the immune system and results in autism. Examination of data collected worldwide affirms that the disorder is not related to immunization programs.

Herd immunity is the immunity engendered to a communicable infectious disease when a major segment of the total population is immune to that disease. The greater the percentage of a total population immune to an infectious agent, the greater the chance that the entire population will be protected. It is not necessary that 100% of a population be immune for a disease to be eliminated. In practice, few vaccines actually confer immunity to all the individuals vaccinated. Generally, herd immunity is conferred if 70% or more of the population is immune. For highly contagious diseases, such as viral influenza, the percentage of the total susceptible population that must be immune to confer herd immunity is close to 95%.

Vaccination of pets for rabies has been effective in controlling this disease in domestic animals, and generally they are not a threat to humans. Unfortunately, there is a large reservoir of the disease in skunks, raccoons, and other carnivores. Rabid animals among these species continue to be a threat to humans.

Antibiotic Resistance

The development of antibiotic resistance in pathogenic organisms is a major public health concern. One somewhat controversial area in this regard is feeding antibiotics to cattle to increase their growth rate. About one-half of the annual production of antibiotics in the United States is incorporated into feed for poultry, swine, and cattle. Tetracycline and neomycin are two antibiotics that have been added to animal feed. Because the animals consuming these antibiotics generally harbor potential pathogens, such as *Salmonella* sp., the indiscriminate use of antibiotics in feed could select for antibiotic-resistant strains. The animal and pharmaceutical industries do not consider this a danger, and it is not restricted. One antibiotic, penicillin, that was added to animal feed in the past has been restricted and is no longer used for this purpose.

The CDC has estimated that 70% of the salmonellosis outbreaks that occurred in the years 1971 and 1983 involved resistant strains that came from food animals. In a documented case, a unique strain of *Salmonella newport*, identified by plasmid profile, was involved. Eighteen cases of salmonellosis were traced to one farm in South Dakota, where the antibiotic-resistant strain apparently originated. Studies in Denmark clearly showed that resistance to four antimicrobials declined following bans on their incorporation into animal feed. For example, resistance to the antibiotic vancomycin in *Enterococcus faecium* strains isolated from chickens declined from 72.7% in 1995 to 5.8% in 2000.

The antibacterial triclosan is a common ingredient of soap and lotions and has also been impregnated into cutting boards, toys, and clothing. The goal is to prevent the spread of pathogens via these materials. However, the result of such widespread nonmedical use is to make us more vulnerable, not less, to infectious agents. It has now been established that bacteria have developed resistance to triclosan, and this in turn engendered resistance to antibiotics such as tetracycline and erythromycin.

A major factor in the development of antibiotic resistant pathogens is the misuse or overuse of available antibiotics worldwide. Surveys indicate that antibiotics have been given clinically far more than advisable. Antibiotics are effective in only 20% to 25% of the cases where they have been used. This misuse promotes the development of resistant strains.

> ### SECTION HIGHLIGHTS
> Control measures for an infectious disease are dictated by its mode of transmission, its reservoirs, and the availability of effective vaccines and antibiotics. Measures that are in wide use are vaccination programs; immunization, quarantine or elimination of animal reservoirs; food inspection and recalls; and water treatment and purification. The rising prevalence of antibiotic resistance may require that we change the ways in which we use antibiotics.

30.7 World Health

The World Health Organization (WHO) was established in 1948 through the United Nations and is now headquartered in Geneva, Switzerland. The WHO is committed to the control of diseases and promoting health for the people in more than 100 member countries and was instrumental in the eradication of smallpox (Box 30.6). The organization is involved with population control, availability of food, and efforts to curtail disease through education. The WHO provides information on developing safe drinking water sources. Worldwide immunization programs for combating such diseases as diphtheria, poliomyelitis, and TB are among the goals of the WHO. It is the intention of the WHO that means be found to bring diseases such as malaria and leprosy under control.

BOX 30.6 *Milestones*

A Success Story: The Eradication of Smallpox

Smallpox is a dreaded disease that was eradicated by a successful worldwide immunization program. Smallpox is an ancient disease, and the millions of deaths caused by this disease are engraved in the 3,000-year record of human history. Through the first quarter of the twentieth century, thousands of cases occurred annually in the United States. The disease was eliminated in the United States by about 1960 through a long-term extensive vaccination program in preschool children. The World Health Organization (WHO) initiated a worldwide eradication program in 1966. As the disease was eliminated in the developed nations, the campaign's thrust turned to India, Africa, South America, and other developing countries. By 1977, the world appeared free of smallpox. In 1980 the WHO declared the disease eradicated. As a result, children are no longer vaccinated for this disease.

There are several reasons that the eradication of smallpox was so successful and that equivalent programs for other diseases might be less so. Among the advantages of the smallpox vaccine were that the vaccine was inexpensive, the vaccine raised a high level of immunity, the vaccine did not require refrigeration, and inoculation was by pinpricks with no need for syringes. Thus, individuals in remote villages could be vaccinated. Not all immunization programs are so amenable.

In 1975, this 8-year-old girl received a cash reward for reporting the last natural case of smallpox in Bangladesh. As the eradication campaign moved into increasingly remote areas, offering cash rewards helped to locate the last remaining smallpox victims. Courtesy of Dr. Stanley O. Foster/CDC/WHO.

Problems in Developing Nations

Diseases that were once prevalent in developed countries but are now controlled by drugs or immunization remain a concern in developing countries. As previously mentioned, infectious diseases cause nearly 50% of the deaths in developing nations, but fewer than 8% of deaths in developed countries. This higher death rate in the developing countries is due to inadequate sanitation, nutritional deficiencies, and substandard housing. The lack of medical care and immunization programs also contribute to a high mortality rate.

Many of the diseases in developing nations result from a lack of adequate supplies of clean water (Table 30.5). Cholera, a serious problem in Southeast Asia and a threat in some South American countries, could be eliminated by providing populations with safe drinking water. Better sanitation practices and safe food handling

TABLE 30.5	A number of the diseases that are a major concern in developing nations
Diseases	**Region Where Disease Occurs**
Cholera	Southeast Asia, Central Africa
Salmonellosis	Most
Shigellosis	Most
Yellow fever	Central & South America
Encephalitis	Most
Typhus	Middle East
Typhoid fever	Middle East, Central & South America
Tuberculosis	Most

would also decrease the incidence of salmonellosis and shigellosis. Typhoid fever is another infectious disease that would be virtually eradicated by the availability of safe water supplies. It is estimated that up to 10% of the population in Latin America are typhoid carriers. Inadequate treatment of sewage is a potential source of typhoid infection from the ever-present carrier reservoir.

Problems in Developed Nations

The developed countries are not without health problems caused by infectious agents. The problems with foot-and-mouth disease in England and the other European countries may well come to the United States. In both developed and developing countries, HIV prevalence is increasing at an alarming rate and is now considered a pandemic (see Figure 30.2). Gonorrhea persists despite the availability of an effective cure for the disease; an estimated 1 of 35 young adults will contract gonorrhea each year. Legionnaires' disease and shigellosis are endemic, and shigellosis is particularly contagious among young children in childcare centers. Lyme disease and Rocky Mountain spotted fever continue to be a concern with no adequate control measures in sight. Vigilance is essential, and the epidemiology programs in the various states and the CDC are doing a commendable job in monitoring communicable diseases in the United States.

Problems for Travelers

When people travel from developed to developing countries, they can encounter potential health problems. Immunizations beyond those routinely given are required or recommended for travel by United States citizens to many areas of the world (Table 30.6). Diseases for which there are no immunizations, such as dengue fever, malaria, and typhus are also prevalent in some developing countries. Travelers should check with health authorities before journeying to countries where communicable diseases not encountered in the United States are a threat.

SECTION HIGHLIGHTS

The World Health Organization (WHO) is an international organization whose mission is to improve public health worldwide. Developing nations often face public health challenges that most developed countries have eliminated, including inadequate sanitation, malnutrition, substandard housing, and lack of medical care and immunization programs. Despite a higher standard of living in developed countries, AIDS, gonorrhea, Legionnaire's disease, shigellosis, and other infectious diseases are still an ongoing problem.

30.8 Clinical and Diagnostic Methods

Early identification of the causative agent of an infectious disease in individuals and the population at large is of utmost importance. Epidemics can be tracked effectively only when the etiological agent is correctly identified. Therapeutic agents, such as antibiotics, can be chosen more efficiently when the infectious agent involved is known and antibiotic resistance and susceptibility have

TABLE 30.6	Immunization recommended or required for travel to developing countries
Disease	**Traveling to**
Vaccination Required	
Cholera	Southeast Asia, Albania, Malta, Central Africa, South Korea
Yellow fever	Central and South America, Africa
Vaccination Recommended	
Plague	Rural Africa, Asia, and South America
Serum hepatitis	Africa, Indochina, Russia, Central and South America
Typhoid fever	Africa, Asia, Central and South America (specific areas)
Most U.S. citizens already immunized	
Diphtheria, polio, tetanus, measles, mumps, rubella	

been determined. Identifying etiological agents can minimize the severity of a disease and shorten the recovery time if proper treatment is immediate and appropriate. In many epidemiological studies the characterization of the microbe must be carried beyond the level of species to the strain involved. Recent technological advances have provided the clinical laboratory with the instruments and techniques that have improved accuracy and shortened the time required for the identification of pathogens.

Searching for Pathogens

There are a number of factors that predispose an individual to the diseased state. Significant among these are the general health of the host, previous contact with the microbe, past medical history, exposure to toxic agents or chemicals, and traumatic or other insults not of microbial origin. These and other elements have considerable bearing on whether an individual will contract a particular disease.

The term *pathogen* can be applied to few organisms if one considers a pathogen to be an organism that *always* causes clinical symptoms. The indigenous microbiota of a human is a varied population, and many of these organisms may cause symptoms of disease under appropriate conditions. Lowered resistance can result in a clinical infection, and it is often difficult to determine which of many organisms present in a diseased patient is the one responsible. Clinical microbiologists are trained to sort through the organisms present in a specimen and make decisions on the most likely cause of clinical symptoms.

Obtaining Clinical Specimens

The accuracy of a diagnosis is dependent on the quality of the specimen delivered to the clinical laboratory. The specimens generally obtained from a patient would be one or more of the following: blood, urine, feces, sputum, biopsy tissue, cerebrospinal fluid, or pus/exudate from a wound. Throat or nasal swabs and fluid aspirated from an abscess may also be submitted to the clinical laboratory for diagnosis. Care must be taken to prevent contamination of the specimen by extraneous microorganisms after it is taken. In addition, the specimen must be sufficiently large that all desired tests can be accomplished on that single specimen.

BLOOD   Presence of bacteria in blood (bacteremia) is generally an indication that there is a significant focus of infection somewhere in the patient's body. If a few bacteria enter the bloodstream, these organisms will probably be cleared quickly by natural host defenses. However, when the number of microorganisms present in blood is significant (several per milliliter) an acute infection is probable somewhere in the body and is shedding microorganisms into the bloodstream. This condition may lead to a general septicemia, which is rapid propagation of pathogens in the bloodstream. Septicemia can result in a fever, chills, and shock. Blood samples for analysis are always taken aseptically with a sterile syringe and the blood should be delivered into a bottle containing a suitable medium and anticoagulant to prevent clots. Clots may entrap bacteria and make isolation of the microbe difficult.

One part of the blood sample is placed in the culture medium to be incubated aerobically, and another anaerobically. These inoculated cultures are placed in an incubator that automatically monitors CO_2 production. Should growth occur, as indicated by generation of CO_2, either aerobically or anaerobically, the microorganism is isolated in pure culture, it is identified, and antibiotic sensitivity is determined. The organisms most commonly associated with blood infections are *S. aureus*, *Streptococcus pyogenes*, *P. aeruginosa*, and enterics.

GENITOURINARY TRACT   Urinary tract infections are frequently caused by gram-negative bacteria similar to those that are part of the natural human microbiota. Because urine itself will support the growth of bacteria, care must be taken to ensure that once a specimen is taken it must not stand for any length of time at room temperature before analysis. If not analyzed immediately, the sample should be refrigerated. Bacteria that are generally involved in urinary tract infections are *E. coli* and species of *Klebsiella*, *Proteus*, and *Enterobacter*.

Urine samples can be analyzed by a direct count of organisms present or by spreading an aliquot over the surface of a MacConkey agar plate and a blood agar plate by using a calibrated loop. The number of colonies that develop on the blood agar plate is a measure of the number of organisms in the urine specimen. If there are 10^5 or more organisms per milliliter of urine, a urinary tract infection is indicated. MacConkey agar is a selective medium for gram-negative bacteria. It contains bile salts and crystal violet that inhibit growth of gram-positive species, the sugar lactose, and neutral red as a dye indicator. *Enterobacter* and *Escherichia* ferment lactose, and the colonies act on neutral red, imparting a reddish color. *Proteus*, *Salmonella*, and *Shigella* do not ferment lactose, and they form white or clear colonies.

Several different media can be employed in characterizing gram-negative bacteria encountered in the clinical laboratory (Table 30.7). These differential media would effectively identify the microorganism involved

TABLE 30.7	Some agar base media that can be used to differentiate clinically significant gram-negative bacteria by colonial appearance					
Medium	*Enterobacter aerogenes*	*Escherichia coli*	*Salmonella* sp.	*Shigella* sp.	*Proteus* sp.	**Gram⁺**
Bismuth sulfate	Mucoid silver sheen	Little growth	Black with metallic sheen	Inhibited to brown	Green	No growth
Eosin-methylene-blue	Pink	Purple with black centers	Colorless	Colorless	Colorless	No growth
Salmonella-shigella	Cream to pink	No growth	Colorless	Colorless	Colorless	No growth
Sodium azide agar	No growth	No growth	No growth	No growth	No growth	Growth
MacConkey agar	Pink to red	Pink to red	Colorless	Colorless	Colorless	No growth

in the infection. The presence of gram-positive staphylococci or streptococci in the urinary tract sample would be determined by inoculating a blood agar plate. These organisms are present less frequently than the gram-negative organisms in urinary tract infections.

Other procedures are followed when the infection might be caused by a sexually transmitted infectious microorganism. Gonorrhea is one of the most common sexually transmitted diseases (STDs) among young adults, and the etiologic agent is *Neisseria gonorrhoeae.* A Gram stain of the purulent urethral discharge in males can be a rapid and reasonably accurate diagnostic procedure. If the symptoms are less obvious, the clinical specimen from males and females must be cultured on selective and nonselective media. A primary medium is the modified Thayer-Martin medium, which contains the antibiotics vancomycin, nystatin, and colistin. These antibiotics inhibit the growth of many microorganisms commonly present in such specimens but not the growth of *N. gonorrhoeae.*

Some pathogenic strains of *N. gonorrhoeae* may be inhibited by vancomycin, so nonselective chocolate agar should also be inoculated. Chocolate agar is prepared from heated blood and is a source of growth factors for *N. gonorrhoeae.* It also absorbs toxic material that may be present in a rich medium that would otherwise inhibit the growth of the gonococcus. The oxidase test is important in identifying *Neisseria* in chocolate agar. The inoculated plates should be incubated in an atmosphere of 5% CO_2. Commercial DNA probes have been developed for a rapid diagnosis of gonorrhea.

Specimens obtained in suspected cases of syphilis would be the exudates from open lesions or material from lymph nodes in the affected regions. The presence of spirochetes as determined by dark-field microscopy is a rapid and generally effective diagnostic tool. Stain-ing is not generally effective, as *Treponema pallidum* stains poorly and is only about 0.2 μm in diameter—near the limit of light resolution. The organism is not readily cultured on laboratory media. Identification can be by fluorescein-labeled antitreponemal antibodies or a slide flocculation test (see later discussion). The flocculation tests are based on the presence of antibody to a specific cardiolipin antigen in the serum of syphilis patients.

The common sexually transmitted pathogen *Chlamydia trachomatis* can be identified by inoculating exudate into cell culture and monitoring for growth of the intracellular pathogens. There are several rapid inexpensive tests for detecting *Chlamydia* in urine samples.

One significant problem in diagnostic tests and treatment of urinary tract infections is that they are often caused by a combination of bacteria. These mixed populations may have differing levels of resistance to antibacterial drugs.

TISSUE/ABSCESS   Samples of tissue from biopsies and material from skin lesions or wounds are streaked on blood agar and other rich media. Duplicates should be prepared for incubation aerobically and anaerobically. Care must be taken in obtaining and handling such specimens, as infections of this sort are frequently caused by strict anaerobes that may be adversely affected by contact with atmospheric oxygen. Exudate from infected areas can be collected in a syringe by aspiration and taken directly to the clinical laboratory for diagnosis. Such samples may be examined by microscopy after staining with fluorescent antibody. The type and the appearance of lesions and the history of the patient may suggest possible etiological agents and indicate which fluorescent antibody should be applied. Anthrax, plague, bubo, and tularemia lesions have distinct characteristics, and confirmation can be made quickly by using serological techniques.

FECAL   The pathogens most often associated with fecal specimens are food- or water-borne microorganisms, and among these are *Vibrio cholerae, Campylobacter jejuni,* and species of *Salmonella* and *Shigella.* Careful handling of fecal specimens is essential, for when left at room temperature for a brief period, the specimens can quickly become acidic. This acidity will kill pathogens such as *Salmonella* or *Shigella.* To overcome this, fecal samples are generally collected in a buffered medium and delivered quickly to the clinical laboratory for diagnosis. The selective medium of choice for fecal specimens is Mac-Conkey or eosin–methylene-blue agar (see Table 30.7). Samples should also be streaked on blood agar to determine whether gram-positive pathogens, such as staphylococcus or streptococcus, might also be present. Special precautions for the identification of *E. coli* O157:H7 are essential. Identification of viral infections must be by immunodiagnosis or nucleic acid hybridization.

SPUTUM   Sputum and material obtained from the upper respiratory tract should be streaked on blood or chocolate agar. Blood agar is useful for culturing *S. pyogenes, Streptococcus pneumoniae,* and *S. aureus. Neisseria meningitidis* and *Haemophilus influenzae* can be detected on chocolate agar. Acid-fast staining of smears and serological tests can be applied where *Mycobacterium tuberculosis* might be involved in the infection.

CEREBROSPINAL   Cerebrospinal fluid specimens are cultured on rich media such as blood and chocolate agar. Organisms involved in meningitis (*N. meningitidis, S. pneumoniae,* and *H. influenzae*) grow on these rich media. Because meningitis is a potentially fatal infection, an immediate diagnosis is mandatory. ELISAs (enzyme-linked immunosorbent assays) (discussed on pp. 998–999) are a major tool in determining the nature of the pathogen that might be involved in infections of this type because the results permit an immediate diagnosis and start of treatment.

SECTION HIGHLIGHTS
Effective treatment, monitoring, and control of an infectious disease all depend on accurate identification of the etiological agent—this is the job of the clinical microbiologist. A variety of techniques have been developed to detect pathogens in the different types of clinical specimens that may be obtained from infected individuals.

30.9 Identifying Pathogens

The major goal of a clinical microbiology laboratory is a prompt, precise identification and characterization of the pathogen involved in an infection. In addition, antibiotic resistance/sensitivity of the causative agent is a major element in selecting a proper treatment. If the infecting strain is resistant to the antibiotic generally employed in such cases, successful treatment depends on administering an alternative antibiotic. From an epidemiological perspective the precise identification of infectious agents is essential in following the dissemination of the disease.

Specimens delivered to the clinical laboratory for microbiological analysis may follow one or more routes:

- The specimen may be examined microscopically following staining or examined directly with dark-field or phase contrast microscopy.

- A specimen can be streaked or cultured on an enrichment, selective, or differential medium (Table 30.8).

- The specimen may be subjected directly to serological, immunofluorescence, ELISA, or other direct diagnostic procedures.

We discuss each of these routes in turn. Direct microscopic examination of stained clinical material can be of value with select specimens. This technique will yield a preliminary diagnosis that can then be confirmed by isolation and identification of the pathogen. In suspected cases of TB, sputum is subjected to the acid-fast stain, a diagnostic tool for identifying bacteria that have a waxy coat. For suspected gonorrhea, cervical scrapings from females and urethral discharge from males may be Gram stained to visualize the typical gonococcus. There may also be observable polymorphonuclear leukocytes that contain the characteristic diplococcal *Neisseria gonorrhoeae* cells. If leprosy is suspected, *Mycobacterium leprae* is identified by direct observation of acid-fast–stained specimens from leprous lesions, since the organism cannot be grown outside the host. A preliminary diagnosis of syphilis is feasible by dark-field microscopy. Examinations for animal parasites in blood, feces, and so on are generally done by direct microscopic examination of properly collected specimens.

Staining and viewing of some specimens microscopically as they are received in the clinical laboratory is not necessarily productive. The specimen from a given source will generally contain an array of microorganisms, and microscopic examination will not indicate which is the agent of infection. In such cases, samples must be inoculated directly into or onto an appropriate growth media. The medium employed will depend on the source of the specimen.

TABLE 30.8	Basic types of media used in clinical microbiology for isolation of bacteria cultures from specimens

Characteristics media—These media test bacteria for specific metabolic activities, enzymes, or growth characteristics.

Citrate agar—Tests ability to utilize sodium citrate as sole carbon source.

TSI (Triple-sugar-iron) agar—Three sugars are lactose, glucose, and sucrose together with sulfates and a pH indicator. Used in slants to determine relative use of each sugar and generation of sulfide.

SIM (Sulfide, indole, motility) agar—Tests for production of sulfide from sulfate, indole from tryptophan and motility.

Differential and **Selective media**—Contain selected chemicals.

Brilliant green agar—A dye that inhibits gram-positive bacteria, thus selecting for gram-negative organisms.

Sodium tetrathionate broth—An inhibitor for normal inhabitants of intestinal tract; favors *Salmonella* and *Shigella* species.

MacConkey agar—Contains bile salts and crystal violet that inhibit gram-positives and many fastidious gram-negative organisms, favoring enterobacteria.

Mannitol salt agar—Contains 7.5% NaCl that is inhibitory to most organisms and favors growth of staphylococci.

Eosin methylene blue—Partly inhibits gram-positive organisms. Eosin gives *Escherichia coli* a metallic greenish sheen. *Enterobacter aerogenes* colonies are pink. Other species are less pigmented.

Enrichment media—Contains blood, serum, meat extract or other nutrients that favor the growth of fastidious bacteria, particularly when present in low numbers. Often employed for clinical specimens such as cerebrospinal fluid.

After examining growth on selective or differential media (see Table 30.8), an experienced clinical microbiologist can select the best pathway to follow in identification. This choice would be influenced by microscopic examination, colony morphology, and specimen source. Microorganisms present in colonies from primary enrichment, selective, or differential media can be subjected to growth-dependent tests for identification (Tables 30.9 and 30.10). The media listed have proved to be accurate in identifying various potential pathogenic microorganisms. These organisms would also be tested for antibiotic sensitivity/resistance.

TABLE 30.9	Growth and colonial characteristics of some commonly isolated organisms on differential agar media

| | Eosin–Methylene-Blue | | Mannitol Salt | |
Organisms	Growth[a]	Color of Colony	Growth of Colony	Color
Enterobacter aerogenes	++	Pink, no sheen	Inhibited	
Escherichia coli	+++	Purple-green metallic sheen	Inhibited	
Klebsiella pneumoniae	++	Green metallic sheen, mucoid colony	Inhibited	
Salmonella typhimurium	++	Colorless	Inhibited	
Staphylococcus aureus	Inhibited		+++	Yellow
Staphylococcus epidermidis	Inhibited		++	Red

[a] ++ good growth; +++ excellent growth.

TABLE 30.10	Major diagnostic tests employed to differentiate pathogenic bacteria in a clinical laboratory	

Test	Purpose	Potential Application
Acid-fast stain	Tests for organisms with high wax (mycolic acid) content in cell surface that stain with hot carbolfuchsin and cannot be decolorized with acid alcohol	*Mycobacterium tuberculosis* is acid-fast
Catalase	Enzyme decomposes hydrogen peroxide $H_2O_2 \rightarrow H_2O + O_2$	*Staphylococcus* from *Streptococcus*
Citrate	Transports iron and is a carbon source	Classification of enteric bacteria
Coagulase	Causes clotting of plasma	*Staphylococcus aureus* from saprophytic staphs
Decarboxylases	Tests for ability to decarboxylate amino acids such as lysine, ornithine, or arginine	Classification of enteric bacteria
Esculin	Tests for cleavage of a glycoside	Separate streptococci
β-Galactosidase	Demonstrates an enzyme that cleaves lactose → glucose + galactose	Separates enterics and identifies pseudomonads
Gelatin liquefaction	Enzymatic hydrolysis of gelatin	Identify clostridia and others
Gram stain	Used as a first and primary differential test	
Hemolysis	Hemolysis of red blood cells; α-hemolysis, indistinct zone of hemolysis, some greenish to brownish discoloration of medium; β-hemolysis, clear, colorless zone around colonies	Pathogenic streptococci
Hydrogen sulfide	Demonstrates H_2S formation by sulfate reduction or from sulfur-containing amino acids	Identify enterics
Indole	Determines hydrolysis tryptophan to indole	Separate enterics
KCN[a] growth	Tests for ability of microorganisms to grow in the presence of cyanide (inhibits electron transport)	Aerobes/anaerobes
Lipase	Presence of the enzyme that cleaves ester bonds of fats yielding fatty acids + glycerol	Separate clostridia
Methyl red	Checks for acid production from glucose via mixed-acid fermentation	Separate enterics
Motility	By microscopic examination or diffusion through soft agar, shows the ability to move	Motile strains
Nitrate reduction	Reduction of $NO_3 \rightarrow NO_2$ (nitrate employed as terminal electron acceptor)	Enterics and others
Oxidase test	Demonstrates presence of cytochrome *c*, which oxidizes an artificial electron acceptor	Enterics from pseudomonads
Phenylalanine deamination	Deaminates the amino acid to phenylpyruvic acid	Proteus group
Protease	Tests for ability to digest casein	Separate *Bacillus* species
Sugar utilization	Growth on pentoses, hexoses, or disaccharides, producing acid and gas	Many
Urease	Detects an enzyme that splits urea to $NH_3 + CO_2$	Separate enterics
Voges-Proskauer	Detects acetoin as product of glucose fermentation and indicates neutral fermentation	Differentiate bacillus species and enterics

[a]KCN, potassium cyanide.

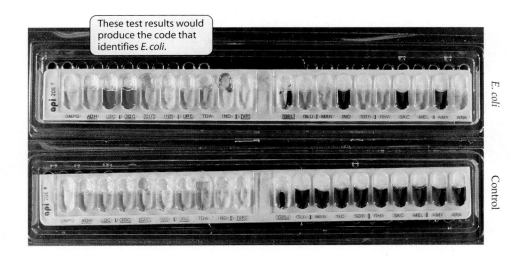

These test results would produce the code that identifies *E. coli*.

E. coli

Control

Figure 30.4 Identification of pathogens
The API 20E system for rapid identification of enteric organisms. There are 21 tests and each yields a positive or negative result. A number is calculated for each group of three tests, and the resulting seven-digit code will identify most enteric bacteria. Photo ©E. Chan/Visuals Unlimited.

Growth-Dependent Identification

Identification of various pathogens can be achieved by growth-dependent methods, and the procedure followed is generally based on preliminary identification of the microorganism involved.

Commercial diagnostic packets containing various media and selected differential tests have been developed. Among the manual rapid test systems is the API 20E system for the identification of enteric bacteria. These systems (Figure 30.4) are compact, they are simple to inoculate, and many clinical isolates can be analyzed in a short period. The sugars and diagnostic tests in these manual systems can be varied and used to confirm the identity of a microorganism based on its presumptive identification. Compact test kits have replaced many of the individual tube tests (Table 30.11) that were routine in clinical laboratories in years past and are still a part of the laboratory section of most general microbiology courses.

Serologic Identification

Culturing of microorganisms including viruses and selected bacteria is not always possible or practical in the clinic laboratory. For rapid diagnosis of many infectious agents, there are commercial test kits that can detect presence of an antigen or antibody. These **serological** or **immunological systems** can identify pathogens in clinical specimens rapidly and accurately without culturing of the infectious agent. Rapid immunologic test kits are available to detect: *H. influenzae* type b, *N. meningitidis* from cerebrospinal fluid, *S. pneumoniae, Helicobacter pylori, Bacteroides fragilis,* Respiratory Syncytial Virus, Herpes simplex virus types 1 and 2, HIV, and other suspected pathogens.

In some cases, diagnosis may rely on the presence of antibody in the serum of a patient, but the detection of antibody does not necessarily distinguish between an active and a previous infection. There must also be sufficient time between the onset of the disease and application of serological tests to permit the formation of circulating antibody, a period of 3 to 7 days.

Automated Identification Systems

Fully automated systems have been developed for the identification of many infectious microorganisms. These systems provide a constant monitoring of growth-dependent reactions and the susceptibility of the microorganism involved to antimicrobial agents. Such identification systems have many advantages over the manual manipulations that have been used for many decades. Among these are:

1. There is a minimalization of sample handling.

2. The chance of human error decreases, as there are fewer manipulations.

3. The system delivers results continuously, and each test specimen is read and recorded separately. There are no batch readings, so results of individual tests are available at the earliest time possible.

4. These systems have broad applicability and can accommodate updating as new infectious agents, chemotherapeutics, and technical capabilities emerge.

5. The systems require very little space.

There are a number of systems available, and the one depicted in Figure 30.5A has been selected as an example. The basic identification module for the system is a white plastic card (Figure 30.5B) that has 64 wells containing dried medium and/or reagent chemicals for differential tests. A number of different test cards are available that differ in the substrates that are present in the

TABLE 30.11	Growth-dependent tests that differentiate members of the enterobacteria

Tests	*Citrobacter freundii*	*Edwardsiella tarda*	*Enterobacter aerogenes*	*Escherichia coli*	*Klebsiella pneumoniae*	*Proteus vulgaris*	*Providencia alcalifaciens*	*Salmonella paratyphi*	*Salmonella typhi*	*Serratia marcescens*	*Shigella dysenteriae*	*Yersinia pestis*
Indole	−	+	−	+	−	+	+	−	−	−	±	−
Methyl red	+	+	−	+	−	+	+	+	+	−	+	+
Voges-Proskauer	−	−	+	−	+	−	−	−	−	+	−	−
Ornithine decarboxylase	−	+	+	±ᵃ	−	−	−	+	−	+	−	−
Motility	+	+	+	+	−	+	+	+	+	+	−	−
Gelatin liquefaction	−	−	−	−	−	+	−	−	−	+	−	−
KCN (growth in)	+	−	+	−	+	+	+	−	−	+	−	−
Glucose acid	+	+	+	+	+	+	+	+	+	+	+	+
Glucose gas	+	+	+	+	+	±	±	+	−	±	−	−
Lipase	−	−	−	−	−	+	−	−	−	+	−	−
NO$_3^-$→NO$_2^-$	+	+	+	+	+	+	+	+	+	+	+	+
Lactose utilization	±	+	±	+	+	−	−	−	−	−	−	−
H$_2$S on TSI	±	−	−	−	−	+	−	−	+	−	−	−
Citrate	+	−	+	−	+	−	+	−	−	+	−	−

ᵃ± most strains positive.

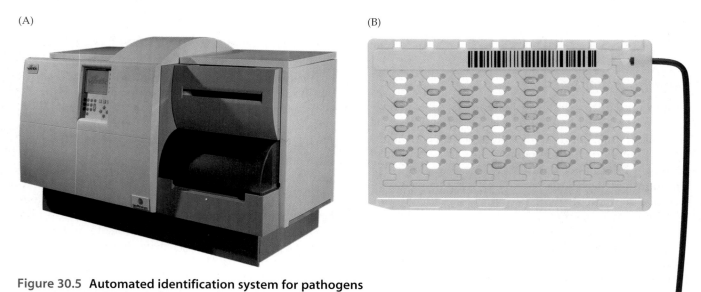

Figure 30.5 Automated identification system for pathogens
(A) The automated VITEK® 2 system for identification of clinical isolates. (B) Sample card that is placed in the automated VITEK® 2 system. The card is clear plastic, measuring 10 cm × 6 cm × 0.5 cm, and with 64 individual wells. Courtesy of bioMérieux, Inc.

TABLE 30.12	Identification cards available for automated identification of commonly encountered organisms	

Test Card	Range	Genera Detected (speciation within the genus is generally determined)
GNI	*Enterobacteriaceae, Vibrionaceae,* glucose nonfermenters	*Achromobacter, Acinetobacter, Citrobacter, Eikenella, Enterobacter, Escherichia, Flavobacterium, Hafnia, Klebsiella, Pasteurella, Proteus, Providencia, Pseudomonas, Salmonella, Serratia, Shigella, Vibrio, Yersinia*
GPI	Coagulase positive and negative, staphylococci, enterococcus, β-hemolytic streptococci, *Corynebacterium, Listeria, Erysipelothrix*	*Corynebacterium, Enterococcus, Erysipelothrix, Listeria, Staphylococcus, Streptococcus* (viridans group)
YBC	Clinically significant yeasts	*Candida, Cryptococcus, Geotrichum, Hansenula, Pichia, Prototheca, Rhodotorula, Saccharomyces, Sporobolomyces, Trichosporon*
ANI	Anaerobic bacteria	*Actinomyces, Bacteroides, Bifidobacterium, Capnocytophaga, Clostridium, Eubacterium, Fusobacterium*
NHI	Fastidious bacteria	*Actinobacillus, Branhamella, Cardiobacterium, Eikenella, Gardnerella, Haemophilus, Kingella, Moraxella, Neisseria*
UID	Detect, identify, and enumerate most common urinary tract pathogens	*Citrobacter, Escherichia, Enterococcus, Klebsiella/Enterobacter, Proteus, Serratia, Staphylococcus, Pseudomonas,* yeast
EPS	Enterics	*Salmonella, Shigella, Yersinia*

(A)

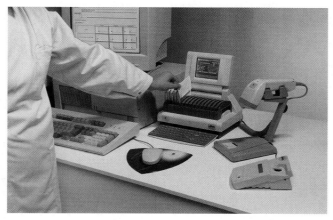

(B)

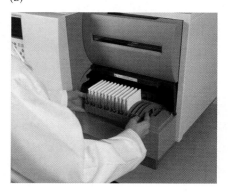

wells. The selection of a test card is based on the presumptive identification of the primary isolate (Table 30.12). This specific instrument is designed so that only the initial microbial suspension is required for identification testing; the instrument will automatically make any dilutions needed for susceptibility tests. For each isolate to be tested, a standard suspension is made in 3 ml of saline from an organism that has been **isolated** in **pure culture** from validated primary enrichment or selective/differential medium. Once the initial suspension is made, the appropriate identification/susceptibility test cards are selected and placed opposite the suspensions on the cassette with their straws inserted into the suspension (Figure 30.6A). The standard cassette holds up to fifteen cards, with the approximate setup time for a single isolate being 90 seconds. Once filled, the cassette is placed into the instrument for testing (the instrument shown can hold up to 60 cards) (Figure 30.6B). As the cassette moves through the instrument, additional dilutions are made (for susceptibility cards only), and cards are filled in the vacuum chamber, sealed, and placed onto a carousel within the incubator module at a selected temperature.

Figure 30.6 Loading samples into the automated identification system
(A) Loading an inoculated sample card into the cassette. (B) Placing the cassette in the incubation chamber. Courtesy of bioMérieux, Inc.

Each cassette is automatically read photometrically once every 15 minutes. Turbidity or color changes are recorded and transmitted to a computer system that analyzes the data. The data points are compared with a known database. When sufficient data are available, the system will print out a most likely and next most likely identification of the microorganism present in the original inoculum. The probability for each is given. The data may also be displayed on a computer screen for reference.

The time required to complete the identification of a clinical isolate varies. Analysis for the enterobacteria takes from 4 to 6 hours, depending on species. The glucose nonfermenting gram-negative bacteria require 6 to 8 hours to identify. The gram-positives such as staphylococci require 6 to 10 hours. Identification of yeasts is accomplished in the same manner, but requires up to 18 hours. A blood culture organism can be identified in 4 to 6 hours, and susceptibility to antimicrobials is run simultaneously with the other identification tests. By use of automated systems of identification and susceptibility, an appropriate treatment can generally be initiated in less than 8 hours.

Susceptibility cards are available to determine the susceptibility to an antimicrobial of a microorganism isolated in primary culture. Test cards contain a standard growth medium with graded levels of different antimicrobials. The instrument is programmed to provide information on the antimicrobial that would be most effective. The effective minimum inhibitory concentration (MIC) of an appropriate antimicrobial is printed out on average in 6 to 8 hours. There are at present seven different standard susceptibility test cards for gram-negative bacteria, and three for gram-positive. Each test card contains 20 to 22 antimicrobials. A nonautomated or manual method for determining the sensitivity of clinical isolates is the Kirby–Bauer disk diffusion assay, which will be discussed in subsequent text of this chapter.

Antigen–Antibody Techniques

Nearly all microbial species and, in many cases, strains of a pathogenic microorganism, will have at least one antigenic component that is unique. Analytical techniques are now available for the recovery and purification of these unique antigens.

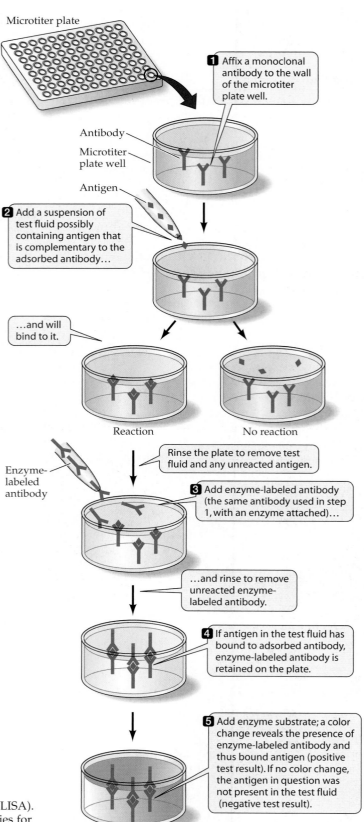

Figure 30.7 Direct ELISA
Direct enzyme-linked immunosorbent assay (ELISA). Using monoclonally produced specific antibodies for detection of an antigen.

A purified antigen can be employed to generate a *monoclonal antibody* that will react specifically with that antigen (see Chapter 27). Monoclonal antibodies to specific antigens are an effective, highly specific, and rapid diagnostic tool for identification.

One of the diagnostic tests that employs monoclonal antibodies is the **direct ELISA** test (Figure 30.7). There are both direct and indirect ELISA tests. The direct ELISA test requires a monoclonal antibody that can be chemically adsorbed onto the walls of a well in a microtiter plate (step 1). A suspension of serum or fluid that is suspected to contain a **specific antigen** is added. If present, the antigen will react with the antibody (step 2) adsorbed to the wall of the well. Binding of the antigen is both specific and strong enough to withstand rinsing that removes unreacted antigen.

A suspension of an enzyme-labeled monoclonal antibody is then added (step 3). This added antibody is equivalent to the antibody adsorbed to the well in step 1 and will react with the same antigen (step 4). This antibody has one modification in that a reporter enzyme is conjugated to it without affecting the capacity of the antibody to react with the specific antigen. Enzymes that are conjugated to antibodies are generally those that will produce a discernible color when the enzyme reacts with its substrate (step 5).

If the antigen is retained by the antibody at step 2, the antibody–enzyme conjugate (step 3) will also be retained specifically and not removed by rinsing. Addition of a substrate (step 5) for the conjugated enzyme will then result in a readily observed color change—a positive test. No color change indicates that no antigen was absorbed at step 2 and that the specimen did not have an antigen present that could react with the antibody—a negative test.

The **indirect ELISA** test is one that determines whether a **specific antibody** is present in a specimen such as serum (Figure 30.8). HIV is one infectious agent that can be identified by the indirect ELISA assay. Because blood transfusion has a potential for spreading HIV, it is essential that the presence of the virus be determined in donated blood. HIV infection induces the host to generate circulating antibody to specific antigens on the HIV particle. These antibodies are formed during the early stages of HIV infection before the immune system is compromised. Detection of these antibodies in the serum of a patient confirms exposure to HIV.

The procedure for an indirect ELISA test is outlined in Figure 30.8. In this case, the HIV antigen itself is adsorbed to the wall of a microtiter plate (step 1).

Serum suspected of containing antibody against this antigen is added to the microtiter well (step 2). Rinsing removes any antibodies not specifically attached to the antigen adsorbed to the well. However, if an antibody attaches to the antigen, it can be detected at step 3 by addition of an antibody–enzyme conjugate suspension. The antibody in this case is one that specifically reacts with human immunoglobulins (generally anti-IgG). Binding of the antibody (step 2) to the antigen results in the formation of a complex (step 4). Addition of a

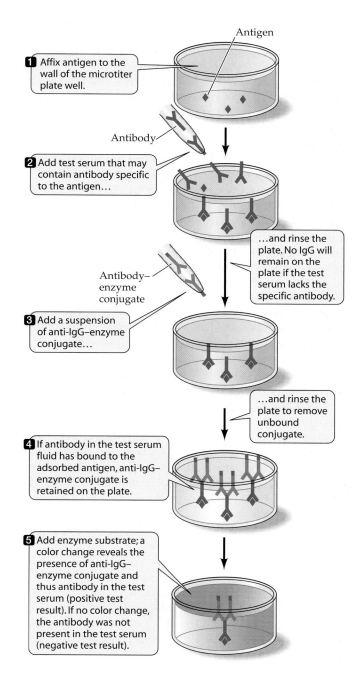

Figure 30.8 Indirect ELISA
Indirect ELISA test to determine the presence of a specific antibody in human serum.

substrate for the enzyme gives a visual indication that the antibody against the original antigen is present. The indirect ELISA can also be employed in diagnosing infections such as salmonellosis, plague, brucellosis, syphilis, tuberculosis, leprosy, and Rocky Mountain spotted fever.

Detection by Immunofluorescence

Antibodies tagged by chemically bonding them to fluorescent dyes are commonly employed in the clinical diagnostic laboratory. A fluorescent dye absorbs light of one wavelength, becomes excited, and emits light at another, longer, wavelength. This technique can be used, for example, in diagnosing the genital tract infection caused by *Chlamydia trachomatis*. To accomplish this, a smear of cervix scrapings would be placed on a slide, then a specific antichlamydial antibody–dye conjugate is applied. The antibody used in detection can be polyclonal or monoclonal and is usually obtained from a select animal species by injecting it with elementary bodies of *C. trachomatis* as antigen. Unreacted antibody–dye conjugate is then removed by rinsing. The slide is viewed with fluorescence microscopy, and the presence of the pathogen is indicated by a green fluorescence against the dark background of counterstained cells. In place of microscopy, a fluorometer can be used that electronically detects the total amount of light emitted at the appropriate wavelength.

A problem in identification of pathogens by immunofluorescence is that some species may share antigens. This is particularly the case with enteric bacteria. The use of monoclonal antibodies that interact solely with an antigen unique to a pathogen lessens the problem with interference from cross-reacting antibodies. Immunofluorescence can be a useful technique for identification of pathogens that grow slowly or are difficult to isolate. For example, mycobacterial, chlamydial, and brucellosis infections are diseases for which isolation of the organism is time-consuming but early treatment is essential. Other specimens that might be examined by immunofluorescence procedures are genital exudates for gonorrhea, feces for cholera, smears from bubonic plague buboes, tularemia lesions, and legionellosis aspirate. The immunofluorescence technique is applicable for cases in which clinical symptoms indicate that one of these pathogens may be responsible for the observed symptoms.

Agglutination

When antigens and antibodies interact, a complex develops that is, quite often, sufficiently large to be visible with the naked eye. This clumping of antigen–antibody is called **agglutination**. Agglutination occurs because antibodies and antigens have two or more binding sites and can form a network of linked antigen and antibody molecules (see Chapter 27). Agglutination has been adapted to the clinical laboratory for the rapid detection of either antigens or antibodies and is a relatively inexpensive and specific test. The agglutination test can be augmented by employing antigen- or antibody-coated latex beads. These beads are spherical and approximately 0.8 µm in diameter (Figure 30.9). Proteins (antigen or antibody) will adhere tightly to the polystyrene surface of the latex beads. When an antigen or antibody is fixed to the bead and a complementary antibody or antigen is added, the agglutination reaction can occur in 30 seconds or less (if positive). A drop of urine, serum, spinal fluid, or other specimen in suspension can be added to the bead–protein complex to determine the presence therein of a specific complementary antigen–antibody. The agglutination test has been used to identify *S. aureus*, *N. gonorrhoeae*, and *H. influenzae*. Yeast infections caused by organisms such as *Cryptococcus neoformans* can be diagnosed by using agglutination tests.

Plasmid Fingerprinting

A plasmid is an extrachromosomal deoxyribonucleic acid (DNA) that replicates autonomously in a bacterial cell. Plasmids are present in many bacterial genera, including pathogenic microorganisms. Bacterial **strains** that are related generally contain the same number of plasmids of equivalent molecular weight. Determining the number and size of plasmids in strains of a species can be a measure of the relatedness. To analyze for plasmids, a cell mass must be lysed to free intracellular components. The plasmid DNA is partially purified and then applied to an agarose gel and subjected to electrophoresis. The migration rate of each plasmid in the gel is inversely proportional to its molecular weight. The DNA can be visualized in the agarose gel by staining with ethidium bromide. The greater the number of bands with equivalent migration rates in the gel, the more accurately the strain can be identified. The number and migration of bands form what is called, a **plasmid profile** or **plasmid fingerprint**. Plasmid fingerprinting is not a definitive test because one strain of a species may have several plasmids and another strain may have few or none. However, if strains have several plasmids in common, it is strong evidence that they are related.

Plasmid profiles are of value in tracing the origin of specific strains that may be involved in a local epidemic such as a nosocomial infection. If a hospital has a series of *Staphylococcus* infections and the plasmid profile of the isolated organisms is similar, this can be indicative of a potential common origin. If the organisms involved

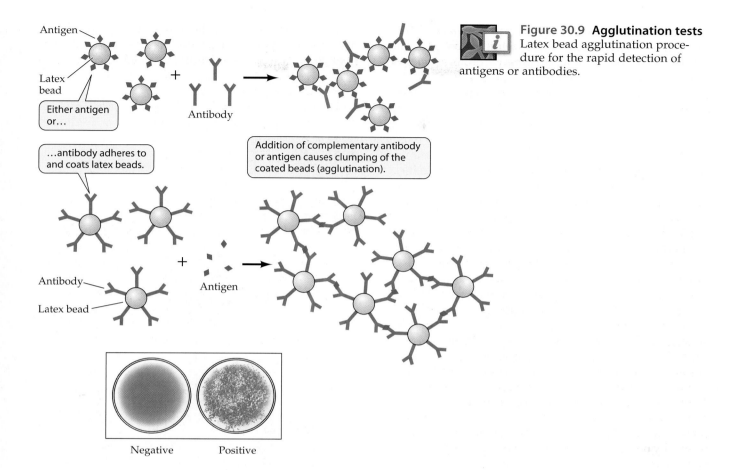

Figure 30.9 Agglutination tests Latex bead agglutination procedure for the rapid detection of antigens or antibodies.

have significantly different profiles, it would suggest that the infections do not have a common origin. Food poisoning cases caused by enterotoxin from coagulase-positive staphylococci may be traced by this technique. The *S. aureus* strain in the nasal passage of a food handler can be subjected to the profile test. If a strain of *S. aureus* from a victim of food poisoning has an identical plasmid profile to that from the food handler, this is evidence that the food handler was the source of the staphylococcal infection.

A related DNA analytical procedure, based on digestion of the entire chromosome by restriction enzymes, has been used successfully in tracing the origin of common-source nosocomial outbreaks. The restriction enzymes yield characteristic DNA fragments that are separated by gel electrophoresis. Patterns of identity based on migration of the fragments indicate relatedness (or unrelatedness).

Phage Typing

Bacterial viruses (bacteriophages) are specific to the species or, in many cases, the bacterial strain they can infect. This specificity is based on the presence of viral re-

ceptors on the surface of the bacterium where the virus attaches to initiate the infectious process. Susceptibility of bacterial strains to viruses can be an indicator of genus, species, and possibly a strain.

Virus susceptibility can be determined by spreading a select bacterium over an agar surface. The plate is squared off by marking with a wax pencil on the bottom of the plate. Then a drop of different viral suspensions is applied to each marked square. After suitable incubation for bacterial growth, the plate is examined for plaques, indicating which of the viruses was capable of attacking that bacterium. Phage typing is useful in tracing localized epidemics such as a case of food poisoning.

Nucleic Acid–Based Diagnostics

Advances in molecular biology have provided techniques for rapid identification of infectious agents. These techniques take advantage of the fact that the nucleotide sequences in one genus or species differ in a measurable way from those in another. A pathogenic microorganism that generates toxins, synthesizes enzymes involved in invasion, or produces virulence factors has unique genetic information. The information for these specific virulence factors resides in the organism's DNA. DNA diagnostics depend on the identification of unique nucleotide sequences that are present in one species or strain but not present in others.

Nucleic acid probes are better than immunological procedures in many ways. Antibodies are sensitive to temperature or other physical conditions that might be necessary to remove interfering material. These harsh conditions that would inactivate an antibody do not harm nucleic acids from pathogens or those in probes.

Diagnostics based on hybridizing or amplifying nucleic acids are gradually being implemented in clinical laboratories. Practical application of these molecular methods has, to date, had some limitations, as these methods may be time-consuming and expensive, and may require the services of experienced technologists. However, these techniques can detect pathogens that available tests cannot and are gaining acceptance, particularly in larger clinical laboratories.

Amplification of target nucleic acids by polymerase chain reaction (PCR) is necessary in using probe assays for the identification of many pathogens. If the disease agent is present in relatively low numbers or isolation of the pathogen is impractical, such as with a viral or rickettsial infection, amplification is a must.

Amplification assays have been developed for many infectious agents including Epstein–Barr virus, enteroviruses, *Ehrlichia, Rickettsia,* Herpes simplex virus, and hepatitis B and C. An amplification assay is also employed to determine the HIV levels in patients with AIDS, particularly for those undergoing chemotherapy.

Pneumonia is an example of a disease for which use of probe assays is impractical. Many microorganisms that are agents of pneumonia are carried in the respiratory tract of asymptomatic individuals. A sensitive amplification assay would be overloaded with organisms that may not be involved in the disease process.

Clinical application of DNA technology relies on the tight binding (hybridization) that occurs between complementary single strands of DNA. A single strand of DNA of a known sequence can be used as a **probe**. If DNA from a clinical isolate has sequences that are complementary to those of a probe, they can hybridize, and these hybrids are detectable.

The DNA from a microorganism obtained from a specimen can be released by treatment with strong alkali, and the DNA rendered single stranded (ssDNA). The source of ssDNA may also be a bacterial colony treated with detergent to release the cellular DNA (double-stranded, dsDNA), and this can be rendered single-stranded by heating. The ssDNA fragments are affixed to a filter (Figure 30.10), and the probe with an attached reporter molecule is added. The temperature and time allowed is adjusted to one at which stable duplexes are formed. This is de-

pendent on the length of the nucleic acid probe and its base composition. After an appropriate period, the unhybridized probe DNA is rinsed through the filter, and

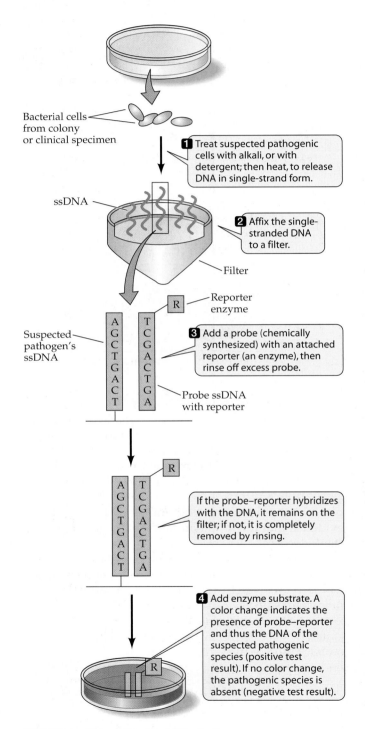

Bacterial cells from colony or clinical specimen

1 Treat suspected pathogenic cells with alkali, or with detergent; then heat, to release DNA in single-strand form.

ssDNA

2 Affix the single-stranded DNA to a filter.

Filter

Reporter enzyme

R

Suspected pathogen's ssDNA

AGCTGACT

TCGACTGA

3 Add a probe (chemically synthesized) with an attached reporter (an enzyme), then rinse off excess probe.

Probe ssDNA with reporter

R

AGCTGACT

TCGACTGA

If the probe–reporter hybridizes with the DNA, it remains on the filter; if not, it is completely removed by rinsing.

R

4 Add enzyme substrate. A color change indicates the presence of probe–reporter and thus the DNA of the suspected pathogenic species (positive test result). If no color change, the pathogenic species is absent (negative test result).

Figure 30.10 DNA probe
DNA probe for detection of a specific microorganism in a clinical specimen.

the amount of reporter remaining is determined. The reporter molecule can be a fluorescent dye, an enzyme, or a radiolabel. A positive test can be determined by fluorescent microscopy, an enzyme dye color change, or radioactivity. There are available DNA probes that will bind to complementary strands of ribosomal RNA (rRNA). A hybrid of rRNA–DNA is more sensitive than the DNA–DNA hybrids, and fewer microorganisms are required for an accurate determination. DNA–DNA probes have been effective in diagnoses of rickettsial and cytomegalovirus infections, among others.

A dipstick probe has been developed for the identification of infectious agents in food and crude clinical specimens (Figure 30.11). In this method, a clinical specimen is treated with alkali to release DNA and generate ssDNA from any microorganism present. The specific single-stranded nucleic acid probe is added. This specific ssDNA probe has incorporated in it a sequence of bases (poly T or poly A) that serve as a capture probe. A low-molecular-weight molecule such as dioxygenin would also be chemically linked to the nucleic acid probe (see D in Figure 30.11). Following incubation of the alkali-treated clinical specimen with the nucleic acid probe, nucleases are added to destroy any single-stranded probe that is unhybridized. A dipstick that bears the capture-probe base sequence would be inserted into the hybridization solution. The capture probe with any hybridized DNA would be exposed to antibodies to the dioxygenin reporter, and these antibodies would be coupled to a fluorescent dye or a detectable enzyme. Whether hybridization occurred would be determined by observing an enzyme reaction or fluorescence by microscopy.

Probes are of particular value in diagnosing *C. trachomatis* and organisms that are involved in respiratory infections. For analysis of crude specimens (sputum, feces, or discharges), a probe is an effective technique because isolation of the organism is not essential. Probes can also be used for identification of viruses and other organisms that may not be grown in culture.

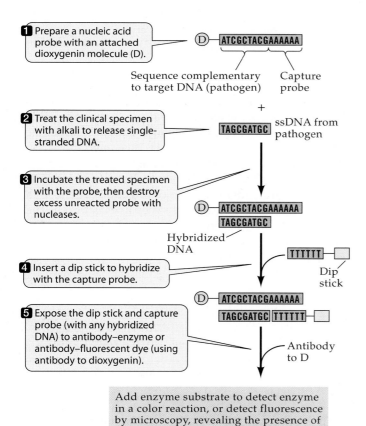

Figure 30.11 Dip stick probe
Dip stick assay procedure for recovery of single-stranded DNA (ssDNA).

> ## SECTION HIGHLIGHTS
>
> The characteristics used by clinical microbiologists to identify pathogens include special growth requirements (detected using selective media), morphology (detected by microscopy), surface structure (detected by stains), unique antigens (detected by serological tests, ELISA, immunofluorescence, and agglutination), the plasmid profile (detected by gel electrophoresis), phage susceptibility (detected by plaque formation), and the presence of specific DNA sequences (detected by hybridization and PCR). Automated identification systems increase the efficiency and consistency of diagnosis.

30.10 Antibiotic Sensitivity

Antibiotics or chemotherapeutics are the treatments available for many bacterial and mycotic infections. As previously mentioned, resistance to these inhibitors is a major concern in the treatment of many infectious diseases. Resistance to one or more antibiotics is now encountered with many infectious microorganisms and is a considerable concern in nosocomial infections. The sooner a patient is treated with an appropriate antibiotic the better.

The susceptibility of a clinical isolate to an antibacterial agent can be determined by the automated system described previously. However, where this technology is not available, a disk diffusion assay can be used. The disk assay widely used was developed by W. Kirby and A. W. Bauer and is termed the Kirby–Bauer method. In this procedure an aliquot of the suspended clinical isolate is added to melted agar and poured into a Petri dish. After solidification, a filter disk impregnated with a measured amount of antibiotic is placed on an agar surface. The disk will absorb moisture from the agar, effecting the diffusion of antibiotic into the agar medium. The distance that the antibiotic moves outward from the disk depends on the solubility, concentration, and other characteristics of the antibiotic. As the antibiotic diffuses outward, a concentration gradient is established with the highest level in the immediate area of the disk.

In practice, a liquid growth medium is inoculated by touching an inoculating loop to a colony of bacteria growing on a primary isolation plate and transferring the inoculum to a liquid medium. The inoculated liquid medium is incubated, and when turbidity is evident, an aliquot is added to melted agar and poured into a Petri plate. After solidification, antibiotic-containing disks are placed on the surface of the agar and the plate is incubated. After a 12- to 18-hour incubation period, **the zones of inhibition** can be measured (Figure 30.12).

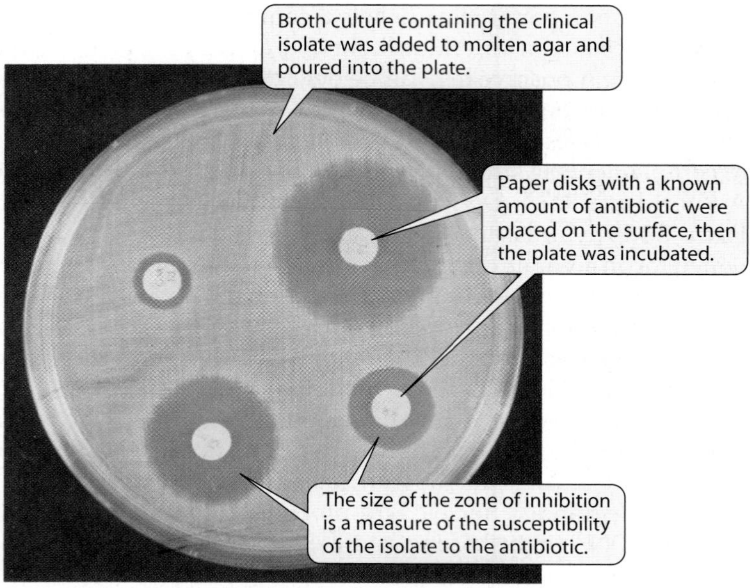

The susceptibility of an organism to an antibiotic is directly related to the diameter of the zone of inhibition. The zone diameters can be interpreted by using a table prepared by Kirby–Bauer (Table 30.13). This table was developed by determining MICs of antibiotics effective against many pathogenic bacteria and relating the MIC to the diameter of an inhibition zone.

MIC values can also be determined as follows: A series of tubes containing an appropriate medium and graded concentrations of antibiotic are inoculated with

Figure 30.12 Disk diffusion assay
Kirby–Bauer procedure for determining the susceptibility of a clinical isolate to antibiotics. Charts are available for estimating the susceptibility of the clinical isolate from the size of the zone of inhibition. Courtesy of J. J. Perry.

equivalent amounts of a suspension of the clinical isolate. The tubes are examined for growth after a suitable incubation period. The lowest concentration of antibiotic that inhibits growth completely is considered the MIC.

Clinical diagnostics has made remarkable advances in the last few years. Automation and rapid specific tests permit accurate identification of infectious agents in a short period. Early treatment saves lives, shortens the illness and convalescence time, and saves money. The epidemiologist and clinical laboratory in concert contribute much to our well-being.

TABLE 30.13	Diameter of the zone of inhibition as a measure of susceptibility to selected antibacterial agents		
Antibacterial	**Amount on Disk (μg)**	**Zone Diameter (mm)**	
		Resistant	**Susceptible**
Ampicillin	10	28 or less	29 or more
Erythromycin	15	13 or less	18 or more
Gentamicin	10	12 or less	15 or more
Tetracycline	30	14 or less	19 or more

SECTION HIGHLIGHTS
The antibiotic chosen to treat an infection depends on the antibiotic sensitivity of the pathogen. Automated identification systems can determine the minimum inhibitory concentration (MIC) for up to a dozen antibiotics simultaneously. When automated systems aren't available, the Kirby–Bauer diffusion assay can be used to determine antibiotic sensitivity.

SUMMARY

- Infectious diseases were the major cause of death worldwide before antibiotics were available. They now are responsible for about 8% of the deaths in the developed countries, but 50% of the deaths in developing countries are due to infectious diseases.

- The science concerned with the prevalence and distribution of diseases in the population is **epidemiology**.

- An **epidemic** occurs when a disease occurs in the population at a higher than normal frequency. A constant low frequency of disease is called the **endemic** level. A worldwide epidemic is a **pandemic**.

- The **incidence** of a **disease** is the number of new cases per unit time in a susceptible population of 100,000. **Mortality** rate is deaths per total infected. **Morbidity** is the number of new cases during a period divided by the total population.

- A **carrier of infection** is an infected individual that transmits a disease to others. A **chronic carrier** is an individual permanently infected and shedding an agent of infectious disease. A **reservoir** is the site where an infectious disease–causing microorganism is maintained between outbreaks.

- **Human contact** diseases and **respiratory** diseases are passed by human interaction such as touching, coughing, or sneezing. Among these are **colds**, **influenza**, and **tuberculosis**.

- **Water-** and **food-borne** diseases are generally caused by organisms that infect the human gastrointestinal tract. They are transmitted through fecal contamination.

- A major public health concern is **sexually transmitted diseases**. HIV infection, chlamydial

infections, and **herpes** are all on the increase, as are **genital warts**.

- Epidemics of **vector-borne** diseases such as **Rocky Mountain spotted fever** and **Lyme disease** occur during summer months, and incidence of Lyme disease in the United States continues to increase at an alarming rate.

- An **infection acquired** in a hospital is termed a **nosocomial infection**. Nosocomial infections affect an estimated 2 million patients each year. The prevalent organism is *Escherichia coli*. **Staphylococcal** and **streptococcal infections** are also a problem.

- Disease **incidence** is monitored in most states by the requirement that physicians report about 60 different diseases to the state division of epidemiology. They in turn report these to the Centers for Disease Control and Prevention (CDC) in Atlanta, Georgia.

- Many diseases are controlled in the United States by **vaccination** of babies and young children. Among the vaccinations are those for **mumps, measles, rubeola, diphtheria, pertussis (whooping cough), tetanus, polio**, *Haemophilus influenzae* b, **chicken pox**, and **hepatitis B. Smallpox** was **eradicated** worldwide by vaccination.

- **Reservoirs** that sustain infectious agents and transmit them to humans are a problem in **epidemiology**, and their elimination is important in controlling disease. This is not possible in all cases.

- **Food inspection** and **water sanitation** are important public health measures.

- A real problem in controlling infectious diseases is development of **drug resistance** among the microbes that cause diseases.

- Rapid identification of pathogens involved in an infection leads to better treatment. A clinical laboratory is charged with this responsibility. Much of the identification has been automated.

- A specimen is the material delivered to the clinical laboratory. It must be handled carefully and may consist of blood, urine, sputum, and so forth.

- Automated procedures are now available that can identify clinical isolates in less than 20 hours and also determine resistance and susceptibility to antibiotics.

- An ELISA (Enzyme-Linked Immunosorbent Assay) is a diagnostic procedure that determines whether a specific antigen or antibody is present in a suspension.

- Antibodies labeled with fluorescent dyes are an important diagnostic tool for identification of pathogens. A specimen is placed on a slide and dye-labeled antibody is applied. Unreacted antibody is rinsed away and the slide is viewed with a fluorescence microscope. Agglutination tests with antibody- or antigen-coated beads, plasmid profiles, and phage typing are other available diagnostic tools.

- **Nucleic acid probes** are a technique for detecting pathogens. The methods take advantage of the complementarity of the nucleotides in a probe and equivalent unique base sequences in the genome of a pathogen. Probes are 60 to 80 nucleotides long and generally are a product of chemical synthesis.

- When small numbers of a pathogen might be present in a clinical specimen the **polymerase chain reaction (PCR)** can be applied to increase the level of DNA present to detectable levels.

- Determination of **antibiotic sensitivity/resistance** of an infectious microorganism is necessary in designing a proper treatment.

 Find more at www.sinauer.com/microbial-life

REVIEW QUESTIONS

1. Define epidemic, endemic, and pandemic.

2. How does a sporadic disease differ from one that is seasonal? Give examples of some seasonal diseases.

3. Two terms employed in disease reporting are *mortality* and *morbidity*. In fact, the Centers for Disease Control and Prevention (CDC) publishes the *Morbidity and Mortality Weekly Report (MMWR)*. What is the significance of these two terms?

4. What is a disease reservoir? Why are diseases that spread from animal reservoirs such a problem?

5. How are human contact diseases spread? Give examples of such diseases. How can they be controlled?

6. Describe how sexually transmitted diseases can be controlled. Which are treatable?

7. Why are nosocomial diseases such a concern? Can they be readily controlled?

8. What are some of the reasons for the doubling of life expectancy in the United States over the last two centuries?

9. We have lessened the incidence of the childhood diseases by vaccination. What are some of these? Why are they still a concern?

10. Name the agency that is responsible for control of diseases worldwide. What major success has it had?

11. What are some major disease problems in developed countries? Developing countries?

12. Why is it important that a disease-causing microbe be isolated and identified?

13. What types of specimens are sent to the clinical laboratory? How are some of these treated once there? What identification procedures are generally used?

14. What is an ELISA? How is it performed?

15. How is immunofluorescence used? Agglutination?

16. How are plasmid fingerprinting and phage typing used in the clinical laboratory?

17. Nucleic acid probes are of growing importance in the clinical laboratory. How are they used?

18. What is PCR? How is it employed in virus identification, and why is it necessary?

SUGGESTED READING

Centers for Disease Control and Prevention. *MMWR Morbidity and Mortality Weekly Reports*. Atlanta: Centers for Disease Control and Prevention.

Centers for Disease Control and Prevention. (Annual) *Surveillance*. Atlanta: Centers for Disease Control and Prevention.

Isenberg, H. D. 1998. *Essential Procedures for Clinical Microbiology*. Washington, DC: ASM Press.

Murray, P. R., editor-in-chief. 2003. *Manual of Clinical Microbiology*. 8th ed. Washington, DC: ASM Press.

Detrick, B., R. G. Hamilton and J. D. Folds. 2006. *Manual of Molecular and Clinical Laboratory Immunology*. 7th ed. Washington, DC: ASM Press.

Salyers, A. A. and D. D. Whitt. 2001. *Bacterial Pathogenesis: A Molecular Approach*. 2nd ed. Washington, DC: ASM Press.

Sussman, M., ed. 2001. *Molecular Medical Microbiology*. San Diego: Academic Press.

Truant, A. L. 2002. *Manual of Commercial Methods in Clinical Microbiology*. Washington, DC: ASM Press.

Tulchinsky, T. H. and E. A. Varavikova. 2000. *The New Public Health*. San Diego: Academic Press.

PART VIII

Applied Microbiology

The objectives of this chapter are to:

◆ Introduce the wide variety of industrial products generated by microbes.

◆ Describe the techniques and equipment used by the fermentation industry.

◆ Outline the approaches used to identify new antibiotics.

◆ Survey some of the important pharmaceutical products now made using genetically engineered microorganisms.

31

Industrial Microbiology

There is no field of human endeavor, whether it be in industry or in agriculture, whether it be in the preparation of foodstuffs or in connection with problems of shelter and clothing, whether it be in the conservation of human and animal health and the combating of disease, where the microbe does not play an important and often a dominant part.
—Selman A. Waksman, 1943

*I*ndustrial microbiology traces its origins to prehistoric times. Fermentation technology was born when early civilizations unwittingly took advantage of the capacity of microorganisms to produce alcoholic beverages as well as leavened bread and cheese. Early Sumerian city–states (fourth millennium BC) produced beer and wine from fermented barley and grapes, respectively. Ancient civilizations also made cheese and salted and fermented meat, thereby preserving these perishable foodstuffs. Modern innovations in food preservation began with the Frenchman François Appert, who developed canning methods in 1809 well before the debunking of spontaneous generation. This was in response to Napoleon's offer of a reward to anyone who developed a means for long-term preservation of food. Preservation of food, particularly in warm weather, would enable his armies to move rapidly and not be dependent on local crops for sustenance.

During the first half of the twentieth century, wine and beer manufacture, vinegar, and bread yeast production moved from the realm of an ancient art to an established science. Large-scale microbial processes for the manufacture of citric and lactic acid were developed. Acetone, butanol, and ethanol were produced in commercial quantities by fermentation, but are now mostly products of chemical synthesis.

The second half of the twentieth century was an era of dramatic change in the fermentation industry. Much of this change originated with the discovery that antibiotics can be an effective weapon against disease. The search for and production of novel antibacterial, antifungal, antitumor, and antiviral agents became major industries in Japan, the United States, and Western Europe. Other products generated by microbes, including vitamins, sterols, organic acids, amino acids, flavoring agents, enzymes, and fermented foods and beverages, gained commercial importance.

Knowledge gained over the last 20 years has markedly broadened the potential for the production of useful compounds by microorganisms. This new field of endeavor is broadly known as **biotechnology**. Biotechnology is mainly based on the use of living systems (usually microbial) to solve technical problems or achieve technological goals. For example, with genetic engineering, human genes for insulin biosynthesis have been introduced into bacteria, thereby enabling bacteria to synthesize this hormone. Thus, diabetics can receive a more effective and safer insulin for their treatment. The synthetic insulin is superior because it does not contain animal derived-material that causes an allergic response in some humans.

Microorganisms had a significant impact on the economic strength of a country. For example, an estimated 5% of Japan's gross national product stems from the broad capabilities of the microbe.

31.1 Microbes Involved in Industrial Processes

The fermentation industry is a collective term applied to highly developed processes that explore and exploit the capacity of microbes to generate useful products. Many of the microbes employed commercially, such as the antibiotic-producing actinomycetes, do not grow anaerobically and therefore do not carry out "ferments" as the term was originally defined (see pages 36, 37). In industry, fermentation is the name applied to production of commercial quantities of a biological product in tanks or containers and generally by aerobic processes. Every microorganism used in the fermentation industry is unique and carefully selected or engineered for the ability to yield maximum quantities of desired product.

Metabolites such as amino acids, B vitamins, or nucleic acid bases may be overproduced by some microbial species in nature and excreted into the proximate vicinity of their microenvironment. Other microorganisms in this environmental niche may assimilate these compounds. Thus, one strain may overproduce the metabolite but require for its growth a metabolite released by another strain. Such harmonious associations occur widely in natural habitats. These microorganisms that overproduce low levels of a metabolite in nature may, when isolated in pure culture and subjected to laboratory growth conditions, produce measurable quantities of a given metabolite.

Techniques are available for the isolation and selection of microbes that excrete selected metabolites such as nucleic acid bases, amino acids, B vitamins, or antibiotics. Once isolated, the microbes can be manipulated to produce significantly higher levels of the desired metabolite. There are many procedures that can be used to accomplish this, including genetic alteration (bioengineering) and selection of progeny that are prolific in generating the selected metabolites. The incorporation of specific precursors or metabolites to the growth medium may increase the yield of valued product at a nominal increase in cost. Through rigorous investigation of various regimens, phenomenal increases in yield can be realized. For example, the original strains of *Penicillium chrysogenum* that yielded 1.2 milligrams of penicillin per liter were manipulated by mutation and selection to produce more than 50,000 milligrams per liter. This led to the availability of penicillin for treatment of disease at a reasonable cost.

Microbial strains developed for the synthesis of commercial products are highly selected specialists and generally unfit to survive outside the laboratory setting. Chosen for their genetic stability during the fermentation process and consistency in production, they are fast-growing organisms that generate useful products in a relatively short period. Such strains are generally harmless to humans and to the environment, as containment of industrial microbes involved in large-scale fermentations would be difficult and expensive.

Primary and Secondary Metabolites

Some metabolites are synthesized and appear in the medium during active growth, and others appear as the organism completes the growth cycle and enters the stationary phase (see Chapter 6). **Primary metabolites** are those produced during active growth, and those synthesized after the growth phase nears completion are designated **secondary metabolites** (Figure 31.1).

A primary metabolite is generally a consequence of energy metabolism, and allows for the continued growth of the microorganism. Recall the discussion of anaerobic metabolism (see Chapter 8) in which a product of glycolysis serves as electron acceptor for the reduced pyridine nucleotide generated by glyceraldehyde-3-phosphate dehydrogenase (see Figure 8.3). In the absence of oxygen, the terminal electron acceptor would be pyruvate and the product of anaerobic energy generation would be lactic acid. The microorganism involved could not generate adenosine triphosphate (ATP) (or grow) without the concomitant production of the primary metabolite, lactic acid, a constituent of yogurt and other food products.

(A)

(B)

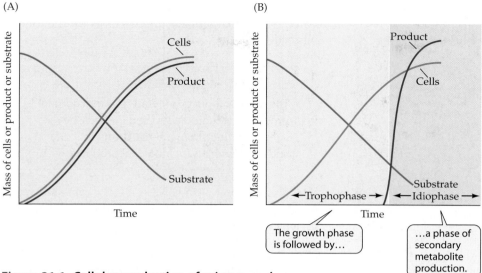

Figure 31.1 Cellular production of primary and secondary metabolites
Time-course for substrate utilization, increase in cell mass (growth), and production of (A) primary and (B) secondary metabolites.

The production of secondary metabolites is a result of complex reactions that occur during the latter stages of primary growth. The hallmark of secondary metabolites is that they are produced by a single or relatively few species. In microorganisms that produce secondary metabolites, the growth stage is called the **trophophase**. The phase involved in production of secondary metabolites is the **idiophase**. In some cases, enzymes involved in trophophase growth shift to the production of metabolite during the idiophase. Citric acid fermentation by *Aspergillus niger* is an example of a product resulting from a metabolic shift. During the trophophase, a substrate such as glucose is mostly used for synthesis and as a source of energy. In generating energy, much of the substrate is oxidized to CO_2. During idiophase, glucose is mostly converted to the secondary metabolite, citric acid, with little respiration (CO_2 production). In the production of antibiotics, which are also secondary metabolites, a different scenario prevails. An actinomycete that produces an antibiotic synthesizes a battery of novel enzymes during late exponential and early stationary phase growth. These are responsible for the synthesis of the antibiotic. The synthesis of these enzymes can be a major undertaking. For example, more than 300 genes are involved in the synthesis of enzymes that generate chlortetracycline by *Streptomyces aureofaciens*. At least 72 intermediates are formed, and only 27 of these have been isolated and characterized. Only a few enzymes involved in chlortetracycline synthesis are required for cell growth.

SECTION HIGHLIGHTS

The fermentation industry exploits specialized microbes, derived from natural microbes by mutation and selection, to make a wide range of useful products. These products can be either primary metabolites, produced during active growth (the trophophase), or secondary metabolites, which are produced as microbes enter stationary phase (the idiophase).

31.2 Major Industrial Products

The diverse industrial products generated by microorganisms are listed in Table 31.1. Bread, cheese, pickles, olives, and mushrooms are common foodstuffs in our diet. Each of these is a direct consequence of microbial activity. Although the role of microbes in the production of cheese and beer is widely known, few are aware that most soft drinks contain citric acid, a product of microbial synthesis.

Flavor enhancers, such as monosodium glutamate (MSG), inosinic acid, aspartate, alanine, and glycine are products of microbial metabolism. Vinegar is a commonly used flavoring agent and preservative. Many of the B vitamins can be produced in significant quantity by selected microbial cultures, but only riboflavin and vitamin B_{12} biosynthesis compete economically with

TABLE 31.1	Major commercial products obtained from microbes

I. Foods, flavoring agents and food supplements, and beverages
  Foods
    Fermented meat
    Cheeses and milk products
    Edible mushrooms
    Leavened bread-baker's yeast
    Coffee
    Pickles, olives, sauerkraut
    Single-cell protein
  Flavoring agents and food supplements
    Vinegar
    Nucleosides
    Amino acids
    Vitamins
  Beverages
    Wines
    Beer, ale
    Whiskey
  Vitamins
    B_{12}
    Riboflavin
II. Organic acids
  Citric acid
  Itaconic acid
III. Enzymes and microbial transformations
  Commercial enzymes
  Sterol conversions
IV. Inhibitors
  Biocides
  Antibiotics
V. Products of genetically engineered microbes
  Insulin
  Human growth factor

Some of the various major commercial products generated by microorganisms are discussed in the following sections.

SECTION HIGHLIGHTS

The useful products generated by microorganisms include ingredients used in foods, beverages, and food supplements, drugs, inocula, and raw materials for industrial products.

31.3 Foods, Flavoring Agents and Food Supplements, and Vitamins

Microorganisms themselves may serve as food, but most often they are used as agents to convert basic foods such as milk or meat to a more desirable product. Microorganisms can also synthesize flavoring agents or flavor enhancers, and these can contribute to the quality and acceptance of foods. Food additives obtained from fermentation can serve as preservative agents or as supplements to improve the nutritional value of a food.

Foods

Many foods are either derived from microbes themselves or processed using steps involving microbes. Examples of such foods, and the microbes used to produce them, are described here.

SINGLE-CELL PROTEIN   Bulk microbial biomass, termed *single-cell protein*, or SCP, can be used as a source of human or animal food. SCP may be valuable where climatic conditions cause agricultural crops to fail, where arable land is scarce, or when populations are under threat of starvation. A potential source of food would be the bulk harvesting of bacterial or yeast cells grown on substrates that in themselves would not support human sustenance (Box 31.1).

Microorganisms are attractive food sources for a number of reasons:

• Rapid growth

• Growth largely unaffected by climate

• Low requirement for land

• Inexpensive substrates for growth

The potential sources of SCP are yeast or bacterial cell mass. Yeasts and filamentous fungi show greater promise as potential food sources for human consumption than do bacteria.

chemical synthesis, the primary source of other B vitamins (biotin, thiamin).

The conversion of low-cost biologically inactive sterols to an activated form is readily accomplished by microorganisms. These substances are important in maintaining human health. Antibiotics synthesized by microbes have had a profound effect on human health and longevity. In the expanding field of biotechnology, microbes are genetically engineered to produce a variety of proteins and peptides including insulin, human growth factor, interferon, blood clotting factors, and vaccines. Production of rhizobia to use as inocula to increase the symbiotic fixation of N_2 in legumes (peanuts, peas) is important in food production.

BOX 31.1   *Milestones*

Bacteria as Food

The consumption of microbes as food has generally been limited to mushrooms (filamentous fungi) and yeasts. Yeasts are eaten as single-cell protein or as a food supplement (see text). Bacteria—as food—have been collected and eaten since ancient times by tribes in the central African republic of Chad. The organism is a spiral-shaped **cyanobacterium** of the genus *Spirulina, Spirulina platensis*, and is collected from the bottom of seasonally dried-up ponds and shallow waters around Lake Chad. The cyanobacterial mats are dried in the sun and cut into small cakes. The cakes are called **dihe**. Dihe is rich in protein (62% to 68% of dry matter) and an important supplement to the diets of the desert nomads. *Spirulina* is grown commercially in Mexico and France. The potential yield of 10 tons of protein per acre dwarfs the yield obtained from wheat (0.16 tons per acre) and beef (0.016 tons per acre). *Spirulina* is now sold as a food supplement in health food stores in the United States. It is marketed as dried cakes or powdered product.

Spirulina platensis, a cyanobacterium used as food. Courtesy of J. T. Staley.

Yeasts grown on a low-nitrogen high-carbon substrate are rich in lipid, but conversely when grown on high nitrogen they have a higher protein content. Yeast cells are rich in essential B vitamins. However, yeast protein is low in sulfur-containing amino acids and is not sufficiently balanced to provide for human or animal needs. Yeast cells also have a high nucleic acid content, and may cause gout or other ailments when consumed in quantity. However, bulk yeast cells can be added to wheat or corn flour to provide B vitamins and selected amino acids and thus yield a product that is more nutritionally balanced.

Mushrooms are filamentous fungi, and several species are already important as human food. The annual consumption of mushrooms in the United States exceeds 300,000 tons. They are typically grown on decaying organic matter in the soil or on rotting wood. The mycelium grows beneath the surface and under favorable conditions will develop the familiar stalked structure. The stalk and cap are actually compact fungal hyphae, and the gills are the site of reproductive spore formation.

Agaricus bisporus and *Lentinus edodes* are two of several mushrooms grown commercially in the United States and commonly sold in grocery stores (Figure 31.2). A generalized flow sheet for growth of *A. bisporus* is presented in Figure 31.3. None of the materials used in preparing the compost on which the mushrooms grow is expensive or has many other uses. The mushrooms are grown at a controlled temperature and humidity and in the dark. The mushrooms appear about 3 weeks after the spawn is inoculated into the trays, and production continues for about 6 weeks. The soil remaining in the trays after growth can be used as topsoil or sterilized, aged, and some of it added back to compost. The shiitake fungus, *L. edodes*, is a more flavorful mushroom that originated commercially in Japan and is now gaining popularity elsewhere. *L. edodes* is a cellulose decomposer that grows best on waste lumber from hardwood trees. Logs or waste lumber are soaked in water and inoculated by placing spawn in holes drilled in the wood. The mushrooms fruit from these holes.

Mushrooms have little nutrient value and are eaten mostly for their distinct flavor and texture. They are low in protein, are essentially fat-free, and provide only low levels of B vitamins.

YOGURT AND CHEESE   **Yogurt** and other fermented milk products are common foodstuffs consumed throughout the world. Commercial yogurt is made from pasteurized low-fat milk thickened by the removal of water or addition of powdered milk. The thickened milk

(A) *Agaricus bisporus*

(B) *Lentinus edodes*

Figure 31.2 Food mushrooms
(A) Commercial production of *Agaricus bisporus*, the white mushrooms most commonly found in grocery stores. (B) *Lentinus edodes*, also known as the shitake mushroom. A, ©Bartomeu Amengual/AGE Fotostock; B, ©Science VU/ Visuals Unlimited.

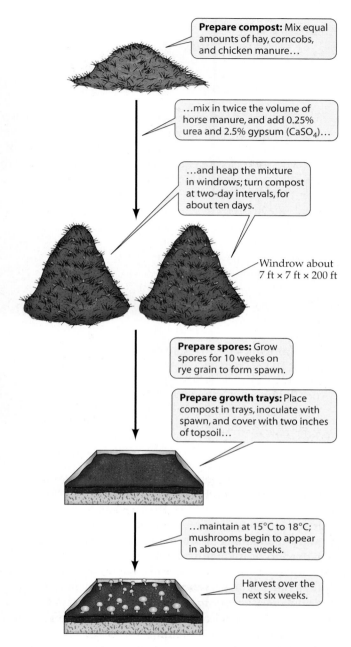

Prepare compost: Mix equal amounts of hay, corncobs, and chicken manure…

…mix in twice the volume of horse manure, and add 0.25% urea and 2.5% gypsum ($CaSO_4$)…

…and heap the mixture in windrows; turn compost at two-day intervals, for about ten days.

Windrow about 7 ft × 7 ft × 200 ft

Prepare spores: Grow spores for 10 weeks on rye grain to form spawn.

Prepare growth trays: Place compost in trays, inoculate with spawn, and cover with two inches of topsoil…

…maintain at 15°C to 18°C; mushrooms begin to appear in about three weeks.

Harvest over the next six weeks.

Figure 31.3 Commercial production of *Agaricus bisporus*
Flow chart for commercial production of *Agaricus bisporus*. After the mushrooms are harvested, the compost is pasteurized and some of it recycled.

is inoculated with a mixture of *Streptococcus thermophilus* and *Lactobacillus bulgaricus* and incubated at 45°C for about 12 hours. Growth of *S. thermophilus* produces lactic acid, and *L. bulgaricus* contributes lactic acid and aromatics that give the product a distinct taste.

Cheese is an ancient food the use of which, in all likelihood, precedes recorded history. Cheese-making originated in the Middle East. From there the practice traveled to Greece and Rome, on to Northern Europe and England, and then to the rest of the world. More than 400 varieties of cheese are made, although many are similar and differ essentially in name only. Cheeses are

TABLE 31.2	The major types of cheese		
Soft	**Semisoft**	**Hard**	**Very Hard**
Unripened	Blue	Cheddar	Parmesan
Cottage	Roquefort	Gouda	
Mozzarella	Limburger	Swiss	
Cream	Muenster		
Ripened			
Brie			
Coulommiers			
Camembert			

generally grouped according to texture, as outlined in Table 31.2.

The manufacture of cheese can be divided into four distinct phases:

• Coagulation of milk to form a curd

• Separation of the curd from the whey

• Shaping of the curd

• Ripening to achieve flavor and texture

The processing of the material at each of these steps and the nature of the milk source have a marked influence on the finished product. Most cheese is made from cow's milk, but many of the varieties produced worldwide are made from the milk of goats, sheep, mares, camels, and water buffalo.

The first step in cheese production is the coagulation of the milk proteins to form a semisolid curd. This curd entraps fat globules and some water-soluble materials. Coagulation is effected by adjusting the pH of the milk to the isoelectric point of casein or adding milk-clotting enzymes. The clotting enzymes (curd-producing) are acid proteases such as rennin, an enzyme originally obtained from calves' stomachs, but now produced by bioengineered microorganisms. Acidification of milk can be accomplished by adding selected cultures of bacteria. Fresh unpasteurized milk has a population of bacteria that will acidify the milk if it is allowed to stand at room temperature. Bacteria that ferment lactose to lactic acid are added to pasteurized milk. These acidifying microorganisms are strains of *Streptococcus lactis, Lactobacillus cremoris, Lactobacillus bulgaricus,* or *Streptococcus bulgaricus.* Some of these bacteria are trapped and remain inside the cheese and become involved in the aging and ripening process.

After the curd is separated from the liquid whey, it is shaped. This process involves adjusting the pH, salting, and physically shaping the curd to ensure proper moisture content and acid production, all of which contribute to the characteristics of the finished cheese.

The ripening process, with or without added microbes, also affects the properties of the cheese product. Fungi or bacteria may be added to the ripening curd, and these add flavors or consistency not contributed by the initial microorganisms. For example, the cheese commonly termed **blue cheese** is inoculated with spores of *Penicillium roqueforti.* If one examines Roquefort cheese in cross section, the "blue veining" where the fungal spores were inoculated are quite evident. For ripening Camembert cheese, *Penicillium camemberti* is added to the surface of the cheese. This fungus produces proteases that transform the cheese to a smooth creamy consistency. Soft cheeses are ripened for a few weeks, semisoft for several months, and a hard cheese like Parmesan for a year or more. Some of the unique features of cheese varieties are the result of bacterial growth. Propionic acid bacteria produce the characteristic holes in Swiss cheese from the CO_2 released by the acid fermentation, and propionic acid is also responsible for the sharp taste of this cheese.

YEAST **Yeast** cells can be a source of single-cell protein and may also be used as a dietary supplement. A number of commercial biochemicals are also obtained from yeast, including enzymes, nucleosides, and natural vitamins. A common component of media for microbial growth is yeast extract, the water-soluble fraction of autolyzed yeast. It is a source of B vitamins, amino acids, and other growth factors.

The production of bakers' yeast annually in the United States exceeds 100,000 tons. The most common use of yeast cells is in the baking industry. Most leavened breads that are marketed throughout the world are products of normal metabolic reactions in yeasts (Box 31.2). Yeast is added to bread dough to cause "rising" or leavening prior to baking. During kneading, the dough is manipulated to aid in the yeast fermentation. Kneading also ensures that the CO_2 bubbles in the bread will be small. As yeast ferments, it converts sugar to ethanol and CO_2. The escaping gas (CO_2) that is trapped in wheat gluten results in an expansion of dough generally referred to as the "dough rises." During baking, the CO_2 and ethanol are driven off, leaving the odd-shaped holes in the baked product.

Bulk yeast used in baking or as a food supplement is grown in large fermentors. A flowchart for yeast growth is presented in Figure 31.4. A small inoculum is transferred batchwise to larger and larger tanks as shown. Before inoculation into the largest fermentor (S-3), the yeast cells are concentrated and added as a thick slurry. Beet or cane sucrose is the source of carbon and energy. The fermentor is vigorously aerated for rapid growth. The addition of the sugar is carefully monitored to ensure that the yeast uses it quickly and that ethanol is not produced rather than more yeast. Yeast is sold for baking as a compressed block, as moist yeast, or as activated dry yeast.

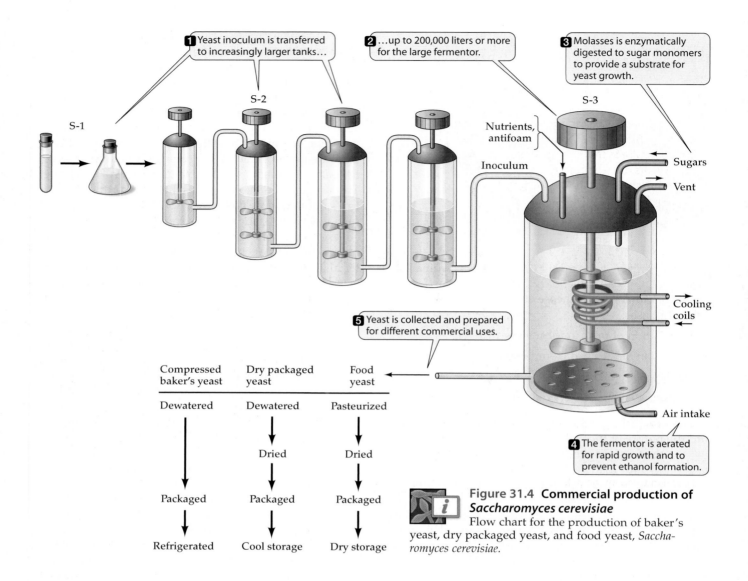

1 Yeast inoculum is transferred to increasingly larger tanks...

2 ...up to 200,000 liters or more for the large fermentor.

3 Molasses is enzymatically digested to sugar monomers to provide a substrate for yeast growth.

S-1

S-2

S-3

Nutrients, antifoam

Inoculum

Sugars

Vent

Cooling coils

5 Yeast is collected and prepared for different commercial uses.

Air intake

4 The fermentor is aerated for rapid growth and to prevent ethanol formation.

Compressed baker's yeast	Dry packaged yeast	Food yeast
Dewatered	Dewatered	Pasteurized
↓	↓	↓
	Dried	Dried
↓	↓	↓
Packaged	Packaged	Packaged
↓	↓	↓
Refrigerated	Cool storage	Dry storage

Figure 31.4 Commercial production of *Saccharomyces cerevisiae*
Flow chart for the production of baker's yeast, dry packaged yeast, and food yeast, *Saccharomyces cerevisiae*.

Yeast cake is about 70% moisture and must be refrigerated to maintain activity. Dry yeast is produced by freeze-drying (lyophilization) to ensure a long shelf life.

FERMENTED MEAT Specialty meat products are manufactured by bacterial fermentation. Among these are summer sausage, Lebanon bologna, salami, and cervelat. The organisms generally involved in meat fermentations are strains of *Pediococcus cerevisiae* and *Lactobacillus plantarum*. Commercial inocula of these microorganisms are available to add to the ground meat. Country cured hams are also a food that gains flavor and texture from fungi that are natural surface contaminants. These fungi are members of the genera *Penicillium* and *Aspergillus*.

OTHER FOOD PRODUCTS Other foods that are products of microbial fermentation are listed in Table 31.3.

Acidification of cabbage by a lactic acid fermentation to produce sauerkraut is an age-old method for long-term storage. Traditionally it is prepared by placing shredded cabbage in a crock or other deep container. Salt (NaCl) is added at 2.2% to 2.8% to restrict the growth of most gram-negative bacteria and to draw out water and nutrients for the fermentation. The crock quickly becomes anoxic, and the natural lactic acid microbiota of the cabbage predominate under these conditions and carry out the fermentation. In today's food industry, *Leuconostoc mesenteroides* and *L. plantarum* are used as inocula in making sauerkraut. The lactic acid produced lowers the pH to a level where little decomposition occurs, and the finished product is stable for extended periods. A typical fermentation requires about 30 days.

Pickles are produced by placing small cucumbers in a salt brine along with dill or other flavoring agents. The

BOX 31.2   *Milestones*

The Staff of Life

The baking of leavened bread is clearly depicted in reliefs from Egyptian tombs that are more than 4,500 years old. The number of individuals and the scale shown in the relief suggest that it was likely a commercial operation. The Old Testament of the Bible affirms that leavened bread was favored:

> *Seven days shall ye eat unleavened bread . . .*
> —Exodus 12:5

Egyptian tomb image of bread-baking. ©Visual Arts Library (London)/Alamy.

Here the consumption of unleavened bread was a sacrifice to be made by the people of Israel. The presence in leavened dough, prior to baking, of an ingredient that was transferable to fresh dough and would transform it to leavened dough was also a part of biblical lore:

> *. . . Know ye not that a little leaven leaveneth the whole lump?*
> —I Corinthians 5:6

It was not until the middle of the nineteenth century that people were aware that the leavening agent was actually a microbe. The early transferable leaven was probably a yeast, a heterofermentative lactic acid bacterium, or a combination of the two.

salt (NaCl) concentration, initially about 5%, is increased to 15.9% over a 6- to 9-week period. The high salt concentration extracts sugar from the cucumbers, and these sugars are fermented to lactic acid. Home production of pickles is a result of fermentation by the indigenous lactic acid biota of cucumbers. Commercial production is accomplished by sterilizing the pickles and adding cultures of *P. cerevisiae* and *L. plantarum* to ensure a uniform product.

Coffee beans develop as berries with an outer pulpy and mucilaginous envelope that must be removed before roasting. The pulp can be removed mechanically, but the pectin that contains the mucilage is removed best by the pectinolytic bacterium *Erwinia dissolvens*. Fungi are also involved in this process. Apparently, microbes contribute little to the flavor or aroma of the finished coffee bean.

TABLE 31.3   Foods produced by microbial fermentation

Product	Fermenting Microorganism
Cocoa beans	*Candida* spp., *Geotrichum* spp., *Leuconostoc mesenteroides*, others
Coffee beans	*Erwinia* spp., *Saccharomyces*
Sauerkraut	*Leuconostoc mesenteroides, Lactobacillus plantarum*
Soy sauce	*Aspergillus soyae, Aspergillus oryzae*
Olives	*Leuconostoc mesenteroides, Lactobacillus plantarum*
Pickles	*Pediococcus cerevisiae, Lactobacillus plantarum*

Chocolate is derived from cacao, a bean that is enclosed in the fruit of a plant that grows in parts of Asia, Africa, and South America. The beans are removed from the fruit and fermented in piles or boxes for several days. The fermentation occurs in two phases: (1) the sugars from the pulp are converted to alcohol and (2) the alcohol is oxidized to acetic acid. The yeasts involved in producing alcohol during the first phase are essential for the development of the chocolate flavor and aroma.

Fermentation of **olives** occurs after the olives have been treated with 1.6% to 2.0% lye at 24°C for 4 to 7 hours. This treatment removes the bitter principal (oleuropein) from the green olive. The lye is removed by soaking, and the olives are placed in oak barrels and 7.0% salt brine is added. The brined olives are inoculated with *L. plantarum* and fermented for 8 to 10 months.

Soy sauce is prepared by mixing soaked and steamed soybeans with crushed wheat and inoculating with *Aspergillus oryzae* or *Aspergillus soyae*. This combination is incubated for several days, and during this period the beans release fermentable sugars, peptides, and amino acids. The fungus-coated beans are then placed in liquid, and salt (NaCl) is gradually added to a final concentration of 18%. The total incubation period is 1 year at room temperature. The addition of salt leads to a replacement of the filamentous fungi by lactic acid bacteria and yeast. The lactic acid bacteria are predominantly strains of *Lactobacillus delbrueckii*, and the yeast is *Saccharomyces rouxii*. The liquid resulting from the long-term fermentation is soy sauce. The solids are pressed and used as food for humans and animals.

Flavoring Agents and Food Supplements

Under appropriate conditions, select bacteria and fungi produce organic compounds that provide a desirable flavor or aroma to food. Many of these organics originally served as preservatives. However, with the advent of refrigeration and other means of preservation these flavorful compounds are no longer necessary as preservatives but are valued as flavoring agents. Sauerkraut, mentioned previously, was originally prepared to preserve cabbage for consumption over the long winter. It is now sold commercially because buyers like the flavor imparted by the bacterial fermentation. The following discussion concerns flavoring agents, flavor enhancers, and supplements such as selected amino acids that are added to food to enhance their nutritional value.

VINEGAR   **Vinegar** making is at least 10,000 years old and originated with the production of wine. The name vinegar comes from the French words vin (wine) and aigre (sour), as it is essentially sour (acetic acid) or spoiled wine. Vinegar has long been valued as a flavoring agent,

preservative, medicine, and consumable beverage. The Romans and Greeks drank dilute vinegar as a beverage and produced it by introducing air into wine casks. An early method of vinegar production was to place an air inlet at the top of a wine cask and draw out vinegar as needed. Fresh wine was then added to replace the vinegar that was removed.

The production of vinegar in the United States amounts to about 160 million gallons per year. Two-thirds of the vinegar produced is used in commercial products such as sauces, in dressings, and in the manufacture of pickles and tomato products. The remainder is used unaltered for domestic purposes. The basic reaction in vinegar formation is as follows:

The organisms employed in commercial vinegar production are *Acetobacter aceti* and *Gluconobacter oxydans*. Along with these two species, hundreds of subspecies or strains of other bacterial species grow in a commercial fermentor, and mixed cultures of all of these bacteria are involved in vinegar production. Two major systems are used for the commercial production of vinegar, the trickling generator and submerged fermentation.

TRICKLING GENERATOR   One of the commercial systems for producing vinegar is the Frings-type trickling generator (Figure 31.5). The generator is constructed of cypress or redwood and packed with curled beechwood shavings. The collection chamber at the bottom holds about 3,500 gallons. A pump circulates the ethanol-water-acetic acid from the collection chamber and sparges it evenly over the upper surface. The temperature is controlled by flow rate at 29°C at the top and 35°C at the bottom. The system is aerated by a blower at the bottom, and the aeration rate is carefully controlled to prevent evaporation of the acetic acid. Ethanol must be fed constantly to sustain the bacterial population. The maximum concentration of acetic acid in solution is 13% to 14%. The life of a well-packed and maintained generator is about 20 years. This is an example of a nonstop **continuous fermentation**.

BATCHWISE FERMENTATION   The other major system for vinegar production is the Frings acetator, which is a submerged, **batchwise fermentation** process (Figure 31.6). The Frings acetator is composed of a stainless-steel tank with a high-speed mixer that constantly stirs

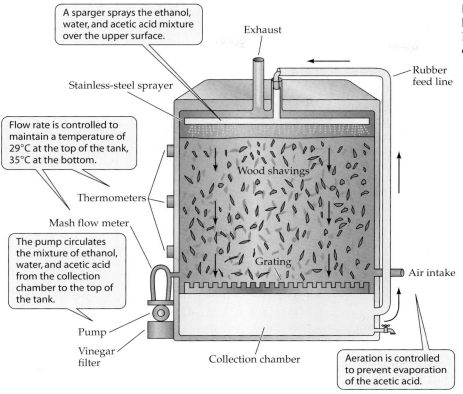

A sparger sprays the ethanol, water, and acetic acid mixture over the upper surface.

Exhaust

Rubber feed line

Stainless-steel sprayer

Flow rate is controlled to maintain a temperature of 29°C at the top of the tank, 35°C at the bottom.

Wood shavings

Thermometers

Mash flow meter

The pump circulates the mixture of ethanol, water, and acetic acid from the collection chamber to the top of the tank.

Grating

Air intake

Pump

Vinegar filter

Collection chamber

Aeration is controlled to prevent evaporation of the acetic acid.

Figure 31.5 Manufacture of vinegar by continuous fermentation
Frings-type trickling generator for the commercial production of vinegar.

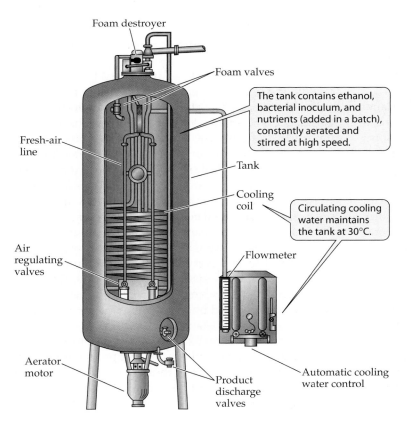

Foam destroyer

Foam valves

The tank contains ethanol, bacterial inoculum, and nutrients (added in a batch), constantly aerated and stirred at high speed.

Fresh-air line

Tank

Cooling coil

Air regulating valves

Circulating cooling water maintains the tank at 30°C.

Flowmeter

Aerator motor

Product discharge valves

Automatic cooling water control

the microbes, air, ethanol, and nutrients to provide a favorable environment for bacterial growth. Small air bubbles are introduced at the bottom to provide aeration. Because aeration can cause foaming, foam produced is removed by a foam destroyer at the top of the tank. The tank is operated at 30°C, and this temperature is maintained by circulation of cooling water. Ethanol, bacterial inoculum, and selected nutrients required for bacterial growth are added to the acetator in a batch. An electronic instrument monitors the ethanol content. When the alcohol level decreases to 0.2% by volume, about one-third of the finished product is removed, and fresh nutrients and ethanol are added. It takes about 35 hours to make 12% vinegar. The acetator is a more efficient system than the trickling generator.

Figure 31.6 Manufacture of vinegar by batch-wise fermentation
Frings acetator, a submerged system for vinegar production.

Several types of vinegar are manufactured and are classified by the origin of the ethanol that the bacterium oxidizes to acetic acid. The basic flavor component in vinegar is acetic acid, but distinct flavors are contributed by the original source of the ethanol. The most widely used types of vinegar include the following:

- Distilled white vinegar—made from distilled ethanol. The origin of the ethanol may be fermentation ethanol or obtained by chemical synthesis

- Cider vinegar—made from the fermented juice of apples

- Wine vinegar—made from a low-quality wine that has been subjected to aerobic oxidation. Either white or red wine can be converted to vinegar

- Malt vinegar—produced from the alcohol obtained by fermenting corn or barley starches that have been treated with enzymes to release sugars for fermentation

NUCLEOTIDES   For centuries the Japanese have used dried seaweed, dried bonito, or other dried fish as flavoring agents. Two of the major ingredients in these dried products that enhance the flavor of food are MSG and the histidine salt of inosinic acid. Inosinic acid is a nucleotide, and the production of this and other nucleotides are discussed herein. MSG fermentations are discussed in the section along with other amino acids. As **flavor enhancers**, the descending order of effectiveness among the nucleotides is guanylic acid (5'-GMP), inosinic acid (5'-IMP), and xanthylic acid. These nucleotides are added to enhance the flavor of soups and sauces and are effective at very low concentrations (0.005% to 0.01%). Production of 5'-IMP and 5'-GMP is about 4,000 tons per year.

Originally nucleotides were produced by enzymatic hydrolysis of yeast ribonucleic acid (RNA). *Candida utilis*, a yeast, is rich in RNA and was one of the sources. Now, more than half of the commercial production of nucleotides is by fermentation. The production of 5'-IMP is by direct fermentation or through the microbial production of inosine that can be chemically phosphorylated. Strains of *Brevibacterium ammoniagenes* that are auxotrophs for adenine and guanine will produce about 30 grams per liter inosine. Some microorganisms can produce 5'-IMP by direct fermentation but must have an altered membrane permeability. Nucleosides (no phosphoryl group) can traverse a normal cytoplasmic membrane but nucleotides (phosphorylated) cannot. Selected permeability mutants of *B. ammoniagenes* can produce about 27 grams per liter 5'-IMP.

The production of 5'-GMP is generally by a combination of fermentation and chemical synthesis. A purine auxotroph of *Bacillus megaterium* is used that accumulates an intermediate of purine biosynthesis, aminoim-

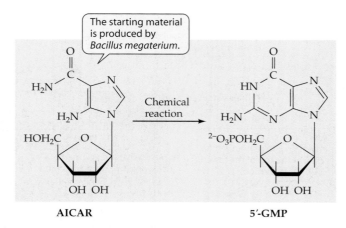

Figure 31.7 Microbiological/chemical production processes
Example of combined microbiological and chemical processes in generating a commercial product. Aminoimidazole carboxamide ribose (AICAR)—an intermediate in purine biosynthesis—is produced microbiologically and transformed chemically to 5'-guanosine monophosphate (5'-GMP).

idazole carboxamide ribose (AICAR). The *B. megaterium* strain will produce 16 grams per liter AICAR (Figure 31.7), and this is chemically transformed to 5'-GMP. Guanosine-excreting mutants are also employed in commercial production of 5'-GMP.

AMINO ACIDS   **Amino acids** are a major industrial product with an annual worldwide production of more than 400,000 tons. Some of the major uses of amino acids are listed in Table 31.4. Amino acids are used to enhance the quality of food, in medicine, or as a starting material in chemical synthesis. The major amino acid produced is glutamic acid at 80% of the total and lysine is second in volume at 10% of the total. The biologically active form of amino acids is the L-form and is the isomer produced by fermentation. Chemical synthesis yields a mixture of the D and L isomers and separation of isomers would be costly.

Amino acids are synthesized in families by microorganisms, so arranged because of the precursor metabolite that serves as the starting material. The synthesis of individual amino acids is tightly regulated, and generally the amount synthesized corresponds to the amount essential for the synthesis of cell protein. Therefore, the initial reactions in the biosynthesis of a particular amino acid may be inhibited by feedback inhibition (see Chapter 13). Repression is also used that curtails the synthesis of an enzyme by an external compound, a repressor. If one wishes to obtain a culture that will produce excessive amounts of an amino acid, then both feedback inhibition and repression (see Chapter 13) must be overcome. Through mutation and selection or bioengineering,

TABLE 31.4	Uses of amino acids
Amino Acid	**Use**
Sodium glutamate (MSG)	Flavor enhancer
L-Lysine	Added to plant-derived foods to improve nutritional quality
L-Methionine	Infusions
L-Threonine	Infusions
L-Tryptophan	Antioxidant in powdered milk
L-Histidine	
Aspartate	Improve flavor of fruit juices
Alanine	Infusions
L-Aspartyl-L-phenylalanine methyl ester (aspartame)	Low-caloric sweetener
Mixed amino acids	Infusion solutions for surgery patients

microorganisms have been developed that are efficient producers of most of the common amino acids.

Most amino acid manufacturing processes originated in Japan. Some strains that are used commercially are shown in Table 31.5. The major glutamic acid–producing organisms have a requirement for biotin and lack significant levels of α-ketoglutarate dehydrogenase. When provided with low levels of biotin, the organism synthesizes leaky and inadequate cytoplasmic membranes (biotin is involved in synthesis of fatty acid). These leaky membranes permit the overproduced glutamic acid to be excreted into the medium. L-Lysine can also be produced in high yield by direct fermentation using a mutant of *Brevibacterium flavum* that requires homoserine and leucine. Lysine is excreted into the medium by active transport.

Vitamins

Mutation and selection can lead to microbial strains that overproduce almost all of the B vitamins. However, only two of the B vitamins, riboflavin and vitamin B_{12}, are produced by fermentation, as chemical synthesis of the others is less costly. The structures of vitamin B_{12} and riboflavin are presented in Figure 31.8. Vitamin B_{12} is a complex molecule that, interestingly, is only synthesized by prokaryotic microorganisms. It is required by humans and supplied in food or is produced by the normal intestinal microbiota. Pernicious anemia can occur in humans who do not obtain sufficient levels of dietary B_{12}.

The annual current production of vitamin B_{12} is about 10,000 kg. Unpurified vitamin B_{12} is added to swine and poultry feed at about 12 mg/ton, and this amount af-

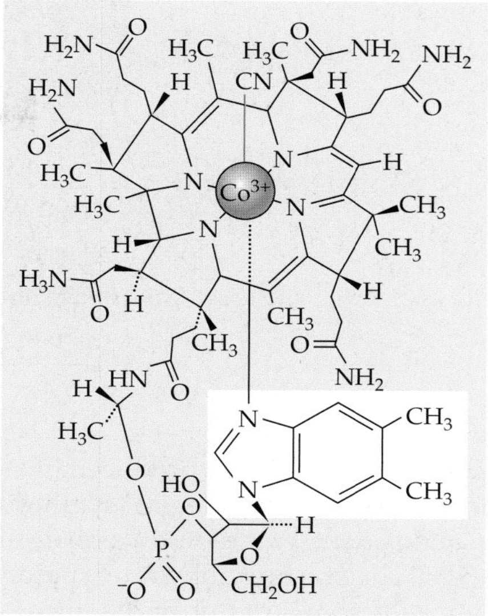

Vitamin B_{12} (cyanocobalamin)

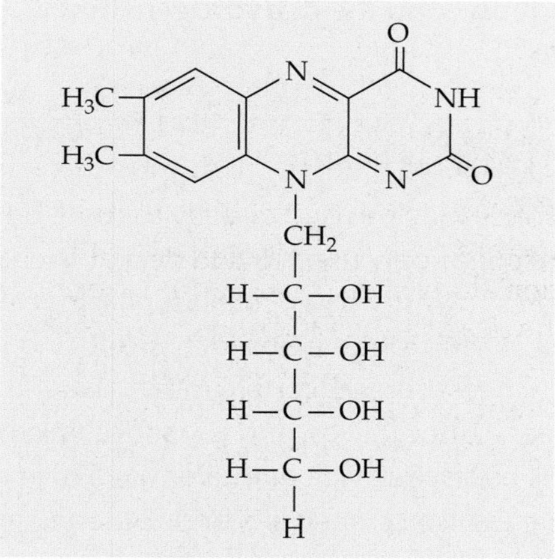

Riboflavin

Figure 31.8 B-complex vitamins
Structures of vitamin B_{12} and riboflavin, two members of the vitamin B complex produced commercially by fermentation.

firms that vitamin B_{12} is active at exceedingly low supplement levels. Vitamin B_{12} is synthesized in bacteria from intermediates common to heme or chlorophyll synthesis. The organisms involved in vitamin B_{12} production are *Propionibacterium shermanii* and *Pseudomonas denitrificans*. The latter produces about 60 milligrams per liter under appropriate growth conditions.

Riboflavin occurs naturally in milk, eggs, meat, and other food. It is added as a supplement to bread, milk, and other foods to improve their nutritional quality. Yeasts such as *Ashbya gossypii are* used in the commercial production of riboflavin, and the yield can be in excess of 7 grams per liter.

TABLE 31.5	Commercially produced amino acids and the microbes involved in their synthesis		
Amino Acid	**Organism**	**Approximate Yield (Grams/Liter)**	**Carbon Source**
L-Glutamate	*Corynebacterium glutamicum*	>100	Glucose
L-Lysine	*Corynebacterium*	39	Glucose
L-Lysine	*Brevibacterium flavum*	75	Acetate
L-Threonine	*Escherichia coli* K$_{12}$	55	Sucrose

Vitamin C (ascorbic acid) is produced by a combination of chemical and microbiological synthesis. It is also used as an antioxidant. World production exceeds 35,000 tons per year. The starting substrate, D-glucose, is reduced chemically to D-sorbitol, and a microbial oxidation step leads from D-sorbitol to L-sorbose, which is chemically converted to ascorbic acid (Figure 31.9). The microbe involved in the dehydrogenation is *Acetobacter suboxydans*.

SECTION HIGHLIGHTS

Microbes may be used directly as food, but are more commonly used to add desirable characteristics to basic foods, such as milk, meat, and produce. Microbes are used in producing yogurt, cheese, bread, certain meats, sauerkraut, pickles, coffee, olives, and soy sauce. Microorganisms also provide additives used to improve the flavor, shelf life, or nutritional quality of food.

31.4 Beverages Containing Alcohol

The alcoholic beverage industry is economically the most significant of all commercial processes that involve microorganisms. Alcohol-containing beverages such as beer and wine are as old as civilization itself. Mead is considered to have been the first consumable alcoholic beverage. Mead is the name given to the product of honey fermentation and is the oldest word associated with drinking; it has been traced to the early Sanskrit language. Quite likely, it originated when honey, which was gathered by the earliest recorded civilizations, became diluted with water, allowing a natural fermentation to take place.

Alcoholic beverages are now produced worldwide from a variety of plant materials. Fruit sugars or grain polysaccharides are the major sources of fermentable substrates (Table 31.6). Wine, beer, and distilled beverages are the major alcoholic beverages produced commercially. Yeasts of the genus *Saccharomyces* are the organisms most involved in alcohol production. A discussion of the processes involved in production of major alcoholic beverages follows.

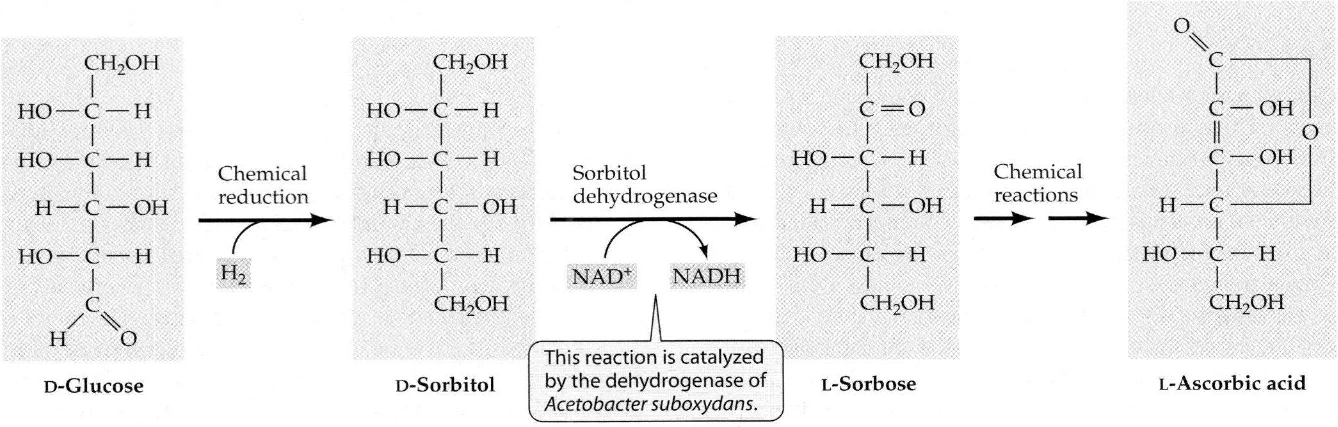

Figure 31.9 Manufacture of vitamin C
Production of vitamin C (ascorbic acid) by a combination of fermentation and chemical reactions.

TABLE 31.6	Major alcoholic beverages and sources of sugars for fermentation
Wine	Fruit sugars
Beer	Barley (sometimes rice or corn)
Whiskey	Rye, corn, barley
Tequila	Agave cactus
Rum	Sugarcane molasses
Mead	Honey
Sake	Rice

Wine

There are many types of wine (such as grape, peach, pear, and dandelion), but by far the most favored is that made from grapes. The leading countries in wine production are Italy, Spain, France, Portugal, United States, Australia, South Africa, and Argentina. Italy and Spain combined produce about 60% of the world total. The best area for wine-producing vineyards is between 30° and 50° North or 30° and 40° South of the equator. In these areas of warm summer and mild winter climate, the grapes ripen slowly and yield superior wines. In the United States, the states of California, New York, Washington, and Oregon have a favorable climate and soil for growth of wine-producing grapes.

Wine occurs in two distinct types—white and red. White wine is made from white grapes or can be made from red grapes provided the skins are removed from the **must** (must is crushed grapes ready for fermentation) prior to fermentation. Red wines gain the red pigmentation (and flavor), as they are partly fermented (3 to 10 days) in the presence of the red skins. During this period, the alcohol formed extracts and solubilizes anthocyanin pigments present in the grape skin. Rosé (pink) wines are made from pink grapes or more often from red grapes but the time of exposure to the skins is for 12 to 36 hours, and much less of the anthocyanin pigments are solubilized. A dry wine is one in which most of the sugar has been fermented to ethanol. A sweet wine has more residual sugar. A fortified wine such as sherry or port contains higher levels of ethanol, and this is attained by supplementing them with ethanol that has been distilled from other wine or from fermented grain.

Yeasts are part of the natural microbiota of grapes. Crushed grapes will undergo a natural fermentation process that eventually produces wine. Natural fermentations are not favored commercially because the natural microbiota can be inconsistent and will not produce sufficient amounts of ethanol. Much of the wine produced by a natural fermentation is not potable. In commercial processes, the grapes are crushed and the must is treated with sulfur dioxide (SO_2) or sometimes pasteurized to destroy the natural microbiota. The must is then inoculated with a proprietary strain of *Saccharomyces ellipsoideus* to bring about the desired fermentation. Laboratory-bred yeasts are favored because they have been selected to tolerate 12% to 15% ethanol, whereas the natural microbiota will cease fermenting when the ethanol content reaches 3.5% to 4%. Quality wine contains 8% to 14% percent ethanol.

A flow chart for red and white wine fermentations is presented in Figure 31.10. White wine fermentations are quite direct. The grapes are de-stemmed, crushed, and skins and solids (pomace) removed. The must is treated with SO_2 and yeast is added. After fermentation, the wine is aged in casks, tanks, or bottles. During a process called racking, red wine is drawn from the primary fermentation vat, placed in casks, and stored at a low temperature.

The aging process is typically much longer for red than white wines. Whereas white wines are ready for consumption within a year of production, most red wines require further aging. During the lengthy aging process used for some of the quality red wines, such as cabernet sauvignon, a second fermentation occurs. This is termed the malo-lactic fermentation. Residual malic acid in the wine is fermented by lactic acid bacteria and converted to lactic acid and CO_2. This lowers the pH of the wine and improves the quality. After aging wines are clarified in a process called fining. Fining can be accomplished by filtering through casein, diatomaceous earth, or bentonite. A few wines are clarified by centrifugation.

Wine fermentations may be accomplished in tanks from 50 to 50,000 gallons, but it is essential that the temperature of the fermentation is maintained at 20°C to 24°C to produce quality wine. Heat generated during metabolism can raise temperatures above the tolerance level of the yeast. Aging is carried out at a lower temperature, as this improves flavor and aroma. Sparkling wines such as champagne are aged in bottles with a secondary yeast fermentation that generates the carbonation (bubbles). Some sparkling wine is carbonated by CO_2 injection after the fermentation. Champagne is aged in bottles that are slanted so that the cork faces down. This allows sediments to gravitate to the neck of the bottle. Periodically the neck is frozen, the bottle uncorked, and sediment removed. Wine lost during this process is replaced before recorking. All wines, including champagne, that are stoppered with a cork should be stored with the bottles lying on their side. This retains moisture in the cork and prevents air from entering. Presence of air allows growth of vinegar bacteria that convert ethanol to acetic acid, thus "spoiling" the product.

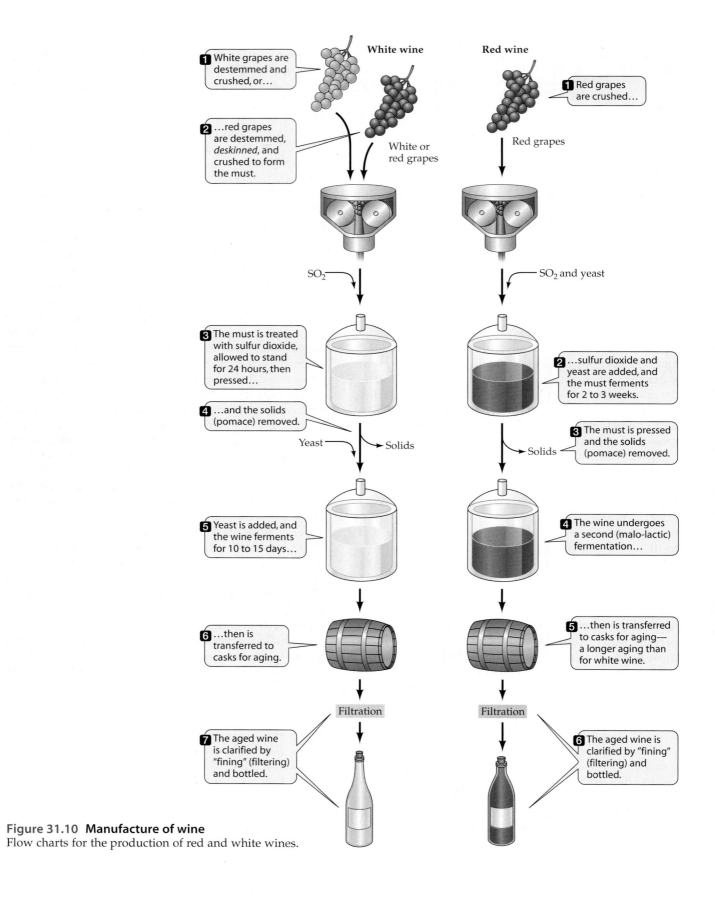

Figure 31.10 Manufacture of wine
Flow charts for the production of red and white wines.

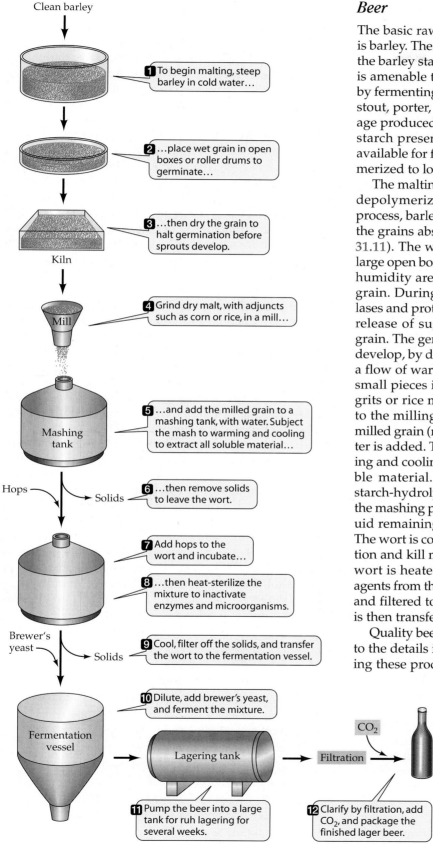

Clean barley

1 To begin malting, steep barley in cold water…

2 …place wet grain in open boxes or roller drums to germinate…

3 …then dry the grain to halt germination before sprouts develop.

Kiln

4 Grind dry malt, with adjuncts such as corn or rice, in a mill…

Mill

5 …and add the milled grain to a mashing tank, with water. Subject the mash to warming and cooling to extract all soluble material…

Mashing tank

Hops — Solids

6 …then remove solids to leave the wort.

7 Add hops to the wort and incubate…

8 …then heat-sterilize the mixture to inactivate enzymes and microorganisms.

Brewer's yeast — Solids

9 Cool, filter off the solids, and transfer the wort to the fermentation vessel.

10 Dilute, add brewer's yeast, and ferment the mixture.

Fermentation vessel

Lagering tank

Filtration

CO_2

11 Pump the beer into a large tank for ruh lagering for several weeks.

12 Clarify by filtration, add CO_2, and package the finished lager beer.

Beer

The basic raw material for brewing of beer worldwide is barley. The barley must be malted, a process whereby the barley starch is converted to a substrate (sugar) that is amenable to fermentation. Malt beverages are made by fermenting the malted barley, and these include beer, stout, porter, and malt liquor. Sake, an alcoholic beverage produced in Japan, is a product of malted rice. The starch present in barley, rice, and other grains is not available for fermentation unless the polymer is depolymerized to low-molecular-weight sugars.

The malting process is a series of manipulations that depolymerize starch to free sugars. In the malting process, barley is soaked (steeped) in cold water so that the grains absorb sufficient water to germinate (Figure 31.11). The wet grain is then placed in rolling tubs or large open boxes in the presence of air. Temperature and humidity are controlled to initiate germination of the grain. During germination, enzymes, including amylases and proteinases, are activated, and these cause the release of sugars and amino acids from the malting grain. The germination process is halted before sprouts develop, by drying the germinated grain in a kiln under a flow of warm air. The dried grain is then broken into small pieces in a grinding mill. Adjuncts such as corn grits or rice may be added at this stage and subjected to the milling process, which exposes the starch. The milled grain (mash) is placed in a mashing tank, and water is added. The mash is subjected to periods of warming and cooling to room temperature to extract all soluble material. During mashing, the proteinases and starch-hydrolyzing enzymes are very active. Following the mashing process, the solids are removed and the liquid remaining is the fermentable portion called wort. The wort is cooked in a brew kettle to halt enzymatic action and kill microorganisms. Hops are added, and the wort is heated to extract tannins and other flavoring agents from the hops. The wort is sterilized by live steam and filtered to remove hops and precipitates. The wort is then transferred to the fermentation vessel.

Quality beer production depends on careful attention to the details involved in the preparation of wort. During these processes, pH, length of heating and cooling

Figure 31.11 Manufacture of beer Flow chart for the production of lager beer.

cycles, removal of precipitates, and other factors are important for the production of a desirable product. Another major concern in beer manufacture is the quality of added water. It must be low in carbonates and calcium and properly balanced in other minerals. The brewing industry developed historically at sites where quality water was available, such as Pilsen, Czechoslovakia; Munich, Germany; and Dublin, Ireland.

After the wort has been cooled and filtered, it is diluted with water. The fermentation is initiated either by a top-fermenting yeast (*Saccharomyces cerevisiae*) or a bottom-fermenting yeast (*Saccharomyces carlsbergensis*). Top fermenters, so-called because they are carried to the top (actually distributed throughout) by the CO_2 produced during the fermentation, are used to produce **ale**. In contrast, the bottom fermenters, which tend to settle to the bottom, are used to produce **beer**. The top fermentation is carried out at 14°C to 23°C for 5 to 8 days. The ale is aged at 4° to 8° C.

The bottom fermentation for beer is carried out at 6°C to 12°C for about the same length of time as the ale. Following fermentation, the beer is pumped into a large tank for maturation. Yeast is added at about 1 million cells per milliliter, and the beer is allowed to "rest" for several weeks at 0.1°C. This process is called the **ruh** (German, meaning "rest") or lagering.

The finished beer is clarified by centrifugation or filtration, CO_2 is added, and the beer is packaged. Draft beer (kegs, bottles, or cans) is generally passed through membrane filters to remove all microorganisms. Prior to the availability of filter systems, bottled beer was pasteurized to prevent spoilage by acetic acid bacteria, and this pasteurization tended to produce off-flavors. Draft beer was considered superior because it was unpasteurized but was kept cold to prevent spoilage. Most European and American beer is lager beer, although ales are increasingly being produced locally by microbreweries in the United States.

Distilled Beverages

Distilled spirits are extensions of the yeast-brewing process discussed in the preceding text. The fermented liquid is placed in a closed vessel with attached cooling coils. The vessel is then heated, and the volatiles are condensed and collected. Ethanol has a boiling point of 78.5°C, so it can be readily separated from the fermentation medium by distillation. The distillate obtained has a high ethanol content and must be diluted before aging and bottling.

Gin, vodka, whiskey, rum, and brandy are examples of distilled alcoholic beverages. Gin and vodka are essentially the distillates from alcoholic fermentation of grain. Gin is produced when the alcohol is refluxed over juniper berries to extract the distinct aroma and flavor of the berries. Rum is manufactured by distilling fermented cane sugar molasses, and brandy is made from distilled wine. Distilled alcoholic beverages are 40% to 43% ethanol by volume (80 to 86 proof). Gin may be somewhat higher.

Whiskey (spelled whisky in Scotland) is defined by the grain employed in the fermentation. Rye whiskey must have 51% or more rye grain in the malting process, and bourbon must have at least 51% corn. There are three types of Scotch whisky: **malt** whisky, **grain** whisky, and **blended** whisky. Malt whisky (single malt) is made from 100% malted barley. Grain whisky is produced from a variety of malted cereals (wheat and maize) that may or may not include malted barley. Blended whisky is a mixture of malt and grain whisky. Most Scotch whisky sold in the United States is blended, but single malt Scotch is growing in popularity. If a Scotch contains any whisky not made from malted barley, it cannot be labeled or sold as malt whisky. It is interesting to note that there are over 300 different single-malt Scotches available in Scotland.

The manufacture of whiskey apparently originated in Ireland or Scotland and moved to the United States (bourbon), Canada, and Japan. Scotch and some bourbons are made by a **sour mash** fermentation. A sour mash is produced by introducing a homolactic bacterium into the mash, and the growth of this bacterium lowers the pH of the mash to about 4.0. The distillate from a malted grain fermentation contains volatile products other than ethanol that are generated during the mashing process or fermentation. These volatile products are responsible for the distinct flavors of the various whiskeys. The distillate from grain fermentation is colorless, and the brownish color of the final product results from the aging process. Whiskey is aged in wooden barrels that have been charred on the inside. Scotch is now aged in barrels previously used for aging bourbon or sherry. During aging, desirable flavors are enhanced and the level of some less desirable substances such as fusel oils may be decreased. Fusel oils are mixtures of propanol, 2-butanediol, amyl alcohol, and other alcohols. Whiskey manufacture is an ancient art that involves minimal science.

SECTION HIGHLIGHTS

Wine, beer, and distilled beverages are the major alcoholic beverages produced commercially, generally by fermentation of various plant materials by yeasts in the genus *Saccharomyces*. Wines are produced from juices, primarily from grapes, and beer from malted grains. Alcohol can be collected from fermented liquids by distillation, and then further processed to produce a variety of distilled beverages.

31.5 Ethanol as Fuel

There are a number of concerns with the use of fossil fuel as energy source. Among these are potential political instability at the source of crude oil, decreasing supply of this finite resource and global warming resulting from petroleum utilization. The United States consumes more than 84 million barrels of oil per day. There is no technology available that will allow for replacement with alternate energy sources in the immediate future. However, microbiologically generated ethanol is a partial replacement source of fuel, and production in the United States is expected to be at 5 billion gallons in 2006. This would represent more than 3% of our gasoline requirement.

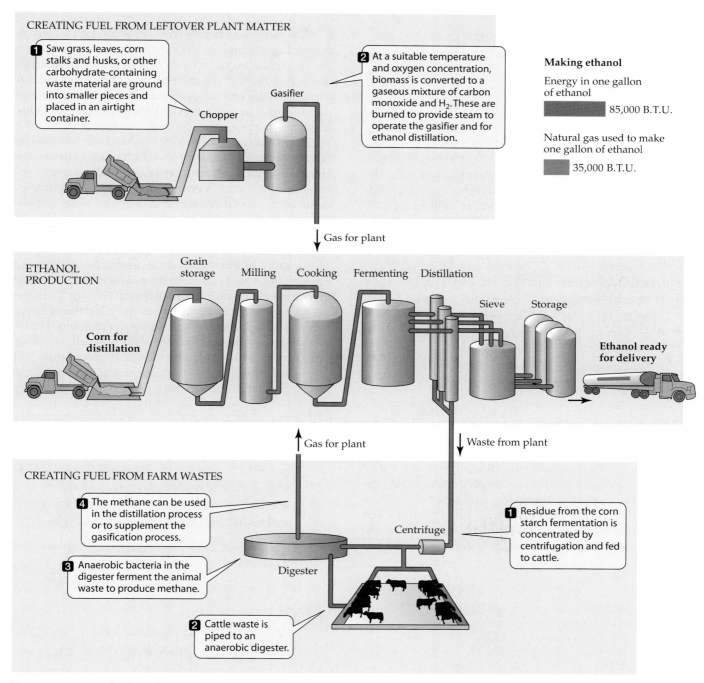

Figure 31.12 Fuel ethanol
Schematic for the production of fuel ethanol with minimal energy input.

Corn and starch from cereal grains have been a major source of sugars for ethanol production. There are several inherent problems with this, as follows:

1. Grain is a potential source of food and beverage. Competition with this use would inevitably increase the price.

2. Energy (steam) is needed to break down cornstarch to a consistency amenable to enzymatic attack. Energy is also required to distill off the ethanol and dry the residual material for cattle food.

3. Fertilizer to grow grain is a made with natural gas. Cultivation and transport are also energy consumers.

4. Energy consumption approaches energy yield as costs of corn for distillation increases.

There are proposals to make the ethanol generating process more energy efficient and a potential operational system is presented in Figure 31.12. In the gasification process, about 90% of the bulk material turns to gas. The variety of materials that can be turned into gas includes wheat straw and sawgrass. This process utilizes the ability of the microbe to ferment grain to ethanol and the generation of methane from animal waste.

> **SECTION HIGHLIGHTS**
>
> Ethanol produced from plant biomass is a potential source of fuel, but there are still several obstacles to the widespread use of this technology, including the cost and availability of raw materials and the energy inputs required.

31.6 Organic Acids

Citric acid is an important organic acid that is synthesized by a microbial fermentation. More than 130,000 tons are produced worldwide annually. Originally, citric acid was extracted from citrus fruit (a lemon is 7% to 9% citric acid). The market was historically controlled by cartels established in citrus-growing areas, leading to prohibitively high prices for citric acid. In 1923 a microbial fermentation was developed with high yields of citric acid, and the price plummeted as availability increased. Approximately two-thirds of the citric acid produced commercially is used in foods and beverages including soft drinks, desserts, candies, wines, frozen fruit, jams and jellies, and ice cream. In the pharmaceutical industry, iron citrate is sold as a source of dietary iron. Citric acid is also used as a preservative for stored blood, tablets, and ointments. Citrate has also been replacing polyphosphates in detergents due to the banning of phosphates, as they are an environmental hazard.

Most citric acid comes from a submerged fermentation in large, stainless-steel fermentors using strains of the fungus *A. niger*. Sucrose is generally the substrate and the citric acid production occurs during the idiophase, as it is a secondary metabolite, and the fermentation follows the pattern presented in Figure 31.1. During trophophase, mycelium is produced and CO_2 is released. During idiophase, glucose and fructose are metabolized directly to citric acid. Little CO_2 is produced. Under optimal conditions, about 70% of the sugar is converted to citric acid. A critical factor in the fermentation is the level of copper, iron, and other minerals present as some of these inhibit citric acid formation. The fermentor must be made of stainless steel or be glass-lined, and copper fittings must be avoided.

A variety of other organic acids can be synthesized by microbes. One example is **itaconic acid**, which is used in methacrylate resins as a copolymer. Adding 5% of this product to printing ink increases the ink's ability to adhere to paper. It also has potential as a detergent. The organism involved in production of itaconic acid is a strain of *Aspergillus terreus*. In submerged fermentation, the yield from a beet molasses medium is as high as 85% of the theoretical (Figure 31.13).

> **SECTION HIGHLIGHTS**
>
> Citric acid and itaconic acid are examples of organic acids produced by microbial fermentation. Citric acid is primarily used as a food additive, but also as a preservative and in detergents. Itaconic acid is used in methacrylate resins.

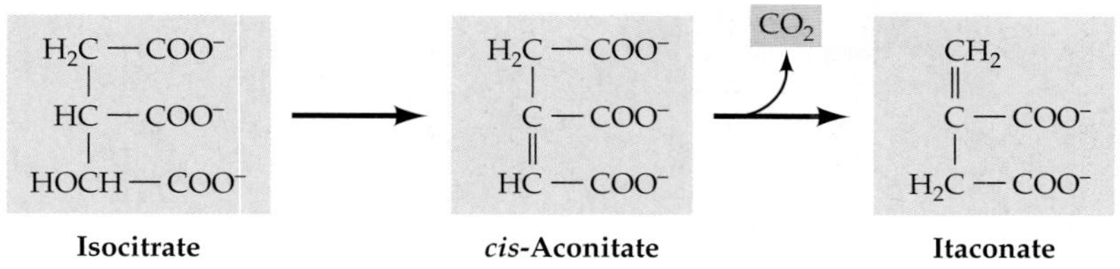

Figure 31.13 Synthesis of itaconic acid
Biochemical reactions in the synthesis of itaconate.

31.7 Enzymes, Microbial Transformations, and Inocula

In a previous section we discussed the enzyme rennin, a coagulant employed in cheese manufacture. This is but one of the many applications of commercially produced enzymes. Selective enzymes from various microorganisms can hydrolyze macromolecules that are of little value to monomers of greater value, at low cost, and under favorable conditions of temperature and pH.

Enzymes

In nature, microorganisms use enzymes to disassemble proteins, starch, pectin, lipids, and other insoluble large molecules, reducing them to monomers that can serve as a carbon or energy source for themselves or others in the environment (see Chapter 12). These hydrolytic enzymes are generally extracellular or on the cell surface and are produced by both bacteria and fungi. As extracellular enzymes, they can be recovered readily from the growth medium. Several commercial applications of microbe-generated enzymes are presented in Table 31.7. The production of bacterial proteases is a leading industrial process both in bulk and value. More than 500 tons of these enzymes are produced each year and are mostly utilized in the manufacture of detergents and cheese (rennin). The enzymes used in detergents are mostly proteases from selected strains of *Bacillus amyloliquefaciens*. They are added to detergents to promote the solubilization of stains present on the material being washed. A significant problem with the addition of pro-teases to detergents has been the allergic response to the bacterial protein. This has been overcome by micro-encapsulation and other techniques that produce dust-less protease preparations. This eliminates the problem with the allergenic dust particles.

Amylases are enzymes that hydrolyze starch. Various microbial amylases hydrolyze starch in different ways to generate short-chain polymers (dextrins) and malt-ose. Other enzymes hydrolyze the dextrins and maltose to glucose. The α-amylases cleave internal α-1,4 glycosidic bonds and are produced commercially by members of the thermophilic, endospore-forming genus *Bacillus*. Fungi such as *Aspergillus* species also produce these enzymes. Glucoamylases split glucose from the nonreducing end of starch. These enzymes are in commercial demand for the production of fructose syrups. Fructose is sweeter than glucose and is the favored sugar in soft drinks and in syrups. *A. niger* is a producer of glucoamylase. Conversion of glucose to fructose can be accomplished with a glucose **isomerase** from bacteria. The most important of the glucose isomerase producers are *Streptomyces* species and *Bacillus coagulans*.

The enzyme Taq polymerase is employed in the **polymerase chain reaction** (see Chapter 16). This enzyme is derived from *Thermus aquaticus*, a thermophilic bacterium that has an optimal growth temperature of 70°C. The polymerase from this organism is heat stable. The enzyme is widely employed in research, diagnostics, and forensic medicine.

Extremophiles are microorganisms that thrive at extremes in temperature (hyperthermophiles/psychrophiles), pH (acidophiles/basophiles) or salt concentra-

TABLE 31.7	Microbial enzymes and their commercial applications	
Enzyme	**Genus of Producer**	**Use**
Bacterial proteases	*Bacillus, Streptomyces*	Detergents
Asparaginase	*Escherichia, Serratia*	Antitumor agent
Glucoamylase	*Aspergillus*	Fructose syrup production
Bacterial amylases	*Bacillus*	Starch liquefaction, brewing, baking, feed, detergents
Glucose isomerase	*Bacillus, Streptomyces*	Sweeteners
Rennin	*Alcaligenes, Aspergillus, Candida*	Cheese manufacture
Pectinase	*Aspergillus*	Clarify fruit juice
Lipases	*Micrococcus*	Cheese production
Penicillin acylase	*Escherichia*	Semisynthetic penicillins
Taq polymerase	*Thermus aquaticus*	Polymerase chain reaction

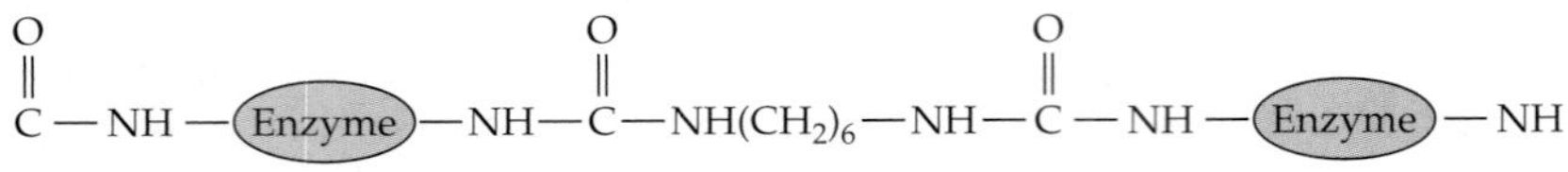

(A) Cross-linking

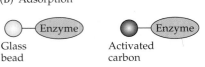

(B) Adsorption

Glass
bead

Activated
carbon

Figure 31.14 Enzyme Immobilization
Immobilization of enzymes: (A) cross-linking with glutaraldehyde, (B) adsorption to inert particles, and (C) encapsulation by trapping in a matrix.

(C) Encapsulation

Penicillin acylases are produced by various bacteria and fungi, but commercial production uses *E. coli* mutants. These enzymes cleave penicillin to 6-amino penicillanic acid and phenylacetic acid (Figure 31.15). Then, the free amine on 6-aminopenicillanic acid can be chemically modified to produce various semisynthetic penicillins (see later discussion of antibiotics).

Microbial Transformations

A dramatic event in the course of medicinal chemistry was the discovery that microorganisms can effectively hydroxylate the steroid nucleus without otherwise altering the molecule. The 3-4 member ring structure com-

tions (halophiles). The enzymes isolated from these microbes (extremozymes) are generally tolerant of the conditions under which the organism can grow. Many of these enzymes have been developed for commercial application and will be exploited much more in the future.

Enzyme preparations have a relatively short shelf life at room temperature and must be stabilized if they are stored for any period. A soluble enzyme can be stabilized in several ways. Adding substrate will stabilize some enzymes; adding low levels of organic solvent such as acetone will stabilize others. Cation additions restrict some enzymes to a tertiary configuration and increase their stability. In general, the most effective means of stabilization is through immobilization. There are three major ways that an enzyme can be immobilized: (1) cross-linking, (2) chemically bonding or adsorption to an inorganic or organic carrier, or (3) encapsulation in a gel or polymer (Figure 31.14).

Cross-linking of an enzyme involves the linkage of functional groups of the enzyme with a cross-linking reagent. A common reagent is glutaraldehyde, and this reacts through the amino groups on the enzyme. Enzymes may be adsorbed on or bonded to glass, activated carbon, starch, or other carriers. They may also be bonded to ionic exchangers. Microencapsulation places the free enzyme (or free cells) inside a semipermeable membrane. Cellulose triacetate can be employed as a fibrous network to entrap an enzyme.

The enzyme **asparaginase** is employed medically in the treatment of some leukemias and lymphomas. The metabolism of these tumor cells requires the amino acid L-asparagine. Asparaginase cleaves asparagine to aspartic acid and ammonia, thereby starving the tumor cell. Commercial production of the enzyme is by selected mutants of *Escherichia coli, Erwinia, Carotovora,* and other species.

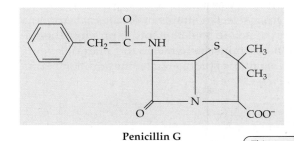

Penicillin G

Penicillin acylase

This reaction is catalyzed by the enzyme produced by *E. coli* mutants.

Phenylacetate

6-Aminopenicillanate

Figure 31.15 Synthesis of precursor of synthetic penicillins
Cleavage of penicillin G (generated by fermentation) to phenylacetate and 6-aminopenicillanate. The latter can be chemically modified to produce effective antibacterials.

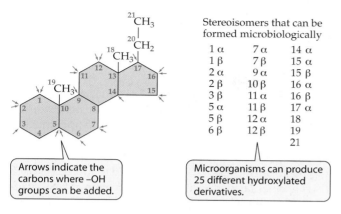

Figure 31.16 Hydroxylation of steroids
Positions at which selected microorganisms can hydroxy-late the sterol nucleus. Two stereoisomers are possible for hydroxylation at many of the positions, designated α and β.

monly present in active sterols is termed the nucleus (Figure 31.16). The anti-inflammatory activity of a steroid compound depends on the presence of an oxygen or hydroxyl in the 11-position. Naturally occurring steroids produced in significant quantities, such as those from yeasts or plants, lack the 11-hydroxylation. Prior to the discovery that microbes could hydroxylate the steroid nucleus and with a virtual stoichiometric yield, the hydroxylation was accomplished chemically and the yield was very low. The simple, low-cost microbial hydroxylation therefore had a marked effect on the availability of biologically active sterols.

An outline of the steps involved in the synthesis of biologically active sterols is presented in Figure 31.17. The starting material, diosgenin, is from a yam grown in Mexico.

Microbial species are available that can hydroxylate the sterol nucleus at any position (see Figure 31.16). Naturally occurring hormones such as progesterone, testosterone, estrone, and those from the adrenal cortex differ in number, type, and location of substituent groups or double bonds on the steroid nucleus. These can now be synthesized inexpensively for medical uses. Other physically active sterols can also be made (Table 31.8). Among the useful products are those that are exceedingly effective in preventing the inflammatory response in asthmatics and other allergic individuals. Some are effective as antitumor agents and for ocular diseases. Oral contraceptives would not have been available at low cost without the microbe-based synthetic process for sterols.

Inocula

Bulk microbial biomass is commercially available as inocula for a number of processes. Lactic acid bacteria

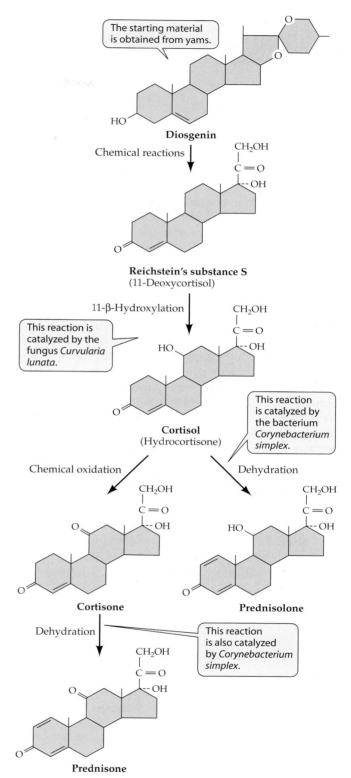

Figure 31.17 Production of medically useful sterols
Conversion of diosgenin, a plant sterol, to medically useful anti-inflammatory agents via chemical and microbiological reactions.

TABLE 31.8	Uses of synthetic sterols

Anti-inflammatory agents
Sedatives
Antitumor agents
Dermatology
Ocular disease
Cardiovascular therapy
Oral contraceptives

are among the microorganisms grown in bulk and sold to fermentation companies involved in the production of cheese and fermented meats. Use of pasteurized milk as a starting material for cheese production is necessary to prevent the spread of milk-borne diseases. However, pasteurization lowers the natural populations of lactic acid bacteria and inoculation with desirable strains to promote cheese production is necessary. Manufacture of summer sausage and other fermented meats is best achieved by inoculating with the appropriate bacterium.

Soils in which leguminous crops are planted often have a low population of symbiotic nitrogen-fixing bacteria. Therefore, planting leguminous crops in these soils may not lead to significant symbiotic nitrogen fixation. Inoculation of seeds with a compatible strain of *Rhizobium* or *Bradyrhizobium* prior to planting can significantly enhance root nodulation and nitrogen fixation. The symbiotic nitrogen fixers are grown commercially and sold as inocula. Field evidence has proved that this is a worthwhile practice.

SECTION HIGHLIGHTS

Microbes are a source of many types of enzymes, including proteases (primarily used in the manufacture of detergents and cheese), amylases (which hydrolyze starch), and enzymes for more specialized uses in cancer treatment, analysis and manipulation of DNA, and antibiotic production. Microbes also perform important steps in the biosynthesis of medically important sterols and are produced in bulk as inocula for food processing and agricultural uses.

31.8 Inhibitors

Microbes can produce catabolites that are antagonistic to other biota. As discussed in the following text, these may be a biocide (toxin) that kills selected insects or other eukaryotes or an antibiotic that inhibits prokaryotic species.

Biocides

The most successful commercial product generated by a microorganism for insect control is a protein synthesized by *Bacillus thuringiensis*. This protein appears as a parasporal crystal during sporulation and is released along with the spore on disintegration of the sporulating bacterium (Figure 31.18). The vegetative cell of *B. thuringiensis* lacks toxicity, but the protein crystal is effective in killing lepidopteran insects. It is most effective in insects that have an alkaline pH in their midgut, as the toxic protein is solubilized at the higher pH. The proteins from ingested spores are cleaved by gut proteases, and the polypeptide toxins destroy gut epithelial cells. The gut contents are released into the blood system of the insect, resulting in paralysis and death. The toxin is harmless to other animals because of the neutral or acid pH of their intestinal tracts.

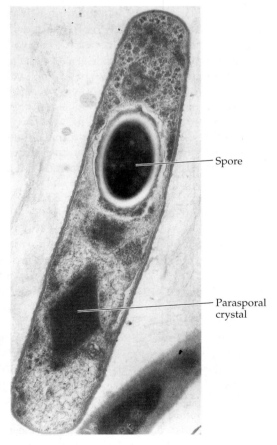

Spore

Parasporal crystal

Figure 31.18 Bacterial insecticide
Spore and diamond-shaped parasporal crystal produced by *Bacillus thuringiensis*. The crystal is an effective insecticide. ©G. B. Chapman/Visuals Unlimited.

B. thuringiensis grows well in submerged fermentation, and under appropriate conditions will produce spores and the parasporal body within 30 hours. After inoculation of the fermentor, growth, and sporulation, the spore/crystals are recovered and incorporated into an inert carrier and marketed as an insecticide to be dusted on plants. The active compound is naturally occurring and therefore degraded by soil microbes. The toxin has been genetically inserted into cotton and corn genomes and now resistance is forming rapidly (much like the antibiotic resistance problems).

Antibiotics

The use of antibiotics in controlling microbial growth has been discussed (see Chapter 7). Approximately 10,000 different **antibiotics** have been characterized, and about 160 of these are of value and in commercial production. Antibiotics are produced mainly by fungi and bacteria. The useful antibiotics are mostly produced by a large group of filamentous soil bacteria known collectively as the actinomycetes. **Streptomyces** is the one genus that has proved to be a most prolific producer, and the search for commercially valuable antibiotics has centered on it and related genera. It is a possibility that other microbial species, not now under investigation, might produce useful antibiotics. Experience, however, has shown that microorganisms that have a life cycle and undergo spore formation (both endospores and other spores), are the most effective in producing useful antibiotic substances. Antibacterial agents are discussed in Chapter 7, but it should be emphasized that agents have also been discovered that are active against some fungi, tumors, and parasites (Table 31.9). Many of the antitumor agents are produced by *Streptomyces* sp., as are agents effective as antifungals. **Ivermectin** is used in veterinary medicine for treatment of intestinal worms.

Many antibiotics are now available for the medical profession, but the search for antibiotics continues. Antibiotics against bacteria, fungi, viruses, and tumors are actively sought. This search includes antimicrobials that are effective against disease-causing microbes now seemingly under control. One of the key problems in the use of antibiotics and chemotherapeutics is the pathogen's development of resistance to the agent. Antibiotic-producing organisms generally bear genetic information that renders them resistant to the antibiotic they produce. This genetic information (an antibiotic resistance gene) is thus available in nature and can pass by horizontal gene transfer through various microbes and, ultimately and inevitably, to disease-causing organisms. Resistance to antimicrobial agents also occurs through mutation and selection (see Chapter 13). Hence, antibiotics are constantly being sought that are not cross-resistant to those in current use. Cross-resistance is the phenomenon that

TABLE 31.9	Antibiotics that are effective against eukaryotic cells
Producing Organism	**Substance Produced**
Antifungals	
Streptomyces griseus	Cycloheximide
Streptomyces noursei	Nystatin
Penicillium griseofulvum	Griseofulvin
Streptomyces nodosus	Amphotericin B
Antitumor	
Streptomyces peucetius	Daunorubicin
Streptomyces antibioticus	Actinomycin C
Streptomyces caespitosus	Mitomycin C
Streptomyces verticillus	Bleomycin
Antihelminth	
Streptomyces avermitoles	Ivermectin

occurs when a bacterium acquires resistance to one antibiotic that spontaneously gives resistance to another. It is clear that new agents effective against bacteria, fungi, viruses, or tumors would be of great value.

Several new approaches have been developed for screening for new antibiotics based on an increased understanding of the unique characteristics of pathogenic microbes. Antimicrobials targeted against specific virulence factors would minimize the emergence of resistant mutants. This would be appropriate for pathogens that develop virulence factors through quorum sensing.

Molecular models of the proteins involved in adherence would permit the design of specific agents that would prevent this important step in establishing infections (see Chapter 26). Sequencing the genomes of pathogens will ultimately identify the gene sequences that code for virulence factors, adherence factors, and toxin production. This can lead to the design of specific chemical agents to counter their activity. Designer drugs such as antisense molecules that react with mRNAs that direct synthesis of proteins involved in pathogenesis are one possibility.

Delving into the internal structure of bacterial ribosomes has led to novel approaches to antibiotic design. The 50S subunit of bacterial ribosomes is composed of rRNA that catalyzes peptide bond formation. Using x-ray detectors scientists have visualized the functional area of the 50S rRNA subunit. This area is highly conserved in most pathogenic bacteria. High-resolution techniques applied to the subunit structure indicate targets for specifically designed antibiotic agents. Drugs are now commercially available such as dorzolamide (Trusopt®) that were derived as a function of structure-based visualizations.

The following discussion presents the industrial approach to the discovery of new and useful antibiotics. Much of the methodology can be applied with some adaptation to the search for biocides, growth factors, or other useful compounds.

SEARCH FOR NEW ANTIBIOTICS   The indispensable factor in any search for novel compounds is the development of an **assay** system. The search is termed a screen, and the detection system for desired activity is the assay. The assay must give accurate, rapid results at a low cost. Whether the screen is for inhibitors, novel enzymes, organic acids, or amino acids, an efficient assay is indispensable. A potential procedure that could be employed in the search for antibiotics is as follows:

- **Identification of potential producers.** In searching for antibiotics, all past experience affirms that actinomycetes are among the best microbes to screen. The best sources for them are neutral to slightly alkaline soils, composts, and peat.

- **Isolation of organisms.** Isolation is usually accomplished by spreading suitable dilutions of the soil on the surface of an agar medium. Petri plates with a medium made up of hydrolyzed casein (animal), soytone (plant), and yeast extract (microbe) are suitable. Sugar content is best kept low to limit the overgrowth of slower-growing actinomycetes by rapidly growing pseudomonads or bacilli. The leathery, compact dry colonies of the actinomycetes are readily apparent to an experienced technician.

- **Preliminary plate test.** The capacity of the isolated microorganism to produce an antibiotic substance can be ascertained by using the plate-test method. The actinomycete is inoculated across an agar plate as depicted in Figure 31.19. *Streptomyces* and other actinomycetes form a compact continuous colony when streaked across a plate. After a few days, the test organisms are streaked perpendicular to the actinomycete. Usually a gram-positive bacterium and a gram-negative bacterium are employed to select for potential agents. A mycobacterium would be used to select for antituberculosis antibiotics and a yeast for antifungals. Rarely is an antibiotic effective against all four of these test organisms unless it is cytotoxic. In the case illustrated, the antimicrobial is effective against *Staphylococcus epidermidis*. If a significant zone of inhibition against any of the test organisms is apparent, the actinomycete passes to the next phase.

- **Assay for novel activity.** With the vast number of known antibiotics, most activity discovered will likely be the result of an antibiotic already described. It is of utmost importance that producers of known antibiotics be eliminated from the screen quickly. This is accomplished by following several possible protocols. First, one grows the selected

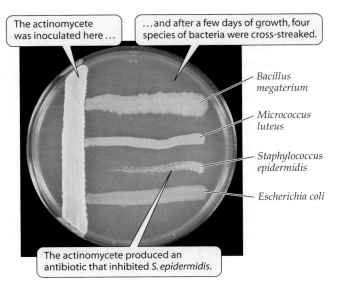

Figure 31.19 Plate test for antibiotic
Production of an antibiotic substance by a *Streptomyces* sp. isolated from soil. The agar plate contains a medium that supports growth of the actinomycete and the four test bacterial species. Courtesy of J. J. Perry.

actinomycete in liquid culture and subjects it to a paper disk plate assay to ensure that antimicrobial activity is repeatable (Figure 31.20). Various tentative identification procedures then might be applied to the growth medium from the initial fermentation including paper chromatography, electrophoresis, pigment production, absorption spectra, and antimicrobial spectrum. Activity

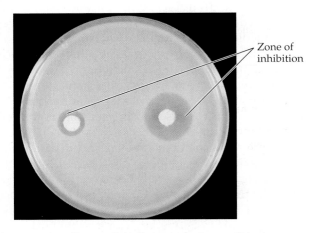

Figure 31.20 Paper disk plate assay for antibiotic
Detection of antibiotic activity by an agar plate assay. Graded amounts of fermentation broth are added to filter-paper disks and placed on the surface of an inoculated agar plate. The size of the zone of growth inhibition is proportional to the amount of antibiotic present. Courtesy of J. J. Perry.

against microbes known to be resistant to antibiotics now in use can also be determined. If the antimicrobial substance appears novel, it would be more extensively characterized.

- **Advanced culture screening.** At this stage, if the activity appears novel, the isolated organism is subjected to a series of procedures to increase the amount produced to levels that permit isolation and characterization. Experience tells us that newly isolated organisms produce only a few micrograms per milliliter of antibiotic and generally synthesize a series of chemically related antimicrobials so that no single one of these would be present in significant amounts. Various parameters are tested for their positive effects on antibiotic production: pH, aeration, growth substrate, length of fermentation, and other conditions. These efforts to increase the amount of antibiotic produced are monitored by assay techniques to ensure that the antimicrobial is the same as that from the original

isolate. Favorable results will lead to production in laboratory-scale fermentors. Chemists are then recruited in efforts to chemically identify the antimicrobial substance. If the compound continues to appear novel, animal testing is warranted.

- **Animal testing.** Mouse tests indicate whether the compound is toxic or destroyed by mammalian enzymes. Such tests determine whether the antimicrobial is effective in vivo. If it proves safe and effective, the antibiotic proceeds to higher animal tests, human tests, and in the rare case, ultimately to the pharmacy.

COMMERCIAL PRODUCTION   The commercial production of antibiotics is accomplished in huge fermentors (Figure 31.21) that vary in size from less than 40,000 to more than 200,000 liters. Size is limited by the ability to provide sufficient aeration for maximum rates of growth. The heat generated during growth is dissipated by cooling coils, as the optimal temperature for a typi-

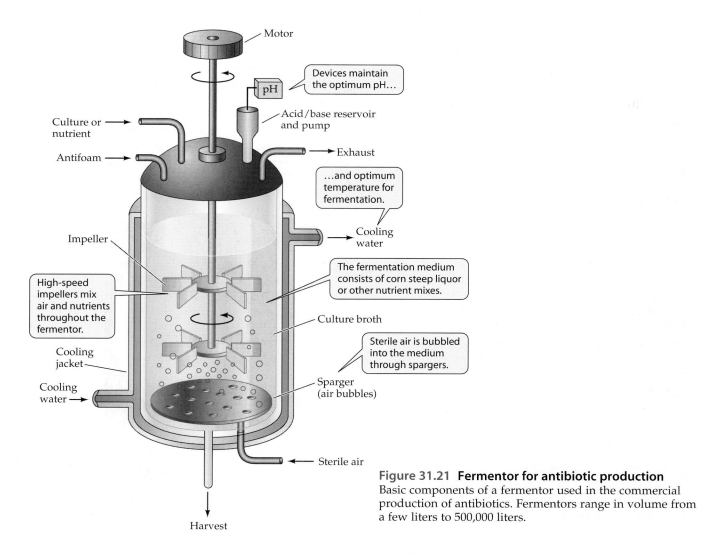

Figure 31.21 Fermentor for antibiotic production Basic components of a fermentor used in the commercial production of antibiotics. Fermentors range in volume from a few liters to 500,000 liters.

cal *Streptomyces* fermentation is generally 26°C to 28°C. Aeration is a significant factor and varies somewhat, but generally requirements are for 0.5 to 1.5 volumes of air per volume of the fermentor per minute. This requires that huge amounts of sterile air be introduced into the fermentor through spargers. Oxygen is poorly soluble in water, and the impellers must turn at more than 100 rpm to ensure adequate air saturation of the growth medium. The pH is monitored to maintain the optimal pH for antibiotic production. Antifoam agents are added to control foaming under these high aeration–mixing conditions. Powerful motors drive the impellers, and the shaft of the impeller passes from the outside to the inside of the fermentor. Contamination of the growth medium is a considerable concern and can cause serious economic loss. The fermentor must be constructed so that it can be sterilized completely, usually with live steam passed through the cooling coils, and every outlet must be constructed to preclude contamination.

The medium in a fermentor varies but generally consists of corn steep liquor, a molasses-like substrate, or various combinations of sugar/yeast extract/soytone.

The addition of substrates as the fermentation progresses is sometimes desirable. Precursors of antibiotics obtained by chemical synthesis may also be added as the fermentation progresses. The time required for a fermentation run is about 4 days.

The fermentation process is outlined in Figure 31.22. The antibiotic-producing stock cultures are carefully maintained either by lyophilization, using liquid nitrogen, or another suitable storage condition. To initiate the fermentation process, the organism is cultivated in a series of fermentation steps, or scale-ups, and at each step the culture is monitored for purity. The final scale-up inoculum is sizable to ensure that the final fermentation occurs quickly and without a lag phase. After growth (trophophase) and antibiotic formation (idiophase), the fermentation medium is filtered to remove mycelia, and the antibiotic is then recovered.

Antibiotics can be classified according to their antimicrobial spectrum, mechanism of action, producer strain, manner of biosynthesis, or chemical structure. A discussion follows on some major antibiotics now employed in medical practice.

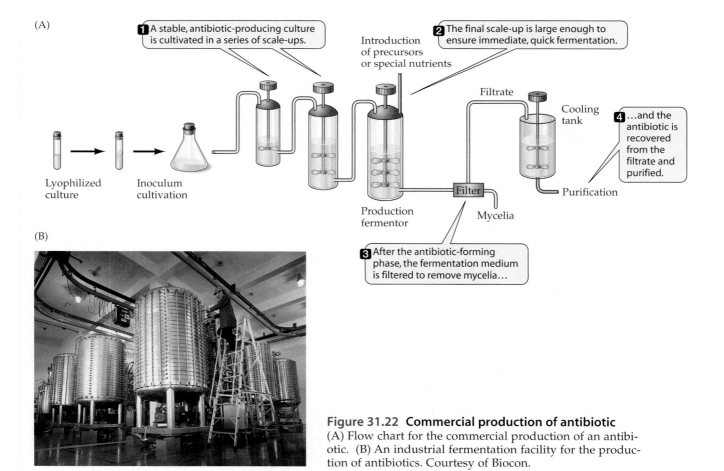

(A)

1 A stable, antibiotic-producing culture is cultivated in a series of scale-ups.

2 The final scale-up is large enough to ensure immediate, quick fermentation.

Introduction of precursors or special nutrients

Filtrate

Cooling tank

4 …and the antibiotic is recovered from the filtrate and purified.

Lyophilized culture

Inoculum cultivation

Production fermentor

Filter

Mycelia

Purification

3 After the antibiotic-forming phase, the fermentation medium is filtered to remove mycelia…

(B)

Figure 31.22 Commercial production of antibiotic
(A) Flow chart for the commercial production of an antibiotic. (B) An industrial fermentation facility for the production of antibiotics. Courtesy of Biocon.

GENERAL STRUCTURE OF A PENICILLIN

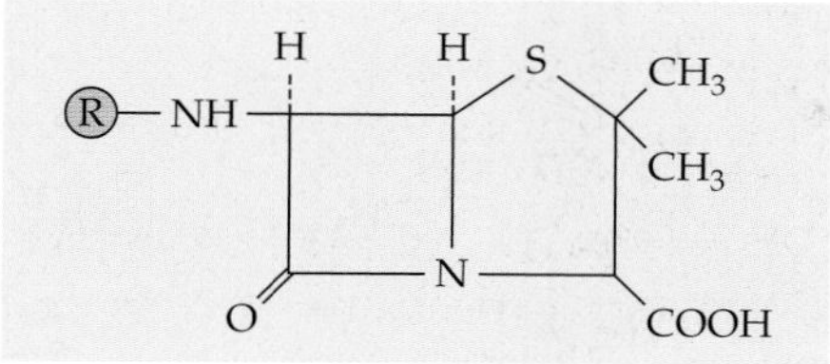

β-Lactam thiazole

R GROUPS

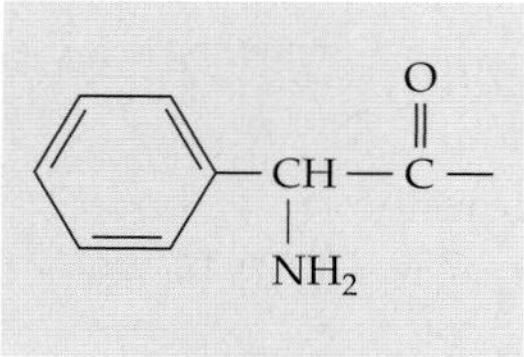

Ampicillin

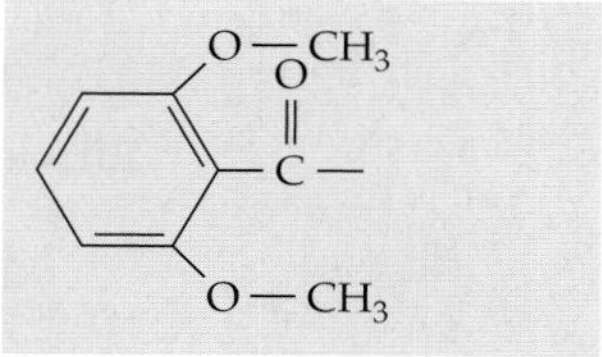

Methicillin

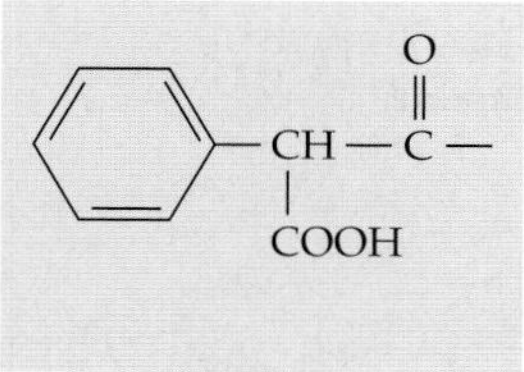

Carbenicillin

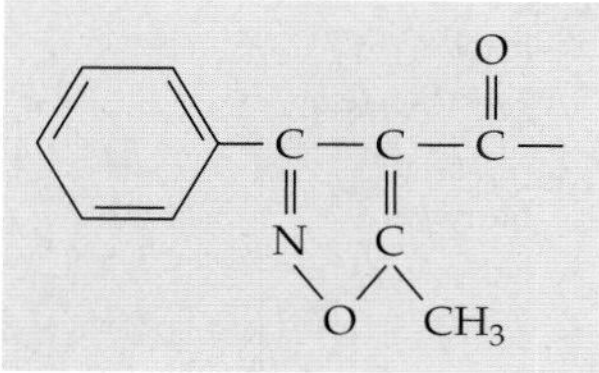

Oxacillin

Figure 31.23 Synthetic penicillins
Chemical derivatives of 6-amino-penicillanate, shown as the R groups substituted in the basic penicillin structure.

GENERAL STRUCTURE OF CEPHALOSPORIN

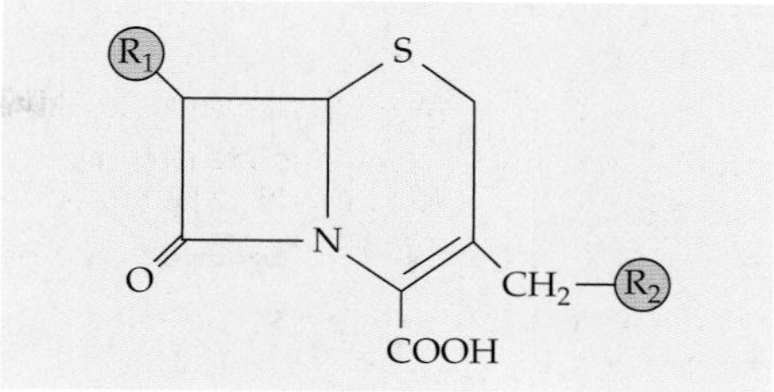

β-Lactam

R GROUPS

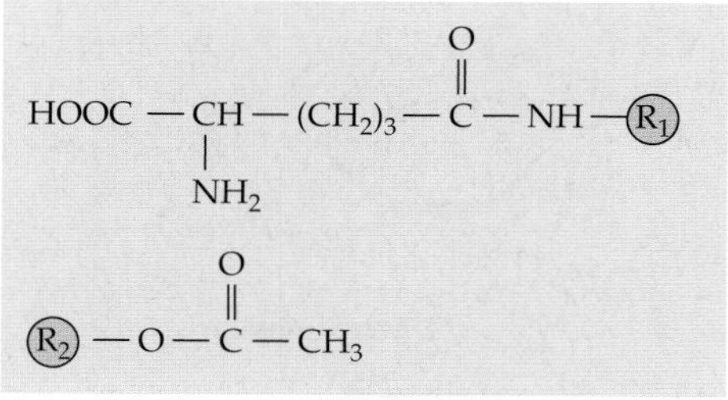

Cephalosporin C

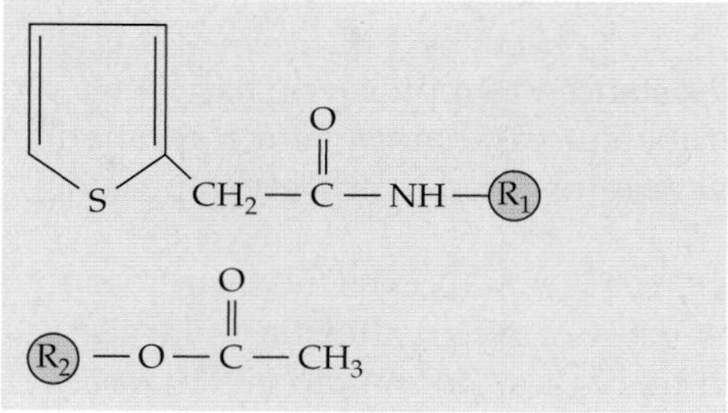

Cephalotin

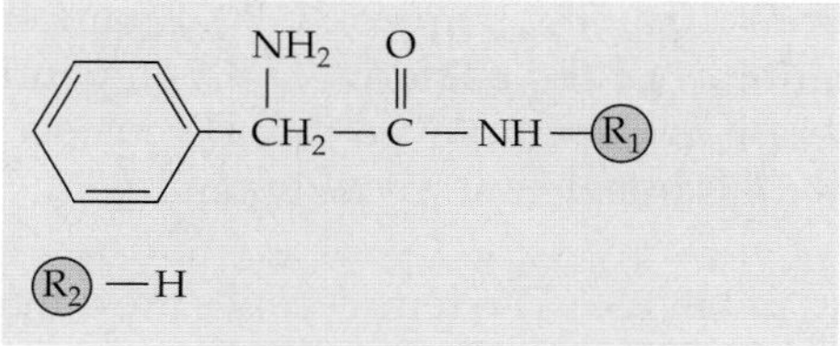

Cephalexin

Figure 31.24 Cephalosporin and derivatives
Cephalosporin C and some semisynthetic derivatives employed in treatment of infections, shown as the R groups substituted in the basic cephalosporin structure.

PENICILLIN Penicillin is produced by *P. chrysogenum* and contains the biologically unusual β-lactam ring. This antibiotic has been in use for more than 50 years, and microbes resistant to it have become a serious concern. Semisynthetic penicillins are more effective against microbes that are resistant to the natural product and are antibiotics of choice clinically. The role of microbial acylase in generating 6-aminopenicillanic acid is shown in Figure 31.15. The structures that can be added to this moiety to form effective antibacterials are illustrated in Figure 31.23.

CEPHALOSPORIN This is also a β-lactam antibiotic that is produced by the fungus *Cephalosporium acremonium*. The cephalosporins have lower toxicity than penicillin and a somewhat broader antimicrobial spectrum. They are also resistant to the penicillinase produced by penicillin-resistant microbes. The cephalosporins are produced by fermentation, and some semisynthetic derivatives are generated by chemical modification. The basic structure of cephalosporin and some semisynthetic derivatives are illustrated in Figure 31.24.

STREPTOMYCIN Many antibiotics are derivatives of sugars, and a major antibiotic in this group is streptomycin (Figure 31.25), which is produced by *Streptomyces griseus*. Streptomycin is composed of amino sugars linked through glycosidic bonds to other sugars. Its discovery was of great medical importance because it was the first effective drug against the scourge of tubercu-

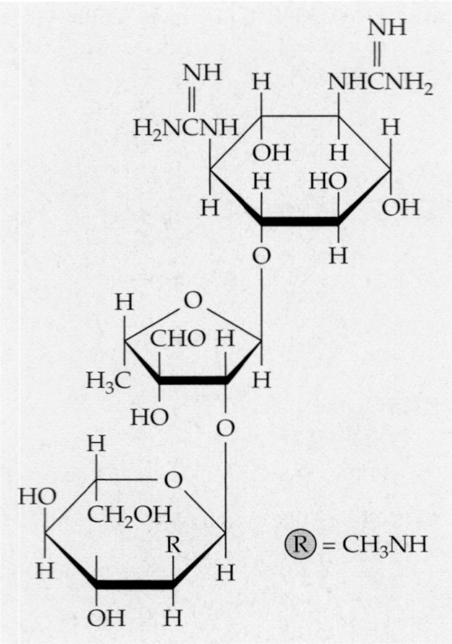

Figure 31.25 Streptomycin
Structure of streptomycin, an aminoglycoside antibiotic.

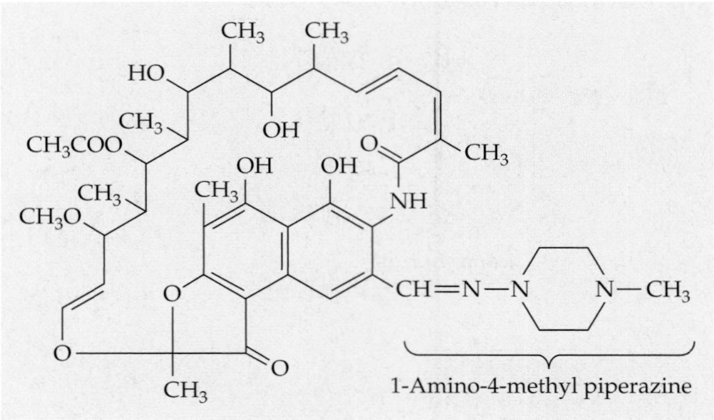

Figure 31.26 Rifampin and other antituberculosis drugs
Structures of some drugs used in treating tuberculosis. Rifampin is synthesized from the antibiotic rifamycin by addition of 1-amino-4-methyl piperazine. Isoniazid, pyrazinamide, and ethambutol are administered in combination with rifampin or streptomycin in tuberculosis treatment.

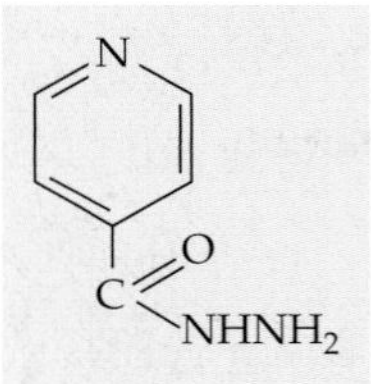

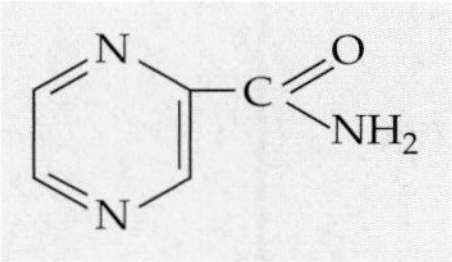

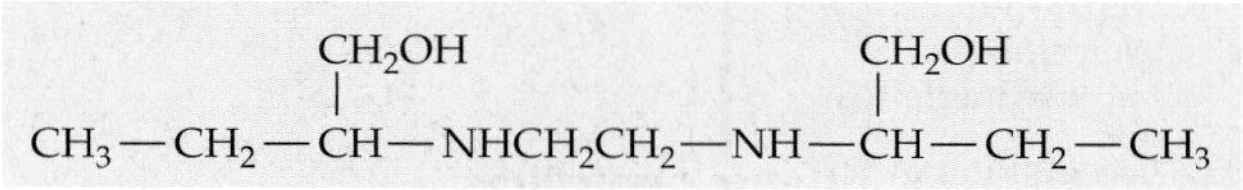

losis. Resistance to streptomycin has been a serious problem in tuberculosis therapy, and combinations of drugs are now administered over a given period.

MACROCYCLIC LACTONES   Macrolides are antibiotics such as erythromycin that have large lactone rings bonded to sugars. Rifamycin is a macrocyclic lactone antibiotic produced by *Nocardia mediterranei*. Rifampin is a semisynthetic macrocyclic lactone derivative synthesized from rifamycin (Figure 31.26). Rifampin is a specific inhibitor of the bacterial DNA-dependent RNA polymerase and is employed in tuberculosis treatment along with isoniazid and pyrazinamide.

TETRACYCLINES   The tetracyclines are a major group of antibiotics effective against both gram-positive and gram-negative bacteria. They are also effective against *Rickettsia, Mycoplasma, Leptospira, Spirochetes,* and *Chlamydia*. In addition, some semisynthetic derivations have been developed to counteract the bacterial drug resistance problem. The tetracyclines have a naphthacene core with a number of added R-groups (Figure 31.27).

Vaccines

Immunity elicited by vaccination is of major consequence in the well-being of humans and many animal species. The production of vaccines is therefore of considerable commercial value. The resistance of microorganisms to antibiotics, and the difficulty in discovering novel and effective antibiotics is of concern. Emerging infections and the lack of effective antiviral agents has generated a considerable interest in the development of vaccines to combat infectious diseases. We now have vaccines that induce immunity to many childhood diseases (see Chapters 28, 29), and others are being developed. Recently a vaccine against the rotavirus (causing disease that kills 600,000 children worldwide each year) was approved.

The economic potential for effective vaccines is huge. Worldwide the vaccine industry generates more than 10 billion dollars each year and it has been predicted that this figure will increase to 24 billion by the end of the decade. For example, a vaccine has been introduced against human papillomavirus (HPV), which causes cervical cancer and kills about 4,000 women in the United States each year (and this number is on the increase). If administered to preteens worldwide, this vaccine would generate about 3.5 billion dollars per year. Vaccines are being sought for genital herpes, meningitis, Epstein-Barr virus (an agent of infectious mononucleosis), AIDS, and other disease agents.

GENERAL STRUCTURE OF TETRACYCLINE

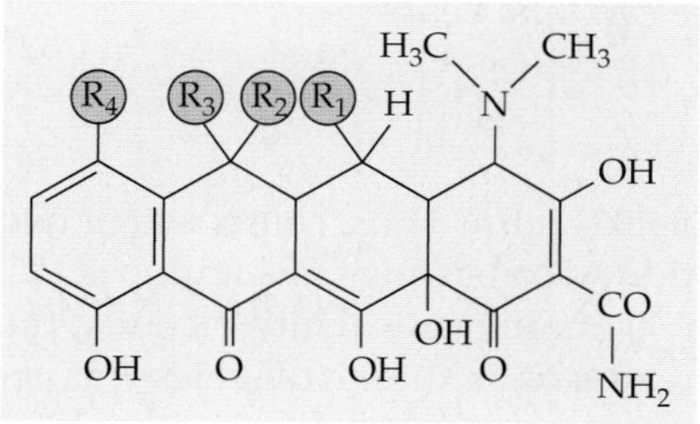

Figure 31.27 Tetracyclines
Structures of the tetracyclines. Some are natural products of *Streptomyces* strains and others are semisynthetic chemical derivatives.

Tetracycline	R_1	R_2	R_3	R_4	Production strain or method
Tetracycline	H	OH	CH$_3$	H	*Streptomyces aureofaciens* (in chloride-free medium); or chemical modification of chlortetracycline
7-Chlortetracycline (Aureomycin)	H	OH	CH$_3$	Cl	*S. aureofaciens*
5-Oxytetracycline (Terramycin)	OH	OH	CH$_3$	H	*Streptomyces rimosus*
6-Demethyl-7-chlortetracycline (Declomycin)	H	OH	H	Cl	*S. aureofaciens* (+ inhibitor)
6-Deoxy-5-hydroxy-tetracycline (Doxycycline)	OH	H	CH$_3$	H	Semisynthetic
7-Dimethylamino-6-demethyl-6-deoxytetracycline (Minocycline)	H	H	H	N(CH$_3$)$_2$	Semisynthetic
6-Methylene-5-hydroxytetracycline (Methacycline)	OH	=CH$_2$		H	Semisynthetic

Genetically engineered antigens are now a reality due to advances in biotechnology. These antigens have many advantages over antigens extracted from pathogenic bacteria or viruses and are generally superior to attenuated bacterial or viral preparations. Antigens generated by bacteria are less costly, more readily purified, and free from contamination by other proteins. One vaccine now obtained from a genetically engineered yeast is that for hepatitis B. Hepatitis B is often transferred person to person via contaminated blood and at risk are intravenous drug users, dialysis patients, and individuals who require multiple transfusions. The hepatitis B virus cannot be grown except in the blood of intentionally infected primates. The blood of individuals chronically ill with hepatitis B virus infection contains a protein particle called HBsAg. The antigen itself is not harmful but can be employed as an effective inducer of anti-hepatitis B antibody. The protein has been cloned into *S. cerevisiae* and is now the approved source of the antigen for human immunization.

This is a promising technique for obtaining antigens related to other infectious agents, particularly when propagation of the causative agent in vitro is difficult.

SECTION HIGHLIGHTS
Microbes are important sources of biocides (toxins) and antibiotics that can be used to kill or inhibit the growth of undesirable organisms. The search for new antibiotics is ongoing and involves several steps: identification of potential producers, isolation of organisms, preliminary plate tests, assays for novel activities, advanced culture screening, and animal testing. Large, industrial fermentors are used to produce antibiotics on a commercial scale. Genetically engineered microorganisms are now used to produce antigens for vaccines.

31.9 Genetically Engineered Microorganisms

Genetically engineered microorganisms contain genes incorporated from a dissimilar organism (see Chapter 15). This approach has been applied to enable other organisms to produce compounds that are commercially useful.

Genetic engineering is being used to develop microorganisms that can produce human-related proteins. Proteins are difficult to synthesize, and extracting clinically useful proteins from natural material is a costly and inefficient process. Blood proteins and hormones are generally present in an animal in limited amounts, and purification of proteins from tissue, glands, or blood is expensive and difficult. Engineered bacteria can grow rapidly on simple nutrients and are potential sources of blood proteins, hormones, and other substances in unlimited amounts.

Several challenges are encountered in the course of producing pharmaceuticals through genetic engineering. Transferring the gene into a foreign organism may be difficult, and once it has been transferred, protecting it from destruction by host nucleases and expressing the gene is not always a certainty. One problem in expressing eukaryotic proteins in bacteria occurs when specific posttranslational modification is required, such as glycosylation. Retrieving the product from the organism and purifying it can also be problematic. The U.S. Food and Drug Administration (FDA) must approve drugs before they can be released for clinical use; their regulations must complied with to gain approval for human use. Proving efficacy through animal and human trials and safety assurance can take years.

Human Insulin

Insulin is a small protein (Figure 31.28) composed of two peptide chains: peptide A, consisting of 21 amino acids, and peptide B, with 30 amino acids. Until recently, the insulin used in the treatment of human diabetes was extracted from the pancreas of slaughtered animals. Porcine- and bovine-derived insulin had two significant shortcomings. The insulin from these animals was not structurally equivalent to human insulin and hence not as effective. In addition, this animal-derived insulin contained animal proteins that caused allergic responses in some individuals with diabetes. By 1984, commercial production of human insulin using genetically engineered *E. coli* was realized.

To engineer a bacterium to produce insulin, a synthetic gene is made for each of the polypeptide chains. The gene for each is incorporated separately into two strains via a plasmid vector. The insulin gene is linked to the end of the gene encoding β-galactosidase so that the insulin polypeptide can be co-produced and secreted along with the enzyme. The two gene-engineered strains, one producing chain A and the other chain B, are grown separately. The peptide produced is separated from β-galactosidase, and the A and B chains are chemically joined to form complete human insulin. More than one-half of the insulin used for diabetes treatment is now obtained from engineered bacteria, and the remainder is recovered from slaughtered animals.

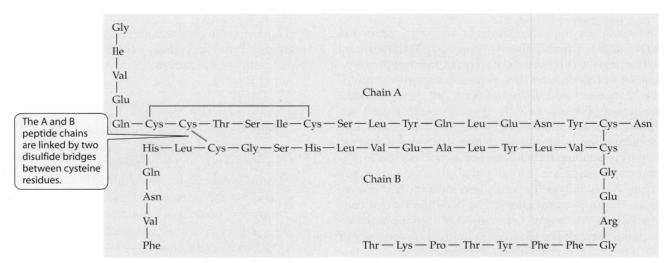

Human insulin

Figure 31.28 Insulin
Amino acid sequence of the hormone insulin. In commercial production of insulin, the two chains are synthesized separately by genetic information cloned into microorganisms and joined chemically to form the active hormone.

Human Growth Hormone

Human growth hormone, HGH, used to treat pituitary deficiencies that result in dwarfism, was previously obtained from the pituitary glands of cadavers. Obtaining enough hormone was both difficult and costly, which severely limited its use. Growth hormone has been recovered from animals but is not effective in humans.

The gene for HGH synthesis has also been engineered into *E. coli* and this is now the commercial source of the hormone. There are more than 30,000 children taking the hormone in the United States each year.

Other Cloned Genetic Systems

The potential for developing cloned genetic systems that will serve as a ready source of medically important compounds is endless. As discussed previously, several of these engineered systems are now available or are in development.

INTERFERONS   Interferons are proteins normally synthesized by cells that interfere with viral propagation. These may also be effective as anticancer agents.

BLOOD PROTEINS   Genetic engineering for the production of important blood proteins is under development. **Tissue plasminogen** activator is involved in dissolving blood clots, especially during wound-healing processes. This protein is useful in dissolving blood clots in the heart or embolisms in other areas of the body. Blood-clotting factors are also essential for hemophiliac patients. These factors are now obtained from pooled blood. With the constant worry about HIV and other viruses in pooled blood, a genetically engineered source would be most welcome.

BONE GROWTH FACTOR   Bone growth protein, found in the intracellular bone matrix, is a hormone used for the treatment of patients with osteoporosis or fractured bones. It induces undifferentiated cells to differentiate into bone-forming cells. As the level of this protein declines with age, it is necessary to provide this hormone to elderly individuals with the factor synthesized by genetically engineered microorganisms.

EPIDERMAL GROWTH FACTOR   This protein stimulates wound healing. A ready source of this factor from genetically engineered bacteria would be a boon to burn and surgery patients.

POTENTIAL FOR FUTURE DEVELOPMENTS   The future of genetically engineered microorganisms capable of generating a wide array of useful substances seems limitless. The purity of proteins, essential to human well-being, is considerably higher than in those extracted from animal tissues. In addition, concerns of contamination by known or unknown human viruses are eliminated. Furthermore, the cost of production is minimized, as microbes can be grown inexpensively and in unlimited amounts. The future for this area of industrial microbiology is indeed very bright.

SECTION HIGHLIGHTS

Microorganisms can be genetically engineered to produce insulin, human growth hormone, interferons, and other clinically useful proteins. Producing these proteins in microbes can significantly increase their availability and lower their cost, as well as providing a purer and potentially safer product.

SUMMARY

- **Industrial fermentations** are used for the commercial production of antibiotics that cure disease, yeast employed in bread making, vitamins, flavoring agents, organic acids, enzymes, and fermented foods and beverages.

- Microorganisms that generate useful industrial products are derived from naturally occurring populations by **mutation** and **selection**.

- **Biotechnology** is a field of research and development that is based on studies with microorganisms.

- Metabolites generated during active growth are considered **primary metabolites**. Those produced after exponential growth are **secondary metabolites**; each type of metabolite is produced by a select group of organisms. The **trophophase** is active growth, and the **idiophase** is involved in stationary growth and results in the production of secondary metabolites.

- Microorganisms may be consumed as food or, more importantly, may **convert foods** such as meat, milk, cabbage, cucumbers, or olives to desirable products. **Mushrooms** are consumed directly.

- **Cheese** is an ancient food that has played a role in the movement of human populations to new territories.

- Bulk **yeast** is used in the baking industry and as a food supplement.

- A lactic acid fermentation of meat produces **Lebanon bologna** and of cabbage yields **sauerkraut**. **Chocolate** and **soy sauce** are products of fermentation, as are olives and pickles.

- **Vinegar** production in the United States amounts to about 160,000,000 gallons per year. Much of this is used in commercial products (sauces, dressings, etc.). Vinegar is a product of oxidation of ethanol to acetic acid.

- **Flavor enhancers** are compounds that add little flavor by themselves but enhance the flavor of the foodstuff to which they are added. Monosodium glutamate and nucleosides are flavor enhancers that are products of industrial fermentation.

- Riboflavin and vitamin B$_{12}$ are **vitamins** produced commercially by microorganisms. Other B vitamins are produced via chemical synthesis. Some steps in the commercial conversion of glucose to vitamin C are carried out by microorganisms.

- **Wine, beer,** and other **alcoholic beverages** are products of fermentation. The starting material in wine fermentation is grapes, and for beer and other alcoholic beverages it is grain. The starch in grain must be depolymerized to yield sugars that are then fermented by yeasts such as *Saccharomyces cerevisiae.*

- **Distilled beverages** (gin, vodka, and whiskey) are essentially distillates from grain alcohol. The flavor is imparted by the volatiles that distill over with the alcohol.

- **Citric acid**, a secondary metabolite produced by *Aspergillus* sp., is added to soft drinks, candy, and other foods and is also used in detergents.

- **Proteases** produced by bacteria such as *Bacillus amyloliquefaciens* are an additive in detergents as stain removers. **Amylases** are employed commercially to convert starch to free sugars.

- **Enzymes** have a short half-life and are stabilized by **immobilization**. The cross-linking of functional groups to reagents, bonding to inert materials, or encapsulation are effective in prolonging the shelf life of enzymes.

- **Biotransformation** of low-cost inactive plant sterols to active forms can be accomplished with microorganisms. Sterols are important medicinally as anti-inflammatory agents.

- A **natural biocide** from a bacterium is the protein crystal formed during sporulation of *Bacillus thuringiensis.* The product is sold as an **insecticide**.

- Approximately 10,000 **antibiotics** have been characterized, with fewer than 100 in commercial production. Most useful antibiotics are produced by actinomycetes. Among the many antibiotics are the **penicillins, cephalosporins, streptomycin,** and **tetracyclines**.

- Microorganisms have been genetically engineered to produce insulin, human growth hormone, vaccines, and other clinically useful compounds.

- Ethanol produced from plant waste is a potential source of fuel (gasohol).

- Production of vaccines is important to human health and a major commercial product.

 Find more at www.sinauer.com/microbial-life

REVIEW QUESTIONS

1. What events occurred in the second half of the twentieth century that had a marked effect on the fermentation industry?

2. How does an industrial microbe differ from a relative that grows in the environment? What are some characteristics of microorganisms that render them potential producers of commercial commodities?

3. Define primary and secondary metabolites and cite examples of each. What do trophophase and idiophase mean in terms of industrial products?

4. What are some foods that are products of microbial fermentations? What is the nature of the microbes involved?

5. There are four major steps involved in cheese manufacture. What is the influence at each of these phases on the final product? How does one cheese differ from another?

6. What is the major use of bulk yeast in the United States? Why is yeast not a quality food for humans?

7. How are microbes utilized in the production of: (a) fermented meat, (b) sauerkraut, (c) pickles, (d) coffee beans, (e) chocolate, (f) olives, and (g) soy sauce?

8. Outline the major processes in vinegar manufacture. How do the major types of vinegar differ?

9. How does a flavoring agent differ from a flavor enhancer?

10. What are some of the major uses of amino acids?

11. How does red wine differ from white wine? Outline the steps in the production of these types of wine.

12. In beer manufacture what are: (a) the organisms involved, (b) malting, (c) wort, (d) top fermentation, (e) bottom fermentation, and (f) lagering?

13. What are distilled spirits? How does bourbon differ from Scotch and rye? How does whiskey differ from whisky?

14. List some uses of citric acid. What organisms are involved in citric acid fermentation?

15. Name three industrial-type enzymes. What are their uses? Why do microorganisms produce these enzymes?

16. What is a biotransformation? How does it apply to steroid manufacture?

17. Outline the steps in the search for a new antibiotic.

18. Genetically engineered organisms are involved in the production of several medical products. What are some of these, and what are their uses?

19. Why is microbial production superior to the isolation of products such as insulin or vaccines from natural sources?

20. What are some problems in production of ethanol as fuel and how might these be overcome?

SUGGESTED READING

Alcamo, E. I. 2001. *DNA Technology: The Awesome Skill.* 2nd ed. San Diego: Academic Press.

Montville, T. J. and K. Matthews. 2004. *Food Microbiology: an Introduction.* Washington, DC: ASM Press.

Demain, A. L. and J. E. Davies. 1999. *Manual of Industrial Microbiology and Biotechnology.* 2nd ed. Washington, DC: ASM Press.

El-Mansi, E. M. T. and C. F. A. Bryce. 2006. *Fermentation Microbiology and Biotechnology.* 2nd ed. Boca Raton, FL: CRC Press.

Glick, B. R. and J. J. Pasternak. 1998. *Molecular Biotechnology: Principles and Applications of Recombinant DNA.* 2nd ed. Washington, DC: ASM Press.

Jay, J. M., M. J. Loessner and D. A. Golden. 2005. *Modern Food Microbiology.* 7th ed. New York: Springer.

The objectives of this chapter are to:

◆ Present the roles that microorganisms play in the treatment of society's waste materials.

◆ Discuss the treatment of sewage.

◆ Describe the treatment of drinking water.

◆ Present practices used in composting and treating solid waste in landfills.

◆ Discuss the degradation of pesticides by microorganisms.

◆ Describe how acid mine drainage results from microbial activities.

32

Applied Environmental Microbiology

*Microorganisms are the premier agents for the degradation
of wastes and toxic substances, and they don't charge a penny
for their services.*
—*Anonymous*

Heterotrophic microorganisms are ideally suited for the disposal of organic waste materials for the following reasons:

- Organic materials serve as the energy and carbon sources of these microorganisms and therefore are not considered waste materials to them.

- Because of the great diversity among microbial heterotrophs as a group, they can grow in a broad range of environmental conditions of temperature, pH, nutrient, and oxygen concentrations, and therefore can utilize organic materials in most places where they exist in the environment.

- Biological processes occur at low temperatures and do not usually produce toxic end products.

- The cost of using microorganisms to dispose of waste organic materials is minimal in comparison to alternatives such as combustion or chemical treatment.

Indeed, it is difficult to stop microorganisms from degrading wastes because it is a major role they play in the recycling of organic materials on Earth. Environmental engineers have exploited the efficient recycling capabilities of microorganisms in the design of specialized microbial bioreactors for the treatment of waste materials. Of course, bioreactors need to be properly designed for the growth of the desired biodegradative microorganisms, and their construction, operation, and maintenance cost money.

Environmental engineers and microbiologists address ecological, environmental pollution, and public-health issues. They are concerned about pathogenic bacteria and their potential impact on the public health, and must ensure that the drinking water supplies are safe. In addition, shellfish and other food products that are obtained from environmental sources should be free of contaminating microorganisms. Environmental engineers and microbiologists are, therefore, concerned about the introduction of pathogenic microorganisms, bacteria, protozoa, and viruses into the environment through contamination by sewage, incomplete treatment of sewage, improper handling of garbage, or by other means.

32.1 Sewage (Wastewater) Treatment

As in natural environments where fecal material from animals is degraded primarily by microbial activities, modern municipal sewage treatment systems utilize microbial degradation as the principal means to degrade these organic materials. However, until the 1900s, cities did not treat sewage. They simply collected the raw or untreated sewage (also termed wastewater) using a sewer system and discharged the raw sewage into a receiving body of water, either a river or the marine environment, depending on the location of the city. However, it soon became apparent that raw sewage discharge in this manner was inadequate for two major reasons:

- Adverse ecological impacts on receiving waters
- Adverse public health impacts, as some of the microorganisms in raw sewage may be pathogenic

Thus, the receiving water cannot be safely used for water contact sports such as swimming and water skiing. A detailed discussion of these issues follows.

Ecological Impact of Raw Sewage on Receiving Water

Imagine a hypothetical river that flows through a city called City A that is located upstream from another city, City B (Figure 32.1A). City A discharges its raw sewage directly into the river, introducing fecal material containing large numbers of bacteria as well as large amounts of organic material into the river (Figure 32.1B). The organic effluent serves as a substrate for the growth of heterotrophic bacteria. Their aerobic respiration activities remove oxygen dissolved in the water. Indeed, if the organic load in the waste is sufficiently great, the resultant bacterial growth removes all of the oxygen and causes anoxic conditions. These anoxic conditions are lethal to fish and other oxygen-requiring animals that live in the river. The resulting "fish kills" will occur because fish require the dissolved oxygen in the water, which they breathe through their gills. In addition to the loss of animal and aquatic plant life, the river becomes a less-desirable recreational area for fishing, boating, and other activities.

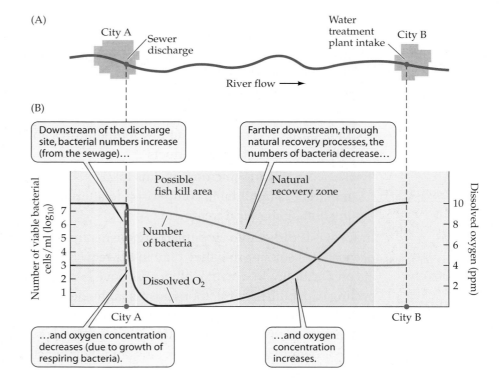

Figure 32.1 Effect of raw sewage discharge
(A) Diagram showing flow of a hypothetical river, with City A discharging raw sewage upstream from City B. (B) Graph showing the change in numbers of bacterial cells and concentration of dissolved oxygen downstream from the site of sewage discharge.

The severity of the impact varies depending on the size of City A, the amount of organic material discharged, the temperature, the volume and flow rate of the river, and a host of other factors. Fortunately, given sufficient time without additional pollutants, rivers have a natural capacity to restore themselves. For example, oxygen is reintroduced into the water by two activities:

- The normal process of oxygenic photosynthesis by cyanobacteria, algae, and aquatic plants restores some of the oxygen
- Turbulence of the water in the river results in the diffusion of oxygen from the overlying air

In addition, the organic material from the raw sewage is eventually degraded by the heterotrophic bacteria in the river and, when it is depleted, the numbers of bacteria will decline, to approach levels comparable to those encountered above City A's wastewater discharge (see Figure 32.1B). The result of these processes is that the stream undergoes a natural recovery process: purifying itself. However, this process may not occur until many miles downstream, perhaps several days' transit from the point of introduction of the wastewater.

Public Health Impact of Raw Sewage Discharge

When very large cities are located on rivers, their discharges can be sufficiently significant that normal recovery processes do not occur before the discharge reaches the next city. As a result, cities located downstream on the river can be adversely affected by the discharges in a number of ways. Of primary importance, the river water cannot be safely used as a source of drinking water because it might carry waterborne infectious disease agents discharged from the feces of infected individuals in the city upstream (Table 32.1). Furthermore, beaches and other recreational areas affected by fecal materials

Figure 32.2 Warning
Sign at a Seattle, Washington, city beach near the discharge point from a sewage treatment plant. The sign indicates that fish and shellfish might be contaminated with microorganisms. Courtesy of J. T. Staley.

cannot be used for contact activities because of the possibility of disease. Commercial shellfish fisheries may be adversely affected as well because clams and oysters, which are filter feeders, ingest particulate materials, including bacteria and viruses, resulting in high concentrations of this matter in their digestive tracts. Regulations restrict harvesting and other uses of shellfish when the numbers of bacteria in shellfish tissues are unacceptably high (Figure 32.2). Although the shellfish themselves may not be affected by ingested pathogenic microorganisms, the contaminated shellfish cannot be marketed for human consumption.

Because of these problems with raw sewage disposal, municipalities in the United States now are required to treat their wastewaters *before* they are discharged into receiving waters. The sewage is collected using a system of sewer pipes, usually by gravity flow, and is sent to the wastewater treatment facility that is located near a river or the marine environment where the treated waste can be discharged. Although wastewater treatment does not completely restore the water to its natural state, it re-

TABLE 32.1	Partial list of waterborne pathogenic microorganisms
Microorganism	**Disease**
Bacteria	
Salmonella typhosa	Typhoid fever
Vibrio cholera	Cholera
Shigella dysenteriae	Bacterial dysentery
Protozoa	
Entamoeba histolytica	Amoebic dysentery
Giardia lamblia	Giardiasis
Naegleria	Meningoencephalitis
Virus	
Infectious hepatitis	Hepatitis

duces the levels of organic matter and the concentrations of bacteria. There are three degrees of wastewater treatment: primary, secondary, and tertiary; these are discussed next.

Primary Wastewater Treatment

The initial step in sewage treatment is termed **primary wastewater treatment** because it involves the first step, namely the settling of the particulate materials and their separation from the dissolved material.

In primary treatment, the sewage coming into the plant is first passed through a screen to remove sticks, plastic bags, and other large pieces of material that might interfere with treatment (Figure 32.3). The wastewater then flows into the primary treatment tank that acts as a settling basin to allow the heavier particulate material (the **primary sludge**) to settle to the bottom. In addition, a skimming bar removes oils that float on the surface of this tank. Thus, primary treatment is simply a physical process whereby the liquid portion of the wastewater (which comprises more than 99% of the waste) is separated from the settleable solids or sludge as well as the floatables.

Secondary Treatment

The typical treatment systems used in the United States are **secondary wastewater treatment** systems. They are called secondary treatment systems because two sequential processes are used in the treatment, a primary treatment stage followed by a secondary treatment stage (see Figure 32.3).

The **secondary treatment** process is a microbiological process. The liquid portion from the primary tank is passed into another tank (the secondary wastewater tank) and is then aerated. Two different types of aerators have been used. An older type is referred to as a **trickling filter**. In this system, the fluid is sprinkled over a bed of rocks that are about the size of a fist. The wastewater is aerated while it passes from the sprinkler and percolates through the bed of rocks. Microorganisms grow as a biofilm (see Chapter 24) on the surface of the rocks and degrade the organic materials that are in the wastewater. Microorganisms, which include protozoa, algae, and bacteria as well as viruses that develop on the surface of the rock biofilms are aerobic, but those underneath are anaerobic fermentative bacteria. Parts of the biofilm are continually detaching from the rocks and are carried into a **secondary clarifier** unit (see subsequent text) along with the fluid wastewater.

The other type of secondary treatment process is referred to as the **activated sludge** process (see Figure 32.3). In this type of treatment, the effluent from the primary treatment tank is aerated by bubbling air through it in the secondary treatment tank. Aerobic microorganisms, particularly bacteria, grow in colonial aggregates or consortia called **flocs** and degrade the organic material in the waste. Among the bacteria that occur here is the genus *Zoogloea*, whose species produce gel-like, flocculent microcolonies due to their capsules (see Chapter 19).

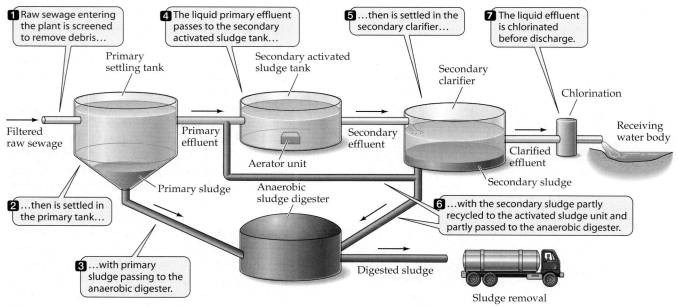

1 Raw sewage entering the plant is screened to remove debris...

2 ...then is settled in the primary tank...

3 ...with primary sludge passing to the anaerobic digester.

4 The liquid primary effluent passes to the secondary activated sludge tank...

5 ...then is settled in the secondary clarifier...

6 ...with the secondary sludge partly recycled to the activated sludge unit and partly passed to the anaerobic digester.

7 The liquid effluent is chlorinated before discharge.

Primary settling tank

Secondary activated sludge tank

Secondary clarifier

Chlorination

Filtered raw sewage

Primary effluent

Secondary effluent

Clarified effluent

Receiving water body

Aerator unit

Primary sludge

Anaerobic sludge digester

Secondary sludge

Digested sludge

Sludge removal

Figure 32.3 Secondary wastewater treatment
Diagram of a typical secondary activated sludge wastewater treatment plant. The end products are clarified, disinfected water, and digested sludge.

As in the trickling filter process, the fluid wastewater from the activated sludge unit, which contains some particulate materials, passes into the secondary clarifier unit. The liquid portion of this material contains much lower levels of organic material and can be satisfactorily discharged into a receiving stream or other body of water. Finally, before discharge, the fluid from the secondary clarifier is chlorinated to kill many of the potentially pathogenic microorganisms that might be present.

A portion of the fluid from the activated sludge digester consists of particulate material, which contains active microorganisms. This settles out in the secondary clarifier and part of this "activated sludge" is recycled back to the aeration tank, which is appropriately named the activated sludge unit. The remainder of the sludge is transferred to the **anaerobic sludge digester**, where it is typically combined with the primary sludge and treated. This is a large-capacity tank in which anaerobic bacteria carry out the final steps of anaerobic degradation of organic materials.

Anaerobic sludge digesters contain fermentative bacteria that produce organic acids, carbon dioxide, and hydrogen gas. Methanogenic bacteria grow on the acetic acid and on the hydrogen and carbon dioxide in these reactors and produce methane gas. The methane produced can be reclaimed and used as a power generator and for heating in the treatment plant.

Anaerobic sludge digestion is a slow process and does not result in a complete conversion of sludge to gases. The digested sludge that remains after treatment is rich in nutrients and must be disposed of elsewhere. It has been used as a fertilizer for agricultural plants that are not used for human consumption because there is concern about viruses and other microbes that might have survived the treatment process.

The organic material in wastewater imparts what environmental engineers refer to as a **biochemical oxygen demand**, or **BOD**, to the waste. The BOD test is used to measure the amount of oxygen demand, or the amount of oxygen required to decompose the organic material in 5 days at 20°C in a wastewater sample. Initially, raw sewage has a very high BOD, but if treatment has been successful, the BOD level is reduced significantly. For example, trickling filters working satisfactorily can reduce the BOD content of raw sewage by about 75%. Activated sludge-treatment units are even more effective when working properly and can reduce the BOD content by at least 85%.

Therefore, secondary treatment effectively removes much of the organic material from wastewater. In fact, the actual BOD reduction is effected by the bacteria and other microorganisms that grow in the secondary treatment units. These microorganisms are doing what they would normally do in nature, but by doing it in a wastewater treatment plant, the process can be monitored and controlled. The microbes carry out these processes effectively and without any cost to society.

Not all microorganisms are removed by settling in the clarifier unit; some will pass to the secondary effluent and be discharged into the receiving waters. Most of these microorganisms have grown on the organic materials of the wastewater. Some of them are coliform bacteria, such as *Escherichia coli*, which come from human feces being treated in the sewage. Because some of the bacteria are waterborne pathogens, it is desirable to kill them. For this reason, the effluent from the secondary clarifier is disinfected, usually with chlorine, before it is finally discharged into the receiving water. Although this process is not 100% effective in disinfection of the effluent, it does dramatically reduce the numbers of coliform bacteria and pathogenic microorganisms that are discharged from the wastewater treatment plant.

Wastewater and septic tanks serve as a source of coliform bacteria discharge into the environment. The coliform bacteria may contaminate beaches and other public areas; therefore, these receiving environments need to be tested by public health microbiologists who are concerned about fecal contamination (see Figure 32.2).

Microbial Treatment Problems

The normal degradative activities occurring in the secondary treatment system depend on the types and status of the microorganisms growing in the reactors. Although it is true that the bacteria carry out their activities "for free," it is essential to ensure that conditions in the bioreactors are satisfactory for the growth of the specific organisms that are best suited for the biodegradative processes needed. Thus, in addition to providing the necessary bioreactor tanks, it is essential that the secondary activated sludge unit has sufficient aeration to ensure that the organic materials can be degraded as completely as possible.

One microbiological problem that may occur in the secondary clarifier units is **bulking**. Bulking refers to the poor settleability of sludge. Ideally, the particulate organic material should be completely removed prior to discharge into a receiving stream or other outlet. Therefore, it is important that microorganisms and other debris settle to the bottom of the clarifier where they can be removed and transferred to the sludge digester. Some species of bacteria are undesirable in secondary clarifiers because they interfere with settling. Filamentous organisms such as *Sphaerotilus natans* and *Thiothrix nivea* are examples of bacteria that interfere with normal settling. Treatment conditions need to be modified to remove them, and this is sometimes difficult to accomplish.

Anaerobic digesters may also encounter problems. Under some circumstances, they "go sour"—they become acidic and cannot support the normal populations of fermenters and methanogens. When this occurs they need to be reseeded with inocula from other active digesters.

Tertiary Wastewater Treatment

Although secondary treatment of wastewaters is effective in removing organic matter from the wastes, it does not remove the inorganic byproducts of the microbial activity. As can be seen from the following overall treatment formula, inorganic substances such as ammonia and phosphate are produced when organic material is oxidized.

$$\text{Organic material} + O_2 \rightarrow CO_2 + NH_3 + SO_4^{2-} + PO_4^{3-} + \text{trace elements}$$

The inorganic products of organic degradation are excellent nutrients for the growth of algae. Discharge of these nutrients into a receiving body of water can cause algal blooms. Such excessive algal growth can lead to eutrophication (Box 32.1). Thus, it is possible to have successfully treated a wastewater by secondary treatment, only to have it cause enrichment in the receiving water.

The major nutrients of concern for enrichment of receiving waters are phosphate and ammonia. These nutrients can be removed from secondary effluent by additional treatment referred to as **tertiary wastewater treatment**. Both chemical and microbiological tertiary treatment processes have been developed. Chemical processes are very expensive and will not be discussed further here.

Typically, ammonia removal is effected biologically through a two-step process. The first step is nitrification, the aerobic oxidation of ammonia to nitrite and nitrate, by nitrifying bacteria (at this time it is not known to what extent archaeal nitrifiers might be involved in wastewater nitrification). Although these are chemolithotrophic bacteria (see Chapter 19), they grow well in properly treated secondary effluent. During this first step, the ammonia is converted to nitrate. In order to remove the nitrate, it is necessary to carry out denitrification. Denitrification results in the anaerobic conversion of nitrate to form N_2O and N_2 gases, which will dissipate into the air. Therefore, during this two-step process nitrogen is removed from the water as a gas and therefore cannot enter and enrich the receiving water. It should be noted that denitrification by organisms such as *Pseudomonas* species requires organic carbon. An effective and inexpensive carbon source, methanol, can be added to serve this purpose.

This two-step process for nitrogen removal can be carried out in normal secondary wastewater treatment systems if they are closely monitored. A period of aeration in the secondary clarifier to enhance nitrification is followed by a period of anaerobic incubation that favors denitrification.

The anammox process (see Chapter 22) is also being tested in pilot studies as a means of removal of fixed nitrogen from wastewaters. Indeed, these bacteria were first discovered in wastewater treatment plants.

Phosphate removal occurs by using bacteria that carry out the uptake of phosphate into their cells. The process is referred to as **enhanced biological phosphorus removal (EBPR)**. Like the anammox bacteria, the microorganism that is responsible for this process has not yet been cultivated in pure culture. However, since it has been well characterized from environmental studies it has been named as a "Candidatus" species *Accumulibacter phosphatis*. Furthermore, a metagenomic study has re-

---

BOX 32.1　*Research Highlights*

Eutrophication of Lake Washington

A classic case of lake eutrophication occurred in Seattle's Lake Washington in the late 1960s. Due to the inflow of secondary effluent from sewage treatment plants around the lake, the lake became increasingly enriched with fixed nitrogen and phosphate. Dr. W. T. Edmondson of the University of Washington noted that phosphate was limiting to algal growth in the lake. Thus, even though the organic material was being satisfactorily removed, the increased growth of algae, due to influx of phospate from the secondary treatment plants, resulted in increased primary production. The resulting abnormal increase in organic material from the growth of the algae in the lake caused lake eutrophication. Based on his recommendation, the metropolitan area took community action and constructed a system of sewers to divert all of the secondary effluent away from the lake. The treated wastewater is now discharged into a much larger body of water, Puget Sound. Since the late 1960s, Lake Washington has returned to its normal mesotrophic status.

cently been performed on organisms from the EBPR process. *A. phosphatis*, which is a member of the *Proteobacteria*, accumulates polyphosphate granules in excess of normal metabolic needs during periods of active growth. The result is an increase in cell density, thereby enabling the bacteria to settle from clarifier units where they can be removed along with the phosphate they contain.

In practice, very few plants in the United States use tertiary treatment. However, this will likely change as the need for protecting our receiving waters increases.

SECTION HIGHLIGHTS

Secondary wastewater treatment reduces the organic pollution of receiving waters and is typically accomplished by activated sludge and anaerobic sludge digesters. The effluent from the secondary clarifier is chlorinated before discharge into the receiving water to reduce the numbers of viable bacteria. Tertiary treatment by microorganisms involves fixed nitrogen removal by either denitrification or anammox, and phosphate removal by the consortium species, *Accumulibacter phosphatis*.

32.2 Drinking Water Treatment

As discussed previously, if a city receives its drinking water from a river, the water may be contaminated with effluent from wastewater treatment plants from cities upstream. In fact, even if there is no city upstream, it is

possible that the water can be contaminated from wild or domestic animals that may harbor infectious waterborne microbial agents such as *Giardia lamblia* or enteric viruses or bacteria. In addition, poorly designed and maintained septic tanks and farm animals in rural areas may also discharge wastewater into the river. Therefore, it is essential for the city to treat its drinking water before it is distributed to its citizens.

Depending on the source of water, the treatment may consist only of disinfection by chlorination or the treatment may be much more extensive. Consider the case of a large city located on a major river such as the Mississippi. Furthermore, consider that the city is downstream of many other cities. In this example, it is important that the city use an extensive treatment system involving four steps:

- Coagulation
- Sedimentation
- Filtration
- Disinfection

These steps are accomplished in a drinking water treatment plant or facility (Figure 32.4).

Coagulation is performed by adding alum (a salt of aluminum sulfate) to the water. When the alum is added to the water, it forms aluminum hydroxide flocs that adsorb particulate materials. The flocs are then settled out by sedimentation in a holding tank. The next step is filtration. The water is passed through a sand filter system to remove the bacteria. Either rapid or slow sand filters are used, and both are effective at removing particulate material, including bacterial cells. If the treatment plant is operating effectively, about 99% of the bacteria present in the raw, untreated water are removed after sand

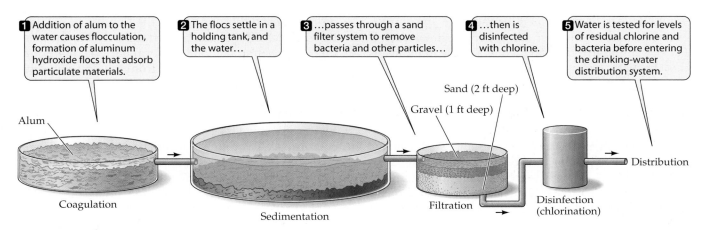

Figure 32.4 Drinking-water treatment system
Diagram of a drinking-water treatment system. Microbiological tests are routinely carried out on the product to ensure potability.

filtration. The final step is **disinfection** of the water. This is usually accomplished by chlorine. The dissolved gas forms a hypochlorite solution (as in household bleach), which effectively disrupts the cell membranes of bacteria, thereby killing them. Disinfection does not kill all of the bacteria, however, so small numbers of bacteria remain in the water.

The drinking water is then passed into a distribution system before it is consumed. In order to ensure that the water is safe for drinking (**potable**), it needs to be tested. Two approaches have been taken to ensure that drinking water is safe or potable. The first approach is to analyze for the **residual chlorine** level. If chlorine is still present in the water at the tap, at a level of 1 ppm, then it may be considered potable. However, as we have stated, even if there is residual chlorine, bacteria may be present. Therefore, microbiological tests are routinely performed to determine water potability.

Current microbiological testing uses an **indicator bacterium**. The standard indicator bacterium used in the United States is *E. coli*. Because *E. coli* is found in the intestinal tracts of all humans and some other warm-blooded animals, its presence in a drinking water is *indicative* of fecal contamination. Why use an indicator bacterium? Why not just look for a particular pathogen? The reason for this is that if a pathogen were present, it would be expected to occur in very low concentrations relative to *E. coli*. This is because the pathogen may be from only one infected individual or carrier out of hundreds or thousands of people who live in the city upstream. Therefore, it is extremely difficult to detect the pathogenic bacterium because so few of them would be released into the sewer system in relation to the numbers of *E. coli*, because *all* individuals carry *E. coli*. Furthermore, even if a test is devised for one pathogenic bacterium, what about all the other pathogens? Because there are many different pathogens, separate tests would have to be performed for each one. In contrast, everyone harbors *E. coli* in his or her intestinal tract. If *E. coli* is found in a drinking water sample, it indicates that the water has been contaminated with fecal material, and this finding alone indicates that it is unsafe for drinking.

Although *E. coli* is the indicator bacterium of choice, it is not simple to identify *E. coli* in natural samples because many other bacteria closely resemble it. It belongs to a group of enteric bacteria called the **coliform bacteria**. Coliform bacteria are defined as non–spore-forming gram-negative rods that ferment lactose to form acid and gas in 24 to 48 hours at 35°C. Thus, coliform bacteria are defined by experimental conditions. Indeed, several different types of tests are used to identify coliform bacteria. A brief description follows of some of the more common tests.

Total Coliform Bacteria Analyses

One group of tests, the **total coliform bacteria** tests, analyze for the concentration of coliform bacteria in a drinking water sample. There are several tests for total coliform bacteria including the most probable number test and membrane filter tests.

MOST PROBABLE NUMBER (MPN) TEST This test is the oldest test for total coliform bacteria and is still used as a standard test. It is performed in three different stages:

- Presumptive
- Confirmed
- Completed

The first stage, called the **presumptive test**, is designed to determine whether coliform organisms might be present in a sample. In this test, a series of lactose broth tubes (lauryl tryptose medium) are inoculated with the drinking water to be tested (Figure 32.5). Typically, fivefold replicates at each of several dilutions are prepared. The first set of five tubes, which are the lowest dilution, is inoculated with 10-ml portions of the water sample to be tested. Because 10 ml is such a large volume of inoculum, 10 ml of the medium is made up double strength, so that it ends up as single strength after addition of the 10-ml inoculum. In the next dilution, the five replicate single-strength tubes are inoculated with 1-ml inocula. Higher dilutions are attained using dilution blanks. The tubes contain lactose, the pH indicator, bromthymol blue, and an inverted vial to detect gas production. They are incubated at 35°C for 24 hours and then read for acid and gas production. If they are positive for gas, they are regarded as presumptive positive. The remaining tubes are incubated an additional 24 hours, and any additional positive tubes are also considered positive presumptive.

The MPN test is quantifiable. Each positive tube is scored at each dilution. Then, the number of positive tubes at each dilution can be used to determine the quantities of presumptive total coliform bacteria. For example, assume that all five tubes from the first two dilutions were positive, three from the third were positive, and only one from the fourth was positive. The results are recorded as 5, 3, and 1 for the final three dilutions that had positive tubes. The final series of tubes had no coliform bacteria and is not being considered. These results can be converted into a quantifiable number taken from a statistical most probable number table (see the American Public Health Association's *Standard Methods for the Examination of Water and Wastewater*). In this particular instance, the number of total coliform bacteria is 1,100 per 100 ml of the original sample. Note that this concentration is given as *per 100 ml* and is called the **coliform index**.

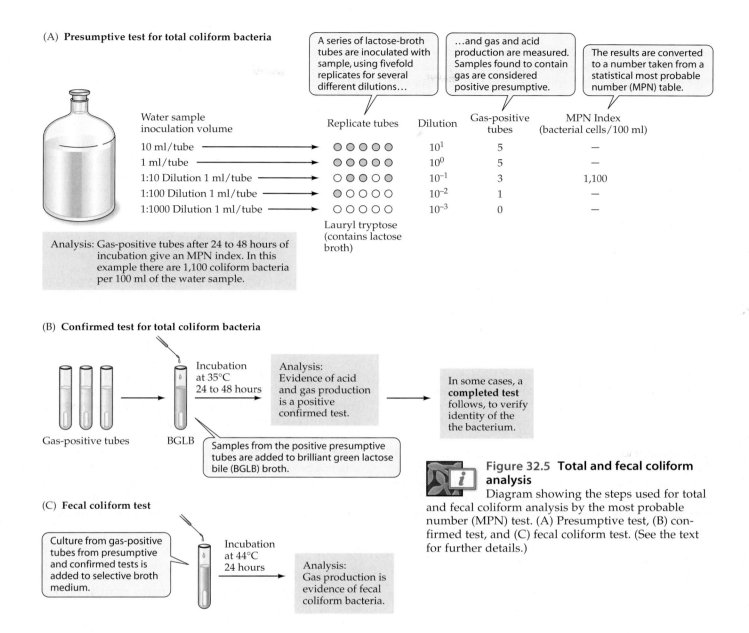

(A) Presumptive test for total coliform bacteria

A series of lactose-broth tubes are inoculated with sample, using fivefold replicates for several different dilutions…

…and gas and acid production are measured. Samples found to contain gas are considered positive presumptive.

The results are converted to a number taken from a statistical most probable number (MPN) table.

Water sample inoculation volume	Replicate tubes	Dilution	Gas-positive tubes	MPN Index (bacterial cells/100 ml)
10 ml/tube	○○○○○	10^1	5	—
1 ml/tube	○○○○○	10^0	5	—
1:10 Dilution 1 ml/tube	○○○○○	10^{-1}	3	1,100
1:100 Dilution 1 ml/tube	○○○○○	10^{-2}	1	—
1:1000 Dilution 1 ml/tube	○○○○○	10^{-3}	0	—

Lauryl tryptose (contains lactose broth)

Analysis: Gas-positive tubes after 24 to 48 hours of incubation give an MPN index. In this example there are 1,100 coliform bacteria per 100 ml of the water sample.

(B) Confirmed test for total coliform bacteria

Gas-positive tubes

BGLB

Incubation at 35°C 24 to 48 hours

Analysis: Evidence of acid and gas production is a positive confirmed test.

Samples from the positive presumptive tubes are added to brilliant green lactose bile (BGLB) broth.

In some cases, a **completed test** follows, to verify identity of the the bacterium.

(C) Fecal coliform test

Culture from gas-positive tubes from presumptive and confirmed tests is added to selective broth medium.

Incubation at 44°C 24 hours

Analysis: Gas production is evidence of fecal coliform bacteria.

Figure 32.5 Total and fecal coliform analysis
Diagram showing the steps used for total and fecal coliform analysis by the most probable number (MPN) test. (A) Presumptive test, (B) confirmed test, and (C) fecal coliform test. (See the text for further details.)

The presumptive test is followed by the **confirmed test**, which provides additional supportive information about the presence of coliform bacteria. At 24 and 48 hours, each positive presumptive tube is used to inoculate a tube of another broth medium, brilliant green lactose bile broth (BGLB). These BGLB tubes are incubated at 35°C for another 24 to 48 hours to *confirm* acid and gas production. If these are positive, they are considered to be confirmed. At this point, it is still not possible to say that coliform bacteria or *E. coli* are present in the sample, only that the presence of coliform bacteria has been confirmed.

However, a confirmed positive is considered adequate for concern about the potability of the water. For drinking water analysis, even a single positive confirmed test is considered serious. Additional testing is required to verify what the bacterium is. This is accomplished by the completed test. In the **completed test**, a loopful of culture from a positive confirmed BGLB tube is streaked on eosin methylene blue (EMB) plates. If coliform bacteria are present, they will produce colonies with a typical metallic green sheen following incubation at 35°C for 24 hours. Cells from these colonies must be Gram stained to determine if they are gram-negative, non–spore-forming bacteria. If gram-negative, non–spore-forming bacteria are found, all the criteria have been met for fulfilling the definition of a coliform bacterium (see Figure 32.5).

It is noteworthy that there are many coliform bacteria other than *E. coli*. Thus, even though this lengthy procedure has been undertaken, it is still not possible to state that *E. coli* is present in the sample. Further tests are needed to verify this. Moreover, some of these coliform bacteria, such as *Enterobacter aerogenes*, are common soil bacteria that do not reside in the intestinal tracts of warm-blooded animals. Therefore, even a positive completed coliform test would not necessarily mean that fecal coliform bacteria are present in a water sample. Thus, the drinking water tests err on the side of safety. Recurring positive results would require that additional tests be performed to determine the identity and source of the coliform organisms that are detected by the completed tests.

MEMBRANE FILTER ANALYSES   In addition, membrane filter tests have been developed for the identification of total coliform bacteria. These tests entail utilizing a sterile 0.45-μm pore size membrane filter to collect cells from the water sample to be tested (see Chapter 6). These filters are then placed on an absorbant pad that contains nutrients for the growth of bacteria. They are incubated for 24 hours at 35°C, and the colonies of total coliform bacteria are identified by their characteristic metallic sheen.

Fecal Coliform Bacteria Analyses

One disadvantage of the total coliform bacteria tests is that they select for a variety of bacteria, such as *E. aerogenes*, that are not indigenous to the intestinal tracts of animals and are therefore not indicative of fecal contamination. Therefore, more specific tests have been designed for fecal coliform bacteria that are more likely to be *E. coli*.

MOST PROBABLE NUMBER (MPN) ANALYSIS   Research has shown that *E. coli* strains can grow at elevated temperatures when compared to soil species such as *E. aerogenes*. Therefore, elevated temperature tests have been developed for identifying fecal coliform bacteria.

For the MPN test for fecal coliform bacteria, positive tubes from the MPN total coliform bacteria test described previously are used to inoculate a selective broth medium for fecal coliform bacteria, which is then incubated at 44°C for 24 hours (see Figure 32.5C). Cultures that grow on this medium and produce gas are regarded as fecal coliform bacteria, and most of them, if identified, turn out to be *E. coli*. As with MPN total coliform bacteria, the actual numbers of fecal coliform bacteria in a sample can be quantified by this technique.

MEMBRANE FILTER ANALYSIS   A membrane filter test has also been developed for fecal coliform bacteria. This test is usually performed directly on water samples. After filtration, the membrane filter is placed on a selective *E. coli* medium (EC medium) and incubated at 44°C.

Colonies of fecal coliform bacteria produce a characteristic blue-colored colony on this medium.

All viable counting procedures for coliform testing have limitations. Among the most serious of these is the fact that the tests take so long to perform. If a city were really concerned about an outbreak of some intestinal disease, it would be at least 2 to 4 days before it would be confirmed that coliform bacteria were present. And, as we have already mentioned, even though coliform bacteria may be found, because they cannot be identified as *E. coli* using any of the tests described, it is really not known whether the water sample has fecal contamination. Therefore, new research is aimed at developing more rapid and specific tests for *E. coli* and other enteric bacteria.

ALTERNATIVE MICROBIOLOGICAL TECHNIQUES FOR DRINKING WATER ANALYSIS   Several different approaches are being taken to develop better procedures for testing the potability of drinking waters. One new approach is the presence/absence test.

PRESENCE/ABSENCE ANALYSIS   The presence/absence test, or P/A test, looks for total and fecal coliform bacteria using a single large volume of the drinking water sample (100 ml) as an inoculum rather than using a series of tubes or membrane filters. In this test, only a positive or negative is recorded for a given sample. Thus, it is not possible to quantify the numbers of total or fecal coliform bacteria, but only to say they are present or absent in a particular 100-ml sample. This type of test is very simple to perform and thereby enables small municipalities serving from 100 to 10,000 citizens to analyze their own water.

COLORIMETRIC AND FLUOROGENIC ANALYSES   Another approach taken to simplify the analysis of total and fecal coliform bacteria in drinking waters, uses **colorimetric** or **fluorogenic compounds**. These compounds, when cleaved enzymatically, release colored end products that can be seen visually, or fluorescent end products that can be detected when illuminated with short wavelength radiation (ultraviolet radiation), respectively. Rapid tests have been developed for total and fecal coliform bacteria using these compounds. For example, one colorimetric test for total coliform bacteria uses orthonitrophenol galactoside (ONPG) as a substrate for the lactose-splitting enzyme, β-galactosidase, found in *E. coli* and other enteric bacteria that ferment lactose. When this enzyme is present, it cleaves the galactoside, freeing the orthonitrophenol, which has a characteristic yellow color. As a result, a yellow color is indicative of the presence of coliform bacteria.

An example of the use of a fluorogenic substrate is 4-methylumbelliferone-β-D-glucuronide, or MUG (Fig-

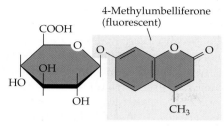

MUG (4-methylumbelliferyl-β-D-glucuronide)

Figure 32.6 Fluorogenic substrate
Chemical structure of 4-methylumbelliferyl-β-D-glu-
curonide (MUG). This fluorogenic compound is not itself
fluorescent, but when cleaved by glucuronidase it produces
the fluorescent product 4-methylumbelliferone.

ure 32.6). Almost all strains of *E. coli,* and few other enteric bacteria, produce glucuronidase and LacZ, two enzymes that cleave this substrate to release the MUG compound that fluoresces when illuminated with ultraviolet
radiation. As in the other tests, these tests require that
the bacteria grow in the vials and that they be metabolically active. In addition, nonenteric bacteria from a sample that have these enzymes may interfere with the test.
However, these latter tests do not require such lengthy
incubations and follow-up inoculations.

Perhaps the future for detection of *E. coli* and water-
borne pathogens lies in the development of probe procedures to look for specific organisms directly in natural samples (see Chapter 24). In addition, procedures
using the polymerase chain reaction (PCR) could greatly
accelerate the detection of fecal contamination.

SECTION HIGHLIGHTS

Drinking water is treated to ensure that the
water has low concentrations of microorganisms and is devoid of pathogens. This is accomplished in many municipalities by use of a
four-step process including coagulation, sedimentation, filtration, and chlorination. To assess whether the water is potable, analyses for
residual chlorine and total and fecal coliform
bacteria are made on the treated water in the
distribution system.

32.3 Landfills and Composting

Microbes are also involved in the degradation of home
refuse and in the process of composting. This is not surprising because these materials contain large amounts

of organic matter, and heterotrophic microorganisms require organic materials for growth.

Landfills

All cities collect garbage and other home refuse, called
solid waste, from private homes and take this to a location away from the city where it is placed in the
ground in a **landfill**. The landfill is excavated from the
ground before receiving waste materials. The landfill is
lined with plastic. The leachate materials produced in
the degradation process can be collected by an underlying piping system and stored in a tank to be treated separately (Figure 32.7). Monitoring wells are used to ensure that the water table does not receive leachate
effluent. When a section of the landfill is filled, it is covered with earth and allowed to decompose. The land-

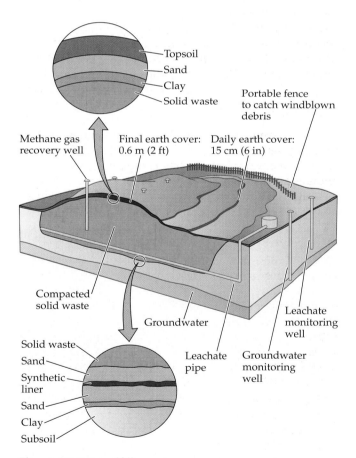

Figure 32.7 Landfill
Diagram of a typical landfill. Solid waste is unloaded into
the landfill, which is first lined with plastic. A drainage system is used to collect leachate to prevent it from contaminating the underlying groundwater. Pipes are installed to
allow methane produced in the degradation to be removed.
When the landfill is full, it is covered with soil. (See text for
details.)

fill is an improvement over the earlier practice of simply dumping refuse in open "garbage dumps."

Landfills contain not only potato peels, tin cans, and discarded cardboard boxes, but also plastic bags, Styrofoam packing materials, broken appliances, and hazardous chemicals. As a result, during the 1960s and 1970s, most old-fashioned garbage dumps became contaminated sites. More recently, new regulations have been imposed to improve the treatment of solid-waste materials. Recycling is practiced widely in the United States. Thus, cardboard, aluminum, and glass are usually treated separately from garbage. Likewise, hazardous materials are also handled separately.

Microbial degradation is the major way in which the organic materials in landfills are degraded. Because the landfilled material is covered with soil, conditions become anoxic. Therefore, anaerobic microorganisms are very important. Methanogens and fermenters, such as *Clostridium* sp. involved in cellulose decomposition, grow in such systems. As a result of the activity of the anaerobic microorganisms, methane is a major gas that is released; therefore, pipes are used to channel its escape. In addition, organic acids and alcohols are produced. These and heavy metals derived from corroding materials may be leached from the landfill and cause deleterious effects on receiving streams (Figure 32.8). Heavy metals tend to accumulate in the environment and have adverse effects on plant and animal life. Therefore, it is important to monitor effluents from landfills to see that they do not adversely impact the watershed.

Composting

Composting is the process whereby plant materials are decomposed. Because plant materials (leaves and stems) are high in organic content but low in nitrogen, they are not as readily degraded as most garbage. In composting, the material is placed in a pile or windrow in which the composting process is allowed to occur. This may be in a container (Figure 32.9) or simply a compost heap on the ground.

Composting typically occurs over a period of weeks. The most important microbial groups involved are the bacteria, in particular the actinobacteria, as well as fungi. The process undergoes several stages, including an initial period of heat generation due to the heat produced by aerobic microorganisms. The heat produced has a pasteurizing effect on mesophilic bacteria, some of which may be plant or animal pathogens. The heat also selects for mildly thermophilic microorganisms, which dominate later on in this successional process (see Chapter 24). Eventually, much of the material being composted is degraded so that the volume is depleted. This material can be used as "mulch," an organic amendment to soils used for gardening. However, it is important to recognize that heavy metals and pesticides or other hazardous materials used in gardening may be concentrated by composting.

The practice of commercial composting is becoming more popular. In this practice, grass clippings, tree trimmings, and other yard waste are collected and composted in large windrows under controlled conditions.

(A)

(B)

Figure 32.8 Leachate effects from a landfill
(A) The small stream (entering from the right) contains organic leachate from a landfill. (B) The gills of this salmon fingerling from a hatchery downstream of the leachate entry show heavy bacterial and fungal growth due to the enrichment effects of the effluent. Courtesy of J. T. Staley and James Huff.

(A)

(B)

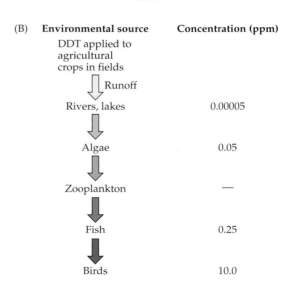

Figure 32.9 Composting
(A) Finished compost on the right, with the starting material on the left.
(B) A barrel for composting plant waste. Courtesy of D. McIntyre.

This practice has certain advantages over landfills, in that the material can be sold back to the public to be recycled as mulch.

SECTION HIGHLIGHTS

Municipalities dispose of home refuse called solid waste by taking it to landfills where it is buried. The organic materials of the solid waste are degraded in the landfill by anaerobic microorganisms. Leachate from landfills can be toxic and may require treatment. Yard clippings and other plant materials are typically degraded in compost piles at home, which, along with commercial compost, can be used as garden mulch.

32.4 Pesticides

Pesticides are chemical substances manufactured by the chemical industry for control of insects (insecticides) or weeds (herbicides). Many of these are manufactured organic compounds, not previously present on Earth, termed xenobiotic compounds. The term xenobiotic is derived from Greek and means literally "stranger (xeno) to the biotic environment."

Dichloro-diphenyl-trichloroethane (DDT) is perhaps the best-known pesticide (**Figure 32.10A**). It interferes with the synthesis of chitin, the tough polymeric material composed of β-1,4-glucosamine found in the ex-

(A)

(B)

Environmental source	Concentration (ppm)
DDT applied to agricultural crops in fields	
⇩ Runoff	
Rivers, lakes	0.00005
Algae	0.05
Zooplankton	—
Fish	0.25
Birds	10.0

Figure 32.10 DDT and its concentration in a food chain
(A) Chemical structure of the pesticide DDT. (B) An example of biomagnification of DDT in an aquatic ecosystem.

oskeletons of insects and crustaceans. Because it is such an effective insecticide, DDT has been used widely all over the world. DDT is an example of a persistent pesticide. Unlike natural organic materials, such as chitin, that are readily degraded by microorganisms in the carbon cycle, some xenobiotic compounds, such as DDT, are not. DDT is recalcitrant to degradation and persists in the environment for long periods (see Chapter 12).

Furthermore, DDT is concentrated in the food chain in a process called **biomagnification** (Figure 32.10B). Although it is usually found in relatively low concentrations in soils and aquatic habitats, it is concentrated in fatty tissues. Thus, as it passes through the food chain, it accumulates in the tissues of animals with longer life spans and that occupy the higher trophic levels. The highest consumers in the food chain—fish, birds, humans, and other mammals—can accumulate high concentrations in their fatty tissues. Indeed, some humans have been found with levels greater than 50 ppb (parts per billion) in their tissues, much higher than the allowed concentration for animals, such as beef, used for human consumption.

DDT has proven to be especially harmful to birds that eat insects or fish that have high concentrations of the pesticide. High levels of DDT in birds interfere with normal eggshell formation. The weak eggs are frequently broken in the nest prior to hatching, affecting bird reproduction rates. In the United States, eagles, ospreys, and other birds of prey have been particularly adversely impacted. For this reason, DDT has been banned for use as a pesticide in the United States. Banning DDT has resulted in increased eagle populations in North America.

Another example of a pesticide that is quite persistent in the environment and has also been banned is 2,4,5-T (Figure 32.11). Although it contains only one additional chlorine atom, it is much more resistant (2 to 3 years in soil to degrade) compared with 2,4-D (3 months to degrade). A mixture of 2,4-D and 2,4,5-T referred to as "agent orange" was used on a large scale as a defoliant in the Vietnam War.

Because of the problems associated with some persistent xenobiotic compounds, only readily degraded pesticides are now permitted for use in the United States. Most of these break down through microbial activity in a period of days or weeks following application.

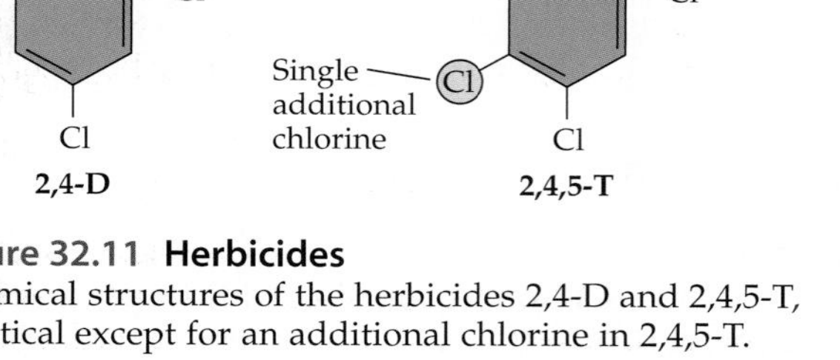

Alternatives to Use of Persistent Pesticides

Another completely different approach that has been taken to control pests is the use of **biological insecticides**. The best example of this is the use of the **Bt protein** produced by *Bacillus thuringiensis* (see Chapter 20). These compounds are produced normally by microorganisms in the environment and therefore can be readily degraded. Furthermore, they are highly specific in their activity, compared with DDT. Bt specifically affects the development of larval stages of insects without having any effect on birds and other animals that might eat the insects. Of course, Bt can have adverse effects due to killing insects that are not specific pest targets. More recently Bt genes have been genetically engineered into some agricultural crops so that undesirable insects that begin to eat plants are killed.

Bt is not only used for the control of plant-eating insects, but for control of other insects as well. For example, some species of *Bacillus* produce effective proteins for control of the *Anopheles* mosquitoes that carry malaria.

SECTION HIGHLIGHTS

Pesticides are used to control insects and weeds. Many of the pesticides are xenobiotic compounds that have been manufactured at chemical plants and resist microbial degradation. More recently biological pesticides such as "Bt" have been produced commercially as an insecticide.

32.5 Bioremediation

The rapid growth of the chemical industry during the last 100 years has significantly altered our lifestyle and improved our standard of living. The "chemical age" has also created problems, including monumental environmental pollution. Today, in the United States alone, more than 50,000 hazardous waste sites have been identified. Of these, the Environmental Protection Agency (EPA) has designated 1,200 as Superfund sites. These particular sites are so hazardous that the federal government has set aside billions of dollars for their cleanup.

Furthermore, it has been estimated that 15% of the 5 to 7 million underground storage tanks containing toxic

Figure 32.11 Herbicides
Chemical structures of the herbicides 2,4-D and 2,4,5-T, identical except for an additional chlorine in 2,4,5-T.

chemicals in the United States are leaking. Most of these tanks, such as those at gasoline stations or for home heating oil, contain petroleum products, but some contain considerably more hazardous chemicals.

Moreover, groundwaters in the United States are often dangerously polluted with an array of toxic chemicals, chiefly commercial solvents, such as trichloroethylene (TCE) used for dry cleaning or cleaning machinery and high technology electronic parts. Groundwater is the source of drinking water for about half of the people in the United States, and it serves as the drinking water supply for more than 95% of rural populations. Among the compounds present in groundwater are those listed in Table 32.2.

In addition, the mismanagement of pesticides and fertilizers, largely in agricultural areas, has created environmental problems even in the most remote rural areas. Not only are our fish and wildlife at risk, but hazards exist for humans as well.

What can be done? A number of chemical and or physical procedures can be used for the removal of hazardous chemicals from environments. These toxic chemicals can then be concentrated and stored safely. They could be chemically oxidized and thereby detoxified by incineration, but this causes air pollution. Potential pollutants from waste streams can be concentrated by absorption on a solid phase, such as activated charcoal. Certain other combinations of absorption and extraction methodologies are also available. However, these processes are usually very expensive and not always applicable, particularly for contamination problems in soils

and other natural environments. A more practical method for remediating environments contaminated by hazardous waste is by **bioremediation**.

Bioremediation is defined as the use of living organisms to promote the destruction of environmental pollutants. The applications of bioremediation methodology are, in many ways, extensions of the technology that has been so effective in the treatment of urban sewage and industrial wastewater. Bioremediation has been applied to the treatment of unusual contaminants including munitions such as TNT (Box 32.2) and even the treatment of radioactive wastes (Box 32.3).

Biodegradative Organisms

Evolution has resulted in a vast array of microbes that have broad and flexible biodegradative capacities. They can survive and destroy or detoxify chemicals in a variety of environmental niches (hot, cold, low pH, with or without oxygen, and so on). Often the organisms best suited for bioremediation are the species that are indigenous to a particular polluted habitat or one that is similar to it. The indigenous microbes have shown that they can survive and grow with the toxic substances. In addition, mixed populations are superior, in many cases, to axenic cultures in the biodegradation of chlorinated aromatic hydrocarbons.

Although some microbial strains have been selected for bioremediation capabilities, they often tend to be less stable and effective than native populations. Furthermore, it is actually *illegal* to introduce *genetically engineered* bacteria into natural environments because of concern about the unknown effects on natural populations and the difficulty of controlling their dispersal to other sites. Therefore, although genetically engineered bacteria may be used for treatment in bioremediation processes in containers, they cannot be inoculated into the polluted environment to degrade the toxic compounds.

Hazardous environments are often formidable to treat in that they usually contain more than one type of toxic substance. For example, it is not uncommon that a site will contain heavy metals such as cadmium, lead, or mercury, as well as chlorinated organic compounds. It is therefore essential that the microbes selected for such sites be able not only to remediate one group of compounds, but also to survive and grow in the presence of other toxic substances. Increased tolerance to acid, base, temperature, salinity, or heavy metals may be essential.

Advantages of Bioremediation

Bioremediation has a number of advantages—such as broad applicability and low cost—over other processes for remediation. Incineration requires that the toxicant is in a

TABLE 32.2	Organic compounds that have been detected in drinking-water wells

Compound	Highest Level Reported (mg/liter)
Trichloroethylene	27.30
Toluene	6.40
1,1,1-Trichloroethane	5.44
Acetone	3.00
Methylene chloride	3.00
Dioxane	2.10
Ethyl benzene	2.00
Tetrachloroethylene	1.50
Cyclohexane	0.54
Chloroform	0.49
Di-*n*-butyl-phthalate	0.47
Carbon tetrachloride	0.40
Benzene	0.33
1,2-Dichloroethylene	0.32

BOX 32.2 *Research Highlights*

Some Bacteria Get a Charge from TNT

Military sites are often contaminated by a variety of toxic compounds. Among the most problematic are munitions such as TNT (trinitrotoluene). This well-known explosive is a common contaminant of soils near munitions storage areas. In some instances, contaminated soils may contain more than 5% TNT!

A group of microbiologists at the University of Idaho led by Ron Crawford has recently discovered that TNT is readily biodegradable by *Clostridium bifermentans* (Figure A). All that is needed is to add starch to the contaminated soils, and the TNT is degraded to harmless compounds. This is an example of cometabolism—the bacteria cannot use TNT as an energy source, but they can degrade it if starch is available as a carbon source for their growth (Figure B). Researchers at the University of Idaho have developed a patented treatment process in which contaminated soils are mixed in large reactor containers with starch and an inoculum of the bacterium. This anaerobic treatment process is now commercially available and is used successfully for the removal of TNT from contaminated sites (Figure C).

(A)

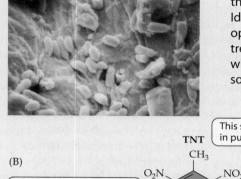

(C)

(B)

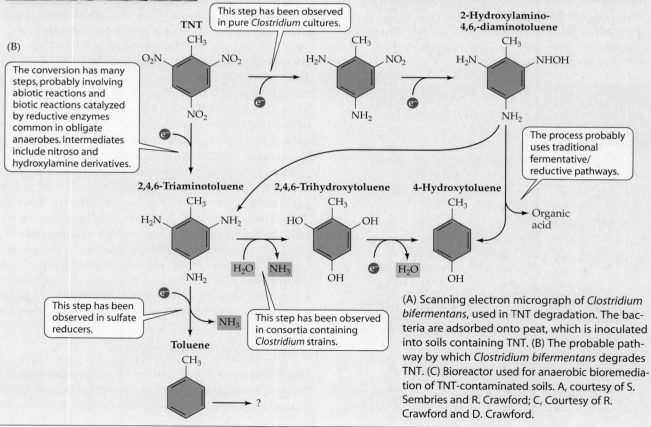

This step has been observed in pure *Clostridium* cultures.

The conversion has many steps, probably involving abiotic reactions and biotic reactions catalyzed by reductive enzymes common in obligate anaerobes. Intermediates include nitroso and hydroxylamine derivatives.

The process probably uses traditional fermentative/ reductive pathways.

This step has been observed in sulfate reducers.

This step has been observed in consortia containing *Clostridium* strains.

(A) Scanning electron micrograph of *Clostridium bifermentans*, used in TNT degradation. The bacteria are adsorbed onto peat, which is inoculated into soils containing TNT. (B) The probable pathway by which *Clostridium bifermentans* degrades TNT. (C) Bioreactor used for anaerobic bioremediation of TNT-contaminated soils. A, courtesy of S. Sembries and R. Crawford; C, Courtesy of R. Crawford and D. Crawford.

BOX 32.3 — *Research Highlights*

Role of Bacteria in Concentration of Radioactive Waste

One of the surprising things about microorganisms is that some can grow in the presence of moderately high concentrations of radioactivity. Of course, they are mutated by radiation, but their high population densities, rapid growth rates, and radiation-repair mechanisms permit many species to survive and grow. Therefore, these microorganisms may in the future play important roles in the management of radioactive wastes. One interesting example concerns the reduction of uranium compounds.

Often, radioactive chemicals occur in very low concentrations and are mixed with other chemicals. Microbiologist Derek Lovley and his colleagues developed and patented a process for concentrating radioactive uranium using bacteria. They use metal-reducing species that oxidize organic compounds while reducing uranium from the U(VI) state to the U(IV) state. Whereas U(VI) is soluble, the

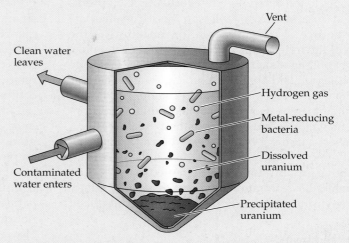

Diagram showing a reactor vessel in which radioactive wastes are concentrated by bacterial reduction and precipitation. Courtesy of Derek Lovley.

reduced form, U(IV), is insoluble and therefore precipitates out from the other wastes. This process thereby concentrates and removes the radioactive uranium from the other waste material (see figure).

One of the bacteria capable of carrying out uranium reduction is the common sulfate-reducing bacterium, *Desulfovibrio desulfuricans*, which uses lactate as a carbon source.

---

burnable form, and exhaust gases must be tested to eliminate toxic emissions. Another advantage to bioremediation is the low risk of exposure to hazardous chemicals during clean up. Chemical–physical excavation methods of remediation require the handling and disposal of the hazardous material. Ideally, sediments in riverbeds, harbors, and lakes should be remediated without disturbing the sediments. When sediments are disturbed, this increases the chance of enlarging the area polluted or of transporting them downstream or to adjacent environments. When bioremediation is successful, there is minimal disturbance of the environment, and the products produced, such as CO_2, H_2O, and chloride, are innocuous.

Problems Associated with Bioremediation

Most chemically synthesized compounds are biodegradable, although many are biodegraded at an unaccept-

ably slow rate. Hence, they remain in the environment for long periods. To enhance bioremediation rates, it is essential that the proper microbe is available along with appropriate nutrients and other conditions—such as pH and O_2—that favor its growth and activity. Contaminated soil and water sites may have to be amended with fixed nitrogen, phosphorus, and other nutrients that may limit microbial growth (see Chapter 6). In many cases, an alternate energy source must be added for rapid biodegradation to occur.

Mineralization to CO_2 is the goal in bioremediation. This does not always occur, and often the pollutant is transformed to a secondary product. This is acceptable when the secondary product is less toxic than the parent compound and can be utilized by other organisms present in the environment. It is unacceptable when the product is more toxic than the parent pollutant and not readily biodegraded.

Methodology of Bioremediation

Bioremediation can be accomplished in situ (without excavation and recovery) or by methods that require recovery and above-ground treatment. In situ treatment of contaminated soil involves the addition of nutrients or microbes to the otherwise undisturbed soil. Nutrients include sources of nitrogen, phosphorus, or an alternate energy source. This augmentation encourages the growth of the indigenous microbes that can catabolize the target pollutant(s). It may also be necessary to sparge air into the polluted area to promote growth of aerobic microorganisms. Microbes that can catabolize selected contaminants may also be inoculated into the environmental niche for in situ bioremediation processes.

There are a number of bioremediation strategies that involve removing or recovering the polluted environment above ground, followed by remediation and replacement after restoration. Some of these methodologies of environmental restoration, which are described next, are:

- Pump and treat
- Bioreactors
- Mound or heap
- Land farming

PUMP AND TREAT   These procedures can be employed effectively in bioremediation of polluted water such as groundwater. The groundwater is pumped to the surface, nutrients are added, and the water is reinjected into the contaminated zone. As an example, for a groundwater pollutant such as TCE, the nutrients added could be methane and O_2, inasmuch as one of the principal organisms responsible for the initial degradation of this compound are the methanotrophic bacteria. Although the methanotrophs do not use TCE as a carbon and energy source, the methane monooxygenase they produce is able to dechlorinate and oxidize this compound in the initial reaction of its degradation. This is an example of **cometabolism**, the degradation of a compound without obtaining energy in the process; the methanotrophs do not derive energy from the degradation of the TCE and still require methane for growth. This is the reason that methane needs to be supplied during this process. Other bacteria and abiotic processes are involved in its complete mineralization. The process of treatment may take months or even years, depending on the degree of contamination of the site.

BIOREACTORS   Bioreactors are often employed to bring together the pollutant and the biodegrading microbe. A bioreactor follows the general principles involved in an industrial fermenter (see Chapter 31). A slurry of soil or groundwater is placed in the reactor, and an inoculum is added. This inoculum may be activated sludge from a sewage treatment plant, an appropriate pure culture or a mixed microbial culture, or an inoculum from a contaminated site. The pure or mixed cultures may be added as a suspension or on a solid support. Supports employed are activated carbon, plastic spheres, glass beads, or diatomaceous earth. The attachment of bacteria to solid support systems was discussed in Chapter 31. Bioreactors can be operated as a continuous culture system, where about 80% of the material is removed periodically. The remaining 20% serves as inoculum for the added contaminated material.

Batch reactors may be placed in sequence so that the microbial population in each has the capacity to biodegrade selected chemicals in the slurry or water. For example, consider the situation in which the toxicant material to be treated contains both chlorinated phenolic compounds and TCE. Bacteria such as *Pseudomonas putida* are able to degrade chlorinated phenolic compounds and use them as a carbon source for growth. However, this species is not able to degrade TCE. As mentioned previously, the methanotrophic bacteria are the preferred organisms for TCE degradation. Furthermore, the conditions necessary for the growth of methanotrophs (methane gas as energy substrate) are not necessary for *P. putida*. Thus, a sequence of two reactors, the first containing *P. putida* with conditions ideal for its growth, followed by a separate reactor for methanotrophs would result in the complete remediation of both chlorinated phenols and TCE.

The sequential reactor arrangement is particularly important when the contaminated material contains a chemical that is generally toxic to microbes. Removal of this chemical by a selected microbe renders the remaining compounds more amenable to bioremediation. Sequential bioreactors also permit the operation of one bioreactor anaerobically and the following one aerobically. This latter arrangement is particularly useful when highly chlorinated compounds are present, such as PCBs (polychlorinated biphenyl compounds). It is known that the most highly chlorinated forms of the PCBs are degraded only anaerobically. Thus, if the first-stage reactor is anaerobic, the more highly chlorinated compounds can be partly dechlorinated and, at least in theory, these less–highly chlorinated products from that reactor could be transferred to an aerobic reactor for further degradation (Box 32.4). However, no one has yet shown complete degradation of highly chlorinated PCBs using this two-step process.

MOUND OR HEAP METHOD   In this procedure, soil is placed in mounds on a plastic liner. Water, microbes, and nutrient are trickled over the soil. The biodegradation of many pollutants including chlorinated aromatics is aug-

BOX 32.4 *Research Highlights*

Degradation of Polychlorinated Biphenyl Compounds (PCBs)

PCBs are double-ringed compounds that are chlorinated (Figure A). Depending on the position and number of chlorine atoms, more than 200 different varieties of PCB, called congeners, exist (Figure B). Like DDT, which is similar chemically, PCBs accumulate in the fatty tissues of animals that consume them. Until recently, the most highly chlorinated congeners of the PCBs were not known to be degraded by microorganisms.

Some of the chemical manufacturers that produced PCBs are located on the Hudson River, which became contaminated with PCBs. In the 1980s, it was discovered that contaminated sediments had lower concentrations of the higher chlori-

nated PCBs than expected, suggesting they were being degraded in the environment. Subsequent laboratory studies at Jim Tiedje's laboratory at Michigan State University confirmed that microorganisms were degrading PCBs. However, this breakdown occurred only under *anoxic* conditions. This was a truly exciting discovery because it proved that these compounds could be degraded by bacteria and therefore could be removed from the environment. Furthermore, it indicated that anaerobic conditions were necessary for degradation of these and possibly other highly chlorinated compounds.

PCB (general structure)

2,3′,4′-Trichlorobiphenyl

(A) General structure of a PCB. Some 200 different congeners are possible, with various locations of the chlorine atoms. (B) One possible PCB congener, 2,3′,4′-trichlorobiphenyl.

---

mented by also adding an energy source that is readily utilizable. The effluent from the process can be collected and recycled and the process operated until the pollutants are degraded.

Land Farming

Selected pollutants can be removed from waste streams by running the waste into a confined soil basin. Nutrients may be added to the soil, and the soil can be tilled to increase mixing and aeration. This method is effective where fertile soil lies over a firm, relatively impermeable clay base. The clay base impedes penetration of the hazardous waste into the groundwater.

The Future of Bioremediation

Bioremediation has a promising future, although it is not the sole answer to our huge problem of hazardous waste in the United States. The low cost and broad applicability to many hazardous waste problems make bioremediation a method of choice. However, more research will be required to develop bioremediation processes so that they can be more widely applied in the field. An example of a successful pilot scale process that would appear

simple to scale up for the field is that for PCE (perchloroethylene or tetrachloroethane) treatment (Box 32.5).

SECTION HIGHLIGHTS

Bioremediation is a process that uses microorganisms to degrade toxic compounds that have been released into the environment. Because of their great metabolic diversity most toxic organic substances can be degraded by one or more microorganisms. A variety of treatment processes have been tested for bioremediation.

32.6 Acid Mine Drainage and Acid Rain

Strip mining practices, in which the surface soil is removed to expose the underlying minerals to be mined, may result in the problem of acid mine drainage. This problem occurs when the minerals that are mined contain significant amounts of sulfide minerals such as pyrite, FeS_2. Strip mining exposes these reduced sulfides

BOX 32.5 *Research Highlights*

Pilot Scale Studies to Remove PCE From Contaminated Soil

(A)

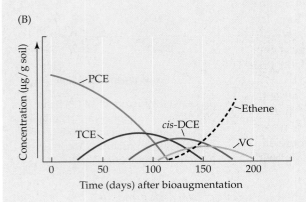

(B)

example of anaerobic respiration. In this process, PCE is used as an electron acceptor by this organism, which grows anaerobically by oxidizing organic acids such as acetate. *D. ethenogenes* successively dehalogenates PCE to TCE, *cis*-DCE, vinyl chloride, and, finally, ethene (Figure A).

Pilot scale studies have been set up in which an injection well is used to inoculate *D. ethenogenes* as well as an organic acid substrate such as acetate into a contaminated site. Another well, termed a sampling well is placed downstream of the groundwater flow. This sampling well is used for the removal and analysis of PCE and its degradation products at various time-intervals. In pilot scale studies, the amount of PCE is seen to decline with time at the sampling well site. As PCE declines, TCE initially builds up, and then *cis*-DCE followed by vinyl chloride (VC) and finally ethene (Figure B).

PCE (perchloroethylene) is used as a dry cleaning fluid and degreasing agent. It is both a teratogen (tumor-causing agent) and carcinogen. When PCE is released into the environment, it and its chlorinated byproducts are therefore of considerable concern, as they may contaminate drinking waters and soil. Soil pilot scale studies have been very successful in converting PCE to the nontoxic gas, ethene. The process includes bioaugmentation

as well as bioremediation. Bioaugmentation refers to the addition of a microbial culture to enhance the removal of a toxic substance. A specific bacterium, *Dehalococcoides ethenogenes* is one of the species of a consortium that is used for bioaugmentation in the treatment of PCE contamination. This species, which is a member of the phylum *Chloroflexi*, degrades PCE anaerobically in a process referred to as halorespiration, an

to oxygen and rainwater. As a result, sulfur-oxidizing members of the *Bacteria* and *Archaea* flourish and oxidize the pyrite to sulfuric acid. The iron can also be oxidized to form iron oxides.

In areas in Appalachia and in certain midwestern states such as Indiana, coal deposits contain large quantities of pyrite. Strip mining has exposed the sulfides and iron to oxidative activity by thiobacilli. Runoff waters leached from these sites can have pH values as low as 2.5 to 4.5—so low that fish are killed, as are the aquatic plants that live in the stream or receiving waters (Figure 32.12).

In addition, acidophilic archaea have been isolated from acid mine waters. These filamentous organisms, which grow as streamers, are *Ferroplasma acidarmanus*. This species obtains energy by the oxidation of pyrite found in the mine sediments. *F. acidarmanus* is capable of growth at pH 0 (Figure 32.13).

There is no simple remedy for acid mine drainage. Therefore, the EPA now requires that strip mining areas be reburied after the mining operation is completed. In this manner, the sulfides are again removed from exposure and oxidation by the bacteria.

Figure 32.12 Acid mine runoff
Acid mine drainage in a historic coal mining area near
Cooke City, Montana. ©Dynamic Graphics Group/
Creatas/Alamy.

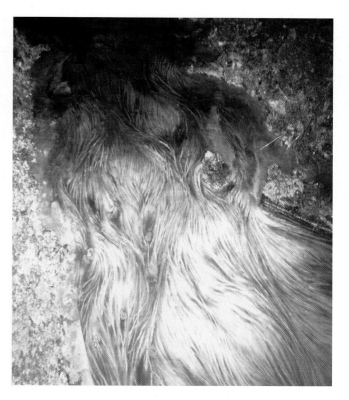

Figure 32.13 *Ferroplasma acidarmanus*
Flowing streamers of *Ferroplasma acidarmanus* growing in
acid runoff from a mine containing pyrite. Photo ©K. J.
Edwards, Woods Hole Oceanographic Institution.

Acid rain is also an environmental problem. Acid rain
is produced when coal and oil that contain large
amounts of sulfur are burned as fuels. When these energy sources are burned, the gas sulfur dioxide is released into the atmosphere. When this reacts with light
and water, it leads to the production of sulfuric acid and
forms acid rain (see sulfur cycle in Chapter 24). Thus, it
is desirable to remove the sulfur from coal and oil before
burning. One way in which this can be accomplished is
by use of thiobacilli to oxidize the sulfur and sulfides
to sulfate, which can be removed in solution before the
fuel is burned.

SECTION HIGHLIGHTS
Acid mine drainage occurs where reduced sulfur minerals such as pyrite have been exposed
to air and rain during the mining process. Bacteria and archaea that oxidize the sulfides
grow aerobically and produce sulfuric acid,
which can decrease the pH of receiving waters
thereby destroying animal and plant life in the
receiving waters.

SUMMARY

• Because of their diverse capabilities of degrading
organic materials, microorganisms, particularly
bacteria, are used in the treatment of **wastewater**
(sewage). Typical sewage treatment plants are
called **secondary wastewater treatment** facilities
because they comprise two successive treatment
systems: **primary treatment**, involving the physical
separation of particulate material (**sludge**) from the
fluid portion of sewage, and **secondary treatment**,
in which organic material is degraded by microbial
communities.

• Most secondary wastewater treatment facilities use
the aerobic **activated sludge process** for reduction
in organic materials, a major component of the biochemical oxygen demand (BOD) in wastewaters.
Anaerobic sludge digesters, in which methanogenesis occurs, are used to reduce the organic content
of sludge. **Biological tertiary treatment** is used to
further treat secondary effluent in some facilities to
remove nitrogen (by nitrification followed by deni-

trification) and phosphorus by enhanced biological phosphorus removal by the "Candidatus" species *Accumilibacter phosphatis*.

- **Drinking water treatment** is necessary to ensure that the public's health is protected from pathogenic bacteria that may contaminate drinking water supplies. Drinking water is treated at filtration plants, which use several process steps including **coagulation** with alum addition; **sedimentation** to remove particulate materials, including bacteria, adsorbed to the alum; and **filtration** through sand filters. The final treatment is **chlorination**, the addition of chlorine gas, which forms hypochlorite solution, to disinfect the water before it is distributed to households and other consumers in the drinking water supply system.

- The safety of drinking water is tested by assaying for **total** and **fecal coliform** bacteria. *Escherichia coli*, which is both a total and fecal coliform bacterium, is used as the **indicator** organism, the presence of which indicates the contamination of a water supply by mammalian feces. Unfortunately, positive total and fecal coliform tests can be caused by other bacteria, such as *Enterobacter aerogenes*, which is a soil organism; therefore, additional tests are needed to validate the presence of fecal contamination by *E. coli*.

- **Landfills** are used to treat **solid waste** materials. Anaerobic bacteria degrade the organic materials placed in landfills. **Composting** involves the degradation of organic materials such as leaves and grass using microorganisms.

- **Bioremediation** is the process whereby toxic materials are degraded by microorganisms, which are capable of degrading almost all organic substances including halogenated compounds such as PCBs, almost all pesticides, aromatic and aliphatic hydrocarbons, and even munitions. **Bioaugmentation** is the practice of inoculating a contaminated site with a culture of a bioremediating organism or a consortium of organisms. **Halorespiration** refers to the process of using halogenated organic compounds as electron acceptors in anaerobic respiration. Bioremediation processes are being used to clean up **EPA superfund sites** scattered about the United States.

- **Acid mine drainage** occurs when mining operations expose reduced sulfur compounds such as pyrite to air and water. *Thiobacillus* **sp**. oxidize the sulfides to produce sulfuric acid and may lower the pH to 2 to 3. This acid mine drainage kills fish and plants in the vicinity. The only solution to this problem is to cover up the mine after the minerals have been removed so that air and water do not reach the sulfides.

 Find more at www.sinauer.com/microbial-life

REVIEW QUESTIONS

1. Differentiate between primary and secondary wastewater treatment.

2. Compare the trickling filter with activated sludge treatment.

3. What is the microbiology involved in tertiary treatment for nitrogen removal? For phosphate removal?

4. What do you believe should be done to with the digested sludge from anaerobic digesters?

5. Why is it so difficult to identify *Escherichia coli* in water samples?

6. Some strains of *E. coli* are pathogenic. Is it still possible to justify the use of this species as an indicator organism?

7. What would be the ideal microbiological test for water potability?

8. Explain the phenomenon of biomagnification and discuss the risk factors for humans living in ecosystems contaminated with DDT.

9. Define and provide an example of cometabolism.

10. Describe how prokaryotic organisms cause acid mine drainage.

SUGGESTED READING

Clesceri, L. S. and A. D. Eaton, eds. 1999. *Standard Methods for the Examination of Water and Wastewater,* 20th ed. Alexandria, VA: Water Environment Federation.

Guthrie, F. E. and J. J. Perry, eds. 1980. *Environmental Toxicology.* New York: Elsevier North Holland.

Hurst, C. J., R. L. Crawford, J. L. Garland, D. A. Lipson and A. L. Mills, eds. 2007. *Manual of Environmental Microbiology.* 3rd ed. Washington, DC: ASM Press.

Jacobson, M. C., R. J. Charlson, H. Rodhe and G. H. Orians, eds. 2000. *Earth System Science.* New York: Academic Press.

Maier, R. M., I. L. Pepper and C. P. Gerba, eds. 2000. *Environmental Microbiology.* New York: Academic Press.

Mitchell, R. 1992. *Environmental Microbiology.* New York: John Wiley & Sons.

Omenn, G. S., ed. 1988. *Environmental Biotechnology: Reducing Risks from Environmental Chemicals through Biotechnology.* New York: Plenum.

Appendix

Taxonomic Outline of the Prokaryotes Release 5.0
Bergey's Manual ™ of Systematic Bacteriology, 2nd Edition

George M. Garrity, Julia A. Bell and Timothy G. Lilburn

Abbreviations: T, the type strain of a species or the type species of a genus; A, on Approved List of species names accepted in 1980; V, a new species officially named (i.e., validly published) after 1980.

Domain *Archaea* VP
Phylum AI. *Crenarchaeota* VP 1
  Class I. *Thermoprotei* VP
    Order I. *Thermoproteales* VP (T)
      Family I. *Thermoproteaceae* VP
        Genus I. *Thermoproteus* VP (T)
        Genus II. *Caldivirga* VP
        Genus III. *Pyrobaculum* VP
        Genus IV. *Thermocladium* VP
        Genus V. *Vulcanisaeta* VP
      Family II. *Thermofilaceae* VP
        Genus I. *Thermofilum* VP (T)
    Order II. "*Caldisphaerales*"
      Family I. "*Caldisphaeraceae*"
        Genus I. *Caldisphaera* VP (T)
    Order III. *Desulfurococcales* VP
      Family I. *Desulfurococcaceae* VP
        Genus I. *Desulfurococcus* VP (T)
        Genus II. *Acidilobus* VP 2
        Genus III. *Aeropyrum* VP
        Genus IV. *Ignicoccus* VP
        Genus V. *Staphylothermus* VP
        Genus VI. *Stetteria* VP
        Genus VII. *Sulfophobococcus* VP
        Genus VIII. *Thermodiscus* VP
        Genus IX. *Thermosphaera* VP
      Family II. *Pyrodictiaceae* VP
        Genus I. *Pyrodictium* VP (T)
        Genus II. *Hyperthermus* VP
        Genus III. *Pyrolobus* VP
    Order IV. *Sulfolobales* VP
      Family I. *Sulfolobaceae* VP
        Genus I. *Sulfolobus* AL (T)
        Genus II. *Acidianus* VP
        Genus III. *Metallosphaera* VP
        Genus IV. *Stygiolobus* VP
        Genus V. *Sulfurisphaera* VP
        Genus VI. *Sulfurococcus* VP
Phylum AII. *Euryarchaeota* VP 3
  Class I. *Methanobacteria* VP
    Order I. *Methanobacteriales* VP (T)

Family I. *Methanobacteriaceae* AL
  Genus I. *Methanobacterium* AL (T)
  Genus II. *Methanobrevibacter* VP
  Genus III. *Methanosphaera* VP
  Genus IV. *Methanothermobacter* VP
Family II. *Methanothermaceae* VP
  Genus I. *Methanothermus* VP (T)
Class II. *Methanococci* VP
  Order I. *Methanococcales* VP (T)
    Family I. *Methanococcaceae* VP
      Genus I. *Methanococcus* AL (T)
      Genus II. *Methanothermococcus* VP
    Family II. *Methanocaldococcaceae* VP
      Genus I. *Methanocaldococcus* VP (T)
      Genus II. *Methanotorris* VP
Class III. "*Methanomicrobia*" 4
  Order I. *Methanomicrobiales* VP (T)
    Family I. *Methanomicrobiaceae* VP
      Genus I. *Methanomicrobium* VP (T)
      Genus II. *Methanoculleus* VP
      Genus III. *Methanofollis* VP
      Genus IV. *Methanogenium* VP
      Genus V. *Methanolacinia* VP
      Genus VI. *Methanoplanus* VP
    Family II. *Methanocorpusculaceae* VP 5
      Genus I. *Methanocorpusculum* VP (T)
    Family III. *Methanospirillaceae* VP
      Genus I. *Methanospirillum* AL (T)
    Genera incertae sedis
      Genus I. *Methanocalculus* VP
  Order II. *Methanosarcinales* VP
    Family I. *Methanosarcinaceae* VP
      Genus I. *Methanosarcina* AL (T)
      Genus II. *Methanococcoides* VP
      Genus III. *Methanohalobium* VP
      Genus IV. *Methanohalophilus* VP
      Genus V. *Methanolobus* VP
      Genus VI. *Methanomethylovorans* VP
      Genus VII. *Methanimicrococcus* VP
      Genus VIII. *Methanosalsum* VP
    Family II. *Methanosaetaceae* VP
      Genus I. *Methanosaeta* VP (T)
Class IV. *Halobacteria* VP
  Order I. *Halobacteriales* VP (T)
    Family I. *Halobacteriaceae* AL

Genus I. *Halobacterium* AL (T)
Genus II. *Haloarcula* VP
Genus III. *Halobaculum* VP
Genus IV. *Halobiforma* VP
Genus V. *Halococcus* AL
Genus VI. *Haloferax* VP
Genus VII. *Halogeometricum* VP
Genus VIII. *Halomicrobium* VP
Genus IX. *Halorhabdus* VP
Genus X. *Halorubrum* VP
Genus XI. *Halosimplex* VP
Genus XII. *Haloterrigena* VP
Genus XIII. *Natrialba* VP
Genus XIV. *Natrinema* VP
Genus XV. *Natronobacterium* VP
Genus XVI. *Natronococcus* VP
Genus XVII. *Natronomonas* VP
Genus XVIII. *Natronorubrum* VP
Class V. *Thermoplasmata* VP
  Order I. *Thermoplasmatales* VP (T)
    Family I. *Thermoplasmataceae* VP
      Genus I. *Thermoplasma* AL (T)
    Family II. *Picrophilaceae* VP
      Genus I. *Picrophilus* VP (T)
    Family III. "*Ferroplasmataceae*"
      Genus I. *Ferroplasma* VP
Class VI. *Thermococci* VP
  Order I. *Thermococcales* VP (T)
    Family I. *Thermococcaceae* VP
      Genus I. *Thermococcus* VP (T)
      Genus II. *Palaeococcus* VP
      Genus III. *Pyrococcus* VP
Class VII. *Archaeoglobi* VP
  Order I. *Archaeoglobales* VP (T)
    Family I. *Archaeoglobaceae* VP
      Genus I. *Archaeoglobus* VP (T)
      Genus II. *Ferroglobus* VP
      Genus III. *Geoglobus* VP
Class VIII. *Methanopyri* VP
  Order I. *Methanopyrales* VP (T)
    Family I. *Methanopyraceae* VP
      Genus I. *Methanopyrus* VP (T)
Domain *Bacteria* VP
Phylum BI. *Aquificae* VP 6
  Class I. *Aquificae* VP

Genus IX. *Gluconobacter* AL
Genus X. *Kozakia* VP
Genus XI. *Muricoccus* VP
Genus XII. *Paracraurococcus* VP
Genus XIII. *Rhodopila* VP
Genus XIV. *Roseococcus* VP
Genus XV. *Rubritepida* VP
Genus XVI. *Stella* VP
Genus XVII. *Teichococcus* VP
Genus XVIII. *Zavarzinia* VP
Order II. *Rickettsiales* AL 34
  Family I. *Rickettsiaceae* AL
    Genus I. *Rickettsia* AL (T)
    Genus II. *Orientia* VP
  Family II. *Anaplasmataceae* AL
    Genus I. *Anaplasma* AL (T)
    Genus II. *Aegyptianella* AL
    Genus III. *Cowdria* AL 35
    Genus IV. *Ehrlichia* AL
    Genus V. *Neorickettsia* AL
    Genus VI. *Wolbachia* AL
    Genus VII. *Xenohaliotis* VP
  Family III. *Holosporaceae* fam. nov.
    Genus I. *Holospora* VP (T)
  Genera incertae sedis 36
    Genus I. *Caedibacter*
    Genus II. *Lyticum* VP
    Genus III. *Odyssella* VP
    Genus IV. *Pseudocaedibacter* VP
    Genus V. *Symbiotes* AL 37
    Genus VI. *Tectibacter* VP
Order III. *Rhodobacterales* ord. nov.
  Family I. *Rhodobacteraceae* fam. nov. 38
    Genus I. *Rhodobacter* VP (T) 39
    Genus II. *Ahrensia* VP
    Genus III. *Albidovulum* VP
    Genus IV. *Amaricoccus* VP
    Genus V. *Antarctobacter* VP
    Genus VI. *Gemmobacter* VP
    Genus VII. *Hirschia* VP
    Genus VIII. *Hyphomonas* VP 40
    Genus IX. *Jannaschia* VP
    Genus X. *Ketogulonicigenium* VP
    Genus XI. *Leisingera* VP
    Genus XII. *Maricaulis* VP
    Genus XIII. *Methylarcula* VP
    Genus XIV. *Oceanicaulis* VP
    Genus XV. *Octadecabacter* VP
    Genus XVI. *Pannonibacter* VP
    Genus XVII. *Paracoccus* VP 41
    Genus XVIII. *Pseudorhodobacter* VP
    Genus XIX. *Rhodobaca* VP
    Genus XX. *Rhodothalassium* VP 42
    Genus XXI. *Rhodovulum* VP 43
    Genus XXII. *Roseibium* VP
    Genus XXIII. *Roseinatronobacter* VP
    Genus XXIV. *Roseivivax* VP
    Genus XXV. *Roseobacter* VP 44
    Genus XXVI. *Roseovarius* VP
    Genus XXVII. *Rubrimonas* VP
    Genus XXVIII. *Ruegeria* VP
    Genus XXIX. *Sagittula* VP
    Genus XXX. *Silicibacter* VP
    Genus XXXI. *Staleya* VP
    Genus XXXII. *Stappia* VP

    Genus XXXIII. *Sulfitobacter* VP
Order IV. *Sphingomonadales* ord. nov.
  Family I. *Sphingomonadaceae* VP
    Genus I. *Sphingomonas* VP (T)
    Genus II. *Blastomonas* VP 45
    Genus III. *Erythrobacter* VP
    Genus IV. *Erythromicrobium* VP
    Genus V. *Erythromonas* VP
    Genus VI. *Novosphingobium* VP
    Genus VII. *Porphyrobacter* VP
    Genus VIII. *Rhizomonas* VP 46
    Genus IX. *Sandaracinobacter* VP
    Genus X. *Sphingobium* VP
    Genus XI. *Sphingopyxis* VP
    Genus XII. *Zymomonas* AL
Order V. *Caulobacterales* AL 47
  Family I. *Caulobacteraceae* AL
    Genus I. *Caulobacter* AL (T)
    Genus II. *Asticcacaulis* AL
    Genus III. *Brevundimonas* VP
    Genus IV. *Phenylobacterium* VP 48
Order VI. *Rhizobiales* ord. nov.
  Family I. *Rhizobiaceae* AL
    Genus I. *Rhizobium* AL (T)
    Genus II. *Agrobacterium* AL
    Genus III. *Allorhizobium* VP
    Genus IV. *Carbophilus* VP
    Genus V. *Chelatobacter* VP
    Genus VI. *Ensifer* VP
    Genus VII. *Sinorhizobium* VP
  Family II. *Aurantimonadaceae* fam. nov. (T)
    Genus I. *Aurantimonas* VP
    Genus II. *Fulvimarina* VP
  Family III. *Bartonellaceae* AL
    Genus I. *Bartonella* AL (T)
  Family IV. *Brucellaceae* AL
    Genus I. *Brucella* AL (T) 49
    Genus II. *Mycoplana* AL
    Genus III. *Ochrobactrum* VP
  Family V. *Phyllobacteriaceae* fam. nov.
    Genus I. *Phyllobacterium* VP (T)
    Genus II. *Aminobacter* VP
    Genus III. *Aquamicrobium* VP
    Genus IV. *Defluvibacter* VP
    Genus V. "*Candidatus* Liberibacter"
    Genus VI. *Mesorhizobium* VP
    Genus VII. *Nitratireductor* VP
    Genus VIII. *Pseudaminobacter* VP
  Family VI. *Methylocystaceae* fam. nov. 50
    Genus I. *Methylocystis* VP (T)
    Genus II. *Albibacter* VP
    Genus III. *Methylopila* VP 51
    Genus IV. *Methylosinus* VP
    Genus V. *Terasakiella* VP
  Family VII. *Beijerinckiaceae* fam. nov. 52
    Genus I. *Beijerinckia* AL (T)
    Genus II. *Chelatococcus* VP
    Genus III. *Methylocapsa* VP
    Genus IV. *Methylocella* VP
  Family VIII. *Bradyrhizobiaceae* fam. nov. 53
    Genus I. *Bradyrhizobium* VP (T)
    Genus II. *Afipia* VP 54
    Genus III. *Agromonas* VP 55
    Genus IV. *Blastobacter* AL 56
    Genus V. *Bosea* VP 57

    Genus VI. *Nitrobacter* AL
    Genus VII. *Oligotropha* VP
    Genus VIII. *Rhodoblastus* VP 58
    Genus IX. *Rhodopseudomonas* AL
  Family IX. *Hyphomicrobiaceae* AL 59
    Genus I. *Hyphomicrobium* AL (T) 60
    Genus II. *Ancalomicrobium* AL
    Genus III. *Ancylobacter* AL 61
    Genus IV. *Angulomicrobium* VP
    Genus V. *Aquabacter* VP 62
    Genus VI. *Azorhizobium* VP 63
    Genus VII. *Blastochloris* VP 64
    Genus VIII. *Devosia* VP 65
    Genus IX. *Dichotomicrobium* VP
    Genus X. *Filomicrobium* VP
    Genus XI. *Gemmiger* AL
    Genus XII. *Labrys* VP
    Genus XIII. *Methylorhabdus* VP 66
    Genus XIV. *Pedomicrobium* AL 67
    Genus XV. *Prosthecomicrobium* AL
    Genus XVI. *Rhodomicrobium* AL 68
    Genus XVII. *Rhodoplanes* VP 69
    Genus XVIII. *Seliberia* AL
    Genus XIX. *Starkeya* VP
    Genus XX. *Xanthobacter* AL 70
  Family X. *Methylobacteriaceae* fam. nov. 71
    Genus I. *Methylobacterium* AL (T)
    Genus II. *Microvirga* VP
    Genus III. *Protomonas* VP
    Genus IV. *Roseomonas* VP 72
  Family XI. *Rhodobiaceae* fam. nov.
    Genus I. *Rhodobium* VP 73
    Genus II. *Roseospirillum* VP 74
Order VII. *Parvularculales* ord. nov.
  Family I. *Parvularculaceae* fam. nov.
    Genus I. *Parvularcula* VP (T)
Class II. *Betaproteobacteria* class. nov. 75
  Order I. *Burkholderiales* ord. nov. (T)
    Family I. *Burkholderiaceae* fam. nov.
      Genus I. *Burkholderia* VP (T)
      Genus II. *Cupriavidus* VP
      Genus III. *Lautropia* VP 76
      Genus IV. *Limnobacter* VP
      Genus V. *Pandoraea* VP
      Genus VI. *Paucimonas* VP
      Genus VII. *Polynucleobacter* VP
      Genus VIII. *Ralstonia* VP 77
      Genus IX. *Thermothrix* VP
      Genus X. *Wautersia* VP
    Family II. *Oxalobacteraceae* fam. nov.
      Genus I. *Oxalobacter* VP (T)
      Genus II. *Duganella* VP
      Genus III. *Herbaspirillum* VP
      Genus IV. *Janthinobacterium* AL
      Genus V. *Massilia* VP
      Genus VI. *Oxalicibacterium* VP
      Genus VII. *Telluria* VP
    Family III. *Alcaligenaceae* VP
      Genus I. *Alcaligenes* AL (T)
      Genus II. *Achromobacter* VP
      Genus III. *Bordetella* AL
      Genus IV. *Brackiella* VP
      Genus V. *Derxia* AL 78
      Genus VI. *Kerstersia* VP
      Genus VII. *Oligella* VP 79

Genus XIII. Thermovenabulum VP
Family II. "Thermodesulfobiaceae"
Genus I. Thermodesulfobium VP
Order III. *Halanaerobiales* VP 154
Family I. Halanaerobiaceae VP
Genus I. Halanaerobium VP (T)
Genus II. Halocella VP
Genus III. Halothermothrix VP
Family II. Halobacteroidaceae VP
Genus I. Halobacteroides VP (T)
Genus II. Acetohalobium VP
Genus III. Halanaerobacter VP
Genus IV. Halonatronum VP
Genus V. Natroniella VP 155
Genus VI. Orenia VP
Genus VII. Selenihalanaerobacter VP
Genus VIII. Sporohalobacter VP
Class II. *Mollicutes* AL
Order I. *Mycoplasmatales* AL (T)
Family I. Mycoplasmataceae AL
Genus I. Mycoplasma AL (T)
Genus II. Eperythrozoon AL
Genus III. Haemobartonella AL 156
Genus IV. Ureaplasma AL
Order II. *Entomoplasmatales* VP
Family I. Entomoplasmataceae VP
Genus I. Entomoplasma VP (T)
Genus II. Mesoplasma VP
Family II. Spiroplasmataceae VP
Genus I. Spiroplasma AL (T)
Order III. *Acholeplasmatales* VP
Family I. Acholeplasmataceae AL
Genus I. Acholeplasma AL
Genus II. "Phytoplasma"
Order IV. *Anaeroplasmatales* VP
Family I. Anaeroplasmataceae VP
Genus I. Anaeroplasma AL (T) 157
Genus II. Asteroleplasma VP 158
Order V. *Incertae sedis*
Family I. "Erysipelotrichaceae"
Genus I. Erysipelothrix AL
Genus II. Bulleidia VP
Genus III. Holdemania VP
Genus IV. Solobacterium VP
Class III. *"Bacilli"*
Order I. *Bacillales* AL 159
Family I. Bacillaceae AL
Genus I. Bacillus AL (T)
Genus II. Amphibacillus VP
Genus III. Anoxybacillus VP
Genus IV. Exiguobacterium VP
Genus V. Filobacillus VP
Genus VI. Geobacillus VP
Genus VII. Gracilibacillus VP
Genus VIII. Halobacillus VP
Genus IX. Jeotgalibacillus VP
Genus X. Lentibacillus VP
Genus XI. Marinibacillus VP
Genus XII. Oceanobacillus VP
Genus XIII. Paraliobacillus VP
Genus XIV. Saccharococcus VP
Genus XV. Salibacillus VP
Genus XVI. Ureibacillus VP
Genus XVII. Virgibacillus VP
Family II. "Alicyclobacillaceae" 160

Genus I. Alicyclobacillus VP
Genus II. Pasteuria AL
Genus III. Sulfobacillus VP 161
Family III. Caryophanaceae AL
Genus I. Caryophanon AL (T)
Family IV. "Listeriaceae"
Genus I. Listeria AL
Genus II. Brochothrix AL
Family V. "Paenibacillaceae" 162
Genus I. Paenibacillus VP
Genus II. Ammoniphilus VP 163
Genus III. Aneurinibacillus VP
Genus IV. Brevibacillus VP
Genus V. Oxalophagus VP
Genus VI. Thermicanus VP
Genus VII. Thermobacillus VP
Family VI. Planococcaceae AL
Genus I. Planococcus AL (T)
Genus II. Filibacter VP
Genus III. Kurthia AL
Genus IV. Planomicrobium VP
Genus V. Sporosarcina AL
Family VII. "Sporolactobacillaceae"
Genus I. Sporolactobacillus AL
Genus II. Marinococcus VP
Family VIII. "Staphylococcaceae"
Genus I. Staphylococcus AL
Genus II. Gemella AL
Genus III. Jeotgalicoccus VP
Genus IV. Macrococcus VP
Genus V. Salinicoccus VP
Family IX. "Thermoactinomycetaceae" 164
Genus I. Thermoactinomyces AL
Family X. "Turicibacteraceae" 165
Genus I. Turicibacter VP (T)
Order II. *"Lactobacillales"*
Family I. Lactobacillaceae AL
Genus I. Lactobacillus AL (T)
Genus II. Paralactobacillus VP
Genus III. Pediococcus AL
Family II. "Aerococcaceae"
Genus I. Aerococcus AL
Genus II. Abiotrophia VP
Genus III. Dolosicoccus VP
Genus IV. Eremococcus VP
Genus V. Facklamia VP
Genus VI. Globicatella VP
Genus VII. Ignavigranum VP
Family III. "Carnobacteriaceae"
Genus I. Carnobacterium VP
Genus II. Agitococcus VP
Genus III. Alkalibacterium VP
Genus IV. Allofustis VP
Genus V. Alloiococcus VP
Genus VI. Desemzia VP
Genus VII. Dolosigranulum VP
Genus VIII. Granulicatella VP
Genus IX. Isobaculum VP
Genus X. Lactosphaera VP 166
Genus XI. Marinilactibacillus VP
Genus XII. Trichococcus VP
Family IV. "Enterococcaceae"
Genus I. Enterococcus VP
Genus II. Atopobacter VP
Genus III. Melissococcus VP

Genus IV. Tetragenococcus VP
Genus V. Vagococcus VP
Family V. "Leuconostocaceae"
Genus I. Leuconostoc AL
Genus II. Oenococcus VP
Genus III. Weissella VP
Family VI. Streptococcaceae AL
Genus I. Streptococcus AL (T)
Genus II. Lactococcus AL
Family VII. Incertae sedis 167
Genus I. Acetoanaerobium VP
Genus II. Oscillospira AL
Genus III. Syntrophococcus VP
Phylum BXIV. *Actinobacteria* phy. nov.
Class I. *Actinobacteria* VP 168
Subclass I. *Acidimicrobidae* VP
Order I. *Acidimicrobiales* VP
Suborder IV. *"Acidimicrobineae"*
Family I. Acidimicrobiaceae VP
Genus I. Acidimicrobium VP (T)
Subclass II. *Rubrobacteridae* VP
Order I. *Rubrobacterales* VP
Suborder V. *"Rubrobacterineae"*
Family I. Rubrobacteraceae VP
Genus I. Rubrobacter VP (T)
Genus II. Conexibacter VP
Genus III. Solirubrobacter VP
Genus IV. Thermoleophilum VP 169
Subclass III. *Coriobacteridae* VP
Order I. *Coriobacteriales* VP
Suborder VI. *"Coriobacterineae"*
Family I. Coriobacteriaceae VP
Genus I. Coriobacterium VP (T)
Genus II. Atopobium VP
Genus III. Collinsella VP
Genus IV. Cryptobacterium VP
Genus V. Denitrobacterium VP
Genus VI. Eggerthella VP
Genus VII. Olsenella VP
Genus VIII. Slackia VP
Subclass IV. *Sphaerobacteridae* VP
Order I. *Sphaerobacterales* VP
Suborder VII. *"Sphaerobacterineae"*
Family I. Sphaerobacteraceae VP
Genus I. Sphaerobacter VP (T)
Subclass V. *Actinobacteridae* VP
Order I. *Actinomycetales* AL
Suborder VIII. *Actinomycineae* VP
Family I. Actinomycetaceae AL
Genus I. Actinomyces AL (T)
Genus II. Actinobaculum VP
Genus III. Arcanobacterium VP
Genus IV. Mobiluncus VP
Genus V. Varibaculum VP
Suborder IX. *Micrococcineae* VP
Family I. Micrococcaceae AL
Genus I. Micrococcus AL (T)
Genus II. Arthrobacter AL
Genus III. Citricoccus VP
Genus IV. Kocuria VP
Genus V. Nesterenkonia VP
Genus VI. Renibacterium VP
Genus VII. Rothia AL
Genus VIII. Stomatococcus VP
Genus IX. Yania VP

Genus I. Planctomyces AL (T)
Genus II. Gemmata VP
Genus III. Isosphaera VP
Genus IV. Pirellula VP
Phylum BXVI. *"Chlamydiae"*
Class I. *Chlamydiae* VP
Order I. *Chlamydiales* AL (T)
Family I. *Chlamydiaceae* AL
Genus I. Chlamydia AL (T)
Genus II. Chlamydophila VP
Family II. *Parachlamydiaceae* VP
Genus I. Parachlamydia VP (T)
Genus II. Neochlamydia VP
Family III. *Simkaniaceae* VP
Genus I. Simkania VP (T)
Genus II. "Rhabdochlamydia"
Family IV. *Waddliaceae* VP
Genus I. Waddlia VP (T)
Phylum BXVII. *Spirochaetes phy. nov.*
Class I. *"Spirochaetes"*
Order I. *Spirochaetales* AL
Family I. *Spirochaetaceae* AL
Genus I. Spirochaeta AL (T)
Genus II. Borrelia AL
Genus III. Brevinema VP
Genus IV. Clevelandina VP
Genus V. Cristispira AL
Genus VI. Diplocalyx VP
Genus VII. Hollandina VP
Genus VIII. Pillotina VP
Genus IX. Treponema AL
Family II. *"Serpulinaceae"* 176
Genus I. Serpulina VP
Genus II. Brachyspira VP
Family III. *Leptospiraceae* AL
Genus I. Leptospira AL (T)
Genus II. Leptonema VP
Phylum BXVIII. *"Fibrobacteres"*
Class I. *"Fibrobacteres"*
Order I. *"Fibrobacterales"*
Family I. *"Fibrobacteraceae"*
Genus I. Fibrobacter VP
Phylum BXIX. *"Acidobacteria"*
Class I. *Acidobacteria* VP
Order I. *Acidobacteriales* VP (T)
Family I. *"Acidobacteriaceae"*
Genus I. Acidobacterium VP
Genus II. Geothrix VP
Genus III. Holophaga VP
Phylum BXX. *"Bacteroidetes"*
Class I. *"Bacteroidetes"*
Order I. *"Bacteroidales"*
Family I. *Bacteroidaceae* AL
Genus I. Bacteroides AL (T)
Genus II. Acetofilamentum VP
Genus III. Acetomicrobium VP
Genus IV. Acetothermus VP

Genus V. Anaerophaga VP
Genus VI. Anaerorhabdus VP
Genus VII. Megamonas VP
Family II. *"Rikenellaceae"*
Genus I. Rikenella VP
Genus II. Alistipes VP
Genus III. Marinilabilia VP
Family III. *"Porphyromonadaceae"*
Genus I. Porphyromonas VP
Genus II. Dysgonomonas VP
Genus III. Tannerella VP
Family IV. *"Prevotellaceae"*
Genus I. Prevotella VP
Class II. *Flavobacteria* VP
Order I. *"Flavobacteriales"*
Family I. *Flavobacteriaceae* VP
Genus I. Flavobacterium AL (T)
Genus II. Aequorivita VP
Genus III. Arenibacter VP 177
Genus IV. Bergeyella VP
Genus V. Capnocytophaga VP
Genus VI. Cellulophaga VP 178
Genus VII. Chryseobacterium VP
Genus VIII. Coenonia VP
Genus IX. Croceibacter VP
Genus X. Empedobacter VP
Genus XI. Gelidibacter VP
Genus XII. Gillisia VP
Genus XIII. Mesonia VP
Genus XIV. Muricauda VP 179
Genus XV. Myroides VP 180
Genus XVI. Ornithobacterium VP
Genus XVII. Polaribacter VP
Genus XVIII. Psychroflexus VP
Genus XIX. Psychroserpens VP
Genus XX. Riemerella VP
Genus XXI. Saligentibacter VP
Genus XXII. Tenacibaculum VP
Genus XXIII. Weeksella VP
Genus XXIV. Zobellia VP
Family II. *"Blattabacteriaceae"*
Genus I. Blattabacterium AL
Class III. *"Sphingobacteria"*
Order I. *"Sphingobacteriales"*
Family I. *Sphingobacteriaceae* VP
Genus I. Sphingobacterium VP (T)
Genus II. Pedobacter VP
Family II. *"Saprospiraceae"*
Genus I. Saprospira AL
Genus II. Haliscomenobacter AL
Genus III. Lewinella VP
Family III. *"Flexibacteraceae"*
Genus I. Flexibacter AL
Genus II. Belliella VP
Genus III. Cyclobacterium VP
Genus IV. Cytophaga AL
Genus V. Dyadobacter VP

Genus VI. Flectobacillus AL
Genus VII. Hongiella VP
Genus VIII. Hymenobacter VP 181
Genus IX. Meniscus AL
Genus X. Microscilla AL
Genus XI. Reichenbachia VP
Genus XII. Runella AL
Genus XIII. Spirosoma AL
Genus XIV. Sporocytophaga AL
Family IV. *"Flammeovirgaceae"*
Genus I. Flammeovirga VP
Genus II. Flexithrix AL 182
Genus III. Persicobacter VP
Genus IV. Thermonema VP
Family V. *Crenotrichaceae* AL
Genus I. Crenothrix AL (T)
Genus II. Chitinophaga VP
Genus III. Rhodothermus VP 183
Genus IV. Salinibacter VP
Genus V. Toxothrix AL
Phylum BXXI. *"Fusobacteria"*
Class I. *"Fusobacteria"*
Order I. *"Fusobacteriales"*
Family I. *"Fusobacteriaceae"*
Genus I. Fusobacterium AL
Genus II. Ilyobacter VP 184
Genus III. Leptotrichia AL
Genus IV. Propionigenium VP
Genus V. Sebaldella VP
Genus VI. Streptobacillus AL
Genus VII. Sneathia VP
Family II. *Incertae sedis* VP
Genus I. Cetobacterium VP
Phylum BXXII. *"Verrucomicrobia"*
Class I. *Verrucomicrobiae* VP 185
Order I. *Verrucomicrobiales* VP (T)
Family I. *Verrucomicrobiaceae* VP
Genus I. Verrucomicrobium VP (T)
Genus II. Prosthecobacter VP
Family II. *"Opitutaceae"*
Genus I. Opitutus VP (T)
Family III. *"Victivallaceae"*
Genus I. Victivallis VP 186
Family IV. *"Xiphinematobacteriaceae"*
Genus I. Xiphinematobacter VP
Phylum BXXIII. *"Dictyoglomi"* 187
Class I. *"Dictyoglomi"*
Order I. *"Dictyoglomales"*
Family I. *"Dictyoglomaceae"*
Genus I. Dictyoglomus VP
Phylum BXXIV. *Gemmatimonadetes* VP
Class I. *Gemmatimonadetes* VP
Order I. *Gemmatimonadales* VP (T)
Family I. *Gemmatimonadaceae* VP
Genus I. Gemmatimonas VP (T)

Glossary

Abiotic Occurring in the absence of living organisms.

Accessory pigment Pigments that trap light energy and transfer it to chlorophyll molecules in reaction centers. Carotenoids and phyco-biliproteins are examples.

Accidental pathogen A microorganism that does not generally cause disease in its normal life cycle. Can cause disease if introduced into a host as a result of a breach in barrier defense.

Aceticlastic A methanoarchaeon that forms methane from the methyl group and CO_2 from the carbonyl group of acetate.

Acetogenic bacteria Species that produce acetic acid via sugar fermentation or from $H_2 + CO_2$ in anaerobic environments.

Acetyl-CoA pathway An autotrophic CO_2 fixation pathway utilized by selected anaerobic bacteria. The assimilation of two molecules of CO_2 results in formation of acetyl-CoA.

Acid dyes Staining compounds that are negatively charged.

Acid-fastness or acid-alcohol fastness A property of some organisms, such as mycobacteria, that have a high lipid content in their outer wall. These organisms retain hot carbol-fuchsin stain when rinsed with acid alcohol.

Acid mine drainage Low pH water emanating from mines.

Acid rain Rainfall that has low pH due to sulfuric acid.

Acidophile An organism that preferentially grows at a pH below 5.4. *Sulfolobus* is one example.

Acne An inflammatory reaction in the sebaceous glands and hair follicles of the skin.

Acquired immunity Most often called adaptive immunity. A specific immunity acquired by exposure of B lymphocytes and T lymphocytes to an antigen.

Acridine orange A fluorescent dye that stains nucleic acids.

Actinobacteria Bacteria from the phylum *Actinobacteria*, some of which produce mycelia.

Activated sludge process An aerobic wastewater treatment process.

Activation energy Energy added that renders molecules capable of chemical interactions.

Activator A protein capable of enhancing transcription of a gene.

Active immunity An immune state generated by challenge with a specific antigen.

Active site The specific area of an enzyme where substrate is bound, forming an enzyme-substrate complex.

Active transport Movement of ions or molecules across the cell membrane at an expenditure of energy.

Acute infection An infection that causes abrupt onset of symptoms and is resolved rapidly by the host, or that kills the host rapidly.

Acute phase response A group of physiological processes occurring soon after the onset of infection, trauma, inflammatory processes, and some malignant conditions. It includes an increase in acute phase proteins in serum many of which arise from the liver, fever, increased vascular permeability, and metabolic and pathologic changes.

Acute stage The stage in the infection cycle where symptoms are most pronounced.

Acyl carrier protein (ACP) A small protein bound to a growing fatty acid during synthesis and released when the chain attains full length.

Acyl-homoserine lactones (AHLs) A class of signaling molecules involved in microbiological quorum sensing.

Adaptive immunity Immune responses characterized by exquisite specificity for the antigen and memory (recall) responses upon secondary encounter with antigen. Such responses are conveyed by lymphocytes known as T cells and B cells.

ADCC Antibody dependent cellular cytotoxicity.

Adenosine-5'-triphosphate (ATP) The energy carrier in all living cells.

Adenylate cyclase toxin A protein toxin produced by *Bordetella pertussis*, catalyzing formation of cyclic AMP in mammalian cells.

Adherence factors Molecules on the surface of pathogens that allow them to attach to host cells and tissues.

Adhesin A cell surface protein that allows pathogenic bacteria to adhere to other cells or a substrate. Adhesins are important in attachment of pathogens to host cells during infection.

ADP ribosylation Transfer of the ADP-ribose moiety from NAD to a protein, a reaction catalyzed by various bacterial toxins.

Aerial mycelium The cellular network of *Actinobacteria* that is produced in the air above the cells on the growth substrate.

Aerobe An organism that utilizes molecular oxygen as terminal electron acceptor in aerobic respiration.

Aerobic respiration A process in which oxygen serves as the electron acceptor in metabolism.

Aerosol Liquid droplets suspended in air.

Aerotolerant An organism that does not utilize molecular oxygen as terminal electron acceptor but is not harmed by O_2.

Affinity maturation The process by which B-cells produce antibodies with increased affinity for antigen. This results from somatic hypermutation of bases encoding the antigen combining site of the B cell antigen specific receptor followed by affinity based selection.

AFM Atomic force microscopy.

Agar A sulfur-containing polysaccharide of marine algal origin that is used as a solidifying agent in culture media.

Agarose gel electrophoresis A method of separating DNA fragments based on size and charge.

Agglutination Sticking together of microbes or cells, often caused by an interaction with an antibody, resulting in an observable clump.

AIDS Acquired immune deficiency syndrome. An infection caused by HIV, the human immunodeficiency virus.

Airborne droplets Liquid suspended in air that may contain pathogens, particularly when of human origin, such as from a sneeze.

Akinetes Resting bodies formed by some cyanobacteria.

Alcoholic fermentation Anaerobic metabolism where alcohol (ethanol, butanol, etc.) and CO_2 are major products.

Algae Photosynthetic eukaryotic aquatic organisms.

Alkaliphiles Organisms that inhabit alkaline lakes and grow at a pH up to 11.5.

Allergic contact dermatitis Allergic response mediated by T cells and activated macrophages.

Allergy Inappropriate immune response to an antigen (allergen).

Allophycocyanin A blue-green colored pigment of cyanobacteria.

Allostery Change in the conformation of a protein resulting from the attachment of a compound at a site other than the reactive site. Generally a reversible inactivation of a protein, such as an enzyme.

Alpha-hemolysis An effect observed on blood agar plates in which bacterial toxins cause a green coloration of the hemoglobin.

Alternative pathway of complement activation Complement activation by generation of a C3 convertase on the surface of pathogens.

Alternative σ factors. A family of sigma subunits of RNA polymerase that are different from the major sigma factor (σ^{70} in *E. coli*).

Amebiasis Infection caused by pathogenic amoebas.

Ames test A procedure for determining the mutagenic potential of a chemical.

Aminoacyl-tRNA synthetase An enzyme responsible for linking amino acids to transfer RNAs.

Ammonia oxidizers Bacteria or archaea that oxidize ammonia as an energy source.

Ammonification Release of ammonia from cell material during decomposition.

Amylases Enzymes that hydrolyze starch or glycogen.

Anabolism See biosynthesis.

Anaerobe An organism that does not employ oxygen as terminal electron acceptor.

Anaerobic respiration Electron transport oxidation where sulfate, sulfur, nitrate, CO_2, or other oxidized compounds are utilized as a terminal electron acceptor.

Anaerobic sludge digester A wastewater treatment process that occurs anaerobically.

Anammox reaction A reaction in which ammonia and nitrite are converted to nitrogen gas.

Anammoxosome A compartment in certain members of the *Planctomycetes* in which the anammox reaction occurs.

Anamnestic response Rapid immune response to antigens to which the host has previously been exposed. Also known as an immunological or recall response.

Anamorphs Asexual growth morphology in fungi.

Anaphylactic shock A destructive reaction between an antigen and antibody. Results in smooth muscle contraction.

Anaphylatoxins Molecules that attract cells to a site of inflammation. Includes complement factors C3a, C4a and C5a.

Anaplerotic reactions Replacement reactions that permit the tricarboxylic acid (TCA) cycle to continue. The replenishing of oxaloacetic acid by carboxylation of pyruvate is a key anaplerotic reaction.

Anastomosis Fusion of cytoplasmic extensions and pseudopodia between the same or different individuals.

Angular aperture The angle through which light enters the objective lens.

Anion A negatively charged atom.

Annotation Notes explaining the putative biological roles of gene products deduced from genome sequencing.

Anoxic An environment that lacks oxygen.

Anoxygenic photosynthesis Photosynthesis in which O_2 is not produced. Employs electron donors other than H_2O. H_2S, H_2, and reduced organic compounds would donate electrons in anoxygenic photosynthesis.

Antenna pigments Light harvesting substances other than chlorophyll *a* used by photosynthetic organisms.

Antheridia One of two types of hyphae in conjugating Ascomycetes. Antheridia transfer haploid nuclei to asconia.

Antibiotic A metabolite produced by a microorganism that inhibits or destroys other microorganisms.

Antibiotic-resistance gene cassettes Segments of DNA, often part of a mobile genetic element such as an integron, encoding antibiotic resistance determinants.

Antibody A glycoprotein present in body fluids that can bind specifically to an antigen. An immunoglobulin.

Anticodon The three-base sequence on tRNA that is complementary to the three-base codon of mRNA.

Antigen A substance recognized by the body as nonself and which elicits the immune response. An immunogen.

Antigen presentation A process wherein certain immune cells (i.e., macrophages, dendritic cells, and B cells) capture and process foreign proteins, display peptide fragments of these on their plasma membrane attached to major histocompatibility antigens, and stimulate T cells, which then mount an immune response.

Antigen-presenting cell (APC) Phagocytic cells that process antigens and present them to T cells in the immune response.

Antigenic determinants Regions of an antigen that elicit an immune response. Also termed *epitopes*.

Antigenic drift (shift) Antigenic variation resulting from alterations in antigen genes.

Antigenic variation See antigenic drift.

Antimetabolite Compound that inhibits metabolism in microorganisms. General structure resembles a natural metabolite and competes for active sites on enzymes.

Antimicrobial agent Anything that inhibits or kills microbial cells.

Antiparallel The strands in double stranded DNA are oriented in opposing directions—one $3' \rightarrow 5'$ the other $5' \rightarrow 3'$.

Antiseptic A chemical generally applied externally that destroys microbes without serious harm to the tissue.

Antiserum A serum that contains specific antibodies.

Anti-sigma factors Proteins that interfere with transcription by sequestering sigma factors.

Antitermination The ability of RNA polymerase to overcome premature termination caused by a hairpin structure called a terminator.

Antiterminator A secondary RNA structure that disrupts the structure formed by a terminator.

Antitoxin An antibody specific for a toxic substance.

Apical complex Set of organelles for host cell invasion found in Apicomplexa.

Apoptosis Elimination of aberrant cells by programmed cell death.

Apothecium Disk-shaped ascocarp.

Aquifex A deep-branching phylum of the *Bacteria*.

ARC Aids Related Complex is a series of symptoms associated with an HIV infection and may lead to an active case of AIDS.

Archaea One of the three domains of living organisms; formerly termed *Archaebacteria*.

Artificial classifications Taxonomies that not based on evolutionary principles.

Artificial competence The ability of bacteria to take up exogenous DNA following treatment with certain chemicals.

Ascocarp In Ascomycetes, hyphae assembled into a single structure that holds the crozier for sexual reproduction.

Ascogonia One of two types of hyphae in conjugating Ascomycetes. Ascogonia are the recipient of haploid nuclei from antheridia.

Ascospores Haploid spores formed inside the ascus of Ascomycetes.

Ascus The terminal hyphae in the ascocarp where meiosis occurs.

Aseptic technique The manipulation of microorganisms such that contamination by undesirable organisms is prevented.

Assimilation Uptake of nutrients and conversion to cellular components.

Assimilatory sulfate reduction Uptake of sulfate to fulfill nutrient requirements of an organism.

Astrobiology The study of early life on Earth and the exploration of life in the Universe.

ATP-Binding Cassette (ABC) transporters Membrane protein complexes that use ATP to transport substrates without alteration of the compound transported.

attB A specific sequence on DNA which serves as a site for integration of bacteriophages or plasmids through recombination with their *attP* sites.

Attenuation (transcriptional) Temporary pause by RNA polymerase during transcription, usually in response to specific signals.

Attenuation (of a microbe) Weakening of a pathogen to be employed in immunization.

Autecology The study of a species in its natural environment.

Autochthonous Indigenous microbial populations.

Autoclave An instrument that uses heat and steam under pressure to destroy all microbes present.

Autoimmune disease A state in which antibodies are formed against self-antigens.

Autoinducers Signaling molecules produced by bacteria for the purpose of coordinating gene expression during quorum sensing.

Autolysin Enzymes in bacteria that form breaks in peptidoglycan, permitting incorporation of newly formed wall units during growth.

Autolysis Process in which peptidoglycan is disrupted, causing cell lysis mediated by autolysins.

Autophosphorylation Modification of an amino acid on an enzyme by self-phosphorylation; often carried out by sensor kinases of the two-component signal transduction systems.

Autotroph An organism that can utilize CO_2 as sole source of carbon.

Auxotroph A mutant that has lost the ability to synthesize a metabolite normally synthesized by the microbial strain.

Avirulent An infectious disease agent that does not have virulence and thus does not cause pathology.

Axenic culture A culture free of contaminating organisms. A pure culture.

Axial filament An endoflagellum produced by members of the phylum *Spirochaetes*.

Axoneme Arrangement of microtubules in the cilium.

Bacillary dysentery A bacterial disease caused by enteric bacteria, esp. *Shigella dysenteriae*.

Bacillus (bacillary) An oblong bacterium; also called a rod; a name for a bacterial genus.

Bacitracin An antibiotic that prevents recycling of the undecaprenol pyrophosphate carrier during peptidoglycan synthesis.

Bacteremia Presence of bacteria in the bloodstream; generally transient.

Bacteria One of the three domains, along with *Archaea* and *Eukarya*.

Bacterial enumeration anomaly The inability to cultivate most of the microorganisms from most natural habitats.

Bactericide Agent that kills bacteria. Mercuric chloride is a bactericide.

Bacteriochlorophyll Chlorophyll pigments of anoxygenic photosynthetic bacteria.

Bacteriocin Protein produced by a bacterial strain that kills related strains.

Bacteriocinogenic plasmids Plasmids carrying genes for bacteriocins.

Bacteriolytic agent An agent that causes dissolution of the bacterial membrane.

Bacteriophage Virus that infects *Bacteria* or *Archaea*.

Bacteriorhodopsin A carotenoid in the purple membranes of halophiles that resembles rhodopsin in the visual cells of animals. It is involved in light-driven ATP synthesis.

Bacteriostatic Inhibits bacterial reproduction or growth but does not kill.

Bacterivores Species that consume bacteria.

Bacteroid An osmotically fragile bacterium that lives within protected areas such as plant root nodules or the intestinal tract.

Bacteroidetes A major phylum of the *Bacteria*.

Bactoprenol A lipid that carries the saccharide units of peptidoglycan or lipopolysaccharide core across the cytoplasmic membrane. Also called undecaprenol.

Baeocyte A reproductive cell produced by some cyanobacteria.

Balanced growth condition When all metabolic intermediates required for growth are available at the appropriate level.

Banded-iron formations (BIFs) Bands of ferric iron deposited in geologic strata during the era when oxygenic photosynthesis evolved.

Barophiles Organisms that thrive at high hydrostatic pressure. Also termed piezophiles.

Barotolerant Can grow under high pressure but generally grows better at normal pressure.

Barrier defense The first level of host defense against infection; includes chemical, mechanical, microbial, and physical components.

Basal body A structure equivalent to the centriole, found at the base of cilia.

Basal structure The component of the flagellum in the cell envelope. Anchors the flagellum to the various layers of the cell wall and includes the flagellar motor.

Base composition Relative percent of the total cellular DNA that is guanine-cytosine (G + C). Remainder is adenine-thymine (A + T).

Base-pairing Hydrogen bond formed between specific bases of DNA and RNA: A with T (or U) and G with C.

Basic dyes Positively charged staining compounds that are most commonly used to stain microorganisms.

Batch culture Growth of a microorganism in a closed vessel, on a suitable medium, at an appropriate temperature, and for a selected time.

B cell (lymphocyte) An antibody-bearing cell that can differentiate during the immune response to secrete antibody (plasma cell) or to form a memory cell.

B-cell receptor Membrane-bound surface immunoglobulin (antibody) on a B lymphocyte; allows it to recognize and react with antigens.

Bdelloplast The spheroidal cell resulting from an infection by *Bdellovibrio*.

Beta (β) A subunit of RNA polymerase.

Beta-hemolysis The lysis of red blood cells due to specific bacterial toxins that produce transparent zones around colonies on blood agar plates.

Beta-hydroxybutyrate A storage polymer produced by many bacteria.

Binary transverse fission Simple symmetrical division of a bacterium to yield two cells.

Binomial system A system for naming organisms with a genus name and a specific epithet.

Bioaugmentation The process of amending contaminated sites with bacteria that are capable of degrading toxic compounds.

Biochemical oxygen demand (BOD) A test in which the oxygen required by microorganisms to catabolize organic matter present in a water sample is measured.

Bioconversion (biotransformation) Biologically induced modification of a substrate to generate a more useful product. The substrate converted is not utilized for growth.

Biodegradation Destruction of organic compounds by microorganisms.

Biodiversity The extent of species richness in habitats.

Biofilms A layer produced by microorganisms on solids in environments that contain fluids.

Biogeochemical cycles The processes whereby chemical materials such as carbon, nitrogen, and sulfur are recycled by biological and geological processes.

Biogeography The distribution of organisms in the environment.

Biological insecticides Organic materials that are produced by organisms and are toxic to insects.

Biological tertiary treatment The processes whereby nitrogen and phosphorus are removed in wastewater treatment.

Bioluminescence Production of light by living organisms.

Biomagnification Accumulation of substances, generally deleterious, by consumer organisms.

Biomarker A chemical substance produced by an organism that serves as a signature for the presence or activity of that organism or group of organisms.

Biomass Total living cellular material in an environment.

Bioremediation Removal of toxic or undesirable environmental contaminants by microorganisms.

Biosynthesis (anabolism) Synthesis of macromolecules from monomers during cell growth.

Biotechnology Use of living organisms to generate useful industrial products. Usually involves genetic manipulation.

Biotic Produced or caused by organisms.

Biotransformation Use of microbes or enzymes to convert a substrate to a modified form.

Biovars A term used to identify a particular subspecies or organism; a biological variety.

Botulism A food-borne disease caused by ingestion of toxin and/or *Clostridium botulinum*.

Bridging immune system A group of defense processes that have some characteristics of both the adaptive and innate immune systems.

Broad-spectrum antibiotic Antimicrobial that is effective against both gram-positive and gram-negative bacteria.

Brucellosis A disease caused by *Brucella* species.

Bt protein A protein produced by *Bacillus thuringiensis* that is toxic to insects.

Bubo Swelling of lymph nodes following infection with *Yersinia pestis*.

Bubonic plague A disease caused by *Yersinia pestis* and characterized by the appearance of buboes; also called the Black Death.

Budding Asexual reproduction in which progeny arise from protuberances on the surface of the parent cell.

Bulking A microbial process that prevents settling in secondary clarifiers and is therefore deleterious to sludge removal.

Butanediol fermentation A specific type of low-acid fermentation typical of some enterics.

Butyric acid bacteria Bacteria that produce butyric acid in sugar fermentation such as some *Clostridium* species.

Cachexia A response to tumor necrosis factor alpha that results in wasting or severe weight loss.

Calorie Amount of heat energy required to raise a gram of water from $14.5°C$ to $15.5°C$.

Calvin cycle The major pathway for CO_2 fixation during photosynthesis in plants, algae, cyanobacteria, and many photosynthetic bacteria. Some chemolithotrophs also utilize this pathway. Sometimes referred to as the Calvin–Benson cycle.

Capsid The protein coat that surrounds viral nucleic acid.

Capsomers Protein subunits that together form an icosahedral capsid.

Capsule An organized layer of biosynthetic origin, often a polysaccharide, that surrounds the outer wall or envelope of the bacterium.

Carboxysomes Inclusion bodies present in autotrophs composed of ribulose-1,5-bisphosphate carboxylase (Rubisco). This enzyme is involved in CO_2 fixation via the Calvin cycle.

Carcinogen An agent, generally a mutagen, that causes cancer.

Carotenoid Red to yellow pigments found in many bacteria. Can serve as accessory pigments in photosynthetic organisms. Also a pigment in many nonphotosynthetic microorganisms.

Carrier An infected individual that transmits disease to others. Usually a carrier has a subclinical infection.

Catabolism Reactions involved in reduction of larger molecules to smaller molecules, generally with the release of energy.

Catabolite repression Curtailment of the synthesis of selected enzymes by the availability of glucose or other metabolites as substrate.

Catalase An enzyme that breaks down hydrogen peroxide to water and oxygen.

Catalyst A compound that increases the reaction rate without itself being altered.

Cationic Having a positive charge.

CD3 A set of glycoproteins on the surface of T cells that provide the signaling component of the T cell receptor for antigen.

CD4 Specific antigen on selected cells such as T cells. $CD4^+$ cells are host for HIV.

CD8 A specific antigen on selected T cells that often have the property of being cytotoxic (kill other cells).

CD28 A costimulatory surface molecule of T cells that must interact with its ligand in order for naïve T cells to be activated by antigen.

cDNA A DNA copied from an RNA template using the enzyme reverse transcriptase.

Cell Basic unit of living matter.

Cell cycle Sequence involved in growth/division of a cell.

Cell envelope The layers that lie outside the cell membrane of microorganisms.

Cell growth Increase in cell size in absence of division.

Cell-mediated immunity T cells and activated macrophages are involved in clearing the host of infection.

Cell sorters Flow cytometers used to count and separate cells.

Cell theory of life The concept that all organisms consists of subunits termed cells.

Cell wall A tough structure surrounding the bacterial cytoplasmic membrane that gives the cell shape and protection.

Cellular differentiation The process whereby cells undergo morphological changes or morphogenesis.

Cellular immunity See cell-mediated immunity.

Centers for Disease Control and Prevention (CDC) A division of the U.S. Public Health Service involved in tracking diseases, providing information on diseases, and advising government agencies on disease prevention.

Centric diatoms Diatoms with radial symmetry.

Centromere Short repeated DNA sequences in eukaryotic chromosomes that bind kinetochore proteins.

Chancre An early lesion that occurs in the primary stages of syphilis.

Chaperones Specific proteins that are involved in the folding and assembly of other proteins.

Chasmolithic Organisms that live in the cracks of rocks.

Chemical mutagens Compounds that modify bases on DNA, resulting in mutations.

Chemical oxygen demand (COD) Amount of chemical oxidant necessary to oxidize organics in water to CO_2.

Chemiosmosis Development of a chemical potential across a cytoplasmic membrane that can drive ATP synthesis. Protons are driven outward by electron carriers or light.

Chemoautotrophic Organisms that obtain their energy from reduced chemical compounds and are capable of synthesizing organic material from carbon dioxide.

Chemoheterotrophic See chemoorganotrophic, also heterotrophic.

Chemolithotroph (chemoautotroph) A microorganism that obtains its energy by oxidation of inorganic compounds and utilizes CO_2 as its carbon source.

Chemoorganotroph (heterotroph) A microorganism that obtains its energy by oxidation of organic compounds and utilizes carbonaceous substrates other than CO_2 for growth.

Chemoreceptors Proteins in the cytoplasmic membrane or periplasm that bind chemicals involved in chemotaxis.

Chemostat A continuous culture apparatus that feeds growth medium, with one substrate at limiting concentration, into a culture vessel and permits removal of cells at a rate that maintains steady state growth.

Chemotaxis Movement of a microorganism toward a chemical attractant or away from a repellent chemical.

Chemotherapy Treatment of disease with chemicals that destroy the agent without serious harm to the host.

Chi site A site on DNA nicked by RecB, RecC, and RecD nucleases during homologous recombination.

Chlamydiae One of the major phyla of *Bacteria*.

Chlorination The process in which chlorine is used to disinfect drinking water and water in swimming pools.

Chlorobi A major phylum of the *Bacteria*.

Chloroflexi A major phylum of the *Bacteria*.

Chlorophyll A tetrapyrrole with a molecule of magnesium in the center that captures light energy during photosynthesis.

Chloroplast The chlorophyll-containing organelle in photosynthetic eukaryotes.

Chlorosome An oblong protein-bound structure in green sulfur bacteria that contains the light-gathering pigments. It is located on the inner surface of the cytoplasmic membrane.

Cholera A water-borne enteric disease caused by *Vibrio cholerae*.

Cholera toxin A secreted toxin produced by *Vibrio cholerae*, responsible for the major symptom (diarrhea) of cholera.

Chromatic aberration Lens defects caused by the differences in wavelengths of white light.

Chromophore The pigmented or colored portion of a dye.

Chromosome The site of cellular DNA. The *Archaea* and *Bacteria* generally have one circular chromosome. One microorganism (*Borrelia burgdorferi*) is known that has a linear chromosome. Eukarya are multichromosomal.

Chromosome mobilizing ability The ability of integrated plasmids to mediate transfer of chromosomal genes.

Chronic infection Long-term infection.

Cilia Projections on the surface of cells that move substances over the cell, or in some cases, move the cell itself.

Classification Placing organisms in groups based on phylogenetic relatedness.

Cleistothecium Spherical ascocarp.

Climax state The stable end-state of a biological succession.

Clonal selection Theory that clones of specific B or T cells are stimulated to reproduce when an appropriate antigen binds to their surface.

Clone Genetically identical progeny of a single parent.

Cloning Generation of multiple copies of the same organism. In genetic engineering, it refers to the capture and propagation of a copy of a gene.

Cloning vector A bit of replicative DNA that can transport inserted foreign DNA into a recipient cell.

Clonotypic population A group of cells that arise from a single progenitor.

Coagulase An enzyme capable of inducing the coagualtion cascade in blood.

Coccus (coccoid) A spherical bacterium.

Codon A sequence of three bases (purines or pyrimidines) in messenger RNA that codes for a specific amino acid.

Coenocytic A filamentous organism that contains nuclei that are not separated from one another by cell septa. Results from repeated nuclear division without cell division.

Coevolution Parallel changes in ecologically interdependent species to maintain an interaction.

Colicins Plasmid-encoded proteins produced by enterics that harm other enteric bacteria.

Coliform bacteria Gram-negative facultative aerobic bacteria that ferment lactose with formation of gas within 48 hours at 35°C.

Colonize To take up residence in a host.

Colony A visible assemblage of microorganisms growing on a solid surface. Generally from reproduction of a single cell.

Colony forming unit (CFU) Single cell that can form a colony on a solid medium.

Colony hybridization Detection of a specific DNA sequence in a bacterial colony by DNA hybridization with a labeled probe.

Cometabolism Transformation of a nongrowth substrate by microorganisms grown or growing on a utilizable substrate.

Commensalism A symbiotic association where one organism benefits and the other(s) is unaffected.

Common-source epidemic One in which all victims are infected from a single source. Food poisoning is an example.

Community A mixed population of microorganisms in a natural habitat or microcosm.

Comparative biochemistry Examination of the universality of equivalent biochemical reactions among all species.

Comparative genomics Elucidation of structure, function, and evolution of genes by comparison of two or more genome sequences.

Compatible solute Organic or inorganic compound that increases the internal osmotic pressure inside a bacterium to that outside.

Competent A bacterial cell that can take up DNA fragments and be transformed.

Competitive inhibitor A chemical analogue that replaces a natural substrate and competes for active sites.

Complement A series of proteins in the blood that act sequentially (cascade) and are involved in the removal of antigen/antibody complexes and microbes by tagging them for phagocytosis (opsonization). Produces pores in cells and microbes, thus killing or inactivating them and attracting phagocytic cells to sites of inflammation.

Complement receptor A cell surface component that binds a fragment of complement.

Completed test The final tests used to identify a fecal coliform bacterium.

Complex medium (undefined) A culture medium in which the chemical composition of the ingredients is not precisely determined.

Composite transposon A transposon consisting of one or more genes (usually antibiotic resistance genes) flanked by two insertion sequences.

Composting Process of degrading plant material, typically in bins.

Compound light microscope A microscope that uses two lens systems, the objective and the ocular, for viewing specimens.

Concatemer Viral genomic material in which individual genomes are linked end-to-end. The linear structure is split to individual genomes upon viral assembly.

Condenser lens A lens used to direct intense light onto the object being viewed.

Confirmed test The second test in the identification of coliform bacteria from natural samples.

Conformation Three-dimensional structure of a folded protein.

Congenital Disease contracted maternally during gestation or birth.

Conidia Hyphae that form asexual dispersal spores.

Conidiospores Aerial spores produced by actinobacteria and fungi; used for dispersal.

Conjugal (sex) pili Organelles used by the donor bacteria to recognize recipients during conjugative transfer of DNA.

Conjugation Transfer of DNA from donor to recipient by direct cell to cell contact.

Conjugative transposon A transmissible genetic element that can excise from the chromosome of a donor cell and transfer to a recipient, where it inserts into a new site.

Consortium An assembly of several bacteria in which all benefit to some degree from the association.

Constitutive enzyme An enzyme not subject to regulation; always expressed during growth.

Consumer An organism that ingests other organisms as a source of nutrients and energy.

Contig A contiguous sequence of DNA generated by assembly of overlapping sequences.

Continuous culture A system where the addition of nutrients and removal of wastes permits long-term active growth.

Contractile vacuole Vacuole system found in many eukaryotes responsible for excretion of excess water and ions by osmoregulation.

Convergent evolution Development of similar characteristics by unrelated organisms because they adapt similarly to an environmental challenge.

Cord factor A component of the cell wall of virulent *M. tuberculosis* consisting of molecules of mycolic acid that are attached to trehalose.

Corepressor A low-molecular-weight compound that functions in enzyme repression.

Cortex Dense area between the endospore coat and core in endospore-forming bacteria.

Coryneform bacteria Bacteria such as the genus *Corynebacterium* that divide by a snapping division process that produces unusual shapes.

Co-translational protein translocation A type of secretion mechanism where the signal peptide of a protein interacts with the Sec secretion apparatus prior to completion of its synthesis by the ribosome.

Covalent bond A chemical bond resulting from sharing of electrons.

Crateriform structures Distinctive pits produced on the outer surface of members of the *Planctomycetes*.

Cristae In-folding membranes in mitochondria that hold some respiratory enzymes.

Critical point drying A process in which water is removed from a specimen as a gas.

Crossbands "Septum-like" structures in caulobacter stalks that are produced during cell division.

Cross-resistance The development of resistance to one antimicrobial agent effects resistance to another.

Crossing-over In meiosis, where adjacent DNA strands are exchanged.

Crown gall A plant tumor produced by *Agrobacterium tumefaciens*.

Crozier Structure that forms at terminal hyphae in Ascomycetes where conjugation occurs.

Crustose Form of a lichen that is compact and grows close to a substrate.

Culture A strain growing on a laboratory medium.

Culture collection A place that holds microbial cultures in a living state so that they can be used by others for comparative purposes.

Culture medium A liquid or solid nutrient on which microorganisms can be grown.

Curing Loss of a plasmid in one of the daughter cells during cell division.

Cyanobacteria The *Bacteria* that can grow photosynthetically via oxygenic photophosphorylation. Morphologically diverse.

Cyanophycin A non-protein polymer consisting of the two amino acids aspartate and arginine.

Cyanosis Blueish appearance of the skin caused by lack of oxygen in the blood.

Cycles of matter Natural processes whereby major constituents of living cells (C, N, S) are recycled.

Cyclic photophosphorylation Cyclical movement of electrons in photosystem I.

Cycloserine An antibiotic that inhibits addition of alanines to UDP-*N*-acetyl muramic acid during early stages of peptidoglycan synthesis.

Cyst Cell temporarily protected inside a cell wall, usually during dormancy.

Cyst (parasites) A sac like structure that forms around parasites.

Cyst (pathology) An abnormal membranous sac containing a gaseous, liquid, or semisolid substance.

Cysticercoid The larval stage of certain tapeworms.

Cytochromes Heme proteins involved in electron transport.

Cytokines A group of proteins produced by leukocytes that direct the functions of cells with specific receptors for them.

Cytokinesis Division of the cytoplasm of a dividing cell.

Cytolethal distending toxin A DNA damaging toxin produced by *Campylobacter jejuni.*

Cytoplasm The liquid contents of a cell. Surrounded by a cytoplasmic membrane.

Cytoplasmic domains The portion of a membrane-bound molecule that is within the cell cytoplasm.

Cytoplasmic inclusion An intracellular storage granule in a bacterium.

Cytoplasmic membrane or **cell membrane** The protein and lipid structure that surrounds the cytoplasm of cell.

Cytoproct A region of the eukaryotic cell cortex specialized in the excretion of undigested food vacuole contents.

Cytoskeleton Network of microtubules and microfilaments that gives a eukaryotic cell its shape and the ability to arrange organelles and move.

Cytostome The oral structure, or specialized feeding area found in many protists.

Cytotoxic A frequent property of CD8 T lymphocytes which allows these cells to kill virus-infected host cells.

Cytotoxin A toxic material that destroys cells. Some pathogens produce these compounds.

Cytotrophs Species that consume eukaryotic cells.

Dark-field microscopy A process for indirect illumination such that the specimen is illuminated against a dark background.

Dark reactions The Calvin cycle of carbon dioxide fixation that accompanies the light reactions of photosynthesis.

Death phase The phase in batch culture in which the viable population is in decline.

Decomposition The degradation or breakup of complex organic material to constituent parts culminating in the production of inorganic materials.

Defensins Host peptides that lyse bacteria and some protozoa.

Defined medium One in which the composition and quantity of all components are known.

Delayed type hypersensitivities A type of host immune response characterized by the presence of T lymphocytes and macrophages and which takes 48 hours to develop. Contact dermatitis is an example.

Deletion Absence of a DNA segment in a mutant, relative to the same region in the chromosome of the wild-type parent.

Denaturation Change in a protein or other macromolecule that destroys activity.

Dendritic cell An antigen-presenting cell; present in lymph nodes and spleen.

Denitrification Reduction of oxidized forms of nitrogen to N_2 gas via anaerobic respiration.

Deoxyribonucleic acid (DNA) A nucleotide polymer composed of deoxyribonucleotides joined by phosphodiester bonds. The genome of an organism is composed of DNA.

Dermatomycosis Fungal skin infections.

Detritus Particulate material that is nonliving.

Dextrans Polymers of glucose or dextrose.

Diamino acid An amino acid such as lysine that has more than one amino group.

Diaminopimelic acid An example of a diamino acid found in the peptidoglycan of some gram-negative bacteria.

Diapedesis Movement between cells, e.g., the process by which a neutrophil migrates from the bloodstream into the tissues.

Diauxie Biphasic growth that can occur when two substrates are available in a culture medium.

Differential media Culture media that permit growth of one bacterial type and inhibit others or permit a microorganism to demonstrate specific biological properties.

Differential stains Stains such as the Gram stain that distinguish one type of bacterium from another.

Differentiation Changes in the structure of a microorganism during the growth cycle.

Dimer Compound formed by joining of two monomers.

Dimorphic life cycle A life cycle in which two different cell types are found.

Diphtheria An infection by *Corynibacterium diphteriae* expressing diphtheria toxin.

Diphteria toxin A protein toxin produced by *C. diphteriae* capable of killing mammalian cells by inhibiting their protein synthesis.

Dipicolinic acid An acid uniquely produced by endospore-forming bacteria.

Diploid A state in eukaryotes in which two copies of each chromosome are found.

Diploid nucleus A nucleus that holds two complete sets of chromosomes.

Disease Disturbance of normal structural or functional capacity of an organism.

Disinfect A way to destroy all microorganisms.

Disproportionation A type of metabolism which requires that part of the substrate is metabolized by one pathway and part is metabolized by another pathway. For instance, a part of the substrate may be oxidized so that the remainder may be reduced.

Dissimilatory sulfate reduction The use of sulfate in anaerobic respiration.

DNA library (gene library) Cloned DNA fragments that represent the genes in the entire genome of an organism.

DNA ligase An enzyme catalyzing the joining of ends of DNA strands.

DNA microarrays A collection of genes or segments of genes, attached to solid support, usually chemically derivatized glass slides.

DNA Pol I, II, III Forms of DNA polymerases used in replicating the bacterial chromosome.

DNA vaccine Immunization by injection of DNA encoding a protective antigen into host cells, resulting in its transcription and induction of an immune response by the transcribed protein.

Domain The highest level of classification of all life based on rRNA analysis. The three domains are *Bacteria, Archaea,* and *Eukarya.* A domain is also used to describe an area of a macromolecule.

Donor A cell that is the source of plasmid during conjugative DNA transfer.

Double-crossover Two homologous recombinations at proximal sites, resulting in the replacement of one DNA segment by another.

Doubling time Time required for a population to double in number.

Duodenum A region of the small intestine.

D value Decimal reduction time. Time required to reduce a population to one-tenth the original.

Early message Messenger RNA formed immediately after viral infection, which is translated to catalytic proteins that disrupt host cell function.

Ecosystem Total organismic community and associated abiotic components of a selected environment.

Ecovars Ecological varieties of species.

Ectomycorrhizae Association between fungi and root tip, in which the fungi form a sheath about the root.

Ectosymbionts Symbionts that reside on the outer surface of organisms.

Edema Accumulated fluid that results in swelling of a tissue.

Electron acceptor The component that accepts electrons during an oxidation-reduction reaction.

Electron donor The component that donates electrons during an oxidation-reduction reaction.

Electron micrograph A photograph taken with an electron microscope.

Electron transport chain Series of electron acceptors and donors that transfer electrons from an electron carrier, such as NADH, to a terminal acceptor, such as O_2. Also functions during photophosphorylation.

Electrophoresis Separation of charged molecules, such as protein or DNA, in an electrical field.

Electroporation Use of an electrical pulse to alter the cytoplasmic membrane, thus promoting uptake of DNA fragments.

Elemental sulfur granules Small spherical deposits of elementary sulfur (S^0) produced by some heterotrophic and photosynthetic bacteria.

Elongation factors EF-G, EF-Ts and EF-Tu Factors responsible for the movement of ribosomes along the mRNA following initiation of protein synthesis.

Embden-Meyerhof pathway (glycolysis) A pathway that catabolizes glucose to two pyruvic acid molecules and generates two molecules of ATP.

Empty magnification Magnification without commensurate increase in resolution.

Enantiomers Molecules whose chemical structures are mirror images of one another.

Encystment The process of forming a cyst.

Endemic disease One that is present at a constant low level in a population.

Endemic species Species, such as kangaroos, that are found in only one geographic area.

Endemic typhus See murine typhus.

Endergonic reaction One that requires energy input to proceed.

Endocarditis Infection of heart valves.

Endocytosis Uptake of a particle, such as a virus, by enclosing it via membrane extension and pinching off of the vesicle formed.

Endoflagella or **periplasmic flagella** Produced by members of the *Spirochaetes*; lie outside the cell membrane but inside the cell wall.

Endogenous pathway of antigen presentation Presentation of antigens by antigen-presenting cells, with the peptides being generated by digesting proteins in the cytoplasm of the host cells. These are normally complexed with MHC class II molecules on the surface of the host cell. They are often derived from pathogens that are living inside the host cell.

Endogenous pyrogen A protein that induces fever in a host.

Endomycorrhizae A fungus–plant root association in which the fungus penetrates into the root cells.

Endophyte A microorganism, often a cyanobacterium, that lives within a plant.

Endospore A heat-resistant dormant cell that forms within the cell of selected bacteria.

Endosymbiosis Growth of a microorganism within another organism that is generally beneficial to both.

Endosymbiotic theory The theory that the mitochondrion, chloroplast, and other organelles arose through an endosymbiotic association of bacteria with eukaryotic ancestors.

Endotoxemia A host response to endotoxin which results in intravascular coagulation of blood, edema, and sometimes death.

Endotoxin The lipopolysaccharide portion of the outer envelope of gram-negative bacteria released by cell lysis or during growth; toxic to animal hosts.

Energy Capacity to do work.

Enrichment culture Addition of a selected substrate to a culture medium to isolate an organism that can grow on that substrate. Physical conditions may also be adjusted to obtain a specific type of organism.

Enteric Intestinal; often used to describe microorganisms associated with the intestinal tract.

Enterocolitis Inflammation of both the small and large intestine.

Enterohaemorrhagic *E. coli* A strain of shiga-like toxin producing *E.coli*.

Enteropathogenic *E. coli* A strain of *E. coli* capable of destruction of intestinal lining.

Enterotoxigenic Enterotoxin-producing strain of *E. coli* causing severe diarrhea.

Enterotoxin A toxin that adversely affects the intestinal tract.

Entner-Doudoroff A nonglycolytic pathway of glucose catabolism that produces pyruvate and glyceraldehyde-3-phosphate.

Envelope The outer structure in gram-negative bacteria of the membranous layer that surrounds the capsid of some viruses.

Enzyme A highly specific protein catalyst.

Enzyme-linked immunosorbent assay (ELISA) A technique to detect specific antigens or antibodies, using enzymes as the detector system.

Eosinophils A type of white blood cell that is in the myeloid lineage and has cytoplasmic granules.

EPA superfund sites Locations in the USA that have been designated by the Environmental Protection Agency as sites that are highly contaminated with toxic materials.

Epidemic A marked increase in the number of individuals affected by an infectious disease during a given period of time.

Epidemic typhus An infection caused by *Ricketsia prowazekii*.

Epidemiology The study of prevalence, incidence, and transmission of infectious diseases.

Epilimnion Aerobic, relatively warm layer of water in a stratified lake; lies above the thermocline.

Epilithic Organism that grows on the surface of a rock.

Episome A plasmid that can integrate into the host genome or function separately.

Epitope A specific area on an antigen that elicits an antibody response.

Erythrogenic toxin A toxin produced by *Streptococcus pyogenes* responsible for the symptoms (rash) of scarlet fever.

Etiologic agent The specific cause of a disease.

Eukarya One of the three domains. Composed of organisms that have a membrane-bound nucleus and division of DNA by mitosis.

Eutrophic An environment, usually aquatic, overly rich in nutrients.

Excystment The process of coming out of a cyst.

Exergonic A reaction that liberates energy.

Exoenzyme An enzyme that is secreted by the organism.

Exogenous pathway Presentation of killed foreign antigens by antigen-presenting cells. Normally these are degraded in the endosomal pathway and the resulting peptides complexed with MHC class II molecules on the surface of the host cell.

Exon Sequences in a split gene that code for messenger RNA.

Exotoxin A toxin released by cells, generally during growth.

Exponential growth Balanced growth in which each microbial cell is dividing during a fixed period of time.

Expression vector Type of plasmid used in recombinant DNA technology, where the expression of the cloned gene(s) is directed from a promoter on the vector.

Extracellular enzyme An enzyme excreted by a microorganism to cleave large molecules to a size transportable into the cell.

Extracellular polymeric substances (EPS) Polymeric materials that are produced outside the cell wall.

Extrachromosomal elements DNA found in bacteria that is not part of the chromosome, such as plasmids.

Extreme Employed with thermophiles or halophiles to indicate that they grow at the highest temperature or salt concentration of microorganisms presently described.

Extremophile Term employed with organisms that live under environmental extremes such as thermophiles or halophiles to indicate that they grow at the highest temperature or salt concentration of microorganisms presently described.

Extrusomes Exocytotic vesicles that contain material to be discharged outside of the cell membrane for defense or offense.

F prime (F′) The F plasmid carrying a segment of chromosomal DNA.

Facilitated diffusion Carrier mediated transport across a cytoplasmic membrane.

Facultative Term applied to aerobes and autotrophs to indicate that the organism can grow either aerobically/anaerobically or autotrophically/heterotrophically under appropriate conditions.

Fall turnover The period in the autumn in which the surface waters of a lake (the epimilinion) of thermally stratified lakes become colder and mix with the deeper waters of the lake hypolimnion.

Fastidious A microorganism with complex growth requirements, generally due to its inability to synthesize requisite monomers.

Fatty acids Organic acids such as acetate and butyrate.

Fc region A fragment of an immunoglobulin composed of the C terminal regions of the heavy chains, which conveys function, such as complement activation, to the antibody molecule. The Fab region combines with a specific antigen.

Fecal coliform Enteric bacteria that can grow at elevated temperature, about 44°C; also see coliform bacteria.

Feedback inhibition Condition in which the end product of a biosynthetic pathway can curtail activity of enzymes at earlier steps in the sequence of synthetic reactions leading to that end product.

Fermentation Production of ATP via catabolic sequences in which organic compounds serve as electron donor and organic intermediates as electron acceptor.

Fermenter An organism that carries out fermentation.

Fermentor A growth chamber used to culture microbes for commercial operations.

Fever An increase in body temperature above the norm.

Filament (flagellar) The external portion of the flagellum, attached to the hook. Consists of thousands of assembled flagellin proteins.

Filamentous A microorganism that, in normal growth, forms long strands.

Filamentous hemagglutin A surface adhesin of *Bordatella pertussis*.

Fimbriae Filamentous structures on bacteria that play a role in adherence or in formation of pellicles or masses of cells.

Fine filaments Fibrous cytoskeletal proteins that assemble into fine filaments.

Fixation (metabolism) The process whereby gases in the atmosphere are fixed into non-gaseous forms, e.g., nitrogen fixation and carbon dioxide fixation.

Fixative (microscopy) A chemical substance, such as formaldehyde or glutaraldehyde, that is used to preserve organisms for observation.

Fixed tissue macrophages Macrophages that were derived from blood monocytes, but which now reside in host tissues and organs permanently.

Flagella Structures involved in motility of bacteria.

Flagellin The principal protein of flagella.

Flavoproteins Riboflavin derivatives that are electron carriers.

Flocs Particulate material that can be suspended in aqueous environments; may or may not be organic.

Flow cytometers Instruments used for the separation of cells; see also cell sorter.

Fluid mosaic model The accepted structure of cytoplasmic membranes with the phospholipids forming a bilayer. Proteins are embedded within the membrane or associated peripherally at either surface.

Fluorescence Emission of light of distinct wavelength after activation with light of a different wavelength.

Fluorescent antibodies Specific antiserum that has been tagged with a fluorescent dye; used for the microscopic identification of organisms and localization of specific proteins.

Fluorogenic compound A chemical substance that becomes fluorescent when cleaved enzymatically.

Flux The rate at which a chemical substance is moved from one place to another.

Foliose Lichens that have a leaflike appearance.

Follicles (within lymphoid organs) Structures in the germinal centers of the spleen and lymph nodes.

Follicular dendritic cells Specialized antigen-presenting cells within follicles of germinal centers.

Fomites Inanimate objects that can harbor pathogens and transmit them to hosts.

Food chain The carbon cycle in which organisms at successively higher levels ingest smaller organisms at lower levels as their source of carbon and energy; photosynthesis is the ultimate source of carbon for consumers in the food chain.

Food infection An infection caused by ingestion of food contaminated with disease-causing pathogens.

Food poisoning Illness caused by toxins present in food due to growth of microorganisms on the foodstuff prior to ingestion.

Food web Interlinked food chains.

Fragmentation Reproduction in actinomycetes in which filaments break into individual cells.

Frameshift mutations Mutations that change the reading frame of a gene, usually by insertion or deletion of one or two (but not three) base pairs.

Free energy Total energy available to do work.

Fruiting body A reproductive structure in some bacteria, such as myxobacteria or actinomycetes. Fungi commonly produce fruiting structures.

Fruticose A lichen that is bush- or treelike.

Functional genomics A study of the biological function of genes and their products.

Fungi Heterotrophic eukaryotes that have rigid walls.

Fungicide An agent that kills fungi.

Fungistatic agent An agent that prevents fungal growth.

G + C The DNA of an organism has guanine paired with cytosine and adenine with thymine. G + C refers to the percent of the total DNA that is guanine-cytosine pairs.

Gametocyte A form of the protozoan *Plasmodium* in infected red blood cells.

Gas vesicles Organelles with protein membranes that fill with gas in aquatic bacteria; serve as flotation devices.

Gastroenteritis Inflammation of the stomach lining or intestines; often caused by food poisoning or infections.

Gene A segment of the genome that codes for a specific polypeptide, protein, tRNA, or rRNA.

Gene replacement Swapping of genes, usually involving homologous recombination.

Gene transfer The process by which genes are transferred from one species to another.

Generalized transduction Transfer of any part of a bacterial genome when packaged randomly in a phage capsid.

Generation time Time required for a population to double in number.

Genetic code Triplet nucleotide sequences that specify a specific amino acid in a protein chain.

Genetic engineering Modification of the genome of an organism.

Genetic map The precise sequence of genes in the genome.

Genome The complete genetic repertoire in a cell or virus.

Genomic island A segment of DNA acquired by horizontal gene transfer.

Genomic library A collection of bacteria carrying segments of DNA, representing the entire genome of an organism.

Genotype Heritable genetic information in a cell.

Genus A grouping of organisms that are closely related phylogenetically.

Geosmin Organic molecules produced by actinomycetes and some cyanobacteria that give soil or water a distinct aroma.

Germicide An agent that kills bacteria.

Germinal center A structure that forms within the spleen and lymph nodes after antigen stimulation and supports developing antibody responses.

Germination Loss of dormancy in an endospore.

Giardiasis Infection of the intestinal tract with *Giardia lamblia*.

Glider An organism that is motile without aid of flagella. Gliders move only when on a semisolid surface.

Gliding motility Motility that occurs without the aid of flagella. Gliding bacteria move on a semi-solid surface such as an agar plate.

Glomerulonephritis Inflammation of kidneys often following bacterial infections.

Glucan A glucose polymer, see dextran.

Glycocalyx Equivalent to capsule, the layer outside the cell envelope or wall.

Glycogen A branched polysaccharide composed of glucose; used as a storage granule.

Glycolysis Anaerobic conversion of glucose to pyruvate via the Embden-Meyerhof pathway, generating ATP.

Glycosidic bonds Covalent bonds between sugars in polysaccharides.

Glyoxylate cycle A modification of the tricarboxylic acid cycle in which isocitrate is cleaved to form succinate and glyoxylic acid. The latter condenses with acetate to form malate. Functional in organisms growing on two-carbon compounds, such as acetate.

Gonorrhea A sexually transmitted disease caused by *Neisseria gonorrhoeae* infection.

Gram stain A differential stain that separates bacteria on the basis of retention of the dye crystal violet. Those with an outer envelope are decolorized by an ethanol wash (gram-negative), whereas those without an outer envelope retain the dye (gram-positive).

Granuloma A nodule that forms in inflamed tissues. This is usually composed of myeloid and lymphoid cells.

Group translocation Transport where a molecule moves across the membrane and is chemically modified during the process.

Growth factor Low molecular weight compounds that must be added to growth media for selected organisms because they cannot synthesize them.

Growth rate The rate at which bacteria reproduce.

Growth yield A measure of cellular mass generated per ATP produced from a substrate.

Gyrase A topoisomerase capable of introducing supercoils into DNA.

Hairpin structure A double-stranded RNA structure formed by the complementary pairing of proximal inverted repeats.

Halophile An organism that requires very high salt (sodium chloride) for growth.

Halorespiration The process of using halogenated compounds as oxidants in respiration.

Halotolerant Able to grow in high salt concentrations, although not requiring it.

Haploid nucleus A nucleus with one complete set of chromosomes.

Hapten A molecule (generally low molecular weight) that cannot elicit an antibody response against self unless coupled with a larger molecule.

Haustoria Specialized tip of a hypha that makes contact with a host cell.

Heat shock proteins Proteins produced under stress, particularly heat, which protect the cell.

α-helix A helical structure present in DNA and some proteins formed by hydrogen bonding.

Helper T cell A T lymphocyte that cooperates with a B cell in initiating the antibody response.

Hemagglutination Coagulation of red blood cells.

Hemagglutinin Adherence molecule on the surface of influenza virus.

Hematopoietic stem cell (HSC) A cell that can mature or differentiate into any of the lymphoid or myeloid cells that make up the host immune system.

Hemolysins Bacterial toxins that can disrupt cytoplasmic membranes of cells. Generally hemolysins are assayed by use of red blood cells. Important in tissue invasion to release substrates for growth of pathogens.

Hemolysis Lysis of red blood cells.

Hemorrhagic fever A fever caused by a virus that leads to hemorrhage and shock. Death is frequently the result.

Herd immunity Immunity to a pathogen in the majority of potential hosts; results in the inability of the pathogen to cause disease.

Heterocyst Specialized cells in cyanobacteria that are sites of nitrogen fixation.

Heterofermentation Fermentation of sugars to a mixture of products.

Heteropolymers Polymers that contain two different types of chemical subunits.

Heterotroph An organism that utilizes preformed organic substrates as its major source of carbon.

Hexose monophosphate shunt (HMS) Metabolism of glucose through 6-phosphogluconate leading to 5-carbon sugars for synthesis of nucleotides and other cellular components.

Hfq A protein chaperone stabilizing the interaction between ncRNAs and target mRNAs.

High endothelial venules Vessels that support the migration of lymphocytes from blood into lymph nodes.

High-energy compound One that yields free energy on hydrolysis.

High-frequency recombinant (Hfr) strain A strain with a self-transmissible plasmid integrated in its chromosome.

Histones DNA binding proteins that package the DNA in eukaryotes.

Holdfast Material produced by some sessile microorganisms that aids in attachment to solid surfaces.

Holoenzyme A complex of RNA polymerase core enzyme with sigma factor.

Homofermentation (homolactic) Fermentation of sugars, mostly to lactic acid.

Homogeneous immersion The procedure in microscopy in which a uniformly high refractive index is provided between the specimen and the objective lens to provide highest resolution in image formation.

Homologous recombination Recombination between nearly identical DNA segments.

Hook Component at the base of the flagellar organelle, which transmits the rotary movement from the shaft to the filament.

Hopanoids Steroid-like compounds produced by cyanobacteria.

Horizontal evolution Gene transfer between distantly related organisms.

Horizontal (lateral) gene transfer Acquisition of new genes from other organisms.

Hormogonia Specialized cells produced by some cyanobacteria.

Host An organism on which a parasite or pathogen can grow.

Human immunodeficiency virus (HIV) The retrovirus that is the etiological agent of AIDS.

Humoral immunity An immune response involving antibodies.

Humus Organic component of soil.

Hybridization Annealing of one single-stranded DNA or RNA to its complementary copy of DNA or RNA.

Hybridoma Fusion of a cancer cell to a specific B cell to generate monoclonal antibodies.

Hydrogen bond A weak, but important, bond between a hydrogen atom and an electronegative atom, such as oxygen or nitrogen. Important in helix formation and other macromolecular interactions.

Hydrogenosomes Organelle evolutionarily derived from mitochondrion, involved in the anaerobic production of about 1 ATP from pyruvate.

Hydrogenotrophic An organism that utilizes or "eats" H_2 as a substrate for growth.

Hydrolysis Cleavage of a compound by addition of water.

Hydrophilic Affinity for, and solubility in, water.

Hydrophobic Lacking affinity for water; mostly insoluble in water.

Hypersensitivity Harmful, exaggerated immune responses, some of which are known as allergies.

Hyperthermophiles *Archaea* or *Bacteria* that have an optimal growth temperature above 80°C.

Hyphae Filamentous cellular extensions from some fungi.

Hyphal growth Filamentous form of growth in many fungi.

Hypolimnion The layer in a stratified lake that has a uniform cold temperature and a low level of oxygen.

Hypolithic Growth of organisms on surface of rocks.

Icosahedral A virus capsid having 20 equilateral triangular faces and 12 corners.

Identification The part of taxonomy that entails the determination of the species to which an unknown organism belongs.

Idiophase End of the exponential growth phase, when secondary metabolites are synthesized in certain organisms, such as antibiotic producing actinomycetes.

Ig class switching The process through which antibody-forming cells couple the antigen-binding parts of an immunoglobulin with different Fc pieces.

Ileum A region of the small intestine.

Immediate hypersensitivity Allergic response characterized by involvement of the antibody class IgE and release of histamine from mast cell granules. The response can result in death within 30 minutes.

Immobilized enzyme One attached to a solid support. May efficiently convert substrate to product.

Immune Resistant to infectious agents, generally due to adaptive immune responses.

Immune response Bodily response to the presence of an antigen; has three components: barrier defense, innate immunity, and adaptive immunity.

Immunization Eliciting of an immune response by introduction of a specific antigen into a host.

Immunodeficiency Inability to generate normal antibody or cellular responses to specific antigens or for phagocytic cells to migrate to areas of inflammation.

Immunogen (antigen) Substance that will elicit an immune response.

Immunoglobulin (Ig) Blood protein fraction that is composed of antibodies.

Immunopathology Host cell, organ, or tissue damage resulting from the host's own immune response.

Immunosuppressant An agent that decreases the immune response.

Incidence Number of cases of a disease in a subset of the general population.

Incompatability group Classification of plasmids based on the inability of closely related plasmids to coexist in the same cell.

Indicator organism A bacterial species that is employed to determine whether an environment is contaminated by human waste. An organism that does not survive for extended periods, but generally outlives enteric pathogens, would be an effective indicator.

Induced phagocytosis Phagocytosis that occurs when the pathogen stimulates the host cells to do so either by binding to receptors on the host cell or inserting bacterial components into the membrane of the host cell.

Inducers Molecules capable of facilitating transcription by their action through transcriptional regulators.

Inducible enzyme Enzyme that is synthesized in the presence of a specific substrate (inducer).

Infection Presence and growth of an organism within a host.

Infection thread A minute tunnel through the roots of a legume, by which rhizobial cells migrate to the host cells that will give rise to a nodule.

Inflammation A response to tissue injury that is generally localized but can become systemic in some cases. The local responses are characterized by pain, swelling, heat, and redness. The host's mechanism for attracting leukocytes, including phagocytic cells, to the site of tissue injury and infection.

Initiation codon The first codon in mRNA that defines the beginning (amino terminus) of a protein.

Initiation site Area of the genome where replication originates.

Innate immunity A group of defense processes that are rapidly deployed after infection and generally involve cells in the myeloid lineage, including granulocytes and phagocytes.

Inoculum Cell mass used to start a microbial culture.

Insertion Placing of a piece of DNA into another sequence of DNA.

Insertion sequence A small transposon that caries the genes for enzymes needed for its integration into a new site in a chromosome or a plasmid.

Integrase An enzyme that promotes recombination between two sequences, resulting in integration of a DNA element into a new site.

Integrase inhibitor A compound that inhibits the incorporation of a DNA sequence into a genome.

Integration Incorporation of a DNA sequence into the genome.

Integrins A family of adhesion molecules that promote stable interactions between cells and their extracellular matrix

Integron A genetic element consisting of an integrase gene and an adjacent *att* site that serves as a site for capturing antibiotic resistance–specifying gene cassettes by site-specific recombination.

Interference Resistance in a lysogenized bacterial cell to invasion by another virus.

Interferons (IFN) There are 3 types of interferons: gamma, alpha, and beta. They have antiviral properties and thus "interfere" with virus replication. Interferon-gamma also activates macrophages for increased anti-microbial activities.

Interspecies hydrogen transfer The process in which hydrogen gas produced by a fermentative bacterium is quickly utilized by a nearby methanogen.

Intimin An *E. coli* surface molecule that allows the bacteria to adhere to host gut cells by interacting with the bacteria-encoded translocated intimin receptor (TIR).

Intracellular pathogens/parasites Microorganisms, including bacteria and protozoa, that survive and replicate inside of host cells.

Intron Noncoding intervening sequences in a split gene. Does not appear in the ultimate RNA product.

In vitro Outside a living organism, such as a test-tube experiment.

In vivo Inside a living organism.

Invasiveness Innate ability of a pathogen to produce substances that aid in dissemination of the organism within the host.

Inverted repeats Two proximal segments of identical DNA sequence that are inverted relative to one another.

Ionizing radiation High-energy radiation that causes loss of electrons from atoms.

Ionophore A compound that disrupts cytoplasmic membranes, resulting in leakage of cytoplasm or transfer of materials (such as calcium) into the cell.

Iris diaphragm An annulus that can be opened and closed to control the amount of light that is focused on the specimen in a compound light microscope.

Iron bacteria Bacteria involved in the oxidation or reduction of iron.

Isomerization Rearrangement of the direction of DNA during recombination.

Isotopes Elements with an increased number of neutrons, but normal electron and proton complement.

Isotopic fractionation The process in which the lighter isotope of an element or compound is preferentially utilized by an organism.

Jejunum A region of the small intestine.
Joule A unit of energy.

Kerogen Organic compounds derived from organisms, but whose degradation over time has rendered the source compounds unidentifiable.

Kilobase pairs 1,000 base pairs in a fragment of DNA.

Kineties Ordered rows of cilia found in ciliates that coordinate the ciliary beat to control swimming.

Kinetochore An assemblage of proteins on the centromere of eukaryotic chromosomes where the mitotic spindle attaches.

Kirby-Bauer An antibiotic disk diffusion assay for testing susceptibility of a clinical isolate to antibiotics or chemotherapeutics.

Koch's postulates Expression of rules for proving the relationship between a specific microorganism and a disease.

Lactic acid bacteria Bacteria that produce lactic acid by fermentation.

Lactic acid fermentation Anaerobic degradation of sugars by glycolysis to produce lactic acid as a major end product.

Lactoperoxidase A peroxidase in milk.

Lag phase Period after inoculation of a culture in which there is no increase in population.

Lagging strand The strand of new DNA that is made during replication by joining short segments of DNA synthesized in the opposite direction as the movement of the replication fork.

Landfill Improved system for treating household food wastes.

Late message Messenger RNA produced sometime after viral infection coding for proteins that are involved in virion synthesis.

Latent virus One present in a host without causing detectable symptoms.

Leaching Release of minerals from ore by microbial action.

Leading strand A DNA strand synthesized during replication that is made in the same direction as the movement of the replication fork.

Lectin Surface protein of a plant cell where microorganisms, such as nitrogen-fixing bacteria, can attach.

Legionellosis (Legionnaires' disease) A respiratory infection by *Legionella pneumonia*.

Legume Plant that can develop nodules for nitrogen fixation.

Leprosy An infection by *Mycobacterium leprae* characterized by severe nerve damage.

Leptospirosis A disease caused by the spirochete genus *Leptospira*.

Lethal dose 50 (LD_{50}) Number of microorganisms or level of toxin that will kill 50% of a test population within a fixed time.

Leukocidin A microbial toxin that will destroy phagocytes.

Leukocyte A white blood cell.

Leukocyte adherence deficiency (LAD) A recessive genetic mutation that prevents leukocytes adhering to blood vessel walls, thus limiting their capacity to migrate from the blood into tissues and consequently reducing the efficacy of inflammatory defenses.

Leukocytosis A decrease in the number and concentration of white blood cells in the host.

Levan Polymer whose subunit is levulose or fructose.

Lichen A beneficial symbiotic association between a fungus and a cyanobacterium or alga.

Light reactions The systems in photosynthesis that covert sunlight into ATP.

Lipopolysaccharide (LPS) Complex structure containing fatty acids and sugars present on the outer envelope of gram-negative bacteria.

Lipoprotein A protein found in either the outer or cytoplasmic membrane of gram-negative bacteria, which is modified by attachment of fatty acids during secretion.

Log phase Exponential phase in growth curve of a culture.

Lophotrichous Tuft of flagella at one or both poles of a rod-shaped bacterium.

Lymph Clear, yellowish fluid that is the interstitial fluid that bathes tissue. Returns to the blood system by flowing through the lymphatic vessels and organs carrying lymphocytes and antigen-presenting cells. Eventually it is dumped into the blood system, although the cells within it may stay in lymphoid tissues along the way.

Lymph node An encapsulated organ that is largely composed of lymphocytes.

Lymphocyte A leukocyte that produces antibodies or is involved in cellular immune responses, including production of cytokines and killing of infected host cells.

Lymphogranuloma venereum A sexually transmitted infection of males by *Chlamydia trachomatis*.

Lymphokine Now generally referred to as a cytokine. Protein secreted by lymphocytes following activation. Mediators in the immune response that transmit signals from cell to cell.

Lyophilization Rapid dehydration of frozen material in a vacuum.

Lysis Physical disintegration of a cell.

Lysogenized cell An archaeal or bacterial cell bearing a prophage.

Lysosome An organelle in eukaryotic cells that contains digestive enzymes.

Lysozyme Enzyme that disrupts the $\beta(1\text{-}4)$ bond in peptidoglycan, causing cell disruption.

Lytic cycle Life cycle of a virus that effects lysis of the host cell.

M cells Cells that are part of the intestinal epithelia lining the gut lumen and through which some microbes pass to access the underlying tissue.

M protein A major surface protein of *S. pyogenes* with antiphagocytic properties.

Macromolecule A large molecule formed by polymerization of small molecules.

Macrophage Phagocytic cell present in blood, lymph, and tissue. These cells destroy pathogens, and some are involved in the immune response by presenting antigen and secreting cytokines.

Magnetosomes Magnetite (Fe_3O_4) particles that are present in certain bacteria. They are tiny magnets that align the bacterium in a magnetic field.

Magnetotactic Bacteria that produce magnetite and orient themselves in a magnetic field.

Maintenance energy Energy required by a cell to remain viable.

Major histocompatability complex (MHC) Cell surface antigens present in all individuals that are the unique marker of self. They are encoded by a family of genes.

Malaria A parasitic infection by members of genus *Plasmodium*.

Marginate To crawl out of the blood through the wall of a blood vessel.

Mast cell Cells that are found in skin tissues, sometimes near blood vessels, and secrete histamine and other active products in the inflammatory response and in hypersensitivity.

Mastigonemes Fine fibrous proteins extending from the cilium membrane, found in many protists.

Mat communities Natural aggregations of microbial species in marine and hot spring environments that form layers of different species.

Mating types Cell membrane proteins in eukaryotes that are responsible for cell–cell recognition and attachment for conjugation. Different alleles in the species provide a diversity of mating types.

Medium The mixture of nutrients employed in the growth of a microorganism.

Meiosis Division of a diploid cell to form two haploid daughter cells typically called the sperm and the egg.

Memory cell Differentiated, specific B cells formed during the immune response that can convert to plasma cells when activated by the presence of a specific antigen.

Meningitis An inflammation of the meninges following infection by any of a number of pathogenic bacteria and viruses.

Mesophile Microorganisms that grow optimally at temperatures between 18°C and 40°C.

Mesosome Cytoplasmic membrane invagination in bacteria.

Mesotrophic A lake with moderate levels of nutrients.

Messenger RNA (mRNA) Single-stranded RNA complementary to template DNA formed via transcription. mRNA generally carries information for one polypeptide and is translated by ribosomes.

Metabolism The sum of all biochemical events in a cell.

Metachromatic granules Dark structures observed in EM thin sections of cells containing polyphosphate.

Metagenomic A term that is used to describe the DNA of organisms in a community when subjected to molecular sequence analyses.

Metalimnion The intermediate layer in a stratified lake, also called the thermocline.

Methanogens Archaea that generate methane in anaerobic environments.

Methanotrophic bacteria Organisms that oxidize methane.

Methylotrophic An organism that utilizes or "eats" compounds that contain methyl carbons bound to O, N, or S as substrates for growth.

Microaerophile Microorganisms that require O_2 but at lower levels than atmospheric pressure.

Microbial ecology The study of the interactions among microorganisms and other organisms as well as the nonbiotic component of their habitat.

Microbial ecosystems Communities composed largely of microorganisms.

Microbiota Microscopic organisms collectively present in an environmental niche.

Microcyst Resting spherical structures formed by some *Cytophaga* and *Azotobacteria*.

Microelectrodes Fine probes used for measuring pH, oxygen concentration, and other factors in the microenvironment.

Microenvironment The small area in which individual microorganisms grow and live in their habitat.

Microfilaments Chains of actin arranged in two parallel helices. Constitutes an important part of the cytoskeleton.

Micropore A coated pit for endocytosis found in Apicomplexa and other Alveolata.

Microtubules Fibrous protein elements that give structure to eukaryotes.

Mineral scales Mineral deposited in an organic matrix in the form of a scale, and secreted from the Golgi apparatus to cover the cell membrane.

Mineral skeleton Deposition of minerals, often in an organic matrix, from the Golgi apparatus, that forms an elaborate system of spines to support the cell.

Mineralization Complete biodegradation of organic matter to CO_2 and other inorganic substances.

Minimum inhibitory concentration (MIC) Smallest amount of an antimicrobial agent that inhibits growth of microorganisms.

Mismatch repair system A system that corrects mismatches in base pairing created during copying of DNA strands.

Mitochondrion The organelle of eukarotic cells that is responsible for respiration and is descended from the *Alphaproteobacteria*.

Mitosis The division of nuclear material in a eukaryote, yielding two nuclei of identical chromosomal composition.

Mitosome A degenerate mitochondrion found in *Giardia*.

Mitotic spindle Microtubular structure that attaches to chromosomes during mitosis to separate the chromosomes.

Mixed acid fermentation One that generates mixed organic acids (lactic, acetic, etc.) as products.

Mixed flagellation An organism that produces two types of flagella.

Mixotroph A microorganism that assimilates organic carbon sources while using inorganic energy sources.

Modification enzymes Enzymes that modify (usually methylate) DNA sequences that

would otherwise be recognized by an organism's own restriction enzymes.

Mol % G + C The proportion of guanine and cytosine to total DNA bases in an organism's DNA, also termed GC ratio.

Molecular mimicry Production of molecules by pathogens that resemble components of the human tissue, thus minimizing recognition by the host immune system.

Molecular pattern receptors These are also known as pathogen pattern receptors and interact with surface molecules on pathogens known as pathogen-associated molecular patterns.

Monoclonal antibody An antibody of a specific type produced by a clone of identical plasma cells.

Monocyte A type of white blood cell that is phagocytic and can migrate to host tissues, where it differentiates into a macrophage. These cells present antigen to T lymphocytes.

Monomer A low-molecular-weight intermediate utilized in the synthesis of cellular macromolecules.

Monophyletic A cluster of organisms that are most closely related to one another.

Monotrichous A bacterium with a single flagellum.

Morbidity rate Incidence of a particular disease in a population during a specified time period.

Mordant A substance that increases the affinity of a cell for a dye.

Morphogenesis Developmental process whereby an organism forms different cell types in its life cycle.

Morphology The shape of an organism or part of it.

Mortality rate Ratio of deaths from a particular disease relative to the total number infected.

Most probable number (MPN) Measure of the number of microorganisms by dilution. End point is highest dilution yielding growth.

Mot proteins The two membrane-embedded components of the flagellum (MotA, MotB) that form the motor of the apparatus and drive flagellar rotation.

Motility Purposeful movement of a microorganism.

mRNA A ribonucleic acid copy of a gene encoded in DNA. Also referred to as a transcript.

Mucociliary escalator Ciliary action that moves mucus towards the pharynx.

Murein See peptidoglycan.

Murine typhus Infection caused by *Rickettsia typhi*. The name refers to rats and squirrels that serve as the reservoir of this organism. Also called endemic typhus.

Mushrooms Filamentous fungi that produce large fruiting structures. Selected species are edible.

Must Crushed fruit, especially grapes, that can be fermented to produce an alcoholic product.

Mutagen A physical agent (radiation) or chemical that induces mutation.

Mutation Heritable change in the genetic makeup of a species.

Mutualism A symbiotic association where both partners will gain.

Mycelium A three-dimensional colony formed from the repeated branching of growing hyphae. Typical of some fungi and members of the *Actinobacteria*.

Mycoplasmas A group of mostly pathogenic bacteria related to the Firmicutes.

Mycorrhiza Fungal-plant root associations.

Myxobacteria Bacteria that have a complex developmental cycle.

N-acetylglucosamine An amino sugar found in peptidoglycan.

N-acetylmuramic acid An amino sugar found in peptidoglycan.

Natural competence The ability of bacteria to take up DNA from their environment.

Natural killer cells (NK) Lymphocytes that are part of the innate immune response and destroy aberrant host cells, such as tumor cells and virus-infected cells, using pattern recognition. They also produce cytokines that influence the responses by other cells, including interferon-gamma.

Near point The closest position an object can be brought to the eye and still be resolved.

ncRNAs Small noncoding regulatory RNAs.

Necrotizing fasciitis A disease characterized by massive destruction of skin and soft tissues following infection by *S. pyogenes*.

Negative chemotaxis Movement of a cell away from a nutrient.

Negative regulation Regulation of gene expression through the action of repressors that interfere with initiation of transcription.

Negative stains Staining procedures that use acidic dyes such as nigrosin which stain the background to distinguish it from the cell.

Negri bodies Viral masses that form in the brain of rabies-infected animals.

Neurotoxin One that harms nerve tissue.

Neutrophils Short-lived leukocytes that are specialized for rapid recruitment from the blood into inflamed tissues, where they phagocytose and destroy pathogens.

N-formylated methionine (fMet) A modified methionine that is the first amino acid incorporated into a polypeptide chain during protein synthesis.

Niche A habitat with all factors necessary for growth of a species.

Nitrification Oxidation of ammonia to nitrate in the environment.

Nitrifiers Members of the *Bacteria* and *Archaea* involved in oxidation of the reduced nitrogen compounds ammonia and nitrite.

Nitrogen fixation Reduction of atmospheric nitrogen to ammonia. A property present in some *Bacteria* and *Archaea*. The enzyme involved is nitrogenase.

Nitrogenase The enzyme that catalyzes nitrogen fixation in microorganisms.

Nodule A structure formed in the roots or stems of symbiotic nitrogen-fixing plants. Filled with the nitrogen-fixing microorganisms.

Nomenclature The aspect of taxonomy that treats the naming of organisms.

Nonprofessional phagocytes Host cells that can be induced to phagocytose microbial organisms. These cells are then colonized by the microbe.

Nonreplicative transposition A mechanism of transposition where the transposon is cut out from the donor site and inserted in the target sequence.

Nonsense codon One that does not code for an amino acid but has a signal to terminate protein synthesis.

Northern blot A technique of hybridization of single-stranded RNA or DNA probes to RNA fragments attached to a matrix.

Nosocomial infection An infection acquired in a health-care facility or hospital.

Nuclear area The area in a bacterium or archaeon that contains DNA.

Nuclear envelope A double membrane surrounding the nucleus and traversed by protein-containing pores.

Nuclear fusion Merging of two haploid nuclei into a single diploid nucleus.

Nucleic acid Polymer composed of nucleotides.

Nucleic acid probe A labeled single strand of nucleic acid that can hybridize with a complementary strand in a crude mixture. Employed in pathogen identification.

Nucleocapsid The basic unit of a virion; nucleic acid surrounded by protein capsid.

Nucleolus Site of transcription of DNA coding for ribosomes (rDNA) to ribosomal RNA (rRNA). Chromosome regions with ribosomal genes aggregated at the nucleolus.

Nucleotide A monomeric unit that can polymerize to form nucleic acid. Composed of sugar (ribose or deoxyribose), phosphate, and a purine or pyrimidine base.

Nucleus The double membrane–enclosed organelle in eukaryotes that contains the genetic material (DNA).

Nuisance blooms Algal blooms that produce odors on the beach and may contain toxic cyanobacteria.

Numerical taxonomy The use of a large number of strains and phenotypic tests to classify organisms.

Nutrient Any substance that is assimilated by a microorganism during growth.

Obligate A term used by microbiologists to indicate an absolute requirement: obligate aerobe, obligate autotroph, obligate intracellular pathogen, etc.

Occam's razor A premise that the simplest explanation is most likely to be the correct explanation.

Okazaki fragments Short fragments of DNA involved in discontinuous replication of DNA on the lagging strand.

Oligotroph A microorganism that can live under low nutrient conditions.

Oligotrophic Term used to describe aquatic habitats that contain very low concentrations of nutrients.

Oncogene A gene that when expressed can convert a normal cell to a tumor cell.

Oocysts Eggs.

Open reading frame (ORF) A contiguous coding sequence between initiation and termination codons.

Operator A specific segment of DNA at the start of a gene where a protein (repressor) can bind to control mRNA synthesis.

Operon A group of adjacent genes organized in such a way that they are transcribed from a single promoter, giving rise to polycistronic mRNA.

Opines Derivatives of the amino acid arginine or sugars that are synthesized in plants transformed by *Agrobacterium tumefaciens*. Utilization of these compounds is restricted to the crown gall–inducing pathogenic agrobacteria.

Opportunist A microorganism that is generally harmless but can cause disease under certain conditions or in an immunocompromised host.

Opsonins Proteins that promote phagocytosis.

Opsonization The process of coating a substance with an opsonin to facilitate phagocytosis.

Oral candidiasis (thrush) Infection of mucous membranes of the mouth by *Candida albicans*.

Organelle A membrane-bound functional structure in a cell.

Organic scales Scales formed inside the Golgi apparatus and secreted to the outside of the cell, usually made of polysaccharides or proteins.

Origin of replication Site of initiation of DNA replication.

oriV Site of origin of replication of plasmids.

oriT Origin of transfer of conjugative plasmids.

Osmophiles Microorganisms that grow best in media of high solute concentration.

Osmosis Movement of water through a membrane from a low solute concentration to one of a higher concentration.

Osmotic balancers Ions or soluble molecules that increase the cytoplasm osmotic potential.

Osmotrophy The uptake of dissolved organic and inorganic nutrients, including minerals, through the cell membrane. This is the only form of nutrient uptake in prokaryotes, and it occurs in many protists.

Outbreak Sudden high incidence of disease in a given population.

Overoxidizer A vinegar bacterium that oxidizes acetic acid to carbon dioxide and water.

Oxidase positive An organism that possesses cytochrome oxidase.

Oxidation Loss of electrons.

Oxidation-reduction (redox) reaction Coupled reaction where one partner donates electrons and the other accepts them. One is oxidized, the other reduced.

Oxidative phosphorylation Employment of the electron transport system to generate a proton motive force that provides energy for ATP synthesis.

Oxygenic photosynthesis Cyanobacterial or plant-type photosynthesis where water serves as the electron donor, resulting in oxygen production.

Pandemic An epidemic worldwide. HIV is now a pandemic.

Paralysis Complete or partial loss of motor function.

Parasite A symbiotic association where one organism benefits and the other is harmed. Used to describe protozoa and worms.

Paraxonemal rod A rod of proteins adjacent to the axoneme inside the cilium.

Paryphoplasm The outlying cytoplasm of a member of the Planctomycetes that is devoid of ribosomes.

Passive diffusion Movement of a molecule from an area of high concentration to one of lower concentration.

Passive immunity A short-lived immunity gained by transferring specific immune antibodies to a nonimmune individual.

Pasteurization Heating a fluid to temperatures that destroy spoilage- or disease-causing organisms.

Pathogen A disease-causing organism.

Pathogenicity A relative term indicating the disease-causing potential of a micro-organism.

Pathology A generic term for tissue damage.

Pathovars Pathogenic varieties of species.

Penicillins Antibiotics that have a β-lactam ring and inhibit bacterial cell wall synthesis by blocking crosslinking of peptide chains in newly synthesized peptidoglycan.

Pennate diatoms Diatoms with a bilateral symmetry.

Pentose phosphate pathway A pathway that oxidizes glucose-6-phosphate to ribulose-5-phosphate. A source of 5-carbon sugars for DNA and RNA synthesis.

Peptic ulcers Lesions of the stomach lining.

Peptide A short linear sequence of amino acids.

Peptide bond A covalent bond, formed by dehydration, between the carboxyl group of one amino acid and the amine of another.

Peptidoglycan (murein) The polymeric cell wall structure present in most *Bacteria*. It is formed by alternating units of *N*-acetylglucosamine and *N*-acetylmuramic acid. The *N*-acetylmuramic acids are cross-linked by amino acids.

Peptones Enzymatically digested proteins that are used in preparation of culture media.

Pericarditis Inflammation of lining that surrounds the heart.

Periplasm (periplasmic space) A space between the cytoplasmic membrane and peptidoglycan layer (gram-positives) and the outer envelope (gram-negative).

Periplasmic flagella Flagella that are located in the periplasm as found in the spirochetes; also called endoflagella.

Perithecium A flask-shaped ascocarp found in some *Ascomycetes* (Fungi).

Peritrichous Having flagella distributed over the surface of a bacterium.

Permease A protein in the cytoplasmic membrane that transports material inward.

Peroxisomes Vesicles that contain oxidative enzymes.

Pertactin One of several *Bordetella pertussis* surface adhesins.

Pertussis toxin A toxin produced by *Bordetella pertussis*; modifies regulators of cyclic AMP synthesis in target cells.

Phagocyte A cell that can capture and digest foreign material, generally macrophages, monocytes, and neutrophils.

Phagocytosis A form of endocytosis where cell membrane invagination engulfs particles such as organic material, bacteria, or protists into phagosomes.

Phagolysosome An intracellular membrane-bound compartment that results from fusion of a phagosome and a lysosome.

Phagosomes Also called food vacuoles, they contain ingested material such as protists or bacteria.

Phagosome (immunology) A membrane-bound vacuole formed in phagocytes by invagination of the cell membrane about a foreign particle.

Phagotrophic Eukaryotic cells that can engulf particulate food and cells.

Pharyngitis Inflammation of the pharynx.

Phase variation A reversible change in the expression of surface molecules on bacteria, particularly fimbriae.

Phenol coefficient Relative strength of a disinfectant compared with phenol.

Phenotype Expressed properties of a microorganism.

Pheromones Molecules secreted to attract individuals of complementary mating types for the purpose of conjugation.

Phosphodiester linkage A bond between 5'-phosphoric acid and 3'-hydroxyl group on neighboring nucleosides in DNA or RNA.

Phospholipid The component part of a cytoplasmic membrane composed of fatty acids, glycerol phosphate, and generally with a polar molecule linked to the phosphate.

Photoautotroph An organism that can utilize light as an energy source and CO_2 as a carbon source.

Photoheterotroph A microorganism that utilizes light energy while assimilating organic compounds as a carbon source.

Photophosphorylation Generation of a proton motive force by use of light energy. This energy drives ATP synthesis from ADP and inorganic phosphate by ATP synthase.

Photoreactivation A process of repair of dimerized pyrimidines by the action of the enzyme photolyase, which is activated by DNA-damaging ultraviolet light.

Photosynthesis Use of light energy to generate chemical energy for cell maintenance and CO_2 assimilation.

Phototaxis Purposeful movement of an organism toward favorable wavelengths of light.

Phototroph Organism that utilizes light as a source of energy.

Phragmoplast Occurs in Charales (Chloroplastida) and higher plants, between the dividing cells at the site of cell wall deposition during cytokinesis.

Phycobiliproteins Phycoerythrin (red) and phycocyanin (blue) pigments that trap light in phycobilisomes.

Phycobilisomes Specialized structures on cyanobacterial membranes involved in light harvesting.

Phylogenetic classifications A classification system based on analysis of sequences of macromolecules, in particular DNA, RNA, and protein.

Phylogenetic trees Figural representations of phylogenetic classifications.

Phylogeny The evolutionary history and genetic relationships between organisms.

Piezophile Organisms that can grow at high hydrostatic pressure, also called barophilic.

Pili Protein filaments, the most common of which are fimbriae, present on bacterial cells that are involved in conjugation.

Pinocytosis A form of endocytosis where invagination of the cell membrane results in external fluid being internalized in small 50–500 nm vesicles.

Pioneer An initial organism in a succession.

Planctomycetes A major phylum of the *Bacteria*.

Plague A disease caused by *Y. pestis* infections.

Plaque A clear area in a lawn of cells on an agar surface resulting from lysis of cells by a virus; or a microbial colony attached to and growing on a tooth.

Plasma The fluid, noncellular portion of blood.

Plasma cell A short-lived differentiated B cell that synthesizes and secretes large quantities of specific antibody.

Plasmid A double strand of circularized DNA that exists and replicates independently in a cell; may integrate into the chromosome. Can carry information for specialized catabolic enzymes or drug resistance.

Plasmodesmata Channels found in higher plants and in Charales (Chloroplastida) that allow communication between the cytoplasm of adjacent cells.

Plasmolysis Loss of water by a bacterial cell placed in high solute concentration, causing shrinkage of the cytoplasmic membrane and potential death.

Platelets Formed elements in blood that are pieces of megakaryocytes without any nuclear material and are important for blood clotting.

Pleomorphic Bacteria that are variable in shape.

Plus strand Viral nucleic acid that is of a base sequence that can serve as mRNA.

Pneumococcus A common name for *Streptococcus pneumoniae*.

Pneumonia Infection and inflammation of the lungs.

Pneumonic plague A severe form of infection by *Yersinia pestis*.

Point mutation Defect in a single base pair in a specific location of a genome.

Polar flagellum The end or both ends of a bacillary bacterium.

Polar mutations Mutations in one gene that affect the expression of an adjacent gene.

Polycistronic mRNA An mRNA molecule containing a number of genes that was transcribed from a single promoter.

Polyclonal B-cell activator A molecule or substance that activates many different populations of B cells.

Poly-β-hydroxybutyrate (PHB) A linear polymer of β-hydroxybutyrate that serves as a storage material or granule in many bacteria.

Poly-hydroxyalkanoates (PHA) Polymers of hydroxy acids such as β-hydroxybutyrate that serve as a storage material in many bacteria.

Polymerase chain reaction (PCR) Amplification of DNA in vitro by synthesis of specific nucleotide sequences from a small amount of template DNA. This technique employs oligonucleotide primers complementary to sequences in the DNA and heat-stable DNA polymerases.

Polymeric substances High-molecular-weight compounds such as DNA, RNA, protein, and polysaccharides found in cells.

Polymorphonuclear leukocyte (PMN) Motile white blood cells that specialize in phagocytosis.

Polyphosphate A storage polymer of phosphate that may serve as a phosphorus or energy source for some bacteria.

Population Mass of cells of the same species.

Porin Proteins that form channels in the outer membrane of gram-negative bacteria for transport of nutrients to the bacteria.

Porters Membrane proteins involved in transport, both inward and outward.

Positive chemotaxis A phenomenon in which bacteria move toward a nutrient.

Positive selection A process that immature T cells undergo in the thymus and which preserves their life.

Posttranslational modification Covalent modification (e.g., phosphorylation or glycosylation) of a protein after its synthesis is complete. May or may not be reversible.

Posttranslational protein translocation A protein secretion pathway where the secreted proteins are first synthesized by ribosomes and then interact with the Sec or Tat secretion machinery.

Potable Water that is safe for drinking.

Prebiotics Food substances that promote the growth of "good bacteria" in the gut.

Precursor metabolites Twelve intermediates that originate in glycolysis, pentose phosphate pathway, or tricarboxylic acid cycle from which all cellular constituents are synthesized.

Presumptive test First test used in identification of coliform bacteria.

Prevalence Total percent of the population infected with a disease at a given time.

Pribnow box A base sequence located about 10 base pairs upstream from the transcription start site. It is the binding site for RNA polymerase.

Primary consumers The first level of consumers in a food chain.

Primary lymphoid organs Those in which lymphoid cells mature, including the bone marrow and thymus in mammals.

Primary metabolites Products secreted during the growth phase. Lactic acid and ethanol are examples.

Primary producer Autotrophic organisms that fix atmospheric CO_2, thus providing sustenance for the ecosystem.

Primary sludge The settleable material in the initial stage of wastewater treatment.

Primary structure of a protein Sequence of amino acids joined in a polypeptide chain.

Primary treatment The first stage in wastewater (sewage) treatment.

Primer A polynucleotide to which the DNA polymerase attaches during DNA replication.

Primordial soup The hypothetical mixture of materials that could have served as the building blocks for the first forms of life.

Prion An infectious proteinaceous particle in which no nucleic acid has been detected.

Probiotics Dietary supplements containing potentially beneficial bacteria.

Prochlorophytes An important group of photosynthetic cyanobacteria found in oceans.

Prokaryote A name applied to all microorganisms that lack a nuclear membrane. Now replaced by *Archaea* and *Bacteria*.

Promiscuous plasmid A plasmid capable of transferring between a wide range of bacterial species.

Promoter The region on DNA at the start of a gene where RNA polymerase binds to initiate transcription.

Proofreading Correction of errors created in DNA during replication.

Prophage State in which a temperate viral genome is integrated into and replicates in concert with the host genome.

Prostaglandin E₂ One of the prostaglandins, a group of hormone-like substances that participate in a wide range of body functions, including the contraction and relaxation of smooth muscle, the dilation and constriction of blood vessels, control of blood pressure, and modulation of inflammation.

Prostheca Extension of the wall and cytoplasmic membrane to form hyphae, stalks, or unusual-shaped bacteria.

Protease An enzyme that cleaves amino acids from a protein.

Protease inhibitor A compound that inhibits a protease by binding to its active site. Certain protease inhibitors are used in the treatment of HIV infections.

Protein A polymer composed of amino acids.

Protein jackets or **S-layers** External cell layers produced by some bacteria and archaea.

Proteobacteria A major phylum of the *Bacteria*.

Proteomics The study of the types and functions of groups of proteins in a cell.

Proteorhodopsin Light-absorbing, retinal-containing protein present in certain bacteria that can effect ATP synthesis.

Proteosome A complex of enzymes in the cytoplasm of host cells that degrades proteins into peptides, some of which are suitable for presentation to T cells as antigens.

Protomer Subunit of a viral capsid.

Proton motive force (pmf) An energized state of a membrane created when an electron transport system transports protons from one side of the membrane to the other. The resulting chemical and electrical gradient can be used to drive ATP synthesis.

Protoplast An osmotically sensitive bacterial or fungal cell resulting from removal of the cell wall. The cytoplasmic membrane remains intact.

Prototroph (wild type) Parent organism of an auxotrophic mutant.

Provirus Viral DNA that has been integrated into the host cell genome.

Pseudomembrane The accumulation of *C. diphtheria* and dead cells in the throat of an individual during onset of diphtheria.

Pseudomembranous colitis The uncontrolled growth of *C. difficile* in the colon, usually caused by perturbation of normal intestinal flora by antibiotic therapy.

Pseudomurein Modified peptidoglycan that is present in the cell walls of some *Archaea*.

Psychrophile A cold-loving organism that can grow at temperatures below 0°C, although 12°C to 15°C is optimal for growth.

Psychrotroph An organism that grows slowly at 0°C but optimally at around 20°C.

Pure culture (axenic) A culture containing a single strain.

Purine Guanine and cytosine.

Purple membrane Areas in cytoplasmic membranes of halophilic bacteria containing bacteriorhodopsin.

Purple nonsulfur bacteria A group of photosynthetic bacteria classified in the *Proteobacteria* that do not deposit sulfur granules in their cells.

Purple sulfur bacteria Photosynthetic bacteria classified as *Proteobacteria* that can deposit sulfur granules in their cells.

Pus A collection of dead host cells and dead microbes at a site of inflammation and infection.

Pyogenic Pus-forming infection.

Pyrimidine Adenine, thymine, and uracil

Pyrogen A fever-inducing molecule.

Pyrogenic Causing a rise in body temperature due to infection or the activity of a toxin.

Quarantine Restriction of movement of individuals to prevent spread of a contagious disease; to isolate animals to ensure that they do not have a disease.

Quaternary structure A complex of more than one protein formed by interactions of amino acid side chains of individual subunits.

Quellung reaction Enlargement of a capsulated microorganism in the presence of antibodies to capsular antigen.

Quoromones Signaling molecules used by bacteria to communicate with each other.

Quorum sensing An ability of a group of bacteria to elicit coordinate responses based on the presence of other bacteria in their immediate vicinity.

R′ A form of antibiotic resistance plasmid carrying a segment of chromosomal DNA acquired following its integration and excision from the bacterial chromosome.

R plasmids Plasmids that carry antibiotic resistance genes. Also referred to as R factors.

Racking Removal of sediments formed in wine bottles during fermentation.

Radioimmunoassay A technique that employs radioisotope-labeled antibody or antigen to detect presence of specific material in body fluids.

Radioisotope An element with a surplus of neutrons that spontaneously decay with emission of detectable radioactive particles.

Reaction center Complex containing multiple bacteriochlorophyll or chlorophyll molecules where photophosphorylation occurs.

Reactive oxygen intermediates Molecules formed within phagosomes of macrophages that have antimicrobial properties. Examples include peroxide and bleach.

Reading frame The sequence of codons in a gene between the initiation and termination signals.

RecA, RecB, RecC, RecD Proteins responsible for distinct steps during homologous recombination.

Real image An image in microscopy that can be projected on a screen.

Recalcitrance A relative term indicating the resistance of a molecule to microbial attack. Generally measurable by the half-life of the compound in the environment.

Receptor-mediated endocytosis Uptake of particles (cells, proteins) bound to a receptor on the surface of mammalian cells though the invagination of the membrane and enclosing the particles in a vesicle.

Recombinant DNA technology Genetic engineering involving the introduction of a gene into a vector such as a plasmid and introducing this into another species to yield a recombinant molecule.

Recombination Combination of genetic material from two separate genomes.

Recombination repair Removal of defects in DNA using the cellular recombination system.

Red beds Geological deposits that appear red due to the oxidized iron forms, such as hematite, they contain.

Reduction potential Tendency of a reduced molecule to donate electrons or an oxidized molecule to accept electrons.

Reductive TCA cycle The reversed form of the TCA cycle that is used by some bacteria to fix carbon dioxide.

Refraction Bending of light as it passes from one medium to another.

Refractory A term used to indicate the inability of microorganisms to readily degrade some chemicals in the environment, such as DDT.

Regulation Sum of the cellular processes that control enzyme function to ensure balanced growth.

Regulon A group of genes or operons controlled by the same regulatory factor.

Relaxase A protein responsible for initiation of conjugative plasmid transfer by nicking the DNA at the *oriT* site and covalently attaching to the free 5′ phosphate group. After transfer, it rejoins the ends of the DNA.

Rep A protein responsible for initiation of rolling circle replication of DNA.

Replica A copy of colonies grown on the surface of a Petri plate, usually generated by overlaying a filter and transferring the attached colonies to a new plate, maintaining the same relative location of the colonies.

Replication Copying of genomic DNA to generate a duplicate copy.

Replication fork A Y-shaped structure in which double-stranded DNA separates, permitting each of the single strands to replicate.

Replicative transposition A mechanism of transposition by a transposon, where the entire transposon is copied during insertion into

a target sequence, followed by resolution of the two copies.

Repressible enzymes The synthesis of these enzymes can be curtailed by the presence of an active repressor protein.

Repressor protein A protein that can bind to an operator to prevent transcription.

Reproduction The process whereby organisms undergo cell division to produce progeny; the process in bacteria and archaea is nonsexual.

Reservoir The site or host in nature where pathogenic microorganisms reside and act as a source of infection for humans or other species.

Resident microbiota Microbial organisms that establish a longterm relationship with the host and which normally do not cause harm.

Residual body Undigested remains in food vacuoles that fuse into a single vacuole that can be excreted.

Residual chlorine The amount of chlorine that is measured in a drinking water sample from a water distribution system.

Resolution The ability of an optical system to identify details of an object.

Resolvase An enzyme that catalyzes recombination between two integrated transposons.

Respiration Energy-yielding catabolic reactions that utilize organic or inorganic compounds as electron donors and acceptors.

Response regulator A component of the two-component signal transduction pathways. It is the recipient of phosphate from an autophosphorylated sensor. Usually a transcriptional regulator.

Restriction endonucleases Enzymes that cleave DNA sequences at specific sites. Probably originated to protect cells against viruses and is now employed in genetic engineering.

Reticuloendothelial system An older term that refers to the presence of phagocytic cells throughout the body, particularly tissues that have a high concentration of such cells, including liver, spleen, and blood.

Retrovirus Viruses that have a single-stranded RNA genome that is replicated by reverse transcription to form a single-stranded complementary DNA copy that is duplicated to form double-stranded DNA. This double-stranded DNA can be integrated into the genome of the host.

Reverse transcriptase An RNA-dependent DNA polymerase that generates DNA from RNA.

Reverse-transcriptase inhibitors Compounds that prevent the transcription of viral RNA to form double-stranded DNA.

Rheumatic fever Inflammation of various tissues, including the heart, during recovery from infection with *S. pyogenes*.

Rhizosphere The area surrounding a plant root.

Rho (ρ protein) A protein that functions to dissociate RNA polymerase after transcription is complete.

Ribonucleic acid (RNA) A polymer composed of ribonucleotides joined by phosphodiester bonds. The bases in RNA are uracil, adenine, guanine, and cytosine.

Riboplasm The part of the cytoplasm of members of the Planctomycetes that contains ribosomes.

Ribosomal RNA (rRNA) The RNA present in a ribosome and involved in protein synthesis.

Ribosome The cellular organelle in which protein synthesis occurs.

Ribosome-binding site (RBS) Also called Shine-Dalgarno sequence. A short sequence on mRNA upstream of the initiation codon that is a recognition site for ribosomes during translational initiation. Shows extensive complementarity to the 3′ end of the 16S rRNA.

Riboswitching A method of regulating transcription by binding of a metabolite to a structure formed by folded mRNA.

Ribozyme RNA with enzyme activity.

RNA polymerase The enzyme that synthesizes mRNA from the DNA template.

Rolling-circle replication A form of DNA replication of circular templates (usually plasmids or bacteriophage genomes) where replication is initiated at a nick on one of the strands. The old strand is displaced as the replication machinery moves around the template.

Root nodule Enlarged structure on a leguminous plant root where nitrogen-fixing endosymbionts live.

Rooted trees Phylogenetic trees that use an outgroup for comparison to the sequences being studied.

Rumen The large organ in a ruminant (herbivore) where cellulosic material is digested anaerobically, generating organic acids that sustain the animal.

Rumposome Flattened vesicle that stores calcium and has been implicated in sensory orientation and taxis in certain chytrids.

Saprophyte An organism that generally grows on decaying organic matter.

Saprotrophy Obtaining nutrients from the digestion of decaying organic matter. Saprotrophs are involved in the decomposition food webs.

SARS (Severe Acute Respiratory Syndrome) A deadly viral infection that emerged in southeast Asia; originated in animals and spread to humans.

Scale-up Gradual increase in size of industrial fermentation to large fermentors for production of useful chemicals.

Scarlet fever Response to infections with *S. pyogenes* characterized by a rash.

SCID Severe combined immunodeficiency.

Scytonemin A UV light absorptive pigment produced by some cyanobacteria.

Sec protein secretion pathway The major pathway for secretion of proteins across the bacterial cytoplasmic membrane.

Secondary clarifier A wastewater treatment unit that separates particulate material (secondary sludge) before discharge or tertiary treatment.

Secondary lymphoid organs Organs in which mature T and B lymphocytes are found in high concentrations. These include the spleen and lymph nodes as well as tonsils and appendix.

Secondary metabolites Products that are synthesized near the end of the exponential growth phase and during early stationary phase.

Secondary structure Product of the initial folding of a polypeptide, generally resulting from hydrogen bonding.

Secondary wastewater treatment A two-step sewage treatment process that removes organic material.

Selectins Adherence molecules on the surface of lymphocytes that facilitate their binding to host cells.

Selection Establishing conditions where organisms with a predetermined desired trait are favored.

Selective medium One that favors growth of selected microorganism(s). The constituents may inhibit others.

Self-transmissible plasmid A plasmid that carries all the genes for transfer between donor and recipient bacteria.

SEM A scanning electron microscope that is used to visualize organisms.

Semiconservative replication A form of replication where the DNA in each daughter cell consists of one old and one newly synthesized strand.

Sense strand The strand of DNA that RNA polymerase copies to produce mRNA, rRNA, or tRNA.

Septicemia Infection of the bloodstream with microorganisms or bacterial toxic products.

Septum In fungi, a cross-wall that separates the filament into compartments.

Serovar An immunological variety or type of a bacterial species.

Serum Fluid portion of blood resulting from removal of blood cells and fibrinogen (normally referred to as "clotting").

Severe combined immunodeficiency (SCID) A defect in both the T cell and B cell parts of the immune system.

Sexually transmitted disease (STD) An infection generally transmitted by sexual contact.

Sheath A tubelike structure that encloses chains of bacterial cells.

β sheet A secondary structure of proteins formed by side-by-side alignment of two or more extended chains.

Shiga toxin A protein toxin produced by *S. dysenteriae*. The toxin kills cells by inhibiting their protein synthesis.

Shiga-like toxin A toxin produced by certain pathogenic strains of *E.coli*. Acts by an identical mechanism as shiga toxin.

Shine-Dalgarno sequence A series of nucleotides on bacterial mRNA that binds to a specific sequence on 16S rRNA in order to properly orient the mRNA on the ribosome.

Siderophore A low-molecular-weight compound that complexes with ferric iron and makes it available for transport into a bacterium.

Sigma (σ) The subunit of the RNA polymerase holoenzyme responsible for promoter recognition.

Signal sequence (peptide) Amino-terminal extension on secreted proteins that serves as the recognition signal for the secretory apparatus.

Signal peptidases Enzymes that remove signal sequences from secreted proteins. Signal peptides on lipoproteins are removed by a different enzyme (SPase II) than the remainder of secreted proteins, which are processed by SPase I.

Simple microscope An optical microscope with a single lens, such as a magnifying glass.

Simple stain Entails the use of a single dye to stain a sample.

Single-cell protein Microbial cell biomass that is primarily a source of protein in animal or human nutrition.

Single-strand binding protein (SSB) A protein whose function is to stabilize single strands of DNA created during replication.

S-layer A structured layer composed of protein or glycoprotein that covers the surface of some bacterial species.

Sleeping sickness A human disease caused by trypanosomatids (Excavata).

Slime layer Diffuse polymeric material exterior to the cell wall of microorganisms.

Sludge The particulate settleable material from primary and secondary wastewater treatment.

Smear The suspension of cells placed on a slide in preparation for staining.

Soil tilth The "health" of a soil for agriculture.

Solid waste The portion of sewage or industrial waste that settles out from the liquid phase.

SOS regulon The sum total of genes whose expression is activated by exposure of bacteria to stress conditions.

SOS response A repair system that is induced in microorganisms that have been damaged.

Southern blot Hybridization of single strands of DNA or RNA to single-stranded DNA immobilized on a matrix.

Speciation The process in which organisms evolve to form new species.

Species Closely related strains that differ from all other strains.

Spherical aberration A defect in the image produced by an optical system that is caused by the inability to produce perfect lenses.

Spheroplast An osmotically fragile cell of a gram-negative bacterium that lacks peptidoglycan.

Spike Projection from the outer envelope of a virus.

Spinae Large pilus-like appendages formed by some bacteria.

Spirochaetes A major phylum of the *Bacteria*.

Spirochete A member of the *Spirocheates*.

Sporozoites The infective forms of malaria parasite.

Spontaneous generation The long-disproved hypothesis that living organisms can arise from inanimate matter.

Sporadic disease A disease that occurs randomly in a human population.

Sporulation The process whereby organisms produce spores, such as bacterial endospores.

Spread plate Isolation of colonies on a solid medium by spreading a dilution of cells over the surface.

SsrA See tmRNA.

Stalk An elongated structure that emanates from a bacterial cell, often used in attaching the cell to a solid surface or another organism.

Starter culture An inoculum consisting of favored microorganisms employed in initiating industrial fermentations.

Stationary phase Phase in batch culture when growth ceases.

Steady state The typical situation in an environment (or a chemostat) in which the concentration of organisms (or a compound) remains constant while the system is dynamic.

Stem cell One capable of extensive proliferation, generating more stem cells, and a large clone of differentiated progeny cells.

Sterile Free of all living things.

Sterilization Destruction of all living things.

Sterols Fat-soluble molecules in cell membranes that increase membrane rigidity. They are found in all eukaryote cell membranes and a few bacteria.

Stickland reaction An ATP-generating reaction employed by some clostridial species where one amino acid is oxidized and another serves as electron acceptor. The products are fatty acids and ammonia.

Stock culture A carefully maintained, stored culture from which working cultures are obtained.

Stop codons Codons UAA, UAG, or UGA in mRNA, specifying the end of a protein coding sequence.

Strain Descendants of a single organism.

Streak plate Isolation of colonies by spreading a culture over an agar surface with an inoculating loop.

Streptococcal pharyngitis Inflammation of the pharynx caused by infection with *S. pyogenes*. Sometimes referred to as strep throat.

Streptolysin O Oxygen-labile membrane-damaging toxin produced by *S. pyogenes*.

Streptolysin S Oxygen-stable membrane-damaging toxin produced by *S. pyogenes*.

Stringent Conditions for a reaction that are more restrictive, as for DNA–DNA hybridization.

Stromatolite A fossilized microbial mat often of cyanobacterial origin. Can be a rounded microbial mat of photosynthetic microorganisms.

Substitution In DNA, replacement of one deoxynucleotide by another. In protein, replacement of one amino acid by another.

Substrate The substance on which an enzyme acts. Also the nutrient(s) on which a microorganism grows.

Substrate-level phosphorylation Generation of a high-energy phosphate bond by reacting an inorganic phosphate with an activated organic compound.

Succession The process of changes in types and concentrations of organisms over time.

Sulfureta Environments in which sulfur cycling processes are occurring, such a sulfur spring.

Superantigen A molecule that is able to bind to T cell receptors outside of the normal antigen-binding pocket. As a result, large numbers of T cells are activated in a nonspecific manner.

Supercoiled DNA Twisted double-stranded circular DNA.

Supercoiling Formation of DNA secondary structures that includes introduction of twists beyond those found in the normal helical structure.

Super-integrons An assemblage of a large number of gene cassettes by sequential integration into a single chromosomal location.

Suppression Reversal of the effect of a mutation by a second mutation at a different site on the chromosome.

Suppressor mutation A mutation that overcomes the effects of another mutation.

Svedberg unit A measure of the sedimentation rate of a macromolecule on centrifugation; dependent on mass and density.

Swarmer cell A motile progeny cell such as in *Caulobacter*.

Symbiosis Different species living together; often used for mutually beneficial interactions between different species.

Synchronous culture A culture in which virtually all cells are at the same stage in the growth cycle.

Synecology A branch of ecology that is involved with the development, structure, and distribution of ecological communities.

Synthetic medium A defined medium of known composition, qualitatively and quantitatively.

Synthrophy A species interaction where metabolic capabilities are shared to permit all to thrive.

Syphilis A sexually transmitted disease caused by *Treponema palidum*.

T-cell receptor The antigen-specific receptor on the surface of T cells that confers specificity to their response.

T lymphocyte (T cell) A lymphocyte that matures in the thymus. T cells are involved in many of the cell-mediated immune responses and in activating specific B cells in the immune response. They have their effect by producing cytokines and by cellular cytotoxicity.

Taxonomy Science of identification, classification, and nomenclature.

TD thymus-dependent A response by T cells is required.

Teichoic acids Glycerol or ribitol polymers joined by phosphates present in cell walls of gram-positive bacteria.

Teleomorph Sexual mycelium of a fungus.

Telomeres Telomeres consist of short repeated sequences of nucleotides that protect the ends of chromosomes.

Temperate virus One that can infect a host without effecting a lytic cycle. The viral genome may be integrated into the host and be replicated along with the host genome.

Template A strand of nucleic acid (DNA or RNA) that can specify the base sequence in a complementary strand.

***Ter* sites** Sites of termination of DNA replication.

Terminator The hairpin loop formed by the nascent mRNA that causes the termination of transcription.

Tertiary configuration Final folding of a polypeptide(s) into an active structure.

Tertiary wastewater treatment Sewage treatment that goes beyond that of secondary treatment.

Thallus Cytoplasmic branches extended by some chytrids (fungi) into the substrate that form a mesh of branching filaments. Some other eukaryotes may also form a thallus.

Thermal stratification The vertical layering of water bodies, particularly lakes in temperate and arctic zones, in which the surface waters are separated from the lower waters by a temperature gradient.

Thermocline The layer of water in a thermally stratified lake where the temperature changes with depth. The thermocline is located between the warmer epilimnion and the colder hypolimnion.

Thermophile A microorganism that grows optimally at temperatures above 45°C with an upper limit of 75°C to 80°C.

Theta (θ) replication An intermediate structure, formed by certain plasmids that utilize a single replication fork.

Thin sections The fine-scale slicing of organisms embedded in resins to produce cross sections showing internal cell structures.

Thylakoid A series of flattened photosynthetic membranes that are impermeable to ions and whose function is to pump out protons to establish a proton gradient essential for ATP synthesis. These membranes contain chlorophyll, electron transport chains, and specific proteins required for photosynthesis.

Th0 A term for uncommitted CD4 T cells.

Th1 Helper T cells that make cytokines such as interferon-gamma that promote inflammation and macrophage activation.

Th2 Helper T cells that make cytokines such as IL-4 and IL-10 that promote antibody responses and inhibit inflammation.

TI T cell–independent. Refers to a type of antigen that stimulates B cells to make antibody in the absence of T cells.

TI-1 A type of TI antigen that is fully T cell–independent.

TI-2 A type of TI antigen that requires only very weak T cell help to stimulate an antibody response.

Ti plasmid Plasmid of *Agrobacterium tumefaciens* carrying the plant tumor–inducing genes. Can be used to transfer foreign genes into plants.

Titer The reciprocal of the highest dilution of antiserum that gives a measurable response by reacting with an antigen in a specific test.

tmRNA A hybrid of tRNA and a small coding RNA, involved in releasing stalled mRNAs from ribosomes due to their loss of proper translational termination signals.

Tolerance Generally refers to the nonresponse of antibodies to self.

Toll-like receptors (TLRs) A family of cell surface molecules that provide a critical link between immune stimulants associated with pathogens and host leukocyte responses.

Topoisomerases Enzymes capable of cutting crossed over DNA strands, passing one strand over the other and resealing the ends.

Total counts The enumeration of all microorganisms in a particular sample using microscopic procedures such as acridine orange staining.

Toxemia Effects of toxins present in the bloodstream.

Toxic shock syndrome A disease caused by production of superantigen toxins during virginal colonization by *Staphyloccous aureus*. The toxins cause massive release of cytokines by immune cells.

Toxigenicity The relative ability of a microorganism to produce toxins.

Toxin A potentially injurious substance, generally protein or lipopolysaccharide, produced by microorganisms.

Toxoid An exotoxin, such as tetanus toxin, that has been modified so that it does not cause damage but will elicit an antibody response.

Tracer A substrate that is radiolabeled to follow its incorporation into a living organism.

Tracer approach The addition of low concentrations of a radiolabeled compound to samples from a habitat in order to follow the metabolism of the compound.

Tracheal cytotoxin A toxin produced by *Bordetella pertussis* consisting of a peptidoglycan fragment.

Trachoma An eye infection caused by *Chlamydia trachomatis*.

Transamination Transfer of an amine from one intermediate to a keto group of another.

Transcription Synthesis of a complementary strand of RNA from a single strand of double-stranded DNA; the enzyme involved is a DNA-dependent RNA polymerase.

Transcriptome The sum total of all RNA transcripts in a cell.

Transcript A messenger RNA.

Transduction Transfer of DNA between bacteria by bacteriophages.

Transfer RNA (tRNA) A small RNA molecule that binds to a specific amino acid and delivers it to a ribosome during translation.

Transformation Uptake and incorporation of exogenous DNA into the bacterial genome.

Transgenic Transfer of foreign genetic information to plants or animals by recombinant DNA.

Transglycosylation Addition of *N*-acetyl glucosamine–*N*-acetyl muramic acid–pentapeptide units to the nascent peptidoglycan chain.

Transition A type of mutation where one purine is replaced with another purine or a pyrimidine with another pyrimidine.

Translation Synthesis of a protein with the aid of a ribosome using the information delivered by mRNA.

Transmissible plasmid Plasmid capable of moving between cells by conjugation.

Transpeptidation Formation of the peptide bridge between the peptides that extend from *N*-acetyl muramic acid of peptidoglycan.

Transposable element Genetic material that can move from one genomic site to another.

Transposase A transposon-encoded enzyme responsible for site-specific recombination between the transposon and the target site.

Transposition Movement of a transposon from one location to another.

Transposon A genetic element capable of moving from one site to another.

Transversion A type of mutation where a purine is replaced with a pyrimidine or a pyrimidine with a purine.

Tricarboxylic acid cycle (Krebs cycle, citric acid cycle) A series of reactions in which a molecule of acetate is completely oxidized to CO_2, generating ATP. Intermediates in the cycle are some of the precursor metabolites involved in biosynthesis.

Trichome A series of bacterial cells in a filament that have close contact with one another. The cells lie parallel lengthwise.

Trickling filter A process sometimes used as a secondary treatment step to remove organic material from wastewater

Trophozoite An active, feeding stage form of pathogenic *Amoeba*.

Tubercle A nodule produced by *Mycobaterium tuberculsosis* and infiltrating macrophages enmeshed in fibrous material.

Tuberculin test A test for exposure to *Mycobacterium tuberculosis.*

Tuberculoid leprosy A mild, self-limiting infection by *Mycobacterium leprae*.

Tuberculosis A respiratory disease caused by infection with *Mycobacterium tuberculosis.*

Tunicamycin An antibiotic that blocks the addition of *N*-acetyl-muramic acid pentapeptide to the undecaprenol phosphate carrier.

Turbidity Measure of the relative opacity of suspended solids in a liquid.

Tus A protein responsible for correct termination of DNA replication.

Two-component signal transduction system A system for relaying signals consisting of two proteins: a sensor and a response regulator. The signal is sensed by a sensor kinase, which first undergoes autophosphorylation and then transfers the phosphate to a response regulator, usually a transcriptional regulator.

Twin arginine protein secretion (TAT) pathway A pathway of secretion for folded proteins, often containing attached cofactors or multiple subunits. Proteins secreted by this pathway are characterized by the presence of two adjacent arginines in their signal sequences.

Two-dimensional gel electrophoresis A method of fractionating proteins based on charge and size.

Type A variety of a species.

Type I–V secretion systems Distinct mechanisms utilized by gram-negative bacteria to secrete proteins into the external medium.

Type strain or **culture** The strain that was first used to describe a new species.

Typhoid fever An infection caused by *Salmonella typhi.*

Ultramicrotome An instrument used to slice thin sections of cells or tissues that have been embedded in resin.

Ultraviolet radiation Wavelengths of light (170–397 nm) that disrupt DNA nucleic acid sequences at the 5′ side of a DNA or RNA molecule.

Undulating membrane A cell membrane in some protists that is continuous with the cilium membrane along part of its length, resulting in a thin connection called an undulating membrane.

Unicellular An organism that grows a single cell.

Unrooted trees Phylogenetic trees in which no outgroup is used for comparison.

UvrABC A complex with nuclease activity that removes lesions in DNA by cutting out the damaged segment**.**

Vaccine A material that can elicit beneficial antibodies to immunize against disease. Employed in vaccination.

Vacuole A general term for large membrane-bound organelles, such as food vacuoles or contractile vacuoles.

Vancomycin An antibiotic that blocks incorporation of the *N*-acetyl glucosamine–*N*-acetyl muramic acid–pentapeptide into the nascent peptidoglycan chain.

Vasoactive Causes vasodilation.

Vasodilation Dilation of blood vessels.

Vector Any of various animals that carry pathogenic organisms from one host to another, often from a reservoir to humans.

Vector (genetics) See plasmid.

Vegetative growth Growth of an organism without production of spores.

Vehicle (fomite) An inanimate object involved in transmission of infectious organisms.

Verrucomicrobia A major phylum of the *Bacteria.*

Vertical inheritance The process whereby the genes from a parent are transferred directly to its progeny.

Vesicle A membrane-bound body. The membrane is usually a lipid bilayer, but in some cases is made from protein.

Viable Alive, but in microbiological terms, able to reproduce.

Viable count Determination of total viable cells in a sample.

Vinegar bacteria Bacteria that produce acetic acid.

Virion A mature virus particle that is the extracellular form of the virus.

Viroid An infectious agent of plants that is a very small single-stranded RNA. It is not associated with protein nor is it transcribed to generate protein.

Virosphere The shell about Earth in which viruses are found.

Virtual image An image from an optical system that cannot be projected onto a screen.

Virulence A relative measure of the invasiveness and pathogenicity of a microorganism. Often measured by LD_{50}.

Virulence plasmids Plasmids of pathogenic bacteria encoding virulence-promoting factors.

Virus An acellular infectious agent composed of DNA or RNA and able to commandeer the synthetic machinery of a host to generate progeny virus.

Volutin A term used to describe polyphosphate granules produced by some bacteria.

Wastewater The liquid portion of sewage or output from a manufacturing process.

Water activity (a_w) A quantitative measure of water availability in an environment.

Wet mounts Cell suspensions used for microscopy.

Whole-genome shotgun sequencing Sequencing of random fragments of DNA to form a genome, resulting in the generation of a large number of overlapping sequences.

Wild type Parental type as isolated from nature.

Winogradsky column A glass cylinder packed with mud bearing photosynthetic bacteria. The lower area is anaerobic and the upper relatively aerobic. Placed in light, all types of photosynthetic *Bacteria* can thrive.

Xenobiotic A synthetic chemical.

Xerophile A microorganism that grows optimally under low a_w conditions.

Yeast A unicellular fungus.

Zoonoses A disease transmitted to humans as a result of their contact with infected animals. It is not transmitted among humans.

Zippering Phagocytosis that occurs by a series of receptors on the phagocyte interacting sequentially with their counter-receptor (or ligand) on the surface of the microbe.

Zygote The fertilized egg or diploid progeny cell produced by a eukaryotic organism. The diploid nucleus results from the conjugation or fusion of two haploid nuclei.

Notes on Part Opener and Chapter Opener Photos

PART I Microorganisms were the first living things on Earth, and after billions of years of evolution they remain the most abundant organisms, both in number of species and in biomass. Courtesy of NASA.

CHAPTER 1 Thylakoid membranes are visible in this TEM of *Dermocarpa* sp., a cyanobacterium. Similarities in nucleic acid sequences and metabolic pathways indicate that the ancestors of eukaryotic chloroplasts were most likely ancient cyanobacteria. Colorized TEM. © Dennis Kunkel Microscopy, Inc.

CHAPTER 2 Antony van Leeuwenhoek (1632–1723) demonstrates his microscopes to Catherine of England in this 1939 portrait by Pierre Brissaud. Courtesy of the National Library of Medicine.

CHAPTER 3 Water is arguably the most important simple molecule for life: Without water, there can be no life as we know it. ©Konradlew/istockphotos.com.

CHAPTER 4 Cocci, bacilli (rods), and helices are the most common shapes for bacteria. Colorized SEM. © Dennis Kunkel Microscopy, Inc.

PART II *Serratia marcescens* (left) and *Kocuria rhizophila* (right) grow on tryptic soy agar in these tube cultures. The slant of the agar medium creates a greater surface area for growth in the narrow tubes. Courtesy of David McIntyre.

CHAPTER 5 *Bacillus anthracis* colonies appear different when grown on bicarbonate agar (left) and blood agar (right). Courtesy of Dr. James Feeley/Centers for Disease Control and Prevention.

CHAPTER 6 CDC laboratory technician Shirley McClinton works with culture plates as part of an epidemiological study to determine the pathogen involved in an outbreak of keratitis, an eye disease, in 2006. Courtesy of James Gathany/Centers for Disease Control and Prevention.

CHAPTER 7 Acidic foods, such as the tomatoes, pickles, corn relish, and salsas shown in this photo, can be sterilized for canning in a boiling water bath. Foods that have a more neutral pH (such as unpickled beans and corn) require a pressure canner to kill the endospores of heat-resistant bacteria. Courtesy of David McIntyre.

PART III Researchers have determined that the flagella on bacteria, such as these *Salmonella* sp., grow from the tip, not the base. The building-block proteins (flagellin) pass through a narrow channel in the lengthening flagellum for assembly at the growing tip. Colorized TEM. © L. Stannard/Photo Researchers, Inc.

CHAPTER 8 The concentration and charge gradient created by protons (white spheres) pumped outside the bacterial cell membrane drives the production of ATP (yellow) from ADP and inorganic phosphate (green) by ATP synthase. Data from Protein Data Bank files 1C17, 1E79, and 1L2P.

CHAPTER 9 The many internal membranes of this cyanobacterium, *Pseudanabaena* sp., are sites of photosynthesis. Colorized TEM. © Dr. Kari Lounatmaa/Photo Researchers, Inc.

CHAPTER 10 This polarized light micrograph shows crystals of the carbohydrate monomer glucose, the most abundant product of living cells. © John Walsh/Photo Researchers, Inc.

CHAPTER 11 This ribbon model of the end view of a bacterial flagellum shows how the flagellin molecules fit together in a helix, leaving a channel at the center for the transport of additional flagellin to the growing tip. Courtesy of Keiichi Namba.

CHAPTER 12 Heat released by biodegradation is most likely responsible for the steam rising from this compost heap. Biodegradation, a function overwhelmingly performed by microorganisms, is essential to the cycle of life in the biosphere. Courtesy of the Pennsylvania Department of Agriculture.

PART IV DNA (yellow) is partitioned equally between two daughter cells of *Salmonella typhimurium* as they divide by binary fission. Colorized TEM. © Dr Kari Lounatmaa/Photo Researchers, Inc.

CHAPTER 13 An unidentified bacterium divides by binary fission into two daughter cells. Colorized TEM. © Scott Camazine/Photo Researchers, Inc.

CHAPTER 14 T4 bacteriophage are complex viruses that attack *Escherichia coli*. When a T4 virus invades an *E. coli* cell, it assumes complete control of the bacterium's reproductive machinery, causing it to replicate DNA at about 20 times its normal rate. Colorized TEM. © Dennis Kunkel Microscopy, Inc.

CHAPTER 15 One cell of *Salmonella enteritidis* transfers DNA to another via a pilus. © David Scharf/Peter Arnold, Inc.

CHAPTER 16 This annotated map of the chromosome of *Synechococcus* sp. includes information about the functional category of genes on the plus and minus DNA strands and identification of "characteristic" genes not present in a related genus. From B. Palenik and 14 others. 2003. "The genome of a motile marine *Synechococcus*." *Nature* 424: 1037–1042.

PART V Diatoms (*Navicula* sp., brown) and cyanobacteria (blue) are small enough to qualify as microbial organisms, but they dwarf other bacteria (visible as tiny cocci and bacilli) in this micrograph. Colorized SEM. © Andrew Syred/SPL/Photo Researchers, Inc.

CHAPTER 17 The structure of ribosomal RNA has been remarkably well conserved among all lineages of organisms; differences in rRNA sequences help to illuminate some of the deepest divisions in the tree of life. This image shows a labeled portion of the small subunit rRNA map in Figure 17.3. Figure by Jamie Cannone, courtesy of Robin Gutell; data from the Comparative RNA Web Site: www.rna.icmb.utexas.edu.

CHAPTER 18 *Pyrococcus furiosus*, a member of the *Thermococcales* order of *Archaea*, prefers temperatures around 100°C; it will "freeze" to death if temperatures drop below 70°C. Colorized SEM. © Eye of Science/Photo Researchers, Inc.

CHAPTER 19 *Campylobacter* sp. are spirilla in the class *Epsilonproteobacteria*. They are the most common cause of food poisoning in the U.S. SEM by De Wood; digital colorization by Chris Pooley. Courtesy of the USDA.

CHAPTER 20 *Clostridium botulinum* is the source of botulinum toxin, a deadly food contaminant and (under the trade name Botox) cosmetic drug. The club-shaped appearance of the cells results from the formation of endospores. © Dennis Kunkel Microscopy, Inc.

CHAPTER 21 These bacteria were collected from the anoxic (oxygen-depleted) lower depths of a lake. The red and green organisms are purple and green sulfur photosynthetic bacteria that have evolved from some of the earliest photosynthetic organisms on Earth. Courtesy of J. T. Staley.

CHAPTER 22 This spirochete, *Treponema pallidum*, is the causative agent of syphilis. Illustration based on an SEM. © Chris Bjornberg/Photo Researchers, Inc.

CHAPTER 23 Two rows of cilia on *Didinium* sp. reveal their membership in the *Ciliophora*. *Didinium* is a predatory ciliate that feeds most often on another ciliate, *Paramecium*. © Aaron Bell/Visuals Unlimited.

PART VI Microbial mat communities can flourish in habitats shunned by all other organisms, such as this thermal pool in Yellowstone National Park. © Fritz Pölking/Visuals Unlimited.

CHAPTER 24 Researchers investigate the unique microbial community of Octopus Spring in Yellowstone National Park. Courtesy of Dave Ward.

CHAPTER 25 Symbiotic colonization by the proteobacterium *Burkholderia tuberum* (green) led to the growth of this root nodule on a purple bushbean (*Macroptilium atropurpureum*). Fluorescence micrograph. Courtesy of Michelle Lum and Ann Hirsch.

CHAPTER 26 *Helicobacter pylori*, a causative agent of gastric and duodenal ulcers and stomach cancer, anchors itself to the stomach wall. Although as much as half the human population carries *H. pylori*, only a small minority develop disease as a result of infection. © Eye of Science/Photo Researchers, Inc.

PART VII A macrophage sends out filopodia in search of *Escherichia coli*. Colorized SEM. © Dennis Kunkel Microscopy, Inc.

CHAPTER 27 A macrophage engulfs the protist that causes leishmaniasis. Phagocytosis is an important defense mechanism of the immune system. Colorized SEM. © Jürgen Berger, Max-Planck Institute/SPL/Photo Researchers, Inc.

CHAPTER 28 *Giardia intestinalis* causes giardiasis, the most prevalent water-borne diarrheal disease in the United States. *G. intestinalis* is a protist with a life cycle that alternates between an infectious cyst form, which can survive outside the host, and a trophozoite form (shown here), which lives and reproduces inside the host. Colorized SEM. Courtesy of Janice Carr/Centers for Disease Control and Prevention.

CHAPTER 29 Viral envelopes are clearly visible in this micrograph of influenza virions. Frequent reshuffling of genetic material in influenza virus populations poses a continual challenge to vaccination efforts. Colorized TEM. Courtesy of Dr. F. A. Murphy/Centers for Disease Control and Prevention.

CHAPTER 30 Insects are vectors for a wide variety of human diseases, from Lyme disease to plague. The *Aedes albopictus* mosquito transmits dengue fever. Courtesy of James Gathany/Centers for Disease Control and Prevention.

PART VIII Bread, cheese, and wine are all products of applied microbiology. © Guido A. Rossi/Photo Researchers, Inc.

CHAPTER 31 Inside these copper fermentation tanks in a German brewery, countless *Saccharomyces cerevisiae* are working tirelessly to turn grain into beer. © imagebroker/Alamy.

CHAPTER 32 Severn Trent's Burton Joyce Works in Nottinghamshire, UK. This was one of the first power stations to generate electricity in conjunction with sewage treatment. Raw sewage passes through the complex from the round settling tanks (at the lower left) toward the gas tanks (top left) through a variety of aeration, settling, and filtering tanks. The decomposing sewage produces methane gas that is burned to generate electricity. © Chris Knapton/Photo Researchers, Inc.

Index

Page numbers in boldface indicate boldface terms in the text; page numbers in italic indicate figures and tables.

Truffles, 705
Trumpet hyphae, 726
Trusopt®, 1033
Trypanosoma brucei
immune evasion and, 834
sleeping sickness, 716–717
toxin, *831*
virulence, 833
Trypanosoma brucei gambiense, 833
Trypanosoma brucei rhodesiensis, 833
Trypanosoma cruzi
as an obligate pathogen, 820
disease caused by, *717, 836*
host cell, *838*
survival in macrophages, *859*
survival in phagocytes, 836, *838*
vector, *821*
virulence, 833
Trypanosoma lewisi, 716
Trypanosomatids, 715–717
Trypanosomes, **45,** *859,* 875, 889
Tryptophan
food supplement, *1021*
negative gene regulation and, 357,
358
pathogens and, 860
structure, *68*
Tryptophan biosynthetic (*trp*)
operon, 369–370
Tsetse flies, 821
TSI agar, *993*
Tsr chemoreceptor, 311
TSST-1 exotoxin, 835
Tube worms, 769, 798–799, *800*
Tuber melanosporum, 705
Tubercle bacillus, 910
Tubercles, **911**
Tuberculin, 888
Tuberculin test, 888, **911,** 976
Tuberculoid leprosy, **912,** 913
Tuberculosis (TB), 627
antibiotic resistance, 912
antibiotics for, 1038
causative agent, *859,* 899, 910
clinical identification, 992
in developing nations, *988*
diagnosis, 911
dormancy of the pathogen, 840
granulomas and, 835
HIV/AIDS and, 820, 980
incidence, 910
increasing number of cases, 979,
980
mechanism of infection, 911
primary reservoir and means of
control, *979*
quarantine and, 986
test for, 888, 911, 976
transmission, 910–911, 978
treatment, 911–912
Tubular particles, 963
Tubulin, 693
microtubules and, 691
phylogeny of, *682*
in the *Verrucomicrobia,* 680–681
γ-Tubulin, 693
Tularemia, *929, 934–935, 979,* **982**
Tumor-inducing (Ti) plasmids,
425–426
Tumor necrosis factor alpha (TNF-
α), 830, 860, *868,* 884
Tumors
benign and malignant, 407
crown gall tumors, 425–426
root tumors, 710
viruses and, 407, *955*
Tunicamycin, 194, **292**
Turbidimetry, **162–163,** *164*
Turbidostats, 168
Turgor pressure, **109**
Turtle grass, 790
Turtles, 920

Tus protein, *345,* **349**
Twin-arginine protein secretion
pathway, *301, 302*
Twitching, 120
Two-component signal transduction
systems, *359, 360*
Two-dimensional gel electrophore-
sis, **480**
Twort, Frederick W., 398–399
Tyler, Stanley, 5
"Tyndalization," **180**
Tyndall, John, 37, **40,** 180
Type 1 cytokine response, 883
Type 1 diabetes, 889
Type 1 (Th1) helper cells, **865,** 866,
867, 868
Type 2 (Th2) helper cells, **865,** 866,
867, 868
Type cultures, **511**
Type I hypersensitivity, 887
Type I methanotrophs, 586
Type I pili, 120, **311**
Type I secretion pathway, 303, *304,*
305
Type II methanotrophs, 586, *587*
Type II secretion pathway, *304,*
305–306
Type III secretion pathway, 306, **309**
Type IV hypersensitivity, 887, 888
Type IV pili, *305,* 306, **311,** *312*
Type IV secretion pathway, *304,*
306–307, 435, **905**
Type strains, **511**
Type V secretion pathway, *304,* 307
Typhoid fever, *1047*
carriers, 976, 977
causative agent, **579,** *918*
in developing nations, *988,* 989
overview of, **920**
primary reservoirs and means of
control, *979,* 980
vaccination, *989*
Typhoid Mary, 976, 977
Typhus, 32
causative agents, 554, 821, 933
in developing nations, *988*
historical significance of, 893
primary reservoirs and means of
control, *979,* 982
public health measures and, 984
transmission pathway, *929*
types of, 933–934
Tyrocidine, *192,* **194**
Tyrosine, *68*

Ubiquinol, 243
Ubiquinones, 211, *218,* **219**
purple nonsulfur bacteria, **243**
redox reactions, 213
UDP-*N*-acetylglucosamine
enolpyruvyl reductase, 470
Ulcers, 917–918
Ultramicrotomes. *See* Dehydration
and embedding
Ultrasound, 294
"Ultrastructure," 92
Ultraviolet (UV) radiation
mutagenesis and, 381
ozone layer and, 24, 235
scytonemin and, 653
sterilization and, **184**
Ulva, 721
Ulvophyceans, 721
Unbalanced growth, **158**
Uncouplers, **220**
Undecarpenol phosphate (Udc), 289,
291, 292
See also Bactoprenol
Underoxidizers, **553**
Undulant fever, 553, *859*
Undulating membranes, **692**
Unicellular organisms, **10**

spherical, **89–91**
sulfur-oxidizing proteobacteria,
572–573, *574*
Uniporters, *135,* 136
United States
ethanol production and, 326
major causes of death in, 974
U.S. Food and Drug Administration
(FDA), 1040
U.S. Public Health Service (USPHS),
983
"Unity of Biochemistry" concept, 50
Unrooted phylogenetic trees, **494**
Unsaturated fatty acids, *277*
"Unweighted pair group method
with arithmetic mean" (UP-
GMA), 495, *496*
Uracil, 69, *70, 274*
Uranium, 1061
Urbani, Carlo, 954
Urea, 328, 611
Ureaplasma, 620
U. urealyticum, 499
Urease, *297,* 328, *329,* **580**
Urease test, *579, 580,* 994
Urediniomycetes, 708–710
Urediniospores, *709,* 710
Uredinium, 710
Urethra, *814,* 818, 924
Urethritis, 924, 927
Uric acid metabolism, 328, *329*
Uridine triphosphate (UTP), 204
Urinary tract infections
cystitis, 895, 924
nosocomial, 983
overview of, 924, 990–991
urethritis, 924, 927
yeast and, 824, 825
Urine, 825, 924
Uroleucon
U. rurale, 801
U. sonchi, 801
Ustilaginomycetes, 710–711
Ustilago
U. avenae, 710
U. maydis, 710, *711*
U. violacea, 710
uvr genes, **383**

V gene, 864
Vaccines/Vaccinations
animals, 987
for anthrax, 915
autism and, 986–987
children, **986**
commercial value and production,
1038–1039
diphtheria, pertusis, tetanus (DPT),
904, *986*
immunological memory and, 861
Edward Jenner and, 38
measles, mumps, rubella (MMR),
945, 946–947, 949, *986,* 987
overview of, 885–887
for pneumonia, 905
for polio, 960, 961, 962
respiratory diseases and, 980
role of, 812
for rubella virus, 886
for smallpox, 988
for travelers, *989*
toxoids and, 831–832
viral diseases and, 944
for whooping cough, 903–904
Vaccinia virus, 410–411
Vacuoles
food, 690
gas, 98, *99,* 100
parasitophorous, 826
Vagina, normal human microbiota,
814, 818
Vaginal infections, 607, 916

Vaginitis, **916**
Valacyclovir, 191
Valence, **58**
Valine, *68*
Valinomycin, **194,** *195*
van der Waals forces, **63**
van Niel, Cornelis B., 14, 20, 21, **50,**
257
Van Niel's postulates, **318–319**
Vanadium, 281
Vancomycin, *192,* 194, **292,** 905, 987,
991
Variable surface glycoprotein (VSG),
833, 834
Varicella-zoster virus (VZV), 945
Varieties, **499**
Variola major, 947
Variola minor, 947
Vasoactive agents, **852**
Vasodilation, **830, 852**
Vector-borne diseases
overview of, 821, 929–935, 982
primary reservoirs and means of
control, *979*
See also specific diseases
Vectors, in vector-borne diseases,
929
Vegetative growth, in endospore-
forming bacteria, **96**
Vegetative propagation, ascomycetes
and, 704
Veillonella, 814
Velocity centrifugation, **294–295**
Venereal warts, *955*
Venezuelan encephalitis, *964*
Venezuelan equine encephalitis, *965,*
966
Venter, Craig, 781
Verrucomicrobia, 11, 508, 680–681,
683
Verrucomicrobium spinosum, 680, *681*
Vertical inheritance, **498**
Vestal, J. Robie, *771*
Vestimentiferan tube worms,
798–799
Viability counts, **162,** *163, 164,* 744,
747
Vibrio (cell shape), **91**
Vibrio
bioluminescence and, 802
characteristics, *575,* 581
chitin degradation and, 324
differentiating from pseudomonads, *583*
electron acceptor, *211*
flagellation, 117
naming of, 38
Vibrio cholerae, 1047
chitin degradation and, 324
cholera, 581, *918,* 921
colonization and, 821
fecal specimens, 992
genome size and genes of unknown function, *471*
hand washing and, 818
horizontal gene transfer, *499*
toxin, *831*
Vibrio fischeri
bioluminesence, 581, 802, 803,
804–805
quorum sensing, 374–375
Vibrio harveyi, 374, 581
Vibrio natriegens, 581
Vibrio parahaemolyticus, 918, 923–924
Vibrionales, 546, 575, 580–582
Vibrionia, 38
Vicia faba, 550
Victoria Land, 771
vif gene, 415
Vinegar, 553, **1018–1020**
Vinegar bacteria, 553
Vinyl chloride (VC), 1064